SYNCHROTRON RADIATION INSTRUMENTATION

To learn more about the AIP Conference Proceedings, including the
Conference Proceedings Series, please visit the webpage
http://proceedings.aip.org/proceedings

SYNCHROTRON RADIATION INSTRUMENTATION

Ninth International Conference on
Synchrotron Radiation Instrumentation

Daegu, Korea   *28 May – 2 June 2006*

PART TWO

EDITORS
Jae-Young Choi
Seungyu Rah
Pohang Accelerator Laboratory,
Pohang, Republic of Korea

SPONSORING ORGANIZATIONS
Korea Ministry of Science and Technology
Korea Synchrotron Radiation Users' Association
The Japanese Society for Synchrotron Radiation Research
Daegu Metropolitan City
Daegu Convention and Visitors Bureau
Korea Tourism Organization
The Korean Physical Society

Organized by:
Pohang Accelerator Laboratory and
Japan Synchrotron Radiation Research Institute

Melville, New York, 2007
AIP CONFERENCE PROCEEDINGS ■ VOLUME 879

Editors:

Jae-Young Choi
Seungyu Rah

Pohang Accelerator Laboratory
San 31 Hyoja-dong
Nam-gu, Pohang, Gyeongbuk, 790-784
Republic of Korea

E-mail: choij@postech.ac.kr
        syrah@postech.ac.kr

L.C. Catalog Card No. 2006939106
ISBN 978-0-7354-0373-4
ISSN 0094-243X

◎ CD-ROM available: ISBN 978-0-7354-0374-1

Printed in the United States of America

CONTENTS

CHAPTER 1

SR-SOURCES AND ADVANCED SOURCES

CHAPTER 2

INSERTION DEVICES

CHAPTER 3

BEAMLINES AND OPTICS

xv

CHAPTER 4

DETECTORS

CHAPTER 5

TIME-RESOLVED TECHNIQUES

CHAPTER 6

MICRO/NANOSCOPY

CHAPTER 7

SR FOR NANO SCIENCE AND TECHNOLOGY

CHAPTER 8

LITHOGRAPHY AND MICROMACHINING

CHAPTER 9

INDUSTRIAL APPLICATIONS

CHAPTER 10

SURFACE AND INTERFACE ANALYSIS

CHAPTER 11

MAGNETISM AND SPINTRONICS

CHAPTER 12

CHEMISTRY AND MATERIALS SCIENCE

CHAPTER 13

LIFE AND MEDICAL SCIENCE

CHAPTER 4
DETECTORS

The Hybrid Pixel Single Photon Counting Detector XPAD

S. Hustache-Ottini[*], J.-F. Bérar[†], N. Boudet[†], S. Basolo[¶], M. Bordessoule[*],
P. Breugnon[¶], B. Caillot[†], J.-C. Clemens[¶], P. Delpierre[¶], B. Dinkespiler[¶],
I. Koudobine[¶], K. Medjoubi[*], C. Meessen[¶], M. Menouni[¶], C. Morel[¶], P. Pangaud[¶],
E. Vigeolas[¶]

[*]Synchrotron SOLEIL, L'Orme des Merisiers, Saint Aubin BP 48, 91192 Gif sur Yvette CEDEX, France
[†]D2AM-CRG/ESRF, 38042 Grenoble and Laboratoire de Cristallographie, CNRS, 38042 Grenoble, France
[¶]CPPM-IN₂P₃, Luminy, 13288 Marseille, France

Abstract. The XPAD detector is a 2D X-ray imager based on hybrid pixel technology, gathering 38400 pixels on a surface of 68*68 mm^2. It is a photon counting detector, with low noise, wide dynamic range and high speed read out, which make it particularly suitable for third generation synchrotron applications, such as diffraction, small angle X-ray scattering or macro-molecular crystallography, but also for small animal imaging. High resolution powder diffraction data and in situ scattering data of crystallization of liquid oxides are presented to illustrate the properties of this detector, resulting in a significant gain in data acquisition time and a capability to follow fast kinetics in real time experiments. The characteristics of the future generation of XPAD detector, which will be available in 2007, are also presented.

Keywords: X-rays, photon counting, pixels.
PACS: 07.85.Fv, 07.85.Qe

INTRODUCTION

On third generation synchrotron radiation sources, diffraction, macro-molecular crystallography and small angle X-ray scattering (SAXS) studies require the simultaneous measurement of both very intense spots and weak diffuse scattering which is often only a few counts above background. The detectors therefore have to offer a very low noise and wide dynamic range, over more than six decades, and furthermore, they should have a fast read-out to allow the study of fast kinetics.

The conventional experimental settings do not fulfill these requirements: slits associated to scintillator+PM allow only a point detection, a lot of scattered photons are lost and the acquisition time is long; large area CCD detectors with phosphor screen and optical demagnification with fibers have a low dynamic range, a slow read-out, a large point spread function and saturated pixels blur their neighbouring.

To overcome these deficiencies, several groups [1-2] develop hybrid pixel counting detectors, which were initially designed for high energy physics. This paper presents the XPAD2 detector, based on this technology, and the results obtained show the potential of such X-ray imagers for synchrotron applications. Therefore, the Laboratoire de Cristallographie de Grenoble, the CPPM/IN₂P₃ and the Synchrotron Soleil are currently developing the next generation, XPAD3, whose characteristics are given in the last section.

THE PRESENT DETECTOR XPAD2

A module of the XPAD2 detector consists of an array of pn diodes (330×330 μm^2 pixels) on a silicon sensor, connected to a symmetrical array of readout channels through a set of microscopic metallic balls, by a so-called "bump bonding" (see Fig. 1.a). The array of readout channels is manufactured with the AMS 0.8μm CMOS technology, in chips of 24×25 pixels (see Table 1.). One module of the detector is made of 8 chips bump bonded to a single 500 μm thick silicon sensor. The electronics is based on the architecture of the first generation XPAD1 [3-4] and runs in counting mode: each channel consists of a preamplifier, shaper, discriminator and 16 bit counter

CP879, *Synchrotron Radiation Instrumentation: Ninth International Conference,*
edited by Jae-Young Choi and Seungyu Rah
© 2007 American Institute of Physics 978-0-7354-0373-4/07/$23.00

(Fig. 1.b). The threshold can be adjusted between 10 and 25 keV. The dynamic range in count rate ranges from 0.01 to 2.10^6 photons per pixel per second.

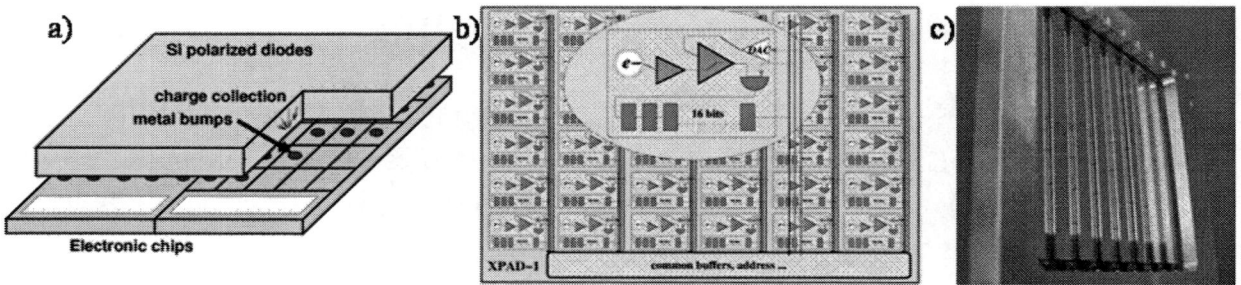

FIGURE 1. a) Principle of a hybrid pixel detector. b) Electronic scheme of the readout chip of the pixels. c) The 8 module X-ray imager (38400 pixels, 68×68 mm²).

The X-ray imager is made of 8 tiled modules (i.e. 8×8 chips), for a total surface of 68×68 mm² (Fig. 1.c). Each chip is wire bonded to a PCB (one PCB per module) and the modules are plugged by parallel wiring to an Altera Nios acquisition card. This parallelism allows to read out the whole detector within 2ms, which is therefore the shortest time between 2 consecutive images. On board memories allow to store 423 images with an exposure time shorter than 10 ms (16 bits), or 233 images with longer exposure (32 bits). Data are transferred to the acquisition PC via a 100 Mb Ethernet link.

POWDER DIFFRACTION AND KINETICS MEASUREMENTS

High resolution data of powder diffraction were recorded with this imager and a conventional setting (slits+ scintillator+ PM) for comparison, on ESRF BM2 beamline, with CeO_2 and X-zeolite samples. Both detectors were located 1m away from the sample, distance where XPAD2 pixel size is 0.02°. On the CeO_2 sample, XPAD2 data were taken at 20 keV, with a step of 1° in 2θ, to be compared with the 0.01° step required with the scintillator. Each Bragg line could be measured 3 or 4 times thanks to the angular covering of the detector, providing the data a high redudancy. Data collection takes 10 minutes, compared to the 3 to 8 hours with the conventional setting, i.e. 18 to 48 times faster with the XPAD 2, with the same data quality. The processing of the images [5] allows to build a Debye Scherrer film, which is shown on Fig. 2.a.

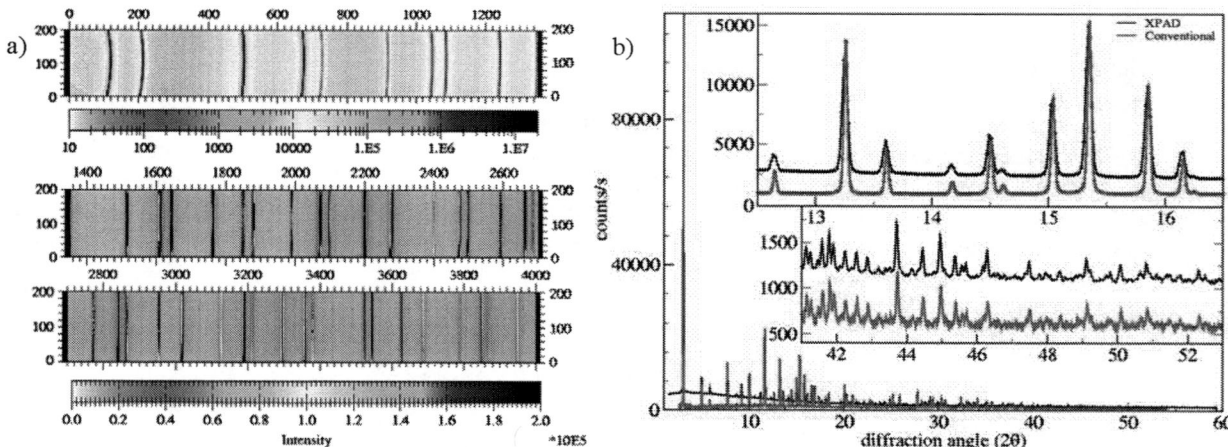

FIGURE 2. a) Debye Scherrer film of CeO_2 reconstructed from XPAD2 data (1m from sample, 20 keV, 1° step in 2θ). b) XPAD2 data (upper curve, 0.1° step) compared to scintillator data (lower curve, 0.006° step) for diffraction on a X-zeolite sample at 16.097 keV. See text for details.

The same technique was applied to collect data on an X-zeolite, at 16.097 keV. The slit aperture in front of the scintillator was 0.35 mm i.e. 0.02°, and 60° were recorded with a step of 0.006° in 2 θ. Data were then taken with the XPAD2, with angular steps of 0.1°, allowing to reduce the acquisition time by 20. Compared data are shown on Fig. 2.b. They were output at a step of 0.01° and scaled to exhibit similar number of counts above the background at wide angle. The diffraction at small angles is detected together with air scattering, which cannot be avoided with the use of 2D detectors. At wide angles, where the signal is weak, XPAD2 data are better than conventional data, even if the background is higher. This increase of background can be explained by a larger integration of the fluorescence of the sample with XPAD2. After Rietveld refinement, both set of data exhibit similar residual factors and refined parameters are in agreement.

To illustrate the potentialities of the XPAD2 detector for the measurements of fast kinetics, data were recorded on ESRF BM2 beamline during the quench of refractive oxides from their liquid state. A small (CaO, 2 Al_2O_3) sphere was heated above its melting point by a laser. The laser was then switched off and data were collected with the XPAD2 during the cooling down. The detector was located at 142 mm from the sample (pixel size of 0.13° and angular aperture of 27°) and the photon beam energy was 17 keV. A movie made of 233 images with an exposure of 20 ms and 2 ms between two consecutive images was recorded. The significant part of the crystallization reaction (4 seconds, 180 images) is displayed on Fig. 3 as a 3D plot.

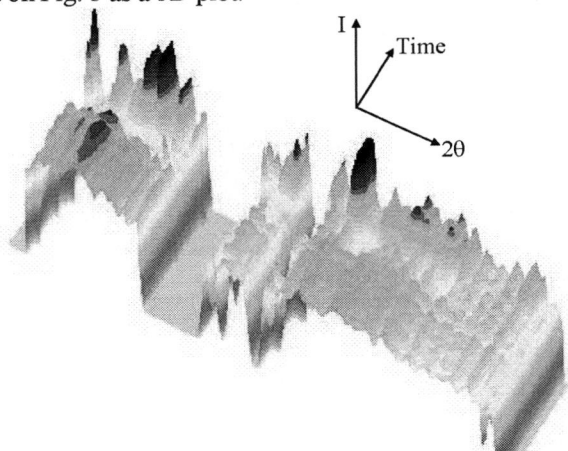

FIGURE 3. 3D plot of 4 seconds of the crystallization process of (CaO, 2 Al_2O_3) reconstructed with 20 ms exposure frames, and a dead time of 2 ms between 2 frames.

Unfortunately, data have been recorded after 1 of the 8 modules was damaged, and this led to the empty band visible in the angular plot. Nevertheless, these data clearly show the capability of the XPAD2 detector to follow fast kinetics.

THE FUTURE DETECTOR XPAD3

Despite the qualities of the X-ray imager presented in the previous sections, the detector has to be improved. The second generation of chips, used in XPAD2, exhibits a wide threshold dispersion, and it was therefore impossible to set the thresholds for the whole detector, leading to a high fraction of useless pixels. This was mainly due to the sensitivity of the design to fluctuations in the CMOS process, within its specifications.

Therefore, it was decided to design a third generation of chips, XPAD3, using the IBM 0.25 μm process, which has the further advantage to be more radiation tolerant. To improve the detector efficiency at high energy, it was also decided to be able to use CdTe as a sensor instead of Si (this is of great interest for small animal imaging, but also for synchrotron applications at energies above 20 keV). Therefore, there will be two types of chip, XPAD3S, for the collection of holes in Si and XPAD3C for the collection of electrons in CdTe. The characteristics of this new generation are summarized in Table 1, together with those of the second generation for comparison.

Since large area and good quality CdTe substrates are not available at the moment, the detectors using this kind of sensor will have a surface of ~2 cm × 1.5 cm. For Si detectors, the mechanical set up will be similar to the XPAD2, with modules of 7 chips tiled to make a large area detector. Smaller detectors can be manufactured, according to the foreseen application, for instance for powder diffraction with crystal analyzer.

In the chip XPAD3C, designed preferentially for CdTe, two thresholds can be adjusted to select incoming photons within an energy window.

TABLE 1. Characteristics of the chips XPAD2 and XPAD3.

	XPAD2	XPAD3S / XPAD3C
CMOS process	AMS 0.8 μm	IBM 0.25 μm
Pixel size	$330 \times 330 \ \mu m^2$	$130 \times 130 \ \mu m^2$
Number of pixels per chip	24×25	120×80
Maximum count rate	10^6 ph/s/mm² (random)	10^7 ph/s/mm² (random)
Energy range	0-25 keV	0-35 keV / 0-100 keV
Threshold range	10-25 keV	4-35 keV (simple) / 4 – 60 keV (window possible)
Read out time	2 ms	2 ms or continuous (exposure > 2ms)

In both chips, it will be possible to reduce the dead time between two images down to less than 1 μs by transferring the 12 bits of the counter to a local memory in the pixel. This memory will be read during the exposure of the next image (this exposure therefore will have to be longer than the read out time). A special care was taken in the design of the readout part of the chip to make it "silent" for the analog part.

XPAD3 is developed in collaboration by the CPPM/IN$_2$P$_3$ for small animal imaging by computed tomography in a PET scanner, the Laboratoire de Cristallographie de Grenoble and SOLEIL for synchrotron applications. The chip has been submitted for fabrication in spring 2006 and the first prototype imager (12×7.5 cm^2, Si) will be available at the very beginning of 2007. The final detector will be ready in summer 2007.

CONCLUSIONS

The results presented here demonstrated that high resolution data can be obtained with the photon counting pixel detector XPAD2. The gain in acquisition time can be more than a factor 50 with respect to conventional settings, with the same quality. The photon counting mode provides a very low noise and a wide dynamic range. The fast readout (2 ms between 2 images) allows to follow fast kinetics in reactions.

The next generation of chips is under design and will offer the same advantages, with some improvements. The submicron CMOS process is more radiation tolerant and makes a smaller pixel size possible. A special care was taken during the design to get a narrower threshold dispersion. To be able to use either Si or CdTe as detection substrate, two chips were designed. With both of them, it will be possible to select a continuous readout mode with a negligible dead time between two consecutive images.

REFERENCES

1. C. Ponchut *et al.*, *Nucl. Instr. and Meth.* **A 484** (2002)396.
2. C. Broennimann et al., *J. Synch. Rad.* (2006) **13**, 120.
3. J.-F. Bérar *et al.*, J. App. Cryst. (2002) **35**, 471.
4. N. Boudet *et al.*, *Nucl. Instr. and Meth.* **A 510** (2003)41
5. J.-F. Bérar *et al.*, to submitted to *J. Synch. Rad.*

HARP: A Highly Sensitive Pickup Tube Using Avalanche Multiplication in an Amorphous Selenium Photoconductive Target

Kenkichi Tanioka

NHK Science & Technical Research Laboratories, 1-10-11 Kinuta, Setagaya-ku, Tokyo 157-8510, Japan

Abstract. We have recently developed a greatly improved version of the HARP pickup tube with a selenium target 25-μm thick. When this advanced pickup tube is operated with an avalanche-mode at a target voltage of 2500 V, its sensitivity is about 60 times(195A/W) as great as the conventional HARP pickup tube with a target 2-μm thick. The HARP handheld camera equipped with the new pickup tubes has a maximum sensitivity of 11 lx at F8. This means that the HARP camera is about 100 times as sensitive as that of a CCD camera. This ultrahigh-sensitivity HARP camera is a powerful tool for reporting breaking news at night and other low-light conditions, the production of scientific programs, and numerous other applications..

Keywords: HARP, pickup tube, amorphous selenium, avalanche multiplication, high sensitivity, imaging device
PACS: 73.50. Pz, 73.61.Jc, 79.20.Hx

INTRODUCTION

To meet the strong demand for a television camera with ultrahigh-sensitivity for broadcasting, we have been studying a very sensitive image sensor since the early 1980s. In 1985, the author found for the first time that an experimental pickup tube with an amorphous selenium photoconductive target exhibited high sensitivity with excellent picture quality because of a continuous and stable avalanche multiplication phenomenon[1]. We named the pickup tube with an amorphous photoconductive layer operating in the avalanche-mode "HARP": High-gain Avalanche Rushing amorphous Photoconductor.

In 1987, we developed a practical pickup tube that consisted of a selenium target doped with impurities 2-μm thick[2]. The tube had sensitivity about 10 times greater than that of ordinary pickup tubes, such as SATICONs.

We have recently developed a greatly improved version of the HARP tube with a selenium target 25-μm thick because sensitivity as a function of the target's electric field increases with the target thickness. This improved version is about 60 times as sensitive as the conventional HARP tube, or about 100 times as sensitive as CCDs. In this article, the target structure and the fundamental characteristics of the newly developed HARP pickup tube and its camera are described.

OPERATIONAL PRINCIPLE OF THE TUBE

An operational representation of the HARP tube is shown in Fig. 1. The light energy absorbed in the selenium target generates an electron-hole pair. The carriers are accelerated by a large electric field, 10^8 V/m, then the hole, which has increased kinetic energy, generates a new electron-hole pair by means of impact ionization. This phenomenon occurs again and again throughout the target. The additional noise produced by the avalanche multiplication is negligible, so that the tube has high sensitivity.

CP879, *Synchrotron Radiation Instrumentation: Ninth International Conference*,
edited by Jae-Young Choi and Seungyu Rah
© 2007 American Institute of Physics 978-0-7354-0373-4/07/$23.00

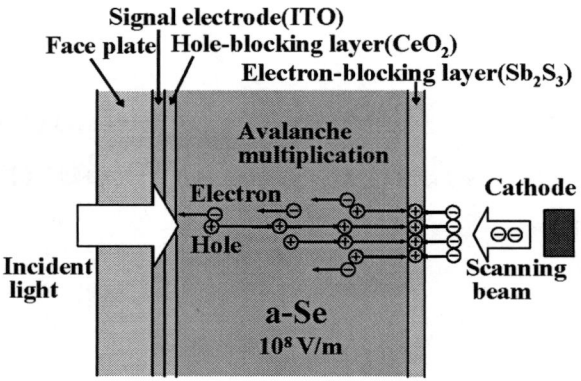

FIGURE 1. Operational representation of the HARP pickup tube.

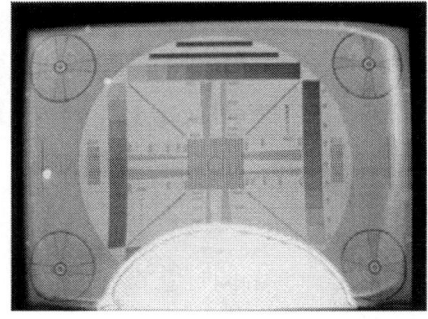

FIGURE 2. Spurious images caused by secondary electron emission.

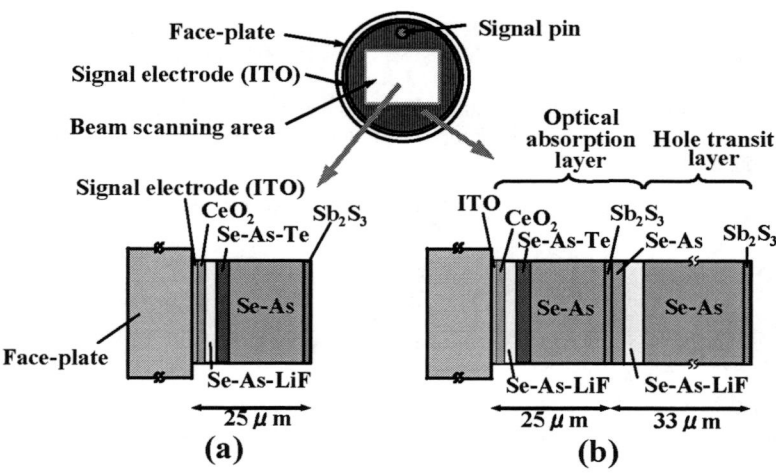

FIGURE 3. Structure of the HARP target.
(a) inside beam scanning area
(b) outside beam scanning area

TARGET STRUCTURE AND FABRICATION

Except for the thickness and the amount of tellurium in the selenium layer, the target structure inside the beam scanning area of the HARP is almost the same as in the conventional HARP target. A thin region of the selenium layer next to the signal electrode was doped with a lithium fluoride to weaken the internal field between the signal electrode and selenium layer by the positive space charge[2]. However, a newly designed structure outside the scanning area of the target stabilizes the beam scanning inside the scanning area even when the applied target voltage is very high. We increased the thickness of the target to 25-µm; therefore, the applied target voltage had to be raised to about 2500 V in order to cause avalanche multiplication. Outside the beam scanning area of the target, the surface potential rose to almost the target voltage. Because of this, a very large difference in surface potential appeared between the outside of the beam scanning area and the inside, where it only rose to about 20 V at most. This distorted the picture quality at the edges as the electron beam was bent toward the higher potential, and the increase in secondary electron emissions from the target caused spurious images to appear as shown in Fig. 2. To solve these problems, the new insensitive target structure was designed and fabricated outside of the beam scanning area. The structure of the insensitive target is compared in Fig. 3 to that of the sensitive target inside the beam scanning area. Basically, the insensitive target (Fig. 3(b)) was fabricated by adding a selenium layer doped with arsenic to the sensitive target (Fig. 3(a)). Lithium fluoride was doped near the boundary surface of the extra layer.

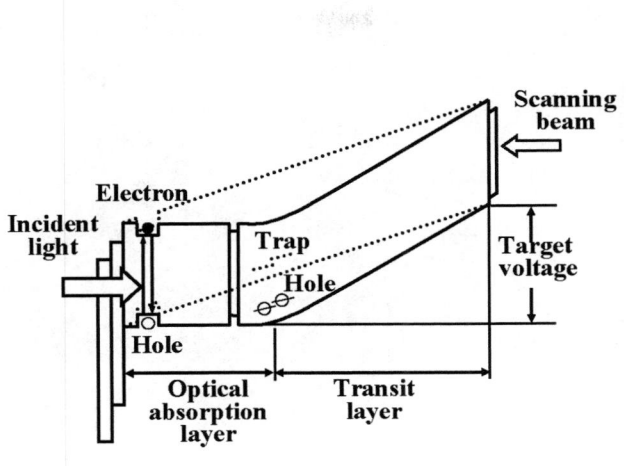

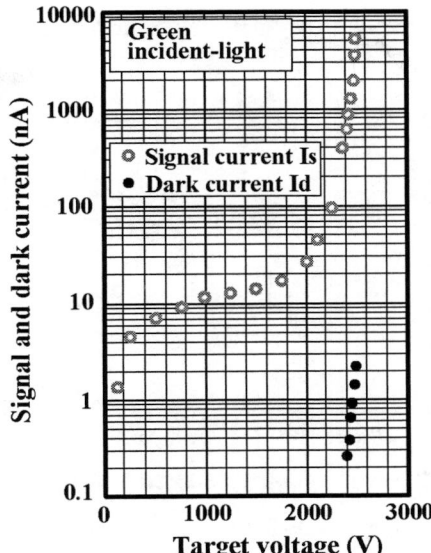

FIGURE 4. Band model of the insensitive target.

FIGURE 5. Signal current and dark current vs. target voltage in the HARP pickup tube.

A band model of this insensitive target is shown in Fig. 4. It consists of an optical absorption layer and a transit layer for holes. Initially, a uniform electric field is applied through the target (the broken line of Fig.4). Once incident light has illuminated the target, positive space charges are formed in the selenium layer doped with LiF in the extra layer. This reduces the internal electric field almost to zero, making the excited carriers disappear through recombination. The electric field of the transit layer for holes is kept less than 8×10^7 V/m by making it about 30 % thicker than the optical absorption layer. No avalanche multiplication effect occurs at this electric field intensity. The sensitivity outside the bean scanning area was measured. It was confirmed that the sensitivity there decreased to $1/10^6$ of the sensitivity inside the beam scanning area. As mentioned above, the new insensitive target introduced outside the beam scanning area made it possible to prevent an increase in surface potential. No picture distortion or spurious images were observed even with 2500 V applied to the target, so stable operation of the HARP tube with a target 25-μm thick is dramatically assured.

As shown in Fig. 3, a thin region of the selenium layer next to the signal electrode was doped with tellurium to increase the quantum efficiency for green and red incident light. The concentration of tellurium in the layer is about 15 % by weight. The thickness of the selenium layer doped with tellurium is about 120 nm.

FUNDAMENTAL CHARACTERISTIC

Figure 5 shows signal current and dark current versus target voltage in the HARP pickup tube. The incident light was green. The signal current rapidly increased at target voltages of more than 1800 V. This phenomenon resulted from avalanche multiplication in the selenium layer of the target. The figure shows that an avalanche multiplication factor of several hundred can be obtained at a target voltage of 2500 V. The sensitivity of the tube rises in proportion to a rise in the multiplication factor because the signal current is proportional to the multiplication factor in the avalanche-mode region. The dark current also increases in the avalanche-mode region. However, at a target voltage of 2500 V(sensitivity; 195A/W), the dark current is as little as approximately 2 nA.

Figure 6 shows the spectral response characteristics of the HARP tube. In order to show the increased quantum efficiency achieved for it by using a selenium layer doped with tellurium, that layer is compared to a selenium layer without tellurium in Fig. 6. The quantum efficiency of the HARP tube for green incident light (wavelength of 540 nm) was found to be double that of the selenium layer without tellurium. We estimated that the signal to shot-noise ratio was improved by 3 dB.

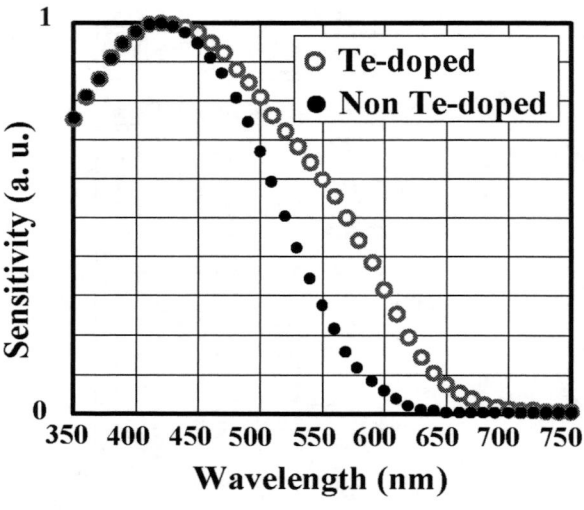

FIGURE 6. Spectral response characteristics of the HARP tube.

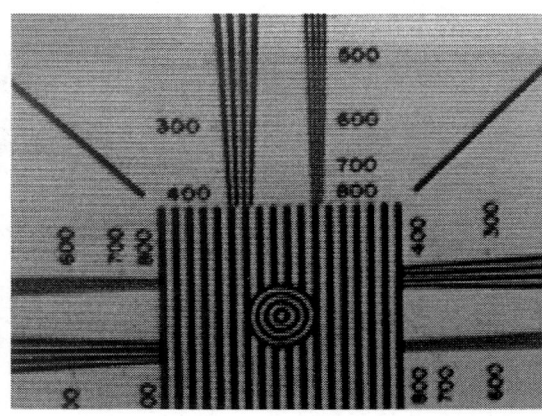

FIGURE 7. Monitor picture of the EIAJ test chart-A reproduced with the HARP tube.

TABLE 1. Specifications of the HARP camera.

Maximum sensitivity	11 lx, F8
Minimum scene illumination	0.03 lx (F1.7, +24 dB)
Signal-to-noise ratio	59 dB
Resolution	800 TV lines
Amplifier gain selection	0, +9 dB, +24 dB
Weight	5 kg
Power consumption	approx. 25 W

FIGURE 8. Appearance of the HARP color camera for NTSC.

The limiting resolution, limited size of the beam, was more than 800 TV lines, as shown in Fig.7. With regard to the lag characteristics, the decay lag in the third field after the incident light was turned off was negligible. This is because the target of the tube has a very small storage capacitance of about 130 pF due to the increased thickness. A thicker target provides a great improvement not only in sensitivity but also in lag.

HARP HANDHELD CAMERA

An ultrahigh-sensitivity HARP camera equipped with the new tubes has been developed. Its appearance is shown in Fig. 8 and its major specifications in Table 1. The target voltages were adjusted to about 2500 V for each channel. Figure 9 (a) shows a monitor picture produced by the three-tube HARP camera. The illumination is 0.3 lx and the lens iris is at F1.7. To illustrate the big difference in sensitivity between the HARP camera and a CCD

(a) Image taken with the HARP camera.

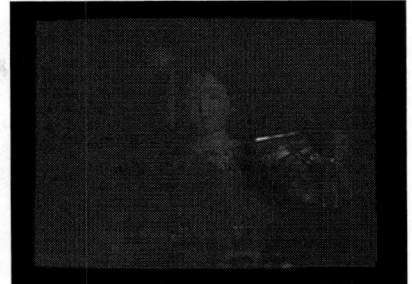

(b) Image taken with a CCD camera (+18dB).

FIGURE 9. Monitor pictures produced by color cameras (NTSC) with HARP tubes and CCDs. Illumination is 0.3 lx and lens irises are at F1.7.

(a) Image taken with the HD HARP camera.

(b) Image taken with an HD CCD camera (+42dB).

FIGURE 10. Monitor pictures produced by HD cameras with HARP tubes and CCDs. Lens irises are at F1.7.

Fig. 9 (b) shows a picture taken under the same conditions with a three-CCD camera (+18 dB). In spite of the dim lighting, the picture produced by the HARP camera is very clear, but a doll in the picture taken with the CCD camera looks like a ghost because of lack of sensitivity. It was confirmed that the HARP camera has a maximum sensitivity of 11 lx at F8. This means that the HARP camera is about 100 times as sensitive as the CCD camera. An ultrahigh-sensitivity HD (High-Definition) HARP camera has also been developed. Fig.10 (a) shows a view of Tokyo at night taken with the HD HARP camera. Fig.10 (b) shows a view taken under the same conditions with an HD CCD camera with a gain of +42dB. The big difference in the noise performance between two cameras can be observed. The result of this camera test reveals that the additional noise of the HARP pickup tube is negligible.

These new HARP cameras can take color pictures of objects under conditions so dark that the objects are imperceptible to the naked human eye. It goes without saying that the sensitivity of the camera can be decreased by decreasing the target voltage, so that the camera is capable of producing excellent picture quality over a wide-range of shooting conditions from daylight to moonlight.

APPLICATIONS IN MEDICAL RESEARCH

The high sensitivity and superior picture quality of HARP cameras has also led to a considerable amount of interest from medical and scientific fields. This section describes how it is applied to research into X-ray medical diagnosis.

A notable example is the potential use of HARP cameras in next-generation X-ray medical diagnosis systems. This research has been done in cooperation with other organizations such as the Tokai University Medical Faculty group and the High Energy Accelerator Research Organization.

The X-ray equipment currently used in hospitals is only able to see large blood vessels with a diameter of at least 0.2–0.5 mm, but this study aims to make it possible to obtain clear images of blood vessels that are several times smaller. It has been said that if narrow blood vessels with a diameter of 0.1 mm or less can be imaged, then it should be possible to detect cancer earlier and make better diagnosis of conditions such as heart attacks and cerebrovascular disorders.

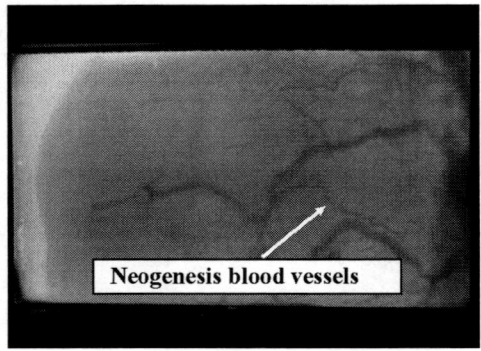

FIGURE 11. An image of minute blood vessels (mouse cancer)

For this purpose, it is necessary to have a special variety of X-rays that are absorbed well by a tiny quantity of contrast medium inside the narrow blood vessels to be imaged, and a TV camera that can clearly reproduce the image formed on a fluorescent screen (placed behind the subject being viewed) due to this absorption. For the special X-rays, we are using monochromatic X-rays with a specific energy obtained from synchrotron radiation. The TV camera is required to have superior sensitivity and resolution. This is because the image on the fluorescent plate is finely detailed and very dark (so as to restrict the exposure of the subject to X-rays).

We have therefore conducted experiments involving the use of an ultra-high sensitivity and high-resolution HD HARP camera in the imaging section of a next-generation X-ray medical analysis system. Figure 11 shows a photograph (obtained using this system) of tiny blood vessels called neogenesis blood vessels that developed in cancerous parts of a mouse. This image shows narrow blood vessels of a characteristic shape with a diameter of 0.1 mm or less, which it has not been possible to see hitherto. This technology is attracting interest as an X-ray diagnosis technique that can lead to the early detection of cancer.

In addition, since a HARP target can convert X-rays into electrons directly, it should be possible to exploit this capability to produce X-ray imaging devices with unparalleled levels of resolution and sensitivity. Consequently, this technology is attracting high levels of interest for applications such as X-ray area detectors for protein crystallographic analyses.

CONCLUTIONS

By increasing the target thickness, we have developed the HARP pickup tube that has better performance than the conventional HARP tube. A special insensitive target structure outside the beam scanning area was added to achieve stable operation at a very high target voltage of 2500 V. The HARP tube is about 100 times as sensitive as CCDs. A handheld camera equipped with the new tubes can serve as a powerful tool for reporting breaking news at night, for the production of scientific programs, and for other applications.

ACKNOWLEDGMENTS

I would like to thank everyone at Hitachi Ltd., Hamamatsu Photonics K.K. and Hitachi Kokusai Electric Inc. for their help in researching and developing the HARP pickup tubes and their cameras for many years.

REFERENCES

1. K. Tanioka et al., "An Avalanche-Mode Amorphous Selenium Photoconductive Layer for Use as a Camera Tube Target," IEEE Electron Device Letters, Vol. EDL-8, No. 9, pp.392-394, Sept. 1987.
2. K. Tanioka et al., "A highly sensitive camera tube using avalanche multiplication in an amorphous selenium photoconductive target." SPIE Vol. 1656 High-Resolution Sensors and Hybrid Systems, pp.2-12, 1992.

Development of Highly Efficient and High Speed X-ray Detectors Using Modern Nanomaterials

Marian Cholewa[a], Shu Ping Lau[b], Gao Xingyu[c], Andrew Thye Shen Wee[c],
Wojciech Polak[d], Janusz Lekki[d], Zbigniew Stachura[d], Herbert O. Moser[a]

[a]Singapore Synchrotron Light Source, National University of Singapore, Singapore
[b]School of Electrical & Electronic Engineering, Nanyang Technological University, Singapore
[c]NUSNNI, Physics Department, National University of Singapore, Singapore
[d]Henryk Niewodniczanski Institute of Nuclear Physics Polish Academy of Science, Krakow, Poland

Abstract. The secondary electron emission (SEE) yield of heterostructures of ZnO nanoneedles coaxially coated with AlN or GaN has been studied for the first time using electron, ion, and X-ray beams. The SEE yield of the heterostructures is enhanced significantly by the intrinsic nanostructure of the ZnO nanoneedle templates as compared to the AlN and GaN thin films on Si substrates. These findings open up a way to develop new universal highly efficient radiation detectors based on the SEE principle by incorporating these one-dimensional (1D) nanostructures as a material of choice.

Keywords: X-ray detector, nanomaterials, secondary electron emission.
PACS: 07.77.-u; 68.55.-a; 79.60.-i; 81.05.-t.

INTRODUCTION

The SEE yield enhancement is attributed to the larger area of the nanostructured surface. These findings open up a way to develop new highly efficient radiation detectors based on the SEE principle by incorporating these one-dimensional (1D) nanostructures as a material of choice. More generally, high SEE yield materials could be also valuable for vacuum devices, such as the protective layer of plasma display [1], cold cathode emitters [2] and electron multipliers [3].

Much work has been done with these new 1D nanostructure materials because of their excellent field emission properties [4]. There is a high potential for developing universal detectors based of these nanomaterials. These detectors will offer higher sensitivity than existing ones because of the higher efficiency in producing secondary electrons. A similar idea was previously developed for carbon foils [5], boron-doped diamond [6] and is now proposed for nanomaterials [4].

EXPERIMENTAL, RESULTS AND DISCUSSION

Sample preparation: ZnO nanoneedles were grown by catalyst-free MOCVD on Si substrates using diethylzinc (DEZn) and oxygen gas in the 400 − 500 °C range [7]. Following the fabrication of ZnO nanoneedles, around 10 nm thick AlN or GaN layers were deposited by low pressure metal-organic vapor phase epitaxy (MOVPE) directly on the ZnO nanoneedles using trimethyl-Ga (TMGa) and trimethyl-Al (TMAl)), and ammonia (NH_3) as precursors, respectively. Details of the growth conditions and structural characterizations of the coaxial heterostructures have been reported elsewhere [8].

CP879, *Synchrotron Radiation Instrumentation: Ninth International Conference*,
edited by Jae-Young Choi and Seungyu Rah
© 2007 American Institute of Physics 978-0-7354-0373-4/07/$23.00

We studied the SEE properties for several materials including Au, Boron-doped CVD diamond, silicon, silicon with coating (e.g. ZnO, GaN, AlN, MgO), carbon nanotubes (CNT), carbon nanotubes with coating (e.g. ZnO/CNT, MgO/CNT, AlN/CNT, GaN/CNT), ZnO nanorods, ZnO nanorods with coating (e.g. AlN/ZnO, GaN/ZnO). A schematic diagram of the experimental set-up is presented in Fig. 1.

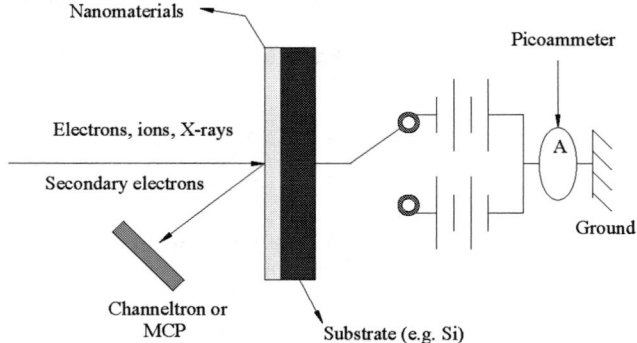

FIGURE 1. Schematic diagram of the experimental set-up and new detector with nanomaterials. Secondary electrons generated by the electrons, ions and X-rays are detected with an electron multiplier (channeltron) or a micro-channel-plate (MCP) detector connected with the proper electronics [9]. SEE yield with electrons was characterized by measuring of the current [10]. The channeltron could offer the rising time of the signal down to 2 ns while the MCP down to 250 ps respectively which will allow to develop a very fast detection system assuming a fast release of secondary electrons from the sample.

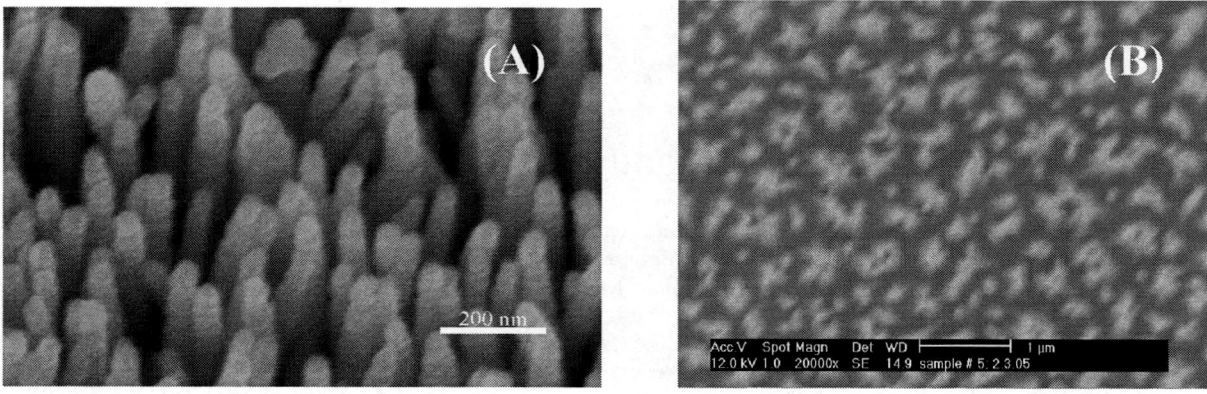

FIGURE 2. SEM image of the GaN/ZnO coaxial heterostructure (Fig. 2A). Secondary electron emission (SEE) image collected under electron microscope (Fig. 2B) shows area of good (light spots) and worse (dark spots) emission properties.

Electrons: In SEE studies [11], an incident electron beam is used to produce secondary electrons inside a sample, and the yield and energy distribution of the secondary electrons emitted from the surface are analyzed to obtain information about the surface properties. The measurement of SEE yield was carried out in a scanning electron microscope (SEM, JEOL-JSM-5910LV) by measuring the specimen current after biasing the specimen in two separate experiments [11]. Figure 2A shows the SEM images of vertically aligned ZnO nanoneedles as well as GaN/ZnO and AlN/ZnO coaxial heterostructures. As compared to the ZnO nanoneedles, the coaxial heterostructures exhibit flat or round tip ends instead of the original sharp tips. Furthermore, it is known that the III-nitride layers are not only deposited on the tips of ZnO nanoneedles but also uniformly along the sidewall of the ZnO core nanoneedles [8]. From samples of AlN and GaN films deposited on flat Si substrates under the same conditions, the thickness of AlN and GaN films is estimated to be 10-13 nm and 7-10 nm, respectively. Therefore, the diameter of AlN/ZnO coaxial heterostructures is expected to be slightly larger than that of GaN/ZnO. Figure 2B shows a 2-dimensional image of the relative SEE yield distribution featuring bright spots of larger yield in darker areas of smaller yield.

Ions: In order to characterize the SEE for ions we have applied the 2.0 MeV H$^+$ beam at the ion microbeam facility at IFJ PAN, Krakow, Poland. In this case the beam was focused to about 15 μm diameter and scanned over 200x200 μm^2 area. We measured the shape of the secondary electron yield by detecting the signal from the channeltron [9]. We

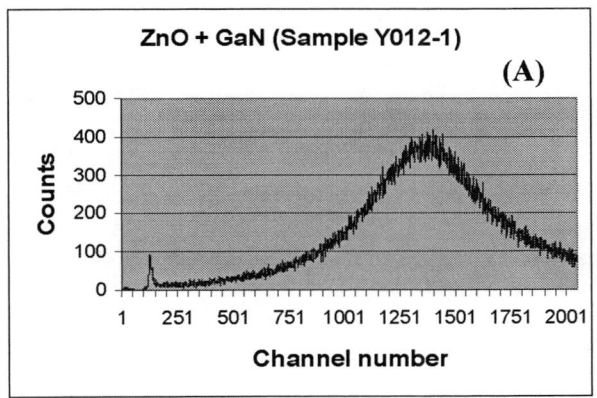
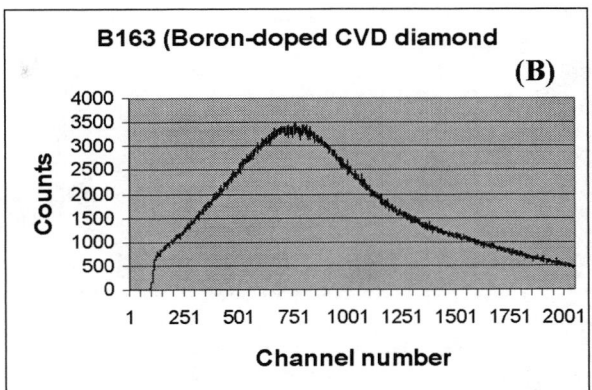

FIGURE 3. Secondary electron spectra for GaN/ZnO (A) and Boron-doped CVD diamond (B) measured with a channeltron detector. Bombarding ions: 2.0 MeV protons. Both spectra have been measured under identical conditions using +2.4 kV operating voltage for the channeltron plus an attracting bias of +180 V. The horizontal axis corresponds in a nonlinear way to the SEE yield from the sample. The vertical axis shows the number of counts registered by the channeltron.

performed experiments on selected samples. The results are presented in for GaN/ZnO (Fig. 3A) and Boron-doped CVD diamond (Fig. 3B). It is clearly visible that the spectrum for the GaN/ZnO heterostructure has a peak at about channel 1,400 indicating a higher SEE yield as compared to Boron-doped diamond which has a peak at about channel 750. The fact that the spectrum for GaN/ZnO extends also to lower channels indicates that the SEE yield from the

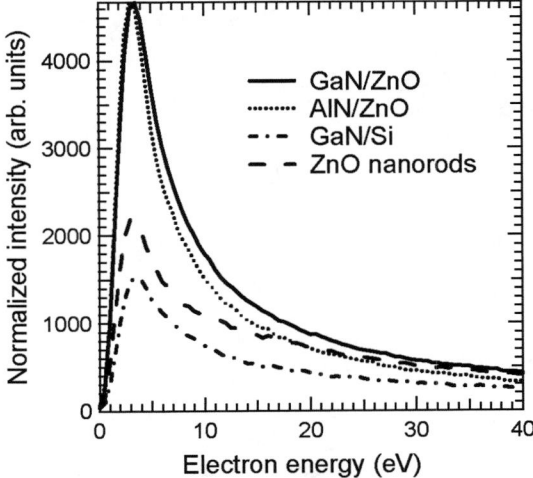

FIGURE 4. XPS (X-ray Photoemission Spectroscopy) spectra of 4 different samples at 1 keV photon energy. The energy of the electrons was measured between 0 and 40 eV. The data for different samples were normalized to the electron beam current of the storage ring, i.e., to the incident photon flux. No bias on the sample.

sample has not been sufficiently high to be separated from the electronic noise of the channeltron. The Boron-doped diamond sample is selected for comparison as it has been described in detail in several publications [9] . It should be noted that the channeltron detector at 2.4 kV voltage is not strictly linear because its amplification tends to saturate at high electron current. This means that the improvement in the SEE yield from GaN/ZnO to Boron-doped diamond may be greater than it appears from Fig. 3A. The channel number (horizontal axis) is related to the SEE yield in a non-linear way and this dependence has not been established for the system at IFJ PAN. The vertical axis corresponds to the number of counts registered by the channeltron detector.

X-rays: In order to investigate the SEE emission for X-rays we used the SINS (surface-, interface- and nanoscience) beamline at the SSLS in Singapore, details of which can be found in ref 12. We performed X-ray photoemission spectroscopy (XPS) using a beam of 1 keV X-rays. The results of the SEE yield are shown in Fig. 4 as a function of the electron energy. It is clearly visible that AlN/ZnO exhibits one of the highest SEE yields.

CONCLUSIONS

Conclusions: After careful analysis of data for many different materials under irradiation of electrons, ions and X-rays we found that AlN/ZnO and GaN/ZnO produced much better SEE yield than other materials investigated. The proposed new nanomaterials are highly efficient emitters of secondary electrons and are much better than diamond or metals (e.g. gold) which have been routinely used in the past as efficient SEE. In conclusion, the AlN/ZnO and GaN/ZnO coaxial heterostructures have a higher SEE yield than AlN and GaN deposited on Si substrates. These results suggest that the enhancement of SEE in the coaxial heterostructure is due to the combined effect of III-nitride layer and nanostructure. However, the understanding of the mechanisms that contribute to its excellent properties is still limited.

As presented in Fig. 1 the secondary electrons generated from the sample are collected either with the channeltron or MCP. Channeltron could offer a 2 ns and MCP down to 250 ps of rising time for the output signal which will make the detector very fast assuming that the act of secondary electron emission from the sample is fast too. These detectors could be used in fast time-of-flight systems where time resolution below 100 ps could be achieved by using only a small fraction of the output signal. However, several issues including uniformity of the surface still need to be addressed. While most of materials used in radiation detectors are good only for detection of one type of radiation (e.g. ions, or electrons or X-rays) these new nanomaterials seems to have higher SEE efficiency for all three types of radiation used in our research and this property alone could make them extremely useful in detectors development.

ACKNOWLEDGMENTS

This work was supported by NTU RGM 17/04. Work partly performed at SSLS under NUS Core Support C-380-003-003-001, A*STAR/MOE RP3979908M and A*STAR 0121050038 grants.

REFERENCES

1. T. J. Vink, R. G. F. A. Verbeek, V. Elsbergen, and P. K. Bachmann, *Appl. Phys. Lett.* **83**, 2285-2287 (2003)
2. J. E. Yater, A. Shih and R. Abrams, J. Vac. Sci. Technol., *A* **16**, 913-918 (1998).
3. W. S. Kim, W. Yi, S. Yu, J. Heo, T. Jeong, J. Lee, C. S. Lee, J. M. Kim, H. J. Jeong, Y. M. Shinm and Y. H. Lee, Appl. Phys. Lett. **81**, 1098-1100 (2002).
4. M. Cholewa, S. P. Lau, G.-C. Yi ,J. K. Yoo, A. P. Burden, L. Huang, X Gao, A. T.S. Wee, H. O. Moser, "Radiation Detector Having Coated Nanostructure and Method", United States Patent Application Serial No.: 11/129,582 (filed on May 13th, 2005).
5. C. Signorini, G. Fortuna, G. Prete, W. Starzecki, A. Stefanini, *Nucl. Instrum. & Meth.* **A224,** 196-223 (1982).
6. T. Kamiya, M. Cholewa, A. Saint, S. Prawer, G.J.F. Legge, J.E. Butler, D.J. Vestyck, *Appl. Phys. Let.,*. **71**, 1875-1877 (1997).
7. W. I. Park, G. C. Yi, M. Kim, and S. J. Pennycook, *Adv. Mater.* **14**, 1841-1843 (2002).
8. S. J. An, W. I. Park, G. C. Yi, Y. J. Kim, H. B. Kang, and M. Kim, *Appl. Phys. Lett.* **84**, 3612-3614 (2004).
9. B.E. Fischer, M. Cholewa, H. Noguchi, *Nucl. Instrum. & Meth.* **B181,** 60-65 (2001).
10. R. Kalish, V. Richter, E. Cheifetz, A. Zalman, P. Yona, *Appl. Phys. Lett.* **73,** 46-48 (1998).
11. S. P. Lau, L. Huang, S.F. Yu, H. Yang, J. K. Yoo, S.J. An, G-C. Yi, Small, Vol. 2, No. 6 (2006) 736-740.
12. Yu XJ, Wilhelmi O, Moser HO, Vidyaraj SV, Gao XY, Wee ATS, Nyunt T, Qian HJ, Zheng HW, J. of Electron Spectroscopy and Related Phenomena **144**, 1031-1034(2005).

The DiagOn : an Undulator Diagnostic for SOLEIL Low Energy Beamlines

K. Desjardins[*], S. Hustache[*], F. Polack[*], T. Moreno[*], M. Idir[*], J.-M. Dubuisson[*], J.-P. Daguerre[*], J.-L. Giorgetta[*], S. Thoraud[*], F. Delmotte[¶], M.-F. Ravet-Krill[¶]

[*] *Synchrotron SOLEIL, L'Orme des Merisiers, Saint-Aubin, BP48, 91192, GIF-sur-YVETTE, France*
[¶] *LCFIO, Institut d'Optique, Centre Scientifique Bât. 503, 91403 Orsay, France*

Abstract. DiagOn is a diagnostic tool intended to precisely define the emission axis of an undulator, that has been developed by SOLEIL synchrotron Optics and Detectors groups. It uses the particular structure of monochromatic undulator emission into a narrow hollow cone centered on the axis. One wavelength is selected by a multilayer mirror to produce a characteristic image of the emission cone on scintillator and CCD camera. We review the emission properties of undulator in order to define the best parameters, K value, photon energy and emission angle, for diagnosis. Then the choice of optimal multilayer coating of the mirror is discussed. We finish by some consideration on mirror cooling and mechanical conception for stable operation. The precise localization of the undulator axis will allow an optimal and fast alignment of the elements located before this detector, particularly adjustable aperture from the head of each of the lines.

Keywords: undulator, diagnostic, multilayer, energy selection, alignment.
PACS: 07.85.Fv, 07.85.Qe

INTRODUCTION

It is well known that in order to reduce the thermal load on the first optical elements of a beamline, it is useful to reduce the collection aperture to a small angle close to the characteristic divergence of the chosen harmonics. In this condition, the collected spectrum is reduced only to narrow peaks corresponding to the on axis harmonics and unwanted energy in between is removed. However photon energy selection not only requires that the undulator is continuously tuned, but also that the aperture is exactly centered on the emission axis otherwise spectral distortions are observed. The aim of the DiagOn device is to experimentally determine the undulator axis and allow optimal and fast alignment of the adjustable diaphragms from the front-end of each of the beamline. DiagOn will equip the first soft X-ray beamlines installed at SOLEIL (Cassiopée, Tempo, Désirs in phase 1 and more in phase 2).

PRINCIPLE OF OPERATION

The spatial distribution of the white beam emitted by an undulator is a broad Gaussian cone. At the scale of the aperture of beamline, there is no position sensitivity to check the emission axis, as can be seen on Fig.1, showing a simulation of the photon flux coming out of an insertion device at 20 m from the source, position of the DiagOn.

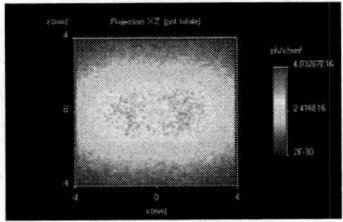

FIGURE 1. Photon flux emitted in undulator HU256, Cassiopée beamline, obtained by simulation (SpotX software[1])

CP879, *Synchrotron Radiation Instrumentation: Ninth International Conference,*
edited by Jae-Young Choi and Seungyu Rah
© 2007 American Institute of Physics 978-0-7354-0373-4/07/$23.00

The position sensitivity required for the determination of the emission axis is obtained by spectral filtering in the DiagOn diagnostic itself (Fig.2). A single wavelength of the white beam is selected by reflection on a multilayer mirror, and directed towards a scintillator. X-rays are then converted into visible light which is observed by a CCD camera with focusing optics.

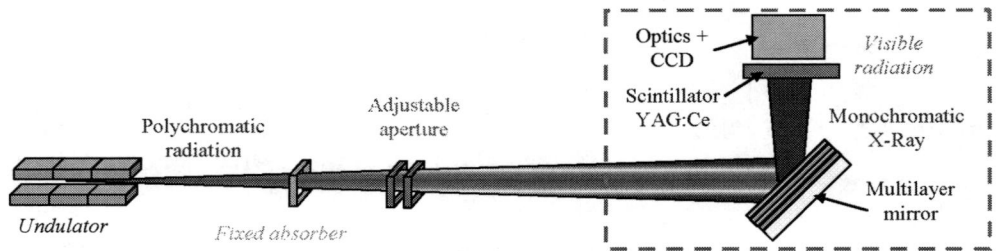

FIGURE 2. Schematic description of the DiagOn's operation principle with the different elements participating to the diagnosis: undulator, adjustable diaphragms and DiagOn: multilayer mirror, scintillator (YAG(Ce)), focusing optics and camera.

Properties of Undulator Emission

To determine precisely the propagation axis, DiagOn uses the undulator's emission properties. In the continuous spectrum emitted, there is a univocal relationship between E, energy of the emitted photons and θ, their angle of emission with respect to the axis. This relationship appears in the fundamental formula for emission in an insertion device [2]:

$$K^2 = n \times \frac{1900\ E_0^2[GeV]}{\lambda_U[cm]\ E[eV]} - 2\ (1 + \gamma^2\theta^2[rad])$$ (1)

where K is the undulator deflection parameter (which is function of the magnetic field between the undulator's jaws), n is the harmonic number, E_0 is the electron beam energy, λ_u is the undulator's period.

For a given K, photons of energy E are emitted on a cone of aperture θ for the nth harmonics of the insertion device, centered on the undulator axis. Consequently, with the selection of one particular wavelength of the emission spectrum, one can extract the corresponding cone. The projection of this cone at the detector position is a ring, whose center is the emission axis. To get the maximum position sensitivity on the position of the center, the energy selection has to be performed within a narrow bandwidth, and this can be achieved by a multilayer mirror.

Optimization of Energy Selection with a Multilayer Mirror

Multilayer coating has a highly chromatic comportment that allows, under certain conditions, to obtain by reflection a quasi monochromatic beam. The wavelength selection follows the Bragg law and depends on two criteria: the beam incidence angle on the mirror θ_m, and the period d of multilayer coating.

$$n_m \lambda = 2d \sin \theta_m$$ (2)

where λ is the wavelength reflected and n_m is the order of the reflection.

DiagOn's multilayer mirror parameters (materials and layer's thickness to get the best reflectivity and bandwidth) have therefore to be optimized with respect to the beamline characteristics, mainly the energy range. In case of a beamline using two insertion devices (Cassiopée and Tempo for instance), the multilayer should have a maximum reflectivity in the range covered by the two sources.

The selection of the multilayer is performed through simulations of the expected photon flux emitted by the undulator folded with the reflectivity of the mirror with different technically feasible multilayer coatings. An example of these optimizations is given in Fig.3, for HU256 on Cassiopée beamline, where energy selection around 180 eV will be performed by a MoB_4C multilayer.

Figure 4 shows two simulations obtained with SpotX software, illustrating what would be observed through the DiagOn, on Cassiopée. The circles corresponding to the first and the second harmonics of the undulator are clearly visible. These circles are defined by n = 1 or 2 in equation (1), which implies the same energy but different angles of emission θ and therefore two distinct rings selected by the multilayer mirror. The emission axis of the insertion device is given by the center of the concentric rings.

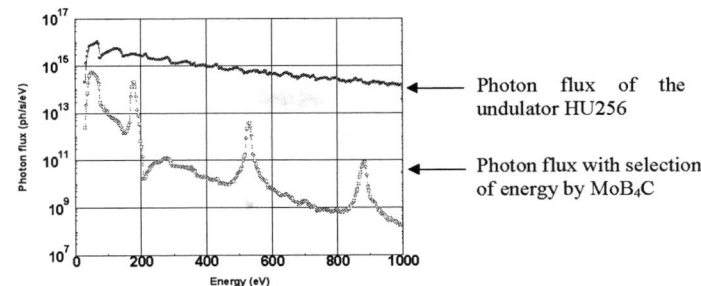

FIGURE 3. Calculation of photon flux as a function of energy of the undulator HU256 on Cassiopée and the same photon flux folded by the reflectivity of a MoB$_4$C (d=25 Å, 50 periods) multilayer calculated with XRV software[1].

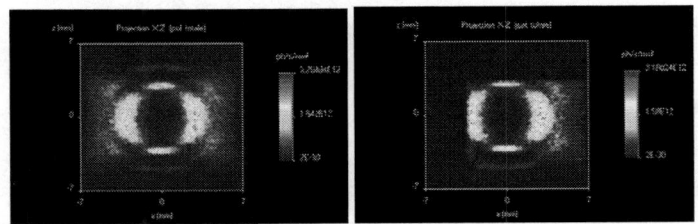

FIGURE 4. Simulations (SpotX software [1]) of photon flux reflected by multilayer mirror at the position of DiagOn (20m from the source) obtained with an aligned (left) and misaligned (right) adjustable aperture in the front end.

PRESENT STATUS OF THE DEVELOPMENT

The different elements of the detector were selected for each phase 1 beamline according to the parameters of the SOLEIL source and the emission properties of the different undulators. The main parameters are summarized in Table 1. Choices of the multilayer mirrors were done in collaboration with the LCFIO which is in charge of their fabrication.

TABLE 1. Results of simulation for phase 1 beamlines

Beamlines/Undulator	DESIRS/HU640	TEMPO/HU80	CASSIOPEE/HU256
Multilayer selected	Mo(50A°)/Si(50A°)	Mo(25A°)/B4C(25A°)	Mo(25A°)/B$_4$C(25A°)
Energy selected	95eV	550eV (3rd order of diffraction)	180eV
Rate of reflectivity	~70%	~5%	~30%
Photon/s incident	$5,55 \times 10^{17}$	$3,14 \times 10^{17}$	$1,38 \times 10^{18}$
Photon/s reflected by multilayer mirror	$5,81 \times 10^{14}$	$1,72 \times 10^{13}$	$6,83 \times 10^{14}$

The first multilayers were deposited and analyzed in grazing incidence with an X-ray Cu tube. Figure 5 shows the first results obtained with a MoB$_4$C multilayer with 50 periods and 25Å thickness per layer.

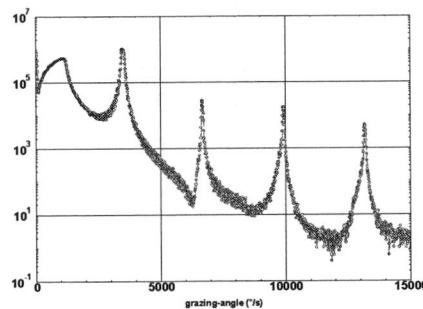

FIGURE 5. Reflectivity of a MoB$_4$C multilayer (50 periods) as a function of grazing angle for E=8keV. Multilayer period deduced from this curve is d= 49.78 Å and γ=0.48.

The mechanical design is complete. Since the device will be one of the first elements of the beamline and will be used in white beam, special care had to be taken for the cooling system. The maximum incident power during functioning of DiagOn will be 50 W on Cassiopée. The temperature of the multilayer mirror must not exceed 100°C, to avoid degradation of the reflecting surface. The Design Engineering group of SOLEIL therefore developed a water cooled holder able to maintain the mirror around 30°C. An indium/gallium interface will be used to ensure a good thermal contact between the silicon substrate of the multilayer and the holder. 3D images of the different parts of the DiagOn are shown in Fig. 6.

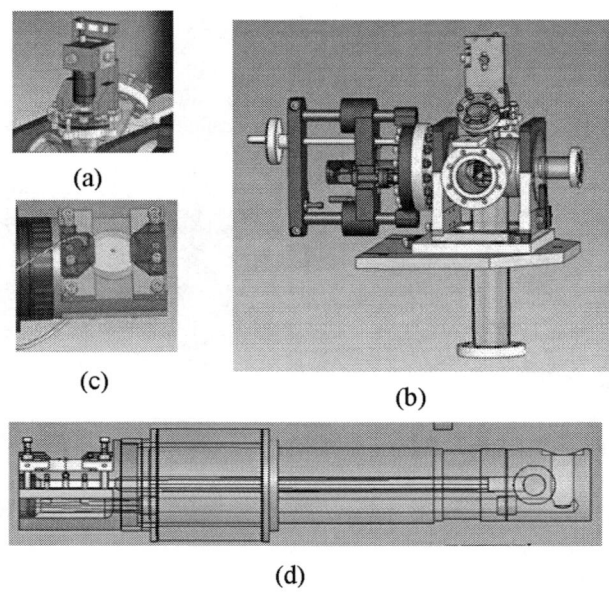

(a)

(c)

(b)

(d)

FIGURE 6. Final drawings of DiagOn (SOLEIL's Design and Engineering group). a) camera and focusing optics, b) general view, c) and d) multilayer mirror holder

SUMMARY

SOLEIL is developing a device for the low energy beamlines (VUV and soft X-rays) Cassiopée, Tempo and Désirs for the optimal and fast alignment of the adjustable apertures located in the front end. This detector, DiagOn, will be used in white beam. It consists of a water cooled multilayer mirror, a scintillator and a CCD camera with focusing optics. The design of this apparatus is complete and the first DiagOn will be installed on Tempo beamline in autumn 2006.

ACKNOWLEDGMENTS

The authors are especially grateful to Thierry Moreno for the access to XRV and SpotX softwares.

REFERENCES

1. Thierry Moreno, SpotX and XRV, private communication
2. K.-J. Kim, *Nucl. Instr. Meth.* **A 246** (1986)67.

Pulsed Neutron Monitoring at High Energy Electron Accelerators with Silver Lined Proportional Counter

P.M. Dighe[1], M.D. Ghodgaonkar[1], M.P. Dhairyawan[2] and P. Haridas[3]

[1]Electronics Division, [2] Personnel Monitoring Section,
Bhabha Atomic Research Centre,
Mumbai 400 085, India
[3] Raja Ramanna Centre for Advanced Technology, Indore, India

Abstract. To meet the challenging requirement of pulsed neutron background measurement, which is present around electron accelerators at the Indus-1 facility of the Raja Ramanna Centre for Advanced Technology (RRCAT) Indore, a silver lined proportional counter with $0.2 cps/n\ cm^{-2}s^{-1}$ thermal neutron sensitivity has been developed. The detector has been tested for its performance in continuous thermal neutron field at Apsara reactor and in pulsed neutron field at Indus-1 facility. The detector shows $\pm 11\%$ signal linearity at various reactor powers and follows the silver decay scheme during reactor scram experiment. Off-line measurements made in pulsed neutron background at the Indus-1 facility compare well with nuclear track detectors (CR-39). For monitoring on-line neutron flux, electronic gating circuit was used that can switch off the scalar counter unit during the prompt X-ray response of the detector taking trigger pulse from the accelerator and experiments showed that the neutron flux measured by the detector is in close agreement with CR-39 values.

Keywords: Pulsed neutron, Proportional counter, Electron accelerator, and Gating circuit.
PACS: 29.40.-n, 29.20.-c, 07.50.-e

INTRODUCTION

Pulsed neutron background is present around electron accelerator facilities, RRCAT, Indore. Here, the pulsed *bremsstrahlung* that is produced due to beam losses and scattering of electrons by the molecules of the residual gas in the vacuum pipes, gives rise to pulsed photoneutron flux [1] that needs to be measured for reasons of personal safety and monitoring the accelerator performance. Various experiments have been carried out with standard neutron-sensitive detectors employing materials such as ^{10}B or ^{3}He in pulsed photo neutron field at electron accelerators in Indus-1 facility and it has been reported that these detectors underestimate dose equivalent rate heavily in pulsed neutron fields of very low duty cycles [2]. Passive pulsed neutron monitoring using foil activation method such as Gold, Indium, Rhodium or Silver foils has been described in literature. On line pulsed neutron monitoring using silver lined or wrapped GM counters also has been described in literature. The GM based systems has certain limitations such as dead time, small resolution time and shorter life limits its use in locations where neutron flux is to be measured in the presence of intense X-ray background. Therefore for this requirement silver lined proportional counter has been developed [3] which is mechanically rugged and has all welded construction. The present paper describes tests and performance of silver lined proportional counter.

DETECTOR DESCRIPTION

A proportional counter with silver lining on the internal diameter of cathode was developed. The thickness of the silver foil was selected as 0.25mm [4]. The main objective of this development was to make a neutron area meter. Therefore, detector dimensions were selected similar to neutron counters used in Andersson and Broun Rem-meter. Table 1 gives the main specifications of the detector. Thermal neutrons interact with silver and produces short-lived beta activity (24.2s, 2.3m), which is detected by the detector.

Thermal neutrons mainly react with silver in following equation:
(i) $^{109}Ag + n \rightarrow \gamma + ^{110}Ag \rightarrow ^{110}Cd + \beta^{-}$, (2.24MeV or 2.82MeV)
(Half-life 24.2s, thermal neutron cross section 110 b)

CP879, *Synchrotron Radiation Instrumentation: Ninth International Conference*,
edited by Jae-Young Choi and Seungyu Rah
© 2007 American Institute of Physics 978-0-7354-0373-4/07/$23.00

(ii) $^{107}\text{Ag} + \text{n} \rightarrow \gamma + ^{108}\text{Ag} \rightarrow ^{108}\text{Cd} + \beta^{-}$, (1.49MeV) (Half-life 2.3min, thermal neutron cross section 30 b)

The energy deposition in the detector is due to beta decay of the silver, therefore spread in the pulse height distribution is observed. The resolving time of the counter is of the order of 0.8μsec and can be estimated from following equation by substituting numerical values. Pulse rise time T is given by [5]

$$T = [a / (a + b)] \{[(b^2 - a^2) \, p \ln (b/a)] / 2 \, \mu \, V\} \qquad (1)$$

Here **a** is the anode radius, **b** is cathode internal radius, **p** is gas pressure, μ is the mobility of ions in the gas and V is applied voltage.

TABLE 1 Main specifications of the silver lined proportional counter

Detector	Cathode (SS)	Anode	Neutron sensing material	Gas-fill	Insulators	Thermal neutron sensitivity
Silver lined P.C.	26mmID x 110mm length	25μ tungsten wire	Silver foil	Ar (90%) + CH$_4$ (10%) at 20 cm Hg	Ceramic to metal feedthroughs	0.2cps/n cm^{-2}s^{-1}

TESTS AND RESULTS

The detector was tested with pulse channel consisting of preamplifier, HV supply, linear amplifier and timer scalar. The neutron sensitivity of the detector was estimated by irradiating it for 15min in 150n cm^{-2}s^{-1} thermal neutron flux to obtain saturation activation and then thermal neutron sensitivity as 0.2cps/n cm^{-2}s^{-1} was derived at maximum counts observed above background divided by the neutron flux on the counter surface. The detector was tested at Apsara thermal column at variable reactor power and counts were measured at each power level waiting for 15m duration to reach the saturation value. Signal linearity of ±11% was observed from 20W to 300W of reactor power and the decay counts were measured at reactor scram follows the silver decay scheme (Fig. 1).

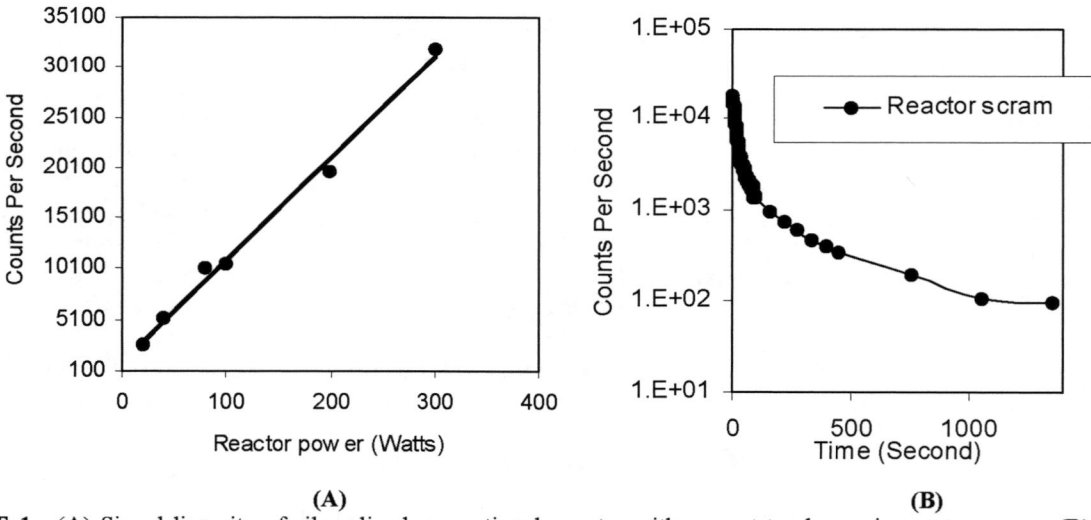

FIGURE 1. (A) Signal linearity of silver lined proportional counter with respect to change in reactor power. (B) Decay signal of silver lined proportional counter during reactor scram

The response of the counter to pulsed neutron background was evaluated in experiments carried out within shielded area at the Indus-1, RRCAT where the photo neutron background is pulsed with 6ns to 1μs duration and 1Hz repetition rate. The neutron pulse repetition rate is very short compared to the beta decay time of silver. The radioactive decay equation is $N = N_0 e^{-\lambda t}$ where N_0 is the initial number of atoms in a given substance, **N** is the remaining atoms at time **t** after radioactive decay and λ is the decay constant. By substituting numerical values it can be derived that more than 97% of the short-lived activity (24.2s) and 99.5% of long-lived activity (2.3m) activity remains in silver before another neutron pulse arrives from the accelerator and therefore the calibration factor derived from continuous radiation has been used to estimate the pulsed neutron field in detector location and all the counts were measured on 0.5V discriminator bias on which neutron sensitivity was derived.

Off line experiments were carried out at the Booster Synchrotron at TL2 location where the 450Mev electron beam was deflected to strike a 30mm thick copper target to produce X-rays and photoneutrons. The detector was placed at different distances from the target and was surrounded with 8cm thick high-density polyethylene annulus as moderator and was 1.2m height from the ground and 1.7m away from the shielding walls to avoid any scattered neutrons from the wall and ground. The counter was irradiated for 15min duration to reach up to saturation activity. During irradiation the counter responded to both X-ray and neutrons. After the accelerator was switched off the detector stopped responding to the X-rays. The counts were taken at regular intervals as the signal decreased (Fig. 2). Neutron flux in the detector location was estimated from the maximum counts observed and derived from the continuous thermal neutron calibration factor. Neutron flux was also estimated by passive CR-39 nuclear track detectors to corroborate the observed values. Theoretical neutron flux estimation is derived by following expression given by Swanson [6].

$$\text{Neutron yield Y (neutrons sec}^{-1}\text{ kW}^{-1}) = 1.21.10^{11}Z^{(0.66)} \qquad (2)$$

The neutron flux estimated by the silver counter is in close estimation with the CR-39 values and theoretically estimated values (Table 2).

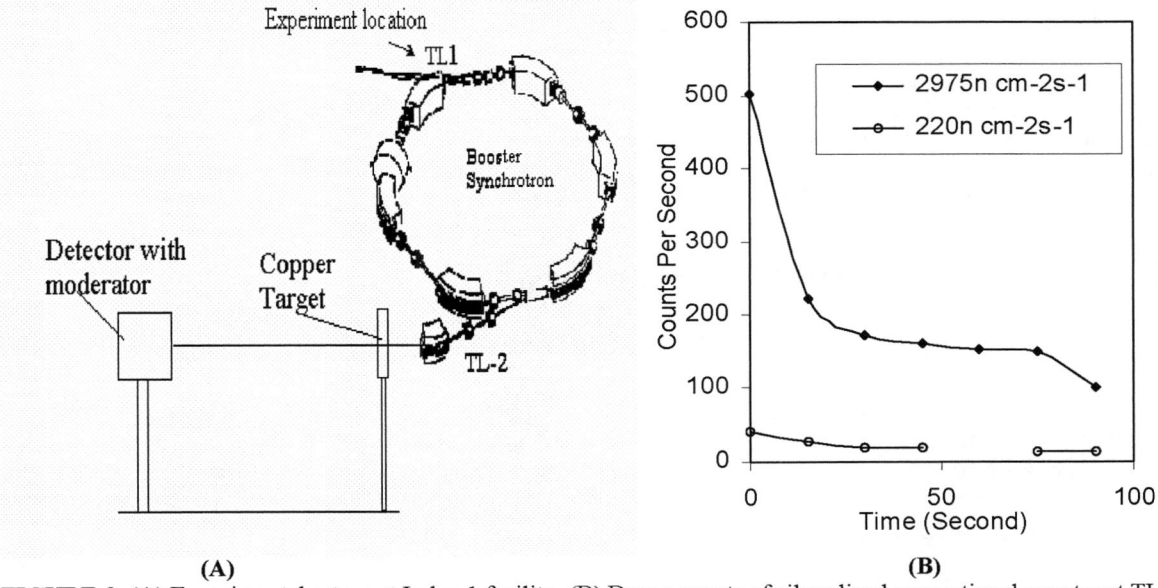

(A)   **(B)**

FIGURE 2. (A) Experimental set-up at Indus-1 facility. (B) Decay counts of silver lined proportional counter at TL2 off line experiment.

To make the detector capable of on-line monitoring of neutron flux electronic gating circuit (Fig. 3) was used that can switch off the scalar counter unit during the prompt X-ray response of the detector taking trigger pulse from the accelerator. The prompt X-ray counts are generated in coincidence with the electron pulse of the accelerator and hence by switching off the scalar counter during this time effectively discriminates the X-ray background counts. This helps to avoid response to prompt X-ray background, which would otherwise interfere with neutron signals. The off time of the timer out-put is controlled by the time delay selector switch and accelerator trigger pulse. The timer is restated by the trigger pulse and the timer remains off for the time selected by the time delay switch. This time can be conveniently selected up to 200μs time, which gives only 0.02%counting losses for 1Hz pulse repetition rate. After the selected time delay the counter starts counting pulses from amplifier and continues to count till next trigger pulse arrives. This process is continued for a selected time. An experiment was conducted at Indus-1 facility to confirm the usability of the gating circuit. The silver counter was placed in HDPE annuals as moderator and was placed at 1m distance from beam line near TL1 bending magnet during normal operation of the accelerator. The neutron flux in the detector location was estimated with CR39. Trigger pulse for gating was obtained from microtron and measurements were taken with and without gating. Before taking measurements the silver counter was first allowed to get irradiated for duration of 15min to reach saturation activity. Very high-count rate (~22times higher compared to counts observed with gating) was observed without gating, which is attributing due to X-ray interaction. Neutron flux was estimated from the counts observed using gating circuit minus background counts. Table 2 gives the results of the experiment carried with gating in pulsed neutron background and it was observed that the neutron flux estimated in the detector location is comparable to CR39 values.

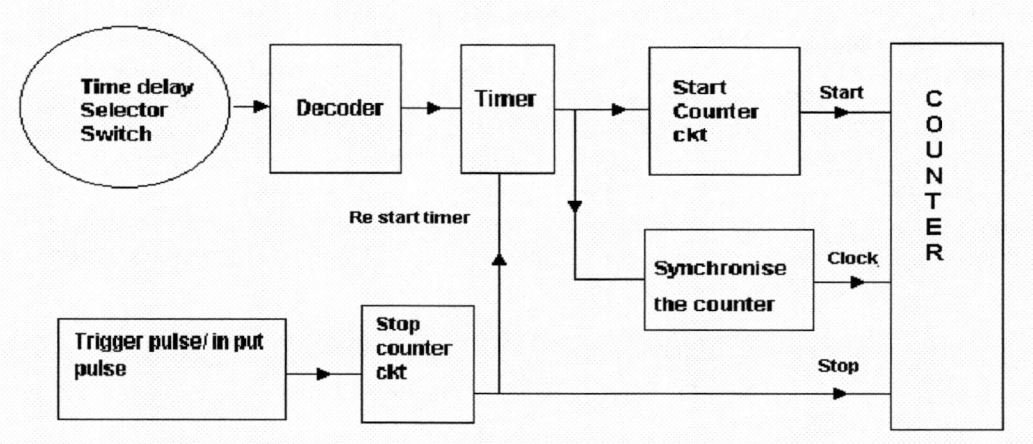

FIGURE 3. Schematic of gating circuit used to test silver lined proportional counter at Indus-1

TABLE 2 Neutron fluence measured with silver lined proportional counter

Detector position	Mode of operation	Silver counter	Neutron flux (n cm^{-2}s^{-1}) CR-39	*Theoretical estimation
At TL2	(OFF –LINE)Position 1	2511	2975	1772
	(OFF –LINE)Position2	207.5	220	-
At TL1	(ON LINE with gating)	50	56.4	-

* The theoretical estimation does not include scattering component.

CONCLUSIONS

The difficulty of monitoring pulsed neutron background has been greatly solved by developing a silver lined proportional counter. The counter is rugged and reliable and studies carried out on the detector shows that it works satisfactorily in continuous as well as in pulsed neutron background. In the pulsed neutron field the calibration of the counter holds good and the experiments carried out at Indus-1 facility indicated that the counter performance is comparable to the data obtained from CR-39 Solid State Nuclear Track Detector.

ACKNOWLEDGMENTS

The authors are grateful to Mr. U.V.Phadnis and his colleagues in RSSD, BARC, Mumbai and Apsara reactor, BARC superintendent and his colleagues for providing tests facilities for thermal neutron experiments. Thanks are also due to M/s Gurnam Singh, R.G. Marathe, K.K. Thakkar & G. Haridas for the experiments at RRCAT, Indore.

REFERENCES

1. K.N.Kirthi, S.D.Pradhan and M.R.Iyer, *Shield design aspects for a synchrotron radiation source facility*, Bulletin of Radiation Protection Vol.13, No.2, April-June 1990

2. Haridas. G et al, *Inter comparison of neutron detectors in pulsed photo neutron field* Radiation protection and environment Vol. 28, No. 1-4 p. 326-328

3. P.M. Dighe, K.R. Prasad and S.K. Kataria, *Silver lined proportional counter for detection of pulsed neutrons*, Nuclear Instruments and Methods in Physics Research (A) **523**, Issues 1-2, 1 May 2004, 158

4. P.R. Zurakowski and E.G. Shapiro, *Pulsed neutron detection system* UCRL-70170 rev. 1, UNCLS June 1971

5. G.F. Knoll, *Radiation Detection and Measurement* Third Edition pp. 181

6. Swanson W.P. (1979) *Improved calculation of photoneutron yields released by incident electrons*, Health physics, **37**, 347-358

Hiresmon: A Fast High Resolution Beam Position Monitor for Medium Hard and Hard X-Rays

Ralf Hendrik Menk[a], Dario Giuressi[a], Fulvia Arfelli[b], Luigi Rigon[c]

[a]Sincrotrone Trieste S.C.p.A, S.S. 14 km 163.5, 34012 Basovizza (TS), Italy
[b]Dept. of Physics - University of Trieste and INFN Via Valerio, 2
34127 Trieste, Italy
[c]The Abdus Salam International Centre for Theoretical Physics
Strada Costiera 11 34014 Trieste, Italy

Abstract. The high-resolution x-ray beam position monitor (XBPM) is based on the principle of a segmented longitudinal ionization chamber with integrated readout and USB2 link. In contrast to traditional transversal ionization chambers here the incident x-rays are parallel to the collecting field which allows absolute intensity measurements with a precision better than 0.3 %. Simultaneously the beam position in vertical and horizontal direction can be measured with a frame rate of one kHz. The precision of position encoding depends only on the SNR of the synchrotron radiation and is in the order of micro meters at one kHz frame rate and 10^8 photon /sec at 9 KeV.

Keywords: Ionization chamber, instrumentation for synchrotron radiation.
PACS: 29.40.Cs, 07.85.Qe

INTRODUCTION

The high resolution beam position monitor (XBPM) described in the following is specifically designed for high flux synchrotron radiation experiments such as x-ray diffraction (protein, SAXS, powder etc), EXAFS, fluorescence and x-ray imaging experiments, etc using medium hard or hard x-rays [from 5keV – 40 KeV]. Those classes of experiments require precise and simultaneous I_0 calibrations in the order of some percent down to some ppm for quantitative measurements. The availability of high intensity micro beams enables time resolved experiments on samples with spatial extensions of some tenth of microns, however results may alter tremendously when fast intensity variation of the beam intensity and / or beam vibration in vertical and horizontal direction occurs. Thus a high-resolution beam position monitor is essential for these classes of experiments.

DETECTOR SETUP

The (XBPM) is based on the principle of a segmented longitudinal ionization chamber [1] and depicted in Fig. 1. In contrast to traditional transversal ionization chambers here incident x-rays are parallel to the collecting field and thus have to cross all 3 electrodes (two high voltage cathodes and the segmented read out anode) and the two windows. The advantage of this configuration is three fold. Firstly in contrast to the conventional setup here the readout current is independent from the beam position, secondly the active conversion volume is well known and thus absolute numbers of absorbed photons can be quoted and thirdly the volume can be kept small which reduces recombination tremendously. As a side effect of the latter the XBPM can be operated with low drift fields and subsequently with low voltage. As mentioned above the x-rays have to cross 3 electrodes and two windows. In order to keep absorption and scatter effects small all high voltage electrodes as well as the windows are made of metallized (two layers of 400 Å Al) Kapton foils (25μm). In case of low energy x-rays thinner foils can be used. As depicted in figure 1 the segmented read out anode, which is arranged in 4-quarter segments, is placed in the middle

CP879, *Synchrotron Radiation Instrumentation: Ninth International Conference*,
edited by Jae-Young Choi and Seungyu Rah
© 2007 American Institute of Physics 978-0-7354-0373-4/07/$23.00

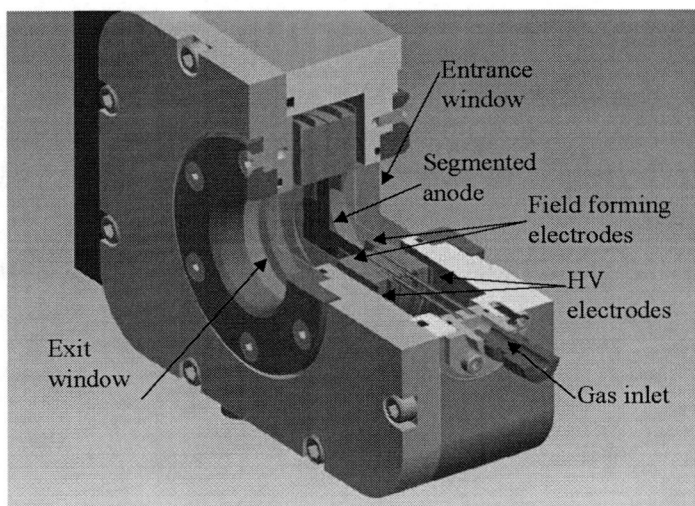

FIGURE 1. Sketch of the XBPM

of the XBPM and hold on virtual ground through the attached amplifiers (DDC112). At a distance of 1 cm the anode is bracketed by two high voltage cathodes. In order to keep the collecting field homogenous half the way between the anode and cathode field forming electrodes are placed. These are hold on half potential utilizing a voltage divider.

In order to avoid deformation of the high voltage electrodes (and subsequently deformation of the collecting volume) caused by Coulomb forces also the windows are metallized and hold on ground potential. Both windows feature the same distance to the HV cathodes as the cathodes to the anode. The XBPM is accommodated in a sealed aluminum vessel and flushed with the conversion gas (preferably N_2). In order to minimize external pick-up the analog part of the readout electronics is placed in the sealed housing as well. The analog readout is connected to the digital part through a sealed feed through. The entire electronics is build around the Burr Brown DDC 112, which is basically an analog integrator, combined with a 20-bit delta sigma converter. Each DDC 112 featured 2 channels and a single serial digital output. Two or more DDC 112 can be daisy chained keeping a single serial digital output. Utilizing thin coax cables the four quarter segments of the readout anode are connected to the four channels of two DDC112 and integrate for a predefined integration time the positive charges released by photo electrical absorption or a Compton Effect in the conversion gas. An on-board micro controller is used as a state machine generating all necessary initialization signals for the DDC 112 and the integration timing (from 0.5 ms to some s). Seven different gains can be set (ranging from 50 pC – 350 pC full scale range) in order to adjust the optimum working conditions to different experiments. Moreover the micro controller handles the communication to the outer world via RS 232 or USB 2. The input noise of the DDC 112 can be quoted with 10 ppm σ_{rms} at the highest gain for the typical input capacity of 200 pF of a quarter segments.

MODE OF OPERATION

As mentioned above the XBPM is an ionization chamber in the classical sense that measures ionization released in the conversion gas after a photoelectrical absorption or by a Compton Effect. As explained later, position encoding here is based on the localization of the barycenter of this ionization and is not based on the centroid of scattered radiation as used in other systems.

Since the XBPM is supposed to remain in the beam path it should be operated with light gases such as N_2 in order to keep absorption for low energy x-rays low. The choice of N_2 features another advantage since it is in ballistic equilibrium with knock-on electrons from the electrodes, which are subsequently released in the active volume and vice versa. Moreover it is a molecule, which can quench avalanche processes due to the emission of UV photons. In principle other gases and air can be used as well. The incident X-rays release after a photoelectrical absorption and thermalization of the photoelectrons typically $n_e = E_\gamma / W_{ion}$ electrons, where E_γ is the x-ray energy and W_{ion} is the mean energy to create a free electron / ion pair in the gas ($\sim$ 30 eV for N_2). Position encoding in the XBPM is based on the determination of the center of gravity of this extended charge distribution. If I_i are the currents measured on the four quadrant segments then

Below is an example equation created with Word 97's Equation Editor. To move this equation, highlight the entire line, then use cut and paste to the new location. To use this as a template, select the entire line, then use copy and paste to place the equation in the new location.

$$\sum I_i = \sum \varepsilon \cdot \frac{E_\gamma}{W_{ion}} \cdot \phi_i \cdot c = \varepsilon \cdot \frac{E_\gamma}{W_{ion} \cdot \tau} \cdot c \sum N_i \qquad (1)$$

is the integrated intensity, where ε is the quantum efficiency ($\sim$ absorption), c the conversion factor between charge an ADC bins, $\phi = N/\tau$ the photon flux in terms of N photons per integration time τ. The beam position in x and y can be found by building the center of gravity:

$$x = \frac{(N_1 + N_4) - (N_2 + N_3)}{\sum N_i} \cdot \Delta x \;\; and \;\; y = \frac{(N_1 + N_2) - (N_4 + N_3)}{\sum N_i} \cdot \Delta y \qquad (2)$$

where Δx and Δy are spatial extensions of the charge cloud in horizontal and vertical direction, respectively, which is a convolution of the beam size with the range of the photoelectrons and the transversal diffusion. It is obvious that (2) can only be applied if (1) is greater than zero. Under the assumption that the beam is in the middle and all segments are calibrated such that they measure the same photon numbers N the error in the position encoding can be calculated taking into consideration Poisson statistics and error propagation as [2]:

$$\sigma_x = \frac{\Delta x}{2} \cdot \sqrt{\frac{\varepsilon \cdot N + \sigma^2}{\varepsilon^2 \cdot N^2}} \;\; and \;\; \sigma_y = \frac{\Delta y}{2} \cdot \sqrt{\frac{\varepsilon \cdot N + \sigma^2}{\varepsilon^2 \cdot N^2}} \qquad (3)$$

$$\Rightarrow \sigma \approx \frac{1}{SNR}$$

where σ is the electronics noise in terms of photons per integration time of the readout electronics attached to a single channel. From (3) it is obvious that the XBPM is working the better, the higher the signal-to-noise ratio SNR and subsequently the photon flux.

MEASUREMENTS OF SPATIAL RESOLUTION

Measurements to determine the intensity precision and the inherent spatial resolution of the XBPM were carried out at the SYRMEP bending magnet beamline during a 2.4 GeV operation of Elettra.

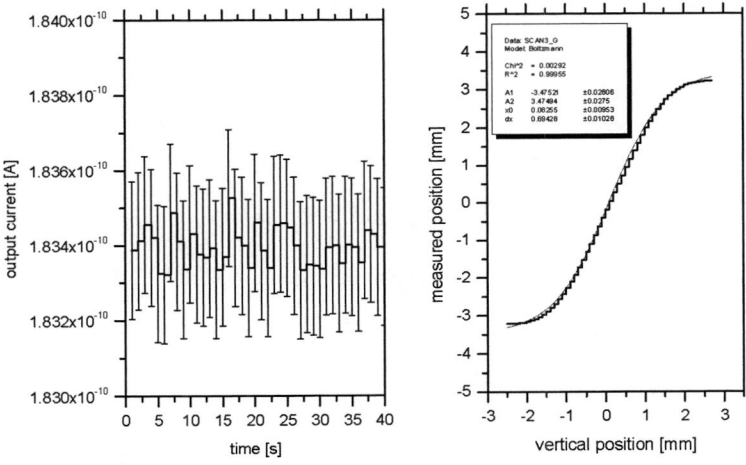

Figure 2. Left: Intensity measurement versus time for 500 samples per point and right: 4mm scan around the center of the XBPM.

The XBPM was mounted on a NEWPORT x-y stage with a nominal resolution of 100 nm two meters downstream a tantalum slit system, defining an aperture of 0.5 mm * 0.5 mm. The x-ray energy ranged from 9 to 35 KeV. The average flux density at 9 KeV can be quoted with $1.4 \cdot 10^8$ photons /s for the aforementioned machine parameters. The integrated current on four sectors is depicted in figure 2 versus time at a constant horizontal position of the beam. Each point is the average of 500 sampling points. The error bar indicated is 0.1% which means a single sampling point precision of 2%. For the resolution measurement the XBPM was scanned in steps of 100 μm downstream the fixed aperture. Zero indicates the middle of the sensitive volume of the XBPM. 500 samples (which corresponds to 1 sec integration time) were taken at each position. The measured sigma in position resolution over 500 samples at that intensity was about 1 μm for the single shot measurement. Since the sigma of the spatial extension of the charge cloud is in the order of 1 mm one would expect a linear position encoding of roughly 4 mm. This is confirmed by the measured position response versus the real position. As expected the curve has sigmoidal shape with a linear response of approximately from –2 mm to 2 mm (Fig. 2). The theoretical resolution as function of the photon flux is shown in Fig. 3 with the energy as a free parameter.

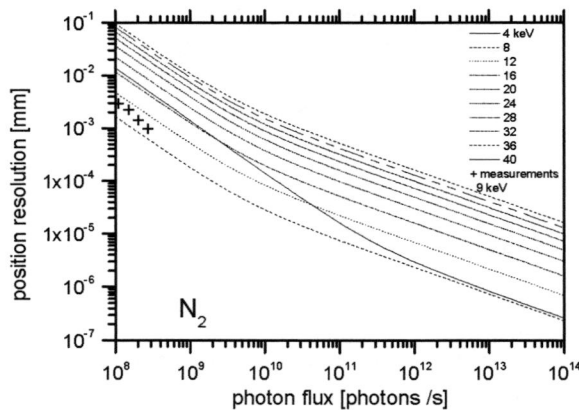

FIGURE 3. Inherent resolution limit of the XBPM for 1 kHz sampling rate

The curves represent the inherent limits for single shot experiments with a repetition rate of 1 kHz. Results of measurements at the bending magnet beamline at 9 KeV are shown as well for photon fluxes between 10^8 and 10^{14} photons / s. According to these curves the inherent limit at the saturation level of the electronics for 1 kHz sampling time should be in the order of some tenth to some hundreds of nano meter.

CONCLUSION

It could be shown that a XBPM on the basis of a segmented longitudinal ionization chamber can provide simultaneous information of the beam intensity with a precision better than 0.3 % and encode the beam position in the sub micron range with 1 KHz sampling rate. The inherent spatial resolution is limited by the SNR of Poisson statistics in case of photon fluxes > 10^{10} photons / sec.

ACKNOWLEDGMENTS

The authors are indebted to Claudio Fava for the mechanical design of the XBPM and are grateful to Giuseppe Cautero for fruitful discussions.

REFERENCES

1. V.K. Myalitsin, H.-J. Besch, H.W. Schenk, A.H. Walenta, *Nuclear Instruments and Methods in Physics Research, Section A: Accelerators, Spectrometers, Detectors and Associated Equipment* **A323**, Issue 1-2, pp 97-103, (1992).
2. H. Wagner, A. Orthen, H.-J. Besch, S. Martoiu, R.H. Menk, A.H. Walenta, U. Werthenbach, *Nuclear Instruments and Methods in Physics Research, Section A: Accelerators, Spectrometers, Detectors and Associated Equipment* **523**, Issue 3, pp 287-301, (2004).

Combination PBPM and Slits at the Pohang Light Source

K. H. Gil, H. C. Lee, G. H. Kim, C. D. Park, C. K. Kim, and C. W. Chung

Pohang Accelerator Laboratory / Pohang University of Science and Technology
San 31, Hyoja-dong, Nam-gu, Pohang, Gyeongbuk 790-784, Korea

Abstract. Three combination photon beam position monitor (PBPM) and slits are in use at the photon transfer line (PTL) of the high flux macromolecular X-ray crystallography (HFMX) beamline at the Pohang Light Source. Each slit, based on a common basic design, drives its two slit plates through a double bellows assembly by two motors. The slits also adopt a turned-grazing inclination of 30° to restrict the passage of scattered photon beams and to enhance cooling efficiency. One of the slits has been successfully applied with cooling function in order to cool the heat load from a multi-pole wiggler. This paper describes the design details and design features of the slits and discusses the results of performance tests for their PBPM function. The combined PBPM function can be used for the analysis of the position behaviors of photon beams at the beamline.

Keywords: combination PBPM and slit, photon transfer line, double bellows assembly, turned-grazing inclination
PACS: 07.85.Fv; 07.85.Qe

INTRODUCTION

A pair of a vertical slit and a horizontal slit is required to exclude the mis-steered photon beams in front of the double crystal monochromator (DCM) and in front of the hutch at the HFMX beamline PTL of the Pohang Light Source, respectively. Since the pair of slits has to be installed together with other beamline equipment in the region from the concrete wall to the DCM, it is not easy to install both the vertical and horizontal slits in this region. Furthermore, since the photon beams in this region are not monochromatic, the slits with cooling capability should be installed.

Thus, a water-cooled vertical slit is installed in this region and a beryllium window ahead of the DCM is modified to function as a fixed horizontal slit. Likewise, both the vertical and horizontal slits without cooling capability are installed for monochromatic beams in front of the hutch.

In this research, for the three slits required in the HFMX beamline PTL, we establish a standardized basic design for the vertical slit, transform this design into one that is for the horizontal slit by altering only directions, and modify the vertical slit into a water-cooled vertical slit by adding cooling lines. In addition, observations for the photon beam position behaviors at the beamline are tried by combining PBPM function with the slits [1].

DESIGN OF SLITS

The vertical slit of the HFMX beamline PTL is shown in Fig. 1. Two tungsten slit plates (1) to be passed by photon beams are bolted to the backside oxygen-free high-conductivity copper (OFHC) upper and lower cooling blocks (2) and (3) with intermediate ceramic insulators (4). Each cooling block is milled into a right triangle of which the hypotenuse is inclined at 30° with respect to the photon beam direction in order to adopt a turned-grazing inclination to prevent photon beams impinging on the slit plates from scattering downstream and to enhance cooling efficiency [2]. Both cooling blocks can translate with three ultrahigh-vacuum ball bushes (5) inserted into holes near the three vertices of the triangle along three guide rods (6), which are mounted on a stainless steel main flange (7). The upper and lower cooling blocks are bolted to upper and C-shaped lower drivers (8) and (9), respectively, which are driven by a double bellows assembly (10). Photoelectrons generated by photon beams impinging on the slit plates flow through two in-vacuum insulated wires (11) and are sensed by a precision current amplifier via two electrical feedthroughs (12) assembled with two elbows welded to the main flange.

CP879, *Synchrotron Radiation Instrumentation: Ninth International Conference,*
edited by Jae-Young Choi and Seungyu Rah
© 2007 American Institute of Physics 978-0-7354-0373-4/07/$23.00

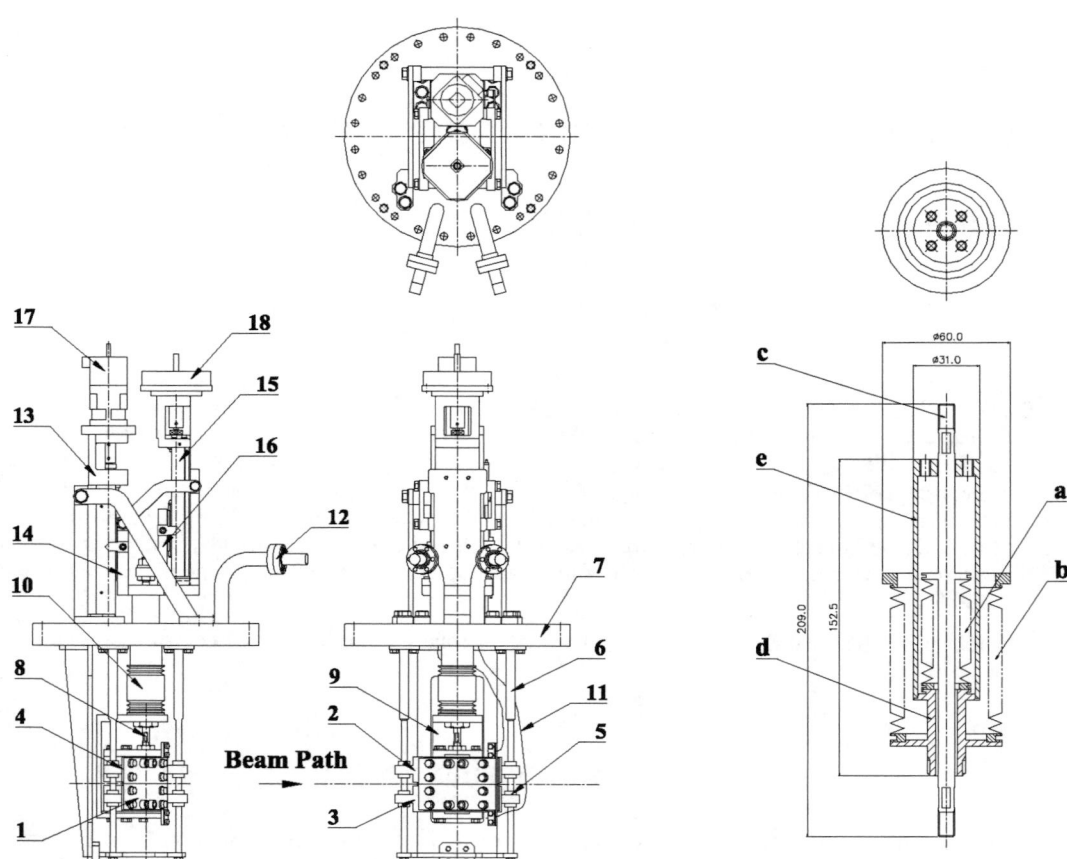

FIGURE 1. General assembly of the vertical slit.

FIGURE 2. Detailed longitudinal sectional view of the double bellows assembly.

Figure 2 shows a detailed longitudinal sectional view of the double bellows assembly. The assembly has a structure that the lower part of an inner bellows (a) is rigidly connected to the lower part of an outer bellows (b). The slide rod (c) welded on top of the inner bellows is used in itself as the upper driver and the connector (d) welded to the bottom of the outer bellows is bolted on top of the lower driver.

FIGURE 3. Cooling line connection of the water-cooled vertical slit.

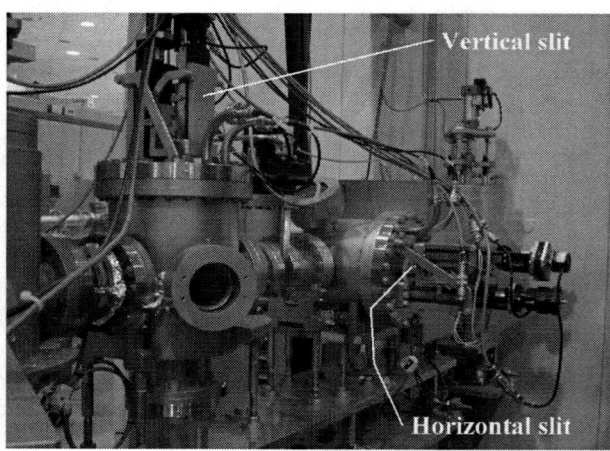

FIGURE 4. Vertical and horizontal slits installed at the HFMX beamline PTL.

The linear table I (13) installed on top of the main flange is connected, through an L-shaped elevation plate I (14), to top of the column (e) welded to the connector of the double bellows assembly and adjusts the position of the slit plates. The linear table II (15) installed on top of the elevation plate I is connected to top of the slide rod of the double bellows assembly through an L-shaped elevation plate II (16) and adjusts the gap of the slit plates. The linear table I and II are driven by a stepping motor (17) and an ultrasonic motor (18), respectively.

The structure of the vertical slit is applied to the horizontal slit and the water-cooled vertical slit as it is. The cooling lines to cool the photon beams from a multi-pole wiggler have been added to the water-cooled vertical slit with the vertical slit structure itself as shown in Fig. 3. Figure 4 shows the vertical and horizontal slits assembled on their vacuum chambers and installed at the HFMX beamline PTL.

MEASUREMENTS AND DISCUSSIONS

The electrode currents of the vertical and horizontal slits have been measured, translating the slit plates of the vertical slit by 20 μm towards the lower position with the horizontal slit fixed. The measured electrode currents are plotted in Fig. 5.

The vertical position index can be formulated from the electrode currents as follows [3]:

$$V.I. = \left(I_{upper} - I_{lower}\right)/\left(I_{upper} + I_{lower}\right) \tag{1}$$

Figure 6 illustrates the vertical position index calculated according to Equation (1) using the electrode currents in Fig. 5. The tilted straight line in the middle of Fig. 6 represents the position response obtained by linearly fitting the vertical position index. Thus, the relation between the vertical position and the vertical position index can be expressed as $y(mm) \propto 0.22 \times V.I.$.

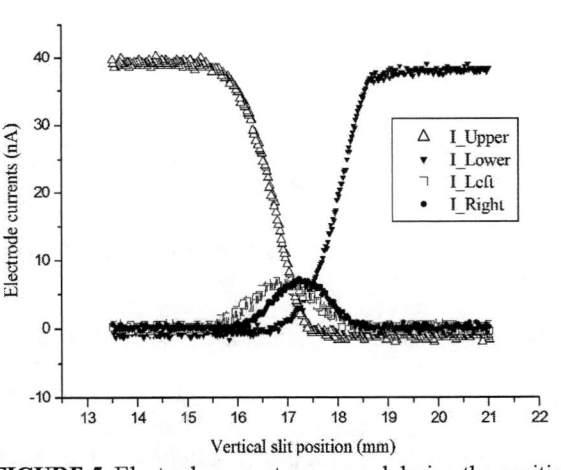

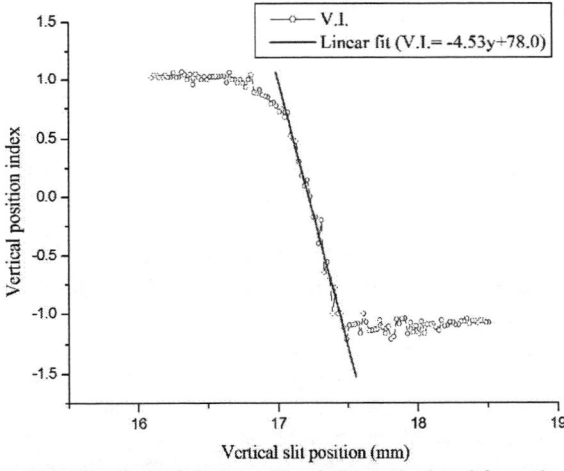

FIGURE 5. Electrode currents measured during the position response test.

FIGURE 6. Vertical position index calculated from the measured electrode currents.

In order to observe the long-duration vertical position drift of photon beams using the result of the previous position response test, a measurement campaign was carried out for 39 hours and 42 minutes. At its first stage, the position of the vertical slit was adjusted so that its upper and lower electrodes yielded evenly distributed current values.

Figure 7 presents four electrode currents of both slits measured during this test. The long-duration vertical position drift of photon beams that resulted from the measured electrode currents is shown in Fig. 8. There were a regular beam injection after 16 hours, an abrupt beam dump followed by an immediate beam injection after 22 hours, and a second regular beam injection for user experiments after 38 hours, since the test started. These incidents are presented in the record of the long-duration vertical photon beam position drift of Fig. 8 as they are.

The position response test mentioned above had been carried out between incidents of the first regular beam injection and the beam dump. After the position response test, the position of the vertical slit was re-adjusted in correspondence with the vertical photon beam position at that time for the next drift test to be resumed.

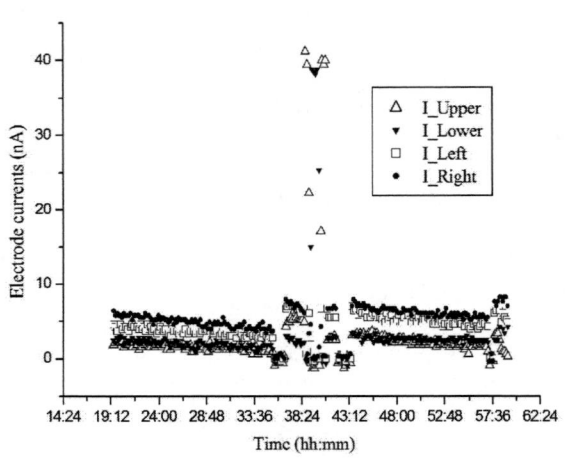

FIGURE 7. Four electrode currents of both slits measured during the long-duration vertical photon beam position drift test.

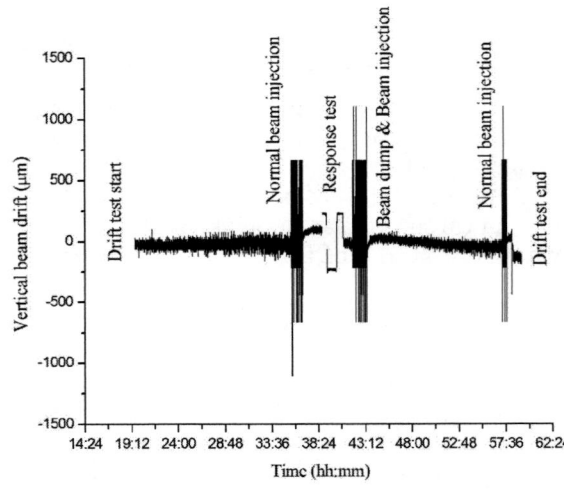

FIGURE 8. Long-duration vertical photon beam position drift.

A tendency is observed that the photon beams in the rear of the HFMX beamline PTL rise up to around 120 μm over an hour just after a beam injection. This seems to be due to transient temperature rises until their surface temperatures reach those of steady state since photon beams started to impinge on the surfaces of optical components such as a collimating mirror and the like.

Analyzing the test period between the beam injection following the beam dump and the second regular beam injection, it can be known that the vertical photon beam position gradually falls down at a rate of 0.12 μm/min and its standard deviation steadily increases from 17.2 μm to around 25.4 μm over 14 hours since it stabilized over an hour.

CONCLUSIONS

The three slits required for the HFMX beamline PTL at the Pohang Light Source have been fabricated based on a common basic design which combines PBPM function. The slits have compact driving mechanisms by using a double bellows assembly.

A water-cooled vertical slit, designed by adding the function of cooling to the basic design, was installed in front of the HFMX beamline PTL and has been in stable operation against the heat load from the multi-pole wiggler of the HFMX beamline.

As a test for the performance of the PBPM function, a long-duration vertical position drift of photon beams was observed using the relation between the vertical position and the vertical position index obtained from a position response test, for a pair of the vertical and horizontal slits installed in front of the hutch of the HFMX beamline. From the test results, some tendencies of the vertical position drift of photon beams at the beamline were grasped.

These combination PBPM and slits can be the basic tools for further research in the near future for the position behaviors of photon beams at the beamline.

ACKNOWLEDGMENTS

This work was supported by the Korean Ministry of Science and Technology.

REFERENCES

1. Y. Xie, T. D. Hu, T. Liu and D. C. Xian, *Nucl. Instr. and Meth. A* **467–468**, 256-259 (2001).
2. D. Shu, C. Brite, T. Nian, W. Yun, D. R. Haeffner, E. E. Alp, D. Ryding, J. Collins, Y. Li and T. M. Kuzay, *Rev. Sci. Instrum.* **66** (2), 1789-1791 (1995).
3. W. Schildkamp and C. Pradervand, *Rev. Sci. Instrum.* **66** (2), 1956-1959 (1995).

Ray-Tracing Analysis of Fresnel-Zone-Plate Optical System as an Electron Beam Profile Monitor

Masami Fujisawa[1], Hiroshi Sakai[1], Norio Nakamura[1], Hitoshi Hayano[2] and Toshiya Muto[2]

[1]*SynchrotronRadiation Laboratory, Institute for Solid State Physics, University of Tokyo, Kashiwanoha, 5-1-5, Kashiwa, Chiba, 277-8581, Japan*
[2]*High Energy Accelerator Research Organization (KEK), Oho 1-1, Tsukuba, Ibaraki, Japan*

Abstract. The analysis of the image distortion made by Fresnel-zone-plate optical system was studied with ray trace simulation and analytical treatment. The tolerable tilt angle depends on the tolerable image size. The distortion appears not only in image size but in image inclination. The simulation and the analysis performed for the optical parameters of the electron beam profile monitor may be useful for the advancement of the X-ray microscope performance.

Keywords: Fresnel zone plate, electron beam profile monitor, ray trace, optical path function.
PACS: 07.60.-j; 07.85.Qe; 41.85.Ew

INTRODUCTION

Among the various transverse beam profile monitors for the diagnosis of the electron beam circulating in an accelerator, the method of observing the image of the radiated light source on the two-dimensional detector made by focusing elements is very useful. The merits of the imaging method are non-destructive and real-time properties. For the third or forth generation synchrotron light source the diffraction limit reaches in the visible range and the electron profile is hidden by the radiated light in the image. It is preferable to observe the image of shorter wavelength light (X-rays) for accurate diagnosis of the electron beam profile because the shorter wavelength of light has the smaller size. Iida et al. [1] recently developed a beam profile monitor based on two Fresnel zone plates (FZPs) and an X-ray CCD, and the first results have been reported. The utility of the optical system as a beam profile monitor has been demonstrated through the diagnosis of the electron beam circulating in an accelerator. However, the system performance may be limited by aberrations due to FZP alignment errors, especially the FZP tilt. They should be estimated with some methods.

Some studies for the aberrations of FZP have already been performed. In the study by Kamiya [2] and Thieme [3], the FZP and the image surface were installed perpendicular to the center axis of the optical ray and no installation errors of the FZP were considered. Another study by Young [4] was for parallel light. In the present study, main subject is the influence of the tilt of FZP on the image and it is necessary to estimate the deformation of a point-source image by the tilt of the FZP in the first place. Then we construct an optical path function for a point source to perform the image analysis on the detector surface perpendicular to the optical axis. It is also needed to determine the zone density parameters which are used in the ray tracing simulations hereafter.

In this article, the optical path function is constructed without installation errors and the zone parameters of the FZP are determined for the ray trace simulation. After that the optical path function is constructed with installation errors of tilt. The errors of tilts are realized by coordinate transformation with rotation. The analytical method based on the optical path function is used for prospect of the tilt effects on the image and confirmation of the ray trace results.

CP879, *Synchrotron Radiation Instrumentation: Ninth International Conference,*
edited by Jae-Young Choi and Seungyu Rah
© 2007 American Institute of Physics 978-0-7354-0373-4/07/$23.00

OPTICAL PATH FUNCTION AND ZONE DENSITY PARAMETERS

When the zone density of the FZPs is expressed by a polynomial of the zone radius r as $D(r) = D_0 + D_1 r^1 + D_2 r^2 + D_3 r^3$, the optical path length $G(r)$ is given as $G(r) = -m\lambda \int_0^r D(r)dr = -m\lambda(D_0 r + \frac{1}{2}D_1 r^2 + \frac{1}{3}D_2 r^3 + \frac{1}{4}D_3 r^4)$. The coefficients $D_i (i = 0-3)$ will be determined later. Here, λ is the photon wavelength, m is its order. The optical path function is constructed under the coordinate definition that the surface of FZP is put on the coordinate axis X, Y and the coordinate axis Z is normal to the FZP surface. For a point on the FZP expressed $P(\xi, \eta, 0)$, the radius r is given by $r = (\xi^2 + \eta^2)^{1/2}$. The source point A and focusing point B are on the Z-axis and can be expressed by $A(0,0,a)$ and $B(0,0,-b)$, respectively. Here $a(>0)$ and $b(>0)$ are the distances between the center point of the FZP and the source point A and focusing point B, respectively. The optical path function $F(\xi, \eta)$ for the point $P(\xi, \eta, 0)$ on the FZP is given by $F(\xi, \eta) = G(r) + r_{AP} + r_{BP}$. Here, r_{AP} and r_{BP} are the geometrical distances between a point $P(\xi, \eta, 0)$ on the FZP and the source point A and focal point B, respectively. They are expressed as $r_{AP} = (\xi^2 + \eta^2 + a^2)^{1/2} = (r^2 + a^2)^{1/2}$, $r_{BP} = (\xi^2 + \eta^2 + b^2)^{1/2} = (r^2 + b^2)^{1/2}$. To rewrite and expand the optical function with r,

$$F(\xi, \eta) = F(r) = G(r) + r_{AP} + r_{BP}$$

$$= (a+b) + \{-m\lambda D_0\}r + \{-\frac{1}{2}m\lambda D_1 + \frac{1}{2}(\frac{1}{a} + \frac{1}{b})\}r^2 + \{-\frac{1}{3}m\lambda D_2\}r^3 + \{-\frac{1}{4}m\lambda D_3 - \frac{1}{8}(\frac{1}{a^3} + \frac{1}{b^3})\}r^4 + \cdots \quad (1)$$

Setting each coefficient of r^i to zero for $\frac{\partial F}{\partial r} = 0$, $D_i(i = 0-3)$ can be determined as, $D_0 = 0$, $D_1 = \frac{1}{m\lambda}(\frac{1}{a} + \frac{1}{b})$, $D_2 = 0$, $D_3 = -\frac{1}{2m\lambda}(\frac{1}{a^3} + \frac{1}{b^3})$. In practice, the FZPs are manufactured neglecting the coefficient D_3. Then also in the present study, the coefficients D_3 are disregarded unless stated otherwise.

When the FZP is tilted around the X axis by an angle T_x and after that tilted around the Y axis by an angle T_y, the coordinate $(\xi, \eta, 0)$ of a point on the FZP changes to $(\xi \cos T_y, \eta \cos T_x - \xi \sin T_y \sin T_x, -\eta \sin T_x - \xi \sin T_y \cos T_x)$ in the XYZ coordinate system. The optical path function $F'(\xi, \eta)$ after tilting of the FZP is expressed as

$$F'(\xi, \eta) = G(r) + r_A + r_B = a + b - \frac{1}{2}(\frac{1}{a} + \frac{1}{b})(\xi \sin T_y \cos T_x + \eta \sin T_x)^2$$

$$+ \frac{1}{2}(\frac{1}{a^2} - \frac{1}{b^2})(\xi \sin T_y \cos T_x + \eta \sin T_x)[-(\eta^2 + \xi^2) + (\xi \sin T_y \cos T_x + \eta \sin T_x)^2]) \quad (2)$$

$$+ \frac{3}{8}(\frac{1}{a^3} + \frac{1}{b^3})(\eta^2 + \xi^2)(\xi \sin T_y \cos T_x + \eta \sin T_x)^2 - \frac{1}{8}(\frac{1}{a^3} + \frac{1}{b^3})(\eta^2 + \xi^2)^2$$

When $T_x, T_y \neq 0$, the terms accompanying $\sin T_x, \sin T_y$ remain as a tilt aberration and contribute to deformation of the image profile. The distortion of image caused by the tilt can be calculated using the following equation [5].

$$\Delta A_x = b\frac{\partial F'(\xi, \eta)}{\partial \xi}, \quad \Delta A_y = b\frac{\partial F'(\xi, \eta)}{\partial \eta} \quad (3)$$

FZP PARAMETERS AND RAY TRACE

Being based on the general description in previous sections, the simulation is performed with the optical parameters of the electron beam profile monitor developed by Iida et al. [1]. The beam profile monitor consists of two FZPs(CZP and MZP) as shown in Fig. 1. The optical arrangement is the same as a con-focal type x-ray microscope. The zone density parameters of both FZPs are evaluated to $D_1 = 2864.584\text{mm}^{-2}$, $D_3 = -0.0012975\text{mm}^{-4}$ for CZP, and $D_1 = 104687.424\text{mm}^{-2}$, $D_3 = -83.001\text{mm}^{-4}$ for MZP for the x-ray of 0.38324nm wavelength. Final magnification of 20 is realized by CZP and MZP with the magnification of 1/10 and 200, respectively.

The computer program for the ray trace analysis is coded with the program language of Visual Basic 6. Diffracted rays are obtained using Snell's Law [6]. In the ray trace simulation, light rays were generated by the Monte Carlo method under the condition that the horizontal and vertical sizes and vertical divergence of the light source are assumed to obey normal distributions, and the horizontal divergence a homogeneous distribution, because the beam profile monitor observes synchrotron radiation emitted from a bending magnet section. The standard deviation of the vertical divergence was set at 126 µrad assuming installation of the monitor at the KEK Accelerator Test Facility [1]. The total width of the horizontal divergence was set at 300 µrad, which is sufficient for light rays to irradiate the entire surface of the CZP and MZP with the radius of 1.5mm and 0.0373mm, respectively.

In order to compare the image on the both screen with the light source easily, data is normalized by being multiplied by 10 at the intermediate screen and by 1/20 at the final screen. Hereafter the image sizes are expressed after such normalization unless stated otherwise.

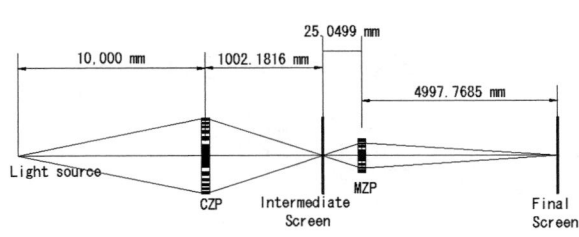

FIGURE 1. Side view of FZP optical system for electron beam profile monitor.

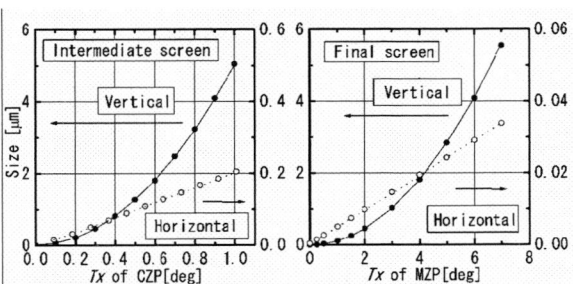

FIGURE 2. Image profile distortions at several tilt angles around X axis. Size is expressed with half the difference between the maximum and minimum values.

RESULTS AND DISCUSSION

The image profile distortions at several tilt angles around X axis are shown in Fig. 2. Here, the size of the distorted image is expressed with half the difference between the maximum and minimum values in order to compare the results of the ray trace with those of the analytical method. The image sizes obtained by ray trace were estimated at the intermediate screen for the CZP and at the final screen for the MZP, where D_3 for the CZP was set at the value obtained in the previous section in the ray trace for the MZP to minimize the effect of the aberration of the CZP. The coincidence of the results demonstrates the validity of both techniques. The amount of distortion made by the MZP tilt is small relative to that by the CZP tilt by about one seventh because of the smaller outer diameter of the MZP. Thus the studies are performed for the CZP tilt hereafter. In order to estimate the image distortion due to simultaneous tilts around X and Y axes, ray trace was performed for a point source with several vertical and horizontal tilt angles simultaneously. The results are summarized in Fig.3 as a contour map with solid and dotted contours for horizontal and vertical distortions, respectively. Here the horizontal and vertical distortions ΔA_x, ΔA_y are expressed in standard deviation. The real size of the image Σ_x, Σ_y are obtained with a convolution of horizontal light source size σ_x and ΔA_x or vertical light source size σ_y and ΔA_y, i.e. $\Sigma_x = (\sigma_x^2 + \Delta A_x^2)^{1/2}$ and $\Sigma_y = (\sigma_y^2 + \Delta A_y^2)^{1/2}$. In order to obtain the image size with the error under 10 percent, distortion ΔA_x, ΔA_y should be under the half of σ_x and σ_y. Namely, $\Delta A_x \le \sigma_x/2 = 20\mu m$, $\Delta A_y \le \sigma_y/2 = 5\mu m$ for the case of $\sigma_x = 40\mu m$ and $\sigma_y = 10\mu m$ [1]. The tolerable tilt angles for the above light source size can be obtained from the area enclosed by the four points expressed by circles in the Fig.3, an origin, an intersection of T_x axis and the 5µm contour (dotted line) of vertical distortion, an intersection of T_y axis and the 20µm contour (solid line) of horizontal distortion and an intersection of these contour. The tolerable vertical tilt angle is 1.5° for $T_y = 0°$. However, it decreases to 0.7° for $T_y = 2.8°$. Ray trace simulation for a light source size of $40\mu m \times 10\mu m$ confirms the insistence as in Fig.4, which shows images on the intermediate screen. The image profile increases not only in size but in (transverse) inclination as indicated in the graph (d) of Fig.4. Here the simulation was performed with ray number of 10,000 points per one trial and the average values and the standard deviations were obtained from ten times trials. It can easily be achieved that the tilt error is kept within tolerable tilt angle.

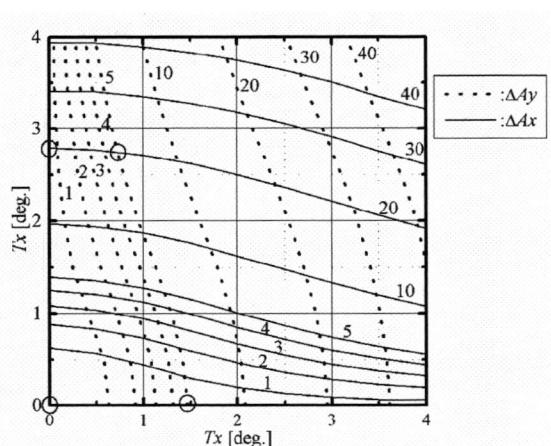

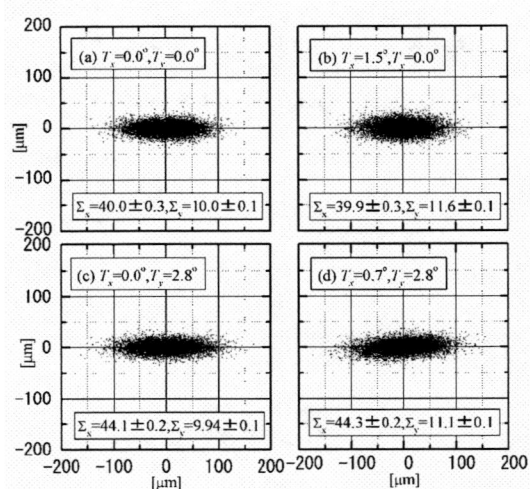

FIGURE 3. Contour map for tilt angles around both X- and Y-axes. Solid contours show the horizontal distortion (ΔA_x) and dotted contours show vertical distortion (ΔA_y). Numerical values in graph show the distortion in μm.

FIGURE 4. Images on intermediate screen obtained by ray trace for light source size of 40×10μm. with several tilt angles (T_x, T_y) of (a) (0.0°, 0.0°), (b) (1.5°, 0.0°), (c) (0.0°, 2.8°) and (d) (0.7°, 2.8°).

CONCLUSION

The performance degeneration of the electron beam profile monitor caused by the FZP tilt is estimated using both ray trace and analytical method. The results are consistent with each other. The distortion appears not only in size but in inclination. For the smaller size of light source, the tolerable tilt angles of FZP become smaller.

Accelerators including synchrotron light sources are still under development and are obtaining lower emittances. For lower emittance accelerators, the sizes of the electron beams or light sources become smaller. Moreover the larger size of FZP may be adopted. In such cases, the tilt error of FZP can be an obstacle in obtaining higher performance. The present simulation and analysis will be useful for estimating the tolerable tilt angles for X-ray microscopes as well as beam profile monitors.

REFERENCES

1. K. Iida, N. Nakamura, H. Sakai, K. Shinoe, H. Takaki, M. Fujisawa, H. Hayano, M. Nomura, Y. Kamiya, T. Koseki, Y. Amemiya, N. Aoki and K. Nakayama: Nucl. Instrum. Methods, A506(2003)41-49.
2. K. Kamiya : Science of Light 12 (1963)35-49.
3. J. Thieme: Theoretical Investigations of Imaging Properties of Zone Plates Using Diffraction Theory in X-ray Microscopy II, Springer Series in Optical Science, ed. by D. Sayre, M. Howells, J. Kirz and H. Rarback (Springer-Verlag, 1988) Vol.56, p70-79.
4. M. Young: J. Opt. Soc. Am. 62(1972)972-976.
5. H. A. Padmore, M. R. Howells and W. R. McKinney: Grazing–Incidence Monochromators for Third-Generation Synchrotron Radiation Sources in Vacuum Ultraviolet Spectroscopy II, Experimental Methods in the Physical Sciences ed. by J. A. R. Samson and D.L. Ederer (Academic Press, 1998), Volume 32, p21-72.
6. G. H. Spencer and V. R. K. Murty: J. Opt. Soc. Am. 52(1962)672-678.

Absolute Photoionization Cross Section with an Ultra-high Energy Resolution for Ne in the Region of 1s Rydberg States

M. Kato[1], Y. Morishita[1], M. Oura[2], H. Yamaoka[2], Y. Tamenori[3], K. Okada[4], T. Matsudo[4], T. Gejo[5], I. H. Suzuki[1] and N. Saito[1]

[1]AIST, NMIJ, Tsukuba, 305-8568, Japan
[2]SPring-8/JASRI, Sayo, Hyogo 679-5198, Japan
[3] RIKEN, Harima Institute, Sayo, Hyogo 679-5148, Japan
[4]Department of Chemistry, Hiroshima University, Higashi-Hiroshima 739-8526, Japan
[5] University of Hyogo, Kamigori, Hyogo 678-1297, Japan

Abstract. The high-resolution absolute photoabsorption cross section with an absolute photon energy scale for Ne in the energy region of 864-872 eV ($1s^{-1}np$ Rydberg states) has been measured using a multi-electrode ionization chamber and monochromatized synchrotron radiation. The natural lifetime width of Ne $1s^{-1}3p$ resonance state has been obtained to be 252 ± 5 meV. The Ne^+ ($1s^{-1}$) ionization potential is determined to be 870.16 ± 0.04 eV by using the Rydberg formula. These absolute values are supposed to be more reliable than those previously reported.

Keywords: photoabsorption, neon, soft x-ray, lifetime
PACS: 32.30.Rj, 32.70Fw, 32.70Cs, 32.80.Hd

INTRODUCTION

The photoabsorption spectra of rare gas atoms are often used for the monochromator calibration or derivation of lifetime width of the corresponding hole state. Furthermore the absolute cross sections of the rare gas atoms in the soft X-ray region are utilized for the measurement of absolute photon intensities [1-2]. For these reasons, the absorption spectra should be converted to photoabsorption cross sections on absolute scale for both the energy values and cross section values, and should be measured with as high resolution as possible.

The $1s \rightarrow np$ Rydberg excitation of Ne as well as the double excitation involving 1s electron has been investigated over 30 years [3-14]. The absolute values of the photoabsorption cross sections provided by some studies are scattered. The measured values of the absolute photoabsorption cross section using synchrotron radiation often suffered from the contamination of impurity photons into the incident beam, for example, stray light and higher order radiation [9,11,15,16]. Therefore it is difficult to obtain the cross sections with an absolute scale correctly. The energies of the $1s^{-1}np$ Rydberg states are not consistent among the reports published [3-8,10]. The natural lifetime width of Ne $1s^{-1}3p$ resonance also distributes among 240 to 259 meV though a narrow bandpass of the photon beam has become available recently [8,10].

In this paper, we measured the photoabsorption cross sections of Ne in the region of $1s^{-1}np$ Rydberg states with ultra-high energy resolution. The energies and the natural lifetime width of the Ne $1s^{-1}np$ Rydberg states, which are often used as reference lines for the absolute energy scale and the energy resolution of the experimental apparatus, respectively are determined precisely. The Ne $1s^{-1}$ ionization potential is also estimated based on the quantum defect method.

EXPERIMENT

The experiment was performed using a high-resolution plane grating monochromator installed on the c-branch of the soft X-ray photochemistry beamline 27SU at SPring-8 [17]. The measurement system and experimental procedure have been described in detail elsewhere [11,15]. The monochromatized soft X-ray beam entered the

CP879, *Synchrotron Radiation Instrumentation: Ninth International Conference*,
edited by Jae-Young Choi and Seungyu Rah
© 2007 American Institute of Physics 978-0-7354-0373-4/07/$23.00

multi-electrode ionization chamber of cylindrical shape, and with a polyimide filter of 1100 Å thickness (LUXEL Co.). The size of the soft X-ray beam was defined using circular aperture of 1 mm diameter. The set of the electrodes consists of six cylinders of 5 mm diameter. Photoion currents from the second to fifth electrodes were utilized for derivation of photoabsorption cross sections. The length of the second and third electrodes is 100 mm and that of forth and fifth 500 mm. The ionization chamber was evacuated with a turbo-molecular pump just before measurements, and supplied with neon gas at about 16-400 Pa in obtaining ion currents. The ion current at each electrode was led to a calibrated picoamperemeter (Keithley 6517) and then the data were transferred to a personal computer. The gas density was detected with a calibrated capacitance manometer (Baratoron 690).

The photoabsorption cross section measured using a double-electrode ionization chamber with ion-collection electrodes having the same length is given in principle as follows [9,15,18]:

$$\sigma = \frac{1}{Lp}\ln\left(\frac{i_1}{i_2}\right) = \frac{1}{L}\frac{\mathrm{d}}{\mathrm{d}p}\left(\ln\left(\frac{i_1}{i_2}\right)\right)$$ (1)

where p denotes the gas density, L the length of the electrodes, and i_1 and i_2 the photoion currents from the second and the third electrodes (or fourth and fifth electrodes), respectively. The gas-density dependence of the σ value obtained from equation (1) dose not show a constant value but show appreciable scatters as we explained in previous paper [14]. These scatters can be ascribed to incompleteness of the spectral purity of the incident photon. We analyzed the pressure dependence of the photoion current using a model, which takes into account the small contribution from stray light and higher orders [11,15]. This analysis has given a photoabsorption cross section for the first order light accurately within an error of 1 % in most photon energies.

The photon energy calibration was performed as follows. The 1s → 3p transition energy of 867.13 eV reported by Wuilleumier [3] was used as the reference for the absolute scale of the incident photon energy. The dispersion of the grating used in this experiment had been determined by measuring the Ne 1s and 2s photoelectron spectra at the photon energies of 2000 eV and 800 eV, respectively, with a high-resolution electron spectrometer (Gammadata SCIENTA SES-2002) [19]. The binding energies of Ne 1s and 2s orbitals used for the dispersion calibration are 870.21 eV and 48.475 eV, respectively [20,21]. Then the dispersion has been determined in extreme high precision.

RESULTS AND DISCUSSION

The photoabsorption cross section of Ne as a function of incident photon energy in the region of the $1s^{-1}np$ Rydberg states is shown in Figure 1. The structures attributed to 1s → 3p, 4p, 5p, 6p transitions were clearly resolved as well as those in the total ion yield spectrum reported by De Fanis et al. [10]. We have analyzed the cross section curve in Fig. 1 with the following expression:

$$\sigma = \sigma_{\mathrm{dir}} + \sum_n \sigma_{1s \to np} + \sigma_{1s}$$ (2)

where σ_{dir} is the nonresonant ionization cross section for outer-shells, $\sigma_{1s \to np}$ the resonant cross sections for 1s→ np transitions represented by Voigt functions, σ_{1s} the 1s ionization cross section represented by an arctangent function. σ_{dir} is considered to be constant in the present energy range. We fitted equation (2) to the experimental photoabsorption cross section in Fig. 1 in order to extract the line strengths, Lorentzian width (w_L), Gaussian width (w_G), and the energies of the resonances. The values of w_L and w_G were assumed to be the same for the resonance peaks in the fitting procedure. The Gaussian width is originated from the bandpass of the incident photon, which was set to be 45 meV from the design parameters in the optics, but the bandpass was larger than the expectation. In the term $\sum_n \sigma_{1s \to np}$, $\sigma_{1s \to np}$ (n = 3-8) were included in the fit for obtaining as a whole a good agreement between the fit and the measured curve although the peaks for n=7 and 8 cannot be resolved in the photoabsorption spectrum. The cross sections contributed from the 1s → np ($n \geq 9$) transitions were included into σ_{1s}. The line strengths of the resonances of $1s^{-1}7p$ and $1s^{-1}8p$ relative to $1s^{-1}6p$ were fixed as the strengths were proportional to n^{-3}. Energies of the resonances for $n \geq 6$ were derived by using quantum defect method. A result of the fit was shown in Fig. 1 with w_L = 252 ±5 meV, w_G = 80 ±8 meV. The natural lifetime widths obtained for the Ne $1s^{-1}3p$ resonance as well as the experimental photon bandpasses are listed in Table 1. Our value for the natural lifetime width for $1s^{-1}3p$ agrees with other reported values within uncertainties, and has the smallest uncertainty among the listed ones in Table 1.

Table 2 shows the energies for the $1s^{-1}np$ (n = 3-8) resonances and the Ne 1s ionization limit compared with some previous works. The Ne $1s^{-1}$ ionization potential of 870.16 ± 0.04 eV has been determined using the Rydberg formula. Our results have given good agreement with the values reported by Wuilleumier [3] for all the resonant peaks and the ionization potential within 10 meV. The energies of the Rydberg series obtained by Coreno et al. [8]

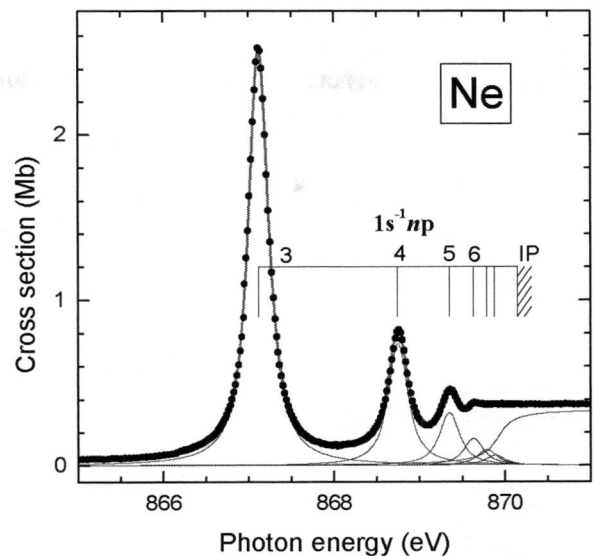

FIGURE 1. Photoabsorption cross section (dots) of Ne as a function of incident photon energy in the region of 1s → np Rydberg transition. The thick curve shows the sum of the Rydberg components (narrow curves) obtained from the results of the fitting procedure (see text). The vertical bars show the $1s^{-1}n$p (n = 3-8) Rydberg resonances and the ion core Ne 1s ionization threshold.

show large discrepancies from the present values and also previously reported values. The other Rydberg energies previously reported are consistent with the present values if the energies are measured relative to the $1s^{-1}3$p peak. The ionization potential measured from X-ray emission experiment [20] also agrees within the uncertainty.

The values of the absolute photoabsorption cross section are compared with the values reported previously in Table 3. At the $1s^{-1}3$p resonance, our value is 2.53 ± 0.01Mb, while the other reported values are 1.6 ± 0.1 Mb [12] and 0.9 Mb [5]. The value at the resonance energy strongly depends on the photon bandpass. Therefore using the natural lifetime width we estimate the peak cross section at infinite resolution as 2.68 Mb, whereas Prince *et al.* [12] reported as 1.94 ± 0.1 Mb. Their value is lower than our estimated value and even our measured value. When the cross sections around the resonance energy were measured, we put Ne gas into the ionization chamber with very low pressure in order to avoid an effect of saturation where the current from the first electrode in the ionization chamber is much larger than that from the second one. The values reported by Prince *et al.* [12] may be suffered from this saturation effect. Below the resonance, the present cross section values are 0.031 Mb at 864 eV and 0.039 Mb at 865 eV. Two values of the cross section were reported previously; One is 0.0228 Mb at 858.8 eV by Suzuki and Saito [9,11], which agrees with the present value reasonably in consideration of photon energy difference. The other value is 0.09 Mb at 865 eV [12], which is significantly larger than the present value. Above the Ne 1s ionization threshold, our value of 0.370 Mb reasonably agrees or is consistent with the previous values by Suzuki and Saito [9,11] and Esteva *et al.* [5], whereas the value of 0.45 Mb reported by Prince *et al.* [12] is significantly larger than our value.

TABLE 1. The natural lifetime width for the Ne $1s^{-1}$ 3p resonance together with the photon bandpass.

References	Natural lifetime width (meV)	Photon bandpass (meV)
This work	252(5)	80(8)
De Fanis *et al* (2002) [10]	240(10)	66
Coreno *et al* (1999) [8]	259(20)	~120
	254(20)	

TABLE 2. The energies of the $1s^{-1}n$p (n = 3-8) Rydberg states of Ne and their ion core Ne$^+(1s^{-1})$. In this work the energies of the $1s^{-1}n$p (n = 6-8) and that of Ne$^+(1s^{-1})$ are determined with the Rydberg formula.

Electronic state	Energy (eV)							
	This work	**Ref [8]**	**Ref [7]**	**Ref [6]**	**Ref [5]**	**Ref [4]**	**Ref [3]**	**Ref [20]**
$1s^{-1}3$p	867.13(2)	867.12(5)	867.13	867.18(2)	867.25	867.05(8)	867.13	
$1s^{-1}4$p	868.76(3)	868.69(4)	868.76(5)	868.85(2)	868.84	868.68(10)	868.77	
$1s^{-1}5$p	869.36(3)	869.27(5)	869.33(5)	869.47(2)	869.5	869.23(15)	869.37	
$1s^{-1}6$p	869.64	869.56				869.63(15)	869.65	
$1s^{-1}7$p	869.80	869.73						
$1s^{-1}8$p	869.99							
$1s^{-1}$	870.16(4)		870.07			870.1(2)	870.17	870.21(5)

The cross section values reported by them is about 0.06-0.08 Mb larger than the present ones above or below the resonance, which suggests that their data are suffered from stray light. Our procedure to obtain the cross section values avoids the effects from the stray light or higher order lights as discussed above.

TABLE 3. The absolute photoionization cross section of Ne 1s ionization region.

Photon energy (eV)	Cross section (Mb)			
	This work	Prince et al. [12]	Saito and Suzuki [9] Suzuki and Saito [11]	Esteva et al. [5]
Below $1s^{-1}3p$ resonance				
858.8			0.0228(10)	
864	0.031(1)			
865	0.039(1)	0.090(5)		
Resonance energy of $1s^{-1}3p$				
measured	2.53(1)	1.6(1)		0.9
estimation at infinite resolution	2.68(1)	1.94(10)		
Above the $1s^{-1}$ threshold				
871	0.370(1)		0.373(3)	
873		0.45(5)	0.371(3)	
879			0.357(3)	0.36

ACKNOWLEDGMENTS

The experiments were performed at SPring-8 with the approval of the program review committee (2004B0369-NSb-np). The work was partially supported by the Grants-in-Aid for Scientific Research (B) from Japan Society for Promotion of Science and by the Budget for Nuclear Research of Ministry of Education, Culture, Sports, Science and Technology, based on the screening and counseling by the Atomic Energy Commission.

REFERENCES

1. N. Saito and I. H. Suzuki, *J Synchrotron Radiat.* **5**, 869 (1998)
2. N. Saito and I. H. Suzuki, *J. Electron Spectrosc. Related. Phenomen.* **101-103**, 33 (1999).
3. F. Wuilleumier, *C. R. Acad. Sci. Paris B* **270**, 825 (1970).
4. A. P. Hitchcock and C. E. Brion, *J. Phys. B: At. Mol. Opt. Phys.* **13**, 3269 (1980).
5. J. M. Esteva, B. Gauthé, P. Dhez and R. C. Karnatak, *J. Phys. B: At. Mol. Opt. Phys.* **16**, L263 (1983).
6. C. M. Teodorescu, R. C. Karnatak, J. M. Esteva, A. El. Afif, and J. -P. Connerade, *J. Phys. B: At. Mol. Opt. Phys.* **26**, 4019 (1993).
7. L. Avaldi, G. Dawber, R. Camilloni, G. C. King, M. Roper, M. R. F. Siggel, G. Stefani, M. Zitnik, A. Lisini, and P. Decleva, *Phys. Rev. A* **51**, 5025 (1995).
8. M. Coreno, L. Avaldi, R. Camilloni, K. C. Prince, M. de Simone, J. Karvonen, R. Colle, and S. Simonucci, *Phys. Rev. A* **59**, 2494 (1999).
9. N. Saito and I. H. Suzuki, *Nucl. Instrum. Mehods A*, **467-468**, 1577 (2001).
10. A. De Fanis, N. Saito, H. Yoshida, Y. Senba, Y. Tamenori, H. Ohashi, H. Tanaka, and K. Ueda, *Phys. Rev. Lett.* **89**, 243001 (2002).
11. I. H. Suzuki and N. Saito, *J. Electron Spectrosc. Related. Phenomen.* **129**, 71 (2003).
12. K. C. Prince, L. Avaldi, R. Sankari, R. Richter, M. de Simone, and M.Coreno, *J. Electron Spectrosc. Related. Phenomen.* **144-147**, 43 (2005).
13. M. Oura, H. Yamaoka, Y. Senba, H. Ohashi, and F. Koike, *Phys. Rev. A* **70**, 062502 (2004).
14. M. Kato, Y. Morishita, F. Koike, S. Fritzsche, H. Yamaoka, Y. Tamenori, K. Okada, T. Matsudo, T. Gejo, I. H. Suzuki and N. Saito, *J. Phys. B: At. Mol. Opt. Phys.* **39**, 2059 (2006)
15. I. H. Suzuki and N. Saito, *J. Electron Spectrosc. Related. Phenomen.* **123**, 239 (2002).
16. I. H. Suzuki and N. Saito, *Rad. Phys. Chem.* **73**, 1 (2005).
17. H. Ohashi, E. Ishiguro, Y. Tamenori, H. Okumura, A. Hiraya, H. Yoshida, Y. Senba, K. Okada, N. Saito, I. H. Suzuki, K. Ueda, T. Ibuki, S. Nagaoka, I. Koyano and T. Ishikawa, *Nucl. Instrum. Methods Phys. Res. A* **467-468**, 533 (2001).
18. J. A. R. Samson and W. C. Stolte, *J. Electron Spectrosc. Related. Phenomen.* **123**, 265 (2002).
19. Y. Shimizu, H. Ohashi, Y. Tamenori, Y. Muramatsu, H. Yoshida, K. Okada, N. Saito, H. Tanaka, I. Koyano, S. Shin, and K. Ueda, J. Electron Spectrosc. Related. Phenomen. **114-116**, 63 (2001).
20. L. Pettersson, J. Nordgren, L. Selander, C. Nordling, K. Siegbahn, and H. Ågren, *J. Electron Spectrosc. Related. Phenomen.* **27**, 29 (1982).
21. V. Schmidt, *Electron Spectrometry of Atoms using Synchrotron Radiation* (Cambridge University Press, 1997), P274.

A Quick and Sensitive Method of Measuring the Proportion of High-harmonic Components for Synchrotron Radiation using Ionization Chambers

Nobuteru Nariyama, Shingo Taniguchi and Yoshihiro Asano

Japan Synchrotron Radiation Research Institute, Beamline Division
1-1-1 Kouto, Sayo, Hyogo 679-5198, Japan

Abstract. The proportion of high-harmonic components included in monochromatized synchrotron radiation was measured with a combination of two free-air ionization chambers. The absolute intensity of the fundamental photons was measured with a chamber having a narrow gap. For the high harmonics, the intensity attenuated through a foil set to remove the fundamental component, was measured with a chamber having a wide gap and converted to the intensity before the foil considering the attenuation. The measurement result was compared with that of an Hp-Ge detector, and close agreement was obtained. Several advantages of this method over the spectrum measurement method were confirmed in the present experiment: namely, a short measuring time, adequate sensitivity even to low high-harmonic proportions and availability at low energies.

Keywords: High harmonic, Ionization chamber, Ge detector, Absolute intensity
PACS: 29.40.Cs, 87.66.Jj, 07.85.Qe

INTRODUCTION

High-harmonic components in synchrotron radiation, which have higher energies than the fundamental and are included inevitably due to Bragg diffraction, are unfavorable in most experiments; hence, the proportion is often lowered by inserting mirrors or detuning the monochromater angles slightly. Because the source intensity is too strong for the pulse measurement, the effect has often been confirmed by measuring the scattered photon spectra. The method, however, requires much time, especially for a small proportion of high harmonics, and cannot be applied to extremely high-intensity photons.

Measurement of the absolute intensity of synchrotron radiation is possible with free-air ionization chambers [1, 2]. The chambers have higher sensitivity to lower-energy photons, and the fundamental component can be easily removed with a foil, leaving the high harmonics. Noticing these points, a quick, sensitive and wide-range measuring method for high harmonics was proposed in this study.

MEASURING METHOD

Use of Ionization Chambers

Figure 1 shows the experimental setup of the proposed and the reference methods. Two parallel-plate free-air ionization chambers were set in tandem: the front one had a narrow gap for high-intensity photons and the back one had a wide gap for high-energy photons. Before the back chamber, a thin foil was set for the purpose of attenuating the fundamental component. To make the foil transparent to high harmonics, a material was chosen that had a K-absorption-edge energy just below the fundamental energy. With the front chamber, the intensity ϕ_1 of the

CP879, *Synchrotron Radiation Instrumentation: Ninth International Conference*,
edited by Jae-Young Choi and Seungyu Rah
© 2007 American Institute of Physics 978-0-7354-0373-4/07/$23.00

fundamental component can be measured because for the high harmonics the intensity is weak, the cross section of air is small, and the possibility of electron loss due to collision with electrodes in the miniature chamber is large.

In the back chamber, the attenuated intensity ϕ'_3 of the high harmonics can be measured. The relation between ϕ'_3 and ϕ_3, the intensity of the high harmonics before the foil, is expressed as follows:

$$\varphi'_3 = B\varphi_3 e^{-\mu t}, \tag{1}$$

where μ is the photon attenuation coefficient in the foil, t the foil thickness and B the buildup factor considering the scattered and fluorescent photons. Collimating the beam just in front of the back chamber was expected to make the value of B almost unity. Thus ϕ_3 is obtained and the proportion of high harmonics is calculated as $\phi_3 / (\phi_1 + \phi_3)$. Only the third-order high-harmonic component, which was dominant, was considered. The advantages of this method are the easy setting, the short measuring time and the possibility of using even high-intensity photons.

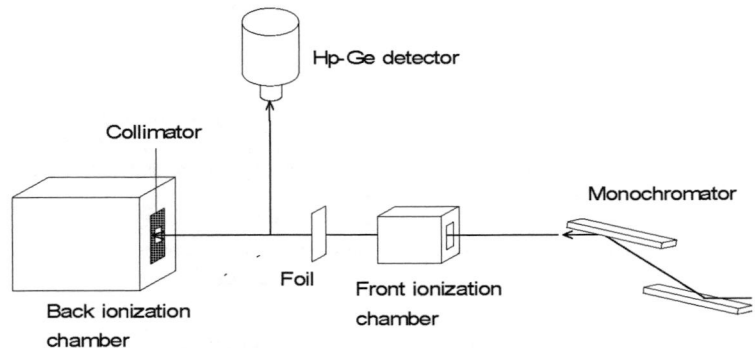

FIGURE 1. Experimental setup

Use of an Hp-Ge Detector

This conventional method measures the scattered photon spectra. For conversion from the scattered spectra to the proportion of high harmonics in the beam, the photon attenuation between the scattered point and the detector, the differential Compton-scattering cross section and the intrinsic full-energy efficiency of the Hp-Ge detector (ORTEC GLP-06165/05-P) [3] were considered.

Compton- and Rayleigh-scattered photons enter the detector. The cross section of Compton scattering is not so energy-sensitive, while the Rayleigh-scattering cross section depends on energy, as shown in Fig. 2 [4]. The differential Compton- and Rayleigh-scattering cross section to linearly-polarized photons are expressed as follows:

$$\frac{d\sigma_C}{d\Omega} = \frac{1}{2}r_0^2 \left(\frac{k}{k_0}\right)^2 \left(\frac{k}{k_0} + \frac{k_0}{k} - 2\sin^2\theta\cos^2\varphi\right) \tag{2}$$

and

$$\frac{d\sigma_R}{d\Omega} = r_0^2 \left(1 - \sin^2\theta\cos^2\varphi\right), \tag{3}$$

where r_0 is the classical electron radius, k the scattered photon energy, k_0 the incident photon energy, θ the polar angle and ϕ the azimuthal angle from the plane including the electric vector. For electron binding correction, the incoherent scattering function for Eq. (2) and the atomic form factor for Eq. (3) need consideration. Thus, the ratios between the fundamental and the high harmonics change after scattering, and the degrees differ between Rayleigh and Compton scattering. As a result, peaks due to Rayleigh and Compton scattered photons need to be separated in the spectra.

Experiment

The experiment was carried out at the BL46XU undulator beamline of SPring-8. The energies of the fundamental photons were 12 to 22 keV. The front ionization chamber had a gap of 4.2 mm [2] and the back chamber had a gap of 85 mm [1]. The back chamber was located 40 cm from the foils. The attenuation foils used were Cu, Zr and Mo, and the thicknesses were all 0.2 mm except 0.05 mm at 12 keV. A collimator made of tantalum with a square hole of 10 mm was set just before the back chamber. For the Ge detector, scattered photons from the air were measured because the overly strong intensity from solid foil induced a pile-up of pulses. For suitable intensity and easy setting, the detector was set at $\theta = \phi = 90°$. A lead collimator covered the head of the Ge detector, and mirrors were not used in the beam. Each measuring time was one minute for the ionization-chamber method and several tens of minutes for the Ge-detector method.

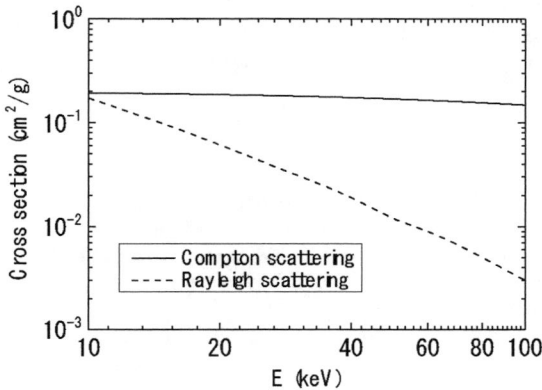

FIGURE 2. Energy dependence of the cross sections of Compton and Rayleigh scattering in air

RESULTS AND DISCUSSION

Figure 3 shows the result of the methods using the ionization chambers and the Ge detector. The two values almost agreed with each other. At 22 keV, the scattered high-harmonic photons were slight in number and a sufficient number of counts could not be obtained for the Ge-detector method in the restricted time; as a result, no value was plotted. On the other hand, the ionizing current could be measured with the ionization chamber because the photons were direct. This sensitivity was the distinguishing merit of the method. The discrepancies at 20 and 21 keV were also due to the small counts from 200 to 1000 of the Ge detector.

At 12 keV, the energy of Compton-scattered photons is 11.7 keV, so that the two peaks by Rayleigh- and Compton-scattered photons could not be separated clearly in the Ge-detector method. On the other hand, the fundamental photons remained even behind the Cu foil of 0.05 mm thickness; their proportion was 5.6%. Owing to their inclusion, the proportion of high harmonics was overestimated by 15% in the ionization-chamber method. Using a thicker foil could have solved this problem.

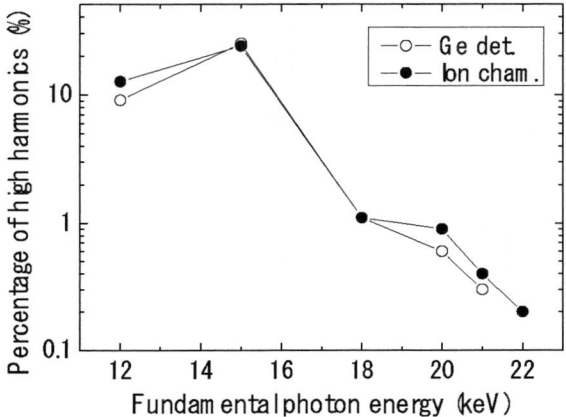

FIGURE 3. Proportions of high harmonics, measured by the ionization-chamber and the Ge-detector methods

In Fig. 3, the proportion of high harmonics was as much as 24% at 15 keV. Even in this unusual case, the overestimation in the ionization-chamber method was only 3.6%, and the measuring range was wide.

When the collimator before the chamber was removed, the current clearly increased. Table 1 represents the ratio with and without the collimator. The increasing degree was large, especially at 12, 20 and 21 keV, which were just above the absorption-edge energies of the foils used: 8.98, 18.0 and 20.0 keV for Cu, Zr and Mo, respectively. That is, this effect owed to the fluorescent photons, rather than the scattered photons, emitted from the foils. Even using the collimator, however, some fluorescent photons may possibly exist; hence, their influence was estimated with the Monte-Carlo photon/electron simulation code EGS4 [5]. Table 1 summarizes the ratio of the current by K-X ray contamination to that by the high harmonics in the same geometry as Fig. 1. The collimator was found to adequately lower the contamination. Because fluorescent photons are emitted isotropically, the flux varies in inverse proportion to the square of the distance; a smaller collimator hole and longer distance between the foil and the back chamber could decrease the influence further.

TABLE 1. Effect of the 10 mm by 10 mm collimator in the ionization-chamber method. The second column denotes the measured ratio of the current without the collimator to that with the collimator. The third column shows the ratio of the current by K-X rays from the foils to that by the high harmonics in the back chamber with the collimator, calculated by photon/electron simulation.

E (keV)	Current increase without the collimator	Current increase by K-X rays	Foil
12	20%	0.48%	Cu
15	2%	0.00072%	Cu
18	<1%	0.16%	Cu
20	17%	1.1%	Zr
21	11%	1.2%	Mo
22	1%	2.8%	Mo

CONCLUSION

The proposed method using two ionization chambers was found to be superior to the spectrum method with respect to the sensitivity even for slightly high harmonics regardless of the short measuring time required. Even at a low fundamental energy of around 10 keV, where the peaks of Compton- and Rayleigh-scattered photons overlapped with each other in the spectrum method, the present method was applicable. Fluorescent photons emitted from the attenuating foils could be removed satisfactorily using a collimator before the ionization chamber.

ACKNOWLEDGMENTS

The experiments were performed at the BL46XU in the SPring-8 with the approval of the Japan Synchrotron Radiation Research Institute (JASRI) (Proposal No. R04B46XU-0031N).

REFERENCES

1. N. Nariyama, N. Kishi and S. Ohnishi, *Nucl. Instru. Meth. A* **524**, 324-331 (2004).
2. N. Nariyama, *Rev. Sci. Instru.* **75**, 2860-2862 (2004).
3. http://www.ortec-online.com/detectors/photon/a5_2.htm
4. http://physics.nist.gov/PhysRefData/Xcom/Text/XCOM.html
5. W.R. Nelson, H. Hirayama and D.W.O. Rogers, SLAC-265 (1985).

Development of the Soft X-ray Intensity Measurement with a Cryogenic Radiometer

M. Kato, A. Nohtomi, Y. Morishita, T. Kurosawa, N. Arai,
I. H. Suzuki and N. Saito

*National Institute of Advanced Industrial Science and Technology (AIST), NMIJ, Tsukuba,
Ibaraki 305-8568, Japan*

Abstract. A cryogenic radiometer has been revised and examined in order to improve the absolute measurement of soft X-ray intensities. The uncertainty of the temperature rise in the cavity absorber has decreased to 0.2 % through an increase in the thermal responsivity and through the suppression of the fluctuation in the back ground temperature.

Keywords: cryogenic radiometer, soft X-ray, primary standard, absolute intensity, synchrotron radiation
PACS: 06.20.F, 07.60.Dq, 29.40.-n

INTRODUCTION

Recent developments of intense light sources such as synchrotron radiation enable us to use brilliant soft X-rays in many research fields. Micro-fabrication and characterization of different materials and micro analyses of biological samples have been investigated using soft X-rays from laser-plasma and synchrotron radiation sources [1]. One of important and fundamental quantities in an efficient utilization of the soft X-ray beam is its absolute intensity. A cryogenic radiometer, which was developed at Physikalisch Thechnische Bundesanstalt (PTB) [2], is used as a primary standard detector for measuring the absolute intensity of the soft X-ray beam. Rabus *et al.* reported that the radiant power can be typically measured using the cryogenic radiometer with a relative standard uncertainty of less than 0.2 % [2]. A multi-electrode ion chamber is another instrument as a primary standard. The absolute intensities of the soft X-rays below 1 keV are measured using the ion chamber with a relative standard uncertainty of about 2 % at National Institute of Standard and Technology (NIST) [3] and AIST [4,5]. However the measurement with ion chambers possesses an advantage that the accuracy of the measurement may be almost constant even if the intensity is low, in comparison with the measurement using the radiometer.

It is important to compare the measurement results obtained using these two methods, which are based on the different principals. The comparison between the measurement result with the cryogenic radiometer and that using the ion chamber is profitable for examining the reliability in the intensity standard of the soft X-rays and for obtaining the correct absolute intensity. We have performed the preliminary measurement of the comparison [5]. The uncertainty of the intensity measured using a previous cryogenic radiometer was large, and we cannot discuss the comparison result in detail and have not yet established the primary standard on a high accuracy. In the present study, the cryogenic radiometer has been redesigned and fabricated in order to improve the performance of the soft X-ray intensity measurement.

EXPERIMENT

Apparatus

The experiment was performed at the beamline 11B at the Photon Factory, Institute of Materials Structure Science, KEK. Synchrotron radiation from the 2.5 GeV storage ring was monochromatized by a double crystal

CP879, *Synchrotron Radiation Instrumentation: Ninth International Conference*,
edited by Jae-Young Choi and Seungyu Rah
© 2007 American Institute of Physics 978-0-7354-0373-4/07/$23.00

monochromator [6-8]. The monochromatized synchrotron radiation with an energy width of about 2 eV was used at the energy of 3300 eV for performance tests of the new radiometer. The photon intensity was monitored with detection of photocurrents from a thin graphite film just after the monochromator. The measurement system and the experimental procedure have been described in detail elsewhere [5]. A schematic diagram of the cryogenic radiometer is shown in Figure 1.

The radiometer was cooled down with liquid helium. The cavity absorber which was made of copper is able to absorb the soft X-ray beam almost completely. The cavity absorber was connected to the reference block using poles and strings. The soft X-ray beam was introduced to the cavity through the entrance aperture of 8 mm in diameter when the gate valve was open.

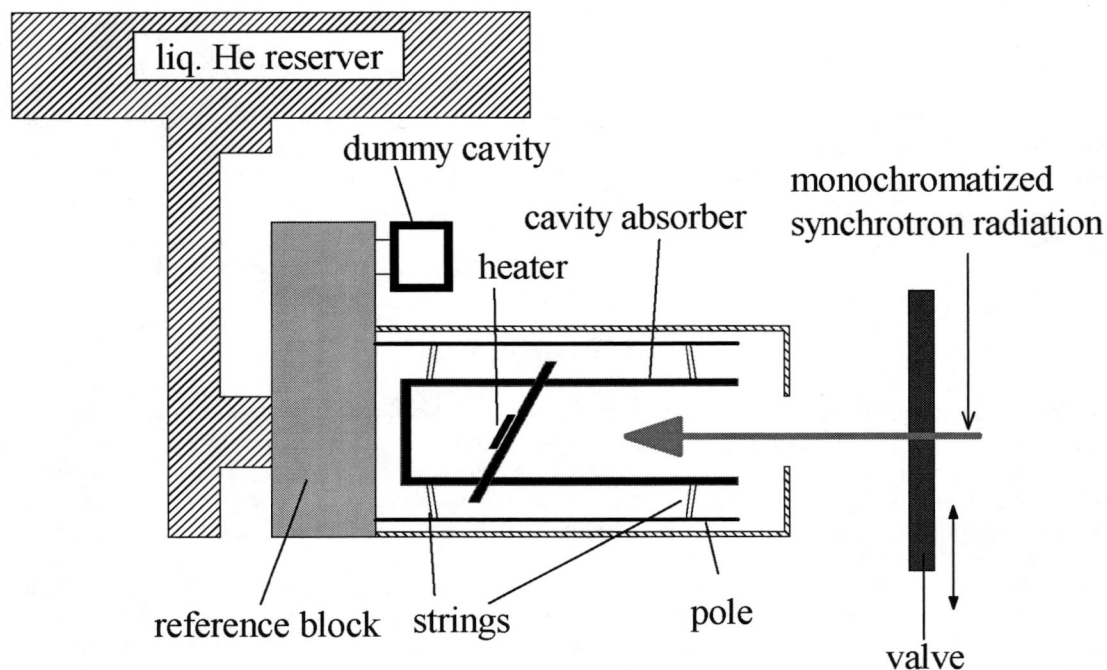

FIGURE 1. Schematic diagram of the cryogenic radiometer.

Procedure for Measurement of the Absolute Soft X-ray Intensity

The soft X-ray intensity represented with the radiant power P (W) is given by

$$P = \Delta T / s, \tag{1}$$

where ΔT (K) denotes the temperature rise in the cavity absorber, and s (K/W) indicates the thermal responsivity of the cavity absorber. Figure 2 shows the time variation in the temperature of the cavity absorber. The term ΔT is obtained as the difference between the temperature with the gate valve being open and that with the valve being closed (see Fig. 1 and Fig. 2). The thermal responsivity s was obtained as a quotient of the temperature rise divided by the input power to heater. For obtaining the correct value of s, we measured the temperature rise as a function of the heater power supplied by an electric circuit.

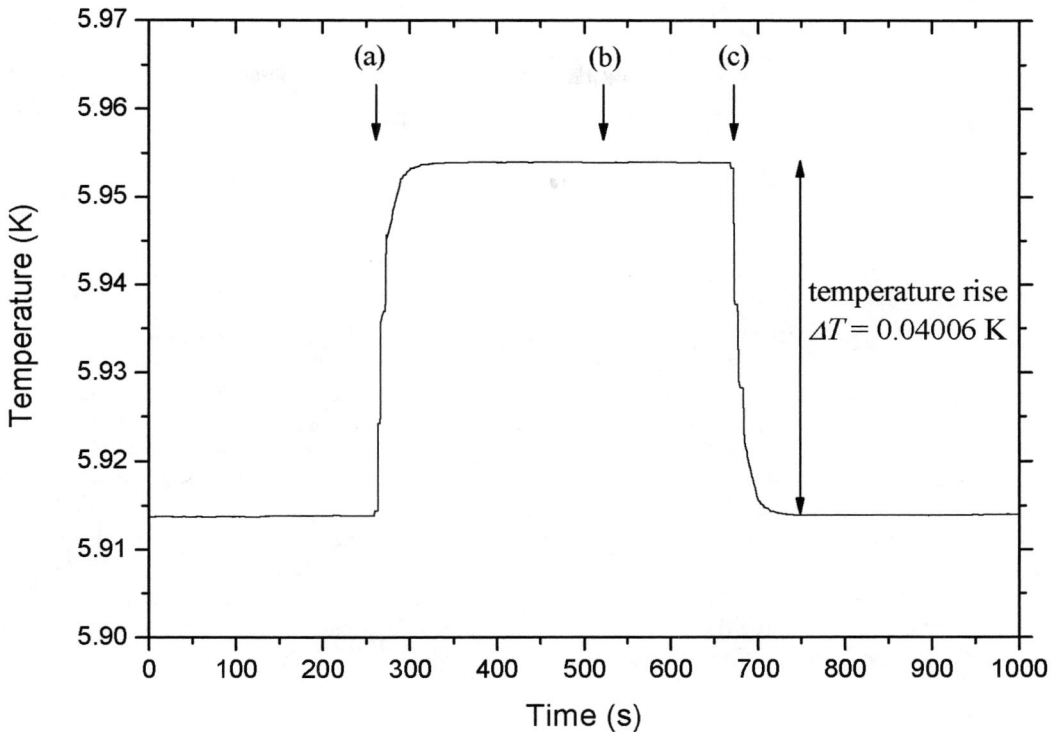

FIGURE 2. Variation of the temperature in the cavity absorber as a function of time lapse. The incident photon energy is 3300 eV. (a) The valve in front of the radiometer (see Fig. 1) opened and the monochromatized synchrotron radiation entered the cavity absorber (at about 260 sec). (b) The temperature of the cavity rose by 0.04006 K and then became steady. (c) The temperature decreased to the background level with the valve closed (at about 660 sec).

Revision in the Cryogenic Radiometer

The cryogenic radiometer has been revised at following three points from the previous instrument [5]: 1) The weight of the cavity absorber has become lighter than that used previously. 2) The method of installing the cavity absorber has been changed, i.e. the cavity absorber is connected to the reference block using poles and strings instead of a 1 mm thickness spacer made of plastic (see Fig. 1). 3) The dummy cavity was set in order to use its temperature as the reference for measuring the cavity absorber temperature. The thermal responsivity s is expected to increase because the heat capacity of the cavity absorber became smaller from the first point than the previous one. The second makes lower the heat conduction efficiency from the absorber to the reference block. The measured temperature of the cavity absorber is expected to be more stable through the third point because the fluctuation in the temperature of the reference block has not made affect to the measured temperature directly.

The performance of the present cavity absorber is summarized in Table 1, together with that of the previous cavity absorber. The present value on the thermal responsivity of the cavity of 5.46 mK/μW is three times as high as the previous value. The background stability improved from 0.1 mK to 0.05 mK. The uncertainty of the temperature rise has been decreased from 10 % to 0.2 % as a result. The performance improvement in the present radiometer clearly becomes visible, as seen in Table 1.

TABLE 1. Thermal responsibility of the cavity absorber, fluctuation in the background temperature and uncertainty of the temperature rise.

	Thermal responsivity (mK/μW)	Fluctuation in the background temperature (mK)[a]	Uncertainty of the temperature rise (%)
Present radiometer	5.46	0.05	0.2
Previous radiometer [5]	1.8	0.1	10

a) The fluctuation in the background temperature for a typical measurement duration of 5 minutes.

Comparison with an Ion Chamber

The absolute photon intensity was measured at the energy of 3300 eV using a multi-electrode ion chamber supplied with dry air, for comparison with the photon intensity measured using the revised radiometer. The experimental apparatus and the procedure for measurements of the photon intensities using the multi-electrode ion chamber were described in detail elsewhere [4,5]. The gas pressure in the ion chamber was 2.6 or 5.3 KPa. The W-value of dry air we used was 33.97 J/C [9]. The ion chamber provided the absolute intensity with a relative standard uncertainty of 1% in this measurement. The absolute intensities measured using the two instruments were normalized with the photocurrents from the graphite film for correction of the attenuation of the incident photon fluence, which changes with the decay of the storage ring current. The ratio between the normalized intensity measured using the ion chamber and that using the radiometer has been obtained to be 0.98 ± 0.01. This deviation from unity seems to be due to the W-value used. The W-value at 3300 eV is probably higher than 33.97 J/C because the energy of 3300 eV is just above the Ar K edge of 3206.26 eV [10] and because the W-value often receives an atomic shell effect of constituent atoms or molecules [11].

CONCLUSION

The cryogenic radiometer has been revised and examined in order to improve the absolute measurement of soft X-ray intensities. The uncertainty of the temperature rise in the cavity absorber of the revised radiometer has been 0.2 %. The photon intensity measured at the energy of 3300 eV using the multi-electrode ion chamber has been slightly lower than that using the radiometer. The W-value of dry air in the energy region just above the Ar K edge is suggested to be higher than that in the other energy regions.

ACKNOWLEDGMENTS

The experiments were performed at KEK with the approval of Photon Factory Program Advisory Committee for proposal No. 2006G217. The work was partially supported by the Budget for Nuclear Research of Ministry of Education, Culture, Sports, Science and Technology, based on the screening and counseling by the Atomic energy Commission. The authors wish to thank the staff group at TERAS and SPring-8 where the preliminary experiments were carried out. The authors are grateful to Dr. Y. Kitajima for valuable advices at BL 11B at Photon Factory. IHS wishes to express an appreciation for support by Dr. K. Ito.

REFERENCES

1. H. Saisho, Y. Gohshi, *Applications of Synchrotron Radiation to Material Analysis*, Elsevier Science, Amsterdam (1996)
2. H. Rabus, V. Persch, G. Ulm, *Appl. Opt.* **36**, 5421 (1997)
3. L. R. Canfield, *Appl. Opt.* **26**, 3831 (1987)
4. N. Saito and I. H. Suzuki, *J. Electron. Spectrosc. Relat. Phenom.* **101-103**, 33 (1999)
5. Y. Morishita, N. Saito and I. H. Suzuki, *J. Electron Spectrosc. Relat. Phenom.* **144-147**, 1071 (2005)
6. M. Funabashi, M. Nomura, Y. Kitajima, T. Yokoyama, T. Ohta and H. Kuroda, *Rev. Scl. Instrum.* **60**, 1983 (1989)
7. T. Ohta, P. M. Stefan, M. Nomura and H. Sekiyama, *Nucl. Instrum. Meyhods A* **246**, 373 (1986)
8. Y. Kitajima, *J. Electron Spectrosc. Relat. Phenom.* **80**, 405 (1996)
9. M. Boutillon and A. M. Perroche-Roux, *Phys. Med. Biol.*, **32**, 213 (1987)
10. M. Breinig, M. H. Chen, G. E. Ice, F. Parente, B. Crasemann and G. S. Brown, *Phys. Rev. A* **22**, 520 (1980)
11. N. Saito and I. H. Suzuki, *Radiat. Res.*, **156**, 317 (2001)

Properties of Wide-dose-range GafChromic Films for Synchrotron Radiation Facility

Nobuteru Nariyama

Japan Synchrotron Radiation Research Institute, Beamline Division
1-1-1, Kouto, Sayo, Hyogo 679-5198, Japan

Abstract. GafChromic films have been used at SPring-8 to detect the intensively irradiated parts and protect them from damage by being covered with shield or moved. To extend the usable dose range more widely, a new type of sensitive film EBT was investigated for the introduction. Calibration curves were obtained irradiated with ^{60}Co γ rays and compared with those of other GafChromic films. For the application, these films were set in the white x-ray hutch and the dose distribution was measured. Ratio of doses given by EBT and XT-R indicated the degree of the photon spectrum hardness, which depended on the positions. As a result, dose range from 50 mGy to 300 kGy became available for dose distribution measurements, and a set of films having different energy responses was found to give information of photon spectra.

Keywords: GafChromic film, Synchrotron radiation, Radiation damage, Radiochromic film, Dose distribution
PACS: 07.89.+b, 87.53.Dq, 87.66.Cd, 87.66.-a

INTRODUCTION

In SPring-8, a radiochromic GafChromic film HD-810 is increasingly used for measuring the dose distribution for radiation damage at storage ring [1,2]. At the ring, intense scattered photons from the crotch absorbers hit components locally because the complicated structures exist around the absorbers. The pinpoint-hit positions have been successfully detected with the thin HD-810 film of 20.3 by 25.4 cm area, which has wide dose range up to 100 kGy [3,4]. Dose at the positions just above the crotch absorbers, however, exceeds this value in one cycle.

Also, in optics and experimental hutches of beamlines, radiation damage has been given to electrical equipments, while the dose level is much weaker than that of the storage ring. The access time is not so limited such as the storage ring that a measurement for short time is possible, that is, low-dose measurement is more convenient. Recently, a new sensitive type of GafChromic film EBT has become available. In this study, properties of EBT and a film for high dose were investigated and applied to the radiation environment of SPring-8.

MATERIALS AND METHODS

The films used were HD-810, MD-55, HS, XT-R and EBT of ISP Technologies. The radiation-sensitive form is based on polydiacetylene and the base is polyester. The thickness is 107.55 μm for HD-810, 278 μm for MD-55, 234 μm for EBT and HS, and 215 μm for XT-R. All the films turn blue when irradiated except XT-R, which becomes orange.

The films were irradiated with ^{60}Co γ rays at 0.02-500 Gy at Koga Institute of Japan Radioisotope Association and Osaka Prefecture University, and the optical density (OD) was measured. The film of XT-R is a reflection type, of which the reading was made with an economical flat-type color scanner (Epson GT-X800) and analyzed with Image-Pro software. The other films, transparent types, were read with a Vidar VXR-16 gray-scale scanner and analyzed with RIT113 software.

CP879, *Synchrotron Radiation Instrumentation: Ninth International Conference*,
edited by Jae-Young Choi and Seungyu Rah
© 2007 American Institute of Physics 978-0-7354-0373-4/07/$23.00

For applications, the films were set at eleven positions in the optics hutch 2 of BL28B2, white X-ray beamline, and irradiated for 170 minutes. Some films near the beam were removed in half or one hour. The ring current was 99.0 mA at top-up mode and no attenuators were used at the beamline. Moreover, HD-810 and GEX B3 film [5] were irradiated directly with white X rays at 120 cm from the beam exit to measure the dose rate. The irradiation period was controlled with a shutter and the open periods were 0.2 to 0.8 sec for HD-810 and 0.2 to 500 sec for B3. The B3 films were read with a Hitachi U-3310 spectrophotometer and the optical density at 554 nm was measured.

RESULTS AND DISCUSSION

Figure 1 shows the relation between the dose and the net optical density. The sensitivities of MD-55 and HS were almost the same. The XT-R, which has the narrow dose range, showed higher sensitivity when red reading was used; in the figure, OD at red reading is shown. The highest sensitivity was obtained for EBT: above 0.05 Gy. The sensitivity, however, depended on the scanning orientation. That is, scanning in the longitudinal orientation produced higher sensitivity than that in the orthogonal orientation. The net OD was the same. To enhance the sensitivity, a sharp-cut optical filter removing shorter wavelengths is effective [6]. Figure 2 shows the comparison with and without a filter SC56 removing the wavelengths below 560 nm. The net OD with the filter became larger than that without the filter; the sensitivity was enhanced.

Figure 3 shows the measured dose-rate distribution at the beam height in BL28B hutch 2; the measurement positions are indicated as the corresponding number in Fig. 4. The doses were high at the positions of 1, 3 and 7, which were along the beam axis. On the other hand, the doses at 4, 5, 6, 9, 11 and 12 were weak, which were far from the axis. The values of EBT, MD-55 and HD-810 agreed with each other and only those of XT-R were larger than the others.

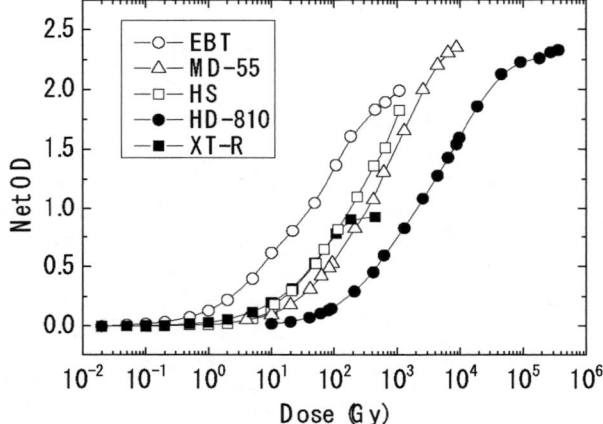

FIGURE 1. Calibration curves at ^{60}Co γ rays

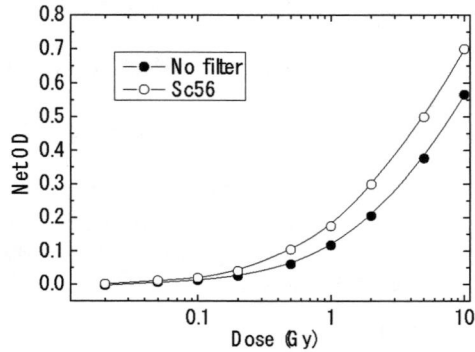

FIGURE 2. Net OD of EBT with and without SC566 optical filter

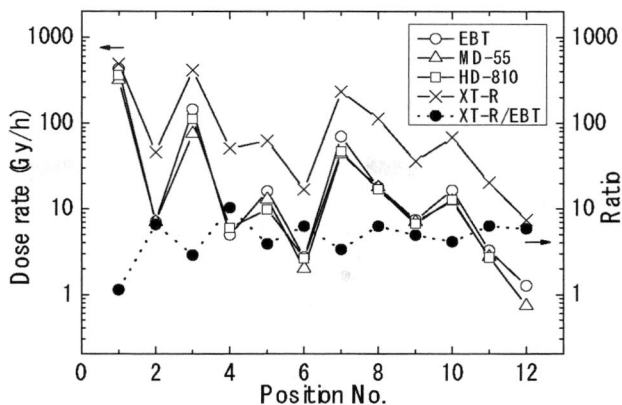

FIGURE 3. Dose rates in BL28B2 hutch 2. Each measurement position is indicated in Fig. 4. The solid circle denotes the ratio of doses given by XT-R and EBT.

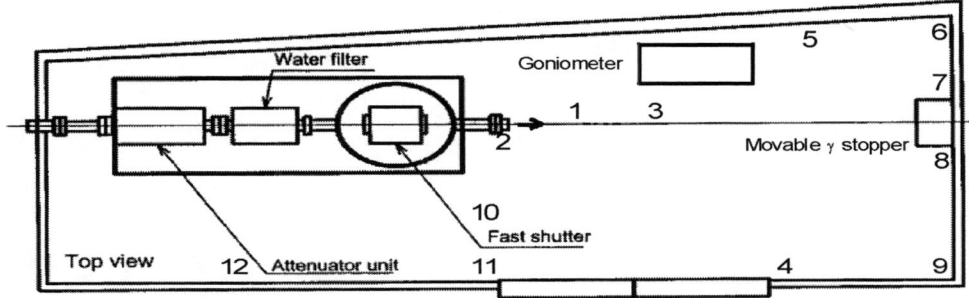

FIGURE 4. Horizontal plan of the BL28B2 hutch 2. The position numbers correspond to those of Fig. 3. Position 2 is low below the beam exit and the other positions were approximately at the beam height.

Energy response of XT-R considerably differs from the others. Figure 5 shows the responses [3,7,8]. Here, the definition of the energy response is described. A dosimeter is often calibrated at ^{60}Co γ rays, in which the signal such as OD is normalized by the absorbed dose in air, $(S/D_{air})_{Co-60}$. For the ideal dosimeter, at any energy the signal per absorbed dose in air, $(S/D_{air})_E$, agrees with $(S/D_{air})_{Co-60}$. The energy response R(E) is defined as

$$R(E) = \frac{(S/D_{air})_E}{(S/D_{air})_{Co-60}} \quad (1).$$

The response of the ideal dosimeter is unity all over the energies. The dosimeter shows D_{air} exactly. Returning to Fig. 5, it is shown that EBT has the most excellent response among the four kinds of films; on the other hand, the response of XT-R is far from the others. This result is related with the components included in the films; XT-R contains heavy elements to enhance the sensitivity.

Because the response ratio of XT-R to EBT varies with energy, the ratio can be used as the index of energy spectrum. Figure 3 shows the ratio at each position. At positions 2 and 4, the ratios are 7 and 10; the ratio of 10 is indicated between 30 and 70 keV in Fig. 5, that is, the spectra are soft. At positions 1, 3, 5 and 7, the ratios are small; the ratio of unity corresponds to 200-300 keV and that of three to 150 keV, which means that the spectra are hard. Along the beam, high-energy components remained but apart from the beam the low-energy components dominated. Thus the combination of films having different response was found to give information of the energy spectra.

In the white X-ray beam, dose rate in the strongest part was 11.3 kGy/s with HD-810. The B3 film showed damage of the paper support from 300 sec. Figure 6 represents the OD of B3 as a function of irradiation period, which increases linearly until 10 sec, 113 kGy, and monotonously until 50 sec, 565 kGy. In the reference [5], the OD is linear up to 20 kGy and saturates at 300 kGy. This film is expected for high-dose measurement.

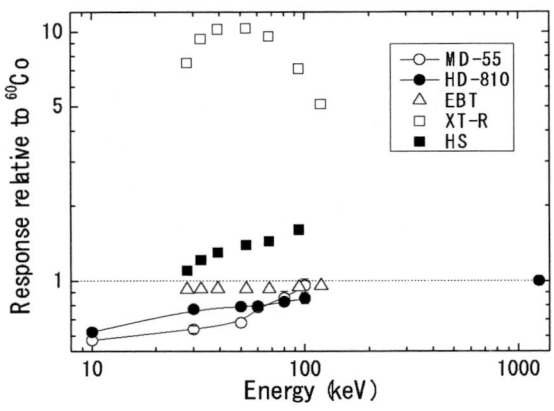

FIGURE 5. Energy responses of the GafChromic films [3,7,8]

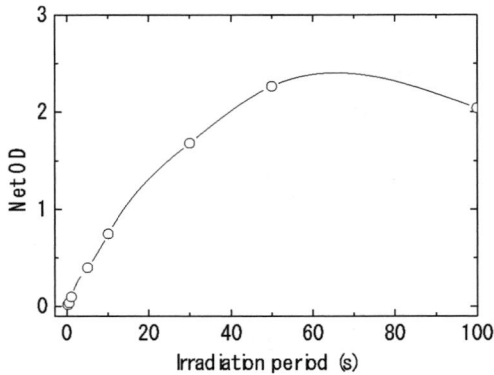

FIGURE 6. OD of B3 film as a function of irradiation period at BL28B2 to white X rays

CONCLUSION

The combination of EBT and HD-810 GafChromic films have made the in-house two-dimensional wide dose measurement possible from 0.05 Gy to 300 kGy. In addition, B3 film could extend the dose range further.

REFERENCES

1. K. Tsumaki, S. Matsui, M. Ohishi and Y. Taniuchi, SPring-8 Document A 2005-002, 122-130 (2005).
2. M. Oishi, T. Yorita, Y. Taniuchi, N. Nariyama, T. Tateishi, H. Yonehara and H. Ohkuma, *J. Vac. Soc. Jpn.* 49, 132-134 (2006) (in Japanese).
3. N. Nariyama, *Appl. Radiat. Isot.* 62, 693-697 (2005).
4. N. Nariyama, T. Magome, M. Ohishi and H. Yonehara, SPring-8 Document A 2005-002, 131-136 (2005).
5. A. Miller, W. Batsberg and W. Karman, *Radiat. Phys. Chem.* 31, 491-496 (1988).
6. http://www.ispcorp.com/products/dosimetry/index.html
7. M. J. Butson, Tsang Cheung and P. K. N. Yu, *Phys. Med. Biol.* 50, N195-N199 (2005).
8. M. J. Butson, Tsang Cheung and P. K. N. Yu, *Appl. Radiat. Isot.* 64, 60-62 (2006).

A Fast Gaseous Integrating Detector to Study Chemical Kinetics in SAXS

K. Medjoubi[*], S. Hustache-Ottini[*], T. Bucaille[*], M. Bordessoule[*], J. Chamorro[*], G. Chaplier[¶]

[*]Synchrotron SOLEIL, 91192 Gif sur Yvette cedex, France
[¶]SATIE ENS CACHAN 61 av du President Wilson 94235 CACHAN Cedex

Abstract. A detector for the study of fast kinetics in chemical reactions with a third generation synchrotron radiation beam is described. Its design makes it able to image, in one dimension, every millisecond X-ray scattering cone from isotropic sample with an efficiency greater than 50 %, few photons sensitivity and a spatial resolution of 800 µm in 8 to 12 keV energy range. Internal structure and electronic read-out of this detector are detailed. Preliminary results obtained with an ^{55}Fe radioactive source and an X-ray generator are presented.

Keywords: SAXS, time resolved, gas detector
PACS: 07.85.Fv, 07.85.Qe

INTRODUCTION

Soleil is a third generation synchrotron radiation source which is being built in France near Paris. Thanks to the high flux delivered, which is many orders of magnitude higher than that emitted by classical sources, and the small beam divergence in both horizontal and vertical directions, many new experiments are possible. A Small Angle X-ray Scattering beamline at Soleil, SWING, takes advantage of these performances to study, with high spatial and time resolutions, phenomena of fast kinetics in chemical reactions. In this kind of experiment, the sample, in solution, is exposed to X-rays with an energy ranging from 8 to 12 keV. The distribution of scattered intensity is isotropic on the azimuthal angle due to the random positions and orientations of the particles inside the solution. The conic diffusion thus obtained must be recorded with a fine angular resolution and resolved in time in the millisecond domain to follow with precision evolution of the chemical reaction.

The detector presented here is especially designed for this kind of experiment. The main requirements are a spatial resolution of 800 µm, a detection efficiency higher than 50 % in the 8-12 keV range, a counting rate more than 100 MHz/detector, a read-out time of 1 ms with 10 % dead time and an X-ray sensitivity of 10 photons at 8 keV/ms/channel.

DETECTOR DESCRIPTION

The detector technology is based on a parallel geometry gaseous detector (Fig. 1) with charge integrating read-out.

Conversion Region

The detector consists of a conversion region, which is 1 cm deep and has a 20 cm diameter, coupled to a thin amplification gap. The detector is enclosed in an aluminum vessel with a 300 µm thick circular carbon fiber entrance window. On the inner side of the window, a 30 µm thick aluminized Mylar foil is used as a drift electrode. The vessel is filled with a Xe/C_2H_6(95 %/ 5 %) gas mixture at the absolute pressure of 1.5 bar. With these characteristics, the calculated quantum efficiency [1] is more than 50 % in the 8-12 keV range. Spatial resolution including parallax error is estimated from an analytical expression based on the convolution of the impulse detector response function

CP879, *Synchrotron Radiation Instrumentation: Ninth International Conference*,
edited by Jae-Young Choi and Seungyu Rah
© 2007 American Institute of Physics 978-0-7354-0373-4/07/$23.00

with the parallax effect. In the centre and at the edge of the detector, estimations of resolutions (FWHM) are respectively ~800 μm and ~1 mm at 12 keV and for a sample-detector distance of 1.25 m (minimum distance).

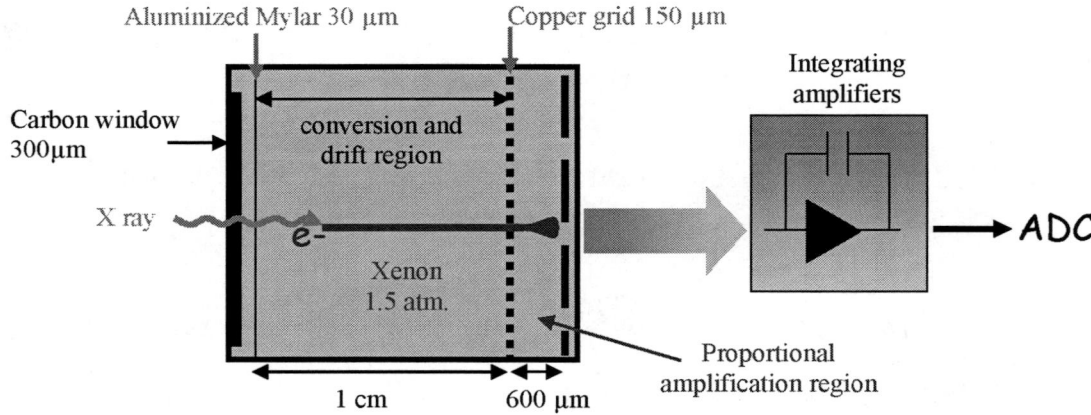

FIGURE 1. Schematic description of the detector.

Amplification Region

The amplification region consists of a thin gap based on CAT technology [2]. It is constituted by a 150 μm thick copper grid with rectangular holes of 200 μm width and pitch. The grid is separated from the anode read-out plane by 600 μm thick ©Kapton spacer (Fig 2.a and 2.b). This structure is mechanically very robust, insensitive to damages that may be caused by sparks and allows to reach high counting rates, up to several MHz locally.

Proportion of inactive areas due to the ©Kapton spacer is estimated, from the drawing, to 20 %. On the other hand, optical transparency of the grid is 40 %.

The read-out electrode is made of 128 concentric copper strips printed on a ©Kapton plane. This geometry takes advantage of the independence on the azimuthal angle of the scattering from isotropic samples. Pitch and strip width are respectively 800 μm and 600 μm.

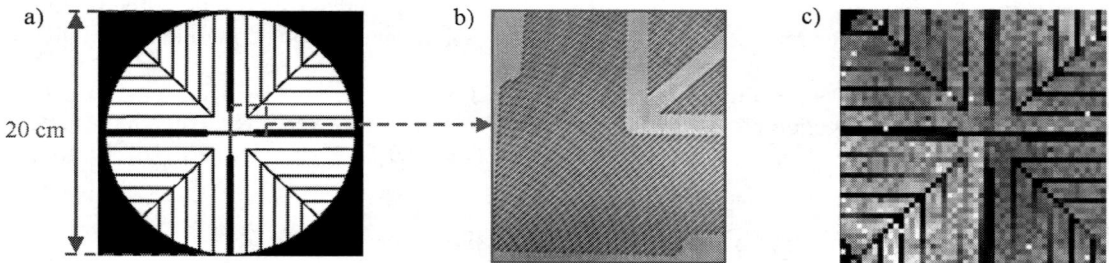

FIGURE 2. a) Drawing of the ©Kapton spacer. b) Picture of a part of the grid. c) X-ray mapping image of the detector realized with a 500 μm collimated beam.

Electronic Read-out

The 128 concentric strips are connected to the input pad of two ASICs [3] controlled by TTL signals, delivered by a Field-Programmable Gate Array. Each ASIC contains 64 low noise charge integrating amplifiers followed by one multiplexer serializing the integrated signals. The sensitivity of the amplifier is determined by its feedback capacitor which can be selected among 4 values: 1 pF, 2 pF, 4 pF or 16 pF. A complete acquisition cycle is adjusted to take 1 ms which is divided in an integration time of 900 μs and a read-out time of 100 μs. Output signal from one ASIC is constituted by a train of 64 pulses, with total a duration of 64 μs, which is repeated every millisecond. The digitization of the entire detector is performed with 2 ADC channels with 12 bits resolution and 1MHz sample frequency.

TABLE 1. Electronic read-out performances.

Integration capacitance	Noise eq. photon @ 8 keV	Maximum signal eq. photon @ 8 keV
1 pF	17	50×10^3
2 pF	24	102×10^3
4 pF	45	206×10^3
16 pF	148	853×10^3

Measured of performances of the electronics with detector disconnected are presented Table 1.

At low integration capacitance and with a gaseous amplification of 100, Table 1 shows that the system can be sensitive to less than one 8 keV photon/ms/channel. At high integration capacitance and with a gas gain of 1, the system should be able to integrate up to 8×10^5 photons @ 8 keV/ms/channel.

EXPERIMENTAL RESULTS

Inactive Areas

X-ray mapping image (Fig 2.c) of the detector has been realized with a 500 μm collimated beam produced by an X-ray tube with a copper anode and perpendicular to the entrance window. Integrated response of the detector was recorded at each beam position on a 15 cm square grid with 3 mm pitch in both directions. The detector was filled with an Ar/CO_2(95 %/ 5 %) gas mixture at the absolute pressure of 1 bar, gas gain was adjusted to ~1 and the high sensitive electronic read-out was selected.

Mapping image, presented Fig. 3, shows clearly the inactive areas due to the ©Kapton spacer. From this image, the proportion of inactive regions is 25 % which is close to the 20 % obtained from the spacer drawing.

Energy Resolution, Gas Gain and Sensitivity

Measurements described below were done with the detector filled with Xe/C_2H_6(95 %/ 5 %) at 1 bar and exposed to a non collimated ^{55}Fe radioactive source.

Energy resolution was measured using a charge pre-amplifier Ortec 142AH connected to a quarter of the copper grid. The pulse height distribution is shown in Fig. 3.a. The energy resolution is measured to be 60 % (FWHM) for a quarter of the detector area.

An image of the source obtained in 1 millisecond is shown on Fig. 3.b. High electronic sensitivity and gas gain of 100 are adjusted. The detector proves its capability to detect very few X-rays.

Figure 3.c shows the measured local gain variation, as a function of the applied grid voltage when detector is exposed to a narrow beam from an X-ray tube with a copper anode. Gain corresponds to the ratio between amplification and ionization currents. The current measurement was performed with a picoammeter Keythley 6485. A gas gain of 100 is safely reached without any spark.

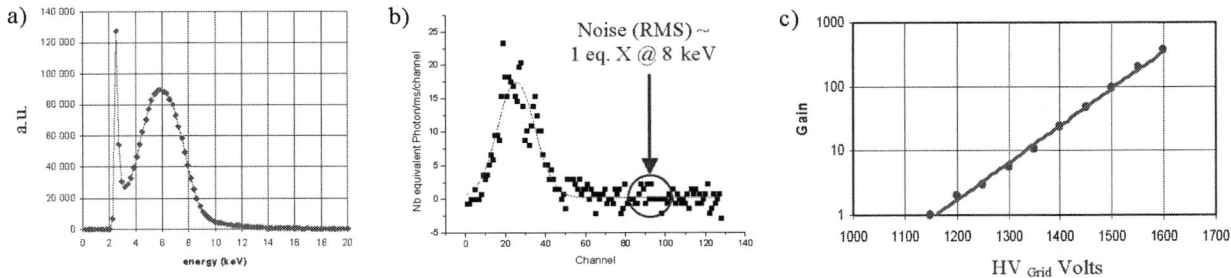

FIGURE 3. a) Pulse height distribution obtained with a non collimated ^{55}Fe source. b) Image of a non collimated ^{55}Fe source acquired in 1ms. c) Detector gain as a function of the grid voltage.

Spatial Resolution

Detector resolution has been measured with a 500 µm collimated beam from an X-ray tube with a copper anode. High electronic sensitivity and gas gain of 30 are adjusted. Detector was filled with Xe/C_2H_6 (95 %/ 5 %) at 1 bar. Detector response is shown in Fig. 4.a and presents a spatial resolution (FWHM) of 1.2 mm. Considering the beam size and the strip pitch, this resolution value is quite good.

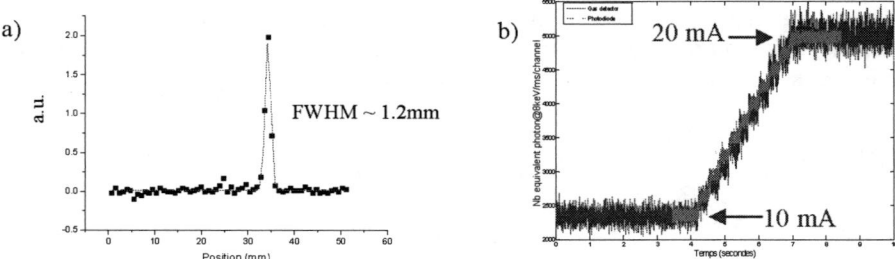

FIGURE 4. a) Spatial response with a 500 µm collimated beam. b) Detector and silicon diode signal vs increasing X-ray tube current

Acquisition

10 000 images were taken in 10 seconds in the conditions described above. During this acquisition, X-ray tube current was increased from 10 mA up to 20 mA. Pixel values corresponding to the peak are plotted every millisecond on Fig. 4.b. Signal is compared with a silicon diode response exposed to the same flux. Detector and diode values are converted into 8 keV equivalent photons. Diodes values are weighted by the ratio between detector and diode efficiencies. The two curves are very similar. Time resolved measurements in the millisecond domain with high rate (exceeding 2 MHz per channel) could therefore be performed with this gas detector.

CONCLUSION

These very preliminary results show that the detector works as expected. Few X-ray sensitivity, 1000 images per second and a 25 % inactive area demonstrates that even at 1 bar pressure, these 3 points are in agreements with required specifications. The next step is to measure its performances under the working pressure i.e. 1.5 bars and to complete its characterization in synchrotron beam

REFERENCES

1. "Compilation of X-ray cross sections" - Lawrence Radiation Laboratory – University of California – Livermore
2. M.Lemonnier et al., *J.Phys.III France* **6** (1996) 337-347.
3. "XL-1 Series 64 Channel variable-gain CMOS buffered multiplexers", Perkin Elmer Optoelectronics.

Methodological Study of a Single Photon Counting Pixel Detector at SPring-8

H. Toyokawa[a], M. Suzuki[a], Ch. Brönnimann[b], E.F. Eikenberry[b],
B. Henrich[b], G. Hülsen[b], and P. Kraft[b]

[a]*Japan Synchrotron Radiation Research Institute, 1-1-1 Kouto, Sayo-cho, Sayo-gun, Hyogo 679-5198, Japan*
[b]*Swiss Light Source, Paul Scherrer Institute, CH-5232 Villigen, Switzerland*

Abstract. PILATUS (Pixel Apparatus for the SLS) is a challenging project to develop a large area single photon counting pixel detector for synchrotron radiation experiments. SPring-8 examined the PLATUS single module detectors in collaboration with the Paul Scherrer Institute. The PILATUS-II single module detector has a desired performance with almost zero defective pixels and a fast frame rate up to 100 Hz using a newly developed PCI readout system on a Linux-PC. The maximum counting rate achieves more than 2×10^6 X-rays/s/pixel.

Keywords: 2D X-ray detector; Pixel detector; Single photon counting; Synchrotron radiation
PACS: 07.85.Fv; 29.40.-n; 29.40.Wk; 29.40.Gx

INTRODUCTION

Position sensitive 2D detectors are powerful devices for use in synchrotron radiation experiments. Imaging plates are representative of them, and CCD-based detectors have become a major tool for protein crystallography recently. These detectors, however, record X-ray intensity by integrating the energy deposited by X-ray photons. Conventional Si, Ge, and NaI detectors are, therefore, still essential instruments, when fluorescence background has to be rejected by energy discrimination. The readout time of CCD is in the second range, and that for imaging plate is minutes. It is often so inefficient and so time consuming that the required measurement becomes almost impossible in practice. In this respect, the single photon counting pixel detector is regarded as a new generation of X-ray detectors. The most important features are the following.

(i) No dark current, no readout noise and energy discrimination, resulting in maximum dynamic range.
(ii) High quantum efficiency.
(iii) Short readout time.

SPring-8 has been investigating such single photon counting pixel detectors in collaboration with the Paul Scherrer Institute (PSI) [1-5]. The developed pixel detector consists of a number of detector modules [4-6]. SPring-8 introduced a single module detector (PILATUS-I SMD) in September 2002 [2,3]. Although the PILATUS-I detector was a prototype with about 5% defective pixels in the readout chip (PILATUS-I chip) due to the DMILL 0.8 μm CMOS technology, it realized a sufficiently high performance to allow the methodological study of its fields of applications. We could study how to operate the pixel detectors in synchrotron radiation applications and how to analyze the data [4-7]. For protein crystallography, the major advantage is that the readout time is much faster than that of a CCD-based detector. In materials science, on the other hand, its major advantage is its ability to discriminate low energy X-rays below a certain threshold and to count a single X-ray photon.

Recently, an advanced pixel detector has been developed by improving the readout chips (PILATUS-II chip) with a 0.25 μm CMOS process, yielding perfect chips with no defective pixel. SPring-8 introduced a PILATUS-II single module detector (PILATUS-II SMD) in October 2005, and it is already open for user experiments. In situ characterization of directional solidification process during welding was carried out using the time resolved X-ray diffraction technique at the BL46XU [8], and a lattice distortion in antiferromagnetic CoO under high magnetic fields up to 40 T was observed at the BL19LXU [9,10].

CP879, *Synchrotron Radiation Instrumentation: Ninth International Conference*,
edited by Jae-Young Choi and Seungyu Rah
© 2007 American Institute of Physics 978-0-7354-0373-4/07/$23.00

PILATUS-I AND PILATUS-II SINGLE MODULE DETECTORS

Figure 1 shows the PILATUS-I and PILATUS-II SMDs. The parameters are listed in Table 1. A module has a single, continuously sensitive, 300 μm thick silicon sensor, which absorbs 100 % incoming X-ray radiation at 8 keV and 75 % at 12 keV. This is superior to the quantum-efficiency of direct-coupled CCD. An array of 2 × 8 custom CMOS readout chips is indium bump-bonded to the sensor. Each pixel contains a charge-sensitive amplifier, a single level discriminator and a counter. An individual pixel is thus capable of being operated in a single photon counting mode. The readout chip is the most important technology in such a hybrid pixel detector development. The PILATUS-I chip with 44 × 78 pixels was fabricated in the radiation tolerant DMILL 0.8 μm CMOS process, and even fully screened chips showed 5 % defective pixels. A subtle design oversight also caused the counters in the pixels to miscount under some circumstances. A redesigned PILATUS-II readout chip with 60 × 97 pixels in the UMC 0.25 μm COMS process has been developed to replace the PILATUS-II detector. Both of above problems have been overcome in the PILATUS-II chip. Major improvements have been the speed of the single photon counting circuit and a much higher yield of working pixels/chip. The PILATUS-II detector achieves 2×10^6 X-rays/s/pixel without counting errors, but which was limited to about 10^4 X-rays/s/pixel in the PILATUS-I. The yield of 'good' chips was about 35% in the DMILL CMOS process, but even good chips still have around 5 % defective pixels. In the 0.25 μm COMS process, on the other hand, the yield of perfect chips was more than 80%. The PILATUS-II SMD has a desired performance with almost zero defective pixels and a fast frame rate up to 100 Hz.

FIGURE 1. PILATUS-I (left) and PILATUS-II (right) single module detectors.

TABLE 1. Properties of PILATUS-I and –II single module detectors

	PILATUS-I SMD	PILATUS-II SMD
Sensor	Si (300 μm)	Si (300 μm)
Pixel size	217 μm × 217 μm	172 μm × 172 μm
Format	157 × 366 = 57,462	195 × 487 = 94,965
Active area	79.4 mm × 34.1 mm	83.7 mm × 33.5 mm
counter	15 bits	20 bits
Counting rate / pixel	$< 10^4$ X-rays/s	$> 2 \times 10^6$ X-rays/s
Readout time	> 6.7 ms	< 3.7 ms
Frame rate	< 2 Hz with VME readout	> 100 Hz with PCI readout
Threshold trimming	4 bits	6 bits

METHODOLOGICAL STUDY

A major purpose of developing a large quantum-limited pixel detector at the SLS is for macromolecular crystallography. The final detector (PILATUS-6M) will be 2463 × 2527 pixels covering 424 mm × 435 mm with 5 × 12 modules. The suitability of the pixel detector for protein crystallography was tested at the BL44B2 of SPring-8. Figure 2 shows 974 × 780 pixel and its zoom images of the diffraction pattern obtained with the PILATUS-II SMD for a themolysin crystal, which is a superposition of images taken at the 2 × 8 different positions. We also measured those for lysozyme and thaumatin crystals with the PILATUS-I SMD, and electron density maps of those have been analyzed successfully.

We have tested a fine ϕ-slicing method, because it has the potential to improve the signal-to-noise ratio (SNR) of the Bragg reflections [4,5,7]. In this mode the crystal is rotated by a fraction of its angular acceptance (0.02–0.2°) during each frame, leading to data sets of as many as 9000 frames for a full 180° rotation. Clearly, such an experiment is very time consuming with a commercial detector as CCD-based detectors. With a fast framing detector the crystal can be rotated continuously in the beam without opening and closing the shutter for each frame. For such an experiment, an electronically gateable detector is indispensable. The PILATUS-6M detector will realize such a desired detection mode.

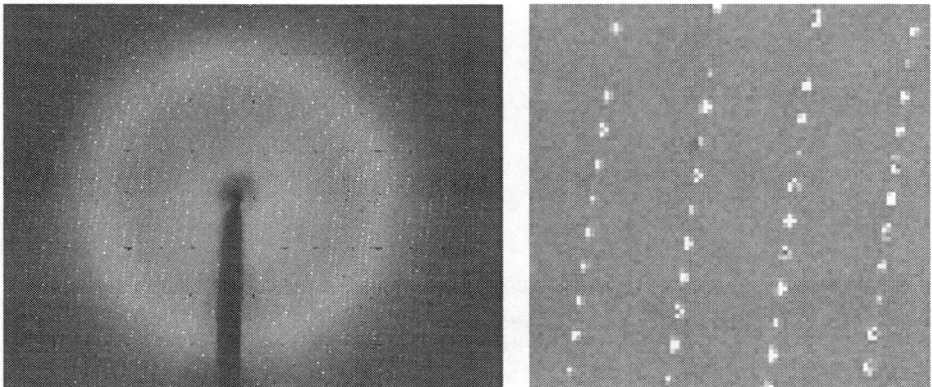

FIGURE 2. 974 × 780 pixel and its zoom images of the diffraction pattern for a themolysin crystal obtained with the PILATUS-II SMD at BL44B2.

Pixels are individually addressable, and each pixel has a 4- or 6-bit threshold trim adjustment DAC, which can be set to minimize threshold dispersion across the chip [1,4,5]. We demonstrated to discriminate low energy fluorescence X-ray below a detective threshold equalized to 15 keV by scattered X-rays with monochromatic beams at the BL46XU using the PILATUS-I SMD. Figure 3 shows diffraction patterns with a [(GaAs)7/(AlAs)3]×100 sample and intensity distributions in the 2θ axis at incident X-rays of 18 keV. Diffraction spots are a Bragg reflection peak and its higher orders. In addition, some powder diffraction circles are obtained from an aluminum base plate. The left figures show the result of the threshold of 10 keV, where week scatterings are smeared out on the interference background. In the 15 keV threshold result described as the right figures, on the other hand, higher order peaks are clearly seen.

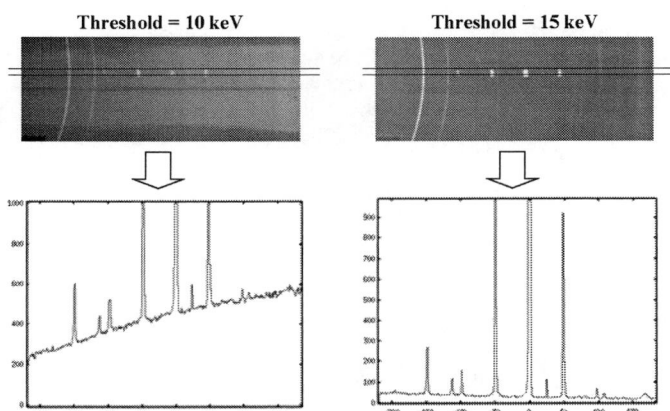

FIGURE 3. Diffraction patterns from [(GaAs)7/(AlAs)3]×100 sample at the 18 keV X-ray beam, and intensity distributions in the 2θ axis obtained with the PILATUS-I SMD at BL46XU.

The PILATUS-II SMD can be operated at the fast frame rate up to 100 Hz using a newly developed PCI readout system on a Linux-PC. In situ characterization of directional solidification process during welding was carried out using the time resolved X-ray diffraction at the BL46XU [8]. Figure 4 shows diffraction patterns for initial states in the frame rate of 20, 50, and 100 Hz. The exposure time was 42 ms, 12 ms, and 2 ms, respectively. The readout time was limited to 8 ms due to the 25 MHz readout clock at this experiment. The intensity of powder diffraction ring during the 2 ms exposure is about one count in one pixel, but it is visible due to the photon counting operation.

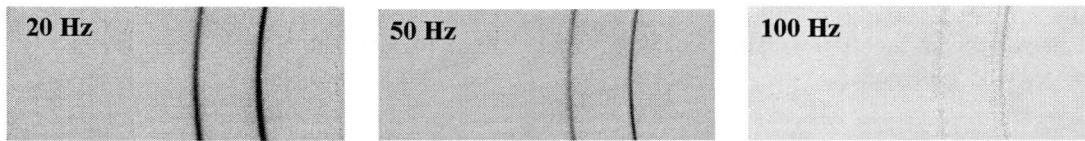

FIGURE 4. Typical fast frame rate diffraction patterns for stainless steel samples obtained with the PILATUS-II SMD at BL46XU. The frame rate was 20, 50, 100 Hz.

CONCLUDONG REMARKS

We have developed the large area single photon counting pixel detectors. Although the PILATUS-I SMD has about 5% defective pixels in the readout chip, we could study how to operate the pixel detectors in synchrotron radiation applications and how to analyze the data. The PILATUS-II SMD realizes a desired performance with almost zero defective pixels and a fast frame rate up to 100 Hz. The PSI is developing the PILATUS-6M detector with the 5×12 modules for the protein crystallography beamline X06SA at the SLS. Its quarter size with the 5×3 modules is currently under fabrication, already showing excellent, and the 6M detector will be completed soon. The PILATUS-6M will realize the fine ϕ-slicing mode with rotating a sample continuously and define the exposure time by the electronically gate. Now we are designing 1M or 2M pixel detectors for small angle scattering and other applications. The PILATUS-1M will be available at SPring-8 in this year.

ACKNOWLEDGMENTS

The authors would like to thank M. Mizumaki, N. Ikeda, S. Kimura, M. Sato, M. Yonemua, T. Otsuki, H. Terasaki, Y. Komizo for their help in performing experiments at BL46XU, N. Kamiya, T. Hikima, H. Nakajima for their help in protein crystallography data collection and analysis at BL44B2. This research was supported in part by the Ministry of Education, Culture, Sports, Science and Technology of Japan.

REFERENCES

1. E.F. Eikenberry, Ch. Brönnimann, G. Hülsen, H. Toyokawa, R. Horisberger, B. Schmitt, C. Schulze-Briese, T. Tomizaki, Nucl. Instr. and Meth. A 501 260-266 (2003).
2. G. Hülsen, E.F. Eikenberry, R. Horisberger, B. Schmitt, C. Schulze-Briese, T. Tomizaki, H. Toyokawa, M. Stampanoni, G.L. Borchert, P. Willmott, B. Patterson, Ch. Brönnimann, Proceedings of the 8th International Conference on Synchrotron Radiation Instrumentation (SRI2003): AIP Conference Proceedings 705, 1009-1012 (2004)
3. B. Schmitt, Ch. Brönnimann, E.F. Eikenberry, G. Hülsen, H. Toyokawa, R. Horisberger, F. Gozzo, B. Patterson, C. Schulze-Briese, T. Tomizaki, Nucl. Instr. and Meth. A 518 436-439 (2004).
4. Ch. Brönnimann, Ch. Bühler, E.F. Eikenberry, R. Horisberger, G. Hülsen, B. Schmitt, C. Schulze-Briese, M. Suzuki, T. Tomizaki, H. Toyokawa, A. Wagner, Synchrotron Rad. News 17, No. 2, 23-30 (2004).
5. Ch. Broennimann, E.F. Eikenberry, B. Henrich, Horisberger, G. Huelsen, E. Pohl, B. Schmitt, C. Schulze-Briese, M. Suzuki, T. Tomizaki, H. Toyokawa, A. Wagner, J. Synchrotron Rad. 13 120-130 (2006).
6. G. Hülsen, Ch. Brönnimann, E.F. Eikenberry, Nucl. Instr. and Meth. A 548 540-554 (2005).
7. Ch. Brönnimann, E.F. Eikenberry, R. Horisberger, G. Hülsen, B. Schmitt, C. Schulze-Briese, T. Tomizaki, Nucl. Instr. and Meth. A 510 24-28 (2003).
8. M. Yonemura, T. Otsuki, H. Terasaki, Y. Komizo, M. Sato, H. Toyokawa, Mater. Trans., to be published.
9. Y. Narumi, K. Kindo, K. Katsumata, M. Kawauchi, Ch. Broennimann, Y. Staub, H. Toyokawa, Y. Tanaka, A. Kikkawa, T. Yamamoto, M. Hagiwara, T. Ishikawa, H. Kitamura, J. Synchrotron Rad. 13 271-274 (2006).
10. Y. Narumi, K. Katsumata, U. Staub, K. Kindo, M. Kawauchi, Ch. Broennimann, H. Toyokawa, Y. Tanaka, A. Kikkawa, T. Yamamoto, M. Hagiwara, T. Ishikawa, H. Kitamura, J. Phys. Soc. Japan., to be published.

Thin Transmission Photodiodes as Monitor Detectors in the X-ray Range

Michael Krumrey, Martin Gerlach, Michael Hoffmann, and Peter Müller

Physikalisch-Technische Bundesanstalt, Abbestr. 2-12, 10587 Berlin, Germany
Corresponding author: Michael.Krumrey@ptb.de

Abstract. Transmission photodiodes with thicknesses between 5 μm and 20 μm have been investigated as monitor detectors for synchrotron radiation experiments in the photon energy range from 1.75 keV to 10 keV. Their responsivity and transmittance has been measured as well as the homogeneity of both quantities. Depending on the photon energy, the optimum diode thickness can be selected.

Keywords: Photodiodes, X-ray detectors, calibration
PACS: 07.85.Fv, 07.85.Qe, 85.60.Dw

INTRODUCTION

Many synchrotron radiation experiments require a monitoring of the incident photon flux using a so-called I_O detector. The ratio of the photon flux in monochromatized radiation to the storage ring current is often not constant, and ionization chambers cannot easily be integrated especially in soft X-ray UHV beamlines. Instead, the photocurrent from a metal mesh, a thin foil or a mirror coating is sometimes used, but the resulting currents are relatively low and often unstable. Silicon photodiodes of appropriate thickness, operated in transmission geometry in air or in UHV, provide high and stable photocurrents as well as high transmittance. With known responsivity and transmittance of the diodes, the transmitted photon flux can be calculated. While the measurement of the transmittance only requires monochromatic radiation of high spectral purity, the determination of the responsivity with low uncertainty also requires a detector standard. In the PTB laboratory at BESSY II, a cryogenic electrical substitution radiometer is available as primary detector standard. Responsivity and transmittance as well as the homogeneity of both quantities over the diode surface have been investigated for photodiodes from different European, Japanese and US manufacturers with thicknesses between about 5 μm and 20 μm.

EXPERIMENTAL

All measurements were performed at the four-crystal monochromator beamline which is one of the detector calibration beamlines in the PTB laboratory at BESSY II [1]. At this beamline, monochromatized synchrotron radiation of very high spectral purity is available in the photon energy range from 1.75 keV to 10 keV [2]. The beam size ranges between 0.2 mm and 0.5 mm. The thin transmission diodes were mounted in a UHV reflectometer in order to investigate the homogeneity of their transmittance and their responsivity.

The transmittance was simply measured by inserting the diode in front of another Si photodiode. The responsivity of the transmission diodes was determined against the other Si photodiode serving here as transfer detector standard. This diode was previously calibrated with relative uncertainties below 1 % against a cryogenic electrical substitution radiometer, serving as primary detector standard [3]. The homogeneity was determined by raster scans over the entire surface. The energy dependence of the transmittance and the responsivity was measured in the diode center as well as on different positions on the surface, if inhomogeneities had been found in the raster scans.

CP879, *Synchrotron Radiation Instrumentation: Ninth International Conference*,
edited by Jae-Young Choi and Seungyu Rah

RESULTS

TABLE 1. Investigated Transmission Photodiodes

Manufacturer	Country of origin	Effective area	Thickness		Measured transmittance at		Measured responsivtiy at	
			nominal	measured	4 keV	10 keV	4 keV	10 keV
		mm²	μm	μm	%	%	mA/W	mA/W
Sintef	Norway	10 x 10	6	6.5	51.3	95.1	113.0	10.9
		10 x 10	12	11.8	29.0	91.1	159.2	19.4
		10 x 10	12	13.0	25.6	90.2	186.1	23.9
Micron Sem.	UK	2.5 x 2.5	20	20.0	12.9	85.8	221.4	35.4
IRD	US	6.5 Ø	5	4.7	61.3	96.5	78.1	8.6
Hamamatsu	Japan	10 x 10	5	5.5	56.4	95.8	110.0	10.1
		10 x 10	10	10.8	33.2	92.2	174.3	20.0
for comparison: Ion chamber, air at ambient pressure, two 50 μm Be windows			50000		53	95	12.3	0.8

The investigated transmission photodiodes are listed in Table 1. The measured thickness was obtained from the transmittance measurements in the diode center. Of course, the thinnest diodes have the highest transmittance, but their responsivity is lower. The optimum thickness depends on the photon energy range and on the available and required photon flux. The photocurrents from ion chambers, which are much larger and more difficult to incorporate in UHV beamlines, are at least one order of magnitude lower, even if operated at atmospheric pressure. At higher photon energies where many experiments are performed in air, the use of ion chambers is easier, but their responsivity decreases rapidly: at 30 keV, the responsivity of the 5 cm long ion chamber mentioned in Table 1 is only 0.02 mA/W while a 100 μm thick transmission diode with 97 % transmittance has a responsivity of 9 mA/W, which is more than two orders of magnitude higher.

The transmittance and responsivity of two different thin transmission diodes is shown in Fig. 1. The calculated transmittance for 5.5 μm and 13 μm Si (in the latter case including 0.3 μm Al for the contact layers) is also shown. Just below the Si K edge at 1.84 keV, the transmittance can again rise above 60 %. The responsivity mainly depends on the diode thickness, especially at high energies. At energies below about 3 keV, the thickness of the radiation entrance layer (silicon oxide or Al) influences the responsivity. It was shown long time ago that Si p on n diodes with a thick oxide layer can degrade under irradiation due to charging effects in the oxide [4], while Si n on p diodes show much better radiation stability [5]. The same behavior is expected for thin diodes. As even the thinnest diodes will not be used in transmission below 1 keV, aluminum contact layers of up to 0.2 μm thickness are also acceptable.

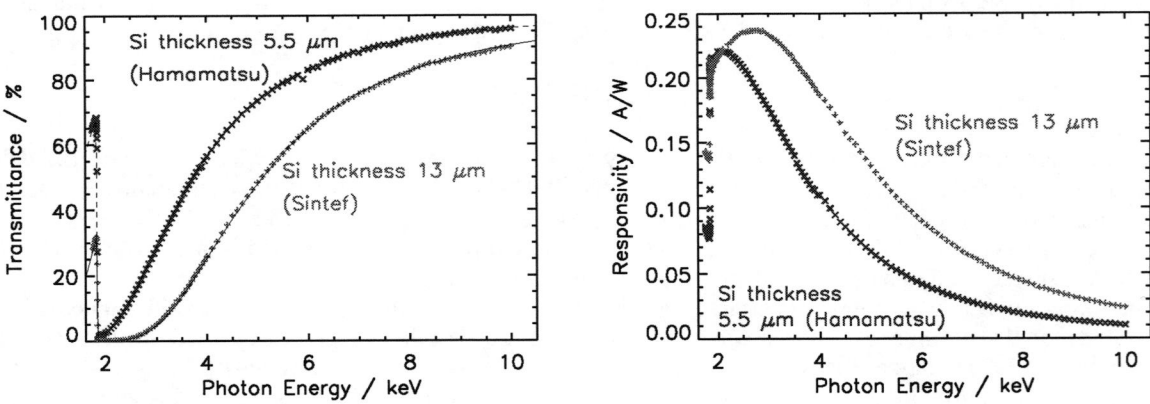

FIGURE 1. Transmittance (left) and responsivity (right) of two different transmission photodiodes. As expected, the thinner diode has a higher transmittance but a lower responsivity.

The homogeneity of the transmittance and the responsivity has been investigated for all diodes at two different photon energies, typically 10 keV and 4 keV. Some examples are shown in Fig. 2 and 3. Several diodes exhibit gradients in the detector thickness over the surface which can be observed in the transmittance and responsivity distribution. This affects the application only if the beam position changes during the experiment. Other diodes showed a homogeneous transmittance, but exhibit structures in the responsivity which can be related to crystal imperfection of this particular device. Further improvements of the homogeneity would be desirable.

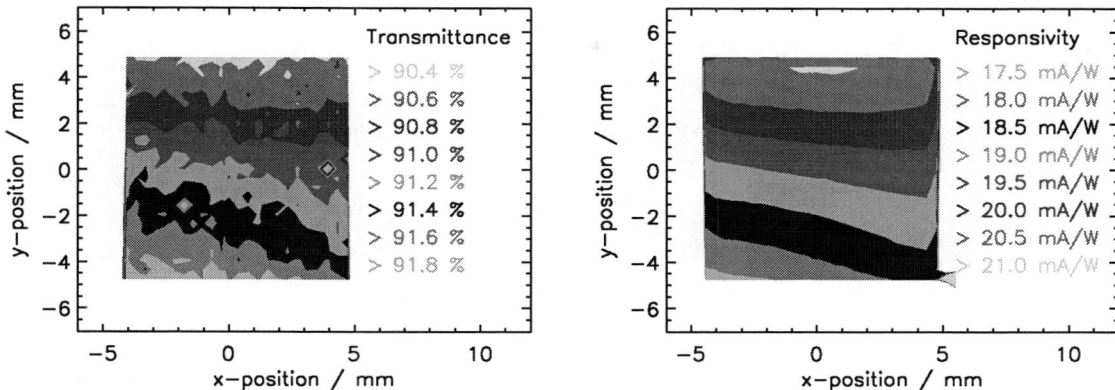

FIGURE 2. Homogeneity of an 11.8 μm thick diode (Sintef) at 10 keV. The gradient in transmittance and responsivity is related to a thickness variation between 11.0 μm and 12.4 μm.

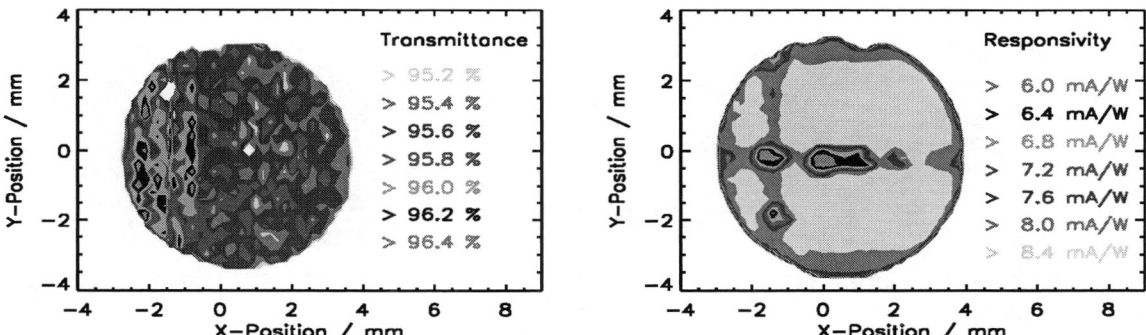

FIGURE 3. Homogeneity of a 4.7 μm thick diode (IRD) at 10 keV. While only a slight gradient is visible in the transmittance distribution (low thickness variation), the responsivity distribution seems to be affected by imperfections in the diode material.

CONCLUSION

Thin transmission photodiodes are well suited as monitor detectors for photon energies above about 4 keV. The appropriate thickness has to be selected depending on the photon energy range. The thinnest investigated diodes are only about 5 μm thick, but not all transmission diodes presented here are currently commercially available. The homogeneity should be further improved by the manufacturers, and the radiation hardness might be critical for diodes with thick oxide layer. For higher photon energies, large area transmission diodes with thicknesses between 100 μm and 500 μm are available and have been tested as well for photon energies up to 60 keV.

ACKNOWLEDGMENTS

The cooperation with John Morse (ESRF) is gratefully acknowledged. We thank all manufacturers for providing the test samples.

REFERENCES

1. A. Gottwald, U. Kroth, M. Krumrey, P. Müller, M. Richter, F. Scholze, and G. Ulm, *Metrologia* **43**, S125 – S129 (2006).
2. M. Krumrey and G. Ulm, *Nucl. Instr. and Meth.* A **467-468**, 1175 - 1178 (2001).
3. M. Krumrey, L. Büermann, M. Hoffmann, P. Müller, F. Scholze, and G. Ulm, *AIP Conf. Proc.* **705**, 861 – 865 (2004).
4. E. Tegeler and M. Krumrey, *Nucl. Instr. and Meth.* A **282**, 701 - 705 (1989).
5. F. Scholze, R. Klein, and T. Bock, *Appl. Opt.* **42**, 5621-5626 (2003).

Development of Si-APD Timing Detectors for Nuclear Resonant Scattering using High-energy Synchrotron X-rays

Shunji Kishimoto[1], Zhang Xiaowei[1], and Yoshitaka Yoda[2]

[1]*High Energy Accelerator Research Organization/CREST*
1-1 Oho, Tsukuba-shi, Ibaraki-ken 305-0801, Japan
[2]*Japan Synchrotron Radiation Research Institute/CREST*
1-1-1 Koto, Sayo-cho, Sayo-gun, Hyogo-ken 679-5198, Japan

Abstract. A timing detector with silicon avalanche photodiodes (Si-APDs) has been developed for nuclear resonant scattering using synchrotron x-rays. The detector had four pairs of a germanium plate 0.1mm thick and a Si-APD (3 mm in dia., a depletion layer of 30-μm thickness). Using synchrotron x-rays of 67.4 keV, the efficiency increased to 1.5% for the incident beam, while the efficiency was 0.76 % without the germanium converters. A measurement of SR-PAC on Ni-61 was executed by using the detector. Some other types of timing detectors are planned for x-rays of E>20 keV.

Keywords: avalanche photodiodes, x-ray detectors, high-energy x-rays, nuclear resonant scattering
PACS: 76.80.+y, 82.80.Ej, 85.60.Dw

INTRODUCTION

A timing detector using a silicon avalanche photodiode (Si-APD) shows a good time resolution of 100 ps to 1 ns, and a wide dynamic range of count-rate, 10^9 -10^{10} [1-3]. However, due to a small cross section of Si (Z=14) with photons of a high energy [4], the detector efficiency decreased to 1% for a 100-μm thick APD at 50 keV, for example. On the other hand, nuclear resonance scattering using synchrotron x-rays is now extending its application to high-energy region of nuclear excited states [5, 6]. Improving efficiency of the timing detector is requested in this field. We can still use some arrangements for Si-APDs in order to increase the efficiency of the Si-APD detectors for a high-energy photon, while keeping a good time resolution. Detecting secondary x-rays, emitting from a converter, with a Si-APD, may be useful for some experiments. A stack of Si-APDs and/or grazing-incidence alignment would be effective in some cases. Here we present the results for the detector using a germanium converter with thin Si-APDs. Another timing detector using a stack of relatively thick Si-APDs is also introduced.

AN APD DETECTOR WITH AN ENERGY CONVERTER

Detector Design

Figure 1 shows a detector consisting of stacked four pairs of a germanium (Ge) plate and a Si-APD. A silicon avalanche photodiode (Hamamatsu SPL3159) was mounted on an aluminum plate 0.6 mm thick. The APD device was 3 mm in diameter and had a depletion layer 30 μm thick. The Ge-plate 0.1 mm thick was attached to the backside of the aluminum plate by silver paste. The distance between the APD tips was 2.4 mm. There were no thick dead-layers in both surfaces of the APD device; each aluminum plate had a through-hole of 3.4 mm in diameter. The beam therefore went through the APDs and the germanium plates. The germanium plate plays a role of the energy converter for the Si-APD. The energy converter absorbs high-energy x-rays and promptly emits fluorescence x-rays, which energy is enough low for absorption in the Si-APD. Moreover, the fluorescence yield should not be too low as the converter. This property depends on the atomic number of the converter material.

CP879, *Synchrotron Radiation Instrumentation: Ninth International Conference*,
edited by Jae-Young Choi and Seungyu Rah
© 2007 American Institute of Physics 978-0-7354-0373-4/07/$23.00

Calculations were given by using a simple model for the metal converters of zirconium (Z=40), niobium (Z=41), zinc (Z=30) and a semiconductor of germanium (Z=32). The self-absorption of the fluorescence x-rays by the converter and the fluorescence yield were considered in the calculation. From the results, germanium is a positive candidate for use of a thin 30-μm APD to obtain a better time resolution. Figure 2 shows the intrinsic efficiency of the APD with a germanium converter for 20, 30 and 50-keV x-rays, as a function of the thickness of the converter. The efficiency was calculated by the simple model of a pair of a Si-APD and a converter, only for the beam direction. Around 100 μm, the efficiency saturated at 1% for 50-keV x-rays. We thus adopted the germanium converter of 100-μm thickness for the Si-APD detector.

FIGURE 1. Inside of the Si-APD detector with germanium converters.

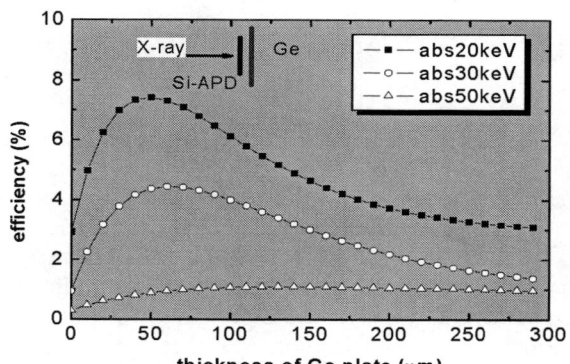

FIGURE 2. Calculation of the intrinsic efficiency of the germanium converter.

Performance For High Energy X-rays

The performance of the detector shown in Fig. 1 was measured by using synchrotron x-rays at beamline BL-14A of Photon Factory (PF). The x-ray energy was set to 67.4 keV. Figure 3(a) is the energy spectrum of the front APD plate (channel 1), measured with a charge-sensitive preamplifier. The fluorescence x-rays, Ge KX-rays, contributed to 44% of the total counts. At a lower bias voltage of the APD, -120 V, the intrinsic efficiency at 67.4 keV was measured and given in Table 1. For four channels of the detector, the efficiency reached 1.53%, two times larger than that of the detector consisting of only four Si-APDs (0.76%). We next measured the pulse-height distribution using a fast amplifier (gain: 800, bandwidth: 100k-1.5GHz) for timing measurements. The pulse-height distribution was measured by scanning the threshold level of a discriminator for outputs from the fast amplifier. The result is shown in Fig. 3(b); the count rate was about 50k s^{-1}. A good resolution was obtained even by a fast pulse of several-nanosecond width, as well as by using the charge-sensitive preamplifier. Time spectra were also measured at E = 50 keV, for the bunch-structure of the multi-bunch mode in PF. The time resolution was 0.36ns (FWHM), determined mainly by the thickness of the APD, not affected by the converters.

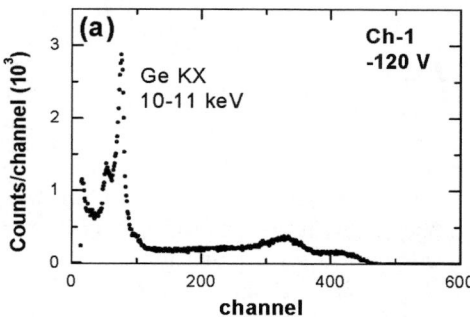

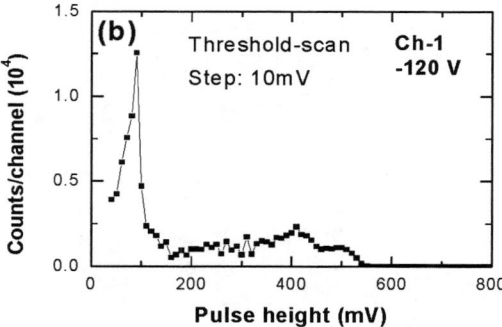

FIGURE 3. Energy spectra measured by (a) a charge-sensitive preamplifier and by (b) a fast amplifier.

TABLE 1. Intrinsic efficiency of each channel. The efficiency was 0.19% for one Si-APD plate.

Channel	Intrinsic Efficiency (%)
Ch-1	0.46
Ch-2	0.41
Ch-3	0.38
Ch-4	0.27
Sum	1.52
Cf. Si 30μm	0.19

Measurement Of Nuclear Resonant Scattering On Ni-61

We tested the detector for the timing measurements of nuclear resonant scattering on Ni-61. The first excited state of Ni-61 exists in 67.412 keV; the half-life is 5.34 ns [7]. By the same set-up of the detector as in the performance test, we measured time spectra for SR-PAC [8]. The experiments were executed at beamline BL09XU at SPring-8. A nickel foil (0.38mm thick, Ni-61: 95% enriched) was put on the detector window, about 2° inclined to the x-ray beam. The time spectra are indicated in Fig. 4. An external magnetic field of 2-3 k gauss, **H**, was applied to the sample in (a) the X- or (b) the Z-direction. Here, the X-direction was parallel to the polarized plane of the synchrotron x-rays; the Z-direction was perpendicular to the polarized plane. The 30-μm Si-APD presented a narrow dead region in time, less than 5 ns, after the prompt radiation. In Fig. 4, the solid curves indicate the exponential decay of a time-constant τ_0, 7.7 ns. Compared with the decay curve, a modulation of anisotropy was seen in (a), but was still lack of statistics in the measurement.

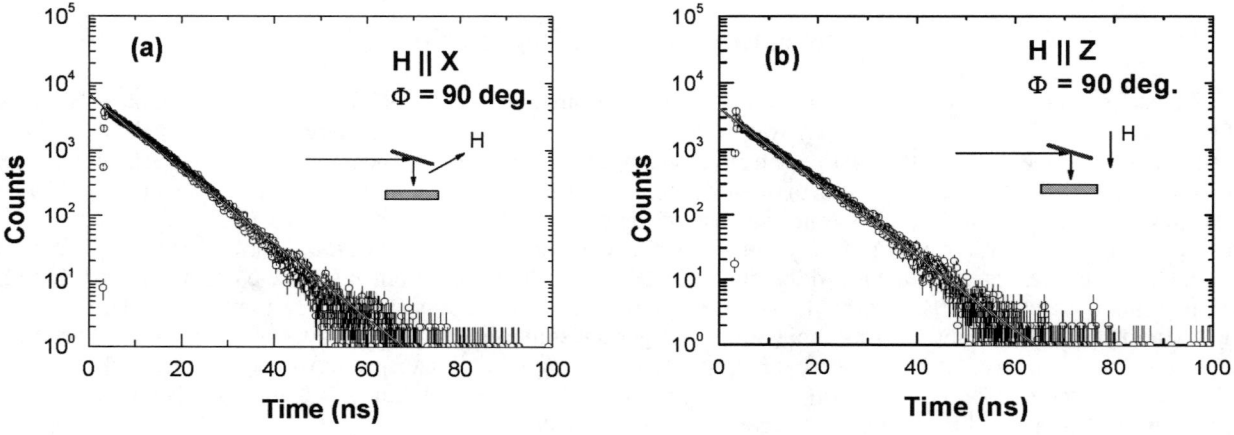

FIGURE 4. SR-PAC time spectra of Ni-61 under an external magnetic field: in the direction (a) parallel or (b) perpendicular to the polarized plane of synchrotron x-rays.

OTHER TYPE OF TIMING DETECTORS

We developed another type of Si-APD detector for a wide dynamic range of count rate [2]. Eight Si-APD plates were installed in an aluminum case of the detector shown in Fig. 5. Outputs from each APD having 3 mm in diameter and a 135-μm thick depletion region, were independently processed by a fast amplifier. This type is also suitable for a timing detector with a high efficiency for high-energy x-rays. We are now preparing a detector having a stack of eight Si-APDs, 45° inclined to the incident-beam direction. Each APD has a thickness of 150 μm, relatively thick, and 3 mm in diameter. The detector is designed to be possible for mounting on the 2θ arm of an x-ray diffractometer. A new fast amplifier is equipped for each APD channel, and is compact, 8×10×50 mm in size. The efficiency will reach 65% at 23.88 keV of the excited level on Sn-119.

If we expect both of high efficiency and a sub-nanosecond time resolution at E >20 keV, to an x-ray timing detector, new fast scintillators, which has a sub-nanosecond scintillation decay, like PbI_2 and HgI_2 [9], may exhibit a good performance for nuclear-resonant scattering experiments. For this application, a low-noise property will be very important for the photomultiplier of the detector.

FIGURE 5. Photograph of the x-ray detector using a stack of eight Si-APD plates.

ACKNOWLEDGMENTS

This work was performed with the approval of Japan Synchrotron Radiation Research Institute (JASRI) (Proposal No.2006A1519) and of the Photon Factory Advisory Committee (Proposal No.2004G031).

REFERENCES

1. S. Kishimoto, *Rev. of Sci. Instr.* **63(1)**, 824-827 (1992).
2. S. Kishimoto, N. Ishizawa and T. P. Vaalsta, *Rev. of Sci. Instr.* **69(2)**, 384-391 (1997).
3. A. Q. R. Baron, S. Kishimoto, J. Morse and J. Rigal, *J. Synchrotron Rad.*, **13**, 131–142 (2006).
4. M.J. Berger, J.H. Hubbell, S.M. Seltzer, J. Chang, J.S. Coursey, R. Sukumar, and D.S. Zucker, XCOM: Photon Cross Sections Database, http://physics.nist.gov/PhysRefData/Xcom/Text/XCOM.html
5. M. Seto, S. Kitao, Y. Kobayashi, R. Haruki, T. Mitsui, Y. Yoda, X. W. Zhang and Y. Maeda, *Phys. Rev. Lett.*, **84** 566-569 (2000).
6. T.Roth, O Leupold, H-C. wille, R. Rüffer, K. W. Quast, R. Röhlsberger and E. Burkel, *Phys. Rev.* **B71**, 140401 (2005).
7. R. B. Firestone, V. S. Shirley, C. M. Baglin, S.Y. F. Chu, and J. Zipkin, *Table of Isotopes,* New York: John Wiley & Sons, Inc., 1998, pp. 1434.
8. A. Q. Baron, A. I. Chumakov, R. Rüffer, H. Grünsteudel, H. F. Grünsteudel, and O. Leupold, *Europhys. Lett.*, **34(5)**, 331 (1996).
9. M. K. Klintenberg, M. J. weber, and D. E. Derenzo, *J. Luminescence*, **102-103**, 287 (2003).

New 36-Element Pixel Array Detector at the ANBF – Choosing the Right Detector for your Beamline

Garry Foran[a], James Hester[a], Richard Garrett[a,b], Pierre Dressler[c], Charles Fonne[c], Jean-Olivier Beau[c] and Marie-Odile Lampert[c]

[a] Australian Synchrotron Research Program, KEK-PF, Oho 1-1, Tsukuba, Ibaraki, Japan 3050801. [b] ANSTO, PMB1, Menai, NSW 2234. [c] Canberra-Eurisys,1, chemin de la Roseraie, Parc des Tanneries, 67380 Lingolsheim, France

Abstract. The Pixel Array Detector for XAFS data collection recently commissioned at the Australian National Beamline Facility is well matched to a busy second-generation bending magnet source where both high and low-flux applications are routinely encountered. In combination with the digital counting chain, throughput has improved by approximately a factor of 5. Detector resolution deteriorates slightly at high count rates.

Keywords: Pixel Array Detectors, Digital Signal Processing, XAFS
PACS: 87.64.Fb,07.50.Qx,07.85.Qe

INTRODUCTION

A new solid state detector system for fluorescence XAS experiments was installed at the Australian National Beamline Facility (ANBF)[1] at the Photon Factory in Tsukuba, Japan in late 2005. The system consists of a 36-element monolithic planar germanium array detector (Canberra-Eurisys), digital signal processing electronics (XIA)[2], a "Zero Boiler" liquid nitrogen recovery system (Canberra Japan KK) and integration and control software written mostly in SPEC (Certified Scientific Software). The complete system was purchased from and is supported by Canberra Japan KK.

The choice of technology (monolithic versus discrete-element array) and specifications of the detector system were matched specifically to the light source and instrumentation at the ANBF which sits on bending magnet port BL-20B at the Photon Factory. On a 2nd generation BM source with no focussing optics, many XAS experiments are "flux-limited". That is, the SSD typically operates within its maximum throughput limit and as a result, data are collected often with the fluorescence detector as close as possible to the sample. In such circumstances, a compact array detector with maximum active area:total area ratio, a total area matched to the existing instrumentation and an optimised pixel size for maximum throughput reaps the most benefit from the available flux. The pixel array detector (PAD) technology (otherwise referred to as monolithic or planar) was deemed the best fit to these criteria[3,4].

The germanium crystal used in the detector is 50mm square and has been surface modified and etched into a 6 x 6 array of pixels in which each pixel is 8mm square. These dimensions produce an array with an active area in excess of 2300 mm^2. The solid angle subtended by the array at a given distance from the sample is very similar to that of the 10-element discreet crystal array formerly in use at the ANBF. The physical size of the detector was determined by the opening angle of controlled-environment instrumentation in use at the beamline (cryostat, furnace etc.) and the need to operate the detector close to the sample. Figure 1 shows an overlay comparison of the new 36-element PAD with the former discrete 10-element and commonly-used 19-element Canberra arrays. The 10-element model had an active area of just 1000 mm^2. Thus, a 2.3 fold increase in active area is achieved with the new system for about the same total area.

Overall detector throughput is also enhanced with this design as each pixel has a smaller surface area than each element of the superseded 10-element model. The larger active area and enhanced throughput, combined with the accurate deadtime correction capabilities of the digital electronics results in a total increase in fluorescence signal throughput of around a factor of 5.

CP879, *Synchrotron Radiation Instrumentation: Ninth International Conference*,
edited by Jae-Young Choi and Seungyu Rah

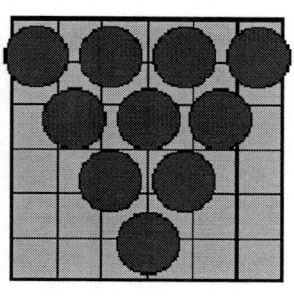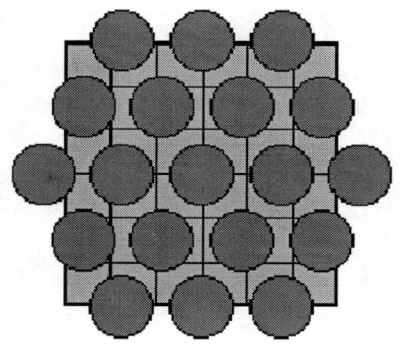

FIGURE 1. Comparison of the physical size and active area of the 36-element PAD with that of the superseded 10-element and the standard 19-element Canberra arrays

DETECTOR CHARACTERISTICS

Results of initial detector characterisation at low count rates are given in Table 1.

TABLE 1. Detector resolution with digital and analog counting chains. All resolution measurements are at 1000 cps, 5.9keV photons.

Characteristic	Analog	Digital
100% rise time	450ns	-
Minimum – maximum - average resolution (eV), 8μs shaping (analog) /17.6μs peaking (digital)	151/175/158	164/183/171
Minimum – maximum – average resolution (eV), 1μs shaping (analog)/2.5μs peaking (digital)	210/260/230	246/277/259

The deadtime of the digital counting chain is determined by the software-configurable minimum peak separation. With filter parameters of 1μs peaking and 0.5μs gap, peaks must be separated by at least 1.5μs, giving a deadtime of at least 3.0μs and a maximum observed count rate of 120,000cps. Fitting of observed count rate curves confirmed that the deadtime was within 40ns of the deadtime calculated from the digital filter parameters.

The energy resolution and peak position were found to vary with count rate. The variation of FWHM for two representative energies is shown in Fig. 2. When a single pixel was exposed, this variation disappeared, suggesting that it is related to charge sharing between pixels. Over the range 0-100kcps, peak position shifted to lower energy by about 90eV (Mn Kα) and 200eV (Mo Kα) (Fig 3). This effect was not present when using an analog counting chain, and is thought to be due to inclusion of signal in the baseline at high count rates. This dependence on count rate can be accommodated by setting appropriately large SCA windows, or by computing appropriate limits from the MCA spectrum at each point in an XAFS scan.

The zero boiler system consists of a metal cap on the dewar with nitrogen recovery hoses connected to a compressor through a vibration isolation system. Tests showed that the detector resolution and signal to noise ratio were not affected by zero boiler operation. However, as the detector signal is sensitive to the formation of ground loops, steps were taken to remove any secondary connection to ground through the nitrogen recovery hoses.

SYSTEM INTEGRATION

Solid-state detectors usually require adjustment of several interrelated parameters for every element: ideally gain, SCA window and shaping time should be optimised for each fluorescence energy and expected maximum countrate. For a large array detector to be efficiently utilised in a busy user facility, maximum automation of this configuration process is essential. As all signal processing parameters of the XIA CAMAC modules are software-configurable, in combination with appropriate control software a high degree of automation is possible.

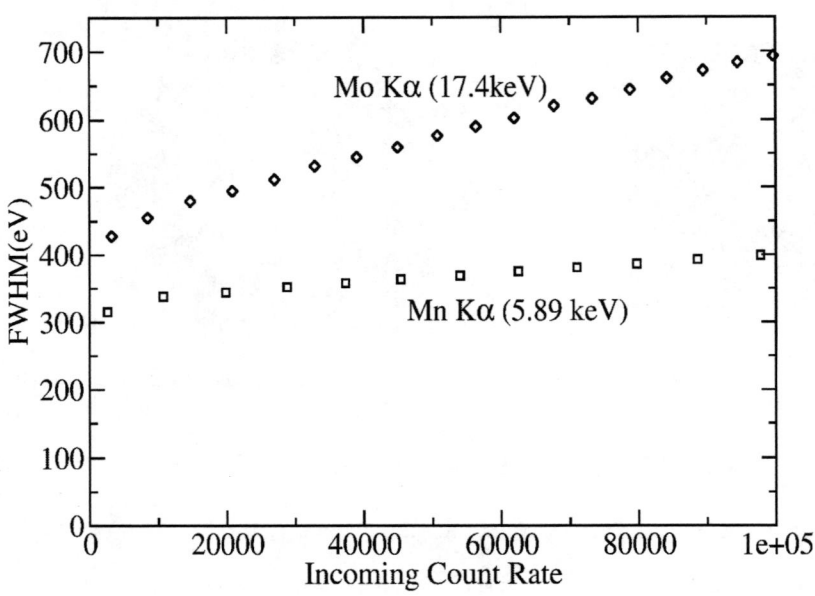

FIGURE 2. Variation of FWHM and peak position with count rate at two energies.

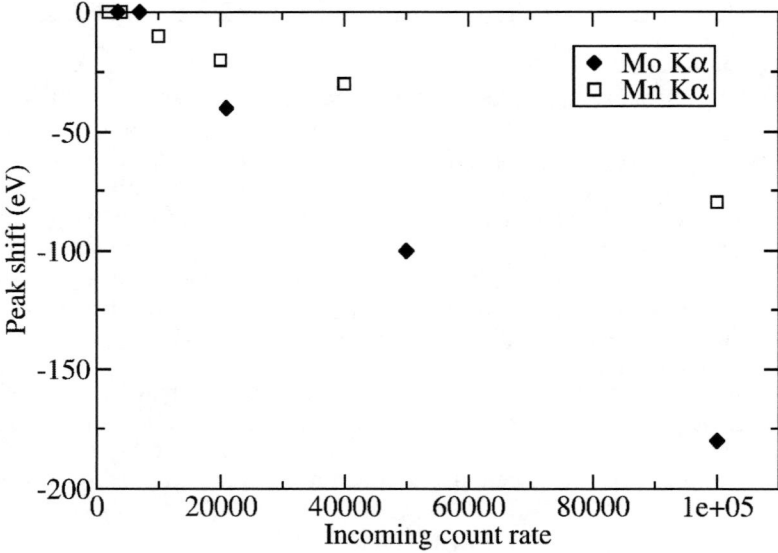

FIGURE 3. Variation of Peak position with Count Rate

The ANBF, in common with many beamlines, operates on a Linux platform, while the standard diagnosis and configuration software bundled with the XIA modules (MESA) runs only on Windows. To minimise time loss due to switching between operating systems we have implemented SPEC macros which duplicate MESA's configuration functionality, and use a Linux-based MCA program to display the contents of SPEC shared memory and interactively set SCA windows. SPEC macros are used to calculate and refine module gains to produce a natural MCA binning factor (10 or 20 eV/bin) and move fluorescence peaks to standard positions. MESA is currently used for its digital oscilloscope diagnostic function, although following anticipated enhancements to SPEC the digital oscilloscope MESA operating mode will also be replaced.

The XIA modules allow accurate automated deadtime correction in the non-linear region of the deadtime curve using the incoming countrate/outgoing countrate ratio reported by the modules. Figure 4 shows the effect of this correction on a Selenium edge collected at medium (40,000 cps) and high (200,000 cps) maximum countrates.

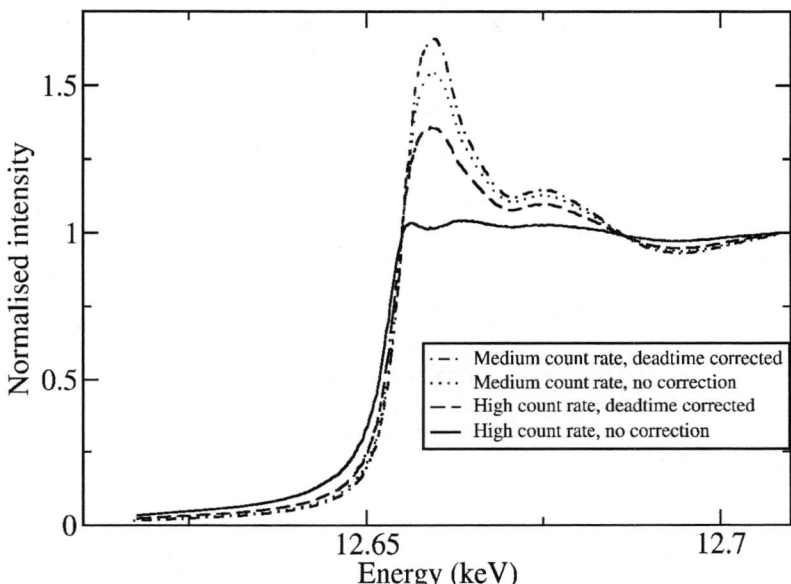

FIGURE 4. Effect of deadtime correction on data collected at the Se edge. The maximum count rate was 200,000 cps and 40,000 cps for the high and medium count rate data respectively. Peaking time 1.0μs, data normalised to the final point.

CONCLUDING REMARKS

After consideration of the various selection criteria involved in matching an XAFS detector system to a given beamline and light source, a novel Ge PAD detector was commissioned. The as-delivered detector exceeds specifications and has already been used by dozens of groups for collection of high-quality XAFS data.

REFERENCES

1. G. J. Foran, D. J. Cookson, and R. F. Garrett, "The Australian National Beamline Facility at the Photon Factory" in *Synchrotron Radiation Facilities in Asia*, Tokyo: Ionics Publishing, pp 119-124 (1994).
2. W. K. Warburton, B. Hubbard and C. Zhou, "Digital spectrometer for automating XAS data collection", XIA, http://www.xia.com (1999)
3. G. Derbyshire, K.-C. Cheung, P. Sangsingkeow and S. S. Hasnain, *J. Synchrotron Rad.*, **6**, pp 62-63 (1999).
4. H. Oyanagi, C. Fonne, D. Gutknecht, P. Dressler, R. Henck, M.-O. Lampert, S. Ogawa, K. Kasai and S. B. Mohamed, *Nucl. Instr. Meth.* **513**, pp 340-344 (2003).

Lanthanum Chloride Scintillator for X-ray Detection

T. Martin and C. Allier and F. Bernard

ESRF, Cyberstar and Saint Gobain

Abstract. In this presentation we describe the testing of a new cerium doped Lanthanum Chloride crystal ($LaCl_3$:Ce), which makes an excellent scintillation material for X-ray counting applications.

Detailed measurements were taken to determine the properties of the scintillator over an energy range of 5 to 60KeV; the results demonstrate that, when used with an appropriate PMT, the crystal sustains high count rates, minimal dead time and good energy resolution. For example an energy resolution of 35% (FWHM) was achieved at 22KeV and count rates of up to 1MHz are possible without dead-time correction.

A comparison of LaCl3:Ce with two conventional scintillation materials, YAP:Ce and NaI(Tl) is also presented, which shows that that LaCl3:Ce offers a good balance of performance parameters for X-ray experiments.

Keywords: Scintillator, LaCl3:Ce, x-ray, detector
PACS: 07.85.-m X and gamma ray instruments, 07.85.Qe Synchrotron Radiation Instrumentation

INTRODUCTION

NaI(Tl) and YAP:Ce counting detectors are readily available and used for a variety of experiments on Synchrotron beamlines. NaI(Tl) based systems demonstrate excellent energy resolution but are limited to count rates of up to 500Kcps, whereas YAP:Ce scintillators can count up to 5Mcps (with counting loss correction). Although YAP:Ce scintillators have high count rate, at energies below 20keV they have poor resolution and so separation of the signal from underlying noise can not be achieved.

Lanthanum Chloride is a new commercial material that seems to offer good energy resolution and fast emission. This manuscript provides an overview of some of the testing carried out on the material over a 5keV to 60keV energy, including measurement of energy resolution and dead time.

Experimental Set-up

These tests were performed using a scintillation counter and processing electronics module available from Oxford Danfysik, produced by Cyberstar.

The scintillation detector head consists of a scintillation crystal from Saint Gobain, a Hamamatsu 10-stage photomultiplier tube (PMT) and a preamplifier. The PMT has a rise time of 2.5ns, a transit time spread of 2.2ns and a sufficiently high anode current limit. The system is operated between 800V and 1000V and the quantum efficiency of the PMT is well matched with YAP:Ce and $LaCl_3$:Ce scintillators; 26% at 370nm.

The processing electronics (X2000 – CBY-2202) is widely used on synchrotron beamlines all over the world. It has a fast preamplifier and fast shaping constants (50ns to 1µs peaking time). The faster shaping times are used for YAP:Ce and $LaCl_3$:Ce scintillation crystals and the slower shaping times for NaI(Tl) scintillators.

Energy Resolution Measurement

In order to measure the detector heads energy resolution it was irradiated with an iron source (^{55}Fe – 5.9 KeV) and a cadmium source (^{109}Cd – 22KeV). The PMT signal was then processed with the X2000 unit detailed above; the spectrum, shown in Fig.1, was recorded with peaking time of 100ns.

CP879, *Synchrotron Radiation Instrumentation: Ninth International Conference*,
edited by Jae-Young Choi and Seungyu Rah
© 2007 American Institute of Physics 978-0-7354-0373-4/07/$23.00

Energy resolution for the 22keV peak was measured at about 35% (FWHM) at room temperature. The peak to valley ratio is excellent at 5.9keV; approximately 20, compared to the ratio of about 5 for YAP. Table 1 shows results for both YAP an LaCl3 scintillators.

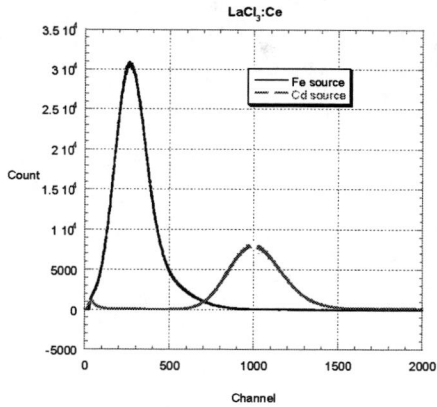

FIGURE 1. Spectrum recorded with LaCl$_3$:Ce scintillator

TABLE 1. LaCl3: Gain 100, ^{55}Fe source (5.9keV)

High Voltage (V)	^{55}Fe source (5.9keV)		^{109}Cd source (22keV)	
	ΔE/E (peaking time=100ns)	ΔE/E (peaking time=300ns)	ΔE/E (peaking time=100ns)	ΔE/E (peaking time=300ns)
850	80%	*130%*	39%	38%
900	75%	*84%*	37%	35%
950	69%	*77%*	34%	35%
1000	70%	73%	25%	34%

Count Rate Measurement

When the distance between the radioactive source and the scintillator is reduced the data input is increased, subsequently modifying the pulse height of the preamplifier. Fig. 2 shows the pulse height of the preamplifier output when excited with a ^{109}Cd source, versus count rate. It is shown that the pulse height is stable until ~900000cps for both YAP and LaCl$_3$ scintillators; the amplitude begins to decrease above count rates of around 1 MHz. This limitation is due to the passive divider network of the PMT, which doesn't allow constant PMT gain at high count rate – this means that for count rates higher than 1Mcps the lower threshold must be very accurately set.

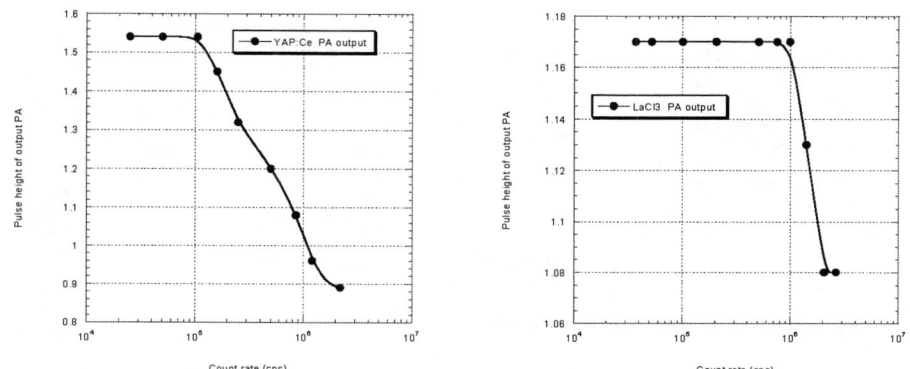

FIGURE 2. Pulse height of the preamplifier output vs. the count rate for YAP (left) and LaCl$_3$ (right)

Dead Time Measurement

Dead time was determined by measuring the intensity of X-rays (from a 60KeV americium source) after passing through a fixed absorber; the true count rate being assumed proportional to the incident count rate and the transmission of the absorber. In this test a single Fe foil was utilised as the absorber.

Data was fitted to both paralyzable and nonparalyzable models, shown in Fig. 3 and Fig. 4 below (LaCl3: Gain=30, peaking time=50ns, HV=1000v and YAP: Gain=30, peaking time=50ns, HV=900v). Table 2 shows a summary of measured dead time for YAP and LaCl3.

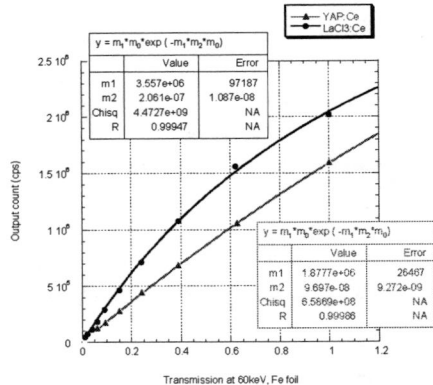

FIGURE 3. Variation of the observed count rate fitted with the nonparalysable model.

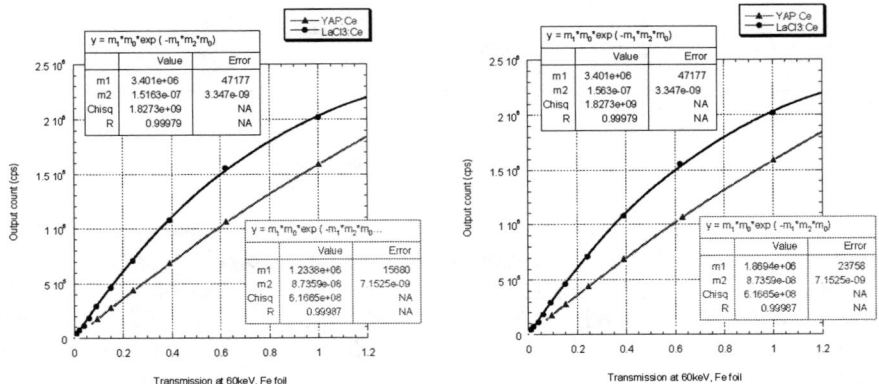

FIGURE 4. Variation of the observed count rate fitted with the paralysable model (left) and paralysable model with corrected absorption (right).

TABLE 2. This table shows a summary of measured dead time

Model type	$\tau_{YAP:Ce}$	$\tau_{LaCl3:Ce}$
Non paralysable	97ns	208ns
Paralysable	87ns	151ns
100% absorption YAP, Paralysable	87ns	151ns

Energy Linearity Measurement

Figure 5 shows measured energies, from obtained peak position, versus the known X-ray energy from a radioactive source, for $LaCl_3$.

The response indicates that $LaCl_3$ is a very linear scintillator over the 5.9keV to 60keV energy range. The highest non-proportionality is about 11% at 8.9keV.

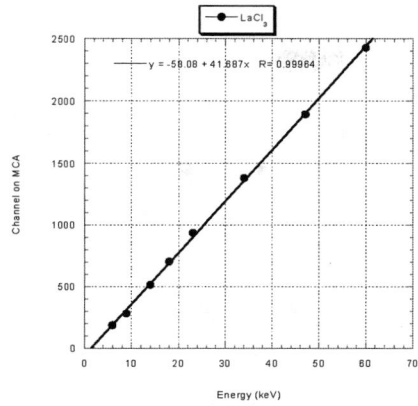

FIGURE 5. Graph demonstrating the energy linearity of LaCl3 (HV = 960v, Gain = 40, peaking time =100ns)

CONCLUSIONS

LaCl3:Ce crystal is a promising crystal for X-ray counting applications; demonstrating a compromise between energy resolution and high counting rate. The scintillator can be used up to 1MHz without dead time correction and the theoretical maximum count rate, using the paralysable model, is 2.5 Mcps. These results were obtained using a radioactive source; future experimentation, to confirm the crystals potential, should be carried out on a Synchrotron beamline.

High quality LaCl3 scintillation detectors, with integrated state-of-the-art pre-amplifiers and ancillary electronics are now available from Oxford Danfysik (shown in Fig. 6 below).

FIGURE 6. Scintillation detector heads and count processing modules available from Oxford Danfysik.

REFERENCES

1. K.S. Shah, J. Glodo, M. Klugerman, L. Cirignano, W.W. Moses, S.E. Derenzo and M.J. Weber, LaCl$_3$:Ce scintillator for Gamma ray detection, Accepted by NIM, to be published.
2. O. Guillot-Noël, J.T.M. de Haas, P. Dorenbos, C.W.E. van Eijk, K. Kramer and H.U. Güdel, Optical and scintillation properties of cerium-doped LaCl$_3$, LuBr$_3$, and LuCl$_3$., Journal of luminescence 85(1999) 21-35 published.
3. M. Harada, K. Sakurai, H. Eba and S. Kishimoto, Performance of the YAP:Ce Scintillation detector, Photon factory activity report 1998, pp292
4. M. Balcerzyk, M. Moszynski, M. Kapusta, Comparison of LaCl$_3$:Ce and NaI(Tl) scintillators in γ-ray spectrometry, NIM A 537 (2005) 50-56
5. Bicron documentation, http:www.detectors.saint -gobain.com

Fiber-Optic Taper Coupled with a Large Format Charge-coupled Device X-ray Detector: Fast Readout and High Duty-Cycle Ratio

Kazuki Ito, Tetsuro Fujisawa and Tadahisa Iwata

RIKEN SPring-8 Center, 1-1-1 Kouto, Sayo, Sayo, Hyogo 679-5148, Japan
Polymer Chemistry Laboratory, RIKEN Central Institute, 2-1 Hirosawa, Wako, Saitama 351-0198, Japan

Abstract. A novel fiber-optic taper (FOT) coupled with a charge-coupled device (CCD) X-ray detector was developed as a basic module of the simultaneous small- and wide-angle X-ray scattering measurement apparatus at SPring-8 BL45XU. The detector consists of two identical units, each comprising a phosphor (CsI:Tl, 80 μm thickness), a 2:1 demagnifying FOT, and an interline transfer-type CCD as an image sensor. The CCD has a 4000 × 2624 pixel format with a 9 μm pitch, resulting in an effective pixel size of 18 μm at the large end of the FOT. The active area size is 72.0 mm × 47.2 mm in each unit. The image stored in the CCD can be readout through the dual readout channels with 12 bits ADC within 220 ms without binning and 65 ms with 8 × 8 binning, respectively. Moreover, this detector is not necessary to stop the incoming X-rays during the CCD readout and can simultaneously execute both exposure and readout on the CCD with a small dead time of ~1 μs so that a duty-cycle ratio of almost 100% is achieved. Therefore, it allows the continuous rotation method, where data is collected without stopping the rotation of a sample crystal in macromolecular crystallography. The continuous rotation method was performed with a total data collection time of 12 s for a range of 180° with 1.5° per frame (120 frames). In this paper, the design, performance characteristics, and verification experiment of the detector will be described.

Keywords: CCD, X-ray detector, SAXS, WAXS, 100% duty-cycle ratio, time-resolved, continuous rotation method, macromolecular crystallography
PACS: 07.85.Qe

INTRODUCTION

The fiber-optic taper (FOT) coupled with a charge-coupled device (CCD) X-ray detector (FOT-CCD X-ray detector) is widely used in both synchrotron macromolecular crystallography and X-ray scattering experiments [1]. In these applications, the FOT-CCD X-ray detector is required for a large active area size, high spatial resolution, high duty-cycle ratio, time-resolved capability, and high sensitivity. In particular, the time-resolved measurement capability is indispensable for its applications in soft-condensed matters such as polymers, surfactants, lipids and liquid crystals. In designing the FOT-CCD X-ray detector, there are several points that should be considered.

Firstly, two parameters (active area size and sensitivity) are important in the detector design. Several research groups [2,3] and companies [4] have developed the detectors comprising an array of FOTs and CCDs. In general, it is difficult to engineer and build an FOT-CCD X-ray detector with both a large active area size and high sensitivity. Owing to the limited size of CCDs (typically 10-50 mm per side), a trade-off between active area size and sensitivity is necessary.

Secondly, with regard to the duty-cycle ratio and time-resolved measurement capability, the readout time of the CCD is an important factor. There are three types of readout schemes in the CCD: full-frame transfer, frame transfer, and interline transfer. The full-frame transfer-type CCD (FFT-type CCD) is widely used in FOT-CCD X-ray detectors such as the commercially available ones [4]. However, the FOT-CCD X-ray detectors, which use the FFT-type CCD, need to stop incident X-rays during the CCD readout. Therefore, the duty-cycle ratio, defined as the ratio of exposure time to the readout time, becomes less than 100%. On the other hand, the FOT-CCD X-ray detectors, which employ the frame transfer-type CCD [5,6] or the framing-mode CCD [7], have a 100% duty-cycle ratio of

---

Corresponding author: kazuki@spring8.or.jp

almost 100% due to the small dead time between recorded frames. However, the sensitivity and readout time was relatively low and slow (2 s). As the third choice, the interline transfer-type CCD (IL-type CCD) can be considered. This is widely used in consumer market devices such as digital video cameras and for machine vision applications, but an FOT-CCD X-ray detector using an IL-type CCD has not yet been manufactured. In general, the surface structure of the IL-type CCD is more complicated than that of the FFT-CCD. Therefore, it is difficult to attach the FOT to the IL-type CCD. Moreover, the IL-type CCD showed a trend of decrease in size with consumer products. At the beginning of the 21st century, the development of a large format IL-type CCD, with over 10 millions pixels had commenced for a high performance digital still camera. Therefore, the large format CCD became commercially available. The increase in the CCD size can reduce the demagnification ratio of the FOT, which also results in an increase in the sensitivity.

Our primary aim was to design a detector for measuring weak X-ray intensities, especially for time-resolved wide-angle X-ray scattering (WAXS) measurements from soft-condensed matters. Moreover, we attempted to increase the data collection efficiency and eliminate the need to control incident X-rays during the CCD readout. In this paper, a brief design of the novel FOT-CCD X-ray detector, which is equipped with an IL-type CCD, will be described. The basic characterization of the detector and verification experiments using a standard powder sample and crystalline biodegradable polymer film will be presented. We will also demonstrate the continuous rotation method of a lysozyme crystal by using this detector to illustrate the small dead time for the CCD readout.

DETECTOR DESIGN

The design goal of this detector is to achieve a time-resolved measurement capability with a time resolution of few tens of milliseconds and high sensitivity in order to discriminate the signals arising due to a few X-ray photons from the system noise. To increase the time-resolved measurement capability, the IL-type CCD was selected as an image sensor. The detector system comprises the following three parts: 1) detector head, 2) camera control unit, and 3) data acquisition system including data processing software. The detector system has been developed in collaboration with Hamamatsu Photonics, Japan.

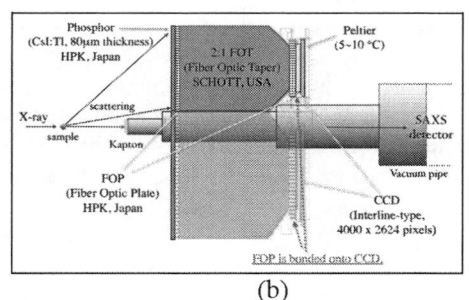

(a)   (b)

FIGURE 1. (a) Photograph and (b) schematic diagram of the detector head.

Detector head

The detector head comprises 1) Thallium doped Cesium Iodide (CsI:Tl) phosphor grown on a fiber-optic plate (FOP), 2) 2:1 demagnifying FOT and 3) an IL-type CCD bonded with an FOP. The photographs of the detector head are shown in Fig. 1(a). A schematic diagram of the detector head is also shown in Fig. 1(b).

The first part of the detector head, which absorbs the incident X-ray photons, is the phosphor. The phosphor material has to be selected such that it matches the spectrum response of the CCD. The central wavelength corresponding to the quantum efficiency of the CCD is around 500 nm. In this detector, the CsI:Tl needle-like crystal (6 μm diameter and 80 μm thickness) grown onto the FOP substrate (Hamamatsu Photonics, Japan), is used. The central wavelength of the emission spectrum of the phosphor is around 560 nm. The 60 nm spectrum difference can be neglected because the spectrum is broad. In addition to the spectrum matching of the phosphor, the thickness of the phosphor is very important. Its thickness is related to the influence of parallax, spatial resolution, and X-ray stopping power, i.e., the sensitivity. For example, the use of thinner screens to optimize the spatial resolution reduces the X-ray stopping power. A low parallax effect and high spatial resolution can be obtained. In general, it is known that CsI:Tl is highly deliquescent, but this characteristic was improved. CsI:Tl also has low persistence and is suitable for time-resolved applications. Therefore, CsI:Tl is selected for this detector.

The second part of the detector head is a 2:1 demagnifying FOT. A FOT is generally more efficient than a combination of lenses in demagnifying geometries [1]. A 2:1 demagnifying FOT of dimensions 72.0 mm × 47.2 mm at the large end (Schott Fiber Optics, U.S.A, fiber material 24A 10 μm with EMA), employs extra mural absorption (EMA) for stray light control in the fiber. The core-clad ratio of the high transmittance type fiber is 70:30. A FOT comprises a coherent bundle of glass-fiber light guides. The bundle may be heated until the glass softens and is stretched so as to produce a FOT to magnify or demagnify an image.

The third part of the detector head is a CCD. The IL-type CCD chip, KAI-11000M, which is manufactured by Kodak, U.S.A., has 4000 × 2624 pixels with 9 μm by 9 μm pixel size. A full well depth is 40,000 e⁻/pixel and has dual readout channels. Each readout channel is operated at a 30 MHz pixel rate and is digitized by the 12 bit analog-to-digital converters (ADCs). The CCD readout gain, analog-to-digital units (ADU) in the ADC, is selectable with 16 steps from 10 e⁻/ADU at the lowest to 2 e⁻/ADU at the highest gain. The full image on the CCD without binning comprises a readout within 220 ms. The minimum readout time is 65 ms with 8 by 8 binning. Temperature of the CCD is stabilized to between 278K (5°C) and 283K (10°C) by a thermoelectrical cooling device (Peltier) attached to the CCD. The readout noise and dark current are 35 e⁻ rms and 5~6 e⁻/pixel/s, respectively. A unique feature of the detector is that the IL-type CCD is used as an image sensor. The IL-type CCD can execute the readout and the exposure with a short interruption (~1 μs) due to its readout structure.

PERFORMANCE CHARACTERISTICS

The conversion gain and dynamic range of the detector were measured in the BL45XU [8] and the spatial resolution was measured in BL44B2 [9] at SPring-8.

The conversion gain in the CCD was 10 e⁻/ADU at the lowest gain setting. The X-ray gain for 13.8 keV X-ray photons was 1.5-1.8 e⁻/xph. For these measurements, the beam was attenuated by a 0.5-mm-thick aluminum sheet and scattered by the glassy carbon, which was positioned at the sample position. The incident X-ray photon flux xph⁻¹ through a 1 mm × 1 mm aperture made by the XY slit was measured with a calibrated Ge solid-state detector; then the intensity through the aperture was recorded by the detector. The intensity was determined by integrating over an area slightly larger than the aperture size of the XY slit.

The dynamic range was measured with different exposure time, under the fixed condition of X-ray intensity and is shown in Fig. 2(a). In Fig. 2(a), three orders of magnitudes are clearly confirmed. However, small fluctuations exist in the lower dose region due to the low sensitivity.

Spatial resolution is the one of the basic characteristics of the X-ray detector. The point spread function (PSF) was measured with an attenuated 12.4 keV X-ray beam and a pinhole collimator of 100 μm thickness Pt with a 14 μm diameter hole. Measured PSF values without binning are shown in Fig. 2(b) in a linear scale and in Fig. 2(c) in a logarithmic scale. The full widths at 50%, 10%, and 1% of the maximum are 69 μm, 150 μm, and 240 μm, respectively. The pixel size at the surface of the phosphor was 18 μm. The PSF deviates from the Gaussian function at around 10% of the maximum.

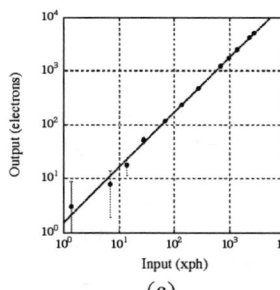
(a)

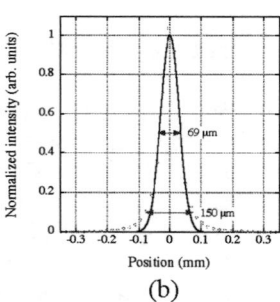
(b)

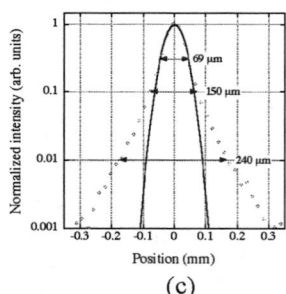
(c)

FIGURE 2. (a) Dynamic range of the detector. Point spread function with linear scale (b) and with logarithmic scale (c).

EXPERIMENTS

An example of the powder diffraction pattern and its circular averaged profile is shown in Fig. 3(a) and Fig. 3(b). Two diffraction images recorded by the detector from the crystalline biodegradable Poly [(R)-3-hydroxybutyrate] [P(3HB)] [10] drawn film and a single crystal of lysozyme are shown in Fig. 3(c) and Fig. 3(d), respectively. All images undergo the image distortion and nonuniform response corrections [11]. We also have performed the continuous rotation method [5,6] of a protein crystal. The speed of the sample rotation was 15° per second and the exposure time in one frame was 0.1 s. The total data collection time for a range of 180° was 12 s in a 4 × 4 binning mode, while a typical FOT-CCD X-ray detector, i. e., with a 2 s readout time, takes about 250 s.

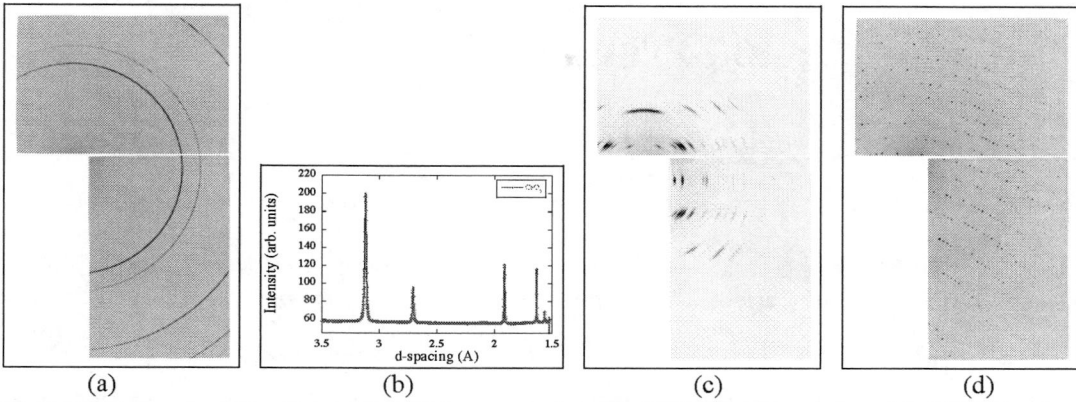

(a)   (b)   (c)   (d)

FIGURE 3. (a) CeO_2 powder diffraction and its circular averaged profile. Diffraction images from (c) a P(3HB) polymer drawn film and (d) a lysozyme single crystal. The sample-to-detector distance and X-ray energy was 123.55 mm and 13.8 keV, respectively. The exposure time was 1 s per frame.

CONCLUSIONS

A novel FOT coupled CCD X-ray detector was developed for synchrotron time-resolved X-ray scattering experiments. This detector is primarily developed as a module of the simultaneous small- and wide-angle X-ray scattering measurement apparatus. The feature of this detector is to combine a demagnifying FOT and a large format interline transfer-type CCD as an image sensor. Therefore, the control of incident X-ray is not needed during the CCD readout. In addition, fast readout (220 ms without binning and 65 ms with 8 × 8 binning) and a duty-cycle ratio of almost 100% are achieved due to a small dead time (~1 μs) between recorded frames. We have demonstrated the basic performance characteristics of the detector and the verification experiments. In the verification experiments, we have obtained the continuous rotation image of a protein crystal. The total data collection time was 12 s for a range of 180° with 1.5° and 0.1 s exposure time per frame.

ACKNOWLEDGMENTS

All experiments were performed at SPring-8 RIKEN Structural Biology beamlines: BL44B2 and BL45XU. This work has been supported by Ecomolecular Science Research provided by the RIKEN Institute. A part of this work has been supported by a Grant-in-Aid for Young Scientists (A) from the Ministry of Education, Culture, Sports, Science and Technology (MEXT) of Japan (No. 15685009). The author (K.I.) would like to thank Drs. Masayoshi Ito (JASRI) and Takaaki Hikima (RIKEN) for their technical assistance in the experiments.

REFERENCES

1. S. M. Gruner *et al.*, *Rev. Sci. Instrum.* **73**, (2002), pp. 2815-2842.
2. N. Allinson, *J. Synchrotron Rad.* **1**, (1994), pp. 54-62.
3. M. W. Tate *et al.*, *J. Appl. Cryst.* **28**, (1996), pp. 196-205.
4. Area Detector Systems Corporation (ADSC), http://www.adsc-xray.com; Mar USA, Inc., http://www.mar-usa.com.
5. K. Ito *et al.*, AIP Conference Proceedings 705 (SRI2003), San Francisco, California, 2004, pp. 913-916.
6. K. Ito, N. Sakabe and Y. Amemiya, (*submitted to J. Synchrotron Rad.*).
7. P. Coan *et al.*, *J. Synchrotron Rad.* **13**, (2006), pp. 260-270.
8. T. Fujisawa *et al.*, *J. Appl. Cryst.* **33**, (2000), pp. 797-800.
9. S. Adachi *et al.*, *Nucl. Instrum. & Method* A **467-468**, (2001), pp. 1209-1212.
10. T. Iwata, *Macromol. Biosci.* **5**, (2005), pp. 689-701.
11. K. Ito *et al.*, *Jpn. J. Appl. Phys.* **44**, (2005), pp. 8684-8691.

A Real-Time Imaging System for Stereo Atomic Microscopy at SPring-8's BL25SU

Tomohiro Matsushita[1], Fang Zhun Guo[1], Takayuki Muro[1], Fumihiko Matsui[2], and Hiroshi Daimon[2]

[1]*JASRI/SPring-8, Kouto 1-1-1, Sayo-cho, Sayo-gun, Hyogo, 679-5198, Japan*
[†]*Graduate School of Materials Science, Nara Institute of Science and Technology (NAIST),8916-5 Takayama, Ikoma, Nara 630-0192, Japan*

Abstract. We have developed a real-time photoelectron angular distribution (PEAD) and Auger-electron angular distribution (AEAD) imaging system at SPring-8 BL25SU, Japan. In addition, a real-time imaging system for circular dichroism (CD) studies of PEAD/AEAD has been newly developed. Two PEAD images recorded with left- and right-circularly polarized light can be regarded as a stereo image of the atomic arrangement. A two-dimensional display type mirror analyzer (DIANA) has been installed at the beamline, making it possible to record PEAD/AEAD patterns with an acceptance angle of ±60° in real-time. The twin-helical undulators at BL25SU enable helicity switching of the circularly polarized light at 10Hz, 1Hz or 0.1Hz. In order to realize real-time measurements of the CD of the PEAD/AEAD, the CCD camera must be synchronized to the switching frequency. The VME computer that controls the ID is connected to the measurement computer with two BNC cables, and the helicity information is sent using TTL signals. For maximum flexibility, rather than using a hardware shutter synchronizing with the TTL signal we have developed software to synchronize the CCD shutter with the TTL signal. We have succeeded in synchronizing the CCD camera in both the 1Hz and 0.1Hz modes.

Keywords: Photoelectron diffraction, Electron holography, photoemission, other methods of structure determination.
PACS: 61.14.Qp, 61.14.Nm , 79.60.-i, 61.18.-j

INTRODUCTION

The photoelectron diffraction technique is an atomic structural analysis method using photoelectrons. The stereo atomic microscope [1] and photoelectron holography [2-4] are based on the photoelectron diffraction technique. The stereo atomic microscope utilizes two photoelectron diffraction patterns excited by left- and right-circularly polarized light. The two patterns can be regarded as stereo image of the atomic arrangement, and no computer processing is required to observe the atomic structure. On the other hand, photoelectron holography utilizes one or more photoelectron diffraction patterns. Recently, new reconstruction algorithms, which can reconstruct a three-dimensional atomic arrangement with about 0.02 nm resolution from a single-energy photoelectron hologram or an Auger-electron hologram, have been proposed [5-7]. The features of these techniques are,

1. An initial model for atomic structure is not required for the atomic structural analysis.
2. Local three-dimensional structure around the target atomic site is observable.
3. Not the electron cloud, but the nuclei position is observable.
4. Perfect long-range order is not required.
5. High surface-sensitivity.
6. Electronic structure (spin etc.) is observable.

These methods are now being applied to the atomic structural analysis of the bulk, the surface, and the local structures around dopants[5-10]. These methods require photoelectron angular distributions (PEAD) of the core-level photoelectron or Auger-electron angular distributions (AEAD) over nearly 2π-steradian (half sphere) with a resolution of about 1 degree. A conventional photoelectron diffraction measurement system is usually composed of a light source (synchrotron radiation or an X-ray tube), a conventional electron analyzer and a sample manipulator

CP879, *Synchrotron Radiation Instrumentation: Ninth International Conference,*
edited by Jae-Young Choi and Seungyu Rah
© 2007 American Institute of Physics 978-0-7354-0373-4/07/$23.00

with two rotation axes. A PEAD/AEAD pattern is recorded by sweeping two rotation axes to give 2π-steradian coverage, and long measurement times (several hours) are required. Real-time measurement is impossible.

We have developed a real-time measurement system of PEAD/AEAD. In addition, a real-time imaging system for circular dichroism (CD) of the PEAD/AEAD has been newly developed. This system proves extremely useful for stereo atomic microscope and photoelectron holography studies.

INSTRUMENT

Figure 1 shows a schematic view of the constructed real-time measurement system for PEAD/AEAD installed at the soft x-ray beamline BL25SU of SPring-8, Japan. The major components are twin helical undulators, a grating monochromator, and a two-dimensional display-type analyzer (DIANA) [11]. The twin helical undulators [12,13] are composed of two helical undulators ID1 and ID2, and a set of kicker magnets. ID1 and ID2 generate left- and right-helicity radiation, respectively. The kicker magnets bump the electron orbit at each undulator, deflecting one radiation component off-axis. By changing the excitation of the kicker magnets periodically, the helicity of the circularly polarized radiation passing to the beamline optics can be periodically switched. Currently a switching frequency of 0.1, 1 or 10 Hz is available. The beamline monochromator is a constant deviation type with varied line-spacing plane gratings (VLSPG) covering an energy region of 0.22 - 2 keV [14,15]. The resolving power of the monochromator is more than 10,000 over the whole energy region. The monochromatic light is incident on a sample, with the resulting emission of photoelectrons. Emitted electrons with kinetic energies corresponding to the pass energy of the DIANA are focused to an aperture, and the electrons that pass through the aperture are projected onto a screen. The two-dimensional PEAD/AEAD pattern directly appears on the screen. The PEAD/AEAD pattern is detected by a CCD camera located outside the vacuum chamber. The acceptance angle covered by the screen is $\pm 60°$. Therefore, DIANA can be used to record PEAD/AEAD patterns in real time.

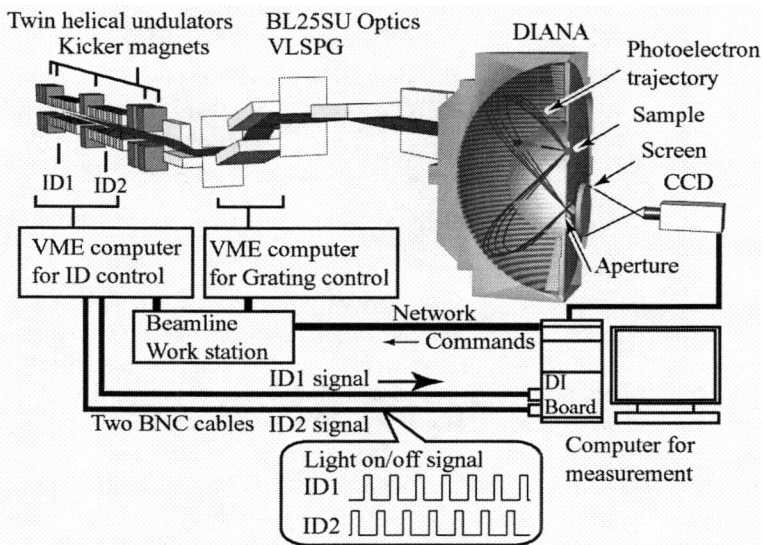

FIGURE 1. A schematic view of the real-time measurement system for the PEAD/AEAD.

In addition, in order to study the CD of the PEAD/AEAD pattern in real time, it was necessary to synchronize the helicity switching and the CCD camera shutter. The sequences for the 0.1Hz and 1Hz modes are shown in Fig.2. In the 1Hz mode, the ID1 light is turned on for 0.3 sec, then both components are turned off for 0.2 sec, the ID2 light is turned on for 0.3sec, and both components are again turned off for 0.2sec. At SPring-8, the VME computers that control the ID and monochromator are controlled over a local network. Synchronization by directly querying the current light helicity over the network, however, is not feasible due to the inherent delays. Therefore, we make a direct connection between the VME computer that controls kicker magnets and the measurement computer with two BNC cables. The two BNC cables correspond to the ID1 light signal and the ID2 light signal. A TTL "high" signal corresponds to the particular ID being "on".

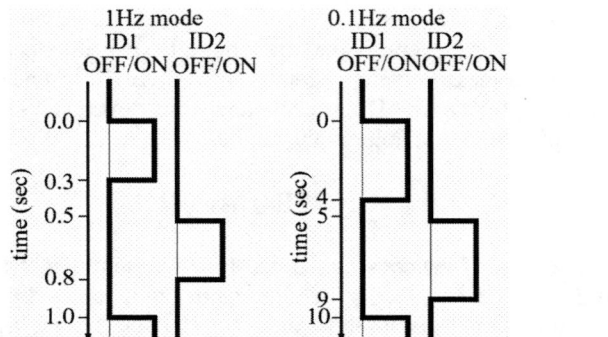

FIGURE 2. A diagram of ID1 and ID2 light of 1Hz mode and 0.1Hz mode.

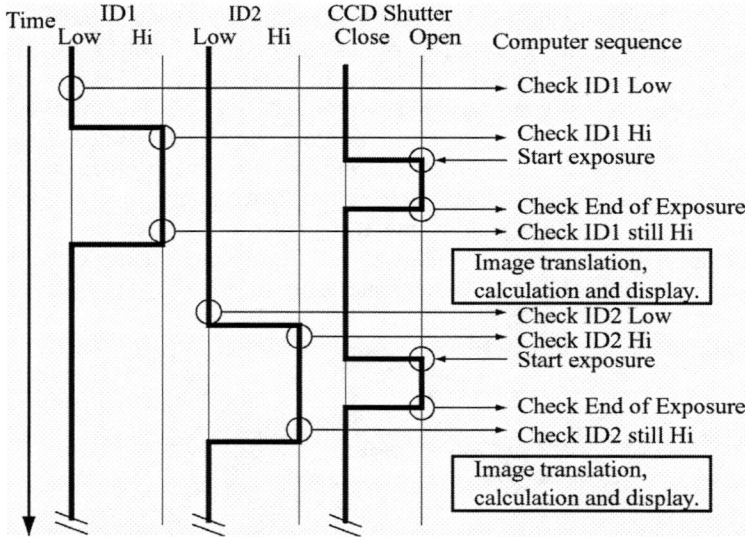

FIGURE 3. A diagram of the computer sequence, in order to synchronize the CCD camera shutter and the helicity switching of ID25.

In order to maximize flexibility for the measurements, we did not develop a hardware shutter synchronizing with the TTL signal, but adopted a software synchronization system. A diagram of the software sequence is shown in Fig. 3. In the case of the 1Hz mode, the time of "Image translation, calculation and display" is quite important, since it must be complete while both lights are turned off. If the calculation time is longer than 0.2sec, the exposure time is reduced, and if it becomes longer than 0.5sec, the sequence collapses. In particular, the translation time of the image of CCD camera is the key to realize this sequence. We have selected for the camera the "Sensicam QE" of "PCO imaging Co.". The camera spec is high-resolution (1376 x 1040 pixels), has a 12-bit dynamic range, and a fast frame rate of 10fbps. In addition, it is necessary to make the calculation and display time as short as possible. Therefore, we did not adopt Labview or other interpreted languages but constructed new original software using the native compiler of "Borland C++ Builder".

PERFORMANCE

The first test was the real-time measurement of AEAD. An LVV AEAD pattern from a Cu(001) sample was successfully recorded with a 0.1sec exposure time at 5~8 frame/sec. Synchronization with the helicity switching was then tested by changing the frequency of the helicity. In the 0.1 Hz mode, an exposure time of about 3.8sec is available. The 1 Hz mode was also tested, and it was possible to synchronize the CCD camera with a 0.16sec exposure time. The loss time is about 0.14sec, mainly due to the two DI readings before and after exposure.

The two PEAD patterns excited with left- and right-circularly polarized light can be regarded as a stereo image of the atomic arrangement. An example of such a stereo atomic microscope image pair is shown in Fig. 4 [8]. These images are sets of Cu (001) PEAD patterns from a Cu (001) surface recorded with a photoelectron energy of 600eV.

The helicity of the photons used for the patterns shown in the left and right images were cw and ccw respectively. These two patterns form a stereo photograph of the atomic arrangements for the left eye and the right eye.

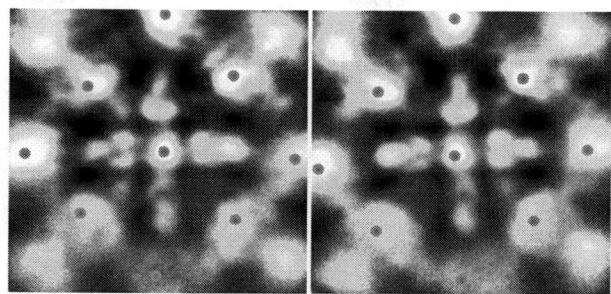

FIGURE 4. Two-dimensional PEAD patterns recorded for Cu(001) 3p excited by cw (left) and ccw (right) helicity light. The kinetic energy is 600eV [8]. A three-dimensional image can be seen by looking at the left image with the left eye, and the right image with the right eye.

CONCLUSIONS

We have constructed a real-time measurement system for PEAD/AEAD at BL25SU of SPring-8. We have confirmed that it is possible to observe AEAD patterns with 0.1sec exposure time at 5~8 frame/sec. In addition we have also constructed a system for synchronizing the periodic photon helicity switching with the CCD camera shutter by using newly developed software. We succeeded in synchronizing at both 1Hz and 0.1Hz modes. We have achieved 0.16sec exposure time at 1Hz mode, and confirmed that it is possible to measure the CD of the AEAD pattern at 1Hz. This system enables real-time atomic stereo microscopy and circular-dichroism photoelectron holography.

REFERENCES

1. H. Daimon, Phys. Rev. Lett., **86**, 2034 (2001).
2. A. Szöke, in *Short Wavelength Coherent Radiation: Generation and Applications*, AIP Conf. Proc. No. 147, 361 (1986).
3. J. J. Barton, Phys. Rev. Lett., **61**, 1356 (1988).
4. J. J. Barton, Phys. Rev. Lett., **67**, 3106 (1991).
5. T. Matsushita, A. Agui and A. Yoshigoe, Europhys. Lett., **65**, 207 (2004).
6. T. Matsushita, A. Yoshigoe and A. Agui, Europhys. Lett., **71**, 597 (2005).
7. T. Matsushita, A. Agui and A. Yoshigoe, J. Electron. Spectrosc. Relat. Phenom., **144-147**, 1175 (2005).
8. F. Z. Guo, F. Matsui, T. Matsushita and H. Daimon, J. Electron. Spectrosc. Relat. Phenom., **144-147**, 1067 (2005).
9. F. Z. Guo, T. Matsushita, K. Kobayashi, F. Matsui, Y. Kato, H. Daimon, M. Koyano, Y. Yamamura, T. Tsuji and Y. Saitoh, J. Appl. Phys., **99**, 024907 (2006).
10. F. Matsui, H. Daimon, F. Z. Guo and T. Matsushita, Appl. Phys. Lett., **85**, 3737 (2004).
11. M. Kotsugi, T. Miyatake, K. Enomoto, K. Fukumoto, A. Kobayashi, T. Nakatani, Y. Saitoh, T. Matsushita, S. Imada, T. Furuhata, S. Suga, K. Soda, M. Jinno, T. Hirano, K. Hattori and H. Daimon, Nucl. Inst. and Meth. in Physics Res. A, **467-468**, 1493 (2001).
12. T. Hara, T. Tanaka, T. Tanabe, X. -M. Marećhel, K. Kumagai and H. Kitamura, J. Synchrotron Rad., **5**, 426 (1998).
13. T. Hara, K. Shirasawa, M. Takeuchi, T. Seike, Y. Saitoh, T. Muro and H. Kitamura, Nucl. Inst. and Meth. in Physics Res. A, **498**, 496 (2003).
14. Y. Saitoh, H. Kimura, Y. Suzuki, T. Nakatani, T. Matsushita, T. Muro, T. Miyahara, K. Soda, S. Ueda, H. Harada, M. Kotsugi, A. Sekiyama and S. Suga, Rev. Sci. Instrum., **71**, 3254 (2000).
15. T. Muro, T. Nakamura, T. Matsushita, H. Kimura, T. Nakatani, T. Hirono, T. Kudo, K. Kobayashi, Y. Saitoh, M. Takeuchi, T. Hara, K. Shirasawa and H. Kitamura, J. Electron. Spectrosc. Relat. Phenom., **144-147**, 1101 (2005).

Bragg Magnifier:
High-efficiency, High-resolution X-ray Detector

Marco Stampanoni*, Amela Groso*, Gunther Borchert[†] and Rafael Abela*

*Swiss Light Source, Paul Scherrer Institut, CH-5232 Villigen, Switzerland
[†]FRM II, Technische Universität München, D-85747, Garching, Germany

Abstract. X-ray computer microtomography is a powerful tool for non-destructive examinations in medicine, biology, and material sciences. The resolution of the presently used detector systems is restricted by scintillator properties, optical light transfer, and charge-coupled-device (CCD) granularity, which impose a practical limit of about one micrometer spatial resolution at detector efficiencies of a few percent. A recently developed detector, called Bragg Magnifier, achieves a breakthrough in this respect, satisfying the research requirements of an efficient advance towards the submicron range. The Bragg Magnifier uses the properties of asymmetric Bragg diffraction to increase the cross section of the diffracted X-ray beam. Magnifications up to 100x100 can be achieved even at hard X-rays energies (>20 keV). In this way the influence of the detector resolution can be reduced accordingly and the efficiency increased. Such a device has been developed and successfully integrated into the Tomography Station of the Materials Science Beamline of the Swiss Light Source (SLS). The device can be operated at energies ranging from 17.5 keV up to 22.75 keV, reaching theoretical pixel sizes of 140 nm.

Keywords: X-ray imaging, synchrotron microtomography, asymmetric Bragg diffraction
PACS: 07.85.Qe , 87.59.Fm , 61.10.Ui , 87.59.Ls

INTRODUCTION

The combination of X-ray microscopy with tomographic techniques as well as the exceptional properties of third-generation synchrotron radiation sources allow to obtain volumetric information of a specimen at micron or sub-micron scale with minimal sample preparation. Microtomographic investigations tend nowadays towards the analysis of millimeter-sized specimens at micrometer resolution within minutes. The requirements on the detectors in terms of spatial resolution and efficiency are therefore very high and tremendous efforts have been made all over the world. The most established detection method consists of converting X-rays into visible light with a scintillator and projecting them onto a charge coupled device (CCD) with the help of suitable optics [1, 2, 3]. It has been shown that the spatial resolution of the visible-light-based approach is intrinsically limited to about 1 μm by scintillation properties, optics efficiency and CCD granularity [4]. Different approaches have been proposed for efficiently exceed the micrometer barrier: Fresnel zone plates [5, 6, 7, 8], parabolic compound refractive lenses [9] and Kirkpatrick-Baez setups [10] have demonstrated to reach submicrometer resolution with sufficient efficiency but with a limited field of view. This article describes a device, called Bragg magnifier, which achieves a breakthrough in this respect, satisfying the requirements for efficient imaging at submicron resolution at high energies. The new instrumentation, inspired by the setup of Kuryiama *et al.* [11], performs two dimensional magnification in the hard X-rays regime (> 20keV) exploiting the well-known principle of asymmetrical Bragg diffraction from two crossed flat crystals: for the prototype described in this work, a pair of Si(220) crystals with an asymmetrical cut of 8° has been used. If the incoming divergence of the beam is kept sufficiently small (< 20 μrad) then a comparison with the relevant rocking curve implies that the first Si(220) crystal accepts 95% of the incoming intensity, which means that almost no flux is lost when passing through the crystals. The magnifying optics collect the intensity distribution of the incoming beam, which contains absorption and phase information about the sample, and magnifies it, acting therefore as a full field microscope. The enlarged image can be converted to visible light in a much more efficient way (thicker scintillator) since the larger spread will be compensated by the X-ray magnification. As a consequence, the efficiency will be enhanced *without* deterring the spatial resolution.

CP879, *Synchrotron Radiation Instrumentation: Ninth International Conference*,
edited by Jae-Young Choi and Seungyu Rah

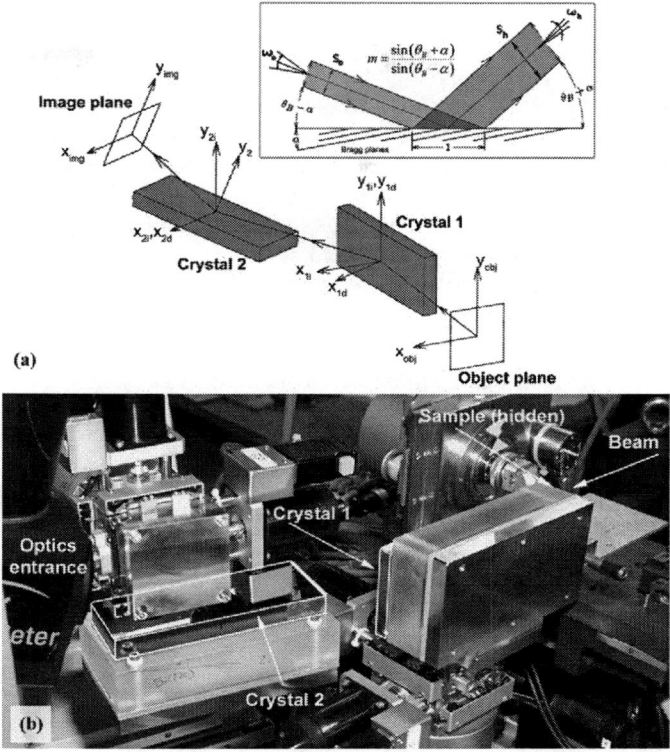

FIGURE 1. (a) Functional principle of the Bragg magnifier depicting both crystals, the object and the image (detector) planes. "*i*" describes the incident beam coordinates, "*d*" the diffracted one. The inset shows the principle of coplanar asymmetric Bragg diffraction: m is the magnification factor, θ_B is the Bragg angle and α is the asymmetry angle. (b) Experimental setup installed at the Materials Science beamline of the Swiss Light Source: visible are both crystals, mounted on their swivels, fixed to the high-resolution goniometers. The beam comes from the right. On the left side, the entrance of the 1:1 optic is also visible.

THEORETICAL BACKGROUND

For σ–polarized X-rays generated by insertion devices of third generation synchrotron facilities, the width of the diffraction pattern for a non-absorbing perfect crystal is given by $\omega_S = \frac{2}{\sin(2\theta_B)} \frac{r_0 \lambda^2}{\pi V} |F_h| e^{-DW}$ where $r_0 = 2.818 \cdot 10^{-15}$ m is the classical electron radius, λ is the wavelength of the incident radiation, V is the volume of the unit cell, θ_B is the Bragg angle, F_h is the crystal structure factor and e^{-DW} is the Debye-Waller factor. Defining the magnification factor as $m = \frac{\sin(\theta_B + \alpha)}{\sin(\theta_B - \alpha)}$ where α is the asymmetry angle, i.e. the angle between the crystal surface and its lattice planes, simple geometrical arguments lead to $S_d = |m| \cdot S_i$, where S_i and S_d are the spatial cross-sections of the incident and diffracted beam. Following Liouville's theorem, it can be deduced that $\omega_i = \sqrt{|m|} \cdot \omega_S$ where ω_i and ω_d are the angular divergences of the incident and diffracted beam respectively. Quantitatively, dynamical theory states that $\omega_i = \sqrt{|m|} \cdot \omega_S$ and $\omega_d = \frac{1}{|m|} \cdot \omega_i$. It follows that if $|m| > 1$ the range of total reflection for the emergent beam is $1/|m|$ times smaller than that of the incident beam, while its spatial cross section is $|m|$ times greater. If two subsequent asymmetrical diffractions with respect to two equal but perpendicular to each other lattice planes occur, we obtain a 2D enlargement of the incoming beam (2D "Fankuchen" effect).

OPERATION WITH THE BRAGG MAGNIFIER

In this paper we summarize the main aspects of the Bragg Magnifier, discussing in particular the properties of the imaging optics and the issues related to the coherence preservation. An extended and more comprehensive theoretical discussion is presented in [12] and lies beyond the aim of this article.

Crystal Preparation

Si(220) crystals with an asymmetry angle of $\alpha = 8°$ have been used as magnifying optics. The crystals have been cut with an angular accuracy of better than 1 arcminute and have been fixed by optical contacting to a precision glass support, the thermal expansion coefficient of which is similar to silicon. The glass support acts as mechanical interface between the silicon crystal and a steel support, which is in turn fixed to a double swivel, ensuring that no mechanical stress is applied to the crystal when the unit is screwed to an high-resolution goniometer, see Fig. 1b. Swivel's pitch and roll accuracy of better than $2''$ as well as goniometer's angular resolution of $0.05''$ perfectly cope with the narrow rocking curves of Si(220), numerically obtained with the help of dynamical X-ray diffraction calculations. Additional details concerning construction and simulations can be found in Ref. [12, 13].

Coherence Preservation and Imaging

The magnified X-ray image, see inset in Fig. 1a, is converted to visible light by a 35x35 mm^2 CsI(Tl) scintillator of 300 micron thickness. The generated light is collected by a high-efficiency 1:1 tandem optic ($f = 150 \; mm$, aperture 1/25, brand "Kinooptik") and projected onto an high-performance CCD camera (THOMSON chip, 2048x2048 pixels,14x14 mum^2 pitch, 14 bit dynamic range, 250 ms full frame readout).

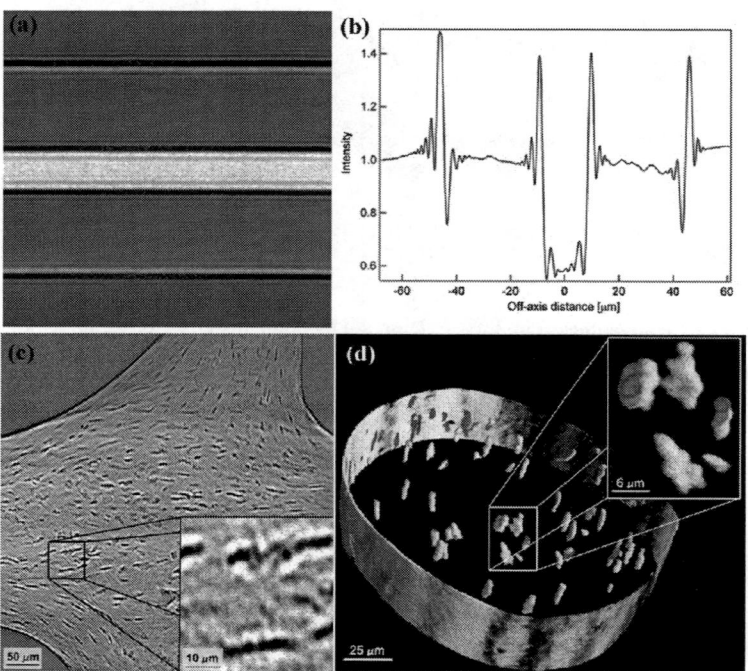

FIGURE 2. X-ray projection (a) and corresponding line profile (b) of a boron fiber with an outer diameter of 100 μm and an inner tungsten core of 14 μm. Edge-enhanced radiographic projection (c) and corresponding 3D rendering (d) of a human bone trabecula.

In a previous experiment [13] a gold mesh with nominal aperture of 11 μm and wire diameter of 5 μm was imaged while tuning the energy from 21.1 keV up to 22.6 keV, producing magnification factors of 20x20 up to 80x80, corresponding to pixel sizes of 700x700 nm^2 down to 175x175 nm^2. Fresnel's diffraction effects were clearly observed, but no quantitative consideration could be grasped from that data. In a more refined experiment [14], it has been shown that the Bragg magnifier optics can form magnified in-line near-field holograms at sub-micrometer resolution, with high magnification and high efficiency, confirming therefore the theoretical predictions of Spal [15]. For this purpose, a boron fiber of 100 μm outer diameter and an inner tungsten core of 14 μm has been imaged at an energy of 22.64 keV, corresponding to a magnification factor of 60x60 and a pixel size of 235x235 nm^2, see Fig. 2a. On a line profile, see Fig. 2b, at least 4 fringes are visible meaning that the optics preserve the phase information carried by the beam,

i. e. can produce phase contrast images of good quality.

Edge-enhancement usually deteriorates the final resolution of a detector but sometime can be exploited in order to detect tiny, low absorbing inclusions in hard materials. As an example we show here a high-resolution, edge-enhanced radiographic projection of a human bone trabecula, see Fig. 2c. Clearly visible are the few microns sized bone lacunaes (osteocytes) within the bone matrix. Performing tomography on the same sample it is possible to obtain the 3D distribution of the lacunaes, see Fig. 2d.

DISCUSSION AND OUTLOOK

The first prototype of the Bragg magnifier has been successfully operated and produced high-resolution, aberration free X-ray microradiographs. Reliability, reproducibility and stability of the X-ray optics have been showed to be excellent [12] and high-resolution tomograms have been recorded. It appears clear however that the optics produce strong edge-enhanced images affecting therefore the final spatial resolution. The effect is well pronounced due to the relatively large distance between the sample and the scintillator mounted on the 1:1 optic (350 mm). This large distance is itself determined by the size of the crystals. For this prototype, 160 mm long crystals have been used, in order to magnify 1x1 mm^2 at least 40x40. Upcoming experiments will try to fully determine the Optical Transfer Function of the system, i.e. to quantitatively characterize the imaging characteristic of the device. Future developments could consider the design of ad-hoc crystals pairs for a given energy or field-of-view.

ACKNOWLEDGMENTS

The authors thank D. Meister and M. Lange from PSI for designing and manufacturing important mechanical parts. We thank Prof. H. Grimmer and Dr. D. Clemens of PSI for their help with the preliminary characterization of the crystals. We appreciated the efficient collaboration with B. Lux and Dr. V. Alex from the Institut für Kristallzüchtung, Berlin. This work has been supported by the ETH Board.

REFERENCES

1. U. Bonse, and F. Busch, *Progress in Biophysics and Molecular Biology* **65**, 133 (1996).
2. B. A. Dowd, G. Campbell, R. Marr, V. Nagarkar, S. Tipnis, L. Axe, and D. Siddons, "Developments in synchrotron x-ray computed microtomography at the National Synchrotron Light Source," in *Developments in X-Ray Tomography II*, edited by U. Bonse, 1999, vol. 3772 of *Proc. SPIE*, p. 224.
3. T. Weitkamp, C. Raven, and A. Snigierv, "An imaging and microtomography facility at the ESRF beamline ID 22," in *Developments in X-Ray Tomography II*, edited by U. Bonse, 1999, vol. 3772 of *Proc. SPIE*, p. 311.
4. A. Koch, K. Raven, P. Spanne, and A. Snigirev, *J. Opt. Soc. Am. A* **15**, 1940 (1998).
5. G. Schneider, D. Hambach, B. Niemann, B. Kaulich, J. Susini, N. Hoffmann, and W. Hasse, *Applied Physics Letters* **78**, 1936–8 (2001).
6. D. Weiss, G. Schneider, B. Niemann, P. Guttmann, D. Rudolph, and G. Schmal, *Ultramicroscopy* **84**, 185–7 (2000).
7. B. Lai, W. Yun, Y. Xiao, D. Legnini, Z. Cai, A. Krasnoperova, F. Cerrina, E. D. Fabrizio, L. Grella, and M. Gentili, *Review of Scientific Instruments* **66**, 2287 (1995).
8. B. Lai, K. M. Kemmer, J. Maser, M. A. Schneegurt, Z. Cai, P. P. Ilinski, C. F. Kulpa, D. G. Legnini, K. H. Nealson, S. T. Pratt, W. Rodigues, M. L. Tischler, and W. Yun, "High-resolution X-ray imaging for microbiology at the Advanced Photon Source," 2000, vol. 506 of *AIP-Conference-Proceedings.*, pp. 585–9.
9. C. Schroer, J. Meyer, M. Kuhlmann, B. Benner, T. Günzler, B. Lengeler, C. Rau, T. Weitkamp, A. Snigirev, and I. Snigireva, *Applied Physics Letters* **81**, 1527 (2002).
10. O. Hignette, G. Rostaing, P. Cloetens, A. Rommeveaux, W. Ludwig, and A. Freund, "Submicron focusing of hard X-rays with reflecting surfaces at the ESRF," 2001, vol. 4499 of *Proc. of SPIE*, pp. 105–16.
11. M. Kuriyama, R. C. Dobbyn, R. D. Spal, H. E. Burdette, and D. R. Black, *J. Res. Natl. Inst. Stand. Technol.* **95**, 559 (1990).
12. M. Stampanoni, G. L. Borchert, R. Abela, and P. Rüegsegger, *Journal of Applied Physics* **92**, 7630 (2002).
13. M. Stampanoni, G. L. Borchert, and R. Abela, "Two-dimensional asymmetrical Bragg diffraction for submicrometer computer tomography," in *Crystals, Multilayers and Other Synchrotron Optics*, edited by T. Ishikawa, A. T. Macrander, and J. L. Wood, 2003, vol. 5195 of *Proc. of SPIE*, pp. 54–62.
14. M. Stampanoni, G. L. Borchert, and R. Abela, *Nucl. Intrum. Meth. Phys. Res. A* **551**, 119–124 (2005).
15. R. D. Spal, *Physical Review Letters* **86**, 3044–3047 (2001).

Adaptation of a Commercial Optical CMOS Image Sensor for Direct-Detection Fast X-ray Imaging

Lyle W. Marschand [*], Xuesong Jiao [†], Michael Sprung [†], Donna Kubik [*], Brian Tieman [†], Laurence B. Lurio [*], and Alec R. Sandy[†]

[*]*Department of Physics, Northern Illinois University, DeKalb, Illinois 60115 USA*
[†]*The Advanced Photon Source, Argonne National Laboratory, Argonne, IL 60439 USA*

Abstract. We have adapted a commercial CMOS optical image sensor for use as a fast x-ray detector. The sensor was used in a mode where the x-rays impinge directly on the sensor. Area detectors can significantly improve the signal-to-noise ratio of acquired data in the low photon count rate situations (even at 3rd generation synchrotron sources) encountered in both small angle x-ray scattering (SAXS) and x-ray photon correlation spectroscopy (XPCS) experiments,. CCD area detectors have been used for these types of experiments, but the relatively slow readout times typical of CCDs limit their use for studying the dynamics and kinetics of many samples. We characterized the performance of a CMOS optical detector for use in XPCS experiments.

Keywords: area detector, x-ray imaging, CMOS camera, small angle x-ray scattering, SAXS, photon correlation spectroscopy, XPCS
PACS: 07.85.Fv, 42.50Ar

INTRODUCTION

Fast area detectors are essential for x-ray photon correlation spectroscopy (XPCS) in a small angle scattering geometry (SAXS) because of the two - dimensional nature of the scattering pattern, and the need to analyze both spatial and temporal scattering fluctuations with high resolution. Area detectors yield significantly improved signal-to-noise ratio (SNR) over equivalent point detector geometries for low photon count rate per image situations encountered in XPCS. Charge-coupled devices (CCDs) have traditionally been used for this application [1,2]. Due to the stringent SNR requirements for XPCS these cameras have been employed in a direct detection mode, where the x-rays directly impinge on the sensor, rather than the more commonly used method of coupling the x-rays to the sensor through a phosphor material. While CCD detectors have excellent spatial resolution, image readout is slow since the charge collected on the camera must be moved to a readout point and read out sequentially. This technology can be sped up through the use of multiple readout points but is generally less flexible than CMOS technology since with CMOS pixels can be individually addressed and read out. Furthermore, CMOS cameras can be manufactured at a much lower cost then CCDs. There are some potential disadvantages to CMOS such as lower overall efficiency, due to thinner depletion layers and inactive areas on the pixels, and higher dark current and noise levels. The objective of this work was to explore the potential uses of CMOS based area detectors for SAXS and XPCS applications.

METHODS AND MATERIALS

Commercially available CMOS cameras are designed for a variety of optical applications. We selected the SI 1280F camera from Silicon Imaging, Inc. because it provided for fast image data acquisition (up to 2000 frames/second) and

CP879, *Synchrotron Radiation Instrumentation: Ninth International Conference,*
edited by Jae-Young Choi and Seungyu Rah

fast data readout (up to 500 frames/second). The image sensor used in this camera is the IBIS-AE manufactured by Cypress Semiconductors, which is a 1024 x 1280 array of 6.7 μm square pixels designed with a 15 μm deep silicon epitaxial for infra-red imaging applications. In order to improve the overall efficiency of the SI 1280F CMOS detector, Cypress Semiconductors developed an "active pixel" architecture, where the charge-to-voltage conversion takes place in each pixel, reducing overall conversion losses.

The SI 1280F also has a second timing board that houses a 12 bit AD converter, system clock (20 MHz – 60 MHz) and Camera Link (100 MB/s) interface to a frame grabber board. We are currently using a Coreco XL-64 frame grabber with 2 GB of onboard frame buffer memory. A vacuum housing with a 1 mil thick aluminized mylar window was designed to enclose the pixel array, timing board and temperature control mechanism. A temperature controlled Peltier cooler was press fit to the back of the image sensor in order to remove fluctuations in dark current caused by thermal fluctuations of the sensor. The sensor housing could be rotated about the horizontal central axis of the pixel array to tilt the sensor at various angles with respect to the incident x-rays. Experiments were performed at the 8-ID beamline at that Advanced Photon Source using an APS undulator-A tuned to 7.49 keV. Calibration data was also taken at the X-ray lab of Northern Illinois University. Two previously calibrated scattering samples were measured: silicon aerogels for static scattering and polystyrene latex spheres suspended in glycerol for dynamic scattering.

We analyzed the sources of potential noise given our planned usage was for low photon counts of <1/pixel and the SI 1280F is designed for optical applications with high photon counts. The IBIS5-AE CMOS imager utilizes correlated double sampling (CDS) as a method to reduce pixel reset noise, which we determined would be the largest source of noise since the pixels reset each time the sensor is read out. The remaining source of noise we wanted to evaluate was dark current (random voltage variations per pixel).

We measured dark current levels and associated rms noise variations at different gains settings and at a constant sensor array temperature of 0° C. We found approximately a 10:1 ratio of dark current levels to rms noise, which was in line with manufacturer specifications, but we had to exclude the rows/ columns near the edge of the image sensor because these pixels had atypical dark and noise statistics.

RESULTS

Noise

In order to estimate the SNR of the camera we took the ratio of the signal from a single 7.49 keV x-ray photon, after subtracting the average dark current, to the root-mean-square (rms) noise per pixel of the sensor dark current. The rms was calculated for each pixel individually from a sequence of 100 dark images and then averaged over all pixels.

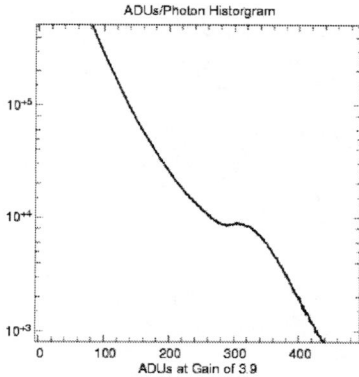

FIGURE 1. Histogram plot of number of pixels vs. ADUs/ Photon using a gain of 3.9, 25 ms exposure and 45 MHz clock speed

The charge deposited per photon is determined from the position of the peak of a histogram of the number of pixels as a function of ADUs per pixel. The peak indicates photon events against the background of dark current and noise. A typical histogram is shown in Fig. 1. ADUs are converted to charge using the specifications of the AD converter. We found SNRs of around 25:1. In the SI 1280F it is possible to adjust the gain of the readout amplifier which measures the charge on each pixel. We found that the SNR did not depend on the amplifier gain, and that the analog-digital-units (ADUs) per photon were linear with amplifier gain, as expected and shown in Fig. 2.

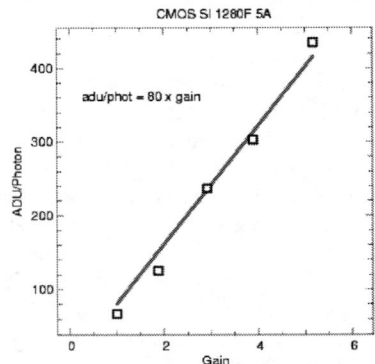

FIGURE 2. Plot of ADU per Photon at 5 different gain settings (1.0, 1.88, 2.92, 3.9, 5.17) with best linear line fit (CMOS imager array at 0° C)

Efficiency

Total efficiency is a measure of the ratio of detected photons to incident photons and is obviously a key measure for the low photon count rate situations typical of many SAXS and XPCS experiments. We calculated the total efficiency by comparing the static structure factors produced by the SI 1280F (Fig. 3.) and a calibrated Princeton Instruments CCD camera for the same sample. Overall efficiency was 11.5% at normal incidence. We measured the efficiency of the camera at different tilt angles, theorizing that the increased "effective" depth of the epitaxial layer might result in a higher efficiency. Results were that the efficiency increased to 16.6% at a tilt angle of 46.0 degrees and 21% at a tilt angle of 56.5 degrees.

Time Scales

We analyzed the dynamics of polystyrene latex spheres in glycerol at various temperatures using standard XPCS analysis methods. The time autocorrelation function (Equation 1) is

$$g_2(\vec{Q},t) = \frac{\left\langle I(\vec{Q},t)I(\vec{Q},t+\tau) \right\rangle_\tau}{\left\langle I(\vec{Q},\tau) \right\rangle_\tau^2} = 1 + A\left[S(\vec{Q},t)/S(\vec{Q}) \right]^2, \tag{1}$$

where I is the intensity of the electric field, A is the optical contrast and $S(Q,t)/S(Q)$ is the normalized intermediate scattering function. This g_2 function is indicative of the dynamic motion of the sample at various wave vectors. Our results are shown in Fig. 4 and compare very favorably with those of the more efficient CCD detectors, but at a faster time constant of 2 msec and with no loss of SNR.

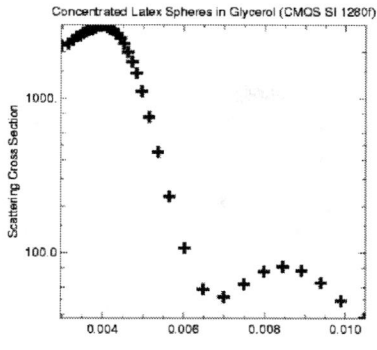

FIGURE 3. Scattering cross section of polystyrene latex spheres in glycerol solution.

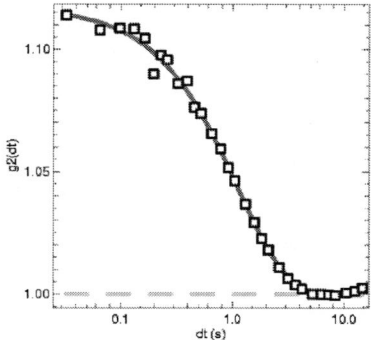

FIGURE 4. Intensity time autocorrelation function, $g_2(Q,t)$, for polystyrene latex spheres in glycerol solution showing contrast of ~ 1.15, a flat baseline of 1.0 and time constants down to ~ 2 msec. Incident photon beam slit size was 50 μm x 50 μm.

DISCUSSION

We were encouraged by the good SNR characteristics of the camera. The fact that we could optimize efficiency levels by tilting the camera housing is very encouraging, resulting in current discussions to manufacture a deeper depletion CMOS sensor for future testing. The high SNR and excellent contrast achieved at an order of magnitude increase in image acquisition speed (2 vs. 30 msec) indicates that CMOS based detectors can be used to study samples that exhibit dynamics in the msec and perhaps even sub-msec range.

ACKNOWLEDGMENTS

Funding for this work was provided by the Department of Education. Use of the APS was supported by the U. S. Department of Energy, Office of Science, Office of Basic Energy Sciences, under Contract No. W-31-109-ENG-38.

REFERENCES

1. D. Lumma, L. B. Lurio, S. G. J. Mochrie, and M. Sutton, Rev. Sci. Instrum. **71**, 3274 (2000)
2. P. Falus, M. A. Borthwick, and S. G. J. Mochrie, Rev. Sci. Instrum. **75**, 4383 (2004)

Detectors for Ultrafast X-ray Experiments at SPPS

D. Peter Siddons[a], Anthony J. Kuczewski[a], Bo Yu[a], John Warren[a], Juana Rudati[b], Paul Fuoss[b], Jerome B. Hastings[c] , Jen D. Kaspar[c], Drew A. Meyer[c] and the SPPS Collaboration[d]

[a]National Synchrotron Light Source, Brookhaven National Laboratory,Upton, New York 11973, USA.
[b]Argonne National Laboratory, Argonne, Illinois 60439, USA
[c]Stanford Linear Accelerator Center, Menlo Park, California 94025, USA
[d]http://www-ssrl.slac.stanford.edu/jbh/

Abstract. The paper describes two detectors designed specifically for the SPPS ultrafast x-ray source, one of which provides 2-D position and intensity information, and the other acts as a low-noise point detector for diffraction experiments. The beam position monitor (BPM) was used as a reference detector for most of the experiments performed there, and the point detector was used for pump-probe experiments involving phonons in bismuth and on photosensitive metal-organic crystals. The schedule for development of these detectors was extremely tight, so as much as possible we used available designs for amplifiers etc.

Keywords: Detector, Position, low-noise
PACS: 07.85.Qe

INTRODUCTION

The SPPS (Sub-Picosecond Photon Source) used the SLAC LINAC in conjunction with a standard APS undulator to generate sub-picosecond pulses of hard x-rays (8 keV). It produced roughly 10^{7} photons per pulse at a repetition rate of 10 Hz. The detector requirements for such a source are already quite challenging, and provided a valuable learning experience in anticipation of the construction of the LCLS project at SLAC. It also produced some truly unique scientific results which serve even more to whet the appetite for the spectacular source which the LCLS will be.

The experiments attempted ranged from those requiring an imaging detector to those requiring a simple point detector. All experiments required per-shot measurements and per-shot beam diagnostics (mainly intensity, position and arrival time).

BEAM POSITION MONITOR

The BPM requirement was for a non-invasive device which could, for every shot, determine the beam intensity and position. A gas-based ionization detector was the obvious choice. Traditionally, such devices have consisted of a parallel-plate chamber, with the beam passing through it, parallel to the plates. One of the plates would be divided in a diagonal manner such that the ion current to each sub-plate would depend on the beam trajectory through the chamber. This is quite satisfactory for a DC-mode device, but we have a pulsed source, and so the signals induced in the collecting electrodes depend on the beam position in both directions [1], and so could not provide a reliable intensity monitor simultaneously with a position readout. Making the plates x-ray transparent and passing the beam perpendicular to them removes the unwanted position dependence since all x-rays traverse the same path, with

CP879, *Synchrotron Radiation Instrumentation: Ninth International Conference*,
edited by Jae-Young Choi and Seungyu Rah
© 2007 American Institute of Physics 978-0-7354-0373-4/07/$23.00

charge generated uniformly all along the beam's track through the detector. Thus the only position dependence comes from the geometrical sharing of the induced charge, as required.

Our BPM consisted of a Plexiglas gas cell with two transparent windows formed from patterned aluminized Mylar foils, on either end of a 12.7 mm diameter cylindrical hole through the cell. The cell was filled with nitrogen or argon gas, depending on the beam intensity at the location of the detector. The aluminum on both foils was lithographically patterned, on one side to form a circular electrode which overfills the hole, and on the other to form a segmented electrode in the form of four quadrants. The incident beam had roughly 10^7 photons per pulse. If we allow our detector to absorb 1% of them, each pulse would generate about 4 pC. Thus, for a centered beam, each readout channel should receive 1 pC. A bias field of 800 V ensured collection of these charges by four identical channels, each containing a hybrid charge-sensitive preamplifier (CSA) of Brookhaven design [2], followed by a simple fixed-gain, fixed time constant shaping amplifier using a set of operational amplifiers for compactness. All of this signal conditioning was contained in the detector housing to ensure low noise and good signal quality.

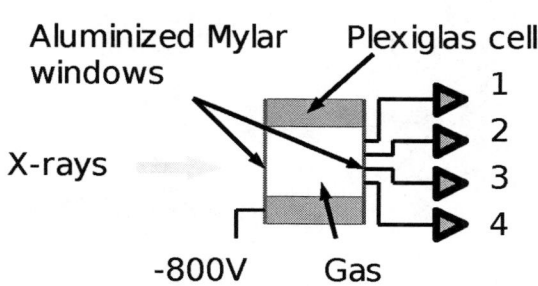 

FIGURE 1. Left: schematic cross-section of gas cell. Left-hand window is fully aluminized, right-hand window is segmented in four quadrants, each of which feed charge-sensitive preamplifier/shapers. Right: photograph of the BPM. The plexiglas cell is on the back side of the PCB. The quadrant patterned mylar window can be seen in the large hole centre-right. The four hybrid charge-sensitive preamplifiers surround the cell, and the shaping amplifiers are to the left.

$$X = \frac{(2+3) - (1+4)}{1+2+3+4}$$

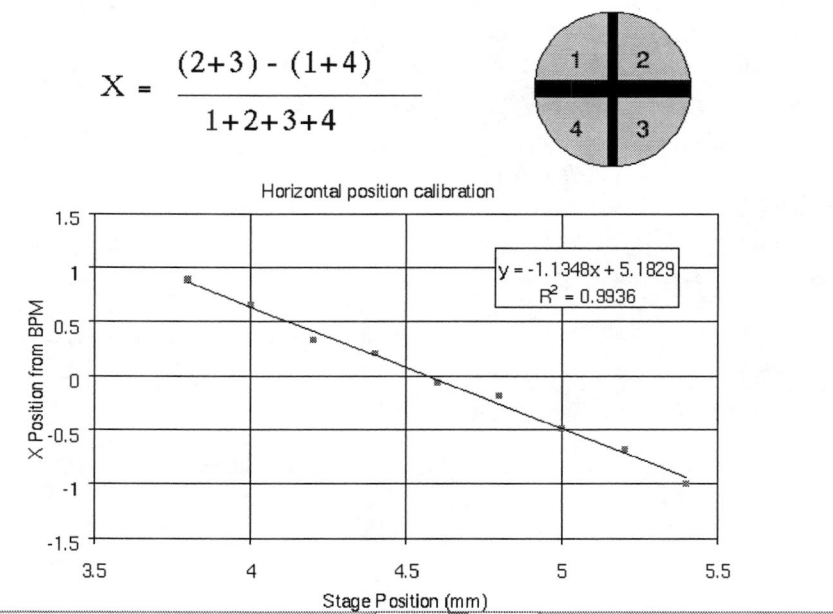

FIGURE 2. The readout scheme for the BPM, showing the position response of the detector in one dimension.

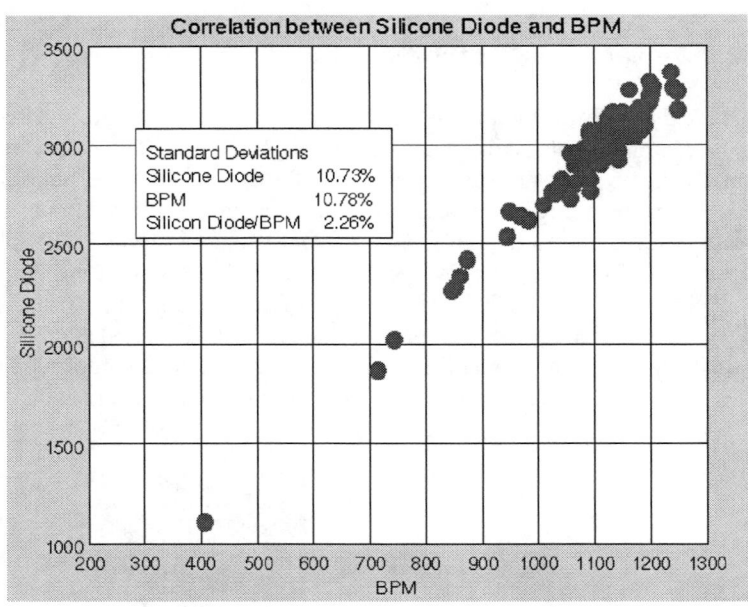

FIGURE 3. Graph showing the degree of correlation between the BPM sum signal and the diode detector, giving confidence that each is providing a reliable signal estimate.

SILICON POINT DETECTOR

The requirements for the point detector were that it should be statistically limited for incident fluxes of more than 2 photons, i.e. its noise should be of order one 8 keV x-ray photon. It should also be capable of handling up to 10000 photons per pulse. This is an extremely challenging specification. Since the project was on an extremely tight schedule, we had to make use of existing designs. The Brookhaven CSA was again pressed into service, in this case a version which was optimized for positive charge collection. It provided a noise level of about 500 electrons RMS. The detector itself was an unpackaged square silicon diode, 2 mm x 2 mm, 0.4 mm thick (actually a test structure

 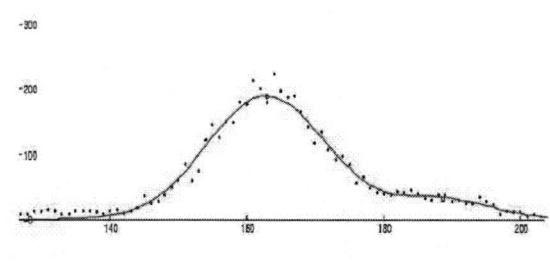

FIGURE 4. Left: Photograph of the silicon diode detector. The diode chip is at the top center, and the hybrid preamplifier to the left. Fine wire bonds can be seen connecting the guard ring and the diode proper. Right: Pulse-height spectrum of the detector output for irradiation by a Barium K emission spectrum, nominally at 32 keV. The fitted curve suggests an energy resolution of 4 keV.

included on most BNL detector wafers). Although a larger device would have been preferred from an experimental point of view, this was quite large from a noise point of view. A larger device would have resulted in both increased capacitance and leakage current. Although the latter could have been significantly reduced by cooling the diode, such a development would have required more time than was available, so we decided on room temperature operation. The diode was directly wire-bonded to the preamplifier input, making every attempt to minimize stray capacitance. At 8 keV, each absorbed photon produces around 2000 electron-hole pairs in silicon. This charge is collected on a 1 pF capacitor, producing a voltage step of 0.3 mV. A full-scale signal of 10000 photons generates a step of 3 V, well within the linear range of the preamplifier. A commercial NIM shaping amplifier was used to condition the signal for digitizing by a nuclear ADC.

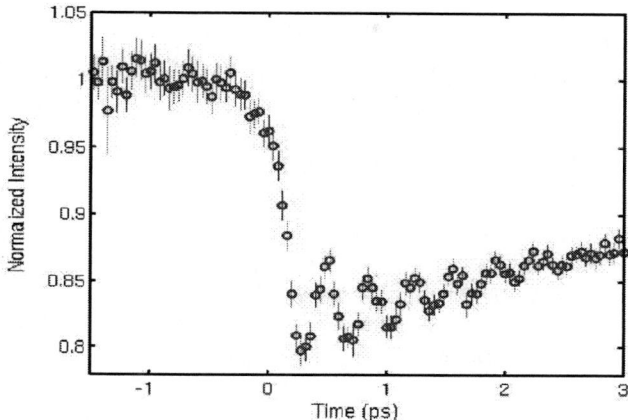

FIGURE 5. Data taken at the SPPS using the silicon diode detector. It shows the time response of a Bragg reflection intensity following optical excitation of a sample of bismuth by an ultrafast laser, probed by the SPPS x-ray beam. SPPS provided the ultrafast x-ray pulse, and our detector allowed the low-noise data above to be acquired.

REFERENCES

1. Glenn F. Knoll, *Radiation Detection and Measurement*, 3rd edition. John Wiley & Sons, Inc. New York 1999.
2. V. Radeka, *Ann. Rev. Nucl. Part. Sci*, 38 (1988) 217-27

New Display-type Analyzer for Three-dimensional Fermi Surface Mapping and Atomic Orbital Analysis

Nobuaki Takahashi[1], Fumihiko Matsui[1,2], Hiroyuki Matsuda[1], Shin Shigenai[1],
Yoshiteru Hirama[1], Yoji Hamada[3], Koji Nakanishi[4], Hidetoshi Namba[4]
Toshiro Kitamura[5], Hiroyoshi Soejima[5] and Hiroshi Daimon[1,2,3]

[1]Graduated School of Materials Science, Nara Institute of Science and Technology, Ikoma, 630-0192, Nara Japan
[2]CREST-JST, Kawaguchi, 332-0012, Saitama Japan
[3]Ritsumeikan SR Center, Kusatsu, 525-8577, Shiga Japan
[4]Department of Physics, Ritsumeikan University, Kusatsu, 525-8577, Shiga Japan
[5]Shimazu Scientific Research, Chiyoda-ku, 101-0054, Tokyo Japan

Abstract. We have developed and installed a new DIsplay-type ANAlyzer (DIANA) at Ritsumeikan SR center BL-7. We measured the angle-integrated energy distribution curve of poly-crystal gold and the photoelectron intensity angular distribution (PIAD) of HOPG to estimate the total energy resolution and to check the condition of the analyzer. The total energy resolution ($\Delta E/E$) is up to 0.78%, which is much higher than the old type. The PIAD of HOPG we obtained was the ring pattern as expected. Therefore, a detailed three-dimensional Fermi surface mapping and an analysis of the atomic orbitals constituting the electron energy bands are possible by combining them with a linearly polarized synchrotron radiation.

Keywords: two-dimensional photoelectron spectroscopy, angle-resolved photoelectron spectroscopy (ARPES), photoelectron intensity angular distribution (PIAD)
PACS: 79.60.-i, 07.81.+a, 71.18.+y, 01.30.Cc

INTRODUCTION

Photoelectron spectroscopy is a direct method used for studying the electronic structure of materials [1]. This method has successfully been applied to understand fascinating properties, such as super-conductivity or charge density wave (CDW), of materials. Conventional analyzers measure photoelectrons in a narrow solid angle. Therefore it takes much time to measure the band structure of a whole Brillouin zone because we have to rotate the sample or the analyzer step by step.

On the other hand, DIsplay-type ANAalyzer (DIANA) [2][3], that is a two-dimensional photoelectron spectrometer, enables us to measure the photoelectron intensity angular distribution (PIAD) in a wide solid angle at once without rotating the sample or the analyzer. PIAD measured by DIANA is an iso-energy cross section of the valence band dispersion at certain k_z. Therefore, by combining it with the energy tunable synchrotron radiation (SR) as an excitation source, three-dimensional Fermi surface and valence band dispersion, which are most important in understanding properties of materials, can be obtained. Furthermore, the configuration of the atomic orbitals constituting each bands is deduced experimentally from the PIADs which are excited by linearly polarized SR. Atomic orbital analysis is the essential key for the development of Orbitronics [4], which is one of the latest frontiers of solid state physics, as well as surface science.

In the case of the investigation of surface electronic state, quick measurement is required, because surfaces are easily contaminated. DIANA can satisfy this requirement, because typical acquisition time for one PIAD is less than 10 minutes.

Up to now, we have succeeded in three-dimensional band mapping of graphite [5] and atomic orbital analysis of Cu [6] at Ritsumeikan SR center using linearly polarized soft x-ray beamline BL-7. However, it was difficult to apply these methods to more complicated band dispersions due to the limited energy resolution of the photon

CP879, *Synchrotron Radiation Instrumentation: Ninth International Conference*,
edited by Jae-Young Choi and Seungyu Rah
© 2007 American Institute of Physics 978-0-7354-0373-4/07/$23.00

monochromator and the old DIANA ($\Delta E/E > 1.5\%$). Hence, we have developed and installed a higher energy resolution type of DIANA. In this paper, we report the points of improvements for New-DIANA

IMPROVEMENTS

New-DIANA has two major improvements. Figure 1 shows a schematic diagram of New-DIANA. Photoelectrons which have escaped from the sample surface excited by incident photon follow ellipsoidal orbits, according to Kepler's first law, in the spherical electric field generated by the obstacle rings, which are used as a low-pass filter. Retarding grids, which are used as a high-pass filter, consist of four concentric-hemispherical grids. Therefore, only electrons having kinetic energies within this energy window can get through the aperture and the retarding grids. These electrons are multiplied by the micro channel plate (MCP) and projected onto the fluorescent screen with the emission angle preserved. PIAD is taken by the CCD camera from outside of the vacuum chamber.

Figure 2(a) shows the transmittance of emitted electrons of the newly designed analyzer. The calculated electrons orbits are depicted in Fig. 2(b). The shape of the outer sphere corresponds to the envelope of the photoelectron orbits of kinetic energy 0.05 % higher than the pass energy. The final shape of the outer sphere was determined so that the transmittance of Fig. 2(a) has the value 0.5 at kinetic energy 1.0005 at all emission angles. As a result, the energy resolution of the obstacle rings ($\Delta E/E$) as a low-pass filter was set to be 0.05%.

The most important factor, to increase the energy resolution, is the shape of the outer sphere and the obstacle rings that generate the spherical electric field. Therefore, at first, the design of the outer sphere was fully optimized. The number of obstacle ring electrodes was increased from 22 to 156. The electrodes were made by engraving ditches on a machinable ceramic, which is used for the outer sphere. By using this outer sphere, the ripple of the transmission coefficient was reduced [3]. This fabrication was a key technology for higher energy resolution and miniaturization of the analyzer. By using ditches and convexities insulated by ditches, the composition accuracy increased. Electrodes were made by varnishing graphite solution on the convexities. In this new design, the detected angle of emitted photoelectrons was increased from ±50° to ±60°.

Another important improvement is the size-variable aperture. The diameter of the aperture of the old DIANA was fixed to 3.0 mm. But the diameter of the new aperture is variable from 0.5 to 17 mm. It is used to choose either high angular resolution or high transmission efficiency. The size is changed from outside of the chamber by a rotary feedthrough. The position of the aperture, MCP and screen are moved by ultra-high vacuum motors in the chamber, which are controlled by a computer.

EXPERIMENTAL

Two experiments were performed to estimate the total energy resolution and the intensity homogeneity in a PIAD.

The first was the measurement of the angle-integrated photoelectron energy distribution curve (EDC) of poly-crystal gold. The total energy resolution was estimated from the Fermi-edge width. Poly-crystal gold was cleaned *in situ* by repeated cycles of Ar$^+$ bombardment. The base pressure was about 1.5×10^{-8} Pa. The photon energy was set to 45 eV. The retarding voltage of the high-pass filter was 200 mV and the diameter of the aperture was set to 3.0 mm. Acquisition time for one point of the EDC data was 10 seconds. All points of the EDC data were average intensity of each PIAD.

The second experiment was the measurement of PIAD of Highly Oriented Pyrolytic Graphite (HOPG) to check intensity homogeneity in the PIAD, which was mainly caused by uneven MCP multiplication and mis-alignment of the analyzer. The cleaved surface of HOPG was cleaned by annealing at 500 °C in ultra-high vacuum for more than half a day. The base pressure is about 6.5×10^{-6} Pa. The photon energy was set to 40 eV. The retarding voltage was 300 mV and the aperture size was set to 3.0 mm. Acquisition time for one PIAD was 10 minutes.

Both of the experiments were performed at room temperature and the incident photon was a linearly polarized SR. The entrance and exit slit size were 0.07 mm.

RESULTS AND DISCUSSIONS

Figure 3 shows raw data of the EDC of poly-crystal gold. The width of the Fermi-edge was 323 meV. Therefore the total energy resolution ($\Delta E/E$) was about 0.78%. This is about twice as good as the total energy resolution of the old DIANA ($\Delta E/E = 1.5\%$). This result agrees with our expectation, because the retarding voltage was 200 mV. The reason why the retarding voltage was 200 mV is that the intensity of the Fermi-edge was very week. Therefore if we can decrease the retarding voltage more, a higher energy resolution can be expected.

Figure 4(a) shows the raw PIAD of the valence band maximum state (π-band) of HOPG and Fig. 4(b) is intensity homogenized PIAD. Intensity homogenization is performed by dividing raw PIAD by different kinetic energy PIAD, which has the energy above valence band maximum state. By this method, uneven multiplication of MCP and other factor are removed.

Single crystal graphite has six-fold π-band around K point made from p_z orbital. However, we obtained a ring pattern. This is due to the summation of PIADs from different domains. Furthermore, because of the linearly polarized excitation, the left and right parts of the ring appeared strongly. This result was consistent with theoretical simulation and indicates that the analyzer was working properly. However there exists uneven intensity in the raw PIADs, which will be removed through further fine-tuning of the analyzer.

CONCLUSIONS

We developed and installed New-DIANA at Ritsumeikan SR center BL-7. The design of the outer sphere and the number of obstacle rings were fully optimized and the size-variable aperture was installed. The total energy resolution ($\Delta E/E$) estimated by the EDC of poly-crystal gold increased from 1.5% to 0.78%.

The PIAD of HOPG was measured. An expected ring pattern was observed. However the intensity of the PIAD was not homogeneous. Therefore we need to tune the analyzer more.

As a future plan, we will install a new electron gun equipped in New-DIANA for measurement of LEED, Augar electron diffraction and SEM.

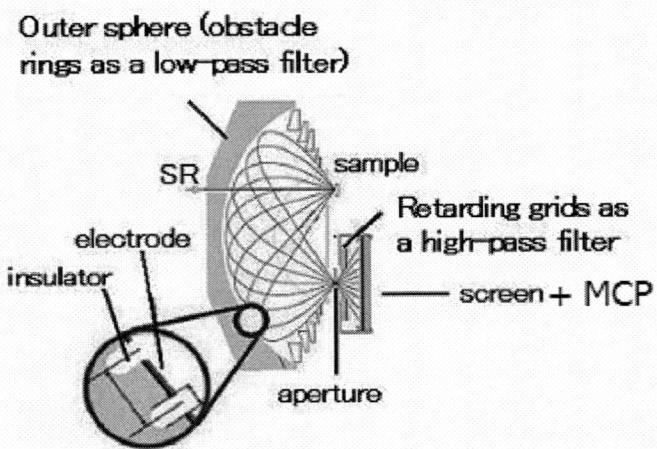

FIGURE 1.   Schematic diagram of New-DIANA. Selected kinetic energy photoelectrons are focused to the aperture.

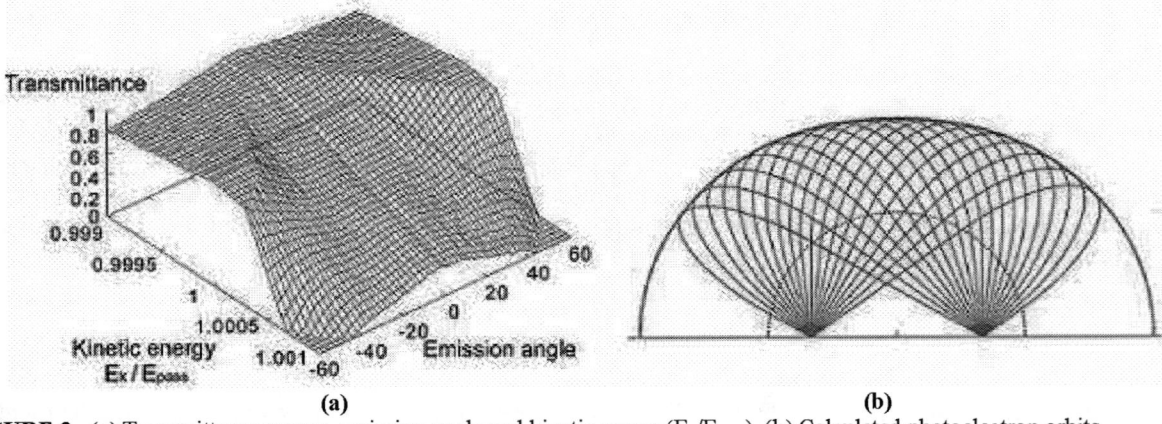

(a)   (b)

FIGURE 2.   (a) Transmittance versus emission angle and kinetic energy(E_k/E_{pass}). (b) Calculated photoelectron orbits.

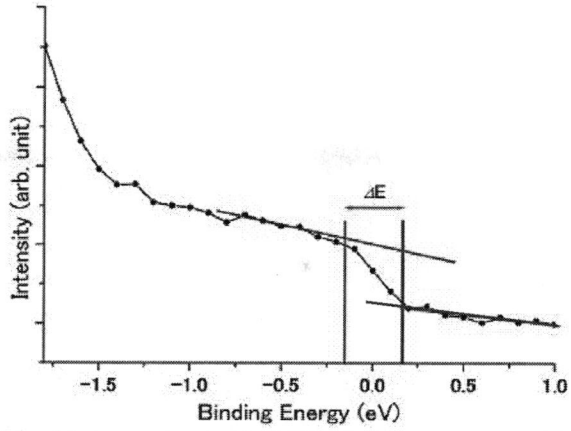

FIGURE 3. The EDC of poly-crystal gold at room temperature. The photon energy was set to 45 eV. The retarding voltage was 200 mV and the aperture size was 3.0 mm.

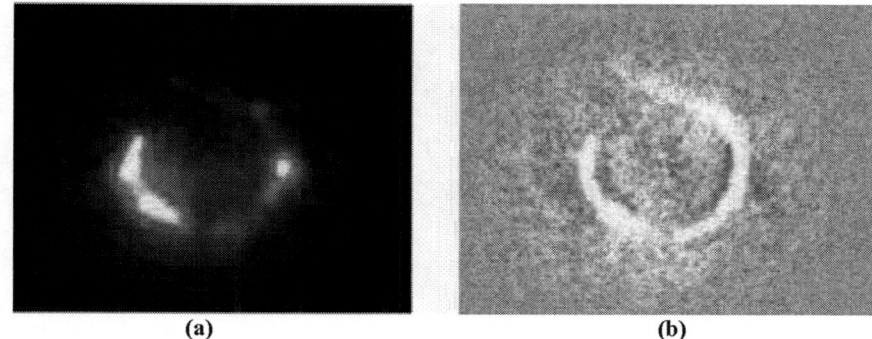

(a)	**(b)**

FIGURE 4. PIADs of the valence band maximum state of HOPG. (a) Raw data. (b) Intensity homogenized pattern.

ACKNOWLEDGMENTS

This work was supported by "Nanotechnology Support Project" of the Ministry of Education, Culture, Sports, Science and Technology (MEXT), Japan.

We acknowledge Dr. T. Matsushita at Japan Synchrotron Radiation Research Institute (JASRI) for letting us use his software of the CCD camera and pass energy of analyzer controlling and data analysis.

REFERENCES

1. S. Hüfner, Photoelectron Spectroscopy (Springer-Verlag, Berlin, 1995).
2. H. Daimon, Rev. Sci. Instrum. 59, 545 (1988).
3. T. Nohno, F. Matsui, Y. Hamada, H. Daimon, *et. al.,* Jpn. J. Appl. Phys. Vol. 42 (2003) 4752-4755.
4. Y. Tokura, N. Nagase, Science, 288, 462 (2000).
5. F. Matsui, Y. Hori, H. Miyata, N. Suganuma, H. Daimon, *et. al.,* Appl. Phys. Lett., 81, 2556 (2002).
6. F. Matsui, H. Miyata, O. Rader, Y. Hamada, Y. Nakamura, K. Nakanishi, H. Daimon, *et. al.,* Phys. Rev. B 72, 195417 (2005).

CHAPTER 5
TIME-RESOLVED TECHNIQUES

Picosecond Diffraction at the ESRF: How Far Have We Come and Where Are We Going?

Michael Wulff[@], Qingyu Kong[@], Marco Cammarata[@], Manuela Lo Russo[@], Philip Anfinrud[#], Friedrich Schotte[#], Maciej Lorenc[%], Hyotcherl Ihee[*], Tae Kyu Kim[*], Anton Plech[&]

[@]European Synchrotron Radiation Facility, 6, rue Jules Horowitz, BP220, Grenoble 38043, France
[#]Laboratory for Chemical Physics, NIDDK, Building 5, National Institute of Health, Bethesda, MD 20892-0520
[*]Korea Advanced Institute of Science and Technology, 373-1 Gu-seong-dong, Yu-sung-gu, Daejeon 305-701, Republic of Korea
[%]Groupe Matière Condensée et Matériaux, Bat 11A, UMR CNRS 6626, Université de Rennes I, 35042 RENNES Cedex France
[&]Fachbereich Physik der Universitaet KonstanzUniversitaetsstrasse 10, 78457 Konstanz, Germany

Abstract. The realization of solution phase pump-probe diffraction experiments on beamline ID09B is described. The pink beam from a low-K in-vacuum undulator is used to study the structural dynamics of small molecules in solution to 100 picosecond time resolution and at atomic resolution. The X-ray chopper and the associated timing modes of the synchrotron are described. The dissociation of molecular iodine in liquid CCl_4 is studied by single pulse diffraction. The data probe not only the iodine structures but also the solvent structure as the latter is thermally excited by the flow of energy from recombining iodine atoms. The low-q part of the diffraction spectra is a sensitive probe of the hydrodynamics of the solvent as a function of time.

Keywords: picosecond diffraction, single bunch exposures, pump-& probe diffraction, structural dynamics, in-vacuum undulators, choppers, toroidal mirror.
PACS: 31.70, 61.10, 82.53, 82.53

INTRODUCTION

Real time observation of temporally varying molecular structures during chemical reactions is a great challenge due to their ultrashort time scales. Ultrafast optical spectroscopy operating on the pico- and femtosecond time scale has provided a wealth of information about the time scales in chemical reactions such as bond breakage, bond formation and electron and proton transfer [1]. Unfortunately the long wavelength of optical light, 300-700 nm, precludes direct structural information at the atomic/molecular level. Another technique is electron diffraction that gives access to the time domain from nano- to picoseconds [2,3]. However the short penetration depth of electrons makes it difficult to apply this technique in condensed matter, which is very important in biological and industrial problems. The third technique is time resolved X-ray diffraction [4] and X-ray spectroscopy [5], which is increasingly important due the many new synchrotrons such as the European Synchrotron Radiation Facility (ESRF) in Grenoble, the Swiss Light Source (SLS) in Villigen, The Advanced Photon Source (APS) in Argonne and Spring8 in Japan. In these facilities, intense and pulsed beams of X-rays are produced by undulators that enhance the intensity 1000 times over conventional bending magnets. This enormous increase leads to much shorter exposure times. The exposure time in a pink beam diffraction experiment on a small unit-cell protein is now in the microsecond range, i.e. comparable to the orbit time for an electron in the storage ring. If now the electrons in the storage ring are concentrated in one bunch, and if the associated X-ray pulse is isolated by a chopper, the exposure time is reduced to the X-ray pulse length, 100 ps, which is about 10^4 times shorter than the open time of the chopper! The high bunch currents in the 4-bunch and 16-bunch mode at the ESRF are excellent for single pulse experiments.

CP879, Synchrotron Radiation Instrumentation: Ninth International Conference,
edited by Jae-Young Choi and Seungyu Rah
© 2007 American Institute of Physics 978-0-7354-0373-4/07/$23.00

The fastest time resolution is realized with pump & probe technology with reversible samples or samples that can be exchanged quickly. A subset of molecules is typically excited by femto or picosecond laser pulses and delayed X-ray pulses probe the evolving structures at that delay. By varying the laser/X-ray delay, snapshots of the moving molecules can be stitched together to form a film. The molecular structures are averaged over the 100 ps (fwhm) X-ray pulse, which is the limiting resolution of a synchrotron.

The time resolution in accelerator based X-ray science was recently dramatically improved at the Sub-Picosecond Pulse Source (SPPS) at Stanford where 80 fs X-ray pulses, produced in a linac, were used to probe an ultrafast solid-liquid phase transition in GaAs [6]. This experiment is a precursor for experiments at future X-ray Free Electron Lasers (XFEL) in the USA, Europe and Japan that are expected to produce ultrashort, coherent and semi-monochromatic (0.1%) X-ray pulses with up to 1×10^{12} photons per pulse.

The purpose of this paper is to give a short description of the ultrafast X-ray diffraction facilities at the ESRF and discuss the recombination dynamics of iodine in liquid CCl_4.

THE PULSED X-RAY SOURCE

Ultrafast time resolved experiments at the ESRF are done on beamline ID09B which is dedicated to pump & probe diffraction experiments. The X-rays are produced by a 236-pole in-vacuum undulator with a 17 mm magnetic period. At the minimum undulator gap, the deflection parameter K is 0.83 and the fundamental energy E_f, is 15.0 keV. The energy of ESRF electrons is 6.03 GeV. The undulator parameters at the fundamental energy are shown in Table 1 and the U17 spectra at 6.0 and 10.4 mm gap are shown in Fig 1.

TABLE 1. Source parameters for beamline ID09 (rms values, low-beta site).

Device	Sy(μm)	Sz(μm)	Sy'(μrad)	Sz'(μrad)	poles	Bmax(T)	Ef(keV)	Pcone(Watt/200mA)
U17	56.5	10.2	87.9	7.5	236	0.54	14.8	386
U20	56.5	10.2	87.9	7.5	200	0.75	8.8	710

The U17 undulator has three important features for time resolved experiments. E_f can be tuned between 15 to 20 keV where diffraction experiments have the best trade-off between detected intensity and radiation damage. Secondly, the spectrum of the fundamental is quasi monochromatic with a δE/E of 3.0 %. This pink beam can often be used without monochromator. Finally the heatload on the toroidal mirror and the chopper is modest, especially in the 4-bunch (40 mA) and the 16-bunch mode (90mA). At closed gap the U17 is used for Laue experiments from proteins. The toroidal mirror can focus 1.0×10^{10} photons per pulse into a 100 x 60 um² spot. When the gap is opened to 10.4 mm, E_f shifts to 19 keV and the second harmonic is reduced to 2% of the fundamental. This open-gap configuration is excellent for measuring transient structures in liquids.

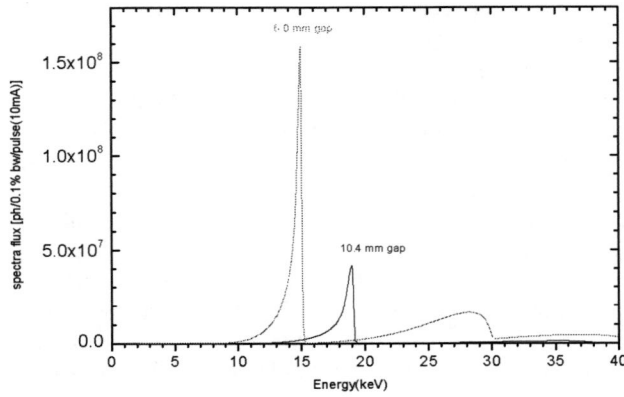

FIGURE 1. Spectral flux. from the U17 undulator. The flux is measured from a single bunch (10 mA) at 6.0 mm gap (Ef = 15 keV) and 10.4 mm gap (19.0 keV).

The circumference of the ESRF storage ring is 844.1 m and it takes an electron 2.826 μs to traverse the ring at the speed of light. The orbit frequency is 355.042 kHz, but the radio frequency cavities (RF) at the ESRF operate at the 992 harmonic. In uniform filling of the ring, the electrons are evenly spaced and separated in time by 2.84 ns. This filling is not suitable for ultrafast experiments with CCD detectors since it impossible to select single pulses with a mechanical chopper from this densely filled pulse train. The solution is to use few-bunch fillings such as 4-bunch, 16-bunch or hybrid modes. In the four-bunch mode, which runs 4% of the time at ESRF, four 10 mA pulses are evenly spaced by 705 ns. A 10 mA bunch produces an intense flash with about 1 x 10^{10} photons per pulse. The 4-bunch mode is excellent for Laue diffraction from proteins where the repeat frequency of the experiment is low due to laser heating. The lifetime is low, 6 hours, due to Coulomb repulsion within the bunch (Touschek Effect). The pulse length is 150 ps (fwhm) at 10 mA and it reduces to 120 ps at the end of a fill at 6 mA. The 16-bunch mode runs 17% of the time at ESRF, and this structure is well suited for stroboscopic liquid experiments where the pulse sequence can be repeated at 1000 Hz by exchanging the sample between shots. At the injection, the current per bunch is 5.5 mA and the lifetime 12 hours. The bunch length starts at 110 ps and ends at 90 ps at 3.5 mA. The longer lifetime improves the stability of the focused beam, in space and in time. The pulse profile of a 16-bunch pulse is shown in Fig. 2.

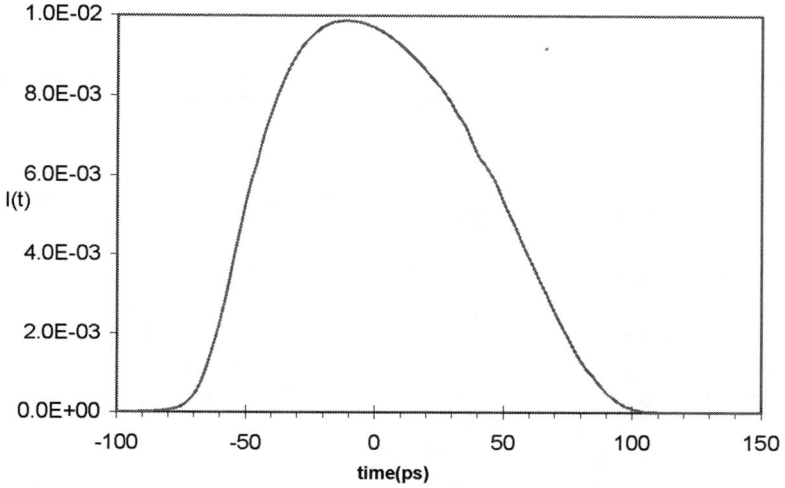

FIGURE 2. Temporal pulse profile in the 16-bunch mode. The current is 4.7 mA and the pulse is 104 ps wide (fwhm). Note that the rising edge is sharper that the falling edge. The measurement was made with the jitter-free streak camera on ID09.

THE FOCUSING MIRROR

The central cone of the undulator is focused by a 1.0 m long cylindrical mirror, which is made from a single crystal of silicon. The mirror is placed 33.1 m from the source where the beam is 6.8 mm wide and 0.6 mm high (fwhm). The mirror is coated by platinum for improved reflectivity at high energy. The incidence angle is 2.668 mrad and energies between 0-26 keV are reflected. The sagittal radius is 71.6 mm. The beam is focused 22.4 m downstream into a spot size of 100 μm x 60 μm spot (fwhm). The mirror body is held at the ends and the shape is controlled by a spring-loaded stepper motor mounted below the mirror. The mirror dimensions are chosen such that gravity curves the mirror into a toroid that focuses the beam 22.4 m downstream (R_m=9.9 km). The measured longitudinal slope-error is as low as 0.7 μrad (rms). The mirror is water cooled through copper plates that are dipped into indium-gallium filled channels along the sides of the optical surface, see Fig. 3. The liquid interface between the mirror body and the cooling system attenuates vibrations from the cooling system. The I_0 stability measured through a 60 μm vertical slit at the focus is a good as 2 x 10^{-3}. The mirror parameters are discussed in depth in [7].

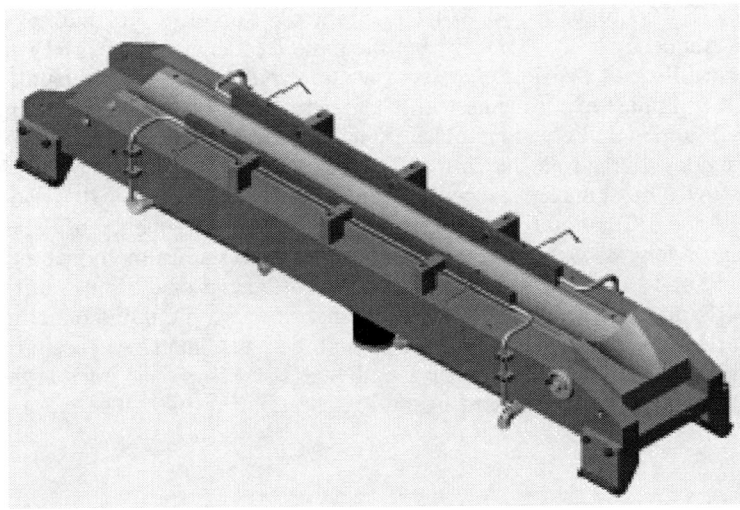

FIGURE 3. The toroidal mirror and its cooling and bending system. The vertical curvature is adjusted by a spring-loaded stepper motor below the mirror.

CHOPPER FOR SINGLE-PULSE SELECTION

One of the challenges in pump & probe experiments at synchrotrons is to control the pulse structure on the sample. The frequency of the X-ray pulses from the 16-bunch mode is 5.7 MHz, which is much faster than commercial femtosecond lasers that typically operate at 1 kHz. IN experiments with integrating CCD detectors, the laser/X-ray pulses have to arrive on the sample in pairs, i.e. the X-ray frequency has to be lowered to that of the laser. The solution is to use a chopper in front of the sample where the chopping efficiency is high due to the small beam size.

The ID09 chopper consists of a flat triangular disk that rotates in magnetic bearings in vacuum about the axis perpendicular to the disk, see Fig.4. It is designed by Bernd Lindenau and Jürgen Räbiger from KFA-Jülich in collaboration with ID09B staff. One of the three edges of the disk has a shallow channel that terminates with small roofs. The channel is thus a semi-open tunnel with slits at the ends. When the X-ray pulse is selected, the tunnel is parallel to the direction of the beam. The advantage of having a tunnel is that the beam envelope is cut from below and above simultaneously which reduces the open-time by two as compared with having the beam cut by the edges of a tunnel-less triangle.

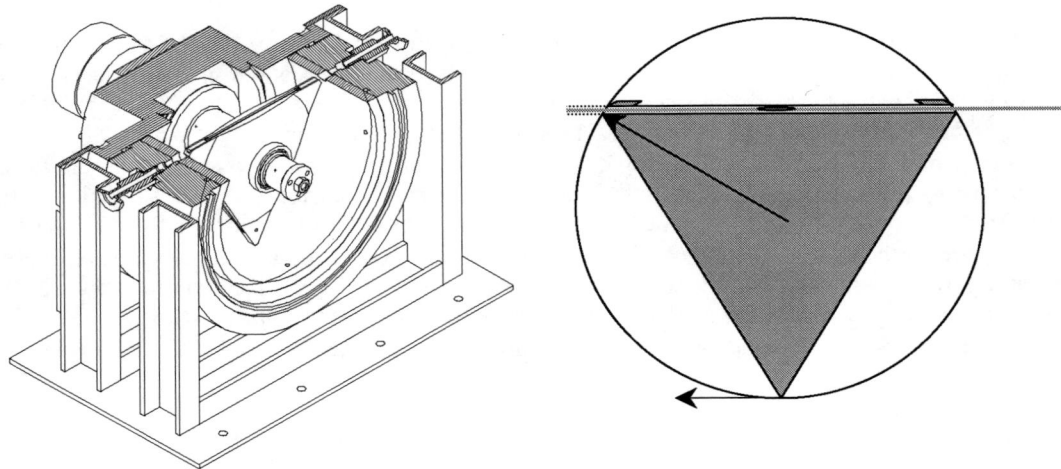

FIGURE 4. Chopper for single pulse selection. The semi-open tunnel geometry is shown on the right (vertical plane).

If the radius of the triangle is r, the height of the tunnel h and the rotation frequency f, the (maximum) open time is

$$\Delta t = \frac{1}{\sqrt{3}} \frac{h}{\pi r f} \qquad (1)$$

The chopper parameters are: r = 96.8 mm, h = 0.12 mm and f = 986.3 Hz giving Δt = 225 ns. In 16-bunch mode where pulses are separated by 176 ns and the 225 ns open window is centered on the given pulse. The clearance to neighboring pulses is thus 63.5 ns. The measured jitter in the rotation is 3-4 ns (rms) which effectively suppresses the neighbor pulses. The open time can be changed slightly by moving the chopper horizontally: the tunnel has a trapezoidal cross section and the tunnel height varies from 0.05 mm to 0.9 mm over its 4.0 mm width. This geometry makes the open time variable from 96 to 1732 ns, which makes it possible to adapt to different bunch structures.

The chopper selects pulses continuously at 986.3 Hz, and produces hence a pulse every 1.014 ms. Protein crystals can't run this fast due to the heatload from the laser. In the case of myoglobin, hemoglobin and PYP, the highest frequency we have used is 3.3 Hz. For these "single-shot" experiments, the x-ray and the laser are gated by a millisecond shutter. The ms-shutter is installed in vacuum upstream the chopper. It consists of a 60 mm long bar with a tunnel along its length. The cross section of the tunnel is trapezoidal: it is 5 mm wide and the height increases linearly from 0.3 to 2.0 mm. The bar is rotated by a stepper motor. The shortest opening time is 0.2 ms, which is obtained by rotating the tunnel from −90 to +90 degree. The stepper motor requires 48 ms to move from −90 degree to the open position at 0 degree and this delay is stored in the software that controls to data collection.

VISUALISING ATOMIC MOTIONS IN LIQUIDS

One of the first elementary reactions probed by X-rays was the recombination of dissociated iodine atoms in liquid CCl_4 [8-10]. A dilute I_2/CCl_4 (1: 360) solution was pumped by a femtosecond green laser pulse (150 fs, 520 nm) that promotes the iodine molecule from its ground state X to the excited states A, A', B and $^1\pi_u$. The emerging picture from numerous laser spectroscopy and X-ray diffraction studies is the following. The excited I_2 molecule dissociates rapidly into two neutral atoms that bounces violently in the solvent cage for a few picoseconds. These hot atoms recombine either geminately (a) or non-geminately (b):

$$I_2 + h\nu \rightarrow I_2^* \ , \ I_2^* \rightarrow I_2(a) \ , \ I_2 \rightarrow 2I \rightarrow I_2(b) \qquad (2)$$

The geminate channel takes 86% of the excited I_2* molecules. Among them 77% recombine to the ground state in 180 ps and the remaining 23% are trapped in the A/A' state for 2.7 ns after which they decay to the ground state in 180 ps. The non-geminate fraction recombines diffusively in 22 ns. During the course of recombination, the solvent cools the solutes. This process is adiabatic for times up to 1 millisecond due to the slow speed of thermal diffusion. The solvent is thus thermally excited and the change in solvent structure is inversely related to the recombination dynamics of iodine.

These atomic motions were probed by X-ray diffraction for twenty time delays between 10 ps to 10 μs. The diffraction spectrum $S(q, \tau)$ depends on the momentum transfer $q = 4\pi \sin(\theta)/\lambda$ and the delay τ. 2θ is the scattering angle and λ is the medium X-ray wavelength. The non-excited spectrum $S(q)$ for the I_2/CCl_4 solution and the laser induced signal is shown in Fig. 5. The feeble difference signal, shown in blue, is typically 1: 1000 of the total signal. Due to this weak signal everything in the experiments has to be stable: the X-ray and laser beam positions and intensities, the liquid jet and the detector. The intensity of the 3% pink beam on the sample was 5 x 10^8 ph/s and the CCD detector was exposed for 10 s per image, limited by saturation of the liquid peak.

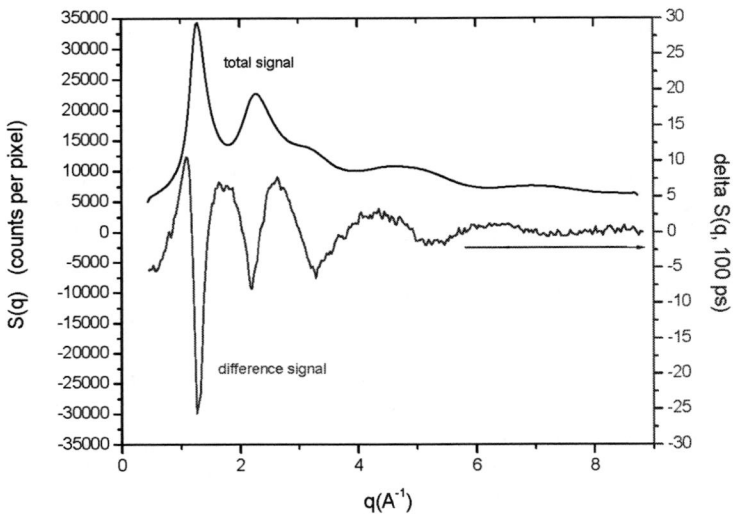

FIGURE 5. The total signal from the I_2/CCl_4 solution is shown in the upper curve. The lower curve is the laser induced signal 100 ps after dissociation. In the q-range above 4 A^{-1}, S(q,100 ps) probes the changes in atom-atom correlations within molecules predominantly. At lower q < 4 A^{-1}, the oscillations stem predominantly from temperature changes in the solvent and changes in the packing of the solvent (cage effect). Note that the CCD records the full diffraction spectrum in one measurement.

The solution-phase diffraction spectrum probes all pairs of atoms in the solution. In fact the spectrum from a molecular liquid is naturally expressed by the atom-atom pair correlation functions $g_{\alpha\beta}$ (r, t) where α and β are the types of atoms, i.e. I, C and Cl for I_2/CCl_4. If the atomic X-ray form factors is denoted by $f_\alpha(q)$, the scattering from the volume V is:

$$S(q,\tau) = \sum_{\alpha\beta} f_\alpha(q) f_\beta(q) \left(N_\alpha \delta_{\alpha\beta} + \frac{N_\alpha N_\beta}{V} \int_0^\infty g_{\alpha\beta}(r,\tau) \frac{\sin(qr)}{qr} 4\pi r^2 \, dr \right) \quad (3)$$

N_α is the number of α atoms in the sample. In the absence of an analytical theory for molecular liquids, the $g_{\alpha\beta}$ functions are usually determined by Molecular Dynamics Simulations (MD), which are based on force-field models. Note that S(q, t) is the formfactor biased sum of the $g_{\alpha\beta}$ functions. In non-resonant X-ray scattering it is thus not possible to extract $g_{\alpha\beta}$ directly. The radial real-space Fourier transform is obtain by [8]:

$$\Delta S[r,\tau] \equiv \frac{1}{2\pi^2 r} \int_0^\infty dq \, [\sum_{\mu\neq} \sum_\nu f_\mu(q) f_\nu(q)]^{-1} \times q \, \Delta S(q,\tau) \sin(qr) \quad (4)$$

which is the quantity used to interpret the data in real-space. It represents a formfactor-biased measure of the change in radial electron density seen by the average excited atom. When a series of snapshots are stitched together, it becomes a film of radial atomic motions during recombination. The $\Delta S[q, 100$ ps] and $\Delta S[r, 100$ ps] curves aer shown in Fig. 6. The first minimum in $\Delta S[r, 100$ ps] at 2.7 A comes from the depletion of the ground state X from the dissociation. The excited molecules then reach the A/A' and higher electronic states, and a maximum appears around 3.2A. In addition, the energy transfer from the solute to the solvent induces a rearrangement of the liquid without thermal expansion (isochoric temperature rise). The minima in $\Delta S[r, 100$ ps] at 4.0 and 6.2 A come from the changes in intermolecular Cl..Cl distances in CCl_4 that generated under isochoric conditions on the picosecond time scale. The transverse gradient in the laser beam leads to gradients in the temperature distribution in the liquid that lead to pressure gradients that eventually release through expansion. For a 100 μm laser beam, the speed of the pressure relaxation is proportional to the time it takes for sound to transverse the laser-illuminated volume, typically 50 ns. From the amplitude of the expansion, which typically leads to an increase in the nearest neighbor distances in the solvent of 1-5 mA and a 1-5 K temperature rise, the conservation of energy can be checked: is the observed temperature rise consistent with the chemical transitions used to explain the high-q data? In almost all reactions studied so far, the expansion is greater than the expected expansion from the observed structural transitions of the solutes. The discrepancy is usually attributed to fast and unresolved geminate recombination that injects energy into the solvent at times below 100 ps.

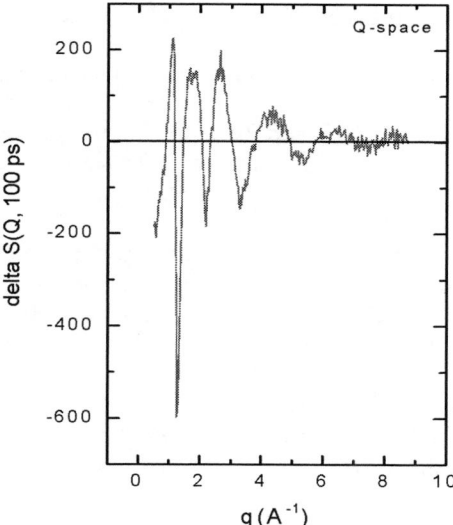 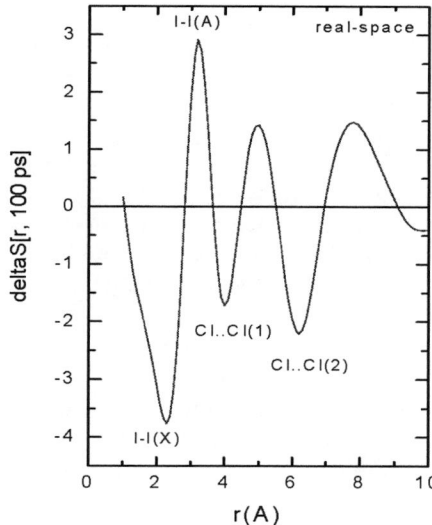

FIGURE 6. The laser induced signal from the dissociation of I_2 in CCl_4. The q-space is shown on the left and r-space on the right. The space curve is a measure of the change in the radial electron density for the average excited atom in the reaction.

Several chemical reactions have now been studied by solution phase X-ray diffraction. A new triangular C-I-C conformation of the C_2H_4I radical was observed in the dissociation pathway for $C_2H_4I_2$ in methanol [11] and a novel linear isomer CH2I-I was observed in the dynamics of CH_2I_2 in methanol [12]. Common to these measurements is that diffraction always measures the sum of two components: the dynamics of the solutes in their cages and the heated solvent. It is impossible to separate the two contributions from one measurement. In some hydrogen containing solvents such a methanol and water, we have used the infrared beam from the femtosecond laser to excite overtones of hydrogen oscillations [13]. This technique does not produce any chemical change, only a transient temperature rise. The differential form of the solvent signal can be expressed as:

$$\Delta S(q,t) = \left(\frac{\partial S}{\partial T}\right)_\rho \Delta T(t) + \left(\frac{\partial S}{\partial \rho}\right)_\tau \Delta\rho(t), \tag{5}$$

a temperature derivative at constant density (or volume) and a density differential at constant temperature. Note that the coefficients are time dependent and determined by the energetics of the chemical reaction (not the structural change of the solutes) and the hydrodynamic parameters of the solvent. Armed with these differentials, the structures of the caged solutes is usually readily determined as discussed in reference [13].

OUTLOOK

The 100 picosecond time resolution of third generation synchrotrons makes is impossible to visualize the primary steps in bond breakage and bond formation. These processes take place on the time scale of molecular vibrations, i.e. in 0.1-1 ps. These times and below will become accessible with the X-ray Free Electron Lasers (XFEL) that are under construction or in an advanced stage of design. Flux comparisons with synchrotron sources indicate that single-shot experiments with 1×10^{12} ph per pulse at 12.4 keV will become possible with unprecedented signal to noise. The actual laser/X-ray time delay will need to be measured for every single shot to compensate for the expected synchronization jitter of 1 ps (rms). However sorting the data after the experiment in time bins followed by averaging should allow to resolve ultrafast phenomena to 100 fs resolution and perhaps below. Note that the propagation velocity of optical light is slower than for X-rays. That implies that future XFEL samples will have to be thin, typically 20 um, to maintain a well-defined time delay during the passage of the two co-linear pulses.

There is an obvious advantage in having ultrafast time resolution. It will become possible to see the *order of events* in chemical and biochemical reactions. In the photo dissociation process of CO in Myoglobin, which might be studied using Laue diffraction from a spontaneous undulator at an XFEL facility, we will see how long it takes for the CO molecule to reach the first docking site. And once there, how and when do neighboring residues respond to the new space constraints?

At synchrotrons the total flux of a given undulator harmonic is essentially proportional to the bunch charge, the number of periods of the undulator and the pulse frequency. At the ESRF it is not possible to increase the bunch charge beyond 15-30 nC since that is detrimental to the emittance and lifetime. There is a potential gain of 3 in using all three sides of the ESRF chopper and such a chopper is being commissioned. Simulations show that if the center of mass and the geometrical center coincide to 10 μm, the thee sides of the rotor can be used to chop the beam at 3 kHz in for time modes with bunch spacing greater than ~ 176 ns (16 bunch mode). A 3 kHz chopper would increase the flux by 3 and hence shorten the exposure time likewise. More importantly, the 3 kHz option will allow us to work at energies where the flux is lower. Finally a heat load chopper is being installed in the optics hutch upstream the monochromator and the mirror. It consists of a 200 mm diameter Cu wheel with five 4 mm high tunnels through the center. When it rotates at 100 Hz, it produces 80 μs pulses at 1 kHz. It promises to lower the heatload on the optics and the chopper by a factor 12.5. The resulting improved stability of the focused beam should make it possible to use longer exposures with improved signal to noise, which will hopefully resolve new fine structure of molecules in action.

ACKNOWLEDGMENTS

The authors are indebted Wolfgang Reichenbach, Laurent Eybert, Laurent Claustre and Rolland Taffut for help with building and running the beamline and to Savo Bratos and Rodolphe Vuilleumier for theoretical assistance. And thanks for Bernd Lindenau and Jürgen Räbiger for engineering the chopper and its electronics.

REFERENCES

1. Y.R. Shen, "The Princibles of Nonlinear Optics": Wiley, New York, 1984.
2. M. H. Pirenne, "The Diffraction of X-rays and Electrons by Free Molecules", Cambridge University Press, 1946,
3. H. Ihee et al., "Direct imaging of transient molecular structures with ultrafast diffraction", Science, 291, 458-462 (2001).
4. J.R. Helliwell and P.M. Rentzepis. "Time-resolved Diffraction". Oxford Series on Synchrotron Radiation, No. 2. Oxford University Press, Oxford, 1997.
5. C. Bressler and M. Chergui. "Ultrafast X-ray absorption spectroscopy". Chem. Rev., 104(4): 1781-1812 (2004).
6. A.M. Lindenberg *et al. Science*, 308, 392 (2005).
7. L. Eybert, M. Wulff, W. Reichenbach, A. Plech, F. Schotte, E. Gagliardini, L. Zhang, O. Hignette, A. Rommeveaux and A. Freund. "The toroidal mirror for single-pulse experiments on ID09B". SPIE Vol. 4782, p 246-257 (2002).
8. Plech, M. Wulff, S. Bratos, F. Mirloup, R Vuilleumier, F. Schotte and P.A. Anfinrud. "Visualising Chemical Reactions in Solution by Picosecond X-ray Diffraction", Physical Review Letters, vol 92, no 12, 125505-1 to 125505-4 (2004).
9. S. Rice, Nature, Vol 429, p 255 (2004).
10. M. Wulff, S. Bratos, A. Plech, R. Vuilleumier, F. Mirloup, M. Lorenc, Q. Kong, H. Ihee. "Recombination of Photo-dissociated Iodine: A time-resolved X-ray Study" Journal of Chemical Physics, 124, 34501-34513 (2006).
11. H. Ihee, M. Lorenc, T. K. Kim, Q. Y. Kong, M. Cammarata, J. H. Lee, S. Bratos, M. Wulff, "Ultrafast X-ray Diffraction of Transient Structures in Solution", Science, 309, 1223-1227 (2005).
12. J. Davidsson, J. Poulsen, M. Cammarata, P. Georgiou, R. Wouts, G. Katona, F. Jacobson, A. Plech, M. Wulff, G. Nyman and R. Neutze, "Structural Determination of a Transient Isomer of CH_2I_2 by Picosecond X-Ray Diffraction", Physical Review Letters 94, 245503 (2005).
13. M. Cammarata, M. Lorenc, T. K. Kim, J. H. Lee, Q. Y. Kong, E. Pontecorvo, M. Lo Russo, G. Schiró, A. Cupane, M. Wulff, H. Ihee. "Impulsive solvent heating probed by picosecond x-ray diffraction". Journal of Chemical Physics, 124, 124504, 2006.

The Advanced Light Source (ALS) Slicing Undulator Beamline

P.A. Heimann[a], T.E. Glover[a], D. Plate[a], H.J. Lee[b], V.C. Brown[a], H.A. Padmore[a] and R.W. Schoenlein[c]

[a]Advanced Light Source, Lawrence Berkeley National Laboratory, Berkeley, CA 94720, USA
[b]Department of Physics, University of California, Berkeley, Berkeley, CA 94720, USA
[c]Materials Sciences Division, Lawrence Berkeley National Laboratory,
Berkeley, CA 94720, USA

Abstract. A beamline optimized for the bunch slicing technique has been construction at the Advanced Light Source (ALS). This beamline includes an in-vacuum undulator, soft and hard x-ray beamlines and a femtosecond laser system. The soft x-ray beamline may operate in spectrometer mode, where an entire absorption spectrum is accumulated at one time, or in monochromator mode. The femtosecond laser system has a high repetition rate of 20 kHz to improve the average slicing flux. The performance of the soft x-ray branch of the ALS slicing undulator beamline will be presented.

Keywords: beamline, electron beam slicing and x-ray absorption.
PACS: 41.50.+h and 78.70.Dm.

INTRODUCTION

Ultrafast x-ray absorption spectroscopy is a powerful technique for studying transient states of matter. A femtosecond (fs) laser pulse induces a phase transition or chemical reaction. Picosecond (ps) synchrotron radiation pulses or femtosecond x-ray pulses generated by the bunch slicing technique are used to obtain the x-ray absorption spectrum. Fs laser pulses can heat a material to high temperature at constant density. Liquid carbon x-ray absorption spectra have been measured. [1] The x-ray absorption near-edge structure of VO_2 has been observed during a photo-induced phase transition. [2]

Fs x-ray pulses may be generated by the bunch 'slicing' technique. [3] The electron bunch is modulated by co-propagating a fs laser pulse with the stored electron bunch through a wiggler. The modulated electrons are spatially separated from the rest of the electron bunch in a straight section with vertical dispersion. Spatial dispersion is employed for bunch slicing at the ALS while angular dispersion is used at BESSY and the SLS. [4,5] Finally, by imaging the radiation from the displaced electrons onto slits, the fs x-rays can be separated from the ps pulse.

ALS beamline 6.0 is dedicated to time-resolved x-ray science. The optical design of this beamline was presented previously in Heimann et al. [6] An in-vacuum undulator / wiggler radiates both soft and hard x-rays. There are two branch lines employing alternatively a grating spectrograph and a crystal monochromator. A 20 kHz fs laser system is located at the end of the beamline. Based on the known storage ring parameters, x-ray pulses of 200 fs duration will be produced. The soft x-ray branch, ALS BL 6.0.2, has been installed and commissioned. Its commissioning will be described here. The hard x-ray branch, ALS BL 6.0.1, is presently being assembled.

COMMISSIONING OF ALS BEAMLINE 6.0.2

The source is an in-vacuum insertion device with a λ_u of 30 mm. At low photon energies, below 2 keV undulator harmonics are used. At high photon energies, the source is wiggler radiation. By using both undulator and wiggler radiation, a wide photon energy range is provided from 120 eV to 10 keV. Figure 1 shows the measured undulator spectrum with the fundamental at 1050 eV. The fundamental has a broad width because of the relatively

CP879, Synchrotron Radiation Instrumentation: Ninth International Conference,
edited by Jae-Young Choi and Seungyu Rah
2007 American Institute of Physics 978-0-7354-0373-4/07/$23.00

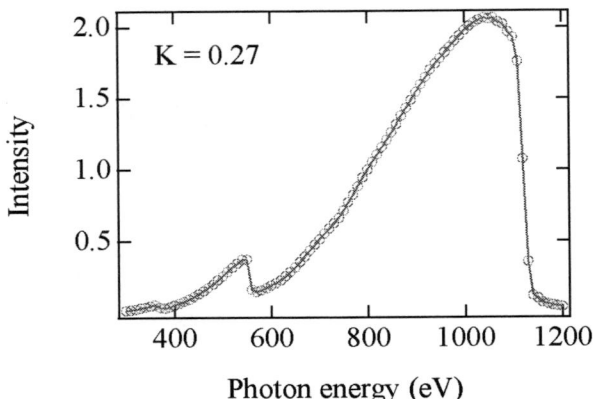

FIGURE 1. The spectrum measured from the in-vacuum undulator at K of 0.27.

large angular acceptance of the beamline, 0.5 x 0.5 mrad^2, and the low K value, 0.27. The photon energy axis was was calibrated using the L_3 and L_2 absorption lines of Co. The peaks at lower photon energies result from the second and third orders of the grating.

The first optical element of both soft and hard x-ray branch lines is a horizontally deflecting toroidal mirror, which produces a ~1:1 image of the insertion device source. The grazing incidence angles of 1.35 degrees and 8 mrad result in energy cut-offs of 2000 eV and 10 keV for the soft x-ray M201 and hard x-ray M101 mirrors respectively. Following the M201 and M101 mirrors are x-ray choppers, which act as power filters and have repetition rates matched to the laser system, 20 kHz. The chopper has not yet been received. As a result, the commissioning has been performed with the undulator at low K. Following the choppers, vertical and horizontal pairs of slits are used to select the femtosecond x-ray pulses and as the entrance slits of the monochromator / spectrometer.

The dimensions of the focus of the M201 mirror was measured to be 80 μm (fwhm) vertically and 540 μm (fwhm) horizontally. The observed intermediate focus is in agreement with ray tracing.

In bunch slicing at the ALS, a fraction of the electrons, ~10^{-4}, are displaced ~200 μm from the center of the bunch. As a result the tails of the vertical focus are important in determining the slicing femtosecond signal / picosecond background. Figure 2 shows both a measured, calculated and ideal Gaussian profile of the vertical focus on a logarithmic scale. The vertical profile was estimated using the Fourier optics program SRW. [7] The radiation is calculated for the undulator source. One dimensional sagittal height data from the mirror metrology is used to generate a phase screen for the mirror. Next a wavefront is propagated from the source to the mirror and then to the image plane. Some present limitation of the calculations are the one dimensional model of the roughness and the exclusion of mirror aberrations. The calculated and measured curves show good qualitative agreement. At the image plane the estimate of the displacement of the x-rays originating from the sliced electrons is 160 μm. These results predict that the slicing femtosecond x-ray pulses will have a signal-to-background ratio of about 1:1.

The grating spectrograph includes a vertical focusing spherical mirror (M202), varied-line-spacing plane gratings (G201 and G202) and a horizontally focusing plane elliptical mirror (M203). An image of the spectrum is produced at the detector. The efficiency of the gratings was measured and compared with calculations based on the Neviere theory of grating efficiency. [8] The measured and calculated efficiencies agree within 10 and 20 % for the two gratings, which indicates the good agreement between predicted and observed performance that is possible for laminar gratings.

The goal of energy resolution is 0.5 eV at 850 eV photon energy determined by the natural linewidth of the nickel L_3 edge. The resolution was optimized by scanning the zero diffracted order across a slit and minimizing the zero order width. The measured zero order implies a resolution of 0.6 eV at 850 eV photon energy. The actual energy resolution for dispersive x-ray absorption spectroscopy may be limited by the spatial resolution of the detector. The M203 mirror produces a 50 μm horizontal focus at the detector.

The flux at 1050 eV photon energy was measured by a silicon photodiode located at the end of the beamline. The observed flux agrees with calculations to within a factor of two.

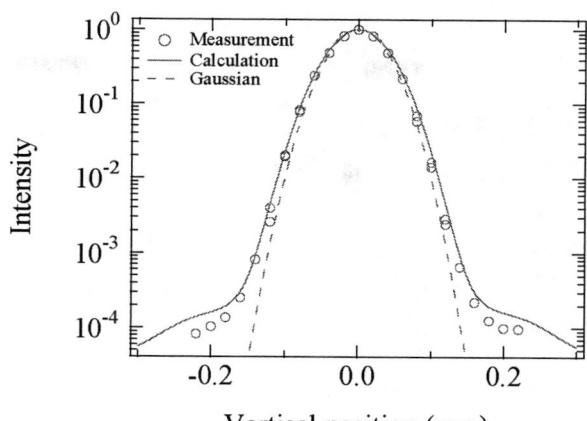

FIGURE 2. The measured (circles), calculated (solid curve) and ideal Gaussian (dashed curve) profile of the intermediate vertical focus.

Dispersive x-ray absorption spectra have been measured with a microchannel plate (MCP) detector, consisting of a MCP, fiber optic and optical CCD. By sending a high voltage pulse to the MCP, the detector can be gated on for individual x-ray pulses from the ALS. An x-ray shutter limited the duration of the exposure to 10 ms. The absorption spectrum of a 60 nm thick Cu foil was measured.

A 20 kHz femtosecond laser system has been constructed with ~1 mJ in both one arm for interacting with the electron beam and a second arm to excite samples. The interaction laser beam has been sent to the storage ring for commissioning studies of the bunch slicing.

REFERENCES

1. S.L. Johnson, P.A. Heimann, A.G. MacPhee, A.M. Lindenberg, O.R. Monteiro, Z. Chang, R.W. Lee, and R.W. Falcone, *Phys. Rev. Letters* **94**, 57407-57411 (2005).
2. A. Cavalleri, M. Rini, H.H.W. Chong, S. Fourmaux, T.E. Glover, P.A. Heimann, J.C. Kieffer, and R.W. Schoenlein, *Phys. Rev. Letters* **95**, 67405-67409 (2005).
3. R.W. Schoenlein, S. Chattopadhyay, H.H.W. Chong, T.E. Glover, P.A. Heimann, C.V. Shank, A.A. Zholents, M.S. Zolotorev, *Science* 287, 2237-2240 (2000).
4. K. Holldack, T. Kachel, S. Khan, R. Mitzner, T. Quast, *Phys. Rev. ST-Accelerators and Beams* **8**, 40704-40711 (2005).
5. G. Ingold, A. Streun, B. Singh, R. Abela, P. Beaud, G. Knopp, L. Rivkin, V. Schlott, T. Schmidt, H. Sigg, J.F. van der Veen, A. Wrulich, S. Khan., *Proceedings of the 2001 PAC* **4**, 2656-2658 (2001).
6. P.A. Heimann, H.A. Padmore and R.W. Schoenlein, in *Synchrotron Radiation Instumentation-2003*, edited by T. Warwick et al., AIP Conference Proceedings 705, American Institute of Physics, Melville, NY, 2004, pp. 1407-1410.
7. O.Chubar, P.Elleaume, S.Kuznetsov, and A.Snigirev, *Proc. SPIE Int. Soc. Opt. Eng.* **4769**, 145 (2002).
8. *Electromagnetic Theory of Gratings*, edited by R. Petit, Springer-Verlag, Berlin, 1980.

Sub-Picosecond Tunable Hard X-Ray Undulator Source for Laser/X-Ray Pump-Probe Experiments

G. Ingold, P. Beaud, S. Johnson, A. Streun, T. Schmidt, R. Abela, A. Al-Adwan, D. Abramsohn, M. Böge, D. Grolimund, A. Keller, F. Krasniqi, L. Rivkin, M. Rohrer, T. Schilcher, T. Schmidt, V. Schlott, L. Schulz, F. van der Veen, D. Zimoch

Swiss Light Source, Paul Scherrer Institut, CH-5232 Villigen, Switzerland

Abstract. The FEMTO source under construction at the μXAS beamline is designed to enable tunable time-resolved laser/x-ray absorption and diffraction experiments in photochemistry and condensed matter with ps- and sub-ps resolution. The design takes advantage of (1) the highly stable operation of the SLS storage ring, (2) the reliable high harmonic operation of small gap, short period undulators to generate hard x-rays with energy 3-18 keV at 2.4 GeV beam energy, and (3) the progress in high power, high repetition rate fs solid-state laser technology to employ laser/e-beam 'slicing' to reach a time resolution of ultimately 100 fs. The source will profit from the inherently synchronized pump (laser I: 100 fs, 2 mJ, 1 kHz) and probe (sliced X-rays, laser II: 50 fs, 5 mJ, 1 kHz) pulses, and from the excellent stability of the SLS storage ring which is operated in top-up mode and controlled by a fast orbit feedback (FOFB). Coherent radiation emitted at THz frequencies by the sliced 100 fs electron bunches will be monitored as on-line cross-correlation signal to keep the laser-electron beam interaction at optimum. The source is designed to provide at 8 keV (100 fs) a monochromized flux of 10^4 ph/s/0.01% bw (Si crystal monochromator) and 10^6 ph/s/1.5% bw (multilayer monochromator) at the sample. It is operated in parasitic mode using a hybrid bunch filling pattern. Because of the low intensity measurements are carried out repetitively over many shots using refreshing samples and gated detectors. 'Diffraction gating' experiments will be used to characterize the sub-ps X-ray pulses.

Keywords: sub-ps X-rays, time-resolved, undulator radiation, fs laser
PACS: 41.75.Ht, 41.60.Ap, 42.65.Re

INTRODUCTION

In the presence of a static magnetic field, an electron can interact with light to either emit or absorb energy. Although the magnetic field of a tightly focused laser beam is much higher than the static one, its coupling to the electron beam is negligible if the electron has relativistic energy ($\beta \simeq 1$) and co-propagates with the laser beam. If the planar magnetic field of a static periodic wiggler is tuned to the wavelength of a horizontally linear polarized laser, the electron will resonantly interact with the optical electric field and—depending on relative phase—will gain or loose energy. The process is described by the theory of a small signal gain FEL amplifier [1]. More recently [2] it has been realized that the same process can be used in a storage ring to impose an energy modulation on the electron beam of several sigma beyond the natural energy spread. By using a fs laser, a sub-ps slice of an electron bunch can be generated after passing a dispersive section in the ring. To achieve sufficient energy modulation to separate the 'sliced' electrons from the core beam, the laser pulse energy should be rather high (3-5 mJ at 2.4 GeV) and the number of wiggler periods matched to the number of optical laser cycles.

The 'slicing' technique first demonstrated at the ALS [3] has recently been implemented at BESSY [4] to generate sub-ps soft x-rays (1-2 keV) with variable polarization. The SLS FEMTO slicing source [5] currently under commissioning at the μXAS beamline of the SLS storage ring (2.4 GeV) is designed to generate x-rays in the range 3-18 keV based on the high harmonic (11./13.) operation of a short period in-vacuum undulator [6].

The FEMTO slicing facility consists of two main parts, the FEMTO laser system and the FEMTO insertion ('slicing spectrometer'). The construction and operation of the facility take advantage of its modular design to allow laser/x-ray pump/probe experiments with time resolution of either 100 ps or 0.1 ps (slicing operation). Hybrid bunch filling ('camshaft' single pulse current: 3-8 mA) requires gated detectors (APDs, Si-μstrip- and pixel- detectors).

So far laser(100 fs)/x-ray(100 ps) pump/probe time-resolved experiments have been performed in absorption and diffraction using both solid and liquid samples. As an example, we present briefly the ps characterization of strain wave dynamics in near-surface heated crystalline InSb.

CP879, *Synchrotron Radiation Instrumentation: Ninth International Conference,*
edited by Jae-Young Choi and Seungyu Rah
© 2007 American Institute of Physics 978-0-7354-0373-4/07/$23.00

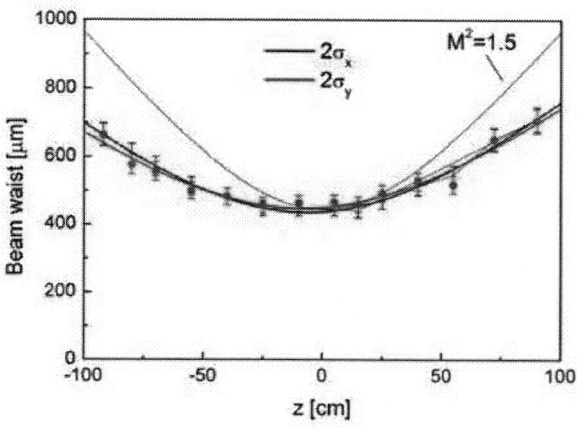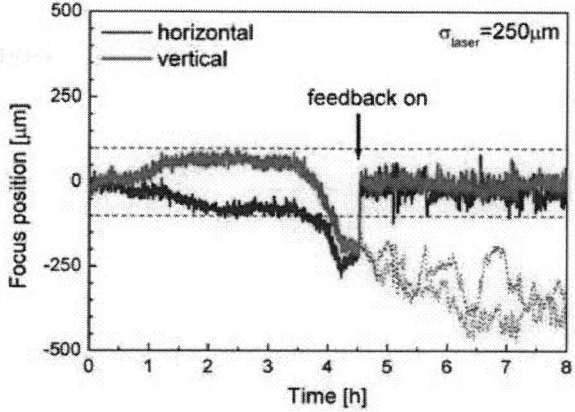

FIGURE 1. Laser amplifier II: Essential for efficient slicing is a near diffraction limited spatial profile of the laser beam and a high pointing stability at the interaction point inside the modulator.

LASER SYSTEM

A high average power femtosecond laser system has been installed near the experimental hutch of the μXAS beamline. The laser system consists of three parts: (1) a fs oscillator (30 fs, 100 MHz, 500 mW, 795 nm) that is synchronized to the rf-master oscillator (500 MHz) with a jitter estimated to be < 1 ps, (2) a 'phase I' regenerative amplifier (115 fs, 1 kHz, 2.1 mJ, 800 nm) for pumping experimental samples, and (3) a 'phase II' amplifier system (consisting of a regenerative amplifier and a symetrically pumped 2-pass amplifier: 50 fs, 1 kHz, 5.2 mJ, 805 nmm) that is used to produce sub-ps x-rays within the storage ring. The Ti:sapphire laser crystal of the amplifier is cryogenically cooled to 45 K to ensure a good mode quality ($M^2 \leq 1.1$) of the outgoing laser beam.

A 5"-aperture magnifying telescope has been set up in the laser hutch to focus the laser in vacuum over an optical distance of $\sim$ 45 m into the modulator where the laser-electron interaction (slicing) takes place. This configuration was chosen to maintain the good spatial and temporal quality of the laser beam by minimizing phase distortions of the intense fs laser pulses, which interact nonlinearly with optical media such as windows, lenses and air.

Special attention was given to the rigidity and accuracy of the mirrors that are placed inside the tunnel to direct the laser into the modulator. These mirrors are placed inside the UHV of the storage ring and are remote controlled. Cameras are used for optical alignment, diagnostics and active beam stabilization (see Figure 1).

SLICING SPECTROMETER

To generate sub-ps undulator radiation above 3 keV at 2.4 GeV, small gap operation of both the modulator and the radiator is needed. To minimize bunch lengthening effects, the FEMTO insertion has been installed in the long straight section 5L of the SLS storage ring. The sequence of magnets act as a 2-stage spectrometer: the energy-modulated satellite electrons ('secondary beam') generated in the modulator—due to the dispersion provided by a chicane—are separated from the core beam inside the radiator, where the x-rays are generated.

A new (FEMTO) quadrupole triplet has been installed to match the beta- functions for the modulator and the radiator (beam stay clear aperture of 8 mm and 5 mm, respectively) installed at the entrance and exit of straight 5L. This break of the 3-fold ring symmetry does not affect the dynamic acceptance of the SLS storage ring.

For sub-ps operation the modulator (hybrid wiggler W138: B_{eff}=1.98 T, g=11 mm, λ_w= 138 mm, 16 periods) is operated at resonance, close to the amplifier II laser wavelength of 805 nm. A 3-dipole chicane provides horizontal spatial and angular dispersion of 0.44 m and 0.1 rad, respectively, to separate the satellite and core electron beams at the center of the radiator. The radiator is a small gap, short period hybrid in-vacuum undulator operated at high harmonics (B_{eff}=0.94 T, g=5 mm, λ_u= 19 mm, 96 periods). For 400 mA (390 bunches) the measured monochromatic flux at 8 keV at the sample is $8 \cdot 10^{12}$ ph/s/0.01% bw (aperture 240 x 54 μrad^2). Using a 5 mA 'camshaft pulse' of 50 ps (FWHM) (using factor 2 bunch shortening by detuning the SC 3rd harmonic cavity), the expected sliced flux at 8

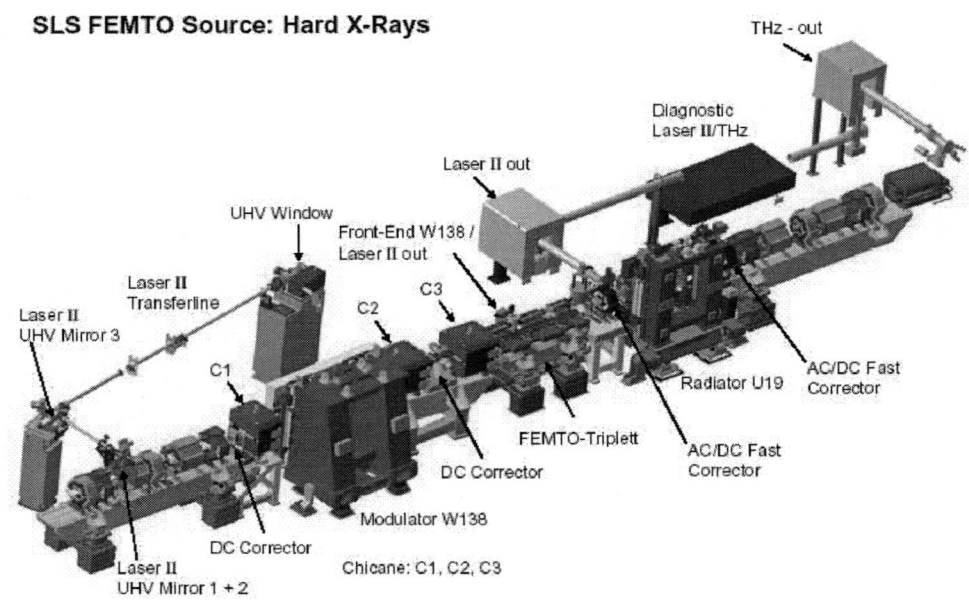

FIGURE 2. Engineering layout of the FEMTO source: Laser II transport line, slicing spectrometer installed in a long (11 m) straight section, and ports for laser and THz diagnostic.

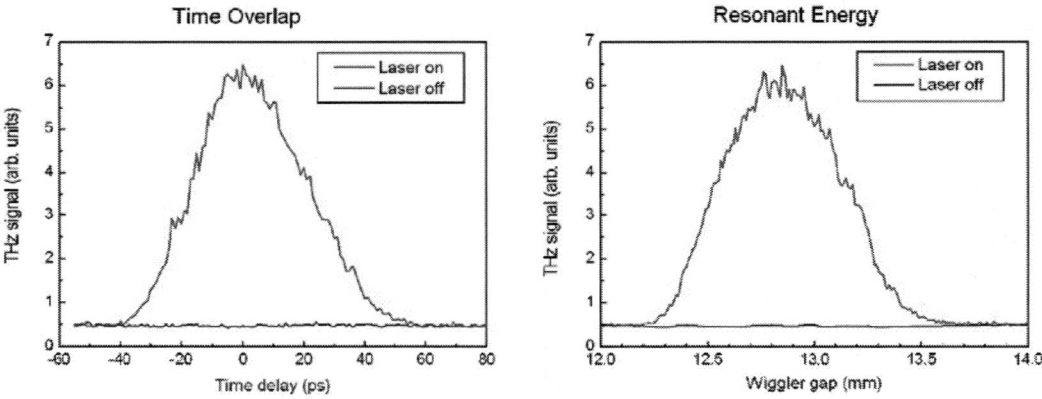

FIGURE 3. Measured coherent THz-radiation emitted by fs laser sliced electrons.

keV at the sample is 10^4 ph/s/0.01% bw (Si crystal monochromator) and 10^6 ph/s/1/5% (multilayer monochromator). Corrector magnets and beam position monitors both for the electron and x-ray beam have been installed to control residual orbit distortions in position and angle to ultimately 1 μm and 1 μrad, respectively, when feedforward (FF) and fast orbit feedback (FOFB) corrections are applied.

To suppress background radiation, the modulator is tilted by 2.4° with respect to the radiator center. Spectrum calculations including all possible background contributions (namely x-rays from the up- and downstream bending magnets, from the chicane dipole magnets C1-C3 and from the modulator) show that x-rays emitted from sliced electrons with energy modulation $\Delta E/E \geq 0.5\%$ can be separated using a slit system installed at the μXAS beamline in 15 m distance from the radiator. Because the x-rays do not pass windows or any optical element in front of the slit, the possibly harmful background is entirely due to the magnetic elements. To achieve a signal-to-background ratio of 10:1, several photon absorbers have been installed along the spectrometer.

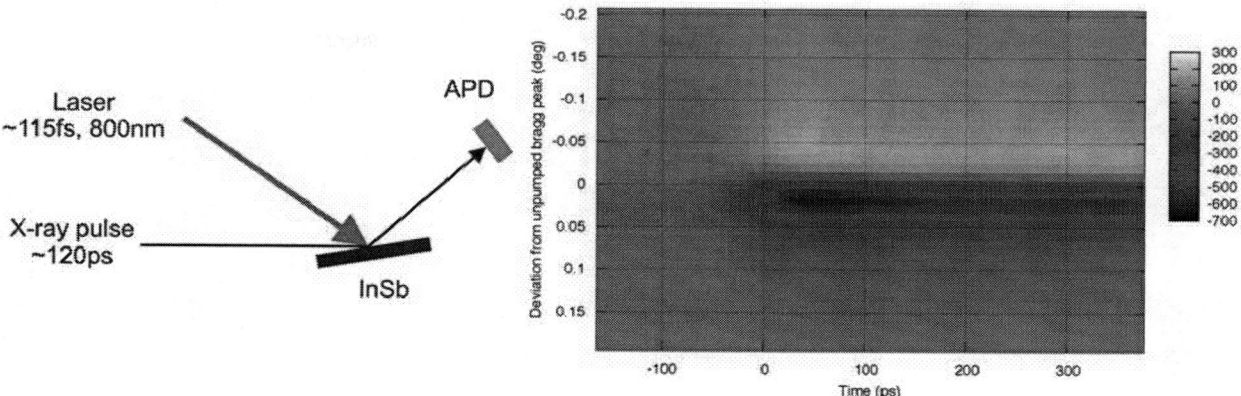

FIGURE 4. Ultrafast X-ray diffraction as a timing diagnostic: ps strain wave dynamics in a surface-heated InSb crystal. The laser strikes the surface at 0 ps, resulting in the formation of an expansion layer over the x-ray probe depth. To resolve the dynamics within the probe depth (< 30 ps), sub-ps X-rays will be used in future experiments.

SUB-PS DIAGNOSTICS

Electrons. Coherent synchrotron radiation (CSR) in the THz spectral range is used as on-line diagnostic to optimize the laser- electron interaction for optimum energy modulation [7]. The CSR is due to the longitudinal charge density modulation on the 100 μm length scale and is extracted at the first bending magnet downstream of the slicing spectrometer. As shown in Figure 3, the THz-intensity measured with an InSb-bolometer (4.2 ° K, rise time 0.3 μs) is clearly correlated when scanning the relative laser-electron timing and matching the resonant energy by changing the magnetic field of the modulator via gap scan.

X-rays. There is no routine method yet to characterize the time structure of sub-ps hard X-rays. For commissioning the timing and pulse length of the slicing source, we are currently preparing an experiment to measure ultrafast diffraction from coherent phonons in InSb and Bi crystals. This experiment [8] is both adequately reversible (i.e. it can be carried out repetitively over hours) and large enough in amplitude to be observable with the sub-ps photon flux expected from this source. In a preliminary experiment we have studied the transient thermal strain in a laser-heated asymmetric cut InSb crystal with ps time resolution shown in Figure 4. "Crossed beam" experimental geometry will allow simultaneous collection of different pump-probe delay times. Gateable 1-D Si μ-strip and 2-D pixel [9] detector modules under development at PSI will be used to resolve the X-ray image.

REFERENCES

1. A. Amir and Y. Greenzweig, *Phys. Rev. A* **34**, 4809 (1986).
2. A. A. Zholents and M. S. Zoloterev, *Phys. Rev. Lett.* **76**, 912 (1996).
3. R. W. Schoenlein et al., *Science* **287**, 2237 (2000); R. W. Schoenlein et al., *Appl. Phys.* **B71**, 1 (2000).
4. S. Khan et al., *Proceedings PAC 2005*, Knoxville, 2309 (2005).
5. G. Ingold et al., *Proceedings PAC 2001*, Chicago, 2656 (2001).
6. G. Ingold et al., *these Proceedings*.
7. K. Holldack et al., *Phys. Rev. Lett.* **96**, 054801 (2006).
8. A. M. Lindenberg et al., *Phys. Rev. Lett.* **84**, 111 (2000); A. Rousse et al., *Nature* **410**, 65 (2001); K. Sokolowski-Tinten et al., *Nature* **422**, 287 (2003).
9. B. Schmitt et al., *Nucl. Instr. and Meth. A* **518**, 436 (2004); Ch. Brönnimann et al., *J. Synchrotron Rad.* **120** (2006).

Dynamic View on Nanostructures: A Technique for Time Resolved Optical Luminescence using Synchrotron Light Pulses at SRC, APS, and CLS

F. Heigl, [1] A. Jürgensen, [1] X.-T. Zhou, [2] S. Lam, [2] M. Murphy, [2] J.Y.P. Ko, [2] T.K. Sham, [2] R. A. Rosenberg, [3] R. Gordon, [4] D. Brewe, [4] T. Regier, [5] L. Armelao [6]

*Canadian Synchrotron Radiation Facility / Synchrotron Radiation Center, University of Wisconsin -
Madison, [1] Chemistry Department, The University of Western Ontario, [2]
Advanced Photon Source, Argonne National Laboratory, [3]
PNC-CAT / Advanced Photon Source, Argonne National Laboratory, [4]
Canadian Light Source / University of Saskatchewan, [5]
Chemistry Department, ISTM-CNR, Padova (Italy) [6]*

Abstract. We present an experimental technique using the time structure of synchrotron radiation to study time resolved X-ray excited optical luminescence. In particular we are taking advantage of the bunched distribution of electrons in a synchrotron storage ring, giving short x-ray pulses (10-10^2 picoseconds) which are separated by non-radiating gaps on the nano- to tens of nanosecond scale - sufficiently wide to study a broad range of optical decay channels observed in advanced nanostructured materials.

Keywords: XEOL, Timing, Synchrotron Radiation, Nanotechnology.
PACS: 07.85.Qe, 78.47.+p, 78.55.-m, 61.46.-w

INTRODUCTION

X-ray excited optical luminescence (XEOL) is a relatively simple photon-in photon-out technique but a promising method to study optical decay channels and their relation to the absorption process [1]. With increasing interest in nanostructured materials and the observation of optical-structural property correlation of light emitting materials in the nanometer scale, XEOL is likely to gain much importance in search of potential candidates for optoelectronic and biosensor applications. Being sensitive to material geometrical constraints on a nanometer scale, photoluminescence is a powerful tool to study crystalline, as well as amorphous low dimensional systems, which are typical for nanostructured materials. A crucial property of luminescence is the variable and sometimes relatively long life times (nanoseconds to milliseconds), which provide a clue to the nature of the electronic states involved in the decay processes.

XEOL AND THE PRINCIPLE OF TIMING

X-ray excited optical luminescence (XEOL) is the emission of optical photons (ultraviolet to infrared) after absorption of X-rays. Figure 1 shows the schematic of the XEOL detection system [2]. The total luminescence yield is collected by a lens system, which focuses on the entrance slit of the monochromator (JY 100). The monochromatic light is detected by a photomultiplier tube (PMT). The tube used in the present setup is a R943-02 (Hamamatsu) type which has a 160 to 930nm spectral range, and a 3ns rise time (anode pulse rise time). The response characteristic is almost flat over the working range [3]. In time-resolved XEOL, interest focuses on the lifetimes of the optical emission [4]. These lifetimes cover a broad range from nanoseconds up to milliseconds. Depending on the fill pattern of the storage ring, the time

CP879, *Synchrotron Radiation Instrumentation: Ninth International Conference*,
edited by Jae-Young Choi and Seungyu Rah
© 2007 American Institute of Physics 978-0-7354-0373-4/07/$23.00

structure (repetition rate) is of the order of tens to hundreds of nanoseconds. In this work we are using the synchrotron time structure for time-resolved XEOL in a 'pump-probe' like experiment.

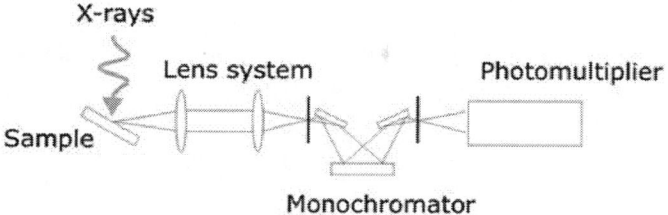

FIGURE 1. Principle setup for X-ray excited optical luminescence (XEOL) using a fast single channel detector (PMT).

In a synchrotron storage ring electrons are bunched by the RF-cavity. As a result, the emitted light is pulsed. The length of the pulses is usually much shorter than the time gap between consecutive bunches. A typical pulse length is in the order of tens of picoseconds (~bunch width/speed of light), whereas typical gaps are in the order of tens to hundreds of nanoseconds. Since luminescent transitions have life times, which are of the order of tens to hundreds of nanoseconds, their decay can be studied within the gap.

At SRC the normal fill consists of 15 equally spaced electron bunches (20ns gap). With one single bunch the gap increases to 300ns. At APS in normal mode the gap between consecutive bunches is already 153ns and thus suitable for most XEOL timing experiments. At CLS with single bunch injection the gap is 570ns. In fact, every spectrum recorded at a synchrotron has the time resolution of the electron bunch pattern in the storage ring. This means that the spectrum shows the response to a repeated excitation integrated over the gap between consecutive bunches. So the spectrum has the time resolution, which corresponds to the width of the gap.

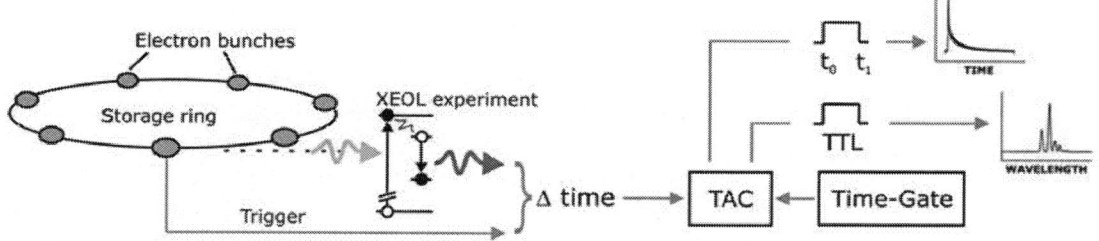

FIGURE 2. Illustration of the principle of time resolved XEOL using the time structure of the electron storage ring. The time difference of signal and trigger is processed by a time-to-amplitude converter (TAC) which can be set to a time gate. The output amplitude over time gives the decay, whereas the true event pulses are used for time-gated spectroscopy.

Figure 2 shows the schematic of time-resolved XEOL. The sample is excited with a synchrotron pulse, the excited state is unstable and decays nonradiatively and radiatively. For light emitting materials, one of the decay channel to release the extra energy is by emission of optical photons. The time elapsed between emission of an optical photon and the trigger is a measure for the decay probability of the optical channel to which the monochromator is tuned. Ideally the trigger is in coincidence with the X-ray excitation, but in practice this is usually not the case. The delay between trigger and X-ray excitation depends on various parameters like position of RF-cavity and beamline, the length of the connecting cables, and the fill pattern in the storage ring. In order to match trigger and X-ray excitation one can use a retarder (basically long cables) to delay the trigger, until excitation and trigger are close. This adjustment is necessary in order to take advantage of the full time gap between consecutive excitations. The trigger can be either the frequency of the RF cavity (revolutions/second) or a bunch clock (some facilities like APS provide this, but its purely generic). A bunch clock indicates each passing of an electron bunch (electron bunches/second). Using the latter, less delay is needed to match trigger and X-ray excitation pulse. In the case of a single electron bunch in the storage ring, the bunch clock frequency and the RF frequency are the same. Another way to match trigger and excitation is to inject the electrons into different buckets.

It is worth noting that this setup uses the 'true event' output from the TAC which is a standard TTL signal, and therefore time gated XEOL data are digitized. The pulse amplitude does not contribute to the signal only the number of counts is accumulated. The advantage is a much reduced noise level since random spikes (dominant at low level signal), which are characteristic for PMT, just add as one count.

There are two possible modes of running the experiment. In Fig. 3a timing is started by the trigger and stopped by the signal, in Fig. 3b the signal starts and the trigger stops. At first glance, there is no difference. It is just a reversal of the time axis. However, technically there is a difference. The number of triggers is about 3 orders of magnitude higher than the number of signals. If started with the trigger and ended with the signal, on average only 0.1% of the measurements are successful. Still the dead-time adds up every time a measurement is started and a good percentage of true events will be missed, reducing the efficiency. So it is better to start with the signal and stop with the trigger to ensure that every start finds a stop.

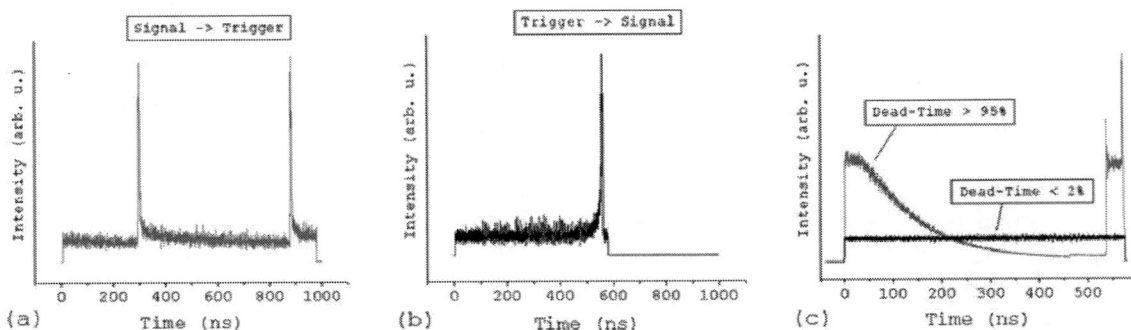

FIGURE 3. Decay of a selected line in the luminescence spectrum using the trigger as start and the signal as stop (a), or reversing the time axis, i.e. using signal as start and trigger as stop (b). Pile-up effect and dead-time of the TAC (c).

Attention: With intense light emitters the multi-channel analyzer (TAC) can be easily overloaded. Being busy with early events, later events will be skipped, suggesting a much faster decay than it actually is. A good reliability measure is the dead-time, which expresses the readiness of the TAC to process an event. If the dead-time is high there are too many events to be processed. One way to reduce the number of events is to reduce the X-ray intensity by closing the entrance slits of the beamline monochromator. As a rule of thumb a dead-time of 2% or less is acceptable for reliable lifetime measurement. This so-called pile-up effect is shown in Fig. 3c. Both curves originate from the same XEOL peak (488nm of $TbCl_3$) but are recorded with different slit opening (20x50 and 5x5 mm). The curve recorded with >95% dead-time shows (fake) 'fast decay' due to pile-up. The curve recorded with <2% dead-time shows the real lifetime, which is in accordance with literature [6,7].

APPLICATION OF TIME-GATED XEOL

We now illustrate the application of time-resolved XEOL using the systems $TbCl_3$ powder, Tb-doped porous Si and in ZnO nanowires. XEOL of terbium compounds involves various f to f optical transitions covering a wide range of lifetimes. Due to their atomic like nature, Tb f-states are mostly indifferent to their environment.

Optical luminescence of Tb (from $TbCl_3$) excited at the giant absorption edge, $N_{4,5}$ (150eV) is shown in Fig. 4a. Typical features from intense $^5D_4$ to ^7F_J (J=0,..,6) and less intense $^5D_3$ to $^7F_{4,5,6}$ transitions are observed [5]. The two curves are recorded at Aladdin storage ring (SRC) using normal 15-bunch mode (20 ns gap) and single bunch mode (300 ns gap). In the single bunch mode the beam current is 12mA at injection, but still the $^5D_4$ to ^7F_J transitions are very intense due to their long lifetime. The different lifetimes of $^5D_4$ to ^7F_J and $^5D_3$ to $^7F_{4,5,6}$ transitions become evident when 547 nm and 382 nm were used to monitor the time response to repeated excitation every 300 ns (Fig. 4b). Assuming single exponential decay the peak at 382nm has a lifetime of 116±5ns, whereas at 547nm the intensity does not show any noticeable decrease over the 300ns gap in accordance with literature [6,7]. A more complete view can be obtained by selecting an appropriate time-window. A comparison of time-gated XEOL at 10-25ns and 40-100ns from Tb in porous silicon is shown in Fig. 4c (APS result). We see that within the 10-25ns window, the fast decay channels still contribute substantially to the total intensity, whereas at 40-100ns the spectrum shows mostly the slow channel contributions.

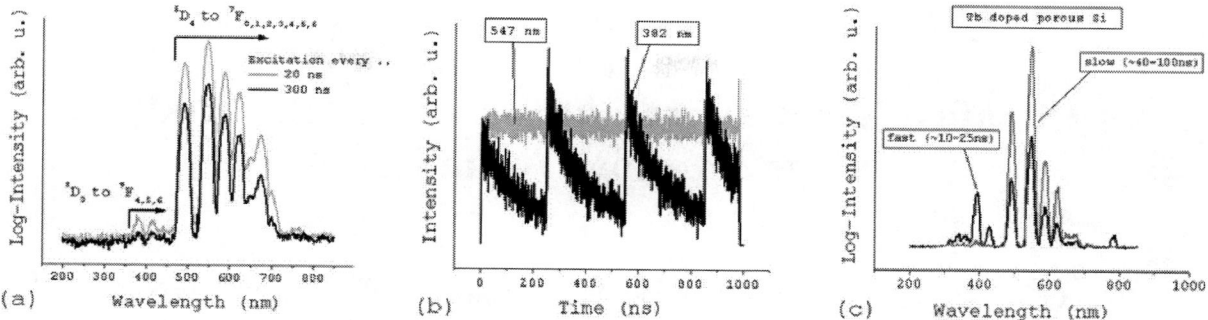

FIGURE 4. Terbium (TbCl₃ powder) XEOL spectrum (hν= 143.9eV) at different excitation rates (a). Time behaviour of intensity peaks at 547 nm and 382 nm (b). Time gated XEOL (hν= 7600eV) of porous silicon doped with Tb^{3+} (c).

Nanostructured ZnO has very pronounced optical properties, which are related to its characteristic size and shape (Fig. 5c). ZnO nanowires are known to emit light in green and in UV corresponding to defect and near band-gap emission, respectively [8]. An example of XEOL of ZnO NW is shown in Fig. 5a, where slow and fast decay channels are separated by the time-gated technique. The partial optical yield XAFS of ZnO NW is shown in Fig. 5b. The fast band-gap emission disappears almost completely in the slow window (20-90ns).

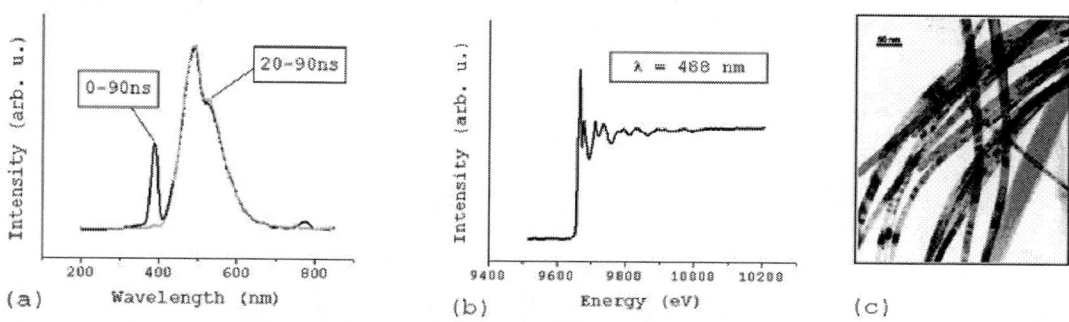

FIGURE 5. Time-gated XEOL from ZnO nanowires (hv = 9665eV), showing the fast (band-gap) and slow (defect) emission (a). Zn K-edge, monitoring the 488nm line (b). TEM image of ZnO nanowires (c).

ACKNOWLEDGEMENTS

We gratefully acknowledge: National Science Foundation, U. S. Department of Energy, Office of Science, Office of Basic Energy Sciences, Canada Foundation for Innovation, Ontario Innovation Trust, National Research Council Canada, Natural Science and Engineering Research Council of Canada. Research at UWO is supported by NSERC, CFI, CRC(TKS) and OIT. CSRF, CLS and the Canadian component at PNC-CAT/APS are supported by NSERC MFA grants. Synchrotron Radiation Center, UW-Madison, is supported by the NSF under Award No. DMR-0084402. The work performed at the Advanced Photon Source was supported by the U.S. Department of Energy, Office of Science, Office of Basic Energy Sciences under Contract No. W-31-109-ENG-38. The Italian Ministry of Education, University and Research (MIUR) is also acknowledged for the FIRB Research Program - RBNE033KMA.

REFERENCES

1. T.K. Sham, D.T. Jiang, I. Coulthard, J.W. Lorimer, X.H. Feng, K.H. Tan, S.P. Frigo, R.A. Rosenberg, D.C. Houghton, and B. Bryskiewicz, Nature (London) **363**, 331 (1993)
2. Ian Coulthard, PhD Thesis, UWO (1998)
3. http://jp.hamamatsu.com
4. R.A. Rosenberg, G.K. Shenoy, F. Heigl, S.-T. Lee, P.-S.G. Kim, X.-T. Zhou, and T.K. Sham, Appl. Phys. Lett. **86**, 263115 (2005)
5. H. Elhouichet, M. Oueslati, N. Lorrain, S. Langa, I.M. Tiginyanu, and H. Föll, phys. stat. sol. (a) **202**, 1513 (2005)
6. http://www.jobinyvon.com/usadivisions/fluorescence/ application_notes.htm
7. M.K. Johansson, R.M. Cook, J. Xu, and K. N. Raymond, JACS **126**, 16451 (2004)
8. X.H. Sun, S. Lam, T.K. Sham, F. Heigl, A. Jürgensen and N.B. Wong, J. Phys. Chem. B **109**, 3120 (2005)

An Ultrafast X-ray Detection System for the Study of Magnetization Dynamics

A. T. Young[1], A. Bartelt[1], J. Byrd[2], A. Comin[1], J. Feng[1], G. Huang[2], J. Nasiatka[1], J.Qiang[2], A. Scholl[1], Hyun-Joon Shin[1,3], Weishi Wan[1], and H.A.Padmore[1]

[1] *Advanced Light Source, Lawrence Berkeley National Laboratory, Berkeley, CA 94720, USA*
[2] *Center for Beam Physics, Lawrence Berkeley National Laboratory, Berkeley, CA 94720, USA*
[3] *Pohang Accelerator Laboratory, Pohang, Korea*

Abstract. We describe the development of a streak camera system optimized for use in time-resolved x-ray absorption studies. Computer simulations of the system characteristics have been performed and a working version of the camera has been developed. The system has been used to perform time-resolved x-ray magnetic circular dichroism (XMCD) experiments on a number of magnetic systems. These experiments demonstrate utility of the technique. Finally, concepts for improving the time resolution to the sub-picosecond level are discussed.

Keywords: ultrafast magnetism, XMCD, streak camera.
PACS: 74.25.Ha; 87.64.Gb

INTRODUCTION

The use of x-rays to probe ultrafast phenomenon is currently a field undergoing intense study, with facilities to produce sub-picosecond long pulses undergoing commissioning (e.g. slicing sources at BESSY II, Berlin and the Advanced Light Source (ALS), Berkeley) or under construction (LCLS, Stanford.) These sources create short pulses of x-rays to provide the ultrafast time resolution desired. Alternatively, by using a time-resolved detector, ultrafast time resolution can be attained with a relatively long pulselength x-ray source. This paper discusses the development of just such an ultrafast x-ray detection system. Centered around a laser-triggered streak camera, sub-picosecond time resolution has been attained with UV illumination using this apparatus. Coupling this detector to an Elliptically Polarizing Undulator (EPU) beamline at the ALS provides the capability to perform time-resolved x-ray spectroscopy. In particular, by utilizing the polarization capabilities of this beamline, magnetic materials can be studied using the technique of x-ray magnetic circular dichroism (XMCD.) Conventional XMCD has proven to be a powerful technique in the study of magnetic materials, and has been applied to many systems. XMCD has demonstrated the ability to independently determine the orbital and spin components of the magnetic moment with monolayer sensitivity and element specificity. Using time-resolved XMCD, fundamental questions about the ultrafast exchange of orbital, spin, and lattice energy can be probed.

EXPERIMENTAL

Synchrotron light sources produce pulses of x-rays. In the case of Advanced Light Source (ALS) at LBNL, these pulses have a duration of 70 pS (FWHM.) To attain a time resolution faster than this pulselength, a streak camera is used. [1] This detector system, analogous to an oscilloscope, translates time to a spatial coordinate, producing a line or streak. The intensity at any point along the streak gives the intensity of the x-rays at a particular time. In a streak camera, photons (either uv or x-rays) strike a photocathode. These photoelectrons are accelerated to ≥10 keV and

CP879, *Synchrotron Radiation Instrumentation: Ninth International Conference*,
edited by Jae-Young Choi and Seungyu Rah
2007 American Institute of Physics 978-0-7354-0373-4/07/$23.00

displaced transverse to the main propagation direction by a small, time-varying electric field. Therefore, different parts of the electron pulse are deflected by different amounts. The electrons are then recorded by a position-sensitive detector, such as a CCD camera or a fluorescent screen.

A schematic diagram of the apparatus is shown in Fig. 1. A laser system producing 100 fs long pulses of 800 nm wavelength is used to both trigger the streak camera (which synchronizes the transverse electric field) and to initiate the demagnetization process. The 800 nm light is also frequency tripled to produce uv light pulses, which are used to calibrate the streaks.

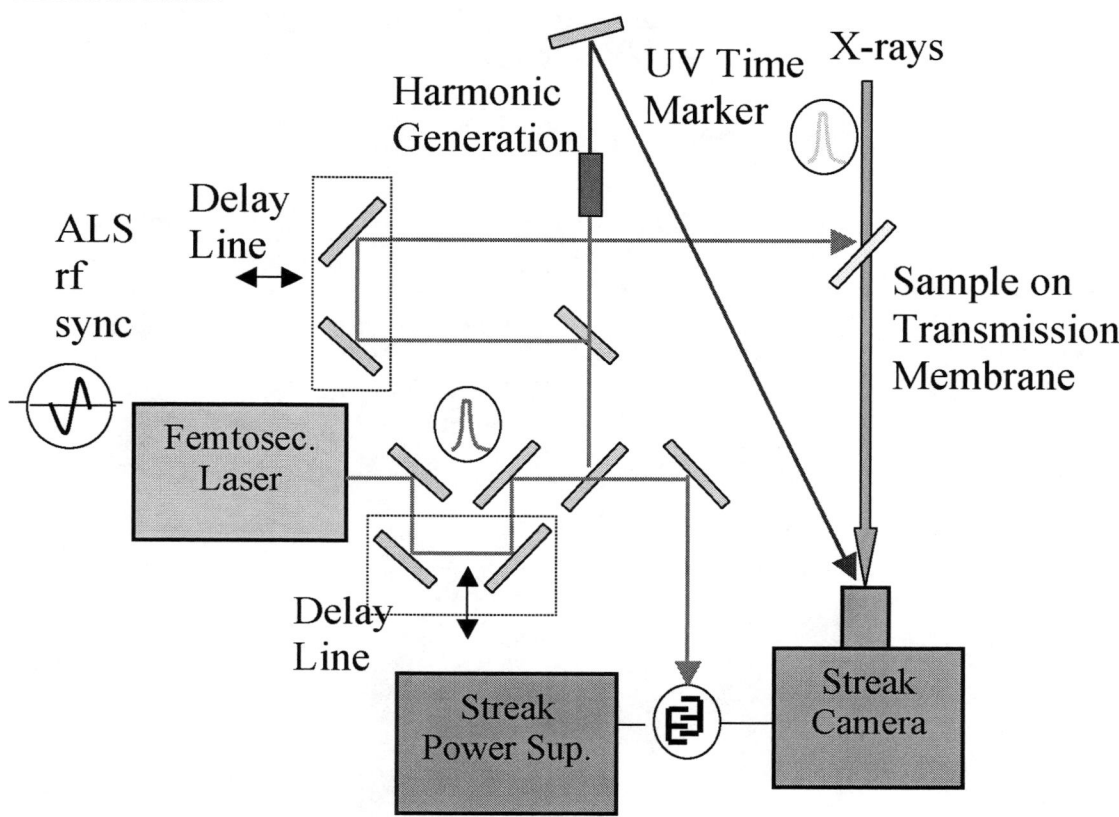

FIGURE 1. Schematic diagram of the time-resolved XMCD experimental setup. System features a picosecond time resolution streak camera and 100 fs laser to initiate magnetization dynamics and to trigger the streak camera.

APPLICATIONS TO STUDIES OF MAGNETIZATION DYNAMICS

Of great recent interest is the study of ultrafast magnetic processes.[2] In addition to providing a better understanding of the basics physics underlying spin and electron dynamics, these studies of ultrafast magnetic processes are technologically important. As the density of magnetic recording media increases, it is vitally important to increase the speed at which the magnetic bits can be written and read. This requires us to have a better understanding of the magnetization/demagnetization processes. Coupling the streak camera to an Elliptically Polarizing Undulator (EPU) beamline, BL 4.0.2 at the ALS, [3] provides the capability to perform time-resolved x-ray spectroscopy on a picosecond and sub-picosecond timescale. In particular, by utilizing the polarization capabilities of this beamline, magnetic materials can be studied using the technique of x-ray magnetic circular dichroism (XMCD.)[4] Conventional XMCD has proven to be a powerful technique in the study of magnetic materials and has been applied to many systems. XMCD has demonstrated the ability to independently determine absolutely the orbital and spin components of the magnetic moment with monolayer sensitivity and element specificity. Using time-resolved XMCD, fundamental questions about the ultrafast exchange of orbital, spin, and lattice energy can be probed.

Figures 2 and 3 illustrate the type of data that can be acquired using the apparatus. Shown in the upper part of Fig. 2 are the time-resolved streaks of the x-ray pulses after they have passed through the sample. The top streak has been obtained with left circularly polarized light, denoted σ⁻, while the center streak was obtained with right

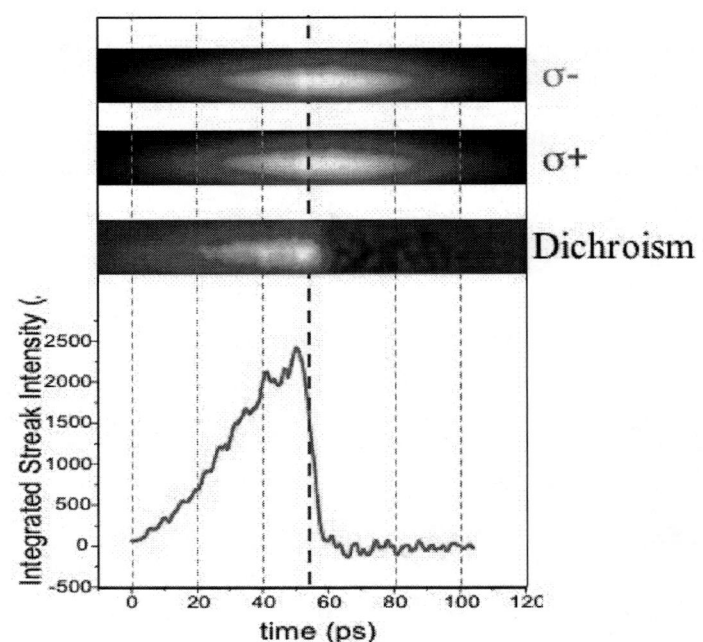

FIGURE 2. X-ray Magnetic Circular Dichroism (XMCD) streaks

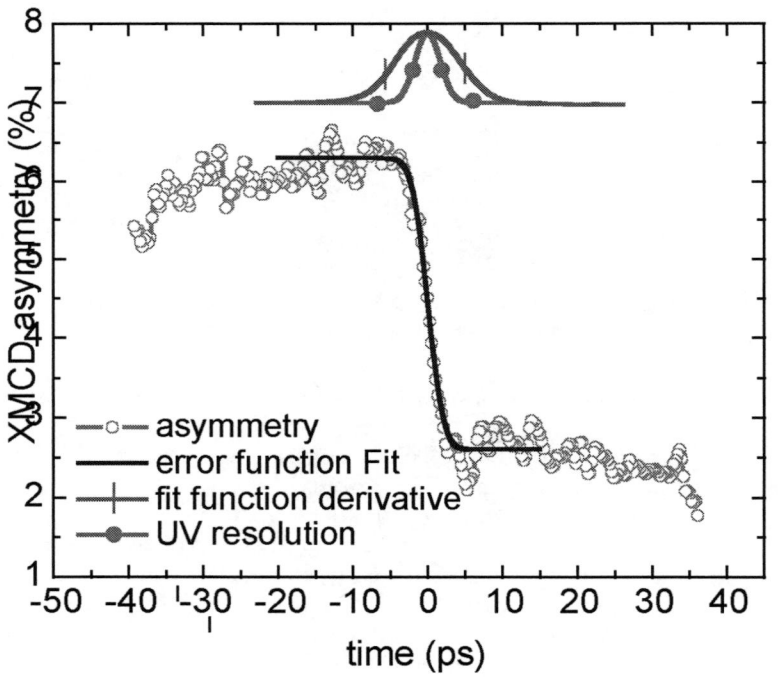

FIGURE 3. Demagnetization time analysis. Shown is the decay of an FeNiPt multilayer demagnetized with a 25 mJ/cm^2 .laser pulse. Decay time is 3.9 pS, which is close to the x-ray time resolution limit.

circularly polarized light, σ^+. The lower streak is the difference between the two, renormalized. At t= 55 ps, a 100 fs IR laser pulse has been directed to the sample, raising the temperature to above the Curie temperature, which should cause the material to demagnetize. As can be seen, the difference streak ("the Dichroism") does go to zero after t=55 ps. The lower part of Fig. 2 shows the intensity of the vertically integrated difference streak, showing the rapid decay of the dichroism and demonstrating that the magnetism has indeed vanished.

Figure 3 shows the detailed analysis for the magnetic multilayer system FeNiPt. The circles represent the experimentally measured XMCD asymmetry, defined as the difference of the data from the two polarization streaks divided by their sum, i.e. $(\sigma^+ - \sigma^-)/(\sigma^+ + \sigma^-)$. To determine the time decay caused by the laser heating, the asymmetry is then fit with the error function, and the FWHM of the derivative of the fit is taken as the time decay constant. For the laser power used in this experiment, 24 mJ/cm², we obtain a time decay of 3.9 ps. Also shown in the figure is the measured time response of the uv time fiducialization pulse, which we measure to be 2.6 ps. Since the uv pulselength is actually approximately 100 fs, we infer that the time resolution for uv detection to be approximately 2.5 ps. Single shot experiments have shown uv time resolutions as high as 0.9 ps.

Although the time constant for the FeNiPt demagnetization is longer than the uv time resolution, it is probable that the 3.9 ps is instrumentally limited. To understand this effect, we have developed a computational model of the streak camera that takes into consideration: (1) the time dependent electric fields in the camera; and (2) the finite energy spread of the photoelectrons emitted from the streak camera photocathode. This model shows that, compared with uv illumination, additional degradation in the time resolution can be expected with the x-ray illumination. The photoelectron energy spread is much wider for the x-rays (>5 ev FHWM) compared to the uv (<0.5 eV.) The spread in electron energies, even after kilovolt acceleration from the photocathode, is thus larger with the x-rays than with the uv. Since the electron optics in the camera can perfectly focus only a small range of electron energies, the larger energy spread leads to imperfect focusing and hence degraded time resolution. We are pursuing a number of techniques to minimize this chromatic aberration effect, including advanced higher gradient acceleration designs. Also being studied is the inclusion of a magnetic dipole-based corrector. A four-magnet set would be used to compensate for the energy spread, bringing all electrons to a common focal point. This system yields a theoretical time resolution of 50 fs.

CONCLUSION

A new time-resolved detector with picosecond time resolution for x-ray detection has been demonstrated. We have coupled this detector to a soft x-ray beamline having an EPU, thus making possible ultrafast measurements of magnetization dynamics. Additional development is progressing, with the goal of producing a sub-picosecond resolution x-ray camera.

ACKNOWLEDGMENTS

The authors would like to acknowledge contributions to this work by Roger Falcone, Dana Weinstein, and Andrew MacPhee, U.C. Berkeley. This work was supported by the Director, Office of Science, Office of Basic Energy Sciences, of the U.S. Department of Energy under Contract No. DE-AC02-05CH11231.

REFERENCES

1. J.Feng, W.Wan , J.Qiang, A. Bartelt, A. Comin, A. Scholl, J.Byrd, R. Falcone, G.Huang, A. MacPhee , J.Nasiatka, K. Opachic, D. Weinstein, T. Young and H.A.Padmore, in Ultrafast X-Ray Detectors, High-Speed Imaging, and Applications, edited by Stuart Kleinfelder, Dennis L. Paisley, Zenghu Chang, Jean-Claude Kieffer, Jerome B. Hastings, Proc. of SPIE Vol. 5920, SPIE, Bellingham, WA, 2005

2. E. Beaurepaire, J.-C. Merle, A. Daunois, and J.-Y. Bigot, PRL 76, 4250 (1996)
3. A.T. Young, E. Arenholz, F. Feng, H.A. Padmore, T. Henderson, S. Marks, E. Hoyer, R. Schlueter, J.B. Kortright, V. Martynov, C. Steier, and G. Portmann Nucl. Instrum. Methods A467/468, 549
4. C.T.Chen, Y.U.Idzerda, H.J. Lin, N.V.Smith, G.Meigs, E.Chaban, G.H.Ho, E.Pellegrin, and F.Sette, Phys.Rev.Lett., 75, (1995)

Studies of Ultrafast Femtosecond-Laser-Generated Strain Fields with Coherent X-rays

Eric M. Dufresne*, Bernhard Adams*, Eric C. Landahl*, Ali M. Khounsary*, David Reis[†], David M. Fritz[†] and SooHeyong Lee[†]

*X-ray Science Division, Argonne National Lab., Argonne, IL 60439
[†]Department of Physics, The University of Michigan, Ann Arbor MI 48109

Abstract. In its 324 bunch-mode of operation, the Advanced Photon Source (APS) has opened new avenues of femtosecond-laser science and techniques. In this new mode, if one uses the tightly focused low-pulse energy (nJ), high repetition rate fs-laser Ti:sapphire oscillator (88 MHz) on beamline 7ID, every laser pulse and X-ray bunch can be overlapped and delayed with respect to each other, resulting in a high-repetition rate pump-probe experiment that uses all the APS X-ray bunches. This paper describes an example of how coherent X-ray experiments may be used to study laser-generated strain fields in semiconductors. With an oscillator beam focused to 6 μm onto GaAs, we have observed coherent X-ray diffraction patterns with a high-resolution camera. We have developed two techniques to observe the strain field, a topographic technique and a coherent diffraction technique. The topographic technique is quite useful to achieve a coarse spatial overlap of the the laser and X-ray beams. The coherent X-ray technique allows one to push the alignment to a few microns. This paper focuses solely on the latter technique. This experiment may help to develop techniques that will be used at the future free electron laser sources, where coherent and pump-probe experiments can be done simultaneously.

Keywords: Ultrafast lasers, coherent X-ray imaging, laser-induced strain fields.
PACS: 61.10.-i, 61.80.Ba, 61.72.Dd, 65.40.De

The future is very bright for coherent X-ray and ultrafast femtosecond (fs) laser experiments at the future X-ray free electron laser (FEL) facilities. The Linac Coherent Light Source (LCLS), for example, will provide 10 orders of magnitude higher instantaneous brightness than the Advanced Photon Source (APS) and a 100 fs pulse duration. It will be transversely coherent, so pump-probe speckle experiments can be conducted routinely. Much work needs to be done to develop new techniques that will open the future science at FEL facilities. With the APS 324-bunch mode, one should be able to use the tightly focused 7ID-D fs-laser oscillator beam (1 nJ/$(5\mu m)^2$ = 4 mJ/cm^2) running at the same frequency of the ring (88 MHz) to develop a laser-pump X-ray-probe experiment, where every X-ray bunch is used. By varying the time delay between the arrival of the laser and X-ray beams, one can study time-resolved diffraction and spectroscopy on a time scale as short as the 100 ps bunch length duration of the APS. Using all the X-rays bunches of the APS allows one to consider experiments where the X-ray cross section is weak or the sample is disordered. Thus provided that significant laser excitations can be produced with a focused nJ per pulse, one can use all the known existing X-ray techniques to study these excitations. In recent work performed on the APS 7ID beamline, an amplified kHz-fs-laser pulse was shown to generate strain fields that modulate Bragg and Laue diffraction [1, 2]. In these studies, laser-pulse energy densities around ten mJ/cm^2 generate coherent acoustical phonons that propagate mostly in one dimension. The work is done in a geometry where the laser beam diameter is much larger than the laser penetration depth, and the excitation is launched impulsively due to the short pulse length. Experiments with a Ti:sapphire laser oscillator will likely be in the weak excitation limit for semiconductors with its peak 4 mJ/cm^2. But using a tightly focused laser beam would allow one to probe a new regime of excitation, where the focused beam diameter is on the same order of magnitude as the laser penetration depth. As shown recently with laser only work, the ratio of the laser beam size over the laser penetration depth influences the strain generation process [3]. In this paper, we show how one can observe laser-generated strain fields in GaAs with coherent X-ray diffraction (CXD).

The 7ID beamline cryogenically cooled Si (111) monochromator was set to 10 keV and detuned by 50% to reduce the third harmonic contamination. The full-width at half maximum (FWHM) of the beam energy bandpass is $\Delta E/E = 0.014\%$. To observe coherent diffraction patterns, the X-ray path-length difference in the sample, Δx, must be smaller than the longitudinal coherence length of the incident X-rays $l_l = \lambda/(\Delta E/E)$, where λ is the X-ray wavelength. In the Bragg diffraction case, one can show that the longitudinal coherence is preserved when $\Delta x = 2l\sin^2\theta_B < l_l$ [4]. Here θ_B is the Bragg angle, and l is the X-ray penetration length, limited by a combination of X-ray absorption and X-ray extinction effects. For perfect crystals of GaAs, we use the X-ray extinction length. For GaAs (111) and (002) at

CP879, *Synchrotron Radiation Instrumentation: Ninth International Conference*,
edited by Jae-Young Choi and Seungyu Rah
© 2007 American Institute of Physics 978-0-7354-0373-4/07/$23.00

FIGURE 1. A) (left) Upstream view of the experimental set up. B) (right) Detector arm set up.

10 keV, the Bragg angles are 10.948 and 12.668 degree respectively, and l is 1.8 and 24.2 μm; thus Δx is 130 nm and 2.3 μm, respectively. Since at 10 keV, $l_l = 0.89$ μm, the longitudinal coherence condition is satisfied for GaAs (111), but not for the (002) reflection, the contrast in coherent diffraction imaging experiment for the (002) reflection will be reduced from unity.

To observe speckle in a coherent diffraction experiment, the illuminated area must also be smaller than the transverse coherence area of the source. An X-ray slit (seen in Fig. 1a) placed 20 cm from the sample can reduce the illumination area to a beam of 3×10^7 coherent X-rays/s within an area of 7 μm by 7 μm. The Ti:sapphire oscillator beam has a nominal wavelength of 800 nm, a pulse length of 30 fs, a repetition rate of 88 MHz, and an average power of 0.45 W. During the span of a week, the output power dropped to about 0.3 W. It was focused down on a GaAs (111) or (001) sample near normal incidence using a long-working-distance Mitutoyo objective with X5 magnification (see Fig. 1a). The transmission of the mirrors in the optical path and the microscope lenses reduced the delivered power to the sample by a factor 0.3, thus the maximum deliverable power on the sample was 135 mW. A set of neutral density filters was used to reduce the power on the sample. The laser focal spot could be displaced on the sample with a stepper-motor-controlled XYZ positioner (three Newport MFN stages). The sample could also be displaced in the center of a 4-circle Huber diffractometer using an Kohzu XYZ stage, controlled by stepper motors.

The diffracted X-rays were absorbed on a YAG:Ce single-crystal screen, generating visible fluorescence at 550 nm, which was imaged using a CoolSNAP HQ CCD camera and a X10 microscope objective (see Fig. 1b). The 12 bit camera is linear and the X-ray resolution of the set up is around 3 μm [5]. The response of the YAG was calibrated against the calculated flux from an ion chamber current following the approach described in Ref. [5]. One analog to digital unit of the CCD corresponds to 51.5 X-rays. The sample to YAG distance was approximately 0.65 m.

The laser focal waist was measured with scans of a razor-blade position mounted on the Kohzu stage, transverse to the laser propagation direction. The signal from a laser power meter recorded the transmitted flux. Figure 2a shows the transmitted laser signal as a function of the blade position at the optimal focal distance. The data are well fit to an error function. The Gaussian beam profile has a FWHM of 6 μm. The position of the best focus was found by displacing the blade along the laser propagation direction. Figure 2b shows a plot of the depth of focus of the microfocused laser beam. The data are well fit to a quadratic function $y = ax^2 + w_0$, where w_0 is the best focal waist, and $a = 510$ μm/mm^2. Defining the depth of focus Δf when $y = 2w_0$, $\Delta f = \sqrt{w_0/a} = 0.108$ mm. The manufacturer specifies a depth of field of 14 μm, which is a factor 8 smaller than our measurement.

Figure 3a shows a coherent diffraction pattern of Si (111) at the peak of its rocking curve. The vertical direction is parallel to the angle 2θ, while the horizontal axis records an horizontal angle $2\theta_H$. The coherence-defining slit was set to $(7 \text{ } \mu\text{m})^2$. From the fringe visibility, it is obvious that beamline 7ID has enough transverse and longitudinal coherence to perform coherent diffraction imaging on low-order Bragg reflections of semiconductors. The imaging detector provides ample resolution to resolve the fringes.

Figure 3b shows three snapshots of the bunch current in 324 bunch mode recorded by the APS accelerator division. Most of the buckets are emptied. Note the large current fluctuations for the filled buckets. We typically measure that the pulse-to-pulse stability of the oscillator power is about one percent. Since we excite our sample at the same pulse

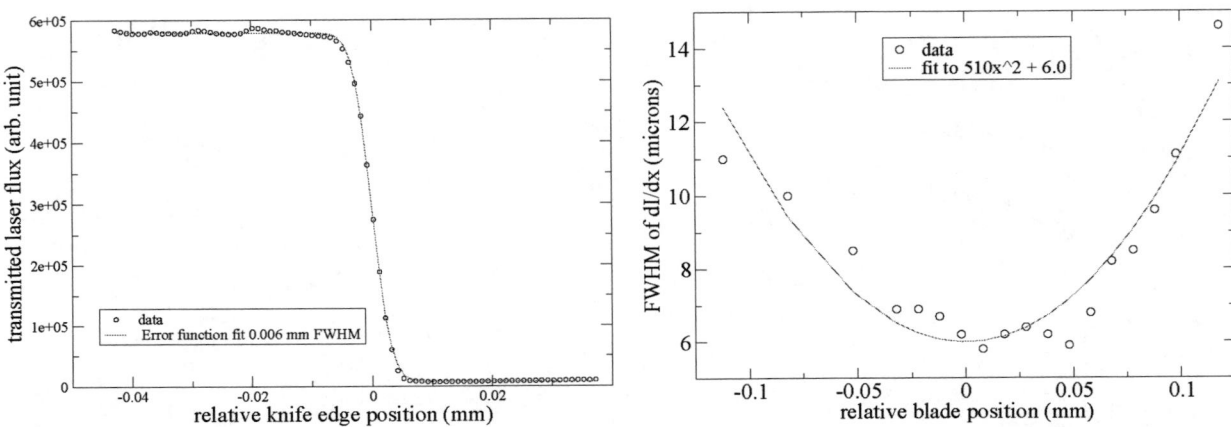

FIGURE 2. A) (left) Focal waist scan versus knife-edge transverse position. B) (right) FWHM focal waist versus knife-edge longitudinal position.

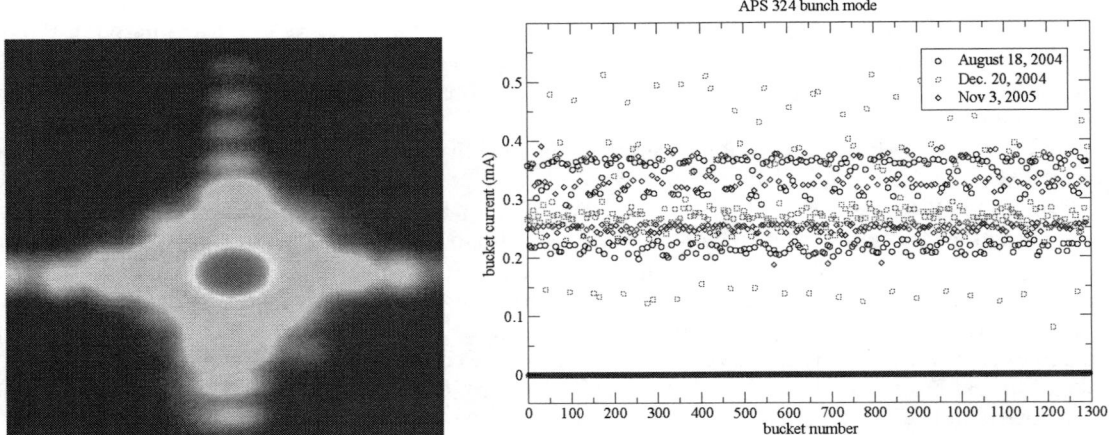

FIGURE 3. A) (left) Fraunhofer diffraction pattern of a coherent beam diffracted by Si (111) with the laser off. B) (right) The bunch current per filled bucket in 324 bunch mode for three fill snapshots.

energy per pulse, we do not expect these current variations to affect our measurements. The spread in bunch current implies a spread in bunch duration through the cube root of the bunch current.

Figure 4 shows the GaAs (002) reflection with laser on and laser off. It is interesting to note that the laser-off pattern does not show any diffraction fringes. This is due to the long optical path difference for this nearly forbidden reflection. Turning the laser on reduces the peak intensity by a factor five. The ratio of integrated intensity between images taken with the laser on and off at the peak of the rocking curve is 38%; thus the laser reduces the integrated intensity by more than a factor 2. The laser-on data were obtained with an incident power on the sample of 68 mW; 71.4% of the power is absorbed in the sample into a 6 μm FWHM Gaussian spot. The incident and absorbed peak energy density per pulse are, respectively, 1.9 and 1.35 mJ/cm^2.

In Ref. [1] and [2], fluences were reported in the range of 5 to 12 mJ/cm^2 with an amplified kHz system. Our data were obtained with a neutral density filter with an optical density of 0.3. Without the neutral density filter, the power was 2.1 times higher, and we noticed a permanent burned-in low-intensity spot on the sample's X-ray diffraction in the location of the laser spot. We did not investigate the structure of the damaged region.

Clearly, the tightly focused oscillator beam generates a strong strain field in GaAs that splits the diffracted beam in three peaks in Fig. 4b and 4c. The angle $\Delta 2\theta_B$ between the two large peaks in Fig. 4c is 75 μrad. The angle corresponds to a strain of $\Delta a/a = \Delta\theta/\tan\theta = 1.7 \times 10^{-4}$. From the thermal expansion coefficient of GaAs (5.73×10^{-6}/K), one deduces a temperature difference of 30 K. A finite element analysis calculation predicts a peak temperature rise of

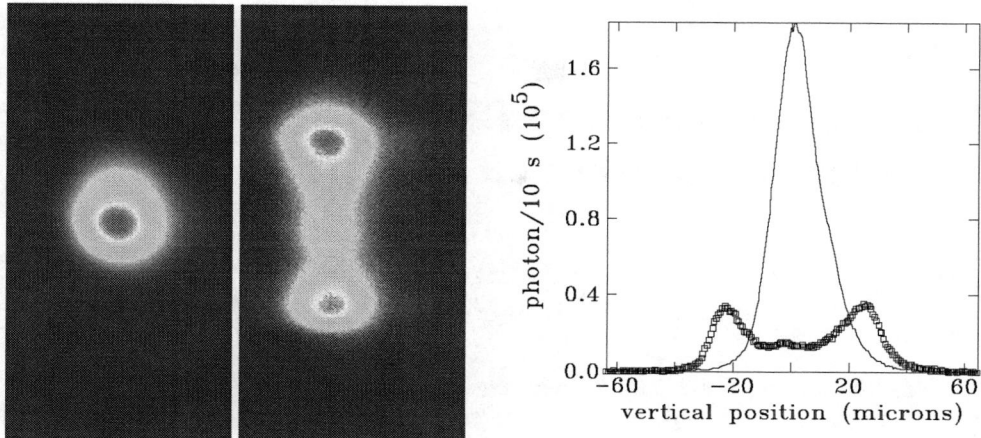

FIGURE 4. A) (left) GaAs (002) coherent diffraction pattern with laser off. The data are displayed with a rainbow color map from 0 to 1.8×10^5 ph/10s. The effective pixel size is 0.65 μm. B) (middle) GaAs (002) diffraction pattern with laser on, displayed with a color map from 0 to 3.7×10^4 ph/10s. C) (right) Vertical slices of the previous two images. The laser-off data are displayed with a solid line, while the laser-on data are shown with squares.

137 K. Since the X-ray beam footprint on the sample is 36 μm, the X-ray probe is much wider than the laser-beam spot size, thus a temperature rise of 30 K is reasonable. We did not find any notable response of the sample to the laser and X-ray time delay or to the laser pulse length.

A quantitative analysis of the microdiffraction patterns is underway to understand the microstructure of the sample. The results from the weak reflection GaAs (002) show that it should be easy to record coherent X-ray diffraction patterns following fs-laser excitation on strong reflections of semiconductors like Ge, and InSb (111) [1]. Future experiments will try to improve the spatial overlap of the X-rays with the laser focus by using microfocused X-rays. Studies of thin films will help to better match the X-ray and laser penetration depth.

ACKNOWLEDGMENTS

Use of the Advanced Photon Source was supported by the U. S. Department of Energy, Office of Science, Office of Basic Energy Sciences, under Contract No. W-31-109-ENG-38. This work was conducted at the MHATT-XOR 7ID beamline at the APS.

REFERENCES

1. D. Reis, M. DeCamp, P. Bucksbaum, R. Clarke, E. Dufresne, M. Hertlein, R. Merlin, R. Falcone, H. Kapteyn, M. Murnane, J. Larsson, T. Missalla, , and J. Wark, *Phys. Rev. Lett.* **86**, 3072 (2001).
2. M. DeCamp, D. Reis, P. Bucksbaum, B. Adams, J. Caraher, C. Conover, E. Dufresne, R. Merlin, V. Stoica, and J. Wahlstrand, *Nature* **413**, 825 (2001).
3. C. Rossignol, J. Rampnoux, M. Perton, B. Audoin, and S. Dilhaire, *Phys. Rev. Lett.* **94**, 166106 (2005).
4. M. Sutton, S. Mochrie, T. Greytak, S. Nagler, L. Berman, G. Held, and G. Stephenson, *Nature* **352**, 608 (1991).
5. E. M. Dufresne, D. Arms, N. Pereira, P. Ilinski, and R. Clarke, "An Imaging System for Focusing Tests of Li Multi-Prism X-ray Refractive Lenses," in *Proceedings of the International Synchrotron Radiation Instrumentation Conference SRI2003 in San Francisco August 2003*, American Institute of Physics, Melville, New York, 2004, vol. 705, pp. 780–783.

Time-Resolved Energy-Dispersive XAFS Station for Wide-Energy Range at SPring-8

K. Kato[1], T. Uruga[1], H. Tanida[1], S. Yokota[1], K. Okumura[2], Y. Imai[1], T. Irie[3], Y. Yamakata[1,4]

[1]JASRI/ SPring-8, 1-1-1 Kouto, Sayo-cho, Sayo-gun, Hyogo 679-5198, Japan
[2]Department of Materials Science, Faculty of Engineering, Tottori University, Koyama-cho, Tottori 680-8552
[3]SES/ SPring-8, 1-1-1 Kouto,Sayo-cho, Sayo-gun, Hyogo 679-5198, Japan
[4]RIGAKU Mechatronics Co., Ltd. Matsubara-cho 3-9-12, Akishima-City, Tokyo 196-8666

Abstract. A time-resolved energy-dispersive XAFS (DXAFS) station has been constructed at the bending magnet beamline BL28B2 at SPring-8 to study the local structural changes of materials during chemical reactions and functional processes. The bending magnet source at SPring-8 has a high photon flux above 50 keV. The purpose of this station is to measure DXAFS spectra in a wide energy range from 7 to 50 keV covering K-edges of lanthanides. Its main components are a polychromator with a bent silicon crystal, a mirror to reject higher harmonics, and a position-sensitive detector (PSD). To correspond to a wide energy range, polychromators for Bragg and Laue geometry were developed for the energy range below and above 12 keV, respectively. The PSD used is CCD coupled with a fluorescent screen and lens system. The fluorescent materials and their thickness were optimized for measurement in the x-ray range. Good quality spectra of Ce K-edge (40.5 keV) were obtained with exposures of 360 ms for the standard samples. The present status of the system and some experimental examples are presented in this report.

Keywords: Energy-dispersive XAFS, Time-resolved
PACS: 07.85.Qe, 61.10.Ht

INTRODUCTION

Structural investigation of transient states undergoing chemical or physical reactions is important for understanding basic scientific phenomena and for further developing functional materials having higher activity. X-ray absorption fine-structure spectroscopy (XAFS) is a powerful tool for investigating the local structures and electronic states of materials in both crystalline and non-crystalline states. XAFS is also useful for studying reactions in that it can provide information about local structure from around a selected element that is a key component of functional materials. Time-resolved in situ measurement is superior to another method that measures an intermediate quenched state during a reaction through off-line treatment because there are cases where the quenching process results in a state different from the actual transient one.

Two major techniques are used in time-resolved XAFS measurements: quick XAFS (QXAFS) and energy-dispersive XAFS (DXAFS). QXAFS is done by quick scanning with a monochromator to obtain XAFS spectra within a short time. QXAFS is superior to DXAFS in that it can be applied to the fluorescence mode and provide spectra with high energy-resolution. DXAFS, on the other hand, has high time resolution. It is capable of a one-shot acquisition time of less than sub-seconds for EXAFS measurement [1-4], which would be hard to achieve with QXAFS.

We constructed a DXAFS station at the bending magnet beamline BL28B2 at SPring-8 in 2002. Taking advantage of a high photon flux light source above several-tens keV, this station is dedicated to DXAFS measurements of up to 50 keV with high-energy resolutions. Especially the energy range from 30 to 50 keV covers K-edge of important elements for catalysts and material science, such as Sb, Te, Ce and La. This station offers ways of studying the K-edge DXAFS spectroscopy of these elements, which can yield significant structural information

CP879, *Synchrotron Radiation Instrumentation: Ninth International Conference*,
edited by Jae-Young Choi and Seungyu Rah
© 2007 American Institute of Physics 978-0-7354-0373-4/07/$23.00

with high reliability and high-spatial resolution that cannot be obtained from L3-edge DXAFS. This report gives an overview of the station's present status and an example of experimental results showing its performance.

BEAMLINE AND INSTRUMENTATION

The light source is a SPring-8 standard bending magnet. The critical energy is 28.9 keV. The BL28B2 contains two experimental spaces: optics hutches 2 and 3 (Fig. 1). Optics hutch 2 is for x-ray diffraction imaging, including white x-ray topography and medical imaging, while optics hutch 3 is dedicated to DXAFS experiments and energy-dispersive diffraction studies of materials under high pressure and high temperature. Some equipment, such as a θ-2θ diffractometer, is used in both types of experiments in optics hutch 3.

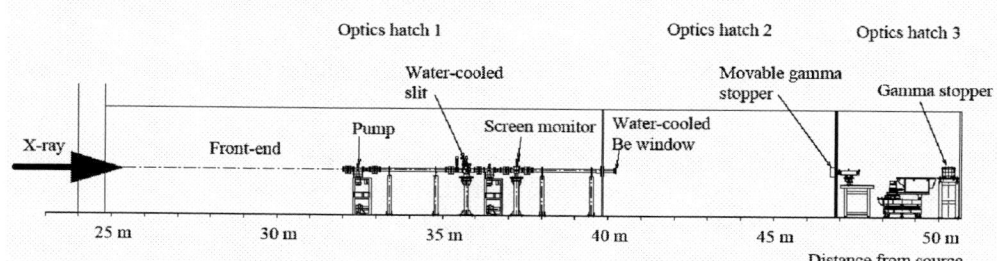

FIGURE 1. Schematic layout of BL28B2 transport channel

Figure 2 shows the configuration of the DXAFS spectroscopy system. The main equipment consists of a polychromator, a higher-harmonic-rejecting mirror, and a position-sensitive detector (PSD). The polychromator, sample and PSD are mounted on a θ-2θ diffractometer. White x-rays coming from the bending magnet source are diffracted by a bent crystal in the polychromator and focused on the sample. The diffracted x-rays have different energies, depending on the diffraction angle. The x-rays are measured with the PSD at the position corresponding to their energy.

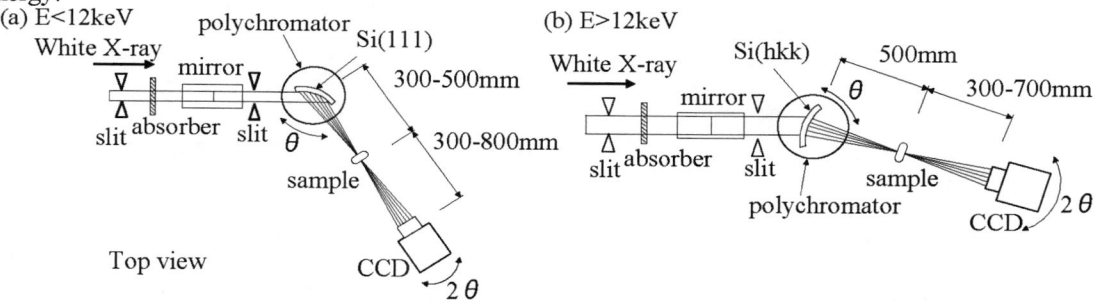

FIGURE 2. Schematic view of DXAFS spectrometer at BL28B2 for (a) Bragg geometry, and (b) Laue geometry

Two types of polychromator were developed. In the Laue configuration, the polychromator is a bender with a fixed radius curvature, R=1000mm. The polychromator crystal has an Si(111) surface, with dimensions of 90(w) x 20(h) x 0.2(t) mm. In the Bragg configuration, the bender is a four-point bender, which can load an independent bending force on each end of the crystal with dimensions of 200(w) x 40(h) x 1(t) mm. In the energy region from 5 to 12 keV, the crystal is positioned using Bragg geometry (Fig. 2(a)). A Laue configuration is used for the net plane Si(hkk), such as Si(422) or Si(311), in the energy region above 12 keV because the energy resolution decreases due to the depth of x-ray penetration into the crystal [5] (Fig. 2(b)).

The PSD is CCD coupled with a fluorescent screen and a lens system, which were developed for x-ray imaging experiments at SPring-8. The CCD used is a C4880-80-24A (Hamamatsu Photonics Co.) comprising 656 x 494 pixels with dimensions of 9.9 x 9.9 μm and a readout time of 6 ms. The CCD detected the beams after the x-rays had been transformed into visible light by a fluorescent screen. The fluorescent materials and their thickness were designed for the measurement x-ray energy region by taking into account x-ray absorption and image blur in the fluorescent screen. The spatial resolution of the x-ray images can be changed by altering the magnification of the optical lens in front of the CCD.

The absorber and mirror are positioned upstream of the polychromator to remove the unwanted components of low and high energy x-rays, respectively. Due to budgetary constraints, a 1000-mm-long vertical-focusing mirror was temporary one, which consists of four pieces of 250-mm-long Si substrate cut from a commercially available 300-mm-diameter Si wafer coated with stripes of Pt and Rh. The bender is a four-point load type.

PERFORMANCE AND EXPERIMENTAL RESULTS

Here we describe the system's performance. The focused beam size in the horizontal direction at the sample is ~110 μm at 25 keV. Figure 3 shows a Pd K-edge (24.3 keV) DXAFS spectrum of Pd foil together with a QXAFS spectrum measured at BL01B1. The minimum time resolution is 6 ms, which is the read-out time of the selected array in the present CCD. For concentrated standard samples, such as foils, good quality spectra can be obtained with exposures on the order of 10 ms; these exposures were shorter by one or two orders of magnitude than those of the QXAFS method. However, problems generally arise with actual samples – such as low concentration, significant x-ray absorption by support elements, and lack of uniformity in the density distribution – which necessitate long exposures to obtain data of a sufficient quality. As a result, a substantial time resolution, on the order of a sub-second or a second, is needed for actual samples. The quality of data obtainable from the DXAFS spectra is also affected by the PSD noise level and by the x-ray specifications (e.g., photon flux detected by the PSD, beam size on the sample, and higher harmonics contamination). The energy resolution was estimated at about 2.5×10^{-4} by convoluting a conventional XANES spectrum taken at BL01B1 with a Gaussian function. This value is slightly worse than the value roughly estimated by calculation. The deterioration in energy resolution may have been caused by unwanted deformation of the polychromator crystal or by x-ray image blur in the fluorescent screen.

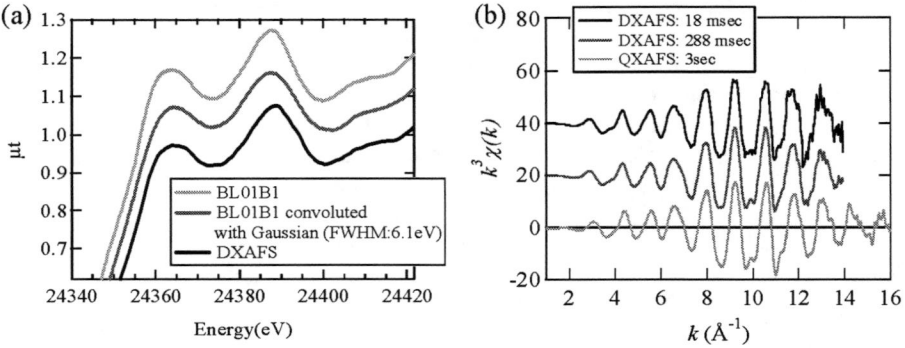

FIGURE 3. Pd K-edge DXAFS spectra (24.3keV) of Pd foil together with QXAFS spectrum measured at BL01B1. (a) XANES spectra (b) $k^3\chi(k)$ spectra

In the high-energy region, both the intensity of x-rays diffracted by the polychromator crystal and their absorption by the fluorescent screen of the CCD were low, so the x-ray measured by the PSD decreased. The energy resolution also deteriorated because of the relation between the energy and diffraction angle of the x-rays. Figure 4 shows Ce K-edge (40.5 keV) DXAFS spectra of a CeO_2 pellet together with QXAFS spectra measured at BL01B1. The DXAFS method provided good quality spectra of the signal to noise ratio at 40 keV with time resolution of about 360 ms for the standard samples. These exposures were shorter by one or two orders of magnitude than those of the QXAFS method. The achieved energy resolution was 3.6×10^{-4} at 40 keV, which was still acceptable.

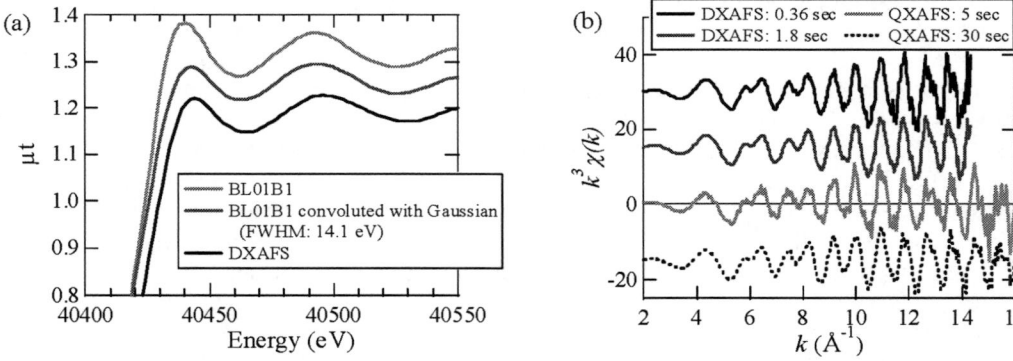

FIGURE 4. Ce K-edge DXAFS spectra (40.5 keV) of CeO_2 pellet together with QXAFS spectrum measured at BL01B1. (a) XANES spectra and (b) $k^3\chi(k)$ spectra

One example of DXAFS experimental data of an actual catalyst is shown below. Figure 5 shows Fourier transform spectra of a 1wt% Rh catalyst loaded onto H-form zeolite during reduction. The sample was heated from room temperature to 773 K at a ramping rate of 5 K/min under an 8% hydrogen flow. The DXAFS spectra were measured with a time resolution of 5 sec for every 10 K. The DXAFS studies revealed that Rh clusters formed inside the pore of zeolites were stable at least up to 773 K.

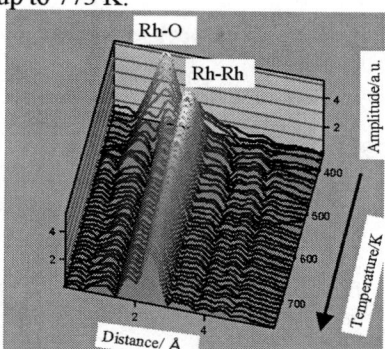

FIGURE 5. Fourier transform DXAFS spectra for 1wt% Rh loaded on to H-beta type zeolite measured in an H_2 flow

CONCLUSION AND FUTURE PLANS

At SPring-8 we have constructed a DXAFS spectrometer with a minimum time resolution of 6 ms from 7 to 40 keV at a bending magnet beamline. Actual samples having low concentration, however, required measurement times on the order of a second to get good quality DXAFS data.

We are improving the devices so we can do DXAFS experiments with a time resolution in the several tens or hundreds of milliseconds over a wide energy range. We plan to increase the photon flux by widening the horizontal beam width incident on the polychromator to be 50 mm. The present CCD system has a fluorescent screen 13 mm in diameter, which limited the EXAFS range and incident beam width. So, we are developing two types of PSD in collaboration with the detector team at SPring-8. One is a 50-mm wide photodiode arrays detector (PDA) with spatial resolution of 25 μm and read-out time of 60 ms. The PDA is coupled with a fluorescent screen and a fiber optic plate. Another is a 50-mm wide one-dimensional Si micro-strip detector. Its spatial resolution and read-out time is 50 μm and 50 μs, respectively. The thickness of the Si strip is 0.3 mm, so the Si micro-strip detector should be useful in an energy region lower than 15 keV. We plan to install a regular long vertical-focusing mirror system at optics hutch 1 to replace the present temporary one.

We are also constructing a handling system for large volumes of harmful or dangerous gases, which are frequently used during in-situ experiments.

ACKNOWLEDGMENTS

We would like to thank Dr. Naoto Yagi and Mr. Kentaro Uesugi of JASRI for the use of CCDs and for their helpful advice. We also thank Dr. Hiroshi Yamazaki of JASRI for providing ingots for polychromator crystals and for helpful comments.

REFERENCES

1. U. Kaminaga, T. Matsushita, and K. Kohra, *Jpn. J. Appl. Phys.* **20**(1981) L355.
2. M. Bonfim, K. Mackay et al., *J.Synchrotron Rad.* (1998), **5**, 750-752.
3. T. Ressler, R.E. Jentoft, J. Wienold, and O. Timpe, *J.Synchrotron Rad.* (2001). **8**, 683-685.
4. T.Shido et al., *Topics in Catalysis* (2002), **18**(1-2), 53-58.
5. M. Hagelstein, C. Ferrero et al., *J.Synchrotron Rad.* (1995). **2**, 174-180.

Development of the Experimental System for Time- and Angle-resolved Photoemission Spectroscopy

Kazutoshi Takahashi, Junpei Azuma, Shinji Tokudomi, and Masao Kamada

Synchrotron Light Application Center, Saga University, Saga 840-8502, Japan

Abstract. Experimental system for the time- and angle-resolved photoemission spectroscopy have been constructed at BL13 in SAGA Light Source, in order to study the electronic non-equilibrium in the surface layer of laser-excited materials The experimental system is very useful for photoemission spectroscopy in the wide temporal and angular ranges. The time- and angle-resolved photoemission spectra can be obtained with using the gate electronics for the MCP detector of the photoemission spectrometer. The gated MCP detector is synchronized with the laser pulse from Ti:sapphire regenerative amplifier with the repetition frequency of 10 to 300 kHz. The time-window of the gated MCP detector can be changed between 10 nano- and 160 micro-second. The time-resolved measurement in pico-second region can be performed with using the pump-probe technique which uses fundamental, second and third harmonics from the Ti:sapphire laser as the excitation source. Using these systems, we can perform the time- and angle-resolved photoemission study for various photo-excited phenomena and surface dynamics.

Keywords: Photoelectron spectroscopy; Surface electronic phenomena; Time-resolved technique
PACS: 79.60.-I; 72.40.+w; 72.20.Jv

INTRODUCTION

Synchrotron radiation (SR) and laser are widely used as useful light sources. Photoemission spectroscopy using synchrotron radiation as the excitation source is one of the most powerful experimental methods to investigate the electronic structures of solids and surfaces. On the other hand, the high brilliance of laser light can produce photo-excited carriers with high density. Various laser-induced phenomena on solids and surfaces, such as electronic non-equilibrium in the surface layer of photo-excited semiconductors [1] have been attracting much interest from the basic scientific point of view and also from the practical applications for photo-electronic devices. Up to now, several groups have been reported photoemission studies for photo-induced phenomena on semiconductor surfaces using the time-resolved technique with the combination of SR and laser. Long *et al.* have reported the temporal surface photo-voltage (SPV) effect on Si(111) surface using photoemission spectroscopy with a temporal resolution of about 20 ns [2]. Marsi *et al.* have studied the SPV effect on several semiconductor surfaces using the combination of a free electron laser synchronized to SR [3]. We have also reported the SPV relaxation on GaAs surfaces in nano- and micro-second range using the pump-probe and TAC-MCA method with SR and laser [4, 5]. Recently, the pump-multi-probe technique with a temporal resolution of 60 ps was used by Widdra *et al.* to study the SPV dynamics of SiO_2/Si (100) interface [6].

In this work, we have constructed new experimental systems for the time- and angle-resolved photoemission spectroscopy, in order to study various laser induced phenomena on solids and surfaces. The experimental system has been developed on the base of a high-resolution angle-resolved photoemission spectrometer. The time- and angle-resolved photoemission spectra in the wide range from nano- to micro-second region can be obtained with using the gate electronics for the MCP. The time- resolved measurement in pico-second region can be performed with using the pump-probe technique which uses fundamental, second and third harmonics from the Ti:sapphire laser.

CP879, *Synchrotron Radiation Instrumentation: Ninth International Conference*,
edited by Jae-Young Choi and Seungyu Rah
© 2007 American Institute of Physics 978-0-7354-0373-4/07/$23.00

EXPERIMENTAL SYSTEM

Experimental system has been constructed at beam line BL13 in SAGA Light Source. The SAGA Light Source has the circumference of 75.6 m, the storage electron energy of 1.4 GeV, critical energy of 1.9 keV and the emittance of 25.5 nm-rad. In order to obtain high energy resolution in a wide energy range from vacuum-ultra-violet to soft X-ray regions, the monochromator system at BL13 is designed to be composed of the grazing incidence mount with two varied line spacing plane gratings (VLSPG) and the normal incidence mount with a multilayer-coated plane grating. Relatively high flux and low higher-order light can be obtained with using the multilayer-coated plane grating in the boundary region between grazing-incidence and normal-incidence mounts. It is found from ray-tracing simulations that the resolving power better than 10,000 and photon flux of 10^{10}-10^{12} photon/sec can be obtained for the synchrotron radiation from the planar undulator of 24 periods [7].

The time- and angle-resolved photoemission spectra can be obtained with using the gate electronics for the MCP detector of the spectrometer. Figure 1(a) shows the schematic diagram of time-resolved measurement with gated detector. Synchrotron radiation from planar undulator is monochromatized with the VLS-PGM. The repetition frequency of synchrotron radiation is 500 MHz in the multi-bunch operation. The laser light from the Ti:sapphire regenerative amplifier (Coherent, RegA9000) which provides the fundamentals around 750-850 nm with the repetition frequency of 10-300 kHz is delivered to the viewport of the main experimental chamber by using an optical fiber. The laser light is then focused onto the sample surface. The hemispherical electron-energy analyzer of high energy and angle resolution type (MB Scientific, A-1) is used to observe the photoemission spectra. The energy resolution of the photoemission spectrometer is about 1.6 meV at the pass energy of 2 eV [7]. The spectrometer is equipped with a 2D detector composed of micro-channel plate (MCP), phosphor screen and CCD camera, which covers the angular range of ±7 degree with the resolution of about 0.5 degree. The gate action is performed with supplying a pulsed-voltage to the MCP. The voltage is synchronized with the laser repetition frequency of the regenerative amplifier by using a TTL clock signal. Both time-window and delay time of the gated MCP detector can be changed from 10 nano- to 160 micro-second. A mechanical shutter is installed in the optical path of the laser, which enables us to measure photoemission spectra with and without laser irradiation in single measurement sequence.

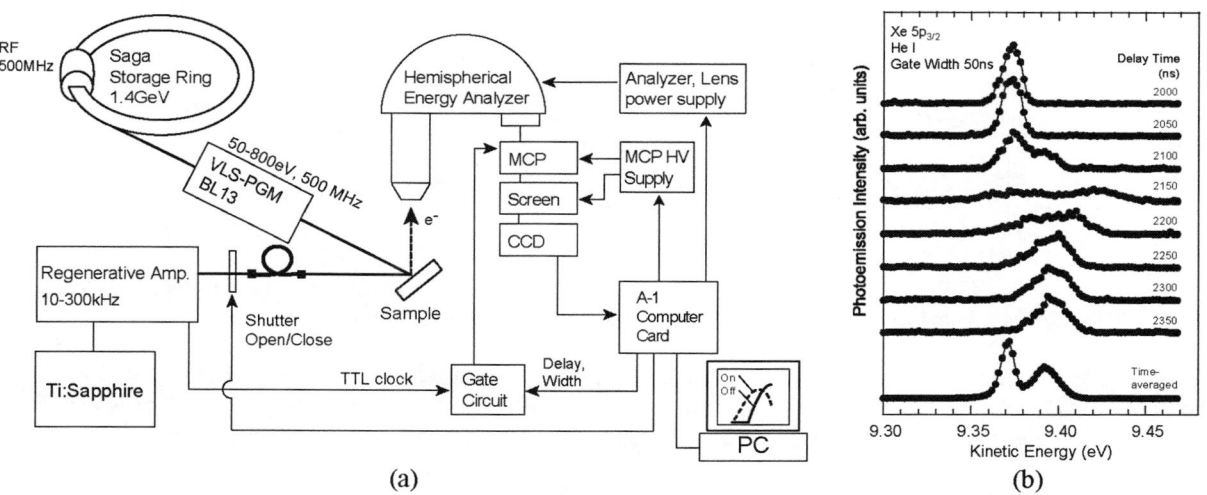

(a) (b)

FIGURE 1. (a) The schematic diagram of time-resolved measurement with gated detector in micro-second region. (b) Xe $5p_{3/2}$ photoemission spectra measured using the present gated-detector with the time-window of 50 ns.

Figure 1(b) shows the results of the preliminary test of the time-resolved measurement with our gated-detector. Xe $5p_{3/2}$ photoemission signal was measured in this test. The He I resonance line was used as the excitation source. The pulsed bias voltage with the repetition frequency of 300 kHz same as that of the regenerative amplifier was supplied to the electrode located at the back side of the Xe ionization cell, in order to change the electrostatic potential at the ionization area of the Xe gas cell. The bottomed spectrum in Fig. 1(b) shows the time averaged spectrum that was measured without gated-detection. The peaks at 9.371 and 9.392 eV in kinetic energy correspond to the photoemission signals without and with voltage supply to the back electrode, respectively. The broadening of the peak at 9.392 eV is due to the slope of the electrostatic potential in the ionization space because the front electrode is connected to the earth. The time-resolved spectra with the gate width of 50 ns show distinct changes at each delay time from the onset of the voltage pulse. Anomalous behavior at the delay time around 2150 ns corresponds to the oscillation of the electrostatic potential due to the electrostatic capacitance of the Xe ionization cell. As shown in Fig. 1(b), it is confirmed that time-resolved measurements with the resolution of 50 ns can be performed with our gated-detector.

The time-resolved photoemission measurement in pico-second region is performed with using the pump-probe technique which uses fundamental, second and third harmonics from the Ti:sapphire regenerative amplifier. Figure 2(a) shows the schematic diagram of time-resolved measurement with using the pump-probe technique. A Ti:sapphire regenerative amplifier (Spectra Physics, Hurricane) provides laser pulse with a photon energy of 1.55 eV, a repetition frequency of 1 kHz, a pulse width of 150 fs, and a pulse energy of 0.8 mJ. The output pulse from the Ti:sapphire laser system is frequency-tripled in a THG unit with BBO crystals. The fundamental and third harmonics from the THG unit are separated using a harmonic separator. The fundamental is used as pump pulse and the third harmonics is used as probe pulse. The frequency-doubled pulse can also be used as pump pulse with changing a optical cut filter. Photoelectrons are generated by the two-photon excitation of probe pulse. The time interval between the pump and probe pulses is controlled by the optical delay line. The time range of the optical delay line is from -100 to 500 ps. A mechanical shutter is also installed in the optical path of the pump pulse, which enables us to measure with and without pump laser irradiation in single measurement sequence.

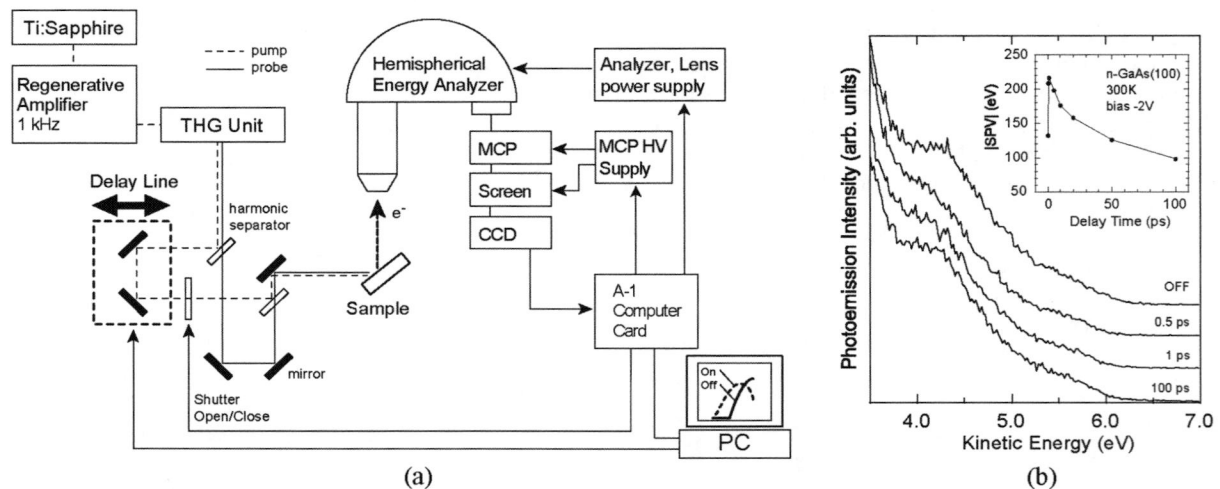

FIGURE 2. (a) The schematic diagram of time-resolved measurement with using the pump-probe technique in pico-second region. (b). Time-resolved photoemission spectra of the n-GaAs(100) surface.

In order to demonstrate the performance of this system, we have observed the temporal changes of the surface photo-voltage (SPV) effect on n-GaAs(100) surface. A Si-doped (3.9-5.0×10^{17} cm^{-3}) n-GaAs(100) was used for the measurements. The sample was cleaned by annealing at 870 K for 60 min in the ultra-high vacuum. The sample cleanliness was checked by XPS and LEED measurement. The fundamental was used as pump pulse in the experiment. The power density of the pump and probe pulses were optimized in order to avoid the space charge effect [8] and the sample damage. The power densities of pump and probe pulses in the present measurement were the 700 MW/cm^2 and 57 MW/cm^2, respectively. The total time resolution was about 300 fs determined by the cross correlation of the pump and probe pulses, which indicates that the duration of the laser pulses were extended by the chirping through the optical path.

Figure 2(b) shows the valence band photoemission spectra in 3.5 eV range from the Fermi level. The topped spectrum shows the photoemission spectrum without pump laser pulse. The time delay of the probe pulse to the pump pulse is also shown in the Fig. 2(b). The valence band shifts to the lower kinetic energy side when the sample is irradiated by pump pulse. This is originating from the SPV effect that the band bending on the semiconductor surface is partially compensated by another electric field caused by spatially separated photo-excited carriers [1]. The temporal SPV determined by photoemission measurement is shown in the inset of Fig. 2(b). One can see that the temporal SPV decreases with increasing the time delay. This results shows that the temporal change of the electronic states near by the Fermi level can be observed in the sub-pico- and pico-second range by using the present system.

In addition to the experimental systems for time-resolved photoemission measurements in micro- and pico-second range described above, the development of the pump-probe measurement system using synchrotron radiation and Ti:sapphire laser is also planned, in order to realize the time-resolved study in nano-second region. The Ti:sapphire laser can synchronize with synchrotron radiation in 1/6 filling operation at SAGA Light Source. Using these systems, we can perform the time- and angle-resolved photoemission study for various photo-excited phenomena and surface dynamics.

SUMMARY

In conclusion, we have constructed the experimental systems for time- and angle-resolved photoemission spectroscopy in the wide temporal range from nano- to micro-second. The time- and angle-resolved photoemission spectra can be obtained with using the gate electronics for the MCP detector of the photoemission spectrometer. The gated MCP detector can be synchronized with the laser from regenerative amplifier with the repetition frequency of 10 to 300 kHz. The time-window of the gated MCP detector can be changed between 10 ns and 160 μs. The time-resolved measurement in pico-second region can also be performed with using the pump-probe technique which uses fundamental, second and third harmonics from the Ti:sapphire laser as the excitation source. The development of the pump-probe measurement system using synchrotron radiation and Ti:sapphire laser is also planned, in order to realize the time-resolved study in nano-second region. The Ti:sapphire laser will be synchronized with synchrotron radiation in 1/6 filling operation at SAGA Light Source. Using these systems, we can perform the time- and angle-resolved photoemission study for various photo-excited phenomena and surface dynamics.

ACKNOWLEDGMENTS

This work was supported by a Grant-in-Aid for Scientific Research from the Ministry of Education and Science, Sports, and Culture, Japan.

REFERENCES

1. Kronik, L., Shapira, Y., Surf. Sci. Rep. **37**, 1-206 (1999).
2. Long, J. P., Sadeghi, H. R., Rife, J. C., and Kabler M. N., Phys. Rev. Lett. **64**, 1158-1161 (1990).
3. Marsi, M., Belkhou, R., Grupp. C., Panaccione, G., Teleb-Ibrahimi, A., Nahon, L., Garzella, D., Nutarelli, D., Renault, E., Roux, R., Couprie, M. E., and Billardon, M., Phys. Rev. B **61**, R5070-5073 (2000).
4. Tanaka, S., More, S. D., Murakami, J., Itoh, M., Fujii, Y., and Kamada, M., Phys. Rev. B **64**, 155308 (2001).
5. Tanaka, S., More, S. D., Takahashi, K., and Kamada, M., J. Phys. Soc. Jpn. **72**, 659-663 (2003).
6. Widdra, W., Bröcker, D., Gießel, T., Hertel, I. V., Krüger, W., Liero, A., Noack, F., Petrov, V., Pop, D., Schmidt, P. M., Weber, R., Will, I., and Winter, B., Surf. Sci. **543**, 87-94 (2003).
7 Takahashi, K., Kondo, Y., Azuma, J., and Kamada, M., J. Electron Spectr, Relat. Phenom., **144-147**, 1093-1096 (2005).
8 Siffalovic, P., Drescher, M., and Heinzmann, U., Europhys. Lett. **60**, 924-930 (2002).

Implementation of Enhanced Quick-scan Technique for Time-Resolved XAFS Experiment at PLS

Nark-Eon Sung, Min-Gyu Kim, Sun-Hee Choi, Jay-Min Lee

Pohang Accelerator Laboratory, Pohang University of Science and Technology, Pohang 790-784, Korea

Abstract. Quick-scan is an important technique for the time-resolved experiments to investigate structural variation in rapid chemical reactions or phase transitions under external perturbations in the material science. Our new quick-scan technique, implemented at BL3C1 of PLS, adopts PCI-bus instead of CAMAC based system which was used in the default quick-scan system. It uses a digital I/O and a counter/timer PCI board, which are connected to DCM encoder and to VFC, respectively. Since the trigger signal is originated for a fixed interval from the counter/timer board operating on 80 MHz internal clock, all channels are synchronized without dead or delay time. When the trigger interval was set as 20 ms for each data point, it took ~100 seconds for each full scan at 7 keV in the range of -200 ~ 1000 eV, and the energy step sizes corresponded to ~0.7 eV. Overall performances of the Quick-XAFS technique will be demonstrated through an *in situ* electrochemical experiment for the Li-ion battery cathode material.

Keywords: Quick-scan, PCI-bus, time-resolved, cathode material, Q-XAFS.
PACS: 39.30.+w

INTRODUCTION

With the advent of synchrotron radiation sources about one fourth century ago, X-ray absorption fine structure (XAFS) has played a crucial role for refining the local structure and electronic states. Up to now, so many XAFS related techniques are developed for special applications. Among these methods, Quick XAFS (Q-XAFS) scan is an essential technique for the instant phase shift or *in situ* chemical reactions of which the normal step-by-step scan is inapplicable [1]. Since Q-XAFS spectra are recorded 'on the fly' by slewing the double crystal monochromator (DCM), so much undesirable mechanical noise, which can occur in step-by-step mode measurement, can be avoided [2]. Q-XAFS for the full scan (~1200 eV) takes time in the range from a few ms to few minutes, which depends on the DCM structure operating without serious vibration or misalignment of crystal parallelism during movement. This quasi-continuous movement can be feasible for the general 'boomerang' type DCM [3]. However, the special electronics system and software are needed to minimize electronic noise and to get the reliable and iterative synchronized absorption coefficients.

CONFIGURATION OF Q-XAFS SYSTEM

We had developed the Q-XAFS technique at BL3C1 at Pohang Light Source (PLS) several years ago [1]. However, the performance has not satisfied user's demands because of too many difficulties in taking Q-XAFS data. For the purpose of making a stable and an easy-to-use system without being interrupts from unexpected noise, we designed a new system which substitutes the PCI-bus for the default CAMAC because the PCI module is rather a compact and cheaper than CAMAC one and has versatile functions. Like our previous Q-XAFS system, we used the broomstick DCM which is similar to the 'boomerang' type.

The main components of DCM to be monitored or controlled by Q-XAFS system are a linear encoder and a harmonic reduction geared motor which rotates a pair of crystals. We use three PCI modules for the new Q-XAFS control system. The first one is a counter/timer (NI PCI-6602) for counting and setting timing-window for each data point. The others are a digital I/O (NI PCI-6533) and a motor controller (NI PCI-7344) for digitizing 23 TTL

CP879, *Synchrotron Radiation Instrumentation: Ninth International Conference*,
edited by Jae-Young Choi and Seungyu Rah
© 2007 American Institute of Physics 978-0-7354-0373-4/07/$23.00

encoder signals and for a continuous DCM movement, respectively. As the counter/timer module can be operated as a counter and/or a timer, two of 8 channels are used to generate the finite periodic pulse trains which can be used as the TTL level trigger signals. Since it uses 80 MHz internal clock, the estimated jitter is within a few ns. The generated trigger signals are distributed as parallel to the other counter/timer channels and digital I/O ports to synchronize all data points. The period of TTL pulses, which depends on the energy step size per data point or motor speed, can be adjusted by software interactively. The minimal period is limited to 1 ms and can be changed in ms unit if necessary. Generally it is set to 20 ms, which corresponds to ~0.5 eV at 7 keV X-ray energy when the slew rate is 2600 pps, for the optimal data points and reasonable data quality. To operate DCM stably we adopt the 'S' curve acceleration and deceleration mode similar to the trapezoidal, and chose a slew rate where the DCM vibration is minimized.

Figure 1 represents the schematic electronics diagram of the newly developed Q-XAFS system. For the time-resolved experiments, all systems are controlled by a code developed using LabVIEW, a commercial software. Since two PCI modules, a counter/timer and a digital I/O, can be operated in cyclic buffered memory mode, these are configured so that each data point can be latched in the corresponding buffers whenever a TTL signal rises up. After the scan if finished, all latched data are read out from buffer to the computer memory though the DMA function and written to a raw data file. As shown in this figure, an electrochemical cell is connected with a potentiostat (PAR 263A) and charged 0.2 mA on constant current mode to confirm the stability and data quality of the Q-XAFS system by measuring *in situ* transmission time-resolved Q-XAFS spectra.

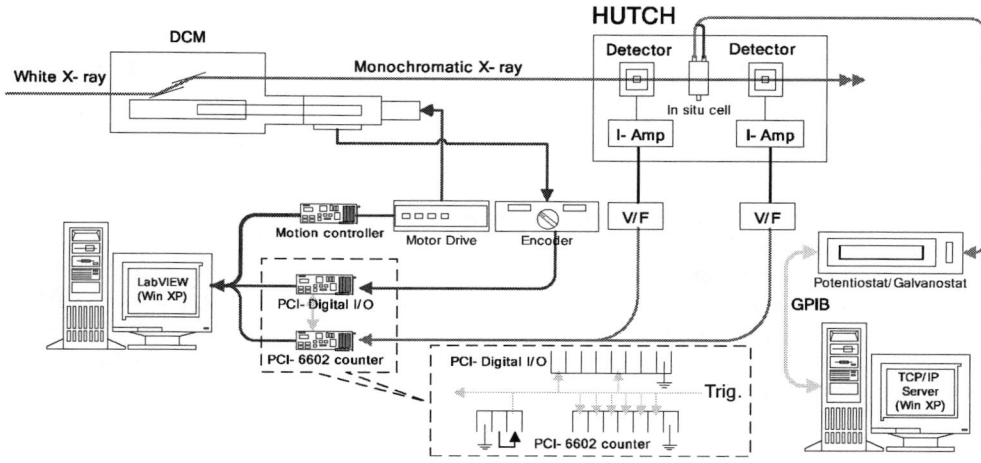

FIGURE 1. The schematic electronics diagram of Q-XAFS control system. Three PCI modules are used for continuous-movement of DCM and synchronizing each data point.

RESULTS AND DISCUSSION

The measurement of time-resolved Mn K-edge Q-XAFS spectra of the *in situ* $LiMn_{0.8}Co_{0.2}O_2$ electrochemical cell, the Li-ion battery cathode material, was performed upon Li-ion deintercalation (charging process) for comparing data quality with that of the step-by-step scan. All Q-XAFS data were recorded in every 5 minutes with a step size of 0.5 eV/point and 20 msec integration per point and took ~35 sec for the scan range 900 eV. On the other hand, the normal step-by-step scan took ~30 min under the same experimental condition. Because too many spectra (over 500) were acquired in a charge/discharge cycle to be analyzed using general method [5], we made a special program, subsidiary software of UWXAFS [6], and processed all measured data in a short time.

Figure 2 shows the normalized X-ray absorption near edge structure (XANES) spectra as a function of Li-ion deintercalation time. A step-by-step scan spectrum of fully discharged sample is shown together as a reference. The overall XANES features are almost identical, on the deintercalation. The energy shift and spectral shape with respect to the Q-XAFS spectra are due to the variation of manganese oxidation state along with the charging process. We note that the Q-XAFS shows comparable spectra even though its energy step is wider 2~3 times than that of the conventional step-by-step scan.

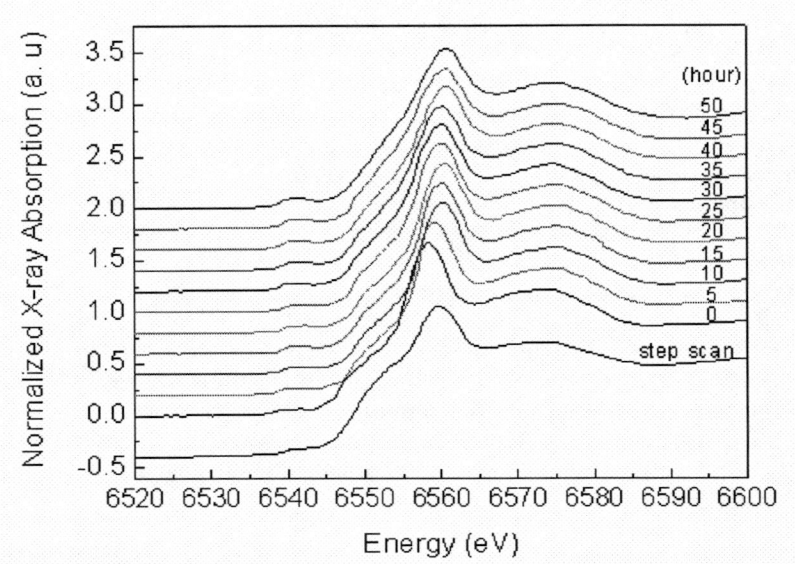

FIGURE 2. XANES spectra of LiMn$_{0.8}$Co$_{0.2}$O$_2$ *in situ* cell at Mn K-edge in Q-XAFS scan mode as a function of Li-ion deintercalation time. The lowest spectrum, taken by the step-by-step scan mode, is shown as a reference. Q-XAFS spectra show a shift to the higher energy side upon Li-ion deintercalation due to the variation of oxidation state of Mn ion. No distinctive quality differences are found between the Q-XAFS spectra and the step-by-step spectrum.

Figure 3 shows the voltage profile and k^2-weighted EXAFS spectra, $\chi(k)$, as a function of Li-ion intercalation time. Figure 3(a) presents a section of the full voltage profile and the voltage range of which Q-XAFS spectra are taken. Figure 3(b) shows Q-XAFS spectra and a step-by-step scan data (bottom line). The difference between the Q-XAFS spectra and step-by-step spectrum is due to the different charged state. In this figure one can observe only a small spectral variation in Q-XAFS spectra even though measured every 5 minutes. It indicates that the oxidation state of Mn ion of this electrochemical cell is not changed seriously in this voltage range. Compare with the conventional scan data, the Q-XAFS data show a good reproducibility and a comparable quality over 12 Å^{-1} in k-space.

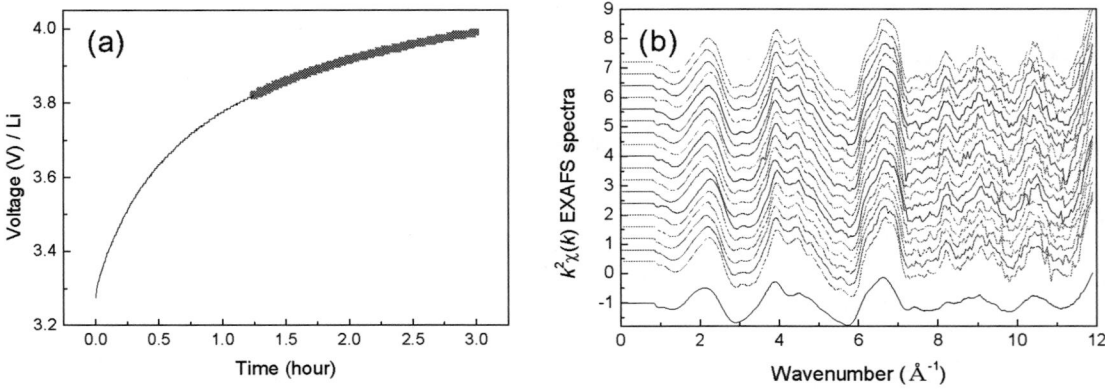

FIGURE 3. A section of the full voltage profile and in situ transmission Q-XAFS spectra of the Li-ion battery cathode material. (a) the voltage profile as a function of charging time. The thick curve indicates the voltage range of which Q-XAFS spectra are taken. (b) Q-XAFS spectra measured in every 5 minutes interval in the voltage range of (a). The bottom spectrum, taken in the normal step-by-step scan, is shown as a reference which is taken at fully discharged state.

CONCLUSION

We designed a new Q-XAFS scan system to replace the old one at 3C1 beamline at PLS using three PCI bus modules. We measured time-resolved *in situ* Q-XAFS transmission spectra of an in situ electrochemical cell at Mn K-edge upon Li-ion intercalation/deintercalation and compared these with a conventional step-by-step scan spectrum in both energy and *k*-space. The number of data points of a Q-XAFS spectrum is ~1800 with an average energy step per data point of ~0.5 eV and 20 ms integration time per point and it took ~35 sec for the scan range of ~900 eV. From the XANES and $\chi(k)$ spectra of both scan methods, we could confirm that the newly designed Q-XAFS system works stably and reliably in time-resolved iterative scan.

ACKNOWLEDGMENTS

The XAFS experiments were performed with approval of authorities concerned of Pohang Light Source (PLS).

REFERENCES

1. L. M. Murphy, B. R. Dobson, M. Neu, C. A. Ramsdale, R. W. Strange and S. S. Hasnain, *J. Synchrotron Rad.* **2**, 64-69 (1995).
2. J. Wong, M. Fröba, R. Frahm, *Physica* **B 208&209** (1995) 249-250
3. M. Ramanathan, P. A. Montano, *Rev. Sci. Instrum.* **66 (2),** February 1995.
4. J. M. Lee, N. E. Sung, J. K. Park, J. G. Yoon, J. H. Kim, M. H. Choi and K. B. Lee, *J. Synchrotron Rad.* **5**, 524-526 (1998).
5. P. A. O'Day, J. J. Rehr, S. I. Zabinsky, and G. E. Brown, *J. Am. Chem. Soc.,* **116**, 2938-2949 (1994).
6. J. J. Rehr, J. M. Leon, S. I. Zabinsky, and R. C. Albers, *J. Am. Chem. Soc.,* **113**, 5135-5140 (1991).

A Simple Short-range Point-focusing Spatial Filter for Time-resolved X-ray Fluorescence

C. Höhr[1], E. Peterson, E. Landahl, D. A. Walko, R. W. Dunford, E. P. Kanter, and
L. Young

Argonne National Laboratory, Argonne, Illinois 60439, USA

Abstract. A simple, compact, point-focusing spatial filter for x-ray fluorescence is presented. This construction maintains the large solid angle and directionality of existing designs but is more easily machined. Combined with a selective absorber, it can be used as an x-ray low-pass filter; this is one common approach taken in inelastic x-ray scattering studies and fluorescence spectroscopy. When combined with a scintillation detector, this device forms a large solid angle energy-selective x-ray detector with ns-scale time resolution that is useful for timing studies.

Keywords: time-resolved x-ray fluorescence, point focusing spatial filter, Argonne Advanced Photon Source
PACS: 61.10.Eq, 61.10.Ht

INTRODUCTION

Ultra-fast photo-induced changes in the local structure surrounding a central atom are typically studied with laser-pump/x-ray probe spectroscopies such as EXAFS (extended x-ray absorption fine structure) and XANES (x-ray absorption near-edge structure) [1]. In such studies, x-ray induced fluorescence detection is typically used due to its elemental specificity. X-ray fluorescence is particularly suitable to study molecules in solution at low concentrations ($\sim$ mM). Due to the large ratio of solvent to solute molecules, the desired signal is weak and there is substantial background from scattering and the possibility of further contamination from auxiliary x-ray lines. Thus, for these studies, a multi-element, energy-dispersive Ge detector is typically used [1] to select the x-ray energy of interest, e.g. the Cu $K\alpha$ line. However, such detectors have slow time response ($\sim$ 2 μs) and are cumbersome and expensive. Therefore, it would be of general interest to have a large solid angle, energy-selective, fast x-ray detector; herein, we report on such a device.

DESIGN

The detector required must be spatially and spectrally resolved to distinguish fluorescence from scattering. At a synchrotron x-ray source, where the laser repetition rate may be much less than the x-ray bunch rate, the detector must be fast enough to reject signals from neighboring (laser-off) x-ray bunches. While an energy-dispersive Si-drift detector can be used to detect emitted x rays, it requires microsecond time spacing between x-ray bunches. We wish to enable work in the Argonne Advanced Photon Source's (APS) 24-bunch mode, where the x-ray pulses are separated by only 154 ns. This mode has the advantage of more available beamtime and a shorter x-ray pulse duration [2]. Faster detectors, in general, have worse energy resolution and often a smaller solid angle. Here, we use a NaI detector with a time resolution of 50 ns and an aperture of 50 mm; the combination of a low-pass absorption filter and a point-focusing spatial filter (PFF) achieves sufficient reduction of unwanted background for studying the near-edge features of Kr. We chose a 41 μm thick Se foil as an absorption filter, corresponding to three absorption lengths for x rays 50 eV above the Se K-edge. We selected a so-called $Z-2$ filter since Br is not practical as a filter material. In order to block the re-fluorescence of the filter, a common method is to combine it with a PFF. The original idea of a PFF was given by Stern and Heald [3] with later improvements by Behne *et al.* [4] and found its application in the widely used Lythle detector [5]. A monolithic PFF is described by Seidler and Feng [6]. The spatial filter permits the fluorescence signal

---

[1] present address: Triumf, Vancouver, British Columbia, Canada

CP879, *Synchrotron Radiation Instrumentation: Ninth International Conference*,
edited by Jae-Young Choi and Seungyu Rah
© 2007 American Institute of Physics 978-0-7354-0373-4/07/$23.00

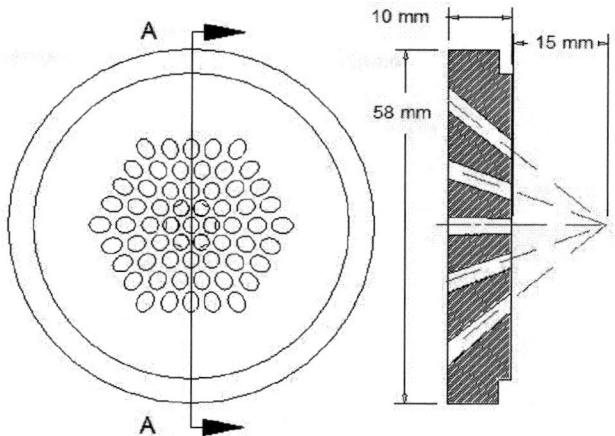

FIGURE 1. Bottom and cross-sectional view of the first design of the point-focusing filter (PFF) made from Al. Each hole has a solid angle of ~ 0.01 sr and all 61 holes together have a total solid angle of ~ 0.61 sr.

photons to travel directly from the target to the detector but blocks most of the Se-filter fluorescence, which its emitted isotropically. The net relative efficiency R of the Se filter and PFF for detecting photons below the Se K-edge relative to those from above depends on the length L of a PFF channel from its entrance to exit aperture and on the area A of one exit aperture:

$$R \propto \frac{L^2}{A} .$$

(1)

For this assumption, direct transmission from the sample is ignored, assuming the filter is thick enough. Also, scattering from the channel walls is neglected. For more details see [4] and the references within.

For a first test and to demonstrate the feasibility, we chose to design a PPF made from a solid Al disc, 10 mm thick and 58 mm in diameter. The 61 holes were drilled using tapered bits available in the local machine shop. This results in a focal length of 15 mm (see Fig. 1). The use of one tapered drill bit instead of several drill bits of different sizes for a step-tapered hole [6] allows a shorter machining time and therefore less cost. It also makes better use of the body material and consequently reaches a bigger solid angle. Each hole has a solid angle of ~ 0.01 sr and all 61 holes together have a total solid angle of ~ 0.61 sr.

TEST AND RESULTS

We tested the PFF at the 7ID beamline of the APS. Monochromatic 14 keV x rays were focused to a FWHM of ~ 10 μm into a Kr gas sample from an effusive gas jet. In addition, a ~ 100 μm FWHM focused laser beam (2.5 mJ energy per pulse, 40 fs pulse width, 800 nm central wavelength, 1 kHz repetition rate) was overlapped in space and time. Kr atoms were ionized by the laser and subsequently probed by the x rays as a function of time delay between pump and probe, space, x-ray energy and laser polarization (for experimental details and discussion of the results, see [7, 8]). In Fig. 2 the background suppression is demonstrated, measured with a Si-drift detector for energy resolution. Spectra (a) - (d) show the fluorescence for different filter combinations. The yields of the different peaks normalized to the corresponding Kr $K\alpha$ peak are listed in Table 1. For the unfiltered spectrum (a) the Kr $K\alpha$ fluorescence, along with background from stainless steel (SS) and Kr $K\beta$ fluorescence which contains elastic scattering background, is seen. As expected, with just the low-pass Se-filter between target and detector, the energy spectrum (b) shows both additional Se fluorescence and an immense reduction for SS and Kr $K\beta$ (which is inseparable from the elastic scattering background). From Fig. 2 (c) it is clear that the PFF alone already eliminates background scattering from the stainless steel of the sample chamber due to its different point of origin. For the same reason, the elastic scattering background mixed with the Kr $K\beta$ peak is significantly reduced. The Kr $K\beta$/Kr $K\alpha$ ratio is now 0.146 and agrees with the literature value [9] of $K\beta/K\alpha = 0.151 \pm 0.005$. Finally, the Se filter and PFF combined (Fig. 2 (d)) purge the spectrum from SS, Se and Kr $K\beta$ fluorescence. In addition, comparing count rates, the advantage of a 50

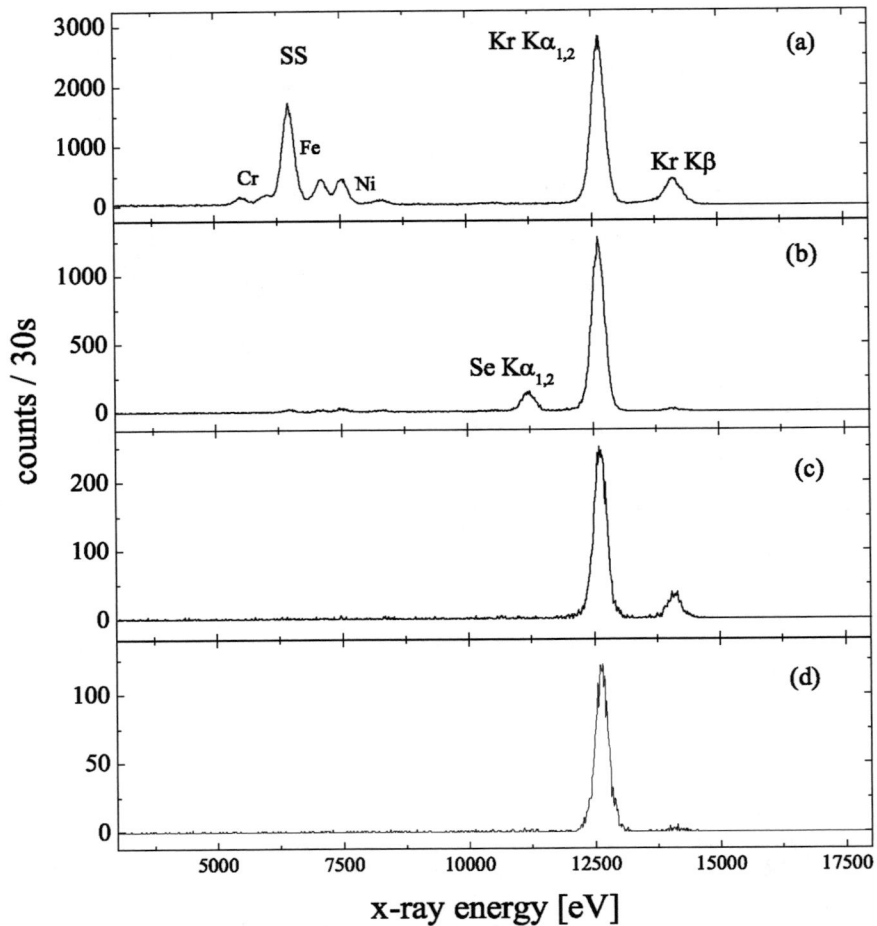

FIGURE 2. X-ray fluorescence spectra of Kr measured with a Si-drift detector and different filter combinations. The raw spectrum (a) shows the Kr $K\alpha$ and $K\beta$ peaks as well as some background at lower energies originating from stainless steel (SS) and under the Kr $K\beta$ peak from elastic scattering. Spectrum (b) shows a suppression of the Kr $K\beta$ and elastic background x rays due to the Se-absorption filter and also a new peak caused by the fluorescence of the Se filter itself. In (c), measuring with the PFF only, the SS is strongly suppressed due to its different point of origin as well as the elastic background underneath Kr $K\beta$. Finally in spectrum (d), with both the Se filter and the PFF in place, a nearly pure Kr $K\alpha$ fluorescence is recorded. Yields normalized to the Kr $K\alpha$ in the corresponding spectrum can be found in Table 1.

mm NaI detector in combination with the PFF over the Si-drift detector is not just the needed time resolution but also a six-fold increase in count rate due to the larger solid angle.

After this successful first test, a second, modified PFF was designed and machined with Mo as the body material, as seen in Fig. 3. Using Mo for the PFF channel walls will result in a more than one order of magnitude smaller contribution from Se fluorescence scattering off the walls than the directly propagated contribution from the Se

TABLE 1. Yields of the peaks in Fig. 2 (a) - (d) normalized to the corresponding Kr $K\alpha$ peak.

	(a) raw	**(b)** Se filter	**(c)** PFF	**(d)** PFF + Se filter
Fe $K\alpha$	0.57 ± 0.01	0.026 ± 0.002	0.002 ± 0.001	-0.002 ± 0.001
Se $K\alpha$	—	0.122 ± 0.002	—	0.005 ± 0.002
Kr $K\beta$	0.21 ± 0.01	0.018 ± 0.002	0.146 ± 0.003	0.017 ± 0.003

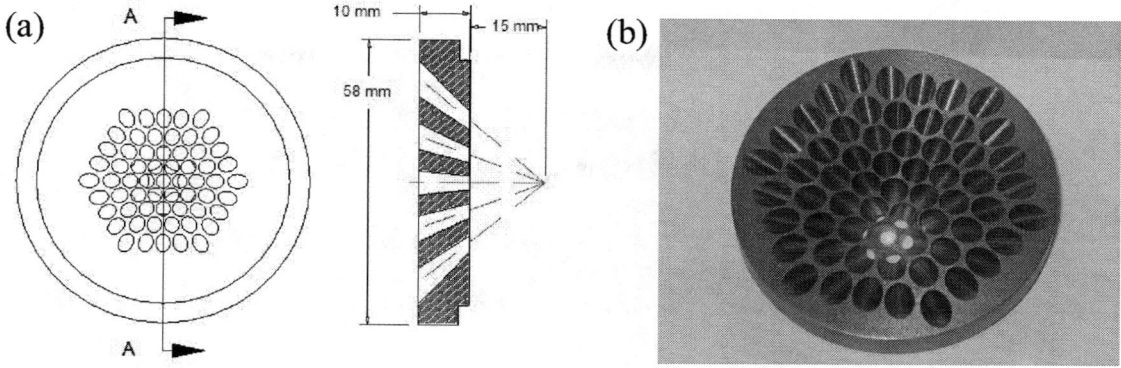

FIGURE 3. Construction drawing (a) and photo (b) of the advanced design made out of Mo with a solid angle of 0.03 sr per hole, with a total solid angle of 1.83 sr for all 61 holes.

filter [4]. The holes are more tapered (solid angle of 0.03 sr) which results in an overall solid angle of 1.83 sr and therefore an expected further improvement of the count rate by a factor of three. Although this new design has a three times bigger channel exit aperture A and, hence from Eq. 1, a three-fold decrease in efficiency R, we still expect a reduction in background sufficient for our needs. A successful test of this new PFF will allow the most common fill pattern of the APS to be used for time-resolved x-ray fluorescence studies.

ACKNOWLEDGMENTS

We are very thankful to Gerry Seidler, Ed Stern, and Steve Heald for fruitful discussions and the loan of an absorption filter. This work was supported by the Chemical Sciences, Geosciences, and Biosciences Division and the Advanced Photon Source by the Office of Basic Energy Sciences, Office of Science, U.S. Department of Energy, under Contract No. W-31-109-Eng-38.

REFERENCES

1. L. Chen, *Angew. Chem. Int. Ed.*, **43**, 2886 (2004).
2. E. Landahl *et al.*, in preparation.
3. E. A. Stern and S. M. Heald, *Rev. Sci. Instrum.*, **50**, 1579 (1979).
4. E. A. Behne, Yejun Feng, and G. T. Seidler, *Rev. Sci. Instrum.*, **72**, 3908 (2001).
5. F. W. Lythle *et al.*, *Nucl. Instrum. Methods*, **226**, 542 (1984).
6. G. T. Seidler and Yejun Feng, *Nucl. Instrum. Methods Phys. Res. A*, **469**, 127 (2001).
7. L. Young *et al.*, *Phys. Rev. Lett.*, accepted.
8. C. Höhr *et al.*, submitted.
9. R. B. Firestone, *Table of Isotopes, CD-ROM Edition*, edited by V. S. Shirley, Wiley Interscience, 1996.

Time-Resolved Dispersive XAFS Instrument at NW2A Beamline of PF-AR

Y. Inada, A. Suzuki, Y. Niwa and M. Nomura

Photon Factory, Institute of Materials Structure Science, KEK, 1-1 Oho, Tsukuba 305-0801, Japan

Abstract. The configuration and performance of the time-resolved dispersive XAFS (DXAFS) instrument, which has been constructed at the NW2A beamline of PF-AR (KEK), are presented. The DXAFS instrument is mainly composed of a polychromator part, a position control part for sample, and a linear detector part. The Bragg- and Laue-type polychromator crystal (Si(111) or Si(311)) is bent using the holder with fixed bending radius, in which the thermostated water is circulated to prevent the temperature change of crystal due to the heat load. The photodiode array (PDA) with and without phosphor screen is used as the linear X-ray detector, and the minimum exposure time is 2 ms for the 1024-element PDA. The phosphor screen on the PDA detector prevents the damage of the chip especially for high energy X-rays but the existence reduces the energy resolution because of the scattering of the visible light converted on the phosphor. The DXAFS instrument was applied to the mechanistic study of the reduction processes of Cu supported on MFI zeolite, and the intermediate Cu(I) states have been successfully observed during the reduction from Cu(II) to Cu(0).

Keywords: time-resolved, dispersive XAFS, mechanistic study, nonreversible process, heterogeneous catalyst
PACS: 61.10.Ht, 82.20.Pm, 82.75.Qx

INTRODUCTION

The X-ray absorption fine structure (XAFS) is the most powerful tool to obtain the radial structural information for the samples without long-range order. The XAFS technique plays an important role for the mechanistic interpretations especially for the heterogeneous catalysis systems [1] because of its useful characteristics, such as the element selectivity, the independency on the sample matrix, the applicability to dilute samples, and so on.

The authors have constructed the time-resolved XAFS instrument at the Photon Factory (KEK) in the viewpoint of the applications for mechanistic investigations of the nonreversible chemical processes. The dispersive optics for the XAFS measurement [2] is an unique solution to achieve the time-resolved observation for the nonreversible processes, because the whole energy range can be covered simultaneously without any mechanical movements and it is not necessary to repeat the reaction processes. The dispersive XAFS (DXAFS) instrument has been initially installed at BL-9C of PF and has been moved to the beamline NW2A [3] with a tapered undullator at PF-AR to improve the time-resolution. The DXAFS instrument has been applied to the mechanistic investigations for many chemical processes and a great success has been achieved for the atomic-scaled mechanistic interpretations about heterogeneous catalysis systems [4–8]. In this paper, the configuration and performance of the DXAFS instrument at NW2A will be presented and the recent results for the mechanistic study will be described about the redox processes of Cu species supported on zeolite.

CONFIGURATION OF DXAFS INSTRUMENT

The NW2A beamline is operated under a white X-ray mode when the DXAFS instrument is used. A flat-bent Rh-coated mirror for vertical focusing and a double mirror system for reducing the high energy components is available in the NW2A beamline. The white X-ray beam is introduced to the experimental hutch with the horizontal size of ca. 15–20 mm, which is limited by the horizontal acceptance of the upstream slit or the double mirror system.

The schematic diagram of the DXAFS instrument is shown in Fig. 1 and is mainly composed of a polychromator part, a position control part for sample, and a linear detector part. The polychromator crystal (Si(111) or Si(311)) is bent using the holder with fixed bending radius, in which the thermostated water at 298.0 ± 0.2 K is circulated to prevent the temperature change of crystal due to the heat load. The crystal and its holder are thermally coupled using liquid InGa and are placed under the flowing He atmosphere. The typical bending radius is ca. 2000–3000 mm for the

CP879, *Synchrotron Radiation Instrumentation: Ninth International Conference*,
edited by Jae-Young Choi and Seungyu Rah
© 2007 American Institute of Physics 978-0-7354-0373-4/07/$23.00

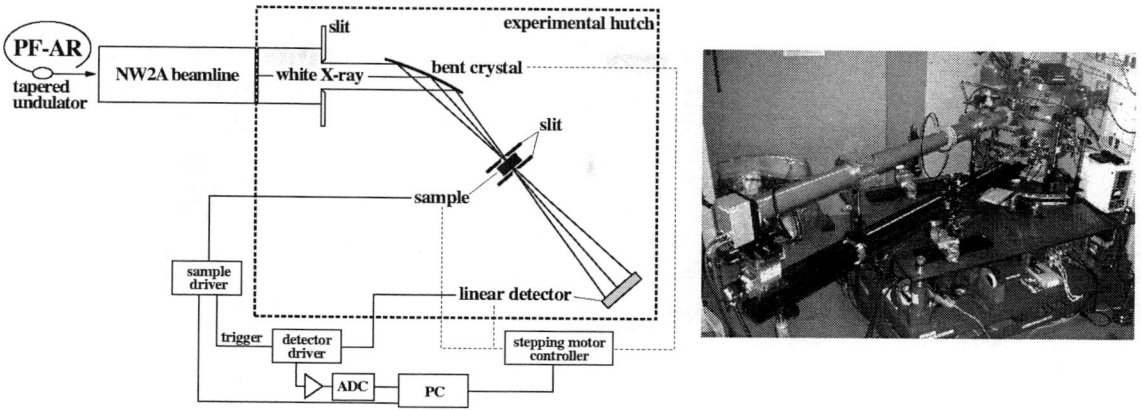

FIGURE 1. Schematic diagram and photograph of the DXAFS instrument installed at the NW2A beamline of PF-AR.

Bragg-type arrangement, while it is ca. 1000 mm for the Laue case. The circulation of water in the holder is important to achieve the good stability of the dispersed X-ray intensity (less than 1 %) and the energy (within one pixel of the linear detector).

The NMOS-type photodiode array (PDA) with 1024 sensing elements is used as the linear detector with and without a phosphor deposited on fiber plate to convert X-ray to visible light. The pixel size of PDA is 20 μm(W) × 2.5 mm(H) and the pitch is 25 μm. The signal output from PDA is digitized by a 14-bit unipolar ADC (525 kHz), and the minimum exposure time is thus 2 ms for one scan. The digitized data are stored in a fast memory module built in the detector driver, in which maximum 256 data sets can be recorded at one run of the measurement. The initiation of the PDA scans can be externally triggered by a TTL signal.

PERFORMANCE

The thickness of a sensing Si channel is 1 μm for the used PDA chip, the sensitivity to the X-rays is potentially low because of the low absorption. The conversion from X-rays to visible lights using a phosphor greatly improves the sensitivity and the use is known to prevent the damage of the PDA chip especially for X-rays with high energy. However, the existence of the phosphor screen reduces the energy resolution because of the scattering of the visible fluorescence. In Fig. 2, the X-ray absorption near edge structure (XANES) of Pt foil measured at the Pt-L_{III} edge using the DXAFS instrument with various phosphor is compared with that measured at the conventional XAFS beamline with an Si(311) double-crystal monochromator. It is clearly found that the energy resolution of the XANES

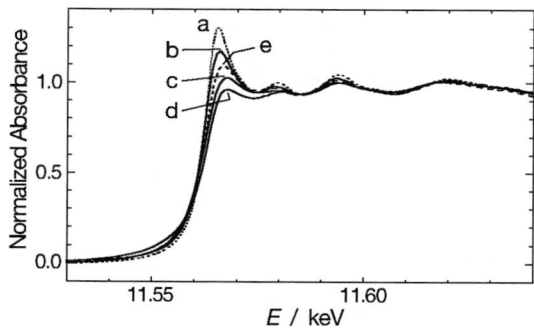

FIGURE 2. XANES spectra of Pt foil at the Pt-L_{III} edge measured using the conventional XAFS beamline with an Si(311) double-crystal monochromator (a) and the DXAFS instrument (b–e). The polychromator crystal and the phosphor were Si(111) and nothing (b), Si(111) and Gd_2O_2S(Tb) (c), Si(111) and CsI(Tl) (d), and Si(311) and Gd_2O_2S(Tb) (e). The exposure time for the DXAFS measurement was 60 ms (b), 500 ms (c), 20 ms (d), and 500 ms (e).

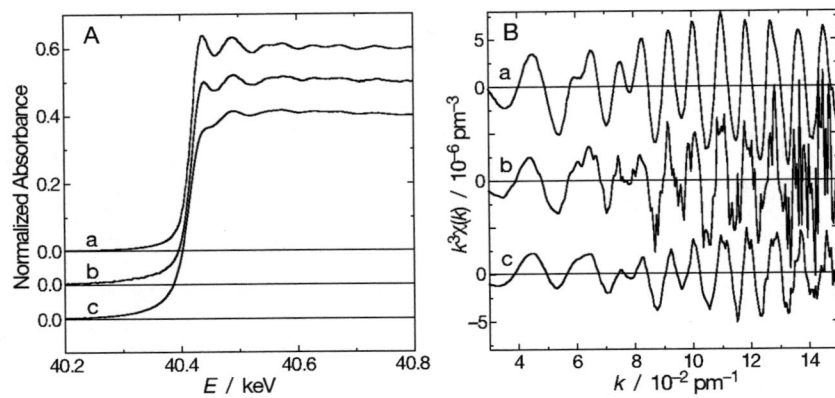

FIGURE 3. XANES spectra (A) and EXAFS oscillations (B) of CeO$_2$ powder at the Ce-K edge measured using the conventional XAFS beamline with an Si(311) double-crystal monochromator (a) and the DXAFS instrument using an Si(311) Laue-type polychromator without phosphor (b) and with phosphor screen of Gd$_2$O$_2$S(Tb) (c). The exposure time for the DXAFS measurement was 50 s (b) and 6 s (c).

spectrum obtained without the phosphor screen (Fig. 2(b)) is almost comparable with that measured at the conventional beamline, while the phosphor reduces the energy resolution. The Gd$_2$O$_2$S(Tb) phosphor (thickness: 20 µm) shows better resolution than CsI(Tl) (150 µm) due to the thinner thickness.

The XAFS measurements at the Ce-K edge were carried out for CeO$_2$ powder diluted with BN using the Laue-type Si(311) polychromator with the bending radius of 900 mm. The XANES spectra and the extracted oscillations in EXAFS region are compared in Fig. 3 with those obtained by the measurement using the conventional XAFS beamline. The results obtained without phosphor screen show good reliability of EXAFS and the Fourier transformation shows excellent agreement with that for the conventional measurement, while very long exposure is necessary to obtain such good statistic result because of the low sensitivity of Si photodiode for the high energy X-rays, although the total measurement time is still much shorter than that at the conventional beamline. The sensitivity is greatly improved by the use of phosphor but the debasement of the energy resolution is very serious especially for the XANES spectrum.

APPLICATION TO COPPER REDUCTION SUPPORTED ON ZEOLITE

The DXAFS instrument was applied to the mechanistic study of the reduction processes of Cu species supported on MFI zeolite, which is regarded to be a key reaction for the deNO$_x$ catalysis [9]. The Cu/MFI powder sample was prepared by the usual ion-exchange method using an aqueous solution of Cu(NO$_3$)$_2$ (pH 5.5), in which the MFI zeolite (HSZ-820NAA, Tosoh, Si/Al molar ratio = 11.9) was suspended at ambient temperature, and the collected Cu/MFI powder was dried at 393 K in air. The Cu loading was 5.1 wt%, which corresponds to 121 % of the conventional ion-exchange molar ratio defined by 2[Cu]/[Al] . The dried powder was calcined at 773 K in air for 1–2 h before all the measurements. It has been confirmed by the static XAFS measurements that the Cu species contained in the calcined powder is quantitatively oxydized as Cu(II) and the local structure parameters around the Cu center suggest that the CuO-like cluster is formed in the supercage of MFI zeolite.

The reduction processes of the Cu(II) species in Cu/MFI were followed after a rapid injection of the CO gas of a known pressure into the evacuated sample cell, in which the Cu/MFI powder was placed at 773 K. The time-resolved XAFS spectra were recorded with the exposure time of 6–10 ms and a suitable time interval (0.1–1.2 s) to follow the entire change from Cu(II) to Cu(0), and the time-resolved XANES spectra are shown in Fig. 4. The X-ray absorption spectrum apparently changes in two phases after injection of CO gas; the early change within 5 s (Fig. 4A) and the later change after 10–20 s (Fig. 4B). The value of $\ln(I_0/I)$ at 8.998 keV, which is the absorption peak energy of the initial Cu(II) species, largely decreases and the absorption threshold synchronously shifts to a lower energy during the early stage. This change is in accordance with the spectral difference between CuO and Cu$_2$O inserted in Fig. 4A taking into account the decreased energy resolution (ca. 5 eV) of the DXAFS instrument with the CsI(Tl) phosphor relative to that at the conventional step-scanning beamline. The similarity in the transient XANES spectra to Cu$_2$O suggests that the Cu(I) state exists as the intermediate species. In the later stage of the reduction, shown in Fig. 4B, a

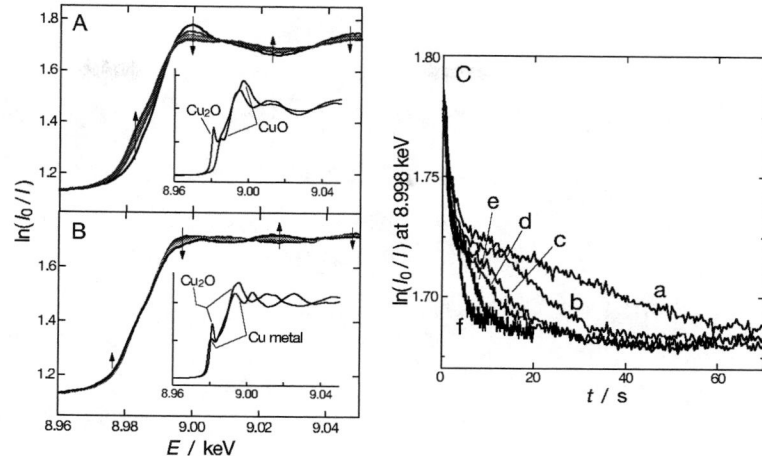

FIGURE 4. Time-resolved XANES spectra during the reduction of Cu/MFI by CO (3.4 kPa) at 773 K. The early (0–8 s) and later (15–40 s) stages are separately drawn in A and B, respectively. The acquisition time of each spectrum was 10 ms. In inset, are shown the XANES spectra of CuO, Cu_2O, and Cu foil recorded at the conventional XAFS beamline using the Si(311) channel-cut monochromator. The changes in X-ray absorbance at 8.998 keV are given in C for the reduction under the CO pressure of 1.8 (a), 3.4 (b), 5.0 (c), 5.3 (d), 6.5 (e), and 13.1 kPa (f) at 773 K.

further decrease in absorbance at ca. 8.995 keV and a slight increase at ca. 8.980 keV are then in agreement with the difference between Cu_2O and Cu metal. Figure 4C shows the changes in the $\ln(I_0/I)$ value at 8.998 keV under various pressure of CO at 773 K. It is found that the transient Cu(I) state is observed at around 5–10 s after the initiation of reaction and that the reaction rate is increased by the increase of the CO pressure. The finding that the $\ln(I_0/I)$ value shows a plateau when the Cu species is existed as Cu(I) strongly indicates at least two Cu(I) states should exist during the reduction process.

ACKNOWLEDGMENTS

This study was financially supported by the Grant-in-Aid for Scientific Research (Nos. 16205005 and 17034022) from the Ministry of Education, Science, Sports and Culture of Japan and was partially supported by the Industrial Technology Research Grant Program in 2005 from New Energy and Industrial Technology Development Organization (NEDO) of Japan. The XAFS measurements were performed under the approval of the Photon Factory Program Advisory Committee (Proposal Nos. 2005G230 and 2005G234).

REFERENCES

1. Y. Iwasawa, *J. Catal.* **216**, 165–177 (2003).
2. T. Matsushita and R. P. Phizackerley, *Jpn. J. Appl. Phys.* **20**, 2223–2228 (1981).
3. T. Mori, M. Nomura, M. Sato, H. Adachi, Y. Uchida, A. Toyoshima, S. Yamamoto, K. Tsuchiya, T. Shioya and H. Kawata, *AIP Conf. Proc.* **705**, 255–258 (2004).
4. A. Yamaguchi, T. Shido, Y. Inada, T. Kogure, K. Asakura, M. Nomura and Y. Iwasawa, *Bull. Chem. Soc. Jpn.* **74**, 801–808 (2001).
5. A. Yamaguchi, A. Suzuki, T. Shido, Y. Inada, K. Asakura, M. Nomura and Y. Iwasawa, *J. Phys. Chem. B* **106**, 2415–2422 (2002).
6. A. Suzuki, Y. Inada, A. Yamaguchi, T. Chihara, M. Yuasa, M. Nomura and Y. Iwasawa, *Angew. Chem. Int. Ed.* **42**, 4795–4799 (2003).
7. A. Suzuki, A. Yamaguchi, T. Chihara, Y. Inada, M. Yuasa, M. Abe, M. Nomura and Y. Iwasawa, *J. Phys. Chem. B* **108**, 5609–5616 (2004).
8. A. Suzuki, Y. Inada and M. Nomura, *Catal. Today* **111**, 343–348 (2006).
9. H. Yahiro, M. Iwamoto, *Appl. Catal. A* **222**, 163–181 (2001).

Energy Dispersive XAFS in the High Energy Region at BL14B1 in SPring-8

Yuka OKAJIMA[1,2], Daiju MATSUMURA[1], Yasuo NISHIHATA[1], Hiroyuki KONISHI[1], and Jun'ichiro MIZUKI[1]

[1]*Japan Atomic Energy Agency (JAEA), 1-1-1 Koto, Sayo-cho, Sayo-gun, Hyogo, 679-5148 JAPAN*
[2]*SPring-8 Service Co., Ltd. 2-23-1 Koto, Kamigori-cho, Ako-gun, Hyogo, 678-1205 JAPAN*

Abstract. An energy dispersive XAFS (DXAFS) system has been constructed in the second optics hutch at BL14B1 in SPring-8. The system is composed of dispersive optics with a position sensitive detector and is distinguished by covering energy region as high as 80 keV. Below 40 keV, two plane mirrors were utilized in optics to eliminate higher order harmonics and to focus X-rays vertically by mechanically-bent devices. The X-rays diffracted by a polychromator have an energy band width of 0.5 ~ 1.5 keV, and its spectrum is measured by a CCD camera. Time-resolved XANES spectra were detected near the La K-edge (38.934 keV) with the exposure time of 200 msec during reduction of $LaCoO_3$ at 600℃. The energy resolution of DXAFS spectra near the La K-edge is as good as the conventional XAFS spectrum, and changing the valence state of La was observed during the decomposition process clearly. The Pt K-edge (78.395 keV) DXAFS spectra for a Pt foil, was observed by summing 100 sets of spectrum whose each exposure time is 500 msec. An EXAFS oscillation was extracted up to wavenumber $k = 14$ Å^{-1}.

Keywords: Energy dispersive XAFS (DXAFS), Time-resolved, High energy region.
PACS: 07.85.Qe

INTRODUCTION

In the conventional XAFS measurement, measuring time of an XAFS spectrum is usually taken at least a few minutes because of a point by point moving monochromator. Even if the energy scanning is carried out with a channel cut monochromator crystal driven by a piezo module, minimum time to measure an XAFS spectrum was not shorter than 500 msec even for an energy range of 600 eV [1]. On the other hand, the energy dispersive XAFS (DXAFS) first proposed in 1981 is a technique for the measurement of time dependent XAFS spectrum [2], and has been developed at synchrotron radiation facility for the two decades [3, 4]. The DXAFS system is composed of dispersive optics with a position sensitive detector which is able to work under high flux conditions. In the dispersive optics, the typical time to measure is a few ten milliseconds because it is not necessary to scan a monochromator. Furthermore, it is possible to measure XAFS spectra accurately because of stable X-rays which are produced by the fixed optical elements. This time resolution makes possible to study dynamics of the local structure change during chemical reactions. Recently, tapered undulators are mostly used as the light source due to more intense light source and moderate spectral range for the DXAFS measurement as time-resolved techniques [3, 5]. However, the energy region which can be measure is limited up to about 30 keV in the case of using the undulators, since the wide and divergent incident X-rays produced by the over bent mirrors are essential for the polychromator to obtain the required energy band width. Accordingly, we have prepared the DXAFS system using a bending magnet source, which is applicable to the measurement of XAFS spectra mainly for the high energy region. This system makes possible to get enough intensity at the sample position by focusing the wide incident X-rays from the bending magnet source.

We report here the compositions of our DXAFS system and its performance mainly in the high energy region. First, we measured time-resolved XAFS spectra near the La K-edge of $LaCoO_3$ for chemical reaction under switching between oxidative and reductive conditions to grasp the performance of this system. Since $LaCoO_3$ is known as a fundamental catalyst and its crystal structure quickly follows atmosphere fluctuation, it is suitable for

CP879, *Synchrotron Radiation Instrumentation: Ninth International Conference*,
edited by Jae-Young Choi and Seungyu Rah

evaluating time-resolution of the measurement. Secondly, we measured XAFS spectra of a Pt foil near the Pt K-edge for the evaluation of performance at higher energies.

OPTICS

The DXAFS system has been constructed in the second optics hutch at JAEA beamline BL14B1, SPring-8. It is distinguished by covering energy region as high as 80 keV. Two water cooled plane mirrors (1000mm in length) were equipped at the first optics hutch at BL14B1 to eliminate higher order harmonics and to make a focus vertically by mechanically-bent devices. In contrast, the mirrors are removed from optical axis over 40 keV. Figure 1 gives the photograph of this system at the second optics hutch. Wide white X-rays (60mm max) from the bending magnet are diffracted by a bent single crystal as a polychromator, and focused horizontally at the sample. The diffracted X-rays have energy band width of 0.5 ~ 1.5 keV. The X-rays penetrate a sample, and the intensities are measured by a position sensitive detector. The polychromator, sample and the position sensitive detector are mounted on the tables, and are arranged in θ-2θposition. The range of 2θ angle is from 0 to 60 degrees. The 2θ rails describe a circle with the radius of 1000mm and 2000mm around the polychromator, and hence the maximum distance about 3000mm is secured between the polychromator and the position sensitive detector. The height of the tables changes as a function of the glancing angle of mirrors. When we use Si(422) for diffraction plane of the polychromator in a transmission geometry (Laue configuration), X-rays are available between 10 and 80 keV. We estimated that the flux near 25 keV was 4×10^9 ph/s (1mrad×50μrad, 0.1% B.W.) at the sample position. The size of the polychromator crystals is 110(w)×25(h)×0.2(t) mm^3. Two types of bent crystal can be chosen according to the energy range of X-rays: radius of curvature, R = 1000mm and 2000mm. The focus size was measured to be less than 0.2mm (FWHM) at the sample position. On the other hand, in the energy region below 10 keV, the diffraction plane of the polychromator is selected to be Si(111) in reflection geometry (Bragg configuration). This crystal should be curved with a parabolic line by a bender in the development stage. The CCD camera (Hamamatsu Photonics, C9300-201) is used as a position sensitive detector. The X-rays are transformed into the visible light on fluorescence screen (20μm thick) consists of P43 (Gd_2O_2S). An XAFS spectrum image is formed through two optical lenses whose focal lengths are both 50mm in the CCD camera system. The number of pixels of this CCD are 640(h)×480(v) and the size of each pixel is 7.4×7.4 μm^2. The typical exposure time was 100 μsec ~ 1 sec in this study.

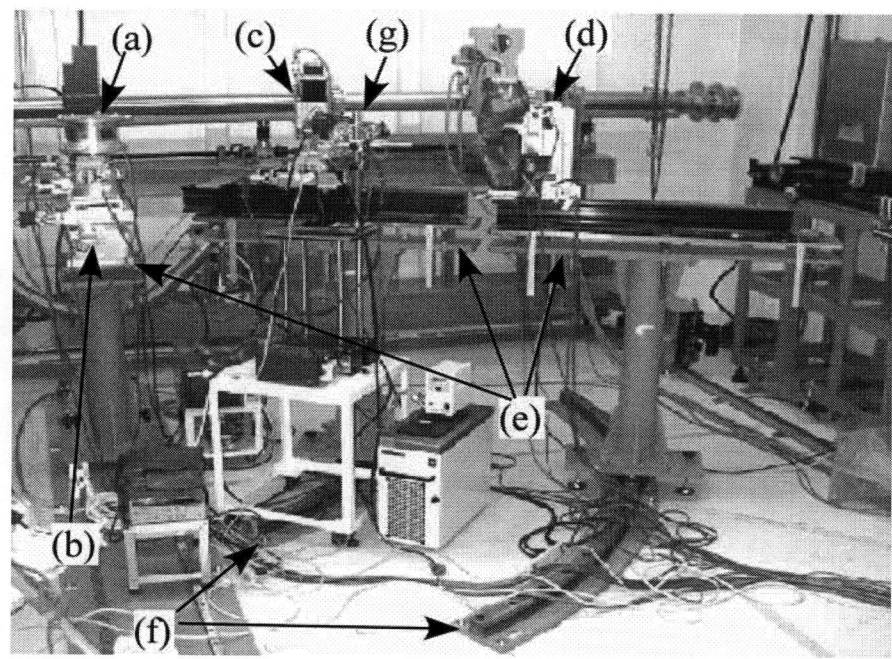

FIGURE 1. A photograph of the DXAFS system at the second optics hutch at BL14B1, SPring-8. The white X-rays are injected from the left side. (a) polychromator chamber, (b) polychromator stage, (c) sample stage, (d) detector stage, (e) tables which can be aligned with the height of incident X-rays, (f) 2θ rails, (g) infrared furnace for heating $LaCoO_3$.

EXPERIMENTAL

Time-Resolved DXAFS Measurement near the La K-edge of LaCoO$_3$

The DXAFS spectrum measured in 200 msec is compared with the conventional XAFS spectrum using a Si(311) double crystal monochromator near the La K-edge for LaCoO$_3$. Figure 2(a) shows the DXAFS spectrum and the conventional XAFS spectrum at room temperature. The energy resolution of the DXAFS spectrum is as good as that measured by the conventional one. An EXAFS function $\chi(k)$ was extracted from this DXAFS spectrum well up to the wavenumber $k = 12$ Å^{-1} (Fig. 2(b)). Fourier transform of the EXAFS function for the DXAFS spectrum also agrees well with that of conventional one (Fig. 2(c)).

Decomposition process of LaCoO$_3$ to La(OH)$_3$ was observed under reductive conditions. The successive phase transformation is known to depend on oxygen pressure. A pellet of LaCoO$_3$ was encapsulated in a glass tube, and was heated to 600°C by an infrared radiant heater. The atmosphere was switched from oxidative gas (O$_2$ 50%, balanced with He) to reductive gas (H$_2$ 50%, balanced with He) with a flow rate of 100 ml/min. The time-resolved measurement in 200 msec was started at the same time of switching gas flow. As shown in Fig. 2(d), XANES spectrum obviously changed at about 20 seconds after switching gas flow, suggesting that the valence state of La was maintained at the initial stage of the decomposition process. We are interested in comparing the local structure change of Co with that of La.

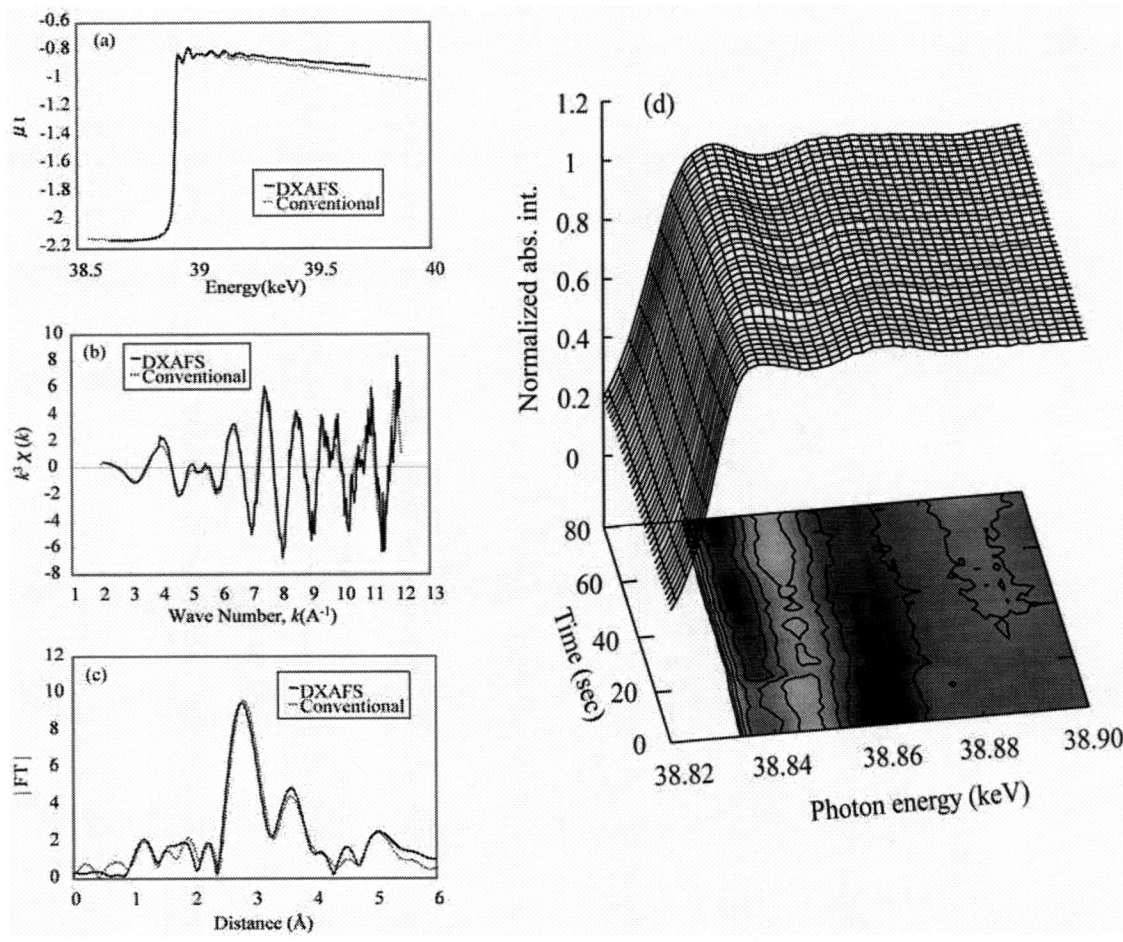

FIGURE 2. (a) DXAFS spectrum and conventional XAFS spectrum at room temperature near the La K-edge for LaCoO$_3$, (b) k^3-weighted EXAFS functions, (c) Fourier transform of the EXAFS functions, (d) time-resolved XANES measured in 200 msec during reduction of LaCoO$_3$ at 600°C.

DXAFS Measurement near the Pt *K*-edge

Measurements of the Pt *K*-edge DXAFS spectra for a Pt foil were carried out at room temperature. We observed EXAFS oscillation by summing 100 sets of DXAFS data each taken with a 500 msec exposure time. Figure 3(a) shows the DXAFS spectrum and the conventional XAFS spectrum using a Si(511) double-crystal monochromator [6]. Figure 3(b) compares the EXAFS functions between $k = 6$ and 14 Å^{-1}. Since the amplitude of EXAFS oscillation for the DXAFS spectrum is a little smaller than that of conventional one, the magnitude of Fourier transformation for the DXAFS spectrum is smaller as shown in Fig. 3(c). These disagreements could be caused by an insufficient energy resolution of the DXAFS spectra. The following refinements regarding the energy resolution in the present DXAFS system are required: bent polychromator, a fluorescence material of the CCD camera which does not contain Gd. Geometrical condition like distance between sample and detector will be also optimized in consideration of energy range and energy resolution.

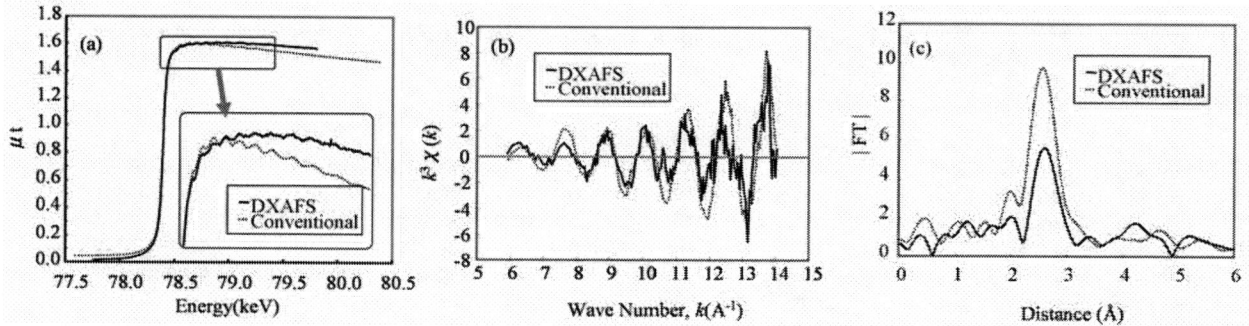

FIGURE 3. (a) DXAFS spectrum and conventional XAFS spectrum near the Pt *K*-edge for a Pt foil, (b) k^3-weighted EXAFS functions, (c) Fourier transform of the EXAFS functions.

SUMMARY

The DXAFS system mainly dedicated to the high energy region has been constructed by making use of the wide and divergent incident X-ray from a bending magnet source. We observed DXAFS spectrum near the La *K*-edge with good energy resolution. Time-resolved XANES spectrum of $LaCoO_3$ during the reduction at 600℃ was detected with the time resolution of 200msec. The change of the obtained spectra suggests that the local structure around La changes suddenly during the decomposition process. An EXAFS oscillation was extracted from the Pt *K*-edge DXAFS spectrum between $k = 6$ and 14 Å^{-1}, however, the energy resolution should be improved by further refinement on the system for higher energies. We are planning to study chemical reaction and structural response of materials to external field, including heavy metals such as lanthanoides, etc.

ACKNOWLEDGMENTS

We would like to thank Dr. T. Uruga, Dr. H. Tanida, Mr. K. Kato for significant discussions, and also Dr. H. Tanaka, Dr. M. Uenishi and Mr. M. Taniguchi for their advice regarding the time-resolved measurement. The powder of $LaCoO_3$ was kindly offered by Hokko Chemical Industry Co., Ltd.

REFERENCES

1. M. Richwin, R. Zaeper, D. Lützenkirchen-Hecht, and R. Frahm, *Rev. Sci. Instrum.* 73, 1668-1670 (2002).
2. U. Kaminaga, T. Matsushita and K. Kohra, *Jpn. J. Appl. Phys. Suppl.* 20, 5, 355-358 (1981).
3. M. Hagelstein, A. Fontaine, and J. Goulon, *Jpn. J. Appl. Phys. Suppl.* 32, 32-2, 240-242 (1993).
4. A. Suzuki, Y. Inada, A. Yamaguchi, T. Chihara, M. Yuasa, M. Nomura, and Y. Iwasawa, *Angew. Chem. Int. Ed.* 115, 39, 4943-4947 (2003).
5. S. Yamamoto, K. Tsuchiya, and T. Shioya, *AIP Conf. Proc.*, 235-238 (2004)
6. Y. Nishihata, J. Mizuki, S. Emura, and T. Uruga, *J. Synchrotron.Rad.* 8, 294-296 (2001)

X-ray Pinpoint Structural Measurement for Nanomaterials and Devices at BL40XU of the SPring-8

Shigeru Kimura, Yutaka Moritomo, Yoshihito Tanaka, Hitoshi Tanaka, Koshiro Toriumi, Kenichi Kato, Nobuhiro Yasuda, Yoshimitsu Fukuyama, Jungeun Kim, Haruno Murayama and Masaki Takata

JASRI/SPring-8, Kouto, Sayo-cho, Sayo-gun, Hyogo 679-5198, Japan & CREST, Japan Science and Technology Agency, Honcho, Kawaguchi, Saitama 332-0012, Japan

Abstract. The pulse characteristic and high coherent x-ray beam of SPring-8 allow us to investigate dynamics of chemical reactions and phase transition of materials caused by applied field. In order to realize such direct investigation, "x-ray pinpoint structural measurement", which is the advanced x-ray measurement technique in nanometer spatial scale and/or pico-second time scale, is being developed at SPring-8. The features of "x-ray pinpoint structural measurement" technique are, 1) spatial resolution: ~ 100 nm, 2) time resolution: ~ 40 ps, and 3) measurement under the photo-irradiation, electric field, magnetic field, high pressure and active devices. Using this technique, we will explore the novel concept and new phenomena for nanomaterials and/or devices, and also demonstrate their validity.

Keywords: pinpoint, nanomaterial, device, synchrotron radiation, femtsecond laser, pump and probe, microdiffraction
PACS: 61.10.Nz, 61.46.-w, 68.37.Yz,

INTRODUCTION

The pulse characteristic and high coherent x-ray beam of SPring-8 allow us to investigate dynamics of chemical reactions and phase transition of materials caused by applied field. In order to realize the direct investigation, "x-ray pinpoint structural measurement", which is the advanced x-ray measurement technique in nanometer spatial scale and/or picosecond time scale, is being developed at the SPring-8.

The scientific targets are "atomic visualization of amorphous-crystal phase change on recording DVD media", "Charge density level visualization of photo-induced phase transition", "Sub-micron single crystal structure analysis", and "Electrode reaction in submicron diffusion zone liquid structure". Nondestructive characterization of running devices in nano-space, pico-second (Fe-RAM etc.) will be planed as an industrial application.

The technical targets in the first stage are set to 100-nm spatial resolution and 40-ps time resolution measurement under photo-irradiation, electric field, magnetic field and high pressure. A new experimental hutch has been constructed at an undulator beam line, BL40XU, to get high brilliance x-ray. The measurement system are composed of an x-ray pulse selector (XPS), a Si(111) channel-cut monochromator, a high-precision microdiffractometer with a CCD detector and an Imaging Plate (IP) detector, a sample cooling system (90 - 300K), and femtosecond Ti-sapphire laser system, which is synchronized to the x-ray pulses.

The main technical problems to be solved are "precise synchronization between timing of both x-ray pulse and applied field on sample", "highly brilliant x-ray pulse generation and proper bunch design" "submicron accuracy diffractometry" and "high brilliance micro-focus technique". The commissioning of the key experimental performances is now going on with a test experiment of photo irradiation time-resolved pump-probe experiment in cooperation with a study on an optimal beam filling pattern by an accelerator group.

"X-ray pinpoint structural measurement" is the part of the CREST projects (2004-2008) funded by Japan Science and Technology Agency (JST).

CP879, *Synchrotron Radiation Instrumentation: Ninth International Conference,*
edited by Jae-Young Choi and Seungyu Rah
© 2007 American Institute of Physics 978-0-7354-0373-4/07/$23.00

RESEACH SUBJECTS AND TIME SCHEDULE

To develop the "x-ray pinpoint structural measurement" technique, five research teams are formed. Research subjects of each team are as follows.

1) Total System Design & Application Team

This team has been constructing the x-ray pinpoint structural measurement system to combine the x-ray microprobe measurement technique, the time-resolved measurement technique, the applied field structural measurement technique, and the sub-micron crystal structure analysis technique. Furthermore, this team will apply the "x-ray pinpoint structural measurement" technique to the nanomaterials and/or nanodevices, and show the validity for industry.

2) Time-Resolved Measurement Team

This team has been developing pico-second time-resolved x-ray measurement technique using a femtosecond pulse laser system. Precise timing control technique of intense femtosecond laser pulses to synchronize synchrotron radiation pulses will be developed.

3) Applied Field Structural Study Team

This team has been developing the synchronization technique of x-ray pulse and many kinds of applied fields in collaboration with Time-Resolved Measurement Team. They will develop technique to measure active devices under photo-irradiation, electric field, magnetic field, and high pressure.

4) Sub-micron Crystal Diffraction Team

This team has been developing structure analysis technique of sub-micron sized single crystal. Decrease of background level, new measurement technique with sub-μm beam, and analysis software will be developed in collaboration with X-ray Microbeam Team.

5) X-ray Microbeam Team

This team has been developing an x-ray focusing system to produce sub-100-nm x-ray beam. They will also develop a high-precision diffractometer system equipped with both the CCD detector and the IP detector to use the sub-100-nm beam. This system will be used for the x-ray pinpoint structural measurement

Time schedule of the research subjects is shown in Table. 1

TABLE 1. Time schedule of the research subjects for the x-ray pinpoint structural measurement.

FY	Total System Design & Application Team	Time-Resolved Measurement Team	Applied Field Structural Study Team	Sub-micron Crystal Diffraction Team	X-ray Microbeam Team
2004	• Pilot study	• Synchronization of x-ray and laser	• Pilot study	• Pilot study	• Pilot study
		Construction of BL40XU experimental hutch 2 Design of x-ray pinpoint structural measurement			
2005-2006	• Analysis of liquid & amorphous	• Install of femto-second laser	• Applied study by combination of sub-micro crystal diffraction and applied field structure measurement		• Development of high-precision μ-diffractometer • Structure measurements of active devices
		Applied study by combination of time-resolved measurement and applied field structure measurement		Establish of nm-size structure analysis	
		• High-intense pulse x-ray source	• Measurement of single-shot phenomena	• Applied study by x-ray microbeam and sub-micro crystal diffraction	
2007	Integrate each technique and experiment				
2008	Explore the novel concept and/or phenomena				

PRESENT STATUS OF SYSTEM DEVELOPMENT

Construction of an New Experimental Hutch and a Laser Booth

A new experimental hutch (EH2) was constructed at downstream of BL40XU experimental hutch 1 in March 2005. The x-ray source of BL40XU [1] is a helical undulator, whose period length is 36 mm and the number of period is 125. Helical undulators were originally used to generate circularly polarized x-rays. One of the specific features of this type of undulator is that the energy of the fundamental radiation is concentrated in the core of the radiation. On the other hand, most of the higher harmonics are emitted off-axis. So, by extracting the central part of the radiation, the fundamental radiation with narrow peak-energy-width, which is treated as a quasi-monochromatic x-ray, can be used without loss of its flux. The undulator gap can be varied so that the fundamental radiation is altered from 8 to 17 keV. The dimensions of the experimental hutch 2 are 5 m long × 4 m wide × 3.3 m high (Fig. 1). At present, the XPS, the Si(111) channel-cut monochromator, the high-precision diffractometer system equipped with both the CCD detector and the IP detector, and the sample cooling system (90 - 300K), are installed in EH2. Also, a femto/picosecond laser system was installed in a laser booth, which was built at the downstream of EH2.

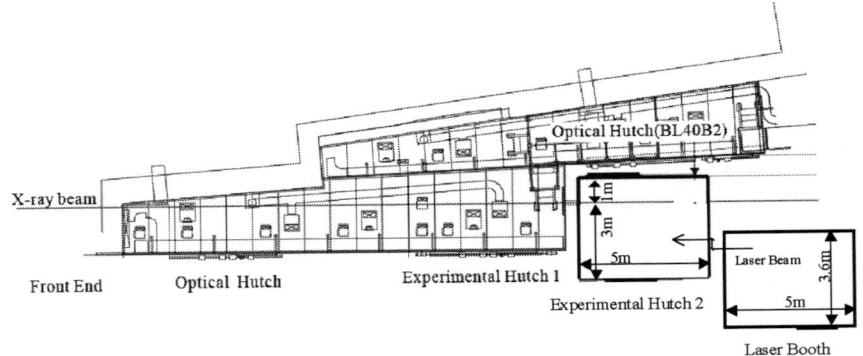

FIGURE 1. Layout of BL40XU experimental hutch 2 and the laser booth.

Femto/picosecond Laser System

Figure 2 shows a photograph and the schematic layout of the femto/picosecond laser system. The laser system consists chiefly of a mode-locked Ti:sapphire laser, a regenerative amplifier, a pulse selector, and an optical parametric amplifier. We have developed a control system to synchronize femtosecond laser pulses to the x-ray synchrotron radiation pulses of SPring-8 [2]. The mode-locked Ti:sapphire laser produces the pulses with a duration of 80 fs, typical energy of 10 nJ/pulse at the wavelength of 800 nm. The output timing of the Ti:sapphire oscillator is synchronized to 84.76 MHz which is 1/6 of a master oscillator of the RF system for storage ring acceleration (508.58 MHz). The synchronization and the control of time delay are key techniques for pinpoint pump-probe (laser + synchrotron radiation) diffraction measurements. The laser pulse width of 80 ps is also available by installing optional optical components in the laser oscillator. Safety interlock system has been completed for laser and synchrotron radiation. The timing-controlled pulse laser system will be applied to studies on fast amorphous-crystal phase change mechanism of DVD media.

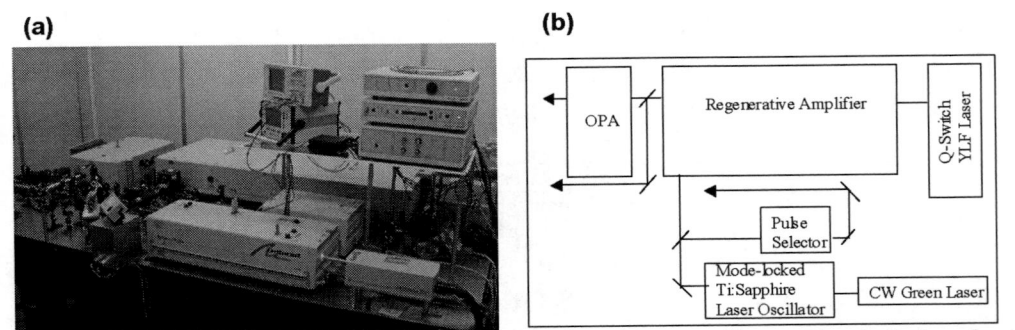

FIGURE 2. (a) Photograph and (b) schematic layout of the femto/picosecond laser system in the laser booth.

High-Precision Diffractometer System

A high-precision diffractometer system (Fig 3) has been developed to use the sub-100-nm x-ray beam. The system mainly consists of a zone plate alignment system, a high-precision goniometer system, an IP detector and a

CCD detector system. The high-precision goniometer system has θ-2θ rotation stages, and 100-nm-resolved XYZ sample positioning stages. Especially, the θ stage uses a high precision air-bearing stage (Canon AB-100R), which realizes very low eccentricity within ±100 nm/360°. We adopt a Rigaku Saturn 70 CCD detector, which is the most sensitive small molecule CCD detector. By using that CCD camera, sub-micron sized single crystal structure can be analyzed. The diffractometer uses a focused beam produced by using a zone plate (ZP). One of two ZPs (ZP1 or ZP2), which were fabricated by NTT-AT [3], can be used for the proper purpose of experiments. The ZP1 has the radius of innermost zone of 5.0 μm, the outermost zone width of 250 nm and the tantalum thickness of 2.5 μm. The thickness was designed to have the highest diffraction efficiency at 15 keV (~26%). The ZP2 has those of 3.0 μm, 75 nm and 750 nm, of which ideal diffraction efficiency at 8 keV is 12%, respectively. Because of thicker tantalum zones, the ZP1 is used mainly for experiments preferring higher energy such as structure analysis for sub-micron sized single crystal. The ZP2 is used for experiments requiring a high spatial resolution.

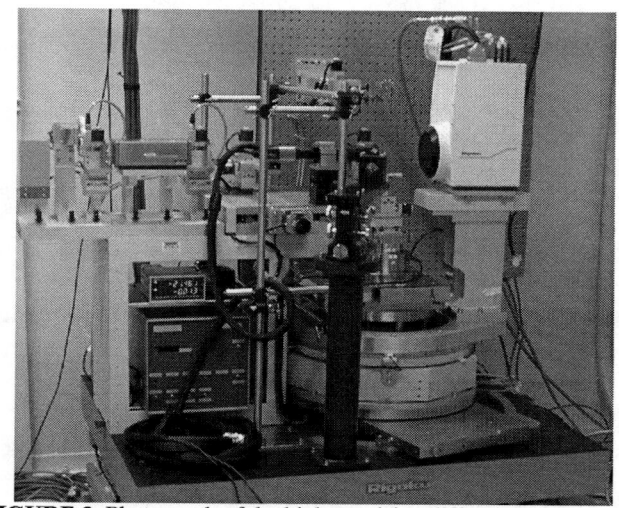

FIGURE 3. Photograph of the high-precision diffractometer system.

SUMMARY

We have been developing the x-ray pinpoint structural measurement technique to investigate the dynamics of chemical reactions and phase transitions of materials caused by applied field. Using this technique, we will explore the novel concept and/or new phenomena for the nanomaterials and/or devices in the near future. We believe the x-ray pinpoint structural measurement shall transfer the role of synchrotron radiation from "to see structure-property relationship of materials" to "to see a function of materials".

ACKNOWLEDGMENTS

We thank all members of the x-ray pinpoint structure measurement group for helpful discussions. This research was conducted at the BL40XU of the SPring-8 with the approval of the Japan Synchrotron Radiation Research Institute (Proposal Nos. 2005A0885-S3-np, 2005B0930 and 2006A1746).

REFERENCES

1. K. Inoue, *Nucl. Instrum. & Methods A*, **467-468**, 674-677 (2001).
2. Y. Tanaka, T. Hata, H. Kitamura and T. Ishikawa, *Rev. Sci. Instrum.*, **71**, 1268-1274 (2000).
3. A. Ozawa, T. Tamamura, T. Ishii, H. Yoshihara and Y. Kagoshima, *Microelectron. Eng.*, **35**, 525-529 (1997).

Opportunities for Time Resolved Studies at the ID24 Energy Dispersive XAS Beamline of the ESRF

O. Mathon, G. Aquilanti, G. Guilera, J.-C. Labiche, P. van der Linden, M.A. Newton, C. Ponchut, A. Trapananti and S. Pascarelli

European Synchrotron Radiation Facility, B.P. 220, F-38043 Grenoble Cedex, France.

Abstract. ID24 is the energy dispersive beamline of the European Synchrotron Radiation Facility dedicated to X-ray Absorption Spectroscopy (XAS). Thanks to the parallel acquisition mode that allows data in a large energy range to be collected simultaneously, XAS using dispersive optics is particularly suited for the study of time dependent processes. The techniques that can be used to study such systems vary according to the timescale of the phenomena under investigation. They take advantage of the temporal structure of the synchrotron radiation in case of time resolution of the order of the intrinsic duration of the x-ray pulse (100 ps), while for time scales above 100 µs or below 100 ps, the x-ray beam can be considered continuous and the time resolution is determined by the different detection systems

Keywords: Dispersive beamline, XAS, time resolved, FReLoN camera, extreme condition.
PACS: 87.64.Fb

DOWN TO THE MS TIMESCALE FOR SINGLE SHOT EXPERIMENTS

Implementation of the Fast Readout Low Noise (FReLoN) CCD Camera

The standard detection scheme of ID24 is based on a CCD detector coupled to the beam through a fluorescent screen and a series of optical components (lenses and a motorized iris) [1]. On a dispersive setup, a quasi one dimensional picture represents the absorption spectra, thus only few lines of the CCD chip are exposed and need to be readout. Up to now, a commercial EG&G Princeton CCD camera was used. In this system 64 (or 32) lines of the camera are exposed. The remaining part of CCD chip is shielded and is used as fast analogical buffer: after exposure of the 64 lines, the image is shifted into the shielded part of the CCD. In the case of a 32 lines image, this analogical shift takes only 77 µs (32x2.4 µs). This operation can be repeated until the CCD chip is filled with successive images (maximum of 32 images = 1024 pixels / 32 lines). The much slower complete readout of the camera starts. It takes about 200 ms and during this period the camera is blind to the evolution of the sample. This is a considerable limitation for time resolved experiments and in particular for non-reversible phenomena like most heterogeneous catalyst process. A second practical limitation of this configuration is the difficulty to pilot the camera and in particular to control and synchronize precisely the timing of the experiment.

To overcome these difficulties we have implemented on ID24 the version 2000 of the Fast Readout Low Noise (FReLoN 2k) CCD camera. The FReLoN detector is a versatile CCD camera developed at the ESRF. Main characteristics combine a low noise, a high dynamic range, high readout speed and an improved duty cycle. Technical details on the camera can be found elsewhere [2]. For time resolved applications, ID24 is particularly interested in the possibilities offered by the FReLoN camera and its control system : (1) to readout quickly only the part of the chip exposed to X-rays. This leads in our configuration to a dead time (shifting time + readout time) as small as 0.8 ms. (2) to repeat this cycle with a quasi infinite number of frames. This mode is called "kinetic pipeline mode". (3) to control very precisely the timing of the camera (input/output of the exposure time, readout time, and eventually shutter closing time through the rear panel of the camera). The camera can operate either in slave or master mode.

CP879, *Synchrotron Radiation Instrumentation: Ninth International Conference,*
edited by Jae-Young Choi and Seungyu Rah

Figure 1 presents an example of a kinetic study obtained with the FReLoN 2k camera during the interaction of NO with reduced 5wt%Rh/Al$_2$O$_3$ powder samples with NO at 573K, with a repetition rate of 62 ms [3]. Full time resolution (exposure + shifting + readout times) of 2 ms are now routinely reached. Down to the ms is possible by compromising on the dynamical range (14 bits instead of the present 16 bits camera) and on the number of exposed lines (for example 32 instead the present 128 lines).

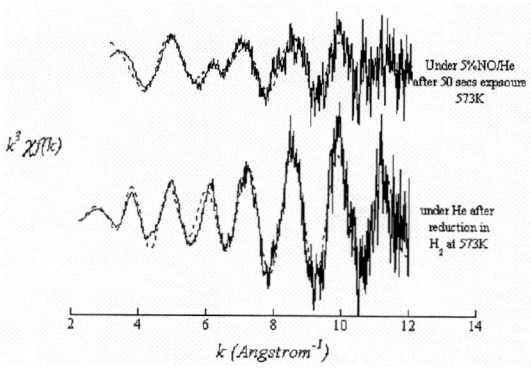
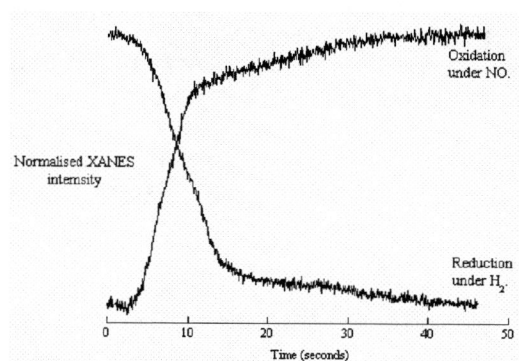

FIGURE 1. Left panel : k^3 weighted Rh K edge XAS spectra from 5wt% Rh/Al$_2$O$_3$ samples in a reduced state and after 50 seconds exposure to 5%NO/He at 573K. Each spectrum is accumulated in 62 msecs. Right panel : kinetic variation in XANES structure observed during oxidation of Rh nanoparticles by 5%NO/He at 573K derived from 750 spectra acquired sequentially in 62 msecs.

Turbo-XAS

In the case of dilute systems or samples not suited for transmission measurements, the ID24 dispersive setup offers the possibility of measurements in fluorescence mode. This mode, called Turbo XAS [4], consists in scanning a thin vertical slit across the polychromatic fan thereby selecting a monochromatic beam. The XAS spectrum is thus recorded by scanning continuously the slit through the polychromatic beam and by recording independently the I$_0$ and the fluorescent yield with standard diodes. Despite the loss of the parallel acquisition, a dispersive setup in Turbo-XAS mode takes advantage of the high flux, of the small focal spot, of the stability and of the relatively fast acquisition time. XANES spectra of about 400 eV at the Pt L$_3$ edge have been recorded with time resolution of about 300 ms on a 1%wt Pt diluted sample [5]. Compared to standard scanning monochromator based beamlines, it takes advantage of the fact that only one slit is scanned instead of two parallel crystals.

PUMP AND PROBE AND PSEUDO TIME RESOLVED EXPERIMENTS

Gated Detector

When reversible phenomena are studied, a stroboscopic data collection method can be applied at a synchrotron taking advantage of the time structure of the beam. In pump-and-probe experiments, the time measurement is given by the delay between the pump (for example a laser or an electric field) and the single X-ray bunch (the probe). The intrinsic time resolution is thus given either by the length of the pump or of the X-ray pulse (typically 100 ps). For such an XAS experiment with a dispersive setup using a CCD camera, the main difficulty is to select the single X-ray bunch that has interacted with the sample with the wanted time delay. This can be done either by using a mechanical X-ray chopper (see for example [6]) or by selecting directly the information inside the counting chain via electronic gating.

With the detection scheme of ID24 based on a CCD camera, we have developed an intermediate solution based on a gating at the level of the visible photons. The unit is inserted between the fluorescent screen and the FReLoN CCD camera. The X-ray converter is a 41 µm thick P47 phosphor screen with decay times of 100 ns for 10% afterglow and 3 µs for 1% afterglow. The gated unit is a 25 mm diameter multichannel plate (MCP) image intensifier with 100 ns minimum gating time from Proxitronic. The fast gating response is achieved by controlling the photocathode potential. The image intensifier is coupled to the phosphor screen and to the CCD camera by high aperture tandem lens systems. The resulting input field width is 51 mm, with 25 µm pixel pitch. The effective spatial resolution (line-spread function FWHM) is of the order of 200 µm, mainly limited by the image intensifier. The internal signal amplification in the MCP provides high sensitivity for efficient detection of short X-ray pulses.

To test the gating performance we have measured the intensity of a continuous visible source as a function of the gating time window. This test has demonstrated good linearity of the MCP gating with windows down to 100 ns. This is sufficient for stroboscopic experiments in 4 bunch mode where individual X-ray bunches are separated by 0.7 µs. The timing capabilities for X-ray experiments of the whole system (fluorescent screen + MCP + camera) has been tested in 4 bunch mode by measuring the integrated intensity on the CCD camera with a gating of 100 ns as a function of the delay between the X-ray pulse arrival and the time window position (stroboscopic technique). The gating window was locked with the arrival of X-ray pulse through the ESRF synchronous RF signal. Figure 2 shows that individual X-ray pulses can be resolved.

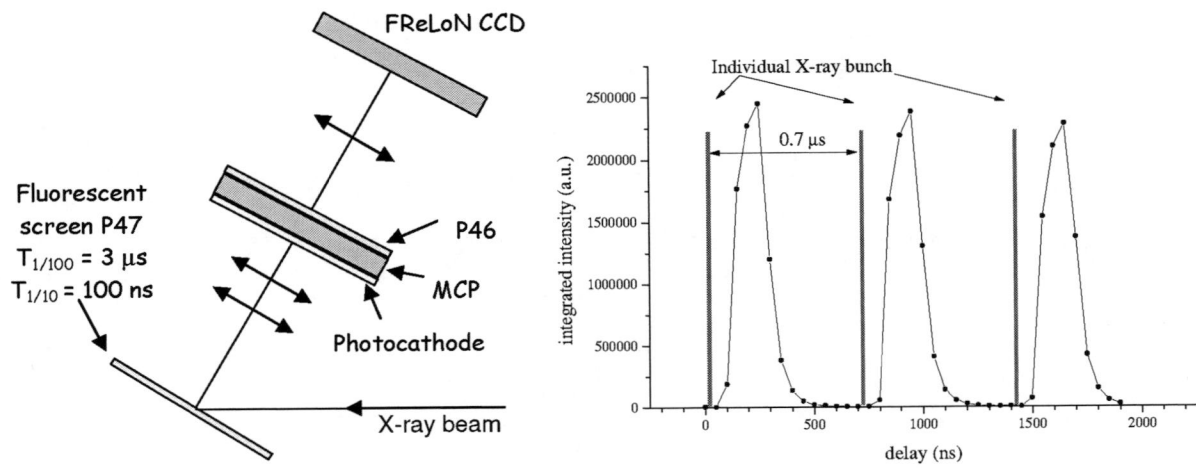

FIGURE 2. Left panel : principle of a gated detector for a dispersive setup. A Multi Channel Plate (MCP) is inserted between the fluorescent screen and the FReLoN CCD camera. Right panel : integrated intensity on the CCD camera as a function of the delay between the arrival of the single X-ray pulse and time position of the gating window. The red line symbolizes the arrival of the individual X-ray pulse. Individual X-ray pulses can be resolved. The FWHM of each pulse is linked to the decay time of the entrance P47 fluorescent screen.

Figure 3 presents Fe-k edge XANES spectra recorded with 10, 50, 500 and 5000 individual X-ray pulses. The individual X-ray pulse signal was obtained by applying a gating window of 0.7 µs. Exploitable signal to noise ratio is obtained already with only 500 pulses and good signal to noise ratio is obtained for 5000 pulses. With the implementation of a fast gating system based on a multi channel plate, it is possible to detect a single X-ray pulse in 4 bunch and 16 bunch modes. The pump and probe technique can be then applied and a time resolution of 100 ps can be achieved. We demonstrate that XAS spectra with reasonable signal-to-noise ratio can be recorded in 500 cycles. This corresponds, for a 1 Hz pump frequency, to a 8 minutes experiment per timing point.

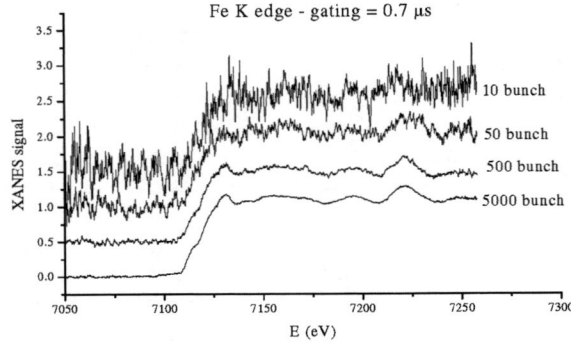

FIGURE 3. Fe-k edge XANES spectra recorded with 10, 50, 500 and 5000 individual X-ray pulse on a Fe foil.

Pseudo Time Resolved Experiment

Pseudo time resolved experiments are carried out using the same stroboscopic technique than pump and probe experiments. The goal is not to study the time evolution of the sample after an excitation but to study the sample under a quasi static excited state that can be obtained only during a short period. This technique can be applied to study systems under extreme conditions such as high pulsed magnetic fields or high pressure shock waves.

The feasibility of this experiment has been demonstrated by recording an X-ray Magnetic Circular Dichroïsm (XMCD) spectrum at the Gd L_3 edge on $GdCo_3$. XMCD is the difference between X-ray absorption spectra obtained with right and left circular polarization. For a finite XMCD signal to be measured the sample must present a net ferromagnetic or ferrimagnetic moment. For symmetry reasons, it is equivalent to record the signal by flipping the magnetic field instead of the circular polarization. We used a 2 ms pulsed magnetic field of 5.5 Tesla produced by the discharge of a 4 kJ (3000V, 1mF) commercially available power supply inside a Cu water cooled coil. The signal was integrated during a 150 μs time window located on the top of the magnetic field pulse using the gating system described in the previous paragraph. A net XMCD signal can be measured for both left and right circular polarization demonstrating the feasibility of this kind of experiment (fig. 4). The goal now is to reach 40 Tesla with the possibility to cool the sample down to 5 Kelvin. For more details on the setup, please report to [7].

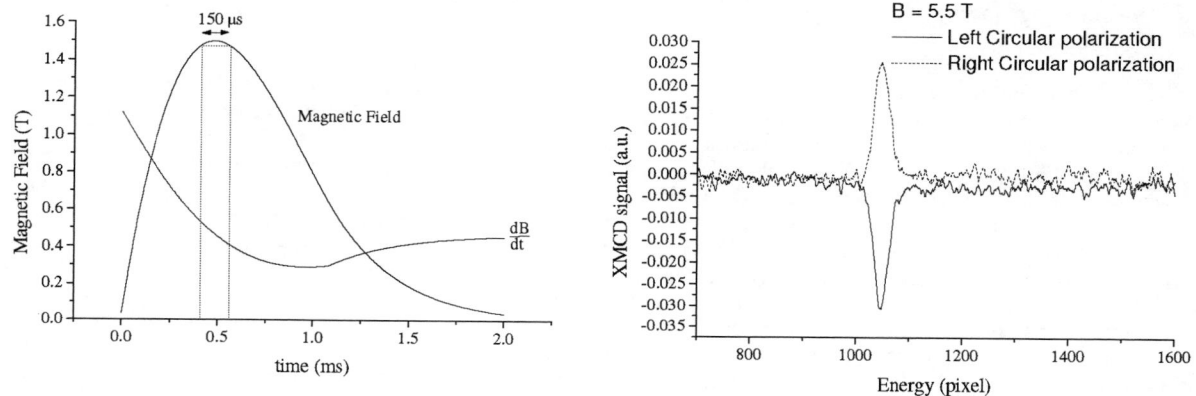

FIGURE 4. Left panel : a typical magnetic pulse. The signal is integrated only during the 150 μs where the field is at its maximum. Right panel : a net XMCD signal with opposite sign for left and right circular polarization is observed demonstrating the feasibility of this experiment

ACKNOWLEDGMENTS

We are very grateful to the ID24 beamline technical staff for the constant and excellent support. In particular we are very grateful to Sébastien Pasternak, Florian Perin, Marie-Christine Dominguez and Trevor Mairs. Rh K edge data was taken on ID 24 under an EPSRC grant (GR/60744/01) to Prof. J Evans (University of Southampton)and Dr A. J. Dent (Diamond Light Source) and a long term proposal allocation from the ESRF.

REFERENCES

1. Koch, M. Hagelstein, A. San Miguel, A. Fontaine and T. Ressler, *SPIE proc.* **2416**, 85-93 (1995).
2. J.-C. Labiche, G. Guilera, A. Hohms, O. Mathon, M.A. Newton, S. Pascarelli and G. Vaughan, "The Fast Readout Low Noise camera as a versatile X-ray detector for time resolved studies.". In preparation (2006).
3. M.A. Newton, S.G. Fiddy, G, Guilera, B. Jyoti and J. Evans, *Chem. Comm.* **118** (2005).
4. S. Pascarelli, T. Neisius and S De Panfilis, *J. Synchrotr. Rad.* **6**, 1044-1050 (1999).
5. G. Guilera, M.A. Newton, O. Mathon, and S. Pascarelli, "State-of-the-art Facilities for Industrial Applications at the ESRF", this conference
6. M. Wulf, F. Schotte, G. Naylor, D. Bourgeois, K. Moffat and G. Mourou, *Nucl. Instrum. Methods* A**398**, 69-84 (1997).
7. P. van der Linden, O. Mathon and T. Neisius, "Pulsed Magnetic Fields For A XAS Energy Dispersive Beamline", this conference.

Time Resolved Detection of Infrared Synchrotron Radiation at DAΦNE

A. Bocci[1,2], A. Marcelli[2], E. Pace[1,3], A. Drago[2], M. Piccinini[2,4], M. Cestelli Guidi[2], D. Sali[5], P. Morini[5], J. Piotrowski[6]

[1]Università degli Studi di Firenze, Dipartimento di Astronomia e Scienza dello Spazio, L.go E. Fermi 2, 50125 Firenze, Italy
[2]INFN-Laboratori Nazionali di Frascati Via E. Fermi 40, I-00044 Frascati, Italy
[3]INFN, Sezione di Firenze, Via G. Sansone 1I, 50019 Sesto Fiorentino (FI), Italy
[4]Università Roma Tre, Dip. Scienze Geologiche, L.go S. Leonardo Murialdo, 1 – 00146 Rome (Italy)
[5]BRUKER Optics s.r.l., Via Pascoli 70/3, 2013 Milano, Italy
[6]VIGO SYSTEM SA, 3 Swietlikow Str., 01-389 Warsaw, Poland, jpiotr@vigo.com.pl

Abstract. Synchrotron radiation is characterized by a very wide spectral emission from IR to X-ray wavelengths and a pulsed structure that is a function of the source time structure. In a storage ring, the typical temporal distance between two bunches, whose duration is a few hundreds of picoseconds, is on the nanosecond scale. Therefore, synchrotron radiation sources are a very powerful tools to perform time-resolved experiments that however need extremely fast detectors. Uncooled IR devices optimized for the mid-IR range with sub-nanosecond response time, are now available and can be used for fast detection of intense IR sources such as synchrotron radiation storage rings. We present here different measurements of the pulsed synchrotron radiation emission at DAΦNE (Double Annular Φ-factory for Nice Experiments), the collider of the Laboratori Nazionali of Frascati (LNF) of the Istituto Nazionale di Fisica Nucleare (INFN), performed with very fast uncooled infrared detectors with a time resolution of a few hundreds of picoseconds. We resolved the emission time structure of the electron bunches of the DAΦNE collider when it works in a normal condition for high energy physics experiments with both photovoltaic and photoconductive detectors. Such a technology should pave the way to new diagnostic methods in storage rings, monitoring also source instabilities and bunch dynamics.

Keywords: fast infrared detectors, time resolved techniques, infrared synchrotron radiation.
PACS: 07.85.Qe.

INTRODUCTION

The availability of fast photon detectors is a fundamental request in many fields of application such as spectroscopy, free space optical communication systems and also for diagnostics at third generation synchrotron radiation (SR) sources and at Free Electron Lasers. To resolve the bunch lengths and detect bunch instabilities X-ray detectors and visible photodiode detectors may be used achieving a few picoseconds response times while the current technology of cooled infrared photovoltaic detectors (PV) realized with HgCdTe semiconductors achieves response times of a few nanoseconds. Infrared photon detectors may be sensitive to a large wavelength range: from approximately 1 micron to about 200 microns, a region that is divided in three main energy domains: nIR (1 – 5 microns), midIR (5 – 20 microns), farIR (20 – 200 microns). Recently new technological capabilities allowed the development of infrared detectors working at room temperature (T≈300 K), optimized to detect nIR and midIR wavelengths. These uncooled detectors may efficiently work with brilliant sources as quantum cascade lasers or synchrotron light sources.

DAΦNE (Double Annular Φ-factory for Nice Experiments) is an electron-positron collider realized at Laboratori Nazionali di Frascati of the INFN for high energy physics experiments [1]. The double ring is designed to work at an energy of 0.51 GeV per beam to produce Φ mesons by e+/e- annihilation. DAΦNE is designed to operate at high current levels and up to 120 bunches. An IR photon beam is extracted from a bending magnet located in the external arc section of the electron ring.

CP879, Synchrotron Radiation Instrumentation: Ninth International Conference,
edited by Jae-Young Choi and Seungyu Rah
© 2007 American Institute of Physics 978-0-7354-0373-4/07/$23.00

SINBAD (Synchrotron Infrared Beamline At DAΦNE) is the beam line dedicated to Infrared Interferometry and Microscopy operational at DAΦNE since 2001 [2]. At the end of the beam line, at the focus of the last toroidal mirror, experiments with both photovoltaic and photoconductive (PC) fast infrared uncooled HgCdTe detectors have been performed to detect the time structure of the IR emission at DAΦNE [3]. We resolved the emission of the time structure of the electron bunches of the DAΦNE collider when it works in a normal condition for high energy physics experiments (105 filled buckets out of the available 120, gap of 15 bunches, bunch distance 2.7 ns, bunch length 100 - 300 ps FWHM with quasi-Gaussian shape, single-bunch current ~20 mA).

EXPERIMENTAL DATA

The IR emission from 105 bunches and the gap of DAΦNE collider was measured by using a single element uncooled PC detector. Sub-nanosecond response times were obtained by a single element uncooled photoconductive VIGO detector – series R005 [4,5]. This is a high-speed photoconductive device operating at room temperature and optimized at 10.6 μm, fabricated on a quaternary HgCd(Zn)Te semiconductor substrate. This device is front-side illuminated, so the absorption of radiation at any wavelength occurs only in the narrow energy gap absorbing layer.

The detector was biased by a current of about 20 mA through a bias tee. The output was amplified by two- linear voltage amplifiers configuration with an input impedance of 50 Ω, bandwidth of 0.05 MHz – 1 GHz, voltage gain of about 20 dB and Noise-Figure (NF) of 3.5 dB and analyzed by an oscilloscope model Tektronix TDS 820 with a bandwidth of 6 GHz and a rise time of 58.3 ps. Detector, amplifiers and scope were connected by double shielded RG223 cables and BNC connectors.

With this setup the IR device was able to discriminate all single pulses of the electron bunches and the gap between the last and first bunch (see Fig. 1). Data in Fig. 1 refer to an average of about 500 acquisitions.

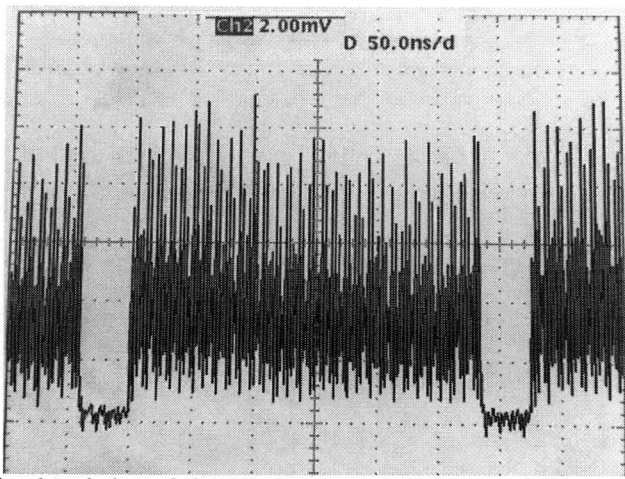

FIGURE 1. The averaged signal emission of the 105 bunches of DAΦNE and the gap obtained with the PC detector (500 acquisitions).

Figure 2 shows the emission of seven single pulses of DAΦNE at high time resolution. The average rise time and fall time of the signal emitted by a single-electron bunch are about 230 ps and 1.24 ns respectively. The response time of the detector, defined as the time that the detector takes to reach the 1/e value of the initial one after the irradiation is switched off, was estimated as 566 ps.

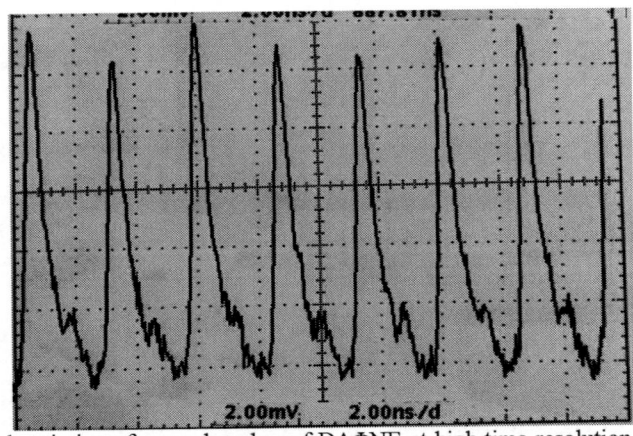

FIGURE 2. The averaged signal emission of seven bunches of DAΦNE at high time resolution obtained with the PC detector (500 acquisitions).

Different measurements were performed also with a VIGO multiple junction photovoltaic detector that is optically immersed, backside illuminated and operating at room temperature [4,6]. This latter is characterized by a ns response time in the 8 to 12 μm range with an optimized sensitivity at 10.6 μm. Its multiple heterojunction PV structure is constituted by a variable gap HgCd(Zn)Te semiconductor optically immersed in a high-refractive-index GaAs (or CdZnTe) hyperhemispherical or hemispherical lens. The main advantage of this fast detector is its capability to operate at room temperature, its compact size due to the absence of any cooling system and the no-bias-voltage operation.

The main drawback is the detectivity that is typically two orders of magnitude less than standard cooled HgCdTe photodiodes as a result of the generation-recombination noise at room temperature [7]. This device was not able to resolve the distance between two single bunches (2.7 ns). In Fig. 3 is reported the emission of one single bunch of DAΦNE with the PV detector.

From the detection of a single pulse emission the fall time of the signal emitted by a single-electron bunch is about 6.5 ns and the response time is about 3 ns. The signal was amplified by two voltage amplifiers and analyzed by an oscilloscope as described in the previous measurements. Actually the detector is not fast enough to resolve the peak to peak signal between two consecutive bunches. The limited response time when compared to the expected value (i.e., < 1 ns) may be due to a long detector RC time constant or to a long collection time of the charge carriers.

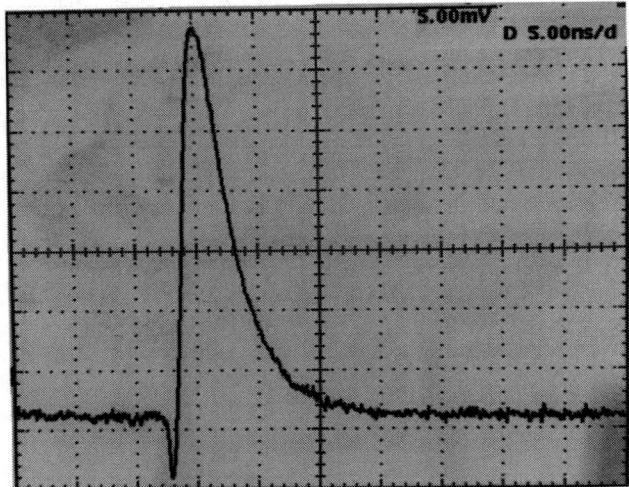

FIGURE 3. The averaged signal emission of a single bunch of DAΦNE obtained with the PV detector (500 acquisitions).

CONCLUSIONS

Technological improvements of the IR detectors manufacturing makes available devices capable to detect fast pulsed long wavelength radiation sources with resolution time in the sub-ns domain. Examples are uncooled HgCdTe photodetectors optimized at wavelengths around 10 μm. At the DAΦNE collider we resolved the emission of the temporal structure of the electron bunches when it works in normal conditions for high energy physics experiments with both photovoltaic and photoconductive uncooled infrared detectors.

The data obtained clearly show that the characterization of the temporal structure of the synchrotron radiation emission at infrared wavelengths at sub-nanosecond resolution time is nowadays possible. New systems and faster configurations are under test and with these devices we expect to resolve in the infrared region the emission shape of a single electron bunch. Moreover, the development of a fast uncooled array detector is also in progress. The device is a bi-linear array with 2x32 pixels: each one is a HgCdTe photoconductive element with a dimension of 50 x 50 microns in size and with an expected response time for pixel of about 500 ps. With such a technology we expect to realize for the first time IR imaging of a synchrotron radiation source with a sub-nanosecond resolution time for each single pixel. It should pave the way to ultra-fast IR imaging of a SR source with a sub-nanosecond resolution time for each single pixel.

ACKNOWLEDGMENTS

A special thank is due to G. Cibin, G. Cinque, A. Grilli, M. Pietropaoli, A. Raco, G. Viviani of the INFN-LNF for their support to the research.

REFERENCES

1. A. Drago, D. Alesini, G. Benedetti, M.E. Biagini, C. Biscari, R. Boni, M. Boscolo, A. Clozza, G. Delle Monache, G. Di Pirro, A. Gallo, A. Ghigo, S. Guiducci, F. Marcellini, G. Mazzitelli, C. Milardi, L. Pellegrino, M.A. Preger, P. Raimondi, R. Ricci, C. Sanelli, M. Serio, F. Sgamma, A. Stecchi, C. Vaccarezza, M. Zobov, Proc. PAC2003, Portland, Oregon, (2003).
2. M. Cestelli Guidi *et al.*, *Jour. Opt. Soc. Amer.* **A22**, 2810 (2005).
3. A. Bocci, M. Piccinini, A. Drago, D. Sali, P. Morini, E. Pace, A. Marcelli, LNF–05-12 (NT), (2005).
4. J. Piotrowski, A. Rogalski, Infrared Phys. Technol. **46**, 115 (2004).
5. http://www.vigo.com.pl/dane/detektory/pc.pdf.
6. http://www.vigo.com.pl/dane/detektory/pvmi.pdf.
7. J. Piotrowski, Opto-Electron. Rev. **12**, 111 (2004).

Soft X-Ray Beamline for fs Pulses
from the BESSY fs-Slicing Source

T. Kachel, K. Holldack, S. Khan[1], R. Mitzner[2], T. Quast, C. Stamm
and H. A. Duerr

BESSY, Albert-Einstein-Str. 15, 12489 Berlin, Germany
[1]Institut für Experimentalphysik, Universität Hamburg, Luruper Chaussee 149, 22761 Hamburg, Germany
[2]Physikalisches Institut der Universität Münster, Wilhelm-Klemm-Str. 10, 48149 Münster, Germany

Abstract. At BESSY II an undulator-based femtoslicing source is in operation. It provides soft X-ray radiation of tunable photon energy and polarization with a pulse length of less than 150 ± 50 fs. In this paper we discuss the beamline layout and our schemes of soft X-ray photon detection with emphasis on special requirements for laser pump – X-ray probe experiments with a time resolution of 150 fs. With our setup we have determined the characteristics of the fs soft X-rays, i.e. their flux, temporal evolution, angular distribution, signal-to-background ratio, and polarization state. Pilot experiments, using fs soft X-rays for absorption spectroscopy are presented.

Keywords: femtosecond, fs, slicing, soft x-rays.
PACS: 07.85 Qe, 78.47 +p, 78.70 Dm

INTRODUCTION

In the near-visible range light sources with pulse durations of 10 fs or less have existed for more than a decade [1]. The most powerful scientific type of application of such short light pulses is the investigation of ultra fast processes in matter. Typically, for such purposes experiments of the pump-probe type are carried out. The pump pulse is used to initialize a fast (and possibly reversible) process on the sample, like, e.g. electronic excitations into ligand states of molecules [2], electronic heating [3] and phonon excitations [4] in solids, or demagnetization of ferromagnets [5]. The probe pulse is used to investigate the dynamical state of matter applying measurements of the reflectivity, absorption, details of light scattering, and/or rotation of polarization vectors in magneto-optical Kerr experiments. The application of fs-lasers using the pump-probe technique has largely stimulated and improved the field of ultra fast dynamical experiments. However, the extension of the probe pulse to the soft X-ray or X-ray range with full tunability of the photon energy and free choice of the polarization offers the large field of investigations that has been exploited using synchrotron radiation sources. Its application allows techniques like photoemission, photon and photoelectron (EXAFS) scattering, X-ray magnetic circular dichroism (XMCD), etc. The possibilities are very promising, so that this aspect plays a major role the in the plans for the installation of Free Electron Lasers (FELs), which are made in several places [6]. FELs will provide high-intensity X-ray pulses of fs duration. A possible first step towards these sources can be made employing the fs-slicing principle [7]. BESSY has developed a fs-slicing source, which provides undulator-based soft X-ray pulses of variable polarization in the photon energy range of 400 – 1200 eV. The layout of the slicing source has been described before in [8].

BEAMLINE LAYOUT

The fs-slicing beamline with its soft X-ray monochromator (see Fig. 1) is close to the standard BESSY plane grating monochromator (PGM) design [9]. However, it is particularly set up with respect to the characteristics of the slicing source. These are the expected low photon flux compared to the standard storage ring operation, the requirement of low pulse broadening at the grating, and a high but adjustable dispersion and photon energy resolution. During the diffraction process every illuminated grating line contributes to the pulse broadening by the

CP879, *Synchrotron Radiation Instrumentation: Ninth International Conference,*
edited by Jae-Young Choi and Seungyu Rah

time that the light requires to proceed the distance of its wave length. In our case, at wave lengths between roughly 1 and 10 nm, this results in a broadening of 3.3 to 33 attoseconds per illuminated grating line. Thus, with 1000 to 10000 lines illuminated, the broadening becomes comparable to the initial pulse width.

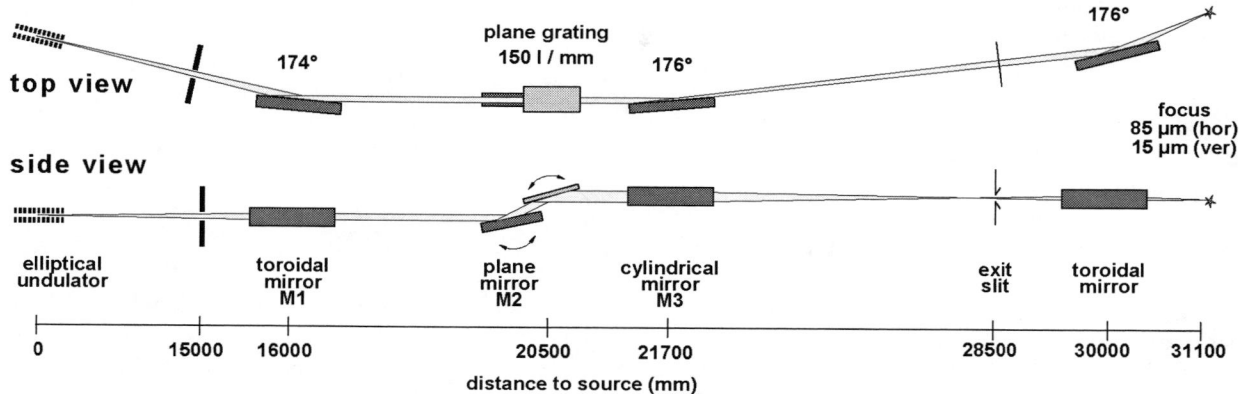

Figure 1. Optical layout of the BESSY femtoslicing beamline.

To circumvent this behavior the monochromator has been equipped with a grating of 150 lines/mm only. Additionally, the monochromator is run with a c_{ff} value of typically 1.2, i.e. a relatively large incidence angle α. The correspondence between pulse broadening, photon energy and c_{ff} value is shown in Fig. 2.

To further enhance the efficiency of the system, the beamline has been equipped with an extractable exit slit. Thus, implementing a one- or two-dimensional photon detector, like a photodiode array or a multi-channel plate, and optimizing the dispersion such that the spectrum matches the detectors field of view, a full spectrum can be accumulated simultaneously. In non-slicing mode we have already been able to obtain such a spectrum for the Fe 2p absorption edge using circularly polarized soft X-rays. The sample was a 50 nm $Fe_{50}Ni_{50}$ Film deposited on a 1 μm Mylar foil. It was mounted on a magnetic yoke to be able to magnetize the film primarily parallel or antiparallel to the light propagation direction. The experiment has been carried out in soft X-ray transmission geometry. As a detector we have used a fluorescent screen imaged by a CCD camera. With this apparatus it is possible to observe

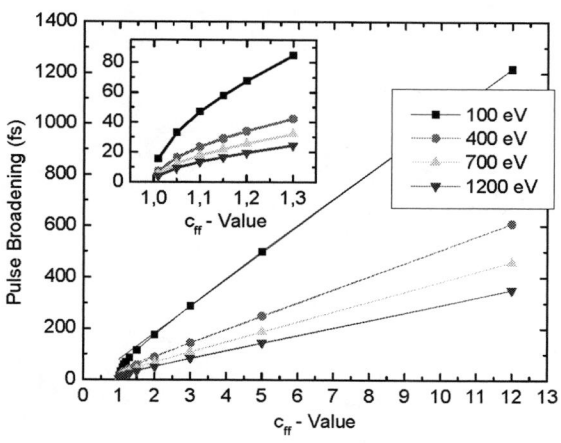

FIGURE 2. Calculated pulse broadening during diffraction at the 150 l/mm grating for selected photon energies as a function of the monochromator c_{ff} - value.

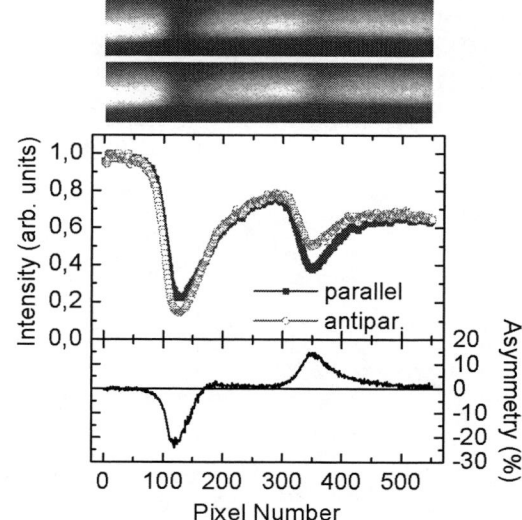

FIGURE 3. Single – shot Fe 2p absorption spectra for opposite magnetizations (raw data). Top: full spectrum images. Middle: integrated line scans of the images. Bottom: asymmetry of the line scans.

the energy-dispersed part of the spectrum as a vertical bright line on the fluorescent screen. Two such images are shown in the top panel of Fig. 3, both displaying the Fe 2p absorption edge. The difference between the two images is the magnetization direction of the Fe film with respect to the light propagation direction, which was antiparallel in the upper and parallel in the lower image. From the line scans in the middle panel, extracted from the CCD images, one can easily identify the Fe 2p absorption spectrum, which is somewhat saturated because of the large sample thickness. In the lower panel we have calculated the XMCD asymmetry

$$A = \frac{I^{\uparrow\uparrow} - I^{\uparrow\downarrow}}{I^{\uparrow\uparrow} + I^{\uparrow\downarrow}}. \tag{1}$$

from the line scans. The slicing experiment will be equipped with a two-dimensional detector based on a multi-channel plate in the near future.

EXPERIMENTAL SETUP AND RESULTS

The experimental setup at the BESSY femtoslicing beamline has to be described in detail elsewhere [10]. In short, an electromagnet with water cooled coils and a Fe yoke supplies a magnetic field of up to 0.5 T parallel or antiparallel to the soft X-ray propagation direction, for typical X-ray magnetic circular dichroism (XMCD) applications. The samples are investigated in transmission geometry under angles between -45° and 45° from normal incidence. They can be prepared in situ under UHV conditions by evaporation from electron beam metal evaporators onto ultra thin SiN or free standing Al or Cu films of 100 to 500 nm thickness.

As soft X-ray detector we use a 1 mm^2 avalanche photodiode (APD) with high gain 1.5 GHz preamplifiers. Gating of the soft X-ray signal is inevitable, to avoid high background signal due to electron bunches which have not undergone the 1 kHz repetition rate slicing procedure. The signal is processed in either Boxcar-integrators [11], a gated single-photon counting device, or by a digital oscilloscope.

A fraction of 10% of the slicing laser pulse is split off and brought to the experiment via a separate laser beamline of roughly 40 m length. Since the pump and probe pulses emerge from the same laser pulse, an intrinsic synchronization is ensured during the experiment. We use a commercial Ti:Sapphire (KMLabs) laser with λ = 780 nm, P_{max} = 2.7 W and a pulse length of typically 50 fs (FWHM) at 1 kHz repetition rate. The time delay between pump and probe pulse can be adjusted by a standard motorized delay unit. Further optical equipment is available to meet the required pump pulse conditions in terms of spot position stabilization, fluence, focusing, and polarization. Available pump laser powers at the experiment reach up to 200 μJ.

As one of the key experiments we were able to record an absorption spectrum of the Fe 2p edge in the slicing mode, i.e. applying pulse lengths with an upper limit of about 150 ± 50 fs. The 20 nm Fe film was evaporated on a 500 nm Al foil. The data were accumulated in the monochromator scan by integrating peak heights of soft X-ray signals on the oscilloscope over 2 s per data point. This corresponds to a photon flux of some 10 photons / pulse.

The lower curve in Fig. 4 shows the same scan, when the laser was switched off. This latter spectrum represents

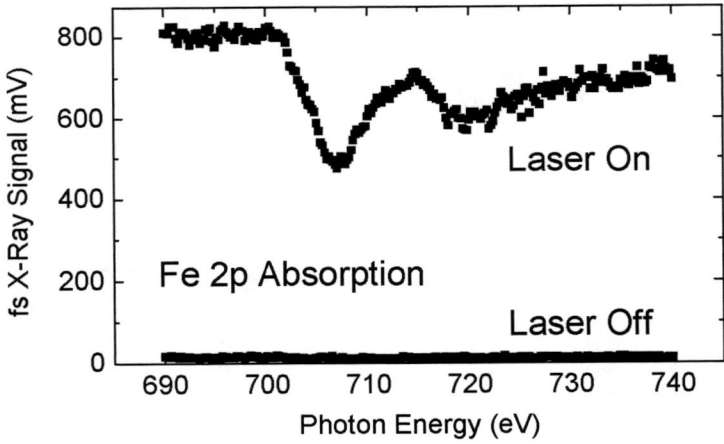

Figure 4. Fe 2p absorption spectrum from a 20 nm thick Fe film measured with fs X-ray pulses in transmission geometry. Accumulation time for each data point was 2 s. The "laser off" spectrum shows that the background contribution from non-sliced pulses is negligible.

the contribution of the non-sliced part of the beam with pulse lengths of 30 – 100 ps (fwhm). Even in our case of horizontal angular separation, that in principle allows for good signal-to-background ratios between sliced and non-sliced soft X-ray pulses, special care has to be taken to suppress the non-sliced background as much as possible [8]. For linearly polarized soft X-rays we are able to obtain signal-to-background ratios close to a factor of 100. The situation becomes more critical, when circularly polarized radiation is used for XMCD measurements. This is due to a broader emission cone in horizontal direction compared to the case of vertical linear polarization. But even in this case we were able to set up conditions for the angular separation that XMCD measurements could be done in the slicing mode with negligible background contribution.

After submission deadline of the current work we have succeeded in accumulating pump-probe delay scans in the slicing mode. They prove the dynamical behavior of electronic excitations [12] and the details of the dynamics of demagnetization in ferromagnetic Ni [10]. These delay scans give a clear indication that an averaged pulse length, including long term delay drifts over several hours, of 150 ± 50 fs (FWHM) can be achieved.

SUMMARY

In the present work we have shown details of the BESSY undulator-based fs-slicing facility. A plane grating monochromator beamline optimized for high transmission and low pulse broadening is in operation. The elliptical undulator provides free choice between linear and circular polarization of the soft X-ray probe pulses, which allows to study magnetization dynamics of ferromagnetic materials. The experimental chamber is laid out for dynamical studies using the laser pump – X-ray probe technique. For the first time we are able to present an absorption spectrum using fs soft X-ray pulses. The achieved photon flux made it possible to acquire the spectrum with 2 s accumulation time per data point only. The available photon energy range, tested until now, is 400 - 1200 eV, which covers many of the relevant absorption edges of the elements. The averaged pulse width of the soft X-rays is proven to be 150 ± 50 fs. At present, most of the experiments have been carried out at the Fe and Ni 2p absorption edges. However, the flexibility of the BESSY slicing source and the beamline will allow studies of fs dynamics in other solids and molecules as well.

ACKNOWLEDGMENTS

We gratefully acknowledge the help of many members of the BESSY staff, namely the undulator and the optics group. Several engineers have contributed with developments of selected components. Technical support has come from our vacuum, electronics, and software groups. Great work has been done by members of the machine group, who made it possible, that the fs-slicing mode can be run during normal user operation without interference to other beamlines at BESSY. The project has been financially supported by the Bundesministerium für Bildung und Forschung and the Land Berlin.

REFERENCES

1. B. B. Hu, E. A. de Souza, W. H. Knox, J. E. Cunningham, M. C. Nuss, A. V. Kuznetsov, and S. L. Chuang, *Phys. Rev. Lett.* **74**, 1689 (1995).
2. C. Bressler, and M. Chergui, *Chem. Rev.* **104**, 1781 (2004) and references therein.
3. W. S. Fann, R. Storz, H. W. K. Tom, and J. Bokor, *Phys. Rev. Lett.* **68**, 2834 (1992).
4. A. M. Lindenberg et al., *Science* **308**, 392 (2005).
5. E. Beaurepaire, J.-C. Merle, A. Daunois, and J.-Y. Bigot, *Phys. Rev. Lett.* **76**, 4250 (1996).
6. J. Arthur et al., *LCLS Conceptual Design Report,* SLAC-R-593 (2002) and
   R. Brinkmann et al. (eds.), *TESLA XFEL Technical Design Report Supplement,* DESY 2002-167 (2002)
   D. Krämer, E. Jaeschke, W. Eberhardt (eds.), *BESSY FEL Technical Design Report,* (2004).
7. A. Zholents, and M. Zoloterev, *Phys. Rev. Lett.* **76**, 912 (1996) and
   R. W. Schoenlein et al., *Appl. Phys.* **B 71**, 1 (2000).
8. K. Holldack, S. Khan, R. Mitzner, and T. Quast, *Phys. Rev. Lett.* **96**, 054801 (2006),
   S. Khan, K. Holldack, T. Kachel, R. Mitzner, and T. Quast, *Phys. Rev. Lett.* accepted for publication and,
   K. Holldack, T. Kachel, S. Khan, R. Mitzner, and T. Quast, *within these proceedings.*
9. R. Follath, *Nucl. Instr. Meth.* **A 467-468**, 418-425 (2001).
10. C. Stamm et al., *to be published.*
11. SR250 from Stanford Research Systems
10. T. Kachel et al., *to be published.*

Ultrafast Time-Resolved Powder Diffraction Using Free-Electron Laser Radiation

C. Blome, Th. Tschentscher,
J. Davaasambuu*, P. Durand*, S. Techert*

Deutsches Elektronen Synchrotron DESY, Notkestr. 85, 22607 Hamburg, Germany
*Max Planck Institut für biophysikalische Chemie, Am Fassberg 11, 37077 Göttingen, Germany

Abstract. Powder Diffraction offers the possibility to collect entire structural information from single profiles. This feature fits well with the possibility and requirement of x-ray free-electron laser sources to collect data on a single shot basis. The requirements for powder diffraction experiments using hard x-ray free-electron laser radiation are discussed in terms of bandwidth, x-ray spot size and detector resolution. Time-resolved measurements are proposed using ultrashort optical laser pulses to excite sample transformation on the 100 femtosecond time scale. Preliminary thoughts on energy absorption in the sample due to optical laser and x-ray beam are discussed.

Keywords: ultrafast, powder diffraction, FEL, damage
PACS: 41.60.Cr, 61.10.Nz, 82.53.–k

INTRODUCTION

Time-resolved x-ray diffraction following photo-excitation constitutes an important class of investigations of atomic resolution structural dynamics in solid crystalline materials. Time-resolution in the sub-picosecond regime will be enabled by using x-ray free-electron laser (FEL) radiation. In this contribution we are discussing the requirements for time-resolved powder diffraction experiments using x-ray FEL radiation. Powder diffraction is particularly interesting since profiles obtained from single pulse exposures provide, in principle, the possibility for complete structure refinement. Such techniques allow to follow structural changes of photo-induced phase transitions and photo-induced structural distortions in solid organic and inorganic materials. To estimate the diffraction signal and detector count rates we have simulated the powder diffraction pattern for methylammonium tetrachloromanganate (II) ($[(CH_3)NH_3]_2MnCl_4$) in the following short MAMC. MAMC and its related compounds are examples of inorganic materials undergoing structural phase transitions with large variation in electronic and magnetic properties [1]. MAMC has three structural phases, namely the tetragonal high temperature (HT) phase (394 K < T), the orthorhombic room temperature (RT) phase (257 K < T < 394 K) and the tetragonal (LTT) (94 K < T < 257 K) and monoclinic (LTM, T < 94 K) low temperature phases. Additionally, MAMC has been shown to be a quasi two-dimensional Heisenberg antiferromagnet with the phase transition to a three-dimensional magnetic order at approx. 45 K [2]. It has furthermore been reported that organic-based manganese compounds are promising candidates for photo-inducing magnetism by light-absorption [3].

EXPERIMENTAL REALIZATION AND REQUIRED RESOLUTION

The requirements for the design of an experimental setup are discussed for the x-ray radiation parameters of the European XFEL facility. The following x-ray radiation parameters have been used [4]: a fundamental photon energy of 12 keV, 10^{12} photons per pulse, a natural bandwidth of x-ray FEL radiation of $\Delta E/E=0.1\%$, a divergence of 0.8 μrad and a resulting beam size at the experiment of 800μm in unfocused conditions. In addition, monochromatization to $\Delta E/E=0.01\%$ and focusing to 100 μm beam size was considered. All values are given as full-width half maximum values. The instrumental resolution in powder diffraction experiments is determined by the x-

CP879, *Synchrotron Radiation Instrumentation: Ninth International Conference,*
edited by Jae-Young Choi and Seungyu Rah
© 2007 American Institute of Physics 978-0-7354-0373-4/07/$23.00

ray beam size, divergence at the sample position, and by the x-ray energy spread. The resulting angular distribution is then sampled by the x-ray detector. In order to be able to observe small changes of diffraction peaks following photo-excitation, both in position and intensity, we investigate the requirements for powder diffraction experiments with instrumental resolutions better than 500 µrad. We have simulated powder diffraction profiles and conclude that collection of scattering data up to angles of 60 deg is useful.

For the above parameters the angular smearing at the detector is determined by the geometrical spreading $\Delta(2\theta)_{geo} = D/L$, due to beam size D and sample-to-detector distance L, and by the bandwidth spreading $\Delta(2\theta)_{bw} = 2 \tan(\theta) \times \Delta E/E$. A contribution to broadening due to the grain size 2r of the powder is taken into account through the Scherrer-equation $\Delta(2\theta)_{grain} = 0.94 \times \lambda/(2r \cos(\theta))$ for the x-ray wavelength λ. In Fig. 1, the total instrumental resolution $\Sigma = (\Delta(2\theta)_{geo}^2 + \Delta(2\theta)_{bw}^2 + \Delta(2\theta)_{grain}^2)^{1/2}$, assuming Gaussian distribution functions, is plotted as a function of scattering angle 2θ for different combinations of D and L and for two grain sizes.

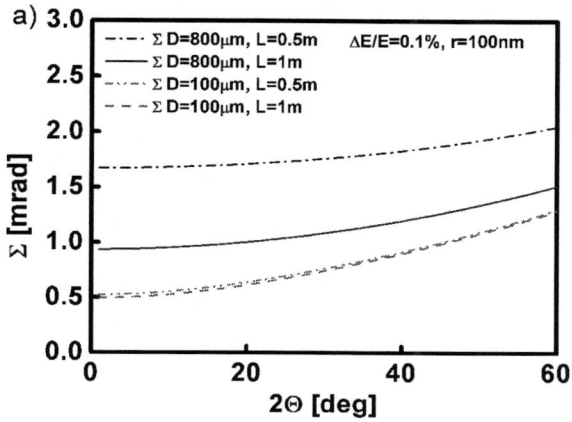 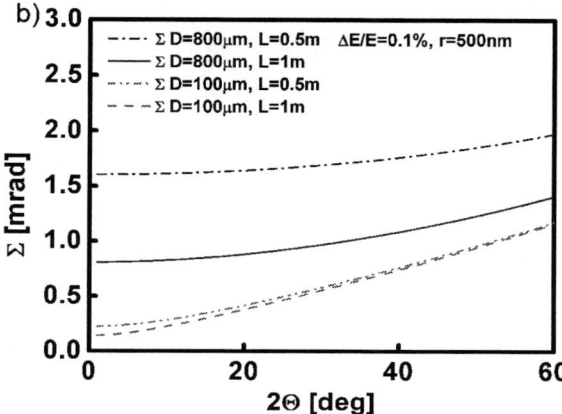

FIGURE 1. Total instrumental resolution Σ for powder diffraction as a function of scattering angle 2θ for various combinations of beam size D and sample-to-detector distance L and a grain size of 200 nm (a.) and 1 µm (b.).

From Fig. 1 one obtains that for the bandwidth of 0.1 % the achievable resolution varies as a function of scattering angle determined by the bandwidth contribution. The resolution is best for forward scattering and degrades increasing the scattering angle. The achievable resolution depends on the geometrical contribution and on the grain size. Performing the same calculations for an improved bandwidth of 0.01% gives total resolution values essentially constant for the angular range of 0 to 60 deg at the identical values achieved in Fig. 1 for forward scattering. In this case the resolution is completely determined by geometrical and grain size contributions. It can be concluded that a total resolution of 500 µrad can be achieved over the entire angular range for monochromatic beam and 100 µm beam size. In the case of unfocused beam size of 800 µm the best achievable resolution is of the order of 1 mrad. For grain sizes larger 200 nm the resolution can be improved to below 500 µrad.

We propose to sample the resulting angular profile by direct detection of the diffracted x-ray beam with a spatially resolving detector. In order to collect the entire profile simultaneously we suggest using a pixel detector covering the entire angular range from 0 to 60 deg, see Fig. 2 (left). The detector pixels are placed on a circle around the sample position. The spatial resolution of the detector Δx relating to the angular broadening Σ must be such that peak profile analysis procedures can be applied. The pixel pitch ideally should be of the order of $0.1 \times L \Sigma$.

INTENSITY DISTRIBUTION AND RELATED ISSUES

In the following we estimate the intensity distribution and the detector count rates per pixel. Following the above findings we assume a beam size of 100µm, a bandwidth of 0.01%, a sample-detector distance of L=0.5m and a powder in a glass capillary with an inner diameter of 50µm and an outer diameter of 150 µm. We assume 100% photo-transformation of the material which is evidentially an overestimation. However, photo-transformations on the order of 20 – 50 % are realistic [5,6]. To evaluate the observable count rate we factorize the various transmission factors. Absorption due to the capillary walls gives $T_{SiO2} = 0.89$, an air path of L = 0.5 m from the sample to the detector $T_{air} = 0.993$ and the absorption by a 50 µm thick MAMC powder $T_{sample} = 0.725$ [7]. The glass capillary forms a hard aperture for an x-ray beam with Gaussian intensity distribution in both lateral dimensions resulting in a transmission factor of $T_{geo} = 0.44$. The restricted bandwidth of the monochromator gives an additional loss of $T_{bw} = 0.094$. For the total transmission T follows: $T = T_{sample} \times T^2_{SiO2} \times T_{air} \times T_{geo} \times T_{bw} = 0.023$. For a cylindrical specimen illuminated by a linearly polarized beam, the integrated number of photons N_{hkl} of a segment of length Δy (see Fig. 2), for a Bragg reflection from hkl-planes recorded at a distance L is:

$$N_{hkl} = \frac{N_o \times T}{A} \left(\frac{e^4}{m_e^2 c^4} \right) \frac{V \lambda^3 m |F_{hkl}|^2}{4\nu^2} \frac{1}{\sin(\Theta)} \frac{\Delta y}{2\pi L \sin(2\Theta)}. \qquad (1)$$

Here, A is the effective area formed by the hard aperture of the capillary, V is the sample volume, ν is the unit cell volume and m is the multiplicity. The other terms have their usual significance for scattering experiments [8].

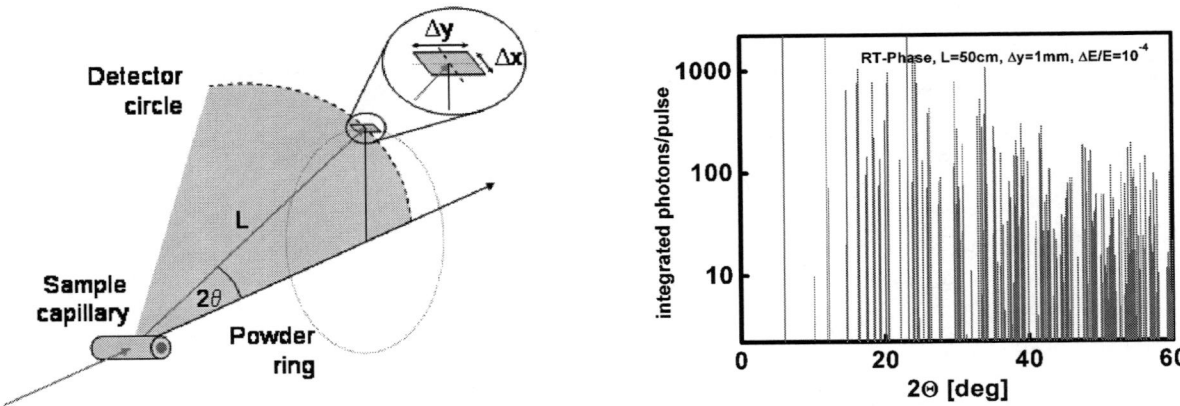

FIGURE 2. Left side: Sketch of the Experimental setup. For clarity only one pixel element of the line detector positioned along the broken line is shown. Right side: integrated photon flux for a transverse pixel size $\Delta y = 1$mm.

The derived formula is valid for a horizontally polarized FEL beam and diffraction in the vertical plane. Figure 2 (right) shows the integrated intensity of powder diffraction rings as a function of 2Θ and integrated over Δy. In the angular regime of 50 - 60 degrees one observes integrated x-ray intensities of $N_{hkl} = 50 - 100$ photons per reflection hkl and pulse. The x-ray intensity increases linearly with the transversal pixel size Δy. Since it is possible to increase this size to 10 mm without degrading angular resolution one could achieve integrated intensities of 500 - 1000 photons per pulse. To observe a statistical accuracy of the order 1 % one would require collecting of the order of 10 - 50 pulses. It should be noted that these data have also to be analyzed and sorted in terms of fluctuations of the experimental parameters (e.g. the time delay, photon energy, beam position). In general, it will be required to collect at least one order of magnitude more pulses to achieve the required statistical accuracy.

To discuss the possibility of using line fitting procedures we calculate the diffracted photons per pixel N_P. We therefore normalize N_{hkl} to the width of the Bragg peak Σ, resulting in the averaged number of photons per radian, and multiply by the angular size of the pixel $\Delta(2\Theta)_{pixel}$:

$$N_P = N_{hkl} \times \frac{\Delta(2\Theta)_{pixel}}{\Sigma}. \qquad (2)$$

As discussed one obtains a nearly constant line broadening of $\Sigma = 0.5$ mrad over the entire range of diffraction angles (D = 50 μm, L = 0.5 m, $\Delta E / E = 0.01\%$, r = 100 nm). With $\Delta(2\Theta)_{pixel} = \Delta x / L = 40$ μrad for $\Delta x = 20$ μm we calculate a constant factor for $\Delta(2\Theta)_{pixel} / \Sigma$ of 0.08. Taking Fig. 2 one thus estimates for the angular regime of 50 - 60 degrees about 4 - 8 photons per shot in a pixel of the size 20×1000 μm, or 40 - 80 per shot for a 20×10000 μm pixel. The significance of line-fitting procedures applied to such intensities for single pulse experiments is uncertain and accumulation of several pulses will be required. To increase the number of photons per pixel without degradation of resolution requires to increase the spectral flux. An attractive possibility here is the application of self-seeding the FEL-process thus narrowing the natural bandwidth to the required 0.01% [9]. Likewise one could achieve of the order of 400 - 800 photons per pixel and pulse.

The coherent illumination using FEL radiation will lead to interference effects in the diffraction pattern. The averaged size δ of the produced speckles scales with the inverse of the illuminated region 1/D, the wavelength λ and the distance between sample and detector L: $\delta = 2\lambda L/D$ [10]. For D=50μm, 12 keV and L=0.5 m one estimates an averaged angular speckle size of 2μm. For the pixel size of 20 μm the instrument will not resolve individual speckles and the coherent illumination should not affect the observable diffraction pattern.

To characterize the damage due to x-ray exposure we have calculated the amount of energy that is deposited in the sample volume. For the experimental parameters discussed above, of the order of 10^{10} photons are absorbed by the sample, corresponding to $\sim$ 10μJ energy deposition per pulse. For the solid-state density of 1.7 g/cm^3 and a sample volume of $\pi \times 100 \times 25^2$ μm^3 a specific energy deposition of 25 J/g follows. This value corresponds to an absorbed energy of 50 meV per molecule and pulse. The corresponding temperature increase is estimated from $\Delta T = E/(n \times c_{mol})$ where E is the deposited energy in the sample volume, n is the mass of the sample volume in mol, and c_{mol} is the heat capacity in J/mol/K. The heat capacity of MAMC was experimentally determined to $\sim$240 J/mol/K for the temperature range 240-350 K [11]. We estimate a temperature increase of 10 K per pulse for an energy absorption of 50 meV per molecule and pulse. Using cryogenic cooling of the relatively small samples we estimate that heat conduction is efficient enough for the overall repetition rate of 10 Hz of the European XFEL facility. Using of the pulse trains with up to 3000 x-ray pulses per 600 μs, however, will lead to severe sample alteration. The use of fast and continuous sample exchange mechanisms must be foreseen in this case. Since the x-ray scattering itself occurs on time scales much shorter than the thermodynamical heating of the sample it is assumed that the sample structure is not affected during the course of the x-ray pulse.

CONCLUSIONS

We have discussed the realization and the requirements for time-resolved powder diffraction experiments using 12 keV FEL radiation provided by the European XFEL facility. Simulations for a relevant powder sample indicate that for the XFEL photon energy the angular range up to 60 deg can be investigated thereby improving the accuracy of structure determination. To allow peak profile analysis the experiments require high instrumental resolution. Analysis of the instrumental resolution shows that it is required to monochromatize the FEL radiation in order to obtain high resolution conditions. Using monochromatic x-ray beam the energy deposition in the sample volume will lead to temperature steps of the order 10 K. Using cryo-cooling this temperature oscillation is expected not to cause severe problems for moderate repetition rates. We have evaluated the use of a one-dimensional, linear array of pixel detectors with a size of 20 times 10000 μm^2, fully covering the scattering angle in a curved, vertical scattering geometry. Calculations of the expected detector count rates have been done for a model compound in a relevant angular range. They show that in the case of typical monochromator bandwidth, a pixel size of 20 $\times$ 10000 μm^2 x-ray intensities per pulse and pixel vary from 40 to 80 photons in the angular regime of 50-60 degrees.

REFERENCES

1. Adams D. M., and Stevens, D. C., *J. Phys. C: Solid State Phys.* **11**, 617 (1978).
2. Achiwa, N., Matsuyama, T. and Yoshinari, T. , *Phase transitions* **28**, 79 (1990).
3. Pejakovic D. et al., *Phys. Rev. Lett.* **88**, 057202 (2002).
4. Altarelli M. et al. (edts.), *XFEL The European X-Ray Free-Electron Laser – Technical Design Report*, DESY Report 2006-097, DESY, Hamburg (2006); also available at http://xfel.desy.de
5. Collet E. et al., *Science* **300** , 612 (2003).
6. Davaasambuu, J. et al., *J. Synchotron Rad.* **11**, 483 (2004).
7. Center for x-ray optics, http://www-cxro.lbl.gov/optical_constants.
8. Warren B.E., *X-ray Diffraction*, Dover Publication (1990).
9. Saldin E.L. et al., *Nucl. Instrum. and Methods Phys. Res.* A**475**, 357 (2001).
10. Svelto O. *Principles of Lasers,* Plenum Publishing Corporation, New York, (1976).
11. White M. et al., *J. Phys. Chem.* **43**, 4 (1982)

Time-Resolved X-Ray Triple-Crystal Diffractometry Probing Dynamic Strain in Semiconductors

Yujiro Hayashi[*,†], Yoshihito Tanaka[†], Tomoyuki Kirimura[†], Noboru Tsukuda[**], Eiichi Kuramoto[**] and Tetsuya Ishikawa[†]

[*]*Interdisciplinary Graduate School of Engineering Sciences, Kyushu Univ., Kasuga, Fukuoka 816–8580, Japan*
[†]*RIKEN SPring–8 Center, Sayo–cho Sayo–gun, Hyogo 679–5148, Japan*
[**]*Research Institute for Applied Mechanics, Kyushu Univ., Kasuga, Fukuoka 816–8580, Japan*

Abstract. Intense synchrotron radiation sources have enabled us to combine time–resolved measurements and triple–crystal diffractometry. The time–resolved triple–crystal diffractometry (TRTCD) determines the time–dependent dilational and shear components of deformation tensor, separately. The TRTCD experiments have been performed at a long undulator beamline of SPring–8. The time-resolved measurement system using pump-probe technique and a fast multi-channel scaler covers a full range of milliseconds with a time–resolution of several tens of picoseconds. The TRTCD with wide time range was applied to the dynamic strain measurement for semiconductor wafers irradiated by a femtosecond pulse laser. We observed a dilational component of acoustic echo pulses to analyze the time–varying pulse shape due to propagation. The lattice motion in the successively induced flexural standing wave has also been observed through a shear component.

Keywords: triple–crystal diffractometry, time–resolved measurement, ultrashort pulse laser, ultrasonic wave
PACS: 61.10.Nz, 62.30.+d, 78.47.+p

INTRODUCTION

When a surface of solid material is irradiated by an ultrashort pulse laser, acoustic wavepackets are produced and propagate in the material [1, 2]. The derivative pulse echo technique using the high–frequency (~ 100 GHz) acoustic pulse is attractive as the investigation tool of underlying structures in materials such as thin layers and defects [3]. The phenomenon has intensively been investigated mainly by ultrafast optical pump–probe technique, which gives integrated effects of deformation such as slope and displacement of the surface, but cannot directly determine strain in materials. On the other hand, x–ray diffraction (XRD) directly detects lattice strain of atomic scale regardless of surface condition. However, few XRD methods have been employed for determining atomic motion in crystals until recently because of the lack of pulsed and brilliant x–ray sources. Significant progress has been made in the development of x-ray sources, which have enabled us to observe dynamic lattice strain even for the optically induced acoustic pulse [4, 5, 6].

Transient lattice strain produced by ultrashort laser pulse light generally includes inclination of lattice planes (i.e. shear) as well as dilation and contraction (i.e. longitudinal strain). In the laser beam deflection method, surface slope caused by optically induced localized longitudinal strain is monitored [7, 8]. The strain with both longitudinal and shear component cannot be quantitatively analyzed from the rocking curve measured by scanning Ewald sphere around a reciprocal lattice point. Triple–crystal diffractometry (TCD) has capability of revealing great details of small lattice deformation in nearly perfect crystals [9, 10]. This is achieved by measuring the two–dimensional distribution of diffracted intensity in a reciprocal space with high momentum resolution. Using parallel and monochromatic incident x–rays to reduce an instrumental blur, Bragg diffracted beam from a sample crystal is angularly resolved by an analyzer crystal, usually made of a highly perfect crystal such as Si, thus obtaining the two–dimensional intensity distribution around the reciprocal lattice point. Highly brilliant x–rays from SR sources have helped enhance the momentum resolution of TCD.

At storage ring facilities, pulsed nature of SR and synchronization technique between SR pulses and ultrashort laser pulses have enabled us to record snap shots of XRD profiles with ~ 100 ps resolution. A few picoseconds resolution has been achieved by using an x–ray streak camera in combination with the synchronization technique. Multichannnel scaling (MCS) technique is advantageous to wide time range measurement with nanosecond resolutions when the repetition rate of SR pulses is much higher than that of pump light pulses [11].

CP879, *Synchrotron Radiation Instrumentation: Ninth International Conference*,
edited by Jae-Young Choi and Seungyu Rah
© 2007 American Institute of Physics 978-0-7354-0373-4/07/$23.00

In this article, we summarize the development of TCD combined with time–resolved measurement using the pump–probe and MCS technique. Since the time–resolved triple–crystal diffractometry (TRTCD) requires more brilliant x–ray sources because of the low efficiency of crystal diffraction, we carried out the measurements at the 27m–long undulator beamline BL19LXU [12] of SPring–8.

EXPERIMENTAL

A triple–crystal diffractometer was set up by using highly perfect Si crystals for a double–crystal monochromator and an analyzer at the beamline BL19LXU of SPring–8 as shown in Fig.1. The monochromatized x–rays were incident to the laser–irradiated point on the surfaces of samples, GaAs (1 0 0) and Si (1 1 1) crystal, through a slit of $\sim 100 \times 100\,\mu\mathrm{m}^2$. The laser beam prepared by a mode–locked Ti:sapphire laser with a generative amplified was guided to the surface center of the samples. The wavelength, the pulse duration, the energy, the repetition rate, and the beam size of the laser were 800 nm, 130 fs, 300 mW, 1.0 kHz, and ~ 1 mm full–width at half–maximum (FWHM) in diameter, respectively. The symmetric reflections from the samples were angularly resolved by the analyzer and detected. The angle ω of the samples and θ of the analyzer were controlled with high–precision goniometers [13]. If we let z and x–axis be parallel to the depth direction and the surface of the sample crystals as shown in Fig.1, $\varepsilon_{33} = \partial w / \partial z$ and $\varepsilon_{31} = \partial w / \partial x$ mean the dilation normal to the surface and the inclination of lattice plane parallel to the surface, respectively, where w is the displacement in the direction to z–axis. When the strain is assumed to be uniform and small within the region illuminated by x–rays, the relationship between ε_{33}, ε_{31} and $\Delta\theta$, $\Delta\omega$ is given by

$$\varepsilon_{33} = -\frac{\Delta\theta}{\tan\theta_B}, \tag{1}$$

$$\varepsilon_{31} = \Delta\omega, \tag{2}$$

where $\Delta\theta$ and $\Delta\omega$ are the profile shift of a ω–2θ scan and a ω scan in the direction parallel and perpendicular to the reciprocal lattice vector, respectively, and θ_B is Bragg angle.

The time–resolved measurement for ω–2θ scans was conducted by MCS with a wide time range and a pump–probe method with a higher time resolution. In the former, the signal pulses from an avalanche photodiode (APD) detector for x–rays were counted as a function of time with the time interval of 0.8 ns for the full time range 1 ms of the laser pulse period by a multichannel scaler (9353, ORTEC), using trigger pulses from an APD detecting laser pulses. Although incident x–rays have pulsed time structure, we can consider them quasi–continuous by accumulating the signals with x–rays and laser pulses asynchronous. In the latter, the laser timing was controlled by phase lock circuits with respect to an electron bunch in the storage ring with the precision of a few picoseconds, which gave the profile snap shots with the time resolution of the x–ray pulse duration (40 ps FWHM) [14]. For ω scans, a Si PIN photodiode detector (S3590–09, Hamamatsu) was used in a photoconductive mode as reported elsewhere [15]. The signal pulses from the PIN photodiode were amplified and averaged by a digital storage oscilloscope of which triggers are from the APD detector for laser pulses. The time resolutions of the pump–probe, MCS, and the PIN photodiode system have been estimated to be 40 ps, 1.2 ns, and 82 ns FWHM, respectively.

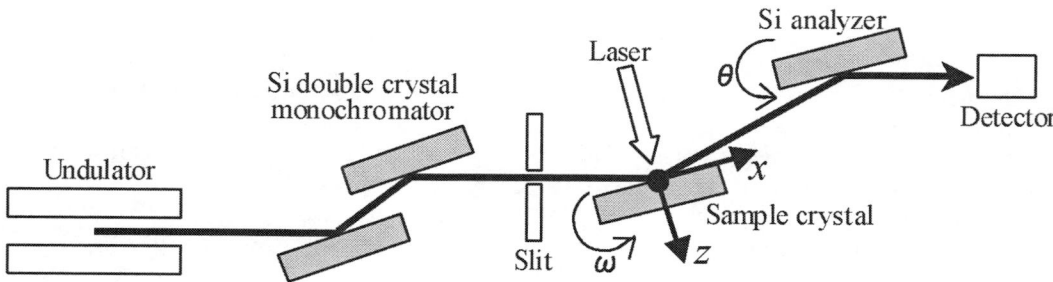

FIGURE 1. Schematic of TRTCD set up.

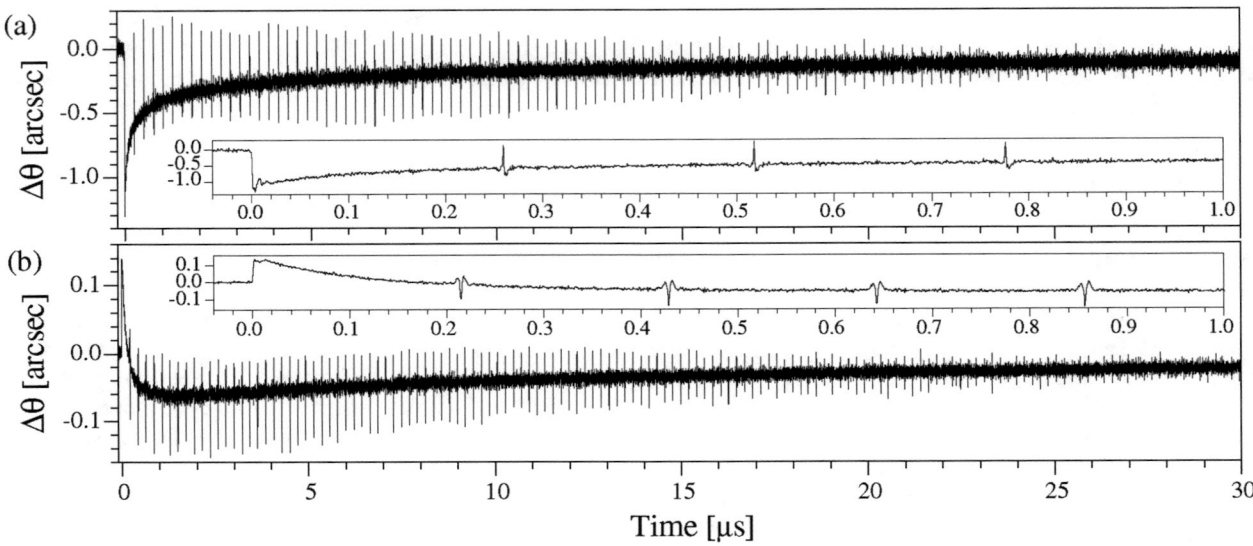

FIGURE 2. The angular shift of the time–resolved ω–2θ scan profiles for GaAs (a) and Si (b) irradiated by a femtosecond pulse laser [16]. Insets: the region around laser irradiation of $t = 0$.

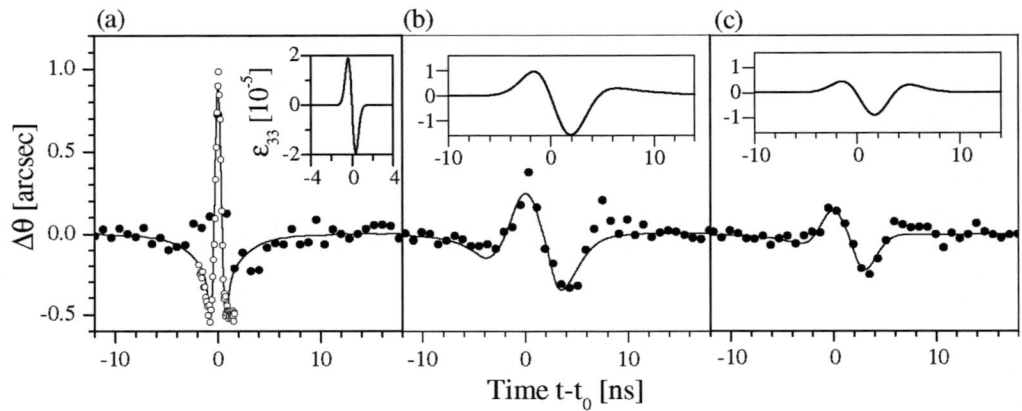

FIGURE 3. The angular shift of the time–resolved ω–2θ scan profile in the part of (a) the 1st, (b) 20th, and (c) 60th pulse echo for GaAs with simulation curves [16]. The time when the profile reaches the largest angle for each echoes is defined as t_0, where $t_0 = 259.2, 5169$, and 15507 ns for the 1st, 20th, and 60th echo, respectively. Insets: the calculated strain ε_{33} for the echoes.

RESULTS AND DISCUSSION

The time–resolved ω–2θ scan profiles of 4 0 0 and 3 3 3 reflection for GaAs and Si were measured using x–rays of 16.2 and 21.9 keV, respectively. The time–dependent $\Delta\theta$ measured by MCS are shown in Fig.2(a) and (b) for the GaAs and Si plate, respectively. The measurement accuracy of ε_{33} has been estimated to be 1×10^{-7} using the time–dependent $\Delta\theta$ from Eq.(1). The moment when $\Delta\theta$ jumps due to the laser irradiation is defined as $t = 0$. In GaAs, the profile shifts towards the smaller angle at $t = 0$ and abruptly shifts to the larger angle at $t = 259$ ns. The profile shift to the smaller and larger angle qualitatively means dilation and contraction, respectively, as seen from Eq.(1). The abrupt shifts occurring at regular time intervals result from multi–reflection of the optically induced acoustic pulse, since the time interval agrees with the round trip time of an acoustic pulse with the longitudinal sound velocity. In Si, the profile shift to the larger angle at $t = 0$, which indicates that optically excited Si is contracted opposite to optically induced dilation of GaAs. Correspondingly, the profile for the pulse echo in Si shifts towards the opposite angle to that of GaAs. The parts of the 1st, 20th, and 60th echo in Fig.2(b) are shown in Fig.3(a), 3(b), and 3(c) by solid circles, respectively.

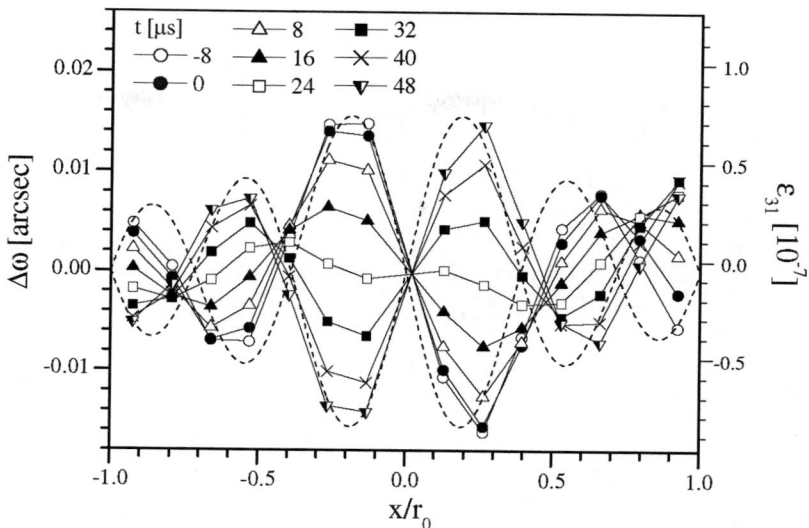

FIGURE 4. The 8 kHz–component of the time–resolved ω scan profile shift as a function of position [15]. Broken lines show the amplitude of the inclination of a flexural standing wave calculated by the classical elasticity theory for thin plates.

For the 1st echo, the pump–probe method reveals fine time structure, which is unavailable by MCS measurements, as shown in Fig.3(a) by open circles. The insets in Fig.3 show the calculated waveforms of each pulse echo. The temporal change in the acoustic pulse waveform due to propagation is clearly observed. Details of the observed dynamic strain have been discussed elsewhere [16].

The time–resolved ω scan profiles of 4 0 0 reflection for the GaAs plate were measured using x–rays of 15 keV. For ε_{31} the measurement accuracy of 4×10^{-8} has been obtained. The experimental result of x–dependence of the time–dependent $\Delta\omega$ shows a transient convex deformation in the normal direction at the surface center irradiated by laser pulses. In addition, the oscillation component of 8 kHz has been significant. This suggests that one of the resonance frequencies of flexural motion should be close to 8 kHz, since the periodic convex deformation leads to the forced oscillations which are the harmonics of 1 kHz. The 8 kHz–component of $\Delta\omega$ is plotted as a function of x/r_0 in Fig.4, where $r_0 = 76$ mm is the radius of the GaAs plate. The result is expressed by a model curve of a flexural standing wave described with Bessel functions as shown in Fig.4 by broken lines.

REFERENCES

1. C. Thomsen, J. Strait, Z. Vardeny, H. J. Maris, J. Tauc, and J. J. Hauser, Phys. Rev. Lett. **53**, 989 (1984).
2. V. E. Gusev and A. A. Karabutov, *Laser Optoacoustics* (American Institute of Physics, Woodbyry, NY, 1993).
3. H. Maris, Sci. Am. **278**, 86 (1998).
4. B. Adams, *Nonlinear optics, quantum optics, and ultrafast phenomena with x–rays* (Kluwer Academic Publishers, Massachusetts, 2003).
5. D. A. Reis, M. F. DeCamp, P. H. Bucksbaum, R. Clarke, E. Dufresne, M. Hertlein, R. Merlin, R. Falcone, H. Kapteyn, M. M. Murnane, J. Larsson, Th. Missalla, and J. S. Wark, Phys. Rev. Lett. **86**, 3072 (2001).
6. M. F. DeCamp, D. A. Reis, D. M. Fritz, P. H. Bucksbaum, E. M. Dufresne, and R. Clarke, J. Synchrotron Rad. **12**, 177 (2005).
7. O. B. Wright and K. Kawashima, Phys. Rev. Lett. **69**, 1668 (1992).
8. O. B. Wright and V. E. Gusev, Appl. Phys. Lett. **66**, 1190 (1995).
9. A. Iida and K. Kohra, Phys. Stat. Sol. A **51**, 533 (1979).
10. P. F. Fewster, J. Appl. Cryst. **22**, 64 (1989).
11. S. Kawado, S. Kojima, T. Ishikawa, T. Takahashi, and S. Kikuta, Rev. Sci. Instrum. **60**, 2342 (1989).
12. T. Hara, M. Yabashi, T. Tanaka, T. Bizen, S. Goto, X. M. Marechal, T. Seike, K. Tamasaku, T. Ishikawa, and H. Kitamura, Rev. Sci. Instrum. **73**, 1125 (2002).
13. T. Ishikawa, Y. Yoda, K. Izumi, C. K. Suzuki, X. W. Zhang, M. Ando, and S. Kikuta, Rev. Sci. Instrum. **63**, 1015 (1992).
14. Y. Tanaka, T. Hara, H. Kitamura, T. Ishikawa, Rev. Sci. Instrum. **71**, 1268 (2000).
15. Y. Hayashi, N. Tsukuda, E.Kuramoto, Y. Tanaka, T. Ishikawa, J. Synchrotron Rad. **12**, 685 (2005).
16. Y. Hayashi, Y. Tanaka, T. Kirimura, N. Tsukuda, E. Kuramoto, and T. Ishikawa, Phys. Rev. Lett. **96**, 115505 (2006).

Ultrafast X-ray Diffraction of Photodissociation of Iodoform in Solution

Jae Hyuk Lee[†], Tae Kyu Kim[†], Joonghan Kim[†], Maciej Lorenc[¶], Qingyu Kong[¶], Michael Wulff[¶], and Hyotcherl Ihee[†*]

[†]Department of Chemistry and School of Molecular Science (BK21), Korea Advanced Institute of Science and Technology (KAIST), Daejeon, 305-701, Republic of Korea
[¶] European Synchrotron Radiation Facility, Grenoble Cedex 38043, BP 220, France

Abstract. We studied structural dynamics in the photodissociation of iodoform (CHI_3) dissolved in methanol by time-resolved x-ray diffraction. A femtosecond laser pulse induces the bond-breaking of an iodine atom from iodoform and an x-ray pulse generated from a synchrotron gives time-dependent diffraction signal which contains the structural information of photoproducts with 100 ps time-resolution and 0.001 Å spatial resolution. CHI_2 radical and I atom are formed by the results of the ultrafast photodissociation of iodoform and these intermediates recombine to form iodoform again via geminate recombination. The iodine atoms which escape from the cages nongeminately recombine to form I_2. Solvent dynamics, heating and solvent expansion, caused by photodissociation, are also explained from time-resolved x-ray diffraction data.

Keywords: Time-resolved solution x-ray diffraction, CHI_3, Hydrodynamics
PACS: 82.53.Uv, 61.10.Eq

INTRODUCTION

Through the advancement of computational methods, the prediction of transient molecular structure during chemical/biological reactions has become relatively easy. In contrast, direct experimental visualization of temporally varying transient structures during ultrafast processes has been recently performed with great technical advances, but it is still difficult [1]. Recent advancements have been achieved in time-resolved optical spectroscopy to correlate frequencies of optical transition of molecules with temporal behavior of molecular structure. Since the optical probe is not able to interfere with all atoms, structural changes of transient species can not be obtained directly, but have to be deduced from potential energy surfaces of molecular species. In this regards, the methodology of time-resolved x-ray diffraction [2-5], in which structural information can be obtained from the diffraction patterns, offers a complementary tool. The reaction is triggered optically as in the conventional spectroscopy, but relevant reactions are probed by the diffraction of ultrashort x-ray pulses. The wavelength of x-ray is typically less than 1 Å and x-ray can track the motions of each atom in the liquid phase. Time-resolved x-ray absorption spectroscopy must also be mentioned as a complementary atomic probe to time-resolved x-ray diffraction [6].

The proof of principle for liquid phase time-resolved diffraction experiments with 100 ps time resolution has been published previously [2-4]. In these works, experimental determination of temporally varying atom-atom pair distribution functions has been achieved. More recently, rather complex structural dynamics of $C_2H_4I_2$ dissolved in methanol has been elucidated [4]. In this approach, the changes in scattering due to photoexcitation were considered as several classes of diffraction modifications: (i) the structural changes in the solutes by the photoreaction, (ii) the changes of solvent cage structure around solute molecules, and (iii) the solvent's structural change due to the energy transfer (such as heating and thermal expansion). The hydrodynamics of the solvents are mathematically correlated with time-dependent solution reaction dynamics, which can transfer energy from photon-absorbing solute molecules

---

[*] Author to whom correspondence should be addressed. E-mail: Hyotcherl.Ihee@kaist.ac.kr

CP879, *Synchrotron Radiation Instrumentation: Ninth International Conference,*
edited by Jae-Young Choi and Seungyu Rah
© 2007 American Institute of Physics 978-0-7354-0373-4/07/$23.00

to the solvent. In the present study, we have investigated structural dynamics of iodoform in methanol solution using synchrotron based time-resolved liquid phase x-ray diffraction.

Following excitation of a nonbonding electron localized on C-I bond to an antibonding (n(I)→σ*(C-I) transition), one of the C-I bond in the iodoform is broken [7]. As results, the iodine atom, which is formed in the cage may escape from the cage, or it may recombine geminately with CHI_2 radical to produce the parent molecule or isomer configuration of iodoform, iso-iodoform (CHI_2-I). Recent investigations using transient resonance Raman spectroscopy and femtosecond pump-probe spectroscopy have shown that initially produced photofragments, iodine atom and CHI_2 radical, recombine within the solvent cage to form the isomer of CHI_2-I and the isomer is stable in the cage up to μs range.[7-8] However, direct evidence of isomer formation can be provided only by using time-resolved x-ray diffraction. Here we determined the primary reaction pathway and subsequent structural dynamics of dissociated CHI_3 in methanol using time-resolved x-ray diffraction. The difference diffraction data were analyzing globally by considering the diffraction contributions from solute, cage, and solvent's hydrodynamic responses.

EXPERIMENTAL

Time-resolved diffraction data were collected on the beamline ID09B at the European Synchrotron Radiation Facility (ESRF) using the optical-pump and x-ray-probe scheme. The detailed experimental geometry and analysis procedure were previously described [2-4]. In brief, 20 mM CHI_3 (Aldrich, 99.5%) was circulated through a high pressure sapphire slit nozzle (300 μm thickness and 1-5 m/s jet speed). Femtosecond laser pulses from a Ti:Sapphire laser system, synchronized with the single pulse of x-ray, were frequency tripled and temporally stretched to 2 ps to avoid multiphoton excitation. After electronically setting the time-delay (~3 ps accuracy), polychromatic x-ray pulses (5×10^8 photons per 100-ps-long pulse, ~ 3% of ΔE/E) were focused onto the sample jet (100×60 μm² spot size). Scattering of x-ray from the sample were collected with an area detector (MARCCD) with a sample-to-detector distance of 43 mm. The diffraction data were averaged over 20 images and were normalized to the total scattering in the region $6.84 \leq q \leq 6.99$ Å⁻¹ where the scattering is insensitive to structural changes. Diffraction data were measured for time-delays of -200 ps, 100 ps, 300 ps, 1 ns, 3 ns, 6 ns, 10 ns, 30 ns, 45 ns, 60 ns, 300 ns, 600 ns, 1μs, and 3 μs. Additional time-delay (-3 ns) was also collected between the images. Difference-diffraction curves ($\Delta S(q,t)$, [$q=(4\pi/\lambda)\sin(2\theta/2)$ where λ is the wavelength of the x-rays and θ is the scattering angle; t is the time delay]) were generated by subtracting the reference data at -3 ns from the data at any other time-delay.

RESULTS AND DISCUSSION

General Information from Difference Diffraction Intensity

The difference diffraction intensities ($q\Delta S(q,t)$) and related radial distribution functions $r\Delta S(r,t)$ are shown in Fig. 1(A) and (B). At -100 ps, which referenced to the time-delay of -3 ns, there is no difference diffraction intensity as expected. At positive time-delays, the difference diffraction patterns emerge and progress with time. More direct information of structural changes can be drawn in the real-space representation of the difference diffraction intensity, $q\Delta S(q,t)$. The sine-Fourier transformation of $q\Delta S(q,t)$ provides one way of the real-space representation, $q\Delta S(q,t)$:

$$r\Delta S(r) = \frac{1}{2\pi^2} \int_0^\infty q\Delta S(q) \sin(qr) \exp(-kq^2) dq \qquad (1)$$

where the constant k ($k=0.03$ Å²) is a damping constant. This difference radial distribution function (RDF), $r\Delta S(r,t)$ is resulted from the average electron density changes by the illumination as a function of interatomic distance, r. Since iodine scatters x-ray more than the other atoms (C, H, and O) in this molecular system, the structural changes of iodine-iodine distance are much pronounced in this representation. The RDFs in Fig. 1(B) exhibit both negative and positive peaks: there is strong depletion of correlation at 3.8 Å with a corresponding rise at 2.9 Å at early times. Since the structural parameters of CHI_3 are well known [8], one can easily assign the negative peak to the bond breakage of C-I in CHI_3 and positive peak to the formation of new C-I bond in CHI_2 radical. However the detailed analysis must include all three contributions: solute-solute, solute-solvent and solvent-solvent which is done in the following section [4,5].

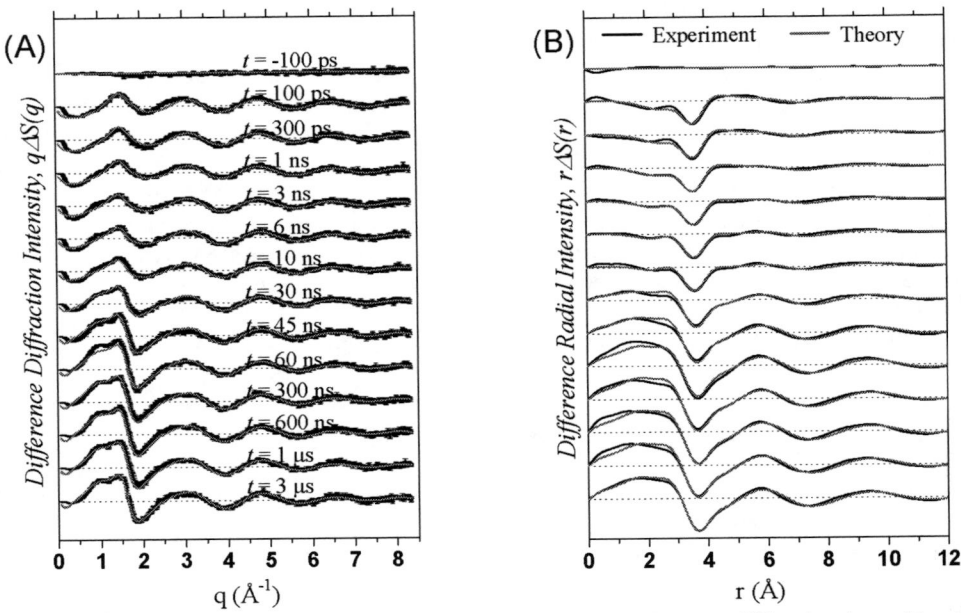

FIGURE 1. Time-resolved diffraction signal for CHI_3 in methanol. (A) Difference diffraction intensities (black), $q\Delta S(q,t)$, excited minus nonexcited. (B) Difference radial distribution functions (black), $r\Delta S(r,t)$ which are sine-Fourier transforms of $q\Delta S(q,t)$ in (A). In both figures, the best fits from a theoretical model are also shown in red curves.

Global Fitting of Difference Diffraction Patterns

To explain and fit the experimental difference diffraction ($q\Delta S(q,t)$), we include all three contributions (solute, solute-solvent cage, solvent's responses). For the solute contribution, we considered all photoreaction pathways. Following photoexcitation, the parent molecules (CHI_3) are dissociated to the CHI_2 and I radicals. These radicals can recombine to the parent molecule (CHI_3) or to the isomer (CHI_2-I). Moreover, the I radical can escape from the cage and recombine to I_2 with I radical from the other cage or make CHI_3 again via nongeminate recombination. In summary, we have five different solute candidates (CHI_3, CHI_2, CHI_2-I, I, and I_2).[7,8] Theoretical scattering curves of these photo-species, which contain the solute and solute-solvent cage structures, were obtained from molecular dynamics (MD) simulations. The pure solvent's contribution is originated from heating and subsequent thermal expansion. Heating is induced by the transfer of energy from the photon-absorbing solutes to the surrounding solvent. Recently we showed that these hydrodynamic effects, i.e. the solvent response to an ultrafast temperature rise ($[\partial S(q,t)/\partial T]_\rho$ and $[\partial S(q,t)/\partial \rho]_T$), can be determined experimentally in pure methanol, without inducing any chemical change, by exciting it with near IR laser pulses that excite over-tones in C-H and O-H vibrations [9].

Once we have obtained all components, the experimental difference diffraction data were fit with a linear combination of these components. The basis set of model functions consists of the solvent terms ($[\partial S(q,t)/\partial T]_\rho$ and $[\partial S(q,t)/\partial \rho]_T$), the difference scattering functions from transitions in the solutes alone, and their caged equivalents. Since hydrodynamic changes (temperature and density changes) of the solvent is mathematically linked by solute reaction dynamics, i.e. time-dependent solute concentration changes, the data of all time-delays are linked and were fitted globally instead of fitting the data at each time-delay separately. The fitting parameters include the reaction constants of all solute chemical reactions, the laser spot size and yields of photodissociation and vibrational cooling processes. As results, this global-fitting process produces time-dependent concentration changes of all putative solute species, and time-dependent temperature and density changes of bulk solvent. More detailed descriptions of global-fitting method can be found in elsewhere [4]. Optimal fits to all experimental data using a kinetic model for $CHI_3 \rightarrow CHI_2 + I$ and subsequent reactions are shown in Figs. 1(A) and 1(B). The fitted theoretical curves successfully reproduce experimental difference curves. The resulting reaction dynamics of solute and hydrodynamics changes of bulk solvent are shown in Fig. 2(A) and 2(B), respectively.

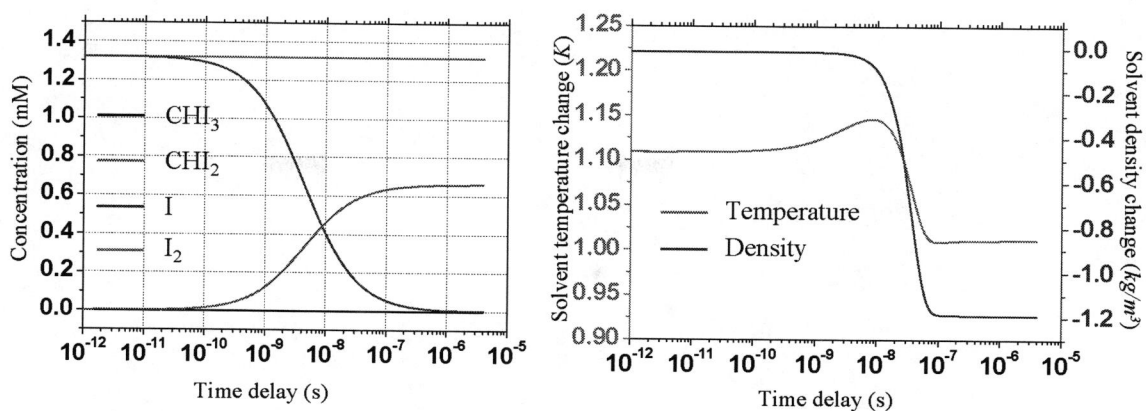

FIGURE 2. (A) The population change of the all photo-species as a function of time-delay. (B) The changes in the solvent temperature (red) and density (black) as a function of time-delay.

Structural Dynamics of Photoreaction of Iodoform

Figure 2(A) shows the population changes of all related chemical species from the global fitting. CHI_2 and I are the dominant species at 100 ps and they originate from prompt dissociation of the parent molecules in the cage. Then most of I radicals recombine non-geminately with iodine atoms that have escaped from another solvent cages to form the I_2 molecules with the rate constant 3.1 ($\pm$0.5) $\times 10^{10}$ $M^{-1}s^{-1}$. The time constant for the non-geminate recombination of I_2 is close to the values found in CCl_4 solutions from optical spectroscopy [10]. After thermal expansion, the density of the solvent decreased by 1.2 kg/m^3 at 1 μs, which corresponds to the temperature change of 1.02 K.

In the spectroscopic results, it was claimed that the isomer was formed after prompt photodissociation and survived up to few μs [7,8]. To check the possibility of isomer formation, we separately performed the global-fitting using the isomer formation reaction. The χ^2 value for the model including the isomer formation process (25.96) was greater than that without it (2.809) as in Fig. 2(A) by 9 folds at 10 ns time-delay. More importantly, if we include the isomer formation channel and let it float in the fitting process, the fraction of isomer formation converged to zero. These findings confirm that the dissociation channel without the isomer formation reproduces the experimental difference diffraction curves with higher reliability than does the model including the isomer formation channel.

ACKNOWLEDGMENTS

This work was supported by a grant from the Nano R&D Program of the Korea Science and Engineering Foundation (2005-02638) to HI.

REFERENCES

1. P. Allen and D. J. Tildesley, *Computer Simulation of Liquids*, Clarendon Press: Oxford, 1989.
2. A, Plech et al., *Phys. Rev. Lett.* **92**, 125505 (2004).
3. M. Wulff et al., *J. Chem. Phys.* **124**, 034501 (2006).
4. H. Ihee et al., *Science* **124**, 1223-1227 (2005).
5. Kim et al., *Proc. Natl. Acad. Sci.* accepted for publication (2006).
6. M. Saes, *Phys. Rev. Lett.* **90**, 047403 (2003).
7. M. Wall et al., *J. Phys. Chem. A* **107**, 211-217 (2003).
8. X. Zheng and D. L. Phillip, *Chem. Phys. Lett.* 324, 175-180 (2000).
9. M. Cammarata et al., *J. Chem. Phys.* **124**, 124504 (2006).
10. S. Aditya and J. E. Willard, *J. Am. Chem. Soc.* **79**, 2680-2681 (1957).

CHAPTER 6
MICRO/NANOSCOPY

Soft X-ray Zone Plate Microscopy to 10 nm Resolution with XM-1 at the ALS

Weilun Chao[1], Erik H. Anderson, Bruce D. Harteneck, J Alexander Liddle and David T. Attwood[1]

Center for X-ray Optics, Lawrence Berkeley National Laboratory, Berkeley, California, CA 94720
[1]University of California, Berkeley, California, CA 94720

Abstract Soft x-ray zone plate microscopy provides a unique combination of capabilities that complement those of electron and scanning probe microscopies. Tremendous efforts are taken worldwide to achieve sub-10 nm resolution, which will permit extension of x-ray microscopy to a broader range of nanosciences and nanotechnologies. In this paper, the overlay nanofabrication technique is described, which permits zone width of 15 nm and below to be fabricated. The fabrication results of 12 nm zone plates, and the stacking of identical zone patterns for higher aspect ratio, are discussed.

Keywords: zone plate microscopy, overlay zone plate fabrication, electron beam lithography, spatial resolution
PACS: 41.50.+h, 07.85.Tt, 42.30.Kq, 42.82.Cr, 42.40.Lx

INTRODUCTION

Soft x-ray zone plate microscopy has proven to be a valuable tool in nanoscience and nanotechnology. Its unique combination of capabilities, including high spatial resolution, elemental and sometimes chemical identification, relatively simple sample preparation, and flexible sample environments for in-situ studies, make the technique complementary to electron, scanning probe, and visible light microscopies. A broad range of physical and life sciences have gained valuable information through the use of this microscopy [1]. To extend the microscopy to nanometer-scale phenomena, resolution improvement has been actively pursued in the past few decades. Resolution of 20 nm has been achieved [2]. Further improvement however was extremely challenging. In this paper, the challenges, the overlay fabrication technique as the solution, and the latest results in resolution and zone plate fabrication by use of the technique are discussed.

CHALLENGES AND SOLUTION TO HIGHER RESOLUTION

The spatial resolution of zone plate microscopy is mainly determined by the quality of the imaging zone plate: the outermost zone width (Δr) and zone placement accuracy [3]. For ideal zone plate lenses, the resolution is approximately equal to Δr. A common way to improve the resolution is to reduce Δr. At present, high resolution zone plates are fabricated by electron beam lithography [4]. Zone Plate lenses with Δr of 20-25 nm have been successfully fabricated [1]. Fabrication process limitations in dense line fabrication, such as electron scattering [5], low e-beam resist contrast and development issues, however, had prevented fabrication of narrower zones. Isolated lines on the other hand, are less susceptible to these problems, and lines of around 10 nm wide can be routinely fabricated. This fact led us to develop a new zone plate fabrication technique, in which a dense zone plate pattern is divided into two (or more) semi-isolated, complementary patterns (Fig. 1), and each pattern is sequentially fabricated and overlaid to the other patterns. The key of success to this overlay technique is the alignment accuracy of the patterns, which for zone plate is needed to be better than one third of the smallest zone width [3]. To obtain this accuracy, before fabrication of any zones, a set of alignment marks is fabricated outside the zone plate area (Fig. 1). Using an internally developed alignment algorithm [6], each zone pattern is fabricated at its desired

CP879, *Synchrotron Radiation Instrumentation: Ninth International Conference*,
edited by Jae-Young Choi and Seungyu Rah
© 2007 American Institute of Physics 978-0-7354-0373-4/07/$23.00

location relative to the alignment marks. The reference to the same alignment mark set in each zone pattern fabrication, as well as extremely accurate (sub-pixel) pattern placement yielded by the algorithm, enables consistent achievement of sub-2 nm zone placement accuracy.

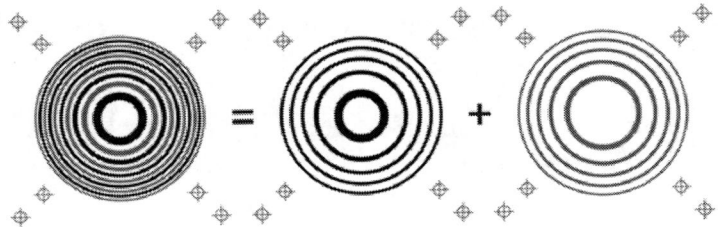

FIGURE 1. Illustration of the overlay fabrication technique. A dense zone plate is divided into two semi-isolated, complementary patterns, which are sequentially fabricated and overlaid using alignment marks (represented by the crosses and circles outside the patterns) to the other patterns in accuracy better than one third of the outermost zone width.

BREAKTHROUGH IN RESOLUTION AND ZONE PLATE FABRICATION MADE POSSIBLE BY THE OVERLAY TECHNIQUE

Using the overlay technique, we have successfully fabricated in our first attempt zone plates of 15 nm outermost zone width at the LBNL's Nanofabrication Laboratory. Figure 2 shows an SEM micrograph of one of the zone plate. The optic, which has a diameter of 30 μm, 500 zones and a gold zone thickness of 80 nm, has near perfect zonal placement. The zone placement accuracy is measured to be 1.7 nm, much smaller than the required accuracy of 5 nm for the 15 nm zone plates. The zone plate was tested at the soft

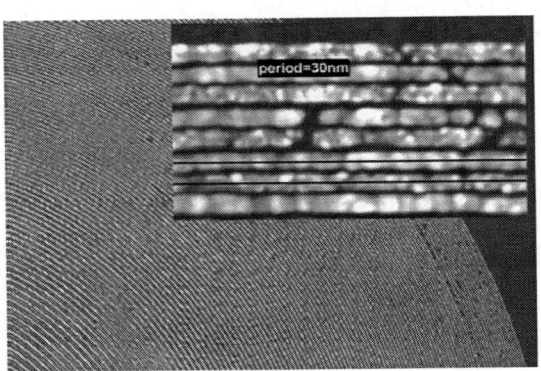

FIGURE 2. A scanning electron micrograph of a 15 nm outermost zone width zone plate fabricated using the overlay nanofabrication technique. The measured zone placement accuracy is 1.7 nm. The zone plate has 500 zones and diameter of 30 μm, with plated gold thickness of 80 nm.

x-ray full-field microscope[7], XM-1, at the Advanced Light Source (ALS). The schematic of the microscope is shown in Fig. 3. Test objects of multilayer coatings in cross section, fabricated using magnetron sputtering and conventional transmission electron microscopy sample preparation techniques [8], were imaged. The images of 19.5 nm and 15.1 nm half-period Cr/Si test object are shown in Fig. 3b and d, respectively. The wavelength of 1.52 nm (815 eV) was used for these images. For comparison, the earlier images of the same objects taken with a zone plate of 25 nm outermost zone width are included in Fig. 4a and c, respectively[2]. For 19.5 nm half period pattern, while good modulation can be seen from the image obtained by the 25 nm zone plate (Fig. 4a), the image quality is drastically improved in the image taken with the new zone plate (Fig. 4b). The improvement by the 15 nm zone plate is particularly evident in the 15.1 nm Cr/Si images, in which the previous zone plate did not yield any modulation in its image due to the theoretical resolution limit of lens (Fig. 4c), whereas the image taken by the new 15 nm zone plate exhibit a clear modulation. Further analysis indicates that sub-15 nm spatial resolution has been achieved with the new zone plate [9]. The result is a significant step forward, permitting application of soft x-ray microscopy to studies that would not be possible with zone plates fabricated using the conventional e-beam lithographic processes. The new zone plate has been used in nano-magnetic studies, in which granular details approaching fundamental length scales were revealed [10].

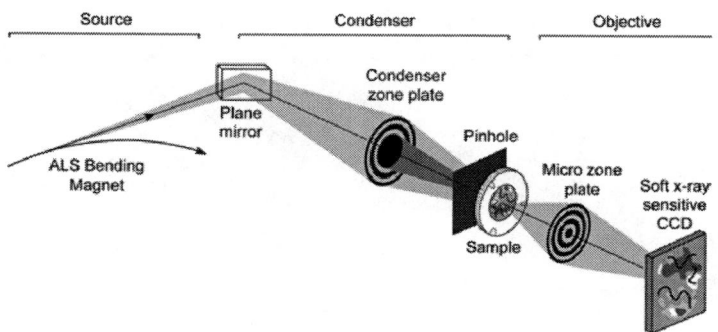

FIGURE 3. A schematic of the soft x-ray full-field microscope, XM-1, at the ALS. The condenser zone plate used in the experiment here has these parameters: $\Delta r = 40$ nm, 56250 zones, 9 mm diameter. The degree of partial coherence, σ, is 0.38.

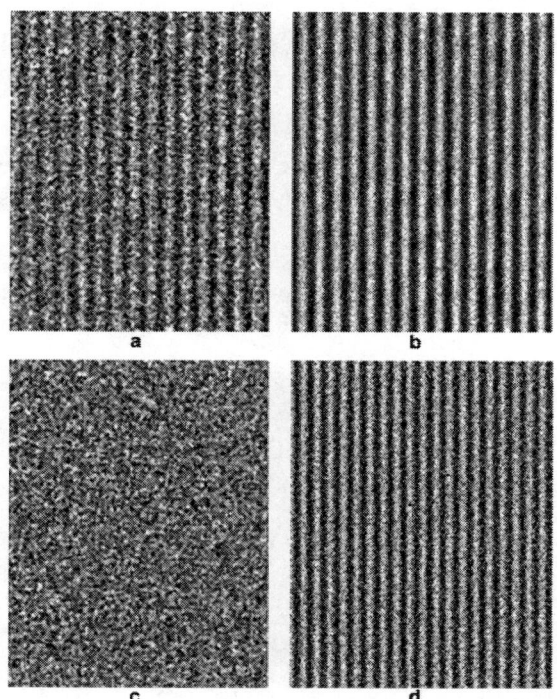

FIGURE 4. Soft x-ray images of 15.1 nm and 19.5 nm half-period Cr/Si multilayer test objects, obtained with zone plates having outer zone widths of 25 nm and 15 nm. (a) Image of 19.5 nm half-period test object obtained previously with a 25 nm zone plate. (b) Image of 19.5 nm half-period object with the 15 nm zone plate. (c) Image of 15.1 nm half-period with the previous 25 nm zone plate. (d) Image of 15.1 nm half-period with the 15 nm zone plate. Images (a) and (c) were obtained at a wavelength of 2.07 nm (600 eV photon energy); (b) and (d) were obtained at a wavelength of 1.52 nm (815 eV). The equivalent object plane pixel sizes for images (a) and (c) are 4.3 nm; the sizes for (b) and (d) are 1.6 nm.

Using the overlay technique, zone plates with 12 nm outermost zone widths have been fabricated. Figure 5 shows an SEM micrograph of one of the zone plates. In our first attempt, zones wider than desired were obtained, as in the case of 15 nm zone plates. This is due to unwanted sidewall etching of the resist in the dry etch process before the formation of the gold zones by electroplating. Breaks can also be seen in the zones, which are contributed by voids in the underlying plating base. Both of these problems can be solved by optimization of the corresponding processes. More importantly, zonal misplacement is revealed in the micrograph. The modest placement accuracy, estimated to be 2-3 nm, is caused by intermittent mechanical failure in isolating vibration in our electron beam lithography system. While we anticipate sub-2 nm placement accuracy to be achieved after replacement of the vibration isolation system, the achieved accuracy is smaller than the required accuracy of 4 nm for the 12 nm zone plates.

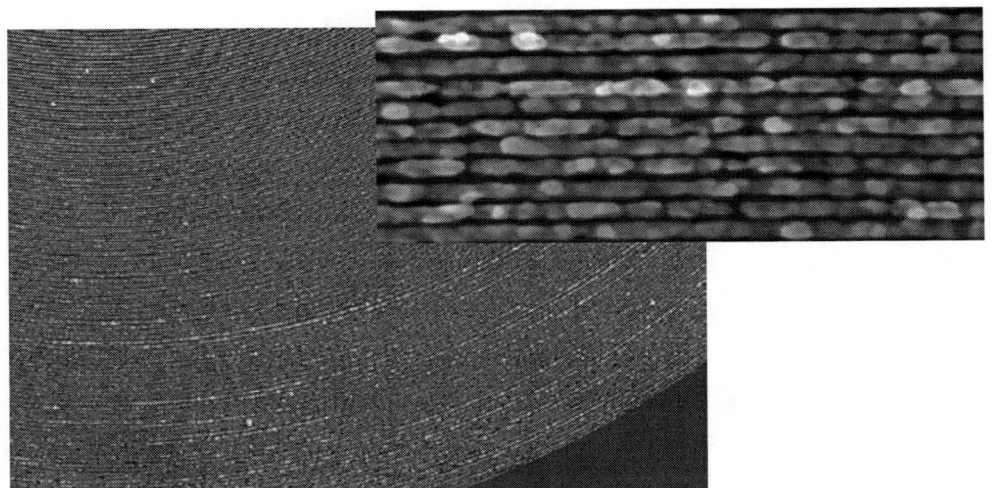

FIGURE 5. A scanning electron micrograph of a 12 nm outermost zone width zone plate fabricated using the overlay nanofabrication technique. The zone placement accuracy is estimated to be 2-3 nm. The zone plate has 500 zones and diameter of 24 μm, with plated gold thickness of 30-40 nm.

While the alternating opaque zones can be fabricated separately and overlaid to achieve narrow zones as in the overlay technique, identical zones can be fabricated and stacked for higher aspect ratio zone plates. Figure 6 shows an SEM micrograph of a supposedly 27 nm outermost zone width zone plate; the zones have a width of 27 nm, separated by an 81 nm space. The zone plate was formed by three identical exposure of one of the zone patterns in the overlay processes. The overlay accuracy is near perfect. The achieved zonal thickness is 70 nm, in comparison to 30 nm obtained in a single exposure.

FIGURE 6. A scanning electron micrograph of a 27 nm outermost zone width zone plate fabricated by overlaying three identical zone patterns in the overlay processes. Only one of the zone patterns in the processes was exposed; the alternating opaque zone pattern is missing. The zones of 27 nm in width are separated by an 81 nm space. The achieved zonal thickness is 70 nm, in comparison to 30 nm obtained in a single exposure.

CHALLENGES OTHER THAN ZONE PLATE FABRICATION TO HIGH RESOLUTION IMAGING

While achieving narrow zones has been the main challenge to high resolution imaging, other issues may arise as the resolution is improved. Both the focal length and depth of field are dramatically reduced with the outermost zone width. For instance, at 2.4 nm wavelength, a zone plate of 10 nm outermost zone width and 500 zones has a focal length of 83 μm and a depth of field of 166 nm. In addition, many samples of interest contains features small in both lateral and longitudinal dimension. A gold sphere of 10 nm diameter has a contrast of only 7% at 2.4 nm wavelength. Phase contrast imaging, if possible, may be needed for enhanced image signal and quality. Furthermore, zone edge roughness may play a more significant role in determining zone plate efficiency, resolution and image contrast.

CONCLUSIONS AND FUTURE WORK

By use of the overlay technique, 15 nm zone plates have been successfully fabricated, and sub-15 resolution is achieved. Preliminary results of 12 nm zone plates clearly show the extensibility of the overlay technique to zone width of 12 nm. We anticipate realization of even 10 nm zone plates using the technique in the near future. By stacking identical zone patterns, higher zonal structures were fabricated. The overlay processes will be optimized for better zonal quality, and dense zone plate patterns will be stacked to achieve higher aspect ratio.

ACKNOWLEDGMENTS

The authors wish to acknowledge financial support from the National Science Foundation's Engineering Research Centre Program, the Department of Energy's Office of Science, Office of Basic Energy Sciences, and the Defense Advanced Research Projects Agency.

REFERENCES

1. J. Susini, D. Joyeux, and P. F., eds., *X-Ray Microscopy VII*, Paris, EDP Sciences, 2003.
2. W. L. Chao, E. Anderson, G. P. Denbeaux, B. Harteneck, J. A. Liddle, D. L. Olynick, A. L. Pearson, F. Salmassi, C. Y. Song, and D. T. Attwood, *Opt. Lett.* **28**, 2019-2021 (2003).
3. A. G. Michette, *Optical Systems for Soft X Rays*, New York, Plenum Press, 1986.
4. E.H. Anderson, D. L. Olynick, B. Harteneck, E. Veklerov, G. Denbeaux, W. L. Chao, A. Lucero, L. Johnson, and D. Attwood, *J. Vac. Sci. & Techn. B* **18**, 2970-2975 (2000).
5. G. Owen and J.R. James, "Electron Beam Lithography Systems," in *Microlithography Science and Technology*, edited by J. R. Sheats and B. W. Smith, New York, Marcel Dekker, 1998, pp. 395-399.
6. E. H. Anderson, D. Ha, and J. A. Liddle, *Microelect. Eng.* **73-74** 74-79 (2004).
7. W. Meyer-Ilse, H. Medecki, L. Jochum, E. Anderson, D. Attwood, C. Magowan, R. Balhorn, M. Moronne, D. Rudolph, and G. Schmahl, Synch. Rad. News 8 29 (1995).
8. W. Chao, E. H. Anderson, G. Denbeaux, B. Harteneck, A. L. Pearson, D. Olynick, F. Salmassi, C. Song, and D. Attwood, *J. Vac. Sci. & Techn. B* **21**, 3108-3111 (2003).
9. W. L. Chao, B. D. Harteneck, J. A. Liddle, E. H. Anderson and D. T. Attwood, *Nature* **435**, 1210-1213 (2005).
10. D. Kim, P. Fischer, W. Chao, E. Anderson, S. Choe, M. Im and S. Shin, *J. Appl. Phys.* **99**, 08H303 (2006).

Hard X-ray Microscopy with sub 30 nm Spatial Resolution

Mau-Tsu Tang[1], Yen-Fang Song[1], Gung-Chian Yin[1], Fu-Rong Chen[1,2], Jian-Hua Chen[1], Yi-Ming Chen[1], Keng S. Liang[1], F. Duewer[3] and Wenbing Yun[3]

[1]National Synchrotron Radiation Research Center, Hsinchu 30076, Taiwan
[2]Department of Engineering and System, National Tsing Hua University, Hsinchu 30043, Taiwan
[3]Xradia Inc., Concord, CA 94520, USA

Abstract. A transmission X-ray microscope (TXM) has been installed at the BL01B beamline at National Synchrotron Radiation Research Center in Taiwan. This state-of-the-art TXM operational in a range 8-11 keV provides 2D images and 3D tomography with spatial resolution 60 nm, and with the Zernike-phase contrast mode for imaging light materials such as biological specimens. A spatial resolution of the TXM better than 30 nm, apparently the best result in hard X-ray microscopy, has been achieved by employing the third diffraction order of the objective zone plate. The TXM has been applied in diverse research fields, including analysis of failure mechanisms in microelectronic devices, tomographic structures of naturally grown photonic specimens, and the internal structure of fault zone gouges from an earthquake core. Here we discuss the scope and prospects of the project, and the progress of the TXM in NSRRC.

Keywords: transmission X-ray microscopy, phase contrast, tomography, synchrotron radiation, zone plate
PACS: 07.85.Tt, 07.85.Qe, 42.79.Ci

INTRODUCTION

Because X-rays have a wave nature, X-ray microscopy has long been expected to be a powerful imaging tool on a nanometer scale. Due to the great depth of penetration of X-rays into matter, X-ray microscopy benefits that research that requires non-destructive and three-dimensional probing *in situ*. The development of X-ray microscopy has been delayed relative to its optical and electron counterparts for two main reasons – the tardy availability of brilliant X-ray sources and the difficulty of fabricating highly performing X-ray optical elements. The modern synchrotron lamp that provides X-rays with brilliance a million times that of an in-house rotating-anode generator has vitally fulfilled the requirement for photons. The invention of highly performing X-ray optical elements began only in the late 1990s, when several focal components such as zone plates[1,2,3], tapped capillaries[4,5], compound refractive lenses[6,7,8], Fresnel-Bragg diffractive lenses[9,10] and Kirk-Baez mirrors[11] were introduced with great progress.

Here we describe the optical design of the nano-Transmission X-ray Microscope (nTXM) newly built in the National Synchrotron Radiation Research Center (NSSRC). Although the optical design of the NSRRC-nTXM inherits the generic design of a full-field TXM, several ambitious optical concepts and advanced optical components have been aggressively adopted for this microscope, such as a tapered-capillary condenser, a high-performance micrometer zone plate (ZP) with outermost zone width 50 nm and aspect ratio 18, the first application of a higher ZP diffraction order to achieve a spatial resolution less than 30 nm, and a Zernike phase-contrast mode for imaging light specimens. Thus the NSRRC-nTXM provides two-dimensional images and three-dimensional tomography with spatial resolution from less than 30 nm to 60 nm, in an energy range 8-11 keV, and with a Zernike phase-contrast mode for imaging of light materials.

LIGHT SOURCE AND BEAMLINE

The radiation for the microscope emanates from a 5-T superconducting wavelength shifter (SWLS), a wiggler-like insertion device to increase the critical energy of the NSRRC 1.5-GeV synchrotron spectrum from 2.14 keV to

CP879, Synchrotron Radiation Instrumentation: Ninth International Conference,
edited by Jae-Young Choi and Seungyu Rah
© 2007 American Institute of Physics 978-0-7354-0373-4/07/$23.00

7.5 keV, with an average photon flux 5×10^{11} photons s^{-1} per 0.1 % bw in an energy range 5-20 keV. The X-rays generated with the wavelength shifter constitute an effective source for a transmission X-ray microscope (TXM) of full field type in that the large illuminating phase space matches efficiently the numerical aperture of the objective zone plate. The beamline 01B at NSRRC has two major optical components – a prefigured focusing mirror (FM) of toroidal shape and a double-crystal monochromator (DCM)[12]. X-rays generated by the SWLS are focused primarily at the charge-coupled detector (CCD) with a FM of focal ratio nearly equal to 1:1. The DCM exploiting a pair of Ge (111) crystals selects X-rays of energy 8-11 keV, with energy resolution better than 1000 to match the zone number of the objective zone plate, in the present case $\sim$ 400. The energy range 8-11 keV is thus chosen to cover the characteristic absorption edges of Cu, Zn, Ga, Ge, As, Ta, W, Au, Hg, Pb etc., which are the most useful elements for semiconductor industries. The beamline construction and commissioning were completed in years 2002 and 2004, respectively.

OPTICAL LAYOUT

Figure 1 shows a schematic optical layout of the NSRRC X-ray microscope. Similar to other transmission-type microscopes, incident X-rays are first condensed onto the sample and then magnified with an objective lens, to form projection images on an area detector.

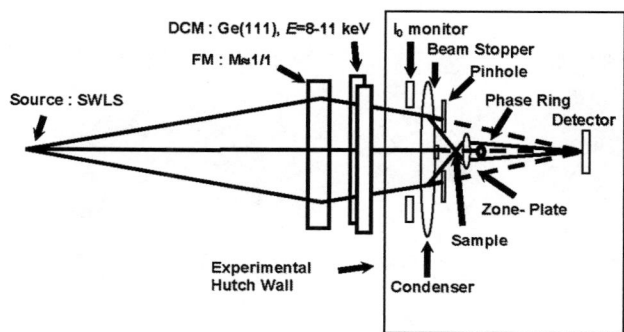

FIGURE 1. Optical layout of NSRRC n-TXM

The condenser is an elliptically shaped capillary that intercepts and focuses (by internal total reflection) the X-rays from the beamline onto the sample. The focusing efficiency of the tapered capillary condenser is $\sim$ 90 %. This focusing efficiency is significantly greater than that of zone-plate condensers used in other existing X-ray microscopes, the focusing efficiency of which is typically less than 10 %. Properly coupled with a beam stop and a pinhole aperture, the condenser delivers a beam to the sample of a hollow-cone shape.

A zone plate is a circular diffraction grating possessing alternating opaque and transparent concentric zones. The zones function in such a way that the phases of the X-rays alter by 2π on each passage through the sequential transparent zones and interfere in phase at the image point. As the numerical aperture of a zone plate is much less than unity in the X-ray case, the zone plate can be adopted like an ordinary refractive lens in an optical microscope so that geometrical optics becomes directly applicable. The spatial resolution δ of the microscope is defined with an objective zone plate according to $\delta = 1.22 \Delta r / m$, in which Δr is the outermost width of the zone plate and $m = 1, 3, 5 \ldots$ is the diffraction order. To improve the spatial resolution, one expects to adopt a zone plate with a finer outermost zone width, and one can achieve m times better spatial resolution using a greater diffraction order m but at the cost of m^2 duration of exposure. In the present microscope, the 50-nm width of the outermost zone delivers a spatial resolution 60 nm in the first diffraction order, and a spatial resolution 20 nm is achievable in principle with the third diffraction order of this zone plate. The challenge to fabricate a highly performing zone plate is set by the large aspect ratio of the outermost zone, in this case ~18. During the test, a weak second-order diffraction image was also observed. The X-ray image is converted into a visible image with a scintillator; this image is further magnified 20x using an imaging system for visible light, producing a total magnification 880x and 2640x for imaging with the first and third diffraction orders, respectively. Over the range 8-11 keV of X-ray energy, the magnification varies between 780x and 880x for imaging with the first-order diffraction through the energy dependence of the zone-plate lens. There are in total six zone plates, each three serving for first-order and third-order diffraction modes.

The interaction of X-rays with a sample is represented with a complex refractive index of the sample, $n(\lambda)=1-\delta(\lambda)-i\beta(\lambda)$, in which λ is the wavelength of incident X-rays, $\delta(\lambda)$ represents the phase shift and $\beta(\lambda)$ the absorptive properties of a sample. The outgoing wave front of X-rays is distorted by a sample in both amplitude and phase. In a conventional X-ray microscope, only the variation of amplitude is recorded, named the absorption contrast mode, whereas the phase term is lost in the imaging process. The phase term is retrievable with Zernike's phase-contrast method that was introduced into optical microscopy from 1942[13] and into X-ray microscopy in 1998 [14]. In the present microscope, the Zernike phase contrast is achieved on placing a gold ring (thickness 3 μm) at the back focal plane of the zone plate, creating a 270° phase retarded from the direct beam, resulting in a record of the phase contrast of the sample on the detector. To cover the energy range 8-11 keV, the first and third diffraction orders, the absorption contrast mode and the phase contrast mode, there are in total three tapered capillaries, six zone plates and 12 phase rings.

PERFORMANCE

The spatial resolution of this microscope was tested on imaging an electroplated gold spoke pattern in first and third diffraction order modes (Fig. 2). The field of view of these images are 15μm×15μm and 5μm×5μm for first and third diffraction order modes, respectively. The visibly resolved 50-nm finest line widths imaged in first diffraction order mode indicates the achievement of the theoretical 60-nm spatial resolution. A modulation transfer function (MTF) test of the third diffraction order mode demonstrates a 25-nm spatial resolution.

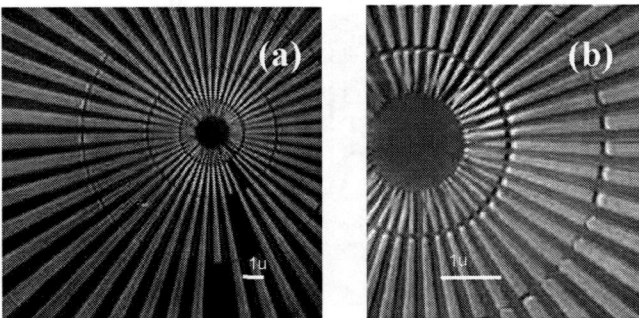

FIGURE 2. Gold spoke pattern imaged in (a) first diffraction order and (b) third diffraction order. The bar scale is 1 μm.

The capability of the Zernike phase contrast was tested on imaging a plastic zone plate. Fig. 3 shows the images of a plastic zone-plate, 1 μm thick, imaged in the absorption mode (a), and in the phase contrast mode (b). The Zernike phase contrast in Fig. 3(b) is calculated according to the criterion of Michelson visibility ($I_{max}-I_{min}/I_{max}+I_{min}$) and is estimated to be ~12 %, which is smaller than the calculated[15] 33 % due to an unoptimized thickness of the phase ring. Cracks formed on the surface of the plastic zone plate are shown clearly in the phase contrast enhancement, but only vaguely in the absorption contrast.

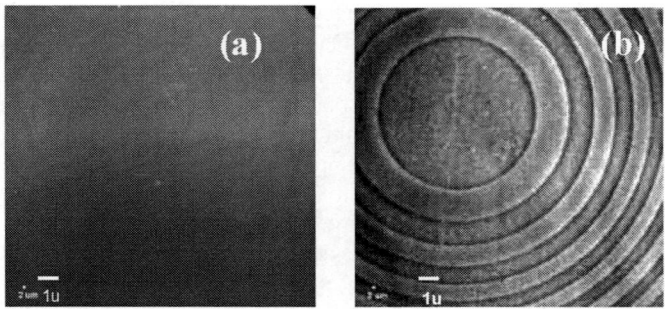

FIGURE 3. A plastic zone plate imaged in (a) absorption contrast mode, and (b) Zernike phase contrast mode. The bar scale is 1 μm.

The tomography of the microscope was tested on imaging an electro-plated gold spoke pattern (Fig. 4). The tomography was reconstructed based on 141 sequential image frames taken in the first diffraction order mode with the azimuthal angle rotating from -70° to 70°. The spatial resolution for the gold spoke pattern is estimated to be 60nm×60nm×80nm.

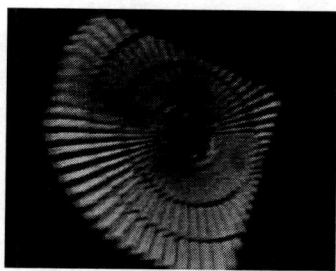

FIGURE 4. Tomographic reconstruction of a gold spoke-pattern with inner line width 50 nm.

CONCLUSIONS

The NSRRC-nTXM has demonstrated two-dimensional images and three-dimensional tomography with spatial resolution better than 60 nm. A spatial resolution better than 30 nm was successfully achieved on using the third diffraction order of an objective zone plate. The Zernike phase contrast has been demonstrated on imaging a plastic zone plate. The Zernike phase contrast has facilitated studies of immuno-labelled proteins in cells and the photonic-like butterfly wing scales (QP-043). Elemental contrast tomography, combining the characteristic absorption of a specific element and tomography, has been applied in investigating the mechanism of failure of an IC (LP-009).

REFERENCES

1. B. Lai, W. B. Yun, D. Legnini, Y. Xiao, J. Chrzas, and P. J. Vidcaro, *Appl. Phys. Lett.* **61,** 1877-1879 (1992).
2. W. Yun and B. Lai, *Rev. Sci. Instrum,* **70** 3537-3541,(1999).
3. W. Yun, B. Lai, Z. Cai, J. Maser, D. Legnini, and E. Gluskin, *Rev. Sci. Instrum.* **70,** 2238-2241 (1999).
4. J. B. Murphy, D. L. White, Alastair A. MacDowell, and Obert R. Wood, *Appl. Optics,* **32,** 6920-6929 (1993)
5. D. H. Bilderback, D. J. Thiel, R. Pahl and K. E. Brister, *J. Synchrotron Rad.* **1,** 37-42 (1994).
6. M. A. Piestrup,a) J. T. Cremer, H. R. Beguiristain, C. K. Gary, and R. H. Pantell, *Rev. Sci. Instrum.* **71,** 4375-4379 (2000).
7. B. Lengeler, C. G. Schroer, M. Richwin, J. Tümmler, M. Drakopoulos, A. Snigirev, and I. Snigireva, *App. Phys. Lett.* **74,** 3924-3926 (1999).
8. B. Lengeler, J. Tümmler, A. Snigirev, I. Snigireva, and C. Raven, *J. App. Phys.* **84,** 5855-586 (1998).
9. P. Dhez, A. Erko, E. Khzmalian, B. Vidal, and V. Zinenko, *Appl. Optics,* **31,** 6662-6667 (1992)..
10. V. V. Aristov, Yu. A. Basov. And A. A. Snigirev, *Rev. Sci. Instrum.* **60,** 1517-1518 (1989).
11. P. Kirkpatrick and A. V. Baez, *J. Opt. Soc. Am.* **38,** 766 (1948).
12. Y. F. Song *et al.,* "X-ray Beamlines on a Superconducting Wavelength Shifter" in *Eighth International Conference on Synchrotron Radiation Instrumentation -2003,* edited by T. Warwick *et al.,* AIP Conference Proceeding 705, American Institute of Physics, Melville, NY, 2004, pp. 412-415 (2003).
13. F. Zernike, *Physica,* **9** , 686 (1942).
14. G. Schneider, *Ultramicroscopy,* **75,** 85 (1998).
15. Max Born and Emil Wolf, *Principles of Optics,* Pergamon Press (1986)

Hard X-ray Microscopic Images of the Human Hair

Jawoong Goo, Soo young Jeon, Tak Heon Oh, Seung Phil Hong, Hwa Shik Yon*, and Won-Soo Lee

Department of Dermatology and Institute of Hair and Cosmetic Medicine, Yonsei University Wonju College of Medicine, Wonju
*Pohang Accelerator Laboratory, Pohang University of Science and Technology, Pohang, Korea

Abstract. The better visualization of the human organs or internal structure is challenging to the physicist and physicians. It can lead to more understanding of the morphology, pathophysiology and the diagnosis. Conventionally used methods to investigate cells or architectures, show limited value due to sample processing procedures and lower resolution. In this respect, Zernike type phase contrast hard x-ray microscopy using 6.95keV photon energy has advantages. We investigated hair fibers of the normal healthy persons. Coherence based phase contrast images revealed three distinct structures of hair, medulla, cortex, and cuticular layer. Some different detailed characters of each sample were noted. And further details would be shown and these results would be utilized as basic data of morphologic study of human hair.

Keywords: Hard X-ray, Microscope, Human hair, PLS
PACS: 07.85.Tt

INTRODUCTION

X-ray microscope is fascinating modality in biologic or medico-clinical research fields and studies on the imaging or computerized tomographic imaging of small organs or cellular level have been performed [1]. Conventionally used methods such as a light microscopy, transmission electron microscopy(TEM) to see cellular architecture or morphology and sub-cellular organelle inevitably needed laborious, time-consuming processing procedures and devitalization of sample so these has limited value. Others, such as an ultrasonography, computerized tomography and magnetic resonance image, have advantages with real-time without devitalization or processing but have limited resolution.

In these respects, highly coherent hard x-ray microscope with KeV photon energy is suitable for studying human or animal organ samples [2] because most of samples from the living organs obtained as a mass or cluster of tissue. Hard x-ray microscope has low absorption contrast in biologic samples. So recently applied Zernike type phase contrast with Fresnel zone plate optics [3] enables us novel morphological studies of small organs to reveal. And it comes to important theme, 'what to see' more than 'how to see'. Hair is one of the most accessible organs from human. However, there have been researches about scattering or spectroscopic of the hair fiber but about the image [4, 5]. So we investigated hairs structure from different body part and races by hard x-ray microscope and some novel images are presented.

EXPERIMENTAL SET-UP

Zernike type phase contrast hard x-ray microscope with Fresnel zone plate was set at Pohang Acceleratory Laboratory, 1B2 beam line (Fig. 1.). The light source is third generation synchrotron from Pohang light source, which is partially coherent x-ray. We used W/B4C multilayer monochromator as a spectral apparatus and 6.95 keV of photon energy. Focus depth longer than 200 μm. Pinhole or center stop is 160 μm diameter, composed of tungsten,

CP879, *Synchrotron Radiation Instrumentation: Ninth International Conference,*
edited by Jae-Young Choi and Seungyu Rah

block on-axis 0th order and higher order diffracted radiation. Other part of hard x- ray microscope is analogous to optical microscope. Condenser zone plate and micro zone plate, composed of Au, is commonly 1.6 μm in thickness

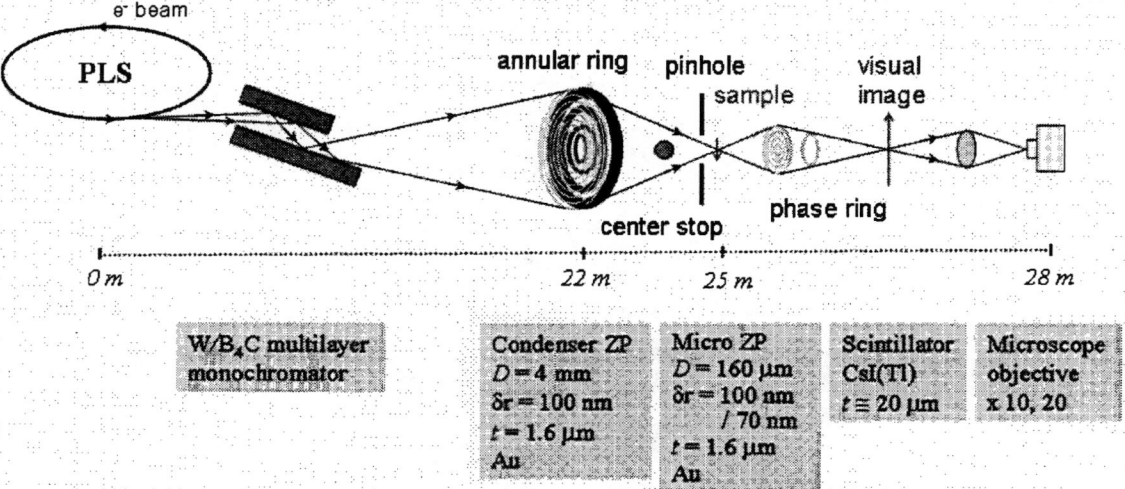

FIGURE 1. Scheme of hard x-ray microscope experimental set-up .

and 4 mm, 160 μm in diameter respectively. Width of the outer most ring zone of these two focusing lenses is 100 nm, so the aspect ratio calculated as 16 and the lenses have at least 20% efficacy. An x-ray images are magnified by 30 times at the objective, micro zone plate and transformed to visual rays by CsI(T1) scintillator. Microscopic objective lens further magnifies visual image by 10 or 20 times and the spatial resolution is sub 100 nm level.

These Fresnel zone plates could not achieve enough image quality with hard x-ray range, 6.95 keV photon energy. So we applied Zernike type phase contrast [2, 6]. Annular ring next to condenser zone plate and phase plate next to micro zone plate are these components. Decreased photon flux to 32.5% by annular ring was compensated with increasing exposure time up to 60 s or more. Outer and inner diameter of 2.1 μm thick Au phase plate is 114.6 μm and 85.9 μm respectively and gives negative phase difference to 3/4λ at 6.95 keV. Sealed with Be vacuum pipe lies x-ray pathway from the micro zone plate to the scintillator so prevents radiation from attenuation by atmosphere. The size of each phase contrast image acquired from CCD camera is 20 x 13.2 μm^2 with 120 s to 150 s exposure time for single shot. An average of hair thickness is about 100 μm. So, the whole hair imaging was taken repeatedly by 5 to 6 frames and put in order.

ANATOMY OF HUMAN HAIR

Human scalp hair is one of the most accessible organs and the diameter is about 100 μm regardless of the races. So this organ is suitable for hard x-ray microscope. Hair is essentially made up of specialized keratinizing cells and keratin fibers. The outermost cuticles usually range 5-10 layers of each 0.3-0.5 μm thickness. Each cuticle mostly has a tapered shape with fringes at their ends. The inner structure of cuticle is cortex consist of macrofibrils, bundle of intermediate keratin filaments. Innermost portion of hair shaft is medulla, which does not exist consistently or continuously. Each hair shaft shows a layered structure consisting of these three major parts, cuticle, cortex, and medulla [7].

RESULTS

Low Power Image

Low magnification image of the Asian hair is shown in Fig. 2. Anatomical location is junction of upper segment of hair follicle (infundibulum) and exposed portion of hair shaft. Diameter of hair shaft was about 100 μm, additionally attached cellular mass which is outer root sheath cell, attached when plugging anagen hair. Translucent zone surrounding hair was assumed as cuticular layers.

FIGURE 2. Low magnification view of scalp hair. Location is between infundibulum and exposed portion of hair shaft. Note that loosely attached outer root sheath(asterisk) and cuticular layer(arrow head) surrounding hair shaft. (bar 50 μm)

High Power Image

Distinct hair layers, cuticular layer, cortex, and medulla, from the outside to inside were clearly seen (Fig. 3, 4, 5). The outermost part of hair fiber and margins of each cuticle circumscribed entire hair shaft was brighter due to edge enhancement or halo effect. The fringe of cuticle homogenously turned toward hair tip. The diameter of each hair fiber ranges from 80-100 μm. Asian shaft was the thickest and American African showed about 90 μm (Fig. 3, 4). Because major axis is bigger than minor axis and curled twisted nature of the African-American hair, diameter of the black man's hair could be varied. So we carefully mounted black man's hair sample, so the side of largest diameter was perpendicular to axis of beam line.

As Fig. 3. shows, Asian hair is well visualized of all three layers. Diameter was about 100 μm, cortex, medulla, and cuticle layer occupied 75-80 μm, 10 μm, and 10-15 μm in thickness, respectively. Cuticular layer showed serrated margins, known that relatively characteristic finding to human hair. Irregularly arranged variable sized cuticle is due to several overlapping image from the front to back side of cuticular layers. Some cuticle was torn out to show wedge shaped defect. The abundant portion of hair shaft, cortex, was seen as vertically oriented bundles of macrofibril, which are most distinguished in the African American hair sample (Fig. 4.). The size of macrofibril is presumably about 0.2-0.3 μm thickness, but limited spatial resolution makes it unable to be evaluated more detailed substructures. The radius of medulla was 10-15 μm, become slender toward hair tip. Some suggested human medulla is filamentous structure but in this study, medulla was obviously amorphous granular structure without evidence of filamentous substructure.

African-American scalp hair (Fig. 4.) measured 90 μm diameter, slightly thinner than that of the Asian. Some cuticles showed coarsely serrated margin and the other cuticles with smooth straight margin showed folding structure. There was an abundant cortex but medulla could barely be seen.

Diameter of axillary hair was about 75 μm and slightly thinner than scalp hair in Fig. 5. Structural components were not different from that of scalp hair. Overlapped cuticles, vertically oriented macrofibril and central medullary portion were clearly identified. Medullary portion showed coarse granular pattern.

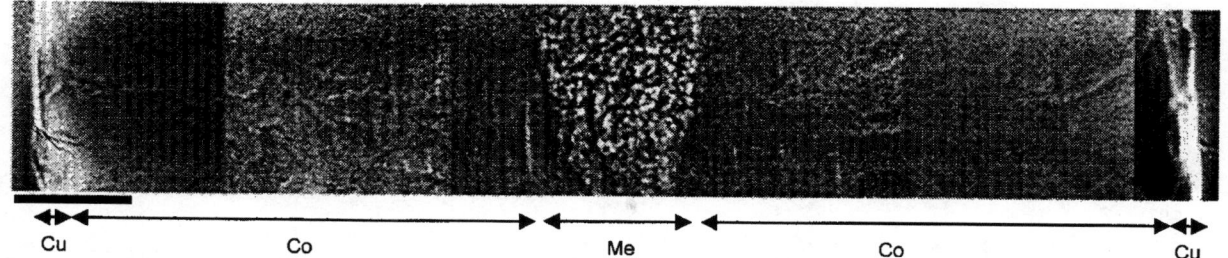

FIGURE 3. Scalp hair from the Asian showed distinct hair compartment of cuticular layer (Cu), Cortex (Co), and medulla (Me) from the outside to inside. Top is proximal and buttom is distal direction. (bar 10 μm, exposure time 120s)

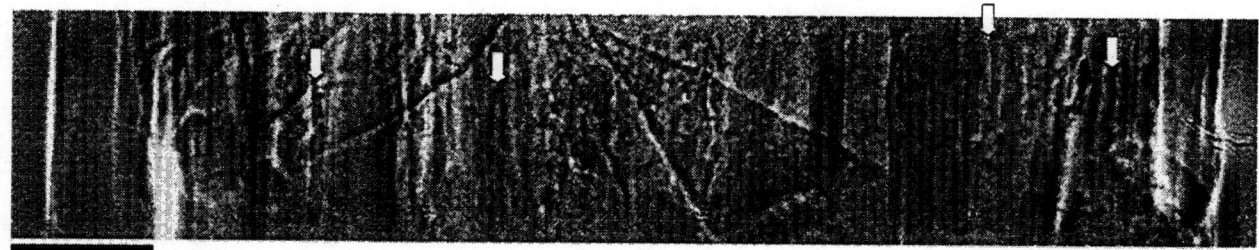

FIGURE 4. Scalp hair from African American. Cuticle showed coarsely serrated or straight margin and some layer is folded. Medulla composed of obvious macrofibrils (white arrow). Medulla could be barely seen. (bar 10 μm, exposure time 150s)

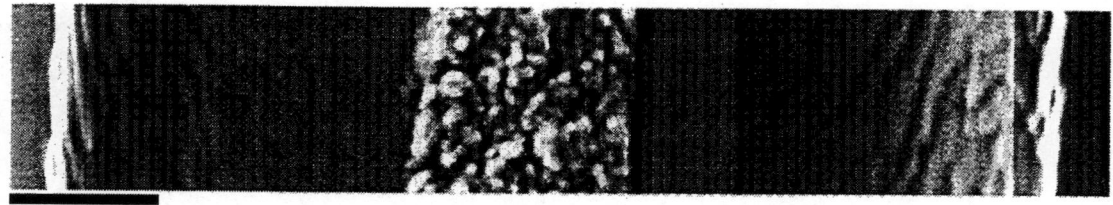

FIGURE 5. Axillary hair shaft from the Asian had 75 μm in diameter. Central medullary portion showed granular structural characteristics. Cuticles oriented obliquely toward hair tip. (bar 10 μm, exposure time 150s)

CONCLUSION

This study shows Zernike type phase contrast hard x-ray microscope is useful tool to investigate morphologic characteristics of human hair and also could be to small organ. Furthermore, there is no need of complex and laborious sample processes by this technique. In particular, generalized application of this result would be difficult but this fundamental pattern could be applicable.

ACKNOWLEDGEMENT

Experiments at PLS were supported in parted by MOST (the Ministry of Science and Technology) and POSTECH.

REFERENCES

1. P. Suortti and W. Thomlinson, *Phys. Med. Biol.* 48, 2-3 (2003)
2. Youn H.S. and Jung S.W., *Phys. Med. Biol.* 50, 5417-5420 (2005)
3. Awaji M., Suzuki Y., Takeuchi A., Takano H., Kamijo N., Tamura S., and Yasumoto M., J. Synchrotron Radiat. 9, 125-127 (2002)
4. Mérigoux C., Briki F., Sarrot-Reynauld F., Salomé M., Fayard B., Susini J. and Doucet J., *Biochim. Biophys. Acta,* 53 , 1619 (2003)
5. Kreplak L., Franbourg A., Briki F., Leroy F., Dallé D. and Doucet J., *Biophys. J.* 82, 2265 (2002)
6. Schmahl G., Rudolph D., Guttmann P., Schneider G., Thieme J. and Niemann B., *Rev. Sci. Instrum.* 66, 1282 (1995)
7. Takami T., Toshihiro T., Seiichi A., Masko O., and Takuma S., Anatomical Record 251, 406-413 (1998)

Trace Element Mapping of a Biological Specimen by a Full-Field X-ray Fluorescence Imaging Microscope with a Wolter Mirror

Masato Hoshino, Norimitsu Yamada, Toyoaki Ishino, Takashi Namiki, Norio Watanabe and Sadao Aoki

Graduate School of Pure and Applied Sciences, University of Tsukuba,
1-1-1 Tennoudai, Tsukuba, Ibaraki 305-8573, Japan

Abstract. A full-field X-ray fluorescence imaging microscope with a Wolter mirror was applied to the element mapping of alfalfa seeds. The X-ray fluorescence microscope was built at the Photon Factory BL3C2 (KEK). X-ray fluorescence images of several growing stages of the alfalfa seeds were obtained. X-ray fluorescence energy spectra were measured with either a solid state detector or a CCD photon counting method. The element distributions of iron and zinc which were included in the seeds were obtained using a photon counting method.

Keywords: X-ray fluorescence microscope, Wolter mirror, element mapping, solid state detector, photon counting
PACS: 07.85.Tt, 78.70.En

INTRODUCTION

Among many analytical methods using X-rays, X-ray fluorescence is a promising probe for qualitative and quantitative analyses. Since X-ray fluorescence has a specific energy for every atom, it can be used to examine the kind of atom and to estimate its quantity. X-ray fluorescence microscopes (XFM) have been expected to be powerful tools in the material science and the biotechnology because they are applicable to investigate bulk samples with non-destruction and high resolution. The most popular type of XFM is a scanning microscope because the optical system is relatively simple and a high quality X-ray microbeam can be easily obtained using Fresnel zone plates or the compound refractive lenses at third generation synchrotron radiation facilities [1,2]. However, spatial resolution suffers from fluctuation of the X-ray source and it takes a very long time to obtain a two dimensional image. A full-field X-ray fluorescence imaging microscope has some advantages over a scanning microscope. The spatial resolution of a full-field imaging microscope is not affected by the fluctuation of the X-ray source and a relatively large field of view can be obtained in one exposure. We are developing a full-field X-ray fluorescence imaging microscope with a Wolter mirror [3]. Although various X-ray energies are detected simultaneously because a Wolter mirror has no chromatic aberration, it is possible to obtain the trace element distribution by means of a photon counting method [4]. In our previous paper, inorganic materials such as a synthesized diamond were observed using an XFM [5].

In this paper, we applied XFM analysis to a biological specimen. We used an alfalfa (*Medicago sativa*) seed as the biological specimen. The X-ray fluorescence images of the alfalfa seeds and the element mappings of metals included in the seeds are shown.

OPTICAL SYSTEM

A schematic diagram of the optical system is shown in Fig.1. A full-field X-ray fluorescence imaging microscope was built at the Photon Factory BL3C2 (KEK). The white beam from the bending magnet of the storage ring was used to generate X-ray fluorescence in order to maximize photons from the sample. The XFM system whose optical length was approximately 2.2 m was constructed normal to the incident X-ray beam to reduce the elastic scattering,

CP879, *Synchrotron Radiation Instrumentation: Ninth International Conference,*
edited by Jae-Young Choi and Seungyu Rah

which is shown in Fig.1. The incident X-ray beam was shaped into a 2 mm×2 mm square by an x-y-slit in front of the experimental hatch. A Wolter Type I mirror whose magnification was 10 was used as an objective device. The length of the mirror was 44 mm and the inner diameter was approximately 10 mm. It was fabricated using a vacuum replication method with a finely polished master mandrel [6]. The detailed mirror parameters are shown in the previous paper [3]. Since the inner surface is coated with Pt, the X-ray energies up to 12 keV can be reflected with a grazing incident angle set to be 7 mrad. Two kinds of charge coupled device (CCD) camera (CCD-1: LCX-1300B, Roper Scientific, CCD-2: C4880, Hamamatsu Photonics) were used as imaging detectors. To prevent the exposure during the readout, a mechanical shutter was set just before the CCD. The X-ray fluorescence energy at the image plane was measured using a solid state detector (SSD). A CCD and an SSD were set on the translation stage to facilitate the exchange of the detectors. The optical path from the end of the Wolter mirror to the detector was evacuated to 1 Pa in order to avoid the absorption of the X-ray fluorescence by air.

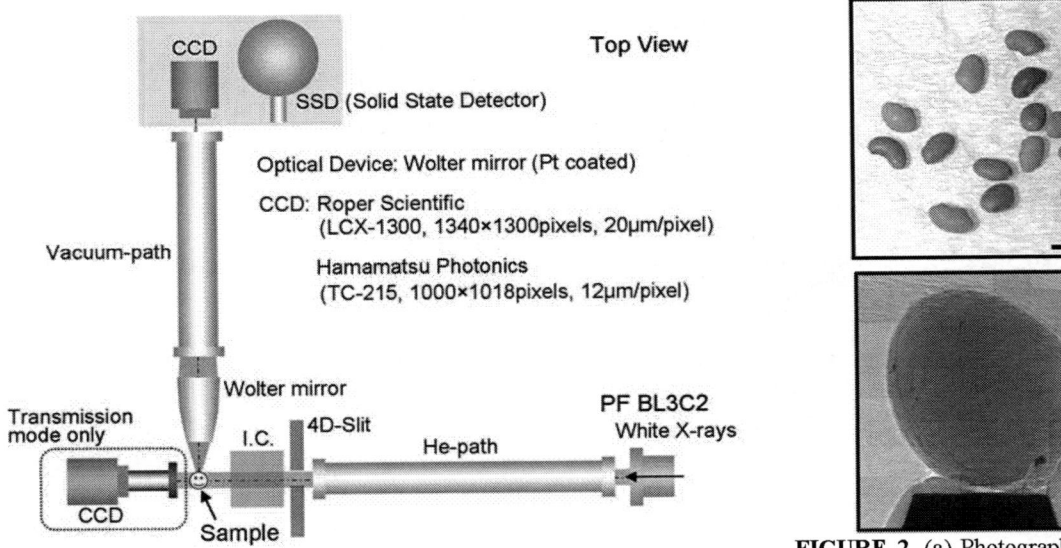

FIGURE 1. Schematic diagram of a full-field X-ray fluorescence imaging microscope with a Wolter mirror.

FIGURE 2. (a) Photograph of alfalfa seeds. Bar: 2 mm. (b) Transmission X-ray image. X-ray energy: 9 keV. Exp: 1.5 s. Bar: 0.2 mm.

X-RAY FLUORESCENCE IMAGING OF ALFALFA SEEDS

Alfalfa is a perennial plant belonging to the beans and it has a unique character that it germinates within 24 hours in a wet environment. An optical photograph and the transmission X-ray image of them are shown in Fig. 2(a) and 2(b), respectively. The size of alfalfa seed is typically 1~2 mm. In this paper, observations of the growth process were tried using an XFM. However, since we could not use one particular growing seed for the observation because of the radiation damages such as the dehydration, we raised many seeds simultaneously. We then chose several seeds at different stage of the growth. The selected seeds were dried sufficiently to avoid the deformation by the dehydration during the exposure. The X-ray fluorescence image of an intact alfalfa seed which is not immersed in water is shown in Fig. 3(a). This is a montage image composed of several images because the field of view of the Wolter mirror used in the experiments was not large enough. In the X-ray fluorescence image, some strong luminous points can be seen. The X-ray fluorescence images of seeds in the growth stage of 15 hours and 30 hours are shown in Fig. 3(b) and 3(c), respectively. All the images of the seeds have strong luminous points, and they show relatively large luminous points near the root embryo. The X-ray fluorescence energy spectrum of the alfalfa seed shown in Fig. 3(a), which was measured with an SSD, is shown in Fig. 4. A lot of iron and zinc were detected and copper, nickel and manganese were also observed. In order to examine the X-ray fluorescence energies of the strong luminous point and other parts, an X-ray microbeam was made by a slit, which was set just before the ion-chamber, and the small area was illuminated. The X-ray fluorescence radiated from this area was detected by the SSD at the image plane. The X-ray fluorescence spectra measured at the strong luminous point and the embryo of the X-ray fluorescence image, which is shown in Fig. 5(a), are shown in Fig. 5(b) and 5(c). From these results, it was found that a lot of iron was included in the strong luminous point and a relatively large amount of zinc was detected at the embryo.

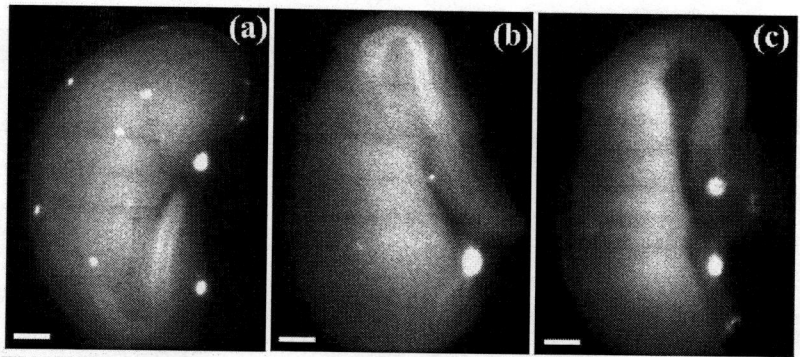

FIGURE 3. X-ray fluorescence images of alfalfa seeds in the stage of (a) 0h (b) 15h (c) 30h. Exp: 30 s×10 integration. Bar: 0.2 mm. (CCD-1)

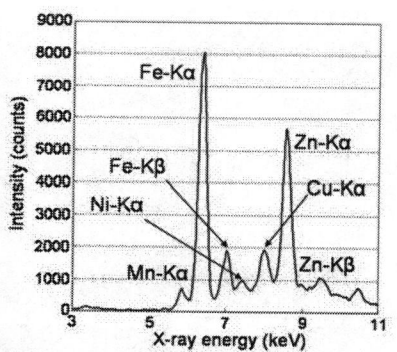

FIGURE 4. X-ray fluorescence energy spectrum measured with an SSD. Exp: 100 s.

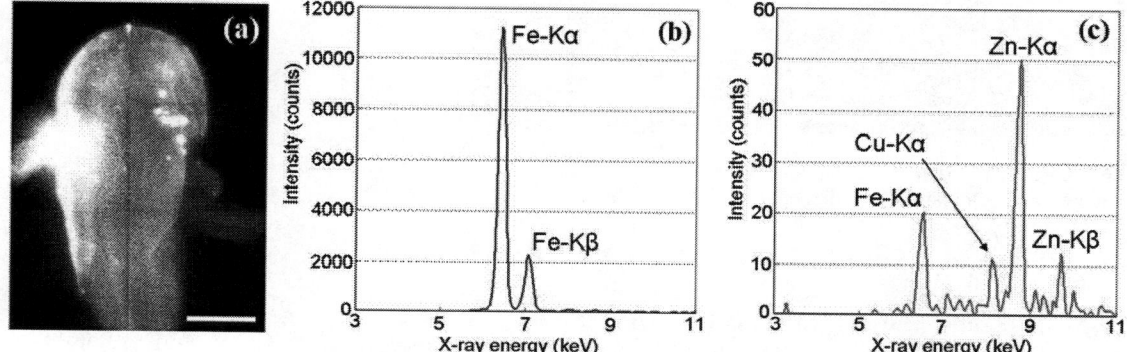

FIGURE 5. (a) X-ray fluorescence image of an alfalfa seed. Exp: 5 m×12 integration. Bar: 0.5 mm. (CCD-2). X-ray fluorescence energy spectra measured at (b) the luminous point and (c) the embryo. Exp: 100 s.

TRACE ELEMENT MAPPING BY PHOTON COUNTING

In the previous section, we examined the distribution of the chemical elements by means of an SSD. In this method, however, we must scan the sample to obtain the element distribution of a whole image. In order to exploit the merit of the full-field imaging microscope, the trace element mapping was carried out using a photon counting method. Although a CCD is usually used as an imaging detector, it can be used as an energy resolvable detector. This is because each pixel of a CCD has the same character as an SSD. In photon counting, we must take an X-ray image under such exposure condition that only one photon enters one pixel of a CCD. The detailed discussion about the photon counting by the full-field X-ray fluorescence imaging microscope is shown in the previous paper [5]. To improve the signal to noise ratio, 2000 images were acquired with an exposure time of 1s. The average of 20 dark images which were taken without X-ray irradiation was used for a background data. The integration image of 2000 photon counting images is shown in Fig. 6(a). To precisely compare the energy spectrum derived from the photon counting dataset with that of the SSD, the photon counting calculation was carried out in a region of interest (ROI) which corresponded to the area measured by the SSD (white square in Fig. 6(a)). The energy spectrum derived from the photon counting calculation is shown in Fig. 6(b). In the calculation, X-rays lower than 3 keV were excluded by the threshold condition. From the photon counting results, good agreements between the SSD and the photon counting spectra can been seen. Furthermore, by extracting the pixels which corresponded to the specific X-ray energy from the photon counting data, the element distribution of iron and zinc were derived. The iron and the zinc images are shown in Fig. 6(c) and 6(d), respectively. The white arrows represent the pixels whose data were absent. They are excluded from the calculation because multiple photons were detected. In this photon counting calculation, it is impossible to derive X-ray energy from the pixel which detects several photons. From the results of element mappings, we found that iron was distributed in the whole area of the seed and showed the luminous points, and zinc was mainly distributed in the embryo.

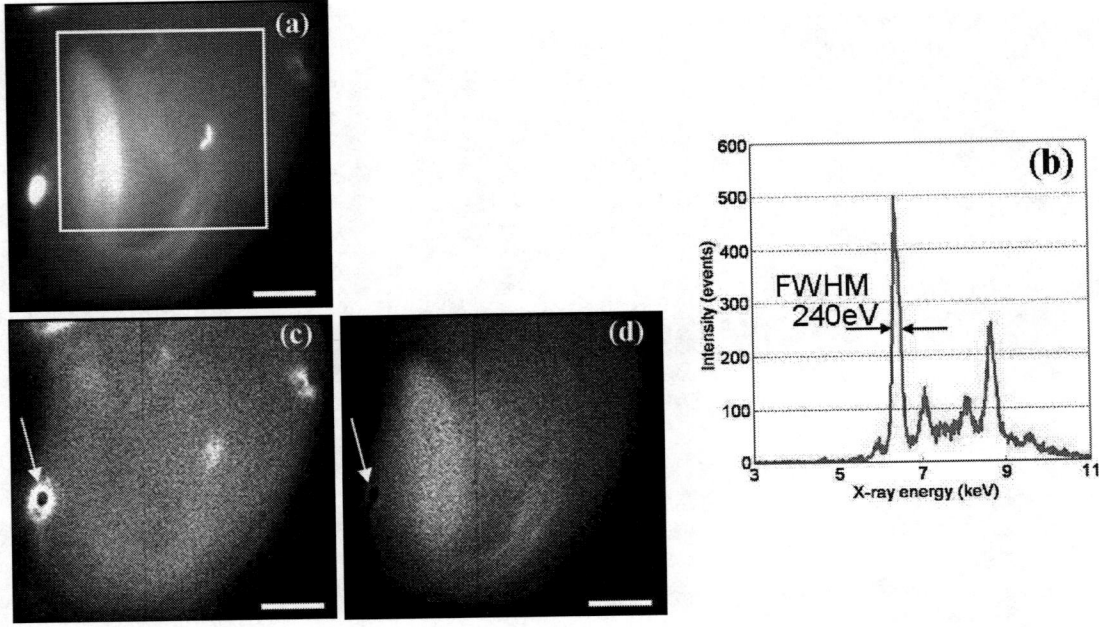

FIGURE 6. (a) X-ray fluorescence image (2000 images integration). Exp: 1 s/image. (CCD-1). (b) X-ray spectrum derived from a photon counting calculation. (c) Element distribution of iron. (d) Element distribution of zinc. Bar: 0.2 mm

CONCLUSION

The X-ray fluorescence images of alfalfa seeds were successfully obtained and the X-ray fluorescence energy spectrum was obtained by a photon counting method with the energy resolution of 240 eV. Furthermore, the element distributions of iron and zinc were derived from the photon counting results. In the X-ray fluorescence images of alfalfa seeds, some strong luminous points were observed. From the SSD and the photon counting analyses, it was found that they included a lot of iron. On the other hand, a relatively large amount of zinc was distributed in the embryo. The quantitative analysis of the alfalfa seed using the X-ray fluorescence microtomography will be shown in the near future. We observed approximately 50 seeds in the experiments and almost all seeds had luminous points. They may have important roles in the growth of a seed. The roles of them are now under investigations.

As the 2-dimensional element distributions were obtained from a photon counting method, a 3-dimensional element mapping will be also possible using a photon counting and a computed tomography. We will also show a 3-dimensional element mapping in the near future.

ACKNOWLEDGEMENTS

This work was partially supported by a Grant-in-Aid for Scientific Research Fund from the Ministry of Education, Culture, Sports, Science and Technology (No. 16206007).

REFERENCES

1. S. Vogt, Y. S. Chu, A. Tkachuk, P. Ilinski, D. A. Walko and F. Tsui, Appl. Surf. Sci. **223**, 214-219 (2004).
2. B. Menez, A. Simionovici, P. Philippot, S. Bohic, F. Gibert and M. Chukalina, Nucl. Instrum. & Methods B **181**, 749-754 (2001).
3. A. Takeuchi, S. Aoki, K. Yamamoto, H. Takano, N. Watanabe and M. Ando, Rev. Sci. Instrum. **71**, 1279-1285 (2000).
4. R. A. Stern, K. Liewer and J. Janesick, Rev. Sci. Instrum. **54**, 198-205 (1983).
5. T. Ohigashi, N. Watanabe, H. Yokosuka, T. Aota, H. Takano, A. Takeuchi and S. Aoki, J. Synchrotron Rad. **9**, 128-131 (2002).
6. Y. Sakayanagi and S. Aoki, Appl. Opt. **17**, 601 (1978).

Optical Tweezers for Sample Fixing in Micro-Diffraction Experiments

H. Amenitsch[*,1], D. Cojoc[**,1], M. Rappolt[*], B. Sartori[*], P. Laggner[*], E. Ferrari[**], V. Garbin[**], M. Burghammer[†], Ch. Riekel[†] and E. Di Fabrizio[**]

[*]Institute of Biophysics and X-ray Structure Research, Austrian Academy of Sciences, Schmiedlstr. 6, 8042 Graz, Austria
[**]CNR-INFM, Lab TASC, Area di Ricerca, 34012 Basovizza (TS), Italy
[†]ESRF, 6 rue Jules Horowitz, BP220, 38043 Grenoble Cedex, France

Abstract. In order to manipulate, characterize and measure the micro-diffraction of individual structural elements down to single phospholipid liposomes we have been using optical tweezers (OT) combined with an imaging microscope. We were able to install the OT system at the microfocus beamline ID13 at the ESRF and trap clusters of about 50 multi-lamellar liposomes (< 10 μm large cluster). Further we have performed a scanning diffraction experiment with a 1 micrometer beam to demonstrate the fixing capabilities and to confirm the size of the liposome cluster by X-ray diffraction.

Keywords: X-ray micro diffraction, Optical manipulation, Phospholipids, Single biological entity
PACS: 87.14.Cc, 87.64.Bx, 82.37.Rs

INTRODUCTION

Since the demonstration of the first optical trapping with dual beams [1] and of the single-beam gradient force optical traps by Asking and co-workers [2], optical tweezers pioneered the field of laser-based optical manipulations. The applications range from manipulation of single atoms, investigation of mesoscopic or colloidal systems up to the manipulation of living cells. Further efforts were devoted to the study of the mechanical properties of single (bio-) polymers like DNA or collagen. A detailed overview of the technical and scientific achievements is given e.g. in [3, 4]. The enormous advantages of optical tweezers are their high positioning precision (down to 5 nm), the fixing of samples non-invasively in their "natural" environment and last the possibility of multiple trapping and subsequent the initiation of particle fusion.

In this work we describe a setup, which is able to fix a single sample entity in a liquid filled capillary and to perform a scanning micro-diffraction experiment on it. As a demonstration experiment - an example for a colloidal system – a small cluster of multilamellar phospholipid vesicles is fixed in the X-ray beam and the diffraction of micron sized vesicle cluster is measured with a micrometer beam.

TWEEZERS AND X-RAY SETUP

The sketch of the setup applied for the micro-diffraction X-ray experiments with optically trapped microparticles in front of the focused X-ray beam is shown in Fig. 1. We use a single mode CW fiber laser (IPG Fibertech PYL-M-10-LP) that generates a collimated linear polarized beam at 1064 nm. The beam is directed to the active area of the spatial light modulator (SLM) (Hamamatsu PPM-X8267) to control the phase of the wave front by means of a pre-calculated diffractive optical element (DOE) which is computer addressed to the SLM. The DOE allows to control the convergence and the shape of the laser beam which is then reflected by the dichroic mirror to the microscope objective (Nikon 60x) and focused inside a thin walled glass capillary of 100 μm diameter (Hilgenberg). Particles immersed in water are fluxed inside the capillary by a syringe pump (TSE Systems GmbH) and trapped at the

---

[1] These authors have contributed equally to the work

CP879, *Synchrotron Radiation Instrumentation: Ninth International Conference*,
edited by Jae-Young Choi and Seungyu Rah

location of the focused laser beam. The working space is imaged by a second microscope objective (40x) on a CCD. The focused X-ray beam is directed orthogonal to the trapped particle in the capillary.

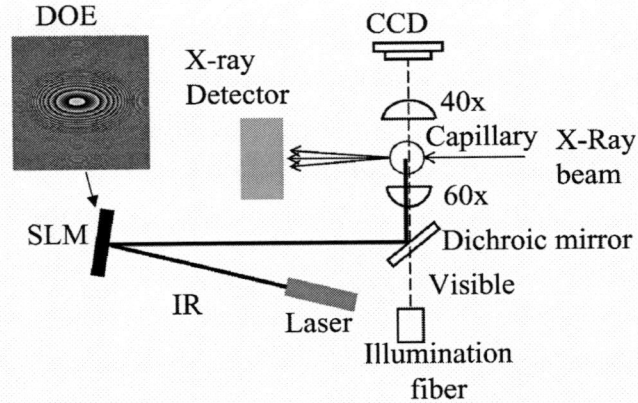

FIGURE 1. Optical setup for trapping and alignment of the microparticles in front of the micro focused X-ray beam. The inlet is an example of diffractive optical element (DOE) displayed on the spatial light modulator (SLM).

Trapping and manipulation of microparticles in solution by means of optical tweezers can be quite easily performed with high numerical aperture (NA) objectives in sample cells with planar walls. Due to specific constraints, we are not able to use these ideal conditions. The view of the optical setup mounted at the X-ray Micro-focus ID13 beamline at ESRF Grenoble is shown in Fig. 2 to demonstrate the spatial constraints in which the trapping setup must be placed.

FIGURE 2. Partial view of the trapping setup at ID13 (laser, SLM, illumination fiber, imaging objective and CCD not shown)

As sample cell we have to choose a capillary because in this way we reduce the volume of water invested by the X-ray beam. Examples of trapping and manipulation of microparticles in samples with particular geometrical configurations are reported in references [5, 6].

Nevertheless, in this experiment we can not use high NA objectives with oil immersion because of the strong scattering of the X-ray beam induced by the oil wetting of the capillary. Moreover, to avoid the vibration of the capillary it should be kept by a holder with a short arm which prevents the use of a dry objective with short working distance and high NA. Therefore we have used an objective with long working distance (2 mm) and moderate NA (0.6). The trapping efficiency with such an objective is low due to the reduced intensity gradient provided. In addition, the cylindrical symmetry of the capillary and the difference between the refractive indexes of air and water make the capillary behave like a cylindrical lens which degrades the shape of the trap and hence its efficiency. In order to compensate for this effect we calculate a diffractive sphero-cylindrical lens (DOE in Fig. 1) and display it on the SLM. The DOE has thus a double role: first to control the convergence of the beam and therefore the position of the trap inside the capillary and second to optimize the shape of the trap.

The trapping setup is installed on a motorized stage of which position can be controlled in Y-Z with micrometric precision. The imaging objective (40x) is mounted on a different suport, to allow the alignment of the trapping setup with respect to the X-ray beam. In a first step, the imaging objective is positioned to get in focus the optical axis of the X-ray beam. In a second step, keeping this objective fixed, the trapping setup is moved in Y-Z until the trapped particle is observed in focus.

The ID13 beamline is set to an X-ray energy of 13 keV and to a sample to detector distance of 348 mm. The diffraction images were recorded with a MAR145 CCD camera (MarResearch), of which the angular calibration was done with silver behenate (D-spacing 5.837 nm). The focus of 1 μm was achieved by using a refractive lens as focusing element.

SAMPLE AND EXPERIMENT

Multi lamellar vesicle dispersions of Palmitoyl-Oleyl-Phosphatidylethanolamine (POPE) (Avanti Polar Lipids) and water (20 w%) were prepared according to standard procedures. Prior to the experiment the stock solution was diluted with a 0.1 mM CaCl$_2$ solution to obtain the final used concentration of 1 w%. This solution was transferred by the remote controlled syringe pump into the capillary held at room temperature. Using the optical tweezers the liposome vesicles were optical sorted until a cluster size of 10 μm had been reached and fixed afterwards in the X-ray beam. The tweezers heating of the sample originated by the laser beam absorption was estimated from the literature to be in the order of 10 °C per W of laser power [7]. Therefore the laser heating is negligible in the phase regime of the POPE sample under investigation. A typical diffraction pattern of the 10 μm large POPE cluster taken with an exposure time of 5 s is shown in Fig. 3. The used exposure time of about 5 s is close to the limit of the radiation damage as tested on a small cluster fixed to the capillary wall. But only single exposures are taken in one position and consequently the radiation damage can be neglected for the used experimental conditions. The measured diffraction ring of the not oriented liquid crystalline phase L$_\alpha$ is demonstrated in Fig. 3a. From the azimuthally integrated 2D diffraction pattern using FIT2D (Fig. 3b) the d-spacing was determined to be 5.29 nm by fitting a Lorentzian onto the 1st order diffraction peak. The d-spacing corresponds to the values usually obtained from diffraction experiments on these samples. The integrated intensity under the peak is a direct measure of the liposome concentration seen by the X-ray beam.

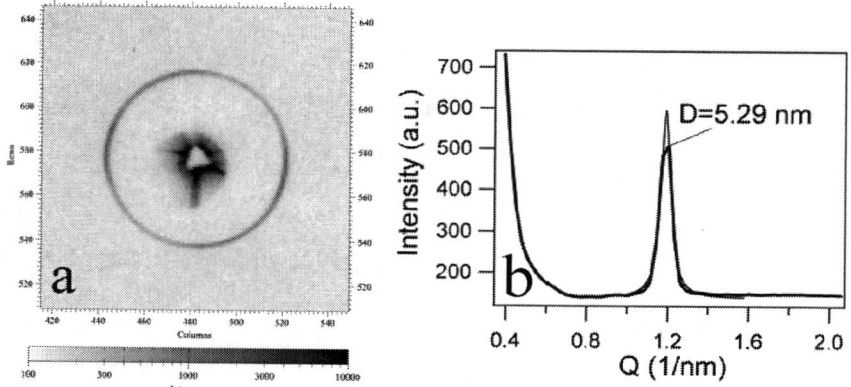

FIGURE 3. 2D Diffraction pattern (a) and plot of the azimuthally integrated intensity (b) of the trapped 10 x 10 x 10 μm^3 large cluster of POPE liposomes. The liposomes are in the liquid crystalline phase having a D-spacing of 5.29 nm as determined from the diffraction experiment.

The trapped sample volume can be imaged by the microscope. For demonstrating the capabilities and to measure the size of the cluster a scanning diffraction experiment was performed with a step size of 2.5 x 5 μm. Figure 4 shows the obtained results. Figure 4a and b show the microscope images of the intense IR laser beam used for trapping and the image taken immediately after turning off the trapping laser to visualize the size of the trapped laser, respectively. From the Fig. 4b a cluster size of 10 x 10 μm^2 has been deduced in the horizontal plane. Additionally some small clusters are visible which are neither in the focal plane nor in the plane of the X-ray beam. Further on the upper half of the images the meniscus of the liquid inside the capillary is seen. The result of the scanning diffraction experiment is given in Fig 4c, which is performed in the vertical plane. Here the integrated intensity under the 1st order reflections is given as a function of the scanning position. The high intensity regime (dark) corresponds to the DOPE cluster, of which the size has been determined with about 10 x 10 μm in vertical plane. The agreement with the cluster dimension measured with microscopy is good.

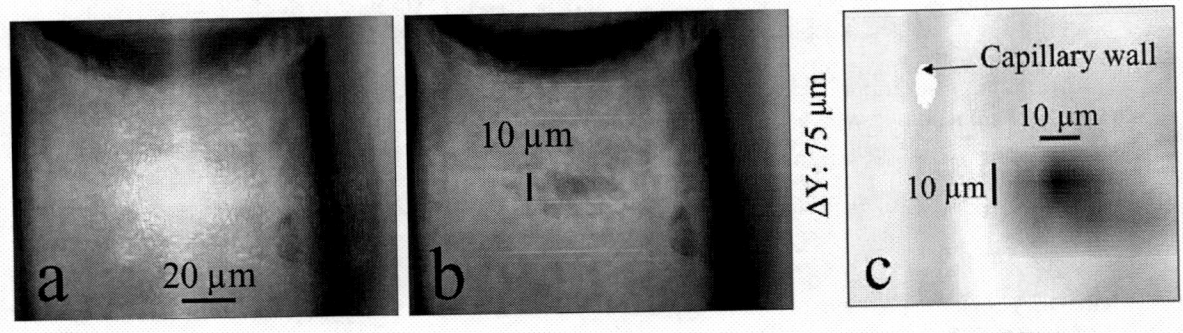

FIGURE 4. Microscope snapshots of the trapping laser (a), of the trapped cluster of POPE liposome immediately after shutdown of the trapping laser (b) and surface plot of the integrated intensity of the POPE versus Y-Z scanning position Y-Z with a scanning increment of 2.5 x 5 μm.

CONCLUSION AND OUTLOOK

Although the setup was found to need further improvement in the future, we have been able to show that: (i) optical tweezers can be used to fix samples in scanning micro-diffraction experiments, (ii) the peak to background ratio over 700 showed that single liposome measurements are feasible (dimension cluster/liposome ~ 10/1).

The improvements include: (i) The installation of multiple trapping tweezers to study e.g. the fusion of single liposomes in different phases (hexagonal/lamellar) and follow their structural transition. (ii) The improvement of the optical imaging with respect to image quality and simultaneous imaging of the trapping laser and trapped sample. (iii) Modification of the sample container i.e. exchange the capillary to a micro fabricated flat cuevette for better optical performance and integrated valves for isolating the observation cell.

Nevertheless these initial experiments have given first insights to the very fascinating new gamble field - to look at single supramolecular assemblies with nm beams.

ACKNOWLEDGMENTS

The authors would like to thank L. Lardiere for his help during the preparation of the experiment at ID13. The ESRF is acknowledged for travel support.

REFERENCES

1. A. Ashkin and J. M. Dziedzic, Appl. Phys. Lett. **19**, 283 (1971).
2. A. Ashkin *et al.*, Opt. Lett. **11**, 288 (1986).
3. D. G. Grier, Nature, **424**, 810 (2003)
4. K. C. Neuman and St. M. Block, Rev. Sci. Instrum., **75**, 2787 (2004)
5. D. Cojoc et al., Optical Trapping and Optical Micromanipulation Conference, Denver USA, *SPIE* Vol **5514** 82-90 (2004)
6. E. Di Fabrizio et al., *Microscopy Research and Technique*, **65** (4-5): 252-262, (2005).
7. K. Dholakia and P. Reece, Nanotoday, **1**, 18 (2006).

X-Ray Microscopy at BESSY: From Nano-Tomography to Fs-Imaging

G. Schneider[*], P. Guttmann[†], S. Heim[*], S. Rehbein[*], D. Eichert[*], and B. Niemann[¶]

[*]BESSY GmbH, Albert Einstein Straße 15, 12489 Berlin, Germany
[†]IRP, c/o BESSY m.b.H., Albert Einstein Straße 15, 12489 Berlin, Germany
[¶]IRP, University of Göttingen, Friedrich-Hund-Platz 1, 37077 Göttingen, Germany

Abstract. The BESSY X-ray microscopy group has developed a new full-field x-ray microscope with glass capillary condenser. It permits tomography and spectromicroscopy of cryogenic as well as heated samples. Correlative light and x-ray microscopy is supported by an incorporated high resolution light microscope. Spectromicroscopy with polarized x-rays from a helical undulator can be performed with $E/\Delta E = 10^4$. With the planned BESSY High Gain Harmonic Generation Free Electron Laser (HGHG-FEL) x-ray imaging with ultra-short pulses and an integral photon flux of about 10^{11} photons/pulse in an energy bandwidth of 0.1% will be possible. Single shot imaging with a full field Transmission X-ray Microscope (TXM) employing a beam shaper as a condenser will be feasible with 20 fs pulses.

Keywords: X-ray imaging, Tomography, Correlative microscopy, Spectromicroscopy, Free Electron Laser.
PACS: 68.37.Yz, 07.85.Tt, 07.85.Qe

INTRODUCTION

X-ray microscopy at 3rd generation synchrotron sources is a powerful imaging method for life, materials and environmental sciences [1]. The penetration depth of x-rays through matter permits to study whole biological cells, fully passivated electronic devices or metal layers in an applied magnetic field, to name modern topics of x-ray microscopy research. Many applications in biology require 3-D information, whereas in other scientific fields temporal resolution is essential, e.g., for magnetic storage devices.

THE NEW BESSY TRANSMISSION X-RAY MICROSCOPE FOR TOMOGRAPHY AND SPECTROMICROSCOPY

The development of a new full-field TXM was motivated by different application fields which we outline first in order to describe their demands on an object stage. Among many scientific questions in life sciences, the cell nucleus which is a vital and complex organelle is still a mystery. How the DNA it contains and its associated proteins are arranged and packaged to fit within this ~10 μm diameter organelle is unknown. Other questions of packaging concern how much "free" space for diffusion is available in the nucleus.

Fluorescence microscopy is an established technique in biophysical investigations of cells and cell nuclei, whereas 3-D x-ray microscopy is a relatively new approach with great potential which enables imaging of whole hydrated cells without chemical fixation, drying or slicing techniques as required in electron microscopy. Conventional optical fluorescence images are diffraction-limited to ~200 nm, whereas current x-ray images can achieve a ten-fold improvement in resolution. The interaction of x-rays is element specific; therefore, x-ray nano-tomography can be used to quantify the packing density of organic material. However, different proteins or molecular structures cannot be distinguished directly in x-ray microscope images. This problem is solved by the availability of specific fluorescent probes detectable by fluorescence microscopy. Thus the two imaging modalities are complementary. Since fluorescence and x-ray microscopy permit analysis of whole cells, it is possible to investigate the same cell in both microscopes. These correlative studies are ideally suited for x-ray microscopy

CP879, *Synchrotron Radiation Instrumentation: Ninth International Conference,*
edited by Jae-Young Choi and Seungyu Rah
© 2007 American Institute of Physics 978-0-7354-0373-4/07/$23.00

because of its ability to image cells in 3-D (see Fig. 1). This enables high throughput imaging of structures larger than a few hundred nanometers, which would otherwise be extremely time-consuming to locate and then serially reconstruct using correlative fluorescence and cryo electron microscopy of thin cell sections. With correlative microscopy, we expect to develop a widely applicable technique that, as applied to nuclear structure, will yield significant new insights.

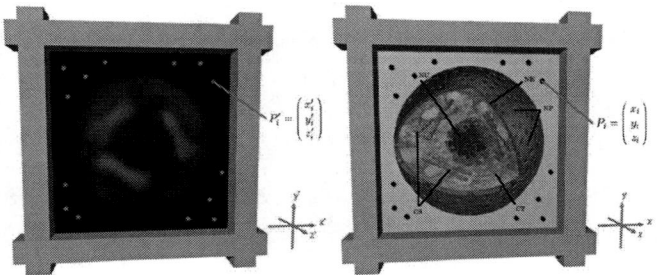

FIGURE 1. Illustration of the 3-D correlative microscopy approach. The fluorescence microscope image (left) will provide information about the location of the labeled structures and the position (x_i', y_i', z_i') of the markers (fluorescent dots outside the cell). The tomographic reconstruction obtained from the data acquired with the x-ray microscope reveals the internal nuclear structures (e.g., chromosomes, nucleolus, nuclear envelope, nuclear pores, and chromatin). It also shows the positions (x_i, y_i, z_i) of the markers by their x-ray absorption. By applying a least-squares fit, the marker positions can be used in a 3-D marker model to align precisely the two coordinate systems of both microscopes.

The current full-field x-ray microscope installed at the undulator U41 was developed for samples in air, e.g., wet cells. Therefore, two pinholes carrying vacuum windows separate the condenser and the x-ray objective chambers. Due to the strong absorption of soft x-rays in air, the distance between these pinholes is typically 300 μm. With such a setup it was possible to perform tomography experiments with samples in a small glass capillary [2,3]. However, many samples, e.g., adherent cells, do not fit into a capillary holder and required a specific setup.

Another application field for x-ray microscopy which is of fundamental interest in materials science is electromigration in advanced copper interconnects buried in low-k dielectric materials [4]. In this case time-resolved x-ray tomography of interconnect structures at about 250°C and at 10^7 A/cm^2 current density has to be performed to determine the exact location of void nucleation and migration as well as to measure quantitatively the mass transport. In order to heat and tilt samples an x-ray microscope with objects in vacuum is advantageous.

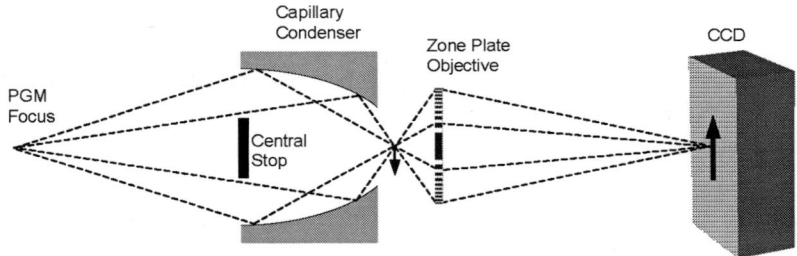

FIGURE 2. X-ray optical setup of the new BESSY full-field TXM with glass capillary condenser.

To meet all these requirements and to permit spectromicroscopy with high energy resolution, the BESSY x-ray microscopy group has developed a new full-field transmission x-ray microscope (see Fig. 2). The new system is based on a tilt stage originally developed for electron tomography which supports automated data collection of cryogenic and heated samples. The stage is able to tilt samples in the x-ray microscope up to ± 80°. Such a large tilt of flat sample holders is impossible on existing x-ray microscopes with bending magnet sources because these require a monochromator pinhole to be positioned close to the specimen. In our new TXM, a plane grating monochromator (PGM) beam line is installed in front of the condenser. The required hollow cone illumination of the object will be generated by an elliptically shaped glass capillary condenser which focuses the x-rays with high reflectivity onto the sample. The TXM can be operated with an energy resolution up to $E/\Delta E = 10^4$. Therefore, the new TXM is well suited for spectromicroscopy studies. The main parameters characterizing the new TXM are listed in Table 1. In comparison with other full-field x-ray microscopes, the features of the new BESSY TXM are unique.

TABLE 1. X-ray optical and mechanical parameters of the new BESSY TXM.

Photon energy range	Soft and hard X-rays
Tomography	Flat sample holder (suited for adherent cells)
	±80° tilt range, 0.01 tilt resolution, 0.01 μm positioning resolution
Object temperature	-180°C up to 500°C, Accuracy: ± 0.1°C suited for cryogenic samples
Spatial resolution	~ 20 nm (depends on objective)
Contrast mechnisms	Amplitude, Zernicke phase, magnetic (with polarized source)
Spectromicroscopy	Energy resolution up to ~10000
Magnification	Adjustable
Source	Compatible with undulators and bending magnets
Objective Stage	Zone plate (XYZ) and phase ring (XYZ) positioning with 0.1 μm resolution
Condenser / Objective aperture	Matched
Software	Computer control of all axes
	Automatic acquisition of tomographic data sets
	Beamline and storage ring interface

THE BESSY HIGH GAIN HARMONIC GENERATION FREE ELECTRON LASER

Free electron lasers are the ultimate tools for time-resolved experiments with x-rays. BESSY has proposed to develop a 2nd generation FEL based on the cascaded HGHG-FEL scheme where the fs x-ray pulse reflects the high-quality pulse shape of the optical fs-seed pulse [5,6]. The proposed BESSY-FEL emits 20 fs pulses with an integral photon flux of about 10^{11} photons/pulse in an energy bandwidth of 0.1% in the photon energy range of the water window. It is known from theoretical calculations and x-ray microscopy experiments at electron storage rings that an x-ray image with 20 nm spatial resolution requires a photon density on the order of several 10^8 photons/μm^2, mainly depending on the contrast produced by the object structures [7]. With its high photon flux in a single x-ray pulse, the BESSY-FEL is therefore an ideal source for single shot x-ray imaging techniques.

An x-ray microscope installed at a soft x-ray FEL enables pump-probe experiments with nanometer spatial resolution. For instance, an optical laser pulse releases biological active molecules and after a well-defined time-delay structural changes are observed by imaging the sample. In this way, structural changes of living cells could be studied on the 20 fs time scale and also correlated with advanced light optical imaging techniques like laser-scan microscopy. In the following, a transmission x-ray microscope for full-field fs imaging is described.

FULL-FIELD X-RAY MICROSCOPE FOR FS IMAGING AT HGHG FEL SOURCE

The diffractive x-ray optics are the key to high spatial and energy resolution x-ray imaging techniques performed at 3rd generation synchrotron sources and probably at the future FEL sources as well. However, for high intensity FEL beams the problem of the thermal stability of the x-ray optics is critical. State-of-the-art transmission zone plates provide about 20 nm resolution below 1 keV photon energy with diffraction efficiencies of 10 - 20%. Such optics are manufactured on thin silicon membranes with about 100 nm thickness. An integral FEL flux of 10^{11} photons/pulse at 0.5 keV photon energy through an aperture of 0.1 and 1 mm diameter would cause a temperature rise of the zone plates of about 1000 and 10 K, respectively. It is known from x-ray microscopy experiments at the BESSY II undulator U41 that transmission zone plates made from silicon under continuous irradiation withstand a temperature increase of about 100 K without damage. Accordingly the illumination has to be chosen such that the maximum temperature rise is less than 100 K.

For single shot experiments it is essential to collect as many photons as possible. Therefore, the diameter of the x-ray condenser has to be matched to the FEL beam diameter and the number of x-ray optical elements in the optical setup has to be minimized. The optical setup of a full-field x-ray microscope for fs imaging requires a condenser which produces a nearly homogeneous illumination of the object field without any movement or rotation of the condenser. A zone plate with 1 mm in diameter would focus the FEL beam onto the object, but at the cost that only a tiny object field of less than 1 μm in diameter is illuminated. To overcome this problem, special x-ray optical devices which influence the wave field by diffraction have to be employed. A specified illumination wave can be converted to a diffracted wave with a desired distribution of its amplitude or phase by a computer generated hologram acting as an x-ray beam shaper. This leads to a homogeneously illuminated object field with a few μm in diameter. With such a beam shaper as a condenser the aperture of the x-ray objective further downstream can be nearly matched and the required hollow cone illumination of the object can be generated by a single x-ray optical element. Such conditions optimize the spatial-frequency transfer of the microscope towards high spatial resolution.

However, the short-pulse duration of only 20 fs and the coherence of the beam have to be taken into account. The beam diameter at the beam shaper is required to be about 1 mm to prevent damage of the optical elements. The optical path difference between the center and the outer diameter of the beam shaper to a point at the sample plane is then about 15 µm which corresponds to a 50 fs propagation time difference of the x-rays. Therefore, the HGHG-FEL offers two major advantages over a SASE-FEL for single shot x-ray imaging. The wave field generated by the inner and outer part of the beam-shaping condenser cannot interfere in the object plane since it arrives at time delays longer than the intrinsic FEL pulse duration (see Fig. 3.a). Second, no monochromator is required in front of the condenser since the monochromaticity of the HGHG FEL-beam is already about $\lambda/\Delta\lambda = 1000$ [5].

To obtain an x-ray image with 20 - 30 nm resolution, at least 10^8 photons/µm^2 are necessary for object illumination assuming an x-ray objective with 100% efficiency. As mentioned above, the integral photon flux of the HGHG-FEL is 10^{11} photons/pulse. Let us assume a diffraction efficiency of 10% for the condenser. Accordingly, about 10^{10} photons/pulse are focused onto the sample. The x-ray objective will be a transmission zone-plate with about 20 nm resolution and about 10% efficiency. As a result 10^9 photons/pulse are available to form an x-ray image in a single pulse exposure.

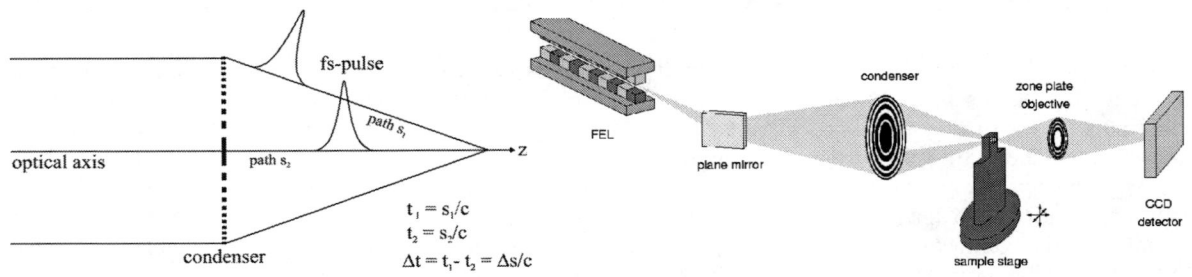

FIGURE 3.a. (left) Illustration of the optical path difference between the central and the outer part of the condenser to the focal spot. **3.b.** (right) X-ray optical setup of the full-field x-ray microscope for single shot imaging. The FEL radiation is collected by a special condenser - a beam shaping diffractive optical device with about 1 mm in diameter - which forms a homogeneous illumination of the object field.

The energy bandwidth of the FEL beam permits to perform x-ray imaging free of chromatic aberration with beam shaping condensers of 1 mm diameter and objectives which are typically 100 µm in diameter. From their diameters and the expected photon flux we estimate a temperature rise in single shots in the beam shaping condenser element of about 10 K and in the much smaller x-ray objective zone plate of ≤ 100 K if both elements are made out of silicon. As discussed above, this temperature rise will not be critical for either x-ray optical element. In Figure 3.b the proposed x-ray optical setup of the full-field TXM in amplitude contrast mode is shown. Basically only the source itself and the condenser are different from an x-ray microscope at 3rd generation synchrotron sources. In summary, we expect with the new x-ray microscopes at BESSY significant steps forward for many scientific and industrial applications.

ACKNOWLEDGMENTS

We are indebted to N. Langhoff and A. Bjeoumikhov (IfG GmbH, Berlin) for providing us the glass capillary condenser. This work was supported in part by the BMBF under Contract No. 05 KS4BY1/7.

REFERENCES

1. J. Susini, D. Joyeux and F. Polack, *X-Ray Microscopy*, edited by J. Susini, et al., Proceedings of the 7th International Conference on X-Ray Microscopy, *J. Phys. IV*, **104**, Paris, France, 2003.
2. D.Weiss, G. Schneider, B. Niemann, P.Guttmann, D. Rudolph and G. Schmahl, *Ultramicroscopy* **84**, 185-197 (2000).
3. G. Schneider, E. Anderson, S. Vogt, C. Knöchel, D. Weiss, M. LeGros, and C. Larabell, ***Surf. Rev. Lett.***, 9, 177-183 (2002).
4. G. Schneider, S. Rudolph, A.M. Meyer, E. Zschech, and P. Guttmann, *Future Fab International* **19**, 115 (2005).
5. The BESSY Soft X-ray Free Electron Laser, in *TDR BESSY March 2004*, edited by D. Krämer et al., BESSY, Berlin, 2004.
6. E. Jaeschke, *Nuclear Instruments and Methods in Physics Research A* **543**, 85 – 89 (2005).
7. G. Schneider, Ultramicroscopy **75**, 85-104 (1998).

Hard X-Ray Nanoprobe based on Refractive X-Ray Lenses

C. G. Schroer*, O. Kurapova†, J. Patommel*, P. Boye*, J. Feldkamp*, B. Lengeler†,
M. Burghammer**, C. Riekel**, L. Vincze‡, A. van der Hart§ and M. Küchler¶

*Institute of Structural Physics, Dresden University of Technology, D-01062 Dresden, Germany
†II. Physikalisches Institut, Aachen University, D-52056 Aachen, Germany
**ESRF, BP 220, F-38043 Grenoble, France
‡Dept. Anal. Chem., Ghent University, Krijgslaan 281 S12, B-9000 Ghent, Belgium
§ISG, Forschungszentrum Jülich, D-52425 Jülich, Germany
¶Fraunhofer IZM, Dept. of Multi Device Integration, Reichenhainer Str. 88, D-09126 Chemnitz, Germany

Abstract. At synchrotron radiation sources, parabolic refractive x-ray lenses allow one to built both full field and scanning microscopes in the hard x-ray range. The latter microscope can be operated in transmission, fluorescence, and diffraction mode, giving chemical, elemental, and structural contrast. For scanning microscopy, a small and intensive microbeam is required. Parabolic refractive x-ray lenses with a focal distance in the centimeter range, so-called nanofocusing lenses (NFLs), can generate hard x-ray nanobeams in the range of 100 nm and below, even at short distances, i. e., 40 to 70 m from the source. Recently, a 47×55 nm^2 beam with $1.7 \cdot 10^8$ ph/s at 21 keV (monochromatic, Si 111) was generated using silicon NFLs in crossed geometry at a distance of 47m from the undulator source at beamline ID13 of ESRF. This beam is not diffraction limited, and smaller beams may become available in the future. Lenses made of more transparent materials, such as boron or diamond, could yield an increase in flux of one order of magnitude and have a larger numerical aperture. For these NFLs, diffraction limits below 20 nm are conceivable. Using adiabatically focusing lenses, the diffraction limit can in principle be pushed below 5 nm.

Keywords: hard x-rays, nanoprobe, refractive x-ray lenses
PACS: 41.50.+h, 07.85.Qe

INTRODUCTION

There is a growing demand for x-ray microscopy techniques in many fields of science, such as physics, chemistry, materials and environmental science, biology, and medical research. One key strength of hard x-ray microscopy is the large penetration depth of hard x-rays in matter that allows one to investigate inner structures of a sample without the need for destructive sample preparation. Combined with tomography, x-ray microscopy can yield the three-dimensional inner structure of a specimen.

In addition, in scanning microscopy contrast can be obtained from hard x-ray analytical techniques, such as diffraction, fluorescence analysis, and absorption spectroscopy. In this way, local structural, elemental, and chemical information can be obtained of a (heterogeneous) specimen. In scanning microscopy, the sample is scanned with a laterally small but intensive beam, recording at each position of the scan the desired x-ray analytical signal. To obtain the desired microbeam, a brilliant x-ray source, such as a synchrotron radiation source of the third generation, and high quality x-ray optics are needed. While hard x-ray beams on the micrometer scale have become routine at modern x-ray sources over the past ten years, only recently beams in the sub-100nm range have become available at these sources, in particular due to developments in x-ray optics. These optics include Fresnel zone plates [1], multilayer Laue lenses [2-4], total reflection and multilayer mirrors in Kirkpatrick-Baez geometry [5, 6], waveguides [7], and refractive lenses [8, 9]. In this contribution, we describe the current state of hard x-ray scanning microscopy based on nanofocusing refractive x-ray lenses (NFLs).

REFRACTIVE X-RAY LENSES

Since their first experimental realization [10] a decade ago, refractive x-ray lenses have been significantly improved, with a large variety of different designs [11-19]. The most important step to making these optics suitable for high resolution imaging and microscopy has been giving them a parabolic (aspherical) shape. In particular the rotationally

CP879, *Synchrotron Radiation Instrumentation: Ninth International Conference*,
edited by Jae-Young Choi and Seungyu Rah
© 2007 American Institute of Physics 978-0-7354-0373-4/07/$23.00

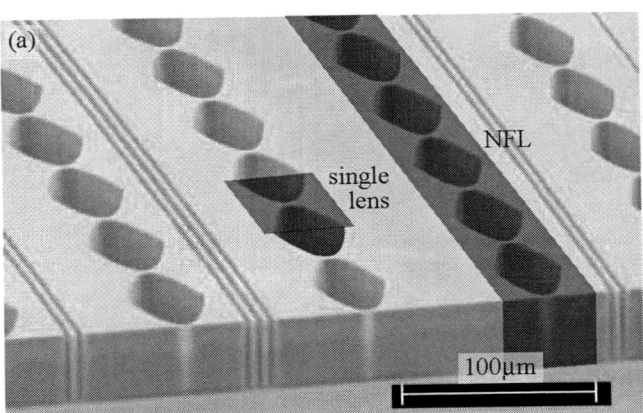
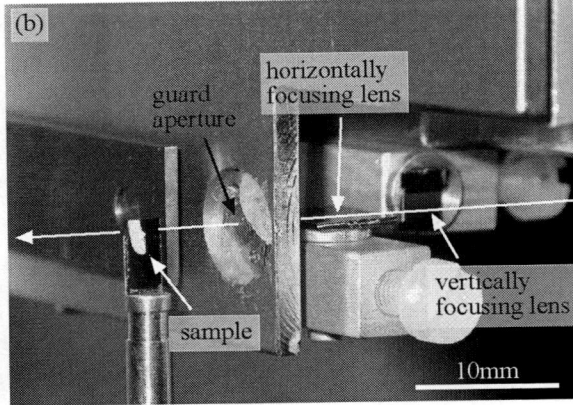

FIGURE 1. (a) Scanning electron micrograph of nanofocusing lenses. A single lens and a nanofocusing lens are outlined by dark shaded areas. (b) Nanoprobe setup: the x-ray beam is focused onto the sample by two crossed nanofocusing lenses.

parabolic lenses allow one to image an object with hard x-rays free of distortion and with sub-micrometer spatial resolution. Using such a lens as objective, a hard x-ray microscope can be built [12, 20, 19]. In combination with tomography, this microscope can yield 3D-reconstructions of an object with sub-micrometer resolution [21].

In addition to full field imaging, rotationally parabolic refractive x-ray lenses are well suited to generate an intensive x-ray microbeam for scanning microscopy. Using refractive optics, a variety of microscopy experiments have been carried out, including fluorescence mapping [22], fluorescence tomography [23-25], tomographic absorption spectroscopy [26], and small angle scattering tomography [27].

For optimal performance in terms of transmission and resolution, refractive x-ray lenses must be made of materials with low atomic number Z. Lenses made of beryllium ($Z = 4$) are very well suited. Rotationally parabolic lenses made of beryllium are available from Aachen University (see [19] for an overview over the current state-of-the-art).

NANOFOCUSING LENSES

With imaging optics, such as refractive lenses, a microbeam is generated by imaging the source onto the sample in the strongly demagnifying geometry. In the absence of aberrations, the beam size is limited by the size of the geometric image of the source and the Airy disc of the optic. While the first depends on the source size and imaging geometry (demagnification ratio), the second is a property of the x-ray optic alone. For refractive optics, both contributions to the focus can be reduced by reducing the focal length f [8]. This is achieved by significantly reducing the radius of curvature of the individual lenses in a refractive lens stack. Fig. 1(a) shows a set of planar parabolic refractive lenses with lens curvatures in the range between 1 and 5 μm made of Si by e-beam lithography and subsequent deep reactive ion etching. These nanofocusing lenses (NFLs) focus hard x-rays in one dimension.

These optics are part of a hard x-ray scanning microscope tested at beamline ID13 of the European Synchrotron Radiation Facility (ESRF). In order to achieve two-dimensional focusing in this microscope, two NFLs are aligned in crossed geometry as shown in Fig. 1(b). The radii of curvature and distances are chosen such that the horizontal and vertical focus coincide at the sample position. A pinhole between the lenses and the sample blocks the radiation falling outside the common aperture of the optics. For a detailed description of the optics and the setup see [8, 28, 9].

We have characterized the hard x-ray nanobeam generated with this lens setup [9]. The nanoprobe setup was located 47 m from the in-vacuum undulator source at beamline ID13 of the ESRF. At $E = 21$ keV a monochromatic nanobeam was generated using a set of lenses with focal lengths $f_h = 10.7$ mm and $f_v = 19.4$ mm for the horizontal and vertical focusing, respectively. Using a fluorescence knife edge technique, a lateral full width at half maximum beam size of 47 nm $\pm$ 9 nm and 55 nm $\pm$ 8 nm was measured in horizontal and vertical direction, respectively. For a storage ring current of 200 mA, the flux in this beam was $1.7 \cdot 10^8$ ph/s with a flux density gain above $2 \cdot 10^4$. The result is in good agreement with the beam size of 43×51 nm^2, predicted for ideal optics. In this geometry, the diffraction limit of 32×45 nm^2 was not reached.

With more transparent lens materials, such as boron or diamond, the flux can be increased by about one order of

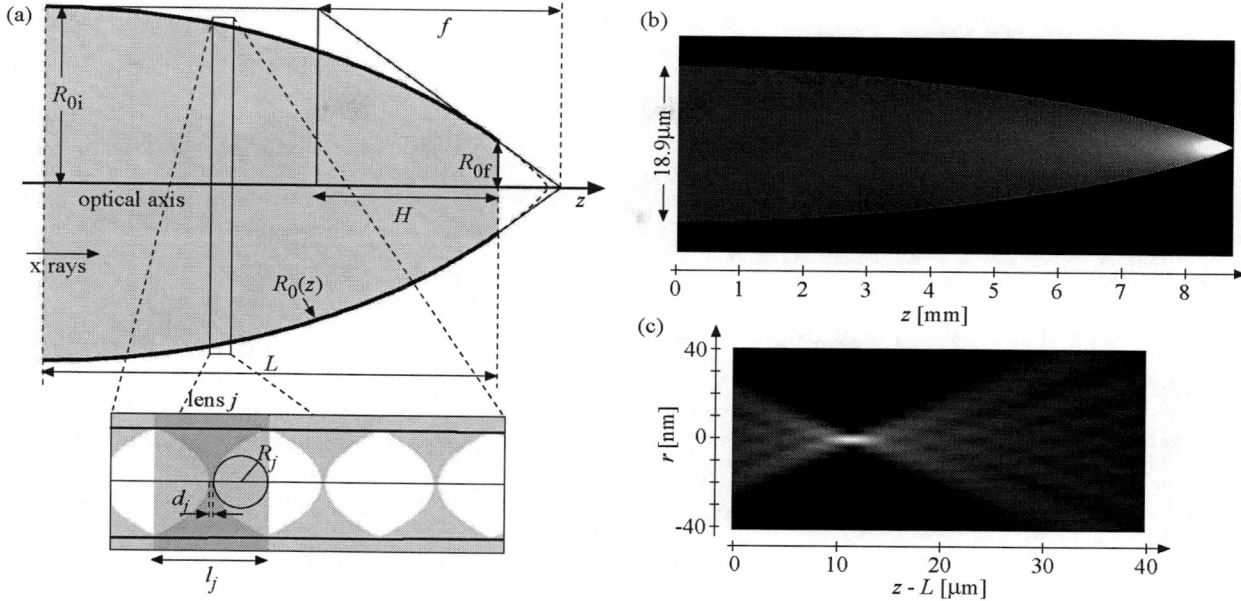

FIGURE 2. (a) Scheme of an adiabatically focusing lens (AFL). The optic is composed of a large number of individual refractive lenses, whose aperture is matched to the size of the converging beam. (b) Calculated intensity distribution inside an AFL and (c) around the focus. The calculated focal spot size is 4.74 nm (FWHM).

magnitude. In addition, the diffraction limit can be pushed below 20 nm [8]. Several experiments have been carried out with a nanobeam based on refractive optics, such as fluorescence tomography [29].

LIMIT OF FOCUSING WITH REFRACTIVE OPTICS

As hard x-rays are focused to ever smaller dimensions, the question arises of what the physical limits to focusing are. For x-ray waveguides, these limits have been determined theoretically in [30]. Bergemann et al. found that the numerical aperture of waveguides is limited by the critical angle of total external reflection $\theta_c = \sqrt{2\delta}$ (with refractive index $n = 1 - \delta + i\beta$). For refractive optics, like compound refractive lenses, the same limit can be deduced from the theory of imaging with thick lenses [31] and for single refractive lenses [32], limiting the Airy disc size to about 15 nm for suitable lens materials [8]. The reason for this limitation lies in the limited refractive power per unit length that can be achieved with a lens of a given aperture [33]. Inside an NFL the aperture and the refractive power are constant. With the focus shortly behind the NFL, the beam converges towards the optical axis along the lens, not fully illuminating the aperture towards the end of the optic. By smoothly (adiabatically) adapting the aperture to the converging beam size, the refractive power per unit length can be increased toward the exit of the optic, leading to an increase of the numerical aperture by typically a factor three and reducing the beam size limit from about 15 nm to about 5 nm. Such a lens is schematically shown in Fig. 2(a). A detailed description of these so-called adiabatically focusing lenses (AFLs) is given in [33]. Figures 2(b) and (c) show the calculated intensity distribution inside an AFL and around the focus, respectively. Using a kinoform lens design, the diffraction limit can in principle be pushed down even further. However, atomic resolution seems to be out of reach.

ACKNOWLEDGMENTS

We thank L. Lardiere for the excellent support during the experiments carried out during the long term project MI-704 at ID13. This work is supported by the German Ministry of Education and Research (BMBF) under grant number 05KS4PA1/9.

REFERENCES

1. XRADIA (????), URL www.xradia.com.
2. C. Liu, R. Conley, A. T. Macrander, J. Maser, H. C. Kang, M. A. Zurbuchen, and G. B. Stephenson, *J. Appl. Phys.* **98**, 113519 (2005).
3. H. C. Kang, G. B. Stephenson, C. Liu, R. Conley, A. T. Macrander, J. Maser, S. Bajt, and H. N. Chapman, *Appl. Phys. Lett.* **86**, 151109 (2005).
4. H. C. Kang, J. Maser, G. B. Stephenson, C. Liu, R. Conley, A. T. Macrander, and S. Vogt, *Phys. Rev. Lett.* **96**, 127401 (2006).
5. H. Mimura, S. Matsuyama, H. Yumoto, K. Yamamura, Y. Sano, M. Shibahara, K. Endo, Y. Mori, Y. Nishino, K. Tamasaku, M. Yabashi, T. Ishikawa, and K. Yamauchi, *Japan. J. Appl. Phys.* **44**, L539–L542 (2005).
6. O. Hignette, P. Cloetens, G. Rostaing, P. Bernard, and C. Morawe, *Rev. Sci. Instrum.* **76**, 063709 (2005).
7. A. Jarre, C. Fuhse, C. Ollinger, J. Seeger, R. Tucoulou, and T. Salditt, *Phys. Rev. Lett.* **94**, 074801 (2005).
8. C. G. Schroer, M. Kuhlmann, U. T. Hunger, T. F. Günzler, O. Kurapova, S. Feste, F. Frehse, B. Lengeler, M. Drakopoulos, A. Somogyi, A. S. Simionovici, A. Snigirev, I. Snigireva, C. Schug, and W. H. Schröder, *Appl. Phys. Lett.* **82**, 1485–1487 (2003).
9. C. G. Schroer, O. Kurapova, J. Patommel, P. Boye, J. Feldkamp, B. Lengeler, M. Burghammer, C. Riekel, L. Vincze, A. van der Hart, and M. Küchler, *Appl. Phys. Lett.* **87**, 124103 (2005).
10. A. Snigirev, V. Kohn, I. Snigireva, and B. Lengeler, *Nature (London)* **384**, 49 (1996).
11. B. Lengeler, J. Tümmler, A. Snigirev, I. Snigireva, and C. Raven, *J. Appl. Phys.* **84**, 5855–5861 (1998).
12. B. Lengeler, C. G. Schroer, M. Richwin, J. Tümmler, M. Drakopoulos, A. Snigirev, and I. Snigireva, *Appl. Phys. Lett.* **74**, 3924–3926 (1999).
13. B. Lengeler, C. Schroer, J. Tümmler, B. Benner, M. Richwin, A. Snigirev, I. Snigireva, and M. Drakopoulos, *J. Synchrotron Rad.* **6**, 1153–1167 (1999).
14. V. Aristov, M. Grigoriev, S. Kuznetsov, L. Shabelnikov, V. Yunkin, M. Hoffmann, and E. Voges, *Opt. Commun.* **177**, 33–38 (2000).
15. V. Aristov, M. Grigoriev, S. Kuznetsov, L. Shabelnikov, V. Yunkin, T. Weitkamp, C. Rau, I. Snigireva, A. Snigirev, M. Hoffmann, and E. Voges, *Appl. Phys. Lett.* **77**, 4058–4060 (2000).
16. B. Lengeler, C. G. Schroer, B. Benner, A. Gerhardus, T. F. Günzler, M. Kuhlmann, J. Meyer, and C. Zimprich, *J. Synchrotron Rad.* **9**, 119–124 (2002).
17. B. Nöhammer, J. Hoszowska, A. K. Freund, and C. David, *J. Synchrotron Rad.* **10**, 168–171 (2003).
18. A. Stein, C. Jacobsen, K. Evans-Lutterodt, D. M. Tennant, G. Bogart, F. Klemens, L. E. Ocola, B. J. Choi, and S. V. Steenivasan, *J. Vac. Sci. Technol. B* **21**, 214–219 (2003).
19. B. Lengeler, C. G. Schroer, M. Kuhlmann, B. Benner, T. F. Günzler, O. Kurapova, F. Zontone, A. Snigirev, and I. Snigireva, *J. Phys. D: Appl. Phys.* **38**, A218–A222 (2005).
20. B. Lengeler, C. G. Schroer, M. Kuhlmann, B. Benner, T. F. Günzler, O. Kurapova, F. Zontone, A. Snigirev, and I. Snigireva, "Beryllium Parabolic Refractive X-Ray Lenses," in *Design and Microfabrication of Novel X-Ray Optics II*, edited by A. S. Snigirev, and D. C. Mancini, 2004, vol. 5539 of *Proceedings of the SPIE*, pp. 1–9.
21. C. G. Schroer, J. Meyer, M. Kuhlmann, B. Benner, T. F. Günzler, B. Lengeler, C. Rau, T. Weitkamp, A. Snigirev, and I. Snigireva, *Appl. Phys. Lett.* **81**, 1527–1529 (2002).
22. S. Bohic, A. Simionovici, A. Snigirev, R. Ortega, G. Devès, D. Heymann, and C. G. Schroer, *Appl. Phys. Lett.* **78**, 3544–3546 (2001).
23. A. S. Simionovici, M. Chukalina, C. Schroer, M. Drakopoulos, A. Snigirev, I. Snigireva, B. Lengeler, K. Janssens, and F. Adams, *IEEE Trans. Nucl. Sci.* **47**, 2736–2740 (2000).
24. C. G. Schroer, *Appl. Phys. Lett.* **79**, 1912–1914 (2001).
25. C. G. Schroer, B. Benner, T. F. Günzler, M. Kuhlmann, B. Lengeler, W. H. Schröder, A. J. Kuhn, A. S. Simionovici, A. Snigirev, and I. Snigireva, "High Resolution Element Mapping Inside Biological Samples using Fluorescence Microtomography," in *Developments in X-Ray Tomography III*, edited by U. Bonse, 2002, vol. 4503 of *Proceedings of the SPIE*, pp. 230–239.
26. C. G. Schroer, M. Kuhlmann, T. F. Günzler, B. Lengeler, M. Richwin, B. Griesebock, D. Lützenkirchen-Hecht, R. Frahm, E. Ziegler, A. Mashayekhi, D. Haeffner, J.-D. Grunwaldt, and A. Baiker, *Appl. Phys. Lett.* **82**, 3360–3362 (2003).
27. C. G. Schroer, M. Kuhlmann, S. V. Roth, R. Gehrke, N. Stribeck, A. Almendarez-Camarillo, and B. Lengeler, *Appl. Phys. Lett.* **88**, 164102 (2006).
28. O. Kurapova, M. Kuhlmann, M. Gather, C. G. Schroer, B. Lengeler, and U. T. Hunger, "Fabrication of Parabolic Nanofocusing X-Ray Lenses," in *Design and Microfabrication of Novel X-Ray Optics II*, edited by A. S. Snigirev, and D. C. Mancini, 2004, vol. 5539 of *Proceedings of the SPIE*, pp. 38–47.
29. C. G. Schroer, T. F. Günzler, M. Kuhlmann, O. Kurapova, S. Feste, M. Schweitzer, B. Lengeler, W. H. Schröder, M. Drakopoulos, A. Somogyi, A. S. Simionovici, A. Snigirev, and I. Snigireva, "Fluorescence Microtomography Using Nanofocusing Refractive X-Ray Lenses," in *Developments in X-Ray Tomography IV*, edited by U. Bonse, 2004, vol. 5535 of *Proceedings of the SPIE*, pp. 162–168.
30. C. Bergemann, H. Keymeulen, and J. F. van der Veen, *Phys. Rev. Lett.* **91**, 204801 (2003).
31. V. G. Kohn, *J. Exp. Theoret. Phys.* **97**, 204–215 (2003).
32. Y. Suzuki, *Japan. J. Appl. Phys.* **43**, 7311–7314 (2004).
33. C. G. Schroer, and B. Lengeler, *Phys. Rev. Lett.* **94**, 054802 (2005).

Polychromatic X-ray Micro- and Nano-Beam Science and Instrumentation

G.E. Ice[1], B.C. Larson[1], W. Liu[1], R.I. Barabash[1], E.D. Specht[1], J.W.L. Pang[1], J.D. Budai[1], J.Z. Tischler[1], A. Khounsary[2], C. Liu[2], A.T. Macrander[2] and L. Assoufid[2]

[1]*Materials Science and Technology Division, Oak Ridge National Laboratory, Oak Ridge TN 37831-6118 USA*
[2]*Advanced Photon Source, Argonne National Laboratory, 9700 S. Cass Ave. Argonne Il 60439*

Abstract. Polychromatic x-ray micro- and nano-beam diffraction is an emerging nondestructive tool for the study of local crystalline structure and defect distributions. Both long-standing fundamental materials science issues, and technologically important questions about specific materials systems can be uniquely addressed. Spatial resolution is determined by the beam size at the sample and by a knife-edge technique called differential aperture microscopy that decodes the origin of scattering from along the penetrating x-ray beam. First-generation instrumentation on station 34-ID-E at the Advanced Photon Source (APS) allows for nondestructive automated recovery of the three-dimensional (3D) local crystal phase and orientation. Also recovered are the local elastic-strain and the dislocation tensor distributions. New instrumentation now under development will further extend the applications of polychromatic microdiffraction and will revolutionize materials characterization.

Keywords: X-ray, microbeam, nanobeam, focusing, diffraction,
PACS: 61.10.Nz, 61.72.Dd

INTRODUCTION AND THEORY

Polychromatic microdiffraction[1] is a rapidly emerging approach to the study of mesoscale materials structure and defect distributions. This approach uses the oldest x-ray diffraction method, Laue diffraction[2], to nondestructively determine local distributions of crystal structure, orientation, elastic deviatoric strain, and dislocation tensors. This information can be mapped with submicron 3D resolution because the diffraction information is determined without the complication of sample rotations. Although the science possible with the world's only 3D X-ray crystal microscope (station 34-ID-E at the APS) is already compelling[3], new instrumentation and technical developments will enable important new applications and will vastly accelerate data collection, analysis and interpretation. In this paper, we briefly survey the current state of the art with respect to instrumentation and science, discuss new instrumentation currently under construction, and speculate on future directions for polychromatic microdiffraction.

Why Use Polychromatic Beams?

The principle motivation for polychromatic microdiffraction is the ability to characterize crystal structure without sample rotations. Max Von Laue first proposed that X-rays could be used to study the atomic structural of materials in 1912. For his work Laue was awarded the Nobel prize in 1914, just 2 years later[2]. As envisioned by Laue, when a collimated polychromatic beam of x-rays impinges on a crystalline lattice, the real-space lattice efficiently reflects x-rays when the scattering from individual atoms constructively interfere at special wavelengths. The wavelengths for constructive interference depend on the lattice and its orientation to the incident beam. This condition is now called the Bragg condition and is conveniently represented by a "so called" Ewald Sphere diagram[4] (Fig. 1).

An Ewald sphere diagram generalized for multiple wavelengths is illustrated in Fig. 1. The real-space crystal lattice has associated with it a reciprocal-space lattice that orients with the real space lattice. For a given wavelength

CP879, *Synchrotron Radiation Instrumentation: Ninth International Conference*,
edited by Jae-Young Choi and Seungyu Rah
© 2007 American Institute of Physics 978-0-7354-0373-4/07/$23.00

(for example the green sphere) the Ewald Sphere indicates the possible momentum transfers for elastic scattering. If the Ewald sphere intercepts one of the reciprocal space lattice points, then a Bragg (or Laue) reflection occurs. With monochromatic diffraction the orientation of the sample is adjusted to pass reciprocal lattice points through the Ewald Sphere. With Laue diffraction a broad bandpass beam is directed at the sample. As shown in Fig. 1, a broad-bandpass beam ensures that *a volume of reciprocal space* is sampled *without* sample rotations. Any lattice point between the maximum and minimum Ewald spere surfaces is efficiently diffracted. This makes it relatively straightforward to determine the local crystal structure and sample orientation.

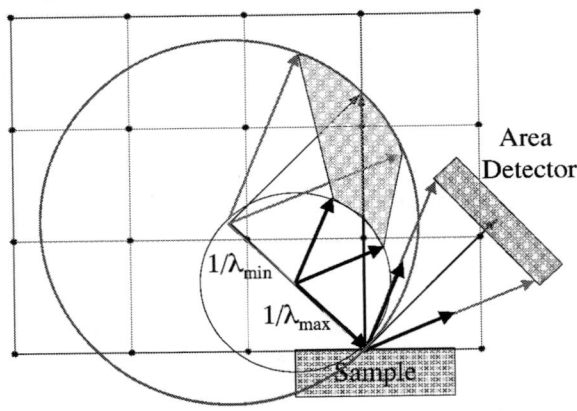

FIGURE 1. Ewald spheres for high energy (large diameter) and low energy (small diameter) limits of a polychromatic x-ray beam. Reciprocal lattice points which lie within the volume defined by the low and high energy limits represent Laue spots (Bragg reflections) excited by the incident beam.

FIGURE 2. Three ways to keep the exit beam coincident with a polychromatic beam: a. Heat one crystal to compensate for the small angular divergence of the straight-through and monochromatized beams; b. a four-crystal monochromator; c. compensate for the slight angular divergence with an upstream pair of mirrors.

The number of reciprocal space points, n_r, that can be detected, depends on the unit cell size a_0, the symmetry of the unit cell, the solid angle subtended by the area detector, Ω, the characteristic wavelength, λ, of the x-ray beam and the bandpass of the radiation $\Delta\lambda$ [5]. For a FCC alloy with the detector centered at 90 degrees to the incident beam, the approximate number of reflections from a crystal is,

$$n_r = \frac{a_0^3 \Omega}{4\lambda^3}\left(\frac{\Delta\lambda}{\lambda}\right)$$ (1)

For a simple FCC metal with $a_0 \sim 3.5$Å, with $\lambda \sim 0.8$ Å, with $\Delta\lambda/\lambda \sim 0.4$ and with a detector solid angle of ~ 1 steradians we expect about 8 reflections. More reflections are detected for larger unit cell size, higher x-ray energy, broader bandpass, or larger solid angle.

Formally a Laue pattern is made up of radial line integrals through reciprocal space weighted by the spectral distribution of the incident beam and the structure factor of the sample volume [6]. This description of Laue diffraction is only realized in the ideal case where the x-ray beam impinges on a large single crystal. In real materials, x-ray beams penetrate through regions with inhomogeneous crystalline structure. Straightforward Laue diffraction from real materials therefore is typically hard to interpret as it contains the overlapping Laue patterns from thousands of volumes with different crystalline phase, orientation or strain.

The problem of overlapping Laue patterns is greatly improved by the use of small x-ray beams; small beams restrict the lateral dimensions of the region probed and hence the number of crystal volumes sampled by the beam. However, x-ray beams penetrate microns to millimeters through materials. For fine-grained low-Z materials this means that the penetrating x-ray beam intercepts tens to thousands of grains. The signal from diffracting volumes along the penetrating beam can be disentangled by a technique called differential aperture microscopy [7].

Differential Aperture Microscopy

Differential aperture microscopy [7] is a knife-edge technique that disentangles the Laue patterns generated along the path of the incident beam. In the differential aperture technique, a smooth high-Z wire is passed near the surface of the sample. For each step of the wire, a "differential" Laue pattern is determined by subtracting the pattern before the step from the pattern after the step. Intensity in the differential pattern arises primarily from rays passing near the leading and trailing edge of the wire. Some rays passing near the leading edge of the wire are occluded *after* the step but not before the step. Some rays passing near the trailing edge of the wire are occluded *before* the step but not after the step. The intensity *in each detector pixel* is fit as a function of wire position to find intensity variations that indicate the changes in the transmission between the sample and the detector. Once a change is identified, the intensity can be assigned to a source point along the incident beam by ray tracing. By following this procedure, the single-crystal-like Laue patterns from each subvolume along the beam can be reconstructed. Because the distance from the sample to the wire is small compared to the distance from the wire to the detector, typical spatial resolution along the wire is submicron even though the CCD pixel size is on the order of 25 microns.[7]

PRACTICAL CONSIDERATIONS FOR A 3D POLYCHROMATIC MICROSCOPE

Station 34-ID-E at the Advanced Photon Source is the first polychromatic microdiffraction facility dedicated to 3D microscopy using differential aperture microscopy. The integrated instrument has been dubbed the 3D X-ray Crystal Microscopy. The key elements of the microscope include, the x-ray source, the nondispersive monochromator [8], the nondispersive Kirkpatrick-Baez total-external-reflection focusing optics [9-11], a precision sample stage, an efficient x-ray sensitive area detector, and advanced software for collecting Laue data and interpreting the patterns [1]. We briefly describe these key elements.

Source (Bend Magnet, Wiggler and Undulator Sources)

Station 34-ID-E utilizes an undulator source, although other polychromatic microdiffraction stations exist on bend magnet sources. The figure-of-merit for microLaue measurements is the integrated source brilliance over an ~ 20-40% bandpass. Intrinsically, the peak brilliance of undulator sources is about four orders of magnitude greater than for bend magnet sources. Even the *wavelength-integrated* brilliance of a type A APS undulator is about 2 orders of magnitude greater than for an APS bend magnet.

Although undulator sources have ultra-high peak brilliance and high average brilliance, spectral peaks, complicate measurements. In particular, small beam motions change the spectral distribution of the beam. These changes complicate measurements of intensity changes during differential aperture microscopy. Large differences in x-ray flux as a function of wavelength also lead to large differences in the intensities of the Laue reflections, which challenges the CCD dynamic range. To reduce these complications, the beam is intentionally steered off axis, which broadens the spectral distribution but nearly conserves the integrated flux in the beam. In addition, a dedicated feedback loop has been installed by the APS to stabilize the beam into the 34-ID beamline.

Monochromator

Although a Laue pattern can quickly determine a number of parameters related to local crystallography, it cannot measure the volume of the unit cell; in order to determine the full strain tensor, the energy of at least one reflection must be determined [1,12]. This information is also useful for determining unknown phases. At 34-ID-E the beam is monochromated using a specially designed microbeam monochromator with a small (~1 mm) displacement. The 34-ID-E monochromator uses a two-crystal design, which allows for smooth and reproducible energy scans. However, because of the beam displacement, there is a small angular difference between the incident beam used for the monochromatic beam and the incident beam used for the polychromatic beam. This difference is corrected by heating the second crystal to adjust the d spacing and Bragg angle (Fig. 2a).

Another nondispersive-monochromator solution is with a four-crystal design as illustrated in Fig. 2b. This design is used by the Advanced Light Source (ALS) microdiffraction beamline 7.3.3 [12]. More recently, the Very Sensitive Probe of Elements and Structure (VESPERS) beamline under construction for the Canadian Light Source has adopted a technique that maintains the simplicity and throughput of the two-crystal approach while avoiding complications introduced with different crystal temperatures. In the VESPERS design two beams, one 0.5 mm below the beam axis and one 0.5 mm above the beam axis are used for the polychromatic and monochromatic options. Different mirrors deflect the two beams at slightly different angles to compensate for the 1 mm offset between the two beams (Fig 2c). These mirrors are also used to focus the beams on an adjustable aperture which allows for simple control of the beam size and flux.

Nondispersive focusing optics

Kirkpatrick-Baez total external reflection mirrors are used to nondispersively focus the polychromatic beam to a submicron spot. High-performance mirrors have been reported based on three distinct approaches: (1) bending of superpolished flat mirrors, [13-15] (2) differential polishing [16] and (3) differential deposition/ differential profile coating [17]. Rapid progress in this area has been reported using all three techniques and mirrors capable of focusing beams to less than 30 nm have been reported. On beamline 34-ID, mirrors made by differential profiling are routinely used. These mirrors have proved to be quite stable and routinely supply beams with less than 500 nm spot size. Beams as small as 70 nm have been delivered under extraordinary conditions [11].

Detectors

The x-ray sensitive area detector is a critical component of polychromatic microdiffraction with requirements considerably different than for protein crystallography and other area-detector intensive techniques. For example, the number of pixels directly correlates into the angular resolution of the detector at a given solid angle. Good spatial resolution is needed to measure peak positions accurately and large solid angles are required to maximize the number of reflections collected and to maximize the angular differences between the reflections. Linearity of the pixel spacing is critical to measurements of elastic strain where errors of 10^{-4} are significant. Variations in pixel sensitivity must also be small to allow for interpolation of peak positions and to allow absolute measurements of integrated intensities. Finally, large dynamic range, fast readout and low point spread functions are all desirable for an ideal detector. For example on 34-ID-E, measurements can be collected in 10 to 100 msec, while our current detector takes ~3-8 seconds to readout.

SOME APPLICATIONS

Good real-space resolution changes everything and opens up countless exciting experimental applications. In particular, x-ray microbeams bridge the gap between their atomic-scale reciprocal-space sensitivity to crystal structure, and their real-space sensitivity to mesoscale (0.1-10 μm) structures. This allows scientists to address key questions of the atomic-scale basis for such familiar emergent behavior as grain growth, deformation and cracking. Polychromatic microbeams have already been deployed to study these and other questions. For example, polychromatic microdiffraction is being used to study grain growth in thin films [18,19] and in 3D [20]. In thin films, the effects of processing conditions, substrate crystallography and/or surface morphology can all be studied. For example, pendeoepitaxial growth of GaN can be studied to quantitatively characterize the number and distribution of defects arising from various processing conditions (Fig. 3a and 3b). Similarly, percolation in high-temperature superconducting polycrystalline films can be determined and related to the growth conditions of the films and the original texture of the substrate (Fig. 4).

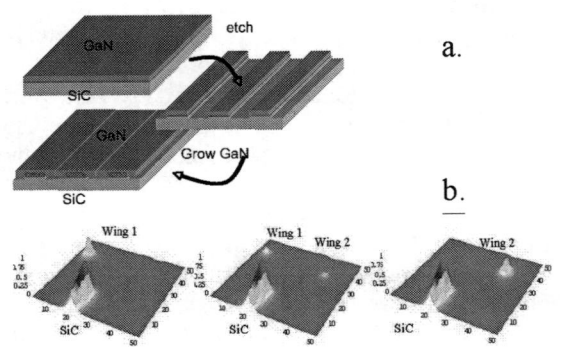

FIGURE 3. (a). Thin GaN film is stress relieved by etching trenches through film and SiC substrate. GaN is overgrown to cover sample. Diffraction maps (b) near a SiC peak and a GaN peak show changes in film orientiation due to position on the underlying SiC pedestal.

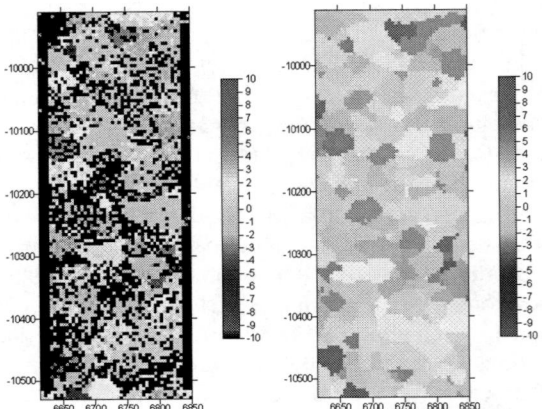

FIGURE 4. The orientation of the superconducting film (left image) can be directly compared to the orientation of the underlying Ni substrate (right image). This allows the eventual electrical percolation to be understood in terms of the starting substrate texture.

Ultra-Small Beams

The increasing importance of nanomaterials requires ultra-small probes of materials properties with spatial resolution to resolve individual nanoparticles or to resolve nanoscale volumes in larger samples. For polychromatic radiation, total-external-reflection Kirkpatrick Baez optics have an intrinsic limit of around 30 nm in spot size which is being approached with ever more perfect optics. We have recently demonstrated beams as small as 70 nm using profile coated optics [11] and have measured beams around 80 nm for polychromatic radiation(Fig. 5). There are currently several strategies emerging to produce even smaller beams. For example, Wolter optics can collect ~4 times larger solid angles. which can theoretically produce beam sizes below 8 nm. Similarly, multilayer optics work at larger grazing incidence angles that similarly allows for beams less than 8 nm. However, Wolter optics are challenging to manufacture for hard x-rays and multilayer optics have reduced bandpass that limits some polychromatic microdiffraction applications. A possible alternative approach-with its own complications- is to use matched pairs of elliptical mirrors to increase the divergence collected in each plane. As illustrated in Fig. 6a if two mirror elliptical surfaces focus to the same image plane, the effective divergence is 4 times larger than is practical for a single mirror. This means that the FWHM of the central beam has a theoretical diffraction limit 4x smaller than with a single elliptical mirror (8 nm). However, because the divergence is not uniform, there will be significant intensity in the region near the central peak (Fig. 6c). We note, that this strategy can be used to resolve very small volumes and the divergence can be further increased with additional reflections (Fig. 6b). The complication of additional maxima can be addressed with deconvolution methods, but is an obvious drawback.

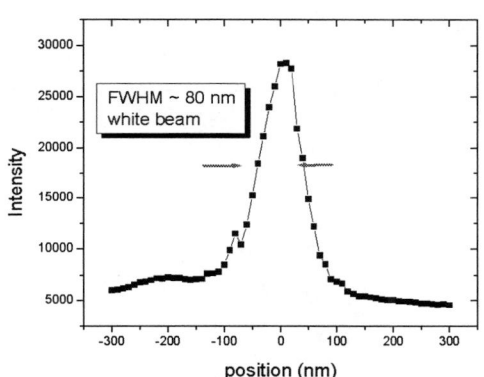

FIGURE 5. Beams below 30 nm have been reported for KB mirrors, and we have recently focused polychromatic beams to less than 80 nm.

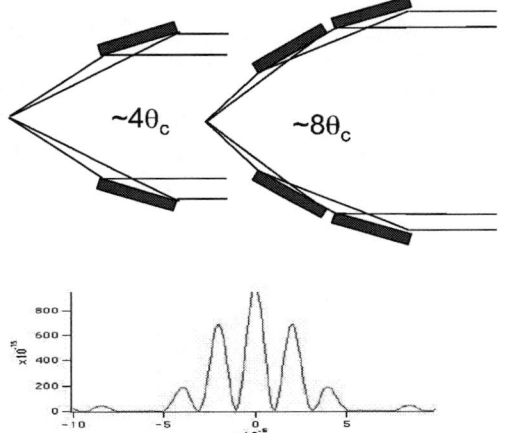

FIGURE 6. (top) Multiple mirrors can be used to increase the divergence in each plane. This reduces the ultimate diffraction limit but leads to structure in the focal plane intensity distribution 6 (bottom).

Diffuse Scattering From Polycrystalline Samples

One potentially exciting application of the instrumentation developed for polychromatic microdiffraction is the study of defects through diffuse scattering. Diffuse x-ray scattering is a powerful tool for determining defects in single crystals with a long history of contributing to our understanding of materials properties. However, best measurements are made on good single-crystal samples that are hard to obtain and as discussed below may even be dangerous. Small polycrystalline sample are however plentiful and easy to prepare for investigations by x-ray microbeams. Because the sample does not need to rotate, thin film samples with grains as small as 10 to 20 μm are sufficient to ensure that the diffraction signal is from a single crystal. This opens up the field of diffuse scattering to a range of samples that would be hard to prepare and invites scientists to address questions that were previously difficult or impossible to answer. For example, we have recently proposed studies of radiation damage in large-grained polycrystalline thin-film samples. Such samples can provide information about primary defects and about factors that influence the survival of these defects. Of great importance is the fact that these samples can be some 5 to 7 orders of magnitude lower in activity than traditional single-crystal samples prepared with the same composition and dose.

Other science also possible includes measurements of local structural fluctuations and measurements of the response of defects to various sinks. Already there have been preliminary measurements in these areas, but the power of nondispersive optics to probe full reciprocal space volumes without changing the real-space sample volume will make these measurements much more sensitive and straightforward.

Accelerated 3D Data Collection

Although 2D maps of thin films and high-density samples can be collected in a few hours, 3D volume maps by differential aperture microscopy are extremely time consuming and virtually impossible for real-time measurements of dynamics. Faster detectors and more advanced data collection can change this situation and efforts are underway to procure 100 hz cameras and to develop new and faster methods of coding the beam position. With a 100 hz camera, images can be collected at about 10-50 hz depending on the sample. This is about 20-500 times faster than with our current detector. At this rate, even straightforward differential aperture microscopy can collect significant volumes with ~500K voxels in 2-5 hours. If multiple apertures or backgammon techniques are employed, an additional gain of about 10 can be expected so that volume information can be collected in about 10-30 minutes.

ACKNOWLEDGMENTS

Work sponsored by the U.S. Department of Energy, Division (DOE), Division of Materials Sciences and Engineering. Oak Ridge National Laboratory (ORNL) is operated by UT-Battelle, LLC, for the U.S. Department of Energy under contract DE-AC05-00OR22725. Experiments were performed on Unicat beamline 34-ID at the Advanced Photon Source Argonne Il. Both 34-ID and the APS are supported by the DOE Office of Basic Energy Science.The submitted manuscript has been authored by a contractor of the U.S. Government under contract No. DE-AC05-00OR22725. Accordingly, the U.S. Government retains a nonexclusive, royalty-free license to publish or reproduce the published or reproduce the published form of this contribution, or allow others to do so, for U.S. Government purposes.

REFERENCES

1. J.S. Chung and G.E. Ice, *J. Appl. Phys.* **86** 5249-5256 (1999).
2. M. von Laue, *Nobel Lecture*, http://nobelprize.org/physics/laureates/1914/laue-lecture.pdf (1915).
3. W. Liu, G.E. Ice, B.C. Larson, W. Yang and J. Tischler, *Met and Mat. Trans. A.* **35A** 1963-1967 (2004).
4. Warren, B. E. *X-Ray Diffraction*, Dover Publications pg. 19 (1990).
5. G.E. Ice, J.-S. Chung, B.C. Larson, J.D. Budai, J.Z. Tischler, N. Tamura, and W. Lowe, *AID Conf. Proc.* **521** 19 (2000).
6. R. I. Barabash, G.E. Ice, B.C. Larson, G.M. Pharr, K.-S. Chung and W. Yang, *Appl. Phys. Lett.* **79** 749 (2001).
7. Larson, B. C., Yang, W., Ice, G. E., Budai, J. D. and Tischler, J. Z., *Nature,* **415,** 887-890 (2002).
8. Ice, G. E., Chung, J. S., Lowe, W., Williams, E. and Edelman, J. *Rev. Sci. Instr.,* **71,** 2001-2006 (2000).
9. P. Kirkpatrick and A.V. Baez, J. Opt. Soc. Am. **38** 766 (1948).
10. Ice, G. E., Chung, J. S., Tischler, J. Z., Lunt, A. and Assoufid, L. *Review of Scientific Instruments,* **71,** 2635-2639 (2000).
11. W. Liu, Gene E. Ice, Jonathan Z. Tischler, Ali Khounsary, Chian Liu, Lahsen Assoufid, and Albert T. Macrander, *Rev. Sci. Instrum.* **76,** 113701 (2005).
12. N. Tamura, A.A. Macdowell, R. Spolenak, B.C.Valek, J.C. Bravman, W.L. Brown, R.S. Celestre, .A. Padmore, B.W. Batterman, J.R. Patel, *J. of Synch. Rad.* **10** 137-143 (2003).
13. P.J. Eng, M.L. Rivers, X. Bingxin and W. Schildkamp *SPIE* **2516** 41 (1995).
14. O. Hignette, P. Cloetens, W. K. Lee, W. Ludwig and G. Rostaing, *Journal de Physique IV* **104** 231-234 (2003).
15. M.R. Howells, D. Cambie, R.M. Duarte, S. Irick, A. Macdowell, H. A. Padmore, T. Renner, S. Rah and R. Sander, *Opt. Eng.* **39** 2748 (2000).
16. K. Yamauchi, K. Yamamura, H. Mimura, Y. Sano, A. Saito, A. Souvorov, M. Yabashi, K. Tamasaku, T. Ishikawa and Y. Mori, *J. Sync. Rad.* **9** 313 (2002).
17. C. Liu, Lahsen Assoufid, R. conley, A. Macrander, G.E. Ice and J.Z. Tischler, *Opt. Eng.* **42** 3622 (2003).
18. R.I. Barabash, G.E. Ice, W. Liu, S. Einfldt, D. Hommel, A.M. Roskowski and R.F. Davis, *Phys. Stat. Sol. A* **5** 732 (2005).
19. J.D. Budai, W.G Yang, N. Tamura, J.S. Chung, J.Z. Tischler, B.C. Larson, G.E. Ice, C. Park, and D.P. Norton. *Nat. Mat.* **2** 487-492 (2003).
20. J.D. Budai, M. yang, B.C. Larson, J.Z. Tischler, W. Liu, H. Weiland and G.E. Ice, *Nat. Sci. Forum* **467-470** 1373-1378 (2004).

Hard X-ray Holographic Microscopy using Refractive Prism and Fresnel Zone Plate Objective

Yoshio Suzuki and Akihisa Takeuchi

Japan Synchrotron Radiation Research Institute (JASRI), SPring-8, Kouto, Sayo, Hyogo 679-5198, Japan

Abstract. Imaging holography in hard x-ray region is realized by combining imaging microscopy with a refractive prism interferometer. The prism is placed behind the back-focal-plane of objective lens in order to configure a wave-front–division interferometer, and a magnified interferogram of object image is generated at an image plane. Spatial resolution of the image hologram is essentially determined by the performance of objective lens. However, speckle noise is a serious problem for fully coherent illumination. We have tried "asymmetric spatial coherence" to reduce the speckle noise. A synchrotron radiation light source with small coupling constant is very suitable for this purpose. The spatial coherence is sufficiently high in the vertical direction to make an interferogram, and low enough in the horizontal direction to suppress the speckle noise. Preliminary experiments at BL20XU of SPring-8 are shown.

Keywords: Holography, Interferogram, Imaging microscopy, Fresnel zone plate, Prism.
PACS: 7.85.Tt, 07.85.Qe, 42.40.-i

INTRODUCTION

Holographic imaging is a unique method for acquiring both amplitude and phase information of objects. Since first report on x-ray holography [1,2], many types of holography optics have been tried in soft and hard x-ray regions [3-6]. Recently, some further progress on x-ray holographic imaging has been reported by using a two-beam interferometer with prism optics [7-11] and image holography using an imaging microscope combined with the prism interferometer [12]. Typical optics system of imaging holography is shown in Fig. 1. The object is illuminated with a coherent parallel beam, and a magnified image of the object is formed at the image plane by the objective lens (A Fresnel zone plate is usually used). A portion of illuminating beam that is also magnified by the objective lens is used as a reference wave. The reference wave is deflected by a prism placed behind the back-focal-plane of the objective, and superimposed on the magnified image of object. Thus, an interference pattern between the image of object and the reference wave is generated on the image plane of microscope. This interference fringe pattern is usually called an interferogram, and the intereferogram includes both the amplitude and quantitative phase information of object. This optical configuration is a well-known system for electron holography [13], but had not been applied to x-ray holography because of absence of practical bi-prism for X-rays.

An advantage of imaging holography is that the spatial resolution is essentially determined by performance of objective lens, while inline holography requires high-resolution imaging-detectors, and a large field of view is needed for Fourier transform holography at high spatial resolution. Recent progress of x-ray optical devices makes it possible to attain nanometer-scale spatial resolution with negligible geometrical aberrations. Now, the spatial resolution of Fresnel zone plate optic is a few ten nanometers even in the hard x-ray region. Therefore, a nm-resolution holographic-imaging is considered to be possible by the imaging holography with FZP objective. However, speckle noise that always appears under the coherent illumination is a serious problem in x-ray holography as well as in laser holography. Different coherence widths in the two orthogonal directions are required in the imaging holography, while two-dimensional coherence is indispensable in most of holographic imaging methods. Therefore, it is possible to suppress the speckle noises by reducing the spatial coherence in a direction horizontal to the interference fringes without any loss of interference fringe visibilities.

In this report, a speckle noise reduction method by utilizing the asymmetric feature of spatial coherence in synchrotron radiation light sources is described. A synchrotron radiation source with small coupling constant is very suitable for this purpose. The coupling constant of SPring-8 storage ring is estimated to be less than 0.1% so that spatial coherence naturally has an extremely asymmetric property, having high spatial coherence in the vertical

CP879, *Synchrotron Radiation Instrumentation: Ninth International Conference*,
edited by Jae-Young Choi and Seungyu Rah
© 2007 American Institute of Physics 978-0-7354-0373-4/07/$23.00

direction and relatively low coherence in the horizontal direction. Therefore, by choosing the beam deflection in the vertical direction, some reduction of speckle noises can be achieved in holographic microscopy without reduction of interference fringe visibility. We have done two experiments with different coherence conditions using a 248 m-long beamline, i.e. fully two-dimensional coherence and one-dimensional coherence, in order to evaluate the feasibility of asymmetric coherence illumination.

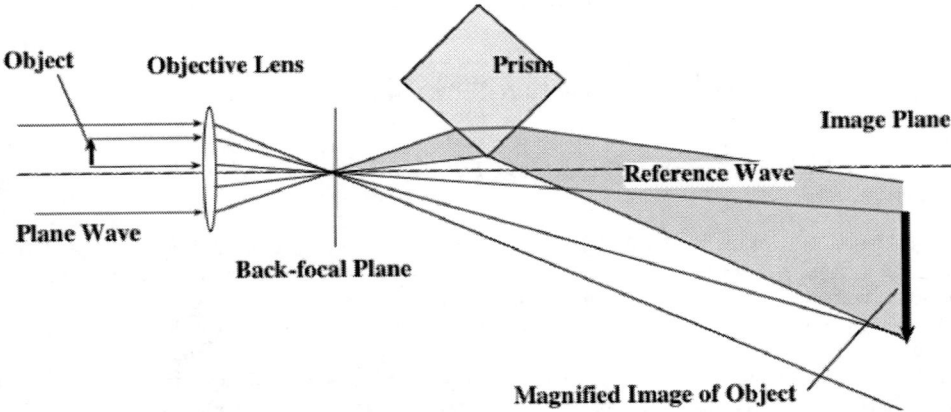

FIGURE 1. Schematic diagram of optical system for imaging holography

EXPERIMENTAL SETUP

Experiments have been done at an undulator beamline BL20XU of SPring-8. The x-ray wavelength of 1.0 Å was used for these experiments. The BL20XU is 248 m long beamline with undulator light source, and there are two experimental stations, one at a distance of 80 m from source point and the other at 245 m from the light source. The storage ring was operated at an emittance of 7 nmrad during the experiment (in the year 2004). So, the vertical size of electron beam was about 20 μm in full width of half maximum (FWHM), and horizontal beam size was about 700 μm in FWWH. High spatial coherence was achieved at the downstream experimental station by using pseudo-point source defined with a quadrant slit located at 200 m from the experimental station. By using a quadrant slit with an opening of 50 μm, the spatial coherence length at the end station is 400 μm in vertical and horizontal direction. Here, the spatial coherent length is defined as $\lambda L/S$ (λ: x-ray wavelength, L: distance from light source, S: source dimension).

Asymmetric coherence condition is instead attained at the upstream experimental station. Here, the vertical coherence is defined by a slit with a width of 10 μm at a distance of 30 m from the experimental station. The horizontal coherence is determined by the synchrotron radiation source. Consequently, the spatial coherent length at the upstream experimental station is estimated to be ~300 μm in the vertical direction, and the horizontal coherence is only about 10 μm. Details of the beamline and spatial coherence at the end station are already reported elsewhere [14,15].

The sample is illuminated by a coherent plane wave (more precisely, a spherical wave from a small point source). One half of incident beam is used for illuminating the sample. The other half of the illuminating beam does not interact with the sample. This directly propagating beam is also magnified by the objective lens. The objective lens used in the experiment is a Fresnel zone plate fabricated at NTT-Advanced Technology, Japan. The zone material is tantalum with a thickness of 1 μm deposited on a Si_3N_4 membrane. The outermost zone width is 0.1 μm, and the diameter is 150 μm. The focal length is designed to be 155 mm at an x-ray wavelength of 1.0 Å. The details of FZP and result of performance tests have already been reported [16]. The FZP has a spatial resolution equal to the theoretical limit, and the diffraction efficiency is 8% at an x-ray wavelength of 1.0 Å. The x-ray prism is a cube of acrylic resin, 15 mm in each dimension. The prism surfaces are finished with a high-precision diamond-cutting machine at RIKEN, and the surface roughness is measured to be about 30 nm root mean square [17]. The beam deflection angle through the prism is usually very small, around a few micro-radian, for normal incidence condition. However, the deflection angle is enhanced by choosing a grazing incidence (or grazing exit) geometry [7]. The glancing angle to the prism surface was set at 0.5° in these experiments, and the beam deflection angle was 0.2 mrad for 1.0 Å X-rays.

The imaging detector used in this experiment is a CCD-based indirect sensing x-ray camera. A thin phosphor screen (Gd-O-S fine powder with stacking thickness of 10 μm) converts the x-ray image to visible light image, and a relay lens system transfers to the CCD (C4742-98-24A, Hamamatsu Photonics, Hamamatsu, Japan) with a magnification of 1/2. The calibrated conversion pixel size is 3.14 μm.

RESULTS AND DISCUSSIONS

Figure 2a shows a conventional bright field image (absorption image) with two-dimensional full-coherent illumination. The sample is a resolution test chart whose finest structure is 0.1 μm line and 0.1 μm space made of 0.5 μm-thick tantalum deposited on a Si_3N_4 thin membrane. A flat field correction (normalization by blank image field) has not been done. The fine structures of object are difficult to be seen, because of the strong artifacts, i.e. concentric ring patterns and grid like patterns. They are considered to be a kind of speckle noise. The concentric ring patterns seem to arise from some imperfection of zone plate structure, because the center of rings coincides with the FZP center. The origin of grid-like patterns is still unknown, but it is apparent that these artifacts are observed only under the coherent illumination. The speckle noises are also seen in hologram, as shown in Fig. 2b. Therefore, it is necessary to reduce speckle noises in order to achieve quantitative holographic imaging, because the flat field correction is invalid for holographic imaging.

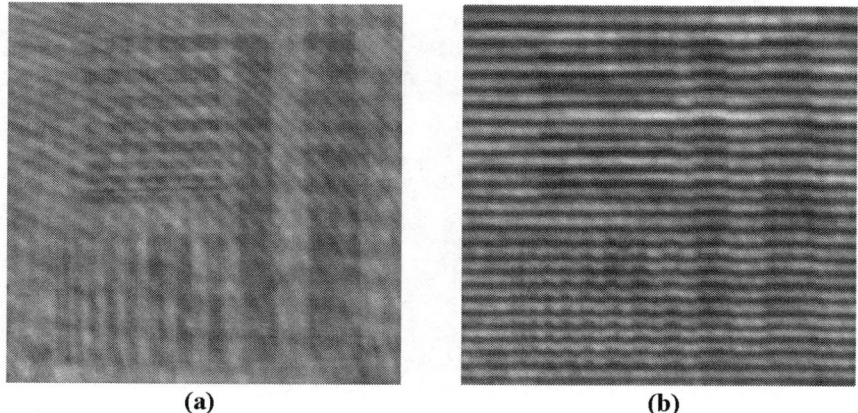

| (a) | (b) |

FIGURE 2. Measured microscopy images under full two-dimensional coherent illumination. Test object is resolution test patterns. x-ray wavelength is 1.0Å. Prism is placed at a distance of 230 mm from FZP (75 mm from back-focal plane), the FZP to detector distance was 6.5 m. Magnification of x-ray microscope is 41. Image size corresponding to the object plane is 23 μm x 23 μm. Exposure time is 100s. (a) Normal bright field image of the resolution test patterns. The prism is removed from x-ray beam. (b) Holographic microscopy image of the resolution test pattern. Beam deflection angle through the prism is 0.2 mrad. The interference fringes are generated by superimposing the reference wave on the object image field.

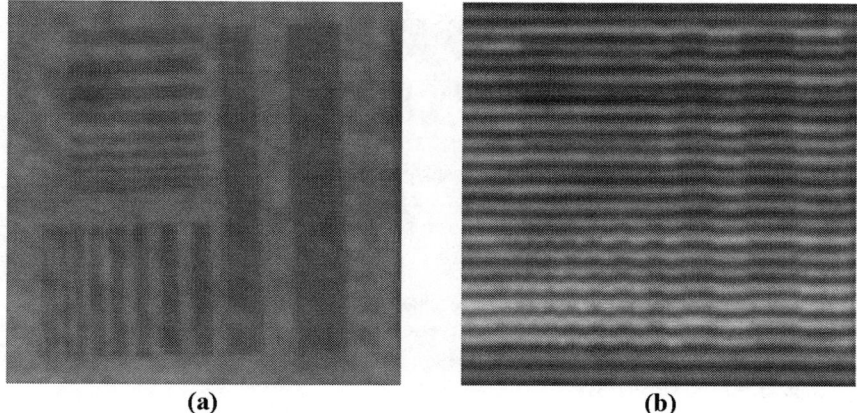

| (a) | (b) |

FIGURE 3. Measured microscopy images under one-dimensional coherent illumination. Test object is resolution test patterns. X-ray wavelength is 1.0Å. Prism is placed at a distance of 230 mm from FZP (75 mm from back-focal plane), the FZP to detector distance was 3.74 m. Magnification of x-ray optics is 23. Image size corresponding to the object plane is 23 μm x 23 μm. Exposure time is 10s. (a) Normal bright field image. The prism is removed from x-ray beam. (b) Holographic microscopy image of the test sample. The beam deflection angle through prism is 0.2 mrad.

Bright field image and holographic image under asymmetric coherence are shown in Fig. 3a and 3b, respectively. The speckle noises are almost erased by the asymmetrically coherent illumination. The fine structures of the resolution test chart are clearly seen both in bright field image and in the interferogram, and the visibility of interference fringes is not affected by the reduction of horizontal coherence. Therefore, it is concluded that the asymmetric coherent illumination is very effective way to reduce speckle noises, and the characteristics of synchrotron radiation light source is appropriate for this purpose.

Figure 4 shows an example of holographic imaging for a pure phase object. The sample is a Kapton foil with a thickness of 50 μm, having a transmissivity of 99 % at 1.0 Å. Although, in the bright field image of Fig. 4a, the transparent object cannot be recognized, it is clearly seen as a positional shift of interference fringes as shown in Fig. 4b. Non-uniformity of the field shown in Fig. 4a is caused by the uniformity of detector's sensitivity (mainly due to non-uniformity of phosphor screen). The measured fringe shift (~2π) agrees well with the calculated phase shift for 1.0Å X-rays in a 50 μm-thick Kapton foil (0.99λ).

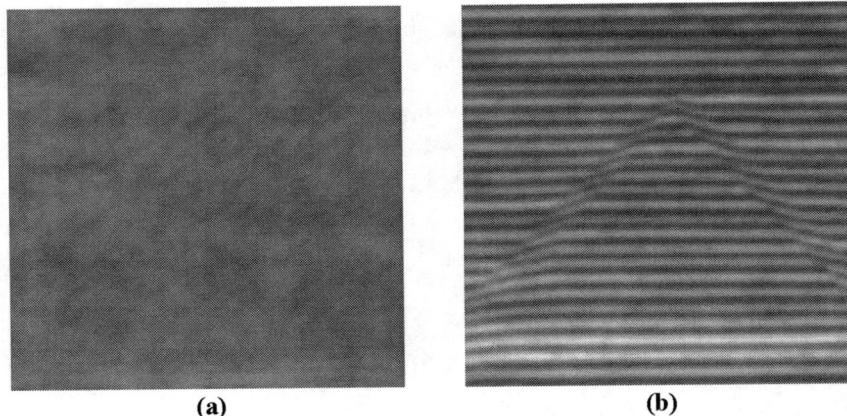

(a)   (b)

FIGURE 4. Measured images of pure phase object. Sample is Kapton film with a thickness of 50 μm. Experimental condition is the same as that in the results shown in Fig. 3. (a) Bright field image. No image contrast can be except for faint edge contrast. (b) Holographic microscopy image. Rectangular edge of the Kapton film is clearly seen as a positional shift of interference fringes.

ACKNOWLEDGMENTS

The authors express appreciation to Dr. Y. Kohmura for preparing the high-quality prism for these experiments. The experiments were performed at SPring-8 under the approval of the Japan Synchrotron Radiation Research Institute (JASRI) (Proposal No. 2004A0172 and 2004B0477).

REFERENCES

1. S. Aoki, Y. Ichihara and S. Kikuta, Jpn. J. Appl. Phys. **11**, 1857 (1972).
2. S. Aoki and S. Kikuta, Jpn. J. Appl. Phys. **13**, 1385 (1974).
3. M. Howells, C. Jacobsen, J. Kirz, R. Feder, K. McQuaid, and S. Rothman, Science **238**, 514 (1987).
4. I. McNulty, J. Kirz, C. Jacobsen, E. H. Anderson, M. R. Howells, and D. P. Kern, Science **256**, 1009 (1992).
5. N. Watanabe, H. Yokosuka, T. Ohigashi, H. Takano, A. Takeuchi, Y. Suzuki, and S. Aoki, AIP Conf. Proc. **705**, 1340 (2004).
6. T. Koyama, Y. Kagoshima, I. Wada, A. Saikubo, K. Shimose, K. Hayashi, Y. Tsusaka, and J. Matsui, Jpn. J. Appl. Phys. **43**, L421 (2004).
7. Y. Suzuki, Jpn. J. Appl. Phys. **41**, L1019 (2002).
8. Y. Kohmura, H. Takano, Y. Suzuki, and T. Ishikawa, J. Appl. Phys. **93**, 2283 (2003).
9. Y. Kohmura, T. Sakurai, T. Ishikawa, and Y. Suzuki, J. Appl. Phys. **96**, 1781 (2004).
10. Y. Suzuki, AIP Conf. Proc. **705**, 724 (2004).
11. A. Takeuchi and Y. Suzuki, Jpn. J. Appl. Phys. **44** (2005) 3293.
12. Y. Suzuki and A. Takeuchi, Rev. Sci. Instrum. **76** (2005) 0937023
13. A. Tonomura, Rev. Mod. Phys. **50**, 639 (1987), and referred therein.
14. Y. Suzuki, et al., AIP Conf. Proc. **705**, 344 (2004).
15. Y. Suzuki, Rev. Sci. Instrum. **75**, 1026 (2004).
16. H. Takano, Y. Suzuki and A. Takeuchi, Jpn. J. Appl. Phys. **43**, L132 (2003).
17. Y. Kohmura, H. Takano, Y. Suzuki, and T. Ishikawa, J. Phy. IV France **104**, 571 (2003).

X-Ray Fluorescence Holographic Study on a Single-Crystal Thin Film of a Rewritable Optical Media

S. Hosokawa, K. Hayashi,[A] N. Happo,[B] K. Horii,[B] T. Ozaki, P. Fons,[C] A. V. Kolobov,[C,D] and J. Tominaga[C]

Center for Materials Research Using Third-Generation Synchrotron Radiation Facilities, Hiroshima Institute of Technology, Hiroshima 731-5193, Japan
[A] Institute of Materials Research, Tohoku University, Sendai 980-8577, Japan
[B] Faculty of Information Sciences, Hiroshima City University, Hiroshima 731-3194, Japan
[C] Center for Applied Near-Field Optics Research, National Institute of Advanced Industrial Science and Technology, Tsukuba 305-8562, Japan
[D] Laboratoire de Physicochimie de la Matière Condensée, Université Montpellier II, 34095 Montpellier Cedex 5, France

Abstract. In this article, we discuss X-ray fluorescence holography (XFH) using a third-generation synchrotron radiation facility through an application to a DVD-RAM material thin film. Three-dimensional atomic images were obtained at 100 K around the Ge atoms in a $Ge_2Sb_2Te_5$ single-crystal thin film by means of XFH technique at the beamline BL37XU of the SPring-8 to clarify the high-speed writing and erasing mechanism of this DVD material. From the obtained XFH images, it was concluded that the single-crystal thin film has a mixture of rocksalt and zinc-blende structures. In addition, the images indicate large distortions associated with the existence of vacancies of the Ge(Sb) site. The present XFH results are in good agreement with the previous XAFS results, which has predicted a phase transition due to an umbrella flip motion of the Ge atoms.

Keywords: Synchrotron radiation, Phase transition; Imaging; Thin film; Chalcogenides
PACS: 61.18.-j; 07.85.Qe; 42.40.-i

INTRODUCTION

X-ray fluorescence holography (XFH) is a technique that allows one to obtain a three-dimensional (3D) atomic image around a specific element [1]. Recent development of fast detectors along with the use of intense X-rays from a third-generation synchrotron facility has enabled measurements to obtain clear atomic images of local structure even up to the 7th neighbor [2], and applied to the local atomic information on several materials [3]. In this paper, we discuss the feasibility of XFH through an application to a DVD-RAM material thin film as an example.

In recent days, rewritable optical media DVD-RAM is widely used for recording massive data or movies. It is well-known that this recording process is governed by a laser-induced crystalline-amorphous phase transition of the media material thin films, such as $Ge_2Sb_2Te_5$, in DVD-RAM. It was, however, difficult to understand the very fast recording and erasing mechanism by a normal idea of laser-induced melting and recrystallizing processes, and the real mechanism was mysterious for about two decades.

Recently, Kolobov and coworkers carried out an X-ray absorption fine structure (XAFS) study on $Ge_2Sb_2Te_5$ thin films [4] to clarify this mechanism in detail. They confirmed that the crystal thin film does not possess a hexagonal structure as in the bulk $Ge_2Sb_2Te_5$, but more likely consists of a distorted rocksalt structure in which the six equal-length bonds of the rocksalt structure are split into two groups of three longer and three shorter bonds [5]. They concluded that the laser-induced amorphization is due to an umbrella-flip motion of the Ge atoms from an octahedral position in the crystal lattice into a tetrahedral position without a rupture of strong covalent bonds.

Since structural information obtained from XAFS measurements is, however, limited to one-dimensional, i.e., a directionally-averaged pair distribution function of mainly first- or second-nearest-neighbors, another probe is

CP879, *Synchrotron Radiation Instrumentation: Ninth International Conference,*
edited by Jae-Young Choi and Seungyu Rah
© 2007 American Institute of Physics 978-0-7354-0373-4/07/$23.00

necessary. For this, we chose XFH to confirm the above unique model. XFH is a technique which can record the amplitude and phase of the wavefronts scattered by atoms relative to direct unscattered wave. FIG. 1 schematically shows the principle of XFH. In normal XFH, as shown in FIG. 1(a), fluorescent X-rays from atoms inside the sample constitute the reference beam, and those scattered by the neighboring atoms act as the object beam. A holographic pattern can be recorded by moving the detector around the sample. The inverse XFH is based on the idea of the optical reciprocity of the normal XFH. As seen in FIG. 1(b), the atoms emitting fluorescent X-rays serve as the detector of the interference field originating from the incident and scattered X-rays, which give the reference and object beams, respectively. The holographic pattern can be obtained by detecting the fluorescent X-rays while changing the sample orientation relative to the incident X-ray beam. Since the incident X-rays can be of any energy above the absorption edge of the emitter element, inverse XFH allows holograms to be recorded at an arbitrary energy, which can suppress the twin-image effect. The present XFH experiment was performed in inverse mode.

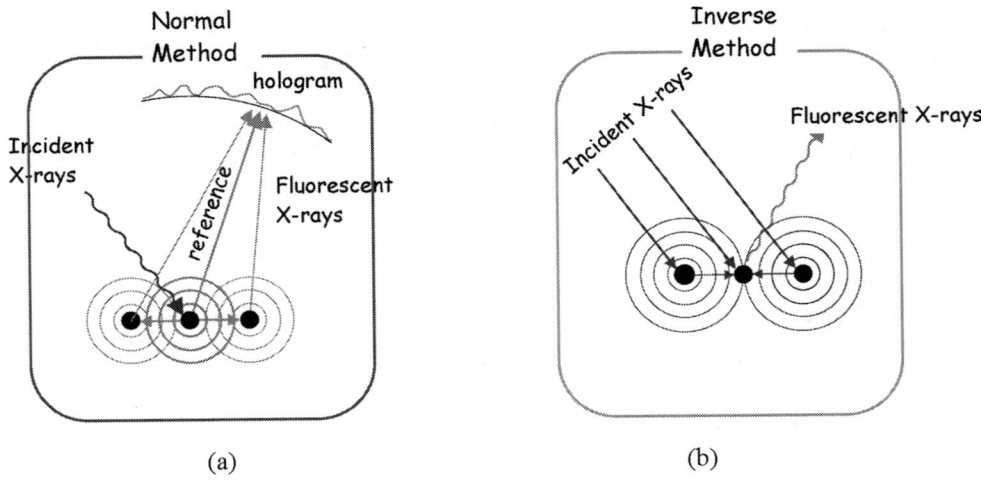

FIGURE 1. Schematic view of the principle of XFH in (a) normal and (b) inverse modes.

EXPERIMENTAL PROCEDURE

The experimental setup is schematically shown in FIG. 2(a). The $Ge_2Sb_2Te_5$ single-crystal thin film sample with a thickness of about 2 μm was obtained by epitaxially growing on an InSb single-crystal substrate having a similar lattice parameter to that of the sample and no Ge content. The sample with the size of about 5 mm square was placed on a two-axes table of a multi-purpose diffractometer installed at the beamline BL37XU of the SPring-8. The measurements were performed by rotating the two axes, $0° \leq \theta \leq 75°$ in steps of 0.5° and $0° \leq \phi \leq 360°$ in steps of about 0.3°, of the sample table, and detecting small intensity variations (~ 0.1 %) of the fluorescent X-rays with angles, which is named as X-ray fluorescence hologram. The incident X-rays were irradiated onto the (001) surface of the sample. The Ge K_α fluorescent X-rays were collected using an avalanche photodiode detector with a cylindrical graphite-crystal energy-analyzer. For each scan, it took about 5.5 hours. The XFH were recorded at eleven incident X-ray energies of 22.0-27.0 keV in steps of 0.5 keV. To suppress the thermal agitations of the atoms, the sample was cooled down to 100 K using a cryostream apparatus.

An extension to the perfect sphere made by a symmetric manipulation was carried out to obtain the hologram patterns. FIG 2(b) shows an example of hologram pattern at 24.0 keV on the (100) plane. Fourfold shadow lines are observed in the hologram pattern, which originate from X-ray standing wave of the (110) direction, indicating that the crystal structure is not a hexagonal structure. From the hologram patterns obtained at different incident X-ray energies, a 3D atomic configuration image was constructed using Barton's algorithm [6]. Since a simple Fourier-transform of the XFH data at a single energy of incident X-rays produces false twin atomic images, images at eleven energies of the incident X-rays were superimposed so that these twin images were suppressed.

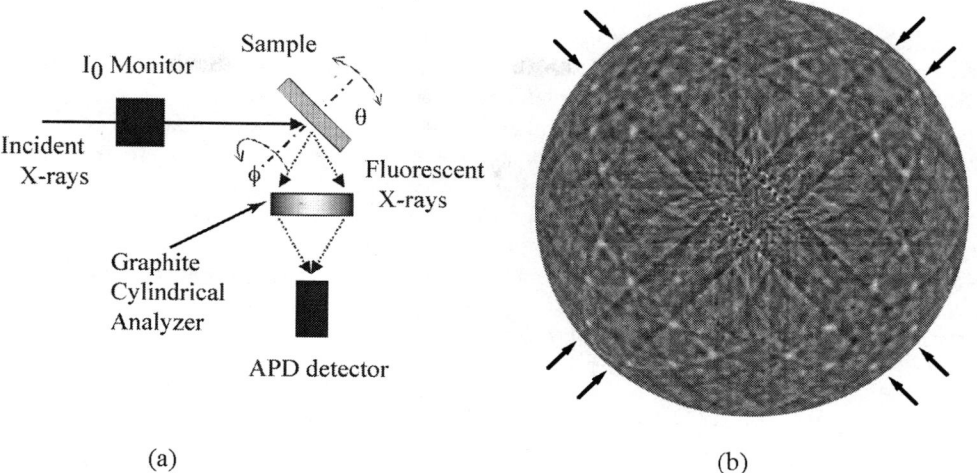

(a)   (b)

FIGURE 2. (a) Schematic view of the XFH experimental setup, and (b) a typical example of the hologram pattern obtained from Ge K_α XFH measurement on $Ge_2Sb_2Te_5$ single-crystal thin film at the incident X-ray energy of 24.0 keV

RESULTS AND DISCUSSION

The obtained atomic image on the (001) plane is given in FIG. 3(a). The cross at the center of the figure indicates the central Ge atom, and the red or black colors represent the position of the neighboring atoms. A clear fourfold atomic structure is visible at each corner of the green square in the figure. This image clearly reveals that the $Ge_2Sb_2Te_5$ single-crystal thin film does not have a hexagonal structure around the Ge atoms. The distance from the central Ge atom to these neighboring atoms is about 0.45 nm, which is much larger than the Ge-Te nearest-neighbor distance of 0.283 nm obtained from the XAFS experiment [4]. Thus, these atomic images should correspond to the second-nearest-neighbor Ge or Sb atoms. As seen in the enlarged figure at the shoulder of FIG. 3(a), these spots are separated into two components, i.e., near and far positions from the central Ge atom. This separation can be understood that the former are composed of the Ge-Te-Ge configurations and the latter the Ge-Te-Sb ones.

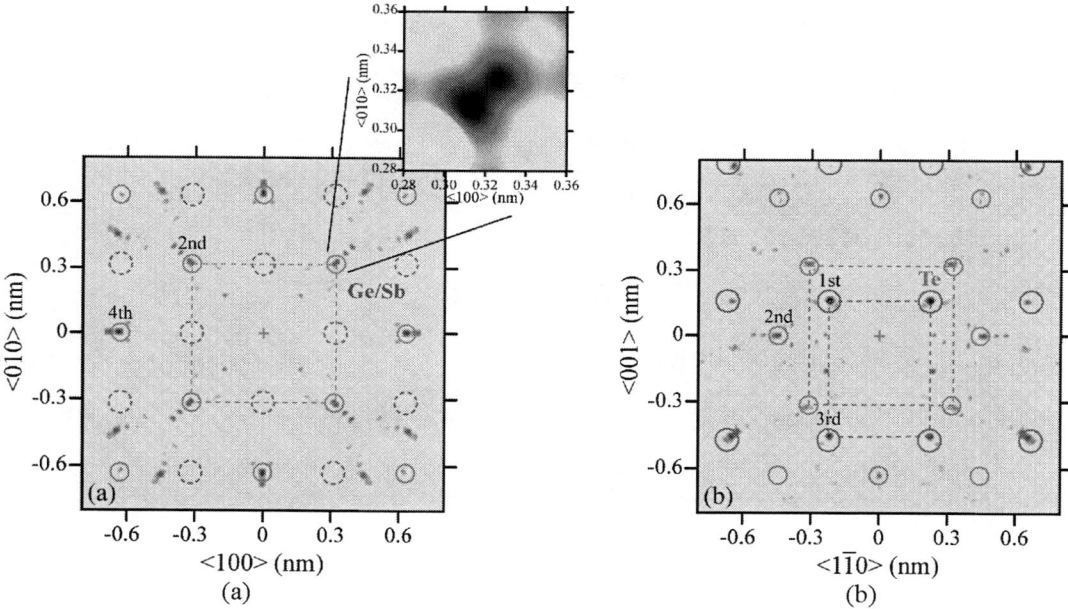

(a)   (b)

FIGURE 3. Atomic images around the central Ge atoms on the (a) (001) and (b) (110) planes. The enlarged image of the second-neighbor atoms is at the shoulder of (a). See the text for details.

Another fourfold atomic images are clearly visible at the symmetric positions of the central Ge atom with respect to each edge of the green square, which may correspond to the forth-nearest-neighbor Ge(Sb) atoms. From these results, it is reasonable to assign that the Ge(Sb) sublattice have a cubic structure. Since the images of sixth-nearest-neighbor Ge(Sb) cannot be observed, the Ge(Sb) sublattice may be highly distorted.

FIG. 3(b) shows the atomic image on the (110) plane. As seen in the figure, two atomic images are seen at the distance of 0.27-0.28 nm from the central Ge atom, as are shown at the upper corners of the red rectangle. Since the bond length is very similar to that obtained from the XAFS experiment, 0.283 nm [4], these atomic images correspond to the nearest-neighbor Te atoms. The bond angle of Te-Ge-Te configuration is about the tetrahedral value of 110°. Thus, it is highly possible that this crystal has a zinc-blende structure. In addition, the second-nearest-neighbor Ge(Sb) atoms can be seen at the distance of ±0.45 nm in the (1-10) direction. The third-neighbor Te atoms are also seen at the lower corners of the red rectangle.

On the other hand, other four atoms are seen at the distance of about 0.45 nm at the corner of green square, which cannot be assigned by the zinc-blende structure. These atoms can be considered as Te atoms of a rocksalt structure. The position of these Te atoms are highly deviated from the original positions of the rocksalt crystal form, but may be shifted due to the existence of vacancies of Ge(Sb) atoms in the rocksalt structure, which was proposed previously [4,5]. The present XFH results are partly inconsistent with the model made of only a distorted rocksalt atomic configuration. The present finding of mixed structure of rocksalt and zinc-blende in the DVD material single-crystal thin film leads to a possibility that an easy crystal-amorphous transformation between octahedra and tetrahedra around the Ge atoms can occur through the zinc-blende in the crystal phase as an intermediate state. These results strongly support an umbrella-flip model on the laser-induced crystal-amorphous transition in $Ge_2Sb_2Te_5$ proposed previously by Kolobov et al. [4].

CONCLUSION

In this article, we discuss the feasibility of recent XFH technique using a third-generation synchrotron radiation facility through an application to a DVD-RAM material thin film. The XFH experiment was performed in inverse mode using a recently developed avalanche photodiode detector with a cylindrical graphite crystal energy-analyzer. A 3D atomic image was obtained at 100 K around the Ge atoms in a $Ge_2Sb_2Te_5$ single-crystal thin film by means of XFH technique at the beamline BL37XU of the SPring-8 to clarify the high-speed writing and erasing mechanism of this DVD material. From the obtained XFH image, it was concluded that the single-crystal thin film has a mixture of rocksalt and zinc-blende structures. In addition, the images indicate large distortions associated with the existence of vacancies of the Ge(Sb) site. The present XFH results are in good agreement with the previous XAFS results, which has predicted a phase transition due to an umbrella flip motion of the Ge atoms.

ACKNOWLEDGMENTS

The authors acknowledge Dr. Y. Terada of JASRI/SPring-8 for the technical help of this XFH experiment. The XFH experiments were performed at the beamline BL37XU of the SPring-8 with the approval of the Japan Synchrotron Radiation Research Institute (JASRI) as a Nanotechnology Support Project of the Ministry of Education, Culture, Sports, Science, and Technology (Proposal. No. 2005B0122).

REFERENCES

1. K. Hayashi, *J. Jpn. Soc. Syn. Rad. Res.* **15**, 267-275 (2002). (in Japanese)
2. M. Tegze, G. Faigel, S. Marchesini, M. Belakhovsky, and O. Ulrich, *Nature* **407**, 38-38 (2000).
3. K. Hayashi, Y. Takahashi, E. Matsubara, S. Kishimoto, T. Mori, and M. Tanaka, *Nucl. Instrum. Met. Phys. Res. B* **196**, 180-185 (2002).
4. A. V. Kolobov, P. Fons, A. I. Frenkel, A. L. Ankudinov, J. Tominaga, and T. Uruga, *Nature Materials* **3**, 703-708 (2004).
5. N. Yamada and T. Matsunaga, *J. Appl. Phys.* **88**, 7020-7028 (2000).
6. J. J. Barton, *Phys. Rev. Lett.* **67**, 3106-3109 (1991).

Beamline Design for a BioNanoprobe: Stability and Coherence

B. Lai, S. Vogt, J. Maser

Advanced Photon Source, Argonne National Laboratory, 9700 S. Cass Ave., Argonne, IL 60439, USA

Abstract. For scanning x-ray microprobes, the angle of the incident beam is required to be stable to better than one microradian during the course of an experiment. This is a very stringent requirement, even more so for micro-XAS measurements when the monochromator energy has to be scanned over hundreds of eV. At the same time, the horizontal emittance of the electron source at most synchrotron facilities is much too large to provide coherent illumination of the microfocusing optics. A beamline design is proposed here that makes use of the large horizontal emittance to provide a very stable beam for the operation of a BioNanoprobe, while also increases the coherence to ensure diffraction-limited resolution in the horizontal direction.

Keywords: Beamline design, nanoprobe, coherence, stability.
PACS: 07.85.Qe, 07.85.Tt, 41.50.+h

INTRODUCTION

Trace metals in biological systems play a vital role in the functioning of cells, and metal deficiencies or metal dysregulation were implicated in a large variety of diseases. High-resolution mapping and quantification of trace metals in cells and tissues has therefore become an important tool at the Advanced Photon Source. To address both the increasing need for trace metal mapping as well as to significantly improve the spatial resolution available, we are designing a new x-ray microscopy beamline. In beamlines dedicated for high-resolution scanning x-ray microprobes, beam stability and coherence preservation are two of the primary design considerations. In particular, the angle of the beam incident on the high-resolution focusing optics must be sufficiently stable so as not to degrade the spatial resolution during the course of measurements. In a diffraction-limited focusing system, the spot size δ is given by $0.6*\lambda/NA$, where λ is the x-ray wavelength and NA is the numerical aperture given by $\sim r/f$, r being the radius and f being the focal length of the focusing optics. It follows that any variation in the beam angle $\Delta\theta$ should be less than δ/f, i.e.

$$\Delta\theta < \frac{0.6\lambda}{r} \tag{1}$$

For $\lambda = 1$ Å and a reasonable optics size of $r = 100$ µm, $\Delta\theta$ needs to be less than 0.6 µrad. In comparison, the Darwin width of a Si(111) crystal is ~ 22 µrad at the same energy. The difficulty in maintaining beam stability is only compounded during microspectroscopy measurements (micro-XAS) when the incident energy is scanned over hundreds of eV. Note that the angular stability requirement (Eq. 1) is almost independent of energy, since r should in principle be matched to the lateral coherence length which is typically proportional to λ.

Another condition that typically exists at synchrotron facilities is that the beam coherence is much higher in the vertical direction than in the horizontal (up to factor of 100). Thus, proper spatial filtering needs to be implemented in the horizontal direction in order to achieve diffraction-limited resolution. We present here a beamline design that will preserve the vertical beam coherence, increase the coherence length in the horizontal direction, maintain angular stability even during XANES scans, and reduce sensitivity to vibration and drift of the monochromator crystals.

CP879, *Synchrotron Radiation Instrumentation: Ninth International Conference*,
edited by Jae-Young Choi and Seungyu Rah
© 2007 American Institute of Physics 978-0-7354-0373-4/07/$23.00

BIONANOPROBE BEAMLINE DESIGN

A BioNanoprobe for life science applications at the Advanced Photon Source with a spatial resolution better than 30 nm has been proposed [1]. The beamline and instrument will be designed for incident energies spanning from the vanadium K-edge at 5.4 keV up to 30 keV, thereby allowing micro-XANES and fluorescence excitation (XRF) of transition and heavy metals of interest. Two collinear undulators with a period of $\lambda_u = 3.0$ cm and a total length of 4.8 m will cover this energy regime continuously using the 1st, 3rd, and 5th harmonic with high brilliance. A schematic of the beamline layout is illustrated in Fig. 1. The first optics will be a horizontally deflecting plane mirror, serving as harmonic rejecter and power filter for downstream optics. It is followed by a double crystal monochromator for micro-XANES and a double multilayer monochromator for micro-XRF, which can interchangeably be inserted into the beam. Both monochromators are horizontally diffracting. They are immediately followed by a horizontal beam-defining aperture (BDA) that acts as the horizontal source for downstream zone-plate-based microfocusing optics.

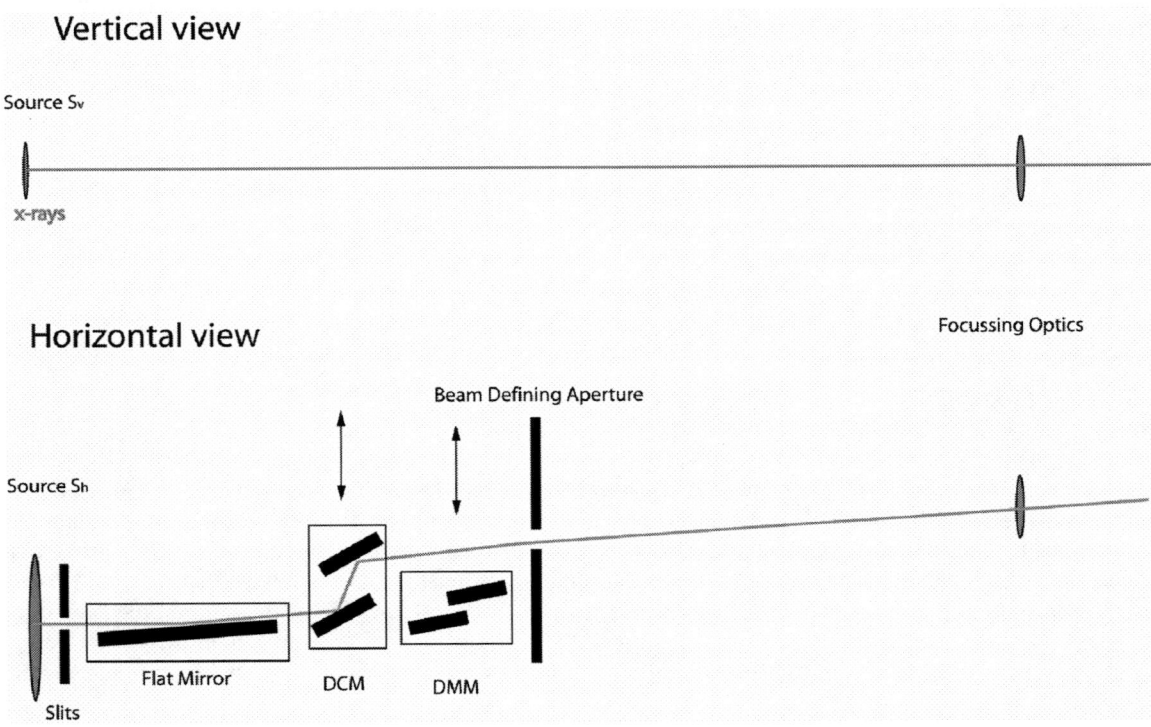

FIGURE 1. Beamline layout for the proposed BioNanoprobe.

Since all beamline optics are flat, they can be fabricated with high precision inexpensively. The all-horizontal deflecting geometry has three advantages. First, vertical coherence will be preserved with minimal degradation. Second, since there are no other optics between the BDA and the downstream focusing optics, the size and position of the focal spot will not be influenced by thermal distortion, drift, and vibration of the crystals, nor imperfect crystal motions during μ-XANES scan. Such imperfections will not affect the spot size or position, but only the beam intensity which can be normalized by proper monitoring. Finally, better stability may be achieved with horizontally instead of vertically diffracting optics due to the absence of gravitational force in the direction of monochromator rotation.

The BDA will be located at 40 m from the source and the microfocusing optics at 70 m. The slit opening of the BDA, Δ_{BDA}, should be sufficiently small to provide a lateral coherence length $\sim 2r$ at the focusing optics. In other words, the geometrically demagnified image of the BDA should be less than δ, i.e.

$$\Delta_{BDA} < \frac{0.6\lambda}{r} * d \qquad (2)$$

where d is the distance between the BDA and the focusing optics. For $d = 30$ m, the aperture size should be < 18 μm which can be reasonably achieved for a slit in monochromatic beam. In comparison, the beam incident on the BDA will have a size of ~ 1.25 mm FWHM based on APS source parameters of $\sigma_x = 276.6$ μm and $\sigma_{x'} = 11.4$ μrad. This provides overfilling of the BDA by ~ 70 times, which is beneficial because intensity variation will be small even when the beam is "walking" drastically on the BDA.

The BDA should not be too far from the source, or else usable x-rays from the source will be lost. Essentially, as illustrated in Fig. 2, the pinhole image of the focusing optics projected by the BDA back onto the source plane should not be larger than the source itself, i.e.

$$d_o < d \frac{\sigma_x}{r} \tag{3}$$

With $\sigma_x = 276.6$ μm, $r = 100$ μm, and a total distance of $d_o + d = 70$ m, the BDA can be located as far as 51.4 m from the source without much loss of usable photons. However, as the distance d between the BDA and focusing optics decreases, the opening of the BDA has to be ever smaller as defined by Eq. 2. Also, the divergence of the beam incident on the focusing optics will increase, which will cause some reflectivity loss if the divergence approaches the Darwin width of the π-polarization rocking curve as discussed below.

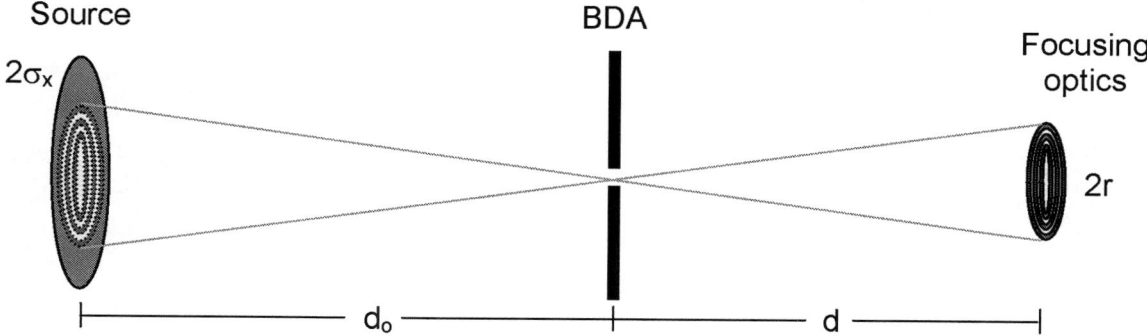

FIGURE 2. Pinhole image of the focusing optics projected back onto the source. The source size is characterized by $2\sigma_x$, and the zone plate focusing optics has an aperture of $2r$.

HORIZONTALLY VERSUS VERTICALLY DIFFRACTING MONOCHROMATOR

The horizontally deflecting geometry and spatial filtering with an appropriate aperture had been employed for x-ray microscopy beamlines at lower energies using grazing incidence optics, such as beamline X1A at the NSLS [2] and 2-ID-B at the APS [3]. In the hard x-ray regime, because of the larger crystal diffraction angle, the disadvantage with a horizontally diffracting double crystal monochromator relative to a vertically diffracting one is potentially lower intensity due to polarization. Figure 3 showed the σ-polarization and π-polarization rocking curve upon reflection from two Si(111) crystals at 10 keV. For measurements that use the full beam, the intensity difference is then determined by the ratio of the integrated reflectivity of the corresponding rocking curves, which is a factor of 0.91 in this example mainly due to the narrower width of the π-polarization rocking curve. However, the angular acceptance of the microfocusing optics is typically smaller than the Darwin width of the rocking curve. In this case, with a 200-μm focusing optics and $d = 30$ m, the angular acceptance is only 6.7 μrad compared to a Darwin width of ~ 25 μrad in π-polarization. Thus, the intensity difference between a horizontally diffracting monochromator and vertically diffracting one is mainly determined by the ratio of the peak reflectivities, which is only a factor of 0.99 at 10 keV. Thus there is essentially no difference in focused intensity at 10 keV caused by the diffraction geometry of the monochromator. At lower energies, the intensity difference is more significant because the peak reflectivity in the π-polarization decreases faster than the σ-polarization due to absorption. For instance, at 5.4 keV the intensities differ by a factor of 0.88. However, these small intensity sacrifice was deemed acceptable for a much more stable beam which is essential for the operation of the BioNanoprobe.

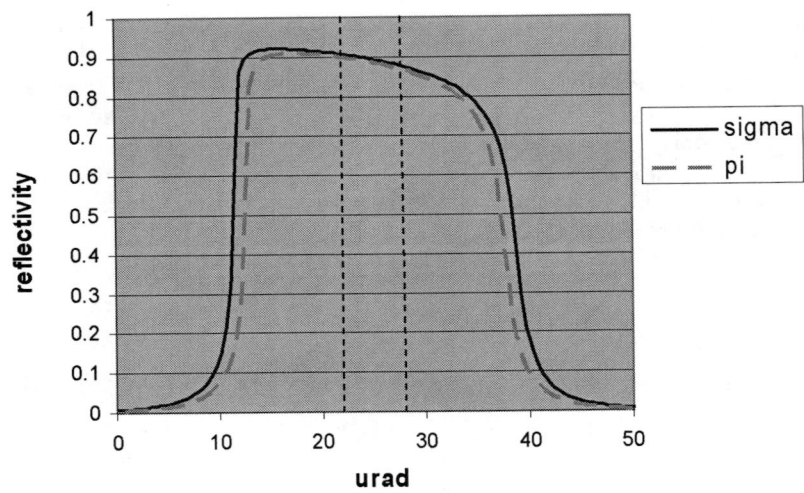

FIGURE 3. Darwin width of Si(111) double-crystal rocking curve at 10 keV in σ-polarization and π-polarization. Angular acceptance of the microfocusing optics is shown between the two vertical dashed lines.

ACKNOWLEDGMENTS

This work was supported by the U.S. Dept. of Energy, Office of Science, Office of Basic Energy Sciences, under Contract No. W-31-109-ENG-38.

REFERENCES

1. G. Woloschak, S. Vogt, *Proposal for a BioNanoprobe* (private communication).
2. H. Rarback, S. Krinsky, P. Mortazavi, D. Shu, J. Kirz, C. Jacobsen, and M. Howells, *Nucl. Instrum Methods Phys. Res.* **A246**, 159-162 (1986).
3. I. McNulty, A. Khounsary, J. Barraza, C. Benson, Y.P. Feng, and D. Shu, *Rev. Sci. Instrum.* **67 (9)**, CD-ROM (1996).

A New Tomography Beamline at a Wiggler Port at the Center for Advanced Microstructures and Devices (CAMD) Storage Ring

Kyungmin Ham*, Heath A. Barnett†, Leslie G. Butler†, Clinton S. Willson**, Kevin J. Morris*, Roland C. Tittsworth[1],* and John D. Scott*

*Center for Advanced Microstructures and Devices, Louisiana State University, Baton Rouge, LA, 70806, USA
†Department of Chemistry
**Department of Civil and Environmental Engineering

Abstract. A new tomography beamline has been built and commissioned at the 7 T wiggler of the CAMD storage ring. This beamline is equipped with two monochromators that can be used interchangeably for X-ray absorption spectroscopy or high resolution X-ray tomography, at best 2-3 μm pixel size. The high-flux double multilayer-mirror monochromator (W-B$_4$C multilayers) can be used in the energy range from 6 to 35 keV with a resolution ($\Delta E/E$) between 0.01- 0.03. The second is a channel-cut Si(311)-crystal monochromator with a range of 15 to 36 keV and resolution of ca. 10^{-4}, this is not yet tested. Tomography has the potential for high-throughput materials analysis; however, there are some significant obstacles to be overcome in the areas of data acquisition, reconstruction, visualization and analysis. Data acquisition is facilitated by the multilayer monochromator as this provides high photon flux, thus reducing measurement time. At the beamline, Matlab© routines provide simple x,y,z fly-throughs of the sample. Off-beamline processing with Amira© can yield more sophisticated inspection of the sample. Standard data acquisition based on fixed angle increments is not optimal, however, new patterns based on Greek golden ratio angle increments offer faster convergence to a high signal-to-noise-ratio image. The image reconstruction has traditionally been done by back-projection reconstruction. In this presentation we will show first results from samples studied at the new beamline.

Keywords: tomography, synchrotron, monochromator
PACS: 81.70.Tx, 87.59.Fm

INTRODUCTION

The CAMD tomography beamline was originally built in 2000 at a bending magnet port supplying filtered x-ray radiation. Recently the beamline moved to one of the 7 Tesla superconducting wiggler ports and was re-installed with a dual monochromator system. The dual monochromator system has channel-cut Si(311)-crystal monochromator for x-ray absorption spectroscopy and a multilayer monochromator for tomography experiment, where the multilayer monochromator is preferred because of higher x-ray flux. Compared to the bending magnet beamline, the new tomography beamline at CAMD can access the energy range from 6 to 35 keV, expanding its application from low x-ray absorbing material to high x-ray absorbing material. Monochromatic light adds the ability to better optimize images based on contrast of internal structures and to evaluate the internal elemental composition of the material.

DESCRIPTION OF CAMD TOMOGRAPHY BEAMLINE

The multilayer monochromator is fabricated with a double crystal cut from a single block of silicon and coated with a number (>200) of W-B$_4$C layers. The W-B$_4$C layers have a d-spacing of 20±1 Å. The multilayer surfaces are 50 mm x 100 mm with reflectivity of single multilayer, R, greater than 65% at 8 keV (Cu-Kα). Energy range is 6 to 35 keV, corresponding to grazing incident angles from 2.96° to 0.5°, respectively. In order to block white beam, which might be transmitted through either monochromator at low grazing incident angles, a beam blocker is installed at the exit of

---

[1] Deceased.

CP879, *Synchrotron Radiation Instrumentation: Ninth International Conference*,
edited by Jae-Young Choi and Seungyu Rah
© 2007 American Institute of Physics 978-0-7354-0373-4/07/$23.00

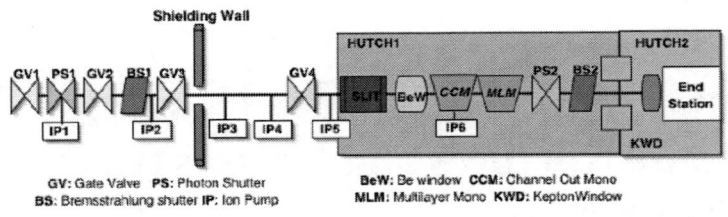 

FIGURE 1. Beamline schematic and photograph of the dual monochromators.

the multilayer monochromator. With the multilayer monochromator, the transmitted beam size is 3 cm wide and 1 mm high and is offset is 4 mm vertically from the incident white beam.

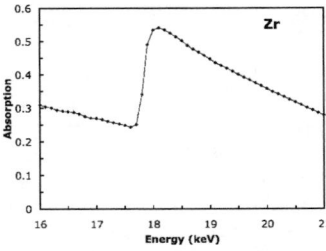

FIGURE 2. Zirconium metal foil K-edge absorption spectrum showing the performance of the high-flux double multilayer-mirror monochromator (W-B_4C multilayers) with an energy resolution ($\Delta E/E$) between 0.01- 0.03. Data were taken at 100 eV increments with ion chambers filled flowing air.

Details of image acquisition are described at [1, 2] and at `http://tomo.camd.lsu.edu`. A conventional tomography sample alignment and rotation stage set is used. We use either YAG(Ce) or CsI(Tl) scintillator, the latter survives several months in the ambient lab atmosphere. We can compare the imaging throughput of the CAMD tomography beamline with that of 13BM at the Advanced Photon Source. At 9 μm pixel size and 20 keV, CAMD achieves 3/4 of CCD maximum counts in 2 s versus an APS shutter speed of about 0.75 s. The near equivalence between CAMD and APS is a result of: (1) the CAMD wiggler versus the APS bending magnet and (2) the CAMD high-throughput multilayer monochromator versus the APS high-resolution double crystal monochromator. APS, however, offers a 5 mm beam thickness versus the current CAMD beam of 1 mm, hence CAMD more often acquires multiple data sets at various z-stage (vertical) settings. One recent innovation has been a change in the time-ordering of the rotation angle settings. Traditionally, tomography has been performed with fixed angle increments for rotation angles between 0° to 180°. If the last portion of rotation angles cannot be accessed, i.e., sample decomposition/dehydration or beam dump, then the reconstruction is distorted. Fortunately, recent mathematical research has considered the data acquisition as a convergence problem and several new rotation angle orderings are now available.[3, 4] At CAMD, used a rotation angle ordering based on modulo angle divided by the Greek golden ratio. With this ordered set of angles, early termination of the experiment still yields a useful reconstruction; the complete angle set is equivalent to that of the fixed angle increment acquisition. MatLab© is used for image reconstruction and the generation of simple x-, y-, and z- axes fly-throughs. Amira is used for 3D visualization.

APPLICATIONS

Biology

Accidents and diseases of cornified end organs, such as equine and bovine hooves, feline claws and psittacine (parrot beaks), are generally accompanied by inflammatory processes with associated lesions and abnormalities of the vascularization. The vascularization of these cornified organs plays a central role in the proper ossification of the underlying bone and the proper nutrition of the cornified tissue.[5, 6] Synchrotron X-ray tomography is being used for imaging these organs. Vascularization patterns in biopsy samples of cornified organs as imaged with a non-destructive diagnostic imaging tool could identify pathological changes at an early stage of the disease. Dehydration of biological samples is one of the biggest challenges for tomography. A humidity chamber built from a sealed aluminum box with Kapton windows is used for this purpose; a large diameter double-race ball bearing keeps the box and windows aligned with the x-ray beam.

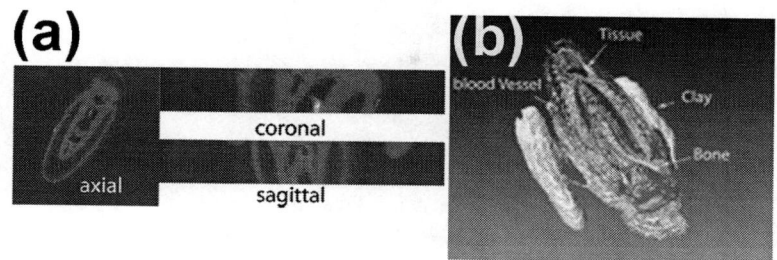

FIGURE 3. A cat claw (front paw) is imaged with 9 μm pixel size in a field of view of 1211x 1211x 192 voxels; projects were acquired at 20 keV x-ray energy for a sample mounted on clay in a 100% humidity chamber. (a) Single slice views of cat claw on axial, coronal, and sagittal planes at various scales. (b) 3D view of reconstructed cat claw volume with isosurfaces for bone, tissue, and blood vessels. Blood vessels are injected with $BaSO_4$ for better contrast with tissues and bones.

Polymer Blends

Our previous work on 3D images of flame retardants blends in polystyrene shows that synchrotron x-ray tomography provides tremendous insight into material structures, in this case, the quality of the blend and the spatial correlation between two additives.[7] From the 3D reconstructed volumes acquired at 3.26 μm pixel size for seven different x-ray energies, we computed the 3D chemical maps of the polystyrene, the flame retardant (a brominated aromatic compound), and antimony(III) oxide, a synergist for the flame retardant. Based on the success of that work, more studies are underway for polymer composite materials and time-evolution of polymer blends.

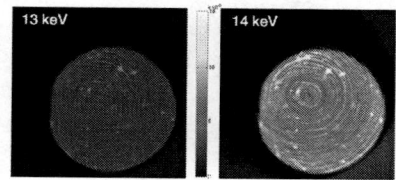

FIGURE 4. A commercial flame retardant (a brominated aromatic) in polystyrene is imaged at 13 keV below Br K-edge energy and 14 keV above Br K-edge energy. This image clearly shows more x-ray absorbance at 14 keV, indicating a heterogenous distribution of the flame retardant, either due to poor initial blending or an unstable blend subject to Ostwald ripening. The field of view is 481x481 with 9 μm pixel size.

Porous media

High-resolution tomography provides the high-quality, 3D images of porous media necessary for mapping pore-scale geometry and topology, i.e., the pore network structure. The tomographic images and the resultant data and information that we get from them provide better understanding of the pore-scale processes that affect single- and multi-phase flow in porous media. Applications include the flow, entrapment, and distribution of immiscible contaminants, e.g., chlorinated solvents, in groundwater systems and of oil and water in petroleum reservoirs. [8, 9, 10]

ONGOING UPGRADES

The present CCD has 85k e$^-$ well depth. Our analysis of the signal-to-noise ratio shows a significant advantage with high-well depth, hence it will be upgraded to the 250k e$^-$ well depth CCD later in 2006. In part, increased well depth

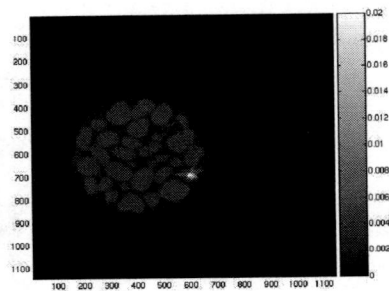

FIGURE 5. A sand grain column held in a plastic tube was imaged with 9 μm pixel size at 25 keV. Field of view is 1101x1101. From images such as these we are now able to extract the distribution of pores and throats, and to then calculate flow parameters. In addition, with selected contrast agents and variable x-ray energies, we are able to separately image air, water, and hydrocarbon solvents.

can be simulated by repetitive imaging, but readout noise reduces the efficacy of this approach.[11] Motorized table alignment (5-axes) will supplement the existing 7-stages of the basic instrument. Height adjustment is a fast way to move the projection across the scintillator for re-alignment and to work around scintillator defects. Also, modification of endstation for phase contrast tomography experiment is in progress.

ACKNOWLEDGMENTS

This project has been supported by NSF IMR-0216875 and the State of Louisiana through the CAMD operational budget.

REFERENCES

1. K. Ham, H. Jin, L. G. Butler, and R. L. Kurtz. A microtomography beamline at the Louisiana State University Center for Advanced Microstructures and Devices Synchrotron. *Review of Scientific Instruments*, 73(3):1521–1523, 2002. Part 2.
2. L.G. Butler, K. Ham, H. Jin, and R.L. Kurtz. Tomography at the Louisiana State University CAMD synchrotorn:applications to polymer blends. *Proceedings of SPIE-The International Society for Optical Engineering*, 4503:54–61, 2002.
3. Thomas Kohler. A Projection Access Scheme for Iterative Reconstruction Based on the Golden Section. *Nuclear Science Symposium Conference Record, 2004 IEEE*, 6(16-22):3961–3965, 2004.
4. Samuel Matej. Iterative Tomographic Image Reconstruction using Fourier-Based Forward and Back-Projectors. *Ieee Transactions on Medical Imaging*, 23(4):401–412, 2004.
5. H. Bragulla and R.M. Hirschberg. Horse hooves and bird feathers: Two model systems for studying the structure and development of highly adapted integumentary accessory organs - The role of the dermo-epidermal interface for the micro-architecture of complex epidermal structures. *Journal of Experimental Zoology Part B-Molecular and Developmental Evolution*, 298B(1):140–151, 2003.
6. H. Bragulla and R.M. Hirschberg. Fetal development of the segment-specific papillary body in the equine hoof. *Journal of Morphology*, 258(2):207–224, 2003.
7. K. Ham, H. Jin, R. I. Al-Raoush, X. G. Xie, C. S. Willson, G. R. Byerly, Larry S. Simeral, M. L. Rivers, R. L. Kurtz, and L. G. Butler. Three-Dimensional Chemical Analysis with Synchrotron Tomography at Multiple X-ray Energies: Brominated Aromatic Flame Retardant and Antimony Oxide in Polystyrene. *Chemistry of Materials*, 16(4032-42), 2004.
8. R.I. Al-Raoush, K. Thompson, and C.S. Willson. Comparison of Network Generation Techniques for Unconsolidated Porous Media. *Soil Sci. Am. J.*, 67:1687–1700, 2003.
9. R. I. Al-Raoush and C.S. Willson. A Pore-scale Investigation of a Multiphase Porous Media System. *Journal of Contam. Hydrology*, 77(1-2):67–89, 2005.
10. R. I. Al-Raoush and C.S. Willson. Extraction of Physically-Representative Pore Network from Unconsolidated Porous Media Systems Using Synchrotron Microtomography. *Journal of Hydrology*, 300(1-4):44–64, 2005.
11. K. Ham, C. S. Willson, M. L. Rivers, R. L. Kurtz, and L. G. Butler. Algorithms for three-dimensional chemical analysis with multi-energy tomography data. *SPIE "Developements in X-Ray Tomography IV"*, 5535:286–292, 2004.

Optomechanical Design of a Hard X-ray Nanoprobe Instrument with Nanometer-Scale Active Vibration Control

D. Shu[2], J. Maser[1,3], M. Holt[3], R. Winarski[1,3], C. Preissner[2], A. Smolyanitskiy[2], B. Lai[3], S. Vogt[3], and G. B. Stephenson[1,4]

[1]*Center for Nanoscale Materials, Argonne National Laboratory, Argonne, IL 60439, U.S.A.*
[2]*APS Engineering Support Division, Argonne National Laboratory, Argonne, IL 60439, U.S.A.*
[3]*X-ray Sciences Division, Argonne National Laboratory, Argonne, IL 60439, U.S.A.*
[4]*Materials Sciences Division, Argonne National Laboratory, Argonne, IL 60439, U.S.A.*

Abstract. We are developing a new hard x-ray nanoprobe instrument that is one of the centerpieces of the characterization facilities of the Center for Nanoscale Materials being constructed at Argonne National Laboratory. This new probe will cover an energy range of 3-30 keV with 30-nm spacial resolution. The system is designed to accommodate x-ray optics with a resolution limit of 10 nm, therefore, it requires staging of x-ray optics and specimens with a mechanical repeatability of better than 5 nm. Fast feedback for differential vibration control between the zone-plate x-ray optics and the sample holder has been implemented in the design using a digital-signal-processor-based real-time closed-loop feedback technique. A specially designed, custom-built laser Doppler displacement meter system provides two-dimensional differential displacement measurements with subnanometer resolution between the zone-plate x-ray optics and the sample holder. The optomechanical design of the instrument positioning stage system with nanometer-scale active vibration control is presented in this paper.

Keywords: X-ray manoprobe, X-ray microscope, X-ray microdiffraction, X-ray imaging
PACS: 07.85.Qe

INTRODUCTION

The Advanced Photon Source (APS) at Argonne National Laboratory (ANL) is a national user facility for synchrotron radiation research. The high-brilliance x-ray beams of this third-generation synchrotron radiation source provide powerful tools for forefront basic and applied research in many field of scientific research. A hard x-ray nanoprobe will be constructed as the centerpiece of the x-ray characterization facilities at the APS for the Center for Nanoscale Materials (CNM) to be constructed at ANL. A dedicated set of source, beamline, and optics will be used to avoid compromising the capabilities of the nanoprobe [1]. This unique instrument will not only be key to the broader nanoscience community, it will also offer diverse capabilities in studying nanomaterials and nanostructures, particularly embedded structures. The combination of diffraction, fluorescence, and phase contrast in a single tool will provide unique characterization capabilities for nanoscience.

This new probe will cover an energy range of 3-30 keV. The working distance between the nanofocusing optics and the sample will typically be in the range of 10-30 mm. With advances in the fabrication of zone-plate optics, and an optimized beamline design, we expect to be able to achieve 30-nm resolution at the nanoprobe. The system is designed to accommodate x-ray optics with a resolution limit of 10 nm, therefore, it requires staging of x-ray optics and specimens with a mechanical repeatability of better than 5 nm.

There are many technical challenges for the development of this new hard x-ray nanoprobe. One of them is to develop a state-of-the-art linear multiple-stage system with nanometer-scale resolution and relative stability with a centimeter-scale dynamic range. The nanoprobe should be functional in a floor environment where typical mechanical vibration amplitude is measured at the 20-nm level in the frequency range of a few hertz. The optomechanical design of the nanoprobe instrument with nanometer-scale active vibration control is presented in this paper.

CP879, *Synchrotron Radiation Instrumentation: Ninth International Conference,*
edited by Jae-Young Choi and Seungyu Rah
© 2007 American Institute of Physics 978-0-7354-0373-4/07/$23.00

GENERAL LAYOUT OF THE NANOPROBE INSTRUMENT

The nanoprobe instrument will combine a scanning probe mode with a full-field transmission mode. The scanning probe mode will provide fluorescence spectroscopy and diffraction contrast imaging. The full-field transmission mode will allow 2-D imaging and tomography. Diffractive optics, such as zone plates, will be used for focusing and imaging in this instrument. High-resolution positioning and scanning is performed using the stage group for zone plate optics in a high-vacuum-compatible instrument chamber. In the same vacuum vessel, a specimen stage group is used for coarse positioning only [2].

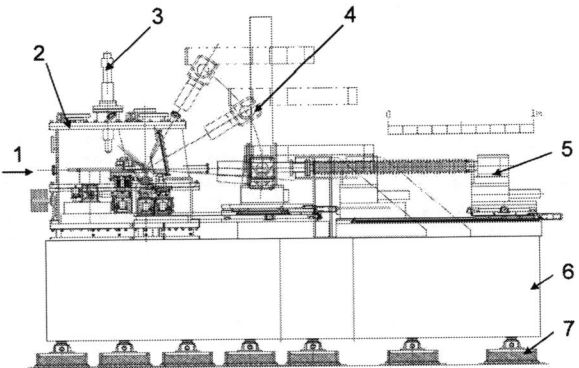

FIGURE 1. Side view of the hard x-ray nanoprobe instrument. (1) Incident beam; (2) Instrument chamber; (3) Optical microscope; (4) Diffraction detector; (5) Transmission imaging detector; (6) Granite base; (7) Isolators.

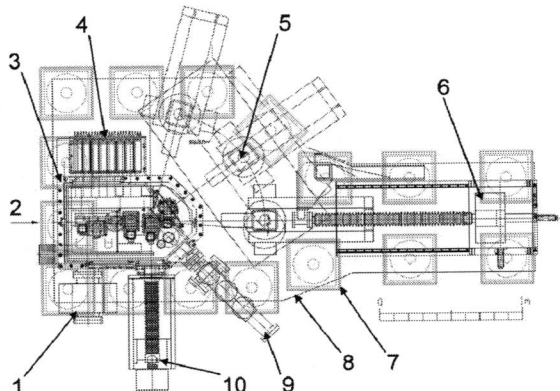

FIGURE 2. Top view of the hard x-ray nanoprobe instrument. (1) Ion pump; (2) Incident beam; (3) Instrument Chamber; (4) Laser head for LDDM; (5) Diffraction detector; (6) Transmission imaging detector; (7) Isolators below base; (8) Granite base; (9) Airlock for specimen exchange; (10) Fluorescence detector.

As show in Fig. 1, the optomechanical structure for the nanoprobe presented here consists of the following major component groups: a nanoprobe instrument chamber (2), a granite base (6) with sixteen sand-box-based isolators (7) to minimize the vibrations excited from the ground, a customized industrial robot-arm-based detector manipulator (4) for microdiffraction applications, an optical microscope (3), and a translation stage system for the transmission imaging detector (5). An airlock for specimen exchange (9) and a fluorescence detector (10) are shown in Fig. 2.

SCANNING STAGE SYSTEM WITH ACTIVE VIBRATION CONTROL

Inside the instrument chamber, as shown in Fig. 3, the positioning stages for the optics are grouped into several subcomponents that are engineered to be moved in or out of the beam to allow configuration of the instrument for either scanning probe mode or full-field transmission mode. These subcomponents are the condensor module (CM) (2), the focusing optics module (FOM) (3), the specimen module (SM) (4), and the imaging optics module (IOM) (6). The CM provides illumination of the specimen in full-field transmission mode. The FOM provides positioning and scanning of the focusing zone plates and order-sorting aperture at high mechanical resolution and accuracy. The SM provides specimen positioning and temperature control. The IOM provides positioning of the objective zone plates and phase plates required for operation in full-field transmission mode. The FOM and SM form a scanning nanoprobe instrument. The CM, SM, and IOM form a transmission x-ray microscope (TXM). The specimen module is shared by both configurations.

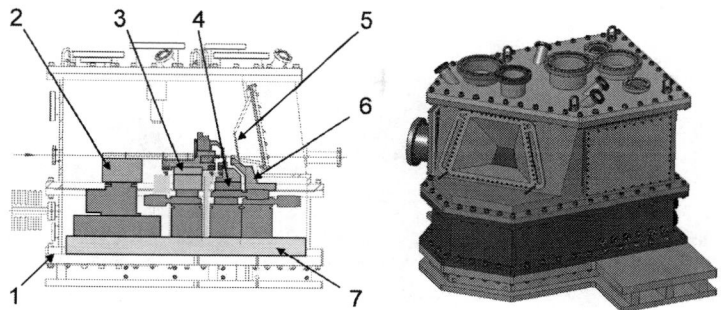

FIGURE 3. Left: A schematic side view of the nanoprobe instrument chamber. (1) Vacuum chamber; (2) CM; (3) FOM; (4) SM; (5) Beryllium window; (6) IOM; (7) Invar reference base. Right: A 3-D model of the nanoprobe instrument chamber.

The laser encoder system consists of a rigid Invar reference frame, a set of LDDMs, and laser optics for resolution extension. The LDDM is based on the principles of the Doppler effect and optical heterodyning. We have chosen a customized LDDM from Optodyne Inc. as our basic system, not only because of its high resolution (2 nm, typically) and fast object speed (2 m/s) but also because of its unique performance independent of polarization, which provides the convenience of creating a novel multiple-reflection-based optical design to attain subnanometer linear resolution [3-5].

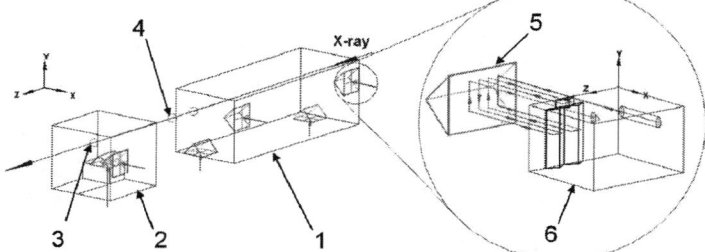

FIGURE 4. Schematic of a six-LDDM encoder system for the two-dimensional differential position encoding between the FOM and SM. (1) Prism holder for zone-plate optics; (2) Prism holder for sample stage; (3) Sample location; (4) Zone-plate optics locations; (5) Prism on the stage; (6) Prism group on the reference frame.

In the self-aligning multiple-reflection optical design for the LDDM system, the heterodyning detector is housed coaxially inside the frequency-stabilized laser source. Unlike a typical single reflection on the moving target, the laser beam is reflected back and forth eight times between the fixed base and the moving target, as shown in Fig. 4 (in circle). The laser beam, which is reflected back to the heterodyning detector, is frequency-shifted by the movement of the moving target relative to the fixed base. With the same LDDM laser source and detector electronics, this optical path provides eight times greater resolution for the linear displacement measurement and encoding. A 0.2-nm resolution was recently reached by a prototype system. A total of six LDDMs will be used to correct possible linear motion trajectory errors. Figure 4 shows the scheme of the six-LDDMs encoder system for the two-dimensional differential position encoding between the FOM and SM. The LDDM encoded travel ranges are 12 mm (X) X 12 mm (Y) X 12 mm (Z) for FOM and 12 mm (X) X 12 mm (Y) X 6 mm (Z) for SM. Figure 5 shows a schematic of differential positioning feedback control in the vertical (Y) direction. The scheme of the differential feedback control in the horizontal (X) direction is similar.

We have developed two sets of piezoelectric-transducer (PZT)-driven horizontal and vertical high-stiffness weak-link stages with different travel range and resolution for FOM fine-motion control. The high-resolution, weak-link stages have a 0.02-nm resolution with a 1/15 motion-reduction mechanism and 1.5-micron travel range [6,7]. For the medium-resolution, weak-link stages, multiple PZT drivers are applied to guarantee the linear motion trajectory accuracy in a specific direction. The resolution of the stage is 0.3 nm with a travel range of 15 microns. Physik Instrumente™ PZT actuators with strain-gauge sensor servo-control modules were used to drive both high- and medium-resolution weak-link stages.

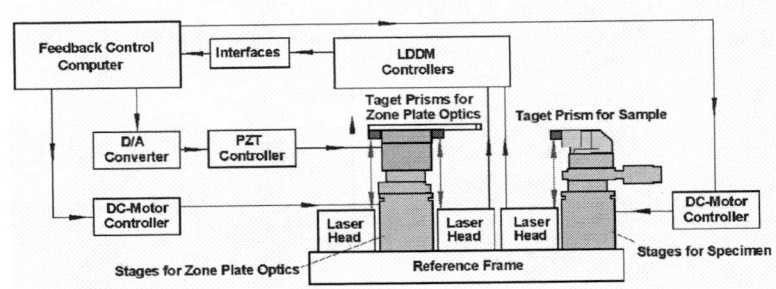

FIGURE 5. Schematic of differential positioning feedback control in the vertical (Y) direction. For the differential feedback positioning control in the vertical (Y) direction, two LDDMs are applied to the FOM and SM in the Y direction. Since the reference frame defines the coordinate system of the nanoprobe, all positions are measured with respect to this frame. To perform a differential measurement between the stage groups for FOM and SM in the Y direction, a DSP-based feedback control computer is used as a control console to collect the position information from the three LDDMs with a positioning update rate of 937 kHz. The DSP computes the position differences between the two stage groups and determines the discrepancy between the actual and desired differential position between the zone-plate optics and sample, and feeds back differential position-correction signals through a proportional-integral-derivative (PID) loop to the PZT-driven weak-link stage on the FOM. Differential scanning motion can be activated by controlling the desired differential position value. In the case of large-range scanning activity, a relay mechanism is implemented into the control software to ensure a smooth transition between the PZT-driven weak-link fine stage and the DC-motor-driven coarse translation stage.

SUMMARY

A new hard x-ray nanoprobe instrument with nanometer-scale active vibration control will be constructed as the centerpiece of the x-ray characterization facilities at the APS for the CNM to be constructed at ANL. The optomechanical design of the nanoprobe instrument is presented in this paper.

To provide initial x-ray microscopy capabilities before the availability of the hard x-ray nanoprobe instrument, we have developed a prototype probe with scanning stage system as an "early user instrument" (EUI). We tested the EUI differential scanning stage system with the high-resolution weak-link stages. A series of 1-nm vertical differential positioning steps (between zone plate optics holder and specimen holder) have been demonstrated with a PC-based closed-loop control. A spatial resolution of 70 nm was obtained with a PC-based closed-loop feedback system for active vibration control. This value is consistent with the value expected for the vertical spot size. During this test, incident x-rays with a photon energy of 7.5 keV were focused by a Fresnel zone plate with an outermost zone width of 50 nm [8]. The experiences we gained from this prototype test will significantly benefit the real CNM nanoprobe instrument design.

ACKNOWLEDGMENTS

The authors would like to thank D. Nocher, and M. Muscia from the Argonne National Laboratory for their help in the EUI experiment. This work was supported by the U.S. Department of Energy, Office of Science, Office of Basic Energy Sciences, under contract No. W-31-109-Eng-38.

REFERENCES

1. J. Maser, G. B. Stephenson, D. Shu, B. Lai, S. Vogt, A. Khounsary, Y. Li, C. Benson, G. Schneider, SRI 2003 Conf. Proc., 705, AIP (2004) 470-473.
2. D. Shu, J. Maser, B. Lai, S. Vogt, M. Holt, C. Preissner, A. Smolyanitskiy, B. Tieman, R. Winarski, and G. B. Stephenson, to be published in the proceedings of X-ray Microscopy 2005, Himeji, Japan, July 2005.
3. LDDM is a trademark of the Optodyne Inc., 1180 Mahalo Place, Compton, CA 90220, U.S.A.
4. U.S. Patent granted No. 5,896,200, D. Shu, 1999.
5. U.S. Patent granted No. 6,822,733, D. Shu, 2004.
6. D. Shu, T. S. Toellner, and E. E. Alp, Nucl. Instrum. Methods A 467-468, 771-774 (2001).
7. U.S. Patent granted No. 6,607,840, D. Shu, T. S. Toellner, and E. E. Alp, 2003.
8. D. Shu, J. Maser, B. Lai, and S. Vogt, SRI 2003 Conf. Proc., 705, AIP (2004) 1287-1290.

Development of a Scanning X-ray Fluorescence Microscope Using Size-Controllable Focused X-ray Beam from 50 to 1500nm

Satoshi Matsuyama[a], Hidekazu Mimura[a], Hirokatsu Yumoto[a], Keiko Katagishi[a],
Soichiro Handa[a], Akihiko Shibatani[a], Yasuhisa Sano[a], Kazuya Yamamura[b],
Katsuyoshi Endo[b], Yuzo Mori[b], Yoshinori Nishino[d], Kenji Tamasaku[d],
Makina Yabashi[c], Tetsuya Ishikawa[c,d] and Kazuto Yamauchi[a]

[a]Department of Precision Science and Technology, Graduate School of Engineering,
Osaka University, Yamada-oka 2-1, Suita, Osaka 565-0871, Japan
[b]Research Center for Ultra-Precision Science and Technology, Graduate School of Engineering,
Osaka University, Yamada-oka 2-1, Suita, Osaka 565-0871, Japan
[c]SPring-8/Japan Synchrotron Radiation Research Institute (JASRI),
Kouto 1-1-1, Mikazuki, Hyogo 679-5148, Japan
[d]SPring-8/RIKEN, Kouto 1-1-1, Mikazuki, Hyogo 679-5148, Japan

Abstract. In scanning X-ray microscopy, focused beam intensity and size are very important from the viewpoints of improvements of various performances such as sensitivity and spatial resolution. The K-B mirror optical system is considered to be the most promising method for hard X-ray focusing, allowing highly efficient and energy-tunable focusing. We developed focusing optical system using K-B mirrors where the focused beam size is controllable within the range of 50 – 1500 nm. The focused beam size and beam intensity can be adjusted by changing the source size, although beam intensity and size are in a trade-off relationship. This controllability provides convenience for microscopy application. Diffraction limited focal size is also achieved by setting the source size to 10 μm. Intracellular elemental mappings at the single-cell level were performed to demonstrate the performance of the scanning X-ray fluorescence microscope equipped with the optical system at the BL29XUL of SPring-8. We will show magnified elemental images with spatial resolution of ~70 nm.

Keywords: X-ray focusing, Kirkpatrick-Baez, X-ray microscopy, Elemental mapping, X-ray fluorescence microscope
PACS: 87.59.-e

SCANNING X-RAY FLUORESCENCE MICROSCOPE

A scanning X-ray fluorescence microscope (SXFM)[1] is an imaging tool capable of visualizing the element distribution of a sample using X-ray fluorescence generated by focusing hard X-ray irradiation onto the sample. Because the excitation beam consists of hard X-rays, there is no need to install the samples under vacuum. From the viewpoint of energy-tunable focusing and sensitivity, the combination of a synchrotron radiation source, with high X-ray brightness and elliptical mirrors (K-B mirrors), which have high focusing efficiencies, represents the most powerful focusing systems[2] for the SXFM. The mirrors employed in this study were designed and fabricated to obtain diffraction-limited focusing at approximately 50 m downstream of an X-ray source. Owing to the distance of 50 m, the X-ray beam size is controllable over a wide range between 30 nm and 1400 nm merely by adjusting the X-ray source size. The beam intensity and size are in a trade-off relationship.

Figure 2 shows a schematic drawing of the SXFM system. A mirror manipulator[3], which was specially developed for these mirrors, enables us to adjust mirror positions with a high degree of accuracy. A silicon drift detector (SDD, Röntec Co., Ltd.) is employed to detect X-ray fluorescence. The detector of an X-ray fluorescence

CP879, *Synchrotron Radiation Instrumentation: Ninth International Conference*,
edited by Jae-Young Choi and Seungyu Rah
© 2007 American Institute of Physics 978-0-7354-0373-4/07/$23.00

spectrum at every point on the scanning area by the SDD makes it easy to carry out, for example, background correction, noise rejection and peak separation after completing the measurements. A linear-encoder-based feedback X–Y stage having a positioning resolution of 1 nm (SIGMA TECH Co., Ltd.) is employed to scan samples finely without backlash. The travel range (±10 mm) is wide enough to select samples from a wide area.

TABLE 1. System specifications for SXFM.

	Vertical direction	Horizontal direction
Mirror aperture (µm)	382	365
Numerical aperture	0.75×10^{-3}	1.20×10^{-3}
Working distance	100 mm	
Depth of focus (µm) [*1]	> 150	> 50
Diffraction limited focal size (nm, FWHM)	48	29
Available X-ray energy	4.4~19keV at BL29XUL of SPring-8	
Minimum step size	1 nm by a piezoelectric stage	
Maximum scan range	20 x 20 mm² by a stepping motor and piezoelectric stage	

[*1] Depth of focus is defined as the tolerance of focal length where the focus-size broadening is within 10%. It depends on source size, so the depth of focus given in the table were calculated under a diffraction-limited condition.

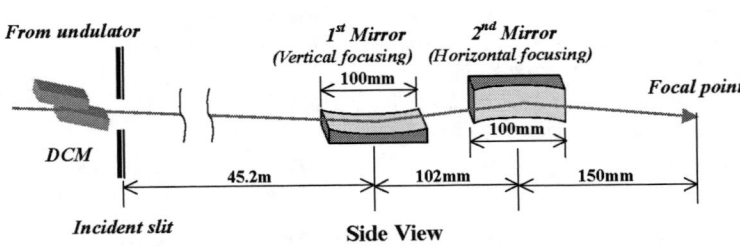

FIGURE 1. Designed optical system for hard X-ray nanofocusing. An incident slit just downstream of a double-crystal monochromator (DCM) is used as a virtual X-ray source to vary source size.

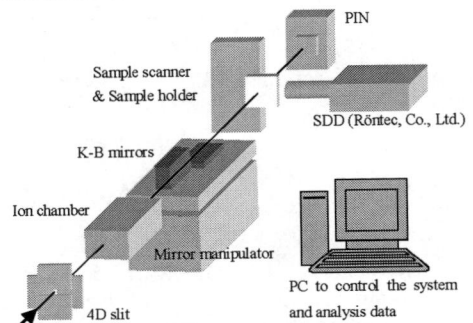

FIGURE 2. Schematic of SXFM system.

TWO-DIMENSIONAL FOCUSING TEST

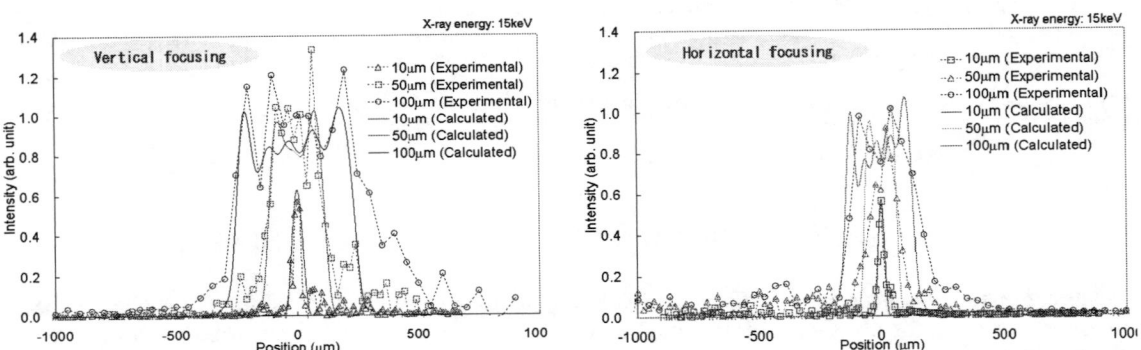

FIGURE 3. Focused beam profiles measured by gold wire (φ200 µm, wire scan method) scanning (E=15 keV). Intensity fluctuation seems to be cased by diffraction from an edge of the slit because K-B mirrors optics doesn't meet Abbe's sin condition.

Focusing tests at an X-ray energy of 15 keV were performed at BL29XUL (EH2) of SPring-8. By finely tuning mirror alignments with the slit size of less than 10 µm, the slit and the sample can be placed at ideal positions of a virtual source and a collecting point. Beam profiles were measured by wire scan method with changing incident slit size (virtual source size). Figure 3 shows the measured beam profiles (dots) and the wave-optically calculated profiles (solid lines). As a result of focusing tests, a minimum beam size of the focus of 30 x 50 nm² (V x H) under diffraction-limited conditions was achieved. Moreover, we could easily control beam size within the range of 30 ~

1400 nm full width at half maximum (FWHM) by changing the virtual source size. Table 2 shows a summary of the relationships between the photon flux and the FWHM of the focused beams obtained in the experiment.

TABLE 2. Relationship between FWHM and photon flux.

	Virtual source size (H x V μm^2)			
	10 x 10	50 x 50	200 x 200	1000 x 1000 (Fully open)
Beam size of focus (H x V nm^2)	30 x 50 **(Diffraction limit)**	150 x 250	600 x 1000	1400 x 1000
Photon flux (photons/s)	6×10^9	3×10^{11}	4×10^{12}	8×10^{12}

OBSERVATION OF ELEMENT DISTRIBUTION IN CULTURED CELLS

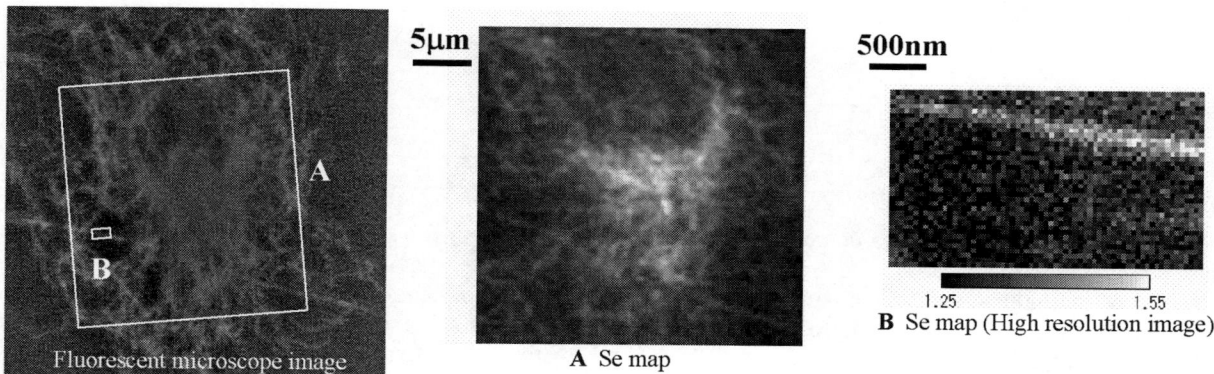

FIGURE 4. SXFM observation of tubulin labeled with CdSe nanocrystals. Because the tubulins were labeled with the CdSe nanocrystals (Quantum Dot Corporation), we can observe intracellular tubulin using both a commercial fluorescence microscope, which utilizes the emission of visible light, and the SXFM, which detects the X-ray fluorescence of the Cd L and Se K lines.

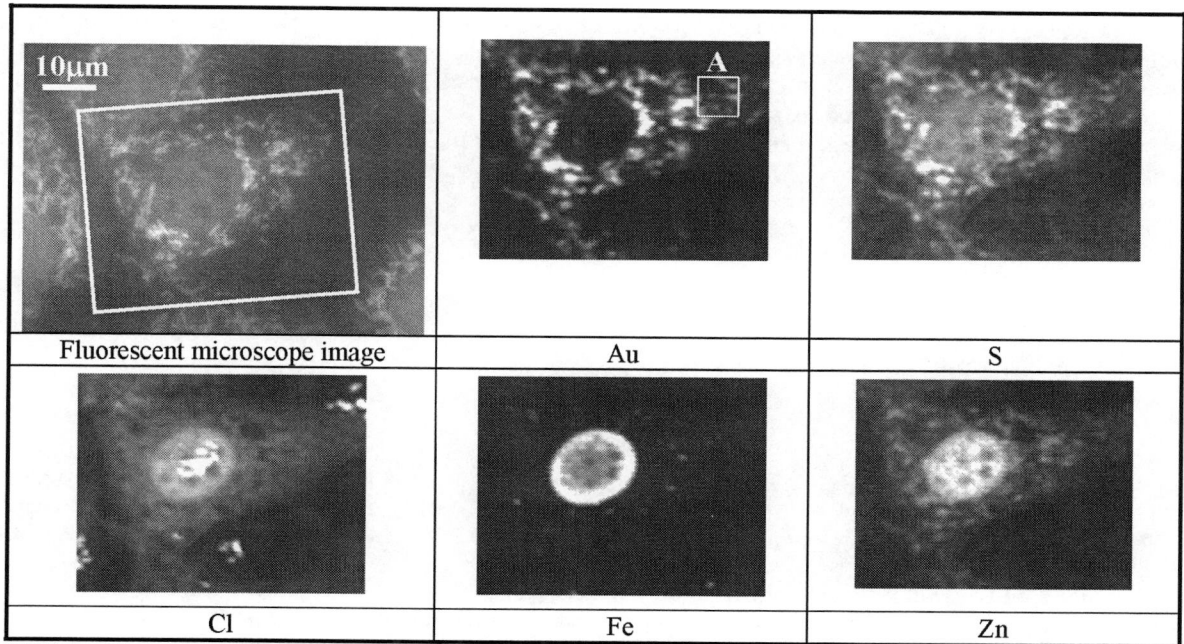

FIGURE 5. Observation of intracellular element distribution using SXFM. The white areas in the Au distribution map show the existence of mitochondria because mitochondria are labeled with gold colloid. In terms of consistency between Au and Zn distributions, we know that mitochondria have relatively high Zn contents, but low Fe contents.

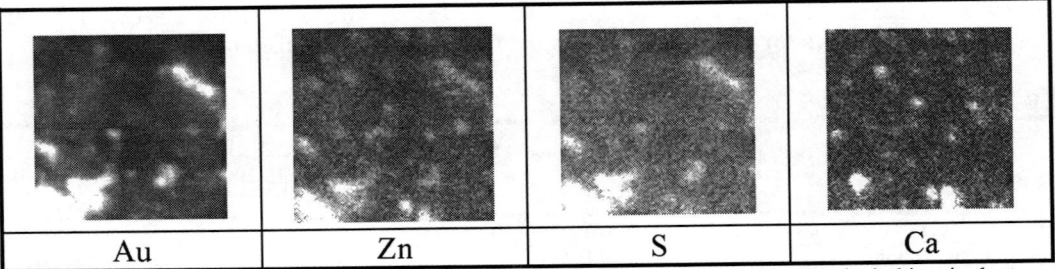

| | Au | Zn | S | Ca |

FIGURE 6. Observation of element distribution in mitochondria at the A (Fig. 5) area. The elliptical object in the top right of Au distribution map shows single mitochondria existence.

TABLE 3. Scan parameters of SXFM.

	Figure. 4 A Scan	Figure. 4 B Scan	Figure. 5 Scan	Figure. 6 Scan
X-ray energy (keV)	15	15	15	15
Beam size (nm, H x V) of focus	1000 x 1000	140 x 140	1000 x 1000	200 x 200
Scanning pitch (nm)	500	70	500	70
Scan area (um, H x V)	40 x 40	3.25 x 1.9	50 x 38	7 x 7
Exposure time (s/pixel)	3	8	1	2

Two samples were prepared to demonstrate the performance of the SXFM. One sample (shown in Fig. 4) is a mouse cell which has tubulins labeled with CdSe nanocrystals by immunostaining. The other sample (shown in Fig. 5 and Fig. 6) is a human cell which have mitochondria labeled with gold colloid and fluorescence dye (FITC). A demonstration of the SXFM was performed at BL29XUL (EH2) at an X-ray energy of 15 keV. The scanning parameters of the SXFM are shown in Table 3. As a result of the demonstration, it was found that tubulins and mitochondria could be observed with a spatial resolution of ~70 nm.

CONCLUSION AND DISCUSSION

We developed a scanning-type X-ray microscope, which makes it possible to visualize trace element distribution in biological samples. The X-ray beam size was controllable within the range of 30 ~ 1400 nm merely by adjusting the source size. As a result of demonstrations, it was found that intracellular element could be visualized with the good sensitivity of ~1 fg. These results suggest that SXFM enables analysis of intracellular trace elements, which contributes to a better understanding of cellular functions. In the near future, the challenge will be to observe frozen and, moreover, living cells to determine in situ intracellular element distributions.

ACKNOWLEDGMENTS

This research was partially supported by Grants-in-Aid for Scientific Research (S), 15106003, 2004 and 21Century COE Research, Center for Atomistic Fabricaton Technology, 2004 from the Ministry of Education, Culture, Sports, Science and Technology of Japan.

REFERENCES

1. M. Shimura, A. Saito, S. Matsuyama, T. Sakuma, Y. Terui, K. Ueno, H. Yumoto, K. Yamauchi, K. Yamamura, H. Mimura, Y. Sano, M. Yabashi, K. Tamasaku, K. Nishio, Y. Nishino, K. Endo, K. Hatake, Y. Mori, Y. Ishizaka and T. Ishikawa, "Element Array by Scanning X-ray Fluorescence Microscopy after Cis-Diamminedichloro-Platinum(II) Treatment", *Cancer Research*, **65** (12) 4998 (2005).
2. H. Mimura, S. Matsuyama, H. Yumoto, H. Hara, K. Yamamura, Y. Sano, M. Shibahara, K. Endo, Y. Mori, Y. Nishino, K. Tamasaku, M. Yabashi, T. Ishikawa and K. Yamauchi, "Hard X-ray Diffraction-Limited Nanofocusing with Kirkpatrick-Baez Mirrors", *Jpn. J. Appl. Phys.*, Part 2, **44** 18 539(2005).
3. S. Matsuyama, H. Mimura, H. Yumoto, H. Hara, K. Yamamura, Y. Sano, K.Endo, Y. Mori, M. Yabashi, Y. Nishino, K. Tamasaku, T. Ishikawa, K. Yamauchi, "Hard X-ray nano-focusing at 40nm level using K-B mirror optics for nanoscopy/spectroscopy", *Proc. SPIE Int. Soc. Opt. Eng.*, **5918** 591804 (2005).

Extension to Low Energies (<7keV) of High Pressure X-Ray Absorption Spectroscopy

J.-P. Itié[1,2], A.-M. Flank[1], P. Lagarde[1], A. Polian[2], B. Couzinet[2], M. Idir[1]

[1] *Synchrotron SOLEIL, L'Orme des Merisiers, Saint-Aubin – BP 48 , 91192 GIF-sur-YVETTE CEDEX (France)*
[2] *Physique des Milieux Denses, IMPMC, CNRS UMR 7590, Université P & M Curie-Paris6, 140 rue de Lourmel, 75015 PARIS (France)*

Abstract. High pressure x-ray absorption has been performed down to 3.6 keV, thanks to the new LUCIA beamline (SLS, PSI) and to the use of perforated diamonds or Be gasket. Various experimental geometries are proposed, depending on the energy of the edge and on the concentration of the studied element. A few examples will be presented: $BaTiO_3$ at the titanium K edge, $Zn_{0.95}Mn_{0.05}O$ at the manganese K edge, KCl at the potassium K edge.

Keywords: Pressure, x-ray absorption spectroscopy.
PACS: 62.50.Ks, 78.70.Dm, 77.84.Dy

INTRODUCTION

High pressure is a growing field, especially in synchrotron radiation centres, whatever the scientific domain (geosciences, physics, chemistry, biology, synthesis of new materials etc.) is. X-ray absorption spectroscopy (XAS) experiments under high pressure have been limited up to now to energy range above 7keV because of the absorption by the diamond anvils used to generate the pressure. The first requirement to lower this limit is to use a very intense and very well focused beam due to the small size of the samples and the complex environment necessary to generate these high pressures. The second one is to reduce the x-ray absorption by the diamonds and/or the gasket. To perform XAS at low energy edges, we have installed on the new LUCIA beamline (SLS, PSI) diamond anvil cells with specific characteristics, depending on the energy of the studied edge and on the nature of the sample (diluted species or not). We first describe the LUCIA beamline, then the high pressure set-up. The newly opened possibilities will be presented with different examples.

THE LUCIA MICRO-XAS BEAMLINE

A new x-ray beamline dedicated to micro absorption and imaging experiments has been developed at the Swiss light source SLS under a collaboration between France (CNRS and Synchrotron SOLEIL) and Switzerland (PSI). The scheme of the beamline is the following. The source is an Apple II type undulator with a 54 mm period, installed on a medium length straight section of the machine; at this location the size of the photon source is 200 (horizontal) x 20 (vertical) μm^2 (FWHM). The beam is first collimated, inside the front end, in order to keep only the central cone of the undulator emission. Then a vertical spherical mirror at an incidence of 0.4° makes an intermediate source which is 80 μm (h) and 760 μm (v). A set of two flat mirrors, with an incidence angle variable from 0.4 to 1.2° acts as a low-pass energy filter to remove the unwanted harmonics of the undulator. The monochromator is a two-crystal where the position of the second crystal is adjusted by a double cam in order to keep the exit beam fixed. In order to cover the full energy range, five crystals (Si(111), InSb(111), KTP(110), beryl(110) and YB_{66}) are set on the same holders and the whole vessel of the monochromator translates perpendicularly to the photon beam. All these optics, as well as the first crystals, are water cooled. The focusing system is located about 3 m downstream of the monochromator; it uses two mirrors in a Kirkpatrick-Baez configuration with an incidence of 0.4°. Each mirror is bent elliptically by a mechanical system with two motorized actuators developed at first at ESRF and adapted here for high-vacuum. More details on the design of the beamline can be found in [1].

CP879, *Synchrotron Radiation Instrumentation: Ninth International Conference,*
edited by Jae-Young Choi and Seungyu Rah
© 2007 American Institute of Physics 978-0-7354-0373-4/07/$23.00

Because of the high quality of the undulator, experiments can use up to the 31th harmonic, therefore a photon energy slightly above 8 keV. Actually the energy domain, from 0.8 to 8 keV, is limited at low energy by the 2d spacing of the beryl crystals, and at the other end by the 0.4° incidence angle of the mirrors, made of silicon covered by 500Å of nickel. At the middle of the energy range, i.e. around 4 keV, the flux at the exit of the monochromator, as measured with a silicon pin diode, is about $3 \ 10^{11}$ ph/sec when the SLS machine is running at 2.4 GeV and 350 mA. This flux decreases by a factor around 3 at the ends of the energy domain, because of the reflectivity of the crystals in one side, and the emission of the undulator at high energy. The transmission of the K-B set-up is of the order of 30%, a value mainly determined by the overall acceptance of the mirrors. Thanks to the topping-up operation of the machine this flux stays constant with time, with no change due to the lifetime of the machine or to the thermal load on the optics.

The spot size has been measured with the knife-edge technique using either the sharp edge of a square titanium dot deposited onto a silicon wafer and monitored by the Ti Kα fluorescence, or by transmission through a copper blade made by electro-erosion. The FWHM's which are indicated in Fig 1 are perfectly in line with the calculations using a ray tracing code which takes into account the actual optical qualities of the mirrors. While these values are of a fundamental importance for imaging experiments, another parameter has to be controlled during x-ray absorption experiments, i.e. the beam spatial stability over the energy domain swept by the monochromator. For that purpose, the roll of the two crystals is first adjusted in order to obtain the horizontal stability, and then the slight misalignment of the pitch of the second crystal when the Bragg angle is changed can be corrected smoothly using a piezo motor which acts on the second crystal. The final result is a stability of ±1 µm in both directions over a full EXAFS energy range. The so-called "undulator gap scan" technique is routinely used to record XAS spectra.

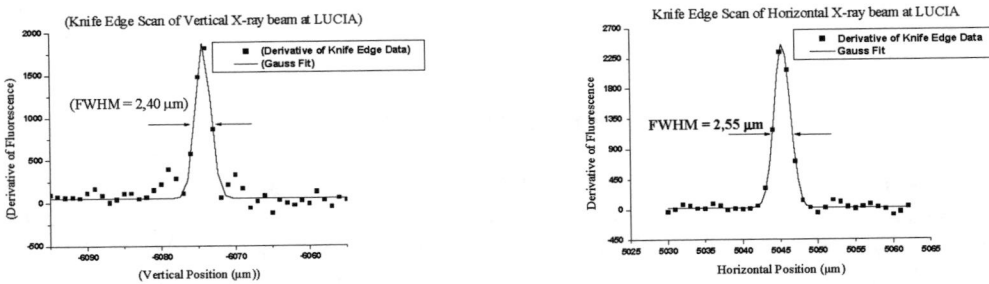

FIGURE 1. Spot size measured at 5 keV on a square titanium dot.

HIGH PRESSURE SET-UP

High pressure XAS experiments use diamond anvil cell. The sample is confined in a hole drilled in a metallic gasket squeezed between the two diamond anvils. Therefore the x-rays have to go through the diamond anvils and/or through the gasket. In order to reduce the absorption of the anvils we used fully perforated diamonds as a support for small diamond anvils (thickness 500 µm) [2, 3]. The total thickness of the anvils is only 1mm, which gives a transmission of $2 \ 10^{-3}$ at 5 keV. Because of the high incident flux ($2 \ 10^{11}$ ph/s/0.1%bw in the x-ray spot), experiments down to the Ti K edge (4.966 keV) are possible in transmission. Cartography of the cell can be done in order to find the sample inside the gasket hole (Fig. 2).

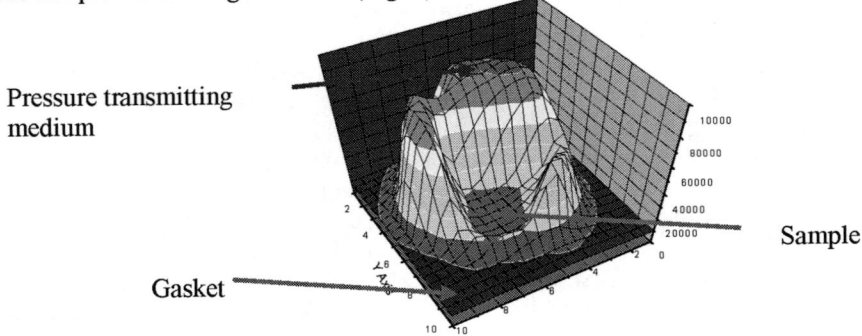

FIGURE 2. Cartography in transmission of the cell at E = 5000 eV (just above the Ti K edge)

For energies below 5 keV, the transmission geometry cannot be used anymore. Two possibilities remain:

i) The incident beam goes through a partially perforated diamond (wall thickness 200 μm) and a fluorescence signal is measured through a Be gasket, at 90° from the diamond axis. This geometry is also very well adapted to diluted elements.

ii) The incident beam goes through the Be gasket and the signal is collected at 90° in the fluorescence mode through the Be gasket. This geometry requires a very small x-ray spot, because of the small gap between the diamonds (less than 20 μm at 20 GPa).

EXAMPLES OF APPLICATION

Titanate Perovskites at the Ti K Edge under High Pressure: the BaTiO3 Case

$BaTiO_3$ is a tetragonal ferroelectric material at ambient condition and undergoes a phase transition above 2 GPa to a cubic paraelectric structure. XAS at the Ti K edge has been performed in the transmission geometry [4]. The feature B (see inset of Fig. 3) remains intense in the cubic phase, just above the structural transition, indicating that the Ti atom remains locally off-centre. Its intensity decreases continuously up to 10 GPa and stays constant above this pressure. This effect has been interpreted as a gradual localization of the Ti atom at the centre of the oxygen octahedron. Above 10 GPa the Ti atom is at the centre and remains there at this position up to the maximum pressure reached during this experiment (19 GPa). Figure 3 shows the spectrum obtained at this pressure.

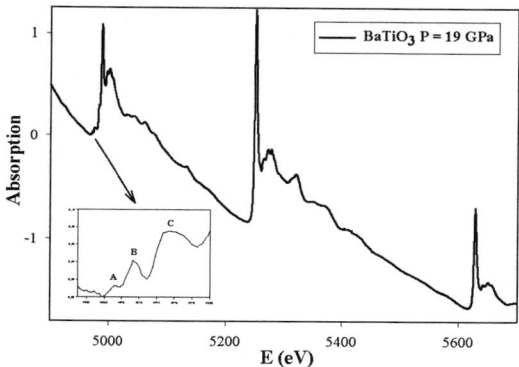

FIGURE 3. X-ray Absorption spectrum of $BaTiO_3$ at 19 GPa. We can observe successively the Ti K edge and the Ba L_{III} and L_{II} edges. The inset shows the pre-edge features of the Ti K edge. The intensity of the feature B has considerably decreased with respect to the room pressure spectrum.

Diluted Samples: High Pressure and Fluorescence Yield

The structural evolution of $Zn_{0.95}Mn_{0.05}O$ thin films [5] with pressure has been studied by X-ray absorption spectroscopy. XAS spectra were recorded beyond the Mn K edge (6539eV) using the diamond gasket geometry. Figure 4 shows XANES spectra of $Zn_{0.95}Mn_{0.05}O$ at various pressures. At ambient conditions, pure ZnO crystallizes in the Wurtzite structure with Zn in a tetrahedral environment. Under pressure its structure changes to Rock-salt at about 9GPa (octahedral Zn environment) [6]. The analysis of the spectra suggests that Mn substitutes Zn, both in the low and high pressure phases, with no evidence of Mn clustering. Signatures of the phase transition are found in the EXAFS part of the spectra (bond lengths and coordination numbers), but also in the intensity of the pre-edge (XANES) which is more intense for a tetrahedral environment (Fig. 4). At this Mn concentration (5%) the phase transition is reversible, although with a noticeable hysteresis.

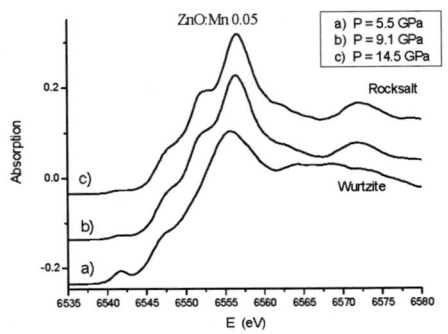

FIGURE 4. $Zn_{0.95}Mn_{0.05}O$ XAS spectra at the Mn K edge in the wurtzite (a) and in the rocksalt (b and c) structures. Spectra have been shifted for clarity.

High Pressure XAS at Very Low Energy: the Potassium K Edge in KCl

Here the gasket-gasket geometry has been used. Spectra at room pressure and at 1.6 GPa have been recorded at the potassium K edge (3608 eV) (Fig. 5). The high pressure spectrum has been rescaled in energy taking into account the pressure-induced variation of the interatomic distances determined from the known KCl equation of state. With this energy scale, the oscillations of the two spectra should have the same period as observed in Fig. 5.

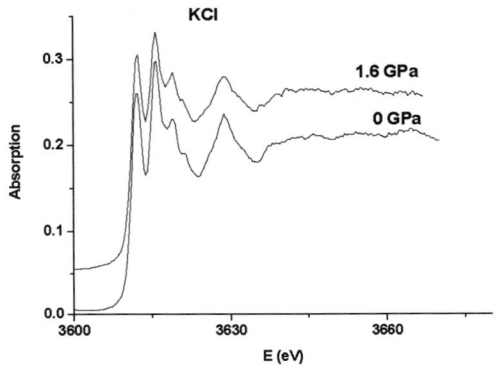

FIGURE 5. X-ray absorption spectrum of KCl at the potassium K edge for room pressure (lower curve) and 1.6 GPa (upper curve). The energy scale of the high pressure spectrum has been modified in order to take into account the bound length variation under pressure.

CONCLUSION

High pressure X-ray absorption at low energy is now possible. Extension to lower energy (down to the Si K edge (1839 eV) will require the gasket-gasket geometry and a modification of the shape of the Be gasket. Such an attempt will be made in the future.

REFERENCES

1. A.-M. Flank, G. Cauchon, P. Lagarde, S. Bac, M. Janousch, R. Wetter, J.-M. Dubuisson, M. Idir, F. Langlois, T. Moreno and D. Vantelon *Nuclear Instruments and Methods B* **246**, 269-274 (2006)
2. A.Dadashev, M. P. Pasternak, G. Kh. Rozenberg and R.D. Taylor, *Rev. Sci. Instr.* **72**, 2633 (2001)
3. J.P. Itié, F. Baudelet, A. Congeduti, B. Couzinet, F. Farges and A. Polian, *J. Phys.: Condens. Matter* **17** S 883 (2005)
4. J.P. Itié, B.Couzinet, A. Polian, A.M. Flank and P. Lagarde, *Europhys. Lett.* **74**, 706 (2006)
5. J.A. Sans, A. Segura, M. Mollar, B. Mari, *Thin Solid Films* **453-454**, 252-255 (2004)
6. F. Decremps, J. Zhang, R.C. Liebermann, *Europhys. Lett.* **51(3)**, 268-274 (2000)

A High Resolution Full Field Transmission X-ray Microscope at SSRL

Katharina Lüning[1], Piero Pianetta[1], Wenbing Yun[2], Eduardo Almeida[3], Marjolein van der Meulen[4]

[1] Stanford Synchrotron Radiation Laboratory, 2575 Sand Hill Road, Menlo Park, CA 94065
[2] Xradia, Inc., 4075A Sprig Drive, Concord, CA 94520
[3] NASA Ames Research Center, Moffett Field, CA 94035
[4] Cornell University, 219 Upson Hall, Ithaca, NY 14853

Abstract. The Stanford Synchrotron Radiation Laboratory (SSRL) in collaboration with Xradia Inc., the NASA Ames Research Center and Cornell University is implementing a commercial hard x-ray full field imaging microscope based on zone plate optics on a wiggler beam line on SPEAR3. This facility will provide unprecedented analytical capabilities for a broad range of scientific areas and will enable research on nanoscale phenomena and structures in biology as well as materials science and environmental science. This instrument will provide high resolution x-ray microscopy, tomography, and spectromicroscopy capabilities in a photon energy range between 5–14 keV. The spatial resolution of the TXM microscope is specified as 20 nm exploiting imaging in third diffraction order. This imaging facility will optimally combine the latest imaging technology developed by Xradia Inc. with the wiggler source characteristics at beam line 6-2 at SSRL. This will result in an instrument capable of high speed and high resolution imaging with spectral tunability for spectromicroscopy, element specific and Zernike phase contrast imaging. Furthermore, a scanning microprobe capability will be integral to the system thus allowing elemental mapping and fluorescence yield XANES to be performed with a spatial resolution of about 1 μm without introducing any changes to the optical configuration of the instrument.

Keywords: Hard X-ray Microscopy, TXM
PACS: 07.85.Tt

INTRODUCTION

High resolution transmission x-ray microscopy in the soft x-ray energy range, in particular in the water window, has been developed over the past several decades into a powerful and versatile technique [1, 2]. It is only recently that advances in hard x-ray focusing optics, have allowed the higher photon energy ranges to be accessed. This development has allowed the study of thicker samples as well as the reduction in the amount of sample preparation. At present, only a few such microscopes exist [3-5] or are under development [6] in the world at synchrotron radiation facilities. We are developing a hard x-ray (5 – 14 keV) imaging facility at SSRL that will provide powerful and unprecedented analytical capabilities for a broad range of scientific areas. This imaging facility will combine the high flux of a SPEAR3 wiggler source, beam line 6-2, with the advanced x-ray imaging technologies developed by Xradia, Inc. This commercially available x-ray microscope from Xradia Inc. is a full-field transmission x-ray microscope (TXM) that has originally been designed for imaging state of the art integrated circuits, routinely achieving spatial resolutions of 60 nm for 2D and 80 nm for 3D images using a rotating anode laboratory x-ray source. At SPEAR3, the wiggler source provides more than 5 orders of magnitude higher usable source brightness than the rotating anode source. This will open up many exciting new possibilities, such as 2D and 3D imaging in third diffraction order for improved spatial resolution, together with spectral tunability for spectromicroscopy, element specific and phase contrast imaging.

CP879, *Synchrotron Radiation Instrumentation: Ninth International Conference*,
edited by Jae-Young Choi and Seungyu Rah
© 2007 American Institute of Physics 978-0-7354-0373-4/07/$23.00

EXPERIMENTAL SETUP

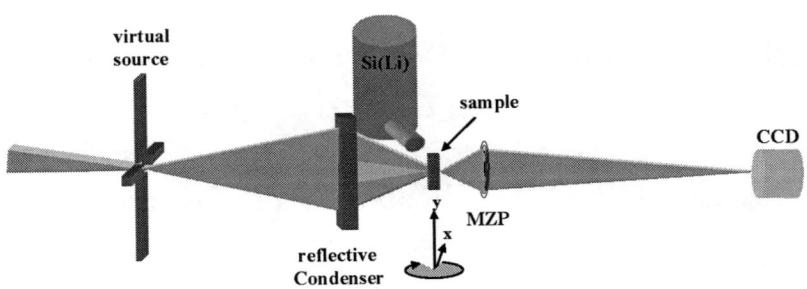

FIGURE 1. Schematic view of the TXM. The virtual source configuration together with the reflective condenser optics will allow operation of this instrument as a fluorescence microprobe.

The general layout of the TXM is schematically shown in Fig. 1. The optical scheme of this x-ray microscope is analogous to a visible light microscope where the incident radiation is condensed onto the object and a magnified image is formed by an objective lens onto a CCD detector. The key optical components of the TXM are the condenser and the zone plate objective lens. An elliptically shaped capillary is used as the condenser [7], which produces a hollow cone illumination at the sample and micro zone plate. A major advantage of this optical concept is that no order-sorting aperture is required since the condenser is not based on a diffractive optical element. The hollow central cone guarantees, intrinsically, that the first positive diffraction order of the objective micro-zone plate is not contaminated by undiffracted zero order x-rays thus leading to improved contrast.

At SSRL a 54 pole wiggler will be used as the x-ray source for the TXM. This provides an x-ray beam of high brightness and flux together with a large phase space as required by this instrument. The optical layout of beam line 6-2 is schematically shown in Fig. 2. This beam line has been designed to accept the full vertical emittance of the wiggler and 1.2 mrad in the horizontal as defined by an aperture located 10 m from the source. The beam is then collimated by a vertically reflecting, Pt coated mirror (M_0) located 13.86 m from the source, which is followed by the liquid nitrogen (LN) cooled, double crystal monochromator (DCM). The DCM is equipped with Si(111) crystals giving an energy resolution $\Delta E/E = 5 \times 10^{-4}$ which meets the requirements of the zone plate objective of the TXM. The tuning range of the monochromator is from 3 to above 20 keV, which covers the energy range of the TXM. Although this monochromator is not specifically designed to have a fixed exit beam height for different photon energies, the vertical motion over the energy range of a typical NEXAFS scan (~100 eV) is only a few microns. This small vertical motion is a consequence of the post-monochromator, torroidal M1 mirror, which is configured to focus the beam onto a slit (S3) and thus reduces the vertical translation of the focused beam significantly. The S3 slit, located 30 m from the wiggler source, will serve as the effective virtual source for the TXM. The beam line is designed to demagnify the wiggler source by a factor of 1.7 in the horizontal and a factor of 1.2 in the vertical. Figure 3 shows the corresponding ray trace result using the XOP version of SHADOW for the horizontal and vertical beam profiles at S3. Assuming perfect optics, the ray trace predicts a photon beam spot size at S3 with a central cone of about 600 microns in the horizontal and 60 microns in the vertical (FWHM). The figure indicates that there is a small contribution from the vertical aberrations of the M1 mirror which extend about 200 microns.

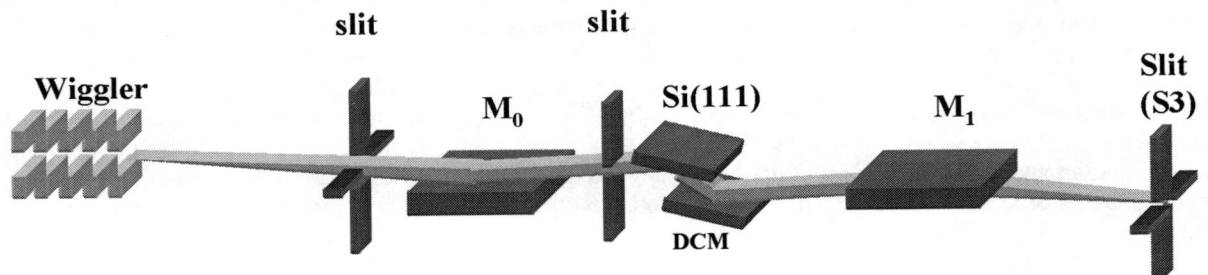

FIGURE 2. Optical configuration of the 54 pole Wiggler beam line 6-2 at SSRL.

Xradia designed the optical components of the TXM to match the beam properties of beam line 6-2. In particular, the proprietary capillary condenser optics is adapted to the beam size and divergence at the virtual source position. The condenser will make a 10:1 demagnified image of the virtual source, defined by the opening of S3. Thus, with the slit setting of the virtual source (S3), the beam spot on the sample can be adjusted from about (1x1) μm^2 to (10x10) μm^2. In addition, the condenser must provide the correct illumination for the numerical aperture of the subsequent zone plate objective. In order to cover the specified photon energy range between 5 keV and 14 keV Xradia will provide 6 zone plates with an outermost zone width of 45 nm and numerical apertures ranging from 1.06 mrad at 14 keV to 2.5 mrad at 5 keV. Since the vertical divergence of the x-ray beam at the virtual source position is only 0.25 mrad, the condenser optics must increase the vertical divergence to match the numerical aperture of the objective zone plate. This will result in an illuminated field at the sample position of about (10 x 10) microns in the higher energy range and about (6 x 6) microns around 5 keV. For applications requiring a larger field of view at the sample position, Xradia will implement the option of vertically scanning the condenser optics in the x-ray beam. For imaging in absorption contrast, the instrument will provide 3 matching elliptical condensers spanning the energy range between 5 and 14 keV. One condenser will be dedicated to the highest resolution imaging mode in third diffraction order at 8 keV. An additional condenser and a Au phase plate will be tailored for Zernike phase contrast imaging at 8 keV, which is well above all the major absorption edges of biological materials. This technique becomes particularly important in the multi-keV photon range where phase contrast is dominating over absorption contrast [8].

It should be noted that this wiggler beam line offers about 5 orders of magnitude higher brightness than the rotating anode source currently being used by Xradia in their commercial instrument. This will allow the use of the positive 3rd order diffraction order of the micro zone plate while compensating for its lower diffraction efficiency. This improves the spatial resolution by a factor of 3 enabling hard x-ray imaging in 2D and 3D with 20 nm spatial resolution.

In addition to high resolution 2D and 3D imaging, the virtual source configuration in combination with the reflective condenser optics will allow the incorporation of a scanning fluorescence microprobe without introducing any changes to the optical configuration of the microscope. As illustrated in Fig.1, when an interesting feature is observed within the (10 x 10) μm^2 field of view of the TXM, it will be possible to bring the feature into the center of the image and reduce the field of view to (1 x 1) μm^2 by closing the virtual source slit and measure the x-ray fluorescence in this region. In fact, fluorescence analysis may be done while imaging at any field of view ranging from (1 x 1) μm^2 up to (10 x 10) μm^2. The fact that the TXM simultaneously provides a high resolution image of the area being analyzed by the fluorescence microprobe will give a definite advantage in identifying the features contributing to the fluorescence yield in a XANES spectrum, for example.

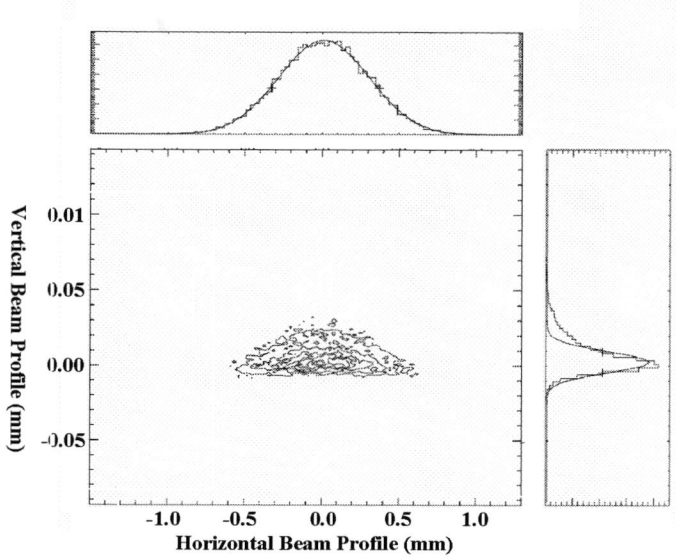

FIGURE 3. Horizontal and vertical beam profiles at the virtual source. The size of the focus is about (600 x 60) μm^2.

OUTLOOK

The TXM will be installed at SSRL in September 2006 and commissioning of the instrument will begin with the start of the next SPEAR3 run in October 2006. We expect to achieve sub 60 nm spatial resolution absorption contrast images of resolution test patterns and bone samples in 2D and 3D by the end of 2006. Our next goal will be to further improve the spatial resolution to 20 nm by employing 3rd order imaging and to implement Zernike phase contrast imaging capabilities.

ACKNOWLEDGMENTS

This program is supported by the National Institutes of Health, National Institute for Biomedical Imaging and Bioengineering, grant number EB004321.

Portions of this research were carried out at the Stanford Synchrotron Radiation Laboratory, a national user facility operated by Stanford University on behalf of the U.S. Department of Energy, Office of Basic Energy Sciences. The SSRL Structural Molecular Biology Program is supported by the Department of Energy, Office of Biological and Environmental Research, and by the National Institutes of Health, National Center for Research Resources, Biomedical Technology Program.

REFERENCES

1. G. Denbeaux, L. Johnson, W. Meyer-Ilse, *Proceedings of the 6th International X-ray Microscopy conference*, 478 – 483 (2000).
2. Niemann, D. Rudoph, G. Schmahl, *NIM* **208**, 367-371 (1983).
3. Kaulich, B.; Neimann, B.; Rostaing, G.; Oestreich, S.; Salome, M.; Barrett, R.; Susini, J., *AIP Conference Proceedings of the Sixth International Conference on X-Ray Microscopy*, **507**, 45-48 (2000).
4. Kaulich, B.; Barrett, R.; Salome, M.; Oestreich, S.; Susini, J., *ESRF Newsletter*; Oct. 2000, **34**, 27-28.
5. M. T. Tang, G. C. Yin, Y.F. Song, J. H. Chen, K. L. Tsang, K. S. Liang, F. R. Chen, F. Duewer, W. Yun, *IPAP Conference Series 7, Proceedings of the 8th International Conference on X-ray Microscopy*, 15 – 17, (2006).
6. J. Maser, R. Winarski,, M. Holt, D. Shu, C. Benson, B. Tieman, C. Preissner, A. Smolyanitskiy, B. Lai, S. Vogt, G. Wiemerslage, G. B. Stephenson, *IPAP Conference Series 7, Proceedings of the 8th International Conference on X-ray Microscopy*, 26-29, (2006).
7. D. H. Bilderback, S. A. Hoffmann, D. J. Thiel, *Science*, **263**, 201-203 (1994).
8. U. Neuhaeusler, G. Schneider, W. Ludwig, M. A. Meyer, E. Zschech, D. Hambach, *J. Phys. D: Appl. Phys.* **36**, A79-A82 (2003).

First X-ray Fluorescence MicroCT Results from Micrometeorites at SSRL

Konstantin Ignatyev[1], Kathy Huwig[2], Ralph Harvey[2], Hope Ishii[3], John Bradley[3], Katharina Luening[1], Sean Brennan[1], Piero Pianetta[1]

[1] Stanford Synchrotron Radiation Laboratory, 2575 Sand Hill Rd, Menlo Park, CA 94025, USA
[2] Case Western Reserve University, Cleveland, OH 44106, USA
[3] Lawrence Livermore National Laboratory, 7000 East Avenue, Livermore, CA 94550, USA

Abstract. X-ray fluorescence microCT (computed tomography) is a novel technique that allows non-destructive determination of the 3D distribution of chemical elements inside a sample. This is especially important in samples for which sectioning is undesirable either due to the risk of contamination or the requirement for further analysis by different characterization techniques. Developments made by third generation synchrotron facilities and laboratory X-ray focusing systems have made these kinds of measurements more attractive by significantly reducing scan times and beam size. First results from the x-ray fluorescence microCT experiments performed at SSRL beamline 6-2 are reported here. Beamline 6-2 is a 54 pole wiggler that uses a two mirror optical system for focusing the x-rays onto a virtual source slit which is then reimaged with a set of KB mirrors to a (2×4) μm^2 beam spot. An energy dispersive fluorescence detector is located in plane at 90 degrees to the incident beam to reduce the scattering contribution. A PIN diode located behind the sample simultaneously measures the x-ray attenuation in the sample. Several porous micrometeorite samples were measured and the reconstructed element density distribution including self-absorption correction is presented. Ultimately, this system will be used to analyze particles from the coma of comet Wild-2 and fresh interstellar dust particles both of which were collected during the NASA Stardust mission.

Keywords: tomography, microprobe.
PACS: 87.59.Fm

INTRODUCTION

Computed tomography (CT) allows reconstruction of 3D information about the inner structure of an object without the need for sample preparation such as sectioning. X-ray absorption CT is well known and is widely used in medicine (CAT scan) as a diagnostic tool. Initially CT measurements were performed on large samples but as technique and technology improved, smaller samples also became a focus of CT studies. MicroCT is a variant of CT that deals with samples which are smaller in size than 1 mm. Recently, a new mode of CT has emerged that employs the detected fluorescence signal of the sample – X-ray fluorescence CT – allowing for the measurement of trace elements. This technique only became feasible with third generation synchrotron sources producing high brightness x-ray beams and the availability of focusing optics [1, 2].

In this work X-ray microCT measurements were conducted on two micrometeorites collected in the Antarctic field. Typically, collected micrometeorites range widely in their internal morphology. MicroCT measurements were used to compare their internal structure non-destructively.

EXPERIMENTAL

The back hutch of the SSRL Wiggler beam line 6-2 was used to collect the data on two Antarctic micrometeorites (100 μm diameter). For the purpose of their easy identification they were named "Mike" and "Olga". Their external morphology and appearance is quite different as can be seen on the optical micrographs in

CP879, *Synchrotron Radiation Instrumentation: Ninth International Conference*,
edited by Jae-Young Choi and Seungyu Rah
© 2007 American Institute of Physics 978-0-7354-0373-4/07/$23.00

Fig. 1. "Olga" has rough external morphology with a lot of bubble-like structures on the surface which suggests a highly vesicular inner structure. Mike on the other hand has a smooth ovoid shape. The meteorites were mounted on top of a silica capillary tube with good mechanical stability. First, X-ray fluorescence spectra were collected at several positions on the micrometeorites with collection times of 200 sec for the purpose of determining their average composition. The incident photon energy was 11 keV. As shown in Fig. 2, an energy-dispersive detector was placed in-plane at 90 degree angle to the incident beam to reduce the scatter contribution. The sample can be rotated and translated in the X-ray beam by a high precision rotation stage mounted on top of a high precision x-y-z translation stage. In addition, the sample can be centered relative to the axis of rotation by a miniature x-y-z stage mounted directly on the high precision rotation stage. The horizontal beam sizes used varied from 2 to 4 μm and vertical beam sizes varied from 3 to 4 μm. The scanning geometry is identical to the one used in 1st generation CT scanners which are based on a pencil beam: the sample is translated first in the direction perpendicular to the beam, rotated a predetermined angle and then the sequence is repeated. The translation step size for all scans was equal to the horizontal beam size. We simultaneously collected the X-ray transmission signal with the photodiode located directly behind the sample. One scan took from 5 to 8 hours depending on the translation step size and number of projections.

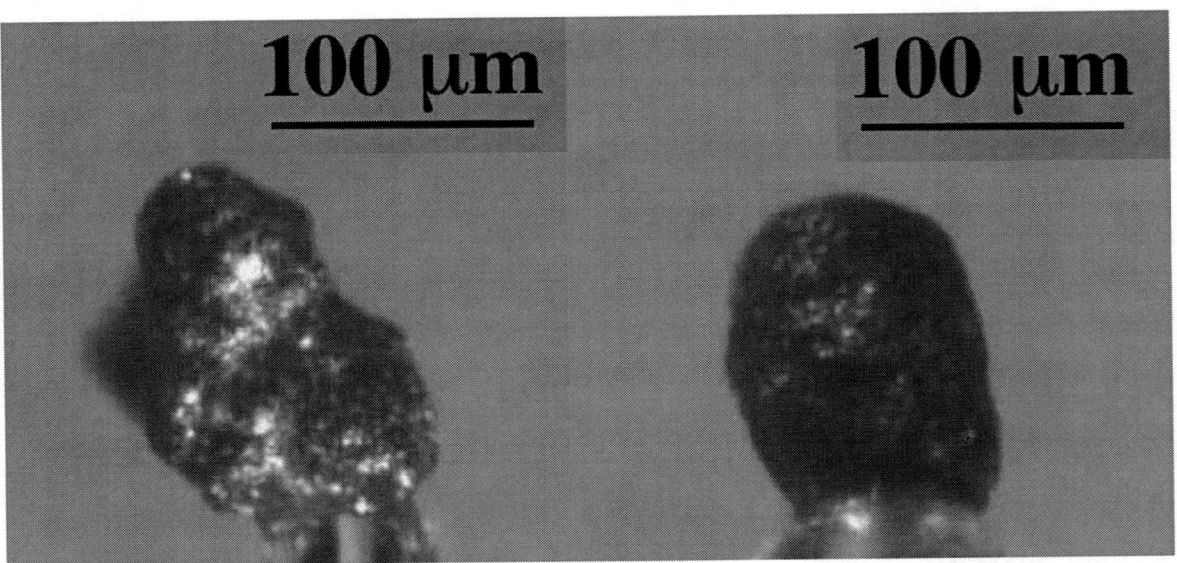

FIGURE 1. Optical micrographs of "Olga" (on the left) and "Mike" (on the right)

RESULTS AND DISCUSSION

For "Mike", fluorescence CT reconstructions were done using the algebraic reconstruction approach as described in [3], employing the Kaczmarz iterative method for solving the system of linear equations. While absorption CT reconstruction is rather straightforward, fluorescence CT is challenging, especially for samples with high X-ray attenuation. X-ray self-absorption in the sample has to be taken into account when measuring highly attenuating samples. The following expression can be used to describe the fluorescence signal excited by the incident X-ray beam in the sample (neglecting second-order effects such as secondary fluorescence):

$$p_{ab}^i = I_0 \xi_{fluor} \varpi^i \int_a^b \mu_0^i(x) \exp\left[-\int_a^x \mu_0(x')dx'\right] \exp\left[-\int_{L(x)} \mu_{fluor}^i(x,y)dy\right]dx$$

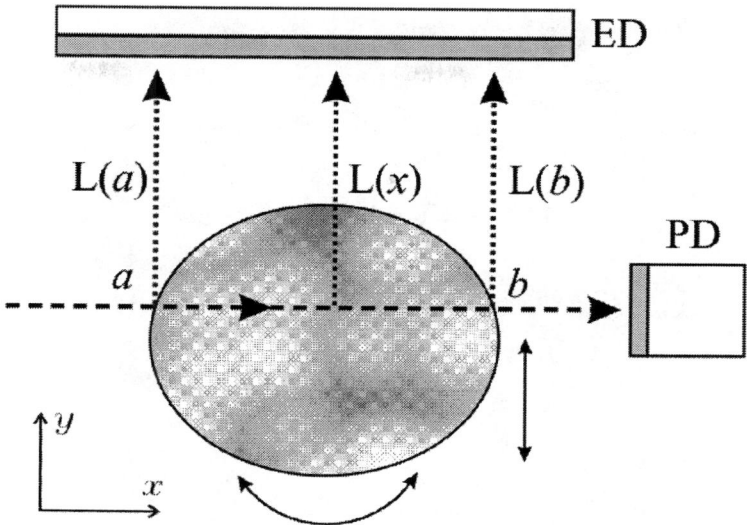

FIGURE 2. Schematic of the X-ray beam passing through a sample with fluorescent radiation measured by the energy dispersive detector ED and transmitted radiation measured by the photodiode PD. L(x) is the line connecting source of the fluorescence radiation at position x with the energy-dispersive detector. Incident beam enters the sample at position $x=a$ and exits the sample at $x=b$. Sample is translated in the horizontal plane in the direction perpendicular to the beam and rotated about the vertical axis.

where superscript i refers to the element of interest, p_{ab} is the measured fluorescence signal for a specific sample position and orientation, I_0 is the intensity of the incident radiation,, ξ_{fluor} is the fluorescence detector efficiency, ω is the fluorescence yield, μ_0^i is the linear attenuation coefficient of the ith element at the X-ray incident beam energy, μ_0 is the linear attenuation coefficient of the sample at the incident energy and μ_{fluor}^i is the linear attenuation coefficient of the sample for the ith element fluorescence. Applying this expression for each beam trace (for each sample translation and rotation) forms a set of equations that can be solved iteratively.

Absorption of the incident radiation for each point and orientation of the sample can be calculated from the absorption map reconstructed from the X-ray transmission signal. The second part of the self-absorption correction dealing with the absorption of the elemental fluorescence signal on its way to the detector is considerably more challenging. One approach exists that allows calculating the absorption of the fluorescence signal without making assumptions about the sample if all major absorbers in the sample fluoresce [4]. Another way to do it is to measure the sample attenuation for all incident beam energies equal to the energies of the fluorescent lines of the elements that are present in the sample, however this is time consuming. For this preliminary study the following assumption was made to facilitate the application of the corrections for self-absorption of fluorescent radiation in the micrometeorites: since Fe is the major constituent in these types of samples, absorption by all the other elements in the sample was assumed to be negligible compared to iron. Thus, the iron distribution was taken to follow the reconstructed attenuation map and could be used to calculate the self-absorption for the fluorescence signal from the elements of interest for all points and orientations of the sample. This approach will introduce errors if the iron distribution is not uniform with respect to the absorption in the sample, however, in our case the reconstructed map of the iron fluorescence showed little deviation from the absorption map in the regions close to the surface (where the absorption correction is small) as well as in the regions closer to the center of the sample.

Figures 3a and 3b show the absorption microtomography reconstructions of two micrometeorites. They were done using filtered backprojection with a Shepp-Logan filter [3]. This dataset was collected with a (2 x 2) μm^2 beam size and the reconstructed pixel size was 2 μm. These images clearly show that although the external morphology is quite different for the two samples, both of them contain large voids and a number of smaller vesicles.

Figures 3c and d show the output of the iterative algebraic reconstruction used on "Mike" (at a different location compared to what is shown on Fig. 3b). Fig. 3c shows the density map of the sample that was used for the self absorption correction. It was used directly for the 11 keV incident beam attenuation correction. In order to correct for the attenuation of the fluorescent radiation it was normalized to the change in the linear attenuation coefficient of iron at the fluorescent energies of Fe Kα and Ni Kα. The reconstructed Ni density map (Fig. 3d) shows a strong Ni contribution in the center of the micrometeorite. This high density grain can also be seen on the absorption map at the same location as the Ni-rich structure.

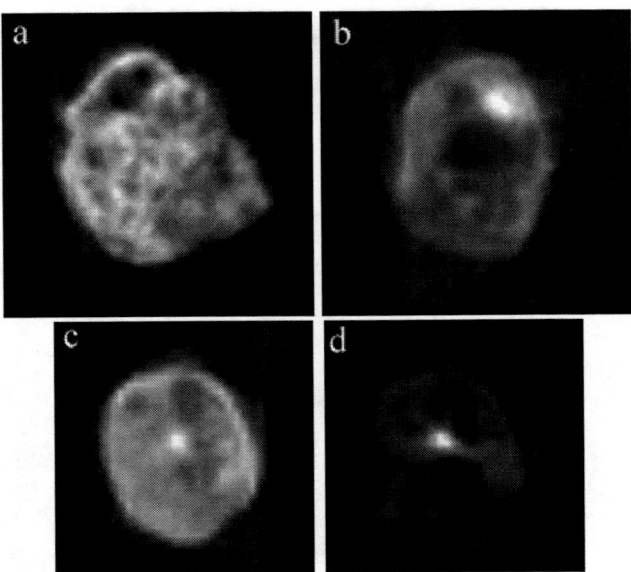

FIGURE 3. Top row - comparison of the density maps of (a) "Olga" and (b) "Mike" obtained with absorption CT reconstruction of a single slice in transmission (beam size 2 μm, angular step size 2° in 180°, number of translation steps 80, integration time 1 s per point, total measurement time 4 hours, resolution 2 μm). Brighter pixels correspond to higher x-ray attenuation by the sample and darker pixels to lower. Bottom row – algebraic reconstruction images of "Mike": (c) absorption and (d) Ni distribution maps (beam size 4 μm, angular step size 4° over 180°, total number of translation steps 50, integration time 4 seconds, total measurement time 3 hours, resolution 4μm). Fluorescent signals for Fe and Ni were ~3x10^4 and 1x10^4 Hz at the peak, respectively. For (c) brighter pixels correspond to higher X-ray attenuation and for (d) brighter pixels correspond to higher elemental concentrations. Field of view of each image is 130 μm.

CONCLUSION

Absorption microCT was used to reconstruct the density cross-section of two micrometeorite samples and it showed that despite differences in external morphology, both of the meteorites had regions of similar internal morphology. Fluorescence microCT was used to determine the distribution of Ni in the cross-section of one of the meteorites. To improve the accuracy of this reconstruction technique, further work will include a modified iterative absorption correction that takes the absorption by all the higher concentration elements in the sample into account.

ACKNOWLEDGMENTS

This research was carried out at the Stanford Synchrotron Radiation Laboratory, a national user facility operated by Stanford University on behalf of the U.S. Department of Energy, Office of Basic Energy Sciences. This research was partially funded by the NASA SRLIDA Program grant number SRL03-0010-0010.

REFERENCES

1. A. Simionovici, B. Golosio, M. Chukalina, A. Somogyi, L. Lemelle, *Developments in X-ray Tomography IV*, SPIE Proceedings, **5535**, 232-242 (2004).
2. D. H. McNear, E. Peltier, J. Everhart, R.L. Chaney, S. Sutton, M. Newville, M. Rivers and D.L.Sparks, *Environ. Sci. Technol.* **39**, 2210-2218 (2005).
3. A. C. Kak and M. Slaney, *Principles of Computerized Tomographic Imaging*, IEEE Press, 1988.
4. M. Chukalina, A. Simionovici, A. Snigirev and T. Jeffries, *X-ray Spectrom.* **85**, 448-450 (2002).
5. IDL absorption CT reconstruction software by Mark Rivers (http://cars9.uchicago.edu/software).

Progress on PEEM3 - An Aberration Corrected X-Ray Photoemission Electron Microscope at the ALS

A.A.MacDowell[1], J.Feng[1], A.DeMello[1], A.Doran[1], R.Duarte[1], E.Forest[2], N.Kelez[1], M.A.Marcus[1], T.Miller[1], H.A.Padmore[1], S.Raoux[3], D.Robin[1], A.Scholl[1], R.Schlueter[1], P .Schmid[1], J.Stöhr[4], W.Wan[1], D.H.Wei[5] and Y.Wu[6]

[1] Advanced Light Source, Lawrence Berkeley National Lab. Berkeley, CA 94720, USA
[2] High Energy Accelerator Research Organization, 1-1 Oho, Tsukuba, Ibaraki,305-0810, Japan
[3] IBM, Almaden Research Center, 650 Harry Road, San Jose, CA 95120, USA
[4] Stanford Synchrotron Radiation Laboratory, PO Box 20450, Stanford, CA 94309, USA
[5] NSRRC, 101 Hsin-Ann Road, Hsinchu 30077, Taiwan
[6] Department of Physics, Duke University, Durham, NC 27708, USA

Abstract. A new ultrahigh-resolution photoemission electron microscope called PEEM3 is being developed and built at the Advanced Light Source (ALS). An electron mirror combined with a much-simplified magnetic dipole separator is to be used to provide simultaneous correction of spherical and chromatic aberrations. It is installed on an elliptically polarized undulator (EPU) beamline, and will be operated with very high spatial resolution and high flux to study the composition, structure, electric and magnetic properties of complex materials. The instrument has been designed and is described. The instrumental hardware is being deployed in 2 phases. The first phase is the deployment of a standard PEEM type microscope consisting of the standard linear array of electrostatic electron lenses. The second phase will be the installation of the aberration corrected upgrade to improve resolution and throughput. This paper describes progress as the instrument enters the commissioning part of the first phase.

Keywords: Photoemission Electron Microscope, PEEM, Aberration Correction.
PACS: 07.85.Qe, 07.78.+s

INTRODUCTION

The photoemission electron microscope (PEEM) has been developed since 1930 [1] to study the surface and thin film properties of various materials. In a PEEM, photons impinging on the sample cause the emission of secondary photoelectrons. These electrons are accelerated to typically 10-30keV and focused to produce a magnified intermediate image by an immersion objective lens. A series of projection lenses are used to magnify this intermediate image further and form a final image on a CCD or other imaging detector. When tunable X-rays from a storage ring are used, different contrast mechanisms such as topographic, elemental, chemical, orientation and magnetic are available. The lateral resolution limit of state-of-the-art PEEMs is about 20nm such as the second-generation PEEM2 instrument [2] now operating on a bend magnet beamline at the Advanced Light Source (ALS). This limit is due to the spherical and chromatic aberrations of the immersion objective lens. By incorporating an electron mirror [3] with its own aberrations but of opposite sign, aberration cancellation can be effected with consequent resolution and throughput improvement. The ALS has embarked on the building of such a PEEM3 instrument that will be stationed on a dedicated EPU insertion device at Beamline 11.0.1 [4].

DESCRIPTION OF THE PEEM3 INSTRUMENT

Figure 1 shows the schematic overview of the electron optics of PEEM3. Photoelectrons from the sample are collected and accelerated by the immersion objective lens to the nominal energy of 20keV. The objective lens is an electrostatic four-electrode lens, in which the sample is also part of the lens and located 2mm away from the second

CP879, *Synchrotron Radiation Instrumentation: Ninth International Conference*,
edited by Jae-Young Choi and Seungyu Rah
© 2007 American Institute of Physics 978-0-7354-0373-4/07/$23.00

grounded electrode. This distance is adjustable to accommodate different sample cases. This objective lens is similar to the PEEM2 objective lens [2], whose electron optical properties have been optimized to have small aberrations [5]. An image with a magnification of 11.23 is formed at the entrance plane of the separator section (indicated by image arrow in Fig.1) between the field lens and the first quadrupole. This field lens makes the field ray parallel upon entering the beam separator. The objective lens, together with the field lens, form a telescopic round lens system. This is necessary for the mirror to run in the so-called symmetric mode in which first-order chromatic distortion and third-order coma can be cancelled [6]. To cope with the inevitable stray DC magnetic fields mechanical misalignments and tolerance errors and we employ dodecapole correctors as indicated in Fig. 1. The choice of dodecapoles was driven by the desire to have not only quadrupole and octapole correctors but also a hexapole component, as three-fold astigmatism has been a problem in PEEM2 which only has octapole correctors. All correctors are dodecapoles for design consistency.

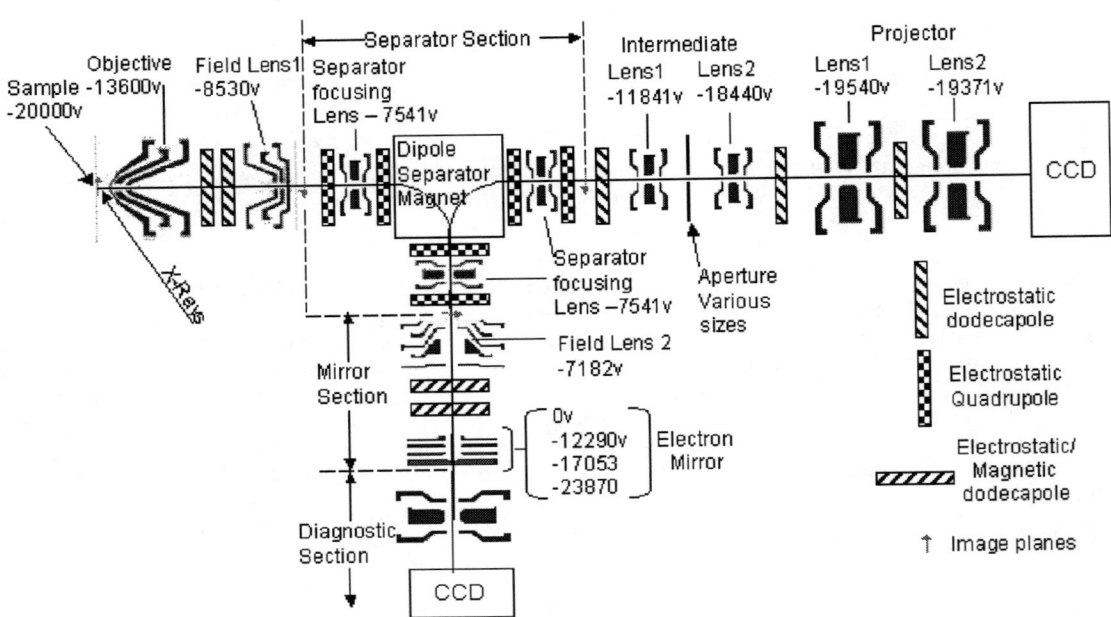

FIGURE 1. Schematic layout of the electrostatic lenses, separator magnet and electron mirror of the PEEM3 microscope. Typical lens voltages are given. As a scale indicator, the distance from the sample to the right hand CCD is 1350mm.

The separator section consists of three Einzel lenses, six quadrupoles and the dipole separator magnet itself [7]. Simple round Einzel lenses were chosen to be the main focusing lenses as the deviation from rotational symmetry caused by the bending magnet and the remaining aberrations of the lenses are small. The weak electrostatic quadrupoles provide the required stigmatic image correction and the 4 degrees of freedom per leg for the electron beam position and angle. On the first pass, the electron beam leaves the separator section chromatically dispersed as it enters the mirror section

The mirror itself consists of four electrodes with the actual mirror reflecting electrode shaped as a spherical segment of radius 5.6mm. The entrance electrode is at ground voltage, while the potentials of other electrodes give three degrees of freedom to determine the focal length, chromatic aberration and the spherical aberrations of the mirror [8]. In order to cancel coma generated by the mirror, the magnification of the mirror is chosen to be -1 and a field lens is placed near the image plane (separator exit on the first pass) to ensure that the linear optics are telescopic for the fundamental ray trajectories. A pair of magnetic/electrostatic dodecapoles effect adjustments of beam trajectory within the mirror section. The set up of the separator and mirror is not going to be trivial. A projector lens and CCD detector are located behind the mirror, which is to be used as a diagnostic PEEM. This PEEM will allow us to independently test and optimize the first half of the beam separator and the incoming beam at the mirror. In this operational mode, the lens voltages are rearranged and the electron beam passes though a 500μm diameter hole in the reflecting electrode before being projected onto the diagnostic CCD.

The mirror section has magnification of −1, so the second pass through the separator cancels the chromatic dispersion of the separator on the first pass. This is the new development for these low energy PEEM microscopes.

A significant amount of design work was put into developing a separator magnet with the initial work directed along the lines of the SMART type separator concept [9]. Here the separator is designed to be aberration free to high order and therefore imposes severe electron optical and engineering challenges. A detailed design was worked out [10] and is very similar to the SMART design being a single magnetic unit wound such that there are 8 dipoles on each pass that employ only edge focusing. The performance of this design depends critically on stringent machining tolerances and is essentially a non-tunable design. By relaxing the chromatic aberration condition for the separator we are now proposing to use, the engineering tolerances are more reasonable and it is also tunable.

The transfer optics and projection system are used to magnify the intermediate image at the exit plane of the magnetic separator onto the CCD detector without distortion. The optical properties of this type of lens with different geometries of individual electrodes have been studied in great detail by Rempfer [11]. Based on this work, the transfer and projection system of PEEM3 consists of four electrostatic uni-potential lenses with the last two lenses having a larger aperture than the first two.

The resolution of the system is dependant on the aberrations of electron optics, the energy and angular spread of the initial electrons and diffraction effects due to the wave nature of the electron. This has been modeled [12] with the resolution defined as the rise-distance of 15% - 85% in intensity of the edge scanning the electron point-spread function. The resolution for 100% transmission is 50nm with the mirror corrector - a significant improvement from that of 440nm without correction. The best predicted resolution is 5nm at 2% transmission, as opposed to 20nm at 0.5% transmission for PEEM2.

PROGRESS WITH PEEM3 CONSTRUCTION

The engineering requirements for this instrument are quite demanding. The magnetic shielding requirements are stringent given that the electrons have very low energy at the sample and electron mirror. It is however the long travel length (1950mm) of the 20KeV electrons that drives the specified AC magnetic field down to a required value of $<2 \times 10^{-7}$ Gauss. In the area where the instrument is to be located, AC field magnitudes of $+-2.5 \times 10^{-4}$ Gauss with a time period of ~1Hz are measured. The source is most likely the injector booster for the main storage ring. Triple mu-metal shielding is used to reduce the fields to the required level. The entire assembly is to be mounted on a single electron optical table that sits on an epoxy granite table with visco-elastic sheet dampers to reduce floor vibrations. The mu-metal shields and vacuum tank are supported separately and wrap around the electron optical table connected only via vacuum bellows. Ultra High Vacuum (UHV) is required for the sample and electron mirror region to reduce surface contamination. Rather than try to evacuate the entire rather large instrument to UHV we have adopted localized UHV sections with the electrons passing in and out of them via small conductance limited holes. The power supplies that supply voltage to the lenses are required to be very stable. An analysis indicated that the electron optics in regions where the electrons have low energy require $<+-2$ parts per million (ppm) stability. Power supplies for the sample, objective lens and the mirror voltages have been custom fabricated.

Constraints imposed by the project complexity and funding have required that the instrument be deployed in two stages. The first stage is in the PEEM2 mode. The schematic of the electron optics is shown in Fig. 2.

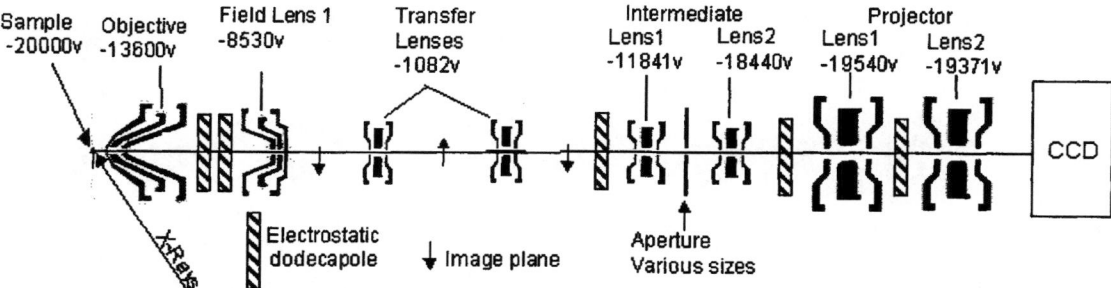

FIGURE 2. Schematic layout of the electrostatic lenses for the PEEM2 (stage 1) mode of the microscope buildup

Here the separator has been replaced with a two-lens (4f) image transfer system. The image from the objective section is relayed to the transfer and projection section. The left panel of Fig. 3 shows the instrument undergoing the final assembly. The electron column is visible along with the sample manipulator system mounted in the UHV sub chamber and surrounded by the triple mu-metal shields. The right panel of Fig. 3 shows a closer view of the UHV non-magnetic sample manipulator and objective lens column. The manipulator is a novel device based on flexures that allows for 5 degrees of freedom of the sample. The degree of freedom missing is rotation of the sample about the electron optic axis, which is not required as the polarization of the x-rays from the EPU can be rotated about this axis. The sample manipulator carries a sample puck that can be inserted and removed via a sample load lock system. The ceramic sample support structure is light and when mounted off the stiff flexures has natural vibration frequencies >200Hz which is high and will help reduce vibration issues. The sample manipulator is mounted on the electron optical bench and provides the required stiff mechanical loop between the sample and objective lens to reduce sample vibration and drift issues. The manipulator's 5 degrees of freedom are driven by non magnetic UHV slides with piezoelectric Nanomotors encoded to 20nm [13]. The stage is designed to be capable of delivering temperature ranges of 60K~1300K to the sample. Commissioning of the instrument and connection to the EPU beamline [4] is underway. The current PEEM2 science program will migrate to this instrument while the separator and mirror section is completed offline. This section should be installed in 2007.

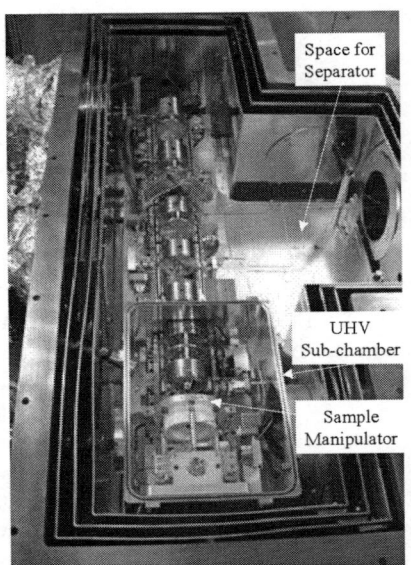 

FIGURE 3. Left – the PEEM2 mode electron column and sample manipulator mounted in the vacuum chamber lined with 3 layers of mu-metal. Right – the objective lens, dodecapole correctors and sample puck mounted on the ceramic sample manipulator.

ACKNOWLEDGMENTS

This work was supported by the Director, Office of Energy Research, Office of Science of the U.S. Department of Energy, under Contract No. DE-AC03-76SF00098.

REFERENCES

1. E.Brucher, Z. Phys., **86, 4**48- (1933).
2. S. Anders et al., Rev. Sci. Instrum, **70**, 3973-3981(1999).
3. D.Preikszas and H.Rose, J.Electro. Microscopy, **1**, 1-9 (1997).
4. T.Warwick et al., "A New VLS grating monochromator beamline with energy stabilization" these proceedings.
5. R.N.Watts, et al., Rev.Sci.Instrum., **68**, 3464-3476 (1997).
6. H.Rose and D.Preikszas, Optik, **92**, 31-44 (1992).
7. W. Wan, J. Feng and H.A. Padmore , "A new separator design for aberration corrected photoemission electron microscopes ", *Nucl. Instr. and Meth. A* in press.
8. W.Wan, et al., Nucl.Instr. Meths., **A519**, 222 -229(2004).
9. R.Fink et.al., J.Elec.Spec Rel Phenom., **84**, 231-250 (1997).
10. Y.K.Wu et al., Nucl.Instr. Meths, **A519**, 230-241 (2004).
11. G.F.Rempfer et al., Ultramicroscopy, **36**, 196-221 (1991).
12. J.Feng, et al., J.Phys: **17**, S1339-S1350(2005).
13. Ibex Engineering, Newbury Park, CA, 91320, USA

In Vivo X-Ray Fluorescence Microtomographic Imaging of Elements in Single-Celled Fern Spores

Yasuharu Hirai*, Akio Yoneyama*, Akiko Hisada*, and Kenko Uchida[†]

Advanced Research Laboratory, Hitachi, Ltd., Hatoyama, Saitama 350-0395, Japan
[†]Central Research Laboratory, Hitachi, Ltd., Kokubunji, Tokyo 185-8601, Japan

Abstract. We have observed *in vivo* three-dimensional distributions of constituent elements of single-celled spores of the fern *Adiantum capillus-veneris* using an X-ray fluorescence computed microtomography method. The images of these distributions are generated from a series of slice data, each of which is acquired by a sample translation-rotation method. An incident X-ray microbeam irradiates the sample with a spot size of 1 μm. The high Ca concentration in the testa and the localized and overlapping Fe and Zn concentrations inside the spore are shown in three-dimensional images. The K concentration is high throughout the cell, and there are localized regions of higher density. The atomic number densities of these elements in the testa and inside the cell in a tomographic slice are estimated with a resolution of about 1 μm.

Keywords: Spore, Testa, X-ray microbeam, X-ray fluorescence computed microtomography, Synchrotron radiation
PACS: 07.85.Qe, 78.70.En, 87.16.Gj, 87.59.Fm

INTRODUCTION

Macronutrients such as Ca, P, and K, and trace elements such as Fe, Zn, and Mn, play important roles in sustaining the homeostasis in a cell. Their distributions are directly correlated with the metabolism of each element. Therefore, several methods to analyze the three-dimensional distribution and density of elements in a living cell have been developed, such as laser-scanning confocal microscopy. This excellent method having enabled dynamic observation, however, requires fluorescent dye molecules that adhere to specific elements such as Ca [1] and Mg [2]; besides molecules that adhere to trace elements are not yet available. In addition, a sample must be transparent to allow detection of these elements using fluorescent light. Thus, an *in vivo* method to detect several elements simultaneously inside an opaque cell will give more information about the distribution and density of those elements and their inter-element interactions. To cope with the above requirements, we have introduced an X-ray fluorescence computed microtomography (or X-ray fluorescence micro-CT) method, which is sensitive at trace-element concentrations with micron spatial resolution. This method was developed by Cesareo *et al.* [3]. Simionovici *et al.* [4] observed images of constituent elements in the root of mahogany plants. The images were obtained as nondestructive "virtual" slices of the plant root.

This paper describes the first *in vivo* observation of three-dimensional distributions of constituent elements, Ca, K, Fe, and Zn, in single-celled spores of the fern *Adiantum capillus-veneris* using an X-ray fluorescence micro-CT system. We have developed this system using the X-ray undulator beamline BL16XU at the SPring-8 facility because an X-ray microbeam with a spot size of 1 μm and an intensity of 10^{10} photons/s was already available by using high-brilliance synchrotron radiation [5]. In the following, we describe the method of the X-ray fluorescence micro-CT and the results of *in vivo* observations of distributions of constituent elements in single-celled spores.

EXPERIMENTS

We irradiated an X-ray microbeam with a 1-μm spot size onto a sample at a photon energy of 11.3 keV. This energy is sufficiently high to observe the interior of a spore sample, which is a quasi-tetrahedron in shape with a

CP879, *Synchrotron Radiation Instrumentation: Ninth International Conference*,
edited by Jae-Young Choi and Seungyu Rah
© 2007 American Institute of Physics 978-0-7354-0373-4/07/$23.00

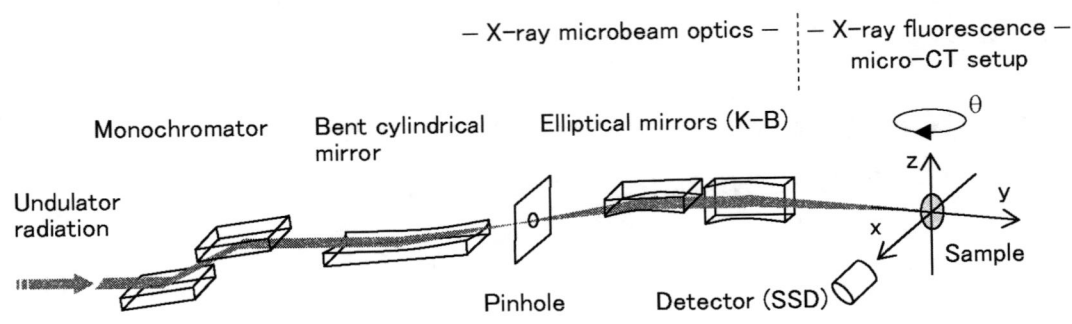

FIGURE 1. X-ray microbeam optics and X-ray fluorescence micro-CT setup.

typical height of 30 – 50 μm, while exciting elements of interest. The beam intensity is on the order of 10^{10} photons/s, which is necessary to detect fluorescence X-rays from trace elements. The experimental setup and procedure are described below.

X-Ray Microbeam Optics

We present the X-ray microbeam optics and the X-ray fluorescence micro-CT setup of the X-ray undulator beamline BL16XU [5] in Figure 1. A monochromatic X-ray beam from a Si 111 double-crystal monochromator is incident on a bent cylindrical mirror that focuses the beam onto a pinhole. This 25-μm high, 40-μm wide pinhole acts as a virtual point source for a Kirkpatrick-Baez (K-B) elliptical mirror system. The first and second mirrors are designed to have vertical and horizontal magnification values of 1/21 and 1/40, respectively, both with a 5-mrad grazing incidence angle. The obtained beam size, which is 1.2-μm high and 1.0-μm wide, is in good agreement with the calculated magnification values. The beam intensity is 8.8×10^9 photons/s at 11.3 keV.

X-Ray Fluorescence Micro-CT Method

The setup of the X-ray fluorescence micro-CT system is shown in Fig. 1 with (x, y, z, θ) coordinates of adjustment degrees of freedom. In this experiment we use three stages corresponding to each of the coordinates, except y. The sample is placed on a vertical rotation axis (z-axis) with rotation θ, where the axis is perpendicular to the incident beam direction. The sample is translated along the x-axis direction perpendicular to both the incident beam and z-axis in 1-μm steps. The sample, then, returns to the original position and is rotated 3°, which is an angle step of θ. Then, the sample is translated again, as described above. The total angular range is 180°. This rotation-translation setup enables horizontally scanning a slice of the sample. A solid-state detector (SSD), which is energy sensitive, is oriented to the sample in a direction parallel to the x-axis. Then, the sample is translated along the z-axis in 1.5-μm steps to obtain a series of slice data.

A quasi-tetrahedral spore with a height of 40 μm, stored in the dark at 4°C before *in vivo* measurement, is glued on the tip of a glass rod with a diameter of 100 μm. A data set of 61 projections is obtained with a linear motion step of 1 μm and an angle step of 3° in air. At each 1-μm step, fluorescence X-ray photons emitted from the sample, Ca-$K\alpha$, K-$K\alpha$, Fe-$K\alpha$, and Zn-$K\alpha$, are detected by the SSD for 1 s. The results of X-ray photon measurements for each element are accumulated in a personal computer, which also controls the X-ray fluorescence micro-CT system. The two-dimensional distribution of each element with a resolution of about 1 μm is reconstructed with a filtered back-projection reconstruction algorithm as if it is obtained from a horizontal slice of the sample.

Next, we translate the sample 14 times in 1.5-μm steps from the tip of the glass rod along the z-axis, obtaining a total data set of 15 slices that are 21 μm high. From this data set of 15 slices, we reconstruct three-dimensional images of Ca, K, Fe, and Zn distributions.

RESULTS AND DISCUSSION

We made the following assumptions to reconstruct the slice images: (a) absorption and scattering of an incident beam by the sample is negligible; (b) self-absorption [6,7] and secondary fluorescence effects in the sample

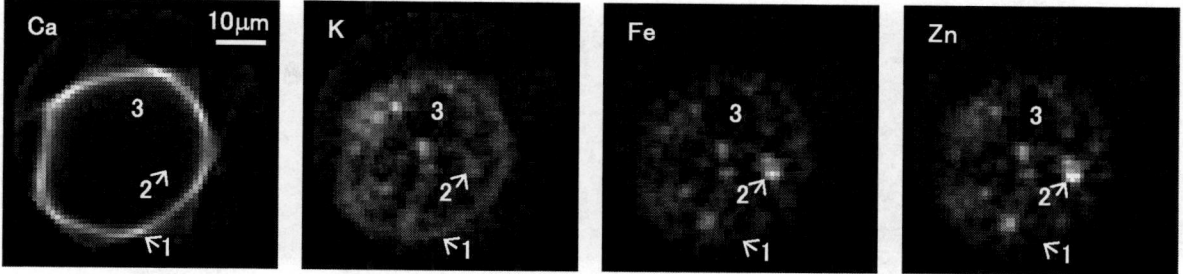

FIGURE 2. Two-dimensional images of Ca, K, Fe, and Zn distributions in single-celled fern spore. Images are reconstructed on 61 × 61 square grids with 1-μm pixel size. Typical pixel points in images are labeled 1, 2, and 3, as described in text.

are negligible; (c) absorption of fluorescence X-rays by air along the path to the detector is constant; (d) the sample is fixed at the same position when seen by the detector. All two-dimensional images are reconstructed on 61× 61 square grids with a 1-μm pixel size. Reconstructed images of Ca, K, Fe, and Zn concentrations at a distance of 16.5 μm from the tip of the glass rod are shown in Figure 2. Three-dimensional images of Ca, K, Fe, Zn, and their total concentrations are shown in Figure 3. These results are summarized as follows: (1) the Ca concentration is very high in the testa (cell wall) [8,9] whose thickness is on the order of 1 – 2 μm or less, (2) the K concentration is almost uniform, though high concentrations are observed at the same spatial regions as those of Ca, Fe, and Zn, (3) Fe and Zn concentrations are quite impressive because of their localizations almost at the same position inside the cell. The sizes of typical localized regions are estimated to be on the order of 1 – 3 μm or less.

These characteristics of each element concentration are represented by three kinds of pixel points, depicted as 1, 2, and 3, in Fig. 2. Point 1 is in the testa, point 2 is in the high-concentration region of Fe and Zn, and point 3 in the low concentration region of all elements. To clarify the differences among these points, we estimate the atomic number densities of elements at each point as follows. The intensity of fluorescence X-ray photons of the element j at detector dF_j is expressed as,

$$dF_j = \gamma_j \sigma_j N_j T_j I_0 \varepsilon \frac{d\Omega}{4\pi} ds \qquad (1)$$

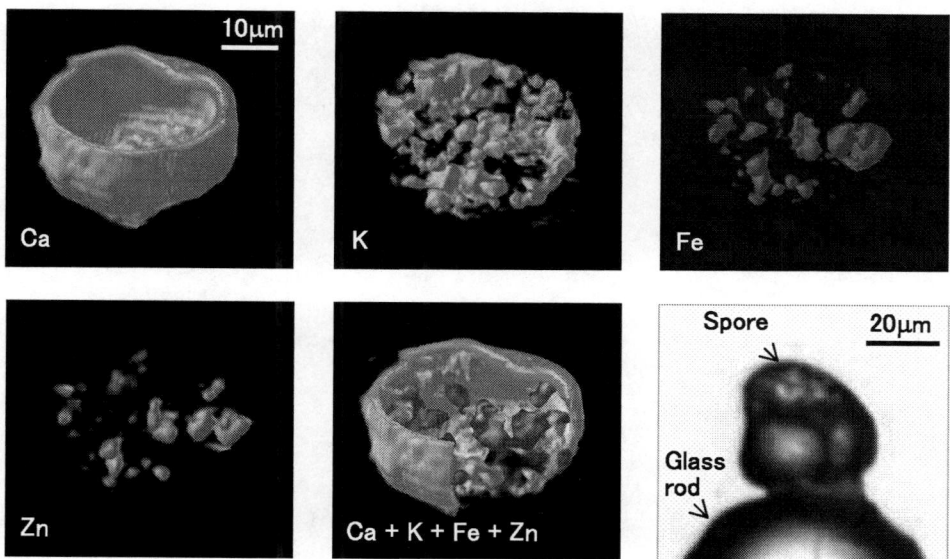

FIGURE 3. Three-dimensional images of Ca, K, Fe, and Zn distributions in single-celled fern spore, and optical microscope image of spore on glass rod.

TABLE 1. Atomic number densities N (atoms/μm^3) of Ca, K, Fe, and Zn, at pixel points 1, 2, and 3 as shown in Fig. 2.

Position	Ca	K	Fe	Zn
Point 1	5×10^8	1×10^8	5×10^6	2×10^6
Point 2	$< 1 \times 10^6$	2×10^8	3×10^7	1×10^7
Point 3	$< 1 \times 10^6$	6×10^7	$< 1 \times 10^6$	$< 1 \times 10^6$

where γ_j is the fluorescence yield of the $K\alpha$-line, σ_j is the atomic photoabsorption cross section at the incident beam energy, N_j is the atomic number density of the element, T_j is the correction factor for the absorption effects of air along the detection path and of the Be window for the detector, I_0 is the intensity of the incident beam, ε is the efficiency of the detector (assumed to be about 100%), $d\Omega$ is the solid angle from the interaction point to the detector surface, and ds is the short path along the incident beam. We define ds as 1 μm. Hence, we can estimate the atomic number density N_j at pixel points 1, 2, and 3, averaged within the volume of 1 μm^3 from Equation (1). In Table 1, the atomic number densities of Ca, K, Fe, and Zn are summarized at three characteristic points in Fig. 2. The peak atomic number densities of Ca and K show values about ten times as much as those of Fe and Zn.

Finally, we evaluate the self-absorption effects from absorption coefficients of the cell at each photon energy of Ca-$K\alpha$, K-$K\alpha$, Fe-$K\alpha$, and Zn-$K\alpha$ from a data set of two projections ($\theta = 0°$ and $180°$). Estimated values of absorption coefficients of Ca, K, Fe, and Zn are 50 cm^{-1}, 75 cm^{-1}, < 10 cm^{-1}, and < 10 cm^{-1}, respectively. Hence, relative transmissions of K, Ca, Fe, and Zn, when the path length through the cell is 40 μm, are 0.74, 0.82, > 0.96, and > 0.96, respectively. Therefore, the self-absorption effects of Fe and Zn are negligible. On the contrary, the atomic number densities for Ca and K in Table 1 have an error of about 20 % because of the self-absorption effects in the sample.

CONCLUTIONS

We have observed three-dimensional distributions of constituent elements in a singe-celled fern spore *Adiantum capillus-veneris* using X-ray fluorescence micro-CT. Each element, Ca, K, Fe, and Zn, exhibits characteristic concentrations that might be related to the metabolism in a cell. More detailed information about concentrations, and atomic number densities of elements, are obtained at each characteristic point in a cell. Relationships between these concentrations and cell structures, however, are not yet clear, and we expect to develop X-ray fluorescence micro-CT with less than 100-nm resolution in the near future.

REFERENCES

1. D. W. Tank, *Science* **242**, 773-777 (1988).
2. H. Komatsu et al., *J. American Chemical Society* **126**, 16353-16360 (2004).
3. R. Cesareo and S. Mascarenhas, *Nucl. Instr. and Meth. A* **277**, 669 (1989).
4. A. Simionovici et al., *Nucl. Instr. and Meth. A* **467-468**, 889-892 (2001).
5. Y. Hirai et al., *Nucl. Instr. and Meth. A* **521**, 538-548 (2004).
6. P. Hogan, R. A. Gonsalves, and A. Krieger, IEEE Trans. Nucl. Sci. **NS-38**, 1721 (1991).
7. B. Golosio et al., *J. Appl.Phys.* **94**, 145-156 (2003).
8. J. D. Cohen and K. D. Nadler, *Plant Physiol.* **57**, 347-350 (1976).
9. M. Kobayashi, H. Nakagawa, T. Asaka, and T. Matoh, *Plant Physiol.* **119**, 119-203 (1999).

Wide-Band KB Optics for Spectro-Microscopy Imaging Applications in the 6-13 keV X-ray Energy Range

E. Ziegler[1], S. de Panfilis[1], L. Peverini[1], P. van Vaerenbergh[1], F. Rocca[2]

1) European Synchrotron Radiation Facility, BP 220, 38043, Grenoble cedex, France
2) Istituto di Fotonica e Nanotecnologie, Centro CNR-ITC di Fisica degli Stati Aggregati, 38050 Povo, Italy

Abstract. We present a Kirkpatrick-Baez optics (KB) system specially optimized to operate in the 6-13 keV X-ray range, where valuable characteristic lines are present. The mirrors are coated with aperiodic laterally graded $(Ru/B_4C)_{35}$ multilayers to define a 15% energy bandpass and to gain flux as compared to total reflection mirrors. For any X-ray energy selected the shape of each mirror can be optimized with a dynamical bending system so as to concentrate the X-ray beam into a micrometer-size spot. Once the KB mirrors are aligned at the X-ray energy corresponding to the barycenter of the XAS spectrum to be performed they remain in a steady state during the micro-XAS scans to minimize beam displacements. Results regarding the performance of the wideband KB optics and of the spectro-microscopy setup are presented, including beam stability issues.

Keywords: Kirkpatrick-Baez, optics, multilayer, monochromator, spectroscopy, microscopy, X-rays, synchrotron
PACS: 07.85.Qe, 78.70.Dm, 78.70.En, 68.37.Yz

INTRODUCTION

While absorption and phase images of large specimens can be obtained through full illumination by an X-ray plane wave and recording with a position-sensitive detector, spectro-microscopy images are generally obtained by raster scanning the sample about a microbeam. Ways of producing a microbeam include Fresnel zone plates, refractive lenses, focusing mirrors, and their combination.[1] While each system comes with its specific load of advantages and disadvantages, one important criterion is the minimum amount of material one is eventually able to detect. The best results achieved so far have been obtained with zone plates for minimum spatial resolution [2] and with Kirkpatrick-Baez optics for maximum irradiance gain.[3] To obtain an image of a sample, either with x-ray fluorescence (XRF) or x-ray absorption spectroscopy (XAS), it is crucial to maintain both the size and the position of the microbeam fixed during the total acquisition time. In the ideal case these variations should remain negligible as compared to the microbeam size. Otherwise, one may attempt to correct for it, providing a mean of monitoring the microbeam variations is available. Stability constraints apply to both micro-XRF and micro-XAS techniques. In practice the problem is more severe in micro-XAS as the positional stability of the microbeam with time is complicated by the fact that the energy of the X-ray beam must be scanned. In practice, the realization of a fixed-exit monochromator that is exempt of vibrations remains a technical challenge. Micro-XAS measurements are also easier to perform when the microfocusing optics is achromatic, which tends to advantage mirror-based optical systems.

A mirror-based solution using a special Kirkpatrick-Baez (KB) optical arrangement has been developed to perform µ-XRF and µ-XAS measurements with a synchrotron bending-magnet source. This paper describes the characteristics and first measurements obtained with such a system.

KIRKPATRICK-BAEZ OPTICS AND WIDE-BAND MULTILAYERS

The KB optics [4] consists of the combination of two cylindrical mirrors set perpendicular to each other, a design that has the advantage of suppressing astigmatism. This is particularly appropriate for operation with hard x-rays where the mirrors are illuminated under grazing incidence. A large demagnification is achieved by minimizing the distance q from the center of each mirror to the focus as compared to the distance p to the source. Depending on the

CP879, *Synchrotron Radiation Instrumentation: Ninth International Conference*,
edited by Jae-Young Choi and Seungyu Rah
© 2007 American Institute of Physics 978-0-7354-0373-4/07/$23.00

experiment, the minimum q value is limited by the presence in XAS measurements of an ion chamber (I_0) at the exit of the KB, and by the sample environment. In our present setup the distance q is set to 0.15 m and the distance p can be set to either 39 m or to 55 m, corresponding to the possible locations of the KB along the beamline. These values, which correspond to geometrical demagnifications of the source dimensions of several hundreds, require off-axis elliptical surfaces. These profiles are achieved by means of an adaptive system composed of a thin (6 mm) mirror of variable width (linear variation) mounted on a 2-moment bending mechanism.[5] Because of the larger horizontal electron emittance of the synchrotron source, the generation of a nearly round spot requires to first concentrate the beam in the vertical direction (mirror M_1, 170-mm long) and to focus the beam horizontally with a shorter mirror (92 mm long) located closer to the focal plane. With perfectly astigmatic mirror shapes, the theoretical spot size expected are 0.7 μm x 1.0 μm for p= 39 m and 0.5 μm x 0.7 μm for p= 55 m, in the vertical and horizontal directions, respectively.

The energy tunability of the synchrotron source allows to perform microanalysis on various kinds of samples by setting the X-ray energy E_i to a given excitation energy in μ-XRF or by scanning the energy around a given absorption edge (μ-XAS). This implies an adjustment of the profile of the KB mirrors prior to the experiment, conditioned by the mean energy of the XAS spectrum and the incident angle θ of the incoming beam that will provide high mirror reflectivity and KB throughput. A fixed θ value is necessary to avoid any mirrors correction during scans, which would unavoidably cause a change in beam size and position. With an undulator source installed at a 3rd generation synchrotron machine it is convenient to use grazing incidence mirrors and to set θ so that all energies concerned are reflected under total external reflection. However, in our case of a bending magnet source, the flux collected would be insufficient to perform spectro-microscopy experiments with a spot size in the micrometer range and below. To circumvent this problem the collection angle was increased by coating both mirrors with multilayers. Because an XAS spectrum spans over an energy range larger than the rocking curve of periodic multilayers, a depth gradient of the period was needed, as already suggested in Ref. [6]. The theory of wide-band multilayers developed in the past years by Kozhevnikov et al.[7] allows designing x-ray reflectors with a bandpass of arbitrary width for a minimum number of layers. Here we aimed at a 15% bandwidth (ΔE/E) for any X-ray energy selected between 6 keV and 13 keV, with a priority on the design for a flat reflectivity response for an energy of 13 keV. The final design and manufacturing were performed at the ESRF multilayer facility where both mirrors were coated with a 35-periods (Ru/B_4C) multilayer.[8] The mean period was 6 nm at the center of the mirror and varied along the longest dimension of the elliptically shaped surface to account for the strongly non-linear variation of the incident angle. The transmission of the overall system (2 reflections) was measured to be 25% and the bandwidth around a mean energy of 12.27 keV was 2.16 keV.[9]

According to the combination of parameters {E_i(θ), p, q} of the experiment the shape of each mirror is modified on-line using the information obtained by sampling the mirror surface with an x-ray pencil beam (slit scanning) and by simultaneous recording of the position of the various impacts at the focal plane with a CCD camera (2.3 μm/pixel). A linear optimization scheme similar to the one used for adaptive telescope mirrors is applied iteratively to infer the bender actuators positions minimizing the focus, as described by Hignette et al.[10]. After optimization of the mirror figure for a mean energy of 9.7 keV a 200-μm Au wire was scanned successively across the beam in the focal plane. For a cross-section of the incoming beam of 1 mm (V) by 0.5 mm (H), spot sizes of 2.0 μm (V) and 2.2 μm (H) were derived after fitting of the measured profiles using an error function.

SPECTRO-MICROSCOPY SETUP AND FIRST MEASUREMENTS

The micro-XAS setup presently installed on the ESRF bending-magnet beamline BM5 [11] is shown on Fig. 1. The incoming beam is monochromatized with a water-cooled double-reflection Si(111) crystal monochromator located at 27.2 m from the source. The two crystals are mounted on a rotary stage to define the Bragg reflection and energy. An additional translation stage of the 2nd crystal maintains the beam at a fixed height of 20 mm above the white beam while a rotation stage (resolution: 1.6 μrad) allows detuning the angle of the incoming beam from the Bragg peak position in order to reduce the contribution of the odd harmonics of higher order. At a distance p of 39.3 m from the source the flux is of $1.6\ 10^{10}$ ph/s/mm^2 for an energy of 25 keV and a ring current of 200 mA; the resolution ΔE/E is of $2.1\ 10^{-4}$. Two pairs of slits separated by a 2 m distance are used to define the incident beam both in direction and in position. A 2D piezoelectric stage mounted on top of a rotation stage is positioned in the focal plane of the KB. The XAS signal in transmission mode can be recorded using two gas-filled ionization chambers placed before (I_0, 30% absorption) and after the sample (I_1, 70%). The fluorescence intensity I_f can also be recorded by tilting the sample at 45° with respect to the beam and positioning a silicon Roentec detector in the horizontal plane at 90° with respect to the beam. The distance to the sample is set to define a dead time of the order of 10%. The fluorescence signal I_f/I_0 normalized for the incoming beam intensity measured after the KB accounts for the intensity fluctuations due to the non-uniform transmission of the device as a function of the energy. It also accounts for the variation of absorption as a function of the energy.

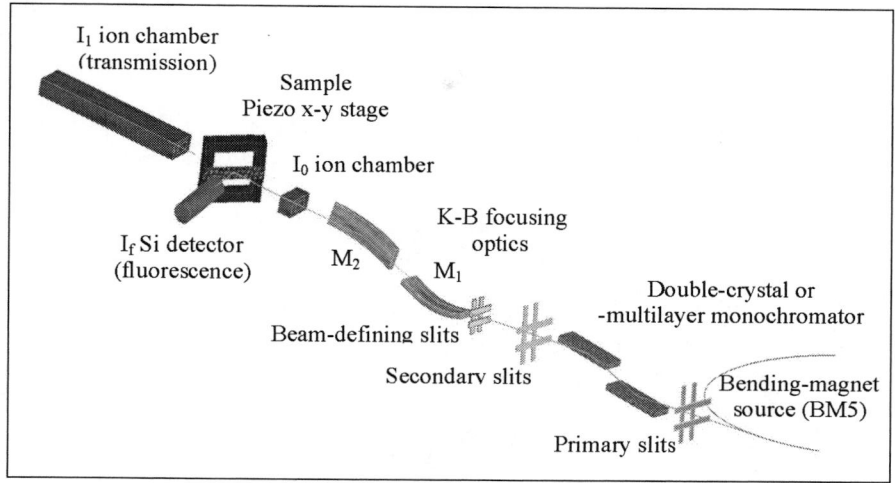

FIGURE 1. Sketch of the spectro-microscopy setup at the ESRF BM05 beamline.

Measurements at various absorption edges (Zn, As, Se) were performed to demonstrate the performance of the system. The measurements at the Zn K-edge are presented in this paper. After aligning the KB at an X-ray energy of 9.60 keV, the energy of the incoming beam was set to 9.67 keV and a ZnO sample was placed in the focal plane. The sample, about 400 nm thick, was produced by dc magnetron sputtering of a metal Zn target onto a Si substrate under Ar gas followed by high temperature oxidation in air at 800°C. Fig. 2-left shows the variation of the Zn fluorescence signal I_f when scanning the sample over a length of 30 μm with a 0.5 μm step. This variation is due to the morphology as observed on the AFM image (see inset in Fig.2). At the position where the Zn fluorescence is maximum a fluorescence XAS spectrum was recorded. Its is compared in Fig. 2-middle with the one of a bulk ZnO powder sample measured at the ESRF beamline BM29 with a standard XAS setup. Notice the overall similarity of the two $\chi(k)$ signals in terms of oscillations frequencies and energy resolution. The smaller amplitude in the XAS oscillations for the ZnO sputtered sample measured on BM05 is attributed both to an increased short range structural disorder and to a greater ratio of the number of atoms present at the surface and in the bulk, due to the intrinsic nature of its morphology. In the Fig. 2-right showing the corresponding Fourier spectra the 2nd peak corresponding to Zn-Zn is strongly reduced for the ZnO sputtered sample measured at BM05 as compared to the bulk powder one, while the magnitude of the 1st peak (Zn-O) is the same for both samples.

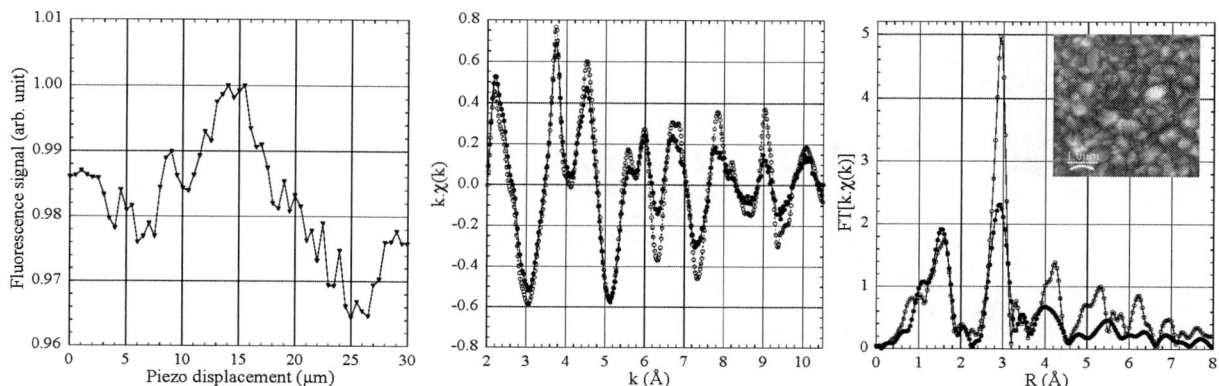

FIGURE 2. (Left):.Zn fluorescence signal when scanning the sample over a length of 30 μm with a 0.5 μm step. (Middle): Comparison of Zn XAS spectra for a bulk powder sample measured with a large beam (empty circles) and with a thin sample and a microbeam (black circles). (Right): Fourier transforms of the spectra shown in Fig. 2-middle. (Inset): AFM image of the ZnO sample produced by sputtering and oxidized in air at 800°C.

When only the micro-XRF needs to be measured, the detection limit can be improved by replacing the Si crystal monochromator with a double-reflection Ru/B_4C multilayer monochromator having a 2.2% bandwidth and located at 28 m from the source. At an energy of 13 keV and with a flux density at the sample of 4.10^{10} ph/s /μm^2, a detection limit below the femtogram was achieved with a 1000 s acquisition time for elements with a Z number higher than 19 and with a 10 s acquisition time for Z $\geq$ 30 (Ref. [9]). The use of a multilayer monochromator provides detection limits comparable to the one obtained with an insertion device and a Si crystals monochromator.

BEAM STABILITY ISSUES

The stability of the focused beam in position and size is an important characteristic of a spectro-microscopy setup. For micro-XRF the stability as a function of time matters. Using a FReLon CCD camera with an equivalent pixel size, through optical magnification, of 0.7 μm the vertical and horizontal beam positions and beam size were recorded. After a stabilization time of 1.8 h, the beam positions varied by 2% and 2.9 % of the respective mean size in the horizontal and vertical directions (standard deviation values) and the beam sizes varied by 7.3% and 3.7 %. This is to be compared to the tolerance defined on the positional stability of the ESRF synchrotron source, given as 10% of its size. For XAS measurements the stability as a function of the energy adds to the one on time. In this case the positions varied by 5.2% and 6.6% of the respective mean size in the horizontal and vertical directions (STD values). A clear oscillation could be seen in the vertical direction, which is also the diffraction plane of the monochromator. The spot size varied by 7.3% (H) and 4.8% (V).

CONCLUSION

In this paper we showed that micro-XAS and micro-XRF can be performed at a bending-magnet beamline using an efficient multilayer-coated KB optics. The wide bandpass allows energy scans over 2 keV with a fixed setup, the mirror shaping being only necessary when defining a new energy of operation. Substantial reduction in focus size can be obtained by operating in the second hutch of the beamline (55 m instead of 40 m). Smaller spot sizes call for improved mirror surface figures, while their use for spectro-microscopy mapping will require better beam stability or positional beam tracking methods. Improvements regarding the sample environment and positioning are now underway.

ACKNOWLEDGMENTS

We wish to thank the ESRF support groups, in particular, the optics group for multilayer and mirror developments, R. Pieritz and O. Svensson for assistance with the alignment routines, L. Claustre for support on beamline control, and the ISG team for detectors. We are also grateful to the staff of the ESRF ID22 beamline for advice and loan of equipment, A. Somogyi from the SOLEIL synchrotron for participation in previous experiments, and A. Kuzmin and R. Kalendarev, from the University of Latvia in Riga, for the fabrication of the ZnO sample.

REFERENCES

1) K. Yamauchi, K. Yamamura, H. Mimura et al., *J. Synchr. Rad.* **9**, 2002, pp. 313; O. Hignette, P. Cloetens, G. Rostaing et al., *Rev. Sci. Instrum.*, **76**, 2005, pp. 063709; W. Chao, B.D. Harteneck, J.A. Liddle, E.H. Anderson, D.T. Attwood, *Nature* **435**, 2005, pp. 1210-3; C.G. Schroer, O. Kurapova, J. Patommel et al., *Appl. Phys. Letters* **87**, 124103 (2005); H.C. Kang, J. Maser, G.B. Stephenson, C. Liu, R. Conley, A.T. Macrander, S. Vogt, *Phys. Rev. Letters* **96**, 2006, pp. 127401
2) W. Chao, B.D. Harteneck, J.A. Liddle, E.H. Anderson, D.T. Attwood, *Nature* **435**, 2005, pp. 1210-3
3) O. Hignette, P. Cloetens, G. Rostaing, P. Bernard, C. Morawe, *Rev. Sci. Instrum*, **76**, 2005, pp. 063709.
4) P. Kirkpatrick, A. V. Baez, *J. Opt. Soc. Am.*, **38**, 1948, pp. 766.
5) E. Ziegler, O. Hignette, M. Lingham, A. Souvorov, SPIE Proc. vol. **2856**, Denver, 1996, pp. 61-7; L. Zhang, R. Hustache, O. Hignette, E. Ziegler, A. K. Freund, *J. Synchr. Rad.*, **5**, 1998, pp. 804-7; E. Ziegler, O. Hignette, C. Morawe, R. Tucoulou, *Nucl. Instr. & Meth.*, **A467-8**, 2001, pp. 954-7.
6) E. Ziegler, I. N. Bukreeva, I. V. Kozhevnikov, A. V. Pirozhkov, E. Ragozin, SPIE-Europto Proc. vol. **3737**, Berlin, 1999, pp. 386-95.
7) I. V. Kozhevnikov, I. N. Bukreeva, E. Ziegler, "Design of X-ray supermirrors", *Nucl. Instr. & Meth.*, **A460**, 2001, pp. 424-43.
8) C. Morawe, C. Borel, E. Ziegler, J.-C. Peffen, SPIE Proc. vol. **5537**, Denver, 2004, pp. 115-126.
9) E. Ziegler, A. Somogyi, et al., 8th International Conference on X-ray Microscopy, edited by S. Aoki et al., IPAP Conference Series 7, Institute of Pure and Applied Physics, Himeji, 2005, pp. 76-8.
10) O. Hignette, A. K. Freund, E. Chinchio, SPIE Proc. vol. **3152**, San Diego, 1997, pp. 188-99.
11) E. Ziegler et al., 8th Int. Conf. on Synchr. Rad. Instrum., edited by T. Warwick et al., AIP Conf. Proc. **705**, 1, San Francisco, 2004, pp. 436-9.

Development and Trial Measurements of Hard X-ray Photoelectron Emission Microscope

T. Taniuchi[1], T. Wakita[2], M. Takagaki[2], N. Kawamura[2], M. Suzuki[2], T. Nakamura[2], K. Kobayashi[2], M. Kotsugi[3], M. Oshima[1], H. Akinaga[4], H. Muraoka[5], and K. Ono[6]

[1]Department of Applied Chemistry, The University of Tokyo, Bunkyo-ku, Tokyo 113-8656, Japan
[2]JASRI/SPring-8, Koto, Hyogo, 679–5189, Japan
[3]Hiroshima Synchrotron Radiation Center (HiSOR), Higashi-hiroshima, Hiroshima 739-8526, Japan
[4]National Institute of Advanced Industrial Science and Technology (AIST), Tsukuba 305-8562, Japan
[5]Research Institute of Electrical Communications (RIEC), Tohoku University, Sendai 980, Japan
[6]Institute of Materials Structure Science, High Energy Accelerator Research Organization (KEK), Tsukuba 305-0801, Japan

Abstract. Photoelectron emission microscope (PEEM) study is performed using hard x-ray illumination. We have successfully obtained images with high spatial resolution of 40 nm with hard x-rays. Spectro-microscopy of Co micro-patterns on Si substrates, which can be applied to XAFS measurements on a minute scale by PEEM. Magnetic imaging has been demonstrated at the Pt L-edges on perpendicular magnetic recording pattern of CoCrPt alloy. These results are the first step toward a new spectroscopic microscopy and magnetic imaging in a hard x-ray region.

Keywords: PEEM, magnetic imaging, perpendicular recording media
PACS: 79.60.-i, 68.35.Ct, 77.55.+f, 85.40.-e

INTORODUCTION

Photoelectron emission microscope (PEEM) is one of the most useful techniques for element-specific magnetic imaging. One of the features of PEEM is that the technique allows us to obtain the spectra on various sizes and positions at the same time in the short measuring time and to produce the spectra of arbitrary areas after the measurement. Recently, there has been a strong interest in techniques of spectro-microscopy using synchrotron radiation with the development of nanotechnology because of the ability of chemical state analysis in a minute area due to energy variability of synchrotron radiation. One of the microscopic techniques commonly used in synchrotron radiation is scanning methods with a focused beam by Fresnel zone plate [1-4] or KB mirrors [5,6], which are mainly used as a hard x-ray microscopy. This method requires an advanced focusing technique in order to improve a spatial resolution, which brings out many studies to achieve the high focusing technique [7,8]. The other is image formation methods with electron lenses such as PEEM as a spectro-microscope [9]. On the contrary PEEM does not require the particular kind of techniques in light source, because PEEM images the photoelectrons from small portion of illuminated sample area by electrostatic lenses or magnetic lenses. PEEM has an advantage of constant focal distance in various photon energies. So far, PEEM is generally utilized in the soft x-ray regions. In this study, we have demonstrated PEEM study as a hard x-ray spectro-microscopy using hard x-ray illumination.

Auger electron emission and fluorescence occur when the specimen is irradiated with x-rays. It is generally known that the probability of these emissions strongly depends on the elements and energy levels to be excited [10,11]. Fluorescence yields represent the probability of a core hole in the K or L shells being filled by a radiative process, in competition with nonradiative processes. Auger processes are the nonradiative process competing with fluorescence especially for K shells. Therefore the excitations of deep energy states tend to emit light, rather than to

CP879, *Synchrotron Radiation Instrumentation: Ninth International Conference*,
edited by Jae-Young Choi and Seungyu Rah
© 2007 American Institute of Physics 978-0-7354-0373-4/07/$23.00

emit electrons in the hard x-ray region. In this experiment we have performed the hard x-ray spectroscopic microscopy by PEEM with a high-intensity hard x-ray light source, demonstrating x-ray absorption spectroscopy (XAS) in a minute area and direct magnetic imaging.

EXPERIMENTAL

We have developed a compact and mobile PEEM system consisting of Elmitec PEEMSPECTOR in which the emitted photoelectrons are magnified by electrostatic lenses [12,13]. The emitted photoelectrons are magnified by electrostatic lens and projected onto a phosphor screen. PEEM images are recorded by a charge-coupled device (CCD) camera (PCO Sensicam). Fields of view (FOV) can be changed from 5 to 500 μm. The sample manipulator has 6-axes with manual XYZ translation, XY tilt and azimuthal rotation. The sample manipulator also has a capability of heating a sample up to 800 °C as well as cooling a sample down to –120 °C using liquid nitrogen. To characterize the performance of our mobile PEEM system, we used a mercury lamp as an excitation source and obtained the spatial resolution of about 35 nm for this microscope.

The experiments reported here were carried out at a hard x-ray undulator beamline BL39XU of the SPring-8 [14] where the PEEM system was installed. The measurements consist of micro-XAS and magnetic imaging by x-ray magnetic circular dichroism (XMCD). This beamline has a helicity-modulation technique with a diamond phase retarder, which provides XMCD signal with a high signal-to-noise (SN) ratio in a shorter measuring time.

To improve SN ratio of magnetic images in the short measuring time, we have developed a new system that can totally control the PEEM system and the beamline optics, and repeatedly operated the acquisition of the images automatically. In the optics system, to minimize the modulation of beam axis during scanning photon energy, we have used Monochromator stabilizer (MOSTAB). In order to minimize the influence of nonuniformity of brightness in the beam spot, we utilized a diffuser. Consequently we have succeeded in improvement of the SN ratio and uniformity in the images using these techniques.

The samples for the estimation of demonstration of micro-XAS were nano-structured patterns of Co on Si substrates formed by electron-beam lithography and liftoff techniques. The sample for the magnetic imaging was a CoCrPt alloy film [15] in which the striped magnetic-recording pattern was formed. The film was deposited by a sputtering method. The directions of magnetization are out-of-plane to the film. All the measurements were performed at room temperature.

RESULTS AND DISCUSSION

We have demonstrated micro-XAFS measurements by obtaining PEEM images with changing the wavelength near the Co K-edge. The field of view is 50 μm. The exposure time to obtain one image is only five seconds. The representative images of the Co patterned sample were shown in Figs. 1(a) and (b), which were acquired at the photon energy of 7.700 keV and 7.730 keV respectively, corresponding to the energy below and above the Co K-shell absorption edge. The contrasts of these images were normalized by the brightness of Si area. The contrast of Co is brighter than that of Si over the K-edge while that of Co is darker than that of Si below the edge. This means that the relative absorption cross sections between Co and Si were inverted in the vicinity of the Co K-edge.

We have plotted the contrast of the square part shown in Fig. 1(a) by changing the wavelength. The area corresponds to 2 μm x 2 μm. We have succeeded in obtaining absorption spectrum with high SN ratio as shown in Fig. 1(c). This result implies that this method can be applied to x-ray absorption near edge structure spectroscopy (XANES) and extended x-ray absorption fine structure (EXAFS) in a minute area. The another plot can be also made at the different Co area of 1 μm x 1 μm from the same data, providing the absorption spectrum with a sufficient SN ratio. One of the features of micro-XAS by PEEM is that the technique allows us to obtain the spectra on various sizes and positions at the same time in the short measuring time and to produce the spectra of arbitrary areas after the measurement [16]. When it is necessary to obtain the spectra with the even higher SN ratio or at the smaller area, we just have to increase the exposure time. Furthermore, our direct imaging has an advantage of availability at various wavelengths without decreasing the spatial resolution and necessity of modulation of focal distance. The focal distance is constant independent of wavelength, because PEEM is a projection type with electrostatic lenses.

We have operated magnetic imaging as one of the demonstrations of spectroscopic microscopy by the hard x-ray PEEM. X-ray magnetic circular dichroism (XMCD) was applied to the magnetic imaging. XMCD in core-level

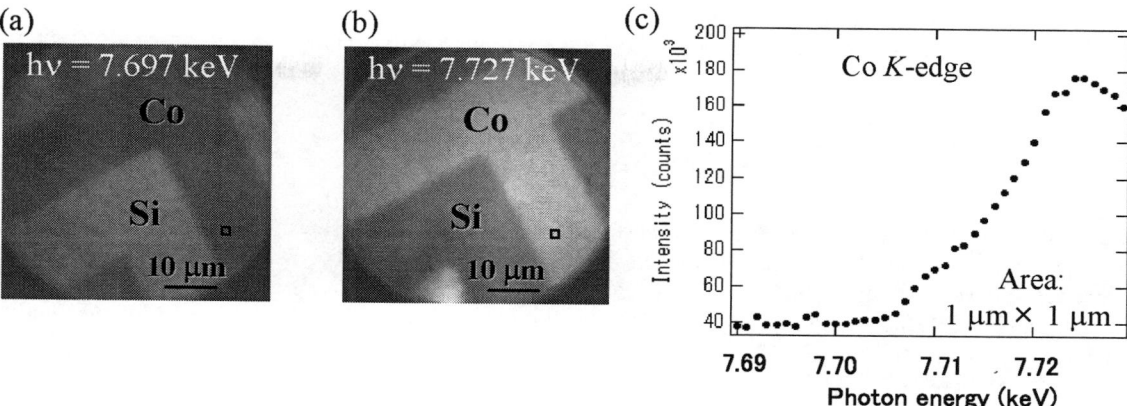

FIGURE 1. The PEEM images of the Co pattern sample. The field of view is 50 μm. The photon energy was corresponding to the Co K-preedge (a) and the K-edge (b). (c) X-ray absorption spectrum in the square area obtained by PEEM. The area corresponds to 2 μm x 2 μm.

absorption detects the magnetization dependence of the x-ray absorption coefficients in the vicinity of element-specific absorption edges of circularly polarized radiation. A diamond x-ray phase retarder was equipped at the beamline for the fast photon-helicity switching. Therefore magnetic images with a higher SN ratio and statistical accuracy of XMCD signal were acquired in the shorter measuring time. It is well known that the XMCD signal in the hard x-ray region is much less than that in the soft x-ray region. In this study we used the system to control the PEEM system and the beamline optics. The CoCrPt alloy film with perpendicular magnetization was used as a sample for magnetic imaging. In this sample, the out-of-plane striped magnetic pattern was formed. Absorption edge to be measured was the Pt L_3 edge (11.56 keV). One cycle of data acquisition consists of the acquisition of a pair of images with different helicity, whose exposure time was 3 seconds per image. The repetition number was 300 times. The magnetic images were obtained by the difference in the images between plus helicity and minus helicity. The measured magnetic images are shown in Fig. 2. The stripe patterns were clearly observed. The result proves that PEEM enable the magnetic imaging with a high SN ratio also in the hard x-ray region.

The spatial resolution of magnetic images in this study was estimated to be about 200 nm, which is much larger than the resolution of the PEEM system. This is due to the larger drift of the sample during measurements which require much longer measuring time. However, this can be improved by the further development of compensation techniques for the sample drifting.

SUMMARY

We have performed PEEM imaging as a new spectroscopic microscopy in a hard x-ray region, whose spatial resolution was 40 nm. We succeeded in micro-spectroscopy, which implies that PEEM can be extended to XANES

$h\nu = 11.56$ keV (Pt L_3 edge) $h\nu = 13.27$ keV (Pt L_2 edge)

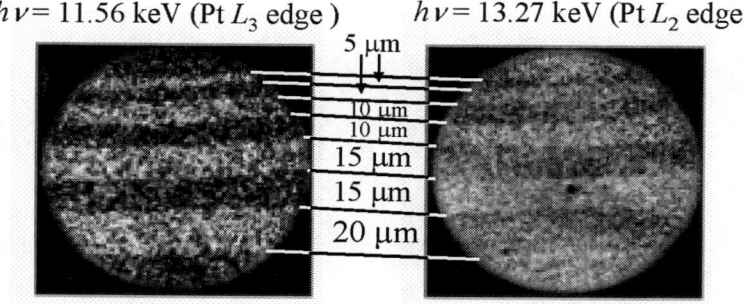

FIGURE 2. The magnetic images of the CoCrPt alloy film with the striping perpendicular magnetic-recording pattern. The field of view is 100 μm. The images were acquired with the photon energies corresponding to the Pt L_2- and L_3-edges. Exposure times are 3000 and 2000 seconds, respectively.

and EXAFS on a minute scale. Magnetic imaging by XMCD has been performed at the Pt *L*-edges by total control of the PEEM system and the beamline optics. This new technique will enable various applications in micro-spectroscopy and spectro-microscopy.

ACKNOWLEDGMENTS

The synchrotron radiation experiments were performed at SPring-8 with the approval of Japan Synchrotron Radiation Research Institute (JASRI) as Nanotechnology Support Project of the Ministry of Education, Culture, Sports, Science and Technology (Proposal No. 2003B0498-NSc-np-Na / BL-39XU). This work was supported by a Grant-in-Aid for Scientific Research (S17101004) from JSPS. One of the authors (T. T.) would like to thank JSPS for financial support (17-11095).

REFERENCES

1. B. Lai, W. Yun, Y. Xiao, L. Yang, D. Legnini, Z. Cai, A. Krasnoperova, F. Cerrina, E. DiFabrizio, L. Grella, and M. Gentili, Rev. Sci. Instrum. **66**, 2287 (1995).
2. M. Awaji, Y. Suzuki, A. Takeuchi, H. Takano, N. Kamijo, S. Tamura, and M. Yasumoto, Nucl. Instrum. Methods Phys. Res. A **467–468**, 845 (2001).
3. Y. Kagoshima, T. Ibuki, Y. Yokoyama, Y. Tsusaka, J. Matsui, K. Takai, and M. Aino, Jpn. J. Appl. Phys., Part 2 **40**, L1190 (2001).
4. Y. Kagoshima, Y. Yokoyama, T. Ibuki, T. Niimi, Y. Tsusaka, K. Takai, and J. Matsui, J. Synchrotron Radiat. **9**, 132 (2002).
5. S. Aoki, in X-Ray Microscopy II, edited by D. Sayre, M. Howells, J. Kirz, and H. Rarback (Springer, Berlin, 1987), p. 102.
6. A. Takeuchi, S. Aoki, K. Yamamoto, H. Takano, N. Watanabe, and M. Ando, Rev. Sci. Instrum. **71**, 1279 (2000).
7. E. Bauer, Rep. Prog. Phys. **57**, 895 (1994).
8. H. S. Youn, S. Y. Baik, and C.-H. Chang, Rev. Sci. Instru. **76**, 023702 (2005).
9. W. Chao, E. Anderson, G. P. Denbeaux, B. Harteneck, J. A. Liddle, D. L. Olynick, A. L. Pearson, F. Salmassi, C. Y. Song, and David T. Attwood, Opt. Lett. **28**, 2019 (2003).
10. M. O. Krause, J. Phys. Chem. Ref. Data **8**, 307 (1979).
11. A. Thompson *et al.*, X-ray Data Booklet, (Lawrence Berkeley National Laboratory, 2001).
12. T. Taniuchi, M. Oshima, H. Akinaga, and K. Ono, J. Electron. Spctro. Rel. Phen. **144-147**, 741 (2005).
13. T. Wakita, T. Taniuchi, K. Ono, M. Suzuki, N. Kawamura, M. Takagaki, H. Miyagawa, F.Z. Guo, T. Nakamura, T. Muro, H. Akinaga, T. Yokoya, M. Oshima, and K. Kobayashi, Jpn. J. Appl. Phys. **45**, 1886 (2006).
14. M. Suzuki, N. Kawamura, M. Mizumaki, A. Urata, H. Maruyama, S. Goto, and T. Ishikawa, Jpn. J. Appl. Phys. **37**, L1488 (1998).
15. T. Shimatsu, H. Sato, T. Oikawa, Y. Inaba, O. Kitakami, S. Okamoto, H. Aoi, H. Muraoka, and Y. Nakamura, IEEE trans. Magn. **20**, 2483 (2004).
16. M. Kotsugi, T. Wakita, T. Taniuchi, K. Ono, M. Suzuki, N. Kawamura, M.Takagaki, M. Taniguchi, K. Kobayashi, M. Oshima, N. Ishimatsu, and H. Maruyama, e-J. Surf. Sci. Nanotech. **4**, 490 (2006).

Observation of a Soft Tissue by a Zernike Phase Contrast Hard X-ray Microscope

Sadao Aoki, Tadahiro Namikawa, Masato Hoshino and Norio Watanabe

Institute of Applied Physics, University of Tsukub
1-1-1, Tennoudai, Tsukuba, Ibaraki, 305-8573, Japan

Abstract. A Zernike-type phase contrast hard X-ray microscope was constructed at the Photon Factory BL3C2 (KEK). A white beam from a bending magnet was monochromatized by a silicon double crystal monochromator. Monochromatic parallel X-ray beam illuminated a sample, and transmitted and diffracted X-ray beams were imaged by a Fresnel zone plate (FZP) which had the outer zone width of 100 nm. A phase plate made of a thin aluminum foil with a pinhole was set at the back focal plane of the FZP. The phase plate modulated the diffraction beam from the FZP, whereas a direct beam passed through the pinhole. The resolution of the microscope was measured by observing a tantalum test pattern at an X-ray energy of 9 keV. A 100nm line-and-space pattern could be resolved. X-ray montage pictures of growing eggs of artemia (plankton) were obtained.

Keywords: X-ray microscope, phase contrast, biological specimen, Fresnel zone plate, phase plate, artemia
PACS: 41.50.+h

INTRODUCTION

X-ray microscopes have been developed mainly in the soft X-ray region [1] because there has been no high resolution optical device which can image hard X-rays. Fortunately recent nanotechnology has made it possible to produce hard X-ray imaging devices such as zone plates, Kirkpatrick-Baez mirrors, Wolter mirros and so on.

With hard X-rays a relatively large and absorptive specimen, for instance, several 100 μm or so in thickness, can be moderately observed, while with soft X-rays it cannot. However, soft tissues like biological specimens are too transparent to observe with hard X-rays. We must manage to enhance the image contrast. In the hard X-ray region, a phase term δ of the complex refractive index of materials is approximately 3 orders of magnitude larger than an absorption term β. Therefore, a phase contrast X-ray microscope, which observes the phase changes as an intensity distribution, enables us to observe weak absorption specimens with relatively high contrast [2].

A Zernike-type hard X-ray phase contrast microscope can be realized by combining a relatively thick tantalum zone plate (1μm thick) and a thin aluminum phase plate. In the previous papers we demonstrated that a simple pinhole made in a thin aluminum foil worked as a phase plate [3,4]. The field of view, however, was limited by the size of the zone plate. In order to expand the effective area of observation we added a scanning sample stage and an image processing technique. For the demonstration of the feasibility of this method we observed growing eggs of artemia (plankton). X-ray montage pictures of growing artemia are shown.

X-RAY MICROSOCPE OPTICS

The optical system of the phase contrast X-ray microscope shown in Fig. 1(a) was constructed at BL3C2 of the Photon Factory (KEK). The white beam from the bending magnet was monochromatized with a Si(111) double-crystal monochromator. X-ray energies from approximately 6 keV to 10 keV were used. From the geometrical restriction the magnifications of the X-ray microscope were from 20 to 30 corresponding to the X-ray energies from 6 keV to 10 keV. The parallel beam was restricted to 100 μm in diameter by a platinum pinhole, which illuminated a

CP879, *Synchrotron Radiation Instrumentation: Ninth International Conference*,
edited by Jae-Young Choi and Seungyu Rah
© 2007 American Institute of Physics 978-0-7354-0373-4/07/$23.00

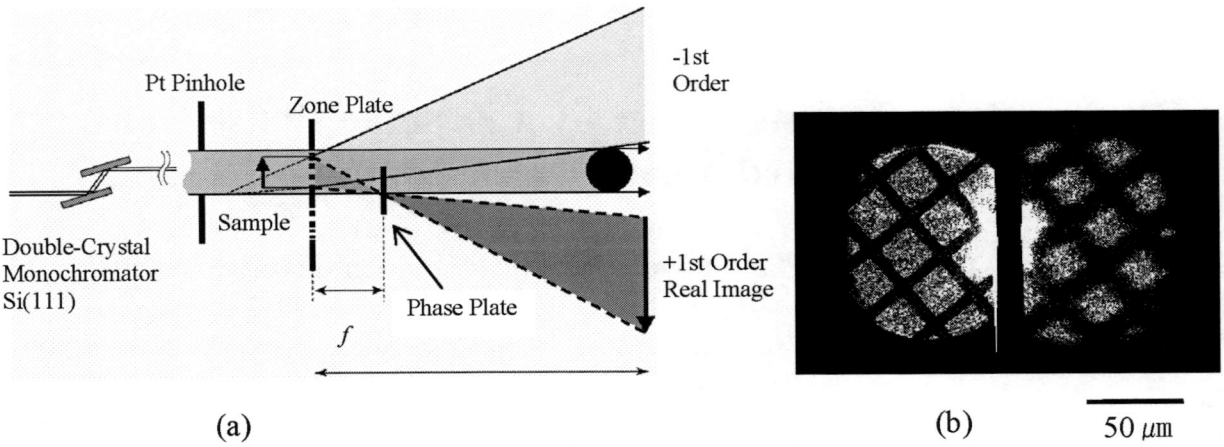

FIGURE 1. (a) The optical system of the phase contrast X-ray microscope.(b) A real (+1st order, left) and a virtual (-1st order, right) absorption contrast images of copper grid (1000lines/inch).

sample and a half of the zone plate. This optical arrangement was used to avoid the overlapping a real image and a virtual image, which is shown in Fig.1(b). An X-ray real image was focused onto a detector by the zone plate. A phase plate was inserted at the back focal plane of the zone plate to modulate the phase of the diffracted or the scattered X-rays. A proper gold wire of several hundreds microns in diameter was put in front of the detector to absorb the zero-th order direct beam.

The parameters of the zone plate were as follows: diameter 155 μm; outermost zone width, 0.1 μm; number of zones, 388; zone material, Ta; pattern thickness, 1μm; substrate, 2.0 μm Si_3N_4, and focal length 113 mm at 9 keV. The phase plate used was a thin aluminum foil which had a pinhole of 5μm in diameter. The thickness of the aluminum phase plate depended on the X-ray energy, which was several micrometers in this experiment. An X-ray CCD camera (Hamamatsu C4880, CCD: TI TC-215, pixel size:12 μm) or a nuclear emulsion plate (Fuji EM C-OC 15) was used as a detector. The latter plate whose resolving power is approximately 1μm was used to obtain a high resolution X-ray image. The optical elements and specimens were set in air and the optical path between the phase plate and the direct beam stop was replaced with helium.

EXPERIMENTS

Resolution Test

The performance of the X-ray microscope was evaluated with a tantalum test pattern fabricated on a 2μm thick Si_3N_4 membrane. The pattern had line structures with line-and-space widths of 0.05 μm, 0.1 μm, 0.2 μm, 0.3 μm and 0.4 μm, which are shown in Fig.2 (b). The thickness of the tantalum test pattern was 0.5 μm, which corresponded to the transmission of 91% for 9 keV X-rays. The 5 μm thick aluminum phase plate having a 5 μm diameter pinhole was used to obtain a phase contrast image. The 5 μm thick aluminum plate corresponded to the transmission of 96% and the phase shift of a quarter wavelength at 9keV. Figure 2(a) shows an X-ray micrograph of the phase contrast image recorded with the nuclear emulsion plate. The line-and-space widths of the image are 0.1 μm, 0.2 μm and 0.3 μm. The exposure time was approximately 10min with the nuclear emulsion plate. From the figure the 100nm line-and-space pattern could be resolved, which was reasonable for the outermost zone width. The thinner pattern of the vertical lines in the image might come from the misalignment of the phase plate or the different spatial coherence between vertical and horizontal directions. The spatial coherence of the vertical direction was better than that of the horizontal direction.

Observation of Biological Specimen (Artemia)

In the present optical system the field of view was limited by either the diameter of the entrance pinhole or the half size of the zone plate. This means that the field of view of the X-ray microscope was approximately 70 μm or so.

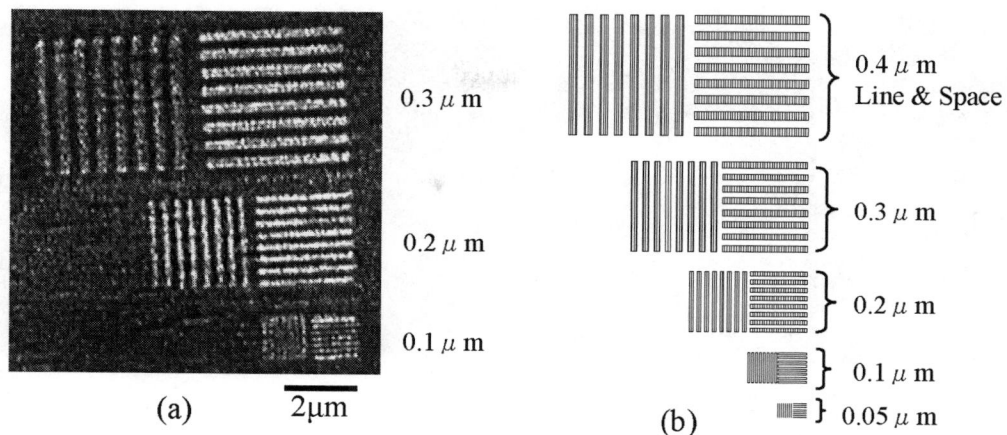

(a) 2μm

0.3 μ m

0.2 μ m

0.1 μ m

(b)

0.4 μ m
Line & Space

0.3 μ m

0.2 μ m

0.1 μ m

0.05 μ m

FIGURE 2. X-ray micrographs of the test pattern. (a) phase contrast image and (b) parameters of line-and-space patterns.

In order to observe the specimen larger than 70 μm in diameter we expanded the effective area of observation by making a montage picture. The montage picture was composed of the multiple images which had been picked up from the fixed rectangular area of a CCD image at each exposure. Only the specimen was scanned for two dimensions. After many exposures each image was displayed for the two dimensions and then a montage picture was obtained. The following pictures were made by this technique.

We selected an egg of an artemia as a specimen to observe several stages of its growth. An artemia is a famous plankton whose egg is able to live in a dry state for a long time. The size of the egg is approximately 300 μm in diameter. When it is immersed in seawater a cyst-like egg hatches within a few hours, and grows to a larva in a day or so. Figures 3(a) and 3(b) are the absorption and the phase contrast X-ray micrographs of an artemia's cyst-like egg. Each exposure was 6 sec at 8 keV. Much larger contrast enhancement was achieved by the phase contrast imaging than that by the absorption contrast one. The bubble-like patterns observed in Fig. 3(a) are parts of the resin bond used to fix the specimen. Discontinuous contrast between adjacent areas may be mainly due to the statistic of the X-ray intensity and the image processing of the CCD camera.

As mentioned above, since an artemia grows to a larva in a day the time lapse observation of its inner morphology is very interesting. Being difficult to keep one egg alive during the exposure we grew several eggs simultaneously for observation. One montage picture was obtained within ten minutes or so. The phase contrast X-ray micrographs of three different stages are shown in Fig. 4. The time lapses of these three images of Fig.4(a), (b) and (c) are 0 h, 2 h and 11 h, respectively. In the dry condition, since the egg might not contain water the shape of the egg was not round at first, which is seen in Fig. 4(a). Soon after having been immersed in water it swelled out due to the water, which is shown in Fig. 4(b).

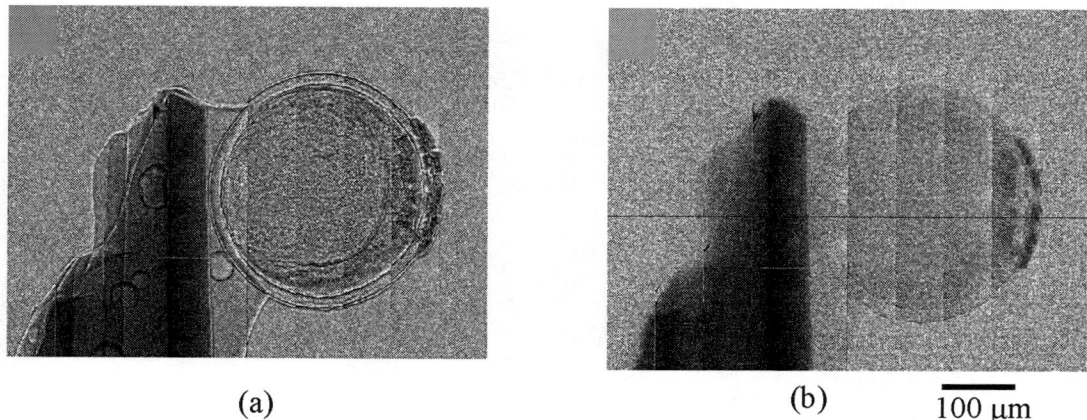

(a)

(b) 100 μm

FIGURE 3. X-ray micrographs of an artemia's cyst-like eggs. (a) phase contrast image and (b) absorption contrast image.

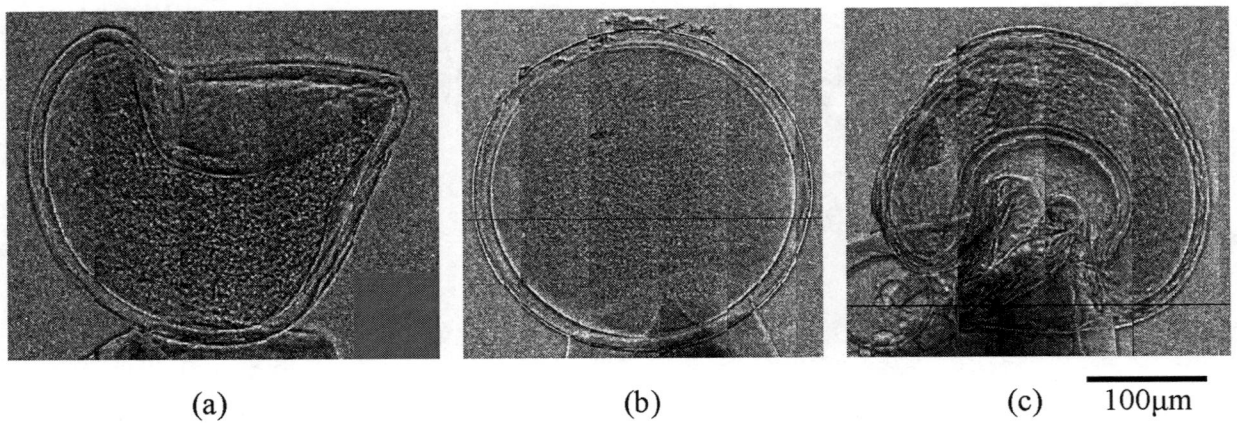

(a)   (b)   (c)   100μm

FIGURE 4. Time lapse X-ray micrographs of an artemia egg. (a) 0 h (b) 2 h and (c) 11 h.

A dramatic change of the form of the cyst-like egg occurred after 11 hours, which is seen in Fig.4 (c). Clear changes of the inner structure could be seen by the observations of every 2 hours. Identification of the parts of the organs is under investigation.

CONCLUSION

The phase contrast X-ray images of the eggs of artemia were successfully obtained by the montage picture. The inner structures of their different growing stages were also observed by hatching several eggs simultaneously. Observation of the growth of one specific egg will be possible in the near future.

ACKNOWLEDGMENTS

This work was partially supported by a Grant-in-Aid for Scientific Research Fund from the Ministry of Education, Culture, Sports, Science and Technology (No. 16206007).

REFERENCES

1. G. Schmahl, D. Rudolph, P. Guttmann, G. Schneider, J. Thieme and B. Niemann, *Rev. Sci. Instrum.*, **66**, 1282-1286 (1995).
2. Y. Kagoshima, T. Ibuki, Y. Yokoyama, Y. Tsusaka, J. Matsui, K. Takai and M. Aino, *Jpn. J. Appl. Phys.* **40**, L1190-L1192 (2001).
3. H. Yokosuka, N. Watanabe, T. Ohigashi, Y. Yoshida, S. Maeda, S. Aoki, Y. Suzuki, A. Takeuchi and H. Takano, *J. Synchrotron Rad.* **9**, 179-181 (2000).
4. H. Yokosuka, N. Watanabe, T. Ohigashi, S. Aoki and M. Ando, "Phase-contrast Hard X-ray Microscope with a Zone Plate at the Photon Factory" in *X-Ray Microscopy*, edited by J. Susini et al., Journal de Physique IV Proceedings, EDP Sciences, Les Ulis, 2003, pp. 591-594.

X-ray Phase Microtomography
by Single Transmission Grating

Yoshihiro Takeda[1], Wataru Yashiro[2], Yoshio Suzuki[3] and Atsushi Momose[2]

[1]*Graduate School of Pure and Applied Sciences, University of Tsukuba,*
1-1-1 Tennodai, Tsukuba, Ibaraki 305-8573, Japan
[2]*Department of Advanced Materials, Graduate School of Frontier Sciences, The University of Tokyo,*
5-1-5 Kashiwanoha, Kashiwa, Chiba 277-8561, Japan
[3]*Japan Synchrotron Radiation Research Institute, SPring-8,*
1-1-1 Kouto, Sayo-cho, Sayo-gun, Hyogo 679-5198, Japan

Abstract. A preliminary experiment of X-ray phase microtomography by a single phase grating is reported. A phase grating was placed behind an object and illuminated by spatially coherent X-rays. At a specific distance from the grating, a periodic intensity pattern caused by the fractional Talbot effect was recorded with a high spatial-resolution image detector. A differential phase map related to the object was retrieved from the deformation in the periodic intensity pattern on the basis of the fringe scanning method. Phase tomograms of a piece of polymer blend were reconstructed and a phase-separation structure in the blend was successfully resolved.

Keywords: Phase, Tomography, Interferometer, Imaging, Grating
PACS: 07.05.Pj, 07.85.Tt, 41.50.+h. 42.25.Hz, 42.30.-d, 42.30.Rx, 42.30.Wb

INTRODUCTION

Because of the penetrating power of X-rays, X-ray absorption imaging makes it possible to observe internal structures of objects nondestructively. However, it is difficult to depict weakly absorbing structures in an object consisting of light elements. When X-rays pass through the object, however, quite a large phase shift occurs. Therefore X-ray phase imaging that measures the phase distribution of the X-rays, can realize high sensitive internal observation nondestructively.

Some methods for X-ray phase imaging was proposed in the middle of 1990s [1-3]. Recently, new X-ray interferometers consisting of X-ray gratings have been proposed and demonstrated [4-8]. The X-ray Talbot effect is used in the interferometers. When a grating is illuminated by spatially coherent X-rays, a periodic intensity pattern appears at a specific distance from the grating. In the X-ray Talbot interferometer [5], an absorption grating is overlaid on the periodic intensity pattern and generated moiré fringes are recorded. The pitch of the absorption gratings is made to be smaller than or comparable to the spatial coherence length of the incident X-rays. At the same time, it is required that the absorption grating is sufficiently thick to absorb the X-rays. This implies that a grating with a high aspect is required, but the fabrication of such a grating is not straightforward. On the other hand, the thickness of an X-ray phase grating can be much smaller than that of an absorption grating. Therefore it is comparatively easy to fabricate X-ray phase gratings. We propose a new method for X-ray phase imaging by a single phase grating that measures the periodic intensity pattern caused by the fractional Talbot effect with a high spatial-resolution detector instead of placing an amplitude grating on it.

PRINCIPLE

Figure 1 shows the experimental set up of the proposed method. A phase grating with a pitch d is placed on the x-y plane ($z = 0$) so that the grooves of the grating are parallel to the y axis. When the phase grating is illuminated by

CP879, *Synchrotron Radiation Instrumentation: Ninth International Conference,*
edited by Jae-Young Choi and Seungyu Rah
© 2007 American Institute of Physics 978-0-7354-0373-4/07/$23.00

spatially coherent X-rays, a periodic intensity pattern appears at the downstream of the grating. Its intensity distribution $I(x,y,z)$ is written by

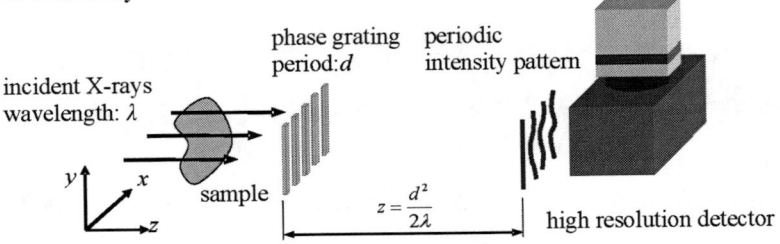

FIGURE 1. Experimental set up of the proposed technique.

$$I(x,y,z) = \sum_n b_n(z)\exp\left(2\pi i \frac{n}{d}x\right) \tag{1}$$

where $b_n(z)$ is the nth Fourier coefficient which depends on the complex transmission function of the phase grating. For a phase grating with a transmission function $T(x,y)=\exp[i\phi(x,y)]$, the periodic intensity pattern at $z = d^2/2\lambda$ is given by [9]

$$I\left(x,y,\frac{d^2}{2\lambda}\right) = 1 + \sin\left[\phi(x,y) - \phi\left(x+\frac{d}{2},y\right)\right] \tag{2}$$

In the case of a $\pi/2$ grating, the visibility of the pattern reaches 100% at $z = d^2/2\lambda$.

If an object with a phase shift $\Phi(x,y)$ is put in front of the grating, $I(x,y,z)$ is given by

$$I(x,y,z) = \sum_n b_n(z)\exp\left(2\pi i \frac{n}{d}\left[x - \frac{\lambda z}{2\pi}\frac{\partial\Phi(x,y)}{\partial x}\right]\right) \tag{3}$$

which indicates that the periodic intensity pattern is distorted by the object. A differential phase map $\partial\Phi(x,y)/\partial x$ can be retrieved by the Fourier transform method or the fringe scanning method from the deformed periodic intensity pattern. By the Fourier transform method, the spatial resolution of the retrieved image is limited by the pitch of carrier fringes; that is, the period of the pattern in this case. On the other hand, the fringe scanning method has no limit with respect to the period in principle, and therefore we adopted the fringe scanning methods.

When the phase grating is displaced by the distance kd/M along the x direction, where k is an integer, the intensity distribution is given by

$$I_k(x,y,z) = \sum_n b_n(z)\exp\left(2\pi i \frac{n}{d}\left[x - \frac{\lambda z}{2\pi}\frac{\partial\Phi(x,y)}{\partial x} + \frac{d}{M}k\right]\right) \tag{4}$$

In the fringe scanning method, the fringe of a sinusoidal profile is normally assumed. Higher orders ($n \geq 2$) in Eq. (4) are therefore the sources of error. However, by adequately selecting a large number for M, the error induced by the harmonics can be reduced [10]. We used 5-step fringe scan since the lowest order that causes error is 9th [7]. Because the magnitudes of such high orders are normally negligible, one can use a relation

$$2\pi\frac{n}{d}\left[x-\frac{\lambda z}{2\pi}\frac{\partial\Phi(x,y)}{\partial x}\right]\approx\arg\left(\frac{\sum\limits_{k=1}^{5}I_k(x,y,z)\sin\!\left(2\pi\dfrac{k}{5}\right)}{\sum\limits_{k=1}^{5}I_k(x,y,z)\cos\!\left(2\pi\dfrac{k}{5}\right)}\right). \tag{5}$$

to calculate a differential phase map $\partial\Phi(x,y)/\partial x$. An X-ray phase map is obtained through operations of unwrapping and integration on the differential phase map. A phase tomogram can be reconstructed from multiple phase maps measured at various angular positions of the sample rotation.

EXPERIMENT

The experiment was performed at BL20XU of SPring-8. X-rays emitted from an undulator was monochromatized to 17.7keV by a Si double-crystal monochromator. We used a phase grating with a gold pattern 8 μm in pitch and 2 μm in height fabricated by UV lithography and electro-plating. The pattern height corresponded to the amount for a π/2 phase shift. We used a CCD camera with a phosphor screen and a coupling lens. The effective pixel size was 1.0 μm and the field of view was about 1.3 mm × 1.0 mm. The camera was placed at $z = d^2/2\lambda$ and generated periodic intensity pattern was measured with it. In front of the grating, a piece of polystyrene (PS) / poly (methyl methacrylate) (PMMA) polymer blend was put as a sample. The sample had phase separation structures consisting of PS-rich phase and PMMA-rich phase. Figure 2(a) shows a periodic intensity pattern obtained by an exposure of 1.5 seconds. An enlarged image of the area indicated with broken lines in Fig. 2(a) is shown in Fig. 2(b), where the distortion of fringes is clearly seen. Five distorted patterns were measured by displacing the phase grating along the x direction by a one-fifth of the pitch. A differential phase map was calculated from the patterns, as shown in Fig. 2(c).

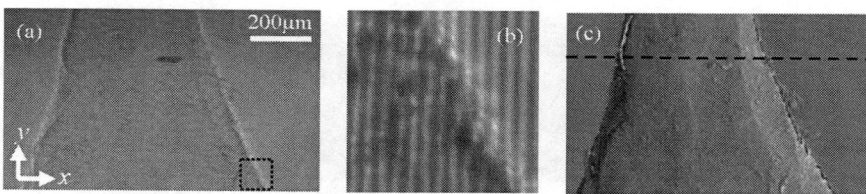

FIGURE 2. Distorted intensity pattern caused by a piece of polymer blend and calculated differential phase map. A close-up image of the region indicated by broken-line square in (a) is shown in (b). (c) is the differential phase map calculated on the basis of the fringe scanning method.

The sample was rotated in a 0.72-degree step over 180 degrees and 250 phase maps were obtained. Figure 3(a) shows a reconstructed phase tomogram at the position of the broken line in Fig. 2(c), and Fig. 3(b) shows a three-dimensional rendering view. The pixel values in reconstructed phase tomogram indicate the refractive index difference between the object and the surrounding air. The refractive index difference between the PS-rich phase and air was 7.5×10^{-7}, and between PMMA-rich phase and air was 8.5×10^{-7}. The detection limit of the refractive index deviation was evaluated to be 1×10^{-8} from the standard deviation of refractive index in the PMMA-rich phase. The spatial resolution in the phase tomogram was evaluated to be 7.5 μm by the FWHM of differential profile across the boundary between the PS-rich and PMMA-rich phases.

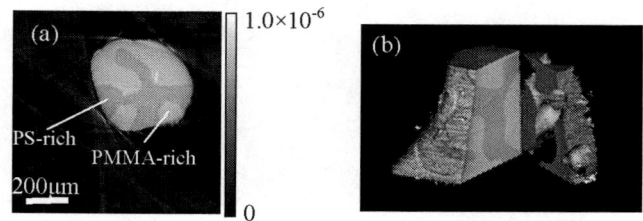

FIGURE 3. Reconstructed phase tomogram of a piece of polymer blend; (a) one of axial sections and (b) a three-dimensional rendering view, where the PS-rich region in the right-side part has been made transparent

DISCUSSION

The detection limit of the refractive index deviation by this technique was worse than that of other techniques [7, 11]. The visibility of the measured periodic intensity pattern was only 8%, which might reflect on the result. The low visibility is considered to be due to the MTF of the detector. Using a detector with a better MTF will raise the visibility of measured pattern and therefore improve the sensitivity to refractive index deviation. If a phase grating with a period larger than that used in this study can be used within the permitted range determined by the spatial coherency, the sensitivity to refractive index deviation will be improved as well.

The spatial resolution of the fringe scanning method is limited by the pixel size in principle. Although presented result exceeded the limit of the sampling theorem in the case of using the Fourier-transform method, the resultant value 7.5 μm was much worse than the pixel size of the image detector used. However, this result was predicted because the edge-contrast as seen in Fig. 2 caused by Fresnel diffraction is unavoidable. The amount of the blur (or the thickness of the edge contrast) is estimated to be $\sqrt{\lambda z} = 5.6$ μm in this case. When the object can be placed downstream of the grating, this blur will be reduced. However, it should be noted that the sensitivity is proportional to distance between the object and the detector, and therefore one has to compromise the decrease in sensitivity in this case. The fractional Talbot effect occurs when the phase grating is illuminated by spherical X-rays. Therefore, another approach for reducing the blur due to Fresnel diffraction is found by combining this method with X-ray imaging microscopy. Provided that an image detector of a wide field of view compatible with a small pixel size in the future, the presented technique would be an approach for attaching a phase-contrast mode in a simple way. In addition, we consider that the combination with an X-ray imaging microscope is an important future direction of the presented method.

CONCLUSION

X-ray phase microtomography with a single phase grating was performed at BL20XU of SPring-8. An Au phase grating was placed behind an object and illuminated by spatially coherent X-rays. Periodic intensity pattern caused by the fractional Talbot effect was measured by a high resolution image detector. A differential phase map was retrieved by the fringe scanning method and phase tomograms were reconstructed. The combination with X-ray microscopy will be a next step, providing a new type differential phase microscopy, with which phase nanotomography would be attainable.

ACKNOWLEDGEMENT

The experiment using synchrotron radiation was performed under the approval of SPring-8 committee 2006A1237-NM-np. This study was financially supported by the project "Development of System and Technology for Advanced Measurement and Analysis" of Japan Science and Technology Agency (JST).

REFERENCES

1. A. Momose, *Nucl. Instrum. Methods* **352**,622-628(1995)
2. T. J. Davis, D. Gao, T. E. Gureyev, A. W. Stevenson and S. W. Wilkins: *Nature* **373**,595-598(1995)
3. A. Snigirev, I. Snigireva, V. Kohn, S. Kuznetsiv and I. Schelov, *Rev. Sci. Instrum.* **66**, 5486-5492 (1995)
4. C. David, B. Nöhammer, H. H. Solak and E. Ziegler: *Appl. Phys. Lett.* **81**, 3287-3289(2002)
5. A. Momose, S. Kawamoto, I. Koyama, Y. Hamaishi, K. Takai and Y. Suzuki: *Jpn. J. Appl. Phys.* **42**, L866-L868(2003)
6. T. Weitkamp, B. Nöhammer, A. Diaz, C. David, F. Pfeiffer, M. Stampanoni, P. Cloetens and E. Ziegler: *Appl. Phys. Lett.* **86**, 054101(2005).
7. A. Momose, S. Kawamoto, I. Koyama and Y. Suzuki, *SPIE Proc.* **5535** , 352-360 (2004)
8. E. Pfeiffer, T. Weitkamp, O. Buck and C. David, *Nature Phys.* **2**, 258-261(2006)
9. J. P. Guigay: *Opt. Acta* **18**, 677-682(1971)
10. K. A. Stetson and W. R. Brohinsky: *Appl. Opt.* **24**, 3631-3637(1985)
11. A. Momose, A. Fujii, H. Kadowaki and H. Jinnai: *Macromolecules* **38**, 7197-7200(2005)

Phase Tomography Using X-ray Talbot Interferometer

A. Momose,[1] W. Yashiro,[1] Y. Takeda,[2] M. Moritake,[1]
K. Uesugi,[3] Y. Suzuki,[3] and T. Hattori[4]

[1]*Graduate School of Frontier Sciences, The University of Tokyo,*
5-1-5 Kashiwanoha, Kashiwa, Chiba 277-8561, Japan
[2]*Graduate School of Pure and Applied Sciences, University of Tsukuba,*
1-1-1 Tennodai, Tsukuba, Ibaraki 305-8573, Japan
[3]*SPring-8/JASRI, 1-1-1 Kouto, Mikazuki, Hyogo 679-5198, Japan*
[4]*Laboratory of Advanced Science and Technology for Industry, University of Hyogo,*
3-1-2 Kouto, Kamigori, Hyogo 678-1205, Japan

Abstract. A biological tomography result obtained with an X-ray Talbot interferometer is reported. An X-ray Talbot interferometer was constructed using an amplitude grating fabricated by X-ray lithography at the LIGA beamline of NewSUBARU and gold electroplating. The pitch and pattern thickness of the grating were 8 μm and 30 μm, respectively. The effective area was 20×20 mm^2, which was entirely illuminated with a wide beam available at the medium-length beamline 20B2 of SPring-8, allowing the acquisition of a three-dimensional tomogram of almost the whole body of a fish. The resulting image obtained with 17.7 keV X-rays revealed organs with bones in the same view.

Keywords: Tomography, Phase, Talbot effect, Talbot interferometer, Grating, High aspect ratio.
PACS: 07.05.Pj, 07.85.Tt, 41.50.+h, 42.25.Hz, 42.30.-d, 42.30.Wb,

INTRODUCTION

Recently, X-ray transmission gratings have been implemented in novel phase-sensitive X-ray imaging [1-6]. Differential phase contrast can be generated by aligning two transmission gratings on an optical axis with a specific separation determined by the pitch of the gratings and X-ray wavelength. The first grating generates self-images that exhibit periodical intensity patterns corresponding to the transmission function of the grating. The self-images are induced by Fresnel diffraction by the grating, which is known as the fractional Talbot effect. When a phase object is placed in the path of X-rays, X-ray refraction at the phase object is reflected on the deformation of the self-images. The second grating, which is of the amplitude type with a pitch almost the same as the average period of a self-image and is placed at the position of the self-image, generates a moiré pattern by superposition with the deformed self-image. This system is known as a Talbot interferometer [7,8].

Because of its use of grating optics, Talbot interferometry has many advantages over other phase-sensitive X-ray imaging methods. One advantage is that the setting up of a Talbot interferometer is easy because gratings can be aligned with an accuracy corresponding to their pitch, which is set to be comparable to the X-ray spatial coherence length typically on the micrometer order. Another advantage is that a cone beam with a broad energy bandwidth (approx. $E/\Delta E > 10$) can be used. In addition, it should be emphasized that the quantitative measurement of a differential phase image is feasible, enabling phase tomography.

In this paper, we describe a biological imaging result obtained by phase tomography using the X-ray Talbot interferometer at SPring-8. An amplitude grating 8 μm in pitch and 30 μm in thickness was fabricated by means of X-ray lithography and gold electroplating. Its effective area was 20×20 mm^2, which was covered with the wide beam of the beamline 20B2.

CP879, *Synchrotron Radiation Instrumentation: Ninth International Conference*,
edited by Jae-Young Choi and Seungyu Rah

X-RAY TALBOT INTERFEROMETER

Figure 1 shows a calculated X-ray intensity downstream from a $\pi/2$ phase grating under plane-wave illumination of X-rays of wavelength λ. The spatial coherence length of the X-rays at the grating was assumed to be almost $3d$, where d is the pitch of the grating. At the positions given by $m\lambda/4d^2$, where m is an integer other than those which are multiples of four, rectangular periodic patterns (self-images) of period d are formed in a process known as the fractional Talbot effect. The visibility of the self-images decreases along the optical axis owing to the finite spatial coherence length.

When a phase object is placed in front of the grating, the self-images are deformed due to the refraction at the phase object. The amount of deformation is proportional to the distance the self-image from the grating. In a Talbot interferometer, an amplitude grating of period d is placed at the position of one of the self-images. Then, the deformation of the self-image is visualized as a moiré pattern induced by the superposition of the self-image and amplitude grating (see Fig. 2). In the experiment presented in this paper, we selected the position corresponding to $m = 2$, where the visibility of the self-image is at its maximum.

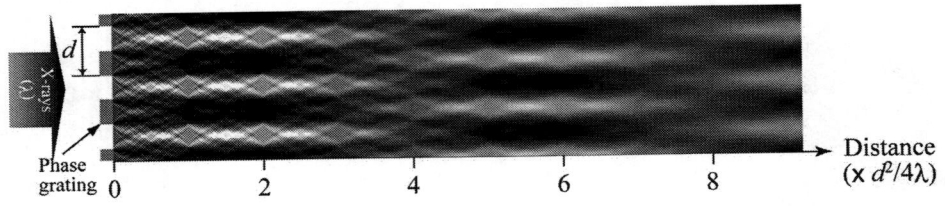

FIGURE 1. Calculation of intensity downstream of $\pi/2$ phase grating induced by fractional X-ray Talbot effect under assumption of illumination of nearly plane-wave X-rays with spatial coherence length of $3d$.

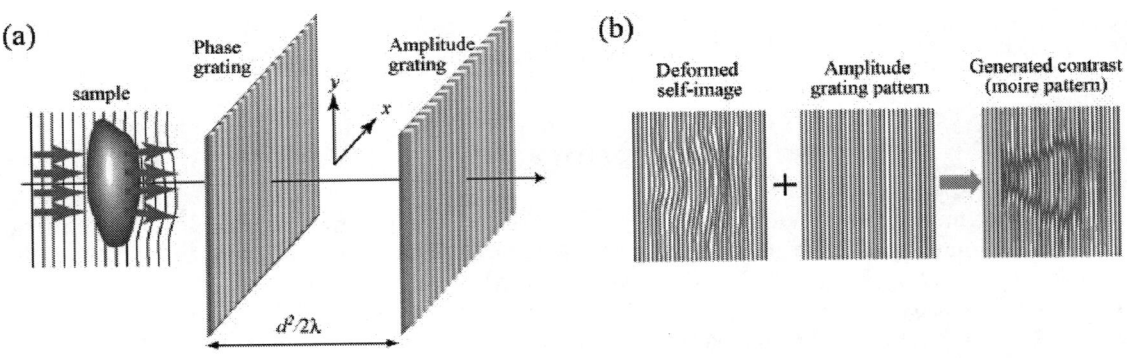

FIGURE 2. Configuration of Talbot interferometer (a) and schematic diagram (b) explaining moiré pattern generation.

The second grating used in the Talbot interferometer must, in principle, be of the amplitude type. The pattern of the grating should therefore be thick enough to absorb X-rays fully. Gold is a suitable material for this grating, considering its absorption coefficient and fabrication process. Nevertheless, a thickness much larger than 10 μm is required. The pitch of the grating should be comparable to or smaller than the spatial coherence length of X-rays to induce the fractional Talbot effect. The spatial coherence length of hard X-rays from a storage ring is typically 10 μm. Therefore, a pattern of high aspect ratio should be fabricated. We used such an amplitude grating fabricated by X-ray lithography and gold electroplating.

A 30-μm-thick X-ray resist film (MAX001, Nagase ChemteX) was spin-coated on a 200-μm-thick Si wafer with a 0.25-μm-thick Ti layer. The beamline 11 of NewSUBARU, Japan, which is dedicated to Lithographie Galvanoformung Abformung (LIGA) fabrication, was used for X-ray exposure through an X-ray mask of an 8-μm-pitch 1:1 line-and-space pattern. After developing, gold lines were formed by electroplating between resist lines, which were left after the electroplating to support the gold lines. The height of the gold lines was nearly 30 μm, and the effective area of the grating was 20×20 mm². The first grating was of the $\pi/2$ phase type for 20 keV X-rays. The basic performance of the set of gratings was reported elsewhere [6]: the visibility of moiré fringes was about 80% at 20 keV; it was greater than 30% at 31 keV.

A wide beam of the beamline 20B2 of SPring-8, Japan through a Si 111 double-crystal monochromator was used 206 m from a bending section, and the entire effective area of the gratings was illuminated by the beam. The spatial coherence length of the X-rays at the interferometer was calculated to be about 20 μm.

A CCD camera (C4742-95HR, Hamamatsu Photonics) lens-coupled with a phosphor screen was used to record moiré fringes. Its effective pixel size was 11.8 μm × 11.8 μm. A cell filled with formalin was placed close to the upstream side of the phase grating, and a sample was immersed in formalin. For tomography, the sample was rotated about the vertical axis. The lines of the gratings were also aligned vertically (parallel to the y-axis in Fig. 2(a)), such that the refraction in the horizontal plane was sensed.

Quantitative measurement of the angular beam deflection induced by the refraction at the sample, that is, a differential phase image, was carried out by observing the change in moiré patterns with the displacement of the amplitude grating in the x-direction (see Fig. 2(a)). The step of the displacement was $d/5$ and five moiré patterns were used to calculate a differential phase image using the fringe scanning method [9]. This measurement was repeated for every angular position of the sample rotation in tomography.

PHASE TOMOGRAM

Figure 3(a) shows the result of the measurement of a differential phase image of a fish (*Hasemania nana*) 3 cm in length obtained with 17.7 keV X-rays. This image was calculated from five moiré patterns, one of which is shown in Fig. 3(b). The contribution of moiré fringes appearing in the background in Fig. 3(b), which was mainly due to the pitch mismatch between the amplitude grating and the self-image, was removed using the data obtained in the absence of the sample.

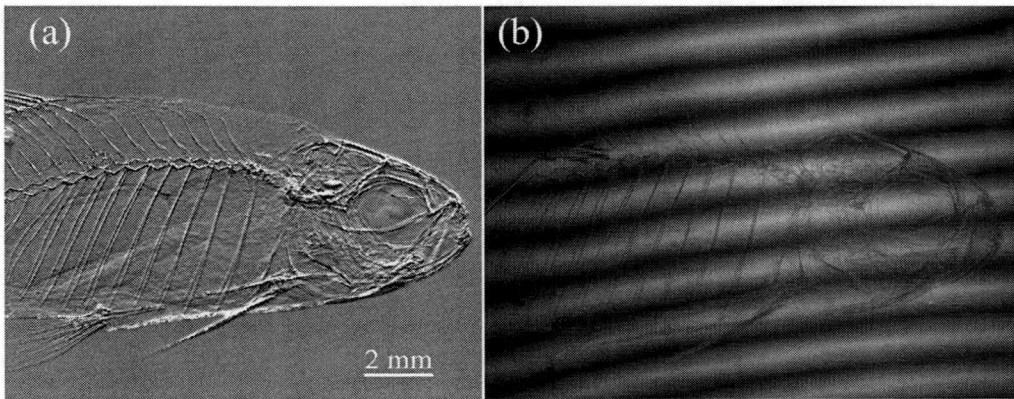

FIGURE 3. Differential phase image (a) of fish (*Hasemania nana*) calculated from a series of moiré patterns, one of which is shown in (b), recorded by fringe scanning. The X-ray energy was 17.7 keV. The vertical direction in these pictures corresponds to the x-direction in Fig. 2(a).

The angular step of the sample for phase tomography was 0.36°, and 500 differential phase images were recorded. The exposure time for obtaining a moiré image was 8.2 s. Before the tomographic reconstruction using the convolution-backprojection algorithm, each differential phase image was converted to a phase image through integration. Figure 4 shows sectional images generated using reconstructed three-dimensional data. Although artifacts remain because strong refractions at the bones and body surface could not be processed correctly within the assumption in the theory of X-ray Talbot interferometry that we used [2,9], organ structures are clearly revealed with bones in the same view, showing a wide dynamic range.

CONCLUSION

Thus, X-ray Talbot interferometry exhibited an excellent sensitivity to soft structures. As for the spatial resolution, the pitch of the gratings gives a limit. Because a cone-beam is available with an X-ray Talbot interferometer, a next possible progress is in combining such an interferometer with a normal X-ray imaging microscope as an attachment for use in a phase-contrast mode. Another possibility of progress, with which X-ray Talbot interferometry will prove its full merit, is the development of an X-ray imaging apparatus for practical applications such as medical diagnosis not only with synchrotron radiation sources but also with compact X-ray sources. For this purpose, we need to move to a higher-energy region, and the thickness and effective area of the amplitude grating should therefore be increased.

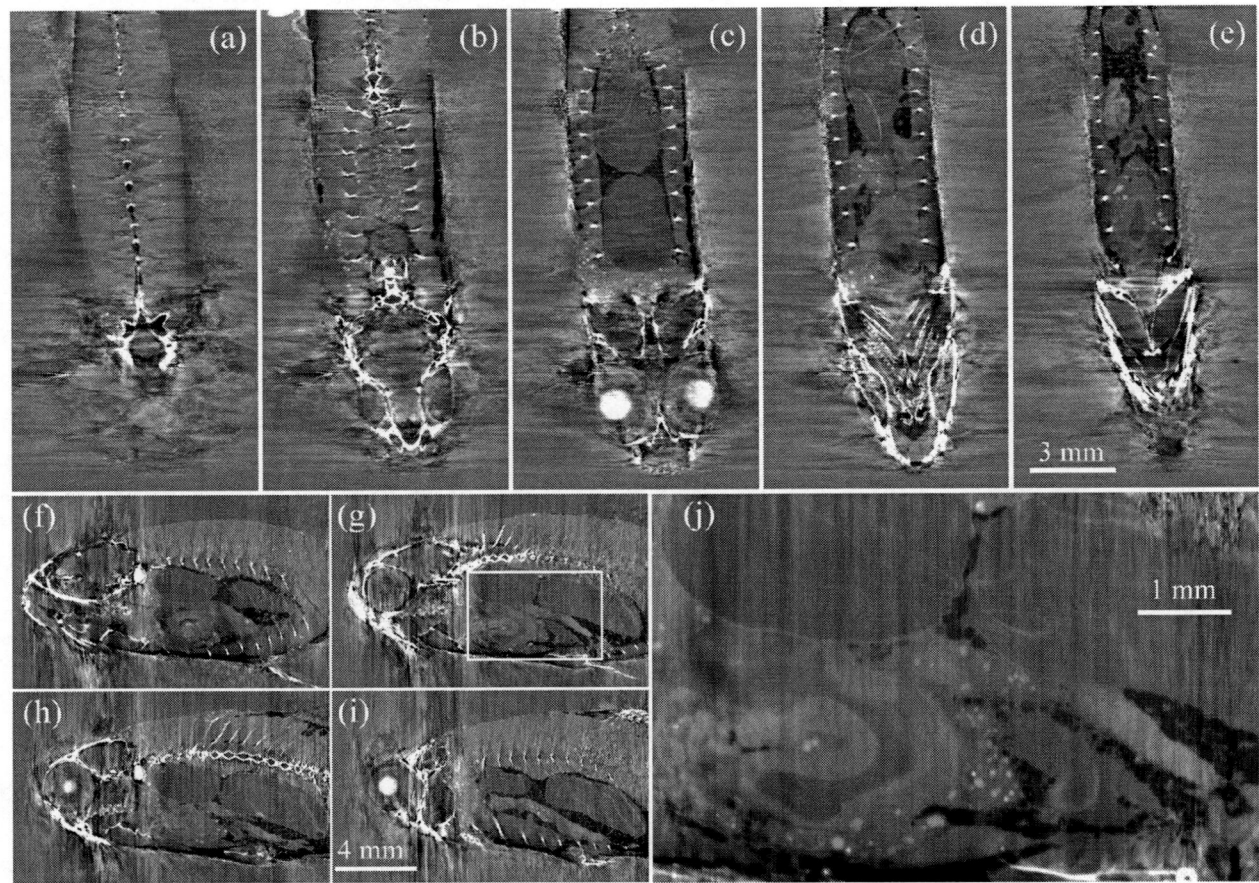

FIGURE 4. Phase tomograms of fish (*Hasemania nana*): coronal views (a)-(e) from back to stomach at 1.18 mm intervals and sagittal views (f)-(i) at 0.59 mm intervals. (j) is a magnified image of the area indicated by a rectangle in (g). Images map the refractive index difference from that of formalin ranging from -1×10^{-7} to 1×10^{-7}.

ACKNOWLEDGMENTS

The experiment using synchrotron radiation was performed under the approval of the SPring-8 committee 2006A1237-NM-np. This study was financially supported by the project "Development of System and Technology for Advanced Measurement and Analysis" of Japan Science and Technology Agency (JST).

REFERENCES

1. C. David, B. Nöhammer, H. H. Solak, and E. Ziegler, *Appl. Phys. Lett.* **81**, 3287-3289 (2002).
2. A. Momose, S. Kawamoto, I. Koyama, Y. Hamaishi, K. Takai, and Y. Suzuki, *Jpn. J. Appl. Phys.* **42**, L866-L868 (2003).
3. T. Weitkamp, B. Nöhammer, A. Diaz, and C. David, *Appl. Phys. Lett.* **86**, 054101 (2005).
4. T. Weitkamp, A. Daiz, C. David, F. Pfeiffer, M. Stampanoni, P. Cloetens, and E. Ziegler, *Opt. Express* **13**, 6296-6304 (2005).
5. F. Pfeiffer, T. Weitkamp, O. Buck, and C. David, *Nature Phys.* **2**, 258-261 (2006).
6. A. Momose, W. Yashiro, Y. Takeda, Y. Suzuki, and T. Hattori, *Jpn. J. Appl. Phys.* **45**, 5254-5262 (2006).
7. S. Yokozeki and T. Suzuki, *Appl. Opt.* **10**, 1575-1580 (1971).
8. A. W. Lohmann and D. E. Silva, *Opt. Commun.* **2**, 413-415 (1971).
9. A. Momose, S. Kawamoto, I. Koyama, and Y. Suzuki, *SPIE Proc.* **5535**, 352-360 (2004).

Three-Dimensional Observation of Polymer Blends with X-ray Phase Tomography

N. Higuchi,[1] A. Momose,[1] W. Yashiro,[1] Y. Takeda,[2]
H. Jinnai,[3] Y. Nishikawa,[3] and Y. Suzuki[4]

[1]*Department of Advanced Materials Science, Graduate School of Frontier Sciences, The University of Tokyo,
5-1-5 Kashiwanoha, Kashiwa, Chiba 277-8561, Japan*
[2]*Graduate School of Pure and Applied Sciences, University of Tsukuba,
1-1-1 Tennodai, Tsukuba, Ibaraki 305-8573, Japan*
[3]*Department of Polymer Science and Engineering, Kyoto Institute of Technology,
Matsugasaki, Sakyoku, Kyoto 606-8585, Japan*
[4]*SPring-8/JASRI, 1-1-1 Kouto, Sayo-cho, Sayo-gun, Hyogo 679-5198, Japan*

Abstract. We performed a quantitative three-dimensional observation of polystyrene (PS)/poly(methyl methacrylate) (PMMA) blends by X-ray phase tomography mainly using the crystal X-ray interferometer at the beamline 20XU of SPring-8, Japan. We observed the time change in the phase-separated structures of PS/PMMA blends annealed at 180°C. We also measured the densities of PS- and PMMA-rich regions in a series of PS/PMMA blends annealed in the temperature range from 220 to 300°C and determined part of the phase diagram of these blends. Finally, we performed a preliminary trial of X-ray phase tomography using an X-ray Talbot interferometer, which will facilitate better control of the environment around a sample than that possible with a crystal X-ray interferometer.

Keywords: tomography, X-ray interferometer, polymer blends, phase separation, phase diagram.
PACS: 07.05.Pj, 07.85.Tt, 41.50.+h, 42.25.Hz, 42.30.-d, 42.30.Wb, 83.80.Tc

INTRODUCTION

To satisfy the increasing needs for new materials with specific properties, polymer blend techniques have been developed. Blending different polymers can create new properties that can never be exhibited by single polymers. Phase separation occurs by blending different polymers with characteristic structures. Thus, it is crucial to observe the three-dimensional phase-separated structures of polymer blends to understand the mechanical properties of these blends. Conventionally, the phase-separated structures of thin polymer blends are observed using an optical microscope or an electron microscope in two dimensions. However, the observation of three-dimensional structures, such as the curvature of an interface, is required to evaluate the strength or elasticity of polymer blends.

Three-dimensional observation techniques of polymer blends, such as laser scanning confocal microscopy [1], transmission electron microtomography [2], and three-dimensional nuclear magnetic resonance imaging [3], have been reported. These techniques require certain treatments for contrast enhancement such as labeling with heavy elements or selective etching. However, such treatments may change the property of polymer blends. Recently, X-ray phase tomography has been used for the three-dimensional observation of polymer blends without treatments for contrast enhancement [4]. X-ray phase tomography exhibits an approximately 1000-fold higher sensitivity for materials consisting of light elements than X-ray absorption tomography, and has been used for biological imaging [5-7]. In addition, X-ray phase tomography enables the measurement of the density of each phase-separated region, and therefore it is considered that the determination of the phase diagram of polymer blends would be feasible if samples annealed at various temperatures are systematically measured.

In this study, first, we analyzed the time change in the phase-separated structures, which were revealed by X-ray phase tomography using a crystal X-ray interferometer [8], of PS/PMMA blends annealed at 180°C. Next, we attempted at determining the phase diagram of a PS/PMMA blend by a composition analysis of PS and PMMA within PS-rich and PMMA-rich regions reconstructed in X-ray phase tomograms. Finally, we performed X-ray

CP879, *Synchrotron Radiation Instrumentation: Ninth International Conference*,
edited by Jae-Young Choi and Seungyu Rah
© 2007 American Institute of Physics 978-0-7354-0373-4/07/$23.00

phase tomography preliminarily using an X-ray Talbot interferometer (an X-ray differential interferometer) [9]. Although the sensitivity of X-ray phase tomography using an X-ray Talbot interferometer is inferior to that using a crystal X-ray interferometer, an X-ray Talbot interferometer provides a wide work space for a sample, enabling the controls of various factors such as temperature, pressure, and stress in the future.

SAMPLE PREPARATION

In this study, we used PS/PMMA blends as samples. PS and PMMA were obtained from Polymer Source, Inc. The weight-average molecular weight (M_w) and coefficient of dispersion (M_w/M_n) of PS were 76 500 and 1.04, respectively. The M_w and M_w/M_n of PMMA were 33 200 and 1.08, respectively. The densities of PS and PMMA were 1.05 and 1.18 g/cm^3, respectively. A solution of 50 vol% PS and 50 vol% PMMA in benzene and containing 5 wt% polymer was freeze-dried to obtain PS/PMMA powder. The PS/PMMA powder was annealed in 2.5-mm-diameter and 5-mm-deep cylindrical holes drilled through a copper plate sandwiched in a melt-press machine (Imoto Co., Ltd.). Then, column PS/PMMA blends were quenched with liquid nitrogen to fix their phase-separated structures. To observe the time change in the phase-separated structures of such PS/PMMA blends, we annealed the samples at 180°C for 3, 5, 7.5, 10 and 15 h. In an attempt at determining a phase diagram, we annealed the samples in the temperature range from 220 to 300°C.

X-RAY PHASE TOMOGRAPHY

The crystal X-ray interferometer has three crystal lamellae with the same spacing, which is cut out monolithically from an ingot of a perfect silicon crystal. When an X-ray beam satisfies the conditions for Bragg diffraction against the lattice planes perpendicular to the surface of the first lamella, the X-ray beam is divided coherently into two beams exiting from the opposite surface of the lamella. The two beams are diffracted at the second lamella in the same manner, and two beams converging onto the third lamella are also diffracted, thereby producing interfering beams. When a sample is placed in one of the paths, an interference pattern is observed. To obtain information on the X-ray phase shift, we use a fringe-scanning technique [10]. For X-ray phase tomography, the X-ray phase shift is measured at different angular positions of sample rotation. A three-dimensional distribution of a refractive index can be reconstructed.

X-ray phase tomography using a crystal X-ray interferometer was performed at the beamline 20XU of SPring-8, Japan. Figure 1 shows the experimental setup. X-rays from an undulator were monochromatized at 17.7 keV by a Si 111 double-crystal monochromator and introduced into an X-ray interferometer. A CCD-based X-ray image detector (Hamamatsu Photonics K.K., C4742-98-24A) of 3.14 μm effective pixel size was employed. The sample was rotated in 0.45° steps in a water-filled cell, and 400 maps of X-ray phase shift were measured.

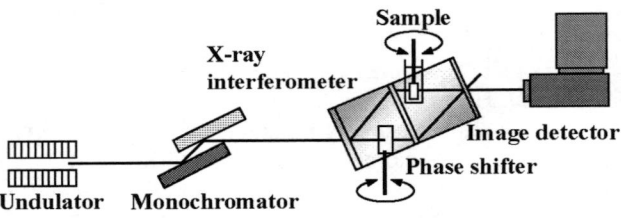

FIGURE 1. Experimental setup for phase tomography using a crystal X-ray interferometer.

RESULTS AND DISCUSSION

Figure 2 shows reconstructed images of the PS/PMMA blends annealed at 180°C. Phase-separated structures were clearly observed even without treatments for contrast enhancement. We identified bright and dark areas in Fig. 2 as PMMA- and PS-rich regions, respectively, from the evaluation of the refractive index differences of PS and PMMA from that of water. Spatial resolution was evaluated to be 10 μm from the contrast profile between the sample and the surrounding water. Density resolution was evaluated to be 3.5 mg/cm^3 from the standard deviation of the reconstructed value in the water region outside the sample.

FIGURE 2. Reconstructed images of PS/PMMA blend annealed at 180℃ for (a) 3, (b) 5, (c) 7.5, (d) 10 and (e) 15 h.

The size of the phase-separated structures of PS/PMMA blends increased with annealing time. We calculated the chord length distribution of the phase-separated structures. The characteristic length of the structures given by the peak of the histogram of the chord length is plotted against annealing time in Fig. 3. The characteristic length was approximately proportional to annealing time up to 7.5 h, as predicted [11] although the values obtained for 10 and 15 h are lower than the prediction. This result may be because the size of phase-separated structures was near the sample size and the speed of the phase separation was slowed down.

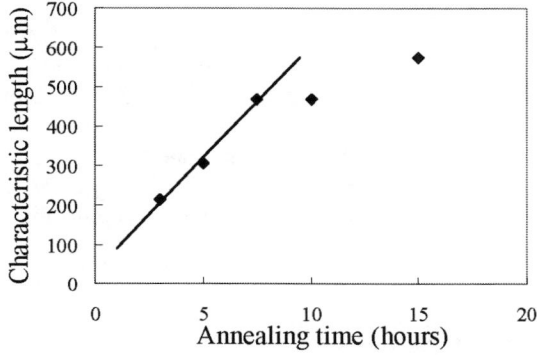

FIGURE 3. Time change in characteristic length of phase-separated structures in PS/PMMA blends annealed at 180℃.

FIGURE 4. Phase diagram of PS/PMMA blend.

Next, we performed a composition analysis of PS and PMMA within PS- and PMMA-rich regions revealed in X-ray phase tomograms. The volume fractions of PS and PMMA were calculated, assuming $\widetilde{\delta}_{PS} = \phi_1 \delta_{PS} + (1-\phi_1)\delta_{PMMA}$ and $\widetilde{\delta}_{PMMA} = \phi_2 \delta_{PS} + (1-\phi_2)\delta_{PMMA}$, where ϕ_1 and ϕ_2 are the volume fractions of PS within PS- and PMMA-rich regions, δ_{PS} and δ_{PMMA} are the refractive index differences of pure PS and PMMA from that of water, and $\widetilde{\delta}_{PS}$ and $\widetilde{\delta}_{PMMA}$ are the averages of 30 × 30 pixels extracted from PS- and PMMA-rich regions, respectively, in the phase tomograms. We performed the calculation on the data obtained for a series of PS/PMMA blend samples annealed in the temperature range from 220 to 300℃. Figure 4 shows a resultant phase diagram. Only part of the phase diagram corresponding to the temperature range from 220 to 300℃ was obtained, but a change in composition with temperature was confirmed. Conventional techniques for determining the phase diagram of polymer blends have been performed using light scattering and light transmission methods [12,13], which determine a cloud point by changing temperature. However, temperature control is a problem in these methods. In X-ray phase tomography, although some composition errors arise, temperature can be kept constant. Moreover, samples are not limited to be thin films as in the methods using light. Thus, X-ray phase tomography can be a new approach to determining the phase diagram of polymer blends.

Because X-ray phase tomography enables the observation of as-grown polymer blends, it is furthermore expected that phase separation or modification of polymer blends can be observed by combination of the apparatuses that apply external force to a sample such as heat or stress with X-ray phase tomography. However, the combination of a crystal X-ray interferometer with other apparatus is not straightforward because the spacing between the lamellae of the interferometer is limited. Therefore, we preliminarily performed X-ray phase tomography with an X-ray Talbot interferometer for the observation of the polymer blends. The X-ray Talbot interferometer consisted of two transmission gratings and provided a wide work space.

Figure 5 shows a reconstructed image of a PS/PMMA blend annealed at 190°C for 2h. The spatial and density resolutions were evaluated to be 17 μm and 5.4 mg/cm^3, respectively. Although the sensitivity of the X-ray phase tomogram using an X-ray Talbot interferometer was inferior to that attained using a crystal X-ray interferometer, phase-separated structures could be thus depicted in Fig. 5. We expect that the X-ray Talbot interferometer can be used for the observation of polymer blends with environmental control.

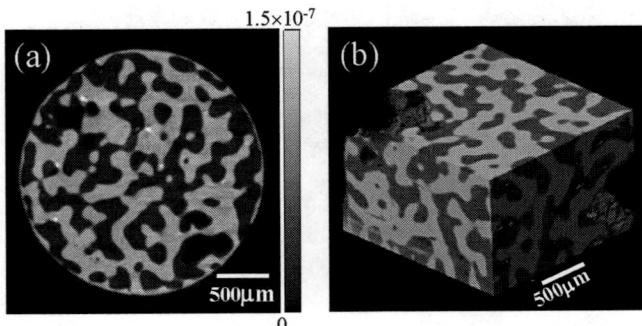

FIGURE 5. Reconstructed images of PS/PMMA blend by X-ray phase tomography using the X-ray Talbot interferometer. (a) Phase tomogram and (b) volume rendering view.

CONCLUSIONS

X-ray phase tomography using a crystal X-ray interferometer was performed for the quantitative three-dimensional observation of PS/PMMA blends at the beamline 20XU of SPring-8, Japan. The time change in the phase-separated structures of the PS/PMMA blend annealed at 180°C was observed. We evaluated the composition of each phase-separated region in a series of PS/PMMA blend samples annealed in the temperature range from 220 to 300°C and determined part of the phase diagram of the PS/PMMA blend. Furthermore, phase tomography using an X-ray Talbot interferometer, which can provide a wide work space, was demonstrated, suggesting new opportunities for research on polymer blends in various controlled environments.

ACKNOWLEDGMENTS

The experiments using synchrotron radiation were performed under the approval (2004B0706-NM-np, 2005A0326-NM-np, 2005B0268-NM-np and 2006A1236-NM-np) of the SPring-8 committee. This study was financially supported in part by the project "Development of System and Technology for Advanced Measurement and Analysis" of Japan Science and Technology Agency (JST).

REFERENCES

1. H. Jinnai, Y. Nishikawa, T. Koga, and T. Hashimoto, *Macromolecules* **28**, 4782-4784 (1995).
2. H. Jinnai, Y. Nishikawa, R. J. Spontak, S. D. Smith, D. A. Agard, and T. Hashimoto, *Phys. Rev. Lett.* **84**, 518-521 (2000).
3. S. Kuroki, S. Koizumi, S. Ymane, and I. Ando, *Kobunshi Ronbunshu* **62**, 458-466 (2005).
4. A. Momose, A. Fujii, H. Jinnai, Y. Nishikawa, and H. Kadowaki, *Macromolecules* **38**, 7197-7200 (2005).
5. A. Momose, *Jpn. J. Appl. Phys.* **44**, 6355-6367 (2005).
6. A. Momose, *Nucl. Instrum. & Methods. A*, **352**, 622-628 (1995).
7. A. Momose, T. Takeda, Y. Itai, and K. Hirano, *Nat. Med.* **2**, 473-475 (1996).
8. U. Bonse, and M. Hart, *Appl. Phys. Lett.* **6**, 155-156 (1965).
9. A. Momose, S. Kawamoto, I. Koyama, Y. Hamaishi, K. Takai, and Y. Suzuki, *Jpn. J. Appl. Phys.* **42**, L866-L868 (2003).
10. J. H. Bruning, D. R. Herriott, J. E. Gallagher, D. P. Rosenfeld, D. E. White, and D. J. Brangaccio, *Appl. Opt.* **13**, 2693-2703 (1974).
11. M. Takenaka, and T. Hashimoto, *J. Chem.Phys.* **96**, 6177-6190 (1992).
12. M. Nishimoto, H. Keskkula, and D. R. Paul, *Polymer* **32**, 272-278 (1991).
13. G. Beaucage, R. S. Stein, T. Hashimoto, and H. Hasegawa, *Macromolecules* **24**, 3443-3448 (1991).

Phase Tomography Reconstructed by 3D TIE in Hard X-ray Microscope

Gung-Chian Yin[1,2], Fu-Rong Chen[1,2,3], Ahram Pyun[4], Jung Ho Je[4], Yeukuang Hwu[5] and Keng S. Liang[1]

[1] National Synchrotron Radiation Research Center, Hsinchu 30076, Taiwan
[2] Department of Photonics, National Chiao Tung University, Hsinchu 30076, Taiwan
[3] Department of Eng. & System Science, National Tsing Hua University, Hsinchu 30076, Taiwan
[4] Department of material science and Engineering, Pohang University of Science and Technology, Pohang, Korea
[5] Institute of Physic, Academic Sinica, Taipei 115, Taiwan

Abstract. X-ray phase tomography and phase imaging are promising ways of investigation on low Z material. A polymer blend of PE/PS sample was used to test the 3D phase retrieval method in the parallel beam illuminated microscope. Because the polymer sample is thick, the phase retardation is quite mixed and the image can not be distinguished when the 2D transport intensity equation (TIE) is applied. In this study, we have provided a different approach for solving the phase in three dimensions for thick sample. Our method involves integration of 3D TIE/Fourier slice theorem for solving thick phase sample. In our experiment, eight sets of de-focal series image data sets were recorded covering the angular range of 0 to 180 degree. Only three set of image cubes were used in 3D TIE equation for solving the phase tomography. The phase contrast of the polymer blend in 3D is obviously enhanced, and the two different groups of polymer blend can be distinguished in the phase tomography.

Keywords: Phase Tomography, Transmission X-ray Microscope.
PACS: 42.30.Rx

INTRODUCTION

Due to the nature of the material, the phase retardation shift is about 2 to 3 orders of magnitude higher than the absorption at hard x-ray regime. In other words, the wave front altered by phase retardation carries stronger information than the attenuation of amplitude of wave front. Therefore, the phase retrieval method has been developed for decades [1-4], and experiments for quantitatively retrieving the phase are also demonstrated [5,6]. The methodology of phase tomography, which is the phase in three dimensions, has been reported [7,8]. The common ways of doing phase tomography is to retrieve the projected phase information from the images of defocus series from all different projected directions. The images of retrieved phase at different angles are then processed by the Filtered-Back Projection (FBP). For the thick sample, the phase is accumulated along the beam path and therefore may wrap around and results in the non-unique value of the phase in two dimensional. It is important to solve the phase in three dimensional to avoid the problem of accumulation and wrapping.

We proposed an alterative method to solve the phase tomography directly from the tomography data set. This method involves integrating the FBP and TIE. The intensity cube of de-focal series deduced from filter-back projection (FBP) method is used as inputs for transportation intensity equation (TIE). The problem is then equivalent to solve 3D TIE. The 3D TIE equation is as written in Eq. (1). Where the φ is the phase in 2D and the ψ is the assistant function for solve the equation, can be written as $\nabla^2 \psi = -k \frac{\partial I}{\partial z}$. The k is the wave factor and $\frac{\partial I}{\partial z}$ is the intensity gradient along the propagation direction. When it was integrated with FBP it can be expressed as Eq. (2). The Eq. (2) can be solved using 3D FFT methods.

---

[*] Corresponding Author, gcyin@nsrrc.org.tw

CP879, *Synchrotron Radiation Instrumentation: Ninth International Conference,*
edited by Jae-Young Choi and Seungyu Rah
© 2007 American Institute of Physics 978-0-7354-0373-4/07/$23.00

$$\nabla^2 \varphi = \nabla(\frac{\nabla \psi}{I}) = \frac{1}{I}\nabla^2 \psi + \nabla(\frac{1}{I}) \bullet \nabla \psi \tag{1}$$

$$\varphi = \frac{-1}{4\pi^2}F^{-1}\{\frac{1}{q^2}F\{(-\frac{k}{I}\frac{\partial I}{\partial Z} + \nabla(\frac{1}{I}) \bullet \nabla(\frac{-1}{4\pi^2}F^{-1}\{\frac{1}{q^2}F\{-k\frac{\partial I}{\partial Z}\}\})\}\} \tag{2}$$

The 3D phase, denoted as φ, is a function of position (x, y, z), which can be solved by the (3D Fast Fourier Transform) 3D FFT by computer. Besides the I, denoted as the intensity, the intensity gradient along the Z direction is also required in this formula. The F is the operator of 3D FFT, and F^{-1} is the operator of Inverse Fast Fourier Transform (IFFT). k is the wave factor and the q is the reciprocal space factor. In 3D-TIE, I in Eq. (2) is the intensity cube solved by tomography of absorption contrast. Hence, data sets of de-focal series in different detector angular positions are reconstructed by FBP and the phase tomography can be solved by Eq. (2).

EXPERIMENT

A polymer blend of PE/PS sample was used to test the 3D phase retrieval method. According to the proposed method, we setup an experiment at BL01A in National Synchrotron Radiation Research Center (NSRRC), as depicted in Fig.1.

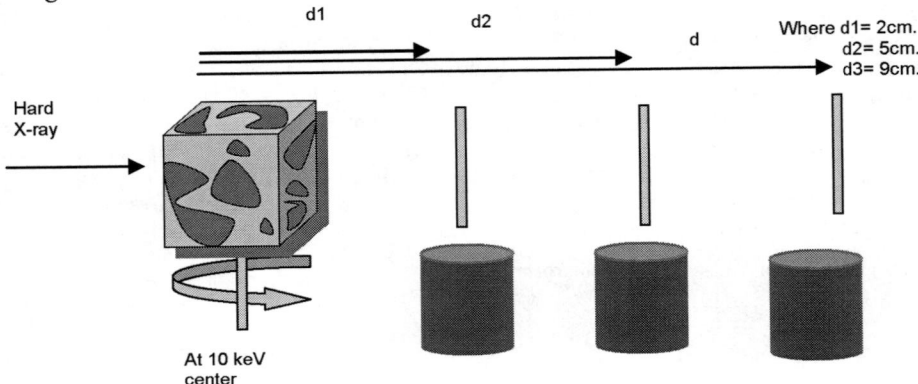

FIGURE 1. The setting up of the phase tomography. The tomography data set was taken at different CCD position. The distances are 2, 5 and 9 cm, and the field of view is about 200 μm. The projection data is first reconstructed and send to 3D-TIE.

The source is broad band directly comes from a superconducting wavelength shifter (SWLS) [9], the energy range is from 6kev to 20 keV, the peak value is around 10 keV. The field of view of image is about 200μm and the detector positions are 2 cm, 5cm, and 9cm. In fact, in our experiment, eight sets of de-focal series image data sets were recorded covering the angular range of 0 to 180 degree. The phase tomography was reconstructed using equation (2). The slice of reconstructed phase tomography and absorption tomography are shown in Fig. 2. The enhancement of the phase contrast is obvious. Two blend of polymer can be identified easily. The 3d rendering of intensity cube of absorption and phase are shown in the Fig. 3(a) and (b).

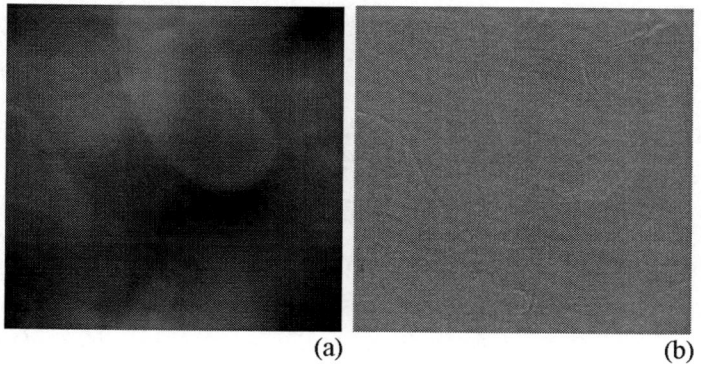

(a)   (b)

FIGURE 2. The slice of (a)phase solved by 3D Tie method and (b) the absorption contrast.

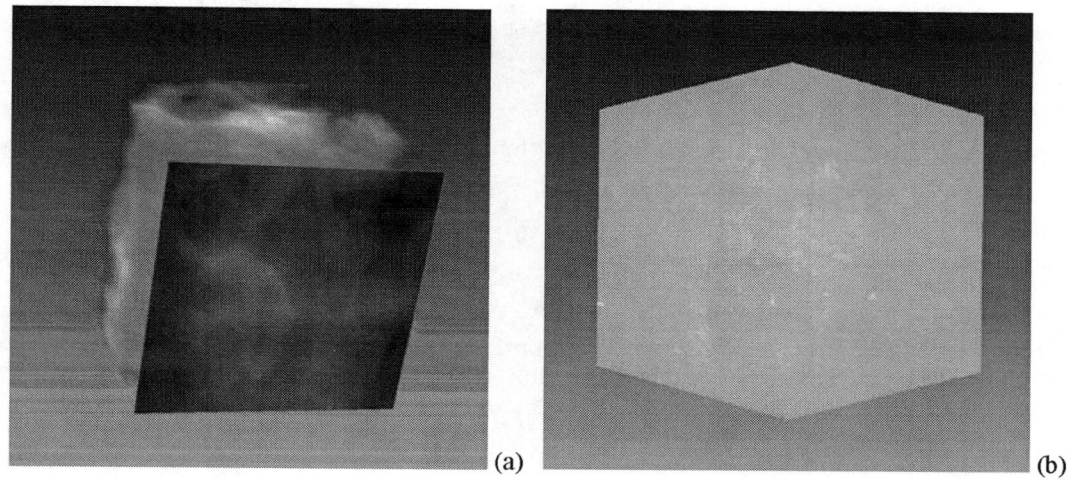

(a)   (b)

FIGURE 3. The 3d rendering of (a) phase contrast and (b) absorption tomogram solved by 3D TIE method.

ACKNOWLEDGMENTS

We thank for the support from the beam-line R&D group at NSRRC.

REFERENCES

1. M. R. Teague, *J. Opt. Soc. Am.* **2**, 2019(1983).
2. T.E.Gureyev, A.Roberts, and K. A.Nugent, *J. Opt. Soc. Am.* **12,**1942(1995).
3. D. Paganin and K.A. Nugent, *phys. rev. lett.,***80**,2586(1998)
4. Andrei V. Bronnikov,, *J. Opt. Soc. Am.*19,472-480 (2002).
5. K. A. Nugent, T. E. Gureyev, D. F. Cookson, D. Paganin, and Z. Barnea, *Phys. Rev.Lett.***77**, 2961-2964(1996).
6. A. G. Peele, F. DeCarlo, P.J. McMahon, B. B. Dhal and K. A. Nugent, *Rev. Sci. Instrum.* 76, 083707 (2005)
7. P. Cloetens, W. Ludwig, J. Baruchel ,D. Van Dyck , J. Van Landuyt, J. P. Guigayb and M. Schlenker *Appl. Phys. lett.* **75**, 2912-2914 (1999).
8. P.J.McMahon, A.G. Peele, D.Paterson, K.A.Nugent, A.Snigirev, T.Weitkamp, C.Rau, *Appl. Phys. Lett.*, 83, 1480-1482 (2003)
9. Y. F. Song, C. H. Chang, C.Y. Liu, L. J. Huang, S. H. Chang, J. M. Chuang, S. C. Chung, P. C. Tseng, J. F. Lee, K. L. Tsang, and K. S. Liang, *SRI2003 proceedings,* (San Francisco, 2003)

Evaluation of In-Vacuum Imaging Plate Detector for X-Ray Diffraction Microscopy

Yoshinori Nishino, Yukio Takahashi, Masaki Yamamoto, and Tetsuya Ishikawa

SPring-8 / RIKEN, 1-1-1 Kouto, Sayo-cho, Sayo-gun, Hyogo 679-5148, Japan

Abstract. We performed evaluation tests of a newly developed in-vacuum imaging plate (IP) detector for x-ray diffraction microscopy. IP detectors have advantages over direct x-ray detection charge-coupled device (CCD) detectors, which have been commonly used in x-ray diffraction microscopy experiments, in the capabilities for a high photon count and for a wide area. The detector system contains two IPs to make measurement efficient by recording data with the one while reading or erasing the other. We compared speckled diffraction patterns of single particles taken with the IP and a direct x-ray detection CCD. The IP was inferior to the CCD in spatial resolution and in signal-to-noise ratio at a low photon count.

Keywords: Imaging Plate, X-ray Diffraction Microscopy, Coherent X-rays, Speckle
PACS: 42.79.Pw, 42.30.Rx, 07.85.Tt

INTRODUCTION

X-ray diffraction microscopy is a novel method to reconstruct electron density distributions from diffraction data with no need of sample crystallization. In experiments, diffraction patterns of sample particles are measured at fine intervals in reciprocal space to satisfy the oversampling condition for solving the phase problem. The phase set is recovered by applying an iterative method developed by Gerchberg, Saxton, and Fienup. The short wavelength and the high penetration power of x-rays offer the unique possibility to achieve high spatial resolution three-dimensional structural analysis also for non-crystalline samples.

In x-ray diffraction microscopy experiments at SPring-8 [1], we have been using PI-LCX 1300 charge-coupled device (CCD) of Princeton Instruments [2]. It directly detects x-rays without using a phosphor screen, and is suitable for the detection of weak diffraction intensities from non-crystalline samples in x-ray diffraction microscopy. A high signal-to-noise (S/N) ratio at a low photon count is achieved, because a single x-ray photon with an energy of, e.g., 5 keV creates about 1300 electron-hole pairs, which is much larger than the noise of typically a few tens of electrons when it is cooled by liquid nitrogen. A deep depletion CCD is adopted to gain high quantum efficiency even for hard x-rays, and it is over 80% for 5 keV x-rays [2]. In addition, high spatial resolution is fulfilled with direct x-ray detection [3]. Despite the above benefits, direct x-ray detection CCDs have a few drawbacks. Due to the direct x-ray detection, the highest photon counts per pixel is as small as about 300 for 5 keV x-rays by creating electrons enough to fill the full potential well. It is therefore required to frequently read data before saturating the detector. It makes, however, measurement inefficient with a relatively slow CCD readout. Another disadvantage is the difficulty in attaining a wide area, which may limit the spatial resolution of the microscope.

Imaging plate (IP) detectors are also popularly used as two-dimensional x-ray detectors [4]. The capabilities of IPs for a high photon count and for a wide area are attracting features in x-ray diffraction microscopy. In this paper, we report on the performance tests of a newly developed in-vacuum IP detector for x-ray diffraction microscopy.

IMAGING PLATE DETECTOR SYSTEM

We developed a large area in-vacuum IP detector system, R-AXIS VIII, in collaboration with RIGAKU [5]. In the detector system, IPs are placed in vacuum, though readers and erasers are in air and they access the IPs through pressure-proof glasses. Table 1 shows a comparison of basic features of R-AXIS VIII and PI-LCX 1300. A distinct

CP879, *Synchrotron Radiation Instrumentation: Ninth International Conference*,
edited by Jae-Young Choi and Seungyu Rah
© 2007 American Institute of Physics 978-0-7354-0373-4/07/$23.00

advantage of IPs is the capability for a wide area. An IP in R-AXIS V III has a total area of 125 mm × 125 mm, which is more than 20 times larger than that of the CCD. A small pixel size is also required in x-ray diffraction microscope for the fulfillment of the oversampling condition with a moderate sample-to-detector distance. R-AXIS VIII has a pixel size (a readout interval length) of 25 μm, which is comparable to the CCD. Here, we note that the spatial resolution of IP detectors is usually determined by IP properties not by the readout interval length. Blue-colored IPs are used in R-AXIS VIII for high spatial resolution. With a large total area and a small pixel size, the total number of pixels became as large as 5000 × 5000.

Another attractive feature of IPs is a high x-ray photon count per pixel. IPs was reported to detect up to the order of 10^5 x-ray photon count per pixel [6], which is far larger than that of direct x-ray detection CCDs, where hundreds of x-ray photons per pixel are enough to saturate it.

A serious drawback of IPs is the slowness in reading and erasing. It takes about five minutes to read a full region-of-interest data, and typically two minutes or more to erase data with R-AXIS VIII. In order to perform efficient measurement, the detector system contains two sets of an IP, a reader, and an eraser, and one can take data with the one while reading or erasing the other. In addition, at the downstream of the IP, a CCD detector can be mounted, which can be used for quick alignment of sample and optical components.

TABLE 1. Comparison of the IP and CCD detectors

	IP detector (R-AXIS VIII)	CCD detector (PI-LCX 1300)
Total Area	125 mm × 125 mm (active area: 115 mm × 115 mm)	26 mm × 26.8 mm
Pixel Size	25 μm	20 μm
Total Number of Pixels	5000 × 5000	1300 × 1340
Max. X-Ray Photon Count per Pixel	~10^5	~10^2

EXPERIMENTAL SETUP

We carried out x-ray diffraction measurement of sample particles to test the performance of the IP detector system in typical condition of x-ray diffraction microscopy experiment. The measurement was conducted at the first experimental hutch (EH1) of BL29XUL in SPring-8. X-rays from an in-vacuum undulator were monochromatized by a silicon 111 double crystal monochromator, and the higher harmonics were reduced by a pair of total-reflection flat-surface mirrors. X-rays were then guided to EH1. Figure 1 shows the experimental setup in EH1. It consists of a main chamber, a beamstop chamber and detectors. In the main chamber, a 30 μm diameter pinhole was placed to limit the x-ray illumination area at the sample, and a silicon guard slit was put just in front of the sample to remove parasitic scatterings from the pinhole. The pinhole, the guard slit, and the sample were mounted on guide rails along the x-ray beam direction, so that the distances among them can be easily modified. In the beamstop chamber, two beamstops can be installed, which can be used to eliminate the missing data region behind a beamstop by performing two diffraction measurements with the beamstops at complementary positions. At downstream of the beamstop, the IP detector system was mounted with a CCD detector at the end. When no IP was at the exposure position, the diffracted x-rays can pass through a 20 mm diameter circular aperture hole on the IP cassette and can reach the CCD. The distances from the sample to the IP and to the CCD were, respectively, 1560 mm and 2274 mm for 5 keV measurement, and 1483 mm and 2196 mm for 10 keV measurement. The chambers were evacuated by turbo molecular and scroll pumps, and the pressure near the pumps was typically of the order of 10^{-3} Pa before cooling the CCD.

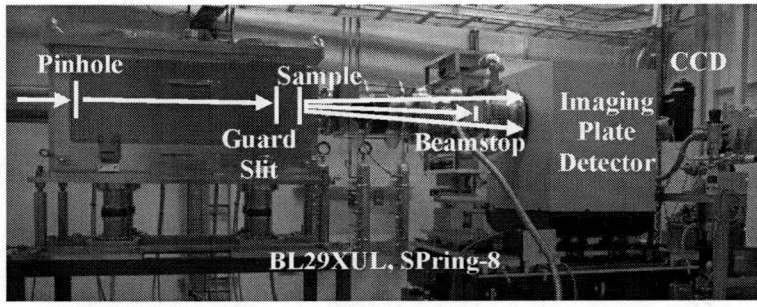

FIGURE 1. Experimental setup for x-ray diffraction microscopy with an imaging plate detector at BL29XUL in SPring-8.

EXPERIMENTAL RESULTS

We first consider the spatial resolution of the detectors, which is important for the fulfillment of the oversampling condition. Because a speckle size is inversely proportional to the sample size, a large sample producing small size speckles is suitable for the purpose. Figure 2 shows a comparison of the diffraction data of a platinum particle about 4 μm in size measured with the IP and CCD detectors. The incident x-ray energy was 5 keV. The CCD data in Fig. 2 (b) was obtained by accumulating 500 diffraction data with 10 ms exposure time, which made the total exposure time 5 s. The accumulation of many short exposure data was necessary in order not to saturate the CCD. The IP data in Fig. 2 (a) was measured with single 10 s exposure.

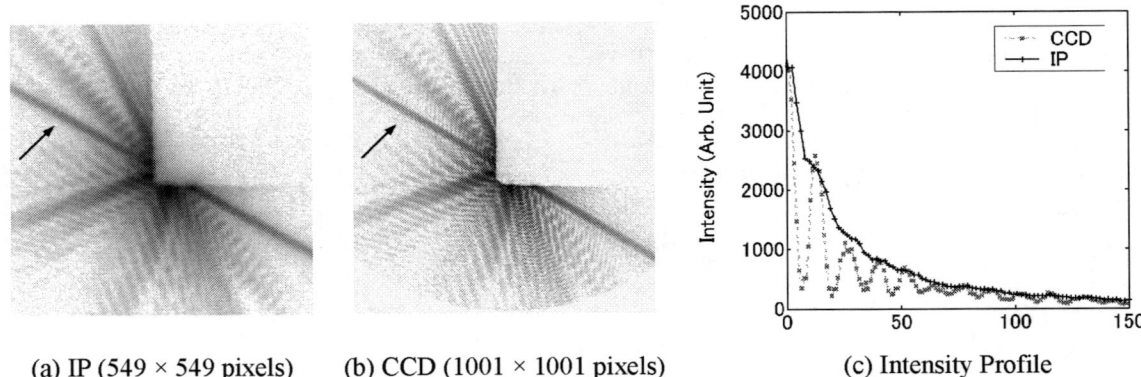

(a) IP (549 × 549 pixels)   (b) CCD (1001 × 1001 pixels)   (c) Intensity Profile

FIGURE 2. Comparison of the spatial resolution of the IP and CCD detectors. The diffraction patterns of a platinum particle about 4 μm in size measured with (a) the IP and (b) CCD detectors are shown together with (c) the intensity profiles along the longest streak indicated by the arrows in (a) and (b). The incident x-ray energy was 5 keV.

It is seen in Fig. 2 (a) and (b) that fine diffraction fringe patterns observed with the CCD were blurred in the image measured with the IP. To visualize the burring more evidently, we showed in Fig. 2 (c) the intensity profiles along the longest streak indicated by the arrows in Fig. 2 (a) and (b). Though a high visibility intensity oscillation was observed with the CCD, it was almost perfectly blurred out with the IP. The periodic length of the intensity oscillation at the IP position was about 178 μm, which was estimated from that of the CCD data and the distance from the sample to the each detector. The periodic length is longer than a reported spatial resolution for blue-colored IPs, which is less than 100 μm [6,7]. The reason of the low spatial resolution may partly be attributed to the reading procedure of the IPs through pressure-proof glasses. The phenomena known as flare, which is observed near an intense spot [6], may also be a reason of the low spatial resolution.

Here, several notes about the diffraction pattern figures in the paper are in order. In the figures, diffraction patterns after background correction were shown in the logarithmic gray scale; diffraction data in about a quarter of the total area were missing behind the beamstop; a part of CCD data was missing due to the circular aperture hole on the IP cassette; a pair of diffraction data with the CCD and IP detectors were plotted in the same scale in reciprocal

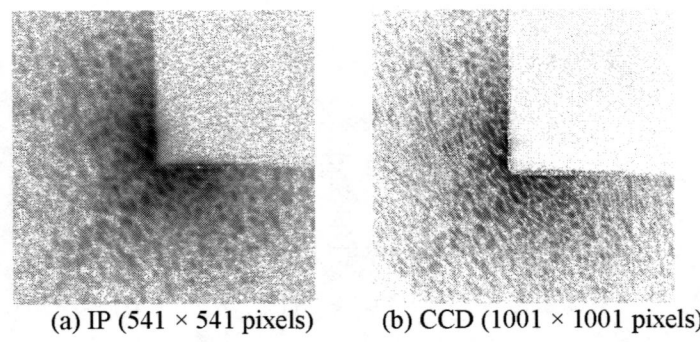

(a) IP (541 × 541 pixels)   (b) CCD (1001 × 1001 pixels)

FIGURE 3. Diffraction patterns of a tungsten particle about 1 μm in size measured with (a) the IP and (b) CCD detectors at an incident x-ray energy of 10 keV. The exposure time was 10 min in the both cases.

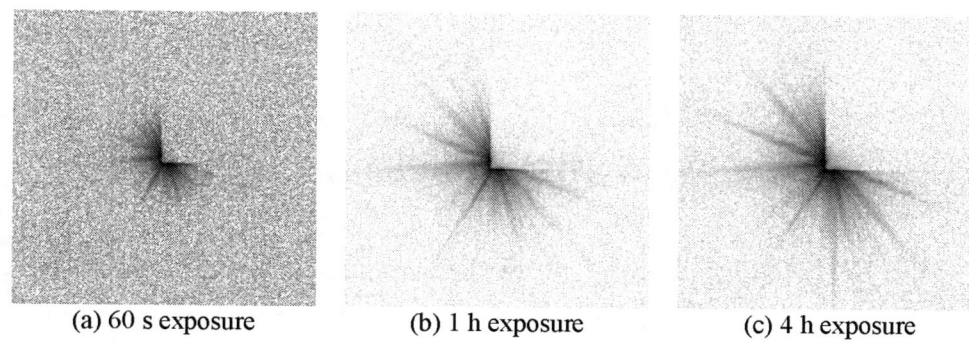

| (a) 60 s exposure | (b) 1 h exposure | (c) 4 h exposure |

FIGURE 4. Diffraction data in the central 2001 × 2001 pixels of the IP with different exposure times. Diffraction patterns of a platinum particle about 1.5 μm in size at an incident x-ray energy of 10 keV are shown.

space. It might be also worth noting that there has been a slight irreproducibility in the IP position of typically about a few pixels long when it was reloaded, and therefore the position alignment of sample and background data was necessary in background correction.

Higher energy x-rays have a potential for higher spatial resolution microscopy. Direct x-ray detection CCD, however, has lower quantum efficiency at higher energies, and it is about 30 % for 10 keV x-rays [2]. As for IPs, S/N ratio has a peak just above the Bromine K absorption edge at about 13.5 keV [8]. In considering different energy dependence of the detectors, we performed the diffraction measurement at 10 keV. Figure 3 shows the diffraction pattern of a tungsten particle about 1 μm in size. The total exposure time was 10 min for the both detectors. It is clear from the noises in the beamstop shadow region that the S/N ratio at a low photon count of the IP is lower than that of the CCD. The efficiencies seem to be similar with the both detectors, though it is difficult to make quantitative analysis due to low spatial resolution and a low S/N ratio at a low photon count of the IP data.

Figure 4 shows diffraction patterns of a platinum particle about 1.5 μm in size measured at 10 keV with the IP detector. The benefit of a large area with increasing exposure time was obviously observed.

SUMMARAY AND OUTLOOK

By comparing diffraction data measured with the IP and direct x-ray detection CCD detectors, we found the disadvantages of the IP in spatial resolution and the S/N ratio at a low photon count. The problem of the low spatial resolution can be circumvented by taking a larger distance between the sample and the detector, but it will reduce the benefit of a large area. Further technical investigations for the improvement of spatial resolution are wanted. A low S/N ratio of the IP at a low photon count may be more problematic, because it will reduce the visibility of speckle patterns and is considered an inevitable feature of IPs. It should be clarified how the S/N ratio affects the image reconstruction before applying the IP detector to x-ray diffraction microscopy.

ACKNOWLEDGMENTS

We thank Dr. Haruhiko Ohashi for valuable advice for vacuum related issues. Thanks are also due to Prof. Yoshiyuki Amemiya and Dr. Kazuki Ito for fruitful discussions about two-dimensional x-ray detectors including imaging plates.

REFERENCES

1. Y. Nishino, J. Miao, Y. Kohmura, Y. Takahashi, C. Song, B. Johnson, M. Yamamoto, K. Koike, T. Ebisuzaki, and T. Ishikawa, *IPAP Conference Series 7: Proceedings of the 8th International Conference on X-ray Microscopy* 386-388 (2006).
2. http://www.princetoninstruments.com/.
3. C. M. Castelli, N. M. Allinson, K. J. Moon, D. L. Watson, *Nucl. Instrum. Methods* **A348**, 649-653 (1994).
4. Y. Amemiya and J. Miyahara, *Nature (London)* **336**, 89-90 (1988).
5. http://www.rigaku.co.jp/.
6. Y. Amemiya, *J. Synchrotron Rad.* **2**, 13-21 (1995).
7. R.H. Templer, N.A. Warrender, J.M. Seddon, J.M. Davis, and A. Harrison, *Nucl. Instrum. Methods* **A310**, 232-235 (1991).
8. M. Ito and Y. Amemiya, *Nucl. Instrum. Methods* **A310**, 369-372 (1991).

Evaluation of Defects inside Beryllium Foils using X-ray Computed Tomography and Shearing Interferometry

Tatsuyuki Sakurai[1], Yoshiki Kohmura[2], Akihisa Takeuchi[3], Yoshio Suzuki[3], Shunji Goto[3] and Tetsuya Ishikawa[2,3]

[1]University of Hyogo, Kouto 3-2-1, Kamigori-cho, Ako-gun, Hyogo, 678-1297, Japan
[2]Harima Institute, RIKEN, Kouto 1-1-1, Sayo-cho, Sayo-gun, Hyogo, 679-5148, Japan
[3]Japan Synchrotron Radiation Research Institute (JASRI) ,Kouto 1-1-1, Sayo-cho, Sayo-gun, Hyogo, 679-5198, Japan

Abstract. When beryllium is used in transmission X-ray optical elements for spatially coherent beams, speckles are usually observed in the transmission images. These speckles seem to be caused by defects either inside or on the surface of beryllium foil. We measured highly polished beryllium foil using two methods, X-ray computed tomography and X-ray shearing interferometry. The results indicate that observed speckle pattern is caused by many voids inside beryllium or inner low-density regions.

Keywords: beryllium, X-ray computed tomography, shearing interferometry.
PACS: 41.50.+h; 42.25.Hz; 42.30.Wb

INTRODUCTION

Defects in beryllium windows cause speckle patterns when the windows are used in coherent X-ray experiments [1]. These speckles affect the accuracy of various experiments, such as coherent X-ray diffraction microscopy and X-ray interferometry. Therefore, the development of defect-free beryllium windows is necessary. Precise evaluation of the defects will help to improve the foil fabrication process.

First, we have estimated the distribution of defects using X-ray computed tomography. We then quantitatively measured the size of defects in the foil using the X-ray shearing interferometry [2]. The result clearly showed that defects detected inside the beryllium foils were voids or inner low-density regions.

EXPERIMENT

Sample

A sample beryllium foil was fabricated by the powder pressing procedure (referred to as O-30) by Brush Wellman Electrofusion Products, U.S.A. The thickness and the purity of the foil were 200 μm and 99.0 %. It was highly polished to a RMS roughness of 0.1 μm.

CP879, *Synchrotron Radiation Instrumentation: Ninth International Conference*,
edited by Jae-Young Choi and Seungyu Rah
© 2007 American Institute of Physics 978-0-7354-0373-4/07/$23.00

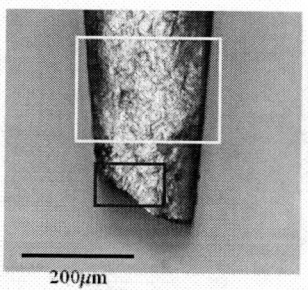

FIGURE 1. An optical microscopic image of the sample. Areas observed by X-ray CT and shearing interferometry are shown by the white and black boxes, respectively.

X-ray Computed Tomography(CT)

We have taken images of a beryllium window at a finite distance from the imaging detector with which the defects are clearly observed using the refraction contrast method. By rotating the sample and using the back-projection method for reconstruction, the 3D distribution of the defects were obtained.

This experiment was performed at BL20XU/SPring-8. Monochromatic X-rays of 12.4 keV (λ=0.1 nm) were utilized. The schematic diagram of the X-ray CT measurement is shown in Fig.2. The distance (L) from sample to detector was 0.16 m. Since high spatial resolution was required, the zooming tube, a modified version of Hamamatsu C5333, was used as the imaging detector. The beryllium foil window of the zooming tube was replaced with Kapton foil. The pixel size and the F.O.V. of the zooming tube were 0.76 μm and 760 μmϕ, respectively. The outline edge enhancement from the sample was effectively reduced by placing the beryllium foil in a water cell, 2 mm in thickness. The exposure time and the step of rotation angle were 10 seconds and 0.72 degree, respectively.

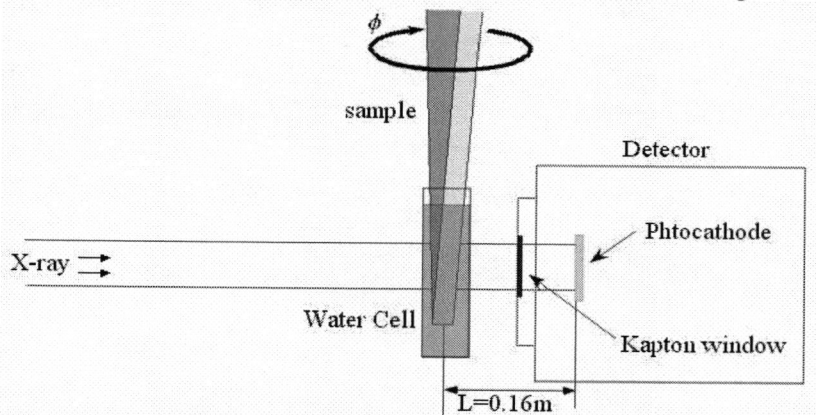

FIGURE 2. Schematic diagram of X-ray computed tomography.

3D defects distribution of beryllium foil by X-ray CT was obtained as shown in Fig.3 (a). A slice image along the white line in Fig.3 (a) is shown in Fig.3 (b). A typical edge enhancement is observed at the edge of the sample with a bright fringe surrounding a dark fringe. The intensity profile around the defect, on the other hand, showed a bright spot at the center. The interpretation of the bright spots will be discussed in the summary. A number of defects are observed inside the beryllium foil with a number density of 2×10^4 /mm^3.

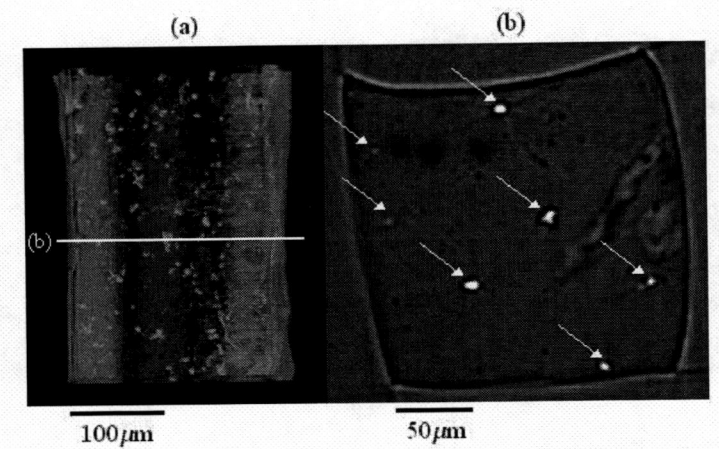

(a)   (b)

100μm   50μm

FIGURE 3. (a) 3D image using X-ray CT. (b) A slice image on the white line of Fig.3 (a).

Shearing Interferomtry

The principle of shearing interferometry is described in reference 2. Shearing interferometry is especially suited for measuring relatively thick samples.

We used an X-ray prism as a two beam wave-front dividing interferometer. The same beamline and X-rays of the same energy were used as with the CT measurement. The size of the slits was set to 20 μm × 20 μm to achieve the coherence length of ~ 1 mm at the sample in both horizontal and vertical directions. The schematic diagram of shearing interferometry is shown in Fig.4. The distance between the X-ray prism and the imaging detector (L_1) and between the sample and the imaging detector (L_2) were 6.3 m and 0.16 m, respectively. The deflection angle of the X-ray was set to 23 μrad ($\Delta\theta$) by setting X-ray glancing angle onto the prism to 4 degree(θ). This deflection angle was chosen to achieve the width of interference region ($L_1\Delta\theta$) of 145 μm, while realizing the detectable fringe spacing of 4.4 μm ($d_f = \lambda/\Delta\theta$). The amount of shear was 3.5 μm ($L_2\Delta\theta$).

We used the same zooming tube as an imaging detector. The pixel size at the sample plane and the F.O.V. of the zooming tube was 0.22 μm and 220 μmϕ, respectively. Due to the detector design, we could not reduce the distance L_2 to less than approximately 0.1 m from the photocathode. The optical thickness relative to the surrounding medium was effectively reduced by placing the beryllium foil in the same water cell.

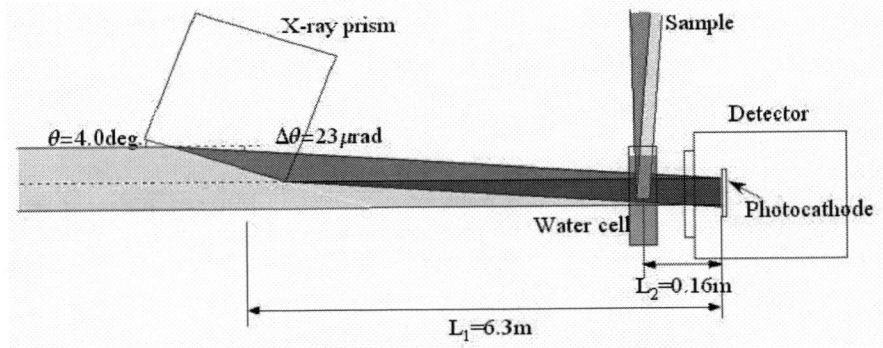

FIGURE 4. Schematic diagram of X-ray shearing interferometry.

A differential phase distribution was calculated by shearing interferometry using fringe scanning method [3], as shown in Fig.5 (a). The phase distribution of sample was obtained by integrating the differential phase distribution in the horizontal direction, as shown in Fig.5 (b). The noise level of the differential phase image was estimated by the standard deviation of phase inside the white box without the sample and was λ/40 (3σ).

The increase of optical thickness along the edge is clearly visible. All of these defects showed a decrease in thickness as evidenced by a reduction in the optical thickness. Since impurity elements included in the sample are heavier than beryllium, defects are likely to be voids or inner low-density regions. Assuming voids, the estimated thickness of three voids in Fig.5 was 4~8 μm, as shown in Table.1.

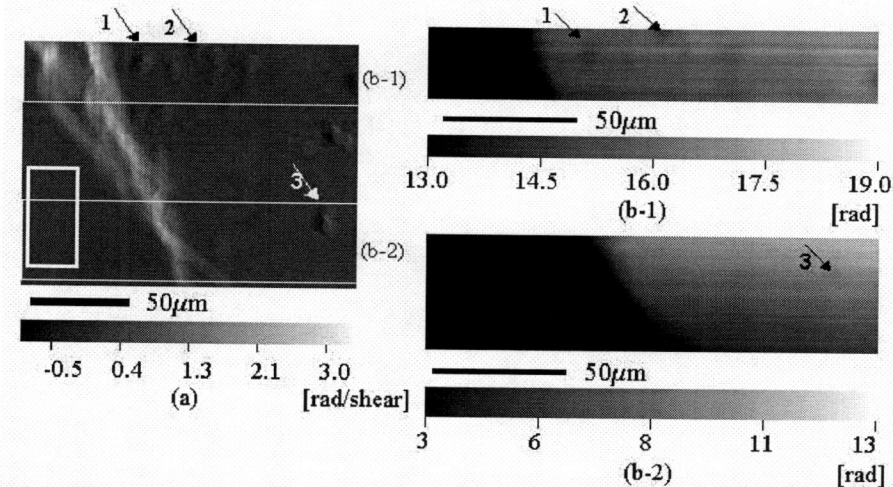

FIGURE 5. The calculated images of the beryllium foil. (a) differential phase distribution and (b) phase distribution inside white boxes in (a).

TABLE 1. Thickness assuming voids in Fig.5.

Void No.	Phase shift [rad]	Thickness of void [μm]
1	1.1	8.2
2	0.5	3.5
3	0.6	4.0

SUMMARY

We have measured the density distribution of defects using X-ray CT method. Bright spots in the image are caused by microfocusing of the beam within the voids or low-density regions at defect site. A number of defects were observed inside beryllium foil. The estimated number density of defects was 2.0×10^4 /mm^3. We have quantitatively measured the size of defects in the foil using the X-ray shearing interferometry. The decrease in thickness was observed at the region of the defects. Our results clearly showed that the beryllium window defects are voids or inner low-density regions.

ACKNOWLEDGMENTS

This work was performed with the approval of the Japan Synchrotron Radiation Research Institute (JASRI). (Proposal No. 2004A0102-CM-np).

REFERENCES

1. S. Goto, M. Yabashi, K. Tamasaku, S. Takahashi and T. Ishikawa, *Eighth International Conference on Synchrotron Radiation Instrumentation, edited by T. Warwick et al., AIP*, 400 (2004)
2. Y. Kohmura, , H. Takano, Y. Suzuki and T. Ishikawa, *J. Appl. Phys.*, **93**, 2283 (2003)
3. J.H. Bruning, D.R. Herriott, J.E. Gallagher, D.P. Rosenfeld, A.D. White, and D.J. Brangaccio, *Appl. Opt.* **13**, 2693 (1974)

CHAPTER 7
SR FOR NANO SCIENCE AND TECHNOLOGY

Developing a Dedicated GISAXS Beamline at the APS

Xuefa Li, Suresh Narayanan, Michael Sprung, Alec Sandy, Dong Ryeol Lee,
Jin Wang

X-Ray Science Division, Argonne National Laboratory, 9700 S. Cass Ave., Argonne, IL 60439, USA

Abstract. As an increasingly important structural-characterization technique, grazing-incidence small-angle scattering (GISAXS) finds vast applications in nanostructures and nanocomposites at surfaces and interfaces for *in situ* and real-time studies because of its probing q-range (10^{-3} – 1 nm^{-1}) and temporal resolution (10^{-3} – 1 s). At the Advanced Photon Source (APS), GISAXS techniques under thin-film waveguide-based resonance conditions were developed to study the diffusion phenomena in nanoparticle/polymer nanocomposites. Also, the kinematics of nanoparticle crystal formation at air/liquid interfaces has been obtained by the similar method in real time during the liquid droplet evaporation. To meet the strong demand from the nanoscience community, a dedicated GISAXS beamline has been designed and constructed as a part of the 8-ID-E beamline at the APS. This dedicated GISAXS setup was developed based on a 4-circle diffractometer so that precise reflectivity of the sample can be measured to complement the GISAXS analysis under the dynamical refection conditions.

INTRODUCTION

Complex nanocomposites are believed to be associated with novel electronic, magnetic and photonic properties of organic and inorganic components. In these nanocomposite systems, although highly ordered structures can often form in a self-assembled fashion, the formation of the structures can be extremely dynamic, far from commonly believed near-equilibrium conditions even at the end of the ordering processes. Therefore, a controlled self-assembling of the nanostructure has to be guided by a thorough understanding of ordering kinetics and nanoparticle dynamics in the complex matrices. For probing the systems involving dynamical structure of surfaces and buried interfaces, many x-ray surface and interfacial characterization techniques provides a unique scientific opportunity to study the principle of formation of ordered nanostructures. Because of its probing q-range (10^{-3} – 1 nm^{-1}) and temporal resolution (10^{-3} – 1 s), GISAXS becomes increasingly important in characterizating nanostructures and nanocomposites and their formation at surfaces and interfaces in real time. At worldwide synchrotron sources, dedicated in-vacuum and *in situ* GISAXS instruments have been setup, facilitating highly sensitive and time-resolved measurement of nanostructure formation. At the APS, we pioneered in using GISAXS techniques under thin-film waveguide-based resonance conditions to study nanoparticle/polymer nanocomposites (Fig. 1) [1, 2] and the kinematics of ordered nanoparticle formation at air/liquid interfaces (Fig. 2) [3]. However, in these experiments, with in-air setup, only a very limited number of systems involving intense scattering particles such as gold nanoparticles can be effectively performed due to limited signal-to-noise ratio mostly induced by background scattering from air, beam-defining and collimating slits and flight-path windows. As a part of strategic plan for upgrading Sector 8, a dedicated GISAXS beamline has been designed, constructed and commissioned in 8-ID-E beamline at the APS targeted to high-throughput experiments for nanoscience research community.

CP879, *Synchrotron Radiation Instrumentation: Ninth International Conference*,
edited by Jae-Young Choi and Seungyu Rah
© 2007 American Institute of Physics 978-0-7354-0373-4/07/$23.00

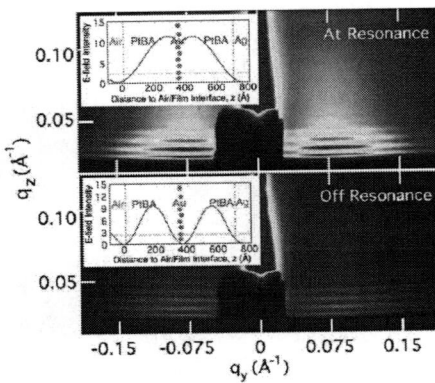

FIGURE 1. Normalized 2D GISAXS patterns of a nanoparticle monolayer embedded in a polymer ultrathin film measured at the first resonance condition (TE0 mode, top panel) and at off-resonance condition (bottom panel), respectively. The insets depict the corresponding calculated E-field intensity distribution.

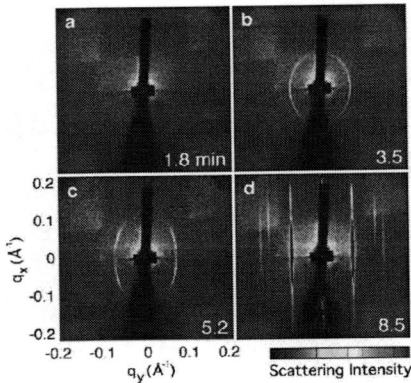

FIGURE 2. In situ SAXS patterns of 2D NCSs formation during the droplet evaporation measured with a narrow x-ray beam. Time in units of minutes is in reference to the deposition of colloid droplet onto the Si3N4 substrate. The exposure time for each frame on the image plate detector was 3 s.

DESIGN OF DEDICATED GISAXS BEAMLINE

Taking advantages of x-ray beam from an undulator, the dedicated GISAXS beamline is designed with both simplicity and flexibility in mind to achieve high resolutions in both reciprocal and real spaces as well as high temporal resolution in measurement. The simplicity is contributed by a fixed photon energy of 7.4 keV with 3 sets of stable upstream in-vacuum slits to ensure a high-throughput and user-friendly operation. The flexibility comes from many aspects: a four-circle diffractometer-based sample holder and the capability of various types of sample environment. More specifically, the samples can be situated in an integrated vacuum chamber on a high-precision heating and cooling stage. The sample chamber can also be isolated from the beamline to allow solvent flows and to accommodating other mechanical systems such as *in situ* dip-coating devices. The setup is shown schematically shown in Fig. 3.

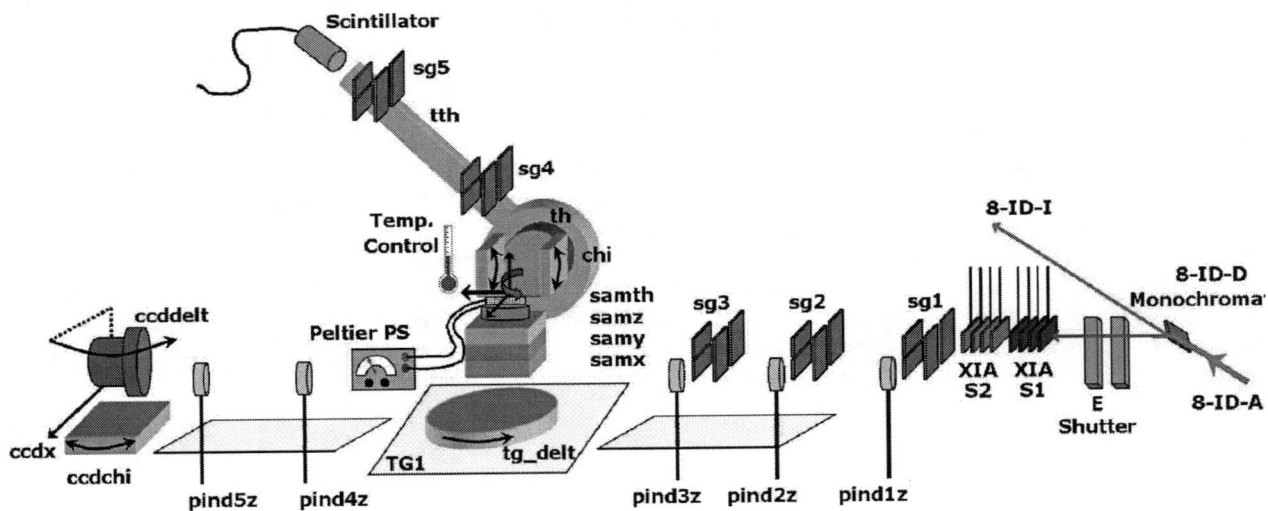

FIGURE 3. Schematic view of the dedicated GISAXS setup at Sector 8 of the APS.

Before the sample chamber, three sets of vacuum-compatible x-y slits (JJ-Xray) are employed in upstream beamline to define the incident x-ray beam size and to eliminate parasitic slit scattering. The first set of slits is 3 m away from sample to initially reduce the incident beam size. The second set is placed at 0.5 m from sample to define the final beam size typically ranging from 20 (horizontal, H) x 20 (vertical, V) to 300 (H) x 300 (V) μm^2 depending on experimental needs. The third serves as guard slits at 0.25 m away from the sample to reduce parasitic slit scattering. The three slits are equipped with intelligent motor system with μm-precision.

The sample chamber is designed based on a 6-way cross with 6-inch-diameter ports. The chamber can be connected to the beamline through bellows, which provides an integrated vacuum to achieve a minimum backgroun. The sample chamber is mounted on x-y-z translation stages and a ϕ-rotation stage coupled with the phi-stage of the 4-cycle Huber diffractometer. The sample-mounting stage is built inside the chamber and has both heating and cooling capabilities ranging from -40° to 600° with 0.1° temperature control precision. The stage is heated by cartridge heaters and cooled by a combination of a Peltier device and a chilled-liquid feed-through.

A pindiode, attached to a motorized vacuum feed-through, is used for sample alignment as well as reflectivity measurement up to 0.35 A^{-1}. A fast NaI scintillation detector (Cyberstar) can also be mounted on the Huber 2θ-arm for high-precision reflectivity beyond 0.35 A^{-1}. Typically, a Mar-165 CCD is used to record 2D GISAXS patterns with a pixel size of 79 microns. The sample to detector distance ranges from 1000 mm to 2500 mm resulting in typical q-range from 0.004 to 0.25 A^{-1}.

COMMISSIONING

This dedicated GISAXS beamline has been commissioned by performing various GISAXS measurements including but not limited to 1) block copolymer thin film ordering kinetics under during heating or/and solvent annealing, 2) nano-crystal formation during solvent evaporation, 3) characterization of quantum dots and nano-lithographic patterns, and 4) *in situ* formation of mesophase nanostructures by dip-coating process. Figure 4 shows the low-q reflectivity data from a 350-nm thick diblock co-polymer thin film using the pin-diode detector fitted with a 300-μm horizontal slit. This measurement demonstrated a q_z-resolution exceeding 0.001 A^{-1}, which is critically needed for analyzing wave-guide-induced resonance-enhanced scattering from the samples in grazing-incidence geometry. This high q_z resolution is achieved by superior beam collimation in the vertical direction.

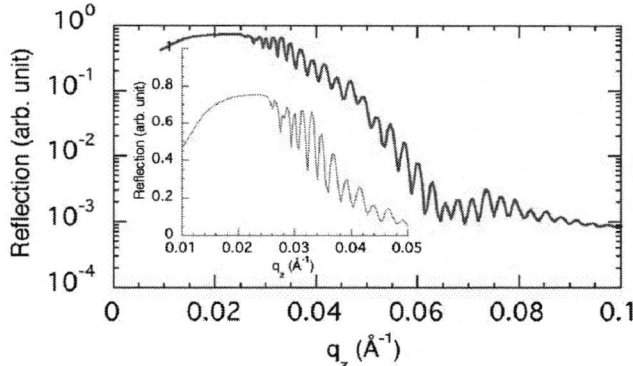

FIGURE 4. Reflection data (not normalized) from a polymer film of ca. 350 nm deposited on Si substrate in low-q (< 0.1 A^{-1}) region. Data in linear scale below and near the critical angle of the Si substrate is shown in the inset. Strong intensity modulations between the critical angles of the polymer film and the substrate indicate the existence of wave-guide-induced resonance enhancement in the film.

Figure 5 shows GISAXS patterns on phase transition in block copolymer thin films. This block copolymer micro-phase separate into spherical morphology. However, depending on thin-film thickness, the spheres organized into different phases such as hexagonal, orthorhombic or body center cubic phases. GISAXS technique is able to precisely determine the ordering structures at different layer thickness [4].

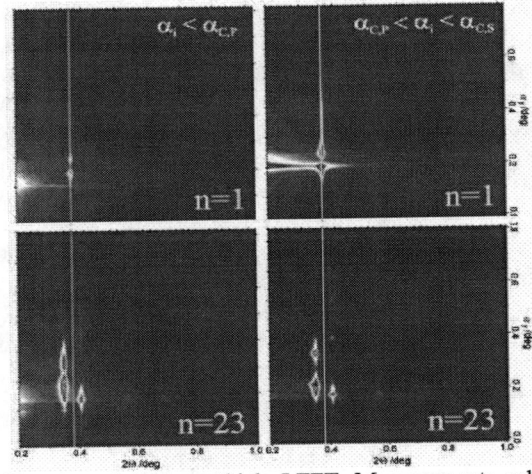

FIGURE 5. GISAXS patterns for films 1- and 23-layer thick. LEFT: Measurements collected below the critical angle of the polymer (top layer of spheres). RIGHT: Measurements collected above the critical angle of the polymer (full film thickness). Yellow lines mark the position of the (10) peak for 2D hexagonal crystalline symmetry.

CONCLUSION

This new and dedicated GISAXS beamline has allowed structural and kinetic characterization of not only nanoparticles of vast varieties but also polymer/polymer nanocomposites that possess only weak scattering contrast. This new capability presents a unique research opportunity in nanoscience and nanoengineering research field. A list of new and exciting research programs has been built around the x-ray measurements such as 1) kinetics of mesoscaled ordered block copolymer thin films during their phase transformation, 2) kinetics of sol-gel processes, 3) formation of organic/inorganic nanocomposite, for example, using organic materials as the templates for inorganic nanocyrstals to form superlattices, 4) preparation and characterization of quantum dots, 5) magnetic nanomaterials and their assembly at surfaces and interfaces, 6) dynamics of surfaces, interfaces and nanoparticles in ultrathin films, 7) dynamics of 2-D nanocrystals and their domain walls: rotational, translational motions, and 8) characteristics of true 2-D systems: phases, phase transition.

ACKNOWLEDGMENTS

This work and the use of the APS are supported by the U.S. DOE under contract W-31-109-ENG-38.

REFERENCES

1. S. Narayanan, D. Lee, R. Guico, S. Sinha, and J. Wang, Phys. Rev. Lett. 94, 145504 (2005).
2. D. Lee, A. Hagman, X. Li, S. Narayanan, J. Wang, and K. Shull, Appl. Phys. Lett. 88, 153101 (2006).
3. S. Narayanan, J. Wang, and X Lin, Phys. Rev. Lett. 93, 135503 (2004).
4. G. Stein, E. Kramer, X. Li, and J. Wang, Abstracts of Am. Chem. Soc. 231: PMSE, March 26-31, 2006.

X-Tip: a New Tool for Nanoscience or How to Combine X-Ray Spectroscopies to Local Probe Analysis

Dhez Olivier[1], Rodrigues Mario[2-3], Comin Fabio[2], Felici Roberto[1], Chevrier Joel[3]

[1]OGG-CNR-INFM, c/o ESRF BP220, 38043 Grenoble cedex, France, [2]ESRF, BP220, 38043 Grenoble cedex, France, [3]LEPES, CNRS, BP 166, 38042 Grenoble Cedex France

Abstract. With the advent of nanoscale science, the need of tools able to image samples and bring the region of interest to the X-ray beam is essential. We show the possibility of using the high resolution imaging capability of a scanning probe microscope to image and align a sample relative to the X-ray beam, as well as the possibility to record the photoelectrons emitted by the sample.

Keywords: Scanning probe microscope, AFM, X-rays, Total Electron Yield.
PACS: 07.79.Lh, 79.60.–i

INTRODUCTION

Our goal is to combine the high lateral resolution imaging capability of Scanning Probe Microscopy (SPM) with the high penetration power and element sensitivity of X-rays. We will show the ability of our home-built microscope to carry out Atomic Force Microscopy (AFM) imaging measurements on a beamline sample holder, align the AFM tip relative to the X-ray beam, and perform total Electron Yield Detection (TEY) via the AFM tip.

We have built a microscope which, in terms of noise and vibrations, is compatible with the relatively high-noise environment of an X-ray beamline. Our microscope is based on a piezoelectric quartz tuning fork oscillator in shear force detection mode [1]. The high-quality factor of the tuning fork allows to perform non-contact experiments in shear mode, even in air. We use a chemically etched tungsten tip glued on one sprung of the tuning fork. The tip apex varies from 10 to 50 nm. In our setup the tip has a fixed position whereas the sample can be adequately moved. A coarse displacement of 4 mm with a resolution of 25 nm is obtained by using a set of Attocube motors. A scanning range of up to 9 μm with 0.02 nm resolution and the tip-to-sample distance regulation are ensured by the microTRITOR from PiezoJena. Microscope electronic control and data acquisition operations are performed by a RHK Technology SPM 1000 Control System.

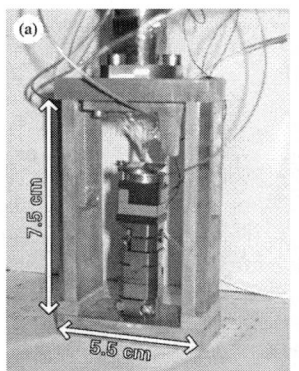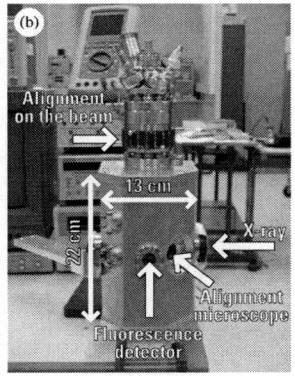

FIGURE 1. (a) View of the microscope composed of 3 Attocube coarse motors and a PiezoJena microTRITOR scanner. (b) General view of the vacuum chamber holding the microscope.

CP879, *Synchrotron Radiation Instrumentation: Ninth International Conference*,
edited by Jae-Young Choi and Seungyu Rah
© 2007 American Institute of Physics 978-0-7354-0373-4/07/$23.00

The whole microscope is enclosed in a vacuum chamber (Fig. 1) that can reach 10^{-8} mbar. The chamber ports allow to measure the X-ray fluorescence signal, the transmitted X-ray and also to place a visual light microscope (VLM) to pre-align the microscope on the beam. The whole microscope can be manually pre-aligned on the beam by using a port-aligner.

THE ATOMIC FORCE MICROSCOPE

The microscope enables imaging analyses to be done on an X-ray beamline. Fig. 2 shows an AFM image, obtained in shear mode of Ge dots on Si. We can achieve a resolution in (X,Y) of 25 to 50 nm depending on the surrounding level of noise and vibrations. Some tests have been made to check the ability of keeping a vertical resolution (tip-sample distance constant) of a few tens of nanometers and move the whole microscope on a standard Huber goniometer.

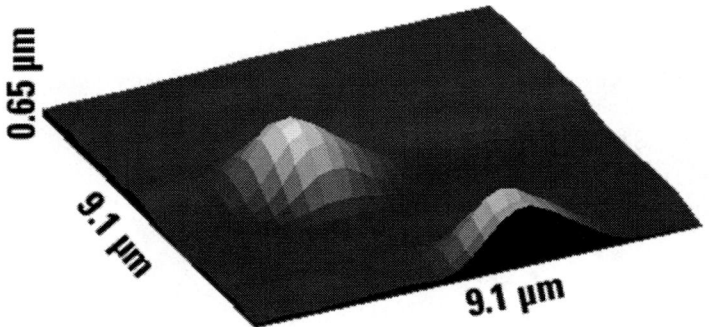

FIGURE 2. AFM image of Ge dots on Si obtained on the ID01 beamline at the ESRF.

We can image and align the AFM metallic tip and the X-ray beam. Different possibilities can be envisaged: measuring the current produced by the interaction of X-rays with the metallic tip, measuring the transmitted intensity or even the fluorescence signal of the tip. In order to measure the tip current we need to use a lock-in amplifier due to the very low current. Fig. 3a shows the current and the phase, "phase" being for us the difference between the measured and the reference signal. On Fig. 3b we show the evolution of the fluorescence signal of the sample (Ge dots in Si), and that of the tip. In this case, the tip was placed at about 500 µm above the sample. When moving the whole chamber vertically one observes first the fluorescence signal of the tip (continuous line) then the fluorescence signal of Ge (dashed line).

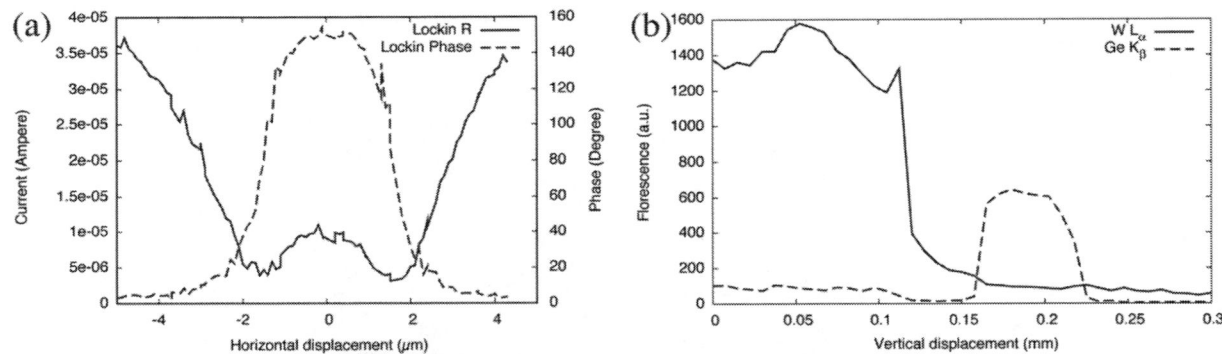

FIGURE 3. (a) Current and phase recorded by the lock-in amplifier through the AFM tip. (b) Fluorescence signal from the W tip and the sample Ge dots.

Figure 4a shows a map of the current measured by the tip, and Fig. 4b a map of the transmitted beam. One can easily argue from these figures that the current detection sensitivity is much higher than the absorption detection sensitivity.

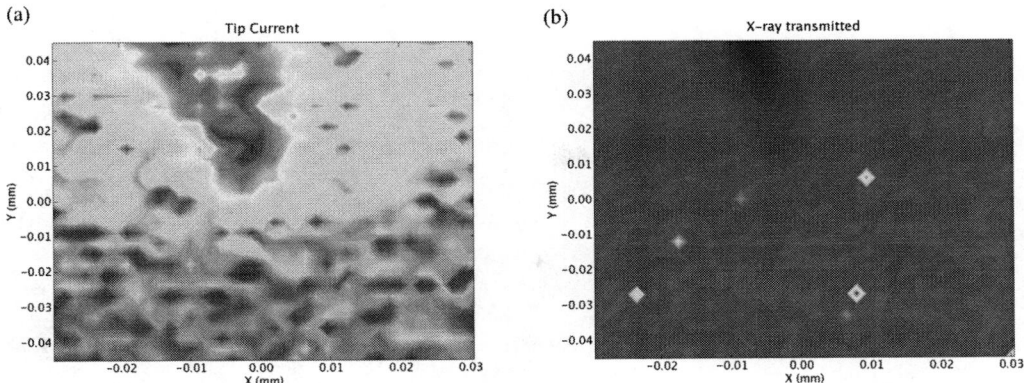

FIGURE 4. (a) Current measured by the metallic tip. (b) Intensity transmitted through the microscope.

After having assessed the ability to obtain good AFM images on a sample table, align the microscope on the X-ray beam and select regions of interest for X-ray studies, we are discussing below the possibility to record TEY signals with the microscope.

TOTAL ELECTRON YIELD DETECTION BY AN AFM TIP

There are several advantages in using a metallic tip as a detector for TEY. Firstly, our device is not sensitive to beam position fluctuations, owing to the size of the tip (the apex dimensions are typically of the order of 50 nm, i.e. they are smaller than the beam focus, which has typically the size of some microns). Secondly, the resolution is independent of the beam size, the area of the electron current collection depending on the tip size.

Figure 5 shows the general setup used to perform TEY experiments as well as the X-ray beam chopper used for the lock-in amplifier.

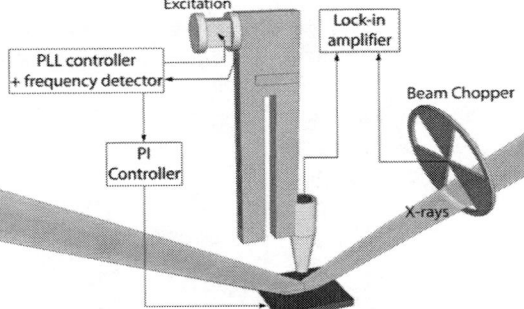

FIGURE 5. Experimental geometry showing the tuning fork and the beam geometry with the locking system.

In order to estimate the expected detectable current, we ran some Monte Carlo simulations using the Penelope code[1]. Figure 6a shows the calculation of the TEY current for a Ge dot of 36 nm radius on a 10 µm thick Si wafer with an X-ray flux of 10^{12} ph/s, assuming that all the X-ray photons are concentrated on the Ge dot. Under these conditions we obtained a jump of 28 pA when crossing the Ge K absorption edge.

---

[1] Penelope code (http://www.nea.fr)

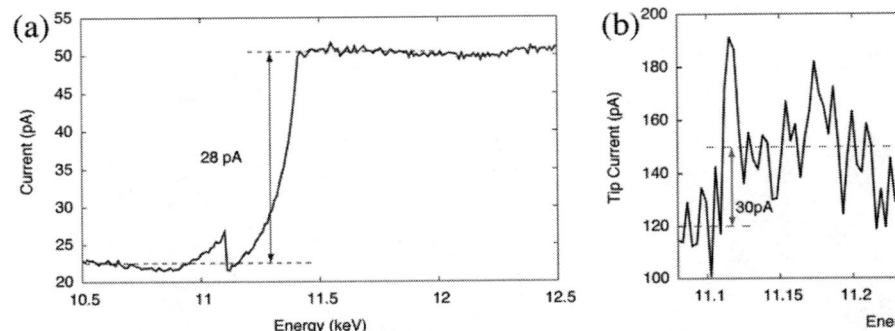

FIGURE 6. (a) Simulation results for the photoelectric current calculated for a Ge dot on Si when crossing the Ge K absorption edge. (b) Current recorded by the tip when crossing the Ge K absorption edge

We shall compare the experimental data shown in Fig. 6b to the theoretical predictions. The experiment was performed at the ESRF beamline ID22, using a microfocused beam (2 x 3 μm^2) with an incoming flux of 6.10^{11} ph/s. The beam grazing angle of incidence on the sample is of the order of 5°. The TEY current is measured by the tip when crossing the Ge K absorption edge. The curve in Fig. 6b indicates that the present measure is quite noisy in spite of the use of a lock-in detection. During the experiment the scanning stage was not mounted and the lateral positioning could not be controlled. As a consequence the position relative to the Ge dot was not actually known. This fact may explain the high level of noise. The difference between the calculated and the measured current may be attributed to the fact that electrons were collected from many dots instead of only one. We can nevertheless recognize a jump at the Ge K absorption edge. The jump is of the same order of magnitude as the Monte Carlo prediction. Further tests are needed, but the use of the TEY detection by an AFM tip, though very challenging, seems to be the right way to go.

CONCLUSION

We have shown that it is possible to obtain AFM topological images in a standard X-ray beamline environment. This opens up the opportunity to use AFM for selecting regions of interest with a nanometric precision and in addition to align it relative to the X-ray beam.

The electron current level is however very low for an acceptable TEY detection. Monte-Carlo simulations show that the current is of the order of several picoamperes The signal-to-noise-ratio may be improved by the use of coaxial insulated tips, which are now under development,.

The use of this kind of microscopes for the purposes of nano-manipulation and interaction with matter on the nanoscale can be prospectively envisaged.

ACKNOWLEDGMENTS

We are greatly indebted to M. Navizet for her technical support. We acknowledge the financial support from the EU's FP6 program which allowed the funding of the STRP 505634-1 X-tip project.

REFERENCES

1. K. Karrai, and R. D. Grober, *Appl. Phys. Lett.* **66**, 1842 (1995)

Microbeam High Angular Resolution Diffraction Applied to Optoelectronic Devices

A. Kazimirov[1], A. A. Sirenko[2], D. H. Bilderback[1], Z.-H. Cai[3], and B. Lai[3]

[1] *Cornell High Energy Synchrotron Source (CHESS), Cornell University, Ithaca, NY, 14853, USA*
[2] *Department of Physics, New Jersey Institute of Technology, Newark, NJ 07102, USA*
[3] *Advanced Photon Source, Argonne National Laboratory, 9700 S. Cass Avenue, IL 60439 USA*

Abstract. Collimating perfect crystal optics in a combination with the X-ray focusing optics has been applied to perform high angular resolution microbeam diffraction and scattering experiments on micron-size optoelectronic devices produced by modern semiconductor technology. At CHESS, we used capillary optics and perfect Si/Ge crystal(s) arrangement to perform X-ray standing waves, high angular-resolution diffraction and high resolution reciprocal space mapping analysis. At the APS, 2ID-D microscope beamline, we employed a phase zone plate producing a beam with the size of 240 nm in the horizontal plane and 350 nm in the vertical (diffraction) plane and a perfect Si (004) analyzer crystal to perform diffraction analysis of selectively grown InGaAsP and InGaAlAs-based waveguides with arc sec angular resolution.
Keywords: microbeam, angular resolution, optoelectronics.
PACS: 61.10.Nz, 68.37.Yz, 68.65.Fg

INTRODUCTION

Recent progress in X-ray focusing optics combined with the high brightness of the third generation synchrotron radiation (SR) sources has made X-ray beams of the sub-micron size routinely available for diffraction experiments (see, e.g. [1]). Some application areas, however, require not only high spatial resolution but high angular resolution, too. Semiconductor technology traditionally depends on high resolution X-ray diffraction (HRXRD) in determining strain, composition, mosaic structure and defect density in thin epitaxial layers [2]. Rapid advances in nanotechnology bring the active device region to a sub-micrometer size while requirements for the angular resolution remain in the arc sec range. For the focused X-ray beam the angular resolution is limited by the angular aperture of the focusing optics, typically of the order of 0.2 deg and more (the smaller the beam the higher is the angular aperture) compromising the angular resolution of XRD technique. The current solutions of this problem are based on (1) a small size pinhole [3,4]; (2) a pinhole of a medium size and compressive crystal optics [5], and (3) a combination of a zone plate and a slit [6]. In [7,8] we proposed to use perfect crystals to condition microbeam for high angular resolution diffraction and scattering experiments. The advantages of using perfect crystal optics are (i) the use of a non-dispersive setup, thus relaxing requirements to the monochromaticity of the incident beam, and (ii) the flexibility of controlling the angular resolution by a proper choice of the collimating crystal.

PERFECT CRYSTAL OPTICS IN MICRODIFFRACTION EXPERIMENTS

At CHESS, during the last few years, different optics schemes have been introduced into microbeam diffraction setup and tested in experiments on (III-V) and (II-VI)-based semiconductor optoelectronic devices. These schemes are shown schematically in Fig. 1. Due to the fact that perfect crystals select only a very narrow angular range out of a converging (diverging) focused beam in a diffraction (typically, vertical) plane, all these arrangements result in a substantial loss of the X-ray flux as a tradeoff for high angular resolution. Out of diffraction plane the focusing is still working effectively offering significant gain in flux in comparison with the pinhole approach. In all experiments at CHESS a one-bounce imaging capillary producing a round microbeam with a typical size of 10 micron has been used as a focusing optic [9]. A one-bounce imaging capillary and a miniature double-bounce Si(004) channel-cut crystal inserted in a limited space between the capillary and the sample was used to perform microbeam X-ray

CP879, *Synchrotron Radiation Instrumentation: Ninth International Conference*,
edited by Jae-Young Choi and Seungyu Rah

standing wave (XSW) measurements [7,8], Fig. 1a. In this experiment an XSW field with high visibility interference fringes has been created within confined areas inside the crystal. In experiments when only the intensity of the diffracted beam has to be measured, a perfect crystal can be placed after the sample assuming the position of the analyzer crystal [10], Fig. 1b. Typically used in a diffraction experiment to select a coherent part of the scattering from the sample and to suppress inelastic contribution, the analyzer crystal in this arrangement is working as an angular filter by selecting a fixed narrow angular region from the focusing aperture and thus providing high angular

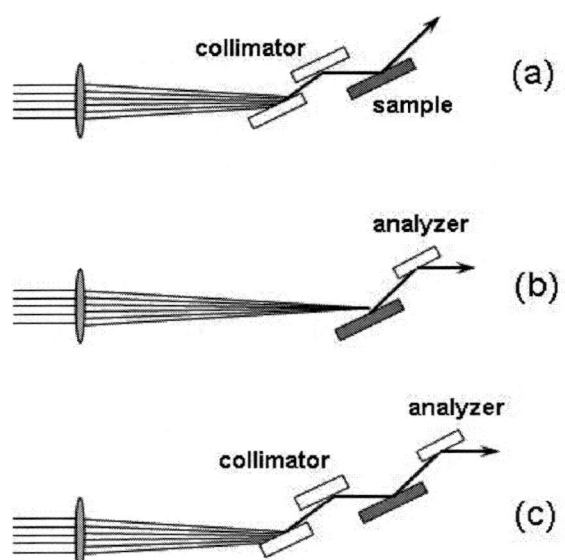

resolution. Both arrangements, with collimating crystal between the focusing capillary and the sample, and with the analyzer crystal after the sample have been used in our recent experiments at to perform reciprocal space mapping with 10^{-5} Å^{-1} resolution [11]. In addition to producing valuable structural information about optoelectronic devices with the a in the range of 15 to 60 microns, our experiments at CHESS served as prototype experiments for sub-micron beams at the third generation sources.

FIGURE 1. Perfect crystal optics in high-resolution microbeam experiments. (a) Perfect collimator crystal is used to condition a microbeam for an XSW experiment. (b) Perfect analyzer crystal is used for high angular resolution diffraction experiments. (c) Both, collimator and analyzer crystals are used for high resolution reciprocal space mapping. Miniature (to fit limited space between focusing optics and the sample) Si or Ge double-bounce channel-cut crystal have been used in (a) and (c) as collimating optics. Both, one-bounce (flat) and three-bounce Si or Ge analyzer crystals have been used in (b) and (c) arrangements.

HIGH ANGULAR-RESOLUTION MICRODIFFRACTION WITH SUB-MICRON BEAM

The experiments were performed at the APS 2ID-D microscope beamline equipped with a phase zone plate [12]. The experimental setup is shown in Fig. 2. The energy of the undulator beam was tuned to 11.890 KeV, above the As-K and below the Au-L edges. A horizontal slit S1 upstream of the Si(111) double-crystal monochromator was used to limit the horizontal source size. The phase Au zone plate with the outer zone of 0.12 μm and the focal distance of 149 mm produced the focal spot size of 0.35 μm (vertical) by 0.24 μm (horizontal). The sample was mounted on a six-circle diffractometer equipped with a precision XYZ-translation stage and a perfect Si(004) analyzer crystal mounted on a detector (2θ) arm. The diffraction curves were measured by performing Δ(θ)-Δ(2θ) scans, where θ and 2θ denote rotation angles of the sample and the detector arm, respectively, and Δ(2θ)=2·Δ(θ) for symmetrical reflection. The angular resolution of our setup was determined by the acceptance range of the analyzer crystal "intrinsic" rocking curve, *i.e.* 2.17 arc sec for Si(004) crystal at 11.89 KeV.

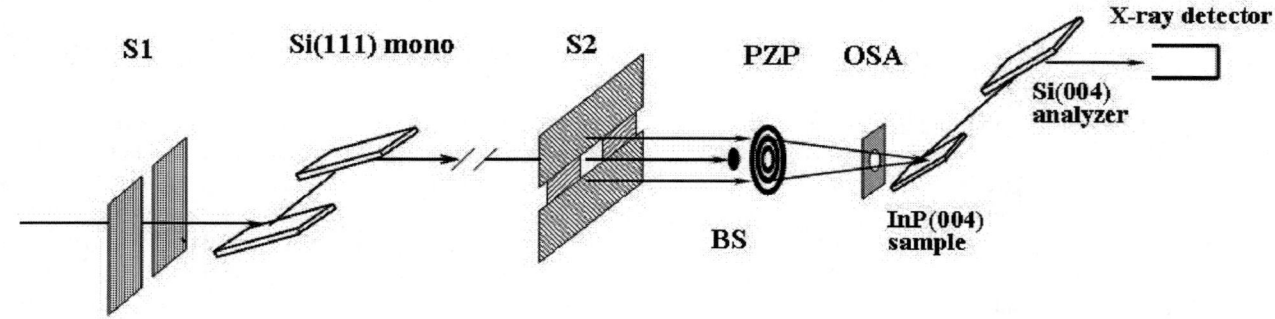

FIGURE 2. Experimental setup for high angular-resolution microdiffraction at the APS 2ID-D beamline. **PZP** - phase zone plate with the outer zone of 0.12 μm and a focal distance of 149 mm; **OSA** – order sorting aperture, **BS** - 30 μm diameter gold beam stop; **S1** – upstream horizontal slit to limit horizontal source size; **S2** – vertical and horizontal slits in the hutch. The InP(004)

sample with optoelectronic structures was mounted on a precision XYZ stage and a single-bounce Si(004) analyzer crystal – on a detector arm of the six-circle diffractometer.

The microbeam HRXRD setup was applied to study compositional and thickness variations in InGaAlAs and InGaAsP-based waveguide arrays grown by Metal Organic Vapor-Phase Epitaxy (MOVPE) in the Selective Area Growth (SAG) regime in narrow openings between SiO_2 stripes. The MOVPE-SAG technique is a modern industrial technology for growing active elements of different optoelectronic devices simultaneously on the same substrate with the properties of these structures effectively controlled by the geometry of the SiO_2 mask [13-15]. The active region of the waveguides was formed by the multiple quantum well (MQW) structures with $N=9$ periods and two 50 nm thick separate confinement layers (SCL) schematically shown in Fig. 3. The length of the waveguides was 600 microns. The microbeam was positioned by monitoring Ga-K and As-K fluorescence and HRXRD curves were measured for different positions of the beam across the waveguide ridge. The commercial RADS-Mercury BEDE software [16] based on the dynamic diffraction theory for x-ray diffraction in layered structures was utilized to determine the period and the strain in the MQW structure [17,18].

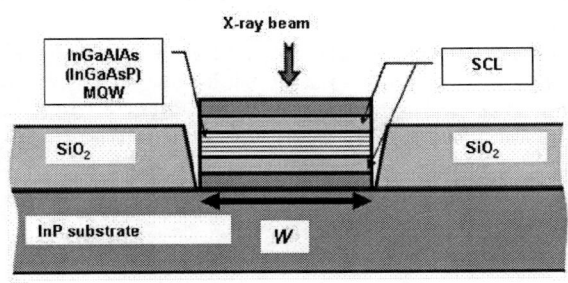

FIGURE 3. Waveguide MQW structure consisting of N=9 periods and two 50 nm thick separate confinement layers (SCL) grown in narrow openings between SiO_2 stripes.

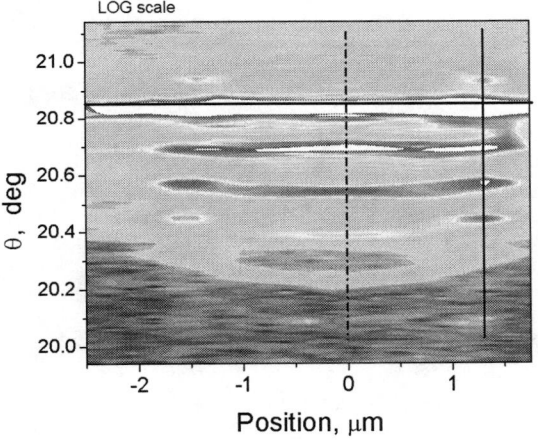 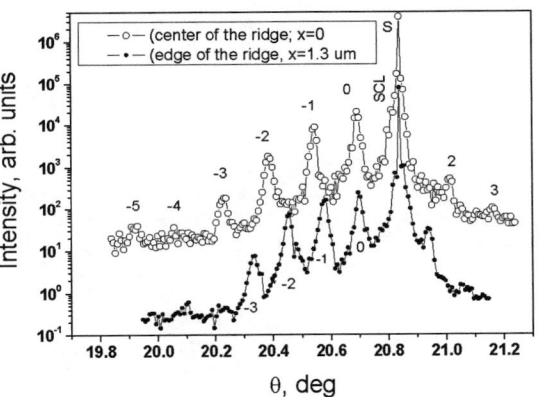

FIGURE 4. Left panel: map representing diffraction pattern for different positions of the microbeam across the InGaAsP waveguide ridge. The horizontal regions of enhanced intensity correspond to the peak from the substrate (at the top) and MQW satellite peaks. Right panel: two diffraction curves corresponding to two vertical lines shown on the left panel: the dashed line runs through the center of the ridge and the solid line at the edge of the ridge. The satellites peaks are marked according to their order on both curves. The shift of the satellite peaks are clearly observed indicating the change in the thickness (period) and the composite strain across the ridge.

As an example, Fig. 4 shows a diffraction intensity map measured from the 1.6 micron wide InGaAsP waveguide ridge. The regions of bright color indicate the angular positions of the satellite peaks. Two diffraction curves corresponding to the center and the edge of the ridge are shown on the right panel. The map clearly reveals significant variations across the ridge: as the microbeam moves away from the center of the waveguide the satellites peaks are getting closer to each other indicating the increase in the MQW period. This is a demonstration of the thickness enhancement at the sidewalls of the ridge, which is well-known for the phosphorous-based SAG structures and usually attributed to the insufficient surface migration of the group-III precursor during growth. The quantitative results were obtained by fitting the diffraction data and determining the MWQ period and the strain across the ridge. The results of the fit are shown in Fig. 5 together with the As-K fluorescence data. The MQW

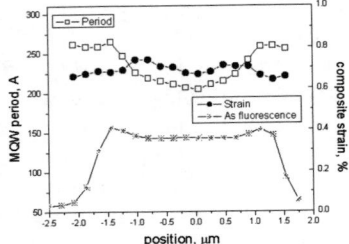

period increases from 205 nm in the center to 260 nm at the edge accompanied by less pronounced changes in lattice mismatch.

FIGURE 5. The MQW period (opened squares, left) and the composite strain (solid circle, right) across 1.6 micron wide InGaAsP waveguide ridge. The As-K fluorescence yield (stars, arbitrary units) is shown for comparison. Both, the MQW period and fluorescence data show thickness enhancement at the edges of the ridge.

A similar analysis performed on InGaAlAs waveguides revealed that the uniformity of the Al-based ridges is superior compared to traditional P-based devices, which is important for further development of a one-growth-step technology of high density integrated optoelectronic devices [17].

In conclusion, an X-ray setup based on a phase zone plate and a perfect analyzer crystal was introduced to perform X-ray diffraction with arc sec angular resolution and sub-micron lateral special resolution. We believe that this approach can be effectively used for structural characterization of future optoelectronic devices based on a variety of III-V materials such as GaN, GaAs, and InP produced by the next generation heteroepitaxial nanotechnology.

ACKNOWLEDGMENTS

Use of the Advances Photon Source was supported by the U.S. Department of Energy, Office of Science, Office of Basic Energy Sciences, under Contract No. W-31-109-ENG-38. The Cornell High Energy Synchrotron Source was supported by the National Science Foundation and the National Institutes of Health/National Institute of General Medical Sciences under award DMR-0225180. The authors thank A. Ougazzaden for the growth of the device structures and S. O'Malley for help with the measurements.

REFERENCES

1. C. Riekel, *Rep. Prog. Phys.* **63**, 233-62 (2000).
2. D. K. Bowen & B. K. Tanner, *High Resolution X-ray Diffractometry and Topography*, London: Taylor and Francis, 1998.
3. Z-H. Cai, W. Rodrigues, P. Ilinski, D. Legnini, B. Lai, W. Yun, E. D. Isaacs, K. E. Lutterodt, J. Grenko, R. Glew, S. Sputz, J. Vandenberg, R. People, M. A. Alam, M. Hybertsen and L. J. P. Ketelsen, *Appl. Phys. Lett.* **75** 100 (1999).
4. D. E. Eastman, C. B. Stagarescu, G. Xu, P. M. Mooney, J. L. Jordan-Sweet, B. Lai and Z. Cai, *Phys. Rev. Lett.* **88**, 156101 (2002).
5. S. Kimura, H. Kimura, K. Kobayashi, T. Oohira, K. Izumi, Y. Sakata, Y. Tsusaka, S. Yokoyama, S. Takeda, M. Urakawa, Y. Kagoshima and J. Matsui, *Appl. Phys. Lett.* **77** 1286 (2000).
6. S. Kimura, Y. Kagoshima, T. Koyama, L. Wada, T. Niimi, Y. Tsusaka, J. Matsui and K. Izumi, *AIP Proceedings* **705** 1275 (2004).
7. A. Kazimirov, D. H. Bilderback, R. Huang and A. Sirenko, *AIP Proceedings* **705** 1027 (2004).
8. A. Kazimirov, D. H. Bilderback, R. Huang and A. A. Sirenko and A. Ougazzaden, *Journal of Physics D: Applied Physics, Rapid Communications* **37** L9 (2004).
9. R. Huang and D. H. Bilderback, *J. Synchrotron Radiation* **13** 74-84 (2006).
10. A. A. Sirenko, A. Kazimirov, R. Huang, D. H. Bilderback, S. O'Malley, V. Gupta, K. Bacher, L. J. P. Ketelsen and A. Ougazzaden, *J. Appl. Phys.* **97** 063512 (2005).
11. A. Kazimirov, et al, under preparation.
12. Z. Cai, B. Lai, and S. Xu, J. Phys. IV, **104**, 17 (2003).
13. M. Gibbon, J. P. Stagg, C. G. Cureton, E. J. Thrush, C. J. Jones, R. E. Mallard, R. E. Pritchard, N. Collis and A. Chew, *Semicond. Sci. Technol.* **8** 998 (1993).
14. M. A. Alam, R. People, E. Isaacs, C. Y. Kim, K. Evans-Lutterodt, T. Siegrist, T. L. Pernell, J. Vandenberg, S. K. Sputz, S. N. G. Chu, D. V. Lang, L. Smith and M. S. Hybersten, *Appl. Phys. Lett.* **74** 2617 (1999).
15. Y. Sakata, Y. Inomoto and K. Komatsu, *Journal of Crystal Growth* **208** 130 (2000).
16. M. Wormington, C. Panaccione, K. M. Matney, and K. D. Bowen, Philos. Trans. R. Soc. London, Ser. A, **357**, 2827-2848, (1999).
17. A. A. Sirenko, A. Kazimirov, A. Ougazzaden, S. O'Malley, D. H. Bilderback, Z.-H. Cai, B. Lai, R. Huang, V. Gupta, M. Chien, S.N.G. Chu, Appl. Phys. Lett. **88** 081111 (2006).
18. A. Kazimirov, A. A. Sirenko, D. H. Bilderback, Z.-H. Cai, B. Lai, R. Huang, A. Ougazzaden, Journal of Physics D: Applied Physics, **39** 1422 (2006).

Poly (methyl methacrylate) Formation and Patterning Initiated by Synchrotron X-ray Illumination

J. Xiao[a], C. H. Wang[b], T. Y, Yang[b], Y. Hwu[b], J. H. Je[a]

[a] Department of Materials Science and Engineering, Pohang University of Science and Technology, Pohang, Korea
[b] Institute of Physics, Academia Sinica, NanKang, Taipei 115, Taiwan

Abstract. A facile radiation method was developed to obtain micro-sized poly (methyl methacrylate) (PMMA) particles and create patterned coating on different substrates by a synchrotron x-ray induced dispersion polymerization. The polymerization of MMA monomer and well defined patterning was successfully realized. The produced PMMA particles and patterning were characterized by Fourier transformation infrared (FTIR), [1]H-Nuclear Magnetic Resonance (NMR), and Scanning Electron Microscope (SEM). The observed patterning contrast essentially derived from a variation of size, density and morphology of particles and the type of substrate materials used.

Keywords: PMMA, patterning, synchrotron x-ray.
PACS: 82.50.Kx

INTRODUCTION

The extremely intense x-rays produced by the third generation synchrotron radiation (SR) facilities possess a variety of unique advantages over conventional light sources and therefore are preferably utilized for the research and development of both fundamental and applied science and technologies. Besides the derived superior characterizing capability, synchrotron x-rays are also utilized to induce photo-chemical reactions for material processing and micro-sized device fabrication. X-ray lithography, deposition and etching account for the most recently developed applications. Novel features associated with these technical approaches include: (1) high flux of energetic x-ray photons enable chemical reaction to be induced and completed in a very short time; (2) very short wavelength that may interfere with devices of micro or sub-micro dimension with high spatial resolution; (3) high energy (especially for hard x-ray) provides deep penetrating lengths and make it possible to induce chemical reactions in a system composing materials of different Z numbers; and (4) the tunability of wavelength of energetic photons opens up the possibility of material-selective or cite- specific photochemistry and patterning. These unique features offered by x-ray deposition demonstrate valuable implication for micro-patterning procedures.

Recently, radiation induced photochemistry has been increasingly utilized to prepare colloidal metallic solutions [1,2] and continuous films composed of nanoclusters [3-5]. Rosenberg and co-workers [4,5] firstly explored the formation of gold and silver coatings induced by synchrotron x-ray irradiation on substrates of various materials (Mo, SiO_2 and polyimide). However, to our knowledge, little work has been reported on depositing polymeric pattern on substrates by a hard x-ray illumination method. This novel polymer patterning technique may find potential applications in microelectronics and nano-fabrication industry. In this work, based on synchrotron x-ray induced dispersion polymerization, we develop a facile radiation chemical approach to form uniform dispersed PMMA particles and create well-defined PMMA patterns on certain substrates.

EXPERIMENTAL

The reaction solutions we employed for PMMA formation were composed of three compositions: the monomer Methyl methacrylate (MMA) (across Inc., US), water/ethanol media and the stabilizer Poly (vinyl pyrolidone) (PVP) (Kant chem. Co. Inc). The monomer MMA was purified by reducing pressure distillation and bubbled with argon gas for 5 minutes to remove the oxygen and then well sealed before use. The MMA containing micro centrifuge tube

CP879, *Synchrotron Radiation Instrumentation: Ninth International Conference*,
edited by Jae-Young Choi and Seungyu Rah

was transferred for x-ray irradiation. The monomer solution was bubbled with argon gas for 5 minutes to remove the oxygen that is a radical scavenger and terminates free radical polymerizations. Unmonochromatized x-ray beam from the 7B2 bending magnet beamline at the 2.5 GeV storage ring of PLS (Pohang Light Source, Pohang, Korea) was utilized throughout the exposure experiments. [6] Bending magnet beamlines produce broadband x-rays and in the case of 7B2 beamline photon energy varies approximately from 5-15 keV.

Suspended PMMA particles were then separated from the solution by centrifuging at 8000 rpm for 10 min and repeatedly rinsed by fresh deionized water for five times to completely remove remnant chemicals attached on PMMA particle surfaces. Such produced samples were studied using a Bruker FT-IR spectrometer at 4 cm^{-1} resolution by mixing with a KBr pellet to confirm the polymerization. Then the products were examined by a proton NMR spectrometer (Bruker DPX (300MHz)) with a CDCl$_3$. To perform the x-ray initiated PMMA patterning, an aluminum mask with ordered circular holes was inserted into the beam path just before the tube. The substrates to be patterned were vertically located in the polymerization solution along the beam tracking path. Two types of substrate were selected in this study: silicon and aluminum coated silicon. The Aluminum coating was done by a sputtering method and the thickness of the aluminum layer is around 500 nm. A series of exposure time range from 15 seconds to 10 minutes was chosen to study the evolution of the pattern formation. The resultant patterns were then observed under optical microscopy (Leica Galen-III, Germany) and SEM (JEOL JSM-6330F, Japan).

RESULTS AND DISCUSSION

The H^1 NMR spectra of MMA, and polymerized MMA particulates formed after different durations of x-ray exposure are given in Fig.1. The observed shifts were specifically assigned to the functional groups of methyl and methacrylate. By comparing the evolution of characteristic shifts of (a)–(c), after 20 min's x-ray shining, the peaks belonging to MMA monomer significantly decreased, indicating the amount of remaining monomer was minimal. Therefore it was found that a radiation of 20 min of x-ray irradiation is sufficient to complete dispersion polymerization of MMA monomer.

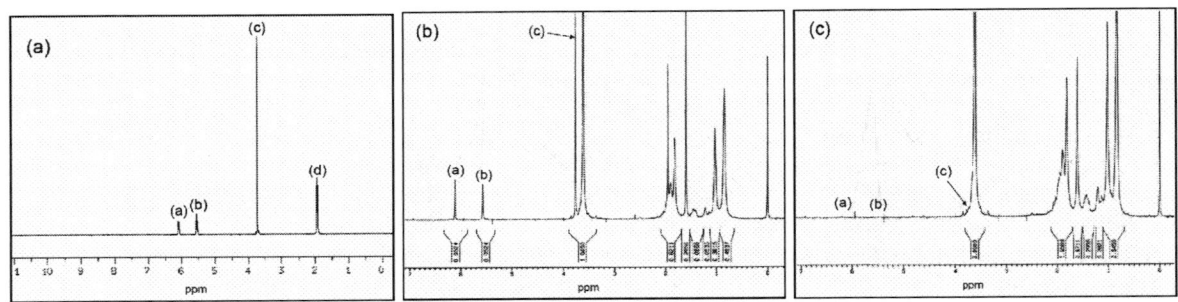

FIGURE 1. H^1 NMR spectrum of (a) standard MMA and PMMA induced by x-ray from purified MMA for different exposure times (b) 10 min; (c) 20 min.

The optical image of PMMA patterns formed on silicon substrate after 3 min x-ray exposure is shown in Fig. 2. The image with higher magnification in Fig. 2 (b) shows the marked region in Fig. 2 (a). A contrast exists between the interfaces of x-ray exposed and non-exposed areas on the substrates, which were derived from the difference of particle density. From Fig. 3, at the early stage of patterning, a distinct difference of the formed particle density can be found between the substrates with beam (Fig. 3 (a)) and without beam (Fig. 3 (b)). The particle size of the beam part was much larger than the non-exposed part while its density was lower. The morphological changes of polymeric particles formed on the beam exposed and non-exposed regions of substrate can also be identified: the particles formed at the beam region were more easily distorted indicating a lower molecule weight. With increasing radiation time, the particle size increased either steadily for beam exposed region or tremendously for the non-exposed region, as shown in Figs. 3 (c) and (d). The influences of exposure time and x-ray beam illumination on the size, density and morphology of deposited PMMA particle on silicon substrate are summarized in Table 1.

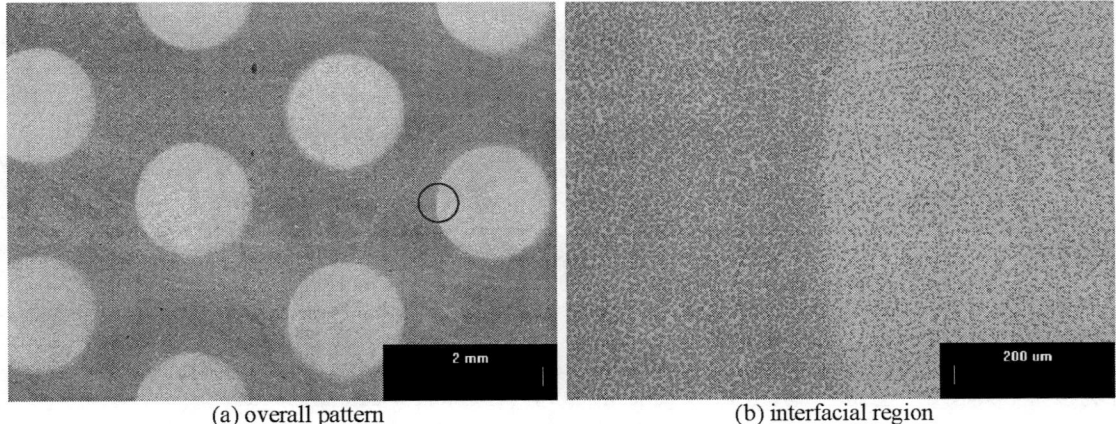

(a) overall pattern   (b) interfacial region

FIGURE 2. Optical image of PMMA pattern formed on silicon substrate for 5 min exposure.

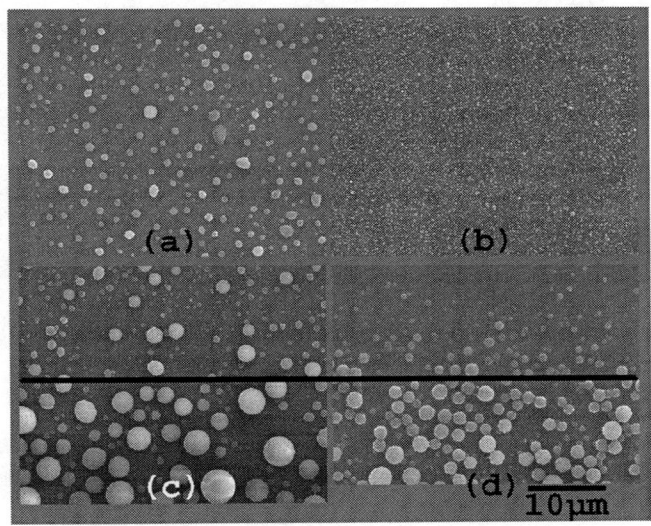

FIGURE 3. SEM micrograph of PMMA deposit on aluminum coated silicon substrate with different radiation time: beam exposed (a) and non-exposed (b) regions for 5min radiation; the intermediate region for 10 min (c) and 20 min (d). The black line roughly marks the boundary of incident x-ray beam path.

TABLE 1. Effect of exposure time and x-ray beam irradiation on the size, density and morphology of PMMA particles.

Radiation time (min)	Region	Particle size* (μm)	Particle density (10^4 number/mm^2)	Rigid shape**
5	Beam	0.8	20	No
	No beam	0.3	400	Yes
10	Beam	1.0	30	No
	No beam	1.2	50	Yes
20	Beam	1.0	10	No
	No beam	1.5	60	Yes

Although with the given radiation time (up to 10 min), a complete polymerization of MMA cannot achieved, we found that a pattern can be formed with x-ray irradiation as short as 15 sec. Furthermore, when the beam exposure was longer than 10 min, the morphology of the obtained patterns got degraded. It is regarded that the interactions between the immerged substrates and incoming x-ray photons at least partly account for the above observations. The effect of substrate materials was further studied by examining the evolution of the size and density of the formed PMMA particles with x-ray radiation length. The proof-of-principle results (data not shown) indicated a significant substrate effect. With longer exposure time, the particle size on the aluminum substrate tended to be kept constant or even decrease a little bit, whereas the particles on the silicon substrate kept growing. Considering the particle density, for aluminum substrate, particles within beam region was much denser than those at non-beam region. However, on the silicon substrate, the density difference existed to a less extent. In this sense, the substrate effect was more pronounced for metallic substrate. However it is not possible to resolve a detail account for this "substrate effect" at this point due to the complicated nature of chemical reactions involved. The process of synchrotron x-ray induced pattern of Polymerized MMA particles essentially involved the following procedures: (1) x-ray radiation induced polymerization of MMA monomers; (2) electrochemical reactions on the surfaces of substrates; and (3) radiation chemical reactions as a consequence of interaction between energetic x-ray photons and various species including solvent and reactants. A special feature of synchrotron x-ray is closely related to the high flux of high energy photons it generated. Therefore, upon the transferring of intense energy to material system, the dissipation of ionization energy is not limited to certain chemical groups that become active specific wavelength, but occurs with all molecules within the incident beam path. Another important consequence of the random dissipation of energy absorption from ionizing radiations is the fact that any substance existed will be radiolyzed and thereby the radiolysis products and intermediate phase contribute a lot to the dispersion polymerization of monomer and the assembly of formed particles upon the immersed substrates. Furthermore, the radolysis of the formed polymer may also lead to the degradation (cross-linking, chain fracture etc.) of polymeric patterns, which further complicated the deposition and patterning procedures. Further systematic research works are required to understand and develop a framework of the chemical events occurred.

CONCLUSIONS

Micro-sized PMMA particles and patterns are formed by a synchrotron x-ray induce dispersion polymerization method. The observed patterning contrast is correlated to the variations of size, density and morphology of particles and type of substrate materials. The complexly nature of reactions involved in the radiation induced polymerization and patterning is highlighted.

ACKNOWLEDGMENTS

This work was supported by the National Science Council (Taiwan), by the BK21 project, by the Korea Institute of Science and Technology Evaluation and Planning (KISTEP) through the National Research Laboratory (NRL), by the SKORE-A projects.

REFERENCES

1. Y. C. Yang, C. H. Wang, Y. Hwu and J. H. Je, *Mater. Chem. Phys.* (in press).
2. C. C. Kim, C. H. Wang, Y. C. Yang, Y. Hwu , S. K. Seol, Y. B. Kwon, C. H. Chen, H. W. Liou, H. M. Lin, G. Margaritondo and J. H. Je, *Mater. Chem. Phys.* (in press).
3. P. H. Borse, J. M. Yi, J. H. Je, W. L. Tsai, Y. Hwu, *Journal of Applied Physics* **95**, 1166 (2004).
4. R. A. Rosenberg, Q. Ma, B. Lai and D. C. Mancini, *J. Vac. Sci. Technol. B* **16**, 3535 (1998).
5. Q. Ma, N. Moldovan, D. C. Mancini, and R. A. Rosenberg, *Applied Physics Letter* **76**, 2014 (2000).
6. S. Baik, H. S. Kim, M. H. Jeong, C. S. Lee, J. H. Je, Y. Hwu and G.. Margaritondo. *Rev. Sci. Instr.* **75**, 4355 (2004).

Design and Performance of the Compact YAG Imaging System for Diagnostics at GMCA Beamlines at APS

Shenglan Xu, Robert F. Fischetti, Richard Benn and Stephen Corcoran

GM/CA CAT, Biosciences Division, Argonne National Laboratory, Argonne, IL 60439, U.S.A

Abstract. A compact YAG (Chromium Doped Yttrium Aluminum Garnet - Cr4+:YAG) imaging system has been designed as a diagnostic tool[1][3] for monochromatic x-rays emanating from the first "Hard" x-ray dual-canted undulator at the Advanced Photon Source at Argonne National Laboratory[4]. This imaging system consists of a flat YAG crystal, right angle prism/mirror, video camera and monitor [2]. A flat YAG crystal with a diameter of 10 mm has been installed in vacuum and positioned downstream of the monochromator of the insertion device beamline. Another 20 mm diameter YAG crystal has been installed in vacuum after the horizontal deflecting mirrors of the second insertion device beamline. CCD cameras are mounted in air close to the window of the vacuum ports to image the fluorescence of the YAG crystals. An additional 25 mm diameter YAG crystal has been used for K-B (Kirkpatrick-Baez) mirror focusing and beamline alignment. These YAG imaging systems have greatly facilitated beamline commissioning as well as sample alignment to the x-ray beam in the macromolecular crystallography endstation. An overview of the optics design, mechanical design and the performance of these devices will be presented in the paper.

Keywords: YAG, diagnostics, alignment.
PACS: : 07.85.Qe

INTRODUCTION

A new macromolecular crystallographic facility developed by GMCA CAT has begun operations at the Advanced Photon Source (APS). The facility consists of three beamlines; the world's first pair of fully tunable "hard" dual-canted-undulator beamlines and one bending magnet beamline. The canted undulator geometry has the potential to double the scientific throughput of synchrotron facilities. The independently tunable ID beamlines are operational, and the bending magnet beamline is being commissioned. A compact YAG (Chromium Doped Yttrium Aluminum Garnet - Cr4+:YAG) imaging system has been designed as a diagnostic tool for monochromatic x-rays emanating from the two canted undulator[4]. These YAG imaging systems have greatly facilitated beamline commissioning as well as sample alignment to the x-ray beam in the macromolecular crystallography endstation.

The beamlines have been design for ease of use and rapid alignment. An imaging system is installed in vacuum after optical component on the ID-lines allowing visual inspection of the beam position shape and intensity. Fig. 1 shows two insertion device beamlines from the dual canted undulators and the locations of the installed YAG image systems.

The location, application, diameter and thickness of the YAG crystals are listed in Table 1. The locations of the YAG beam monitors are summarized in column 2. Three sets of YAG imaging system were used for beamline commissioning, five sets were installed on in vacuum beam position monitors (BPMs), and two sets are used for sample alignment at ID experiment stations. The YAG system can be used as tools to help align mirrors to the X-ray beam. In this application, it is beneficial to be able to see the beam before the mirrors intercept the beam and after they deflect it. Therefore, 20 mm diameter YAG crystal has been placed in the vacuum system after the horizontal deflecting mirrors and the Kirkpatrick-Baez (K-B) focusing mirrors. Larger 25 mm diameter YAG crystals are incorporated in the compacted set of YAG imaging system.

CP879, *Synchrotron Radiation Instrumentation: Ninth International Conference,*
edited by Jae-Young Choi and Seungyu Rah
© 2007 American Institute of Physics 978-0-7354-0373-4/07/$23.00

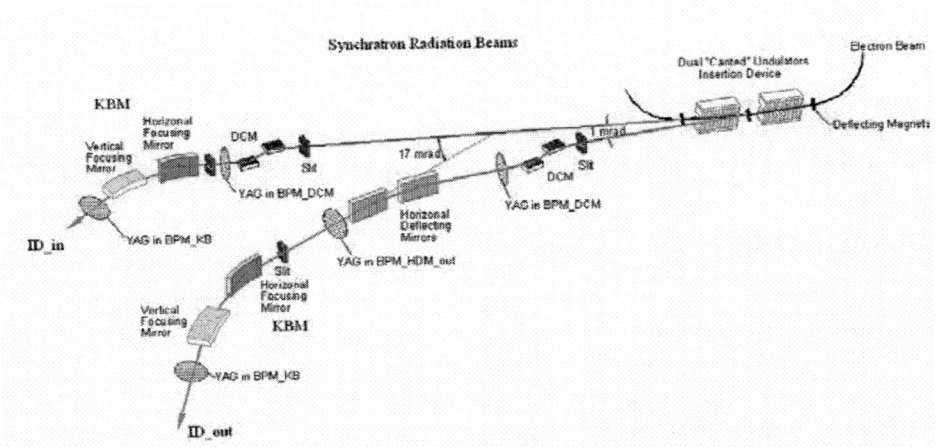

FIGURE 1. Schematic of canted ID beamlines and YAG imaging system at GMCA CAT at the Advanced Photon Source.

TABLE 1. List of YAG image systems installed at the three beamlines of GMCA CAT at APS

YAG	Location	Vacuum or in Air
Ø10mm X 100µ	BPM_DCM of ID_in Beamline	UHV
Ø20mm X 200µ	BPM_K-B, of ID_in Beamline	UHV
Ø10mm X 100µ	BPM_DCM of ID_out Beamline	UHV
Ø20mm X 200µ	BPM_HDM of ID_out Beamline	UHV
Ø20mm X 200µ	BPM_K-B of ID_out Beamline	UHV
Ø20mm X 200µ	ID_in Station for commissioning	In Air
Ø20mm X 200µ	Sample position of ID_in for alignment	In Air
Ø20mm X 200µ	ID_out Station for commissioning	In Air
Ø20mm X 200µ	Sample position of ID_out for alignment	In Air
Ø20mm X 200µ	BM Station for commissioning	In Air

OPTICS DESIGN OF THE YAG IMAGING SYSTEM

The YAG imaging system consists of a flat YAG crystal, right angle front surface mirror or flat mirror, video camera and monitor is shown in Fig.2. Two types of YAG imaging systems that have been used for commissioning and alignment of beamlines are shown in Figure 3. The set up on the left is a very compact system using just a CCD camera, lens and a block to hold the YAG crystal and a mirror. The system sits on a miniature rail for easy transportation and can readily be mounted on a platform to monitor the X-ray beam. The overall dimensions of the YAG system are 150 mm x 38 mm x 76 mm. All components are connected together by a block that has C-mount threads. The 25mm diameter YAG is glued into a C-mount ring that is then screwed into the block. A C-mount collar screws in to the block perpendicular to the YAG. The lense of the camera slips into the collar and is held with set screws. The CCD camera is directly connected with the objective lens A protected aluminum-coated first surface mirror (35mm x 35 mm x 3 mm) is positioned in the housing at 45 degrees relative to the incident beam. The compact YAG imaging system mounted on XY stages for ID beamlines commissioning and alignment.

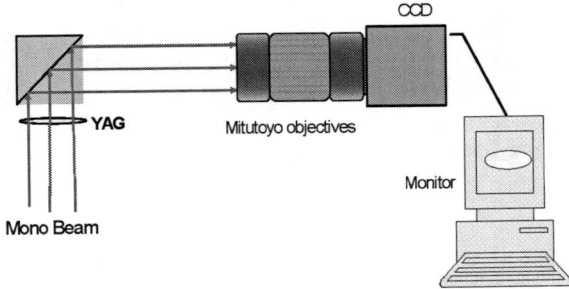

FIGURE 2. Schematic of YAG imaging system.

The right-hand picture in Figure 3 shows a set up used for the bending magnet beamline commissioning. The unfocused monochromatic beam is almost 50 mm wide in the experimental station. It is desirable to see the entire beam so a 50 mm diameter YAG is used here. A large mirror prism (50 mm X 50 mm X 71 mm) servers as a front surface mirror so the camera can view the beam from the side, thus protecting the lens form radiation damage.

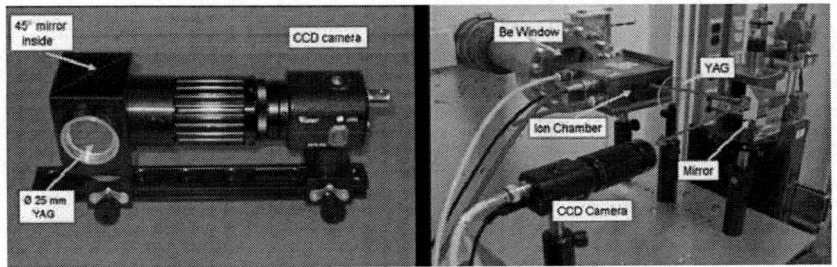

FIGURE 3. Left-hand of figure 3 is a compact imaging system for ID beamlines commissioning and alignment. The Ø25 mm YAG crystal, flat mirror, video camera were mounted on the PRL-6 rail. Right-hand of figure 3 is a set-up for bending magnet beamline commissioning and alignment. The Ø50 mm YAG crystal, right angle mirror and camera were mounted on the base plate which installed on PXY stages

MECHICAL DESIGN OF THE YAG IMAGING SYSTEM IN VACUUM

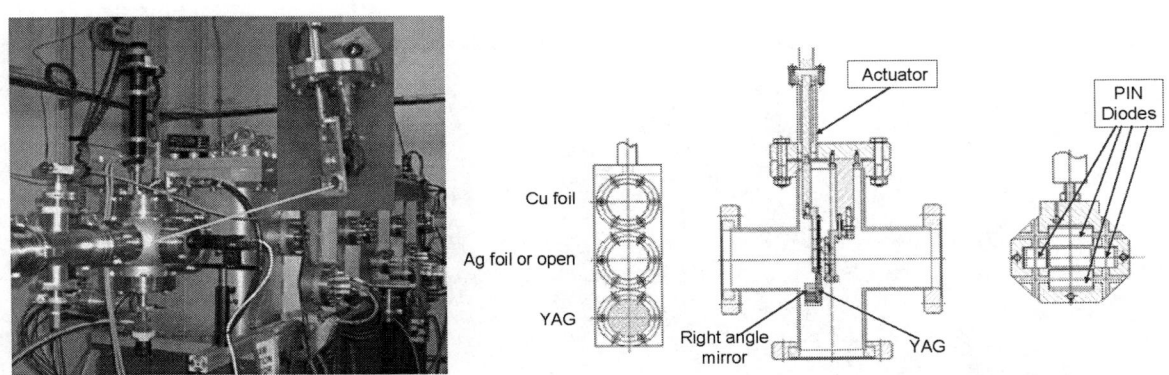

FIGURE 4. YAG crystals of various diameters have been installed in the UHV vacuum system of the two ID-beamlines just downstream of the major optical components (monochromators, horizontal deflecting mirrors and K-B focusing mirrors). A CCD camera mounted outside the vacuum system monitors the fluorescence form the YAG crystal through a viewport. The right drawing shows a flat Ø10 mm YAG crystal and a 10 mm X 10mm X 14 mm right angle mirror have been installed in the vacuum chamber

The YAG systems are attached to the beam position monitors that providing intensity and positional read outs. The beam position monitor (BPM) has to comply with high vacuum requirements because it is integrated into high vacuum sections. Four photodiodes are arranged around the beam facing downstream (Fig. 4 right picture). The beam passes through a thin foil fluorescence or scattering that is continuously monitored by the photodiodes. The photo current from each diode is analyzed to extract the beam position. The device is based on a NW 63 CF 6-way cross (standard) see Fig. 4 (middle picture) and on a NW 100 CF cross needed for the BPM-DCMs and the BPM-HDM-1. In general, two ports of the cross are for beam entrance and beam exit, one port will serve as access for the diode holder and the target foil translation, and additional ports are for survey of the diode. Copper and silver foils are mounted on the first and second positions of the holder. A YAG image assembly was installed on the bottom port of the holder (see Fig 4 left picture of right drawing) which on the actuator will provide enough yields. The distance between the two ports is 30.5 mm. The stroke of the actuator allows the foil assembly and the YAG assembly to be pulled out of the direct beam. These YAG imaging systems have greatly facilitated beamline commissioning

The requirements are similar for the monitor after the HDM, for which both beams, ID-in and ID-out, have to be accounted. In addition, when using the monitor for the HDM alignment, both, the diode holder and the foil translator, have to accommodate a beam travel range of the deflected ID-out beam from 0 mrad to 16 mrad. The 20 mm diameter YAG has installed for accepting the beam.

Fig.5 shows the simultaneous observation of monochromatic X-rays from the dual-canted undulators. The picture shows the images of monochromatic X-ray from the first "Hard" dual-canted undulators on the YAG crystals (February 17, 2005). In the left image the YAG shows monochromatic X-rays from the downstream

FIGURE 5. Simultaneous Observation of Monochromatic X-rays from the 1st "Hard" Dual-Canted Undulator

YAG FOR SAMPLE ALIGNMENT AND CENTERING

These YAG imaging systems are routinely used to align the X-ray beam and the sample. The end station goniometer has a high magnification optical system for viewing and centering the sample. The optical system has a hole along the optical axis through which the X-ray beam passes. One can align the X-ray beam and the optical axis of the camera by placing a YAG on the goniometer at the sample position and centering the image of the beam on the cross hairs as indicated in Fig. 6. In the upper right picture the X-ray beam is misaligned. The lower right picture shows a beam properly aligned to the cross hairs. Once the ebam and the cross hairs are aligned the sample can be mounted instead of the YAG and easily aligned to the X-ray beam by centering it on the cross hairs.

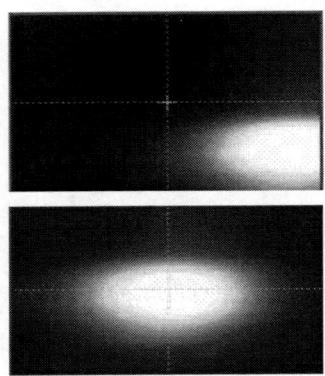

FIGURE 6. The picture on the left shows a Ø10 mm X 100 μm YAG crystal mounting at sample position. The top right and lower left pictures are images of the beam captured with a video frame grabber.

ACKNOWLEDGMENTS

GM/CA CAT has been funded in whole or in part with Federal funds from the National Cancer Institute (Y1-CO-1020) and the National Institute of General Medical Sciences (Y1-GM-1104). Use of the Advanced Photon Source was supported by the U.S. Department of Energy, Basic Energy Sciences, Office of Science, under contract No. W-31-109-ENG-38. We thank W. Smith, D. Yoder and R. Sanishvili for helpful discussions and support.

REFERENCES

1. S. Xu, Z. Cai and B. Lai "Mechanical design for an x-ray diffraction microprobe at the Advanced Photon Source" 2004 MEDSI meeting in ESRF (May 24-27)..
2. S. Xu, R.F. Fischetti, D. Yoder, R. Benn, S. Corcoran, R. Sanishvili and W.W. Smith. "Compact system utilizing YAG crystals to image the monochromatic beam" for APS 2006 User meeting. In Argonne National Laboratory (May 1-5, 2006).
3. Reinhard Pahl "Toys and tools" TWG Meeting in APS (August 15, 2002)
4. R.F. Fischetti "First light from the second GM/CA – CAT canted undulator"APS Science 2005, P160-P161

Hard X-ray Photoemission Spectroscopy using Excitation Energies of up to 10 keV for Materials Science

J. J. Kim[1], E. Ikenaga[1], M. Kobata[1], M. Yabashi[1], and K. Kobayashi[1], Y. Nishino[2], D. Miwa[2], K. Tamasaku[2], and T. Ishikawa[2]

[1] *SPring-8/JASRI Kouto 1-1-1, Sayo-cho, Sayo-gun, Hyogo 679-5198, Japan*

[2] *SPring-8/RIKEN Kouto 1-1-1, Sayo-cho, Sayo-gun, Hyogo 679-5198, Japan*

Abstract. Hard X-ray photoemission spectroscopy has been shown to be a powerful tool for investigating the bulk sensitive chemical and electronic states of solid materials. High energy resolution and high throughput have been achieved using the highly brilliant undulator X-rays available at SPring-8 along with an electron analyzer optimized for high electron kinetic energy. Here, we report the performance and characteristics of hard X-ray photoemission spectroscopy of photoelectron excitation energies of up to 10 keV.

Keywords: Photoemission spectroscopy, Hard X-ray, Bulk sensitive, Electronic structure
PACS: 07.81.+a; 07.85.Qe; 71.20.-b

INTRODUCTION

Nanoscience continues to make rapid process, and demands experimental setups capable of differentiating between surface and bulk properties. Photoemission spectroscopy is an extensively used method in materials science for investigating the chemical and electronic states of target materials. However, conventional photoemission spectroscopy using ultra-violet, Al and Mg K_{α}, and soft X-ray light sources is known to be surface sensitive, as the inelastic mean free paths of the photoelectrons in the solid are in the range of 0.5-3 nm [1]. This means that ultra high vacuum (UHV) is necessary to maintain the surface free from adsorbates during the measurement, and that surface effects must be taken into account in the interpretation of the resulting spectra. Therefore, hard X-ray photoemission spectroscopy (HX-PES), which can extend the probing depth to greater than ca. 5-10 nm, is greatly needed. We have confirmed that HX-PES probes the intrinsic bulk electronic structure [2, 3] and have applied this method to various solid materials including high-k HfO_2/interface/Si CMOS gate dielectric structures [2], III-V and II-VI compound semiconductors [4, 5], diluted magnetic semiconductor systems of Cr-doped and Mn-doped GaN [6-8], and strongly correlated electron materials [9, 10] at SPring-8 over the past four years.

There are two decisive factors in achieving a successful HX-PES experiment: One is a highly brilliant radiation source and the other is an electron analyzer optimized for high electron kinetic energy. The former is necessary to compensate for the weak signal intensity due to the rapid decrease of the atomic subshell photoionization cross section with increasing excitation energy and the decrease in the spectral transmissivity of lens systems in electron analyzers with increasing photoelectron kinetic energy. The latter is necessary to achieve low photoelectrons pass energies for high instrumental energy resolution. We have succeeded in HX-PES experiments with the photoelectron excitation energies of up to 10 keV with both high throughput and high energy resolution by using the highly brilliant undulator X-rays available at SPring-8 and the electron analyzer optimized for high electron kinetic energy of Gammadata Scienta R4000-10keV. Here we report the developed performance and characteristics of the HX-PES.

EXPERIMENTS

HX-PES experiments have been performed at the planar undulator beam lines, BL47XU and BL29XU of SPring-8. Photoelectron excitation energies of 6, 8 and 10 keV were used. X-rays monochromatized with a Si (111) double

CP879, Synchrotron Radiation Instrumentation: Ninth International Conference,
edited by Jae-Young Choi and Seungyu Rah
© 2007 American Institute of Physics 978-0-7354-0373-4/07/$23.00

crystal were directed onto samples mounted in a measurement chamber. A channel-cut post-monochromator with (nnn) reflection planes of a Si single crystal (n=3 for 6 keV, n=4 for 8 keV, and n=5 for 10 keV) was used to reduce the energy bandwidth. The bandwidth of the X-rays could be reduced to 60 meV-90 meV depending on the configuration of the X-ray optics. High throughputs are realized with a resolution of 75 meV-250 meV for the characterization of samples.

We have used two different instrumental configurations for the HX-PES experiments, a "room temperature mode (RT mode)" and a "low temperature mode (LT mode)", according to experimental aims. In the RT mode, an analyzer, a motorized xyzθ stage for a sample manipulator, two turbo molecular pumps (TMP's), and two charge-coupled device (CCD) cameras are equipped in the measurement chamber. The vacuum of the analyzer and measurement chamber was ~10^{-5} Pa during measurements. This level of vacuum is would be completely unsuitable for conventional PES experiments, which require UHV condition even better than 10^{-8} Pa. However, the large escape depths at high photoelectron excitation energies allow us to determine bulk electronic states owing to negligibly small contributions from the surface in the low vacuum condition. In the RT mode, we can mount several samples in the measurement chamber very easily, only breaking the vacuum of the measurement chamber and setting the samples in the manipulator. Thirty minutes is enough to reach a vacuum of ~10^{-5} Pa. In the LT mode, the measurement chamber is equipped with a preparation and a load-lock chamber. The vacuum of the measurement chamber in the LT mode is maintained at UHV condition better than 10^{-8} Pa. Therefore, we can measure samples at low temperatures without worrying about the adsorption of moisture at sample surface.

RESULTS AND DISCUSSIONS

The first HX-PES experiment in SPring-8 took place in June 2002 at BL29XU with an excitation energy of 5.95 keV by a group of collaborators from JASRI/SPring-8, RIKEN/SPring-8, and HiSOR. We succeeded in measuring an Au Fermi edge with a total resolution of 240 meV at 300 K and an adequate throughput, even though the instrumentation was not optimized [2, 3]. We have continued to improve the instrumentation and optics. At present, we can perform HX-PES experiments at excitation energies of up to 10 keV with high throughput and high energy resolution. Before presenting the performance, we introduced an improved spectral signal to noise ratio (S/N) by changing CCD camera. Figure 1(a) shows the schematic drawing of the detection system in the photoelectron analyzer. The system consists of a micro channel plate (MCP) detector, phosphorus screen, and CCD camera. The d1 and d2 distances indicate the distances of one capillary to the nearest neighbor capillary in the MCP (15μm) and that of one pixel to the nearest neighbor pixel in the CCD camera, respectively. Figure 1(b) shows the background spectrum above the Fermi edge of Au measured using the first HX-PES instrumental setup. We noticed a large modulation in the background spectrum. Since the distance ratio of d2/d1 was ca. 0.9 in the first instrumental setup, we supposed the modulation could be caused by Moire fringe, given by the equation of (d1d2)/(|d1-d2|). We replaced the old CCD camera with a high resolution CCD camera with a which has the pixel distance of 6.45 μm, below half of the d1 (d2/d1 is ca. 0.4). As a result of the changing CCD camera, the modulation was drastically decreased, as shown in Fig.1(c).

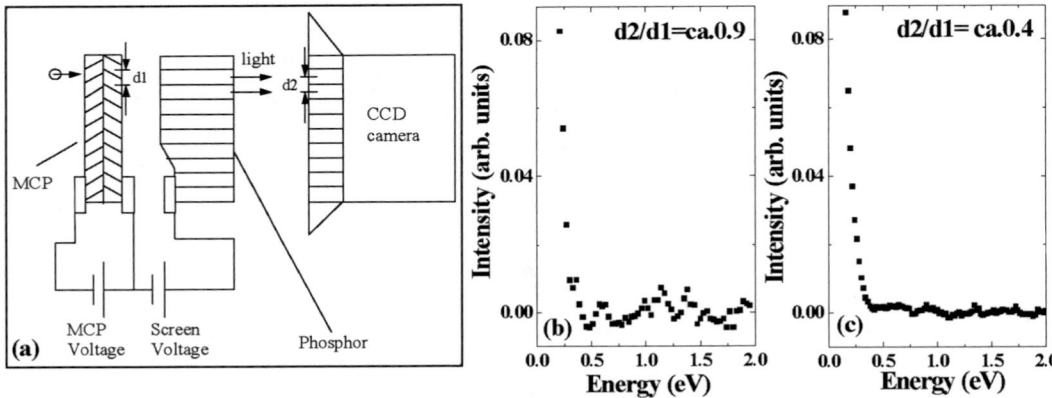

FIGURE 1. (a) Schematic drawing of the detection system in photoelectron analyzer. The detection system consists of MCP, screen, and CCD camera. d1 and d2 are the distances of one capillary to the nearest neighbor capillary in MCP and that of one pixel to the nearest neighbor pixel in CCD camera, respectively. (b) Background spectrum above the Fermi edge of Au when the d2/d1 is about 0.9. There is a large modulation in this spectrum. (c) Background spectrum above Fermi edge of Au when the d2/d1 is about 0.4. The modulation was drastically decreased.

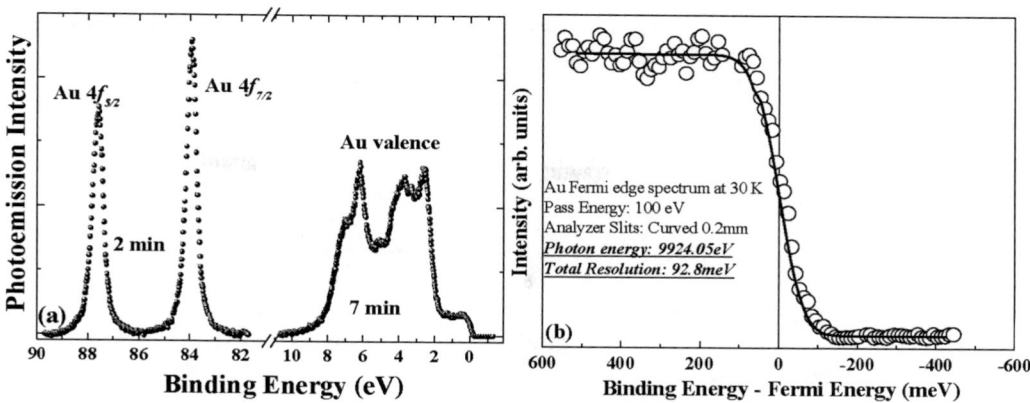

FIGURE 2. (a) Au 4f core level and valence band spectra of a gold plate measured at 7.942 keV with the pass energy of 200 eV. Measurement temperature was 300K. (b)Au Fermi edge spectra measured at 9.924 keV with the pass energy of 100 eV.

Figure 2(a) and (b) show the state-of-the-art performance of the present HX-PES experiment, high throughput and high energy resolution, examined by measuring the Au 4f core level and valence band spectra of a gold plate with a photoelectron take-off angle (measured from the sample surface) of 80°. Figure 2(a) shows the observed spectra with accumulation times of 2 min for the Au 4f spectrum and 7 min for the Au valence band spectrum measured at the excitation energy of 7.942 keV with the RT mode configuration. Despite the fact that no surface cleaning was carried out before introducing the Au plate into the analyzer chamber, a very clear Au valence band spectrum could be observed with only a short measuring time. Figure 2(b) shows the Au Fermi edge measured at 9.924 keV and the instrumental energy resolution including X-ray band width was determined to 92.8 meV at a measuring temperature of 30 K with the pass energy of 100 eV. Analysis of the curve fit gives an estimated analyzer resolution of 23.8 meV.

Until the successful HX-PES experiment with the excitation energies of up to 10 keV, we had suffered from the discharge problem between the deceleration lens and the slit of the photoelectron analyzer. This is because a high voltage must be applied to the deceleration lens to achieve high resolution in HX-PES, unlike for conventional PES. We have overcome this discharge problem and achieved a very stable high voltage supplying up to 10 kV in collaboration with VG Gammadata Scienta. Figure 3(a) shows the full range photoemission spectrum of Cu metal from the most intense Cu 1s (binding energy (BE) = 8979 eV) core level to the valence band at a photoelectron excitation energy of 9.924 keV. The inset of Fig. 3(a) shows the enlarged valence band spectrum of Cu metal. The full range spectrum gives us the experimental photoionization cross section ratio of Cu core level spectra in Cu metal at the excitation energy. It should be mentioned that long collection times were necessary to record the valence band spectrum at the excitation energy of 9.924 keV. This indicates that the electron analyzer has good stability under the applied voltage of 10 kV.

Successful HX-PES experiments at 10 keV were also applied to copper oxides. Previous studies on the copper oxides have concentrated on the investigation of Cu 2p core level spectrum to understand the valence band states. However, the Cu 2p spectrum is difficult to interpret due to the very complex multiplet splitting of the d^9 peak, the splitting between $2p_{3/2}$ (BE = 933 eV) and $2p_{1/2}$ (BE = 952 eV) by the spin-orbital interaction, and the interference with the Bi 4s core level spectrum (BE = 939 eV) in case of Bi-related Cu oxides. The exchange integral of Cu 1s (0.06 eV from the atom) with Cu 3d is very small compare to those of Cu 2p (2.9 eV from the atom) and Cu 3s (19 eV from the atom) with Cu 3d and a negligible multiplet splitting in Cu 1s is expected. Therefore, the Cu 1s spectrum should provide simple information for the valence band states. Figure 3(b) shows the Cu 1s, Cu 2p and Cu 3s core level spectra for CuO. We have clearly observed very intense and simple Cu 1s spectrum while the Cu 2p and Cu 3s spectra are weak and overlapped with a complex satellite structure. Figure 3(c) shows the Cu 1s spectra for some Cu oxides including Cu metal. It was possible to determine the charge transfer states of the copper oxides in just a few minutes by recording the intense Cu 1s core level spectra. HX-PES makes it possible to study a wide range of core level spectra, and extending further the probing depth. It is also possible to study the core level and valence band spectra together at excitation energies of up to 10 keV. This cannot be achieved without very stable high-voltage supplies for the electron analyzer.

The bulk sensitive photoelectron spectra which can be studied using the HX-PES technique provide us with a lot of important information necessary for investigating the electronic structure of solid state materials. We have been successful in performing HX-PES experiments with excitation energies of up to 10 keV with high throughput and high energy resolution. The method has proved itself as a powerful bulk-sensitive tool for investigating the chemical and electronic states of solid materials. It is anticipated that combined core level and valence band studies will continue to produce new information on the electronic properties of various materials.

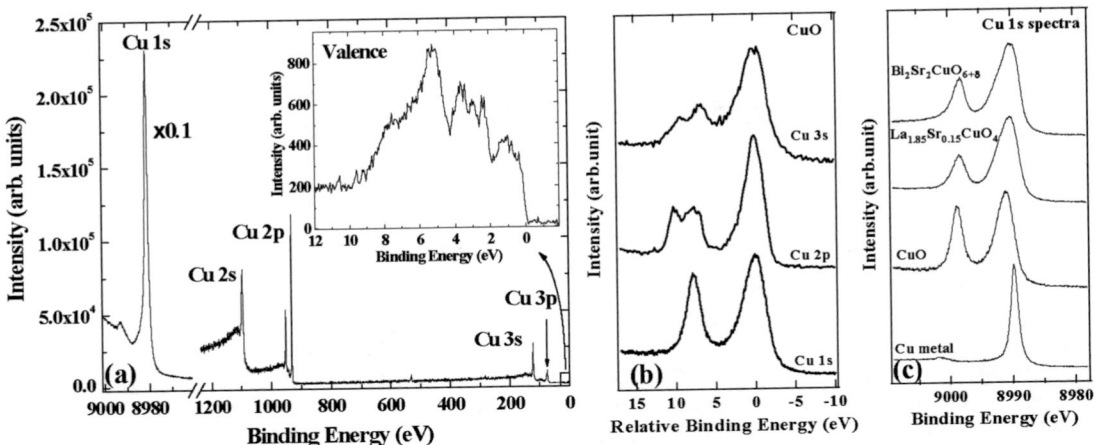

FIGURE 3. (a)Full range photoemission spectrum measured from Cu $1s$ to valence band spectra of Cu metal. (b) Cu $1s$, Cu $2p$, and Cu $3s$ core level spectra of CuO (c) Cu $1s$ core level spectra of some Cu oxides including Cu

ACKNOWLEDGMENTS

We would like to thank Mr. H. Anzai and Dr. A. Ino for offering the results of Cu oxides shown in Fig. 3(b) and (c). We would like to thank Dr. J. Harries for the examination of this manuscript. This work was partially supported by the through a Grant-in-Aid for Scientific Research (A) (No. 15206006, Principal Investigator: Takeo Hattori) and also partially supported by a Nanotechnology Support Project of The Ministry of Education, Culture, Sports, Science and Technology.

REFERENCES

1. The electron inelastic-mean-free-paths were estimated using NIST Standard Reference Database 71, "NIST Electron Ineleastic-Mean-Free-Path Database: Ver. 1. 1." It is distributed via the Web site http://www.nist.gov/srd/nist.htm, and references therein.
2. K. Kobayashi, M. Yabashi, Y. Takata, D. Miwa, T. Ishikawa, H. Nohira, T. Hattori, Y. Sugita, O. Nakasuka, A. Sakai, and S. Zaima, *Appl. Phys. Lett.*, 85, 1005-1007 (2003).
3. Y. Takata, K. Tamasaku, T. Tokushima, D. Miwa, S. Shin, T. Ishikawa, M. Yabashi, K. Kobayashi, J. J. Kim, T. Yao, T. Yamamoto, M. Arita, H. Namatame, and M. Taniguchi, *Appl. Phys. Lett.*, 84, 4310-4312 (2004).
4. K. Kobayashi, Y. Takata, T. Yamamoto, J. J. Kim, H. Makino, K. Tamasaku, M. Yabashi, D. Miwa, T. Ishikawa, S. Shin, and T. Yao, *J. J. Appl. Phys.*, 43, L1029-L1031 (2004).
5. J. J. Kim, H. Makino, K. Kobayashi, P. P. Chen, E. Ikenaga, M. Kobata, A. Takeuchi, M. Awaji, T. Hanada, M. W. Cho and T. Yao, *Phys. Stat. Sol. (c)*, 3, 1846-1849 (2006).
6. J. J. Kim, H. Makino, P. P. Chen, T. Hanada, T. Yao, K. Kobayashi, M. Yabashi, Y. Takata, T. Tokushima, D. Miwa, K. Tamasaku, T. Ishikawa, S. Shin, and T. Yamamoto, *Materials Science in Semiconductor Processing*, 6, 503-506 (2003).
7. J. J. Kim, H. Makino, K. Kobayashi, Y. Takata, T. Yamamoto, T. Hanada, M. W. Cho, E. Ikenaga, M. Yabashi, D. Miwa, Y. Nishino, K. Tamasaku, T. Ishikawa, S. Shin, and T. Yao, *Phys. Rev. B*, 70, 161315(R) (2004).
8. J. J. Kim, H. Makino, T. Yao, Y. Takata, K. Kobayashi, T. Yamamoto, T. Hanada, M. W. Cho, E. Ikenaga, M. Yabashi, D. Miwa, Y. Nishino, K. Tamasaku, T. Ishikawa, and S. Shin, *J. Electron. Spectrosc. Relat. Phenom.*, 144-147, 561-564 (2005) .
9. K. Horiba, M. Taguchi, A. Chainani, Y. Yakata, E. Ikenaga, D. Miwa, Y. Nishino, K. Tamasaku, M. Awaji, A. Takeuchi, M. Yabashi, H. Namatame, M. Taniguchi, H. Kumigashira, M. Oshima, M. Lippmaa, M. Kawasaki, H. Koinuma, K. Kobayahsi, T. Ishikawa, and S. Shin, *Phy. Rev. Lett.* 93, 236401 (2004).
10. H. Tanaka, Y. Takata, K. Horiba, M. Taguchi, A. Chainani, S. Shin, D. Miwa, K. Tamasaku, Y. Nishino, T. Ishikawa, E. Ikenaga, M. Awaji, A. Takeuchi, T. Kawai, and K. Kobayashi, *Phys. Rev. B*, 73, 094403 (2006).

Characterization of Nanoscale Domain Structures in Epitaxial Ferroelectric PbTiO$_3$ Capacitors by Reciprocal Space Mapping

Kilho Lee[1], Hee Han[1], Hyunjung Yi[2], Yong Jun Park[1,3,], Jae-Young Choi[3,], and Sunggi Baik[1]

[1]*Dept. of Materials Science & Engineering, Pohang University of Science and Technology*
[2]*Nano Device Research Center, Korea Institute of Science and Technology*
[3]*Pohang Accelerator Laboratory*

Abstract. In order to test a critical lateral dimension in two dimensional (2D) planar ferroelectric, epitaxial PbTiO$_3$ discrete islands were fabricated by both lithography process and chemical solution deposition process with different lateral size. The evolution of 90° domain structures as a function of film thickness and lateral dimensions was characterized extensively by reciprocal space mapping using the Huber six-circle diffractometer of the synchrotron x-ray diffraction beam line at Pohang Light Source. This method is useful for considering a reciprocal lattice point that is tilted out of the growth plane and has many advantages for the investigation of imperfect heterostructures. As the pattern size decreases in micron scale, the substrate clamping effects is significantly reduced and thus the misfit strain in the films could be relaxed further. As the lateral size of patterns decreases to about 100 nm by e-beam lithography, the quite different domain structures were observed especially in the strain distribution as well as tilting of *a*-domains. The new type of *a*-domains appeared in the nanoscale PbTiO$_3$ patterns in the center of the reciprocal space map of tilted PbTiO$_3$ (100) that has been observed in the continuous films. These new *a*-domains are aligned along the substrate normal without the tilting angle required to maintain an *a/c* twin boundary relationship. Equilibrium domain structures in the PbTiO$_3$ thin film islands are also analyzed by the finite element simulation and found to be consistent with the experimental observation. In the case of nano islands formed by chemical solution deposition, *c*-domain abundance increases as both size and thickness decrease identically, which implies that if the size can be reduced below a certain critical size there exists only single *c*-domain. Compared to the case of patterns with lateral dimension of 100 nm and thickness of 100 nm which has a tendency of reduced *c*-domain abundance due to the reduction of compressive strain, islands with thickness of 30 nm and 120 nm lateral size have larger *c*-domain abundance.

Keywords: Epitaxial PbTiO$_3$ Capacitor, Domain Structure, Reciprocal Space Mapping
PACS: 61.46.-w, 68.03.Hj, 68.65.-k

INTODUCTION

Ferroelectric thin films continue to draw considerable attention due to their potential applications in microelectronic, electro-optic, and nonvolatile memory devices [1]. In order to achieve very high-density devices, it is crucial to downscale the lateral dimensions of the ferroelectric capacitors. As the ferroelectric thin films are patterned into discrete islands, ferroelectric and piezoelectric properties greatly change due to the increased free surface and reduced substrate clamping effect [2,3]. Moreover, because the domain structure of patterned films should be different from that of continuous films, understanding of such domain structures and their formation mechanism is essential to control and achieve desirable ferroelectric properties for practical applications [4]. In this study, size effects of domain structures in epitaxial Pb(Zr,Ti)O$_3$ (PZT) thin films were investigated systemically from the view point of strain relaxation mechanism. The effects of lateral dimension and film thickness were extensively analyzed by reciprocal space mapping using synchrotron x-ray as well as finite element simulation.

CP879, *Synchrotron Radiation Instrumentation: Ninth International Conference*,
edited by Jae-Young Choi and Seungyu Rah
© 2007 American Institute of Physics 978-0-7354-0373-4/07/$23.00

EXPERIMENTAL PROCEDURES

Equilibrium domain structures commonly observed in epitaxial Pb-based ferroelectric thin films are analyzed by finite element method (FEM) using a commercial package, ABAQUS. Structures of periodic 90°-domains in epitaxial PbTiO₃ thin films on cubic single crystalline substrates are analyzed as a function of decreasing temperature in order to simulate cooling process after the film deposition at elevated temperature (T_G). The degree of c-axis orientation (α) is determined as a function of temperature below the Curie temperature and compared to the experimental results. It is then possible to calculate the magnitude of misfit strain during film growth and its relaxation due to dislocation generation. FEM simulation is performed with the assumption that the major driving force for such domain formation is thermo-elastic strains arising from the film-substrate interaction and the cubic-tetragonal phase transformation. The FEM analysis also suggests that initial misfit stress at T_G is not fully relaxed and the residual stress is inversely related to final c-domain abundance.

Evolution of 90° domain structures was characterized in situ using a high temperature stage installed on the Huber six-circle diffractometer of the 3C2 synchrotron x-ray diffraction beam line at the Pohang Light Source (PLS). The beam size at the focal point is typically less than 1 mm². A scintillation detector was used to record the diffracted beam intensities. Complete structural characterization was performed as a function of lateral size and thickness using two-dimensional HKL-mesh scans in reciprocal space for the PbTiO₃ (001) and (100) reflections ($a = 3.899$ □, $c = 4.153$ □). The MgO substrate ($a = 4.213$ □) was used as an internal standard, assuming that it is unstrained. Figure. 1 shows a portion of the reciprocal space together with the adjacent Ewald sphere. Major crystal planes having indices (hkl) are represented by points located in the reciprocal lattice. Mapping of the reciprocal lattice space is conducted by keeping one of the Miller indices, taking HL lattice plane as an example, k in the reciprocal lattice fixed, and varying h by $\pm\Delta h$ every increment Δl in l. Instrumentally, a number of ω-2θ scans along the reciprocal lattice vector $q_\perp$ normal to reflecting planes are measured for a sequence of different angular positions χ (ω for HK space map, where ω and χ-scan direction are perpendicular to $q_\perp$) of the sample. Quantitative c-domain abundance was estimated by the ratio of the integrated intensities of HK-plane maps in reciprocal space for (001) and (100) reflections,

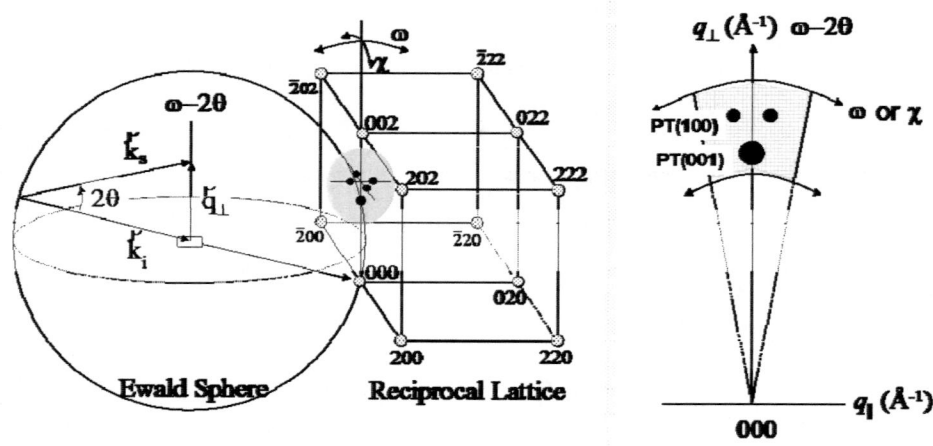

FIGURE 1. Schematic illustration of a portion of the reciprocal lattice together with the adjacent Ewald sphere. k_i : incident wavevector, k_s : scattered wavevector, θ : Bragg angle. The black solid circles indicate the reciprocal points corresponding (001) and (100) reflections of PbTiO₃ with their spatial positions exaggerated.

RESULTS AND DISCUSSION

In the epitaxial PbTiO₃ thin films directly grown on MgO (001) single crystal substrate, the lattice constant of PbTiO₃ film is smaller than that of the substrate at the growth temperature, which induces a large tensile misfit strain in the PbTiO₃ thin films. As the lateral 2D planar size in PbTiO₃ patterns decreases, some of the a-domains turned into c-domains due to the relaxation of residual misfit tensile strain as shown in Fig. 2. In the PbTiO₃ / Pt(001) / MgO (001) system, in contrast, PbTiO₃ film matches with underlying Pt electrode not with MgO substrate and the lattice constant of PbTiO₃ film is larger than that of the Pt electrode at the growth temperature. Thus the introduction of epitaxial Pt electrode layer changes the polarity of misfit strain from tensile to compressive, which enhances c-domain formation [5]. As the pattern size decreases, the c-domain abundance is also reduced because of relaxation of residual compressive misfit strain. The solid lines in Fig. 2 represent the estimations based on our experimental observation that c-domain abundance is linearly dependent on the strain relaxation in PbTiO$_3$ continuous films [6].

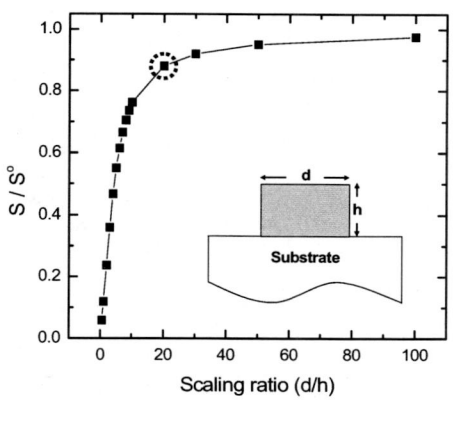 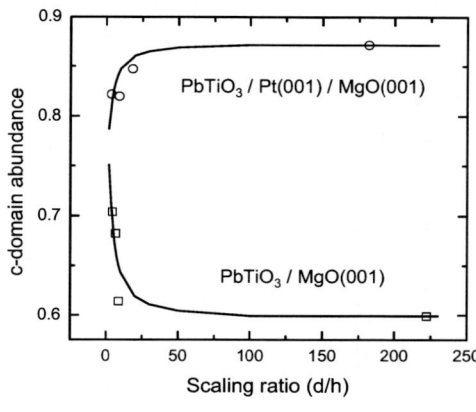

(a)   (b)

FIGURE 2. (a) The average in-plane strain across the entire islands is calculate by ABAQUS simulations with Visual FORTRAN 6.1 subroutines and normalized with respect to the in-plane strain of continuous film as a function of scaling ratio. (b)The synchrotron XRD results on c-domain abundance of epitaxial PbTiO$_3$ thin films on MgO (001) and on Pt (001) / MgO (001) as a function of scaling ratio. The solid lines represent the estimations based on our experimental observation that c-domain abundance is linearly dependent on the strain relaxation in PbTiO$_3$ continuous films

In order to investigate the lateral size effect in smaller scaling ratio about 1, the epitaxial nanoscale patterns with the lateral size of 100 nm were fabricated by e-beam lithography. As shown in Fig. 3, the a-domains in the nanoscale PbTiO$_3$ patterns were quite different from those in continuous films especially in the untilted a-domains in the center of the reciprocal space mapping of PbTiO$_3$ (100). These untilted a-domains could arise from the almost fully relaxed misfit strains in the upper part of the patterns predicted by the finite element simulation. The calculated c-domain abundances of the continuous thin film and nanoscale patterns were 0.85 and 0.68. The large amounts of a-domains are observed in 100 nm-lateral size patterns and this discrepancy between the expected and experimental values could be come from the ultilted a-domains in the upper part of the nanoscale patterns.

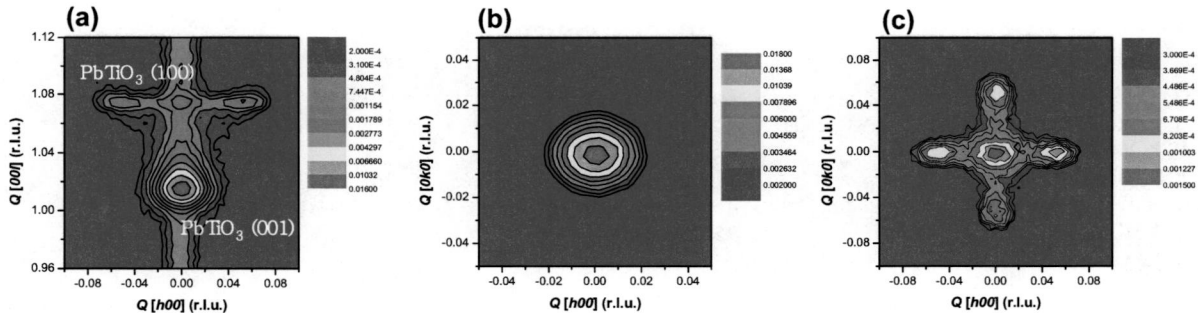

FIGURE 3. The reciprocal space mapping of epitaxial nanoscale PbTiO$_3$ thin films with lateral size of 100 nm; (a) HL scans, (b) HK scans of PbTiO$_3$ (001), (c) HK scans of PbTiO$_3$ (100).

Ferroelectric PbTiO$_3$ nano islands have been also fabricated on Pt (001)/MgO (001) single crystal substrate by using chemical solution deposition which is based on the instability of ultrathin film at high temperature and the evolution of their equilibrium domain structures as a function of size of nano islands was characterized. The size of the ferroelectric nano structures could be changed by varying initial thickness of as-deposited films, annealing time and temperature. The largest size of islands and the smallest size of islands were about 450 nm and 55 nm, respectively. Figure 4 shows the degree of *c*-axis orientation (*c*-domain abundance) as a function of lateral size. With the decrease of the lateral size, the degree of *c*-axis orientation, α increased gradually and this trend implies that if the size can be reduced below a certain critical size there exists only single *c*-domain. The new type of *a*-domains as mentioned above was observed in this case too. The amount of defective *a*-domain which aligns along substrate normal direction increased as the lateral size as well as thickness of nano islands decreased.

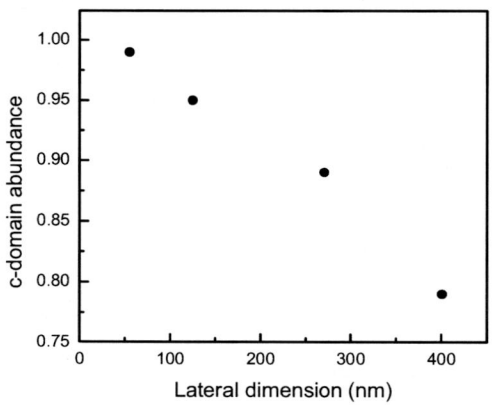
(a)

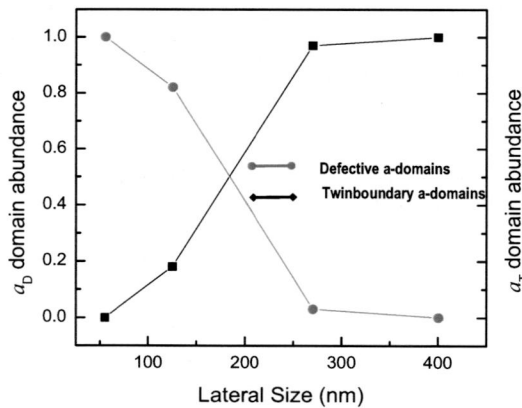
(b)

FIGURE 4. (a) *c*-domain abundance as a function of lateral size of PbTiO$_3$ nano islands. As the size of islands decreases the *c*-domain abundance increases. (b) The abundance change of defective *a*-domains and *a*-domains with twin boundary as a function of lateral size. In the smaller size of islands, defective *a*-domain structures are more favorable.

CONCLUSIONS

We have observed the size effect of ferroelectric thin films, especially on domain structures. When the ferroelectric thin films are patterned into micron size capacitors, the substrate clamping effect decreases and strain relaxation occurs. In the nano scale patterns, however, quite different domain structures are obtained. New type of *a*-domains which align along the substrate normal direction without tilting angle have been observed. Moreover, amount of the defective *a*-domains increases as the thickness as well as the lateral size of nano islands decrease.

REFERENCES

1. O. Auciello, J. F. Scott, and R. Ramesh, *Phys. Today*, **51**, 22 (1998).
2. A. L. Roytburd, S. P. Alpay, V. Nagarajan, C. S. Ganpule, S. Aggarwal, E. D. Williams, and R. Ramesh, *Phys. Rev. Lett* **85**, 190 (2000).
3. T. M. Shaw, S. Trolier-McKinstry, and P. C. McIntyre *Annu. Rev. Mater. Sci.,* **30**, 263 (2000).
4. K. S. Lee, J. H. Choi, J. Y. Lee, and S. Baik *J. Appl. Phys*, **90**, 4095 (2001).
5. Y. K. Kim, K. Lee, and S. Baik, *J. Appl. Phys*, **95**, 236 (2004).
6. K. Lee and S. Baik, Annu Rev. Mater. Res. **36**, 81 (2006).

Compact Resonant Inelastic X-Ray Scattering Equipment at BL19LXU in SPring-8

A. Higashiya[*], S. Imada[†], T. Murakawa[†], H. Fujiwara[†], A. Yamasaki[†],
A. Sekiyama[†], S. Suga[†], M. Yabashi[‡], and T. Ishikawa[*‡]

[*]SPring-8/RIKEN, Kouto 1-1-1, Sayo-cho, Hyogo 679-5148, Japan
[†]Graduate School of Engineering Science, Osaka University, Toyonaka, Osaka 560-8531, Japan
[‡]SPring-8/JASRI, Kouto 1-1-1, Sayo-cho, Hyogo 679-5198, Japan

Abstract. Resonant inelastic x-ray scattering (RIXS) has been carried out at BL19LXU in SPring-8 with use of a compact setup. RIXS was measured for a Cu-O one-dimensional (1D) strongly correlated insulator system $Sr_{14}Cu_{24}O_{41}$ with the incident photon energy ($h\nu$) near the Cu $1s$ absorption edge (K-edge) as a test of the equipment. Dependences of the spectral features on $h\nu$ were systematically studied.

Keywords: resonant inelastic x-ray scattering, one-dimensional cuprate
PACS: 78.70.Ck, 71.20.-b, 71.27.+a, 75.10.Pq

INTRODUCTION

Electronic structure of many transition metal compounds with strong electron correlation has intensively been studied from both theoretical and experimental points of view. Angle-resolved photoelectron spectroscopy and angle-resolved inverse photoemission spectroscopy have been very useful for probing occupied and unoccupied electronic states, respectively, of metallic as well as semiconducting materials.

Resonant inelastic X-ray scattering (RIXS) has recently been attracting much attention as a tool to probe the momentum dependence of low-energy excitations in solids [1, 2, 3]. In RIXS, an electron is excited from the occupied state to the unoccupied state, and the energy difference and the momentum difference between the occupied and unoccupied states are measured. In the general inelastic X-ray scattering, the excitation of valence electrons is very weak and thus difficult to be distinguished from other excitation processes with good statistics. Therefore, resonance enhancement near the core excitation threshold is utilized in RIXS.

Since RIXS is a light-in and light-out method, it is applicable to insulators, which can neither be studied by photoemission nor by inverse photoemission. Moreover, RIXS is much more bulk sensitive than photoemission and inverse photoemission, in which electrons are detected and incident to the sample, respectively. Another advantage of RIXS measurement is the freedom to change photon energy (hv) near the core absorption threshold, resulting in different intermediate states [4].

In this paper, we introduce the experimental setup of RIXS at BL19LXU in SPring-8 and show an example of its application to the study of a one-dimensional insulator cuprate.

EQUIPMENT

The experiment was carried out in the third hutch of the beam line BL19LXU [5] of SPring-8 with a 27m long X-ray linear undulator of in-vacuum type, where the period length is 32 mm and the number of periods is 780. In Fig. 1 is shown the experimental setup optimized for RIXS in the Cu 1s absorption region around 9 keV. All the components shown in the 'Top view' part of the figure were set up on a single optical bench with a size of 1m × 2m.

CP879, Synchrotron Radiation Instrumentation: Ninth International Conference,
edited by Jae-Young Choi and Seungyu Rah
© 2007 American Institute of Physics 978-0-7354-0373-4/07/$23.00

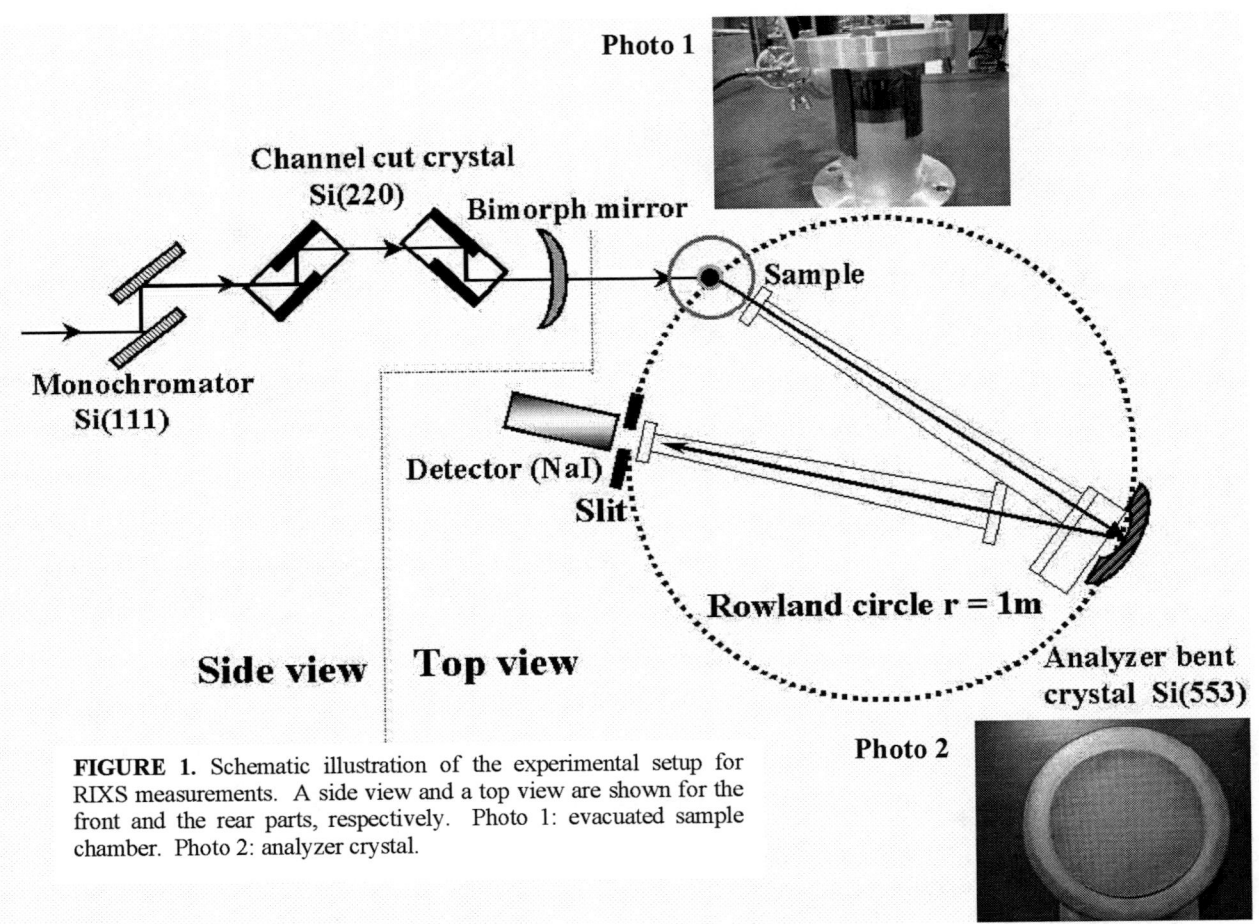

Photo 1

Channel cut crystal
Si(220)

Bimorph mirror

Sample

Monochromator
Si(111)

Detector (NaI)

Slit

Rowland circle r = 1m

Side view **Top view**

Analyzer bent
crystal Si(553)

Photo 2

FIGURE 1. Schematic illustration of the experimental setup for RIXS measurements. A side view and a top view are shown for the front and the rear parts, respectively. Photo 1: evacuated sample chamber. Photo 2: analyzer crystal.

The linearly polarized undulator radiation was tuned to a proper monochromatic hv by two Si (111) crystals. It was further monochromatized by two channel-cut Si (220) crystals. Next, the light was focused by the bimorph mirror [6] with a focusing length of about 1m. The horizontal size of the light was smaller than 100 μm on the sample. The full width at half maximum (FWHM) of the energy resolution of the excitation photons was set to 360 meV. The monochromatic light was incident onto the polished surface of the sample, which was kept at room temperature in an evacuated chamber with polyimid windows as shown in Photo.1.

Horizontally scattered radiation was analyzed by a setup constructed on an arm rotating around the sample position. The angle of this arm determines the scattering vector and therefore the momentum transferred to the excited electron. The energy of the scattered radiation was analyzed by a spherically bent Si (553) crystal show in Photo 2, which was mounted on the biaxial goniometer enabling both energy analysis and adjustment of the tilting angle. The diameter of the Rowland circle was set to 1 m. The analyzed light was focused on the NaI scintillation detector. The size of the slit in front of the detector was set to 0.4 mm. The slit and the detector were put on the same plate, which was moved horizontally so that the slit followed the Roland circle. The total energy resolution, determined from the width (FWHM) of the quasi-elastic scattering peak, was about 440 meV. Most parts of the optical path from the sample position to the detector were evacuated using chambers with polyimid windows, in order to minimize the loss due to the scattering by air.

In order to set the scattering vector Δq, the sample was rotated around the vertical axis and the arm holding the analyzer crystal and the detector was rotated. At each Δq, the energy of the scattered x-ray was calibrated using the quasi-elastic scattering peak. Finally, the inelastic scattering spectrum was taken by rotating the analyzer crystal and moving the detector simultaneously. The signals from the detector were analyzed by a single-channel analyzer in order to distinguish the real signals from the noises.

RIXS measurement for $Sr_{14}Cu_{24}O_{41}$

$Sr_{14}Cu_{24}O_{41}$ has attracted much attention in the past decade. This compound contains the Cu_2O_3 planes with two-leg ladders together with the planes containing weakly coupled CuO_2 chains [7-10]. The unit cell is very large along the c-axis due to the mixture of both CuO_2 chains and Cu_2O_3 ladders.

Figure 2 shows the Cu 1s-4p XAS spectrum measured by means of the fluorescence yield (a) and hv-dependence of RIXS spectra (b), (c) for $Sr_{14}Cu_{24}O_{41}$. The labels in XAS indicate the incident photon energies used for RIXS measurement shown in Figs. 2 (b) and (c). In the Fig. 2 (a), the Cu 1s-3d quadrupole peak is located at hv = 8.981 keV. The shoulder feature located around hv = 8.990 mainly reflects the Cu 1s-4pπ transition. The main peak is located near hv = 9.000 keV and the Cu 1s-4pσ absorption is dominant in this region. Figures 2 (b) and (c) show the RIXS spectra at $\Delta q = 3.0\pi$ for the chain layer and the ladder layer, respectively. The lattice parameters along the c direction are about 2.73 Å and 3.92 Å for the chain and ladder layers, respectively [11]. Namely Δq = 3.0π for the ladder layer corresponds to $\Delta q = 2.1\pi$ for the chain layer.

Figures 2(b) and 2(c) are shown in the same scale. In these figures, the vertical bars show the positions of visible structures. At hv = 8.981 keV (A) corresponding to the quadrupole excitation, two features are observed near 2 and 5 eV in both (b) and (c). At hv=8.99 keV, however, these features are not clearly observed in $\Delta q = 3.0\pi$ of the chain layer as shown in Fig. 2 (b). On the other hand, two structures are observed around 3.5 eV and 5.7eV in Fig. 2 (c). For the main peak excitation near hv=9.000 keV (C) the peak structure around 5.5 eV is prominent in both (b) and (c). Additionally, a new shoulder structure near 6.5 eV is recognized in both cases. Although the RIXS peak near 5.5 eV stays for hv=9.006 keV (D) in $\Delta q = 3.0\pi$ of the chain layer, the shoulder structure becomes separated from this structure and is observed around 6.8 eV as shown in Fig. 2 (b).

In order to understand the physical mechanisms underlying the RIXS spectra in detail, we must measure the momentum (Δq) dependence of RIXS spectra and further compare these spectra with the theoretical calculation.

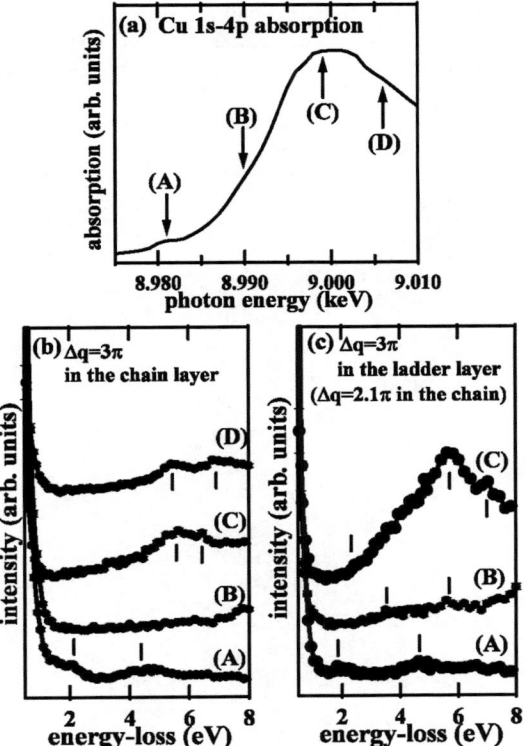

FIGURE.2. (a): XAS spectrum of $Sr_{12}Cu_{24}O_{41}$. (b): hv-dependence of RIXS spectra in the chain layer of $Sr_{12}Cu_{24}O_{41}$. (c): hv-dependence of RIXS spectra in the ladder layer of $Sr_{14}Cu_{24}O_{41}$. The labels (A)-(D) indicate the excitation energies (A) hv= 8.981 keV, (B) 8.99 keV, (C) 8.999 keV, and (D) 9.006 keV.

CONCLUSIONS

We performed resonant inelastic x-ray scattering (RIXS) measurements at BL19LXU in SPring-8. Our setup for RIXS was very compact and was built on an optical bench with a size of 1m × 2m. We demonstrated the performance of the experimental setup by showing the results of RIXS measurements on $Sr_{14}Cu_{24}O_{41}$. Further improvement of the Si (220) channel cut crystal and Si (553) analyzer crystal is expected to lead to a better energy resolution.

ACKNOWLEDGMENTS

We thank Mr. Daigo Miwa for his technical support. This experiment was performed at BL19LXU in SPring-8 under the proposal of 2004B0469-ND3d-np.

REFERENCES

1. M. Z. Hasan, E. D. Isaacs, Z. -X. Shen, L. L. Miller, K. Tsutsui, T. Tohyama and S. Maekawa, Science **288**, 1811 (2000).
2. M. Z. Hasan, P. A. Montano, E. D. Isaacs, Z-X. Shen, H. Eisaki, S. K. Sinha, Z. Islam, N. Motoyama, and S. Uchida , Phys. Rev. Lett. **88**, 177403 (2002).
3. S. Suga, S. Imada, A. Higashiya, A. Shigemoto, S. Kasai, M. Sing, H. Fujiwara, A. Sekiyama, A. Yamasaki, C. Kim, T. Nomura, J. Igarashi, M. Yabashi, and T. Ishikawa, Phys. Rev. B **72** 089101(R) (2005).
4. N. F. Mott, Metal-insulator transitions. Second Edition (Taylor & Francis, London, 1990).
5. M. Yabashi, K. Tamasaku, and T. Ishikawa, Phys. Rev. Lett. **87** 140801 (2001).
6. R. Signorato, T. Ishikawa and J. Carre, SPIE Proceedings, Vol. **4501** 76-87 (2001).
7. L. -C. Duda, T. Schmitt, A. Augustsson and J. Nordgren, J. Alloys & Compounds **362**, 116 (2004).
8. E. M. McCarron, M. A. Subramanian, J. C. Calabrese, and R. L. Harlow, Mater. Res. BNull. **23** 1355 (1988).
9. R. S. Roth, C. J. Rawn, J. J. Ritter, and B. P. Burton, J. Am. Ceram, Soc. **72**, 1545 (1989).
10. P. Abbamonte, G. Blumberg, A. Rusydi, A. Gozar, P. G. Evans, T. Siegrist, L. Venema, H. Eisaki, E. D. Isaacs, and G. A. Sawatzky, Nature **431**, 1078 (2004).
11. J. Etrillard, M. Braden, A. Gukasov, U. Ammerahl, and A. Revcolevschi, Physica C **403**, 290-296 (2004).

Direct Observation of a Gas Molecule (H$_2$, Ar) Swallowed by C$_{60}$

H. Sawa[1,2], T. Kakiuchi[2], Y. Wakabayashi[1], Y. Murata[3], M. Murata[3], K. Komatsu[3], K. Yakigaya[4], H. Takagi[4.5], and N. Dragoe[6].

[1]Photon Factory, KEK, Tsukuba 305-0801, Japan,
[2]Department of Materials Structure Science, The Graduate Univ. for Advanced Studies, Tsukuba 305-0801, Japan,
[3]ICR, Kyoto University, Uji, Kyoto 611-0011, Japan,
[4]AMS, Univ. of Tokyo, Kashiwa 277-8561, Japan
[5]RIKEN, The Institute of Physical and Chemical Research, Wako, 351-0198, Japan,
[6]LEMHE, Univ. Paris XI – ICMMO, 410 - 91405 Orsay Cedex – France;

Abstract. Various types of endohedral fullerene complexes are known to date. The well known metallofullerenes are generally produced by arc-discharge method, but the use of such extremely drastic conditions is apparently not suitable for encapsulation of unstable molecules or gases. We recently succeeded in incorporation of a H$_2$ molecule or an Ar atom in 100% into a C$_{60}$. In order to observe the endohedral gas molecule directly, the X-ray diffraction analysis using synchrotron radiation were carried out. We observed a gas molecule encapsulated in each fullerene cage using structure analysis and the maximum entropy method. These gas molecules are floating inside of the hollow cavities and are completely isolated from the outside

Keywords: endohedral fullerene, structure analysis, charge density analysis
PACS: 61.48.+c

INTRODUCTION

In order to research into molecular capsules for use as the hydrogen storage materials or as nano scale medicinal applications, a technique to insert gas or unstable molecules into a molecular cage is a significant problem. In this regard, the exploration of a methodology of organic synthesis as well as the investigation of the encapsulated molecular state can become an important issue. Fullerene, C$_{60}$ is a typical cage molecule, and the endohedral C$_{60}$ is a exotic material. We succeeded in incorporation of a H$_2$ molecule in 100% into a derivative of an aza-thia open-cage fullerene (ATOCF).[1,2] This compound can be regarded as a nano-sized container for a single hydrogen molecule,

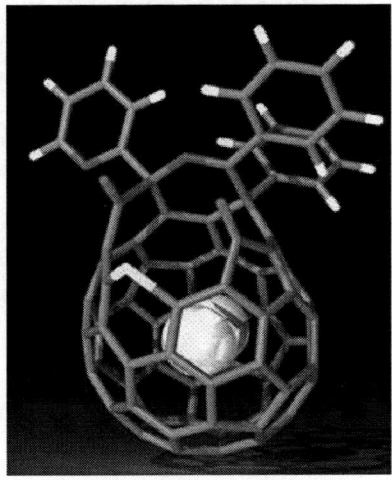

FIGURE 1. Molecular structure of H$_2$@open-cage C$_{60}$. The encapsulated H$_2$ molecule is shown as a space-filling model.

CP879, *Synchrotron Radiation Instrumentation: Ninth International Conference,*
edited by Jae-Young Choi and Seungyu Rah
© 2007 American Institute of Physics 978-0-7354-0373-4/07/$23.00

with which we can control hydrogen storage with pressure and temperature (See Fig.1). By X-ray diffraction analysis on a single crystal of this fully H_2-encapsulating molecule, we have succeeded in direct observation of a floating single H_2 molecule inside of the hollow cavity of an ATOCF molecule. This result demonstrates that this H_2-encapsulating compound is the first material which can allow us to argue the physical properties of single floating H_2 molecule.[3]

On the other hand, the wealth of novel properties and functions produced by the capture of different atoms inside the fullerene cage has led to a great deal of interest in endohedral fullerenes. Endohedral C_{60} is usually intercalated with non-ionic-interacting elements such as nitrogen, phosphorus or noble gas atoms and therefore is inert. In this sense, endohedral C_{60} compounds may be considered as novel van der Waals derivatives. We may expect then that some of the unique characteristics of C_{60}, for example the superconductivity of doped C_{60}, are only modestly modified in "inert" endohedral compounds. We have now modified the previous method to achieve a large scale synthesis. We also report the result of the charge density analysis of $Ar@C_{60}$ which was synthesized using a hot isostatic pressure vessel. [4]

EXPERIMENT AND RESULTS

In the case of single crystal analysis for ATOCF, an X-ray diffraction study on the crystals of H_2 containing ATOCF ($H_2@ATOCF$) and of empty ATOCF as the exactly matching reference was conducted. We obtained accurate X-ray diffraction data by synchrotron radiation with a Weissenberg type imaging-plate detector at BL-1A at the Photon Factory, KEK, Japan. The X-ray wavelength was 1.0 Å. The intensity of the Bragg reflections were measured in a half-sphere of reciprocal space in the range $2\theta<120$ degrees. The sample was cooled to 200K by N_2-gas flow type refrigerator. The Rapid-Auto program by MSC Corporation was used for the two dimensional image processing. Sir2003 program was used for the direct method.

The number of observed reflections with $I > 5\sigma(I)$ was 9449 for the $H_2@ATOCF$ crystal and 9302 for the empty ATOCF crystal. The Shelx97 program was used for refinements. After full refinement, the reliability(R) factor was 9.0% for the $H_2@ATOCF$ crystal and 8.7% for the empty one. Both results showed good agreement with a previous report on the crystal structure of ATOCF.[2] It is not possible to specifically locate the electron-density of such a small species as H_2 molecule by the conventional least-square refinement. However, we were able to picture an image, which enabled us to construct a three-dimensional electron density map, by the maximum entropy method (MEM) using the diffraction data. The MEM analysis was carried out by the use of the Enigma program[5] at a resolution of $128 \times 128 \times 128$ pixels. The R-factors of the final MEM charge density were 2.8% and 2.4% for $H_2@ATOCF$ and empty ATOCF, respectively. The contour maps of the final MEM charge densities of $H_2@ATOCF$ and empty ATOCF are shown in Fig. 2. The obtained MEM charge density maps clearly demonstrate the presence of a piece of electron density, floating at the center of the cage in the $H_2@ATOCF$ molecule. In sharp contrast, no electron density was observed in the case of the empty ATOCF.

The synthesis of $Ar@C_{60}$ is similar to that described previously.[6] Eventually, we obtained about 0.6 mg of $Ar@C_{60}$ with purity better than 95%. Structural information of solid state samples, C_{60} and $Ar@C_{60}$, was obtained by the powder X-ray diffraction patterns measured at BL02B2 in SPring-8 as a function of temperature between 300K and 100K. The rotational disorder at room temperature (RT) makes it possible to treat the cage as a spherical electronic distribution. Therefore, scattering factor of C_{60} can be expressed by a Bessel function. Rietveld analysis was carried out at RT. The refined fractional occupancy of the Ar atom was 97.8%. On the other hand, the fit is in good agreement at 100K using Pa3 model. The $Ar@C_{60}$ unit cell parameter compared to that of empty one is slightly larger at RT but smaller at 100K. This result indicates that the inserted atom has an influence to the lattice parameter.

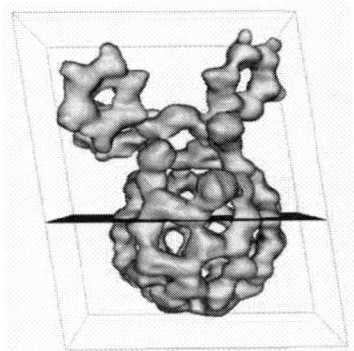 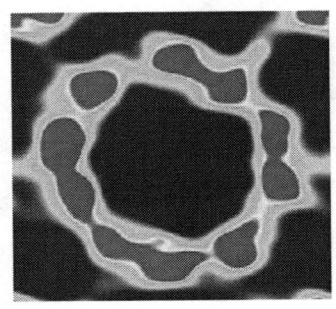 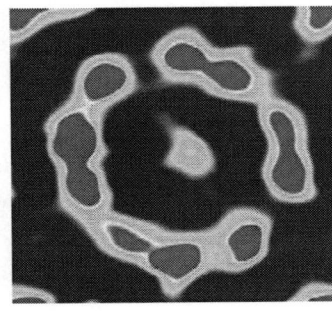

FIGURE 2. A left figure shows the positions of horizontal division of ATOCF molecule. The MEM electronic density distributions of empty ATOCF (center) and H2@ATOCF (right). The contour maps are drawn from $0.01e/Å^3$ to $0.11e/Å^3$.

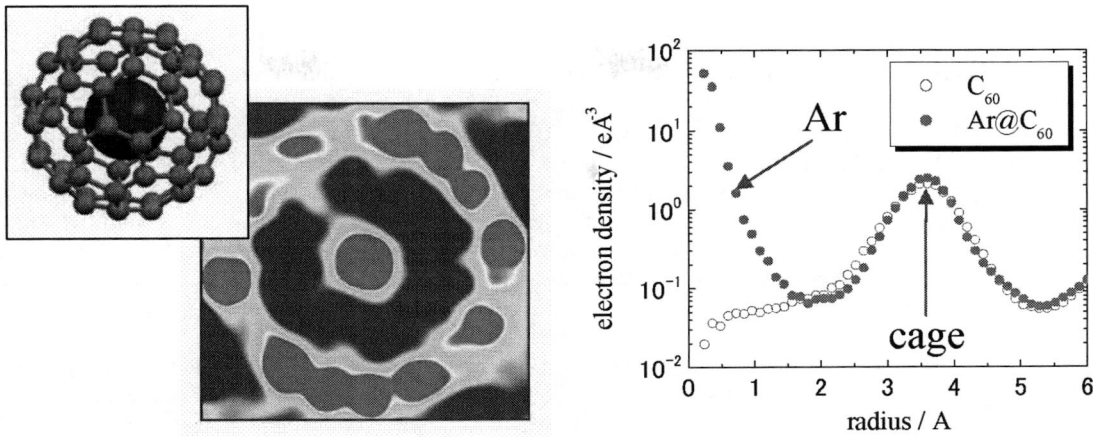

FIGURE.3 Left side: Electron density contour map of Ar@C_{60} using MEM analysis with the structure model. Right side: Dependence of electronic density upon the radius from the center of the cage for C_{60} molecules. The filled circles and open circles represent the profiles for endohedral and empty, respectively. The electron density peak for Ar@C_{60} shows an Ar.

The precise electron density profile was obtained by the MEM analysis (Fig.3 left side). The charge density of an Ar atom is located at the center of the cage.

DISCUSSION

In both incorporated endohedral fullerenes, the main charge density of encapsulated matter is condensed almost at the center of the cage and much lower density seems to exist in the area between the center and the inside wall of the cage (Fig. 2,3).

In the hydrogen contained case, it is difficult to determine the figure of the encapsulated charge density to be either spherical or elliptical. This ambiguity is most probably caused by the molecular motion of H_2, because there is no chemical bond between the trapped H_2 and the carbon cage. M. Carravetta et al. investigated the rotational motion of the encapsulated H_2 in H_2@ATOCF by means of low-temperature solid-state NMR measurements.[7] Their results show that the motional anisotropy of H_2 inside ATOCF is very small and the orifice in ATOCF only slightly perturbs the rotational motion. There is no observed electron density along the neck of ATOCF, indicating that the H_2 molecule is electronically disconnected from the outside of the cage. Therefore, this H_2 molecule is considered to be completely isolated from the outside. In that sense, the environment of this encapsulated H_2 is very similar to that of the H_2 molecule incorporated in the pristine C_{60} itself, i.e., H_2@C_{60}.

Is there any charge transfer interaction between the encapsulated H_2 molecule and the cage? In the case of metallofullerenes M@C_{82}, it is reported that the presence of significant charge transfer interaction between the encapsulated metal and the cage, which also causes the selection of cage symmetry according to the stability of different isomers.[8,9] In striking contrast, our results show observed no difference in cage structure between H_2@ATOCF and empty ATOCF. Thus, we conclude that there exists no appreciable charge transfer between the encapsulated H_2 and the cage. The number of electrons belonging to the encapsulated H_2 estimated by integration from the center of the cage to the minimum point of electron density was 2.0. This result exactly corresponds to the presence of one H_2 molecule at the center of the hollow cage of the ATOCF molecule in excellent agreement with the ^{1}H NMR result.[1]

On the other hand, having established that pure Ar@C_{60} was obtained, we explored possible superconductivity by doping it with potassium obtained by the thermal decomposition of KN_3. The potassium doped fullerene K_3C_{60} is a superconductor having critical temperature (T_C) of 19.2K. The T_C of K_3Ar@C_{60} was investigated by magnetization measurement under a field of H=10Oe, and found to be 17.8K. We measured X-ray powder diffraction data for C_{60}, Ar@C_{60}, K_3C_{60} and K_3Ar@C_{60}. The peculiarities of the solid state structure and dynamics of C_{60} and related materials are well documented. Rietveld analysis allowed us to conclude that the solid state behavior of Ar@C_{60} is similar to that of C_{60}, that is, there is a rotational disorder at room temperature. The refined fractional occupancy for the Ar atom was 97.8%.[4]

Unfortunately, Rietveld analysis was not possible for the data of K_3C_{60} and K_3Ar@C_{60} because of impurities, disorder and stacking faults common in these systems. The lattice parameters of them were obtained at room temperature. The lattice parametere of K_3Ar@C_{60} is smaller than K_3C_{60} by 0.07 Å. It may be caused by sample non-stoichiometry, inhomogeneities or possibly the effect of the inserted Ar atom on the lattice. In order to clarify the effect of inserted Ar atom, the precise electron density profile obtained by the present MEM analysis is visualized by the dependence of the electron density on the radius (r), or distance from the center of the cage, as shown Fig. 3(right side). The solid circles and open circles represent the profiles for $Ar@C_{60}$ and pristine C_{60}, respectively. The maximum peaks at around $r \sim 3.6$Å correspond to the cage frame of the C_{60} molecule itself. The electron density of $Ar@C_{60}$ appears to show a maximum peak at the center ($r = 0$ Å) whereas there is no peak in the case of C_{60}. The number of electrons belonging to encapsulated Ar estimated by integration from the center of the cage to the minimum point of electron density was 18.0. It is unlikely, in such a van der Waals system where no or little charge transfer occurs, that the electronic density states at the Fermi level are directly modified. While a subtle change in the intermolecular interactions triggered by the inserted atom should alter the band width. We therefore propose that the changes in the critical temperatures observed for these compounds are related to indirect changes in the density of states together with, to a lesser extent, changes in the phonon spectra intermolecule.

To summarize, the molecule of $H_2@ATOCF$ is regarded as an excellent system to examine the important issue characteristic to a single isolated H_2 molecule, e.g. the ortho-H_2 to para-H_2 conversion, quantum motion of H_2 molecule at low temperature, and so on. On the other hand, further work is required to determine the change in the superconductivity of endohedrals, particularly for smaller atom like hydrogen and/or larger atoms like krypton. The possibility that indirect changes of the density of states at the Fermi level are induced by the inserted atoms suggests that an increase in the T_C of these materials is possible.

ACKNOWLEDGMENTS

This work was supported by a Grant-in-Aid for Creative Scientific Research (13NP0201), TOKUTEI (16076207) from the Ministry of Education, Culture, Sports, Science and Technology of Japan.

REFERENCES

1. Y. Murata, M. Murata, K. Komatsu, *J. Am. Chem. Soc.* **2003**, 125, pp. 7152-7153.
2. Y. Murata, M. Murata, K. Komatsu, *Chem. Eur. J* **2003**, *9*, 1600-1609.
3. H. Sawa, Y. Wakabayashi, Y. Murata, M. Murata, K. Komatsu, *Angew. Chem. Int. Edit.* **44** (2005) pp. 1981-1983
4. A. Takeda, Y. Yokoyama, S. Ito, T. Miyazaki, H. Shimotani, K. Yakigaya, T. Kakiuchi, H. Sawa, H. Takagi, K. Kitazawa and N. Dragoe, *Chem. Comm. 2006, pp.912-914*
5. H.Tanaka, M. Takata, E. Nishibori, K. Kato, T. Iishi, M. Sakata, *Journal of Applied Crystallography* **35** (2002) pp 282-286
6. Y. Chai, T. Guo, C. Jin, R. E. Haufler, L. P. F. Chibante, J. Fure, L. Wang, J. M. Alford and R. E. Smalley, *J. Phys. Chem.*, **95**, (1991), 7564.
7. M.Carravetta et al., *J. Am. Chem. Soc.* **124** (2004) 4092-4093.

NEXAFS Study of the Annealing Effect on the Local Structure of FIB–CVD DLC

Akihiko Saikubo[1,2], Yuri Kato[1,2], Jun–ya Igaki[1,2], Reo Kometani[1,2,3], Kazuhiro Kanda[1,2] and Shinji Matsui[1,2]

[1]*Graduate School of Science, LASTI, University of Hyogo, 3–1–2 Kouto, Kamigori–cho, Ako–gun, Hyogo 678–1205, Japan*
[2]*CREST JST, Kawaguchi Center Building, 4–1–8 Honcho, Kawaguchi–shi, Saitama 332–0012, Japan*
[3]*JSPS, 8 Ichibancho, Chiyoda–ku, Tokyo 102–8472, Japan*

Abstract. Annealing effect on the local structure of diamond like carbon (DLC) formed by focused ion beam–chemical vapor deposition (FIB–CVD) was investigated by the measurement of near edge x–ray absorption fine structure (NEXAFS) and energy dispersive x–ray (EDX) spectra. Carbon K edge absorption NEXAFS spectrum of FIB–CVD DLC was measured in the energy range of 275–320 eV. In order to obtain the information on the location of the gallium in the depth direction, incidence angle dependence of NEXAFS spectrum was measured in the incident angle range from 0° to 60°. The peak intensity corresponding to the resonance transition of $1s \rightarrow \sigma^*$ originating from carbon–gallium increased from the FIB–CVD DLC annealed at 200°C to the FIB–CVD DLC annealed at 400°C and decreased from that at 400°C to that at 600°C. Especially, the intensity of this peak remarkably enhanced in the NEXAFS spectrum of the FIB–CVD DLC annealed at 400°C at the incident angle of 60°. On the contrary, the peak intensity corresponding to the resonance transition of $1s \rightarrow \pi^*$ originating from carbon double bonding of emission spectrum decreased from the FIB–CVD DLC annealed at 200°C to that at 400°C and increased from that at 400°C to that at 600°C. Gallium concentration in the FIB–CVD DLC decreased from $\approx 2.2\%$ of the as–deposited FIB–CVD DLC to $\approx 1.5\%$ of the FIB–CVD DLC annealed at 600°C from the elementary analysis using EDX. Both experimental results indicated that gallium atom departed from FIB–CVD DLC by annealing at the temperature of 600°C.

Keywords: near edge x–ray absorption fine structure, incident angle dependence, diamond like carbon, focused ion beam–chemical vapor deposition, annealing effect, gallium concentration
PACS: 61. 10. Ht

INTRODUCTION

Focused ion beam–chemical vapor deposition was known to be suitable for fabrication of three dimensional nano materials [1]. In this method, focused ion beam of gallium was irradiated to silicon substrate under the phenanthrene gas atmosphere using as a source gas and carbon material was deposited by dissociation of phenanthrene absorbed on the surface of silicon substrate by secondary electrons, which were emitted in the collision of gallium ion with silicon substrate surface. It was known that the fundamental structure of carbon material formed by FIB–CVD is diamond like carbon (DLC) from the measurement of Raman spectrum and near edge x–ray absorption fine structure (NEXAFS) spectrum of carbon K edge absorption [1,2]. In addition, the gallium using as a source gas was reported to remain in the FIB–CVD DLC and to depart from the FIB–CVD DLC by the annealing at 600°C [3]. The material properties of FIB–CVD DLC, such as hardness and conductivity, are considerable to be strongly dependent on the concentration and position of gallium atoms. These material properties are expected to control by varying annealing condition.

NEXAFS technique has great advantages in evaluation of the electronic state with high accuracy because the peak corresponding to the $1s \rightarrow \pi^*$ resonance transition can be observed isolatedly [4–6]. In our previous work, it was found that the residue gallium in the FIB–CVD DLC moved toward the neighboring surface and subsequently departed from the surface by annealing [7]. In the present work, the local structure of three samples formed by FIB–CVD with a different annealing condition was investigated by the measurement of incident angle dependence of

CP879, Synchrotron Radiation Instrumentation: Ninth International Conference,
edited by Jae-Young Choi and Seungyu Rah
© 2007 American Institute of Physics 978-0-7354-0373-4/07/$23.00

NEXAFS spectrum of carbon K edge absorption. In addition, amount of gallium in the FIB–CVD DLC was measured using energy dispersive x–ray (EDX) analyzer for the confirmation of gallium movement by the annealing.

EXPERIMENT

The measurement of NEXAFS spectra of FIB–CVD DLC was performed at BL8B1 of UVSOR, which is synchrotron radiation facility with 0.75 MeV electron storage ring [8]. X–ray beam in the desired energy range was extracted using monochromator equipped with a 540 lines/mm laminar grating which had a 15 m radius. The carbon K edge absorption NEXAFS spectra were measured in the energy range of 275–320 eV. The energy resolution of x–ray beam was estimated to be ≈ 0.5 eV in full width at half maximum with slit width of 20 μm. The reading of monochromator was calibrated against the pre–edge resonance corresponding to the carbon $1s \rightarrow \pi^*$ transition appeared at 285.3 eV in the NEXAFS spectrum of graphite. The detection of electrons coming from the sample was performed in the total electron yield mode. The intensity of incident x–ray was measured by detecting the photocurrent from a gold film. The NEXAFS spectrum was given by the ratio of the photocurrent from the sample to that from the gold film. In the present work, incident angle dependence of NEXAFS spectrum was measured in the angle region from 0° to 60° by 15° step in order to observe the distribution of gallium atom in the depth direction.

Elementary analysis of gallium in the FIB–CVD DLC was performed using EDX (JEOL Ltd.; JSM–6700F). The emission spectrum of FIB–CVD DLC was measured using 3 kV electron at 10 μA. Concentration of gallium in the FIB–CVD DLC was estimated from the peak intensity of gallium Lα line.

FIB–CVD DLC film was deposited on silicon substrate using focused ion beam apparatus (SII NanoTechnology Inc.; SMI 9200). Gallium ion beam accelerated to 30 kV was focused to 50 nm at 90 pA beam current. Phenanthrene using as a source gas was introduced into the chamber at the pressure of $\approx 10^{-5}$ Pa. The FIB–CVD DLC film was formed with the vertical size of 2 mm and the horizontal size of 10 mm. The film thickness of FIB–CVD DLC was estimated to be ≈ 9 nm using atomic force microscope. For the purpose of investigation of annealing effect on the local structure of FIB–CVD DLC, the samples were annealed for an hour at 200°C, 400°C and 600°C, respectively.

RESULTS AND DISCUSSION

The incident angle dependence of NEXAFS spectra of carbon K edge absorption for FIB–CVD DLC annealed at 200°C, 400°C and 600°C is shown in Figs. 1–3, respectively. In these NEXAFS spectra of FIB–CVD DLC, sharp peak located at 285.3 eV is ascribable to the $1s \rightarrow \pi^*$ resonance transition originating from carbon double bonding. Broad peak located at the energy range higher than 293 eV is ascribable to the $1s \rightarrow \sigma^*$ resonance transition. In the NEXAFS spectrum of FIB–CVD DLC, the novel peak appears at 289.0 eV, which is ascribable to the $1s \rightarrow \sigma^*$ resonance transition originating from carbon–gallium bonding [2]. NEXAFS spectra of FIB–CVD DLC annealed at 200°C for an hour did not almost depend on the incident angle, as shown in Fig. 1. On the other hand, in the NEXAFS spectrum of FIB–CVD DLC annealed at 400°C for an hour (see Fig. 2), the peak intensity located at 289.0 eV increased remarkably with the incident angle. This indicated that the gallium concentration increased at the neighboring surface. In other words, the gallium atoms, existed in the FIB–CVD DLC, moved toward surface by annealing at 400°C. On the other hand, the peak due to the $1s \rightarrow \pi^*$ transition at 285.3 eV originating from carbon double bonding disappeared. These experimental results show that the gallium, moving to the neighboring surface, decoupled the carbon double bonding and generated newly carbon–gallium bonding. In the Fig. 3, obvious incident angle dependence of the peak intensity of the $1s \rightarrow \sigma^*$ transition of at C–Ga site was not observed. This elimination of incident angle dependence is considered that the gallium was departed from the FIB–CVD DLC surface by annealing at 600°C for an hour. On the other hand, the peak intensity of $1s \rightarrow \pi^*$ transition of at C=C site increased to almost same as that of Fig. 1. This is interpreted to the recoupling of the carbon double bonding by departure of the gallium from the FIB–CVD DLC.

For the purpose to understand the variation in the amount of gallium in the FIB–CVD DLC, EDX spectra of FIB–CVD DLC without and with 600°C annealing for an hour were observed. The gallium concentration decreased drastically from 2.2 to 1.5% for FIB–CVD DLC annealing at 600°C by compared to as–deposition. This result supported the conclusion derived from NEXAFS study, that gallium departed from the FIB–CVD DLC by the annealing at the substrate temperature of 600°C.

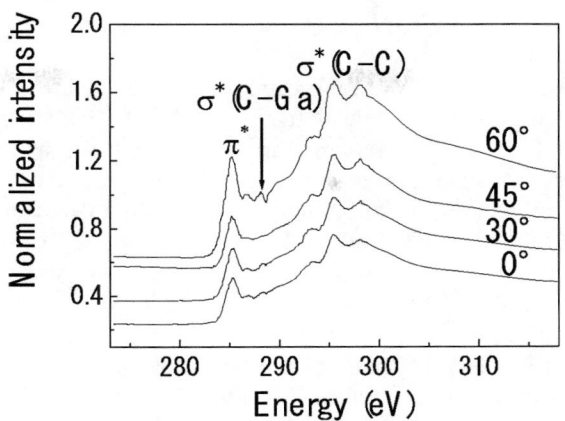

FIGURE 1. Incident angle dependence of NEXAFS spectra for FIB–CVD DLC annealed at 200°C.

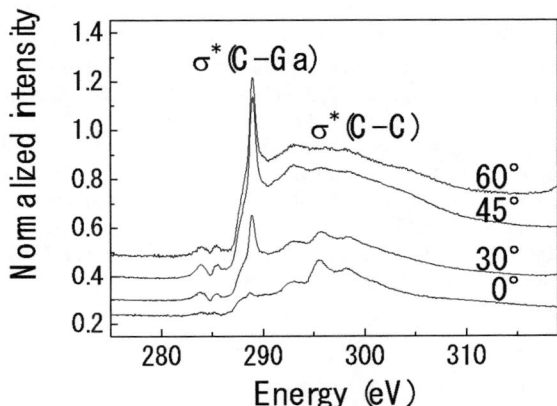

FIGURE 2. Incident angle dependence of NEXAFS spectra for FIB–CVD DLC annealed at 400°C.

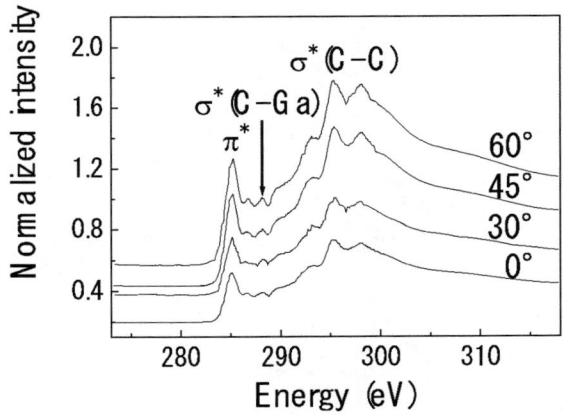

FIGURE 3. Incident angle dependence of NEXAFS spectra for FIB–CVD DLC annealed at 600°C.

CONCLUSION

Incident angle dependence of carbon K edge NEXAFS spectrum was measured for FIB–CVD DLC annealed for an hour at 200°C, 400°C and 600°C in order to investigate the difference of local structure of FIB–CVD DLC. In addition, emission spectra of gallium Lα line were measured in order to confirm the movement of gallium in the FIB–CVD DLC by annealing. Both experimental results showed that the gallium moved toward the surface of FIB–CVD DLC by annealing. For annealing temperature less than 400°C, gallium remained in the FIB–CVD DLC. However, for annealing temperature higher than 400°C, the gallium departed from FIB–CVD DLC.

ACKNOWLEDGMENTS

The authors appreciate the support of CREST Japan Science and Technology Co. This work was performed at BL8B1 of UVSOR with the support by the Joint Studies Program of Institute for Molecular Science under the proposal number of 17–813. The authors thank UVSOR staff, especially for Dr. Nakamura, for the NEXAFS measurement at BL8B1 of UVSOR.

REFERENCES

1. S. Matsui, T. Kaito, J. Fujita, M. Komuro, K. Kanda and Y. Haruyama, *J. Vac. Sci. Tech. B* **18**, 3181 (2000).
2. K. Kanda, J. Igaki, Y. Kato, R. Kometani, A. Saikubo and S. Matsui, *to be published in Rad. Phys. Chem.*
3. J. Fujita, M. Ishida, T. Ichihashi, T. Sakamoto, Y. Ochiai, T. Kaito and S. Matsui, *Jpn. J. Appl. Phys.* **41**, 4423 (2002).
4. C. Lenardi, P. Piseri, V. Briois, C. E. Bottani, A. Li Bassi and P. Milani, *Jpn. J. Appl. Phys.* **85**, 7159 (1999).
5. K. Kanda, T. Kitagawa, Y. Shimizugawa, Y. Haruyama, S. Matsui, M. Terasawa, H. Tsubakino, I. Yamada, T. Gejo and M. Kamada, *Jpn. J. Appl. Phys.* **41**, 4295 (2002).
6. K. Kanda, Y. Shimizugawa, Y. Haruyama, I. Yamada, S. Matsui, T. Kitagawa, H. Tsubakino and T. Gejo, *Nucl. Instrum. Methods Phys. Res. B* **206**, 880 (2003).
7. A. Saikubo, J. Igaki, Y. Kato, R. Kometani, K. Kanda and S. Matsui, *to be published in Jpn. J. Appl. Phys.*
8. A. Hiraya, E. Nakamura, M. Hasumoto, T. Kinoshita and K. Sakai, *Rev. Sci. Instrum.* **66**, 2104 (1995).

Synchrotron X-Ray Induced Gold Nanoparticle Formation

Y. C. Yang[a], C. H. Wang[b], T. Y. Yang[b], Y. Hwu[b], C. H. Chen[c], J. H. Je[d], and G. Margaritondo[e]

[a]Dept. Mater. Min. Reso. Eng., National Taipei Univ. Tech., Taipei, Taiwan 106, R. O. C.
[b]Institute of Physics, Academia Sinica, Nankang, Taipei, Taiwan 115, R. O. C.
[c]Dept. Mater. Sci. Eng., Pohang University of Science and Technology, Pohang, Korea
[d]China Steel Corporation, 1 Chung-Kang Road, Kaohsiung, Taiwan 812, R. O. C.
[e]Ecole Polytechnique Fédérale de Lausanne (EPFL), CH-1015 Lausanne, Switzerland

Abstract. We reported a simple approach to generate gold nanoparticles from $HAuCl_4$ containing aqueous solution by synchrotron x-ray irradiation at room temperature. The gold colloidal were investigated by a variety of characterization methods including Transmission Electron Microscope (TEM), Scanning Electron Microscope (SEM), Fourier transformation infrared (FTIR), Ultraviolet and Visible (UV-VIS) spectrometer and the effects of variables including pH value, radiation time were examined.

Keywords: gold, nanoparticle, synchrotron x-ray.
PACS: 81.40.Wx

INTRODUCTION

The novel features of extremely intense x-rays produced by the third generation synchrotron radiation (SR) facilities are increasingly employed to trigger photo-chemical reactions for material processing and micro-sized device fabrication. Specifically the very high flux of energetic x-ray photons enables chemical reactions to be induced and completed in a very short time, which is favorable for nanomaterial fabrication. The high energy (especially for hard x-ray) provides deep penetrating lengths and makes it possible to perform photosynthesis within embedded materials system. The possibility of synchrotron x-ray source for the direct reduction of gold precursor solutions has been firstly explored by Rosenberg and co-researchers.[1,2] In their reports, using a commercial electroplating solution, an irradiation for ~11 min at 75 mA beam current could generated particulate gold with a mean size of 18 nm (as decided from x-ray diffraction). Even though detailed information on thus produced gold nanoparticles and the influences of processing parameters are not available in these reports, the promise of utilizing synchrotron x-rays to synthesis colloidal gold solutions has been demonstrated.

Recently, an irradiation method utilizing synchrotron x-rays to form gold nanoparicles in aqueous has been proved to be an alternative way to obtain dispersed nanoparticles.[3] Compared to other radiation methods to obtain metallic nanoparticles including γ-ray [4,5] and in-house x-ray,[6] this method is unique with terms of: (1) free of reducing and stabilizing agent; (2) short radiation time needed. However, the effects of experimental parameters on the structure and morphology of such produced gold nanoparticles are unknown. In this work, we studied the dependence of particle size and dispersion on variables such as pH value, beam condition, irradiation time. Under optimal conditions, the precipitated nanoparticles were well dispersed and with uniform size, ≈10 nm. The mechanism involved in the formation of colloidal gold was also concisely discussed.

EXPERIMENTAL

Gold nanoparticles were synthesized from aqueous hydrogen tetrachloroaurate trihydrate ($HAuCl_4 \cdot 3H_2O$, Aldrich) solutions by synchrotron x-ray irradiation. Centrifuge tube (2 ml) were used as the solution container. The pH value of precursor solution was adjusted to different pH values by adding NaOH solution. Deionized water was used for solution preparation throughout the experiment. The experiments were performed on the 7B2 beamline of

CP879, *Synchrotron Radiation Instrumentation: Ninth International Conference*,
edited by Jae-Young Choi and Seungyu Rah
© 2007 American Institute of Physics 978-0-7354-0373-4/07/$23.00

2.5 GeV PLS storage ring (Pohang Light Source, Pohang, Korea).[7] We used unmonochromatized ("pink") x-ray beams with no optical elements except one set of beryllium and Kepton windows. In addition to a CCD camera installed to monitor the color changes of the solution, the formation of the particles were also monitored by synchrotron x-ray microscopy in solution and in real time by analyzing the local change of the metal concentration.

The pH value was measured with an Orion 720A pH-meter. UV and visible light absorption spectra were taken with a Shimadzu UV-160 spectrometer with 1 cm quartz cuvette. The particle morphology, structure and size were measured with a JEOL JSM-6330F field emission scanning electron microscope (FESEM) and with a JEOL JEM 2010 F FEEM (200 kV) transmission electron microscope (TEM). The samples for TEM measurements were prepared by placing droplets of nanoparticle-containing solution on carbon-coated Cu grids and allowing them to dry at ambient atmosphere.

RESULTS AND DISCUSSION

Figure 1 shows SEM images of Au nanoparticles with different pH values after 90 seconds of irradiation under 165 mA electron beam current. For pH = 3, the nanoparticles agglomerated forming large porous particles within 10 minutes after irradiation (Fig. 1(a)). When the pH value was increased to 6 (Fig. 1(b)), an extended structure composed of large Au particle was formed. Figures. 1(c) to 1(d) show precipitated nanoparticles from neutral and alkaline solutions. The particle sizes ranged from 5nm to 40nm and the nanoparticles were well dispersed. The corresponding solution did not show any agglomeration even after a long period of time at room temperature.

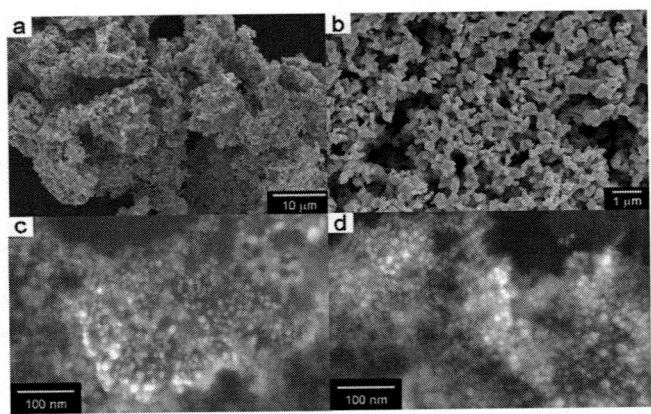

FIGURE 1. Scanning electron micrographs (SEM) of particles derived from precursor solutions with various pH values: (a) pH=3, (b) pH=6, (c) pH=7, (d) pH=8.

The TEM micrographs of gold particles produced from precursor solutions with different pH values are shown in Fig. 2. As shown in the Fig. 2 (a), the particles produced under acidic condition were easily aggregated and interlinked with each other. Moreover, both larger particles (>30 nm) with irregular shapes and smaller spheres (< 10 nm) were observed. When the pH of precursor solution increased to neutral and acidic conditions, the formed gold particles became well dispersed with sphere gold particles. The size of formed gold particles decreased drastically with increasing pH value: 15-20 nm for pH 7 and 10-15 nm for pH 9.

Free radical H• and •OH are known to develop by dissociation of water molecules exposed to ionization radiation.[4,5] The active hydrogen free radical H• acts as an electron donor (reducing agent) and thus $HAuCl_4$ is reduced to metallic gold. It is well-known that the formed gold cluster may experience agglomeration in the absence of stabilizing agent. However, in our approach, no stabilizer was introduced beforehand. The drastic pH dependence indicated that the OH^- groups present in the precursor solution probably adsorbed on the gold nanoparticle surfaces and prevented the created particles from agglomeration and further precipitation. To examine this speculation, FTIR measurement on the precipitated gold nanoparticles after evaporating the solvent was performed (data not shown). A very broad absorption peak was observed at 3400 cm^{-1}, which could be unambiguously attributed to the vibrating mode of hydroxyl and therefore supported the above argument. The UV-VIS spectra of gold nanosols produced from solutions of different pH values are shown in Fig. 3. The characteristic absorption (surface Plasmon resonance, SPR) peaks of gold nanoparticles (522 nm for pH 7, 524 nm for pH 8 and 528 nm for pH 9) were clearly observed. The evolutions of absorption maximum and intensity are commonly related to the variations of particle size, shape and dispersion. By correlating the variations of SPR features to particle morphology demonstrated in Fig. 2, the red-shift of SPR with increasing pH values cannot be contributed to the decreased gold particles, which normally resulted in blue-shift of SPR peaks. The origin of this abnormal red-shift was still unclear. It was also found that the SPR intensity decreased with increasing pH value indicating less metallic gold was produced.

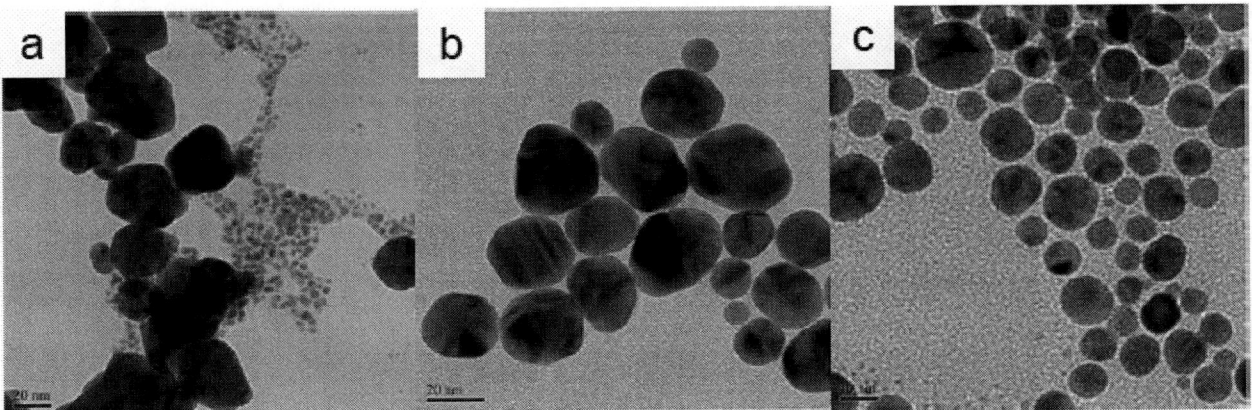

FIGURE 2. TEM images of gold nanoparticles from precursor solutions with different pH values, produced by synchrotron irradiation: (a) pH=5; (b) pH=7; (c) pH=9.

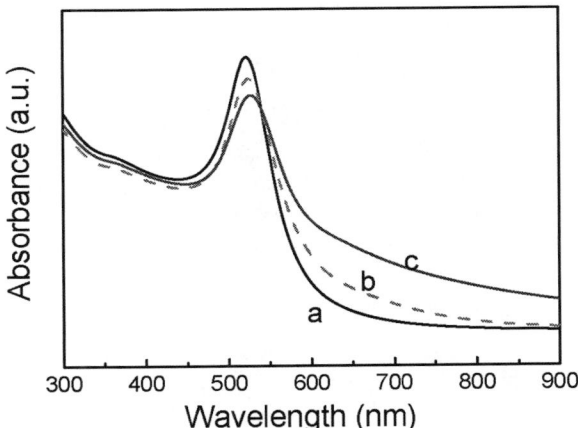

FIGURE 3. UV-visible absorption spectra of gold nanoparticles from solutions with different pH values, produced by synchrotron irradiation: (a) pH=7; (b) pH=8; (c) pH=9.

Experimental parameters such as beam current and radiation time were found to influence the formation of gold nanoparticles as well. The evolution of gold production as indicated by the absorbance at 520 nm was plotted against the beam current and radiation time. For this exposure, the precursor solutions were contained in a 15 ml centrifuge and transferred to x-ray exposure. As shown in Fig. 4, for longer x-ray radiation time, the influence of exposure time became negligible while for short radiation time, the influence of beam condition (deriving from the decay operating mode) should be taken into account.

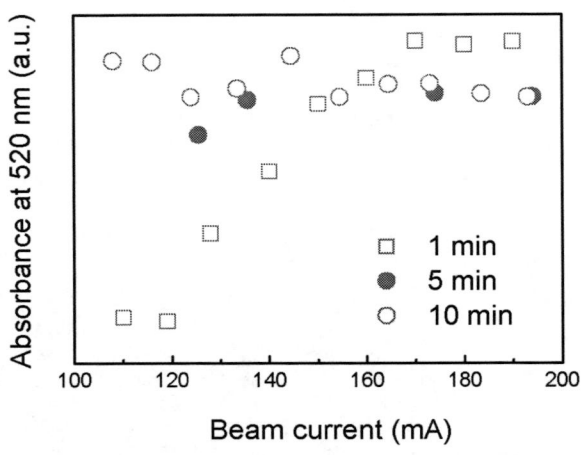

FIGURE 4. Evolution of absorbance (520 nm) of gold nanosol as function of beam current and radiation time.

CONCLUSIONS

We demonstrated that gold nanoparticles can be conveniently prepared by synchrotron x-ray irradiation. The size, shape and dispersion of formed Au particles depended largely on the pH value of the precursor solutions and were also influenced by the exposure parameters such as beam condition, radiation dose etc. Well dispersed Au nanoparticles with ≈10 nm diameter were obtained without stabilizers such as surfactants and polymers. This unique synthesis route is simple, clean and very fast without unreacted residue left and promising for biomedical applications.

ACKNOWLEDGMENTS

We wish to thank G. M. Chow for the stimulating discussion in characterizing the samples. This work was supported by the National Science Council (Taiwan), by the Academia Sinica (Taiwan), by the BK21 Project, by the Korea Institute of Science and Technology Evaluation and Planning (KISTEP) through the National Research Laboratory (NRL) and SKORE-A projects, by the Swiss Fonds National de la Recherche Scientifique and by the EPFL.

REFERENCES

1. R. A. Rosenberg, Q. Ma, B. Lai, D. C. Mancini, *J. Vac. Sci. Technol. B* **16,** 3535 (1998).
2. Q. Ma, N. Moldovan, D. C. Mancini, R. A. Rosenberg, *Appl. Phys. Lett.* **76,** 2014 (2000).
3. Y. C. Yang, C. H. Wang, Y. Hwu and J. H. Je, *Mater. Chem. Phys.* (in press).
4. E. Gachard, H. Remita, J. Khatouri, B. Keita, L. Nadjo, J. Belloni, *New J. Chem.* **22,** 1257 (1998).
5. J. Belloni, M. Mostafavi, H. Remita, J. L. Marignier, M. O. Delcourt, *New J. Chem.* **22,** 1239 (1998).
6. F. Karadas, G. Ertas, E. Ozkaraoglu, S. Suzer, *Langmuir* **21,** 437 (2005).
7. S. Baik, H. S. Kim, M. H. Jeong, C. S. Lee, J. H. Je, Y. Hwu and G.. Margaritondo. *Rev. Sci. Instr.* **75,** 4355 (2004).

Nanoparticle Decoration of Carbon Nanotubes by X-Ray Irradiation

T. Y. Yang[*], C. H. Wang[*], Y. Hwu[*], C. J. Liu[*], H. M. Lin[†], J. H. Je[¶] and G. Margaritondo[‡]

[*]Institute of Physics, Academia Sinica, 128 Academia Rd, Nankang, Taipei 115, Taiwan

[†] Department Material Engineering, Tatung University, Taipei 104, Taiwan

[¶]Department Material Science Engineering, Pohang University of Science and Technology Pohang, Korea

[‡] Ecole Polytechnique Fédérale de Lausanne (EPFL), CH-1015 Lausanne, Switzerland

Abstract. Titanium oxide (TiO_2) nanoparticles were immobilized onto the surfaces of multiwall carbon nanotubes (MWNTs) by a facile synchrotron X-ray exposure method. The influence of processing parameters such as beam dose and duration on the structures and properties of nanoparticle/nanotube hybrid were investigated by X-ray Absorption Spectroscopy (XAS), Raman spectroscopy, Transmission Electron Microscopy (TEM). The influences of synchrotron x-ray exposure on the sol-gel process and structure of produced TiO_2 nanoparticles were also investigated. Furthermore, the locations of functional groups between TiO_2 and MWNTs were also probed. This new approach is envisaged to be quite promising for the synthesis of novel nanocomposite materials.

Keywords: TiO_2, MWNTs, Synchrotron X-ray.
PACS: 61.46.-w

INTRODUCTION

Carbon nanotubes (CNTs) continuously attract wide interests in many areas of science and technology due to their unique structure-dependent optical, electronic and mechanical properties [1]. In order to optimize the application of carbon nanotubes, it is often necessary to attach other nanostructures or functional groups onto their surface. The synthesis of carbon nanotubes with other nanocrystals has been used to enhance performance in applications such as catalysts, gas sensors, field emission displays, biomedical.[2] It is also possible to immobilize metal oxides such as TiO_2 [3], SnO_2 [4], ZnO [5], SiO_2 [6] onto the surfaces of CNTs. Among them, titanium oxide is one of the most important metal oxide semiconductors due to its high photocatalytic properties in elimination of organic pollutants. It has unique catalytic activity, stability and biocompatibility when compared to other wide band gap metal oxide semiconductors. Several routes have been studied to increase photo-efficiency of titanium oxide by doping noble metal. To take of the advantage on the conductivity of MWNTs, we expect that to synthesis MWNTs and titanium oxide may further enhance the photocatalytic properties of titanium oxide and to alter charge transfer behavior. Because of their material stability and chemical inertness, carbon nanotubes can be used as potential hosts for titanium oxide in environmental and medical benefic applications.

In this paper, we propose a facile synchrotron X-ray exposure method to surface coverage MWNTs with titanium oxide and performed a systematic study of the properties of the resulting compound materials. The novelty of this study is that titanium oxide nanoparticles were reduced and precipitated onto MWNTs by electrostatic interaction. The fact that adopting $TiCl_4$ as starting material, compared to use organometallic compound, makes the control of the hydrolysis reaction easy . Our finding demonstrates that x-ray exposure method was indeed convenient, fast, versatile. It does not require any reducing agent and can be scale down to nanosize dimensions. It is easy to control

CP879, *Synchrotron Radiation Instrumentation: Ninth International Conference,*
edited by Jae-Young Choi and Seungyu Rah

the particle size and the whole process is performed at room temperature. The process is also kept at high cleanliness deposition environment. The process also allow an easy "dissecting" of the CNTs to short lengths. The above benefits prove that this simple one-step X-ray exposure method indeed is effective in immobilizing nanosize TiO_2 on the multiple-walled carbon nanotubes.

EXPERIMENTAL

Multi-wall carbon nanotubes (MWNTs) used in this study was commerce material provided by CNT co. Ltd from Korea. TiO_2/MWNTs hybrid materials were synthesized by sol-gel method and X-ray exposure. In a typical hybrid material synthesis experiment, 0.06g MWNTs and 1.2g Titanium tetrachloride (*Showa 2021-8250*) were added into 200ml of Isopropyl Alcohol (*Wako 166-04836*) solution and sonicated for 30 minutes and then stirred for 30 minutes at room temperature to obtain well-dispersed suspensions, By irradiating a solution using synchrotron X-rays to synthesize the hybrid material. The exposure times are 5, 10 and 15 minutes; other parameters are 1, 1.5 and 2 hours, respective.

The experiment was performed at 01A beamline of NSRRC (Nation Synchrotron Radiation Research Center, Taiwan), operating at 1.5 GeV. We used unmonochromatized ("white") beam with no optical elements except beryllium and silicon windows. The beam size was controlled as 0.6×1.3 cm^2 by tungsten slits.

Raman spectra were recorded using Raman spectrometer (RENISHAW inVia Raman Microscopy, 0.5mW) and 514 nm as the excitation source with a CCD detector. The final spectrum presented is an average of 10 spectra recorded at different regions over the entire range of the sample. The morphology and structure of the synthesized products were observed through a Field emission electron microscopy (FE-TEM, JEOL, JEM-2100F) operated at 200 KV. Transmission electron microscope observation was performed to compare the microstructure of the samples before and after X-ray exposure. This treatment provides a critical prerequisite for attaching TiO_2 nanoparticles onto the nanotube.

RESULTS AND DISCUSSION

Raman spectroscopy was first used to characterize the chemical composition of the TiO_2/MWNTs nanostructure. Figure 1 (a) shows a typical Raman spectrum of MWNTs indicating two characteristic peaks. The *G*-band peak at 1577.33 cm^{-1} corresponds to the high-frequency, Raman-active, E_{2g} mode of graphite. The strong and broad *D*-band at 1350.12 cm^{-1} and a weak *D'*-band at around 1621 cm^{-1} were attributed to disorder-induced carbon features arising from finite particle size effect or lattice distortion. First-order Raman spectra have been measured at room temperature in anatase TiO_2 of D_{4h} point group. The 6 Raman- active fundamental vibrations, predicted by group theory to have symmetries A_{1g}, B_{1g}, and E_g were observed. The spectrum exhibited strong lines of A_{1g}, E_g, and B_{1g} symmetries. Figure 1 (b) shows the Raman spectrum of anatase TiO_2. The results indicated that the strong lines of A_{1g} and E_g symmetries, a sharp low-frequency line of B_{1g} symmetry, they are characteristic peaks of anatase TiO_2 of Raman spectra. Figure 1 (c) shows the Raman spectra vibration modes of the TiO_2/MWNTs at different exposure times. The peaks of TiO_2/MWNTs samples in the Raman spectra can be divided into two groups at different X-ray exposure times. One group is the vibration modes of MWNTs. These Raman shifts are 1350 and 1577 cm^{-1}, which are consistent with the *D*-band and *G*-band vibration modes, respectively. While for the Raman spectra of MWNTs coated with TiO_2 (Fig. 1 (c) a, b, c, d, e and f), peaks at around 146, 198, 395, 519 and 633 cm^{-1} are attributed to the E_g, B_{1g} and A_{1g} vibration modes of TiO_2, respectively. The results demonstrated that nanoparticles of TiO_2 can be successfully coated onto the surface of MWNTs.

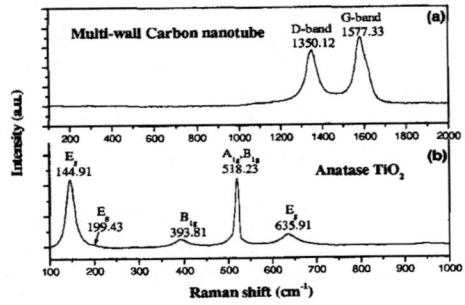
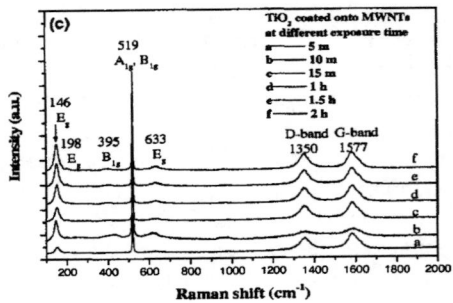

FIGURE 1. A typical Raman spectrum of (a) Multi-wall carbon nanotube; (b) Anatase TiO_2; (c) TiO_2/ MWNTs at different exposure times.

High resolution TEM micrograph reveals the interface between TiO_2 nanocrystallines and MWNTs. The HRTEM studies indicate that the MWNTs are coated with TiO_2 nanoparticles and they appear undamaged by the X-ray exposure. A representative HRTEM image of the MWNTs is shown in Fig. 2 (a) and (b), which shows that NWNTs were almost the only species in the purified products with a mean outer diameter of 15-20 nm, a mean inner diameter of 3-5 nm and lengths up to several micrometers. For TiO_2-coated surface of MWNTs, about 5 nm TiO_2 nanoparticles attach to the outer layer of the nanotubes after 5 minutes X-ray exposure as shown in Fig. 3 (a) and (b). Apparently X-ray irradiation method could form a TiO_2 and adsorption layer around the outer wall of MWNTs. It seems the TiO_2 nanooparticles were dispersed well on the MWNTs. Figure 3 (c) and (d) present MWNTs coated with nanoparticles of TiO_2 after X-ray exposure during 1h. It is obvious that the nanosized TiO_2 particles were coated onto the surface of the MWNTs. The amount of TiO_2 is increase with long exposure time. The thickness of TiO_2 layer on the MWNTs was found to be around about 5-15 nm on average. Moreover, the TiO_2 coated onto MWNTs might lap over together and from layer which is thicker than the particle size of the TiO_2.

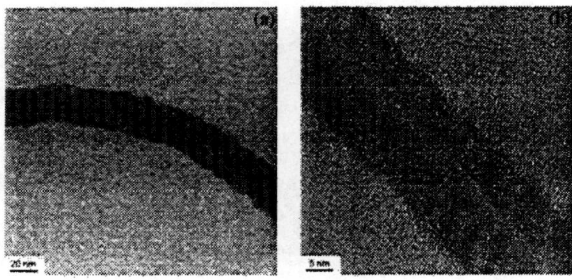

FIGURE 2. HRTEM images of MWNTs (a) before and (b) after expose by X-ray.

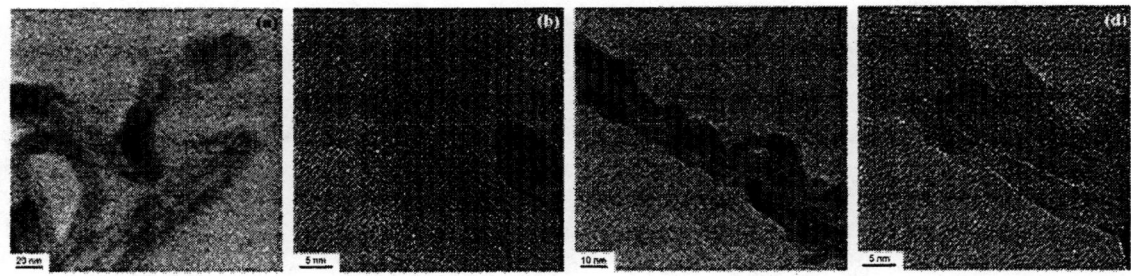

FIGURE 3. HRTEM micrographs (a) of MWNTs coated with TiO_2 after 5 minutes X-ray exposure (b) the detailed microstructural of TiO_2/ MWNTs (c) of MWNTs coated with TiO_2 after 1 hour X-ray exposure (d) the detailed microstructural of TiO_2/ MWNTs.

Figure 4 (a) and (b) summarizes the O K-edge and Ti L-edge features of TiO_2 at difference X-ray exposure times. The NEXAFS spectra were recorded using the fluorescence-yield (or sample current) method to assure the measurements of nanoparticles stoichiometries of the titanium oxides. Figure 4 clearly indicates that both the O K-edge and Ti L-edge features are very sensitive to the oxidation states of titanium. The local coordination of oxygen in TiO_2 is octahedral for these oxides. As the oxidation state of Ti changes, the energy difference of $\Delta E(e_g - t_{2g})$ in the O K-edge decreases about 2.7 eV for TiO_2. In the Ti L-edge, the four separate L_3-t_{2g}, L_3-e_g, L_2-t_{2g} and L_2-e_g features of TiO_2. The X-ray induced reaction method of TiO_2 is clearly indicated by the following spectral in the O K-edge features: (1) the $\Delta E(e_g - t_{2g})$ in the O K-edge have 2.7eV for stoichiometric TiO_2 at different X-ray exposure times. (2) The O K-edge features at 531.3, 534.0, 539.3 and 544.8eV were observed at different X-ray exposure times. Furthermore, the X-ray induced reaction method is also supported by the following spectral in the Ti L-edge features: (1) the conversion of the four well separated L_3-t_{2g}, L_3-e_g, L_2-t_{2g} and L_2-e_g features of TiO_2 to four broad peaks and (2) the subsequent gradual intensity of the peak is increase as the X-ray exposure time increases. The characteristic changes in the O K-edge and Ti L-edge regions clearly indicated a conversion of TiO_2 to oxides with lower oxidation states as result of X-ray exposure method. In the XAS results, it is clearly that TiO_2 can attach to MWNTs successful by X-ray exposure method.

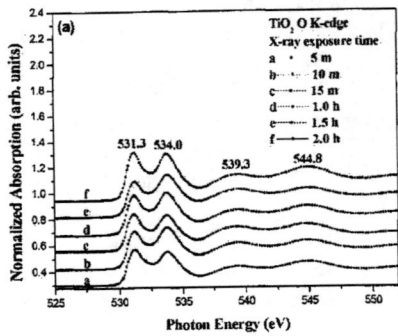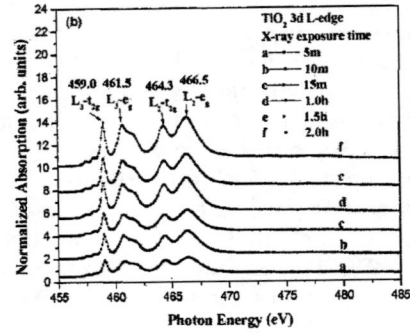

FIGURE 4. Comparison of O K-edge (a) and Ti L-edge (b) features for TiO_2. NEXAFS measurements were carried out by measuring the fluorescence yield to assure nanoparticle for TiO_2.

This process consists of a precursor organometallic or suspension followed by hydrolysis and polycondensation reaction resulting in the final coating. In this study titanium tetrachloride ($TiCl_4$) was performed to make nanocrystalline TiO_2. The hydrolysis and polycondensation of titanium $TiCl_4$ precursor according to the following scheme:

$$TiCl_{4(l)} + 2H_2O_{(l)} \rightarrow TiO_{2(s)} + 4HCl_{(l)} \tag{1}$$

On X-ray irradiation of water, the following radicals are produced:

$$H_2O \rightarrow (H_2O^+, e^-_{aq}) + (H^\bullet, OH^\bullet) + (H_2, O_2) + (H_2O^\bullet, HO_2)$$

(2)

Where H and e^-_{aq} are reducing agents, the e^-_{aq} has higher potential energy of reducing (e^-_{aq} = -2.77eV). The e^-_{aq} reacts with metal ions to produce Ti^+ at room temperature in the solution. During nucleate and grow metal ions are forming the nanoparticles of TiO_2 that condensed onto the surface of MWNTs.

CONCLUSIONS

By applying intense synchrotron x-rays, we successfully fabricate with high-rate nanocomposite of TiO_2/MWNT at room temperature, without a reducing agent. By adjusting parameters such as the irradiation time and the concentration of $TiCl_4$ precursor in the synthesis system ultimal properties can be obtained. This method is attractive since it eliminated a high-temperature heat treatment step normally required in chemicall synthesis process. X-ray exposure can transform the amorphous TiO_2 formed into a crystalline phase on the MWNTs and make this one stop method advantageous in many applications.

ACKNOWLEDGMENTS

This work was supported by the National Science Council (Taiwan), and National Synchrotron Radiation Research Center (NSRRC).

REFERENCES

1. C.N.R. Rao, B.C. Satishkumar, A. Govindaraj, M. Nath, *Chem. Phys. Chem.* 2 (2001) 78.
2. Alberto Bianco, Kostas Kostarelos, Charalambos D. Partidos, *Chem. Commun.* (2005), 571–577.

3. K. R. Lee, S. J. Kim, J. S. Song, J. H. Lee, Y. J. Chung and S. Park, *J. Am. Ceram. Soc.*, 2002, 85, 341.
4. Jining Xie, Vijay K. Varadan, *Materials Chemistry and Physics* 91 (2005) 274–280.
5. Hansoo Kim and Wolfgang Sigmunda, *Appl. Phys. Lett.*, Vol. 81, No. 11, 9 September 2002.
6. Q. Fu, C. Lu, J. Liu, *Nano Lett.* 2 (2002) 329.

Microstructural Changes upon Milling of Graphite in Water and Subsequent MWCNT Formation During High Temperature Annealing

Adriyan Milev, Nguyen Tran, G.S. Kamali Kannangara and Michael Wilson

College of Health and Science, University of Western Sydney,
Locked Bag 1797, Penrith South DC 1797, Australia.

Abstract. The method of preparing carbon nanotube (CNT) by milling of graphite particles in water followed by high temperature annealing is proposed and the mechanism discussed. Transmission electron microscopy (TEM) and X-ray diffraction (XRD) line broadening analysis reveal that cleavage of the graphite particles occurs preferentially along the out-of-plane π bonds. Carbon K-edge near edge X-ray absorption fine structure (NEXAFS) of the milled graphite shows an increased sp^3 character of the C=C bonds, but no major bonds rupture in the graphene sheets. The annealing at 1400 °C for 4 h of the milled graphite in argon results in formation of multiwalled carbon nanotubes accompanied with a number of coiled and twisted stacks of graphene sheets. The increased structural disorder of the milled graphite and presence of iron contaminations facilitate the rolling up of the cleaved graphene sheets during annealing.

Keywords: Carbon nanotube, Graphite, Milling, Annealing, NEXAFS.
PACS: 81.05.Uw, 81.07.De, 81.20.Wk, 81.40.Gh, 82.80.Ej.

INTRODUCTION

Graphite consists of stacks of parallel two-dimensional graphene sheets with the carbon atoms arranged in hexagonal rings through localized in-plane $2s$, $2p_x$ and $2p_y$ (sp^2) orbitals, whereas the individual sheets are weakly bonded in the third dimension by delocalized out-of-plane $2p_z$ orbitals. Spontaneous rolling up of individual graphene sheets in freshly cleaved graphite crystals has been reported [1]. The rolled graphene sheet has tubular shape resembling carbon nanotube. However, the $2p_z$ orbitals of the carbon atoms overlap most effectively when they are parallel, thus the graphene sheet has lowest energy when it is completely flat. Therefore, the rolling up of a graphene sheet is energetically unfavorable process.

Chen and co-workers [2] reported that ball milling of graphite develops reactive nanoporous carbons with high surface areas, which upon annealing produce multiwalled carbon nanotubes. They attributed the formation process to be a low temperature solid-state crystal growth mechanism. Hiura and co-workers [1], however, suggested that the driving force for graphene sheet folding could be a certain degree of disorder in form of structural defects and broken bonds that may modify the highly delocalized C=C structure. To compensate the increased energy, graphene sheets are folded. However, no experimental evidence to show the formation of defects and change of C=C bond character was provided.

Our previous work showed that excessive destruction of the two-dimensional hexagonal graphene sheet structure has a detrimental effect on the possibility of carbon nanotube formation upon annealing [3]. The aim of this work is to study the role of the disruption of C=C delocalization and hence formation of nanotubes. The microstructural changes of graphite during processing are investigated by Transmission electron microscopy (TEM), X-ray diffraction and near edge X-ray absorption fine structure (NEXAFS) spectroscopy.

EXPERIMENTAL

Sample Preparation. Graphite powder was ball milled in a Pulverizette 6 (Fritsch) planetary mill. The 2g samples were loaded in a stainless-steel container with volume of 80 ml containing 25 ml water. Then 300 stainless

CP879, *Synchrotron Radiation Instrumentation: Ninth International Conference,*
edited by Jae-Young Choi and Seungyu Rah
© 2007 American Institute of Physics 978-0-7354-0373-4/07/$23.00

steel balls with diameter of 5 mm were added. The milling was carried out for 100 h at 400 r.p.m. The milled powders were dried at 100 °C for 12 h in air and then were annealed in an alumina tube furnace at 1400 °C for 4 h in argon (99.999 % purity) flowing at 15 – 20 ml/min.

Transmission Electron Microscopy. A Philips Biofilter-120 TEM operating at 120 kV was used for imaging. The specimens were sonicated in ethanol in order to separate aggregates. A drop of the suspension was transferred onto lacey carbon foils supported on copper grids for examination.

X-ray Diffraction. The X-ray diffraction patterns were acquired by a Philips PW 1825/20 powder diffractometer (Cu K_α, 40 kV, 30 mA, 15 – 65 deg., 0.02° step, 10 sec/step) on background free silicon wafer sample holder. Longer acquisition times (240 sec/step) were required to obtain good signal-to-noise ratio for (*110*) reflections. The diffraction lines were corrected for instrumental broadening. The out-of-plane (L_c) crystallite sizes were determined by the Debye-Scherrer formula applied to (*002*) reflection.

Near Edge X-ray Absorption Fine Structure. The carbon *K-edge* NEXAFS spectra were acquired at the wide range, bending magnet beam line 24A at the National Synchrotron Radiation Research Centre (NSRRC) in Hsinchu, Taiwan. Prior to the measurements, samples were heated in the ultra high vacuum analysis chamber at ~200 °C for 24 h in order to minimize the adventitious surface contaminants. All spectra were acquired *via* bulk sensitive fluorescence yield mode [4] and were normalized to reference beam intensity measured on a gold grid monitor. The incident X-ray energy was calibrated using the π^* of precursor graphite at 285.4 eV.

RESULTS AND DISCUSSION

TEM. The examination of the morphological and structural transformation of graphite milled in water shows that the graphite particle morphology on micrometer length scale is preserved (Fig. 1 a).

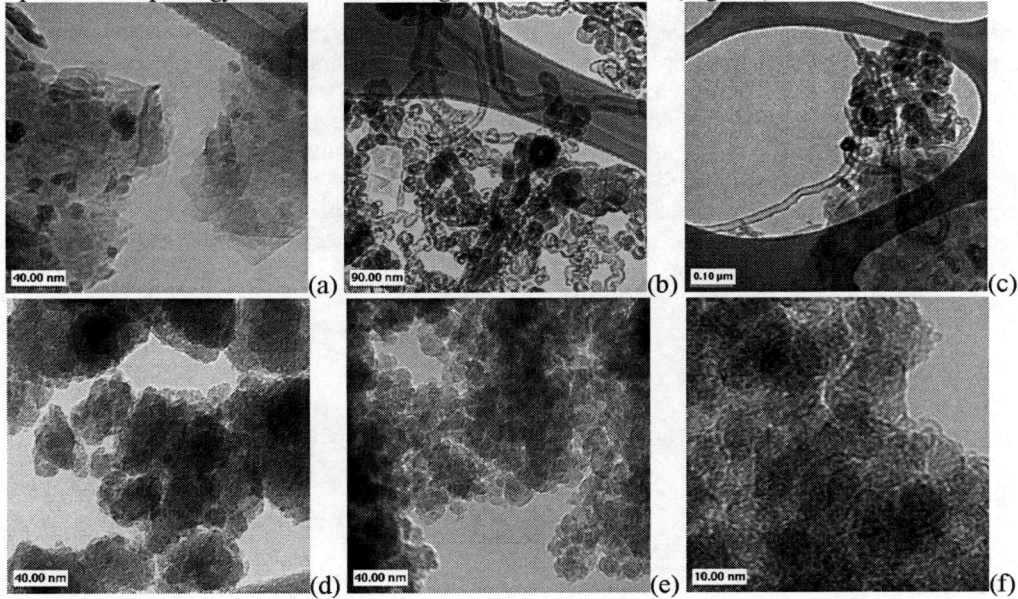

FIGURE 1. TEM micrographs of processed graphite. (a) Milled in water, (b-c) Annealed after milling in water, (d) Dry-milled, (e-f) Dry milled and annealed.

For comparison, graphite sample was dry-milled in argon under the same conditions (Fig. 1 d). The dry-milled graphite consists of thick agglomerates without specific morphology. Clearly, the water protects to great extent the two-dimensional morphology of the graphite particles. Annealing at 1400 °C for 4 h in argon of water-milled graphite changed the sheet shape (Fig. 1 b and c). Several groups of MWCNTs with rather irregular shapes and outer diameters ranging from 20 to 60 nm along with a number of bent and coiled ribbons are present. The annealed dry-milled graphite shows no significant morphological developments (Fig. 1 e, f).

The X-ray diffraction patterns of milled in water and annealed graphite are compared with that of the graphite precursor. Two strong *001* diffraction peaks can be distinguished in the precursor pattern (Fig. 2 A). These correspond to the lamellar organization of the graphene sheets with an interplanar distance d_{002} = 0.336 nm. Milling in water reduces the peak intensity and broadens the (*002*) reflection considerably (Fig. 2 b). It is noteworthy to mention that reflections due to magnetite (Fe_3O_4) appeared, which is attributed to abrasion of the mill container and balls followed by a chemical reaction of the abrasion products with the water.

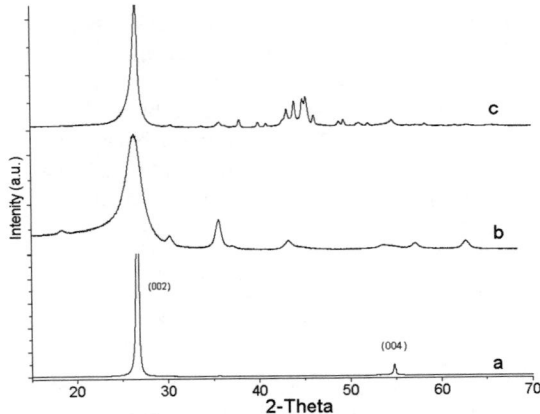

FIGURE 2. XRD patterns of processed graphite. (a) Precursor, (b) Milled in water graphite, (c) Annealed after milling graphite.

Annealing slightly narrows the main (*002*) reflection (Fig.2 c). Several non-graphitic reflections are also detected in the 2θ interval of 35 to 63 deg., which are identified as iron carbide phase (Fe_3C). This phase is probably formed by reduction of the iron oxide phase by the graphitic carbon during annealing. The changes of the width of the major (*002*) reflection indicate that the out-of-plane crystallite size (crystallite thickness) is changed during processing. The average crystallite thickness (L_c) decreases from ~ 170 nm of the precursor, to ~ 5 nm upon milling, and then increases to ~ 15 nm of the annealed material.

NEXAFS. Normalized fluorescence yield X-ray absorption spectra, recorded above the carbon *K*-edge for the precursor and processed graphite samples are compared in Fig. 3. The samples are characterized by excitations at 285.4 eV, and at 291.8 eV, known to originate from C 1s→ π* and C 1s→ σ* transitions of sp^2 carbon atoms in the aromatic graphene rings. If the order in the rings is disturbed, the electron charge is reorganized, which causes a chemical shift of the NEXAFS features or appearance of additional transitions. In particular, the σ* excitation is known to be very sensitive to structural and electronic local changes [5]. Relatively broad excitation features in 287 – 289 eV spectral region are present in all samples. Such excitations have been assigned to π* states originating from free-electron-like interlayer states in graphite [4] or bonding between carbon and heteroatoms [6].

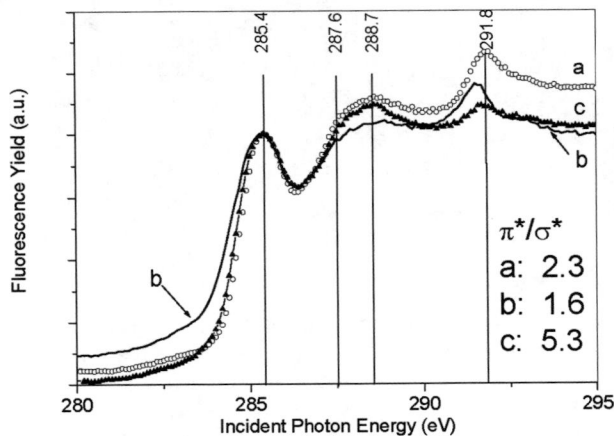

FIGURE 3. Normalized C *K-edge* NEXAFS spectra of processed graphite. (a) Precursor graphite (b) Milled in water graphite, (c) Annealed after milling graphite. The insert gives π*/σ* index.

The spectral features of the processed graphite are similar to those of the precursor graphite, however the peak relative intensities and peak widths differ (Fig. 3 b and c). The broader π* resonance, the weaker peaks in the 287 –

289 eV excitation interval and the 0.3 eV lower energy shift of the structure sensitive σ^* transition all indicate that the local order is changed during milling. The strong peak at 285.4 eV, however, shows that the sp^2 C=C hexagonal network is not greatly affected. Apart from the reduced intensity of the σ^* transition, all spectral features of the annealed material are very similar to those of the precursor graphite. This indicates that annealing restores the local order to great extend. The assignment of the excitations in 287 – 289 eV interval to heteroatoms, however, cannot explain why these resonances are stronger of the precursor graphite and of the annealed sample, than of the milled in water sample. It seems that the features in this spectral region arise from a modification of π-electron system of graphite, which may be caused not only by chemical as previously suggested [6] but also by structural factors. The π^*/σ^* relative intensity ratio can be used as an index to estimate the aromatic fraction of carbon in processed graphite [7]. The π^*/σ^* indexes inserted in Fig. 3 show lowest value of the milled graphite, which indicates an increased sp^3 bond character. Annealing seem to restore the local order and sp^3 bond character decreases, and hence the π^*/σ^* ratio increases.

Under ball milling conditions, at least in the reactors and conditions described here, water facilitates the graphite particle thinning, and to great extent protects the hexagonal graphene sheet structure from the severe impact forces. The energy of the system increases due to development of crystallite defects, changes of the bond character and probable formation of dangling bonds. During annealing, the system is driven to minimize its free energy by restoration of the sp^2 bond character, and elimination of the dangling bonds. This opens the opportunity for bending of the graphene sheets and formation of new C=C bonds between sheets with various mutual orientations. The presence of iron contaminations may also contribute to the rolling up of the graphene sheets as suggested previously [2]. It is noteworthy that once the hexagonal network of graphene sheets is destroyed it cannot be recovered by annealing at 1400 °C [3]. Such restoration may require much higher temperatures where the formation of flat two-dimensional graphene structure is kinetically and thermodynamically favorable process.

CONCLUSIONS

1. Milling of graphite in water protects the lamellar particle morphology on micrometer scale level. Yet, milling changes the order on nanometer and sub-nanometer level and increases the sp^3 character of the C=C bond.

2. Thick multi walled carbon nanotubes (MWCNTs) are produced by annealing at 1400 °C in argon of the milled graphite. The high temperature, iron contaminants and the restoration of the sp^2 bond character seem to be the driving forces for the rolling up of the graphene sheets.

ACKNOWLEDGMENTS

This work was produced as part of the activities of the Centre for Functional Nanomaterials funded by the Australian Research Council under the ARC Centers of Excellence program. Part of this work was performed at the Australian National beamline facility with support from Australian Synchrotron Research program.

REFERENCES

1. H. Hiura, T. W. Ebbesen, J. Fujita et al., Nature (London, United Kingdom) 367 (6459), 148-151 (1994).
2. Y. Chen, M. J. Conway, J. D. Fitzgerald et al., Carbon 42 (8-9), 1543-1548 (2004).
3. C. Marshall and M. Wilson, Carbon 42 (11), 2179-2186 (2004); D. E. Smeulders, A. S. Milev, G. S. Kamali Kannangara et al., Journal of Materials Science 40 (3), 655-662 (2005).
4. D. A. Fischer, R. M. Wentzcovitch, R. G. Carr et al., Physical Review B: Condensed Matter and Materials Physics 44 (3), 1427-1429 (1991).
5. R. Schloegl, V. Geiser, P. Oelhafen et al., Physical Review B: Condensed Matter and Materials Physics 35 (12), 6414-6422 (1987).
6. U. Dettlaff-Weglikowska, V. Skakalova, R. Graupner et al., Journal of the American Chemical Society 127 (14), 5125-5131 (2005).
7. S. D. Berger, D. R. McKenzie, and P. J. Martin, Philosophical Magazine Letters 57 (6), 285-290 (1988).

CHAPTER 8

LITHOGRAPHY AND MICROMACHINING

A MEMS-Based Micro Biopsy Actuator for the Capsular Endoscope Using LiGA Process

Sunkil Park[1], Kyo-in Koo[1], Gil-sub Kim[1], Seoung Min Bang[2], Si Young Song[2], Chong Nam Chu[3], Doyoung Jeon[4], and Dongil "Dan" Cho[1*]

[1]School of Electrical Engineering and Computer Science, ASRI, NBSRC, ISRC, NAVRC, Seoul National University, Korea, [2]Department of Internal Medicine, Yonsei University College of Medicine, Korea, [3]School of Mechanical and Aerospace Engineering, Seoul National University, Seoul, Korea, [4]Department of Mechanical Engineering, Sogang University, Seoul, Korea,

Abstract. This paper presents a LiGA (German acronym for LIthografie, Galvanoformung, Abformung) based micro biopsy actuator for the capsular endoscope. The proposed fabricated actuator aims to extract sample tissues inside small gastric intestines, that cannot be reached by conventional biopsy. The actuator size is 10 mm in diameter and 1.8 mm in length. The mechanism is of a slider-crank type. The actuator consists of trigger, rotational module, and micro biopsy tool. The core components are fabricated using the LiGA process, for overcoming the limitations in accuracy of conventional precision machining.

Keywords: Capsular endoscope, actuator, LiGA, biopsy, x-ray
PACS: 61.43.Dq, 68.03.Hg, 68.90.+g

INTRODUCTION

It is becoming increasingly common to find research relating to capsular endoscopes replacing conventional wire endoscopes. PillCam (Given-Imaging, Israel), OMOM (Jinshan, China), and Endo Capsule (Olympus, Japan) were commercialized in July of 2001, March of 2005 and October of 2005, respectively. MiRO was developed by the Intelligent Microsystem Center (Seoul, Korea) in March of 2005. In addition, Norika3 (RFsystems, Japan) and SmartPill (Smartpill, US) are developing similar products. These capsular endoscopes can transfer images in all gastro-intestines and cause little discomfort to patients when swallowing the capsules. Conventional wire endoscopes, are characteristic of continuous discomfort and limited diagnostic scope. Even though capsular endoscopes have these advantages over conventional wire endoscopes, they are not expected to completely replace conventional wire endoscopes, because of their single function as an image transport.

In this paper, a MEMS (Microelectromechanical Systems) based micro biopsy actuator is developed for the multi-function capsular endoscope. This actuator was designed for integration with a capsular endoscope. To overcome the limitation of conventional precision machining, the actuator was fabricated using the LiGA (German acronym for LIthografie, Galvanoformung, Abformung) process.

MECHANISM AND DESIGN

The MiRO has a diameter of 10 mm and height of 26 mm. A space of diameter 10 mm and height 2 mm is permitted for a micro biopsy actuator, because of other components of the capsular endoscope such as CCD camera, image transport module, telemetry system, and battery. The actuator (diameter 10 mm and height 1.8 mm) was designed and fabricated using the LiGA process, which is smaller than the space permitted.

A Si micro spike was developed as a micro biopsy tool without an actuator [1]. The design of the Si micro spike is applied to a biopsy component of the actuator. The slider-crank mechanism was utilized for spike actuating, as

---

*Corresponding author: dicho@snu.ac.kr

CP879, *Synchrotron Radiation Instrumentation: Ninth International Conference,*
edited by Jae-Young Choi and Seungyu Rah
© 2007 American Institute of Physics 978-0-7354-0373-4/07/$23.00

shown in Fig. 1. The actuating force is stored at a torsional spring (Diameter 0.14 mm). The spring triggering is performed by a melting polymer string. The polymer string is melted by allowing current to heat the SMA (Shape Alloy Memory), which coils the polymer string. Then, the micro spike strokes the target tissue. All the dimensions of the actuator's components and specification data are extracted from preliminary research [1, 2].

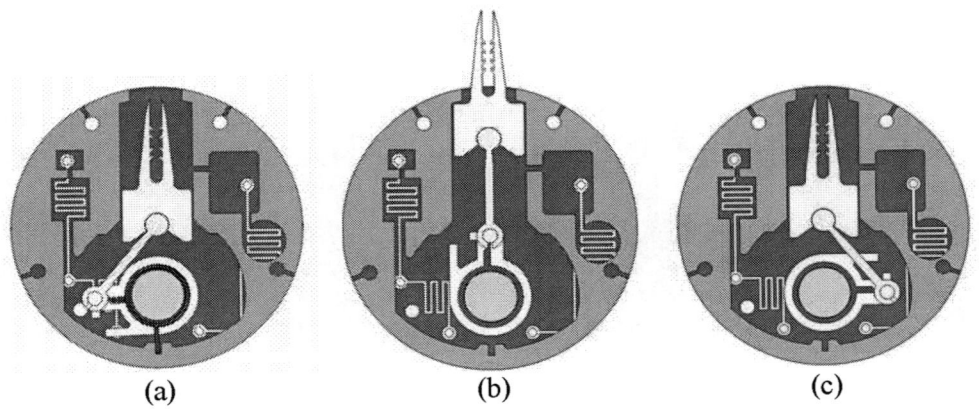

(a) (b) (c)

FIGURE 1. Schematics of tissue sampling mechanism of micro biopsy actuator (Before actuation (1), actuation mode (2), after actuation (3))

FABRICATION OF MICRO BIOPSY ACTUATOR

To integrate the actuator to the permitted space (diameter 10 mm and height 2 mm), the actuator component cannot be fabricated by conventional precision machining. Therefore, the LiGA process is applied to making precise dimension fabrication with fine sidewall roughness and high aspect ratio structure. The recent development of the LIGA process, which consists of x-ray lithography, electroplating and electromoulding, has led to the high-volume fabrication of various plastic MEMS components. DXRL (Deep x-ray lithography), the first step in the LIGA process, is considered an important process step, because it determines the quality and accuracy of the final product. In addition, DXRL can provide a final product several millimeters high, with an aspect ratio exceeding 100:1.

Fabrication of X-ray Mask

A polyimide film of thickness 125 μm is bonded on the wafer using dry photoreist, and then the Ti /Au (300 Å / 1500 Å) seed layer is deposited by a metal sputter on the polyimide film for electroplating. TOK PMER P-LA 900PM photoresist is coated on the deposited seed layer with 1000 rpm, 35 seconds using spin coater. The PMER photoresist is uniformly deposited with 27 μm thickness. The wafer is exposed and developed sequentially. Gold electroplating is performed to manufacture the X-ray absorber structure. Figure 2 shows the process flow of x-ray mask fabrication.

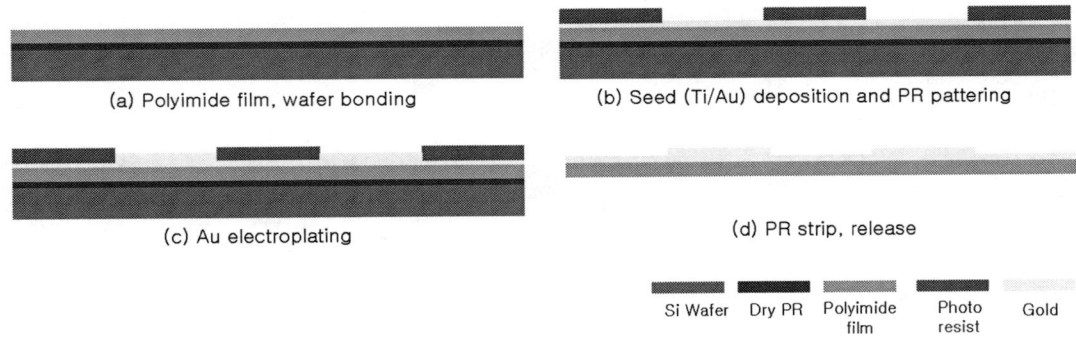

(a) Polyimide film, wafer bonding   (b) Seed (Ti/Au) deposition and PR pattering

(c) Au electroplating   (d) PR strip, release

Si Wafer   Dry PR   Polyimide film   Photo resist   Gold

FIGURE 2. Process flow of x-ray mask fabrication using polyimide film

X-ray Irriadiation & Electroplating Process

The LiGA process flow is presented in Fig. 3. A PMMA bonded on a titanium substrate is irradiated using the fabricated X-ray mask. The irradiated PMMA is developed at room temperature with a specific organic developer, commonly known as GG developer (2-(2-butoxy-ethox) ethanol 60 %, Morpholine 20 %, Ethanolamine 5 %, DI water 15 %). Nickel electroplating is performed on the developed PMMA mold. Finally, the core components are fabricated by removing the mould parts, as shown in Fig. 4 [3, 4].

Irradiation is performed at the PLS (Pohang Accelerator Laboratory) in Pohang, Korea. It operates at 2.5 GeV with an electron current between 110 and 190 mA. Among many beam lines, the 9C1 beam line is equipped with the 250 μm Be filters and a 50 μm polyimide filter. The distance in air from the polyimide filter to the irriadiation vacuum jig is 18 cm. The theoretical calculation of the critical wavelength is 2.0 Å at the end of the beam line. The x-ray beam size was 180 x 15 mm^2, the PMMA samples were irradiated under a scanning velocity of 3 cms^{-1}. For fast and effective development of irradiated polymer samples, the bottom dose of the sample must be greater than 4 kJcm^{-3}. In addition, the maximum dose under the membrane must be less than 20 kJcm^{-3}. For making the components of the micro biopsy actuator, the PMMA as an x-ray photoresist is exposed to 4.0 kJcm^{-3} of bottom dose [3, 4, 5].

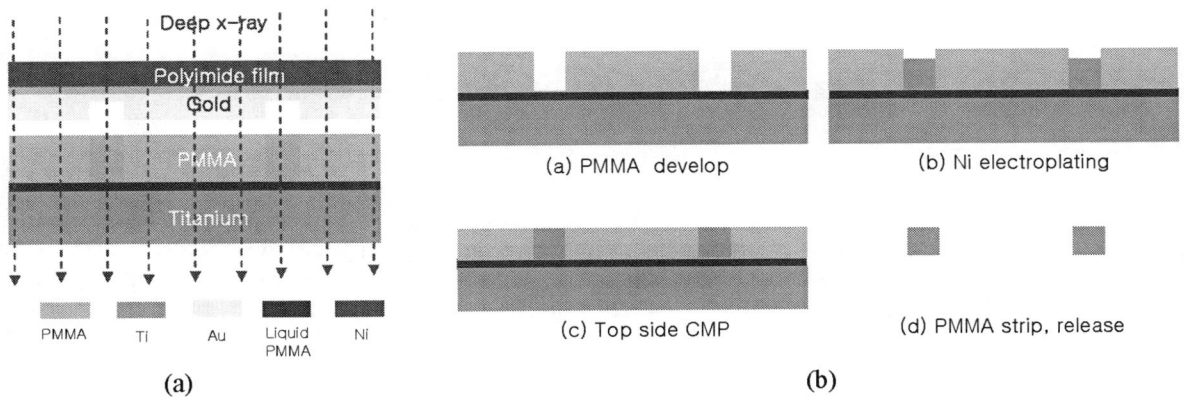

FIGURE 3. LiGA process flow (X-ray irriadiation diagram (a), electroplating process (b))

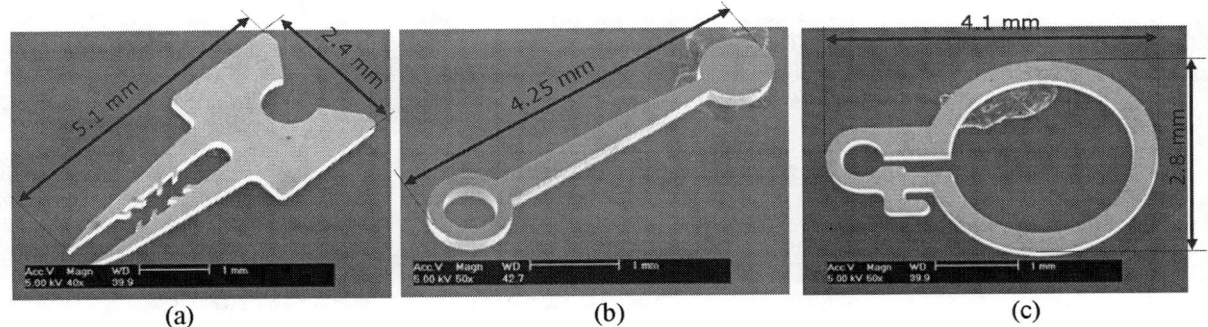

FIGURE 4. FE-SEM photograph of fabricated actuator components using LiGA process (Micro spike (a), connecting bar (b), spring fixation pin (c))

EXPERIMENT & ANALYSIS OF MICRO BIOPSY ACTUATOR

Figure 5 (a) presents the assembled micro biopsy actuator with LiGA processed components, PCB and torsional srping. Firstly, for triggering the micro biopsy actuator, the SMA of diameter 0.0015" is connected to a 3 V button cell battery on the backside of actuator. Then, the spring fixation pin mounting torsional spring is rotated 180° anticlockwise, for readying tissue samples. A spring fixation pin is then fixed with polymer wire of diameter 0.1 mm. When the power switch is turned on, the polymer string is melted by heated SMA wire, then the spring is rotated clockwise 180° and the micro biopsy tool moves back and forth at the same time. The micro biopsy actuator is tested with the force measurement system, when the micro biopsy tool moves forward. Figure 5 (b) presents the experimental setting which consists of a load-cell with a strain gauge, micro-biopsy actuator, XYZ stage, and notebook connected indictator with a serial (RS-232) port [6, 7]. Figure 5 (c) presents the measured force when the spike moves forward. The measured force is 0.66 N, which is greater than the tissue penetration force of 0.22 N with micro-biopsy tool. The tissue penetration force of tissue is extracted from the preliminary tissue sampling experiment with micro biopsy tools [1, 2].

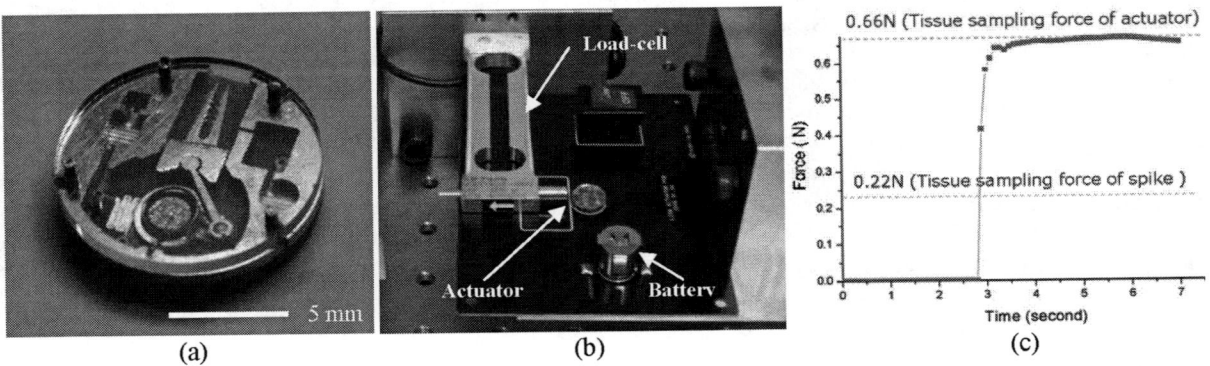

(a) (b) (c)

FIGURE 5. Photograph of assembled actuator of diameter 10 mm (a), setup photograph of force measurement with load-cell, indicator, and PCB (b) and graph of measured force of micro biopsy actuator with load-cell (c)

CONCLUSION

In this paper, a micro biopsy actuator for capsular endoscope integration, is presented. This actuator has a diameter 10 mm and height 1.8 mm. The core components are fabricated using the LiGA process to overcome the limitations of conventional precision machining. It reliably performs with a 3 V button cell battery. The tissue biopsy experiment is currently being performed using a pig's small intestine *ex vivo*, with the fabricated micro biopsy actuator.

ACKNOWLEDGMENTS

This research has been supported by the Intelligent Microsystem Center(IMC; http://www.microsystem.re.kr), which carries out one of the 21st century's Frontier R&D Projects sponsored by the Korea Ministry Of Commerce, Industry and Energy, by a grant of the Korea Health 21 R&D Project, Ministry of Health & Welfare, Republic of Korea (A05-0251-B20604-05N1-00010A), by the SRC/ERC program of MOST/KOSEF (grant # R11-2000-075-01001-0), and by pohang accelerator laboratory (PAL; http://pls.postech.ac.kr)

REFERENCES

1. S.W, Byun, J.M. Lim, S.J. Paik, A.R. Lee, K.I. Koo, S.K. Park, J.H. Park, B.D. Choi, J.M. Seo, K.A. Kim, H. Chung, S.Y. Song, D.Y. Jeon, S.S. Lee, and D. I. Cho, "Novel Barbed Micro-Spikes for Micro-Scale Biopsy," IOP Journal of Micromechanics and Microengineering, vol. 15, no. 6, pp. 1279-1284, 2005
2. S.K, Park, A.R. Lee, M.J. Jeong, H. M. Choi, S.Y. Song, S.M. Bang, S. J. Paik, J. M. Lim, D.Y. Jeon, S.K. Lee, C.N. Chu, and D.I. Cho, "A Disposable MEMS-Based Micro-Biopsy Catheter for the Minimally Invasive Tissue Sampling," IEEE/RSJ International Conference on Intelligent Robots and Systems, Edmonton, Alberta, Canada, Aug. 2-6, 2005
3. J.T, Kim and C.G, Choi, "Absorber embedded x-ray mask for high aspect ratio polymeric optical components," IOP Journal of Micromechanics and Microengineering, vol. 15, pp. 615-619, Jan 2005
4. S.J, Moon and S.S, Lee, "A novel fabrication method of a microneedle array using inclined deep x-ray exposure," IOP Journal of Micromechanics and Microengineering, vol. 15, pp. 903-911, Mar 2005.
5. S. Griffiths, "Fundamental limitations of LiGA x-ray lithography: sidewall offset, slope and minimum feature size," IOP Journal of Micromechanics and Microengineering, vol. 14, pp. 999-1011, May 2004
6. M. Rentschler, J. Dumpert, S. Platt, D. Oleynikov, S. Farritor, K. Iagnemma, "Mobile *In Vivo* Biopsy Robot," Proceedings of the 2006 IEEE International Conference on Robotics and Automation, Orlando, Florida, May, 2006
7. T.K. Podder, D.P. Clark, J. Sherman, D. Fuller, E.M. Messing, D.J. Rubens, J.G. Strang, Y.D. Zhang, W.O.Dell, W.S. Ng, and Y. Yu, "EFFECTS OF TIP GEOMETRY OF SURGICAL NEEDLES: AN ASSESSMENT OF FORCE AND DEFLECTION," The 3rd European Medical and Biological Engineering Conference EMBEC'05, Prague, Czech Republic, Nov. 20 – 25, 2005

A Transdermal Drug Delivery System Based on LIGA Technology and Soft Lithography

Marco Matteucci[1], Frederic Perennes[1], Benedetta Marmiroli[2] and Enzo Di Fabrizio[3]

[1]Sincrotrone Trieste, Area Science Park, I-34012 Basovizza -Trieste, Italy
[2]INFM-TASC-CNR S.S.14 km 163,5 in Area Science Park, 34012 Basovizza-Trieste, Italy
[3]Universita' Magna Graecia, Camups Germaneto-BIONEM lab, viale europa 88100, Catanzaro, Italy

Abstract. This report presents a transdermal drug delivery system based on LIGA fabricated microparts. It is a portable device combining a magnetically actuated micro gear pump with a microneedle array. The fluidic behaviour of the system is analyzed in order to predict its performance according to the dimension of the microparts and then compared to experimental data. The manufacturing process of both micropump and microneedle array are described.

Keywords: LIGA, DXRL, Transdermal drug delivery, microneedles, microgear pump.
PACS: 85.85.+j

INTRODUCTION

In recent years a major effort has been done on optimization of drug delivery, due to increasing demand in disease targeting and precision of clinical diagnostic tools. The general aim is to deliver in a more localized way smaller quantities of prescribed drug in precise times in order to improve therapeutic effects and avoid negative side-effects linked to uncontrolled mass delivery. At present, a large number of invasive and non-invasive techniques are being developed [1]. Although most of these have proven to be effective especially in delivering substances composed of large molecules, a complete control of delivered drug quantities together with minimum invasiveness have yet to be achieved. In this context, transdermal delivery has the potential to significantly improve the treatment of certain diseases as well as patient comfort.

The outermost 20 µm of skin, called stratum corneum, is primarily composed of dead cells. The viable epidermis, up to 50-100 µm below the stratum corneum, contains living cells, but it has no blood vessels and contains few nerves. Therefore, in order to obtain effective transdermal delivery, the drug solution must cross the stratum corneum and the viable epidermis and reach the dermis where the outermost nerves and blood vessels are located. Once the drug solution has reached the dermis, the interstitial fluid can absorb the drug and then diffuse into the blood stream through capillaries. Microneedles able to reach depths between 150 and 200 µm under the skin surface, have the advantage of enhancing effective drug delivery and, at the same time, lower the chance of causing pain through nerve pinching.

At present, transdermal drug delivery systems based on microneedle arrays offer a wide range of delivery rates compared to other systems, together with the possibility of having disposable items, minimum invasiveness and high patient comfort. Silicon is a widely used construction material for microneedles [2],[3]. However, due to its brittle nature, metal and hard polymer materials have recently gained considerable interest, especially when they show high biocompatibility [4] and can have large scale production at low costs. At this regard, the combination of Deep X-ray Lithography (DXRL), electroforming, and molding techniques [5], [6] offers the possibility to choose from a wide range of polymer compounds with biocompatible properties. Moreover these techniques allow the fabrication of microneedles with bevel tips ensuring an easy penetration into the skin [7]. A pumping system is also needed to transfer the fluid from a reservoir through the system of needle channels; microgear pumps can be efficiently fabricated with DXRL with very small tolerances.

CP879, *Synchrotron Radiation Instrumentation: Ninth International Conference,*
edited by Jae-Young Choi and Seungyu Rah

In this work a complete system was realized with DXRL and successfully tested. A short description of the fabrication process is here reported together with the first performance measurements.

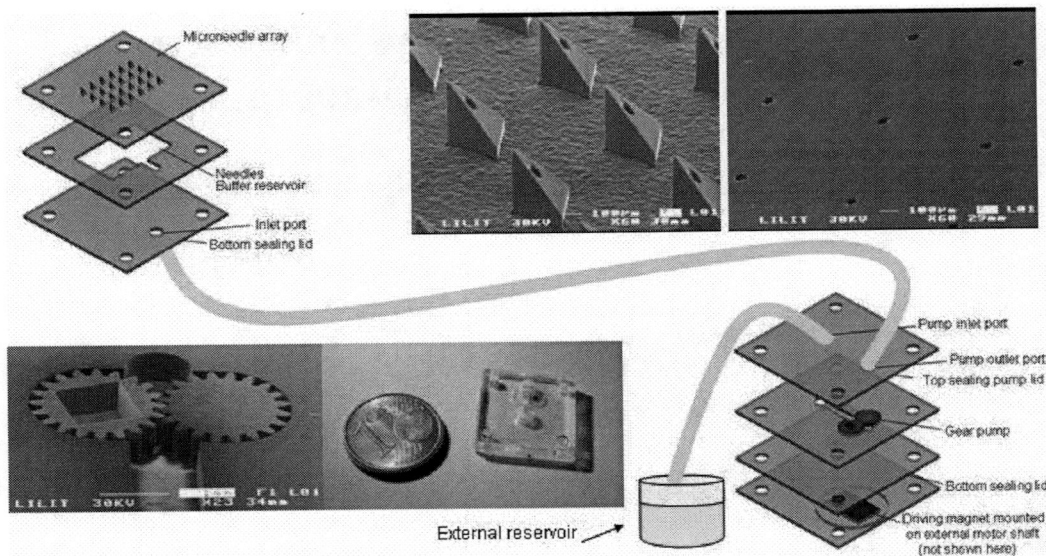

FIGURE 1. The microfluidic system: drug delivery is obtained by connecting through a microfluidic system a set of microneedles (top) with a micro gearpump (bottom).

SYSTEM FABRICATION

The transdermal delivery system consists of a microgear pump connected with an array of microneedles as shown in Fig. 1. A large number of different micropumps have been developed up to now and reported in literature [8]; microgear pumps in PMMA able to work with fluids with a wide range of viscosities can be successfully fabricated with tight tolerances by DXRL [9]. The micropump actuation is achieved through the coupling of a permanent magnet embedded into one of the gears with a second magnet mounted on a micromotor. This kind of magnetic actuation allows complete sealing of the pump case, thus avoiding possible leaks and eliminating the need for a complicated dynamic shaft seal assembly [10]. Since the measurement of the rotating gears speed is quite complex, the speed taken into account is that of the motor shaft, assuming an ideal magnetic coupling [9]. A SEM photograph of an assembled microgear pump with perfect vertical sidewalls is shown in Figure1 together with a sealed micropump. A detailed description of the pump fabrication and assembly is reported elsewhere [9]. The connection with the external reservoir and the microneedles has been made possible via two holes on the top lid of the micropump.

Hollow microneedle arrays entirely made of PMMA are obtained through multiple DXRL exposures, electroplating and a soft polymer casting [11]. The hollow microneedle pattern used for the first flow rate measurements has 121 needles and is shown in Fig. 1. The final structures have a tip angle of 40 degrees and are 400 µm high, the microneedle channel has an 84 µm internal diameter. The buffer reservoir and the bottom sealing lid were both made by mechanical milling of 500 µm PMMA sheets. The reservoir consists of a 1mm sheet with a circular hole of 1 cm diameter. The three parts were then sealed through solvent bonding. The microfluidic system was completed by connecting the pump and the needles with a Tygon™ tube 22 cm long and with an 800 µm internal diameter. A similar connection was used between the pump inlet and the external reservoir. To have a more compact system, a pump housing that allows bonding the microneedles to the micropump and a connection to an external reservoir has been built and is now being tested.

THEORETICAL CALCULATION

In order to predict the capabilities of such a microfluidic system, a numerical calculation of the volumetric flow rate was made considering the rate of delivery of the micropump and the flow through all the microfluidic elements.

According to previous calculations [12], the total flow rate determined by the micropump is given by:

$$Q_g = bn\pi \left(\frac{d_h^2}{2} - \frac{a_k^2}{2} - \frac{m^2\pi^2}{6}\cos^2\alpha_0 \right) - \frac{wh^3}{12\eta d_h}\Delta p \qquad (1)$$

Where h is the height difference between the case and the gears, n the rotational speed of the gears, η the fluid dynamic viscosity, w the assembly slit width, h the height mismatch between housing and gear and Δp the pressure difference created by the micropump. All gears have a thickness b of 500 μm. Other quantities used for calculations are listed in table I. Most drugs prior to administration are diluted in water-based fluid that was therefore chosen as working medium for the calculation.

The flow rate through the microchannels with a length l and radius R, is given by the Hagen-Poiseuille equation,

$$Q = \frac{\pi}{8}\frac{R^4}{\eta l}\Delta p \qquad (2)$$

Where Δp is the pressure difference between channel input and output, l the microchannel length and R its radius.

To be able to use Hagen-Poseuille equation, a regime of laminar flow has to be assumed. Even though the reservoir radius is quite large, considering a maximum flux of 10 ml/min, we obtain a Reynolds number of 477. Although this number is significant, it is still more than 4 times lower than the limit fixed for turbulent flow.

Measurements of the subcutaneous tissue pressure through a simple direct manometric method in normal patients give an average value of 3 hPa with a range extending between 0.5 and 8 hPa (pressure values are given relatively to the atmospheric pressure = 10^3 hPa [13]. In order to test the capabilities of the system, calculations and measurements of flow rate were performed at 5, 10 and 20 hPa head pressure.

TABLE 1. Quantities used in calculation. The height mismatch h was measured before sealing the micropump parts.

Module (m, μm)	136	Contact Angle (α_0, °)	20	Gear thicknes (b, μm)	500
Pitch diameter (a_k, μm)	2727	External diameter (d_h, μm)	3000	Height mismatch (h, μm)	20

EXPERIMENTAL RESULTS

Theoretical calculations and first experimental data are shown in Fig. 2. Experiments show values of flux that go from 1 to 6.5 ml/min for head pressures that vary between 5 and 20 hPa.

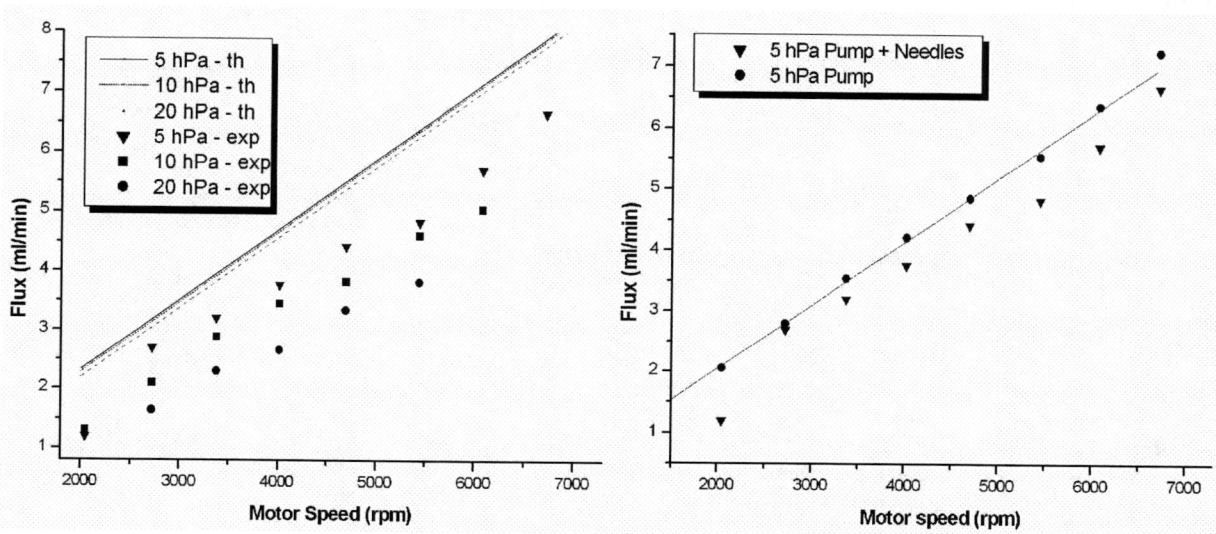

FIGURE 2. Comparison between theoretical and experimental data of flux versus rotational speed of the actuating magnet (left). The flux in the microfluidic system is driven by a micropump with an external diameter of 3 mm, 20 teeth and a height mismatch of 20 μm. Comparison between system flux and flux of the micropump versus speed (right) at 5 hPa head pressure.

The measurements at different head pressures were performed by changing the height difference between the output (microneedles) and the input (reservoir water level). This height difference was assumed to enhance the head pressure as the micopump input and output were aligned with reservoir water level. The difference between theory and experiment can be explained either with higher height mismatch caused by bottom lid bending as in previous measurements [8] or with significant localized pressure drops in connections between microfluidic elements.

More accurate measurements of the height mismatch could improve theoretical prediction of flux. Nevertheless, the values obtained are of great interest for drug delivery: even if the presence of skin can lower the flow rate through a single microneedle of a few orders of magnitude [14], flow rates of a few ml/min obtained without insertion into skin, can be enough to infuse fluid into the skin at rates up to 1 ml/h that is sufficient for many applications. Efficiency of the microfluidic system can depend on the difference in dimensions between the internal diameter of the microneedles and the internal dimensions of the other microfluidic components (greater by one order of magnitude). To check the effect of the microneedle internal diameter on the micropump flux, the data of output from the system were compared with the data of pump flux (Fig. 2) at 5 hPa head pressure. The similar trend of the two curves, allows us to say that the small microneedle internal diameter causes no significant decrease of overall flux.

CONCLUSIONS

First measurements of a miniaturized portable transdermal drug delivery system were performed. The final goal is to produce a versatile and compact system capable of delivering precise amount of drug solution on daily basis. The microgear pump parts geometry and actuation velocity can be adjusted in order to supply the required drug quantity over the desired period. Most elements of the system array can be fabricated with very high precision in one single DXRL exposure from which a mould can be manufactured for future mass-production. DXRL on a pre-form PMMA foil allows the fabrication of beveled tip microneedles that can be replicated through casting process. Further miniaturization of the system will be provided by sealing the microneedle array on top of the micropump output. The magnetic actuation of the gearpump still remains to be investigated in detail in order to predict how much the dimensions can be scaled down to achieve the flow rates required for drug delivery.

ACKOWLEDGEMENTS

This work is supported by MIUR (Ministero dell' Istruzione, dell' Università e della Ricerca) through the project FIRB RBNE01XPYH_003.

REFERENCES

1. S. L. Tao, T. A. Desai, *Advanced Drug Delivery Reviews* 55 (2003) 315-328.
2. S. Henry, D. V. McAllister, M. G. Allen, M. R. Prausnitz, *J. of Pharmaceutical Sciences*, vol.87, No.8, 922-925 (1998).
3. B. Stoeber. D. Liepmann, *Proc. of S-S Sens. and Act. Workshop* (2002) Hilton Head.
4. Jung Hwan Park, M. J. Allen and M. R. Prausnitz *J of Controlled Release* 104 (2005) 51-66
5. K. Kim, D.S. Park, H. M. Lu, W. Che, K. Kim, J. B. Lee, H. Ahn , *J. Micromech. Microeng.*, 14, p597-603, (2004).
6. S.J. Moon, S.S. Lee, *J. Micromech. Microeng* 15 (2005) 903-911.
7. P.M. Wang, M.G. Cornwell, and M.R. Prausnitz, Proc. of the Second Joint EMBS/BMES Conf., Houston, TX, USA, Oct.2002, 506-507.
8. D. J. Laser, D. J. Santiago, *J. Micromech. Microeng.* 14 (2004) R35-R64.
9. M. Matteucci et al, *Microelectronic Engineering*, Volume 83, Issues 4-9, Pages *1288-1290* (April-September 2006)
10. K. Deng, A.S. Dewa, D.C. Ritter, C. Bonham, H. Guckel, *Microsystem Technologies* 4, 163-167, (1998).
11. F. Pérennès et al, *J. Micromech. Microeng.* 16 473-479 (2006)
12. J. Döpper, M. Clemens, W. Ehrfeld, S. Jung, K-P Kämper and H. Lehr, *J. Micromech. Microeng.*, 7, p230-232, (1997).
13. H. S. Wells, J. B. Youmans and D. J. Miller JR, *J Clin Invest*. 1938 July; 17 (4): 489–499.
14. W. Martanto et al. *Pharmaceutical research* Vol 23, No 1, January 2006 p104-112

Replication of Metal-Based Microscale Structures

J. Jiang[1], W.J. Meng1, G.B. Sinclair[1,] C. O. Stevens[2], E. Lara-Curzio[2]

[1]Mechanical Engineering Department, Louisiana State University, Baton Rouge, Louisiana 70803, USA
[2]High Temperature Materials Laboratory, Oak Ridge National Laboratory, Oak Ridge, Tennessee 37831, USA

Abstract. Technologies capable of economical mass production of metal-based high-aspect-ratio microscale structures (HARMS), with structural heights of several hundred μm offer the potential to realize many metal-based microdevices. In this paper, our recent research on high-temperature microscale compression molding of metal-based HARMS from surface-engineered, microscale, refractory metal inserts is summarized.

Keywords: LiGA technology, microfabrication, replication, metal-based HARMS
PACS: 85.40.Hp, 87.80.Mj, 83.50.Uv, 85.85.+j, 87.15.La

INTRODUCTION

Economical mass production of metal-based high-aspect-ratio microscale structures (HARMS), with a typical height of several hundred microns, may lead to the realization of many metal-based micromechanical and microchemical devices. These devices can be passive, such as micro heat-exchangers [1, 2], or active, such as electromagnetic micro-relays [3], and have functionalities not achievable using Si-based micro-electro-mechanical systems (MEMS). One important fabrication strategy for metal-based HARMS is LiGA (Lithographie, Galvanoformung, Abformung) and LiGA-related approaches [4], combining deep X-ray-/UV- lithography, fabrication of primary high aspect ratio mold inserts by electrodeposition or other means, and replication by compression molding. Our recent research has concentrated on advancing the technology of replicating metal-based HARMS. Replication has been successfully demonstrated in Pb and Zn [5], Al [6], and Cu [7], using surface- and bulk- engineered microscale mold inserts. The mechanics of microscale compression molding has been studied through modeling based on contact mechanics principles [8] and preliminary finite element analysis (FEA) [9].

In what follows, we outline a hybrid protocol for fabricating high aspect ratio mold inserts out of elemental refractory metals as well as alloys with high strengths at elevated temperatures by combining UV-LiGA with micro-electrical-discharge-machining (μEDM), and summarize results from instrumented Al molding experiments.

EXPERIMENTAL PROCEDURES AND RESULTS I: INSERT FABRICATION

Fabrication of Ta Inserts with a Meandering Microprotrusion Pattern

Initial definition of Ni micropatterns was accomplished with UV-LiGA with SU-8/100 resist [10]. Ni HARMS were fabricated via electrodeposition into lithographically defined recesses in SU-8 after development. The remaining SU-8 resist after Ni electrodeposition was removed by heating the specimen in air at ~750°C for 1h, followed by ultrasonic cleaning in acetone and methanol. The Ni HARMS serve as the electrode for subsequent μEDM process. Further details on UV-lithography and Ni electrodeposition were given elsewhere [11]. An electrodeposited Ni HARMS with one single meandering microchannel is shown in Fig.1(a). The width and depth of the microchannel in Ni are about 200μm and 1100μm, respectively.

Parallel μEDM was carried out using a SARIX High Precision Micro-Erosion Machine (SR-HPM-B). Electrodeposited Ni HARMS served as electrodes for machining arc-melted Ta (99.9at.%) blanks, with a square active area of 9500μm×9500μm and a height of 2000μm. The top surface of the Ta active area was mechanically

CP879, *Synchrotron Radiation Instrumentation: Ninth International Conference*,
edited by Jae-Young Choi and Seungyu Rah
© 2007 American Institute of Physics 978-0-7354-0373-4/07/$23.00

polished with SiC abrasive papers down to 1200 grit size prior to μEDM. During μEDM, the discharge parameters of frequency, on-time width, and maximum voltage were set respectively at 120kHz, 5μsec, and 90V for pattern transfer from Ni HARMS to Ta. A hydrocarbon-based dielectric fluid, IonoPlus/3000®, was used as the dielectric medium for μEDM. As shown in Fig.1(b), the μEDM process transforms the meandering microchannel on the Ni HARMS into a meandering rectangular microprotrusion on the Ta blank. Figure 1(c) shows that surfaces of as-machined Ta blanks contain numerous microscale nodules and cracks. This altered surface layer (ASL) forms during μEDM, as a result of the reaction between Ta and the dielectric fluid. As shown in Fig.1(d), this ASL was removed by electrochemical polishing (ECP) in a mixed acid solution of $H_2SO_4(98\%)/HF(49\%)$ [7]. Further details on the ECP process were described elsewhere [7, 11]. The final width and height of the rectangular Ta microprotrusion is about 105μm and 450μm, respectively. Figure 1 shows the feasibility for transferring complex micropatterns onto high-temperature compatible materials with the hybrid LiGA/parallel μEDM fabrication strategy.

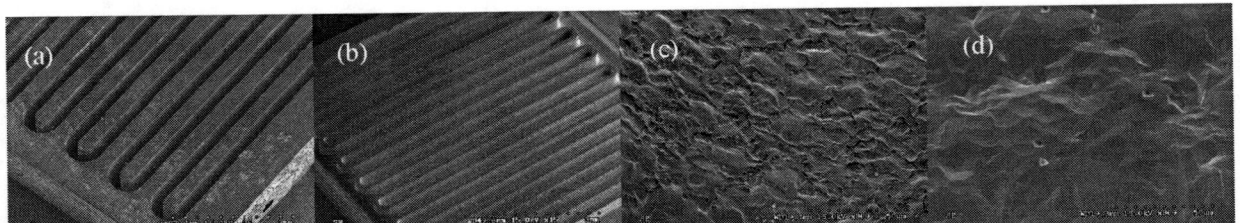

FIGURE 1. Hybrid UV-LiGA/parallel-μEDM fabrication of a high aspect ratio mold insert with a meandering microchannel pattern in elemental Ta: (a) primary micropattern defined by UV-LiGA and Ni electrodeposition; (b) micropattern transferred to Ta by parallel μEDM; (c) typical surface morphology of as-machined Ta surface; (d) typical morphology of Ta insert surfaces after ECP.

Fabrication of Ta and Inconel Mold Inserts with Parallel Microprotrusions

Using Mo blade electrodes 170μm in width, a series of parallel microchannels were machined into elemental Ta (99.9at.%) and Inconel X-750® (Ni 70wt.% min., Cr 14.0-17.0wt.%, Fe 5.0-9.0wt.%, Ti 2.25-2.75wt.%, with C, S, Si, Cu, Co, and Al impurities) blanks by μEDM. The Inconel X-750® alloy is engineered for high strengths at high temperatures [12]. The active area dimensions for Ta and Inconel X-750® blanks are 9500μm×9500μm and 15000μm×15000μm, respectively. To remove the ASL on Ta and Inconel X-750® resulting from the μEDM process, ECP of Ta and Inconel X-750® was carried out in mixed acid solutions of $H_2SO_4(98\%)/HF(49\%)$ and 1:1 $HClO_4(70\%)/CH_3COOH(100\%)$, respectively [7, 11]. Figure 2(a) shows a typical Inconel X-750® insert after ECP, which successfully removed the microscale nodules and cracks due to μEDM. The insert contains a series of parallel, rectangular, microprotrusions on the active surface. X-ray energy dispersive spectroscopy (EDS) data were obtained from the insert surface after ECP. A typical EDS spectrum is shown in Fig.2(b), and is substantially similar to that obtained from as-received Inconel X-750® material. The width and height of the microprotrusions are about 180μm and 500μm, respectively. The center-to-center spacing between microprotrusions is about 750μm. Figure 2(c) shows a typical Ta insert with a series of parallel, rectangular, microprotrusions on the active surface. Surface engineering of this insert included ECP followed by conformal deposition of a Ti-containing hydrogenated carbon (Ti-C:H) coating. Figure 2(d) shows a magnified view of one typical rectangular microprotrusion on the Ta insert after Ti-C:H deposition. The microscale roughness on the protrusion sidewall resulted from grain orientation dependent etch rate, and is typical of ECP processed materials as evidenced in Fig.2(a) and 2(d). The width and height of the microprotrusions are about 150μm and 500μm, respectively. The center-to-center spacing between microprotrusions is about 750μm. Figure 2 shows that, by combining μEDM with ECP and conformal coating deposition, microscale mold inserts can be fabricated out of high-temperature compatible elemental metals and alloys not achievable with electrodeposition.

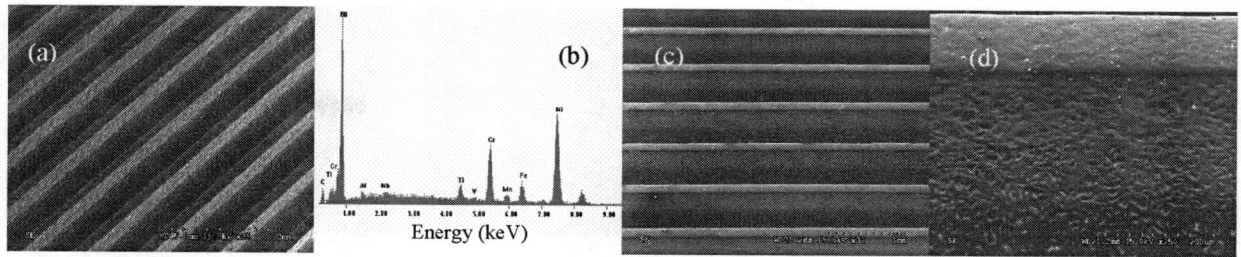

FIGURE 2. μEDM fabrication of high aspect ratio mold inserts made of Inconel X-750® and Ta: (a) an Inconel X-750® insert with a series of parallel, rectangular, microprotrusions after ECP; (b) an EDS spectrum obtained from the surface of the Inconel X-750® insert after ECP; (c) a Ta insert with a series of parallel, rectangular, microprotrusions after ECP and conformal deposition of a Ti-C:H coating; (d) a magnified view of a typical rectangular microprotrusion on the Ta insert. The presence of a thin Ti-C:H coating does not alter the surface roughness

EXPERIMENTAL PROCEDURES AND RESULTS II: MICROSCALE COMPRESSION MOLDING OF AL

Procedures for Al Molding

Circular Al (1100H14, 99at.% Al) disks, 35mm in diameter and 6.4mm in thickness, were molded at different temperatures using the Ti-C:H coated Ta insert shown in Fig.2(c). The top surface of the Al disks were mechanically polished with SiC abrasive papers down to 1200 grit size prior to being molded. Microscale compression molding was carried out in a high-vacuum, high-temperature, instrumented, compression molding apparatus, shown in Fig.3. Molding and demolding occurred isothermally, with insert and Al temperatures controlled to be within 10°C of each other. The total axial force on the insert and the total axial displacement of the actuator, onto which the mold insert was mounted, were continuously measured with a load cell and a linear variable displacement transducer (LVDT), both located outside the high-vacuum molding chamber [13].

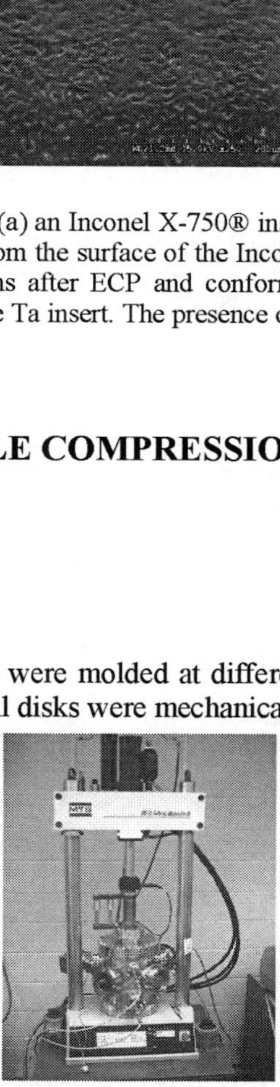

FIGURE 3. A custom built, high-vacuum, high-temperature, instrumented molding apparatus.

Al Molding Response and Molded Al Structures

Response for molding Al with the Ti-C:H coated Ta insert shown in Fig.2(c) was measured at 360°C, 400°C, 450°C, and 500°C. Multiple molding runs were conducted at the same temperature. To obtain the true molding response in terms of the relationship between the axial force on the insert and the actual depth of penetration of the rectangular microprotrusions into the molded metal, the stiffness of the molding apparatus was obtained by measuring force-displacement curves of a Ta blank with the same active area as the actual insert compressing on Al disks. The stiffness calibration runs occurred over a similar range of force as that used in actual molding experiments. The true molding response was obtained by subtracting measured system stiffness contribution from measured axial-force versus total-displacement curves.

Figure 4(a) shows measured true molding response as a function of the molding temperature. The reproducibility of measured molding response at one temperature appears to be about 10% in terms of the molding force at the same depth. The initial molding response appears to be stiff, with molding force rising rapidly with increasing indentation depth. The maximum molding force decreases by more than a factor of 2.5 as the molding temperature increases from 360°C to 500°C. The top view of a typical molded Al structure is shown in Fig.4(b), in which a series of parallel microchannels result from indentation by the series of parallel rectangular microprotrusions on the Ta insert. The cross-sectional view of the molded Al microchannels, shown in Fig.4(c), shows that the sidewalls of the microchannels are close to vertical. The top surface of the molded Al appears rounded, due to indentation-induced pileup.

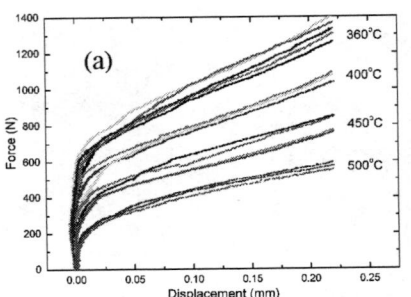 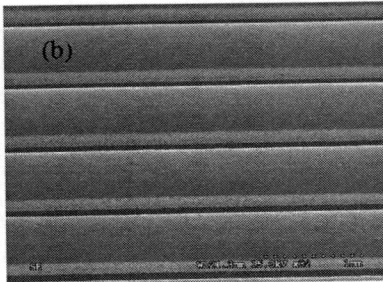 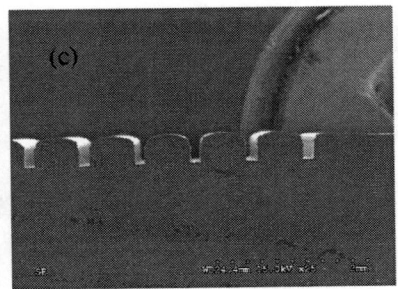

FIGURE 4. Al molding response and molded Al structures: (a) stiffness-corrected molding force – indentation depth curves for Al micromolding at different temperatures; (b) top view of a typical molded structure in Al; (c) cross-sectional view of a typical molded structure in Al.

High Temperature Tensile Testing of Al and Data Interpretation

To understand measured Al molding response, the mechanical response of Al at the molding temperatures needs to be known. Tensile testing of Al was therefore conducted at each of the four molding temperatures. Because the actual strain rate during Al micromolding is unknown, Al tensile testing was conducted at several strain rates.

Al rod (1100O, 99at.% Al) specimens with a gauge section of 6.35mm in diameter and 31.75mm in length were custom made and tested with an Instron 4507 single-axis system in a 45cm long, three-zone furnace with three independent temperature controls. Specimen temperatures at the two ends and the mid-point of the gauge section were controlled to be within 3°C of each other. Specimen extension at high temperatures was determined using a custom-built extension stage. The two ends of the extension stage were rigidly fastened to the two ends of the Al rod specimen. The relative displacement between the two ends was measured by two separate LVDTs whose outputs were averaged to calculate the mean specimen extension and axial strain. At each temperature, tensile tests were conducted at three different strain rates. At all temperatures and strain rates tested, measured stress-strain curves approximate that of an elastic-ideally plastic solid at strains < 10%. The measured yield stresses as a function of strain rate and temperature are shown in Table 1. The tensile testing results are consistent with power-law creep.

TABLE 1. Yield stress of Al at different temperatures and strain rates.

Strain Rate (1/sec)	Flow Stress (MPa)			
	360°C	400°C	450°C	500°C
3×10^{-4}	15.01	11.83	9.05	7.3
1×10^{-3}	18.16	14.06	10.79	7.36
5×10^{-3}	22.18	15.95	11.96	9.1
3×10^{-4}	15.24	-	-	-
1×10^{-3}	-	-	10.05	
5×10^{-3}	-	16.04	-	8.8

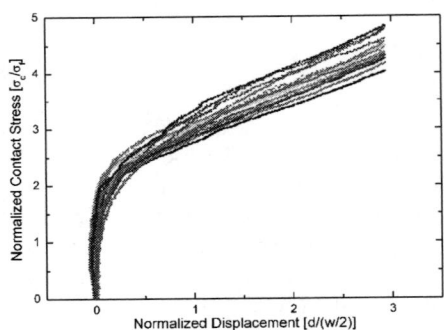

FIGURE 5. Normalized molding response.

The contact stress σ_c, obtained by dividing the measured axial force by the nominal contact area, was normalized by the measured yield stress at the corresponding temperature. The indentation depth is likewise normalized by a characteristic width, i.e., the half-width of the rectangular microprotrusion on the Ta insert. As shown in Fig.5, this normalization combines the measured molding response at different temperatures into what might be viewed as a universal molding response curve. In Fig.5, the variation in the normalized contact stress at the maximum depth is within a factor of 1.2, in marked contrast to Fig.4(a) where a force variation of more than a factor of 2.5 is observed. Furthermore, this combining of responses for different temperatures into a single response curve appears to be not extremely sensitive to the particular strain rate selected (from Table 1) for the yield stress. The present normalization is consistent with our previous micromolding results on Pb, and lends further support to our mechanics-based indentation model [8].

ACKNOWLEDGMENTS

Partial project supports from NSF through grants DMI-0400061/DMI-0556100 and Louisiana Board of Regents through contract LEQSF(2004-07)-RD-B-06 are gratefully acknowledged.

REFERENCES

1. F. Arias, S. R. J. Oliver, B. Xu, R. E. Homlin, G. M. Whitesides, Fabrication of metallic heat exchangers using sacrificial polymer mandrils, J. MEMS 10, 107 (2001).
2. C. Harris, K. Kelly, T. Wang, A. McCandless, S. Motakef, Fabrication, modeling, and testing of micro-cross-flow heat exchangers, JMEMS 11, 726 (2002).
3. J. D. Williams, W. Wang, Microfabrication of an electromagnetic power relay using SU-8 based UV-LIGA technology, Microsystem Technologies 10, 699 (2004).
4. M. Madou, Fundamentals of Microfabrication (CRC Press, Boca Raton, Florida, 2000).
5. D. M. Cao, D. Guidry, W. J. Meng, K. W. Kelly, Molding of Pb and Zn with microscale mold inserts, Microsystem Technologies 9, 559 (2003).
6. D. M. Cao, W. J. Meng, Microscale compression molding of Al with surface engineered LIGA inserts, Microsystem Technologies 10, 662 (2004).
7. D. M. Cao, J. Jiang, W. J. Meng, J. C. Jiang, W. Wang, Fabrication of high-aspect-ratio microscale Ta mold inserts with micro-electrical-discharge-machining, Microsystem Technologies, s00542-006-0198-8 (2006).
8. W. J. Meng, D. M. Cao, G. B. Sinclair, Stresses during micromolding of metals at elevated temperatures: pilot experiments and a simple model, J. Mater. Res. 20, 161 (2005).
9. D. M. Cao, J. Jiang, W. J. Meng, G. B. Sinclair, D. M. Cao, J. Jiang, W. J. Meng, G. B. Sinclair, Metal micromolding: further experiments and preliminary finite element analysis, Microsystem Technologies, s00542-006-0200-5 (2006).
10. J. D. Williams, W. Wang, Using magasonic development of SU-8 to yield ultra-high aspect ratio microstructures with UV lithography, Microsystem Technologies, 10, 694-698 (2004).
11. D. M. Cao, J. Jiang, R. Yang, W. J. Meng, Fabrication of high-aspect-ratio microscale mold inserts by parallel μEDM, Microsystem Technologies, s00542-006-0131-1 (2006).
12. http://www.hightempmetals.com/techdata/hitempInconelX750data.php (accessed May 18, 2006).
13. D. M. Cao, W. J. Meng, K. W. Kelly, High-temperature instrumented microscale compression molding of Pb, Microsystem Technologies 10, 323 (2004).

Instrumentation for Microfabrication with Deep X-ray Lithography

F.J. Pantenburg

Forschungszentrum Karlsruhe GmbH, Institut für Mikrostrukturtechnik, Postfach 3640, D- 76021 Karlsruhe, Germany, E-Mail: Pantenburg@ fzk.imt.de

Abstract. Deep X-ray lithography for microfabrication is performed at least at ten synchrotron radiation centers worldwide. The characteristic energies of these sources range from 1.4 keV up to 8 keV, covering mask making capabilities, deep X-ray lithography up to ultra deep x-ray lithography of several millimeters resist thickness. Limitations in deep X-ray lithography arise from hard X-rays in the SR-spectrum leading to adhesion losses of resist lines after the developing process, as well as heat load due to very high fluxes leading to thermal expansion of mask and resist during exposure and therefore to microstructure distortion. Considering the installations at ANKA as an example, the advantages of mirrors and central beam stops for DXRL are presented. Future research work will concentrate on feature sizes much below 1 μm, while the commercialization of DXRL goes in the direction of massive automation, including parallel exposures of several samples in a very wide SR-fan, developing and inspection.

Keywords: LIGA, Deep X-ray lithography, heat load, adhesion, Mirror, commercialization
PACS: 85.40.HP; 85.85.+j;89.20.BP

INTRODUCTION

LIGA is one of several microfabrication technologies used for the realization of plastic and metallic microparts and microsystems. It uses deep X-ray lithography with synchrotron radiation to produce plastic molds with freestanding features in the micrometer range, with high aspect ratio (up to 100), smooth sidewalls (R_a<30 nm) and sub-micron details in sidewalls [1]. These outstanding values result from the homogeneity and parallelism of synchrotron radiation together with the high selectivity of the resist/developer system and an highly refined X-ray mask technology.

The basic process is the transfer of an X-ray mask by shadow printing into an X-ray sensitive polymer using synchrotron radiation. Polymethylmethacrylate (PMMA) with a molecular weight of more than $2*10^6$ g/mol is a typical resist material. This material requires a minimum dose of 3 kJ/cm³ to become completely soluble in GG-developer, while the surface dose should not exceed 12-18 kJ/cm³ to avoid foaming of the resist during exposure. To avoid attacking of unexposed polymer areas, the dose deposition underneath the Au-absorbers structures of the X-ray mask should be less than 0.1 kJ/cm³.

SU8 is a negative resist used very often in DXRL. Nazmov et al. [2] found optimized exposure values for minimal size deviations and minimal sidewall roughness. In this regime, SU-8 requires only 50-100 J/cm³, but the ratio of top to bottom dose should be in the range of 1÷1.7. Those exposure strategies are strongly dependent on the preparation of the SU-8 resist. Therefore, Becnel [3] found slightly different exposure conditions. What remains is that the resist material is hardly to remove after the post exposure back, making PMMA still the standard resist material.

Structure Limitations due to Adhesion Problems

During DXRL, the resist is mounted typically on a Ti/Cu/Ti or TiOx-layer on top of silicon, ceramic or metallic substrate to allow electroforming after development. The adhesion of positive resist lines to the metallic substrate is influenced by secondary electrons produced inside the plating base during the irradiation process [4]. The high

CP879, *Synchrotron Radiation Instrumentation: Ninth International Conference,*
edited by Jae-Young Choi and Seungyu Rah

energy part of the synchrotron radiation spectrum penetrates the Au-absorber on the X-ray mask and generates secondary radiation. This radiation degrades the resist at the interface to the substrate and causes adhesion failure after the development process: For this reason, the reemitted dose from the plating base should be kept below 0.1 kJ/cm³.

Different techniques have been used to decrease this secondary radiation dose: increasing the gold absorber thickness on the X-ray mask, using low-Z substrates [5], introducing a buffer layer [6] or a soft X-ray spectrum [7]. These will be discussed in more detail below:

UV-lithography is used to structure 50 µm thick resist layers and to fabricate X-ray masks with sufficient thick Au-absorbers in one process step for the LIGA process. The diffraction of light (λ=248 nm) at the chromium mask limits the resolution to 4 µm for 50 µm thick resist layers. The post exposure bake may introduce striations of several 100 nm in the sidewalls, due to the shrinkage of negative resist during crosslinking. Those striations are transferred to the Au-absorber structure on the X-ray mask during electroforming and into the PMMA-mold by DXRL. Decreased sidewall roughness of the PMMA molds and increased demolding forces of electroplated molds are the consequence.

An optimized mask fabrication process uses e-beam lithography into a 3 µm thick resist layer. After the developing and Au-electroforming process these intermediate masks are copied by X-ray lithography with synchrotron radiation into a thicker PMMA-resist layer. Minimum features sizes of 0.2 µm in the sidewalls and freestanding sub-µm lines have been fabricated. Striations in the sidewall are much below the 20 nm range and arise from e-beam patterning.

FIGURE 1. An optimized mask making technology combines electron beam and soft X-ray lithography. Primary X-ray mask fabricated by e-beam lithography with gold absorber thickness of 2.2 µm (left) and working mask with 22 µm thick absorbers produced by shadow printing of the primary mask with DXRL (right).

Alternative low-Z substrates (like C and Al) have been used for processing in DXRL. However, better quality is obtained by the above mentioned plating bases, due to the higher starting point density for the electroforming process and better adhesion of the metallic micro parts to the substrate. Anodized aluminum substrates [5] developed at Sandia Nat. Lab. seems to be the substrates for DXRL in the future.

A polyamide buffer layer was introduced by De Carlo [6] to absorb the electrons form the substrate. This layer is not attacked by the developer and is removed by an O_2-plasma step after the development process, but small and narrow trenches are difficult to clean out while preserving the sidewall roughness.

Megtert et al. [7] used an X-ray mirror to reduce the amount of high energy photons in the SR-spectrum. They showed significantly increased adhesion of small resist lines to a TiOx plating base in the mask making regime.

Structure Limitations due to Heat Load

During the exposure of mask and resist, X-rays are absorbed not only in the resist itself, but also in the mask membrane, the gold absorber structures and the substrate [8]. Especially at high flux sources, this results in thermal expansion of mask and substrate during exposure and leads to a structure smearing of several micrometers. Additionally, the dose smearing in the sidewalls lead to micro cracks in the resist sidewalls. Even foaming of the resist was observed at beamline X-14B at NSLS/BNL. Gas molecules are produced during the exposure in the resist. If the resist warms up, the mobility of these molecules is significantly increased leading to a foaming of the resist. Mrowka [9] used a rotating cylinder chopper to reduce the flux by a factor of 14, to reduce the flux and to produce microstructures with sufficient accuracy and quality at NSLS/BNL.

Considering the installations at ANKA as an example, the instrumentation for microfabrication with deep X-ray lithography is presented:

INSTRUMENTATION AT THE MICROFABRICATON LABORATORY AT ANKA

High resolution X-ray masks are fabricated by a combination of e-beam lithography and DXRL. At Forschungszentrum Karlsruhe, a high resolution 100kV electron beam writer (EBPG 5-HR) is used to pattern a resist layer of 3 μm on a Ti membrane, to achieve an Au-absorber thickness of 2.2 μm on the primary mask (Fig. 1). This mask is transferred by DXRL into PMMA resist layers of 50 to 70 μm using soft X-rays to achieve Au-absorber thickness up to 40 μm. This working X-ray mask is used to structure resist layers in the range of 100 to several 1000 μm thickness. This process flow allows the fabrication of very smooth sidewalls (R_a = 20nm) and smallest freestanding resist lines below 1 μm in PMMA resist.

Mirror Installation for Adhesion Improvements

In the past, we used the synchrotron radiation source ELSA of Bonn university at different electron energies. X-ray masks were structured at a characteristic energy of 0.835 keV (electron energy: E_e=1.6 GeV), structures up to 800 μm were realized at 2.5 keV (E_e=2.3 GeV) and ultra deep structures up to 3 mm thickness at 4 keV (E_e=2.7 GeV) [10].

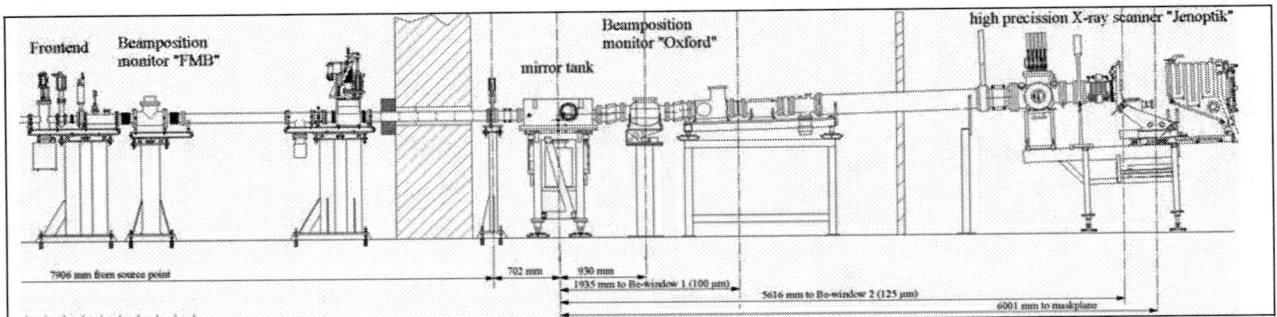

FIGURE 2. DXRL-beamline layout from source point to experimental station at ANKA. Beamposition monitors before and behind the x-ray mirror control the position and constant reflection angle of the SR-beam.

Since July 2001, we started to establish these production capabilities at the new synchrotron radiation source ANKA. In a transitional phase, one beamline at ANKA was used at different electron energies (1.3 GeV, 1.8 GeV and 2.5 GeV) to cover the whole production sequence for research and industrial projects. Today three beamlines are up and running; a high resolution X-ray working mask production beamline (up to 100 μm resist), one beamline for resist structures up to 800 μm and an ultra deep X-ray lithography beamline for structures up to several millimeters. They are operated now at the nominal electron energy 2.5 GeV of ANKA and two of them are equipped with a cooled, flat mirror to provide a high energy cut-off. The layouts of these beamlines are very similar (Fig. 2). Beam position monitors are installed before and after the mirror chamber, and high precision scanners are used as exposure end stations. The installations differ mainly in the mirror coating and reflection angle (Table 1).

The mirror parameters are optimized for the specific tasks of each beamline. Depending on the resist thickness, the mask membrane (2.5 μm Ti) and vacuum windows (225 μm Be), the spectral distributions have been chosen to achieve a typical top to bottom dose ratio inside the resist of 4. This requires a lower energy cut-off for the mask fabrication beamline than for the standard deep X-ray lithography station. The spectral distributions after the Be-windows are shown in Fig. 3 for the different exposure conditions with and without mirror and at reduced electron energy. The higher energetic part of each distribution may pass the gold absorber and produce secondary radiation in the metallic substrate. If the reemitted radiation produces more than 0.1 kJ/cm³ in the resist interface, adhesion failures of the resist lines might be observed. The spectral distributions with the lowest component of high energy photons should lead to the smallest gold absorber thicknesses. Corresponding calculations with the computer code LEX-D show that the reduction in gold absorber thickness is approximately a factor of 2 for the installed beamlines and the exposure time decreased by a factor of 3 to 8, with mirror, at 2.5 GeV compared to the corresponding spectrum at different electron energies (Table 1). This is a result of the well defined cut-off behavior of the installed mirrors and the higher amount of low energetic photons at 2.5 GeV in general.

TABLE 1. Mirror characteristics of DXRL beamlines at ANKA

Beamline	optical componet	Cut-off energy	Minimum absorber thickness		Exposure time [min/cm]	
			mirror @ 2.5 GeV	White beam	mirror @ 2.5 GeV	White beam
X-ray mask fabrication (f.e. 50 µm resist)	flat Si-mirror /120 nm Cr	2.8 keV @ 15.4 mrad	1.7 µm	3 µm @ 1.3 GeV	8.4	54.6 @ 1.3 GeV
Deep X-ray lithography (f.e. 300 µm resist	flat Si-mirror /120 nm Ni	6.9 keV @ 8.65 mrad	7.5 µm	17.5 µm @ 1.8 GeV	9	24 @ 1.8 GeV
ultra deep X-ray lithography	Central beam stop	none	dependant on structure and beam stop height		dependant on structure and bam stop height	

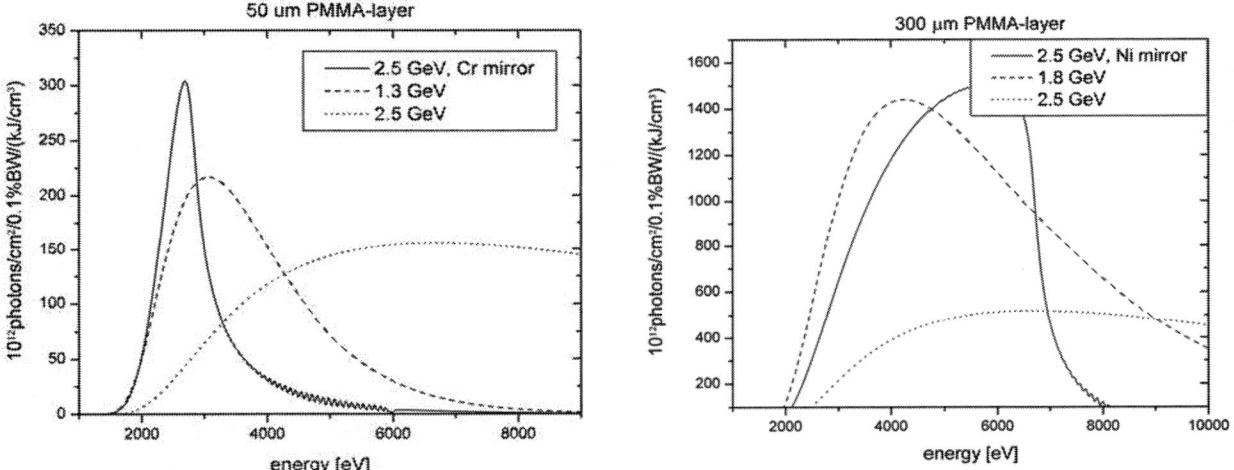

FIGURE 3. Synchrotron radiation distribution behind the vacuum windows to deposit a dose of 1 kJ/cm^3 in the resist bottom layer as a function of electron energy and mirror parameters at ANKA for different resist thicknesses behind the last Be-window.

The reduced gold absorber thickness for the primary mask at the mask fabrication beamline will push the limits of DXRL to freestanding submicron structures with increased linewidth accuracy. As the necessary gold absorber is reduced by a factor of 2, thinner resist layers might be used for e-beam writing. Electron beam broadening and dose smearing in the resist layer will be decreased significantly. Additional advantages of this optimized installation will be increased process stability with the existing process flow at IMT/FZK.

Structure Limitations due to Heat Load

Litho 3 beam line at ANKA is used for Ultra Deep X-ray lithography to structure resist layers above 1 mm and needs high energy photons. Consequently no mirror is installed inside the beam line, instead the white beam of the ANKA bending magnet is used for exposure. The spectrum is adapted to the resist layer only by preabsorbers which are mounted in the scanner filter chamber. Using a 1500 µm resist layer, a preabsorber of 450 µm thick carbon foil is used, besides the 225 µm thick Be-vacuum window. In this regime, an X-ray power of 1.3 W/cm² is still hitting the mask and most of the radiation is absorbed inside the Au-absorbers on the 650 µm thick Be-membrane. Part of the energy is transferred to the cooled mask frame and a much smaller part is transferred over the resist layer to the cooled substrat holder. Consequently, mask and substrate warms up and expand during exposure. Concentric circles with a diameter of 40 μm in the outermost zone of the design field on the Be-mask appear as eggs in the resist with a deviation from a circle of at least 3 μm.

Additional preabsorbers would lead to an unwanted hardening of the SR-spectrum, resulting in adhesion problems of the resist due to additional photoelectrons from the substrate. Therefore, we have installed a central beam stop to block the central part of the Gaussian synchrotron radiation fan. A changeable densimet rod of 14 cm length and rectangular cross-section is mounted directly on the Be-window inside the exposure chamber. The rod thickness is 5 mm while the blocking height might be chosen from 1 to 5 mm. The rod is fixed at one end in the holder, while the other end is able to move horizontally and pulled in line by a spring. This will prevent the rod from being bend if thermally loaded by the X-ray beam. The influence of the beamstop on the exposure conditions is presented in Fig. 4. A power reduction up to factor of 22 is possible with a beam stop height of 1 to 5 mm. For example a beamstop height of 4 mm will result in a power reduction of a factor of approx. 10, additionally it will soften the X-ray spectrum. The amount of photons above 12.5 keV is significantly decreased using a 4 mm beamstop. This is a result of the naturally smaller divergence of the high energetic photons of synchrotron radiation spectra. Therefore the minimum Au-absorber thickness is decreased from 36 μm to 29 μm using a 4 mm beam stop.

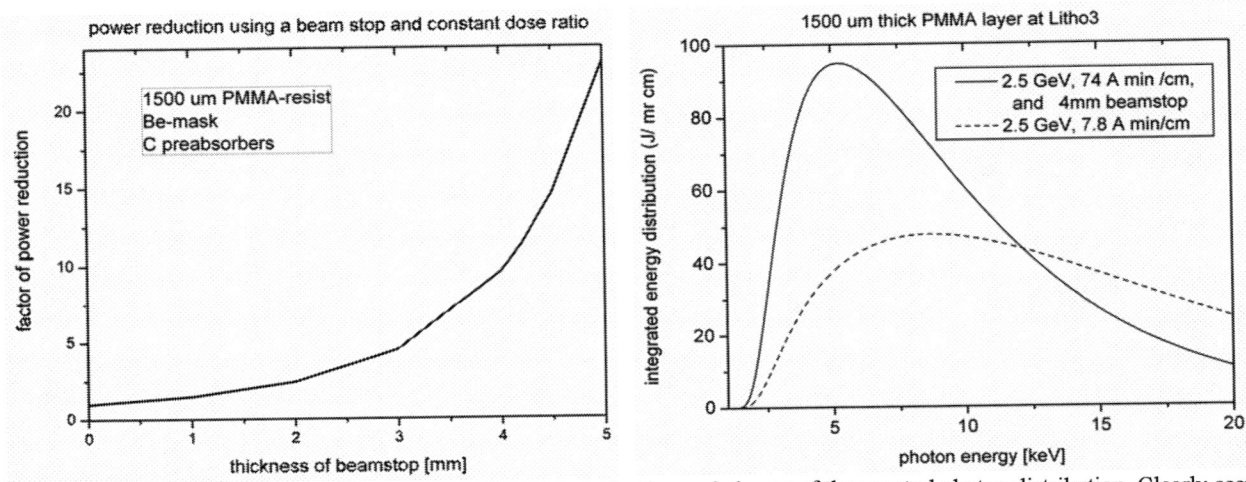

FIGURE 4. Influence of a central beam stop on the power reduction and change of the spectral photon distribution. Clearly seen is the reduction of high energy photons above 12.5 keV for a 4 mm wide beam stop..

The deviation from a circle inside the developed resist for the above mentioned resist-mask system is less then 0.2 μm using a 4 mm beamstop, which represents the measurement accuracy of our optical microscope setup. It should be mentioned that a temperature rise of 50 °C have been measured at the frame of the beamstop. The temperature rise in the cooling circulation of mask and substrate were decreased by 1.5°C to 0.2°C during exposure.

The Exposure Stations

All installed exposure stations [10] are equipped with a vacuum chamber to allow exposure under a controlled 100 mbar He-atmosphere. Special filter foils are mounted in a filter chamber for further adaptation of the exposure spectrum. The scanning stages are equipped with a precision tilting device to allow for tilted exposures. Aligned exposures are done via a special alignment stage in combination with an optical microscope. An overlay accuracy of 0.2 μm is possible. The scanners have been fabricated by the company Jenoptik/Mikrotechnik in close cooperation with IMT/FZK.

Outlook on Commercialization

In the past, researcher in LIGA thought that it would be sufficient to increase the X-ray flux on mask and sample or to expose samples stack wise [12] to increase the throughput of samples. The increased flux on mask and resist lead to heat load, thermal expansion and microstructure smearing. Besides structure inaccuracies of more than several micrometer, micro cracks in the smooth sidewalls are the result.

A consortium of manufacturers for process equipment and industrial microstructure users are building under the leadership of research center Karlsruhe a fabrication line for massive parallel manufacturing of LIGA microparts [13]. This line is planned as a fully automated process flow for exposure, development, electroforming and inspection. The exposures will be done in parallel on a 60 cm radiation fan from a tunable bending magnet at ANKA. The line will start commissioning in 2008.

CONCLUSION

The new installation at ANKA for deep x-ray lithography have been optimized with respect to resist adhesion and structure accuracy. The three beamlines allow the parallel manufacturing of x-ray working masks, deep and ultra deep microstructures with optimized exposure conditions at standard operation conditions of ANKA. No special special shifts at ANKA are required throughout the manufacturing process. This will shorten the process cycle time to less than 2 weeks starting from the design to the final DXRL-product, which is very essential for developing prototype production capabilities for industrial customers and their products. Besides the existing standard exposures for resist layers between 50 μm and 3 mm it allows the fabrication of even higher microstructures and smaller line widths with higher structure accuracies. Freestanding sub-micron microstructures will be part of the future standard process at Forschungszentrum Karlsruhe.

New steps for commercialization of LIGA are on their way. A fully automated fabrication line for exposure, developing, electroforming and inspection will decrease the cost production costs for LIGA-microstructures.

ACKNOWLEDGMENTS

The reemitted dose values into the resist by secondary electrons in Table 1 were calculated with the computer code LEX-D developed at SANDIA National Laboratories, California. Special thanks to Stewart Griffiths who implemented X-ray mirrors to the program code.

REFERENCES

1. E.W. Becker, W. Ehrfeld, P. Hagemann, A. Maner, and D. Münchmeyer, "Fabrication of microstructures with high aspect ratios and great structural heights by synchrotron radiation lithography, galvanoforming and molding", Microelectronic Eng. **4**, pp. 35-56, 1986
2. V. Nazmov, E. Reznikova, J. Mohr, A. Snigirev, I. Snigireva, S. Achenbach, V. Saile, "Fabrication and preliminary testing of X-ray lenses in thick SU-8 resist layers", Microsystem Technologies **10** (2004) 716–721
3. Ch. Becnel, Y. Desta, K. Kelly, "Ultra-deep x-ray lithography of densely packed SU-8 features: II. Process performance as a function of dose, feature height and post exposure bake temperature", J. Micromech. Microeng. **15** (2005) 1249–1259
4. F.J. Pantenburg, J. Chlebek, A. El-Kholi, H. L. Huber, J. Mohr, H. K. Oertel, J. Schulz: "Adhesion problems in deep-etch lithography caused by fluorescence radiation from the plating base", Microsystem engineering **23**, (1994) 226-226
5. S.K. Griffiths et al.,: "Resist substrate studies for LIGA microfabrication with application to a new anodized aluminum substrate", J. Micromech. Microeng. **15** (2005) 1700-1712
6. F. DeCarlo, J. J. Song, D. C. Mancini, "Enhanced Adhesion Buffer Layer for Deep X-ray Lithography Using Hard X-rays," J. Vac. Sci. Technol. B **16**, November, 3539-3542, (1998)
7. S. Megtert, F.J. Pantenburg, S. Achenbach, R. Kupka, J. Mohr, M. Roulliay: „Preliminary results on the use of a mirror-system for LiGA-process", SPIE, Volume **3680** (1999), pp. 917-923
8. M. Neumann, F.J. Pantenburg, M. Rhode, M.Sesterhenn: „Heat transport in masks for deep X-ray lithography during the irradiation process", Microelectronics Journal 28 (1997) 349-355
9. Stanley Mrowka, Sandia National Laboratory, Livermore Ca, USA, private communication
10. F.J. Pantenburg, J. Mohr, "Deep X-ray lithography for the fabrication of microstructures at ELSA", Nuclear Instruments and Methods A **467-468** (2001) 1269-1273
11. F. Perennes, F.J. Pantenburg, "Adhesion improvement in the deep X-ray lithography process using a central beam-stop, Nucl. Instrum. Meth B, **174** (2001), pp. 317-321
12. H. Guckel, K. Fischer, E. Stiers, B. Chaudhuri, S. McNamara, M. Ramotowski, E. D. Johnson, C. Kirk: "Direct, high throughput LIGA for commercial applications: a progress report", Microsystem Technologies 6/3 (2000), pp. 103-105
13. Arendt M., Meyer P., Saile V.: "Launching into a golden age (1) - marketing strategy for distributed LIGA- fabrication." Commercialization of Micro and Nano Systems (COMS 2005): 10th Internat.Conf., Baden-Baden, August 21-25, 2005 Albuquerque, N.M. : MANCEF, 2005 S.547-51

Towards 3D Electromagnetic Metamaterials in the THz Range

B. D. F. Casse, H. O. Moser, M. Bahou, P. D. Gu, L. K. Jian, J. R. Kong, S.B. Mahmood and Li Wen

Singapore Synchrotron Light Source (SSLS),
National University of Singapore (NUS), 5 Research Link, Singapore 117603.

Abstract. SSLS has been using its lithography-based micro/nanofabrication facility LiMiNT (Lithography for Micro and Nanotechnology) and its infrared spectro/microscopy facility ISMI to develop and characterize the first electromagnetic metamaterials having their spectral response in the THz range. Derived from Pendry's nested-split-ring resonator design, these structures require micro/nanofabrication in order to have resonances in the THz range. They exhibit a negative refractive index and hold promise of sub-diffraction limit imaging. Besides the reduction of the size of the resonating structures to extend the spectral range towards the visible, outstanding issues include the production of high-aspect-ratio resonators that are sensitive for the magnetic field in any direction (3D sensitivity) and the capability to produce copious amounts of the electromagnetic metamaterials with a good yield. In this paper, we shall report on first results of 3D EM^3 structures made by inclined exposures.

Keywords: Electromagnetic Metamaterials, EM3, THz, Left-Handed Materials, 3D structures, Microfabrication, LIGA
PACS: 42.70.Qs, 41.20.Jb, 73.20.Mf, 78.20.Ci

INTRODUCTION

Electromagnetic Metamaterials (EM^3) refer to artificially engineered structures having simultaneously negative permittivity ε and permeability μ. EM^3 represent a new class of composite materials capable of exhibiting a negative index of refraction and possessing *"superlenses"* properties [1].

The significant achievements, from a micro/nanofabrication point of view of EM^3, are: 1) theoretical investigation of left-handed materials and prediction of their exotic properties by Veselago in 1967 [2]; 2) Pendry's recipes for manufacturing $\varepsilon_{eff} < 0$ [3] and $\mu_{eff} < 0$ [4] by a combination of wire arrays and split-ring resonators (SRRs) respectively; 3) EM^3 operating in the GHz range; 4) the first microfabricated EM^3 operating in the THz range [5]; 5) negative μ square split rings in the THz range [6]; 6) the first nanofabricated negative μ materials operating around 100 THz [7] and various EM^3 structures beyond the 100 THz regime [8] [9] [10] [11]. EM^3 structures produced so far in the terahertz range have been mostly two-dimensional (2D), and are therefore highly anisotropic. By anisotropy, it is inferred that the response of the system depends on the direction of illumination. For instance the split ring resonators behave as an LC circuit when incoming electromagnetic radiation has magnetic field components polarized parallel to the rings' axis. Gay-Balmaz and Martin [12] were among the first to propose more three-dimensional (3D) EM^3 by assembling planar EM^3 structures such that they offer full coupling for the incident electric and magnetic fields in two or three orthogonal directions. Although this is seemingly the most logical way of obtaining isotropic structures, the technique of assembling planar unit cells becomes highly impractical when it comes to μm-size structures and below.

Recently, we proposed schemes [8] [13] to produce more isotropic structures, within the same matrix, via tilted X-ray exposures that were introduced in the LIGA[1] process years ago [14] [15] [16] where LIGA stands for the German words LIthographie (Lithography), Galvanoformung (Electroplating) and Abformung (Molding). In this paper, we present first results of microfabrication of nearly 3D EM^3 structures for the THz range. For the sake of rapid prototyping, we limit ourselves to the fabrication of $\mu_{eff} < 0$ rings as the implementation of the additional rod structure is obvious.

CP879, *Synchrotron Radiation Instrumentation: Ninth International Conference*,
edited by Jae-Young Choi and Seungyu Rah
© 2007 American Institute of Physics 978-0-7354-0373-4/07/$23.00

EXTRUDING THE 2-D PENDRY'S NESTED RINGS TO PRODUCE 3D STRUCTURES

The split ring resonator of Sir John Pendry [4] shown in Fig. 1 forms the backbone of practically all metamaterials. It is designed to have a very strong magnetic response and potentially one that can lead to a negative effective permeability. The idea of achieving a magnetic response from conductors comes from

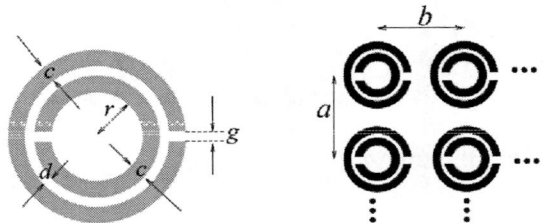

FIGURE 1. Plan view of a split ring showing geometric parameter definitions (left). Periodic arrangement the split rings (right).

the basic definition of a magnetic dipole moment $\overline{m}$.

where j is the current density. From the definition in (1), we can see that a magnetic response can be obtained if local currents can be induced to circulate in loops. Now if one introduces a resonance into the element, a negative effective permeability can be achieved at resonance because of the phase shift between exciting and resonating fields. The split ring resonator can be viewed as an LC circuit where a time-varying magnetic field applied parallel to the axis of the rings induces an emf in the plane of the element, driving currents within the 'split rings'. The lower and upper limit of the frequency interval over which $\mu_{eff} < 0$ was calculated from Pendry's analytical formula [4]

$$v_0 = \frac{1}{2\pi}\sqrt{\frac{3dc_0^2}{\pi^2 r^3}} < v_{mp} = \frac{v_0}{\sqrt{1 - \pi r^2/ab}} \tag{2}$$

where c_0 is the speed of light *in vacuo*.

In X-ray deep lithography, the angle of incidence can be varied to obtain inclined exposures as shown in Fig. 2 (left). The stack could even be rotated either continuously or to different positions between exposures. This tilting, rotating and even wobbling was proposed several years ago [14] [15]. By exploiting the full potential of LIGA, we can produce EM³ structures in which the axes of the SRRs would cover two (or even three) intersecting or perpendicular directions so that the incident field can always couple

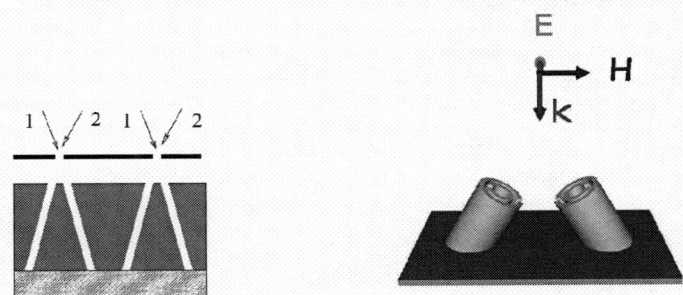

FIGURE 2. Double angle X-ray exposures [1=+θ , 2=−θ to the normal] to produce 3D inclined SRR structures within the same matrix (left). For structures inclined at an angle θ° to the normal, an incident beam would always have a component $H \sin\theta$ along the rings' axis (right).

efficiently to the SRRs as shown in Fig. 2 (right). For structures inclined at an angle θ to the normal, an incident beam would always have a component $H \sin\theta$ along the rings' axes. The planar case corresponds to θ = 0°.

MICROFABRICATION OF THE 3D EM³ STRUCTURES

To produce the X-ray mask for the inclined EM³ structures, a two-stage lithography process has been developed: In the first stage, an intermediate optical mask was generated by direct laser writing. This mask is made up of 0.09 inch soda lime glass covered with a first layer of 800 Å thick chromium and a 0.5 μm thick AZ 1518 resist layer on top of the Cr. An AutoCAD design file containing split rings (with parameters $r = 15$ μm, $c = 15$ μm, $d = 10$ μm, $g = 10$ μm) was generated in 2 × 2 cm² arrays and transferred into the AZ 1518 photoresist by direct laser writing with the DWL 66 (registered trademark of Heidelberg Instruments Mikrotechnik, Germany). The DWL 66 is equipped with a 20 mW HeCd laser of 442 nm wavelength. The 20 mm write-head, allowing minimum feature sizes of 4 μm, was used in a double pass exposure. The positive photoresist was then developed in the AZ 400K developer followed by an immersion of the optical mask in a chromium etch for 2 min.

The second stage consisted of producing a graphite mask with a gold absorber. SU8 2025 was spin coated onto a graphite wafer such that a thickness of 20 μm was achieved. Pattern transfer from the optical mask to the graphite mask was done by deep UV exposure using a Karl Suss MA8 contact UV mask aligner. A gentle oxygen plasma etch was applied to the graphite mask to remove residual resist in the developed areas. The graphite mask was brought into a gold electroplating bath in order to obtain nested split rings on the graphite membrane. Electroplating was carried out at a current density of 0.1 A/dm² and temperature of 55 °C.

In order to produce EM³ materials for transmission experiments, the rings have to be embedded in a material that features high transparency in the relevant THz frequency range. SU8 2100 was chosen because of its good transparency in the far infrared region, short exposure time even for thick SU8 samples, and sufficient ruggedness to survive mechanical handling. SU8 2100 was spin coated onto a 4 inch silicon wafer, which had been sputtered with 40 nm of titanium and 100 nm of gold on top of the Ti layer, until a thickness of 200 μm was achieved. X-ray exposures were performed at angles of ± 10°, ± 30° and ± 45° to the normal of the mask-substrate stack to show the expected deterioration of the mask contrast at higher incidence angles. A secondary slit-type absorber was used, on alternate rings columns to prevent exposing open areas twice, i.e., the slits exposed odd columns of rings to the X-rays and blocked even ones for the first exposure, and vice versa for the second exposure. The final step was to postbake the SU8, resist development, descumming, and nickel electroplating at 0.4 A/dm². The SU8 chips were then peeled off the silicon wafer by stressing the plating and cleaving the back side of the wafer.

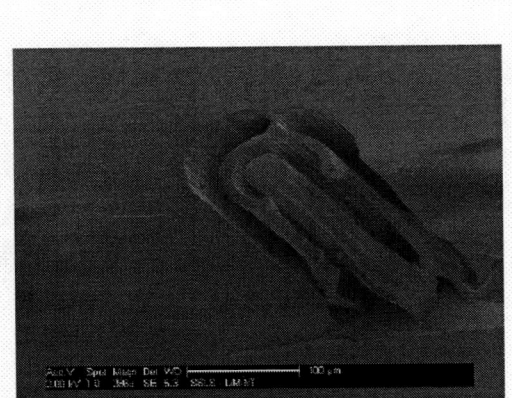
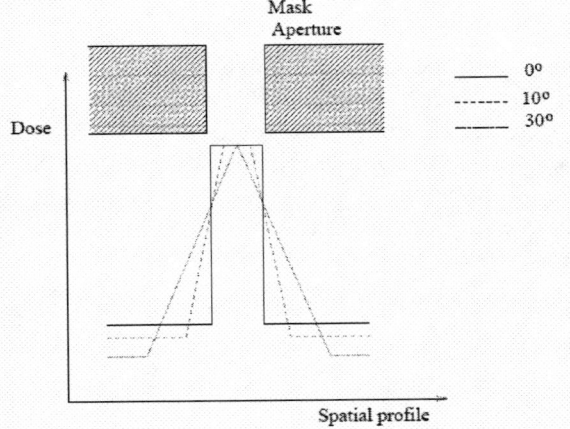

FIGURE 3. Cross section of resist template at 30° (left) exposure to the normal [scale bar 100 μm]. Lacking mask contrast prevents the formation of good structures. Qualitative dose deposition profile for various irradiation inclinations (right).

At 45° exposures the rings are deformed because the X-rays cannot penetrate the small inner ring gap ($d = 10$ μm) at this angle to expose the volume in between the rings. At 30° to the normal (Fig. 3 left), X-rays can just penetrate the nested rings gap. The top part of the template looks acceptable, but the bottom part is still deformed. The ideal dose profile is rectangular, corresponding to perpendicular irradiation (Fig. 3 right). At 10°, the profile becomes trapezoidal, while at 30°, the dose profile is triangular. To obtain the best inclined structure, a 10° inclination was chosen, in order to have a dose profile as close as possible to the ideal one. Figure 4 (left) shows the cross section of a resist template that has been exposed at 10° to the normal. For this configuration, we observe a well-defined structure with straight sidewalls. Clean double angle exposures at $\pm 10°$ to the normal are shown in Fig. 4 (right).

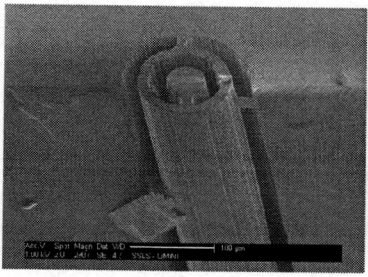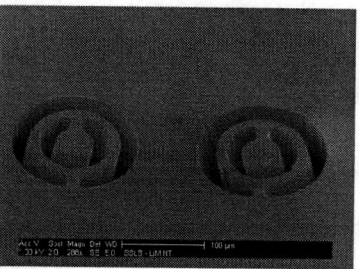

FIGURE 4. Cross section of resist template at 10 ° exposure to the normal (left). Double angle exposures at $\pm 10°$ to the normal (right). [Scale bar 100 μm].

CONCLUSION

Using the LIGA process, nearly three-dimensional (3D) split ring resonators (SRRs) were produced within the same resist matrix. By making the structures more isotropic, we remove the H polarization constraint placed on the impinging electromagnetic waves by the media. Basic geometry indicates that such 3D nested rings could lead to an improvement of the coupling of the H vector and thus ease future implementations of EM^3 in real-life applications. To produce inclined structures facing each other in the matrix of the SU8 2100 negative resist, a secondary slit-type absorber was used which prevents exposing the same areas twice. The slits are not necessary if we were dealing with a positive resist (e.g. PMMA), but the known drawback of PMMA is the long exposure time for high-aspect-ratio structures. Work is under way in SSLS to characterize the 3D composite structures by means of infrared spectroscopy.

ACKNOWLEDGMENTS

The authors would like to thank Joe Wing Lee for his contribution to the microtechnology processes. This work was performed at the Singapore Synchrotron Light Source (SSLS) under A*STAR/MOE RP3979908M, A*STAR 0121050038, and NUS Core Support C-380-003-003-001 grants.

REFERENCES

1. J. B. Pendry, Phys. Rev. Lett. 85, 3966 (2000).
2. V. G. Veselago, Sov. Phys. Usp. 10, 509 (1968).
3. J. B. Pendry, A. J. Holden, W. J. Stewart, and I. Youngs, Phys. Rev. Lett. 76, 4773 (1996).
4. J. B. Pendry, A. J. Holden, D. J. Robbins, and W. J. Stewart, IEEE Trans. Microwave Theory Tech. 47, 2075 (1999).
5. H. O. Moser, B. D. F. Casse, O. Wilhelmi, and B. T. Saw, Phys. Rev. Lett. 94, 063901 (2005).
6. T. J. Yen, W. J. Padilla, N. Fang, D. C. Vier, D. R. Smith, J. B. Pendry, D. N. Basov, and X. Zhang, Science 303, 1494 (2004).
7. S. Linden, C. Enkrich, M. Wegener, J. Zhou, T. Koschny, and C. M. Soukoulis, Science 306, 1351 (2004).
8. H. O. Moser, B. D. F. Casse, O. Wilhelmi, and B. T. Saw, "Electromagnetic Metamaterials over the whole THz range — achievements and perspectives, Proceedings of the ICMAT 2005 Symposium R (Electromagnetic Materials)," World Scientific, Singapore, 2005, pp. 55–58, ISBN 981-256-411-X(pbk).
9. B. D. F. Casse, H. O. Moser, O. Wilhelmi, and B. T. Saw, "Micro- and Nano-Fabrication of Electromagnetic Metamaterials for the Terahertz range, Proceedings of the ICMAT 2005 Symposium R (Electromagnetic Materials)," World Scientific, Singapore, 2005, pp. 18–25, ISBN 981-256-411-X(pbk).
10. B. D. F. Casse, H. O. Moser, M. Bahou, L. K. Jian, and P. D. Gu, "NanoElectromagnetic Metamaterials Approaching Telecommunication Frequencies", NanoSingapore 2006, 2006 IEEE Conference on Emerging Technologies — Nanoelectronics, pp 328–331, ISBN 0-7803-9357-0, Singapore 2006.
11. S. Zhang, W. Fan, N. C. Panoiu, K. J. Malloy, R. M. Osgood, and S. R. J. Brueck, Phys. Rev. Lett. 95, 137404 (2005).
12. P. Gay-Balmaz, and O. J. F. Martin, J. Appl. Phys. 81, 939–941 (2002).
13. B. D. F. Casse, H. O. Moser, L. K. Jian, M. Bahou, O. Wilhelmi, B. T. Saw, and P. D. Gu, Journal of Physics: Conference Series 34, 885–890 (2006).
14. H. O. Moser, W. Ehrfeld, M. Lacher, and H. Lehr, "Fabrication of Three-dimensional Microdevices from Metals, Plastics and Ceramics," Institut des Microtechniques de Franche-Comté, Besançon, France, 1992.
15. W. Bacher, P. Bley, and H. O. Moser, Optoelectronics Interconnects and Packaging, SPIE Critical Reviews of Optical Science and Technology CR62, 442–460 (1996).
16. G. Feiertag, W. Ehrfeld, H. Freimuth, H. L. H. Kolle, M. Schmidt, M. M. Sigalas, C. M. Soukoulis, G. Kiriakidis, T. Pedersen, J. Kuhl, and W. Koenig, Appl. Phys. Lett. 71, 1441 (1997).

Using E-Beam and X-Ray Lithography Techniques to Fabricate Zone Plates for Hard X-ray

T. N. Lo[a], Y. T. Chen[a,c], C. J. Liu[a], W. D. Chang[a,b], T. Y. Lai[a,b], H. J. Wu[a,b], I. K. Lin[a], C. I. Su[a,b], B. Y. Shew[d], J. H. Je[e], G. Margaritondo[f] and Y. Hwu[a,b]

[a]Institute of Physics, Academia Sinica, Taipei 115, Taiwan
[b]Institute of Optoelectronic Sciences, National Taiwan Ocean University, Keelung 202, Taiwan
[c]Department of Materials Engineering, Tatung University, Taipei 104, Taiwan
[d]National Synchrotron Radiation Research Center, Hsinchu 300, Taiwan
[e]Department of Materials Science and Engineering, Pohang University of Science and Technology, Pohang, Korea
[f]Ecole Polytechnique Fédérale, CH-1015 Lausanne, Switzerland

Abstract. A high-resolution zone plate for focusing and magnifying hard-x-rays require very high fabrication precision and high aspect ratio metal structures and therefore post perhaps the most challenging task to the nanofabrication. We present a nanofabrication strategy to fabricate such devices for x-ray applications which takes advantage of the state-of-the-art x-ray lithography and electroplating processes. The substrate used to fabricate the x-ray mask is first prepared by depositing Si_3N_4 film on a Si wafer. Electron beam lithography capable of writing small lines (<10nm on photoresist) is then used to create desired zone plate pattern. An Au layer is electrodeposited into nano-paterned photoresist structure of sufficient thickness and the resist is then removed to expose the Au zone plate structure. Finally, the KOH is used to etch the Si wafer down to the Si_3N_4 film and to produce the desired x-ray hard mask. This hard mask defines the zone plate pattern with the x-ray lithography method. We demonstrate that the mask prepared by this method is of very high precision—30nm outermost zone—can be obtained. X-ray lithography process is attempted with convincing results.

INTRODUCTION

The high penetration is the most important characteristics of x-rays and has been used as a radiology tool ever since its discovery. In spite of the very short wavelength (<1nm), x-ray microscopy, or radiology failed to produce diffraction limited image mainly because of the poor quality of the x-ray optics compared to other microscopy. One of the most efficient x-ray focusing and imaging devices, the x-ray phase zone plate, cannot produce image with resolution better than the smallest zone. On the other hand, to produce the required phase shift between adjacent zones, the thickness of the zone plate would have to be larger than ~500nm if Au is used. Therefore, to achieve an image resolution of <50nm, it is necessary to fabricate a zone plate device with a smallest zone <50nm and an Au structure of an aspect ratio >10. Either characteristics, the small zone width and the high aspect ratio, can be difficult to achieve, the simultaneous requirement make fabrication of this device one of most challenging task in nanofabrication.

We carefully evaluated the available nanofabrication and conclude that the only feasible approach is to use electron beam lithography for structures of a thickness of less than 500nm and x-ray lithography for structure thicker than 500nm. In the latter case, a 1-to-1 mask of smallest zone width <50nm for x-ray lithography is required which can only be fabricated by e-beam lithography. We therefore designed a fabrication strategy of hard-x-ray phase zone plate by the combined use of e-beam and x-ray lithography.

A typical zone plate has concentric circles structure with alternating filled and empty zones. The width of the zone, ΔRn, decreases when the diameter of the outmost zone, OD, increases following the simple rule: $f = (OD \times \Delta Rn)/\lambda$, which results in the desired phase shift at the focal length f and to produce a focusing effect. If a coherent source is used to illuminate the zone plate, the size of the focal point depends critically on the dimension of the smallest zone. For example, to achieve a 50nm resolution when the zone plate device is used as a magnifying lens or a focusing device, the width of the smallest (the out most) zone would have to less than 50nm [1].

Zone plate devices of this level of performance are difficult to fabricate by the conventional ultraviolet

CP879, Synchrotron Radiation Instrumentation: Ninth International Conference,
edited by Jae-Young Choi and Seungyu Rah

lithography process. In order to carry out the zone plate nanofabrication, the electro beam lithography (EBL) is applied as the principal approach according to its direct writing and nano dimensions resolution [2]. However, the EBL is a rather time consuming process and the high cost to definite the patterns that will result in difficulty on the mass production. Meanwhile, for zone plate structure thicker than 500nm, the focusing effect of electron beam would not be able to produce sharp and vertical structure and x-ray lithography become a viable approach.

The main difficult in this latter process is to prepare x-ray hard masks for deep X-ray lithography (DXL) which is fabricated with the same resolution as the final product desires. In this work, we will demonstrate the feasibility of combined EBL and DXL techniques to fabricate zone plate. For zone plate of a thickness less than 500nm EBL will be used to write directly the zone plate structure and later fill the etched trench with Au by electrodeposition. When a thick zone plate >500nm is desired, the same EBL process will be used to fabricate the x-ray mask whose pattern will be transferred to a thick photoresist.

EXPERIMENTAL PROCEDURE

The silicon wafer with 1 μm thick Si_3N_4 is spin coated a layer of PMMA resist with thickness about 300 nm. After baking, the pretreated samples are patterned by e-beam writer. After developing, a gold layer about 200 nm is deposited on the patterned samples by electroplating. Finally, a lift-off process is applied to the samples to remove the residual resist. The finished pattern is then used as the hard mask for X-ray exposure.

The X-ray exposure process is applied with BL18B beamline at the National Synchrotron Radiation Research Center in Taiwan. This beamline provides photon beams with energies 500-1500 eV from the bending magnet. Before X-ray exposure, Si_3N_4 is deposited on the Si wafer substrate and back-side etched to create an opening window. Subsequently, the wafer will be spin coated a layer of PMMA resist about 300 nm. Figure 1 shows the schematic diagram of the combined EBL and DXL process. The final zone plate product of the EBL and DXL processes is examined by scanning electron microscope (SEM).

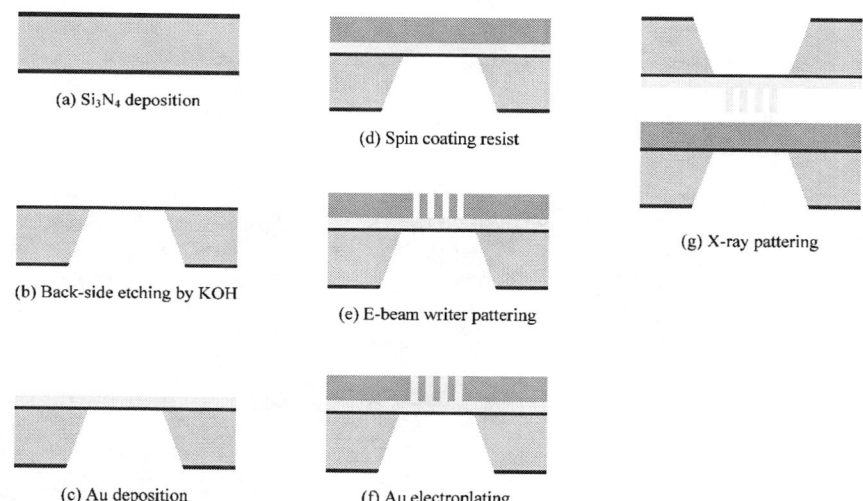

(a) Si_3N_4 deposition

(b) Back-side etching by KOH

(c) Au deposition

(d) Spin coating resist

(e) E-beam writer pattering

(f) Au electroplating

(g) X-ray pattering

FIGURE 1. The e-beam lithography for zone plate fabrication and X-ray lithography process.

RESULTS

Figure 2 shows the SEM images of hard mask with zone plate pattern, which indicates that the EBL process is able to fabricate zone plate with high precision. It also reveals that the smallest line width of the outermost zone is about 50nm, meaning that the aspect ratio can be accomplished is already over 6. Although the EBL is successfully pattern the very small line width of the outermost zone, the yield is rather low and requires very expensive e-beam writer time.

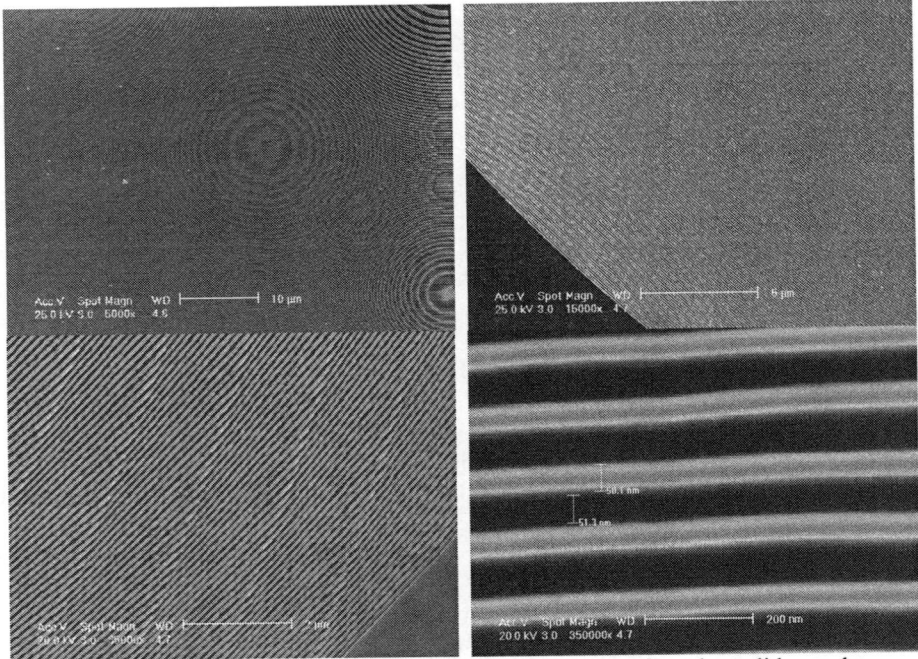

FIGURE 2. The SEM observation of the zone plate pattern after e-beam lithography.

Figure 3 shows images of the zone plate pattern after Au electroplating. These SEM micrographs indicate that the Au structure is continuous and with good precision resulted from the excellent EBL process. It also reveals that the Au structure has the aspect ratio about 4 sufficient for the DXL process as the hard mask. The surface of the zone band is not as smooth as desired but we expect that its variation will be reduced with the DXL process rather than amplified. We also expect the surface roughness can be improved by controlling the current density and electrolyte concentration during plating which is underway.

FIGURE 3. The SEM observation of Au electroplating on etched zone plate pattern.

Figure 4 is an SEM micrograph of the photoresist patterned by a DXL process. The ring pattern can be clearly identified. However, due to the high aspect ratio, electroplating of Au is not successful yet and further improvement in optimize the electroplating condition is underway.

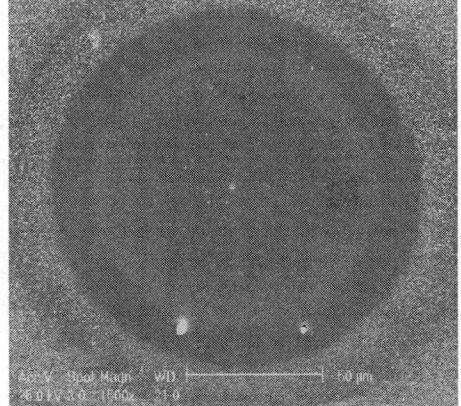

FIGURE 4. The SEM observation of the zone plate pattern after X-ray lithography.

CONCLUSION

We demonstrate that a <50nm phase zone plate of a thickness less than 500nm can be fabricated by EBL and Au electrodeposition. If a thicker structure is desired, a DXL would be a good candidate to transfer designed pattern to thick resist. However, careful selection of suitable photoresist and the capability to electroplate thick Au layer into small trenches would dictate the quality of the final device.

ACKNOWLEDGMENTS

The authors are grateful for the assistance of Mr. C. H. Wu and Y. H. Tsai in operating the BL18B beamline at National Synchrotron Radiation Research Center, Hsinchu, Taiwan.

REFERENCES

1. D. Attwood, Soft X-Rays and Exteme Ultraviolet Radiation: Principles and Applications, Chapter 9, Cambridge University Press (1999).
2. C. David, R. Medenwaldt, J. Thieme, P. Guttmann, D. Rudolph and G. Schmahl, J. Optics (Paris) **23**, 255-258 (1992).

Beneficial Photoacid Generator for CA Resist in EUVL

Takeo Watanabe[1,3], Hideo Hada[2], Yasuyuki Fukushima[1,3], Hideaki Shiotani[1,3], Hiroo Kinoshita[1,3], and Hiroshi Komano[2]

[1]Laboratoy of Advanced Science and Technology for Industry, University of Hyogo,
3-1-2 Kouto, Kamigoori-cho, Akou-gun, Hyogo 678-1205, Japan.
[2]Advanced Material Development Division 1, Tokyo Ohka Kogyo Co., Ltd.,
1590 Tabata, Samukawa-cho, Koza-gun, Kanagawa 253-0114, Japan.
[3]CREST-JST, Kawaguchi Center bldg., 4-1-8 Hon-cho, Kawaguchi-shi, Saitama 332-0012, Japan.

Abstract. We succeeded in developing beneficial photoacid generator (PAG) based on onium salts for extreme ultraviolet lithography resist. The CA resist employing this beneficial PAG has E_0 sensitivity of 1.1 mJ/cm^2. We confirmed that the distinctive acid production reaction is occurred under EUV exposure in comparing under EB exposure. As results of the time dependent mass spectroscopy and the Fourier Transform Infrared Spectroscopy (FT-IR), it is confirmed that multiple acids are generated from cyclo(1,3-perfluoropropanedisulfone) imidate employed as an anion of PAG under EUV exposure.

Keywords: EUVL, lithography, resist, high sensitivity, chemically amplified resist, onium salts
PACS: 85.40.Hp

INTRODUCTION

Extreme ultraviolet lithography (EUVL) [1] is now planned to address the 32 nm node technology below. The lithographic performances of variety of resist types have been previously evaluated [2-7] including chemistries of both a CA resist[8-15] and a non-CA resist under EUV exposure.[8, 10, 11] However, at the industrial standing point, the current EUV resist is not yet mature enough to achieve some of the required specifications, such as line edge roughness (LER), sensitivity and outgassing. The LER specification was reduced to 1.5 nm (3σ) to maintain the electronic-device performance. Since the required LER specification is as same as a size of resist molecule, this target of LER is challenging. Furthermore, it is especially challenging when satisfying both LER of 1.5 nm (3σ) and sensitivity of 2~5 mJ/cm^2.[16] Since the wavelength of 13.5 nm using in EUVL is about 1/20 shorter than that using for conventional optical lithography, shot noise in EUV exposure influences LER controllability. Therefore, focusing on CA resist, quantum yield of PAG generation has to increase under EUV exposure to satisfy the specification of LER. It is reported that the resist using a sulfonium salt on the basis of a new photo acid generator (PAG) could achieve higher E_0 sensitivity in comparison with a conventional PAG under EUV exposure.[17, 18] However, there is not effective under EB and 248-nm exposures. This indicates that there is a distinctive photolysis process occurring in the resist film matrix under EUV exposure.

This paper focused on a study of beneficial PAG for EUV resist. We will discuss the chemical reaction of a onium salts under EUV exposure.

EXPERIMENTS

The samples were formulated with a polymer, a solvent and 10wt% onium salt employed as a PAG against the base polymer. We employed poly(hydroxystyrene-co-t-butylacrylate) as a base polymer. Resists A and B employ triphenylsulfonium cyclo(1,3-perfluoropropanedisulfone) imidate (TPS-IMIDATE) and triphenylsulfonium perfluorobutanesulfonate (TPS-PFBS) as PAGs, respectively. Resists C and D employ bis-4-(tert-butyl)phenyliodonium cyclo(1,3-perfluoropropanedisulfone) imidate (BPI-IMIDATE) and bis-4-(tert-butyl)phenyliodonium perfluorobutanesulfonate (BPI-PFBS) as PAGs, respectively. Propyleneglycol

CP879, Synchrotron Radiation Instrumentation: Ninth International Conference,
edited by Jae-Young Choi and Seungyu Rah
© 2007 American Institute of Physics 978-0-7354-0373-4/07/$23.00

monomethyletheracetate (PGMEA) was employed as a solvent. The detail of resist sample formulations and structures are shown in Table 1.

TABLE 1. Resist samples which employed in the experiments.

Sample	Resist A	Resist B	Resist C	Resist D
Resin				
PAG	(triphenylsulfonium cation) O₂S–CF₂ / O₂S–CF₂ with N⁻	(triphenylsulfonium cation) $C_4F_9SO_3^-$	(bis-phenyl iodonium cation) F_2C–SO_2 / F_2C–SO_2 with N⁻	(bis-phenyl iodonium cation) $C_4F_9SO_3^-$
Solvent	Propylene glycol monomethylether acetate (PGMEA)			

The sample films were spin-coated on a silicon wafer and were baked on a hot plate at a temperature of 130°C for 90 s. The film thickness was fixed to be 100 nm and determined by using Nanospec model 6100A manufactured by NANOmetrics Co., Ltd. After exposure, post exposure bake (PEB) are carried out at a temperature of 110°C for 90 s. The puddle development was performed with NMD-3 at 23°C for 60 s, which was mainly containing of 2.38% tetra-methylammonium hydroxide (TMAH) aqueous solution. The rinse was performed with deionized water at a temperature of 23°C for 60 s. Then, the remaining thickness was measured to obtain the sensitivity curve. The outgas experiment was carried out at the BL3 beamline in NewSUBARU. The light source of this beamline is a bending magnet. The storage ring operates at electron energy of 1.0 GeV. Using he resist evaluation system [19] which can simulate a six-mirror imaging optics, SR light was mono-chromated to 13.5 nm. The outgassing species was measured by HAL511/3L quadrupole mass spectrometer (Hiden Analytical Ltd.) connected to the resist evaluation chamber. The atmospheric pressure in the resist test chamber is maintained at less than 3.0×10^{-5} Pa.

RESULTS AND DISCUSSIONS

E_0 sensitivity is defined as a threshold dose to completely develop. E_0 sensitivities of resists A, B, C, and D are 1.1 mJ/cm², 3.8 mJ/cm², 2.0 mJ/cm², and 4.0 mJ/cm², respectively. In comparing resists A and B, sensitivity of resist A is four times higher than that of resist B. Both resist A and B employ TPS as a cation of PAGs. The different points were only anions. Resists A and B employ IMIDATE and PFBS as anions of PAGs, respectively. Furthermore, in comparing resists C and D, sensitivity of resist C is twice higher than that of resist D. Both resist C and D employed BPI as a cation of PAGs. The different points were only an anion structure. Resists C and D employ IMIDATE and PFBS as anions of PAGs, respectively. As results, IMIDATE employed as an anion of PAG is more sensitive than PFBS employed as an anion of PAG under EUV exposure. Since an E_0 sensitivity tendency in iodonium salts shows slightly lower than that in sulfonium salts, it is considered that sensitivity caused by an inhibition effect of iodonium salts under EUV exposure. E_0 sensitivity measurements of resists A and B were carried out under EB exposure to compare E_0 obtained under EUV exposure. EB exposure was carried out by using an electron beam direct writing tool which has an accelerated voltage of 70 kV. Resist thickness was 150 nm and resist components and process condition were the same as using under EUV exposure. Table 2 shows E_0 sensitivities for resists A, B, C and D under EUV and EB exposures. As a result, E_0 sensitivities of resists A and B under EB exposure are almost the same. E_0 sensitivity of resist C and B under EB exposure are almost same. However, E_0 sensitivity of resist A is higher than that of resist B under EUV exposure. E_0 sensitivity of resist C is twice higher than that of resist D under EUV exposure. It is confirmed that a selection of an anion of PAG under EUV exposure is much more effective to achieve a high quantum yield of acid generation rather than that of PAG under EB exposure.

FT-IR-spectra were measured to study the decomposed mechanisms of resists A and B. The spectra were measured to compare the cases before and after EUV exposures for both resists A and B. FT-IR spectra before and after exposures of resists A are shown in Fig. 1. For resist B, the absorbance peaks of the spectra did not show any significant change before and after exposures. However, there is an absorbance peak change at approximately 1156 cm^{-1} in resist A. The absorbance at 1156 cm^{-1} increases after exposure. FT-IR spectra of PAG of resists A is shown in Fig. 2. As a result of

TABLE 2. E_0 sensitivities under EUV and EB exposures

	E_0 (mJ/cm^2) under EUV exposure	E_0 (μC/cm^2) under EB exposure
Resist A	1.1	14.3
Resist B	3.8	14.6
Resist C	2.0	14.0
Resist D	4.0	15.0

electron orbital calculation by Gaussion03 [20], the peak at approximately 1156 cm^{-1} in Fig. 1 corresponds to a C-F bonds of an anion of TPS-IMIDATE. It indicates that IMIDATE decomposed under EUV exposure. As a result, the outgassing measurements and the FT-IR spectra measurements show that the EUV-induced reaction of TPS-IMIDATE occurred more efficiently than that of TPS-PFBS. It is considered that since the IMIDATE employed as the anion of PAG decomposed under EUV exposure to produce a larger amount of acid than PFBS did, the large amount of sulfonic acid is induced by the photodecomposition reaction of IMIDATE. Therefore, IMIDATE which is employed as the anion of PAG, achieves the fast photospeed characteristics under EUV exposure.

The outgassing characteristics are measured to estimate the chemical reaction mechanism of PAG under EUV exposure. The outgassing species was measured by quadruple mass spectrometer. In order to avoid any potential contamination of the sample through outgassing of the filament in the mass spectrometer, the resist sample was moved into the chamber 600 s prior to starting the outgas measurement. The transient mass spectra measurement time for each resist sample was a total of 189 s. The first 63 s is for background measurement. After that EUV light is then irradiated to measure outgassing by mass spectrometer for 63 s. The last 63 s is for background re-measurement, which corresponds to a dose of 23.9 mJ/cm^2. However, the content of PAG is almost 10wt% for the base polymer. Thus the outgassing species from PAG is very weak. Therefore, to increase the signal noise ratio of ion counts of the outgassing species from PAG, the integrated ion counts over the exposure time is calculated based on the measurement results of the time dependency of ion counts. As a result of outgassing measurement of resist A, outgassing

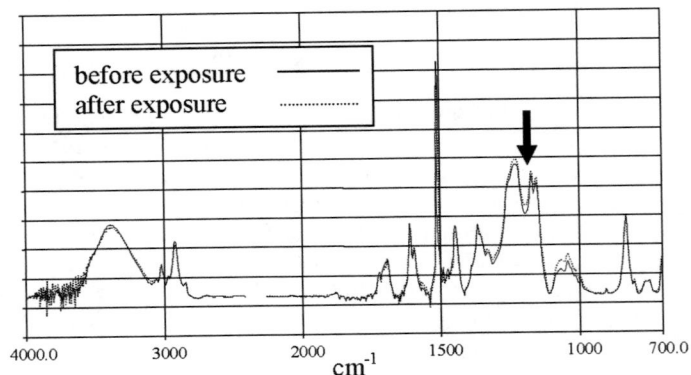

FIGURE 1. FT-IR spectra of resist A before and after exposure.

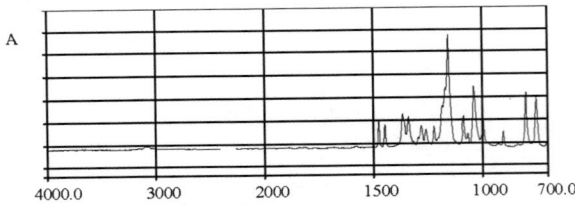

Wave number

FIGURE 2. FT-IR spectra of PAGs of resists A

species of mass numbers of 64, 80, 128, 142, 150 are observed as the photodecomposition species from PAG. For resist B, outgassing species of mass number of 64 and 80 are observed as the photodecomposition species from PAG. As shown in Fig. 3, the integrated ion counts over 60 s exposure for each outgassing species which have mass number of are obtained by integrating ion counts over 60 s. The outgassing species which has mass number of 64, 80, 128, 142, and 150 are assigned as SO$_2$, SO$_3$, CF$_2$SO$_2$N, SO$_2$NO$_2$S, and (CF$_2$)$_3$, respectively. Furthermore, for resist B, outgassing species of mass numbers of 63 and 64 are observed as the photodecomposition species from PAG. Since the ion counts are very small for other species from TPS-PFBS, the ion counts are too small to integrate over 60 s. The outgassing species which has number of 64, 80, are assigned as SO$_2$, SO$_3$, respectively. Resist A contains PAG anion of imidate derivatives, which carried out photodecomposition reaction of PAG anion under EUV exposure. This reaction will be expected to generate many acidic species, which has the potential of becoming a catalyst for the de-protecting reaction. This mechanism is useful for a resist design to obtain a high sensitivity EUV resist comparison with resist B. As a result of time dependency of the outgassing species from resist C, large amountof inspecting species which has a mass number less than 80 was observed. There is a possibility of solvent, PAG decomposition and de-protect species induced outgassing and fragmentation from large mass number compounds. Thus many fragment species of decomposed PAG anions were observed from resist C. Large amount of m/z 64, which related SO$_2$ was observed. It is confirmed that IMIDATE employed as an anion of PAG, which carried out distinctive photodecomposition-reactions under EUV exposure. Thus these reactions can induce the higher sensitivity in resist C than resist D.

As results of the time dependency mass spectroscopy and the FT-IR spectroscopy, the distinctive photodecomposition reactions of IMIDATE of an anion of PAG will be expected to generate many acidic species, which has the potential of becoming a catalyst for a de-protecting reaction. This mechanism is very useful for the resist design to develop a high sensitivity EUV resist for EUVL.

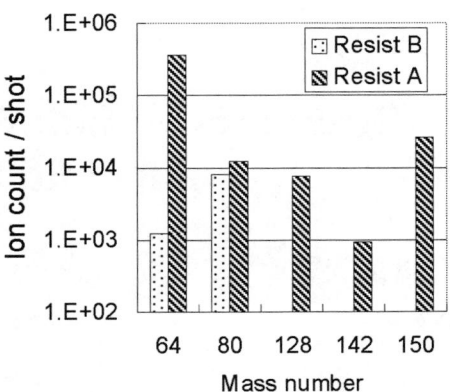

FIGURE 3. Decomposition fragments of anions of PAGs.

CONCLUSIONS

We succeeded in developing high sensitivity CA resist with using beneficial PAG to achieve E_0 sensitivity of 1.1 mJ/cm^2. In a high annealing CA type resist system of which poly-hydroxystyrene employed as a base resin, it is found that cyclo(1,3-perfluoropropanedisulfone) imidate as a anion of PAG of onium salts are more sensitive rather than those which employed perfluorobutanesulfonate as a anion of PAG of onium salts under EUV exposure. However, the sensitivities were different between under EUV and EB exposures. It is confirmed that the distinctive acid production reaction is occurred under EUV exposure in comparison with EB exposures. As results of the time dependency mass spectroscopy and the FT-IR spectroscopy, the distinctive photolysis reactions of IMIADTE of an anion of PAG will be expected to generate many acidic species, which has the potential of becoming a catalyst for a de-protecting reaction. Furthermore, it is confirmed that the EUV-induced reaction of IMIDATE employed as an anion of PAG occurred more efficiently than that of perfluorobutanesulfonate employed as an anion of PAG. Therefore, IMIDATE which is employed as an anion of PAG is high sensitive under EUV exposure.

ACKNOWLEDGMENTS

The authors would like to thank Mr. T. Hirayama, Mr. D. Shiono, Mr. S. Matsumaru, Mr. T. Ogata, Mr. R. Takahashi, Mr. J. Onodera, and Ms. S. Samejima of Tokyo Ohka Kogyo for their guidance and encouragement.

REFERENCES

1. H. Kinoshita, K. Kurihara, Y. Ishii and Y. Torii: J. Vac. Sci. & Technol. B**7** (1989) 1648.
2. H. Ito, C. G. Willson, and J. M. J. Frechet: Digest of Tech. Papers 1982 Symp. VLSI Tech., (1982) 86.
3. H. Ito and C. G. Willson: Polym. Eng. Sci. **23** (1983) 1012.
4. H. Ito, G. Breyta, D. Hofer, R. Sooriyakumaran, K. Petrillo, and D. Seeger: J. Photopolym. Sci. Technol. **7** (1994) 433.
5. Handbook Mark A. MaCord, Michael J. Rooks: Microlithography, Micromachining, and Microfabrication Volume 1; Microlithography, ed. P.Rai-Choudhury, p.208.
6. T. Watanabe, H. Kinoshita, A. Miyafuji, S. Irie, S. Shirayone, S. Mori, E. Yano, S. Okazaki, H. Hada, K. Ohmori, and H. Komano, Proc. SPIE **3997** (2000) 600.
7. T. Watanabe, H. Kinoshita, H. Nii, Y. Li, K. Hamamato, T. Oshino, K. Sugisaki, K. Murakami, S. Irie, S. Shirayone, Y. Gomei and S. Okazaki, J. Vac. Sci. & Technol. B**18** (2000) 2905.
8. T. Watanabe, H. Kinoshita, H. Nii, K. Hamamoto, H. Hada, H. Komano, and S. Irie, B**19** (2001) 736.
9. K. Hamamoto, T. Watanabe, H. Tsubakino, H. Kinoshita, T. Shoki and M. Hosoya: Photopolym. Sci. Technol. **14** (2001) 567.
10. T. Watanabe, K. Hamamoto, H. Kinoshita, H. Tsubakino, H. Hada, H. Komano, M. Endo and M. Sasago, Photopolym. Sci. Technol. **14** (2001) 555.
11. K. Hamamoto, T. Watanabe, Hideo Hada, Hiroshi Komano and Hiroo Kinoshita: J. Photopolym. Sci. Technol. **15** (2002) 361.
12. H. Hada, T. Watanabe, K. Hamamoto, H. Kinoshita and H. Komano, Proc. SPIE **5374** (2004) 686.
13. T. Watanabe, H. Kinoshita, K. Hamamoto, H. Hada and H. Komano, J. Photopolym. Sci. Technol. **17** (2004) 362.
14. T. Watanabe, K. Hamamoto, H. Kinoshita, H. Hada and H. Komano, *Jpn. J. Appl. Phys.* **43** (2004) 3713.
15. H. Hideo, T. Hirayama, D. Shiono, J. Onodera, T. Watanabe, S. Y. Lee and H. Kinoshita, Jpn. J. Appl. Phys. **44** (2005) 5824.
16. W. Yueh, H. Cao, M. Chandhok, S. Lee, M. Shumway and J. Bokor: Proc. SPIE. **5376** (2004) 434.
17. T. Watanabe, H. Hada, S. Y. Lee, H. Kinoshita, K. Hamamoto and H. Komano, Jpn. J. Appl. Phys **44** (2005) 5866.
18. H. Hada, T. Watanabe, H. Kinoshita and H. Komano, J. Photopolym. Sci. Technol. **18** (2005) 475.
19. T. Watanabe, H. Kinoshita, N. Sakaya, T. Shoki and S. Y. Lee, Jpn. J. Phys. **44**, (2005) 5556.
20. Gaussian03 electron orbital calculation software is informed in URL:http://www.gaussian.com/home.htm.

Beamlines on Indus-1 and Indus-2 for X-ray Multilayer Optics and Micro Fabrication Research

G.S. Lodha, V.P. Dhamgaye, M.H. Modi, M. Nayak, A.K. Sinha, R.V. Nandedkar

Synchrotron Utilisation and Material Research Division,
Raja Ramanna Centre for Advanced Technology, Indore-452 013, India

Abstract. A soft X-ray/extreme ultra violet (EUV) reflectometry beamline is operational at Indus-1 synchrotron source. The beamline is used for the characterization of multilayer optics for EUV lithography. A soft/deep X-ray lithography beamline is being set up on Indus-2, for undertaking research activities on micro electro mechanical systems (MEMS) and sub micron X-ray lithography structures. Present status of these beamlines is presented.

Keywords: Synchrotron Radiation Instrumentation, X-ray Lithography, X-ray Mirrors, X-ray Optics.
PACS: 07.85.Qe, 85.40.Hp, 07.85.Fv, 41.50.+h

INTRODUCTION

Indus-1 (450 MeV) is a soft X-ray/EUV synchrotron radiation (SR) source (3 to 100 nm) operational at our Institute. Indus-2 (2.5 GeV) is a third generation SR source [1], currently undergoing commissioning. At present 2 GeV electron beam has been stored at low currents (few mA). The machine is expected to be operational in a few months and will be available to users for installation of various beamlines on bending magnet sources. In this paper we report the details of the reflectometry beamline operational at Indus-1 and soft/deep X-ray lithography beamline, being setup at Indus-2.

SOFT X-RAY REFLECTOMETRY BEAMLINE ON INDUS-1

Soft X-ray/ EUV reflectometry beamline is operational on a bending magnet source of Indus-1 [2]. The first optical element is a torroidal mirror (pre mirror), which focuses the source onto the entrance slit of the torroidal grating monochromator (TGM). The TGM covers 4 to 100 nm wavelength range using three interchangeable torroidal gratings. The monochromatic focused image at the exit slit of the TGM is imaged at the sample point using a torroidal mirror (post mirror). For suppressing higher order contamination, a filter wheel mechanism, with a provision to install eight filters, is incorporated between the exit slit of the monochromator and the post mirror. Beam viewers are installed at various locations in the beamline to see the beam size and shape. EUV/soft X-ray photodiode detector is installed between the TGM exit slit and post mirror. This detector is used for periodically determining photon flux and resolution at various wavelengths, for maximizing photon flux at the experimental station. The beamline operates in a windowless mode in an ultra high vacuum environment using sputter ion pumps. Residual gas analyzers are installed before the pre mirror and before the experimental station to check the quality of vacuum. The sub sections of the beamlines are isolated using a combination of pneumatic and manual gate valves. Fast closing vacuum interlocks are provided for safety. To improve the beamline safety further, a comprehensive interlocking scheme is implemented. Operational status of the vacuum valves, fast closing shutter, sputter ion pumps, pneumatic pressure line, vacuum gauges, residual gas analyzer etc. is provided on a LABVIEW platform. At the experimental station, a beam spot of ~1 mm² with high photon flux (~10^{11} photons/sec) and moderate wavelength resolution ($\lambda/\Delta\lambda$ ~ 200-500) is obtained.

A soft X-ray reflectometer has been installed on this beamline [3]. A two axes goniometer is assembled using coaxially mounted two high vacuum compatible rotary stages of 160 mm diameter, and with an angular resolution of 0.0025^0. A spring-loaded sample holder is mounted on a motorized linear translation stage, which in turn is mounted

CP879, *Synchrotron Radiation Instrumentation: Ninth International Conference,*
edited by Jae-Young Choi and Seungyu Rah
© 2007 American Institute of Physics 978-0-7354-0373-4/07/$23.00

on one rotation stage. Sample holder reference surface is aligned to coincide with the rotation axes of the goniometer. The linear translation stage enables the measurement of direct beam intensity, as sample can be retracted from the beam path. A EUV photo diode detector is mounted on the other axis of the goniometer, with a detector arm of 200 mm. A gold pinhole of 2 mm diameter is inserted in the incident SR beam, just before the sample, and photocurrent from the pinhole is constantly monitored during measurement, for normalizing the incident SR intensity. Four segment soft X-ray photodiode, with a central aperture to pass the SR beam is installed before the reflectometer, to monitor the incident beam shape and position. The goniometer is mounted in a high vacuum cylindrical chamber. A glass window gate valve is mounted in the SR entrance port of the reflectometer station. When this valve is closed, the sample can be positioned and aligned at atmospheric pressure, using visible portion of the SR radiation. The reflectometer station is evacuated using a turbo molecular pump. Due to the presence of goniometer, stepping motors, electrical wires etc. inside the chamber, the vacuum in the reflectometer station is limited to $\sim 5 \times 10^{-7}$ mbar. To interface this with ultra high vacuum requirement of the beamline, a differential pumping system is installed between the post mirror and the reflectometer chamber. The data acquisition, analysis and hardware control are done on LABVIEW platform. Angle and wavelength dependent reflectivity measurements from grazing to near normal incidence are possible on the reflectometer station.

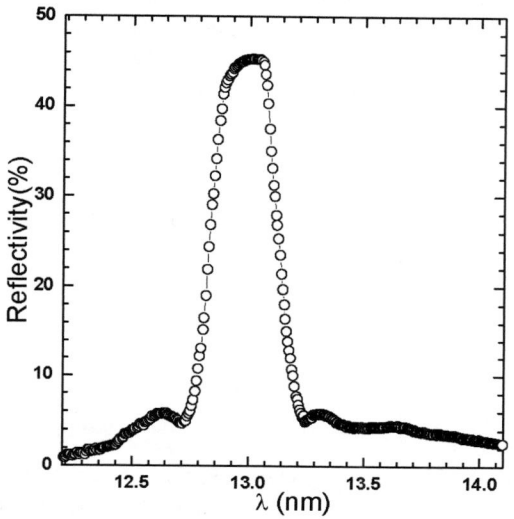

FIGURE 1. The wavelength scan of Mo/Si ML at 45^0 **FIGURE 2.** Peak reflectivity of Mo/Si ML (N=30)

The beamline is used for a variety of research activities in X-ray optics e.g. determination of optical constants [4,5], surface interface characterization [6], testing of X-ray multilayer performance [7]. Studies on Mo/Si multilayers near the silicon L absorption edge are done, with an aim to develop optics for EUV lithography and polarimetry. A typical reflectivity scan of Mo/Si ML at an incidence angle 45^0 is shown in Fig. 1. Peak reflectivity from 11.5 to 16 nm region is shown in Fig. 2. This multilayer structure is planned to be coated on an AXUV 100 photodiode, with the aim of doing polarization studies. Recently another experimental station on this beamline has been setup for studying photo ionization and related processes in gases targets [8]. A vacuum ultra violet polarimeter has been designed for installation at this beamline [9] and this is now under fabrication.

SOFT/DEEP X-RAY LITHOGRAPHY BEAMLINE ON INDUS-2

A dedicated beamline for soft and deep X-ray lithography (SDXRL) is being built for research on (i) high aspect ratio, three-dimensional MEMS structures using hard X-rays and (ii) high spatial resolution (<100nm) structures using soft X-rays. SDXRL beamline is designed to operate between 1.5 keV to 20keV energy range. The main design considerations are wide energy range, high flux in the appropriate X-ray range, and a homogeneous intensity profile (± 3%) in the horizontal plane. Penumbral blur and runout error are optimized in order to achieve minimum feature size of few tens of nanometers. The beamline is designed for 1.5keV where highest spatial resolution can be achieved due to reduced photoelectron blur. The selection of various energy domains is achieved with combinations of filters and mirrors. The beamline design allows continuous change of spectral range of interest [10, 11].

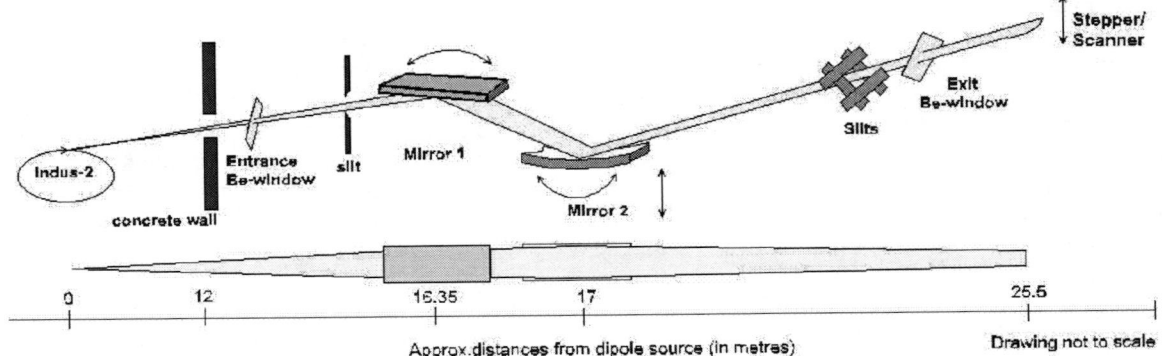

FIGURE 3. Optical design layout of SDXRL beamline.

Optical design of SDXRL beamline is shown in Fig. 3. The salient features of the beamline are given in Table 1. Mirror 1 is a plane mirror, which acts as a power and high energy (harmonic) filter. Mirror 2 is a torroidal, which is used to collimate beam horizontally and to focus beam in vertical direction. Beamline accepts 5 mrad horizontal and 0.83mrad (@1.5keV) vertical. Mirror 1 undergoes pitch motion and mirror 2 undergoes pitch motion plus vertical translation. The vertical translation of mirror 2 is required to keep the beam height fixed with respect to undeviated optic axis at the experimental station. The distance between the two mirrors is minimized, so that both the mirrors can be housed in a single vacuum chamber.

TABLE 1. Salient features of SDXRL beamline.

Energy range	1.5 keV – 20 keV
Reflecting mirrors	Two mirrors (1 plane @ 16.35m and 1 torroidal @ 17m)
	Pt coated Si-substrate with side-cooling arrangement
Mirror sizes	100mm x 650mm
Beamline acceptance	Horizontal 5 mrad and Vertical 0.83 mrad 4σ (@1.5 keV)
Beam size at sample	55 (Horizontal) x 2 (Vertical) mm^2
Angular Range	0.16 – 2.8 degrees
Filters	Be, C, Al etc.

Taking into account the figure errors, thermal slope and misalignment errors of the optical elements, detailed ray tracing simulations are done using RAY [12], SHADOW and SHADOW VUI [13, 14]. Figure 4 shows the point diagram of the SR beam at experimental station. The size of the beam is 2 mm vertical and 55 mm horizontal, with horizontal intensity uniformity within ±3%. For exposing photo-resist at mask-wafer stage a wide beam with high spectral power is required. Figure 5 gives the energy-power spectra for various mirror settings and Be-window

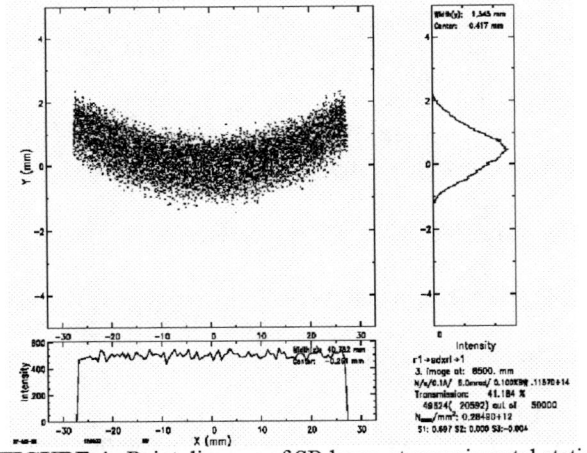

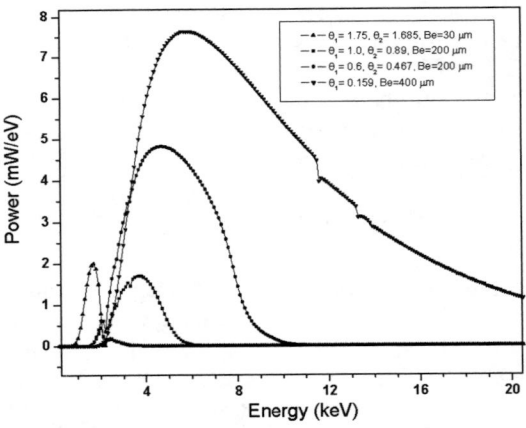

FIGURE 4. Point diagram of SR beam at experimental station at 1.5 keV mirror settings.

FIGURE 5.   Energy-Power spectrum offered by SDXRL beamline.

1476

thicknesses at 300mA beam current. Slope error values of 5 mrad (meridional) and 25 mrad (sagittal) for mirrors give acceptable beam quality at the experimental station.

Thermal bump is introduced on the mirror surface due to high absorbed heat load. On mirror 1 and 2 the maximum absorbed power densities are 770 and 0.2 mW/mm^2 respectively. Side cooling arrangement is selected due to ease in mechanical design. Furthermore, the side cooling offers the advantage that thermal gradient is along the sagittal direction. For silicon mirror substrates of thickness 50mm, analysis for temperature rise and thermal bump is done using finite element methods (FEM). For mirror 1, temperature rise is ~3 ^0C. Effect of thermal bump on the beamline performance is simulated. The rise in temperature on mirror substrate does not produce any appreciable effects on image shape. FEM analyses are also done for the Beryllium windows and edge filters [15].

The beamline will be operated in high vacuum (~ 1 x 10^{-8} mbar) and mask wafer stage will be kept in He-environment to avoid heat load on the mask-resist. Storage ring vacuum will be protected with the help of fast closing shutter and gate valves and acoustic delay line.

Initially the beamline will be operated at energies above 4 keV. This mode will allow the use of 200 micron Be-windows for isolating the storage ring vacuum. This would allow research on fabricating deep etch structures. In the first phase of the utilization of this beamline, the experimental station will be assembled using high precision rotary and linear stages. The experimental station will be housed in a clean room. Near the experimental station, a scanning electron microscope and an optical microscope will be installed. Support facilities like electron beam writer, reactive ion beam etching, plasma enhanced chemical vapor deposition, and electroforming etc. will be setup in phases.

To summarize, reflectivity beamline on Indus-1 is a powerful facility for developing optical elements for EUV lithography and for undertaking a variety of other research studies. SDXRL beamline will be useful for undertaking research programs on the development of MEMS structures using hard X-rays and high spatial resolution structures using soft X-rays.

ACKNOWLEDGMENT

K.J.S. Sawhney (DLS, UK) was actively associated with the reflectivity beamline on Indus-1 and for planning of SDXRL beamline. We acknowledge his contributions.

REFERENCES

1. R.V. Nandedkar and G. Singh, *Synchrotron Radiation News* **16(5)**, 43-48 (2003).
2. R.V. Nandedkar, K.J.S. Sawhney, G.S. Lodha, A. Verma, V.K. Raghuvanshi, A.K. Sinha, M.H. Modi and M. Nayak, *Current Science* **82(3)**, 298-304 (2002).
3. G.S. Lodha, M.H. Modi, V.K. Raghuvanshi, K.J.S. Sawhney, and R.V. Nandedkar, *Synchrotron Radiation News* **17(2)**, 33-35 (2004).
4. Pragya Tripathi, G.S. Lodha, M.H. Modi, A.K. Sinha, K.J.S. Sawhney and R.V. Nandedkar, *Optics Communications* **211**, 215-223 (2002).
5. Mohammed H. Modi, Gyanendra S. Lodha, Kawal Jeet S. Sawhney, and Rajendra V. Nandedkar, *Applied Optics* **42(34)**, 6939-6944 (2003).
6. M.H. Modi, G.S. Lodha, M. Nayak, A.K. Sinha and R.V. Nandedkar, *Physica B* **325**, 272-280 (2003).
7. Mohammed H. Modi, G.S. Lodha, S.R. Naik A.K. Srivastava and Rajendra V. Nandedkar, *Thin Solids Films* **503**,115-120(2006)
8. B. Bapat, R.K. Singh, K.P. Subramanian and G.S. Lodha, *Radiation Physics and Chemistry* **74**, 71-75 (2005).
9. S.R. Naik and G.S. Lodha, *Nuclear Instruments and Methods in Physics Research A* **560**, 211-218 (2006).
10. E.Di. Fabrizio, A Nucara, M. Gentilli and R. Cingolani, *Rev. of Sci. Inst.* **70(3)**, 1605-1613 (1999).
11. V. P. Dhamgaye and G. S. Lodha, RRCAT Internal Report, 12-2004 (2004).
12. F. Schafers, BESSY Internal Report, TB 202/96 (1996).
13. B. Lai and F. Cerrina, *Nuclear Instruments and Methods in Physics Research A* **246**, 337(1986).
14. M. Sanchez Del Rio and R.J. Dejus, *SPIE proceedings* **3448**, 340-345 (1998)
15. V.P. Dhamgaye, A.K. Sinha and G.S. Lodha, SUMRD-RRCAT Internal Report (2005).

Actinic Mask Inspection using EUV Microscope

Hiroo Kinoshita [1,4], Kazuhiro Hamamoto [1,4], Yuzuru Tanaka [1,4],
Noriyuki Sakaya [2,4], Morio Hosoya [2,4], Tsutomu Shoki [2,4], Donggun Lee [3]
and Takeo Watanabe [1,4]

[1] *Laboratory of Advanced Science and Technology for Industry, University of Hyogo,3-1-2Kouto,
Kamigori-cho, Ako-gun, Hyogo 678-1205,Japan*
[2] *HOYA Corporation Electro-optics Company R&D Center 3-3-1 Musashino, Akishima-shi, Tokyo 196-8510, Japan*
[3] *Samsung Electronic Co.,Ltd*
[4] *CREST, JST Kawaguchi Center bldg., 4-1-8 Hon-cho, Kawaguzhi-shi, Saitama, Japan*

Abstract. We constructed an EUV microscope (EUVM) for actinic mask inspection which consists of Schwarzschild optics (NA0.3, 30X) and an X-ray zooming tube.This system has been used to inspect both complete EUVL masks as well as plain Mo/Si-coated glass substrates. Based on imagery of a 250-nm width pattern, the resolution of the EUVM can be estimated to be 50 nm or less. The EUVM has also been used to inspect programmed bump phase defects in the EUV mask. Programmed phase defects with widths of 90 nm, 100 nm, 110 nm, a bump of 5 nm and a length of 400 m have been clearly observed. Moreover, the EUVM resolved programmed pit phase defects of 100 nm-wide and 2 nm in depth. The EUVM described here has enabled the of topological defects within a multilayer film. These results show that it is possible to image the internal reflectance distribution of a multilayer using an EUV microscope, without being dependent on surface figure.

Keywords: EUV Lithography, EUV microscope, mask, defect
PACS: 85.40.Hp

INTRODUCTION

Extreme ultraviolet lithography (EUVL) has been proposed as a next generation lithography at the 32 nm node around 2009 [1]. Defect-free mask fabrication is one of the critical issues to lead EUVL into production. Based on the ITRS [2], defect widths of less than 20 nm are required at the 32 nm node.

There are two types of defects in an EUVL mask. One is amplitude defect and another is a phase defect. Amplitude defects are either particles on a surface of the multilayer or flaws in the multilayer. These defects can be detected directly by measuring the intensity of EUV light which is scattered from the defects. On the other hand, phase defects are produced when the multilayer is deposited over a bump or pit on the substrate. These phase defects are swellings and depressions on the surface of the multilayer.

There are two techniques of detecting a small swelling on the surface: using deep ultraviolet (DUV) light and using EUV light at the exposure wavelength. Mask defect inspection using DUV light is a conventional method in optical lithography [3, 4]. However, this inspection is difficult to detect defects on a EUVL mask because defects to be detected are much smaller than the inspection wavelength. So, it is extremely difficult to detect defects of 50 nm or less using DUV light. Furthermore, since the characteristics of phase defects depend on exposure wavelength, it is necessary to observe phase defects with the same wavelength as the exposure wavelength.

CP879, *Synchrotron Radiation Instrumentation: Ninth International Conference,*
edited by Jae-Young Choi and Seungyu Rah
© 2007 American Institute of Physics 978-0-7354-0373-4/07/$23.00

EUV MICROSCOPE

EUV Microscope

Figure 1 shows the configuration of the actinic EUV microscope installed at the BL-3 beamline in the NewSUBARU SR facility. It consists of Schwarzschild optics, a Mirau interferometer for phase-shift interference measurement, an X-Y sample stage, a focus detector, an X-ray zooming tube connected to a CCD camera, and an image processing computer. It is installed in a vacuum chamber evacuated down to $1 \times 10\text{-}5$ Pa, and the vacuum chamber is set on a vibration isolation table.

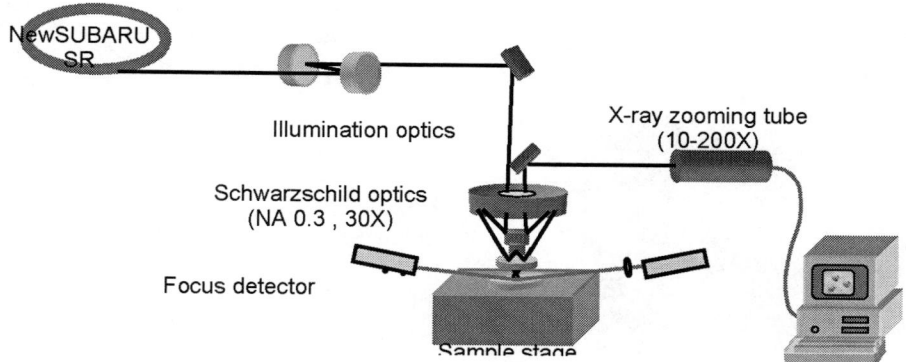

FIGURE 1. Configuration of the EUV microscope

Schwarzschild Optics

The system employs Schwarzschild optics with 0.3 NA and 30X magnification. Our simulation predicted that the system can resolve 10-nm-wide isolated lines under practical illumination conditions. The figure error of the mirrors was less than 0.4 nm and the mid-frequency surface roughness was less than 0.15 nm. These mirrors, made of Zerodur, were fabricated by ASML Tinsley. Mo/Si multilayers were coated on these optics by an X-ray instruments company in Russia. D-space matching of less than 0.01 nm has been achieved at the wavelength of 13.5 nm. The wave-front error of the Schwarzschild optics after assembly was measured with a Fizeau interferometer (ZYGO GPI) and found to be about 2 nm (rms). This optics was installed in an optical housing made of Invar to prevent the thermal expansion effect.

X-ray Zooming Tube

The X-ray imaging system is composed of an X-ray zooming tube, a CCD camera, and an image processing computer. The mask image is projected to an X-ray zooming tube (Kawasaki Heavy Industries Co., Ltd.) with electromagnetic lenses that can be tuned to vary the magnification in the range from 10 to 200. Therefore, considering the magnification of the Schwarzschild optics, the total magnification of the microscope system is from 300X to 6000X. The resolution of the X-ray zooming tube is 0.3 m on a CsI photocathode. Defects of 10 nm in size are magnified by the Schwarzschild optics to 300 nm. Thus, the resolution of the X-ray zooming tube is sufficient to enable the detection of 10 nm defects and to produce diffraction-limited images of mask patterns. The X-ray zooming tube has a field of view of 1.5 mm x 1.5 mm at the CsI photocathode, which corresponds to 50 m x 50 m on the sample surface. The magnified EUV microscope images are taken into a CCD camera and displayed on the screen of the image processing computer, and the image data can be stored in the computer.

EXPERIMENTAL

We observed the finished EUVL mask and mask blanks with programmed substrate defects using the EUV microscope without the Mirau interferometer.

Finished Mask Observation [5-8]

A finished EUVL mask was observed using the EUV microscope. The mask with a 6025-format substrate (ULE glass, Corning Inc.) was fabricated by HOYA Corporation. Figure 2 shows an image of the mask with 300 nm-wide isolated lines. The white part is a Si surface of the Mo/Si multilayer and the dark part is the absorber material of TaBN. Clearly, the EUV microscope is capable of resolving 300 nm-wide absorber patterns on a mask, which corresponds to 75 nm-wide patterns on the wafer, on the assumption of 1/4 magnification for commercial exposure tools. The detected resolution of 50 nm is estimated from the contrast at the edge of the pattern (see Fig. 3).

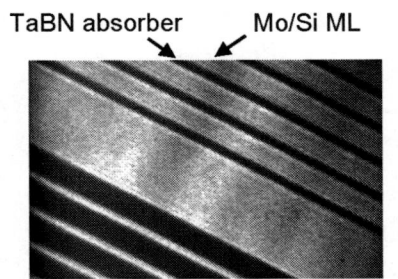

FIGURE 2. EUV microscope image of 300 nm-wide absorber pattern.

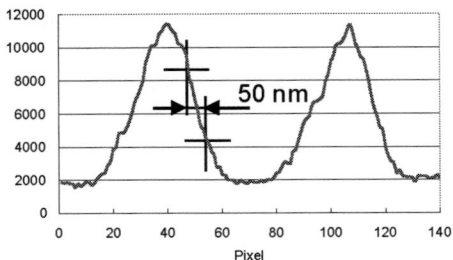

FIGURE 3. Intensity distribution at the edge of the absorber.

Programmed Phase Defect Observation

Bump Defect [9]

Mask blanks with programmed phase defects on the substrates (ULE grass) were made. The programmed bump defects were fabricated on CrN layer deposited on grass substrates. And Mo/Si multilayer is fabricated on CrN later which has bump defects.

Figure 4 shows a EUV microscope image of program phase defects of 1 m-wide dots. In spite of astigmatic aberration of the illumination optics, phase defects of dots with 1 m-wide and 5 nm-high can be confirmed.

Figure5 shows EUV microscope images of phase defects of 90 nm, 100 nm and 110 nm-wide isolated lines. The height and the length of the lines are 5 nm and 500 m, respectively. It is thought that the phase defect of 5 nm height was printable.

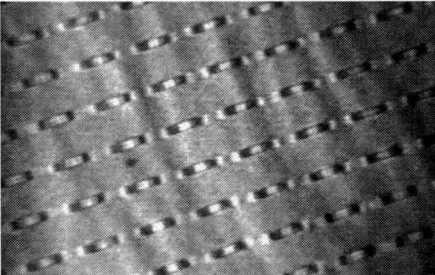

FIGURE 4. EUV microscope image of programmed phase defect of 1 m dot.

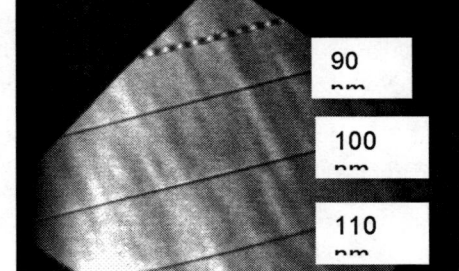

FIGURE 5. EUV microscope image of programmed phase defect of 90 nm, 100 nm, 110 nm isolated lines.

Bump defects are shown in previous section. But it is thought that bump defect will be removed by cleaning technology. Pit defects are significant defect of mask substrate. Pit defects were covered by Mo/Si multilayer coating after dry etching of Si wafer substrate. In this case, there is no buffer layer under the multilayer. Several depth of the defects were fabricated. Figure 6 and 7 show atomic force microscope (AFM) image of Si substrate pattern of 500 nm-wide lines and spaces and 100 nm-wide isolated line of pit defects with 5 nm-depth before multilayer coating and EUV microscope image of after multilayer coating, respectively. Phase defects of 5 nm-depth have the capability of printable.

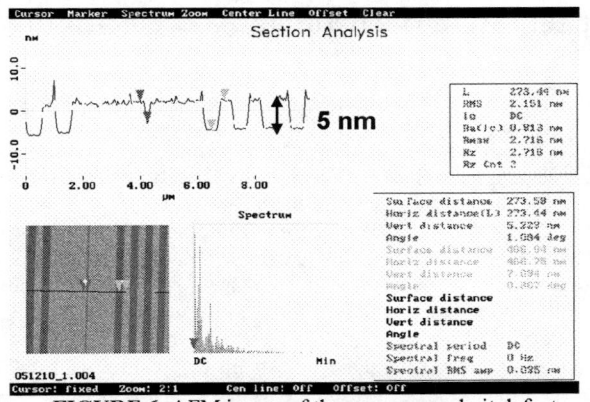
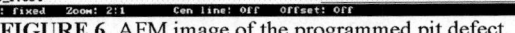

FIGURE 6. AFM image of the programmed pit defect.

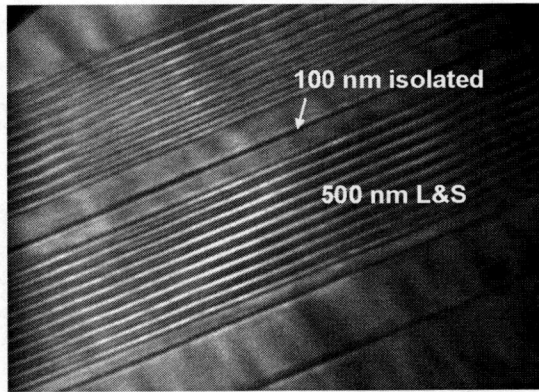

FIGURE 7. EUV microscope image of programmed defect.

Thus, we observed a programmed defect using the EUV microscope, and confirmed to be able to observe intensity distributions according to the change of the reflectivity in the defect.

CONCLUSION

We have constructed an EUV microscope system for at-wavelength areal image mask inspection. Using the EUV microscope, the image of the phase defect due to a programmed bump defect of 90 nm wide and programmed pit defect of 100 nm wide were observed using the EUV microscope without a Mirau interferometer. Furthermore, for programmed defects of broad size, the edge of the program defect is observed. Using the EUV microscope, it succeeded in observation of the topological defect structure inside the multilayer. These results show that our system has a capability to detect internal reflectivity distributions, without depending on surface perturbation.

REFERENCES

1. T. Watanabe, H. Kinoshita, K. Hamamoto, M. Hosoya, T. Shoki, H. Hada, H. Komano and S. Okazaki: Jpn. J. Appl. Phys. 41 (2002) 4105.
2. Semiconductor Industry Association, "International Technology Roadmap for Semiconductors: 2004 update: Lithography" Austin TX, International SEMATECH.
3. T. Hashimoto, H. Yamanashi, S. Miyagaki and I. Nishiyama: Proc. SPIE 5446 (2004) 860.
4. D. Pettibone and S. Stokowski: Proc. SPIE 5567 (2004) 807.
5. T. Haga, H. Takenaka and M. Fukuda: J. Vac. Sci. Technol. B 18 (2000) 2916.
6. T. Haga , H. Kinoshita, K. Hamamoto, S. Takada, N. Kazui, S. Kakunai, H. Tsubakino and T. Watanabe: Jpn, J. Appl. Phys. 42 (2003) 3771.
7. H. Kinoshita, T. Haga, K. Hamamoto, S. Takada, N. Kazui, S. Kakunai, H,Tsubakino, T. Shoki, M. Endo and T. Watanabe: J. Vac. Sci. Technol. B 22 (2004) 264.
8. K. Hamamoto, Y. Tanaka, H. Kawashima, S. Y. Lee, N. Hosokawa, N. Sakaya, M. Hosoya, T. Shoki, T. Watanabe and H. Kinoshita: Jpn. J. Appl. Phys. 44 (2005) 5474.
9. K. Hamamoto, Y. Tanaka, S. Y. Lee, N. Hosokawa, N. Sakaya, M. Hosoya, T. Shoki, T. Watanabe and H. Kinoshita: J. Vac. Sci. Technol. B 23 (2005) 2852.

High-Precision Multilayer Coatings and Reflectometry for EUV Lithography Optics

Stefan Braun, Peter Gawlitza, Maik Menzel, Stefan Schädlich, Andreas Leson

IWS Dresden, Fraunhofer-Institut für Werkstoff- und Strahltechnik, 01277 Dresden, Winterbergstr. 28, Germany

Abstract. The topic of this paper is the fabrication and characterization of EUV reflective coatings based on molybdenum/silicon (Mo/Si) multilayers. For the fabrication of such nanometer structures, the technologies of magnetron sputter deposition (MSD) and ion beam sputter deposition (IBSD) are used in IWS Dresden. The main challenges for extreme ultraviolet (EUV) optics are high reflectance, precise thickness profiles, low internal stress and long-term stability. Reflectances > 70 %, uniformities > 99.9 % and overall internal stresses < 20 MPa have been reached. In addition to sophisticated deposition technologies, precise metrology tools are mandatory for the characterization of the coatings. Together with a number of partners, IWS Dresden has developed a stand-alone EUV reflectometer that makes it possible to measure EUV reflectance R and peak position λ on substrates with diameters of up to 500 mm. Current improvements of the reflectometer resulted in differences compared to calibrated measurements at PTB/BESSY of ΔR = -0.2...-0.6 % and $\Delta\lambda_{50}$ = +4...+9 pm.

Keywords: extreme ultraviolet (EUV) lithography, optics, Mo/Si multilayers, MSD, IBSD, reflectometry
PACS: 07.60.Hv, 07.85.Qe, 41.50.+h, 61.10.Kw, 68.35.Ct, 68.60.-p, 68.65.Ac, 81.15.Cd

INTRODUCTION

According to the semiconductor roadmap, the density of structures on integrated circuits has to be doubled every 18 month (Moores law). In order to follow this forecast, in a few years it is necessary to change the operation wavelength used for the illumination process from 193 nm to the EUV range with λ = 13.5 nm [1]. Since no materials with high EUV transmittances are available, mirrors instead of lenses have to be used in the optical system. Such mirrors consist of superpolished substrates and high-reflective Mo/Si multilayer coatings.

Since the first attempts to grow Mo/Si multilayers for EUV reflection coatings [2], within the last 20 years remarkable progress has been made in improving and understanding structure and growth mechanisms of the nanometer structures. One of the main goals has been to improve the reflectance of the coatings close to the theoretical limit of R = 75 %. Because of the fact that approximately ten reflections are necessary in a realistic EUV illumination system any reflectance improvement of every single mirror remarkably improves the overall throughput of the optical system. Independent from the deposition technology used for the production of the Mo/Si multilayers reflectances close to R = 70 % can be reached (evaporation [3,4], MSD [5-8], IBSD [9]). In this paper we will give a short review about the current status of EUV reflection coatings made by MSD and present first results obtained with a new large-area IBSD machine.

In connection with the development of EUV reflection coatings IWS Dresden together with a number of different partners (Carl Zeiss SMT AG Oberkochen, Physikalisch-Technische Bundesanstalt PTB Berlin, BESTEC GmbH Berlin, Max-Born-Institut Berlin, AIS GmbH Dresden) has built an EUV reflectometer suitable for the characterization of optics with diameters of up to 500 mm [10]. The working principle is based on set-ups that have been published in the literature already several years ago [11,12]. In this paper we will summarize the short-term reproducibility of the measured peak positions and reflectances. Additionally we will show the comparison of reflectance spectra of multilayers with different EUV peak positions between 13 and 14 nm measured at PTB/BESSY II and our reflectometer.

CP879, *Synchrotron Radiation Instrumentation: Ninth International Conference,*
edited by Jae-Young Choi and Seungyu Rah
© 2007 American Institute of Physics 978-0-7354-0373-4/07/$23.00

EUV REFLECTION COATINGS

As mentioned above several deposition techniques can be used to coat substrates with highly reflective Mo/Si multilayers. In the past we intensively studied pulsed laser deposition (PLD) and MSD. With both technologies tiny barrier layers were applied between the molybdenum and silicon single layers in order to improve the EUV reflectance [6]. Concerning highest possible reflectances MSD has been shown to be the most suitable technology with R = 70.1 % (angle of incidence α = 1.5 °, peak wavelength λ = 13.3 nm) and R = 71.4 % (α = 22.5 °, λ = 12.5 nm). The layer thickness uniformities that can be obtained with MSD are > 99.9 % on substrates with diameters of 150 mm. By applying stress compensation layers the overall stress of the coatings could be reduced down to $|\sigma|$ < 20 MPa [13]. Long-term measurements made from PTB at BESSY II show no reflectance altering and no EUV peak shift within a time period of 2 years if barrier layers are used. Without barrier layers the center wavelength decreases by 10-15 pm [14].

While MSD is the favorite option for the deposition of EUV projection optics, IBSD will probably be applied for the deposition of Mo/Si multilayers on mask substrates and on collection optics. The advantages of IBSD compared to MSD are the lower defect levels and the lower sensitivity of the multilayer reflectance to substrate roughness [9]. The new approach of the IBSD machine that we have developed together with the company Roth&Rau Wüstenbrand, Germany, is that linear sources with a length of 400 mm are used (Fig. 1). This concept ensures the scalability of the process to even larger substrates. Today, we are able to handle substrates with diameters of up to 200 mm via the load-lock. Larger substrates with lengths of up to 500 mm or with diameters of up to 450 mm have to be introduced in the deposition chamber via the front door. First reflectance and uniformity results of IBSD Mo/Si multilayers are shown in Fig. 2.

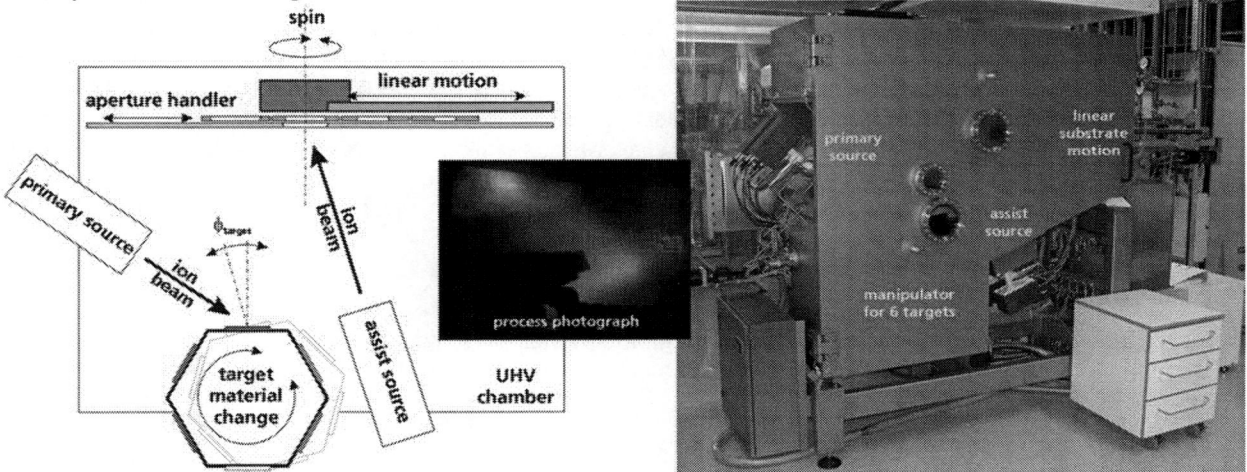

FIGURE 1. Front view of the principle of the large-area IBSD machine developed by the company Roth&Rau Wüstenbrand and IWS Dresden. The chamber is equipped with two linear ECR (electron cyclotron resonance) ion beam sources of 400 mm length. Substrates with diameters of up to 200 mm can be handled via the load-lock at the backside of the deposition chamber. Larger substrates with lengths of up to 500 mm or diameters of up to 450 mm have to be introduced via the front door.

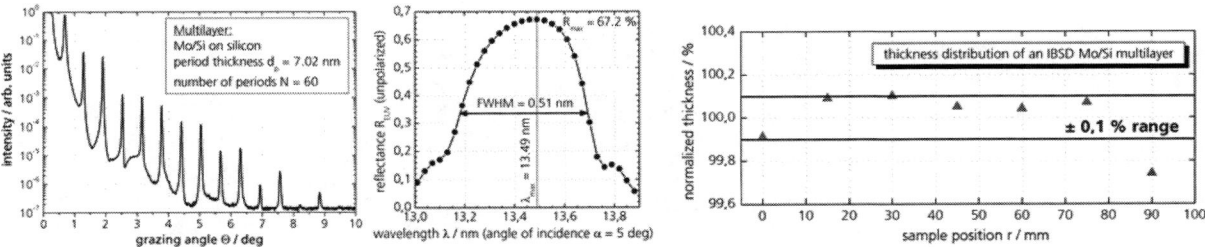

FIGURE 2. Initial results of IBSD Mo/Si multilayers without barrier layers: left hand side: Cu-Kα and EUV reflectographs, right hand side: uniformity obtained on substrates with diameters of 200 mm.

EUV REFLECTOMETRY

Simultaneously to the development of EUV reflective coatings the corresponding characterization methods like reflectometry have to be developed. The availability of EUV light at synchrotron beamlines offers direct access to reflectance measurements [15,16]. However, with the commercialization of EUV components it became necessary to develop laboratory reflectometers that are independent of synchrotrons and directly available next to the coating machine. One successful approach in replacing synchrotron light was to use laser pulse plasma (LPP) sources with gold or tungsten targets [11,12]. This concept was applied to install a laboratory reflectometer that can be used to characterize large-scale optics with diameters of up to 500 mm and masses of up to 30 kg [10].

Short-term reproducibility

Due to the fact that the LPP source of the laboratory reflectometer has not the same stability as a synchrotron source, it requires a lot more effort to perform absolute measurements of reflectance R and peak wavelength λ with the necessary precision. Practically it is much easier to perform EUV reflectance measurements on calibrated samples with known values of R and λ and to consider possible shifts of R and λ for the measurements of unknown samples. The accuracy that can be obtained by this method is then limited by the total uncertainty of the calibrated measurements at PTB/BESSY II and by the short-term reproducibility of the laboratory reflectometer. The carefully analyzed values of the total relative uncertainty values from PTB are 0.14 % for the peak reflectance and 0.014 % for the center wavelength λ_{50} [15].

With the laboratory reflectometer we have to take into account an additional uncertainty arising from the short-term reproducibility of the measurements. In order to quantify these values we regularly perform repeatability experiments where a calibrated sample is measured five times on the same position. A typical result of this procedure is shown in Fig. 3. The relative standard deviations are $\sigma_{relative, R} = 0.29$ % and $\sigma_{relative, \lambda50} = 0.00085$ %. Therefore the total relative uncertainties of the laboratory reflectometer are increased compared to the PTB values to 0.32 % for the peak reflectance and 0.014 % for the center wavelength λ_{50}. However these values are only valid for flat samples. With curved samples the reflectance of convex and concave surfaces is under- and overestimated, respectively. The quantification of the error contributions depending on the radius of curvature is still under investigations.

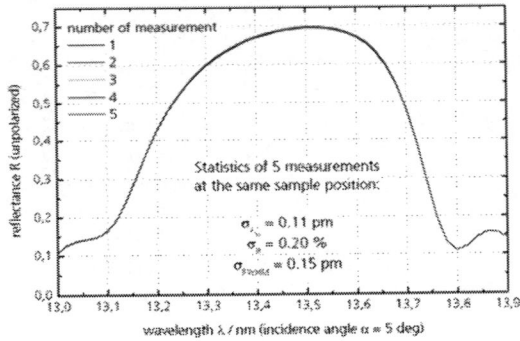

FIGURE 3: Short-term reproducibility of EUV reflectance spectra measured with the laboratory reflectometer. Five identical measurements at the same sample position show standard deviations of the wavelength λ_{50}, reflectance R and half width FWHM of $\sigma_{\lambda50} = 0.11$ pm, $\sigma_R = 0.20$ % and $\sigma_{FWHM} = 0.15$ pm.

Comparison of measurements from PTB/BESSY II and IWS

In order to get accurate information about the peak reflectance R and center wavelength λ_{50} of unknown samples using the laboratory reflectometer, reflectance spectra of PTB-calibrated samples have to be determined directly before or after the measurements of the unknown samples. Due to the fact that reflectance and wavelength differences between synchrotron and laboratory reflectometry generally also depend on the center wavelength itself we have fabricated multilayers with different period thicknesses. Their reflectance peaks are between 13 and 14 nm (Fig. 4). By interpolating $\Delta\lambda$ and ΔR we are able to calculate functions $\Delta\lambda = \Delta\lambda(\lambda)$ and $\Delta R = \Delta R(\lambda)$. Of course these functions have to be updated regularly, especially after hardware changes of the source (i.e. target exchange). The observed differences of the half widths of the BRAGG peaks can be explained with the different polarization states of the light. With perpendicular polarized EUV light larger FWHM values are expected than with unpolarized light.

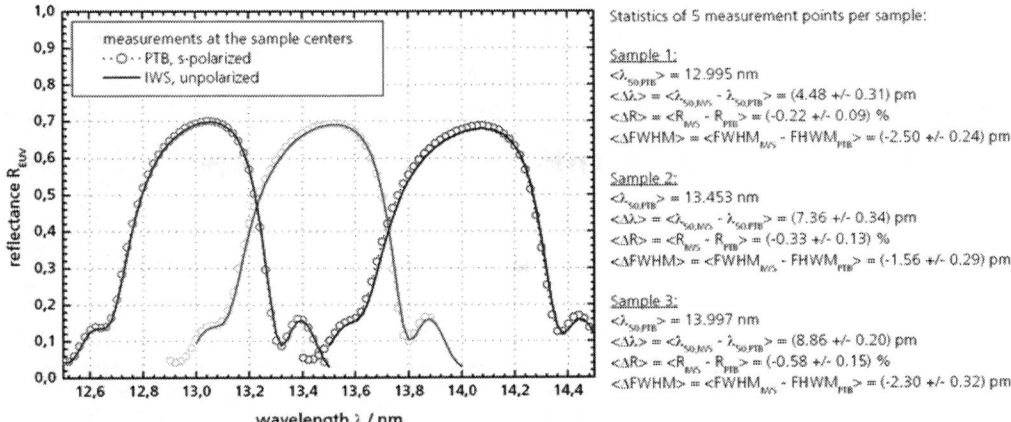

FIGURE 4. Comparison of reflectance measurements made at PTB/BESSY II and IWS with the laboratory reflectometer. Three multilayer samples with different EUV peak positions have been measured at 5 positions per sample. The mean values of the differences $\Delta\lambda$, ΔR and $\Delta FWHM$ between the measurements have been compared.

In addition to the absolute differences of λ_{50}, R and FWHM between PTB and IWS measurements, the standard deviations of the individual differences should also be noticed. From these values an estimation can be derived how close IWS measurements can predict PTB measurements. In agreement with the reproducibility tests the largest deviation is observed for the differences of the reflectance values. The standard deviation of the absolute differences between PTB and IWS values are between 0.09 and 0.15 % and are predominantly caused by the reproducibility of the IWS measurements. Much smaller deviations occur for the differences of λ_{50} and FWHM. Both are in the order of 0.20-0.35 pm. Hence, the current prediction limits of PTB measurements of λ_{50} and FWHM are given by relative standard deviations of approximately 0.003 % and 0.06 %, respectively.

ACKNOWLEDGMENTS

The EUV reflectometer has been financially supported by the Bundesministerium für Bildung und Forschung and Carl Zeiss SMT AG, Oberkochen. We would like to thank M. Nestler (Roth&Rau) for the fruitful cooperation during the IBSD machine development and F. Scholze (PTB) for the EUV reflectance measurements.

REFERENCES

1. D. Atwood, these proceedings, session K3: Lithography and Micromachining
2. T. W. Barbee, S. Mrowka, M. C. Hettrick, Applied Optics 24, 1985, pp. 883
3. E. Louis, A. E. Yakshin, P. C. Görts, S. Abdali, E. L. G. Maas, R. Stuik, F. Bijkerk, D. Schmitz, F. Scholze, G. Ulm, M. Haidl, Proceedings of SPIE 3676, 1999, pp. 844
4. U. Kleineberg, T. Westerwalbesloh, W. Hachmann, U. Heinzmann, J. Tümmler, F. Scholze, G. Ulm, S. Müllender, Thin Solid Films 433, 2003, pp. 230-236
5. S. Bajt, J. Alameda, T. Barbee Jr., W. M. Clift, J. A. Folta, B. Kauffman, E. Spiller, Optical engineering 41, 2002, pp. 1797
6. S. Braun, H. Mai, M. Moss, R. Scholz, A. Leson, Japanese Journal of Applied Physics 41, 2002, pp. 4074-4081
7. M. Shiraishi, N. Kandaka, K. Murakami, Proceedings of SPIE 5374, 2004, pp. 104-111
8. T. Feigl, S. Yulin, N. Benoit, N. Kaiser, Optics and Precision Engineering 13, 2005, pp. 421-429
9. E. Spiller, S. L. Baker, P. B. Mirkarimi, V. Sperry, E. M. Gullikson, D. G. Stearns, Applied Optics 42, 2003, pp. 4049-58
10. L. van Loyen, T. Böttger, S. Braun, H. Mai, A. Leson, F. Scholze, J. Tümmler, G. Ulm, H. Legall, P. V. Nickles, W. Sandner, H. Stiel, C. Rempel, M. Schulze, J. Brutscher, F. Macco, S. Müllender, Proceedings of SPIE 5038, 2003, pp. 12-21
11. D. L. Windt, W. K. Waskiewicz, Proceedings of SPIE 1547, 1991, pp. 144
12. E. M. Gullikson, J. H. Underwood, P. C. Batson, V. Nikitin, Journal of X-ray Science and Technology 3, 1992, pp. 283
13. M. Moss, T. Böttger, S. Braun, T. Foltyn, A. Leson, Thin Solid Films 468, 2004, pp. 322-331
14. F. Scholze, C. Laubis, C. Buchholz, A. Fischer, S. Plöger, F. Scholz, H. Wagner, G. Ulm, 3rd International EUVL Symposium, 1.-4.11.2004, Miyazaki, Japan
15. J. Tümmler, H. Blume, G. Brandt, J. Eden, B. Meyer, H. Scherr, F. Scholz, F. Scholze, G. Ulm, Proceedings of SPIE 5037, 2003, pp. 265-273
16. E. M. Gullikson, S. Mrowka, B. B. Kaufmann, Proceedings of SPIE 4343, 2001, pp. 363-373

Optical and Thermal Comparative Study of a Soft X-Ray Lithography Beamline for the Australian Synchrotron

Matteo Altissimo[1*], David Wang[2] and Steve Wilkins[1]

[1] *CSIRO Manufacturing and Materials Technology, Gate 5 Normanby Road, 3168 Clayton (VIC), Australia.*
[2] *Australian Synchrotron Project, 800 Blackburn Road, 3168 Clayton (VIC), Australia.*

Abstract. In the last years we witnessed an increasing trend in miniaturization of electronic, mechanical, optical and magnetic components. Currently, the ultimate Critical Dimension (CD) of such components is rapidly approaching the 10 nm length scale, with a growing interest in integrating different components in complex systems. Advanced fabrication techniques are thus required to rapidly pattern substrates on the 10-100 nm length scale, and soft X-Ray Lithography (XRL) has proven its patterning ability down to 40 nm, with the possibility to be further extended. The construction and commissioning of the Australia Synchrotron, that will be operational in 2007, gives the possibility to build such a facility on a powerful third-generation synchrotron source. This work is a comparative study of three possible beamlines devoted to XRL, one of which could be installed at the Australian Synchrotron. The optical layout and the thermal load have been studied, under the constraints given by the storage ring, the ultimate patterning ability, the current mask technology and two possible exposure stations, aiming at replicating structures in the sub-100 nm level. Ray tracing simulations have been performed in order to assess the optical layout and the performances of the beamlines, which in principle can cover a wide lithographic window between 1 and 8 keV incoming photon energy.

Keywords: X-Ray Lithography, optical layout, thermal load.
PACS: 85.40.Hp

INTRODUCTION

X-Ray Lithography (XRL) was invented at the beginning of the 1970's [1] as a fast and simple parallel patterning technique. In the past 3 decades, it has shown its ability in patterning down to the 40 nm level and with possibilities to extend to the 25 nm node [2-4]. It has produced working DRAM at the 70 nm level [5], is currently used to fabricate photonic crystals both by industry [6] and by researchers [7], and also X-Ray Optics [8]. With the current trend in miniaturization, and with the above-mentioned possibilities in patterning ability, XRL is a valuable tool in nano-fabrication, especially as a possible bridge between top-down and bottom-up techniques. Usually, the first is considered to be patterning of structures starting from bulk materials, and the second is thought as assembling structures atom by atom or molecule by molecule.

Here we report on an XRL beamline design based on the one proposed by [9], which in our opinion provides a reasonable trade-off between cost and performance. The source chosen for the beamline is 6 mrad of the orbit from a Bending Magnet (BM) of the Australian Synchrotron, which will be relatively easy to handle in terms of thermal load on the beamline optical elements. These will be two beryllium windows, and two mirrors, one plane and the other toroidal. The first mirror will be plane and the second toroidal if the exposure station (stepper) will have a scanning stage, whereas the toroidal mirror will be installed upstream the plane one if the stepper will require a scanning beam.

In the second section, we will describe the optical setup of the beamline, and in the third we will calculate the thermal load on each optical element. The last section will provide a summary of these results.

---

for correspondence: matteo.altissimo@csiro.au

CP879, *Synchrotron Radiation Instrumentation: Ninth International Conference*,
edited by Jae-Young Choi and Seungyu Rah
© 2007 American Institute of Physics 978-0-7354-0373-4/07/$23.00

OPTICAL LAYOUT

The optical layout of the beamline is defined taking into account the difference in height between the e-beam orbit in the storage ring (1.3 m from the floor level), the exit port of the steppers (1.021 and 1.024 m from floor for the scanning stage and the scanning beam respectively), the closest possible distance to source (14m) and the maximum mask size (30x30 mm^2). These geometrical constraints have to be coupled with the constraints about the passing band, which should be centered around 1.7 keV, with a width of a few hundreds of eV. This energy window provides the best resolution in replicas [10,11]. To ensure the highest possible flux at the experimental station, Pt mirrors are preferred, because of the dependency of the critical angle for total external reflection on the material's density, which improves the mirror's reflectivity at the high glancing incidence angles needed for soft- XRL applications.

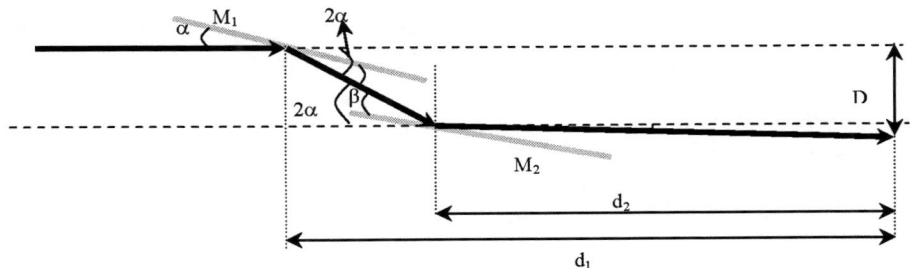

FIGURE 1. Optical layout of the beamline. D is the distance of the image from the optical axis, α and β are the glancing incidence angles on the first (M$_1$) and the second (M$_2$) mirror respectively; d$_1$ and d$_2$ are the distances of the first and second mirror from the image plane. The photon path is represented by the black, solid arrows.

Figure 1 shows the schematic of the beamline. M$_1$ and M$_2$ are first and second mirror, respectively; d$_1$ and d$_2$ are their relative distance from the wafer plane, α and β are the glancing incidence angles on first and second mirror and D is the distance of the beam from the optical axis at the wafer plane. Given the geometry in the figure, it is possible to see that:

$$\beta = \alpha + \frac{1}{2}\arctan\left[\frac{(d_1 - d_2)\tan(2\alpha) - D}{d_2}\right] \tag{1}$$

We set the mirrors distances from the wafer plane to 11 m and 9 meters, for first and second mirror respectively, and chose to design the beamline with the beam lowered by D=0.1 m from the orbit height. With this geometry, and according to Equation 1, a variation in α will define a different energy passing band. A corresponding β variation will let the beam through the beamline at the wafer, allowing in this way to perform both soft- and deep-XRL on the same beamline.

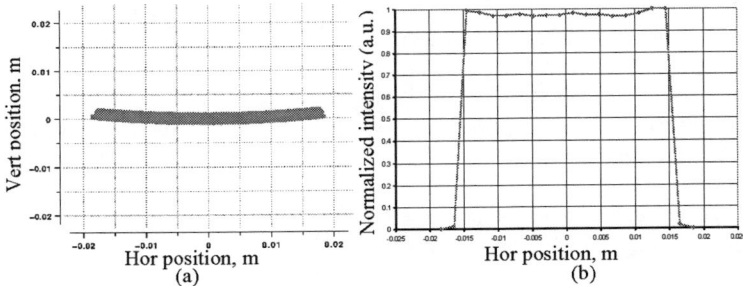

FIGURE 2. Results of ray tracing at the wafer plane for the fix-mirror configuration: (a) is an image of the beam at the wafer plane and (b) is a crosscut of (a), normalized to unity, to determine the beam uniformity, which is around ± 1.5%.

The ray tracing simulations have been performed with SHADOW [13], using the beamline parameters discussed throughout the text. Figure 2 shows the results for the configuration with fix mirrors. The toroidal mirror radii ρ and R are 0.499 m and 422.209 m respectively, and the beam size is around 2x34 mm (VERTxHOR) at wafer plane, as required by mask fabrication standards. The intensity distribution varies by ± 1.5% around its average value in the region between -0.015 and +0.015 m, thus giving a uniform exposure window 30 mm in size. The horizontal divergence of the beam (called *runout*) is ± 4.5 mrad, as given by SHADOW. In the vertical direction, the beam divergence will be averaged out by the scanning mechanism.

In the case of a scanning beam, we decided to analyze two different beamline configurations as described in Table 1, where ρ and R are the toroidal mirror radii of curvature for the two chosen glancing incidence angles, d_1 and d_2 are the distances between the first/second mirror to the wafer plane, while β_1 and β_2 are the extreme grazing incidence angles on the second mirror needed to scan the beam across a 50 mm writing field.

TABLE 1. Beamline configurations for the scanning mirror.

	ρ (m)	R (m)	d_1 (m)	d_2 (m)	β_1 (deg)	β_2 (deg)
α = 1.5 deg	0.4353	575.55	11	7.5	1.818	1.62
α = 1.8 deg	0.522	479.649	11	9	1.8822	1.7167

The toroidal mirror has been designed to provide uniform runout across the whole scanning distance, and its parameters have been calculated following the procedure outlined above. Figure 3 refers to the beam image at the wafer plane for two of the four cases reported in Table 1: Fig. 3(a,b) refers to the configuration with α = 1.5°, and α = 1.8° respectively, while Fig. 3(c) shows the average power impinging onto the mask for the two considered configurations, the solid line referring to α=1.5° and the dotted line referring to α=1.8°. The integrated power delivered to the mask turns out to be 712 mW for α=1.5°and 465 mW to α=1.8°. Our simulations have shown that there is no appreciable difference in the beam shape at the wafer plane. Furthermore the runout of the two configurations is the same, and is equal to ± 2.5 mrad on the 30 mm writing field over the whole vertical scanning range.

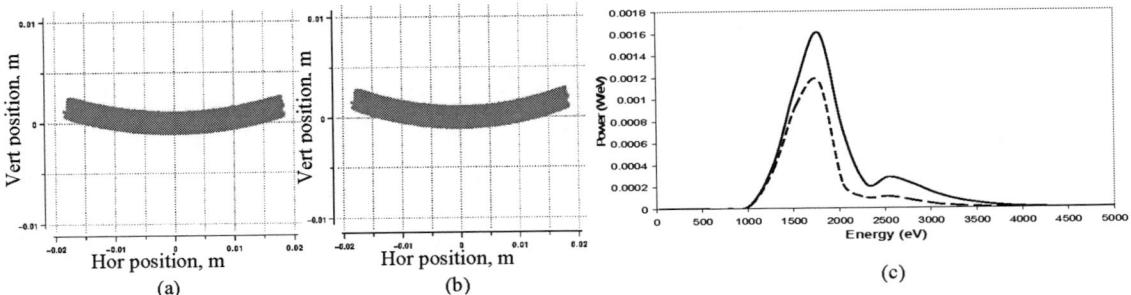

(a)   (b)   (c)

FIGURE 3. Results of ray tracing at the wafer plane for the scanning mirror configuration for α = 1.5° (a) and α = 1.8° (b); (c) shows the average power impinging on the mask

THERMAL LOAD

The analysis of the radiation-induced thermal load is necessary in order to understand the cooling requirements for each optical element in the beamline. Using one of SHADOWS utilities, we computed the power absorbed by the beamline's optical elements, which in turn allow us to calculate the equilibrium temperature of the Be windows and the thermal distortions on the first mirror. We will concentrate only on the first mirror, since the power absorbed by the second one is two orders of magnitude less, and thus thermal effects on it can be safely neglected.

The equilibrium temperature of the first Be window can be evaluated in the framework of grey-body theory. In the absence of cooling, the only thermal exchange mechanism is irradiation. Therefore, if the window has a surface A_1 with an emittivity of ε_1, and is irradiating a surrounding box with surface A_2 (emittivity ε_2), the power density exchanged by A_1 and A_2 is given by the following relation [9]:

$$Q = \frac{A_1 \sigma (T^{*4} - T_2^4)}{\frac{1}{\varepsilon_1} - \frac{A_1}{A_2}\left(\frac{1}{\varepsilon_2} - 1\right)} \cong A_1 \sigma \varepsilon_1 (T^{*4} - T_2^4) \tag{2}$$

where σ is the Stefan-Boltzmann constant, and T^* and T_2 are the two equilibrium temperature. The right-hand side of Equation 2 can be justified by observing that in the case of the beryllium window located at the front end, A_2 is much bigger than A_1 (it is the ratio of the inner surface of the beamline compared to the Be window area), and therefore (2) can be taken as the limiting case where $A_1/A_2 \rightarrow 0$. If Q = 19.77 W, ε_1 = 0.82, and assuming a room temperature of 293.15 K, then T^*=811 K. For the second beryllium window, one gets T^*= 395 K for the fixed-mirror configuration, and T^* = 419 K and T^* = 400 K for the configuration with α = 1.5 degrees and α = 1.8

degrees respectively. These values are to be compared with the melting temperature of beryllium, which is 1560 K. It is therefore advisable to cool both the windows.

TABLE 2. First mirror distortions induced by the impinging radiation.

Static mirrors configuration		
R_{th} (m)	28000	2485
H(μm)	0.22	2
$\Delta\theta$ (μrad)	0.804	7.3
Scanning mirror, configuration 1, α= 1.5 degrees		
R_{th} (m)	19900	1760
H (μm)	0.31	2.84
$\Delta\theta$ (μrad)	1.61	14.74
Scanning mirror, configuration 2, α= 1.8 degrees		
R_{th} (m)	19500	1731
H (μm)	0.32	2.9
$\Delta\theta$ (μrad)	1.64	15

The radiation-induced effects on the mirrors can be described as a thermally-induced bump of height H, a curvature of radius R_{th} and a slope error $\Delta\theta$ in the mirror figure related to R_{th}. Reference [14] shows how it is possible to relate H, R_{th} and $\Delta\theta$ to the absorbed heat Q, the mirror thickness t, and the thermal properties of the material constituting the mirror. Setting t = 10 cm, and considering silicon and fused silica as possible mirror materials, we can compute the thermal distortion values, reported in Table 2. We see that R_{th} is one order of magnitude larger and H is one order of magnitude smaller for Si than for SiO_2. Therefore silicon is a better choice than silicon dioxide for our case, due to its thermal characteristics.

CONCLUSIONS

We have proposed three different possible configurations for a soft-XRL beamline to be installed at the Australian Synchrotron, and analyzed their optical layout and the thermal load on each element. Further considerations will have to be made, like allowable slope error in mirror fabrication, roughness on mirror surfaces, and effects of the thermal distortions on the image, which will most probably affect the shape of the beam at the wafer plane, but not its energy distribution. These considerations will be carried out in a subsequent work, as soon as the specific exposure station will be selected. The design has been carried out with the aim of replicating structures at the sub-100 nm level, which is confirmed by simulations, given the low divergence of the beam at the wafer plan.

ACKNOWLEDGMENTS

M. Altissimo would like to acknowledge Prof. E. Di Fabrizio for precious advices and many useful discussions.

REFERENCES

1. D.L Spears and H. I. Smith, *Electron. Lett*, July (1972).
2. M. Khan, et al, *J. Vac. Sci. Technol.* **B 19**(6), (2001).
3. E. Toyota, et al, *J. Vac. Sci. Technol.*, **B 19**(6), *(2001)*.
4. Y. Vladimirsky, et al, *J. Phys. D: Appl. Phys* **32** L114 (1999).
5. Y. Iba et al., *Jpn. J. Appl. Phys.*, vol. 39, 2000.
6. www.brl.ntt.co.jp/E/research/pdf/pamphlet_E_04.pdf
7. L. Businaro, et al, *J. Vac. Sci. Technol.* **B 21**, 748 (2003)
8. E. Di Fabrizio, et al, *J. Vac. Sci. Technol.* **B 12**(6) 3979 (1994).
9. E. Di Fabrizio, et al, *Review of Scientific Instruments*, Vol 70 (3), 1999.
10. M. Feldman, J. Sun, *J. Vac. Sc. Technol.* **B 10** (6) 1992.
11. M. Khan, et al, *J. Vac. Sci. Technol.* **B 17** (6) 1999.
12. S. Goto, et al, *J. Vac. Sc. Technol.* **B 11**(2) 1993.
13. Welnak, et al, *Nucl. Instrum. Methods Phys. Res.* **A347** 1994.
14. R.K. Smither, *Nucl. Instr. Meth. in Phys. Res.* **A291,** 1990.

Soft X-ray Lithography Beamline at the Siam Photon Laboratory

P. Klysubun, N. Chomnawang, and P. Songsiriritthigul

National Synchrotron Research Center, 111 University Ave., Muang District, Nakhon Ratchasima 30000, Thailand

Abstract. Construction of a soft x-ray lithography beamline utilizing synchrotron radiation generated by one of the bending magnets at the Siam Photon Laboratory is finished and the beamline is currently in a commissioning period. The beamline was modified from the existing monitoring beamline and is intended for soft x-ray lithographic processing and radiation biological research. The lithography exposure station with a compact one-dimensional scanning mechanism was constructed and assembled in-house. The front-end of the beamline has been modified to allow larger exposure area. The exposure station for studying radiation effects on biological samples will be set up in tandem with the lithography station, with a Mylar window for isolation. Several improvements to both the beamline and the exposure stations, such as improved scanning speed and the ability to adjust the exposure spectrum by means of low-Z filters, are planned and will be implemented in the near future.

Keywords: deep x-ray lithography (DXRL), lithography beamline, LIGA process, micromachining
PACS: 0785

INTRODUCTION

In the last several years, miniaturization of combined electronics and mechanical systems has become increasingly more prevalent in the same manner as the miniaturization of electronics happened earlier. This advance in microengineering is evidenced by the introduction of new and more complex microelectromechanical systems (MEMS) and microoptoelectromechanical systems (MOEMS) in a variety of applications. A few examples are microoptical components for communication applications [1], sensor devices, for e.g. microaccelerometers for automotive industrial applications [2], microfluidic devices for liquid handling systems [3], microactuators, micromotors, microconnectors for microelectronics applications [4], and miniaturized chemical analysis devices known as lab-on-a-chips (LOC) for clinical and healthcare applications [5].

One of the prominent and most promising processes for microstructure fabrication is LIGA (Lithographie, Galvanoformung, and Abformung) [6,7]. The technique begins with the fabrication of three-dimensional microstructures by deep x-ray lithography (DXRL), followed by electrodeposition and molding processes. There are several advantages offered by the LIGA process over other microfabrication techniques. First and foremost is the flexibility of fabricated materials. Metals, metal compounds, plastics, polymers, ceramics can be utilized as final products. The technique uses highly collimated synchrotron x-rays for lithographic process, which results in high accuracy with feature sizes in submicron range. Moreover, the high penetration depth of x-rays makes the fabrication of high-aspect-ratio microstructures (HARMS) possible. In addition, the cost of fabrication is relatively low, making the process suitable for commercial mass production. These benefits make LIGA technique one of the most attractive microfabrication processes. Several LIGA beamlines have been constructed at synchrotron facilities around the world [8-11].

Siam Photon Laboratory (SPL) is a synchrotron radiation research facility located at the National Synchrotron Research Center (NSRC) in Nakhon Ratchasima, Thailand. The Siam Photon Source (SPS) is a synchrotron accelerator complex with a 1.2-GeV, 41-nm•rad electron storage ring. Currently there are two beamlines in operation. One is a vacuum ultraviolet (VUV) beamline for photoelectron spectroscopy (BL4) [12]. The other is a soft x-ray beamline for x-ray absorption fine-structure spectroscopy (BL8) [13]. Both beamlines utilize synchrotron radiation from the 1.44 T bending magnet. Now, the SPL has just completed the construction of the third bending magnet beamline (BL6). The beamline is an x-ray lithography beamline intended primarily for microfabrication and is currently in a commissioning period.

CP879, *Synchrotron Radiation Instrumentation: Ninth International Conference*,
edited by Jae-Young Choi and Seungyu Rah
© 2007 American Institute of Physics 978-0-7354-0373-4/07/$23.00

SOURCE CHARACTERISTICS

As mentioned above, the Siam Photon Source operates at electron beam energy of 1.2 GeV with a natural emittance of 41 nm•rad. The source of BL6 is one of the bending magnets in the SPS storage ring. All the bending magnets have equal bending radius of 2.78 m. Synchrotron radiation is extracted from the 10-degree port of bending magnet number six (BM6). At this position the electron beam Gaussian widths are 320 microns and 47 microns in the horizontal and vertical directions, respectively. The critical photon energy, given by

$$\varepsilon_c[keV] = 0.665E^2[GeV^2]B[T], \tag{1}$$

is 1.38 keV. The synchrotron radiation spectrum generated by the SPS bending magnet is shown in Fig. 1.

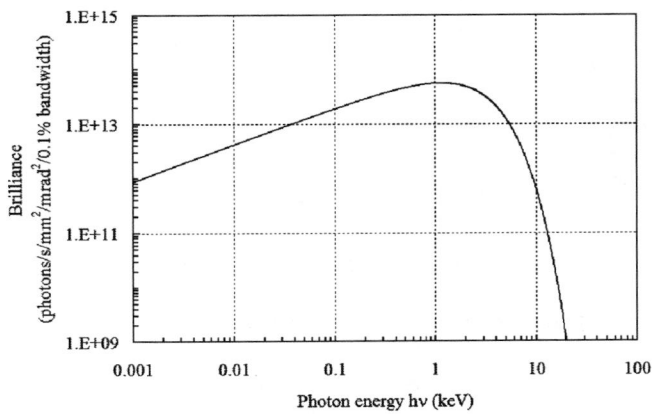

FIGURE 1. Synchrotron radiation spectrum from the SPS bending magnet

BEAMLINE AND END STATION

The x-ray lithography beamline was modified from the existing electron beam monitoring beamline. The beamline is a 'white light' beamline with a source point-to-scanner distance of 17.44 m. A water-cooled beryllium window was installed in the upstream portion of the beamline just outside the radiation shielding wall. The size of the Be window is 40 mm (h) × 8 mm (v), which corresponds to the opening angles of 5.3 mrad (h) × 1.1 mrad (v). The high-purity beryllium foil is 100-microns thick and is coated on the atmosphere (downstream) side to protect the beryllium surface from oxidation and contamination. A compromise has been made in the selection of the Be foil thickness, that is, the Be window has to be able to withstand the pressure difference between the storage ring and the exposure chamber with a safety margin to avoid accidental window breakage. On the other hand, thicker window means reduction in the x-ray intensity. This is especially true for a low energy synchrotron machine such as the Siam Photon Source. A few modifications have been made to the front-end of the beamline. The previously installed beam size aperture (Mask 1 in Fig. 2) with angular acceptance of only 1.9 mrad (h) × 6.05 mrad (v), was first rotated by 90°, and later replaced altogether with a new aperture with larger opening. The existing second beam aperture (Mask 2 in Fig. 2) with opening angles of 5.0 mrad (h) × 5.0 mrad (v) is kept and can be replaced later if there is a need for larger horizontal beam size. The layout of the beamline is shown in Fig. 2.

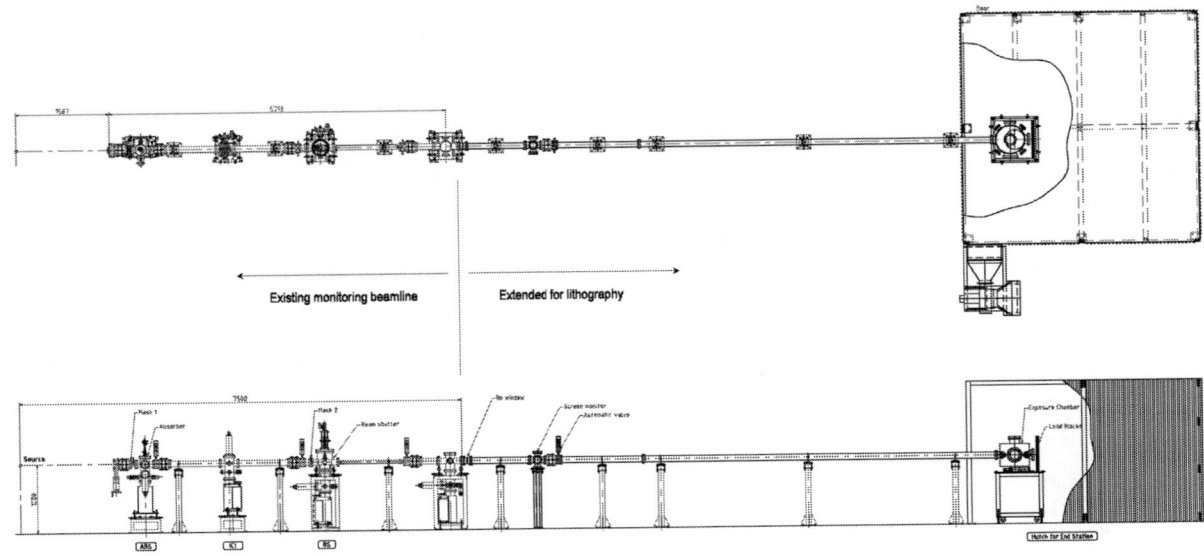

FIGURE 2. Layout of the x-ray lithography beamline

An exposure chamber is installed in a clean room at the end of the beamline and is equipped with a vertical scanning stage. The stage has a 100 mm travel range and the current maximum scanning speed is only one millimeter per second. During exposure the chamber is filled with helium gas for cooling the mask/substrate assembly. The beam size at the exposure plane is 87.2 mm (h) × 7.4 mm (v). The beamline characteristics can be summarized as listed in Table 1.

TABLE 1. Characteristics of the SPS x-ray lithography beamline (BL6)

Source	BM6
Optics	No optics
Monochromator	– (White spectrum)
Window	100 μm Be window
Transmitted bandpass spectrum	1.5 – 6 Å (2 – 8 keV)
Source to exposure plane distance	17.44 m
Beam size at exposure plane	87.2 mm (h) × 7.4 mm (v)
Horizontal acceptance	5.0 mrad
Maximum scanning length	10 cm
Heat load on mask	720 mW/1 cm (h) (@ 100 mA)

FIGURE 3. Beamline and the clean exposure area

FUTURE IMPROVEMENTS

As mentioned above, the current scanning speed of the vertical scanner is only 1 mm/s, which is too slow. In order to avoid uneven exposure and heating of the x-ray mask and photoresist, the scanning speed will be increased to around 100 – 200 mm/s. After increasing the scanning speed, a filter chamber will be installed in front of the exposure chamber to harden the radiation for exposure of thicker photoresist. A number of filters that are in preparation include aluminum, graphite, and kapton of varying thicknesses. Afterward, another exposure chamber will be installed in tandem with the lithography chamber. This radiation biology chamber will be used for studying the effects of radiation on biological samples. In addition, user interface of the control system will be further improved for easier operation by researchers.

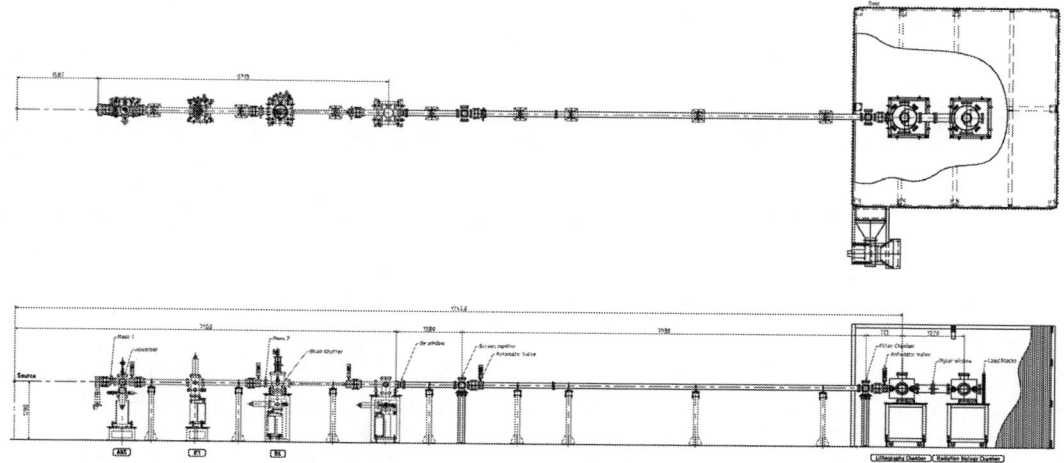

FIGURE 4. Final layout of the lithography/radiation biology beamline

SUMMARY

An x-ray lithography beamline with an exposure station at the Siam Photon Source has been modified from the existing electron beam monitoring beamline and is currently in a commissioning stage. Deep x-ray lithography will be conducted for fabrication of microstructures and HARMS. Several improvements to both the beamline and the exposure station were planned and will be carried out in the future.

ACKNOWLEDGMENTS

The authors would like to express their thanks to NSRC staff and especially P. Sombunchoo, S. Boonsuya, P. Janphuang, and P. Prawatsri for their helpful assistance and support.

REFERENCES

1. J. Gottert and J. Mohr, *SPIE Proceedings – Micro-Optics II* **1506**, 170-178 (1991).
2. C. Burbaum, J. Mohr, P. Bley, and W. Ehrfeld, *Sensors and Actuators* **A25-27**, 559-563 (1991).
3. W. K. Schomburg *et al.*, *J. Micromech. Microeng.* **4**, 186-191 (1994).
4. K. Miura *et al.*, *SEI Tech. Rev.* **60**, 30-35 (2005).
5. T. Vo-Dinh, B. Cullum, and J. Fresenius, *J. Anal. Chem.* **366**, 540-551 (2000).
6. E. W. Becker *et al.*, *Microelectronic Engineering* **4**, 35-56 (1986).
7. W. Ehrfeld *et al.*, *J. Vac. Sci. Technol. B* **6**, 178-182 (1988).
8. E. Burattini *et al.*, *Rev. Sci. Instrum.* **60**, 2133-2136 (1989).
9. E. Di Fabrizio, A. Nucara, M. Gentili, and R. Cingolani, *Rev. Sci. Instrum.* **70**, 1605-1613 (1999).
10. Y. Cheng, B.-Y. Shew, C.-Y. Lin, and D.-H. Wei, *Proc. Natl. Sci. Counc. ROC(A)* **23**, 537-543 (1999).
11. D. C. Mancini, F. DeCarlo, Y. S. Chu, and B. Lai, *Rev. Sci. Instrum.* **73**, 1550 (2002).
12. P. Songsiriritthigul, W. Pairsuwan, T. Ishii, and A. Kakizaki, *Surf. Rev. Lett.* **8**, 497-500 (2001).
13. W. Klysubun *et al.*, these proceedings.

The Study of Deep Lithography and Moulding Process of LIGA Technique

Yuhua Guo, Gang Liu, Ya Kan, Yangchao Tian

National Synchrotron Radiation Laboratory, University of Science and Technology of China, Hefei, Anhui, 230029, P.R.China

Abstract. The knowledge of the development behavior, especially the development rate, is of primary importance for the study of deep x-ray lithography in LIGA technique. In the first part of this paper, we have measured the development rates of crosslinked PMMA foils irradiated in NSRL covering a wide dose range (bottom dose in the range:2.5-8.0 kJ/cm^3). After the exposure, we use a so-called period-development method (to dip development in GG-developer for 20 minutes and clean in rinse solution for 40 minutes as a development period). For processing the experiment data, we get the KD^β model to describe our PMMA/GG-developer system. The aim of this work is to find out a stable experiment condition for deep X-ray lithography and development. The result shows that in small amount of dose (bottom dose range: 2.5-4 kJ/cm^3), this model is very stable. While in large amount of dose (bottom dose range: 5-8kJ/cm^3), the model becomes very sensitive and even unavailable. To verify the conclusion validity, the fixed dose range (bottom dose range: 3.5-4 kJ/cm^3) is applied on PMMA microstructures. And the result shows an effective development process. In the following procedure, mold inserts can be produced by micro-electroforming and plastic replicas can be mass produced by hot embossing. To emboss high-aspect-ratio microstructures, the deformation of microstructures usually occurs due to the demolding forces between the sidewall of mold inserts and the thermoplastic (PMMA). To minimize the friction force the optimized experiment has been performed using Ni-PTFE compound material mold inserts. Typical defects like pull-up and damaged edges can be greatly reduced.

Keywords: LIGA, deep x-ray lithography, development, moulding, hot embossing
PACS: 01.30.Cc

INTRODUCTION

The LIGA process, which combines deep X-ray lithography with electroplating and moulding, is a main fabrication method for producing MEMS [1]. The LIGA station which is operated at National Synchrotron Radiation Laboratory (NSRL) of the University of Science and Technology of China (USTC) has been tested in May of 2002. It is the beamline for a general purpose of Synchrotron Radiation (SR) exposing system to perform MEMS R&D. In this paper, recent researches including deep x-ray lithography and moulding technologies are introduced.

THE STUDY OF DEEP LITHOGRAPHY

A fine microstructure is of priority for the fabrication of metal mold insert. In the first step (deep X-ray lithography) of LIGA, the absorber structure of an X-ray mask is transferred into resist layers by using synchrotron radiation. Usually, poly-methylmethacrylate (PMMA) is used as resist material. Through exposure, the molecular weight of the PMMA decreases and therefore becomes soluble in an organic developer. The GG-developer is the standard developer for PMMA in deep X-ray lithography. The achievable quality of microstructures irradiated in deep X-ray lithography is decisively defined by the development process. The knowledge of the development behavior, especially development rate, is of primary importance for the LIGA technique. The system material/development (PMMA/GG-developer) needs to be characterized, optimized, and stabilized. In this part of work, the measurement of the development rate of crosslinked PMMA foils irradiated in NSRL is reported. The dose covered a wide range (bottom dose in the range: 2.5-8.0 kJ/cm^3) and a so-called period-development method have

CP879, *Synchrotron Radiation Instrumentation: Ninth International Conference*,
edited by Jae-Young Choi and Seungyu Rah
© 2007 American Institute of Physics 978-0-7354-0373-4/07/$23.00

been used. In the experiment data processing, we get the KD^β model [2] to describe our PMMA/GG-developer system which helps us to find out a stable experiment condition for deep X-ray lithography and development.

Experiment

In all experiments, Si wafers covered with Ti/Cu/Ti (50nm/100nm/100nm) layers and a layer of spin-on PMMA (1-2 μm) are used as substrate. And commercially available PMMA sheets (Molecular weight: 6.0×10^6 g mol^{-1}, crosslinked) are used as resist layer. The thickness of the sheets is 1mm. Sheets are cut into pieces (35x35 mm) and annealed in hot oven (15 ℃/h up to 80 ℃, 1 h at 80 ℃ and cool to room temperature). To glue the prefabricated PMMA foils, a resin (MMA) is applied to the substrate. After gluing, the samples are relaxed at room temperature for 24 h. Finally, the PMMA is flied cut to desired thickness. To measure the development rate of irradiated PMMA, we have used a gold mask with one circle pattern (the diameter is about 1 cm). The gold is thick enough for absorbing x-ray.

The exposure has been performed at the LIGA station of National Synchrotron Radiation Laboratory (NSRL) of the University of Science and Technology of China (USTC) working at 800MeV (magnetic field: 6T; the e-beam current: 150-200 mA). A 200-μm thick beryllium window separates the vacuum of the beam line and the working chamber. The distance source-sample is 12.3 m. The scanning speed is 8 mm s^{-1}. In all experiments, we have used several 100 μm -KAPTON polyimide films to obtain the desired dose profile. The exposure is performed in a 10 kPa Helium gas condition.

After the exposure, we use a so-called period-development method. At the development temperature of 37℃, we define "dip development in GG-developer for 20 minutes and clean in rinse solution for 40 minutes without any agitation" as a development period. In the subsequent data processing, the latter 40-minute of every period is not counted in development time. The well-known GG-developer is a mixture of water and three different organic solvents (15 vol. % water, 60 vol. % butoxyethoxyethanol, 20 vol. % tetrahydro-oxazine, and 5 vol. % aminoethanol). And the rinse solution is composed of water and other organic solvent (20 vol. % water and 80 vol. % butoxyethoxyethanol). After every development period, the sample is dried with nitrogen gas. The depth-etched by the developer into the irradiated region of the resist is measured using a profilometer (Alpha 500 Surface profiler from Tencor Instruments) at the same place. After several development periods, the development process is completed and the experimental data between the depth and time are also obtained.

Model and Calculation of the Development Rate

The development rate $R(D)$ as a function of the deposited dose D is approximated by the empirical formula:[2]

$$R(D) = KD^\beta \tag{1}$$

K and β are assumed constants related to the resist and developer. The parameter β, is usually used to describe the contrast of the developer (a higher contrast indicates a smaller attack of non-irradiated resist areas for a given developed depth). And

$$K = C \times \left(\frac{G}{100 \times 1.6 \times 10^{-19} \times N_A \times \rho} \right)^\beta \tag{2}$$

C is characteristic constant of the resist and developer; N_A is the Avogrado's number; ρ is the density of PMMA; the G value is defined as the number of effective scission events per 100 eV. The calculation of the rate, using the experimental data depth h versus dose, is based on the following relationship:

$$h_i = \int_0^{t_i} R(D_{h(t)})dt \text{ and } D_h \approx D_s e^{-\alpha h} \tag{3}$$

with D_s being the absorbed dose at the top of the resist, h the distance from the surface, $h(t)$ the distance from the surface at time t. α represents the average linear absorption coefficient of the resist, base on the fact that the PMMA absorption spectrum is narrow for the energy range used in the LIGA process. Using formulas (1) and (2), one obtains:

$$h = \frac{1}{\alpha\beta} Ln(1 + KD_s^\beta \alpha\beta t) = aLn(1 + bt) \tag{4}$$

with:

$$a = (\alpha\beta)^{-1} \text{ and } b = KD_s^\beta \alpha\beta \qquad (5)$$

And finally we can get:

$$R(t) = \frac{dh}{dt} = \frac{ab}{1+bt} \qquad (6)$$

Relation (4) is used to fit the experiment results, depth versus time for a set of dose profiles. So the parameters a and b can be obtained. Subsequently the corresponding dose and the rate using Equation (6) are calculated. Using relation (5), K and β can be calculated. Thus we get the KD^β model to describe our PMMA/GG-developer system.

Results

We have performed several experiments using a range of dose profile; the fitting data and exposure parameters are summarized in Table 1.

After obtaining parameters a and b, using relation (4), K and β can be calculated. Thus we get the KD^β model to describe our PMMA/GG-developer system. And the calculated parameters (K-β) in all samples are given in Table 2.

In the KD^β model, K and β are assumed constants related to the resist and developer. The result shows that in small amount of dose (bottom dose range: 2.5-3.5 kJ/cm³), the value of K doesn't change. And in normal dose range (bottom dose range: 4-4.5 kJ/cm³), the value of K is relatively stable. While in large amount of dose (bottom dose range: 5-8 kJ/cm³), the K-value becomes very sensitive; small variations of bottom dose induce significant of the K. It means that excessive dose range (in this case, above 5 kJ/cm³) causes the model to be unstable and may have influenced the character behavior of the system PMMA/GG-developer. In other words, the KD^β model is not available in excessive dose range. [3]

The result also shows that in small amount of dose, the value of β is bigger and more stable than that in large amount of dose. It means that the contrast of the system PMMA/GG-developer is reduced by excessive dose. That is to say, excessive amount of dose will induce a great attack of non-irradiated resist areas for a given developed depth and reduce the quality of microstructures even make microstructures fall-off.

The aim of this work is to find out a stable experiment conditions for deep X-ray lithography and development. An unstable development process will lead to unpredictable development time. Result shows that appropriate amount of dose is good for setting up a stable experiment conditions. To verify the conclusion validity, the fixed dose range (bottom dose range: 3.5-4 kJ/cm³) is applied on PMMA microstructures. The development process of the profiles completes in predictable time and turns out good qualities. The scanning electron microscope (SEM) pictures of the profiles are given in Fig. 1.

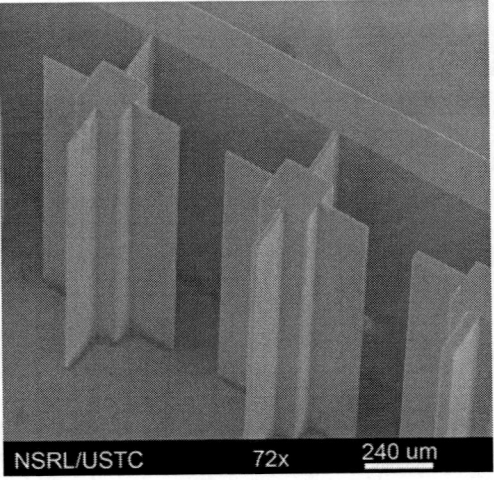 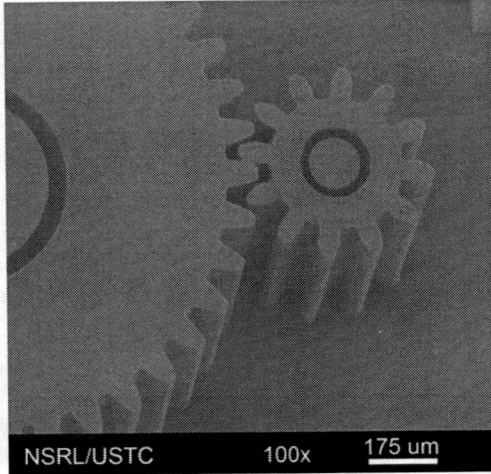

FIGURE 1. Structures irradiated in deep x-ray lithography at LIGA beamline of NSRL, (a) PMMA structures with minimum width of 5 μm and height of 1 mm (bottom dose: 3.5 kJ/cm³); (b) PMMA structures with height of 500 μm (bottom dose: 4.0 kJ/cm³).

TABLE 1. The exposure parameters and the fitting results

Sample	Dose D (kJ/cm³)	Parameter a	Parameter b
NSRL01	2.5-11.59	144.197±2.994 (2.076%)	0.0470±0.0023 (4.836%)
NSRL02	3.0-13.91	149.784±2.444 (1.632%)	0.0716±0.0030 (4.255%)
NSRL03	3.5-16.23	153.567±3.157 (2.056%)	0.1067±0.0063 (5.944%)
NSRL04	4.0-18.55	178.173±4.317 (2.423%)	0.1871±0.0071 (3.795%)
NSRL05	4.5-20.87	191.283±8.258 (3.153%)	0.1726±0.0086 (4.983%)
NSRL06	5.0-23.19	235.240±7.957 (3.383%)	0.1171±0.0095 (8.121%)
NSRL07	6.0-27.82	293.556±6.179 (2.105%)	0.0973±0.0047 (4.806%)
NSRL08	7.0-32.46	347.711±13.29 (3.822%)	0.0900±0.0077 (8.544%)
NSRL09	8.0-37.10	537.611±38.54 (7.168%)	0.0494±0.0064 (12.92%)

THE STUDY OF MOULDING PROCESS

After obtaining fine resist microstructures, metal mold inserts can be produced using micro-electroforming. In the following procedure, plastic replicas can be mass produced by hot embossing, which is one of main processing techniques for polymer microfabrication. Most problems in polymer micro molding are not caused by the filling of the mold, but by the demolding process. During demolding, microstructures may be torn apart, deformed or destroyed. Demolding also very much affects the wear of mold inserts. If the microstructure is not designed properly or unsuitable process parameters are chosen, some delicate parts of the mold insert cannot survive in the first single molding process. Typical defects are shown in Fig.2.

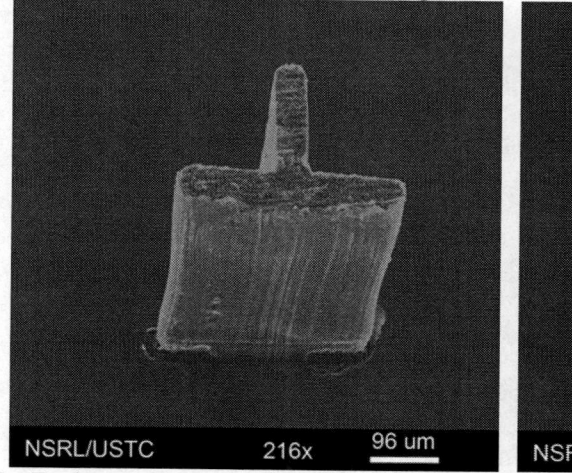 

FIGURE 2. Typical defects when embossing microstructures, deformed in the left, pull-off in the right, the aspect ratio is 16.5.

Our previous study shows that adhesion force between surfaces is the main reason of the deformation especially when demolding high aspect ratio microstructures. [5] For minimizing the effect of adhesion force, one effective way is to use Ni-PTFE compound material mold inserts which are electroformed in the nickel bath added with PTFE particles. And the detail of fabrication method can be referred in our previous work [5-8]. The demolding effects of two ways are shown in Fig.3. Typical defects like pull-up and deformed edges can be greatly reduced in this way.

TABLE 2. The calculated parameters (K-β) in all samples

Sample	K	β	R=KD$^\beta$
NSRL01	0.0010	3.6157	R=0.0010D$^{3.6157}$
NSRL02	0.0011	3.4809	R=0.0011D$^{3.4809}$
NSRL03	0.0012	3.3951	R=0.0012D$^{3.3951}$
NSRL04	0.0065	2.9263	R=0.0065D$^{2.9263}$
NSRL05	0.0083	2.7257	R=0.0083D$^{2.7257}$
NSRL06	0.0260	2.2164	R=0.0260D$^{2.2164}$
NSRL07	0.0777	1.7761	R=0.0777D$^{1.7761}$
NSRL08	0.1695	1.4995	R=0.1695D$^{1.4995}$
NSRL09	0.7986	0.9698	R=0.7986D$^{0.9698}$

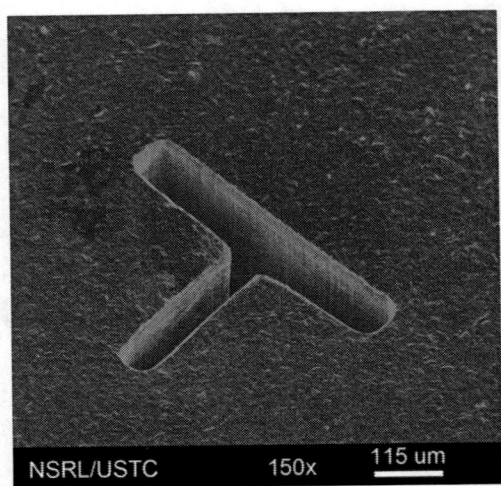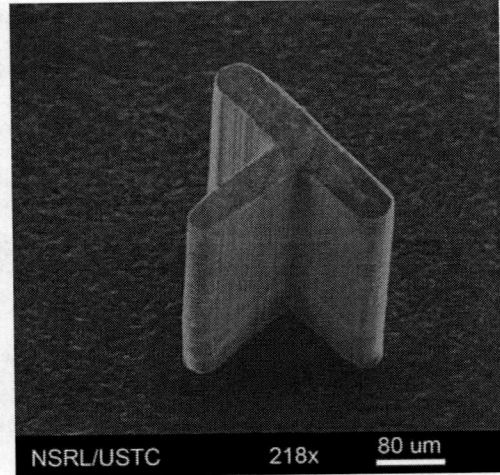

FIGURE 3. Optimizing the demolding process using Ni-PTFE compound material mold insert, Ni-PTFE LIGA mold insert in the right, and replica in the left, and the minimum width is 20 μm and the max aspect ratio is 16.5.

CONCLUSION

As a foundry of LIGA, basic research of deep x-ray lithography has been well studied. By measuring the development rate of PMMA irradiated in deep x-ray, it has been shown that the character behavior of the system PMMA/GG-developer is influenced by deposited dose and appropriate amount of dose makes a stable model. A stable development system results in a predictable development process and better structure quality in deep x-ray lithography. And the study of the moulding process is becoming increasingly important which plays a key role in commercial manufacturing of polymer replicas using LIGA technology. The friction force caused by surface adhesion between the mold and polymer are main source of the demolding forces. To improve the surface adhesion property is important for reducing the influence of the friction force. And optimized experiment using Ni-PTFE compound material mold inserts are proposed and proved effective.

ACKNOWLEDGMENTS

This work is supported by National Natural Science Foundation of China (No.10375058).

REFERENCES

1. E. W. Becker, W. Ehrfeld, A. Maner and D. Munchmeyer, *Microelectronics*, **4**, 35-36 (1986).
2. P. Meyer, A.El-Kholi and J. Schulz, *Microelectronic Engineering*, **63**, 319-328 (2002).
3. Yuhua Guo, Gang Liu, Xuelin Zhu and Yangchao Tian, The 4th International Workshop on Microfactories, Shanghai, China, 2004, pp. 321-325.
4. H. M. Pollock, D. Maugis and M. Barquins, *Appl. Phys. Letters* **33**, 798 (1978).
5. Yuhua Guo, Gang Liu, Xuelin Zhu and Yangchao Tian. The High Aspect Ratio Micro Structure Technology Workshop (HARMST 2005), Gyeongju, Korea, 2005, pp.238-239.
6. Y Tian, P Zhang, G liu and X Tian, *Microsystem Technologies* **11**(4-5), 261-264 (2004).
7. Zhang P, Liu G, Tian Y and Tian X, *Sensors and Actuators A* **118**, 338-34 (2005).
8. Yuhua Guo, Gang Liu, et al, *Journal of Physics: Conference Series* **34**, 870-874 (2006).

Nickel Micro-spike for Micro-scale Biopsy using LiGA Process

Gilsub Kim[1], Sunkil Park[1], Kyo-in Koo[1], Hyun-Min Choi[1], Myeong-Jun Jung[1], Si-Young Song[2], Seoung-Min Bang[2] and Dongil "Dan" Cho[1*]

[1]*School of Electrical Engineering and Computer Science, ASRI, ERC-NBS, ISRC, Seoul National University, San 56-1, Shillim-dong, Kwanak-ku, Seoul 151-744, Korea, [2]Department of Internal Medicine, Yonsei University College of Medicine, 134, Shinchon-dong, Seodaemun-ku, Seoul, Korea*

Abstract. In this paper, biopsy tools are developed for minimally invasive tissue sampling using the LiGA (Lithographie Galvanoformung Abformung) process. The micro-spike is composed of two barbed-shanks and a body. The shank of the micro-spike is between 2 mm ~ 3 mm and the opening gap is approximately 350 μm between the shanks. The micro-spike is integrated with the conventional catheter, for medical diagnostics. Tissue samples were extracted from the anesthetized pigs using biopsy catheters *in vivo*, and observed with hematoxylin and eosin (H&E) staining. The amount of extracted sample is sufficient to diagnose abnormal cells.

Keywords: Micro-spike, LiGA, Catheter
PACS: 61.43.Dq, 68.03.Hg, 68.90.+g

INTRODUCTION

Biopsy catheter is essential equipment for pathological testing, medical examination and health monitoring. Due to the size and sampling mechanism, conventional biopsy catheter forceps often cause significant discomfort, infectious risk, and injury to patients. According to the Ministry of Health, Labor and Welfare of Japan, gastric intestine punctures occurred 66 times during biopsy tests from May to November, 2003. The protruding and spiking mechanism with silicon micro-spikes, developed by this research, could solve these problems of conventional biopsy tools [1]. Even though there were no damaged silicon micro-spikes in several animal experiments, silicon micro-spikes are suspected to be brittle [2].

In this paper, the nickel micro-spike is designed, fabricated, and examined for diagnostic examination. The fabrication of the micro-spike employed the use of the LiGA process. The shank length of the micro-spike is between 2 ~ 3 mm and the opening size between shanks is approximately 350 μm. The fabricated micro-spike was mounted on the conventional catheter for the *in vivo* test. The mounted micro-spike was tested in the small intestine of a pig. Animal experiment results with the micro-spike were pathologically evaluated using hematoxylin and eosin (H&E) staining. The pathological test results show that the amount of tissue extruded by micro-biopsy tools is sufficient for diagnostics. Although it is known that nickel is not bio-compatible, it is easily electroplated. Therefore, for the clinical usage, we will coat the nickel micro-spike with 2 μm parylene C which is known to be bio-compatible.

TISSUE SAMPLING MECHANISM

The micro-spike contains barbs for minimally invasive tissue sampling. Barbs protruding from the micro-spike shanks facilitate the biopsy procedure by tearing off the samples from the target tissue and retaining these samples. In addition, a micro-spike promotes tissue sampling by trapping samples between shanks and holding them, as

---

* Corresponding author : dicho@snu.ac.kr

CP879, *Synchrotron Radiation Instrumentation: Ninth International Conference*,
edited by Jae-Young Choi and Seungyu Rah
© 2007 American Institute of Physics 978-0-7354-0373-4/07/$23.00

shown in Fig. 1. The micro-biopsy spike can extract a small amount of tissue samples in a minimally invasive way. Therefore, micro-spikes not only simplify surgical procedures, but also lower patient risk and pain.

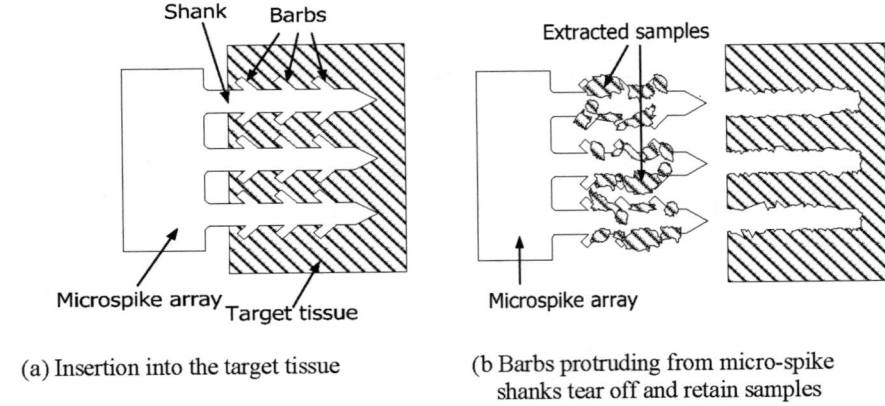

(a) Insertion into the target tissue

(b Barbs protruding from micro-spike shanks tear off and retain samples

FIGURE 1. The schematic diagram of the biopsy procedure with a micro-spike [2].

Design

The dimensions of micro-spike shanks are designed considering the thickness of mucous membranes of the digestive organs as well as mechanical stability during the insertion of the micro-spike to animal organs [3]. The shank length of the designed micro-spike is between 2~3 mm and the opening size between the shanks is approximately 350 μm. Figure 2 shows the basic design of the micro-spike.

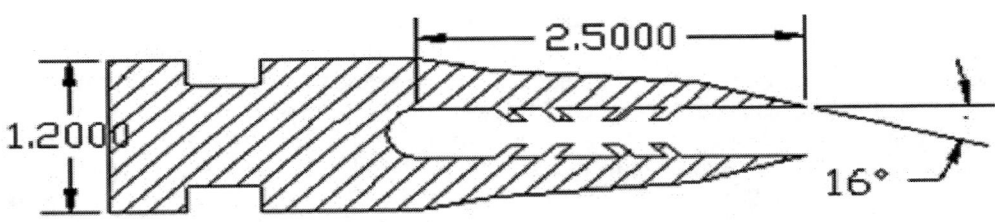

FIGURE 2. The basic design of the micro-spike.

FABRICATION

LIGA is a fabrication process for high aspect ratio micro-structures using Deep X-ray Lithography. This process is composed of the x-ray mask fabrication process and electroplating process for device fabrication. First, we fabricated X-ray mask using silicon wafer for LiGA process. Sequentially, PMMA bonded on a titanium substrate is irradiated using the fabricated X-ray mask. Electroplating, developing and polishing process are performed for achieve micro-spike after irradiation.

X-ray Mask

Figure 3 shows the process flow of the X-ray mask. A Ti /Au (300 Å / 1500 Å) seed layer is deposited by a metal sputter on the silicon wafer for electroplating. Photo resist (PMER) is coated on the deposited seed layer at 1000 rpm, 35 seconds before using spin coater. PMER photo resist is uniformly deposited with 27 um thickness. Sequentially, PMER is patterned using conventional UV lithography equipment for Au electroplating. Gold electroplating is performed to create the X-ray absorber structure. The thickness of gold absorber is 15 μm which is sufficient for absorbing deep X-ray. Because silicon also absorbs the deep X-ray, the Au electroplated Si substrate is etched away to the 350 μm thickness.

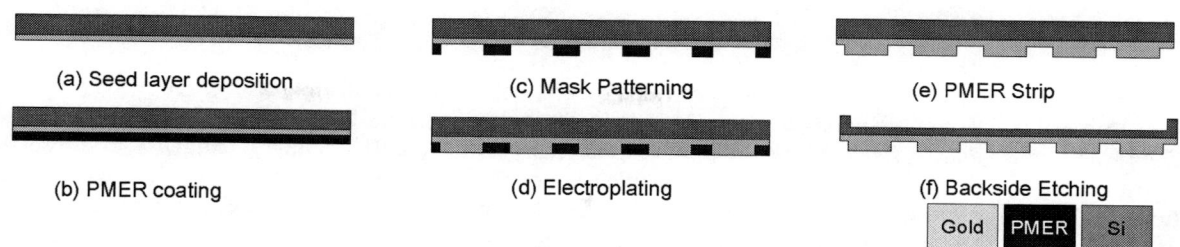

(a) Seed layer deposition

(c) Mask Patterning

(e) PMER Strip

(b) PMER coating

(d) Electroplating

(f) Backside Etching

| Gold | PMER | Si |

FIGURE 3. Process flow of x-ray mask.

X-ray Exposure & Electroplating Process

LiGA process flow is shown in Fig. 4. PMMA bonded on a titanium substrate is irradiated using the fabricated X-ray mask. Irradiated PMMA is developed at room temperature with a specific organic developer, commonly known as GG developer (2-(2-butoxy-ethox) ethanol 60 %, Morpholine 20 %, Ethanolamine 5 %, DI water 15 %). Nickel electroplating is performed on the developed PMMA mold. Finally, the core components are fabricated by removing the mould parts. The FE-SEM photographs of the nickel micro-spike are shown in Fig. 5. Irradiation was performed at the Pohang Accelerator Laboratory (PAL) in Pohang, Korea. This irradiation operated at 2.5 GeV with an electron current between 110 and 190 mA. Among many beam lines, the 9C1 beam line is equipped with 250 μm Be filters and a 50 μm polyimide filter. The distance in air from the polyimide filter to the exposure vacuum jig is 18 cm. The theoretical calculation of the critical wavelength is 2.0 Å at the end of the beam line. The x-ray beam size was 180 x 15 mm^2, PMMA samples were irradiated under a scanning condition of 3 cms^{-1}. For fast and effective development of irradiated polymer samples, a bottom dose of 4 kJm^{-3} for exposed sample was chosen. In addition, the maximum dose under the membrane must be less than 20 kJm^{-3}.

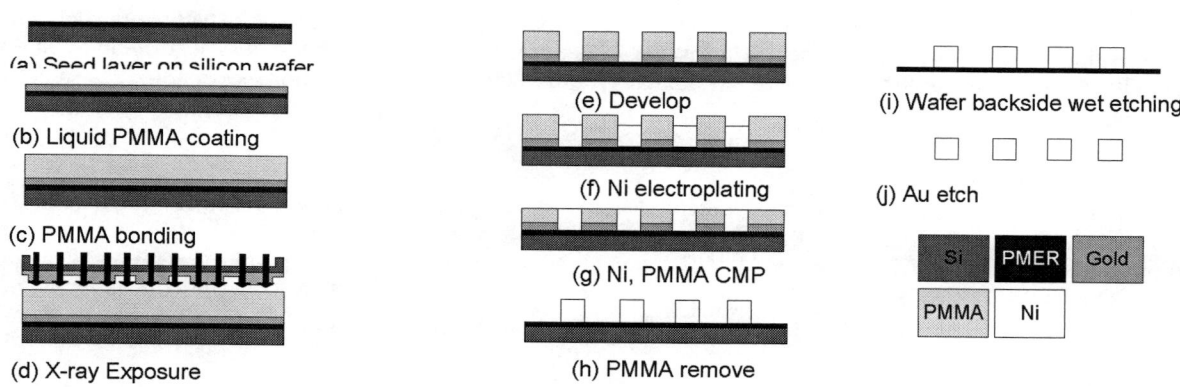

(a) Seed layer on silicon wafer

(b) Liquid PMMA coating

(c) PMMA bonding

(d) X-ray Exposure

(e) Develop

(f) Ni electroplating

(g) Ni, PMMA CMP

(h) PMMA remove

(i) Wafer backside wet etching

(j) Au etch

| Si | PMER | Gold |
| PMMA | Ni | |

FIGURE 4. LiGA process flow

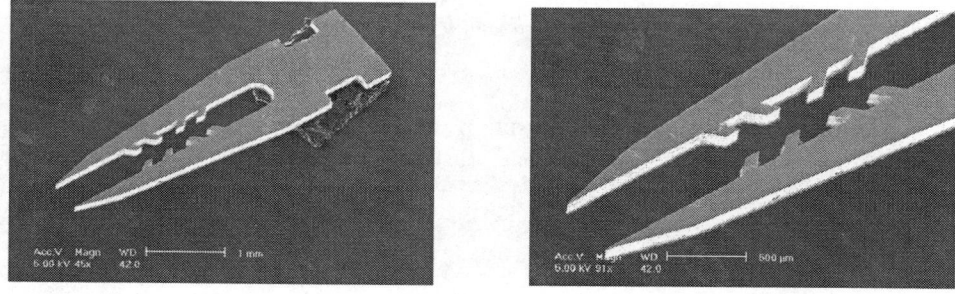

FIGURE 5. FE-SEM photographs of fabricated nickel micro-spike using LiGA process.

IN VIVO BIOPSY TEST

Several *in vivo* animal biopsy tests were performed on the small intestine of a pig using a conventional biopsy catheter. Several *in vivo* tests showed the repeatable results of tissue extraction. The extracted samples were rinsed in distilled water for 3 minutes, fixed in 70 % ethanol for 3 minutes, strained in Rhodamine B (Sigma-Aldrich) solution for 10 minutes, and finally rinsed again in the distilled water for 1 minute. The micro-spike can easily extract sufficient tissue samples from the animal small intestine to observe the tissues and cells. Figure 6 shows the H&E (Hematoxylin and Eosin) stained image after sampled tissue paraffin block. Figure 6 shows clinical diagnosis evaluation.

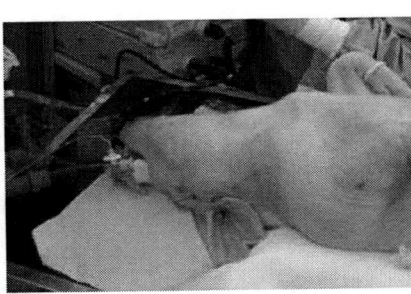 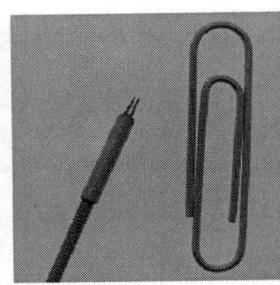 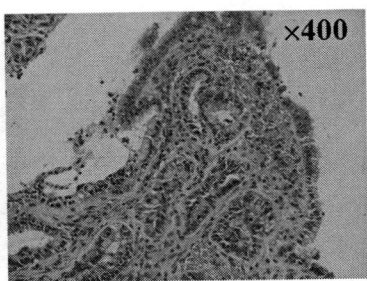

FIGURE 6. *In vivo* animal biopsy tests on small intestine of a pig (left), Biopsy catheter using micro-spike (middle), Biopsy results of H&E stained images after paraffin section (right)

CONCLUSION

A micro-spike for minimally invasive tissue sampling was fabricated and evaluated for pathological tests *in vivo*. Micro-spike was designed and fabricated using the LiGA process. Several animal biopsy tests were demonstrated with the porcine's small intestine *in vivo*. Several tests results showed repeatability for successful tissue sampling.

ACKNOWLEDGMENTS

This research was supported by the Intelligent Microsystem Center(IMC; http://www.microsystem.re.kr), which carries 21st century Frontier R&D Projects sponsored by the Korea Ministry Of Commerce, Industry and Energy, through grants from the Korea Health 21 R&D Project, Ministry of Health & Welfare, Republic of Korea (A05-0251-B20604-05N1-00010A), SRC/ERC program of MOST/KOSEF (grant # R11-2000-075-01001-0), and 9C1 X-ray Nano/Micromachining(LIGA) beamline of Pohang Accelerator Laboratory (PAL; http://pls.postech.ac.kr).

REFERENCES

1. Park, S. K., Lee, A. R., Jeong, M. J., Choi, H. M., Song, S. Y., Bang, S. M., Paik, S. J., Lim, J. M., Jeon, D. Y., Lee S. K., Chu, C. N., and Cho, D. I., "A Disposable MEMS-Based Micro-Biopsy Catheter for the Minimally Invasive Tissue Sampling," *IEEE/RSJ International Conference on Intelligent Robots and Systems*, Edmonton, Alberta, Canada, Aug. 2-6, 2005.
2. Byun, S. W., Lim, J. M., Paik, S. J., Lee, A. R., Koo, K. I., Park, S. K., Park, J. H., Choi, B. D., Seo, J. M., Kim, K. A., Chung, H., Song, S. Y., Jeon, D. Y., Lee, S. S., and Cho, D. I., "Novel Barbed Micro-Spikes for Micro-Scale Biopsy," *IOP Journal of Micromechanics and Microengineering*, **15**, no. 6, pp. 1279-1284, 2005.
3. Paik, S. J., Byun, S. W., Lim, J. M., Park, Y. H., Lee, A. R., Chung, S., Chang, J. K., Chun, K. J., and Cho, D. I., "In-plane single-crystal-silicon micro needles for minimally invasive micro-fluid system," *Sensors and Actuators A*, **114**, pp. 276-284, Sep. 2004.

Deep X-Ray Lithography in the Fabrication Process of a 3D Diffractive Optical Element

S. P. Heussler[1,2], H. O. Moser[1], C. G. Quan[2], C. J. Tay[2], K. D. Moeller[3], M. Bahou[1] and L. K. Jian[1]

[1] Singapore Synchrotron Light Source, National University of Singapore, 5 Research Link, Singapore 117603
[2] Department of Mechanical Engineering, National University of Singapore, 10 Kent Ridge Crescent, Singapore 119260
[3] Department of Electrical and Computing Engineering, New Jersey Institute of Technology, Newark, NJ 07102-1982

Abstract. We present first results of the fabrication process of a diffractive optical element (DOE) using deep X-ray lithography. The DOE forms the core of our proposed fast parallel-processing infrared Fourier transform interferometer (FPP FTIR) that works without moving parts and may allow instantaneous spectral analysis only limited by detector bandwidth. Design and specifications of the DOE are discussed. A fabrication process including deep X-ray lithography (DXRL) on stepped substrates is introduced.

Keywords: Diffractive Optical Element, Fourier Transform Spectroscopy, X-ray lithography.
PACS: 42.82.Cr, 42.25.Fx, 07.57.Ty, 85.40.Hp, 78.47.+p

INTRODUCTION

In 1994, two of us (HOM and KDM) proposed a novel three dimensional diffractive optical element (DOE) [1] as a basis of a Fast Parallel-Processing Fourier Transform Interferometer (FPP FTIR) without moving parts. The FPP FTIR is a stationary lamellar grating interferometer which may allow measurements of transients in a broad spectral range. The instantaneous recording of spectra is of great importance when either the light emitting object or the detector are moving fast or when the emitter changes its spectral distribution and intensity rapidly. Such spectrometers may be useful for measuring transient emission sources such as explosions or plasmas. They may also be of particular interest for satellites, missiles and robots, when one can assume that the input to the spectrometer is constant only during a brief moment.

THE DIFFRACTIVE OPTICAL ELEMENT

The core of the FPP FTIR is the diffractive optical element. It consists of N x N binary grating cells with varying grating depth arranged in a square array in a plane. Its schematic is shown in Fig. 1a. Figure 1b illustrates the function of a single grating cell. A perpendicularly incident plane wave is split into two coherent beams by front and back facet mirrors and dispersed in an angle α. Wave fronts reflected by either front or back facet mirrors travel a different optical path and produce a phase shift φ. Upon interference, the phase shift results in a modulation of the reflected amplitude. The intensity pattern of the dispersed light is calculated by means of equation (1) [2].

$$I \propto \left[\frac{\sin \pi p \sin \alpha / 2\lambda}{\pi p \sin \alpha / 2\lambda} \right]^2 * \left[\frac{\sin N \pi p \sin \alpha / \lambda}{\sin \pi p \sin \alpha / \lambda} \right]^2 * \cos^2 \left[\underbrace{\frac{1}{2} * \frac{2\pi}{\lambda} * (\frac{p}{2} \sin \alpha + d(1 + \cos \alpha))}_{\varphi} \right] \qquad (1)$$

CP879, *Synchrotron Radiation Instrumentation: Ninth International Conference*,
edited by Jae-Young Choi and Seungyu Rah

where p is the grating period, λ is the wavelength, α the diffraction angle, N the number of illuminated lamellae, and d the grating depth.

By changing the grating depth d from cell to cell, the optical path difference and in consequence the phase difference between the two interfering beams is shifted and the amplitude alters accordingly. Since the DOE comprises a multitude of grating cells of constantly changing grating depth, collimated light reflected from the DOE is split into N x N sub beams. Each of these beams is modulated in amplitude according to the specific grating cell. The zeroth diffraction order of the sub-beams is recorded at a certain position on a pixel array detector. The recorded intensity signal with respect to the optical path difference constitutes the interferogram Fourier analysis of which yields the spectrum.

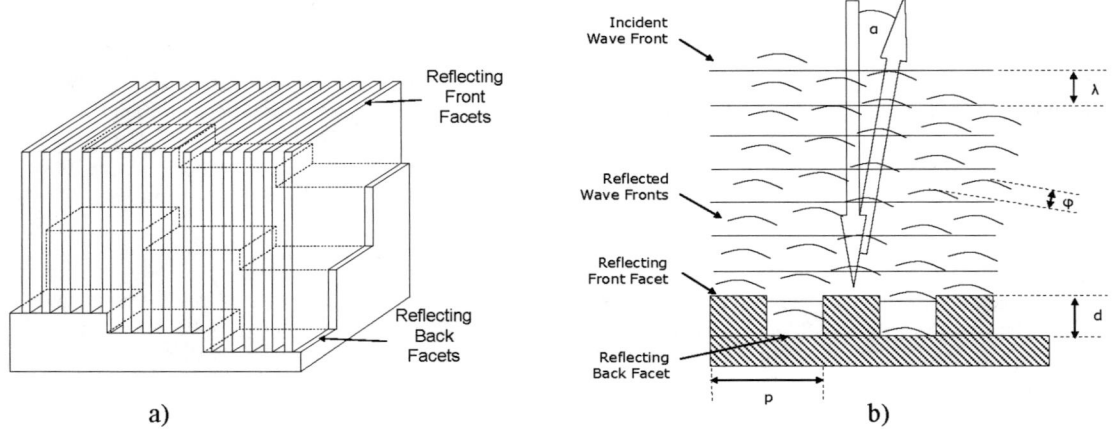

FIGURE 1. a) Schematic drawing of a diffractive optical element consisting of 3 x 3 lamellar grating cells. b) Schematic of the wave front division and diffraction, p is the grating constant, d the grating depth, λ the wavelength of the incident light, α the diffraction angle and φ the phase difference due to the optical path difference of rays reflected by the front and by the back facet mirrors.

Design Requirements

Lamellar grating interferometers work in the zeroth diffraction order of the diffracted beam and higher diffraction orders must be prevented from entering the exit slit of the interferometer [2]. The efficiency of these interferometers drops when the first diffraction order enters the aperture of the detector. Since the shortest wavelength in the instrument are dispersed the least, a short wavelength limit is placed on the lamellar grating interferometer. Therefore, at shorter wavelengths a small grating period is essential. Figure 2a exemplifies the relation of wavelength to the angle of first diffraction order. As the wavelength reduces from 20 μm to 15 μm and to 10 μm at constant illuminated lamellae number and grating period, the first order diffraction pattern moves towards the zeroth diffraction order and upon entering the detector slit, depending on the aperture size and distance to the detector, modulates the zeroth diffraction order pattern. However, as the grating period reduces, the walls of the lamellae form a cavity which polarizes the long wavelength radiation passing through. This results in a change of the group velocity of the passing beam and a subsequent error in the effective optical path difference. The effect is known as the cavity effect and limits the long wavelength, minimal grating period as well as the maximum grating depth [3].

The resolution of a spectrometer measures its ability to distinguish two spectral peaks that are very close to each other [4]. For Fourier transform interferometers, the resolution is inversely proportional to the maximum optical path difference between the interfering waves [3]. Its dependence on the grating depth with the incidence angle as a parameter is demonstrated in Fig. 2b. A resolution of 5 cm^{-1} can be achieved by a maximum grating depth of 1 mm and perpendicular incident light. This is a resolution value commonly used in routine operations of medium resolution Fourier transform interferometers in the infrared spectrum. A larger grating depth is within reach of the process when a higher resolution is required.

The step height of the back facet mirrors and consequently the number of cells of the DOE depend on the smallest wavelength of the spectrum under examination and on the available number of pixels of the array detector. Nyquist criterion states that the minimum sampling frequency must be at least twice the smallest frequency of the signal. For perpendicularly incident white light with a minimal wavelength component λ_{min}, constructive interference occurs at grating depths equal to multiples of $\lambda_{min}/2$. Thus, the stepping height of the grating depth must at least equal $\lambda_{min}/4$.

Further requirements on the DOE are reflective surfaces of mirror quality with a high degree of parallelism, highly reproducible mirror steps and a surface roughness (R_a- value) of the mirror facets of $< \lambda_{min}/20$.

The study of a spectrum from 500 cm⁻¹ to 0 would require a sampling interval of 5 μm steps. A detector of 16 x 16 pixels would provide 256 sampling points. An interferogram of 256 points would then correspond to a maximum grating depth of 1.275 mm and a resolution of 3.92 cm⁻¹.

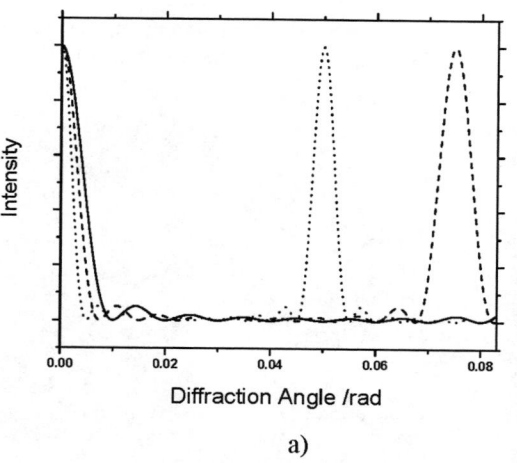

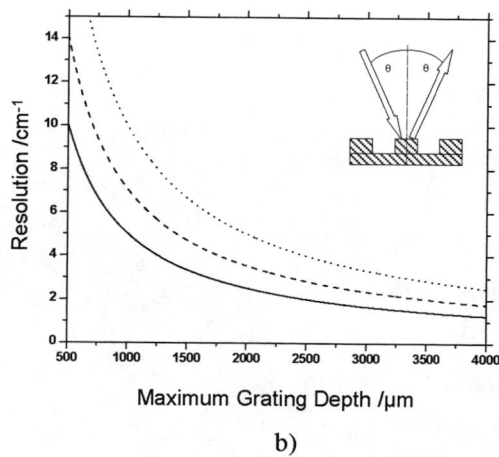

a) b)

FIGURE 2. a) Influence of wavelength on the pattern of the zeroth and first order diffraction angle for 5 illuminated grating periods of 200 μm for 20 μm (solid), 15 μm (dashed) and 10 μm (dotted) wavelengths. b) Dependence of resolution on the maximum grating depth d for incidence angles θ of 0° (solid), 45° (dashed) and 60°.

Fabrication of Lamellar Structures via Deep X-ray Lithography

From the discussion above it can be seen that the design of the lamellae plays a prominent role for the performance of the FPP FTIR. Period and maximum height of the lamellae determine the obtainable spectrum and resolution of the interferometer. The unique low divergence of intense hard synchrotron radiation provides a powerful tool for deep X-ray lithography to fabricate high-aspect-ratio lamellar structures. Jian et al. [5] demonstrate the capability of the technique for the fabrication of precise high aspect ratio structures. Using negative SU-8 photoresist, they were able to fabricate micro components of 3.6 mm height with an aspect ratio of 360.

In our work on the DOE, we cast thick negative SU8 photoresist on a stepped substrate and transfer the pattern of the lamellae by deep X-ray lithography. Then, a graphite mask carrying a lamellar gold pattern is placed on top of the substrate-SU-8-stack which is subsequently exposed by perpendicularly incident X-ray radiation, thus transferring the absorber pattern by a 1:1 shadow projection. During exposure, X-rays passing through the clear areas of the mask irradiate the underlying photoresist, whereas the gold structure prevents the radiation from penetration. During post exposure bake, the exposed photoresist areas crosslink and become, contrary to their unexposed counterpart, insolvable for the resist developer. Figure 3 illustrates the process. With a lamellae surface trimming procedure prior to the resist development, the SU8 structure itself can be used as sub-structure for the front facet mirrors. The transparency of SU8, however, necessitates a gold-reflection coating. Alternatively, a mould for reproduction of the element by hot embossing can be fabricated by the complete LIGA process cycle.

For the exposure of a stepped substrate with a constant exposure dose, the exposure dose profile has to be taken into consideration. Due to the changing photoresist thickness, the exposure dose deposited at the bottom of the structure varies. For the exposure of 1 mm thick SU-8 resists in comparison to 0.05 mm thick resists, the bottom exposure dose varies by approximately 8 times. Furthermore, the amount of SU8 shrinkage varies due to the changing SU8 volume. Both effects on the structural dimensions are currently under examination.

Figure 4 a and b show SEM images of 0.7 mm tall SU-8 lamellae fabricated via X-ray lithography. With a minimum grating constant of 125 μm, the lamellae feature an aspect ratio of 12.

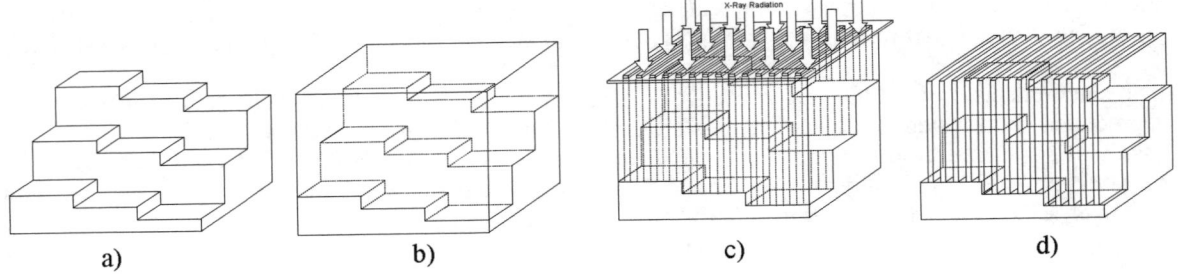

a)   b)   c)   d)

FIGURE 3. Schematic fabrication process of a diffractive optical element. a) Stepped metal substrate. b) Casting of substrate in SU-8. c) Deep X-ray lithography for the fabrication of lamellae. d) SU-8 lamellae on top of the stepped substrate after resist development.

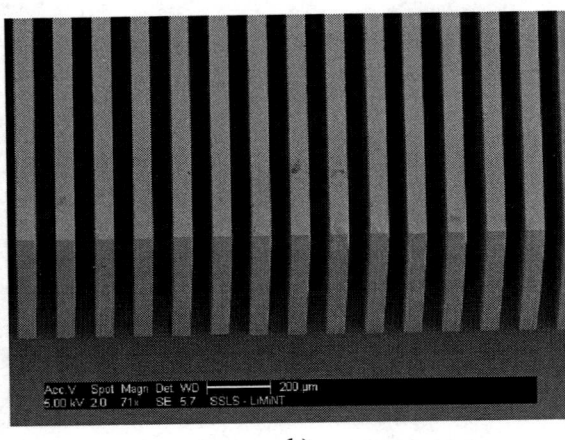

b)

FIGURE 4. SEM pictures of 700 μm tall SU-8 lamellae fabricated via X-ray lithography. Pictures were taken under an inclined angle of 30°. a) Lamellae of 700 μm height and grating constant of 200 μm, b) lamellae of 700 μm height and grating constant of 125 μm.

CONCLUSION

Based on the above results, we believe that the fabrication, by DXRL and the LIGA process, of the 3D diffractive optical element as it is required for a fast parallel-processing infrared Fourier transform interferometer is a viable approach. Although the lithography on stepped surfaces and the shrinkage of crosslinked SU8 photoresist need further study, it is felt that fly-cutting and polishing techniques may lead to the required flatness and surface roughness of the lamellae.

ACKNOWLEDGMENTS

This work was performed at the Singapore Synchrotron Light Source (SSLS) under NUS Core Support C-380-003-003-001, A*STAR/MOE RP 3979908M and A*STAR 12 105 0038 grants.

REFERENCES

1. H. O. Moser and K. D. Moeller, EP No. 0 765 488 B1, (1994).
2. J. Strong and G. A. Vanasse, *J. Opt. Soc. Am.* **50**, 113-118 (1960).
3. R. J. Bell, "Lamellar Grating Interferometers," in *Introductory Fourier Transform Spectroscopy,* New York and London: Academic Press, 1972, pp. 200-230.
4. V. Saptari, "Principles of Interferometer Operation" in *Fourier-Transform Spectroscopy Instrumentation Engineering,* edited by A. R. Weeks, Jr., Bellingham: SPIE, 2004, pp. 17-35.
5. L. K. Jian, B. Loechel, H. U. Scheunemann, M. Bednarzik, Y. M. Desta and J. Goettert, *ICMENS*, pp. 10-14 (2003).

X-ray Energy Filter using Compound Refractive Lenses

S.S. Chang[1], J.H.Kim[1], G.B.Kim[2] ,J.P.Lee[2]and S.J.Lee[2]

[1]*Pohang Accelerator Laboratory, POSTECH, San 32, Hyoja-Dong, Nam-Gu, Pohang, 790-784, Korea*
[2] *Department of Mechanical Engineering, POSTECH, San 32, Hyoja-Dong, Nam-Gu, Pohang, 790-784, Korea*

Abstract. We suggest a novel X-ray energy filter based on LIGA fabricated compound refractive lenses, and describe the results that we have been carrying out a feasibility study on our proposal. The filter consists of both of lenses array and slits, which are exactly aligned to optical axis by a lens-guide built on the same plan with slit. Its fine resist microstructure is obtained through an exposure/develops process to be characterized, optimized, and stabilized. In addition, the resist structure of slits is filled as Cu by using electroforming with accurate replication after the developing process. We describe a manufacturing process of the filter and describe some quantitative results to prove that our proposal is feasible.

Keywords: LIGA, Deep etch X-ray lithography, CXRL, PMMA
PACS: 41.50.+h, 42.15.Eq, 42.30.Va, 42.70.JK, 42.79.Bh

INTRODUCTION

Many applications in basic science and technology at synchrotron radiation (SR) sources require devices which generate not only a small focal spot but also photons of successive energy. X-ray focusing to get a small focal spot has been mainly used well-known reflective optics such as Fresnel Zone plates, bent crystals and multilayer mirrors, planar waveguides and capillary lenses. Meanwhile, Focusing X-rays by means of refractive lenses have been considered as very difficult or even impossible, because X-rays are hardly refracted (refractive index decrement for hard X-rays: $\delta \sim 10^{-5} - 10^{-6}$) but strongly absorbed in matter. However, it is experimentally proved that focusing by X-ray lenses is possible if the radius of curvature R of the lenses is chosen to be small, if many lenses are stacked behind one another in a row, and if a lens material with low Z, like aluminum, is chosen[1,2]. Generally, a collecting lens with radius r for X-rays region can produce a point focus at a distance $f=r/2\delta$. For example, the simple lens of 300 μm radius will focus the parallel X-ray beam at distance of 54 m, , if $\delta = 2.8 \times 10^{-6}$ for Al at E = 14 keV. To shorten the focal distance, a compound X-ray refractive lens (CXRL) with a series of N lenses is used and then focal length is reduced by 1/N. For the same condition, a CXRL with 30 holes brings the focal length into a range acceptable for many micro-focus experiments (f=1.8 m)

Since the lenses machined as a row of holes of 1 mm in diameter in a block of aluminum was firstly proposed for focusing high-energy X-rays in 1996, many articles have been published to describe the methods of CXRL fabrication [3]. A simple way to produce X-ray lenses is to drill a linear array of cylindrical holes inside a low Z material (such as carbon, beryllium, aluminum and polymers). The advantage of cylindrical profile lenses is the easiness of fabrication and their shortcomings are the high roughness of the machined surface, spherical aberration and very small optimal aperture of about a few tens of micrometers. For the higher accuracy, some scientific institutes have applied a lithographic technology to fabricate lenses such as the methods of deep X-ray lithography. The method of X-ray lithography has high manufacturing precision and geometrical variety. Using this method, the lens profile which has accuracy better than 0.1 μm and its surface roughness is less than 10 nm is feasible [4].

Meanwhile the refractive index of X-rays in CXRL depends on a wave length (photon energy). It means that a focal point is changed with photon energy and placed in a row along an optical axis. When a slit with reasonable hole size is located a point on optical axis, only photons focused at the point will pass a slit. Therefore it is possible to obtain selectively photons with a narrow band energy spectrum, if we can assemble a CXRL and an applicable slit. For energy resolved experiment, this is much more compact scheme then a conventional mirror system controlled by mechanical manipulator and more useful.

CP879, *Synchrotron Radiation Instrumentation: Ninth International Conference,*
edited by Jae-Young Choi and Seungyu Rah
© 2007 American Institute of Physics 978-0-7354-0373-4/07/$23.00

In this work, we propose the novel X-ray filter capable of obtaining photons with consecutive energy and describe feasible study. We also describe fabrication of CXRLs based on LIGA fabrication and the result about transmission characteristic of X-rays focused by a series of cylindrical unit lenses fabricated in PMMA

REFRACTIVE X-RAY LENSES AND ENERGY FILTER

To realize X-ray focusing optics by refractive lenses, there are points to be duly considered. Since the real part of the refractive index is smaller than one for hard X-rays in matter, focusing is done by concave lenses. In addition, materials consisting of light atoms allow to reduce X-ray absorption and to increase the gain in the focal spot of an X-ray lens compare to heavier materials. Besides, the exact patterning of the lens element form with the possibility of a coaxial arrangement of the elements is necessary in order to obtain the minimal size of a focal spot. A high roughness of the surface of the lens elements needs also to be avoided because it reduces the gain in the focal spot. Depending on these limitations, LIGA technique would be a suitable process to pattern these lens elements.

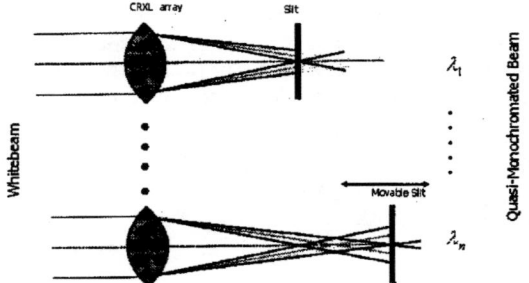

FIGURE 1. Scheme view of the energy filter using energy dependency of the refractive index in matter.

Due to the very weak refraction of hard X-rays in matter, a CXRL is composed of many lenses stacked behind each other. It is known for visible optics that the focal length of a series of N lenses in contact with each other is reduced by $1/N$. Thus, X-ray CXRL with short lengths have been made by aligning an array of N closely packed X-ray focusing bi-concave lenses with radius of curvature R made of low Z materials with resulting focal length given by[2]

$$f = R/2N\delta \qquad (1)$$

where the complex refractive index of the CXRL material is expressed by $(n = 1 - \delta - i\beta)$

We design consecutive holes with 3 different radius of curvature ($R \approx 50$ µm, 100 µm, 150 µm). Therefore the each lens array with 200 elements (momentarily neglecting absorption effects) have a focal length of 30 cm, 60 cm and 90 cm respectively, which is reasonable for most SR source experimental configuration. To avoid strong absorption, the bridge distance of between adjacent concave is made as thin as possible ($d \approx 20um$).

The refractive index decrement δ decreases by increase of photon energy. Therefore the refractive index of X-rays in matter depends on photon energy. It means that the focal length in matter is changed as photon energy changes and consecutively placed on optical axis. In the synchrotron radiation with white energy spectrum, using a CXRL and a movable slit with reasonable width (Fig. 1), we build up the X-ray energy filter which can obtain the continuous photon energy spectrum. It is consists both of a lens and slit, and patterned them to be located on the same plane for the easy alignment. After individual process, CXRL is inserted through lens-guide and then aligned with an optical axis

PROCESSING AND FABRICATION

Micro-fabrication using X-ray lithography in PLS (Pohang Light Source) is a well established process. Lens material use PMMA, which is the most popular resist in the conventional deep X-ray lithography process, because it can provide high-resolution capability and excellent sidewall quatity. X-ray exposure of the PMMA is performed at the beamline 9C1 which is dedicated for micromachining. The main influence of X-ray lithography processing on structure quality, namely solvent content after pre-exposure baking,, value of absorbed radiation in the exposed volume of the resist layer and temperature time regime for post-exposure baking were optimized on the basis of the characteristics of PMMA resist.

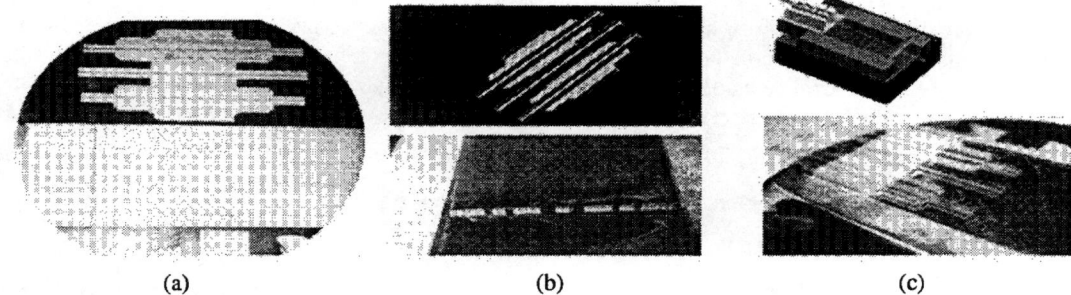

(a)	(b)	(c)

FIGURE 2. (a) X-ray mask with lenses array(top) and slit pattern with lens guider (bottom), (b) PMMA structure of lenses array (top) and metal structure of slit (bottom), (c) assembling of lenses array and slit.

In deep X-ray lithography, a mask patterning Au absorber to cut off selectively X-ray needs. The substrate used for mask is 4 inches silicon wafer with 150 μm thick. It is deposited with a thin Cr/Au layer of 30, 150 nm, respectively by e-beam evaporation. Final substrate comprised of a 180 μm thick membrane spin-coating the SU-8 photoresist of 30 μm thick on it. To create the X-ray masks, the substrate is first patterned by UV-exposure and developed, and then resist mold is filled as gold. As shown in Fig.2 (a), the resultant X-ray mask plated a gold absorber of 25 μm minimal thickness on Si substrate is used for the formation of PMMA lenses with element of cylindrical profile. A masking pattern is divided two parts. One is a lenses array and the other is a slit pattern with lens-guide. A mask is fitted with small gap ($\approx$ 2mm) on the substrate to minimize the divergence effect.

By shadow printing, the layout of the mask absorbers is copied into the resist. In the case of PMMA resist, the molecular weight is locally reduced by 3 orders of magnitude, rending those areas soluble in an organic developer. The GG-developer is used for dip development at 20 °C. Best results are achieved if the dose at the irradiated resist top does not exceed 20 kJ/cm³ and decays almost exponentially to 4 kJ/cm³ at the irradiated resist bottom. Using this optimum conditions for deep X-ray lithography process the designed PMMA lens and slit structure could be exposed and developed as shown in Fig.2 (b). In the case of lens, we use a bare resist of 1 mm thick to avoid under cut which is generated by secondary radiation from boundary layer of substrate. For slit we use PMMA resist based on Cu substrate. Hence, after developing, the negative-structure of PMMA is filled by Cu.

Figure 4 shows SEM photograph of structure of lenses and slit. Figure 4 (a) is lens array with diameter of 100 μm and 200 μm (top). The surface of opened sidewall studied by SEM photographs and Laser surface measurements is obtained identical features to design and reasonable roughness values. The overall r.m.s. roughness value is in the range from 10 to 15 nm. We analogize the rough of lens wall might be the same value. In the lower SEM photograph, it is clearly shown that the gap distances 20 μm between neighbouring circular holes is fabricated by deep X-ray lithography. Figure 4 (b) shows metal structures of oval shaped slit with 3 sizes of widths (10 μm, 20 μm and 30 μm). The oval shape is more effective to collect beam focused by lens array and filter out noisy beam than other shape such as rectangular or circular type. As shown in Fig.4 (b) the sidewall in slit is very rough and inhomogeneous, but it goes work well for our use to filter out extra-energy beam and cuts off noisy beam.

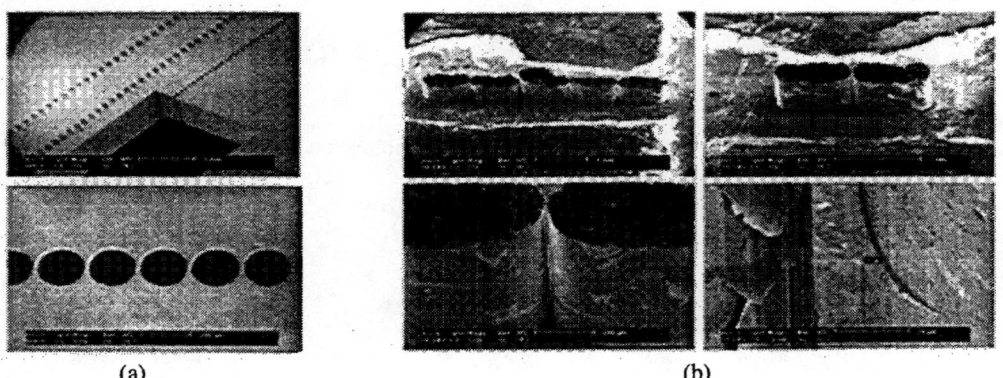

(a)	(b)

FIGURE 3. (a) X-ray mask with lenses array(top) and slit pattern with lens guider (bottom), (b) PMMA structure of lenses array (top) and metal structure of slit (bottom), (c) assembling of lenses array and slit.

EXPERIMENTAL RESULTS

Figure 4 is the experiment set up used to appraise focusing capability and filter characteristic in relation to a CRXL capable of focusing in one dimension (top). It is carried out at the 7C "X-ray Imaging beamline" in PLS. The lens is placed a distance 14 m from source and transfers an image onto a screen at a distance L_4 behind the slit. Here the translatable slit is used to determine a corresponding energy to the spot size at the focal point.

The SR source has the fine size in practice. Thus the image of the source is formed at the distance D, which is determined by the equation $D=FL/(L-F)$ [4], where $L(\sim L_1 + L_2)$ is a distance from the SR source to the lens and $F(\sim L_1)$ is the total focal distance for N identical lenses. For the nominal photon energy of 8 keV, the focal point F is

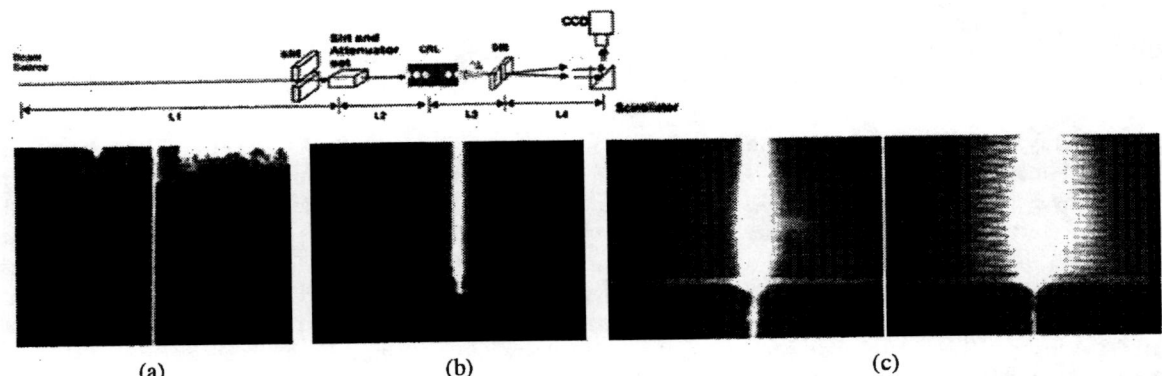

(a) (b) (c)

FIGURE 4. The experimental set up used to test performance of CXRL capable of focusing in one direction (top): Left side mages obtained through silt in case of without lenses (a) and with lenses (b). (c) are images of a free standing W mesh (20 µm x 20 µm with 20 µm in diameter.) obtained in case of lenses with 200 µm (left) and 300 µm.(right) in diameter

obtained 6 cm from theoretical calculations. We perform all of our experimental to this energy.

Figure 4 (a) shows the image of beam obtained from the slit when lens isn't on the beam path. Meanwhile (b) is the image from a slit with CXRL, which is clearly shown that the lens is capable of focusing in one dimension.

We assume that the image in the focal plane of the lens is a scalar sum of intensities from all points of the source. Then, using the law of geometrical optics, we can write down the size of the expected image:

$$S_f = \frac{D}{L} S_{SR} \qquad (2)$$

where S_f is the size of the SR source image in the CXRL focal plane, and S_{SR} is the SR source size.

Figure 4(c) depicts the image of a two dimensional W mesh, whose periodicity is 40 µm in both directions. It isn't an exact transcription of mesh in observation plan (screen). But the image of mesh shows is magnified toward horizontal direction. Note that image of object is transferred, illustrating the possibilities of the imaging reduction and magnification in X-ray optics.

DISCUSSIONS AND CONCLUSION

The fabrication and quantitative analysis for LIGA fabricated CXRL are performed. The lenses are shown to be suitable for use in SR in the energy range of 8 keV around. The combination of both a lens and a silt provides a possibility of energy selectable filter. These results could be applicable to basic science such as a nano-patterning, 2D imaging and an energy resolved experiment in the X-ray range. To further improve these results, high quality and precision of the fabrication would be added.

REFERENCES

1. A. Snigirev , V. Kohn, I. Snigireva and B. Lengeler *Nature (London)* **384** 49 (1996)
2. B. Lengeler, J. Tummler J, Snigirev A, Snigireva I and Raven C *J. Appl. Phys.* **84** 5855–61 (1998)
3. P. Elleaume, Nucl. Instr. and Meth. A 412 (1998)

4. V.A. Chernov, K.E. Kuper et.al, Nucl. Instr. and Meth. A 543, 326-332 (2005)

X-ray-based Micro/Nanomanufacturing at SSLS — Technology and Applications

L.K. Jian, B.D.F. Casse, S.P. Heussler, J.R. Kong, H.O. Moser,
Y.P. Ren, B.T. Saw, Shahrain bin Mahmood

Singapore Synchrotron Light Source (SSLS)
National University of Singapore (NUS), 5 Research Link, Singapore 117603

Abstract. Deep and high-aspect-ratio micro/nanostructures can be accurately patterned in a variety of resists by proximity lithography using high energy, intense, parallel beams of X-rays from synchrotron radiation sources. Using so called LIGA technology, high-aspect-ratio micro/nanostructures can be produced with vertical dimensions ranging from micrometers to millimeters and horizontal dimensions as small as microns. SSLS has set up its LiMiNT facility (Lithography for Micro/Nanotechnology) and is running it partly as a foundry, partly as a research lab. Under the foundry aspect, work is done for customers in various fields of applications. SSLS' own research is focusing on the development of devices and artificial composite materials such as electromagnetic metamaterials. In this paper, the technology capabilities of the LiMiNT facility and application examples are presented.

Keywords: Synchrotron radiation, micro/nanomanufacturing, LIGA, X-ray lithography
PACS: 07.85.Qe, 41.20.Jb, 81.07.-b

INTRODUCTION

The LIGA (an acronym for the German words for lithography, electroplating, and molding) process utilizes X-ray synchrotron radiation as a lithographic light source. Deep X-ray lithography has its advantages when it comes to the fabrication of tall, high aspect ratio microstructures due to the shorter wavelength, higher penetration depth and larger depth of focus of the X-ray photons [1,2]. Based on LIGA, high-aspect-ratio micro/nanostructures can be produced with vertical dimensions ranging from micrometers to millimeters and horizontal dimensions as small as microns. The LiMiNT facility (Lithography for micro/nanotechnology) at SSLS is a one-stop shop for micro-nanomanufacturing using the full LIGA process and is run partly as a foundry, partly as a research lab [3-8]. Under the foundry aspect, work is done for customers in various fields of applications. SSLS' own research is focusing on the development of devices and artificial composite materials such as electromagnetic metamaterials for the THz spectral range including the near infrared that is relevant for telecommunication, near infrared photonic devices, X-ray and infrared microoptics, and mechanical nanofilters. In this paper, the technology capabilities of the LiMiNT facility and application examples are presented. To demonstrate its performance, latest results including THz electromagnetic metamaterials, plastic templates, and mould inserts applications are reported.

TECHNOLOGY CAPABILITIES IN MICRO-NANOMANUFACTURING

SSLS is operating the compact 700 MeV electron storage ring Helios 2 that produces synchrotron radiation from two superconducting dipoles which run at a magnetic flux density of 4.5 T. The characteristic photon energy is 1.47 keV. The useful spectral range extends from about 15 keV to the far infrared. Beam current after injection is up to 500 mA. Lifetime ranges between 11 and 17 hours depending on beam current.

The LiMiNT (Lithography for micro/nanotechnology) beamline is connected to Dipole 2 of Helios 2 and the end station is an Oxford Danfysik scanner (Fig. 1). The LiMiNT beamline provides reasonable photon flux for (deep) X-ray lithography. The useful spectral flux at the sample covers a bandwidth from 2 keV to 10 keV.

CP879, *Synchrotron Radiation Instrumentation: Ninth International Conference*,
edited by Jae-Young Choi and Seungyu Rah
© 2007 American Institute of Physics 978-0-7354-0373-4/07/$23.00

The currently available equipment at SSLS allows complete prototyping using the integral cycle of the LIGA process for producing micro/nanostructures. It extends from mask writing via either laser direct writing or e beam lithography over X-ray irradiation, development, to electroplating in Ni, Cu, or Au, and, finally, hot embossing in a wide variety of plastics as one of the capabilities to cover a wide range of application fields and to go into higher volume production. The process chain also includes plasma cleaning and sputtering as well as substrate preparation processes including metal buffer layers, plating bases, and sacrificial layers, as well as spin coating, polishing, and dicing. Furthermore, metrology using scanning electron microscopy (SEM), optical profilometry, optical microscopy, and X-ray phase contrast imaging and tomography is available. Except for dicing and polishing, these processes are all located in a class 1000 cleanroom.

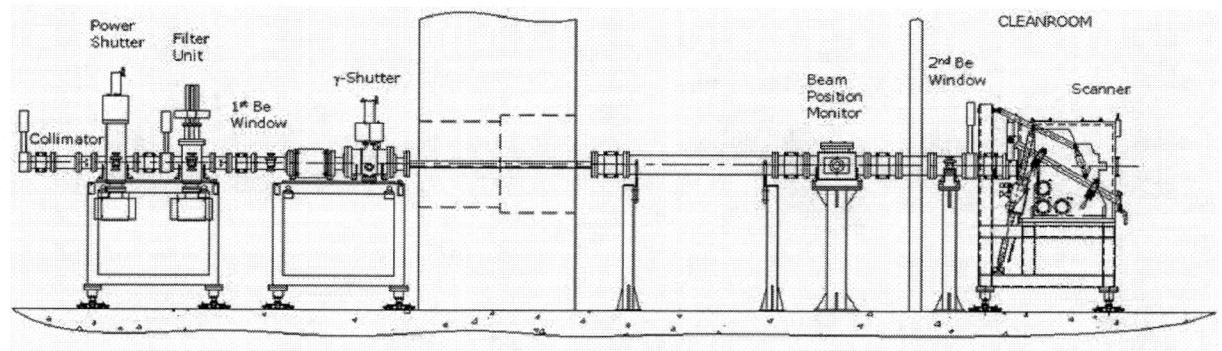

FIGURE 1. Schematic layout of the LiMiNT beamline for micro/nanofabrication at SSLS.

Figure 2 shows some of the key equipment used to realize the micro-nanomanufacturing capability at SSLS.

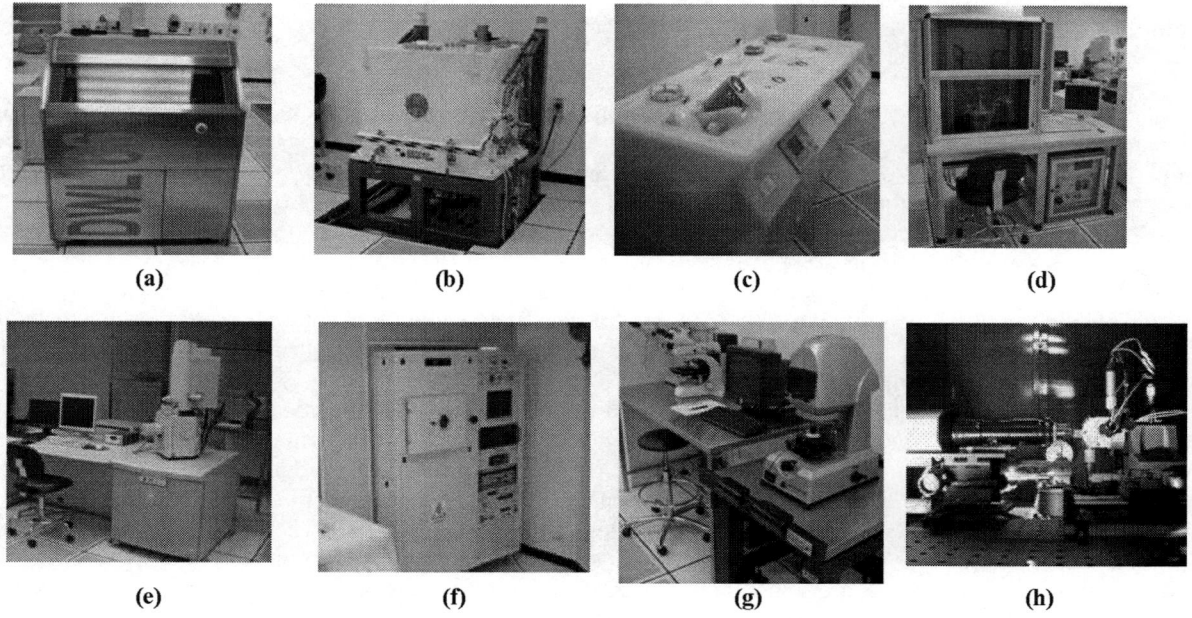

FIGURE 2. Key equipment at SSLS. **(a)** Heidelberg Instruments DWL 66 direct-write laser system for mask fabrication. **(b)** Oxford scanner, designed for 4-inch-mask complying with the NIST-standard. **(c)** µGalv plant (M-O-T) with process circuits dedicated to Au, Cu and Ni plating. **(d)** HEX 01 hot embossing system from Jenoptik-Mikrotechnik. **(e)** Sirion scanning electron microscope (FEI) which also serves as platform for electron beam lithography with Nabity NPGS. **(f)** NSP 12-1 (Nanofilm Technologies International) magnetron sputtering system with DC and RF magnetron sputtering guns. **(g)** Wyko NT 1100 profiler for measurement of height and surface roughness of microstructures. **(h)** PCIT station for the analysis of hollow high aspect-ratio structures by phase contrast imaging and tomography.

EXAMPLES OF MICRO-NANOMANUFACTURING

Applications pursued under SSLS' own research include electromagnetic metamaterials (EM³), near infrared photonic bandgap devices, X-ray multielement optics, and infrared diffractive optical elements (DOE).

SSLS has produced the first THz EM³ by microfabrication with resonance frequency extended from 1.5 to 2.4 THz [6]. Meanwhile, SSLS has succeeded to extend the accessible spectral range by two more orders of magnitude by means of nanofabrication of such EM³ to resonance frequencies approaching the near infrared spectral range as used in telecommunication [8]. Figure 3 shows a resonance curve and a picture of such an EM³ at 187 THz that is close to the 193 THz that correspond to 1.55 µm wavelength used in fiber optical telecommunication.

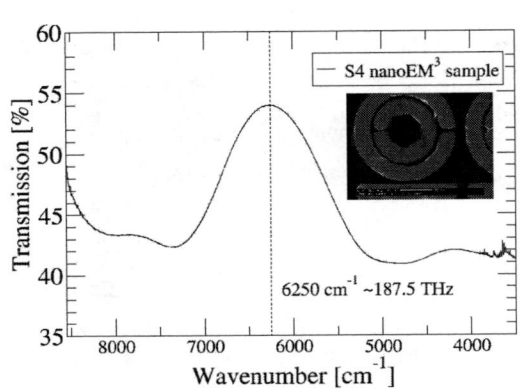

FIGURE 3. Measured infrared transmission spectrum of a rod-split-ring EM³ as shown in the inset. The outer diameter of the rings is 620 nm.

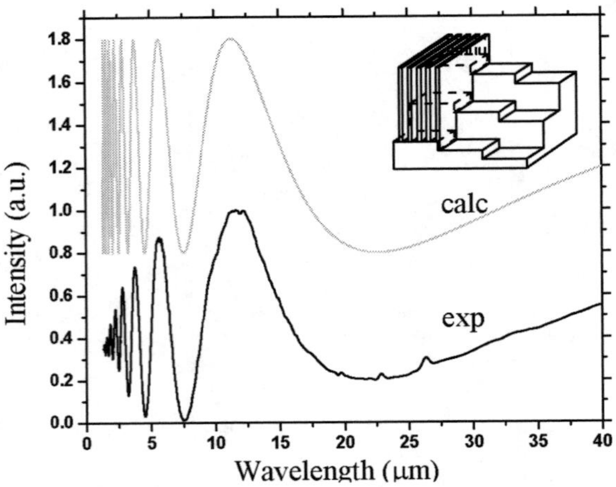

FIGURE 4. Spectral response of a single DOE cell with a depth of 8 µm and a grating period of 200 µm.

Another example is the development of an infrared diffractive optical element (DOE) [7] that aims at setting up a fast parallel-processing infrared Fourier transform interferometer. Intended applications include high-bandwidth non-periodic phenomena like explosions or combustion processes that are not readily amenable to standard Fourier transform interferometers that usually feature mechanically scanning mirrors. Figure 4 (inset) illustrates the principle of the diffractive optical element that is the heart of the system as well as the simulated and measured interferogrammes of such a diffractive optical element.

SSLS is increasingly active for commercial customers. It has produced a wide variety of devices covering resist structures for testing and electroplating, and Ni molds for injection molding and hot embossing. For demonstration purposes, high aspect ratio nickel hot embossing mould inserts have been fabricated for MiniFab Pty Ltd representing a group of customers.

The fabrication of the required hot embossing mould inserts comprises the process steps of mask layout design, optical mask fabrication, X-ray mask fabrication, substrate preparation, X-ray exposure, resist development, nickel electroplating, polishing, dicing and metrology. AutoCAD is used in the mask layout design. The pattern area is usually inscribed in an 80 mm diameter circle. The optical mask fabrication is carried out all in-house with our Heidelberg Instruments DWL 66 direct-write laser system with 4 µm resolution. The X-ray mask was fabricated with a graphite wafer of 200 µm thickness as membrane material and 20 µm electroplated gold as absorber. The finished optical and X-ray masks are shown in Fig. 5. X-ray lithography was performed at the LiMiNT beamline of SSLS with an SU-8 X-ray lithography process [9].

For the fabrication of the metal mould for hot embossing or injection molding, there are two process routes to form the base of the mould insert. One is to form the base of the mould insert through overplating which can be time consuming. An alternative is to use a thick metal plate as the base on top of which the microstructures are plated. In the present application, the second option was selected using a 2 mm thick nickel plate as the base of the mould insert and plating the microstructures on top (Fig. 6).

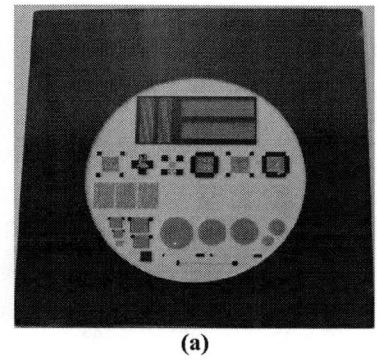 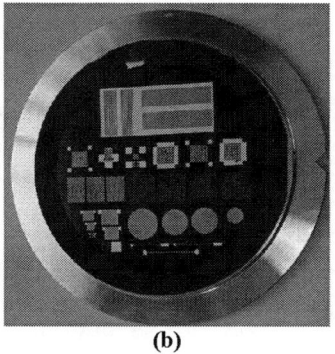

(a)   (b)

FIGURE 5. Masks used for the fabrication of mould inserts for the multi-project wafer contract. **(a)** Optical mask. **(b)** X-ray mask.

 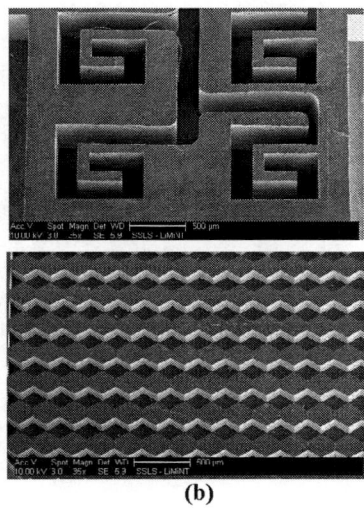

(a)   (b)

FIGURE 6. Nickel hot embossing mould inserts fabricated by LIGA process at SSLS **(a)** Ni plated microstructures before polishing **(b)** SEM image of Ni mould insert.

ACKNOWLEDGMENTS

Work performed at SSLS under NUS Core Support C-380-003-003-001, A*STAR/MOE RP 3979908M and A*STAR 12 105 0038 grants.

REFERENCES

1. E.W. Becker, W. Ehrfeld, P. Hagmann, A. Maner and D. Muenchmeyer, *Microelectron. Eng.* **4**, 35 (1986).
2. J. Hruby, *MRS Bulletin*, Vol. 26, No.4 (2001).
3. H.O. Moser, L.K. Jian, B.D.F. Casse, S.P. Heussler, M. Bahou, M. Cholewa, J.R. Kong, B.T. Saw, Shahrain bin Mahmood and P. Yang, *Journal of Physics: Conference Series* **34**, 15-21 (2006).
4. L.K. Jian, B.D.F. Casse, S.P. Heussler, J.R. Kong, B.T. Saw, Shahrain bin Mahmood and H.O. Moser, *Journal of Physics: Conference Series* **34**, 891-896 (2006).
5. B. D. F. Casse, H. O. Moser, L. K. Jian, M. Bahou, O. Wilhelmi, B. T. Saw and P. D. Gu, *Journal of Physics:Conference Series* **34**, 885-890 (2006).
6. H.O. Moser, B.D.F. Casse, O. Wilhelmi and B.T. Saw, *Phys. Rev. Lett.* **94**, 063901 (2005).
7. S. P. Heussler, H. O. Moser, C.G. Quan, C. J. Tay, K. D. Möller, M. Bahou and L.K. Jian, *this conference*
8. B.D.F. Casse, H.O. Moser, M. Bahou, L.K. Jian and P.D. Gu, *IEEE Transactions on Nanotechnology*, (2006), in press.
9. L.K. Jian, Y. Desta, J. Goettert, M.Bednarzik, B. Loechel, Y. Jin, G. Aigeldinger, V. Singh, G. Ahrens, G. Gruetzner, R. Ruhmann and R. Degen, *Proc. SPIE* Vol. 4979, 2003, pp. 394-401.

Study of LOR5B resist for the Fabrication of Hard X-ray Zone Plates by E-beam Lithography and ICP

Y. T. Chen[a, b], I. K. Lin[a], T. N Lo[a], C. I. Su[a,c], C. J. Liu[a], J. H. Je[d], G. Margaritondo[e] and Y. Hwu[1,3]

[a] *Institute of Physics, Academia Sinica, Taipei 115, Taiwan.*
[b] *Department of Materials Engineering, Tatung University, Taipei 104, Taiwan.*
[c] *Institute of Optoelectronic Science at National Taiwan Ocean University, Keelung 202, Taiwan*
[d] *Department of Materials Science and Engineering, Pohang University of Science and Technology, Pohang, Korea*
[e] *Ecole Polytechnique Fédérale, CH-1015 Lausanne, Switzerland*

Abstract. We used an approach combining Inductively Coupled Plasma (ICP) etching process and high resolution electron beam lithography and successfully fabricate high aspect ratio zone plate for hard-x-ray applications. The electron beam lithography defines the pattern with outmost zone dimension smaller than 100nm while the consequently ICP produced high aspect ratio structures. Both chacteristics, high resolution patterning and high aspect ratio are required to produce zone plate devices for multi-keV x-rays. We demonstrated that a zone plate with a 60nm outmost zone and a thickness of 500nm is achievable by this approach.

Keywords: ICP, Zone Plate, Hard X-ray
PACS: 61.10.Kw

INTRODUCTION

The thick nanopatterned zone plate of metallic structure is required to create sufficient phase change to the high energy x-ray pass through different part of the zone plate. In addition to the thick metallic structure small outmost zone width determines the imaging or focusing resolution when used in an x-ray microscopy. With the advanced electron beam lithography, large patterns (> 500 μm area) with precision better than 10nm can be defined with state-of-the-art electron beam writer. However, it is not so trivial to produce high aspect ratio structure of such precision which is the primary reason that even the best achieved lateral resolution of a transmission x-ray microscope are of the order of 60nm. Specifically, electron beam writer loose its placement accuracy when the thickness of the resist become too thick, due to the sharp focusing of the electron beam. We used 30KeV cold field emission scanning electron microscope (SEM Hitachi S-4200) to be our e-beam writer. The etching development process also produces further uncertainties due to the difficulty in controlling etching time. With a successfully demonstrated process of fabricating zone plate of 30nm line with an aspect ratio of 10, we seek complementary process which can help to extend the aspect ratio further than currently limited by the process of e-beam lithography alone.

In this work, we used Elionix EIS-700 Inductively Coupled Plasma Etching system (ICP) etching method to replicate a thin metallic structure patterned by the e-beam lithography. This combined approach is to seek circumvent the problem in producing thick structures by e-beam lithography and wet etching by a dry etching process. ICP as a well know dry etching process is superior in many ways to another popular dry etching process, the reactive ion etching (RIE). Although reactive ion etching (RIE) has been widely used in zone plate fabrication, ICP obtains the higher etch rate and more precise vertical profile than is possible with RIE. In RIE, only one r.f.

CP879, *Synchrotron Radiation Instrumentation: Ninth International Conference,*
edited by Jae-Young Choi and Seungyu Rah

power is applied to regulate both plasma density or ion energy (or d.c. bias) and it is impossible to control them separately. Although the etch rate increases with r.f. power, the increased r.f. power can cause severe sputtering by energetic ions at the sample surface, leading to increased damage on the sample and decreased etch selectivity. In this regard, high density plasma etching using an inductively coupled plasma (ICP) source offers an attractive alternative over conventional RIE [1]. ICP etching uses a secondary r.f. power (ICP power) to generate a high density plasma without correlating to the ion energy. Thus, ICP etching produces low surface damage while achieving high etching rates. In this study, ICP etching of LOR 5B has been conducted to improve the etch rates. The etch rate, vertical profile, and sidewall roughness of the side wall of zone plate have been studied as a function of ICP power and r.f. power

FABRICATION PROCESS

In Fig. 1, there is the zone plate fabrication process. We prepare a Si (100) wafer and the thickness of wafer is 525um. Low stress SiN membrane is deposited on Si wafer by LPCVD. Then we coat photo resist (PR) S1813 on back side of Si wafer and we define the pattern by Mask Aligner. We use this pattern to etch SiN membrane with CF_4 by RIE. This wafer would put into KOH to etch Si and we can get a SiN window on Si wafer. And we deposit Au to be conduction layer by thermal evaporation. And we coated PR LOR 5B with 2000 rpm by spin coater and baking time is 5min with $180^\circ C$.

In order to make hard mask on LOR 5B, we deposit Cr layer (20nm) to be hard mask by thermal evaporation. To define the pattern on Cr layer, we use PMMA to define the zone plate pattern by E-beam writer [2]. And we etch Cr layer by Cr etching solution with 30 seconds. And we etch LOR 5B by ICP with ICP power from 250W to 350W and DC bias from 50V to 100V to get high aspect ratio structure. We deposit Au on our structure by electroplating to get high aspect ratio zone plate structure.

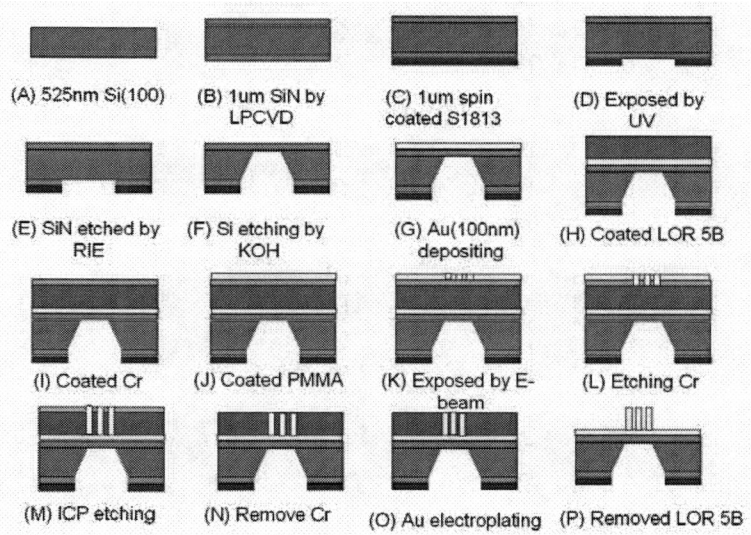

FIGURE 1. Zone plate structure fabrication process.

ION COUPLE PLASMA (ICP) ETHCING

ICP etching of LOR 5B was carried out by varying the ICP power while other parameters were DC bias from 50V to 100V, 5 mTorr, and 30 sccm O_2. As the ICP power increased from 250 W to 350 W, the cross section morphologies of LOR 5B are different and the etching rate is increasing via the ICP power increased.

In Fig. 2, the under cut damage is occurred with ICP power of 250w. With increasing the etching time, the structure would be damaged and the Cr layer is lifted off from the LOR 5B. When the ICP power is increasing from 250w to 350w, the under cut damage would be reduced but the ion bombing effect would be increased. In Fig. 3, the side wall of LOR 5B is attacked by ion bombing and appear wave wall morphologies. The outmost line width is about 60nm and the thickness of PR is about 500nm, we can get the aspect ratio is about 8:1.

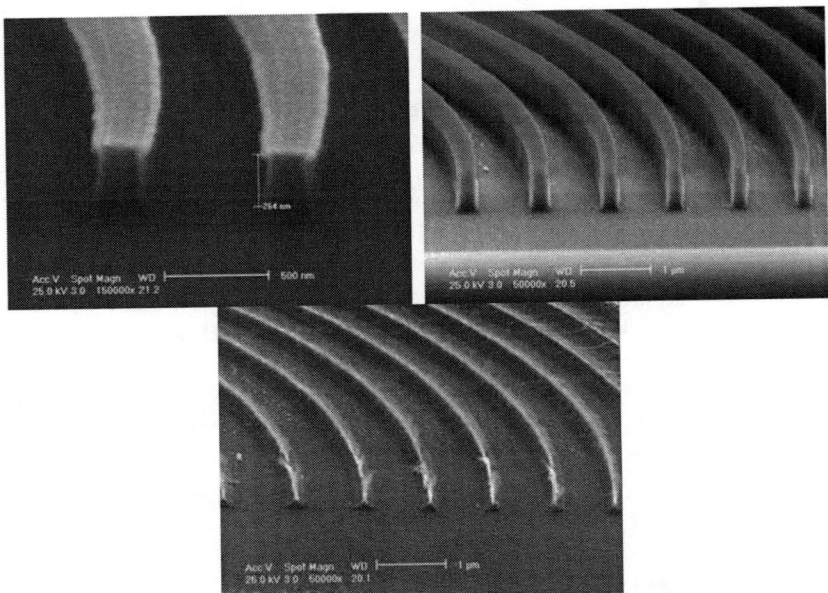

FIGURE 2. SEM images of the etching profiles of LOR5B at ICP power of 250 W and DC bias of 50V (a)60s (b)120s (c)240s.

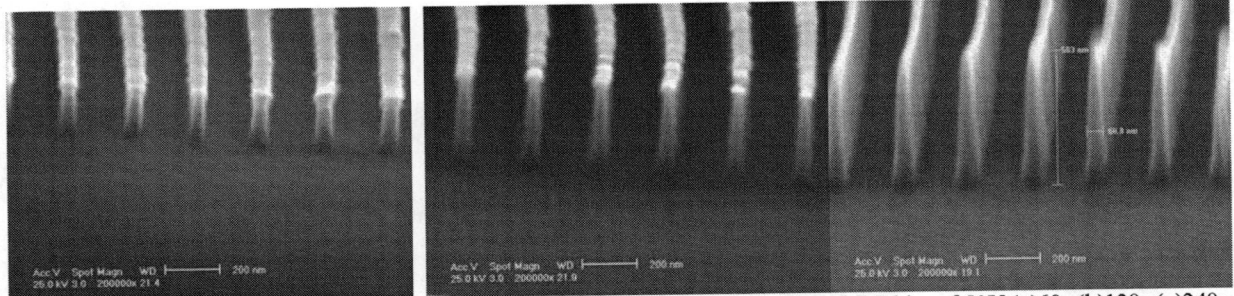

FIGURE 3. SEM images of the etching profiles of LOR5B at ICP power of 300 W and DC bias of 50V (a)60s (b)120s (c)240s.

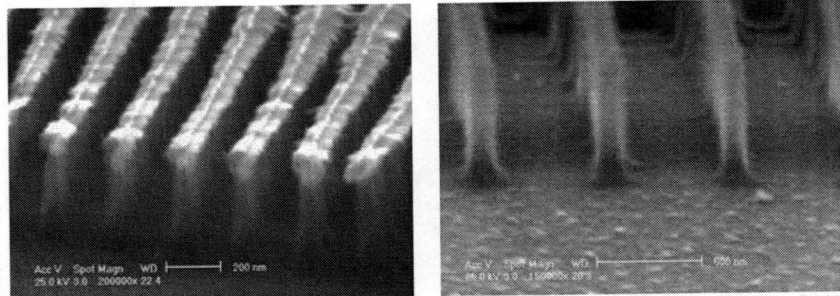

FIGURE 4. SEM images of the etching profiles of LOR5B at ICP power of 350 W and DC bias of 50V (a)60s (b)120s.

In Fig. 5 and Fig. 6, we use the ICP power of 300W with various DC bias from 75V to 100V. The etching rate of DC bias for 100V and 75V is faster than that for 50V. It is well know that increasing the substrate DC bias would increase the etching rate. In the etching time of 60s, the etching thickness of 50V in DC bias is about 250nm but the 75V and 100V is over than 500nm.

Although increasing the DC bias would improve the etching rate, the ion bombing damage would also be increased. In Fig. 5(c) and Fig. 6(c), the Cr layer had be damaged to form rough surface. These rough structures on the surface would cause the ICP efficiency and the electron field would condense on the tip of these rough surfaces. It would increase the ion bombing damage in our hard mask to change the original shape in our pattern.

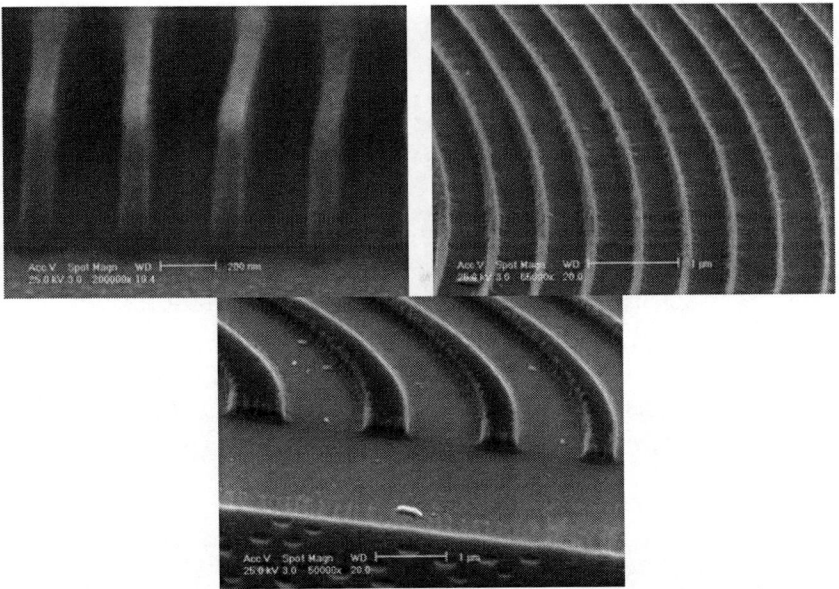

FIGURE 5. SEM images of the etching profiles of LOR5B at ICP power of 300 W and DC bias of 75V (a)60s (b)120s (c)120s.

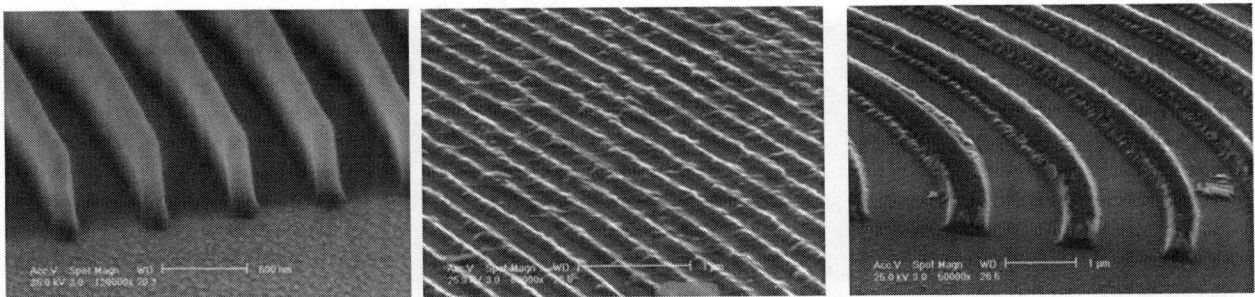

FIGURE 6. SEM images of the etching profiles of LOR5B at ICP power of 300 W and DC bias of 100V.(a)60s (b)120s(c)120s.

SUMMARY

In this study, we have investigated the effect of ICP power and substrate bias on the side wall structure after ICP etching. Although increasing the ICP power and substrate bias can get the high vertical structure of the sample, ion bombing damage in the side wall was observed simultaneously. The optimum ICP power and DC substrate bias investigated in this study are 300 W and 50 V, respectively, to obtain a high aspect ratio (8:1) structure and good side wall morphology. Once in combination of ICP with optimum parameters and subsequent electroplating process, a high resolution x-ray zone plate can be obtained.

REFERENCES

1. A. Holmberg , S. Rehbein, H.M. Hertz, Nano-fabrication of condenser and micro-zone plates for compact X-ray microscopy, Microelectronic Engineering **73–74** (2004) 639–643.
2. A. Ozawa, T. Tamamura, T. Ishii, H. Yoshihara and T. Kagoshima, Application of x-ray mask fabrication technologies to high resolution, large diameter Ta Fresnel zone plates, Microelectronic Engineering 35 (1997) **525-529**.
3. T. Schliebe, Generation of large area condenser zone plates with smallest zone width below 40nm by electron beam lithography, Microelectronic Engineering 41/42 (1998) **465-468**.

Wavefront Metrology for EUV Projection Optics by Soft X-ray interferometry in the NewSUBARU

Masahito Niibe, Katsumi Sugisaki,[1] Masashi Okada,[1] Seima Kato,[1] Chidane Ouchi,[1] and Takayuki Hasegawa[1]

Laboratory of Advanced Science &Technology for Industry, University of Hyogo,
3-1-2 Kouto, Kamigoori, Ako-gun, Hyogo 678-1205 Japan
[1]Extreme Ultraviolet Lithography Association (EUVA)
Nishimotokousan Bldg., 3-23 Kanda Nishiki-cho, Chiyoda-ku, Tokyo 101-0054 Japan

Abstract. Precise measurement of the wavefront errors of projection optics with 0.1 nm RMS accuracy is necessary to develop extreme ultraviolet (EUV) lithography. To accomplish this, an experimental EUV interferometer was developed and installed at the NewSUBARU SR facility, with which various types of interferometry experiments can be carried out by replacing optical parts easily. The wavefront error of a Schwarzschild-type test optics was measured by several methods. Finally, reproducibility below 0.045 nm RMS was achieved with the point diffraction interferometer (PDI) method, and the residual systematic error was reduced to 0.066 nm RMS excluding axial symmetric aberration.

Keywords: Optical testing, interferometry, EUV lithography, multilayer mirror, projection optics.
PACS: 41.50.+h, 42.15.Dp, 42.25.Hz, 42.62.Cf, 42.62.Eh, 42.82.Cr, 42.87.Bg

INTRODUCTION

EUV lithography (EUVL) is a promising exposure technology for next-generation semiconductors in which the patterns on a reflection mask are reductively projected on the surface of a silicon wafer through mirror projection optics. The projection optics is made of plural Mo/Si multilayer mirrors that have high reflectivity of light at a wavelength of 13.5 nm and at normal incidence. The fabrication of the projection optics for EUVL requires superfine finishing of aspherical mirror surfaces and precise assembly. For example, the wavefront error of the projection optics for 32 nm in half-pitch needs to be less than 0.45 nm RMS. Therefore, the wavefront of the optics has to be measured within an accuracy of 0.1 nm RMS. However, conventional visible light interferometers using reference mirrors are unsuitable because it is difficult to get a reference mirror with such a precise surface. Moreover, a wavefront measured by visible light is not strictly the same as that measured by EUV because of the phase shifts originated in a multilayer reflection. Therefore, it is necessary to measure the wavefront of the optics with EUV light whose wavelength is the same as that of the actual exposure tools (at-wavelength).

To realize such high-precision wavefront metrology, several at-wavelength interferometry techniques have been proposed in which a precisely fabricated reference mirror is not used. The energetic development of such metrology started in the mid-1990s at the Lawrence Berkeley National Laboratory [1]. Point diffraction interferometer (PDI) and lateral shearing interferometer (LSI) are known as types of at-wavelength interferometry using no reference mirror. Several interferometric methods were proposed that overcome the defects of the preceding technologies [2]. To estimate how feasible these methods are, an experimental EUV interferometer (EEI) was developed; using the EEI, various interferometry methods can be tested by simply changing the optical parts. A bright and stable light source with high coherency was needed for the EEI system. The EEI was then installed in the NewSUBARU SR facility at the University of Hyogo, which is equipped with an 11-m long undulator [3]. The purpose of this study is the development and selection of the best method of wavefront metrology for the actual projection optics for EUVL. In this paper, we report mainly the results of the measurement of Schwarzschild test optics and the estimation of the measurement accuracy obtained by rotating the optics.

CP879, *Synchrotron Radiation Instrumentation: Ninth International Conference,*
edited by Jae-Young Choi and Seungyu Rah

EXPERIMENTAL EUV INTERFEROMETER (EEI)

In this paper, the PDI method is mainly described as interferometric experimental method. The principle of PDI is based on the physics concept according to which the wavefront of a light forms an ideal sphere after passing through a small pinhole and the resulting spherical wave can be used as a reference wavefront. The pinhole size must be less than λ/NA, where λ is the wavelength and NA is the numerical aperture. A schematic diagram of the PDI for the EUV wavelength region is shown in Fig. 1. An aberration-free light emitted from the initial pinhole located at the object plane of the test optics is passed though a binary grating and introduced to the test optics. The grating functions as a beam splitter. After passing through the test optics, diffracted lights of the 0th and 1st order are focused at slightly separated points on the image focal plane of the test optics. A mask on which the second pinhole and a window are bored is placed at the focal plane. The 0th-order diffracted light becomes again an ideal spherical wave after passing through the 2nd pinhole. The 1st-order diffracted light passes through the window while conserving the aberration of the test optics. Both lights passing through the mask were interfered with and formed fringes on the CCD camera. Any aberration in the test optics can be determined by analyzing the fringes.

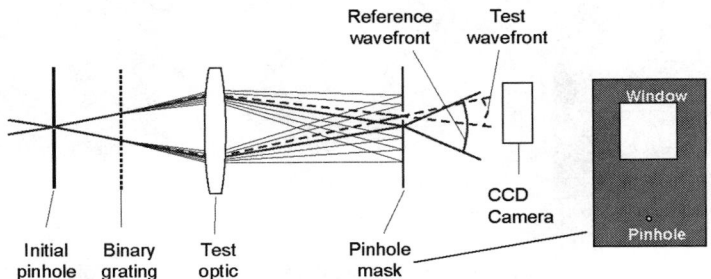

FIGURE 1. Schematic diagram of the point diffraction interferometer (PDI) at the EUV wavelength region.

Figure 2 is a schematic diagram of the EEI operating at the NewSUBARU SR facility. The light source is the long undulator (10.8-m length, 200 periods, principal harmonics: λ=13.5 nm at a 35-mm gap, Ee=1.0 GeV). An illumination optics including upstream beamline optical components is optimized, as its NA matches the test optics. Six different interferometric methods that are classified as PDI and LSI can be examined by simply replacing the optical components, such as the pinhole mask and grating [4]. The test optics was a Schwarzschild-type optics with NA=0.2 and demagnification M=1/20. The test optics was assembled with an aberration of 1.06-nm RMS by visible light PDI equipment that was developed individually for another project [5] and installed in the EEI. In this experiment, the plugging of the pinhole originated by an EUV irradiation-induced carbon deposition became a problem. The pinholes are subjected to an oxygen flow to suppress the plugging.

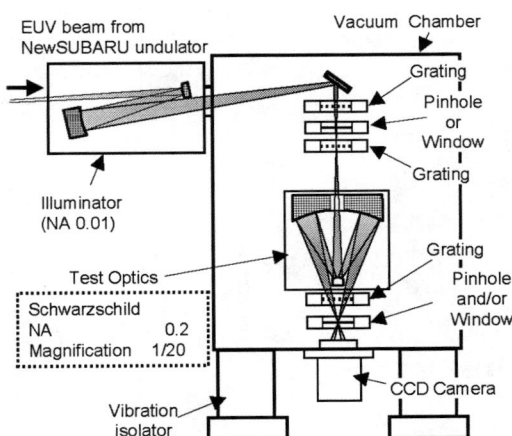

FIGURE 2. Schematic diagram of the experimental EUV interferometer (EEI).

EXPERIMENTAL RESULTS

A sample interferogram of the test optics obtained by the PDI method is shown in Fig. 3(a). The contrast of the fringes is sufficient to retrieve the wavefront. Figure 3(c) shows the retrieved wavefront of the test optics obtained by the EEI. For comparison, the wavefront obtained when assembling the optics by the visible light PDI method [5] is shown in Fig. 3(b). All obtained wavefronts were evaluated in terms of the RMS of the wavefront errors approximated by using 5th to 36th terms of the Zernike polynomial [6,7]. The RMS values of the wavefront errors for EUV-PDI and visible light PDI were 1.10 nm and 1.06 nm, respectively, and both wavefront shapes agreed well with each other. In Schwarzschild optics, there is little differences in wavefronts of visible light and EUV light, since the incidence of the major light on the mirror plane is nearly normal.

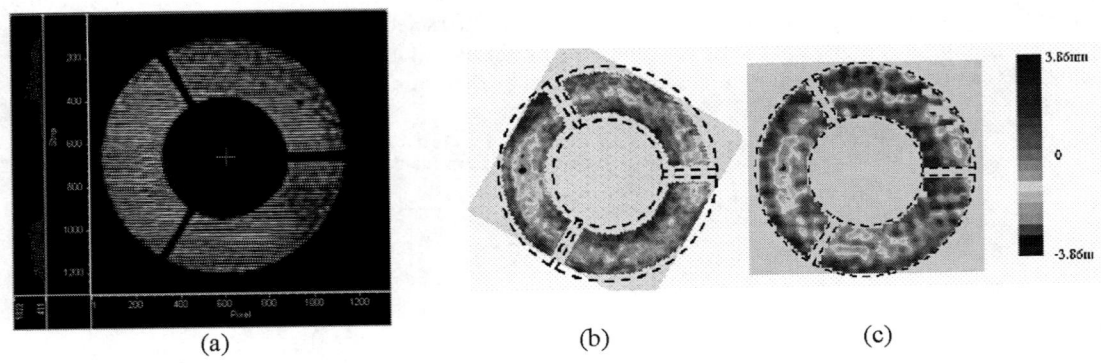

(a)　　　　　　　(b)　　　　　　　(c)

FIGURE 3. Interference fringes and wavefront map of the Schwarzschild test optics obtained by the PDI method.

Reproducibility

The alignment error of the 2nd pinhole proved to affect the measurement accuracy significantly. Therefore, the alignment was conducted very carefully, and, finally, we achieved an alignment accuracy that was better than 10 nm in the image plane and 15 nm along the optical axis by analyzing the wavefront of the preliminary measurement. To evaluate the measurement reproducibility, measurements were carried out 7 times while adjusting the alignments, and the deviation from the mean value of those Zernike coefficients was obtained. The results indicated that the reproducibility of the PDI measurement was estimated to be 0.045 nm RMS.

Systematic Error

The obtained wavefront includes the systematic error of the measurement in addition to the real aberration of the test optics. To estimate the systematic error, the measurements were repeatedly carried out by rotating the optics from 0° to 90°, 180° and 120° around the optical axis. The optics was rotated with breaking the vacuum of the EEI chamber, by manually lifting up and rotating the optics within an interval of 2 – 3 weeks.

The wavefront maps of the test optics obtained by the PDI method are shown in Fig. 4 at the rotation angles of (a): 0°, (b): 90° and (c): 180°. The images were rotated back again for the sake of comparison.

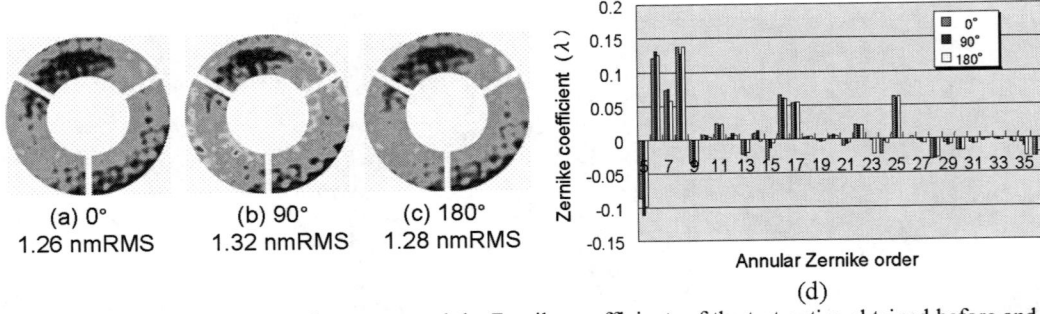

(a) 0°　　　　(b) 90°　　　(c) 180°
1.26 nmRMS　1.32 nmRMS　1.28 nmRMS

(d)

FIGURE 4. Comparison of the wavefront maps and the Zernike coefficients of the test optics obtained before and after the rotation of the optics (a) 0°, (b) 90°, (c) 180°, and (d) Zernike coefficients (unit = λ : 13.5 nm).

The three measured maps were very similar to each other. In addition, in Fig. 4 (d), the values of the 36 Zernike coefficient terms of the three wavefront maps are in good agreement with each other. The systematic error of the measurements was estimated by the difference in the wavefronts measured before and after the optics rotation. As a result, the systematic error was evaluated to be as small as 0.066 nm RMS except for the axial symmetrical components such as $z9$, $z19$, $z25$, and $z36$, which cannot be separated by rotation.

EUV Wavefront Metrology System (EWMS)

Continuing the success of the EEI, we are extending our research to the construction of the EUV wavefront metrology system (EWMS) by March 2006; using this system, the six-mirror projection optics with NA=0.25 for EUVL will be evaluated. Fig. 5 shows a full view of the EWMS installed in NewSUBARU. Its development is based on the techniques acquired with the EEI. The EWMS is a wavefront metrology standard, and it will be used for the calibration of a non-EUV wavefront sensor that will be used in the mass-production of EUV tools.

FIGURE 5. Full view of the EUV wavefront metrology system (EWMS) in NewSUBARU.

CONCLUSION

We developed metrological techniques to evaluate EUV lithographic optics using the EEI and achieved 0.045 nm RMS of reproducibility and 0.066 nm RMS of systematic error by the PDI.

ACKNOWLEDGMENTS

This work was performed under the management of EUVA as a research and development program of the New Energy and Development Organization (NEDO), Japan. The authors thank the members of operation stuff of NewSUBARU for their support and help with stable operation of the storage ring.

REFERENCES

1. K. A. Goldberg et al., *Extreme Ultraviolet Lithography*, F. Zernike and D. T. Atwood (eds.): OSA, Washington, D.C., 1995, pp. 134-141.
2. K. Murakami et al., *Proc. SPIE* **5037**, 257-264 (2003).
3. M. Niibe et al., in *Synchrotron Radiation Instrumentation*, edited by T. Warwick et al., AIP Conf. Proc. 705, 2004, pp. 576-579.
4. K. Sugisaki et al., *Proc. SPIE* **5374**, 702-709 (2004).
5. K. Otaki et al., *J. Vac. Sci. Technol.* **B20**, 2449-2458 (2002).
6. M. Born and E. Wolf, *Principle of Optics, 5th ed.*, Pergamon press, Oxford 1975.
7. S. R. Restaino et al., *Opt. Eng.* **42**, 2491-2495 (2003).

Reflection Passband Broadening by Aperiodic Designs of EUV/Soft X-ray Multilayers

Toshihide Tsuru

Research Center for Soft X-ray Microscopy, IMRAM, Tohoku University,
2-1-1 Katahira, Aoba-ku, Sendai, Miyagi 980-8577, Japan

Abstract. By using three conventional optimization algorithms, we have developed computer programs of layer thickness designing for reflection passband broadening of EUV/soft X-ray multilayers. Three programs with optimization by Simplex, quasi-Newton and Gradient methods were found to be effective to search for aperiodic Mo/Si multilayers of a leveled reflectance of 35% within a wavelength region between 13 nm and 15 nm, though the thickness structures were considerably different. For much shorter wavelengths in the water window, solutions of Cr/Sc multilayers were also found for a leveled 30% reflectance between 3.14 nm and 3.16 nm, and also for a 15% reflectance between 3.13 nm and 3.17 nm. The bandwidth $\lambda/\Delta\lambda$ of these designs were improved from 286 of periodic multilayer to 137 and 66, respectively. Two practical design solutions were used to fabricate aperiodic Mo/Si multilayer mirrors by our ion beam sputtering system. The samples show EUV reflectance of more than 15% between 13 nm and 15 nm.

Keywords: multilayer mirror, optimization method, aperiodic structure
PACS: 41.50.+h, 42.79.Bh

INTRODUCTION

In the extreme ultraviolet (EUV) and soft X-ray wavelength regions, multilayer mirror optics are actively utilized for various normal incidence X-ray imaging optics. Particularly for the reflection at a wavelength of 13.5 nm, Mo/Si multilayer mirrors are adopted as projection optics for the next generation lithography tool [1]. At much shorter wavelengths of soft X-ray in the "water window" region between 2.4 nm and 4.4 nm, Cr/Sc multilayer mirrors were good candidates for biological and medical applications [2].

In these multilayer mirrors, several tens to hundreds layers should be stacked at high accuracy, which is inversely proportional to the number of layers [3]. Because of the large number of layers, the reflection peak is of a narrow bandwidth with $\lambda/\Delta\lambda$ being a few tens to hundreds. This causes difficulties to attain high throughput in the imaging optics composed of multiple multilayer mirrors since the error of period thickness control should be less than 1%, which is extremely difficult to achieve by deposition rate stabilization. To increase the tolerance of error, reflection passband broadening can be a practical solution.

Recently, Wang *et al.* successfully achieved the broadening with aperiodic structures of Mo/Si multilayers at a wavelength range between 13 nm and 19 nm by sophisticated computer routine [4]. The reflection passband broadening method is expected to be crucial not only for the wavelength matching but also for the spreading applications using EUV/soft X-ray multilayer optics. Thus, we have tried to use simple procedure with several conventional optimization algorithms [5] for the reflection passband broadening for our application of microscope development.

In this study, we briefly describe the procedure of computer designing with three commercial optimization algorisms for reflection passband broadening by aperiodic EUV/soft X-ray multilayer mirrors. Then we present the results of design examples at two wavelengths with theoretical reflectance spectra. Finally, experimental results of reflectance spectra of Mo/Si multilayer mirrors we fabricated with the designed thickness structure are shown for demonstration of a practical use.

CP879, *Synchrotron Radiation Instrumentation: Ninth International Conference,*
edited by Jae-Young Choi and Seungyu Rah
© 2007 American Institute of Physics 978-0-7354-0373-4/07/$23.00

PASSBAND BROADENING PROCEDURE

For computer simulation and designing of the EUV and soft X-ray multilayers, we have used Berning's formula based on the Fresnel formulae of optical multilayers [6]. With the formula, layer-by-layer optimization of the thickness for the most effective increase of reflectance can be calculated as the most smooth variation of amplitude reflectance in the complex plane plots [6]. For given optical constants of a pair of materials and a substrate, the optimum thickness structure starts with a specific optimal 1st layer thickness defined by the amplitude reflection coefficients of its boundaries. After the initial a few to several layers appear as an aperiodic structure, the optimum thickness structure varies smoothly into a periodic one as the reflectance increase is saturating at several tens to hundreds layers. Then, the optimum thickness of the top terminating layer departs clearly from the periodic structure since the next material outside is environmental medium, which is vacuum in our case. It should be noted that in the optimum aperiodic structute, the thickness values of the 1st and the last layers appear as specific odd values.

We have used the periodic thicknesses of the pair at the reflectance saturation as the initial set to be used in the next computer optimization routine. The reflection bandwidth with the optimum periodic structure was taken as the start value to be compared since the bandwidth gain by introducing aperiodicity is small enough to be ignored.

In order to optimize layer thicknesses numerically for a desired reflectance profile at a specific wavelength range, a merit function (MF) is defined as;

$$MF = \frac{1}{m} \sum_{k=1}^{m} I_k^{\,2} (R_k - R^T_{\,k})^2 , \qquad (1)$$

where m is the total number of wavelength sampling with an integer k representing the position of the sampling equally spaced. At every wavelength λ_k, a reflectance difference between the calculated R_k and the target $R^T_{\,k}$ multiplied by the irradiation I_k to the multilayer structure is calculated to sum up the residuals. In this paper, an unified irradiation was assumed by settting I_k=1.

For minimization of Eq. (1), we have employed commercially available Simplex, quasi-Newton (Variable) and Gradient methods [5] to treat multiple variables of layer thicknesses. We have found that the result of layer thickness distribution varies depending on the initial layer structure, the leveled target reflectance, the wavelength region, the wavelength sampling interval, and so on. Therefore, the results of the optimum periodic structure described above were used as a common starting structure for comparison. As the target reflectance and the sampling interval, we have tried several values till we obtain reasonably flat spectrum as shown in the following examples. The target wavelength regions for broadening were chosen as preferable values between ×1.5 and ×4.0 for practice. The optical constants of materials used for calculation were taken from a web site of the Center for X-ray Optics, Lawrence Berkeley National Laboratory [7].

DESIGN EXAMPLES AT AN EUV REGION

In this example, we tried passband broadening of a Mo/Si multilayer mirror suited for application in the EUV wavelength region. As the initial structure for optimization, a periodic 40 pairs of Mo and Si layer optimized for the maximum peak reflectance at an angle of incidence of 5° was used with their thicknesses set at 2.693 nm and 4.512 nm, respectively. This structure shows the maximum s-reflectance of 70% at a wavelength of 14.0 nm. The target reflectance was set at 35% between a wavelength region of 13.0 nm and 15.0 nm. A wavelength interval of a 0.01 nm was used.

When the iteration numbers for Simplex, quasi-Newton and Gradient methods reached 73587, 113 and 820, the solutions of the thickness distributions were found as shown in Fig. 1 (a), (b) and (c), respectively. From the viewpoint of residual deviations, the Gradient method was the best with 0.740 compared to the Simplex and the quasi-Newton with 2.488 and 1.780, respectively. Note that the iteration in Simplex may include deterioration and the residuals may be increased because of the random nature of the Simplex method [5]. As shown in Fig.1, passband broadening was successfully achieved by all three optimization methods used. Although the thickness structure obtained by Simplex method is of random nature, quasi-Newton and Gradient methods gave more systematic structures. Period thicknesses of each layer indicated by plus marks distribute around the period thickness used for initial structure. The theoretical calculations of the s-reflectance using optimum thickness distributions show good broadened and leveled reflectance profiles as shown in Fig. 1 (d). At the target reflectance of 35%, a slight reflectance oscillation is still remaining at shorter wavelengths. This oscillation can be made much smaller if we set a smaller target value. Actually, this oscillation was used to judge the optimal condition to set the largest target reflectance for the widest region. In the Gradient case, the optimum layer thickness distribution consists of five layer blocks as shown in Fig. 1 (c). The initial four blocks to 54th layers contribute to the reflection at around 13.5 nm, whereas the fifth final block contributes to around 14.5 nm. This thickness distribution should help avoiding an absorption loss in total by placing the layer blocks for shorter wavelengths at the bottom.

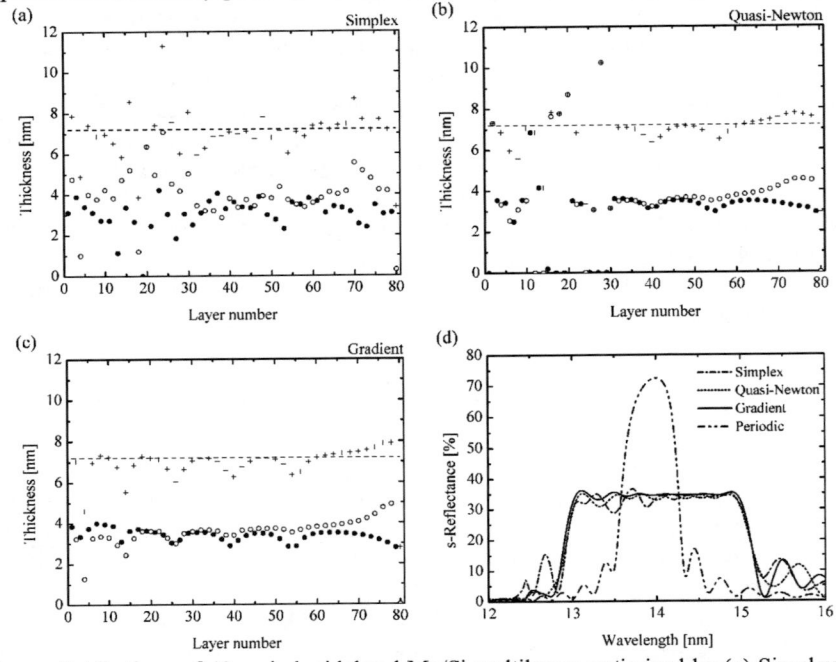

FIGURE 1. Thickness distributions of 40 period wideband Mo/Si multilayers optimized by (a) Simplex, (b) quasi-Newton and (c) Gradient methods. An angle of incidence was set at 5 deg. Solid, open and plus marks indicate the layer thickness of odd (Mo), even (Si) layers and period, respectively. The thickness of periodic multilayer used as an initial structure are indicated by dotted lines. (d) Broadened and leveled s-reflectance of Mo/Si multilayers calculated by optimum thickness distributions.

DESIGN EXAMPLES AT THE WATER WINDOW REGION

As an example for the passband broadening of multilayers at the water window region, materials of Cr and Sc were selected. We firstly broadened and leveled the reflectance at normal incidence between 3.14 nm and 3.16 nm by Gradient method. The target reflectance was set at 30%. Periodic Cr (0.624 nm)/Sc (0.953 nm) composed of 500 period was used as the initial structure. Then, using the result of this optimum thickness distribution as the initial structure, 15% broadband Cr/Sc multilayer was derived at a wavelength region between 3.13 nm and 3.17 nm by the same optimization methods.

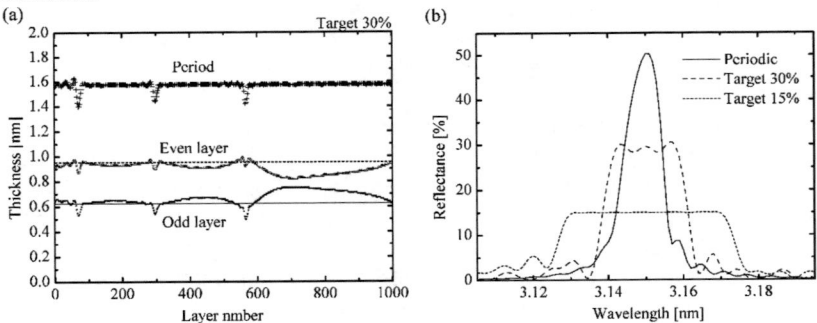

FIGURE 2. (a) Thickness distribution for a target reflectance of 30% optimized by Gradient method. Solid and dotted lines indicate the layer thickness of odd (Cr) and even (Sc) layers of the periodic multilayer, respectively. Solid, open and plus marks indicate the layer thickness of odd, even layers and period, respectively. (b) Broadened and leveled reflectance at normal incidence of Cr/Sc multilayers calculated by optimum thickness distributions.

The optimum thickness distribution having a leveled wideband reflectance of 30% is shown in Fig. 2 (a). Odd and even layers show thickness variations with the period being almost constant except for a few periods. Since the most of odd layer thickness is larger than that of periodic multilayer, the optimum thickness distribution would be within a technical limit to realize. As shown in Fig. 2 (b), the theoretical reflectance spectra for the design targets of 30% as well as a case of 15% proved wide enough profiles. Compared with the resolution $\lambda/\Delta\lambda=286$ of the periodic multilayer, bandwidth broadening factors of these designs are 286/137=2.09 and 286/66=4.3, respectively.

FABRICATION OF BROADENED AND LEVELD MO/SI MULTILAYER MIRRORS

To test the experimental feasibility, the wideband Mo/Si multilayers designed by the Simplex and the Gradient methods for the target reflectance of 35% between 13.0 nm and 15.0 nm were fabricated by our ion beam sputtering (IBS) system [8]. The deposition durations of each layer were controlled to the optimum thickness structures derived. The designed thickness of null is treated by adding the layer thicknesses on both sides into a thick equivalent layer since the materials of the layers are the same. This treatment is correct theoretically though the total effective number of layers is reduced by two.

EUV reflectance of the wideband Mo/Si multilayer mirrors fabricated were measured at BL-12A, Photon Factory, KEK with a reflectometer. As shown in Fig. 3, s-reflectance more than 15% between 13.0 nm and 15.0 nm was successfully achieved. The reduction of the reflectance level could be attributed to the thickness controlling error during fabrication. Optimization methods with conventional algorithms were found to be effective to search for aperiodic EUV multilayers of a broadened and leveled reflectance. The use of passband broadened multilayer mirrors should help relieving the severe engineering tolerances of the wavelength matching for relaying reflecting optics including imaging optics.

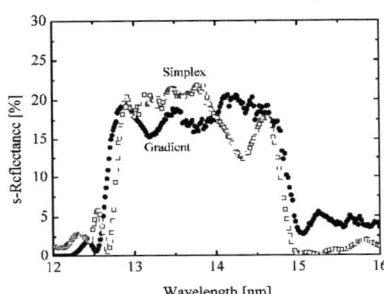

FIGURE 3. The measured s-reflectance of two broadened and leveled Mo/Si multilayer mirrors.

ACKNOWLEDGMENTS

This research was partially supported by Giant-in-Aid for Specially Promoted Research, No. 15002001 from the Ministry of Education, Culture, Sports, Science and Technology, Japan. The author thanks to Mr. T. Harada and Dr. T. Hatano for the soft X-ray reflectance measurements. Thanks are also due to Prof. M. Yamamoto for critically reading the manuscript and to Dr. T. Komiya, Dr. K. Murakami and Prof. Z. Wang for useful discussions.

REFERENCES

1. H. Kinoshita, K. Kurihara, Y. Ishino and Y. Torii, *J. Vac. Sci. Technol* **B7**, 1648-1651 (1989).
2. K. Sakano and M. Yamamoto, "Development of Soft X-ray Multilayer Mirrors for a Wavelength of 3 nm" in *EUV, X-Ray, and Neutron Optics and Sources*, edited by C. A. MacDonald et al., Proceedings of SPIE 3767, Denver, Colorado, 1999, pp. 238-241.
3. T. Tsuru, T. Tsutou, T. Hatano and M. Yamamoto, "Thickness monitoring of nm period EUV multilayer fabrication by ellipsometry" in *Synchrotron Radiation Instrumentation*, edited by T. Warwick et al., AIP Conference Proceedings 705, American Institute of Physics, San Francisco, USA, 2004, pp. 732-735.
4. A. G. Michette and Z. Wang, *Opt. Commun.* **177**, 47-55 (2000).
5. W. H. Press, S. A. Teukolsky, W. T. Verrerling and B. P. Flannery, Numerical Recipes in C, New York, Cambridge University Press, 2002, pp. 394-455.
6. M. Yamamoto and T. Namioka, *Appl. Opt.* **31**, 1622-1630 (1992).
7. http://www.cxro.lbl.gov/optical_constants/
8. T. Tsuru, T. Tsutou, T. Hatano and M. Yamamoto, *J. Electron Spectrosc. Relat. Phenom.* **144-147**, 1083-1085 (2005).

CHAPTER 9
INDUSTRIAL APPLICATIONS

State-of-the-Art Facilities for Industrial Applications on BM29 and ID24 at the ESRF

Gemma Guilera, Mark Newton, Olivier Mathon, Bernard Gorges and Sakura Pascarelli

European Synchrotron Radiation Facility, BP 220, 38043 Grenoble CEDEX, FRANCE

Abstract. BM29 and ID24 are two independent but complementary beamlines at the ESRF dedicated to X-ray Absorption Spectroscopy (XAS). The implementation of new state-of-the-art facilities, specially on ID24, devoted to homeogeneous and heterogeneous catalysis, material science and solid-state chemistry has attracted industries and academic users from all over the world. Here we present some of the activities carried out on these beamlines.

Keywords: XAS, complementary beamlines, time resolved techniques, multi-technique approach, catalysis
PACS: 30, 33.20.Rj, 80, 81.16.Hc, 82.20.-w

INTRODUCTION

BM29 and ID24 are two independent beamlines at the European Synchrotron Radiation Facility (ESRF) dedicated to X-ray Absorption Spectroscopy (XAS). On one hand, BM29 is a standard energy scanning XAS beamline with a bending magnet X-ray source; on the other hand, ID24 has an energy dispersive setup using radiation from two tapered undulators.

The excellent signal-to-noise ratio, versatility, reliability and high automation level of BM29 together with the natural features of the ID24 energy dispersive setup such as parallel and fast acquisition, stability during the measurement, small focal spot and high flux makes these two beamlines very complementary and complete, and therefore very attractive for industries to pursue entire projects.

Continuous collaboration between the different specialized groups at the ESRF, such as Sample Environment and Detector Pool, coupled with the constant feed-back and challenging requests from academic and industrial users has permitted targeted instrumental developments and a wide range of sample environments which are readily accessible to the entire scientific community.

Herein we present some of the experimental facilities available on BM29 and ID24 and some related examples.

BM29, THE STANDARD XAS BEAMLINE

BM29 is a standard XAS beamline installed on a bending magnet [1]. Accordingly, structural and electronic information on the short-range environment around selected atomic species in condensed matter can be obtained. This poses an advantage over other techniques for the study of amorphous and highly disordered solids, as well as for liquid solutions.

The strengths to which BM29 operates arise from the intrinsic properties of the ESRF synchrotron. Principally, it has a very large operational energy range (4 to 74 keV); high energy resolution; high spectral signal to noise ratio (above $7.0 \cdot 10^4$ for well prepared samples); high beam stability; and a high level of automation which allows to perform non-conventional scanning measurements, like single energy temperature scans and energy scanning x-ray diffraction. These characteristics are reinforced by the diverse sample environments (e.g. cryostat working between 3 and 400K, furnace reaching 3000°C, high pressure device –Paris-Edinburgh press– up to 15 GPa working also at high temperatures and cells for liquids) and detection modes (transmission, fluorescence and total electron yield).

The scope of samples analyzed on this beamline is countless. Examples vary from biological enzymes, ionic liquids and catalysts to magnetic, semiconductor and superconductor materials [2].

CP879, *Synchrotron Radiation Instrumentation: Ninth International Conference,*
edited by Jae-Young Choi and Seungyu Rah
2007 American Institute of Physics 978-0-7354-0373-4/07/$23.00

ID24, THE DISPERSIVE XAS BEAMLINE

ID24 is a very specialized beamline with dispersive optics for XAS measurements. The scheme of the optical setup is shown in Figure 1. Further technical information can be found elsewhere [3]. From this setup can be highlighted the high flux achievable thanks to the two tunable and tapered undulators, the high rejection of harmonics coming from the two coated mirrors, the highly focusing polychromator based on elliptically bent silicon crystals (111, 311, 220) in Bragg configuration, the small focal spot that can be reached after additional vertical refocusing, and the remarkable stability related to the dispersive setup where none of the optical components of the beamline are submitted to mechanical movement during the entire measurement. Energy range of operation is 5 to 27 keV.

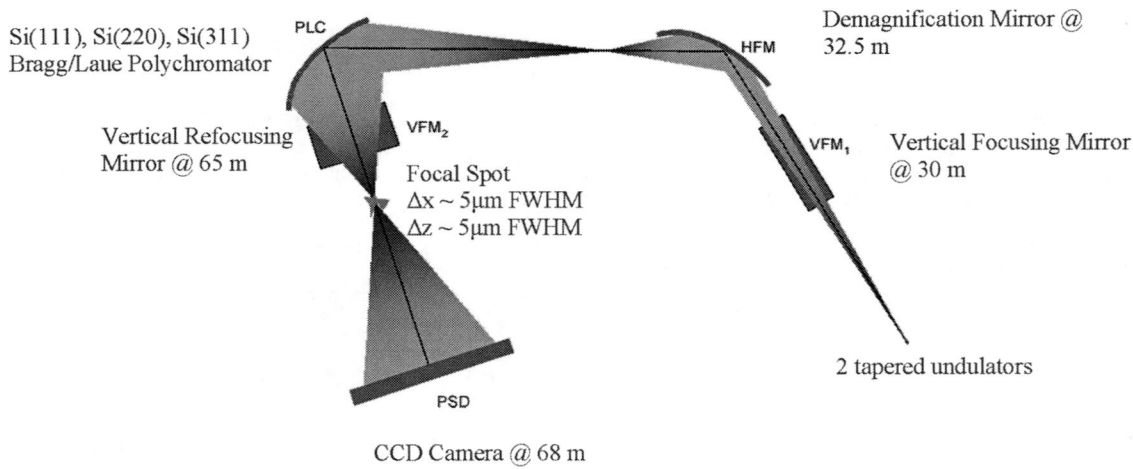

FIGURE 1. Schematic drawing of the ESRF dispersive XAS beamline ID24.

This beamline is optimized to pursue XAS experiments in transmission. The data acquisition is obtained in parallel over a large energy window that covers a complete XAS spectrum with a FReLoN2k detector (Fast Readout Low Noise), a CCD camera developed at the ESRF [4]. This arrangement provides an ideal configuration for performing time resolved studies. With the current setup we can obtain up to 400000 images, each of these being an entire EXAFS spectrum, with a time resolution of 2 ms; the only limitation being the detection system itself [5].

During the past few years, ID24 has attracted the interest of industrial and academic communities in the field of catalysis, material science and solid state chemistry aiming to study chemical processes occurring at relatively fast rates *dynamically* and *in-situ*. Catalytic exhaust converters, H-storage nanomaterials and homogeneous catalysts for Heck chemistry are only a few examples of areas where ID24 can give answers for [6].

If we closely look at the statistics given for industrial activities in developed countries we would realize that about 80% of the industrial processes use catalysts to manufacture all sorts of products. For this reason, and because of the stricter environmental regulations, chemical industries need to develop new, more active, more selective, more efficient, cleaner and cheaper catalysts. Knowledge of how catalysts act and behave while the reaction takes place is essential for such developments. This information cannot be obtained through static measurements.

Because of the unique potential of ID24 to study catalysts under *operando* conditions we have developed a set of facilities in which several techniques are coupled and synchronized at a millisecond regime. These are: Energy Dispersive EXAFS (EDE)/Diffuse Reflectance Infrared Spectroscopy (DRIFTS)/Mass Spectrometry (MS) for the study of heterogeneous systems in the solid-gas phase and Stopped-flow/UV-Visible Spectroscopy (UV-Vis)/EDE for the study of homogeneous systems in the liquid phase. In this way, direct structuro-kinetic information and correlation can be attained. These facilities are readily available for all users and can be utilized independently from the beamline when collaboration with our scientific staff is established.

(a) 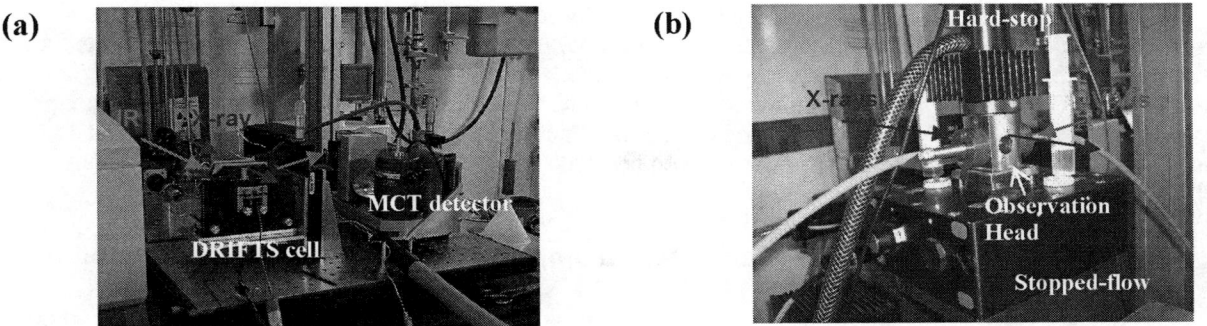 (b)

FIGURE 2. **(a)** Picture of the EDE/DRIFTS/MS experiment. **(b)** Picture of the Stopped-flow/UV-Vis/EDE experiment.

EDE/DRIFTS/MS Facility for Heterogeneous Solid-Gas Systems

Several prototypical catalytic exhaust converters have been studied with EDE/DRIFTS/MS on ID24. These sorts of catalysts are typically nanoparticles of Pt, Rh and/or Pd dispersed on a solid support such as alumina or silica. Other additives (e.g. zirconia, ceria, etc.) are generally included to enhance certain properties of the catalysts such as oxygen storage capacity, activity and lifetime, among others. The experiment consists on flowing through the sample continuous cycles of oxidative and reductive atmospheres, which are automatically switched by a highly precise electro-valve. The sample is presented as loosely packed powder placed in a costum built DRIFTS cell with minimal dead volume. This can be heated up to 673K. The gas-solid interactions occurring during the reaction are monitored via an IR-FT. This techniques is synchronized with the in-situ EDE measurements at the millisecond regime while a quadrupole spectrometer continuously measure the composition of the gas phase. The prominence of the facility lies on the possibility to obtain simultaneous and direct information about structure, functionality, reactivity and selectivity of the chemical species at high time resolution.

Figure 3 shows an example of changes observed in the Rh K edge XANES along with those observed for the Rh(NO$^+$) species in DRIFTS during oxidation by NO and subsequent reduction using 5%H$_2$/He at 573K on a sample containing 5wt% Rh supported on γ-Al$_2$O$_3$. EXAFS and IR data were collected at ca. 64 ms time resolution during 50 s [7].

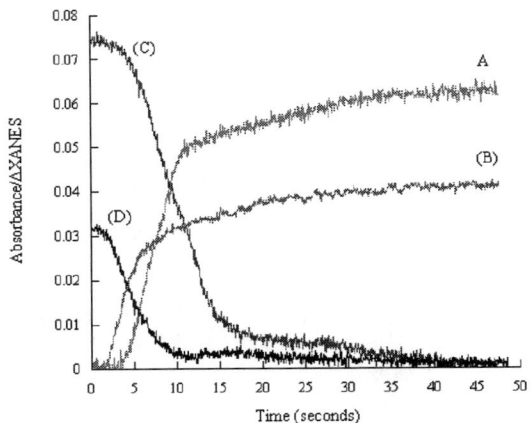

FIGURE 3. Graph showing the synchronously measured temporal variations observed at Rh-edge XANES and at 1910 cm^{-1} in DRIFTS for the oxidation/reduction of 5wt% Rh/Al$_2$O$_3$ at 573K. (A) XANES during NO exposure; (B) DRIFTS during exposure to NO; (C) XANES during exposure of NO reacted sample to 5% H2/He; (D) during exposure of NO reacted sample to 5% H2/He.

For those cases where the loading of metal particles is lower than 1wt%, or the solid support is a highly absorbing matrix, measurements in fluorescence are required. ID24 has developed a Turbo-XAS mode that allows performing such measurements [8]. This is achieved by inserting a slit on a tray downstream of the polychromator that can rapidly move back and forth; thus, being able to monochromatise the x-ray beam. In this configuration it is possible to obtain XANES spectra down to 300 ms in fluorescence. Two different cells are adapted for such measurements: a plug-flow reactor (ready for November 2006) and a Fluo cell that can reach temperatures up to 600°C and 850°C respectively under gas flow conditions.

Stopped-flow/UV-Vis/EDE Facility for Homogeneous Systems in the Liquid Phase

The stopped-flow/UV-Vis/EDE service is one of the few synchrotron multi-technique facilities that permits the study, *dynamically* and *in-situ*, of chemical reactions in the liquid phase down to the millisecond regime. Furthermore, this multi-technique approach not only has the advantage of gaining correlated and complementary structural, electronic and kinetic information and identifying short-lived intermediates, but also of proving the reliability of the experiment by cross-checking of spectroscopies. This last aspect is of prime importance when working at high-flux third generation sources, where X-ray induced local heating or sample degradation can seriously undermine the reliability of the conclusions drawn [9]. The recent implementation of this setup has widened enormously the possibility to study a myriad of chemical systems under different temperature and solvent conditions. These include a fast diode array multi-wavelength UV-Vis spectrometer (range 200-1000 nm), a series of different cuvettes with path lengths ranging from 10mm to 1.5mm, new stopped-flow components made of materials resistant to the majority of solvents and a thermostatic bath that permits examination of reactions at temperatures between 5°C and 180°C.

Remarkable results and high quality spectra have been obtained regarding the study of the oxidative addition process of phenyl iodide on Pd-based catalysts -involved in the formation of C-C and C-N bonds- in toluene solution sand at 40°C [10]. EDE spectra were collected with the FReLoN camera working in kinetic mode and with time resolution of 500 ms.

ACKNOWLEDGMENTS

We are very grateful to ID24 and BM29 technical staff for the constant and excellent support. In particular we are very grateful to Florian Perrin, Trevor Mairs, Marie-Christine Dominguez, Alejandro Homs and Sebastien Pasternak. We also acknowledge EPSRC for funding (grant GR/60744/01) given to Prof. J Evans (University of Southampton) and Dr A. J. Dent (Diamond Light Source) and a long term proposal allocation from the ESRF.

REFERENCES

1. A. Filipponi, M. Borowski, D. T. Bowron, S. Ansell, A. Di Cicco, S. De Panfilis, and J-P. Itiè, *Rev. Sci. Instr.* **71**, 2422-2432 (2000).
2. *e.g.* G. Ciatto, H. Renevier, M. G. Proietti, A. Polimeni, M. Capizzi, S. Mobilio, and F. Boscherini, *Phys. Rev. B* **72**, 085322(1-8) (2005).
3. S. Pascarelli, O. Mathon and G. Aquilanti, *J. All. Comp.* **363**, 33-40 (2004).
4. O. Mathon , G. Aquilanti, G. Guilera, M.A. Newton, A. Trapananti and S. Pascarelli, "Opportunities for time resolved studies at the Energy Dispersive X-ray Absorption Spectroscopy beamline of ESRF", this conference.
5. J.-C. Labiche , G. Guilera, A. Homs, O. Mathon, M.A. Newton, S. Pascarelli and G. Vaughan. "The Fast Readout Low Noise (FReLoN) camera as a versatile X-ray detector for time resolved studies at the ms time scale". In preparation.
6. *e.g.* M. A. Newton, S. G. Fiddy, G. Guilera, B. Jyoti and J. Evans, *Chem. Commun.*, 118-120 (2005).
7. M. A. Newton, A. J. Dent, S. G. Fiddy, B. Jyoti and J. Evans, *Catalysis Today*, submitted.
8. S. Pascarelli, T. Neisius and S De Panfilis, *J. Synchrotr. Rad.* **6**, 1044-1050 (1999).
9. J. G. Mesu, A. M. J. van der Eerden, F. M. F. de Groot and B. M. Weckhuysen, *J. Phys. Chem. B* **109**, 4042-4047 (2005).
10. G. Guilera, M. A. Newton, C. Polli, S. Pascarelli, M. Guinó and King Kuog (Mimi) Hii, *Chem. Commun.*, submitted.

Ultrafast X-ray Imaging of Fuel Sprays

Jin Wang

X-Ray Science Division, Argonne National Laboratory, 9700 S. Cass Ave., Argonne, IL 60439, USA

Abstract. Detailed analysis of fuel sprays has been well recognized as an important step for optimizing the operation of internal combustion engines to improve efficiency and reduce emissions. Ultrafast radiographic and tomographic techniques have been developed for probing the fuel distribution close to the nozzles of direct-injection diesel and gasoline injectors. The measurement was made using x-ray absorption of monochromatic synchrotron-generated radiation, allowing quantitative determination of the fuel distribution in this optically impenetrable region with a time resolution on the order of 1 μs. Furthermore, an accurate 3-dimensional fuel-density distribution, in the form of fuel volume fraction, was obtained by the time-resolved computed tomography. These quantitative measurements constitute the most detailed near-nozzle study of a fuel spray to date. With high-energy and high-brilliance x-ray beams available at the Advanced Photon Source, propagation-based phase-enhanced imaging was developed as a unique metrology technique to visualize the interior of an injection nozzle through a 3-mm-thick steel with a 10-μs temporal resolution, which is virtually impossible by any other means.

INTRODUCTION

High-pressure, high-speed sprays are an essential technology in many industrial and consumer applications, including fuel injection, inkjet printers, liquid-jet cutting and cleaning systems. In particular, liquid fuel sprays and their atomization and combustion processes have numerous technological applications including energy sources for propulsion and transportation systems including internal combustion engines. In fuel-spray applications, diesel and gasoline direct-injection systems aim to achieve better fuel efficiency and control of emissions. Both objectives bring more impetus for in- and near-nozzle characterization of the fuel flow and the spray formation to optimize transient injection and spray characteristics. However, high-speed fuel sprays are optically opaque due to dense liquid droplets generated by the sprays so that the detailed structure of the sprays cannot be resolved by conventional optical means. Other challenges arise from the transient nature of the sprays, frequently requiring images on μs time scales. X-rays are highly penetrative in materials composed of extremely dense droplets composed of low-Z materials, which makes x-rays a suitable tool for fuel-spray studies designed to overcome the limitations of visible light. With the advent of synchrotron radiation sources, extremely brilliant monochromatic x-ray beams are now available to reveal many transient characteristics of fuel sprays quantitatively and unambiguously that were never previously known and/or that could not be measured. The measurement allows quantitative determination of the fuel distribution in the optically impenetrable region. For monochromatic x-radiography and -tomography, the fuel distribution can be determined by $I/I_0 = e^{-\mu_M \cdot M}$, where μ_M, the mass absorption coefficient, can be measured accurately for the absorbing medium, and M the amount of fuel in the beam path, can be easily calculated from the transmission, I/I_0. Even with the x-radiography technique, however, the highly transient fuel sprays have never been visualized or reconstructed in a true three-dimensional (3D) manner. The intent to qualitatively visualize and measure fuel-spray characteristics, such as internal structure, density distribution, and flow dynamics, requires the development of an ultrafast x-tomography technique capable of capturing the transient nature of the sprays [1-5]. In addition, taking advantage of the high energy and high brilliance of the x-ray beams produced by third-generation synchrotron sources, such as the Advanced Photon Source (APS), propagation-based phase-enhanced imaging was developed as a unique metrology technique to visualize the internal structure of high-pressure fuel-injection micronozzles. We have visualized the micrometer-scale machining and finishing defects inside a 200-μm fuel-injection micronozzle in a 3-mm-thick steel housing using phase-enhanced x-ray imaging. Because of the phaseenhancement, this new microimaging-based metrology technique has paved the way to directly study highly transient fluid dynamics in micronozzles *in situ* and in real time, which is virtually impossible by any other means.

CP879, *Synchrotron Radiation Instrumentation: Ninth International Conference*,
edited by Jae-Young Choi and Seungyu Rah

X-RADIOGRAPHY OF DIESEL JETS

For the first time, by using wide-bandpass synchrotron x-ray beams and a novel fast x-ray framing detector developed by Gruner's group at Cornell University [6,7], time-resolved x-radiography clearly captures propagation of the diesel spray-induced shock waves in a gaseous medium [2]. It also allows quantitative analysis of the thermodynamic properties of the shock waves, which has been impossible with optical imaging methods. We used a high-pressure common-rail diesel-injection system, typical of that in a passenger car with a specially fabricated single-orifice nozzle. The injection pressure was set at between 20 and 135 MPa, and the injection was perform into an ambient condition of 0.1 MPa of SF_6, which is a heavy gas (molecular weight of 146), and was used to simulate the relatively dense ambient gas environment in a diesel engine during the adiabatic compression part of the engine cycle when the diesel fuel is normally injected. The sonic speed in SF_6 at room temperature (30°C) has been measured to be 136 m/s [8], considerably less than the 330 m/s speed of sound in air at the same temperature.

Figure 1 shows a series of x-radiographs of the fuel spray for times ranging from 38 to 192 μs after the beginning of the injection process, taken at the D-1 beamline of the Cornell High Energy Synchrotron Source (CHESS). The imaged area shown in the largest panel is 61.7 mm (horizontal) by 17.5 mm (Verticle) with data corrected for the divergence of the x-ray beam. The exposure time per frame was set to 5.13 μs (twice the CHESS synchrotron period) with subsequent images taken after an additional 2.56 μs delay. Each position shown is the average of images from 20 fuel-injection cycles. In this measurement, the fuel-injection pressure was set to 135 MPa, resulting in maximum leading-edge speeds of 345 m/s. The leading-edge speed exceeds the sonic speed upon emergence. The shockwave front, or the so-called Mach cone, is clearly observed as emanating from the leading edge of the fuel jet soon after emergence with an x-ray absorption of up to 3% in the shock front. The false-color levels of the images have been set to accentuate small differences in the x-ray intensity arising from the slightly increased (ca. 15%) x-ray absorption in the compressed SF_6 gas. This characteristic has been quantitatively determined for the first time. This shockwave generation process was also simulated, for the first time, by a computational fluid dynamics (CFD) model. The agreement between the measurement and the simulation is remarkably good.

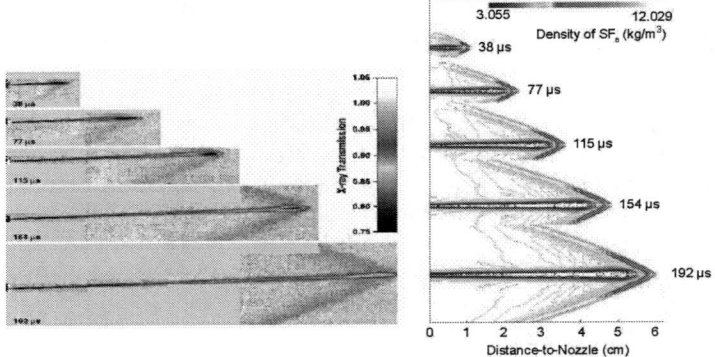

FIGURE 1. Shockwaves generated by high-pressure high-speed diesel fuel sprays imaged with x-radiography (left panels) and illustrated with CFD simulations (right panel).

ULTRAFAST X-TOMOGRAPHY OF SPRAYS

Using the ultrafast x-ray detector and intense x-ray beams from synchrotron radiation, the interior structure and dynamics of the direct injection gasoline spray from a multihole direct injector were elucidated for the first time by a newly developed, ultrafast computed microtomography technique [9]. Many features associated with the transient liquid flows are readily observable in the reconstructed spray. Furthermore, an accurate 3D fuel-density distribution was obtained as the result of the computed tomography in a time-resolved manner. These results not only reveal the near-field characteristics of the complex fuel sprays with unprecedented detail but will also facilitate realistic computational fluid dynamics simulations in highly transient, multiphase systems. The key component in the setup is the integrated tomography fuel-chamber system that includes spray injection chamber, rotation and translation stages. The injection chamber is intended to provide an environmental enclosure for the fuel sprays. There are two identical x-ray-transparent windows situated symmetrically on the chamber with a 120° x-ray viewing angle. The injector is mounted on the top of the chamber. The environment in the spray chamber is maintained at a pressure of

0.1 MPa and at room temperature. The spray chamber is designed to rotate and to translate in precise steps, while the x-ray source and the detector are stationary. The spray is triggered at 1.15 Hz, and a series of frames is taken at various delay times. The exposure time per frame is set to 10.25 μs with an interval between frames of 25.6 μs. Each image is obtained by averaging 20 fuel-injection cycles.

The images of the multi-orifice gasoline direct injection sprays recorded by the Cornell pixel array detector (PAD) at selected projection angles and time instances are shown in Fig.2 (left panels). These images show the progression of the spray with unprecedented details. Different phases of the transient spray characteristics, including the "sac", streak, and "bounce", can be readily observed. The representative instances selected here are 1349 μs, when the spray tip just appeared at the nozzle; 1457 μs, when the sac portion was exiting the nozzle; 1601 μs, when the spray cone fully opened and the sac portion started to break up with the main spray; 1925 μs, when the spray became stabilized; 3113 μs, just after the nozzle was closed; and 3257 μs, when the first bounce occurred.

From these projection images, the spray cross section can be reconstructed in 3D by the computerized tomography technique. The principles of transmission tomography show that the linear attenuation coefficient distribution of the spray cross-section, $\mu_L(x,y)$, can be reconstructed from values of its line integrals provided the x-ray energy is monochromatic [5]. For our case, the line integrals of $\mu_L(x,y)$ can be easily resolved from the radiography images as shown in Fig.2 (left panels). With these line integrals (or sinogram), $\mu_L(x,y)$ is computed by several numerical methods based on filtered backprojection, algebraic iteration, and the Fourier transform method. The Fourier transform method was selected as our working algorithm due to its computation efficiency and relatively easy implementation. Finally, the fuel-density distribution, $\rho(x,y)$, can be derived based on the following simple relation, $\rho(x,y) = \mu_L(x,y)/ \mu_M$, the 3D fuel density distribution is, then, built upon all the reconstructed cross sections at different locations of z, as shown in Fig.2 (right panels).

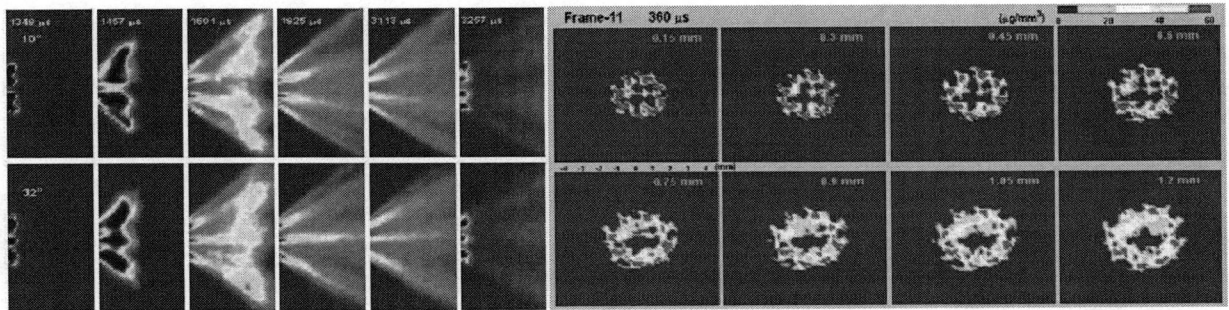

FIGURE 2. X-radiography images of 8-hole nozzle spray at 2 selected projection angles at 10° and 32° and 6 time instances at 1349, 1457, 1601, 1925, 3113 and 3267 μs (left panels) and cross-sectional view of reconstructed fuel mass distribution in the near-nozzle region (from 0.15 to 1.2 mm from the nozzle exits) and 360 μs after the fuel emerged from the nozzle (right panels).

MICROIMAGING OF DIESEL SPRAY NOZZLES

Direct measurements involving direct imaging of the internal structure of a real diesel injection nozzle, made of mm- or cm-thick steel, even in a static condition, will greatly complement and expand the current understanding of the fuel-spray process. Nondestructive visualization of the internal structures has thus far been difficult, if not impossible. With high-energy and high-brilliance x-ray beams available at the APS, propagation-based phase-enhanced imaging can be used to develop a unique metrology technique to visualize the injection process in a high-pressure fuel-injection system with high temporal resolution, which is virtually impossible by any other means. The feasibility study was performed by imaging static nozzles with a high-energy monochromatic x-ray beam. The experiments were carried out at the XOR 1-ID-C beamline at the APS [10]. Great details, which are of great significance for understanding the fundamental principle in liquid breakup, have been clearly revealed in unprecedented spatial resolution.

In the experiment, a 1 (horizontal, H) x 1 (vertical, V) mm^2 beam with 70 keV photon energy was used to illuminate the sample. The source-to-sample distance was about 60 m, and the sample-to-detector distance was varied from 0.01 to 3 m. A 5-μm-thick Ce-doped YAG single-crystal film grown on a YAG substrate or a CdWO$_4$ crystal was used to convert x-rays to visible light. The images were magnified and recorded with a cryogenically cooled CCD (1024x1024 pixels, 19.5 μm pixel size) detector, resulting in an effective pixel size of 0.98 μm. For clarity, a specially fabricated single-orifice minisac nozzle was initially used in this study. The structure of a single-hole minisac nozzle near the orifice exit is schematically shown in Fig.3a. Figures 3b and 3c show the absorption-based and phase-enhanced radiographs, respectively. The field of view of the images is 0.98 x 0.98 mm^2. Calibrated with a sample the diameter of the orifice is measured to be approximately 200 μm. The dramatic difference between the absorption and phase-enhanced images is readily observed. The phase-enhanced images contain edge enhancements that make it easy to discern the edges of the needle, the narrow openings, the sac and the orifice. To demonstrate the usefulness of the method, we are also able to image a production 6-hole nozzle as shown in Fig.3d where the 6 orifices can be clearly seen.

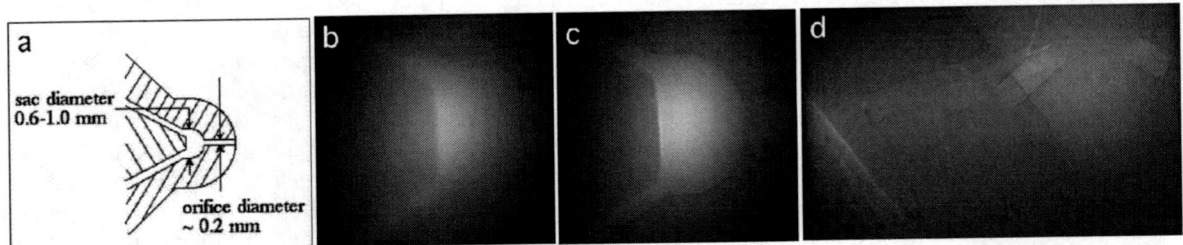

FIGURE 3. Phase-enhanced radiography of diesel injector nozzles: a) Schematic of the single-hole nozzle, b) absorption-based radiograph, c) phase-contrast radiograph of the single-hole nozzle, and d) phase-contrast radiograph of a 6-hole production nozzle, without the needle. The 6-hole nozzle was oriented 90° from that in b and c with a much different magnification.

ACKNOWLEDGMENTS

This work and the use of the APS are supported by the U.S. DOE under contract W-31-109-ENG-38. CHESS is funded by the U.S. NSF and NIH under award DMR-9713424. The results are obtained through continuing collaborations between the fuel spray research group at the ANL and Sol Gruner's group of Cornell University, Ming-chia Lai's group of Wayne Station University, David Hung and James Winkelman of Visteon Corp., Johannes Schaller and Jochen Walther of Bosch GmbH, and Wah-Keat Lee and Kamel Fezza of the APS. Use of the APS was supported by the U. S. Department of Energy, Office of Science, Office of Basic Energy Sciences, under Contract No. W-31-109-Eng-38.

REFERENCES

1. C.F. Powell, Y. Yue, R. Poola and J. Wang. J. Synchrotron Rad. 7, 356 (2000).
2. A.G. MacPhee, *et al.* Science 295, 1261 (2002).
3. W. Cai, *et al.* Appl. Phys. Lett. 83, 1671 (2003).
4. J. Wang, X-ray Vision of Fuel Sprays, J. Synchrotron Rad., 12, 197-207 (2005)
5. X. Liu, *et al.* SAE Paper 2006-01-1-41 (to be published in SAE Transaction: J. Engines)
6. S.L. Barna, *et al.* IEEE Transactions Nuclear Science 44, 950 (1997).
7. G. Rossi, *et al.* J. Synchrotron Rad. 6, 1096 (1999).
8. J.J. Hurly, *et al.* International J. Thermophysics 21, 739 (2000).
9. X. Liu, *et. al.*, Proc. SPIE, 5535, 21 (2004).
10. W.-K. Lee, K. Fezzaa, and J. Wang, Appl. Phys. Lett., 87, 084105(3) (2005).

Nanotechnology and Industrial Applications of Hard X-ray Photoemission Spectroscopy

K. Kobayashi [1,2], E. Ikenaga [1], J. J. Kim[1], M. Kobata[1], and S. Ueda [1]

[1] SPring-8/JASRI Kouto 1-1-1, Sayo-cho, Sayo-gun, Hyougo 679-5198, Japan
[2] SPring-8/NIMS Kouto 1-1-1, Sayo-cho, Sayo-gun, Hyougo 679-5198, Japan

Abstract. Since the first test experiments of the high resolution hard X-ray photoemission spectroscopy by the JASRI-RIKEN-HiSOR collaboration at BL29XU in 2002, we have continued efforts to widen the application field of this novel method. We have conducted collaborations on Si-LSI related materials, diluted magnetic semiconductors, and strongly correlated electronic device materials with several leading edge groups to test the feasibility of the method in these areas. Based upon these preliminary efforts, we began to accept public user proposals under the framework of the Nanotechnology Support Project, at BL47XU from 2004. In 2005, industrial application subjects were also introduced. The present status of applications to both nano-materials science and technology, and to industrial research and development is described. Typical examples of users' experimental outcomes are illustrated.

Keywords: hard X-ray photoemission spectroscopy, high resolution, nanotechnology, industrial application.
PACS: 72., 73., 78.40.Fy

INTRODUCTION

Increasing the excitation photon energy to the hard X-ray region greatly enhances the versatility of the photoelectron spectroscopy (PES) technique, provided that the signal loss due to the rapid decrease in photo-ionization cross-sections [1] is compensated for [2, 3]. The greater probing depth afforded by hard X-rays frees the measurements from sample surface conditions, and also removes the need for careful surface cleaning methods, thus enabling the study of laboratory prepared nano-scale thin films and clusters. We have developed the high-resolution, high-throughput hard X-ray photoelectron spectroscopy (HX-PES) technique through collaboration among SPring-8/JASRI, SPring-8/RIKEN, and the Synchrotron Radiation Center of Hiroshima University [3-5]. The first test experiment was performed at BL29XU in June 2002. Since then, we have developed an analyzer optimized for high electron kinetic energies under the support of the Ministry of Education, Science, Sports and Culture (MEXT) through a Grant-in-Aid for Scientific Research (A) (No. 15206006, Principal Investigator: Takeo Hattori) in collaboration with the Musashi Institute of Technology. The results of various instrumental developments and improvements have been already reported by Takata et al [4, 5]. It has recently been verified that the valence band spectrum of Si can be recorded with an overall energy resolution of 0.2 eV and a sufficiently high S/N ratio with 300 sec accumulation time at a photon energy of 6 keV [6].

On the basis of these efforts, our JASRI HX-PES group has concentrated on pioneering applications of the technique, especially in the applied physics and industrial fields. During the 2004B (Sep. 2004-Dec. 2004) SPring-8 operations period, we offered beam line at BL47XU to users under framework of the nanotechnology support project of MEXT. For the 2005A period (Mar. 2005-Jul. 2005), the status of BL47XU was changed from an internal R&D beam line to a full public use beam line, and about a half of the beam time has been opened to HX-PES users. From the same period we also started to accept industrial proposals under the frameworks of the MEXT project "The Program for Strategic Use of Advanced Large-scale Research Facilities", and the Hyogo Prefecture project "Collaboration of Regional Entities for the Advancement of Technological Excellence". In this report, we describe the present status and achievements to date in nanotechnology and industrial applications of HX-PES at SPring-8.

CP879, *Synchrotron Radiation Instrumentation: Ninth International Conference,*
edited by Jae-Young Choi and Seungyu Rah
© 2007 American Institute of Physics 978-0-7354-0373-4/07/$23.00

PRESENT STATUS OF HX-PES PUBLIC USE

HX-PES studies at SPring-8 mainly make use of one of two 10 keV analyzers (VG SCIENTA, model R4000-10kV) for the general, nanotechnology support, industrial, long term, and proprietary subjects, mainly at BL47XU, and partly at BL39XU. Some of the nanotechnology support subjects have been carried out with the support of RIKEN at BL29XU, a RIKEN beam line. Both horizontal and vertical focus mirrors have already been installed at BL29XU, and will also be installed at BL47XU. Operation at 10 keV has been confirmed with a total resolution of 93 meV by the measurement of the Au Fermi edge at 30 K, as reported by Kim et al. [7].

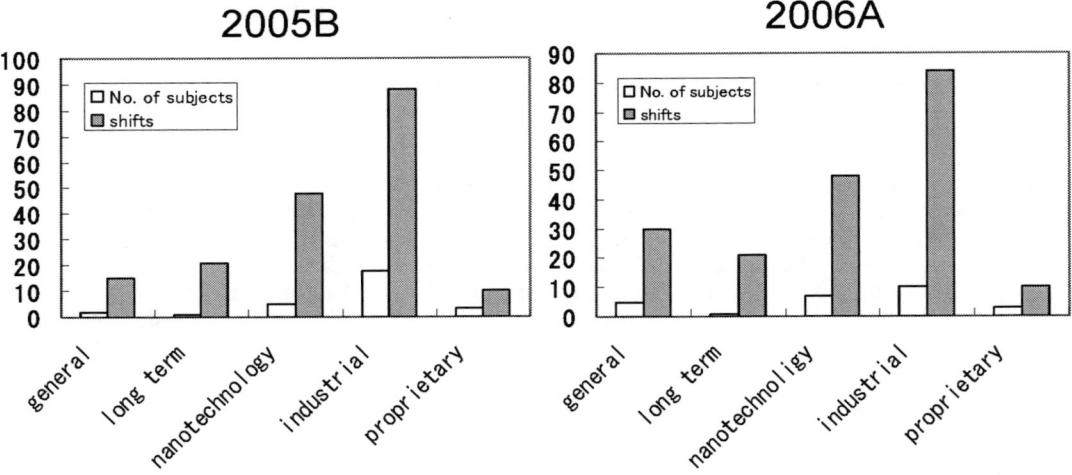

FIGURE 1. Statistics of HX-PES user subjects during the 2005B and 2006A periods.

Statistics of public use in 2005B (Sep. 2005-Dec. 2005) and 2006A are shown in Fig. 1. Proposal in the "general" category are few compared to the nanotechnology and industrial subjects. However, if we also take into account research activities by the SPring-8/RIKEN group at BL29XU, and by the Osaka University group at BL19LXU, the balance between basic and applied physics-industrial subjects is good at the present stage. A long term project has been running on Si-ULSI related research since 2005A. This was proposed by a Nagoya University-Musashi Institute of Technology-Hiroshima University joint group. Proprietary subjects are proposed by several groups in order to secure the beam time without the need for proposal assessment. This is a sign of beam time demand outstripping beam time availability.

EXAMPLES OF USER ACTIVITIES

Si ULSI Related Applications

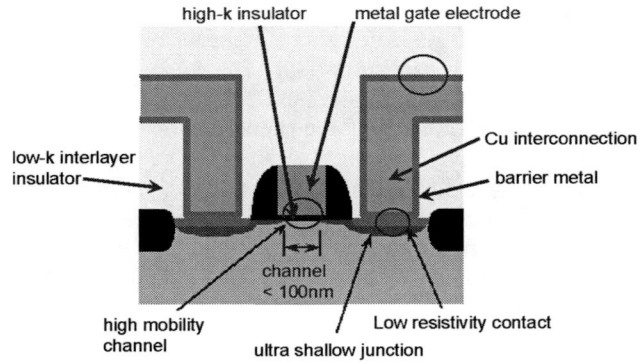

FIGURE 2. Schematic drawing of a Si-ULSI structure surrounding a MOS transistor.

Figure 2 is a simplified schematic drawing of a structure surrounding a Si-ULSI MOS transistor. A channel width of around 90 nm is already well established, and the sizes of 60 nm, and then 45 nm are imminent [8]. Si-ULSIs are in effect typical top-down-type nano-devices, employing various kinds of nano-scale thin layers such as high-k dielectrics, gate metal electrodes, low-k interlayer insulators, Cu interconnection, and barrier-metals for preventing Cu diffusion into the Si and low-k interlayer insulators. High concentration doping is necessary in the drain and the source regions which have extremely shallow depth profiles. All these nano-structures and their interfaces are required to be stable against annealing during the fabrication process. Photoemission spectroscopy has been used to investigate interface problems due to any such instabilities, however, the typical thicknesses of the nano-scale structures in real devices are far beyond the probing depth of the conventional photoemission techniques. We have shown that HX-PES can be used to probe buried layers 20-35 nm below the surface [2, 4, 5], presenting a very promising method for the investigation of nano-scale advanced materials and devices.

The on-current I_d(on) of the MOS transistor is proportional to the dielectric constant k of the gate insulator and the area of the channel S, and inversely proportional to the thickness of the gate insulator T_{ox}. High-speed performance requires a reduction of S. Thus it is necessary to decrease T_{ox} in order to maintain I_d(on) at a certain level. This causes an increase in leak current. The value of T_{ox} already approaches the material limit of 1 nm (only four monolayers) for SiO_2. To increase the T_{ox}, the quest for so-called high-k gate insulators is one of the urgent issues for the future development of Si-ULSI. Because the high-k materials and Si substrate interfaces are less clean and less stable than the SiO_2/Si interface, much effort is being dedicated to solve the problem.

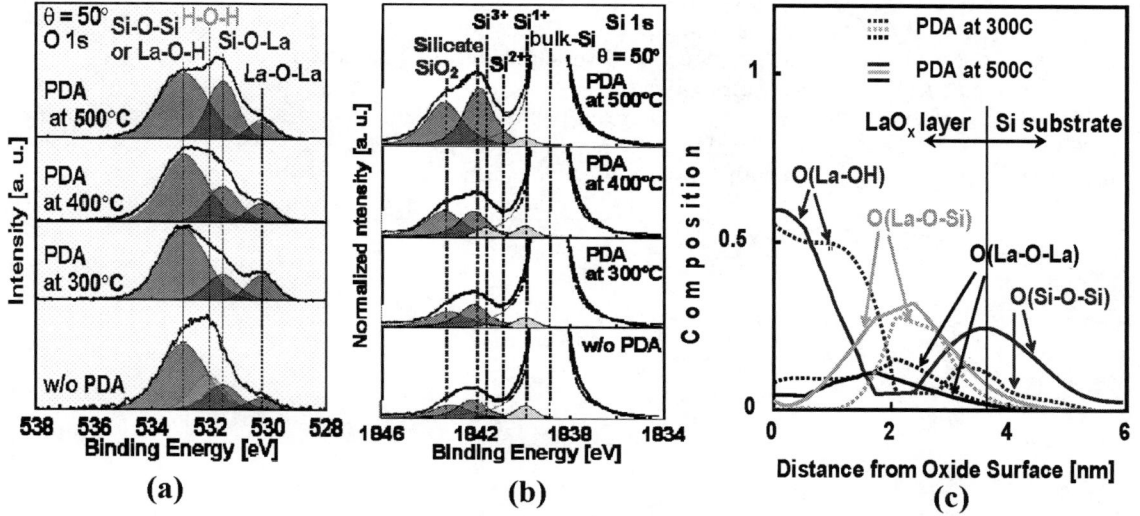

FIGURE 3. (a) O 1s and (b) Si 1s spectra of as-deposited and annealed La_2O_3/Si(100) samples. (c) Depth profiles of chemical states in La_2O_3/Si (100) interfaces annealed at 300 °C, and 500 °C.

Among the various high-k dielectrics, La_2O_3 has recently been investigated as one of the most promising candidates [9-12]. Nohira et al. have recently performed an HX-PES analysis of the La_2O_3/Si interface thermal stability [12]. Figures 3(a) and 3(b) show the O 1s and Si 1s spectra of as-grown and post-deposition annealed (PDA) La_2O_3(4 nm) /Si(100) samples as recorded with an 8 keV incident X-ray energy. The deposition was carried out at room temperature and PDA in an atmospheric pressure of N_2 gas flow at 300 °C, 400 °C, and 500 °C for 10 min each. The O 1s spectra decomposed into essentially three components, which can be assigned to originate from Si-O-Si(or La-O-H), Si-O-La, and La-O-La bonds. Room temperature deposition alone already causes La silicate formation with the amount of silicate increasing with increasing PDA temperature. This enhancement is also evident in the Si 1s spectra shown in Fig. 3(b). The take-off-angle (TOA) dependence of O1s, Si 1s and Hf 3d spectra has been analyzed using the Maximum Entropy Method [13, 14] to obtain the depth profiles of the chemical states as shown in Fig. 3(c) for PDA samples at 300 °C and 500 °C [12]. It is clearly seen that whereas the Si atoms diffuse into the La oxide layer to form silicate, La atoms do not diffuse into Si substrate. It is worth noting that interface Si oxide layer formation becomes more prominent as the PDA temperature increases.

Stability at gate metal/gate insulator interfaces is another important problem. Control of the work function at the interface is also an urgent requirement in establishing complementary MOS (CMOS) structures, the most essential components in Si-ULSI. Attempts to apply HX-PES to these problems have been already carried out at SPring-8.

Ikenaga et al. found that Poly Si layers are oxidized near the poly Si /HfO$_2$ interfaces [15]. In this study test MOS devices were fabricated from the same wafers, from which samples for HX-PES measurements were prepared. HX-PES results were compared with measurements of threshold (V_t) of transistor action, and it was found that V_t is progressively stabilized as the interface oxidation proceeds. This phenomenon can be reasonably understood by applying Shiraishi's theory [16] on the interface oxidation by itinerant electron transfer from the HfO$_2$ side to the poly Si side through doubly charged oxygen vacancy states. This oxidation proceeds until the Fermi level position is stabilized at the interface. The mechanism of the work function control by doping of the full silicide gate has also been investigated by the Toshiba Corporation group [17]. They argued that the work function is modified by the dipoles formed at the interface due to the difference in electro negativity between Ni and B, based upon the measured chemical shift of the B 1s peak. Another successful application was the investigation of the shallow junctions in the drain and the source regions formed by plasma doping. The activations of the doped carriers by several different annealing techniques were deduced from the Si 1s spectra [18].

Hard Disc Lubricant Analysis

As the area density of hard disc (HD) memory devices increases, the bit size decreases. Accordingly a reduction in the spacing between the read-write head and the disc surface is necessary to maintain the signal level. In current devices this spacing has already reached less than 10 nm. A typical HD has a structure of DLC (3-5 nm)/ CoCrTaPt multi layers (20-40 nm)/NiP layer (10-20 nm), where DLC stands for diamond like carbon. On top of this layer a lubricant layer is formed to avoid damage of the HD structure due to head-scraping or crashing by the head. MORESCO (Matsumura Oil Research Corporation) recently developed a new type of lubricant for the next generation of HD devices. This advanced lubricant contains chain molecules with the formula X-CH$_2$CF$_2$O(CF$_2$CFO$_2$)$_m$(CF$_2$O)$_n$CF$_2$CH$_2$OH, where X stands for a phosphorzene based molecular assembly. The MORESCO group prepared test samples without the magnetic memory multilayers with 1-2 monolayers of lubricant

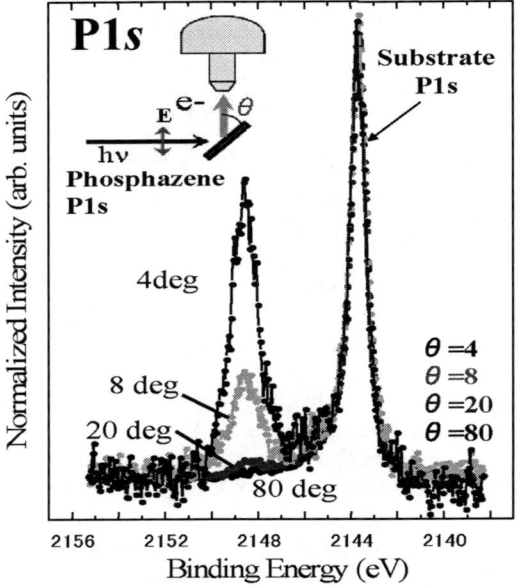

FIGURE 4. P 1s spectra recorded at take-off angles (TOA) of 4, 8, 20 and 80 degree from the sample surface.

molecules adsorbed on to the DLC covered NiP layers. HX-PES P 1s spectra at various TOA s were recorded, and these are shown in Fig. 4 [18]. The intensities are normalized to have the same intensity at the lower binding energy peak, which is assigned to come from the NiP layer, in this figure. The P 1s peak intensity of phosphorzene is plotted as a function of TOA in Fig. 5(a). It is clearly recognized that P 1s photo electrons from the phosphorzene can only be detected in a small angle cone around the surface normal. The phosphorzene P 1s spectra consist of a single Lorentzian component convoluted with a single Gaussian with the instrument width. This shows that the three P atoms of the phsoperzene molecule are equivalent. The C 1s spectra however were found to exhibit two groups of chemical shift components. The weak components at high binding energies are chemically shifted components from

O-CF$_2$-O and C-CF$_2$-O in the chain. The main components, which appear at lower binding energy, were assigned as originating from DLC. The main peak can be decomposed into sp^2, sp^3, and sp components, of which the sp^2 and sp^3 components are attributed to DLC. The sp component was thought to come from P-C, N-C bonding at the lubricant/DLC interface. Based on all of the above results a model structure for the lubricant molecule onto the DLC covered HD surface was deduced, and as shown in Fig. 5(b) [18]. It is not clear as yet whether the –OH end is free or terminated onto the DLC surface.

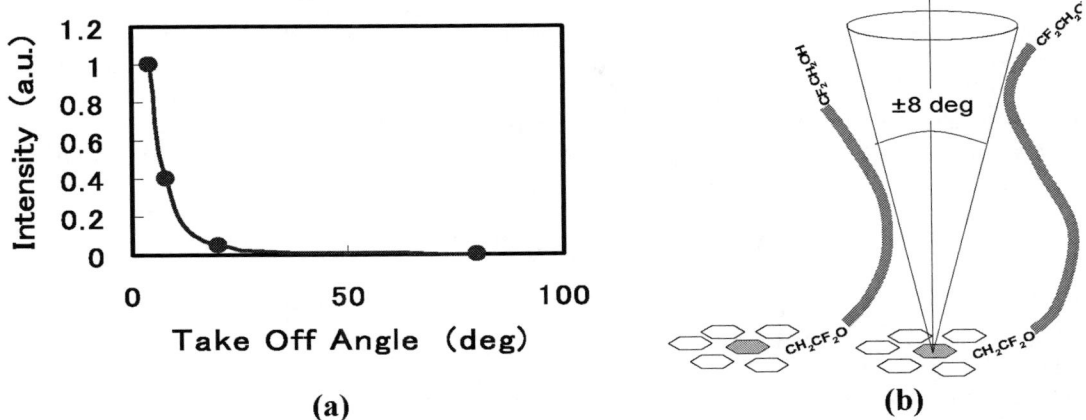

FIGURE 5. (a) Take-off angle dependence of P 1s intensity. (b) Adsorption model of lubricant molecule.

Analysis of Phase ChangeMemory Disc Materials

Toshiba Corporation has recently been developing a dual-layer rewritable HD DVD (High Density Digital Versatile Disc) media. Figure 6(a) shows the structure of the recording layers, which consist of Ge-Bi-Te alloy-films, are sandwiched by oxide interface layers to accelerate phase changes. Samples with the structure shown in Fig. 6(a) were prepared at the Toshiba laboratory. Figure 6(b) shows HX-PES valence band spectra of the samples

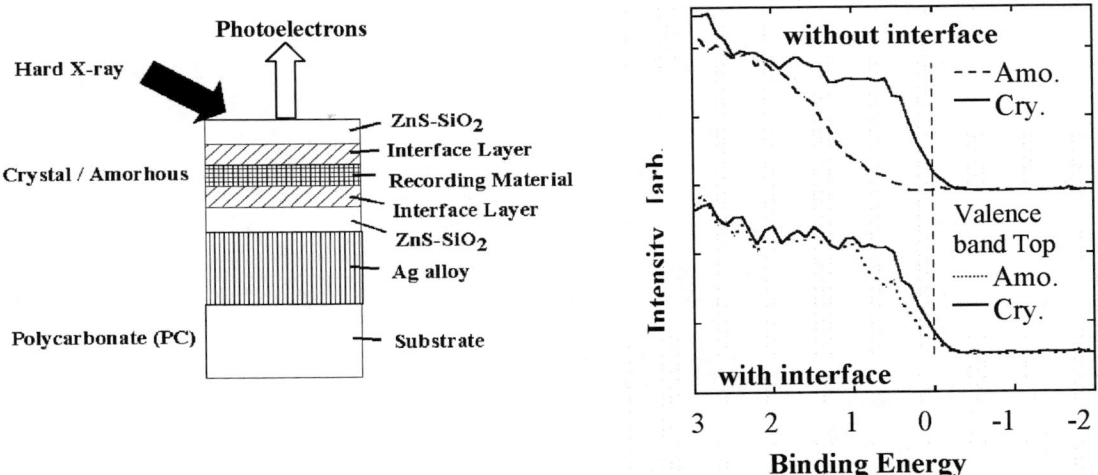

with **FIGURE 6. (a)** Cross sectional structure of HD DVD memory layer. (b) Valence band HX-PES spectra with and without interface layers.

(lower spectrum) and without (upper spectrum) interface the layers [20]. It should be noticed that the valence band which spectra of the buried memory layers are visible through the ZnS-SiO$_2$ and the interface layers. The state density of the crystalline phase steeply rises at the Fermi level in the sample without the interface layers, indicating the metal-like characteristics of this material. On the other hand, the amorphous phase shows a lack of the state density around the Fermi level. This result is consistent with the fact that the resistivity of the amorphous phase is much larger than that of the crystalline phase. With the interface layers, the valence band spectra near the Fermi level resemble each other for both the amorphous and the crystalline phases. Both show nearly metallic features. Nevertheless, the optical reflectivity measured (at 405 nm wavelength) showed a large difference between the two phases. The resemblance of the valence band structure is considered to be closely related to the mechanism of the fast rewriting capability of their recording media.

SUMMARY

High-Resolution Hard X-ray Photoemission Spectroscopy has been established to be a vital tool for the investigation of nano-scale thin solid films. The number of SPring-8 proposal submissions from the applied physics and advanced industrial R&D fields is rapidly increasing at Spring-8. We believe that this novel method will continue to open up a wide range of new applications hitherto impossible in solid-state physics, nano-scale materials science and technology, and also industrial research and development, which has been impossible without this HXPES.

ACKNOWLEDGMENTS

We are grateful to Drs. K. Tamasaku, Y. Nishino, D. Miwa, M. Yabashi, and T. Ishikawa for their help in construction and improvements of the beam line optics and X-ray monochromators at BL47XU. We are also grateful for their help during measurements at BL29XU. Dr. M. Suzuki is appreciated for his help in the measurements at BL39XU. Collaborations with Y. Takata and his colleagues are much appreciated on developments and improvements of the high energy analyzer and photoemission measurements. One of the analyzers was introduced with support of the Ministry of Education, Science, Sports and Culture (MEXT) through a Grant-in-Aid for Scientific Research (A) (No. 15206006, Principal Investigator: Takeo Hattori). Cooperation with VG SCIENTA AB. has also been essential. This work is partly supported by the "Nano Technology Support Project", and "The Program for Strategic Use of Advanced Large-scale Research Facilities", and also partly supported by the Hyogo Prefecture project "Collaboration of Regional Entities for the Advancement of Technological Excellence". The examples of HX-PES activities as typically introduced here are the results of the collaborations with many public users – not named here but deeply thanked.

REFERENCES

1. J. J. Yeh and I. Lindau, At. Nucl. Data Tables 32, 1. (1985)
2. K. Kobayashi et al. Appl. Phys. Lett. 83, 1005. (2003)
3. Y. Takata, K. Tamasaku, T. Tokushima, D. Miwa, S. Shin, T. Ishikawa, M. Yabashi, K. Kobayashi, J. J. Kim, T. Yao, T. Yamamoto, M. Arita, H. Nama.tame, and M. Taniguchi, *Appl. Phys. Lett.* **84**, 4310 (2004).
4. K. Kobayashi, Nucl. Instrum. Methods A 547, 98 (2005).
5. Y. Takata,M. Yabashi, K. Tamasaku, Y. Nishino, D. Miwa, T. Ishikawa, E. Ikenaga, K. Horiba, S. Shin, M. Arita, K. Shimada, H. Namatame, M. Taniguchi, H. Nohira, T. Hattori, S. Södergren, B. Wannberg, K. Kobayashi, *Nucl. Instrum. Methods.* **A 547**, 50 (2005).
6. Y. Takata et al., invited talk of this conference.
7. J. J. Kim, E. Ikenaga, M. Kobata, M. Yabashi, K. Kobayashi, Y. Nishino, D. Miwa, K. Tamasaku, and T. Ishikawa, presented at this conference.
8. International technology roadmap for semiconductorw: http://public.itrs.net/.
9. H. Iwai et al., IEDM Tech. Dig. 625 (2002), and reference therein.
10. T. Hattori et al. Microelectronic Engineering, 72. 283 (2004)
11. H. Nohira et al., Appl. Surf. Sci. 234, 493 (2004).
12. H. Nohira et al. ECS Transaction, Vol. 1, No. 1, The Electrchemical Society. Pennington, NJ.(2005), pp. 87-95.
13. G. C. Smith and A. K. Liversey, Surf. Interface Anal., 19, 175 (1992)
14. S. Shinagawa, H. Nohira, T. Ikuta, M. Hori, M. Kase, and T. Hattori, 80, 98 (2005).
15. E. Ikenaga et al., J. Electron Spectrosc. And Relat. Phenom., 144-147, 491 (2005).
16. K. Torii, K. Shiraishi, S. Miyazaki, K. Yamabe, M. Boero, T. Chikyow, K. Yamada, H. Kitajima, and T. Arikado, , Tech. Digest of 2004 IEEE International Electron Device Meeting.
17. Y. Tsuchida, M. Yoshiki, M. Koyama, A. Kinoshita, and J. Koga, IEDM2005 Tech. Dig., 637 (2005).
18. C. G. Jin et al., presented at 6th International Workshop on Junction Technology (IWJT), May 15-16, 2006, (Hotel Equitorial, Shanghai, Chaina). to be published.
19. Y. Sakane, A. Wakabayashi, and T.Hirano,presented at International Tribology Conference KOBE 2005, May 29-June 2,
20. T. Nakai, M. Yoshiki, and N. Ohmachi, presented at ODS06, Montreal, Canada, and to be published in SPIE.

Dislocation Elimination in Czochralski Silicon Crystal Growth Revealed by White X-ray Topography Combined with Topo-tomographic Technique

Seiji Kawado*, Satoshi Iida[a], Kentaro Kajiwara[b], Yoshifumi Suzuki[c], and Yoshinori Chikaura[c]

*Rigaku Corporation, 3-9-12 Matsubara-cho, Akishima-shi, Tokyo 196-8666, Japan
[a]Faculty of Science, Toyama University, 3190 Gofuku, Toyama 930-8555, Japan
[b]Japan Synchrotron Radiation Research Institute, SPring-8, 1-1-1 Kouto, Sayou, Hyogo 679-5198, Japan
[c]Graduate School of Engineering, Kyushu Institute of Technology, Tobata-ku, Kitakyushu 804-8550, Japan

Abstract. We have examined the neck of a large-diameter [001]-oriented Czochralski silicon crystal by synchrotron white X-ray topography combined with a topo-tomographic technique in order to explain the mechanism of dislocation elimination due to Dash necking in industrial-scale crystal growth. In the portion where the grown crystal was transformed from a dislocated region to a dislocation-free region, dislocation half loops were first generated at the dislocation tangles. These loops then expanded on the {111} glide planes and then terminated inside the crystal. In some cases, they reached the side of the crystal. A new mechanism for the elimination of dislocations is proposed based on the fact that dislocations in the neck are not accompanied by the solid-melt interface during the crystal growth, and they proceed in the crystal after the movement of the interface.

Keywords: X-ray topography, silicon, crystal growth, Czochralski method, dislocation elimination
PACS: 61.72.Ff, 81.05.Cy

INTRODUCTION

We have demonstrated that a combination of synchrotron white X-ray topography and a topo-tomographic technique is useful for the determination of three-dimensional structures of individual dislocations in silicon, i.e., the direction of dislocation, its Burgers vector, and the glide plane [1]. In the previous study [2], we examined the three-dimensional structure of the dislocations in the neck of a 2-inch diameter silicon crystal that was grown by the Czochralski method in the laboratory. The result of the structural analysis revealed that the elimination of dislocations due to Dash necking was caused by the termination of the expansion of the dislocation half loops inside the crystal and by pinning dislocations on the side of the crystal. Recently, we have examined the neck of a large-diameter [001]-oriented Czochralski silicon crystal in the same manner in order to clarify the mechanism of dislocation elimination due to Dash necking in industrial-scale crystal growth.

In this paper, we first describe the ability of synchrotron white X-ray topography combined with a topo-tomographic technique to determine the three-dimensional structure of dislocations in silicon. This combined technique was developed at beamline BL28B2 of SPring-8. Further, we elucidate the detailed structure of the residual dislocations in the neck at the final stage of dislocation elimination in the large-diameter Czochralski silicon crystal growth. Lastly as an alternative to the traditional mechanism, we propose a new mechanism based on this dislocation structure with regard to the dislocation elimination process due to Dash necking.

WHITE X-RAY TOPOGRAPHY COMBINED WITH A TOPO-TOMOGRAPHIC TECHNIQUE

Figure 1 shows the experimental set-up at the experimental hutch of beamline BL28B2 of SPring-8. The sample crystal was about 6 mm in diameter, and it was cut from the neck of an 8-inch diameter CZ-Si crystal. The crystal

CP879, Synchrotron Radiation Instrumentation: Ninth International Conference,
edited by Jae-Young Choi and Seungyu Rah
© 2007 American Institute of Physics 978-0-7354-0373-4/07/$23.00

was fixed by positioning its growth axis [001] nearly parallel to the ω-axis on the sample holder of the subsidiary goniometer that was mounted on the swivel stage of the main diffractometer.

From an in situ observation of the transmission Laue patterns, the orientation of the sample was adjusted such that the (110) plane was perpendicular to the incident X-ray beam [about 8 mm (H) × 5 mm (V)] and the [$\overline{1}$10] orientation was horizontal. This angular position was designated as ω = 0°. Therefore, when ω = 45°, 90°, and 135°, the plane perpendicular to the incident X-ray beam corresponded to (100), (1$\overline{1}$0), and (0$\overline{1}$0), respectively. The sample was inclined at 4.36° (Bragg angle) using the R_x stage of the main diffractometer so that the 004 Laue spot could be formed by 60 keV X-rays. Finally, the growth axis [001] of the sample crystal was precisely adjusted so as to make it parallel to the ω-axis. After the adjustment was completed, the 004 Laue spot was recorded by rotating the ω-axis at intervals of 3° between 0° and 180°. The data on the Laue spot was stored on the hard disk of a personal computer connected to a cooled CCD camera. In addition, several sets of Laue patterns were recorded on X-ray films at ω-intervals of 45°. Two representative examples of the Laue pattern are shown in Fig. 2. The details of the dislocation structure were examined by enlarging the image of each Laue spot.

The advantage of this method lies in its ability to acquire information regarding the configuration of the dislocations from the variation in their features observed in a specific Laue spot—in this case, the 004 spot—by the tomographic technique. In addition, the technique can acquire information on the image contrast of the dislocations observed in several Laue spots by conventional white X-ray topography in order to determine their Burgers vectors.

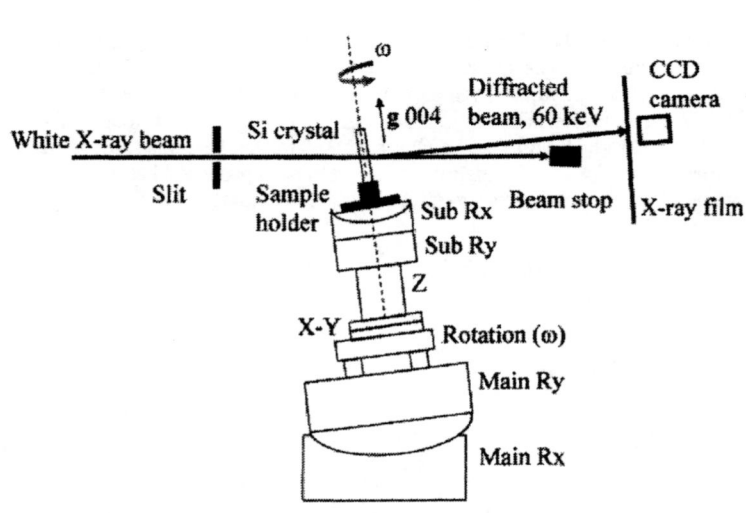

FIGURE 1. Schematic illustration of the experimental set-up (side view). The subsidiary goniometer comprises a sample holder, R_x, R_y, z, x-y and ω-rotation stages. Only the R_x and R_y swivel stages are shown for the main diffractometer.

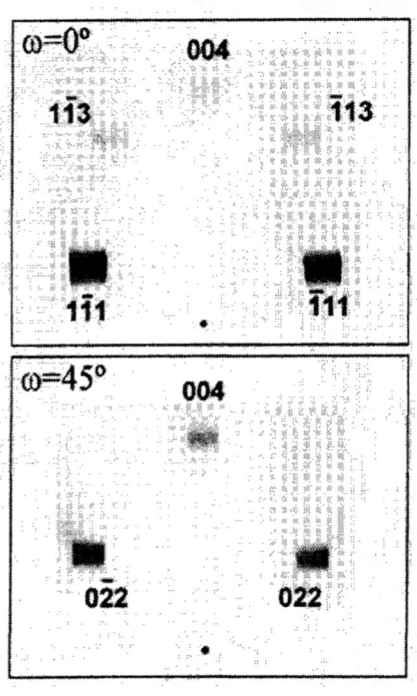

FIGURE 2. Examples of two representative Laue patterns observed at ω = 0º and ω = 45º.

OBSERVATION OF RESIDUAL DISLOCATIONS IN THE NECK

Figure 3 shows a series of enlarged images in the 004 spot at ω = 0°, 45°, and 90° that were obtained from a portion where the grown crystal transformed from a dislocated region to a dislocation-free region. The images reveal that the dislocations were multiplied by a spiral mechanism proposed by Frank and Read [3–5]. Furthermore, the dislocation half loops were generated at tangled dislocations. They then expanded on the {111} glide planes and terminated inside the crystal (e.g., dislocations A, B, and C). In some cases, they reached the side of the crystal (e.g., dislocations D and E).

Figure 4 shows the enlarged images of the 1$\overline{1}$1 and $\overline{1}$11 spots obtained at ω = 0° and those of the $\overline{1}$$\overline{1}$1 and 111 spots obtained at ω = 90°. These images correspond to the upper portions in Fig. 3 and were used to determine the

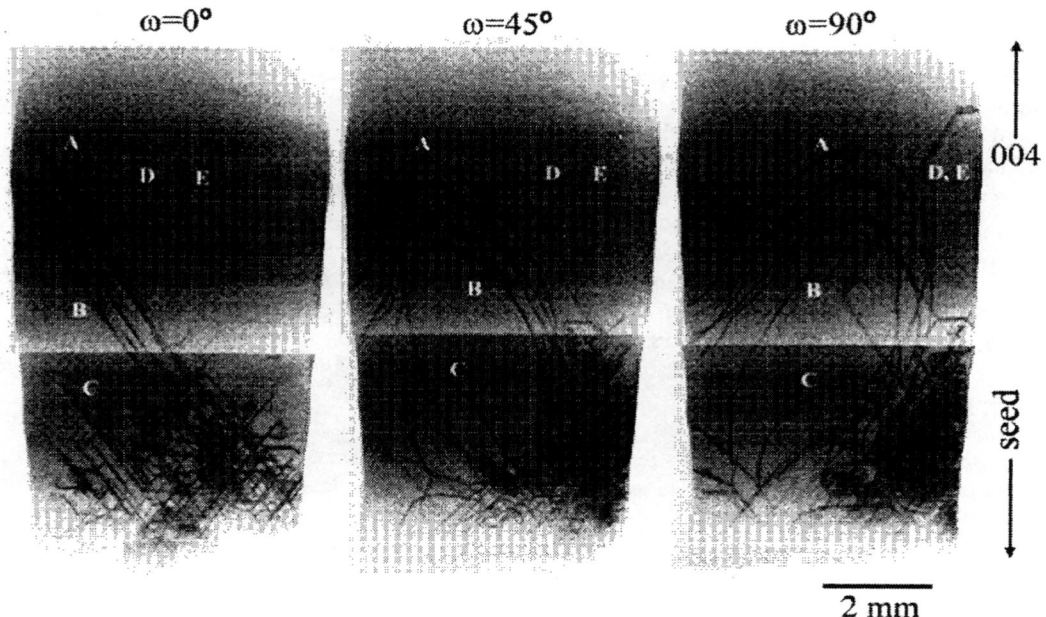

$\omega=0°$ $\omega=45°$ $\omega=90°$

004

seed

2 mm

FIGURE 3. Enlarged images of the neck crystal in the 004 spot obtained at $\omega = 0°$, $45°$, and $90°$. From the variation in their spatial configuration, dislocations A, B, and C were found to be on ($\bar{1}11$) and dislocations D and E on ($\bar{1}\bar{1}1$). The image contrast of dislocations D and E was very weak at $\omega = 0°$ but it gradually improved with the ω-rotation. This is probably attributable to the anisotropy of the strain field around the dislocation.

Burgers vector **b** of individual dislocations. We have used the rule that an approximate invisibility often arises when **g**· **b** = 0 for all types of dislocations, where **g** is a diffraction vector [6]. The Burgers vector of dislocations A and B was determined to be **a**/2 [101] and that of dislocations D and E was **a**/2 [1$\bar{1}$0], where **a** was a translation vector [7].

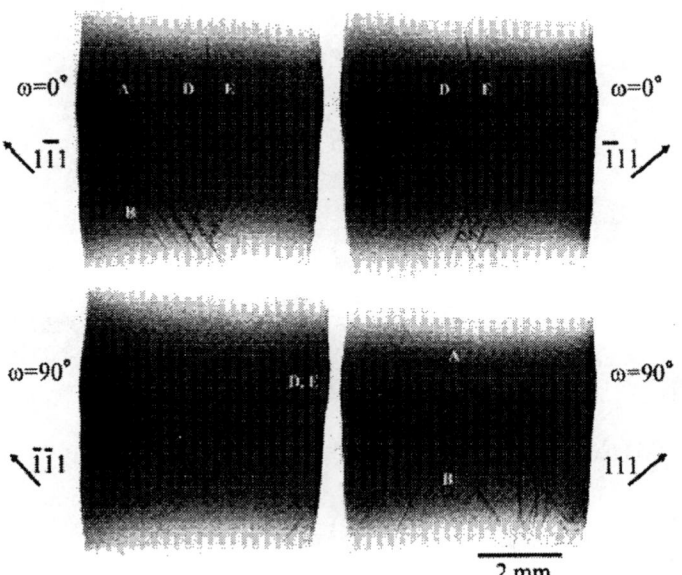

$\omega=0°$

$\bar{1}1\bar{1}$

$\omega=90°$

$\bar{1}\bar{1}1$

$\omega=0°$

$\bar{1}\bar{1}1$

$\omega=90°$

111

2 mm

FIGURE 4. Enlarged images of the neck crystal in the $1\bar{1}1$ and $\bar{1}11$ spots obtained at $\omega = 0°$ and those in the $\bar{1}\bar{1}1$ and 111 spots obtained at $\omega = 90°$. The Burgers vector of dislocations A and B was determined to be **a**/2 [101] since these dislocations showed a weak contrast in the $\bar{1}\bar{1}1$ and $\bar{1}11$ spots. The Burgers vector of dislocations D and E was **a**/2 [1$\bar{1}$0] since they showed a weak contrast in the 111 and 004 spots (see Fig. 3).

MECHANISM OF DISLOCATION ELIMINATION

It has been long believed that the residual dislocations proceeding in contact with the solid-melt interface during the crystal growth play an important role in the dislocation elimination during Dash necking in the Czochralski crystal growth. When the diameter of the grown crystal is reduced, the shape of the solid-melt interface becomes convex with respect to the melt; consequently, these dislocations are elongated in the direction perpendicular to the interface and they finally exit the crystal, as shown in Fig. 5 [8].

However, based on the presence of dislocation half loops, the present study has revealed that the residual dislocations in the neck are not accompanied by the solid-melt interface, and they proceed in the grown crystal after the movement of the interface during the crystal growth. This result is consistent with the previous observation [2]. We propose a new mechanism of the dislocation elimination during Dash necking as follows:

1) Many dislocations are generated by a thermal shock at the seed-melt interface in the early stage of the crystal growth and they penetrate into the grown crystal. This is followed by their impingement on the crystal side and the formation of tangled dislocations inside the crystal, as shown in Figs. 1 and 5 of ref. 2. The dislocations also multiplied by the Frank-Read spiral mechanism; this was followed by the formation of tangled dislocations, as shown in Fig. 3.

2) Some segments of the tangled dislocations begin to expand on the {111} planes forming dislocation half loops. The driving force is assumed to be the stress caused by the thermal gradient in the crystal. This mechanism is similar to that where a dislocation half loop is expanded on the glide plane from the segment of the dislocation networks when a shear stress is applied in the crystal [9].

3) The final stage of dislocation elimination is caused by the termination of the expansion of the dislocation half loops inside the crystal and by the pinning dislocations on the crystal sides (see Fig. 6) when the driving force is reduced.

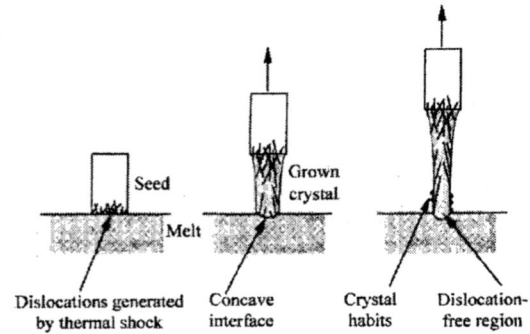

FIGURE 5. Traditional model of dislocation elimination during Dash necking. The original figure was given by T. Abe in ref. 8.

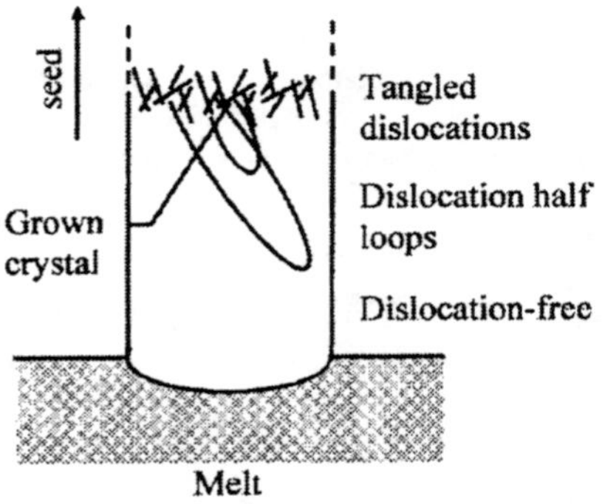

FIGURE 6. Schematic illustration of the final stage of dislocation elimination in the new model.

ACKNOWLEDGMENTS

The authors would like to thank Dr. K. Kashima of Toshiba Ceramics Co. for providing the test samples, and Messrs. K. Fukuda and A. Sato for their assistance in this experiment. The experiment was performed at SPring-8 with the approval of the Japan Synchrotron Radiation Research Institute (JASRI) (Proposal No. 2004B0532-ND3c-np).

REFERENCES

1. S. Kawado, T. Taishi, S. Iida, Y. Suzuki, Y. Chikaura, and K. Kajiwara, *J. Synchrotron Rad.* **11**, 304–308 (2004).
2. S. Kawado, T. Taishi, S. Iida, Y. Suzuki, Y. Chikaura, and K. Kajiwara, *J.Phys. D: Appl. Phys.* **38**, A17–A22 (2005).
3. F. C. Frank and W. T. Read, *Phys. Rev.* **79**, 722–723 (1950).
4. W. C. Dash, in *Dislocations and Mechanical Properties of Crystals*, edited by J. C. Fisher et al., New York: John Wiley & Sons, 1957, pp. 57–68.
5. A. Authier and A. R. Lang, *J. Appl. Phys.* **35**, 1956–1958 (1964).
6. B. K. Tanner, *X-Ray Diffraction Topography*, Oxford: Pergamon, 1976, chap. 4.
7. J. P. Hirth and J. Lothe, *Theory of Dislocations*, New York: McGraw-Hill, 1968, chap. 11.

8. T. Abe, in *Dynamics of Melt Growth*, edited by S. Miyazawa, Tokyo: Kyoritsu, 2002, p. 153. (in Japanese)
9. M. Kato, *Introduction to the theory of dislocations*, Tokyo: Shokabo, 1999, p. 67. (in Japanese)

Analysis of Dislocation Structures in Bulk SiC Single Crystal by Synchrotron X-ray Topography

Satoshi Yamaguchi, Daisuke Nakamura, Itaru Gunjishima and Yoshiharu Hirose

Toyota Central R&D Labs,. Inc. 41-1 Yokomichi, Nagakute,Nagakute-machi,Aichi,480-1192, Japan

Abstract. The detailed properties of the dislocations of SiC crystals were analyzed by means of transmission X-ray topography using ultrahigh-quality substrates manufactured by RAF (repeated a-face) growth method. From this analysis, we revealed the detailed features of one type of basal plane dislocations and two types of threading dislocations. The basal plane dislocations were screw type with Burgers vector were parallel to the $<11\bar{2}0>$ direction. One of the threading dislocations was mixed type close to screw dislocation parallel to the growth direction with Burgers vector of $1c+na$ ($n=0$, 1, 2, ...). Another was the edge type parallel to the c-axis, which lay between two basal plane dislocations. Moreover, these dislocations were found to be connecting with each other, constituting large network structures.

Keywords: X-ray Topography, SPring-8, BL20B2, BL16B2, Silicon Carbide, single crystal, dislocation.
PACS: 61.72.Ff

INTRODUCTION

Although some rough sketches of dislocation structures in bulk SiC crystals have been reported [1], detailed features were not yet revealed by transmission X-ray topography. In conventional SiC crystals, it was difficult to obtain a clear image of a single dislocation by the transmission X-ray topography, because each dislocation image included some superpositions from neighboring ones due to the high density of dislocations. In the meanwhile, we have developed a new method of bulk growth process for ultrahigh-quality SiC single crystals. The RAF (repeated a-face) growth method [2] was designed to reduce crystal defects in the SiC single crystals. The RAF substrate allowed us to analyze detailed properties of dislocations in the SiC crystal using transmission X-ray topography with reduced superpositions of dislocation images.

EXPERIMENT

Two types of in-house 4H-SiC substrates were prepared using the RAF growth method. Sample A was sliced into a plate with 400μm thickness, whose slicing plane was perpendicular to the crystal growth direction, i.e. <0001> direction with 8 degrees off-axis. Sample B was sliced with the slicing plane parallel to the growth direction. The surfaces of the substrates were carefully polished and treated to eliminate the mechanically damaged layer on both sides.

The observations of high-resolution synchrotron monochromatic X-ray topography were performed at SPring-8, the second hutch of BL20B2 and BL16B2. X-ray topographs were taken with transmission geometry and reflection geometry. The former shows the defects inside the bulk of samples, whereas the latter is suitable for investigating the surfaces layer of samples. The Sample A was observed using $11\bar{2}0$ and $1\bar{1}00$ diffractions with transmission geometry, and 0004 diffraction in reflection one. 0004 and $22\bar{4}0$ transmission topographs of the Sample B were taken. The experimental conditions are shown in Table 1. All the topographs were recorded on nuclear plates, Ilford L4, at a high resolution.

CP879, *Synchrotron Radiation Instrumentation: Ninth International Conference*,
edited by Jae-Young Choi and Seungyu Rah
© 2007 American Institute of Physics 978-0-7354-0373-4/07/$23.00

TABLE 1. The measurement conditions of this topographic experiment

Sample	Diffraction plane	Transmission or Reflection	X-ray Energy [keV]	Observation depth	Beamline (SPring-8)
A	11$\bar{2}$0	Trans.	16.09	bulk	
	1$\bar{1}$00	Trans.	16.71	bulk	BL20B2
	0004	Ref.	11.94	15μm	
B	0004	Trans.	14.20	bulk	BL20B2
	22$\bar{4}$0	Trans.	17.68	bulk	BL16B2

RESULTS

Figure 1 shows the X-ray topographs for Sample A taken in the same areas with various diffraction conditions. Two sorts of defect images are observed in Fig. 1, which are line shaped and dotted ones. The dotted ones seem to localize in the end of the line shaped ones, as shown in Fig. 1 (a). Furthermore, dotted images are observed under both conditions of g=11$\bar{2}$0 and 0004, and disappeared in g=1$\bar{1}$00. Those facts indicate that the dotted images observed in Fig. 1 are threading mixed dislocations with Burgers vector of 1c+na (n=0, 1, 2, ...), where component "a" is parallel to the [11$\bar{2}$0] direction. On the other hand, the line shaped images parallel to three crystallographically equivalent directions of [11$\bar{2}$0], [1$\bar{2}$10] and [$\bar{2}$110] are suggested to be the basal plane dislocations. The Burgers vector of the dislocations parallel to the [11$\bar{2}$0] direction is perpendicular to [1$\bar{1}$00], because the dislocation images disappeared in the topograph with g=1$\bar{1}$00.

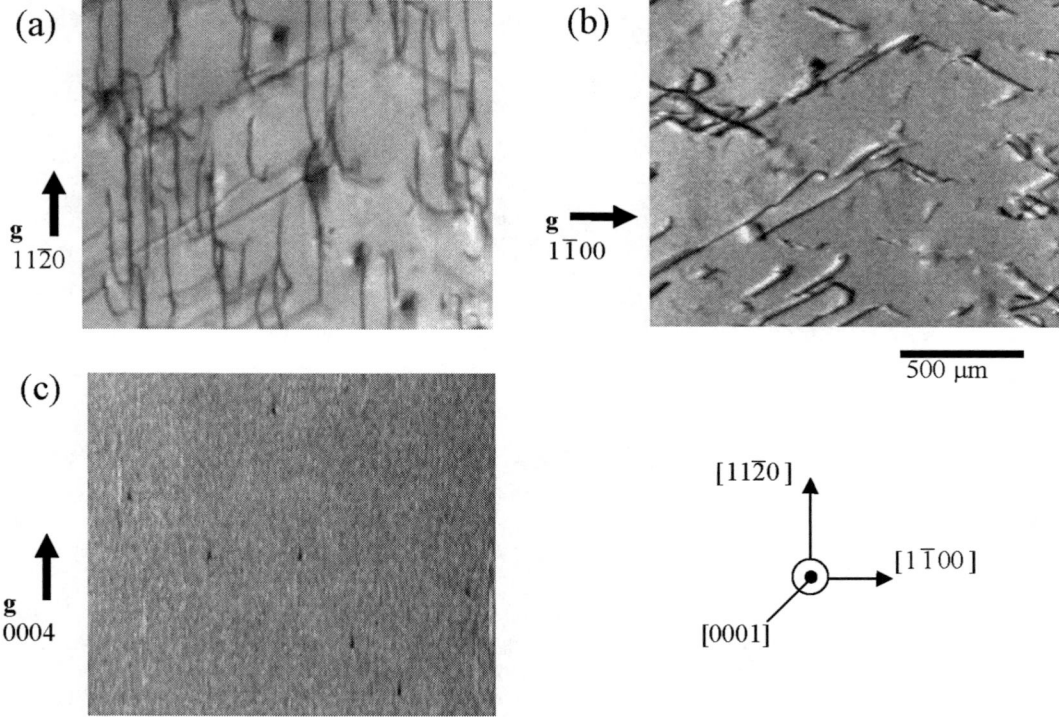

FIGURE 1. The X-ray topographs for Sample A, (a) 11$\bar{2}$0 and (b) 1$\bar{1}$00 transmission topographs, (c) 0004 reflection topograph. In 11$\bar{2}$0 and 0004 topographs, dotted defect images are observed. These defects correspond to screw or mixed dislocations parallel to growth direction. The line shaped images can be seen in transmission topographs, which are the basal plane dislocations.

X-ray topographs for the same areas of the Sample B with g=0004 and g=$22\overline{4}0$ are shown in Fig. 2 (a) and (b), respectively. Figure 2 shows the trace of propagation behaviors of the dislocations during the crystal growth. In these topographs, three sorts of defect images can be seen, i.e., long threading dislocations approximately parallel to the growth direction, line shaped dislocations lying on the basal plane, and short threading dislocations parallel to the c-axis. The first dislocation is the same sort of defect as the dotted images in Sample A. These defects were observed in the same positions of both topographs with g=0004 (Fig. 2 (a)) and g=$22\overline{4}0$ (Fig. 2 (b)). Most of the long threading dislocations are suggested to be mixed dislocations close to screw type with Burgers vector of 1c+na (n=0, 1, 2, ...), as mentioned above, connecting some basal plane dislocations. Remarkably, these long threading dislocations propagated parallel to the crystal growth directions, rather than the exact c-axis. The second, line shaped dislocation lying on the basal plane is equivalent to line shaped images in topograph for Sample A. Taking into account the result of Sample A, those basal plane dislocations can be easily understood as screw ones, whose Burgers vectors and dislocation lines are parallel to the $<11\overline{2}0>$ directions. The last, short threading dislocation parallel to the c-axis, are observed under the condition of g=$22\overline{4}0$ as shown in Fig. 2 (b), lying between two basal plane dislocations. Image annihilation of these dislocations in the 0004 topograph indicates that the short threading dislocations are edge dislocations. These short threading dislocations can not be observed obviously in the Sample A.

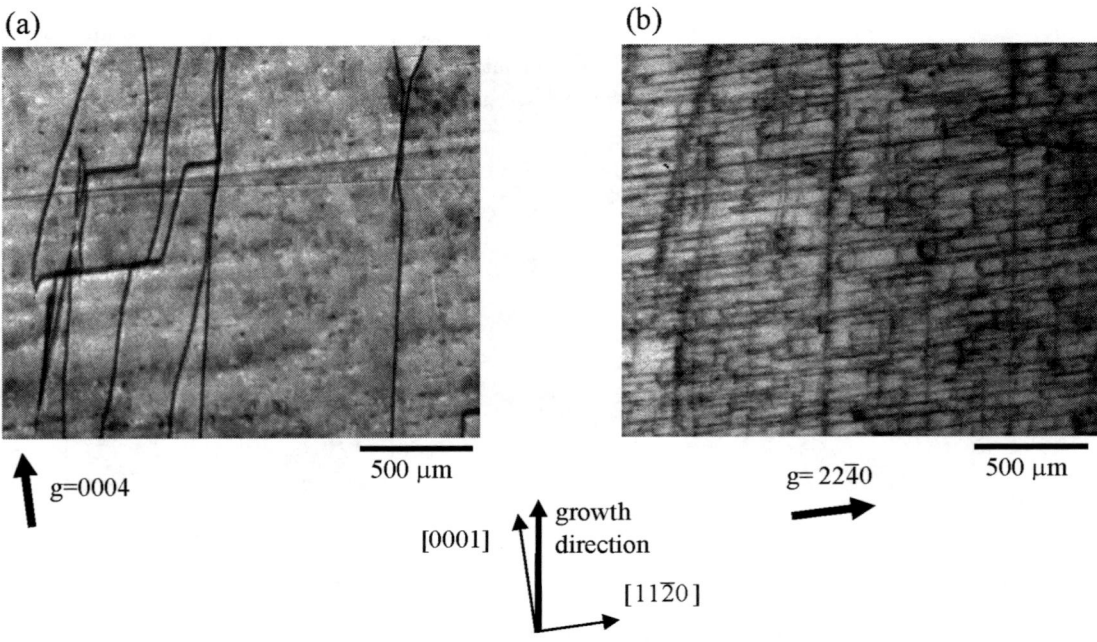

(a)　　　　　　　　　　　　　　　　　(b)

g=0004　　　　　500 μm　　　　　g=$22\overline{4}0$　　　　500 μm

[0001]　growth direction

[11$\overline{2}$0]

FIGURE 2. Transmission topographs of (a) 0004 and (b) $22\overline{4}0$ diffraction for Sample B. In the $22\overline{4}0$ topograph, we can see defect images, which are lying on the basal plane and with Burgers vector parallel to the basal plane. Furthermore, the short dislocations parallel to the c-axis, and the long ones, which are running approximately parallel to the growth direction, are observed clearly.

SUMMARY

By means of the X-ray topographic observation with synchrotron radiation, the characteristics of dislocations in 4H-SiC single crystals were successfully observed using the ultrahigh-quality SiC single crystal. The dislocations were identified as follows; long threading mixed dislocations with Burgers vectors of $1c+na$ (n=0, 1, 2, ...) which propagated parallel to the crystal growth direction, the basal plane screw dislocations parallel to $<11\bar{2}0>$, and short threading edge dislocations. Moreover, these dislocations were found to be connecting to each other, constituting large network structures.

ACKNOWLEDGEMENT

The X-ray topographic experiments were performed at the SPring-8 with the approval of the Japan Synchrotron Radiation Research Institute (JASRI).

REFERENCES

1. M. Dudley, S. Wang, W. Huang, C.H. Carter Jr., V. F. Tsvetkov and C. Fazi, *J.Phys.D*, **28**, A63-A68 (1995)
2. D. Nakamura, I. Gunjishima, S. Yamaguchi, T. Ito, A. Okamoto, H. Kondo, S. Onda and K. Takatori, *Nature*, **430**, 1009-1012 (2004)

Characterization of 3D Trench PZT Capacitors for High Density FRAM Devices by Synchrotron X-ray Micro-diffraction

Sangmin Shin[1,2], Hee Han[3], Yong Jun Park[3,4], Jae-Young Choi[4], Youngsoo Park[1,2], and Sunggi Baik[3,4]

[1]Nano Devices Laboratory
[2]Samsung Advanced Institute of Technology
[3]Dep. of Materials Science & Engineering, Pohang University of Science and Technology, San 31, Hyoja-dong, Pohang, Gyungbuk, 790-784 Korea
[4]Pohang Accelerator Laboratory, Pohang University of Science and Technology, San 31, Hyoja-dong, Pohang, Gyungbuk, 790-784 Korea

Abstract. 3D trench $PbZr_xTi_{1-x}O_3$ (PZT) capacitors for 256 Mbit 1T-1C FRAM devices were characterized by synchrotron X-ray micro-diffraction at Pohang Light Source. Three layes, Ir/PZT/Ir were deposited on SiO_2 trench holes with different widths ranging from 180 nm to 810 nm and 400 nm in depth by ALD and MOCVD. Each hole is separated from neighboring holes by 200 nm. The cross sectional TEM analysis for the trenches revealed that the PZT layers were consisted of columnar grains at the trench entrance and changes to polycrystalline granular grains at the lower part of the trench. The transition from columnar to granular grains was dependent on the trench size. The smaller trenches were favorable to granular grain formation. High resolution synchrotron X-ray diffraction analysis was performed to determine the crystal structure of each region. The beam was focused to about 500 μm and the diffraction patterns were obtained from a single trench. Only the peaks corresponding to ferroelectric tetragonal phases are observed for the trenches larger than 670 nm, which consist of fully columnar grains. However, the trenches smaller than 670 nm showed the peaks corresponding the pyrochlore phases, which suggested that the granular grains are of pyrochlore phases and non-ferroelectric.

Keywords: PZT capacitor, FRAM device, synchrotron x-ray
PACS: 85.50.GK, 61.10.Nz

INTRODUCTION

One important issue on future ferroelectric random access memory (FRAM) devices is the scalability of ferroelectric capacitor, which is closely related with the memory density [1]. In Al2 level, assuming the remnant polarization $2Pr = 20$ μC/cm^2 from the cell capacitor, a 3-dimensional (3D) trench capacitor structure would be required with lateral hole size of 0.25 μm and depth of 0.4 μm, and the resultant sensing margin would be 270 mV via 200 fF bit line capacitance [2]. When we employ 0.13 design rule (D/R) and 9F^2 cell size, the memory density will reach 256 Mbits [3]. This is a major breakthrough in commercializing high density FRAM. In this article, we report on the PZT grain structures with trench hole size ranging from 180 nm to 810 nm observed by synchrotron x-ray micro-diffraction.

EXPERIMENTAL PROCEDURE

The substrate with the structure of diffusion barrier /SiO_2/TiAlN/Si was prepared. The total thickness of the insulating layers was about 400 nm. We prepared various-sized trench holes in order to observe the size dependence of PZT grain growth along the trench sidewall. To identify the grain structure, TEM analysis and high resolution

CP879, *Synchrotron Radiation Instrumentation: Ninth International Conference,*
edited by Jae-Young Choi and Seungyu Rah
© 2007 American Institute of Physics 978-0-7354-0373-4/07/$23.00

synchrotron x-ray micro-diffraction were employed. Trench hole array from 180 nm to 810 nm of diameter were formed on a 400 nm thick SiO_2 substrate. Each hole is separated from neighboring holes by 0.2 μm spacing. As shown in Fig. 1 (a), three layers, bottom Ir (20 nm), PZT (60 nm), top Ir (20 nm), were sequentially deposited in the trench holes. Atomic layer deposition (ALD) was used for top and bottom Ir electrodes, and metal organic chemical vapor deposition (MOCVD) for PZT layer. After depositing PZT layer, the top part of PZT layer was etched off by using ion milling to observe the grain structure difference between the top part and the sidewall part of trench holes. For two kinds of samples, the trench holes with different sizes ranging from 180 nm to 810 nm were fabricated on each sample. In order to focus the incident x-ray beam on the each region with different size, we reduced the beam size down to 500 μm. The array of same sized trench holes is aligned with incident beam direction, in order to avoid the x-ray scattering from another sized trench holes. Figure 1 (b) shows the region of 250 nm trench hole.

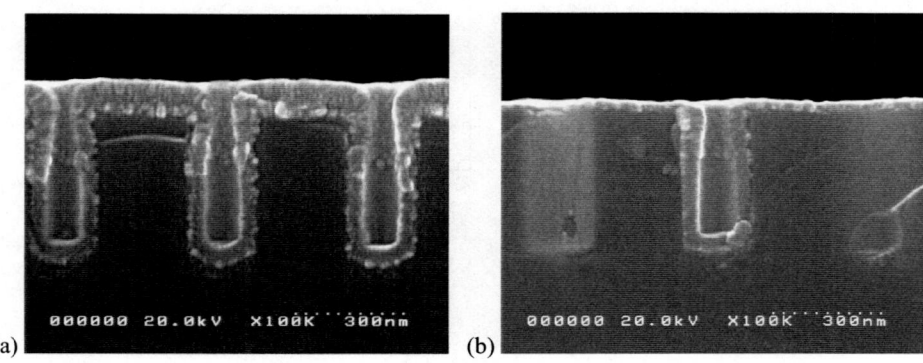

FIGURE 1. (a) SEM image of cross section of deposited Ir/PZT/Ir on SiO_2 trench holes. (b) Top PZT layers were etched off by ion-milling.

RESULTS AND DISCUSSION

Figure 2 (a) is a TEM image showing the grain growth difference on planar Ir and trench side wall Ir. On planar Ir, columnar grains were grown, while granular grains were formed on the side wall. As shown in Fig. 2 (b) columnar grains extended downward along the side wall as the hole size became larger. Therefore, the transition from columnar to granular grains was dependent on the trench size. The smaller trenches were favorable to granular grain formation.

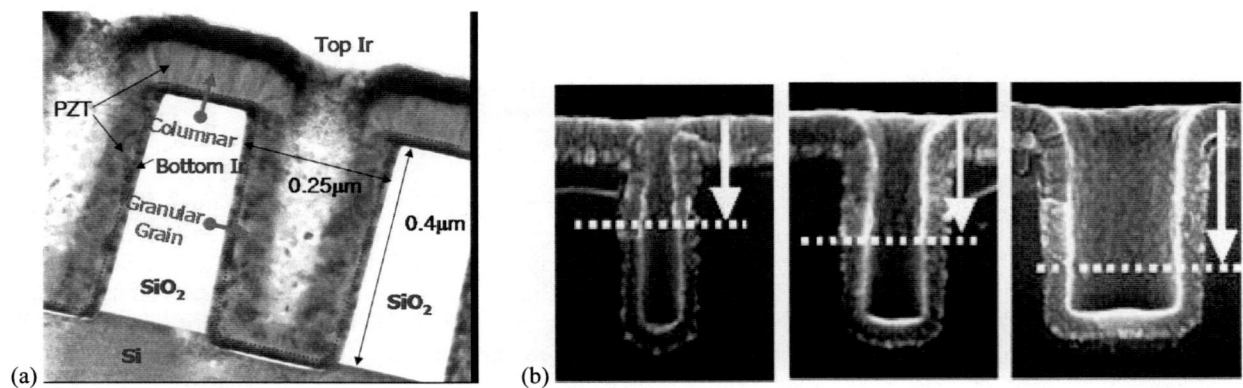

FIGURE 2. (a) TEM image of cross section of deposited PZT/Ir on SiO_2 trench holes. (b) Size dependence of columnar grain region extension along the sidewall of various-sized trench holes. 180 nm, 250 nm, 400 nm of diameter, respectively from the left.

In order to explain that both columnar and granular grains were truly PZT, synchrotron x-ray micro-diffraction analysis was performed. In the case of the sample with top PZT layer, all trench holes showed the highly (001) and (100) oriented structure, and as the trench hole size decreased, PZT (110) peak was presented as shown in Fig. 3 (a). In all trench holes, no peaks corresponding to pyrochlore phase were observed. However, in the sample with removed planar PZT by ion milling, pyrochlore phase was obtained at the smaller trench hole which means that the grains grown on side wall are not PZT but pyrochlore as shown in Fig. 3 (b). Trench holes with the size of only larger than 670 nm which consist of fully columnar grains showed the peaks corresponding to ferroelectric phase. Most sidewall region consists of granular grains in smaller than 670 nm trench holes and the granular grains are confirmed as pyrochlore phase.

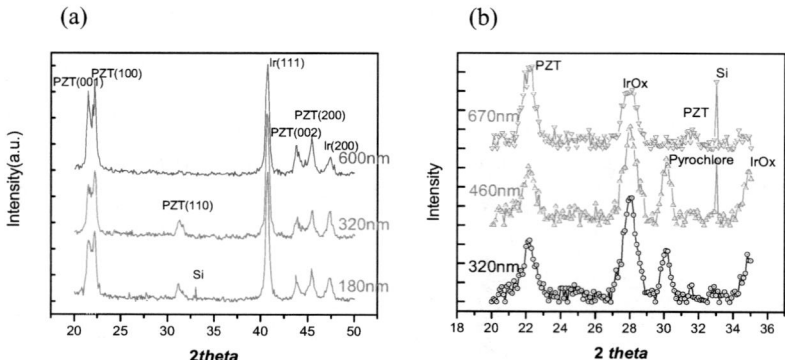

FIGURE 3. Synchrotron x-ray micro-diffraction analysis of (a) PZT layer grown on bottom Ir electrode. (b) removed planar PZT layers and only sidewall PZT remained

CONCLUSIONS

We observed the size dependence of PZT grain growth along side wall and planar region by using synchrotron x-ray micro-diffraction. PZT layers which are columnar grains grown on planar region are ferroelectric, in contrast, PZT layers which are granular grains grown on side wall are non-ferroelectric phase. The transition from ferroelectric to non-ferroelectric phase was dependent on the trench size and the critical trench hole size is 670 nm.

REFERENCES

1. D. C. Yoo, B. J. Bae, J. –E. Lim, D. H. Im, S. O. Park, H. S. Kim, U. –In. Chung, J. T. Moon and B. I. Ryu, "High Reliable 50nm-thick PZT Capacitor and Low Voltage FRAM Device using Ir/SrRuO$_3$/MOCVD PZT Capacitor Technology", *Symp. On VLSI Tech. Dig.*, pp. 100-101, 2005
2. K. Kim : "Future Emerging New Memory Technology" in plenary session, *International Symposium of Integrated Ferroelectrics*, Shanghai, China, 2004
3. The International Technology Roadmap for semiconductors : Process Integration, Devices and Structures, pp. 22, 2004

Preliminary Hard X-ray Micro-spectroscopic Investigations on Thin-Film Ta-and-W Based Diffusion Barriers for Copper Interconnect Technology

James M. Ablett[1], Joseph C. Woicik[2] and Zsolt Tokei[3]

[1] *Brookhaven National Laboratory, National Synchrotron Light Source, Upton, New York, 11973, USA.*
[2] *National Institute of Standards and Technology, Gaithersburg, Maryland, USA.*
[3] *IMEC, Kapeldreef 75, B-3001 Leuven, Belgium.*

Abstract. Within the microelectronics industry, the requirement for reducing device dimensions for increased circuit performance and lower manufacturing costs has led to many avenues of research in advanced materials and fabrication processes. One of the most important challenges in ultra-large scale integrated technology is the fabrication of thin-film diffusion barriers that prevent copper interconnect lines diffusing through the barrier material and into the neighboring silicon layers. In this paper, we present preliminary synchrotron x-ray spectroscopy measurements as a tool for studying the properties of these buried barrier layers and consider the opportunity of applying the spatial resolution of an x-ray microbeam in probing different regions of the barrier material.

Keywords: Microelectronics, Diffusion Barrier, X-ray Spectroscopy, Synchrotron.
PACS: 82.80.Ej, 85.40.-e.

INTRODUCTION

As the width of Cu interconnect lines falls beyond 90 nm and is anticipated to reach 45 nm within the next few years, the need for forming ultra-thin highly-conformal barrier layers along the entire surface area of the interconnect trench is required for reliable device operation. Current research is focused on finding the optimal fabrication processes and barrier materials as devices continue to scale down in size. Barrier properties depend on a wide range of parameters such as growth processes, anneal temperatures, amorphous/polycrystalline structure, composition and thickness.[1-9] Barrier liners, comprising of refractory materials such as Ta, W, Ti and their nitrides, have been studied in their effectiveness for preventing Cu diffusion through the barrier layer and into the nearby silicon.[1-9] The major barrier reliability concern is that Cu impurity atoms are extremely mobile and therefore detrimental to active silicon transistor performance. Good adhesion of the barrier layer to the substrate and the copper interconnect is another important factor to consider. The ability to determine the local bonding environment of thin barrier layers using x-ray spectroscopy would be extremely beneficial as it could reveal important information governing the overall performance of the barrier material, which can be amorphous and therefore not measurable by x-ray diffraction. In this paper, we show through preliminary investigations, that x-ray micro-spectroscopy is a powerful analytical tool for studying these barrier systems.

EXPERIMENT

Feasibility studies were performed on tantalum and tungsten based barriers layers, grown by physical vapor deposition (PVD) and atomic layer deposition (ALD), with ALD identified as the likely candidate for growing highly conformal barrier films in sub 65 nm interconnect lines.[7] Both blanket barrier films, with and without Cu overlayers, and passivated Cu interconnect samples fabricated using damascene and chemical mechanical planarization processes, with varying line spacing, widths and barrier thicknesses were investigated. All samples were grown on Si(100) wafers, which had a SiO_2 insulating layer on the top. The experiments were performed at the

CP879, *Synchrotron Radiation Instrumentation: Ninth International Conference,*
edited by Jae-Young Choi and Seungyu Rah
© 2007 American Institute of Physics 978-0-7354-0373-4/07/$23.00

National Synchrotron Light Source on the X27A micro-focus bending-magnet beamline, which has a 6 µm [vertical] x 17µm [horizontal] *fwhm* spot size.[10] The samples were mounted upright and at ~ 45° to the incident x-ray beam, on a motorized x-y-z stage for accurate positioning. A 13-element liquid nitrogen cooled germanium detector and associated digital processing electronics were used for collecting and analyzing the fluorescence x-rays emitted from the sample. For x-ray absorption near-edge structure (XANES) and extended x-ray fine structure (EXAFS) measurements, the incident x-ray energy was tuned across a particular absorption-edge of interest using a Si(111) channel-cut monochromator as the change in intensity of the various fluorescence line(s) of interest were recorded. More details on the X27A micro-focus experimental beamline and capabilities can be found in [10].

RESULTS AND DISCUSSION

One of the difficulties in applying standard x-ray fluorescence spectroscopy to these barrier layers is the proximity of the core-level x-ray fluorescence energies of the associated elements at the L-edges of Ta and W, and at the K-edge of the Cu interconnect/overlayer. With a typical energy-dispersive detector, with an energy resolution of ~ 2%, i.e. 200 eV *fwhm* at 10 keV, peak overlap between the Ta/W $L\alpha$ and Cu $K\alpha$ lines is problematic. However, probing the L_2 absorption edge of Ta and W opens up additional decay channels that can be used to investigate these species of interest. To highlight this, Fig. 1. shows (plotted on a log scale) the MCA spectra taken above (diamond symbol) and below (straight line) the Ta L_2 absorption edge on a tantalum-based barrier film with a 1000 nm Cu overlayer. The barrier layer was prepared by PVD and the system was ordered as; Si(100) / SiO_2 / Ta_2N(5nm) / Ta(10nm) / Cu(1000 nm). As can be seen, the Ta $L\beta_1$ emission (L2-M4) is well separated from the neighboring fluorescence lines and the Compton/elastic scatter peaks.

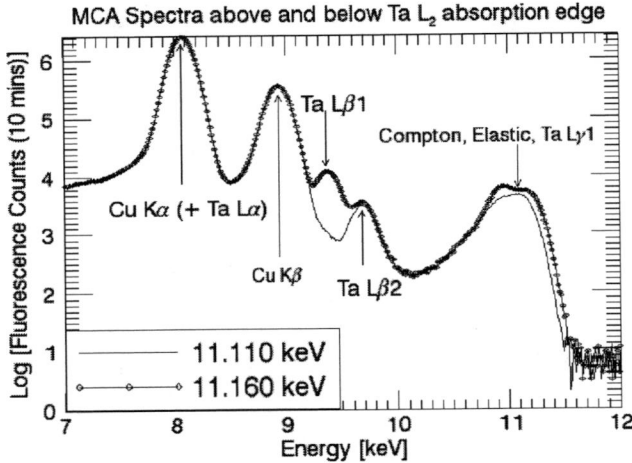

FIGURE 1. MCA Spectra from a Ta-based barrier liner with a Cu overlayer (blanket film), above and below the Ta L_2 absorption edge (11.136 keV). The peaks are labeled according to the main contributions from the various decay channels.

The ability of obtaining chemical and bonding information from the near- and- extended regions of the x-ray absorption spectra on Ta-based barrier systems is now realized. This is confirmed in Fig. 2., where Ta $L\beta_1$ fluorescence is plotted as a function of incident x-ray energy, across the Ta L_2 absorption edge (left), which is a combination of 40 scans, lasting ~ 40 hours in total. The corresponding EXAFS spectra is shown on the right, and although this preliminary data are somewhat noisy, better statistics will be available from real-world systems that have much lower copper coverage. We believe it is possible, through analyzing our other preliminary data, that good EXAFS spectra on < 5nm thin barrier films can be obtained under typical < 200 nm thick Cu interconnect lines within a reasonable amount of measurement time i.e. a few days.

Furthermore, another diffusion barrier system candidate, WNC/Ta has been examined for potential x-ray spectroscopic analysis. These barrier layers were grown by ALD and the system was ordered as; Si(100) / SiO$_2$ / WNC(6nm) / Ta(10nm) / Cu(50-100 nm). Figure 3. (left and center) shows the MCA spectra taken above (diamond symbol) and below (straight line) the W L$_2$ absorption edge. The W Lβ1 fluorescence was identified as being a possible applicant for measuring XANES/EXAFS on these systems, and Fig. 3. (right) shows W L$_2$ XANES measurements on ALD 60 cycle (4nm) and 90 cycle (6nm) WNC/Ta- based barrier layers using this emission channel.

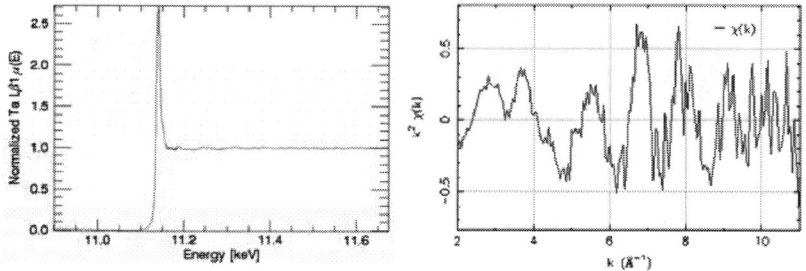

FIGURE 2. Ta L$_2$ absorption edge scan, obtained by recording the Ta Lβ$_1$ fluorescence intensity and corresponding EXAFS data taken on a PVD Ta$_2$N (5nm) / Ta (10 nm) barrier with a 1000 nm Cu overlayer.

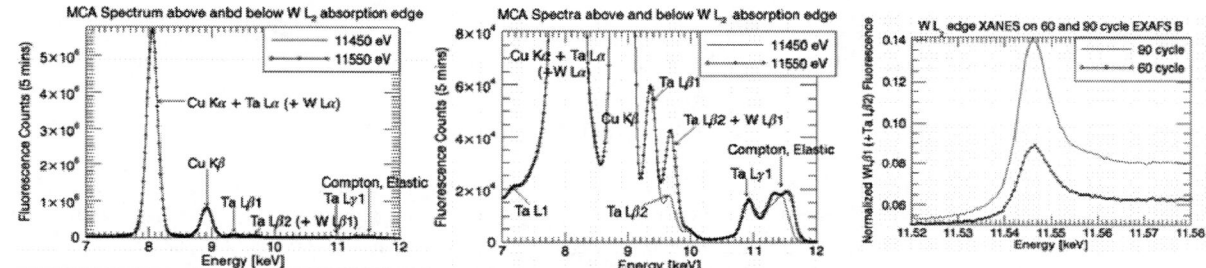

FIGURE 3. Left and Center: MCA Spectra from a WNC/Ta-based barrier layer with a Cu overlayer (blanket film), above and below the W L$_2$ absorption edge (at 11.544 keV). The peaks are labeled according to the main contributions from the various decay channels. Right: W L$_2$ absorption-edge XANES scan on 60 (4nm) and 90 (6nm) cycle WNC/Ta based barrier layers with a Cu overlayer.

The spatial resolution offered by x-ray micro/nano probes allows for the possibility of examining the chemistry/bonding properties of these buried barrier layers at various locations around and beneath the Cu interconnect lines. Preliminary measurements on passivated 10 μm wide 120 nm thick Cu interconnects, separated by 40 μm and 90 μm 'field' regions, and with a 4 nm thin ALD Ta$_3$N$_4$ barrier layer were performed at X27A. Cu Kα fluorescence recorded at an incident x-ray energy of 9.5 keV, and Ta Lβ$_1$ fluorescence recorded at the peak of the Ta white-line at 11.142 keV, are shown in Fig. 4a. as the 10 μm wide interconnect lines, spaced by 40 μm, are scanned across the x-ray microbeam. The Cu and Ta fluorescence increases across the interconnect lines, but do not fall to zero off a line. The reason for this was that sub-micron copper line 'dummies' were inserted within the 'field' regions for these particular test structures. Figure 4b. similarly shows the variation in Cu fluorescence across 10 μm wide, 90 μm spaced interconnect lines and Fig. 4c. shows a Cu fluorescence map across an area that contains two copper pad structures. Clearly, the application of x-ray microprobes for exploring the chemistry and bonding of the barrier layers at various locations in device structures is evident. Figure 5. shows the Ta L$_2$ XANES on a PVD Ta$_2$N(5nm)/Ta (20nm) blanket film (straight line A14) and a 4nm ALD Ta$_3$N$_4$ barrier used in a copper interconnect test structure (diamond and asterisk symbols). As is shown, the nitrogen content within the Ta$_3$N$_4$ barriers is clearly evidenced by the spectral signature of the near-edge region and the shift in the absorption edge towards higher x-ray energies, compared to the Ta$_2$N(5nm)/Ta(20nm) blanket film which is of predominantly Ta character.

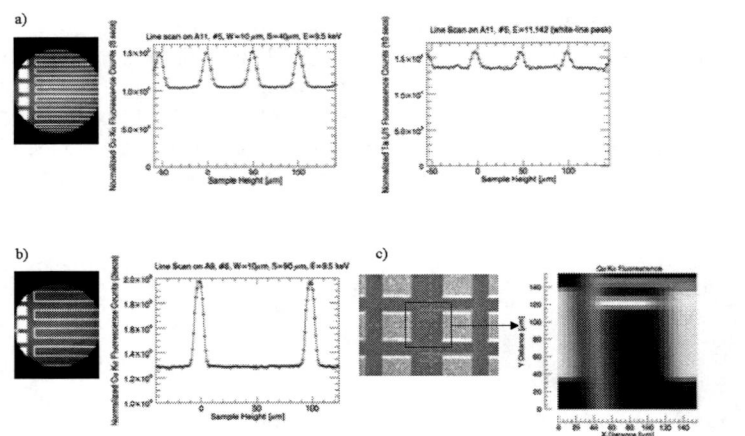

FIGURE 4. X-ray fluorescence scans across copper interconnect a) & b) and pad structures c).

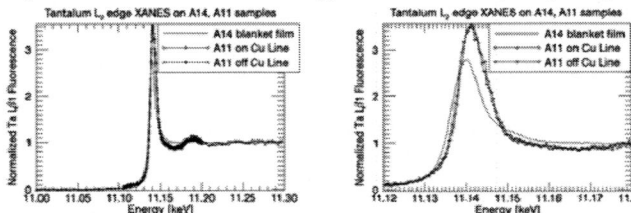

FIGURE 5. Ta L_2 absorption-edge XANES on Ta_2N (5nm)/Ta (20nm) blanket film (straight line A14) and the Ta_3N_4 barrier on a Cu interconnect patterned sample, both on and off the interconnect lines (diamond and asterisk symbols A11).

CONCLUSION

In this paper, we have shown the prospective use of x-ray spectroscopy for studying buried thin-film diffusion barriers for copper interconnect technology. By accessing the L_2 absorption edges of Ta and W, these barrier layer systems can be investigated, with the potential of revealing the chemistry and bonding of these species through XANES and EXAFS measurements. In addition, measurements on test structure devices can be performed using x-ray micro-spectroscopic techniques to probe different regions of the barrier material, such as the sidewalls and trench bottoms of Cu interconnect lines. Furthermore, with the increasing availability of synchrotron x-ray nano-probes with spatial resolutions below 100 nm, these instruments open up new opportunities for studying sub-100 nm device structures for real-world microelectronics applications.

ACKNOWLEDGMENTS

Research carried out at the National Synchrotron Light Source, Brookhaven National Laboratory, which is supported by the U.S. Department of Energy, Office of Basic Energy Sciences and Division of Materials Sciences under Contract No. DE-AC02-98CH10886.

REFERENCES

1. Chung, H.C. and C.P. Liu, Surface and Coatings Technology **200**, 3122 (2006).
2. Hübner, R., et al., Thin Solid Films **500**, 259 (2006)
3. Jiang, L., et al., Jap. Journ. Appl. Phys. **41**, 6525 (2002).
4. K.-L., O., et al., Journ. Vac. Sci. Techn. B **23**(1), 229 (2005).
5. Kim, H. and S.M. Rossnagel, Thin Solid Films **441**, 311 (2003).
6. Kim, H., et al., Journ. Appl. Phys. **98**, 014308-1 (2005).
7. Kim, H., Surface and Coatings Technology **200**, 3104 (2006).
8. Li, N., D.N. Ruzic, and R.A. Powell, Journ. Vac. Sci. Techn. B **22**(6), 2734 (2004).
9. Wang, H., et al., Appl. Phys. Lett. **81**(8), 1453 (2002).
10. Ablett, J.M., et al., Nucl. Instrum. and Meth. in Phys. Res. A **562**, 487 2006.

Magnified 3D-CT System Using the Portable Synchrotron "MIRRORCLE-6X"

M. Sasaki[1,2], Y. Oda[1], J. Takaku[1], T. Hirai[1], H. Yamada[1,2], T. Nitta[3], M. Takahashi[3], and K. Murata[3]

[1] *Faculty of Science and Engineering Ritsumeikan University, 1-1-1 Nojihigasi, Kusatsu, Shiga, 525-8577, Japan*
[2] *Synchrotron Light Life Science Center, Ritsumeikan University, 1-1-1 Nojihigasi, Kusatsu, Shiga, 525-8577, Japan*
[3] *Department of Radiology, Shiga University of Medical Science, Seta-Tsukinowa-cho, Otsu, Shiga, 520-2192, Japan*

Abstract. The portable synchrotron MIRRORCLE-6X is a novel x-ray source suitable for hard x-ray imaging. Highly brilliant x-rays are generated by MIRRORCLE-6X in the shape of cone from a source with micron size. We obtained images with high magnification and high resolution by these x-rays, and found that they can be utilized for medical imaging. Recently, we started to perform ideal 3-dimentional computed tomography (3D-CT) based on the Feldkamp algorithm, which is the most popular reconstruction algorithm for a cone beam CT, using a flat panel detector. Therefore, we developed a magnified 3D-CT system of high quality using MIRRORCLE-6X.

Keywords: magnified imaging, computed tomography, hard x-ray
PACS: 07.85.Tt

MAGNIFIED IMAGES USING MIRRORCLE-6X

The portable synchrotron MIRRORCLE-6X is a novel x-ray source suitable for hard x-ray imaging [1,2,3,4]. Highly brilliant x-rays are generated by MIRRORCLE-6X in the shape of cone from a source with micron size. We obtained images with high magnification and high resolution by these x-rays, and found that they can be utilized for medical imaging, nondestructive inspection and so on [5,6]. Figure 1 shows a magnified image of a chest phantom, in which an urethane ball of 8 mm diameter was implanted as an imitation of a tumor. The magnification rate was 5.4 times and the detector was an imaging plate (*FUJIFILM, IMAGING PLAATE ST-IV*, 150 μm pixel size). We see not only a shape of the urethane ball, but narrow blood vessels surrounding the ball through ribs. Such magnified images with high resolution cannot be obtained by conventional medical x-ray tubes, which have sources with a few millimeters sizes. Using x-rays generated by MIRRORCLE-6X and a flat panel detector, we can perform an ideal 3-dimentional computed tomography (3D-CT) based on the Feldkamp algorithm, which is the most popular reconstruction algorithm for a cone beam CT. Therefore, we developed a magnified 3D-CT system of high quality using MIRRORCLE-6X.

MAGNIFIED 3D-CT SYSTEM BASED ON MIRRORCLE-6X

Figure 2 shows the magnified 3D-CT system based on MIRRORCLE-6X. Flat panel detector, *PAXSCAN2520* made by *VARIAN* medical systems, is used in this system. This detector has an irradiation area of 244 mm wide x 195 mm height and 127 μm of the pixel size. Samples are set at 0.5m from the source point, and the magnification rate is determined by the position of the detector. When the detector is located at 5.4 m from the source point, the projection data can be obtained with maximum magnification rate of 10.8 times for samples with width smaller than 23 mm. We developed an original program to control the rotation stage and the detector by using the software, *LabView* and *Visual C++.NET*. From the obtained projection data, we derived the 3D-images using the Feldkamp reconstruction algorithm.

CP879, *Synchrotron Radiation Instrumentation: Ninth International Conference*,
edited by Jae-Young Choi and Seungyu Rah
© 2007 American Institute of Physics 978-0-7354-0373-4/07/$23.00

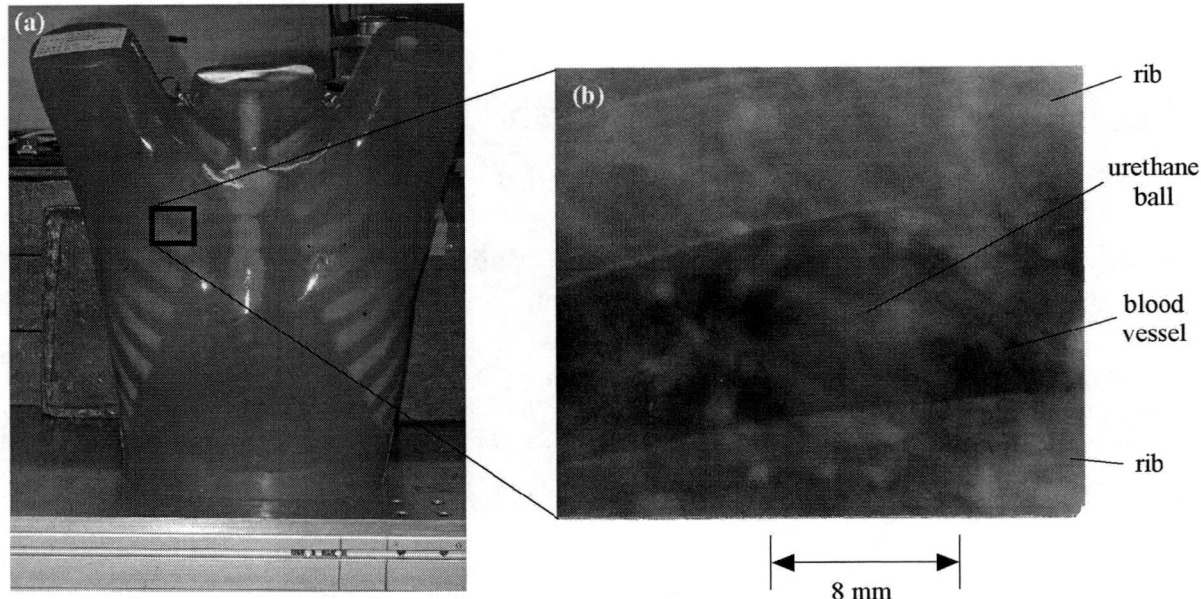

FIGURE 1. Magnified image using MIRRORCLE-6X. (a) Photograph of a chest phantom, and (b) magnified image near the urethane ball as an imitation of a tumor. Magnification rate was 10.8 times and a diameter of the ball was 8 mm.

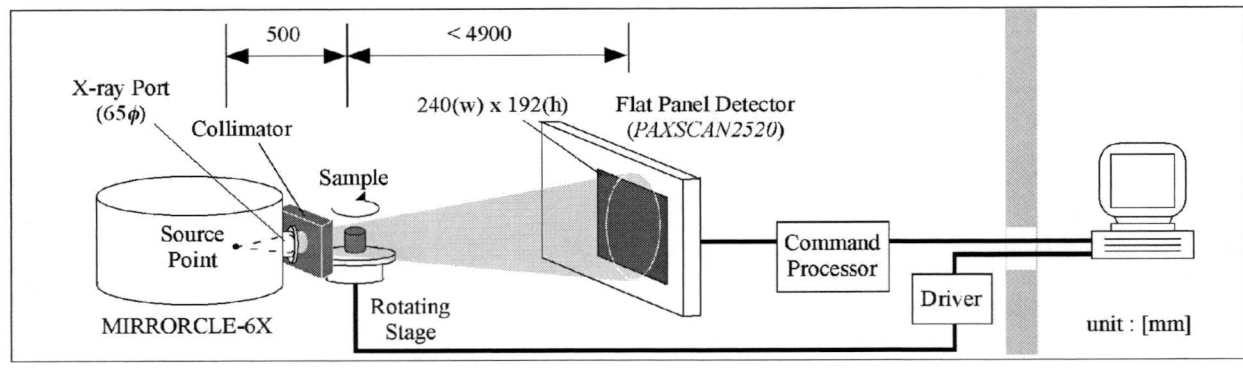

FIGURE 2. Magnified 3D-CT system based on MIRRORCLE-6X.

RESULTS

In order to investigate this system, we performed the magnified 3D-CT with low magnification rate. The samples were a connector and a transistor, and magnification rates were 2 times and 4 times, respectively. For these samples, projection data were taken by rotation angle of 2 degree and 2.5 degree, respectively. The size of the source point was 100 μm in diameter.

Figure 3 shows cross sectional images and volume renderings for the connector by the magnified 3D-CT. From the cross sectional images (Fig. 3-(b), (c) and (d)), we see that the shapes of pins are different in accordance with the depth. In addition, we also see a screw thread and a pawl of the connector from the volume renderings (Fig. 3-(e)).

In Fig. 4, the inner structure of the transistor is shown. We see a semiconductor plate in the transistor. But, we do not see a bonding wire between the plate and a terminal with a diameter of several tens μm.

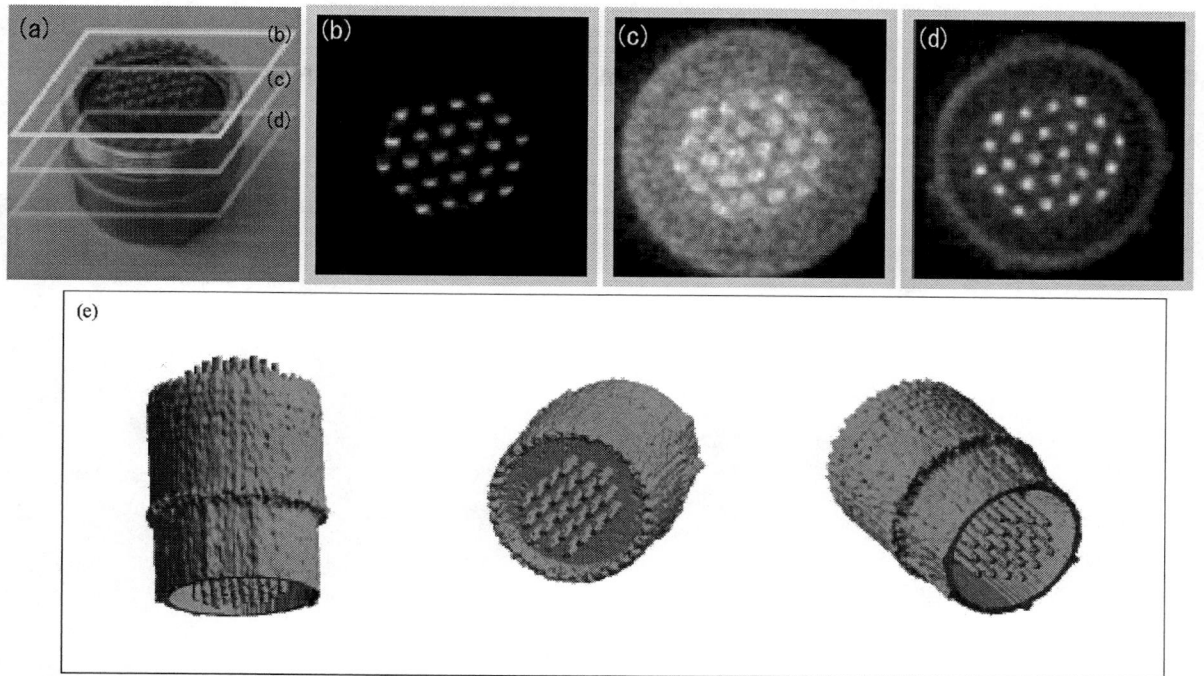

FIGURE 3. 3D-CT image of a connector. (a) Photograph, (b)(c)(d) cross sectional images, and (e) volume renderings.

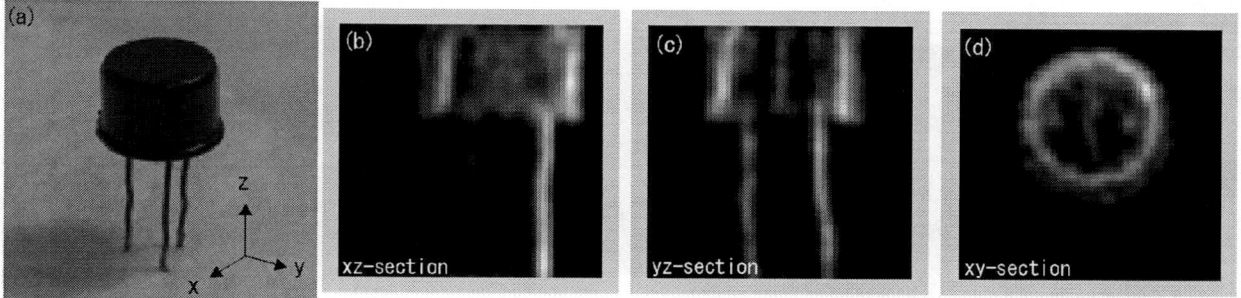

FIGURE 4. 3D-CT image of a transistor. (a) Photograph, and (b)(c)(d) cross sectional images.

SUMMARY

We developed the magnified 3D-CT system using MIRRORCLE-6X, and we succeeded to obtain the 3D images with the magnification rate of 2 ~ 4 times. All projection data for the transistor were obtained for 1.3 minutes. From this fact, it was found that it takes about 8 minutes to perform the magnified 3D-CT with the maximum magnification. Even if very high density samples are used, the exposure time is expected to be about 20 minutes.

For the maximum magnification rate, spatial resolution of about 100 μm should be achieved in this system. The spatial resolution will be improved by using a smaller source and by using a flat panel detector with smaller pixel. After these improvements, we will try to perform the 3D-CT with the magnification rate of 10 times. Consequently, the magnified 3D-CT based on MIRRORCLE-6X is useful for nondestructive inspection, as well as for biological study of small animals.

REFERENCES

1. H. Yamada, *Journal of Synchrotron Radiation* (1998), pp.1326-1331
2. H. Yamada, *Nucl. Instr. and Meth.* **A467-468** (2001), pp.122
3. H. Yamada, *Nucl. Instr. and Meth.* **B199** (2003), pp. 509-516
4. H. Yamada, *AIP Proc. of Int. Sympo. on Portable Synchrotron Light Sources and Advanced Applications* (2004), pp.12-17
5. T. Hirai, et al., *Proc. Int. Conf. Synchrotron Radiation Instrumentation, San Francisco* (2003), pp.24-29
6. T. Hirai, et al., *AIP Proc. of Int. Sympo. on Portable Synchrotron Light Sources and Advanced Applications* (2004), pp. 132-134

Using Fluorescence XANES Measurement to Correct the Content of Hexavalent Chromium in Chromate Conversion Coatings Determined by Diphenyl Carbazide Color Test

Junichi Nishino, Toshikazu Sekikawa*, Haruka Otani*, Hironori Ofuchi, Yosuke Taniguchi, Tetsuo Honma, and Akio Bando*

Japan Synchrotron Radiation Research Institute (1-1-14 Kouto, Sayo-cho, Sayo-gun, Hyogo 679-5198 Japan)
*Mihara sangyo co., ltd. (15-26, 2chome, Mikuriyasakaemachi, Higashiosaka 577-0036 Japan)

Abstract. The Restriction of the use of certain Hazardous Substances (RoHS) directive will take effect on July 1 of this year. From that date, the use of chromate conversion coatings containing hexavalent chromium will not be permitted. By comparing the concentration of $Cr6+$ determined by the diphenyl carbazide color test and by fluorescence XANES (X-Ray Absorption Near Edge Structure) measurement, we can correct for the $Cr6+$ content of the color test. This will enable the use of the diphenyl carbazide color test to check product shipments in compliance with the RoHS directive.

Keywords: XANES, chromate conversion corting, hexavalent chromium, SPring-8, RoHS.
PACS: 61.10.Ht

INDUSTRIAL BACKGROUND

Chromate conversion coatings have been widely used for the surface layer of Zn plating on steel products. Recently, $Cr3+$ conversion coating was developed in preparation for the RoHS directive [1], because of its durability, strong adhesion to steel and so on.

The diphenyl carbazide color test is commonly used because of its simplicity and low cost. But, this test has the following problems.

· During the dissolution of $Cr6+$ into water, $Cr3+$ sometimes changes by reacting to $Cr6+$.

· The whole coating layer does not always dissolve into water in the limited time allowed (several minutes).

· Discoloration sometimes occurs because of the presence of $Fe3+$, $Mo4+$, and $V5+$.

To avoid these problems, XANES (X-Ray Absorption Near Edge Structure), is measured $Cr6+$ directly. But, this method requires high cost equipment, making it impractical for checking shipments of low cost products such as bolts, nuts, washers, etc. A practical method of checking products using the diphenyl carbazide color test as corrected by XANES needs to be established.

RoHS Directive

The key point of the RoHS directive is as follows; "Member States shall ensure that, from 1 July 2006, new electrical and electronic equipment put on the market does not contain lead, mercury, cadmium, hexavalent chromium, polybrominated biphenyls (PBB) or polybrominated diphenyl ethers (PBDE)". The phrase "does not contain" means must contain less than 1000 ppm (weight)/unit. The major electric and electronic companies in Japan will adopt the interpretation of "unit" as the weight of the Zn plating layer of one piece of the product (bolts, nuts, washers, and so on) for chromate conversion coatings [2].

CP879, *Synchrotron Radiation Instrumentation: Ninth International Conference,*
edited by Jae-Young Choi and Seungyu Rah

EXPERIMENT

Nomura demonstrated the method for correcting the diphenyl carbazide color test by using fluorescence XANES measurement at a characteristic peak of Cr6+ at 5993 eV of the chromate conversion coating [3]. The contents of Cr6+ in layers of conversion coating were determined by measuring the Cr6+ peak height of Cr-Kα fluorescence. We applied the same method for very low Cr6+ content of Cr3+ coating layers using a 19-channel solid-state detector (19ch-SSD) with precise measurement. The experiments were done at the BL19B2 at SPring-8.

Samples

As samples we used washers that were 10.0 mm in outer diameter, 3.0 mm in inner-hole diameter, and 1.0 mm thick. The washers were prepared with three different treatments. Series-A washers had a Zn plated layer treated by nitric acid solution containing Cr2O3 (Cr3+) and CrO3 (Cr6+). The Cr3+/Cr6+ ratios of the solutions varied from 0 to 10 weight %. Series-B washers had a Zn plated layer treated by LUSTER M-200A and M-200B [4] solutions with added CrO3(Cr6+). Series–C washers had a Zn plated layer treated by M-200A solution only and added CrO3(Cr6+). Figure 1 shows a cross section SEM image and a schema of a series-B sample. Chromate conversion coatings of series-B washers consisted of a chromate layer covered with a glass layer which protect a chromate layer. The total thickness of the conversion coatings ranged from about 100 nm to 200 nm; the thickness of the glass layer was not able to be measured by cross-section SEM image. Series-A and series-C were not covered with a glass layer.

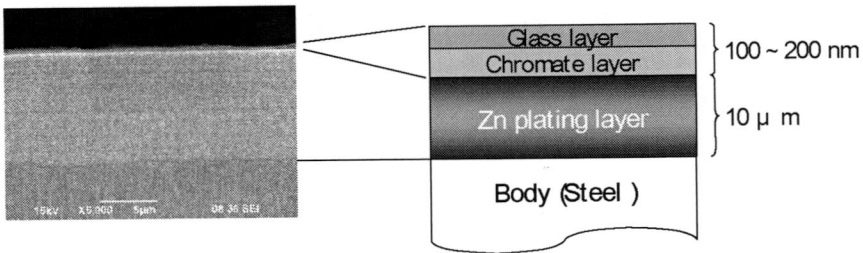

FIGURE 1. Cross- section SEM image and schema of a series-B sample.

The concentration of Cr6+ in the series-A samples was measured by the diphenyl carbazide color test using the UV-mini1240 Spectrometer [5] and the Pack-test [6] for Cr6+, these are widely used in Japan. Table1 shows Cr6+ contents of samples.

Sample number	CrO3 concentration of treat solution	Cr6+ contents by Diphenyl Carbazide color test
A-1	10000ppm	777ppm
A-2	1000ppm	361ppm
A-3	500ppm	185ppm
A-4	200ppm	73ppm
A-5	100ppm	6.5ppm
A-6	50ppm	0.8ppm
A-7	20ppm	Under detection limit
A-8	None added	Under detection limit

TABLE 1. Cr6+ contents of samples.

Experimental Configuration and Measurement Conditions

Figure 2 shows the layout of the fluorescence XANES measurement using 19ch-SSD. The sample was set at 45 degrees to incident SR and 19ch-SSD was set at 90 degrees. BL19B2 is a typical bending magnet beamline with a double crystal monochromator of Si(111) and (311) and a double mirror coated with Rh. The 19ch-SDD is made by Canberra. The peak area was used to calculate Cr6+ contents for taking large total yield.

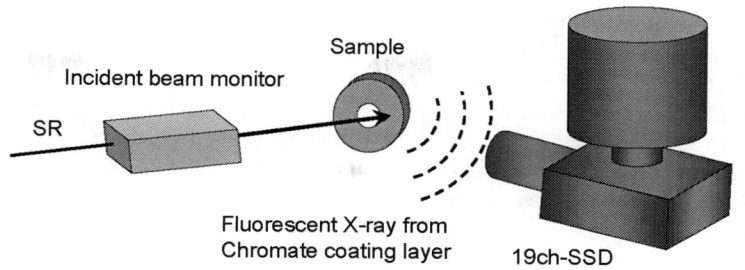

FIGURE 1. Layout of fluorescence XANES measurement using 19ch-SSD.

RESULT AND DISCUSSION

Figure 3 (a) shows spectra of fluorescence XANES of A-series samples. The height of the Cr6+ characteristic peak is increased by increasing Cr6+ contents of the conversion layers. Figure 3 (b) shows spectra of B-series and C-series samples. There was no clear difference between series-B and C, so that it is possible to measure the real products that are covered glass layer by fluorescence XANES method, directly.

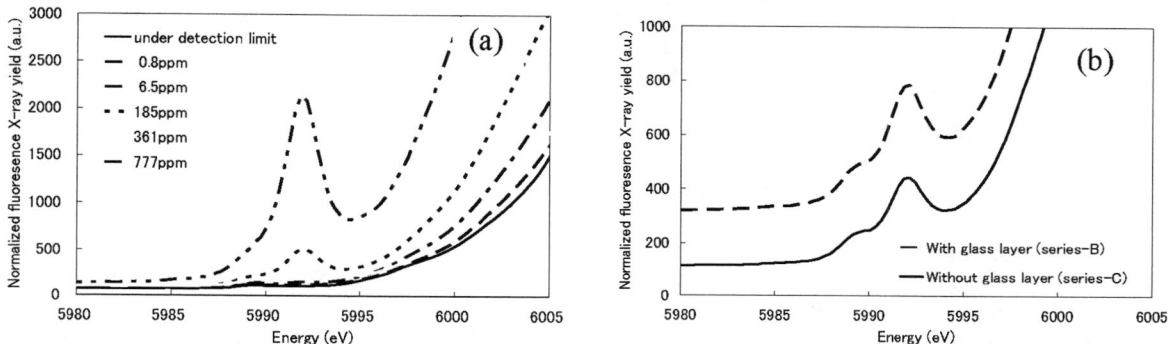

FIGURE 3. Spectra of fluorescence XANES of sample A-series (a) and sample B and C-series (b).

Correction of Diphenyl Carbazide Color Test

Figure 4 shows the correlation between the Cr6+ contents of the Diphenyl Carbazide color test and the fluorescence XANES measurement using A-series samples. Cr6+ contents as measured by fluorescence XANES were derived from peak areas of spectra. There was good linearity down to concentrations of 100 ppm of Cr6+ making a correction equation and a coefficient of the target products possible in compliance with the RoHS directive.

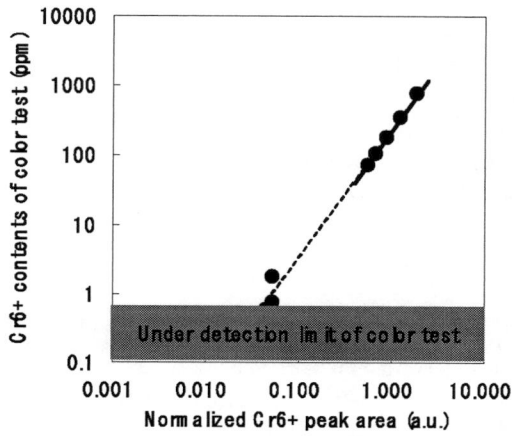

FIGURE 4. Correlation of color test using A-series samples.

Difference of Spectra Among Detectors

There were clear differences among the signals of each of the 19 channel detectors. Figure 5 shows an example of spectra of raw data of fluorescence XANES for the A-series samples. The yields of the lower energy side of the Cr-K absorption edge were higher in the channel 12 and 16 detectors than in the channel 10 detector. We assume that the background increased in specific channels is because of diffracted X-rays from the Zn plating layer. There is broad background in X-ray diffraction of the sample of same making condition. Bragg angle of this background at 5.8 keV of incident X-ray energy is around 110 degrees. This is almost the same angle as the direction to the channel 12 and 16 detector. To avoid incidence of diffraction, it is needed to take long distance from sample to SSD.

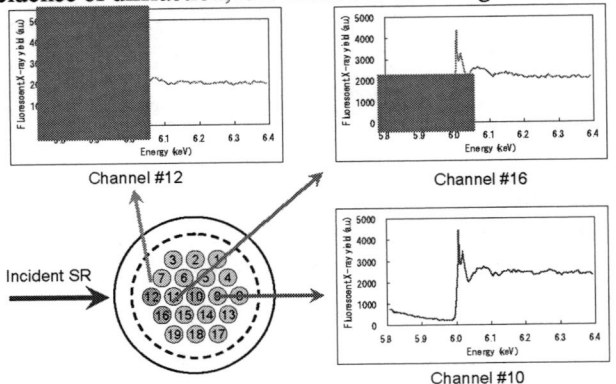

FIGURE 5. Difference in background level of each detector.

Approach to Decreasing Lower Detection Limit

With concentrations less than 100 ppm, more precise measurement will be required. For fluorescence XANES measurement, either more measurement time or increased detection efficiency of fluorescence X-rays is required. One possible method for decreasing the lower detection limit is glazing incidence configuration. For this configuration a test piece with a flat plate shape is required. The other possible method is to use an insertion device (undulator) beam line. Using these it will be possible to check shipments of products in compliance with the RoHS directive.

SUMMARY

Correlations between the diphenyl carbazide color test and the fluorescence XANES measurement down to concentrations of 100 ppm Cr6+ in chromate conversion coatings were obtained. This is a sufficient level for the diphenyl carbazide color to test product shipments in compliance with the RoHS directive. For future progress, direct measurement of real products will be required. We will be able to measure products directly using an insertion device beamline.

ACKNOWLEDGMENTS

The synchrotron radiation experiments were performed at the BL19B2 at SPring-8 with the approval of the Japan Synchrotron Radiation Research Institute (JASRI) (Proposal No. 2005B#0986). The authors thank Dr. Sugiura, Dr. Komiya, and Dr. Watanabe for their valuable help

REFERENCES

1. DIRECTIVE 2002/95/EC OF THE EUROPEAN PARLIAMENT AND OF THE COUNCIL of 27 January 2003.
2. T. Sekikawa (private communication).
3. K. Nomura et al., Oral presentation at 2005 Autumn meeting of The Japan Society for Analytical Chemistry (in Japanese)
4. Products of Mihara sangyo, web site, http://www.mihara-sangyo.co.jp/index.htm (in Japanese).
5. Product of SHIMADZU corporation, web site http://www.shimadzu.co.jp/ (in Japanese).
6. Product of Kyoritsu Chemical-Check Lab., Corp., web site http://kyoritsu-lab.co.jp/ (in Japanese).

Annealing-induced Interfacial Reactions between Gate Electrodes and HfO$_2$/Si Gate Stacks Studied by Synchrotron Radiation Photoemission Spectroscopy

H. Takahashi[1], J. Okabayashi[1], S. Toyoda[1], H. Kumigashira[1], M. Oshima[1],
K. Ikeda[2], G. L. Liu[2], Z. Liu[2], K. Usuda[2]

[1]*Department of Applied Chemistry, The University of Tokyo, Bunkyo-ku, Tokyo 113-8656, Japan*
[2]*Semiconductor Technology Academic Research Center, Kohoku-ku, Kanagawa 222-0033, Japan*

Abstract. We have investigated the interfacial reactions between gate electrodes (polycrystalline-Si and TiN) and HfO$_2$/Si gate stacks by annealing in ultrahigh vacuum using synchrotron radiation photoemission spectroscopy. Hf $4f$ high-resolution photoemission spectra have revealed that a Hf-silicide formation starts at temperature 900 °C for a HfO$_2$/Si structure without a gate electrode. On the other hand, in the case of a polycrystalline-Si/HfO$_2$/Si gate stack structure, the silicidation occurs at as low temperature as 700 °C. It is derived from the Si sub oxide which is formed at the polycrystalline-Si/HfO$_2$ interface by the oxygen diffusion from HfO$_2$ to polycrystalline-Si. In the case of a TiN/HfO$_2$/Si gate stack structure, Hf nitridation occurs at 700 °C, and HfN is formed selectively by further annealing above 800 °C.

Keywords: photoemission spectroscopy, high-k, poly-Si, metal gate
PACS: 79.60.-i, 68.35.Ct, 77.55.+f, 85.40.-e

INTORODUCTION

For continuous improvements in ultra-large-scale integrated (ULSI) device performance, the device size has been shrinking, resulting in the limitation of the SiO$_2$ gate dielectrics due to the direct tunneling leakage current [1]. Among various alternative high dielectric constant (high-k) materials, HfO$_2$ is one of the most promising materials because of its thermal stability, wide band gap, and high dielectric constant [2-4]. However, one of the current problems is thermal stability at the interface between conventional polycrystalline-Si (poly-Si) gate electrodes and high-k gate dielectrics during dopant activation annealing. For example, the silicidation at the interface between gate electrodes and high-k gate dielectrics leads to the increase of the leakage current [5,6] and the flat-band voltage shift by Fermi-level pinning [7,8]. Furthermore, in the case of poly-Si gate electrodes, there are some problems like the gate depletion and the dopant penetration [9]. To solve these problems, the introduction of metal gate electrodes is required. As the candidates of metal gate electrodes, high melting point metals, metal silicides, and metal nitrides have been proposed [10]. However, the interfacial reactions may degrade device performances for metal gate electrodes just like the poly-Si gate electrodes. In order to improve device performances, the interfacial reactions have to be clarified explicitly and suppressed. In previous reports, we have investigated the silicidation between poly-Si gate electrodes and HfO$_2$ gate dielectrics in ultrahigh vacuum (UHV) annealing [11]. In this paper, we have chosen TiN as a metal gate electrode because of its thermal stability and have clarified the difference in the interfacial reactions for poly-Si/HfO$_2$/Si and TiN/HfO$_2$/Si gate stack structures in UHV annealing.

CP879, *Synchrotron Radiation Instrumentation: Ninth International Conference*,
edited by Jae-Young Choi and Seungyu Rah
© 2007 American Institute of Physics 978-0-7354-0373-4/07/$23.00

EXPERIMENTAL

We deposited 2 nm HfO₂ gate dielectrics on clean p-type Si (001) substrates. Poly-Si and TiN electrodes were deposited on HfO₂/Si gate stacks by magnetron sputtering. The thickness was estimated by the elipsometry to be 3 and 5 nm for poly-Si and TiN electrodes, respectively. An interfacial layer (SiO₂ rich) with the thickness of 0.5 nm exists on Si substrates. Synchrotron radiation photoemission spectroscopy measurements were carried out at an undulator beam line BL-2C of the Photon Factory in High-Energy Accelerator Research Organization (KEK). In the case of poly-Si/HfO₂/Si and TiN/HfO₂/Si gate stack structures, sample surface was etched in a HF solution to remove surface native oxides and contaminations just before loading into the vacuum chamber. Annealing was performed in UHV for 10 min at each temperature by the direct current flowing method through samples. In addition, we fabricated two HfSiON (3 nm)/p-type Si (001) samples which have different nitrogen concentration. After UHV annealing, we performed photoemission spectroscopy.

RESULTS AND DISCUSSION

Figure 1 shows the annealing-temperature dependence of Hf 4f photoemission spectra for HfO₂/Si structures. Each spectrum is normalized at the peak height. The components appearing around the binding energy of 18 eV with spin-orbit splitting are derived from HfO₂ [12]. Until the 800 °C annealing, there is no peak component besides HfO₂ peak component. By the 900 °C annealing, a new peak component appears at the binding energy of 14.5 eV, which can be assigned as a Hf-silicide component from its binding energy [12].

The annealing-temperature dependence of Hf 4f photoemission spectra for poly-Si/HfO₂/Si structures is shown in Fig. 2. We performed normalization with the peak height like as Fig. 1. Angular dependence with increasing the surface sensitivity is also shown. By the 700 °C annealing, the Hf-silicide component appears at the binding energy of 14.5 eV. Furthermore, the intensity ratio of the Hf-silicide component to the HfO₂ component increases in the

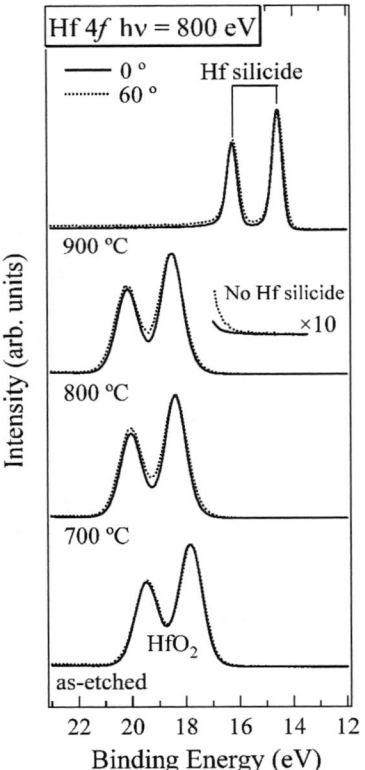

FIGURE 1. Hf 4f photoemission spectra for HfO₂/Si depending on the annealing temperature. Solid curves show spectra taken at the emission angle of 0 °. Dashed curves show those of 60 °.

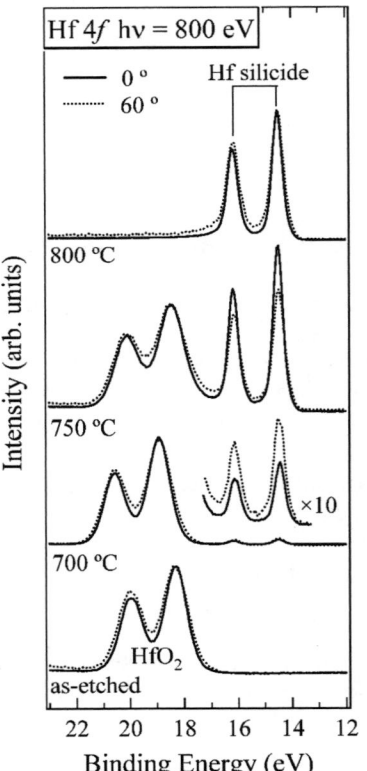

FIGURE 2. Hf 4f photoemission spectra for poly-Si/HfO₂/Si depending on the annealing temperature. Solid curves show spectra taken at the emission angle of 0 °. Dashed curves show those of 60 °.

surface-sensitive measurements at the emission angle of 60 °, suggesting that the Hf silicide is formed at the "upper interface" between poly-Si electrodes and HfO₂ gate dielectrics. On the other hand, by the annealing at 750 °C the intensity ratio decreases in the surface-sensitive measurements. This is probably because the silicidation may propagate into the lower interface between HfO₂ gate dielectrics and Si substrates. By the annealing at 800 °C, the HfO₂ component completely disappears and only the Hf-silicide component remains.

Compared to HfO₂/Si structures, the silicidation occurs at lower temperatures for poly-Si/HfO₂/Si structures. It is because of the difference between the upper interface and the lower interface. Before the annealing, a stoichiometric SiO₂ interfacial layer exists at the lower interface. On the other hand, we have detected a Si sub oxide at the upper interface before the silicidation by Si 2p photoemission spectra (not shown). By the annealing, oxygen atom diffuses from HfO₂ into poly-Si and the nonstoichiometric Si sub oxide (SiO$_x$) is formed at the upper interface. It is related to the report by Shiraishi *et al.*, HfO₂ with ionic bonds easily forms oxygen vacancies [13]. The nonstoichiometric SiO$_x$ reacts with poly-Si more easily than the stoichiometric SiO₂ at the lower interface and accordingly SiO gas is produced. The SiO gas reacts with HfO₂ and Hf silicide is formed consequently. Therefore, the SiO$_x$ at the upper interface can trigger the silicidation [11]. To suppress the silicidation, the interfacial layer as the stoichiometric SiO₂ should be formed at the upper interface.

Next, we have investigated the interfacial reactions for the TiN/HfO₂/Si structure. Figure 3 shows the annealing-temperature dependence of Hf 4f photoemission spectra for the TiN/HfO₂/Si structure. We have performed peak fitting with 5 components, HfO₂, Hf₃N₄, HfN, Hf metal, and O 2s component. The O 2s component is derived from the Ti oxide on TiN gate electrodes. In the as-etched sample, there are Hf nitrides and Hf metal at TiN/HfO₂ interface. By the annealing at 700 °C, Hf₃N₄ and HfN components increase because the Hf nitridation occurs. By further annealing above 800 °C, only the HfN component becomes dominant. This is probably because the annealing is performed in UHV and Hf₃N₄ has been changed into HfN and N₂ in the following reaction [14],

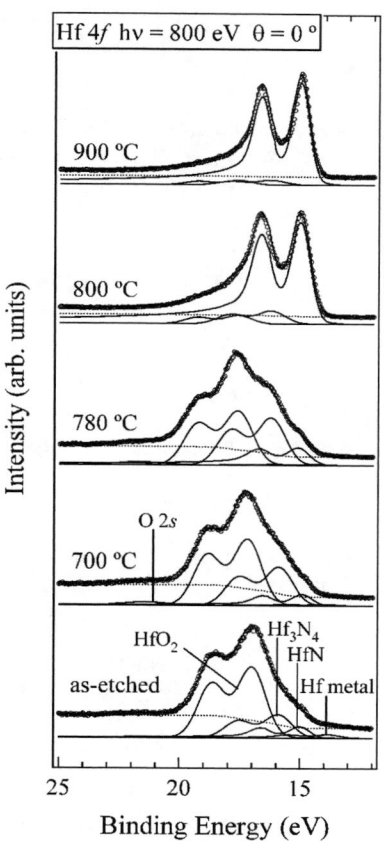

FIGURE 3. Hf 4f photoemission spectra for TiN/HfO₂/Si depending on the annealing temperature.

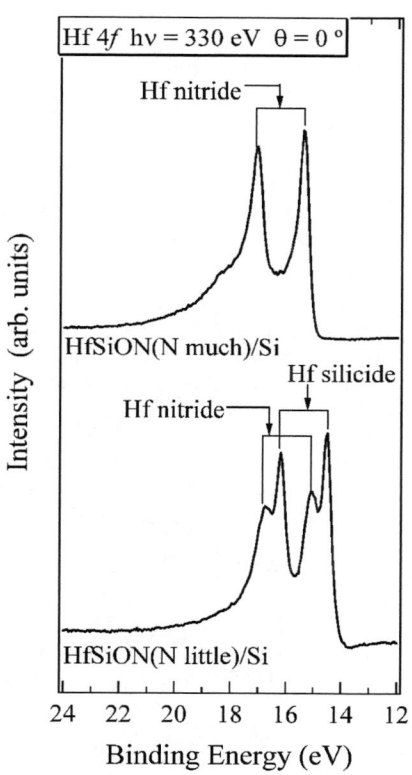

FIGURE 4. Hf 4f photoemission spectra for HfSiON/Si depending on the nitrogen concentration.

$$Hf_3N_4 \rightarrow 3HfN + \frac{1}{2}N_2. \tag{A}$$

Furthermore, in the case of the TiN/HfO$_2$/Si structure, the silicidation does not occur compared to poly-Si/HfO$_2$/Si and HfO$_2$/Si structures. Therefore, Hf nitride is formed more easily than Hf silicide.

We performed additional annealing experiments with two HfSiON (3 nm)/Si samples to confirm the above interfacial reactions. Each sample has different nitrogen concentration. We show the Hf $4f$ photoemission spectrum for each sample annealed by 1000 °C for 10 min in Fig. 4. In the case of the low nitrogen concentration sample, both Hf silicide and Hf nitride are formed. On the other hand, in the case of the high nitrogen concentration sample, only Hf nitride is formed. Therefore, it is concluded that the larger the nitrogen concentration is, the more the amount of the Hf nitride is. In the case of the TiN/HfO$_2$/Si structure, Hf nitride is formed selectively since TiN gate electrodes contain nitrogen.

SUMMARY

We have investigated the interfacial reactions between gate electrodes and HfO$_2$/Si gate stacks for poly-Si/HfO$_2$/Si and TiN/HfO$_2$/Si gate stack structures. In the case of the poly-Si/HfO$_2$/Si gate stack structure, Hf $4f$ high-resolution photoemission spectra have revealed that a Hf-silicide formation starts at temperature 700 °C at the upper interface. However, in the case of the samples without poly-Si gate electrodes, the silicidation occurs at as high temperature as 900 °C. The difference can be attributed to the nonstoichiometric Si sub oxide which is formed at the poly-Si/HfO$_2$ interface by the oxygen diffusion from HfO$_2$ to poly-Si. In contrast, in the case of the TiN/HfO$_2$/Si gate stack structure, Hf silicidation does not occur. Alternatively, Hf nitridation occurs at 700 °C, and HfN is formed selectively by further annealing above 800 °C. This is because TiN gate electrodes contain nitrogen.

ACKNOWLEDGMENTS

This work was supported by Semiconductor Technology Academic Research Center (STARC). Synchrotron-radiation photoemission measurements were performed under the project 02S2-002 at the Institute of Materials Structure Science in KEK.

REFERENCES

1. ITRS Report 2005 version.
2. D. C. Gilmer, R. Hegde, R. Cotton, R. Garcia, V. Dhandapani, D. Triyoso, D. Roan, A. Franke, R. Rai, L. Prabhu, C. Hobbs, J. M. Grant, L. La, S. Samavedam, B. Taylor, H. Tseng, and P. Tobin, Appl. Phys. Lett. **81**, 1288 (2002).
3. S. K. Kang, S. Nam, B. G. Min, S. W. Nam, D. -H. Ko, and M. -H. Cho, J. Appl. Phys. **94**, 4608 (2003).
4. P. S. Lysaght, P. J. Chen, R. Bergmann, T. Messina, R. W. Murto, and H. R. Huff, J. Non-crystalline Solids **303**, 54 (2002).
5. K. Muraoka, J. Appl. Phys. **96**, 2292 (2004).
6. K. Y. Lim, D. G. Park, H. J. Cho, J. J. Kim, J. M. Yang, I. S. Choi, I. S. Yeo, and J. W. Park, J. Appl. Phys. **91**, 414 (2002).
7. D. Y. Kim, J. Kang, and K. J. Chang, Appl. Phys. Lett. **88**, 162107 (2006).
8. C. C. Hobbs, L. R. C. Fonseca, A. Knizhnik, V. Dhandapani, S. B. Samavedam, W. J. Taylor, J. M. Grant, L. G. Dip, D. H. Triyoso, R. I. Hegde, D. C. Gilmer, R. Garcia, D. Roan, M. L. Lovejoy, R. S. Rai, and E. A. Hebert, IEEE Trans. Electron Devices **51**, 978 (2004).
9. H. P. Tuinhout, A. H. Montree, J. Schmitz, and P. A. Stolk, IEDM Technical Digest pp. 631 (1997).
10. J. K. Schaeffer, S. B. Samavedam, D. C. Gilmer, V. Dhandapani, P. J. Tobin, J. Mogab, B. Y. Nguyen, B. E. White, Jr., S. Dakshina-Murthy, R. S. Rai, Z. X. Jiang, R. Martin, M. V. Raymond, M. Zavala, L. B. La, J. A. Smith, R. Garcia, D. Roan, M. Kottke, and R. B. Gregory, J. Vac. Sci. Technol. B **21**, 11 (2003).
11. H. Takahashi, S. Toyoda, J. Okabayashi, H. Kumigashira, M. Oshima, K. Ikeda, G. L. Liu, Z. Liu, and K. Usuda, Appl. Phys. Lett. submitted.
12. J. H. Lee, Thin Solid Films **472**, 317 (2004).
13. K. Shiraishi, K. Yamada, K. Torii, Y. Akasaka, K. Nakajima, M. Konno, T. Chikyow, H. Kitajima, and T. Arikado, Jpn. J. Appl. Phys. **43**, L1413 (2004).
14. P. Kroll, J. Phys.: Condens. Matter **16**, S1235(2004).

Characterization of Amorphous High-k Thin Films by EXAFS and GIXS

Momoko Takemura*, Hideyuki Yamazaki*, Hirobumi Ohmori*, Masahiko Yoshiki*,
Shiro Takeno*, Tsunehiro Ino*, Akira Nishiyama*, Nobutaka Sato [†],
Ichiro Hirosawa [¶], and Masugu Sato [¶]

*Corporate R&D Center, Toshiba Corporation 1, Komukai Toshiba-cho, Saiwai-ku, Kawasaki 212-8582, Japan
[†]Toshiba Nanoanalysis Corporation, 1,Komukai Toshiba-cho, Saiwai-ku, Kawasaki 212-8583, Japan
[¶]JASRI/SPring-8, 1-1-1, Kouto, Sayo-cho, Sayo-gun, Hyogo 679-5198, Japan

Abstract. Silicon and nitrogen incorporated Hf oxide (HfSiON) is considered to be a promising alternative gate insulator for next-generation MOSFETs. EXAFS and GIXS (Grazing Incidence X-ray Scattering) have been applied to the characterization of amorphous HfSiON films at SPring-8. Novel cluster models have been suggested based on the analogy to the ordered states for the Zr-O-N ternary system.

Keywords: EXAFS, GIXS, gate insulator, high-k, hafnium oxide, oxynitride, amorphous, SPring-8
PACS: 73.61.Ng, 61.10.Ht, 61.10.-I

INTRODUCTION

Recently, the shrinkage of semiconductor devices has led to the CMOS gate dielectric becoming so thin that SiO_2 no longer acts as a good insulator. A thicker layer of high dielectric constant (K) oxides, which are called high-k materials, is being investigatd with a view to its application as a replacement for SiO_2. Among many high-k candidates, Hf, Zr,and La oxides have been investigated for the purpose. The high-k candidates should preferably be amorphous and able to withstand high temperature up to 1000 C.

Hf oxide incorporating silicon and nitrogen (HfSiON) is considered to be a promising alternative gate insulator for next-generation CMOS. Incorporation of silicon in Hf oxide prevents crystallization, and furthermore, incorporation of nitrogen prevents not only crystallization but also phase separation, and minimizes interfacial layer with silicon substrate when it is heated during the device manufacturing process [1,2]. Although microstructure of high-k films has an important bearing on the physical properties of semiconductor devices, only a few methods are available for the structural analysis of amorphous high-k thin films. We have successfully applied EXAFS and GIXS (Grazing Incidence x-ray Scattering) to the characterization of amorphous HfSiON films. EXAFS is a well-known method for analysis of amorphous structure. GIXS, which was developed at SPring-8 [3,4], is another effective method for the structural analysis of amorphous thin films. EXAFS and GIXS can be complementary to each other, since GIXS is advantageous about information for medium-range ordering, on the other hand EXAFS analysis is advantageous about information for elements.

EXPERIMENTAL

The sample films listed in Table 1 have been deposited on Si (100) wafers by co-sputtering of Hf and Si targets under oxygen, nitrogen and argon gas flow. Nitrogen and oxygen contents in the films have been adjusted by changing gas flow rates. All sample films are without any thermal treatments after deposition. The intended Hf content (Hf/(Hf+Si)) is 80 atomic%. The sample name Nxx in Table 1 indicates that the intended nitrogen content is

CP879, *Synchrotron Radiation Instrumentation: Ninth International Conference*,
edited by Jae-Young Choi and Seungyu Rah

xx atomic%. N00 contains no nitrogen. The intended film thickness is 100nm. Film thicknesses are measured on the SEM image and listed in Table 1. The compositions in Table 1 have been determined by Rutherford backscattering spectrometry. Exceptionally, the film thickness of N20t10 is set to 10nm.

Hafnium oxide (HfO_2)-powder, which is a commercial product, has been used as a reference for EXAFS analysis.

The x-ray in-plane diffraction data has been taken from the sample films using a RIGAKU ATX-G, and as shown in Figure 1, does not show any signal that can be ascribed to crystals.

TABLE 1. Samples

Sample Name	Thickness (nm)	Composition (atomic %)			
		Hf	Si	O	N
N00	72	24	7	68	0
N10	86	25	6	58	9
N20	92	31	8	40	20
N35	97	32	9	23	35
N50	102	30	8	9	51
N20t10 *	10	31	8	40	20

*The composition of N20t10 is inferred from the value determined for N20.

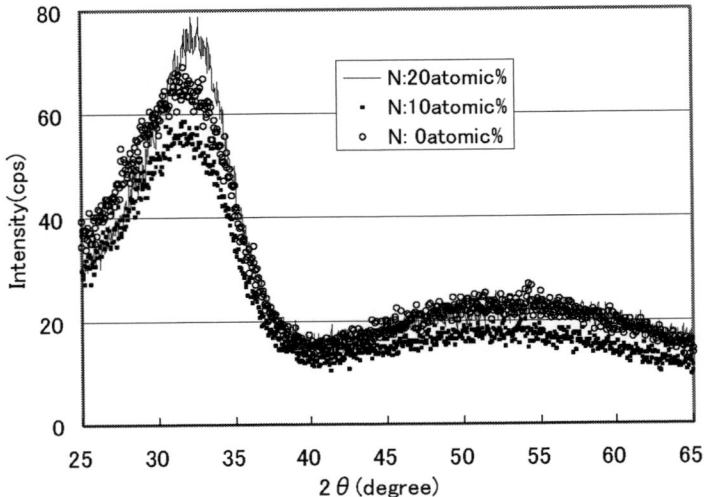

FIGURE 1. In-plane x-ray diffraction diagram for HfSiON films.

EXAFS experiments were performed at SPring-8, BL16B2 (Industrial Consortium Beamline). The synchrotron radiation at BL16B2 was monochromatized by the Si(311) planes. X-ray absorption spectra around the L3 edge of Hf were obtained by the TEY method using a vacuum chamber. EXAFS analyses have been done using the REX program of RIGAKU.

A GIXS experiment was performed for sample N20t10 with x-rays of 15 keV, monochromatized by Si(111), using a multi-axis diffractometer at BL46XU/SPring-8. The incident grazing angle on a sample was adjusted to be 0.1 degrees.

RESULTS AND DISCUSSION

As shown in Fig. 2, the Fourier transforms of Hf L-EXAFS for HfSiON films of 100 nm thickness show that the anion coordination number decreases remarkably with increased nitrogen content, although the change of distance between Hf (Si) and anions is small.

Although the previous reports concerning hafnium oxynitrides are few, we can refer to literature concerning investigations of zirconium oxynitrides since hafnium compounds are remarkably similar to zirconium compounds in terms of both structure and chemistry. Following discussion is based on the viewpoint of comparison of coordination number between amorphous HfSiON films and a series of zirconium oxynitride crystals.

Anion substitutions of O^{2-} by N^{3-} bring about a series of zirconium oxynitrides, which have anion vacancies according to $ZrO_{2-x}N_{2x/3}V_{x/3}$. ("V" denotes vacancy.) M. Lerch et al. [5] and S. J. Clarke et al. [6] showed that they are composed of ordered anion vacancies, and reported several ordered phases denoted as β, β', β" and γ [5~7].

The oxynitride details cited from the literature [5,6] are listed in Table 2, where Zr atoms are substituted by Hf atoms. The β, β' and β" have the same component, $Hf_7O_8N_4V_2$, which is named a Bevan cluster [5]. The β' and β" phases are formed as superlattices with stacking coherency of Bevan clusters and a zirconium oxide unit (Zr_7O_{14}). Calculated nitrogen content and anion coordination numbers of hafnium oxynitrides (or oxide) are listed in Table 2. Additionally, experimental results for amorphous HfSiON series are listed in the right columns of Table 2. It is noteworthy that the tendency of decreasing feature in coordination number with increasing nitrogen content for both calculated and experimental ones is well coincident.

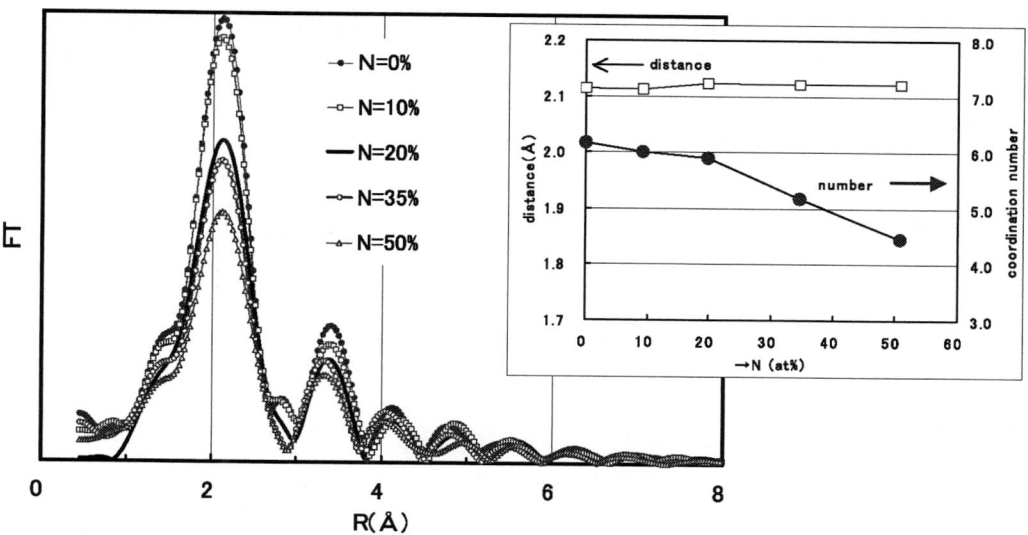

FIGURE 2. Magnitude of Fourier transform* of Hf L-EXAFS for HfSiON films. The result of EXAFS analysis is shown in the inset figure. (*The phase shift has been taken account for the horizontal axis.)

TABLE 2. Hf-O-N crystal data and experimental results for HfSiON.

Hf-O-N crystal series cited from [5].						Experimental results		
Composition	Unit cell	Phase	N content (atomic %)	Anion coordination number *1		Sample name	N content (atomic %)	Anion coordination number
HfO_2	Hf_4O_8 Cubic, tetragonal or monoclinic		0.0	7~8	→	N00	0.0	6.2
$Hf_7O_{11}N_2$	Hf_7O_{14} + $Hf_7O_8N_4V_2$ Hexagonal	β'	10.0	7.3	→	N10	9.3	6.0
$Hf_7O_{9.2}N_{3.2}$	Hf_7O_{14} + 4($Hf_7O_8N_4V_2$) Hexagonal	β"	16.5	7.1				
$Hf_7O_8N_4$	$Hf_7O_8N_4V_2$ Hexagonal	β	21.1	7.0	→	N20	19.7	5.9
Hf_2ON_2	$Hf_{32}O_{16}N_{32}V_{16}$, Cubic	γ	40.0	6.0	→	N35	34.5	5.2
						N50	50.7	4.5

*1 The numbers are calculated based on assumption that the anion coordination number of a hafnium atom for HfO2 is 7.5 (=average value of 7 and 8).

Figure 3 shows a Fourier transform of the GIXS oscillation for N20t10 sample film (details are in Table 1). The peaks correspond to the atomic pairs in HfSiON. The first peak corresponds to the shortest Hf-O bond. Other peaks are considered to correspond to various Hf-Hf bonds, because the scattering from Hf-O should be so weak that only the nearest Hf-O can be clearly observed. It was reported that the average distance of the shortest Hf-Hf pairs is 3.4 angstrom, and the average of the second shortest Hf-Hf is 4.0 angstrom for monoclinic HfO_2, and the ratio of numbers of them is 7 vs. 2 [8]. The distances and the ratio coincide well with our result shown in Fig. 3. More distant pairs of Hf-Hf are observed in the range up to 10 angstrom as shown in Fig. 3. As these distant pairs cannot be explained in terms of single Bevan cluster unit, several cluster units may possibly be combined in amorphous HfSiON film.

The coincidence between our results and oxynitride crystal structures seems to suggest that amorphous HfSiON films have some piece of microstructure including some units of the crystal structure.

CONCLUSION

We have concluded that hafnium oxynitride films accumulated on Si substrate by sputtering, although they are amorphous in terms of x-ray diffraction characterization, are composed of certain clusters similar to the Bevan cluster, which is known as a unit of Zr-O-N ternary crystal system, and includes anion vacancies.

EXAFS and GIXS analyses using brilliant synchrotron radiation at SPring-8 are successfully applied to the characterization of amorphous films for the next-generation MOSFET.

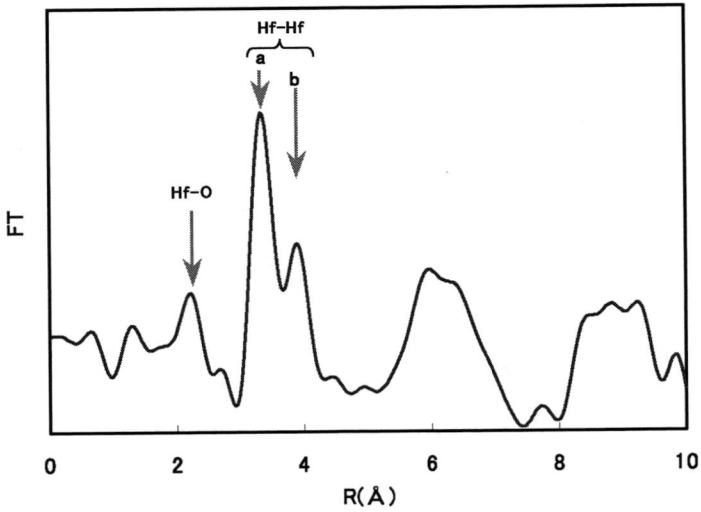

FIGURE 3. FT of GIXS oscillation for an HfSiON film of 10nm thickness.

ACKNOWLEDGMENTS

The authors are grateful to Dr. Norimasa Umesaki of JASRI for the valuable advice. The experiments were performed at SPring-8 with the approval of the Japan Synchrotron Radiation Research Institute (JASRI) (Proposals No.C04A16B2-4060-N and No.2004B0493-NI-np-TU.)

REFERENCES

1. M. Koike, T. Ino, Y. Kamimuta, M. Koyama, Y. Kamata, M. Suzuki, Y. Mitani, A. Nishiyama, and Y. Tsunashima, *International Electron Devices Meeting Technical Digest*, 107-108 (2003).
2. T. Ino, Y. Kamimuta, M. Suzuki, M. Koyama, and A. Nishiyama, *Jpn. J. Appl. Phys.* 45, 2908 (2006).
3. I. Hirosawa, H. Kitajima, and K. Torii, *Trans. Mat. Res. Soc. Japan* 30, 221-224 (2005).
4. M. Sato, T. Matsunaga, T. Kouzaki, and N. Yamada, *Mat. Res. Soc. Symp. Proc.* 803, 245 (2004).
5. M. Lerch, F. Krumeich, and R. Hock, *Solid State Ionics* 95, 87-93 (1997).
6. S. J. Clarke, C. W. Michie, and M. J. Rosseinsky, *J. Solid State Chem.* 146, 399 (1999).
7. G. Van Tendeloo, L. Anders and G. Thomas, *Acta metall.* 31, 1619-1625 (1983).
8. J. Wang, H. P. Li and R. Stevens, *J. Mat. Sci.* 27, 5397-5430 (1992).

Non-destructive Measurement of Residual Stress Depth Profile in Laser-peened Steel at SPring-8

Masugu Sato[a], Yuji Sano[b], Kentaro Kajiwara[a], Hirotomo Tanaka[c], Koichi Akita[c]

[a] Japan Synchrotron Radiation Research Institute, Sayo, Hyogo, 679-5198 Japan
[b] Power and Industrial Systems Research and Development Center, Toshiba Corporation, Isogo-ku, Yokohama, 235-8523 Japan
[c] Dept. of Mechanical Systems Engineering, Musashi Institute of Technology, Setagaya-ku, Tokyo, 158-8557 Japan

Abstract. We investigated the residual stress depth profile near the surface of steel treated by laser peening without coating using X-ray diffraction at SPring-8. This investigation was carried out using a constant penetration depth $\sin^2\psi$ method. In this method, the $\sin^2\psi$ diagram is measured controlling both the ψ angle and the X-ray penetration depth simultaneously with a combination of the ω and χ axes of the 4-circle goniometer. This method makes it possible to evaluate the residual stress and its depth profile in material with a stress gradient precisely and non-destructively. As a result, we confirmed that a compressive residual stress was successfully formed all over the range of the depth profile in the steel treated properly by laser peening without coating.

Keywords: laser peening, residual stress, constant penetration depth $\sin^2\psi$ method.
PACS: 42.62.Cf

INTRODUCTION

Laser peening without coating (LPwC) is a new surface enhancement technology for metallic materials [1]. This treatment forms a compressive residual stress on the surface by irradiating laser pulses onto the material in an aqueous environment. Since the process is simple and remotely operable due to the absence of reactive force against laser irradiation, LPwC has been used to mitigate stress corrosion cracking (SCC) in operating nuclear-power reactors [2, 3]. To check the effect of this treatment, it is necessary to measure the depth profile of the residual stress in the laser-peened material using X-ray diffraction. Such measurements have previously been carried out using destructive methods such as successive electrolytic polishing. However, destructive methods cannot be used to precisely evaluate the durability of the treatment because they cannot use the same samples to measure the dependence on the external loading. Moreover, electrolytic polishing inevitably influences the residual stress depth profile due to the removal of the surface layer. Therefore, we investigated the residual stress depth profile in laser-peened steels non-destructively using the constant penetration depth (CPD) $\sin^2\psi$ method at BL19B2 of SPring-8.

EXPERIMENT

The $\sin^2\psi$ method is generally used in the measurement of residual stress with X-ray diffraction, where ψ is the angle between the scattering vector of the X-ray diffraction from the sample and the normal vector of the sample surface. When a residual stress σ remains on the surface of a polycrystalline sample, angles 2θ of diffraction from the arbitrary crystal plane of a crystal grain of the sample change with ψ, as follows,

$$2\theta = \left(-2\tan\theta_0 \cdot \frac{1+\nu}{E} \cdot \sigma \right) \sin^2\psi + \left(\tan\theta_0 \cdot \frac{\nu}{E} \cdot \sigma + 2\theta_0 \right) \tag{1},$$

where ν is the Poisson ratio, E is the elastic constant and $2\theta_0$ is the diffraction angle of the crystal plane without any strain. Here, it is supposed that the penetration depth of the X-ray is shallow enough that the direction of the observable residual stress is in-plane. When 2θ is plotted as the function of $\sin^2\psi$ (the so-called $\sin^2\psi$ diagram), σ is estimated from the slope of the $\sin^2\psi$ diagram, $d(2\theta)/d(\sin^2\psi)$, as follows,

CP879, *Synchrotron Radiation Instrumentation: Ninth International Conference*,
edited by Jae-Young Choi and Seungyu Rah

$$\sigma = K \cdot \frac{d(2\theta)}{d(\sin^2 \psi)} \tag{2}.$$

where K is called X-ray stress constant and written as the following equation.

$$K = -\cot\theta_0 \cdot \frac{E}{2(1+\nu)} \tag{3}.$$

The slope is positive in the case of compressive stress, while it is negative in the case of tensile stress. The ψ angle can be controlled using the ω or χ axis of the 4-circle diffractometer as shown in Fig. 1. The relation among ω, χ and ψ is

$$\cos\psi = \cos(\theta - \omega) \cdot \cos\chi \tag{4}.$$

When the ψ angle is changed by ω or χ, the X-ray penetration depth D of the sample surface changes as follows,

$$D = \frac{1}{\mu} \cdot \frac{\cos\chi}{1/\sin\omega + 1/\sin(2\theta - \omega)} \tag{5},$$

where μ is the X-ray absorption coefficient.

When the $\sin^2\psi$ diagram is measured using the conventional $\sin^2\psi$ method, the ψ angle is controlled with either ω or χ. In iso-incline mode, the ω axis is used ($\psi = \theta - \omega$; $\chi = 0$), and in side-incline mode, χ axis is used ($\psi = \chi$; $\omega = \theta$). In both modes, the X-ray penetration depth varies along with ψ: $D = (1/\mu)/(1/\sin(\theta - \psi) + 1/\sin(\theta + \psi))$ in the iso-incline mode and $D = (1/\mu) \cdot \cos\psi \cdot \sin\theta / 2$ in the side-incline mode. Because the strain is estimated from the shift of the diffraction peak that is the average weighted by the X-ray absorption ratio over the penetration depth, a precise evaluation of residual stress in the material with the stress gradient is difficult to obtain using the conventional $\sin^2\psi$ method. In the CPD $\sin^2\psi$ method, which has been developed at BL19B2 of SPring-8, the ψ angle and the X-ray penetration depth are simultaneously controlled with a combination of the ω and χ axes [4]. When the diffraction signal with the diffraction angle 2θ is detected with arbitrary values of the ψ angle and the X-ray penetration depth D, ω and χ can be calculated by using equations (4) and (5),

$$\omega = \theta \pm \arccos\left(\frac{\cos\theta}{\sqrt{1 - \mu D \dfrac{2\sin\theta}{\cos\psi}}}\right) \tag{6},$$

$$\chi = \arccos\left(\frac{\cos\psi}{\cos\theta}\sqrt{1 - \mu D \frac{2\sin\theta}{\cos\psi}}\right) \tag{7}.$$

The $\sin^2\psi$ diagram is obtained by changing the ψ angle with a fixed X-ray penetration depth D as in the equations (6) and (7). Using this method, one can precisely evaluate the residual stress and its depth profile in material with a stress gradient.

Experimental Procedure

We prepared two samples of high tensile-strength steel (SHY685) treated by LPwC. One was treated normally to form a compressive stress from the surface deeper into the sample (sample A). The other was treated differently to form a steep gradient of the residual stress near the surface (sample B). Sample B was prepared to investigate the ability of the CPD sin2ψ method. The sample preparation procedure is shown in Fig. 2. In the LPwC method, when one laser pulse is shot onto a surface of the material, compressive residual stress is formed around the irradiated spot, but tensile residual stress is induced on the irradiated area by the thermal affect of direct laser irradiation on the material without coating. In order to eliminate the tensile residual stress, the laser pulses are shot onto the surface so densely as to overlap each other as shown in Fig. 2. In the preparation of sample B, the number of laser pulses per unit area was much less than that in the normal preparation process, so that tensile residual stress remained near to the surface. The conditions of LPwC are shown in Table 1.

The experiments were carried out using the multi-axis diffractometer installed in BL19B2 at SPring-8. The energy of the X-ray we used was 25 KeV. The diffraction peak we detected was (521) of α-phase of SHY685. The diffraction angle was approximately 56.3 deg. The X-ray stress constant K for this peak was -2431 MPa/degrees. The maximum penetration depth was estimated to be 22 μm. A soller slit with a divergence angle of 0.2 degrees was placed between the sample and a detector to improve the parallel beam optics. Widths of divergence and receiving slits were tuned to keep the detection area on the sample surface. An example of peak profiles is shown in Fig. 3. The diffraction angle of each peak was estimated by profile fitting with Lorentzian function.

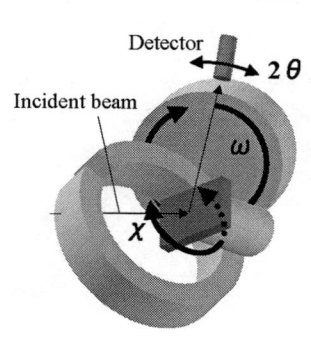

FIGURE 1. Schematic view of the 4-circle diffractometer

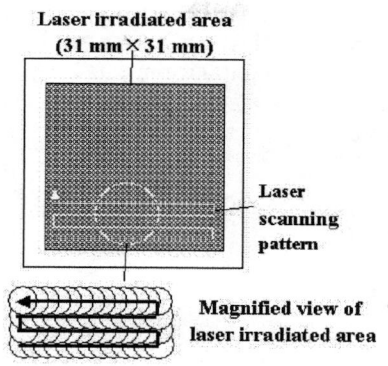

FIGURE 2. Schematic view of laser irradiation on a sample.

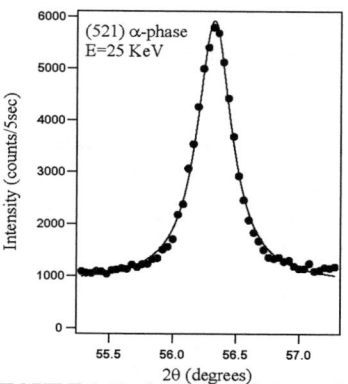

FIGURE 3. Peak profile of (521) of α-phase of SHY685. The solid line is the calculated intensity of fitted Lorentzian function.

TABLE 1. Condition of laser peening without coating

Sample	Sample A	Sample B
Pulse energy	200 mJ	215 mJ
Spot diameter	Φ 0.8 mm	Φ 1.0 mm
Irradiation density	10000 pulse/cm^2	1019 pulse/cm^2

RESULTS & DISCUSSION

At first, we measured the residual stress in sample B using both the CPD sin2ψ method and the conventional sin2ψ method and compared the results in order to investigate the effectiveness of the CPD sin2ψ method. Figure 4 shows the sin2ψ diagram for the (521) peak of sample B measured using the conventional sin2ψ method in side-incline mode. The slope of the sin2ψ diagram changed significantly from positive to negative as sin2ψ increased, and the X-ray penetration depth shown by the dashed line decreased as sin2ψ increased. This suggests that the mean residual stress in the X-ray penetration depth changed from tensile to compressive as the X-ray penetration depth became deeper. Because of the bend of the sin2ψ diagram, it is difficult to exactly evaluate the residual stress in the sample with the stress gradient using the conventional sin2ψ method. The measurement taken using the CPD sin2ψ method is shown in Fig. 5. We measured sin2ψ diagrams changing the X-ray penetration depth from 17.6 μm to 4.4 μm by a step of 2.2 μm. The figure shows that the sin2ψ diagram for each penetration depth was linear to sin2ψ due to the precise control of the X-ray penetration depth. This made it possible to properly evaluate the mean residual strain in each X-ray penetration depth from the sin2ψ diagram. The slopes of the sin2ψ diagrams decreased as the X-ray penetration depth became shallower and changed from positive to negative at around D = 6.6 μm. These results indicate that the CPD sin2ψ method can be used to precisely evaluate the residual stress depth profile in samples with a steep stress-gradient.

Figure 6 shows the sin2ψ diagram of sample A measured using the CPD sin2ψ method. All slopes of the sin2ψ diagrams were positive and their dependence on X-ray penetration depth was unremarkable. Figure 7 compares the residual stress depth profiles of the two samples estimated from the measurements taken using the CPD sin2ψ method. These data represent the mean residual stress in the X-ray penetration depth converted from the slopes of the sin2ψ diagrams shown in Figs. 5 and 6. These data clearly show that the residual stress is compressive over the full range of the depth profile in sample A, while it changes steeply from tensile to compressive as the X-ray penetration depth increases in sample B.

CONCLUSION

We used the CPD $\sin^2\psi$ method to investigate the residual stress depth profile in high tensile-strength steel (SHY685) treated by LPwC. The results showed that the method prevented the distortion of the $\sin^2\psi$ diagram due to the stress gradient and could evaluate the mean residual stress over the X-ray penetration depth precisely. It was also confirmed that the LPwC successfully formed a compressive residual stress on the surface of steel that was properly treated. These results proved that the CPD $\sin^2\psi$ method is applicable and extremely useful for non-destructive evaluation of surface enhancement technologies.

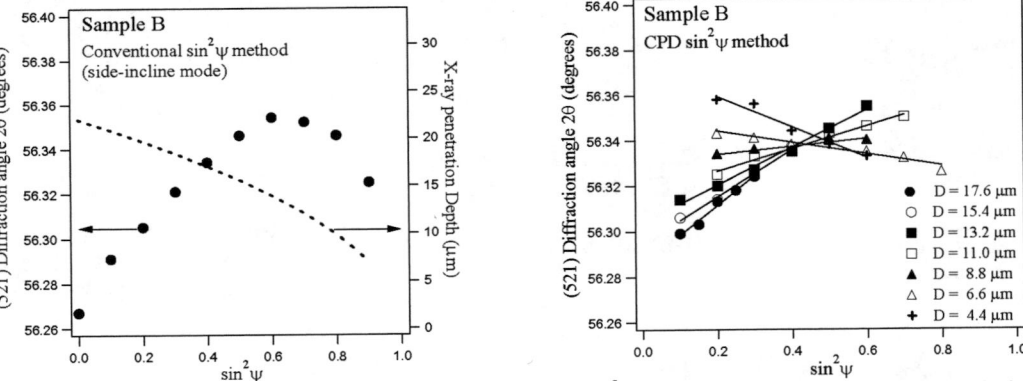

FIGURE 4. $\sin^2\psi$ diagram of sample B measured using conventional $\sin^2\psi$ method in side-incline mode. The dashed line is the dependence of the X-ray penetration depth on $\sin^2\psi$ (left side).
FIGURE 5. $\sin^2\psi$ diagram of sample B measured using CPD $\sin^2\psi$ method (right side).

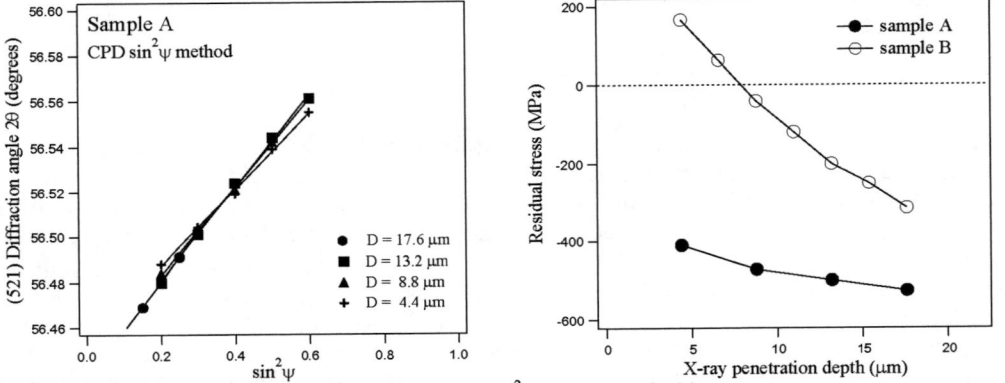

FIGURE 6. $\sin^2\psi$ diagram of sample A measured using CPD $\sin^2\psi$ method (left side).
FIGURE 7. Residual stress depth profiles of samples A and B measured using CPD $\sin^2\psi$ method (right side).

ACKNOWLEDGMENTS

The authors would like to acknowledge Dr. I. Hirosawa for his technical support and helpful discussions.

REFERENCES

1. Y. Sano, M. Obata, T. Kubo, N. Mukai, M. Yoda, K. Masaki and Y. Ochi, *Mater. Sci. Eng.* A **417**, pp. 334-340 (2006).
2. Y. Sano, M. Kimura, K. Sato, M. Obata, A. Sudo, Y. Hamamoto, S. Shima, Y. Ichikawa, H. Yamazaki, M. Naruse, S. Hida, T. Watanabe and Y. Oono, Proceedings of the Eighth International Conference on Nuclear Engineering (ICONE-8), Baltimore, 2000 (paper no: ICONE-8441).
3. I. Chida, M. Yoda, N. Mukai, Y. Sano, M. Ochiai, T. Miura, R. Saeki, Proceedings of the 13th International Conference on Nuclear Engineering (ICONE-13), Beijin, 2005 (paper no: ICONE13-50334).
4. S. Tsuchiya, T. Oshika, F. Tsuchima and A. Nishiyama, Transactions of the Materials Research Society of Japan, **29** pp. 425-428 (2004).

Strain Measurements using High Energy White X-rays at SPring-8

T. Shobu[1,2], H. Kaneko[1,2], J. Mizuki[1], H. Konishi[1],
J. Shibano[3], T. Hirata[3] and K. Suzuki[4]

[1] Japan Atomic Energy Agency, Koto, Sayo-cho, Sayo-gun, Hyogo, 679-5148, Japan
[2] SPring-8 Service Co. Ltd, Koto, Sayo-cho, Sayo-gun, Hyogo, 679-5148, Japan
[3] Kitami Institute of Technology, Kouen-cho, Kitami, Hokkaido, 090-8507, Japan
[4] Niigata University, Igarashi-2no-cho, Niigata, 950-2181, Japan

Abstract. The strain in the bulk of a material was evaluated using high energy white X-rays from a synchrotron radiation source at SPring-8. The specimen, which was a 5 mm thick austenitic stainless steel sample (JIS-SUS304L), was subjected to bending. The internal strain was measured using white X-rays, which ranged in energy from 60 keV to 125 keV. Highly accurate internal strain measurements were accomplished by simultaneously using strain data from several lattice planes of α-Fe. Furthermore, utilizing diffracted beams with a high energy, a high peak count, and a profile similar to a Gaussian distribution decreased the error of the strain measurement. The results indicated that high energy white X-rays can effectively measure the internal strain at a millimeter depth.

Keywords: residual stress, high energy white X-rays, energy dispersive diffraction, JIS-SUS304 austenitic stainless steel
PACS: 81.40.Jj; 61.10.Nz;

INTRODUCTION

High energy X-rays can penetrate deep below the surface. Therefore, information on the interplanar spacing from the surface to the deeper layers can be obtained. Third generation synchrotron X-ray sources such as the European Synchrotron Radiation Facility (ESRF), the Advanced Photon Source (APS), and SPring-8 have made very intense beams of extremely high energy X-rays available for the first time. Techniques, which use an angular divergent X-ray diffraction method with high energy synchrotron radiation to non-destructively measure internal residual stress, are well established and can provide engineers, as well as material scientists, with valuable information [1]. Compared to these techniques, energy dispersive diffraction techniques, which use white radiation, provide several advantages. For example, the multitudes of reflections recorded in one spectrum offers additional information that can be used to evaluate a stress gradient [2]. G. Brusch *et al.* [3] and W. Reimers *et al.* [4] have reported residual stress measurements in materials using high energy white X-rays to transmit diffracted X-rays through the specimen. Although measurements using high energy white X-rays are useful to non-destructively evaluate stress in the bulk of a material, this method has yet to be established. The purpose of this study is to measure the strain in the bulk of materials using high energy white X-rays from a synchrotron radiation source and to develop this technique.

ENERGY DISPERSIVE METHOD

Because a white X-ray beam has a wide energy range, an interplanar spacing, d, of a crystal should be measured under a fixed diffraction angle 2θ. Based on Bragg's law, the fundamental equation of energy dispersive method is expressed as equation 1.

CP879, *Synchrotron Radiation Instrumentation: Ninth International Conference,*
edited by Jae-Young Choi and Seungyu Rah
© 2007 American Institute of Physics 978-0-7354-0373-4/07/$23.00

$$d = \frac{hc}{2\sin\theta}\frac{1}{E_n},\tag{1}$$

where h is Planck's constant, c is the velocity of light and E_n is the energy value of the diffracted X-rays. Differentiating equation 1 gives equation 2 because the Bragg angle, θ, is constant in this method

$$\Delta d = -\frac{hc}{2\sin\theta \cdot E_n^2}\Delta E_n.\tag{2}$$

Combining of equation 2 with equation 1 shows that

$$\frac{\Delta d}{d} = -\frac{\Delta E_n}{E_n} = \frac{E_n^0 - E_n}{E_n},\tag{3}$$

where E_n^0 is the peak energy value of the diffracted beam profile of a non-strained specimen and E_n is that of a strained specimen. Hence, the strain, ε, is obtained from equation 3 using the diffracted X-ray energy.

Equation 3 demonstrates that the accuracy of a strain measurement increases as the X-ray energy increases and that the accuracy is independent of the diffracted angle, 2θ. These characteristics are major differences between the energy dispersive method and the angle dispersive method.

EXPERIMENTAL

Measurements were conducted in BL14B1 at SPring-8 using high energy white X-rays from a synchrotron radiation source. The sizes of the white X-ray beams were 50 μm (vertical), and 300 μm (horizontal). The patterns were collected by a Ge Solid State Detector (SSD) mounted on the arm of a two-axes diffractometer behind two sets of secondary collimating slits. The Bragg angle, 2θ, of the SSD was set at 10 degrees. A Gaussian equation was used to determine the peak energy of the diffracted beam profile in each lattice plane.

Figure 1 (a) shows the specimen, which was austenitic stainless steel (JIS-SUS304L). The grain size was 30 μm or more. The specimen beam, which was 5 mm thick, was bent by the clamp load of a bolt. The surface strains of the beam were measured by strain gauges and the maximum strain of the tension side was 770×10^{-6}, while that of compression side was -740×10^{-6}. The interplanar spacing of the crystal in the measurement position indicated in Figs. 1 (b) and (c) were measured using diffracted X-rays transmitted through the beam of the specimen. The bending strain was calculated by equation 3 using the peak energy obtained from the non-strained and the strained specimens.

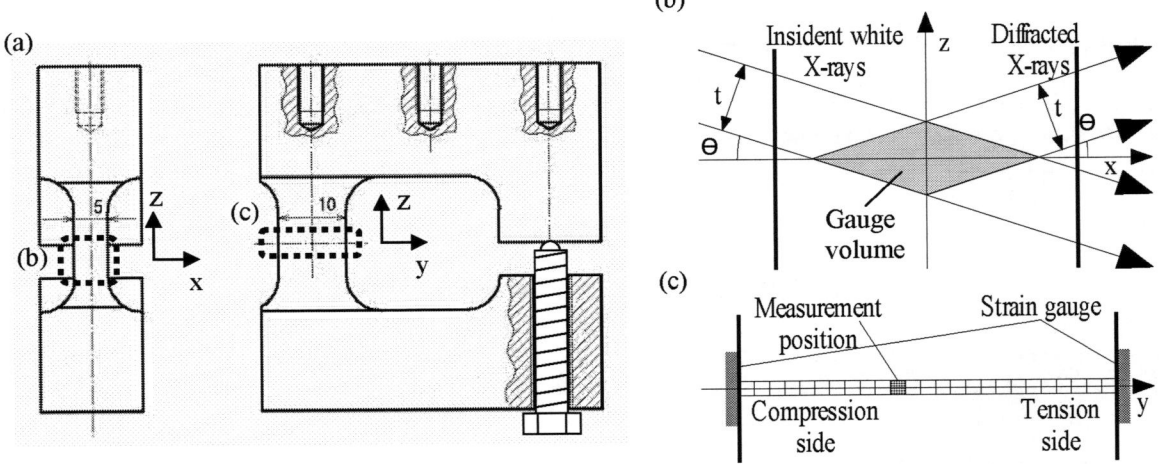

FIGURE 1. (a) Specimen configuration, (b) schematic of gauge volume using transmitted diffracted X-rays and (c) schematic of measurement positions and strain gauges.

RESULT AND DISCUSSION

A translated oscillation of ±2 mm along the x direction and a rotated oscillation of ±2 degree in the xz plane of the specimen were simultaneously performed because the diffracted X-ray peak counts at each measurement position fluctuated due to coarse grains and preferred orientation or texture. Figure 2 shows the diffracted X-ray profiles at all measurement positions. It was confirmed that the interplanar spacing of the crystal in the 5 mm thick specimen can be measured using white X-rays, which range in energy from 60 keV to 125 keV. In this experiment, W-K α 1 radiation was detected from a collimator, which consisted of tungsten steel. The measured peak energy values of the diffracted X-rays must be modified so that the W-K α 1 values coincide with the theoretical value due to the constant energy value of the white X-rays.

Figure 3 (a) depicts the strain distribution of the specimen using the strain data of the α -Fe (620) lattice plane. The strain distribution had large errors compared to the applied strain. This phenomenon also appeared in the strain distribution obtained by other lattice planes. When the peak count and the determination coefficient of the Gaussian fitting of the diffracted beam profile were low, the measurement accuracy decreased. Therefore, strain data from several lattice planes of α -Fe were simultaneously used to determine the strain distribution. Furthermore, the diffracted beam with a high energy, a high peak count, and a profile closed to a Gaussian distribution was used to calculate the measurement strain. Figure 3 (b) shows the strain distribution obtained from strain data from four lattice planes, which shows that the applied and measured strains are consistent. These results show that high energy white X-rays can effectively measure the internal strain of a material at a depth, which is on the millimeter order.

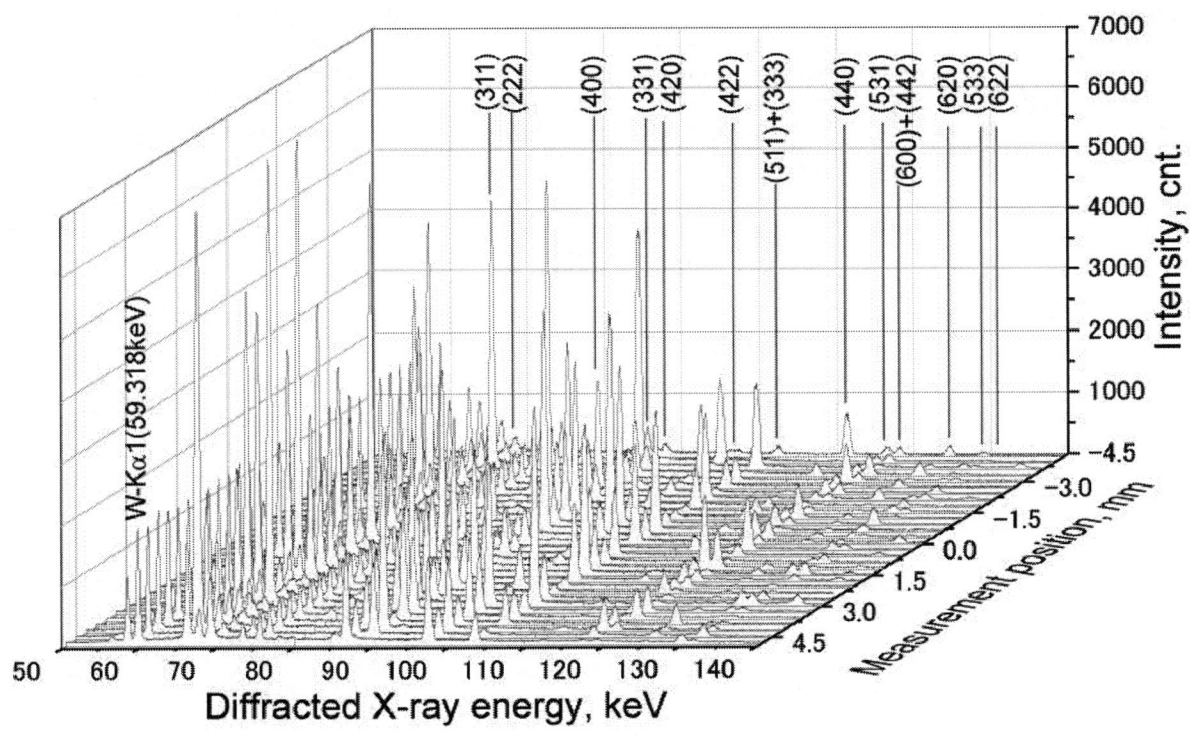

FIGURE 2. Distribution of SUS304L diffraction profiles by white X-rays.

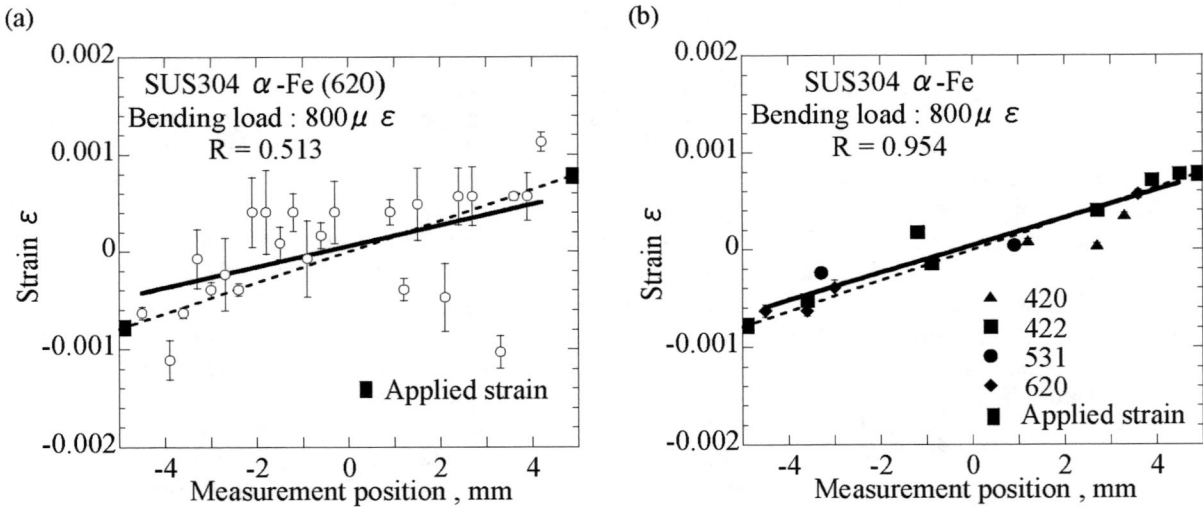

FIGURE 3. Strain distribution of SUS304L under a bending load obtained from the strain data of (a) an α-Fe (620) lattice plane and (b) four lattice planes.

CONCLUSION

A 5 mm thick austenitic stainless steel specimen was subjected to bending. High energy white X-rays from a synchrotron radiation source of SPring-8 was used to measure the internal strain of this specimen. Consequently, the internal strain was evaluated using white X-rays, which range in energy from 60 keV to 125 keV. Simultaneously using the strain data from several lattice planes increased the accuracy of the measured strain. In addition, using the diffracted beam with a high energy, a high peak count, and a profile similar to a Gaussian distribution decreased the strain measurement error. Hence, it was confirmed that the high energy white X-rays can effectively measure the internal strain at a millimeter depth in the material.

ACKNOWLEDGMENTS

We would like to thank Dr. Utsumi, Dr. Nishihata and Miss Okajima of JAEA (Japan Atomic Energy Agency) for their technical support during the experiments.

REFERENCES

1. P. J. Withers, et al., Materials Science Forum Vols. 404-407 (2002), pp.1-12.
2. A. M. Korsunsky, et al., J. Synchrotron Radiat. 9 (2002), pp.77-81.
3. G. Brusch and W. Reimers, The Fifth International Conference on Residual Stresses, Vol.1(1997), pp.557-562.
4. W. Reimers, M. Broda, G. Brusch, D. Dantz, K. D.Liss, A. Pyzalla, T. Schmackers and T. Tschentscher, J. Nondestructive Evaluation, Vol.17-3(1998), pp.129-140.

CHAPTER 10
SURFACE AND INTERFACE ANALYSIS

High-Resolution Core Level Photoemission of Mg:Ag Deposited on Tris(8-hydroxyquinolato) Aluminum Probed by Synchrotron Radiation

Tun-Wen Pi*, Hsiao-Hsuan Lin[†], Hsin-Han Lee[†] and J. Hwang[†]

*National Synchrotron Radiation Research Center, Hsinchu 30076, Taiwan, R.O.C.
[†]Department of Materials Science and Engineering, National Tsing Hua University, Hsinchu 30013, Taiwan, R.O.C.

Abstract. Deposition of magnesium on tris(8-hydroxyquinolato) aluminum (Alq_3) precovered with a thin silver dopant was investigated with high-resolution core-level photoemission via synchrotron radiation. First, the noble-metal dopant that make contact with the molecules reside at the vicinity of the pyridyl ring, similar to the case of the alkaline earth metals on Alq_3. Further, a fit to this Alq_3-derived Ag component delivers a non-zero Doniach-Sunjic singularity index of 0.061, suggesting that the incorporated Ag dopant behaves as a metallic cluster. Upon Mg adsorption on the Ag/Alq_3 surface, the size of the Ag cluster remains intact, but its binding energy now appears lower than that of the bulk. The charge added onto the clusters comes from part of Mg which is mixed with Ag. As to the other part of the Mg atoms, they gather about the chelated oxygen. This is certainly in contrast to the case of Mg on Alq_3 where Mg accumulates at the vicinity of the nitrogen atoms.

Keywords: OLED, Alq_3, Mg:Ag, interface
PACS: 73.90.+f, 79.60.Jv, 81.07.Nb

INTRODUCTION

The interfacial electronic structure of tris(8-hydroxyquinolato) aluminum (Alq_3) with a low work-function metal remains a topic of interest in the realm of basic researches in organic light-emitting devices (OLEDs), in spite of the vast number of applications currently availble. In the very first demonstration of Alq_3 as an electroluminescent organic material for OLEDs by Tang and VanSlyke [1], a mixture of magnesium and silver (Mg:Ag) with an atomic ratio of 10:1 was used as the electrode material. To the best of our knowledge, existing work in photoemission [2, 3, 4, 5, 6] has addressed the issue of the alkaline cathode interface with Alq_3 only under conditions in which the silver deposit is absent, the lack of which leaves the question of the necessity of introducing Ag as part of the cathode for good light emission unanswered. Whether the knowledge obtained from the study of the Mg/Alq_3 interface is transferable to the $Mg:Ag/Alq_3$ interface remains uncertain.

In this report, we present a high resolution photoemission study of Mg deposited on a Alq_3 surface precoated with a thin layer of silver. Our study indicates that the Mg deposit, which prefers a pyridyl ring in Alq_3 without the presence of Ag, was found to localize at a phenoxide ring, since the Ag deposit preoccupied the pyridyl ring. Nevertheless, some Mg remain mixed with Ag, whose product hinders subsequent Mg from diffusing into the Alq_3 bulk and therefore provides a basin for the fast growth of metallic Mg on top of it. Moreover, we found that Mg caused the Ag deposit to phase change from bulk to clusters.

EXPERIMENTAL

Our photoemission experiments were performed at the National Synchrotron Radiation Research Center (NSRRC) in Taiwan, Republic of China. Monochromatic radiation was provided by either a low-energy spherical grating monochromator (LSGM) or a wide-range spherical grating monochromator (WR). We prepared samples by vacuum deposition of purified Alq_3 powder onto a clean Si(001)-2×1 surface within an ultrahigh-vacuum photoemission chamber with a base pressure better than 8×10^{-11} Torr. Evaporation of Alq_3 was controlled below a pressure of 1.5×10^{-9} Torr, and the film was estimated to be 70 Å thick. A thin layer of silver was first deposited on top of this film, followed by a series of Mg depositions. Pressure during evaporation of the metals was below 5×10^{-10} Torr. Photoelectrons were collected with an 125-mm hemispherical analyzer (OMICRON Vakuumphysik GmbH).

CP879, Synchrotron Radiation Instrumentation: Ninth International Conference,
edited by Jae-Young Choi and Seungyu Rah
© 2007 American Institute of Physics 978-0-7354-0373-4/07/$23.00

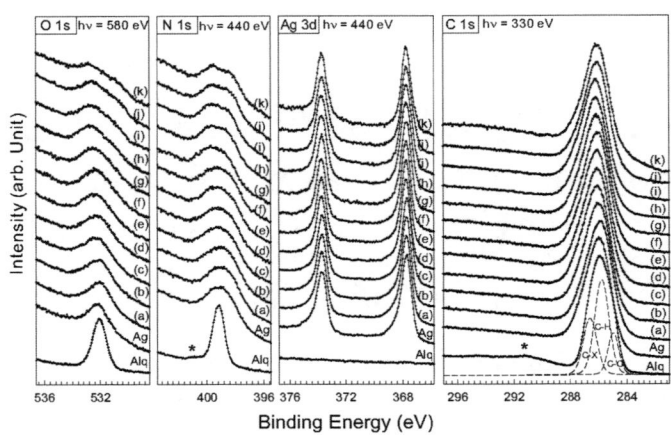

FIGURE 1. The O 1s, N 1s, Ag 3d, and C 1s core-level spectra of Mg deposited on 2 Å Ag on an Alq$_3$ film. The thickness is in unit of angstrom, t: (a) t = 0.08; (b) t = 0.25; (c) t = 0.5; (d) t = 0.75; (e) t = 1; (f) t = 1.5; (g) t = 2.5; (h) t = 4.0; (i) t = 5.5; (j) t = 8.0; (k) t = 12. The mark * in (C) stands for the $\pi \rightarrow \pi^*$ transitions.

RESULTS AND DISCUSSION

Figure 1 displays O 1s, N 1s, Ag 3d and C 1s core-level spectra, taken with photon energies of 580, 440, 440 and 330 eV, respectively, in normal emission for the clean Alq$_3$ surface covered first with 2 Å Ag, followed by a sequential deposition of various Mg thicknesses. Kinetic energies of electrons ejected by these photons fall between 20 to 50 eV, thereby giving rise to a 3 - 5 Å inelastic mean-free-path. High surface sensitivity of the presented spectra is thus emphasized. Nevertheless, the photon energy for taking the Ag 3d core level spectra is a bit high so as to reduce the high rising background at small kinetic energies. The bottommost curve in each panel of Fig. 1 reveals the pristine Alq$_3$ film prior to incorporation of foreign atoms. Analysis of these line shapes can be found in detail elsewhere [7], however, we briefly mention the results here. For the O 1s and N 1s core levels, each shows a single peak at binding energies of 532.10 (O) and 399.40 (N) eV, reflecting only one bonding configuration for these atoms. For the C 1s core level three components are extracted, which originate from three different bonding environments for carbon in the structure of the 8-quinolinol ligand; that is, a bridged carbon connected to only other carbons (the C-C bond), an outer carbon that bonds to both carbon and hydrogen (the C-H bond), and an inner carbon attached to nitrogen or oxygen (the C-X bond). The binding energies of the three components are 285.05, 285.90 and 286.77 eV for the C-C, C-H and C-X components, respectively. All these core level spectra are manifests of loss structures marked as *, the absence of which (on atoms' adsorption) suggests their strong surface origin.

Right above the bottommost curves are the spectra with deposition of about 2 Å Ag. As is discernable in these curves, the silver deposit has caused changes in the line shapes of all the 1s cores so that new components develop on the low binding energy side of the original components. This is rather peculiar, since it is commonly understood in the study of core-level spectra of Alq$_3$ that an induced component that lies above the original component is cased by a charge donation from the deposited atoms. Applying this model to the present Ag/Alq$_3$ interface, we could then say that the deposited Ag has donated charge to the Alq$_3$ molecules.

Upon Mg deposition, development of the 1s cores reveal a few differences from those shown in the case of the Mg/Alq$_3$ interface where silver is absent at the surface. For O 1s, the induced component continues to gain strength in the low-energy position, while for the Mg/Alq$_3$ interface the induced component develops at a high-energy position [6]. With regard to the N 1s core, the line shape remains virtually unchanged, in contrast to that in Mg/Alq$_3$ where the induced component increases in intensity with increasing Mg coverage [6]. With regard to the C 1s core-level spectra, a line narrowing is clearly seen, which is mainly due to an intensity reduction of the C-X component. On the contrary, the C 1s counterpart in the Mg/Alq$_3$ interface exhibited little change with various Mg depositions.

In order to derive detailed information from the measured curves, we have analyzed them by shaping each component with a Voigt function by four adjustable parameters, line position, intensity, gaussian width and lifetime width. The background function is denoted with a power-law form. The residual plot beneath each spectrum shows only statistical fluctuations, indicating that the model function properly represents all line shapes.

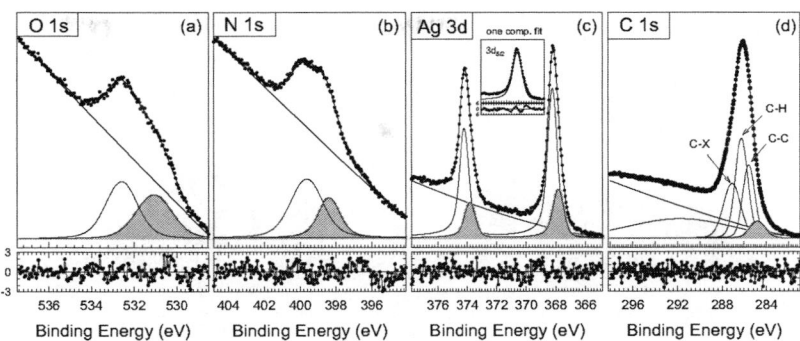

Mg:Ag on Alq$_3$

FIGURE 2. A fit to the Ag 3d (a), C 1s (b), N 1s (c), and O 1s (d) core-level spectra for 8 Å of Mg deposited on the Ag/Alq$_3$ surface.

In the case of Mg deposited on the Ag-covered Alq$_3$, a representative fit was exercised on a given Mg thickness of 8 Å and the results are shown in Fig. 2. For the N 1s core, a fit delivers an area ratio of 0.5 between the induced and original components. This magnitude stays the same throughout all the N 1s curves with various Mg thicknesses. Nevertheless, the induced component becomes narrowed and its gaussian width is 1.06 eV, a 25% reduction from that in the Ag/Alq$_3$ surface. Furthermore, it separates more from the original component, now by 1.25 eV. This indicates that additional charge has been added onto the attached nitrogen by Mg. In regard to the O 1s core, the induced component grows considerably in strength upon Mg deposition. In regard to the C 1s core-level spectrum, the induced component gains slightly in intensity due to a farther weakening of the C-X component. Accordingly, we then understand that some of the Mg atoms have also gathered about the phenoxide rings.

The analysis at the Ag 3d core-level spectrum is much more involved. Examining the measured data, one can see that the dotted curve becomes less skewed towards higher binding energies upon Mg deposition. An analytical fit to this curve is unsatisfactory if it is represented by only one spin-orbit splitting state, as exhibited in the inset of Fig. 2(c), where an unwanted structure appears in the difference curve (χ^2 (1.67)). However, the addition of a second spin-orbit splitting state to the model function alleviated the structural fluctuation such that χ^2 is reduced to 1.08. As a consequence, the lifetime width and the singularity index become reasonable values of 0.28 eV and 0.061, respectively. In a fit, the second component is only about 0.40 eV above the first component, suggesting that the core-hole screening is enhanced at the corresponding Ag. The source of charge for the enhancement is the Mg atoms that are affiliated with these silver atoms.

Figure 3 shows the development of the Mg 2p core-level spectra plotted with increasing deposition time from bottom to top. In the Ag-free interface (Fig. 3(a)), three components are resolved, marked as B, S and Z, respectively. The low-energy S+B region is attributed to the metallic Mg [4], while the high-energy component Z to the Mg clusters [6]. The binding energy of the B component is 49.60 eV, which originates from the metallic Mg. Component S is shifted $+200 \pm 10$ meV from component B, whose sign and magnitude matches nicely with that of the surface core-level shift of a crystalline Mg surface.

Upon farther examination of Fig. 3, we found that the metallic S+B region appears late after a certain amount of Mg has already deposited onto the Alq$_3$ film. Furthermore, the S+B region actually consists of two components, where the lower energy one starts to increase after the higher energy one has reached a certain strength. The delayed appearance suggests then that the Mg metal grows on top of the contacted interface. An off-angle spectrum which is more surface sensitive than the normal-emission spectrum shows an enhanced S+B region, thus confirming this statement. As can be seen in Fig. 2, the growth of the Mg metal is even sooner when silver is present at the interface. In other words, the Ag buffer effectively blocks the Mg atoms from diffusing down to the Alq$_3$ film, and allows them to grow fast as a metal.

For the Mg:Ag/Alq$_3$ interface, a fourth component, denoted as C, was resolved, in addition to the B, S and Z' components, although the broadening of components C and Z' made them hard to be differentiable at small Mg

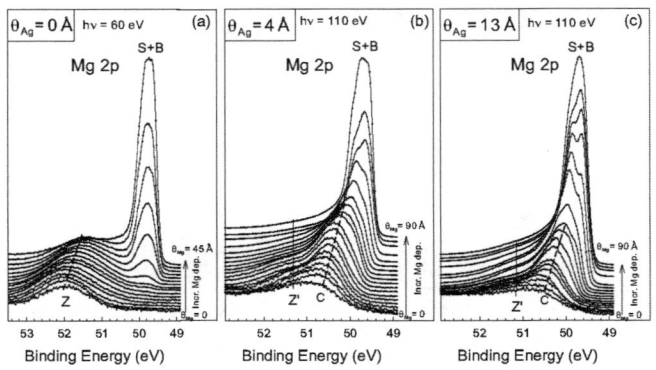

FIGURE 3. Mg 2p core-level spectra without (a) and with (b) and (c) with the presence of Ag in the Alq$_3$ surface.

exposures. An analytical fit to all the measured curves showed that the S, B and Z' components remain fixed in energies, while the C component exhibits a gradual shift towards smaller binding energies with increasing Mg thicknesses. It is suggestive, then, that enlargement in the size of Mg clusters with deposition is now manifest in the C component in the Mg:Ag/Alq$_3$ interface. Since the C component only appears when silver is present in the interface, its development should correlate closely with the electronic structure of the noble metal. As revealed in the valence band spectra (not shown), the increase in Mg deposition has rendered the Ag 4d band narrower and its centroid shifted towards greater binding energies. In other words, the noble metal has endured a phase change from bulk-like to cluster-like. Since Mg increases in size and Ag decreases, we surmise that those Mg atoms landing at the pyridyl rings and giving rise to component C are becoming mixed with Ag. This exceptional result is further justified by the fact that the valence band spectra of the mixed MgAg phase closely resembles that of the Mg$_x$Ag$_y$ compounds in the previous literature.[8] Thus, the Alq$_3$ surface has endured a change from an Ag-rich phase to an Mg-rich phase.

We attribute the Z' component of the Mg 2p core-level spectra that appears without a shift at about 51.0 eV Eb, as having originated from the Mg atoms that hit on the Ag-free region, or the phenoxide rings, since the pyridyl rings have already preoccupied by Mg that has mixed with Ag. This argument is supported by Fig. 1(a), in which the O 1s cores exhibit excess charge with Mg deposition.

In summary, we have described the interfacial chemistry of Mg deposited on the Ag/Alq$_3$ surface probed by high resolution core-level photoemission via synchrotron radiation. The electronic structure of the Mg:Ag/Alq$_3$ interface is intrinsically different from that of the Mg/Alq$_3$ interface. For the latter, the Mg dopant is prone to gather around the pyridyl ring and donate charge to the nitrogen atoms. However, the Ag deposit present at the interface prior to the Mg deposition quickly captures the site of the pyridyl ring, and thus repels the subsequent incoming Mg atoms at the site of the phenoxide ring. At the same time, some Mg begins to mix with the Ag clusters at the pyridyl ring. At large exposures, Mg grows fast as a metallic layer on top of the Mg:Ag/Alq$_3$ surface layer.

ACKNOWLEDGMENTS

This project is sponsored by the National Science Council under the contract no. NSC-94-2112-M-213-002.

REFERENCES

1. C. W. Tang and S. A. VanSlyke, Appl. Phys. Lett. **51**, 913 (1987).
2. A. Rajagopal and A. Kahn, J. Appl. Phys. **84**, 355 (1998).
3. P. He, F. C. K. Au, Y. M. Wang, L. F. Cheng, C. S. Lee, and S. T. Lee, Appl. Phys. Lett. **76**, 1422 (2000).
4. C. Shen, A. Kahn, and J. Schwartz, J. Appl. Phys. **89**, 449 (2001).
5. M. G. Mason, *et al.*, J. Appl. Phys. **89**, 2756 (2001).
6. T. -W. Pi, C. -P. Ouyang, T. C. Yu, J. -F. Wen, and H. L. Hsu, Phys. Rev. B **70**, 235346 (2004).
7. T.-W. Pi, T. C. Yu, C.-P. Ouyang, J.-F. Wen, and H. L. Hsu, Phys. Rev. B **71**, 205310 (2005).
8. M. Davies, P. Weightman, and D. R. Jennison, Phys. Rev. B **29**, 5318 (1984).

Synchrotron Radiation Studies of the Interface Morphology in Co/Pd Multilayers

Amir S. H. Rozatian

Department of Physics, University of Isfahan, Isfahan 81746-73441, Iran

Abstract. Relating the interface morphological parameters to the magnetic properties of thin films and multilayers has been a field in which x-ray scattering has played an important role over the last decade. As the bit size in magnetic storage media has continued to reduce, it was always clear that, at some point, a switch would occur from longitudinal to perpendicular recording. Seagate announced its first perpendicular recording drive in January 2006. Co/Pd magnetic nano-multilayers show perpendicular anisotropy with the interface anisotropy playing a key role. The interface morphology is believed to be important in determining the magnitude of the anisotropy. In this paper, x-ray scattering studies, which measure the statistical interface parameters directly, will be presented. Grazing incidence x-ray specular and diffuse scattering experiments were carried out on station 2.3 at the SRS (Daresbury). A few measurements were made at the XMaS beamline, BM28, at the ESRF, Grenoble. Layer and interface structural parameters were deduced by fitting the specular and transverse diffuse data to profiles simulated from model structures using the Bede REFS and REFS Mercury codes.

Keywords: Perpendicular magnetic anisotropy, Co/Pd multilayers, Grazing incidence x-ray scattering.
PACS: 61.10Kw.

INTRODUCTION

The evolution of the interface morphology during the growth of thin films and multilayers has a major impact on the overall magnetic and transport properties. In particular, the mechanism of roughness propagation from the substrate to the surface has been the focus of a number of experimental and theoretical studies. The model of Holý and Baumbach assumes a conformal roughness component that propagates completely between interfaces, a separate component of uncorrelated roughness being added sequentially at successive interfaces. The model of Spiller *et al* assumes that roughness components with larger in-plane length scales propagate preferentially. Depending on the values of the various parameters, the overall roughness amplitude may increase or decrease with the number of repeats in a multilayer. The Ming model assumes that the roughness is self-affine and that propagation does not depend on the in-plane length scale. The roughness propagation is characterised by a single out-of-plane correlation length. Controlled ballistic deposition leads to a self-affine surface morphology whose scaling behaviour corresponds to the Kardar-Parisi-Zhang (KPZ) equation, which is one of the most general local models of growth [1].

EXPERIMENTAL AND DATA ANALYSIS DETAILS

Four series of {Pd 30Å / (Co xÅ / Pd 30Å)*N} multilayer films were grown on single crystal (001) oriented silicon using the magnetron sputtering technique at the University of Leeds. The number of bilayers, N, was varied between 2 and 30. Grazing incidence x-ray reflectivity experiments were carried out on station 2.3 at the SRS (Daresbury, UK). Grazing incidence diffuse x-ray scattering experiments were performed at the ESRF XMaS beamline, BM28, at the ESRF (Grenoble, France) using a MAR CCD detector. All synchrotron radiation measurements were made with a wavelength 1.3 Å. Structural parameters were obtained by fitting the observed x-ray scattering distribution to that simulated from a self-affine fractal model of the interface introduced by Sinha *et al*

CP879, *Synchrotron Radiation Instrumentation: Ninth International Conference,*
edited by Jae-Young Choi and Seungyu Rah
© 2007 American Institute of Physics 978-0-7354-0373-4/07/$23.00

[2]. The results for series 3 samples can be seen in Fig. 1 and Table 1. Further results relating to series 1 and 2 may be found in [3, 4, 5]. Results relating to series 4 samples will be published later.

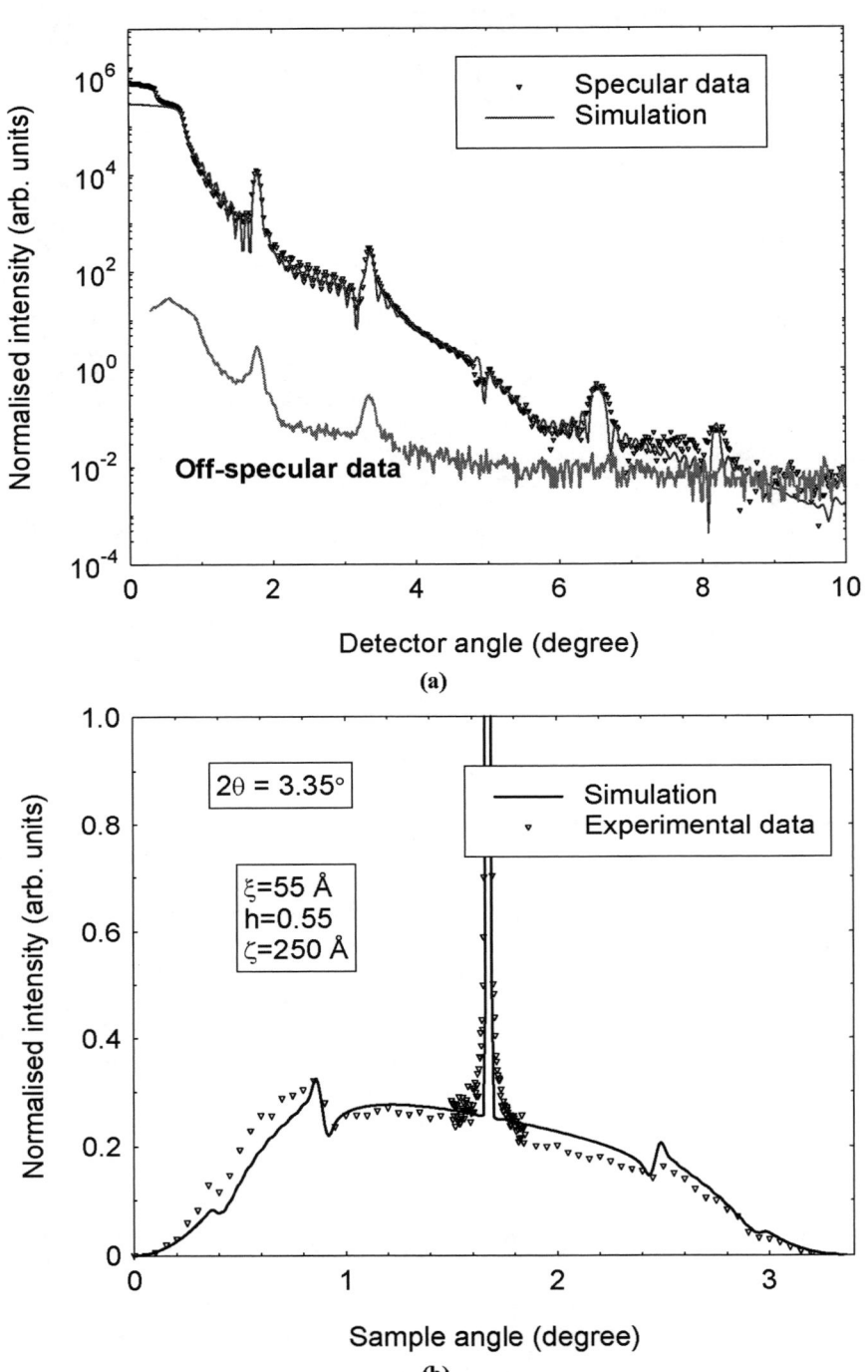

FIGURE 1. **(a)** Specular and Off-specular data and simulation for the 12-bilayer sample. **(b)** Transverse diffuse scan and simulation for the 12-bilayer sample, synchrotron data with $\lambda = 1.3$Å.

TABLE 1. Multilayer structure parameters for series 3 samples, determined from x-ray scattering simulations (σ is the interface roughness and t is the layer thickness).

N	t_{Buffer} ±0.3Å	t_{Co} ±0.3Å	t_{Pd} ±0.3Å	σ_{Buffer} ±0.3Å	σ_{Co} ±0.3Å	σ_{Pd} ±0.3Å	ζ (Å) ±10%	ξ (Å) ±10%	h ±10%
8	33.0	14.4	29.2	3.9	3.8	3.7	200	18	0.20
10	32.8	14.6	31.0	8.0	2.6	9.1	250	60	0.20
12	35.5	14.9	30.9	3.8	3.9	3.6	250	70	0.24
16	32.0	15.4	30.1	4.8	4.2	4.2	250	120	0.15
26	32.3	13.0	30.9	7.5	4.8	8.4	250	300	0.50

RESULTS

As the bilayer number is increased, the off-specular Bragg peak in the diffuse data remains, indicating an out-of-plane correlation over several bilayer repeats is always retained. However, the off-specular Kiessig fringes, which correspond to roughness correlations extending over the entire thickness of the sample, disappear as the stack increases (Fig. 2). No significant interdiffusion could be measured at the interfaces; almost all of the interface width determined from fitting the specular scatter was identified as topological roughness. As the number of bilayers is increased, a broad peak at low in-plane wavevector appears in the diffuse scatter which corresponds to an increase in the in-plane correlation length. The integrated intensity of the diffraction satellites associated with the multilayer initially rises with bilayer number in sample series 1, but saturates at N = 20, where the largest PMA per layer is found.

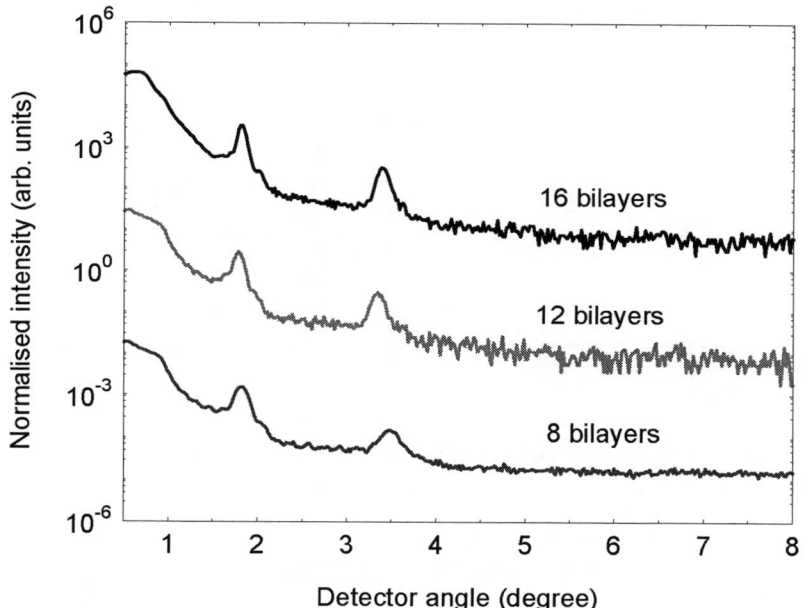

FIGURE 2. A series of Co/Pd Off-specular scans (sample off-set -0.1°, λ = 1.3Å).

Scaling behaviour is observed in all the Co/Pd multilayers examined for large out-of-plane scattering vectors (Figs. 3 and 4(a)). The scaling exponent rises with increasing number of multilayer repeats, there being reasonable agreement in the asymptotic limits with the TAB and KPZ models of film growth. As the number of repeats rises, the length scale to which fractal behaviour is observed falls, although the correlation length defined within the growth models and obtained by fitting the Sinha model rises. The length scale over which the height-to-height correlation is retained thus rises, but on the corresponding length scale, fractal behaviour is not observed. Caution must therefore be exercised in applying the Sinha model as it does not describe the differing changes in the conformal and random roughness frequency components. Both approaches show that as the number of bilayer repeats increases, the interfaces become more two-dimensional in character, as the Hurst fractal parameter, h, rises (Fig. 4(b)) [3, 4, 5].

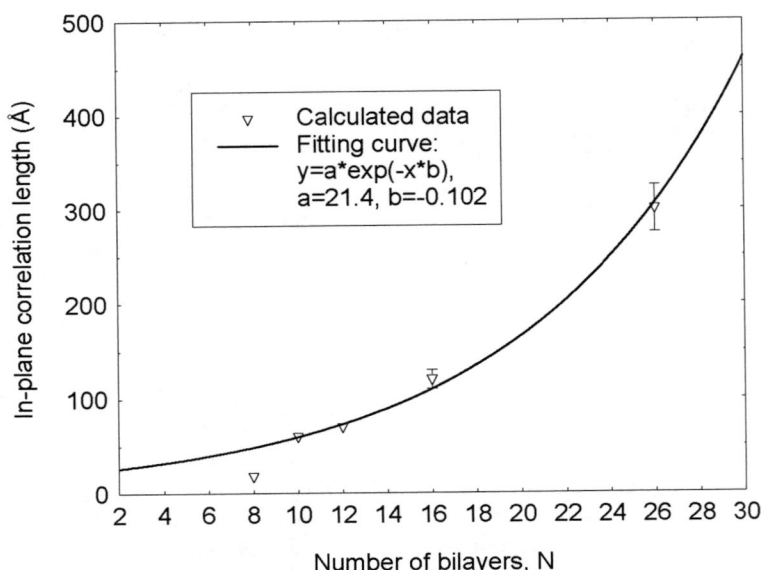

FIGURE 3. Relation between the in-plane correlation length, ξ and number of bilayers.

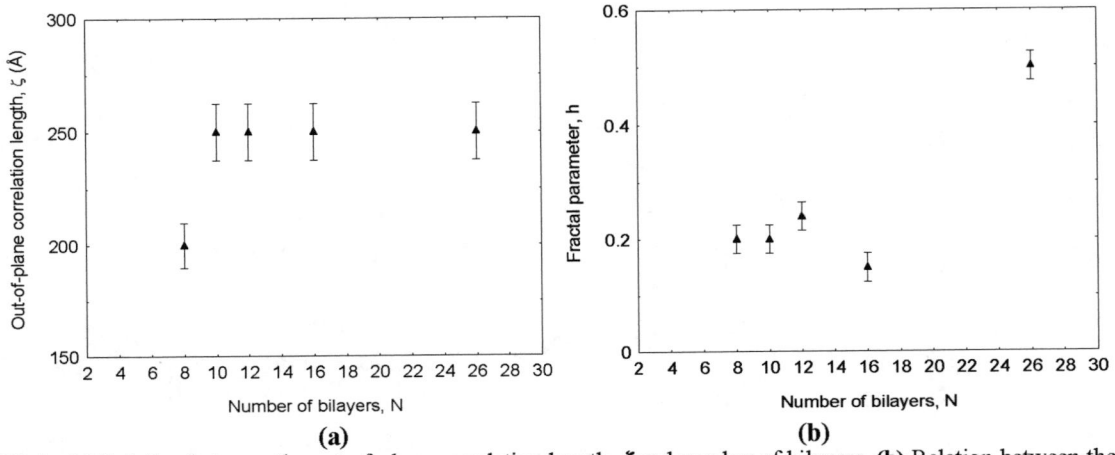

(a)   **(b)**

FIGURE 4. **(a)** Relation between the out-of-plane correlation length, ζ and number of bilayers, **(b)** Relation between the fractal parameter, h and number of bilayers.

ACKNOWLEDGMENTS

Thanks are expressed to Prof. Brian Tanner for helpful discussion and the University of Isfahan for financial support during the period at Durham University where most of the work was performed. Thanks are also given to Dr. C. H. Marrows of the University of Leeds for specimen preparation, the staff at Daresbury, SRS beamline 2.3, and BM28 of the ESRF, Grenoble.

REFERENCES

1. A.-L. Barabasi and H. E. Stanley, *Fractal Concepts in Surface Growth*, Cambrige University Press, 1995.
2. S. K. Sinha et al, Phys. Rev. B **38 (4)**, 2297-2311 (1988).
3. A. T. G. Pym et al, *Physica B* **357**, 170-174 (2005).
4. A. T. G. Pym et al, *J. Phys.: D* **38**, A190-A194 (2005).
5. A. S. H. Rozatian et al, *J. Phys.: C* **17**, 3759-3770 (2005).

High Resolution Hard X-ray Photoemission Spectroscopy at SPring-8: Basic Performance and Characterization

Yasutaka Takata[1]*, Koji Horiba[1], Masaharu Matsunami[1], Shik Shin[1], Makina Yabashi[2,3], Kenji Tamasaku[2], Yoshinori Nishino[2], Daigo Miwa[2], Tetsuya Isikawa[2,3], Eiji Ikenaga[3], Keisuke Kobayashi[3], Masashi Arita[4], Kenya Shimada[4], Hirofumi Namatame[4], Masaki Taniguchi[4], Hiroshi Nohira[5], Takeo Hattori[5], Sven Södergren[6], Björn Wannberg[6]

[1]Soft X-ray Spectroscopy Laboratory, RIKEN SPring-8 Center, 1-1-1 Kouto, Sayo-cho, Sayo-gun, Hyogo 679-5148, Japan
[2]Coherent X-ray Optics Laboratory, RIKEN SPring-8 Center, 1-1-1 Kouto, Sayo-cho, Sayo-gun, Hyogo 679-5148, Japan
[3]JASRI/SPring-8, 1-1-1 Kouto, Sayo-cho, Sayo-gun, Hyogo 679-5198, Japan
[4]HiSOR, Hiroshima University, Kagamiyama 2-313, Higashi-Hiroshima 739-5826, Japan
[5]Department of Electrical and Electronic Engineering, Musashi Institute of Technology, Tamazutsumi 1-28-1, Setagaya-ku, Tokyo 158-8557, Japan
[6]VG Scienta AB, P.O. Box 15120, SE-750 15 Uppsala, Sweden

Abstract. Photoemission Spectroscopy (PES) is a powerful method to investigate electronic structure of materials. However, conventional vacuum ultraviolet (VUV) and soft x-ray (SX) PES is surface sensitive because of the short inelastic-mean-free-paths of photoelectrons. This requires us to prepare clean surface and sometimes obscures intrinsic bulk states. In order to realize bulk-sensitive or surface-insensitive PES, we have developed hard x-ray (HX) PES using high-brilliance synchrotron radiation at BL29XU in SPring-8. Large probing depth of high energy photoelectrons enables us to probe intrinsic bulk states almost free from surface condition. A combination of x-ray optics and an electron energy analyzer dedicated for HX-PES achieved the total instrumental energy resolution of 63 and 55 meV (FWHM) at 5.95 and 7.94 keV, respectively. A special arrangement of an analyzer and a sample was employed and increased photoelectron intensity drastically. We describe present performance of our apparatus and characterize HX-PES by showing typical spectra.

Keywords: Photoemission Spectroscopy, Hard X-ray, Electronic Structure.
PACS: 79.60.-i

INTRODUCTION

Photoemission spectroscopy (PES) has been used extensively to experimentally determine electronic structure of core levels and valence bands (VBs)[1]. However, conventional PES is surface sensitive because of short inelastic-mean-free-paths (IMFPs) [2]. In order to attain larger probing depths of VBs than that in vacuum ultraviolet (VUV) spectroscopy, soft x-ray (SX) VB-PES using synchrotron radiation (SR) has recently become attractive[3]. However, it is obvious that SX-PES is still surface sensitive, because, for example, the IMFPs of a valence electron are only 1.3 and 2 nm for Au and Si at a kinetic energy (KE) of 1 keV, respectively[2]. In the case of core levels, smaller KEs than those of VBs enhance surface sensitivity, make it rather difficult to probe bulk character, using SX-PES.

In contrast to the above-mentioned surface sensitive PES techniques, the IMFP values of a valence electron for Au and Si increase to 5.5 and 9.2 nm, respectively at 6 keV[2], which lies in the range of hard x-rays. The straightforward way to realize an intrinsic bulk probe is to increase KE of photoelectrons by use of hard x-rays. The first feasibility test of hard x-ray (HX)-PES was done by Lindau et al. in 1974 using a 1st generation SR source[4].

CP879, *Synchrotron Radiation Instrumentation: Ninth International Conference*,
edited by Jae-Young Choi and Seungyu Rah
© 2007 American Institute of Physics 978-0-7354-0373-4/07/$23.00

However, the feeble signal intensity even of Au 4f core level excluded possibility of studies of VBs. What has prevented HX VB-PES is the rapid decrease in subshell photo-ionization cross section (σ). The σ values for Au 5d ($1\times10-5$ Mb) and Si 3p ($3\times10-5$ Mb) at 6 keV are only 1-2% of those at 1 keV [5].

In order to realize HX-PES with high-energy-resolution and high-throughput, both high-brilliance SR and a high performance electron energy analyzer are required. After 2nd generation SR became available, there have been a few reports on core level photoemission and resonant Auger spectroscopy using several keV x-rays[6]. In the last few years, unprecedented high-flux and high-brilliance SR at third generation facilities such as ESRF, APS and SPring-8 has stimulated us to develop HX-PES with high-energy-resolution and high-throughput. The results of the feasibility test at the excitation energy of 6 keV done at SPring-8 in 2002 demonstrated the capability to probe intrinsic bulk electronic structure of both core levels and VBs [7, 8]. HX-PES has also been developed at ESRF [9-11]. All these experimental achievements indicate that HX-PES will contribute significantly in the study of electronic structure of solids. The wide range of applications includes depth-resolved electronic structure, buried layers, interfaces, ultrashallow junctions and the bulk electronic structure of strongly correlated electron systems. Here, we describe present performance of our apparatus and characterize HX-PES by showing typical spectra.

EXPERIMENTAL

The essential problem to overcome and realize HX-PES is weak signal intensity due to small σ values as pointed out above. Of course, intensity of x-rays and detection efficiency of an electron energy analyzer are critical factors. In addition to these, configuration of the experimental setup also influences the signal intensity. Figure 1 shows IMFPs up to KE of 10 keV for several materials[2]. IMFPs at the KE of 6 keV range from 4 to 15 nm and are almost 5 times larger than those at 1 keV. However, these values are much shorter than the x-ray attenuation length (30 μm for Si and 1 μm for Au at 6 keV). In order to avoid wasting x-rays in the region deeper than the electron escaping depth, grazing incidence of x-rays relative to the sample surface is desirable.

The detection angle of photoelectron relative to the polarization vector of x-rays also plays a role in gaining photoelectron intensity. When we use linearly polarized light as an excitation source, photoelectrons from free atoms show angular distribution depending on the asymmetry parameter β (see Eq. (5) in Ref. [5]) as shown in Fig. 2. For HX-PES, typical photon energy is above 6 keV, and almost all subshells have positive β values [5]. In this case, photoelectron intensity has a maximum along the direction parallel to the polarization vector. This behavior is considered applicable even to solids. On the other hand, to achieve large probing depth, photoelectrons should be detected along the direction close to the normal of the sample surface.

Following these considerations, an HX-PES apparatus with the configuration shown in Fig. 3 has been constructed at an x-ray undulator beamline BL29XU [12] in SPring-8. The lens axis of the analyzer is placed perpendicular to the x-ray beam and the incidence angle relative to the sample surface is typically set to about 1° for samples with a flat surface[13]. The first version electron energy analyzer (a modified SES-2002) has recently been

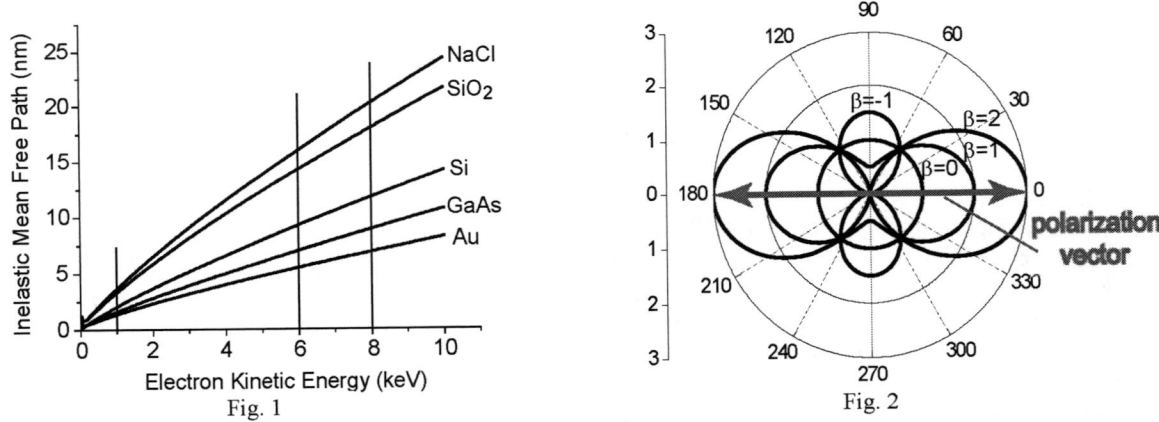

Fig. 1

Fig. 2

FIGURE 1. Inelastic mean free paths for electron kinetic energies up to 10 keV, for Au, GaAs, Si, SiO$_2$, and NaCl[2].

FIGURE 2. Angular distribution of photoelectrons from free atoms. For positive asymmetry parameter β, the intensities have a maximum along the direction of the polarization vector.

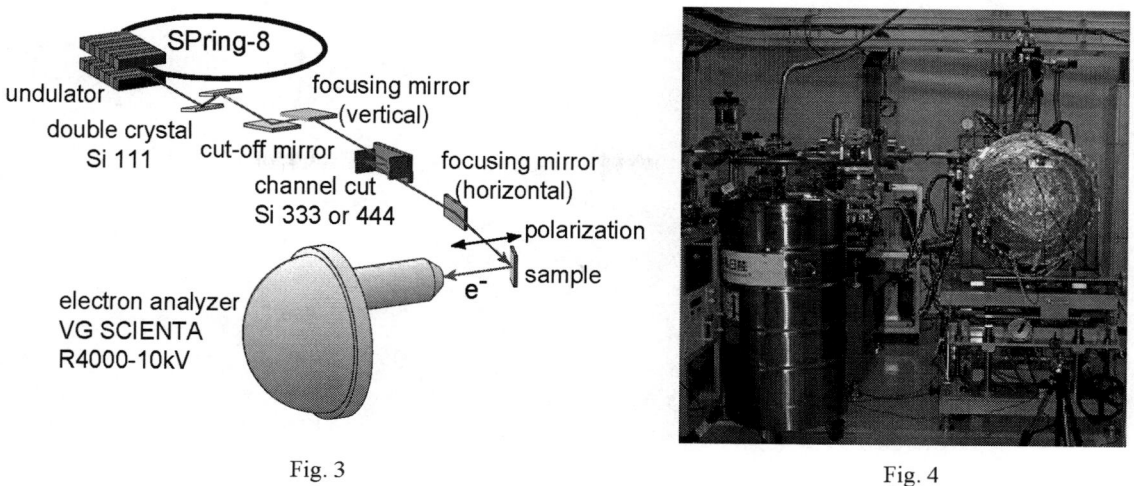

Fig. 3   Fig. 4

FIGURE 3. Schematic of experimental setup including x-ray optics at the beamline BL29XU in SPring-8.

FIGURE 4. Apparatus for HX-PES placed in the experimental hutch at BL29XU in SPring-8.

replaced by a newly developed one, R4000-10kV (VG SCIENTA Co.), and measurable kinetic energy has been extended from 6 to 10 keV. In addition to the analyzer, a sample manipulator with a motorized XYZΘ stage, a flow-type He cryostat for sample cooling, two turbo molecular pumps, and CCD cameras are equipped on the measurement chamber. The whole system including load-lock and preparation chambers is mounted on a position adjustable stage. The design is made as compact as possible so as to carry the apparatus into the experimental hutch (see Fig. 4). The vacuum of the measurement chamber is 10^{-8} Pa, and the lowest sample temperature achieved is 20 K. In order to realize high-energy-resolution and high-throughput, x-ray optics dedicated for HX-PES is essential. Figure 3 shows the schematic of the optics at BL29XU [12]. X-rays from an undulator are premonochromatized with a Si 111 double-crystal monochromator. A channel-cut Si monochromator is placed downstream to realize high energy resolution. The Bragg angle is fixed at 85°, and Si 333 and 444 reflections are used for 5.95 and 7.94 keV x-rays, respectively. The incident energy band width is less than 60 meV. Horizontal and vertical focusing mirrors are installed and the spot size at the sample position is 50 (vertical) × 35 (horizontal) μm^2 with the x-ray intensity of 10^{11} photons/sec. Details of x-ray optics for HX-PES is described by Ishikawa et al.[14]. The fine focus considerably increases photoelectron intensity because the lens system of the analyzer magnifies the spot size on the sample surface by 5 times at the entrance slit of the hemispherical analyzer.

PERFORMANCE AND CHARACTERISTICS

In this section, typical spectra are shown to demonstrate the high-energy resolution and high- throughput of HX-PES and also to characterize this method. It should be noted that all the HX-PES spectra presented in this section were measured without surface treatment in the vacuum.

High Throughput and High Energy Resolution

Figure 5 shows HX-PES spectra of a Au plate measured at 20 K with the excitation energy of 7.94 keV. In both, the Au 4f core level and valence band spectra measured with the analyzer pass energy (Ep) of 200 eV, the signal-to-noise ratio is very good even with the short accumulation times of 5 and 30 sec, respectively. The total instrumental energy resolution including the x-ray band width was 230 meV in this conventional setting. With the lowest Ep of 50 eV and the narrowest slit setting (0.1 mm width), we have checked the total energy resolution by measuring Fermi-edge spectra of Au at 20 K. Figure 6 shows the experimentally obtained spectrum. By fitting this profile, the total instrumental energy resolution at 7.94 keV is determined to be 55±5 meV (E/ΔE=140000). At 5.95 keV, the highest resolution of 63±3 meV was realized. At ESRF, Torelli et al. also achieved the energy resolution of 71±7 meV at 5.93 keV[11].

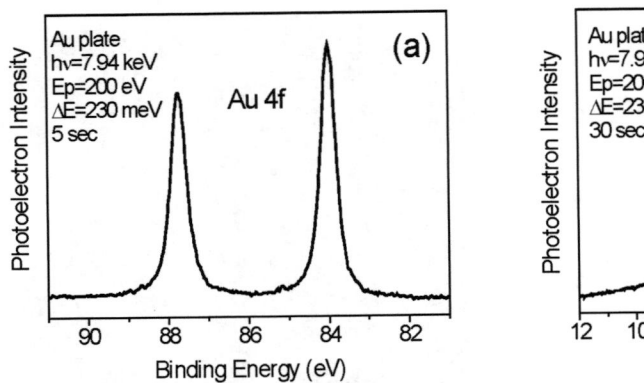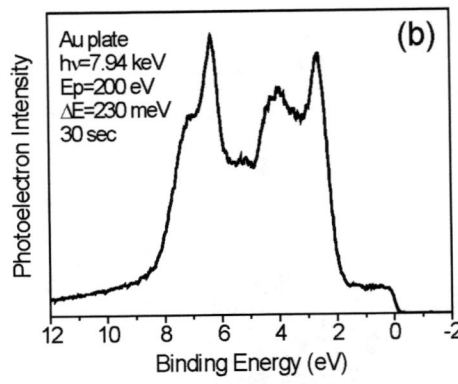

FIGURE 5. (a) Au 4f and (b) valence band spectra of a Au plate measured at 20 K with 7.94 keV excitation. High quality spectra with the total instrumental energy resolution ΔE of 230 meV are obtained within a short acquisition time of 5 and 30 sec.

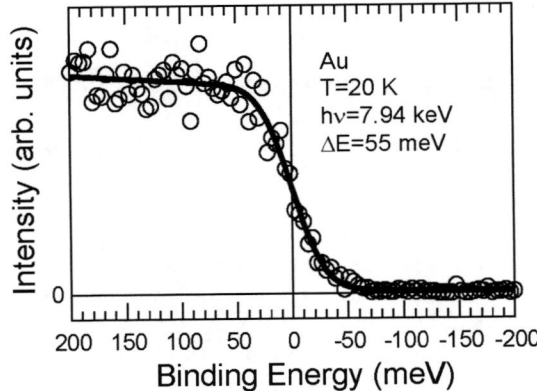

FIGURE 6. Fermi-edge spectrum of a Au plate measured with the lowest Ep of 50 eV and the narrowest slit setting (0.1 mm width) at 7.94 keV. By fitting the profile, the total instrumental energy resolution at 7.94 keV is determined to be 55±5 meV (E/ΔE=140000).

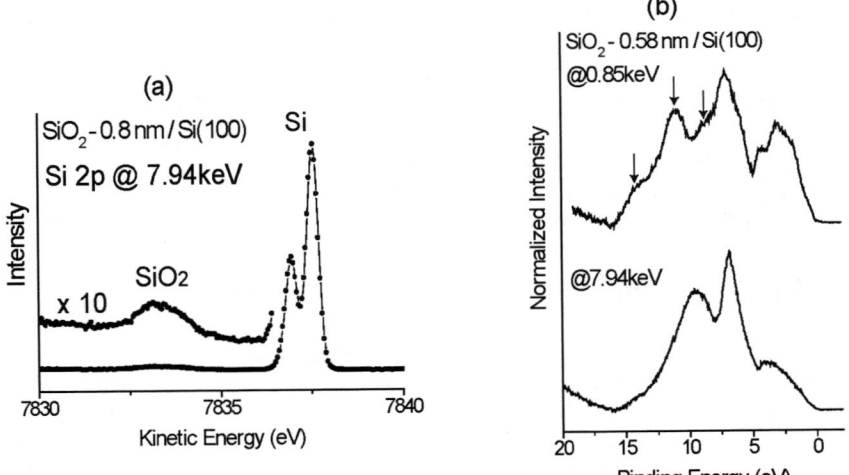

FIGURE 7. (a) Si 2p and (b) valence band spectra of a Si(100) surface with a thin-SiO_2 layer measured at 7.94 keV. SX valence band spectrum measured at 0.85 keV is also shown as a reference. Arrows indicate the structure originating from the surface SiO_2 layer.

Surface Insensitivity

The capability of HX-PES to probe bulk states of reactive surfaces is demonstrated by measuring core and VB spectra of a Si(100) surface with a thin-SiO_2 layer. Figure 7 (a) shows the Si 2p (binding energy, BE~100 eV) spectrum of 0.8 nm-SiO_2/Si(100) measured at 7.94 keV. The peak intensity of the surface SiO_2 layer is negligibly small (~2%) in comparison with that of substrate Si. Negligible surface contribution is also confirmed for the VB spectrum in Fig. 7(b). Comparing the HX spectrum with the SX (0.85 keV) spectrum, the structures marked by arrows in the SX spectrum are due to the 0.58 nm-surface SiO_2 layer. These features almost vanish in the HX spectrum. The "surface insensitivity" of HX-PES enables us to investigate the intrinsic bulk state of thin films which are beyond the reach of "surface sensitive" PES. This is because surface sensitive PES necessarily requires surface cleaning and preparation procedure. It should be noted that these Si 2p and VB spectra can be obtained within short acquisition times of 30 and 300 sec, respectively.

Large Probing Depth

In addition to "surface insensitivity", the large probing depth of HX-PES extends the applicability to embedded layers and interfaces in nano-scale buried layer systems. Figure 8 shows the Sr $2p_{3/2}$ spectra of a bare Sr-TiO_3 (STO) substrate and the substrate covered with a thin layer (20 nm) of $La_{0.85}Ba_{0.15}MnO_3$ (LBMO) measured at 5.95 keV [15]. The Sr $2p_{3/2}$ (BE=1940 eV) photoelectrons from the substrate with the KE of 4010 eV are still observable through the 20 nm thick overlayer. The small KE difference between these two samples is attributed to band bending. From the intensity variation, the IMFP value of electrons with KE=4010eV in LBMO is estimated as 4 nm. Concerning the probing depth of HX-PES, Sacchi et al. have recently determined the effective attenuation length over the KE range from 4 to 6 keV in Co, Cu, Ge and Gd_2O_3, and showed the use of HX-PES for studying buried layers and interfaces[16]. Dallera et al. have recently reported a study on AlAs layer buried under different thickness of GaAs, emphasizing the role of HX-PES in non-destructive analysis of buried layers[10].

SUMMARY

HX-PES with high-throughput and high-energy-resolution has been realized for core-level and VB studies using high-energy and high-brilliance SR at the beamline BL29XU in SPring-8. In addition to x-ray optics and an electron analyzer dedicated for HX-PES, optimized experimental configuration, such as photoelectron detection along the direction parallel to x-ray polarization and grazing incidence of well-focused x-ray beam, strongly improved photoelectron intensity. When we set the total instrumental energy resolution to be about 230 meV, not only core-level but also VB spectra of Au and Si can be measured within several tens of seconds to several minutes, demonstrating the high throughput of the experimental system. The most important characteristics of HX-PES, i.e.,

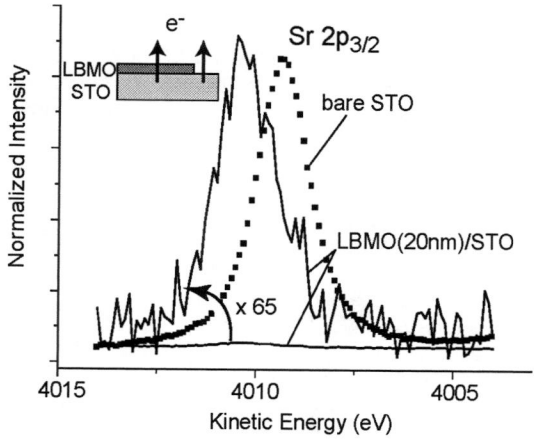

FIGURE 8. Sr $2p_{3/2}$ core level spectra of bare $SrTiO_3$ (STO) substrate and the substrate covered with a thin layer (20 nm) of $La_{0.85}Ba_{0.15}MnO_3$ (LBMO) measured at 5.95 keV [15]. The Sr $2p_{3/2}$ (BE=1940 eV) photoelectrons from the substrate with KE of 4010 eV are still observable through the 20 nm thick overlayer.

surface insensitivity and large probing depth, were confirmed. The highest total energy resolution achieved is 63 meV at 5.95 keV, and 55 meV at 7.94 keV.

HX-PES has been applied to studies of Si high k dielectrics[7], diluted magnetic semiconductors[8,17], f-electron systems with valence transition[18], and 3d transition metal compounds with strong electron correlation[19-21]. All the results confirm capability of HX-PES to probe depth-resolved electronic structure, buried layers, interfaces, ultrashallow junctions and the bulk electronic structure of strongly correlated electron systems. We believe that HX-PES will become a standard method to study electronic properties of various materials in the field of basic science and technologies.

Concerning future perspectives, improvement of the energy resolution down to 30 meV can be expected to be done soon. Further improvement down to 10 meV is a challenge because the development of a new power supply system based on alternative technology will be necessary. It should be noted that x-ray band width of 120 μeV has been achieved at 14.41 keV [14]. Angle resolved PES of VB is also very attractive even with the present resolution of ~60 meV. For this purpose, high angular resolution less than 0.1° is required because first Brillouin zone shrinks at several keV. It is also interesting to develop spin polarized HX-PES.

This work was partially supported by the Ministry of Education, Science, Sports and Culture through a Grant-in-Aid for Scientific Research (A) (No. 15206006) .

REFERENCES

1. S. Hüfner, *Photoelectron Spectroscopy*, Berlin-Hidelberg: Springer-Verlag, 1995.
2. The electron inelastic-mean-free-paths were estimated using NIST Standard Reference Database 71, "NIST Electron Inelastic-Mean-Free-Path Database: Ver. 1.1". It is distributed via the Web site http://www.nist.gov/srd/nist71.htm, and references therein.
3. A. Sekiyama, T. Iwasaki, K. Matsuda, Y. Saitoh, and S. Suga, *Nature* **403**, 396 (2000).
4. I. Lindau, P. Pianetta, S. Doniach, and W. E. Spicer, *Nature* **250**, 214 (1974)..
5. J. J.Yeh and I. Lindau, *At. Data Nucl. Data Tables* **32**, 1 (1985).
6. W. Drube, Th. Eickhoff, H. Schulte-Schrepping, and J. Heuer, *AIP conference proceedings* **705**, 1130 (2002).
7. K. Kobayashi, M. Yabashi, Y. Takata, T. Tokushima, S. Shin, K. Tamasaku,D. Miwa, T. Ishikawa, H. Nohira, T. Hattori, Y. Sugita, O. Nakatsuka, A. Sakai, and S. Zaima,*Appl. Phys. Lett.* **83**, 1005 (2003).
8. Y. Takata, K. Tamasaku, T. Tokushima, D. Miwa, S. Shin, T. Ishikawa, M. Yabashi, K. Kobayashi, J. J. Kim, T. Yao, T. Yamamoto, M. Arita, H. Namatame, and M. Taniguchi, *Appl. Phys. Lett.* **84**, 4310 (2004).
9. S. Thiess, C. Kunz, B.C.C. Cowie, T.-L. Lee, M. Renier, and J. Zegenhagen, *Solid State Commun.* **132**, 589 (2004).
10. C. Dallera, L. Du`o, L. Bricovich, G. Panaccione, G. Paolicelli, B. Cowie, J. Zegenhagen, *Appl. Phys. Lett.* **85**, 4532 (2004).
11. P. Torelli, M. Sacchi, G. Cautero, M. Cautero, B. Krastanov, P. Lacovig, P. Pittana, R. Sergo, A. Tommasini, A. Fondacaro, F. Offi, G. Paolicelli, G. Stefani, M. Grioni, R. Verbeni, G. Monaco, G. Panaccione, *Rev. Sci. Instrum.* **76**, 23909 (2005).
12. K. Tamasaku, Y. Tanaka, M. Yabashi, H. Yamazaki, N. Kawamura, M. Suzuki, T. Ishikawa, *Nucl. Instrum. Methods.* **A 467/468**, 686 (2001).
13. Y. Takata,M. Yabashi, K. Tamasaku, Y. Nishino, D. Miwa, T. Ishikawa, E. Ikenaga, K. Horiba, S. Shin, M. Arita, K. Shimada, H. Namatame, M. Taniguchi, H. Nohira, T. Hattori, S. Södergren, B. Wannberg, K. Kobayashi, *Nucl. Instrum. Methods.* **A 547**, 50 (2005).
14. T. Ishikawa, K. Tamasaku, and M. Yabashi, *Nucl. Instrum. Methods.* **A 547**, 42 (2005).
15. H. Tanaka, Y. Takata, K. Horiba, M. Taguchi, A. Chainani, S. Shin, D. Miwa, K. Tamasaku, Y. Nishino, T. Ishikawa, E. Ikenaga, M. Awaji, A. Takeuchi, T. Kawai, and K. Kobayashi, *Phys. Rev.* B **73**, 094403 (2006).
16. M. Sacchi, F. Offi, P. Torelli, A. Fondacaro, C. Spezzani, M. Cautero, G. Cautero, S. Huotari, M. Grioni, R. Delaunay, M. Fabrizioli, G. Vankó, G. Monaco, G. Paolicelli, G. Stefani, G. Panaccione, *Phys. Rev.* B **71**, 155117 (2005).
17. J.J. Kim, H. Makino, K. Kobayashi, Y. Takata, T. Yamamoto, T. Hanada, M.W. Cho, E. Ikenaga, M. Yabashi, D. Miwa, Y. Nishino, K. Tamasaku, T. Ishikawa, S. Shin, T. Yao, *Phys. Rev.* B **70**, 161315 (2004).
18. H. Sato, K. Shimada, M. Arita, K. Hiraoka, K. Kojima, Y. Takeda, K. Yoshikawa, M. Sawada, M. Nakatake, H. Namatame, M. Taniguchi, Y. Takata, E. Ikenaga, S. Shin, K. Kobayashi, K. Tamasaku, Y. Nishino, D. Miwa, M. Yabashi, T. Ishikawa, *Phys. Rev. Lett.* **93**, 246404 (2004).
19. A. Chainani, T. Yokoya, Y. Takata, K. Tamasaku, M. Taguchi, T. Shimojima, N. Kamakura, K. Horiba, S. Tsuda, S. Shin, D. Miwa, Y. Nishino, T. Ishikawa, M. Yabashi, K. Kobayashi, H. Namatame, M. Taniguchi, K. Takada, T. Sasaki, H. Sakurai, E. Takayama-Muromachi, *Phys. Rev.* B **69**, 180508 (2004).
20. K. Horiba, M. Taguchi, A. Chainani, Y. Takata, E. Ikenaga, D. Miwa, Y. Nishino, K. Tamasaku, M. Awaji, A. Takeuchi, M. Yabashi, H. Namatame, M. Taniguchi, H. Kumigashira, M. Oshima, M. Lippmaa, M. Kawasaki, H. Koinuma, K. Kobayashi, T. Ishikawa, S. Shin, *Phys. Rev. Lett.* **93**, 236401(2004).
21. M. Taguchi, A. Chainani, K. Horiba, Y. Takata, M. Yabashi, K. Tamasaku, Y. Nishino, D. Miwa, T. Ishikawa, T. Takeuchi, K. Yamamoto, M. Matsunami, S. Shin, T. Yokoya, E. Ikenaga, K. Kobayashi, T. Mochiku, K. Hirata, J. Hori, K. Ishii, F. Nakamura, and T. Suzuki, *Phys. Rev. Lett.* **95**, 177002 (2005).

Crossover in the Scaling Behavior of Ion-sputtered Pd(001)

T. C. Kim[1], Y. Kim[1], D. Y. Noh*[1], B. Kahng[2], and J.-S. Kim[3]

[1]*Department of Materials Science and Engineering, GIST, Gwangju, 500-712, Korea*
[2]*School of Physics and Center for Theoretical Physics, Seoul National University, Seoul, 151-747, Korea*
[3]*Department of Physics, Sook-Myung Women's University, Seoul, 140-742, Korea*

Abstract. We investigate morphological evolution of Ar^+ ion sputtered Pd(001) by *in situ* real-time x-ray reflectivity and grazing incidence small angle x-ray scattering (GISAXS) experiments. Surface roughness W, and its dynamic scaling behavior show a definite crossover across a crossover time t_c. Before t_c, growth exponent β, varies from 0.20 to 0.49 depending on substrate temperature, T. After t_c, β drops to ~0.1, irrespective to substrate temperature. Satellite peaks in GISAXS indicating laterally ordered structure develop as the growth time approaches t_c, which become clear with further sputtering. We think that the crossover behavior near t_c indicates the reduction of non-linear effect and the scaling behavior would follow the Edwards-Wilkinson model.

Keywords: x-ray reflectivity, ion-sputtering, scaling theory
PACS: 68.55.-a, 05.45.-a, 64.60.Cn, 79.20.Rf

INTRODUCTION

Recently, there has been increasing interest in surface modification by ion sputtering. Fabrications of self-organized ordered nano-structures are demonstrated by simply adjusting physical parameters involved in ion sputtering such as ion beam energy, flux, incidence angle, and substrate temperature [1-3]. Numerous continuum models and simulations [4-9] have successfully elucidated morphological changes during sputter-erosion such as roughening and coarsening.

Theoretical investigations of sputter-eroded surface such as hydrodynamic [9] and Sigmund's theory [8], showed that a new higher order nonlinear term $\nabla^2 (\nabla h)^2$ is required to describe the evolution of the surface height h,

$$\frac{1}{c}\frac{\partial h}{\partial t} = -1 - \nu \nabla^2 h - D \nabla^4 h + \lambda_1 (\nabla h)^2 + \lambda_2 \nabla^2 (\nabla h)^2 + \eta(x,y,t) \tag{1}$$

where c is the surface recession rate gives by a material constant times ionic flux, ν the effective surface tension generated by the erosion process, D the effective diffusion coefficient, λ_1 and λ_2 tilt-dependent erosion rates, and η uncorrelated white noise with zero mean, mimicking the randomness resulting from the stochastic nature of ion arrival at the surface. This equation explains self-ordering as well as coarsening by incorporating redeposition and viscous flow effects [8, 9]. However, the scaling behavior of the surface in long time limits is still unresolved questions [10].

EXPERIMNET

Sputtering of the Pd(001) surface was carried out with Ar^+ ion beam incident normal to the sample surface. Ion energy, ε was 0.5 keV and ion flux, $f = 0.5 \times 10^{13}$ ions/cm^2/s. The ion beam was defocused in order to irradiate the

CP879, *Synchrotron Radiation Instrumentation: Ninth International Conference*,
edited by Jae-Young Choi and Seungyu Rah

sample surface uniformly. To avoid possible contamination of the sample during sputtering, fresh Ar gas (purity of 99.999 %) was continuously flown through the chamber under a pressure maintained at 2.0×10^{-5} Torr.

RESULTS AND DISCUSSION

Figure 1(a) shows that XRR decreases more sharply as the sputtering proceeds, which indicates that the surface roughness W increases as the sputtering proceeds. W is determined by fitting of the experimental XRR curve according to the Parratt's formalism. Figure 1(b) summarizes the evolution of $W(t)$ thus obtained as a function of sputtering time t at various sample temperatures. The linear dependence of $W(t)$ on sputtering time t in a double logarithmic scale plot implies a power-law $W(t) \sim t^{\beta}$, where β is the growth exponent. In between the two temperature 306 and 440 K, well defined crossovers are observed in the dynamic scaling behavior of $W(t)$: After the crossover time t_c, $W(t)$ grows much slower than before t_c, which is also manifested in a small value of $\beta \approx 0.1$ after the crossover time. To examine the lateral structure formed during sputtering, GISAXS is made along the in-plane, close packed [110] direction $q_{\parallel}$ across the specular rod at $q_z = 0.1769$ Å^{-1} (Fig. 2(a)). We observe a specular peak at the center and two satellite peaks on its sides. It may reflect the formation of a laterally ordered structure and its coarsening during the sputtering process. The lateral characteristic length ξ is obtained from the satellite peak positions $\Delta q_{\parallel}^{peak}$ as $\xi = 4\pi / \Delta q_{\parallel}^{peak}$. Figure 2(b) shows the temporal evolution of ξ at various substrate temperatures. $\xi(t)$ is well described by a power law $\xi \sim t^{1/z}$, where $1/z$ is the coarsening exponent. It is noteworthy that ξ does not show any crossover behavior, in contrast to $W(t)$ in Fig. 2(b).

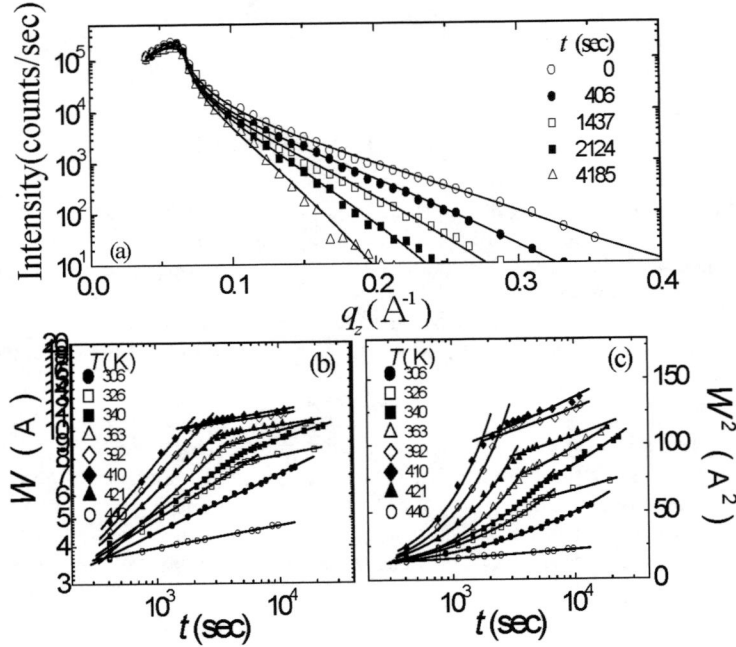

FIGURE 1. (a) Specular x-ray reflectivity (symbols) as a function of the out-of-plane momentum transfer, q_z for increasing t at 421 K. (b)$W(t) \sim t^{\beta}$ plot, and (c)$W^2(t) \sim \log t$ plot at various temperatures.

(a)

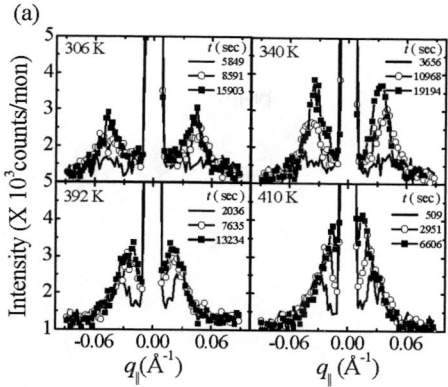

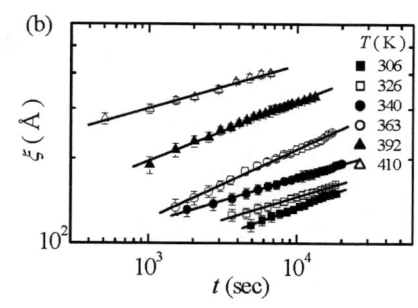

FIGURE 2. (a) Evolution of GISAXS profiles at various temperatures. (b) Lateral characteristic length ξ as a function of sputter time t for different substrate temperatures (symbols). Solid line represent the curves best fit by $\xi(t) \sim t^{1/z}$.

The roughness evolution shown in Fig. 1(b) exhibits distinguished behavior before and after t_c, giving very different β's. While the growth exponent β has relatively large value ~0.2-0.5 before t_c, after which β is reduced to $\beta \sim 0.1$, irrespective of the substrate temperature. Such a small value of β suggests that the surface evolution may belong to the class of the Edwards-Wilkinson (EW) model, in which the square of the surface width W^2 depends on time in logarithmic way. In Fig. 1(c), we find that the experimental data are also nicely fit to the presentation of EW model, $W^2 \sim \log t$.

The crossover behavior thus implies that the nonlinear effect such as the Kardar-Parisi-Zhang (KPZ) term, which manifests itself in the form of large β values, disappears for $t > t_c$. Recently, E. S. Tok *et al.* [11] has performed a kinetic Monte-Carlo simulation for the evolution of sputter-eroded surface based on Sigmund's theory. They also find that $W(t)$ exhibits a crossover similar to what we observe in the experiment. The crossover behaves to the EW scaling in the long time limit. In their model, the EW behavior mainly results from the balance between the erosion due to ion bombardment and the redeposition of sputtered atoms onto the surface. As sputtering proceeds, the surface become rough, and then the sputtered atoms have more chance to be captured at the hill than in the case where the surface is flat. Thus, as the surface becomes rougher, the balance can be achieved more easily. The partial volume of vacancy island generated by ion impact is usually smaller than the mass actually removed, since the vacancy island exposed to vacuum likely encloses vacancies beneath it. If the apparent volume of vacancy island and the redeposited volume are balanced, then the nonlinear term such as the KPZ term $(\nabla h)^2$ vanishes, and the surface is reduced to the EW scaling.

SUMMARY

We have investigated the asymptotic scaling behavior of ion sputtered Pd(001) surface employing *in situ*, real-time x-ray reflectivity (XRR) and grazing incidence small angle x-ray scattering (GISAXS). We have found a definite crossover of surface roughness W and growth exponent β with increasing sputter time t. The morphological evolution is controlled by the mass redistribution mechanism, and the asymptotic behavior of the scaling shows that the surface is well described by a linear model, the Edwards-Wilkinson model.

ACKNOWLEDGMENTS

This work is supported from National Research Laboratory program on synchrotron x-ray technology in GIST. All the experiments were performed at Pohang Accelerator Laboratory (PAL).

REFERENCES

1. S. Facsco *et al.* ,Science **285**, 1551 (1999).
2. G. Costantini *et al.* , Surf. Sci. **416**, 245 (1998).
3. S. Rusponi *et al.* , Phys. Rev. Lett. **81**, 2735 (1998).; S. Rusponi et al. , Phys. Rev. Lett. **78**, 2795 (1997).
4. R. M. Bradly and J. M. E. Harper, J. Vac. Sci. Technol, A **6**, 2390 (1988).
5. Y. Kuramoto, Chemical Oscillations, Waves and Turbulence (Springer, Berlin, 1984); G. I. Sivashinsky and D. M. Michelson, Prog. Theor. Phys. **63**, 2112 (1980).
6. R. Cuerno *et al.* Phys. Rev. Lett. **74**, 4746 (1995).
7. S. Park *et al.* Phys. Rev. Lett. **83**, 3486 (1999).
8. T. C. Kim *et al.* Phys. Rev. Lett. **92**, 246104 (2004).
9. M. Castro *et al.* Phys. Rev. Lett. **94**, 016102 (2005).
10. C. Jayaprakash *et al.* Phys. Rev. Lett. **71**, 12 (1993); V. S. L'vov *et al.* ,Phys. Rev. Lett. **69**, 3543 (1992).
11. E. S. Tok *et. al.* , Phys. Rev. E. **70**, 011604 (2004).

Circular Dichroism in Photoelectrons from Adsorbed Chiral Molecules

J. W. Kim[*], J. H. Dil[†], Th. Kampen[†], and K. Horn[†]

[*]Korea Research Institute of Standards and Science, 1 Doryong-dong, Daejon 305-340, Korea
[†]Fritz-Haber-Institut der Max-Planck-Gesellschaft, Faradayweg 4-6, Berlin 14195, Germany

Abstract. The study of chiral molecules on two dimension is important for an analysis of enantio-selectivity in heterogeneous catalysis aiming to measure and control various biochemical functions. Here we show that such molecules can be identified through circular dichroism in core level photoemission from the stereoisomers of 2,3-butanediol molecules adsorbed on Si(100) and tartaric acids on Cu(110), using circularly polarized x-rays. The dichroic signal changes sign with going from the (R,R) to the (S,S) enantiomer; it is absent in the achiral (R,S) isomer, demonstrating that the dichroism is caused by the chiral molecular environment of the photoionized atoms. This observation demonstrates the possibility of determining molecular chirality in the adsorbed phase.

Keywords: Circular dichroism, Chiral molecules, Adsorption, Photoemission spectroscopy
PACS: 79.60.Dp, 33.55.Ad

INTRODUTION

Chirality in chemistry has an analogy with electron spin in physics. While the electron spin governs magnetic properties of materials, the molecular chirality controls many biochemical functions. This is why the most modern drugs are made of single enantiomers, and their synthesis or separation has been a key technology in chemistry and biology. In particular, the measurement and control of such chiral molecules on two-dimensional system attract great attention on enantio-selective heterogeneous catalysis. In search of a method to control the enantio-selectivity in chemical reactions, recent interest has turned to adsorbed chiral molecules [1]. The detection of chirality in adsorbed molecules is difficult, however, having so far been mostly demonstrated in scanning tunneling microscopy (STM) studies [2-4]. The interaction between polarized light and chiral molecules leads to optical rotation and circular dichroism, and both methods have been widely used in the visible and near UV region, e.g. for structural analysis of biomolecules [5]. The use of higher photon energies, in an analysis of circular dichroism in valence and core level absorption or electron emission, has been less actively pursued. Now, with the availability of high-intensity circularly-polarized tunable x-rays, interest has returned to this topic since it may provide access to specific atomic or molecular orbitals. However, an application of photoelectron spectroscopy to adsorbed molecules has been hindered by the realization that the detection of "true" circular dichroism, i.e. induced by the chiral nature of the molecular environment, may be hampered by the competing effect of circular dichroism in angular distribution (CDAD) brought about by the handedness of the experimental geometry [6].

Here are demonstrated two examples on the studies of adsorbed chiral molecules, i.e. 2,3-butanediol (H_3C-CHOH-CHOH-CH_3) on Si(100) [7] and tartaric acid (HOOC-CHOH-CHOH-COOH) on Cu(110), using circular dichroism in core level photoemission. We have successfully distinguished between circular dichroism from the chirality of molecules, and that induced by the chiral nature of the experimental geometry, through a comparison of dichroism observed under different experimental geometries.

EXPERIMENTALS

All measurements were performed at beam line UE56/2 PGM-1 of BESSY (Berliner Elektronenspeicherring GmbH) in Berlin. The CD measurement has been performed using right and left circularly-polarized (RCP and LCP)

CP879, *Synchrotron Radiation Instrumentation: Ninth International Conference*,
edited by Jae-Young Choi and Seungyu Rah
© 2007 American Institute of Physics 978-0-7354-0373-4/07/$23.00

light at the UE56/2-PGM1 beam line at BESSY II. The degree of circular polarization of the LCP and RCP light was above 98 % at all photon energies used here. Photoelectron signals ejected from C $1s$ core levels were recorded using a PHOIBOS 100 electron energy analyzer equipped with a 2D CCD detector (Specs GmbH) permitting to record photoelectrons over a 14° angular range with about 0.1° angular resolution. Special attention was paid to the influence of the geometrical CDAD effect, which occurs when the direction of the photon beam (**q**), surface normal (**n**), and electron emission (**k**) in Fig. 1(a) are not coplanar, *i.e.* have an "experimental handedness". All data were collected only in a coplanar geometry within an error of 0.5°. The angle between photon beam and electron emission was 54 ± 7°.

The clean Si(100) surface was obtained by flashing a few times to 1400 K after several hours of outgassing at 800 K. The 2,3-butanediol was adsorbed on the Si(100) surface to a saturated amount at room temperature. A Cu(110) crystal was cleaned by repeated cycles of Ar$^+$ sputtering (800 eV) and annealing to 500 K. The cleanliness and surface ordering of the initial substrates were confirmed repeatedly by the absence of impurities in the core level spectra and sharp LEED patterns corresponding to the clean surfaces. Three different optical isomers of tartaric acids (Chiron AS, 99 %) were evaporated from a home-made Knudsen cell at 355 K in ultrahigh vacuum onto the clean Cu(110) surface.

RESULTS AND DISCUSSION

Adsorption of 2,3-butanediols on Si(100)

As a Si(100) offers the well-known dimer reaction site, the adsorption of the 2,3-butanediol on the Si(100) makes a six-membered ring by the formation of two Si-O bonds after O-H bond breaking [8,9]. This interaction is analogous to the (4+2) cycloaddition or Diels-Alder reactions. Since 2,3-butanediol has two identical -(CHOH-CH$_3$) units, the molecule is expected to form with C$_2$ rotational symmetry on the Si(100)-2×1 surface.

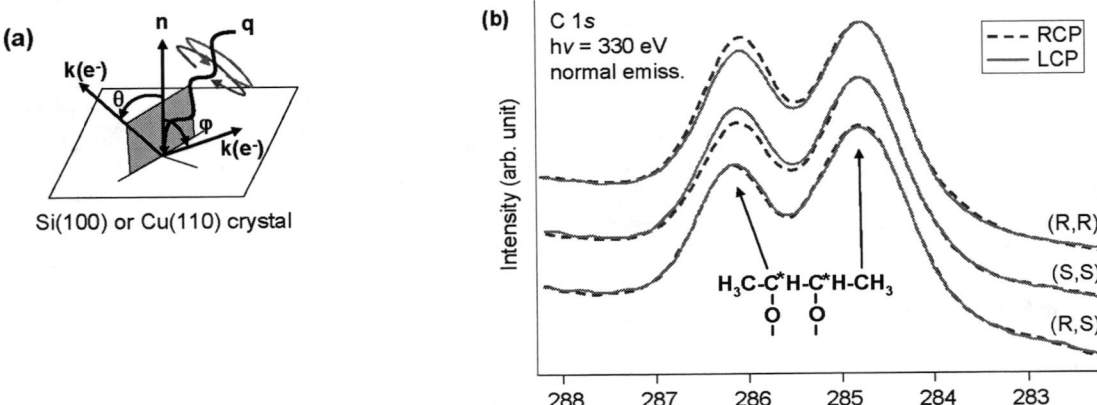

FIGURE 1. (a) Schematic display of measurement geometry. (b) Photoelectron spectra of the C $1s$ core levels for the adsorbed stereoisomers of 2,3-butanediol excited by right (dashed) and left-circularly polarized (solid line) light. Dichroism is observed on the higher-binding energy component attributed to emission from the carbon atoms in the chiral center from the (R,R) and (S,S) forms. The dichroic effect changes sign either with photon polarization or molecular handedness. No dichroism is observed in the achiral (R,S) form.

The two types of carbon atoms in 2,3-butanediol, *i.e.* the one bonded to the oxygen atom, and another in the CH$_3$ group, are easily distinguished since they exhibit a difference in C $1s$ binding energies of 1.3 eV, on account of their different chemical environment. This provides the opportunity to measure an atom-specific CD effect. The data for the C $1s$ core level from (R,R)-2,3-butanediol under RCP (blue) and LCP (red) light are shown in the topmost spectra of Fig. 1(b). These spectra were recorded in normal emission at a photon energy of hv = 330 eV with an incident angle of 54° from the surface normal. A sizeable dichroic effect occurs in the peak arising from the carbon atom bonded to oxygen. As seen from the middle spectra, an asymmetry of equal magnitude but opposite sign occurs when the (S,S) enantiomer, the mirror image of the (R,R)-form, is investigated. Finally, no asymmetry occurs when the achiral (R,S) isomer is adsorbed on the surface as shown in the bottom spectra. These observations are a clear proof that this dichroic signal is caused by the chiral nature of the environment of the respective carbon atom as discussed in more detail below. The C $1s$ core level is unlikely to feel the chiral nature of its molecular environment; the asymmetry then arises from a transition of the photoexcited electron from the core level into an unoccupied or continuum state. The influence of the latter determines the sign and magnitude of the circular dichroism [10]. Considering the matrix element governing the photoemission intensity, the strongest transitions will occur into those unoccupied levels that are localized near the core ionized carbon atom, *i.e.* those that exhibit chirality as a result of the electronic charge distribution around them. The effect is proportional to the interference of pairs of dipole matrix elements which differ by the signs of all projections of orbital momenta. For non-chiral molecules having a plane of symmetry, these differences are equal to zero. In fact, no dichroism is found in the achiral enantiomer [bottom spectra of Fig. 1(b)] where the emission from the R and S forms of carbon atoms in a molecule cancels.

Angular Distribution in Circular Dichroism from Tartaric Acids on Cu(110)

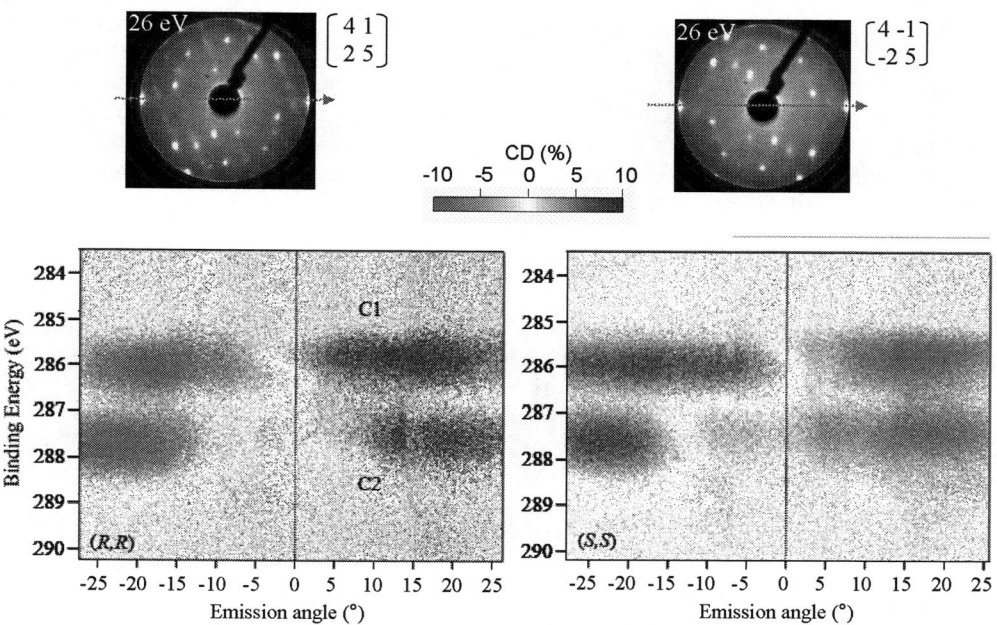

FIGURE 2. Circular dichroism images from the C $1s$ region along the substrate [001] direction, from the high coverage monotartrate phase, for (R,R) on the left and (S,S) tartaric acid on the right. The respective LEED patterns for the (4 1, 2 5) (R,R) and (4 -1, -2 5) (S,S) are shown on above. The asymmetry values scaled on the color table bar are defined like as $(I^{RCP} - I^{LCP})/(I^{RCP} + I^{LCP})$.

There create two kinds of stable phases upon adsorption of tartaric acid on Cu(110) at an elevated temperature: a low coverage bitartrate phase (with a {9 0, 1 2} structure in matrix notation of the LEED), and a high coverage monotartrate phase with {4 1, 2 5} structure [3]. In this report, only the results for high coverage phase are presented. In the tartaric acid molecule, there are two different types of carbon atom from a chemical point of view: one in the carboxylic group, and one in the alcohol group. Structural studies have demonstrated that tartaric acid is bound to the Cu surface atoms through the carboxylic oxygen atoms after deprotonation. The different carbon atoms in the adsorbed molecule can be readily identified in core level photoemission where two peaks are observed at 286.60 (C1) and 288.15 eV (C2), respectively, similar to a previous study on glycine adsorbed on Cu(110) [11].

By utilizing the parallel detection capabilities of our photoelectron energy analyzer, we obtain images of photoelectron intensity rather than separate spectra. This provides a graphic representation of changes induced by a reversal of photon helicity. The magnitude of circular dichroism is defined as the asymmetry $A = (I^{RCP} - I^{LCP})/(I^{RCP} + I^{LCP})$, where I is the measured intensity under the respective light helicity. Photoemission images in an angular range of $\pm7°$ are recorded with RCP and LCP light and are subtracted from one another after a background normalization. From a set of images at different angle settings, images such as in Fig. 2 are assembled, where blue means a positive and red a negative value as indicated in the color table bar. The binding energy range encompasses the C1 and C2 features mentioned above, clearly seen by the two horizontal bars of intensity. In Fig. 2 we compare the asymmetry values in the two C $1s$ components for the (R,R) and (S,S) enantiomers of tartaric acid in the high coverage monotartrate (4 1, 2 5) phase (left and right side, respectively). This structure gives rise to the LEED patterns shown on above each asymmetry plot; they are mirror images of one another, brought about by the interplay between the molecular structure of the adsorbate and the Cu(110) surface geometry as demonstrated through STM [3]. Two effects are immediately evident: emission from both carbon atoms exhibits circular dichroism, and the asymmetry reverses sign with emission angle in the plane spanned by the light and the surface normal. The asymmetry is rather small in normal emission, but reaches values of up to 10 % at emission angles around 20° and beyond. A striking observation comes from a comparison of circular dichroism from the (R,R) and (S,S)-forms, however. A comparison of circular dichroism from the (R,R) and (S,S) forms of tartaric acid adsorbed on Cu(110) yields the clear result that, within a good approximation, the intensity patterns (blue and red contrast) of Fig. 2 are mirror images of one another with respect to the 0° emission angle line. This is a most important result since it demonstrates that the observed dichroism is due to the chiral nature of the adsorbed molecule, and is not related to the geometrical CDAD. This is because in our experiment, in which the three axes (**q**, **n**, and **k**) are all coplanar [Fig. 1(a)], the CDAD will be zero if that plane coincides with a mirror plane of the (nonchiral) surface as stated in the previous section. Within this mirror plane, the observed dichroism is therefore a direct consequence of the chirality of the adsorbed molecule. This is also clear from the reversal of asymmetry between the (R,R) and (S,S) forms of tartaric acid. The above results are readily understood if one considers that in optical transitions at low photon energies, parity-violating processes are negligible. The result of an experiment involving chiral molecules must therefore be invariant upon an application of the parity operation which changes both the photon helicity and the handedness of the molecule. This invariance causes the interaction of RCP light with a "right-handed" enantiomer to yield an identical result as LCP light wit a "left-handed" enantiomer.

CONCULSION

We have observed a clear circular dichroism in core-level photoelectrons from two kinds of enantiomer pairs of 2,3-butanediol adsorbed on Si(100) and tartaric acids on Cu(110) surfaces. We are able to distinguish this effect from a dichroism induced by the handedness of geometric arrangement of the incident electromagnetic radiation and the outgoing photoelectron path, since the asymmetry changes sign with enantiomer and photon polarization. In view of the increasing importance of an analysis of biologically active surface species, our observation may provide a path to analyzing chiral centers in complex adsorbed molecules using photoemission spectroscopy.

ACKNOWLEDGMENTS

This work is supported by EU under the ARI program. J.W.K. acknowledges supports by the Alexander von Humboldt Foundation and by grant No. R01-2006-000-10920-0 from the Basic Research Program of the Korea Science & Engineering Foundation.

REFERENCES

1. R. Raval, *Cattech* **5**, 12–28 (2003).
2. A.Kühnle, T. R.Linderoth, B.Hammer, and F. Besenbacher, *Nature* **415**, 891–893 (2002).
3. M. Ortega Lorenzo, C. J.Baddeley, C.Muryn, R. Raval, *Nature* **404**, 376–379 (2000).
4. R. Fasel, M. Parschau, and K.-H. Ernst, *Nature* **439**, 450–452 (2006).
5. N. Berova, K. Nakanishi, R. W. Woody, "Circular Dichroism - Principles and Applications", 2nd edition, New York: Wiley-VCH, 2000.
6. G. Schönhense, *Physica Scripta* T. **31**, 255–275 (1990).
7. J. W. Kim, M. Carbone, J. H. Dil, M. Tallarida, R. Flammini, M. P. Casaletto, K. Horn, and M. N. Piancastelli, *Phys. Rev. Lett.* **95**, 107601 (2005).
8. J. W. Kim, M. Carbone, M. Tallarida, J. H. Dil, K. Horn, M. P. Casaletto, R. Flammini, and M.N. Piancastelli, *Surf. Sci.* **559**, 179–185 (2004).
9. K. Seino and W.G. Schmidt, Surf. Sci. **585**, 191–196 (2005).
10. N. Chandra, *Phys. Rev.* A **39**, 2256–2257 (1989).
11. J. Hasselström, P. Karis, M. Weinelt, N. Wassdahl, A. Nilsson, M. Nyberg, L. G. M. Pettersson, M. G. Samant, and J. Stöhr, *Surf. Sci.* **407**, 221–236(1998).

A High-Resolution Soft X-Ray Photoemission Apparatus Combined with a Laser Molecular-Beam Epitaxy System at SPring-8 BL17SU

K. Horiba*, R. Eguchi*, N. Kamakura*, K. Yamamoto*, M. Matsunami*, Y. Takata*, Y. Senba[†], H. Ohashi[†] and S. Shin*,**

*RIKEN SPring-8 Center, Sayo-cho, Sayo-gun, Hyogo 679-5148, Japan
[†]JASRI/SPring-8, Sayo-cho, Sayo-gun, Hyogo 679-5198, Japan
**Institute for Solid State Physics, The University of Tokyo, Kashiwa, Chiba 277-8581, Japan

Abstract. We have constructed a high-resolution synchrotron-radiation photoemission apparatus combined with a laser molecular-beam epitaxy (laser MBE) system in order to investigate the electronic structure of thin films. The system is installed at the newly-built soft x-ray undulator beamline BL17SU of SPring-8. Single crystal thin films fabricated by laser MBE can be transferred quickly into the photoemission chamber under ultra-high vacuum condition. The photoemission spectrometer is equipped for high-throughput and high energy-resolution angle-resolved photoemission measurements. High energy-resolution and stability of the beamline optics enable us to achieve the best total energy resolution of 51 meV at 867 eV excitation. The performance and characteristics of the system is demonstrated by showing results on $LaNiO_3$ thin films.

Keywords: photoemission spectroscopy, laser molecular beam epitaxy
PACS: 79.60.-i, 71.20-b, 73.21.-b

INTRODUCTION

Photoemission spectroscopy (PES) is a powerful experimental technique for determining the electronic structure of transition metal oxides, especially with a layered structure. Recent progress in high-resolution angle-resolved PES (ARPES) enables us to address the band structure near the Fermi level of such materials and its relation to physical properties [1, 2, 3]. In contrast, there are few PES studies on strongly-correlated oxides of a three-dimensional crystal structure with no cleavable plane, which nonetheless exhibit fascinating electrical and magnetic properties such as colossal magnetoresistance, metal-insulator transitions, and spin-charge ordering [4, 5]. This is simply because it is difficult to obtain well-ordered surfaces of such materials by conventional surface preparation such as sputtering and annealing procedures or cleavage of single-crystal samples.

Laser molecular-beam epitaxy (laser MBE), or the pulsed laser deposition (PLD) technique with *in situ* reflection high energy electron diffraction (RHEED) monitoring, is one of the best methods for fabricating the oxides with well-ordered surfaces [6]. A wide variety of high-quality thin films have been successfully grown by this technique. Therefore, *in situ* PES measurements on transition metal oxides grown by the laser MBE technique are indispensable for paving the way to detailed investigations of the electronic structure of three-dimensional strongly-correlated materials. Recently, "*in situ* ARPES - Laser MBE" system, comprising of a high-resolution angle-resolved photoemission apparatus and laser-MBE equipment in ultra high vacuum (UHV), was constructed at a vacuum ultraviolet (VUV) beamline for ARPES measurements on three-dimensional transition metal oxides [7].

In addition, recent progress in the development of high photon flux soft x-ray (SX) undulators at third-generation synchrotron-radiation (SR) light sources enables us to carry out ARPES study using high photon energy SX [8]. PES studies of transition metal compounds using SX light sources have advantages over using VUV light sources for the following reasons: (i) The probing depth is large, owing to large mean free paths of the emitted high kinetic energy electrons. Recently, correlation-induced changes at the surface of $3d$ valence electron systems has been reported and questions arose as to the reliability of surface-sensitive PES spectra for addressing the bulk electronic structure of strongly-correlated materials[9, 10, 11]. (ii) We can directly observe the transition-metal $3d$ band structure which is directly related to the strongly-correlated properties, since the photoionization cross section of transition metal $3d$ orbitals are much larger than those of ligand $2p$ orbitals at SX energies [12].

Thus, it is important to perform the *in situ* SX-ARPES measurements on strongly-correlated thin films fabricated

CP879, *Synchrotron Radiation Instrumentation: Ninth International Conference*,
edited by Jae-Young Choi and Seungyu Rah
© 2007 American Institute of Physics 978-0-7354-0373-4/07/$23.00

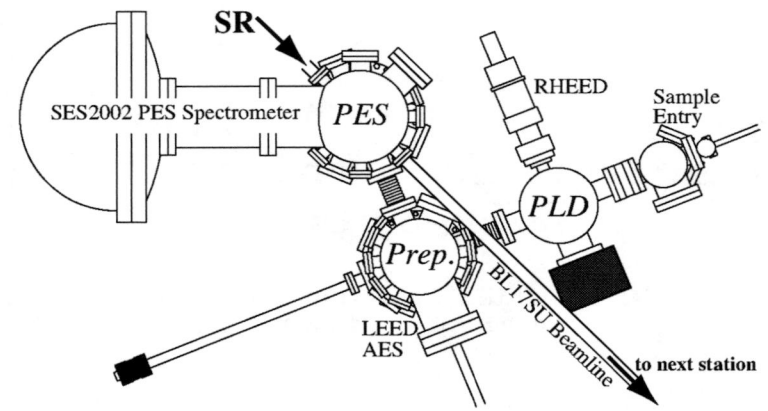

FIGURE 1. A schematic drawing of the "*in situ* SX-ARPES - Laser MBE" system.

by laser MBE for revealing the electronic structure of three-dimensional strongly-correlated materials. In order to realize this, we have constructed a high-resolution angle-resolved photoemission apparatus, which is connected to the laser-MBE equipment in UHV. The system is installed at a SX undulator beamline BL17SU of SPring-8 as an end-station.

SYSTEM DESCRIPTION

System Overview

Figure 1 shows a schematic view of the "*in situ* SX-ARPES - Laser MBE" system [13]. This system consists of four interconnected chambers: a sample entry chamber, a laser MBE chamber, a preparation chamber, and a photoemission chamber. The photoemission chamber is connected to a soft X-ray undulator beamline BL17SU at SPring-8. These chambers are connected to each other in UHV and each chamber can be isolated by gate valves. Typical sample growth and measurement sequence is as follows. At first, a substrate mounted on a sample holder is loaded into the sample entry chamber and transferred to the laser MBE chamber. Thin films are grown on the substrate by pulsed laser deposition while observing the RHEED intensity oscillations. The thin film samples are then transferred to the sample characterization chamber where their surface structure and surface cleanness can be characterized by low energy electron diffraction (LEED) and Auger electron spectroscopy. After surface characterization, the sample is moved with a transfer rod into the photoemission measurement stage. The sample transfer is carefully operated under UHV of $< 10^{-8}$ Pa in order to avoid the sample surface contamination during the transfer.

Laser MBE Apparatus

A combinatorial laser MBE chamber (PASCAL: Mobile Combi-PLD) is attached to the system. The chamber was specially designed for the use at the synchrotron beamlines. The chamber includes a rotatable sample stage, a four-target carousel for ablation targets, two movable masks for combinatorial or high-throughput thin film library fabrication, and a differentially-pumped RHEED system. A compact pulsed Nd-YAG laser with maximum pulse energy of 100 mJ at the third harmonic (355 nm) is used for ablation. For sample heating, we use a continuous-wave (CW) semiconductor diode laser. Light from the laser is delivered to the chamber through a 600 μm core diameter fiber and focused on the back side of the sample holder [14]. The laser operates at 808 nm and can deliver up to 100 W of CW power. The high-power CW laser allows us to heat the substrate up to 1400 °C while keeping the heater outside of the vacuum system. Film growth can be monitored with a real-time RHEED specular intensity monitor, which allows the deposited film thickness to be controlled with submonolayer accuracy. All aspects of the deposition system are computer controlled, including all the sample and target control motors and the substrate temperature. Temperature feedback is obtained from an optical pyrometer.

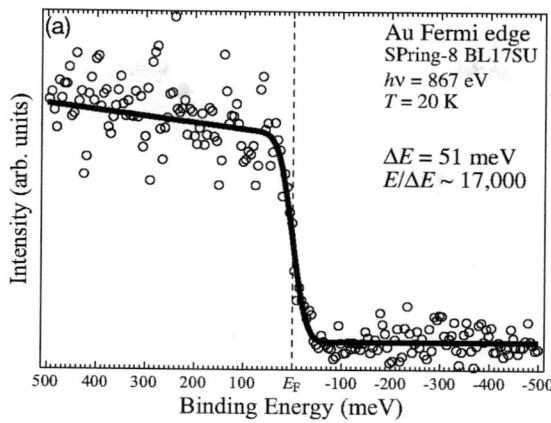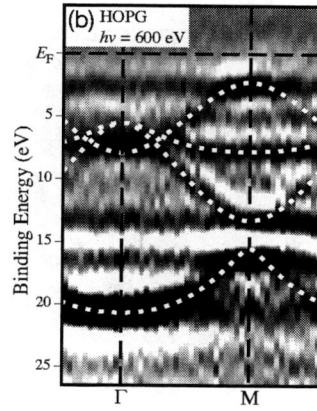

FIGURE 2. (a) (open circle) High-resolution photoemission spectrum near E_F of Au measured with the photon energy of 867 eV. (solid line) Fitting result of the Au Fermi edge using a Fermi-Dirac function convoluted with a Gaussian function of 51 meV representative of instrumental (photon + analyzer) resolution function. (b) Experimental band structure of highly oriented pyrotytic graphite obtained from the ARPES measurements with photon energy of 600 eV at room temperature.

Photoemission Apparatus

For the electron energy analyzer, a GAMMADATA SCIENTA SES-2002 photoelectron energy analyzer with a multi-channel detector is adopted to achieve a high energy- and angular-resolution as well as high count rates of photoelectrons. The angle between the incident direction of SR and the photoemission spectrometer is set to 45°. In order to optimize the energy resolution of our instrument, the analyzer has been set in a μ-metal chamber for screening of external magnetic fields. In order to especially minimize the influence of a leakage magnetic field to photoelectrons, an additional cylindrical μ-metal magnetic shield inside the chamber surrounds the electronic lens of the analyzer and is tightly coupled with the μ-metal shield of the analyzer. The photoemission spectra have been recorded using soft X-rays at BL17SU of SPring-8. The available photon-energy is in the range from 200 eV to 2000 eV. The solid sample can be cooled to a temperature below 20 K. The temperature is measured with an accuracy of ~ 1 K using a calibrated Si-diode sensor placed beside the sample base plate. The sample temperature is regulated by optimizing a proportional-integral-derivative (PID) controlled ceramic heater. The sample temperature can be controlled from 20 K to 400 K. The present performance is demonstrated in Fig. 2 (a) and (b) by the Fermi edge profile of Au and ARPES spectra of graphite, respectively. For photon energy of 867 eV, we obtained a total energy resolution (analyzer and photon) of 51 meV by fitting of Au Fermi edge. Clear dispersive bands in the spectra of graphite with photon energy of 600 eV demonstrate that SX-ARPES measurements can be reliably carried out using this system.

EXPERIMENTAL RESULTS

In this section, we demonstrate the capabilities of this system by *in situ* SX-PES results of LaNiO$_3$ (LNO) thin films. LNO thin films were grown epitaxially on SrTiO$_3$ (STO) substrates. Sintered stoichiometric LNO pellets were used as ablation targets. A Nd: YAG laser was used for ablation in its frequency-tripled mode ($\lambda = 355$ nm) at a repetition rate of 1 Hz. The wet-etched STO (001) substrates were annealed at 900 °C at an oxygen pressure of 1×10^{-4} Pa before deposition. The substrate temperature was set to 650 °C and the oxygen pressure was 10 Pa during the deposition. The fabricated LNO thin films were subsequently annealed at 400 °C for 30 minutes in atmospheric pressure of oxygen to remove oxygen vacancies. After cooling the sample to below 100 °C and evacuating the growth chamber, the surface morphology and crystallinity of the fabricated LNO thin films was checked by *in situ* observation of reflection high-energy electron diffraction (RHEED) pattern. Sharp streak patterns shown in Fig. 3 (a) indicate the smooth and high-quality single crystal surface of the fabricated LNO thin films.

Figure 3 (b) shows *in situ* PES and x-ray absorption (XAS) spectra of the fabricated LNO thin film. In contrast to the previous PES results on polycrystalline LNO surfaces [16], *in situ* PES spectra show sharp structure and strong intensity of a peak crossing E_F, reflecting the metallic ground state and large effective electron mass of LNO [17]. This result suggests the importance of *in situ* PES measurements for revealing the intrinsic electronic structure of

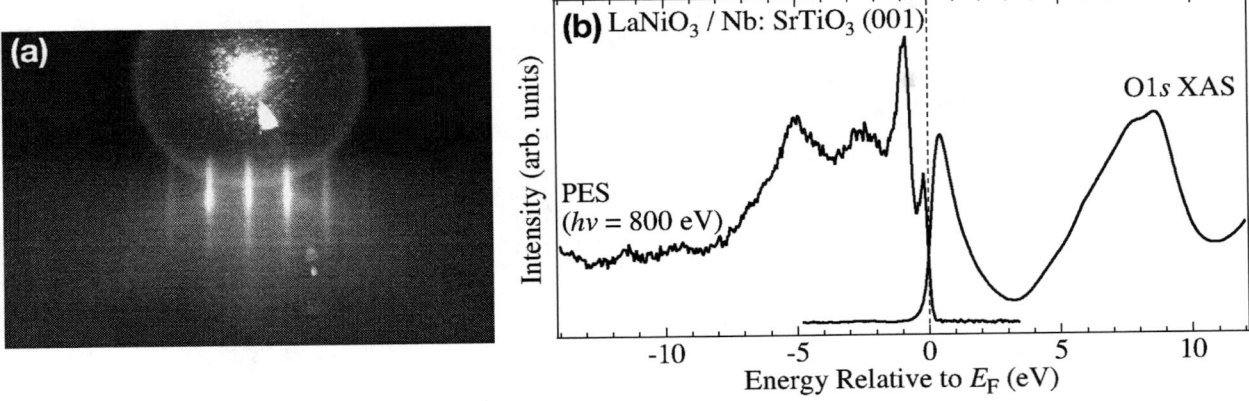

FIGURE 3. (a) *In situ* RHEED pattern and (b) *in situ* PES and XAS spectra of LNO thin films.

strongly-correlated oxides.

CONCLUSION

In order to investigate the electronic structure of strongly-correlated materials with no cleavable plane, we have constructed a high-resolution synchrotron-radiation photoemission apparatus combined with a laser MBE system. The system is installed at the newly-built soft X-ray undulator beamline BL17SU of SPring-8. Single crystal thin films fabricated by laser MBE can be transferred quickly into the photoemission chamber under ultra-high vacuum condition. High energy-resolution and stability of the beamline optics enable us to achieve the best total energy resolution of 51 meV at 867 eV excitation. The results of LNO demonstrate the importance of *in situ* PES measurements for revealing the intrinsic electronic structure of strongly-correlated oxides.

REFERENCES

1. Z. -X. Shen and D. S. Dessau, *Phys. Rep.* **253**, 1 (1995).
2. A. Damascelli, Z. Hussain, and Z. -X. Shen, *Rev. Mod. Phys.* **75**, 473 (2003).
3. Y. -D. Chuang, A. D. Gromko, D. S. Dessau, T. Kimura, and Y. Tokura, *Science* **292**, 1509 (2001).
4. M. Imada, A. Fujimori, and Y. Tokura, *Rev. Mod. Phys.* **70**, 1039 (1998).
5. *Colossal Magnetoresistive Oxides*, Advances in Condensed Matter Science, Vol. 2, edited by Y. Tokura (Gordon and Breach, Amsterdam, 2000).
6. H. Koinuma *et al.*, *Appl. Surf. Sci.* **109/110**, 514 (1997).
7. K. Horiba *et al.*, *Rev. Sci. Instrum.* **74**, 3406 (2003).
8. N. Kamakura, Y. Takata, T. Tokushima, Y. Harada, A. Chainani, K. Kobayashi, and S. Shin, *Europhys. Lett.* **67**, 240 (2004).
9. A. Sekiyama, T. Iwasaki, K. Matsuda, Y. Saitoh, Y. Onuki, and S. Suga, *Nature* **403**, 396 (2000).
10. K. Maiti, P. Mahadevan, and D. D. Sarma, *Phys. Rev. Lett.* **80**, 2885 (1998).
11. S. -K. Mo *et al.*, *Phys. Rev. Lett.* **90**, 186403 (2003).
12. J. J. Yeh and I. Lindau, *At. Data Nucl. Data Tables* **32**, 1 (1985).
13. K. Horiba *et al.*, *J. Electron Spectrosc. Relat. Phenom.* **144-147**, 1027 (2005).
14. S. Ohashi, M. Lippmaa, N. Nakagawa, H. Nagasawa, H. Koinuma, and M. Kawasaki, *Rev. Sci. Instrum.* **70**, 178 (1999).
15. Y. Aiura, H. Bando, T. Miyamoto, A. Chiba, R. Kitagawa, S. Maruyama, and Y. Nishihara, *Rev. Sci. Instrum.* **74**, 3177 (2003).
16. S. R. Barman, A. Chainani, and D. D. Sarma, *Phys. Rev. B* **49**, 8475 (1994).
17. K. Sreedhar *et al.*, *Phys. Rev. B* **46**, 6382 (1992).

Study on Formation of ZnO/SiC Interface by SRPES

P.S. Xu[*], C.W. Zou, B. Sun, Y.Y. Wu, F.Q. Xu, H.B. Pan

NSRL, University of Science and Technology of China, Hefei, 230029, China

Abstract. The adsorption and the thermal oxidation of Zn on 6H-SiC surface and the interface formation of ZnO/SiC have been investigated by SRPES and XPS. With increasing of Zn coverage, the surface exhibits metallic characteristic. When the deposited Zn film is annealed at 180°C in oxygen flux, it could be partly oxidized. While it is annealed at 600°C, the total deposited Zn atoms could be oxidized to form ZnO and a thin layer of silicon oxide exists at the interface of ZnO/SiC.

Keywords: SiC, ZnO, interface, synchrotron radiation, photoemission
PACS: 73.20. –r, 73.40. Lq, 79.60. –i, 79.60. Jv

INTRODUCTION

In the recent decades, ZnO has attracted much attention due to its high thermal and chemical stability, direct wide band gap, large exciton binding energy and its potential application of photoelectron devices [1-3].

Many techniques have been employed in the preparation of ZnO thin films [4]. Recently, an easy and useful method for ZnO preparation by thermal oxidation of metallic Zn has been applied and obtained great success [5]. This method can prepare high-quality ZnO thin film with excellent crystallinity and strong UV emission. However, the initial absorption, oxidation and interface formation of ZnO and substrate of this method have not been investigated.

In this paper, we study the adsorption as well as the thermal oxidation of Zn on 6H-SiC surface and the interface formation of ZnO/SiC by using synchrotron radiation photoelectron spectroscopy (SRPES) and X-ray photoemission (XPS).

EXPEIMENTAL

The experiment is performed at the surface physics station of NSRL. The experimental station is mainly composed of VG ARUPS10 angle resolved photoelectron spectrometer system. The beamline covers the energy range from 10 to 300eV and the energy resolution (E/ΔE) is better than 1000. More details of the experiment station and the related beamline are described elsewhere [6]. The substrate of 6H-SiC single crystal in this experiment is from Cree. To obtain clean surface, the sample is treated chemically using the conventional method. Then the sample is introduced into the UHV chamber followed by degassing at 550°C for about five hours. After above treatment, the clear patterns of substrate surface are observed in low energy electron diffraction (LEED). High purity Zn (~99.99%) granules are evaporated by K-cell at 200°C in the deposition chamber and the evaporation rate is about 0.1nm per 16 seconds measured by quartz crystal thickness monitor. The total film thickness is around 2.5nm. Then the Zn films are exposed in oxygen flux of 2.0×10^{-6} mbar and annealed at 180°C and 600°C for 10 min respectively. At each stage the sample is transferred to the analysis chamber for SPRES and XPS measurements.

RESULTS AND DISCUSSION

Figure 1 shows the valence band (VB) spectra with the photon energy of 32 eV at different stages during the adsorption and the thermal oxidation of Zn on 6H-SiC surface. When metallic Zn of 0.5nm is deposited on 6H-SiC

---

[*] Corresponding author: Tel. 0551-3602037, Fax. 0551-5141078, e-mail. psxu@ustc.edu.cn

CP879, *Synchrotron Radiation Instrumentation: Ninth International Conference*,
edited by Jae-Young Choi and Seungyu Rah
© 2007 American Institute of Physics 978-0-7354-0373-4/07/$23.00

surface, a broad peak located at −10.0eV appears, which is regarded as Zn3d peak. When the deposition of Zn increases up to 2.5 nm, the intensity of Zn3d peak enhances obviously. In contrast, the VB of substrate is depressed

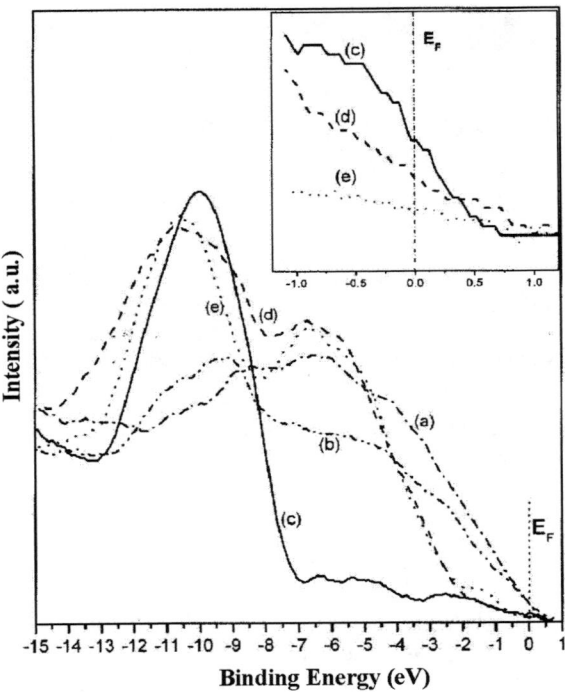

FIGURE 1. The valence band spectra with the photon energy of 32 eV at different stages (a) 6H-SiC after chemical treatment (b) 0.5nm Zn deposition (c) 2.5nm Zn deposition (d) annealing at 180 °C in oxygen flux (e) annealing at 600 °C in oxygen flux

greatly. After the deposited Zn film is annealed at 180°C in oxygen flux of 2x10-6mbar, the Zn3d peak becomes broader, which includes the components of metallic Zn as well as oxidized Zn and another peak at −6.6eV appears,

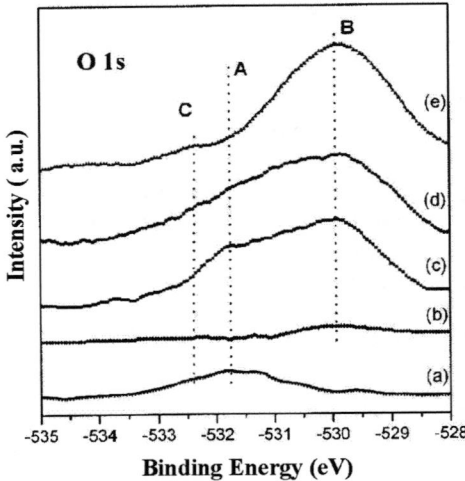

FIGURE 2 The O1s XPS spectra at different stages (a) 6H-SiC substrate; (b) 2.5nm metallic Zn deposition; (c) exposed in oxygen flux; (d) annealed at 180°C and (e) annealed at 600°C in oxygen flux.

which is originated from the contribution of O2p of ZnO. When the annealing temperature is increased up to 600°C, the Zn3d peak becomes sharper, which shows that it is mainly composed of the component of ZnO. The VB appears the characteristic of ZnO [7], which indicates that the ZnO film is formed on the surface. The inset of this figure shows the Fermi edges for three stages. When 2.5nm Zn film has been deposited onto the substrate, the Fermi edge has a very clear step, which shows the high density of states (DOS) at Fermi edge and reflects the metallic characteristic of surface. After annealing at 180°C in oxygen flux, the height of the step becomes low. It means that the DOS at Fermi edge decreases, which indicates that the metallic surface is partly oxidized and the metallic characteristic is quickly weakened. When the annealing temperature increases to 600°C in oxygen flux, the step at Fermi edge almost disappears. It means that the DOS at Fermi edge is very low, which indicates that the metallic Zn

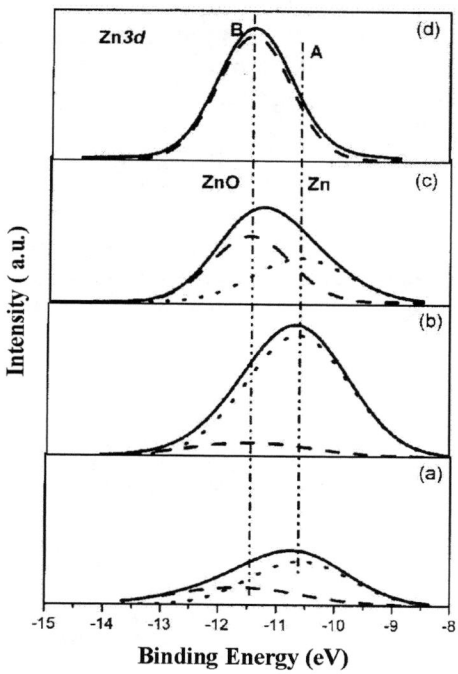

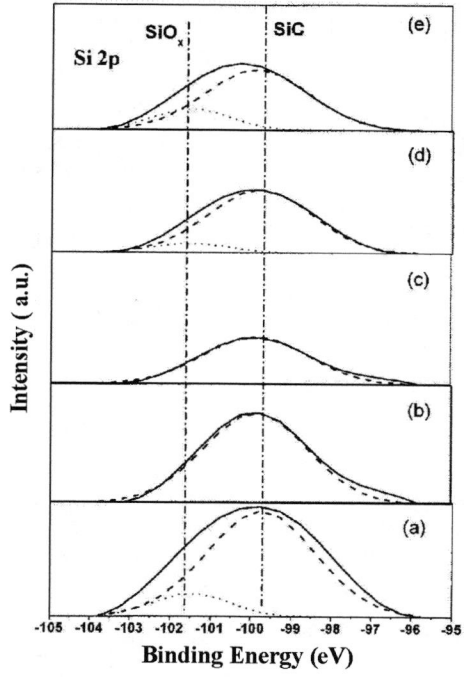

FIGURE 3 The Zn3d spectra at different stages (a) 0.5nm Zn film deposition (b) 2.5nm Zn film deposition (c) annealing at 180°C in oxygen flux (d) annealing at 600°C in oxygen flux.

FIGURE 4 The Si2p spectra at different stages (a) 6H-SiC after chemical treatment (b) 0.5nm Zn film deposition (c) 2.5nm Zn film deposition (d) annealing at 180°C in oxygen flux (e) annealing at 600°C in oxygen flux.

film is oxidized completely to form ZnO.

O1s XPS spectra with X-ray of Mg Kα at each stage are shown in Fig.2. From Fig.2 (a), we can see a broad peak A at 531.7eV, which is from Si oxides and hydroxides remained on SiC after chemical treatment. When metallic Zn of 2.5 nm is deposited on 6H-SiC surface, the peak A disappears, while another broad and very weak peak B at 529.8eV appears. It can be explained that the deposited Zn atoms capture and bond with the oxygen atoms remained on the substrate to form ZnO clusters. Therefore, the peak B is from ZnO. When the deposited Zn is exposed in oxygen flux, the intensity of peak B increases fast, which illustrates that more Zn atoms bond with O atoms in oxygen ambience. At the same time, the intensity of peak A also increases, which indicates that more oxygen is adsorbed on the surface. After annealing at 180°C in oxygen flux, the intensity of peak B continue to increase while peak A decreases, which illustrates that more O atoms including adsorbed oxygen bond with Zn atoms to form ZnO. Besides, we can also find a new weak peak C at 532.4eV, which is from Si oxides. It indicates that during annealing in oxygen flux, not only covered Zn atoms could be oxidized to form ZnO, but the substrate of SiC could also be oxidized slightly to form Si oxides. Further annealing at 600°C in oxygen flux, the intensity of peak B increases very fast, while peak A disappears completely. It illustrates that after annealing at high temperature in oxygen flux, all Zn atoms combine with O atoms to form ZnO and the spectrum is mainly composed of the component of ZnO. We can also find that the intensity of peak C enhances, which indicates that the substrate is more oxidized.

Zn3d spectra with photon energy of 32 eV at different stages are shown in Fig. 3. The peak A and peak B are Zn3d emission from metallic Zn and ZnO respectively. When metallic Zn of 0.5nm is deposited on 6H-SiC surface, the peak is mainly composed of A component, but B component can still be seen, which illustrates that the metallic Zn has been a little oxidized by the adsorbed oxygen atoms remained on the substrate. With increasing of Zn deposition, the intensity of peak A increases rapidly, but the weak peak B still exists. After annealing at 180°C in oxygen flux, the peak B increases quickly. It indicates that the Zn film is partly oxidized. However comparing with the spectrum in 3(b), the total intensity decreases unexpectedly. It might be because of the evaporation of metallic Zn due to its low evaporation temperature in UHV. Further annealing at 600°C in oxygen flux, just B component can be seen, while A component almost disappears, which illustrates that Zn film has been completely oxidized to form ZnO.

In order to study the effect of adsorption and thermal oxidation of Zn on substrate, Si2p spectra with photon energy of 150 eV at different stages have been taken as shown in Fig.4. The Si2p peaks of SiC and SiOx have been displayed in the figure respectively. There is a weak peak of SiOx on 6H-SiC substrate after chemical treatment, which shows that a few amount of Si oxides remained on the substrate. When metallic Zn of 0.5nm is deposited on 6H-SiC surface, Si 2p peak of SiOx almost disappears, which indicates that the deposited Zn atoms capture the oxygen atoms remained on the substrate. When the deposition of Zn increases, the peak intensity decreases due to more Zn coverage. However the peak position has no change. While annealing at 180°C in oxygen flux, a very weak peak of SiOx appears, which illustrates that the substrate could be oxidized after annealing. However, comparing Fig.4 (d) with Fig.4 (c), the total peak intensity increases, which is because of part of Zn atoms escaping from surface due to its low evaporation temperature in UHV, which enhances the intensity of substrate. When the annealing temperature increases to 600°C, the peak intensity of SiOx increases, which indicates that more SiOx have been produced.

CONCLUSION

The adsorption and the thermal oxidation of Zn on 6H-SiC surface and the interface formation of ZnO/SiC have been investigated. With increasing of Zn coverage, the surface exhibits metallic characteristic. When the deposited Zn film is annealed at low temperature in oxygen flux, it could be partly oxidized. While it is annealed at high temperature, the Zn film could be oxidized to form ZnO completely and a thin layer of silicon oxide exists at the interface of ZnO/SiC.

ACKNOWLEDGMENTS

This work is supported by the Specialized Research Fund for the Doctoral Program of Higher Education (Grant No. 20030358054) and the National Natural Science Foundation of China (Grant No. 50532070).

REFERENCES

1.   R.F. Service, *Science* **276**, 895-897 (1997).
2.   D. Bagnall, Y.F. Chen, Z. Zhu *et al.*, *Appl. Phys. Lett.* **70**, 2230-2232 (1997).
3.   D.C. Look, *Mater. Sci. Eng.* B **80**, 383-387 (2001).
4.   R. Triboult, J. Perriere, *Progress in Crystal Growth and Characterization of Materials* **47**, 65-138 (2003).
5.   S. Cho, J. Ma, Y. Kim *et al*, *Appl.Phys. Lett.* **75**, 2761-2763 (1999).
6.   C. Zou, B. Sun, W. Zhang *et.al.*, *Nucl. Instrum. Meth.* A **548**, 574-581 (2005).
7.   R.R. Gay, M.H. Nodine, V.E. Henrich *et.al*, *J Amer. Chem. Soc.* **102**, 6752-6761 (1980).

Measurement of Compositional Grading at InP/GaInAs/InP Hetero-interfaces by X-ray CTR Scattering Using Synchrotron Radiation

Y. Ohtake[1], T. Eguchi[1], S. Miyake[2], W. S. Lee[1], M. Tabuchi[3], and Y. Takeda[2,3]

[1]*Department of Materials Science and Engineering,* [2]*Department of Crystalline Materials Science,*
[3]*Venture Business Laboratory, Nagoya University, Furo-cho, Chikusa-ku, Nagoya 464-8603, Japan*

Abstract. Compositional grading of group-III and V elements in InP/GaInAs/InP structures grown by OMVPE (organometallic vapor phase epitaxy) with various V/III ratios was investigated by utilizing X-ray CTR (crystal truncation rod) scattering. The results showed that the compositional grading both of As and Ga decreased by the increase of the V/III ratio. The compositional grading of Ga arised by the exchange with In in the InP cap layer, since the total amount of Ga showed almost no change with the increase of the V/III ratio. On the other hand, for As, the accumulation of precursors on the reactor wall and/or the adsorption of atoms on the surface and the susceptor mainly caused the compositional grading when V/III ratio was less than about 20. The exchange of atoms mainly caused the compositional grading of As when the V/III ratio was greater than about 20.

Keywords: InP/GaInAs/InP, OMVPE, X-ray CTR scattering, hetero-interface, compositional grading
PACS: 68.35.Ct

INTRODUCTION

The optical and electronic properties of quantum well structures are strongly influenced by the quality of their hetero-interfaces. With today's crystal growth technology, it is generally expected that the abruptness of the hetero-interfaces can be achieved at the scale of 1 monolayer. However, when the heterostructures are grown by OMVPE (organometallic vapor phase epitaxy) which is one of the most advanced crystal growth technology, such completely abrupt hetero-interfaces cannot be obtained and compositional grading arises at the hetero-interfaces unintentionally.

We have utilized X-ray CTR (crystal truncation rod) scattering measurement[1-3] to investigate the compositional grading at the scale of 1 monolayer[4-6]. In these works, it was shown that the compositional grading of group-V elements was able to be suppressed by optimizing the source-gas purging time (growth interruption time)[7]. On the other hand, the compositional grading of group-III elements was not suppressed by the growth interruption, but was suppressed by decreasing the growth temperature[8]. However, the lower growth temperature might degrade the crystal quality.

In this work, the compositional grading of the group-III and V elements in InP/$Ga_{0.47}In_{0.53}As$(15ML)/InP structures grown by OMVPE with different V/III ratios was investigated using the X-ray CTR scattering. For the group-III elements, it was reported that the compositional grading decreased with the increase of the V/III ratio[9]. For the group-V elements, it was expected that the compositional grading increased with the increase of the V/III ratio, since the accumulation of group-V precursors or elements was expected to increase with the increase of the V/III ratio.

EXPERIMENTS

InP/$Ga_{0.47}In_{0.53}As$(15ML)/InP structure samples were grown by OMVPE. The InP substrates were exactly [001] oriented. The sample structure is schematically shown in Fig. 1. The OMVPE growth was performed in a vertical reactor at a low-pressure (76.0 Torr). The V/III ratio was varied from 5 to 50 by changing the flow rates of group-V

CP879, *Synchrotron Radiation Instrumentation: Ninth International Conference,*
edited by Jae-Young Choi and Seungyu Rah
© 2007 American Institute of Physics 978-0-7354-0373-4/07/$23.00

precursors while keeping the flow rates of group-III precursors.   Precursors were triethylgallium (TEGa), trimethylindium (TMIn), tertiarybutylarsine (TBAs) and tertiarybutylphosphine (TBP). Hydrogen was used as a carrier gas.   Total flow rate of the hydrogen was the same for all the samples. Growth rate was set at 0.4ML/s and growth temperature was 620°C for all the samples. Ga composition in GaInAs layer was designed to be 0.47.

The X-ray CTR scattering measurement was performed at the BL6A of the Photon Factory at the High Energy Accelerator Research Organization in Tsukuba.   Wavelength of the X-ray was set at 1.0 Å by a triangular Si (111) mo-nochromator. A Weissenberg camera was used to record the X-ray CTR scattering intensity around 002 Bragg diff-raction spot of InP by utilizing CCD camera as a detector.

Measured X-ray CTR spectra are shown in Fig. 2.   The spectra were measured along [00 ℓ] direction in reciprocal space. The direction was normal to the surfaces of the samples. The peaks observed at ℓ=2 were 002 Bragg peaks of InP, although they were truncated near the Bragg point since the peaks were too high.   As shown in Fig. 2, clear oscillations were observed in the range of 1.8 < ℓ < 2.2 for all the samples. The oscillation was caused by the interferences between reflected X-ray by InP and GaInAs layer.

The X-ray CTR spectra were analyzed by comparing them with theoretically calculated spectra based on a model structure shown in Fig. 3. CTR spectrum is very sensitive to layer thickness and composition profile of each atom. These parameters were changed independently, so we can define the real structure by fitting of CTR spectrum. Distributions of Ga and As at hetero-interfaces were assumed to be represented by a function as follows:

$$x(n) = x_h \exp\left(-\frac{n}{d}\right) \qquad (1)$$

where $x(n)$ is composition in a layer n[ML] away from a interface, x_h is peak Ga or As composition in GaInAs layer, and d indicates degree of compositional grading. The best-fitted spectra are also shown in Fig. 2.

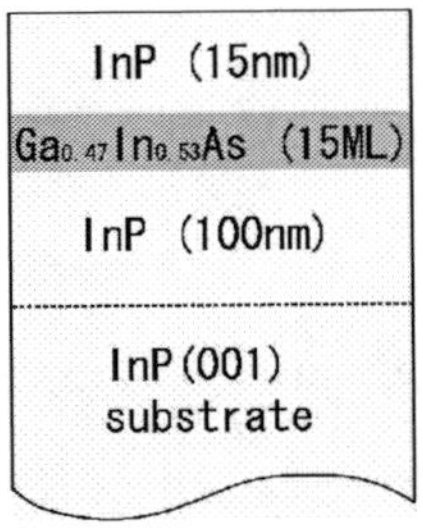

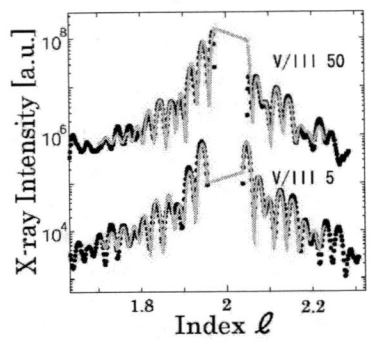

FIGURE 1. Sample structure grown by OMVPE. The V/III ratio was changed from 5 to 50. The growth rate was 0.4ML/s and the reactor pressure was fixed at 76.0 Torr for all samples.

FIGURE 2. Measured and theoretically calculated X-ray CTR scattering spectra. InP 002 Bragg peaks were at ℓ =2.0 although they were truncated since X-ray intensity was too high.

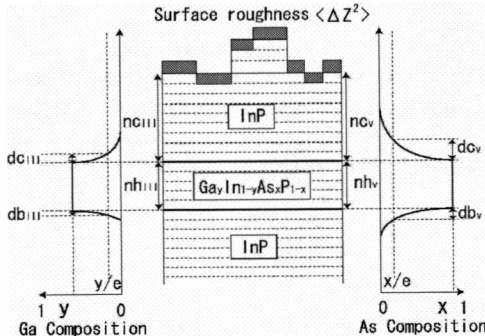

FIGURE 3. A model structure to calculate CTR scattering spectra. Measured X-ray CTR spectra were analyzed by comparing them with the theoretically calculated spectra.

RESULTS AND DISCUSSIONS

Figure 4 shows distribution profiles of As and Ga calculated at the best-fit. It was observed that when the V/III ratio was low, As and Ga atoms considerably incorporated into the InP cap layer. In other words, unintentional GaInAsP layers were formed at the upper interfaces. By increasing the V/III ratio, almost abrupt interfaces were obtained. Figure 5 shows the relationship between the V/III ratio and the degree of the compositional grading of As and Ga. As shown in Fig. 5, the compositional grading both of As and Ga decreased with the increase of the V/III ratio.

To understand the reason why the compositional grading decreased with the increase of the V/III ratio, we considered two mechanisms which caused the compositional grading. One was accumulation of precursors on the reactor wall and/or adsorption of excess atoms on the surface and the susceptor. The other was exchange of atoms in growing surface layer and in grown subsurface layer. When the accumulation and/or the adsorption is the main reason of the compositional grading, the total amount of As and Ga should increase with the increase of the compositional grading, since the incorporation of the excess atoms cause the grading. On the other hand, when the exchange is the main reason of the compositional grading, the total amount of As and Ga should not change while changing the compositional grading. Figure 6 shows the relationship between the compositional grading and the total amount of As and Ga contained in the samples which were obtained by the X-ray CTR scattering measurement. In Fig. 6, Ga and As showed different dependence on the degree of compositional grading. The total amount of Ga was almost the same while changing of the V/III ratio. This result suggests that the compositional grading of Ga arised by the exchange of atoms. On the other hand, for As, the accumulation and/or the adsorption occurred mainly when the degree of compositional grading was large, in other words, when the V/III ratio was low, since the total amount of As increased with increase of the compositional grading. When the degree of compositional grading was small, in other words, when the V/III ratio was high, the exchange mainly caused the compositional grading, since the total amount of As kept constant while changing the compositional grading.

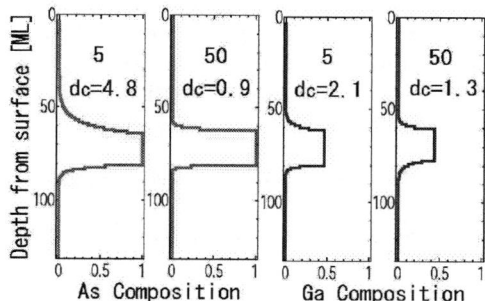

FIGURE 4. Examples of composition profiles of As and Ga analyzed by the X-ray CTR scattering measurements. When the V/III was high, the compositional grading decreased for both As and Ga. 'dc' indicates degree of compositional grading at the interface between GaInAs and InP cap layers.

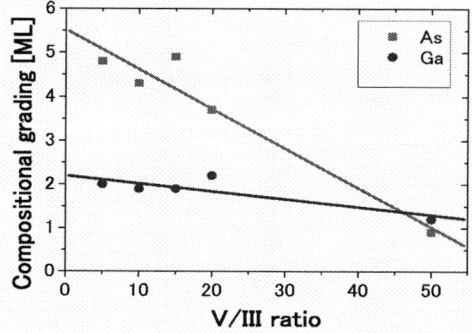

FIGURE 5. Relationship between V/III ratio and the degree of compositional grading. With the increase of the V/III ratio, the compositional grading both of Ga and As decreased.

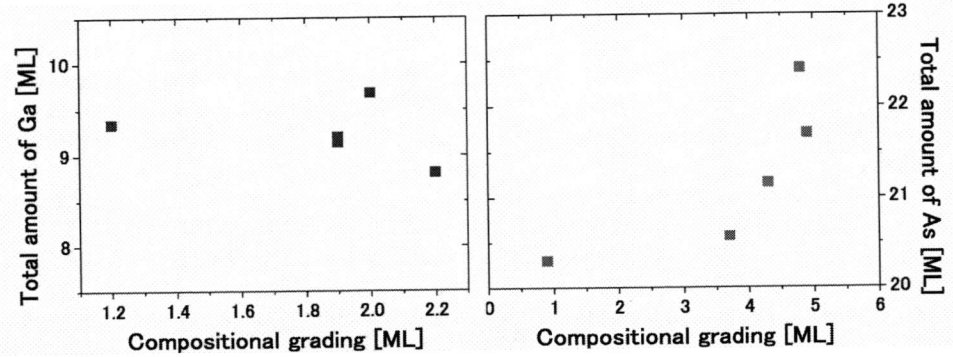

FIGURE 6. Relationship between the degree of compositional grading and the total amount of atoms. The total amount of Ga was almost the same while the increase of the compositional grading. On the other hand, the the total amount of As changed obviously when the compositional grading was large, in other words, the V/III ratio was high.

CONCLUSIONS

We have investigated the compositional grading of Ga and As elements in InP/GaInAs/InP structures with various V/III ratios by utilizing X-ray CTR scattering. Although, it was reported that when the V/III ratio was high, the composition grading of group-III was suppressed[9], our results showed that the compositional grading both of As and Ga decreased by increasing of the V/III ratio. From the dependence of the compositional grading on the total amount of atoms, we could identify the main mechanism of the composition grading of Ga and As, respectively. The compositional grading of Ga arised by the exchange of atoms between Ga and In consisting InP cap layer. On the other hand, for As, the accumulation of precursors on the reactor wall and/or the adsorption of atoms on the surface and the susceptor mainly caused the compositional grading when V/III ratio was less than about 20. The exchange of atoms mainly caused the compositional grading of As when the V/III ratio was greater than about 20.

ACKNOWLEDGMENTS

This work was performed as a part of the projects 2004G222 and 2005G155 accepted by the Photon Factory Program Advisory Committee.

REFERENCES

1. I. K. Robinson, *Phys. Rev.* B **33** (1986) 3830
2. I. K. Robinson, D. J. Tweet, *Rep. Prog. Phys.* **55** (1992) 599
3. T. Shimura, and J. Harada, *J. Appl. Cryst.* **26** (1993) 151
4. M. Tabuchi, H. Kyozu, M. Takemi, Y. Takeda, *Appl. Surf. Sci.* **216** (2003) 526-531
5. Y. Takeda, Y. Sakuraba, K. Fujibayashi, M. Tabuchi, T. Kumamoto, I. Takahashi, J. Harada, and H. Kamei, *Appl. Phys. Lett.* **66** (1995) 332
6. I. Yamakawa, R. Oga, Y. Fujiwara, Y. Takeda, A. Nakamura, *Appl. Phys. Lett.* **84** (2004) 4436-4438
7. M. Tabuchi, R. Takahashi, M. Araki, K. Hirayama, N. Futakuchi, Y. Shimogaki, Y. Nakano, Y. Takeda, *Appl. Surf. Sci.* **159** (2000) 250
8. M. Tabuchi, S. Hisadome, Y. Ohtake, W. S. Lee, and Y. Takeda, "REALIZATION OF SQUARE QUANTUM-WELL STRUCTURE InP/ GaInAs/InP BY ORGANOMETALLIC VAPOR PHASE EPITAXY" in *Indium Phosphide & Related Materials-2005*, May 7-11, 2005, Glasgow, UK
9. K. Muraki, S. Fukatsu, Y. Shiraki, R. Ito, *Appl. Phys. Lett.* ,**61**, 557 (1992)

Energy-Level Alignment at Interfaces between Gold and Poly(3-hexylthiophene) Films with two Different Molecular Structures

Yeong Don Park, Jeong Ho Cho, Do Hwan Kim, Wi Hyoung Lee, Kilwon Cho

Department of Chemical Engineering / Polymer Research Institute, Pohang University of Science and Technology, Pohang, 790–784, Korea

Abstract. The electronic structures of the interfaces between Au and poly(3-hexylthiophene) (P3HT) films with two different molecular orientations and orderings were investigated using synchrotron radiation photoemission spectroscopy. We found that, depending on whether thermal treatment was used, the P3HT thin film adopts two different molecular orientations—parallel and perpendicular to the silicon substrate—which result in different values of the vacuum level shift and hole-injection barrier. Thus the molecular orientation and ordering of the P3HT material strongly affect the energy level alignment at the P3HT/Au interface.

Keywords: Energy-level alignment, poly(3-hexylthiophene), molecular structure

INTRODUCTION

In recent years, there has been increasing interest in organic field effect transistors (OFETs) [1]. Poly(3-hexylthiophene) (P3HT) is an organic semiconductor which has received considerable attention because of its high mobility values when used as a semi-conducting material in OFETs [1]. The use of this polymer enables simple, low-cost processing because it is soluble and conducting.

In top contact devices, gold is usually deposited as a source and drain electrode on a P3HT thin film surface. In such devices, one of the factors that determine device function and performance is charge carrier injection from the electrode into the organic material [2-10]. The interface between the polymer semiconductor and the metal contact is expected to play a prominent role in the operation of such devices. The electrical properties of these devices are much affected by the choice of the metallic contact used as a hole injector. However, few studies of the interface between the conjugated polymer and the metal electrode [10-12] have been reported.

In the present study, we investigated the interface between Au and P3HT by measuring the secondary electron cutoff and the valence band maximum using synchrotron radiation techniques. We determined the electronic structure of the interface between Au and P3HT with different molecular orientations (i.e., edge-on and face-on) with a low-coverage evaporation technique and synchrotron radiation photoemission spectroscopy. From the viewpoint of investigations of metal-organic interfaces [4, 5, 8], the differences between edge-on and face-on P3HT provide an excellent opportunity for examining the basic issues in the formation of interfaces between metal electrode and spin-coated polymer semiconductors [13, 14].

EXPERIMENTAL

Two samples were prepared by spin coating 0.5 wt% regioregular P3HT solution in chloroform ($CHCl_3$) onto silicon substrates in the glove box (thickness of film : 32 nm). One of the P3HT films was kept at 240°C for 20 min in closed jar under inert Ar conditions, and then cooled slowly. Au was deposited in several steps in an ultrahigh vacuum system (base pressure $<1\times10^{-10}$Torr) at submonolayer coverages on the P3HT substrates. At each step, the interface between Au and P3HT was characterized by measuring the valence band spectrum and secondary electron cutoff using synchrotron radiation photoemission spectroscopy at the 2B1 and 4B1 beam lines of the Pohang Accelerator Laboratory, Korea. The Au coverage was determined at each step with a thickness monitor.

CP879, *Synchrotron Radiation Instrumentation: Ninth International Conference*,
edited by Jae-Young Choi and Seungyu Rah
© 2007 American Institute of Physics 978-0-7354-0373-4/07/$23.00

RESULTS AND DISCUSSION

In order to investigate the inner structure of the P3HT thin films, grazing-incidence X-ray diffraction (GIXD) was used [15] at the 3C2 and 8C1 beamlines at the Pohang Accelerator Laboratory. Figure 1 shows the out-of-plane X-ray diffraction patterns of the P3HT films before and after thermal treatment above the order-disorder transition temperature of P3HT. In the case of the as-prepared P3HT thin film, the (100) reflection due to lamellar layer structure is weak, but the out-of-plane (010) reflection due to π-π interchain stacking (3.8Å) is strong. However, in the case of the thermally treated P3HT thin films, the second Bragg peak (200) and the first Bragg peak (100) were easily detected, but the (010) reflection was not observed. From these results, we conclude that the thermally treated P3HT film has an edge-on structure with a preferential orientation of its (100)-axis normal to the P3HT film, and the as-prepared P3HT film has a face-on structure with its (010)-axis normal to the P3HT film.

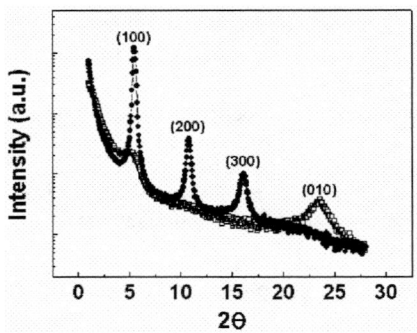

FIGURE 1. Out-of-plane grazing incidence angle X-ray diffraction intensities as a function of the scattering angle 2θ for regioregular P3HT thin films on SiO₂/Si substrates: ● thermally treated P3HT, □ as-prepared P3HT.

Figures 2 show the secondary electron cutoffs and valence band maxima obtained from our study of Au deposition on the as-prepared and thermally treated P3HT films respectively. Deposition of 2 Å Au produced abrupt upward shifts (0.3 eV for as-prepared P3HT, and 0.5 eV for thermally treated P3HT) in the position of the vacuum level, corresponding to the formation of an interface-dipole, as shown in Figs. 2(a) and 2(c). Accompanying this shift was the appearance of spectral features characteristic of the Au overlayers. Further deposition resulted in slight increases in the vacuum level shift, which saturated at 0.44 eV and 0.69 eV at 16 Å respectively. The large upward shift, 0.69 eV, of thermally treated P3HT compared to that of as-prepared P3HT, 0.44 eV, is due to the interface-dipole that arises from the structural ordering of P3HT [11].

Figures 2(b) and 2(d) show the changes in the valence band spectra of the Au/P3HT samples with increases in the thickness of deposited Au. The spectra were recorded using a photon energy of 80 eV. From the spectra obtained for P3HT/Si before Au deposition, the HOMO levels were found to be 0.8 eV and 0.95 eV for as-prepared and thermal treated P3HT, respectively. With increases in the thickness of the Au layer, the HOMO level shifted to the Fermi level. In other words, the top of the HOMO was 0.8 eV below the Fermi level for as-prepared P3HT, and 0.95 eV below the Fermi level for thermally treated P3HT. In this analysis, it was assumed that the gap between the HOMO level and the Fermi level is the hole-injection barrier.

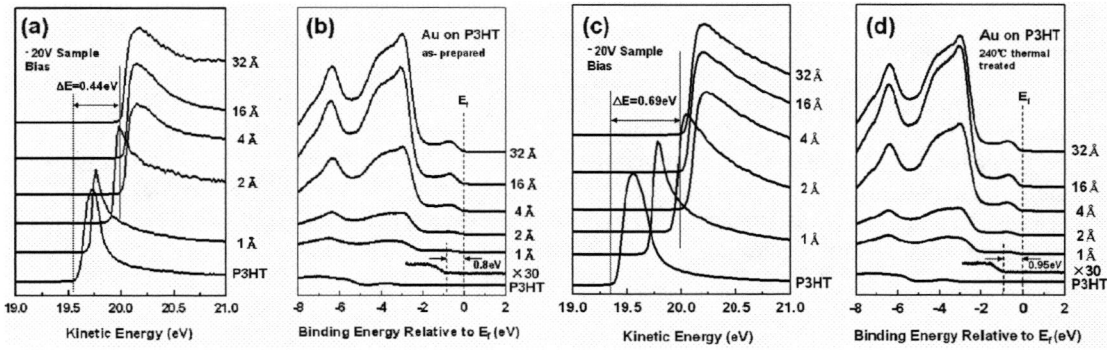

1624

FIGURE 2. X-ray Photoemission Spectra for Au deposited on P3HT films. (a) Secondary electron emission spectra of an as-prepared P3HT film for various amounts of Au deposition. (b) Changes in the valence band spectra of the as-prepared P3HT film as a result of Au deposition. (c) Secondary electron emission spectra of a thermally treated P3HT film for various amounts of Au deposition. (d) Changes in the valence band spectra of the thermally treated P3HT film as a result of Au deposition.

CONCLUSIONS

In summary, we have reported the difference between the energy-level alignments at the interfaces of gold and P3HT films with two different molecular orientations and ordering. P3HT films that have undergone thermal heating above the melting point and a subsequent cooling process have an ordered structure and an edge-on orientation. The interfaces between P3HT and Au of the samples were investigated with synchrotron photoemission spectroscopy. The interface-dipole and hole-injection barrier at interfaces between well-ordered P3HT with an edge-on orientation and Au are larger than those of face-on oriented P3HT, which indicates that the energy level alignment is altered by changes in the molecular orientation and ordering of P3HT.

REFERENCES

1. H. Sirringhaus, N. Tessler, and R. H. Friend, *Science*, **280**, 1741 (1998).
2. H. Ishii, K. Sugiyama, E. Ito, and K. Seki, *Adv. Mater.,* **11**, 605 (1999).
3. D. Cahen, and A. Kahn, *Adv. Mater.*, **15**, 271 (2003).
4. I. Hill, A. Rajagopal, and A. Kahn, *J. Appl. Phys.,* **84**, 3236 (1998).
5. N. Koch, A. Kahn, J. Ghijsen, J.-J. Pireaux, J. Schwartz, R.L. Johnson, and A. Elschner, *Appl. Phys. Lett.,* **82**, 70 (2003).
6. S. Kim, J. Lee, K. Kim, and Y. Tak, *Appl. Phys. Lett.*, **86**, 133504 (2005).
7. K. Ihm, T. Kang, K. Kim, C. Hwang, Y. Park, K. Lee, B. Kim, C. Jeon, C. Park, K. Kim, and Y. Tak, *Appl. Phys. Lett.,* **83**, 2949 (2003).
8. N. Watkins, L. Yan, and Y. Gao, *Appl. Phys. Lett.,* **80**, 4384 (2002).
9. N. Koch, A. Elschner, J. Rabe, and R. Johnson, *Adv. Mater.,* **17**, 330 (2005).
10. P. Dannetun, M. Boman, S. Stafström, W. R. Salaneck, R. Lazzaroni, C. Fredriksson , J. L. Brédas, R. Zamboni, and C. Taliani, *J. Chem. Phys.,* **99**, 664 (1993).
11. B. Boer, A. Hadipour, M. M. Mandoc, T. Woudenbergh, and P. W. M. Blom, *Adv. Mater.,* **17**, 621 (2005).
12. I. H. Campbell, J. D. Kress, R. L. Martin, D. L. Smith, N. N. Barashkov, and J. P. Ferraris, *Appl. Phys. Lett.,* **71**, 3528 (1997).
13. G. R. Hutchison, M. A. Ratner, and T. J. Marks, *J. Phys. Chem. B*, **109**, 3126 (2005).
14. E. Bundgaard, and F. C. Krebs, *Macromolecules*, **39**, 2823 (2006)
15. D. H. Kim, Y. D. Park, Y. Jang, H. C. Yang, Y. H. Kim, D. G. Moon, S. Park, T. Chang, C. Chang, M. Joo, C. Y. Ryu, and K. Cho, *Adv. Funct. Mater.*, **15**, 77 (2005).

Studies of Mn/ZnO (000$\bar{1}$) Interfacial Formation and Electronic Properties with Synchrotron Radiation

C.W. Zou[1], P.S. Xu[1][*], Y.Y. Wu[1], B. Sun[1], F.Q. Xu[1],
H.B. Pan[1], H.T. Yuan[2], X.L. Du[2]

[1] NSRL, University of Science and Technology of China, Hefei, 230029, China
[2] Beijing National Laboratory for Condensed Matter Physics, Institute of Physics,
The Chinese Academy of Sciences, Beijing 100080, China

Abstract. The initial growth, interfacial reaction and Fermi level movement of Mn on the O-terminated Zn (000$\bar{1}$) surface have been investigated by using synchrotron radiation photoelectron spectroscopy (SRPES) and X-ray photoemission (XPS). Mn is found to be grown on the surface in the layer-by-layer (Frank-van der Merwe) mode and be quite stable on the O-terminated surface at room temperature. With increasing the coverage of Mn, a downward Fermi level movement in band structure measurement of SRPES is observed and the resultant Schottky Barrier Height (SBH) is calculated to be about 1.1eV. Annealing behavior of the interface is investigated and we find that annealing at 600 ^{0}C induces a pronounced Mn-Zn atoms exchange reaction at the interface.

Keywords: Mn film, ZnO, interface, synchrotron radiation, photoemission.
PACS: 73.20. –r, 73.40. Sx, 79.60. –i, 81.15.-z

INTRODUCTION

ZnO has attracted much attention for its potential applications of opto-electronic devices such as light emitting diodes and laser diodes, which benefit from its wide direct band gap and large exciton binding energy [1-3]. The interface property of metal/ZnO is essential for the performance of all kinds of devices. With the achievement of ZnO high quality films by MBE, MOCVD and other methods, the interfacial structures and Ohmic or Schottky properties of metal/ZnO have been investigated recently. However, many aspects of the Metal/Semiconctor contact, such as the interface reaction, the Fermi level movement and the effects of annealing still have not been well understood.

In this paper, we study the initial growth, interfacial reaction and Fermi level movement of Mn on the O-terminated Zn (000$\bar{1}$) surface by using synchrotron radiation photoelectron spectroscopy (SRPES) and X-ray photoemission (XPS).

EXPEIMENTAL

The experiment is performed at the surface physics station of NSRL. The experimental station is mainly composed of VG ARUPS10 photoelectron spectrometer system. The beamline covers the energy range from 10 to 300eV and the energy resolution (E/ΔE) is better than 1000. More details of the experimental station and the related beamline are described elsewhere [4].

Unipolar ZnO film with O-terminated surface is prepared on c-plane sapphire substrate by radio frequency plasma-assisted molecular beam epitaxiy [5]. After being degreased in acetone and rinsed in DI (deionized) water, the ZnO/α-Al$_2$O$_3$ sample is blown dry and quickly introduced into the UHV chamber. Then it is annealed at 600 ^{0}C in oxygen ambient with the pressure of 2x10^{-5}mbar for 20min. After above treatments, clear low energy electron

---

[*] Corresponding Author, E-mail: psxu@ustc.edu.cn, Fax: +0086-551-5141078

diffraction (LEED) patterns of the ZnO surface are observed. High purity Mn (~99.99%) granules are evaporated by K-cell at 830 °C in the deposition chamber. The growth rate is about 0.18nm/s which is measured by quartz crystal thickness monitor. After deposition the sample is transferred to the analysis chamber for SPRES and XPS measurements.

RESULTS AND DISCUSSION

Figure 1 shows the valence band photoemission spectra of Mn/ZnO sample by using the photon energy of 25eV with different Mn coverages. It is clearly shown that the Fermi edge emission appears when the nominal thickness of metallic Mn is 0.1 nm and becomes pronounced with the increasing coverage of metallic Mn, which reflects the metallic characteristic of ZnO surface. At the same time the intensity of Zn3d peak located at -10.8eV decreases rapidly. However, after the sample is annealed at 600°C, the clear Fermi edge emission almost disappears. It is obvious that interfacial reaction near the interface greatly weakens the metallic properties of Mn/ZnO surface. This

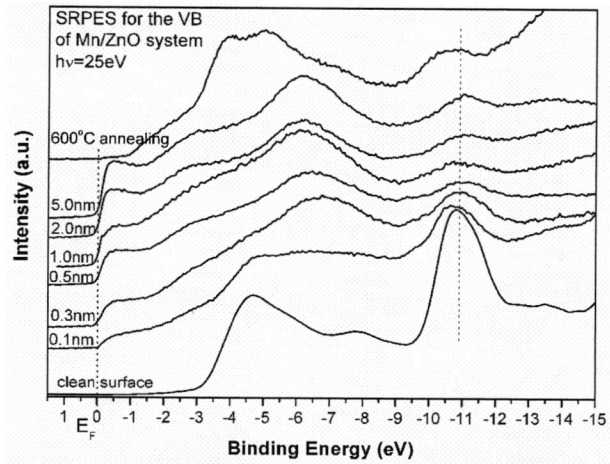

FIGURE 1. The valence band photoemission spectra excited by synchrotron radiation with photon energy of 25eV.

chemical reaction will be discussed in detail later.

Figure 2 (a), (b) and (c) shows the XPS (Mg Kα, 1253.6eV) spectra of Mn2p, O1s and Zn2p at different stage, respectively. According to the XPS spectra, the intensity ratio of $Mn2p_{3/2}$ to $Zn2p_{3/2}$ versus the Mn coverage at RT is calculated and shown in Fig. 2 (d). It is clear that the $Mn2p_{3/2}$ intensity increases quickly with the increasing of Mn deposition, while for $Zn2p_{3/2}$, it decreases. This behavior indicates a layer-by-layer growth mode. Actually many similar results have been reported for this kind of metal growth on wide-band semiconductor, such as Ni or Pd deposition on GaN surface at RT [6, 7].

From Fig. 2 (b) and (c) we can observe that during the Mn deposition, the energy positions of Zn2p and O1s shift toward low binding energy direction by 0.72eV and 0.60eV, respectively, following the same trend. Therefore, the binding energy shifts are assumed to be rigid Fermi level shifts and corresponding to band bending at the interface. The little disagreement value (~0.12eV) is perhaps originated from the interfacial reaction at RT, although this reaction is quite weak. Actually in Mn2p spectra in Fig. 2 (a), Mn2p shows broad peak when the coverage is 0.1nm, which indicates a weak interfacial reaction occurs. However, when more metallic Mn film is grown on the surface layer-by-layer, the reaction is suppressed immediately. In addition, ordered ZnO(000$\bar{1}$) surface can be obtained by annealing in oxygen ambient and this annealed clean surface has great chemical stability at room temperature, and it keeps metallic properties when Mn film is deposited..

Figures 3 (a) and (b) show the band bending and Schottky Barrier Height (SBH) evolution process with different Mn coverages. Due to the band bending, a downward Fermi level movement occurs and reaches stable state. Thus the final SBH is calculated to be 1.1eV.

After 5nm Mn film is grown on the surface, Mn/ZnO sample is annealing at 600°C in UHV chamber. Figure 2 shows the O1s peak has little shift (less than 0.2eV) after annealing, and this shift may be originated from the difference of binding energies in ZnO and the Mn oxides. While Mn2p peak shift about 2.5eV toward high binding

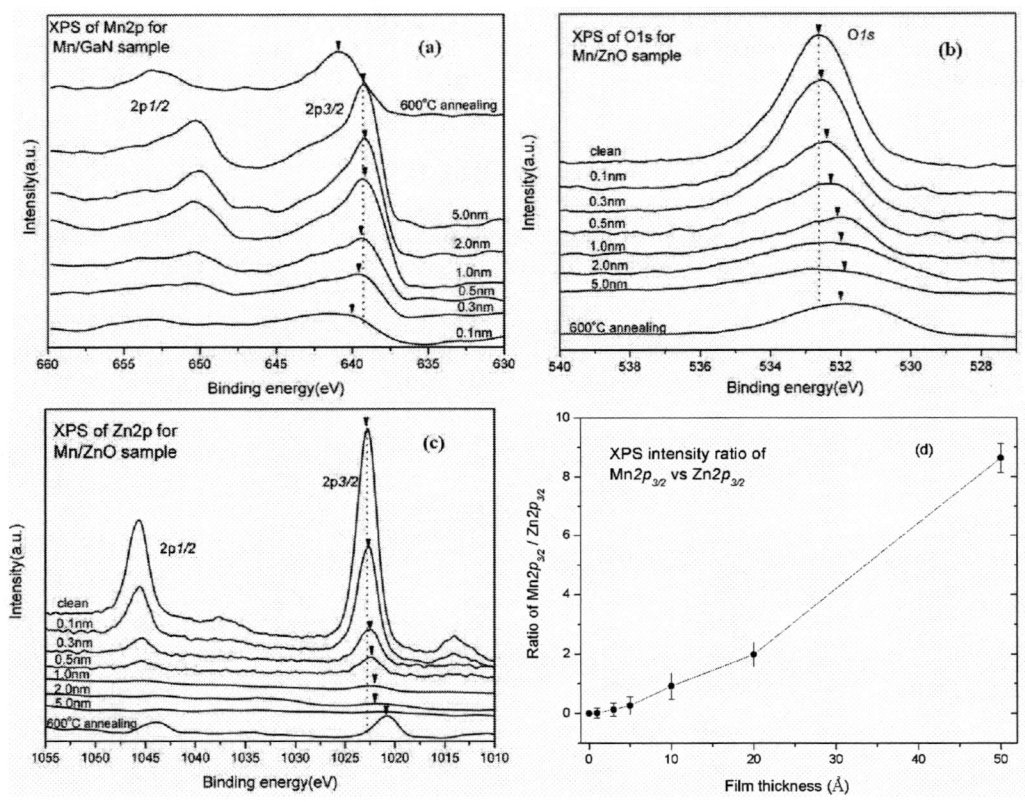

FIGURE 2. The XPS spectra of Mn2p (a); O1s (b) and Zn2p (c) at different stages excited by Mg $K\alpha$ (1253.6eV); (d) the XPS intensity ration of Mn2p$_{3/2}$ to Zn2p$_{3/2}$ as a function of Mn coverage at room temperature.

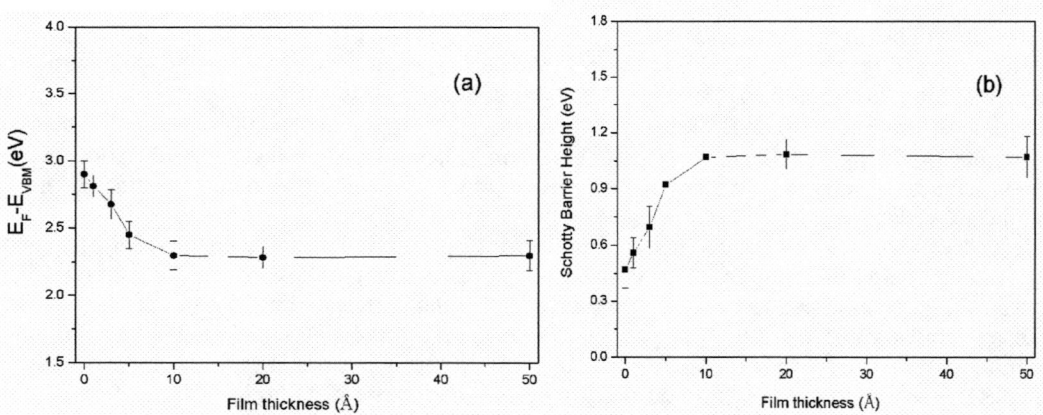

FIGURE 3. Band bending (a) and SBH (b) as a function of Mn coverage.

energy direction to 641.2eV, and obviously this binding energy is related to the oxidized state of Mn atoms [8]. Simultaneously, the Zn2p$_{3/2}$ peak shift to opposite direction for about 0.4eV. All the above spectra shifts indicate that high temperature annealing treatment induces violent interfacial chemical reaction according to the following rule: $Mn + ZnO \rightarrow MnO_x + Zn$. Thus metallic zinc atoms are formed while Mn atoms are oxidized near the interface while annealing at high temperature. Another point to notice is that after annealing process, the Zn2p signal intensity increases, which indicates zinc atoms are segregated onto the topmost surface.

CONCLUSION

The Mn/ZnO(000$\bar{1}$) interface and its property are investigated by SRPES and XPS methods. Mn atoms are quite stable on the O-terminated surface and Mn film can be grown in the layer-by-layer mode. Prominent M/S contact is obtained at room temperature and the resultant Schottky Barrier Height (SBH) is calculated to be about 1.1eV. However, when we anneal the Mn/ZnO in high temperature, the interfacial reaction between Mn and ZnO happens violently. As a result, Mn oxides are formed and metallic znic atoms are segregated onto the topmost surface. From the above analysis, we can conclude that the clean ZnO(000$\bar{1}$) surface, which is prepared by annealing the sample in oxygen ambience, has great chemical stability, metallic Mn atoms can not react with the atoms on the clean surface at room temperature. But annealing treatment with high temperature will break the chemical stability, and then pronounced interfacial reaction occurs.

ACKNOWLEDGMENTS

This work is supported by the Specialized Research Fund for the Doctoral Program of Higher Education (Grant No. 20030358054) and the National Natural Science Foundation of China (Grant No. 50532070, 50532090, 60476044, 60376004).

REFERENCES

1. A.B. Djurisic, Y. Chan, E.H. Li, *Mater. Sci. Eng. B* **38**, 237-242 (2002).
2. D.M. Bagnall, Y.F. Chen and Z. Zhu et.al. *Appl. Phys. Lett.* **70**, 2230-2232 (1997).
3. K. Ellmer, *J. Phys. D: Appl. Phys.* **34**, 3097-3102 (2001).
4. C. W. Zou, B. Sun, W.H. Zhang, et.al., *Nucl. Instrum. Meth.* A **548**, 574-581 (2005).
5. Y. Wang, X.L. Du, Z.X. Mei et.al., *Appl. Phys. Lett.* **87**, 051901-051903 (2005).
6. V.M. Bermudez, R. Kaplan, M.A. .Khan et.al., *Phys.Rev* .B **48**, 2436 -2443(1993).
7. M. Eycjeler, W. Monch, T.U.Kampen et.al., *J.Vac.Sci.Technol* .B **16**, 2224 -2230(1998).
8. J. F. Moulder, W. F. Stickle, P.E. Sobol et.al., *Handbook of X-ray Photoelectron Spectroscopy*, edited by J.Chastain, Perkin-Elmer Corporation, Eden Prairie, Minnesota,1992, pp78-79

SR-Excited Angle-Resolved-Ultraviolet-Photoelectron-Spectroscopy Study of One-Dimensional Electronic State on Ni(332) Stepped Surface

Koji Ogawa, Nobuyuki Fujisawa, Koji Nakanishi, Hidetoshi Namba

Department of Physical Sciences, Ritsumeikan University
1-1-1 Nojihigashi, Kusatsu, Shiga 525-8577 Japan

Abstract. We have investigated the surface electronic structure of stepped Ni(332)[=6(111)×(111)] in the expectation of observing one-dimensional surface state by angle-resolved ultraviolet photoelectron spectroscopy using synchrotron radiation(SR-ARUPS). In SR-ARUPS spectra, we observed two surface-sensitive peaks at ~0.5 and ~0.1 eV below Fermi level E_F revealed by O_2 adsorption. The former peak disperses downwards away from E_F parallel and perpendicular to the steps, which can be attributed to the (111) terrace-derived two-dimensional surface state. Though the latter peak disperses upwards parallel to the steps, on the other hand, it has almost no dispersion in the perpendicular direction. This flat dispersion is the clear evidence that the peak just below E_F stems from an electronic state localized at the steps. Considering the flat and upwards dispersions perpendicular and parallel to the steps, this step-localized state is concluded to be one-dimensional electronic state.

Keywords: Surface states, Clean metal stepped surface, Nickel, Photoemission, Synchrotron radiation
PACS: 73.20.At, 79.60.Bm, 71.20.Be, 79.60.-i

INTRODUCTION

The properties and nature of surface states on metals, which are localized in the very vicinity of the surfaces, strongly depend on the atomic structure of the surfaces where they are induced. On planar surfaces such as the closest packed surfaces of Cu(111) and Ni(111), the surface states have a high in-plane isotropy being two-dimensional[1,2]. The two-dimensional nature of such surface states is evident from identical energy dispersions in two orthogonal directions and circular Fermi surfaces obtained by means of angle-resolved ultraviolet photoelectron spectroscopy(ARUPS). When the isotropy in the surface structure is hindered by, for example, a regular array of surface steps, the anisotropy must have some influence on the surface states. On stepped Ni(7 9 11) which has a regular array of (110) steps, ARUPS spectra revealed that the two-dimensional surface state retains the nature and more importantly that a new anisotropic state is induced[3,4]. The new state on Ni(7 9 11) shows almost flat dispersion perpendicular to the steps. Moreover, alkali metals adsorbed at the steps more sensitively reduced the intensity of the new-state peak than that of the two-dimensional one in the ARUPS spectra. So the new state was concluded to localize at the steps. Similarly one-dimensional step state is found at (100) steps on Ni(755)[5].

These results pose an important question about how the step-localized states are induced on the stepped surfaces. One possibility is that we observe surface states only on (110) and (100) steps. This is partially denied since we observe the step states both on Ni(7 9 11) and Ni(755). The other is whether a termination of in-plane periodicity at the steps induces the step-localized states just as surface states on planer surfaces. Possibly, therefore, there is no step dependency. By observing the step state on Ni stepped surfaces with (111) steps, we can conclusively confirm this point.

In order to clarify these points, in the present study, we determined to investigate the surface state on Ni(332) whose step is (111) different from Ni(7 9 11) and Ni(755). We report in the present paper the results of ARUPS measurement on Ni(332) using synchrotron radiation for excitation.

CP879, *Synchrotron Radiation Instrumentation: Ninth International Conference*,
edited by Jae-Young Choi and Seungyu Rah

EXPERIMENTAL

Ni(332)[=6(111)×(111)] is inclined by 10.02° with respect to (111) towards [11$\bar{2}$]. The nominal average terrace width is 11.5 Å. The steps run along the [1$\bar{1}$0] direction. The samples were cut from a (111)-oriented single-crystal rod and polished mechanically. We cleaned the samples by repeated cycles of Ar$^+$ ion sputtering at 0.5 keV and annealing at 1080 K in a sample-preparation chamber whose base pressure is below 1×10^{-10} Torr. We confirmed the surface cleanliness and crystallinity by Auger electron spectroscopy and low-energy electron diffraction (LEED) patterns (Fig. 1). We can observe clear split spots in Fig. 1 which are characteristic of stepped surfaces.

Measurements of SR-ARUPS spectra were carried out at the SORIS Beamline(BL-8), a compact VUV beamline for high-resolution photoelectron spectroscopy at the SR Center, Ritsumeikan University[6]. The available photon energy is from 5 to 700 eV. SR-ARUPS spectra were measured at room temperature using a high-resolution electron energy analyzer[7]. The total energy resolution of the monochromator and the energy analyzer was estimated to be about 30 meV for the present measurements. The acceptance angle of the energy analyzer for photoelectrons was set to be ±2°. The angle between the incident photon and the axis of the energy analyzer was fixed to be 55°. The incident photon is p-polarized in oblique incidence. In the present study, the samples were rotated for the measurements of angular distribution of photoelectrons. The incident angle of photons linearly polarized in the horizontal plane is |55°-θ_e| and varied with θ_e, where θ_e is the emission angle measured from [111]. The azimuth angle is changed also by rotating samples around the terrace normal of [111] in vacuum. When the steps are parallel and perpendicular to the plane of incidence, the setup is referred as the parallel and perpendicular configurations, respectively.

RESULTS AND DISCUSSION

Figure 2 is the angular dependence of the photoemission spectra on stepped Ni(332) taken around the $\bar{\Gamma}$ point of the surface Brillouin zone at $h\nu$=10 eV in the perpendicular (a) and parallel (b) configurations. The abscissa is the binding energy and the ordinate the intensity normalized by the storage-ring current. The emission angles of photoelectron are indicated in the figure. The binding-energy region just below the Fermi level is shown where the surface state on planar Ni(111) is observed.

In Fig. 2 we observed a single broad peak with some structures from bulk and surface components. In order to extract surface components from these spectra we adsorbed O_2 on Ni(332) since surface components are very sensitive to the condition of the surface and decrease its peak intensity upon surface contamination more rapidly than bulk components. Figure 3(a) shows the change in the spectra taken at the emission angle of +10° in the perpendicular configuration upon O_2 adsorption at various exposures. The change is clearer in the difference (Adsorbed - Clean) spectra in Fig. 3(b). Upon O_2 adsorption one can see two peaks at ~0.7 and ~0.1 eV below the Fermi level E_F decreased their intensity rapidly, revealing that these peaks are surface component.

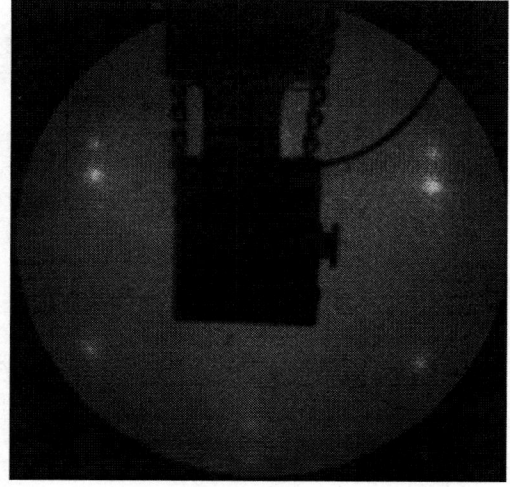

FIGURE 1. LEED pattern from stepped Ni(332) at E_k=$-$118 eV.

The difference spectra at all observed emission angles of photoelectron are collected in Fig. 4. We can see that the peak at higher binding energy disperses downwards away from ~0.5 eV at the $\bar{\Gamma}$ point in both the parallel and perpendicular configurations. This isotropic dispersion is very similar to the surface state on planar Ni(111) and thus the peak at the higher binding energy is assigned to be the (111) terrace-derived surface state. On the other

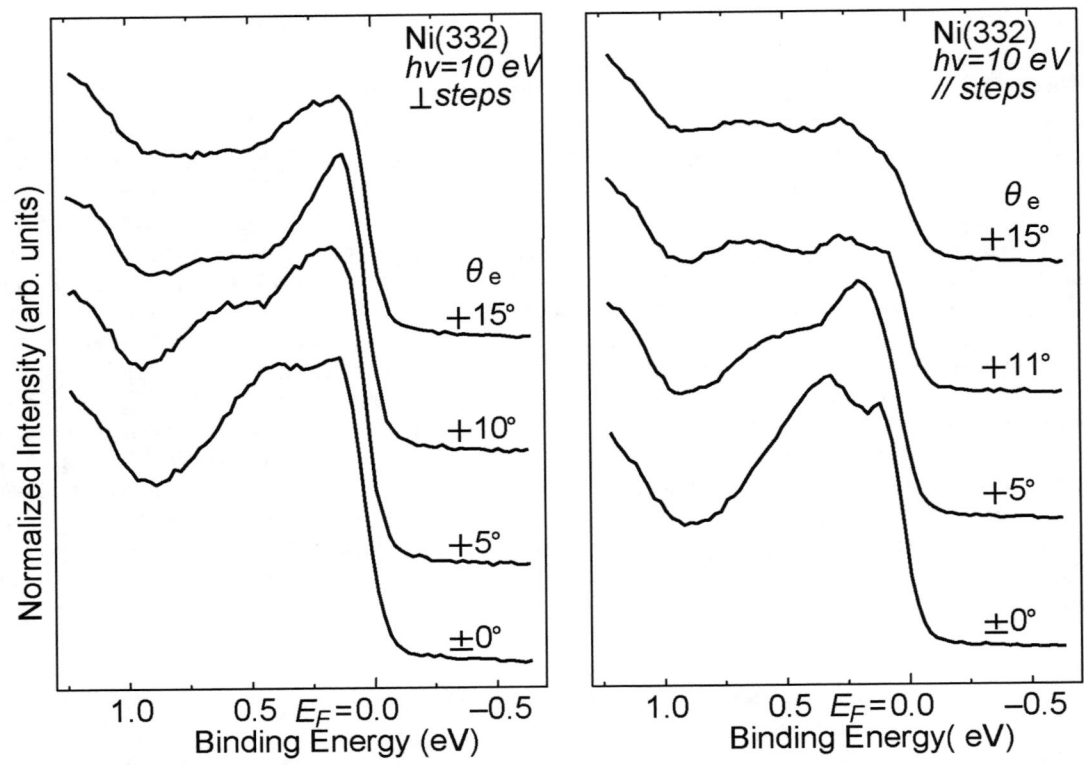

FIGURE 2. ARUPS spectra taken at $h\nu$=10 eV in the perpendicular (a) and parallel (b) configurations on stepped Ni(332).

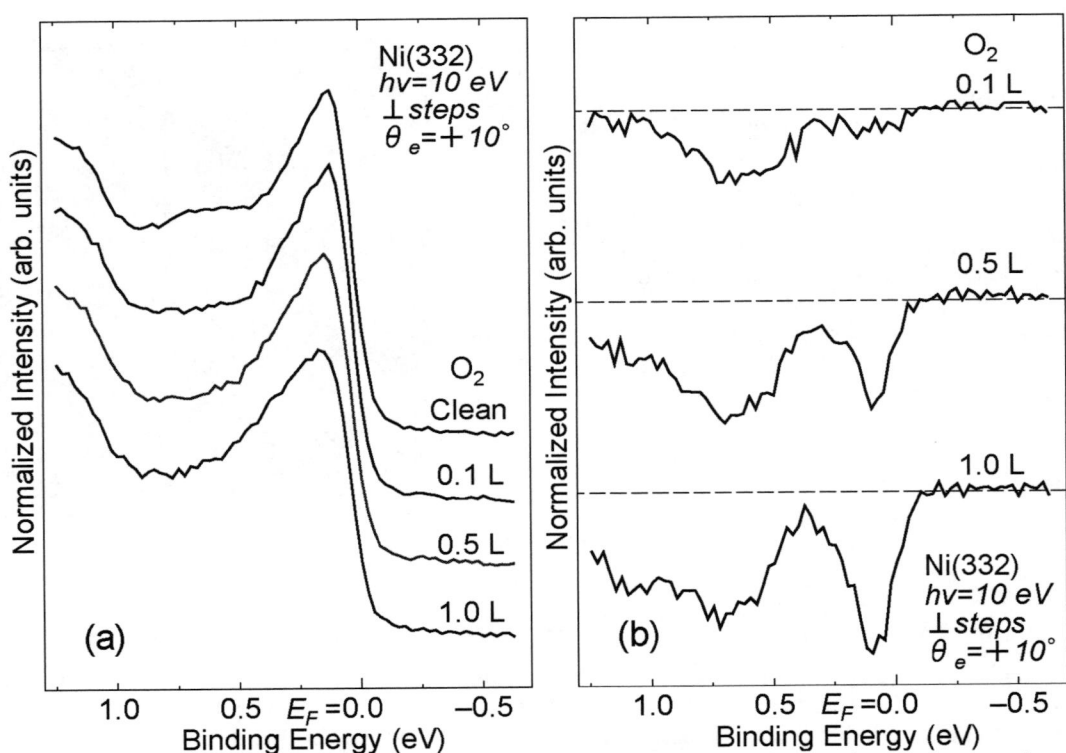

FIGURE 3. ARUPS spectra taken at $h\nu$=10 eV in the perpendicular configuration after O_2 adsorption at several exposures on stepped Ni(332) (a) and the difference (Adsorbed-Clean) spectra (b).

hand, the peak just below E_F disperses anisotropically. The peak shows almost flat dispersion perpendicular to the steps, whereas in the parallel configuration the peak slightly disperses upwards. This flat dispersion perpendicular to the steps is the clear evidence that this electronic state is localized at the steps and that there is almost no interaction between the steps. The small but finite dispersion parallel to the steps indicates that the electronic state is one-dimensional.

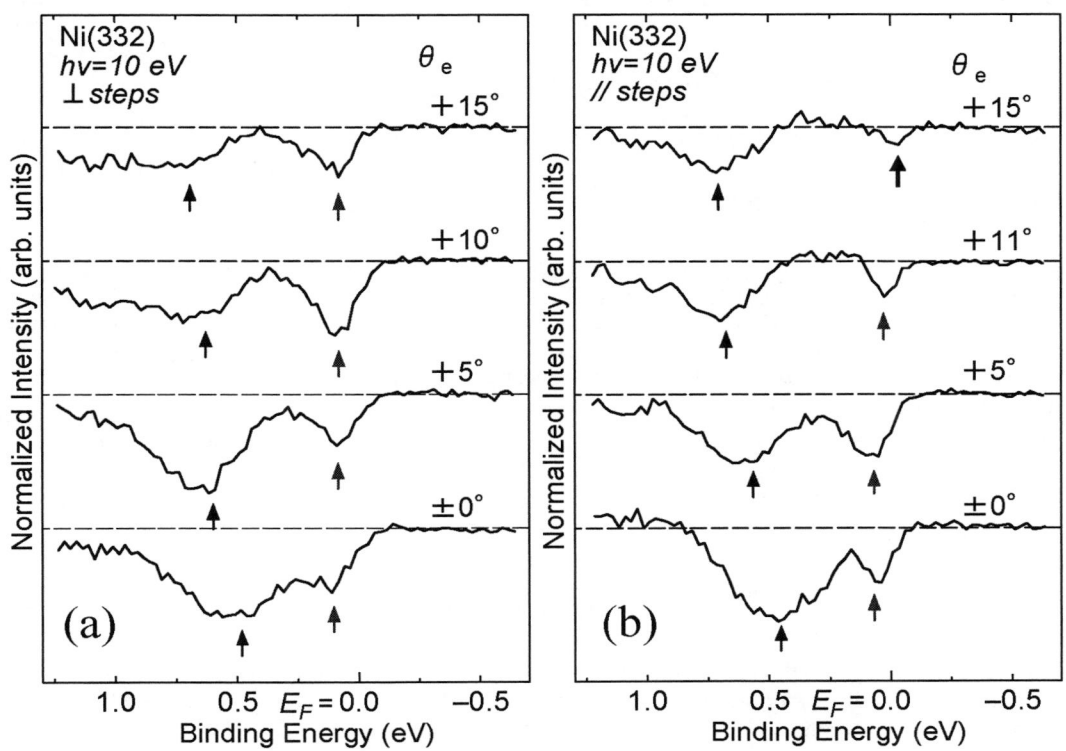

FIGURE 4. Difference (Adsorbed-Clean) spectra taken at hv=10 eV in the perpendicular (a) and parallel (b) configurations after O_2 adsorption of 1.0 L on stepped Ni(332). The arrows indicate the position of surface-sensitive peaks revealed by O_2 adsorption.

SUMMARY

We have investigated the surface electronic structure of stepped Ni(332). In SR-ARUPS spectra, we observed two surface-sensitive peaks at ~0.5 and ~0.1 eV below Fermi level E_F revealed by O_2 adsorption. The former peak disperses downwards away from E_F parallel and perpendicular to the steps, which can be attributed to the (111) terrace-derived two-dimensional surface state. Though the latter peak disperses upwards parallel to the steps, on the other hand, it has almost no dispersion in the perpendicular direction. This flat dispersion is the clear evidence that the peak just below E_F stems from an electronic state localized at the steps. Considering the flat and upwards dispersions perpendicular and parallel to the steps, this step-localized state is concluded to be one-dimensional electronic state.

REFERENCES

1. J. Kutzner, R. Paucksch, C. Jabs, H. Zacharias and J. Braun, *Phys. Rev. B* **56**, 16 003-16 009 (1997).
2. F. Reinert, G. Nicolay, S. Schmidt, D. Ehm and S. Hüfner, *Phys. Rev. B* **63**, 115415-1-7 (2001).
3. H. Namba, N. Nakanishi, T. Yamaguchi and H. Kuroda, *Phys. Rev. Lett.* **71**, 4027-4030 (1993).
4. H. Namba, N. Nakanishi, T. Yamaguchi, T. Ohta and H. Kuroda, *Surf. Sci.* **357-358**, 238-244 (1996).
5. H. Namba, K. Yamamoto, T. Ohta and H. Kuroda, *J. Electron Spectrosc. Relat. Phenom.* **88-91**, 707-710 (1998).
6. H. Namba, M. Obara, D. Kawakami, T. Nishimura, Y. Yan, A. Yaghishita and Y. Kido, *J. Synchrotron Rad.* **5**, 557-558 (1998).
7. H. Iwai, H. Namba, Y. Kido, M. Taguchi and R. Oiwa, *J. Synchrotron Rad.* **5**, 1020-1022 (1998).

Photoelectron Spectroscopy with Phases of Standing Waves Observed by Total Electron Yield and Reflection Spectra

T. Ejima, A. Yamazaki, K. Sato, and Y. Nakamura

Institute of Multidisciplinary Research for Advanced Materials, Tohoku University
2-1-1 Katahira, Aoba-ku, Sendai, 980-8577 JAPAN

Abstract. Photoelectron spectroscopy using standing waves generated by a reflection multilayer is an effective method for investigating electronic and/or chemical structures at interfaces. Changes in photoelectron spectra are difficult to measure when the standing wave phase is changed, because no known method exists to measure phases in situ. In this study however, we present a method to measure the phases of standing waves in situ by combining the reflection and total electron yield spectra. The photoelectron spectra of Fe/Si/Fe trilayers deposited on a Mo/Si reflection multilayer were demonstrated by the phase results.

Keywords: Standing wave, total electron yield, reflectance, photoemission, multilayer, phase, thickness of uppermost layer, Fe/Si/Fe trilayer
PACS: 68.49.Uv, 73.20.-r, 79.60.-i, 75.47.De

INTRODUCTION

Interfaces are believed to play an essential role in the behavior of current semiconductor devices, which generally consist of layered structures. As one example of such layered structures in magnetism, the giant magnetoresistance (GMR) exhibits enhanced resistance in multilayered structures, consisting of alternating magnetic and non-magnetic layers, following exposure to an external magnetic field. In the case of Fe/Si multilayers, the magnetic behavior is initially ferromagnetic, and then antiferromagnetic, and finally non-coupling with an increase in the non-magnetic Si layer [1]. In spite of many works to investigate this property [2, 3], the origin of the enhancement, and the identity of the material which mediates the interlayer magnetic coupling, are still contentious issues. At present, the electronic structure and material of the interface are roughly classified into either metallic compounds such as Fe-Si compounds [2], or semiconductor materials, such as amorphous Si [3].

One of the most effective methods to investigate the electronic structures of interfaces is photoemission spectroscopy, which uses standing waves generated by Bragg reflections from a reflection multilayer [4]. The node and anti-node positions of the standing waves can move easily within the top layers of the multilayer, therefore photoelectrons excited in these uppermost layers can be effectively scanned through the interface of the layers. Recently, total electron yield (TEY) spectra, measured simultaneously with the reflection spectra, and the TEY peak intensity, varies according to changes in the uppermost layer of a reflection multilayer around the reflection peak [5]. These TEY spectra can be reproduced by reflection spectra using a phase which contains the reflection phase and the thickness of the uppermost layer [6]. In this study, Fe/Si/Fe trilayers are deposited on a Mo/Si reflection multilayer, and the TEY spectra of these multilayers measured in situ prior to measuring the photoelectron spectra. The TEY spectra are expected to clarify that the phase alters according to changes in the incident photon energy and thicknesses of the uppermost Fe/Si/Fe trilayers. Combining this phase information with the photoelectron spectra for the valence band, electronic structures of the Fe/Si interfaces will be investigated.

EXPERIMENTS

Mo/Si multilayers were simultaneously deposited onto $20 \times 10 \text{mm}^2$ Si wafers using a magnetron sputtering system (ANELVA SPL-500). Then, Fe/Si/Fe trilayers were deposited onto the Mo/Si multilayer, where the thickness

CP879, *Synchrotron Radiation Instrumentation: Ninth International Conference*,
edited by Jae-Young Choi and Seungyu Rah

of the central Si layer was changed from 0.7 to 2.2 nm in 0.5 nm steps, while the thickness of the Fe layers were fixed at 3.0 nm. The deposition of Mo and Si was conducted in an Ar atmosphere at a pressure of 2.0×10^{-3} Torr, following pre-evacuation of the chamber at a base pressure of 7.5×10^{-7} Torr. The effective deposition rates of Mo and Si (for the Mo/Si multilayer fabrication), Si (for the Fe/Si multilayer fabrication), and Fe were 1.74 nm/min, 1.71 nm/min, 1.43 nm/min, and 1.49 nm/min, respectively. The periodic lengths of the Mo/Si multilayers were approximately equal to each other (7.63 nm). The thicknesses of the uppermost Fe/Si/Fe trilayers were estimated from the deposition rate and the deposition times. The magnetization measurements suggest that the ferromagnetic behavior in the Fe (3.0nm)/Si (1.7nm)/Fe (3.0 nm) trilayer is consistent with previous results [1].

The TEY and reflection spectra of the multilayers were measured using the reflectometer on the BL5B beamline at the Ultraviolet Synchrotron Orbital Radiation Facility in the Institute for Molecular Science [7]. The incident wavelength was calibrated according to the position of the Si-$L_{2,3}$ absorption edge. The resolving power of the wavelength $\lambda/\Delta\lambda$ (estimated from the width of the Si-$L_{2,3}$ absorption edge) was determined to be approximately 200 at 12.5 nm. The reflectance of the Mo/Si multilayer was 0.61 at an angle of incidence of 10°. The reflectance peak was located at 14.5 nm, and did not change following deposition of the Fe/Si/Fe trilayers.

The photoelectron spectra were measured alongside the corresponding TEY spectra using the photoelectron spectrometer of the BL-18A beamline at the Photon Factory in the High Energy Accelerator Research Organization. The resolving power of the wavelength $\lambda/\Delta\lambda$ was approximately 200 at 12.5 nm. Photoelectrons were measured using VG ADES 500, which has a spherical condenser-type analyzer, with an acceptance angle of photoelectrons of about 1°. The angles of the incident light and emitted photoelectrons were 10° to the sample normal. Clean sample surfaces were obtained by Ar^+ etching the substrates under an Ar pressure of 3.0×10^{-5} Torr. The photoemission measurements were made within 7 hours of cleaning the sample. S-polarized incident light was used in all measurements, and the drain current was measured in the TEY spectra.

RESULTS AND DISCUSSION

The TEY spectra of the [Si/Mo]×20+/Fe/Si/Fe multilayers are shown in Fig. 1(a). The intensities of the TEY spectra were normalized to 1 when the corresponding reflectance is equal to 0, and the intensity is $1 + R + 2\sqrt{R}$ when $\cos\theta = 0$. A peak with satellite peaks was observed at around the 14.4 nm wavelength (Bragg reflection peak position) in the 0 min spectrum, and the peak position was moved from 14.7 nm to 14.1 nm with an increase in the Ar^+ etching time from 0 min to 50 min, respectively. At 60 min, the peak position increased suddenly to 14.9 nm, and then decreased down to 14.8 nm.

Thickness changes in the Fe/Si/Fe trilayer were obtained as follows. At first, to obtain the reflection phase δ, the reflection spectrum of the Mo/Si reflection multilayer before deposition of the Fe/Si/Fe trilayer was measured and the spectral shape of the reflection spectrum reproduced using a computer program [8]. In the spectrum, the values of the optical constants were used in the program [9, 10]. The periodic length obtained from the calculation was 7.63 nm (Si 4.47 nm/Mo 3.16 nm).

When the reflectance R and the reflection phase δ are known, the electron yield intensity I_{TEY} is proportional to:

$$1 + R + 2\sqrt{R} \cos(\delta - 2\xi(d + L)) \tag{1}$$

using thickness d of the uppermost layer, propagation vector ξ, and the attenuation length L of the emitted photoelectrons [10]. In the reproduction procedure of TEY spectral shapes, the term $2\xi(d+L)$ was treated as a fitting parameter. The positions of the local minimum values (represented by arrows in Fig. 1(a)) were reproduced, because

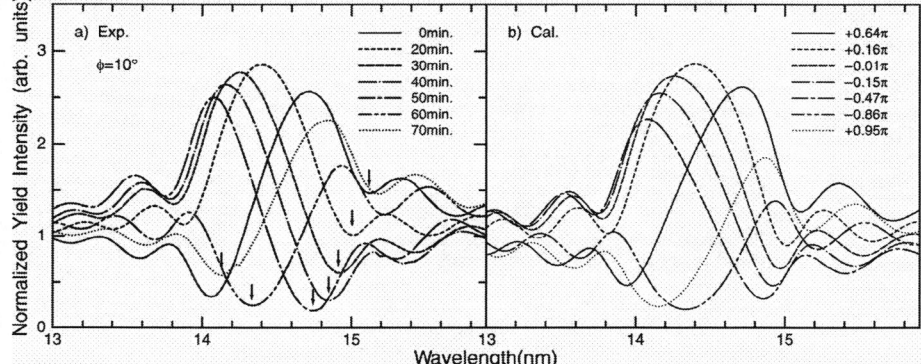

FIGURE 1. Total electron yield spectra changing the Ar^+ etching time (a) and the calculation results changing the phase term (b).

the wavelength of these positions was used to excite photoelectrons. The reproduced spectral shapes are presented in Fig. 1(b) and the parameters obtained from the fitting procedure are listed in Table 1. In the table, the column "phase" represents the term $2\xi(d+L)$, while the column "phase diff. (rad)" corresponds to the phase difference between the adjacent lines. The column "phase diff. (nm)" is transformed into the film thickness from the column "phase diff. (rad)", assuming that the propagation vector ξ corresponds to that observed in a vacuum, and the column "Total Etching Thickness" represents the summation of etched thicknesses of the Fe/Si/Fe trilayer surface before Ar^+ etching. The obtained thicknesses suggest that the etching rate was not proportional to the etching time and Ar pressure.

The photoemission spectra measured after each etching is presented in Fig. 2. The photoelectron spectra from 50min to 70min show the same spectral shape as that observed for iron. According to the spin resolved photoemission study of the Fe (100) surface [11], the Fe bulk valence band has two features: the first is a single band structure at about 0.3 eV below E_F, while the second displays two structures located at 0.7 and 2.3 eV, respectively. These peaks are assigned as minority and majority spin Fe d-bands states, $\Delta_{5\downarrow}$ at 0.3 eV, $\Delta_{1\uparrow}$ at 0.7 eV, and $\Delta_{5\uparrow}$ at 2.3 eV, respectively. These peaks are designated as α, β, and γ, respectively (Fig. 2).

The Fe d-band states are emitted from the substrate-side Fe layer of the Fe/Si/Fe trilayer etched for 60 min and 70 min, and the total etching thicknesses are located at this layer. After etching for 50 min, the total etching thickness is 3.96 nm, indicating that the surface of the trilayer is now located at the middle Si layer. However, the spectral shape resembles that of iron, indicating that the photoelectrons are emitted from the bottom Fe layer. This is due to the difference in the node position of the standing wave excited in the multilayer. The wavelength used to excite the photoelectrons had a local minimum value in the TEY spectra at around 14.5 nm. In the local minimum value, the node of the standing wave is located at the surface of the multilayer, and the incident light energy absorbed in the trilayer is increased according to the direction from the surface to the substrate. Therefore, photoelectrons will be emitted from the deepest region of the trilayer. The attenuation length is estimated at about 1 nm, which is evaluated from both the nominal thicknesses at deposition and the total etching thicknesses.

To investigate the differences in the photoelectron spectra with respect to the node positions, the photoelectron spectra were measured while the incident wavelength was changed from the local minimum value to the local maximum value in the TEY spectrum. The measurement results are shown in Fig. 3, and the exhibited spectra were excited at 15.0 nm (local minimum value) and at 14.4 nm (local maximum value) in the same 20 min etched trilayer. The node position is located in the surface at the 15.0 nm wavelength; therefore the photoelectron spectrum excited by this wavelength reflects the substrate side of the sample. At the 14.4 nm wavelength, the antinode position is located in the surface, such that the photoelectron spectrum reflects the surface side of the sample. In Fig. 3, the spectra were normalized with respect to the peak height around the Fermi level, and the difference spectrum between the two is designated by a solid curve. The two peaks observed in the difference spectrum at 6 and 10 eV correspond to peaks c' and d' in Fig. 2. The spectral shape of the difference spectrum resembles that of amorphous Si [12]. The same spectral shapes are also observed in the difference spectra of the 30min and 40min etched trilayers. Concerning the Total Etching Thickness, these results suggest that the amorphous Si layer will exist at the center of the middle Si layer.

The other spectroscopic properties are summarized as follows. Similar sharp peaks α and β are observed at around the Fermi level in the corresponding 20-40 min etched trilayer spectra. This similarity suggests that peaks α and β in the 20-40 min etched trilayer originate from the Fe d-band states. In addition to these peaks, the small and broad peaks of a and b are observed at 1.3 and 2.8 eV, respectively, while the clear and wide peaks c and d, are observed at 7.0 and 11.0 eV, respectively. Peak a is considered to originate from the Fe3d and/or Si3d states of the Fe-Si compounds located at the interface of the Fe/Si layers. Peaks b, c, and d are deemed to originate from the Si3p and 3s states, on account of the peak positions and the thicknesses of Fe/Si/Fe trilayers.

TABLE 1. Phase values obtained from the TEY spectra and the thicknesses of the Fe/Si/Fe trilayers.

Etching Time (min)	Phase (rad)	Phase Diff.(rad)	Thickness Diff.(nm)	Total Etching Thickness (nm)
0	$+0.64\pi$	----	----	0.00
20	$+0.16\pi$	0.48π	1.71	1.71
30	-0.01π	0.17π	0.61	2.32
40	-0.15π	0.14π	0.50	2.82
50	-0.47π	0.32π	1.14	3.96
60	-0.86π	0.39π	1.39	5.35
70	$+0.95\pi$	0.19π	0.68	6.03

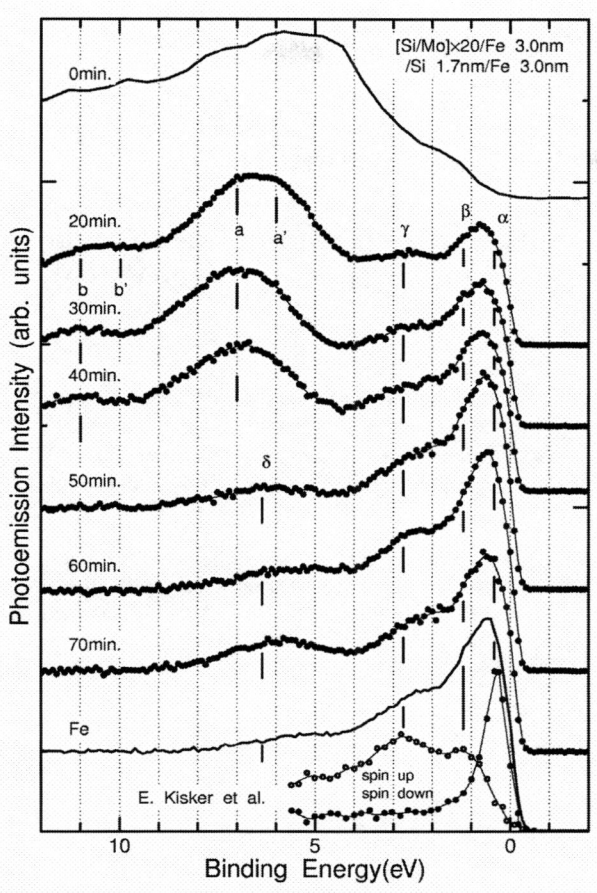

FIGURE 2. Photoelectron spectra with increase of Ar⁺ etching time.

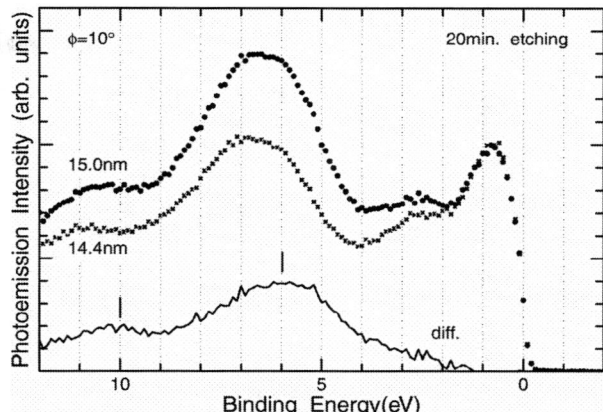

FIGURE 3. Photoelectron spectra with the difference of incident wavelength..

SUMMARY

The thicknesses of Fe/Si/Fe trilayers deposited on Mo/Si reflection multilayers were obtained using both reflection and total electron yield spectra. The photoelectron spectra were obtained with the corresponding depth profiles of the uppermost Fe/Si/Fe trilayer. These results suggest that the present method for measuring the thickness of the uppermost layers on a reflection multilayer will enable us to investigate the electronic structures at the interfaces of a layered structure.

REFERENCES

1. Eric E. Fullerton, J. E. Mattson, S. R. Lee1, C. H. Sowers, Y. Y. Huang, G. Felcher, S. D. Bader, and F. T. Parker, J. Magn. Magn. Mater. **117**, L301(1992).
2. P. Bruno and C. Chappert, Phys. Rev. Lett., **67**, 1602(1991), W. Baltensperger and J. S. Helman, Appl. Phys. Lett., **31**, 2954 (1990), and Z.-P. Shi, P. M. Levy, and J. L. Fry, Phys. Rev. B **49**, 15159 (1994).
3. D. M. Edwards, J. Mathon, R. B. Muniz, and M. S. Phan, Phys. Rev. Lett. **67**, 493(1991), M. D. Stiles, Phys. Rev. B **48**, 7238 (1993), and P. Bruno, J. Appl. Phys. **76**, 6972 (1994).
4. S.-H. Yang, B.S. Mun, N. Mannellal, S. -K. Kim, J. B. Kortright, J. Underwood, F. Salmassi, E. Arenholz, A. Young, Z. Hussain, M. A. Van Hove, C. S. Fadley, J. Phys., Condens. Matter. **14**, L407 (2002).
5. T. Ejima, A. Yamazaki, T. Banse, and T. Hatano, J.Electron Spectrosc. Relat. Phenom., **144-147,** 897 (2005).
6. T. Ejima, T. Harada, A. Yamazaki, Appl. Phys. Lett., **89**, 021914 (2006).
7. M. Sakurai, S. Morita, J. Fujita, H. Yonezu, K. Fukui, K. Sakai, E. Nakamura, M. Watanabe, E. Ishiguro, and K. Yamashita, Rev. Sci. Instrum. **60**, 2089(1989).
8. D. L. Windt, Comp. Phys., **12**, 360 (1998).
9. Hand book of Optical Constants of Solids I, II, and III, edited by E. D. Palik, (Acad. Press. Inc., New York & London, 1985, 1991, and 1998).
10. B. L. Henke, E. M. Gullikson, and J. C. Davis, At. Data Nucl. Data Tables **54**, 181 (1993) at http://www-cxro.lbl.gov/optical_constants/.
11. M. Eddrief, M. Marangolo, V. H. Etgens, S. Ustaze, F. Sirotti, M. Mulazzi, G. Panaccione, D. H. Mosca, B. Lepine and P. Schieffer, Phys. Rev. B **73**, 115315 (2006).
12. R. Cimino, F. Boscherini, F. Evangelisti, F. Patella, P. Perfetti, and C. Quaresima, Phys. Rev. B **37**, 1199 (1988).

XAFS Measurement System for Nano, Bio and Catalytic Materials in Soft X-ray Energy Region

Shinya Yagi, Toyokazu Nomoto, Takaki Ashida, Kazuya Miura, Kazuo Soda, Kazue Yamagishi[*], Noriyasu Hosoya[#], Ghalif Kutluk[+], Hirofumi Namatame[+], and Masaki Taniguchi[+]

School of Engineering, Nagoya University, Furo-cho, Chikusa-ku, Nagoya, 464-8603 Japan
[*]*School of Dentistry, Tokyo Medical and Dental University, Yushima, Bunkyo-ku,Tokyo, 113-8549 Japan*
[#]*Department of Physiology, Tsurumi University, 2-1-3 Tsurumi, Tsurumi-ku, Yokohama, 230-8501 Japan*
[+]*Synchrotron Radiation Center, Hiroshima University, Kagamiyama, Higashi-Hiroshima, 739-8526 Japan*

Abstract. In soft X-ray energy region, there are really many absorption edges for the constituents of the useful materials. We have reported the powerful XAFS measurement system at soft X-ray energy beamline on small or compact storage ring HiSOR and also recent NEXAFS spectra data about nano, bio and catalytic materials. The XAFS measurement system has a Be window, which can separate the ultra-high vacuum (UHV) and the atmospheric pressure condition. Moreover the XAFS spectra can obtain by yielding a fluorescent X-ray emitted from the samples.

Keywords: XAFS measurement system, He path, soft X-ray energy region, HiSOR
PACS: 07.85.Qe, 78.70.Dm

INTRODUCTION

A XAFS measurement system for nano, bio and catalytic materials in soft X-ray region has been reported. These materials are usually using under atmospheric pressure and/or in water environment. Moreover these materials do not have a good electrical conductivity. Therefore the XAFS measurement system with hard X-ray energy region is effective to investigate the surface reaction on those materials. Because we do not need to consider a vacuum condition or a charge-up, we can easily obtain the XAFS spectra with using hard X-ray. In soft X-ray energy region of 1800-3200 eV, however, there are many important elements in the catalytic and bio materials field. Those elements have K- or L-absorption edge in soft X-ray region. In previous study, we have constructed the XAFS measurement system under atmospheric pressure with He path [1]. In case using He path, the soft X-ray can easily reach at the sample without decreasing X-ray photons and the emitted fluorescence X-ray from the sample can also reach at the X-ray detector. Thus we can carry out the XAFS measurements for the surface and interface reactions. In this paper we introduce some XAFS spectra with using synchrotron light of the soft X-ray energy region about the catalyst made from the noble metal nanoparticle, the bio materials and the molecular adsorption reaction at the interface between solid surface and solvent molecule.

XAFS MEASUREMENT SYSTEM

All XAFS measurements were carried out at soft X-ray beamline on Hiroshima Synchrotron Radiation Center (HiSOR: 700 MeV) [2,3]. Figure 1 (a) and (b) show the XAFS measurement system under atmospheric pressure connected with the surface XAFS measurement system under UHV condition with the Be window (thickness of 20 μm) [1]. And also synchrotron light (SL) comes through the Be window. A soft X-rays within the range of 1800-3000 eV can penetrate the Be window by 60 % or more. All XAFS spectra were obtained by yielding the fluorescent X-rays with an UHV-compatible gas-flow type proportional counter (FPC) with P-10 gas (10% CH_4 in Ar).

CP879, *Synchrotron Radiation Instrumentation: Ninth International Conference*,
edited by Jae-Young Choi and Seungyu Rah
© 2007 American Institute of Physics 978-0-7354-0373-4/07/$23.00

FIGURE 1. Photographic views of (a) Be window part and (b) whole XAFS system under atmospheric pressure.

EXAMPLE DATA OF NEXAFS MEASUREMENTS AND DISCUSSIONS

Pd-L₃ Edge Nexafs Measurement of Pd Nanoparticle

Figure 2 shows the Pd-L$_3$ edge NEXAFS spectra of a Pd nanoparticle and bulk Pd. The Pd nanoparticle was fabricated by the gas evaporation method (GEM). A measurement samples were prepared by the GEM under low pressure of He (about 10^4 Pa) [4,5]. The Pd nanoparticle was deposited on a polycrystalline nickel substrate at 300 K. After deposition, the sample was brought out from the GEM chamber and set in the XAFS measurement system. Therefore the Pd nanoparticles were exposed to the air at 300 K. There are mainly two peaks at near the absorption edge. Those peaks observed at 3177.5 eV (a) and 3179.7 eV (b) are assigned to the bulk Pd and the oxide Pd phases, respectively. These results imply that the Pd nanoparticles are oxidized by the air, nevertheless the bulk Pd is not easily oxidized by the air at room temperature.

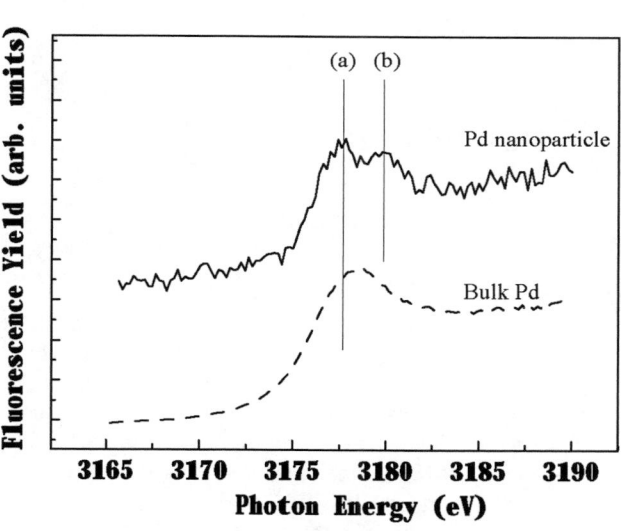

FIGURE 2. Pd-L3 edge NEXAFS spectra for Pd nanoparticle and bulk Pd.

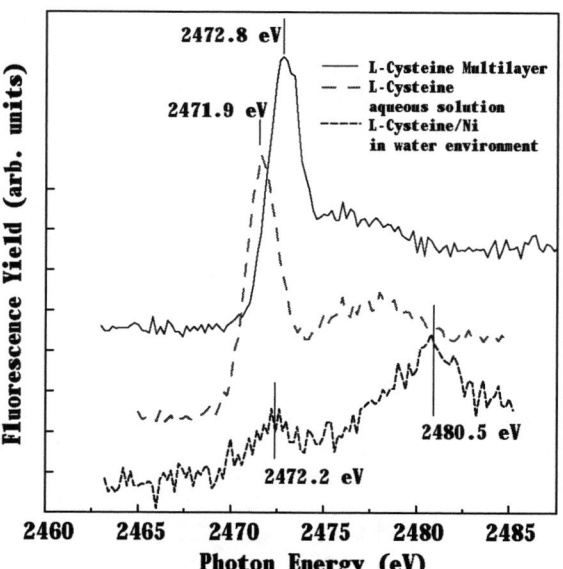

FIGURE 3. S-K edge NEXAFS spectra for L-Cysteine multilayer, aqueous solution and adsorbed on nickel substrate under water environment.

S-K Edge Nexafs Measurement of Bio Amino Molecule (Dry and Wet Samples)

Figure 3 shows the S-K edge NEXAFS spectra for L-Cysteine multilayer (dry sample), an aqueous solution (wet sample, 1 mmol) and interface between L-Cysteine and Ni substrate under water environment (wet sample). The L-Cysteine multilayer was prepared by the thermal evaporation method on a polycrystalline copper substrate under high vacuum condition. The L-Cysteine aqueous solution was set inside of the liquid cell with using a polyvinylidene chloride $(CH_2CCl_2)_n$ window [1]. For the interface sample, L-Cysteine molecule adsorbed on the polycrystalline nickel surface inside of the L-Cysteine aqueous solution and the substrate was rinsed by the distilled water. There are three peaks at around 2472 eV. All those peaks are assigned to the excitation from sulfur 1s to σ^* (S-C) orbital. The peak positions show that the intra molecular S-C bond length for the sample in the aqueous solution is qualitatively elongated. Since the peak observed at 2480.5 eV for the interface sample is almost same as the peak of K_2SO_4 (2481.7 eV), the sulfur atom of L-Cysteine seems to have a coordinate bond with some water molecules.

P-K Edge Nexafs Measurement of Dental Root Surface

Figure 4 shows the P-K edge NEXAFS spectra for dental root surface. The dental root was extracted from a socket. There are a strong peak and a shoulder structure in the NEXAFS spectra. Those structure positions are same as that of K_3PO_4 spectrum. This result indicates that the dental root surface is formed by the phosphoric salt component.

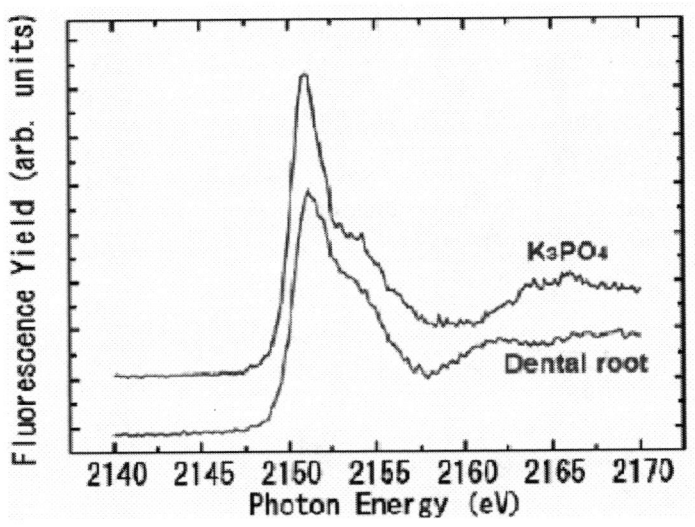

FIGURE 4. P-K edge NEXAFS spectra for dental root surface and K_3PO_4 powder.

Rh-L$_3$ Edge and Cl-K Edge Nexafs Measurements of Rh(PVP) Nanoparticle

We have fabricated a Rh nanoparticle with using PVP(poly-vinyl-pyroridone) as a surfactant molecule, which prevents an aggregation each other. The diameter of the fabricated Rh(PVP) nanoparticle has been controlled by the molecular weight of the PVP [6]. But the fabricated Rh(PVP) nanoparticle has some chlorine atoms inside and/or surface of the nanoparticle. The origin of the chlorine atom is a start material of $RhCl_3$. Figure 5 shows both (a) Rh-L$_3$ edge and (b) Cl-K edge NEXAFS spectra, and TEM (transmission electron microscope) image as inset (c). Both Rh-L$_3$ and Cl-K edge NEXAFS spectra were obtained before/after annealing at 400 °C. For the Cl-K edge results after annealing, both the edge-jump and the peak intensity decrease in comparison with the results before annealing. And the peak position shifts to higher energy side. This result represents the chlorine impurities desorb from the Rh(PVP) nanoparticle and change its chemical condition by annealing. On the contrary, the Rh-L$_3$ edge spectrum after annealing only changes the peak position, which shifts to lower energy side, compared to before annealing. This indicates that the chemical condition of the most Rh atom only changes into the reduction state.

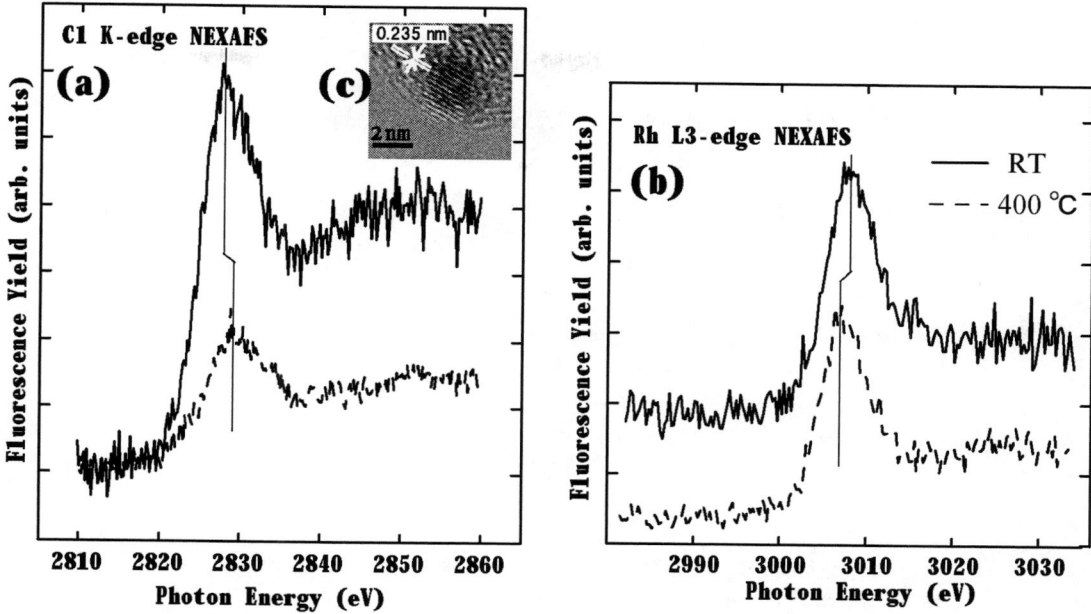

FIGURE 5. (a) Cl-K and (b) Rh-L3 edge NEXAFS spectra for Rh(PVP) nanoparticle before/after annealing at 400 °C. The inset (c) is a TEM image of Rh(PVP) nanoparticle.

SUMMARY

We have reported the XAFS measurement system at soft X-ray energy beamline on HiSOR and recent NEXAFS results about the nano, bio and catalytic materials. The XAFS measurement system under atmospheric pressure is actually easy to use for all different kinds of bulk materials and interfaces. It is found that the NEXAFS measurements using He path for any materials of conductor, semiconductor, insulator, liquid can be executed, if there are absorption edge (for example K-, L- or M-edge) in the covering photon energy.

ACKNOWLEDGMENTS

The authors are grateful for the financial supported of a Grant-in-Aid for Scientific Research from the Ministry of Education, Science and Culture, Japan (No.15360358), 21-Century COE "Isotope Science and Engineering from Basics to Applications" and an Innovation Plaza Hiroshima of JST (Japan Science and Technology Agency) "Taniguchi-MAZDA Project". This work was performed under the approval of HSRC Program Advisory Committee (No. A05-15).

REFERENCES

1. S. Yagi, Y. Matsumura, K. Soda, E. Hashimoto and M. Taniguchi, *Surf. Interface Anal.* **36**, 1064-1067 (2004).
2. M. Taniguchi and J. Ghijsen, *J. Synchrotron Radiat.* **5**, 1176-1179 (1998).
3. S. Yagi, G. Kutluk, T. Matsui, A. Matano, A. Hiraya, E. Hashimoto, and M. Taniguchi, *Nucl. Instrum. Methods Phys. Res.* **A 467-478**, 723-726 (2001).
4. S. Yagi, H. Sumida, K. Miura, T. Nomoto, K. Soda, G. Kutluk, H. Namatame, M. Taniguchi, *e-J. Surf. Sci. Nanotech.* **4**, 258-262 (2006).
5. K. Miura, T. Nomoto, S. Yagi, K. Soda, H. Namatame, M. Taniguchi, in preparation.
6. T. Ashida, K. Miura, T. Nomoto, S. Yagi, K. Soda, H. Namatame, M. Taniguchi, in preparation.

Mechanism of Exfoliation of Clays

Nguyen Tran*, Michael Wilson*, Adriyan Milev*, Gary Dennis*,
G S Kamali Kannangara*, Robert Lamb¶

*School of Natural Sciences, University of Western Sydney, Locked Bag 1797, Penrith South DC 1797, Australia
¶School of Chemistry, University of New South Wales, Sydney 2052, Australia

Abstract. The exfoliated structures of lamellar clays offer potentials as precursor for formation of nano-structured materials. We explored the Synchrotron Radiation soft X-ray techniques and nuclear magnetic resonance to study the exfoliation of phyllosilicate clays by polymers. Experiments were carried out in dispersions containing approximately 1% weight phyllosilicate in 5% aqueous solution of poly(acrylic acid) at different temperatures. The clays were exfoliated as the reaction was performed at 85°C. X-ray photoemission spectroscopy indicated that the exfoliated structures were consisted of virtually pure silica nano-plates. ^{29}Si nuclear magnetic resonance and oxygen K-edge near edge X-ray absorption fine structure indicated that the surface of the plates was terminated by high concentrations of the silanol groups, which created structural branches. The formation of the branches created a steric effect that inhibited the stacking of the plates, which eventually resulted in the exfoliation.

Keywords: XPS, NEXAFS, NMR, clay, exfoliation.
PACS: 81.05.-t, 82.80.Pv

INTRODUCTION

Phyllosilicate clays are assemblies of nano-plates of silicate with the dimension of height to length being 1 : 100 nm or more. The preferential face-to-face stacking of these silicate plates via van der Waals forces leads to formation of the lamellar, layered structures. These relatively inexpensive, readily available clays with low dimensional strutures have been used in the formation of nano-structured materials such as the exfoliated hybrid nanocomposites, which exhibit vastly improved properties compared to the pristine, host materials [1].

Extensive research of the exfoliation of clays prior to the formation of the nanocomposites has been carried out. For example, treatments of the clays using polymers have resulted in the formation of the exfoliated nano-plates of silicate or silica, which were incorporated in a polymer or dispersed in a liquid phase [2]. We present an alternative, one-step procedure for exfoliation of clays and characterise the exfoliated products using photoemission spectroscopy, near edge X-ray absorption fine structure and ^{29}Si nuclear magnetic resonance.

RESULTS AND DISCUSSION

Figure 1 shows the XRD patterns of the untreated Lucentite and the solid products recovered from the reactions of Lucentite with an aqueous solution of poly(acrylic acid). The main peak in the diffraction pattern of Lucentite (2θ approx. 6.5°) was attributed to the interlayer spaces formed by regular stacking of the silicate plates along the c-axis or out-of-plane direction. The interlayer distance estimated from the 2θ value was 1.3 nm. The main peak of the diffraction pattern of the solid product from the reaction of poly(acrylic acid) and Lucentite at 20°C was recorded at 2θ ~5.8°, corresponding to an increase of the interlayer distance to 1.5 nm. This was interpreted as due to the intercalation of fully extended chains of poly(acrylic acid) within the interlayer spaces between the silicate plates.5,6 The out-of-plane peaks were not detected in the diffraction pattern of the solid product from the reaction at 85°C. This indicated that the layered structures were exfoliated.

CP879, *Synchrotron Radiation Instrumentation: Ninth International Conference*,
edited by Jae-Young Choi and Seungyu Rah
© 2007 American Institute of Physics 978-0-7354-0373-4/07/$23.00

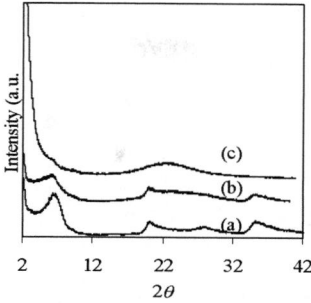

FIGURE 1. Low angle X-ray diffraction patterns of Lucentite (a) and products recovered from reactions of Lucentite and aqueous solution of poly(acrylic acid) at (b) room temperature (20°C) and (c) 85°C

In the high resolution Si 2p XPS scans (Fig. 2), the peak at 102.9 eV for Lucentite and the product at 20°C were primarily due to the Si-O-Mg species. The Si 2p peak from the product at 85°C was recorded at 103.5 eV, in agreement to that of an established silica reference [3]. For the O 1s XPS scan of Lucentite, the peak at 532.0 eV was due to the oxide components. For the O 1s scan of the product at 20°C, the relatively broad peak at 533.0 eV with a shoulder at higher binding energy was mainly due to the Si-O-Mg species of silicate and carboxylic groups of the intercalated poly(acrylic acid). The symmetric O 1s peak for the scan of the product at 85°C was recorded at 532.6 eV, in agreement to that of silica reference. The combined O 1s and Si 2p XPS results indicated that silicate was converted into silica as the reaction was carried out at 85°C.

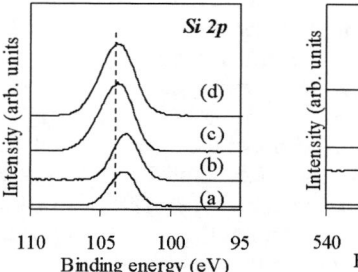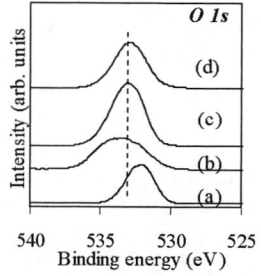

FIGURE 2. High resolution XPS spectra of Si 2p and O 1s photoelectrons in (a) Lucentite, reaction products at (b) 20°C, (c) 85°C and (d) silica reference

Figure 3 showed the ^{29}Si NMR spectra of Lucentite and the solid products recovered from the reactions at varying temperatures. In the spectrum of Lucentite, the main Q3 peak (approx. -94.8 ppm) was interpreted as primarily due to the binding between the tetrahedral SiO_4 units in the lattice and three other SiO_4 tetrahedra, while the Q2 peak (approx. -86.0 ppm) was due to the binding between the tetrahedra at the edges and two tetrahedral [4]. In the spectrum of the reaction product at 20°C, the relative intensities of the Q2 and Q3 peaks reduced significantly compared to those of the spectrum of Lucentite. The presence of the Q4 peak at approx. -110.9 ppm was due to the fully co-ordinated SiO_4 tetrahedral units [5, 6]. This spectrum therefore showed evidence of a partial conversion from silicate into silica plates during the reaction at 20°C. In addition, the presence of the Q3 peak at approx. -101.5 ppm was interpreted as due to silanol -SiOH groups on the plate surfaces [5, 6]. Formation of the silanol groups confirmed that the poly(acrylic acid) chains were intercalated within the interlayer spaces via ion exchanges between the -SiONa components and the protons from the acidic polymer [7]. In the ^{29}Si NMR spectrum of the reaction product at 85°C, the relative intensities of the Q4 and the silanol Q3 peaks were increased more significantly, compared to those of the spectrum of the intercalated product. This confirmed the XPS results and indicated that the exfoliated silica plates had relatively high concentrations of the silanol groups. In addition, the small peak at approx. -91.9 ppm was probably due to the presence of residual silicate, which was not detectable by the long - range XRD measurements.

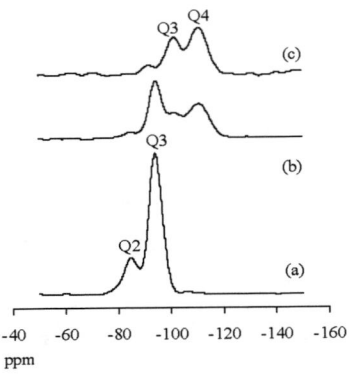

FIGURE 3. ^{29}Si nuclear magnetic resonance spectra of (A) Lucentite, reaction products at (B) 20°C and (C) 85°C

Figure 4 shows the oxygen K-edge NEXAFS spectra of Lucentite, silica and the reaction products at 85°C. The O K-edge spectra of silica and Lucentite were in agreement with those of previously reported [8]. In particular, these spectra were consisted of three regions including the pre-edge (~530 eV), near-edge (≤ ~540 eV) and multiple - scattering regions (> 540 eV). According to the dipole selection rule (ie an electron transition requires a change in angular momentum quantum number of $\Delta l = \pm 1$) [9], the near-edge regions were due to the transition of photoelectrons from $1s$ orbitals to the nearest empty p-states of $2p$ orbitals, similar to those of carbon K-edge electronic transitions [10]. The multiple - scattering regions were primarily due to scattering between electrons and neighbouring atoms. The slight differences of these spectra at the high energy ranges were related to the multiple - scattering of O $1s$ electrons with atoms either in the neighbouring Si tetrahedral or Mg octahedral structures.

The pre-edge peak was previously suggested as due to the surface contamination such as oxygen residues [8]. However, theoretical calculations of O K-edge NEXAFS using different cluster sizes of SiO_2 suggested that this pre-edge peak was primarily due to the formation of branched structures on the cluster surfaces [11]. In our experiments, the pre-edge peak in the O K-edge spectrum of Lucentite (Fig. 4A) was interpreted as being due to the branches associated with the Q2 and Q3 groups on the surfaces and at the edges of silicate plates [12]. The small pre-edge in the spectrum of the silica reference (Fig. 4C) was due to the branches formed by the Q3 silanol groups on the silica surfaces. The main resonances in the O K-edge spectrum of silica were reproduced in that of the product from the reaction at 85°C (Fig. 4B). The significance of the O K-edge NEXAFS measurements remained in the increase in the relative intensity of the pre-edge peak in the spectrum of the product at 85°C, compared to that of the silica reference. This indicated that the concentration of the branches increased with increasing the concentration of the silanol groups on the surfaces of the silica plates.

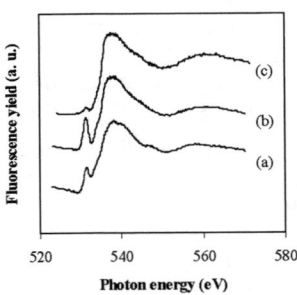

FIGURE 4. O K-edge NEXAFS of (A) Lucentite, (B) reaction product at 85°C and (C) silica reference

Our combined results allowed the development of a simple model, which involves the polymer intercalation, to explain the exfoliation of the lamellar structures. Under acidic conditions, these Q3 and Q2 groups within the lattice and at the edge of silicate plates were converted into Q4, in agreement with Zhu *et al* [4]. The conversion probably involved the replacements of protons with Mg ions at the edges and the tetrahedral SiO_4 network was formed via condensation process [5, 6]. In addition, the polymer intercalation resulted in the formation of further Q3 silanol groups on the surfaces [7]. As the temperatures increased, the mobility of the previously intercalated poly(acrylic acid) chains was also increased and therefore allowed the chains penetrating into the bulk. Within the bulk, the chains may have reacted with octahedral Mg (or Li) ions leading to formation of -COOMg (or -COOLi) complexes, subsequently removed by washing. The removal of magnesium or lithium would create new surfaces consisting of more Q3 silanol groups [6]. Condensations of these Q3 would lead to Q4 and subsequently silica plates. The remaining Q3 silanol groups would increase the concentration of the branches on the surfaces of the silica plates.

The polymer intercalation and the removal of cations by the intercalated polymers or protons would significantly reduce the electrostatic, van der Waals forces between the plates. Also, the increased concentration of the branches on the surfaces would create a steric effect that inhibited the face-to-face stacking of the plates. Together, these resulted in the exfoliation.

ACKNOWLEDGMENTS

This work was funded by University of Western Sydney via Greater Western Post-doctoral Fellowship Scheme. Access to the wide range BL24A, National Synchrotron Radiation Research Centre, Hsinchu, Taiwan was supported by the Australian Synchrotron Research Program. We thank Dr B Gong for XPS support, Drs Y Yang and L Fan for NEXAFS support and Ms V Mellors for providing clay samples.

REFERENCES

1. S. S. Ray and M. Okamoto, *Prog. Polym. Sci.* **28**, 1539 (2003)
2. J. Lin, C. C. Chou and T. Y. Juang, U.S. Patent No. US 2004/0071622 (2004); J. J. Lin and C. C. Chu, U.S. Patent No. 2005/0080180 (2005)
3. A. Cros, R. Saoudi, C. A. Hewett, S. S. Lau and G. Hollinger, *J. Appl. Phys.* **67**, 1826 (1990)
4. H. Y. Zhu, J. A. Orthman, Y. J. Li, J. -C. Zhao, G. J. Churchman and E. F. Vansant, *Chem. Mater.* **14**, 5037 (2002)
5. K. Kosuge, K. Shimada and A. Tsunashima, *Chem. Mater.* **7**, 2241 (1995)
6. K. Okada, A. Shimai, T. Takei, S. Hayashi, A. Yasumori and K. J. D. MacKenzie, *Micropor. Mesopor. Mater.* **21**, 289 (1998)
7. N. Tran, G. Dennis, A. Milev, K. Kannangara, M. Wilson and R. Lamb, *J. Colloid Interface Sci.* **290**, 392, (2005)
8. B. T. Poe, C. Romano and G. Henderson, *J. Non-Cryst. Solids* **341**, 162 (2004)
9. A. Bianconi and A. Marcelli, "X-ray Absorption Near-Edge Structure: Surface XANES", in *Synchrotron Radiation Research, Advances in Surface and Interface Science*, edited by R. Z. Bachrach, New York, Plenum Press, 1992, pp. 63-109
10. J. G. Chen, *Surf. Sci. Rep.* **30**, 1 (1997)
11. V. Nazabal, E. Fargin, G. Le Flem, V. Briois, C. Cartier dit Moulin, T. Buffeteau and B. Desbat, *J. Appl. Phys.* **88**(11), 6245 (2000)
12. 12K. A. Carrado, R. Csencsits, P. Thiyagarajan, S. Seifert, S. M. Macha and J. S. Harwood, *J. Mater. Chem.* **12**, 3228 (2002)

Effect of Phase State of Self-Assembled Monolayers on Pentacene Growth and Thin Film Transistors Characteristics

Hwa Sung Lee, Do Hwan Kim, Jeong Ho Cho, Minkyu Hwang, Kilwon Cho

Department of Chemical Engineering / Polymer Research Institute, Pohang University of Science and Technology, Pohang, 790–784, Korea

Abstract. With the aim of investigating the effect of alkyl chain orientation of self-assembled monolayers (SAMs) on the pentacene growth mode and the performances of organic thin-film transistors (OTFT), pentacene films were deposited on octadecyltrichlorosilane (ODTS) with different alkyl chain orientation. The alky chain orientation of ODTS was controlled by adjusting the reaction temperature. Pentacene films on ordered ODTS were found to have higher crystallinity and better device performances than those on the disordered ODTS, and their differences increased with the substrate temperature. These results can be explained by (1) lattice matching effect between the pentacene crystals and ODTS, and (2) temperature-dependent alkyl chains mobility of ODTS.

Keywords: Self-assembled monolayer, Phase state, Pentacene, OTFT, Chain mobility, Lattice matching.
PACS: 73.50.-h

INTRODUCTION

Among all organic semiconductors, pentacene has been shown to have the highest mobility reported to date [1, 2]. To improve pentacene-TFTs performance, the dielectric surface in organic thin film transistors (OTFTs) is generally modified by self-assembled monolayers (SAMs) [3, 4]. Among various SAM materials, an octadecyltrichlorosilane (ODTS) has been widely used for a simple and effective surface treatment due to the uniform and reproducible surface properties. The effect of surface treatment using ODTS on pentacene structures and device performances was focused on the variation of pentacene grain size, molecular structure, and charge carrier density induced by the change of surface energy [5, 6]. However, the phase states of alkyl chains were affected by the reaction temperature because it has a transition temperature. In spite of this phenomenon, the effects of the conformation of alkyl chains on the pentacene growth mode and the device performance are neglected.

In this work, in order to investigate the alkyl chain orientation of ODTS on pentacene growth mechanisms and its electrical properties, we have controlled the alkyl chain orientation of ODTS. To investigate the structural and morphological properties of pentacene films deposited on dielectrics, X-ray diffraction (XRD) was used. The relationship of these properties to the electrical performance of pentacene TFTs has also been investigated.

EXPERIMENTAL

For the fabrication of the OTFTs, a highly doped p-Si wafer with a 300 nm oxide layer was used as a substrate. The wafer serves as a gate electrode and the oxide layer acts as a gate insulator. To fabricate ordered and disordered alkyl chain structures, ODTS [Gelest] was formed with a dipping method at 4 oC and 65 oC. After ODTS treatment, pentacene (Aldrich Chemicals) films of 50 nm thick were deposited at 0.2 Å/s. Gold was used as the source and drain electrodes. The channel length and width were fixed at 100 μm and 800 μm, respectively.

To investigate the structure of pentacene films, the x-ray diffraction (XRD) was measured at the 8C1 and 3C2 beam at the Pohang Accelerator Laboratory, Korea. It was obtained at scanning intervals of 2θ between 3o and 33o.

In our study of the current-voltage characteristics of the prepared devices, the OTFTs were operated in accumulation mode by applying a negative gate bias, where the source electrode was grounded and the drain

CP879, *Synchrotron Radiation Instrumentation: Ninth International Conference*,
edited by Jae-Young Choi and Seungyu Rah
© 2007 American Institute of Physics 978-0-7354-0373-4/07/$23.00

electrode was negatively biased. All the measurement results were obtained by using Keithley 2400 and 236 source/measure units.

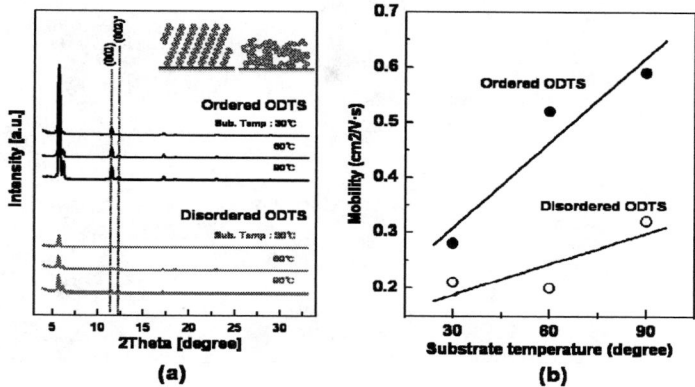

FIGURE 1. (a)X-ray diffraction (XRD) patterns of pentacene films deposited on the ordered (black) and disordered ODTS (with various substrate temperatures (30, 60, 90 °C). (Inset) Ordered and disordered ODTS. (b) Mobilities of TFTs fabricated on the ordered and disordered ODTS with the various substrate temperatures

RESULTS AND DISCUSSION

To investigate the crystalline phase of the pentacene films, synchrotron X-ray diffraction (XRD) measurements were performed. The X-ray patterns of pentacene films on the ordered and disordered ODTS at various substrate temperatures (30, 60, and 90oC) were shown in Fig. 1(a). XRD patterns indicate the presence of two distinct crystalline phases, the "thin-film phase" and the "bulk phase," which are characterized by different d001-spacings, 15.5 ± 0.1 Å and 14.5 ± 0.1 Å respectively. The intensity of the diffraction peaks increased with substrate temperature. Moreover, pentacene films on the ordered ODTS were found to have the higher peak intensity rather than those on the disordered ODTS. That is, these results indicate that the crystallites of pentacene film on ordered ODTS significantly increased against those on disordered ODTS.

The mobilities of pentacene-TFT fabricated on the ordered and the disordered ODTS with the various substrate temperatures were shown in Fig. 1(b). We found that the field-effect mobilities of pentacene films on the ordered ODTS are higher than those on the disordered ODTS. Furthermore, the mobilities of pentacene films on the ordered ODTS significantly increase with the substrate temperature. However, those of pentacene films on the disordered ODTS were not noticeably improved with the substrate temperature.

Figure 2 shows the selected AFM topographs and height profiles of 30 Å thick pentacene films on the ordered and disordered ODTS at various substrate temperatures. On the ordered ODTS (Fig. 2(a), (b), and (c)), the laterally isotropic and flat pentacene domains are formed in the cases of all substrate temperatures. These heights of domains exhibit 2~3 layer thicknesses comparable to the long axis of the pentacene molecule (The thickness of pentacene monolayer is about 1.5~1.6 nm). However, on the disordered ODTS (Fig. 2(d), (e), and (f)), the conglomerated island-type pentacene domains are formed, and their height is elevated. From these result, it can be expected that the conglomerated island-type pentacene domains may have the fewer crystal phase with their (00l) planes parallel to the gate dielectric than those of the laterally isotropic and flat pentacene domains, which results in the lower crystallinities of pentacene films as shown in Fig. 1.

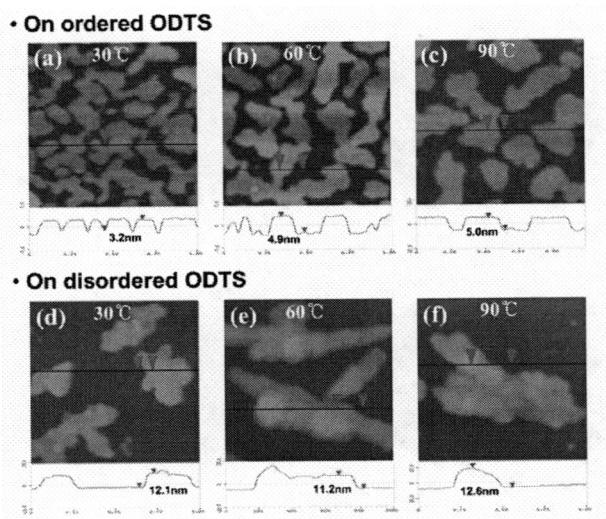

FIGURE 2. AFM images and height profiles of 30 Å thick pentacene film on the ordered [(a), (b), and (c)] and disordered ODTS [(d), (e), and (f)] with various substrate temperatures (30, 60, and 90 ˚C). All images are 1 μm x 1 μm in size.

CONCLUSIONS

To demonstrate the effects of alkyl chain orientation of ODTS on the pentacene growth mode and the pentacene-TFT performances, we used ODTS with different alkyl chain orientation. In our experiments, we found that pentacene films on the ordered ODTS were found to exhibit higher crystallinity and better mobility than those on the disordered ODTS, the their differences are increased with the substrate temperature. Although we do not have unequivocal evidences, these results can be explained as (1) lattice matching effect, and (2) temperature-dependent alkyl chain mobility. However, additional studies and experiments were necessary to understand these phenomena because the detailed mechanism is not clarified yet.

REFERENCES

1. G. Horowitz and M. E. Hajlaoui, *Adv. Mater.* **12**, 1046-1050 (2000)
2. M. A. Loi, E. D. Como, F. Dinelli, M. Murgia, R. Zambon, F. Biscarini, and M. Muccini, *Nat. Mater.* **4**, 81-85 (2005)
3. S. Kobayashi, T. Nishikawa. T. Takenobu, S. Mori, T. Shimoda, T. Mitani, H. Shimotani, N. Yoshimoto, S. Ogawa, and Y. Iwasa, *Nat. Mater.* **3**, 317-322 (2004)
4. I. H. Campbell, J. D. Kress, R. L. Martin, D. L. Smith, N. N. Barashkov, and J. P. Ferraris, *Appl. Phys. Lett.* **71**, 3528-3530 (1997)
5. A. S. Killampalli and J. R. Engstrom, *Appl. Phys. Lett.* **88**, 143125 (2006)
6. C. D. Dimitrakopoulos, P. R. L. Malenfant, Adv. Mater, **14**, 99-117 (2002)

1-D Molecular Chains of Thiophene on Ge(100)

Seok Min Jeon[†], Soon Jung Jung[†], Hyeong-Do Kim[¶], Do Kyung Lim[†], Hangil Lee[¶] and Sehun Kim[†]

[†]Department of Chemistry and School of Molecular Science (BK21), Korea Advanced Institute of Science and Technology, Daejeon 305-701, Republic of Korea
[¶]Beamline Research Division, Pohang Accelerator Laboratory (PAL), POSTECH, Pohang, Kyungbuk 790-784, Republic of Korea

Abstract. The adsorption geometry of thiophene on Ge(100) have been studied by high-resolution core-level photoemission spectroscopy (HRPES) using synchrotron radiation and scanning tunneling microscopy (STM). From the analysis of the Ge $3d$, S $2p$, and C $1s$ core-level photoemission spectra, we found three different adsorption geometries, which were assigned to a dative bonding feature, a [4+2] cycloaddition reaction product, and a desulfurization reaction product. Furthermore, we investigated that the ratio of the components induced by three adsorption geometries changed depending on the molecular coverage and the annealing temperature. At low coverage, the kinetically favorable dative bonding features favorably form 1-D molecular chains. Increasing the molecular coverage, the energetically more stable [4+2] cycloaddition reaction products are additionally created.

Keywords: Thiophene, Ge(100), 1-D molecular chain, HRPES, STM
PACS: R73. 22. -f

INTRODUCTION

Recently, the organic-semiconductor hybrid structure has attracted the great attention of a significant number of science groups because various profits of organic molecules can be incorporated to existing microelectronic technology [1,2]. Among diverse organic molecules, thiophene has widely studied and utilized in a variety of research fields because it is used as a building block of conducting polythiophene. In addition, group IV semiconductors; Si, Ge, and C(diamond), especially have been a subject of great interest not only from a scientific perspective but also due to its applications to the semiconductor industry.

Because a thiophene molecule contains a sulfur atom and has aromaticity, it can adsorb onto semiconductor (100) surfaces through a variety of surface reactions. On one hand, if the conjugated diene participates in a reaction, thiophene undergo a [4+2] cycloaddition reaction similar to that observed for benzene adsorbed on a Ge(100) surface [3]. On the other hand, if only one of the double bonds takes part in the reaction, a [2+2] cycloaddition reaction may occur. Furthermore, if the lone pair electron of the thiophene molecule is donated to the electron-deficient down Ge atom, a Lewis acid-base reaction is anticipated to occur like the case of pyridine on Ge(100) [4]. To date, several research groups have reported about the adsorption of thiophene on a Si(100) surface [5-8], but a few have been published about that on a Ge(100) surface [8,9]. They recently concluded that the conjugated diene of a thiophene molecule underwent the [4+2] cycloaddition reaction with a surface dimer of the Si(100) surface to form 2,5-dihydrothiophene-like species at room temperature [5-8]. Due to the similarity of the reconstructed surface structure and surface reactions between Si(100) and Ge(100), we could predict that the similar reaction would be able to occur on the Ge(100) surface.

In the present study, we have investigated the adsorption of thiophene onto a Ge(100) surface at room temperature by core-level photoemission spectroscopy (HRPES) using synchrotron radiation and scanning tunneling microscopy (STM).

CP879, *Synchrotron Radiation Instrumentation: Ninth International Conference*,
edited by Jae-Young Choi and Seungyu Rah
© 2007 American Institute of Physics 978-0-7354-0373-4/07/$23.00

RESULTS AND DISCUSSIONS

Based on earlier studies for thiophene on Si(100) [5-8] and Ge(100) [8,9] as well as our DFT calculations [9], we considered four most possible adsorption geometries for thiophene on Ge(100), a dative bonding state, [2+2] cycloaddition bonding state, [4+2] cycloaddition bonding state, and decomposed bonding state. In the dative bonding state, the sulfur atom of thiophene donates its lone pair electrons to the electron-deficient down Ge atom. In this case, the Ge-S bonding is so weak that this bonding state is expected to be developed at low temperature and be kinetically favorable. Second, in the [2+2] cycloaddition bonding state, one of double bonds of a thiophene molecule perform cycloaddition reaction with a Ge(100) dimer. The similar bonding state for the adsorption of the thiophene molecule onto Si(100) was theoretically compared with a [4+2] cycloaddition adduct by Lu et al. using the hybrid density functional (B3LYP) method in combination with a cluster model approach [6]. They concluded that the more thermodynamically and kinetically favorable adsorption structure between a [4+2] adduct and a [2+2] adduct was the former one. Similarly, after we performed analogous calculations about thiophene on Ge(100) systems, we led to the same result, namely, the [4+2] cycloaddition reaction is more preferred both thermodynamically and kinetically [10]. Therefore, we exclude the possibility to form the [2+2] cycloaddition product. Third, the [4+2] cycloaddition bonding state is formed via a Diels-Alder cycloaddition reaction which results in strong chemical bonding between the α-carbons of a thiophene molecule and a reconstructed dimer of a Ge(100) surface. Lastly, the decomposed bonding state can be created by desulfurization of the [4+2] adduct via S atom migration to a neighboring Ge dimer to form Ge_2S and $C_4H_4Ge_2$.

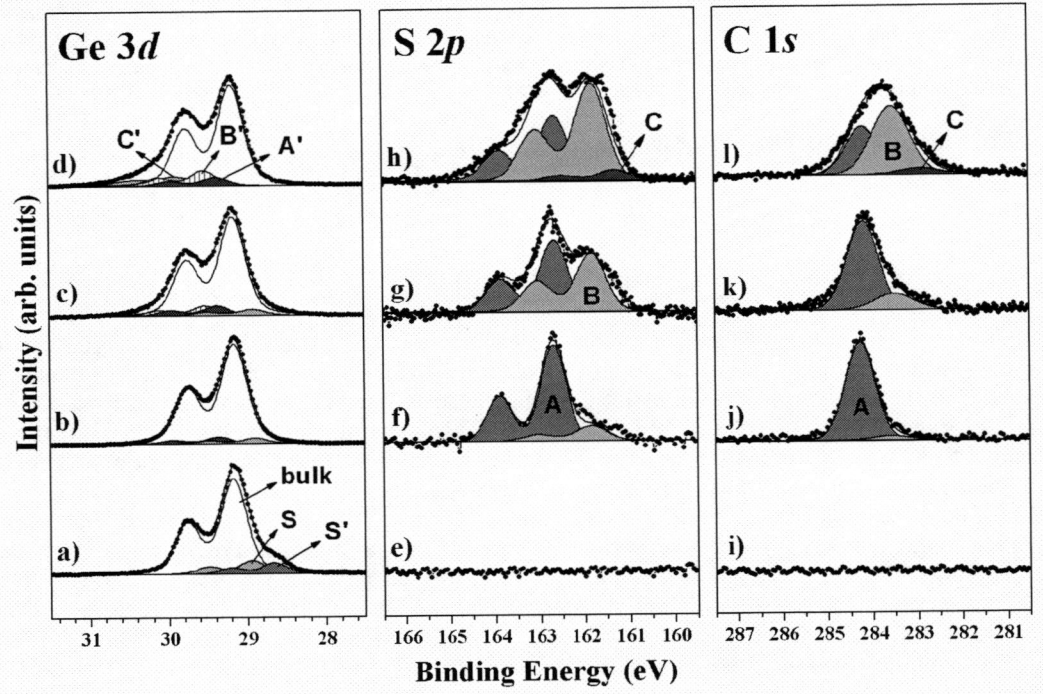

FIGURE 1. Three sets of core-level spectra for the room-temperature adsorption of thiophene on Ge(100) at various coverages. The Ge 3d (left panel), S 2p (middle panel), and C 1s (right panel) core level spectra of the clean (a, e, and, i), 25 L (b, f, and, j), 50 L (c, g, and, k), and 200 L (d, h, and, l) thiophene systems are displayed from bottom to top, respectively. The photon energies were set at hv = 100 eV (Ge 3d), 500 eV (S 2p), and 500 eV (C 1s). The overall resolution was 0.03 eV for the Ge 3d spectrum and 0.25 eV for the S 2p and C 1s spectra. The dots are experimental values and the solid lines represent the results of peak fitting.

Figure 1a shows a clean Ge 3d core level spectrum, which is resolved into three well-defined peaks; bulk, subsurface (S), and up-dimer surface (S'). The binding energies are assigned to 29.2 eV (bulk), 29.0 eV (subsurface), and 28.7 eV (up-dimer surface), which are well consistent with the previous result [10]. We concurrently measured S 2p (Fig. 1e) and C 1s (Fig. 1i) core-level spectra to confirm the cleanness of the Ge(100) surface. As shown in these spectra, we could not identify any peaks caused by impurities or residual gases. After confirming the clean surface, we exposed the clean surface to various amounts of thiophene at room temperature and tracked variations of the bonding states representing adsorption structures of thiophene on Ge(100). Fig. 1b illustrates the Ge 3d peak after exposing a clean Ge(100) to 25 L of thiophene at room temperature. As shown in this figure, we note the remarkable change in this peak. First, the surface state (S′) is fully disappeared and subsurface state (S) is being decreased, which means that thiophene molecules start to be adsorbed on the Ge(100) surface. Second, the new peak (A′) with a small portion starts to come out, which locates at 0.2 eV higher than the bulk Ge 3d peak. We elucidate that a small amount of thiophene causes the appearance of this peak (A′) at binding energy of 29.4 eV, which we assign to the dative bonding state. Fig. 1f and j display S 2p and C 1s peaks, respectively after exposing of 25 L of thiophene at room temperature. In the S 2p core level spectra, we resolve two bonding states; one is located at 162.7 eV (A) and the other is located at 161.8 eV (B). We clearly assign these two peaks to the dative bonding state and the [4+2] cycloaddition bonding state. As shown in this peak, two bonding states coexist even though the dative bonding state is dominant at this coverage. The C 1s peak displays a similar trend. Specifically, the C 1s core level spectrum can be resolved into two components; one is located at 284.2 eV (A) and the other is at 283.5 eV (B). Similar to S 2p core level spectrum, these two components can be assigned to the dative bonding state and [4+2] cycloaddition bonding state with the dative bonding state dominating. Therefore, through the variation of bonding states for three types of core-level spectra, we elucidate that dative bonding state is dominant at 25 L exposure of thiophene. To more precisely observe the changes of intensities among bonding states as a function of molecular coverage, we continually increased the exposure of thiophene. Fig. 1c,g and k display three core-level spectra (Ge 3d, S 2p, and C 1s) after exposing a clean Ge(100) surface to 50 L of thiophene at room temperature. In Fig. 1c, we find a new peak (B′), which is located at the binding energy of 29.5 eV. Considering this shift toward the higher binding energy region, we expect that this bonding feature is more strongly bonded than the dative bonding state. Accordingly, we consider this feature as the species arising from the [4+2] cycloaddition reaction between a Ge(100) dimer and the conjugated diene of a thiophene molecule. Fig. 1g and k show S 2p and C 1s core-level spectra recorded after exposing to 50 L of thiophene at room temperature. Comparing these spectra with the corresponding spectra for the 25 L exposure, we clearly recognize that the intensity of the [4+2] cycloaddition bonding state (B) is more enhanced relative to the dative bonding state (A) when thiophene exposure increases from 25 L to 50 L, even though the intensity of the later bonding state is still larger at this coverage. To clarify the change of the two bonding states as a function of molecular coverage and to verify whether any other feature appears, we increased exposure of thiophene up to 200 L at room temperature (Fig. 1d, h, and l). As shown in Fig. 1d, the subsurface feature (S) disappears and new bonding state is manifestly shown with a binding energy of 29.8 eV (C′) that is 0.6 eV higher than the bulk peak. From previous reports, the binding energy shift of Ge$_2$S against the bulk peak is 0.66 eV [11] and 0.67 eV [12], which are similar to the shift of 0.6 eV investigated here for the C′ bonding feature. We assign this state (C′) to the strong bonding state between a Ge(100) dimer and a sulfur atom. Therefore, we expect that the sulfur atom of the thiophene molecule directly link to the surface Ge atoms through a desulfurization reaction. To see more precise change for each component, we examined the S 2p (Fig. 1h) and C 1s (Fig. 1l) core level spectra. In these spectra, we identify a new bonding peak (C′) is positioned at 161.2 eV (S 2p) and 282.9 eV (C 1s). We speculate that this peak corresponds to the decomposed bonding feature, although the peak for this state has the slightest intensity among the three adsorbed bonding features.

We summarize coverage dependent behavior of thiophene on Ge(100) at room temperature as follows. Three bonding states (dative bonding state, [4+2] cycloaddition bonding state, and decomposed bonding state) coexist at room temperature even though they show different intensities as a function of thiophene exposure. When the thiophene exposure is 25 L, the dative bonding state dominates, indicating that it is kinetically favored. However, the ratio of the [4+2] cycloaddition bonding state to the dative bonding state grows up when the exposure of thiophene increases up to 50 L. Further increase of molecular exposure to 200 L results in the appearance of the decomposed bonding state. The binding energies of each component are summarized in Table 1.

TABLE 1. Curve fitting results for the predicted structures shown in Fig. 1.

Component	Predicted Geometry	Binding Energy (eV)		
		S 2p	C 1s	Ge 3d
A	dative bonding	162.7 ± 0.1	284.2 ± 0.1	29.4 ± 0.1 (A′)
B	[4+2] cycloaddition	161.8 ± 0.1	283.5 ± 0.1	29.5 ± 0.1 (B′)
C	decomposition	161.2 ± 0.1	282.9 ± 0.1	29.8 ± 0.1 (C′)

Considering the bonding states investigated in HRPES, we observed filled-state STM images at a variety of thiophene exposures at room temperature. Fig. 2 show filled-state STM images of Ge(100) surfaces that have been exposed to 2 L and 50 L of thiophene at room temperature, respectively. After exposure to 2 L of thiophene (Fig. 2a), the thiophene molecules create 1-D molecular chain structures (Feature A) on the Ge(100) surface. Carefully analyzing the position of the bright protrusion of feature A, which is located on the middle of neighboring Ge atoms along dimer row direction, we conclude that the feature is assigned to the dative bonding state. In order to confirm the existence of [4+2] cycloaddition bonding feature, we increase thiophene exposure up to 50 L (Fig. 2b). The resulting surfaces show not only the molecular chain structure on every second dimer row but also additional bright and round-shaped structures (Feature B) between the chains. Based on the evidence that feature B is located on top position of the every second Ge dimer along dimer row direction and is symmetric against the mirror plane halving Ge dimers along dimer row direction, we assign the feature to the [4+2] cycloaddition bonding state.

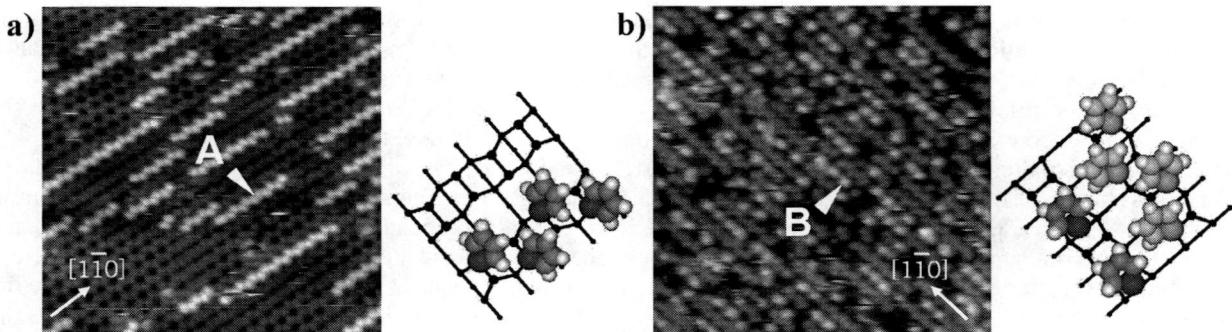

FIGURE 2. Filled-state STM images (20×20 nm^2, $V_s = -2.0$ V, $I_t = 0.1$ nA) and schematic models of clean Ge(100) surfaces that have been exposed to (a) 2 L (0.13 ML) and (b) 50 L (0.26 ML) of thiophene at room temperature.

We conclude that at room temperature, adsorption of thiophene at low coverages lead to the growth of 1-D molecular chains of thiophene on Ge(100) via kinetically favorable Lewis acid-base reaction. On increasing the coverage to over 0.25 ML, however, a thermodynamically stable [4+2] cycloaddition reaction product additionally appeared between adjacent molecular chains.

ACKNOWLEDGMENTS

This research was supported by the Brain Korea 21 project, the SRC programs (Center for Nanotubes and Nanostructured Composites and the Center for Strongly Correlated Material Research) of MOST/KOSEF, the National R & D Project for Nano Science and Technology, Korean Research Foundation Grant No. KRF-2005-070-C00063, and the Basic Research Program of KOSEF Grant No. R01-2006-000-11247-0.

REFERENCES

1. M. A. Filler and S. F. Bent, *Prog. Surf. Sci.* **73**, 1 (2003).
2. J. T. Yates *Science* **279**, 335 (1998).
3. A. Fink, D. Menzel, and W. Widdra *J. Phys. Chem. B* **105**, 3828 (2001).
4. Y. E. Cho, J. Y. Maeng, S. Hong, and S. Kim *J. Am. Chem. Soc.* **125**, 7514-7515 (2003).
5. M. H. Quao, Y. Cao, F. Tao, Q. Liu, J. F. Deng, and G. Q. Xu *J. Phys. Chem. B* **104**, 11211-11219 (2000).
6. X. Lu, X. Xu, N. Wang, Q. Zhang, and M. C. Lin *J. Phys. Chem. B* **105**, 10069-10075 (2001).
7. M. Shimomura, Y. Ikejima, K. Yajima, T. Yagi, T. Goto, R. Gunnella, T. Abukawa, Y. Fukuda, and S. Kono *Appl. Surf. Sci.* **237**, 75-79 (2004).
8. G. B. D. Rousseau, V. Dhanak, and M. Kadodwala *Surf. Sci.* **494**, 251-264 (2001).
9. S. M. Jeon, S. J. Jung, D. K. Lim, H. -D. Kim, H. Lee, and S. Kim *J. Am. Chem. Soc.* accepted (2006).
10. E. Landmark, C. J. Karlsson, L. S. Johansson, and R. I. G. Uhrberg *Phys. Rev. B.* **49**, 16523-16533 (1994).
11. T. Weser, A. J. Bogen, B. Konrad, R. D. Schnell, C. A. Schung, and W. Steinmann *Rhys. Rev. B* **35**, 8184-8188 (1987).
12. G. W. Anderson, M. C. Hanf, P. R. Norton *Appl. Phys. Lett.* **66**, 1123-1125 (1995).

CHAPTER 11
MAGNETISM AND SPINTRONICS

Pulsed Magnetic Fields for an XAS Energy Dispersive Beamline

Peter van der Linden[1], Olivier Mathon[1], Thomas Neisius[2]

[1] *European Synchrotron Radiation Facility, B.P.220, F-38043 Grenoble CEDEX, France*
[2] *CRMCN-CNRS, Campus de Luminy CASE 913, F-13288 Marseille CEDEX 09, France*

Abstract. Pulsed magnetic fields constitute an attractive alternative to superconducting magnets for many x-ray techniques. The ESRF ID24 energy dispersive beamline was used for pulsed magnetic field room temperature XMCD measurements on GdCo$_3$. The signal has been measured up to a magnetic field of 5.5 Tesla without signs of deterioration.

Keywords: pulsed magnetic field, XMCD, energy dispersive
PACS: 75.25+z; 07.55.Db; 07.85.-m

INTRODUCTION

High magnetic fields up to 15 Tesla for synchrotron radiation research have until recently only been generated by superconducting magnets. To overcome this field limit there is a growing interest in the generation of pulsed magnetic fields at synchrotron radiation sources. Unlike 30 Tesla continuous field resistive magnets these systems only need a modest investment and show low power consumption. They are transportable from one beamline to another, the physical dimensions are well suited for SR experiments and the pulse time structure allows for many different experiments to be designed.

Very high magnetic fields in the 10 to 50 Tesla range can be reached in pulsed mode using three different technologies depending on the pulse duration. At the European Synchrotron Radiation Facility, diffraction experiments with a 30 Tesla (25 ms duration, 110 kJ power supply) have already been reported [1]. As the standard demand for homogeneity ($\Delta B/B < 10^{-4}$ over a 1 cm ball volume) can be relaxed considerably due to the small X-ray focal spots available on highly focused insertion device beamlines (typically below 20 x 20 µm) , the magnet size can be reduced. Thus, a second technology, based on Portable Pulsed Fields developed by Nojiri [2] allows to reach 40 Tesla (<1 ms pulse duration, 1 kJ power supply) with a repetition rate of 17 Hz (at 4 Tesla) on a millimeter size sample. Finally, the Cu microcoil technique can also be used and currently produces magnetic field intensities of 50 Tesla (200 ns pulse duration, mJ power supply) with a repetition rate of 1 Hz on a sample of 150 µm [3].

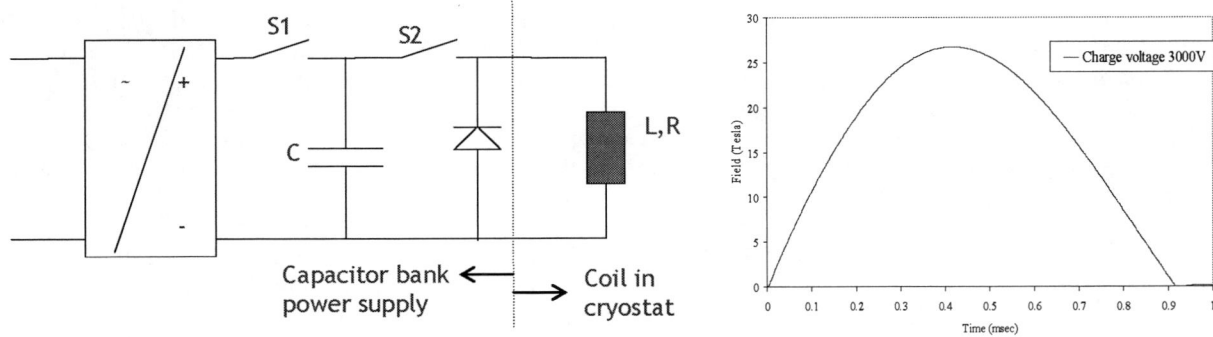

FIGURE 1. Principle of generation of a pulsed magnetic field, and a test pulse of 27 Tesla

CP879, *Synchrotron Radiation Instrumentation: Ninth International Conference,*
edited by Jae-Young Choi and Seungyu Rah
© 2007 American Institute of Physics 978-0-7354-0373-4/07/$23.00

The technique of field generation remains the same in all of these cases (see Fig.1): a DC power supply charges a capacitor through switch S1 which is then discharged over the high field coil. Using a thyristor as S2, the switch closes at zero current yielding a half sine pulse shape. Inserting a crowbar diode antiparallel to the coil yields a damped pulse shape as in Fig. 2.

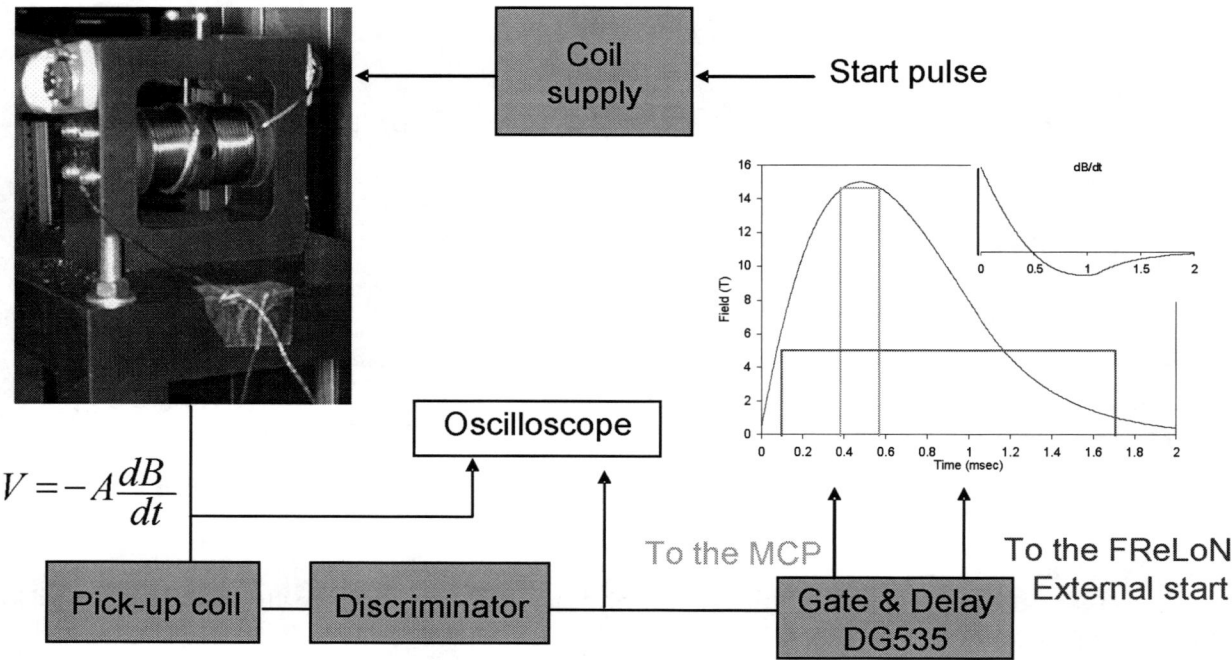

$$V = -A\frac{dB}{dt}$$

FIGURE 2. Timing scheme of data acquisition.

COIL AND PULSE TIMING

The setup presented here has been developed by the ESRF Sample Environment Support Service, a group which was created to provide the beamlines with extreme sample conditioning such as high and low temperatures, high pressure and high magnetic field. Concerning the latter, the ESRF is collaborating with the CNRS-LNCMP (Toulouse, France) to develop a pulsed field installation fixed on one beamline; the in house development of miniature coils targets the use of different beamlines and X-ray measurement techniques.

Unlike solid state properties such as magnetoresistance which can be measured in a single high field pulse, x-ray optical techniques may require the averaging of many pulses. This effect takes a special importance in an XMCD measurement where hundreds of pulses are to be averaged. It is therefore of prime importance to optimise the repetition rate of the pulsed field coil and power supply combination. Thus, the design philosophy is to optimally cool the coil using liquid nitrogen while keeping the inner diameter of the coil sufficiently large to place a small sample cryostat inside the bore of the coil. For this first experiment the copper wound coil dimensions were maintained at $2r_i$=16 mm, $2r_o$=32mm, l=30mm but the coil was water cooled to simplify the setup as the sample was measured at room temperature. Using a 4 kJ (3000V, 1mF) commercially available power supply [10] the coil was calculated for a maximum field of 20 Tesla, but actually used up to 8 Tesla. The half sine shaped magnetic field pulse has a duration of about 3 ms. At the maximum field the applied power supply repetition rate of about one pulse per second was higher than the coil cooling allowed for, resulting in a coil failure. Therefore no data measured at the highest field reached can be presented.

Data acquisition timing was done using two oppositely wound pickup coils giving a pickup voltage $V=AdB/dt$, A being the surface area of the coil (see Fig. 2). The very sharp pickup voltage rise at the start of the pulse allows to precisely set the timing parameters. The two voltages were fed into a discriminator and delay generator to separately control the detector and multi channel plate. In this way the length of the data acquisition window can be traded off against the acceptable error in dB/B for a given measurement.

FEASABILITY EXPERIMENT ON ID24

ID24 is the energy dispersive beamline of the European Synchrotron Radiation Facility dedicated to X-ray Absorption Spectroscopy. The innovative optical scheme of ID24, based on an undulator source coupled to a polychromator through a Kirkpatrick-Baez (KB) mirror system, has natural features such as parallel and fast acquisition, stability during the measurement, high flux and small focal spot. Applications are in a variety of different fields, ranging from the measurement of tiny atomic displacements [4] to time resolved techniques [5], from measurements under extreme conditions [6-8] to micro-XAS studies [9]. The three techniques to generate pulsed magnetic fields described above are fully compatible in terms of photon flux, sample size, pulse duration and repetition rates with an X-ray Absorption measurement on an energy dispersive beamline like ID24 equipped with a gateable detector.

The feasibility experiment reported here proves the possibility to do XAS measurements under pulsed high magnetic fields on an energy dispersive beamline. We have measured the room temperature XMCD spectrum of a GdCo3 sample at the Gd L3 edge in a miniature 8 Tesla pulsed magnet. XMCD is the difference between X-ray absorption spectra obtained with right and left circular polarization. For a finite XMCD signal to be measured the sample must present a net ferromagnetic or ferrimagnetic moment. For symmetry reasons, it is equivalent to record the signal by flipping the magnetic field instead of the circular polarization. The signal was integrated during a 150µs time window located around the maximum of the magnetic field pulse using the gating system described in the previous paragraph. Figure 3 shows the XMCD signal at 5.5 Tesla for left and right circular polarised light.

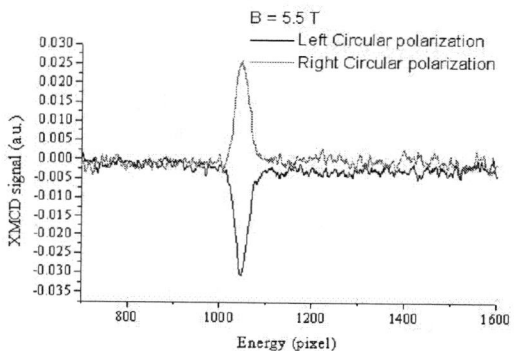

FIGURE 3. XMCD signal for a magnetic field of 5.5 Tesla and left and right polarised light

FUTURE PERSPECTIVES

For the near future the use of a miniature liquid Nitrogen cryostat for the coil with a Helium flow cryostat insert for the sample is planned (see Fig. 4). The coil cryostat is actually under test, the sample cryostat is in construction. The flexible design allows the use of this combination for XAS measurements along the field direction and diffraction perpendicular to the field. Principal characteristics targeted include a maximum field strength of 40 Tesla and sample temperature of 5 Kelvin. At this stage the use of a standard solenoid is planned, in a later stage the split pair coils will be developed. Further development plans include an increase in field strength, lowering of the sample temperature and optimisation of the repetition rate.

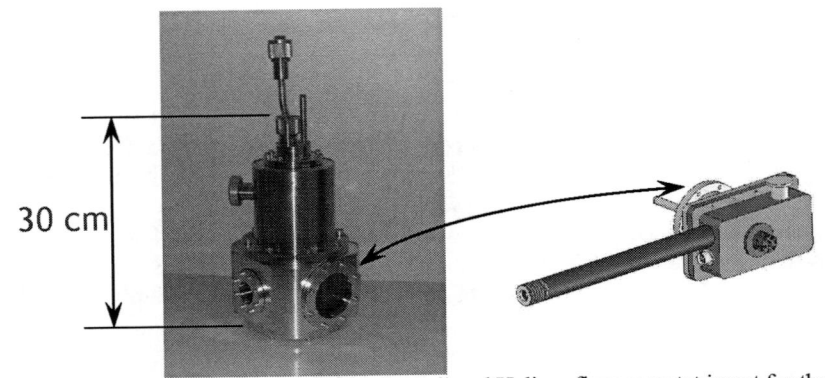

FIGURE 4. Miniature nitrogen cryostat for the coil and Helium flow cryostat insert for the sample

ACKNOWLEDGMENTS

The authors would like to thank W. Joss (Grenoble High Field Magnet Laboratory), P. Frings, G. Rikken, and F. Lecouturier, O. Portugall, N. Ferreira (CNRS - Laboratoire National de Champs Magnétiques Pulsés, Toulouse) and Carsten Detlefs (ESRF).

REFERENCES

1. P. Frings et al., "Synchrotron X-ray powder diffraction studies in pulsed magnetic fields", submitted to *Rev Sci Instr.*
2. Nojiri, H. et al., *IEEE Trans. Appl. Supercond.* **10**, 534 (2000).
3. Mackay, K. et al., *J. Appl. Phys.* **87**, 1993 (2000)
4. R.F. Pettifer, O. Mathon, S. Pascarelli, M.D. Cooke, M.R. Gibbs, *Nature.* **435**, 78 (2005)
5. M.A. Newton, S.G. Fiddy, G, Guilera, B. Jyoti and J. Evans, *Chem. Comm.*, 118 (2005).
6. O. Mathon, F. Baudelet, J.-P. Itié, A. Polian, M. d'Astuto, J.-C. Chervin and S. Pascarelli, *Phys Rev Lett* **93**, 25503 (2004).
7. E. Duman, M. Acet, E. F. Wassermann, J.-P. Itié, F. Baudelet, O. Mathon and S. Pascarelli, *Phys. Rev. Lett.* **94**, 077502 (2005).
8. G. Aquilanti and S. Pascarelli, *J. Condens. Matter* **17**, 1811 (2005).
9. S. Pascarelli, O. Mathon, M. Munoz and J. Susini, "Energy dispersive absorption spectroscopy for micro-XAS applications", submitted
10. www.metis.be, CDMM 4kJ 3kV Capacitive Discharge Magnetizer

Electronic Structures of Hexagonal Manganites HoMnO₃ Studied by X-ray Absorption Near-edge Structure

K. Asokan[1,2], C.L. Dong[2], C.W. Bao[2], H.M. Tsai[2], J.W. Chiou[2], C.L. Chang[2], W.F. Pong[2], P. Duran[3], C. Moure[3], and O. Peña[4]

[1]*Inter-University Accelerator Centre, Aruna Asaf Ali Marg, New Delhi-110 067 India*
[2]*Department of Physics, Tamkang University, Tamkang, Taiwan, ROC*
[3]*Electroceramics Department, Instituto de Ceramica y Vidrio, Campus de Cantoblanco, c/Kelsen 5, 28049 Madrid, Spain*
[4]*Sciences Chimiques de Rennes, UMR-CNRS 6226, Université de Rennes 1, 35042 Rennes, Cedex, France*

Abstract. X-ray absorption near-edge structure (XANES) spectroscopy is sensitive to the chemical environments and the geometrical arrangement of the atoms surrounding the absorbing atom. This aspect has been used to understand the electronic structure of a non-perovskite hexagonal prototype manganites, HoMnO₃. While the pre-edge spectral features observed at O K-edge arise due to the crystal field effects and results in five non-degenerate Mn 3d- O 2p derived orbitals, the Mn K-edge spectra show that Mn ions are in the geometrically frustrated triangular lattices. Above results may provide the possible reasons for the coexistence of magnetism and ferroelectric nature in these manganites.

Keywords: XANES, HoMnO₃, Electronic structure, multiferroics.
PACS: 78.70 Dm; 77.84.-s;75.50.-y

INTRODUCTION

Current research efforts are in the direction of understanding the factors that determine the electronic and structural properties and phase transitions in transition-metal oxide materials. Magnetic and transport properties of these materials are closely related to their charge, orbital, and spin orderings [1, 2]. In recent years, perovskite rare-earth manganites have become the focus of many studies, since they are the parent compounds of the hole-doped oxides showing colossal-magnetoresistance properties [1, 2]. The large-sized R^{3+} cations (for example R= La, Ce, Pr, Nd, Sm, Eu, Tb, or Dy) adopt an orthorhombic structure and exhibit magnetic ordering. Some of the manganites interestingly exhibit two different types of properties: ferroelectric and magnetic, which are known to be mutually exclusive [3, 4]. The RMnO₃ compounds (for R = Ho, Er, Tm, Yb, Lu, Y) with smaller ionic radius crystallize in hexagonal non-perovskite structure and belong to a class of materials, popularly known as multiferroic, characterized by the coexistence of magnetic and ferroelectric orderings [4, 5]. The hexagonal HoMnO₃ system is a prototype multiferroic and its ferroelectric transition occurs around 600-1000 K and the antiferromagnetic transition occurs below 70- 130 K [3-6]. Apart from this, in contrast to RMnO₃ perovskites, no Jahn-Teller distortion takes place in the hexagonal RMnO₃ [4]. A stable hexagonal HoMnO₃ can be described as dense oxygen ion packing (ABCACB) with Mn^{3+} ions having 5-fold trigonal bipyramidal coordination and R^{3+} ions in 7-fold monocapped octahedral coordination. Since the hexagonal structure is formed by trigonal bipyramids, MnO₅, results in change in the local environment or crystal field around Mn ions and hence has direct consequence in the degeneracy of the Mn 3d orbitals that are hybridized with O 2p orbitals [7]. X-ray absorption near edge structure (XANES) is very sensitive spectroscopic tool to understand the chemical environments and the geometrical arrangement of the atoms surrounding the absorbing atom and hence it is an ideal technique to see the electronic structure of these materials [8]. In this communication, we report the electronic structure of hexagonal manganites (HM): HoMnO₃ using the XANES spectra at O, and Mn K-edges with specific references to Mn 3d-O 2p hybridization and identify the

CP879, *Synchrotron Radiation Instrumentation: Ninth International Conference*,
edited by Jae-Young Choi and Seungyu Rah
© 2007 American Institute of Physics 978-0-7354-0373-4/07/$23.00

unoccupied Mn $3d$ orbitals that may be responsible for multiferroic nature in HM. We also include $YMnO_3$ which is also known to belong to the same class for consistency of their spectral features for generalization.

EXPERIMENTAL DETAILS

Manganites were synthesised by standard solid-state reaction from stoichiometric mixtures of the corresponding oxides Ln_2O_3 (Ln=Ho, and Y), and MnO. Samples were characterized by x-ray diffraction (XRD) and magnetic measurements using SQUID susceptometer. The XANES measurements at the O K-edges were carried out using high-energy spherical grating monochromator (HSGM) beamline at the National Synchrotron Radiation Research Center (NSRRC), Hsinchu, Taiwan, operating at 1.5 GeV with a maximum stored current of 200 mA. The spectra were obtained using the sample drain current mode at room temperature and the vacuum in the experimental chamber was in the low range of 10^{-9} Torr. The typical resolution of the HSGM beamline was better than ~ 0.2 eV. The wiggler beamline BL17C of NSRRC was used for the XANES measurements at the Mn K-edge in the transmission and fluorescence mode and the resolution was ~ 0.5eV. All measurements were done at room temperature.

RESULTS AND DISCUSSION

From the structural details derived from XRD show that $HoMnO_3$ posses hexagonal structure with lattice parameters of a=6.134 Å and c=11.406 Å and $YMnO_3$ with a = 6.136 Å, and c = 11.387 Å. These samples were free from impurities since MnO was used. The magnetic moments of Mn ions of $YMnO_3$ and $HoMnO_3$ were found respectively to be 4.81 and 4.74μ_B. These are reported elsewhere [9-11]. These results are consistent with previous studies [3-6]. Figure. 1(a) shows the normalized XANES spectra of $HoMnO_3$ and $YMnO_3$ at O K-edge. For the sake of clarity, the spectra of reference oxides measured during this experiment are not shown in the figure. It may be noted that some results have been discussed elsewhere with different perspective [9]. All these spectra were normalized in the energy range between 550 and 560eV (not fully shown in the figure) after subtracting the background. The O K-edge XANES spectra basically reflect transitions from the O 1s core state to the unoccupied O 2p-derived states. To the first order, one can view the resulting O K-edge spectrum as an image of the oxygen p projected unoccupied density of states. The spectral features, near the threshold i.e., below 538 eV, are attributable dominantly to the hybridization with the Mn 3d orbitals and above threshold correspond to Ho -5d4f and Mn 4sp bands [4, 9,12,13]. For transition metal cations, the shape and occupation of the d-orbitals become important for an accurate description of the properties of the compounds. It has been shown that in a distorted octahedral coordination, the O 2p- and Mn 3d-orbitals hybridize to give crystal field split levels of t_{2g} and e_g bands [12,13]. The d orbitals of the Mn^{3+} ion in an ideal triangular bipyramid of five O^{2-} ions are split by the crystal field into three components: two doubly degenerate states, e_1 (d_{xz}, & d_{yz}), e_2 (d_{xy}, & $d_{x^2-y^2}$), and a_1 ($d_{3z^2-r^2}$) states in the order of increasing energy [7]. The latter orbital has the highest energy as the Mn apical O bond lengths are shorter than in-plane distances. The in-plane d_{xy}, and $d_{x^2-y^2}$ orbitals are strongly hybridized with in-plane oxygen p orbitals. In the ground state one would expect that the four electrons of Mn^{3+} ion occupy four lowest orbitals leaving $3d_{3z^2-r^2}$ orbital unoccupied [7,9]. However, in a non-ideal or frustrated triangular-bipyramidal crystal field environment, these states may non-degenerate states resulting in five discrete Mn 3d levels. As evident from the Fig.1, such a situation is observed in these hexagonal manganites. There are five distinct spectral features below 538 eV with their energy positions centered with respect to $YMnO_3$ at 530, 531.4, 532.8, 534.4, and 535.2 eV (labeled respectively 'a' to 'e') and correspond to five Mn 3d orbitals in the order of increasing energy respectively, $3d_{xz}$, $3d_{yz}$, $3d_{xy}$, $3d_{x^2-y^2}$, and $3d_{3z^2-r^2}$. All above features are distinct but the intensity of 'd' is significantly high (at least four times) compared to other peaks. To identify the occupied and unoccupied Mn 3d orbitals in $HoMnO_3$, one needs to consider the interaction of Mn 3d orbitals with the ligands and the crystal field geometry wherein the lobes of $3d_{xz}$, $3d_{yz}$ and $3d_{xy}$ orbitals are low in energy compared to $3d_{x^2-y^2}$ and $3d_{3z^2-r^2}$ orbitals. Hence out of four electrons from Mn^{3+} ion, three electrons occupy the $3d_{xz}$, $3d_{yz}$ and $3d_{xy}$ orbitals correspondingly the spectral features in O K-edge are with very low intensities. Larger intensity of spectral feature, 'd' in the O K-edge implies that $3d_{x^2-y^2}$ orbital is unoccupied indicating that the fourth electron occupies $3d_{3z^2-r^2}$ orbital whose intensity is very low but comparable with other occupied orbitals. It is intriguing to note that electron occupies a higher energy orbital leaving the lower energy orbital unoccupied. This is accountable by considering the factors: (i) the shape of the low energy occupied orbitals and their repulsive forces on $3d_{x^2-y^2}$ and $3d_{3z^2-r^2}$ orbitals and, (ii) the $3d_{x^2-y^2}$ and $3d_{3z^2-r^2}$ orbitals differ in their shapes and point directly towards the O ligands. Such an environment results in more repulsive force on $3d_{x^2-y^2}$ orbital compared to $3d_{3z^2-r^2}$. The feature C_1 at 538.6 eV corresponds to Ho-5d4f (Y-3d) –O 2p hybridized states[9,12,13]. As expected, an energy shift of ~ +0.5 eV for HoMnO₃ is observed due to the large difference in the ionic radii [9]. The peak labeled D_1 corresponds primarily to Mn $4sp$ –O $2p$ derived states [9,12]. This feature is divided into two distinct bands, (541.4 and 543 eV) and this feature is broader in HoMnO₃ [9]. It may be noted that in spectral features 'a' to 'c' merge into t_{2g} orbitals (A_1) and 'd' & 'e' into e_g orbitals (B_2) if the structure is transformed into perovskite structure [9,12,13].

The normalized XANES spectra at Mn K-edge of the hexagonal HoMnO₃ and YMnO₃ are shown in Fig.1 (b). The energy was calibrated by using a Mn metal foil and the oxides of Mn: MnO, Mn₂O₃ and MnO₂ were also measured as standard references for different valence states of Mn. From the position of main edge of Mn K-edge spectra of HoMnO₃ and YMnO₃, it is found to be in trivalent state consistent with the expected valence state for those compounds. The spectral features are marked as A_2 to E_2. It may be noted that these spectral features are similar to that of C.T. Wu *et al* reported for YMnO₃ [14]. However, the spectral features reported here appears to be somewhat distinct especially features C_2 to E_2. There is a pre-edge feature (A_2) centred at ~6540.5 eV is assigned to 1s-3d transition which is normally forbidden by dipole selection rules, but it becomes allowed with p-d orbital mixing [13,14]. A closer view of this feature is shown as inset. This can occur in a site without a center of symmetry or distortion of a centrosymmetric site. In other words, it reflects that the coordination environment of Mn ion is non-centrosymmetric, there is vacancy in d orbitals, and strong covalency with the metal-ligands. The feature B_2 is attributed to an increased p-character in the lowest empty Mn 3d-like electronic band [13]. The main-edge spectral signatures marked C_2 -E_2 arise due the multiple-scattering effects of the ejected photoelectron wave from the central absorbing atom among its nearest and next-nearest neighbours [8, 14]. These effects also are related to different overlapping of the mixed orbitals and consequently to the main trigonal-bipyramid geometry of the absorbing atom, Mn. Apart from this, we can also infer that Mn 4p states that are highly delocalized and extended over near-neighbors, at least three near-neighbors of Mn ions. From structural consideration, it is known that this compound contains a triangular arrangement of Mn³⁺ cations. If these three Mn ions are in geometrically perfect, one would expect the photoelectron scattering resonances, i.e, intensities of peaks C_2, D_2 and E_2 should have been of same strength due to the multiple scattering processes involved in the XANES region [8]. Since intensities of these peaks are different, it thus shows that the Mn ions are in geometrically frustrated environment.

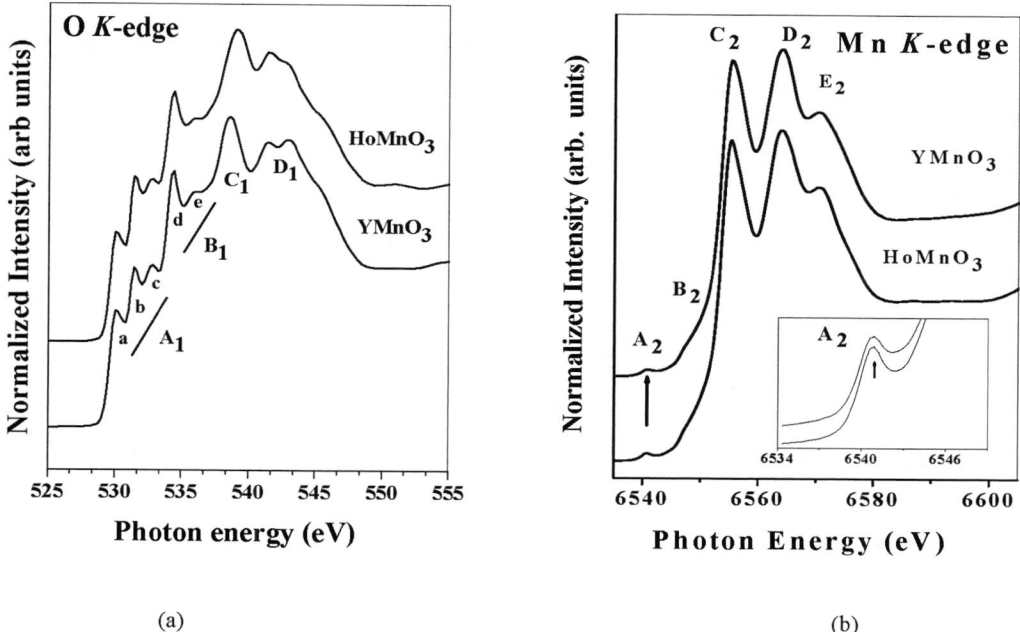

(a)   (b)

FIGURE 1. shows the normalized XANES spectra at (a) O K-edge and (b) Mn K-edges of hexagonal manganites: YMnO₃ and HoMnO₃ and inset figure is the closer view of the pre-edge feature A_2. Prominent spectral features are marked discussed in text.

Above details from XANES at O K-edges are not supported by recent theoretical studies concerning the electronic structures of manganites mainly because of assumptions made in their first principles calculations [15,16].

Theoretically, one would expect for the hexagonal symmetry, the four d electrons of the Mn^{3+} ion entirely occupy the two orbital doublets $(d_{xy}\uparrow, d_{x^2-y^2}\uparrow)$ and $(d_{xz}\uparrow, d_{yz}\uparrow)$, leaving d_z^2 orbital which is the highest in energy, empty. Hence Medvedeva et al and Filippetti et al assumed that $\mathbf{a}_1$ (Mn $3d_{z^2-r^2}$) is totally unoccupied [17,18]. This inconsistency has been highlighted by Kalashnikova and Pisarev while understanding their experimental optical spectra [19]. In other words, holes are present, predominately in $3d_{x^2-y^2}$ orbital. Presence of partially filled $3d$ orbital is necessary for magnetic ordering [20]. J.-S. Kang et al. studied multiferroic $YMnO_3$ and $ErMnO_3$ using both XANES and photoemission measurements and arrived at different conclusions [21].

Previous studies indicated that the magnetic unit cell coincides with the chemical cell and also the magnetic moments of the Mn^{3+} are oriented within the c-plane and makes a triangular arrangement, where the moments of three neighboring Mn^{3+} ions in the same MnO plane and aligned in such a way as to make an angle of 120° each other [5, 22, 23]. Above XANES measurements at Mn K-edges demonstrate that the Mn ions are not in perfect in triangular arrangement. This view is consistent with the findings from other studies [20,22,23]. Such an arrangement is likely to lead to generation of a ferroelectric polarization [20]. Recently, based on XANES measurements at Ho M_5-edge, it was reported that Ho is in trivalent state and does not show additional features after the main peak that can be associated with the mixed valence nature [9].

CONCLUSION

The electronic structures of prototype hexagonal manganites, $HoMnO_3$, studied by x-ray absorption near-edge structure at O and Mn K-edges. Since the environment of Mn ion in HM phase is not an ideal crystal symmetry of trigonal bipyramid, but distorted, all levels non-degenerate into five distinct orbitals and the Mn $3d_{x^2-y^2}$ orbitals are unoccupied indicating that the presence of holes in these orbitals. The XANES spectra at Mn K-edge display that Mn^{3+} ions are in geometrically frustrated environment. Above study demonstrates that the holes in Mn $3d_{x^2-y^2}$ O 2p orbitals and geometrically frustrated environment of Mn ions may be responsible for the coexistence of ferroelectric and magnetic nature in these materials.

ACKNOWLEDGMENTS

One of the authors (K.A) thanks Dr. A. Roy, Director, Inter University Accelerator Centre for encouragement and ICTP, Trieste for short term visiting fellowship, and NSRRC for technical support during measurements.

REFERENCES

1. For review and references: Y. Tokura (ed.) Colossal Magnetoresistance Oxides, Gordon and Breach, The Netherlands 2000.
2. A.P. Ramirez, J. Phys. Condens. Matter 9, (1997) 8171-8199.
3. M.N. Iliev et al., Phys. Rev.B 57 (1998) 2872-2877; Z. J. Huang et al., Phys. Rev. B56 (1997). 2623-2626.
4. B.B. van Aken, "Structural Response to Electronic Transitions in Hexagonal and Ortho-manganites", Ph.D Thesis, Reiksuniversity, Groningen, 2001.
5. T. Lottermoser, et al., Nature 430 (2004) 541-544.
6. A. Munoz et al., Inorg. Chem. 40 (2001) 1020-1028.
7. A.B. Souchkov, et al., Phys. Rev. Lett. 91 (2003) 027203
8. J. Stohr, NEXAFS Spectroscopy, Springer, Berlin, 1992.
9. K. Asokan et al., Solid State Communications,134 (2005)821-826.
10. C. Moure, et al., J. Mater Sci. 34, (1999)2565.
11. O. Peña (private communication).
12. J.-H. Park, T. Kimura, and Y. Tokura, Phys. Rev. B58 (1998) R13330-R1334.
13. K. Asokan etal., J. Phys. Conds. Matter 16 (2004) 3791-3799
14. C.T. Wu et al., Physica B 329-333 (2003) 709-710.
15. M. Qian, J. Dong, and D.Y. Xing, Phys. Rev. B63 (2001) 155101.
16. T. Katsufuji, et al., Phys. Rev. B64 (2001) 104419.
17. J. E. Medvedeva et al., J. Phys: Condens. Matter 12 (2000) 4947.
18. A. Filippetti, and N.A. Hill, Phys. Rev. B65 (2002) 195120.
19. A.M. Kalashnikova, and R.V. Pisarev, JETP letters, 78 (2003) 143.
20. M. Fiebig, J. Phys. D; Appl. Phys. 38 (2005) R123-R152.
21. J.-S Kang et al., Phys. Rev.B 71 (2005) 092405.
22. H. Sugie et al., J. Phys. Soc. of Japan, 71 (2002) 1158-1564.
23. O.P. Vajk et al., J. Appl. Phys. 99 (2006) 08E301.

Element Selective X-ray Detected Magnetic Resonance

J. Goulon*, A. Rogalev*, F. Wilhelm*, N. Jaouen*, C. Goulon-Ginet*, G. Goujon*,
J. Ben Youssef† and M.V. Indenbom†

*European Synchrotron Radiation Facility (ESRF), B.P. 220, 38043 Grenoble Cedex, France
†Laboratoire de Magnétisme de Bretagne, CNRS FRE 2697, UFR Sciences et Techniques, 29328 Brest Cedex
03, France

Abstract. Element selective X-ray Detected Magnetic Resonance (XDMR) was measured on exciting the Fe K-edge in a
high quality YIG thin film. Resonant pumping at high microwave power was achieved in the nonlinear foldover regime and
X-ray Magnetic Circular Dichroism (XMCD) was used to probe the time-invariant change of the magnetization ΔM_z due to
the precession of *orbital* magnetization densities of states (DOS) at the Fe sites. This challenging experiment required us to
design a specific instrumentation which is briefly described.

Keywords: XDMR,XMCD,FMR
PACS: 76.20.q, 76.50.g, 78.20.Ls, 78.70.Dm

INTRODUCTION

X-ray Magnetic Circular Dichroism (XMCD) benefits of the unique advantage to be element/edge selective and
became particularly attractive when magneto-optical sum-rules made it possible to resolve the contributions of spin
and orbital moments at different sites [1]. We discuss below how XMCD can be used to probe *locally* the resonant
precession of orbital magnetization components under the influence of a strong microwave pump field [2]. X-ray
Detected Magnetic Resonance (XDMR) is then a transposition into the X-ray regime of Optically Detected Magnetic
Resonance (ODMR) [3]. An alternative approach to the same physics was explored by Bailey *et al.* who combined
time-resolved soft X-ray reflectometry with Pulsed Induction Magnetometry (PIM) [4]. XDMR spectra recorded in
the frequency domain, however, offer better prospects to extract very weak signals from noise whereas resonance
frequencies up to the THz range could potentially be reached at the ESRF.

As shown in Figure 1, two different XDMR geometries can be envisaged. In the *longitudinal* geometry, the
wavevector $\mathbf{k}_x$ of the incident, circularly polarized (cp) X-rays is nearly parallel to the static magnetic *bias* field $\mathbf{H}_0$

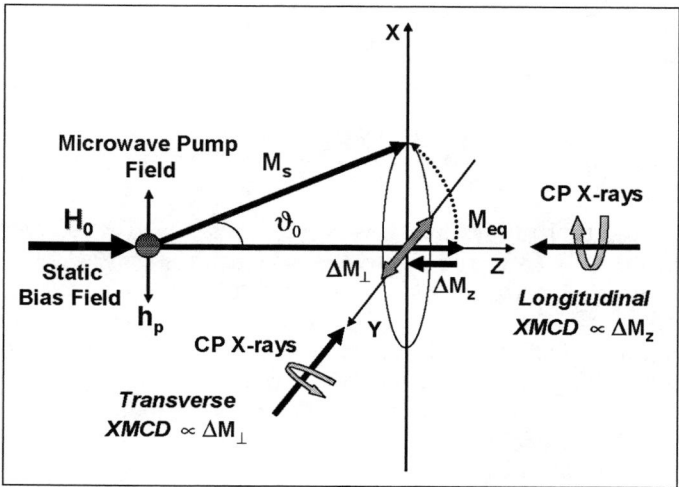

FIGURE 1. XDMR detection in *Longitudinal* and *Transverse* geometries. Note the time-invariant character of ΔM_z.

CP879, *Synchrotron Radiation Instrumentation: Ninth International Conference*,
edited by Jae-Young Choi and Seungyu Rah
© 2007 American Institute of Physics 978-0-7354-0373-4/07/$23.00

whereas the microwave pump field $\mathbf{h}_p$ oscillates in a direction perpendicular to $\mathbf{H}_0$. If one assumes that the length of the equilibrium magnetization (M_s) remains invariant in the precession, there should be along the direction of $\mathbf{H}_0$ a time-invariant change of the magnetization ΔM_z that could be probed by XMCD. In the *transverse* geometry, the wavevector $\mathbf{k}_x$ of the incident cp X-rays would be set perpendicular to both $\mathbf{H}_0$ and $\mathbf{h}_p$: what should now be measured is a stronger XMCD signal proportional to the transverse magnetization $\Delta M_\perp$ but oscillating at the microwave frequency. In both cases, the XMCD/XDMR signal is recorded in the X-ray fluorescence excitation mode.

What makes the longitudinal geometry more attractive is the argument that no fast X-ray detector is required because the XDMR signal is proportional to the microwave power that can be amplitude modulated at low frequency. There is, nevertheless, a price to be paid: since the precession opening angle θ_0 is small, ΔM_z is only a 2nd order effect in comparison with $\Delta M_\perp$ and the signal is always very small. To enhance sensitivity, it is desirable to increase the microwave power at the expense of running into the nonlinear foldover regime of FMR [5]. Fortunately, it was also recognized long time ago by Bloembergen and co-workers [6] that the longitudinal geometry was much less sensitive to magnon-magnon scattering processes and could provide us with a much higher saturation limit regarding the microwave pump power.

XDMR SPECTROMETER

A spectrometer has been designed to record XDMR spectra at high microwave pumping power over the 1-18 GHz microwave frequency range. This spectrometer, which has the capability to detect XDMR spectra both in the longitudinal and in the transverse geometries, is permanently installed on the ESRF beamline ID12 [7]. With its powerful helical undulator sources covering the 1.8-18 keV energy range, this beamline was optimized to record high quality XMCD spectra at the K-edges of all $3d$ transition metals as well as at the L-edges of the rare-earths and all $4d$-$5d$ transition elements.

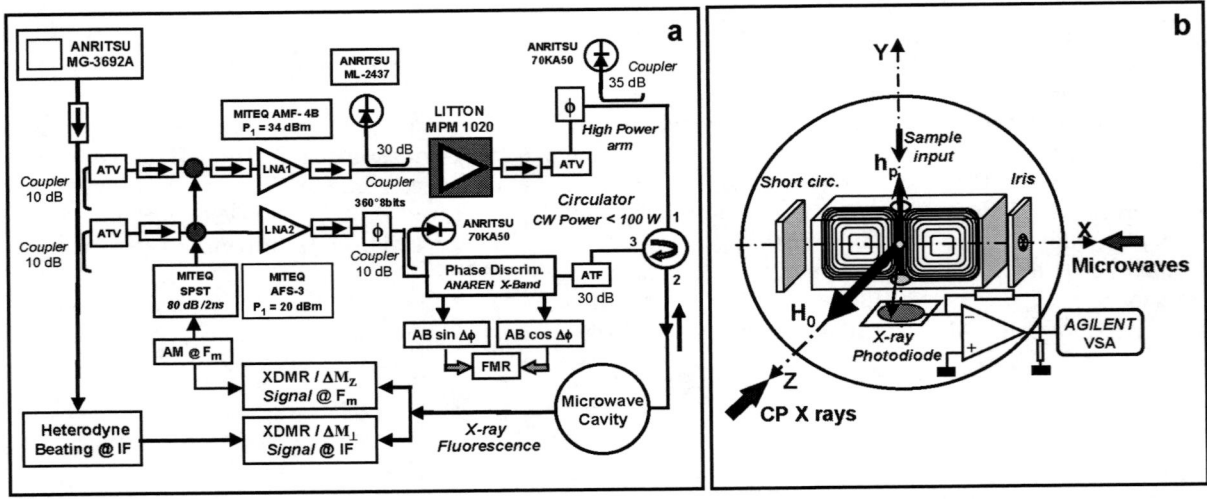

FIGURE 2. a) Arrangement of the microwave X-band bridge. Boxes with arrows are for ferrite isolators; ATV: Variable attenuator; ATF: Fixed attenuator; b) Schematic representation of the XDMR TE_{102} rectangular cavity. The large circle figures out the poles of the electromagnet generating the bias field $\mathbf{H}_0$.

The arrangement of the X-band microwave bridge is shown in Fig. 2a. The microwave source is a wide-band generator with a very low phase noise (Anritsu MG-3692). Two amplification stages make it possible to handle the desired microwave pumping power: (i) a low noise solid state amplifier (LNA-1: Miteq AMF-4B) can deliver up to 32 dBm which was enough to excite our YIG thin film; (ii) a microwave power module (Litton MPM-1020) based on a micro-TWT device can deliver up to 50dBm. For the experiments carried out on YIG which we report in this paper, the 2nd amplification stage was simply by-passed. A circulator (Channel Microwave Inc.), that can accommodate up to 50W(CW) reverse power, allows us to extract the signal reflected back from the microwave cavity. This reflected microwave signal, together with the amplified signal of the reference arm (LNA-2: Miteq AFS-3) are fed into a phase discriminator (Anaren 20758) in order to record routinely *in situ* FMR spectra. For XDMR experiments in the longitudinal geometry, the microwave power is square-modulated using a fast switch (Miteq: SPST 124796) featuring over 80dB isolation with a very short rise/fall time (≤ 2 ns).

Samples ($2x2 mm^2$) are glued on low loss sapphire rods (⌀ 4mm) terminated by a flat surface slightly tilted from the rod axis. The sample holder is inserted in a rectangular TE_{102} X-band cavity (Fig.2b). The resonance frequency of the empty cavity was $F_{cav} \approx 9.450$ GHz with $Q_L \leq 4300$. This home-made cavity had to be modified with respect to commercial EPR cavities: (i) a small hole (⌀3 mm) let the X-ray beam enter the cavity throughout a thin Be window and propagate along the axis of the external modulation coils; (ii) there is at the bottom of the cavity an X-ray detector collecting the X-ray excited fluorescence in as large a solid angle as possible. This detector is a PNN^+ Si photodiode optimized for us by Canberra-Eurisys to keep a low capacitance (≤ 11 pF) for an active area of 80 mm^2 [8]. The detector is carefully shielded by an ultra-thin Be window which also prevents any direct detection of microwave that could result in unwanted artefacts. The detector readout electronics combines a home-made, magnetically shielded, ultra-low noise preamplifier with a multichannel Vector Spectrum Analyzer (Agilent Technologies Inc. VSA 89600-S) exploiting 23-bits digitizers. The dynamic range of the detection system was checked to exceed 126 dBc.

EXPERIMENTS

The sample was a high quality thin film of YIG (8.9 μm thick) grown by liquid phase epitaxy on a (111) GGG substrate. The FMR spectra exhibited a rich pattern of very sharp lines assigned to magnetostatic modes. However, the resonance of the uniform mode vanished due to a very high radiation damping effect. The signal could be recovered by overcoupling the cavity ($Q_L \approx 800$) and off-setting the microwave frequency from F_{cav} by as much as 50 MHz . In a true perpendicular magnetization geometry, the linewidth of the uniform mode was found to be quite narrow: $\Delta H_{fwhm} = \sqrt{3}\Delta H_{pp} = 1.13$ Oe. A Gilbert damping factor of $6.0 \ 10^{-5}$ was extracted from a detailed conventional FMR study: such a low value of the damping factor can only be obtained from high quality thin films.

For our XDMR experiment, the sample was inclined by: $\beta_N = -30°$ in order to prevent the X-ray fluorescence from being heavily reabsorbed. This resulted in a broader FMR linewidth: $\Delta H_{fwhm} = 3.64$ Oe but also in a smaller microwave frequency offset (5 MHz). The incident microwave power could be increased up to 30 dBm: there is no ambiguity that resonant pumping occurred then in the non-linear foldover regime with fully distorted lineshapes. The resonance field ($\mathbf{H}_0$) was carefully scanned down to the onset of the critical foldover jump at $\mathbf{H}_{C2} = 3980.6$ Oe, whereas the X-ray monochromator was independently tuned to 7113.74 eV, *i.e.* to the maximum of the XMCD signal in the pre-peak of the XANES spectrum of the YIG film.

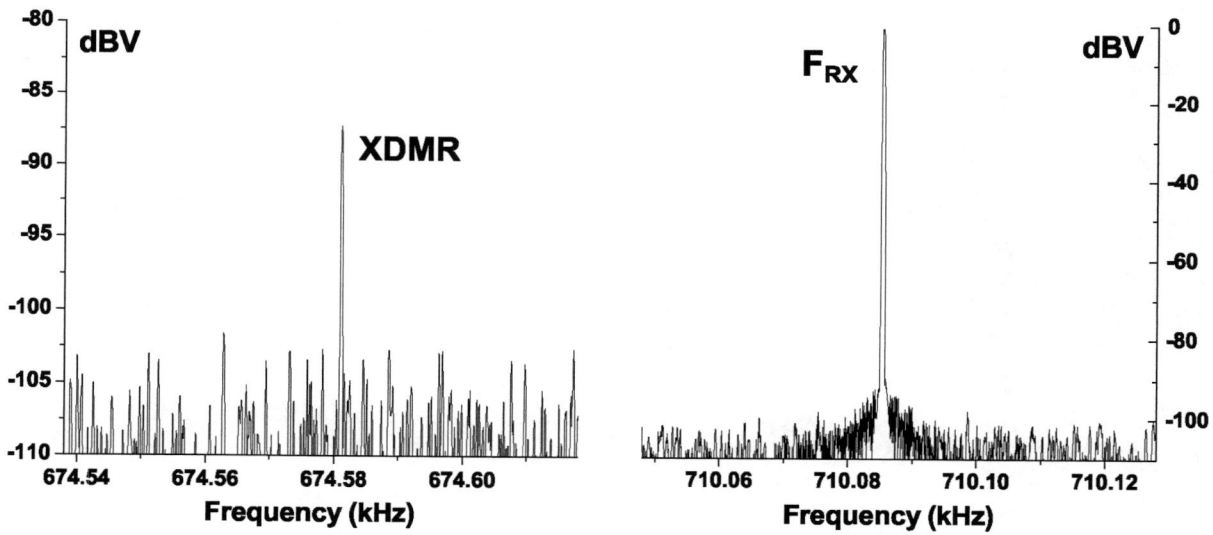

FIGURE 3. Low-frequency side-band of F_{RX} due to XDMR. Note the high dynamic range of the detection system.

Since the ESRF storage ring was run in the 2x1/3 filling mode, the X-ray beam was modulated at the macrobunch repetition frequency ($F_{RX} = 710.084$ kHz) whereas the microwave amplitude modulation was triggered at $F_m = F_0/10 = 35.5042$ kHz in which $F_0 = RF/992$ is the revolution frequency of the ESRF storage ring. Data acquisition was performed in the synchronous time-average mode of the VSA using an external triggering signal at F_m. Under such conditions, the XDMR signature should show up as modulation sidebands at $F = F_{RX} \pm F_m$. The XDMR signal displayed in Fig. 3 is typically the modulation sideband at $F_1 = 674.58$ kHz: it is peaking *ca.* 20 dBV above the noise floor, which is, however, well above the intrinsic detector noise and is dominated by the statistical noise of the X-ray

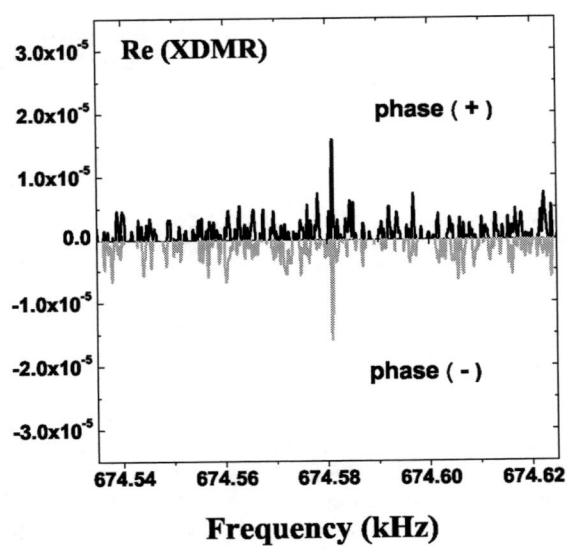
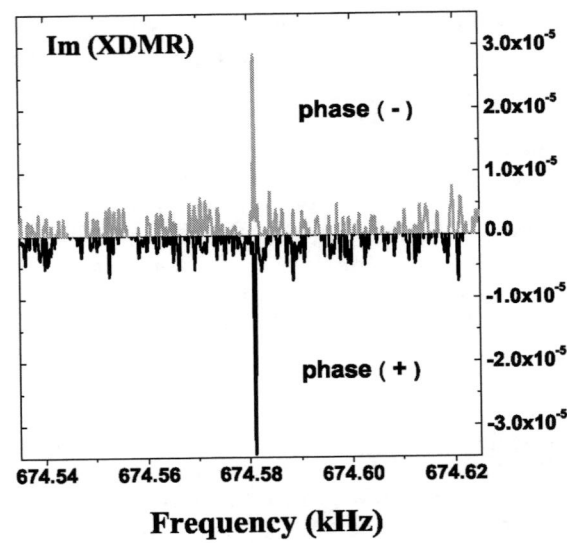

FIGURE 4. Inverted XDMR signatures when the X-ray helicity is switched from left to right, phase (+) and phase (-), respectively.

beam. The signal at F_{RX} was used only for intensity monitoring and data renormalization. This led us to calculate a very small differential cross-section normalized to the edge jump: $\Delta\sigma_{XDMR} \approx 1.34x10^{-5}$. The real and imaginary parts of the XDMR signal are also reproduced in Fig. 4: note that the XDMR, just like XMCD get nicely inverted when the phase of the X-ray helical undulator is inverted, *i.e.* when the helicity of the photons is changed from left to right.

DISCUSSION

On combining properly XDMR with *static* XMCD measurements performed under similar conditions, one could determine quite reliably the average opening cone θ_0 of the precessing moments. In the longitudinal detection geometry ($\mathbf{k}_x = \mathbf{k}_\parallel$), one may write [9]:

$$[\Delta\sigma_{XDMR}(k_\parallel)]/[\Delta\sigma_{XMCD}(k_\parallel)] \simeq -1/2\tan^2\theta_0$$

in which $\Delta\sigma_j$ refers to the relevant differential X-ray absorption cross-sections. Under the experimental conditions described above, the precession cone near critical foldover was found to be: $\theta_0 \simeq 3.5°$. Recall that the effective operators responsible for XMCD at a K-edge are of pure *orbital* nature [1]. This holds true for electric dipole ($E1$) as well as for electric quadrupole ($E2$) transitions. Thus, our XDMR experiment at the Fe K-edge produced for the first time a clear evidence of forced precession of local, orbital magnetization DOS [2]. In this respect, XDMR appears as a unique and very promising tool to investigate the precession dynamics of orbital magnetization components [9].

ACKNOWLEDGMENTS

Technical assistance by S. Feite and P. Voisin during the construction phase of the XDMR spectrometer and during the experiments is warmly acknowledged.

REFERENCES

1. A. Rogalev *et al.*, "X-ray Magnetic Circular Dichroism", in *Magnetism: A Synchrotron Radiation Approach*, edited by E. Beaurepaire *et al.*, *Lectures Notes in Physics* **697**, Berlin: Springer Verlag, 2006, pp. 71–94.
2. J. Goulon *et al.*, *JETP Letters* **82**, 791-796 (2005).
3. K. Gnatzig *et al.*, *J. Appl. Phys.* **62**, 4839–4843 (1987).
4. W.E. Bailey *et al.*, *Phys. Rev. B* **70**, 172403 (2004).
5. A. G. Gurevich, G. A. Melkov "Magnetization Oscillations and Waves", Boca Raton: CRC Press Inc., 1996.
6. N. Bloembergen, S. Wang, *Phys. Rev.* **93**, 72-83 (1954).
7. J. Goulon *et al.*, *J. Synchrotron Rad.* **5**, 232-38 (1998).
8. J. Goulon *et al.*, *J. Synchrotron Rad.* **12**, 57-69 (2005).
9. J. Goulon *et al.*, *Eur. Phys. J. B, submitted* (2006).

Combinatorial *in situ* Growth-and-Analysis with Synchrotron Radiation of Thin Films for Oxide Electronics

M. Oshima[1], H. Kumigashira[1], K. Horiba[1], T. Ohnishi[2], M. Lippmaa[2], M. Kawasaki[3], and H.Koinuma[4]

[1] *School of Engineering, The University of Tokyo, Tokyo 113-8656, Japan*
[2] *Institute for Solid State Physics, The University of Tokyo, Kashiwa 277-8581, Japan*
[3] *Institute for Materials Research, Tohoku University, Sendai 980-8577, Japan*
[4] *National Institute for Materials Science, Tsukuba 305-0047, Japan*

Abstract. We have developed a combinatorial *in situ* growth-and-analysis system combining thin film growths by laser molecular beam epitaxy (MBE) with synchrotron radiation photoemission spectroscopy and X-ray absorption spectroscopy to investigate electronic structures of strongly correlated oxide systems, especially $La_{1-x}Sr_xMnO_3$ (LSMO) for tunneling magnetoresistance device applications. For combinatorial high-throughput analysis, we have developed a computer-controlled manipulator. The usefulness of combinatorial *in situ* analysis was verified by resonant photoemission and angle-resolved photoemission spectroscopy of Mn $3d$ e_g derived majority band of compositional spread LSMO with various Sr compositions and thickness-dependent core-level shifts reflecting band bending.

Keywords: photoemission spectroscopy, oxide electronics, laser MBE, combinatorial growth and analysis
PACS: 72.80.Ga, 79.60.-I, 71.20.-b, 73.20.-r

INTRODUCTION

Strongly correlated transition metal oxides [1] have attracted much attention because of their interesting magnetic and electronic properties such as high-Tc superconductivity, colossal magnetoresistance, half-metallicity [2] and metal-insulator transition. Angle-resolved photoemission spectroscopy is a very powerful tool to reveal electronic structures of these materials. However, atomically flat surfaces are strongly required for these measurements. So far there have been few ARPES studies on transition metal oxides with a three dimensional perovskite structure like $La_{1-x}Sr_xMnO_3$ (LSMO). In order to solve this problem, it is necessary to combine high-resolution synchrotron radiation photoelectron spectroscopy (SRPES) with combinatorial laser molecular beam epitaxy (MBE) technique [3, 4]. By the growth-and-analysis using laser-MBE and SRPES, it is also possible to reveal the epitaxial stress effect and superlattice effect on electronic structures, resulting in development of new materials such as nanodots and nanowires. Furthermore, high-throughput screening of materials by means of synchrotron radiation analysis of combinatorial libraries with various compositions and structures would be possible, since the synchrotron radiation beam is only 0.1mm times 0.5 mm in size at the undulator beamline BL-2C of the Photon Factory.

In this study, we have developed a sample manipulator for synchrotron radiation combinatorial high-throughput analysis of strongly correlated transition metal oxide films. Then, we have characterized electronic structures of $La_{1-x}Sr_xMnO_3$ (LSMO) films with different Sr compositions and thickness on $SrTiO_3$ (STO) substrates.

EXPERIMENTAL

We have developed an *in situ* SRPES system combined with a combinatorial laser MBE system [5]. Figure 1 shows a sample manipulator with θ and ϕ rotation and X,Y and Z translation which are computer-controlled on the LabVIEW software for the combinatorial *in-situ* growth-and analysis system with synchrotron. After growing combinatorial libraries, samples are transferred to the preparation chamber for LEED observation, and then moved

CP879, *Synchrotron Radiation Instrumentation: Ninth International Conference,*
edited by Jae-Young Choi and Seungyu Rah

to the photoemission chamber. Samples used in this study were $La_{1-x}Sr_xMnO_3$ thin films grown on Nb-doped STO substrates using various $La_{1-x}Sr_xMnO_3$ targets with different Sr compositions. SR beam is irradiated onto the different sample position to investigate the compositional dependence of electronic structure. Samples can be rotated along the axis θ and the in-plane axis φ.

A Nd: YAG laser was used for ablation in its frequency-tripled mode (λ = 355 nm) at a repetition rate of 1 Hz. LSMO films with a thickness of about 400 Å were deposited on the wet-etched TiO_2-terminated STO substrates [6] at a substrate temperature of 950 °C at an oxygen pressure of 0.1 mTorr. The intensity of the specular spot in the reflection high-energy electron diffraction (RHEED) pattern was monitored during the deposition to determine the surface morphology and the film growth rate. The epitaxial growth of LSMO thin films on STO substrates was confirmed by the observation of clear oscillations due to the layer-by-layer growth mode. The LSMO thin films were subsequently annealed at 400 °C for 45 min in the atmospheric pressure of oxygen to remove oxygen vacancies. The surface morphology of the measured films was analyzed by atomic force microscopy (AFM) in air.

RESULTS AND DISCUSSION

First we measure photoelectron spectra as an example for combinatorial photoemission spectroscopy of Sr $3d$ core levels of five samples grown on a STO substrate using a combinatorial mask. As shown in Fig. 2, by descending the sample manipulator, five different Sr $3d$ photoelectron spectra for the TiO_2-terminated STO substrate, the 5 ML SrO layer on STO, 1 ML SrO layer on STO, the LSMO layer on 1 ML SrO/STO and the 5 ML LSMO /STO were appearing one by one, showing the usefulness of this combinatorial growth-and-analysis technique. From the higher binding energy to the lower binding energy, the Sr $3d$ peaks can be assigned to be NaCl-type SrO, perovskite-type STO, (La,Sr)O-terminated LSMO, and MnO_2-terminated LSMO, respectively [7].

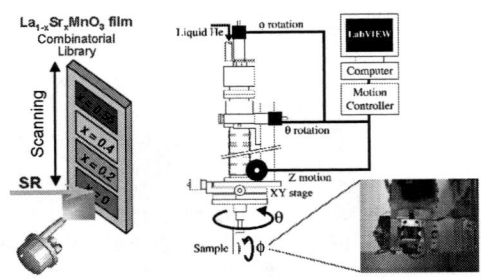

FIGURE 1. A sample manipulator with θ and φ rotation and X,Y and Z translation which are computer-controlled on the LabVIEW software for synchrotron radiation combinatorial high-throughput analysis.

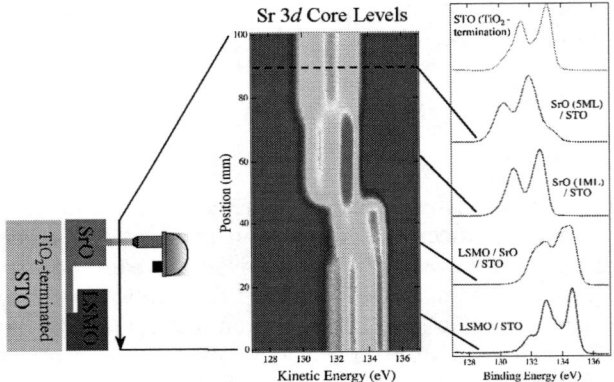

FIGURE 2. An example for combinatorial photoemission spectroscopy of Sr $3d$ core levels of five samples grown on a STO substrate using a combinatorial mask.

We also fabricated a compositional spread $La_{1-x}Sr_xMnO_3$ series with x=0, 0.2, 0.4 and 0.55 on a STO substrate, and measured valence band spectra using the photon energy of 644 eV corresponding to Mn $2p$-$3d$ resonant photoemission [8, 9]. We found the unique evolution of near Fermi level features, showing that the Mn $3d$ e_g peak at 0.8 eV of binging energy is decreasing with increasing the Sr composition, while the Mn $3d$ t_{2g} peak intensity at 2.1 eV remains unchanged. This is a clear evidence for hole doping into the Mn $3d$ e_g states. Then, we performed combinatorial ARPES band mapping of LSMO[10, 11]. Angle-resolved photoemission spectroscopy has long played an important role in band structure study on layered perovskite crystals such as high-T_C superconductors. However, for 3D perovskite structures there have been few ARPES studies simply because they do not have any cleavage plane. Thus, it is quite important to prepare clean and well-ordered surfaces for ARPES measurements. Based on the ARPES spectra of LSMO along the Γ-X direction, we find that there exists a clearly dispersive feature near the Fermi level, as shown in Fig. 3 where we compared the measured band structure with the LDA+U band calculation. The observed electron pocket is originated from the Mn $3d$ e_g x^2-y^2 majority band. Thus, *in-situ* ARPES measurements have revealed the existence of electron Fermi surface in LSMO. Furthermore, we investigated the band structures depending on the Sr composition. It is found that for the antiferromagnetic insulator with x=0 the electron pocket is not seen, while with increasing hole doping the electron pocket is becoming more prominent and shifting upward. Thus, almost all features can be explained by the rigid band shift except for the electron pocket.

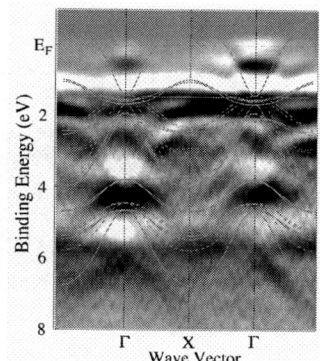

FIGURE 3. ARPES band map of LSMO(x=0.4) in the second derivative form.

Next example is combinatorial analysis for interfacial properties which are very important for TMR elements for MRAM and HDD head applications. TMR elements consisting of half metal should show infinite magnetoresistance ratio based on the Julliere's equation [12]. However in fact, almost no TMR ratio was obtained at room temperature. This phenomenon may be due to a kind of interface degradation. We have already found that with increasing the Sr composition, holes are doped into Mn $3d\, e_g$ state, and the e_g peak intensity decreases, as shown in Fig. 4 from x=0.4 to x=0.55. When a $La_{1-x}Sr_xFeO_3$ (LSFO) film of 1 ML was deposited on LSMO, the e_g peak drastically decreased down to the case of x=0.55. With the 2 ML LSFO overlayer, the e_g peak almost disappeared [13]. This feature can be explained by charge transfer from LSMO Mn $3d\, e_g$ (about 0.8 eV) to LSFO Fe $3d\, e_g$ (about 1.3 eV). The similar reduction of e_g states has been observed in STO/LSMO multilayers [14, 15].

Finally, we investigated the thickness dependence of LSMO. We grew a kind of staircase-like LSMO library and analyzed photoelectron spectra. Figure 5 shows Ti $2p$, Mn $2p$ and O $1s$ core level spectra depending on the LSMO thickness. The Ti $2p$ shift of about 1.3 eV can be explained by the upward band bending of the Nb-doped STO substrate at the LSMO/STO interface, that is a kind of Schottky contact formation of metal/n-type semiconductor. Since the valence band maximum of the Nb-STO substrate whose Fermi level is almost equal to the conduction band minimum is measured to be 3.1 eV, the Schottky barrier height for the LSMO/STO is calculated to be 1.3 eV which is saturated at 5 ML of LSMO. The small Mn $2p$ shift toward lower binding energy can be attributed to the degraded quality of LSMO film at the interface probably due to the strain from the substrate. The large shift of O $1s$ core level as a function of LSMO thickness up to 1.7 eV can be attributed to both the chemical shift change of O $1s$ from STO to LSMO, and the upward band bending (1.3 eV). It is also revealed that the valence band structure is changing from that of STO with about 3.1 eV band gap to an emerging LSMO valence band. The valence band maxima of STO are moving toward lower binding energy with increasing the LSMO thickness, which is in good agreement with the upward band bending observed in Ti $2p$.

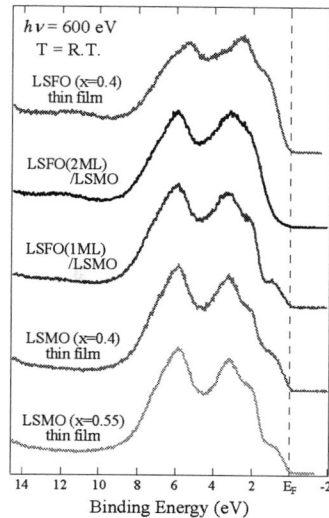

FIGURE 4. Photoelectron spectra of valence band from $La_{1-x}Sr_xMnO_3$ films with x= 0.4 and 0.55 grown on a $SrTiO_3$ substrate, LSFO films (1 ML and 2 ML) grown on LSMO/STO, and LSFO grown on STO.

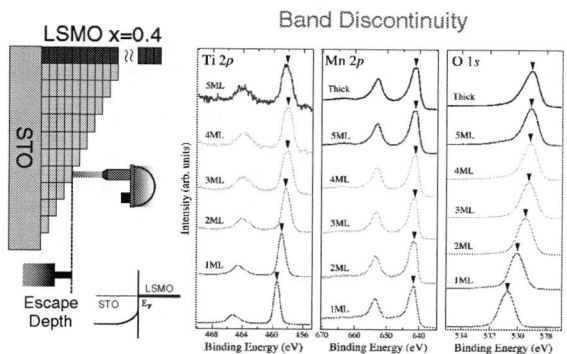

FIGURE 5. Thickness dependence of Ti $2p$, Mn $2p$ and O $1s$ core levels for LSMO/STO. Photon energy was 600 eV.

SUMMARY

We have developed a combinatorial *in situ* growth-and-analysis system combining thin film growths by laser molecular beam epitaxy (MBE) with synchrotron radiation photoemission spectroscopy and X-ray absorption spectroscopy to investigate electronic structures of strongly correlated oxide systems, especially $La_{1-x}Sr_xMnO_3$ (LSMO) for tunneling magnetoresistance device applications. For combinatorial high-throughput analysis, we have developed a computer-controlled manipulator. The usefulness of combinatorial *in situ* analysis was verified by resonant photoemission spectroscopy and angle-resolved photoemission spectroscopy of Mn $3d$ e_g derived majority band of compositional spread LSMO with various Sr compositions. Furthermore, thickness-dependent valence band features and core-level shifts have been analyzed in a combinatorial manner.

ACKNOWLEDGMENTS

This work was supported by a Grant-in-Aid for Scientific Research (S17101004) from JSPS. Synchrotron radiation photoelectron spectroscopy measurements were performed under the project 02S2-002 at the Institute of Materials Structure Science in KEK.

REFERENCES

1. M. Imada, A. Fujimori, and Y. Tokura, Rev. Mod. Phys. **70**, 1039 (1998), and references there in.
2. J. –H. Park, E. Vescovo, H. -J. Kim, C. Kwon, R. Ramesh, and T. Venkatesan, Nature **392**, 794 (1998).
3. H. Koinuma and I. Takeuchi, Nature Materials 3, 429 (2004).
4. T. Ohnishi, D. Komiyama, T. Koida, S. Ohashi, C. Stauter, and H. Koinuma, A. Ohtomo, M. Lippmaa, N. Nakagawa, and M. Kawasaki, T. Kikuchi and K. Omote, Appl. Phys. Lett. **79**, 536 (2001).
5. K. Horiba, H. Ohguchi, H. Kumigashira, M. Oshima, K. Ono, N. Nakagawa, M. Lippmaa, M. Kawasaki, and H. Koinuma, Rev. Sci. Instrum. **74**, 3406 (2003).
6. M. Kawasaki, K. Takahashi, T. Maeda, R. Tsuchiya, M. Shinohara, O. Ishiyama, T. Yonezawa, M. Yoshimoto, and H. Koinuma, Science **266**, 1540 (1994).
7. H. Kumigashira, K. Horiba, H. Ohguchi, K. Ono, M. Oshima, N. Nakagawa, M. Lippmaa, M. Kawasaki, and H. Koinuma, Appl. Phys. Lett. **82**, 3430 (2003).
8. K. Horiba, A. Chikamatsu, H. Kumigashira, M. Oshima, N. Nakagawa, M. Lippmaa, K. Ono, M. Kawasaki, and H. Koinuma, Phys. Rev. B **71**, 155420 (2005).
9. H. Kumigashira, K. Horiba, H. Ohguchi, D. Kobayashi, M. Oshima, N. Nakagawa, T. Ohnishi, M. Lippmaa, K. Ono, M. Kawasaki, and H. Koinuma, J. Electron Spectr. Rel. Phenom. **136**, 31-36 (2004).
10. A. Chikamatsu, H. Wadati, M. Takizawa, R. Hashimoto, H. Kumigashira, M. Oshima, A. Fujimori, N. Hamada, T. Ohnishi, M. Lippmaa, K. Ono, M. Kawasaki, and H. Koinuma, J. Electron Spectr. Rel. Phenom. **144-147**, 511-514 (2005).
11. A. Chikamatsu, H. Wadati, H. Kumigashira, M. Oshima, A. Fujimori, N. Hamada, T. Ohnishi, M. Lippmaa, K. Ono, M. Kawasaki, and H. Koinuma, Phys. Rev. B 73, 195105 (2006).
12. M. Julliere, Physics Letters A, **54** 225 (1975).
13. H. Kumigashira, D. Kobayashi, R. Hashimoto, A. Chikamatsu, M. Oshima, N. Nakagawa, T. Ohnishi, M. Lippmaa, H. Wadati, A. Fujimori, K. Ono, M. Kawasaki, and H. Koinuma, Appl. Phys. Lett. **84**, 5353 (2004).
14. H. Kumigashira, R. Hashimoto, A. Chikamatsu, M. Oshima, T. Ohnishi, M. Lippmaa, H. Wadati and A. Fujimori, K. Ono, M. Kawasaki, and H. Koinuma, *J. Appl Phys.* **99**, 08S903 (2006).
15. H. Kumigashira, A. Chikamatsu, R. Hashimoto, M. Oshima, T. Ohnishi, M. Lippmaa, H. Wadati, A. Fujimori, K. Ono, M. Kawasaki, and H. Koinuma, *Appl. Phys. Lett., in press*.

Magnetic Circular Dichroism in X-ray Fluorescence Cascade Processes

A. Rogalev*, J. Goulon*, F. Wilhelm*, N. Jaouen*,† and G. Goujon*

*European Synchrotron Radiation Facility (ESRF), B.P. 220, 38043 Grenoble Cedex, France
†Synchrotron Soleil, Saint-Aubin - BP 48, 91192 Gif-sur-Yvette Cedex, France

Abstract. The excellent performances of our 35-element Silicon Drift Diode detector over a very wide energy range (≥ 0.5 keV), allowed us to investigate the existence of magnetic circular dichroism in X-ray fluorescence cascade processes. We report the results of a test experiment performed at the Tb L_{III}-edge using an amorphous $TbCo_2$ thin film which was magnetically saturated with an external magnetic field of 0.5 T. In a single energy scan, two XMCD spectra were obtained by monitoring separately the integrated intensities of L_α and M_β emission lines of the Tb atoms.

Keywords: X-ray Magnetic Circular Dichroism, X-ray fluorescence, Rare-earth elements
PACS: 78.20.Ls, 78.70.En, 75.20.En

INTRODUCTION

The combination of element and orbital selectivity in X-ray absorption spectroscopy and magneto-optical sum rules make X-ray Magnetic Circular Dichroism (XMCD) a very powerful tool in magnetism[1]. However, the interpretation of the L-edges XMCD spectra of rare-earth atoms is still a matter of debate. Their spectral shape and amplitude could not be explained by a simple model and the presence of both strongly localized $4f$ shell and delocalized $5d$ electrons is at the origin of difficulties. It is therefore highly desirable to develop another spectroscopic approach that could give access to the magnetic properies of the $4f$ and the $5d$ states. We tried to investigate the existence of Magnetic Circular Dichroism in X-ray fluorescence cascade processes in a rare-earth ion excited at the $L_{II,III}$ absorption edges.

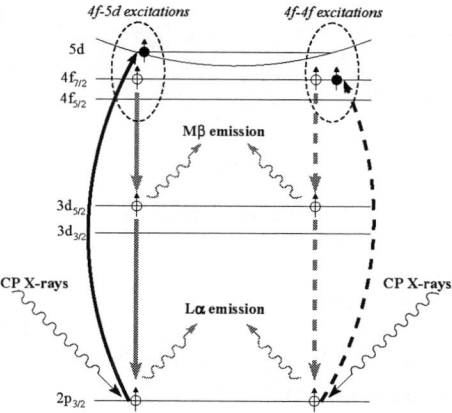

FIGURE 1. The diagram of the X-ray fluorescence cascade in a rare-earth ion following the dipolar (solid line) or quadrupolar (dashed line) excitation at the L_{III} absorption edge.

Our strategy relied on the assumption that, in an XMCD experiment carried out at the L_{III} edge of a rare-earth atom, the hole created in the $2p_{3/2}$ core level should be spin-polarized. Since the primary L_α X-ray fluorescence emission ($3d \rightarrow 2p$ dipolar transitions) should itself create a spin-polarized hole in the $3d$ level, one may expect the latter to recombine radiatively (M_β fluorescence) with one electron of the strongly magnetic $4f$ shell. Therefore, the excitation spectra recorded at the L_{III} edge should exhibit XMCD in both the L_α and M_β emission channels. An

CP879, *Synchrotron Radiation Instrumentation: Ninth International Conference,*
edited by Jae-Young Choi and Seungyu Rah
© 2007 American Institute of Physics 978-0-7354-0373-4/07/$23.00

one-electron picture of this cascade process is shown on Fig. 1. Moreover, the dipolar ($2p_{3/2} \rightarrow 5d$) and quadrupolar ($2p_{3/2} \rightarrow 4f$) XMCD signals detected with M_β fluorescence should provide us with complementary information, since the final states are different and correspond to two different excitations, $4f - 5d$ and $4f - 4f$, respectively. The present contribution deals with the first XMCD measurements in X-ray fluorescence cascade that were performed at the ESRF beamline ID12[2].

EXPERIMENTAL STATION

The experimental station used for our experiment was originally developed for XMCD studies on ultrathin films and is illustrated with Fig. 2. The key element is a 35-element silicon drift diode (SDD) detector array[3] developed in collaboration with Eurisys-Mesures (now Canberra Eurisys). The detector shown in Fig. 2 consists of an array of 7 x 5 cylindrical Si drift-diodes with an active area of 10 mm^2 for each diode. The detector is cryogenically cooled to the optimum temperature ($T \simeq 143K$). The anode diameter (200 μm) accommodates an external J-FET (EuriFET) featuring a very low input capacitance (0.9 pF). This results in a very small readout noise: the FWHM energy resolution of the individual diodes measured with a ^{55}Fe source is as good as 129 eV using a standard pulse processing time of 12 μs, whereas the peak-to-background ratio is in excess of 1000. Under normal operating conditions, the peaking time can be reduced to 0.5 μs in order to maximise the counting rate (10^5 cps), however with some deterioration of the energy resolution.

FIGURE 2. (a) Photograph of the 35-element SDD array now installed on the ESRF beamline ID12. (b) Photograph of the experimental station dedicated to XMCD measurements on ultrathin films.

A well performing multichannel digital pulse processing electronics (XDS boards) has been developed at the ESRF. Digital pulse processing offers a number of significant advantages: the possibility to implement optimal filters; the insensitivity to pick-up noise after digitization; the flexibility to reconfigure or recalibrate the whole system with simple software procedures. Each XDS board designed in the VXI c-size accommodates in a compact environment 4 readout channels and is fully controllable by software. The output of the charge preamplifier is first differentiated and noise whitening is performed; a second amplifier stage which includes an anti-aliasing filter was added to benefit from the full dynamic range of the ADC. Digitization is performed with a 12-bit ADC (10 MHz). The digitized signal is then transferred to a Programmable Digital Signal Processor (Plessey) which carries out the finite impulse response filtering with up to 128 coefficients. Any kind of filter shape can be synthesized with peaking times ranging from 500 ns to 6.4 μs. Powerful FPGAs (XILINX) are used to trigger the pile-up rejection, to restore the baseline according to various algorithms, to detect the pulse maxima and to histogram the pulse heights into memory buffers which are read periodically. Further local 'on-line' pre-processing of the data is also possible. This includes the definition of regions of interest in the spectra, count-rate integrations over given regions of interest etc... As of today, XDS system offers nearly the same energy resolution as a standard analog processing system. However, it is envisaged to upgrade the

ESRF XDS system hardware with faster ADC (16-bit/25 MHz) and larger FPGAs in order to push the performances of this detector to its ultimate limits.

To record XMCD spectra under optimal conditions, this detector was operated in the standard UHV environment of beamline ID12. We have designed a compact sample chamber that was inserted between the poles of a 0.7 T electromagnet. The detector is housed in a special, UHV-compatible chamber separated from the sample chamber by a gate valve. Special care has been taken to maximize the solid angle of collection of the fluorescence photons. The detector is systematically operated windowless: this allowed us to extend the operation of the detector down to the soft X-ray range where scattering is a major limitation.

EXPERIMENTS

For our test experiment, the sample was a 4-μm-thick film of amorphous TbCo$_2$ alloy deposited on a Kapton substrate. It is ferromagnetic at room temperature and a modest magnetic field of 0.5 T is sufficient to get it saturated magnetically. A typical emission spectrum of the sample excited by X-ray photons of energy above the L$_{III}$-edge of Tb (≈ 7.56 keV) is reproduced in Fig. 3. This spectrum recorded using only one single Si drift diode channel is dominated by the Tb L$_\alpha$ emission line ($E \simeq 6.28$ keV) whereas the signal associated with Tb M emission lines ($E < 1.7$ keV) is nearly two orders of magnitude smaller. We also detected additional fluorescence liness of about the same intensity which were due to small amounts of titanium and argon in the sample, or to potassium impurities in the kapton film. Under normal operating conditions, the energy resolution of the detector was found excellent: ≈ 130eV for the Tb L$_\alpha$ fluorescence and ≤ 100eV for the Tb M$_{\alpha,\beta}$ fluorescence. This illustrates the excellent performances of the Si drift diode detector over a very wide spectral range.

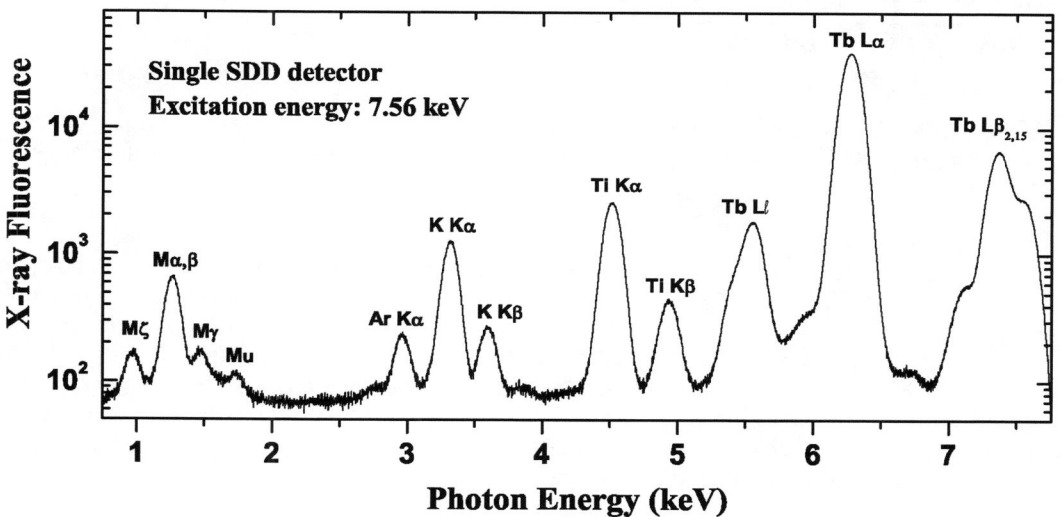

FIGURE 3. X-ray emission spectrum of 4-μm-thick TbCo$_2$ film deposited on kapton. The energy of exciting photons was just above the Tb L$_{III}$ absorption edge: 7.56 keV.

In standard X-ray fluorescence analyses, it is not trivial to disentangle true cascade processes from other uncorrelated events involving either the re-absorption of a primary fluorescence photon, or even independent excitations directly caused by the incident photons. The SDD detector offers a remarkable possibility to select two regions of interest in the emission spectrum corresponding to Tb L$_\alpha$ and M$_\beta$ emission lines and to recors two excitation spectra in a single energy scan across the Tb L$_{III}$ absorption edge. The two XANES spectra, reproduced in the top panels of Fig. 4, look fairly similar. This result suggests that the M$_\beta$ fluorescence signal detected in the emission spectrum is mainly due to a cascade process and is not due to the direct excitation of the 3d shell. Note, that in the latter case one should observe an anticorrelated spectrum[4].

As for XANES spectra, one can record two XMCD spectra in a single energy scan by flipping the direction of applied magnetic field at every energy point. To make sure that the measured XMCD spectra were free of artefacts the spectra were recorded with two helicities of the incoming X-rays. Whereas the XMCD spectrum recorded with L$_\alpha$ fluorescence is very similar to the one measured previously[5], the M$_\beta$-XMCD spectrum exhibits distinct differences. Surprisingly, the feature corresponding to quadrupolar excitation channel is absent in the M$_\beta$-XMCD spectrum. What

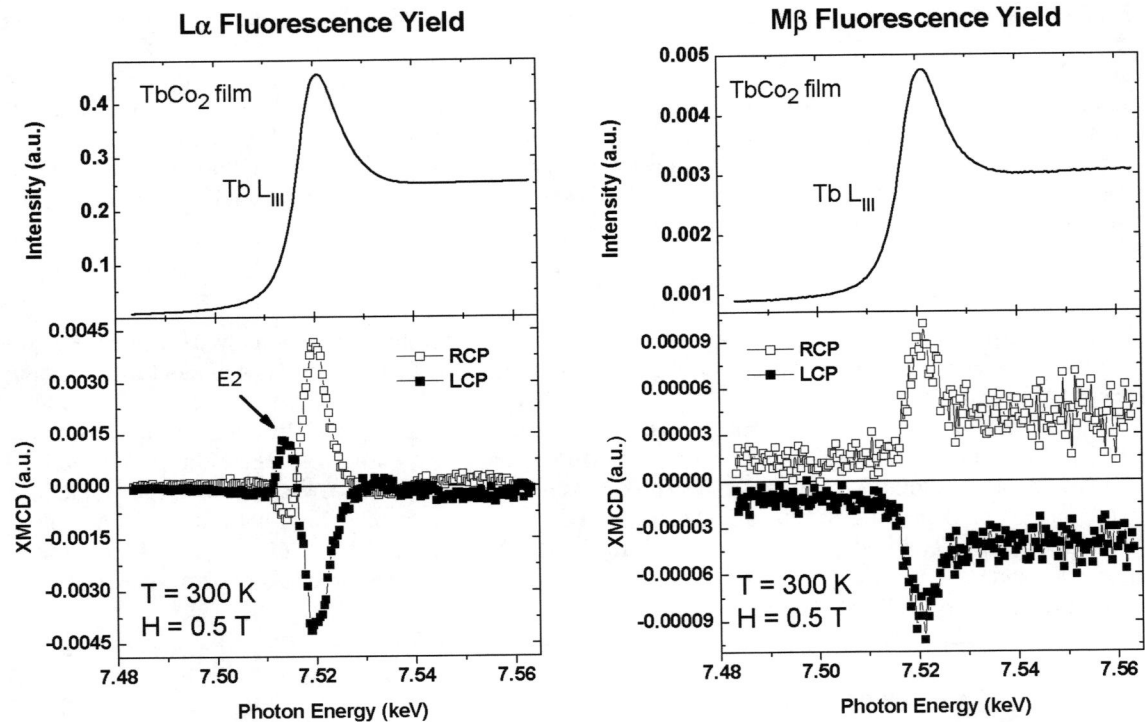

FIGURE 4. Tb L_{III} XANES (top) and XMCD (bottom) spectra in $TbCo_2$ amorphous film recorded by monitoring the integrated intensities of L_α (left panel) and M_β (right panel) fluorescence. XMCD spectra recorded with right and left circularly polarized X-rays are shown.

is even more remarkable is the presence of a non-zero XMCD signal at excitation energies well above and below the Tb L_{III} absorption edge. Unfortunately there is as yet no theory available for the magnetic effects in polarization dependence of the cascade processes: this would be a pre-requisite to understand how these spectra are related to the magnetic properties of the $4f$ and $5d$ states and their relaxation dynamics. Work is in progress to clarify such a theory.

In conclusion, our SDD detector has revealed excellent performances over a very wide energy range and allowed us to produce the first experimental evidence of magnetic circular dichroism in X-ray fluorescence cascade process in the rare-earth atom excited at the L_{III}-edge. Whether or not one could use this technique to access to magnetic properties of the $4f$ and the $5d$ states remains an open question.

ACKNOWLEDGMENTS

Technical assistance by S. Feite and P. Voisin is warmly acknowledged. We wish also to express our gratitude to C. Gauthier and E. Moguiline for excellent work in the early stage of the SDD project.

REFERENCES

1. A. Rogalev *et al.*, "X-ray Magnetic Circular Dichroism", in *Magnetism: A Synchrotron Radiation Approach*, edited by E. Beaurepaire *et al.*, *Lectures Notes in Physics* **697**, Berlin: Springer Verlag, 2006, pp. 71–94.
2. A. Rogalev *et al.*, "Instrumentation Developments for Polarization Dependent X-ray Spectroscopies" in *Magnetism and Synchrotron Radiation*, edited by E. Beaurepaire *et al.*, *Lecture Notes in Physics* **565**, Springer Verlag, Berlin, 2001, pp. 60–86.
3. J. Goulon *et al.*, *J. Synchrotron Rad.* **85**, 57–69 (2005).
4. C. Gauthier *et al.*, *NIM A* **382**, 524–532 (1996).
5. J. Goulon *et al.*, *J. Synchrotron Rad.* **5**, 232–38 (1998).

Hard X-ray Linear Dichroism Using a Quarter Wave Plate for Structural Characterization of Diluted Magnetic Semiconductors

F. Wilhelm[1], E. Sarigiannidou[2], E. Monroy[2], A. Rogalev[1], N. Jaouen[1], H. Mariette[2] and J. Goulon[1]

[1]European Synchrotron Radiation Facility, 6 rue Jules Horowitz, B.P. 220, 38043 Grenoble, France
[2]CEA-CNRS-UJF, Laboratoire de Spectrométrie Physique, Université Joseph Fourier and DRFMC/SP2M/PSC, CEA-Grenoble, 17 Avenue des Martyrs, 38054 Grenoble cedex 9, France

Abstract. Diluted magnetic semiconductors are actually of high interest from the viewpoint of basic and applied physics due to their potentially for spintronic applications. However, it is fundamental to characterize in detail the structure since the magnetic properties depends strongly on the sample quality. We will show that hard x-ray linear dichroism is a well suitable characterization tool to probe the local environment (bonding and symmetry) of the transition metal atoms diluted in the semiconductor crystal. With the help of theoretical simulation, it is possible to distinguish the crystallographic site which the diluted transition metals atoms occupy. Finally, it allows detecting a possible non-negligible presence of secondary phases or metallic cluster.

Keywords: XLD, DMS.
PACS: 75.50.Pp, 75-25.+z, 78.70.Dm

INTRODUCTION

Much efforts are devoted to incorporate $3d$ transition metals in semiconductors in order to obtain so-called *diluted magnetic semiconductor* (DMS) showing ferromagnetic properties [1]. Although many groups already succeed to grow such materials, a controversy still exists regarding their magnetic properties. Those discrepancies may be attributed to the presence of parasitic magnetic secondary phases that mask the intrinsic magnetic properties. Prior to magnetic characterization, it is therefore necessary to characterize in detail the local structure around the diluted $3d$ transition metal. Very few techniques with a high sensitivity make it possible to probe the local structure and symmetry around a given atom. We will demonstrate that hard x-ray linear dichroism (XLD) provides complementary informations to standard laboratory techniques (e.g. x-ray diffraction) by offering in addition to its higher sensitivity, the element and shell selectivity. In case of non-cubic structure DMS, XLD is a suitable technique to explore the local structure and symmetry by probing the anisotropy of the electronic structure. We will show that XLD experiments offer a very high sensitivity to detect the presence of secondary phases. Finally, using both XANES and XLD, it is possible to distinguish which crystallographic sites are occupied by the diluted transition metal atoms.

X-RAY QUARTER WAVE PLATE

XLD measurements were done at the ESRF ID12 beamline [2] and performed at the K-edge of the $3d$ transition metals and the $4p$ metal on wurtzite DMS. The K-edge absorption spectra offers the advantage to probe the directionality of the p final states and add new information regarding chemical bonding environment in selected directions. In the case of the $L_{2,3}$-edges, the angular dependence is more complicated due to the presence of two final state components ($l=0$ and $l=2$). Moreover, the K-edge spectra are often much more structured and make possible a refined comparison with theoretical calculations. XLD could be recorded using directly the linearly polarized x-rays

CP879, *Synchrotron Radiation Instrumentation: Ninth International Conference*,
edited by Jae-Young Choi and Seungyu Rah
© 2007 American Institute of Physics 978-0-7354-0373-4/07/$23.00

delivered by the undulator source, but we prefer to use a quarter wave plate (QWP) to convert circularly polarized x-rays into linearly polarized. X-ray phase plates exploit the birefringence of perfect crystals under the conditions of Bragg diffraction: this implies that the refraction indices are different for σ and π polarization components [3] and the key point that there is a phase shift ϕ between the σ and π wave components transmitted by the crystal. The latter phase shift ϕ is inversely proportional to the angular offset $\delta\theta$, i.e. the difference between the true angle of incidence and the angle corresponding to the center of the reflection profile:

$$\phi = -\frac{\pi}{2} \cdot \frac{A.t}{\delta\theta.\cos(\psi)} \qquad (1)$$

where t is the crystal thickness and ψ the angle between the incident beam and the normal to the crystal surface. When the phase shit is set to be equal $\pm\pi/2$, then the crystal acts as a quarter wave plate and the incoming circularly polarized x-ray beam will be converted into a linearly polarized beam.

The QWP set-up described by J. Goulon *et al.* [4] is suitable for spectroscopic studies. The QWP is made of a 0.9mm thick diamond <111> crystal exploiting the asymmetric (11-1) reflection. The typical offset value resulting in the desired $\pm\pi/2$ phase shift is of the order of ±140arcsec at the Mn K-edge and ±25arcsec at the Ga K-edge (Fig. 1). The angular offset is controlled by a digital piezoactuator which offers the advantage to flip rapidly the linear polarization from horizontal to vertical. The polarization can be flipped several times for every energy point. The XLD signal at a given edge is recorded by varying together the Bragg angle of the monochromator and the angle corresponding to the center of the reflection profile keeping the angular offset constant in such a way that the condition to flip the polarization is always satisfied. There are several advantages to perform XLD with a QWP: (i) the polarization can be flipped several times at every energy point of the scan, (ii) the beam footprint, the sample orientation and the solid angle detection remain constant, (iii) the measured XLD signal is insensitive to eventual low frequency beam displacements. The energy range of our QWP is: 4.5keV up to 15keV.

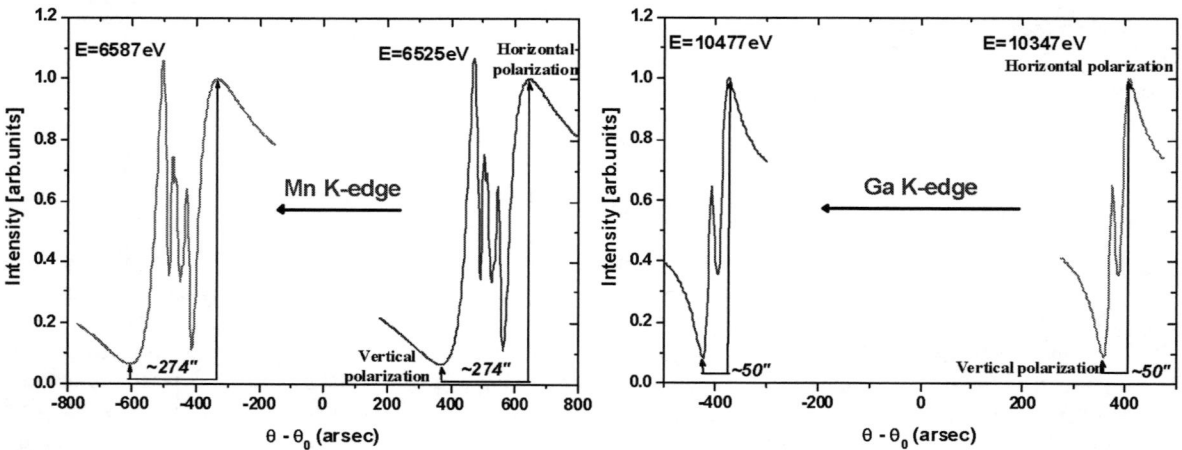

FIGURE 1. Intensity profiles recorded below and above the Mn K-edge (left) and Ga K-edge (right). They were measured on varying the angular offset of the diamond QWP.

EXPERIMENT

We have investigated a GaMnN epilayer film containing 6.3 at.% of Mn. This film was prepared by plasma-assisted molecular beam epitaxy (PAMBE) in a chamber equipped with standard effusion cells for Ga and Mn. Active nitrogen was supplied by a radio-frequency plasma cell. During the growth process, the surface quality was monitored *in situ* by reflection high-energy electron diffraction (RHEED). The structural quality of the films is first investigated by x-ray diffraction (XRD), and within the detection limits of the method, no secondary phases are observed. XLD experiments were performed at both the Mn and Ga K-edges. The circularly polarized x-ray photons were delivered by an apple-II type HU-38 undulator. The 0.9 mm thick diamond QWP was located downstream with respect to the fixed exit monochromator which was equipped with a pair of Si(111) crystals. The polarization was flipped at a frequency of 1Hz for every energy point. The total fluorescence yield was detected using 8 photodiodes in the backscattering geometry. In front of each photodiode, we inserted a 5μm thick Cr filter to minimise diffuse scattering effects. The experiment was carried out at an incidence angle of 80° with respect to the **c**-axis of the wurtzite crystal. XLD was obtained by combining the x-ray absorption cross-section measured with two orthogonal linear polarizations of the x-ray beam (oriented perpendicular and parallel to the **c**-axis for wurtzite symmetry). This XLD spectrum contain information on the anisotropy of the unoccupied electronic density of state of the $4p$-shell projected perpendicularly or parallel to the **c**-axis. The angular dependence of the XLD spectra was discussed in ref. [5].

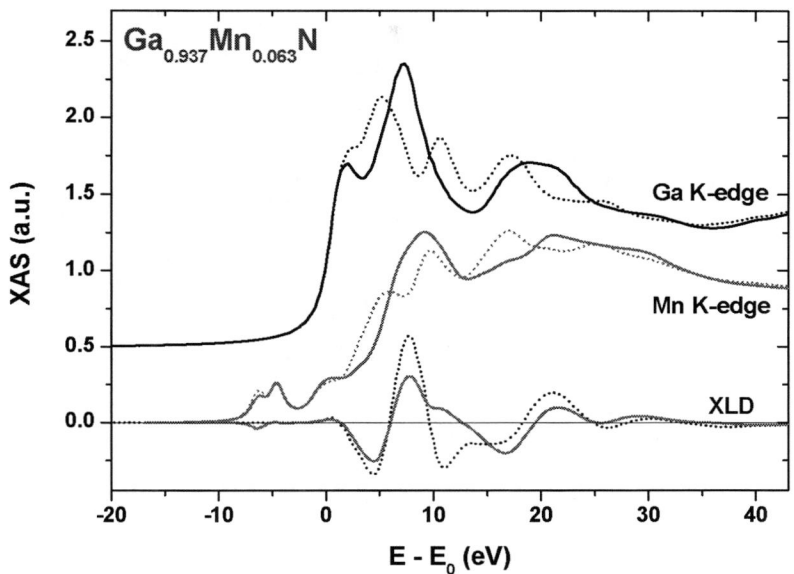

FIGURE 2. XLD signal (bottom) recorded at the Mn (full line) and Ga (dotted line) K-edges for a GaMnN epilayer film containing 6.3 at. % of Mn. The normalized XANES spectra (top) are shown for two orthogonal orientation of the film: the polarization of the x-rays was either perpendicular (dotted line) and parallel (full line) to the **c**-axis of the crystal. Both Mn and Ga XANES spectra have been normalized to the edge jump of unity far form edge. Note that the relative energy positions of the XLD signal maxima for both Mn and Ga coincide very well.

Figure 2 reproduces the XLD spectra recorded at the Mn and Ga K-edge for GaMnN with 6.3 at.% of Mn. A strong XLD with an amplitude of 58% compared to the edge jump was found at the Ga K-edge. It is identical in shape and amplitude to the XLD of pure wurtzite GaN epilayers (not shown). It suggests that the local symmetry at the Ga atoms in our GaMnN sample is the same as in GaN. At a first sight, we did not find any signature of secondary phases such as $GaMn_3N$. At the Mn K-edge, there is a strong XLD signal with an amplitude of 30% compared to the edge. The XANES spectra at the Mn K-edge also exhibit two peaks in the pre-edge region which are absent in the Ga K-edge spectra. Only the first pre-peak exhibits a strong XLD signal.

DISCUSSION

Our dichroism spectra suggest that the Mn atoms have a wurtizte symmetry. The XLD spectral shape of Mn is very similar to the XLD of Ga but nearly twice smaller in amplitude. The fact that the XLD maximal position in relative energy coincide perfectly indicates already that the Mn atoms may have a wurtizte symmetry. To support our observation, we have simulated XANES and XLD spectra using the FDMNES code [6]. The simulation was performed using a wurtzite supercell (i) $Ga_{15}MnN_{16}$ in which one Mn atom is in substitution with one Ga atom, (ii) $Ga_{16}MnN_{15}$ in which one Mn atom is in substitution with one N atom and (iii) $Ga_{16}MnN_{16}$ in which one Mn atom is an interstitial site. All simulations were consistent with a uniformly doped GaN semiconductor with 6.25at.% of Mn, just as in the real sample. The radius of the cluster was 8.5Å. The lattice parameter was chosen to be that of bulk GaN. The simulations were performed using the multiple scattering approach within the muffin tin approximation. The electronic configuration state of Mn that reproduces best in amplitude and shape the experimental XANES and XLD spectra is intermediate between $3d^4 4s^2$ and $3d^{4.5} 4s^2$. This would support the conclusion that Mn is rather $3d^4$ (Mn^{3+}) than $3d^5$ (Mn^{2+}) as deduced from x-ray magnetic circular dichroism [7]. The FDMNES simulations are reproduced in figure3 and support the interpretation that the Mn atoms are substituted with Ga. The ~1:2 amplitude ratio for Mn:Ga XLD signal is also well reproduced by our calculations for Mn substituted with Ga atoms in GaN wurtzite crystal [7]. Finally, since the simulations agree rather well with the amplitude of the Mn XLD experimental spectra, we may exclude the presence of secondary phases such as Mn_4N or Mn metallic clusters. Since these phases have anyhow a different crystal symmetry (e.g. perovskite, cubic), we would expect XLD to be a very sensitive tool to probe their existence in our sample. If such phase were present, we should have observed a significant change of the XLD signal which we do not find.

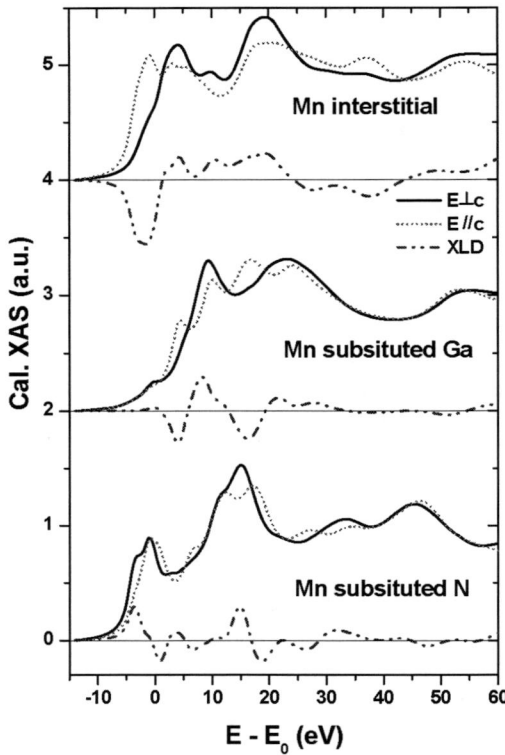

FIGURE 3. XANES and XLD Mn K-edge simulations for a wurtzite supercell (top) $Ga_{16}MnN_{16}$ in which a Mn atom is an interstitial site, (middle) $Ga_{15}MnN_{16}$ in which a Mn atom is in substitution with one Ga atom, (bottom) $Ga_{16}MnN_{15}$ in which a Mn atom is in substitution with one N atom.

In conclusion, we have demonstrated that XLD in the hard x-ray range is a very valuable tool to characterize in detail the local environment, symmetry and crystallographic site occupancy of transition metals atoms diluted in wurtzite semiconductors. This technique is extremely promising to study the structural quality of diluted magnetic semiconductor at concentrations down to a few ppm.

ACKNOWLEDGMENTS

We thank E. Bellet-Amalric, R.M. Galera and J. Cibert for x-ray diffraction and magnetic characterizations.

REFERENCES

1. T Dietl, H. Ohno, F. Matsukura, J. Cibert and D. Ferrand, Science **287**, 1019-1022 (2000).
2. A. Rogalev, J. Goulon, C. Goulon-Ginet and C. Malgrange, "Instrumentation Developments for Polarization dependent X-ray Spectroscopies" in *Magnetism and Synchrotron Radiation*, edited by. E. Beaurepaire, F. Scheurer, G. Krill and J.-P. Kappler, *Lecture Notes in Physics* **565**, Berlin: Springer Verlag, 2001, pp. 60-86.
3. V.E. Dmitrienko and V.A. Belyakov, *Sov. Phys. Uspekhi.* **32**, 697-719 (1989).
4. J. Goulon, C. Malgrange, C. Giles, C. Neumann, A. Rogalev, E. Moguiline, F. De Bergevin and C. Vettier, *J. Synchrotron Rad.* **3**, 272-281 (1996).
5. C . Brouder, *J. Phys. : Cond. Matter* **2**, 701-738 (1990).
6. Y. Joly, *Phys. Rev. B***63**, 125120 (2001).
7. E. Sarigiannidou, F. Wilhelm, E. Monroy, R.M. Galera, E. Bellet-Amalric, A. Rogalev, J. Goulon, J. Cibert and H. Mariette, *Phys. Rev. B***74**, 041306(R) (2006).

Development of a Novel Piezo Driven Device for Fast Helicity Reversal Experiments on the *XMaS* Beamline

L. Bouchenoire[a,b], S.D. Brown[a,b], P. Thompson[a,b], M. G. Cain[c], M.Stewart[c], M.J. Cooper[d]

[a] *XMaS, The UK-CRG, ESRF, BP220, F-38043 Grenoble CEDEX, France*
[b] *Dept of Physics, University of Liverpool, Oliver Lodge Laboratory, Oxford Street, Liverpool L69 7ZE, United Kingdom*
[c] *Functional Materials, National Physical Laboratory, Hampton Road, Teddington, Middlesex TW11 0LW, United Kingdom*
[d] *Dept of Physics, University of Warwick, Gibbet Hill Road, Coventry, CV4 7AL, United Kingdom*

Abstract. The XMaS diamond phase retarder has been combined with a phase sensitive detection method to probe weak signals buried in high backgrounds. The system was tested in x-ray magnetic circular dichroism (XMCD) studies. Fast reversal of the photon helicity was effected by a new piezo driven device operating at 10Hz and higher frequencies. The XMCD signal was detected through the use of a lock-in amplifier. This technique was successfully applied at the L_2 and L_3 edges of a $GdCo_2$ foil.

Keywords: x-ray phase retarder, XMCD, polarisation modulation, piezo actuator
PACS: 7.85.Qe; 7.50.Qx; 32.55.Ad; 41.50.+h

INTRODUCTION

X-ray magnetic circular dischroism (XMCD) is a useful technique to investigate the magnetic properties of complex systems due to its element specificity [1-3]. XMCD is defined as the difference in absorption between right and left handed circularly polarised x-ray beam of a sample magnetised along the photon wave vector. Circularly polarised x-rays can be produced in three different way: i) by extracting the bending magnet synchrotron radiation from above or below the orbital plane of the electron beam [4,5], ii) by using exotic insertion devices such as helical undulators [6,7], iii) by inserting a quarter phase-plate [8,9].

Most previous XMCD measurements have been made using a fixed helicity and switching the direction of the sample's magnetisation (field reversal method, FR) [10-12]. However, this method is limited because it requires a long integration time between each field orientation to get good quality data and relies on extremely good stability of the incident x-ray beam over the same period. Also it is only viable to saturated or hysteresis-free system. Another way of measuring a dichroic signal is to keep the direction of the magnetic field constant and to reverse the photon helicity (helicity reversal method, HR). As the FR method is not employed, the sample environment remains unchanged during the measurements even with different experimental conditions (e.g. temperature, external magnetic and electric fields with different strengths, pressure). X-ray phase retarders have been intensively utilised for that purpose [9,13] as it is relatively easy to reverse the polarisation of the beam by changing the sign of the offset angle from the Bragg condition. However, the quality of XMCD spectra was essentially identical to those obtained with the FR method taken in comparable time scale, and also suffered from beam instability. Very weak signals can be extracted from high backgrounds by combining fast reversal of the photon helicity with a phase-sensitive detection [14,15]. This method results in extremely high quality data taken in a very short period [16,17].

The oscillation stages employed to date have been driven by piezoelectric translators [14,15] of relatively large size which could not be implemented into the XMaS beamline due to space restrictions. In this paper, we present a novel compact piezo driven device and we report its efficiency in a XMCD experiment performed with a phase-sensitive detection method.

CP879, *Synchrotron Radiation Instrumentation: Ninth International Conference*,
edited by Jae-Young Choi and Seungyu Rah

INSTRUMENTS FOR FAST HELICITY REVERSAL

The XMaS beamline located on the soft end of the bending magnet 28 (BM28) at the ESRF is dedicated to studies of magnetic materials mainly using x-ray diffraction techniques. A complete description of the beamline equipment is reported elsewhere [18]. In brief, the white beam is monochromated with a Si(111) double-bounced water-cooled monochromator and focused down to the sample position by a toroidal rhodium-coated single crystal silicon mirror. The linear polarisation of the synchrotron beam can be converted to circular polarisation with a diamond phase plate. XMaS has three diamond phase-plate crystals which have different thicknesses [12] to cover an energy range between 3 to 11 keV.

The novel piezo driven oscillation stage, the so-called flipper (Fig.1), was designed by the Functional Material Group of the National Physical Laboratory (UK). The device uses two pairs of multilayer piezoelectric stacks (P in Fig.1) mounted on opposite sides of, and coupled to, an aluminium plate (A2) which is free to rotate via two weak links. They are driven in opposite directions to provide displacement of up to 350 arcseconds. This specification is considerably larger than the displacement obtained by piezoelectric translators used in reported helicity modulation techniques [14,19] which allows each diamond to be used in a wider energy range. A second advantage is that this flipper is very compact compared to existing oscillation stages and allows easy mounting on the phase-plate Huber 410 circle. The diamond (D) is attached to a small copper goniometer (G) which screws into an aluminium adapter plate (A1). The resulting block is then mounted onto the aluminium plate with weak links (A2). Two manual linear translations (T) allow positioning of the diamond at the centre of rotation of the phase-plate assembly.

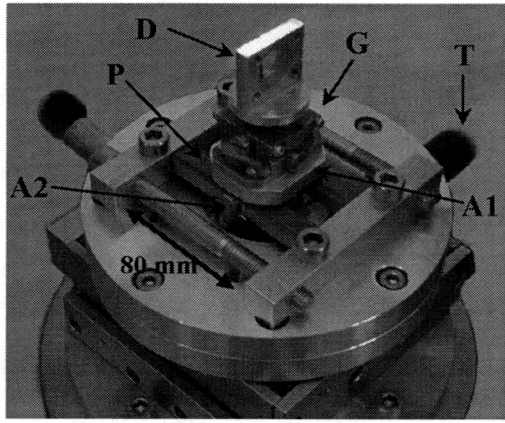

FIGURE 1. XMaS flipper assembly. **D**: diamond phase-plate. **G**: copper goniometer. **A1**: aluminium adapter plate. **A2**: aluminium plate with two weak links. **P**: (four) multilayer piezoelectric stacks. **T**: (two) manual translations.

The performance of the flipper was tested under real conditions by measuring the XMCD signals of a GdCo$_2$ foil at the L$_2$ and L$_3$ edges of gadolinium. The foil was mounted at 45° from the direction of the incident beam and was magnetised with an electromagnet. The linearly polarised x-rays were converted into circular polarised x-rays with a 0.8 mm thick (110) diamond phase plate crystal used in Laue transmission geometry with the (111) reflection planes inclined by 45° with respect to the polarisation plane of the incoming beam. The diamond was first driven to the correct Bragg angle with the Huber 410 circle. The diamond was then oscillated around the Bragg conditions by an offset angle $\pm\Delta\theta$ using the flipper driven at 10Hz by a sinusoidal function generator. Alternate right and left-handed circularly polarised x-rays were incident on the sample. In order to improve the signal to noise ratio, the novel flipper was combined with phase sensitive detection. The incident and transmitted intensities were measured with ionisation chambers and the resulting currents were amplified and converted to voltages. A log divider measured the normalised intensity from the foil to that transmitted through the diamond. The resultant signal was then fed to the lock-in amplifier. The dichroic signal was measured from the lock-in output synchronised to the flipping of the helicity.

XMCD RESULTS AND DISCUSSION

The dichroic signal was measured as a function of energy. For each energy, while the oscillation of the flipper was fixed, the new Bragg angle was automatically recalculated and the diamond was driven by the Huber circle to the correct position to satisfy the Bragg conditions. In order to remove any residual non-magnetic signal deriving mainly from the different diamond absorption between right (+Pc) and left (-Pc) helicity, the measurements were performed for both magnetic field directions (+B and -B). The resulting dichroic signal was obtained by calculating the half-difference between these two measurements. The results are presented in Fig.2. The dichroic signal measured with the flipper (closed circles) was recorded in ~2 hours and is compared to the standard method (open circles) which has been multiplied by -1 for clarity. The standard method has been described by Pizzini et al. [11] and involves four independent "static" measurements, (+Pc+B, +Pc-B, -Pc-B, -Pc+B), in order to subtract any non-magnetic signal. Note that the L_3 data taken with this method (open circles) were obtained in about 10 hours during a different experiment reported elsewhere [12]. They are only shown here to demonstrate first of all that the size of the effect is similar to that obtained with the lock-in technique and that to get similar statistics the counting time must be multiplied by at least a factor of five. The XMCD signal measured at the L_2 edge with the standard method (open circles) was also recorded in ~2 hours and during the same experiment as the flipper (the time taken to perform one single static measurement at the L_2 edge was equivalent to two scans using the flipper). As expected, the signal to noise ratio is considerably lower using the lock-in detection. Another advantage over the standard method is that the XMCD signal equal to $\Delta\mu/\mu$ is obtained much more easily and more reliably as no background subtraction nor rescaling of μ to tabulated values has to be done. Also, the range of the energy scans must be larger (100eV) with the standard method than with the lock-in detection (50eV) in order to fit a correct background which is therefore time consuming. Note that the sign of the dichroic signal (positive and negative for the L_3 and L_2 edges, respectively) is also preserved with the lock-in method.

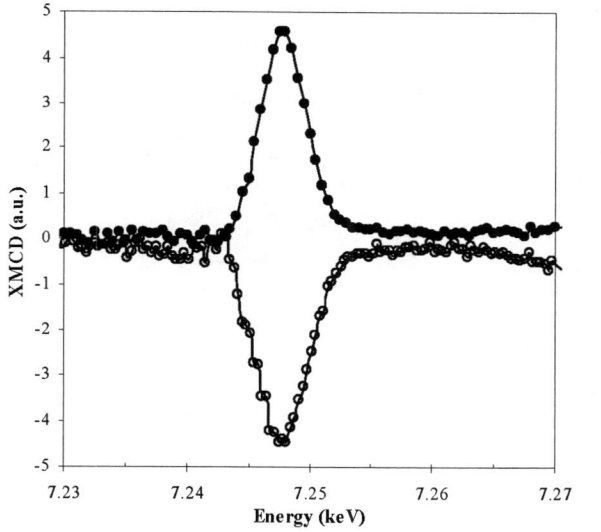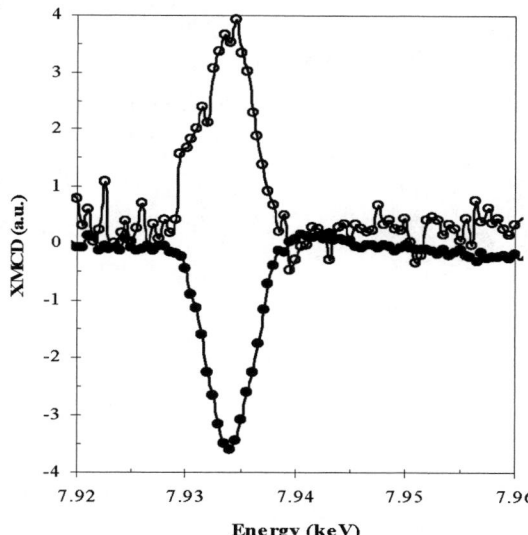

FIGURE 2. XMCD measurements performed on a $GdCo_2$ foil at the Gd L_3 (left) and L_2 (right) edges with the flipper (closed circles) and the standard method (open circles). The dichroic signal obtained with the standard method was multiplied by -1 for clarity. Note that the open circles data for the L_3 edge were not measured during this experiment, see reference [12].

To conclude, we have demonstrated that the novel piezo driven device designed to rapidly reverse the helicity performed as expected. Combined with lock-in detection, it considerably improved the signal to noise ratio of XMCD spectra measured in shorter period.

The XMaS beamline will strongly benefit from the advantages brought by this novel piezo driven device as it will open up new fields. XMaS has recently purchased a 4T superconducting magnet but the magnetic field can only be reversed every 20min. Combined with the new piezoelectric flipper, the study of magnetic disorders at the interfaces of magnetic multilayers in strong fields will be possible for example. The device can also operate under vacuum so that experiments (e.g. XMCD, x-ray resonant magnetic reflectivity measurements, etc...) can be carried out in the low energy range of XMaS, e.g. at the M edges of actinides, for which absorption (air, kapton, diamond) is large.

REFERENCES

1. G. Schütz, R. Frahm, P. Mautner, R. Wienke, W. Wagner, W. Wilhem, P. Kienle, Phys. Rev. Lett., 62, 2620-2623 (1989).
2. J. C. Lang, G. Srajer, C. Detlefs, A.I. Goldman, H. König, X. Wang, B.N. Harmon, Phys. Rev. Lett., 74, n° 24, 4935-4938 (1995).
3. H. Wende, Z. Li, A. Scherz, G. Ceballos, K. Baberschke, A. Ankudinov, J.J. Rehr, F. Wilhwm, A. Rogalev, D.L. Schlagel and T.A. Lograsso, J. Appl. Phys., 91, n° 10, 7361-7363 (2002).
4. D. Laundy, S.W. Collins, A.J. Rollason, J. Phys. Condens. Matter, 3, 369-372 (1991).
5. D. Laundy, S.D. Brown, M.J. Cooper, D. Bowyer, D. Paul, W.G. Stirling, J. Synchrotron Rad., 5, 1235-1239 (1998).
6. H. Kawata, T. Miyahara, S. Yamamoto, T. Shioys, H. Kitamura, S. Sato, N. Kananaya, A. Lida, A. Mikiuni, M. Sat, T. Iwazumi, Y., Kitajima, M. Ando, Rev. Sci. Instrum., 60, 1885-1888 (1989).
7. P. Elleaume, J. Synchrotron Rad, 1, 19-26 (1994).
8. K. Hirano, K. Izumi, T. Ishikawa, S. annaka, S. Kikuta, Jpn. J. Appl. Phys. Vol. 30, n°3A, L407-L410 (1991).
9. C. Giles, C. Malgrange, J. Goulon, F. de Bergevin, C. Vettier, E. Dartyge, A. Fontaine, C. Giorgetti, S. Pizzini, J. Appl. Cryst., 27, 232-240 (1994).
10. J.P. Rueff, R.M. Galera, S. Pizzini, A. Fontaine, L.M. Garcia, Phys. Rev. B, 55, n°5, 3063-3070 (1997).
11. S. Pizzini, M. Bonfin, F. Baudelet, H. Tolentino, A San Miguel, K. Mackay, C. Malgrange, M. Hagelstein, A. Fontaine, J. Synchrotron Rad., 5, 1298-1303 (1998).
12. L. Bouchenoire, S.D. Brown, P. Thompson, J.A. Duffy, J.W. Taylor, M.J. Cooper, J. Synchrotron Rad., 10, 172-176 (2003).
13. C. Giles, C. Malgrange, J. Goulon, F. de Bergevin, C. Vettier, A. Fontaine, E. Dartyge, S. Pizzini, Nucl. Instrum. & Methods, A349, 622-625 (1994).
14. K. Hirano, T. Ishikawa, S. Koreeda, K. Fuchigami, K. Kanzaki, S. Kikuta, Jpn. J. Appl. Phys. Vol. 31, L1209-L1211 (1992).
15. J. C. Lang and G. Srajer, Rev. Sci. Instrum. 66, 1540-1542 (1995).
16. M. Suzuki, N. Kawamura, M. Mizumaki, A. Urata, H. Maruyama, S. Goto, T. Ishikawa, Jpn. J. Appl. Phys. Vol. 37, L1488-L1490 (1998).
17. M. Suzuki, N. Kawamura, M. Mizumaki, A. Urata, H. Maruyama, S. Goto, T. Ishikawa, J. Synchrotron Rad, 6, 190-192 (1999).
18. S.D. Brown, L. Bouchenoire, D. Bowyer, J. Kervin, D. Laundy, M.J. Longfield, D. Mannix, D.F. Paul, A. Stunault, P. Thompson, M.J. Cooper, C.A. Lucas, W.G. Stirling, J. Synchrotron Rad., 8, 1172-1181 (2001).
19. J. Lang (private communication).

A 4 Tesla Superconducting Magnet Developed for a 6 Circle Huber Diffractometer at the *XMaS* Beamline

P.B.J.Thompson[1,2], S.D. Brown[1,2], L. Bouchenoire[1,2], D. Mannix[1,2], D.F. Paul[1,3], C.A. Lucas[2], J. Kervin[2], M.J. Cooper[3], P. Arakawa[4], G. Laughon[4]

[1] *XMaS, The UK-CRG, ESRF,BP220, F-38043 Grenoble CEDEX, France.*
[2] *Dept of Physics, University of Liverpool, Liverpool, United Kingdom,*
[3] *Dept of Physics, University of Warwick, Gibbet Hill Road, Coventry, CV4 7AL, United Kingdom,*
[4] *American Magnetics Inc, P.O. Box2509, 112 Flint Road, Oak Ridge, TN 37831-2509, U.S.A.*

Abstract. We report here on the development and testing of a 4 Tesla cryogen free superconducting magnet designed to fit within the Euler cradle of a 6 circle Huber diffractometer, allowing scattering in both the vertical and horizontal planes. The geometry of this magnet allows the field to be applied in three orientations. The first being along the beam direction, the second with the field transverse to the beam direction a horizontal plane and finally the field can be applied vertically with respect to the beam. The magnet has a warm bore and an open geometry of 180°, allowing large access to reciprocal space. A variable temperature insert has been developed, which is capable of working down to a temperature of 1.7 K and operating over a wide range of angles whilst maintaining a temperature stability of a few mK. Initial ferromagnetic diffraction measurements have been carried out on single crystal Tb and Dy samples.

Keywords: Superconducting Magnet, Magnetism, Sample environment
PACS: 07.55.Db, 84.71.Ba

INTRODUCTION

The *XMaS* beamline [1,2] at the European Synchrotron Radiation Facility (E.S.R.F.) has been designed to perform high-resolution diffraction and magnetic scattering and has been operational since April 1998. The beamline is sited on the soft end of dipole D28, which has a critical energy of 9.8 KeV. The primary optics consist of a fixed exit double crystal Si(111) monochromator, followed by a toroidal mirror which focuses the beam down to a size of less than 1mm^2. The experimental hutch is equipped with a six circle Huber diffractometer that allows scattering in both the vertical and horizontal planes. A comprehensive description of the beamline has been given by Brown *et al.* (2001) [2]. Currently the beamline sample environments range in temperature from 1k to 800 K and a 1 Tesla conventional electromagnet is also available. Recently our magnetic field sample environment has been augmented by the arrival of a novel cryogen free 4 Tesla superconducting magnet, supplied by American Magnetics Inc.

The *XMaS*/AMI superconducting magnet has been designed to fit within the Euler cradle of the Huber diffractometer and allows three field orientations. The geometry allows the magnet to be turned along the vertical axis through 90 degrees, facilitating application of magnetic fields both along and transverse to the incident beam direction. Thus, both transverse and longitudinal fields may be applied. This allows the separation of spin and orbital contributions to magnetic scattering signal in non-resonant ferromagnetic studies [3-5]. It may also be mounted to provide a vertical field allowing additional contrast in resonant magnetic scattering studies. It can deliver a field of 4 T in a large 40 mm opening warm bore with a 180° scattering aperture. An efficient yoke configuration occupies the lower half of the vertical scattering plane, leaving the other half open for the cryostat and scattered beam. The geometry of the coil former has been optimized to allow for a maximum number of turns within the geometrical constraints of the diffractometer. The combination of the new 4 Tesla magnet and low temperature insert at *XMaS* will also enable detailed studies of complex field and low temperature (~2 Kelvin) phase diagrams. The 180° scattering aperture allows large access to reciprocal space, so that important information on wave-vector dependence can be readily obtained. As this magnet can be used in vertical or horizontal scattering geometry, which will allow

CP879, *Synchrotron Radiation Instrumentation: Ninth International Conference*,
edited by Jae-Young Choi and Seungyu Rah
© 2007 American Institute of Physics 978-0-7354-0373-4/07/$23.00

the experimentalist to take advantage of either incident σ-polarised or π-polarised photons from the ESRF storage ring. Additional information on the scattering process may also be obtained when this magnet is used in conjunction with the *XMaS* polarization analyzer [6], which determines if the incident x-rays are scattered with the same or rotated linear polarization. The magnet is ideal for investigations of the multipolar ordering found in rare-earth intermetallic compounds. In these systems, the applied field perturbs the spontaneous magnetic order and stabilizes the onset of quadrupole order. Diffraction experiments in applied fields using resonant x-ray scattering will deepen our understanding of these complex ordering phenomena and magnetic structures.

DESIGN CONCEPTS AND CONSTRUCTION

The primary design requirements for this magnet were that it should be able to operate in any angular geometry, thus necessitating a cryogen free design. It was also decided that the magnet should be able to fit within the Euler cradle of the Huber diffractometer. Another important criterion for the design was a 180° open warm bore access for the scattered x-ray beam, allowing access to a very large area of reciprocal space. The three versatile geometries of operation of the magnet mounted within the Huber diffractometer are shown in Fig.1.

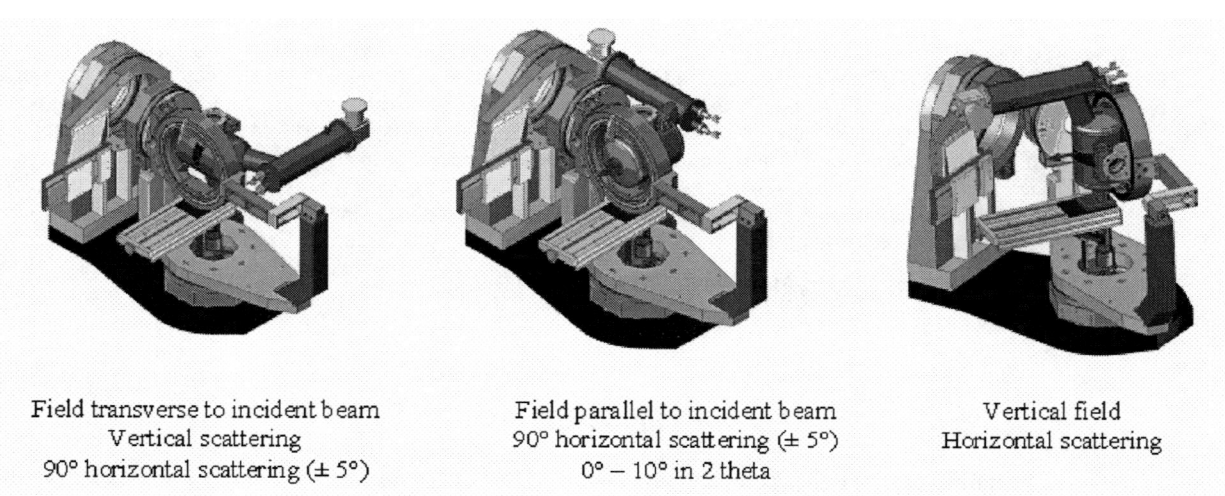

Field transverse to incident beam	Field parallel to incident beam	Vertical field
Vertical scattering	90° horizontal scattering (± 5°)	Horizontal scattering
90° horizontal scattering (± 5°)	0° – 10° in 2 theta	

FIGURE 1. The *XMaS*/AMI superconducting magnet mounted within the Euler cradle on the Huber diffractometer, illustrating the three possible scattering geometries.

The superconducting magnet, shown in Fig.2, is wound of twisted multifilamentary Niobium-Titanium (NbTi) superconductor embedded in a copper matrix. Twisted filaments maximize magnetic stability and minimize magnetic hysteresis. The former for the magnet coil is constructed of non-magnetic titanium alloy. Quench protection diodes are mounted within the magnet. The magnet system has been optimized to allow for a maximum number of ampere-turns within the geometrical constraints of the diffractometer. One of the most demanding requirements was the provision for 180° of clear room temperature radial access to the 4 Tesla central magnetic field, while maintaining a fixed, compact outer vacuum vessel which fits within the various Huber diffractometer configurations. At 4 Tesla, the split coil magnet produces a force of approximately 68,000 N, which acts to collapse the 180° gap. The magnet former was designed to support these forces and be rigid enough to minimize coil movement, which could cause a premature quench (i.e., below 4 Tesla). The field magnitudes are shown in Fig.3, with the plots in the zr plane. The z-axis is along the main axis of the split coil and the r-axis is along the radial axis of the split coil. Although the central field at $z=0$, $r=0$ is 4 Tesla, the peak fields within the windings are approximately 8.5 T. This is approaching the critical field limit of about 9 Tesla for the Nb-Ti wire used in this magnet, with some margin for conduction cooling. A commercial Sumitomo closed cycle refrigerator is used to cool the magnet system and AMI high temperature superconducting feedthroughs are installed to energise the magnet, whilst minimizing the heat load on the refrigeration system. The requirement that the refrigerator should be located remotely from the magnet, due to the need for multiple orientations within the Huber diffractometer also caused a number of design concerns. The vacuum vessel of the magnet needed to support the

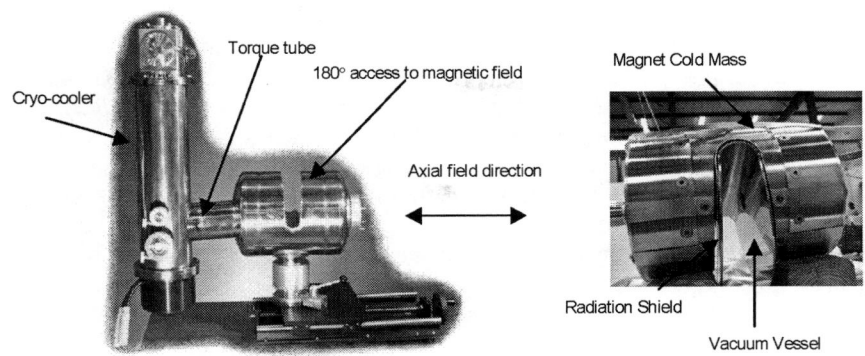

FIGURE 2. A view of the completed magnet is shown (to the left), showing cryocooler and open room temperature access to the magnetic field. A partially assembled view is also shown (to the right) illustrating the minimal spacing between the magnet cold mass, thermal radiation shield and the vacuum vessel.

weight of the refrigerator and it was also to ensure a good thermal conduction path between the magnet and the cryocooler. Heat transfer between the magnet and the cryocooler is accomplished by thermal links located within the torque tube. Differential thermal contraction is mitigated though flexible joints on both the first and second stage thermal links from the cryocooler. Another problem associated with multiple magnet orientations was locating the cold (< 4 K) magnet assembly within the radiation shield and vacuum vessel. Internal supports made from glass-fibre reinforced resin were designed to support the full weight of the magnet (some 34 Kg) in radial and axial loading conditions, whilst simultaneously limiting the heat loads to the magnet to less than 200mW. These supports also have to tolerate the dimensional changes due to the thermal contraction of dissimilar materials whilst locating the magnet in its correct position. In some areas there is less than 4mm between the cold magnet and the vacuum vessel at room temperature, as shown in Fig.2.

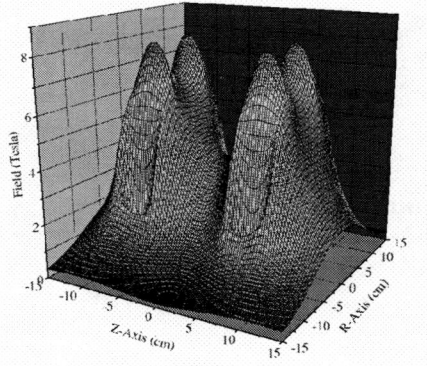

FIGURE 3. The 3D-magnetic field profile when the system is energized to 4.0 Tesla. Peak fields of ~8.5 Tesla can be seen within the windings at ~±7cm along the z axis.

The variable temperature insert is based around a RICOR 2/9 two-stage displex, capable of reaching 10K. However, a third stage has been developed by the cryogenics group at the I.L.L. in Grenoble [7]. This novel device is capable of operating down to 1.7 K using ^{4}He gas to within a few mK stability. It may also be operated over a wide range of angles without degradation of the base temperature.

FIRST RESULTS ; FERROMAGNETIC SCATTERING FROM TERBIUM AND DYSPROSIUM

Terbium and Dysprosium have hcp structures and become ferromagnetic at low temperatures with magnetic moments lying in the basal-plane. In order to investigate the dependence of the electronic band structure on crystallographic direction, the 4 Tesla XMaS/AMI superconducting magnet was employed to align a component of the moment along the c-axis. The ferromagnetic measurements were performed scattering horizontally through scattering angles of around 90° at the Tb and Dy L_{III} edges. The 4 Tesla magnetic field was applied in the vertical direction and reversed every 20 minutes. The asymmetry ratio, defined as the fractional change in intensity after reversing the samples magnetisation direction, was measured across the resonant energy and is shown in Fig.4 (light) along with the results obtained with a conventional 1 Tesla magnet where the moments were aligned in the basal plane (dark). The results show that the line shapes of both the Tb and Dy magnetised along the c-axis were similar to those recorded for other heavy-rare earths magnetised along the same crystallographic direction. These spectra consist of a weak pre peak and a stronger peak ~5 eV higher in energy. Note that when these samples were magnetised along the easy direction, i.e. in the basal plane, the intensities of these two peaks were similar. The separation and size of these two peaks is very important to obtain a qualitative and quantitative description of electronic structure and magnetism in the heavy rare-earths.

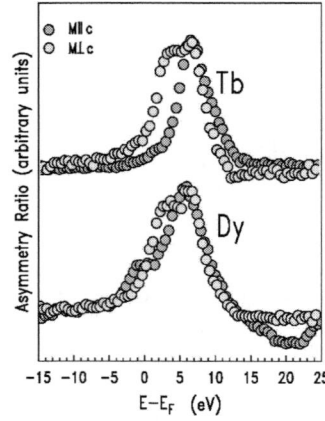

FIGURE 4. Charge-magnetic interference scattering measured in Tb and Dy at 2K. The 4 Tesla magnetic field applied vertically along the c-axis was reversed every 20 min.

CONCLUSIONS

A 4 Tesla supercoducting magnet has been designed and built to fit into a standard Huber diffractometer in both vertical and horizontal scattering geometries. Ferromagnetic diffraction measurements on single crystal Tb and Dy have been made at a base temperature of 2K in a 4 Tesla field, illustrating one application of this versatile magnet.

ACKNOWLEDGMENTS

This work was performed on the EPSRC funded ***XMaS*** beamline at the ESRF. We gratefully acknowledge American Magnetics Inc. USA, Huber Diffraktionstechnik GmbH, Germany and A.S. Scientific Ltd, U.K. for this collaboration. Also we thank S. Beaufoy from The University of Warwick, for additional support.

REFERENCES

1. D.F. Paul, M. J. Cooper, W. G. Stirling, Rev. Sci. Instrum. **66** (1995) 1741
2. S.D. Brown, L. Bouchenoire, D. bowyer, J. Kervin, D. Laundy, M. Longfield, D. Mannix, D. Paul, A. Stunault, P. Thompson, M. J. Cooper, C. A. Lucas and W. G. Stirling, J. Sync. Rad
3. D. Laundy, S.P. Collins, A.J. Rollason, J. Phys. Condens. Matter 3 (1991) 369
4. S.P. Collins, D. Laundy, A.J. Rollason, Philos. Mag. B 65 (1992) 37.
5. S.P. Collins, D. Laundy, G.Y. Guo, J. Phys.: Condens. Matter 5 (1993) L637.
6. S.D. Brown, P. Thompson, M. J. Cooper, J. Kervin, D. F. Paul, W. G. Stirling and A. Stunault. "Proceedings of 7th International Conference on Synchrotron Radiation Instrumentation", Part 2 NIM A, 467-468 (2000), pp 727-732
7. I.L.L. Annual report, 2001

A Low Temperature Two-Axis Goniometer for Azimuth Dependent Studies

Flora Yakhou*, Alistair Harris[†], Pascal Bernard*, Jean-Paul Valade*, Pascale P. Deen*,** and Gérard Lapertot[‡]

*European Synchrotron Radiation Facility - BP 220, 38043 Grenoble Cedex 9, France
[†]Etudes et Conception Mécaniques - Les Coings, 38210 Montaud, France
**presently at Institut Laue-Langevin - BP 156 , 38042 Grenoble Cedex 9, France
[‡]Commissariat à l'Energie Atomique, Département de Recherche Fondamentale sur la Matière Condensée,
SPSMS, 38054 Grenoble Cedex 9, France

Abstract. A novel insert was developed for top-loading liquid helium cryostats that allows a combined $\pm 90°$ sample tilt with a full $360°$ azimuthal rotation. This sample stick is operational down to 1.5 K and was primarily designed for X-ray scattering in reflecting geometry. The device not being intended for scanning but positioning purposes, motion precision, resolution and repeatability specifications were relaxed to $0.1°$ and readily achieved. This setup was initially designed for a specific cryostat and implemented at beamline ID20 of the ESRF but it can be easily adapted to similar top loading cryostats including cryomagnets. Initial results on the low temperature phases of CeB_6 are presented.

Keywords: x-ray diffraction, cryostat, low-temperature goniometer, azimuthal dependence, cerium hexaboride, orbital ordering
PACS: 61.10.-i, 07.20.Mc, 07.85.Jy, 71.27.+a

MOTIVATION

Condensed matter studies often require such parameters as temperature, pressure or magnetic field to be varied and set to rather extreme values. Temperatures below 5 Kelvin and magnetic fields of a few Tesla are of particular interest and now are readily achieved through heavy sample environment equipment such as cryostats and cryomagnets. In addition, recent experimental works in the field of strongly correlated electronic systems have outlined the prime importance of azimuthal dependencies of scattered intensities [1].

However, measuring such a dependence requires the sample and its environment to be rotated about the scattering vector of interest. In a standard four-circle geometry [2], this can in principle be achieved with a combined motion of the three sample rotation axes *theta*, χ and ϕ. In practice, the exercise is much more difficult when taking into consideration the volume, weight and tubing required for low temperature cooling devices.

A few years ago a dedicated setup was developed on beam line ID20 [3] at ESRF which allowed the rotation of a closed-cycle refrigerator (T > 10 K) about an additional axis on the existing diffractometer. The compact nature of closed-cycle refrigerators allows a large range of movements. Nevertheless, geometrical and off-axis load constraints limit this setup to scattering vectors close to specularity (with respect to the sample surface) and relatively low Bragg angles.

To date it has been only possible to perform an azimuthal dependence using closed-cycle refrigerators[1]. Heavy equipment such as liquid helium cryostats and/or cryomagnets cannot be tilted more than a couple of degrees from the horizontal scattering plane and a full rotation remained unfeasible. A dedicated sample stick has therefore been developed for the so-called "orange cryostat" [4] that accommodates a sample tilt supported by a full $360°$ azimuthal rotation stage (ρ axis). Any scattering vector in the horizontal plane can hence be set along the ρ axis by means of the sample tilt and the azimuthal dependence measured by a simple rotation about the axis.

---

[1] Recent developments have allowed to reach lower base temperatures of about 1.2 K

CP879, *Synchrotron Radiation Instrumentation: Ninth International Conference*,
edited by Jae-Young Choi and Seungyu Rah
© 2007 American Institute of Physics 978-0-7354-0373-4/07/$23.00

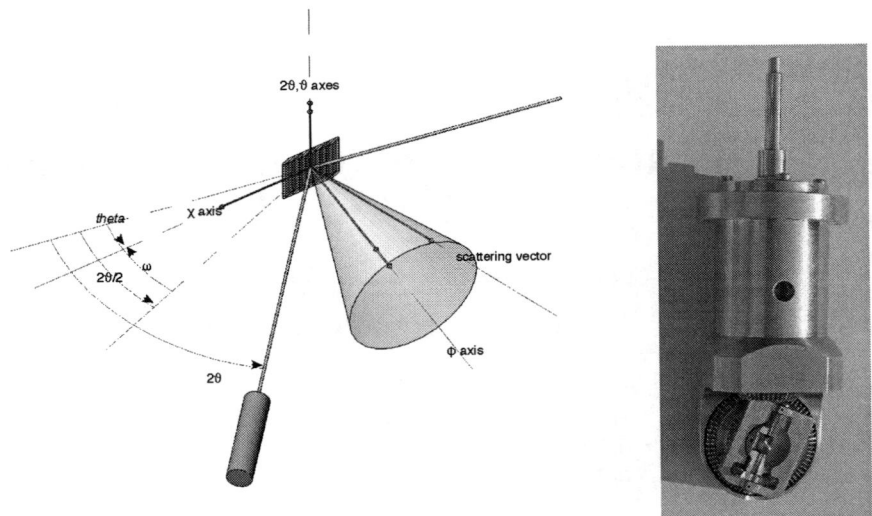

FIGURE 1. Scattering geometry and lower part of the insert

DESIGN AND MANUFACTURING

Specifications and Constraints

The design of the setup had to accommodate the following specifications: low temperature (1.5 K) operation, restricted Φ50 mm sample chamber diameter with full clearance for reflection geometry (hence 25 mm available only), 360°azimuthal rotation with an additional perpendicular axis that allows a minimum ±45°sample tilt.

The χ and ϕ axes of a standard four-circle diffractometer are restricted to ±3°when using the orange cryostat on beamline ID20. The effective angles at which the crystal plane of interest can be set into the horizontal scattering plane depend on the precise a priori knowledge of the crystal orientation and the way it is glued on its support. It is usually not better than a couple of degrees. The precession cone of the scattering vector is then defined by the deviation $\omega = theta - 2\theta/2$ from the bisecting geometry ($theta = 2\theta/2$) and the effective χ angle (see Fig. 1). For a given scattering vector this cone would be set within the accessible range in χ by means of the sample tilt and the usual diffractometer axes used for precise scanning. The motion precision, resolution and repeatability of both the azimuth and tilt axes were therefore relaxed to 0.1°.

Due to the above mentioned spatial constraints the sample size was limited to 6 mm in diameter with a maximum thickness of 2 mm. For the very same reasons, no sample adjustment with respect to the device axes was considered. The initial centering of the sample is hence critical and would require careful attention.

Solutions Chosen

In the solution retained the two motions are transmitted concentrically over 1m from the top flange of the sample chamber; a relative motion of the two concentric axes controls the sample tilt in the cradle, a joint motion rotates the cradle about the azimuthal axis. Special care was taken in the design and manufacturing processes to avoid any internal friction that would induce a relative motion of the two axes once the tilt is set.

The parts were manufactured and assembled by the company Elements Robert Pennacchiotti in Grenoble [5]. The size of the parts and the small (0.5) modulus gears required proven expertise in precision mechanics. Axes perpendicularity or concentricity were within the requested 100 μm of machining precision.

Mechanical plays at room temperature were optimized to accommodate the thermal contraction and to meet the specifications at low temperature. All the parts were machined in the same batch of bronze to minimize differential contraction effects. Bearings were excluded from the design for vacuum and low temperature compatibility. All surfaces were treated with molybdenum disulphide for self-lubrication (Lubodry®). Figure 1 shows the lower part of the insert before the surface treatment was applied. The thickness of the stainless steel tubes was optimized to minimize the amount of heat transmitted to the lower (colder) part of the insert. Baffles were included to prevent the natural convection of the exchange gas. A similar design to the standard orange cryostat sample stick was adopted with a close copper to copper contact in the exchanger zone of the cryostat.

FIRST TESTS

The actual commissioning of the setup presented a major difficulty. Because of the loose plays at room temperature, any test had to be performed in low temperature conditions where no optical access is available. All tests were therefore performed in-situ with a known sample using the x-ray beam. CeB_6 was chosen as the test sample being the leading case for the development of the insert.

Cryogenic tests consisted in measuring the base temperatures reached on the insert where a Si diode sensor is permanently installed and at the sample position where a test diode was installed. Comparison with the cryostat exchanger sensor temperature gives indication on the overall cooling performance of the insert and its inertia. Sample cooling within an orange cryostat is achieved through both conduction and exchange gas. Cooling cycles were hence performed with and without exchange gas. The results are fairly similar in terms of base temperature and cooling time. The thermal inertia after the initial cooling is equivalent to that of the standard sample stick and is negligible. However the ultimate cooling performance is given by the temperature effectively reached on the sample illuminated hence heated by the x-ray beam. It depends *in fine* on the sample conductivity and thermal contact to the holder.

The mechanical performances of the 360°azimuthal rotation were checked using the two blind zones at 90°and 270°. The sample tilt was tested through its capacity to compensate for a known variation of the *theta* angle once set in Bragg reflecting conditions. Repeated scans were used to check for resolution and reproducibility. A straightforward conclusion is that low temperature operation is definitely achieved. The results also show that the specifications in terms of motion precision, resolution and reproducibility are met.

In a typical experiment for measuring azimuthal dependences, the diffractometer angles for the scattering vector of interest are optimized at the starting azimuthal position; subsequent angles for a different azimuth are then derived from the so-called orientation matrix that relates the crystal coordinate system to the diffractometer angles. The accuracy of the calculation depends on the accuracy of the matrix it-self and the proper alignment of the diffractometer axes with respect to the beam. The accuracy that is effectively required depends in turn on the sample mosaicity and detector (or analyser) acceptance for the *theta* and χ angles respectively. Typical values are in the range 0.05°. The next section will show that these conditions can be met and a proper azimuthal dependence measured even in the case of very low scattering intensities.

APPLICATION TO A SCIENTIFIC CASE: CEB$_6$

CeB_6 is an example of a strongly correlated electronic system displaying an unusual magnetic field H versus temperature T phase diagram. It is still the subject of intensive investigations to explain its peculiar low temperature properties. In particular, a mysterious ordered phase II was identified by several techniques with a transition temperature T_Q=3.2 K at zero field. Several interpretations were given that could not reconcile all the experimental results until a recent non-resonant x-ray diffraction work was performed [6]. It is shown qualitatively and quantitatively that both quadrupole and hexadecapole moments order in phase II while octupole moments are further induced by magnetic fields.

Whilst non-resonant x-ray measurements do allow easier quantitative interpretation of the data, resonant x-ray scattering techniques directly probe the local symmetries and are element and electronic state selective. As a symmetry probe, the azimuthal dependence of the scattering intensity is the ideal tool to discriminate between various ordering models but is still being developed. A set of data on an otherwise "known" structure is of major interest to validate the technique.

In preliminary investigations at the Cerium L_{III} absorption edge, several reflections corresponding to the wave vector (1/2 1/2 1/2) were measured. According to its energy lineshape, the (5/2 1/2 1/2) reflection is likely to include both magnetic and multipolar ordering components and was chosen for the first tests. Typical count rates were in the 10 cts/s range with an incident beam attenuated by a factor 200 with respect to the full power of the beamline ID20. The attenuation was needed to lessen the effects of beam heating. It is worth noting that the same attenuation factor had to be used in similar conditions with the standard sample stick, thus showing the very good overall cryogenic performances of the new insert.

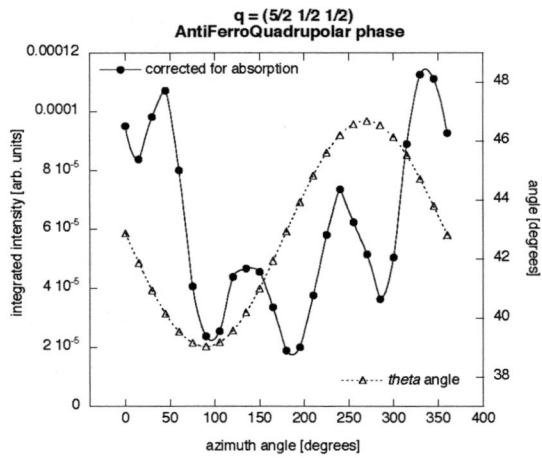

FIGURE 2. Azimuthal dependence of the (5/2 1/2 1/2 reflection) in phase II

Within the available crystal geometry, this reflection displays an $\omega = 15°$ off-specularity. Figure 2 shows the azimuthal dependence in the multipolar ordered phase II. Once the diffractometer and azimuthal stick were properly aligned, a complete such set could be measured within four to five hours. A lithium fluoride crystal was used as polarization analyzer. The analyzer crystal acceptance and sample mosaicity were about 0.05° and 0.01° respectively. This imposed stringent conditions on the accuracy of the angle calculations. The observed and calculated angles are plotted on Fig. 2. The difference remains small enough to be able to recover the total intensity within a few alignment scans. It demonstrates that this novel insert is reliable enough for routine azimuthal measurements of small signals in a short amount of time.

CONCLUSIONS

A novel insert with two perpendicular axes was developed for the so-called "orange cryostat". It is operational at the base temperature of the cryostat (1.5 K) and allows positioning of the respective angles within 0.1°. Full 360° azimuthal dependencies were recorded in the low temperature phases of CeB$_6$ on an off-specular reflection as a first application of this new setup. UHV operation should in principle be possible with slight modifications. Adaptation to the neutron case in transmission geometry can be envisaged as well as modifications for compatibility with high magnetic fields. In a more general way, the additional degrees of freedom offered by this setup make it a perfect in-situ low temperature crystal alignment tool.

ACKNOWLEDGMENTS

Fruitful discussions with Serge Pujol (ILL) and Riccardo Steinmann (ESRF) are gratefully acknowledged

REFERENCES

1. D. Mannix, Y.Tanaka, D. Carbone, N. Berboeft and S. Kunii *Phys. Rev. Lett.*, **95** 117206, (2005) and references within.
2. W.R. Busing and H.A. Levy, *Acta Cryst.* **22**, 457–464 (1967).
3. http://www.esrf.fr/UsersAndScience/Experiments/XASMS/ID20/
4. AS Scientific Products Ltd: http://213.107.68.166/asscientific/
5. Elts Robert Pennacchiotti, 35 Bd des Alpes 38240 Meylan, France
6. Y. Tanaka et al., *Europhys. Lett.* **68**, 671–677 (2004).

Upgrade of X-ray Magnetic Diffraction Experimental System and Its Application to Ferromagnetic Material

Kosuke Suzuki[A], Masahisa Ito[B], Naruki Tsuji[A], Hiromitu Akiyama[A], Kensuke Kitani[B], Hiromichi Adachi[C] And Hiroshi Kawata[C]

[A] *Graduate School of Engineering, Gunma University, Tenjin-cho 1-5-1, Kiryu, Gunma 376-8515, Japan.*
[B] *Faculty of Engineering, Gunma University, Tenjin-cho 1-5-1, Kiryu, Gunma 376-8515, Japan.*
[C] *Institute of Materials Structure Science, KEK, Oho 1-1, Tsukuba, Ibaraki 305-0801, Japan.*

Abstract. We have performed X-ray magnetic diffraction (XMD) experiment of ferromagnets at the Photon Factory (PF) of the High Energy Accelerator Research Organization (KEK) in Tsukuba. In this study, we have upgraded the XMD experimental system in order to apply this method to as many samples as possible. Upgrade was made for (1) the X-ray counting system and related measurement program, (2) the electromagnet, and (3) the refrigerator. The performance of the system was enhanced so that (1) the counting rate capability was improved from 10^4cps to 10^5cps, (2) the maximum magnetic field was increased from 0.85T to 2.15T, and (3) the lowest sample temperature was reduced from 15K to 5K. The new system was applied to an orbital ordering compound of $YTiO_3$, and we obtained spin magnetic form factor for the reflection plane (010) perpendicular to the b axis. The magnetic field of 2T was needed to saturate the magnetization of $YTiO_3$ along the b axis. These are the first data with the magnetization of $YTiO_3$ saturated along the b axis by the XMD.

Keywords: X-ray magnetic diffraction (XMD)
PACS: 601.10.Nz, 74 .25.Ha

INTRODUCTION

X-ray magnetic diffraction (XMD) for ferromagnets is an experimental technique that uses elliptically polarized synchrotron radiation. With this method, we can measure spin and orbital magnetic form factors, separately. [1,2] As magnetic form factor is related to magnetic moment density distribution through the Fourier transform, we can observe spin and/or orbital magnetic moment density distribution by the XMD.

We have performed the XMD experiment of ferromagnets at the Photon Factory (PF) of the High Energy Accelerator Research Organization (KEK) in Tsukuba. The experimental condition was that the X-ray counting rate capability was 10^4cps, the maximum magnetic field was 0.85T and the lowest sample temperature was 15K. We upgraded the XMD experimental system in order to apply this method to as many materials as possible and to perform the experiment more effectively. The upgrade was made for the X-ray counting system, the electromagnet and the refrigerator, and the experimental condition was improved for the counting rate capability, the maximum magnetic field, and the lowest sample temperature.

The new experimental system was applied to an orbital ordering compound $YTiO_3$. Though there is already a report of neutron diffraction experiment of this compound, [3,4] we would like to provide an alternative method with X-rays for measuring spin magnetic form factor. Final goal of this experiment is to observe three dimensional spin density distribution in real space.

DETAILS OF UPGRADE AND EXPERIMENTS

Details of upgrade are as follows.
(1) X-rays counting system: In the previous system, we used a spectroscopy amplifier, an ADC (analog to digital converter), and hardware-type MCA (multi channel analyzer) for electronic counting system. In the upgraded system, we use a high count-rate type Ge-SSD, a DSP module (digital signal processor, CANBERRA Model 9660), and software-type MCA (CANBERRA Genie 2000). By this upgrade, the count rate performance was expected to be

CP879, *Synchrotron Radiation Instrumentation: Ninth International Conference*,
edited by Jae-Young Choi and Seungyu Rah

elevated from 10^4cps to 10^5cps. In order to control the MCA in the XMD experiment we made the measurement program by using Visual Basic.

(2) Electromagnet: In the previous system, we used relatively small electromagnet which was operated at 15A and produced 0.85T. The upgraded electromagnet (Tamakawa Ltd) can be operated at 60A and is designed to produce magnetic field over 2T. Precise magnetic field measurement showed that the maximum field was 2.15T. Approximate size of the new electromagnet is 280mm in diameter and 300mm in width.

(3) Refrigerator: In the previous system, we used a relatively low-power refrigerator in which the sample temperature was as low as 15K. In the upgraded system, we use a more powerful refrigerator (Iwatani Ltd. HE05) in which the sample temperature is as low as 5K.

We performed the following experiments to check the performance of the upgraded system, and then we applied it to ferromagnetic YTiO$_3$.

Count Rate Performance

Preliminary experiment was made to check the count rate performance of the upgraded system. The sample was a single crystal of Fe. White beam of elliptically polarized synchrotron radiation was irradiated on the sample. The sample crystal was arranged so that the Bragg angle of (110) plane was 45 degree. The diffracted X-ray intensities of (220), (330), and (440) reflection planes were measured by the SSD and the related counting system. The X-ray intensity incident on the SSD was changed by inserting aluminum foils of various thicknesses between the sample and the SSD. Total count rate was varied from 1.7×10^4cps to 2.1×10^5cps.

The obtained data was fitted with the following conventional equation (1),

$$I = I_0 e^{-\mu t},$$
(1)

where I_0 is the incident X-ray intensity on the Al foils, and I is the transmitted X-ray intensity through the Al foils, μ is the absorption coefficient of Al and t is the thickness of Al foils.

Count Rate Dependence of the Flipping Ratio

We examined count rate dependence of the flipping ratio (magnetic effect) of Fe 220 diffraction intensity. The diffraction intensity was changed by changing the size of the incident X-ray beam on the sample crystal from 90μm $\times$ 90μm to 200μm $\times$ 200μm. Then, the X-ray intensity was varied from 2.0×10^4cps to 1.1×10^5cps and the flipping ratio of Fe 220 diffraction was measured.

Application to Ferromagnetic YTiO$_3$

The upgraded system was applied to the XMD experiment of YTiO$_3$. YTiO$_3$ is ferromagnetic below 30K, and measurement was at 15K. The previous magnetization measurement showed that the magnetization of YTiO$_3$ was saturated with magnetic field 2T even along the hard axis (the b axis). White X-ray beam was irradiated on the sample and the Bragg angle at the sample was fixed to 45 degree. The magnetic field of 2.15T was applied and the magnetization was aligned along the scattering vector. This configuration gives us spin magnetic form factor and is called S-configuration.

The flipping ratio R is given by the following equation (2),

$$R = \frac{(I_+ - I_-)}{(I_+ + I_-)},$$
(2)

where, I_+ is the diffracted X-ray intensity when the magnetic field was applied along one direction, and I_- is the diffracted X-ray intensity when the magnetic field was applied along reversed direction. The flipping ratio of the S-configuration is expressed by the following equation (3) according to the X-ray magnetic scattering theory. [1,2]

$$R = \frac{\gamma f_p \mu_s(k)}{\sqrt{2} n(k)}.$$
(3)

Here, $\gamma = \hbar\omega / mc^2$, where, $\hbar\omega$ is energy of X-rays and mc^2 is the electron rest mass energy. And, $f_P = P_c/(1-P_l)$, where P_c is the degree of circular polarization and P_l is the degree of linear polarization. $\mu_s(k)$ is spin magnetic form factor and $n(k)$ is crystal structure factor. Here, $k = \sin\theta/\lambda$, where λ is the X-ray wavelength and θ is the Bragg angle. By measuring the flipping ratio R of the diffracted X-ray intensity, we can observe the spin magnetic form factor $\mu_s(k)$. Spin density distribution can be obtained by the Fourier transform of the observed $\mu_s(k)$.

RESULTS AND DISCUSSION

Count Rate Performance

The result of the experiment of count rate performance is shown in Fig.1. In Fig.1 the measured X-ray intensities of 220, 330 and 440 diffraction of Fe were plotted against the thickness of Al foils. The data of solid circles, open squares, and crosses are corresponding to the diffraction of 220, 330, and 440, respectively. X-rays energies of 220, 330 and 440 diffraction were 8.65keV, 12.98keV and 17.3keV, respectively, as the Bragg angles for the diffraction was fixed to 45 degree. Because of larger absorption coefficient for lower X-ray energy observed slope was steeper for the lines of lower X-ray energy. These observed data were expressed well by the lines that were calculated with the equation (1). Without Al foils, the total diffracted X-rays intensity amounted to 2.1×10^5cps. It was confirmed that upgraded X-ray counting system is able to measure the X-ray intensity up to 2.1×10^5cps which is ten times higher than the maximum count rate of the previous system.

We measured FWHM (full width at half maximum) of the peak profile of Fe 220 diffraction X-rays of the upgraded system and compare it with the one of the previous system. The result is shown in Fig.2. In Fig.2, normalized intensity is plotted against X-ray energy. The solid circles are for the upgraded system at the total counting rate of 7×10^4cps, and the open circles are for the previous system at the total counting rate of 7×10^3cps. The FWHM was 0.38keV for the present system and 0.31keV for the previous system. This result shows that we achieved the count rate performance ten times higher than the previous system at the sacrifice of degradation of the FWHM by 20%.

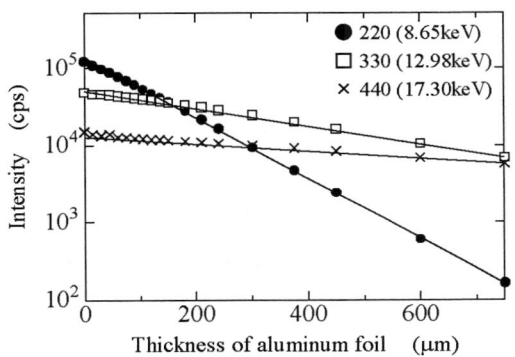

FIGURE 1. Measured X-ray intensity of 220, 330 and 440 diffraction of Fe for various thicknesses of Al foils. The energy of each diffraction is shown in parenthesis.

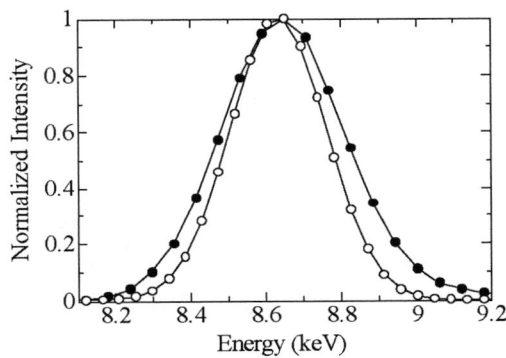

FIGURE 2. Comparison of FWHM between the upgraded system at 7×10^4cps (solid circles) and the previous system at 7×10^3cps (open circles).

Count Rate Dependence on the Flipping Ratio

We examined count rate dependence of the flipping ratio of Fe 220 diffraction. The result is shown in Fig.3. In Fig.3, flipping ratio is plotted against the total intensity of diffracted X-rays. Because of limited measurement time we have fairly large error bars. The observed flipping ratio keeps constant within the estimated error bars in the range of observed count rate. Therefore, it is concluded that the upgraded system is able to measure flipping ratio up to the count rate of 10^5 cps.

Measurement of Spin Magnetic form Factor of YTiO₃

We applied the upgrade system to the XMD experiment of $YTiO_3$. Reflection plane was (010) and Bragg angel was 45 degree. The magnetic field of 2.15T was applied along the [010] direction, which is parallel to the scattering vector. We measured the flipping ratio of the reciprocal lattice points of 060, 080, 0 10 0, 0 12 0, 0 14 0 and 0 16 0. Because the X-ray energies for 020 and 040 were too low to be detected, this system was not able to observe them. The peak profile of 0 10 0 was contaminated with fluorescent X-rays of YK_α. The X-ray energies of 0 10 0 diffraction and fluorescent X-rays of YK_α were 15.4keV and 14.9keV, respectively. We obtained the peak profile of the 0 10 0 diffraction by subtracting fluorescence profile from the contaminated profile. The polarization factor was determined by the same method as was reported previously. [5,6] The spin magnetic form factor was obtained from the observed flipping ratio R by using the equation (3). The observed spin magnetic form factors are shown in Fig.4.

The open circles are the data obtained previously. [7-9] The solid circles are obtained in this study. In Fig.4, $\mu_s(k)$'s for another points of 042, 084, 620, 820, 510, 10 2 0, 530 are shown, which were obtained also in this study. These data will be used to 3D reconstruction of the spin density distribution in the real space by using the maximum entropy method.

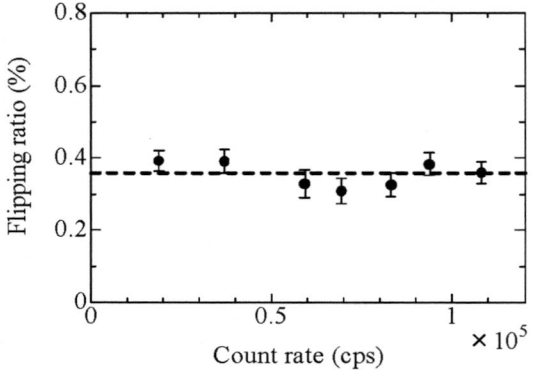

FIGURE 3. Count rate dependence of the flipping ratio of Fe 220 diffraction.

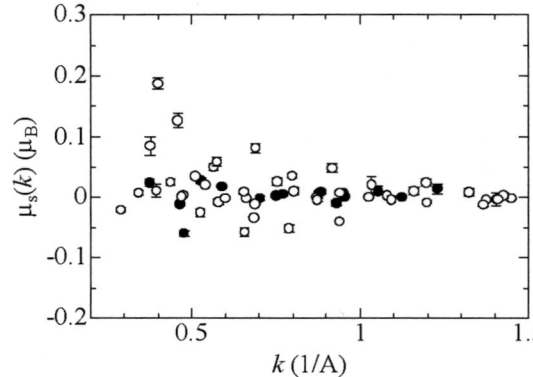

FIGURE 4. Spin magnetic form factor of YTiO$_3$. The solid circles are obtained in this study. The open circles are the data obtained previously. [7-9]

CONCLUSION

We performed upgrade of the XMD experimental system at the PF of the KEK. The upgrade was made for (1) the X-rays counting system, (2) the electromagnet, and (3) the refrigerator. We checked the performance of the upgraded system, and we confirmed that the counting rate capability was 10^{5}cps, the maximum magnetic field was 2.15T, and the lowest sample temperature was 5K. The experimental system was applied to ferromagnetic YTiO$_3$, and we obtained the spin magnetic form factor of YTiO$_3$ with the magnetization saturated along the b axis for the first time in the XMD.

ACKNOWLEDGMENTS

We acknowledge Dr. N. Nakao, Prof. Y. Murakami, Prof. Y. Taguchi and Prof Y. Tokura for providing the single crystal of YTiO$_3$ and useful discussion. The present research was partly supported by a Grant-in-Aid for Scientific Research from the Ministry of Education, Science, Sports and Culture (16340098). The XMD measurements were performed in the KEK-PAC proposal of 2005G005 and 2005G109.

REFERENCES

1. M. Blume: J. Appl. Phys. **57** 3615-3618 (1985).
2. S. W. Lovesey: J. Phys. C**20** 5625-5639 (1987).
3. H. Ichikawa, J. Akimitsu, M. Nishi and K. Kakurai: Physica B**281&282** 482-484 (2000).
4. J. Akimitsu, H. Ichikawa, N. Eguchi, T. Miyano, M. Nishi and K. Kakurai: J. Phys. Soc. Jpn. **70** 3475-3478 (2001).
5. M. Ito, H. Kawata, Y. Tanaka, A. Koizumi, T. Ohata, T. Mori, N. Sakai, N. Shiotani, M. Matsumoto, and S. Wako: Photon Factory Activity Report #13 38 (1995)
6. M. Ito, T. Fujii, H. Hashimoto, T. Nakamura, H. Kawata, T. Mori, N. Sakai, F. Itoh, and S. Nanao: Photon Factory Activity Report #13 41 (1995)
7. M. Ito, N. Tsuji, F. Itoh, H. Adachi, E. Arakawa, K. Namikawa, H. Nakao, Y. Murakami, Y. Taguchi and Y. Tokura: J. Phys. Chem. Solids **65** 1993-1997 (2004).
8. M. Ito, N. Tsuji, H. Adachi, H. Nakao, Y. Murakami, Y. Taguchi and Y. Tokura: Nucl. Instrum. Methods Phys. Res. B**238** 233-236 (2005).
9. N. Tsuji, M. Ito, H. Akiyama, K. Suzuki, H. Adachi, H. Nakao, Y. Murakami, Y. Taguchi, and Y. Tokura: Acta Cryst. A**61** C429-C430 (2005).

Synchrotron X-ray Powder Diffraction Studies in Pulsed Magnetic Fields

C. Detlefs*, P. Frings†, J. Vanacken†,**, F. Duc†, J. E. Lorenzo‡, M. Nardone†,
J. Billette†, A. Zitouni†, W. Bras§ and G. L. J. A. Rikken†

*European Synchrotron Radiation Facility, B.P. 220, F-38043 Grenoble, France
†Laboratoire National des Champs Magnétiques Pulsés, 143, avenue de Rangueil, F-31400 Toulouse, France
**Pulsveldengroep, Institute for Nanoscale Physics and Chemistry, Celestijnenlaan 200 D, B-3001 Leuven,
Belgium
‡Laboratoire de Cristallographie, CNRS, B.P. 166X, F-38043 Grenoble, France
§Netherlands Organisation for Scientific Research (NWO), DUBBLE CRG @ ESRF, BP 220, F-38043 Grenoble,
France

Abstract. X-ray powder diffraction experiments under pulsed magnetic fields were carried out at the DUBBLE beamline (BM26B) at the ESRF. A mobile generator delivered 110 kJ to the magnet coil, which was sufficient to generate peak fields of 30 T. A liquid He flow cryostat allowed us to vary the sample temperature accurately between 8 K and 300 K.

Keywords: Synchrotron instrumentation, high magnetic fields, magneto-mechanical properties, Jahn-Teller effect
PACS: 07.55.Db, 07.85.Qe, 75.80.+q, 71.70.Ej

INTRODUCTION

X-ray diffraction is the by far dominant method for the determination of crystal structures. Nevertheless, it has until recently [1, 2, 3] only been combined with moderate (low) steady fields produced by split-pair superconducting (resistive) magnets. The same limitations apply also for neutron diffraction, which is also sensitive to magnetic structures. Electron diffraction is not compatible with magnetic fields, due to the charge of the electron. Consequently, at present there is no experimental tool for structure determinations under magnetic fields higher than 17 T, the highest magnetic field available at synchrotron or neutron facilities.

For our pilot studies we have chosen to combine synchrotron x-ray powder diffraction with pulsed magnetic fields. Compared to neutron facilities, synchrotron x-ray sources offer much larger flux (monochromatic beams up to 10^{13} photons/s) so that a diffraction pattern can be recorded within a few milliseconds. Compared to resistive steady magnetic field installations, pulsed field setups are much smaller and do not require specialized mains connections or high power cooling systems. They can therefore be adapted for installation on existing synchrotron beamlines without major infrastructure investments.

EXPERIMENTAL SETUP

The experiment was performed on the DUBBLE CRG-beamline [5] (BM26B) at the European Synchrotron Radiation Facility. The radiation hutch EH2 of this beamline is large enough to hold the entire setup, including the generator and the high field magnet/sample cryostat assembly. Magnetic fields up to 30 T were generated in the magnet coil. A more detailed description of the experimental configuration can be found in Ref. [4].

The pulsed field generator used for the experiments presented here was developed at the LNCMP, Toulouse. It consisted of three subunits: Two identical storage units contained the capacitor banks, crowbar diodes and -resistors, and inductive current limiters. The third (control) unit housed the charger, thyristor stack, dump resistors and -relays, and the current and voltage monitors. The generator could be charged up to 130 kJ. Charging up to full voltage (16 kV) took less than one minute. The energy of the capacitor bank was released into the coil through an optically triggered thyristor switch.

The assembled generator had dimensions of ($h \times d \times w$) $1.25 \times 1.30 \times 2.85 \, \text{m}^3$ and weighed approximately 2800 kg. It was connected to the high field magnet by a high voltage-high current coaxial cable. The magnet coil was designed

CP879, Synchrotron Radiation Instrumentation: Ninth International Conference,
edited by Jae-Young Choi and Seungyu Rah

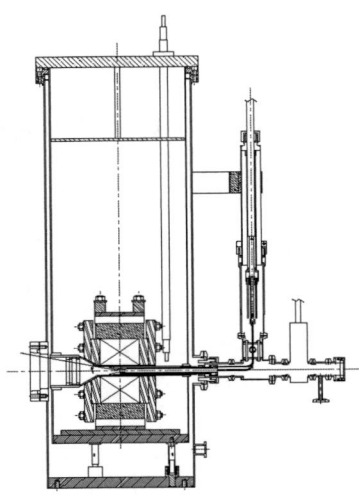

FIGURE 1. Cross section of the magnet/sample cryostat assembly.

and manufactured at the LNCMP in Toulouse. Its bore was Ø22 mm, the external diameter of the coil was 124 mm, and the height was 74 mm.

The coil was horizontally mounted into a cryostat with separate magnet and sample spaces (Fig. 1). The beam axis was along the bore of the magnet (Faraday geometry). In normal operation the magnet was immersed in liquid nitrogen, which reduced the electrical resistance of the coil and also acted as a thermal screen for the sample volume. Apart from this thermal screening effect the sample and magnet temperatures were completely decoupled.

A He flow cryostat, custom designed and built at the LNCMP in Toulouse, was inserted from the upstream side into the bore of the magnet (Fig. 1). Samples were loaded and removed *in situ*, using a load lock and transfer stick. Downstream of the sample the assembly was optimized for the largest possible optical access. However, x-ray scattering angles were limited to 14 degrees in order to stay with a well proven classical coil design. The sample space was evacuated. Kapton foil (120 μm thickness) was used for the entrance and exit windows.

Samples were prepared and mounted as follows: Single crystals were ground into a fine powder and embedded in low molecular weight polyvinylpyrolidon (PVP) in order to inhibit field-induced motion of the powder particles while at the same time improving thermal contact. The resulting pellets were mounted in a sample holder that was screwed into the cryostat by means of an actuator.

The time structure of the magnetic field pulse is shown in Fig. 2. The field rose from zero to maximum within 4 ms. This time constant was governed mainly by the capacitance of the generator, and the added inductances of the magnet coil and the current limiters integrated into the generator. During the decaying half cycle the magnet was short circuited by the crowbar resistor and -diode. The resulting time constant was 4.1 ms for maximum field. The full width at half maximum (FWHM) of the magnetic fields pulses was 18 ms.

Due to the restrictions of magnetic pulse length and X-ray beam intensity the experiment requires the accumulation of some tens of magnetic field shots to acquire statistically relevant data. After several tests we have run the experiment in the fast-pulse mode, triggering 10–20 pulses as fast as the charger permitted before allowing the coil to cool down back to liquid nitrogen temperature. The detector was read out after such a series of pulses. During these sequential pulses the magnet heated up gradually so that the cool-down period after a series of pulses had to be extended – after 15 pulses at the maximum field of 30 T the coil had heated up to 260 K, and 45 minutes of cool down time were required. Furthermore, the electrical resistance of the wire material increased with temperature and a progressively larger part of the energy dissipated into the coil. Consequently the peak field and the pulse length obtained for a given energy both decreased, as shown in Fig. 2. At this point of our studies the decrease of the peak field is not problematic. The effect can be controlled by limiting the number of pulses per series before cool-down, or by adapting the capacitance of the generator such that the pulse length corresponds to the maximum admissible temperature increase in one single pulse (see below).

The DUBBLE beamline is situated at the dipole port BM26 of the ESRF. Photons emitted from the 0.8 T bending magnet source were monochromatized by a Si(111) double crystal monochromator tuned to 21 keV ($\lambda = 0.59$). The

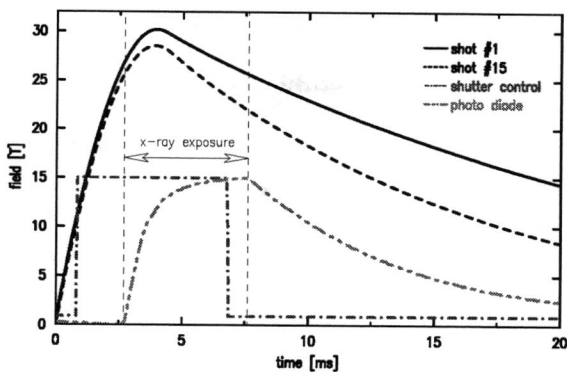

FIGURE 2. Time structure of the experiment. A common trigger signal triggers the magnetic field pulse and the x-ray shutters. The delays are adjusted such that the x-ray shutters expose the detector so that the average magnetic field is within 90% of its maximum value. In the fast-pulse mode (see text) the coil progressively heats up, increasing its resistance and thus leading to reduced field strength and faster decay. In the first pulse the magnetic field varied between 25.6 and 30.2 T (∅28.2 T), whereas in the 15th pulse it varied from 21.9 to 28.5 T (∅25.8 T). Due to the delay between the control pulse and the opening of the shutter it was not possible to use the magnet current to control the shutter.

limited opening angle (14 deg) of the cryostat restricts the available Q-space to 2.6^{-1}, i.e. Bragg reflections having d-spacings larger than 2.4 . In order to obtain the best possible resolution the x-ray beam was focused onto the detector.

In the absence of a large area detector that is fast enough to follow the evolution of the magnetic field pulse (see Fig. 2) we were forced to record a time-integrated signal over a short x-ray pulse synchronized with the field generator.

The timing of the x-ray pulse was generated by two mechanical x-ray shutters. The x-rays then passed through the cryostat and the sample. Downstream of the sample the direct beam was intercepted by a lead beam stop equipped with a photodiode and a small fluorescent screen providing a second means of controlling the time structure of the x-ray beam (see Fig. 2). Scattered radiation impinged on the 2D detector, an on-line image plate (MAR Research model MAR345).

The opening of the X-ray beam shutter must be synchronized with the magnetic field pulse. The shutter opened and closed after predefined delays (Fig. 2), adjusted such that the average magnetic field during the exposure is within 90% of the maximum. This corresponds to time frames of 4.9 ms. To obtain enough diffraction counts in the detector, sequences of pulses were required. Each pulse added 4.9 ms measuring time for a spectrum at one given field value and the detector integrates the x-ray signal over this exposure time. This process was repeated until the data were statistically relevant. In this mode of data acquisition a separate series of field pulses is required for each value of the magnetic field.

The diffracted intensity as function of scattering angle, 2θ, was obtained by angular integration of the Debye rings.

OBSERVATION OF A FIELD INDUCED PHASE CHANGE IN TBVO$_4$

Terbium ortho-vanadate, TbVO$_4$ is a textbook example for a cooperative Jahn-Teller transition (JT) mediated by quadrupolar interactions between the $4f$ moments [6]. At high temperatures TbVO$_4$ crystallizes in the tetragonal zircon structure. Upon lowering the temperature through $T_{JT} = 33$ K it undergoes a cooperative JT transition: The crystal spontaneously distorts along the [110] direction to orthorhombic symmetry.

Single crystals of TbVO$_4$, grown by the flux method, were kindly provided by P.C. Canfield of Ames Laboratory (USA). Powder samples were prepared as described above.

Powder x-ray diffraction diagrams were recorded as described above, using an x-ray photon energy of 21 keV. Figure 3 shows the field dependence of the X-ray spectrum, as observed at $T = 7.5$ K (left) and $T = 39$ K (right) for different magnetic fields. The graphs are shifted vertically for clarity.

We first consider the zero field measurements at 39 K and 7.5 K. In the high temperature phase we observe two Bragg reflections with Miller indices (using the high temperature tetragonal unit cell, subscript tet) $(211)_{tet}$ near 11.8 deg and $(112)_{tet}$ near 12.6 deg. The low temperature distortion due to the Jahn-Teller effect is clearly visible as a splitting of these peaks: $(211)_{tet} \rightarrow (311)_{ortho} + (131)_{ortho}$ and $(112)_{tet} \rightarrow (202)_{ortho} + (022)_{ortho}$, where the subscript

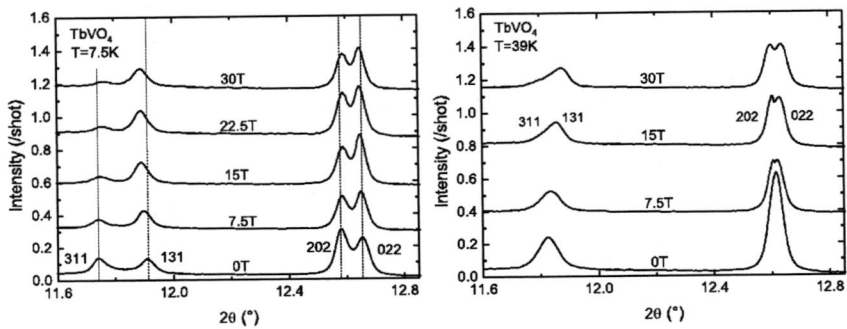

FIGURE 3. Field dependence of the diffraction pattern at $T = 7.5\,\mathrm{K}$ and $T = 39\,\mathrm{K}$.

ortho signifies indexing with respect to the orthorhombic unit cell.

Next, we investigate the evolution of the spectra as function of applied field at low temperature, $T \approx 7.5\,\mathrm{K}$ (Fig. 3, left). Immediately visible is a change of the relative amplitude of the high- and low angle partners. We attribute this to the preferential population of domains, due to an in-plane (a_{ortho} vs. b_{ortho}) magneto-crystalline anisotropy of TbVO$_4$. The splitting decreases with increasing field, indicating that the Jahn-Teller effect is suppressed.

In the high temperature spectra (Fig. 3, right) we observed that upon applying the magnetic field a splitting ($(202)/(022)_{\mathrm{ortho}}$ pair) appears, and that a preferential domain population develops ($(311)/(131)_{\mathrm{ortho}}$ pair), i.e. the magnetic field *induces* the Jahn-Teller state.

At $B = 30\,\mathrm{T}$ the magnitude of the distortion is similar for both temperatures. At all temperatures the peaks broaden as the field is applied. This is due to the strong dependence of the phase diagram on the angle between the magnetic field and the sample's c-axis.

Our experiment thus shows that the magnetic field pulse modified the Jahn-Teller distortion in TbVO$_4$, and that these changes can be detected using x-ray powder diffraction at a synchrotron source. A quantitative analysis of these data will be published elsewhere [7].

ACKNOWLEDGMENTS

The authors acknowledge the NWO/FWO Vlaanderen and ESRF for granting the beamtime for these experiments, and thank the staff of DUBBLE CRG and the ESRF for help in setting up these experiments. F. Meneau, D. Detollenaere, and J. Jacobs provided much needed technical assistance. The safety issues of the equipment were addressed with the help of ESRF Safety Group. P. van der Linden's active support with the cryogenics was greatly appreciated. The MAR345 online image plate detectors were lent to us by the ESRF detector pool and by T. Buslaps (ESRF, ID15 beamline). The samples of TbVO$_4$ were kindly provided by P.C. Canfield.

Part of this work was supported by the European Community under the FP6 contracts DeNUF (RIDS-CT-2005-011760) and EuroMagNET (RII3-506239).

REFERENCES

1. Y. H. Matsuda et al, Physica B **346–347** 519 (2004).
2. Y. H. Matsuda, T. Inami, K, Ohwada, Y. Murata, H. Nojiri, Y. Murakami, H. Ohta, W. Zhang, and K. Yoshimura, J. Phys. Soc. Jpn, **75**, 024710 (2006).
3. Y. Narumi et al, J. Syn. Rad. **13**, 271 (2006).
4. P. Frings et al, Rev. Sci. Instr. **77**, 063903 (2006).
5. W. Bras et al, J. Appl. Cryst. **36**, 791 (2003).
6. G. A. Gehring and K. A. Gehring, Rep. Prog. Phys. **38**, 1 (1975).
7. J. Vanacken et al, in preparation.

Element-Specific Hard X-ray Micro-Magnetometry to Probe Anisotropy in Patterned Magnetic Films

Motohiro Suzuki*, Masafumi Takagaki*, Yuji Kondo†, Naomi Kawamura*, Jun Ariake†, Takashi Chiba†, Hidekazu Mimura** and Tetsuya Ishikawa‡,*

*SPring-8/JASRI, 1-1-1 Kouto, Sayo, Hyogo 679-5198, Japan
†Akita Research Institute of Advanced Technology, 4-21 Sanuki, Araya, Akita 010-1623, Japan
**Department of Precision Science and Technology, Graduate School of Engineering, University of Osaka, 2-1 Yamada-oka, Suita, Osaka 565-0871, Japan
‡SPring-8/RIKEN, 1-1-1 Kouto, Sayo, Hyogo 679-5148, Japan

Abstract. We developed a technique for element-specific magnetic hysteresis measurements with a 2.5-μm spatial resolution using a focused hard X-ray beam to study magnetic films with lateral microstructures. This technique was applied to a 100-nm-square magnetic dot array fabricated by focused ion beam lithography in an 8×8 μm area on a CoPt film. Measured magnetic hysteresis curves illustrated the notable change in magnetic reversal process of the dots, induced by cutting a continuous film into isolated dots.

Keywords: X-ray magnetic circular dichroism, microscopy, magnetic recording, patterned media
PACS: 68.37.Yz, 07.55.Jg, 75.50.Ss, 87.64.Ni

INTRODUCTION

Patterned magnetic recording media is one of the key technologies for ultra-high density information storage in the next decade [1]. Recording on isolated magnetic dots would greatly improve the thermal stability and would reduce the media noise compared to continuous granular media presently used. Patterned media with a ~ 25 nm dot pitch adopted in the perpendicular recording scheme can achieve recording density of 1 Tbit/in^2. In this context, magnetization measurements with a lateral resolution have become increasingly important to characterize the magnetic properties, in particular magnetic anisotropy, of those media. A magnetic force microscope (MFM) offers a sub-100 nm spatial resolution and can visualize the magnetization of an individual recording bit. However, MFM is not sensitive to in-plane magnetization; therefore, full magnetometry (measurements for various magnetization direction) to determine the anisotropy is impossible.

X-ray magnetic circular dichroism (XMCD) magnetometry is an effective means for characterizing magnetic films, offering element specificity and monolayer sensitivity. Using this technique element-specific magnetization curves [2] are measured in a magnetic field with arbitrary strength and direction to a sample to study the magnetic anisotropy. XMCD magnetometry would be a desirable tool for characterizing patterned magnetic films when a microscopic capability is properly added.

In this paper, we report the development element-specific hard X-ray micro-magnetometry technique at an undulator beamline BL39XU [3] in SPring-8. Basis of the technique is XMCD measurements in a variable magnetic field, using an X-ray microbeam generated by Kirkpatrick and Baez (KB) mirror optics [4, 5]. An application to patterned magnetic media fabricated by focused ion beam (FIB) lithography is described to demonstrate the effectiveness of the technique.

INSTRUMENTATION

The instruments for X-ray micro-magnetometry consist of the scanning X-ray magnetic microprobe [5] developed in 2004 and equipments currently introduced: (i) a dedicated electromagnet and (ii) stages to incline the sample at an arbitrary angle with respect to the applied magnetic field. Figure 1(a) shows an illustration of the instruments. The KB focusing mirrors are made of silicon. The surfaces are elliptic cylindrical shaped by the plasma chemical vaporization machining and the elastic emission machining [6, 7]. The focal lengths are 300 mm for the vertical and 150 mm for

CP879, *Synchrotron Radiation Instrumentation: Ninth International Conference*,
edited by Jae-Young Choi and Seungyu Rah
© 2007 American Institute of Physics 978-0-7354-0373-4/07/$23.00

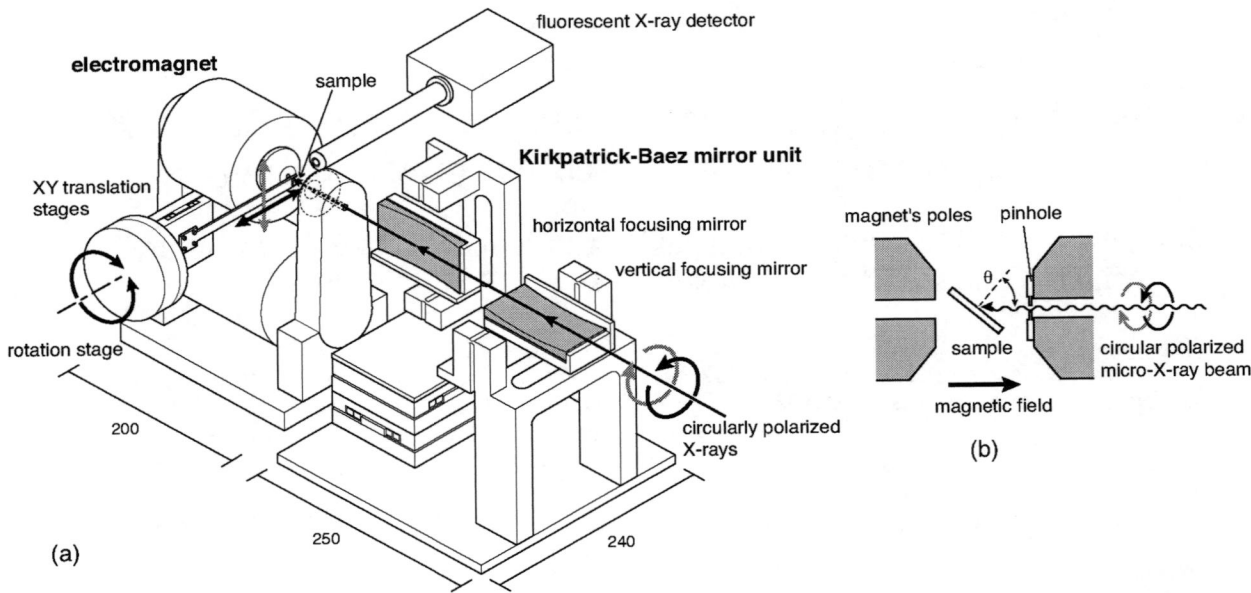

FIGURE 1. (a) Setup of the hard X-ray magnetic microprobe. (b) Configuration around the sample.

the horizontal mirrors. The glancing angles are 1.4 and 1.8 mrad for the vertical and horizontal mirrors, respectively. The detail of the KB mirror unit and the focusing performance are described elsewhere [4, 5].

The electromagnet is downsized ($200D \times 150W \times 250H$ mm) such that the sample can be placed at the focal point, just 100 mm from the horizontal mirror tail. The design with an asymmetric coil arrangement offers a large open space around the poles, allowing a detector head access the sample as close as possible. Available magnetic field is 8 kOe, which is sufficient for saturating magnetization of most hard magnetic films. Field can be applied horizontally, along the X-ray beam. A focused X-ray beam is introduced onto the sample placed between the magnet poles through a $\phi 3$ mm hole drilled at the center of the poles and a $\phi 20$ μm pinhole attached to the upstream pole [Fig. 1(b)]. The pinhole works as a diaphragm to cut stray X-ray intensity on the tail of the focused beam profile to improve the signal-to-background ratio in XMCD measurements.

High-precision pulse-motorized XY stages are used to translate a sample in vertical and horizontal directions within the sample film plane. The minimum step is 0.125 μm/pulse. The sample is rotatable along the horizontal axis so that the angle between sample surface and X-ray beam (also magnetic field) is adjustable from $\theta = 0°$ (normal incidence) to 90° (grazing incidence).

X-ray fluorescence yields from the sample are measured using a silicon drift detector (RÖNTEC, XFlash Detector 1000) placed at the side of the sample. The detector is operational with a total photon input of 10^5 counts/s and offers moderate energy resolution of ≤ 300 eV.

X-RAY MICRO-MAGNETOMETRY IN PATTERNED MAGNETIC FILMS

The technique was applied to CoPt alloy thin films with a dot array structure fabricated in a limited area of the film. This application aimed to detect changes in magnetic properties of total magnetization induced by cutting a continuous film into isolated dots despite our technique offering element specificity. Element-specific magnetization hysteresis measurement were performed at the Pt L_3 edge based on the assumption that XMCD amplitude of Pt is proportional to the total magnetization of the alloy films.

The magnetic dot array sample was prepared by patterning a continuous film using FIB lithography [8]. The original film was $Co_{80}Pt_{20}(at.\%)(15$ nm)/Au(6 nm)/Ti(5 nm), deposited on a glass substrate by magnetron sputtering at room temperature. Operation condition of an FIB apparatus (SMI2050, SII NanoTechnology Inc.) was Ga ions of 30 kV acceleration voltage and 1 pA sample current. The original film was prepared on the whole area (5×5 mm^2) of the substrate, whereas the dot array were placed only in an 8×8 μm area in the center of the sample. Figure 2 shows

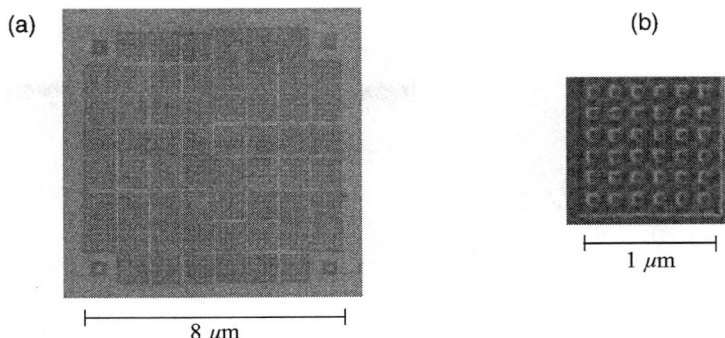

FIGURE 2. Scanning ion microscope images of the patterned magnetic film: (a) whole dot array, (b) one array element with larger magnification.

images of the dot array, taken by a scanning ion microscope. Whole dot array shown in Fig. 2(a) consists of 60 pieces of elemental array of 1×1 μm square shown in Fig. 2(b). The elemental array includes 6×6 matrix of 100 nm square isolated dots with a 100 nm spacing. Magnetic properties of the original film were characterized by alternating gradient force magnetometry (AGM).

Element-specific micro-magnetometry measurements were performed using the setup shown in Fig. 1. X-ray energy was tuned to 11.560 keV at which the XMCD amplitude for the Pt L_3 resonance takes a maximum. A magnetization curve was determined by monitoring the XMCD amplitude, $\Delta\mu(H) = \mu^-(H) - \mu^+(H)$, where $\mu^+(H)$ [$\mu^-(H)$] denotes the fluorescence yields when the incident photon momentum is parallel [antiparallel] to the X-ray wave vector, as a function of external magnetic field H. The focused beam size was 2.4 (H)×2.5 (V) μm^2 in full width at half maxima. Magnetization curves were taken at particular area on the sample, *i. e.* for the unpatterned area (continuous film) and the patterned area (dot array), by changing the sample position so that the X-ray beam was positioned at the target area. The beam size was rather large compared to a single dot, and measured magnetization signals were averaged over ~ 120 dots included in the X-ray beam spot. Moreover, for dot array magnetization curves for various direction of external magnetic field, from out-of-plane to nearly in-plane, were obtained at different sample orientations ($\theta = 0$–70°). Photon flux within the X-ray beam spot was 3×10^{10} estimated from an ionization chamber current. Resulting Pt L_α fluorescence intensity was ~ 10000 counts/s for continuous film and ~ 3000 counts/s for dot array, which ensures a $< 1\%$ statistical accuracy for an acquisition time of ten seconds per points.

RESULTS AND DISCUSSION

Figure 3(a) compares magnetization curves of continuous film and dot array for out-of-plane magnetization ($\theta = 0°$). This result demonstrate that magnetic properties of the film was significantly modified by FIB patterning. The XMCD hysteresis curve for continuous film shows the linear feature with a small coercive field, $H_c = 200$ Oe and remanent magnetization ratio, $M_r/M_s = 0.1$, which are consistent with the total magnetization curve recorded using AGM. For dot array, the hysteresis is rather square shape; coercivity value was enhanced to $H_c = 1600$ Oe, and remanent magnetization was increased to $M_r/M_s = 0.7$. These results suggest that the magnetization reversal process changes from the domain wall motion type in the original continuous film to different one with narrower switching field distribution in the patterned dot array. Nevertheless, size of the dots is a few times larger than the exchange length so that coherent magnetization rotation is unlikely to occur. Further investigation would be needed to correctly understand the reversal process in the patterned dot array [8].

Vertical axis of Fig.3 denotes the XMCD amplitude relative to the polarization-averaged X-ray fluorescence yield, $[\mu^+(H) + \mu^-(H)]/2$, so that XMCD values plotted are proportional to the magnetic moment of a Pt atom. For dot array the saturation value of XMCD amplitude for $H > 3000$ Oe is half of that for continuous film. The decrease in XMCD amplitude represents net reduction of magnetization, which is probably due to damage from Ga ion irradiation in the FIB process [9, 10].

Figure 3(b) shows hysteresis curves of patterned dot array in out-of-plane ($\theta = 0°$) and inclined ($\theta = 60°$) magnetization. As changing the magnetic field direction from out of plane to in plane, shape of the hysteresis curves changed from square to linear, decreasing the coercivity. These results directly show strong perpendicular anisotropy of the dot

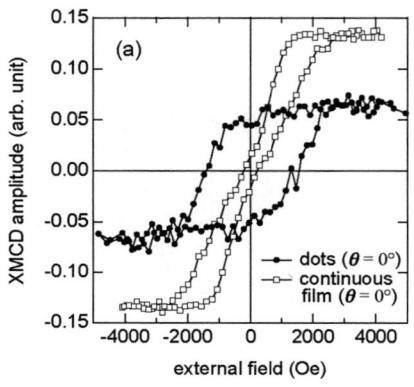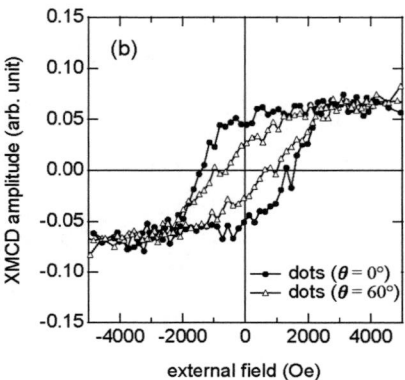

FIGURE 3. Element-specific magnetization curves of a $Co_{80}Pt_{20}$(at.%) film. (a) Out-of-plane magnetization hysteresis obtained for areas of continuous film (open square) and dot array (solid circle). (b) Hysteresis curves of out-of-plane ($\theta = 0°$) and inclined ($\theta = 60°$) magnetization for dot array.

array.

CONCLUSION

Effectiveness of the X-ray micro-magnetometry technique has been demonstrated by the magnetic hysteresis measurement of the 100-nm dot array in an 8×8 μm area. Change in magnetization reversal process induced by FIB patterning was clearly observed. Full magnetometry measurements for magnetic materials of such a small size are unattainable neither using a superconducting quantum interference device nor MFM. This technique will be a powerful tool for characterization of patterned magnetic media or magnetic random access memories, those are candidate for a future magnetic recording device of extreme high density.

ACKNOWLEDGMENTS

The experiments were performed at BL39XU in SPring-8 with the approval of the Japan Synchrotron Radiation Research Institute (JASRI) (Proposal No. 2005B0785). This work was partially supported by Akita Prefecture Collaboration of Regional Entities for the Advancement of Technological Excellence, JST.

REFERENCES

1. C. A. Ross, H. I. Smith, T. Saras, M. Schattenburg, M. Farhoud, M. Hwang, M. Walsh, M. C. Abraham, and R. J. Ram, *J. Vac. Technol.* **B17**, 3168–3176 (1999).
2. V. Chakariana, Y. U. Idzerda, G. Meigs, E. E. Chaban, J.-H. Park, and C. T. Chen, *Appl. Phys. Lett.* **66**, 3368–3370 (1995).
3. H. Maruyama, M. Suzuki, N. Kawamura, M. Ito, E. Arakawa, J. Kokubun, K. Hirano, K. Horie, S. Uemura, K. Hagiwara, M. Mizumaki, S. Goto, H. Kitamura, K. Namikawa, and T. Ishikawa, *J. Synchrotron Rad.* **6**, 1133 (1999).
4. K. Yamauchi, K. Yamamura, H. Mimura, Y. Sano, A. Saito, K. Endo, A. Souvorov, M. Yabashi, K. Tamasaku, T. Ishikawa, and Y. Mori, *Jpn. J. Appl. Phys.* **42**, 7129–7134 (2003).
5. M. Takagaki, M. Suzuki, N. Kawamura, H. Mimura, and T. Ishikawa, "Proceedings of the 8th International Conference on X-ray Microscopy," IPAP Conference Series 7, 2006, p. in press.
6. K. Yamauchi, H. Mimura, K. Inagaki, and Y. Mori, *Rev. Sci. Instrum.* **73**, 4028–4033 (2002).
7. K. Yamamura, K. Yamauchi, H. Mimura, Y. Sano, A. Saito, K. Endo, A. Souvorov, M. Yabashi, K. Tamasaku, T. Ishikawa, and Y. Mori, *Rev. Sci. Instrum.* **74**, 4549–4553 (2003).
8. Y. Kondo, T. Keitoku, S. Takahashi, N. Honda, and K. Ouchi, *J. Magn. Soc. Jpn.* **30**, 112–115 (2006).
9. T. Rettner, S. Anders, T. Thomson, M. Albrecht, Y. Ikeda, M. E. Best, and B. D. Terris, *IEEE Trans. Magn.* **38**, 1725–1730 (2002).
10. Y. Kondo, J. Ariake, T. Chiba, M. Suzuki, M. Takagaki, N. Kawamura, H. Mimura, and T. Ishikawa (in preparation).

Soft X-ray Magnetic Circular Dichroism of Ce(Fe$_{0.8}$Co$_{0.2}$)$_2$

M. Tsunekawa[*], S. Imada[*], A. Matsumoto[*], A. Yamasaki[*], H. Fujiwara[*], S. Suga[*], B. Schmid[*†], H. Higashimichi[*], T. Nakamura[¶], T. Muro[¶], and H. Wada[‡]

[*]Graduate School of Engineering Science, Osaka University, Toyonaka, Osaka 560-8531, Japan
[†]Experimentalphysik II, Universität Augsburg, D-86135, Augsburg, Germany
[¶]Japan Synchrotron Radiation Research Institute, Mikazuki, Hyogo, 679-5198, Japan
[‡]Graduate School of Science, Kyushu University, Higashi-ku, Fukuoka 812-8581, Japan

Abstract. Soft x-ray magnetic circular dichroism study was performed at the $L_{2,3}$ edges of iron and cobalt and the $M_{4,5}$ edge of cerium in the ferromagnetic Ce(Fe$_{0.8}$Co$_{0.2}$)$_2$ compound. The spin and orbital angular magnetic moments of Fe and Co were separately estimated by applying the sum-rules to the magnetic circular dichroism spectra. The estimated magnetic moment of Co is much smaller than that of Fe, suggesting that the small magnetic moment of Co is associated with the suppression of the magnetization in this system. We have also estimated the Ce $4f$ electron number from the absorption spectrum.

Keywords: Magnetic circular dichroism, soft X-ray absorption, Synchrotron radiation, Rare earth, transition metal.
PACS: 75.50.Cc, 78.70.Dm

INTRODUCTION

Soft x-ray magnetic circular dichroism (SXMCD) has attracted much interest as a useful tool to investigate ferromagnetic and ferrimagnetic states. The spectra at the $L_{2,3}$ edges of transition metal compounds directly reflect the net magnetic moment, because the core electron is excited into the $3d$ state responsible for the magnetic order. Special interest has been focused on the magnetic properties of Ce(Fe$_{1-x}$Co$_x$)$_2$ systems ranging from the superconductor CeCo$_2$ to the ferromagnet CeFe$_2$ with a Curie temperature $T_C \approx 235$ K [1]. Although the temperature dependence of the magnetization of CeFe$_2$ shows a transition from the high-temperature paramagnetic phase to the ferromagnetic phase around T_C with decreasing temperature, the substitution of Fe by Co results in the appearance of the second transition (first-order-transition) from the ferromagnetic to antiferromagnetic phase at $T_S \approx 72$ and 69 K in the Ce(Fe$_{1-x}$Co$_x$)$_2$ systems with $x = 0.1$ and 0.2, respectively [2]. On the other hand, such a first-order transition disappears for $x \geq 0.3$ [2]. The current understanding is that magnetic properties for the Ce(Fe$_{1-x}$Co$_x$)$_2$ systems depending on the substitution of Fe are related with both the deformation of the transition metal 3d band [3] and the itinerancy of the Ce $4f$ electrons [4]. In order to reveal detailed magnetic properties of the Ce(Fe$_{1-x}$Co$_x$)$_2$ system, SXMCD for Fe-$L_{2,3}$ and Co-$L_{2,3}$ absorption edges has been measured for polycrystalline Ce(Fe$_{0.8}$Co$_{0.2}$)$_2$. Magnetic moments of the Fe and Co can be separately estimated by our results due to the element selectivity of SXMCD, by applying the sum rules [5,6] to the spectra. Additionally, SXMCD at the Ce-$M_{4,5}$ edges was carried out.

EXPERIMENT

Ce(Fe$_{0.8}$Co$_{0.2}$)$_2$ samples were prepared by argon arc-melting of pure constituents of at least nominal 99.9 % purity, followed by annealing at 850 °C in an evacuated quartz tube for 1 week [2]. SXMCD was carried out at BL25SU of SPring-8 by using a SXMCD spectrometer system with a water-cooled electromagnet of 1.9 T, combined with the photon spin (helicity) switching of highly circularly polarized light [7,8]. Circularly polarized light was supplied from twin-helical undulator with which almost 100 % polarization was obtained at the peak of first-harmonic radiation. The helicity modulation method with the frequencies of 1 and 10 Hz was available for the SXMCD at BL25SU by using the kicker magnets distributed around the twin-helical undulator [7]. The frequency was set to 1

CP879, *Synchrotron Radiation Instrumentation: Ninth International Conference*,
edited by Jae-Young Choi and Seungyu Rah

Hz in the present study. SXAS spectra were measured in total electron yield detection mode by detecting the photocurrent. Magnetic field of ±1.9 T was applied so as to be perpendicular to the sample surface [8]. The angle between the magnetic field and soft x-ray wave vector was fixed to 10 ° in the horizontal plane. In order to compensate possible artifacts, the averaged absorption intensities I_+ and I_- are defined as $(I_+^+ + I_-^-)/2$ and $(I_-^+ + I_+^-)/2$, respectively, where I_+^+, I_+^-, I_-^- and I_-^+ represent the absorption intensities in the configuration as shown in Table 1. The SXMCD spectrum is defined as $I_+ - I_-$. The samples were cooled down to 100 K. The clean surfaces were obtained by *in situ* cleaving. The base pressure was 9.0×10^{-8} Pa during the measurement.

TABLE 1. Definitions of the absorption intensities with the direction of the helicity and the magnetic field.

Helicity	magnetic field + 1.9 T	magnetic field − 1.9 T
Positive	I_+^+	I_+^-
Negative	I_-^+	I_-^-

RESULTS AND DISCUSSION

The SXAS spectra in the Fe and Co $L_{2,3}$ regions of $Ce(Fe_{0.8}Co_{0.2})_2$ at 100 K are shown in Fig. 1 (a) and (b), respectively. They are normalized to the incident photon flux. The vertical axis in Fig.1 (b) is set to the same scale of that in Fig.1 (a). Two sharp peak structures L_3 and L_2 are observed in Fig. 1 (a (b)). It is found in both Fig. 1 (a) and (b) that the absorption intensity I_+ is stronger than that of I_- in the L_3 region, whereas the intensity I_+ is weaker than that of I_- in the L_2 region. As a result, the SXMCD spectra have positive and negative intensities in the L_3 and L_2 regions, respectively. These results indicate that the directions of the magnetic moments of Fe and Co are parallel to that of the magnetic field. One should note that Ce $4f$ moment of $CeFe_2$ has been under debate [9]. By comparing the overall shape of Ce-$M_{4,5}$ SXMCD spectrum shown in Fig. 1 (c) with that of the theory [10,11], it is expected that the Ce $4f$ spin moment is also parallel to the magnetic field. However, estimation of the orbital magnetic moment needs further analysis. The M_5 and M_4 intensities of the SXMCD spectrum of Ce in Fig. 1 (c) is ~0.8 and ~0.4 times of the value smaller than those so far reported for $CeFe_2$ [12], respectively. It follows that the magnetic moment of Ce for the $Ce(Fe_{0.8}Co_{0.2})_2$ is expected to decrease with increasing the concentration of Co.

Here, we discuss the spin and orbital angular momentum ($<S_z>$ and $<L_z>$) of the Fe and Co $3d$ states. The spin and orbital angular momentum can be obtained by using the sum rules [5,6] for the Fe and Co $2p$ SXMCD as shown in Table 2 under assumption that the contribution of the magnetic dipole moments for Fe and Co is neglected because the sample is polycrystalline. The Fe and Co magnetic moments per formula unit (f.u.) can be estimated as $\mu_{Fe} \sim 1.8$ μ_B/f.u and $\mu_{Co} \sim 0.2$ μ_B/f.u. when the $3d$ electron number of $n_{3d} = 6.66$ and 7.45 for Fe and Co are assumed, respectively [13]. According to the magnetization measurements, the total magnetic moment for the $Ce(Fe_{1-x}Co_x)_2$ systems decreases with increasing the concentration of Co [2]. The estimated magnetic moment of Fe per atom for $Ce(Fe_{0.8}Co_{0.2})_2$ (1.13 μ_B) is comparable with that for $CeFe_2$ (1.17μ_B) [14] and much larger than that of Co (μ_{Co}/μ_{Fe} ~0.11). This is quite in contrast with Fe-Co binary alloys. As the peak is seen in the Slater-Pauling curve for Fe-Co binary alloys, the magnetic moment of Fe shows a linear increase from 2.2 to 2.9 μ_B as Co content increases up to 50 % [15,16]. However, the magnetic moment of Fe does not change obviously for the $Ce(Fe_{1-x}Co_x)_2$ systems as Fe is substituted by Co. Additionally, as shown in Table 2, the magnetic moments of Co for $Ce(Fe_{0.8}Co_{0.2})_2$ is also much smaller than those of Co for Fe-Co binary alloys, ~1.8 μ_B independently of the Co concentration [16]. The magnetic moments of Fe and Co for the $Ce(Fe_{1-x}Co_x)_2$ system and Fe-Co binary alloys show different behaviors with the concentration change of Co. Therefore, the Ce $4f$ and $5d$ electronic and magnetic states should be investigated in order to understand the ferromagnetic instability of the $Ce(Fe_{1-x}Co_x)_2$ systems. The SXMCD measurements have hence qualitatively revealed element specifically the effect of Co-doping related to the ferromagnetic instability in the $Ce(Fe_{1-x}Co_x)_2$ systems [2].

Next we attempt to estimate the $4f$ electron number (n_f) of $Ce(Fe_{0.8}Co_{0.2})_2$ from the Ce-M_5 SXAS spectrum at 100 K. Since the n_f for CePdSb is almost unity according to the non-crossing approximation (NCA) based on the single-impurity Anderson model (SIAM) calculation [17], CePdSb can be regarded as one of the localized $4f$ systems (Ce^{3+}). Then we obtained the difference spectrum by subtracting the Ce-M_5 SXAS spectrum of CePdSb from that of $Ce(Fe_{0.8}Co_{0.2})_2$ in the higher energy side of the Ce-M_5 absorption region as seen in Fig. 2. Assuming that the difference consists of the $3d^9 4f^1$ and the $3d^9 4f^2$ final state components, we reproduce the difference spectrum as seen in Fig. 2. The $4f$ occupancy is expected to correspond to the ratio of the weight of the estimated $3d^9 4f^1$ final state component to that of all in the M_5 region. The weight of the $3d^9 4f^1$ and the $3d^9 4f^2$ final state components are additionally obtained by the similar way of convoluting the difference spectra in the case of CeRu$_2$ and CeRu$_2$Si$_2$ (not shown here). Then we can estimate $n_f \sim 0.78$ for Ce(Fe$_{0.8}$Co$_{0.2}$)$_2$ by referring to the n_fs of CeRu$_2$ and CeRu$_2$Si$_2$ calculated by the NCA [18-19]. The so far reported n_f of CeFe$_2$ (0.76-0.78) [20,21] is a little larger than that of CeCo$_2$ (0.74-0.76) [21,22]. The estimated $n_f \sim 0.78$ for Ce(Fe$_{0.8}$Co$_{0.2}$)$_2$ is closer to that for CeFe$_2$, which suggests that the Ce $4f$ occupancy in Ce(Fe$_{0.8}$Co$_{0.2}$)$_2$ is similar to that in CeFe$_2$. However, the band structure, the f-d hybridization and the magnetic moment of Ce could be modified in the Ce(Fe$_{1-x}$Co$_x$)$_2$ because the lattice parameters become smaller with increasing the Co concentration [23]. The n_f for Ce(Fe$_{0.8}$Co$_{0.2}$)$_2$ should be more accurately estimated by applying the SIAM to the hard x-ray high-energy Ce $3d$ core-level and the valence band photoemission spectra with the high energy resolution in near future in order to precisely compare the results for CeFe$_2$,CeCo$_2$, and the other Ce(Fe$_{1-x}$Co$_x$)$_2$ systems.

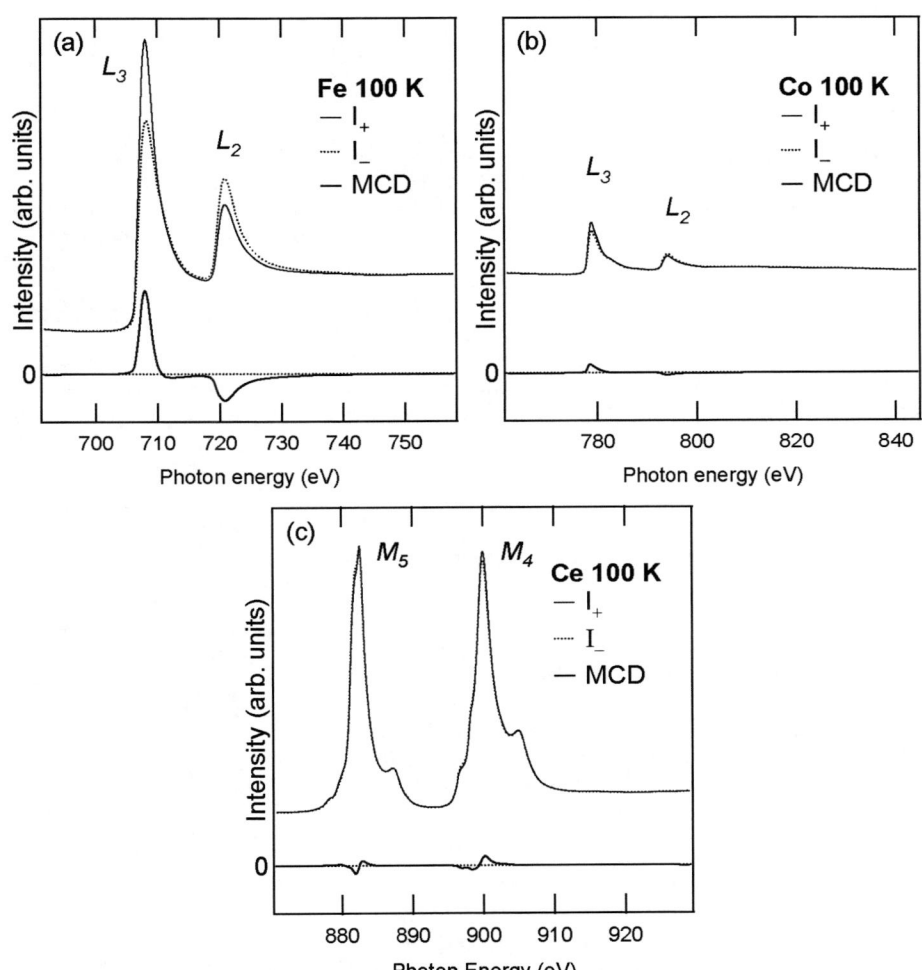

FIGURE 1. Fe (a) and Co (b) *2p-3d*, and Ce (c) *3d-4f* SXAS spectra of Ce(Fe$_{0.8}$Co$_{0.2}$)$_2$ at 100 K. Solid lines and dots correspond to the absorption spectra I_+ and I_-, respectively. The bottom panel shows the SXMCD spectrum.

TABLE 2. Values of μ_{Fe}/atom and μ_{Co}/atom for the Ce(Fe$_{0.8}$Co$_{0.2}$)$_2$ compounds derived from the SXMCD spectra.

Element (T = 100 K)	$<L_z>$	$<S_z>$	$\mu = <L_z>+2<S_z>$	$<L_z>/<S_z>$
Fe	0.06 ± 0.01	0.54 ± 0.01	1.13 ± 0.03	0.09 ± 0.01
Co	0.08 ± 0.01	0.23 ± 0.01	0.54 ± 0.03	0.36 ± 0.04

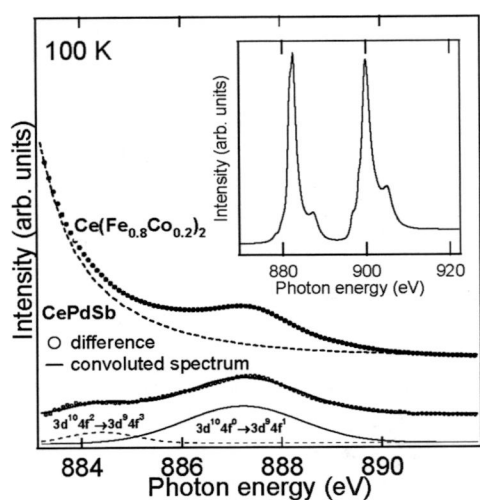

FIGURE 2. Ce-M_5 SXAS spectrum of Ce(Fe$_{0.8}$Co$_{0.2}$)$_2$. The dots and the thick broken lines correspond to the spectra of Ce(Fe$_{0.8}$Co$_{0.2}$)$_2$ and CePdSb, respectively. The linear backgrounds were subtracted from these SXAS spectra. In the lower panel, the open circles (the thick solid line) represent the difference between the SXAS spectrum for Ce(Fe$_{0.8}$Co$_{0.2}$)$_2$ and that for CePdSb (the convoluted spectrum which is the sum of the $3d^9 4f^1$ (—) and the $3d^9 4f^3$ (– –) final state components). Inset: Ce-$M_{4,5}$ SXAS spectrum at 100 K.

ACKNOWLEDGMENTS

We thank A. Sekiyama and T. Shirai for supporting the experiment. This work was supported by a Grant-in-Aid for 21st century COE "Core research and Advanced Education Center for Materials Science and Nano Engineering" from the Ministry of Education, Culture, Sports, Science and Technology, Japan. The SXMCD measurements were performed in BL25SU of SPring-8 under the approval of the Japan Synchrotron Radiation Research Institute (2005A0206).

REFERENCES

1. G. E. Fernández, M. Gormez Berisso, O. Trovarelli, and J. G. Sereni, *J. Alloys Compd.* **261**, 26-31 (1997).
2. J. Chaboy, C. Piquer, L. M. García, F. Bartolomé, and H. Wada, et al., *Phys. Rev. B* **62**, 468-475 (2000).
3. H. Wada, M. Nishigori, and M. Shiga, *J. Phys. Soc. Jpn.* **62**, 1337-1345 (1993).
4. P. K. Khowash, *Phys. Rev. B* **43**, 6170-6173 (1990).
5. B. T. Thole, P. Carra, F. Sette, and G. van der Laan, *Phys. Rev. Lett.* **68**, 1943-1946 (1992).
6. P. Carra, B. T. Thole, M. Altarelli, and X. Wang, *Phys. Rev. Lett.* **70**, 694-697 (1993).
7. T. Muro, T. Nakamura, T. Matsushita, and H. Kimura, et al., *J. Electron Spectrosc. Relat. Phenom.* **144-147**, 1101-1103 (2005).
8. T. Nakamura, T. Muro, F. Z. Guo, T. Matsushita, and T. Wakita, et al. *J. Electron Spectrosc. Relat. Phenom.* **144-147**, 1035-1038 (2005).
9. A. P. Murani, and P. J. Brown, *Europhys. Lett.* **48**, 353-354 (1999).
10. T. Jo, and S. Imada, *J. Alloys Compd.* **193**, 170-174 (1993).
11. S. Imada, and T. Jo, *J. Phys. Soc. Jpn.* **59**, 3358-3373 (1990).
12. A. Delobee, A.-M. Diask, M. Finazzi, L. Stichauer, and J.-P. Kappler, *EuroPhys. Lett.* **43**, 320-325 (1998).
13. D. S. Wang, R. Q. Wu, L. P. Zhong, and A. J. Freeman, *J. Magn. Magn. Mater.* **140-144**, 643-646 (1995).
14. S. J. Kennedy, P. J. Brown, and B. R. Coles, *J. Phys. Condens. Matter* **1**, 5169-5178 (1993).
15. G. G. Low, *Adv. Phys.* **XVIII**, 371-400 (1968).
16. M. F. Collins and J. B. Forsyth, *Phil. Mag.* **8**, 401-410 (1963).
17. T. Iwasaki, A. Sekiyama, A. Yamasaki, M. Okazaki, and K. Kadono et al., *Phys. Rev. B* **65**, 195109-1-9 (2002).
18. J. W. Allen, S.-J. Oh, O. Gunnarsson, K. Schönhammer, and M. B. Maple, et al., *Adv. Phys.* **35**, 275-316 (1986).
19. A. Sekiyama, K. Kadono, K. Matsuda, T. Iwasaki, and S. Ueda, et al., *J. Phys. Soc. Jpn.* **69**, 2771-2774 (2000).
20. T. Konishi, K. Morikawa, K. Kobayashi, T. Mizokawa, and A. Fujimori, et al. *Phys. Rev. B* **62**, 14304-14312 (2000).
21. L. Braicovich, N. B. Brookes, C. Dallera, M. Salvietti, and G. L. Olcese, *Phys. Rev. B* **56**, 15047-15055 (1997).
22. W.-D. Schneider, B. Delley, E. Wuilloud, J.-M. Imer, and Y. Baer, *Phys. Rev. B* **32**, 6819-6831 (1985).
23. H. Fukuda, H. Fujii, H. Kamura, Y. Hasegawa, and T. Ekino, et al., *Phys. Rev. B* **63**, 054405-1-9 (2001).

Magnetism of Impurities Probed with X-ray Standing Waves

N. Jaouen[1,2], F. Whilhelm[1], A. Rogalev[1], and J. Goulon[1]

[1]*European Synchrotron Radiation Facility (ESRF), rue J. Horowitz, 38043 Grenoble, France*
[2]*Synchrotron SOLEIL, l'Orme des merisiers 91192 Gif-sur-Yvette, France*

Abstract. We report on the measurement of a Magnetic X-ray Standing Wave in the hard X-ray range which is the most interesting energy range for diffraction-like experiment. X-ray standing waves produced by a Bragg interference combined with element and orbital selectivity of magnetic circular dichroism have been measured at the L edges of Pt diluted in a $Ni_{90}Pt_{10}$ crystal. By comparing with XMCD experiments done on the same sample, our result suggest that there is two different magnetic type of Pt atoms. These results demonstrate the capability to directly probe the magnetic properties of dilute impurities located in bulk *3d* metals and suggest a broad applicability of such standing wave measurements in the hard x-ray range to site-selective study of magnetic properties in systems lacking long-range order.

Keywords: XMCD, X-ray standing wave
PACS: 75.70.Cn, 75.25.+z

INTRODUCTION

The growing interest on magnetic and electronic properties of impurities and alloys is driven by both fundamentals and technological interests. In recent industrial materials, such as magnetic semiconductors, the bulk properties are related to the impurities atoms which are known, from theoretical studies, to be strongly affected by hybridization with the local surrounding environment. Up to now only few experimental techniques could give information on the magnetic properties of the dilute impurity with an element and site selectivity. Regarding this great sensitivity to the local chemical order, the quest for experimental techniques allowing exploring the link between local order and magnetic properties is highly relevant.

Recent advance in synchrotron radiation sources allow using spectroscopic technique such as X-ray Magnetic Circular Dichroism (XMCD) to give element specific information's on the induced magnetic properties of very dilute *4d* and *5d* impurities in a *3d* host [1]. Although in very dilute system with a random distribution this approach is suitable, this will be not longer the case for more concentrated system or in presence of non-equivalent crystallographic site. Nevertheless the lack of long range order in disordered phase prevent the use of the additional site selectivity normally arising from a resonant scattering experiment [2] which can only distinguish the atoms located at the topmost surface by using grazing incidence geometry [3].

In this paper we report on an experimental technique explicitly able to probe magnetic properties of impurity atoms depending on their surrounding environment. This experiment is based on the so-called X-ray Standing Wave (XSW) techniques, which is a well established tool for obtaining structural information [4] and to yield information on the impurity location [5]. A disordered $Ni_{90}Pt_{10}$ alloy was used as prototype systems to demonstrate the sensibility of the XSW to magnetic properties of Pt atoms located in the bulk Ni crystal at regular crystallographic position. The important separation in energy between the Ni K (8.33 keV) and the Pt $L_{2,3}$ (13.27 and 11.56 keV respectively) absorption edges, allow us to measure the Pt $L_{2,3}$ XMCD spectra while strong X-ray standing waves are created by rocking the crystal through the intense Ni (111) Bragg peak.

EXPERIMENTAL

The measurements have been performed at the ESRF beamline ID12 [6]. Incident x-rays were tuned to the Pt L edges and have been used to generate the standing wave field by rocking the crystal in angle θ through the Darwin curve of the strong Ni (111) Bragg reflection. Both the scattered intensity and the fluorescence yield were recorded

CP879, *Synchrotron Radiation Instrumentation: Ninth International Conference*,
edited by Jae-Young Choi and Seungyu Rah
© 2007 American Institute of Physics 978-0-7354-0373-4/07/$23.00

simultaneously using two Si photodiodes (mounted behind a rectangular aperture for the one dedicated to measure scattered intensities and the other one being always parallel to the sample surface) using the compact reflectometer available on ID12 [7]. Circularly polarised X-rays from a HU38 helical undulator with measured degree of polarization P_c=0.99 were used with an applied magnetic field oriented along the incident beam direction and in the diffraction plane. Rocking curves were recorded by scanning θ through the (111) reflection at several fixed hv. For the standing wave measurement, the same experimental procedure was used, but we measure simultaneously the resulting fluorescence intensity. Conventional x-ray absorption (XAS) and XMCD spectra in a backscattering geometry were recorded in order to aid in normalizing the angular dependent scans far from the Bragg peak.

RESULTS

In Fig.1 we show normalized x-ray absorption (XAS) and XMCD spectra at the $L_{2,3}$ edges of Pt for the studied $Ni_{90}Pt_{10}$ crystal. The ratios L_3/L_2 was normalized to 2.21/1 for the XAS spectra according to [8]. The existence of large XMCD signal confirms that Pt atoms have acquired a net induced magnetic moment. To assure that the XMCD spectra are free of experimental artefacts both the helicity of the incidents x-rays and the directions of the applied magnetic field were reversed. After correction from self-absorption effects, the magneto-optical sums rules [9] have been applied in order to extract the average spin and orbital induced magnetic moments of Pt atoms. Our analysis yields a value of 0.341 μ_B for the spin moment and of 0.095 μ_B for the orbital moment per Pt atom.

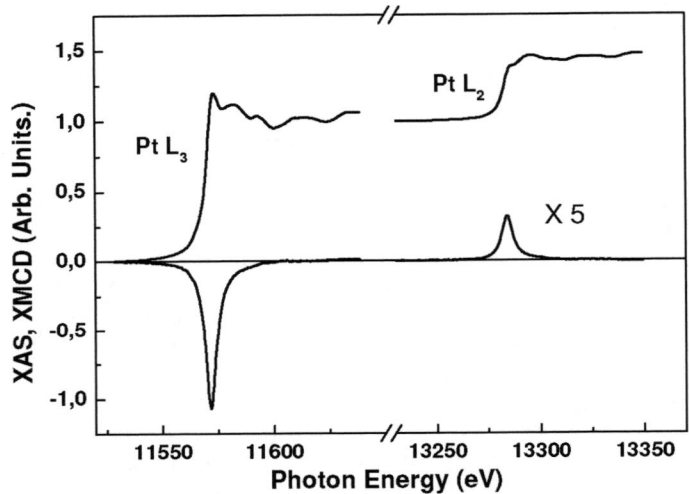

FIGURE 1. XAS (top) and XMCD (bottom) spectra measured at the $L_{2,3}$ edges of Pt in the $Ni_{90}Pt_{10}$ crystal. For better illustration the XMCD spectra have been multiplied by 5.

In Fig. 2 the fluorescence yield, when going through the rocking curve is shown for two energies, 11.564 and 13.27 keV, which are respectively close to the Pt L_3 and L_2. The diffracted intensities are reported in the same panel as a reference for the Bragg position. In both case, the characteristic line shape of the fluorescence curve [4], with a large angular range necessary for the low-angle tail to approach background and an increase of fluorescence in the high angle clearly indicate that a strong SW resonance occurs under Bragg peak position. This also suggests that the Pt L edges are sufficiently far from the Ni K edge to ensure that the presence of Pt atoms in the Ni crystals could be considered as a negligible perturbation of the standing wave field. In order to remove the SW modulation away from the Pt L_2 and L_3 edge the data were renormalized using modification to the standard techniques in accordance with Refs. [10, 11]. This leads to assume that the SW field is constant at fixed wave vector q except for the small perturbation resulting from Pt absorption. In this approach the pre-edge background is extrapolated and subtracted, and the result normalized according to [8] in the post edge region. Considering the edges are separated by more than 1500eV, this approach appears to be a reasonable hypothesis. The normalized helicity dependent spectra are displays in the low part of each panel in Fig. 3.

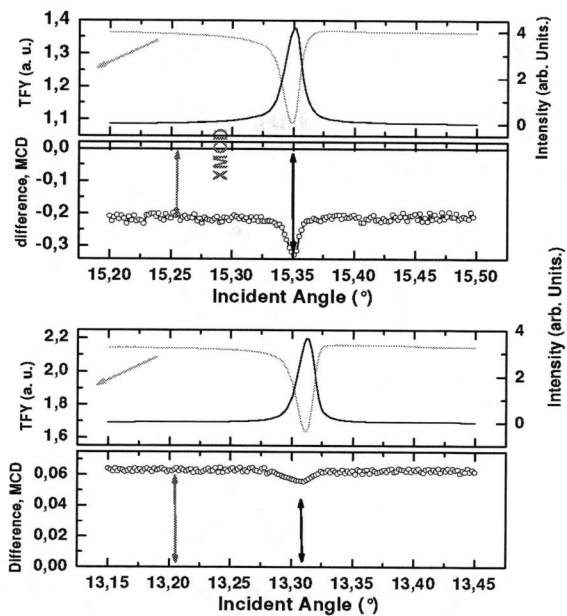

FIGURE 2. Angular line shape of the diffracted intensity for the Ni (111) reflections at the Pt L_3 (top in black) and L_2 (bottom) respectively. The corresponding fluorescence yield is presented in the same picture in grey line. Bottom part presents the difference in fluorescence obtained by subtracting the fluorescence yield for two opposite helicity of the incoming beam.

The excellent signal to noise ratio, related to the high photon flux and degree of polarization delivered by third generation synchrotron radiation facility [7], allow one to clearly identify a structure in the helicity dependent measurements. This modulation of the XMCD signal, when the Bragg condition are satisfied, results from sensibility to existence of different Pt magnetic moments probe at the SW resonance. The crucial point here is to keep in mind that a 10% Pt concentration corresponds to more than one Pt atom by lattice cell. The basic idea of our experiment was to combine these two experiments in order to separe the Pt really in impurity to those having other Pt atoms as neighbour.

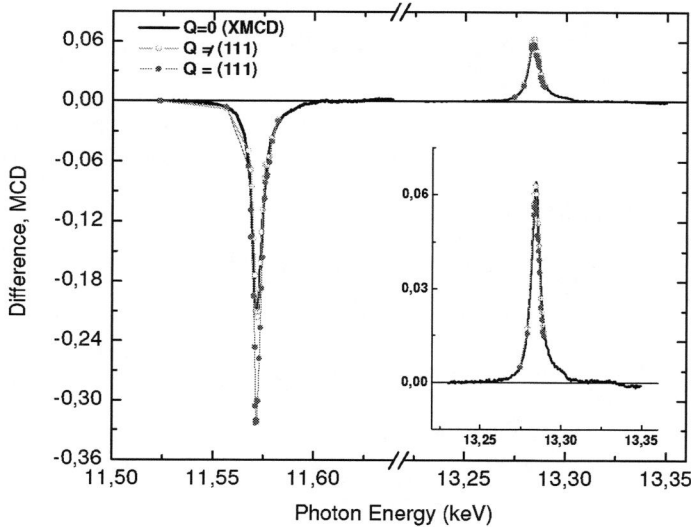

FIGURE 3. Energy line shape of the XMCD measured in backscattering geometry (strait line) outside from the Bragg peak position as described in text (cross) and of the MXSW (circle).

Nevertheless to do such quantitative analysis, one needs to measure the energy dependence of the magnetic X-ray Standing Wave (MXSW) spectra. This has been done by repeating the experiment for several energies close to the Pt $L_{2,3}$, the MXSW being the sum of two term. The first one will be estimate by integrating the difference in fluorescence for opposite helicity of x-rays in a rocking scan (see Fig. 2), after subtraction of a constant value. After proper normalization to the edge jump [10, 11] this constant signal, which corresponds directly to the XMCD far from the Bragg condition, was added to the first term to give the full MXSW. Figure. 3 shows the energy dependence of the MXSW, measured at q=(111), together with the constant signal which correspond to the XMCD measure at non-zero q value but far from the (111) and the XMCD previously displayed in Fig. 1.

One can directly see that the XMCD and MXSW signal have different amplitude and L_3/L_2 ratio. Assuming that we can apply the sum-rules, we find that MXSW give us an equivalent moment of $0.415\mu_B$ and $0.12 \mu_B$ for the spin and orbital moment respectively. This different have to be assign to the different weight in the scattering factor of non-equivalent Pt atoms. At this stage to go further and extract the magnetic moments of these non-equivalent Pt atoms, one needs first to know the distribution of the non equivalent atoms and to develop a theory for Magnetic standing wave which is far beyond the scope of our paper.

CONCLUSION

Present experiments have shown that observable effect, dependent on magnetization, can be measured in fluorescence yield under x-ray standing wave field in dilute ferromagnetic systems. In Ni crystals with a dilute amount of Pt atoms, a comparison between XMCD to MXSW energy dependence of the magnetic contribution have been done. The modulation of the XMCD when the crystal is rock in Bragg condition allow to probe selectively the Pt atoms located in bulk position despite the absence of long range order. A theoretical development for analyzing Magnetic Standing Wave experiment, which includes full X-rays polarization properties, is mandatory to pass from an experimental demonstration to the development of a technique largely used. Finally and beyond these point, this experimental evidence of magnetic contribution in a hard X-ray standing wave experiment open new opportunities to probe with a element and site selective technique the magnetic properties of atoms in impurity in bulk materials, such as ferromagnetic semi-conductors, to deposited atoms on a surface.

ACKNOWLEDGMENTS

The authors would like to thank L. Ortega for providing us access to conventional X-ray diffractometer used for pre-alignment of the sample. We also would like to thanks O. Robach and Y. Jugnet for lending us the crystal.

REFERENCES

1. F. Whilhelm, *et al.*, private communication.
1. F. de Bergevin, M. brunel, R.M. Galera, C. Vettier, E. Elkaïm, M. Bessière, and S. Lefèbvre, Phys. Rev. B 46, 10772 (1992).
2. S. Ferrer, P. Fajardo, F. de Bergevin, J. Alvarez, X. Torrelles, H.A. van der Vegt and V.H. Etgens, Phys. Rev. Lett. 77, 747 (1996).
3. B.W. Batterman, Phys. Rev. 113, A759 (1964).
5. B.W. Batterman, Phys. Rev. Lett. 22, 703 (1969).
6. J. Goulon , A. Rogalev, C. Gauthier, C. Goulon-Ginet, S. Paste, R. Signorato, C. Neumann, L. Varga, and C. Malgrange, J. Synchr. Rad. 5, 232 (1998).
7. N. Jaouen, F. Whilelm, A. Rogalev, J. Goulon and J.M. Tonnerre , AIP Conf. Proc. 705, 1134 (2004).
8. B.L. Henke, E.M. Gullikson, and J.C. Davis, At. Data Nucl. Data Tables 54, 181-342 (1993).
9. B.T. Thole, *et al.*, Phys. Rev. Lett. 68, 1943 (1992); P. Carra, *et al.*, Phys. Rev. Lett. 70, 694 (1993).
10. Sang-Koog Kim and J.B. Kortright, Phys. Rev. Lett. 86, 1347 (2001).
11. See-Hun Yang, *et al.*, J. Phys.: Condens. Matter. 14, L407-L420 (2002).

Orbital Ordered Structure of a Manganite Thin Film Observed by K-edge Resonant X-ray Scattering

Y.Wakabayashi[1], D.Bizen[2], H.Nakao[2], Y.Murakami[2,3], M.Nakamura[4], Y.Ogimoto[5], K.Miyano[6] and H.Sawa[1]

[1]Photon Factory, KEK, Tsukuba 305-0801, Japan,
[2]Department of Physics, Tohoku University, Sendai 980-8578, Japan,
[3]Synchrotron Radiation Research Center, JAERI, Sayo, 679-5148, Japan,
[4]Department of Applied Physics, University of Tokyo, Tokyo 113-8586, Japan,
[5]Devices Technology Research Laboratories, SHARP Corporation, Nara 632-8567, Japan,
[6]Research Center for Advanced Science and Technology, University of Tokyo, Tokyo 153-8904, Japan

Abstract. The low-temperature orbital ordering structure of $Nd_{0.5}Sr_{0.5}MnO_3$ thin film on $SrTiO_3$ (011) substrate ($NSMO/STO_{011}$) has been clarified by resonant x-ray scattering. This thin film is the first example of the manganite thin film that has sharp metal-insulator transition, and synchrotron x-ray diffraction reveals the structure of the low-temperature orbital ordered phase. The orbital order structure was found to be the same as the structure of bulk $Nd_{0.5}Sr_{0.5}MnO_3$, while the tilt/rotation of the MnO_6 octahedra were different.

Keywords: Mn oxide, orbital ordering, resonant x-ray scattering
PACS: 75.47.Lx

INTRODUCTION

Charge order and orbital order (CO/OO) are the characteristic phenomena of correlated electron systems. Manganese oxides are good examples of such systems because of their various ground states and complicated roles of charge, spin and orbital degrees of freedom. A number of theoretical and experimental studies [1] on CO/OO in $RE_{1-x}AE_xMnO_3$ (RE: rare earth metals; AE: alkali earth metals) have been conducted in the vicinity of $x=0.5$ in order to understand the mechanism of the ordering and the resulting electronic properties. One fascinating achievement of these studies is controlling the OO state by means of the thin film technique [2]. It utilizes the strong coupling between the orbital degree of freedom and the lattice distortion, which can be controlled by this technique.

Orbital ordering is classified into two groups: ferro orbital order and antiferro orbital order. Konishi *et al.*[2] demonstrated to control the ferro OO by using tensile strain and compressive strain given by cubic perovskite (001) substrates. However, the strong coupling between orbital and lattice has prohibited them from having sharp phase transition to the OO phase for long years[3-6]. In such circumstances, it was reported[7] that $NSMO/STO_{011}$ shows sharp first-order phase transition around 170K (T_{MI}). We have studied[8] the OO in this new system and reported that it has an antiferro OO state below 150K and ferro OO state between 150K and 170K. In this study, we clarified the structure of the antiferro OO by means of resonant and non-resonant x-ray scattering technique.

EXPERIMENT AND RESULTS

The x-ray diffraction experiments were carried out at BL-4C and BL-16A2 at the Photon Factory, KEK and at X22C at the National Synchrotron Light Source, BNL. The beamlines are equipped with standard four-circle diffractometers connected to closed-cycle refrigerators. Epitaxial films were grown by the pulsed laser deposition method. The thickness of the sample was 80nm.

CP879, *Synchrotron Radiation Instrumentation: Ninth International Conference*,
edited by Jae-Young Choi and Seungyu Rah
© 2007 American Institute of Physics 978-0-7354-0373-4/07/$23.00

Figure 1 shows the temperature dependence of the lattice parameters in pseudo cubic notation; throughout this paper, we use this notation. The lattice parameters at room temperature were a=3.905Å, b=c=3.824 Å, α=90.5°, and β=γ=90.3°, and those at 10K were a=3.896 Å, b=3.867 Å, c=3.761 Å, α=90.4°, β=90.1°, and γ=90.6°. The lattice parameter a, which is locked to the lattice constant of the substrate, was nearly unchanged while the b and c lattice parameters changed drastically at T_{MI}.

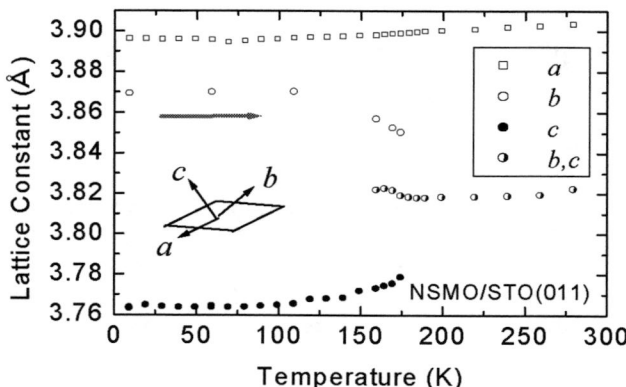

FIGURE 1. Temperature dependence of the lattice parameters in pseudo cubic notation in a heating run. Sharp first order transition was observed at 170K.

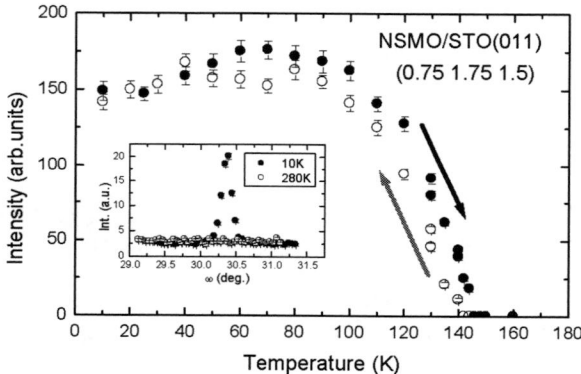

FIGURE 2. Temperature dependence of the (0.75 1.75 1.5) superlattice reflection intensity. Inset shows the rocking curve.

Next we searched for superlattice reflections in order to find out the correct lattice periodicity in the insulator phase. Superlattice reflections with modulation vector q_m=(1/2 1/2 1/2) were observed at room temperature; this wavevector is different from bulk $Nd_{0.5}Sr_{0.5}MnO_3$ structure, which has $\sqrt{2} \times \sqrt{2} \times 2$ structure at room temperature[9]. These reflections are caused by the displacement of A-site (Nd and Sr) ions and MnO_6-octahedra rotation. At 10K, superlattice reflections characterized by the wavevector (1/4 1/4 0), (1/2 1/2 0), (1/4 1/4 1/2) and (0 0 1/2) were observed; one example is shown in the inset of Fig. 2. In ref. 8, we have reported that the orbital order in this film has no (1/4 1/4 0) modulation judged from the result of the peak search in the region $0.5 \leq h \leq 1$ and $1.5 \leq \{k, l\} \leq 2$. Here, we found superlattice reflections having wavevector (1/4 1/4 0) in a wide-range peak search; this finding changes the result of the analysis. As a result, $\sqrt{2} \times 2\sqrt{2} \times 2$ structure was confirmed at a low temperature below 150K. Figure 2 shows the (0.75 1.75 1.5) superlattice reflection intensity as a function of temperature. The superlattice reflection disappeared at 150K in the heating run, which is significantly lower than T_{MI}. In the intermediate temperature region, there were no superlattice reflections except for those having q_m=(1/2 1/2 1/2).

In order to clarify the orbital order structure, we conducted Mn K-edge resonant x-ray scattering measurements. Figure 3 shows the energy spectrum of the (0.25 2.25 2), (0.25 1.25 0.5), (2.25 2.25 0) and (0.5 1.5 2) superlattice reflection intensity. A striking feature in these spectra is no intensity anomaly at Mn K-edge at the (0.25 1.25 0.5) superlattice reflection; this means the Mn ions do not have structural modulation having wavevector (1/4 1/4 1/2) while Nd ions have. The structure of the orbital ordering will be discussed in the following section.

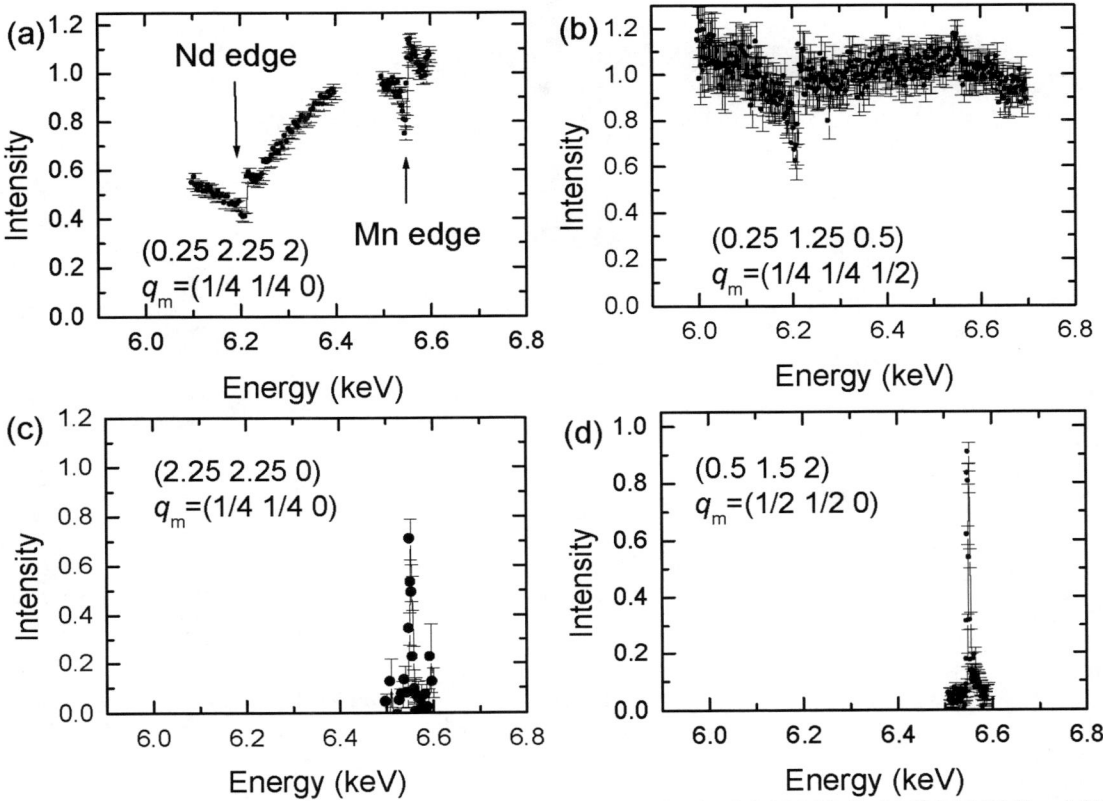

FIGURE 3. Incident energy dependence of the intensity of (a) (0.25 2.25 2), (b) (0.25 1.25 0.5), (c) (2.25 2.25 0) and (d) (0.5 1.5 2) superlattice reflections whose modulation wavevectors are (1/4 1/4 0), (1/4 1/4 1/2), (1/4 1/4 0) and (1/2 1/2 0), respectively.

DISCUSSION

Judged from lattice parameters, lack of superlattice reflections and reported material properties, this thin film has A-type antiferromagnetic x^2-y^2 ferro OO state between 150K and 170K[8]. Here, we discuss the structure of the low-temperature antiferro OO state. The energy spectrum shown in Fig. 3(d), the sharp intensity enhancement of a *charge-ordering* peak, supports the existence of the checker board type charge ordering. Assuming the checkerboard type charge ordering in cubic perovskite structure, there are only two types of orbital ordering that has $\sqrt{2} \times 2\sqrt{2} \times 2$ structure: CE-OO and AP-OO (antiphase orbital order) shown in Fig. 4.

As mentioned above, the Mn ions do not have structural modulation having wavevector (1/4 1/4 1/2). This information gives strong clue of the structure of the orbital ordering in this film because antiferro OO structure in a MnO_2 plane involves transverse mode Mn displacement. This implies the superlattice reflection from OO contains the scattering amplitude from Mn ions. Since AP-OO is characterized by the wavevector (1/4 1/4 1/2), this structure is denied. In contrast, CE-OO having the wavevector (1/4 1/4 0) is supported by the (2.25 2.25 0) resonant peak.

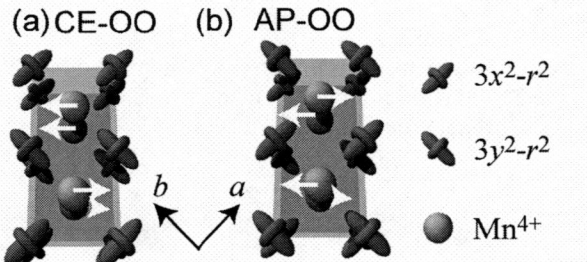

FIGURE 4. Possible orbital arrangement having checkerboard type charge ordering: (a) CE-OO and (b) AP-OO. White arrow shows concomitant Mn^{4+} displacement.

ACKNOWLEDGMENTS

The authors are grateful to Prof. T.Arima and Dr. J.P.Hill for fruitful discussions. This work was supported by a Grant-in-Aid for Creative Scientific Research (13NP0201), TOKUTEI (16076207) and Support for long-term visit from the Ministry of Education, Culture, Sports, Science and Technology of Japan, JSPS KAKENHI (15104006), and Yamada Science Foundation. Financial support to M. N. by the 21st Century COE Program for ``Applied Physics on Strong Correlation" administered by Department of Applied Physics, The University of Tokyo is also appreciated.

REFERENCES

1. Y.Tokura and N.Nagaosa, *Science* **288**, 462 (2000).
2. Y.Konishi, Z.Fang, M.Izumi, T.Manako, M.Kasai, H.Kuwahara, M.Kawasaki, K.Terakura and Y.Tokura, *J. Phys. Soc. Jpn.* **68** 3790 (1999).
3. W.Prellier, A.M.Haghiri-Gosnet, B.Mercey, Ph.Lecoeur, M.Hervieu, Ch.Simon, and B.Raveau, *Appl. Phys. Lett.* **77**, 1023 (2000).
4. W.Prellier, Ch.Simon, A.M.Haghiri-Gosnet, B.Mercey, and B.Raveau, *Phys. Rev. B* **62**, R16337 (2000).
5. Y.Ogimoto, M.Izumi, T.Manako, T.Kimura, Y.Tomioka, M.Kawasaki, and Y.Tokura: *Appl. Phys. Lett.* **78**, 3505 (2001).
6 A.Biswas, M.Rajeswari, R.C.Srivastava, T.Venkatesan, R.L.Greene, Q.Lu, A.L.de Lozanne, and A.J.Millis: *Phys. Rev. B* **63**, 184424 (2001).
7. Y.Ogimoto, M.Nakamura, N.Takubo, H.Tamaru, M.Izumi, and K.Miyano *Phys. Rev. B* **71**, 060403(R) (2005).
8. Y.Wakabayashi, D.Bizen, H.Nakao, Y.Murakami, M.Nakamura, Y.Ogimoto, K.Miyano and H.Sawa, *Phys. Rev. Lett.* **96**, 017202 (2006).
9. H.Kawano, R.Kajimoto, H.Yoshizawa, Y.Tomioka, H.Kuwahara and Y.Tokura, *Phys. Rev. Lett.* **78**, 4253 (1997).

Resonant X-Ray Magnetic Scattering Study of $BaFe_{12}O_{19}$ and $BaTiCoFe_{10}O_{19}$

Seiji Ohsawa, Maki Okube, Takahiro Aki, Shoichi Sakurai, Norio Shimizu, Koichi Ohkubo, Takeshi Toyoda*, Takeharu Mori** and Satoshi Sasaki

Materials and Structures Laboratory, Tokyo Institute of Technology, Nagatsuta 4259, Yokohama 226-8503, Japan
*Industrial Research Institute of Ishikawa, Kuratsuki 2-1, Kanazawa 920-8203, Japan
**Photon Factory, Institute of Materials Structure Science, KEK, Oho 1-1, Tsukuba 305-0801, Japan

Abstract. Synchrotron diffraction studies have been carried out for single crystals of ferrimagnetic M-type ferrites at the BL-3A of the Photon Factory. Resonant x-ray magnetic scattering (RXMS) and x-ray magnetic circular dichroism (XMCD) at the Fe K edge were utilized to examine the magnetic order of $3d$ states through the hybridized $4p$ states. A magnetic satellite reflection of 0 0 8+(2/3) was observed for $BaTiCoFe_{10}O_{19}$ with the resonant enhancement. The magnetic anomalous scattering factor f''_m of $BaFe_{12}O_{19}$ was estimated as 0.23 at E = 7128.2 eV in the least-squares procedure to use the difference of asymmetrical ratios. The crystal structure with spin arrangement has been determined for $BaFe_{12}O_{19}$, which coincides with the Gorter model.

Keywords: magnetic structure determination, resonant x-ray magnetic scattering, RXMS, ferrimagnetic Ba ferrite.
PACS: 75.25.+z, 78.20.Ls, 78.70.Ck, 78.70.Dm

INTRODUCTION

High flux, tunability and linear polarization of synchrotron radiation make possible to increase the experiments on various magnetic systems by x-ray scattering. Resonant x-ray magnetic scattering (RXMS) has attracted much interest as a useful tool to determine the magnetic structures associated with specific electronic states such as $3d$-$4p$ interactions. The RXMS experiment in ferromagnetic Ni was first made at the Ni K edge [1]. Large resonant exchange scattering was predicted in L and M absorption edges to dominate the magnetic scattering [2]. The Bragg intensity was estimated with the resonant enhancement between charge and magnetic scatterings at the L edge of the rare-earth systems [3]. Various resonant experiments have been carried out for the x-ray energies near the M_{IV} edges of actinides and near the L_{III} edges of rare-earth and transition-metal compounds [4]. There are not so many reports on the K edge of $3d$ transition metal because of relatively weak enhanced intensity. However, it is now known that magnetic resonance arises from the spin-orbit coupling at the K edge through the process of either strongly polarized $3d$ band or interatomic exchange between $4p$ and $3d$ sates [5, 6].

M-type barium hexaferrite is ferrimagnetic below T_c = 723 K [7, 8]. The magnetic structure was proposed with the spin collinearity on the basis of analysis of magnetic properties, where all the magnetic moments are ordered parallel or antiparallel to c axis [8,9]. Although $BaFe_{12}O_{19}$ has the strong uniaxial magnetic anisotropy [10], the substitution of Fe^{3+} by a pair of Ti^{4+} and Co^{2+} results the reduction of the anisotropy. The magnetic anisotropy was investigated by the single-crystal neutron diffraction, where magnetic reflections of type 0 0 $2l\pm\tau$ (τ = 2/3 or 0.575) were observed for the crystals, having x ≥ 0.8 in $BaTi_xCo_xFe_{12-2x}O_{19}$ [11].

In this study, the magnetic structures of M-type $BaFe_{12}O_{19}$ and $BaTiCoFe_{10}O_{19}$ have been examined to confirm the potential ability of RXMS in the relatively complicated system, having five independent Fe sites such as tetrahedral $4f_1$, bipyramidal $2b$, and octahedral $2a$, $4f_2$ and $12k$ sites. The quite large resonant effect at the K edge makes it possible to determine both crystal structure and spin arrangement for a tiny single crystal.

CP879, *Synchrotron Radiation Instrumentation: Ninth International Conference*,
edited by Jae-Young Choi and Seungyu Rah
© 2007 American Institute of Physics 978-0-7354-0373-4/07/$23.00

EXPERIMENTAL

Powder crystals of $BaFe_{12}O_{19}$ and $BaCoTiFe_{10}O_{19}$ were synthesized by the conventional solid state reaction using starting materials of $BaCO_3$, $2CoCO_3 \cdot 3Co(OH)_2$, Fe_2O_3 and TiO_2 [12, 13]. Cell dimensions of $BaFe_{12}O_{19}$ and $BaCoTiFe_{10}O_{19}$ are $a = 5.8888(3)$, $c = 23.268(3)$ Å (hexagonal, $P6_3/mmc$) and $a = 5.8955(3)$, $c = 23.205(2)$ Å, respectively. The crystal sizes are $0.22 \times 0.10 \times 0.06$ mm and $1.0 \times 0.57 \times 0.25$ mm, respectively. The chemical composition of $BaCo_{1.00}Ti_{0.90}Fe_{10.11}O_{19}$ was analyzed by HOROBA XGT2000 fluorescence spectrometer.

Preliminary diffraction experiments were carried out at the BL-10A station of the Photon Factory. XMCD and RXMS experiments were performed at the Fe K absorption edge at the BL-3A. The horizontally polarized white x-rays were monochromatized by the Si(111) double-crystal monochromator. The incident beam was guided into a synthetic single crystal of (001) diamond with a thickness of 0.492 mm in order to produce circularly polarized x-rays. The scattering plane of the diamond crystal was inclined by 45° from the vertical plane in order to balance the σ and π components of the transmitted beam. The diamond crystal was set near the 111 Bragg condition in the asymmetric Laue case. A standard transmission setup was used for XMCD measurements with the Faraday arrangement, using rare-earth magnets in a magnetic field of 0.4 T. A four-circle geometry at the BL-3A was used for the RXMS study at the Fe K edge at temperatures of 100 or 296 K. Low-temperature experiments were performed with the Oxford Cryostream Cooler, where cold and dry nitrogen gas is directly blown onto the crystal.

XMCD AND 3D STATES MIXING

XMCD spectra at the K edge of transition-metal compounds are considered to reflect the magnetic order by absorption from the core electron to $4p$ states. Since the state is not directly given to construct the magnetic order, the XMCD has been analyzed at the Fe K edge using the ferrimagnetic ferrite and the magnetized sample. The absorption is split into a spin-independent global electronic part and a spin-dependent one. When I and I_0 are the transmitted and incident intensities for the parallel (+) and antiparallel (-) configurations of the x-ray wave vector and the spin direction of the sample, respectively, the spin-dependent to spin-independent ratio, $\Delta\mu/\mu_0$ gives a thickness-free quantity:

$$\frac{\Delta\mu}{\mu_0} = \frac{\ln(I_0^+/I^+) - \ln(I_0^-/I^-)}{\frac{1}{2}[\ln(I_0^+/I^+) + \ln(I_0^-/I^-)]} = \frac{\mu^+ - \mu^-}{\mu^+ + \mu^-}. \tag{1}$$

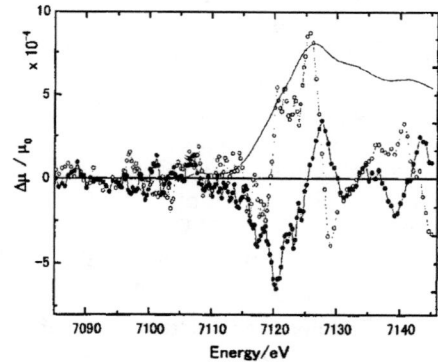

FIGURE 1. XMCD spectra of $BaFe_{12}O_{19}$ at the Fe K edge. (a) Ferrimagnetic powder crystals (solid circle) and (b) the sample magnetized parallel to the c axis (open circle). The fine solid line displays the XANES spectra in arbitrary unit.

The Fe K-edge XMCD and XANES spectra of $BaFe_{12}O_{19}$ are shown in Fig. 1. The XMCD signal is characterized by a dispersion-type profile, having a negative peak close to the edge ($E = 7121$ eV) and a positive one above 7 eV. The profile is similar to that of the Fe K main-edge XMCD in $NiFe_2O_3$ (Ni ferrite, spinel structure) [14, 15]. Therefore, the Fe $3d$ states in $BaFe_{12}O_{19}$ is similar in the electronic structure to those of $NiFe_2O_4$, which has the A-O-B superexchange interaction. The hybridization is between $3d$ and $4p$ orbits through the $2p$ orbits of neighboring oxygen atoms [16]. The XMCD intensity is the order of 5×10^{-4} to XANES one. The dichroic signal of the magnetized sample is also shown in open circle in Fig. 1. At least three types of spin inversion can be observed: (1) a negative peak at $E = 7120$ eV changes to be positive, (2) a positive peak at $E = 7128$ eV is turned into negative one at $E = 7129$ eV, and (3) a positive peak appears between the previous two peaks, by breaking the counterbalance of spins. Thus, the $4p$ transition would be useful to examine the magnetic structure by RXMS at the Fe K edge.

OBSERVATION OF MAGNETIC SATELLITE REFLECTION

X-ray diffraction experiments for $BaCoTiFe_{10}O_{19}$ were made at a wavelength of $\lambda = 1.7406$ Å ($E = 7122.8$ eV) of the Fe K edge, based on the characteristic XMCD signals for the crystals. A magnetic reflection appears at 100 K (Fig. 2), which does not obey the extinction rules of the $P6_3/mmc$ symmetry and is described as 0 0 8+(2/3). Although the satellite reflection associates with the diffuse intensity, the sharpness of peak is characteristic of a long-range magnetic order. The appearance of the peak coincides with the result by the neutron diffraction for $BaCo_{0.8}Ti_{0.8}Fe_{10.4}O_{19}$, which is interpreted in terms of a magnetic helix propagated along the hexagonal c axis [11, 16]. Thus, the Fe K edge can excite resonantly but indirectly the $3d$ states and the magnetic resonant enhancement is sufficiently large to study the magnetic order of Fe compounds.

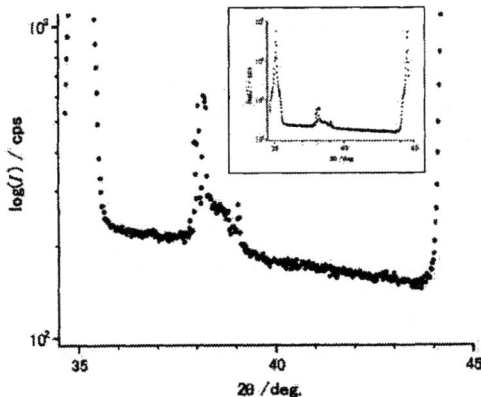

FIGURE 2. Scans between 008 and 0 0 10 reflections of $BaCoTiFe_{10}O_{19}$. Beside the fundamental reflections, 0 0 8+(2/3) is observed at $T = 100$ K. The inset shows the same measurements on a full scale. The satellite reflection disappeared at $T = 200$ K.

MAGNETIC STRUCTURE ANALYSIS BY XRMS

In order to examine the direction of the majority spin, the interference between the charge and magnetic exchange scattering has been investigated in the ferrimagnetic $BaFe_{12}O_{19}$. X-ray scattering amplitude including magnetic terms is given for the $1s$ to $4p$ dipole transition [2, 19]. Having the scattering intensities Y^+ and Y^- for magnetization parallel and antiparallel, respectively, which is perpendicular to the scattering and polarization plane, an asymmetrical ratio ΔR [$= (Y^+ - Y^-) / (Y^+ + Y^-)$] can be approximated as [3, 18]

$$\Delta R \cong 2 \tan 2\theta \frac{(F_0 + F')F_m'' - F''F_m' - F''F_{0,m}}{|F_c|^2}. \qquad (2)$$

In Eq. (2), F is the structure factor given by $F(hkl) = \Sigma f \exp 2\pi i(hx+ky+lz) \exp(-W)$ and h, k, l = reflection indices, x, y, z = atomic coordinates, W = Debye-Waller factor and c = charge scattering. The atomic scattering factor f is given by $f = f_0 + f' + if'' + i(f_{0,m} + f'_m + if''_m)$, where f_0 = Thomson scattering factor, f' and f'' = real and imaginary parts of anomalous scattering factor, $f_{0,m}$ = non-resonant magnetic scattering factor, f'_m and f''_m = real and imaginary parts of resonant magnetic scattering factor. The f'' term was estimated from the XANES spectra. The f' term was calculated from f'' by the program DIFFKK of the Kramers-Kronig's transformation [19, 20]. The values obtained are $f' = -6.721$ and $f'' = 5.165$ at $E = 7128.2$ eV. The $f_{0,m}$ value is taken from the reported one [18].

Intensity measurements were made by an ω-step-scan technique (10 sec/step, 92 steps/scan) at $\lambda = 1.7398$ Å ($E = 7128.2$ eV) of the Fe K edge. The observed asymmetrical ratio ΔR_{obs} was obtained in line with the following procedures: (1) background correction, (2) estimation of R^+ and R^- from the circularly-polarized integrated intensity, (3) estimation of ΔR_{obs} from R^+ and R^-, (4) construction of magnetic structure models, (5) calculation of structure factors corresponding to charge and magnetic scatterings, (6) calculation of ΔR_{calc} in Eq. (2), and (7) calculation of residual factors of $\Sigma(\Delta R_{obs} - \Delta R_{calc})^2$ for each magnetic model, where the summation is over all reflections used.

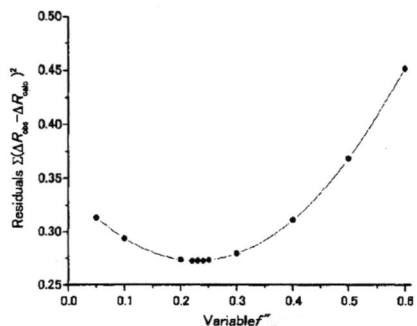

FIGURE 3. Changes in residual factors of $\Sigma(\Delta R_{obs} - \Delta R_{calc})^2$ as a function of magnetic resonant scattering factor f''_m in the least-squares calculation ($E = 7128.2$ eV). There is only one minimum at $f''_m = 0.23$.

The estimation of resonant magnetic scattering factors was made with the structure-refinement procedure. A full matrix least-squares program RADY was used for 27 independent reflections such as 200, 300, 110, 220, 211, 102, 302, 112, 203, 104 and 304 [21]. Based on the Gorter model, the residual factors of $BaFe_{12}O_{19}$ were calculated changing a parameter f''_m [8]. Figure 3 shows a variation of residual factors as a function of f''_m values, which has a minimum at $f''_m = 0.23$. Similar estimation gave a value for f'_m close to zero. Then, the spin orientations in five kinds of Fe sites were evaluated to calculate the asymmetry ratio ΔR_{calc} and the difference of $\Delta R_{obs} - \Delta R_{calc}$ for all 16 possible magnetic structures of $BaFe_{12}O_{19}$. The residual factors range between 0.16 and 0.32. The most plausible magnetic structure, having the smallest value of 0.16 among the residual factors, coincides with the Gorter model [8] and the neutron diffraction results [9, 10]. The Fe sites are as follows: Fe1 site (Wyckoff notation = 2a, coordination number = 6, block = S, magnetic moment direction = up), Fe2 (2b, 5, R, up), Fe3 (4f_1, 4, S, down), Fe4 (4f_2, 6, R, down) and Fe5 (12k, 6, R-S, up), having the stacking along c axis as RSR*S* (asterisk = 180°-rotation block).

We thank K. Yamawaki, S. Ishikawa, A. Ohno, M. Tada, T. Ohno and S. Hara of our group for their help. This study was performed under the auspices of the Photon Factory (PAC No. 2003G183 and 2005G131).

REFERENCES

1. K. Namikawa et al., *J. Phys. Soc. Jpn.* **54**, 4099-4102 (1985).
2. J. P. Hannon et al., *Phys. Rev. Lett.* **61**, 1245-1248 (1988).
3. P. Carra et al., *Phys. Rev. B* **40**, 7324-7372 (1989).
4. D. Gibbs, *J. Magn. Magn. Mater.* **104-107**, 1489-1495 (1992).
5. W. Neubeck et al., *Phys. Rev. B* **60**, R9912-R9915 (1999).
6. J. Igarashi and M. Takahashi, *J. Phys. Soc. Jpn.* **69**, 4087-4094 (2000).
7. P. B. Braun, *Philips Res. Rep.* **12**, 491-548 (1957).
8. E. W. Gorter, *Proc. IEEE B* **104**, 225-260 (1957).
9. O. P. Aleshko-Ozhevskii et al., *Sov. Phys. Crystallogr.* **14**, 367-369 (1969).
10. A. Collomb et al., *J. Magn. Magn. Mater.* **62**, 57-67 (1986).
11. J. Kreisel et al., *J. Magn. Magn. Mater.* **224**, 17-29 (2001).
12. T. Toyoda et al., *J. Ceram. Soc. Jpn.* **112**, PacRim 5 Spec. Issue, S1455-S1458 (2004).
13. T. M. Robinson and M. Labeyrie, *IEEE Trans. Magn.* **23**, 3727-3729 (1987).
14. F. Saito et al., *Physica B* **270**, 35-44 (1999).
15. K. Matsumoto et al., *Jpn. J. Appl. Phys.* **39**, 6089-6093 (2000).
16. L. Kalvoda et al., *J. Magn. Magn. Mater.* **87**, 243-249 (1990).
17. M. Blume and D. Gibbs, *Phys. Rev. B* **37**, 1779-1789 (1988).
18. K. Kobayashi et al., *J. Synchr. Rad.* **5**, 972-974 (1988).
19. J. O. Cross et al., *Phys. Rev. B* **58**, 11215 (1998).

20. S. Sasaki, *KEK Report*, National Laboratory for High-Energy Physics, Jpn., 88-14 (1989).
21. S. Sasaki, *KEK Internal*, National Laboratory for High-Energy Physics, Jpn., 87-3 (1987).

CHAPTER 12
CHEMISTRY AND MATERIALS SCIENCE

Application of High Energy- and Momentum-Resolution ARPES with Low Energy Tunable-Photons to Materials Science

Masaki Taniguchi*

*Hiroshima Synchrotron Radiation Center and Graduate School of Science, Hiroshima University, Higashi-Hiroshima 739-8526, Japan

Abstract. Recently, resolutions of energy and momentum in ARPES experiments have been greatly improved to 600~700 μeV and 4×10^{-3} Å^{-1}, respectively, at excitation photon energy of hv~8 eV on helical undulator beamline connected to compact light source; HiSOR. During the experiments, temperature of specimens on multiaxis cryogenic manipulator can be controlled down to 4.8 K. The system has been applied to investigations of quasiparticles in high T_c superconductor and 3d transition metals.

Keywords: electronic structure, quasi-particle, ARPES, high T_c superconductor, transition metal
PACS: 71.18.+y, 74.25.Jb, 74.72.Hs, 75.25.+z, 79.60.-i

INTRODUCTION

Low-energy excitation photons have advantages in precise Angle-Resolved Photoemission Spectroscopy (ARPES) experiments, with respect to high energy- and momentum-resolution, deep probing depth and selectivity of final states of excitations.

The energy resolution due to a combination of those of excitation photon and photoelectron energy analyzer reaches to ~1 meV or below it, by use of highly monochromatic and intense excitation-photons in UV~VUV region on undulator beamline and high-precision photoelectron energy analyzer. The in-plane momenta $k_{//}$ of photoelectrons are magnified into large emission angles, and higher momentum resolution is realized by a given instrumental angular resolution. In addition, the momentum $k_\perp$ perpendicular to the sample surface is well specified by a final-state wave function of a long mean-free-path λ, $\Delta k_\perp = 1/\lambda$. The effect of possible surface imperfection and contamination is minimized and the spectra are highly sensitive to bulk properties, because λ for the low-energy ARPES (e.g. excitation photon-energy, hv~8 eV) is by one order of magnitude longer than that, λ ~10 Å, for the conventional ARPES (hv~20 eV). Tuning the photon energy to the final state is also critically important to enable the photoelectron transition, since low-energy final states have a structure unlike free electrons.

In this paper, we present recent applications of precise ARPES with low energy tunable-photons from compact light source to investigations of quasiparticles in high T_c superconductor and 3d transition metals.

EXPERIMENTALS

The compact light source at Hiroshima Synchrotron Radiation Center (HiSOR operated at 700 MeV) accommodates two straight sections for insersion devices; helical undulator with perfect circular polarization in photon energy range of 4~40 eV and linear undulator for 30~300 eV [1, 2].

On helical undulator beamline, a 3m off-plane Eagle monochromator is installed to cover 4~40 eV range with resolving power of 12000~30000 [3, 4]. The end station with R4000 analyzer (Gammadata-Scienta), multiaxis cryogenic manipulator dedicated to ARPES and preparation chamber is always fixed to the beamline, to keep and

CP879, Synchrotron Radiation Instrumentation: Ninth International Conference,
edited by Jae-Young Choi and Seungyu Rah
© 2007 American Institute of Physics 978-0-7354-0373-4/07/$23.00

improve the performance. The multiaxis cryogenic manipulator realizes to scan the emission angle of photoelectrons and visualize electronic structures in the wide range of Brillouin zone [5].

Within the practical use, the best resolutions of energy and momentum in ARPES experiments are 600~700 μeV and 4×10^{-3} Å^{-1}, respectively, at hν ~8 eV on helical undulator beamline. During the experiments, temperature of specimens on a multiaxis cryogenic manipulator can be controlled down to 4.8 K.

For ordinary use, photon flux on the sample is set to ~5×10^{11} photons /sec with energy resolution of 3~4 meV at hν~8 eV. Details on ARPES experiments on linear undulator beamline are described in Refs. [6-8].

QUASI-PARTICLES NEAR THE FERMI LEVEL

Quasi-Particle Dynamics in High-T_c Superconductor $Bi_2Sr_2CaCu_2O_{8+\delta}$

In a d-wave superconducting state, quasi-particles survive solely at the node of the superconducting gap. The nodal quasi-particles govern the thermodynamic properties at low temperatures. So far, ac transport studies such as microwave experiments have shown that the quasi-particle scattering rate is dramatically suppressed in the superconducting state [9]. However, the transport data are velocity-weighted averaged over the momentum space, and the analyses require the Drude or extended Drude model for interpretation. Therefore, direct observation of the single quasi-particle excitation has been attempted by ARPES experiments. Resolution in the energy-momentum space is, then, the key to trace the energy dependence along the nodal direction and to resolve the contributions of two proximate bilayer-split nodes in $Bi_2Sr_2CaCu_2O_{8+\delta}$. In ARPES experiments so far, however, the residual spectral width has been broader than the nodal bilayer splitting, and the discontinuous change at T_c has been unidentified in the temperature dependence of the nodal spectral width in contrast to the transport studies [10]. In this section, we will report on an application of the low-energy tunable photons to the ARPES study of a bilayer cuprate superconductor $Bi_2Sr_2CaCu_2O_{8+\delta}$, and show that this technique provides us the single quasi-particle scattering rate, which is well resolved in the energy-momentum space [11].

The ARPES results obtained using low-energy excitation photons of hν=7.57 eV is shown in Fig. 1. The samples are nearly optimally doped $Bi_2Sr_2CaCu_2O_{8+\delta}$ of T_c=86 K. The spectra are collected around the nodal Fermi-surface crossings along the $(0,0)$-(π, π) direction at T=9 K in the superconducting state. The spectral images demonstrate the performance of the low-energy ARPES. First, Fig. 1(a) clearly shows the presence of a characteristic energy of $|\omega|$~70 meV, at which the nature of quasi-particle excitations dramatically changes. Compared to the behavior observed around the "kink" so far, the quasi-particle peak is strikingly sharpened on crossing $|\omega|$~70 meV towards the Fermi level ω=0 in good correlation with the abrupt deceleration of the quasi-particle group velocity v_k=dω_k/dk. Second, as shown in Figs. 1(b)-(e), the peak becomes so sharp near ω=0 that the small nodal splitting ~0.0075 Å^{-1} is clearly resolved. The momentum width, Δk=0.0065 Å^{-1} at ω=0, is much sharper than that of the conventional APRES before, Δk≥0.02 Å^{-1}. The nodal quasi-particles travel for a length longer than 1/Δk~150 Å without scattering in the superconducting state.

As a result of photon-energy-dependent study, the peaks of the doublet in Fig. 1(e) are identified as the bonding and antibonding bands of two proximate CuO$_2$ layers. Such bilayer splitting has previously been resolved in the antinodal region, and ascribed to the intrabilayer hoppings via Cu 4s orbitals [12]. In the nodal direction, on the other hand, the hybridization between Cu 3d$_{x2-y2}$ and 4s orbitals vanishes as $\Box$ $|\cos(k_xa)-\cos(k_ya)|^2$, and the effect of residual nodal bilayer splitting has been controversial. In the present result, the nodal splitting is finite, k^b-k^a=0.0075±0.0015 Å^{-1}, but much smaller than that, 0.015 Å^{-1}, reported by conventional ARPES [13]. In contrast, only single band dispersion is recognized in the ARPES spectra collected with 6-eV laser [14], even though the momentum resolution is similar to the present study.

The apparent inconsistency has been resolved by our investigation of the photon-energy dependence of the low-energy ARPES spectra. We have found that the photoelectron intensity ratio between the bonding and antibonding bands remarkably changes when hν is changed by several hundred meV. Here, the splitting width is almost constant with changing hν. In particullar, the bonding band intensity rapidly decreases on going from hν=7.8 to 7.0 eV, and is invisible for hν<7.0 eV. The results make it clear that the tuning hν is so critical to the ARPES experiments, and the careful investigation of the hν-dependence using synchrotron radiation is required to exclude unexpected final-state effect. In this work, we tuned the photon energy to hν=7.57 eV so that both the bonding and antibonding band are distinctly observed. Both the bonding and antibonding-band scattering rates have a linear energy-dependence near $\omega=0$, and the antibonding-band scattering rate increases more rapidly with energy $|\omega|$ than the bonding one [11].

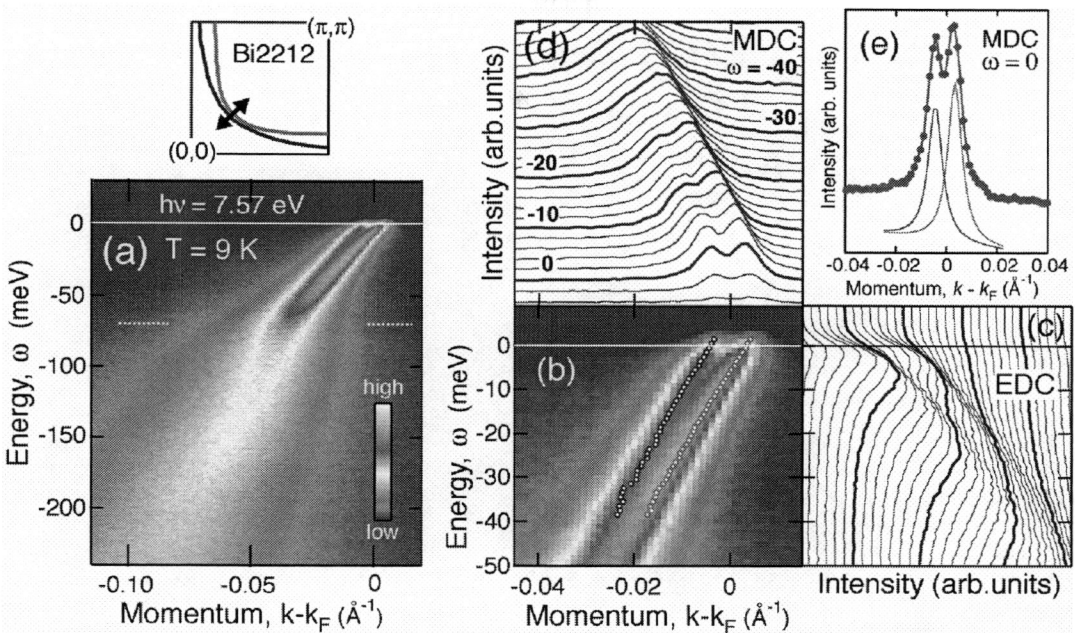

FIGURE 1. Result of low-energy ARPES ($h\nu$=7.57 eV), taken in the nodal direction of a superconducting bilayer cuprate $Bi_2Sr_2CaCu_2O_{8+\delta}$ at T=9 K. (a) Overview of the spectral-intensity map in the energy-momentum space. White dotted lines indicate the characteristic energy $|\omega|$~70 meV of the renormalization. (b) Enlarged view of the spectral-intensity map around the Fermi surface crossings. Red and blue open circles denote the peak positions of the momentum distribution curves (MDCs), corresponding to the bonding and antibonding bands of two proximate CuO_2 layers, respectively. (c) Energy distribution curves (EDCs) at each 0.0012 Å⁻¹. (d) MDCs at each 2 meV. (e) MDC at ω=0. The small nodal bilayer splitting, k^b-k^a=0.0075 Å⁻¹, is clearly resolved.

Figures 2(a) and 2(b) exhibit the quasi-particle renormalization energy and scattering rate, i.e. the real and imaginary parts of the self-energy $\Sigma(\omega)$, respectively [15]. We performed the fitting analysis of all the momentum distribution curves (MDCs), regarding splitting parameters as constants, and obtained the energy dependence of the momentum width and dispersion averaged between the bilayer-split quasi-particles. The real part, $Re\Sigma'(\omega)$, has been deduced from the dispersion deviation from a straight line. The imaginary part, $Im\Sigma(\omega)$, is directly given by the momentum width, namely the inverse scattering length, at the scaling factor of a constant unrenormalized velocity. As shown in Fig. 2(c), the causal consistency of the real and imaginary parts of $\Sigma(\omega)$ is established by the comparison between the width-derived $Im\Sigma(\omega)$ and the Kramers-Krönig transformation $Im\Sigma_{KK}'(\omega)$ of the dispersion-derived $Re\Sigma'(\omega)$. The step features of $Im\Sigma(\omega)$ and $Im\Sigma_{KK}'(\omega)$ agree excellently in the position $50\leq|\omega|\leq90$ meV and in the height ~35 meV. The bump feature of $Im\Sigma(\omega)$ at $|\omega|$~90 meV is cancelled out in the difference $Im\Sigma(\omega)$-$Im\Sigma_{KK}'(\omega)$. The remaining featureless background may be interpreted as the electron-electron scattering rate, because $Re\Sigma'(\omega)$ includes no ω-linear background. Both $Re\Sigma'(\omega)$ and $Im\Sigma(\omega)$ consistently indicate that the scattering rate abruptly increases from $|\omega|$=50 to 90 meV due to the coupling with bosonic modes.

The quasi-particle scattering rate in the form of the inverse lifetime, $1/\tau(\omega)$, is given by the energy width ΔE, which is also deduced from the MDC analysis by the relation, $1/\tau(\omega)=\Delta E=v_k\Delta k$. Fig. 2 (d) exhibits comparison between the nodal single-particle scattering rate obtained by ARPES, and the transport scattering rate derived from the optical conductivity [16]. The quantitative agreement indicates that the extrinsic broadening of the ARPES spectral peak is minimized by using the low-energy excitation photons.

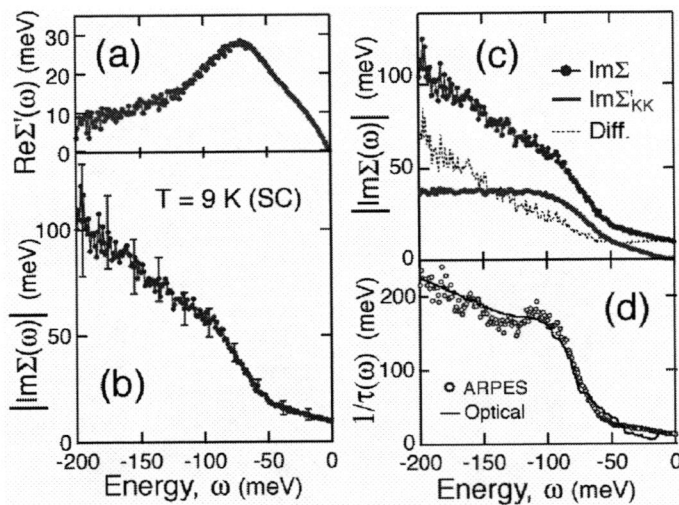

FIGURE 2. Self-energy $\Sigma(\omega)$ of the nodal quasi-particles, obtained by the MDC-fitting analysis regarding the splitting parameters as constants. (a) Real part of self-energy $Re\Sigma'(\omega)$. The energy shift of quasi-particle excitation has been determined from the the dispersion by $Re\Sigma'(\omega_k)=\omega_k-\varepsilon_k^{0'}$. (b) Imaginary part of self-energy. The quasi-particle scattering rate has been determined directly from the momentum width Δk (FWHM) by $\Sigma(\omega)=-(1/2)v_0\Delta k$, where the scaling factor is $v_0=2.8$ eVÅ. (c) Comparison between the width-derived $Im\Sigma(\omega)$ (filled circles) and the Kramers-Krönig transformation $Im\Sigma_{KK}'(\omega)$ (solid lines) of the dispersion-derived $Re\Sigma'(\omega)$ for T=95 and 9 K. The bump at $|\omega|\sim90$ meV is common between them, and cancelled out in the difference, $Im\Sigma(\omega)$-$Im\Sigma_{KK}'(\omega)$ (black dotted lines). (d) Inverse lifetimes, deduced from ARPES by $1/\tau(\omega)=v_k\Delta k$ (open circles) and from optical conductivity (line) [16].

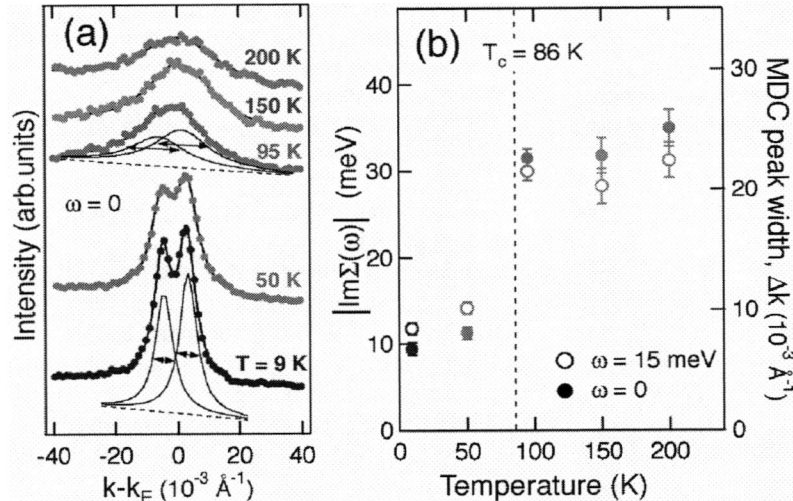

FIGURE 3. (a) Temperature dependence of the MDC at ω=0. The spectral peaks are dramatically sharpened in the superconducting state. (b) Temperature dependence of the scattering rate, $-Im\Sigma(\omega)=(1/2)v_0\Delta k$, directly determined from the momentum width Δk (FWHM) at ω=0 (filled circles) and 15 meV (open circles).

Figure 3 represents the temperature-dependence of the MDC and scattering rate at ω=0. As shown in Fig. 3(b), we have found that the momentum width of the nodal quasi-particle is abruptly suppressed by ~0.01 Å^{-1} (60-70 % drop) upon the superconducting transition (T_c=86 K). This observation reconciles the ARPES results with transport studies [9]. The residual scattering rate in the superconducting state is lower than the linear extrapolation of the normal-state scattering rate for T→0, suggesting that the reduction of the scattering rate at T_c occurs not only for the inelastic part but also for the elastic part. In the normal state, a zero-energy electron is elastically scattered into the other segment of the Fermi surface. The opening of the d-wave superconducting gap would close these scattering channels except for those into the other nodal points. Consequently, the nodal quasi-particles are hardly scattered by the impurities in the superconducting state, even though the superconducting gap is closed there.

As shown in Fig. 2(b), the nodal scattering rate appears to be linear in energy $|\omega|$ at low energies, $|\omega|<50$ meV. Knowing that the nodal scattering rate is strongly affected by the opening of the d-wave gap, the energy dependence of the scattering rate may also be governed by the d-wave gap. In the d-wave superconducting state, the low-energy electronic density of states is reduced to the amount proportional to the energy $|\omega|$. Consequently, the elastic scattering rate may have a linear ω-dependence, while the inelastic scattering rate depends on $|\omega|$ as an order higher than the first order and is relatively negligible near $\omega=0$. A possible candidate is the elastic forward scattering by the antisite defects or excess oxygens in the BiO and/or SrO layers, as proposed recently [17]. In contrast to the transport scattering rate, the single-particle scattering rate observed by ARPES may be affected by such elastic forward scattering [18].

Energy Band and Spin-Dependent Many-Body Interaction in Ferromagnetic Ni

The ARPES measurements have revealed that electron correlation is important in the spin-polarized Ni 3d bands [19]: the 3d band width is narrowed by ~25%, and the exchange splittings are reduced by ~50% compared with band-structure calculations with the local spin-density approximation (LSDA). A spin-polarized 6 eV satellite also exists in the core-level and valence-band photoemission spectra [20]. These unusual spectral features have been explained in terms of the electron correlation effect [21].

In order to clarify the energy-band and the spin-dependent many-body interactions on the quasi-particles near E_F, we carried out ARPES measurements on Ni(110) at low temperature [7]. Figures 4(a) and 4(c) show energy distribution curves (EDC's) and intensity plot for the $\Sigma_{2\downarrow}$ bands, respectively. The evaluated peak positions are indicated by open circles in the intensity plots. A kink structure or a sudden change of the group velocity is apparent at ~-40 meV in the $\Sigma_{2\downarrow}$ band. Figure 5(b) exhibits the intensity plot for the $\Sigma_{1\sigma}$ ($\sigma=\uparrow, \downarrow$) band and Figs. 5(a) and 5(c) the bands on the different points of the Fermi surface. A weak kink structure in the minority-spin bands but is much less clear in the majority-spin bands.

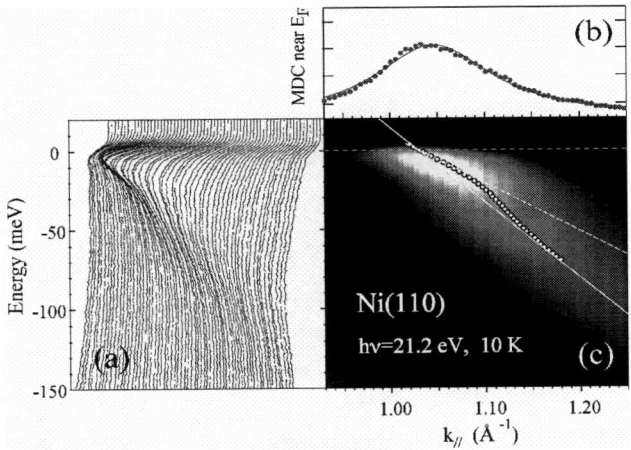

FIGURE 4. The ARPES results of Ni(110) at 10 K for hv=21.2 eV. (a) shows EDCs, and (c) intensity plot for the $\Sigma_{2\downarrow}$ band. The circles and solid line respectively indicate evaluated peak positions, and linear dispersion (ε_k^0) without the kink structure. The dashed line shows the gradient at E_F.

Figure 6(a) provides the evaluated $|\mathrm{Im}\Sigma^\sigma|$ of the $\Sigma_{2\downarrow}$, $\Sigma_{1\uparrow}$ and $\Sigma_{1\downarrow}$ bands [15]. $|\mathrm{Im}\Sigma^\downarrow|$ of the $\Sigma_{2\downarrow}$ and $\Sigma_{1\downarrow}$ bands decreases for the energy of $\omega >-40$ meV, which implies that the kink structures originate from the many-body interaction and not from the energy dispersion. Since the energy scale of the kink structures coincides well with the Debye temperature, $\Theta_D=450$ K ($k_B\Theta_D=39$ meV), it is reasonable to assume that the structure is derived from the electron-phonon interaction.

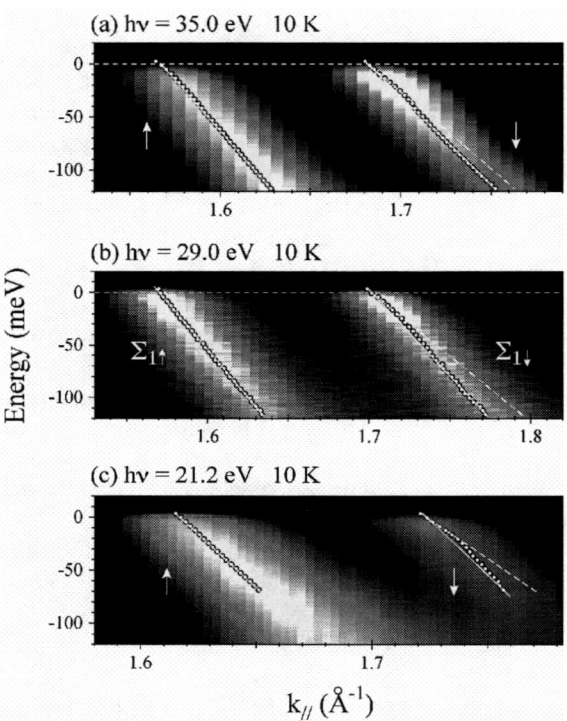

FIGURE 5. The ARPES intensity of Ni(110) at 10K taken at hν = (a) 35.0 eV, (b) 29.0 eV and (c) 21.2 eV. (b) shows the $\Sigma_{1\uparrow}$, and $\Sigma_{1\downarrow}$ bands. The circles and solid lines indicate the evaluated peak positions, and linear dispersions ($\varepsilon_k^{0\prime}$) without the kink structure, respectively. The dashed lines show the gradient of the down-spin bands at E_F [7].

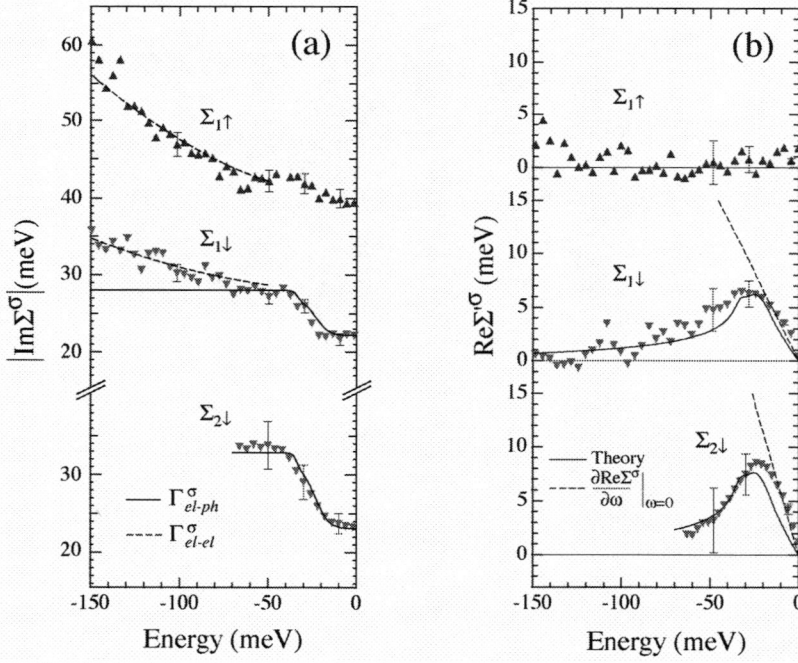

FIGURE 6. Experimentally obtained imaginary (a) and real (b) parts of the self-energy of the $\Sigma_{2\downarrow}$, $\Sigma_{1\uparrow}$, and $\Sigma_{1\downarrow}$ bands. The symbols represent the observed imaginary/real parts of the self-energy. The solid and dashed lines in (a) exhibit the theoretical $\Gamma_{el\text{-}ph}$ and $\Gamma_{el\text{-}el}$, respectively. The theoretical $\Gamma_{el\text{-}ph}$ and $\Gamma_{el\text{-}el}$ in (a) are shifted by the offset Γ^σ_0. The solid and dashed lines in (b) indicate the theoretical $Re\Sigma^{\downarrow}$, and the gradient of experimental $Re\Sigma^{\uparrow}$ near E_F, respectively.

The solid lines in Fig. 6(a) represent the calculated lifetime broadening due to the electron-phonon interaction (Γ_{el-ph}; $\Gamma^\sigma=2|Im\Sigma^\sigma|$) for the $\Sigma_{2\downarrow}$ and $\Sigma_{1\downarrow}$ bands, which explain well the observed $|Im\Sigma^\sigma|$ especially the drop for $\omega>$-40 meV. The theoretical $Re\Sigma^\sigma$'s (Solid lines in Fig. 6(b)) also explain well the observed ones. These results confirm that the kink structure is produced by the electron-phonon interactions.

The electron-phonon coupling constant ($\lambda=|\partial Re\Sigma'^\sigma/\partial\omega|_{\omega=0}$) was evaluated from the dashed lines in Fig. 6(b). We obtained λ=0.57±0.06 and 0.33±0.05 for the $\Sigma_{2\downarrow}$ and $\Sigma_{1\downarrow}$ bands, and $\lambda\sim$0 for the $\Sigma_{1\uparrow}$ band, respectively. We have found that λ is dependent on energy-band and spin-direction. The value of λ has been theoretically evaluated using the phase shifts of the wave function scattered by the core potential [22]. In the case of Ni, the d-f scattering contributes most strongly to the coupling constant [22].

Assuming that the d weight given by the LSDA calculation is a measure of the coupling constant, the $\Sigma_{2\downarrow}$ band is purely d-like, while the d weight at E_F for the $\Sigma_{1\downarrow}$ band, ~90%, is smaller due to s-d hybridization. Although the $\Sigma_{1\uparrow}$ and $\Sigma_{1\downarrow}$ bands have the same symmetry, the d component in the $\Sigma_{1\uparrow}$ band at E_F, ~80%, is slightly smaller than that in the $\Sigma_{1\downarrow}$ band, due to exchange splitting. It seems that this explanation works well in a qualitative way, the difference, ~10%, however, looks rather too small to account for the large spin-dependence of λ in the $\Sigma_{1\uparrow}$ bands. One should take electron correlation into account when considering the spin-dependent spectral-weight distribution.

In order to see the electron correlation effects, we have evaluated the group velocity, $v_F^{ARPES}=2\pi/h(d\varepsilon_k^{0}/dk)$. Obtained v_F^{ARPES} values are smaller than those given by the LSDA calculation, v_F^{LSDA}, by a factor of v_F^{ARPES}/v_F^{LSDA} ~ 47-69 %. On the other hand, the ratio of the Fermi wave numbers k_F^{ARPES}/k_F^{LSDA} was estimated to be 0.9-1.0. Thus the significant deviation from the LSDA is in the Fermi velocity. The mass enhancement factor due to electron correlation can be evaluated by $\eta=m/m_b=(k_F^{ARPES}/k_F^{LSDA})(v_F^{LSDA}/v_F^{ARPES})$, where m_b represents an electron mass given by the LSDA calculation.

The overall effective mass enhancement can be expressed by $m^*/m_b=(1+\lambda)\eta$ derived from both the electron-phonon interaction and electron correlation. The m^*/m_b ratio for the d-like $\Sigma_{2\downarrow}$ band ($m^*/m_b\sim$2.8, $\eta\sim$1.8, $1+\lambda\sim$0.57) is larger than those for the $\Sigma_{1\uparrow}$ band ($m^*/m_b\sim$1.9, $\eta\sim$2.2, $1+\lambda\sim$1) and $\Sigma_{1\downarrow}$ band ($m^*/m_b\sim$2.2, $\eta\sim$1.4, $1+\lambda\sim$1.3). The present results for the $m^*/m_b\sim$1.9-2.8 agree well with those from the de Haas-van Alphen measurement $m^*/m_b\sim$1.8-2.3 [23]. Note that the m^* enhancement is mainly derived from electron correlation in the $\Sigma_{1\uparrow}$ band, while both the electron-phonon interaction and electron correlation contribute to the m^* enhancement in the $\Sigma_{1\downarrow}$ band.

The lifetime broadening due to the electron-electron interaction $\Gamma_{el-el}^\sigma\sim2\beta^\sigma[(\pi k_B T)^2+\omega^2]$ [24] can be measured by a coefficient $2\beta^\sigma$. By the fit of the lower energy side of $|Im\Sigma^\sigma|$, we evaluated $2\beta^\uparrow\sim$1.4 eV^{-1} and $2\beta^\downarrow\sim$0.6 eV^{-1} for the $\Sigma_{1\uparrow}$ and $\Sigma_{1\downarrow}$ bands, respectively. These results indicate that the quasi-particles with an up-spin are strongly scattered compared with those with a down-spin.

Now we are examining the wave-vector dependence of the many-body interactions, namely, the anisotropy of the coupling constants. The ARPES study of Pd(110) [25] and Cu(110) which are located next to Ni in the periodic table have been done, which provides us a systematic view on the many-body interactions leading to different physical properties or magnetism in these elements. Recently, it has been reported that a kink structure is derived from the electron-magnon interaction for the surface states of Fe thin film [26]. We have carried out high-resolution ARPES on Fe(110) bulk single crystal and have evaluated magnitudes of the electron-phonon, electron-magnon, and electron-electron interactions for the bulk derived energy bands [27].

ACKNOWLEDGMENTS

The author would like to thank all the staff members of Hiroshima Synchrotron Radiation Center and scientists of graduate school of science, Hiroshima University, for promoting cooperative researches on materials science in the field of solid state physics.

REFERENCES

1. K. Yoshida, T. Takayama and T. Hori, *J. Synchrotron Rad.* **5**, 345 (1998).
2. A. Hiraya, K. Yoshida, S. Yagi, M. Taniguchi, S Kimura, H. Hama, T. Takayama and D. Amano, *J. Synchrotron Rad.* **5**, 445 (1998).
3. T. Matsui, H. Sato, K. Shimada, M. Arita, S. Senba, H. Yoshida, K. Shirasawa, M. Morita, A. Hiraya, H. Namatame and M. Taniguchi, *Nucl. Instrum. Method Phys. Res. A* **467-468**, 537 (2001).
4. M. Arita, K. Shimada, H. Namatame and M. Taniguchi, *Surf. Rev. Lett.* **9**, 535 (2001).

5. Y. Aiura, H. Bando, T. Miyamoto, A. Chiba, R. Kitagawa, S. Maruyama and Y. Nishihara, *Rev. Sci. Instrum.* **74**, 3177 (2003).

6. K. Shimada, M. Arita, T. Matsui, K. Goto, S. Qiao, K. Yoshida, M. Taniguchi, H. Namatame, T. Sekitani, K. Tanaka, K. Shirasawa, N. Smolyakov and A. Hiraya, *Nucl. Instrum. Method Phys. Res. A* **467-468**, 504 (2001).

7. M. Higashiguchi, K. Shimada, K. Nishiura, X. Y. Cui, H. Namatame and M. Taniguchi, *Phys. Rev. B* **72**, 214438 (2005).

8. X. Y. Cui, H. Negishi, S. G. Titova, K. Shimada, A. Ohnishi, M. Higashiguchi, Y. Miura, S. Hino, A. M. Jahir, A. Titov, H. Bidadi, S. Negishi, H. Namatame, M. Taniguchi and M. Sasaki, *Phys. Rev. B* **73**, 085111 (2005).

9. D. A. Bonn, R. Liang, T. M. Riseman, D. J. Baar, D. C. Morgan, K. Zhang, P. Dosanjh, T. L. Duty, A. MacFarlane, G. D. Morris, J. H. Brewer, and W. N. Hardy, C. Kallin and A. J. Berlinsky, *Phys. Rev. B* **47**, 11314 (1993); D. A. Bonn, P. Dosanjh, R. Liang, and W. N. Hardy, *Phys. Rev. Lett.* **68**, 2390 (1992); A. Hosseini, R. Harris, S. Kamal, P. Dosanjh, J. Preston, R. Liang, W. N. Hardy, and D. A. Bonn, *Phys. Rev. B* **60**, 1349 (1999).

10. T. Valla, A. V. Fedorov, P. D. Johnson, B. O. Wells, S. L. Hulbert, Q. Li, G. D. Gu, N. Koshizuka, *Science* **285**, 2110 (2001); T. Valla, A. V. Fedorov, P. D. Johnson, Q. Li, G. D. Gu, and N. Koshizuka, *Phys. Rev. Lett.* **85**, 828 (2000); Z. M. Yusof, B. O. Wells, T. Valla, A. V. Fedorov, P. D. Johnson, Q. Li, C. Kendziora, *Phys. Rev. Lett.* **88**, 167006 (2002).

11. T. Yamasaki, K. Yamazaki, A. Ino, M. Arita, H. Namatame, M. Taniguchi, A. Fujimori, Z.-X. Shen, M. Ishikado, and S. Uchida, cond-mat/0603006.

12. D. L. Feng, N. P. Armitage, D. H. Lu, A. Damascelli, J. P. Hu, P. Bogdanov, A. Lanzara, F. Ronning, K. M. Shen, H. Eisaki, C. Kim, Z.-X. Shen, J.-i. Shimoyama and K. Kishio, *Phys. Rev. Lett.* **86**, 5550 (2002).

13. A. A. Kordyuk, S. V. Borisenko, A. N. Yaresko, S.-L. Drechsler, H. Rosner, T. K. Kim, A. Koitzsch, K. A. Nenkov, M. Knupfer, J. Fink, R. Follath, H. Berger, B. Keimer, S. Ono, and Y. Ando, *Phys. Rev. B* **70**, 214525 (2004).

14. J. D. Koralek, J. F. Douglas, N. C. Plumb, Z. Sun, A. V. Fedorov, M. M. Murnane, H. C. Kapteyn, S. T. Cundiff, Y. Aiura, K. Oka, H. Eisaki, and D. S. Dessau, *Phys. Rev. Lett.* **96**, 017005 (2006).

15. The major photoemission spectral features are given by $A(k,\omega)$ related to the imaginary part of the single-particle Green's function, $A(k,\omega)=-(1/\pi)ImG(k,\omega)=-(1/\pi)Im[1/\{\omega-\varepsilon_k^0-\Sigma(k,\omega)\}]$, where ε_k^0 represents the energy of the non-interacting band. The imaginary part ($Im\Sigma$) and the real part ($Re\Sigma$) of the self-energy can be evaluated from the spectral width ($\Delta E=2|Im\Sigma|$) and the energy shift from the non-interacting band ($Re\Sigma=\varepsilon-\varepsilon_k^0$), respectively (See S. Hüfner, *Photoelectron Spectroscopy* 3rd Ed., Springer-Verlag, Berlin, 2003). We have assumed that the dispersion of ε_k^0 is linear without the kink structure and put $Re\Sigma'=\varepsilon-\varepsilon_k^{0'}$ using the linear $\varepsilon_k^{0'}$.

16. A. V. Puchkov, D. N. Basov, and T. Timusk, *J. Phys.* **8**, 10049 (1996).

17. L. Zhu, P. J. Hirschfeld, and D. J. Scalapino, *Phys. Rev. B* **70**, 214503 (2004); T. Dahm, P. J. Hirschfeld, D. J. Scalapino, and L. Zhu, *Phys. Rev. B* **72**, 214512 (2005).

18. E. Abrahams and C. M. Varma, *PNAS.* **97**, 5714 (2000).

19. F.J. Himpsel, J.A. Knapp and D.E. Eastman, *Phys. Rev. B* **19**, 2919 (1979), and references therein.

20. S. Hüfner and G.K. Wertheim, *Phys. Lett.* **51**, 299 (1975); R. Clauberg, W. Gudat, E. Kisker, E. Kuhlmann and G.M. Rotheberg, *Phys. Rev. Lett.* **47**, 1314 (1981), and references therein.

21. D.R. Penn, *Phys. Rev. Lett.* **42**, 921 (1979); A. Liebsch, *Phys. Rev. Lett.* **43**, 1431 (1979).

22. D.A. Papaconstantopoulos, L.L. Boyer, B.M. Klein, A.R. Williams, V.L. Morruzi, J.F. Janak, *Phys. Rev. B* **15**, 4221 (1977).

23. E.I. Zornberg, *Phys. Rev. B* **1**, 244 (1970).

24. T. Valla, A.V. Federov, P.D. Johnson, and S.L. Hulbert, *Phys. Rev. Lett.* **83**, 2085 (1999).

25. M. Higashiguchi *et al.*, *Physica B* (2006) in press.

26. J. Schäfer, D. Schrupp, E. Rotenberg, K. Rossnagel, H. Koh, P. Blaha, R. Claessen, *Phys. Rev. Lett.* **92**, 097205 (2004).

27. X. Y. Cui *et al.*, in preparation.

Hard X-Ray Photon-In-Photon-Out Spectroscopy with Lifetime Resolution – of XAS, XES, RIXSS and HERFD

P. Glatzel[a], M. Sikora[a], S. G. Eeckhout[a], O. V. Safonova[a], G. Smolentsev[b], G. Pirngruber[c], J. A. van Bokhoven[c], J.-D. Grunewaldt[c], M. Tromp[d]

[a]*European Synchrotron Radiation Facility (ESRF), BP22, 6 rue Jules Horowitz, 38043 Grenoble, France*
[b]*Faculty of Physics, Rostov State University, Rostov-on-Don, Russia*
[c]*Institute for Chemical and Bioengineering, ETH Zurich, 8093 Zurich, Switzerland*
[d]*University of Southampton, School of Chemistry, Southampton, SO17 1BJ, United Kingdom*

Abstract. Spectroscopic techniques that aim to resolve the electronic configuration and local coordination of a central atom by detecting inner-shell radiative decays following photoexcitation using hard X-rays are presented. The experimental setup requires an X-ray spectrometer based on perfect crystal Bragg optics. The possibilities arising from non-resonant (X-Ray Emission Spectroscopy - XES) and resonant excitation (Resonant Inelastic X-Ray Scattering Spectroscopy – RIXSS, High-Energy-Resolution Fluorescence Detected (HERFD) XAS) are discussed when the instrumental energy broadenings of the primary (beamline) monochromator and the crystal spectrometer for x-ray emission detection are on the order of the core hole lifetimes of the intermediate and final electronic states. The small energy bandwidth in the emission detection yields line-sharpened absorption features. In transition metal compounds, electron-electron interactions as well as orbital splittings and fractional population can be revealed. Combination with EXAFS spectroscopy enables to extent the k-range beyond unwanted absorption edges in the sample that limit the EXAFS range in conventional absorption spectroscopy.

Keywords: Emission Spectroscopy, Coordination Chemistry, Electronic Structure, EXAFS
PACS: 61.10.Ht, 87.64.Fb, 87.64.Gb, 71.70.Ch

INTRODUCTION

Photon-in-photon-out techniques are targeted at applications where the sample environment cannot be chosen freely, i.e. UHV conditions suitable for photoemission experiments are not possible, or when a bulk sensitive probe is desired. The preferred X-ray spectroscopy technique to study element specifically electronic structure and local coordination is X-ray absorption spectroscopy (XAS).[1,2] The near edge structure (XANES) is mainly used to obtain oxidation states, even though XANES also contains information on the local geometry and coordination. A detailed analysis of the XANES structure is a complex task because of the numerous interactions that contribute to its shape. The spectroscopy using the extended range (EXAFS) is well developed theoretically and experimentally but the technique has its inherent limitations (e.g. differentiation of elements close in atomic number Z) and the ideal experimental conditions (e.g. sample thickness, homogeneity) for a correct EXAFS analysis are not always given. It is thus desirable to introduce other techniques on the beamline that either provide a mean to verify the results obtained from XAS or yield additional information on the sample.

An X-ray spectrometer based on perfect crystal Bragg optics opens up new possibilities for X-ray spectroscopy. Detecting the emitted X-rays with an instrumental energy bandwidths on the order of the core hole lifetime broadening enables to resolve fine structure in the X-ray emission spectrum. This fine structure contains information on the electronic configuration and chemical environment of the emitting atom that is complementary to what can be obtained in XAS.[3] Such a secondary monochromator provides an additional tunable energy detection to the primary monochromator of the synchrotron radiation source. X-ray emission spectrosocpoy (XES) thus adds a dimension to XAS. The techniques arising from such an experimental setup have been named non-resonant XES, resonant XES or resonant inelastic X-ray scattering spectroscopy (RIXSS) and high-energy-resolution fluorescence

CP879, *Synchrotron Radiation Instrumentation: Ninth International Conference*,
edited by Jae-Young Choi and Seungyu Rah
© 2007 American Institute of Physics 978-0-7354-0373-4/07/$23.00

detection (HERFD). They will be discussed here. Another technique, non-resonant X-ray Raman scattering, will be left out to be discussed by others.[4]

X-RAY EMISSION DETECTION WITH LIFETIME RESOLUTION

Non-resonant XES

With the excitation energy set well above (> 30 eV) an absorption edge the X-ray emission spectrum is independent of the incident energy. The emitted X-rays are generally referred to as fluorescence lines. The chemical dependence of the Kβ lines in 3d transition metals has been discussed by many authors.[3] The K$\beta_{1,3}$ lines reflect the valence shell spin state and thus enable to detect oxidation state changes as well as high-spin-low-spin transitions. The Kβ satellite lines occur right below the Fermi level and thus show a very strong chemical dependence. In particular, they show a strong sensitivity to the type of ligand. Unlike XAS, the XES shows distinct signatures for different ligands that enable to identify the quality of the chemical environment much more exact than in XAS. For example, nitrogen and oxygen ligands can be clearly distinguished.[5] Also, ligand protonation yields spectral changes that can be assigned to e.g. OH$^-$ or H$_2$O groups (Fig. 1). The spectral signature does not depend strongly on the bond distances in contrast to the X-ray absorption features. It is thus easier to assign a spectral feature to a change in the ligand environment. However, bond distances cannot be obtained accurately using non-resonant XES and the technique has thus to be seen as a complementary tool to EXAFS.

The Kβ satellite spectral features can be calculated using the FEFF code or density functional theory. The latter has the advantage that the molecular orbitals can be visualized and assigned to spectral features. This opens a new realm for spectral interpretation that will help to understand the effect of changes in the ligand environment on the electronics structure. When interpreting the spectra it is important to be aware of possible multi-electron transitions. Their intensities depend on the incident energy and it is thus possible to identify them in the Kβ satellite spectrum.[6]

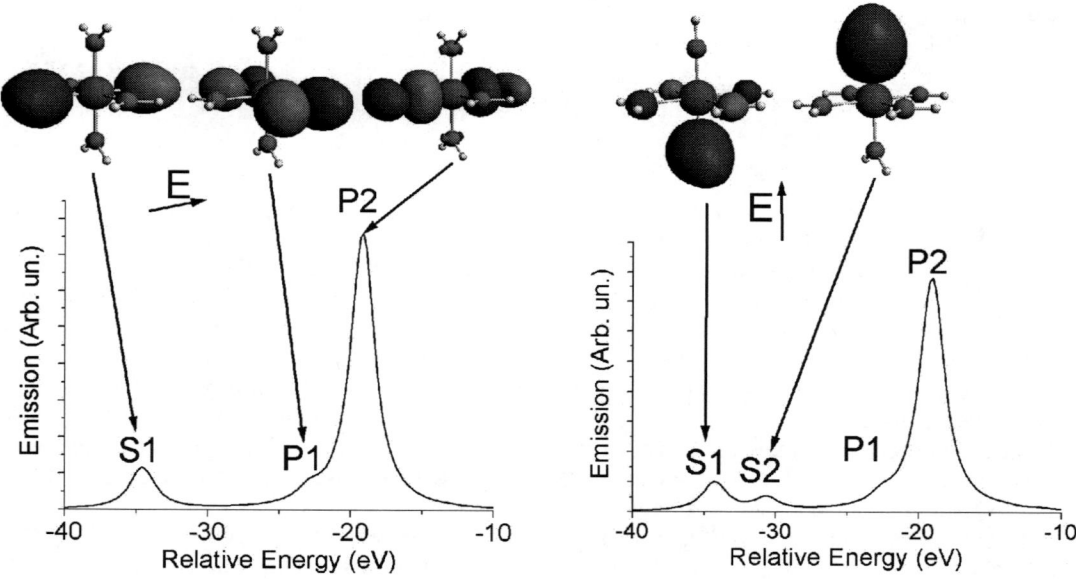

FIGURE 1. Kβ satellite lines for model systems [Mn(H$_2$O)$_6$]$^{2+}$ (left) and [Mn(H$_2$O)$_5$OH]$^+$ (right). Ligand deprotonation lowers the symmetry and yields an additional peak (S2) that can be assigned to a molecular orbital localized on the OH ligand.

HERFD XAS

Fluorescence detected absorption spectroscopy was developed in order to be able to measure dilute samples that do not absorb enough photons to obtain a XAS spectrum in transmission mode. A XAS in fluorescence mode is detected either without energy resolution (e.g. photodiode) or with energy resolution (e.g. Ge detector). The better the energy resolution, i.e. the smaller the energy bandwidth around the fluorescence line, the better will be the signal to background ratio. The limit for improvement is set by the core hole lifetime broadening because a smaller bandwidth in the emission detection will not significantly improve the signal to background ratio anymore. We thus define high-energy-resolution fluorescence detected absorption spectroscopy (HERFD XAS) as detecting the fluorescence line with an energy bandwidth on the order or below the core hole lifetime broadening. Apart from the advantage of having the best possible signal to background ratio, emission detection with lifetime resolution also yields line-sharpened absorption features. This effect considerably helps to analyze absorption features. For 3d transition metals it has been used to separate the K absorption pre-edges from the strong main edge. Figure 2 shows that the effect is very pronounced for high-Z materials (e.g. Pt, Au, ...). The reduced line-broadening can be readily simulated in the FEFF code. The technique has been used to study catalysts under working conditions and it successfully helped to identify the bonding configuration of CO on Pt nano-particles or the actication of O_2 over Au nano-particles.[7]

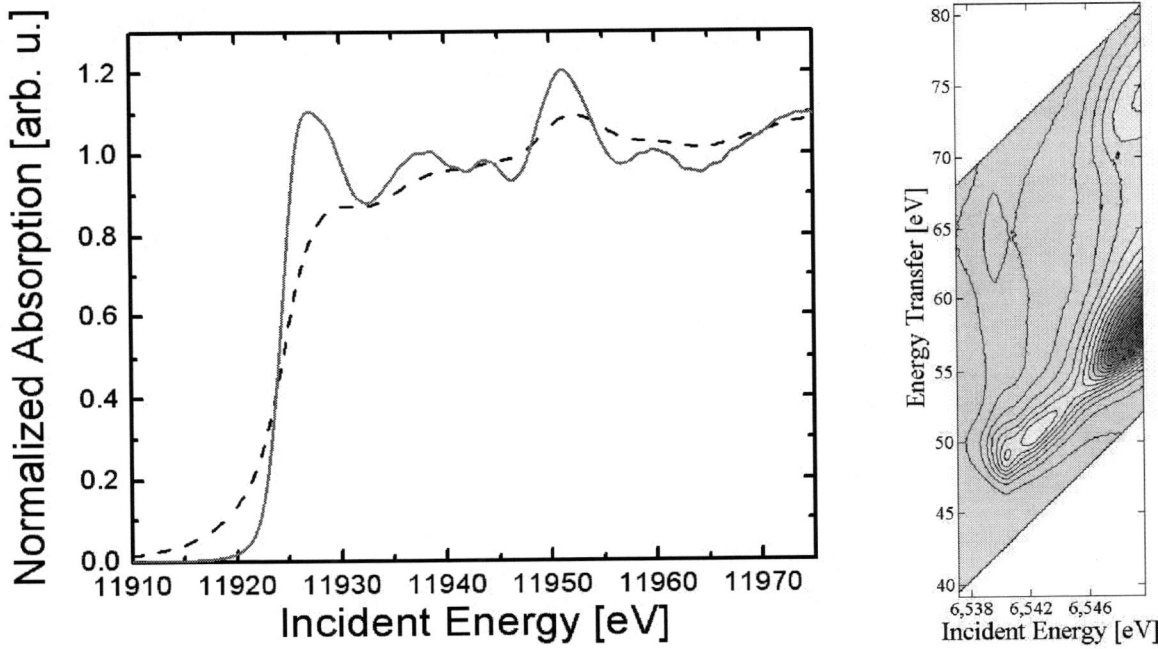

FIGURE 2. left: The L_3 absorption edge of a Au foil detected in transmission mode (dotted) and using the HERFD technique (solid). Right: 1s3p RIXS plane of Mn_2O_3. The spin-down and spin-up directions occur at around 50 eV and 65 eV energy transfer, respectively. The spectra were recorded on beamline ID26 of the European Synchrotron Radiation Facility (ESRF).

The fact that HERFD XAS fully separates most fluorescence lines can be used to record extended absorption scans beyond unwanted edges that occur either in transmission or medium energy resolution fluorescence detection mode. The undesired edge may arise either from another element present in the sample or actually from the same element that is being studied. An example for the former is Fe in protein samples where the active site containing Mn is studied. The Fe K-edge occurs in the Mn EXAFS range and limits the available k-range for EXAFS analysis. Using HERFD XAS the Fe K-edge can be eliminated and the Mn EXAFS analysis greatly improved. The technique was successfully applied to study the oxygen evolving complex in the photosynthesis multi-protein complex photosystem II (PS II).[8] The same principle was applied for rare earths to eliminate the L_1-edge in the L_2 EXAFS.[9] However, here the pathways of possible electronic transitions are more complex and difficulties were encountered caused by cascade decays. We note that the constraint of having a dilute sample applies to HERFD as it applies to standard fluorescence detected XAS.

RIXSS

The incident energy can be tuned to an absorption feature near the Fermi level that arises from an excitation into a localized state. If the radiative decay of this resonantly excited state is detected with lifetime resolution one can refer to resonant inelastic x-ray scattering spectroscopy (RIXSS). Figure 2 shows the 1s3p RIXS plane of Mn_2O_3. A 1s electron is resonantly excited into the K absorption pre-edge and the subsequent 3p to 1s decay is monitored using the emission spectrometer.[3]

For a meaningful analysis of the RIXS spectra it is very helpful to record a full RIXS plane, i.e. cover the full range of possible combinations of incident and emitted energies as shown in Fig. 2. A line plot will only show the intensity variations in one direction. However, for the RIXS process intensity variations in two directions form a peak or spectral feature and interpretation of simple line plots is very difficult.

Experimentally the RIXS plane is recorded by fixing the emission energy and performing fast or continuous scans with the incident energy monochromator. This corresponds to diagonal scans in the 1s3p RIXS plane. To build up the RIXS plane the emission energy is changed step wise. A full RIXS plane can thus be recorded in one hour on a concentrated sample. In case the sample shows damage due to the X-ray beam it is possible to change the beam position on the sample with every change of emission energy. It is then necessary to correct the intensity for changes in the sample concentration or thickness.

RIXS spectroscopy greatly facilitates analysis of the K absorption pre-edges in transition metals because the lifetime of the final state and not the 1s excited state limit the resolution of the spectral features. It can be used to analyze valence shell spin and oxidation states as well as a diagnostic of the local symmetry.[10,11] The pre-edge can also be deconvoluted into spin-up and spin-down excitations which provides a stringent test of the theoretical model that aims to simulate the spectral shape.[12]

CONCLUSIONS

X-Ray emission detection with lifetime resolution is a technique complementary to XAS. An experimental setup should therefore ensure that both techniques can be applied simultaneously where the XAS is measured either in transmission or in fluorescence detection mode. Sample characterization on an X-ray beamline will then be more complete and the data analysis will be more robust because the different spectra will also serve to countercheck each other. X-ray emission spectrometers are currently being developed at various synchrotron radiation sources. It is desirable that these endstations will be installed permanently on a XAS beamline in order to minimize setup time and to make the technique readily available for users. With increasing user-friendliness in the operation, more and more user groups will appreciate the benefits of XES and it is realistic to assume that XES will become a standard technique.

ACKNOWLEDGMENTS

We would like to thank Dr. Frank F. M. de Groot for fruitful discussions. The ESRF support groups are gratefully acknowledged.

REFERENCES

1. *EXAFS Spectroscopy. Techniques and Application*, edited by B. K. Teo and D. C. Joy, New York, Plenum Press, 1981
2. *X-ray Absorption: Principles, Applications, Techniques of EXAFS, SEXAFS, and XANES*, edited by D. C. Koningsberger and R. Prins, New York, John Wiley & Sons, 1988
3. P. Glatzel and U. Bergmann, Coord. Chem. Rev. **249**, 65 (2005).
4. U. Bergmann, O. C. Mullins, and S. P. Cramer, Anal. Chem. **72** (11), 2609 (2000).
5. U. Bergmann, C. R. Horne, T. J. Collins et al., Chemical Physics Letters **302** (12), 119 (1999).
6. P Glatzel, U Bergmann, F. M. F. de Groot et al., in *X-Ray and Inner-Shell Processes*, edited by A. Bianconi and A. Marcelli (American Institute of Physics, Rome, 2002), Vol. 652, pp. 250.
7. J. A. van Bokhoven, C. Louis, J. T. Miller et al., Angewandte Chemie - International Edition in English **45**, 4651 (2006).
8. J. Yano, Y. Pushkar, P. Glatzel et al., J. Am. Chem. Soc. **127** (43), 14974 (2005).
9. P. Glatzel, F. M. F. de Groot, O. Manoilova et al., Phys. Rev. B **72** (1) (2005).
10. P. Glatzel, U. Bergmann, W. W. Gu et al., J. Am. Chem. Soc. **124** (33), 9668 (2002).
11. P. Glatzel, U. Bergmann, J. Yano et al., J. Am. Chem. Soc. **126** (32), 9946 (2004).
12. X. Wang, F. M. F. deGroot, and S. P. Cramer, Physical Review B **56** (8), 4553 (1997).

Microreactor Cells for High-Throughput X-ray Absorption Spectroscopy

Angela Beesley [a], Nikolaos Tsapatsaris [a], Norbert Weiher [a], Moniek Tromp [b], John Evans [b], Andy Dent [c], Ian Harvey [d], Sven L. M. Schroeder [a,e*]

[a]School of Chemical Engineering and Analytical Science, The University of Manchester, Sackville Street, P.O. Box 88. Manchester, M60 1QD, UK.
[b]School of Chemistry, The University of Southhampton, Highfield, Southampton, SO17 1BJ, UK.
[c]Diamond Light Source Ltd., Diamond House, Chilton, Didcot, Oxfordshire, OX11 0DE, UK.
[d]Synchrotron Radiation Source (SRS), Daresbury Laboratory, Warrington, Cheshire, WA4 4AD, UK
[e]School of Chemistry, The University of Manchester, Sackville Street, P.O. Box 88. Manchester, M60 1QD, UK.

Abstract. High-throughput experimentation has been applied to X-ray Absorption spectroscopy as a novel route for increasing research productivity in the catalysis community. Suitable instrumentation has been developed for the rapid determination of the local structure in the metal component of precursors for supported catalysts. An automated analytical workflow was implemented that is much faster than traditional individual spectrum analysis. It allows the generation of structural data in quasi-real time. We describe initial results obtained from the automated high throughput (HT) data reduction and analysis of a sample library implemented through the 96 well-plate industrial standard. The results show that a fully automated HT-XAS technology based on existing industry standards is feasible and useful for the rapid elucidation of geometric and electronic structure of materials.

Keywords: High-throughput, XAS, XANES, EXAFS, catalysis.
PACS: 61.10.Ht, 81.16.Hc, 82.65.-s.

INTRODUCTION

High throughput (HT) approaches can potentially increase research productivity considerably. HT methods developed originally in the context of pharmaceutical research [1] are currently extended to other research areas [2], and have become accepted in the catalysis community in recent years [3-4]. Detailed studies of catalytic materials using high-throughput spectroscopic techniques such as Infrared (IR) [5], Raman [6], X-ray Fluorescence (XRF) [7], Fluorescence Microscopy [8], Imaging Polarimetry [9] and Nuclear Magnetic Resonance (NMR) [10], have been reported. In catalysis it is crucial to establish novel routes that facilitate the discovery of new catalysts and the correlation of reaction outcomes with the atomic structure of the chemical species involved. Investigating the nature of catalytic precursors and their configuration with the support materials is the first step for determining catalyst stability, activity and selectivity in a particular chemical reaction [11-13]. X-ray absorption spectroscopy (XAS) is a synchrotron-based spectroscopic technique that provides this information by probing the local coordination geometry and electronic structure around X-ray-absorbing atoms [14]. It can investigate samples under in-situ reaction conditions and in any aggregation state and therefore is suitable for high-throughput experimentation.

We have developed XAS end-station instrumentation that allows the rapid elucidation of the metal local structure in a library of catalytic precursors based on a 96-wellplate sample array. This library contains 91 supported catalyst precursors in a ternary system Cu/Pt/Au prepared by impregnation of γ-Al$_2$O$_3$ with aqueous CuCl$_2$, PtCl$_2$ and HAuCl$_4$ solutions. Here are presented the results from the automated HT data reduction and analysis of XAS spectra taken at the Cu K-edge. Results from the data analysis of spectra taken at Au L$_{III}$ and Pt L$_{III}$ will be extensively described in a forthcoming publication [15]. This HT spectral analysis was carried out using a set of scripts utilizing the IFEFFIT library, which provided values of coordination number, first shell distances and Debye-Waller factors

CP879, Synchrotron Radiation Instrumentation: Ninth International Conference,
edited by Jae-Young Choi and Seungyu Rah
© 2007 American Institute of Physics 978-0-7354-0373-4/07/$23.00

for all samples. Phase changes were identified yielding valuable knowledge on the ternary catalyst precursors system.

A feasibility study on the inclusion of 2D correlation analysis of XAS spectra [16] as a feedback mechanism to the automated data acquisition system is also presented.

EXPERIMENTAL

Library Preparation

Catalyst precursors were prepared by wet impregnation of $CuCl_2$ (Aldrich, 99.995%), $PtCl_2$ (Aldrich, 98%) and $HAuCl_4$ (Riedel-de Haën, 51% Au) on $\gamma-Al_2O_3$ (D1011, BASF) support (200-450 μm particle size). A standard 96 well plate was used to prepare and store the samples. The library comprises all possible combinations of Cu, Pt and Au in the concentrations 0, 0.1, 1 and 5 wt% leading to a total of 64 elements; 27 additional elements were added as permutations of Cu, Pt and Au in concentrations of 0.1, 1 and 10 wt%, resulting in a library of 91 elements. Each well containing the metal solution (200 μl) was filled with 200 mg of $\gamma-Al_2O_3$ corresponding to the pore volume of the support. The samples were then left to dry at room temperature. Figure 1a shows the Cu concentration in wt % for each of the library members in the 96 well-plate.

High-Throughput XAS Data Acquisition and Analysis

XAS experiments were carried out at station 9.3 of the SRS in Daresbury Laboratory, UK [17]. Cu K-edge spectra were collected in fluorescence-yield mode using a Canberra 13 element Ge solid state detector. The EXAFS region was scanned in the k-scan mode from 2 to 14 $Å^{-1}$ using k-steps of 0.04 $Å^{-1}$. In the pre-edge region, data were recorded on an energy grid of 22 eV/step and with 1.0 eV/step in the XANES region. These parameters resulted in data acquisition times of 9 min per spectrum. An energy calibration of the beam was performed by collecting the transmission spectra of Cu reference foil at the Cu K-edge. An in-house built XYZ translation stage (Parker) was used for sample positioning. Data acquisition and automatic movement of the XYZ stage as well as synchronization with the station main computer through the TCP/IP protocol were achieved using a LabVIEW control system. The software allows the scan of a variety of library sizes, in several positions and angles with respect to the fluorescence detector.

Four individual spectra containing 5% wt of Au, Pt and/or Cu samples were initially analyzed and suitable starting parameters for background subtraction, normalization, k-weighting, Fourier transformation, edge-position, coordination number (N), nearest neighbour distance (R) and Debye-Waller factor (σ^2) were determined. These values were then used as a priori information to the HT analysis software. This software is based on a series of Linux shell scripts which reads sequentially all the spectra in a subdirectory, applies the IFEFFIT background subtraction routine and performs first-shell fitting analysis according to a prescribed physical model. Theoretical amplitudes and phase functions used in the fitting procedure were calculated with FEFF6 [18]. All the fits were performed using multiple k-weightings of 1, 2 and 3. The script reports the shift on the edge position, the edge-step (ES) values, the normalized data, the $\chi(k)$ data, the Fourier transformed data, the fitted curves and the resulting fitted values for N, R and σ^2. This software routine performs single-shell EXAFS analysis on 96 samples at three different edges (i.e., a total of 288 spectra in approximately 20 minutes. In order to improve the versatility of the HT system, a feasibility study was carried out based on the inspection of 2D correlation maps from the collected spectra. These maps were built based on 2D XAS correlation analysis reported by Haider [16] and using the 2Dshige software [19].

RESULTS AND DISCUSSION

Some of the results from the HT-XAS analysis of spectra taken at the Cu K-edge are shown in Figs. 1b and 1c. These figures present a map of the Cu for one particular XAS parameter, namely the edge-step (Fig. 1b) and coordination number (Fig. 1c). The color black in the figures indicates either the absence of the metal under study or spectra with an extremely low S/N ratio. All results of the automated analysis were additionally cross-checked individually, and it was found that the information provided by the maps was reliable.

Most samples follow the expected strong correlation between concentration (Fig. 1a) and edge step (Fig. 1b), which supports the reliability of the HT analysis. In most cases, fits for the Cu K-edge are in line with the 4-fold

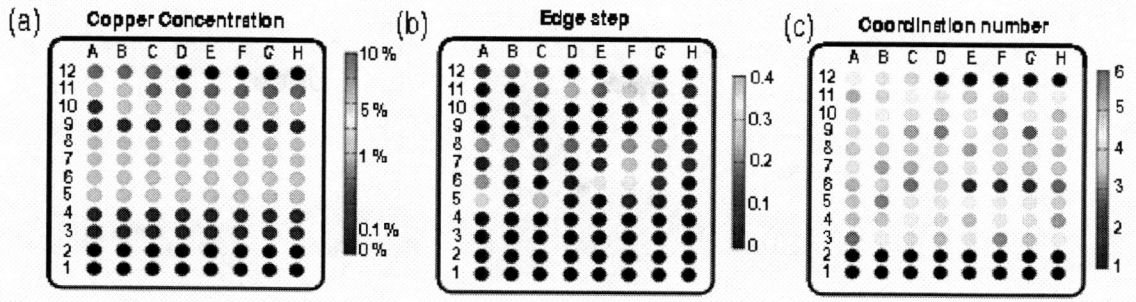

FIGURE 1. Color maps representing (a) the known copper loading (wt%) of the catalysts, (b) the edge step values and (c) the coordination numbers found from the HT-XAS analysis.

coordination of Cu found in CuCl$_2$ [20]. From immediate visual inspection one can observe high ES values (Fig. 1b) for samples E6 and F6 associated to 1 wt% Cu metal. This may be due to an uneven distribution of Cu salt on the support or to irregularities in the synthesis procedure. In addition the N color map (Fig. 1c) clearly shows that E6, F6 and G6 present extremely low coordination numbers with non-physical significance. These samples were further analyzed individually leading to a metal local structure in line with the 12-fold coordination found in Cu metal. Figure 2a shows the XANES spectra of sample A7 corresponding to CuCl$_2$ as expected from the impregnation process, sample E6 corresponding to reduced Cu structure and Cu foil shown for comparison purposes. The reduction is believed to take place at the expense of Pt(II), which oxidizes to Pt(IV), as indicated by an increase in the intense near-edge feature ("white line") in the Pt LIII XANES spectra [21]. This is illustrated in Fig. 2b, which compares spectra from samples E6 and E2; the latter contains only PtCl$_2$ impregnated on alumina and therefore the XANES spectrum presents a reduced white line feature. Figure 2b also shows the Pt foil XANES spectrum for comparison.

The selected results described above demonstrate that HT XAS analysis is an efficient tool for rapid structural phase identifications and may be used in the future for the determination of structural changes under reaction conditions.

Figure 3 shows the 2D XAS synchronous correlation maps for the normalized spectra under discussion. Figure 3a corresponds to the synchronous correlation spectra of samples A7 and E6. One can identify peaks corresponding to spectral variations between the two Cu K-edge spectra. This is in line with the difference in spectral features associated to different local structures i.e. Cu metal and CuCl$_2$. Figure 3b shows the 2D XAS synchronous correlation spectra for samples E6, F6 and G6. In contrast to Fig. 3a, spectral correlations are found that appear as a blurry image with "crossing lines" that are associated with noise in the data [22]. This sole presence of noise in this figure is expected, as all three samples presented the same 12-fold Cu metal structure. Finally Fig. 3c shows the correlation map of two identical spectra (both from sample A7) as an illustration of total correlation. These results show that using 2D XAS correlation analysis as a feedback mechanism may, with minimum effort, (i) improve the rapid data acquisition by discriminating between spectral noise levels and (ii) provide a strategy for the rapid determination of possible phase changes under reaction conditions.

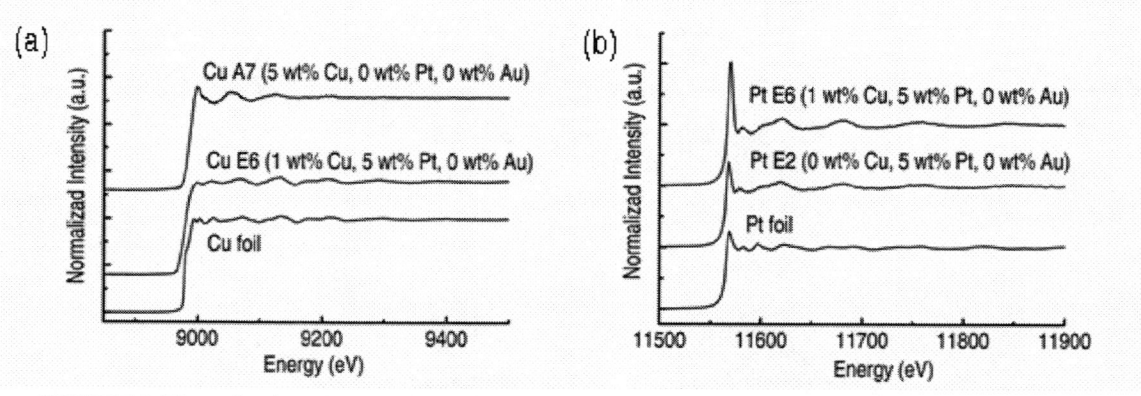

FIGURE 2. Normalized XANES spectra of selected samples taken at the (a) Cu K-edge and (b) Pt L$_{III}$-edge.

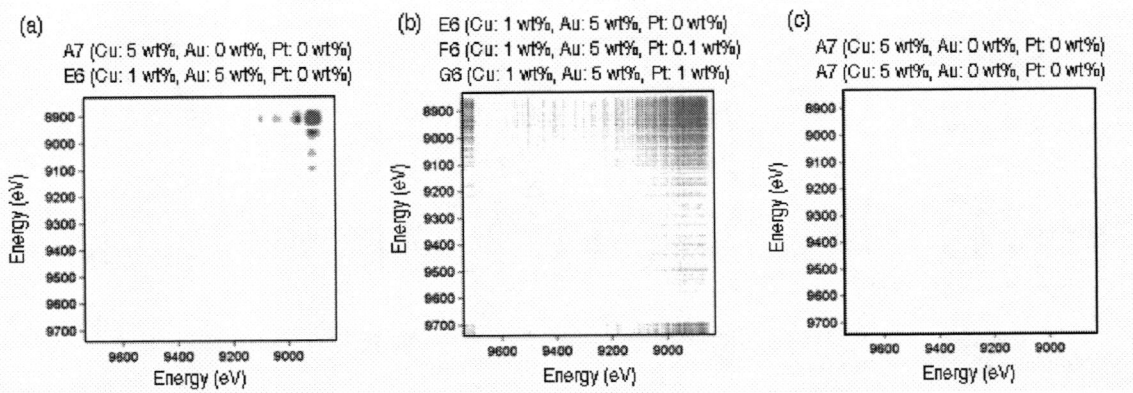

FIGURE 3. 2D XAS correlation spectra taken at the CuK-edge for samples: (a) A7 and E6, (b) E6, F6 and G6 and (c) A7.

CONCLUSIONS

We have demonstrated that HT-XAS can significantly reduce the amount of time needed to acquire and analyze XAS data as compared to standard XAS experimentation, yielding rapid information about correlations between structural parameters and providing fast and reliable structural phase identification.

ACKNOWLEDGMENTS

This work was supported through EPSRC research grants GR/S85801/01 and GR/S85818/01. The CCLRC Synchrotron Radiation Source (SRS) at Daresbury Laboratory is acknowledged for providing beamtime at station 9.3 under award number 44135.

REFERENCES

1. R. A. Houghten, C. Pinilla, S. E. Blondelle, J. R. Appel, C. T. Dooley and C. T. Cuervo, *Nature* **354**, 84-86 (1991).
2. R. J. Hendershot, C. M. Snively and J. Lauterbach, *Chem. Eur. J.* **11**, 806-814 (2005).
3. T. Zech, G. Bohner, O. Laus, J. Klein and M. Fischer, *Rev. Sci. Instrum* **76**, 62215-1-8 (2005).
4. C. Klanner, D. Farrusseng, L. Baumes, C. Mirodatos, F. Schuth, *QSAR. Comb. Sci.* **22**, 729-735 (2003).
5. A. Venimadhav, K. A. Yates, M. G. Blamire, *J. Comb. Chem.* **7**, 85-89 (2005).
6. A. Leugers, D. R. Neithamer, L. S. Sun, J. E. Hetzner, S. Hilty, S. Hong, M. Krause, K. Beyerlein, *J. Comb. Chem.* **5**, 238-244 (2003).
7. T. C. Miller, G. Mann, G. J. Havrilla, C. A. Wells, B. P. Warner, R. T. Baker, *J. Comb. Chem.* **5**, 245-252 (2003).
8. M. M. Taniguchi, R. A. Farrer, J. Fourkas, *J. Comb. Chem.* **7**, 54-57 (2005).
9. P. R. Gibbs, C. S. Uehara, P. T. Nguyen, R. C. Willson, *Biotechnol. Prog.* **19**, 1329-1334 (2003).
10. R. A. Kautz, W. K. Goetzinger, B. L. Karger, *J. Comb. Chem.* **7**, 14-20 (2005).
11. C. L. Bianchi, P. Canton, N. Dimitratos, F. Porta, L. Patri, *Catalysis Today* **102-103**, 203-212 (2005).
12. J. Słoczyński, R. Grabowski, A. Kozlowska, P. Olszewski, J. Stoch, J. Skrzypek, M. Lachowska, *Appl. Catal. A* **278**, 11-23 (2004).
13. M. Ritcher, M. Langpape, S. Kolf, G. Grubert; R. Eckelt, J. Radnik, M. Schneider, M. -M. Pohl, R. Fricke, *Appl. Catal. B* **36**, 261-277 (2002).
14. J. J. Rehr, A. L. Ankudinov, S. I. Zabinsky, *Catal. Today* **39**, 263-269 (1998).
15. N. Tsapatsaris, A. Beesley, N. Weiher, M. Tromp, J. Evans, A. Dent, I. Harvey, S. Hayama, S. L.M. Schroeder to be submitted to *J. Comb. Chem.*
16. P. Haider, Y. Chen, S. Lim, G. L. Haller, L. Pfefferle and D. Ciuparu, *J. Am. Chem. Soc.* **127**, 1906-1912, (2005).
17. Dent, A. J.; Derst, G.; Van Dorssen, G. E.; Mosselmans, J. F. M. Daresbury Synchrotron Radiation Source (SRS), Manual Station 9.3. http://www.srs.dl.ac.uk/srs/stations/station9.3.htm.
18. S. I. Zabinsky, J. J. Rehr, A. Ankudinov, R. C. Albers, M. Eller, *Phys. Rev. B* **52**, 2995–3009 (1995).
19. 2Dshige (c) Shigeaki Morita, Kwansei-Gakuin University, 2004-2005.
20. P. C. Burns, F. C. Hawthorne, *Am. Mineral.* **78**, 187-189 (1993).
21. R. Ayala, M. E. Sanchez, S. Diaz-Moreno, V. A. Sole, A. Munoz-Paez, *J. Phys. Chem. B.* **105**, 7588-7593 (2001).
22. M. A. Czarnecki, *Appl. Spectrosc.* **52**, 1583-1590 (1998).

High Throughput *In Situ* EXAFS Instrumentation for the Automatic Characterization of Materials and Catalysts

Nikolaos Tsapatsaris[¶], A. M. Beesley[¶], Norbert Weiher[¶], Moniek Tromp[‡], John Evans[‡], A. J. Dent[§], Ian Harvey[†], Sven L. M. Schroeder*[,¶¶]

[¶]*School of Chemical Engineering and Analytical Science, Molecular Materials Centre, The University of Manchester, Sackville Street, P.O. Box 88. Manchester, M60 1QD, UK.*
[‡]*School of Chemistry, The University of Southhampton, Highfield, Southampton, SO17 1BJ, UK.*
[§]*Diamond Light Source Ltd., Diamond House, Chilton, Didcot, Oxfordshire, OX11 0DE, UK.*
[†]*Synchrotron Radiation Source (SRS), Daresbury Laboratory, Warrington, Cheshire, WA4 4AD, UK*

Abstract. An XAS data acquisition and control system for the *in situ* analysis of dynamic materials libraries under control of temperature and gaseous environment has been developed. It was integrated at the SRS in Daresbury, UK, beamline 9.3, using a Si (220) monochromator and a 13 element solid state Ge fluorescence detector. The core of the system is an intelligent X, Y, Z, θ positioning system coupled to multi-stream quadrupole mass spectrometry analysis (QMS). The system is modular and can be adapted to other synchrotron radiation beamlines. The entire software control was implemented using Labview and allows the scan of a variety of library sizes, in several positions, angles, gas compositions and temperatures with minimal operator intervention. The system was used for the automated characterization of a library of 91 catalyst precursors containing ternary combinations of Cu, Pt, and Au on γ-Al_2O_3, and for the evaluation and structural characterization of eight Au catalysts supported on Al_2O_3 and TiO_2. Mass spectrometer traces reveal conversion rate oscillations in 6wt % $Au/\gamma Al_2O_3$ catalysts. The use of HT experimentation for *in situ* EXAFS studies demonstrates the feasibility and potential of HT *in situ* XAFS for synchrotron radiation studies.

Keywords: High throughput, Combinatorial chemistry, XAS, Catalysis, Au, γ-Al_2O_3, SBS standard.
PACS: 07.85.Qe, 82.65.–s, 61.10.Ht, 81.16.Hc

INTRODUCTION

The development of intelligent and modular instrumentation is imperative for the identification of catalysts and materials with new target functionalities. Integration of many analytical techniques under a single high throughput (HT) experiment enables detailed screening of numerous candidates and reduces the time scale of the experiments [1]. Parallel experimentation increases the probability of significant discoveries by revealing trends in complex data sets. It increases efficiency and allows for significant reduction of research costs. HT methods have recently become more accepted in the catalytic community, with combinatorial experimentation a promising research avenue for the discovery of new catalysts and optimization of their yield and selectivity [2-4]. The analysis of catalytically active materials requires the use of various scientific methods and experimental setups. Discovering a promising catalyst requires synthesis and *in situ* screening of sufficiently large, statistically significant libraries. Each catalyst candidate can be defined by a multitude of descriptors (e.g., its constituents, structure, synthetic parameters). which relate to its activity and selectivity and the reaction mechanism.

Various catalyst screening methods have been reported in the literature. Besides kinetic reaction studies, methods such as Infrared spectroscopy [5], Raman spectroscopy [6], X-ray Fluorescence [7], Fluorescence Microscopy [8], Imaging Polarimetry [9], Nuclear Magnetic Resonance spectroscopy [10] and XRD [11] have been incorporated as part of high-throughput setups. These techniques are valuable for characterizing large sample libraries but have limitations, e.g. difficult *in situ* realization (NMR) or the restriction to crystalline materials (XRD). In catalysis and materials research, EXAFS is used routinely for characterization of the electronic and local

CP879, *Synchrotron Radiation Instrumentation: Ninth International Conference*,
edited by Jae-Young Choi and Seungyu Rah
© 2007 American Institute of Physics 978-0-7354-0373-4/07/$23.00

geometric structure and for revealing changes in the chemical environment during catalytic reactions [12]. In contrast to the expanding use of HT X-ray diffraction (XRD) [13], the flexibility and inherent advantages of X-ray absorption spectroscopy (XAS) have not been utilized in HT research and present new and promising research ground [14]. XAS characterisation is usually divided into an interpretation of two prominent X-ray absorption fine-structure (XAFS) regions in the spectra. The X-ray absorption near-edge structure (XANES) probes the local electronic structure at the site of the X-ray absorbing atoms, whereas the extended X-ray absorption fine-structure (EXAFS) provides structural information, including bond-lengths, coordination numbers as well as static and thermal disorder.

MATERIALS AND METHODS

Sample Preparation

The catalysts investigated in this system contained different concentrations of Au on Al_2O_3 and TiO_2 supports. Au/Al_2O_3 was prepared using a modified incipient wetness method, 100 mg of $HAuCl_4$ were dissolved in 5.1 ml distilled water, 10 g of sieved and dried Al_2O_3 (Condea, 0.125-0.250 μm) were added and the system was vigorously shaken for one hour. The resulting gray powder was poured into 200 ml of distilled water and heated to 340 K. The pH was kept at 8 using 1M NaOH. After two hours of stirring, the solid was filtered, washed with hot water and dried in vacuum at room temperature for 48 hours. Elemental analysis reported a chlorine content below the detection limit (<0.2 wt%). Au/TiO_2 was prepared by deposition-precipitation. 400 mg $HAuCl_4$ were dissolved in 200 ml distilled water. The pH was adjusted to 7 using 1M NaOH. 10 g of TiO_2 (P25, Degussa) were added and the suspension was heated to 340 K. After two hours of stirring, the solid was filtered and washed with hot water. The resulting powder was dried in vacuum at room temperature for 48 hours. Elemental analysis reported a chlorine content below the detection limit (<0.2 wt%).

HT EXAFS Screening

Our HT XAS data acquisition and control system comprises of an XYZθ–positioning stage that allows precise positioning of highly compact arrays (> 1000 discrete materials/cm^2). Individual gas flows are supplied via independent mass flow controllers (MFC); the effluent of each cell is monitored by quadrupole mass spectrometry (MS). XAS experiments were carried out at station 9.3 at the SRS in Daresbury, UK [15]. The XANES spectra were collected in fluorescence mode using a 13 element Ge solid state detector. Acquisition times were 15 minutes per XANES spectrum. The energy axis was calibrated with a transmission Au L_3-edge spectrum of a Au reference foil. The system was initially tested (without mass spectrometric measurements) using a library of 91 candidates consisting of ternary Al_2O_3 supported powder catalyst precursors containing Cu, Pt, and Au [16]. For the present experiments a custom made 8-fold microreactor was used that permitted additional investigations while monitoring the individual conversions using a QMS through 1/16" connections (Omnifit) Au L_{III} XANES spectra were obtained from the catalysts during temperature programmed reduction (TPR) under CO (Purity 3.8, BOC) atmosphere. Subsequently the samples were exposed to various $CO:O_2$ (99.99%): He (99.99%) mixtures and spectra were acquired *in situ*. Data reduction was performed using the IFEFFIT library and the corresponding front ends ATHENA & ARTEMIS [17].

RESULTS & DISCUSSION

Figure 1(left) demonstrates interrelations between the different components of the HT setup and the beamline. Positioning control was accomplished using a ±10 μm step 4 degree of freedom mechanical apparatus composed of three linear stages (Parker) and a 1 circle segment (Huber). The stages were stacked in three vertical planes providing range of movement of 400x200x300 mm (x, y, z) and ±10° rotation around the y axis. The mechanical setup was mounted on the existing optical table of the SRS station and required virtually no setup changes. The sample library was positioned in such way so that the procedure of loading samples is more efficient by instructing the software to position the *in situ* cell closer to the operator. Additional adjustments were made to the design to accommodate for space restrictions in the beamline, for example the small distance of the hutch wall to the X-ray beam (125 mm). The gas and MFC manifold (Bronkhorst) permitted parallel control of four gases (CO, O_2, He, H_2)

to within 1% at the maximum flow 100 ml/min/gas through the *in situ* cell. Stainless steel tubing (1/4") was used for interconnecting the MFCs and gas cylinders. Various protocols, including the industry-standard Profibus were incorporated for flexibility and expandability (Fig. 2 right). Combining the RS232 and CAN protocols and a custom made driver allowed communication with the positioning stages and the rest of the components of the HT setup. Data acquisition cards were used to provide the necessary analogue and digital lines to the MFCs and solenoid valves. Synchronization with the station control computer was provided via a TCP/IP network connection with a custom made handshake protocol. The protocol contains commands for the start, stop and mode of EXAFS scans, adjusting monochromator position, scan downloading and also more detailed calibration routines. The main control computer hosts an in-house LABVIEW program that is responsible for automatic positioning of the 8 well array in any position or angle with respect to the X-ray beam, allowing accurate measurements of XAFS data for each individual catalyst or material. The program is built around libraries responsible for the individual control of every component such as multiple temperature control, valve manifolds, mass flow controllers, yet maintaining compatibility with the various instrumentation infrastructures present at different synchrotron radiation beamlines.

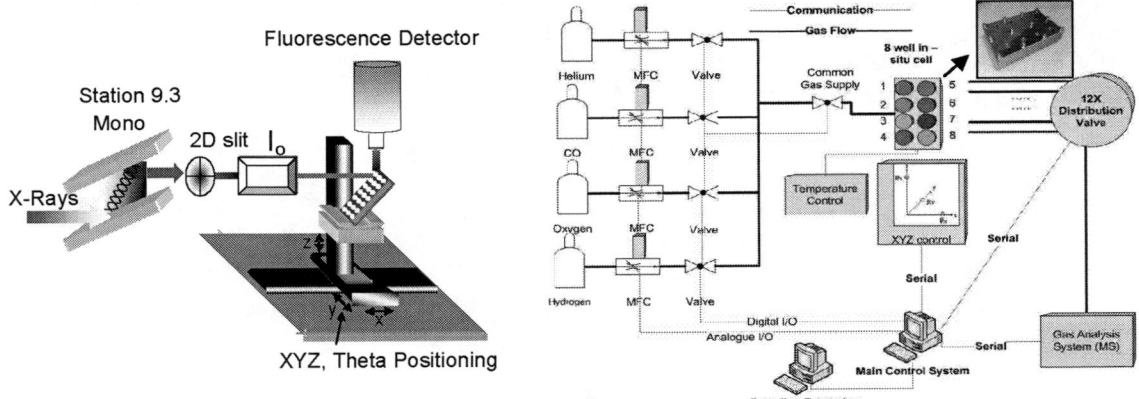

FIGURE 1. Integration of HT *in situ* EXAFS instrumentation in Station 9.3 of SRS Daresbury. Left: Robotic stage with respect to the station detector and X-ray beam, Right: Main system components and communication interface with beamline. The inset shows the custom made 8 well reactor.

Initial examination of the XANES spectra of Au catalysts revealed saturation effects at high metal concentrations. A linear correlation between edge step and nominal Au concentration was found up to a metal loading of approximately 4 wt%. The spectral distortions at higher Au concentrations can be attributed to self-absorption effects and/or a non linear response of the detector. The development of the chemical state of the Au component during temperature programmed reduction (TPR) of one of the catalysts in the 8 well reactor can be seen on Fig 2 (left). It is seen how Au^{3+} was reduced in an (1/1) CO/He stream, as indicated by a decrease of the near-edge feature at 11925 eV [18]. It shall be noted that the material was already partially reduced before exposure to CO as indicated by the low whiteline intensity. Fig 2 (right) shows MS transients of O_2, CO, He and CO_2 during CO

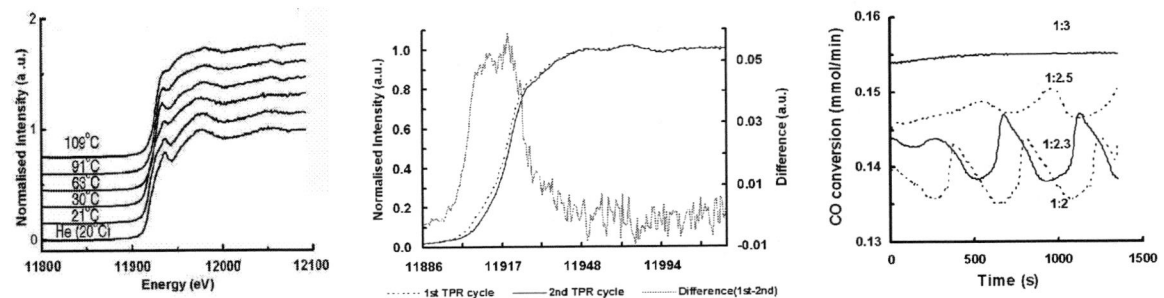

FIGURE 2. Left: Au L_{III} XANES series of Temperature Programmed Reduction of 4% Au/Al_2O_3 on a CO:He (1:1) stream. Centre: XANES Spectra of 4% Au/Al_2O_3 during CO oxidation with 1:2 $CO:O2$ after two reduction cycles. Oscillations on this sample were observed only after the first reduction cycle. Right: Non linear CO conversion response of 6% Au/Al_2O_3. Sustained oscillations were observed at CO/O_2 ratios between 0.4 and 0.625.

oxidation over a 1% Au/Al₂O₃ catalyst. An oscillatory response of CO conversion during CO oxidation was first identified on a 4% Au/Al₂O₃ catalyst (Fig. 2, center) under a CO:O 1:2 gas mixture and after a deactivation period of 6 hours. Reproduction of the short-lived oscillatory response was not successful after a 2nd TPR cycle. The difference XANES spectrum shown in Fig. 2 (center) indicates the presence of a -1 eV edge shift between the oscillating and steady state catalysts. The appearance of a shoulder at -3 eV could be attributed to weak oxidation by adsorbed gas species and is in agreement with previous observations [19]. Conversion oscillations similar to those observed here were previously observed in single crystal transition metal catalysts, e.g. for Pt, Rh and Ir [20-24]. Usually, oscillations are related to reduction/oxidation cycles, resulting, for example, in non linear rate phenomena arising from oxide or sub-surface layers that deactivate slowly the reactive surface [24]. Preliminary XANES analysis indicates that such oxidation/reduction cycles maybe present on the Au component of our catalysts (Fig. 2, center). The 15 min delay in the acquisition of XANES spectra caused by the acquisition electronics of the fluorescent detector as well as during the experiments did not allow for resolving the chemical state of the catalyst within an oscillation cycle.

In summary, integration of the HT system in the synchrotron station facilitated the rapid acquisition of *ex situ* and *in situ* XAS data with parallel MS, gas-flow and temperature control. This parallel control system has already facilitated the discovery of CO oxidation rate oscillations over Au catalysts. The results demonstrate the advantages of HT experimentation when used *in situ* EXAFS studies.

ACKNOWLEDGMENTS

This work was supported through EPSRC research grants GR/S85801/01 and GR/S85818/01. The CCLRC Synchrotron Radiation Source (SRS) at Daresbury Laboratory is acknowledged for providing beamtime at station 9.3 under award number 44135.

REFERENCES

1. R. Hoogenboom, F. Wiesbrock, M A.M. Leenen, M A.R. Meier, U S. Schubert, *J. Comb. Chem* **7**, 10-13 (2005)
2. R.A. Potyrailo, W.J. Morris,R.J. Wroczynski, P.J. McCloskey, *J. Comb. Chem* **9**, 869-873 (2004)
3. A. Potyrailo, R. J. Wroczynski, J. P. Lemmon, W. P. Flanagan, O. P. Siclovan, *J. Comb. Chem* **5**, 8-17 (2003)
4. R. Malhotra, Ed. Combinatorial Approaches to Materials Development; *ACS Symposium Series* **814**; American Chemical Society: Washington DC, 2002.
5. A. Venimadhav, K.A. Yates, M. G. Blamire, *J. Comb. Chem* **7**, 85-89 (2005)
6. A. Leugers, David R. Neithamer, Larry S. Sun, Jack E. Hetzner, Sean. Hilty, Sam Hong, Matthew Krause,Kenneth. Beyerlein, *J. Comb. Chem* **5**, 238-244 (2003)
7. T. C. Miller, Grace Mann, George J. Havrilla, Cyndi A. Wells, B. P Warner, Tom R, Baker, *J. Comb. Chem* **5**, 245- 252 (2003)
8. M. M. Taniguchi, R.A. Farrer, J. Fourkas, *J. Comb. Chem* **7**, 54-57 (2005)
9. P. R.Gibbs, C. S. Uehara, P. T. Nguyen, R.C. Willson, *Biotechnol. Prog.* **19**, 1329-1334 (2003)
10. R.A. Kautz, W.K Goetzinger, B.L. Karger, *J. Comb. Chem* **7**, 14-20 (2005)
11. J.H. Reibenspies, N.S.P. Bhuvanesh, *Journal of Pharmaceutical and Biomedical Analysis* **37**, 611-614 (2005)
12. J. J. Rehr, A. L. Ankudinov, S. I. Zabinsky, *Catal. Today* **39**, 263 (1998).
13. A. E. Russell, A. Rose, *Chemical Reviews* **104**, 4613 (2004)
14. A. Corma, J. M. Serra, P. Serna, M. Moliner, *Journal of Catalysis* **232**, 335 (2005)
15. A. J. Dent, G. Derst, G. E. Van Dorssen, J. F. M. Mosselmans, Daresbury Synchrotron Radiation Source (SRS), Manual Station 9.3. http://www.srs.dl.ac.uk/srs/stations/station9.3.htm.
16. N. Tsapatsaris, A. Beesley, N. Weiher, M. Tromp, J. Evans, A. Dent, I. Harvey, S. Hayama, S. L.M. Schroeder to be submitted to *J. Comb. Chem*.
17. M. J Newville, *J. Synchrotron Rad* **8**, 322 (2001)
18. N. Weiher, E. Bus, L Delannoy, C. Louis, D. E. Ramaker, J. T. Miller, J. A. van Bokhoven, *Journal of Catalysis* **240**, (2), 100-107 **(**2006)
19. J. A. van Bokhoven, C. Louis, J. T. Miller, M. Tromp, O. V. Safonova, P. Glatzel, *Angew. Chem. Int. Ed.*, DOI: 10.1002/anie.200123456 in press
20. P. A. Carlsson et al, *Journal of Catalysis* **226**, (2), 422-434 (2004)
21. R. Danielak, A. Perera, M. Moreau, M. Frankowicz, R. Kapral *Physica A* **229**, (3-4), 428-443 (1996)
22. V. V. Gorodetskii, V. I. Elokhin, J. W. Bakker, B. E. Nieuwenhuys, *Catalysis Today* **105**, (2), 183-205 (2005)
23. L. Li, W. Wlodarski, K. Galatsis, D. A. Powell, *Sensors and Actuators B: Chemical* **96**, (3), 610-614 (2003)
24. C. D. Lund, C. M. Surko, M. B. Maple, S. Y. Yamamoto, *Surface Science* **459**, (3), 413-425 (2000)

Anomalous X-Ray Scattering Using Third-Generation Synchrotron Radiation

S. Hosokawa[A,B] and J.-F. Bérar[C]

[A] Center for Materials Research Using Third-Generation Synchrotron Radiation Facilities, Hiroshima Institute of Technology, Hiroshima 731-5193, Japan
[B] Institut für Physikalische-, Kern-, und Makromolekulare Chemie, Philipps Universität Marburg, D-35032 Marburg, Gemany
[C] Laboratoire de Cristallographie, CNRS, F-38042 Grenoble Cedex, France

Abstract. In this paper, we discuss the recent development of anomalous X-ray scattering (AXS) technique as a tool of investigating local structures of non-crystalline materials using a third-generation synchrotron radiation facility, ESRF. In order to obtain differential structure factors with a high statistical quality, it is necessary to acquire scattering data with a good energy resolution to discriminate elastic signals from fluorescence and Compton scattering intensities, as well as with a sufficient number of scattered X-ray photons. For this we chose a single-crystal graphite energy-analyzer with a long detector arm. In order to show the feasibility of this detecting system, we describe in detail examples of our recent AXS results on As_2Se_3 chalcogenide glass and $(As_2Se_3)_{0.4}(AgI)_{0.6}$ superionic glass.

Keywords: Intermediate-range order; Short-range order; Local structure; Synchrotron radiation; Chalcogenide glass
PACS: 61.10.Nz; 61.43.Fs; 66.30.Hs

INTRODUCTION

More than two decades have already passed through since anomalous X-ray scattering (AXS) using synchrotron radiation was expected as a promising tool for studying partial atomic structures in non-crystalline materials [1]. A similar method using synchrotron X-ray sources, X-ray absorption fine structure (XAFS), was started to be improved at almost the same period, and already becomes a well-established technique for studying local structures around a specific element. On the contrary, AXS technique is still not widely used for the partial structure analysis, although compared to the XAFS, this technique has a large advantage that information on intermediate-range atomic correlations is included. However, the corrections for the Compton and fluorescence contributions are complex and time-consuming, and the anomalous terms are hard to experimentally determine correctly.

Unfortunately, recent developments in synchrotron radiation sources have not helped to overcome the above difficulties in carrying out the AXS experiments. We believe that the main reason is only on the detecting system. Pure Ge solid-state-detector (SSD) has commonly been used as a detector for AXS experiments using old-generation synchrotron radiation facilities. This detector is very sensitive, and can collect energy-resolved data with the resolution of some hundred eV, which unfortunately requires the above-mentioned complicated corrections. Moreover, the SSD detector has a long blind time of some microseconds. For this reason, high flux incident X-rays from third-generation synchrotron radiation facilities never help in improving the statistic quality of AXS data, or even bring other new and complicated corrections.

Instead of the SSD detector, we chose a single-crystal graphite energy-analyzer with a long arm for a scintillation detector. Using this improved detecting system, we have recently carried out the AXS experiments at the beamline BM02 of the ESRF on several chalcogenide glasses [2,3], room-temperature superionic glasses [4,5], and bulk metallic glasses [6]. In this paper, we describe in detail the new detecting system, and show examples of our recent AXS results on As_2Se_3 chalcogenide glass and $(As_2Se_3)_{0.4}(AgI)_{0.6}$ superionic glass.

CP879, *Synchrotron Radiation Instrumentation: Ninth International Conference*,
edited by Jae-Young Choi and Seungyu Rah
© 2007 American Institute of Physics 978-0-7354-0373-4/07/$23.00

EXPERIMENTAL PROCEDURES

The AXS technique utilizes an anomalous change of an atomic form-factor of a specific element near an absorption edge of the element. An atomic form-factor of an element is given as

$$f(Q,E) = f_0(Q) + f'(E) + if''(E)$$

where f_0 is the usual energy-independent term, and f' and f'' are the real and imaginary parts of the anomalous contributions, respectively. In general, f of an element is governed by the Q-dependent $f_0(Q)$ in a normal X-ray scattering process. Near the absorption edges of the constituent element, however, the energy-dependent anomalous terms, f' and f'', become prominent. Fig. 1 shows examples of theoretical f' and f'' functions of As and Se elements at energies near the K absorption edges [7]. As seen in the figure, f' has a large negative minimum and f'' has an abrupt jump near the corresponding absorption edge energies of each element. If two X-ray scattering experiments are carried out at energies far from and very close to the absorption edge of i-th element as seen by circles in the figure, a differential structure factor, $\Delta_i S(Q)$, can be obtained by the difference of the scattering intensities, which highly enhances the contribution of the i-th element-related partial structures and suppresses the other partials.

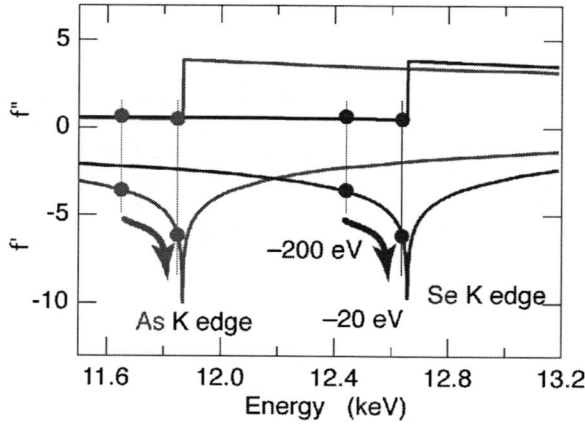

FIGURE 1. The real and imaginary parts of anomalous terms in atomic form-factors of As and Se near the K absorption edges obtained by theory [7].

As mentioned in the introductory section, in order to obtain $\Delta_i S(Q)$ with a high statistical quality, two requirements should be fulfilled: 1) A good energy resolution to discriminate the elastic signal from the fluorescence and Compton scattering contributions, and 2) a sufficient number of scattered X-ray photons within a reasonable data collection time. Due to the relatively bad energy resolution and the long blind time, the SSD is not enough feasible as a detector for the AXS experiments using third-generation synchrotron radiation facilities. The use of single-crystal energy-analyzer, such as Ge and InSb, is another candidate for the detecting system, having a very good energy resolution of some eV. However, the count rate is very weak and it takes more than one week for each Q scan.

Instead, we are using a graphite crystal energy-analyzer with a long detector arm for a scintillation detector. Figure 2(a) shows the schematic view of setup for measuring AXS on non-crystalline mixtures. The graphite crystal provides a good Bragg reflection of the scattered X-ray photons. The distances from the analyzer crystal to the sample and to the receiving slits for the detector were equivalent and long enough (40 cm) to allow a good resolution of 60 eV full-width at half-maximum. Using this detecting system, the elastic signal can be discriminated from the Compton and fluorescence contributions. Typically more than one million counts can be acquired at the $S(Q)$ maximum for at most eight hours per scan. In addition, the typical duration of the data analysis for one sample is much less than one day, while that using the old SSD system takes more than two weeks.

The solid curves in Fig. 2(b) show energy-scan curves obtained from this detector system at incident X-ray energies close to the Se K edge (−20 and −200 eV) by simultaneously changing the θ-2θ angles of the energy analyzer crystal and the detector arm. For this scan, the energy resolution was about 60 eV full-width at half-maximum. The dashed curves are enlarged ten times to clearly show the Se K_β fluorescence and Compton scattering intensities. As seen in the figure, these contributions can be estimated to be less than 0.2 % at energies where the elastic spectra were measured (arrows in Fig. 2(b)). Nevertheless we performed such energy scans at several Q values to estimate these contributions for the data reduction. Then, we directly derived $\Delta_i S(Q)$ spectrum by taking the difference between two diffraction data sets around each absorption edge using the procedure given in the literature [2]. Incident X-ray energies were calibrated using the absorption edge of each element.

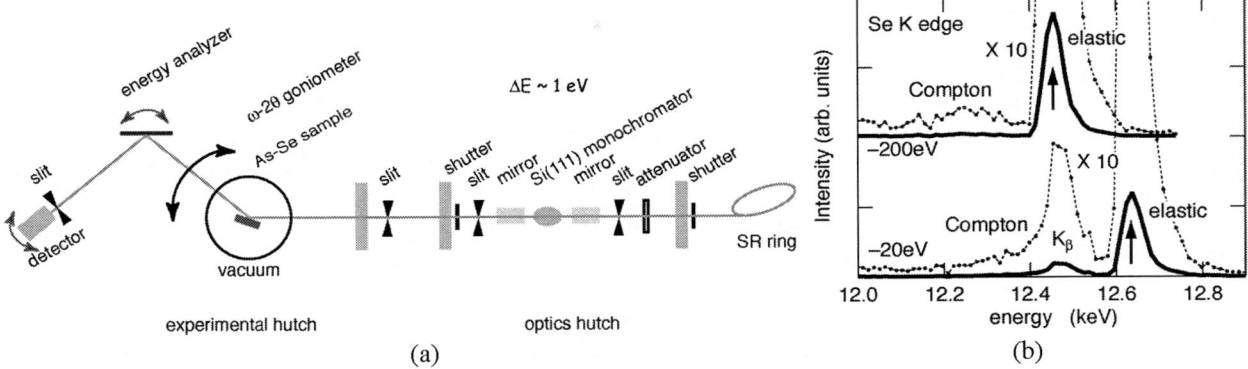

(a)   (b)

FIGURE 2. (a) Schematic diagram of the AXS diffractometer, and (b) energy-scan curves of the present detector system at incident X-ray energies near the Se K edge (-20 and –200 eV). From T. Usuki et al. [4].

RESULTS AND DISCUSSION

Figure 4(a) shows $\Delta_i S(Q)$ spectra of As_2Se_3 chalcogenide glass measured at energies close to the As (crosses) and Se (circles) K edges [3]. For the comparison, $S(Q)$ measured at 11667 eV (200 eV below the As K edge) is also shown by the solid curve at the bottom of the figure. Clear contrasts are seen between $\Delta_i S(Q)$ and $S(Q)$, which are very prominent near the prepeak position of $Q \sim 12$ nm^{-1}, the prepeak being the clear evidence of the existence of intermediate-range order (IRO) in this glass. $\Delta_{As} S(Q)$ has a distinct and sharp prepeak, the position of which coincides with that in $S(Q)$. Any structures at this Q position can be barely visible in $\Delta_{Se} S(Q)$, while it shows a shoulder at the higher Q value of 14.5 nm^{-1}, where $\Delta_{As} S(Q)$ has no indication. As explained above, $\Delta_{As} S(Q)$ is composed of mainly $S_{AsAs}(Q)$ and $S_{AsSe}(Q)$ partial structure factors, while $\Delta_{Se} S(Q)$ consists mainly of $S_{AsSe}(Q)$ and $S_{SeSe}(Q)$. Therefore, it appears plausible that the main part of the prepeak at about 12 nm^{-1} originates from the As-As partial correlation, and its higher Q side around 14.5 nm^{-1} from the Se-Se correlation [3].

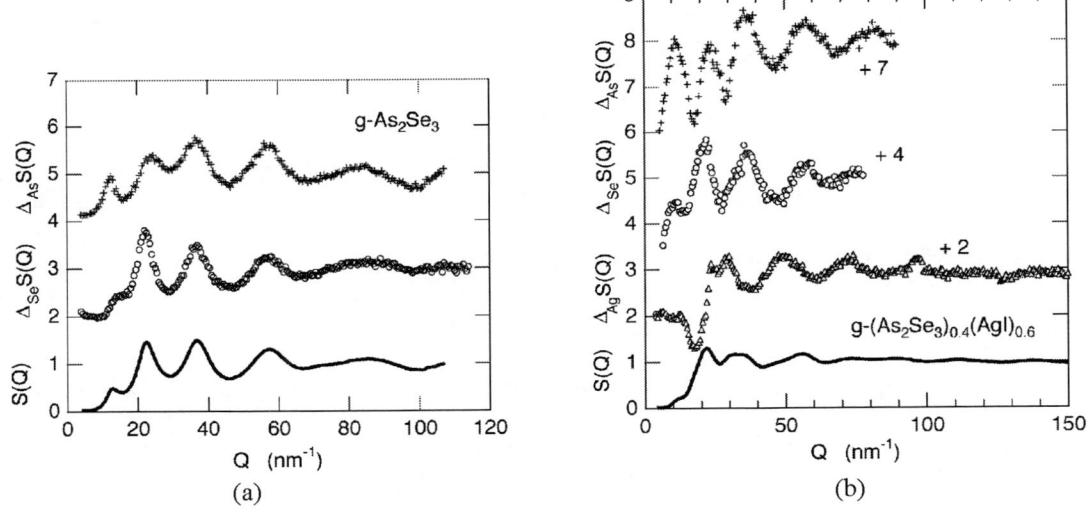

(a)   (b)

FIGURE 3. $\Delta_i S(Q)$ spectra of glassy (a) As_2Se_3 and (b) $(As_2Se_3)_{0.4}(AgI)_{0.6}$ together with $S(Q)$. From S. Hosokawa et al. [3] and T. Usuki et al. [5].

Figure 4(b) shows $\Delta_i S(Q)$ spectra of glassy $(As_2Se_3)_{0.4}(AgI)_{0.6}$ measured at energies close to the As (crosses), Se (circles), and Ag (triangles) K edges. The $\Delta_I S(Q)$ spectrum has not yet measured due to the energy limitation of the BM02/ESRF beamline. For the comparison, $S(Q)$ measured at 25315 eV (200 eV below the Ag K absorption edge) is also shown by the solid curve at the bottom of the figure. As shown at the bottom of Fig. 3(b), the prepeak in $S(Q)$ of the mixture of As_2Se_3 and AgI becomes much smaller in intensity than that in $S(Q)$ of *pure* As_2Se_3. In addition, the Q position of the prepeak slightly shifts towards a higher Q value by adding the AgI salt. Moreover, $S(Q)$ changes also in the higher Q range. Thus, a previous interpretation of this $S(Q)$ results was that the IRO in As_2Se_3 drastically changes by mixing with AgI [8].

As seen in Fig. 3(b), clear contrasts are seen between $\Delta_i S(Q)$s and $S(Q)$. The differences are very prominent at the whole Q range measured. In the Q range near the prepeak around 12 nm^{-1}, $\Delta_{As} S(Q)$ has a distinct prepeak, where a small shoulder locates in $S(Q)$. $\Delta_{Se} S(Q)$ also has a small peak at the same Q position. However, $\Delta_{Ag} S(Q)$ shows no indication of the prepeak.

Surprisingly, the $\Delta_{As} S(Q)$ and $\Delta_{Se} S(Q)$ results look very similar to those in pure glassy As_2Se_3 except the prepeak position, although the $S(Q)$ spectra are very different from each other. Therefore it is plausible that the SRO of this glassy superionic conductor around the As and Se atoms are very similar to that in glassy As_2Se_3, and the IRO around these elements would be modified with AgI component. The feature of $\Delta_{Ag} S(Q)$ is completely out-phase, and looks likely molten salt as is $\Delta_{Cu} S(Q)$ in glassy $(As_2Se_3)_{0.4}(CuI)_{0.6}$ [4], although the partial information on liquid AgI is lacking so far.

Differential pair-correlation function, $\Delta_i g(r)$, were obtained from the Fourier transform of $\Delta_i S(Q)$, which were shown in Fig. 3 of ref. [5]. The results clearly indicate that a pseudo-binary mixture of the As_2Se_3 network matrix and AgI-related ion conduction pathways would be a good structural model for this superionic glass, and the correlation between the Ag atoms and As_2Se_3 network can hardly be seen in $\Delta_{Ag} g(r)$. This experiment strongly revealed that the present AXS technique enables one to clarify the partial structural information in even such a complex four-component glass experimentally.

CONCLUSION AND PERSPECTIVE

In this paper, we discuss the recent development of AXS technique as a tool of investigating local structures of non-crystalline materials using a third-generation synchrotron radiation facility, ESRF. In order to obtain $\Delta_i S(Q)$s with a high statistical quality, it is necessary to acquire scattering data with a good energy resolution to discriminate elastic signals from fluorescence and Compton scattering intensities, as well as with a sufficient number of scattered X-ray photons. For this we chose a single-crystal graphite energy-analyzer with a long detector arm. In order to show the feasibility of this detecting system, we describe in detail examples of our recent AXS results on As_2Se_3 chalcogenide glass and $(As_2Se_3)_{0.4}(AgI)_{0.6}$ superionic glass. This technique will be a break-through for the developing of the AXS experiment using a high-flux X-ray source from third-generation synchrotron radiation facilities, especially using insertion device optics in the near future.

ACKNOWLEDGMENTS

The authors would like to thank Dr. D. Raoux for several suggestions of the AXS technique at the beginning phases of this project. The glassy As_2Se_3 and $(As_2Se_3)_{0.4}(AgI)_{0.6}$ samples were provided by Prof. P. Boolchand group and Prof. T. Usuki, respectively. The AXS experiments were performed at the beamline BM02 of the ESRF (Proposal No. HS1562, HS1860, HS2184, HS2348, HS2798, and ME1002).

REFERENCES

1. P. H. Fuoss, P. Eisenberger, W. K. Warburton, and A. Bienenstock, *Phys. Rev. Lett.* **46**, 1537-1540 (1981).
2. S. Hosokawa, Y. Wang, J.-F. Bérar, J. Greif, W.-C. Pilgrim, and K. Murase, *Z. Phys. Chem.* **216**, 1219-1238 (2002).
3. S. Hosokawa, Y. Wang, W.-C. Pilgrim, J.-F. Bérar, S. Mamedov, and P. Boolchand, *J. Non-Cryst. Solids*, (2006) in press.
4. T. Usuki, S. Hosokawa, and J.-F. Bérar, *Nucl. Instrum. Met. Phys. Res.* B **238**, 124-128 (2005).
5. T. Usuki, S. Hosokawa, and J.-F. Bérar, *J. Non-Cryst. Solids*, (2006) in press.
6. S. Hosokawa et al., to be published in *Mater. Trans.*
7. S. Sasaki, *KEK report 1989* (Nat. Lab. High Energy Phys., Tsukuba, 1989), p. 1-136.
8. T. Usuki, S. Saito, K. Nakajima, O. Uemura, Y. Kameda, T. Kamiyama and M. Sakurai, *J. Non-Cryst. Solids* **312-314**, 570-574 (2002).

A New, Rapid 3D Tomographic Energy Dispersive Diffraction Imaging System for Materials Characterization and Object Imaging (Rapid TEDDI)

K. H. Khor and R. J. Cernik

Manchester Materials Science Centre, University of Manchester, Grosvenor Street, Manchester M1 7HS, UK

Abstract. Tomographic energy dispersive diffraction imaging (TEDDI) using white beam synchrotron X-rays has been demonstrated to yield images of the interior features of solid objects non-destructively[1]. However, to image a solid object in three-dimensional (3D) via the currently available TEDDI technique is a very time consuming process. As a consequence, an array solid state detector has been developed together with a high resolution X-ray collimator array. This system is capable of reducing the time for object scanning from hours to minutes. A wide range of diverse applications will benefit from this new TEDDI system which will provide rapid 3D density maps with structural and chemical information at each voxel of the image.

Keywords: Energy-dispersive diffraction; Synchrotron radiation; Tomography; X-ray collimator
PACS: 87.59-e

INTRODUCTION

In 1998 Hall *et al* [1] demonstrated that the interior features of solid objects could be very effectively and simply imaged in a non-destructive manner using synchrotron energy dispersive diffraction. The sample volumes were approximately 125 cm^3 scanned with spatial resolutions down to tens of microns. They chose to demonstrate the new technique on two materials, the first was a semi-crystalline polymer (PEEK) with a test pattern of pre-formed features; the second sample was a limestone plug with a varying mineralogical composition. They were able to demonstrate a big improvement over the measurements obtained by Harding [2] who used conventional rotating anode X-ray sources. However both papers demonstrate the huge potential of TEDDI applied to soft tissue; ceramics; minerals; large scale engineering components; and high density metals using either laboratory or synchrotron X-rays.

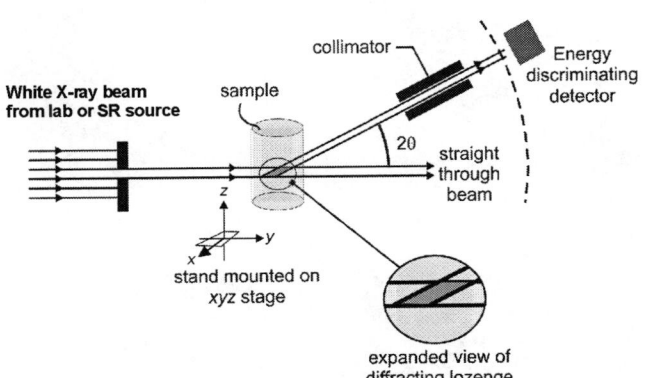

FIGURE 1. One dimensional TEDDI experimental arrangement.

The TEDDI experimental arrangement is shown in fig 1. A white beam from synchrotron or laboratory X-ray source is collimated to the desired spatial resolution. The small diffracting sample volume (called the diffracting

CP879, *Synchrotron Radiation Instrumentation: Ninth International Conference*,
edited by Jae-Young Choi and Seungyu Rah
© 2007 American Institute of Physics 978-0-7354-0373-4/07/$23.00

lozenge) is defined by the track of the incident and scattered beams through the sample and the angle subtended by the collimator aperture. A solid-state detector collects the diffraction pattern from each diffracting lozenge. By step-scanning the sample in X, Y and Z direction, a 3D structural map can be built up. However, the speed of conventional TEDDI method (as described above) is limited by current technology. The main drawback of the current system is that the tomographic data have to be read-out, point by point, from volume elements (microns to mm in size) within the study object. Assembling two-dimensional (2D) or 3D-images is extremely slow and commonly taking 14 to 16 hours even with synchrotron radiation. This makes the method impractical for a laboratory based analytical tool and certainly rules out medical application such as in-vivo studies. As a consequence, a new rapid TEDDI technique was developed in this study with the aim to increase the speed of data collection, and hence, further expand the potential of TEDDI across different disciplines.

RAPID TEDDI EXPERIMENTAL CONFIGURATION

Figure 2 shows a schematic representation of the rapid TEDDI conceptual layout that can collect data on large 3D objects in seconds rather than hours. Figure 2(a) shows the conventional TEDDI technique with one detector. Figure 2(b) shows the situation where there are four detectors stacked vertically. If the collimation in front of the detectors is kept parallel each detector will define a different diffracting lozenge arranged in sequence along the X-axis. Figure 2(c) shows the situation with a number of detectors arranged along the Y direction. Again with perfectly parallel collimation each detector will see the scattering from a different lozenge this time spatially arranged in Y. Ideally we would have a two-dimensional array of detectors such that all diffracting lozenges within a defined spatial resolution and a defined value of Z are collected simultaneously, this is shown in fig 2(d). In order to collect a complete set of diffraction patterns for all voxels in the sample we need to scan the sample in the Z direction through the fan beam (vertical beam ~ 50 μm and horizontal beam ~ 10-20 mm) as shown in fig 2(e).

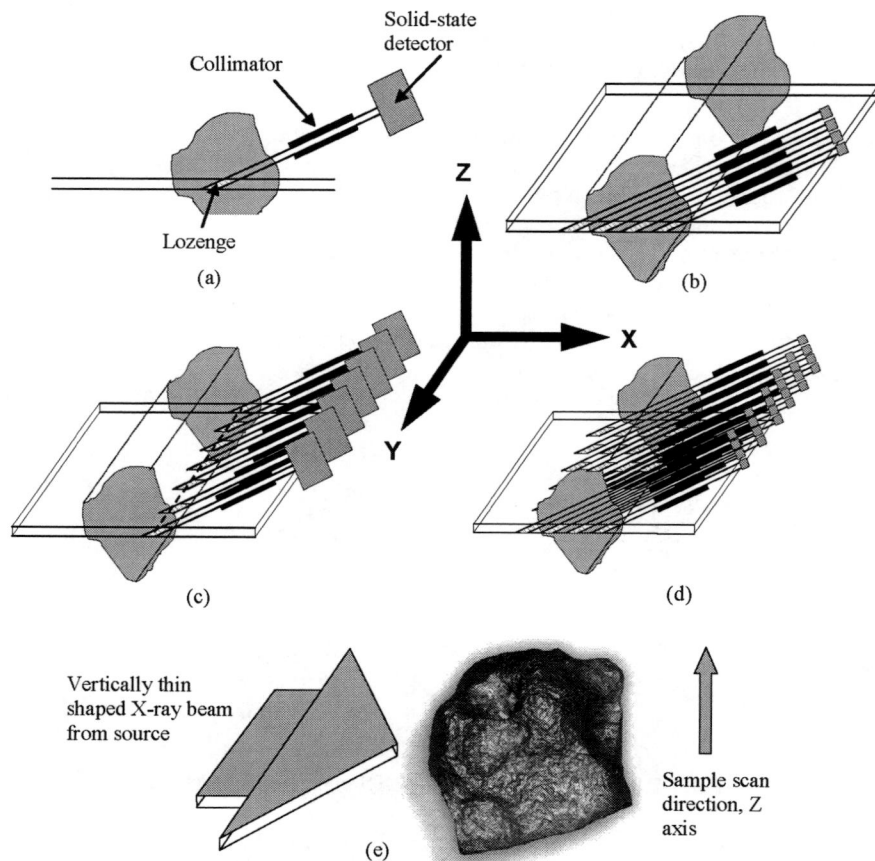

FIGURE 2. Rapid TEDDI experimental configuration.

COLLIMATOR

Constructing the array of X-ray collimator for rapid TEDDI is a challenging process. A novel manufacturing principle (*patent pending*) was established to manufacture the collimator and is described elsewhere [3]. Figure 3 shows the current prototype of collimator. It is laminar in design, and consists of a series of 0.1 mm tungsten plates stacked together at specific distances to achieve the desired full angle divergence beam at the collimator end. There are 9 tungsten plates aligned in this collimator, each of these plates has an array of 16 x 16, 75 μm diameter holes with a pitch distance of 300 μm drilled by femtosecond pulse duration laser. This collimator has a length of 400 mm and gives a full angle divergence beam of ∼ 0.01°. Each of the tungsten plates has to be carefully aligned relative to each other over the full length of the collimator in order to give satisfactory alignment of all the 75 μm holes.

FIGURE 3. X-ray collimator with 16 x 16 highly collimated holes.

Synchrotron X-ray tests involved placing the collimator on a goniometer stage on Station 7.6 at Daresbury Laboratory. The collimator was tilted horizontally and vertically on the stage in order to maintain a parallel position to the incident X-ray beam. Figure 4(a) shows measured Gaussian shape X-ray intensity profile from this trial. The extinction is excellent around the transmission area. Each hole in the collimator was then tested individually. Figure 4(b) shows a map of the X-ray intensity from the individual holes of the full array of this collimator. It shows that all holes have been correctly drilled and aligned between plates. However, only a 14 x 16 array of hole profiles were observed as opposed to the designed 16 x 16 array. Two horizontal lines of holes were missing. This was found to be caused by a programming error at some stage during the laser hole drilling. These results show the effectiveness of short pulse duration laser engineering in manufacturing the 2D collimator with very high precision (<5 micron).

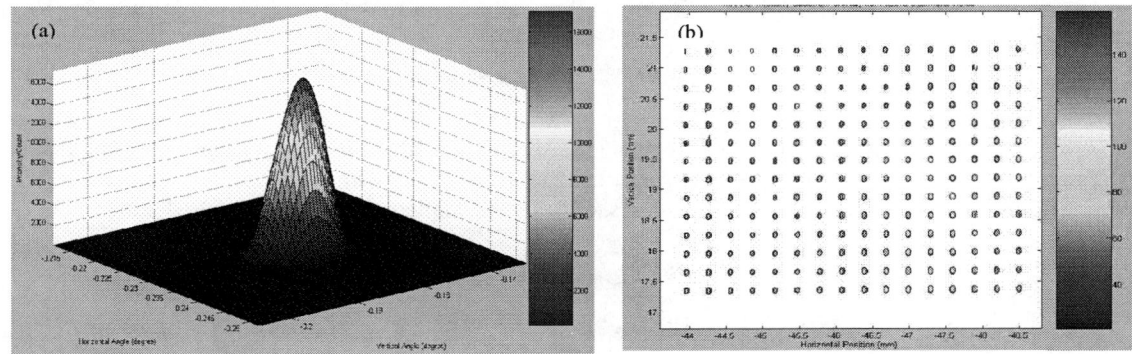

FIGURE 4. (a) The transmission functions from the full length collimator. (b) Individual hole transmission function.

MULTI PIXEL SOLID-STATE ENERGY RESOLVING DETECTOR

To collect individual collimated diffraction patterns from each hole of the 2D X-ray collimator, a multi pixel array solid-state energy resolving detector was developed [4]. The detector consists of a 16 x 16 array of 300 μm² pixels where a totally active Si surface is bump bonded to an electronic read out chip behind each pixel. Figure 5(a) shows the size of the module, the active square detector area (4.8 x 4.8 mm) and the associated read out electronics.

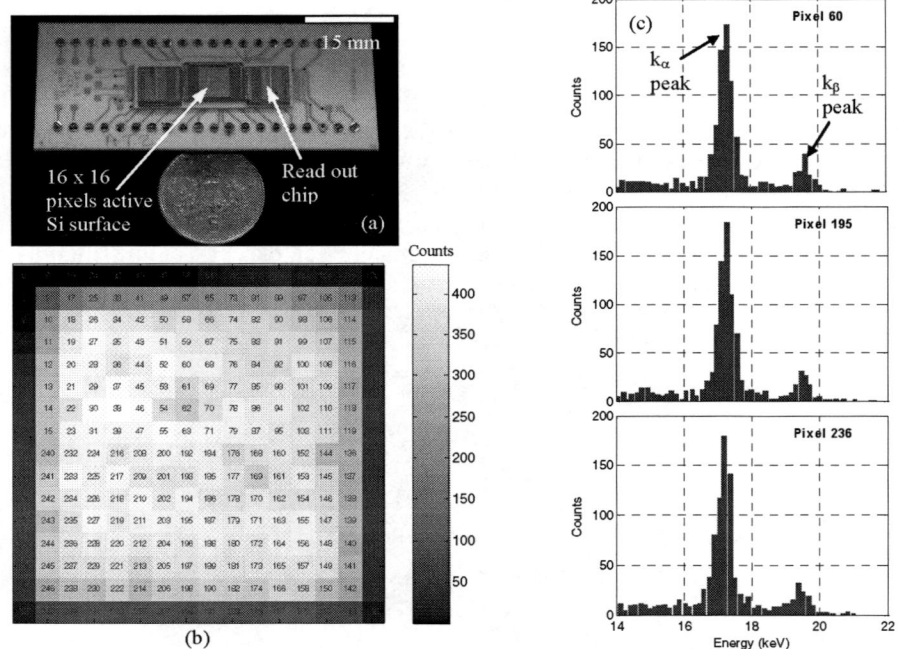

FIGURE 5. (a) Multi pixel array solid-state energy resolving detector. (b) Detector pixel mapping. (c) Energy spectra of three individual pixels.

We have tested the Si detector using a radioactive variable-energy X-ray source. The source consists of a sealed ceramic primary 370MBq [241]Am alpha-particle source which can excite characteristic X-rays from any one of its six targets (Cu, Rb, Mo, Ag, Ba and Tb) with different energy levels. Figure 5(b) shows a typical map of a single detector pixel under a Mo k_α and k_β target source at 17.44 and 19.63 keV, with all pixels being individually numbered. Except all the edge pixels which were not functioning properly (due to bump bonding non uniformity) all interior pixels appeared to show a fairly uniform count rate. Figure 5(c) shows typically measured Mo k_α and k_β spectra from three individual pixels, with all of them being able to detect both energy peaks at the correct energy levels. The measured full width half maximum (FWHM) energy resolution across all working pixels was fairly consistent ranging from 230 eV to 470 eV, with an average value of 350 eV. Our target design value is 250 eV.

CONCLUSIONS AND FUTURE WORK

We have shown that short pulse laser engineering is suitable method for manufacturing the challenging 2D collimator necessary for rapid TEDDI. The first stage testing of the 2D multi pixel energy resolving detector using variable-energy X-ray source shows promising results. The next phase of this project is to couple the 2D detector with the collimator to perform rapid TEDDI with a series of test sample using synchrotron radiation X-rays.

ACKNOWLEDGMENTS

The authors wish to thank EPSRC and CCLRC for their funding and support in this project. The University of Cambridge IMRC is also acknowledged for their help in laser machining.

REFERENCES

1. C. Hall, P. Barnes, J. K. Cockcroft, S. L. Colston, D. Hausermann, S. D. M. Jacques, A. C. Jupe and M. Kunz, *Nuclear Instruments & Methods in Physics Research Section B-beam interactions with materials and atoms*, **140** (1-2), pp. 253-257 (1998).
2. G Harding, M. Newton and J. Konsanetzky, M. P. Brown, *Physics in Medicine and Biology*, **35** (1), pp. 33-41 (1990).
3. L. Tunna, P. Barclay, R. J. Cernik, K. H. Khor, W. O'Neill and P. Seller, *Journal of Measurement Science and Technology,* in press, (2006).
4. P. Seller, et. al. "Silicon pixel detector for X-ray spectroscopy.", SPIE Vol 3445, EUV, X-ray and Gamma-ray Instrumentation for Astronomy IX, July 1998.

In-Situ White Beam Microdiffraction Study of the Deformation Behavior in Polycrystalline Magnesium Alloy During Uniaxial Loading

P. A. Lynch[*,†], A. W. Stevenson, D. Liang, D. Parry, S. Wilkins[†], I. C. Madsen[*], C. Bettles[**] and N. Tamura, G. Geandier[‡]

[*]*Commonwealth Scientific and Industrial Research Organisation (CSIRO), Minerals, Private Bag 33, Clayton South MDC, 3169, Australia*
[†]*CSIRO, Manufacturing and Infrastructure Technology.*
[**]*ARC Centre of Excellence for Design in Light Metals, Department of Materials Engineering, Monash University, Clayton, 3800, Victoria, Australia*
[‡]*Lawrence Berkeley National Laboratory, 1 Cyclotron Road, Berkeley, CA 94720, U.S.A.*

Abstract. Scanning white beam X-ray microdiffraction has been used to study the heterogeneous grain deformation in a polycrystalline Mg alloy (MgAZ31). The high spatial resolution achieved on beamline 7.3.3 at the Advanced Light Source provides a unique method to measure the elastic strain and orientation of single grains as a function of applied load. To carry out in-situ measurements a light weight ($\sim$0.5kg) tensile stage, capable of providing uniaxial loads of up to 600kg, was designed to collect diffraction data on the loading and unloading cycle. In-situ observation of the deformation process provides insight about the crystallographic deformation mode via twinning and dislocation slip.

Keywords: white beam microdiffraction, plastic deformation, magnesium.
PACS: 07.85.Qe, 62.20.Fe, 61.10.-i, 61.72.Dd

INTRODUCTION

As a non-destructive test, X-ray diffraction techniques are well suited to provide microstructural information during a dynamic process, as encountered in the standard engineering mechanical strength test. To gain a greater understanding of the bulk specimen microstructure, in-situ X-ray diffraction represents a routine method for measuring the residual stress and texture as a function of applied tensile or compressive load, see for example [1, 2]. Most recently, high-spatial resolution microdiffraction synchrotron beamlines have been used to study the dynamic behavior of individual grains in a polycrystalline material under applied tensile load. Using a high energy focused monochromatic beam, Margulies et al.[3] have measured the lattice rotation of an individual grain under applied load. This technique has been further refined to study strain tensor development from individual [4] and multiple grains [5] in a polycrystalline sample under tensile load. In a different approach, using a highly focused synchrotron white beam, Joo et al. [6, 7] have used in-situ tensile measurements to study the deformation of individual copper grains. An underpinning requirement of these in-situ studies is the necessity to load the sample in incremental steps to enable data collection, either by specimen rotation or raster scanning the sample under the focussed incident beam. Both of these approaches provide a unique set of microstructural information from individual grains. Using focused synchrotron white or pink beam and appropriate sample positioning loading device, real time stress-strain microstructural information can be realised. The importance of this further development is that it will enable critical features of an engineering stress test, such as strain rate, to be measured on a grain and sub-grain level. Furthermore, the ability to monitor a region of interest during a loading or unloading cycle will provide a measurement scenario for pseudo-elastic deformation studies.

This article will outline the progress towards development of a real-time dynamical white beam microdiffraction experimental approach to measure the deformation of individual grains under load. The principle requirements of the deformation stage for performing high-spatial resolution microdiffraction experiments are discussed. This is followed by in-situ uniaxial tensile data collected for an Mg alloy sample at the microdiffraction beamline 7.3.3 at the Advanced Light Source.

CP879, *Synchrotron Radiation Instrumentation: Ninth International Conference*,
edited by Jae-Young Choi and Seungyu Rah
© 2007 American Institute of Physics 978-0-7354-0373-4/07/$23.00

EXPERIMENTAL

Deformation Stage Design

To accommodate scanning microdiffraction requirements the deformation stage had to meet two key requirements. Firstly, due to weight constraints introduced by the high spatial resolution translation stages the total weight of the stage had to be restricted (< 0.5kg). Secondly, to ensure the incident beam is in close proximity to the sample, the physical size of the stage had to fit into a small volume of ~20cm length, 5cm wide x 5cm height. Consequently, standard commercially available tensile stages were inappropriate. In view of the experimental demands and sample requirements, a deformation stage capable of producing a tensile load of up to 6kN was designed and manufactured. The stage with Mg alloy 'dog bone' sample is shown in Figure 1.

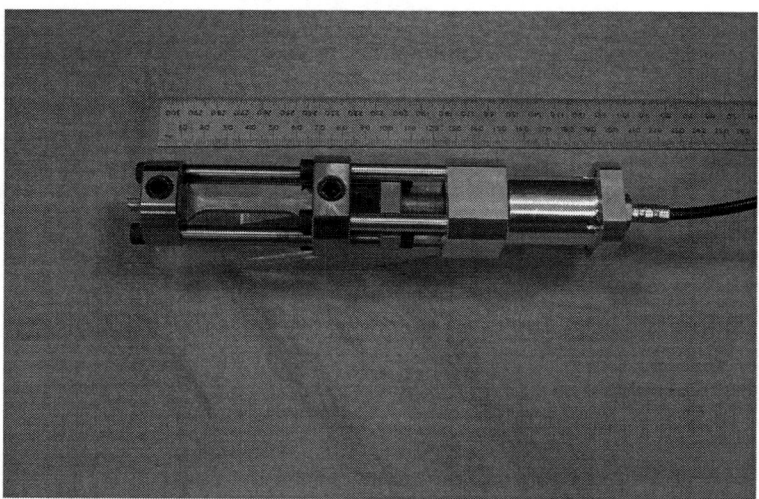

FIGURE 1.   Hydraulic deformation stage configured for uniaxial tensile experiments.

The stage and hydraulic control system was designed as a fully portable device to allow for transfer to a synchrotron beamline end-station. For the current experimental requirements, the loading system configuration operates via a single-action hydraulic cylinder that provides a maximum tensile load of up to 6.0kN. Sample loading and control is achieved by an input and output transducer connected to the hydraulic intensifier. The intensifier requires house or compressed air which is scaled according to the required sample load. All input and output transducer control is achieved with a portable analog pc card which uses a USB connection, enabling control with a laptop personal computer. Finally, loading cycles and feedback are achieved with a windows GUI interface.

Microdiffraction Results and Analysis

In-situ experiments were performed on the high-resolution synchrotron microdiffraction beamline (7.3.3) at the Advanced Light Source. For the present experiments the white beam from the bend magnet is focused using Kirkpatrick Baez mirrors to a spot size of ~$1\mu m^2$ onto the sample. For a detailed description of the beamline characteristics the reader is referred to MacDowell et al. [8]. Measurements were collected from a MgAZ31 (96%Mg, 3%Al, 1%Zn) sample rolled to a thickness of 3.6mm. After a post-rolling anneal the sample was cut into the form of a regular 'dog bone' shape for tensile testing.

To conduct in-situ experiments the tensile stage was mounted on an XYZ translation stage, the sample was then positioned in a symmetric diffraction geometry with the incident beam angle ω= 45° and CCD detector set at 90° 2θ. To test the stability of the tensile stage, diffraction data was collected using a raster scan approach; the sample was translated across the beam in $8\mu m$ steps, generating an XY diffraction map from a specific region in the sample. A total of 7 diffraction maps were collected in load increments of 0.36kN, ranging from 0 to 2.16kN.

Indexation of the collected white beam patterns was achieved using the software package XMAS [9]. Figure 2 shows an example of the in-situ grain orientation maps collected for the MgAZ31 sample at load increments of 0.36kN,

0.72kN and 1.8kN. Fig. 2a, c, e represents the out-of-plane orientation of [0002] relative to the surface normal and Fig. 2b, d, f shows the in-plane orientation of [10$\bar{1}$0] relative to the scanning X direction, where the X and Y directions are parallel and perpendicular to the tensile axis respectively. The dashed box in each figure highlights the same grain region for each data set. By tracking this region in each load increment the deformation of 5 grains could be recorded.

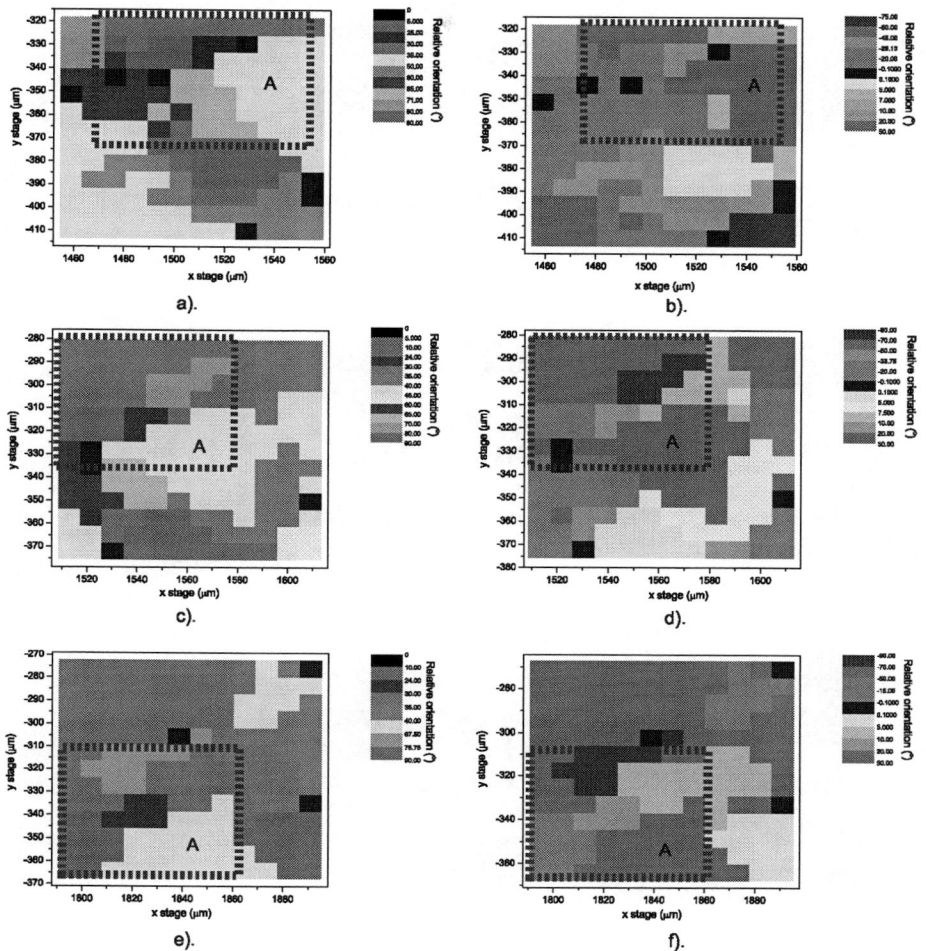

FIGURE 2. Grain orientation maps of the MgAZ31 sample at load increments of 0.36kN (a, b), 0.72kN (c, d), and 1.80kN (e, f). The dashed box for each figure is used to highlight the same region on the sample at each load increment.

As an example, the deviatoric strain in the X and Y directions and the in-plane rotation angle have been plotted for Grain A in Figure 3. As illustrated in Fig. 3a the applied load is observed as an increase in the tensile strain parallel to the load axis and compressive strain perpendicular to the load axis. For the in-plane rotation angle (Fig. 3b) only a relatively small shift between the [10$\bar{1}$0] and tensile axis was seen. In similar in-situ microdiffraction tensile experiments performed on FCC structures [6] an in-plane lattice rotation of up to 10° was reported. Grain lattice rotation was attributed to the active slip system in the FCC structure. In this study it is proposed that the limited slip geometry in the hcp structure may constrain the in-plane lattice rotation and consequently induced lattice deformation is relieved via deformation twinning. To test this hypothesis an algorithm is presently under development which will provide information regarding the nature and content of deformation twinning in the sample.

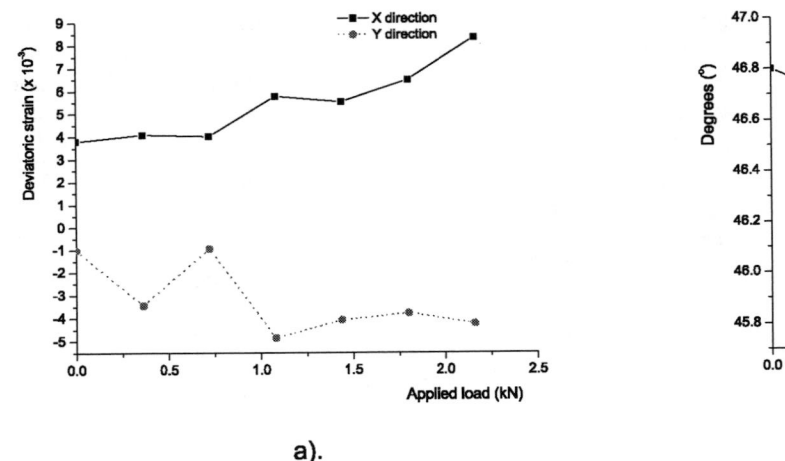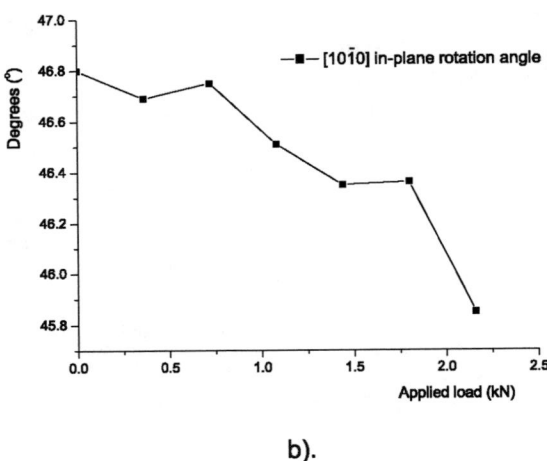

| a). | b). |

FIGURE 3. The observed deformation of grain A as a function of applied load. a). Deviatoric strain parallel and perpendicular to the load axis. b). Change of the in-plane angle between the [10$\bar{1}$0] and load axis.

CONCLUSION

In this investigation a light weight strain stage has been successfully used to study the in-situ deformation of individual grains during uniaxial tensile loading. Using a raster scan approach it was possible to monitor the grain orientation and deviatoric strain as a function of applied load. The outcome of this preliminary work lends strong support for the future development of real time dynamic deformation studies based on white beam microdiffraction. This dynamic process can be realised with grain tracking strain stage development, currently in progress.

ACKNOWLEDGMENTS

We would like to thank K. Venkatesan and M. Gibson for scientific discussions and assistance with sample preparation. This work was supported by the Victorian Centre for Advanced Materials Manufacturing (VCAMM). The Advanced Light Source is supported by the Director, Office of Science, Office of Basic Energy Sciences, of the U. S. Department of Energy under Contract No. DE-AC02-05CH11231.

REFERENCES

1. S. R. Agnew, M. H. Yoo, and J. A. Horton, *Magnesium Alloy* pp. 119– 124 (2000).
2. D. W. Brown, S. R. Agnew, M. A. M. Bourke, T. M. Holden, S. C. Vogel, and C. N. Tome, *Mat. Sci. Eng. A* **399**, 1–12 (2005).
3. L. Margulies, G. Winther, and H. F. Poulsen, *Science* **291**, 2392–2394 (2001).
4. L. Margulies, T. Lorentzen, H. F. Poulsen, and T. Leffers, *Acta Mat.* **50**, 1771–1779 (2002).
5. R. V. Martins, L. Margulies, S. Schmidt, H. F. Poulsen, and T. Leffers, *Mat. Sci. Eng. A* **387-389**, 84–88 (2004).
6. H. D. Joo, J. S. Kim, K. H. Kim, N. Tamura, and Y. M. Koo, *Scripta Mat.* **51**, 1183–1186 (2004).
7. H. D. Joo, J. S. Kim, C. W. Bark, J. Y. Kim, Y. M. Koo, and N. Tamura, *Mat. Sci. For.* **475-479**, 4149–4152 (2004).
8. A. A. MacDowell, R. S. Celestre, N. Tamura, R. Spolenak, B. Valek, W. L. Brown, J. C. Bravman, H. A. Padmore, B. W. Batterman, and J. R. Patel, *Nucl. Inst. Meth. Phys. Res. A* **467-468**, 936–943 (2001).
9. N. Tamura, A. A. MacDowell, R. Spolenak, B. C. Valek, J. C. Bravman, W. L. Brown, R. S. Celestre, H. A. Padmore, B. W. Batterman, and J. R. Patel, *J. Syn. Rad.* **10**, 137–143 (2003).

Johann Spectrometer for High Resolution X-ray Spectroscopy

Pavel Machek*, Edmund Welter[†], Wolfgang Caliebe[†], Ulf Brüggmann[†], Günter Dräger[¶] and Michael Fröba*

*Institut für Anorganische und Analytische Chemie, Justus-Liebig-Universität Gießen
Heinrich-Buff-Ring 58, 35392 Gießen, Germany
[†]Hamburger Synchrotron Strahlungslabor (HASYLAB) am Deutsches Elektronen Synchrotron (DESY)
Notkestraße 85, 22607 Hamburg, Germany
[¶]Martin-Luther-Universität Halle-Wittenberg, Fachbereich Physik
Friedemann-Bach-Platz, 06108 Halle(Saale), Germany

Abstract. A newly designed vacuum Johann spectrometer with a large focusing analyzer crystal for inelastic x-ray scattering and high resolution fluorescence spectroscopy has been installed at the DORIS III storage ring. Spherically bent crystals with a maximum diameter of 125 mm, and cylindrically bent crystals are employed as dispersive optical elements. Standard radius of curvature of the crystals is 1000 mm, however, the design of the mechanical components also facilitates measurements with smaller and larger bending radii. Up to four crystals are mounted on a revolving crystal changer which enables crystal changes without breaking the vacuum. The spectrometer works at fixed Bragg angle. It is preferably designed for the measurements in non-scanning mode with a broad beam spot, and offers a large flexibility to set the sample to the optimum position inside the Rowland circle. A deep depletion CCD camera is employed as a position sensitive detector to collect the energy-analyzed photons on the circumference of the Rowland circle. The vacuum in the spectrometer tank is typically 10^{-6} mbar. The sample chamber is separated from the tank either by 25 μm thick Kapton windows, which allows samples to be measured under ambient conditions, or by two gate valves. The spectrometer is currently installed at wiggler beamline W1 whose working range is 4-10.5 keV with typical flux at the sample of 5×10^{10} photons/s/mm^2. The capabilities of the spectrometer are illustrated by resonant inelastic experiments on 3d transition metals and rare earth compounds, and by chemical shift measurements on chromium compounds.

Keywords: X-ray spectrometer, curved crystal, inelastic X-ray scattering, X-ray fluorescence.
PACS: 07.85.Nc, 78.70.En, 71.20.Be, 71.20.Eh

INTRODUCTION

A new focusing spectrometer has been build and installed at Hamburger Synchrotron Strahlungslabor (HASYLAB). The design of the spectrometer is adapted for the use of a large spot size at the DORIS III storage ring and it has proved as a suitable tool for measurements of resonant inelastic x-ray scattering (RIXS), and for high resolution x-ray fluorescence spectroscopy. In the past, some experiments with focusing analyzers have been performed at HASYLAB by Schülke ([1], scanning geometry) and Dräger (dispersive geometry with position sensitive detector), respectively, but wide access to these instruments was not possible. The installation of the new spectrometer is permanent, and access to the spectrometer is open to the whole community. Therefore, some emphasis in the construction of the spectrometer was laid on easy operation, and on high flexibility. However, polarization dependent measurements are not possible, because the focusing plane of the spectrometer is fixed in the plane of the storage ring.

CP879, Synchrotron Radiation Instrumentation: Ninth International Conference,
edited by Jae-Young Choi and Seungyu Rah
© 2007 American Institute of Physics 978-0-7354-0373-4/07/$23.00

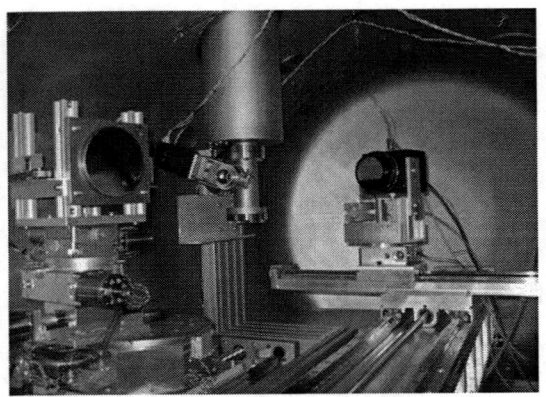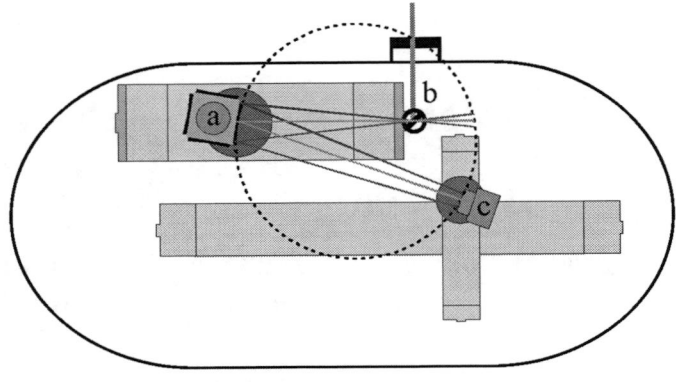

FIGURE 1. View into the opened vacuum tank and schematic set-up of the spectrometer in the dispersive geometry: (a) crystal holder, (b) sample, (c) CCD camera

THE SPECTROMETER

The instrument has been described in detail in [2] and therefore only a short summary is given here.

The essential elements of the spectrometer are spherically and cylindrically bent crystals. The spherical crystals have a diameter of up to 125 mm, and were prepared by pressing and gluing thin wafers with appropriate crystal orientation into spherical substrates. The cylindrical crystals are bent by a 4-point bender [3]. The standard radius of curvature of both kinds of crystals is 1000 mm, however the measurements with smaller and larger bending radii are also feasible with some limitation, though.

A basic set of spherically (Si(111) and Si(311)) and cylindrically bent crystals (Si(111), Si(110), Si(100), and Si(311)) is currently available to cover the working range of the beamline monochromator at W1 (4-10.5 keV) without using Bragg angles < 55° on the analyzer. Up to four of these crystals are mounted on a revolving crystal changer, and thus measurements with different crystal orientations are possible without breaking the vacuum.

The angle of incidence of the emitted photons from the sample on the crystal is adjusted by means of a goniometer on which the crystal holder is mounted. The distances between the sample and the crystal and the position of the detector are set up by means of three large high load linear slides. Since the spectrometer is predominantly designed for measurements in dispersive mode, the spectrometer offers a large flexibility to choose the optimum position of the sample inside the Rowland circle.

The scattered photons are detected by a deep depletion CCD camera (Princeton Instruments) with a working range of 2-15 keV. The CCD chip has 1024 pixels x 256 pixels, each sized 26 um x 26 um. The design of the camera head enables the application of the camera in vacuum as well under ambient conditions. That is useful for alignment, when the tank is vented to atmosphere. For measurements where the actual shape of the fluorescence spectra is not important, the use of other kind of detector, e.g. PIN diode, is possible.

While the mechanical parts (slides, rotation stages) of the spectrometer are controlled by the Hasylab *Online* software on a Linux-PC, the data acquisition with the CCD camera is controlled by *WinSpec* (running under Windows XP) that can be further extended by programs written in Microsoft Visual Basic. During measurements, the synchronization between the CCD controller and *Online* is realized by means of 8 bit TTL input and output signals, and the data transfer by means of *WinSock* connections, respectively. The full control of the experiment is thus possible using only *Online*.

To minimize air absorption in the low energy range, and, at the same time, to minimize the amount of air scattering, the spectrometer is installed inside a large cylindrical vacuum tank. The samples are then mounted on a holder inside a separate chamber. That chamber is either separated from the tank vacuum by 25 μm thick Kapton windows, or it is part of the tank vacuum. In the first case, Kapton windows allow to measure samples under ambient conditions. In the other case, two gate valves are closed for sample change so that only the relatively small volume of the sample chamber has to be evacuated after sample change. It is planned to use cryostats for cooling samples to cryogenic temperatures, and other special sample environments.

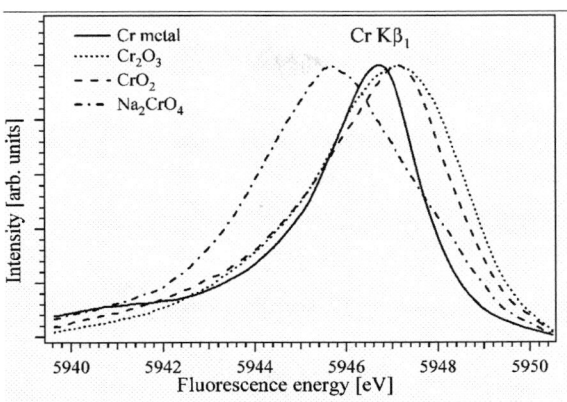

FIGURE 2. Measurements of the Cr Kβ₁ fluorescence lines of Cr, Cr₂O₃, CrO₂ and Na₂CrO₄ at the excitation energy E=6100 eV.

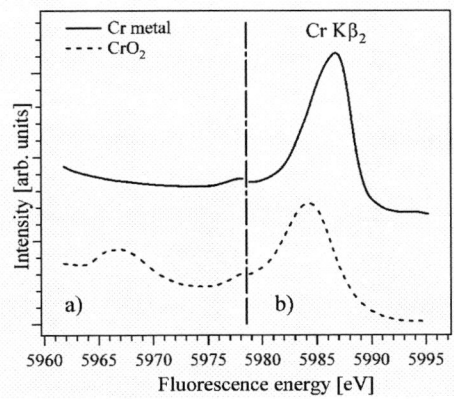

FIGURE 3. Measurements of the Cr Kβ₂ fluorescence lines of Cr and CrO₂ at the excitation energy E=6100 eV. The spectrometer was set to the energy a) 5972 eV and b) 5988 eV.

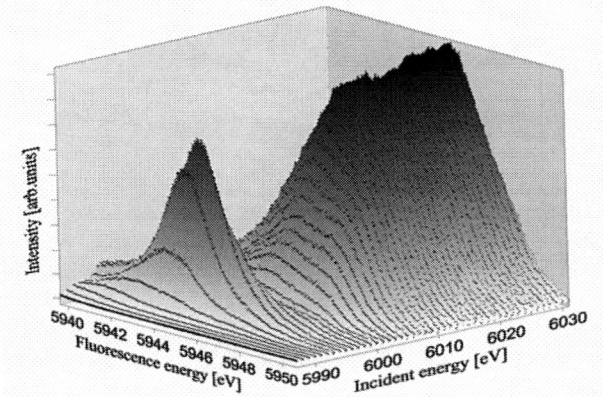

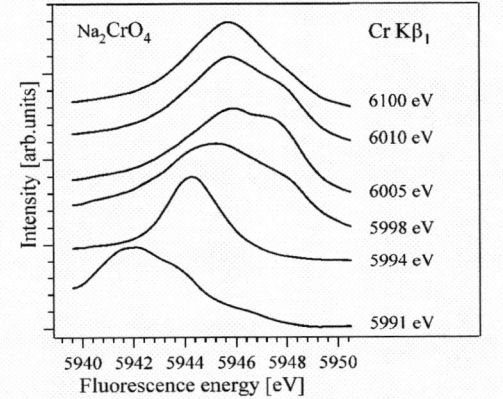

FIGURE 4. Measurements of 1s3p RIXS on Na₂CrO₄: (a) 3D plot and (b) the fluorescence spectra at several excitation energies.

EXPERIMENTAL

The spectrometer has been installed at the wiggler-beamline W1 at the storage ring DORIS III. The available working range of the beamline monochromator is 4-10.5 keV, which covers K absorption edges of the 3d-transition metals and L absorption edges of the rare-earth elements. The energy resolution is about 2 eV at 9 keV and the monochromatic flux is 5×10^{10} photons/s/mm². The approximately size of the beam on the sample is 2x4mm² (vxh).

To measure one spectrum on the CCD camera, typical exposure time of 0.2s and 200 accumulations are used, which is the basic loop. This loop is, depending on the intensity of the fluorescence signal at a given excitation energy, repeated as many times as necessary. At the beginning of each loop the intensity of the incident beam is measured in an ionization chamber, which is mounted at the entrance window of the vacuum tank. This value is used for the intensity scaling of the fluorescence signal.

The energy scale is usually calculated from the known energy dispersion of the crystal, and from the geometry of the experiment. Furthermore, the spectrometer can be calibrated by measuring elastic peaks, or in special cases from the measurements of two known emission lines (e.g. Kα₁ and Kα₂) well above threshold. Whenever possible, at least two of these methods are compared in order to verify energy scaling.

The energy resolution of a spectrometer in Johann geometry is in principle best for Bragg angles close to 90 degrees [4]. A detailed discussion of different contributions to the energy resolution that are specific for the presented spectrometer is given in [2]. According to the discussion, one can estimate that the energy resolution ranges between 1.0 eV (for Cu Kα₁ and Si(444) crystal, θ=79.3°) and 4.9 eV (for Cu Kβ₁ and Si(444) crystal, θ=62.6°). For geometrical reasons we can achieve at the most 86 degrees (for Cr Kβ₁ and Si(333) crystal).

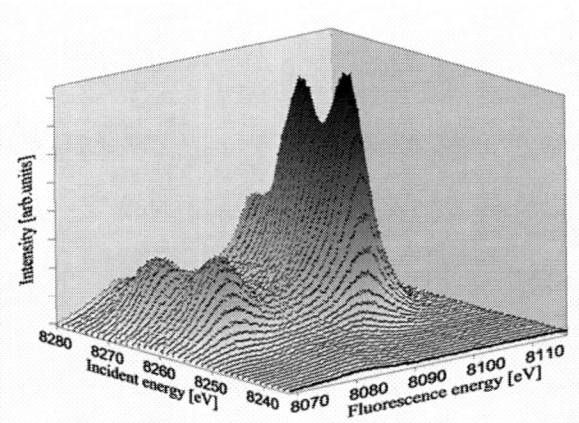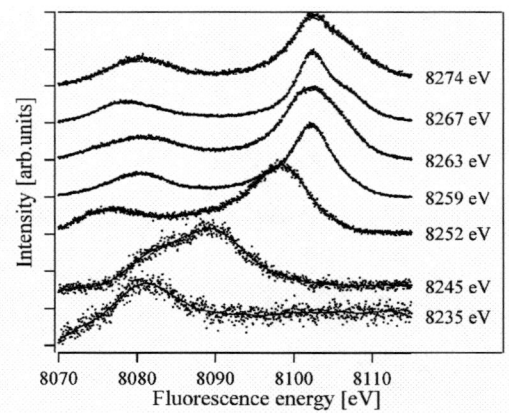

FIGURE 5. Measurements of 2p4d RIXS on Tb_2O_3: (a) 3D plot and (b) the fluorescence spectra at several excitation energies

The example of a measurement at high Bragg angle is shown in Fig.1. Chemical shifts of Cr $K\beta_1$ lines of Cr_2O_3, CrO_2 and Na_2CrO_4 with respect to Cr $K\beta 1$ of chromium metal were measured at the excitation energy of 6100 eV. The observed shifts between the compounds and the metal are ΔE=-1.08 eV for Na_2CrO_4, ΔE=0.38 eV for CrO_2 and ΔE=0.44 eV for Cr_2O_3.

Figure 4 shows experimental data of 1s3p RIXS of Na_2CrO_4, which was performed with the same experimental setup as the previous measurement. In the left hand side, one can observe the strong resonance at the excitation energy of 5994 eV. The other resonance can be identified at 6005 eV, which is shown in the right hand side of Fig.4.

The disadvantage of the high resolution measurements at the angles close to 90 degrees is that the energy dispersion of the crystal is very small, and the CCD camera can pick up only a small part of the fluorescence spectra. In such a case, the spectrometer has to be tuned subsequently to several energies in order to cover the larger energy region of the spectra. The measurements of the fluorescence spectra of chromium metal and CrO_2 in the vicinity of Cr $K\beta_2$, where the spectrometer was set to 5972 eV and 5988 eV, respectively, are presented in Fig.3. The measurements exhibit in addition to the main Cr $K\beta_2$ fluorescence lines also a satellite line in the case CrO_2, which can be attributed to oxygen ligand. However, because the spectra were recorded at the different positions of spectrometer, we could not simply glue them together. It would be advisable to complete this high resolution measurements with the measurement at the low Bragg angle (by means of Si(422) crystal in the presented case) in order to verify both the energy and intensity scaling.

The other possibility how to overcome the problem with the low energy dispersion is to use the spectrometer in the stepping mode. In this mode, tuning energy of the spectrometer varies according to the excitation energy of the monochromator. This method is useful at the measurements of the RIXS spectra when we have intense fluorescence signal even at the excitation energies well bellow the absorption edge.

Finally, 2p4d RIXS of Tb_2O_3 in the Fig.5 represents the measurements of comparatively weak L fluorescence line $L\gamma_1$ (the intensity of $L\gamma_1$ is about 11% of $L\alpha_1$). The measurement shows a pronounced splitting of the main as well as the satellite peaks and a weak resonance at the excitation energy 8245 eV.

ACKNOWLEDGMENTS

The project is funded by the federal German ministry of research and education (BMBF): „Weiterentwicklung des hochauflösenden Fluoreszenzdetektorsystems mit einstellbarer Energie-auflösung auf der Basis eines fokussierenden Johannspektrometers für Energien von 2-25 keV " (Project No. 05KS4RGA/5)

REFERENCES

1. W. Schülke, A. Kaprolat, T. Fischer, K. Höppner and F. Wohlert, *Rev. Sci. Instrum.* **66**, 2446-2453 (1995)
2. E. Welter, P.Machek, G. Dräger, U.Brüggmann and M. Fröba, *J. Synchrotron Rad.* **12**, 448-454 (2005)
3. K. Läuger, Ph.D. Thesis, Ludwig-Maximilians-Universität München, 1968
4. U. Bergmann and S. P. Cramer, *Proc.* **3448**, 198-209 (1998)

The Infrared Microspectroscopy Beamline at CAMD

O. Kizilkaya[1], V. Singh[1], Y. Desta[1], M. Pease[1], A. Roy[1], J. Scott[1], J. Goettert[1], E. Morikawa[1], J. Hormes[1], and A. Prange[1,2]

[1]*Center for Advanced Microstructures and Devices, Louisiana State University, Baton Rouge, Louisiana 70806 USA*
[2]*Microbiology and Food Hygiene, Niederrhein University of Applied Sciences, Germany*

Abstract. The first infrared microspectroscopy beamline at the Louisiana State University, Center for Advanced Microstructures and Devices (LSU-CAMD) has been constructed and dedicated to investigation of samples from various disciplines including chemistry, geology, biology, and material sciences. The beamline comprises a simple optical configuration. A planar and toroidal mirror pair collects 50 and 15 mrad synchrotron radiation in horizontal and vertical directions, respectively, and focuses the beam through a diamond window located outside of the shielding wall. This focus acts as a new source point for the rest of the optical systems. The synchrotron beam spot size of 35 μm and 12 μm is measured in the x and y direction of the sample stage position of the microscope. This small beam spot has a superior brightness compared to conventional IR sources and allows spatially resolved measurements with very good signal/noise ratio. Compared to a conventional thermal source, synchrotron radiation provides 30 times better intensity and a two orders of magnitude greater signal/noise ratio when measuring with microscope aperture size of 15 x 15 μm^2. The results of the studies on the fungus-plant interaction with its resultant effects on the healthy leaves, and bacterial growth process in the crystallization of gordaite, a mineral, are presented.

Keywords: Infrared microspectrocopy, signal to noise ratio, bean rust, gordaite mineral
PACS: 07.85.Qe, 91.65.An, 78.30.Jw

INTRODUCTION

The advances in the instrumentation of the Fourier transform infrared (FT-IR) spectrometer, in the last century, have made this technique one of the most fundamental and unique analytical tools to investigate the chemical components of samples from various research disciplines. IR microscopes utilized with spectrometer and conventional thermal sources have developed over the last thirty years. The implementation of IR microscope to image and resolve the chemical structure of samples has catapulted the application of FTIR technique into a wide range of materials from animal tissues to geological samples. A conventional thermal source confronts signal-to-noise (S/N) ratio limitation when small size samples are measured (<40 μm). Since the thermal source has a large emittance, only a fraction of radiation illuminates the sample when aperture of the microscope, which confines the beam into the area of interest of the sample, is close to 40 μm or less. When the thermal source was substituted with synchrotron radiation, the performance of IR microscopes for small size samples, 5-40 μm, greatly improved due to high brightness (photon flux emitted from unit source area and unit solid angle) of the synchrotron radiation source. The synchrotron infrared radiation source, which has 100-1000 times higher brightness, makes possible the investigation of chemical variations across the small size samples and also small heterogeneous regions of large samples with high spatial resolution. However, as the aperture size is decreased for the measurements of small samples, the diffraction limit becomes effective on the IR spectrum. The diffraction limit for confocal microscope is obtained to be ~λ/2 [1]. Here, we will present the CAMD's first IR beamline configuration, its optical elements and performance. In addition, the results acquired from this beamline from the studies of fungi growth on bean leaves and a geological sample of gordaite will be discussed.

CP879, *Synchrotron Radiation Instrumentation: Ninth International Conference*,
edited by Jae-Young Choi and Seungyu Rah
© 2007 American Institute of Physics 978-0-7354-0373-4/07/$23.00

BEAMLINE DESIGN AND PERFORMANCE

CAMD's IR beamline has been designed to implement the radiation emitted only from the homogeneous magnetic field of a bending magnet. The current dipole chamber permits radiation exraction of 70 mrad in the horizontal and 15 mrad in the vertical direction. Although these openings facilitate measurements in mid-IR region, the vertical opening limits the performance in the far IR region. Therefore, to utilize the beamline in the far-IR region as well, a new dipole chamber with 50×50 mrad^2 opening angle will replace the current chamber in the near future. As shown in Fig. 1, the IR beamline is a relatively short beamline; therefore, it comprises a compact design and carries a simple optical configuration. A planar/torodial mirror pair is designed to collect 50 mrad of radiation in both horizontal and vertical direction. Synchrotron radiation is reflected back and down to the toroid mirror by the plane mirror. Since the incidence angle of radiation on each mirror is 19°, the most of the radition is absorbed by the plane mirror; therefore, this mirror is water-cooled. The toroid mirror focuses synchrotron beam onto a diamond window, which is located outside the shield wall. This focus serves as a new source for the rest of the optical system. The beam passed through the diamond window is reflected by a small off-axis gold-coated parabolic mirror which collimates the synchrotron beam. Then the beam is directed to FTIR spectrometer by employing two gold-coated plane mirrors. The thermo Nicolet Continuum microscope, used at the beamline, is currently equipped with two objectives (15X and 32X), a condenser (15X) and a mercury-cadmium-telluride (MCT) detector which covers the mid-IR spectral range of 11700-400 cm^{-1}. The beam collimating optics, spectrometer and microscope are all purged with dry nitrogen gas.

The advantage of synchrotron radiation over thermal source stems from its brightness, and this advantage tremendously enhances the S/N ratio. In Fig. 2a, the measured S/N ratio is plotted as a function of microscope's aperture size for synchrotron and thermal sources. It is explicit that for the aperture size smaller than 15 μm, the synchrotron radiation provides a two orders magnitude greater S/N ratio compared to the one provided by the thermal source. This superiority is also verified in the 100% transmission line plotted by taking the ratio of two successive background spectra. Figure 2b shows the comparison of 100% transmission line measured at 10×10 μm^2 for both sources. For this aperture size, diffraction, apparent as deterioration in 100% transmission line, dominates the spectrum below 1000 cm^{-1}. The 100% transmission line obtained with synchotron radiation also indicates that there is no significant noise components in the measured frequency range due to physical vibrations.

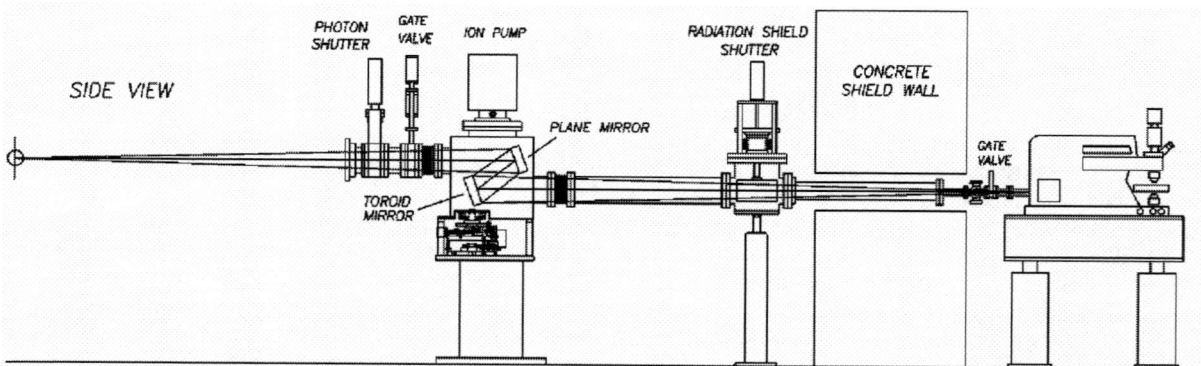

FIGURE 1. Schematic diagram of the infrared microspectroscopy beamline at CAMD. The infrared beam is collected and focused on the diamond window by a pair of mirrors, a plane (16×16 cm^2) and a toroid (17×17 cm^2).

EXPERIMENTAL RESULTS

The combination of IR microspectroscopy with synchrotron radiation has greatly improved the capability of infrared measurements of small areas in biological samples, specifically, for detection and classification of diseases. For example, aggregates of misfolded proteins in the brain tissue of Alzheimer's disease and spectral evidence of bone

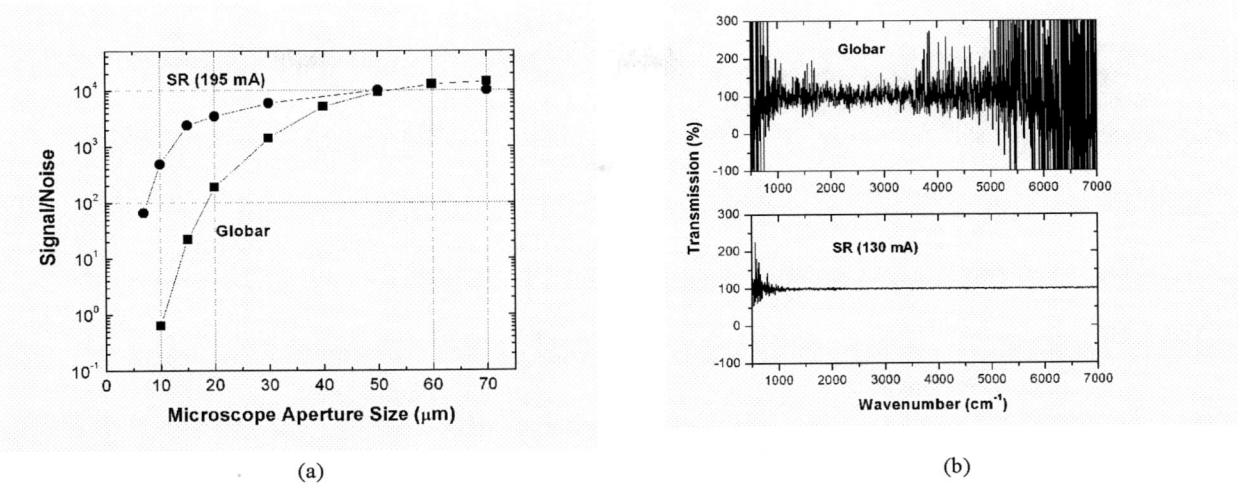

(a)   (b)

FIGURE 2. a) Signal to noise ratio as a function of the microscope aperture size for synchrotron and thermal sources. b) The ratio of two successive background spectra (100% lines) measured in transmission mode with 10x10 μm^2 aperture size.

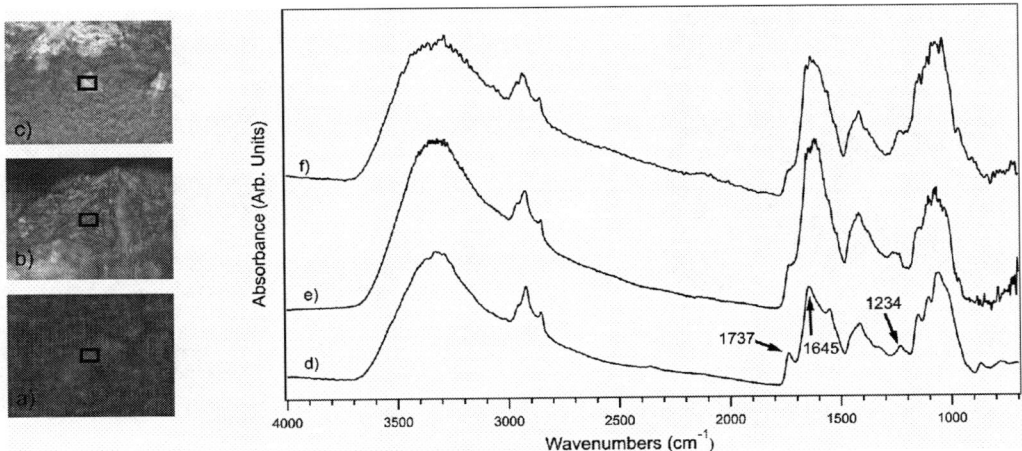

FIGURE 3. The optical image of healthy bean leaf (a), less infected (b), and highly infected (c) parts of infected bean leaf. Infrared spectra of healthy, less and highly infected samples are shown in (d), (e) and (f), respectively.

disease have been identified [2]. In this regard, we have performed synchrotron radiation based IR microspectroscopy to investigate and characterize the host-pathogen interaction of bean rust *Uromyces appendiculatus* with bean leaves. The bean rust *Uromyces appendiculatus* was cultivated on planta under greenhouse conditions at INRES-Phytomedicine, Bonn University. Samples were used directly for the measurements without any pretreatment. Figure 3 shows the optical image of a healthy bean leaf (a), less infected (b) and highly infected (c) parts of an infected bean leaf. The transmission mode with 40x30 μm^2 aperture size (indicated as the rectangle in the optical images) was employed for the collection of IR spectra of these samples. IR spectra acquired from healthy, less and highly infected samples are plotted in Fig 3d, 3e and 3f, respectively. There are no noticeable changes observed in the IR features of lipids (stretching vibrational mode of CH_2 and CH_3, 2900-3000cm^{-1}) between IR spectrum of the healthy and infected bean leaves. A sharp peak located at 1737 cm^{-1} due to lipid ester (C=O) in the healthy bean leaf appears as a shoulder in the infected leaf. Furthermore, evident changes are elucidated in the features of proteins (~1650 cm^{-1}) and carbohydrates (~1100 cm^{-1}). A well resolved protein Amide I band at 1645 cm^{-1} becomes very broad peak in the IR spectra of infected leaf (Fig. 3e and 3f). This change may arise as a result of differences in the protein structure of healthy and infected bean leaves. The band located at 1234 cm^{-1}, which is attributed to C-O stretching vibrational mode of carbohydrates, shifts to lower wavenumbers. In addition, the well resolved bands from the deformational vibrational mode of carbohydrates in the region 1000-1200 cm^{-1} turns into one broad band. Another stark contrast is the observation of the enhanced band intensity of protein features (region of 1500-1650 cm^{-1}) in the less infected leaf compared to the intensity of the features of carbohydrates (1000-1200 cm^{-1}).

The superior S/N ratio of synchotron based IR microspectroscopy also became apparent in an corrosion experiment. In order to test the efficacy of a copper flake embedded epoxy coating to inhibit bacterial growth, epoxy-coated steel plates were immersed in artificial seawater. Within 30 days a several mm thick crust grew on the surface of the plates. An optical image of the crust in Fig. 4a shows black nodules (ca. 2mm across) which are filled with white translucent crystals. The crust scraped from the steel surface was placed on a gold-coated slide for infrared microspectroscopy. The reflection mode of the IR microscope and 15x15 μm^2 aperture size were employed for dark nodules and colorless crystal sample. The IR spectra acquired from these samples are shown in Fig. 4b, which reveals that the crystal is gordaite, a zinc chloro-sulfate (NaZn$_4$(SO$_4$)(OH)$_6$Cl.6H$_2$O), first described in 1997 by Schluter et al. [3]. In Fig. 4b, the bands from streching vibrations of OH group appear in the region 3300-3550 cm^{-1} and the bands between 600-1200 cm^{-1} is ascribed to vibration of the sulfate group. Very similar measured IR spectral features between the dark nodules and colorless crystal suggest that both have similar chemical compositions, though morphologically different.

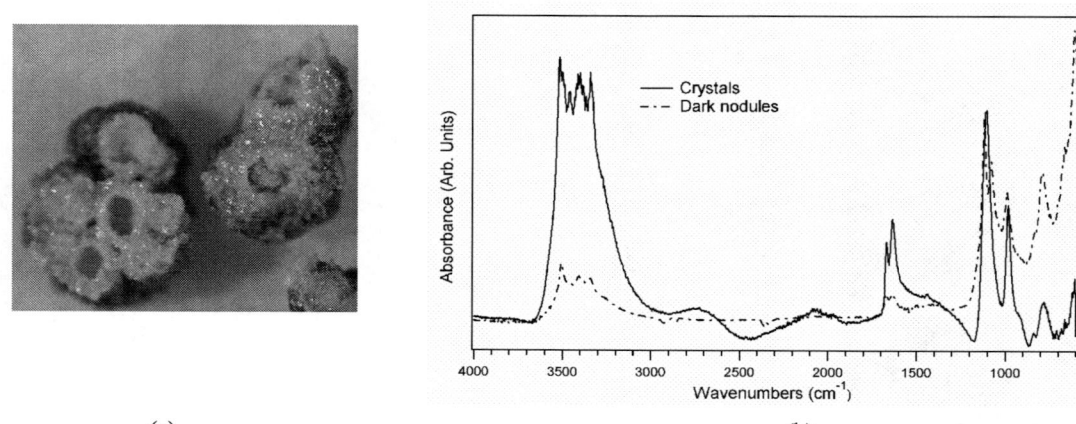

(a) (b)

FIGURE 4. a) Optical image of the goadaite and b) IR spectrum of crytal and bulk powder of Gordaite.

ACKNOWLEDGMENTS

The Louisiana Educational Quality Support Fund (LEQSF) and the Louisiana Board of Regents are acknowledged and thanked for the funds to build most of this beamline and purchase the microscope and detectors.

REFERENCES

1. G.L.Carr, *Rev. Sci. Instrum.* **72**, 1613-1619 (2001).
2. L. M. Miller, P. Dumas, J. L. Teillaud, J. Miklossy, L. Forro, *Rev. Sci. Instrum.* **73**, 1357-1360 (2002).
3. J. Schluter, K.H. Klaska, *Neues Jahrbuch Fur Mineralogie-Monatshefte* **4**, 155-162 (1997).

New HMI hard X-ray Diffraction Beamlines at BESSY

I.A. Denks, C. Genzel, E. Dudzik, R. Feyerherm, M. Klaus, G. Wagener

Hahn-Meitner-Institut Berlin c/o BESSY, Bereich Strukturforschung,
Albert-Einstein-Straße 15, 12489 Berlin, Germany

Abstract. Since April 2005 the Hahn-Meitner-Institute is operating two new beamlines for energy dispersive diffraction experiments (EDDI) and for (resonant) magnetic scattering (MAGS) at BESSY. The source for both beamlines is a superconducting 7 T multipole wiggler which provides hard X-ray photons with energies between 4 and 150 keV. The EDDI beamline uses the white beam and is intended for residual stress measurements on small samples as well as heavy engineering parts. The MAGS beamline delivers a focussed monochromatic beam with photon fluxes in the 10^{12} (s 100 mA 0.1 % bandwidth)$^{-1}$ range at energies from 4 to 30 keV. It is equipped for single crystal diffraction and resonant (magnetic) scattering experiments as well as for the study of thin films, micro-, and nanostructures in materials science.

Keywords: beamline; X-ray diffraction; energy dispersive diffraction; residual stress analysis; resonant scattering; magnetic scattering
PACS: 61.10.Nz

INTRODUCTION

Both X-ray and neutron diffraction are powerful tools for structural investigations in different research fields ranging from basic solid state physics, such as the study of charge, orbital, and magnetic ordering, to materials science, e.g. residual stress analysis. To complement its existing neutron instrumentation, the Hahn-Meitner-Institute Berlin has set up two new beamlines for hard X-ray diffraction at BESSY which are now in user operation. Both beamlines use parts of the radiation fan from the same superconducting multipole wiggler source. Because the two beamlines are separated by only 13 mrad, they had to be designed and built simultaneously. The construction and installation was carried out by Accel Instruments.

The energy dispersive diffraction (EDDI) beamline is designed for white beam experiments in materials science such as residual stress, texture and microstructure analysis. Compared to angle-dispersive (AD) diffraction, energy-dispersive (ED) diffraction provides major advantages, as the ED diffraction patterns are recorded with both sample and detector in fixed positions relative to the primary beam, so that no scan is necessary. A multitude of diffraction lines is recorded simultaneously, each of which delivers additional information in depth resolved residual stress, texture and phase analysis [1,2].

The monochromatic (MAGS) beamline provides a high intensity focussed beam with energies from 4 to 30 keV (up to 70 keV using the Si(333) reflection of the monochromator). It is intended for single crystal diffraction and resonant (magnetic) scattering experiments as well as the study of thin films, micro-, and nanostructures in materials science. Different cryostats are available which allow sample temperatures from 4 to 800 K. Recently, a small angle X-ray scattering (SAXS) detector arrangement has been set up as a further experimental option.

PHOTON SOURCE AND BEAMLINE LAYOUT

The 7 T multipole wiggler was designed jointly by BESSY and the HMI, and built by the Budker Institute in Novosibirsk (Russia). It consists of a superconducting 7 Tesla wiggler with 13 full poles and 3/4 and 1/4 end poles with a period length of 140 mm. The wiggler source thus is an array of point sources separated horizontally by 1.22 mm and by half the period length along the beam direction. The beam from the wiggler is about 92 % linearly polarized in the orbit plane. The wiggler critical energy is 13.5 keV at the current BESSY operation conditions of 1.7 GeV. The 7 T multipole wiggler has been described elsewhere [3,4] in detail.

CP879, *Synchrotron Radiation Instrumentation: Ninth International Conference,*
edited by Jae-Young Choi and Seungyu Rah
© 2007 American Institute of Physics 978-0-7354-0373-4/07/$23.00

As shown in Fig.1 the 7 T multipole wiggler supplies the two neighbouring beamlines, which are separated by 13 mrad. The MAGS beamline is built +1 mrad off axis and thus the array of source points appears as two large source spots when seen from the beamline. The EDDI beamline is situated at −12 mrad leading to a separation of the two beamlines of 390 mm at the EDDI experiment (30 m from the source). The two beamlines have separate radiation safety hutches and are operated fully independently.

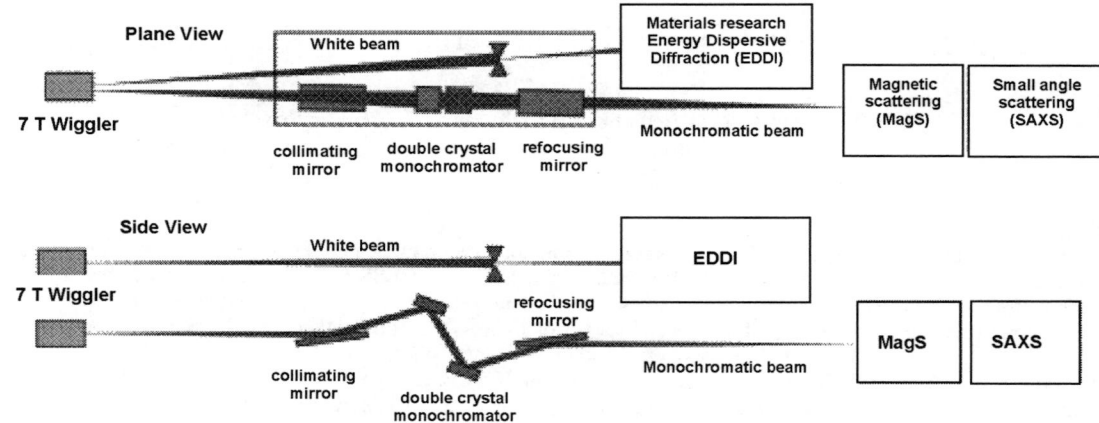

FIGURE 1. Schematic beamline layout

THE ENERGY DISPERSIVE DIFFRACTION (EDDI) BEAMLINE

The EDDI beamline uses the white beam as delivered by the wiggler. Figure 2a shows the energy spectrum measured directly (i. e. without any attenuators in the beam) at low ring current of 110 pA and then extrapolated to 250 mA. We found the beam conditions to be homogeneous within a cross-section of 4x4 mm².

The white synchrotron beam is collimated in both the axial (horizontal) and equatorial (vertical) direction by a slit system at 29.5 m from the source. A filter system consisting of Al-foils of different thicknesses is to attenuate low energy photons < 10keV in order to reduce the heat load on the sample.

In order to specify the energy range that can be used for diffraction experiments, we performed measurements in reflexion and transmission geometry at different scattering angles $2\theta_d$. Due to the basic relation in ED diffraction, $E(hkl) = hc/[2\sin\theta_d\ d(hkl)]$ (h - Planck's constant, c - speed of light), larger angles θ_d lead to a shift of the diffracted spectrum towards smaller energies and vice versa. As shown in Fig. 2b diffraction lines of reasonable intensity are available at energies as high as 150 keV. In transmission, we were able to penetrate steel samples up to a total thickness of 15 mm.

The EDDI diffractometer system in the experimental hutch was delivered by GE Inspection Technologies (formerly Seifert). It consists of a θ-θ-diffractometer MZ VI as basic unit, which operates in the vertical plane, and two sample positioners to handle small as well as large and heavy samples. The 5-axes positioner is based on an Eulerian cradle segment with integrated x-y-z translation. Even under extreme conditions (maximum load 5 kg) the sphere of confusion is smaller than 90 μm. The 4-axes positioner is designed for heavy samples up to 50 kg and operates in the Ω-mode. A laser adjustment system with integrated CCD camera is used for precise sample positioning. Two Canberra solid state germanium detector systems are available for data recording. Their intrinsic resolution is specified as 150 eV at 5.9 keV and 500 eV at 122 keV.

As the sample dimensions on the diffractometer are limited, a separate goniometer for large samples and industrial parts is being constructed and will be available soon.

The EDDI beamline components are controlled by the SPEC software from Certified Scientific Software which runs under Linux. Although it is rather simple to prepare and to run SPEC macros for any kind of ED measurements, we provide our users a special Mathematica software package, which enables them to prepare and to carry out investigations on EDDI and to evaluate the measured diffraction spectra immediately after having finished the experiment. This quasi 'in-situ' assessment of the achieved results is helpful in various respects: On the one hand, starting with calibration measurements performed at stress-free samples, systematic diffraction line shifts may be minimised by optimising the geometrical diffraction parameters (beam cross-section, slit apertures, filters etc.). On the other hand, prompt data evaluation helps to avoid logistic problems in handling the enormous amount of data accumulated in one beamtime, which may reach some thousand diffraction spectra. A more detailed description is given in [5].

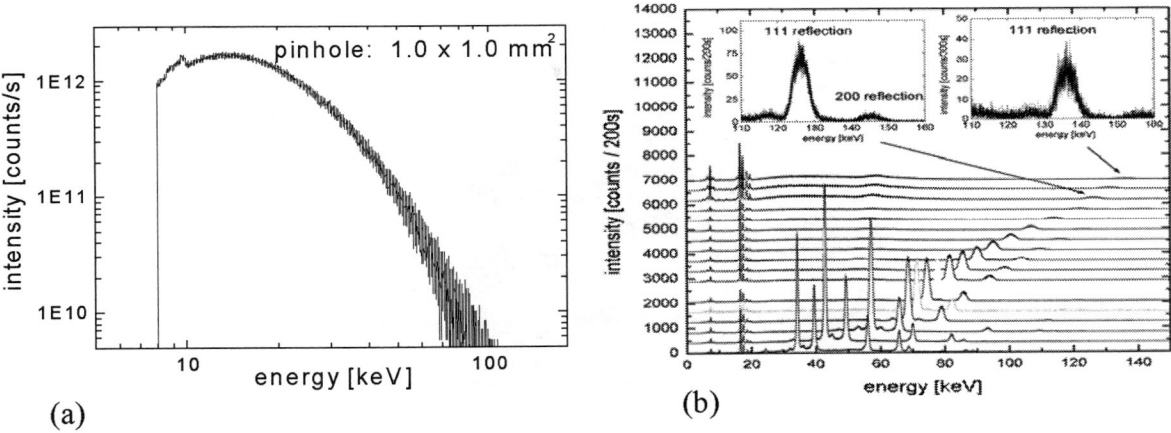

FIGURE 2. (a) Energy spectrum of the 7 T multipole wiggler (photon flux through a pinhole of 1x1 mm² 30 m behind the source), scaled up to a ring current of 250 mA. (b) Inconel (IN 718) diffraction spectra measured in reflexion geometry after stepwise variation of $2\theta_d$ between 2.5° (top spectrum) and 10° (bottom spectrum).

THE MAGNETIC SCATTERING (MAGS) BEAMLINE

The main beamline optics consist of the monochromator unit and two mirrors (as indicated in Fig. 1). The mirrors and the mirror benders were built by SESO, France. Both mirror substrates are Si single crystals, 1200 mm long, 200 mm wide and 50 mm thick, with the reflecting surfaces polished to 3 Å rms roughness. The mirrors have two tracks, one coated with Rh, the other is the uncoated Si substrate. For both mirrors the pitch angle can be varied between 0 and 5 mrad. The variable pitch and the different coatings are used to match the mirror reflectivity cut-off to the photon energy for harmonic rejection. The collimating mirror at 20.5 m from the source is mounted on a 4-cylinder SESO bender and has a variable bending radius to adapt to changes in pitch and heat load, with bending radii between 40 and 4 km. The mirror is indirectly water-cooled, with cooling pipes running through grooves cut into the mirror substrate. An eutectic liquid (Galinstan) serves as heat conductor between the substrate and the cooling pipe. The cooling was designed for heat loads up to 2000 W. The refocusing mirror at 25.5 m is mounted on a SESO U-bender allowing variable bending radii between ∞ and 2.3 km. This mirror needs no cooling.

The fixed exit double crystal monochromator at 23 m from the source covers the energy range from 4 to 30 keV with a pair of Si(111) crystals, and can reach photon energies up to approximately 70 keV using the Si(333) reflection. With the collimating mirror in the beam the energy resolution at the Cu K-edge is 2 eV; without mirrors the energy resolution is limited by the vertical beam divergence of 0.3 mrad. The monochromator crystals are mounted on a goniometer, with the first crystal surface at the centre of rotation. The goniometer can be moved vertically by 50 mm to adjust to the changing first mirror pitch. The first crystal is indirectly water-cooled and furnished with a counterbending mechanism to compensate the thermal bump from the wiggler beam [6]. The second crystal is mounted on an ESRF-type sagittal bender [7] with a variable bending radius between 1 and 80 m. It provides 2:1 horizontal focussing at all photon energies. A vertical crystal translation of 50 mm range and a horizontal translation of 200 mm provide a fixed beam offset of about 25 to 30 mm (this can be chosen to some extent). There is a coarse and a fine pitch adjustment for the second crystal, along with a roll and a yaw movement. Figure 3 shows the possible focus settings of the beamline.

There are white beam slits and four graded absorbers upstream from the first mirror to reduce the heat load on the optics. Behind the monochromator there is a second set of slits, as well as an I_o which consists of a scattering foil and a photodiode.

The last section of the beamline is inside the experimental hutch. It contains a set of graded attenuators, followed by horizontal and vertical slits and another I_o (again using a scattering foil and a photodiode). The diffractometer (positioned at 36.5 m) is a 6+3-circle Huber diffractometer with a standard 512 Eulerian cradle. The diffractometer can be used both in the vertical and the horizontal scattering geometry. It has an optional polarisation/energy analyzer with a full set of analyzer crystals in the energy range from 3.5 keV to 20 keV. Two detector systems are available: a liquid nitrogen cooled Canberra low energy Ge diode detector with a Canberra digital signal processor and a multichannel analyzer provides an easy way of separating fluorescence, higher order light etc. An Oxford Danfysik Cyberstar scintillation detector serves as a robust everyday detection system, and is also small enough to be used with the analyzer. Two closed cycle Displex based cryostats are available for sample temperatures from 6 to 800 K, and from 4 to 300 K, respectively. Another closed cycle cryostat for temperatures down to 1.5 K will be available in the near future; a superconducting 5.5 T magnet is under development. Beamline and experiment are controlled by two Linux-PCs with the SPEC software from Certified Scientific Software. More details about the beamline and first successful experimental results will be published soon [8,9].

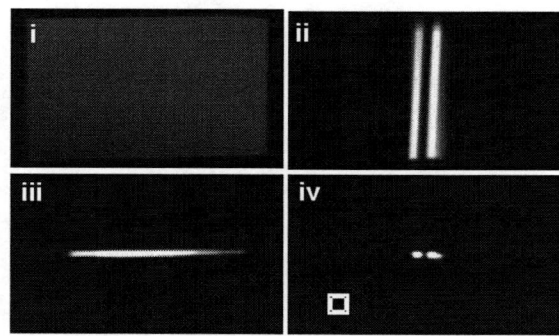

FIGURE 3. i) completely unfocussed beam, ii) horizontal focussing (the two stripes are the image of the two source spots in the wiggler), iii) vertical focussing, iv) both horizontal and vertical focussing. The focus size is about 1.2×0.4 mm^2. All figures are to the same scale indicated by the white 1 mm^2 square in iv).

ACKNOWLEDGMENTS

The building of this beamline has been supported by the BMBF through the HGF Vernetzungsfonds, grant No. 01SF0005 and 01SF0006, as well as the European Union. We also wish to thank the BESSY staff for their support.

REFERENCES

1. H. Ruppersberg and I. Detemple, *Mat. Sci. Eng.* **A 161**, pp. 41-44 (1993)
2. Ch. Genzel, C. Stock and W. Reimers, *Mat. Sci. Eng.* **A 372**, pp. 28-43 (2004)
3. D. Berger, E. Weihreter, N. Mezentsev and V. Shkaruba, *8th European Particle Accelerator Conference*, pp. 2595-2597 (2002)
4. D. Berger, H. Krauser, M. Rose, V. Duerr, E. Weihreter and S. Reul, *8th European Particle Accelerator Conference*, pp. 2598-2600 (2002)
5. Ch. Genzel, I. A. Denks, M. Klaus, *Proceedings of the Seventh European Conference on Residual Stress (ECRS 7)*, in press
6. R. Zaeper, M. Richwin, R. Wollmann, D. Lützenkirchen-Hecht and R. Frahm, *Rev. Sci. Instrum.* 73, pp. 1564-1567 (2002)
7. A. K. Freund, F. Comin, J.L. Hazemann, R. Hustache, B. Jenninger, K. Lieb and M. Pierre, *SPIE* Vol. 3448, pp. 144-155 (1998)
8. E. Dudzik, R. Feyerherm, W. Diete, R. Signorato and C. Zilkens, to be published
9. R. Feyerherm, E.Dudzik, N. Aliouane, D. N. Argyriou, *Phys. Rev.* **B** (2006), in press

A Guinier Camera for SR Powder Diffraction: High Resolution and High Throughput.

D. Peter Siddons[a], Steven L. Hulbert[a], Peter W. Stephens[b]

[a]National Synchrotron Light Source, Brookhaven National Laboratory, Upton, New York 11973, USA.

[b]University of New York at Stony Brook, Stony Brook, New York 11794, USA

Abstract. The paper describe a new powder diffraction instrument for synchrotron radiation sources which combines the high throughput of a position-sensitive detector system with the high resolution normally only provided by a crystal analyser. It uses the Guinier geometry [1] which is traditionally used with an x-ray tube source. This geometry adapts well to the synchrotron source, provided proper beam conditioning is applied. The high brightness of the SR source allows a high resolution to be achieved. When combined with a photon-counting silicon microstrip detector array, the system becomes a powerful instrument for radiation-sensitive samples or time-dependent phase transition studies.

Keywords: Powder diffraction, High Resolution.
PACS: 07.85.Qe

GUINIER GEOMETRY

The Guinier geometry [1] has been used with laboratory sources for many years. It is capable of providing high-resolution powder patterns from large samples using low-brightness sources such as a standard x-ray tube. Figure 1 shows the geometry. It can take two forms, one (a) optimized for low diffraction angles, and the second (b) for high angles. A production instrument would allow rapid transition between these two modes, requiring primarily a simple translation of the circle along the incident beam, and provision of a second sample holder.

OPTICS FOR SYNCHROTRON USE

Synchrotron radiation beams are typically highly collimated or slightly convergent, depending on the intended setup at the beamline. Powder diffraction instruments are installed at both types of beamline, usually in some form of the Debye-Scherrer geometry, with the sample at the center of the diffractometer and the detector system rotating about that point. In parallel-beam instruments the detector is usually a combination of an analyzer crystal and a point detector. In modern designs this single detector may be augmented by a set of several identical systems mounted on the detector arm and rotating together to increase the throughput of the instrument. Beamlines with focusing optics typically use a position-sensitive detector and generally achieve lower resolution than the analyzer instruments, but much higher throughput.

The instrument described here tries to provide both high resolution and high throughput. It is a strongly focusing device, identical to the classical Guinier geometry, but scaled up to allow angular resolutions at the 10^-4 level or better. Its intrinsic resolution is determined by the size of the focal spot at the Rowland circle and the detector spatial resolution.

CP879, *Synchrotron Radiation Instrumentation: Ninth International Conference*,
edited by Jae-Young Choi and Seungyu Rah
© 2007 American Institute of Physics 978-0-7354-0373-4/07/$23.00

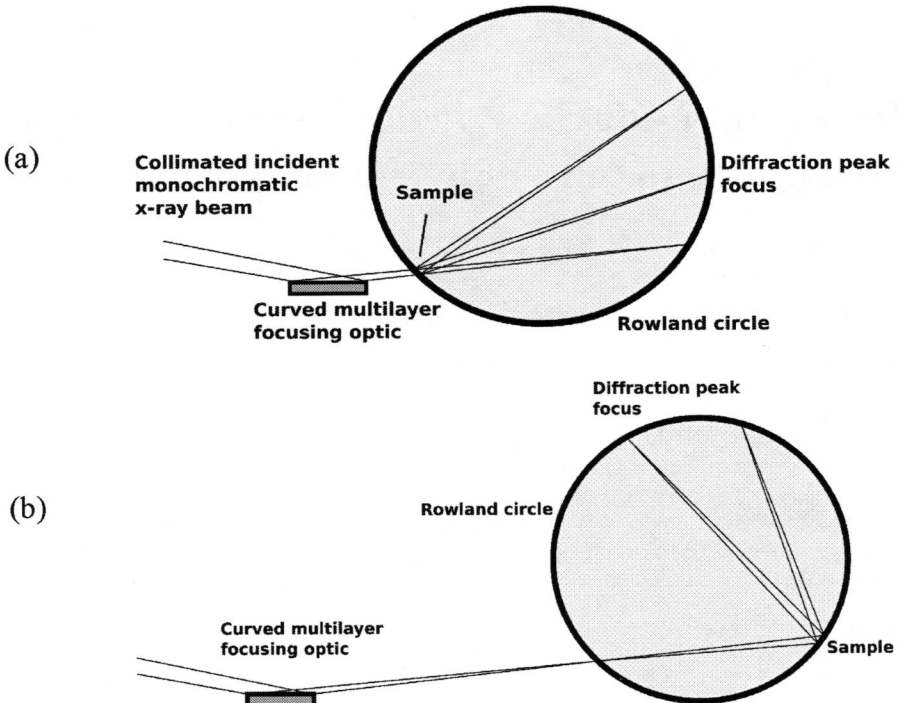

FIGURE 1. The Guinier geometry is a Rowland circle construction which can take two forms. (a) is optimal for low Bragg angle peaks and operates in transmission. (b) is optimized for high Bragg angle peaks and operates in reflection.

THE SILICON STRIP DETECTOR

The advantage of this geometry is only properly realized with an efficient, low-noise, high-resolution position sensitive detector. We have developed such a detector based on BNL's HERMES integrated circuit [2] and silicon microstrip arrays produced at BNL. It consists of a monolithic array of silicon diodes coupled to a set of application-specific integrated circuits designed at BNL. This allows each diode to detect x-rays with good energy resolution and hence low noise. The sensitive thickness of the diodes is 0.4mm, allowing efficient detection up to about 12keV. The array used for the tests reported here had 384 strips each 0.125 x 4 mm^2, read out by 12 chips. Each chip has 32 full counting channels. Each channel consists of charge-sensitive preamplifier, shaping amplifier and a set of three discriminators and 24-bit counters. A choice of four shaping time constants is provided, from 0.5 to 4 microseconds. Two gain setting are provided, one having a maximum photon energy of 20keV, and one with 40keV maximum. Since the photon energy used for these experiments was 8keV, we used the high gain setting. Also, the detector was operated near room temperature and under these conditions the optimum performance is achieved with the shortest shaping time of 0.5 microseconds. It is capable of operating down to below -20 degrees Celsius if required for energy resolution reasons. This was not necessary for these experiments. It became clear from these experiments that there would be some advantage in having a wider detector. The line shapes measured show minimal asymmetry. Since the effects of asymmetry on the lineshape are well-known it would probably be of benefit to use a sufficiently wide detector such that some asymmetry is observed in order to increase the intensity detected. A factor of 2 or three would probably not be unreasonable.

FOCUSING OPTIC

If high resolution is to be achieved, we need an optical system which provides a good energy resolution, and at the same time provides a strongly demagnified image of the source at the detector. We would like the beam size at the sample to be large, to ensure as many powder particles are intercepted by the beam as possible, and we need a small focal spot (or line) to provide the required good Bragg angle resolution. While asymmetric crystal

FIGURE 2. The silicon diode microstrip array. The aluminum frame is water-cooled to prevent overheating of the readout chips, and to provide a sink for the heat produced by the Peltier coolers when the detector is operated below ambient temperature.

monochromators can provide reasonable energy resolution and a small focus, they cannot provide a large beam cross-section at the sample in a synchrotron context because of the large asymmetry necessary to maintain the Cauchois geometry and hence good energy resolution. We chose to use the standard beamline 2-crystal monochromator to provide the high energy resolution, and to use a secondary optic to provide the strong focusing. We need to take the collimated beam from the beamline monochromator (1.5mm x 4mm) and focus it down to as small a line focus as possible. The ultimate resolution of this instrument depends on the size of this focus and the detector spatial resolution. To perform this focusing, the beam downstream of the focusing optic will have a convergence of about 2 mr. This implies a range of incidence angles on the optic of 1 mr. A grazing-incidence mirror could do the job, since typical grazing angles are 3 mr or more (depending on the mirror coating). Assuming an average grazing angle of 2 mr, the mirror would need to be almost 1m long. This was not practical for us, since the instrument would then not fit in the available space. We chose to use a multilayer mirror to increase the grazing angle and maintain a high reflectivity. The multilayer used was a W/SiC multilayer grown on a standard silicon device wafer, and had a d-spacing of around 4nm, providing a grazing angle of 1 degree at 8keV. It was mounted in a simple spring-loaded bender, illustrated below, and bent to the correct radius by inspection of the focal spot using a YAG scintillator viewed by a video camera.

FIGURE 3. The multilayer mirror mounted on its bender. The bender is a simple 'U' bender, actuated by stretching a tension spring using a motorized translation stage. The feet of the 'U' ride on steel balls against a glass plate, minimizing any stiction. The mount is kinematic to minimize any undesired twist.

Test data for the system was collected using Maltose as a test material. It makes a good test, since it has many lines, even at low angles, and its structure is well-known. The system was operated at 8keV, primarily because the available bendable multilayer could accept a significant beam aperture at that energy. Unlike a crystal analyzer, the momentum resolution of the instrument is energy-dependent since it offers a constant angular resolution, whereas a crystal analyzer offers constant momentum resolution. The data was taken using the 48mm long array illustrated above, and stepping through the required angular range in 1.5 degree steps (the detector covered 2.6 degrees). Obviously a full instrument would have a complete position-sensitive detector covering the entire range. Below we show the data, together with a Reitveld fit using a Voigt line profile. We also show a crystal analyser scan of the same material for comparison. The crystal scan was collected using 17keV radiation and so it's angle scale is roughly half that for the Guinier data. It consisted of 3274 points of 1 second each, i.e. about 60 minutes for the scan. The Guinier instrument covered a similar d-spacing range in 14 exposures of 50 seconds each, collecting roughly 5 times the counts. If we had the full detector array, this would represent a speed increase of 300 for equivalent counting statistics, a worthwhile improvement for many applications. If that detector had 3 times wider diodes (a practical possibility), then the speed gain would approach 1000. The ideal instrument would have an improved multilayer optic, with improved figure error and coating quality, smaller d-spacing and would be much longer to allow operation at higher energies without losing beam aperture. It would have a full detector array of about 7000 elements covering 90 degrees of the Rowland circle (45 degrees in 2-theta). The detector would be mounted in such a way that it could switch between the two modes of operation (as shown in fig. 1) quickly and easily. Such an instrument would be a powerful tool for chemistry and materials science.

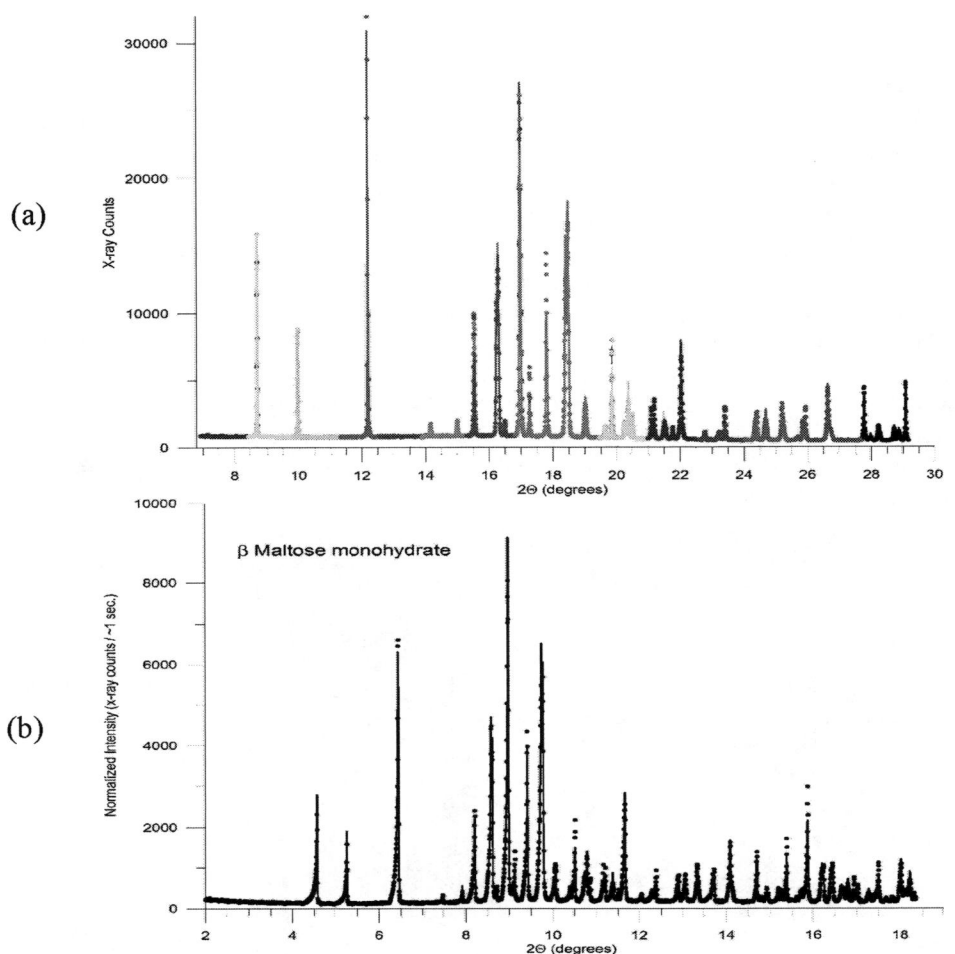

FIGURE 4. Data collected using (a) the Guinier camera and (b) a crystal anayzer instrument.

REFERENCES

1. A. Guinier, 'X-ray Crystallographic Technology', Hilger & Watts, London, 1952.
2. G. De Geronimo et al, IEEE Trans. Nucl. Sci, 50 (2003) 885-891

A New Attachment of the Large Debye-Scherrer Camera at BL02B2 of the SPring-8 for Thin Film X-ray Diffraction

Keiichi Osaka, Shigeru Kimura[a], Kenichi Kato[a] and Masaki Takata[a,b]

Japan Synchrotron Radiation Research Institute (JASRI), 1-1-1 Kouto, Sayo-cho, Sayo-gun, Hyogo 679-5198, Japan
[a]CREST/JST, Honcho, Kawaguchi, Saitama 332-0012, Japan
[b]RIKEN, Kouto, Sayo-cho, Sayo-gun, Hyogo 679-5198, Japan

Abstract. We have developed a new attachment for thin film x-ray measurements equipped with the large Debye-Scherrer camera at BL02B2 of the SPring-8. It is quite easy to handle and control the attachment using user-friendly computer programs. It is also notable that the attachment realizes both low-glancing-angle incident beam condition and in-plane measurement condition. The attachment enables us to measure high precision x-ray diffraction data from many kinds of thin films with the Debye-Scherrer camera. Using this attachment we try to analyze some dimensionally controlled thin film structures.

Keywords: thin film, nanomaterials, Debye-Scherrer camera, in-plane diffraction.
PACS: 61.10.Kw, 68.05.Cf, 68.65.-k

INTRODUCTION

For many applications of functional materials, fabrication of the high quality thin films is a key issue. In order to fabricate thin films with a good property, it is important to determine the crystallographic structure of the thin films. However, it is difficult to measure x-ray diffraction from the thin film using a conventional x-ray diffractometer with laboratory x-ray source because of their poor crystallinity and divergence of the incident beam. We have therefore developed a new attachment for thin film x-ray measurements equipped with the large Debye-Scherrer camera [1,2] installed at BL02B2 of the SPring-8 which was developed for the x-ray powder diffraction measurements using high-intense synchrotron radiation source. It is quite easy to handle and control the attachment using user-friendly computer programs. The attachment enables us to measure high precision x-ray diffraction data from many kinds of thin films with the Debye-Scherrer camera. It is also notable that the measurements with both low-glancing-angle incident beam and in-plane condition can be performed without any changes in the setup of the attachment. Using this attachment we try to analyze some dimensionally controlled thin film structures and contribute to the improvement of thin film devices.

SYSTEM ARRANGEMENT AND PERFORMANCE

The large Debye-Scherrer camera has only one motor–driven stage (ω-axis) that rotates a sample filled in a fine glass capillary. The attachment (Fig. 1) has three translation stages (X, Y, Z), and two tilt stages (RY, RZ), where Y means the direction of the incident x-ray beam, Z the vertical direction and X the normal direction of YZ plane, respectively. Table 1 shows the specification of the stages. Each stage is driven by a stepping motor. The attachment has the dimension of 16 cmϕ x 30 cm and the weight of only 3 kg. Therefore it is easily installed on the ω-axis of the large Debye-Scherrer camera, and the system combined with the large Debye-Scherrer camera and the attachment has six motor-driven stages (see Fig. 2(a)).

CP879, *Synchrotron Radiation Instrumentation: Ninth International Conference,*
edited by Jae-Young Choi and Seungyu Rah
© 2007 American Institute of Physics 978-0-7354-0373-4/07/$23.00

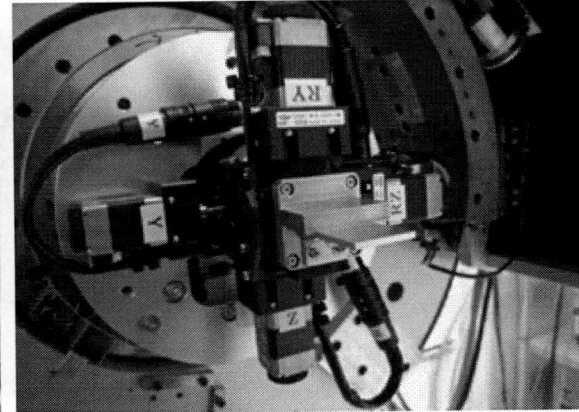

FIGURE 1. Photographs of the new attachment for thin film x-ray measurements equipped with the large Debye-Scherrer camera at BL02B2 of the SPring-8: (a) a overview and (b) a close-up image around the attachment.

TABLE 1. Specification of the attachment for thin film diffraction installed at BL02B2 of the SPring-8. Schematic representation of the behavior of the stages is shown in Fig. 2(a).

Name of Axis		Stroke	Resolution	Behavior
Translation stage				
	X	16 mm	2 μm	Normal to YZ plane (Identical with the direction of ω-axis)
	Y	16 mm	1 μm	Parallel to the incident beam
	Z	17 mm	1 μm	Vertical
Tilt stage				
	RY	17°	1.5×10^{-3} °	Rotation around Y axis
	RZ	12.6°	1.1×10^{-3} °	Rotation around Z axis
	RX	No limitation	2.0×10^{-3} °	Rotation around X axis (Identical with ω-axis)

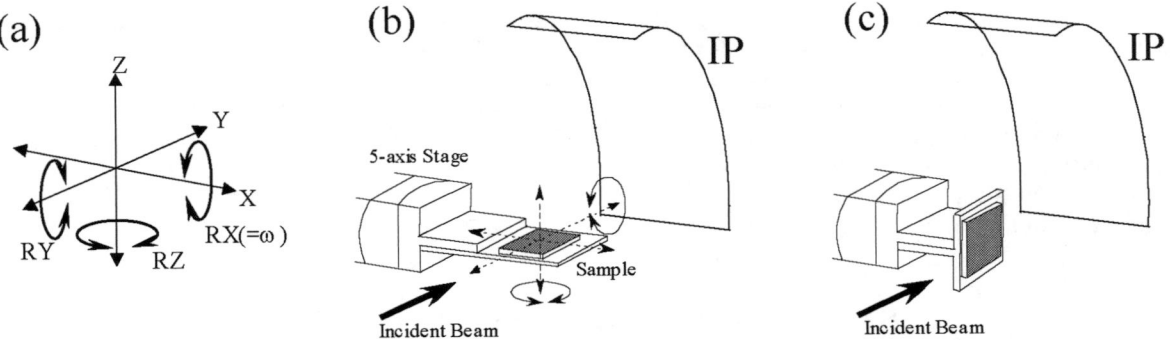

FIGURE 2. Schematic figures around the position of the sample. (a) Behavior of the stages of the attachment, (b) low-glancing-angle incident beam condition and (c) in-plane measurement condition.

The attachment realizes both low-glancing-angle incident beam condition (Fig. 2(b)) and in-plane measurement condition (Fig. 2(c)). These two conditions are effective to measure thin film x-ray diffraction. Two types sample holders made of aluminum are designed:

1. plane plates for low-glancing-angle incident beam condition (Fig. 2(b)),
2. T-shaped plates for in-plane condition (Fig. 2(c)).

These holders are designed for holding the flat sample with the size of 5 to 20 mm, and easily mounted on the sample position on the attachment. Instead of them, even glass plates which are commercially available can be mounted.

For control of the attachment, a set of the computer programs has been newly developed using *LabVIEW*, which is the programming code produced by National Instruments Corporation. These programs can control all motor-driven stages of the attachment and also the Debye-Scherrer camera. We can therefore record the diffraction patterns automatically on the imaging plate detector of the Debye-Scherrer camera. Using this system, we have succeeded to obtain some fine diffraction patterns from thin films of the nano-porous coordination compounds with thickness of less than 1 nm.

EXPERIMENTAL RESULTS

For examples, diffraction patterns for UHV-CVD grown 1,4-Fc$_2$Aq thin films on SiO$_2$/Si substrates[3] are shown in Fig. 3. The thin films have nominal thickness of about 40 to 50 nm. Two-dimensional diffraction patterns recorded on the imaging plates and their intensity profiles for (a) as-deposited, (b) THF-absorption-treated and (c) 120 °C-annealed samples are shown, respectively. These patterns were recorded under the low-glancing-angle incident beam condition (see Fig. 2(b)). X-ray with the wavelength of 0.1 nm was used as incident beam. The beam size was 0.2 mm in the vertical and 3.0 mm in the horizontal direction, respectively. The glancing angle at the

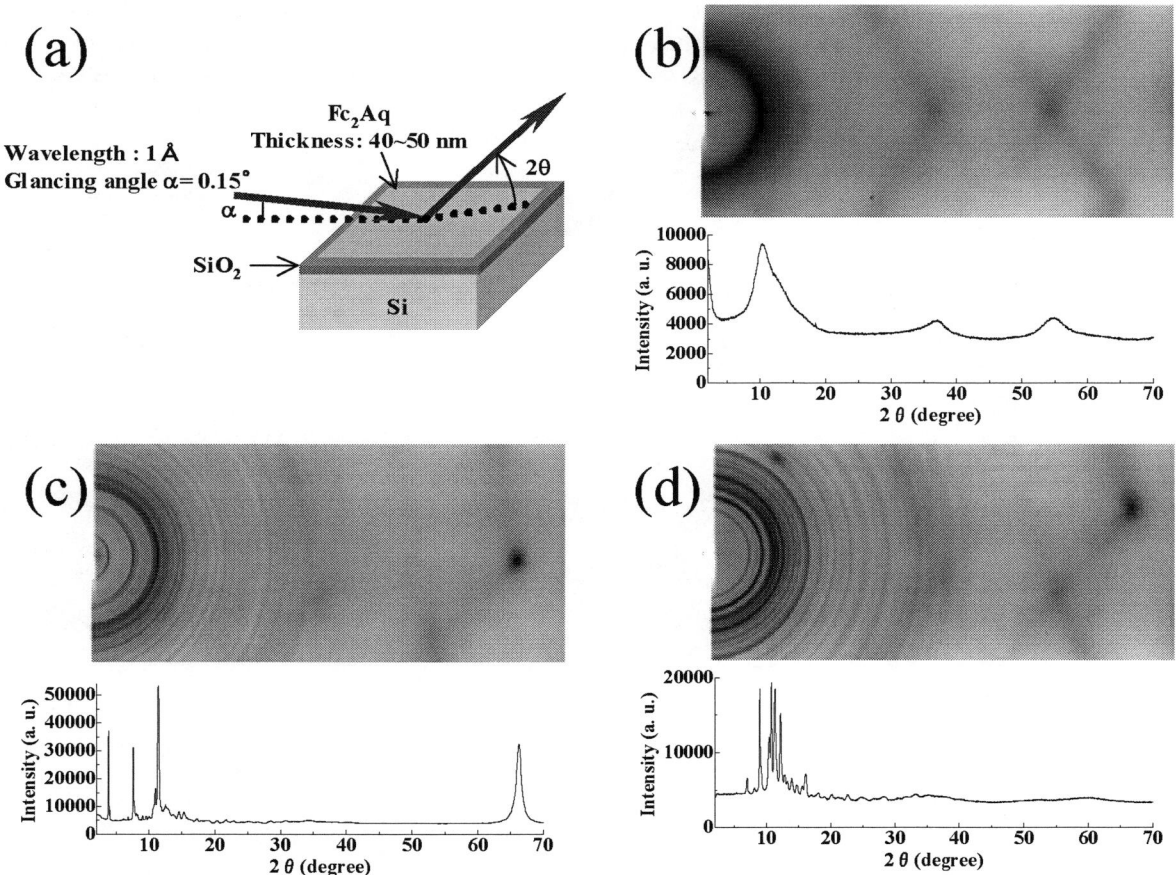

FIGURE 3. Diffraction results for UHV-CVD grown 1,4-Fc$_2$Aq tin films on SiO$_2$/Si substrates: (a) experimental and sample condition, Diffraction patterns for (b) as-deposited sample, (c) THF-treated sample, and (d) 120 °C-annealed sample. Diffracted intensity vs. diffraction angle 2θ is also plotted for each diffraction pattern.

surface of the sample was 0.15°, which is near the total reflection condition. Exposure time was 10 minutes.

In Fig. 3(b), a broad Debye ring is seen around $2\theta=10°$, which is the diffraction peaks due to the thin film. No sharp ring was observed, which means the structure of the thin film is amorphous. After THF-absorbing treatment (Fig. 3(c)), one can recognize that fine Debye rings become dominant and broad ones reduce their intensity. This means the structure of the thin film changes from amorphous to crystal with the THF absorption. It is clearly seen in Fig. 3(d) that additional annealing at 120 °C results in a change in the crystallographic structure.

CONCLUDING REMARKS

We developed a new attachment for thin film x-ray measurements equipped with the large Debye-Scherrer camera at BL02B2 of the SPring-8. The attachment enables us to measure high precision x-ray diffraction data from many kinds of thin films with the large Debye-Scherrer camera.

Further modification of the thin film measurement has been in progress. At the start of it, we are getting started to prepare a collimator system which has aperture as narrow as 0.02 mm. Using the system, it is expected that back ground intensity should be suppressed and weak intensity from the thin film should be clearly observed. We are also ready to provide a high temperature system up to 1000 K. This will enable us to perform *in situ* diffraction analysis of thin films such as electronic devices used in the high temperature environment.

ACKNOWLEDGMENTS

This work was supported by the Grant-in-Aid for Scientific Research in a Priority Area 'Chemistry of Coordination Space' (No. 17036073 and No. 18033062) from the Japanese Ministry of Education, Culture, Sports, Science and Technology. The authors also thank to Ms. M. Kondo, Dr. M. Murata and Prof. H. Nishihara for providing samples used in this experiment. The thin film diffraction experiments were performed with the approval of JASRI (Proposal No. 2005B0442 and 2006A1793) at BL02B2 of the SPring-8.

REFERENCES

1. E. Nishibori, M. Takata, K. Kato, M. Sakata, Y. Kubota, S. Aoyagi, Y. Kuroiwa, M. Yamakata, N. Ikeda : *Nucl. Instrum. Methods Phys. Res. A* (2001), A467-468, 1045-1048.
2. M. Takata, E. Nishibori, K. Kato, Y. Kubota, Y. Kuroiwa and M. Sakata : *Advances in X-ray Analysis* (2002), 45, 377-384.
3. The thin film diffraction experiment was performed with the approval of JASRI (Proposal No. 2005B0442) at BL02B2 of the SPring-8.

High-Resolution Anti-Parallel Double-Crystal Spectrometer at BL15XU in SPring-8

H. Oohashi*, A. M. Vlaicu*, D. Horiguchi[†], K. Yokoi[†], H. Mizota[†], S. Sakakura[†], Y. Ito[†], T. Tochio**, H. Yoshikawa*, S. Fukushima[‡] and T. Shoji[§]

*National Institute for Material Science, Harima-office, SPring-8, 1-1-1 Koto, Sayo, Hyogo 679-5198, Japan
[†]Laboratory of Atomic and Molecular Physics, ICR, Kyoto University, Uji, Kyoto 611-0011, Japan
**Keihanna Institution Plaza Inc., Seika-cho, Soraku-gun, Kyoto 619-0237, Japan
[‡]National Institute for Material Science, 1-2-1 Sengen Tsukuba Ibaraki 305-0047, Japan
[§]X-ray Research Laboratory, RIGAKU Co,. 14-8 Akaoji, Takatsuki, Osaka 569-1146, Japan

Abstract. A simple double-crystal spectrometer with an anti-parallel setting was developed in the third generation synchrotron facility BL15XU, SPring-8. This is derived by modifying the commercial RIGAKU Spectrometer (3580E). Although the resolving power of the two-crystal x-ray spectrometer is largely dependent on the perfection of the crystals, the x-ray spectrometer secures high resolving power, intensity, and reproducibility. This system is convenient for the experiment of Resonant Inelastic X-ray Scattering (RIXS). As an example, RIXS data of several elements in ferroelectric materials $LiTaO_3$ and $BaTiO_3$ and XAS spectra of Cu metal and Cu oxide are presented.

Keywords: two-crystal x-ray spectrometer, anti-parallel setting, high resolution
PACS: 07.85.Nc, 07.85.Qe, 32.30.Rj, 78.70.En, 78.70.Dm

INTRODUCTION

Fluorescent X-rays emitted from excited atoms provides us with a lot of information about materials. Especially, the shape of emission spectra and the behaviour of them under several conditions are clearly affected by the core electronic state and the valence or conductive band. So we can study about the chemical state and local structure of a sample. Using hard X-ray beam, bulk sensitive X-ray emission spectroscopy can be performed.

We quoted an Introduction of the two-crystal spectrometer from the late prof. L. G. Parratt [1]; "The development of double-crystal x-ray spectrometry has provided the researcher with an invaluable tool in making more precise measurements of x-ray diffraction ···."

In order to examine the detail of the emission and absorption spectra using the high brightness of the third generation synchrotron radiation, we modified and improved the commercial RIGAKU Spectrometer (3580E) [2] to be used in SPring-8. We settled a high-resolution double-crystal spectrometer (HRDCS) with a single axis in BL15XU. The theoretical possibilities of the two crystal method were explored by Tochio et al. [3]. With this spectrometer, not only usual X-ray emission spectroscopy (XES) but also X-ray absorption fine structure spectroscopy (XAFS), resonant inelastic X-ray scattering spectroscopy (RIXS) partial fluorescence yield spectroscopy (PFY) are available. Furthermore, this spectrometer has considerably high resolution, so fine analysis of X-ray emission spectra can be obtained. There are some ways to search the same phenomenon of electron's transitions in an atom such as X-ray photoelectron spectroscopy, transmission mode spectroscopy, and total fluorescence yield (TFY) spectroscopy, mössbauer spectroscopy. However, they have their own strict conditions in device settings and for the sample state. Compared with above methods, HRDCS supplies the simple settings and doesn't need a high vacuum environment.

In this work, about the apparatus feature of HRDCS using synchrotron radiation, several useful points are described in detail.

CP879, *Synchrotron Radiation Instrumentation: Ninth International Conference,*
edited by Jae-Young Choi and Seungyu Rah
© 2007 American Institute of Physics 978-0-7354-0373-4/07/$23.00

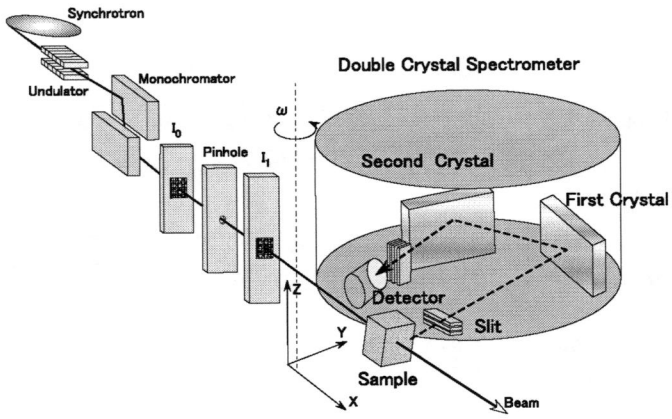

FIGURE 1. A schematic apparatus of the double-crystal spectrometer and entering synchrotron beam. Pinhole shrinks the beam right spotted to the sample. The beam intensity can be measured as I_0 and I_1 by meshed Au detector. The beam monitor can be set behind the sample. We can adjust this system by Y, Z, ω and the diameter of the pinhole.

EXPERIMENT

Beamline and Spectrometer

Figure 1 shows the schematic view of the experiment. At SPring-8 BL15XU, there is an undulator beam source and Si (111) crystals as a monochrometer before the spectrometer.

The key point of the resolution problem is in $(++)$ configuration (anti-parallel) of two crystals and their integrity. Double convolution process between the two instrumental functions of rocking curve and slit function can result in the effective high resolution. Further analysis and numerical treatment are discussed [1, 3].

Compared with Johann-typed X-ray spectrometer, the instrumental function, which is the convolution in rocking curves of the crystal, and slit function of a soller slit in the system, becomes simple. In many cases, it can be neglected and the FWHM of the line widths in the X-ray spectrum is practically the natural line width of the sample. Because the only wavelength satisfies the Bragg's law on both of the two crystals, the first crystal takes an important role as a slit. Therefore, scattered beam can be allowed to enter the first crystal in every direction with natural strong intensity, which enables measurement of high counts of X-ray.

HRDCS has either a F-PC (a gas flow proportional counter) or S-PC (a sealed proportional counter) detector. Measurable 2θ angle ranges from $20°$ to $147°$ only due to the mechanical reason. Scanning steps of 2θ angles for use are $0.0005°$, $0.001°$, $0.002°$, $0.005°$ and $0.01°$. The base stage of HRDCS has two translation axes, y and z, and one rotation axis ω. These axes can be controlled from the outside of the experimental hutch in order to align HRDCS for the synchrotron radiation beam. Sample can be just a bulk form and even liquid can be available. Selectable atmospheres are low vacuum (a few Pa), air, and He gases. There are two soller slits; one (100 mm $\times$ 1 mm layer) restricts the beam's vertical divergence behind the entrance of emitted X-ray, and the other (15 mm $\times$ 0.45 mm layer) is the sub-soller slit in front of the detector.

Si(400), Si(220), Si(111), InSb(111), Ge(111), ADP(101) and TAP(001) crystals are available in this system. Though mixed using of different two crystals needs two axes, only one axis is set in this spectrometer for the convenience of the operation in a vacuum. The shape of two crystals is a rectangular parallelepiped form, 25 mm $\times$ 40 mm $\times$ 5 mm in each size. Piezo $\delta\theta$ fine adjustment for tuning the most suitable alignment in two crystals is done before the measurement.

Sample

Measurable elements are roughly from N to Mo for K lines and from Cu to U for L lines. $2d$ value of each crystal and the spectrometer's movable range of 2θ ($20° \sim 147°$) determine each crystal's energy scan range. Energy range of the measured X-ray and the usable range of the angle of the crystal allow the most suitable crystal.

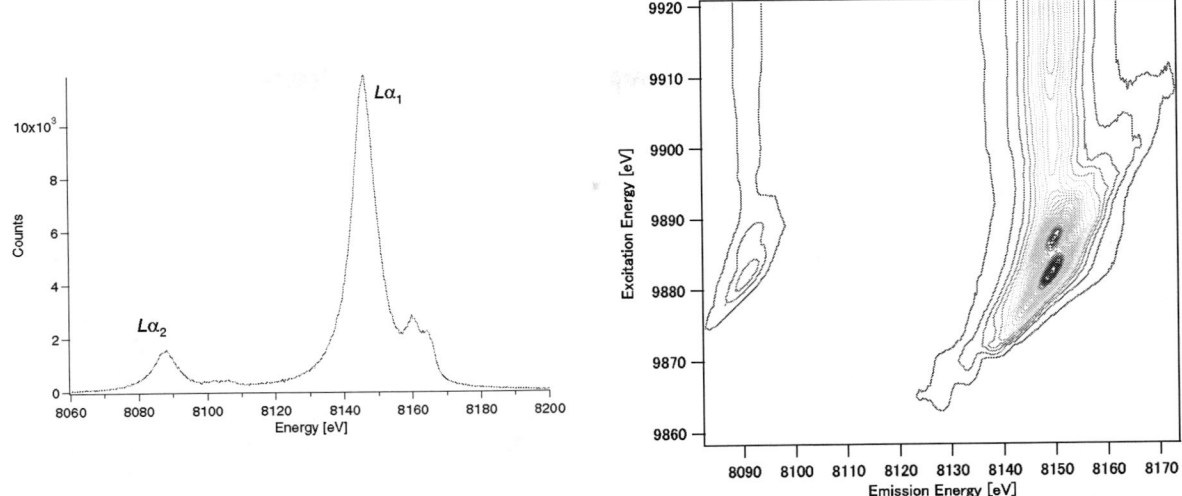

FIGURE 2. (left):Ta $L\alpha$ emission spectra in LiTaO$_3$ at excitation energy 9900 eV. (right):RIXS contour map of Ta $L\alpha$ in stoichiometric LiTaO$_3$.

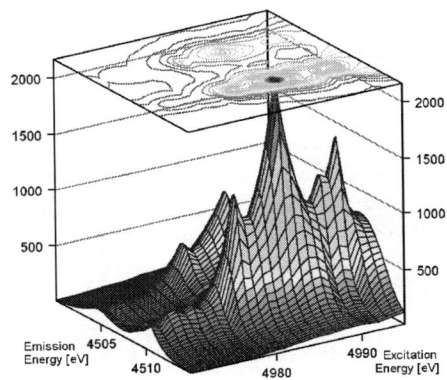

FIGURE 3. RIXS map of Ti K-edge in BaTiO$_3$ at Ti $K\alpha$.

In this study, we measured Cu metal and oxide, and the famous and focused ferroelectric materials, LiTaO$_3$ and BaTiO$_3$. We used a 5-μm-thick Cu foil, CuO powder spread on the both sides of the tape uniformly, pellets of LiTaO$_3$ and BaTiO$_3$ for the measurement.

RESULTS AND BRIEF DISCUSSION

Figure 2(left) shows the Ta $L\alpha$ emission spectra in LiTaO$_3$ at excitation energy 9900 eV. This is measured by this HRDCS in SPring-8. RIXS components can be seen on the higher energy side of each diagram line. Measuring time is 2.5 second per one point and about 30 minutes in total, using Ge (111) crystal.

Figure 2(right) shows RIXS spectra (XANES region) contour map made from Ta $L\alpha$ emission data of each energy. Using this method, XAFS and EXAFS signals can be obtained. Long measurement to gather all fluorescent X-rays can make TFY spectroscopy.

Next sample is BaTiO$_3$ as one of the Ba and Ti compounds. Figure 3 shows a RIXS map of Ti K-edge in BaTiO$_3$ at Ti $K\alpha$ peaks. Some resonance points can be seen clearly.

The third sample is Cu, metal and oxide CuO. The absorption spectra by PFY at Cu $K\alpha_1$ peak and transmission

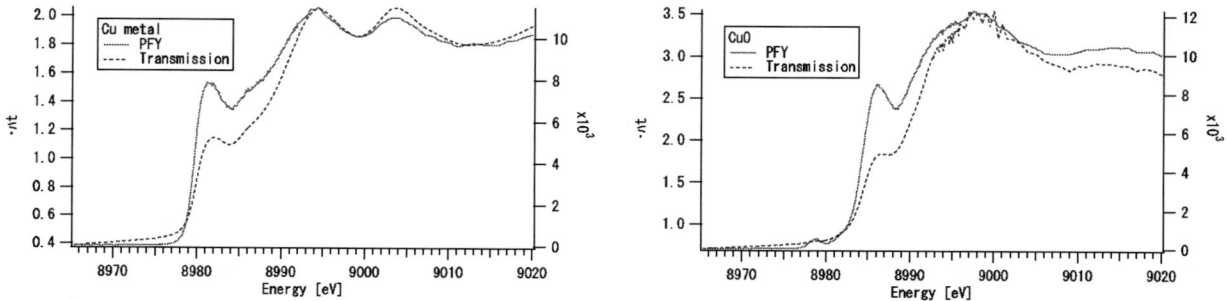

FIGURE 4. The Cu *K*-edge absorption spectra of Cu metal and CuO by transmission and PFY method at Cu $K\alpha_1$ peak.

of each sample are shown in Fig. 4. The beam intensities in front of the sample (I_0) and behind the sample (I_1), and Cu $K\alpha_1$ peak intensity analyzed by HRDCS (I_{emit}) were measured simultaneously at each excitation energy. I_{emit}/I_0 are ploted as PFY and $-\ln(I_1/I_0)$ are ploted as transmission in Fig. 4. We can confirm pre-edge more clearly by PFY method than by transmission method. Especially, the pre-peak of CuO (about at 8979 eV) can be found only by PFY. This peak is due to 1s–3d resonance.

Additionally, temperature device was introduced in this spectrometer during the public beam time at 2006A. So dynamic changes of material states under various temperatures (15 K~ 300 K) can be investigated using this spectrometer. More detailed data analysis of the RIXS in the ferroelectric materials will be in progress for a submission.

CONCLUSION

HRDCS, the spectrometer with the anti-parallel double-crystal setting was settled in synchrotron facility, SPring-8. The measurement of absorption spectra and emission spectra in a Cu metal, a Cu oxide, and two ferroelectric compounds was carried out. At the same time, HRDCS was evaluated in energy, intensity and the resolution with the system of synchrotron light source. Referring to intensity and measuring time, about a couple of days are needed for enough counts. HDRCS has a potential to make a more detailed study of fluorescent X-rays and its application.

ACKNOWLEDGMENTS

The experiments are carried out at SPring-8 under proposals of No. 2004B0654-Nxa-np, No. 2004A0474-Nxa-np, and No. 2005A0457-Nxa-np. This project is supported by Asahi Röntgen Corporation and RIGAKU Corporation, and performed under the approval of the proposal No. C05A15XU-2001N and No. C03A15XU-2004N, and Kyoto Prefecture Collaboration of Regional Entities for the Advancement of Technological Excellence, JST.

REFERENCES

1. L. G. Parratt, *Phys. Rev.*, **41**, 553 (1932).
2. Y. Gohshi, *Appl. Spectr.*, **36**, 171 (1982).
3. T. Tochio, Y. Ito, and K. Omote, *Phys. Rev. A*, **65**, 042502 (2002).

Application of the Quantitative-Phase and Crystal-Structure Simultaneous Analysis to the X-ray Diffraction Data Obtained by Synchrotron Gandolfi Camera System

Masahiko Tanaka[1], Tomoki Nakamura[2] and Takaaki Noguchi[3]

[1] WEBRAM, SPring-8, National Institute for Materials Science,
Kouto Sayo-cho 1-1-1, Sayo-gun, Hyogo 679-5198, Japan
[2] Department of Earth and Planetary Sciences, Faculty of Sciences,
Kyushu University, Hakozaki, Fukuoka 812-8581, Japan
[3] Department of Materials and Biological Sciences, Ibaraki University,
Bunkyo 2-1-1, Mito, Ibaraki 310-8512, Japan

Abstract. Quantitative-phase and crystal-structure simultaneous analysis method has been successfully developed by applying the multi-phase Rietveld refinement to the diffraction data obtained by synchrotron Gandolfi camera system. This newly developed analytical method has been applied to diffraction experiments with micro-amount extraterrestrial samples, which the diameter is a hundred micrometers or less, and the mineral composition ratio of crystalline phase included in the sample has been successfully determined. As the crystal structure is also refined in Rietveld analysis by this method, the structural information of each mineral phase are simultaneously obtained. This method is a totally non-distractive quantitative-phase analysis method and will be a new and unique analytical method for planetary material science.

Keywords: Gandolfi camera, Powder diffraction, Crystal structure analysis, Quantitative phase analysis
PACS: 07.85.Qe

INTRODUCTION

The Gandolfi camera is a unique method to obtain a pseudo-powder-diffraction pattern from a single crystal [1]. This camera gives semi-random rotation to the sample and enables to collect a pseudo-powder-diffraction pattern. Since the development of the imaging plate, this camera gives new attention as a unique method for mineralogical or material-scientific analysis [2]. We have developed a synchrotron Gandolfi camera system [3] at BL-3A beamline of Photon Factory and have applied it to the characterization of many micro-amount extraterrestrial samples [3-5].

Combining this synchrotron Gandolfi camera and multi-phase Rietveld analysis, we developed a new quantitative-phase analysis method for micro-amount extraterrestrial sample, which the diameter is a hundred micrometers or less. The powder diffraction data collected from extraterrestrial samples by using the synchrotron Gandolfi camera is then analyzed by the multi-phase Rieveld refinement method. As the amount of crystalline mineral phase is proportional to the scale factor obtained by Rietveld refinement [6], the mineral composition ratio is determined by using this scale factor. Formerly, only mineral amount ratio between two mineral phases could be determined by comparing the powder diffraction peak intensity characteristic for mineral phases. However, by this method, the mineral composition ratio of many mineral phases can be calculated and determined at once. The crystal structures are also refined in the Rietveld cycle, and the simultaneous analysis of quantitative-phase and crystal-structure is achieved.

The effective and simple method to determine the mineral composition ratio did not exist, in spite of the importance of the mineral composition information to investigate planetary material origin and evolution. However the quantitative-phase analysis by Gandolfi camera can determine the mineral composition ratio from micro-amount samples non-destructively and will become a unique method in planetary material science.

CP879, Synchrotron Radiation Instrumentation: Ninth International Conference,
edited by Jae-Young Choi and Seungyu Rah
© 2007 American Institute of Physics 978-0-7354-0373-4/07/$23.00

This method has merits in this field of analysis as follows. A) Since extraterrestrial materials are usually an agglomerate of small single crystals and are close to powder samples, the Gandolfi camera method is suitable to obtain better quality powder diffraction data. B) High intensity synchrotron radiation enables to collect powder diffraction data from small specimens in practical measurement time. C) High angular resolution caused by low divergence synchrotron X-rays enables to decompose the overlapped diffraction peaks from multi phase samples with ease. D) This method is completely none-destructive and thus the same samples can be characterized by different methods for elemental and isotopic abundances.

It is important that quantitative-phase analysis by the Gandolfi camera is a non-destructive analytical method, for extraterrestrial materials are usually high-cost to obtain and we should gather as much information as possible. We plan to apply this method to characterize mineral compositions of cometary particles that will be brought back by the NASA stardust mission.

DESCRIPTION OF THE SYNCHROTRON GANDOLFI CAMERA

The synchrotron Gandolfi camera system was constructed by modifying a Gandolfi camera produced by Tenno Co. Ltd., Italy. Figure 1 shows the photograph of the Gandlofi camera. The camera is set on the diffractometer at BL-3A experimental station [7] (Fig.1-a). The Gandolfi camera has two axes to rotate the sample randomly (Fig.1-b). The main axis is placed at the center of the sample chamber, which is normal to the incident X-ray beam. The sub-axis crosses at 45 degrees to the main-axis. While the main axis rotates once a minute the sub axis rotates several times at a chaotic ratio around the main axis. The powder diffraction pattern is recorded on the imaging plate attached inside of the chamber. The feature of the Gandolfi camera is tabulated in Table 1.

FIGURE 1. Photographs of the Gandolfi Camera set on the diffracotometer at BL-3A of Photon Factory (a), and the sample rotation mechanism of Gandolfi Camera (b).

TABLE 1. Main feature of the synchrotron Gandolfi camera and its experimental conditions

Camera radius	r = 57.3 mm	X-ray Source	Photon Factory, BL-3A Beamline
Detector	Fiji Film Imaging Plate	Ring operation energy	2.5GeV
IP size	360 * 50 mm	Monochromator	Si(111) double crystal monochromator with saggital focusing system
2θ degree to mm relation	1°(2θ) = 1mm	Mirrors	Pseudo-parabolic collimating and focusing mirrors
Observed 2θ range	12 – 160 °	Beam size at sample	0.3 * 0.3 mm^2
Read out system	Fiji Film BAS-2500	Wavelength (Å)	2.161 +/- 0.002 Å
Pixel resolution	50*50 μ m^2	Exposure time	10 – 180 minutes

EXPERIMENTS

The diffraction experiments are carried out at BL-3A experimental station of Photon Factory, Tsukuba, Japan [7]. The experimental conditions are also summarized in Table 1. The synchrotron radiation from a bending magnet was monochromatized and focused by using the 3A beam line optics and finally cut to 0.3*0.3 mm by a collimator. The higher harmonics are also reduced by the mirror system. In order to decrease background noise, the sample chamber was kept in a vacuum while measurement. The X-ray wavelength used for the diffraction experiment was 2.161 +/- 0.002 Å. Exposure time was 10 to 180 minutes, depending on the size of the sample particle. The recorded diffraction pattern was read out by BAS-2500 IP reader (Fuji Film Co.) and translated to 2theta-intensity format by the program, GanCon2, which we have developed.

The samples were a series of extraterrestrial materials, including interplanetary dust particles and cometary particles brought back by NASA stardust mission. The diameter of each sample particle is about a hundred micrometers or less. Sample particles were mounted on a thin glass fiber with 5 μm in diameter without any pre-treatments after sampling.

ANALYSIS

The determinations of mineral phases were carried out by calculating the d-value of the observed diffraction peaks. The mineral phases included in these extraterrestrial materials are generally well known and the mineral phases were easily determined. These mineral phases were re-confirmed by electron microscope analysis after diffraction experiments.

The quantitative-phase analysis by Rietveld method [8] is carried out after determination of the included mineral phases. In the multi-phase Rietveld refinement, the obtained scale factors are proportional to the amount of included crystalline phases. The mass weight ratio is calculated by the following formula,

$$x_n = (s_n Z_n M_n V_n) / (\Sigma_i s_i Z_i M_i V_i)$$

Where x is mass weight ratio, s is the scale factor, Z is the chemical number, M is the molecular weight and V is the unit cell volume. Thus the mass weight ratio is determined by the evaluation of the scale factor values.

The Rietveld structure refinement program, PFLS [9], was used for Rietveld analysis. In our target samples various mineral phases are included. If all of the parameters are varied in the Rietveld refinement cycles, refinement is hard to converge. Only selected necessary parameters are refined and other parameters are fixed throughout refinements. The crystal structure models for each mineral were obtained from references. Isotropic temperature factors were also adopted from the same references. The chemical compositions of mineral phases, which are determined by an electron microscope after diffraction experiments, are used and assumed to be homogeneous. Absorption correction carried out by using the absorption coefficient estimated from the peak intensity of olivine $((Mg,Fe)_2SiO_4)$, which is the major component of this series of samples. The same peak profile function is used for all mineral phases. In principle, when quantitative-phase analysis is carried out by the Reitveld method, all of the compositions in the sample should be included in calculation, but in this analysis the minor minerals or amorphous phases were neglected.

RESULTS AND DICUSSION

Figure 2 shows the observed diffraction images recorded on the imaging plate. The Debye-Scherrer rings are homogeneous and diffraction spots from large crystalline particles are not observed. Our target samples are usually an agglomerate of fine size single crystal of silicate and iron-related minerals and such samples are suitable for obtaining good quality powder diffraction pattern by the Gandolfi method. Calculating the d-values of each Debye-Scherrer ring, the mineral phases included in the samples are determined, as enumerated in Table 2. Only the major mineral phases are shown in the table, but occasionally small amount of accessory minerals such as iron oxide and iron sulfide exist.

FIGURE 2. Observed diffraction images recorded on the imaging plate.

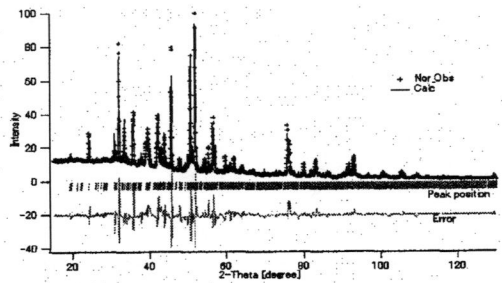

FIGURE 3. A result of Rietveld analysis. This sample (2AF6) includes 4 minerals, olivine, clino-pyroxine, feldspar and troilite. The weight percents ratio is determined as 70.5, 17.1, 11.5 and 0.9 %, respectively. Some errors are still remained around strong diffraction peaks but the observed and calculated diffraction patterns are basically corresponded.

Multi-phase Rietveld refinements ware carried out using the crystal structure models of four minerals tabulated in Table 2. As for the three major silicate minerals, the scale factors, the lattice constants, the atomic positions and the peak profile parameters were refined and converged properly. For troiliete, FeS, the scale factor was only refined. Figure 3 shows a result of the Rietveld analysis. Some errors are still remained around strong diffraction peaks but the observed and calculated diffraction patterns basically correspond.

From the scale factors obtained by the Rietveld analysis, the mass weight ratios of ten kinds of samples were calculated. Table 2 summarizes the result of the quantitative-phase analysis. Considering that the samples used for this experiment did not have any pre-treatment to produce good quality powder sample after sampling, the powder diffraction quality observed are satisfactory and the result of mineral composition ratio determination is acceptable. In order to get further precise analysis result, the following factors should be considered; the variation of peak profile function for mineral phases, the effect of minor minerals or amorphous phase, variation of chemical composition of mineral phases, absorption coefficient and large size crystalline particle effect.

TABLE 2. The result of quantitative phase analysis by Rietveld method. [wt%]

Sample No.	Olivine [(Mg,Fe)2SiO4]	Clino Pyroxine [(Mg,Fe)SiO3]	Plagioclase [NaAlSi3O8]	Troilite [FeS]
2AF1	60.8	24.0	13.1	2.1
2AF2	46.3	31.6	18.1	4.0
2AF3	61.6	23.2	14.9	0.3
2AF4	58.8	26.4	12.6	2.3
2AF5	52.2	32.6	13.8	1.5
2AF6	70.5	17.1	11.5	0.9
2AF7	50.5	30.5	16.5	2.6
2AF8	54.3	29.2	14.6	1.8
2AF9	56.5	27.2	14.0	2.3
2AF10	52.0	32.8	12.3	2.9

SUMMARY

A new quantitative-phase analysis method for micro-amount samples has been successfully developed by a combination of the synchrotron Gandolfi camera and multi-phase Rietveld analysis. Using the multi-phase Rietveld analysis, both information of mineral amount ratio and crystal-structure was obtained simultaneously. This method was applied to analysis of micro-amount extraterrestrial materials and successfully obtained the mineral composition ratio. This method has many advantages for extraterrestrial material analysis. The Gandolfi method is suitable to observe powder diffraction patterns from an agglomerate of small single crystals like extraterrestrial materials. High intensity synchrotron X-rays enable to collect high quality powder diffraction data from micro-amount specimens in practical measurement time. High angular resolution of synchrotron X-rays enables to decompose the overlapped diffraction peaks with ease. This method is completely none-destructive and thus the same samples can be characterized by other methods for elemental and isotopic abundances.

ACKNOWLEDGMENTS

This study was performed under the auspices of the Photon Factory (PAC No. 2001G241, 2003G201 and 2005G125). We thank Mr. Mori for technical assistance during experiments.

REFERENCES

1. Gandolfi G., Miner. Petrogr.Acta, 13, 67-74 (1967).
2. Otto H.H., Hofmann W. and Schroder K., J. Appl. Cryst. 35, 13-16 (2002).
3. Nakamura T., Noguchi T., Zolensky M. E., and Tanaka M., Earth and Planetary Science Letters 207, 83-101 (2003).
4. Nakashima D., Nakamura T., and Noguchi T., Earth and Planetary Science Letters, 212, 321-336 (2003).
5. Nakamura T., Earth and Planetary Science Letters, 242, 26-38 (2006).
6. Werner,P.E., Salome,S. and Malmros, G., J. Appl. Cryst. 12, 107-109 (1979).
7. Sasaki, S., et al, Rev. Sci. Instrum. 63, 1047-50 (1992); Kawasaki, K., et al., Rev. Sic. Instrum. 63, 1023-1027 (1992).
8. Rietveld, H.M., J. Appl. Crystallogr. 2, 65-71 (1969).
9. Toraya H., J.Appl.Cryst.,31,333-343 (1998).

Far infrared Spectroscopy with FTIR Beam Line of MIRRORCLE 20

Nobuhiro Miura[1] Ahsa Moon[2], Hironari Yamada[1,2] Kishi Nishikawa[2],
Toshimichi Kitagawa[2], and Nobuhiko Hiraiwa[2]

[1]Synchrotron Light Life Science Center, Ritsumeikan University,
Noji-higashi 1-1-1, Kusatsu, Shiga 525-8577, Japan,
[2]Department of Photonics, Faculty of Science and Engineering, Ritsumeikan University,
Noji-higashi, 1-1-1, Kusatsu, Shiga 525-8577, Japan

Abstract. A beam line for far infrared spectroscopy using Fourier Transform Infrared Spectrometer (FTIR) has been developed as a facility of tabletop synchrotron MIRRORCLE 20 in Ritsumeikan University and has been utilized to study liquid structure through analysis of intermolecular vibration in aqueous solutions. We report recent developments in the system and the examples of measured spectra.

Keywords: Tabletop synchrotron, far infrared, liquid structure, FTIR
PACS: 61.20.-p

INTRODUCTION

It is well known that water has the network structure constructed by intermolecular hydrogen bonding [1]. This kind of liquid structure is different from a crystal structure in a mineral. The liquid structure is unstable and has only short-distance order since molecular arrangement is changed by micro-Brownian motion continuously. Water molecule interacts with the solute through hydrogen bonding in aqueous solution, and the structure is more complex than pure solvent. Therefore, the liquid structure in aqueous solution has not been clarified completely yet although investigation has been continued for more than thirty years.

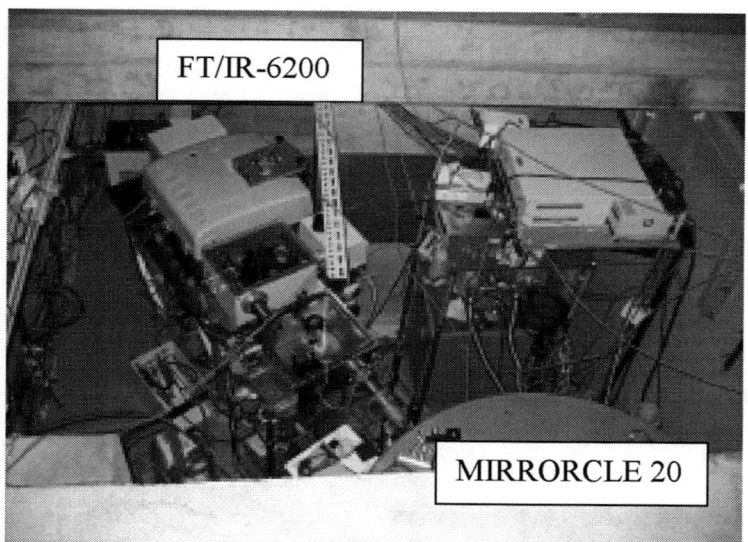

FIGURE 1. FTIR beam line of MIRRORCLE 20.

CP879, *Synchrotron Radiation Instrumentation: Ninth International Conference*,
edited by Jae-Young Choi and Seungyu Rah
© 2007 American Institute of Physics 978-0-7354-0373-4/07/$23.00

Far infrared spectroscopy is a suitable technique to investigate the water structure. We can observe two intermolecular vibration modes attributed to the libration of molecule around 700cm^{-1} and the stretch of hydrogen bonding around 200cm^{-1} in water.[1] These are lattice modes in ice and expected to reflect a condition of hydrogen bonding. However, it is thought as a special technique in the water-containing sample since there is no appropriate light source, and water has large absorption. To overcome this difficulty, the FTIR beam line is prepared to utilize the synchrotron light from MIRROCLE 20 as light source of far infrared [2-5]. We report the present condition and the far infrared spectra.

FTIR Beam Line of MIRRORCLE 20

Far infrared synchrotron light is generated by tabletop MIRRORCLE 20, which has a 30cm diameter electron orbital [2,3]. The light from the whole orbit is focused by an optical system with a quasi-ellipsoidal mirror and a circular mirror [4]. Critical wavelength of synchrotron light is 10 μ m, and the infrared light with range from 2 μ m to 500 μ m is available for the spectroscopy. Figure. 1 shows the FTIR beam line to transfer synchrotron light to FT/IR-6200. The diameter of the body of MIRRORCLE 20 is 1.2m. The beam line is under the ground.

The ceramic heater is commonly utilized as the source for infrared spectroscopy. The calculated spectrum of SR in MIRRORCLE 20 was compared with that of black body radiation in Fig. 2. The SR spectrum is much broader than that of black body radiation in the lower frequency region. Therefore, there is the significant advantage using SR in far infrared.

Absorption spectroscopy was performed by FT/IR-6200 (JASCO Corp., Japan), which equipped step scanning system for a moving mirror and a function of time resolution measurements. Available frequency is in the range between 7800cm^{-1} and 20cm^{-1}. The FTIR main body can be exhausted to avoid water vapor and carbon dioxide on optical pass. Detector of silicon bolometer (Infrared Lab., USA) was used.

We prepared liquid sample cells made of crystal silicon and polyethylene for transmission measurements. The silicon and polyethylene were flat plates of 3mm thickness and 25mm diameter. Samples were put between two plates. Sample thickness were controlled by the Teflon film of thickness of 25 μ m and 50 μ m. The cell was sealed by o-rings. A sample holder was equipped with Peltier elements to control sample temperature between -10.0℃ and 90.0℃.

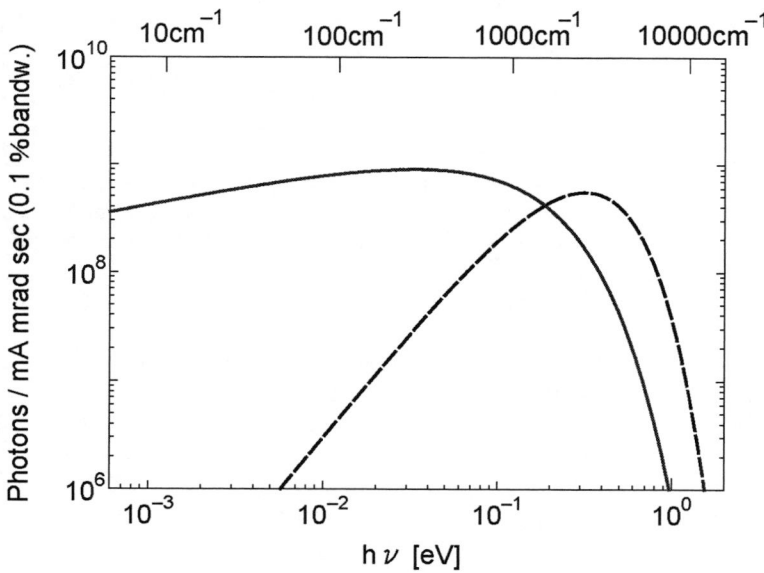

FIGURE 2. Calculated spectra. Solid line: SR light in MIRRORCLE 20. Break line: Blackbody radiation at 1300K. Intensity of the SR light is much higher than that of blackbody radiation in far infrared region.

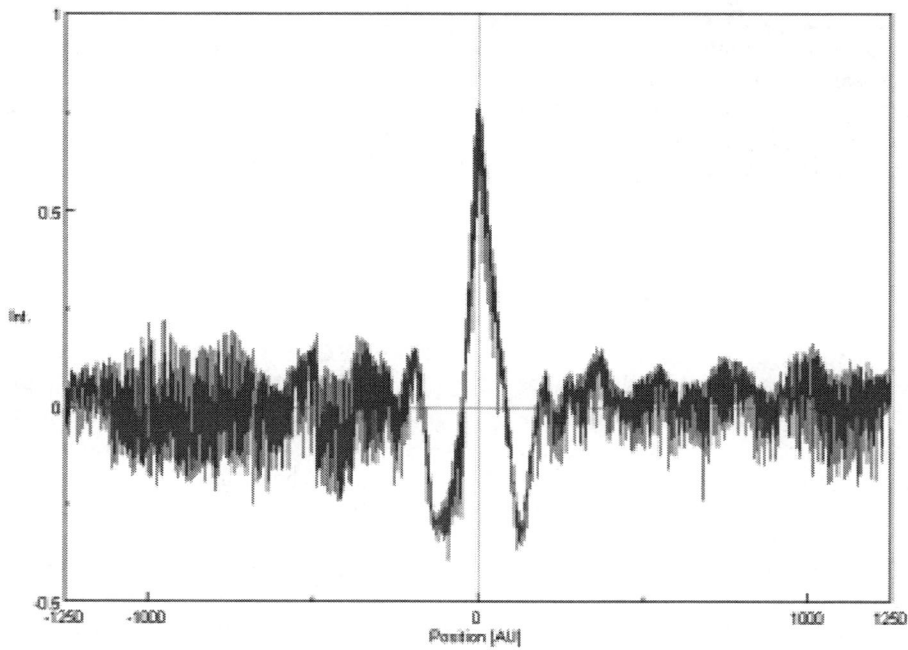

FIGURE 3. Interferogam measured by FTIR beam line. Injection frequency of electron beam was 50Hz, and beam current was 80mA.

Life time of the electron beam in the synchrotron was about 10msec. The rapid scan type of a moving mirror drive mechanism in the interferometer is not appropriate because of time dependence of light source. The FTIR with step scanning system and time resolution function was employed for the beam line. Figure. 2 shows an example of interferogram obtained by the beam line. A single point in interferogram was measured by single beam injection.

FAR INFRARED SPECTROSCOPY

We performed absorption spectroscopy for water in far infrared region. Spectra of water at 20.0℃ were shown in Fig. 4. Closed circles are for experimental data of FTIR beam line measurements using synchrotron light, and the solid line is for off line experiment data of the FTIR. Error bars for closed circles correspond to standard deviation. The spectrum obtained by FTIR beam line was connected with the off line data continually at $100cm^{-1}$. Opened symbols were bibliographic data [6-10]. The experimental data are in good agreement with the bibliographic data. The apparent peak around $200cm^{-1}$ is due to the stretch of hydrogen bonding, which corresponds to the translational lattice mode in ice [1].

Knowledge about intermolecular hydrogen bonding between solute and water is the important factor to understand the aqueous solution properties such as liquid structure, hydration structure and concentration fluctuation. It is known that adding primary alcohol changes water structure drastically, but the interaction with water and alcohol even in such a simple mixture has not been clarified yet. We expect that the far infrared spectroscopy is a useful tool to investigate the interaction in the various systems like biopolymer aqueous solutions, gels, hydrate, and living matter. These basic researches may promote practical applications to utilize the far infrared synchrotron light.

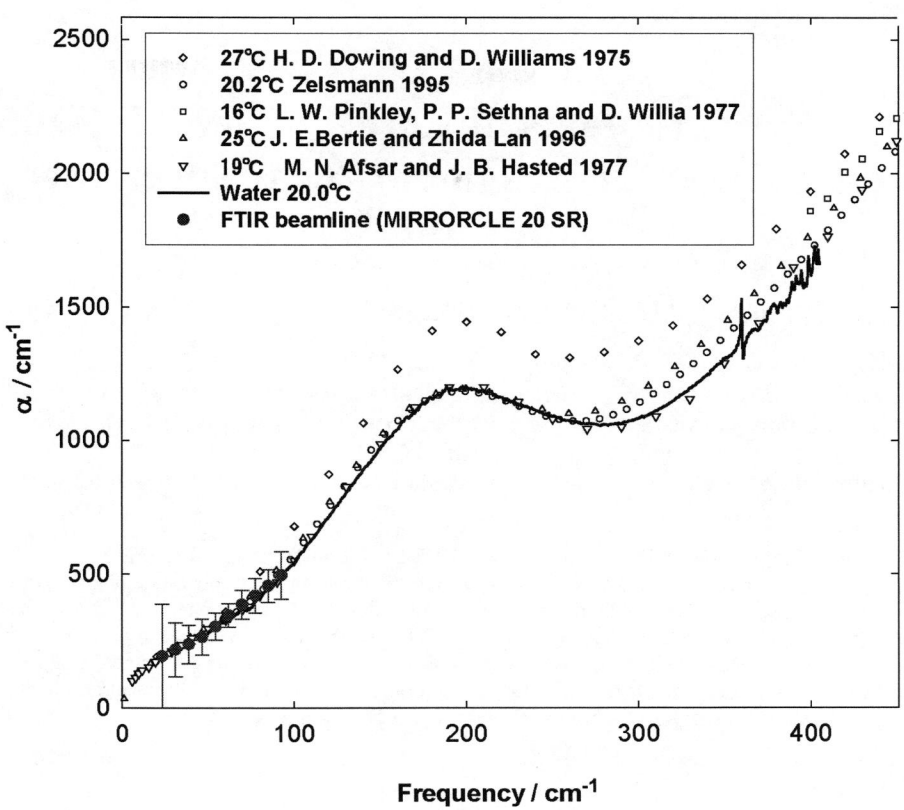

FIGURE 4. Far infrared spectra in water. The experimental data, which obtained by FTIR beam line using MIRRORCLE 20, was shown by plots of closed circle, and the data of off line experiments was shown by solid line. Bibliographic data [6-10] were shown by opened symbols.

REFERENCES

1. D. Eisenberg and W. Kauzmann, "The structure and properties of water", Oxford, 1969.
2. H. Yamada, Adv. Colloid Interface Sci. 71-72, 371-392 (1997).
3. H. Yamada, Proceedings of International Symposium on Portable Synchrotron Light Sources and Advanced Applications, 12-17 (2004).
4. A. Moon, Y. Nakamura, T. Toma, and H. Yamada, Proceedings of International Symposium on Portable Synchrotron Light Sources and Advanced Applications, 167-170 (2004).
5. N. Miura, H. Yamada, A. Moon, and K. Nishikawa, Proceedings of International Symposium on Portable Synchrotron Light Sources and Advanced Applications, 171-174 (2004).
6. H. R. Zelsmann, J. Mol. Struct. 350, 95-114 (1995).
7. J. E. Bertie and Z. Lan, Appl. Spectrosc. 47, 1047-1057 (1996).
8. M. N. Asfer and J. B. Hasted, J. Opt. Soc. Am. 67, 902-904 (1977).
9. H. D. Dowing, and D. Williams, J. Geophys. Res. 80, 1656-1661 (1975).
10. L. W. Pinkley, P. P. Sethna, and D. Williams, J. Opt. Soc. Am. 67, 494-499 (1977).

Coincidence Measurements of Core-Excited Molecules and Clusters Using TOF Fragment-Mass Spectroscopy

Kiyohiko Tabayashi[a,c], Tomoé Maruyama[a],
Kenichiro Tanaka[b,c], Hirofumi Namatame[b,c], and Masaki Taniguchi[b,c]

[a]Department of Chemistry, Graduate School of Science, Hiroshima University, Higashi-Hiroshima 739-8526, Japan
[b]Department of Physical Science, Graduate School of Science, Hiroshima University, Higashi-Hiroshima 739-8526, Japan
[c]Hiroshima Synchrotron Radiation Center (HSRC), Hiroshima University, Higashi-Hiroshima 739-0046, Japan

Abstract. In order to investigate photo-induced processes and dynamics of photochemical reactions involving molecular complexes and clusters in the vacuum ultraviolet and soft X-ray region, a cluster beam-photoreactive scattering apparatus has been designed and constructed at the HiSOR facility in Hiroshima University. The apparatus has a basic arrangement of a cluster beam source and a linear time-of-flight mass spectrometer, where it was designed to give strong cluster beam intensity enough to perform a variety of photoelectron and photoion coincidence measurements. Here, we present spectroscopic and coincidence measurements of the photo-induced processes of atomic and molecular clusters in the soft X-ray region to make a fundamental assessment for the present cluster beam-photoreactive scattering apparatus.

Keywords: Molecular cluster, Core-level excitation; TOF mass spectroscopy; Charge separation mechanism; PEPIPICO
PACS: 36.40; 82.30 Nr; 82.33 Fq.

INTRODUCTION

Molecular clusters have received significant attention from the view points that it is concerned with the problems of intermolecular reaction and energy transfer processes, and the solvent effects on relaxation dynamics of electronically excited molecular systems. So far a number of studies on the dissociation of molecular clusters have been performed up to the vacuum ultraviolet (VUV) energy region [1], however, core-level excitation and induced processes of molecular clusters have been less frequently studied using soft X-ray radiation [2]. This is generally due to low photon intensities available for synchrotron radiation (SR), if it is compared with UV and VUV laser sources, when the most convenient supersonic expansion technique is employed to produce molecular clusters. In our research project for studying SR photochemistry of molecular clusters, we have designed and constructed [3] a cluster beam-photoreactive scattering apparatus (CBPRSA) that assures the strong cluster-beam intensity and effective detection to perform a variety of photoelectron and photoion coincidence measurements in the soft X-ray region.

Here, we conducted core-level excitation studies on the dissociation and ionization of molecular clusters made up of atoms or simple molecules to make a fundamental assessment for the present CBPRSA cluster machine.

EXPERIMENT

Cluster Beam-Photoreactive Scattering Apparatus

Studies on the SR photochemistry of molecular clusters are carried out using the CBPRSA apparatus that can be placed at an end-station of the beamlines of the HiSOR facility in Hiroshima University. A cross sectional view of

CP879, *Synchrotron Radiation Instrumentation: Ninth International Conference*,
edited by Jae-Young Choi and Seungyu Rah

the apparatus is shown in Fig. 1. It is a modified version of the one designed previously [4]. Basically, it consists of three vacuum chambers: a cluster-beam source chamber, a main scattering chamber, and a pumping chamber. The rectangular shaped main chamber (with inner wall dimensions of 640 mm×640 mm×570 mm) is divided into three regions: a differential pumping region, a time-of-flight (TOF) mass detector region, and a scattering region. The apparatus has been designed so that the following three axes cross with each other: (a) the axis of the cluster beam, (b) that of incident SR light, and (c) the centerline of the TOF spectrometer. The main scattering chamber has several flange holes along the axes. The top hole is used to place a rotatory lid table where the TOF mass detector region is furnished. One of the side flange holes is used to mount the cluster-beam source chamber and another one to connect a gate valve where the SR light from a beamline-monochromator is introduced. The axes of the above flange holes are chosen perpendicularly crossed. The cluster-beam source chamber is pumped with a 3300 L/s turbo-molecular pump whereas the three regions (differential pumping, TOF detector, and scattering regions) in the main chamber are pumped with turbo-molecular pumps at the rates of 1200, 300, and 3300 L/s, respectively. The three regions can be pumped down below 5×10^{-6} Pa.

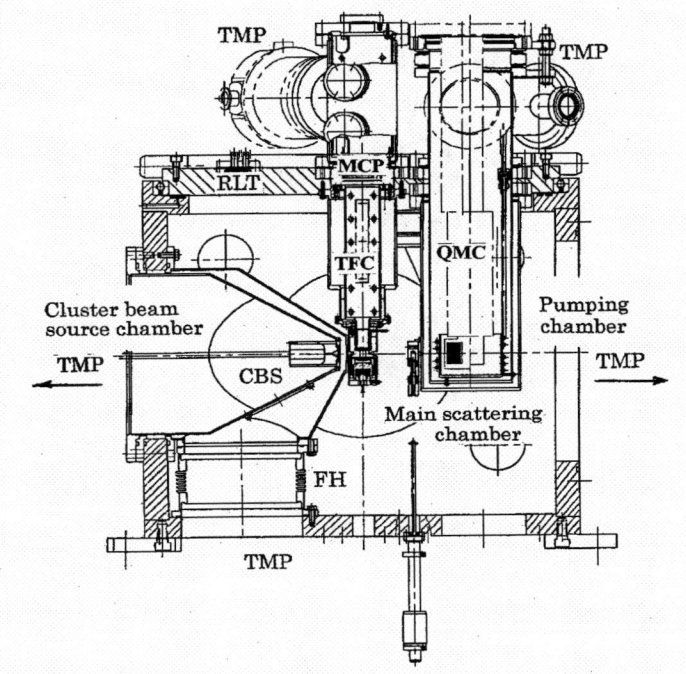

FIGURE 1. Cross sectional side view of the cluster beam-photoreactive scattering apparatus. CBS: Cluster beam source, RLT: Rotatory lid table, TFC: TOF detector chamber, MCP: Multi-channel plate detector, QMC: Quadrupole mass detector chamber, TMP: Turbo-molecular pump, FH: Flange hole

Molecular complexes and clusters are generated by supersonic expansion of sample mixtures from a nozzle of a temperature-controlled beam source. The nozzle is replaceable but the one made of Pt aperture with an opening of $\phi 50$ μm I.D. is typically used at stagnation pressures up to 0.5 MPa. The source is mounted in a holder made of copper block that can be cooled down to 243 K by circulation of coolant. An effusive beam is also produced by introduction of a neat sample gas at a low stagnation pressure through the nozzle with a large aperture (typically $\phi 200$-μm nozzle hole). Either beam is skimmed with a $\phi 1.0$-mm skimmer into the differential pumping region where it is further collimated with 1×4 mm^2 slit. The collimated beam is allowed to enter the main scattering chamber of the CBPRSA apparatus, where it is crossed with a beam of monochromatized SR-light at an ionization zone of the TOF spectrometer of Wiley-McLaren-type [5]. The nominal distance between the nozzle hole and the ionization zone of the TOF spectrometer is ~55 mm.

Product cations resulting from photoionization processes are extracted to a drift tube (L=300 mm) of the TOF spectrometer with a $\phi 47$-mm multichannel plate (MCP) detector, whereas electrons are guided to the opposite direction and immediately detected with another MCP. A continuous electrostatic field across the ionization zone is applied at 525 V/cm in the usual TOF measurements. The axis of the spectrometer is directed perpendicular to the polarization vector of the incident SR-photon beam. With the present optical arrangement and parameters, most of singly charged ions with kinetic energies up to ~15 eV are estimated to be detected without angular discrimination.

Core Level Excitation and Coincidence Measurements

The experiments were performed at the beamline BL-6 of HiSOR [6]. The SR-light from an electron storage ring was monochromatized with a varied-line-spacing plane grating monochromator and focused at the ionization zone of the CBPRSA machine. Core-level excitation spectroscopy and fission mechanism can be analyzed by using TOF fragment-mass measurements. It is generally considered, for example, that resonant Auger relaxation in the pre-edge regime predominantly produces singly charged molecules (atoms) and clusters that may then partially autoionize to form dication depending on the final-state energies. The coincidence techniques such as photoelectron-photoion-coincidence (PEPICO) and photoelectron-photoion-photoion-coincidence (PEPIPICO) were advantageously applied to the core-level excitation studies. Partial ion yields (PIY) spectroscopy by using the PEPICO method is of particular importance to analyze "cluster-specific" excitation measurements when the product cations originated from clusters can be monitored.

The PEPICO and PEPIPICO signals were recorded using a multichannel scaler (Ortec Turbo MCS) and a fast multi-hit digitizer (FAST ComTec 7886). The time resolution was typically set to 2-5 ns. Energy resolution of the soft X-ray radiation was better than $E/\Delta E = \sim 500$ at the energy corresponding to the oxygen K-edge. The photon energy was confirmed using the O($1s$) resonance transitions of CO_2 determined by Prince et $al.$ [7].

RESULTS AND DISCUSSION

Total and Partial Ion Yield Spectra of Ar$_n$ Clusters

Total and partial ion yield (TIY, PIY) spectra of Ar$_n$ clusters in the Ar($2p$) excitation region have been measured using TOF mass spectroscopy by the PEPICO technique. Figure 2 compares typical TIY and PIY (Ar$_2^+$ yield)

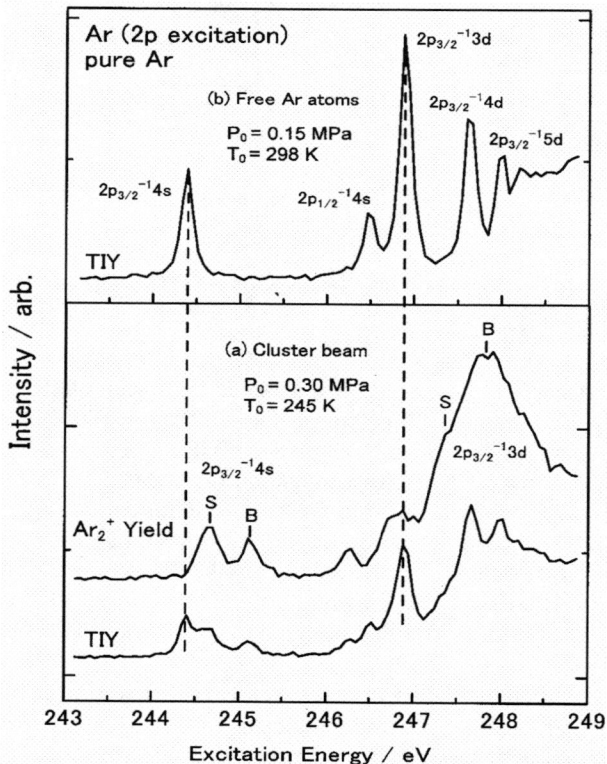

FIGURE 2. Comparison of Ar ($2p$) pre-edge spectra of Ar$_n$ cluster beam with that of free atomic Ar beam.

spectra of (a) Ar_n cluster beam with a TIY spectrum of (b) atomic Ar beam. The cluster beam produced under the stagnation condition at P_0=0.3 MPa with T_0=245 K still contains a certain amount of uncondensed free atoms, but "cluster-specific" excitation spectra could be recorded without atomic contributions by monitoring the PIY for fragment Ar_n^+ cations. The band peaks indicated by **S** and **B** in the Ar_2^+ yield are those of cluster "surface" and "bulk" excited states for the $2p_{3/2}^{-1}4s$ and $2p_{3/2}^{-1}3d$ transitions, respectively. The average cluster size of the present Ar_n cluster beam was estimated to be $<N>$= 20-30 from comparison with the relative intensities of the "surface"/"bulk" cluster bands observed in the soft X-ray absorption (XA) spectra [8]. We have also calculated a scaling parameter Γ^* for condensation given by Hagena [9]. The Γ^* value for the present beam conditions was found to correspond to an average cluster size of $<N>$= ~20 when the experimental ($<N>$ vs. Γ^*) correlation by Wörmer *et al.* [10] was employed. The average cluster size could be thus consistently characterized by both XA-spectroscopic and scaling-law analyses of Ar_n cluster beam experiments.

Dissociative Double Ionization of $(CO_2)_n$ Clusters by PEPIPICO Measurements

C(1s) core-level excitation of $(CO_2)_n$ clusters has been investigated by zero-kinetic-energy electron spectroscopy by Dietrich *et al.* [11]. Although the classification of correlated cation pairs originated from the $(CO_2)_n$ clusters was studied, the details of charge separation and fragmentation dynamics have not been examined. Here, we studied dissociative double ionization mechanism of $(CO_2)_n$ clusters in the O K-edge region, from shape analysis of the contour plots obtained by the PEPIPICO measurements.

The $(CO_2)_n$ cluster beam was generated by supersonic expansion of 10%CO_2 in He or Ar mixture under the beam conditions of P_0=0.50 MPa and T_0=272 K. In the PEPICO spectroscopic measurements, we identified series of mixed fragment-cluster cations, $(CO_2)_mO^+$ and $(CO_2)_mCO^+$ with $m \leq 3$ as well as O_2^+ product characteristic of ion-molecule reaction within the clusters, besides the intact cluster cations $(CO_2)_n^+$ with $n \leq 10$. Intensities of correlated cation pairs were also accumulated by the PEPIPICO technique. The PEPIPICO intensities for the cluster beam with 10%CO_2/He revealed that fragment/cluster (such as O^+, $CO^+/(CO_2^+)_n$) cation pairs were more intense than the cluster/cluster $(CO_2)_n^+/(CO_2)_m^+$ cation pairs. When a cluster beam was generated using more efficient diluent Ar for the cluster formation, i.e. 10%CO_2/Ar, intensity of the cluster/cluster channels increased from 31 to 66 % out of the total PEPIPICO intensities from core-excited clusters, whereas that of the fragment/cluster channels decreased, accordingly.

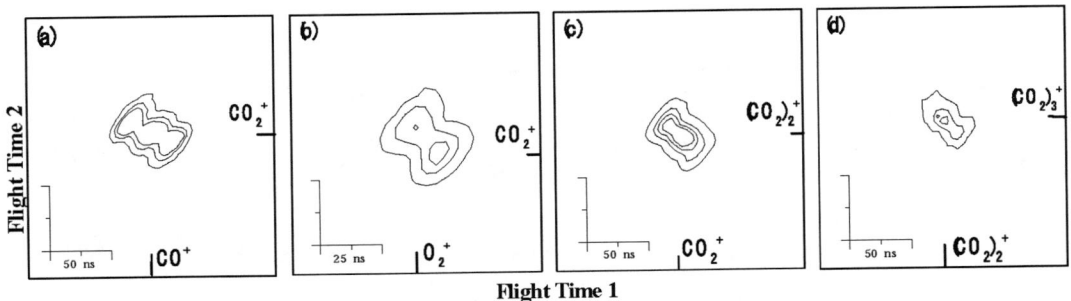

FIGURE 3. PEPIPICO contour plots of $(CO_2)_n$ clusters recorded with O1s$\rightarrow\pi^*$ excitation at 535.4 eV. (a) CO^+/CO_2^+ region; (b) O_2^+/CO_2^+ region; (c) $CO_2^+/(CO_2)_2^+$ region; (d) $(CO_2)_2^+/(CO_2)_3^+$ region.

Figure 3 shows the PEPIPICO contour plots for the predominant channels recorded at 534.5 eV in a cluster beam (10%CO_2/He, P_0=0.50 MPa, T_0=272 K). The contours are analyzed based on the dynamics of multi-body dissociation of core-excited dication [12,13]; when the flight times, (t_1 and t_2), of correlated cations with single charge are plotted, the slopes of the contour allow us to examine the dissociation mechanism, and their lengths provide kinetic energy releases (KERs) of the dominant fragmentation steps. The results are summarized in Table 1.

The contour plots indicate that the dissociation of the prominent cation pairs such as $(CO_2)_n^+/(CO_2)_m^+$ and O^+, CO^+/CO_2^+ takes place via the secondary decay (SD) mechanism where charge separation proceeds prior to the secondary releases of neutral moieties. From analysis of the charge separation mechanism, we found relatively small KERs (2.5-3.5 eV) with charge separation distances (CSDs) of 4.0-5.5 Å that correspond to the fission of double charges located separately at the nearest-neighbor molecular distance.

Further details of the above results will be presented elsewhere. In conclusion, the CBPRSA apparatus constructed for studying the SR photochemistry of molecular clusters has showed the present examples of satisfactory achievement for both spectroscopic and coincidence measurements of core-excited atomic and molecular clusters.

TABLE 1. Charge separation mechanism in core-excited $(CO_2)_n$ clusters

Correlation	Intensity	Reaction Mechanism	Slope		KER	CSD
			Expt.	Model	(eV)	(Å)
CO^+/CO_2^+	102	$(CO_2)_4^{2+} \rightarrow CO_2^+ + (CO_2)_3^+$ $CO_2^+ \rightarrow O + CO^+$	-0.51	-0.52 (SD)	5.5	2.6
O_2^+/CO_2^+	35	$(CO_2)_3^+ \rightarrow CO_2^+ + 2CO_2$ $(CO_2)_7^{2+} \rightarrow (CO_2)_3^+ + (CO_2)_4^+$ $(CO_2)_3^+ \rightarrow O_2^+ + 2CO + CO_2$	-1.09	-1.03 (SD)	3.1	4.6
$CO_2^+/(CO_2)_2^+$	100	$(CO_2)_4^+ \rightarrow CO_2^+ + 3CO_2$ $(CO_2)_7^{2+} \rightarrow (CO_2)_2^+ + (CO_2)_5^+$ $(CO_2)_2^+ \rightarrow CO_2^+ + CO_2$	-0.83	-0.80 (SD)	2.6	5.4
$(CO_2)_2^+/(CO_2)_3^+$	9.3	$(CO_2)_5^+ \rightarrow (CO_2)_2^+ + 3CO_2$ $(CO_2)_{14}^{2+} \rightarrow (CO_2)_5^+ + (CO_2)_9^+$ $(CO_2)_5^+ \rightarrow (CO_2)_2^+ + 3CO_2$ $(CO_2)_9^+ \rightarrow (CO_2)_3^+ + 6CO_2$	-0.85	-0.83 (SD)	1.4	10.4

ACKNOWLEDGMENTS

The authors thank the staff of HSRC for stable operation of the synchrotron radiation source. The present study was performed under the Cooperative Research Programs of HiSOR, in HSRC, Hiroshima University.

REFERENCES

1. For example, see *Vacuum Ultraviolet Photoionization and Photodissociation of Molecules and Clusters*, edited by C.-Y. Ng (World Scientific, Singapole 1991).
2. E. Rühl, *Int. J. Mass Spectrom.* **229**, 117-142 (2003).
3. K. Tabayashi, K. Yasunaga, J. Aoyama, S. Wada, K. Tanaka, A. Hiraya, and K. Saito, *Proceedings of the Fourth Hiroshima International Symposium on Synchrotoron Radiation*, HSRC, Hiroshima University, 2000, p. 333.
4. K. Tabayashi and K. Shobatake, *J. Chem. Phys.* **84**, 4919-4929 (1986).
5. W.C. Willey and I.H. McLaren, *Rev. Sci. Instr.* **26**, 1150-1157 (1955).
6. H. Yoshida, Y. Senba, T. Goya, Y. Azuma, M. Morita, and A. Hiraya, *J. Electron Spectrosc. Relat. Phenom.* **144-147**, 1105-1108 (2005).
7. K.C. Prince, L. Avaldi, M. Coreno, R. Camilloni, and M. de Simone, *J. Phys. B* **32**, 2551-2657 (1999).
8. O. Björneholm, F. Federmann, F. Fössing, and Möller, *Phys. Rev. Lett.* **74**, 3017-3020 (1995).
9. O. F. Hagena, *Z. Physik* D **4**, 291-299 (1987).
10. J. Wörmer, V. Guzielski, J. Stapelfeldt, G. Zimmerer, and T. Möller, *Phys. Scri.* **41**, 490-494 (1990).
11. H.-J. Dietrich, R. Jung, E. Waterstradt, and K. Muller-Dethlefs, *Ber. Bunsenges. Phys. Chem.* **96**, 1179-1183 (1992).
12. M. Simon, T. Lebrun, R. Martins, G.G.B. de Souza, I. Nenner, M. Lavollee, and P. Morin, *J. Phys. Chem.* **97**, 5228-5237 (1993).
13. K. Tabayashi, S. Tada, J. Aoyama, K. Saito, H. Yoshida, S. Wada, A. Hiraya, and K. Tanaka, *J. Electron Spectrosc. Relat. Phenom.* **144-147**, 179-182 (2005).

Development of Auger-Electron–Ion Coincidence Spectrometer to Study Decay Dynamics of Core Ionized Molecules

T. Kaneyasu, Y. Hikosaka and E. Shigemasa

UVSOR Facility, Institute for Molecular Science, Nishigonaka 38, Myodaiji, Okazaki 444-8585, Japan

Abstract. A new electron-ion coincidence spectrometer for studies on decay dynamics of core excited/ionized molecules has been developed. The coincidence spectrometer consists of a double toroidal electron analyzer (DTA) and a three-dimensional ion momentum spectrometer, which are followed by time- and position-sensitive detectors (PSDs). The Auger electron is energy analyzed by the DTA, while the ions are extracted from the interaction region by a pulsed electric field triggered by the detection signal of the electron, and the time-of-flight of the ions and their position information on the ion PSD are obtained. The design and performance of the coincidence spectrometer are reported.

Keywords: Toroidal electron analyzer; Ion momentum spectrometer; Energy and angle correlations; Auger electron; Coincidence technique
PACS: 07.81.+a; 32.80.Hd; 33.80.Eh

INTRODUCTION

The understanding of the dynamics induced by inner-shell ionization of molecules using synchrotron radiation has been considerably progressed in the last two decades, thanks to the rapid progress of synchrotron radiation instrumentation and related experimental technique. The core-ionized molecules generally relax by emitting Auger electrons and fragment ions. Coincidence detection and correlation analysis among the ejected particles during the decay process bring a deeper insight into the decay dynamics of the core-ionized molecules. The first Auger electron–photoion coincidence (AEPICO) experiment has been performed, to investigate the state-to-state dissociation dynamics of defined molecular ion states [1]. Recently, an improved AEPICO technique has been applied for studying the photofragmentation of some complex molecules, and has highlighted the role played by the lone-pair electrons localized on the target atom in selective fragmentation [2,3].

Another important aspect for the AEPICO technique lies in vector correlation measurements between the Auger electron and fragment ion. For investigating the molecular photoionization dynamics, angular distribution measurements of fixed-in-space molecule photoelectrons, which have been realized by the vector correlation measurement between the photoelectron and the fragment ion, have been successfully performed [4]. However, the vector correlation measurement between the Auger electron and fragment ion has been rarely obtained [5,6], since it is still difficult to observe momentum vectors of fast Auger electrons and ionic fragments simultaneously with sufficient resolution.

In order to overcome such experimental difficulties, and to assess further information on the decay dynamics relevant to the core-hole creation, we have developed an AEPICO spectrometer. This spectrometer enables us to make correlation analysis through the measurement of three-dimensional momentum vectors of ionic fragments, and energy- and angle-resolved Auger electrons. An overview of the new setup and results of photoelectron-ion coincidence measurements as a performance test of the spectrometer are described.

CP879, *Synchrotron Radiation Instrumentation: Ninth International Conference,*
edited by Jae-Young Choi and Seungyu Rah
2007 American Institute of Physics 978-0-7354-0373-4/07/$23.00

ELECTRON-ION COINCIDENCE SPECTROMETER

A schematic drawing of the AEPICO spectrometer which we have newly developed is illustrated in Fig. 1. The spectrometer is composed of the double toroidal electron analyzer (DTA) [7] and an ion momentum spectrometer, both of which are followed by time- and position-sensitive detectors (PSDs). Ionization light crosses the gas beam effusing from the 0.2φ aperture on the repeller plate (see Fig. 1(b) for details).

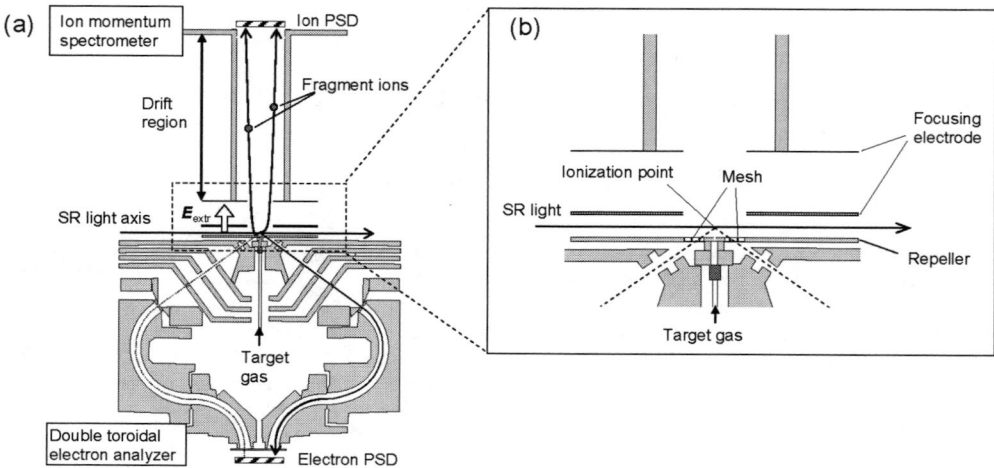

FIGURE 1. Schematic drawing of the electron-ion coincidence spectrometer

On the observation of Auger electrons, the interaction region is kept to be field-free, which is essential to achieve high energy- and angular resolutions. The DTA for the Auger electron observation is basically the same as that Céolin *et al.* have developed [8]. The DTA consists of a four element conical electrostatic lens, two toroidal shaped deflectors and the PSD. The lens system defines the central detection angle as 54.7° with respect to the analyzer symmetry axis, and the acceptance solid angle is 5% of 4π sr. Auger electrons emitted in the acceptance angle are focused by the electrostatic lens onto the entrance slit, and then dispersed in energy by the toroidal deflectors. The Auger electrons within an energy range 10% of the pass energy can pass through the toroidal deflectors and arrive at the PSD. From the arrival positions' information, the energies and emission angles of the Auger electrons can be determined. The energy and angular resolutions are estimated by a numerical simulation to be <1% of the pass energy and 5-7°, respectively.

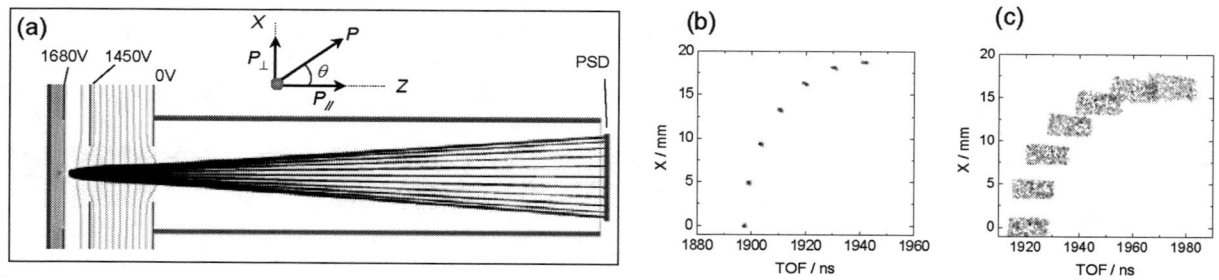

FIGURE 2. (a) Ion trajectories of N^+ ions and equipotential surfaces in the ion momentum spectrometer. 500 N^+ ions with kinetic energy of 5 eV are ejected from the random points inside the source volume, at every 15° from 0° to 90° relative to the Z axis. *P* denotes the initial momentum vector of an emitting N^+ ion, and $P_{//}$ and $P_\perp$ are projections of *P* onto the Z and X axes, respectively. (b) Contour plot of X position and TOF for each ion trajectory shown in (a) with focusing condition. (c) Contour plot under defocusing condition.

In response to the detection of each Auger electron, a pulsed electric field is applied to the interaction region. The ions relevant to the Auger electron are thus extracted into the flight tube of the ion momentum spectrometer. From the time-of-flights (TOFs) of the ions and the arrival positions on the PSD, one can reconstruct the initial momentum vectors of the ions. Here, the momentum resolution is limited by the volume of the interaction region, because a large source volume leads to both the blurring in images and the temporal structures in TOFs. Therefore it is essential for achieving a high momentum resolution to cancel out the blurring induced by the source volume [8-11]. The electrostatic lenses for our ion momentum spectrometer, which consist of open-face electrodes and make an inhomogeneous electric field in the interaction region, are designed to fulfill both the image and time focuses at the same time. Figure 2 shows a simulation result of ion trajectories in the electric field formed by the designed electrostatic lenses, where N^+ ions with the kinetic energy of 5 eV are assumed to be emitted from the 2 mm × 2 mm source. Since the present setup has a cylindrical symmetry around the collinear axis of the two spectrometers, two-dimensional numerical simulations are well valid. The chosen coordinate system is indicated in Fig. 2(a). The emission angles of the ions from the source are set to be at every 15° from 0° to 90° relative to the Z axis. It is seen that the trajectories for ions, emitted at the same direction but from different source positions, concentrate into a small area on the PSD. The simulation gives a positional resolution of <0.2 mm (FWHM) as a result of the image focus. The TOFs of the ions under this image focus condition are plotted in Fig. 2(b) as a function of X position (radius measured from the PSD center). Depending on the seven different emission angles, the corresponding seven small areas appear on the contour plot of TOF and X. From the intensity distributions in each area in Fig. 2(b), the temporal resolution is estimated to be 2.5 ns (FWHM). When we detune the potentials slightly from the image focus condition, the concentrations become rather worse (see Fig. 2(c)). These results clearly imply that image and time focuses are achieved at the same time. At such focus conditions, the spectrometer assures a quasi-linear relationship between X position on the PSD and $P_\perp$ and that between TOF and $P_{//}$, similar to the case for using a homogeneous electric field. The quasi-linear relationships are useful for ion momentum analysis. Concerning the pulsed electric field extraction of ions, a new high voltage pulse generator with the electronic time delay of 160 ns and rise time of 20 ns has been introduced.

PERFORMANCE TEST OF THE SPECTROMETER

The performance test of the coincidence spectrometer has been done at the beamline BL8B1 of UVSOR. The beamline is equipped with a constant-deviation constant-length spherical grating monochromator. The cylindrical axis of the AEPICO spectrometer lies along the electric vector of the monochromatized light. A raw image from the observation of Kr MNN Auger electrons in the kinetic energy range of 48 - 58 eV is presented in Fig. 3(a), where the pass energy of the DTA was set to 80 eV. The concentric circles corresponding to individual Auger lines are clearly seen on this image. The anisotropic intensity distributions along the circumferences of the circles result from the unevennesses of the analyzer transmission efficiency and the detection efficiency of the PSD, since the azimuthal angle distribution with respect to the electric vector should be isotropic. Conversely, we can calibrate the transmission and the detection efficiencies, by using the present image data. The radial intensity distribution of the image, which corresponds to the Auger electron spectrum, is shown in Fig. 3(b). The Auger structures exhibited can be interpreted by referring the reported Auger energies [12]. A fitting with Gauss functions results in a mean FWHM of 1 eV for the Auger lines. The observed energy resolution is reasonable in comparison with the simulation result.

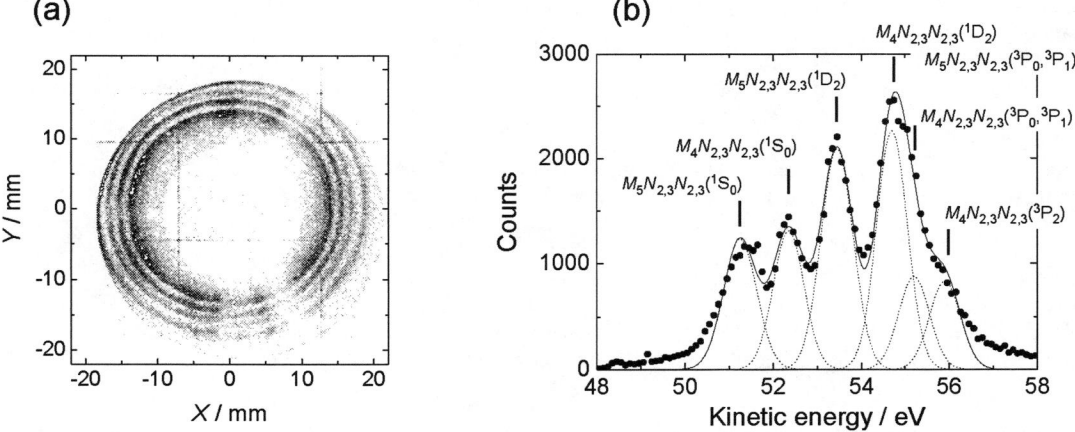

FIGURE 3. (a) Two-dimensional image of Kr MNN Auger electron following the Kr $3d$ photoionization. The DTA was operated at the pass energy of 80 eV. (b) Auger electron spectrum deduced from the two-dimensional image.

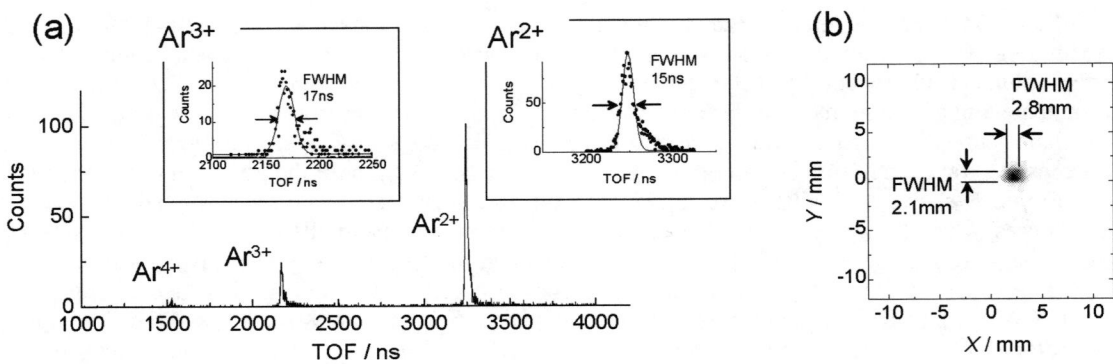

FIGURE 4. (a) Time-of-flight spectrum of Ar ions following the 2*p* photoionization at *h*ν = 335 eV. (b) Two-dimensional image of Ar ions on the PSD.

The performance of the ion momentum spectrometer has been investigated by measuring Ar ions associated with Ar 2*p* photoionization. The TOF spectrum and the two-dimensional image of the Ar ions detected in coincidence with the Ar 2*p* photoelectrons are denoted in Fig. 4. The different charge states of the Ar ions, produced by the Auger decays from the Ar 2*p* hole states, are exhibited in the TOF spectrum. The relative intensities of the product ions are 84%, 14%, and 2% for Ar^{2+}, Ar^{3+} and Ar^{4+}, respectively. These values agree with the previously reported ones [13], which suggests that the detection efficiencies of the ion PSD for different charge states are constant. The narrow peaks in the TOF spectrum and the small spot on the image indicate that the time and image focuses are established simultaneously. Practically, we observe the temporal resolution less than 17 ns (FWHM) and the positional resolution of 2.8 mm (FWHM). The observed focuses are, however, somewhat worse than the expected resolutions by the numerical simulation, which may result from imperfection of the repelling electric field induced by the pulse generator.

ACKNOWLEDGMENTS

We are grateful to the UVSOR staff for the stable operation of the storage ring and their excellent support. This work has been supported by Grant-in-Aid for Scientific Research from JSPS.

REFERENCES

1. W. Eberhardt, E. W. Plummer, I. W. Lyo, R. Carr, and W. K. Ford, *Phys. Rev. Lett.* **58**, 207 (1987).
2. C. Miron, M. Simon, N. Leclercq, D.L. Hansen, and P. Morin, *Phys. Rev. Lett.* **81**, 4104 (1998).
3. D. Céolin, C. Miron, K. Le Guen, R. Guillemin, P. Morin, E. Shigemasa, P. Millié, M. Ahmad, P. Lablanquie, F. Penent, and M. Simon, *J. Chem. Phys.* **123**, 231303 (2005).
4. For example, E. Shigemasa, J. Adachi, M. Oura, and A. Yagishita, *Phys. Rev. Lett.* **74**, 359 (1995).
5. R. Guillemin, E. Shigemasa, K. Le. Guen, D. Ceolin, C. Miron, N. Leclercq, P. Morin, and M. Simon, *Phys. Rev. Lett.* **87**, 203001 (2002).
6. Th. Weber, M. Weckenbrock, M. Balser, L. Schmidt, O. Jagutzki, W. Arnold, O. Hohn, M. Schöffler, E. Arenholz, T. Young, T. Osipov, L. Foucar, A. De Fanis, R. Díez Muiño, H. Schmidt-Böcking, C. L. Cocke, M. H. Prior, and R. Dörner, *Phys. Rev. Lett.* **90**, 153003 (2003).
7. C. Miron, M. Simon, N. Leclercq, and P. Morin, *Res. Sci. Instrum.* **68**, 3728 (1997).
8. D. Céolin, C. Miron, M. Simon, and P. Morin, *J. Electron. Sepctrosc. Relat. Phenom.* **141**, 171 (2004).
9. R. Dörner, H. Bräuning, J. M. Feagin, V. Mergel, O. Jagutzki, L. Spielberger, T. Vogt, H. Khemliche, M. H. Prior, J. Ullrich, C. L. Cocke, and H. Schmidt-Böcking, *Phys. Rev. A* **57**, 1074 (1998).
10. Y. Hikosaka, and J. H. D. Eland, *Chem. Phys.* **281**, 91 (2002).
11. M. Lebech, J. C. Houver, and D. Dowek, *Rev. Sci. Instrum.* **73**, 1866 (2002).
12. L. O. Werme, T. Bergmark, and K. Siegbahn, *Phys. Scripta.* **6**, 141 (1972).
13. S. Brünken, Ch. Gerth, B. Kanngießer, T. Luhmann, M. Richter, and P. Zimmermann, *Phys. Rev. A* **52**, 052708 (2002).

Ar·NO Neutral and Cationic van der Waals Complexes Study: Experiment and Theory

Sisheng Wang[†], Ruihong Kong[†], Xiaobin Shan[†], Yunwu Zhang[†], Liusi Sheng[†], Zhenya Wang[¶], Liqing Hao[¶], and Shikang Zhou[¶]

[†] *National Synchrotron Radiation Laboratory, University of Science and Technology of China, Hefei, 230029, People's Republic China*
[¶] *Laser Spectroscopy Laboratory, Anhui Institute of Optics and Fine Mechanics, Chinese Academy of Sciences, Hefei, 230031, People's Repubic China*

Abstract. Photoionization of Ar·NO complex is studied with photoionization mass spectroscopy (PIMS) and Photoionization efficiency spectroscopy (PIES) by synchrotron radiation, supersonic molecule beam and reflection time of flight mass spectrum (RTOF-MS). In the PIES of Ar·NO, an intensive semi-resonance structure is observed. The ionization energy (IE) of Ar·NO and dissociation energy (DE) of Ar·NO$^+$ and Ar·NO are calculated using Gaussian-2 method calculation. The equilibrium geometry and harmonic vibrational frequencies of Ar·NO$^+$ and Ar·NO complexes are calculated using a variety of basis sets and different theories.

Keywords: Ar·NO, Photoionization, Quantum chemical calculation, van der Waals
PACS: 36.40.Mr, 36.40.-C, 31.15.Ar

INTRODUCTION

The spectroscopy of rare-gas and open-shell complexes has been a rich area of research in recent years. Weakly bound complexes of open-shell molecules with rare-gas atoms are of fundamental interest[1]. These complexes act as prototypes for the study of weak intermolecular interactions and spectroscopic and theoretical investigations provided important information about the intermolecular energy surface, equilibrium geometry and harmonic vibrational frequencies.

The adiabatic ionization energy of Ar·NO was originally determined by Kimura and co-workers to be 73783 ± 40 cm^{-1}[2]. Recently, a more precise value for the adiabatic ionization energy of Ar·NO has been determined by Takahashi using a threshold photoelectron technique similar to that employed in this study[3]. A two-color (2+1') REMPI scheme was used and the ionization energy of Ar·NO was determined to be 73869 ± 6 cm^{-1}. From the ionization energies of Ar·NO and free NO, together with the dissociation energy of the neutral ground state, the dissociation energy, D_0, of the $X^1\Sigma^1$ state of Ar·NO was determined to be 951 cm^{-1}. Two regular vibrational progressions were identified in the threshold photoelectron spectra and these were assigned to transitions to excited van der Waals stretching and bending levels of the Ar·NO$^+$ cation. A simple analysis of the observed vibrational spacings yielded fundamental frequencies for the two vibrational modes of 94 and 79 cm^{-1}, respectively. Franck–Condon calculations based on the relative intensities observed when exciting through different intermediate levels were used to predict that the structure of the Ar·NO$^+$ ion is a skewed T-shape. Ar·NO$^+$ has also been the subject of extensive high-level theoretical studies[4,5,6,7]. The *ab initio* calculations of Wright and co-workers suggest that the ground state of the ion is skewed T-shaped with the argon atom lying closest to the nitrogen atom.

However, thermochemial properties of Ar·NO are little known. Above the ionization threshold of Ar·NO, it is not reported about the excited states of Ar·NO$^+$, Rydberg states and intermolecular energy transfer. This paper, experimental investigation in photoionization is performed above the ionisation threshold of Ar·NO. Furthermore, the equilibrium geometry and harmonic vibrational frequencies of Ar·NO$^+$ and Ar·NO complexes are calculated using a variety of basis sets and a variety of theories. The ionization energy of Ar·NO and dissociation energy of

CP879, *Synchrotron Radiation Instrumentation: Ninth International Conference*,
edited by Jae-Young Choi and Seungyu Rah

Ar·NO$^+$ and Ar·NO are calculated using Gaussian-2 method calculation.

EXPERIMENTS AND CALCULATIONS

This experimental study was performed at atomic and molecular physics experiment station at the National Synchrotron Radiation Laboratory (NSRL) in Hefei, China. Synchrotron radiation from undulator of the 800MeV electron storage ring is monochromized by using a SGM monochromator equipped with three gratings (1250, 740 and 370 lines·mm^{-1}) covering the wavelength range from 10 to 177 nm. The dispersed radiation is focused by a toroidal mirror into the photoionization chamber. This experiment adopts the 370 lines·mm^{-1} grating, covering the energy range from 7.5 to 22.5 eV, resolving power E/ΔE is above 5000 and energy error is below 0.002 eV . This grating is alignmented by the ionization threshold of Ar, Kr and Xe and so on. The photon flux was above 10^{12} photons/second.

The molecular beam is produced by expanding the sample gas through a 70 μm nozzle into the beam source chamber and then through two skimmers into the photoionization chamber. It intersects the monochromatized synchrotron radiation beam. The mass spectrograph is installed above the intersecting point. The Photoionization, differential and beam source chambers are pumped by turbomolecular pumps respectively. The background pressure is 10^{-4}, 10^{-3} and 10^{-2} Pa respectively when the molecular beam is on. The setup is equipped with a computerized control and data acquisition system.

In this experimental study, the purities of sample Ar and NO are both 99.9%. The sample gas used in the present work was a 10% mixture of NO in Ar and the stagnation pressure is 1-3 bars.

The total energies of molecule (neutral and ionic state) are calculated with G2 method[8,9] by using the Gaussian 03 program[10] . The equilibrium geometry and harmonic vibrational frequencies of Ar·NO$^+$ and Ar·NO complexes are also calculated using a variety of basis sets by HF、MP2、QCISD、QCISD(T),CCSD,CCSD(T), B3LYP and BPW91 levels of theory.

RESULTS AND DISCUSSION

Photoionization Efficiency Spectroscopy

When photon energy is 17 eV, the mass spectrum is obtained. In this spectrum, the monomer ions: NO$^+$, Ar$^+$, and the dimmer ions: (NO)$_2^+$, ArNO$^+$ and (Ar)$_2^+$ are observed. The photoionization efficiency spectroscopy can be obtained when the mass of molecules is fixed and the wavelengh of synchrotron radiation is scanned. Figure 1 shows PIES of different energe range. In the photoionization efficiency spectroscopy of Ar·NO, an intensive semi-resonance structure is observed in the energy range of 11.5-12.0 eV. From analysis, this resonance structure comes from the Ar atom transition spectrum line. In the theory, Ar atom has two transition spectrum lines in the resonance transition of 3p^6(^{1}S)→ 4s(1/2, 3/2), counterpart $\lambda_{1/2}$ = 104.82 nm (11.83 eV), $\lambda_{3/2}$ = 106.66 nm (11.62 eV), respectively. When the photon energy is equal to Ar atom resonance transition energy, Ar atom of Ar·NO cluster is subjected resonance excitement. In interior of Ar·NO cluster, there is energy transfer between excited Ar and NO, which is made Ar·NO ionized. This semi-resonance may be expressed by below equation.

$$\text{Ar·NO} + h\nu \rightarrow \ (\text{Ar}^*\text{·NO}) \ \rightarrow \ (\text{Ar·NO}^+) + e^- \tag{1}$$

This process is named intermolecular penning ionization[11]. In this experiment, blue shift is discovered in the semi-resonance transition.

An intensively spectral structure due to NO$^+$ is detected at the energy 13.94 eV. A relatively intensive widened spectral structure starting from 15.88 eV is also observed. There were some relatively intensive peas in the Fig. 1 (b). We conside that these peaks were mainly contributed to the spectral structure of NO ion in the Ar·NO complex.

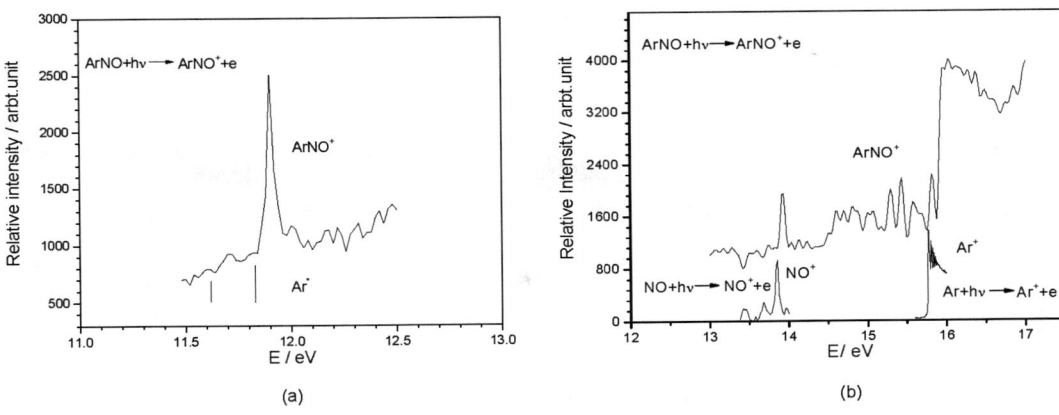

FIGURE 1. (a) The photoionization efficiency spectrum of Ar·NO cluster at the energy range 11.5−12.5 eV. (b) The photoionization efficiency spectrum of Ar·NO cluster at the energy range of 13.0−17.0 eV

Results of Quantum Chemical Calculations

The calculated results for Ar·NO$^+$ are shown in Table 1. Through calculations, the neutral or cationic complexes are calculated to have a shewed T-shaped structure. The harmonic vibrational frequencies are mainly calculated to confirm that the calculated geometries are minima, and not saddle points. We note that the QCISD method except density functional methods gives a good value for the NO$^+$ and NO vibrational frequencies, which is expected to be almost identical with the uncomplexed value of 2374.31 and 1904.20 cm^{-1}[12] respectively; also the NO$^+$ and NO bond length are very close to the experimental value of 1.06434 and 1.15077 Å[13] respectively. About the Ar·NO$^+$ cluster, the *ab initio* harmonic frequeuencies results from QCISD, QCISD(T), CCSD and CCSD(T) levels of theory are about 100 and 75 cm^{-1}, which are fairly close to the experiment values of Takahashi[3]. From the density functional results, the bond lengths are clearly predicted to be significantly different to the *ab initio* values. In the Ar·NO complex, both the intermolecular stretch and bend are too low compared with resuls from QCISD, QCISD(T), CCSD and CCSD(T) theory; nevertheless, in the Ar·NO$^+$ complex, it may be seen that both the intermolecular stretch and bend are quite high compared with experimental or *ab initio* results. Because the intermolecular density of Ar and NO is too small, density functional calculations are underestimating the amout of electron density between Ar atom and NO molecule. On the other hand about Ar·NO$^+$, for the *ab initio* calculations, the charge is mainly on the argon atom <0.002e; however, for the density functional methods, the charge on the argon atom is > 0.1e. Thus the density functional calculations are overestimating the amount of electron density between Ar atom and NO$^+$ molecule.

An adiabatic ionization energy of Ar·NO complex can be obtained as the difference energy between its neutral molecule and ion in their electronic ground state:

TABLE 1. Calculated geometries and harmonic vibrational frequencies for the Ar·NO$^+$ complex.

Method	r(N=O) (Å)	R(Ar-N) (Å)	θ (Ar-N-O) (°)	V$_1$(bend) (cm^{-1})	V$_2$(stretch) (cm^{-1})	V$_3$(N-O$_{str}$) (cm^{-1})
hf/6-31g*	1.04059307	3.27251327	106.74490819	46.3972	73.5664	2852.1610
mp2/6-311g*	1.08618154	3.05859654	109.23502721	62.3903	100.9022	2164.2528
mp2/cc-pvtz	1.0827087	2.97548077	105.51668867	76.2186	112.6837	2159.3126
qcisd/6-311g*	1.06730793	3.11252552	109.00031381	57.4938	93.2404	2426.6704
qcisd/cc-pvdz	1.07485216	3.07067674	109.04982332	60.7354	102.0641	2416.5908
qcisd(t)/cc-pvdz	1.08025898	3.03916221	109.49954658	62.6261	102.2684	2358.4856
ccsd/cc-pvdz	1.07205149	3.07502763	109.04369439	61.0752	102.2508	2465.9138
ccsd(t)/cc-pvdz	1.07949629	3.03965647	109.48199517	62.4578	101.9088	2368.1830
b3lyp/6-31g*	1.07933216	2.68321375	114.85884377	123.9020	242.1102	2398.8246
bpw91/cc-pvtz	1.07949164	2.64181253	115.56247622	123.5415	257.3400	2257.5985
b3lyp/cc-pvdz	1.07295262	2.68281118	115.06724545	124.3740	242.0351	2413.5603

$$ArNO + IE \rightarrow ArNO^+ + e \tag{2}$$

$$IE(ArNO) = E_0(ArNO^+) - E_0(ArNO) = 9.177 \text{ eV} \tag{3}$$

E_0 is the total energy from G2 calculation method at 0 K, which contains electron energy and zero point vibrational energy. This result is accord with the ionisation threshold from experiment[2,3].

Binding energy (BE) of neutral Ar·NO complex can be deduced from below equation.

$$Ar \cdot NO + D_0 \rightarrow Ar + NO \tag{4}$$

$$BE(Ar \cdot NO) = E_0(Ar) + E_0(NO) - E_0(Ar \cdot NO) = 0.010 \text{ eV} \tag{5}$$

This result is in agreement with 0.013 eV by resonant multiphoton ionization spectroscopy[2].

If cationic $ArNO^+$ complex receive some energy, DE, they can be dissociated into Ar and NO^+:

$$ArNO^+ + D_0 \rightarrow Ar + NO^+ \tag{6}$$

$$DE(Ar \cdot NO^+) = E_0(Ar) + E_0(NO^+) - E_0(Ar \cdot NO^+) = 0.114 \text{ eV} \tag{7}$$

This result is accord with the energy determined by Kimura and co-workers and by Takahashi using a threshold photoelectron technique[2,3].

CONCLUSTION

This experiment using PIMS and PIES by synchrotron radiation is performed above the ionisation threshold of Ar·NO. An intensive semi-resonance structure is observed. On the other hand, ionization energy of cluster, binding energy of neutral complex and dissociation energy of cationic complex are obtained theoretically by using G2 method. These results predicted from quantum chemical calculation are in good agreement with those got from experiments. Furthermore, The equilibrium geometry and harmonic vibrational frequencies of Ar·NO^+ and Ar·NO complexes are calculated and compared using a variety of basis sets and a variety of theories.

ACKNOWLEDGEMENTS

This work is supported by the National Natural Science Foundation of China (Grant No.10374084).

REFERENCES

1. M.C..Heaven, *Annu.Rev.Phys.Chem.* 43, 283-310(1992); *J.Phys.Chem.* 97, 8567-8577 (1993).
2. K. Sato, Y.Achiba and K.Kimura, *J. Chem. Phys.* 81,57-62 (1984).
3. T. Masahiko, *J. Chem. Phys.* 96, 2594-2599 (1992).
4. A. M. Bush, T. G. Wright, V. Spirko and M. Jurek, *J. Chem. Phys.* 106, 4531 – 4535 (1997).
5. T. G. Wright, *J. Chem. Phys.* 105, 7579 – 7582 (1996).
6. T. G. Wright, V. Spiko and P. Hobaza, *J. Chem. Phys.* 108, 5403-5410 (1994).
7. E. P. F. Lee, P. Soldan and T. G. Wright, *J. Phys. Chem. A*, 102, 6858 – 6864 (1998).
8. L. A. Curtiss, K. Raghavachari, G. W. Trucks and J. P. Pople, *J. Chem. Phys.*, 94, 7221-7230 (1991).
9. L. A. Curtiss, K. Raghavachari and J.P. Pople, *J. Chem. Phys.*, 98, 1293-1298 (1993).
10. Gaussian 03,Revision B.03, Gaussian, Inc., Pittsburgh PA, 2003.
11. E. Ruhl, P. Bisling, B. BrutschyK, O. Leisen and H. Morgner, *Chem. Phys. Lett.* 128, 512 (1986).
12. K. P. Huber and G. Herzberg, *Constants of Diatomic Molecules*, Van Nostrand Reinhold: New York, 1977, pp.474,476,782.
13. K. P. Huber and G. Herzberg, *Molecular Spectra and Molecular Structure IV:Constants of Diatomic Molecules*, van Nostrand: New York,1979.

Determination of Dissociation Barrier Heights of C_2H_3 and C_2H_2F Radicals Using Photofragment Translational Spectroscopy and Synchrotron VUV Ionization

Shih-Huang Lee[a],* and Yuan T. Lee[b]

[a] *National Synchrotron Radiation Research Center (NSRRC)*
101 Hsin-Ann Road, Hsinchu Science Park, Hsinchu 30076, Taiwan
[b] *Institute of Atomic and Molecular Sciences (IAMS), Academia Sinica*
P.O. Box 23-166, Taipei 106, Taiwan

Abstract. Using synchrotron radiation as a light source for ionization of reaction products allows us to detect all the photofragments, eight species in total, upon photodissociation of vinyl fluoride (CH_2CHF) at 157 nm. Four primary dissociation pathways leading to products C_2H_2 + HF, C_2HF + H_2, C_2H_2F + H, and C_2H_3 + F are identified. Spontaneous decomposition of internally hot C_2H_3 and C_2H_2F fragments are observed and the barrier heights of $C_2H_3 \rightarrow C_2H_2$ + H and $C_2H_2F \rightarrow C_2H_2$ + F are determined to be 38 and 42 kcal mol^{-1}, respectively.

Keywords: vinyl fluoride, vinyl, fluorovinyl, synchrotron radiation, photofragment translational spectroscopy.
PACS: 34.50.Lf

INTRODUCTION

For years, photofragment translational spectroscopy (PTS) coupled with electron-impact ionization has been widely used to investigate the photodissociation dynamics of molecules in gas phases. Due to the severe dissociative ionization and the large detection background, some photoproducts are undetectable using a conventional electron ionizer. Photoionization with tunable vacuum ultraviolet (VUV) radiation from a synchrotron has been successfully applied to the investigation of chemical reaction dynamics at the Advanced Light Source [1–3] and the Taiwan Light Source [4–6]. Typically, photofragments, even though atomic H and H_2, are all detectable using the state-of-the-art experimental apparatuses [4–6]. The *universal* detection of reaction products allows us to understand the whole picture of photodissociation dynamics of a molecule.

Photolysis has been widely used to generate a radical R from a precursor R–X. Since the X-atom carries no internal energy, the distribution of internal energy of radical R is derivable from a distribution of total kinetic energy of the R-radical and its counterpart X-atom detected using the PTS technique. Spontaneous decomposition of the radical can take place provided that internal energy of the radical is greater than its dissociation barrier. Upon the photolysis of acetyl chloride (CH_3COCl) at 248 nm, the internally hot CH_3CO (acetyl) radical was observed to dissociate to CH_3 + CO following the C–Cl bond fission using the PTS and electron-impact ionization [7]; the dissociation barrier height of CH_3CO was determined to be 17.1 kcal mol^{-1} from the kinetic-energy distribution of the surviving CH_3CO. Recently, we have determined dissociation barrier heights of several radicals produced from photolysis of supersonically cooled molecules using the PTS and VUV ionization. The barrier height of $CH_2CHCH_2 \rightarrow CH_2CCH_2$ + H was determined to be 55 kcal mol^{-1} upon the photolysis of propene (CH_3CHCH_2) at 157 nm [4]. The dissociation barrier height of CH_3O (methoxyl) produced from photolysis of methanol (CH_3OH) at 157 nm was determined to be 25 kcal mol^{-1} to form CH_2O + H [5]. Moreover, the barrier heights of HCO $\rightarrow$ H + CO and FCO $\rightarrow$ F + CO were evaluated to be 22±2 and 35±1 kcal mol^{-1}, respectively, upon the photolysis of formyl fluoride (HFCO) at 193 nm [6].

The significant elimination mechanism of hydrogen halide from halo-olefin attracts interests of scientists in the photodissociation of vinyl fluoride [8–11]. In contrast, less work was dedicated to single bond breaking channels, $C_2H_3F \rightarrow C_2H_3$ + F and $C_2H_3F \rightarrow C_2H_2F$ + H. Sato *et al.* [12] and Tu *et al.* [13] have investigated the

CP879, *Synchrotron Radiation Instrumentation: Ninth International Conference*,
edited by Jae-Young Choi and Seungyu Rah
© 2007 American Institute of Physics 978-0-7354-0373-4/07/$23.00

photodissociation dynamics of vinyl fluoride at 157 nm using PTS and electron-impact ionization. Sato *et al.* only reported the time-of-flight (TOF) spectra of three photofragments. Tu *et al.* extensively reported the TOF spectra of seven fragments and identified three primary dissociation channels except the reaction $C_2H_3F \rightarrow C_2H_3 + F$. Following optical excitation at 157 nm, vinyl fluoride can produce C_2H_3 and C_2H_2F radicals with a loss of F- and H-atom, respectively, in addition to $C_2H_3F \rightarrow C_2H_2 + HF$ and $C_2H_3F \rightarrow C_2HF + H_2$. In the present work, we measured TOF spectra of all photofragments of vinyl fluoride using tunable VUV ionization.

EXPERIMENTS

The experimental apparatus has been described in detail elsewhere [4–6]; only a brief description is given here. The experimental apparatus includes a molecular beam end-station, an F_2-excimer laser, and a VUV light source. Shown in Fig. 1 is the molecular beam apparatus that comprises two rotating source chambers, a main chamber, and a detection chamber. A source chamber equipped with an Even-Lavie valve and two successive skimmers served to generate a collimated beam of vinyl fluoride with a stagnation pressure of 300 Torr. The pulsed valve was heated to 115 °C to avoid formation of clusters. The F_2–excimer laser served to generate radiation of wavelength 157 nm for photolysis of vinyl fluoride. The photolysis light was introduced into the main chamber and focused into an area of 2×7 (W×H) at the interception region with the molecular beam. After free flight along a path of length 100.5 mm, photoproducts were ionized with the VUV radiation from an undulator. Electrons in the storage ring emitted photons after being wiggled by the U9-undulator. The Chemical Dynamics beamline delivered VUV photons from the undulator to the detection chamber for ionization of products. Since only the fundamental frequency was desired in the present work, a windowless gas cell (called harmonic suppresser) filled with noble gas (e.g., Ne, Ar and Kr) served to absorb photons of frequencies at high harmonics. The fundamental frequency of the VUV photons is readily tunable by adjusting the gap between two rows of magnets of the undulator. The detection chamber contains ion optics, a quadrupole mass filter (QUAD), and a Daly-type ion counter. Following photoionization the ion optics extracted cations into the quadrupole mass filter for selection of species at a ratio (*m/z*) of mass to charge. The Daly-type detector counted ions and a multichannel scaler sampled ions into 4000 bins of 250 ns. We obtained the TOF spectra of neutral products after subtracting the flight interval of cations from the total flight duration.

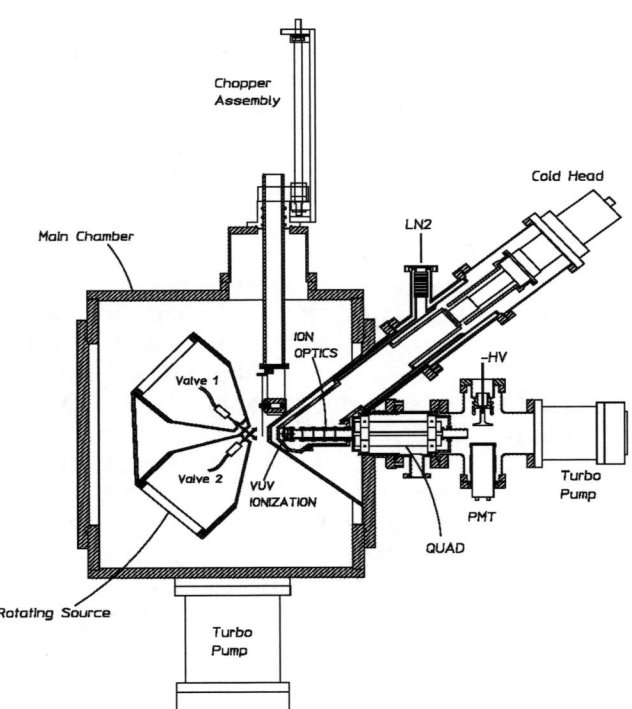

FIGURE 1. Schematic representation of the molecular beam apparatus

RESULTS AND DISCUSSION

Upon the photolysis of vinyl fluoride at 157 nm, we observed eight photoproducts, e.g., C_2H_2F, C_2H_3, C_2HF, C_2H_2, HF, H_2, F and H, associated with four primary dissociation channels $C_2H_3 + F$, $C_2H_2F + H$, $C_2HF + H_2$, and $C_2H_2 + HF$; their branching ratios are 0.04, 0.32, 0.05, and 0.59, respectively. A fraction of the C_2H_3 and C_2H_2F products were observed to undergo secondary dissociation. In the proceedings we exhibited only the TOF spectra of C_2H_2F, C_2H_3, F-atom, and H-atom to reveal the dissociation thresholds of the former two radicals. A computer program PHOTRAN serves to mimic the TOF spectra of two momentum-matched fragments with a trial kinetic-energy distribution $P(E_t)$. After iterative forward convolution, we can obtain a satisfactory $P(E_t)$ of the two momentum-matched products from the best fit to the experimental TOF spectra. E_t is the total kinetic energy including two momentum-matched products.

Figure 2 (a) and (b) depict the experimental TOF spectra of C_2H_3 (vinyl) and atomic F, respectively, along with their own TOF simulations. Due to a minor channel, signals of the two momentum-matched products C_2H_3 and F-atom are small. A signal of the ^{13}C isotopic variant of C_2H_2 was observed in Fig. 2 (a) and fitted with a dotted line. The surviving and dissociating C_2H_3 correlate with the solid-line and dashed-line components of atomic F, respectively. The dash-dotted-line component of F-atom originates from dissociative ionization of product HF. Dissociation to $C_2H_3 + F$ has available energy (E_{ava}) 55 kcal mol^{-1} [12] and product C_2H_3 has an internal energy of $E_{ava}-E_t$. The $P(E_t)$, shown in Fig. 2 (c), indicates that a fraction (45%) of C_2H_3 with internal energy greater than 38 kcal mol^{-1} spontaneously decomposes to $C_2H_2 + H$; the evaluated barrier height is in good agreement with a theoretical value 38.3 kcal mol^{-1} [14]. Product C_2H_2 has two features in the TOF distribution, not shown here; the rapid feature is due to elimination of H_2 and the slow feature is due to secondary dissociation $C_2H_3 \rightarrow C_2H_2 + H$. The secondary H-atom might contribute to the small bump of atomic H in flight time around 13 μs (*vide infra*).

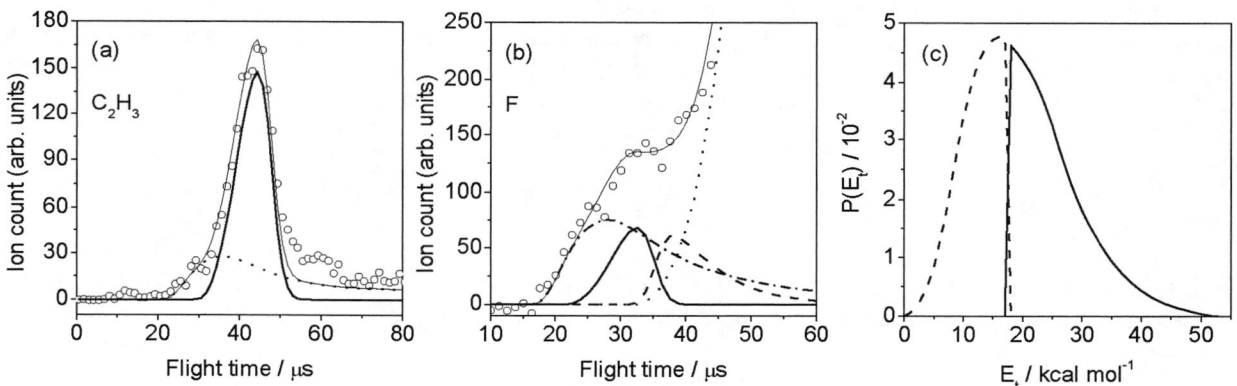

FIGURE 2. (a) A TOF spectrum of C_2H_3 detected at a scattering angle 30° using photoionization energy 11.1 eV; (b) a TOF spectrum of atomic F detected at a scattering angle 20° using photoionization energy 18.3 eV; (c) the corresponding kinetic-energy distributions for the surviving C_2H_3 (solid line) and the dissociating C_2H_3 (dashed line).

Analogously, Fig. 3 (a) and (b) depict the TOF spectra of C_2H_2F (fluorovinyl) and atomic H, respectively, along with their own TOF simulations. Shown in Fig. 3 (c) is the primary $P(E_t)$ for the $C_2H_2F + H$ channel. C_2H_2F has a sharp feature near 100 μs and a broad feature around 200 μs; the rapid and slow features correspond to the forward and backward parts of C_2H_2F in the center-of-mass frame, respectively, and thus have the same $P(E_t)$. The surviving C_2H_2F correlates only with the leading part of atomic H, indicating that a large fraction (93%) of the C_2H_2F product that correlates with the dashed-line component of H-atom undergoes spontaneous decomposition. From analysis of product yields, we deduce that C_2H_2F dissociates to $C_2H_2 + F$ and $C_2HF + H$ with a branching ratio of 95:5. The large second feature of atomic F peaking at ~80 μs and the small bump of atomic H around 13 μs are attributed to the decomposition of C_2H_2F. Because the reaction $C_2H_2F \rightarrow C_2H_2 + F$ has a smaller enthalpy of reaction and a larger yield than the reaction $C_2H_2F \rightarrow C_2HF + H$, we propose that the breakpoint at 25 kcal mol^{-1}, as shown in Fig. 3 (c), reflects the dissociation barrier height of the F-leaving channel. Since the reaction $C_2H_3F \rightarrow C_2H_2F + H$ has available energy 67 kcal mol^{-1} [12], the dissociation barrier height of $C_2H_2F \rightarrow C_2H_2 + F$ is determined to be 42 kcal mol^{-1}.   The barrier height of the reaction $C_2H_2F \rightarrow C_2HF + H$ cannot be unraveled in the present work.

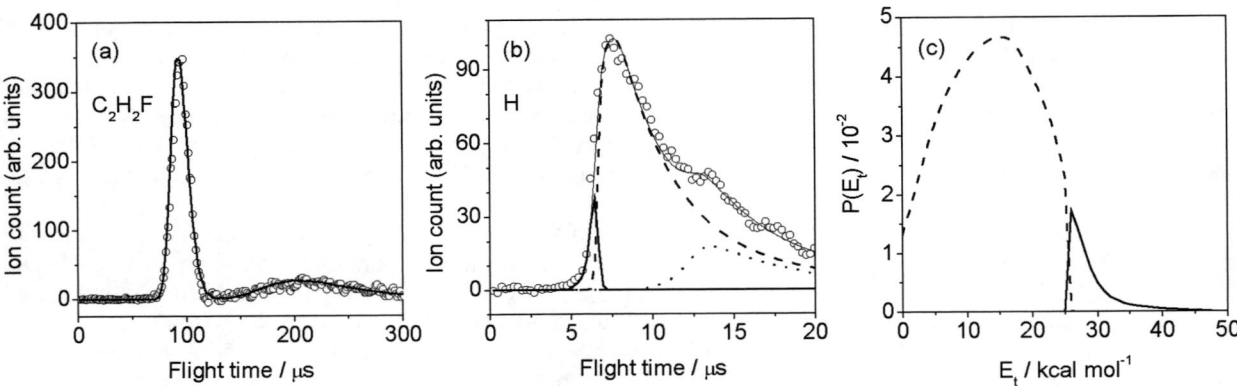

FIGURE 3.   (a) A TOF spectrum of C_2H_2F detected at a scattering angle $10°$ using photoionization energy 11.1 eV; (b) a TOF spectrum of atomic H detected at a scattering angle $30°$ using photoionization energy 17.1 eV; (c) the corresponding kinetic-energy distributions for the surviving C_2H_2F (solid line) and the dissociating C_2H_2F (dashed line).

In summary, it is the first time to identify the $C_2H_3 + F$ dissociation pathway of vinyl fluoride photolyzed at 157 nm.   All the photoproducts, eight species in total, are detectable using selective photoionization, which allows us to determine total branching ratios of seven (four primary and three secondary) dissociation pathways.   The dominant reaction $C_2H_3F \rightarrow C_2H_2 + HF$ proceeds on the ground potential energy surface, indicating that internal conversion from the photo-excited state π-π^* to the ground electronic state is more efficient than direct dissociation.   From the cutoffs in kinetic-energy distributions of products C_2H_3 and C_2H_2F, we determine the dissociation barrier heights of $C_2H_3 \rightarrow C_2H_2 + H$ and $C_2H_2F \rightarrow C_2H_2 + F$.   The tunneling effect is neglected here.

ACKNOWLEDGMENTS

We thank the National Synchrotron Radiation Research Center, Academia Sinica, and the National Science Council of Taiwan (Grant No. NSC94-2113-M-213-004) for financial supports.   S.-H. Lee thanks Prof. Xueming Yang and Prof. Jim J. Lin for establishment of the molecular beam apparatus.

REFERENCES

1.   X. Yang, J. Lin, Y. T. Lee, D. A. Blank, A. G. Suits, and A. M. Wodtke, Rev. Sci. Instrum. **68**, 3317-3326 (1997).
2.   N. Hemmi and A. G. Suits, J. Chem. Phys. **109**, 5338-5343 (1998).
3.   F. Qi, O. Sorkhabi, and A. G. Suits, J. Chem. Phys. **112**, 10707-10710 (2000).
4.   S.-H. Lee, Y.-Y. Lee, Y. T. Lee, and X. Yang, J. Chem. Phys. **119**, 827-838 (2003).
5.   S.-H. Lee, H.-I Lee, and Y. T. Lee, J. Chem. Phys. **121**, 11053-11059 (2004).
6.   S.-H. Lee, C.-Y. Wu, S.K. Yang, and Y.-P. Lee, J. Chem. Phys. **123**, 074326 (2005).
7.   S. W. North, D. A. Blank, and Y. T. Lee, Chem. Phys. Lett. **222**, 38 (1994).
8.   J. F. Caballero and C. Wittig, J. Chem. Phys. **82**, 1332-1337 (1985).
9.   E. Martínez-Núñez and S. A. Vázquez, Chem. Phys. Lett. **332**, 583-590 (2000).
10.   E. Martínez-Núñez and S. A. Vázquez, J. Chem. Phys. **121**, 5179-5182 (2004).
11.   S.-R. Lin, S.-C. Lin, Y.-C. Lee, Y.-C. Chou, I-C. Chen, and Y.-P. Lee, J. Chem. Phys. **114**, 7396-7406 (2001).
12.   K. Sato, S. Tsunashima, T. Takayanagi, G. Fijisawa, and A. Yokoyama, Chem. Phys. Lett. **242**, 401-406 (1995).
13.   J. Tu, J. J. Lin, Y. T. Lee, and X. Yang, J. Chem. Phys. **116**, 6982-6989 (2002).
14.   A. H. H. Chang, A. M. Mebel, X. Yang, S. H. Lin, and Y. T. Lee, J. Chem. Phys. **109**, 2748-2761 (1998).

Measurements of Molecular Alignment in an Intense Laser Field by Pulsed Undulator Radiation

Takahiro Teramoto[1,2], Jun-ichi Adachi[1,2], Kaoru Yamanouchi[2], and Akira Yagishita[1,2]

[1] Photon Factory, IMSS, High Energy Accelerator Research Org., Tsukuba 305-0801, JAPAN
[2] Department of Chem., Graduate School of Sci., The University of Tokyo, Tokyo 113-0033, JAPAN

Abstract. We have initiated a new experiment using pulsed undulator radiation to study geometrical and electronic structures of free molecules in an intense laser field. Our experimental system consists of a coincidence velocity map spectrometer, a nanosecond-pulse Nd:YAG laser, a optical system for the laser, a delayed-coincidence logic circuit, and a data acquisition system. We have measured the angular distributions of photofragment ions due to the C 1s ionization, induced by soft x-ray undulator radiation, of CS_2 molecules in the intense laser field of ~0.5 TW/cm^2. The angular distributions have been slightly different from those without the laser field. However, the anisotropy parameter of the photofragments from the molecules in the laser field has been much smaller than the expected. Further improvements on the spatial overlap between the focused undulator radiation and laser beams are in progress.

Keywords: Intense laser field, Alignment of molecules, Soft x-ray undulator radiation, Coincidence velocity map spectrometer
PACS: 33.20.Ni, 33.90.+h, and 39.90.+d

INTRODUCTION

Controlling molecules with intense laser fields attracts a lot of interest of molecular scientists, because the manipulation of molecules by the intense laser fields can be widely applicable not only to spectroscopy but also to control of molecular reactions [1]. It has been demonstrated that a strong, non-resonant, linearly polarized laser field is capable of aligning molecules. For example, Iwasaki et al. have observed CS_2 molecules aligned in a strong laser field, measuring fragment ions due to Coulomb explosion induced by a probe ultrashort intense laser [2]. And Hoshina et al. have observed pulsed electron diffraction patterns from CS_2 molecules aligned in an intense laser field [3]. The purpose of our present study is to observe CS_2 molecules aligned in a strong laser field, using inner-shell photoionization processes induced by soft x-ray undulator radiation (UR).

Recently we have started our new project to investigate molecules aligned in intense laser fields using soft x-ray UR. Our first attempt is to probe spatially aligned CS_2 molecules which are produced by interaction with the intense laser fields. Linear molecules in the moderate intense laser fields of 0.1-1 TW/cm^2 are ensembles in 'pendular' states and can be aligned along the electric vector of the laser [1-6]. The degree of alignment of CS_2 molecules in the laser fields could be determined from the angular distributions of fragment ions produced in the inner-shell ionization of the sample molecules by the soft x-ray UR, because the direction of the emitted fragments coincides with the molecular axis at the moment of the photoionization [7,8], reflecting the fragmentations following the inner-shell ionization being much faster than molecular rotational periods.

EXPERIMENTAL

Our experimental system consists of a coincidence velocity map spectrometer (CO-VIS), a nanosecond-pulse Nd:YAG laser, a laser optical system, a delayed-coincidence logic circuit, and a data acquisition system. For the detection of the photoelectrons and photoions and the data acquisition system, the products of RöntDek Handels GmbH [9] have been used. The system has two-dimensional detectors with delay-line anodes and a time-to-digital

CP879, *Synchrotron Radiation Instrumentation: Ninth International Conference*,
edited by Jae-Young Choi and Seungyu Rah
© 2007 American Institute of Physics 978-0-7354-0373-4/07/$23.00

converter. A time reference signal from the micro channel plate and delayed signals from both ends of each delay line have been recorded their time information in the list format. The detail of the data acquisition and analysis has been described in Ref. 10. The CO-VIS, developed for studies of inner-shell photoionization [10,11], has been modified for the present experiment. The modifications are as follows; increasing target density in a supersonic molecular beam, introducing a pin-hole for monitor of light beam positions in the experimental chamber, and setting a four-way slit to reduce the focus size of the soft x-ray UR between the beam line and experimental chamber (see Fig. 1).

We have used a fundamental output (1064 nm, ≤ 1 J/pulse, 30 Hz) from the nanosecond pulse Nd:YAG laser (Spectra-Physics GCR-290) [12] to yield the intense laser fields. The laser system is equipped with a seeder, so that time jitter of lasing out can be less than 1 ns. The laser beam was once expanded to reduce its divergence, then was focused by a lens with f = 750 mm onto the collision center to yield the intense laser fields with the power density of about 0.47 TW/cm^2.

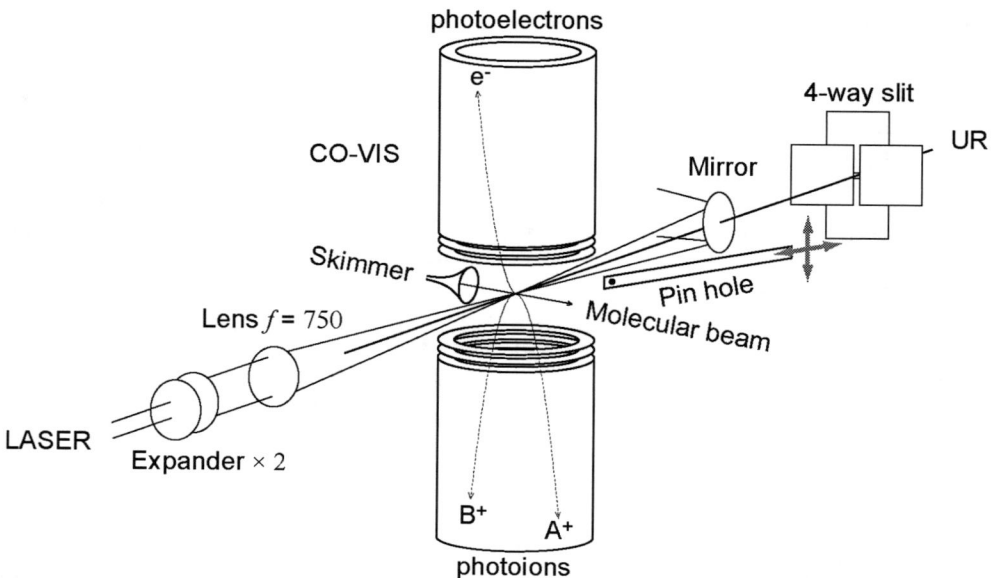

FIGURE 1. A schematic drawing of the experimental set-up. The laser and UR are introduced coaxially in the opposite directions. The laser is dumped away the beam line by the mirror with a central hole, through which the UR has come into the collision region.

Experiments have been performed under the single bunch operation of the storage ring at the Photon Factory. We have used the zero-th order light of the monochromator at beam line 2C [13,14] to use intense soft x-rays under the condition of an undulator fundamental energy of about 320 eV (pulse width of ~150 ps). On the other hand, we introduced the focused laser (λ = 1064 nm, duration ~8 ns) from the nanosecond pulse Nd:YAG laser. The UR and laser beams were overlapped both in time (less than ± 1 ns) and in space (less than ± 0.1 mm) at the collision center. The fragment ions were detected with the CO-VIS to obtain their angular distributions with respect to the polarization vector of the laser, which is parallel to that of the UR. The vapor of CS$_2$ molecules with a purity of 99 % has been used as a sample gas. The CS$_2$ gas diluted about 10 % in He gas with a stagnation pressure of about 3 atm was expanded from the aperture of 0.05 mm into the source chamber that was evacuated by the turbo molecular pumps (the total evacuation rate was nominally 1600 L/s). Then, the supersonic molecular beam was formed through a skimmer into the main chamber.

In order to overlap the UR with laser beams spatially, at first we determined the position of the UR beam by the two-dimensional scan of the pin hole with the diameter of 0.1 mm. Then we adjusted the laser steering mirrors for the attenuated laser beam to pass through the pin hole. Here, laser power was reduced to be the order of 100 µW by an attenuation optics, which was consisted of three splitters of the polarized beam and three $\lambda/2$ plates, to avoid ablation of a pin-hole plate.

As shown in Fig. 2, a divider module converted a RF signal of 500.1 MHz into a NIM fast signal (t_{RF}) with a frequency of 1.6 MHz, the frequency of single bunch. Another divider module converted a t_{RF} signal into a signal (t_{laser}) with a frequency of 30 Hz (*i.e.*, the repetition of the laser oscillation) with negligible jitter in time. Then, a signal t_{laser} was used to generate a delayed signal ($t_{delayed}$) from a digital delay module (DG535), which was used as a trigger pulse for the laser. A 'gate' pulse was generated by a photoelectron signal t_e, which was used as an 'AND gate' for a signal t_{RF}. The signal (t_e) of photoelectrons from the CS_2 molecules ionized by the UR, without the laser irradiation, was measured to obtain the time correlation between t_{RF} and t_e by a time-to-amplitude converter. Then, the signal t_e of photoelectrons from the CS_2 molecules ionized by the fundamental output of the Nd:YAG laser, without the UR, was also measured to obtain the time correlation. Finally, the $t_{delayed}$ was tunned to coincide the peak of the time correlation of t_{RF}-t_e due to the laser with the peak of that of t_{RF}-t_e due to the UR (see Fig. 2).

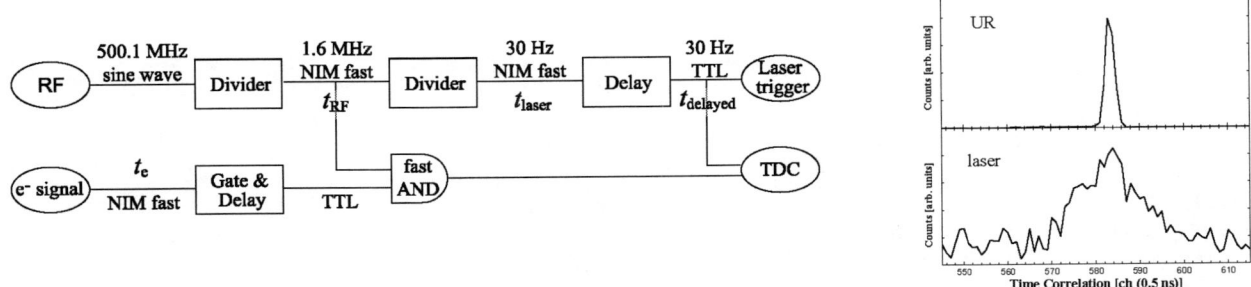

FIGURE 2. Diagram of the synchronization circuit and the time correlation t_{RF}-t_e. The figures on the right side show the time correlation in the case only on UR (upper) and only on the laser (bottom). The peak width in the case only on UR corresponds to the distributions of the flight time of the electrons, but the peak width in the case only on the laser to its pulse width.

RESULTS AND DISCUSSION

In order to estimate the degree of alignment of CS_2 molecules, we have calculated it using a program code [15] on the basis of the theory of Friedrich and Herschbach [5]. Figure 3 shows the theoretical estimation of the degree of alignment with the rotational constant B of 0.10910 cm^{-1} and the polarizability difference $\Delta\alpha$ of 8.8 Å^3. If the rotational temperature of the sample molecules lowers to 10-20 K, which is the expected value under the present experimental condition, the degree of alignment of CS_2 molecules in the laser field of 0.47 TW/cm^2 is about 0.45 at least, as shown in Fig. 3(b).

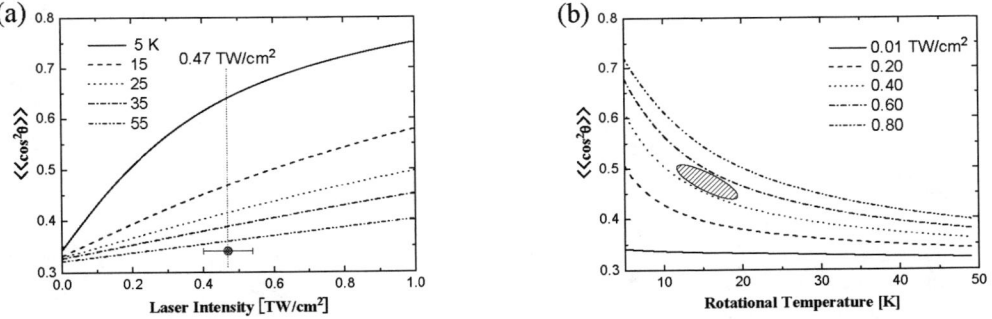

FIGURE 3. Theoretical estimation for the degree of alignment of CS_2 molecules in intense laser fields. (a) shows the dependence of the degree of alignment, $\langle\langle\cos^2\theta\rangle\rangle$, on power densities of the laser fields. θ is the angle between the molecular axis and the electric vector of the laser. (b) shows the dependence of the degree of alignment on rotational temperatures. In (a), the vertical dotted line indicates the value corresponding to the present condition (0.47 TW/cm^2).

Figure 4 shows that coincidence momentum images of S^+ and CS^+ ion pairs from the C 1s photoionization of CS_2 molecules by the soft x-ray UR and polar plots of relevant images; left without the laser fields and right with. Without the laser fields, the angular distribution of the fragment ions is isotropic as expected. Comparing the two results, one can say that CS_2 molecules in the intense laser fields are slightly aligned along the electric vector of the laser fields. For quantitative discussion, we have determined the anisotropy parameter β from the angular distribution of the fragment ions shown in the polar plots; $\beta = -0.002 \pm 0.015$ for the condition without the laser fields and $\beta = 0.070 \pm 0.082$ with the laser fields. As the degree of alignment is quantified by $\langle\langle\cos^2\theta\rangle\rangle$, we have calculated it from the measured angular distributions characterized by the β; the former value gives $\langle\langle\cos^2\theta\rangle\rangle \sim$

0.33 and the latter $<<\cos^2\theta>> \sim 0.34$. For randomly oriented molecules, $<<\cos^2\theta>>$ becomes 1/3. Therefore, the result of $<<\cos^2\theta>> \sim 0.34$ implies that CS_2 molecules in the intense laser fields are slightly aligned along the polarization vector of the laser fields.

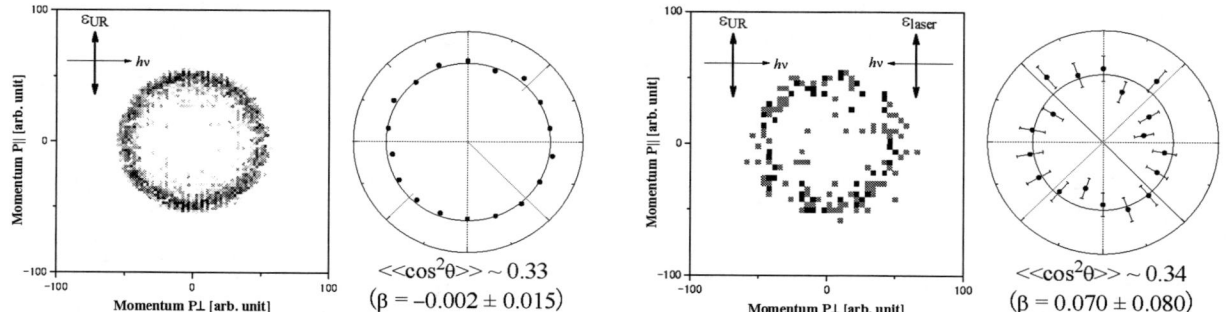

FIGURE 4. Coincidence momentum images and polar plots for the angular distributions of the fragment ions. The two figures on the left show the results without the laser irradiation, the two figures on the right show the results with the laser.

However the result of $<<\cos^2\theta>> \sim 0.34$ is quite smaller than the expected value of 0.45. This discrepancy suggests that spatial overlap between the laser and UR beams may not be perfect under the present experimental condition. Another possible origin of the discrepancy may be caused by high rotational temperature, *i.e.*, the actual rotational temperature may be higher than 10-20 K which we expected. If the rotational temperature was reached at 10-20 K, the degree of alignment $<<\cos^2\theta>>$ should be about 0.45-0.50 (see Fig. 3 (b)).

To establish this new experimental approach, further improvements, that is, finer adjustment of the spatial overlap between the undulator radiation and focused laser beams and decrease of the rotational temperature of sample molecules, are in progress.

ACKNOWLEDGMENTS

We grateful to Dr. Watanabe and Prof. Takahashi for the advices on the coincidence measurement, and Prof. K. Hoshina and Dr. T. Okino for really useful advices on the laser experiment. The study was supported by a Grant-in-Aid for Scientific Research on Priority Areas from the MEXT of Japan and by Grant-in-Aid for Scientific Research (B) 1430126 from JSPS.

REFERENCES

1. K. Yamanouchi, *Science* **295**, 1659-1660 (2002).
2. A. Iwasaki, A. Hishikawa and K. Yamanouchi, *Chem. Phys. Lett.* **346**, 379-386 (2001).
3. K. Hoshina, K. Yamanouchi, T. Ohshima, Y. Ose and H. Todokoro *J. Chem. Phys.* **118**, 6211-6221 (2003).
4. B. Friedrich and D. Herschbach, *Phys. Rev. Lett.* **74**, 4623-4626 (1995).
5. H. Sakai, C.P. Safvan, J.J. Larsen, K.M. Hilligsoee, K. Hald and H. Stapelfeldt, *J. Chem. Phys.* **110**, 10235-10238 (1999).
6. H. Stapelfeldt and T. Seideman, *Rev. Mod. Phys.* **75**, 543-557 (2003).
7. A. Yagishita, K. Hosaka, and J. Adachi, *J. Electron Spectrosc.* **142**, 295-312 (2005); and references therein.
8. J. Adachi, N. Kosugi and A. Yagishita, *J. Phys.* **B 38**, R127-R152 (2005); and references therein.
9. RöentDek Handels GmbH, http://www.roentdek.com/.
10. K. Hosaka, J. Adachi, A.V. Golovin, M. Takahashi, N. Watanabe and A. Yagishita, *Jpn. J. Appl. Phys.* **45**, 1841-1849 (2006).
11. K. Hosaka, "Molecular inner-shell photoionization dynamics studied by momentum imaging spectroscopy" (in Japanese), Doctor Thesis, University of Tokyo, 2005.
12. Spectra-Physics, http://www.spectra-physics.com/.
13. M. Watanabe, A. Toyoshima, J. Adachi and A. Yagishita, *Nucl. Instrum. Methods*, **A 467–468**, 512-515 (2001).
14. M. Watanabe, A. Toyoshima, Y. Azuma, T. Hayaishi, Y. Yan and A. Yagishita, *Proc. SPIE* **3150**, 58 (1997).
15. K. Hoshina and T. Okino (private communication).

Photofragment Imaging Apparatus for Measuring Momentum Distributions in Dissociative Photoionization of Fullerenes

Bhim P. Kafle*, Hideki Katayanagi*¶, and Koichiro Mitsuke*¶

*Department of Structural Molecular Science, Graduate University for Advanced Studies, Myodaiji, Okazaki 444-8585, Japan
¶Institute for Molecular Science, Myodaiji, Okazaki, 444-8585, Japan

Abstract. Description is made on a design of a new version of photofragment imaging spectrometer which will be applied to observe the momentum distributions of ionic fragments from large molecules, clusters, and fullerenes. The apparatus consists of several components: a three-element velocity focusing lens system, a time-of-flight drift tube, a potential-switcheable mass gate, an ion reflector, and a position sensitive detector. The velocity focusing lens system of Eppink-Parker type [Eppink and Parker, *Rev. Sci. Instrum.* **68**, 3477 (1997)] realizes high-resolution photofragment images. Moreover, the mass gate is incorporated inside the tube in order to separate fragment ions with a particular cluster size (e.g. C_{58}^+) from those with other sizes (e.g. C_{60}^+ and C_{56}^+). The optimum arrangement and dimensions of the components are determined from the results of ion trajectories of C_{56}^+, C_{58}^+, and C_{60}^+ simulated by using the SIMION software. The calculated images of C_{58}^+ ions show that kinetic-energy resolution of 10 meV is achievable. It is expected that useful information on reaction mechanism of highly-excited fullerene ions can be derived from the momentum distributions of the fragments.

Keywords: Momentum Imaging, Ion trajectories, Fullerenes, Photodissociation, Fragmentation.
PACS: 39.30.+w; 34.30.+h, 33.20.Ni; 33.80.Eh; 36.40.Qv; 36.40.Wa.

INTRODUCTION

Much experimental work has been devoted to the study of fragmentation of solitary fullerenes, C_{60} and C_{70}, in the extreme UV region by using energy-controlled electron [1,2], synchrotron radiation [3,4], laser [5,6], and fast heavy particles [7]. It is well documented that decomposition of vibrationally-hot C_{60}^{z+} and C_{70}^{z+} formed from C_{60} and C_{70} leads to various carbon clusters C_{60-2n}^{z+} and C_{70-2n}^{z+} ($n \geq 1$, $z \geq 1$) with even-numbered atoms, though dynamical aspects of energy partitioning and fragmentation are not fully elucidated. Very recently dissociative photoionization of C_{60} and C_{70} have been studied by measuring the yield curves of C_{60-2n}^{z+} and C_{70-2n}^{z+} in a wide excitation energy range [3,4]. The behavior of the yield curves can be interpreted in terms of the stepwise mechanism, i.e. internal conversion of the electronically excited states, statistical redistribution of the excess energies, and consecutive ejection of C_2 units [4]:

$$C_{60}^+ \rightarrow C_{58}^+ + C_2, \quad C_{58}^+ \rightarrow C_{56}^+ + C_2, \quad \cdots, \quad C_{60-2n+2}^+ \rightarrow C_{60-2n}^+ + C_2. \tag{1}$$

There are only a few experimental studies of product analysis of the fragments. Several groups have measured the translational energy distribution of C_{60-2n}^+ to gain insight into the energetics and mechanism of fragmentation. Hertel and co-workers [5] evaluated the average kinetic energies of C_{60-2n}^+ ($1 \leq n \leq 14$) produced by laser multiphoton ionization of C_{60} by making use of a time-of-flight method. Later Märk and co-workers [2] fulfilled electron impact ionization of C_{60} and reported a value of ca. 0.45 eV as the total average kinetic energy release in the decomposition of C_{60}^+ into C_{60-2n}^+ ($1 \leq n \leq 8$). These authors suggested that not only sequential C_2 ejection of process (1) but also single-step two-fragment fission of the parent C_{60}^+ ions

$$C_{60}^+ \rightarrow C_{60-2n}^+ + C_{2n} \tag{2}$$

are possible mechanisms for the formation of C_{60-2n}^+.

CP879, *Synchrotron Radiation Instrumentation: Ninth International Conference*,
edited by Jae-Young Choi and Seungyu Rah
© 2007 American Institute of Physics 978-0-7354-0373-4/07/$23.00

In the present study we will develop a new version of momentum imaging spectrometer to obtain reliable velocity distribution of the fullerene fragments. This paper describes a basic design of this spectrometer. From photofragment images, we will be able to decide on which mechanism dominates fragmentation of fullerene ions, because three-dimensional velocity distributions are expected to considerably differ for different mechanisms. Moreover, closer inspection of the images may allow us to directly probe the properties of transition states correlated to the dissociation channels [8].

Momentum imaging technique was first introduced to the field of molecular reaction dynamics in 1987 by Chandler and Houston [9]. In 1997 a new velocity focusing lens system was invented by Eppink and Parker and a great advance was made in improving the resolution of the image [10]. Using momentum imaging technique photofragmentation processes of fundamental molecules have been studied extensively. In general neutral fragments under study are ionized selectively by resonance-enhanced multiphoton ionization method (REMPI) [11] to avoid the interference from imaging of other fragments. In contrast, it is difficult to obtain clear images of the ionic fragments which are free from signals of unwanted species. Some authors applied a pulsed high voltage to the front plate of the microchannel plate electron multipliers (MCP) just on time for the arrival of the ionic fragments with particular mass-to-charge ratio m/z [12]. This method of MCP switching is not so advantageous to large molecules because time of flights of many kinds of fragments come close to one another. Furthermore, application of the pulsed high voltage field brings about distortion of equipotential surfaces inside a drift tube which reduces the number of ion trajectories reaching the detector. To overcome these difficulties we incorporate a mass gate and an ion reflector inside a drift tube of our photofragment imaging spectrometer, which may provide uncontaminated images of particular ionic fragments from large molecules, clusters, or fullerenes.

BASIC CONCEPT AND DESIGN OPTIMIZED BY SIMULATIONS

Figure 1 illustrates a schematic design of our spectrometer. All the electrodes and ion drift tube are cylindrically symmetrical. A velocity focusing lens system contains three electrodes: repeller, extractor, and the entrance electrode of the drift tube. Pulsed voltages are applied to the former two electrodes to extract ionic fragments from C_{60}, while the drift tube is connected to a ground throughout our trajectory calculations. Every electrode of the three-element lens system is made of circular plates 1 mm in thickness and 50 mm in outer diameter, separated by 15 mm from each other. Extractor and the entrance electrode of the drift tube have central hole of 20 mm in diameter. Such open-hole structure of the two electrodes allows us to bend the equipotential surfaces by simply manipulating the extractor voltage and to achieve excellent focusing of momentum image on a position sensitive detector PSD [10,13]. Near the end of the drift tube there are a cylindrical mass gate and an ion reflector having the inner diameter of 40mm. Thin meshes with high transmittance are so fixed to both ends of the mass gate that distortion of equipotential surfaces due to fringe effect becomes negligibly small. The ion reflector is comprised of three electrodes with fine meshes. Its central electrode is floated to a high positive voltage of 320 V, while the other two electrodes are grounded. The latter electrodes, located 5mm apart from the central electrode, can keep equipotential surfaces flat and parallel near the ion reflector. As long as the mass gate is kept grounded, all fragments are repelled by the ion reflector and do not impinge against the PSD (effective size = 40 mm in diameter). When an entire bunch of the fragments having an expected m/z arrives inside the mass gate, a pulsed voltage is applied there. Thereby, the potential energies of the ions in this bunch are suddenly elevated, so that only these ions can pass through the ion reflector and reach the PSD.

We performed ion trajectory simulations utilizing the SIMION 3D (ver. 7.0) software [14] to optimize the dimensions of the electrodes in Fig. 1. The grid size of the simulations of 0.5 or 1mm was adopted to keep a good scale factor. Here, dissociative ionization of C_{60} is considered to take place within a region of rectangular

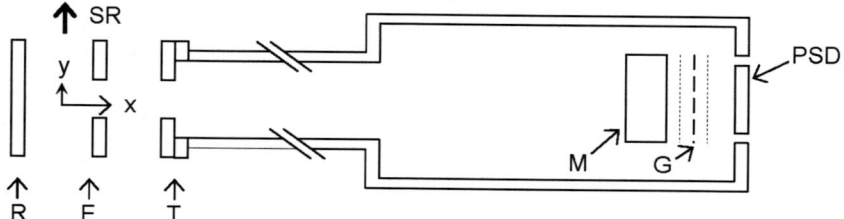

FIGURE 1. Schematic view of the momentum imaging spectrometer in combination with the mass gate M and ion reflector G. The dimensions of all the electrodes are determined from the SIMION 3D software. R, repeller, E, extractor, T, entrance electrode of a drift tube; SR, synchrotron radiation.

parallelepiped $\Delta x \Delta y \Delta z = 1 \times 3 \times 1$ mm³ as depicted in Fig. 2. Since the y-direction is assigned to the passage of synchrotron radiation, the y-coordinate of this region is made to range from -1.5 to +1.5 mm. In ion trajectory simulations the eight corners and center of the ionization region were chosen for the starting points of the trajectories. From each point 171 trajectories were generated in the elevation angle range of -90° to +90° at intervals of 22.5°, and in the azimuth angle range of 0° to +180° at intervals of 10°. The definition of the two angles is given in Fig. 2. The optimum distance from the ionization region to the center of the mass gate was 335.5 mm and that to the entrance of the PSD was 360.5 mm. The length of the mass gate (= 10 mm) was so chosen as to accommodate all fragment ions with a particular m/z ratio inside the mass gate.

Application of the pulsed voltages to repeller, extractor and the mass gate was realized by means of a "user program" of SIMION [14]. The amplitude and duration of the pulsed voltage to repeller were 300 V and 7 µs, respectively. Simultaneously a similar pulsed voltage was applied to extractor. The ratio of voltages

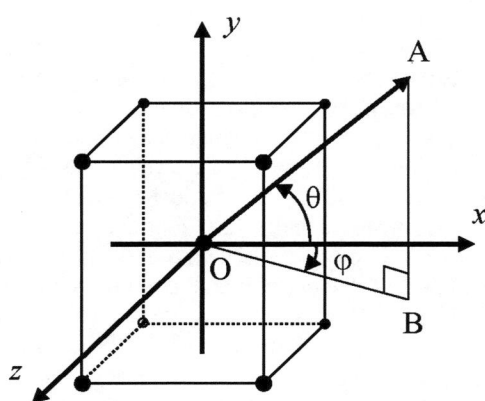

FIGURE 2. Ionization volume and coordinate system defined for simulations. OA, ion emission direction; OB, projection of OA on the x-z plane; θ, elevation angle; φ, azimuth angle.

between that applied to extractor and that to repeller was set to be constant at 0.714. Such applications permit the ions produced during the past ~ 13 µs to be guided into the drift tube. The duration of 7 µs was long enough for the ions to escape from the effect of equipotential surfaces near the three-element lens system. The rising edge of the pulsed voltages for repeller and extractor preceded that for the mass gate by 44.5 µs. At the mass gate the spread of the time of flight was estimated to be 0.65 µs which arises from the finite volume of the ionization region and distribution of the kinetic energy of the fragment ions. Thus, the duration and amplitude of the pulsed voltage to the mass gate were chosen to be 1 µs and 120 V, respectively.

Panel (b) of Fig. 3 shows simulated trajectories of C_{58}^+ at initial kinetic energy of 0.1 eV, the ion whose momentum image we wish to measure, and Panels (a) and (c) present the trajectories of unwanted ions, C_{60}^+ and C_{56}^+, respectively. Though the trajectories of both C_{60}^+ and C_{56}^+ are reflected completely, most of the trajectories of C_{58}^+

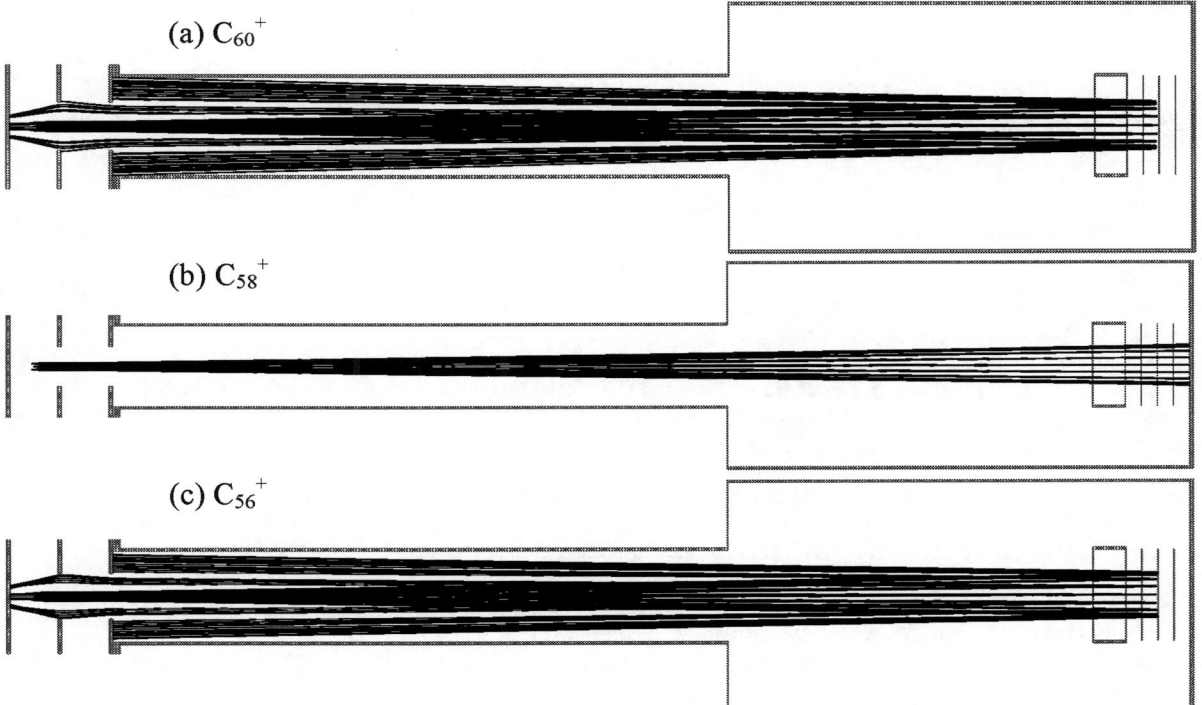

(a) C_{60}^+

(b) C_{58}^+

(c) C_{56}^+

FIGURE 3. Simulated trajectories of (a) C_{60}^+, (b) C_{58}^+ and (c) C_{56}^+ at initial kinetic energies of 0.1 eV. The trajectories of C_{60}^+ and C_{56}^+ are found to turn around at the ion reflector, travel in the opposite direction and terminate at repeller.

are found to go beyond the ion reflector and reach the PSD. This observation provides direct evidence for exclusive detection of the image of C_{58}^+ without interference by C_{60}^+ and C_{56}^+ having the same kinetic energies.

Figure 4 shows the simulated momentum images of C_{58}^+ ions on the PSD at the kinetic energies of 0.1 eV (triangles) and 0.11 eV (circles). It should be noted that these images result from momentum distributions of the ions in the laboratory system. We took into account the ion trajectories generated in the elevation and azimuth angle ranges of 0° to +90° and 0° to +180°, respectively, which cover only one quarter of the full three-dimensional trajectories over the 4π solid angle. The trajectories with a given elevation angle form a horizontal stripe, and the envelope of all the stripes makes an arc, which clearly demonstrates that scattering distribution in spherical symmetry can be successfully projected on an image plane. It is likely that C_{58}^+ fragment ions with kinetic energy difference of 0.01eV are almost separable. Comparison between the simulations with and without the ion reflector confirmed that the images are not distorted in the presence of the ion reflector. The present momentum imaging spectrometer will be constructed and installed in the end station of beam line 2B in the UVSOR facility.

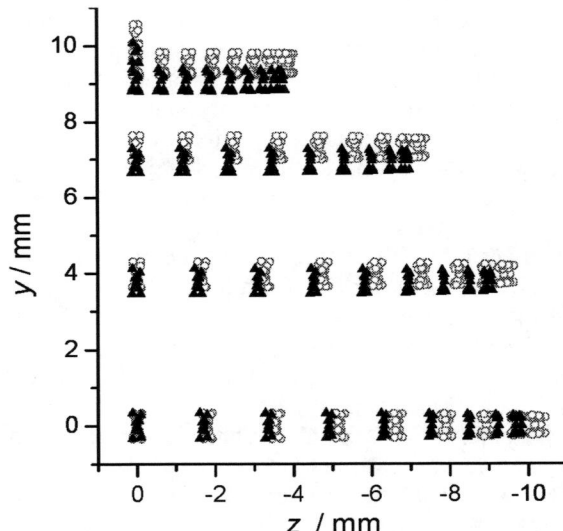

FIGURE 4. Simulated image of C_{58}^+ ions at the kinetic energies of 0.1 (▲) and 0.11 eV (○). The three-dimensional scattering distribution of the ions is projected on the PSD.

ACKNOWLEDGMENTS

This work has been supported by a Grant-in-Aid for Scientific Research (Grant No. 14340188,18350016) from the Ministry of Education, Science, Sports, and Culture, Japan, and by a grant for scientific research from Research Foundation for Opto-Science and Technology.

REFERENCES

1. M. Foltin, M. Lezius, P. Scheier, and T. D. Märk, *J. Chem. Phys.* **98**, 9624-9634 (1993); P. Scheier, B. Dünser, R. Wörgötter, M. Lezius, R. Robl, and T. D. Märk, *Int. J. Mass Spectrom. Ion Proc.* **138**, 77-93 (1994); S. Matt, B. Dunser, M. Lezius, H. Deutsch, K. Becker, A. Stamatovic, P. Scheier, and T.D. Mark, *J. Chem. Phys.* **105**, 1880-1896 (1998).
2. D. Muigg, G. Denifl, P. Scheier, K. Becker, and T.D. Mark, *J. Chem. Phys.* **108**, 963-970 (1998).
3. A. Reinköster, S. Korica, G. Prümper, J. Viefhaus, K. Godehusen, O. Schwarzkopf, M. Mast, and U. Becker, *J. Phys. B* **37**, 2135-2144 (2004).
4. J. Kou, T. Mori, Y. Kubozono, and K. Mitsuke, *Phys. Chem. Chem. Phys.* **7**, 119-123 (2005); K. Mitsuke, H. Katayanagi, J. Kou, T. Mori, and Y. Kubozono, *Am. Inst. Phys. CP* **811**, 161-166 (2006).
5. H. Gaber, R. Hiss, H.G. Busmann, and I.V. Hertel, *Z. Phys. D* **24**, 302 (1992).
6. P. Wurz and K.R. Lykke, *J. Phys. Chem.* **96**, 10129-10139 (1992); D.Ding, R.N. Compton, R.E. Haufler, and C.E. Klots, *ibid.* **97**, 2500-2504 (1993); J. Laskin, B. Hadas, T.D. Märk, and C. Lifshitz, *Int. J. Mass Spectrom.* **177**, L9-L13 (1998).
7. B. Walch, C.L. Cocke, R. Voelpel, and E. Salzborn, *Phys. Rev. Letters* **72**, 1439-1442 (1994); T. LeBrun, H.G. Berry, S. Cheng, R.W. Dunford, H. Esbensen, D.S. Gemmell, E.P. Kanter, and W. Bauer, *ibid.* **72**, 3965-3968 (1994).
8. If dissociation channels involve passage over an exit barrier, the distribution of the translational energy in the center-of-mass system shows a peak at some fraction of the barrier height. In contrast, if little or no exit barrier exists on dissociation channels, the distribution shows peak at close or equal to zero. References: D.H. Mordaunt, D.L. Osborn, and D.M. Neumark, *J. Chem. Phys.* **108**, 2448-2457 (1998); A.E. Faulhaber, D.E. Szpunar, K.E. Kautzman, and D.M. Neumark, *J. Phys. Chem. A* **109**, 10239-10248 (2005).
9. D.W. Chandler and P.L. Houston, *J. Chem. Phys.* **87**, 1445-1447 (1987); P.L. Houston, *J. Phys. Chem.* **100**, 12757-12770 (1996).
10. A.T.J.B. Eppink and D.H. Parker, *Rev. Sci. Instrum.* **68**, 3477-3484 (1997).
11. M. Ito and M. Fujii, "Advances in Multi-Photon Processes and Spectroscopy," ed. S.H. Lin, World Scientific, Singapore 1988, Vol. 4, pp. 1-68.
12. F. Aguirre and S.T. Pratt, *J. Chem. Phys.* **118**, 6318-6326 (2003).
13. B. Tang and B. Zhang, *Chem. Phys. Lett.* **412**, 145-151 (2005).
14. D. A. Dahl, "SIMION v7.0," Idaho National Engineering and Environmental Lab., Idaho Falls 2000.

A Two Magnetron Sputter Deposition Chamber Equipped with an Additional Ion Gun for *in situ* Observation of Thin Film Growth and Surface Modification by Synchrotron Radiation Scattering

Norbert Schell, Johannes von Borany, and Jens Hauser

Forschungszentrum Rossendorf, P.O. Box 510119, 01314 Dresden, Germany

Abstract. We report the design of a sputter deposition chamber for the *in situ* study of film growth and modification by synchrotron x-ray diffraction and reflectivity. The chamber is sealed with four Be-windows allowing unhindered scattering access of –2 up to +50 degrees off-plane and –2.9 up to +65 degrees in-plane, respectively. The chamber fits into a standard six-circle diffractometer from HUBER which is relatively widespread in synchrotron laboratories. Two commercial miniature magnetrons with additional gas inlets allow for the deposition of compound films and multilayers. Substrate heating up to 950°C and different substrate bias voltages are possible. An additional ion gun up to 6 keV and 10 μA allows post-deposition ion irradiation with light atoms or energetic ion bombardment during sputter deposition. The performance of the chamber was tested with the deposition of MAX phase Ti_2AlN and with the off-sputtering of a thin Pt film.

Keywords: X-ray diffraction, Thin film structure and morphology, Deposition by sputtering.
PACS: 61.10.Nz, 68.55.-a, 81.15.Cd.

INTRODUCTION

The technological importance of thin films has led to an unabated interest in the detailed characterization of their structure, morphology, and their interfaces. A real understanding of the underlying growth mechanisms and their microstructural development requires sophisticated *in situ* techniques. X-ray diffraction is such a well established and powerful technique: Small angle scattering informs about nucleation, time-dependent crystal truncation rod scattering about the growth mode (i.e. layer-by-layer growth, island formation or step flow mode), specular reflectivity gives the density (*via* critical angle), film thickness (*via* Kiessig fringes) and roughness (*via* intensity drop), large angle scattering the lattice parameter, i.e. phase, and a detailed analysis of peak profiles and intensities reveals texture, grain size and microstrain. A combination of in-plane and off-plane scattering geometries allows a 3-dimensional stress analysis, and finally, all techniques performed time dependently *in situ* reveal the dynamics of the microstructural development *during* film growth.

Numerous examples of corresponding investigations can be found in the literature, not only with x-rays but also other probes like electrons, ions and neutrons. For one widespread layer deposition technique, i.e. magnetron sputtering, however, only x-ray techniques are suitable as a routine and versatile investigative tool for structural characterization, the reason being magnetic stray fields and vacuum requirements in the Pa-range. Again, while the literature is rich in examples of special equipment, there exist few devices allowing all above mentioned scattering techniques at the same time (see e.g. [1] and references therein). Based on our experiences with a precursor versatile sputter deposition chamber with just two magnetrons [1] and in view of the potential of combining the deposition with energetic ion bombardment for growth assistance or amorphization, we commissioned an improved process chamber with two magnetrons for balanced and unbalanced DC operation, several gas inlets for reactive sputtering and an additional ion gun allowing post-deposition ion irradiation or ion bombardment during sputter deposition. This contribution describes its design and shows first results of test experiments.

CP879, *Synchrotron Radiation Instrumentation: Ninth International Conference*,
edited by Jae-Young Choi and Seungyu Rah
© 2007 American Institute of Physics 978-0-7354-0373-4/07/$23.00

DESCRIPTION OF THE CHAMBER

Like its predecessor, the new chamber (Fig. 1) is equipped with two commercial balanced cylindrical magnetrons [2] each with a target diameter of 1 in as deposition source. Their axes are tilted 30° away from the substrate surface normal at a target to substrate distance of 100 mm. Chimneys and pneumatic-driven shutters in front of each magnetron avoid cross-contamination and allow a precise determination of deposition times. Two additional gas inlets for different sputter gases allow reactive sputter deposition of multicomponent films or multilayer structures. A BORALECTRIC heater [3] isolated by ceramic supports (Fig. 2) allows sample temperatures up to 950°C and a bias voltage of ±10 kV. Temperature control is performed with chromel-alumel thermocouples (one for heating control, the other for surface temperature calibration). The ion current is measured in a Faraday cup positioned on the substrate holder in the line of the ion gun axis. With two turbo-molecular pumps and a liquid nitrogen trap inside the chamber the base pressure reaches a vacuum better than 2×10^{-7} mbar. The additional ion gun IQ 100 [4] operating under a fixed incidence angle of 20° to the substrate normal can deliver energetic ions of ≤6 keV and up to 10 μA current. It has a sub-mm focus and a maximum scanning field of 10 x 10 mm^2. Separate differential pumping allows its operation independent from the working pressure of 0.1–1 Pa characteristic for magnetron sputtering.

The chamber is equipped with a quartz view port for visual inspection and with four 400 μm thick Be windows (protected inside by Al-foils on protruding bolts) allowing unhindered scattering access of –2° up to +50° off-plane and –2.9° up to +65° in-plane, respectively, and thus the application of all above mentioned characterizing x-ray techniques under large and in-plane grazing angles up to 10°. The size and total weight of <15 kg is restricted such that the chamber fits into the 6-circles HUBER diffractometer [5] on a separate adapter flange for the φ circle (inner diameter 80 mm), itself inside the closed Eulerian cradle (modified model 512.1) of the multipurpose materials research experimental station BM20 at ESRF (ROBL-CRG) without loss of the high precision of the goniometer settings of 0.001°.

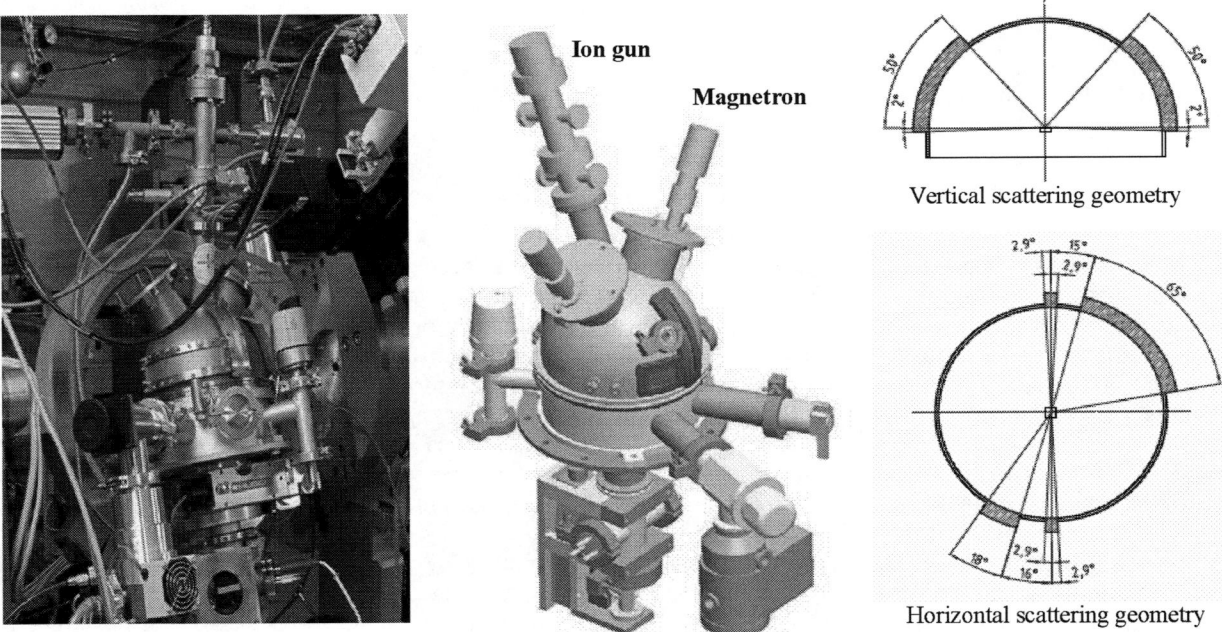

Vertical scattering geometry

Horizontal scattering geometry

FIGURE 1. New sputter deposition chamber: (left) inserted into the 6-circles HUBER goniometer of ROBL viewed from the detector side with only the ion gun and its gas supply connected; (middle) design study from the beam entrance side with symmetric magnetrons, ion gun, viewport between Be entrance windows, one turbo pump (70 l/s) at a corner valve which can be used for throttling and substrate alignment and handling devices below; (right) accessible angles in vertical and horizontal scattering geometries.

At substrate temperatures >500°C, the chamber walls are cooled with air fans. All vacuum flanges are sealed with Viton® O-rings. The substrate is changed by disconnecting only the substrate carrier unit at the bottom flange of the chamber. Flooding the chamber with dry nitrogen during this change reduces considerably the contamination of the inner walls with moisture or oxygen.

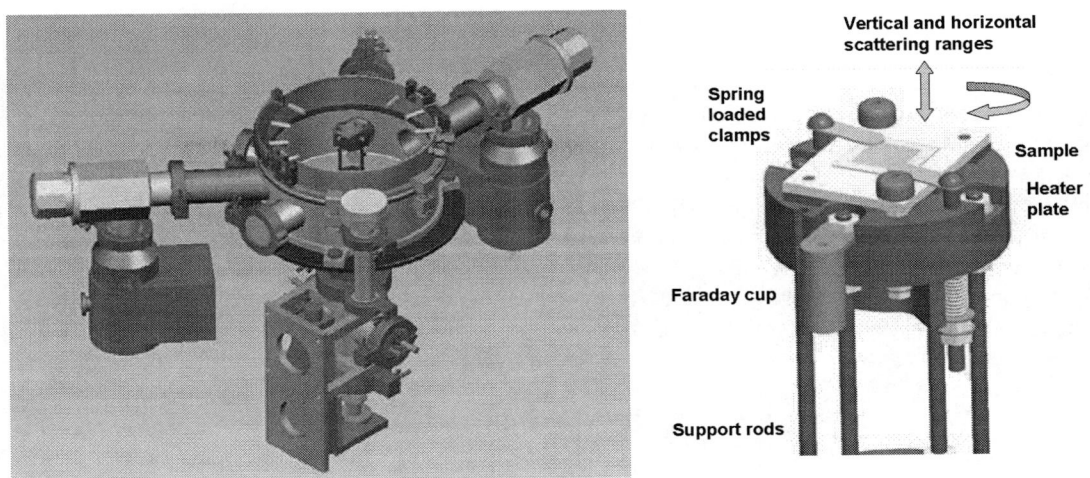

FIGURE 2. Interior of the chamber with alignment stage (left) and substrate holder with sample (right).

TEXTURE DEVELOPMENT AND PHASE STABILITY OF MAX Ti₂AlN THIN FILMS

As an improved modification of a successful pilot deposition chamber, the new process chamber comprises all technical possibilities of our previous experiments, allowing e.g. to follow the texture development [6] and the growth mode of TiN hard coatings [7], the grain development of nano-crystalline Au [8], or the real-time structural design of smart materials [9]. For our up-to-now most demanding deposition, i.e. pseudo-epitaxially grown MAX phase Ti₂AlN – a high-temperature perovskite-similar alloy combining ceramic and metallic properties – we were at the limit concerning heating and base pressure, leading to tilted basal plane growth prohibiting most likely the technical use of such thin films [10]. With the new chamber, i.e. a base pressure by nearly one order of magnitude better and 200°C higher temperatures possible, we could not only succeed in real basal plane growth but were also able to fully investigate the phase stability between growth start and final dissolution by interface spinel formation (*to be published by M. Beckers et al. in JAP and APL*). Figure 3 shows a comparison from the relevant publications.

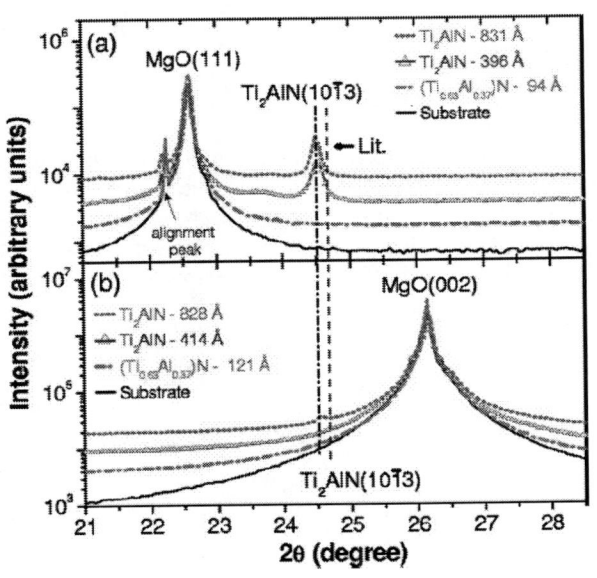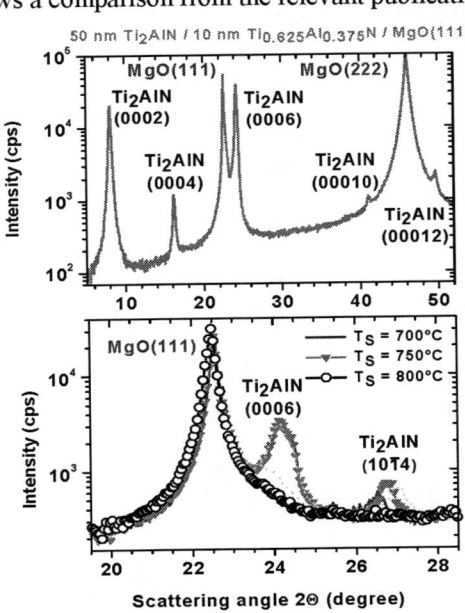

FIGURE 3. Comparison of the sputter deposition of MAX phase Ti₂AlN thin films onto MgO(111) substrates with a thin fcc TiAlN seed layer in the old chamber [10] (left) and the new one (right). The larger windows allow more reflections to be seen, the higher achievable temperature allows basal plane growth (recognizable by the multiplicity 0001 peaks) and a determination of the overall phase stability with temperature.

OFF-SPUTTERING OF PT THIN FILMS

The off-sputtering of a 55 nm thick Pt film deposited on SiO_2(1.5 μm)/Si(100) has been investigated by XRD in Bragg-Brentano-geometry (λ = 1.54 Å). The film was irradiated with Ar ions of 5 keV energy under an incidence angle of 20°; the total ion current of 2.72 μA within a scanning field of 4.2 x 4.2 mm² corresponds to an ion flux density of 9.6×10^{13} ions/cm²s. Figure 4 shows scans around the dominating Pt(111) peak recorded after consecutive sputtering steps as well as corresponding TRIDYN code simulations [11]. Three features are clearly observed: (i) the integrated peak area decreases as expected due to the film thickness reduction from sputtering, (ii) a second component appears at a lower scattering angle (2θ = 39.15°) reflecting an expanded Pt lattice due to Ar incorporation, and (iii) at the end of the sputtering process the scan peaked with a very low intensity at the initial position (2θ = 39.60°) which can be interpreted in terms of a very thin Pt film or even Pt islands (size <5 nm) with a relaxed lattice as Ar becomes volatile.

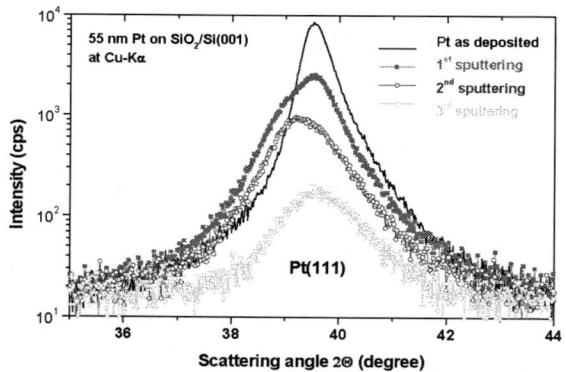 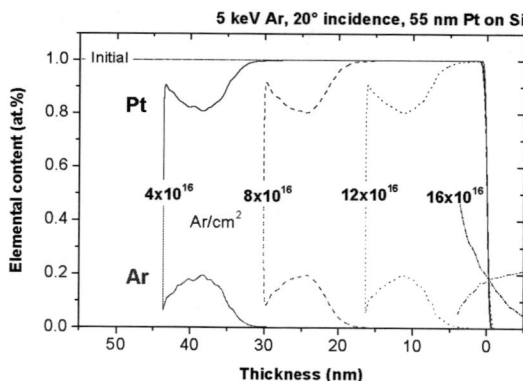

FIGURE 4. XRD scans in Bragg-Brentano-geometry around the dominating Pt(111) peak after three consecutive off-sputtering steps lasting 13, 8, and 8 min, respectively (left); Pt off-sputtering and Ar incorporation into the residual film according to TRIDYN calculations for different Ar fluences (right): The reduction of the Pt film thickness with increasing Ar ion fluence is obvious, the sputter yield amounts to 2.8 at/ion. The calculated loss of the Pt areal density due to off-sputtering agrees quantitatively very well with the measured decrease of the integrated Pt(111) peak area. The Ar ions modify a near-surface region of the Pt films of approximately 10 nm; the maximum Ar content saturates at about 20 at.%.

CONCLUSION

A process chamber with two magnetrons and one ion gun, heating and bias voltage, equipped with four Be windows for various x-ray scattering techniques off-plane and in-plane has been commissioned at a synchrotron radiation source. First experiments demonstrate its possibilities for *in situ* x-ray investigations of the microstructural development during growth of thin films or their modification by additional ion bombardment. Future experiments will be dedicated to study changes of film properties by ion irradiation *during* deposition. Energy and flux of the ions delivered from the gun can be varied independently from the ions and atoms from the magnetron targets.

REFERENCES

1. W. Matz et al., *Rev. Sci. Instrum.* **72**, 3344-3348 (2001).
2. AJA International, P.O. Box 246, 809 Country Way, North Scituate, MA 02060, U.S.A., http://www.ajaint.com
3. http://www.tectra.de/heater.htm#BORALECTRIC%AE
4. Kremer Vakuumphysik, Chorbuschstr. 13, 50259 Pulheim, Germany, http://www.kremer-vakuumphysik.de
5. HUBER X-ray Diffraction Equipment, 83253 Rimsting, Germany, http://www.xhuber.com
6. N. Schell et al., *J. Appl. Phys.* **91**, 2037-2044 (2002).
7. J. Bottiger et al., *J. Appl. Phys.* **91**, 5429-5433 (2002).
8. K.P. Andreasen et al., *Mat. Res. Soc. Symp. Proceedings* **788**, 49-54 (2004).
9. N. Schell et al., *Appl. Phys. A* **81**, 1441-1445 (2005).
10. M. Beckers et al., *J. Appl. Phys.* **99**, 034902-1-034902-8 (2006).
11. W. Möller and M. Posselt, Tridyn-FZR User Manual, Wiss.-Tech. Berichte, FZR-317, 2001.

Photodissociation of Butyl Cyanides and Butyl Isocyanides in the Vacuum UV Region

Kazuhiro Kanda[1], Koichiro Mitsuke[2], Kaoru Suzuki[3] and Toshio Ibuki[4]

[1] LASTI, University of Hyogo, Koto, Kamigori, Ako, Hyogo 678-1205, Japan
[2] Institute for Molecular Science, Myodaiji, Okazaki 444-8585, Japan
[3] Faculty of Foreign Studies, Tokoha Gakuen University, Sena, Shizuoka 420-0911, Japan
[4] Faculty of Education, Kyoto University of Education, Fushimi-ku, Kyoto 612-8522, Japan

Abstract. Photodissociation process to produce the electronically excited $CN(B\Sigma^+)$ fragment has been studied for four structural isomers, n-C_4H_9CN, t-C_4H_9CN, n-C_4H_9NC and t-C_4H_9NC by using synchrotron radiation. Photoexcitation spectra for the $CN(B^2\Sigma^+ \rightarrow X^2\Sigma^+)$ transition were measured in the excitation wavelength range of 85-165 nm (7.5-14.5 eV) and the $CN(B^2\Sigma^+ \rightarrow X^2\Sigma^+)$ emission spectra were dispersed at several wavelengths in the range of 57-140 nm. Quantum yields for the production of the CN(B) state from photodissociative excitation of the isocyanides were larger than those from the cyanides. The quantum yields from the molecules consisting of *tertiary*-butyl group were larger than those from the *normal*-butyl compounds.

Keywords: photoexcited fluorescence spectrum, photoemission spectrum, synchrotron radiation, cyanide, isocyanide
PACS: 32.50.+d

INTRODUCTION

Photodissociation of the CN-containing compounds in the vacuum ultraviolet region is known to produce CN radicals in the $B^2\Sigma^+$ state. Extensive works have been performed by the photon impact study of cyanides. However, photodissociation of isocyanides has not been investigated sufficiently. Internal energy distributions of the produced $CN(B^2\Sigma^+)$ state produced by collision with rare gas metastable atoms were reported to be much different between in the case of cyanides (n-C_4H_9CN and t-C_4H_9CN) and that of isocyanides (n-C_4H_9NC and t-C_4H_9NC) [1]. Deeper insight into the dissociation process requires the knowledge of the highly excited states of the parent molecules in the VUV region and information on the photodissociation via these excited states. In our previous study, the highly excited states of these four kinds of structural isomers, n-C_4H_9CN, t-C_4H_9CN, n-C_4H_9NC and t-C_4H_9NC were investigated by the measurements of the photoabsorption spectra in the wavelength range 85-165 nm using synchrotron radiation (SR) and HeI photoelectron spectra [2]. In the present study, photodissociation of these structural isomers has been investigated in the extreme vacuum ultraviolet region using SR. Absolute cross sections of the $CN(B^2\Sigma^+ \rightarrow X^2\Sigma^+)$ emission for these molecules have been determined over the excitation wavelength range 85-165 nm (7.5-14.5 eV). In addition, the $CN(B^2\Sigma^+ \rightarrow X^2\Sigma^+)$ emission spectra produced from photodissociation of these molecules at several wavelengths were measured in the range of 57-140 nm in order to confirm the emitting species and to observe for the internal energy distributions of the CN(B) fragments.

EXPERIMENTAL

The experiments were carried out using SR provided by the 0.75 GeV storage ring of the Synchrotron Radiation Facility (UVSOR) at the Institute for Molecular Science. Commercial *normal*-butylcyanide:n-C_4H_9CN, *tertiary* butylcyanide: t-C_4H_9CN, *normal*-butylisocyanide: n-C_4H_9NC and *tertiary* butylisocyanide: t-C_4H_9NC were used as sample after a trap-to-trap cycle in order to eliminate the possible impurities.

CP879, *Synchrotron Radiation Instrumentation: Ninth International Conference*,
edited by Jae-Young Choi and Seungyu Rah
© 2007 American Institute of Physics 978-0-7354-0373-4/07/$23.00

Measurement of Photoexcitation Spectra

The photoexcitation spectra were measured in the wavelength range 85-165 nm on the fluorescence apparatus for vapor phase photochemistry at BL2A stage [3]. The SR from bending magnet was dispersed by a 1 m Seya-Namioka monochromator and was introduced into a 12.3 cm-length reaction flow cell. The spectral resolution was 0.3 nm fwhm. The fragment emissions were isolated by a band-pass filter (Toshiba C-39A), whose transmittance wavelength region was 360-440 nm, and detected by a photomultiplier tube (Hamamatsu R585). Absolute values for the emission cross sections were scaled by a comparison with the intensity of the CN(B→X) emission produced in the photodissociation of HCN, for which the absolute emission cross section has been reported [4]. The emission cross section determined in the present study corresponds to the cross section for the production of CN(B), σ_B, because the radiative decay can be regarded as the dominant deexcitation process for the CN formed in the B state. The relative uncertainty of σ_B was estimated to be ≈10%.

Measurement of $CN(B^2\Sigma^+-X^2\Sigma^+)$ Emission Spectra

The CN(B→X) emission was dispersed in the 330-460 nm wavelength region at BL3A2. The light source was an undulator, which has 24 periods with a period length of 80 mm. Details of the experimental apparatus was described in ref. [5]. Briefly, the fundamental light of the undulator radiation was dispersed by a monochromator of a constant deviation grazing-incidence type with a 2.2 m focal length. A spectral resolution of SR was 0.24 nm (fwhm, 83 meV at 20.85 eV) with slit widths of 300 μm. The fluorescence was collected by an optical detection system and introduced to the 300 mm focal-length imaging spectrograph (Roper Scientific, SpectroPro-300i) with a 600- and 1200-grooves mm^{-1} gratings, being equipped with a liquid nitrogen cooled CCD array detector. The emission from photodissociation of the cyanides and isocyanides was dispersed at 9.03, 10.2, 11.8 and 21.7 eV.

RESULTS AND DISCUSSION

Panels (a)-(d) in Fig. 1 depict the photoabsorption spectra (thin lines) taken from ref. [2] and relevant photoexcitation spectra for the CN(B→X) emission (thick lines) of n-C_4H_9CN, t-C_4H_9CN, n-C_4H_9NC and t-C_4H_9NC, respectively, in the wavelength range of 65-165 nm. Solid arrows in each panel indicate the first and the second ionization potentials of the molecule [2]. Broken arrows in the panels (b) and (d) indicate the photon energies for the dispersed emission spectra in Fig. 2. Panels (a) and (b) in Fig.2 are the emission spectra of t-C_4H_9NC and t-C_4H_9CN excited at the 10.4 eV, respectively, being dispersed in the wavelength region from 320 nm to 460 nm with a low resolution. Figures 2c and 2d show the CN(B→X) emission spectra of 1-0 sequence for t-C_4H_9NC excited at 9.03 and 10.4 eV, respectively, being dispersed in the wavelength region from 410 nm to 425 nm with a high resolution.

In the absorption spectra of n-C_4H_9NC and t-C_4H_9NC, numerous sharp peaks converging to the first ionization potential have been assigned to the excitations to the Rydberg states [2]. On the other hand, sharp peaks attributable to Rydberg transitions were not observed in the absorption spectra of n-C_4H_9CN, t-C_4H_9CN. The overall intensity of absorption became stronger towards the shorter wavelengths and absolute value of absorption cross sections increased from >10 Mb at 165 nm to ≈120 Mb at 100 nm for four molecules. The absolute cross sections for the production of CN(B→X) emission are, however, very different for the molecules. In order to confirm the emitting species in Fig. 1, the emission spectra were dispersed. The dominant band at ≈388 nm for t-C_4H_9NC in Fig. 2a is the 0-0 sequence of CN(B→X) emission and the small bands at ≈358 and ≈410 nm correspond to the 1-0 and 0-1 sequences of CN(B→X) emission, respectively. Thus, no emission except the CN(B→X) was observed for the photodissociation of t-C_4H_9NC in the excitation wavelength range of 85-165 nm. In other words, the large emission cross section for t-C_4H_9NC was confirmed to reflect only the CN(B→X) emission. On the other hand, a small peak at 390 nm for t-C_4H_9CN in Fig. 2b corresponds to the 0-0 sequence of $N_2^+(B^2\Sigma^+\to X^2\Sigma^+)$ emission due to the higher harmonic light of undulator radiation. Thus, N_2^+(B→X) emission due to the impurity nitrogen molecule in the vacuum was observed in the case that the σ_B was small. In the excitation wavelength region shorter than 100 nm, where the emission cross section increased toward shorter wavelength, N_2^+(B→X) emission was merely observed from the photodissociation of t-C_4H_9CN, n-C_4H_9NC, and n-C_4H_9NC. Namely, the emission cross sections in this excitation wavelength region for these molecules reflect dominantly the N_2^+(B→X) emission.

The structure in the CN(B) excitation spectrum mostly corresponds to the structure in the absorption spectrum in the excitation wavelength region shorter than ≈140 nm for each cyanides and isocyanides. Therefore, the excited photodissociative states of the parent molecule mainly responsible for the CN(B) production are the Rydberg states. The σ_B was observed to increase with excitation energy and decrease at the excitation wavelength corresponding to the opening of the ionization channel in the photoexcitation spectra of n-C_4H_9NC (Fig. 1c) and t-C_4H_9NC (Fig. 1d). This decrease was not observed in the photoexcitation spectra for n-C_4H_9CN and t-C_4H_9CN. This result supports that the emission cross sections for the n-C_4H_9CN and t-C_4H_9CN are dominantly due to the N_2^+(B-X) emission in this excitation wavelength region. Overall emission cross section in the wavelength range of 120-140 nm is t-C_4H_9NC $>>>$ n-C_4H_9NC > t-C_4H_9CN >> n-C_4H_9CN. The 6 Mb observed for t-C_4H_9NC at the 142 nm, which was assignable to the $2\pi \rightarrow 3p\sigma$ transition, was remarkably large as the emission cross section from the photodissociation of this size molecule. As the absorption cross section of the four molecules showed mostly resemble values, quantum yields for the production of the CN(B) fragment from the photodissociation of the isocyanides were larger than those from the cyanides and that those from the molecules consisting of *tertiary*-butyl group were larger than those from the *normal*-butyl compounds.

As shown in Fig. 2 (c) and (d), the vibrational excitation up to v=3 and 4 were observed in the emission spectra of t-C_4H_9NC at 9.03 and 10.4 eV, respectively. This vibrational excitation was remarkably high in the photodissociation of this size molecule. For other three molecules, the emission spectrum could not measured with a high resolution, because of small emission cross section, however, the vibration excitation was found to be small by comparison of band shape of 0-0 sequence from t-C_4H_9CN with that from t-C_4H_9NC as shown in Fig. 2b and 2a. This tendency agreed with the dissociative excitation by the collision with rare gas metastable atom [1]. The internal energy distribution and emission cross section were considerable to be affected by the flexibility of molecule. In other word, the available energy dissipated to alkyl group can be estimated to increase with decreasing the energy given to the electronical and rovibrational excitations of CN fragment. The *tertiary*-butyl group has more rigid structure than *normal*-butyl group. From the molecular orbital calculation, the highest occupied molecular orbital

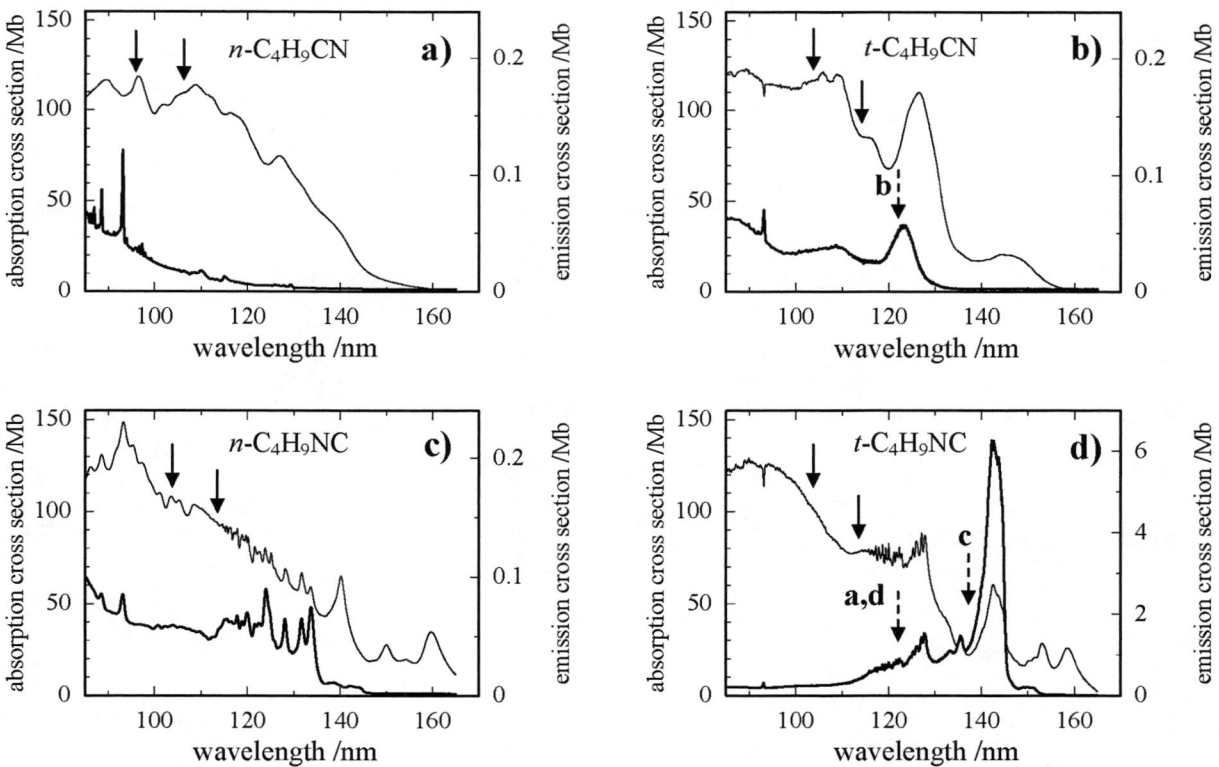

FIGURE 1. Absolute cross-section for absorption (thin line) and the $CN(B^2\Sigma^+$-$X^2\Sigma^+)$ emission (thick line) from the photodissociation of a) n-C_4H_9CN), b) t-C_4H_9CN, c) n-C_4H_9NC and d) t-C_4H_9NC. Solid lines indicate the first and second ionization potentials. Broken arrows indicate the excitation photon wavelengths for the dispersed emission spectra in Fig.2.

found to be localized on the CN atoms for the cyanides, however, to be extended to the CNC atoms for the isocyanides [2]. As a result, the rigid structure of t-C_4H_9NC gives the large quantum yield and high vibrational excitation for the photodissociation of this molecule.

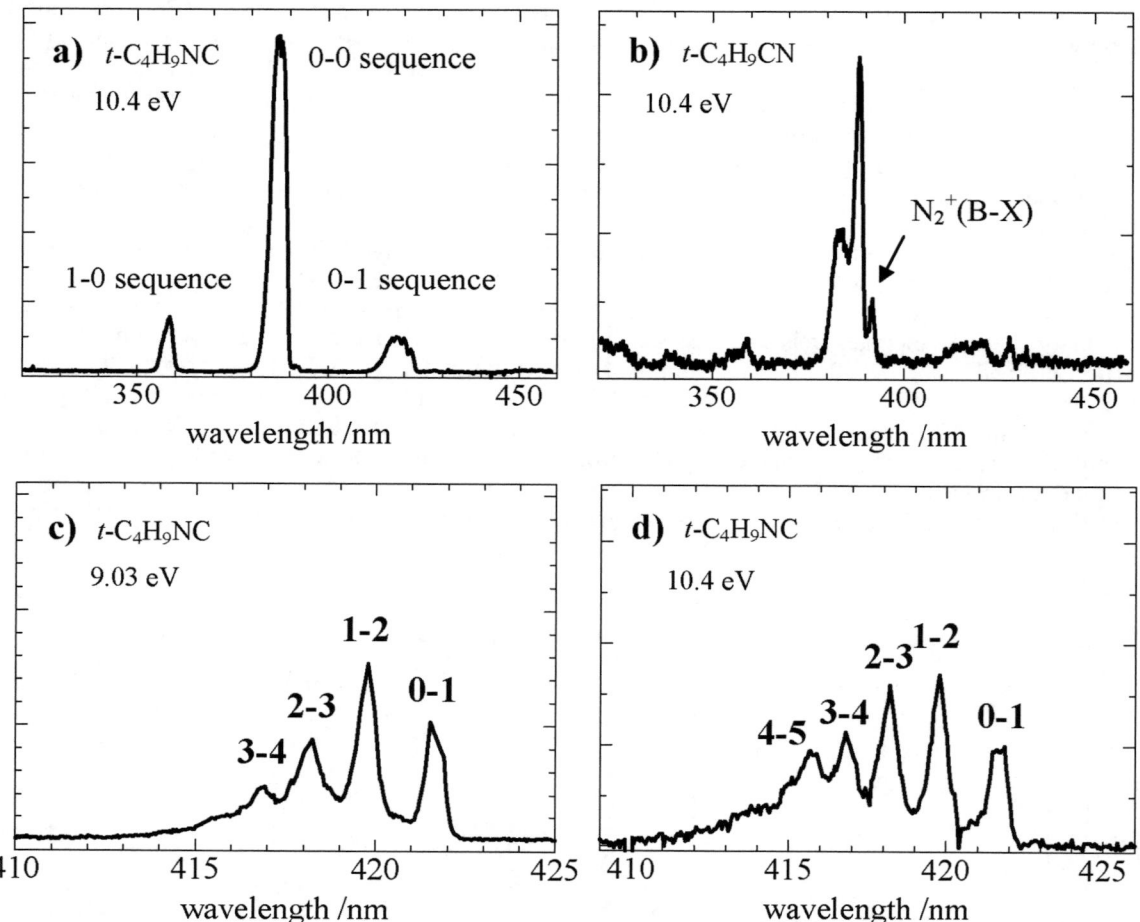

FIGURE 2. CN(B-X) emission spectra produced from the photodissociation of t-C_4H_9CN (a) and t-C_4H_9NC (b) at the 10.4 eV with a low resolution. Emission spectra of the $\Delta v = -1$ sequence in the main bands of the CN($B^2\Sigma^+$-$X^2\Sigma^+$) emission from the photodissociation of t-C_4H_9NC at c) 9.03 eV and d) 10.4 eV with a high resolution.

ACKNOWLEDGMENTS

The authors thank Professor S. Matsui for his interest and encouragement in the present study. This work was supported by the Joint Studies Program of the Institute for Molecular Science.

REFERENCES

1. K. Suzuki and T. Kondow, *Atomic Collision Research in Japan, Progress Report*, **16**, 56-59 (1990).
2. K. Kanda, K. Suzuki, E. Matsumura, N. Kishimoto, K. Ohno and T. Ibuki, The 2005 Pasifichem Meeting Abstract, Program number 697 and under preparation for submitting to Bull Chem. Soc. Jpn.
3. K. Kanda, S. Katsumata, T. Nagata, Y. Ozaki, T. Kondow, K. Kuchitsu, A. Hiraya and K. Shobatake, Chem. Phys. **175**, 399 (1993).
4. L.C. Lee, Chem. Phys., **72**, 6414 (1980).
5. K. Mitsuke and M. Mizutani, Bull. Chem. Soc. Jpn., **74**, 1193 (2001).

Electronic Structure of
Pd- and Zr-based Bulk Metallic Glasses Studied by Use of
Hard X-ray Photoelectron Spectroscopy

Kazuo Soda, Shunji Ota, Takaharu Suzuki, Hidetoshi Miyazaki, Manabu Inukai,
Masahiko Kato, Shinya Yagi, Tsunehiro Takeuchi[*], Masashi Hasegawa[#],
Hirokazu Sato[+] and Uichiro Mizutani[$]

*Department of Quantum Engineering, Graduate School of Engineering, Nagoya University,
Furo-cho, Chikusa-ku, Nagoya 464-8603 Japan.*
[*]*Ecotopia Science Institute, Nagoya University, Furo-cho, Chikusa-ku, Nagoya 464-8603 Japan.*
[#]*Institute for Materials Research, Tohoku University, Katahira, Aoba-ku, Sendai 980-8577 Japan.*
[+]*Aichi University of Education, Hirosawa, Igaya-cho, Kariya 448-8542 Japan.*
[$]*Toyota Physical and Chemical Research Institute,
41-1 Aza-yokomichi, Oaza-nagakute, Nagakute-cho, Aichi 480-1192 Japan.*

Abstract. The hard x-ray photoelectron spectroscopy has been applied to the investigation of the electronic structures and the chemical states of the constituent atoms of the binary Zr-TM (TM = Ni and Cu) and ternary Zr-TM-Al metallic glasses and a $Pd_{42.5}Ni_{7.5}Cu_{30}P_{20}$ bulk metallic glass. The Zr $2p$ and Al $1s$ spectra as well as the O $1s$ line show that the Zr-based glasses are covered with a thick oxide surface layer, while the core level spectra other than the P $1s$ and $2s$ levels in $Pd_{42.5}Ni_{7.5}Cu_{30}P_{20}$ reveal almost no oxide components, indicating that the oxide layer is essentially negligible. The bulk valence-band electronic structures of the Zr-based glasses can be successfully estimated by measuring the detection angle dependence of the valence-band and Zr $2p$ core level spectra. In Zr-Cu, the Cu $3d$ band is shifted toward the high binding energy with the Cu concentration increased, which is consistent with the chemical shifts of the bulk Zr and Cu $2p$ levels observed in this study. These core level shifts indicate an increase in the charge transfer from Zr to Cu with increasing the Cu concentration. In $Pd_{42.5}Ni_{7.5}Cu_{30}P_{20}$, the bands at the binding energy of 0.7, 1.9, 2.7 and 4.0 eV are ascribed to the Ni $3d$-, P $3p$-, Cu $3d$- and Pd $4d$-derived states, respectively. The deep Pd $4d$-band may be indicative of the strong interactions between Pd and its surrounding elements.

Keywords: hard x-ray photoelectron spectroscopy, Pd-based bulk metallic glass, Zr-based bulk metallic glass, electronic structure, glass formation ability.
PACS: 79.60.-i, 71.23.Cq, 82.80.Pv

INTRODUCTION

Bulk metallic glasses, bulky multi-element amorphous alloys, have attracted much attention as new materials possessing useful engineering properties such as high mechanical strength, high corrosion resistance, good shaping ability, and soft-magnetic properties [1]. In spite of their thermodynamically metastable phase, they exhibit very high resistance against the crystallization of the super-cooled melt and show a clear glass transition. In order to understand the origin of their large glass forming ability from the microscopic point of view, we have investigated the electronic structure of the bulk metallic glass experimentally by means of the photoelectron spectroscopy [2,3]. In this study, we have applied a hard x-ray photoelectron spectroscopy with use of 8-keV photons as an excitation source to Zr-based (bulk) metallic glasses, binary Zr-TM (TM = Ni and Cu) and ternary Zr-TM-Al, and a Pd-based one, $Pd_{42.5}Ni_{7.5}Cu_{30}P_{20}$. The large probing depth of the hard x-ray photoelectron spectroscopy may enable us to investigate the bulk electronic structures of the amorphous metals for wide composition range without any special surface treatment. This is very important for the study of specimens which can be prepared only in a ribbon shape.

CP879, *Synchrotron Radiation Instrumentation: Ninth International Conference*,
edited by Jae-Young Choi and Seungyu Rah
© 2007 American Institute of Physics 978-0-7354-0373-4/07/$23.00

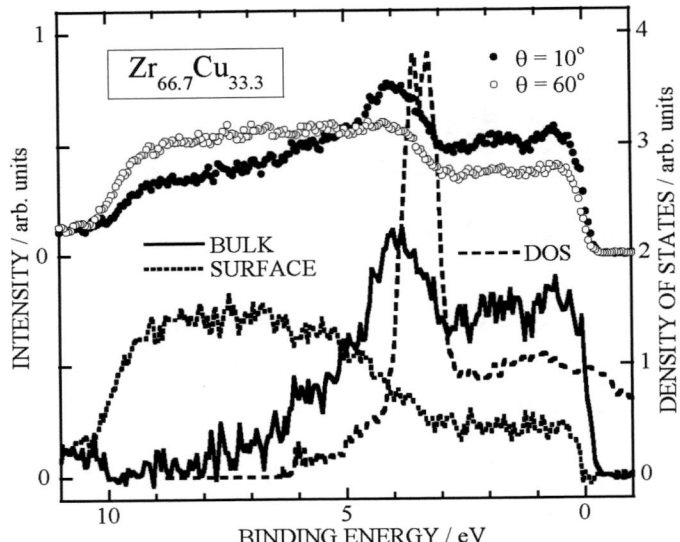

FIGURE 1. Valence-band spectra of metallic glass $Zr_{66.7}Cu_{33.3}$ recorded at two detection angles θ and its bulk and surface components. Calculated electronic density of states (DOS) is also shown for a metallic glass Zr_6Cu_3.

EXPERIMENTAL

The hard x-ray photoelectron measurement was performed at the beamline 47XU of SPring-8 at the Japan Synchrotron Radiation Research Institute. Synchrotron light from an undulator was monochromatized by a double-crystal monochromator and emitted photoelectrons were measured at room temperature with a high-resolution analyzer (GAMMADATA-SCIENTA R4000). The excitation photon energy $h\nu$ and the total energy resolution ΔE including the thermal broadening were estimated by measuring the Fermi edge of Au to be 7941 eV and 0.3 eV, respectively. The origin of the binding energy E_B, i.e. the Fermi energy E_F, was also determined from the Au edge.

Specimens of the Zr-based metallic glasses were typically 1 mm wide and 30 μm thick ribbons prepared by a single roller melt-spinning technique [4], and they were cut into pieces of ~5 mm length. The $Pd_{42.5}Ni_{7.5}Cu_{30}P_{20}$ bulk metallic glass prepared by a water quenching method [5] and cut into a size of 1x2x5 mm³ was polished with a rapping paper before the measurement. These specimens were attached on a holder with carbon adhesive tape.

RESULTS AND DISCUSSION

In the upper part of Fig.1, we show the valence-band spectra of a metallic glass $Zr_{66.7}Cu_{33.3}$ recorded at the detection angles θ, i.e. the angle between the surface normal and the lens axis of the electron analyzer, of 10° (closed circles) and 60° (open circles). Here, the spectral intensity is normalized with respect to the intensity integrated up to $E_B \sim 11$ eV. For the surface-sensitive spectrum of $\theta = 60°$, the enhancement and suppression, relative to the spectrum of $\theta = 10°$, due to the oxide or contaminated surface layer is clearly seen in the regions between $E_B \sim 5$ and 10 eV and from $E_B \sim 5$ eV to E_F, respectively. While the Cu $2p$ spectra reveal no chemically-shifted components, the Zr $2p$ spectra show the oxidation peak as shown in Fig.2, where two prominent peaks are attributed to the bulk and surface Zr $2p_{3/2}$ components in an increasing order of the binding energy. Taking into account of the observation of the O $1s$ line and the large photoelectron escape depth in this study, we are led to conclude that the Zr-based specimen is covered with a considerably thick Zr-oxide layer. In other Zr-based glasses studied, the Ni and Cu $2p$ spectra show no chemically-shifted (oxidation) components but the Zr $2p$ and Al $1s$ ones do.

Assuming both the valence-band and Zr $2p_{3/2}$ spectra to consist of two components, i.e. the surface and bulk ones, and their intensity ratio to be the same between the valence and core spectra, we have decomposed the valence-band spectrum into its bulk and surface components, $I_B(E_B)$ and $I_S(E_B)$, as follows:

$$I_B(E_B) = \frac{a_S(60)I(E_B;10) - a_S(10)I(E_B;60)}{a_S(60)a_B(10) - a_S(10)a_B(60)}, \quad I_S(E_B) = -\frac{a_B(60)I(E_B;10) - a_B(10)I(E_B;60)}{a_S(60)a_B(10) - a_S(10)a_B(60)}. \tag{1}$$

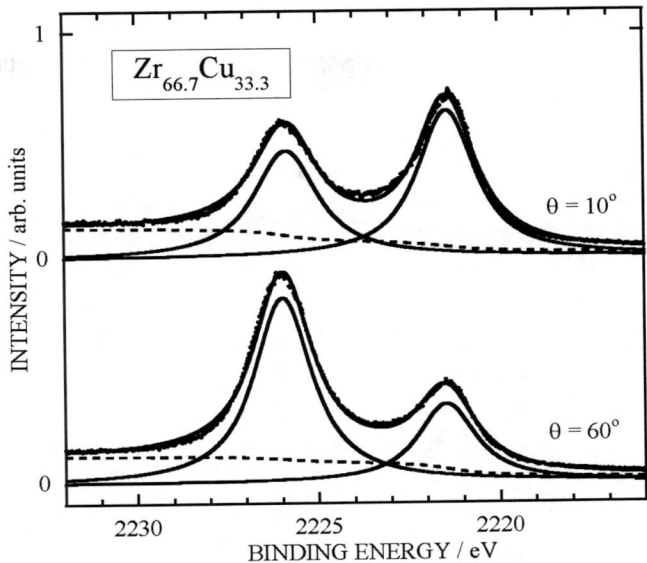

FIGURE 2. Zr $2p_{3/2}$ core level spectra of metallic glass $Zr_{66.7}Cu_{33.3}$ recorded at the detection angles $\theta = 10°$ and $60°$.

Here, $a_B(\theta)$ and $a_S(\theta)$ are the fractions of the bulk and surface components, respectively, in the spectrum recorded at θ, and $I(E_B;\theta)$ is the measured valence-band intensity at E_B for θ. These θ-dependences are expected to arise from the difference in the effective probing depth. Then the fractions may be determined by decomposing the Zr $2p_{3/2}$ spectrum into two bands shown by the Voigt functions (solid curves in Fig.2) and a background (broken one);

$$a_B(\theta) = \frac{I_B(\theta)}{I_B(\theta) + I_S(\theta)}, \quad a_S(\theta) = \frac{I_S(\theta)}{I_B(\theta) + I_S(\theta)} = 1 - a_B(\theta), \tag{2}$$

where $I_B(\theta)$ and $I_S(\theta)$ stand for the integrated intensity of the bulk and surface Zr $2p_{3/2}$ components, respectively. Thus the bulk and surface valence-band spectra, $I_B(E_B)$ and $I_S(E_B)$, are obtained as shown in the lower part of Fig.1 by solid and dotted curves, respectively.

In the bulk valence-band spectrum, a Cu $3d$ band is located at $E_B \sim 4$ eV and the Zr and Cu sp states extend from 7 eV to E_F. A small hump at ~ 0.5 eV might be attributed to the Zr $4d$ states; the relative Zr $4d$ intensity is expected to be fairly low because of its relatively small ionization cross section at $h\nu = 8$ keV [6]. In Fig.1, we also show a tentative result on the theoretical calculation by the recursion method for a giant atomic cluster, where the atomic positions are determined by the structural analysis of the metallic glass $Zr_{66.7}Cu_{33.3}$ [7]; a broken curve in the figure represents the sum of the calculated local electronic density of states (DOS) for six Zr and three Cu atoms, which are selected at random among 1000 atoms in the cluster. The overall feature agrees fairly well with the present photoemission results. It is also found that the Cu $3d$ band in Zr-Cu is shifted toward the high binding energy side with the Cu concentration increased. This is consistent with the observed chemical shifts of the bulk Zr and Cu $2p$ levels, indicating the charge transfer taking place from Zr to Cu with increasing the Cu concentration. In the surface valence-band spectrum, we recognize the low intensity at E_F, indicating the insulating nature of the surface layer, as well as the O $2p$- or contamination-derived states at the high binding energy region.

In Fig.3, we compare the valence-band spectra of $Pd_{42.5}Ni_{7.5}Cu_{30}P_{20}$ with the spectrum at $h\nu = 40$ eV, which was recorded at UVSOR of the Institute for Molecular Science and will be reported in detail elsewhere. The present spectra show the features at $E_B \sim 0.7$, 1.9, 4.0 and 6.5 eV, while the spectrum at $h\nu = 40$ eV reveals two peaks at $E_B \sim 2.7$ and 4.0 eV and a faint shoulder around $E_B \sim 6$ eV. The θ-dependence of the spectrum shows the very weak reduction of the main band and the faint enhancement of the band around 6.5 eV, the latter being indicative of the O $2p$-derived states in the surface layer. Since neither the O $1s$ line nor oxidized lines are observed in the investigated Pd, Ni, and Cu core level spectra but the P $1s$ and $2s$ ones and a very weak O $2s$ line is recognized, it might be attributed to the P $3s$-derived states or P oxides. Taking into account of the $h\nu$-dependence of the ionization cross sections [6], we have ascribed the features at $E_B \sim 0.7$, 1.9, 2.7 and 4.0 eV to the Ni $3d$, P $3p$, Cu $3d$ and Pd $4d$ states, respectively. The Pd $4d$-band is considerably deepened in comparison with the crystalline Pd, which may be taken as indication for the presence of the strong interactions between Pd and its surrounding elements.

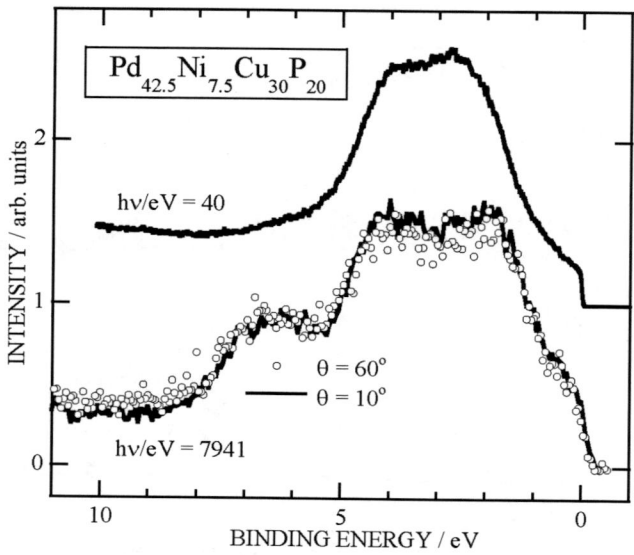

FIGURE 3. Valence-band spectra of $Pd_{42.5}Ni_{7.5}Cu_{30}P_{20}$. The excitation photon energy $h\nu$ and the detection angle θ are indicated.

SUMMARY

We have studied the electronic structures of the binary Zr-TM, ternary Zr-TM-Al and $Pd_{42.5}Ni_{7.5}Cu_{30}P_{20}$ bulk metallic glasses and the chemical states of their constituent elements by the hard x-ray photoelectron spectroscopy. The Zr $2p$ and Al $1s$ spectra as well as the O $1s$ line suggest a thick Zr oxide overlayer in the Zr-based metallic glasses, while the observation of very weak O-related lines and small oxidation components only for P implies that an oxide layer is fairly thin for $Pd_{42.5}Ni_{7.5}Cu_{30}P_{20}$. Using the detection angle dependence of the Zr $2p_{3/2}$ and valence-band spectra, we have successfully extracted the bulk valence-band spectrum for the specimens of the Zr-based metallic glasses in a ribbon shape without any special surface treatment for the photoemission measurement. Detailed results, including the chemical shift of the core levels and the theoretical calculation for the Zr-based metallic glasses with various compositions, will be reported elsewhere.

ACKNOWLEDGMENTS

We would like to thank Dr. H. Kato and Dr. N. Nishiyama for the preparation of the metallic glass samples. The hard x-ray photoelectron measurement was performed at the SPring-8 with the approval of the Japan Synchrotron Radiation Research Institute (Proposal No. 2005B0072). We would like to appreciate the staff members, in particular Dr. E. Ikenaga and Dr. K. Kobayashi for their kind support during the measurements. This work was partly supported by a Grant-in-Aid for Scientific Research on Priority Areas "Materials Science of Bulk Metallic Glasses" from the Japanese Ministry of Education, Culture, Sports, Science and Technology.

REFERENCES

1. M. Telford, *Materials Today* **7**, 36-43 (2004).
2. K. Soda, K. Shimba, S. Yagi, M. Kato, T. Taketomi, U. Mizutani, T. Zhang, M. Hasegawa and A. Inoue, *J. Electron Spectrosc.Relat. Phenom.* **144-147**, 585-587 (2005).
3. T. Suzuki, H. Miyazaki, K. Soda, T. Takeuchi, M. Hasegawa, H. Sato and U. Mizutani, *J. Jpn. Soc. Powder Powder Metallurgy* **53**, 107-110 (2006).
4. A. Inoue, D. Kawase, A. P. Tsai, T. Zhang and T. Masumoto, *Mat. Sci. Eng.* **A178**, 255-263 (1994).
5. N. Nishiyama and A. Inoue, *Appl. Phys. Lett.* **80**, 568-570 (2002).
6. J. J. Yeh and I. Lindau, *Atomic Data and Nucl. Data Tables* **32**, 1-155 (1985).
7. T. Fukunaga, K. Itoh, T. Otomo, K. Mori, M. Sugiyama, H. Kato, M. Hasegawa, A. Hirata, Y. Hirotsu and A.C. Hannon, *Intermetallics* **14**, 893-897 (2006).

Pressure Induced Phase Transition in PbTiO₃ Studied by X-ray Absorption Spectroscopy at the Ti K edge.

A.C. Dhaussy[1], N. Jaouen[2,3], J.P. Itié[3,4], A. Rogalev[2], S. Marinel[1] and A. Veres[1]

[1]CRISMAT, ENSICAEN, Bd Mal Juin, 14050 Caen, France
[2]European Synchrotron Radiation Facility, rue J. Horowitz, 38043 Grenoble, France
[3]Synchrotron SOLEIL, l'Orme des merisiers 91192 Gif-sur-Yvette, France
[4]Physique des Milieux Denses, IMPMC, Université P & M Curie, 75015 Paris, France

Abstract. The Ti-K edge X-ray Absorption Near Edge Structure (XANES) for CaTiO₃ and PbTiO₃ have been measured under high pressure in a diamond anvil cell at room temperature. Despite the huge absorption from the diamond cell and the sample high quality XANES allows us to observe that in CaTiO₃ no change occurs when applying pressure, at the opposite of PbTiO₃ in which the pre-edge features vary strongly. It allows studying the phase transition from ferroelectric to paraelectric phase in PbTiO₃ from the local point of view. Under pressure the change in intensity of the pre-edge indicates qualitatively that the Ti atom is moving toward the centre of the oxygen octahedron along the c-axis.

Keywords: High Pressure, X-ray absorption spectroscopy.
PACS: 81.40.Vw, 78.70.Dm, 77.84.-s

INTRODUCTION

X-ray spectroscopy is now a powerful technique routinely employed to probe the local structure of the absorbing atom under very high pressure but mainly for elements with rather high atomic number ($Z > 26$) at incident X-ray energies ≥ 7 keV, and it is only recently that XANES has been recorded at the Ba L3 edge [1] and at the Ce L edge [2] under extreme conditions. Taking into account of the wide $2 - 16$ keV energy range provided by the ESRF-ID12 beamline, we have investigated the effect of pressure on the local structure at the Ti – K edge (4.9 keV) in titanium oxides i.e. CaTiO₃, PbTiO₃ and PZT (not reported in this paper).

Recent papers [3,4] show that Ti K edge XANES is a powerful approach for studying the small local displacements occurring at the ferro to paraelectric phase in oxide perovskites. In particularly, the pre-edge features give very sensitive information on the Ti position with respect to the surrounding oxygen. All these peaks refer to dipolar and/or quadrupolar transitions in the absorption cross-section. In a simple way, the first and third peaks correspond to interactions between Ti – Ti atoms although the second peak is related to the Ti-O bonds. In that sense, the later strongly depends on structural and ferroelectric features and its intensity increases both with the displacement of Ti atom off the centre of the cell and with ferroelectricity [3,4].

The ATiO₃-type perovskites (A = Ca, Ba, Sr, Pb) form one of the most important classes of ferroelectric materials with nonlinear electro-optical properties and they can be employed in several applications in electronic technology. The majority of these materials show a ferroelectric to paraelectric phase transition induced by temperature or pressure effect. In particular, PbTiO₃ shows the tetragonal-ferroelectric to cubic-paraelectric phase transition at about 770 K at ambient pressure and at about 12 GPa at room temperature. It has been also reported that the temperature of the phase transition decreases with increasing the pressure [5]. The wide temperature stability of the ferroelectric phase, which is the highest among the ferroelectric perovskites, is one of the major interest of the extensively studies reported in last decades. Under ambient pressure and room temperature, the ferroelectric PbTiO₃ phase is reported to have a tetragonal structure (space group P4mm, c/a = 1.06) and the cation shifts along the ferroelectric axis are markedly larger than those of other ferroelectric perovskites. Such feature can be explained by the existence of the $6s^2$ lone pair of Pb^{2+} ion and the orbital hybridization between the Pb $6s$ states and O $2p$ states which play crucial roles in the tetragonal distortion [6,7]

CP879, *Synchrotron Radiation Instrumentation: Ninth International Conference*,
edited by Jae-Young Choi and Seungyu Rah
© 2007 American Institute of Physics 978-0-7354-0373-4/07/$23.00

CaTiO$_3$ presents an orthorhombic structure (Pbnm) at ambient temperature. Previous studies on high temperature phase transitions evidence considerable differences in the number of phase transitions, transition temperature and structural changes. Indeed, CaTiO$_3$ undergoes a number of phase transitions above 1000 °C from orthorhombic to tetragonal followed by a tetragonal to cubic Pm3m phase transition. On the pressure side, HP energy dispersive X-ray diffraction experiments carried out on CaTiO$_3$ at room temperature revealed no evidence of any structural transformation up to 45 GPa [8].

In the present paper, we will concentrate on a presentation of some key experimental aspects which will underline the challenge of measuring Ti K edge XANES in lead-based compounds. In the second section, we propose to start with a presentation of our results for the CaTiO$_3$ sample. Unlike we don't observe significant change at the local scale; this study will serve to illustrate the extreme sensibility one can obtain on the ID12 beamline when the X-rays are not strongly absorbed (only by the diamond from the high pressure cell). In the second subsection, we will present and qualitatively discuss the results obtained by replacing Ca by Pb.

EXPERIMENTAL

The samples have been previously synthesized by solid-state chemistry in air and fully characterized in term of purity and structure by XRD experiments previous to the synchrotron beamtime.

A Le Toullec-type gas-driven membrane diamond anvil cell has been used for the experiments. In order to reduce the absorption of the diamonds, specific anvils with perforated diamonds have been adapted [9,10]. Such specificity reduces the total thickness of the diamond anvils to 1 mm leading to a 1.10^{-3} transmission value at the Ti K edge. Silicon oil was used as pressure transmitting medium for all the experiments. Pressure has been measured by the ruby fluorescence technique. The samples were load in inconel gasket with typically a 100 μm diameter hole and a 50 μm thickness. Considering the air absorption as negligible at 5 KeV in our setup configuration (~ 2cm air thickness before and after the cell) all the experiments have been performed in air. As ID12 bealine [11] is not dedicated to High Pressure experiments, a special set-up has been mounted especially for this experiment. With two precise translations and a rotation around the vertical axis it allows respectively aligning the cell and removing contributions of diffraction peaks from the diamond. XAS spectra were measured in transmission geometry by using a unique photodiode. The incoming X-rays delivered by the Apple II HU38 undulator (in linear mode) has been focused vertically using the V2FM down to 20 μm FWHM and collimated to ~100 μm horizontally using the beamline slits.

In addition to the necessity to close the beamline slits and the diamond absorption our experiment was complicated by the sample itself. Due to the strong absorption by Pb atoms in the sample (transmission of 4.10^{-14} through 50 μm of PbTiO$_3$!), the powdered samples have been diluted in BN powder. The relative concentrations of sample and BN have been optimized (PbTiO$_3$-BN in a 1:4 weight ratio leading to a ~12μm of PbTiO$_3$) to reach a final transmission value of 10^{-6} (diamonds + sample). As well as air absorption, the 38μm BN absorption have been neglected here for simplicity. Nevertheless, despite the low transmission through the setup (diamonds + sample), high quality XANES spectra (see Fig1-2-3) have been successfully recorded at several pressures for all samples thanks both to the high stability and the high quality detection scheme available at the ID12 beamline.

RESULTS

CaTiO$_3$ Powdered Sample

XANES spectra have been recorded in the pressure range 0 – 8.4 GPa (Fig.1). The global behavior of the pre-edges can be compared to those reported in the literature for the ATiO$_3$ perovskite type [12]. All the XANES spectra revealed no change in the local order symmetry when increasing the pressure This first result is directly in relation with the absence of phase transition under pressure detected by X-ray diffraction measurement that have been carried out to 9.7 GPa [13]. Nevertheless, the shortening of the interatomic distances with pressure is clearly visible in the EXAFS regime where one can observe shifts of the oscillations toward high energy.

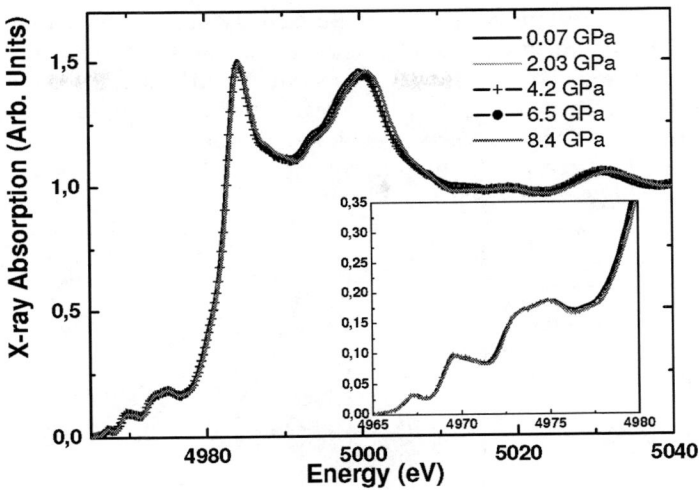

FIGURE 1. Experimental Ti K-XANES for CaTiO$_3$ polycrystalline sample at various pressures from 0 to 8.4 GPa. Inset shows pre-edge features in details.

PbTiO$_3$ Powdered Sample

In PbTiO$_3$, absorption spectra measured from 0 to 13 GPa clearly evidence a phase transition towards a local higher symmetry which seems to occur between 10.6 and 13 GPa with gradual change in the shape of the A, B and C peaks of the pre edge zone with applied pressure (Fig.2). The pre edge-considerations are associated with a shift to higher energy in the XANES region (4080-5000eV) and confirm the pressure induced phase transition which can be partly correlated with the X-ray diffraction measurements evidencing the tetragonal to cubic phase transition around 11 GPa [5].

More in details, taking into account of the examination of the pre-edge peaks in term of shape and intensity, PbTiO$_3$ behavior can be directly correlated with BaTiO$_3$ [14]. Indeed, both compounds exhibit the similar tetragonal structure and ferroelectric properties at ambient conditions. In comparison with CaTiO$_3$ spectra, the ferroelectric feature related to the distortion of the Ti atom off the center of the octahedral site is clearly evidenced since the intensity of the B peak is drastically larger and sharper than that in the Ca phase.

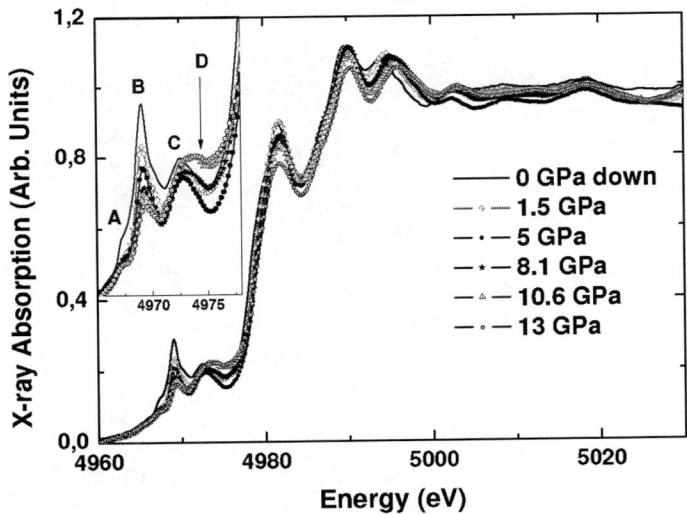

FIGURE 2. Experimental Ti K-XANES spectra for PbTiO$_3$ polycrystalline sample at various pressures from 0 to 13 GPa. Inset shows pre-edge features in details.

With increasing pressure, displayed spectra clearly reveal a gradually decrease of the peaks intensity in the investigated pressure range (Fig. 3 left). Such feature is correlated with a displacement of Ti atoms toward the central site along the c axis in the case of $PbTiO_3$. However it is worth noting that the decrease of intensity still remains between 10.6 and 13 GPa, which means that the Ti atoms are still slightly distorted and the mentioned tetragonal to cubic phase transition has not been completely achieved. Remarkably, one can observe the appearance of the D peak under pressure starting from 8 GPa but clearly appearing from 10.6 GPa. This feature will be discussed in details in future paper using modern calculations [15].

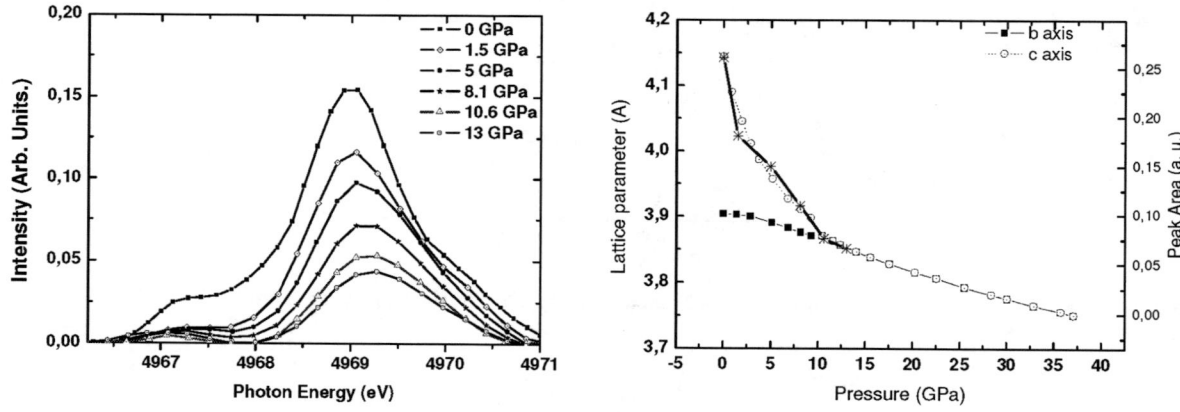

FIGURE 3. Left: Experimental details of the major pre-edge features (background subtracted) of $PbTiO_3$ polycrystalline sample at the Ti K edge for various pressures from 0 to 13 GPa. Right: Evolution of the $PbTiO_3$ cell edges with external pressure from [3]. The black dashed line shows the variation of the peak area versus the pressure.

In Fig. 3 (Right), we arbitrally scaled the intensity of the B peak (left) at the two extreme pressures to the variation of c lattice parameters taken from Ref. [5]. The common behavior of both curves indicates a correlation between the Ti position and the c-lattice parameter.

CONCLUSION

Even under the difficult experimental conditions described in this paper, high quality X-ray absorption spectra have been measured at the Ti K edge under high pressure. This work indicates that recording X-ray absorption at energy below 7 keV under HP is feasible at third generation synchrotron source. We show that in $CaTiO_3$ there is no change in the local structure with pressure up to 8.4 GPa. In the case of $PbTiO_3$, X-ray absorption indicates that the tetragonal to cubic transition in $PbTiO_3$ is analogous in a general way to the temperature-induced transition.

REFERENCES

1. A. San Miguel, A. Merlen, P. Toulemonde, T. Kume, S. Lefloch, A. Aouizerat, S. pascarelli, G. Aquilanti, O. mathon, T. Le Bihan, J.P. Itié, and S. Yamanaka., Europhys. Lett., 69(4), 556 (2005).
2. J.P. Rueff *et al.*, Phys. Rev Lett. 96, 237403 (2006).
3. Y. joly, *et al*, Phys. Rev. Lett. 82, 2398 (1999)
4. R.V. Vedrinski, *et al*, J. Phys. : Condens. Matter 10, 9561 (1998)
5. A. Sani, M. Hanfland, and D. levy, J. Phys.: Condens. Matter 14, 10601 (2002).
6. R.E. Cohen, Nature 358(6382), 136 (1992).
7. Y. Kuroiwa, S. Aoyagi, A. Sawada, J. Harada, E. Nishibori, M. Takata, M. Sakata, *Phys. Rev. Lett.* 2001, 87, 217601.
8. X. Wu, Acta Physica Sinica, 53 (6), 1967 (2004).
9. A. Dadashev, M.P. Pasternak, G. Kh. Rozenberg, and R.D. Taylor, Rev. Sci. intrum., 72, 2633 (2001).
10. J.P. Itié, F. Baudelet, A. Congeduti, B. Conzinet, F. Farges and A. Pollian, J. Phys.: Condens. Matter., 17(S), 883 (2005)
11. A. Rogalev, J. Goulon, C. Goulon-Ginet and C. Malgrange, in Magnetism and Synchrotron Radiation, eds E. beaurepaire, F. Scheurer, G. Krill and J.P. Kappler, Lectures Notes in Physics, Vol. 565 (2001) Springer.
12. F. Farges, G.E. Brown, and J.J. Rehr, Phys. Rev. B 56, 1809 (1997).
13. N.L. Ross and R.J. Angel, American Mineralogist, Vol 84(3), 217 (1999).
14. J.P. Itié, B. Couzinet, A. polian, A.M. Flanck, and P. Lagarde, Europhys. Lett., 74 (4), 706 (2006).
15. A.C. Dhaussy, N. Jaouen, J.P. Itié, A. Rogalev, S. Marinel and Y. Joly. Inpreparation.

Advanced Structural Analyses by Third Generation Synchrotron Radiation Powder Diffraction

M. Sakata, S. Aoyagi, T. Ogura* and E. Nishibori

Department of Applied Physics, Nagoya University, Nagoya, 464-8603 Japan
* Present address: Isuzu Motors Limited, Fujisawa, Kanagawa, 252-0806 Japan

Abstract. Since the advent of the 3rd generation Synchrotron Radiation (SR) sources, such as SPring-8, the capabilities of SR powder diffraction increased greatly not only in an accurate structure refinement but also *ab initio* structure determination. In this study, advanced structural analyses by 3rd generation SR powder diffraction based on the Large Debye-Scherrer camera installed at BL02B2, SPring-8 is described. Because of high angular resolution and high counting statistics powder data collected at BL02B2, SPring-8, *ab initio* structure determination can cope with a molecular crystals with 65 atoms including H atoms. For the structure refinements, it is found that a kind of Maximum Entropy Method in which several atoms are omitted in phase calculation become very important to refine structural details of fairy large molecule in a crystal. It should be emphasized that until the unknown structure is refined very precisely, the obtained structure by Genetic Algorithm (GA) or some other *ab initio* structure determination method using real space structural knowledge, it is not possible to tell whether the structure obtained by the method is correct or not. In order to determine and/or refine crystal structure of rather complicated molecules, we cannot overemphasize the importance of the 3rd generation SR sources.

Keywords: 3rd generation Synchrotron Radiation, powder diffraction, Genetic Algorithm
PACS: 61.10.Nz

INTRODUCTION

Advent of third generation Synchrotron Radiation (SR) has greatly improved the capabilities of X-ray diffraction. Powder diffraction is no exception. A large Debye-Scherrer camera [1] as installed at BL02B2, SPring-8 in order to carry out advanced structural studies by SR powder diffraction. Many SR powder data collected by the camera were analyzed [2] by the advanced analytical method [3], which is the combination of Maximum Entropy Method (MEM) [4] and Rietveld refinements. In this method, Rietveld refinements are performed as a preliminary analysis. In other word, MEM provides an analytical method of crystal structure, which can progress further than Rietveld refinements. As products of such an advanced structural analysis, one can obtain a MEM charge density map. It is consistent with the observed integrated Bragg intensities included in the SR powder data and least biased with unobserved integrated Bragg intensities. If one could measure a SR powder data very precisely with enough resolution, a MEM charge density map derived from the data could be very accurate. It may be said that the advanced analytical method by the combination of MEM and Rietveld refinements utilizing 3rd generation SR powder data is more or less established.

The advantage of *ab initio* structure determination based on 3rd generation SR data seems not very well developed. A part of reason may be due to the fact that *ab initio* structure determination by powder diffraction often stops before very accurate structure refinement is performed presumably assuming any further refinement is not necessary since it is structure determination but not refinement. Because of such an attitude, *ab initio* structure determination is often carry out by laboratory X-ray sources, which can not provide details of whole powder pattern due to the luck of intensities of incident X-ray photons and angular resolution. One has to admit information included X-ray powder pattern collected by an ordinary laboratory X-ray source is much less than SR powder pattern.

The purpose of this study is to demonstrate the importance of 3rd generation SR not only for structure refinements but also for *ab initio* structure determination in the case of relatively complicated organic materials. In order to show

CP879, *Synchrotron Radiation Instrumentation: Ninth International Conference*,
edited by Jae-Young Choi and Seungyu Rah
© 2007 American Institute of Physics 978-0-7354-0373-4/07/$23.00

the practical problems, *ab initio* structure determination and refinement of Prednisolone Succinate ($C_{25}H_{32}O_8$) powder specimen is described.

EXPERIMENTAL DATA

The power specimen of Prednisolone Succinate was sealed in a capillary of 0.4 mm diameter. Then, a SR powder pattern was collected by the Large Debye-Scherrer camera installed at BL02B2, SPring-8 under ambient temperature. The wavelength of incident X-ray was 1.0014 Å and the exposure time was 145 min. The homogeneity of Debye ring was confirmed on Imaging Plate (IP), which is the detector for the camera. Since IP is two-dimensional detector in nature, it is very easy to check the homogeneity of Debye rings. This sometime gives additional advantage to eliminate peaks, which come from impurities. Debye rings of impurities often are very spotty, because impurities exist much less quantities compared with the original sample. The collected data, which will be shown in a later section as Fig. 1 together with the fitting results of Rietveld refinement, shows very sharp independent peaks at low angle regions and details of Bragg intensities undulations at higher angle region. The structural information is included in these undulations.

To have some sharp independent peaks is extremely important to determine unit cell parameters, which is the first step of *ab initio* structure determination. It is possible to perform *ab initio* structure determination of powder specimens by laboratory X-ray sources. But they are all limited for materials with simple structures, for which some independent peaks could be observed in a whole powder pattern collected by laboratory X-ray sources. At this point, there is no doubt that 3rd generation SR source has essential importance. It is obvious that the correct structure can interpret details of Bragg intensities undulations at higher angle region. This will be shown in the next section.

AB INTIO ATRUCTURE DETERMINATION AND REFINEMENT

At first, *ab initio* structure determination of Prednisolone Succinate was done by four steps. First step is cell parameters determination, which was done by *DICVOL04* [5]. The obtained parameters by *DICVOL04* were refined by Le Bail [6] fitting. Prednisolone Succinate is monoclinic and cell parameters are a = 21.1400(2), b = 9.16066(9), c = 24.5891(3) [Å] and β = 98.1456(7) [°]. At second stage, space group is fortunately determined as $P2_1$ unambiguously by observing extinction rule. In many cases, space group could not be uniquely determined at this stage. In such a case, structure determination has to be done for all candidates of space groups. Third step is structure determination process. In the present study, Genetic Algorithm (GA) [7] is adopted. The model structure used in GA is constructed based on relatively similar molecule, which is 6α-methylprednisolone. At fourth step, the crystal structure obtained GA is refined by Rietveld method. Figure 1 shows the fitting results of Rietveld refinement. The R-factors are R_{wp} = 8.56 % and R_I = 19.2 %. The refined structure is shown in Fig. 2. The value of R_{wp} is less than 10 %, which may be normally regarded satisfactory. On the other hand, the value of R_I seems a little too big. It is well know that all the structural information is included in the integrated Bragg intensities. R_I is evaluated based on the integrated Bragg intensities. The discrepancy of two R-factors might suggest that there would be better solution.

In order to study such a possibility, we did further analysis using a kind of MEM analysis, which is slightly different from the ordinary MEM. To have MEM charge densities, the phases of structure factors are calculated from a structure. In the calculation the phases of structure factors, several atoms are intentionally omitted. By omitting these atoms, model bias in the phase calculation, which comes from these atoms, can be partly excluded. MEM charge density distribution obtained in this way will be called omit-MEM map in this study. Omit-MEM map still shows charge densities, which corresponds to the atoms omitted. The atomic positions shown in omit-MEM map are not always same as the structure refined by Rietveld method. Sometime it shows rather different positions. Then, the omitted atoms are placed at the position shown by the omit-MEM map on the viewer program, such as *PyMOL* [8]. In this way, it is possible to perform much more flexible search for a better solution. At the next stage, Rietveld refinement is done to adjust the central position and the direction of the molecule more precisely than the viewer program. The R-factors become R_{wp} = 3.74 % and R_I = 8.15 %, which is much smaller than that of Fig. 1. The observed intensity undulation around 20 ° is now very well fitted by the newly obtained structure and the value of R_I became well below 10 %. At the final stage of the refinement, restrained Rietveld refinement is carried out to adjust atomic positions very slightly. Eventually, the R-factors become R_{wp} = 2.26 % and R_I = 3.47 %. The fitting results of Rietveld program done by this process are shown in Fig. 3. It is concluded that the fitting is satisfactory at all angle regions and that any further improvement should not be expected. In the consequence, it is reasonable to consider that the structure obtained is correct.

The final structure is shown in Fig. 4, which is quite different from the structure shown in Fig. 2 in both aspects, i.e. crystal structure and molecular structure.

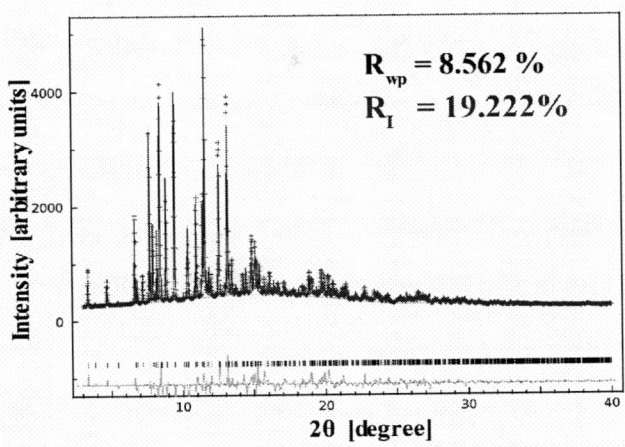

FIGURE 1. Fitting result of Rietveld refinement based on the structure model obtained by GA.

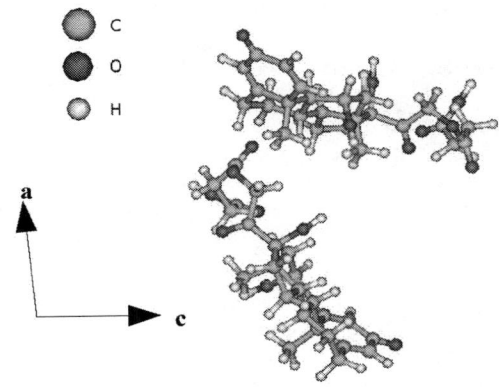

FIGURE 2. The asymmetric unit of the refined structure by Fig. 1.

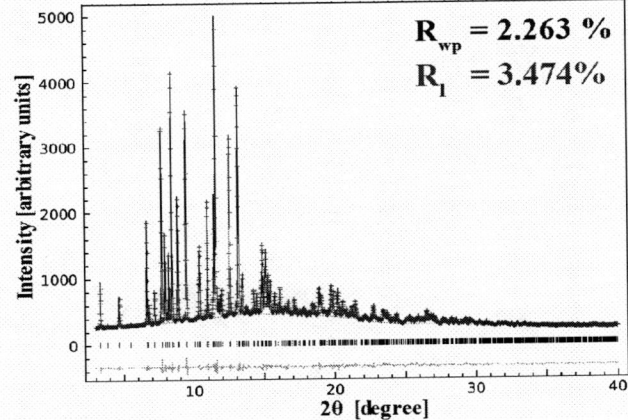

FIGURE 3. The final fitting result of Rietveld refinement.

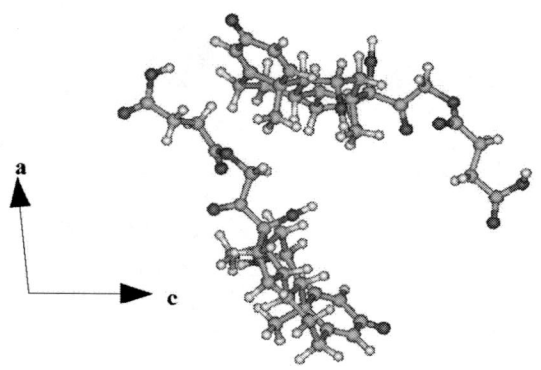

FIGURE 4. The finally refined crystal structure of Prednisolone Succinate (asymmetric unit).

DISCUSSION AND CONCLUSION

The implication of the present study seems rather serious in *ab initio* structure determination by X-ray powder diffraction. Obviously the experimental data taken by 3rd generation SR include much more structural information than the data taken by laboratory X-ray or even 2nd generation SR sources. This is a big advantage of *ab initio* structure determination by 3rd generation SR source. It has to be noted that the crystal and molecular structures, which explain 3rd genaration SR powder data at less than 10 % level in R_{wp} was still not quite right in the present case. It leaves a very difficult problem in *ab initio* structure determination by X-ray powder diffraction. That is how far the experimental data should be analyzed. It of course depends on how complicated structure to be solved. In the present case, the correct structure is obtained when R_I becomes less than 10 %. In R_{wp}, which should be influenced by many factors, such as background level, the value happened to be less than 4 %.

One thing is certain. Until the satisfactory refinement of an accurate experimental powder data, which include enough structural information, is done, it is not possible to tell whether the structure obtained is correct or not. Generally speaking, structure refinement process, such as omit-MEM or Rietveld, is much more time consuming compared with structure determination process, such as GA. There is no guarantee that the structure obtained so-called structure determination process is basically correct. Therefore, an accurate structure refinement utilizing, for example, omit-MEM has to be done to make R_I as small as possible, though an accurate refinement is time consuming and may not be suitable for automatic analysis by a computer program at the present stage.

The present study seems suggest that *ab initio* structure determination of complicated organic materials, such as medicine, by X-ray powder diffraction can be done at least under two conditions. Firstly, a very accurate powder diffraction data, which include enough structural information, has to be measured. Secondly, the data has to be refined extremely well. In this context, an advanced structure analyses by 3rd generation SR has essential importance in both *ab initio* structure determination and accurate refinements by X-ray powder diffraction.

REFERENCES

1. E. Nishibori *et al.*, *Nucl. Inst. Method Phys. Res.* **A467-468**, 1045-1048 (2001).
2. For example, Y. Kuroiwa *et al.*, *Phys. Rev. Lett.* **87**, 217601 (2001).
3. M. Takata *et al.*, *Z. Kristallogr.* **216**, 71-86 (2001).
4. M. Sakata and M. Sato, *Acta Cryst.* **A46**, 263-270 (1990)
5. A. Boultif and D. Louër, *J. Appl. Cryst.* **37**, 724-731 (2004).
6. A. Le Bail, H. Duroy and J. L. Fourquet, *Mat. Res. Bull.* **23**, 447-452 (1988).
7. B. M. Kariuki, H. Serrano-Gonzalez, R. L. Johnston and K. D. M. Harris, *Chem. Phys. Lett.* **280**, 189-195 (1997).
8. W. L. Delano, "The PyMOL Molecular Graphics System", 2002, on World Wide Web http://www.pymol.org

Versatile Collimating Crystal Stage for a Bonse-Hart USAXS Instrument

J. Ilavsky, D. Shu, P.R. Jemian, and G.G. Long

Advanced Photon Source, Argonne National Laboratory, Argonne, IL 60439 USA

Abstract. An advanced ultra-small-angle X-ray scattering (USAXS) instrument, using the Bonse-Hart design and installed at APS, is a robust and reliable instrument, providing a scattering vector (q) range of nearly 4 decades (0.00015 to 1 Å^{-1}), an intensity dynamic range of up to 9 decades, standard-less absolute intensity calibration, and USAXS imaging capabilities. This type of instrument typically uses channel-cut crystals in both the collimating (before sample) and analyzing (after sample) stages. The optical surfaces of these crystals are finished by etching processes, which leave an orange-peel surface texture, which would compromise the USAXS imaging quality. Therefore optics with highly polished surfaces using separated crystals in both collimating and analyzing stages were developed. A novel design of the optics and mechanical stage uses a fixed gap between the two separated collimating crystals in which a triangular section of the first crystal is removed, allowing for a variable number (1, 2, 4, 6, or 8) of crystal reflections for X-ray energies between 7 and 19 keV. The number of reflections is selected by lateral translation of the collimating crystal pair. Rotational alignment of the second crystal in the pair by an artificial channel-cut crystal mechanism, implemented with a novel high-stiffness weak link actuated by both a picomotor and a piezo-electric transducer, provides the capability to align or adjust an assembly of crystals to achieve the same performance as a single channel-cut crystal with integral weak link. The arrangement of both crystals is held on a removable base that can be remounted with precision within the Si(111) rocking curve on a three-point kinematic mount. Additional tilt adjustments are also provided for initial alignment. This monochromator has proven to be highly robust with respect to motions and vibrations, as well as flexible with respect to selection of number of reflections, and its performance directly resulted in the highly reliable performance of the whole USAXS instrument.

Keywords: USAXS optics, artificial channel cut, the high-stiffness weak-link
PACS: 41.50.+h (X-ray optics)

INTRODUCTION

Small-angle X-ray scattering (SAXS) is the premiere technique for the size (typically 1 nm to 100 nm) characterization of materials on the nanoscale, with ultra-small-angle X-ray scattering (USAXS) instruments extending the range of sizes measured upwards past a micrometer [1,2]. Advanced USAXS instruments installed at synchrotron facilities were able to extend the large-size end of the microstructural range measured without compromising their ability to measure nanoscale structures, making them unique instruments bridging the gap between the regular SAXS and light scattering [3,4].

Addition of the USAXS imaging technique [5] adds unique capabilities for materials characterization-but puts more stringent requirements on the optics used. Therefore, it is necessary to use highly polished optics without deep etching, which do not have an "orange-peel" surface, as this surface is directly imaged by the instrument and overlaid with the sample image.

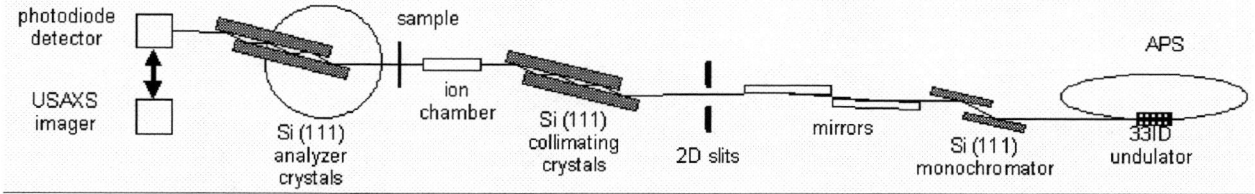

FIGURE 1. The 33ID beam line and USAXS standard 1-D collimation geometry (side view).

CP879, *Synchrotron Radiation Instrumentation: Ninth International Conference,*
edited by Jae-Young Choi and Seungyu Rah
© 2007 American Institute of Physics 978-0-7354-0373-4/07/$23.00

The Advanced Photon Source (APS) USAXS instrument is currently installed at Sector 33, which is equipped with an undulator A X-ray source with a double-crystal Si (111) monochromator to select the energy of the incident X-ray beam, and a pair of vertically reflecting mirrors in a (1, -1) orientation to reject harmonics. At the sample position, 11 keV photon intensities of order 10^{13} ph s^{-1} are incident on a 0.4 x 2.5 mm^2 area.

Figure 1 shows a layout of the USAXS instrument in its 1D collimated geometry. A crystal collimator, discussed in this paper, is placed after the 2D entrance slits that are used to control the size of the X-ray beam. This collimator is an artificial channel-cut Si (111) optic with a triangular section removed from the first crystal (Fig. 2). The collimator is mounted on novel high-stiffness weak-link mechanism stage discussed in next section. This geometry enables us to change the number of reflections from 2 to 8 (we typically use 6) and at the same time (without changing the optic) allows the use of energies from 7 to 19 keV. We have previously reported on this design of the USAXS collimator as single-piece channel cut [6,7]. In this paper we report further advances in this design, where the general geometry of the collimator is preserved, but, using a novel high-stiffness weak-link stage, we can use separated crystals to create an artificial channel-cut crystal. These separated crystals can be highly polished, individually, to provide necessary surface quality for imaging applications.

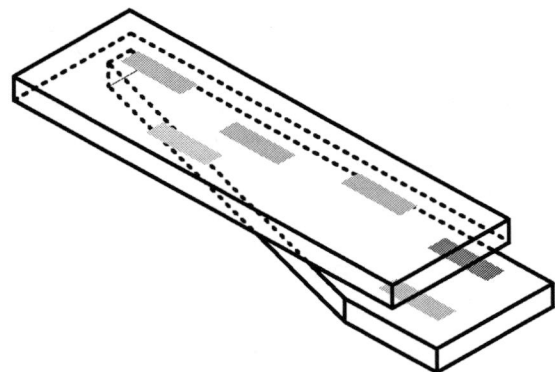

FIGURE 2. Schematics of the channel-cut optics with paths indicated for 2 and 4 reflection beam paths.

Further downstream, the beam passes through a windowless ionization chamber, which serves as the beam intensity monitor used for normalization of the detected signal. After the sample there is an analyzer crystal pair that uses a separated-function pair of crystals with changeable gap. Stable and accurate alignment of the second crystal in the pair with respect to the first crystal in the pair is achieved by means of a similar high-stiffness rotary stage as reported previously [7,8]. Finally, a photodiode detector, linear over 10 decades of operation, measures the scattered beam. The photodiode detector can be replaced by a high-resolution X-ray imager.

STAGE DESIGN

To develop a mechanism that allows us to align or adjust an assembly of crystals to achieve the same performance as a single channel-cut crystal, we have developed a novel high-stiffness weak-link mechanism. In this "artificial channel-cut crystal" design, we have chosen overconstrained mechanisms to optimize the system stiffness. The precision of modern photochemical machining processes using lithography techniques makes it possible to construct a strain-free (or strain-limited) overconstrained mechanism on a thin metal sheet [9]. By stacking these thin metal weak-link sheets with alignment pins, we can construct a solid complex weak-link structure for a reasonable cost. The test results show that the contribution of the angular drift of two crystals attached to each other with the high-stiffness weak-link mechanism is less than 25 nrad per hour [10].

Figure 3 shows the design of the artificial channel-cut crystal mechanism for the multiple-reflections monochromator. There are two sets of stacked thin-metal weak-link modules used in the driving mechanism: one is a planar-shaped, high-stiffness, high-stability weak-link mechanism acting as a planar rotary shaft (2, numbers refer to Fig. 3), and the other is a weak-link mechanism acting as a linear stage (7) to support a coupling plate between a PZT (6) (mounted on the base plate) and a Picomotor^{TM} [1] (8) (mounted on the sine-bar). Both weak-link mechanisms have two modules mounted on each side of the base plate (1). The sine-bar (5) is installed on the center

---

[1] Picomotor^{TM} is a trademark of Newfocus Co. California.

of the planar rotary shaft for the pitch alignment between the second crystal (3) and the first crystal which is not shown in this figure. There are two linear drivers to adjust the sine-bar serially. The rough adjustment is performed by the Picomotor[TM] [2] (8) with a 20 - 30 nm step size. The Physik Instrumente[TM] [3] closed-loop controlled PZT (6) with strain sensor provides 1 nm resolution for the pitch fine alignment. A pair of commercial flexure bearings (9) is mounted on the second crystal holder, and a Picomotor[TM]-driven structure (4) provides the roll alignment for the second crystal.

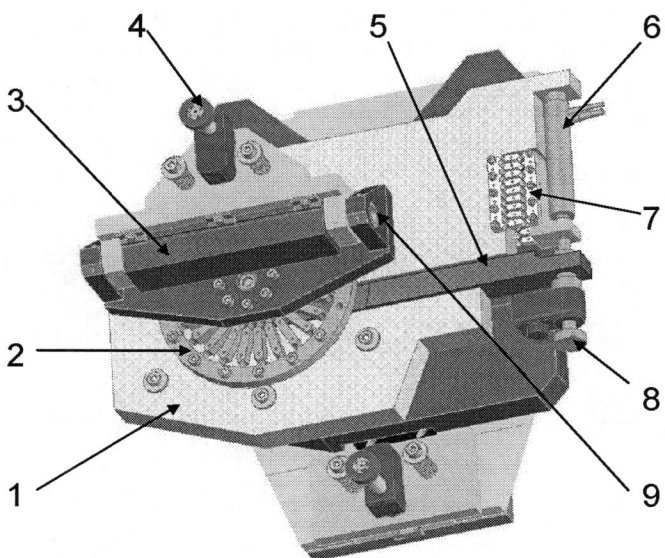

FIGURE 3. The design of the artificial channel-cut crystal mechanism which includes: (1) base plate; (2) weak-link module acting as a planar rotary shaft; (3) second crystal; (4) Picomotor[TM] actuator; (5) sine-bar; (6) PZT actuator; (7) weak-link module acting as a linear stage; (8) Picomotor[TM] actuator; (9) flexure bearing. The first crystal and its holder are not shown in this figure.

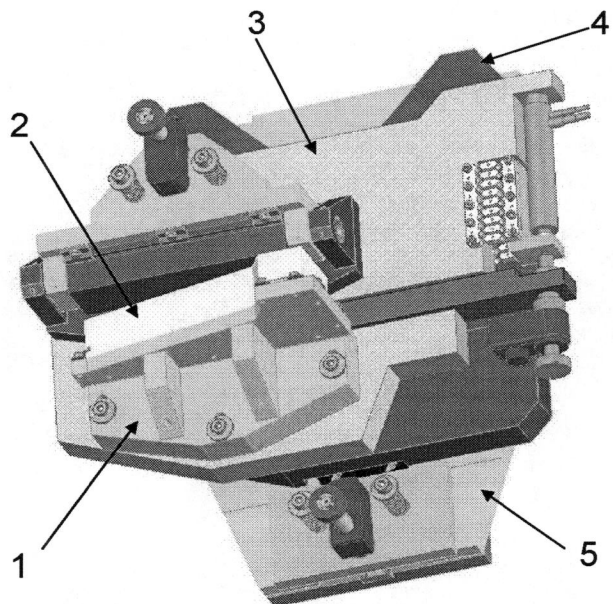

FIGURE 4. The design of the artificial channel-cut crystal mechanism with the first crystal and its holder. (1) first crystal holder; (2) the first crystal; (3) base plate; (4) magnetic couplers; (5) Picomotor[TM]-driven roll alignment structure.

---

[2] Picomotor[TM] is a trademark of Newfocus Co. California.
[3] Physik Instrumente[TM] is a trademark of Physik Instrumente Inc. Germany.

As shown in Fig. 4, a triangle-shaped first crystal (2) is mounted on its holder (1) fixed on the base plate (3). The entire artificial channel-cut crystal mechanism is kinematically mounted on a Picomotor™-driven roll alignment structure (5) through three commercial magnetic couplers (4).

CONCLUSIONS

The new artificial channel-cut crystal stage enables using two highly polished crystals in the first crystal collimator while preserving the necessary high stability and robustness. This novel design enables one to use the same collimating system for both imaging, as well as for regular USAXS, operations. It has been found to offer the stability, robustness and reliability comparable to a single-piece channel cut, as well as the advantages of separated crystals. While this design has been found to be somewhat more complicated to align initially, it has resulted in significant improvement of functional properties of the USAXS instrument.

The collimator X-ray throughput is routinely better than 50% of the incident X-ray beam depending on the quality of the alignment of the double-crystal monochromator and the mirror pair upstream. This compares exceptionally well to previous experience with the single-crystal channel-cut stage used previously, which exhibited comparable throughput.

ACKNOWLEDGMENTS

The APS is supported by the U.S. DOE, Basic Energy Sciences, Office of Science under contract No. W-31-109-ENG-38.

REFERENCES

1   U. Bonse and M. Hart, *Appl. Phys. Lett.* **7** (9), 238-240 (1965).
2   G. G. Long, P. R. Jemian, J. R. Weertman, et al., "High-Resolution Small-Angle X-Ray-Scattering Camera for Anomalous Scattering," *J. Appl. Cryst.* **24**, 30 - 37 (1991)
3   G. G. Long, A. J. Allen, J. Ilavsky et al., "The Ultra-Small-Angle X-Ray Scattering Instrument on UNICAT at the APS," in *Synchrotron Radiation Instrumentation, 11th Us National Conference Proceedings (Sri'99)*, edited by P. Pianetta (American Institute of Physics, 2000), Vol. 521, pp. 183-187.
4   A. J. Allen, P. R. Jemian, D. R. Black et al., *Nuclear Instruments & Methods In Physics Research Section A-Accelerators Spectrometers Detectors and Associated Equipment* **347** (1-3), 487-490 (1994).
5   L. E. Levine and G. G. Long, *Journal of Applied Crystallography* **37**, 757-765 (2004).
6   J. Ilavsky, A. J. Allen, G. G. Long et al., *Review of Scientific Instruments* **73** (3), 1660-1662 (2002).
7   I. Ilavsky, P. Jemian, A. J. Allen et al., "Versatile USAXS (Bonse-Hart) Facility for Advanced Materials Research," in *Synchrotron Radiation Instrumentation* (2004), Vol. 705, pp. 510-513.
8   D. M. Shu, T. S. Toellner, and E. E. Alp, *Nuclear Instruments & Methods in Physics Research Section A-Accelerators Spectrometers Detectors and Associated Equipment* **467**, 771-774 (2001).
9   D. Shu, T. S. Toellner, and E. E. Alp., "Novel Miniature Multi-Axis Driving Structure with Nanometer Sensitivity for Artificial Channel-Cut Crystals," in *Synchrotron Radiation Instrumentation, 11th Us National Conference Proceedings (SRI'99)*, edited by P. Pianetta (American Institute of Physics, 2000), Vol. 521, pp. 219-223.
10 D. Shu, T. S. Tollner, and E. E. Alp, USA Patent No. 6,607,840 (2003).

Large Solid Angle Spectrometer for Inelastic X-ray Scattering

F. Gélebart[*], M. Morand[*], Q. Dermigny[*], P. Giura[*], J.-P. Rueff[†], A. Shukla[*]

[*]*Institut de Minéralogie et de Physique des Milieux Condensés, Campus Boucicaut, 140 rue de Lourmel, 75015 Paris, France*
[†]*Synchrotron SOLEIL, L'Orme des Merisiers, BP 48 Saint Aubin, Gif-sur-Yvette, France*

Abstract. We have designed a large solid angle spectrometer mostly devoted to inelastic x-ray scattering (IXS) studies of materials under extreme conditions (high pressure / temperature) in the hard x-ray range. The new IXS spectrometer is designed to optimize the photon throughput while preserving an excellent resolving power of ~10000 in the considered energy range. The spectrometer consists of an array of up to 4 spherically bent 0.5 m radius analyzer crystals and a solid-state detector positioned on the Rowland circle. The four analyzers can cover a solid angle more than one order of magnitude larger than conventional spectrometers. The spectrometer is to be installed on the GALAXIES beamline at SOLEIL in the near future.

Keywords: Inelastic x-ray scattering, Instrumentation, Spectrometer.
PACS: 07.85.Nc

INTRODUCTION

Inelastic x-ray scattering (IXS) has proven to be a powerful technique of investigation of electronic properties of materials. IXS can be simply sketched as a two-step process consisting of the absorption of incident photons followed by the emission of secondary photons which are analyzed in energy. As an all-photon technique and thanks to the high penetration depth of photons in the hard x-ray region (above ~5 keV), IXS is well adapted to constrained sample environments such as high pressure cell or catalytic chamber. Unique information about low-energy excitations, local magnetism or structure, valence properties or density of states can then be obtained. Of particular interest is the possibility to use IXS in the non resonant regime for probing the K absorption edges of light elements with hard x-rays. This is the case when the energy transferred by the incident photon is sufficient to excite 1s core electrons. The absorption resonance is then observed in the energy loss spectra, typically few hundred eV above the quasi-elastic line. Such measurements have been carried out in the last years in particular at the C and O K-edges under high pressure conditions [1,2].

The poor signal statistics has however limited so far the extension of this spectroscopic technique beyond case studies. IXS is a second order probe, intrinsically of weak cross section. This limitation can be partly overcome thanks to the resonant enhancement of the scattering cross section when the incident photon energy is chosen close to an absorption edge. This is no longer the case when one has to deal with non-resonant processes such as in the energy loss spectra as described above. Additionally, the sample chamber windows may also absorb part the scattered light, which further degrades the signal statistics and this independently of the resonant conditions. Increasing the signal / noise ratio then becomes necessary especially for high pressure experiments in a diamond anvil cell. A simple yet efficient way to improve the count rate consists of enlarging the spectrometer acceptance. In the Rowland circle geometry, it is defined by the angular aperture of the analyzer which serves to select the photon energy and focuses the emitted light onto the detector. In that perspective, we have designed an IXS spectrometer which covers a solid angle larger by one order of magnitude than conventional setups. The new instrument is more particularly devoted to photon-hungry experiments. It is to be installed on the IXS-beamline (GALAXIES) at SOLEIL (France).

CP879, *Synchrotron Radiation Instrumentation: Ninth International Conference*,
edited by Jae-Young Choi and Seungyu Rah
© 2007 American Institute of Physics 978-0-7354-0373-4/07/$23.00

IXS SPECTROMETER SETUP

Technical Specifications

Spectrometer Geometry

The IXS spectrometer which we propose is based on the Rowland circle geometry. In this geometry, the source, analyzer and detector are positioned on a circle of diameter ρ (cf. Fig. 1). The Johan approximation to the Rowland condition imposes that the circle diameter corresponds to the bending radius of analyzer crystal. The exact focusing condition is then only fulfilled at the point of the analyzer surface which intercepts the Rowland circle. Deviation from this is negligible providing the analyzer size is small compared to ρ.

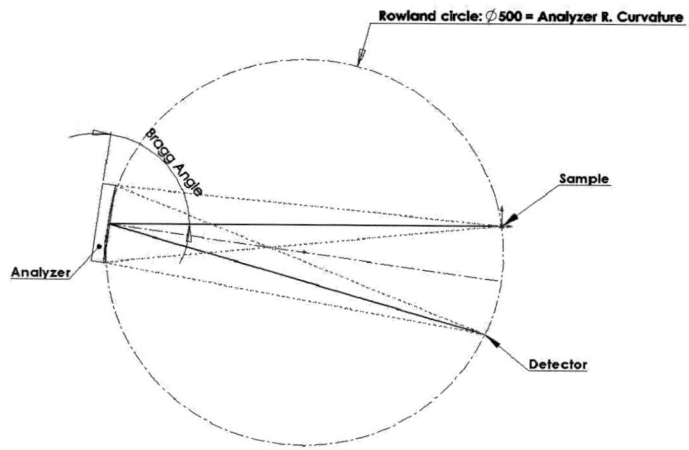

FIGURE 1. Johann geometry with Rowland circle

Analyzer Crystals

The analyzer serves both to select the energy of the scattered radiation by Bragg reflection in a narrow bandwidth and to focus the beam onto the detector position, thus maximizing the spectrometer throughput. The spectrometer therefore benefits from large solid angle while the experimental resolution is conserved. The analyzers which are envisaged for the spectrometer are prepared from Si wafers spherically bent onto a glass substrate. In order to minimize strain due to the gluing process, we have recently developed at the IMPMC a novel technique of preparation of analyzers [3] where the wafer is maintained on the substrate by anodic bonding, without resin. Analyzers prepared by this method show superior quality in terms of reflectivity and resolving power. We have obtained an intrinsic resolution of 200 meV at 8 keV in a recent series of experiments with a 2-m analyzer [4]. It is now possible to routinely make analyzers with a bending radius of 0.5 m and above.

Technical Requirements

The spectrometer design results from a compromise between energy resolution, and experimental details such as sample handling, scanning procedure, and solid angle. Our main aim is to realize a large solid angle IXS spectrometer by combining an array of multiple analyzers. We will use an assembly of up to 4 spherically bent analyzers of Silicon (111) with radius of curvature of 500 mm and a diameter of 100 mm. Note that the solid angle which is covered by four analyzers is equivalent to that of sixteen analyzers of 1 m bending radius and same diameter, as mounted in conventional RIXS spectrometers. The advantage of using 500 mm bent radius analyzer is in fact manifold: i) firstly, the large solid angle (0.125 sr with four analyzers) implies higher photon count rate on the detector, thus improved experimental statistics; ii) the reduced weight and the compact design of the spectrometer makes this instrument easily portable; an important aspect for the testing phase during which the instrument is to be moved on various experimental stations; iii) finally, a significantly reduced cost.

The main shortcoming of the large solid angle setup resides in the difficulty to precisely define the source size. This is more particularly problematic when the sample is embedded in a constrained environment such as in a high pressure cell. Then, large part of the incident beam is scattered by the immediate sample surrounding, hence degrading the signal to background ratio. One way to prevent unwanted scattered radiation to reach the detector is to install a Soller slit device near the sample position. Implementation of sample slits will be considered in a second phase.

In a first step, the spectrometer will be devoted to non-resonant IXS experiments at fixed Bragg angle. No movement of the analyzers is necessary during the measurements, and the analyzers only need to be aligned before the experiment. In order to obtain a total resolution of 0.5 eV at incident energy of 8 keV and a Bragg angle about 83°, we must ensure a precision on the analyzer rotation better than 0.029°.

Experimental Details

Since the spectrometer technical specifications are not compatible with commercial solutions, it was decided to develop a "home-made" design, including full motorization of the 4 analyzers and mounting.

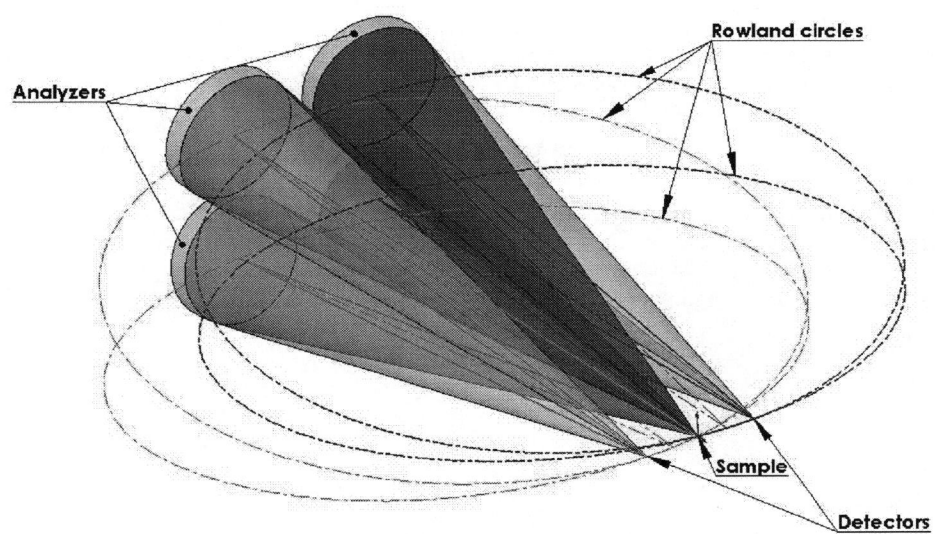

FIGURE 2. Positions between sample, detectors and analyzer matrix in Rowland circle geometry.

Each analyzer crystal will have three degrees of freedom: Two rotations θx and θy, to select the Bragg angle and center the reflected beam onto the detector, and one Z translation in order to adjust the analyzer centre of rotation on the Rowland circle. As illustrated in Fig. 2, each analyzer in fact defines its own Rowland circle. This requires that each analyzer to be aligned independently.

For mechanical reasons a pivot-axe has been preferred to spherical ball which induces more friction. θx and θy movements are motorized with an angular amplitude of +/- 10° by a set of two precision actuators (cf. Fig. 3), which push the analyzer support along the two orthogonal directions. Each actuator is mounted at the back of the analyzer, perpendicularly to the plane of rotation. The originality of this solution consists of using a unique traction spring to permanently maintain the contact between the analyzer and both actuators. The one-axis accuracy of the actuators is 10 μm which translates into an accuracy on the rotation of $\Delta\theta x$ of 0.008° and $\Delta\theta y$ of 0.013°, well within the specifications. The Z translation will be motorized by a step-motor with a precision of 0.02mm. The main parts of spectrometer are machined from aluminium alloys in order to reduce global weight (around 5 kg for one analyzer device, cf. Fig. 3).

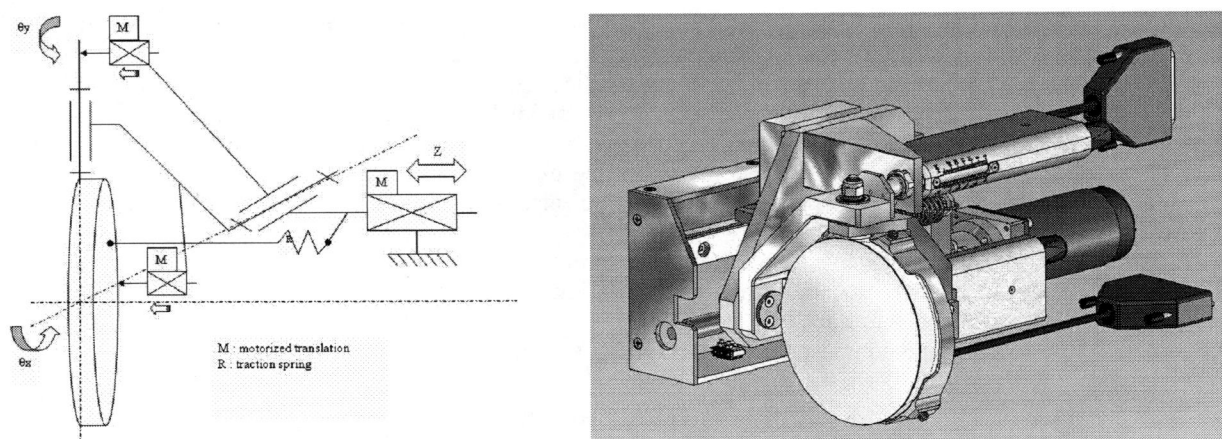

FIGURE 3. (Left) Cinematic design of analyzer motorized motions and (right) mechanical device for one analyzer.

The chosen motor controller is an XPS-Newport high performance 1-8 axes integrated motion controller/driver. It offers high speed 10/100 Base-T Ethernet communication interface and powerful programming functionality from almost any operating system. The controller is fully compatible with the TANGO device server structure developed at SOLEIL. Proprietary driver modules allow the XPS to control many motion devices as stepper motors, DC brushless motor and piezoelectric actuators. An intuitive object oriented TCL (Tool Command Language) can be used for high speed process control and multi-trajectories calculation.

Planning and Installation

The instrument will be installed on the inelastic x-ray scattering beamline GALAXIES at the new SOLEIL synchrotron light source (France). The beamline should be operational during the second semester of 2008. The beamline energy range covers 3-12 keV in an adaptable energy bandwidth. It will be equipped with a versatile 2 m IXS spectrometer in addition to the compact spectrometer. Figure 4 shows an artist's view of the spectrometer, once installed on the beamline.

FIGURE 4. Overview of the future large solid-angle IXS spectrometer on the GALAXIES beamline at SOLEIL.

REFERENCES

1. W. L. Mao et al., *Science* **302**, 425 (2003).
2. Y. Q. Cai et al., *Phys. Rev. Lett.* **94**, 025502 (2005).
3. E. Collart et al., *J. Synchrotron Rad.* **12**, 473-478 (2005).
4. E. Collart et al., *Phys. Rev. Lett.* **96**, 157004 (2006).

Dispersive XAFS Image Radiograph by Parametric X-ray Radiation

*Akira Mori[1], Yasushi Hayakawa[2], Isamu Sato[2], Toshinari Tanaka[2], Ken Hayakawa[2], Takao Kuwada[2], Koji Kobayashi[1], Hisashi Ohshima[1]

[1]Pharmacy, Nihon University, Narashinodai, Funabashi, Chiba, 274-8555, JAPAN
[2] Laboratory for Electron Beam Research and Application, Nihon University, Narashinodai, Funabashi, Chiba, 274-8501, JAPAN

Abstract. The parametric X-ray (PXR) generator system at Laboratory for Electron Beam Research and Application (LEBRA) in Nihon University is a variable-wavelength and quasi-monochromatic X-ray source, which was developed as one of the advance applications of the 125-MeV electron linear accelerator. Since the first observation of the X-rays generated by the system in April 2004, application studies have been performed using the PXR beam in the region from 6.0 to 20keV. The PXR beam extracted from the fixed output port of the generator has characteristic energy dispersion. The theoretical energy spread at the output port with an inner diameter of 98mm changes approximately from 300eV to 2keV depending on central X-ray energy from 7keV and 20keV. The dispersion occurs linearly only in horizontal direction, with a high energy resolution. The characteristics of the PXR beam from the generator suggest a possibility of for the kind of energy dispersive X-ray absorption fine structure (DXAFS) measurement using density distribution in the radiographs of materials. Using the uniform film of the sample materials, DXAFS can be deduced from the measurement of the horizontal density distribution in the radiograph due to the characteristics of the PXR beam. Since the PXR generator system is based on the S-band liner accelerator, it has the potential for the time-resolved XAFS measurement with several ten pico-second resolutions.

Keywords: DXAFS, Parametric X-ray, Energy distribution
PACS: 41.60.-m, 41.50.+h, 29.17.+w, 61.10.Ht

INTRODUCTION

At the Laboratory for Electron Beam Research and Application (LEBRA), Nihon University, a tunable monochromatic X-ray source was developed based on Parametric X-ray radiation (PXR) [1,2]. The PXR system was constructed with a dedicated beam line connected to the 100MeV - class electron linear accelerator of LEBRA. Since the LEBRA-PXR source serves a tunable X-ray beam involving liner energy dispersion, X-ray absorption fine structure (XAFS) measurement is possible using an imaging device. This method is a kind of energy-dispersive XAFS (DXAFS) which has recently been studied using bent-crystal polychrometers at several large synchrotron facilities. In the case of the LEBRA-PXR, however, the energy dispersion of the X-ray beam is obtained using flat-plane crystals and has a good linearity to the emission direction of X-rays. In a sample setup, one can obtain XAFS spectra using the PXR beam from the LEBRA system, although the measurement requires large uniform samples.

Considering the large area of irradiation field of the PXR beam, imaging devices for radiography, such as an imaging plate (IP), are useful for the detector in the DXAFS experiment using the PXR beam. We therefore carried out preliminary experiments of DXAFS using the PXR beam and imaging devices such as an IP and an X-ray CCD. This paper reports the results and discusses the properties of the LEBRA-PXR system with respect to XAFS measurements.

CP879, *Synchrotron Radiation Instrumentation: Ninth International Conference*,
edited by Jae-Young Choi and Seungyu Rah
© 2007 American Institute of Physics 978-0-7354-0373-4/07/$23.00

TABLE 1. Specifications of LEBRA-PXR system.

Typical electron energy	100 MeV
Average beam current	≤ 5 μ A
PXR generator target (first crystal)	0.2mm thick Silicon (1 1 1)
PXR energy range (Silicon (1 1 1))	4-20 keV
PXR beam size (Output window)	98 mm (diameter)
Distance between PXR source to output window	7.2 m
Range of elements for XAFS measurement	Ca to Mo (K-edge) Sb to Cf (L-edge)

ENERGY DISPERSIVE XAFS MEASURMENT USING PXR

The LEBRA-PXR system is a monochromatic X-ray source on the base of the 125-MeV electron liner accelerator. PXR is a kind of radiation phenomena which result from the interaction between a relativist charged particle and crystal medium. The radiation has high directivity and unique energy distribution.

If both the electron vector v and the reciprocal lattice vector g of the target crystal are along the horizontal plane, one can express the PXR energy $\hbar\omega$ in the case of the Bragg angle θ, which is the incident angle of the electron into the crystal plane, following,

$$\hbar\omega = \frac{\hbar c^* |g| \sin\theta}{1 - \beta\cos\phi} , \qquad (1)$$

where c^* is the light speed in the target medium and $\beta = |v|/c^*$; ϕ is the angle between the electron velocity and the X-ray direction. According to eq.(1), the PXR energy depends almost linearly on the angle $\phi - 2\theta$ for fixed if β is approximately equal to 1 and ϕ is close to 2θ [3-5]. For $\phi = 2\theta$, i.e., the Bragg condition in eq. (1), the PXR energy is approximately equal to the Bragg energy $\hbar\omega_B = \hbar c^* |g| / 2\sin\theta$.

Around the Bragg condition, the PXR energy shift due to slight difference $\Delta\theta$ of the angle ϕ is approximately written by

$$\hbar\omega' = \hbar\omega_B + \Delta\hbar\omega \approx \hbar\omega_B(1 - \frac{\Delta\theta}{\tan\theta}) , \qquad (2)$$

within the approximation $\beta \approx 1$.

Considering the reflection of X-rays by second crystal and the geometry of the system as shown in Fig.1(a), eq.(2) is replaced by the function of horizontal position x,

$$\hbar\omega' \approx \hbar\omega_B(1 + \frac{1}{\tan\theta} \cdot \frac{x}{L}) , \qquad (3)$$

where L is the distance between the PXR source and the observation plane in the Fig.1(a). Figure 1(b) shows the PXR energy as a function of the horizontal position which is accompanied with the results of the X-ray energy

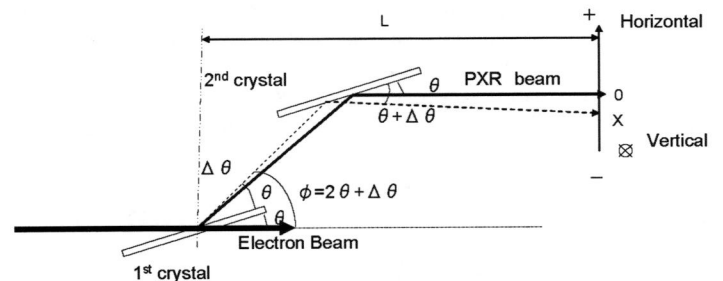

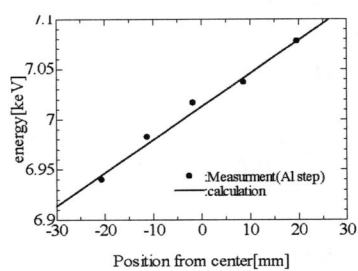

(a) Arrangement (b) energy distribution (Center energy is 7keV)
FIGURE 1. PXR generated system with double crystal system

measurement using the absorption due to the Aluminum step. Using the distribution property, the XAFS spectrum of the samples can be directly observed using imaging device. In the case of the LEBRA PXR system, the irradiation field area has rather large area, of which diameter 98mm at the output port. Therefore the imaging plate (IP) is useful for imaging device.

The tunability of the PXR beam in the ranges of 4 keV to 20 keV makes possible XAFS measurement of elements from Ca to Mo for K–absorption edge and from Sb to Cf for L –absorption edge. The dynamic ranges of the energy distribution are estimated to be 90 eV and 2.7 keV when the center energies are 4 keV and 20 keV. When there are uniform samples as long as the PXR beam diameter, EXAFS can be measured.

EXPERIMENTAL RESULTS AND DISCUSSION

To demonstrate DXAFS measurement using the PXR beam, several samples were prospered and these absorption images were taken in the simple setup as shown in Fig.2(a). Figure 2(b) shows the typical images of an Yb_2O_3 X-ray filter, Cu-Ni alloy foil, a pure copper foil and $CuSO_4$ gel In this case, the PXR energy was turned around 9 keV and an IP (Yosida) with 50 um square area was used as the imaging device, the measurement conditions were L1 of 600 mm and L2 of 20 mm. The time that had been required to take image was 1800 second. The absorption edge of each sample was clearly defined and striped pattern were observed in the higher energy region than the absorption edge. Figure 3(a) shows absorption spectra obtained from Fig.2(b), where the X-ray energy was calculated by applying eq. (3). In the spectra of pure Cu and Cu - Ni alloy, the oscillators that seem to be a part of EXAFS of copper were observed. On the other hand, energy shifts of the absorption edges and sharp peaks were observed in the spectra of $CuSO_4$ and Yb_2O_3.These seems to be typical XANES spectra. In addition, the strong absorption due to the L-absorption edge of Yb_2O_3 was also observed. The energy difference of Yb_2O_3 L -absorption edge from Cu K-absorption edge obtained in the experiment was almost equal to the value of the "Table of Isotopes". These results suggest that the calculation can be easily performed by measured sample simultaneously. The range irradiation field is one of the advantages of the DXAFS measurement using the PXR beam.

In other hand, Fig.3(b) shows the absorption spectra obtained using an X-ray CCD with an element size of 24 um square and an active area of12 mm square. The time that had been required to take image was 900 second. The DXAFS measurements for pure copper foil and $CuSO_4$ gel were carried out under the condition that L1 and L2 were 480 mm and 180 mm, respectively. The energy range covered by the X-ray CCD is approximately 60 eV at 9 keV because of the X-ray CCD action area. Although the use of the X-ray CCD is not efficient for wide-range EXAFS measurement, the device is suitable for the XANES measurement become of the high position resolution.

In the case of the PXR source, both the electron beam size on the target and the position resolution of the imaging device restrict the energy resolution of DXAFS measurement. In particular, the electron beam size effect is significant since it degrade the energy dependence itself. Assuming the diameter of the electron beam is 750μm. L is 7650 mm at 9 keV, the energy resolution is estimated to be 3.93 eV. In order to improve the energy resolution, smaller electron beam or larger distance L is required. The well forward electron beam has an advantage over the long distance L because of the photon density and the compact measurement system. Perhaps, the destruction of the

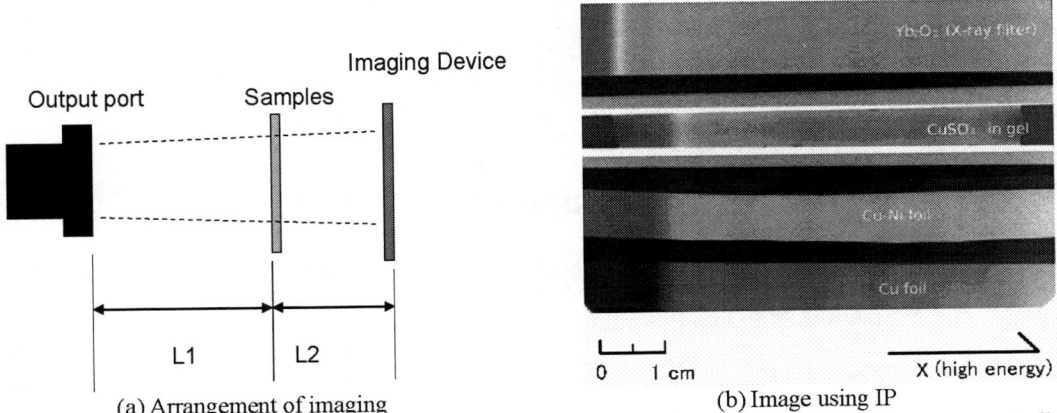

(a) Arrangement of imaging   (b) Image using IP

FIGURE 2. X-ray absorption image(The Yb_2O_3 X-ray filter, alloy ribbon of copper and nickel, the pure copper ribbon and $CuSO_4$. gel)

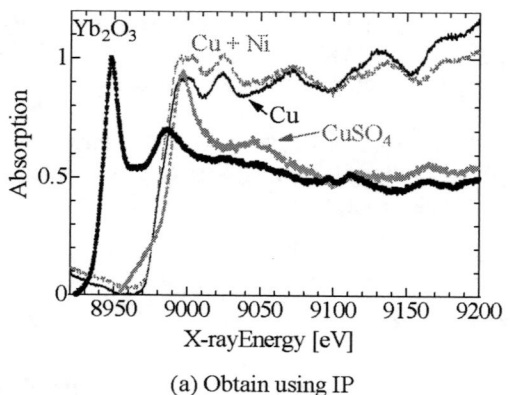

(a) Obtain using IP

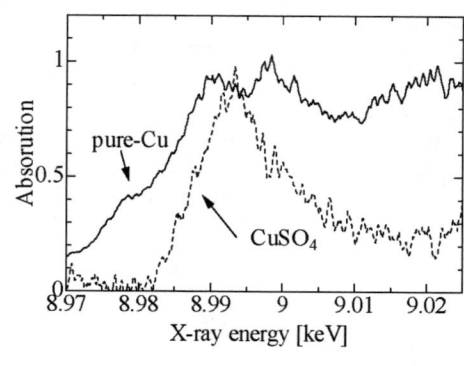

(b) Obtain using X-ray CCD

FIGURE 3. The absorption spectra

target crystal due to the electron beam may limit the energy resolution of the PXR dispersion.

Because the spatial resolution of the IP and the X-ray CCD are about 50μm and 24 μm, the energy resolution attributed to the position resolution of the imaging device are estimated to be 1.16 eV and 0.56 eV respectively. In the spec, the IP and the X-ray CCD are unquestionable for the energy resolution because the resolutions these imaging devices are higher than the PXR source. Moreover, the resolution of the IP is more poor that of the CCD (Especially, XANES region). This is due to the problem of the device when IP is developed. It is necessary to use wide view CCD used in synchrotron radiation facilities to request more detailed resolution.

The energy resolution when IP was used was worse than the result of using X-rays CCD. It is measured that this deterioration is not caused by the spatial resolution of the IP but by the distortion in the amplifier linearity of the IP reader. Therefore, the improvement of the imaging device is also important for DXAFS measurement using radiography techniques.

SUMMARY

There is a possibility that energy-dispersive XAFS (DXAFS) is obtained by taking the penetration image using ribbon material without PXR generator scanning, because PXR has a straight line energy distribution. This method can observe two or more material and authentic sample at the same time because the energy distribution in the vertical direction is constant. As a result of the experiment, the spectra pattern and the chemical shift was observed. However, the energy resolution is decided the electron beam diameter and the imaging device.

ACKNOWLEDGMENTS

We express gratitude for Dr. T.Sakai, Dr. K.Nogami and Dr. K.Nakao to control the accelerator. The development of this X-ray source has been supported by the "Academic Frontier" Project for Private Universities: matching fund subsidy from MEXT, 2000-2004 and 2005-2007. Part of the work was also supported by MEXT.KAKENHI(17760059).

REFERENCES

1. Y. HAYAKAWA, et. al., in: proc. Of 12th Symposium on Accelerator Science and Technology, Wako, Japan, 1999, p. 391
2. Y. Hayakawa et. al., Nucl. Instrum. Methods A 483 (2002) 29.
3. Y. Hayakawa,et.al., Nucl. Instrum. Methods B 227 (2005) 32.
4. M. L. Ter-MIkaelian, "High-energy electromagnetic processes in condensed media", Wiley-Interscience, New York, 1972
5. Y.Hayakawa et.al. "Tunable Monochromatic X ray Source Based on Parametric X ray Radiation at LEBRA, Nihon University", *in these proceedings.*,2006

Using Synchrotron-based X-ray Absorption Spectrometry to Identify the Arsenic Chemical Forms in Mine Waste Materials

Vitukawalu P. Matanitobua[*], Barry N. Noller[*], Barry Chiswell[*], Jack C. Ng[*], Scott L. Bruce[*], Daphne Huang[*], Mark Riley[Ψ], and Hugh H. Harris[¶]

[*]National Research Centre for Environmental Toxicology (EnTox), University of Queensland, 39 Kessels Road, Coopers Plains, Queensland 4108, Australia
[Ψ]Chemistry Department, School of Molecular and Microbial Sciences, University of Queensland, St Lucia, Queensland 4067, Australia
[¶]School of Chemistry, University of Sydney, New South Wales 2006, Australia

Abstract. X-ray Absorption Near Edge Spectroscopy (XANES) gives arsenic form directly in the solid phase and has lower detection limits than extraction techniques. An important and common application of XANES is to use the shift of the edge position to determine the valence state. XANES speciation analysis is based on fitting linear combinations of known spectra from model compounds to determine the ratios of valence states and/or phases present. As(V)/As(III) ratios were determined for various Australian mine waste samples and dispersed mine waste samples from river/creek sediments in Vatukoula, Fiji.

Keywords: arsenic speciation, selective extraction technique, mine wastes, XANES, As(V)/As(III) ratio.
PACS: 80.87.87.64.Gb.89.60.-k.92.40.Gc

INTRODUCTION

Arsenic speciation in the solid phase is difficult to perform directly. Traditionally arsenic species are identified using time-consuming, indirect methods that determine the identity of the compound through chemical manipulation. Thus different selective extraction techniques have been used to demonstrate classes of arsenic species in each extracted fraction, for example, using oxalic acid and sodium dithionite to release arsenic bound by metals [1]. Pre-treatment of samples may alter the chemical form of the arsenic. Inductively-Coupled Plasma Atomic Emission Spectroscopy is used to determine the total amount of arsenic in a sample acid digest but cannot distinguish particular arsenic species.

X-ray absorption spectroscopy is capable of providing detailed chemical and structural information about a specific absorbing element in situ with minor or no pre-treatments [2]. There are two regimes in X-ray Absorption Structure (XAS) Spectroscopy: X-ray Absorption Near-Edge Spectroscopy (XANES) and Extended X-ray Absorption Fine Structure (EXAFS) Spectroscopy; these contain diverse information about an element's local coordination and chemical state. The absorption edge energy in XANES spectra is sensitive to the oxidation state of arsenic, and the position increases with an increase in oxidation state. XANES is applicable to solution or solid-phase samples, is element and oxidation state specific, and sensitive to parts per million [3]. XANES gives chemical identity empirically by comparing sample spectra to known compounds. This paper illustrates the usefulness of the XANES technique to identify distinguishing features of arsenic species in mine wastes and dispersed sediments from various mines in Australia and Fiji.

METHOD

"Total" mine waste samples and their respective fractions were obtained via selective chemical extraction (SCE) technique based on the sequential series of 8 selective extraction reagents [1] with increasing extraction power. A modification of the 8-step sequential procedure was developed for mining waste solid-phase samples containing arsenic [4]. Eight fractions of arsenic were separated from mine waste material and dispersed creek/river sediment

CP879, *Synchrotron Radiation Instrumentation: Ninth International Conference*,
edited by Jae-Young Choi and Seungyu Rah
© 2007 American Institute of Physics 978-0-7354-0373-4/07/$23.00

by SCE [1]. The oxidation states and the ratio of arsenic species present in these samples, as well as some selected fractions (those with highest As concentrations Fraction 3 Al-As; Fraction 4 Fe-As; Fraction 5 Ca-As; & Fraction 6 Fe occluded As) were determined using XANES.

Mine Sites

Tailings were collected from four Australian mines: Gympie Gold Mine (GHT 100 yr old), the rehabilitated Jibbinbar Arsenic Mine (JAsMT), Kidston Gold Mine (the acidic (KAT) and neutral (KNT) tailings) and at the Red Dome Heap Leach (RDHL). Dispersed contaminated and upstream sediments were collected in Fiji from Vunisina Creek (8SdsVsc-02), Dakavono Creek (15SdsDvc-02) and the Nasivi River (23SdsNsr-01) upstream from Emperor Gold Mine at Vatukoula.

Model Arsenic Compounds

Standard arsenic model compounds were used to identify suspected arsenic chemical forms in the mine waste materials, and in their sequentially extracted fractions. The model inorganic arsenic compounds used in this study (Table 1) were classified as either As(-I), As(III) or As(V) compounds, according to their formal oxidation states. The synthetic arsenic compounds (Table 1) calcium arsenite, aluminium arsenate, calcium arsenate and iron arsenate were prepared by precipitation reactions of suitable soluble salts using standard methodologies. The arsenopyrite used was a mineralogical sample from Jibbinbar analysed to show % arsenic present. Crystalline reagent grade sodium arsenite tetrahydrate, $NaAsO2.4H2O$ and disodium orthoarsenate heptahydrate, $Na2HAsO4.7H2O$ (BDH, Australia), arsenic(III) sulfide and arsenic(V) sulfide, (Aldrich Chemical Company, Inc., WI, USA) were also used as model compounds. They were all diluted to about 1000mg As/kg using the "neutral" (<10ppm As) 23SdsNsr-01 upstream sediment of Nasivi River by adding 5g of each model compound with 5g of 23SdsNsr-01 prior to grinding to fine powder (<75μm) in a zirconia tema swing mill (N.V. Tema, Germany) and stored in clear capped clean 5mL clear plastic vials prior to analysis. At the synchrotron facility, approximately 2g of the model compounds/sand mixtures and samples were evenly loaded into 1mm thick aluminium holders. Both sides of the holder were encapsulated with X-ray transparent Kapton tape. The valency and individual quantity of arsenic present in each mine waste sample were calculated after scanning using XANES.

TABLE 1. Model arsenic compounds used and their class

As(-1)	As(III)	As(V)
Arsenopyrite, FeAsS	Arsenic sulfide, As_2S_3	Aluminum arsenate, $AlAsO_4.8H_2O$
	Calcium arsenite, $CaAsO_3H$	Arsenic sulfide, As_2S_5
	Sodium arsenite, $NaAsO_2.4H_2O$	Calcium arsenic, $Ca_3(AsO_4)_2$
		Iron arsenate, $FeAsO_4$
		Sodium arsenate, $Na_2HAsO_4.7H_2O$

Spectroscopic Technique/XANES Data Analysis

Arsenic K-edge XAFS spectra were collected at the Australian National Beamline Facility (ANBF BL-20B) Photon Factory, Tsukuba Science City, Japan over the energy range 11840–11940eV (ring conditions: 2.5GeV, 300-400mA). BL–20B was equipped with a channel-cut Si (111) monochromator which was detuned 50% to reject harmonics. The monochromator step size was reduced to 0.25eV per step in the XANES region (11845-11895eV) to collect high-resolution spectra. XAFS data for the above-mentioned samples and model compounds were collected at ambient temperature and pressure in fluorescence, with the simultaneous collection of an As reference foil for energy calibration (the first peak of the first derivative of the spectrum of elemental As was assumed to be 11867.0eV). A 10-element germanium fluorescence detector was used to collect the data. The detection limit for adequate speciation was approximately 6mg/kg arsenic (see Fig. 1).

Data analysis was processed using the EXAFSPAK suite of programs [5]. XANES spectra of model compounds and samples were background subtracted and normalised to edge jump (normalised to the absorbance value of the spline at 11885eV). The XANES analysis consisted of fitting linear combination of model spectra to sample spectra using the program DATFIT [5]. The precision of this fit procedure was determined to be ~10% based on analyses of control mixtures of model compounds.

RESULTS

The results are tabulated (see Tables 2 & 3) and shown in Figs. 1 and 2 below.

TABLE 2. The XANES fitting analysis of the oxidation states and ratio of arsenic species identified

(a) Original mine waste samples from various Australian mine sites

Sample Description	Total [As] (mg/kg)	As(III) sulfide	Fe arsenate	Arseno-pyrite	Al arsenate	Ca arsenate	Ca arsenite	Fit Residual
GHT	750		~100%					0.006%
KAT	1200	9.30%	60.70%		32.10%			0.002%
KNT	320	7.30%	27.30%	63.50%				0.01%
RDHL	520		~100%					0.02%
JAsMT	2600		15.90%		54.40%	27.50%		0.001%

(b) Sequentially extracted fraction

Sample Description	Total [As] (mg/kg)	As(III) sulfide	Fe arsenate	Arseno-pyrite	Al arsenate	Ca arsenate	Ca arsenite	Fit Residual
KAT Desorption Fr. 4 (Fe-As) incl.	1100		63.70%	16%	13.20%	9.10%		0.001%
KAT Desorption Fr. 5 (Ca-As) incl.	1100		72.40%	11.90%	3.10%	14.30%		0.003%
KAT Desorption Fr. 6 (Fe Occl.-As) incl.	610	12.80%	78.90%		8.50%			0.002%
JAsMT Desorption Fr. 3 (Al-As) incl.	2500		78.70%	23.40%				0.003%
JAsMT Desorption Fr. 4 (Fe-As) incl.	2000		67.30%			35.40%		0.002%
JAsMT Desorption Fr. 5 (Ca-As) incl.	860	2.20%	59.70%		22.70%	19.20%		0.002%

TABLE 3. The XANES fitting analysis of arsenic species identified and determined on dispersed mine waste samples from creek sediments together with their sequentially extracted fraction in Vatukoula, Fiji

Sample Description	Total [As] (mg/kg)	As(III) sulfide	Fe arsenate	Arseno-pyrite	Al arsenate	Ca arsenate	Ca arsenite	Fit Residual
8SdsVsc-02	300		40.60%	7.80%	56.30%			0.0008%
8SdsVsc-02 Desorption Fr. 4 (Fe-As) incl.	190	3.30%	84.20%	12.70%				0.0009%
8SdsVsc-02 Desorption Fr. 5 (Ca-As) incl.	120	8.30%	87%	4.60%				0.0010%
15SdsDvc-02	690	5.20%	89.40%	6.30%				0.0004%
15SdsDvc-02 Desorption Fr. 4 (Fe-As) incl.	520		~100%					0.0010%
15SdsDvc-02 Desorption Fr. 5 (Ca-As) incl.	380		75.40%		9.20%	17.80%		0.0010%
23SdsNsr-01	10		16.70%		69.40%		21.30%	0.0070%

DISCUSSION AND CONCLUSIONS

XANES analysis indicates that As(V) comprising the arsenates of iron, aluminium, and calcium is the dominant oxidation state in most mine waste samples, including the ones that were dispersed as creek sediments. Mixed oxidation states; supposedly As(V) (comprising the arsenates of iron, aluminium, and calcium), As(III) (consisting of arsenic sulfide), and As(-I) (comprising of arsenopyrite only) were observed to be present in most of the total mine waste samples and their respective selective extraction fractions. The XANES analysis technique is a direct measurement of the composition of mine waste components in relation to the oxidation states and chemical forms (species) of arsenic that may be present in original sample and selected fractions. Thus, the usefulness of the XANES technique to measure arsenic speciation in the solid phase of mine wastes has been demonstrated.

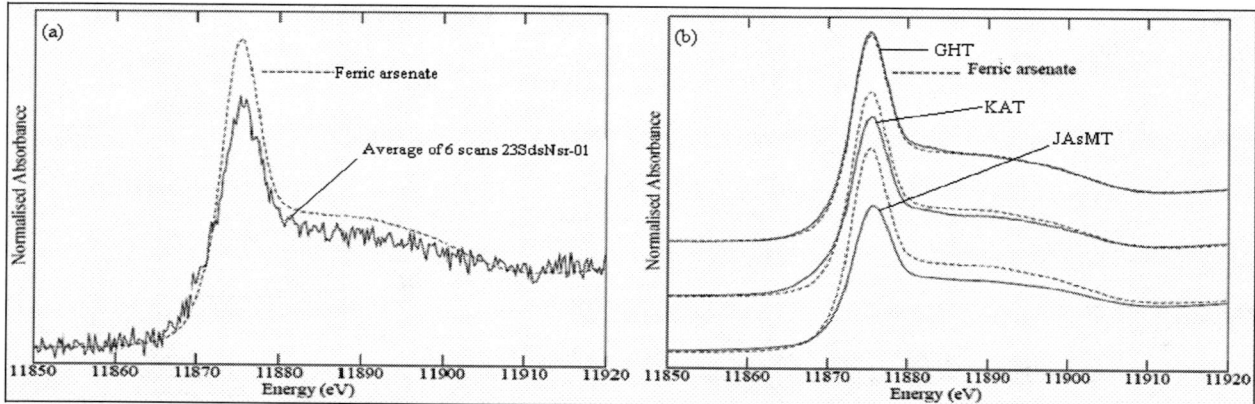

FIGURE 1. Comparison of sample spectra and the model ferric arsenate compound spectrum: (a) average spectra of 23SdsNsr-01 upstream sediment of Nasivi River, Fiji revealing a low level of arsenic, compared with ferric arsenate model spectrum; and (b) GHT with almost 100 % ferric arsenate showing good fit to the model spectrum, KAT - about 61 % ferric arsenate and JAsMT – approx. 16 % ferric arsenate. Table 2 gives the % composition of arsenic species in these samples.

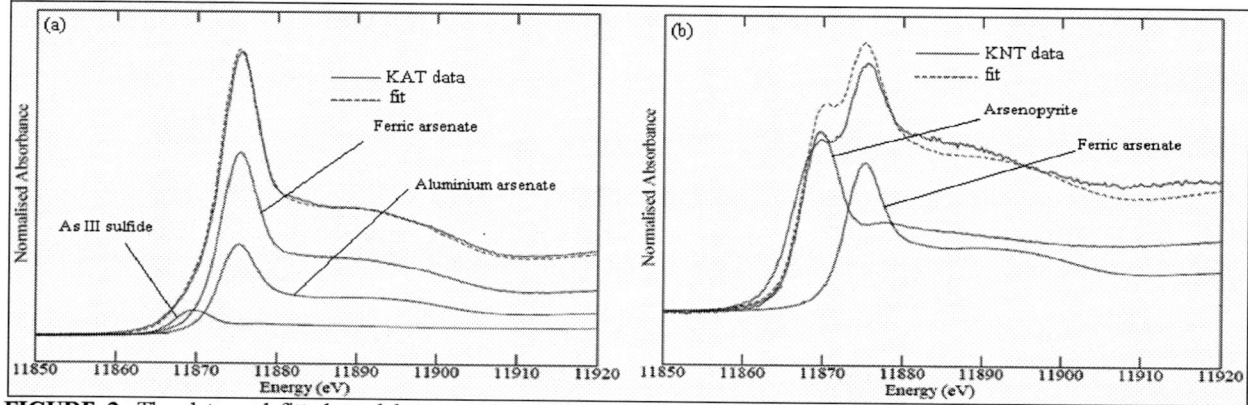

FIGURE 2. The data and fitted model spectra (normalized vs % composition) for two types of tailings collected from the Kidston Gold Mine: (a) KAT - acid tailings, mainly As V arsenates; and (b) KNT – neutral tailings showing the presence of As(-I) and As(III) in the neutral tailings together with ferric arsenate. Table 2 gives the ratio of arsenic species present in these two samples.

ACKNOWLEDGMENTS

This work was performed at the ANBF with support from the Australian Synchrotron Research Program, which is funded by the Commonwealth of Australia under the Major National Research Facilities Program. EnTox is funded by Queensland Health, the University of Queensland, Griffith University and Queensland University of Technology. V. P. Matanitobua received financial support from the Ministry of Fijian Affairs for PhD studies. Thanks to Garry Foran at ANBF for assistance.

REFERENCES

1. Chunguo, C. and Zihui, L., Chemical speciation and distribution of arsenic in water, suspended solids and sediments of Xiangjiang River, China. The Science of the Total Environment, 1988. 77: p. 69-82.
2. Bertsch, P.M. and Hunter, D.B., Elucidating fundamental mechanisms in soil and environmental chemistry: The role of advanced analytical spectroscopic and microscopic methods. Soil Sc. Soc. Am. Spec. Publ., 1998. 55: p. 103-122.
3. Cutler, J.N., Jiang, D.T., and Remple, G., Chemical Speciation of Arsenic in Uranium Mine Tailings by X-ray Absorption Spectroscopy. Can. J. Anal. Sci. Spectrosc., 2001. 46(4): p. 130-135.
4. Noller, B.N., Parry, D., and Eapaea, M.P., Arsenic dispersion and retention at mine sites in the Northern Territory Australia. in 8th International Conference on the Biogeochemistry of Trace Elements (ICOBTE). 2005. Adelaide, Australia. pp 668-669.
5. George, G.N. and Pickering, I.J., EXAFSPAK: A Suite of Computer Programs for Analysis of X-ray Absorption Spectra. 2000, Stanford Synchrotron Radiation Laboratory: Stanford, CA.

Experimental Study of Polarization Clusters in 0.72Pb(Mg$_{1/3}$Nb$_{2/3}$)O$_3$-0.28PbTiO$_3$ Relaxor Ferroelectrics by means of Synchrotron Radiation X-ray Diffraction

Zhi Guo[1], Renzhong Tai[1*], Hongjie Xu[1], Chen Gao[2], Haosu Luo[3], Guoqiang Pan[2], Chuansheng Hu[2], Di Lin[3], Rong Fan[2], Ruipeng Li[2], and Kazumichi Namikawa[4]

[1]*Shanghai Synchrotron Radiation Facility, Shanghai Institute of Applied Physics, Chinese Academy of Sciences, P.O. Box 800-204, Shanghai 201800 China.*
[2]*National Synchrotron Radiation Laboratory, University of Sciences and Technology of China, Hefei 230029 China.*
[3]*Shanghai Institute of Ceramics, Chinese Academy of Sciences, Dingxi road 1295, Shanghai 200050 China.*
[4]*Department of Physics, Tokyo Gakugei University, 4-1-1 Nukui-Kita Machi, Kogannei-Shi, Tokyo 184-8501 Japan.*

Abstract. X-ray diffraction has been conducted to study the microscopic-scale structures for 0.72Pb(Mg1/3Nb2/3)O3-0.28PbTiO3 relaxor ferroelectrics in a high external DC field during phase transition. Clear quasi-periodic structures were observed along <111> and <1-11> directions near Tc induced by the high external DC field. The formation of these periodic structures are interpreted as a type of Coulomb interaction among adjacent polar clusters. The cluster size was estimated to be 17nm. It was also found that the dominating interaction direction among clusters (periodic direction) were changeable among <111> and <1-11> as temperature changed.

Keywords: polarization clusters; relaxor ferroelectric; x-ray diffraction
PACS: 41.50.+h, 77.80.Bh, 77.84.-S, 77.84.Dy

INTRODUCTION

Relaxor ferroelectrics have been extensively studied for more than 40 years since Pb(Mg$_{1/3}$Nb$_{2/3}$)O$_3$(PMN) was first synthesized by Smolenskii and Agranovska [1-5]. The typical relaxor ferroelectrics of perovskites lead-based oxides (Pb(Mg$_{1/3}$Nb$_{2/3}$)O$_3$-PbTiO$_3$(PMN-PT) exhibits ultrahigh piezoelectric(with d_{33} over 1500pC/N)and electromechanical responses (with k_{33} over 90%). They have been commercialized and have many potential applications in medical imaging, telecommunication and ultrasonic devices, cyberetics, artificial intelligence and neutral networks[2]. The observed properties of relaxor are strongly affected by the reorientation of polar clusters[5].

Some theoretical models for polar clusters have been proposed [5, 6]. Such as the theory of the reorientation of the clusters proposed by B. E. Vugmeister [5] and spherical random-bond-random-field (SRBRF) model proposed by R. Pirc [6], Recently some experiments had been conducted to characterize these clusters [7-9]. P. M. gehring used the neutron inelastic scattering measurements to study the nanometer-sized polar regions, which was speculated to be 3.1nm [7]. Hyun M Jang studied the same question using the Raman scattering measurement, and the size of the polar clusters of PbTiO$_3$-based relaxor ferroelectrics in tetragonal symmetry was estimated to be 1.2nm [8]. Another similar experiment was also conducted with Raman scattering method, where the size of polar region was estimated to be 1.6~2.0nm, and the size was independent of temperature [9]. A. A. Bokov also proposed the size of polar clusters to be about 10nm [10].

However, to our best knowledge, there is no direct observation reported on the behavior of polarization clusters in relaxors to date. In this paper, we attempted to observe the nano-sized polarization clusters in relaxors by means of small-angle x-ray diffraction using a partially coherent synchrotron radiation x-ray beam. As a preliminary result, obvious quasi-periodic structures are found to exist from Curie temperature T_c to several hundreds of degrees above

---

*Corresponding author: tairenzhong@sinap.ac.cn

CP879, *Synchrotron Radiation Instrumentation: Ninth International Conference,*
edited by Jae-Young Choi and Seungyu Rah
© 2007 American Institute of Physics 978-0-7354-0373-4/07/$23.00

T_c. These appearances of the structures manifested a type of long-range coherence of polar clusters. This experiment provides a more straight-forward way to observe the polarization clusters within relaxors just as that performed on prototype ferroelectrics [11, 12].

EXPERIMENTS AND RESULTS

The 0.72Pb(Mg$_{1/3}$Nb$_{2/3}$)O$_3$-0.28PbTiO$_3$ (0.72PMN-0.28PT) single crystal was grown from its melt by using the Bridgman method [13]. Its Curie temperature T_c is 401K, the piezoelectric factor is 1200 pC/N, and the electromechanical coupling factor is 76%. The sample was in (001) cuts, with dimensions of 5×5×0.06 mm^3. Surfaces of the specimen were polished to diminish the diffuse scattering of surface. For PMN single crystal, there is no phase transition from relaxor to ferroelectric phase and similar to glassy behaviour, transformation from relaxor state to ferroelectric state is induced by PT-content(x) [14].

Experiments were carried out on the Diffraction and Scattering station at National Synchrotron radiation laboratory (NSRL), as shown in Fig. 1. The light source is a superconducting wiggler. A photon energy of 9.6 keV (0.129nm) was chosen by a double crystal monochromator. The maximum photon flux is about 1.3×10^9 photons/s of this source at energy resolution of 5×10^{-4}. A 400(H)×100(V)μm^2 slit was placed in the front of the sample to improve the spatial coherence of the incoming beam. The photon flux incident on the sample was estimated to be 1×10^5 photons/s. The source-to-slit, slit-to-sample and sample-to-detector distances are 16m, 100 mm and 1050 mm, respectively. The pixel size of the CCD camera is 70μm horizontally and 100μm vertically and with pixel number 2048×2048. A lead beamstop was used to suppress the direct transmission beam (central peak) and to improve the signal to noise ratio. Transmittance of this thin sample is about 0.4%. The exposing time was 5400s. The partially coherent x-ray beam transmitted the sample in a normal way. Temperature of the sample could be controlled with a precision of ±1K from room temperature (291K) to 473K, and a high DC bias field (2.3kV/cm) was applied along <100> direction, parallel to the x crystallographic axis.

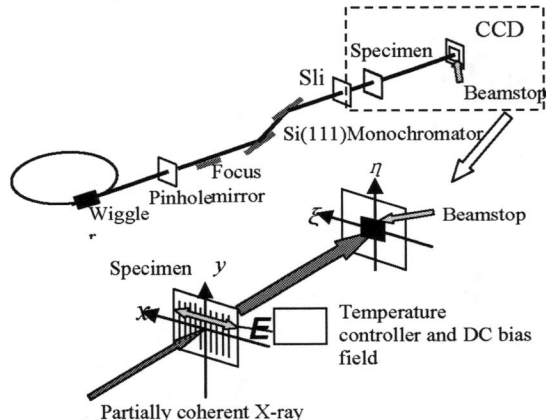

FIGURE 1. Schematic diagram of the experimental setup

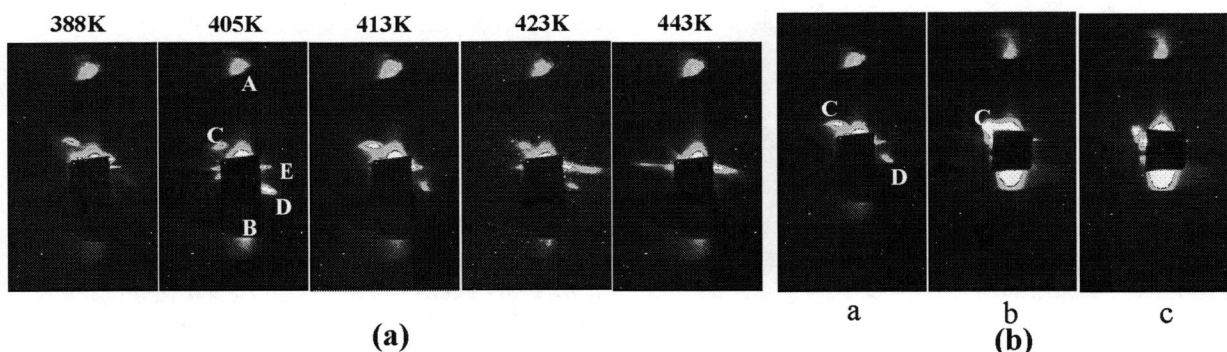

(a)

a   b   c

(b)

FIGURE 2. (a)Diffraction patterns measured at 388K, 405K, 413K, 423K, 443K and 413K with the high DC bias field. (b) Diffraction patterns when the DC bias field was removed. a: 1 hour, b: 2 days, c: 2 days, after the removal of the high DC bias field at 413K, 405K ,423K respectively.

The data are shown in Fig. 2(a). Clear differences of those diffraction patterns were observed during phase transition on a heating process. The joint line between C and D forms an angle of 45° from x crystallographic axis, which is ascribed to the diffraction of a structure along <111> direction. The C and D are relatively weak below T_c (401K), and reached the maximum at 413K, then decrease gradually with the increasing of temperature, finally vanish completely at 443K. The E along x crystallographic axis is diffuse-like, implying another kind of structure along <100> direction. Since there is no clear temperature dependence for those spots, A and B, they are believed to originate from some other trivial reasons independent on polar clusters. Experiment also shows that after the removal of DC bias field, this symmetric feature gradually disappear, even though it could be observed within 2-3 hours, as shown in Fig. 2(b).

ANALYSIS

The scattering patterns along <100> and <111> directions as shown in Fig. 2 must correspond to some particular types of structures which appears accompanying the phase transition. Defining the scattering vector $\mathbf{q}$ as $\mathbf{q}=k\xi/z\ \mathbf{e}_\xi$, where ξ is the coordinate on CCD camera along <100> direction with a unit vector $\mathbf{e}_\xi$, z is the distance between the sample and CCD camera. The ±1st-order spots, that is C and D, appeared at $q=\pm2\pi/d$. The distance between the symmetric spots C and D $(2|\xi|)$ is 95 pixels (7.5 mm), so the period length is about 36.1nm along <111> direction. The spots along <100> direction are analogous to a diffuse scattering, implying that there is no strict period distribution along <100> as that along <111> direction. A strong fluctuation for this period exists along this direction.

There exists a large abundance of polarization clusters in relaxor ferroelectrics [3, 5, 6]. These clusters should exhibit property of clear temperature dependence, since they emerge from Burn temperature and grow gradually into ferroelectric domain at T_c [15]. Therefore, it is reasonable to think that the diffraction patterns observed in our experiment arise from those polarization clusters.

A simple and reasonable model to interpret the experiment is to describe the structure of clusters as a phase grating. Those clusters exhibit a quasi-periodic distribution along <100> and <111> respectively. The cluster size can be estimated from the intensity distribution. Supposing that the periods number along <100> direction is N, the length of each period is d and the scattering size is a, the intensity of scattering pattern can be described according to the Frauhofer diffraction theory [16]:

$$I(\xi,\eta)=\left(\frac{A_\sigma}{\lambda z}\right)^2\left[a^2 sinc^2\,(af_x)\right]\frac{sin^2\left(\pi Nf_x d\right)}{sin^2\left(\pi f_x d\right)}\Bigg|_{f_x=\frac{\xi}{\lambda z}} \qquad (1)$$

where, A_σ is the amplitude of the incident x-ray, $sinc^2(af_x)$ is the diffraction of a single scattering element (cluster), while $sin^2(\pi Nf_x d)/sin^2(\pi f_x d)$ denotes the interference of x-rays diffracted by the multi-elements. The intensity of the zero-order diffraction (cut by the beam-stopper) was estimated to be about 1.7×10^6 counts in a same exposing time, 5400 seconds. The spot C was 330 counts at 140℃. The size of cluster could be deduced from the relationship of $I(\pi dz,0)/I(0,0) = sinc^2(\pm a/d)$. Substituting the clusters' period, 36.1 nm estimated on the above, the cluster's size is 17nm along <111> direction.

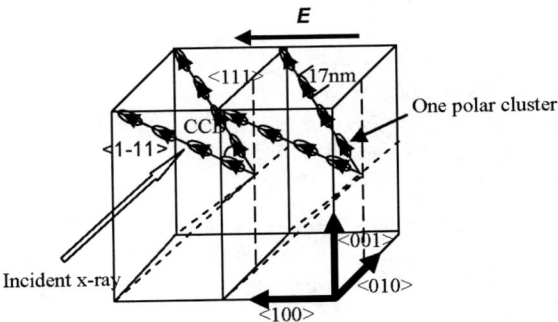

FIGURE 3. Illustration for the periodic structure of polar clusters within the sample when a DC bias field is applied along <100>. Both <111> and <1-11> will contribute to the diffraction pattern. <111> component would enhance the intensity C, while <1-11> will enhance the intensity of D.

DISCUSSIONS

Park [17] and Yin [18] showed that a <001> external-field-poled relaxor had four possible degenerated polar directions: <111>, <-111>, <1-11> and <-1-11>. As shown in the Fig. 3, the periodic structures along <111> and <1-11> for the clusters would contribute to the diffraction patterns shown in Fig. 2. The periodic structure should originate from the strong Coulomb interaction among clusters with their polar axis along same direction.

The dissymmetry of intensity for the spots C and D also give information about which direction (<111> or <1-11>) dominates at a certain temperature, since the angle formed by the incident x-ray and the polar axis of ach cluster are complementary, as shown in Fig. 3. If we consider that the periodic structure along <1-11> direction dominate, then the intensity of spot C is weaker than that of spot D because this angle disfavour photon scattering to C, and this is the case of the diffraction pattern observed at 405K; On the contrary, a periodic structure along <111> direction would result in a strong intensity for C, and this is the case as observed at 413K; If these periodic structures were distributed along either <111> or <1-11> with equal possibility, then no significant intensity difference would be observed for C and D, as the pattern observed at 423K. As a summary, when a DC bias field is applied along <100> direction during phase transition near T_c, the cluster's orientation and their interaction direction that dominate at a certain temperature is changeable on a heating process. This feature could contribute in some way to the ultrahigh electromechanical coupling and strain under DC bias field.

In Fig. 2(b), when the DC bias field was removed, the original symmetric diffraction pattern as shown in Fig. 2(b)a degenerates, only spot C exists. The reason might be that this interacting direction still survive among polar clusters (periodic structure) along <111> direction, while another direction along <1-11> disappears, which favours the scattering of x-ray to spot **C** in Fig. 2(b). The different shape around the central transmission spots between figures in Fig. 2(b) is probably due to the little change of illuminated position on the sample.

CONCLUSIONS

X-ray diffraction experiment has been conducted to study the microscopic-scale structures for 0.72PMN-0.28PT relaxor ferroelectrics in a high external DC field during phase transition. Clear quasi-periodic structures were observed along <111> and <1-11> directions near T_c induced by the high external DC field. The formation of these periodic structures are interpreted as a type of Coulomb interaction among adjacent polar clusters.The cluster size was estimated to be 17nm. It was also found that the dominating interaction direction among clusters (periodic direction) were changeable among <111> and <1-11> as temperature changed.

ACKNOWLEDGEMENT

One of the authors (Chen Gao) would like to thank the support from NSFC (50421201).

REFERENCES

1. G. A. Smolenskii and A L. Agranovska, *Sov. Phys. Sol. State* 1 (1959) 1429
2. Vinod K. and Wadhawan. *Material Science and Engineering B.*, **120**, 199 (2005).
3. V.S.Tiwari, G. Singh, V.K. Wadhawan. *Solid State Commuciation.*, **121**, 39-43 (2002).
4. B. Noheda. D. E. Cox, G. Shirane. *Phys Rev Lett.*, **86**, 3891 (2001).
5. B.E.Vugmeister. H. Rabitz, *Phys Rev B.*, **57**, 7581-7584 (1998).
6. R.Pirc. R. Blinc, *Phys. Rev. B.*, **60**, 13470-13478 (1999).
7. P.M. Gehring, S.-E. Park, and G. Shirane. *Phys Rev Lett.*, **84**, 5216-5219 (2000).
8. Hyun M.Jang, Tae-Yong Kim, *Solid State Commu.*, **127**, 645-648 (2003).
9. E.A.Rogache, *Physica B.*, **291**, 359-361 (2000).
10. A. A. Bokov, *Ferroelectrics.*, **190**, 197 (1997)
11. R.Z.Tai.K.Namikawa, *Phys Rev Lett.* **89**, 257602-1.-157602-4 (2002).
12. R. Z. Tai. K. Namikawa. M. Kishimoto, *Phys. Rev. Lett.*, **93**, 087601-1-087601-4 (2004).
13. Guisheng Xu, Haosu Luo, Haiqing Xu and Zhiwen Yin, *Phys. Rev. B.*, **64**, 020102-1-020102-3 (2001)
14. A. A. Bokov and Z.-G. Ye, *Phys. Rev. B.*, 66, 064103 (2002).
15. G. Burns and F. H. Dagol, Ferroelectrics, **104**, 25(1990)
16. Born, M. and Wolf, E., *Principle of Optic*s, Pergamon, Oxford, 1965, pp.405.
17. Seung-Eek park. Thomas R. Shrout, *J. Appl. Phys.*, **82**, 1804-1811 (1997).
18. Jianhua Yin and Wenwu Cao, *J. Appl. Phys.*, **87**, 7438-7441 (2000).

The Gas Pressure-dependent Contact Angle of Diamond-like Carbon Surface by SR Exposure Under Perfluorohexane(C$_6$F$_{14}$) Gas Atmosphere

Noriko Yamada, Yuri Kato, Kazuhiro Kanda, Yuichi Haruyama
and Shinji Matsui

Graduate School of Science, Lasti, Univercity of Hyogo
3-2-1Koto, Kamigori, Ako, Hyogo, 678-1205, Japan

Abstract. The contact angle on a diamond-like carbon (DLC) surface was controlled by synchrotron radiation (SR) under perfluorohexane (C$_6$F$_{14}$) gas atmosphere. It was found that the fluorocarbon group bonded to the DLC surface in the modified hydrophobic area from the measurement of X-ray photoelectron spectrum. The contact angle of a DLC surface was succeeded to increase from 73° to 116° by the SR irradiation with 400 mA·h at C$_6$F$_{14}$ gas pressure higher than 1.2 Pa.

Keywords: synchrotron radiation process, gas pressure-dependence, surface wettability, fluorine coating diamond-like carbon, X-ray photoelectron spectroscopy
PACS: 81.15-z

INTRODUCTION

Diamond-like carbon (DLC) has attractive properties, such as high hardness, low friction coefficient and biocompatibility as industrial materials [1,2]. Furthermore, chemical modification of a DLC surface has been explored by several methods in order to expand application field of a DLC [3-6]. Recently, the coating of F-containing DLC on the artificial cardiac and vascular products are interest for the medical field because of its high hydrophobicity and antithrombogencity [7,8]. On the other hand, the application of soft X-ray (SR) process has increased in utility in the surface modification, because soft X-ray irradiation is a low temperature process and causes an inner shell excitation, which it is expectable to a selective large reaction cross-section [9-11]. In the previous study, our group attempted to modify the DLC surface by SR exposure under C$_6$F$_{14}$ gas atmosphere and succeeded to increase the surface wettability of a DLC film [12]. In the present study, we discussed on the reaction process on the DLC surface by SR exposure under C$_6$F$_{14}$ gas atmosphere from X-ray photoelectron spectroscopy (XPS) analysis and measured C$_6$F$_{14}$ gas pressure-dependence of the contact angle of a DLC surface.

EXPERIMENT

The SR exposure of the DLC film was carried out at BL6 of NewSUBARU. The details of the experimental apparatus at BL6 are described in ref [11]. The electron energy of the NewSUBARU ring was 1.0 GeV during this experiment. The white radiation, which has the photon energy range of 50 – 1000 eV, is irradiated to the BL6 sample stage [13]. The SR dose is represented by the integrated value of ring current and exposure time. Three differential pumping systems were mounted at BL6. The base pressure in the irradiation chamber of BL6 was 2×10^{-6} Pa. For the purpose of irradiation experiment under the gas atmosphere, the sample stage was surrounded by a 105 mmϕ × 212 mm gas cell. SR was introduced to a sample stage through the hole window of 5 mm in diameter at the side wall of the gas cell. DLC films with the thickness of ≈500 nm were formed on silicon wafer by ion plating

CP879, *Synchrotron Radiation Instrumentation: Ninth International Conference*,
edited by Jae-Young Choi and Seungyu Rah
© 2007 American Institute of Physics 978-0-7354-0373-4/07/$23.00

(NANOTEC) and were used as a sample after cleaning with acetone. For the surface modification, C_6F_{14} (Wako Pure Chemical Industrials, purity 99%) gas was introduced into the gas cell. The SR irradiation to the DLC film was performed at C_6F_{14} gas pressure of 0.3, 0.6, 1.2 and 2.4 Pa. The sample substrate was not heated during SR exposure. After exposure to SR, modified DLC sheets were kept in air atmosphere. Wettability of the DLC was evaluated from a contact angle of a water drop against the DLC film surface. Contact angle was measured with a contact angle meter (Kyowa Interfac, Drop Master 500) with a water drop of 1 μl. XPS spectra were measured using the conventional photoelectron spectroscopy apparatus, which was mounted with a CL150 (VSW) hemispherical electron energy analyzer. The Mg Kα line ($h\nu$=1253.6 eV), used as X-ray source, was incident at 45° with respect to the emission angle of 45° to the surface normal. The base pressure in the photoelectron analysis chamber was 2×10^{-8} Pa.

RESULTS AMD DISCUSSION

In the previous work, we reported the DLC surface was modified by SR exposure under C_6F_{14} gas pressure and confirmed that SR was necessary for this modification [12]. Figure 1 shows the photograph of the DLC surface after SR irradiation of 400 mA·h under C_6F_{14} gas atmosphere of 1.2 Pa. The surface color was changed in the circle at the center of the sample, where SR was irradiated directly through a 5 mmφ hole of the gas cell. Contact angle increased with SR dose at the irradiated area and not-irradiated area, where was the surrounding area of the irradiated circle. The presence of F atom was confirmed by the measurement with energy dispersive X-ray in the both areas. From these results, the increase of the contact angle was concluded to be ascribable to the fluorine groups on the modified DLC surface introduced by SR exposure under C_6F_{14} gas atmosphere. On the other hand, the contact angle of the irradiated area was found to be 10° smaller than that of the not-irradiated area. This decrease of contact angle at the irradiated area was considered to occur by the secondary reaction of deposited fluorocarbon materials with incident SR. The contact angle of a fluorocarbon polymer was known to decrease by the deconstruction of CF_2 structure due to SR irradiation in the soft X-ray region [9].

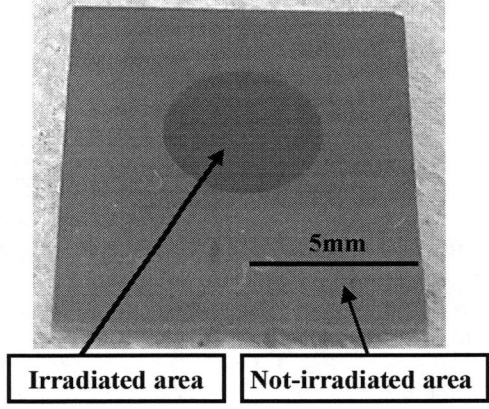

FIGURE 1. Image of the DLC film after SR exposure of 400 mA·h dose under C_6F_{14} gas atmosphere of 1.2 Pa [12].

The XPS Analysis of DLC Surface

The chemical composition of the DLC surface was measured by XPS. Figure 2 shows the XPS spectra of wide scan (a) and the carbon 1s region (b) of the DLC surface exposed to SR under C_6F_{14} gas atmosphere at the irradiated area and the not-irradiated area with the spectrum of the DLC film before SR irradiation for comparison. In the spectrum of the DLC film before irradiation of Fig. 2 (a), the C1s and C Auger peaks were observed at 286 and 1005 eV, respectively. The peaks observed at 533 and 748 eV corresponded to the O1s and O Auger electrons, which were emitted from oxygen atom, adsorbed on the DLC surface. In the spectrum of the DLC surface in the not-irradiated area, the peak intensities originated from carbon and oxygen decreased drastically and the novel peaks, which were assignable to F1s and F Auger, were appeared at 693 and 606 eV, respectively. In the spectrum of the DLC surface in the irradiated area, the peak intensities due to oxygen increased by the comparison with that of the not-irradiated area. In the carbon 1s region spectrum of the DLC before irradiation, the peak corresponding to the C-C component was merely observed. O-containing compounds were estimated to absorb dominantly on the DLC surface physically, because peaks corresponding to the C-O and C=O components were not observed in the C1s XPS spectrum in spite of a great amount of O atom in Fig. 2 (a). In the spectra of the DLC surface in the irradiated area and not-irradiated area, the CF_3, CF_2 and CF peaks were confirmed and the peak at ≈ 287 eV consisted of the C=O, C-CF and C-O components. At the not-irradiated area, the chemical component on the surface consisted mainly of CF_2. The large amount of the CF_2 component on the modified DLC surface was ascribable to that C_6F_{14} molecule consists of a certain mount of the CF_2 component. At the irradiated area, amount of the CF_2 component decreased and amounts of the C=O, C-CF and C-O components increased by compared to the not-irradiated area. This alternation can be interpreted that the CF_2 component on the DLC surface was deconstructed by SR and coupled subsequently with residual O atom in the vacuum and/or O_2 molecule in the air. This secondary process was reported on the surface modification of a PTFE film by SR exposure and the O-containing component on the surface was known to lead to decrease contact angle [9]. Therefore, it was considered that the O-containing component on the DLC surface leaded to decrease the contact angle at the irradiated area. As a result, a hydrophobic surface formed at the not-irradiated area was ascribable to the fluorine group, which was introduced to the DLC surface by the dissociation of C_6F_{14} molecules in the gas phase by SR and subsequent deposition of fragments of C_6F_{14} onto the DLC surface. The relative intensities of each component estimated from XPS spectra are summarized in Table 1.

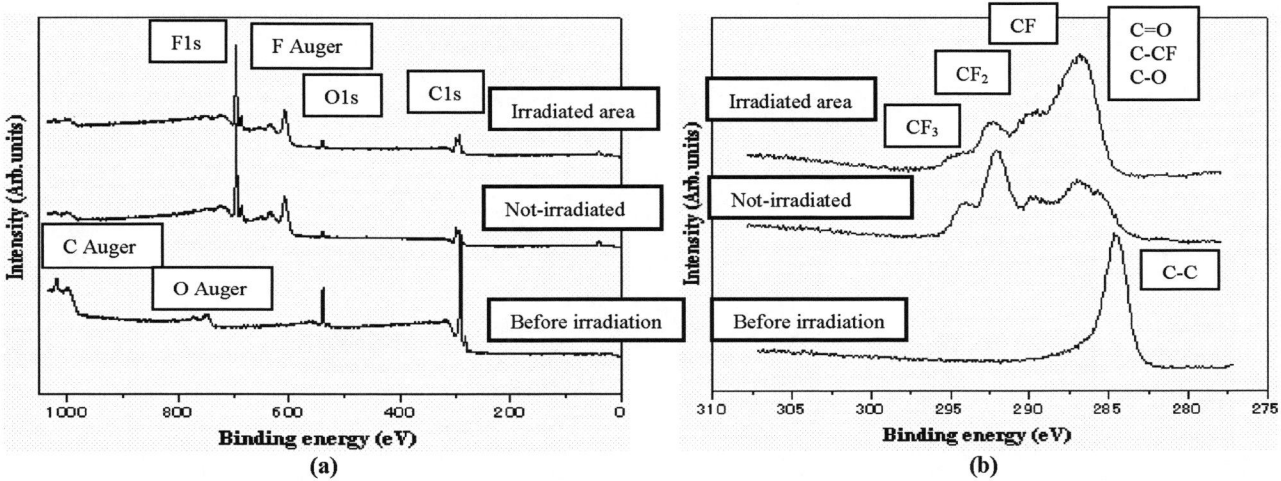

FIGURE 2. The XPS spectra of wide scan (a) and the carbon 1s region (b) of the DLC surface irradiated by SR under C_6F_{14} gas atmosphere at the irradiated area, not-irradiated area and before irradiation.

TABLE 1. Relative intensities of each component estimated from XPS spectra of the wide scan (1) and the carbon 1s region (2) of the DLC surface at the area, not-irradiated area and before irradiation.

(1)   (%)

Wide scan	C Auger	O Auger	F1s	F Auger	O1s	C1s
Irradiated area	37.4	—	35.4	19.3	2.4	5.5
Not-irradiated area	31.8	—	38.4	26.3	1.2	2.3
Before irradiation	76.9	5.2	—	—	5.9	12.0

(2)

C1s region	CF_3	CF_2	CF	C=O,C-CF	C-C
Irradiated area	11.0	4.1	42.2	42.6	—
Not-irradiated area	6.5	40.8	14.3	29.1	9.2
Before irradiation	—	—	—	—	100

The Gas Pressure-dependent Contact Angle of DLC Surface

For the deeper insight into the surface modification process on the DLC film using SR under C_6F_{14} gas atmosphere, the contact angle was measured at various C_6F_{14} gas pressures. As the contact angle at the irradiated area was found to decrease by the secondary reaction as described above, the SR dose dependence of contact angle at only the not-irradiated area of the DLC surface was measured under various C_6F_{14} gas pressures. Measured contact angle was plotted against SR dose at each C_6F_{14} gas pressure in Fig. 3. The closed circle represents the contact angle of the DLC surface before irradiation, 73°. The cross, open square, open triangle and open circle show the contact angle of the DLC surface irradiated by SR under C_6F_{14} gas atmosphere at 0.3, 0.6, 1.2 and 2.4 Pa, respectively. DLC films were irradiated with 1, 100, 250 and 400 mA·h SR dose at each C_6F_{14} gas pressure. Contact angle of the DLC surface was found to increase with SR dose and converge to the certain angle in the high SR dose region at each pressure. The contact angle increased with C_6F_{14} gas pressure in the region from 0.3 to 1.2 Pa, however it did not change in the pressure region > 1.2 Pa. These indicated that sufficient C_6F_{14} molecules for the DLC surface modification were provided at C_6F_{14} gas pressure of 1.2 Pa in the present experimental condition. The contact angle took a maximum value of 116° at 400 mA·h SR dose at C_6F_{14} gas pressure of 1.2 and 2.4 Pa.

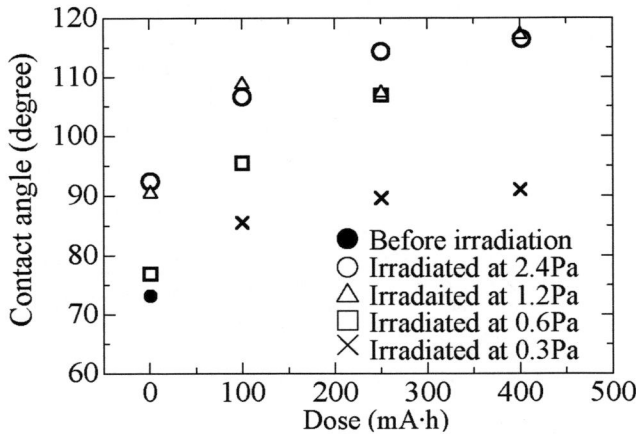

FIGURE 3. The dose dependence of the contact angle of the DLC surface by SR exposure under various C_6F_{14} gas pressure at the not-irradiated area.

CONCLUSIONS

The surface modification process of DLC films by SR exposure under C_6F_{14} gas atmosphere was investigated. It was confirmed that the hydrophobic surface in the not-irradiated area of the DLC surface was formed by the fluorocarbon group on the DLC surface due to the dissociation of C_6F_{14} molecule in the gas phase, and that the decrease of contact angle in the irradiated area occurred by the coupling of oxygen to the fluorocarbon surface due to the secondary reaction. Increasing rate of contact angle increased with C_6F_{14} gas pressure in the region of 0 – 1.2 Pa, however, it did not depend on C_6F_{14} gas pressure in the region higher than 1.2 Pa.

REFERENCES

1. A. Grill, Diamond Relat. Mater. 8 (1999) 428-434.
2. L. Valentini, E. Braca, J, M. Kenny, L. Lozzi and S. Santucci, J. Vac. Sci. Thechnol. A19 (2001) 2168-2173.
3. L. G. Jacobsohn, D. F. Franhceschini, M. E. H. Maia da Costa and F. L. Freire, Jr, J. Vac. Sci. Technol. A18 (2000) 2230-2238.
4. T. Watanabe, K. Yamamoto, Y. Koga and A. Tanaka, Jpn. J. Appl. Phys. 40 (2001) 4684-4690.
5. T. Nakamura, T. Ohana, M. Suzuki, M. Ishihara, A. Tanaka and Y. Koga, Surf. Sci. 580 (2005) 101-106.
6. M. K. Fung, K. H. Lai, C. Y. Chan, N. B. Wong, I. Bello, C. S. Lee and S. T. Lee, Diamond Relat. Mater. 9 (2000) 815-818.
7. T. Saito, T. Hasebe, S, Yohena, Y. Matsuoka, A. Kamijyo, K. Takahashi and T. Suzuki, Diamond Relat. Mater. 14 (2005) 1116-1119.
8. T. Hasebe, Y. Matsuoka, H. Komada, T. Saito, S. Yohena, A. Kamijyo, N. Shiraga, M. Higuchi, S. Kuribayashi, K. Takahashi and T. Suzuki, Diamond Relat. Mater. 15 (2006) 129-132.
9. K. Kanda, T. Ideta, Y. Haruyama, H. Ishigaki and S. Matsui, Jpn. J. Appl. Phys. 42 (2003) 3983-3985.
10. Y. Kato, K. Kanda, Y. Haruyama and S. Matsui, Jpn. J. Appl. Phys. 43 (2004) 3938-3940.
11. Y. Kato, K. Kanda, Y. Haruyama and S. Matsui, J. Elect. Spectrosc. Rel. Phenom. 144-147 (2005) 413-415.
12. N. Yamada, Y. Kato, K. Kanda, Y. Haruyama and S. Matsui, Jpn. J. Appl. Phys. in press.
13. J. Taniguchi, K. Kanda, Y. Haruyama, S. Matsui, M. Tkunaga and I. Miyamoto, Jpn. J. Appl. Phys. 41 (2002) 4304-4306.

Structure Refinement Based on Inverse Fourier Analysis in X-Ray Fluorescence Holography

K. Hayashi

Institute of Materials Research, Tohoku University, Sendai 980-8577, Japan

Abstract. A new reconstruction technique for X-ray fluorescence hologram data was proposed based on extractions of holographic oscillations from single scatterers within a sample. The extractions were iteratively carried out by the inverse Fourier transformation of selected atomic images, which were obtained by the Fourier transformation of one-dimensional hologram averaged over azimuth about a given polar axis in **k**-space. The refinement of the real space reconstruction was performed using the measured holograms and the extracted holographic oscillations. I applied this data processing to the theoretical holograms of fcc Au cluster at 12.0, 12.5 and 13.0 keV, and successfully obtained clear atomic image without artifacts.

Keywords: X-ray holography, Single crystal; Imaging; Local structure; X-ray fluorescence
PACS: 41.40.-i; 61.10.-i; 61.50.-f

INTRODUCTION

X-ray fluorescence holography (XFH) provides 3D atomic image around a specified element. Thus, this technique has attracted much attention as a tool for evaluating a local structure around dopant in single crystal. XFH has two main experimental setups, one in the normal mode and one in inverse mode. The inverse XFH applied here uses fluorescing atoms within the sample to detect the interference between the incident beam reaching such atoms either directly or after being scattered by nearby atoms.[1] Since the XFH can determine the 3D atomic arrangement around a specified element, it has been expected as a promising tool for the determination of local structures such as the extended X-ray absorption fine structure (EXAFS). Although a number of applications have been reported such as the structural analysis of the environment around dopants[2], quasicrystal[3] and thin films to date, the number of reported applications has been limited. One of reasons for this is that no novel data analysis technique has been proposed since the development of Barton reconstruction algorithm.[4] If we could determine atomic positions within an accuracy of ±0.01 Å, local lattice distortion around dopants could be evaluated quantitatively and the mechanisms responsible for material properties would be elucidated.

I proposed the inverse Fourier analysis for the X-ray fluorescence hologram data, which uses the inverse Fourier transformation of a selected atomic image like the EXAFS analysis.[5] Using this method, the interatomic distances of neighboring atoms around an emitter were estimated from 16 experimental holograms of a Au single crystal and most of them were in good agreement with the actual values within error of 0.3%. However, this technique needs over 8 holograms at different X-ray energies. Therefore, I advanced this technique to the new reconstruction algorithm, which is a combination of the iterative extraction of holographic oscillations from single scatterers within a sample using the inverse Fourier transformation and a real space reconstruction from the hologram data with the extracted holograms. In the present study, I described the concept of this new data processing and application to theoretical holograms of Au cluster.

CONCEPT OF REAL SPACE RECONSTRUCTION

Let's consider some basic imaging concepts for XFH using calculated holograms at 12.0, 12.5 and 13.0 keV for 16726 fcc Au cluster, whose radius is about 40 Å. The pattern at 12.0 keV is depicted in Fig. 1(a). The calculations

CP879, *Synchrotron Radiation Instrumentation: Ninth International Conference,*
edited by Jae-Young Choi and Seungyu Rah
© 2007 American Institute of Physics 978-0-7354-0373-4/07/$23.00

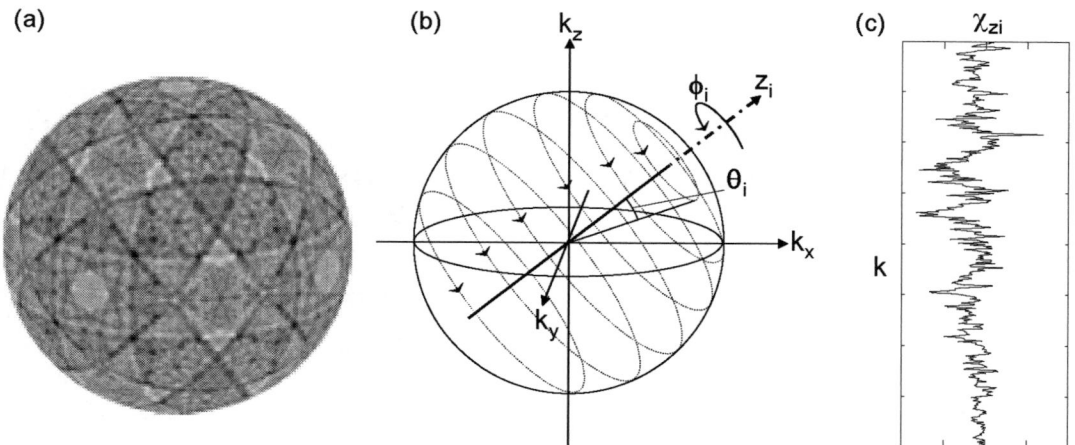

FIGURE 1. Illustrations of procedure for averaging hologram one-dimensionally. (a) X-ray hologram $\chi_n(\mathbf{k})$. (b) The average process. (c) Projection on z_i axis of the hologram after averaging and summing.

were carried out on the assumption of the inverse XFH experimental mode with unpolarized X-ray beams. In the data processing of the XFH, we ordinary use Barton algorithm, which is a kind of 3D Fourier transformation. A real space image at (001) plane was reconstructed from the three holograms by the Barton algorithm, as shown in Fig. 2 (a). Cross points of the grid lines indicate the original atomic positions of the Au cluster used for the calculations. The atomic image has a lot of artifacts, and true atomic images cannot be easily distinguished from false ones. In particular, it is found that the artifacts exist in the vicinity of the origin, at which the emitter atom is located. Moreover, first neighbor atoms are spilt and second neighbor atoms shift toward the origin. This problem can be resolved by adding several holograms at different X-ray energies. However, this needs long time for the experiment.

I proposed a new holographic reconstruction method by utilizing the data processing as described in Ref. 5. As shown in Fig. 1(b), the calculated hologram $\chi_n(\mathbf{k})$ was averaged over azimuth ϕ_i about a given polar axis z_i, and the averaged intensity variation is projected on to the z_i axis. This procedure is quite similar to one process in a reconstruction method proposed by Matsushita et al.[6] The averaged holograms were summed up as shown in Fig. 1(c), then the EXAFS like oscillation was obtained. These processes are represented by the following equation,

$$\overline{\chi}_{zi}(k) = \sum_n \frac{\int_S \chi_n d\phi_i}{\int_S d\phi_i} . \tag{1}$$

Here, the minimum values of k_x of these holograms are defined to be 0. The real space reconstruction along the z_i direction is obtained by a simple Fourier transformation. Before applying the Fourier transformation, I doubled the k-range by defining $\chi_{zi}(-k) = \chi_{zi}(k)$, because the phase of the holographic cosine-like curve at $k = 0$ from any scatterer is fixed at approximately $-\pi$. This increases the resolution of the peaks of the atomic images. By carrying out this procedure for any directions, atomic image can be obtained as well as Barton algorithm. The real space reconstruction at (001) plane from the calculated three Au holograms is shown in Fig. 2(b). This image seems to be sharp as compared with the image in Fig. 2(a). However, its main feature is same, such as strong artifacts in the vicinity of the origin, spilt of the first neighbor atoms and shift of the second neighbor atoms. These are also resolved by adding holograms similarly to the Barton algorithm. However, the reconstruction method presented here gives no advantage at this stage.

ITERATIVE REFINEMENT OF ATOMIC IMAGE

As described in the preceding session, the reconstruction techniques based only on Fourier transformation could not refine the atomic images from the three Au holograms. In the previous work, I found that holographic oscillation from atoms at single site can be extracted from the measured holograms using the inverse Fourier transformation from selected atomic image. I utilized the extracted oscillations for the refinement of the atomic image. Solid line in Fig. 3(a) represents the cross section along [110] direction in Fig. 2(b). Strong artifacts are seen in addition to the ½½0 and 110 atomic images. The peak at ½½0 atom position was inverse Fourier transformed, for

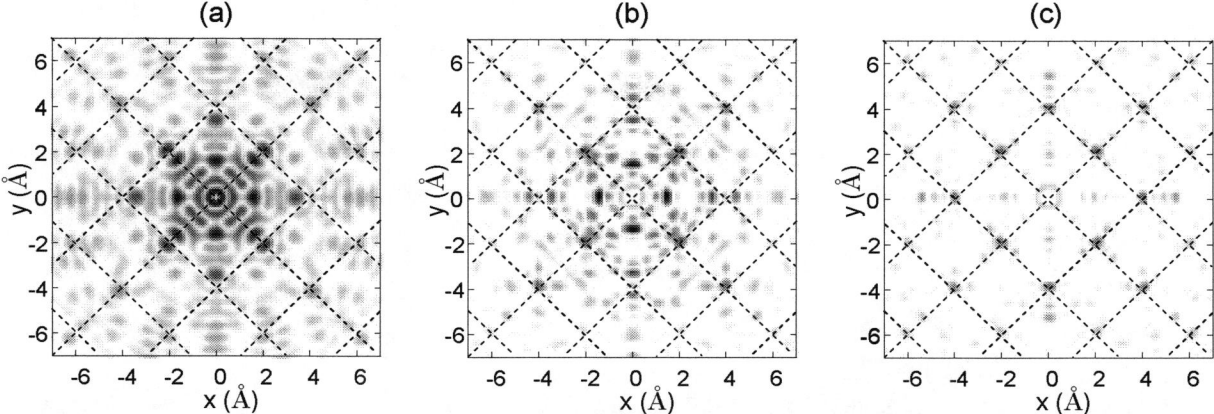

FIGURE 2. Real space reconstructions at (001) plane of Au fcc cluster from the holograms at 12.0, 12.5 and 13.0 keV. (a) Barton algorithm. (b) Fourier transform after averaging of holograms. (c) Modified method of (b). Cross points of grid lines indicate original atomic positions.

which the a R-range of the filter window was between 2 and 3.5 Å. This R-range adopted here is wide enough to reproduce original holographic oscillation if noise signals do not exist. The obtained oscillation is plotted as solid line in Fig. 3(b), which corresponds to a hologram only from the ½½0 atom. However, the present oscillation does not well coincide with true holographic oscillation, which is also plotted as dotted line in Fig. 3(b), because the ½½0 peak is heavily contaminated by the strong artifacts. To process the structure refinement, I extracted the holographic oscillation χ_{nj} from each atom up to 7th coordination shell (134 atoms), using the inverse Fourier transformation, where j means the jth atom. From the χ_{nj} and direction of z_i axis, the hologram of jth atoms in **k**-space $\chi_{nj}(\mathbf{k})$ can be calculated.[5] Furthermore, the hologram of the small cluster $\chi_{ns}(\mathbf{k})$ can be regenerated from the $\chi_{nj}(\mathbf{k})$ as

$$\chi_{ns}(\mathbf{k}) = \sum_j \chi_{nj}(\mathbf{k}). \tag{2}$$

However, difference between the χ_{ns} and theoretical hologram of 134 atoms cluster is not small, because the first extracted holograms are less accurate.

At this point, I turned to the extraction of hologram from jth atom using

$$\overline{\chi}_{zi,s,j}(k) = \sum_n \frac{\int_S (\chi_n - \chi_{ns} + \chi_{nj})d\phi_i}{\int_S d\phi_i}. \tag{3}$$

All the holographic oscillations except of that of jth atom obtained at the preceding stage were subtracted from the hologram χ_n. Using $\overline{\chi}_{zi,s,j}$, we can extract the holographic oscillation of jth atom without the effect of holograms of nearby atoms. Break lines in Figs. 3(a) and (b) show the reconstructed intensity along [110] direction obtained from the $\overline{\chi}_{zi,s,j}$ and the extracted holographic oscillation, respectively. In the reconstructed intensity, the peak of 110 atom and artifacts around ½½0 atom are diminished. The extracted hologram of ½½0 atom becomes close to the true oscillation. I again obtained each χ_{nj} up 7th coordination shell, and calculated new profile of $\overline{\chi}_{zi,s,j}$. The every step is the extractions of holograms of single scatterers using Eq. (3). I carried out five iterations of this processing in the present study.

I reconstructed the (001) plane from the finally obtained χ_{nj}. Here, instead of the $\overline{\chi}_{zi}$, $\overline{\chi}_{zi,s,j}$ was used for the reconstruction 1 Å around jth atom. The coordinate of jth atom was estimated from the direction of z_i axis and the period of cosine curve in χ_{nj}.[5] For the reconstruction of other area, $\overline{\chi}_{zi,s}$ was used, which is expressed as

$$\overline{\chi}_{zi,s}(k) = \sum_n \frac{\int_S (\chi_n - \chi_{ns})d\phi_i}{\int_S d\phi_i}. \tag{4}$$

The $\overline{\chi}_{zi,s}$ is simply a one-dimensional hologram generated from the cluster excluding central 134 atoms cluster. Fig. 3 (c) shows the atomic image resulted from the present data processing. Strong artifacts in the vicinity of the origin

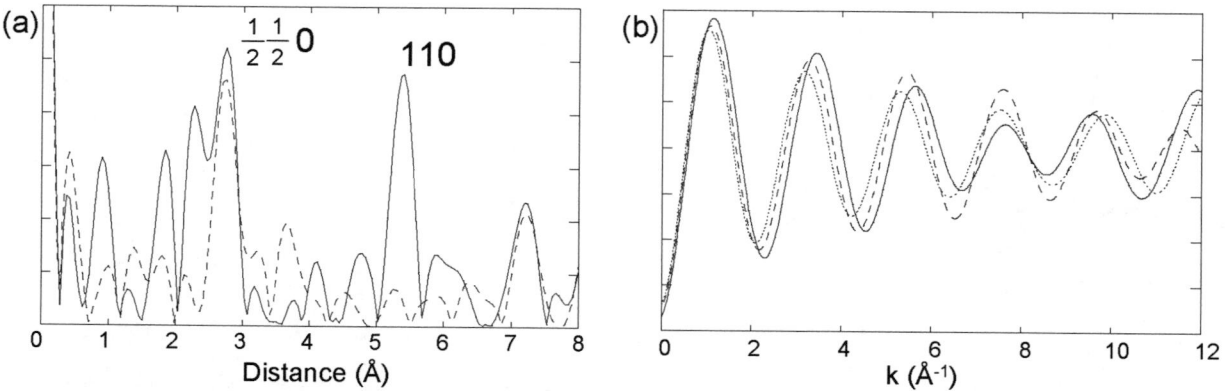

FIGURE 3. (a) Reconstructed .intensities along [110] direction and (b) extracted holographic oscillations from ½½0 atom. Solid and break lines indicate the plots obtained from $\bar{\chi}_{zi}(k)$ and $\chi_{zi,j}(k)$, respectively. Dotted line in (b) indicates true hologram of ½½0 atom.

are strongly suppressed and all atomic images appear at accurate atomic positions. This structure refinement was successfully performed at the present study. Moreover, through the present study, I understood that most artifacts come from overlaps of holograms from nearby atoms. Weak artifacts still remain along x and y axes. This would be due to interference of strong X-ray standing wave lines.

CONCLUSION

I proposed a new real space reconstruction technique based on the iterative extractions of holographic oscillations from single scatterers and the imaging by the Fourier transformation of the projection of multiple energy hologram on any polar axis in **k**-space. The image obtained from three theoretical holograms of Au cluster is clear without artifacts and atomic images appear at accurate positions. This result reveals that the present data processing greatly advances the structure refinement of the crystal structure obtained from XFH data. The present article reports the result only on the theoretical holograms. I also applied it to the experimental holograms of Au single crystal and relatively good image was obtained as compared with the image obtained by Barton algorithm. The result concerning to the experimental data will appear elsewhere. Using the XFH technique, I and my coworkers are now evaluating crystal structures of dilute magnetic semiconductor $Zn_xMn_{1-x}Te$ and rewritable DVD RAM material $Ge_2Sb_2Te_5$, which are very difficult to seek by conventional X-ray diffraction. The present data processing will help these structural analyses.

ACKNOWLEDGMENTS

This study was supported by the Industrial Technology Research Grant Program in 00' from the New Energy and Industrial Development Organization (NEDO) of Japan. Part of this work was financially supported by a Grant-in-Aid for Scientific Research (B) (18360300) and for young scientists (15686025) from the Ministry of Education, Culture, Sports, Science and Technology.

REFERENCES

1. T. Gog, P. M. Len, G. Materik, D. Bahr, C. S. Fadley and C. Sanchez-Hanke, *Phys. Rev. Lett.* **76**, 3132-3135 (1996).
2. K. Hayashi, M. Matsui, Y. Awakura, T. Kaneyoshi, H. Tanida and M. Ishii, *Phys. Rev. B* **63**, R41201 (2001).
3. S. Marchesini, F. Schmithüsen, M. Tegze, G. Faigel, Y. Calvayrac, M. Belakhovsky, J. Chevrier and A. Simionovici, *Phys. Rev. Lett.* **85**, 4723 (2000).
4. J. J. Barton, *Phys. Rev. Lett.* **67**, 3106-3109 (1991).
5. K. Hayashi, *Phys. Rev. B* **71**, 224104 (2005).
6. T. Matsushita, A. Agui and Y. Yoshige, Europhys. Lett. **65**, 207-213 (2004).

In-situ Stress Measurements on SUS316L Stainless Steel in High Temperature Water Simulated Boiling Water Reactor

A. Yamamoto[1], S. Nakahigashi[2], M. Terasawa[1], T. Mitamura[1], Y. Akiniwa[3], T. Yamada[1], L. Liu[1], T. Shobu[4] and H. Tsubakino[1]

[1] Graduate School of Engineering, University of Hyogo,2167 Shosha, Himeji, Hyogo 671-2201, Japan
[2] Japan Power Engineering and Inspection Corp., Japan, 14-1 Benten-cho, Tsurumi-ku, Yokohama 230-0044, Japan
[3] Graduate School of Engineering, Nagoya University, Furo-cho, Chikusa-ku, Nagoya 464-8601, Japan
[4] Synchrotron Radiation Research Center, Japan Atomic Energy Agency, 1-1-1 Kouto, Sayo-cho, Sayo-gun, Hyogo 679-5148, Japan

Abstract. An in-situ straining device has been developed, which enables one to apply a load of 240 N to a specimen in hot water at 561 K and a pressure of 8 MPa, simulating the environment in a boiling water reactor (BWR). The device is equipped with sapphire glass windows for a light path, that is, the device can be used for dynamic measurements of stress induced in the specimen using a synchrotron radiation facility. In-situ stress measurements have been carried out at SPring-8 (BL02B1) on a specimen prepared from SUS316L stainless steel. Inhomogeneity in stress distribution and time-dependent changes in stress were successfully measured.

Keywords: In-situ stress measurement, SUS316L, hot water with high pressure, Synchrotron radiation, SPring-8..
PACS: Stress measurement, synchrotron radiation, high temperature techniques and instrumentation.

INTRODUCTION

It has been generally recognized that stress corrosion cracking (SCC) can be suppressed by lowering carbon contents in stainless steels, therefore, the usage of SUS316L, low carbon content stainless steel, can solve the problems on SCC phenomena. Cracking in reactor core shrouds made with SUS316L stainless steel have recently been reported on a BWR in service in Japan. The phenomena occurring seemed to be similar to those observed in general stress corrosion cracking, except the fact that cracks were initiated in grain interiors and portions where cracks were formed showed higher hardness compared with the surrounding area.

Hand-grinding finishing of the surface after welding is supposed to increase the hardness of the surface, which could also cause a severe layer deformation including high dislocation density and surface roughening. The former would accelerate the precipitation of carbides, and the latter would induce stress concentrations. Consequently, introducing heterogeneous residual stress distributions. The following research is important to clarify this type of cracking; (1) reproducing the cracking in an environment simulating the conditions of BWR, (2) measuring a two-dimensional stress distribution of residual stress on the ground surface, and (3) investigating the precipitation at low temperatures for prolonged aging.

The present study is concerned with the first area to measure the stress induced in the specimen under the simulated conditions. Synchrotron radiation (SR) can provide a way of measuring the stress for such a situation, that is, sufficient intensity of a SR beam can be achieved using the in-situ stage described above. Moreover, small sizes of the SR beam, a few tens micron meter in area, can allow the measurement of stress distributions in the specimen.

The present authors have reported on a newly developed in-situ straining device used for SR diffraction measurements in a previous paper [1] in which preliminary results were shown, that is, clear diffraction patterns could be measured where the $\sin^2\psi$ plot results in good linearity with small deviations.

In this paper, results of dynamic stress measurements carried out at SPring-8 (BL02B1) using the same device is reported.

CP879, *Synchrotron Radiation Instrumentation: Ninth International Conference*,
edited by Jae-Young Choi and Seungyu Rah
© 2007 American Institute of Physics 978-0-7354-0373-4/07/$23.00

EXPERIMENTAL PROCEDURES

The details of the in-situ straining device have been reported in a previous paper [1]. The specimen chamber is shown in Fig. 1 (a). The specimen is set between the fixed rod and pulling rod. The rods can be rotated by 90 degrees, which enables one to carry out two types of measurement. In the first geometry shown in (b), a reflected beam is used for diffraction measurements, in which a relatively thin surface layer contributes to the diffracted intensity, where the irradiated area is dependent upon the incident angle. In the second geometry shown in (c), the transmitted beam is used for diffraction measurement, in which the irradiated area can be kept to a small size. We carried out another experiment using this device with the second geometry at SPring-8 (BL22XU). Strain distribution in an area of 1 mm by 1mm was measured using a beam of 0.2 mm by 0.3 mm at a beam energy of 67 keV. In the present study, the reflection geometry shown in (b) was adopted.

The other end of the pulling rod is free except for friction due to an O-ring used for sealing. When the specimen chamber is filled with hot and high pressure water, the pulling rod is pushed out, but the specimen restricts the displacement of the rod, that is, the specimen is stretched depending on the pressure of water.

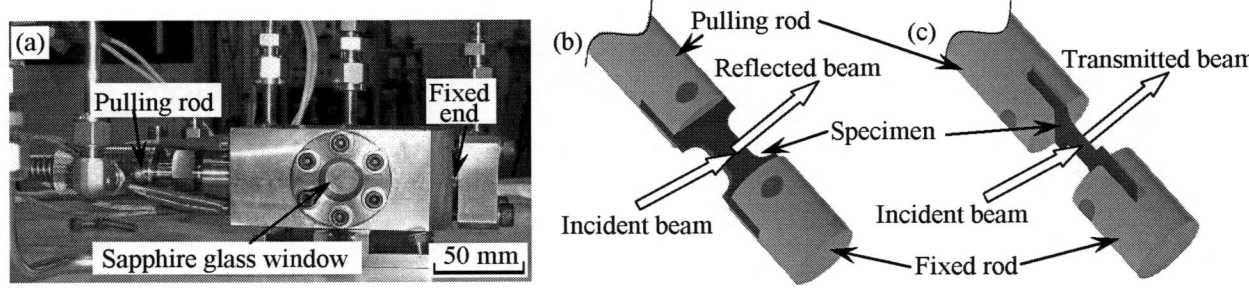

FIGURE 1. In-situ straining device for SR diffraction measurements. (a) Appearance of the device, and schematic illustrations used to measure the reflected beam and transmitted beam, (b) and (c), respectively.

The maximum pressure of the water is 8 MPa at 561 K, simulating the BWR environment, which produces a load of about 240 N. The gauge sizes of the specimen were 1 mm length and 1×1 mm^2 cross section, a stress of 240 MPa is induced in the specimen. An additional load of 120 N can be applied by a spring attached to the pulling rod.

The chemical composition of the SUS316L stainless steel used is listed in Table 1. The tensile specimens used for the in-situ experiments were prepared by the following procedures: specimens were solution heat treated at 1323 K for 0.9 ks, surfaces of the specimens were mechanically polished using an emery paper of #1000, and then one surface was ground by using a #100 grinding disk, pre-deformation was applied to the specimen by using a conventional slow strain rate testing (SSRT) machine. The conditions for the pre-deformation were as follows; strain rate: 10^{-3} s^{-1}, temperature: 561 K, pressure: 8 MPa, stretching: up to 80 % of the time to failure which was previously measured using other specimens.

TABLE 1. Chemical composition of the specimen (mass%).

C	Si	Mn	P	S	Ni	Cr	Mo	N	Fe
0.011	0.89	1.05	0.024	0.006	12.09	17.60	2.03	0.026	bal.

Micrographs of a specimen after surface grinding are shown in Fig. 2. The grinding direction was parallel to the longitudinal axis of the specimen as shown in Fig. 2 (a). Microscopic crack-like grooves are seen in Fig. 2 (b). Residual stress measurements on the surface ground specimen showed that tensile stress was induced along the longitudinal axis by grinding [2]. The specimen surface due to stretching by the SSRT test up to 80 % of the failure time is shown in Fig. 3. A number of slip bands are observed to be formed, moreover many voids/micro-cracks are observed.

The in-situ stress measurement was carried out at BL02B1 at SPring-8. Diffraction of {113} planes in austenite was selected for the measurements. The two theta angle for the planes under the beam energy of 72 keV was about 9.11 deg. The beam sizes were 0.3 to 1.0 mm and 0.2 mm in the horizontal and vertical directions, respectively. Stress was evaluated using the reflected geometry (Fig. 1 (b)) changing the inclination angles in the range of 0 to 5.6 degree. When the size of the beam in horizontal direction was 0.3 mm, stresses of three portions in the gauge part of the tensile specimen can be individually measured, whereas when the size was 1 mm, stress on the whole of the gauge part can be measured.

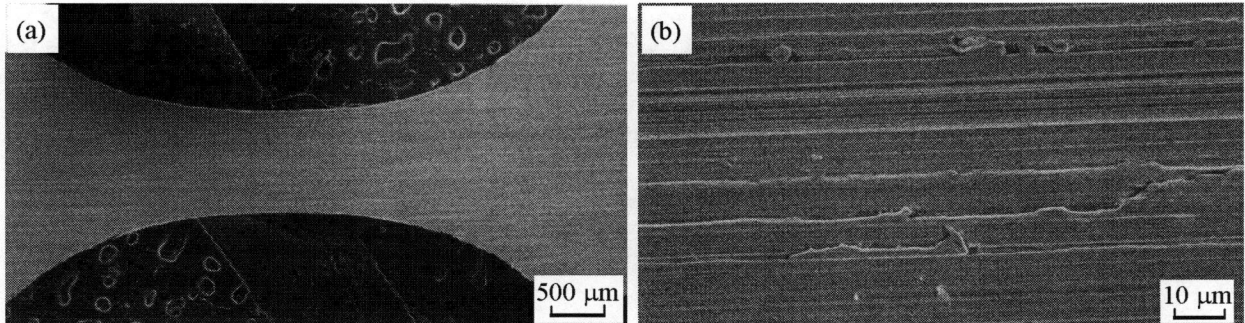

FIGURE 2. SEM micrographs showing the specimen surface after surface grinding at low and high magnifications, (a) and (b), respectively.

FIGURE 3. Specimen after stretching by a conventional slow strain rate test up to 80 % of the time to failure.

RESULTS AND DISCUSSION

The in-situ measurement lasted for about three days by repeating the reflection geometry measurement. The measured stresses are plotted in Fig. 4. In the early stage of the experiment, all of the gauge area of the specimen was irradiated by SR beam with a beam size of 1 mm (horizontal) x 0.2 mm (vertical) under a load of 240 N. Stress approximately decreased with time. In the second stage, the horizontal size of the beam was reduced to 0.3 mm, and the stress was measured at three positions, pulling side, middle portion and fixed side, under the same load. The obtained stresses were different to each other as shown in Fig. 4, and were higher than those obtained with the beam of 1.0 mm. In the last stage, additional load was applied by using the spring attached to the pulling rod; the total load was 330 N. The stress measurement at the middle position was skipped in order to shorten the measuring cycle. The stresses obtained at the pulling side and fixed side showed different values, although the change in stress were similar to one another as shown in Fig. 4.

The measured stress is a sum of the applied stress due to a tensile load and the residual stress due to surface grinding and stretching by the SSRT. The applied stress is caused by the high pressure and high temperature water, which were controlled at constant values during measurements. Thus, the decrease and increase in stress shown in Fig. 4 is supposed to be due to changes in dislocation substructures and diffusion of solute elements. Interpretation of the details of the change in stress seems to be difficult, but the following important features were revealed; (1) stress induced in the specimens exposed in the BWR conditions can be directly measured by using the device developed by the authors utilizing the SR diffraction technique, (2) stress is not uniformly distributed in the specimen, relatively high stress was detected when the beam size was reduced, which implies the necessity of two dimensional stress measurements for clarifying the cracking phenomena in the core shroud.

Inhomogeneous microstructures composed of micro-cracks and slip bands are formed by surface grinding [2]. Grain boundaries do not play an important role in such a situation. Cracking would occur at a highly stress concentrated position, even in grain interiors. The beam sizes used in the present study were relatively large, because BL02B1 is a bending magnet type of beam line; beam intensity is not so high. For example, in the BL22XU, undulator type of beam line, smaller beam can be provided. Smaller the beam size, higher the image resolution in two-dimensional stress mapping.

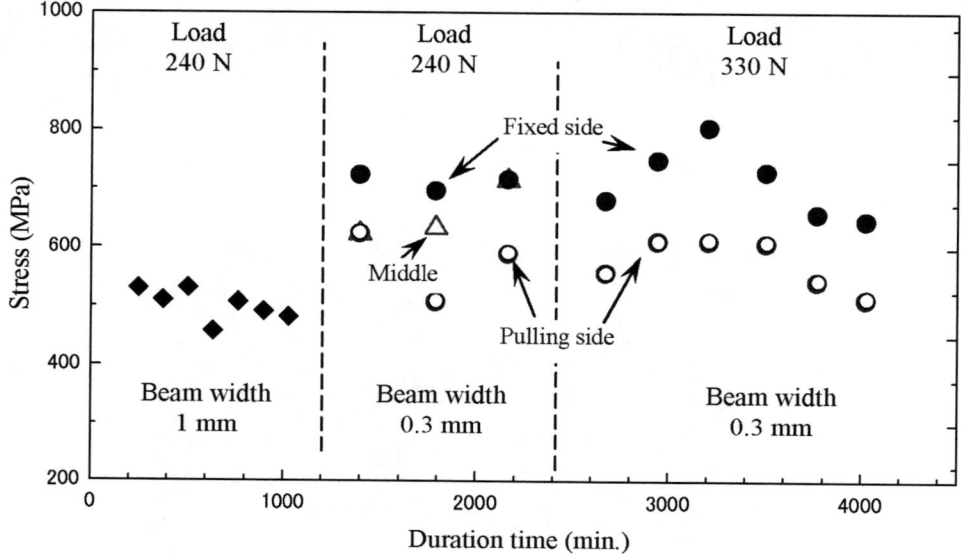

FIGURE 4. Changes in stress induced in the specimen in high temperature and high pressure (561 K, 8 MPa) water with tensile load obtained by using the in-situ straining device.

SUMMARY

The in-situ stress measurement on the specimen in high temperature, 561 K, and high pressure, 8 MPa, water was successfully carried out for about 240 ks. Of course the time is too short to clarify the SCC, but the results showed the change in stress with time occurred in the range of about 100 MPa. The magnitude of the stress became obvious when the beam size was reduced, which implies the importance of two dimensional stress measurements.

REFERENCES

1. A. Yamamoto, S. Nakahigasi, M. Terasawa, T. Mitamura, Y. Akiniwa, T. Yamada, L. Liu, T. Shobu and H. Tsubakino, *J. Syncrotron Rad.* **13**, 14-18 (2006).
2. A. Yamamoto, T. Yamada, S. Nakahigashi, L. Liu, M. Terasawa and H. Tsubakino, *ISIJ Inter.* **44**, 1780-1782 (2004).

Orientation-Dependent Structural Properties and Growth Mechanism of ZnO Nanorods

H.-J. Yu[1], E.-S. Jeong[1], S.-H. Park[2], S.-Y. Seo[2], S.-H. Kim[2], S.-W. Han[1]*

[1]*Division of Science Education and Institute of Proton Accelerator, Chonbuk National University, Jeonju 561-765, Korea*
[2]*Department of Materials Science and Engineering, Pohang University of Science and Technology, Pohang 790-784, Korea*

Abstract. We present the local structural properties of ZnO nanorods studied by using extended x-ray absorption fine structure (EXAFS). Vertically aligned ZnO nanorods were fabricated on Al_2O_3 substrates by a catalyst free metal organic chemical vapor deposition (MOCVD). The polarized EXAFS measurements on the ZnO nanorods were performed at Zn K-edge. The polarized EXAFS study revealed that the nanorods had a wurtzite structure, and that there were substantial amount of structural disorders in Zn-O pairs in the beginning of the nanorod growth. The EXAFS measurements revealed that the orientation-dependent disorders of the Zn-O pairs were directly related to the growth mechanism and crystal quality of the ZnO nanorods.

Keywords: EXAFS, ZnO, nanorod, growth mechanism, MOCVD
PACS: 61.60.Ht, 61.46.Hk, 61.43.Dq

INTRODUCTION

Semiconducting nanostructures have attracted considerable attention for their practical applications to nanometer-scale electronics and photonics as well as fundamental academic research. ZnO is a candidate material for room temperature UV applications due to 3.36 eV band gap energy and 60 meV exciton binding energy at room temperature. In addition, it is relatively easy to fabricate vertically aligned ZnO nanorods with various fabrication techniques [1-5]. The ZnO nanorods were intensively studied for its practical applications, AFM cantilevers [6], gas sensors [7,8], nanotransistors [9], logic gate device [10], and UV light emitting diodes [11]. Many previous studies attempted to understand the growth mechanism of ZnO nanorods [12-15]. However, the growth mechanism is still unclear. We investigated if structural stress due to lattice mismatch between the ZnO and substrates play a role in the ZnO nanorod growth. We employed polarized extended x-ray absorption fine structure (EXAFS) which can describe the angles and distances of neighboring atoms from a probe atom [16], to investigate the orientation-dependent residual strain in ZnO nanorods. Two independent EXAFS measurements from the vertically aligned ZnO nanorod arrays with different lengths were made with the x-ray polarizations parallel and perpendicular to the nanorod length direction. Both sets of EXAFS data were *simultaneously* fitted with the same parameters and the orientation-dependent structural properties of the ZnO nanorods were obtained.

EXPERIMENTAL

ZnO nanorod arrays were fabricated using metal-organic chemical vapor deposition (MOCVD) on $Al_2O_3(0001)$ substrates. The substrate temperature was maintained at 300 – 450 °C and the growth rate was about 10 nm per minute. To investigate the structural residual strain in ZnO nanorods, we synthesized the ZnO nanorods with average length of 0.1, 0.2, 1.0 µm on the Al_2O_3 substrates. The average diameters of the ZnO nanorods were 50 - 100 nm. Scanning electron microscope (SEM) images revealed vertically well-aligned ZnO nanorods. High-resolution x-ray diffraction (XRD) and high-resolution transmission electron microscope (TEM) measurements demonstrated that the ZnO nanorods were well-ordered single crystals. The EXAFS measurements at Zn K edge (9659 eV) with fluorescence and transmission modes for ZnO nanorods and powder, respectively, were made at

---

* Author to whom correspondence should be addressed; Email: swhan@chonbuk.ac.kr.

CP879, *Synchrotron Radiation Instrumentation: Ninth International Conference*,
edited by Jae-Young Choi and Seungyu Rah
© 2007 American Institute of Physics 978-0-7354-0373-4/07/$23.00

beamline 3C1 of Pohang Light Source at room temperature. Incident x-ray energy was selected with a three-quarters tuned Si(111) double monochromator.

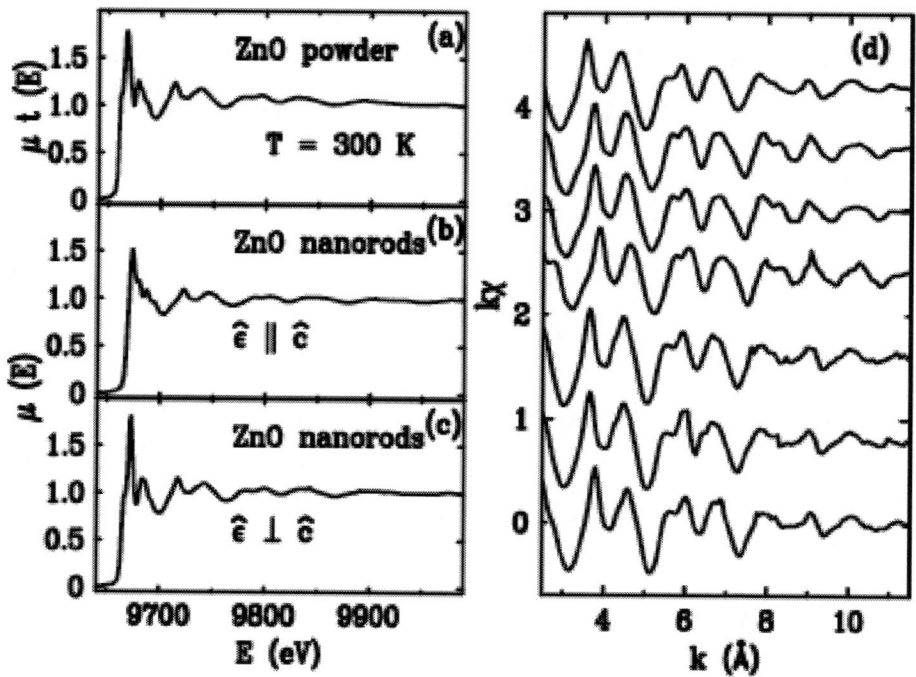

FIGURE 1. Total x-ray absorption from ZnO powder and (b), (c) orientation-dependent x-ray absorption coefficient of ZnO nanorods near Zn K-edge as a function of incident x-ray energy. For the orientation dependent EXAFS measurements, the electric field vector of the incident x-ray was aligned (b) parallel ($\varepsilon \parallel c$) and (c) perpendicular ($\varepsilon \perp c$) to the ZnO nanorod length direction. (d) EXAFS ($k\chi$) from ZnO powder (1st) ZnO nanorods with lengths of 0.1 μm (2nd, 5th), 0.2 μm (3rd, 6th) and 1.0 μm (4th, 7th) as a function of photoelectron wave vector described in the text. The 2nd-4th are for $\varepsilon \parallel c$ and the last three are for $\varepsilon \perp c$.

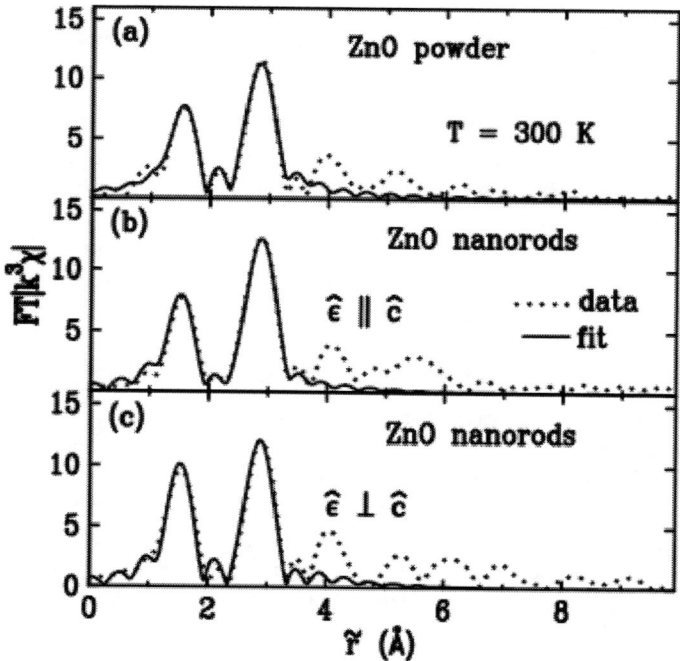

FIGURE 2. Magnitude of Fourier transformed EXAFS from the data of ZnO nanorods with 1.0 μm length as a function of distance from a zinc atom. For the Fourier transformation, a Hanning window with a windowsill width of 0.5 Å^{-1} was used. Data in the range of r = 1.2 - 3.3 Å were fitted.

TABLE 1. Results of EXAFS Analysis

specimen	length	N	d	σ^2	N	d	σ^2	N	d	σ^2	N	d	σ^2
			Zn-O(1)			Zn-O(2)			Zn-Zn(1)			Zn-Zn(2)	
powder		1	1.903(9)	0.003(1)	3	1.98(2)	0.003(1)	6	3.206(1)	0.008(4)	6	3.246(13)	0.008(2)
nanorods	0.1 μm	1	1.939(5)	0.0052(5)	3	2.018(7)	0.0045(6)	6	3.209(3)	0.0088(4)	6	3.255(4)	0.0085(4)
nanorodsx	0.2 μm	1	1.939(6)	0.0049(6)	3	2.024(7)	0.0027(6)	6	3.209(5)	0.0089(4)	6	3.259(2)	0.0090(3)
nanorods	1.0 μm	1	1.939(5)	0.0055(5)	3	2.018(6)	0.0028(6)	6	3.206(5)	0.0085(4)	6	3.257(4)	0.0085(4)
nanorods	0.5 μm	1	1.903(6)	0.003(1)	3	1.98(1)	0.003(1)	6	3.216(10)	0.004(1)	6	3.236(1)	0.004(1)

Coordination number (N), bond length (d) and Debye-Waller factor (σ^2) of ZnO powder (top), ZnO nanorods/Al$_2$O$_3$ (middle three) and ZnO nanorods/GaN/Al$_2$O$_3$ (bottom). The EXAFS results of ZnO nanorods/Al$_2$O$_3$ at 300 K were determined by polarized-EXAFS measurements whereas those of ZnO powder at 300 K and ZnO nanorods/GaN/Al$_2$O$_3$ at 30 K were obtained from unpolarized EXAFS. S$_0^2$ of 0.95(5) was used in the data fits. The R-factors of a goodness fit were 0.005, 0.006, 0.009, 0.012, and 0.007 for the specimens from the top, respectively.

RESULTS AND DISCUSSION

Figures 1 show the normalized total x-ray absorption and x-ray absorption coefficient from the ZnO powder and nanorods near the Zn K edge as a function of the incident x-ray energy. The near edge structures from the nanorods obviously show the dependence of the crystal orientations, and are very comparable with previous reports [17-19]. The EXAFS data were analyzed with the UWXAFS package [20] using standard procedures [16,18,19]. After the atomic background absorption μ_0 was determined using AUTOBK (part of the UWXAFS package), the EXAFS function, $\chi = \mu(E)/\mu_0(E) - 1$, was obtained. Figure 1 (d) shows the EXAFS from ZnO nanorods at the Zn K edge for $\varepsilon \parallel c$ and the last three are for $\varepsilon \perp c$, as a function of the photoelectron wave vector, $k = \sqrt{2m_e(E - E_0)}/\hbar$, where m_e is the electron rest mass, E is the incident photon energy and E_0 is the edge energy. To minimize uncertainty only the EXAFS data in the k-range of 2.5 - 10.5 Å^{-1} were used for further analysis. Figure 2 shows the magnitudes of the Fourier transformed EXAFS data from the ZnO nanorods with different lengths. It should be noted that the peaks shifted by about 0.4 Å on the r-axis from their true bond lengths due to the phase shift of the back-scattered photoelectrons. The EXAFS data Fourier transformed to r-space were *simultaneously* fitted to the theoretical EXAFS calculations [21]. The fits included single- and multi-scattering paths. The data were fitted with a fully occupied model of a wurtzite structure, varying the bond lengths and Debye-Waller factors (σ^2, including thermal vibration and static disorder). The polarized-EXAFS data can independently determine the distances and σ^2s from the probe zinc atom to one O(1) located just below the Zn atom along the c-axis, three O(2)s located about 19° off from the Zn ab-plane, six Zn(1)s located at ~55° off from the ab-plane and six Zn(2)s located in the ab-plane. The fit results are summarized in Table 1.

From the EXAFS data analysis, we observed that the σ^2s of Zn-O pairs in the ZnO film, with thickness of about 0.1 μm, which was constantly observed in the beginning of the ZnO nanorod growth [22], had a substantial amount of disorders, compared with ZnO powder and high quality ZnO nanorods [18]. As ZnO formed into nanorods, the extra disorder in the Zn-O(2) pairs disappeared. However, the extra disorder in the Zn-O(1) pairs was observed in even 1 μm nanorods. EXAFS revealed no extra disorders existing in all Zn-Zn pairs, comparing to ZnO powder counterpart. EXAFS results strongly suggested that the structural strain of Zn-O pairs in the ab-plane due to the lattice mismatch should be first relaxed for the ZnO nanorod growth. This result strongly suggested that the strain relaxation in ab-plane was the ZnO nanorod seeds of the Stranski-Krastanov growth mode. We also investigated the structural properties of the ZnO nanorods grown on Al$_2$O$_3$ with GaN interlayer. The EXAFS analysis results of the ZnO nanorods/GaN/Al$_2$O$_3$ are summarized in Table 1. The EXAFS measurements of ZnO nanorods grown on Al$_2$O$_3$ substrates with GaN interlayers demonstrated that the extra disorder of Zn-O(2) pairs as well as Zn-O(1) pairs was disappeared as the ZnO formed into nanorods. This provides more evidence that the strain relaxation of Zn-O(2) pairs are critical in the formation of ZnO nanorods. The XRD measurements (not shown here) of the ZnO nanorods showed that the lattice constant of the ZnO crystals was about 5.206 Å which corresponded well with that of ZnO bulk [18]. Interestingly, the full widths at half maximum (FWHM) of ZnO (0002) diffraction peak of ZnO nanorods grown on Al$_2$O$_3$ substrates was about three times larger than that of the ZnO nanorods grown on Al$_2$O$_3$ substrates with GaN interlayers. Comparing with EXAFS results, we concluded that the extra broadening of the diffraction peak in ZnO nanorods/Al$_2$O$_3$ was mainly affected by the disorder of Zn-O(1) pairs instead of Zn-Zn pairs.

CONCLUSIONS

The polarized EXAFS technique was employed to investigate the growth mechanism of ZnO nanorods. Vertically aligned ZnO nanorods with various lengths were fabricated on Al_2O_3 substrates with and without GaN interlayers by MOCVD. The orientation-dependent structural properties of the ZnO nanorods were determined by the EXAFS at Zn K-edge. Polarized EXAFS measurements of ZnO nanorods grown on Al_2O_3 substrates revealed that a substantial amount of structural strain existed in the Zn-O pairs located in the *ab*-plane and that the strain disappeared as the ZnO formed into nanorods. Our observations present the strong evidence that the structural stress due to the lattice mismatch between nanorods and substrates dominantly influenced ZnO nanorod formation. We observed extra disorder existing in Zn-O(1) pairs of ZnO nanorods grown on the Al_2O_3 substrates, comparing with the ZnO nanorods grown on the GaN/Al_2O_3 substrates.

ACKNOWLEDGEMENTS

This work was supported by the Korea MOEHRD (KRF-2005-042-C00055), the Korea Basic Research Program (KOSEF-R01-2006-000-10800-0), the Korea MOCIE through the New Technology R&D Program, and the Korea MOST through PEFP User Program as a part of the 21C Frontier R&D Program. The EXAFS data were collected at 3C1 beamline of the Pohang Light Source.

REFERENCES

1. M. H. Huang, S. Mao, H. Feick, H. Yang, Y. Wu, H. Kind, E. Weber, R. Russo and P. Yang, *Science* **292**, 1897 (2001).
2. W. I. Park, D. H. Kim, S.-W. Jung and Gyu-Chul Yi, *Appl. Phys. Lett.* **80**, 4232, (2002).
3. Lionel Vayssieres, *Adv. Mater.* **15**, 464 (2003).
4. D. Whang, S. Jin, Y. Wu and C. M. Lieber, *Nano lett.* **3**, 1255 (2003).
5. Q. Wen, Q. H. Li, Y. J. Chen, T. H. Wang, X. L. He, J. P. Li, and C. L. Lin, *Appl. Phys. Lett.* **84**, 3654 (2004).
6. William L. Hughes and Zhong L. Wang, *Appl. Phys. Lett.* **82**, 2886 (2003).
7. Q. Wan, Q. H. Li, Y. J. Chen, T. H. Wang, X. L. He, J. P. Li, C. L. Lin, *Appl. Phys. Lett.* **84**, 3654 (2004)
8. Q. H. Li, Y. X. Liang, Q. Wan, T. H. Wang, *Appl. Phys. Lett.* **85**, 6389 (2004).
9. Xiangfeng Duan, Yu Huang, Yi Cui, Jiangfang Wang, and Charles M. Lieber, *Nature* **409**, 66 (2001).
10. Won Il Park, Jin Suk Kim, Gyu-Chul Yi, and Hu-Jong Lee, *Adv. Mater.* **17**, 1393 (2005).
11. Won Il Park and Gyu-Chul Yi, *Adv. Mater.* **16**, 87 (2004).
12. H. Q. Le, S. J. Chua, Y. W. Koh, K. P. Loh, Z. Chen, C. V. Thompson and E. A. Fitzgerald, *Appl. Phys. Lett.* **87**, 101908 (2005).
13. Wen-Jun Li, Er-Wei Shi, Wei-Zhuo Zhong and Zhi-Wen Yin, *J. Cryst. Growth* **203**, 186 (1999).
14. Xiang Liu, Xiaohua Wu, Hui Cao and R. P. H. Chang, *J. Appl. Phys.* **95**, 3141 (2004).
15. G. W. Cong, H. Y. Wei, P. F. Zhang, W. Q. Peng, J. J. Wu, X. L. Liu, C. M. Jiao, W. G. Hu, Q. S. Zhu and Z. G. Wang, *Appl. Phys. Lett.* **87**, 231903 (2005).
16. S.-W. Han, E. A. Stern, D. Hankel and A. R. Moodenbaugh, *Phys. Rev. B.* **66**, 94101 (2002).
17. H. Chik, J. Liang, S. G. Cloutier, N. Kouklin and J. M. Xu, *Appl. Phys. Lett.* **84**, 3376 (2004).
18. S.-W. Han, H.-J. Yoo, Sung Jin An, Jinjyoung Yoo and Gyu-Chul Yi, *Appl. Phys. Lett.* **88**, 111910 (2006).
19. S.-W. Han, H.-J. Yoo, Sung Jin An, Jinkyung Yoo and Gyu-Chul Yi, Appl. Phys. Lett. **86**, 21917 (2005).
20. E. A. Stern, M. Newville, B. Ravel, Y. Yacoby and D. Haskel, *Physica B* **208\&209**, 117 (1995).
21. A. L. Ankudinov, B. Ravel, J. J. Rehr and S. D. Conradson, *Phys. Rev. B.* **58**, 7565 (1998).
22. J. Y. Park, Y. S. Yun, Y. S. Hong, H. Oh, J.-J. Kim, and S. S. Kim, *App. Phys. Lett.* **87**, 123108 (2005).

CHAPTER 13
LIFE AND MEDICAL SCIENCE

SPring-8 Structural Biology Beamlines / Automatic Beamline Operation at RIKEN Structural Genomics Beamlines

Go Ueno[a], Kazuya Hasegawa[b], Nobuo Okazaki[b], Raita Hirose[c], Hisanobu Sakai[b], Takashi Kumasaka[a, d] and Masaki Yamamoto[a, b]

[a]RIKEN SPring-8 Center, 1-1-1 Kouto, Sayo-cho, Sayo-gun, Hyogo, 679-5148, JAPAN
[b]SPring-8/JASRI, 1-1-1, Kouto, Sayo-cho, Sayo-gun, Hyogo, 679-5198, JAPAN
[c]PharmAxess Inc., 3-1-1 Kouto, Kamigori-cho, Ako-gun, Hyogo, 678-1205, JAPAN
[d]Tokyo Institute of Technology, 4259 Nagatsuta-cho, Midori-ku, Yokohama, Kanagawa, 226-8501, JAPAN

Abstract. RIKEN Structural Genomics Beamlines (BL26B1 & BL26B2) at SPring-8 have been constructed for high throughput protein crystallography. The beamline operation is automated cooperating with the sample changer robot. The operation software provides a centralized control utilizing the client and server architecture. The sample management system with the networked database has been implemented to accept dry-shipped crystals from distant users.

Keywords: automatic data collection, sample changer, mail-in data collection, high-throughput protein crystallography
PACS: 61.10.Nz

INTRODUCTION

Structural genomics research progressing worldwide is aiming at accumulating the information on a significant number of protein structures and functions corresponding to the genetic sequencing analyses [1]. In order to deal with a vast amount of protein crystals, development of high-throughput facilities for macromolecular structure analysis had been required. RIKEN Structural Genomics Beamlines I & II (BL26B1 & BL26B2) at SPring-8 [2] have been constructed to execute the rapid data collection to contribute to the structural genomics research. The beamline operation is automated cooperating with the sample changer robot named SPACE (SPring-8 Precise Automatic Cryo-sample Exchanger) [3], to improve the beam-time efficiency. The beamline optics adopted the standard design for SPring-8 bending magnet beamlines, which is suitable for automation. In the end station, two types of area detectors (CCD and IP) were installed, which can be automatically exchanged one another by an equipment stage. The beamline operation software BSS (Beamline Scheduling Software) [4] provides the intuitive GUI and unified control of beamline instruments with the networked client-server architecture. Since October 2003, BL26B2 has been continuously operated with the sample changer robot. Protein crystals are stored and delivered using the special sample trays for SPACE, which can be sent using a commercially available dry shipper. In July 2005, a networked database D-Cha (Database for Crystallography with Home-lab. Arrangement) has been implemented for distant users to edit the experimental schedules for sample trays they have sent. The beamline has been constantly accepting dry-shipped sample trays from distant users, and executing diffraction measurements according to the experimental conditions specified *via* D-Cha. Presently, mail-in data collection service for industrial users utilizing the automation system is prepared.

BEAMLINE DESIGN AND CONTROL SYSTEM

The light source and optics of the RIKEN Structural Genomics Beamlines adopted the standard design for bending magnet beamlines in the SPring-8, which is suitable for automation [5]. The X-rays emitted from bending magnet are truncated by a water-cooled slit and monochromatized by a fixed-exit Si double-crystal monochromator

CP879, *Synchrotron Radiation Instrumentation: Ninth International Conference,*
edited by Jae-Young Choi and Seungyu Rah
© 2007 American Institute of Physics 978-0-7354-0373-4/07/$23.00

[6]. Available photon energy range is from 6 to 17 keV, which is suitable for MAD (Multi-wavelength Anomalous Diffraction) experiments with heavy atoms typically used in protein crystallography. The focusing mirror is a vertically bent cylindrical mirror glancing in downward direction, which enables the two-dimensional focusing of the X-ray beam. The beam size at the sample position is 150 μm FWHM diameter. The energy resolution $\Delta E/E$ of the monochromatic X-rays is on the order of 10^{-4} for available energy range. The total photon flux at the sample position is about 10^{11} photons/sec at 12 keV photon energy.

In the experimental hutch, all devices are mounted on the remote-controllable equipment stage (Fig. 1). Optic devices such as slits, an X-ray shutter, and attenuators are placed in vacuum housings. The crystal goniometer is equipped with κ-axis that makes it possible to align the crystal orientation for efficient data collection. The goniometer head can be positioned by a remote translation stage for sample centering. The sample changer robot SPACE is located facing to the goniometer. Two types of area detectors, a mosaic CCD (Charged Coupled Device) detector (Jupiter 210, Rigaku co.) [7], and a large IP (Imaging Plate) detector (R-AXIS V, Rigaku co.) [8] are installed

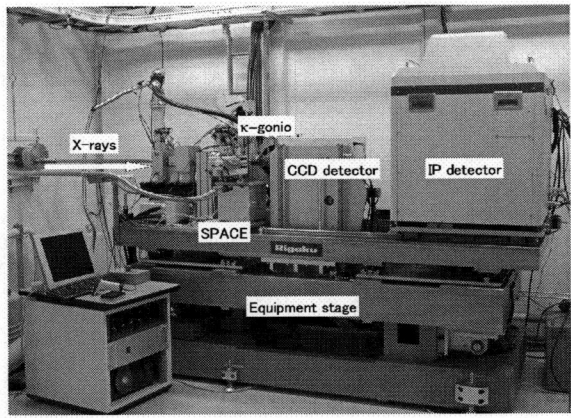

FIGURE 1. End station of the SPring-8 RIKEN Structural Genomics Beamline II (BL26B2).

The operation software of the RIKEN Structural Genomics Beamlines adopted the client and server architecture, on which beamline devices are controlled by individual server programs (Fig. 2(a)). The server program for beamline common components such as monochromator, beam shutter, goniometer and so on, adopted the SPring-8 standard control program [9]. BSS is the GUI based client software, which realizes the central control of the beamline devices *via* the network (Fig. 2(b)). The main feature of the BSS is the scheduling function of successive data collections for multiple sample crystals. The schedule can be registered to BSS as a series of various measurements; crystal check, single-wavelength data collection, MAD data collection or XAFS (X-ray Absorption Fine Structure) spectrum measurement. Beamline can be operated within the unified environment of BSS GUI.

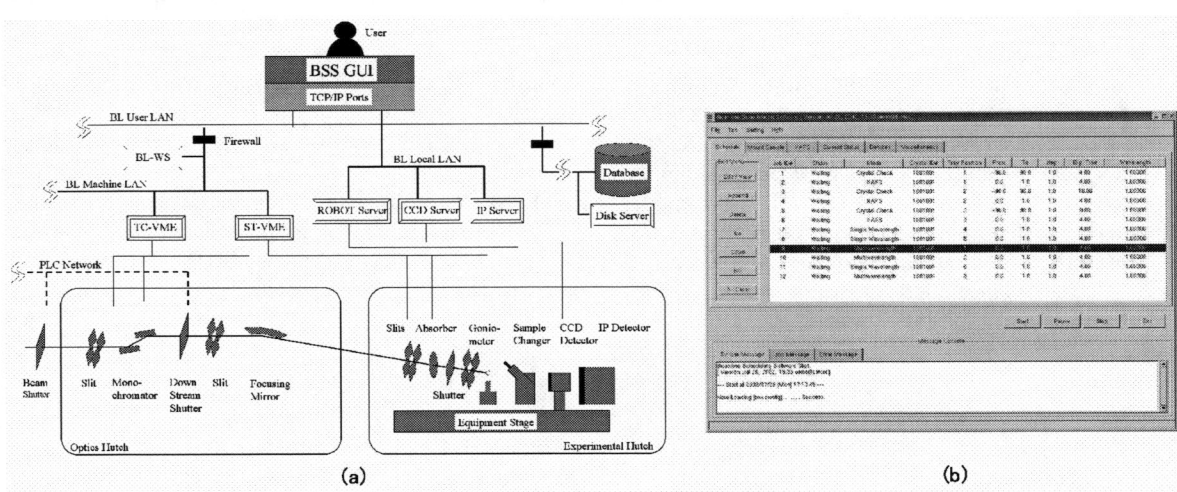

FIGURE 2. Beamline control system of the SPring-8 RIKEN Structural Genomics Beamlines. (a) Network and hardware configuration. (b) BSS main window.

SAMPLE MANAGEMENT SYSTEM

The automation of beamline is based on the sample management system with a new-concept sample pin and sample changer SPACE. The system covers whole sample-handling procedures from the laboratory to the beamline. SPACE robots are installed in both laboratory and beamline, and those facilities are connected to the sample management database D-Cha through the Internet (Fig. 3). Users store crystals into sample tray by using SPACE at laboratory. The sample tray is identified by barcode sticker, and the individual crystal information and experimental schedule is uploaded to D-Cha. D-Cha manages the massive sample information by tray identification number, and provides the GUI to edit the experimental schedule of each tray on a web browser. Access to D-Cha is authenticated by user account and password so that only valid owners can access the contents of the tray information. At the beamline, the information of delivered tray is extracted from D-Cha, and registered to BSS. The SPACE installed at the beamline sequentially exchanges the crystals according to the experimental schedule. The acquired data is uploaded to D-Cha, and the diffraction images can be browsed or downloaded by users *via* the Internet.

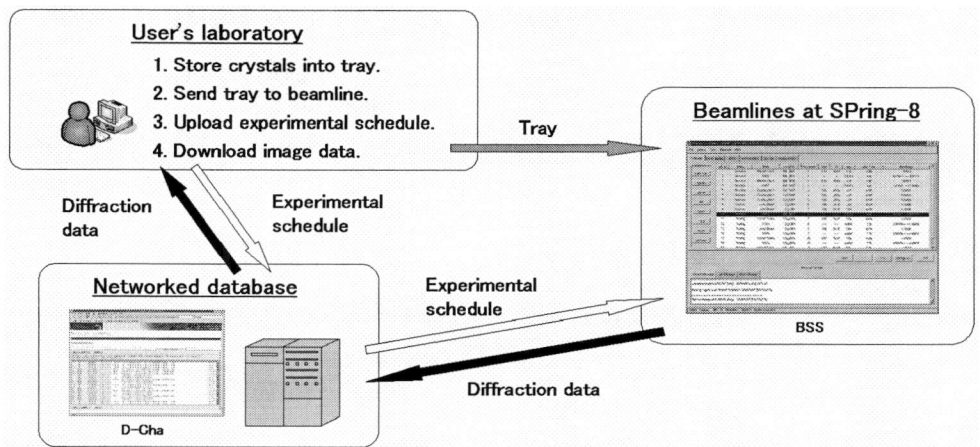

FIGURE 3. Sample management system at SPring-8 RIKEN Structural Genomics Beamlines

The sample pin and tray adopted a special design for SPACE (Fig. 4 (a) and (b)). The pin is equipped with screw threads to be attached to the goniometer. The mounting method using the screw thread is the key feature of SPACE, which enables to mount crystals with high positional reproducibility (< 10μm). SPACE hardware consists of three-axis robotic arm and X-Y translation stage for the sample storage (Fig. 4(c)). The sample pin is mounded on or dismounted from the goniometer by revolving it with SPACE robotic arm. The sample tray is located in the liquid nitrogen bath of the storage. The X-Y stage aligns the position of the sample tray so that the robotic arm can pick up an arbitrary sample. The level of the liquid nitrogen bath is monitored and filled with an automatic supplying system.

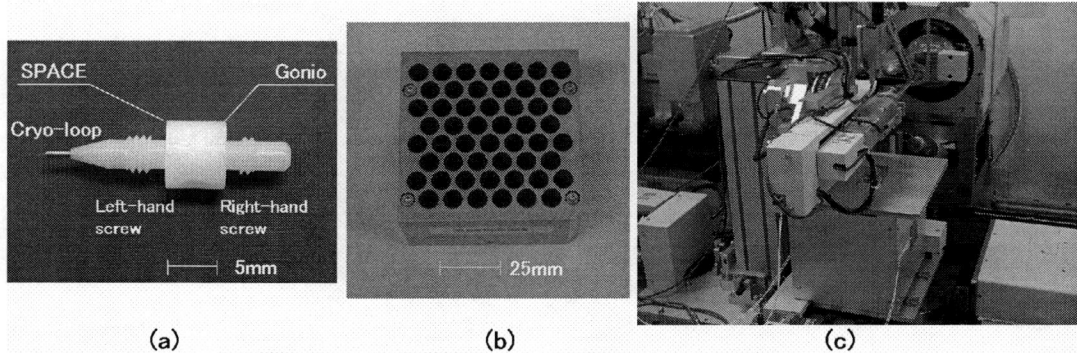

FIGURE 4. Robotic system developed and implemented at SPring-8 RIKEN Structural Genomics Beamlines. (a) Sample pin. (b) Sample tray. (c) SPACE.

CURRENT STATUS OF AUTOMATIC BEAMLINE OPERATION

The automatic beamline operation is executed by taking two experimental steps named "two-mode beamline operation" as follows. In the first mode, all crystals in the tray are successively mounted on the gonoimeter to collect the screening data, which is used to identify the crystal quality. During this mode, beamline operators assure the centering of each crystal. The diffraction data and the record of the goniometer translation are registered to D-Cha as the sample-specific information. Since the screening data for a crystal can be obtained within ten minutes, the first-mode experiment for a sample tray (maximum fifty-two crystals) finishes within daytime. Users qualify the crystals by using diffraction image browser of D-Cha, and make the experimental schedules for selected crystals. In the second mode, automatic overnight data collections without any attendant at the beamline are conducted. Samples are automatically mounted and centered utilizing the record of the goniometer translation, taking advantage of the screw-type sample pin with SPACE. During the interval of data collections, the beam intensity at sample position is automatically optimized by scanning the monochromator axis and equipment stage, with monitoring an ionization chamber signal located at upstream of the sample. RIKEN Structural Genomics Beamline II (BL26B2) has been continuously operated with this method since 2003. The summary of automatic beamline operation in 2005 is shown in Table 1.

With the automation system, RIKEN Structural Genomics Beamlines have achieved the acceleration of crystal screening and efficient use of the beam-time. Presently, the mail-in data collection service for industrial users, based on the beamline automation system, is prepared.

TABLE 1. Summary of automatic beamline operation at BL26B2, in 2005.

Total operation period	70 days
Total samples screened	1,526 (average: 21/daytime, maximum: 51/daytime)
Screening rate	6 min/sample (diffraction measurement only)
	10 min/sample (diffraction and XAFS measurement)
Total data sets collected	486 (average: 6.9/night, maximum: 21/night)

ACKNOWLEDGMENTS

We gratefully acknowledge the cooperation of Atsushi Nisawa, Hironori Murakami and other beamline staffs. Yukito Furukawa at SPring-8/JASRI is greatly appreciated for providing the control system and device servers. We also appreciate useful advice and cooperation of Shunji Goto and Kunikazu Takeshita at SPring-8/JASRI. We also thank Tetsuya Ishikawa and Masashi Miyano at RIKEN SPring-8 Center for valuable supports and discussions. A part of this work was supported by National Project on Protein Structural and Function Analysis by MEXT of Japan.

REFERENCES

1. D. Baker & A. Sali, *Science*, **294**, 93-96. (2001).
2. G. Ueno, H. Kanda, R. Hirose, K. Ida, T. Kumasaka and M. Yamamoto, *J. Struct. Funct. Genomics*, (2006), published online.
3. G. Ueno, R. Hirose, K. Ida, T. Kumasaka, and M. Yamamoto, *J. Appl. Cryst.* **37**, 867-873. (2004).
4. G. Ueno, H. Kanda, T. Kumasaka, and M. Yamamoto, *J. Synchrotron Rad.* **12**, 380-384. (2005).
5. S. Goto, M. Yabashi, H. Ohashi, H. Kimura, K. Takeshita, T. Uruga, T. Mochizuki, Y. Kohmura, M. Kuroda, M. Yamamoto, Y. Furukawa, N. Kamiya, and T. Ishikawa, *J. Synchrotron Rad.* **5**, 1202-1205. (1998).
6. M. Yabashi, H. Yamazaki, K. Tamasaku, S. Goto, K. Takeshita, T. Mochizuki, Y. Yoneda, Y. Furukawa, and T. Ishikawa, *Proc. SPIE* **3773**, 2-13. (1999).
7. M. Suzuki, M. Yamamoto, T. Kumasaka, K. Sato, H. Toyokawa, I. F. Aries, P. A. Jerram, and T. Ueki, *Nucl. Instrum. Meth. A.* **436**, 174-181. (1999).
8. M. Yamamoto, T. Kumasaka, H. Yamazaki, K. Sasaki, Y. Yokozawa, and T. Ishikawa, *Nucl. Instrum. Meth. A.* **467-468**, 1160-1162. (2001).
9. T. Ohata, H. Konishi, H. Kimura, Y. Furukawa, K. Tamasaku, T. Nakatani, T. Tanabe, N. Matsumoto, & T. Ishikawa, *J. Synchrotron Rad.* **5**, 590-592. (1998).

Development of Auto-Mounting System for Screening Protein Crystals

G.H. Kim[*], C.G, Ryu[*], K.J. Kim[*], H.Y. Kim[*], C.J. Yu[*], K.W. Kim[*], S.N. Kim[*], M.J. Kim[*], J.H. Kim[*], S.Y. Kwon[*], Y.J. Choi[*], G.W. Song[§], Y.J. Jang[†], H.S. Lee[*]

*Beam line Division, Pohang Accelerator Laboratory, Pohang, Kyungbuk, Korea.
§Department of Electronic and Electrical Engineering, Pohang University of Science and Technology, Pohang, Kyungbuk, Korea.
†Department of Information & Communication, Dong-Guk University, Kyung-Ju, Kyungbuk, Korea.

Abstract. We are developing a robotic system for auto-mounting protein crystals from a sample cassette in liquid nitrogen to a goniometer head. A small industrial six joint axis robot is adopted for this system with a custom built actuator. It will be installed at a wiggler protein crystallography beam-line in Pohang Accelerator Laboratory. Our design goal is to make the system applied to every type of sample pins such as a screw or magnetic types of pins with 18mm length. The sample cassette is designed to contain 10 crystals, which is similar to ALS[1] and ACTOR[2] systems. The diameter of a hole is 17 mm for several type caps. It's bigger than others. A sample Dewar with two type sensors for monitoring liquid nitrogen level can contain 5 cassettes at a time. We designed two types of actuator for holding the cap containing a protein crystal. These whole hardware and software are under development in PAL. Our system and some test results will be introduced in this paper.

Keywords: robot, auto-mounting, actuator, cassette, PAL.
PACS: 01.30.Cc

INTRODUCTION

Pohang Accelerator Laboratory (PAL) has presently three beam-lines for macromolecular crystallography. One is a multipole wiggler source beam-line and the other two are bending magnet source beam-lines. The wiggler source beam-line is our first insertion device beam-line in PLS as an X-ray source. Therefore, we met several problems coming from heat load by the wiggler source, because we have had no experience for operating the wiggler source beam-line. The multipole wiggler has 28 poles with 14 cm period. The characteristics of it are described in Table 1.

It is important to use beam time efficiently in this kind of insertion device source beam-line. Several institutes have been trying to develop an auto-sample mounting system[3]. We also began to develop our auto-mounting system from October 2003. We will describe our analysis of beam time used by domestic crystallographers and then introduce our auto-mounting system (PLS type).

ANAYSIS OF BEAM TIME USED BY USERS

Recently we have installed a program which can monitor the pattern of beam time used by general users. So we made an analysis with accumulated data from March 3rd 2006 to May 16th 2006. We set a criterion of data collection time as 5 minutes in this analysis. Data collection time less than 5 minutes is considered as a crystal screening time. This data collection time involves crystal centering and fluorescence scanning for multi-wavelength anomalous diffraction (MAD) experiment. Table 2 shows the results of beam time analysis. The averaged value of crystal screening time does not include the time for sample centering and exposure. It is only the time for sample

CP879, *Synchrotron Radiation Instrumentation: Ninth International Conference*,
edited by Jae-Young Choi and Seungyu Rah

mounting, demounting and accessing to the hutch. We know that the averaged off time for screening and collection data is longer than the averaged off time for screening. Most users are, on the average, spending more than 5 minutes for crystal screening. One data set can be, on the average, obtained in every 37.3 minutes in this beam-line.

TABLE 1. Analysis results of beam time used by users

Total used time (hour)	453.5
Averaged off time for screening (sec)	308
Averaged off time (sec)	356
Averaged time for taking data (min.)	31.4
No. of data	516
No. of screening	874

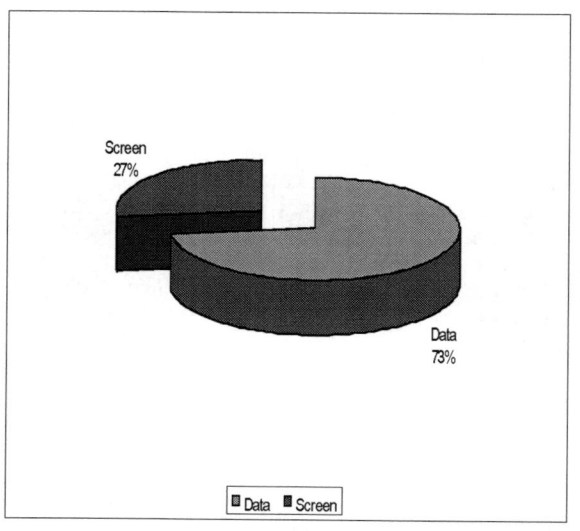
(a)

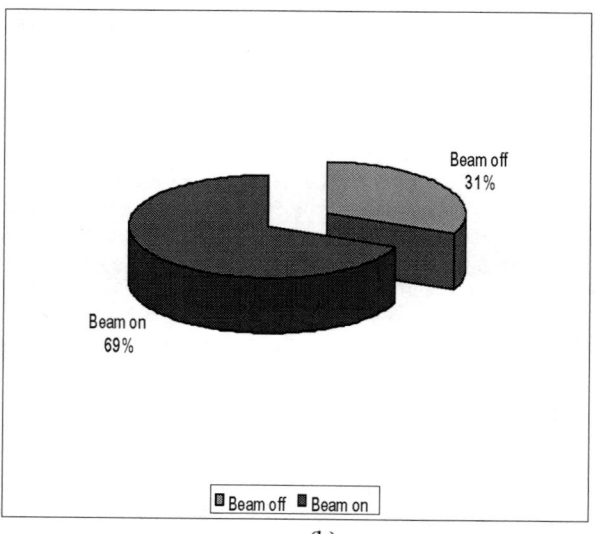
(b)

FIGURE 1. (a) : Ratio time used for screening sample and taking data, (b) : Ratio between wasted and used time

And we have analyzed the beam time to know how users are using their beam time efficiently. Figure1 (a) explains that users are spending 27 % of their beam time for screening samples. From Fig.1 (b), we know that 31% beam time is dumped for mounting and demounting samples, and to access the hutch. From this results, we can know that about 25 % beam time could be saved if the averaged off time were reduced from 356 seconds to less than 60 seconds by using the auto-mounting system. Users could use more beam time in this beam-line and get much more high quality data than that obtained from a bending magnet source beam-line. This is the main reason why we are trying to develop and install the auto-mounting system.

AUTO-MOUNTING SYSTEM

Auto-mounting Robot

There are several types of auto-mounting system in the world. We tried to make a comparison among these types and then decided to use a commercial robot for our auto-mounting system through collecting opinions from a committee organized with users and PLS staffs. The robot (FARA AT2 model made in Samsung (Fig.2 (a)) has six joints and working area of $36 \times 36 \times 36$ cm^3. It has position repeatability of ± 20 μm

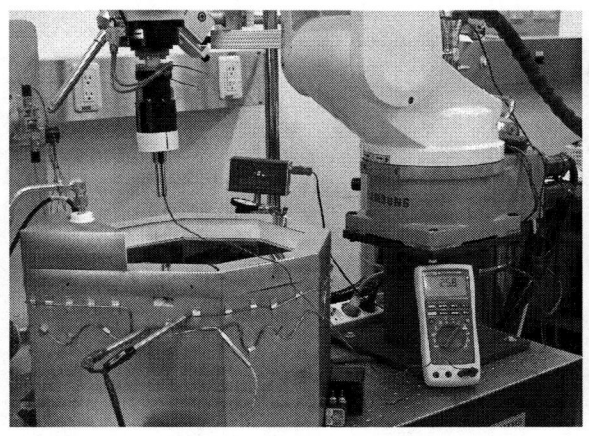 

(a)   (b)

FIGURE 2. (a) This is the picture of the robot attached a finger. (b)A Dewar filled with liquid nitrogen and installed sample cassettes

Finger, Dewar and Cassette

Finger is the core of the auto-mounting system because it decides cassette type and Dewar shape. It should keep the temperature of samples and prevent the crystal from icing caused by air moisture while it carries the sample from a cassette installed in the Dewar filled with liquid nitrogen to a goniometer head. Most of fingers developed already have restricted the type of caps for application. Therefore, we have tried to develop a new type of finger which is applicable to all types including a screw and magnet types of pins. A new finger was developed after all, which is applicable to all types. But our cassette has larger diameter holes, compared with other similar cassettes. Figure2 (b) is the picture of the Dewar filled with liquid nitrogen and installed cassettes

The finger, so called PLS type, shields a crystal from air moisture completely and keeps the temperature of the sample more stable. We measured temperature variations of the shield inside to find the optimum condition for a normal operation. First step is that the finger is dipped in liquid nitrogen for 30 seconds. Second, the finger is put out from liquid nitrogen. Third, a thermocouple is put inside the shield and the temperature of inside is measured. We recorded the time reaching to low temperature of the thermocouple inside the shield. And inside temperature variation of the shield was measured with 30 seconds interval. These procedures were repeated several times. The measured results are displayed in Fig. 3.

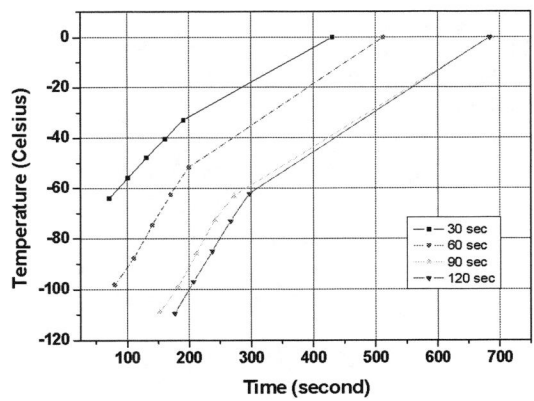

FIGURE 3. Temperature variations of the shield inside vs. time

Control Program

We are using three computers for beam-line control and experiment. The first one is for beam-line optics such as mirrors, double crystal monochromator, slits, and beam-line diagnosis like scanners and screens. A program for controlling these components is developed with Visual C++ language based on Window 2000. The second is for CCD (Quantum 210) operated on LINIUX. Third is for the robot. These all have to be integrated with a concept that users could handle the program easily. Last year, we developed an auto fluorescent scanning program which can perform automatic scanning and selecting a absorption peak and an absorption edge from selected metal atom, and auto X-ray beam alignment program which can automatically align X-ray at the sample position by clicking a button. Figure4 (a) displays a scanned result. So far, we have been developing an integrated program. Figure4 (b) shows a main control screen for combined operation.

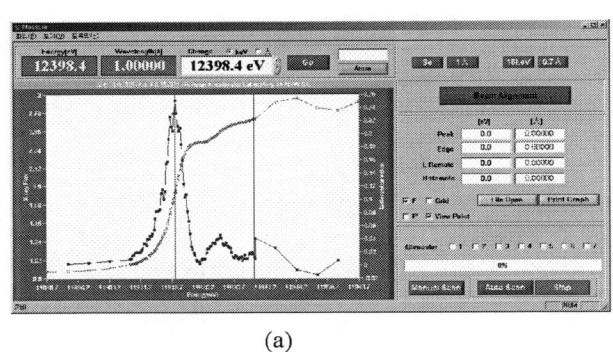

(a)

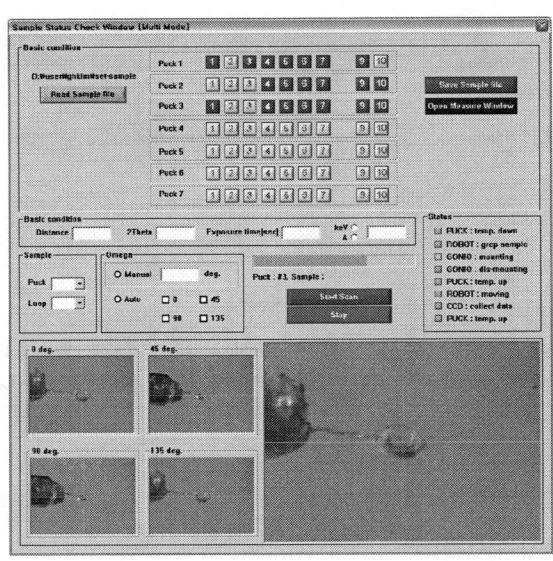

(b)

FIGURE 4. (a) Auto scanning program with a graph scanned at Se absorption edge. (b)A program for integrated operation under developing

SUMMARY

We developed the finger which can grip any types of sample pins. And ten holes cassettes are fabricated. Several tools and a sensor for protecting the finger is under development. Integrated software is under development. All system will be installed and examined from July 2006.

ACKNOWLEDGMENTS

We are grateful to thank to Dr. Soichi Wakatsuki in Photon Factory for his kind advice and assistance, and SSRL staffs who gave us their drawings and explaining for their system. This study was supported by a grant of the Automation of BT Beam-line Project, Ministry of Science and Technology.

REFERENCES

1. G. Snell, C. Cork, R. Nordmeyer, E. Cornell, G. Meige, D. Yegian, J. Jaklevic, J. Jin, R. Stevens, and T. Earnest, *Structure 12*, 537-545 (2004).
2. Web Site of ACTOR: http://www.rigaku.com/protein/actor.html
3. Enrique Abola, Pete Kuhn, Thomas Earnest, and Raymond C. Stevens, *Nature Structural Biology* 7, 973-977 (01 Nov. 2000) Progress.

Radiation Damage of Protein Crystal in Various X-ray Energies

Nobutaka Shimizu[*], Kunio Hirata[†], Kazuya Hasegawa[*], Go Ueno[†]
and Masaki Yamamoto[*,†]

[*]Structural Biology Group, Research and Utilization Division, Japan Synchrotron Radiation Research Institute, 1-1-1 Kouto Sayo, Sayo-gun, Hyogo 679-5198, Japan, and [†]Division of Synchrotron Radiation Instrumentation, RIKEN SPring-8 Center, Hyogo, Japan

Abstract. The radiation damage of crystals is the most serious problem for obtaining accurate structure in protein crystallography. In order to compare the effect of radiation damage among various X-ray energies, we collected 12 to 15 data sets of tetragonal lysozyme crystals at nine different X-ray energies at 6.5, 7.1, 8.3, 9.9, 12.4, 16.5, 20.0, 24.8 and 33.0 keV, using BL41XU at SPring-8. The processing results of each data set became worse in all nine energies as the measurement progressed. However, it was shown that the change of these parameters depends upon the absorbed dose, expressed in Gray (Gy). Disruption of the disulphide bonds due to the radiation damage was observed on the electron density map even in the highest photon energy (33 keV). Therefore, these results suggest that the radiation damage in such an energy region should be discussed based on the absorbed dose.

Keywords: Protein crystallography, Synchrotron radiation, Radiation damage
PACS: 61.10.Nz

INTRODUCTION

The highly brilliant X-rays from synchrotron radiation have become the powerful tool for modern protein crystallography. However, it often gives serious problem arising from radiation damage for structural analysis. As X-ray exposure is accumulated in a protein crystal, broadenings of diffraction spots as well as decrements of diffracted intensities are often observed. For example, the crystallographic statistics, the mosaicity and R_{merge}, are also increasing as the radiation damage progresses. For a cryo-cooled crystal, the theoretical limit of absorbed dose per unit mass is known as Henderson limit, 2×10^7 Gy (1 Gray (Gy) = 1 J/kg) [1]. Various techniques to avoid this phenomenon in protein crystallography were designed and reported. One of them is the helium gas cryostream to lower the sample temperature (~ 40 K). The effect of low temperature around 40 K against the radiation damage has been reported by several works [2, 3]. In addition to this, utilization of the high energy X-ray may be one of the other methods. Many protein crystallographers generally consider that the radiation damage would be decreased at higher X-ray energy from the view point of the absorbance [4]. In order to compare the effect of radiation damage among various X-ray energies, hence, we collected 12 to 15 data sets utilizing 9 different X-ray energies (6.5, 7.1, 8.3, 9.9, 12.4, 16.5, 20.0, 24.8 and 33.0 keV).

MATERIALS AND METHODS

Lysozyme crystals with tetragonal form ($P4_32_12$) were adopted as a sample for this study. The crystallization was performed by hanging-drop vapor diffusion at 20 °C, pH 4.2 with 100 mM acetate buffer including 1.25 M NaCl. Crystals having the approximate size of $100 \times 150 \times 150$ μm^3 were selected for the diffraction measurement. Paratone-N (Hampton Research) was selected as a cryoprotectant for the measurement at 100 K in nitrogen gas stream. The diffraction experiments were performed at BL41XU of SPring-8. In order to compare the radiation damage on the basis of the absorbed dose (Gy = J/kg), we calculated the quantity using *RADDOSE* [5]. The incident

CP879, *Synchrotron Radiation Instrumentation: Ninth International Conference*,
edited by Jae-Young Choi and Seungyu Rah
© 2007 American Institute of Physics 978-0-7354-0373-4/07/$23.00

photon flux needed for estimating the absorbed dose was calculated from the ionization current measured with the PIN photodiode installed at the sample position. In order to adequately calibrate the PIN photodiode, the observed flux at 12.4 keV was compared with a reference value, which was observed by an ion chamber (pure nitrogen gas flow type). Photon flux at other photon energies were obtained by theoretical calculation based on the conversion efficiency and absorbance of the detector materials. The beam size at sample position was measured by scanning with 30 μm pinhole. Two sets of the beam shaper slits were adjusted so that the beam size at the sample position was 100×100 μm^2 (F.W.H.M.). Table 1 shows the experimental conditions for the nine energies studied. One data set includes 180 images with an oscillation step of 1.0 °. A Quantum 315 (ADSC) was used as the detector. The shortest available detector distance was 155 mm in BL41XU. The exposure time was adjusted in the range from 0.3 to 5 seconds on the basis of the diffraction intensity. All of the 130 data sets were processed and scaled using *HKL2000* [6]. The processing results of the first data set of each energy are listed in Table 2. Molecular replacement was carried out with *AMoRe* [7]. Structural refinement of the data at 8.3, 12.4, 24.8 and 33.0 keV was carried out using *REFMAC 5* [8].

TABLE 1. Experimental condition at each energy.

X-ray Energy (keV)	Detector distance (mm)	Exposure time (sec.)	Irradiated time per 1data (sec.)	Number of datasets	Photon flux (photons/sec)	Dose per 1data (Gy)
33.0	450	5.0	900	13	1.6×10^{11}	2.2×10^4
24.8	310	5.0	900	12	1.6×10^{11}	1.2×10^4
20.0	270	1.0	180	15	3.6×10^{11}	3.2×10^3
16.5	180	0.3	54	15	4.0×10^{11}	1.9×10^3
12.4	155	0.3	54	15	5.0×10^{11}	8.3×10^2
9.9	155	0.5	90	15	1.3×10^{11}	2.0×10^3
8.3	155	1.0	180	15	8.8×10^{10}	2.1×10^3
7.1	155	1.5	270	15	5.5×10^{10}	2.5×10^3
6.5	155	1.6	288	15	5.0×10^{10}	2.3×10^3

TABLE 2. The result of the first data set.

X-ray Energy (keV)	Resolution range (Å)	a=b (Å)	c (Å)	R_{merge}[a]	$<I/\sigma(I)>$[a]	Completeness (%)	Mosaicity
33.0	25.0 - 1.35	78.56	37.21	0.081 (0.295)	31.7 (11.5)	100.0	0.120
24.8	25.0 - 1.21	78.50	37.25	0.072 (0.294)	35.7 (6.0)	98.2	0.143
20.0	25.0 - 1.24	78.59	37.26	0.061 (0.294)	44.6 (6.9)	99.9	0.178
16.5	25.0 - 1.25	78.55	37.24	0.064 (0.295)	36.6 (6.0)	99.5	0.117
12.4	25.0 - 1.33	78.61	37.20	0.048 (0.310)	50.3 (7.0)	99.8	0.196
9.9	35.0 - 1.50	78.58	37.25	0.063 (0.228)	43.2 (6.9)	99.5	0.160
8.3	50.0 - 1.85	78.64	37.25	0.068 (0.162)	41.9 (13.3)	99.4	0.181
7.1	50.0 - 2.06	78.46	37.29	0.067 (0.106)	40.0 (15.7)	99.0	0.150
6.5	50.0 - 2.33	78.70	37.18	0.063 (0.120)	44.5 (16.9)	98.9	0.207

[a]Values in parentheses are for the highest-resolution shell.

RESULT AND DISCUSSION

In general, it is necessary to conduct diffraction data collection in a wide energy range to evaluate energy dependent effects in protein crystals. The high flux X-ray that can cause the radiation damage in short time is suitable for the repetitive measurement. Since the light source of BL41XU at SPring-8 is the in-vacuum undulator [9], not only the high flux but also the wide X-ray energy range (6.5 − 35.0 keV) are available. Therefore, BL41XU is considered to be a suitable beamline to conduct such energy dependency comparison. Although the detector distances were changed based on the maximum resolution of the diffraction, the diffraction intensity in a wider angle could not be recorded at energies lower than 9.9 keV in the present study because of the limitation of the detector distance. The exposure time of 33.0 and 24.8 keV is longer than that of the others. The exposure time of 5 seconds or longer was necessary for obtaining the data with enough statistical accuracy at these energies. The reason is presumably that the phosphor of CCD detector had been optimized for 1 Å photons. The absorbed doses were calculated with *RADDOSE* by using various parameters, for example, the X-ray energy, the crystal size, the solvent contents for the crystallization, and the beam size, etc. Since the exposure time was also one of parameters taken into account, the long exposure time at two high energy experiments should not seriously influence the following discussions.

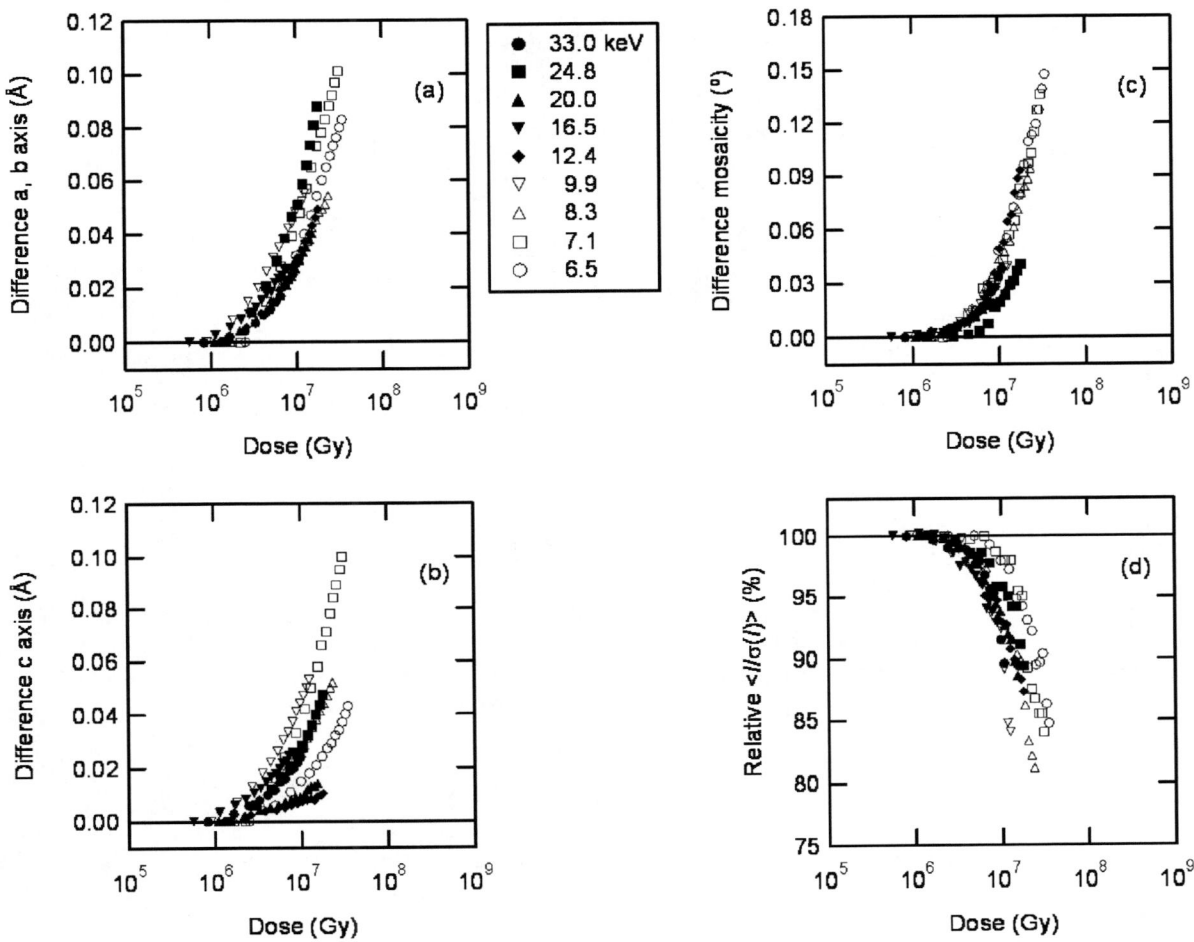

FIGURE 1. Comparison of the effect of the radiation damage among nine energies. These parameters obtained by processing and scaling each data set were plotted against the absorbed dose calculated by *RADDOSE*.

As shown in table 2, the lattice constants of the first data set are the almost same among all energies. For data sets measured at 12.4 keV or higher photon energy, the resolution cut off was defined in such a way that R_{merge} in the highest resolution shell did not exceed 30%. On the other hand, the resolution cut off for the 9.9 keV data set or lower photon energy was determined by the detector edge, because of the limitation of the detector distance. In both cases, the resolution range at each energy was fixed on the first set and applied to the analysis of all data sets. It was also shown that these crystals kept good mosaicities, although they were not identical. In Fig. 1, the changes of the lattice constants, the mosaicity and $<I/\sigma I>$ are compared among nine energies. These parameters are plotted against the absorbed dose. In all data sets, remarkable radiation damages are observed after the absorbed dose exceeded 10^6 Gy. Furthermore, the plotted curve for all the parameters in all X-ray energies are almost superimposing. The result reveals that the effect of the radiation damage depends primarily on the absorbed dose. It is found that the level of the radiation damage can be discussed on the basis of the absorbed dose at least between 6.5 and 33.0 keV.

In order to evaluate the change in the electron density map, structural refinements were performed using the 8.3, 12.4, 24.8 and 33.0 keV data sets. Figure 2 shows the electron density map calculated at 1.9 Å resolution in 8.3 and 33.0 keV as representative data. Breakage of the disulphide bond due to radiation damage was observed on the electron density map not only at 8.3 keV but also at 33.0 keV (the highest energy used in this study). The same breakages were also observed in the other two energies (data not shown). It is, therefore, suggested that the radiation damage occurred in the high energy experiments as the same manner as observed at lower energy.

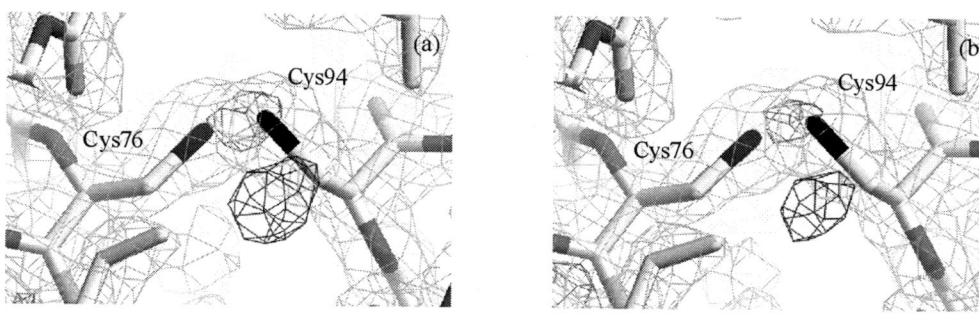

FIGURE 2. The electron density map of the last data set measured at 8.3 keV (a) and 33.0 keV (b). Carbon, oxygen, nitrogen and sulfur atoms are colored white, light grey, dark grey and black, respectively. Both 2Fo-Fc maps contoured at 1.0 σ (light grey) and Fo-Fc maps contoured at 3.0 σ (black) or -3.0 σ (dark grey) are superimposed on the disulphide bond between Cys76 and Cys94. These figures were drawn using CueMol (http://www.cuemol.org).

CONCLUSION

A systematic study of the radiation damage on the lysozyme crystals has been conducted for various X-ray energies at SPring-8 BL41XU. In this study, it was confirmed that the radiation damage depends on the absorbed dose and that no dependence on X-ray energy was observed. This study suggests that optimizing absorbed doses in crystals for diffraction data collection is critical at any X-ray energy to obtain accurate protein structures.

ACKNOWLEDGMENTS

We are grateful to Dr. Garman of the Oxford University for valuable advice and discussion. We also thank all the member of both the structural biology group in JASRI/SPring-8 and the synchrotron radiation instrumentation group in RIKEN SPring-8 Center.

REFERENCES

1. Henderson, R., *Proc. R. Soc. London Ser. B* **241**, 6-8 (1990).
2. Nakasako, M., Sawano, M., and Kawamoto, M., *The Rigaku Journal*, 47-53 (2001).
3. Hanson, B. L., Harp, J. M., Kirschbaum, K., Schall, C. A., DeWitt, K., Howard, A., Pinkerton, A. A. and Bunick, G. J., *J. Synchrotron Rad.* **9**, 375-381 (2002).
4. Jan Drenth, *Principles of Protein X-ray Crystallography*, Springer-Verlag, Berlin; Heidelberg; New York, 1994.
5. Murray, J. W., Garman, E. F. and Ravelli, R. B. G., *J. Appl. Cryst.* **37**, 513-522 (2004).
6. Otwinowski, Z. and Minor, W., Processing of X-ray Diffraction Data collected in Oscillation Mode, *Methods Enzymol.*, 276, edited by Carter, C. W. Jr. and Sweet, R. M., New York: Academic Press, 1997, pp. 307-326.
7. Navaza, J., *Acta Crystallogr.* **A50**, 157-163 (1994).
8. Murshudov, G. N., Vagin, A. A. and Dodson, E. J., *Acta Crystallogr.* **D53**, 240-255 (1997).
9. Kitamura, H., *J. Synchrotron Rad.* **7**, 121-130 (2000).

X-ray Crystallography at High Pressure to Probe Conformational Fluctuations in Biological Macromolecules.

Eric Girard[1], Richard Kahn[2], Anne-Claire Dhaussy[3], Isabella Ascone[1,4], Mohamed Mezouar[5] and Roger Fourme[1]

[1] Synchrotron-SOLEIL, BP48 Saint Aubin, 91192 Gif sur Yvette, France
[2] IBS, 41 rue Jules Horowitz, 38027 Grenoble Cedex, France
[3] CRISMAT ENSICAEN, Bd Maréchal Juin, 14000, Caen, France
[4] Physics Department, Bldg E. Fermi, Piazzale A. Moro, University of Rome La Sapienza, Italy
[5] ESRF, BP220, 38027 Grenoble, France.

Abstract. Macromolecular crystals can be compressed hydrostatically at room temperature in a diamond anvil cell. The quality of diffraction data recorded on the ESRF ID30/ID27 beamlines using a parallel X-ray beam of ultra-short wavelength can meet usual standards. The 3D structures of proteins (monomeric, dimeric and tetrameric) and of a virus have been refined both at atmospheric and at high pressure. High pressure is a way to explore the high energy landscape of macromolecular systems, from the fully folded state to the unfolded state. High energy conformers of biological significance can be selected and trapped under high pressure.

Keywords: high pressure, macromolecular crystallography, protein, virus, fluctuations, energy landscape, diamond cell.
PACS: 87-15

INTRODUCTION

The combination of macromolecular crystallography and pressure perturabation was pioneered nearly 20 years ago [1] and other publications have been very scarce [2] until recently. Progress in instrumentation including the use of diamond cells [3] and ultra-short wavelength synchrotron radiation was the basis of a renewal in the last few years [4]. High pressure macromolecular crystallography (HPMX) can now be considered as mature and prospects are bright, as discussed in a recent mini-review [5].

INTEREST OF HPMX

According to Le Chatelier's principle, a perturbed system shifts its equilibrium state so as to counteract the effect of the disturbance. Consequently an increase in pressure favours reduction of the volume of the system, with several important consequences. Sufficiently high pressure has a denaturing effect on protein because the volume is smaller in the unfolded state than in the folded state. When a solution of macromolecules consists at atmospheric pressure of a series of subensembles differing in volumes, increasing pressure will populate subensembles of increasingly smaller volume [6]. Pressure increases the population of higher-energy conformers, as shown by high pressure NMR [7-9]. By choosing an appropriate pressure, an intermediate conformer can be stably trapped in local free energy minimum. Proteins have evolved to sample multiple defined conformations that are critical for function, and pressure may allow us to explore such conformations. Indeed, beyond classical studies such as the molecular basis of life adaptation to extreme conditions, exploring the energy landscape of a macromolecule - and accordingly most of its biologically relevant conformational space - is probably the major interest of combining pressure perturbation with structural methods.

A macromolecular crystal is sufficiently plastic to accommodate not only elastic compression of the lower energy subensemble, but also subensembles with relatively large conformational changes. With respect to solution

CP879, *Synchrotron Radiation Instrumentation: Ninth International Conference,*
edited by Jae-Young Choi and Seungyu Rah
© 2007 American Institute of Physics 978-0-7354-0373-4/07/$23.00

studies by NMR or other techniques, the crystalline state may have distinct advantages in the study of high energy conformers: MX can reach atomic or quasi-atomic resolution on systems of arbitrary complexity; further, we have suggested, on the basis of thermodynamic arguments, that the crystalline state may act as a "conformation filter" [5] favouring monodispersity and paving the way for high resolution studies of kinetic intermediate states of biological relevance.

MATERIALS

Data collection has been performed at the ESRF (Grenoble) with a setup [4] initially installed on the high pressure beamline ID30 then on ID27 [10]. Main components of ID27 are the following:

- X-ray source and beam optics: ID27 features two in-vacuum U23 undulators, which are very brilliant sources of X-rays even at ultra-short wavelengths. The radiation is monochromatized by a (Si 111) channel cut crystal. The multilayer Kirkpatrick-Baez optics is generally not used, as a parallel beam is required for HPMX experiment.

- high pressure device is diamond anvil cell (DAC) of the cylinder-piston type where the thrust applied to the piston is generated by a toroidal metallic membrane inflated by helium. Two coaxial diamonds are mounted on metal supports. The first is part of the cell body and the second one is attached to the piston. Two new DAC providing useful apertures around 90° (coll. JC Chervin and B. Couzinet, IMPMC Paris; JP Itié, SOLEIL) are currently under test, one with thin diamonds, the other with thicker conical diamonds adapted to the higher pressure range (Girard et al., to be published).

- monitoring pressure and viewing sample. The pressure in the compression cavity is monitored by means of the wavelength shift of the fluorescence emission of a ruby chip. The optical system, placed on the X-ray beam path between the DAC and the detector, is also used to image the sample. It is removed out of the X-ray beam path during data acquisition.

- goniometer: we use now a three-circle goniometer mounted on a stack of three accurate orthogonal translations.

- detector: data were collected with a MAR345 imaging plate. The wavelength was adjusted at 0.0331 nm (on the high energy side of the Ba K-absorption edge) where the detector efficiency is maximized for elastic scattering and reduced for softer Compton scattering, thus improving the signal-to-noise ratio. A MAR 555 with direct conversion of photons in a Se layer has also been tested at the same wavelength. Attractive characteristics of this new detector for HPMX, include reasonable efficiency, short readout time, large area and small PSF.

METHODS

A metal gasket is placed between the diamonds and indented. A cavity is produced by electro-erosion at the centre of the indentation. Macromolecular crystals are prepared at atmospheric pressure by conventional techniques (e.g. hanging drop). The compression medium is generally the mother liquor used for crystallization, with in some cases a slightly modified composition. A crystal is gently transferred to a drop of this liquor deposited on the cavity. As compression must be hydrostatic, the ultimate pressure is set by solidification of liquor (~ 1 GPa for pure water at room temperature and somewhat higher for water mixtures). In the elastic compression regime (and of course below the pressure of protein denaturation), macromolecular crystals withstand pressure extremely well (figures 1and 2). Due to the plasticity and annealing capacity of the crystalline phase, packing contraction as well as relatively large conformational changes in macromolecules are in many cases accomodated without loss of 3D order.

Data collection is performed by the rotation method, recording "frames" (figure 2) over small contiguous rotations of the cell around the vertical axis of the goniometer. In order to get high quality electron density maps, it is crucial to collect high completeness diffraction data devoid of blind zones in reciprocal space. A large oscillation domain is accesible thanks to new large aperture DAC. Crystals with a very anisotropic shape (e.g. thin plate) tend to orient always in the same way with respect to culet tips. In such cases, different orientations can be obtained by depositing the crystal on a diamond chip (transparent to X-rays) introduced into the cavity.

The beam cross section at the sample location is ~50x50 microns. The crystal is moved every few frames in order to irradiate successively fresh zones of the crystal.

These particular experimental conditions ensure a high signal to noise ratio [4,11], which explains why the quality of data is good in spite of the small irradiated volume. Indeed, the quality (resolution, R_{sym}, completeness) of HPMX data can meet usual standards [11-14] and lead to refinements with low R and Rfree factors.

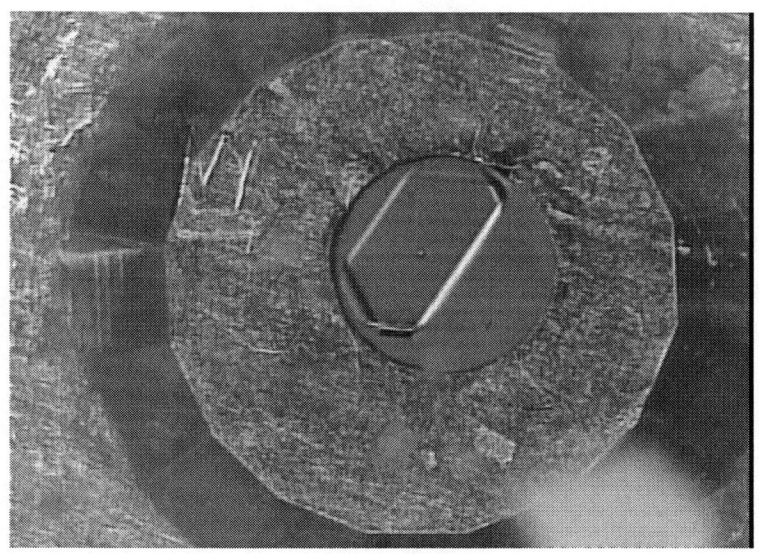

FIGURE 1. Urate oxidase crystal compressed at 140 MPa (3D structure in [14]).

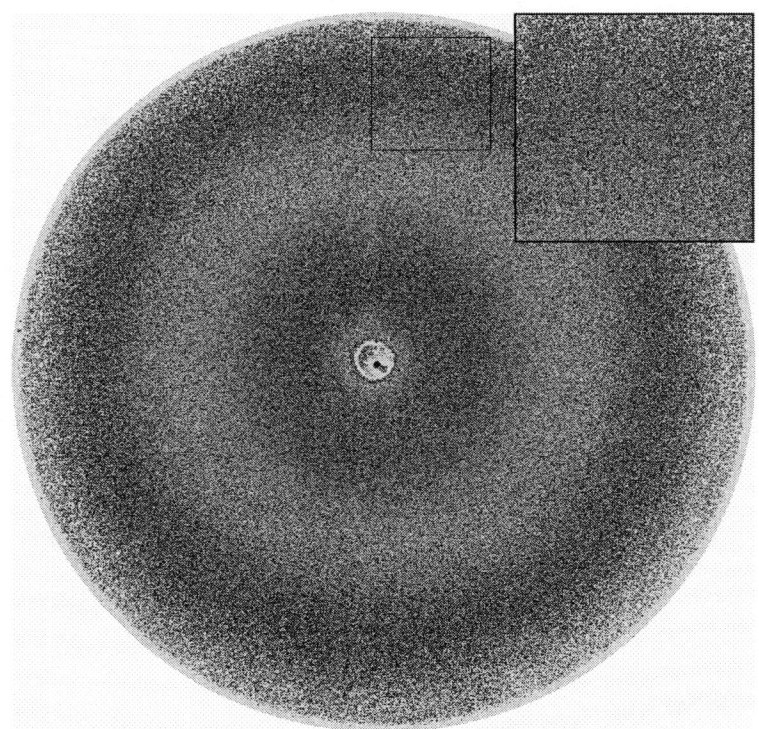

FIGURE 2. Diffraction picture of Cowpea Mosaic Virus crystal at 330 MPa. Resolution 0.27 nm (3D structure in [13]).

Conventional MX (i.e. at atmospheric pressure) could take advantage of techniques developed for HPMX [11]. On the one hand, by using ultra-short wavelengths and parallel beam geometry. On the other hand by taking advantage of moderately high pressure to improve order in pre-existing crystals. These issues are part of investigations under way in the context of a Long Term Project at ESRF on ID27.

RESULTS

System	Aminoacids in asymmetric unit	Goals	Status
Hen egg-white Lysozyme (HEWL)	129	Structural evolution under HP.	Data collected at 700 and 650 MPa with refinement. New data to be collected at various pressures. Published in preliminary form [12].
Cowpea mosaic virus (CpMV)	5 x (189 + 369) = 2790	Disorder-to-order packing transition. Evolution of viral capsid proteins under HP.	Structure refined with 5-fold NCS at 330 MPa and AP. Published [13].
Urate oxidase from *A. flavus*	295	Evolution of 3D structure and tetrameric association under HP.	Structure refined at 140 MPa and AP. Onset of tetramer dissociation captured in crystal. Published [14].
Cellulase from *P. Haloplanktis*	2 x 293 = 586	Molecular basis of life adaptation to low temperature and HP.	3D-structures refined at 175 MPa and AP. *Coll. N. Aghajari, IBCP Lyon, France*
Bovine Cu,Zn superoxide dismutase (SOD)	2 x 150 = 300	Evolution of 3D structure and dimeric association under HP. Phase transition at the active site.	3D structure refined at 540 MPa .
Green Fluorescent Protein (GFP), wild type	230	Evolution of 3D structure and chromophore environment with pressure.	Data collection planned 2006. *Coll. P. Oger, ENS Lyon, France*
Ubiquitin	76	Trapping and 3D structure of high energy conformer.	Data collection planned 2006. *Coll. K. Akasaka., Kinki Univ., Japan*
Oligonucleotides	--	Evolution of 3D structure with pressure.	Data collection planned 2006. *Coll. T. Prangé, Paris 5 Univ., France*

REFERENCES

1. C. E. Kundrot and F. M. Richards, *J. Mol. Biol.* **193**, 157-170 (1987).
2. P. Urayama, G.N. Phillips and S.M. Gruner, *Structure* **10**, 51-60 (2002).
3. A. Katrusiak and Z. Dauter, *Acta Cryst. D* **52**, 607-608 (1996).
4. R. Fourme, R. Kahn, M. Mezouar, E. Girard, C. Hörentrup, T. Prangé and I. Ascone, *J. Synchr. Rad.* **8**, 1149-1156 (2001).
5. R. Fourme, E. Girard, R. Kahn, A-C Dhaussy, M. Mezouar, N. Colloc'h and I. Ascone, *BBA* **1764**, 384-390 (2006).
6. K. Akasaka, *Biochemistry* **42**, 10875-10885 (2003).
7. R. Kitahara, S. Sareth, H. Yamada, K. Akasaka, F. Ohmae, K. Gekko and K. Akasaka, *Biochemistry* **39**, 12789-12795 (2000).
8. K. Inoue, H. Yamada, K. Akasaka, C. Hermann, W. Kremer, T. Maurer, R. Doeker and H.R. Kalbitzer, *Nat. Struct. Biol* **7**, 547-550 (2000).
9. R. Kitahara, S. Yokohama and K. Akasaka, *J. Mol. Biol.* **347**, 277-285 (2005).
10. M. Mezouar, W.A. Crichton, S. Bauchau, F. Thurel, H. Witsch, F. Torrecillas, G. Blattmann, P. Marion, Y. Dabin, J. Chavanne, O. Hignette, C. Morawe and C. Borel, *J. Synchr. Rad.* **12**, 659-664 (2005).
11. R. Fourme, E. Girard, R. Kahn, I. Ascone, M. Mezouar, A. C. Dhaussy, T. Lin and J. E. Johnson, *Acta Cryst. D* **59**, 1914-1922 (2003).
12. R. Fourme, I. Ascone, R. Kahn, E. Girard, M. Mezouar, T. Lin and J. E. Johnson, in *New trends in macromolecular crystallography at high hydrostatic pressure*, edited by R. Winter, Heidelberg : Springer, 2003, pp 161-170.
13. E. Girard, R. Kahn, M. Mezouar, A-C Dhaussy, T. Lin, J. E. Johnson and R. Fourme, *Biophysical J.* **88**, 3562-3571 (2005).
14. N. Colloc'h , E. Girard, A.-C.Dhaussy , R. Kahn, I. Ascone, M. Mezouar and R. Fourme, *BBA* **1764**, 391-397 (2006).

Synchrotron Radiation Computed Tomography Station at the ESRF Biomedical Beamline

C. Nemoz[1], S. Bayat[1], G. Berruyer[1], T. Brochard[1], P. Coan[1], G. Le Duc[1],
J. Keyrilainen[1], S. Monfraix[1], M. Renier[1], H. Requardt[1], A. Bravin[1], P. Tafforeau[1,6],
J.F. Adam[2], MC. Biston[2], C. Boudou[2], AM. Charvet[2], S. Corde[2], H. Elleaume[2],
F.Estève[2], A. Joubert[2], J. Rousseau[2], I. Tropres[2], M. Fernandez[3], L. Porra[3],
P. Suortti[3], S. Fiedler[1,4], W. Thomlinson[1,5]

[1]European Synchrotron Radiation Facility, ESRF, BP 220, 38043 Grenoble, France
[2]INSERM U647, c/o ESRF, BP 220, 38043 Grenoble, France
[3]University of Helsinki Central Hospital, POB 340, FIN-00029 HUS, Helsinki, Finland
[4]EMBL c/o DESY, Notkestr. 85, D-22603 Hamburg, Germany
[5]CLS, 101 Perimeter Road Saskatoon, SK., Canada. S7N 0X4
[6]CNRS, LGBPH, Université de Poitiers, 40 avenue du Recteur Pineau 86022 Poitiers Cedex, France

Abstract. The different tomography imaging modalities of the ESRF Medical Beamline are described and research applications are presented

Keywords: Medical, Imaging, Tomography
PACS: 07.85.Qe

INTRODUCTION

Synchrotron radiation is now applied to virtually all areas of biomedical sciences using ionizing radiation. The availability of intense monochromatic X-ray beams over a wide energy range differentiates these sources from standard clinical and research tools and gives access to *in vivo* and *in vitro* research not possible otherwise. At the European Synchrotron Radiation Facility (ESRF) a dedicated biomedical beamline (ID17) has been built to perform coronary angiography clinical trials [1]. The protocols took place between years 2000 and 2003. In the meantime, a powerful Synchrotron Radiation Computed Tomography (SRCT) system has been developed. Thanks to the availability of different monochromators and detectors, the system can be used either for *in vivo* high temporal resolution experiments or *static* high spatial resolution experiments. SRCT was then one of the main activities of the beamline during 2002-2005. Other activities are radiotherapy and planar imaging.

The main SRCT scientific topics addressed are Brain Imaging (25 % of SRCT beam time), Mammography and Cartilage imaging with Diffraction Enhanced Imaging (DEI) (25 %), Bronchography in Lungs Physiology Research (20 %), Fossils Imaging in Paleontology (15 %) and Cerebral Angiography (10 %).

BEAMLINE COMPONENTS

The complete beamline design can be found in [2]. The X-ray beam produced by the wiggler (typically: 0.7 T, gap: 50 mm - critical energy: 16.5 keV) travel across slits and attenuators before impinging on the monochromator systems located 150 m from the source. This long distance permits to obtain a wide fan beam (up to 150 mm width, 5 mm height). Then the monochromatized beam delivered in the imaging hutch cross the sample located on a rotating stage and the attenuation profiles are recorded by the detector systems.

CP879, *Synchrotron Radiation Instrumentation: Ninth International Conference,*
edited by Jae-Young Choi and Seungyu Rah
© 2007 American Institute of Physics 978-0-7354-0373-4/07/$23.00

Monochromator Systems

The optics hutch is equipped with two monochromator systems.

The first one called TOMO is based on double-crystal bent Laue scheme and provides a fixed-exit beam over an energy range from 17 keV to about 90 keV. This monochromator has two advantages for *in vivo* tomography: the produced beam is horizontal in such a way the sample rotates in a horizontal plane, and changing beam energy is very fast (< 2 minutes). Typical photon flux is 4.10^8 photons/s/mm^2.

The second system, ANGIO (it has been originally developed for the planar coronary angiography imaging protocols) is a single crystal monochromator, operated in the bent Laue transmission mode. It can provide two separated inclined beams with respective energy bracketing the K-edge absorbing threshold of a contrast agent and is used for the K-Edge Digital Subtraction method (KEDS). The energy can be selected in the range 17 to 80 keV. Typical photon flux is 6.10^{11} photons/s/mm^2. As the ANGIO system is non standard, it is displayed on Fig 1.

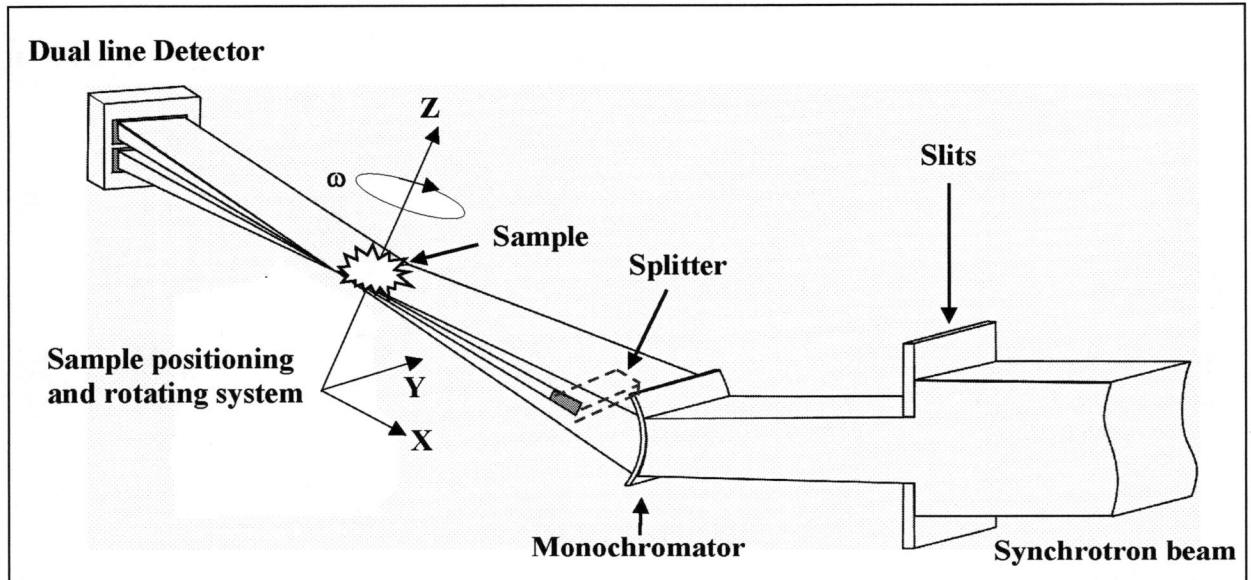

FIGURE 1. Experimental setup in K-edge mode with the ANGIO monochromator and the Germanium detector

Detector Systems

The imaging hutch is equipped with two detectors which can be chosen according to the experimental needs.

The first one (Germanium) is a high purity solid-state detector (Eurisys Mesures). It is a monolithic germanium crystal 150 mm long with two rows of 432 parallel strips, having a pitch of 350 μm. Each row can be acquired simultaneously in KEDS mode. The minimum time between two projections recording is 1 ms, so the typical time between two images can be less than 0.5 s. In computed tomography, angular projections acquisitions during 180° rotation is enough. However, by shifting the rotation center near the side of the detector and by doing acquisition on 360°, the field-of-view can reach 300 mm.

The other detector (FReLoN, for Fast REadout LOw Noise) is a Peltier-cooled 2048x2048 pixel CCD-based camera, developed at the ESRF [3]. It is coupled, via tapered fiber optics, to an exchangeable fluorescent screen from different thickness (40-200μm). The field-of-view of the taper optics is 94x94mm2 with a corresponding pixel size of 46x46μm2. As in the previous case, by redundant acquisition the field-of-view can be increased up to 180 mm. Because the incident beam is 1D-like, only a horizontal region of interest is readout. As the data transfer across the CCD dominates the readout time, it has been necessary to synchronize the acquisition with a beam chopper [4] also developed at ESRF to avoid beam exposure and to save dose during readout. This chopper consists in 12 planetary stainless-steel blades mounted on two large discs. The blades are kept permanently parallel between them by a series of gears, which also allow changing the orientation of the blades, and thus allow changing the duty cycle.

Rotating Devices

The Medical Beamline was designed to permit clinical trials in such a way a Patient Positionning System (PPS) has been built to vertically scan the patient chest for angiography protocol planar imaging. The PPS is also able to rotate the Patient for tomography or tomotherapy purposes. This rotation can handle heavy sample and is very precise at 180°/s speed. However it cannot be inclined and so it cannot be used with the ANGIO monochromator. As this rotation is reliable at high speed due to position and speed feedback regulation, this system cannot be used with the FReLoN camera too. Thus K-edge or FReLoN camera acquisitions are based on small stepper motor rotating stages.

CONTROL AND ANALYSIS SOFTWARE

The combinations of these beamline components require that the control system and the reconstruction software's are very flexible. The control system has to handle the detectors and the rotating motors in such a way a very precise synchronization between the data acquisition and the angular rotation can be achieved. The synchronization is even more challenging when the chopper, the contrast agent injector or the vertical movement must be included in the chain. The control system is built on a client-server basis using SPEC (Certified Scientific Software) as the client and Linux drivers running the equipment as servers. The sinogram treatment is handled with IDL (Research Systems) software. This software is parameterized in function of the imaging modality to permit image subtraction either in KEDS, temporal or DEI mode. Some tools to define rotation center, to filter and clean the sinograms have been included. The tomography back-projection reconstruction algorithm is either Snark (Pennsylvania University) or HST, developed at ESRF.

RESULTS AND DISCUSSION

Avoiding beam-hardening effects, absolute concentration measurements of a contrast agent are achievable with an optimal combination of signal to noise ratio and X-ray dose parameters. Regarding the broad variety of applications by combining the different beamline components as summarized in Table 1, the achievements of the ID17 experimental tomography station are numerous and challenging. For some applications the system must be as fast as possible in order to achieve dynamic studies as Cerebral Blood Volume (CBV) and Cerebral Blood Flow (CBF) assessment in Cancer Research, and Bronchography, whereas high-spatial resolution is more important in DEI and Paleontology. The rotating devices permit SRCT imaging of a large variety of samples, ranging from small rodents to pigs. In combination with the precise vertical motion of the PPS, fast helical tomography can be achieved to record volumes.

The image acquisition is very flexible. It can be operated in the KEDS mode where the two beams with slightly different energies tuned around the K-edge of the contrast agent are recorded at the same time. This mode is mandatory for moving objects like lungs. On the other hand, the temporal digital subtraction mode, which quickly records images before and after contrast agent injection, will provide better quantitative results but is applicable only for static objects like a rodent brain.

Figures 2 and 3 illustrate the very different types of results one can obtain with this system. Figure 2 shows a typical image of a rat brain where the tumor is perfused by some contrast agent (1 s data acquisition) whereas Fig. 3 shows the volume reconstruction of a precious fossil (24 h data acquisition). Images acquired with the ANGIO mono in KEDS mode may be found in [5].

TABLE 1. Typical Tomography Imaging Modalities

Monochromator \ Detector	Germanium	FReLoN
TOMO monochromator	Low spatial resolution, fast acquisition *In vivo* temporal subtraction possible to use PPS rotation **brain perfusion imaging**	high spatial resolution, slow acquisition Volume reconstruction **Diffraction Enhanced Imaging** **Paleontology**
ANGIO monochromator	Low spatial resolution, fast acquisition *In vivo* K-edge subtraction **lungs imaging**	high spatial resolution, slow acquisition *In vivo* K-edge subtraction **cerebral angiography**

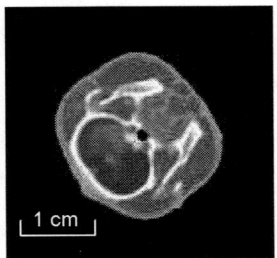

FIGURE 2. Tomographic slice of a rat bearing a brain tumor highlighted here by some iodinated contrast agent

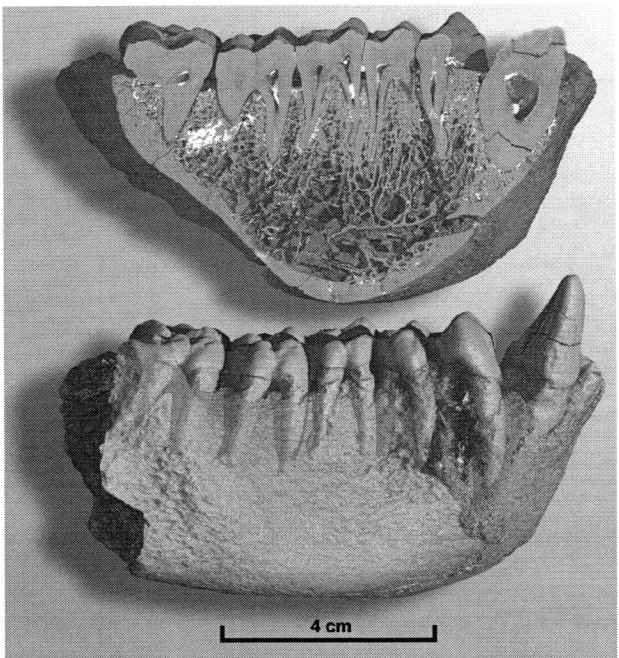

FIGURE 3. Mandible of an ancestor of modern orang-utans from Thaïland. Right teeth have been virtually pulled off.

ACKNOWLEDGMENTS

The authors gratefully acknowledge MC. Dominguez, D. Fernandez, H. Gonzalez, G. Goujon, JC. Labiche, A. Mirone and M. Perez from ESRF instrument and computing support groups for their high implication in the development and commissioning of the apparatus.

REFERENCES

1. B. Bertrand et al, Comparison of synchrotron radiation angiography with conventional angiography for the diagnosis of in-stent restenosis after percutaneous transluminal coronary angioplasty - *European Heart Journal* - Vol.26 - pp.1284-1291 - 2005
2. H. Elleaume et al, Instrumentation of the ESRF medical imaging facility. *Nucl Instr and Meth A* 428, 513-527, 1999
3. P. Coan et al, Evaluation of image performance of a taper optics CCD FReLoN camera designed for medical imaging. *Journal of Synchrotron Radiation* 13, 260-270 (2006).
4. M. Renier et al, A mechanical chopper with continuously adjustable duty cycle for a wide X-ray beam, *MARS 2004 Proceedings, Nucl Instr and Meth A* 548, 111-115, 2005
5. S. Monfraix et al, Quantitative measurement of regional lung gas volume by Synchrotron Radiation Computed Tomography – *Phys. Med. Biol* 50, 1-11 (2005)

Synchrotron X-ray PIV Technique
for Measurement of Blood Flow Velocity

Guk Bae Kim[1,3], Jung Ho Je[2,3] and Sang Joon Lee[1,3]

[1]Dept. of Mechanical Eng., Pohang University of Science and Technology, Pohang 790-784, Korea
[2]Dept. of Material Eng., Pohang University of Science and Technology, Pohang 790-784, Korea
[3]Systems Bio-Dynamics Research Center, Pohang University of Science and Technology, Pohang 790-784, Korea

Abstract. Synchrotron X-ray micro-imaging method has been used to observe internal structures of various organisms, industrial devices, and so on. However, it is not suitable to see internal *flows* inside a structure because tracers typically employed in conventional optical flow visualization methods cannot be detectable with the X-ray micro-imaging method. On the other hand, a PIV (particle image velocimetry) method which has recently been accepted as a reliable quantitative flow visualization technique can extract lots of flow information by applying digital image processing techniques However, it is not applicable to opaque fluids such as blood. In this study, we combined the PIV method and the synchrotron X-ray micro-imaging technique to compose a new X-ray PIV technique. Using the X-ray PIV technique, we investigated the optical characteristics of blood for a coherent synchrotron X-ray beam and quantitatively visualized real blood flows inside an opaque tube without any contrast media. The velocity field information acquired would be helpful for investigating hemorheologic characteristics of the blood flow.

Keywords: X-ray micro-imaging, PIV, X-ray PIV, Blood flow, Quantitative velocity field
PACS: 87.59.-e, 07.85.Qe, 87.57.Ce

INTRODUCTION

Most *flow visualization techniques* using visible light are unsuitable to visualize opaque human blood flow. Only for the case of micro-flows, we can visualize blood flows inside a microchannel of *several tens* micrometer in depth by tracing RBCs or tracer particles using such a micro-PIV technique[1]. However, because most important hemodynamic phenomena related with vascular diseases occur in the blood vessels of several millimeters or even larger in diameters, such conventional flow visualization methods can not be used for direct visualization of blood flows. On the other hand, most *medical diagnosing instruments* are capable of observing the internal *structures* of opaque objects, but they have not been developed yet to the level of visualizing internal *flows* of opaque objects. Therefore, direct visualization of blood flow is also not easy for conventional medical techniques.

Therefore, there is a long term demand for developing a new advanced measurement technique that can extract quantitative flow information of blood flows. To resolve these limitations encountered in conventional visualization techniques, we developed an X-ray PIV technique[2] and found its feasibility for visualizing blood flow[3]. In this study we investigated the optical characteristics of blood for a coherent synchrotron X-ray more systematically and analyzed the non-Newtonian flow characteristics of blood moving in a circular tube.

METHODS

The high coherence of the synchrotron light source offers various approaches to radiology[4,5]. Several imaging techniques utilizing coherent light sources, such as holography and interferometry, have been studied and some phase contrast imaging methods also have been developed[6]. We expect these high coherent characteristics of synchrotron light source would work powerfully to visualize blood flows. In this study, to visualize blood flow inside an opaque conduit and analyze the flow characteristics quantitatively, we combined a synchrotron X-ray micro-imaging technique and a PIV technique. We used an unmonochromatic beam (about 10-60 keV) at 7B2 beam

CP879, *Synchrotron Radiation Instrumentation: Ninth International Conference*,
edited by Jae-Young Choi and Seungyu Rah
© 2007 American Institute of Physics 978-0-7354-0373-4/07/$23.00

line of Pohang Light Source (Pohang, Korea). The size of source in the vertical and horizontal directions is 45 μm and 120 μm, respectively. The beam size at the position of sample is closed up to the image size by using slits for avoiding unnecessary exposure of X-rays to samples. The detailed beam characteristics and performance of this beamline are described in Je *et al*[7].

We developed a new combined technique, which can visualize quantitatively an opaque fluid, but still have problems to apply this technique to blood flows. At first, we should extract velocity information from the blood flow itself without adding any tracer particles. Next, because the biological specimens composed of low-density elements are transparent to hard X-rays, moreover, the optical properties of RBCs and plasma are so similar, it is not easy to discriminate blood cells clearly even using the phase-contrast X-ray imaging method.

We solved these problems by finding speckle patterns of blood flow induced by synchrotron X-ray beam. We could obtain patterns of blood sample by using the propagation-based phase-contrast enhancement via optimizing the sample(blood)-to-scintillator distance d and the interference-based enhancement via increasing the sample thickness t. Figure 1 shows X-ray images of blood flow according to the phase-contrast/interference-based enhancement method. When d is larger than 8 cm, the speckle pattern is recognizable for all thickness tested in this study. When the distance d is over 40 cm, however, the blood patterns begin to be blurry slightly due to excessive phase-contrast enhancement. Nevertheless, the blood pattern images are still reasonably recognizable inside the solid-line box, suitable for applying PIV algorithm to extract velocity field information. Actually, whether a certain speckle pattern image is suitable for PIV algorithm depends on how apparent the pattern's intensity variations are. In the viewpoint of interference enhancement, all patterns of blood sample with thickness t = 0.3~10 mm are recognizable, if d is selected properly.

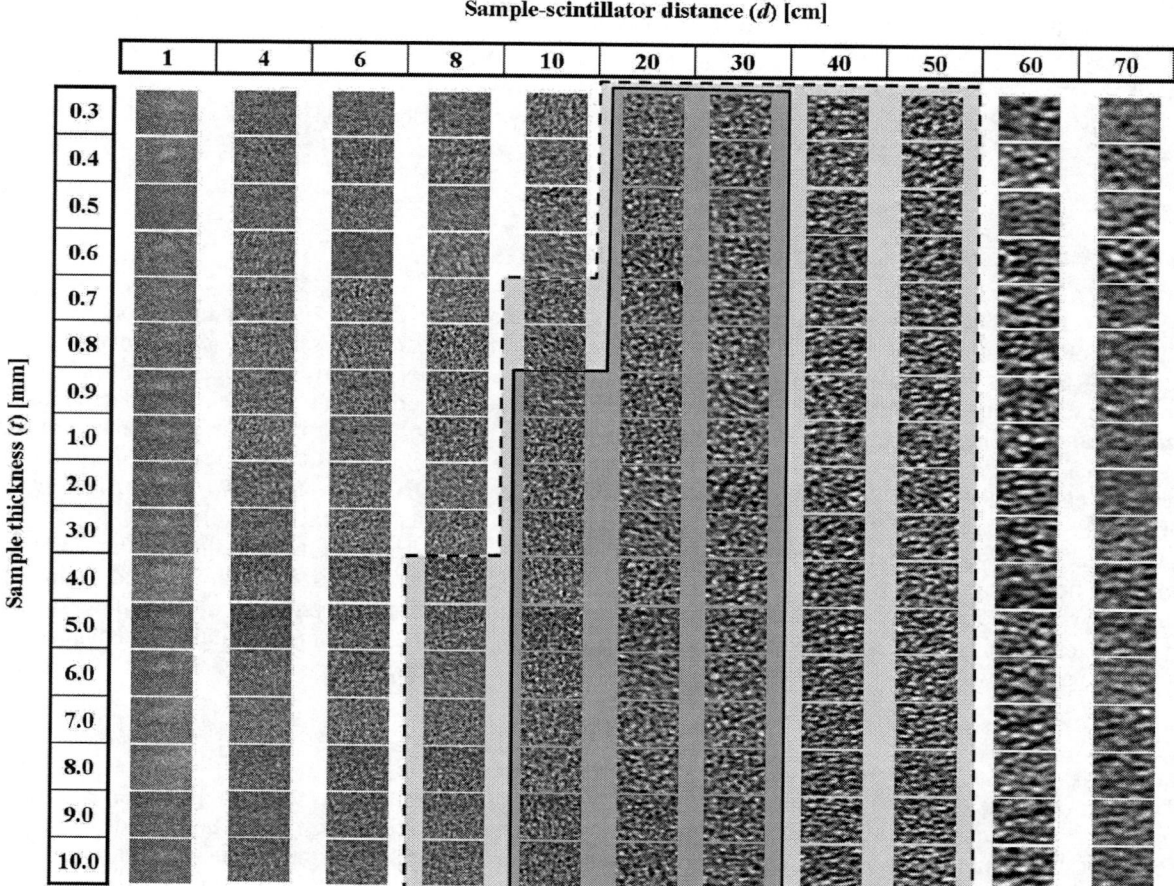

FIGURE 1. X-ray images of blood captured using the phase-contrast/interference-based enhancement method

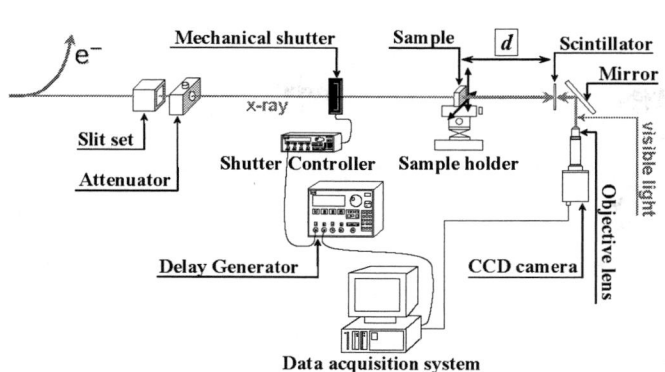

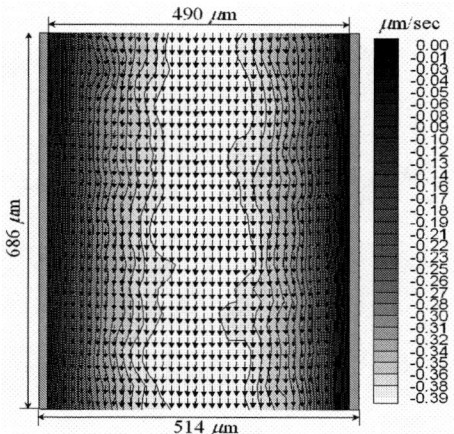

FIGURE 2. Schematic diagram of x-ray PIV system

FIGURE 3. Streamwise mean velocity field of blood flow measured in an opaque microchannel

Using the phase-contrast and interference-based edge enhancement methods of synchrotron X-ray micro-imaging, it is possible to directly visualize blood flow in an opaque conduit without seeding any tracer particles or contrast materials. A schematic diagram of the experimental setup is shown in Fig. 2. Sample images were captured by a cooled charge coupled device (CCD) camera with 1280×1024 pixels resolution and 6.7×6.7 μm^2 in physical pixel size, after converting X-rays to visible lights by passing then through a thin $CdWO_4$ scintillator crystal. Field of view (FOV) of a whole image is 642×514 μm^2 by adapting a $10\times$ objective, therefore, the pixel size of x-ray images are 0.502 μm/pixel. Because the X-ray beam was supplied continuously, we installed a mechanical shutter to make a pulse-type beam. A delay generator was used to synchronize the mechanical shutter and the CCD camera. A syringe pump was employed to supply blood.

At first, we captured X-ray images of blood flow in a rectangular-shaped opaque microchannel of width 490 μm and depth 1390 μm under the optimized conditions of $d = 40$ cm and $t = 1390$ μm. By applying a two-frame cross-correlation PIV algorithm to the X-ray images of the blood flow, we could obtain instantaneous velocity fields. Each X-ray image was divided into many small interrogation windows of 13×13 μm^2 in physical size. The mean velocity field was obtained by ensemble averaging 200 consecutive instantaneous velocity fields statistically. The measured streamwise mean velocity field is shown in Fig. 3. The flow speed increases with going toward the channel center from the channel wall. This velocity field is similar to the velocity distribution typically observed in a macro-sized channel of a rectangular cross-section. This quantitative velocity field result shows the reliability of X-ray PIV method for measuring blood flows.

We also investigated blood flow inside an opaque tube with a circular cross-section. This is a realistic application of the X-ray PIV method for analyzing the non-Newtonian characteristics of real blood flows. Inner diameter of the opaque tube was 2.77 mm and blood was injected by a syringe pump at a flow rate of 50 μl/min. Figure 4 shows a typical streamwise mean velocity profile extracted from the mean velocity field data along a horizontal line. From the results of mean velocity profiles of real blood flow, we compared the experimental velocity profile with hemorheologic models of blood flow. The measured velocity profile is well agreed with the velocity profile suggested by Casson model[8,9]. The typical parabolic velocity profile of Newtonian flow has some discrepancy with the experimental result. The diameter of center potential region in which no velocity gradient exists due to the yield stress of blood is $r_c / R = 0.044$. Conclusively, we found that the X-ray PIV method can be used for revealing various hemodynamic phenomena experimentally.

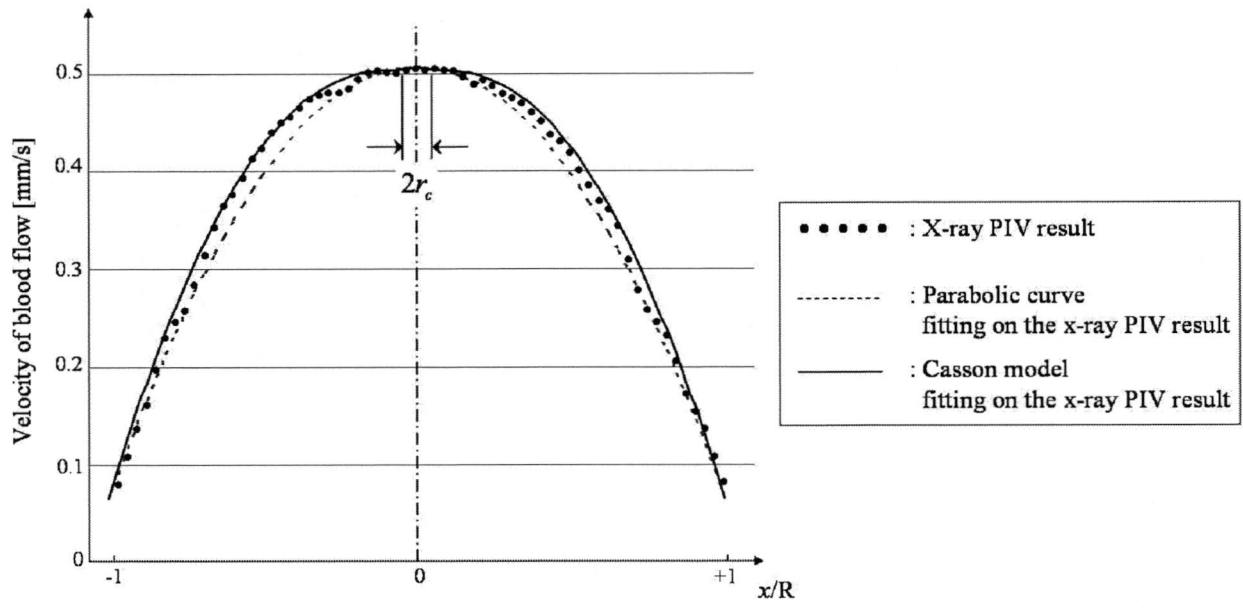

FIGURE 4. Comparison of streamwise mean velocity profile of a blood flow inside a circular tube.

CONCLUSION

In this study, we optimized the experimental condition such as sample(blood)-to-scintillator distance and sample(blood) thickness to acquire suitable X-ray images of blood flow for PIV measurements. The X-ray PIV technique was also applied to blood flows in an opaque rectangular microchannel and a circular tube for analyzing its non-Newtonian flow characteristics. The X-ray PIV results are well matched with the Casson's hemorheological model. The X-ray PIV method has a strong potential for visualizing blood samples non-invasively to obtain detailed flow information such as flow rate, spatial distributions of velocity, and shear stress. We hope the developed X-ray PIV method can be used for investigating various hemodynamic phenomena for which the hemodynamics and the pathology of vascular diseases play a key role.

ACKNOWLEDGMENTS

Experiments at the 7B2 beamline of PLS were supported in part by MOST and POSTECH. This work was also supported by MOST (KOSEF) through grant no. (R01-2004-000-10500-0) from the Basic Research Program and Systems Bio-Dynamics Research Center.

REFERENCES

1. C. W. Park, G. B. Kim and S. J. Lee, *Int. Symp on Biomedical Optics*, San Jose, CA, 2004, #5325-09.
2. S. J. Lee and G. B. Kim, *J. Appl. Phys.* **94**, 3620–3623 (2003).
3. S. J. Lee and G. B. Kim, *J. Appl. Phys.* **97**, 064701 (2005).
4. A. Snigirev et al., *Rev. Sci. Instrum.* 66, 5486–5492 (1995).
5. D. Chapman et al., *Phys. Med. Biol.* **42**, 2015–2025 (1997).
6. Z. H. Hu, *Nature* **392**, 690–693 (1998).
7. J. H. Je, *Rev. Sci. Instrum.* **75**, 4355–4358 (2004).
8. O. Syoten, *Cardiovascular Hemorheology*, Cambridge University Press, 1981, pp. 40–43.
9. C. W. Macosko, *Rheology Principles, Measurements and Applications*, Wiley-VCH, 1994, pp. 95–96.

Synchrotron Radiation Mammography: Clinical Experimentation

Fulvia Arfelli[1], Alessandro Abrami[2], Paola Bregant[3], Valentina Chenda[2], Maria A. Cova[4], Fabio de Guarrini[3], Diego Dreossi[1], Renata Longo[1], Ralf-Hendrik Menk[2], Elisa Quai[2], Tatjana Rokvic[1], Maura Tonutti[4], Giuliana Tromba[2], Fabrizio Zanconati[4], Edoardo Castelli[1]

[1]Department of Physics, University of Trieste and INFN, Via A. Valerio 2, 34127 Trieste, Italy
[2]Sincrotrone Trieste SCpA, S.S. 14 km 163.5, 34012 Basovizza, Trieste, Italy
[3]Health Physics, Hospital, Via Pietà 19, Trieste, Italy
[4]Department of Radiology, University and Hospital, St. di Fiume 447, 34139 Trieste, Italy

Abstract. For several years a large variety of *in-vitro* medical imaging studies were carried out at the SYRMEP (Synchrotron Radiation for Medical Physics) beamline of the synchrotron radiation facility ELETTRA (Trieste, Italy) utilizing phase sensitive imaging techniques. In particular low dose Phase Contrast (PhC) in planar imaging mode and computed tomography were utilized for full field mammography. The results obtained on *in-vitro* samples at the SYRMEP beamline in PhC breast imaging were so encouraging that a clinical program on a limited number of patients selected by radiologists was launched to validate the improvements of synchrotron radiation in mammography. PhC mammography with conventional screen-film systems is the first step within this project. A digital system is under development for future applications. During the last years the entire beamline has been deeply modified and a medical facility dedicated to *in-vivo* mammography was constructed. The facility for PhC synchrotron radiation mammography is now operative in patient mode. The system reveals a prominent increase in image quality with respect to conventional mammograms even at lower delivered dose.

Keywords: X-ray Imaging, Phase Contrast, Mammography
PACS: 87.59.-e, 87.59.Ek, 07.85.Qe

INTRODUCTION

Synchrotron radiation is an attractive X-ray source for radiography, in particular for mammography, because of the wide energy spectrum and the intrinsic laminar geometry of the beam [1]. The use of a tunable monochromatic beam has the advantage that the optimal energy for the object under examination can be chosen in order to achieve the best image quality i.e. to maximize the signal-to-noise ratio minimizing at the same time the delivered radiation dose. The peculiar vertical collimation of the beam brings a strong scattered radiation reduction without the aid of the anti-scattering grid with a consequent improvement of the image contrast. In addition the high degree of coherence of the source allows the applications of phase sensitive techniques, in particular of Phase Contrast Imaging. This method can give a further improvement on the visibility of very low contrast details.

The bending magnet beamline SYRMEP, dedicated to medical imaging, was built at the synchrotron radiation facility ELETTRA [2] and it is under operation since 1996. Here a large variety of studies on phantoms and *in-vitro* tissue samples have been carried by our group, a collaboration among the University of Trieste, INFN, the Società Sincrotrone Trieste and the Trieste Hospital [3]. Innovative phase techniques, such as Phase Contrast (PhC) radiography and Diffraction Enhanced Imaging (DEI), have been investigated [4] and tomography [5] has been implemented. These techniques yield numerous results in different fields of application [6]. Pioneering pre-clinical feasibility studies have been performed applying low dose PhC on human breast tissue samples [7].

CP879, *Synchrotron Radiation Instrumentation: Ninth International Conference*,
edited by Jae-Young Choi and Seungyu Rah
© 2007 American Institute of Physics 978-0-7354-0373-4/07/$23.00

The successful *in-vitro* imaging studies for full field mammography carried out at SYRMEP have encouraged the *in-vivo* synchrotron mammography project. Phase contrast mammography with commercial screen-film systems is the first step of the program. A digital system is under development for future applications including breast tomography.

The aim is a clinical trial on a limited number of patients selected by radiologists, after routine examinations, on the basis of BI-RADS (Breast Imaging Reporting And Data System) classification [8], according to the research program approved by the local Ethical Committee.

For this purpose the beamline layout has been modified in order to realize a clinical facility. It has been extended for accommodating the radiologist room and the patient examination hall. An exhaustive set of tests on the whole system has been performed to check the safety system, the exam procedure and to assure a correct automatic film exposure with a real-time dose control. After the final assent from the Italian Ministry of Health the first clinical examinations have been started in March 2006 with encouraging results.

MATERIAL AND METHODS

The source is one of the bending magnets of ELETTRA. The beamline, described in Ref. [9,10], comprises a fixed exit Si(111) double crystal monochromator capable of tuning the energy from 8.5 keV to 35 keV with an energy resolution about 0.2%. In our protocol for clinical mammography the energy range is restricted to 16-22 keV. The maximum available flux is of the order of 10^8 photons/(s*mm^2) and can be reduced by means of a set of calibrated aluminum filters.

During the last years the beamline has been substantially modified for the construction of the dedicated *in-vivo* mammography medical facility. The new part of the beamline accommodates the patient examination room and the radiologist room. A layout of the beamline is depicted in Fig.1.

Special care was drawn on the patients' movement device, the patients' security and the dosimetry system. The former is necessary since planar images are obtained by a vertical movement of the patient through the fixed laminar beam. The patient lays prone on a high precision movement support (Fig.2), which comprises a special opening for the breast alignment. Size and shape of the opening are consistent with the chest anatomy. A compression system, similar to the clinical one, is used for equalizing thickness and stretching tissues. The vertical velocity is constant within few percent during the entire scan. Besides the vertical movement the support can move horizontally for positioning and can rotate for the oblique breast projection and for tomography.

The screen-film system is mounted on a two meters long rail to allow the optimization of the distance for PhC imaging and it is moved synchronized with the patient support.

In order to fulfill the severe security requirements the safety system follows redundancy criteria and a *fail safe* philosophy. The experimental hutch was expanded to accommodate the beam control system and the dosimetry system (Fig.1): a slit system, a first ionization chamber, two safety shutters, an imaging shutter and a second ionization chamber.

The beam geometry is defined vertically by a slit with a fixed aperture of 3 mm and horizontally by a motorized slit equipped with absolute encoders. The beam cross section at the breast position, about 30 m from the source, is about 21 cm horizontal and 3.4 mm vertical.

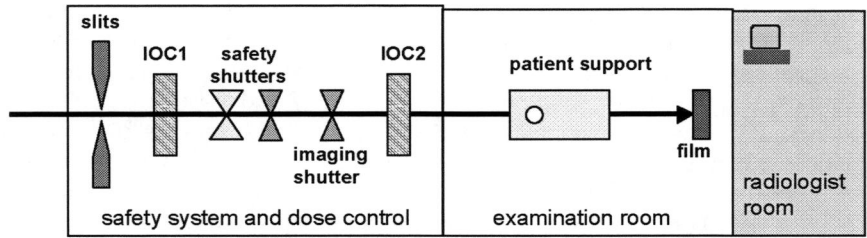

FIGURE 1. The layout of the beamline for clinical mammography

The two safety shutters are designed with different mechanics. The first one, always kept in open position, is capable of closing in 10 msec in a case of emergency. The second one opens only during the examination and can close in about 40 msec. The imaging shutter, which has the same design of the second safety shutter, is synchronized with the patient movement and defines the duration of the exam.

Two custom-made high precision ionization chambers (IOC) including the readout electronics were especially constructed for the laminar beam geometry. In this geometry the beam impinges perpendicular to the electrodes which are made of very thin (50 μm) aluminized Mylar foils. The IOCs entrance window, which defines the sensitive area, is 23.4 x 2 cm^2, while the absorption thickness is 25 mm. They are placed about 27 m from the source and 3 m upstream the patient and they are air-filled: the readings are corrected according to the atmosphere pressure and humidity monitors. The first IOC is utilized as beam monitor in order to measure the radiation flux before the patient shutters opening. Both chambers are then used for the dose control during the examination. The electronics noise was reduced accommodating the readout in the IOC housing. The fast 20-bit digital output is sent via optical fiber simultaneously to the exam control workstation and to the safety system.

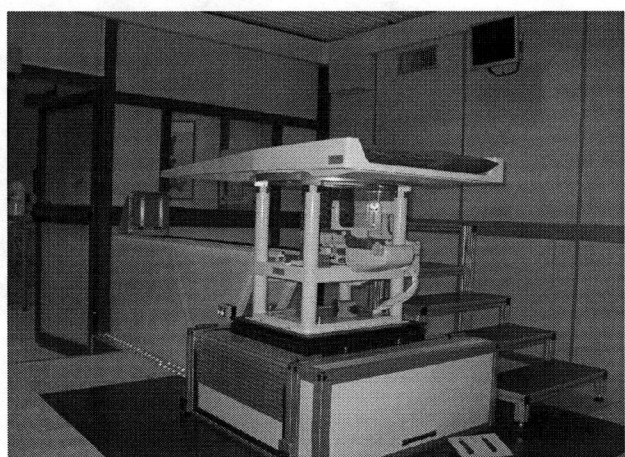

FIGURE 2. The examination room with the patient support.

Furthermore the IOCs evaluate in real time the dose delivered to the patient during the examination and in combination with a custom-made exposimeter system they define the parameters of the entire exam. The exposimeter system consists of an array of 4 solid state photodiodes read out by the same electronics of the IOCs. It is placed behind the screen-film cassette and it is illuminated by the beam only during the pre-scan procedure. Due the scanning mode applied to perform the exam, it is necessary to define *a priori* the exposure time that is correlated to the velocity of the vertical movement of the screen-film (Kodak MinR 2000). For this purpose during the pre-scan the exposimeters measure the real average absorption of the breast under examination with a very low dose irradiation. From the measurement of the radiation flux and the breast absorption, together with the previously evaluated film response, the scan parameters are completely defined.

RESULTS

Before starting the examination on patients all the devices connected to the safety and dose control system have been tested to check if they fulfill all the stringent requirements. The ionization chambers have demonstrated high long term stability and a very uniform response over the entire sensitive area. They have been calibrated against the national standard ionization chamber. The exposimeters have been calibrated against the IOCs over the energy range 16-22 keV. They have shown a high sensitivity that allows a very low delivered dose during the pre-scan phase.

One of the major issues was how to define an automatic procedure of the film exposure. The exhaustive study of the film exposure dependence on the X-ray energy and flux as well as on the scan velocity has brought to an analytical formulation in terms of simple fit parameters of the film characteristic curve (HD curve) [11]. This formulation is also able to deal with the *reciprocity law* failure [12].

Breast samples from mastectomy surgery have been studied using the complete final set-up in order to test the overall control system as well as the security and the image quality protocols. Figure 3 shows a breast tumor sample image obtained with our synchrotron system at 18 keV compared to that from the hospital mammography system.

The entire automatic procedure has obtained always the best exposure on the film for each energy and scan velocity. Moreover the procedure has been demonstrated to be capable of selecting automatically the optimal energy, for a dose lower or comparable to that delivered in the conventional exam, according to the compressed breast thickness and the glandularity estimation. Up to now nine examinations on patients have been performed. The synchrotron images have been compared to the images obtained at the mammography unit of the Trieste Hospital (GE Senographe 2000D). Our system has revealed a prominent increase in image quality with respect to conventional mammograms even at lower dose delivered to the breast. The preliminary results are encouraging and a complete evaluation of the clinical impact of the new method is in progress.

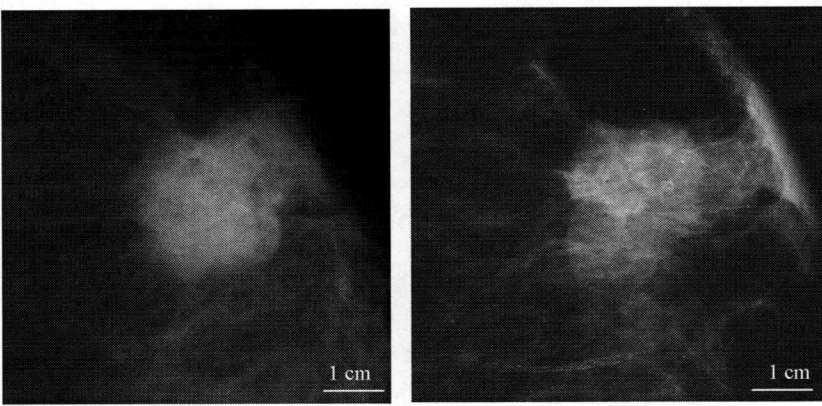

FIGURE 3. Image of a breast tumor obtained with the hospital mammography unit (left) and with synchrotron at 18 keV (right).

CONCLUSIONS

In the last years several experiments on phantoms and *in-vitro* samples have been carried out at the beamline SYRMEP at ELETTRA in order to assess the potential of Phase Contrast (PhC) Imaging in synchrotron-based mammography. In particular it has been demonstrated the possibility to achieve an enhanced image quality using a clinical screen-film receptor with a dose comparable or lower than that delivered in the clinical practice. The beamline has been completely modified in order to implement a synchrotron medical facility according to the standard requirements on safety, image quality protocols and dose control.

After the successful tests on the recently installed medical equipment the clinical mammographic program has been started in March 2006. The first nine patients, with an uncertain diagnosis after the examination at the hospital, have been imaged with our synchrotron mammographic unit. The images obtained have shown an improved contrast and better definition of small details. In some cases the medical doctors could clarify some ambiguities observed in the previous diagnosis. Our protocol plans about 100 patients in two years in order to achieve a good statistics of cases and to provide a final assessment on the synchrotron radiation PhC mammography.

REFERENCES

1. P. Suortti, W. Thomlinson, Phys. Med. Biol. 48, pp. R1-R35 (2003).
2. F. Arfelli, G. Barbiellini, V. Bonvicini et al., Physica Medica 13 Suppl.1 , pp. 25-30 (1997).
3. E. Castelli et al., "Clinical mammography at the SYRMEP beam line", to be published on the Proceedings of X Pisa Meeting on Advanced Detectors, May 21-27, 2006, La Biodola, Isola d'Elba (Italy).
4. F. Arfelli, V. Bonvicini, A. Bravin et al., Radiology 215, pp.286-293 (2000).
5. S. Pani, R. Longo, D. Dreossi et al., Phys. Med. Biol. 49, pp.1739-1754 (2004).
6. A. Abrami, F. Arfelli, R.C. Barroso et al., Nucl. Instr. and Meth. 548, pp. 221-227 (2005).
7. F. Arfelli, M. Assante, V. Bonvicini et al., Phys. Med. Biol. 43, pp.2845-2852 (1998).
8. American College of Radiology. "Illustrated Breast Imaging Reporting And Data System (BI-RADS) -4th Edition" Reston (VA), ACR (2003).
9. F. Arfelli, G. Barbiellini, V. Bonvicini et al., IEEE T. Nucl. Sci. 44, pp. 2395-2399 (1997).
10. F. Arfelli, V. Bonvicini, A. Bravin et al., Proceedings SPIE 3770, pp.2-12 (1999).
11. C.J. Vyborny, Med. Phys. 6, pp.39-44 (1979).
12. A. De Almeida, W.T. Sobol, G.T. Barnes, Med. Phys. 26, pp. 682-688 (1999).

2D and 3D Refraction Based X-ray Imaging Suitable for Clinical and Pathological Diagnosis

Masami Ando[1,2,3], Hiroko Bando[4], Zhihua Chen[5], Yoshinori Chikaura[6], Chang-Hyuk Choi[7], Tokiko Endo[8], Hiroyasu Esumi[9], Li Gang[10], Eiko Hashimoto[3], Keiichi Hirano[2], Kazuyuki Hyodo[2], Shu Ichihara[11], SangHoon Jheon[12], HongTae Kim[13], JongKi Kim[13], Tatsuro Kimura[14], ChangHyun Lee[15], Anton Maksimenko[2], Chiho Ohbayashi[16], SungHwan Park[17], Daisuke Shimao[18], Hiroshi Sugiyama[2,3], Jintian Tang[19], Ei Ueno[4], Katsuhito Yamasaki[20], and Tetsuya Yuasa[21]

[1] Institute of Science and Technology, Tokyo Univ. of Science, Yamasaki 2641, Noda, Chiba 278-8510, Japan,
[2] Photon Factory, IMSS, KEK, Oho 1-1, Tsukuba, Ibaraki 305-0801, Japan,
[3] Dept. of Photo-Science, GUAS, Shonan, Hayama, Kanagawa 240-0193, Japan,
[4] Dept. of Breast-Thyroid-Endocrine Surgery, Univ. of Tsukuba, Ibaraki 305-8573, Japan
[5] China-Japan Friendship Hospital, Yinhua Donglu, Beijing 100029, P.R.China,
[6] Dept. of Material Science, Kyushu Institute of Technology, Kitakyushu-shi, Fukuoka, 804-8550, Japan,
[7] Dept. of Anatomy, School of Medicine, Catholic Univ. of Daegu, Daegu 705-718, Korea
[8] Dept. of Radiology, Nagoya Medical Center, National Hospital Organization, Naka-ku, Nagoya 460-0001, Japan,
[9] National Cancer Center Research Institute East, Kashiwanoha 6-5-1, Kashiwa, Chiba 277-8577, Japan,
[10] Beijing Synchrotron Radiation Facility, IHEP, CAS, Yuchuang Lu, P.O.Box 918, Beijing 100039, P.R.China,
[11] Dept. of Pathology, Nagoya Medical Center, National Hospital Organization, Naka-ku, Nagoya 460-0001, Japan,
[12] Thoracic and Cardiovascular Surgery, Seoul National Univ., Gyeonggi-do 463-707, Korea
[13] Dept. of Diagnostic Radiology and Biomedical Engineering, Catholic Univ. of Daegu, Daegu 705-718, Korea
[14] Dept. of Environmental Medicine, Kobe Univ., Kobe, Hyogo 650-0017, Japan,
[15] Dept. of Radiology, Seoul National Univ. Hospital, 110-744 Seoul, Korea
[16] Dept. of Pathology, Kobe Univ., Kusunoki-Cho 7-5-2, Kobe, Hyogo 650-0017, Japan,
[17] Dept. of General Surgery, Catholic Univ. of Daegu, Daegu 705-718, Korea
[18] Dept. of Radiological Sciences, Ibaraki Prefectural Univ. of Health Sciences, Ibaraki 300-0394, Japan,
[19] Department of Engineering Physics, Tsinghua University, Beijing 100084, P.R.China,
[20] Institute of Genome, Kobe Univ., Kusunoki-Cho 7-5-2, Kobe, Hyogo 650-0017, Japan,
[21] Faculty of Engineering, Yamagata Univ., Jonan 4-3-16, Yonezawa, Yamagata 992-8510, Japan,

Abstract. The first observation of micro papillary (MP) breast cancer by x-ray dark-field imaging (XDFI) and the first observation of the 3D x-ray internal structure of another breast cancer, ductal carcinoma in-situ (DCIS), are reported. The specimen size for the sheet-shaped MP was 26 mm x 22 mm x 2.8 mm, and that for the rod-shaped DCIS was 3.6 mm in diameter and 4.7 mm in height. The experiment was performed at the Photon Factory, KEK: High Energy Accelerator Research Organization. We achieved a high-contrast x-ray image by adopting a thickness-controlled transmission-type angular analyzer that allows only refraction components from the object for 2D imaging. This provides a high-contrast image of cancer-cell nests, cancer cells and stroma. For x-ray 3D imaging, a new algorithm due to the refraction for x-ray CT was created. The angular information was acquired by x-ray optics diffraction-enhanced imaging (DEI). The number of data was 900 for each reconstruction. A reconstructed CT image may include ductus lactiferi, micro calcification and the breast gland. This modality has the possibility to open up a new clinical and pathological diagnosis using x-ray, offering more precise inspection and detection of early signs of breast cancer.

Keywords: X-ray refraction, X-ray dark-field imaging (XDFI), DEI, breast cancer, DCIS, pathological diagnosis, clinical diagnosis, ductus lactiferi.

PACS: 87.59.-e, 87.59.Ek, 87.59.Fm, 87.59.Hp, 87.61.Pk, 87.62.+n, 87.64.Bx

CP879, Synchrotron Radiation Instrumentation: Ninth International Conference,
edited by Jae-Young Choi and Seungyu Rah
© 2007 American Institute of Physics 978-0-7354-0373-4/07/$23.00

INTRODUCTION

Throughout the world, breast cancer has indicated such a rapid growth rate that it is now extremely important to develop a precise visual technique to detect the early signs of breast cancer. Absorption contrast, as used in current mammography, can barely visualize breast cancer except for calcification. Because of this, therefore, using the current method of mammography hardly seems appropriate for early check-ups. Two other visual techniques, phase-inference contrast (PIC)[1] and refraction contrast, could have higher capability of visualizing a phase object such as breast cancer. Thus far, a variety of imaging schemes using refraction contrast[2-5], diffraction-enhanced imaging (DEI)[6-8], and x-ray dark-field imaging (XDFI)[9] have been proposed. In order to visualize breast cancer, following the pioneering work on the imaging of breast cancer by Burattini's group[10], further trials to reveal breast-cancer tissue were performed by PCI[11, 12], DEI[13-15], PIC[16], super magnification imaging (SMI)[17], XDFI[18,19] and x-ray fluorescence (XRF)[20]. These were trials of 2D imaging, and more recently trials of 3D reconstruction have been carried out[21-23]. Maksimenko et al. recently proposed a novel tomographic imaging protocol based on a physico-mathematically defined reconstruction algorithm[24, 25] with a paraxial-ray approximation in the domain of geometrical optics. Meanwhile, Zhu et al. also presented a detailed description of the algorithm[26]. Our algorithm has been successfully applied in visualizing ductal carcinoma in situ (DCIS) in a 3D mode with high contrast and high resolution[27]. Thus, in this report we would like to introduce a newly developed visual technique for the early inspection of signs of breast cancer[18-20, 27].

EXPERIMENT

XDFI was first proposed in 2002. The XDFI illustrated in Fig. 1 utilizes an asymmetric-cut[28] monochromator-collimator (MC) to provide an extremely parallel incident beam onto a specimen. Located beyond the specimen is a transmission-type angular analyzer (TAA) that has a specified thickness so that only refraction components from an object can pass through it. This is due to the nature of forward diffraction, while the x-rays that do not change their direction will be diffracted. As a result, it is possible to obtain an x-ray image with a very high contrast of soft tissue due to refraction. Furthermore, due to asymmetric diffraction at the MC, a large field of view (FOV) suitable for an x-ray photo in a single shot is provided.

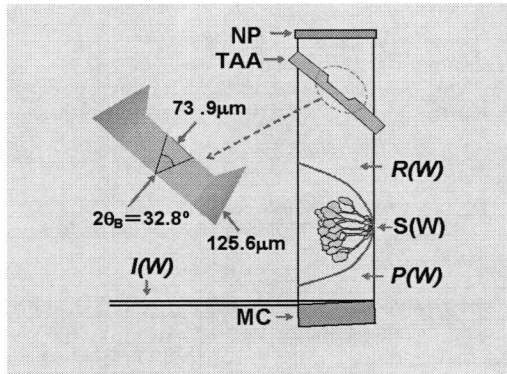 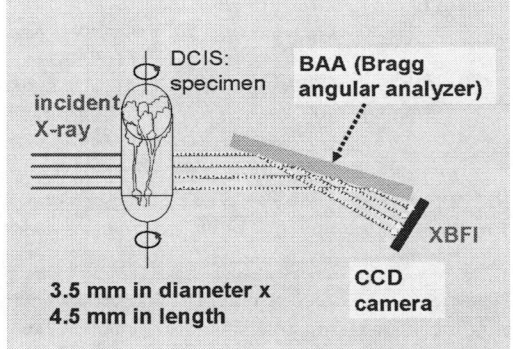

FIG. 1. X-ray optics for x-ray dark-field imaging (XDFI). The refraction component from an object may pass through towards the direction of forward diffraction by the help of a specified thickness transmission angular analyzer (TAA). The x-ray image was captured on film.

FIG. 2. X-ray optics diffraction-enhanced imaging (DEI) for CT data acquisition. The sample was rotated 180 degrees at intervals of 0.2 degrees around the rotation axis. Each image was captured by a CCD camera.

The x-ray optics DEI[5] with Bragg-case angular analyzer (BAA) at the x-ray energy of 11.7 keV was used because the FWHM of the rocking curve available by a double-crystal monochromator was relatively broad, and thus the system was relatively stable. The sample was remotely rotated at intervals of $\Delta \Theta = 0.2°$. The MC of Si(220) was cut with the angle of $\alpha = 9.5$, where α is the angle between the surface and the (220) diffracting planes. The specimen used in this experiment was a rod-shaped DCIS, and the number of sample rotations for data acquisition was 900. The time for data acquisition of each frame was 200 msec ~ 1 sec, and it took 2-5 seconds to transfer this data to a PC,

1 sec for sample rotation, and an additional 1 sec for stabilizing the system free of vibration from the motor. Every ten data acquisitions, the x-ray intensity was measured by CCD without the specimen for background subtraction. An air-cooled CCD camera, X-FDI 1.00:1, with a view size of 8.7 mm (h) x 6.9 mm (v) was supplied by Photonic Science. This is compatible with 16-bit, and provides 1392 x 1040 pixels with a pixel size of 6.3 μm x 6.6 μm. Data transfer was carried out through a FireWire (IEEE 1394) port. Following the initial series of measurements, the angular position of the analyzer crystal was changed to continue another series of measurements. In total, data acquisition time was approximately 3 hours.

RESULT

XDFI was applied to a breast cancer micro papillary specimen[18] and successfully revealed breast-cancer cells and nests as well as stroma as depicted in Fig. 3. A refraction-based algorithm was also applied to another breast cancer specimen, DCIS, and this revealed a reconstructed image (Fig. 4a)[26] that corresponds well to a pathological view (Fig. 4b).

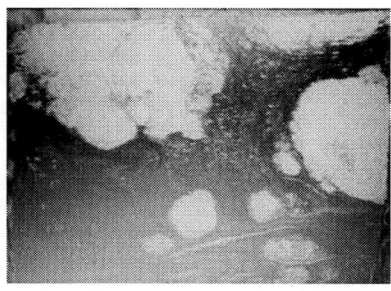

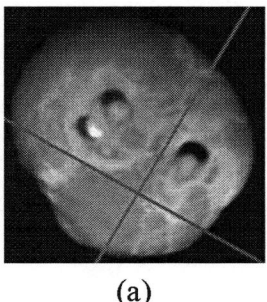
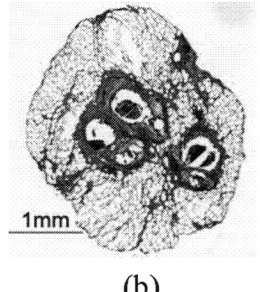

(a)   (b)

FIG. 3. Micro papillary taken with XDFI at 35 keV. The specimen size is 26 mm horizontal, 22 mm vertical and 2.8 mm in depth. Breast-cancer nests and large cells with a size of around 100 μm are clearly seen. White corresponds to an excess of x-rays in order to adjust the normal contrast due to absorption in the medical-imaging film.

FIG. 4. (a) Reconstructed 3D image of DCIS breast cancer. In the three circular shapes ductus lactiferi calcification is discovered. Further outside of these we can find the mammary gland showing reasonable contrast. (b) depicts the pathological view of the same location as (a). Both are in very good agreement, and each one is approximately 3.5 mm in diameter.

CONCLUSION

The current technique we have developed can provide the following specifications as described in Table 1. It is possible to apply this technique to (1) mammography for early inspection of breast cancer with an x-ray energy of ~35 keV and (2) pathological diagnosis. The skin radiation dose at 35 keV could be dramatically improved, at least by a factor of 30 compared to the classical one.

TABLE 1. Achievement of Field of View and Spatial Resolution

Type of View	X-ray Energy	Field of View	Spatial Resolution
2D	35 keV	90 mm x 90 mm	140 μm
2D	17.5 keV	10 mm x 10 mm	25 μm
3D	11.7 keV	5 mm x 5 mm x 5 mm	200 μm

ACKNOWLEDGMENTS

The authors would like to thank the program advisory board at the Photon Factory, KEK where use of synchrotron radiation was performed under the program numbers #2002G045 and #2005G085. This research was performed by the following financial support to which authors would like to highly appreciate: Grant-in-Aid for Exploratory Research #15654042, Grant-in-Aid for JSPS Fellows #1705335, Grant-in-Aid for Scientific Research

(A) #18206011 from the Ministry of Education, Culture, Sports, Science and Technology (MEXT), Suzuken Research Fund and the 2006 Horiba Special Award.

REFERENCES

1. A . Momose, *Nuclear Instruments and Methods in Physics Research*, **A 352**, 622-628 (1995).
2. K. M. Podurets, V. A. Somenkov and S. Sh. Shil'shtein: *Physica*, **B156 & 157**, 691-693 (1989).
3. V. N. Ingal and E. A. Beliaevskaya: *J. Phys. D: Appl. Phys.* **28**, 2314-2317 (1995).
4. T. J. Davis, D. Gao, T. E. Gureyev, A. W. Stevenson and S. W. Wilkins, *Nature* **373**, 595-598 (1995).
5. D. Chapman, W. Thomlinson, R. E. Johnston, D. Washburn, E. Pisano, N. Gmur, Z. Zhong, R. Menk, F. Arfelli and D. Sayers,: *Phys. Me. Biol.* **42**, 2015-2025 (1997).
6. N. Yagi, Y. Suzuki, K. Umetani, Y. Kohmura and K. Yamasaki, *Med. Phys.* **26**, 2190-2193 (1999).
7. K. Hirano, A. Maksimenko, H. Sugiyama and M. Ando, *Jpn. J. Appl. Phys.* **41**, L595-L598 (2002).
8. A. Maksimenko, H. Sugiyama, K. Hirano, T. Yuasa and M. Ando, *Meas. Science and Technology*, **15**, 1251-1254 (2004).
9. M. Ando, H. Sugiyama, A. Maksimenko, W. Pattanasiriwisawa and K. Hyodo, *Jpn. J. Appl. Phys.* **41**, L1016-L1018 (2002).
10. E. Burattini, E. Cossu, C. Di Maggio, M. Gambaccini, P. L. Indovina, M. Marziani, M. Pocek, S. Simeoni and G. Simonetti, *Radiology*, **195**, 239-244 (1995).
11. R. E. Johnston, D. Washburn, E. Pisano, C. Burns, W. C. Thomlinson, L. D. Chapman, F. Arfelli, N. F. Gmur, Z. Zhong and D. Sayers, *Radiology*, **200**, 659-663 (1996).
12. D. Chapman, E. Pisano, W. Thomlinson, Z. Zhong, R. E. Johnston, D. Washburn and D. Sayers, *Breast Disease*, **10**, 197 (1998).
13. M. Di Michiel, A. Olivo, G. Tromba, F. Arfelli, V. Bonvicini, A. Bravin, G. Cantatore, E. Castelli, L. Dalla Palma, R. Longo, S. Pani, D. Pontoni, P. Poropat, M. Prest, A. Rashevsky, A. Vacchi and E. Vallazza, *Medical Applications of Synchrotron Radiation*, edited by M. Ando and C. Uyama, Tokyo: Springer-Verlag, 1998, pp. 78-82.
14. F. Arfelli, V. Bonvicini, A. Bravin, G. Cantatore, E. Castelli, L. Dalla Palma, M. Di Michiel, M. Fabrizioli, R. Longo, R. H. Menk, A. Olivo, S. Pani, D. Pontoni, P. Poropat, M. Prest, A. Rashevsky, M. Ratti, L. Rigon, G. Tromba, A. Vacchi, E. Vallazza and F. Zanconati, *Radiology*, **215**, 286-293 (2000).
15. M.O. Hasnah, Z. Zhong, O. Oltulu, E. Pisano, R.E. Johnson, D. Sayers, W. Thomlinson and D. Chapman, *Med. Phys.* **29**, 2216-2221 (2002).
16. T. Takeda, J. Wu, Y. Tsuchiya, A. Yoneyama, T. Lwin, Y. Aiyoshi, T. Zeniya, K. Hyodo and E. Ueno, *J. J. Appl. Phys.* **43**, 5652-5656 (2004).
17. F. Toyofuku, Y. Higashida, K. Tokumori, A. Yoshida, M. Matsumoto, T. Ideguchi, K. Hyodo and M. Ando, *J. J. Med. Phys.* **23**, **Supp.3** 127 (2003).
18. M. Ando, K. Yamasaki, F. Toyofuku, H. Sugiyama, C. Ohbayashi, G. Li, L. Pan, X. Jiang, W. Pattanasiriwisawa, D. Shimao, E. Hashimoto, T. Kimura, M. Tsuneyoshi, E. Ueno, K. Tokumori, A. Maksimenko and Y. Higashida, *Jpn. J. Appl. Phys.* **44**, L528-L531 (2005).
19. M. Ando, H. Sugiyama, S. Ichihara, T. Endo, H. Bando, K. Yamasaki, C. Ohbayashi, Y. Chikaura, H. Esumi, A. Maksimenko, G. Li, *Jpn. J. Appl. Phys.* **45**, L740-L743 (2006).
20. M. Ando, K. Yamasaki, C. Ohbayashi, H. Esumi, G. Li, A. Maksimenko and T. Kawai, *Jpn. J. Appl. Phys.* **44**, L998-L1001 (2005).
21. F. A. Dilmanian, Z. Zhong, B. Ren, X. Y. Wu, D. Chapman, I. Orion and W. C. Thomlinson, *Phys. Med. Biol.* **45**, 933-947 (1999).
22. E. Pagot, S. Fielder, O. Cloetens, A. Bravin, P. Coan, K. Fezzaa, J. Baruchel and J. Hartwing, *Phys. Med. Biol.* **50**, 709-724 (2005).
23. T. Sera, K. Uesugi and N. Yagi, *Med. Phys.* **32**, 2787-2792 (2005).
24. A. Maksimenko, M. Ando, H. Sugiyama and T. Yuasa, *Appl. Phys. Lett.* **86**, 124105-1~124105-3 (2005).
25. A. Maksimenko, M. Ando, H. Sugiyama and E. Hashimoto, *Jpn. J. Appl. Phys.* **44**, L633-L635 (2005).
26. P. P. Zhu, J. Y. Wang, Q. X. Yuan, W. X. Huang, H. Shu, B. Gao, T. D. Hu and Z. Y. Wu, *Appl. Phys. Lett.* **87**, 264101-1~264101-3 (2005).
27. M. Ando, A. Maksimenko, T. Yuasa, E. Hashimoto, K. Yamasaki, C. Ohbayashi, H. Sugiyama, K. Hyodo, T. Kimura, H. Esumi, T. Akatsuka, G. Li, D. Xian and E. Ueno, *Bioimages*, **13**, 1-7 (2006).
28. K. Kohra, *J. Phys. Soc. Jpn.* **17**, 589-590 (1962).

Dynamic Studies of Lung Fluid Clearance with Phase Contrast Imaging

Marcus J Kitchen[1], Rob A Lewis[2], Stuart B Hooper[3], Megan J Wallace[3],
Karen K W Siu[1,4], Ivan Williams[1], Sarah C Irvine[1], Michael J Morgan[1],
David M Paganin[1], Konstantin Pavlov[2], Naoto Yagi[5], Kentaro Uesugi[5].

[1]*School of Physics, Monash University, Victoria 3800, Australia.*
[2]*Monash Centre for Synchrotron Science, Monash University, Victoria 3800, Australia.*
[3]*Department of Physiology, Monash University, Victoria 3800, Australia.*
[4]*Department of Medical Imaging and Radiation Science, Monash University, Victoria 3800, Australia.*
[5]*SPring-8/JASRI, Mikazuki, Hyogo 679-5198, Japan.*

Abstract. Clearance of liquid from the airways at birth is a poorly understood process, partly due to the difficulties of observing and measuring the distribution of air within the lung. Imaging dynamic processes within the lung *in vivo* with high contrast and spatial resolution is therefore a major challenge. However, phase contrast X-ray imaging is able to exploit inhaled air as a contrast agent, rendering the lungs of small animals visible due to the large changes in the refractive index at air/tissue interfaces. In concert with the high spatial resolution afforded by X-ray imaging systems ($<100~\mu$m), propagation-based phase contrast imaging is ideal for studying lung development. To this end we have utilized intense, monochromatic synchrotron radiation, together with a fast readout CCD camera, to study fluid clearance from the lungs of rabbit pups at birth. Local rates of fluid clearance have been measured from the dynamic sequences using a single image phase retrieval algorithm.

Keywords: Lung; Aeration; Fluid Clearance; Phase Contrast; X-ray Imaging
PACS: 87.59.-e; 42.25.Gy; 87.59.Hp; 07.85.Qe

INTRODUCTION

The factors regulating lung aeration at birth are poorly understood, largely due to difficulties in distinguishing between liquid-filled and air-filled airways and the inability to quantitatively assess local rates of aeration. Without a detailed understanding of this process, the ventilation strategies employed to assist breathing in newborn infants suffering respiratory failure, especially preterm infants, can injure the developing lung tissue. Propagation-based phase contrast X-ray imaging (PBI) provides a method of visualizing the lungs of small animals with high contrast resolution and high spatial resolution ($<100~\mu$m) [1-7]. Alternative imaging techniques, including MRI, CT and X-ray bronchography, cannot offer these possibilities without the use of contrast agents or a high dose of ionizing radiation. PBI renders phase changes visible upon propagation of a partially coherent X-ray beam through an object, with the propagating waves interfering to provide edge-enhancing contrast. For these experiments the object is imaged in the regime between geometrical and wave optics, such that a single fringe highlights the boundaries of the airways, which become apparent when aerated [2]. The airways become visible as they fill with air, creating a speckled intensity pattern that is hypothesized to arise from a focusing effect - where overlapping airways serve as aberrated compound refractive lenses [4].

Combining the method of phase contrast with the high flux provided by synchrotron radiation and a high-speed digital detector enables the clearance of lung fluid to be imaged at approximately one frame per second; in principle, video frame rates are also possible. By digitally recording the image sequence it is possible to numerically extract localized rates of lung aeration using an appropriate phase retrieval algorithm.

CP879, *Synchrotron Radiation Instrumentation: Ninth International Conference*,
edited by Jae-Young Choi and Seungyu Rah

MATERIALS AND METHODS

Lungs of rabbit pups delivered by caesarean section were imaged *in vivo* on beamline 20B2 of the Biomedical Imaging Centre at the SPring-8 synchrotron radiation research facility, Japan (proposal no. 2005A0064-NL3-np). Experimental protocols were established in accordance with regulations set by the SPring-8 Animal Care and Use Committee. Approval was also obtained from the Monash University Animal Ethics Committee. Anaesthesia of pregnant rabbits at 30–31 days of gestation was induced by a 12 mg kg^{-1} bolus and maintained by a 100 mg kg^{-1} h^{-1} infusion (i.v.) of Rapinovet (10 mg ml^{-1} propofol, Schering).

A 25 keV X-ray beam was used to provide good phase and absorption contrast (see [7]). Rabbit pups were positioned approximately 210 m downstream of the synchrotron source in the anterior/posterior position. The detector was placed a further 2.0 m downstream to record phase contrast images (Fig. 1). To capture the entire chest cavity on the image detector a demagnifying (×0.52) tandem lens system and thin phosphor screen (P43, 10 μm thickness) were coupled to the CCD camera (Hamamatsu, C4742-95HR), providing an effective pixel size of 22.47 μm (2 × 2 binning mode) and an active area of 44.94(H) × 29.48(V) mm^2. Following image acquisition, custom software was used to correct for dark current and the effects of non-uniform beam intensity. Rabbit pups were imaged every 0.8 s for several minutes immediately after birth. The active area of the detector was reduced to 24.45(H) × 20.85(V) mm^2 to capture only the chest cavity, whilst enabling a short exposure time of 83 ms to minimise motion blur during respiratory efforts. A square wave pulse was used to trigger both the detector exposure times and a pre-object shutter to avoid unnecessary irradiation of the pup.

Pups were positioned in a water-filled plastic vessel to prevent the skin/air interface producing unwanted phase contrast and to aid the measurement of lung aeration. The head of each pup was supported by a rubber seal and as the pups breathed, fluid was expelled from this sealed vessel via an external tube. Aeration of the lung was quantified by measuring the relative local rate of fluid removed from the vessel. We assume that the entire object (including vessel and pup) is comprised of water, in which case the phase and absorption information present in each image is taken to be directly proportional to the projected thickness (T) of water, i.e., the phase is given by $\varphi = 2\pi\delta T/\lambda$, where δ is the refractive index decrement for water and λ is the X-ray wavelength, whilst the intensity is $I = I_o \exp(-\mu T)$, where I_o is the intensity incident on the sample and μ is the attenuation coefficient. The projected thickness is calculated using the method of Paganin et al. [8] and the volume is found by summing the projected thickness over the region of interest. The difference in volume from the first image of the sequence to every other frame provides a relative volume measure. The absolute volume of liquid removed is equal to the aerated volume. In these studies, μ for water at 25 keV was taken as 50.7 m^{-1} [9] and δ was computed as 3.81 × 10^{-7} using the method outlined in [4].

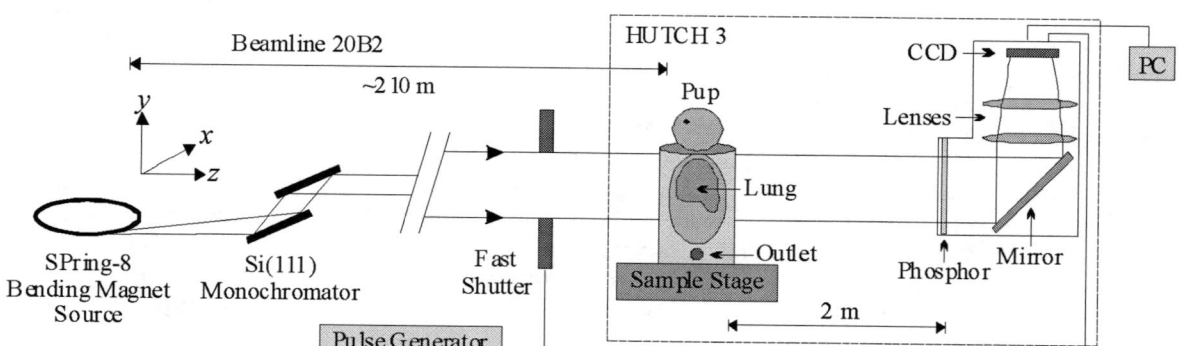

FIGURE 1. Schematic of the experiment for imaging lung aeration at birth with synchrotron-based phase contrast imaging.

RESULTS

Figure 2 shows a sequence of images acquired for a single pup within the first eight minutes after birth. The first image reveals that the pup had inhaled before imaging commenced, as is evident in the appearance of the trachea, major bronchi and some of the smaller airways (Fig. 2(a)). The phase contrast images reveal the branching structure with remarkable clarity. Within the first minute following birth smaller airways became apparent, either individually or as overlapping segments (most probably lobules) giving rise to a speckled pattern. Over time the larger airways became less visible as the speckles dominate the images. Note that the smaller right-hand lung appears to have aerated less quickly than the left lung. Five minutes after birth the lungs appeared well inflated and the diaphragm and the shadow of the heart became evident.

Aeration rates of the individual lungs were determined by separating the images at the pup's midsection along the spinal column. Figure 3 shows the aeration pattern for both lungs. Sudden jumps in air volume correspond to deep inspiratory efforts as evidenced by motion blur in the corresponding images. For this pup the pattern of aeration follows the same overall trend for both lungs, yet the difference in lung size led to different levels of aeration.

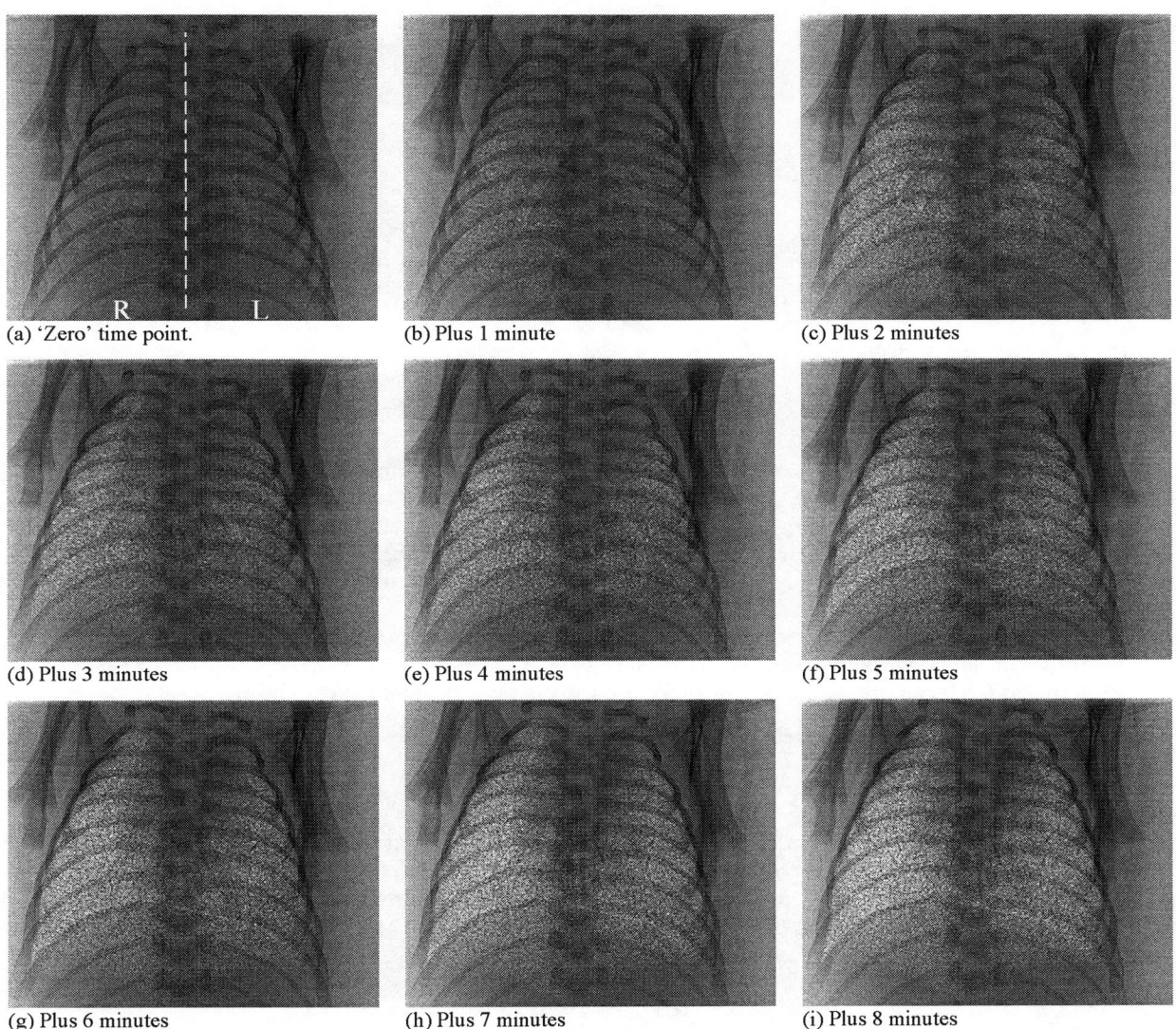

(a) 'Zero' time point. (b) Plus 1 minute (c) Plus 2 minutes

(d) Plus 3 minutes (e) Plus 4 minutes (f) Plus 5 minutes

(g) Plus 6 minutes (h) Plus 7 minutes (i) Plus 8 minutes

FIGURE 2. Images acquired at one minute time intervals after the first image was captured following the birth of a single pup. Exposure time: 83 ms. Image size: 24.45(H) × 20.85(V) mm^2. The dashed line in (a) distinguishes the right (R) and left (L) lungs.

DISCUSSION

The short exposure times required for such near real-time sequences can lead to poor signal-to-noise ratios. Strong phase contrast at the air/tissue interfaces can provide an increase in contrast of up to an order of magnitude over absorption contrast (see [7]). This enables the airways to be seen clearly under relatively poor illumination. Figure 2 clearly indicates that, with a sufficiently bright X-ray source and a high speed, high resolution detector, lung aeration at birth can be revealed with high spatial and contrast resolution using phase contrast imaging. For lung tissue we have previously shown (see [6]) that the enhanced signal-to-noise ratio provides a more reliable reconstruction of the projected thickness of the lung tissue than would be possible if the absorption image alone had been recorded.

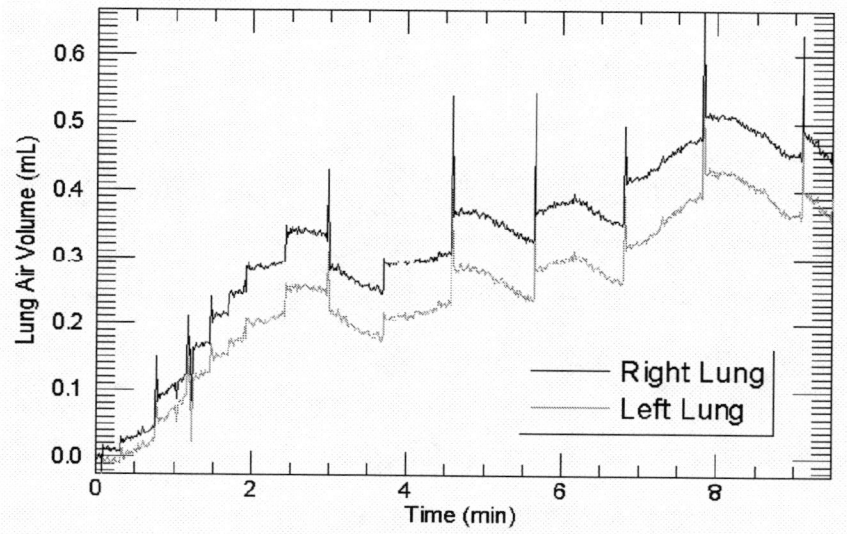

FIGURE 3. Aeration measured for both lungs seen in Fig. 2.

It is interesting that following the first couple of breaths lung volume continued to increase between breaths (Fig. 3), whereas after the lung had largely aerated, lung volume tended to decrease between breaths. It is possible that the increase in lung volume between the first couple of breaths reflects expiratory braking and the maintenance of a positive airway pressure, which helps to establish a functional residual capacity. The decrease in lung volume between breaths may reflect an increase in the surface area of the air/liquid interface resulting in increased surface tension and increased lung recoil.

Future studies will attempt to elucidate aeration rates of the lung lobes; however, this will likely involve more complex image tracking algorithms (see [7]). We aim to build a larger body of evidence to reveal the variations in lung aeration and the factors that influence this process, namely body positioning and the effects of corticosteroids and the presence of lung surfactant.

ACKNOWLEDGMENTS

We acknowledge the Access to Major Research Facilities Program (managed by the Australian Nuclear Science and Technology Organization) for supporting this work and funding the overseas visits of the Australian coauthors to conduct the experiments. We thank the Japan Synchrotron Radiation Research Institute for the use of the SPring-8 facility. M. J. Kitchen acknowledges the receipt of a Monash University Postgraduate Publications Award. R. A Lewis, M. J. Morgan and S. B. Hooper acknowledge the Australian Research Council for a Discovery Project Grant.

REFERENCES

1. Yagi, N., Suzuki, Y., Umetani, K. et al., *Med. Phys.* **26**, 2190-2193 (1999).
2. Suzuki, Y., Yagi, N., and Uesugi, K., *J. Synchrotron Rad.* **9**, 160-165 (2002).
3. Lewis, R. A. Hall, C. J., Hufton, A. P. et al., *Br. J. Radiol.* **76**, 301-308 (2003).
4. Kitchen, M. J., Paganin, D., Lewis, R. A. et al., *Phys. Med. Biol.* **49**, 4335-4348 (2004).
5. Kitchen, M. J., Lewis, R. A., Yagi, N. et al., *Br. J. Radiol.* **78**, 1018-1027 (2005).
6. Kitchen, M. J., Paganin, D., Lewis, R. A. et al., *Nucl. Instrum. Meth.* A **548**, 240-246 (2005).
7. Lewis, R. A., Yagi, N., Kitchen, M. K. et al., *Phys. Med. Biol.* **50**, 5031-5040 (2005).
8. Paganin, D., Mayo, S. C., Gureyev, T.E. et al., *J. Microsc.* **206**, 33-40 (2002).

9. Chantler, C. T., Olsen, K., Dragoset, R. A. et al., Online: *National Institute of Standards and Technology*, http://physics.nist.gov/PhysRefData/XrayMassCoef/cover.html (2005).

Synchrotron X-Ray Synthesized Gold Nanoparticles for Tumor Therapy

C. C. Chien[a], C. H. Wang[a], T. E. Hua[a,b], P. Y. Tseng[a], T. Y. Yang[a],
Y. Hwu[a,c], Y. J. Chen[d], K. H. Chung[e], J. H. Je[f], G. Margaritondo[g]

[a]Institute of Physics, Academia Sinica, Nankang, Taipei, Taiwan 115, R. O. C.
[b]School of Life science, National Tsing-Hua Univiersity, Hsinchu, Taiwan, R. O. C.
[c]Institute of Optoelectronic Science, National Taiwan Ocean University, Keelung, Taiwan, R. O. C.
[d]Department of Radiation Oncology and Medical Research, Mackay Memorial Hospital, Taipei, Taiwan, R. O. C.
[e]Restorative Dentistry, School of Dentistry, University of Washington, Seattle, Washington 98195-7456, U. S. A.
[f]Department of Materials Science and Engineering, Pohang University of Science and Technology, Pohang, Korea
[g]Ecole Polytechnique Fédérale de Lausanne (EPFL), CH-1015 Lausanne,Switzerland

Abstract. Highly concentrated gold nanoparticles (20 ± 5 nm) were produced by an x-ray irradiation method. The particles were then examined for the interactions between gold and tumor cells under x-ray radiation conditions. The biological effects of gold nanoparticles were investigated in terms of the internalization, cytotoxicity and capability to enhance x-ray radiotherapy. The results of this investigation indicated that x-ray derived gold nanoparticles were nontoxic to CT-26 cell line and immobilized within cytoplasm. The irradiation experiments provided further evidence that gold nanoparticles were capable of enhancing the efficiency of radiotherapy.

Keywords: Gold nanoparticle, synchrotron x-ray, tumor therapy
PACS: 87.50.-a, 87.50.Gi

INTRODUCTION

Colloidal gold particles possess a wonderful blend of unique physical, chemical and biological properties that enable them to be used in a wide spectrum of applications including bio-imaging and tumor treatment [1-3]. Nano-sized gold particle enabled photo-thermal effects in cancer treatment have been validated recently by employing the plasmonic resonant feature of gold based particles. However, the above practice is not without limitations when implemented in medical application. For example, the IR and near IR light sources lack sufficient penetrating and focusing capability especially for deep embedding tumors and therefore require fiber-optics or additional surgery in some case. To tackle these intrinsic drawbacks associated with treatments using laser sources, x-rays may offer advantages with its high penetration and focusing abilities.

The effectiveness of sub-micro sized gold particles in enhancing x-ray radiotherapy has been reported by Hainfeld and his colleagues [4]. In their study, a tremendously high concentration (up to 2.7g Au/kg mice body weight) of 1.9 nm gold particles was intravenously administered prior to x-ray therapy. Compared to mice treated with x-ray or gold alone, this combined treatment produced significantly prolonged survival of mice with tumors. The results indicated a drastic improvement in x-ray radiotherapy to the local buildup of gold within tumors that rendered a tumor-to-normal-tissue ratio of gold of approximately 8:1 during x-ray exposure. Although the therapeutic effect is promising, the treatment remains impractical for clinical applications due to limited availability and high cost of large amounts of gold particles with size less than 2 nm in diameter. Herold et al [5]. investigated the efficacy of larger gold particles (1.5-3μm) both in vivo and in vitro. The results of their investigation revealed an average enhancement of 1.43 in cell killing in the presence of 1% gold particles. However, the lower suspension stability of micro-sized gold particles caused serious problem of particle flocculation and precluded the possibility to evaluate the efficacy of gold particles with higher concentrations. Recently our group developed a novel x-ray assistant synthesis method to prepare nano-sized gold particles with high

CP879, *Synchrotron Radiation Instrumentation: Ninth International Conference*,
edited by Jae-Young Choi and Seungyu Rah
© 2007 American Institute of Physics 978-0-7354-0373-4/07/$23.00

concentration, controllable size and sufficient colloidal stability. These characteristics are essential for future in vitro and in vivo applications. In this work, we present the preliminary results of in vitro experiments.

EXPERIMENTAL

Gold nanoparticles were synthesized via a simple approach by x-ray induced reduction of gold precursor aqueous solution [6]. In a typical experiment setup, 1ml 1mM $HAuCl4 \cdot 3H2O$ (Aldrich) solution was mixed with appropriate amount of NaOH (Showa) solution and carried into 15 ml plastic vial for exposure experiment. The exposure experiments were performed at the beamline 01A at National Synchrotron Radiation Research Center (NSRRC), Hsinchu, Taiwan. The parameters of storage ring were 1.5 GeV and 200 mA. Un-monochromatized "white" x-ray beam was utilized throughout the exposure. UV and visible light absorption spectra were taken with a Shimadzu UV-160 spectrometer. The particle morphology, structure and size were measured with a JEOL JSM-6330F field emission scanning electron microscope (FESEM) and with a JEOL JEM 2010 F FEEM (200 kV) transmission electron microscope (TEM). The samples for TEM measurements were prepared by placing droplets of nanoparticle-containing solution on carbon-coated Cu grids and allowing them to dry at ambient atmosphere.

CT26 cells were obtained from N-nitroso-N-methyl urethane-induce mouse colorectal adenocarcinoma cells of Balb/C origin.[7] Cells were cultured in RPMI (GIBCO) medium containing 10% fetal bovine serum (FBS), 1% antibiotics (penicillin at 100 U/mL and streptomycin at 100 μg/mL) and L - glutamine at 37℃ in a humidified 5% $CO2$ incubator. Cells grew to confluence and were detached by trypsin (0.5 g porcine trypsin and 0.2 g EDTA · 4Na per liter of Hanks' Balanced Salt Solution) (Sigma, Saint Louis, Missouri, USA). The cytotoxicity of gold nanoparticles was assessed by MTT ((3-(4,5-Dimethylthiazol-2-yl)-2,5-diphenyltetrazolium bromide)). CT-26 cells were seeded within 24-well culture dishes and were treated with gold nanoparticles of different concentrations (0, 0.125 mM, 0.25 mM, 0.5 mM, 1 mM, and 2 mM) for 3 days. The MTT reagent was purchased from Sigma Inc. Absorbance was recorded at 570 nm using a plate reader (Sunrise, Tecan). Colonogenic cell survival assay was evaluated as follows: 150 CT-26 cells were seed in 6-well culture dish for 24 hours, and then cocultured with gold nanoparticles for 24 hours and subsequently irradiated by the electrons from linear accelerator (Clinac 1800, Varian Associates, Inc., PaloAlto, CA; dose rate 2.4Gy/min) with electron beam energy of 6 MeV and various doses (0, 0.5, 1, 2, and 3 Gy) in a single fraction. Full electron equilibrium was ensured for each fraction by a parallel plate PR-60C ionization chamber (CAPINTEL, Inc., Ramsey, NJ). The radiated cells were further incubated for 10-14 days. Finally the cells were stained by 0.4% crystal violet and colonies were counted. The survival curve was fitted linear quadratic model.[8] The established procedures for TEM sample preparation [9] were followed. The embedded cells were sectioned to 90-100 nm in thickness using a Leica Ultracut R ultramicrotome and the images were taken by using a Hitachi H-7500 operating at 100 keV. Hitachi H-7500 operating at 100 keV.

RESULTS AND DISCUSSION

The results of characterization of gold nanoparticles produced with synchrotron x-ray are shown in Fig.1. The average size of gold nanoparticles was 20 ± 5 nm with reasonable dispersion. The data of UV-VIS data indicated the formation of nano-sized gold particles with the characteristic surface plasmon resonance maximum at 522 nm. Also note that all the gold ions in precursor were reduced to metallic gold due to the absence of absorption peak at 300-320 nm for $HAuCl_4$.

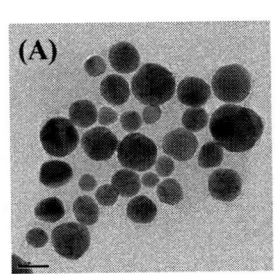 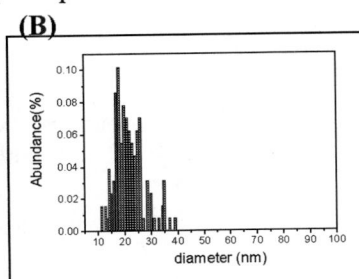 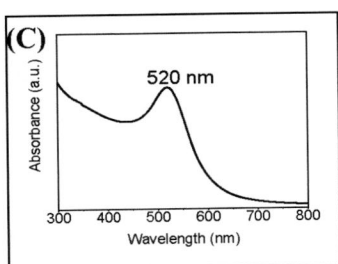

FIGURE 1. Characterization of x-ray synthesized gold Nanoparticles: (a) TEM image (scale bar 50 nm); (b) Histogram showing the size distribution; (c) UV-VIS absorption spectrum.

TEM observation may provide direct evidence to confirm the presence of gold particles within tumor cells. As displayed in Fig.2, some cluster or aggregate composed of gold nanoparticles were observed and located in the vesicles within cytoplasm. It validated the successful internalization of gold particles. No gold particles were found inside the nucleus entity and the nuclear pore structure was completely integrated. The cytotoxicity assay of gold nanoparticles was also performed as shown in Fig.3 and the MTT data supported that gold nanoparticles with the described size were toxic to tumor cells. This finding is controversial with other researchers [10, 11].

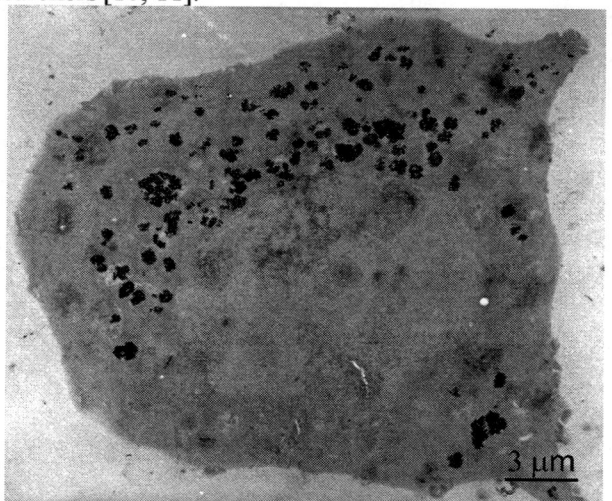

FIGURE 2. TEM micrograph of a cell containing gold nanoparticles.

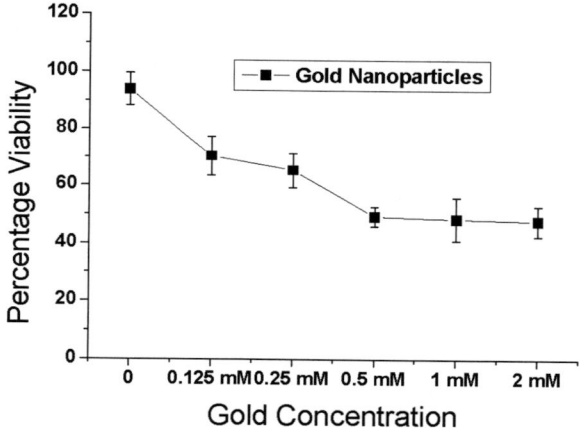

FIGURE 3. Cell viability of CT-26 cell line with respect to the concentration of gold nanoparticles.

The positive effect of gold nanoparticle-enhanced tumor cell killing is described in Fig.4. The result indicated that under identical radiation doses, cells exposed to nanoparticles showed lower survival percentages with increasing amounts of gold. It was also found that within the dosages attempted, the enhanced cell inhibition was more pronounced at higher radiation doses. For example, at a dosage of 3 Gy, the survival percentages were 30.3% and 17% for control cell and 2000 μM gold mediated cell, respectively. As mentioned above, the effectiveness of gold nanoparticle enhanced radiotherapy is essentially dependent on the preferably accumulation of large amount of gold particles within tumor cell/tissue environment. The nano-sized gold particles tested in this report showed the cytotoxicity depended on the concentration of the gold nanoparticles. Furthermore, the unique synthesis route of gold nanoparticles by synchrotron x-ray exposure possesses numerous advantages including the lack of need for any reducing agent and stabilizer, high reproducibility, and easy operation and scaling up. The availability of suitable gold nanoparticles facilitates our future work, especially on animal model experiment, which is underway in our group.

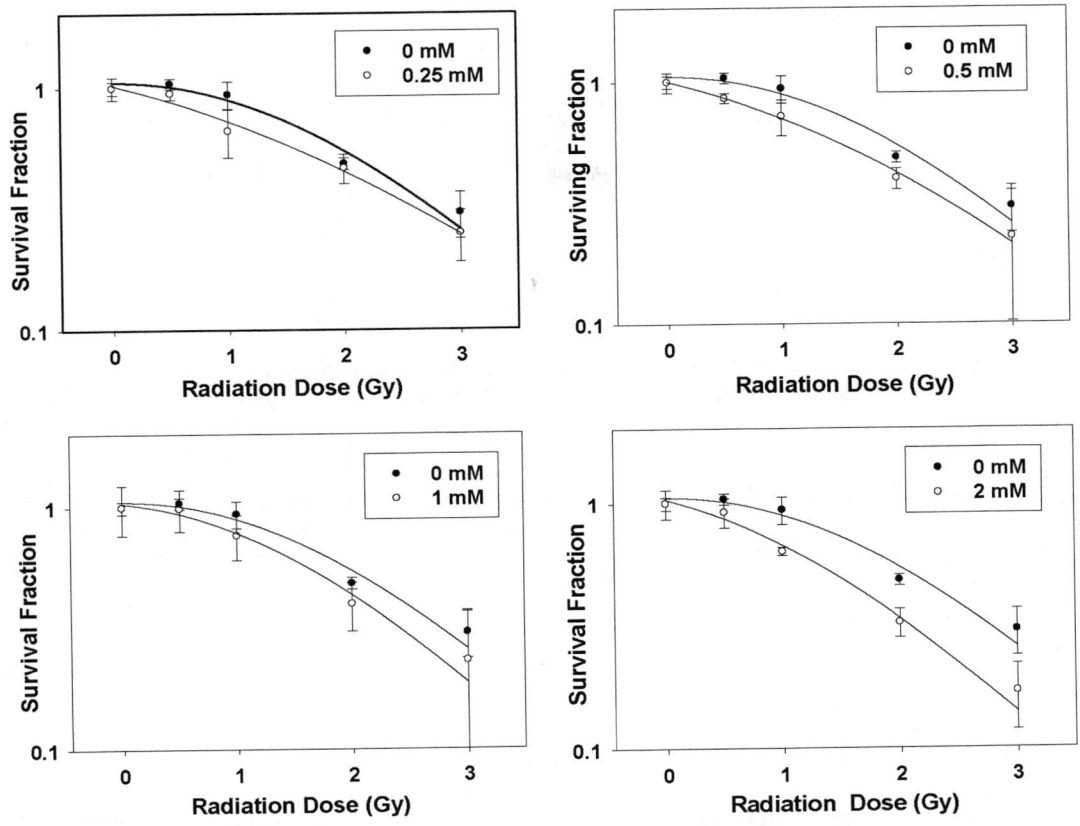

FIGURE 4. Radiation survival curves for CT-26 cell line exposed to a variety concentration of gold nanoparticles: (a) 250 μM; (b) 500 μM; (c) 1000 μM; (d) 2000 μM.

CONCLUSIONS

Gold nanoparticles are successfully fabricated by synchrotron x-ray induced reduction reaction. The synthesis method produced Au nanoparticles with high concentration, controllable sizes and high colloidal stability. In vitro study on the synchrotron x-ray synthesized gold nanoparticles were conducted with the anticipation that gold particles may enhance radiotherapy. The preliminary investigation indicated evident efficiency of radiotherapy.

ACKNOWLEDGMENTS

This work was supported by the National Science Council (Taiwan), by the Academia Sinica (Taiwan) and by the Swiss Fonds National de la Recherche Scientifique and by the EPFL.

REFERENCES

1. T. M. Lee, A. L. Oldenburg, S. Sitafalwalla, D. L. Marks, W. Luo, F. Jean-Jacques Toublan, K. S. Suslick, and S. A. Boppart, Optic. Lett. 28, 1546 (2003).
2. Q. Sun, Q.Wang, B. K. Rao, and P. Jena, Phys. Rev. Lett. 93, 186803 (2004).
3. C. Loo, A. Lowery, N. Halas, J. West, and R. Drezek, Nano Lett. 5, 709-711 (2005).
4. J. F. Hainfeld, D. N. Slatkin and H. M. Smilowitz, Phys. Med. Biol. 49, N309-N315 (2004).
5. D. M. Herold, I. J. Das, C. C. Stobbe, R. V. Iyer, and J. D. Chapman, Int. J. Radiat. Biol. 76, 1357-1364 (2000).
6. Y. C. Yang, C. H. Wang, Y. K. Hwu and J. H. Je, Mater. Chem. Phys. (in press).
7. Y. J. Chen, H. F. Liao, T. H. Tsai, S. Y. Wang and M. S. Shiao, Int. J. Radiat.Oncol. Biol. Phys. 63, 1252 (2005)
8. A. Joubert, M. C. Biston, C. Boudou, J. L. Ravanat, T. Brochard, A. M. Charvet, F. Esteve, J. Balosso and N. Foray, Int.J.Radiat.Oncol.Biol.Phys. 62, 1486 (2005).
9. B. D. Chithrani, A. A. Ghazani and W. C. W. Chan, Nano. Lett. 6, 662 (2006).
10. E. E. Connor, J. M.wamuka, A. Gole, C. J. Murphy and M. D. Wyatt, Small 1, 325 (2005).
11. R. Shukla, V. Bansal, M. Chaudhary, A. Basu, R. R. Bhonde and M. Sastry, Langmuir 21, 10644 (2005).

A New Method for Phase Contrast Tomography

A. Groso, R. Abela, C. David and M. Stampanoni

Paul Scherrer Institut, CH-5232 Villigen PSI Switzerland

Abstract. We report how a method which requires a simple experimental setup, in principle equivalent to the one for doing classical absorption tomography, can yield phase information in three dimensions in only one single step. As an alternative to two steps approaches (phase retrieval and reconstruction of the object function applying a conventional filtered backprojection algorithm), A.V. Bronnikov suggested an algorithm which provides direct 3D reconstruction of the refractive index from the intensity distributions measured in a single plane (pure phase object) and from the intensity distributions measured in two planes (mixed phase-amplitude object). We adapted the original algorithm in order to retrieve the 3D reconstruction of the refractive index from the intensity distributions measured in a *single plane* also for the case of mixed weakly absorbing-phase object. The results show that the contrast is increased, while keeping dose minimal and spatial resolution equivalent to the conventional absorption based technique.

Keywords: x-ray imaging, phase contrast, tomography
PACS: 81.70.Tx, 42.30.Wb

INTRODUCTION

Conventional x-ray computed tomography (CT) is based on the difference in x-ray absorption by different materials and is nowadays a standard technique with spatial resolution around one micron [1-4]. On the other hand, a wide range of samples studied in materials science, biology and medicine show very weak absorption contrast, yet producing significant phase shifts of the x-ray beam. The use of phase information can therefore increase the contrast substantially.

As far as phase tomography (quantitative 3D reconstruction of the phase or the refractive index from 2D phase images) is concerned, several attempts were made by using interferometric [5-6] and non-interferometric phase retrieval methods [7-13]. All these methods are based on a two step approach: first, the projections of the phase are determined in the form of Radon projections and then the object function, i.e. the refractive index δ is reconstructed applying a conventional filtered backprojection algorithm.

Alternatively, the reconstruction algorithm suggested by Bronnikov in his theoretical works [14, 15] presents an approach which requires no intermediate step of 2D phase retrieval and provides a direct 3D reconstruction of refractive index from the intensity distributions measured in a single plane (pure phase object) and from the intensity distributions measured in two planes in the case of mixed phase-amplitude object.

In this work we applied the quantitative phase contrast CT method presented in [14, 15] to real experimental data. We adapted the original algorithm in order to retrieve the 3D reconstruction of the refractive index from the intensity distributions measured in a *single plane* also in the case of *mixed weakly absorbing-phase* object. In this way the experimental setup can be kept very simple and the radiation damage is kept to the minimum.

THEORETICAL BACKGROUND

Let consider a monochromatic plane wave with wavelength λ that propagates along the positive z-axis and that impinges upon a thin mixed phase-amplitude object, which is characterized by the linear absorption coefficient $\mu(x_1, x_2, x_3)$ and the real part of the index of refraction $\delta(x_1, x_2, x_3)$. The intensity distribution $I(x, y)$ at a distance $z = d$ and angle of rotation θ can be expressed by [9]:

CP879, *Synchrotron Radiation Instrumentation: Ninth International Conference*,
edited by Jae-Young Choi and Seungyu Rah
© 2007 American Institute of Physics 978-0-7354-0373-4/07/$23.00

$$I_{\theta,z=d}(x,y) = I_{\theta,z=0}(x,y)\left[1 - \frac{\lambda d}{2\pi}\nabla^2\phi_\theta(x,y)\right] \qquad (1)$$

with $\phi_\theta(x,y) = -\dfrac{2\pi}{\lambda}\displaystyle\int_{\Re^2}[\delta(x_1,x_2,y)\times\widetilde{\delta}(x - x_1\cos\theta - x_2\sin\theta)]dx_1 dx_2$ being the phase function of the object and

$\widetilde{\delta}$ denoting the Dirac function. Expression (1), which corresponds also to the Transport of Intensity Equation (TIE) in its simplified form [6-11], is valid (i) in the near-field Fresnel region ($d << a^2/\lambda$, where a is the transversal size of the smallest structure in the object) and (ii) for a mixed phase-amplitude object with weak and almost homogeneous absorption (i.e. $\partial\mu/\partial x$, $\partial\mu/\partial y \approx 0$).

The goal of quantitative phase tomography is to reformulate Eq.1 to obtain $\delta(x_1,x_2,x_3)$ from the knowledge of $I_{\theta,z=d}(x,y)$ for $\theta \in [0, \pi]$. Expressing Eq.1 as $\nabla^2\phi_\theta(x,y) = -\dfrac{2\pi}{\lambda d}g_\theta(x,y)$ with $g_\theta(x,y) = \dfrac{I_{\theta,z=d}(x,y)}{I_{\theta,z=0}(x,y)} - 1$, applying then the 3D Radon transform (denoted by the symbol $\wedge$), and calculating finally the second derivative with respect to the variable $s = x\sin\omega + y\cos\omega$, one gets: $\dfrac{\partial^2}{\partial s^2}\overset{\wedge}{\delta}(s,\theta,\omega) = -\dfrac{1}{d}\hat{g}_\theta(s,\omega)$.

Above expression is a theorem which states that from the 2D Radon transform of the measured value g, one can directly find the 3D Radon transform of δ [15]. The computation of the two dimensional Radon transform and its backprojection can be combined into a single step and, defining $q(x,y) = \dfrac{|y|}{x^2 + y^2}$, one obtains:

$$\delta(x_1,x_2,x_3) = -\frac{1}{4\pi^2 d}\int (q**g_\theta)d\theta \qquad (2)$$

where the stars indicate a 2D convolution. This convolution integral can be computed in the Fourier domain by taking the two dimensional Fourier transform. In the Fourier domain, equation $q(x,y) = \dfrac{|y|}{x^2 + y^2}$ has a low-pass filter form given by [15]:

$$q(\xi,\eta) = \frac{|\xi|}{\xi^2 + \eta^2} \qquad (3)$$

Expressed in this form (Eq. 2), the approach becomes very interesting since in case of a pure phase object (i.e. $I_{\theta,z=0}(x,y) = 1$) the 3D distribution of the refractive index can be recovered from only one single tomographic data set. This is a significant improvement since (a) the experimental setup can be kept extremely simple (it is actually the same as for standard, absorption based tomography) and (b) the radiation damage is kept to the minimum, which is of special importance for biological specimens.

RESULTS AND ANALYSIS

In a previous work [16] we simulated intensity projection data, implemented the aforesaid reconstruction algorithm and successfully reconstructed the object refractive index. In a further step, the original algorithm [14,15] has been used to reconstruct phase tomograms from experimental data sets.

Experiments were performed at the Tomography Station of the Materials Science beamline at the Swiss Light Source [3]. As already pointed out, the experiment setup is equivalent to the one for standard absorption tomography. Nevertheless, two additional conditions have to be satisfied: (i) sample-detector distance (SDD) is now increased from the minimum SDD=0 used in absorption tomography, to the near field Fresnel region where $SDD << a^2/\lambda$ (a is the transversal size of the smallest structure in the object and λ is the x-ray wavelength) and (ii) the photon energy was set to the maximum attainable at the experimental station (25 keV) in order to satisfy as much as possible weak and almost homogeneous absorption condition ($\partial\mu/\partial x$, $\partial\mu/\partial y \approx 0$).

As a test, one millimeter thick polyacrylate, starch and cross-linked rubber matrix sample was investigated. There is a rather weak contrast when absorption tomography is used [Fig. 1a)]. In order to apply phase tomography method in pure phase object approximation ($I_{\theta,z=0}(x,y) = 1$) a single data set ($I_{\theta,z=d}(x,y)$) of 721 angular projections at d of 80 mm and 25 keV photon energy was acquired. Fig. 1b) presents a slice through the sample reconstructed using Eq.

(3) and (2). Even though the absorption level for this object is calculated to be in the range of only 3 %, pure phase object condition is not satisfied – the image is severely corrupted by the residual absorption artifact (see cap in the center of Fig. 1b)) meaning that in principle also intensity measurement at 'zero' distance ($I_{\theta,z=0}(x,y)$) is required.

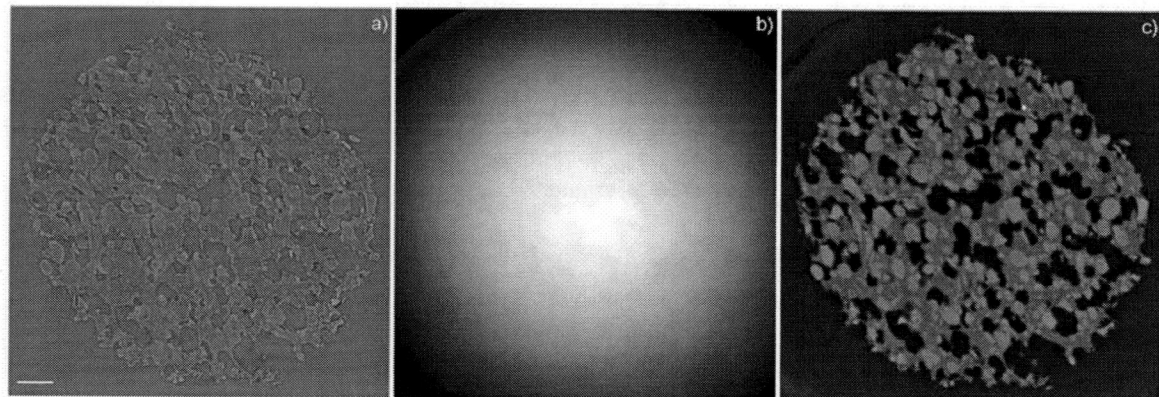

FIGURE 1. Tomographic reconstruction of a polymer sample: (a) absorption contrast, (b) phase contrast obtained with uncorrected filter, Eq. 3 (c) phase contrast obtained with correct filter, Eq. 4. The length of the scale bar is 100 μm.

We have adopted another approach - the reconstruction artifacts (observe Fig. 1b)) are corrected by amending the original method [15] leading to a modified version of Bronnikov's algorithm (MBA). The correction consists of adding, in the denominator of the low-pass filter given in Eq. (3), an absorption dependent correction factor α_{exp}. Consequently, the new filter has the following form:

$$q(\xi,\eta) = \frac{|\xi|}{\xi^2 + \eta^2 + \alpha_{exp}}$$

(4)

The values of α_{exp} to be used are found by using a semi empirical (simulations-experiment) approach. The details are given in [17]. Applying the MBA method, we reconstructed the same data set (rubber matrix) and the (artifacts free) result is presented in the Fig. 1c).

In order to validate our approach, we have compared its results with the ones obtained using an established differential phase contrast method (DPC) [6]. To make a direct evaluation, experiments using the two techniques have been performed on the same samples and with the same detector effective pixel size (1.4 microns).

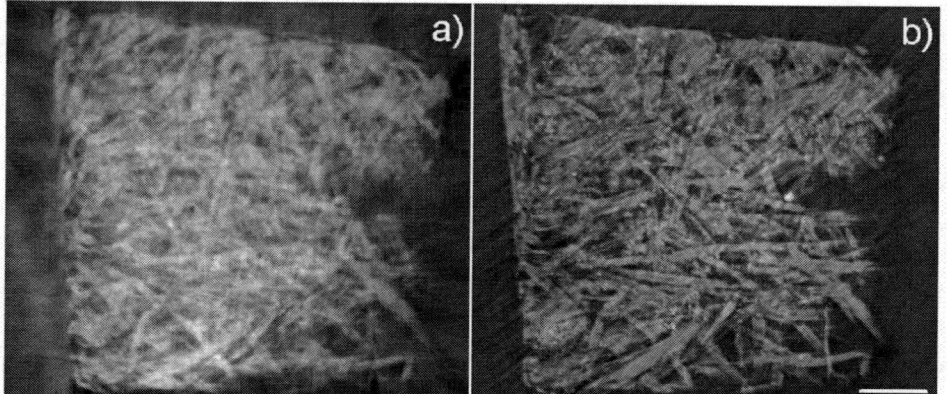

FIGURE 2. Validation of the MBA method: Phase tomographic reconstruction of paper obtained (a) using DPC and (b) using MBA. The length of the scale bar is 100 μm.

Figure 2a) presents the result obtained using the DPC technique. 501 angular projections with 4 phase steps for every angular projection were acquired at a photon energy of 17.5 keV. Total scan time was 6.5 hrs and total exposure time 3.3 hrs. For every angular position the projection of the phase is determined and then the phase in 3D is reconstructed using conventional filtered backprojection algorithm. Figure 2b) shows the same sample reconstructed with the MBA technique. One tomographic data set (501 angular projections) was acquired at SDD of

22 mm (using 25 keV photon energy) and reconstructed applying Eq. (4) and Eq. (2). Total scan time was 1.5 hrs and total exposure time 50 minutes.

DISCUSSION

We have demonstrated that the 3D distribution of the phase (refractive index) of a weakly absorbing object can, alternatively to the combined phase retrieval-backprojection methods (two steps) presented up to now, be directly reconstructed from a single tomographic data set (one step). In order to reduce the residual absorption artifact the method presented in Ref. [15] has been amended leading to what we named the Modified Bronnikov Algorithm (MBA). The results show that the contrast is increased, while keeping experimental setup simple and radiation dose minimal.

The MBA was validated using the Differential Phase Contrast Imaging technique. In general, the two methods are complementary. The full 3D phase imaging approach (MBA) is experimentally simple, very fast and it is ideally suited for small objects when resolution around 1 micron is needed. Deposited dose is equivalent (or lower, since higher photon energy is used) to the one deposited in conventional tomography. The DPC is more demanding in terms of instrumentation and acquisition time (therefore also deposited dose) but it is more sensitive and can be scaled up to large fields of view.

Artifacts/non perfect reconstruction using the MBA method are due to the fact that estimated absorption (prior to the filtering step) is approximated as constant over entire object. Additionally, detector sensitivity/noise and beam instability also induce low frequencies problem, deteriorating the reconstruction quality.

Future developments will focus on the fine tuning of the filtering step and on the streamlining of the whole process in order to allow high-resolution and high throughput imaging of soft biological and materials science specimens.

ACKNOWLEDGMENTS

We are grateful to A. Bronnikov and O. Bunk for fruitful discussions and to R. Rudolf for providing the samples. The help of P. Schneider, S. Linga and R. Müller in performing simulations and F. Pfeiffer in DPC reconstructions is gratefully acknowledged. Part of this work was supported by the Swiss National Science Foundation (FP-620-58097.99 and PP-104317/1).

REFERENCES

1. U. Bonse and F. Bush, Prog. Biophys. Molec. Biol., 65, 133-169 (1996)
2. B.A. Dowd, G.H. Campbell, R.B. Marr, V.V. Nagarkar, S.V. Tipnis, L. Axe and D.P. Siddons, Developments in X-Ray Tomography II, Proc. SPIE 3772, 224-236 (1999)
3. M. Stampanoni, G. Borchert, P. Wyss, R. Abela, B. Patterson, S. Hunt, D. Vermeulen P. Rüegsegger, Nucl Instrum Meth. A 491 (1-2), 291-301 (2002)
4. T. Weitkamp, C. Raven and A. Snigirev, Developments in X-Ray Tomography II, Proc. SPIE 3772, 311-317 (1999)
5. A. Momose, Nuclear Instruments and Methods A 352, 622-628 (1995)
6. T. Weitkamp, A. Diaz, C. David, , F. Pfeiffer, M. Stampanoni, P. Cloetens, E. Ziegler, Op. Express 13, 6296-6304 (2005)
7. L. J. Allen, W. McBride, N.L. O'Leary, and M. P. Oxley, Ultramicroscopy 100, 91-104 (2004)
8. W. K. Hsieh, F.R. Chen, J. J .Kai, A.I. Kirkland, Ultramicroscopy 98, 99-114 (2004)
9. S. C. Mayo, P.R. Miller, S.W. Wilkins, T.J. Davis, D. Gao, T. E. Gureyev, D. Paganin, D. J. Parry, A. Pogany and A. W. Stevenson, Journal of Microscopy, 207, 79-96 (2002)
10. T.E. Gureyev, A. Pogany, D. M. Paganin and S.W. Wilkins, Optics Communications 231, 53-70 (2004)
11. A. G. Peele, F. DeCarlo, P.J. Mahon, B.B. Dhal and K.A. Nugent, Rev. Sci. Instr. 76, 083707 (2005)
12. T. E. Gureyev and K. A. Nugent, Journal of the Optical Society of America A-Optics Image Science and Vision 13, 1670-1682 (1996)
13. P. Cloetens, W. Ludwig, J. Baruchel, D. Van Dyck, J. Van Landuyt, J.P. Guigay and M. Schlenker, Appl. Phys. Letters 75, 2912-2914 (1999)
14. A.V. Bronnikov, Optics Communications, 171(4-6), 239-244 (1999)
15. A.V. Bronnikov, J. of the Opt. Soc.of America A 19 (3), 472-480 (2002).
16. A. Groso, M. Stampanoni, R. Abela, P. Schneider, S. Linga and R. Müller, Appl. Phys. Lett. (2006) (to be published)
17. A. Groso, R. Abela, C. David, F. Pfeiffer and M. Stampanoni, submitted

Beamline for Biological Macromolecular Assemblies (BL44XU) at SPring-8

Masato Yoshimura[1*], Eiki Yamashita[1*], Mamoru Suzuki[1], Masaki Yamamoto[2], Shinya Yoshikawa[3], Tomitake Tsukihara[1] and Atsushi Nakagawa[1]

[1]Institute for Protein Research, Osaka University, 3-2 Yamadaoka, Suita 565-0871, Japan
[2]RIKEN/SPring-8, 1-1-1, Kouto, Sayo, Sayo, Hyogo 671-5198, Japan
[3]Graduate School of Life Science, University of Hyogo, 3-2-1, Kouto, Kamigohri, Hyogo 678-1297, Japan

Abstract. To reveal the function of the biological macromolecular assemblies at atomic level, three-dimensional structure of the complex molecules is essential. The beamline including detectors is designed to collect high resolution and high quality diffraction data from macromolecular assembly crystals with large unit cells. This beamline uses a standard undulator of SPring-8 as a light source. X-rays from the undulator are monochromatized by a liquid nitrogen cooled double crystal. The monochromatized X-ray beam is collimated or focused reflected by a Rhodium-coated mirror, and it is also used for elimination of higher-order harmonics. At downstream of the Beryllium window, we place a fast shutter system, two quadrant slit systems, and an ionization chamber in this order. Size of the X-ray-beam can be changed by the slit systems from 1 μm to fully-open. Typical photon flux is about 10^{11} photons/sec. with the slit size of 0.07×0.07 mm^2 at 0.9 Å. A sample is cooled either to 90K by a nitrogen or to 30K by a helium cryo-stream system. A DIP6040, which is a hybrid-type of image plates and a CCD, is used for data collection. The DIP6040 has six image plates with individual readout systems, and effective area of each image plate is circular with 400 mm in diameter. A 185 mmφ CCD detector can be used both for data collection and screening of the crystals. Frame rate of the image plate system can be achieved to 120 images per hour at maximum.

Keywords: X-ray beamline; protein crystallography.
PACS: 61.10.Nz

INTRODUCTION

Biological macromolecular assemblies play significant roles in many biological reaction systems. To reveal the function of the macromolecular assemblies at atomic level, three-dimensional structure of the complex is essential. Crystals of biological macromolecular assemblies such as the complex of membrane protein, the complex of proteins, the complex of protein and nuclear acid and the virus, are often extremely weak diffraction, large in unit cell size, small in crystal size. Furthermore, many macromolecular crystals are weaker than normal protein crystals in x-ray radiation damage. The beamline for biological macromolecular assemblies beamline (BL44XU) at SPring-8 was designed for the high precision diffraction data collection of biological macromolecular crystals. The beamline is operated by the Institute for Protein Research, Osaka University. The description of the beamline design, X-ray optics, detector, data processing and some crystallographic results are presented.

BEAMLINE AND DETECTOR

Since crystals of biological macromolecular assemblies are often X-ray radiation sensitive and extremely weak diffraction power, it is essential to use high-brilliance and highly paralleled synchrotron radiation for diffraction data collection. Schematic layout of the beamline is shown in Fig. 1.

---

[*] Equally contributed

CP879, *Synchrotron Radiation Instrumentation: Ninth International Conference*,
edited by Jae-Young Choi and Seungyu Rah
© 2007 American Institute of Physics 978-0-7354-0373-4/07/$23.00

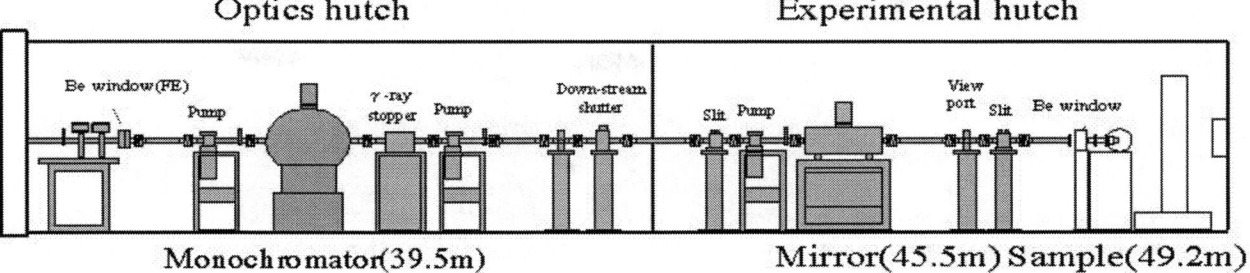

FIGURE 1. Schematic layout of the beamline. Main components are followed with the distance from the light source

Optics

Optics is designed to optimize brilliance and divergence of the X-ray beam. An undulator is used as light source. The main optical components are the monochromator and the mirror. Typical photon flux is about 10^{11} photons/sec. with the slit size of 0.07×0.07 mm^2 at 0.9 Å. X-ray path to about 2.5 m downstream from the mirror are evacuated by turbo molecular pump systems. For crystallographic techniques such as multi- and single-wavelength anomalous diffraction wavelength (MAD/SAD), new GUI was developed for easy use of changing wavelength and optimizing the wavelength for the MAD/SAD experiments.

Light Source

The SPring-8 storage ring runs at electron energy of 8 GeV with nominal current of 100 mA in its top-up operation mode. The beamline uses a SPring-8 standard-type in-vacuum undulator as a light source. The undulator is 4.5 m in length and 32 mm in period length, so that, number of period is 140. Light from the undulator are masked with the front-end slit. Typical aperture is 0.5 mm in horizontal and 0.5 mm in vertical, and is located at 32.2 m from the light source.

Monochromator and Mirror

X-ray photons from a light source are monochromatized with a liquid nitrogen cooled double-crystal monochromator. The monochromator is located at 39.5 m from the light source and in the optics hutch. Using a silicon crystal in 111 orientation, design allows access to photon energies between 5.8 and 23 keV. However, it is typically operated at energies between 7 and 21keV. The energy resolution ($\Delta E/E$) is about 1×10^{-4}. To prevent from silicon crystal distortion by the high power density of the undulator, both crystals are sandwiched and are cooled with the two copper blocks in which liquid nitrogen (LN$_2$) streams. Because of LN$_2$ cooling power, simple geometry of crystals without inclination is allowed. The mirror is set at 45.5 m from light source and in the experimental hutch. The mirror is horizontal focusing type and is Rh-coated mirror, which is mainly used for elimination of multiple harmonics.

Goniometer Stage

Down-stream from the Be window which keeps vacuum for upstream side, there are ion-chamber, the fast shutter system, a 1st quadratic slit system, a 2nd quadratic slit system, a beam intensity monitor and a collimator of 0.3 mm in diameter in that order. To reduce a background from air in X-ray path, the path between the Be window and the collimator which is next to a sample is filled with He gas. And these components are located within 1 m region.

Shutter System, Slit Systems, Beam Intensity Monitor

A fast shutter system, of which opening/closing edge time is less than 1 msec., is used to minimize systematic error in intensity measurements of partial reflections. Two quadrant slit systems are used to define shape of the X-ray beam in horizontal and vertical directions. Using these slit systems, the size of X-ray beam can be defined in the step of 1 μm by the control computer. After defining the X-ray beam sizes by the slits, the X-ray intensity is always monitored with an ionization chamber. Typical size of X-ray beam used is 0.05×0.05 mm^2.

Goniometer, CCD camera, Sample Cooling system

On the goniometer stages, there are the slit systems and the ionization chamber described above, CCD camera for sample centering and a beam stopper movable in photon direction. This stage can be controlled and adjusted in the four-axis directions. A sample is mounted on the goniometer head. The sample position can be adjusted by a controller in the hutch or by the computer outside the hutch. To reduce the radiation damage by cooling samples, a cryo-cooler is used. Using with this cryo-cooler at He gas mode, the temperature of the sample can be controlled and reached to 30 K at lowest. Typically, the sample is kept about 100 K with Nitrogen mode.

Detector and Data processing tools

For highly precise data taking for crystal of heavily decaying diffraction with higher resolution or of very large unit cell, large area detector and large dynamic range is the key point of choice of the detector. We mainly use large area image plates detector system. For checking diffraction data or fast processing data, we serve the analyzing and processing tools besides the experimental hutch.

DIP6040

As the two-dimensional detector, we use DIP6040, which is hybrid type of CCD detector and imaging plates (IPs). There are six independent image plates (IP) accompanies with the readout systems and one CCD detector (SMART6500, Bruker-AXS) in the DIP6040 system. The size of each IP is 400 mm in diameter, data of which are read with two photo-multipliers by spiral readout method. Demerit of using IP is its long readout time. To reduce this dead time, readout time for one IP is divided into exposure time for the other five IPs. If exposure time is about 1 second, 120 frames per hour can be taken.

TABLE 1. Performance of the detector DIP6040

Performance	Imaging Plates	CCD
Sample to detector distance	234-1200 mm	160-1200 mm
Detection area (in diameter)	400 mmφ	165 mmφ
Number of IPs/CCD	6	1
ADC	16 bit	16 bit
Dynamic range	1048576	65536
Pixel size	0.1 mm	0.085 mm
Dead time from exposure to exposure	30 sec (exposure time < 1s)	10 sec

FIGURE 2. Photograph of overview of the DIP6040. The CCD detector is placed at 30 cm down stream of the front black panel.

Analyzing Tools

There are five or more Linux machines besides the experimental hutch as analyzing tools. Every taken image, its analyzed highest diffraction resolution and time-dependent resolution change are automatically displayed. Many users use this graph of the resolution change as a radiation damage monitor of sample crystal. The data processing and reduction software for protein crystallography such as HKL2000, Mosflm are offered to users.

OBTAINED RESULTS

From the beginning, many solved crystal structure by the beamline is larger than 3Å resolution and larger than 70 Å² of the averaged temperature factor. For the experiment of such crystals, long exposure time was required. To taking the advantage of the dynamic range of the IP, better counting statistics of diffraction at higher resolution was obtained without saturating that of lower one.

Large Particles

As one of the very large particles, the diffraction result for rice dwarf virus (RDV)[1] is shown in Fig. 3., which crystallized in *I222* with unit-cell dimensions of about $a = 760$ Å, $b = 780$ Å, $c = 350$ Å. With long exposure time of 60 seconds, 3.24 Å resolution data for RDV was obtained. Large dynamic range of the IP allows us to obtain high-resolution data without saturation of the intensities of strong diffractions.

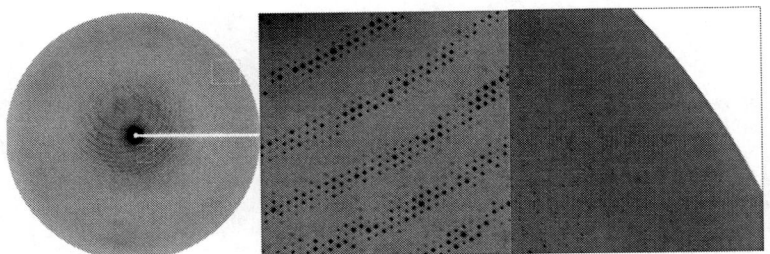

Condition	
wavelength	0.9 Å
exposure time	60 sec.
camera length	700 mm
oscillation angle	0.3 deg.

FIGURE 3. Diffraction data for the crystal of RDV are shown. Left image shows overall image, Middle image is the enlarged picture for lower resolution part. Right image is that for higher part. Spots of diffraction near the edge (3.24 Å) of IP are seen. Measurement conditions for the image data are summarized in the text box.

Membrane Proteins

Cytochrome c Oxidase

The structure of cytochrome *c* oxidase firstly was solved in 1995. Since then, the mechanism of proton pump of the enzyme is one of the hottest topics in the field of bioenergetics. As the first priority subject of the beamline, the diffraction data for the enzyme elaborately have been taken [2].

Other Selected Membrane Proteins

The structures of the other membrane proteins solved using with the beamline are listed below, Bacterial multidrug export, Structure of *E. coli* protein AcrB[3], Structural changes in the calcium pump accompanying the dissociation of calcium[4], The Cytochrome b₆f complex of oxygenic photosynthesis[5], Monoamine Oxidase A[6], Membrane fusion protein, MexA, of the multidrug transporter in *pseudomonas aeruginosa* and the drug discharge outer membrane protein, OprM [7]

REFERENCES

1. A. Nakagawa et al., *Structure* **11**, 1227-1238 (2003).
2. T. Tsukihara et al., *Proc. Natl. Acad. Sci. U.S.A* **100**, 15304-15309 (2003). E. Yamashita et al. *Acta Crysta. Sec. D*, **61**, 1373-1377 (2005)
3. S. Murakami, R. Nakashima, E. Yamashita, and A. Yamaguchi, *Nature*, **419**, 587-593 (2002).
4. C. Toyoshima and H. Nomura, *Nature*, **418**, 605-611 (2002).
5. G. Kurisu et al., *Science*, **302**, 1009-1014 (2003).
6. J. Ma et al., *J. Mol. Biol.*, **338**, 103-114 (2004).
7. H. Akama et al., *J. Biol. Chem.* **279**, 25939-25942 (2004). *J. Biol. Chem.* **279**, 52816-52819 (2004).

SPring-8 Structural Biology Beamlines / Current Status of Public Beamlines for Protein Crystallography at SPring-8

Masahide Kawamoto*, Kazuya Hasegawa*, Nobutaka Shimizu*, Hisanobu Sakai*, Tetsuya Shimizu[†], Atsushi Nisawa[†], and Masaki Yamamoto*[†]

*SPring-8/JASRI, 1-1-1, Kouto, Mikazuki, Sayo-gun, Hyogo 679-5198, Japan
[†]RIKEN SPring-8 Center, 1-1-1, Kouto, Mikazuki, Sayo-gun, Hyogo 679-5148, Japan

Abstract. SPring-8 has 2 protein crystallography beamlines for public use, BL38B1 (Structural Biology III) and BL41XU (Structural Biology I). The BL38B1 is a bending magnet beamline for routine data collection, and the BL41XU is an undulator beamline specially customized for micro beam and ultra-high resolutional experiment. The designs and the performances of each beamline are presented.

Keywords: Protein crystallography, undulator, mail-in data collection, micro beam, Ultra-high resolution, SPring-8
PACS: 61.10.Nz

INTRODUCTION

There are 9 beamlines for protein crystallography at SPring-8 as of December 2005. Two beamlines, BL38B1 (Structural Biology III) and BL41XU (Structural Biology I), out of them are dedicated for public uses. The BL38B1 is designed based on a standard SPring-8 bending magnet beamline, and used for routine protein crystallography. The BL41XU is an undulator beamline using the SPring-8 standard in-vacuum undulator as the light source. Several challenging themes, such as measurement in ultra-high resolution and data collection from micro-crystals, are executed with the high brilliant undulator beam. In this paper, we report the optics, experimental stations and current status of each beamline.

BEAMLINE OPTICS

The BL38B1 is designed based on the standard SPring-8 bending magnet beamline [1]. The white X-rays generated by the bending magnet are monochromatized using a fixed-exit double crystal monochromator, which has a pair of Si (111) crystals that are cooled by water directly for first crystal and indirectly for second one. This monochromator provides X-rays in an energy range from 6.5 to 17.5 keV. The monochromatized X-ray beam is focused into the size of 0.2 mm in the horizontal and vertical directions at the sample position with the 1-m-long rhodium-coated bent-cylindrical mirror. In order to obtain the optimized focus beam size at the sample position and to reject higher-order harmonics, the glancing angle of the mirror is set to 3.3 mrad. The photon flux at the sample position is 1×10^{11} photons/sec at 12.4 keV. The flux is estimated by using Si-PIN photodiode.

The BL41XU is an undulator beamline [2]. The SPring-8 standard in-vacuum-type undulator [3] is employed as a light source, whose magnetic periodicity is 3.2 cm and a total length is 4.5 m. The magnetic gap is variable from 9.6 to 50 mm. A corresponding energy range of fundamental emission is from 6.3 to 18 keV, and that of third harmonics from 19 to 54 keV. The undulator beam is monochromatized using a rotated-inclined double crystal monochromator. The monochromator has a pair of Si (111) crystals, first crystal has a "pin-post" structure for direct water cooling [4], and second one is cooled by water indirectly. The monochromator provides highly brilliant X-ray in a wide energy range from 6.5 to 17.5 keV with fundamental emission and from 19 to 37.5 keV with third

CP879, Synchrotron Radiation Instrumentation: Ninth International Conference,
edited by Jae-Young Choi and Seungyu Rah
© 2007 American Institute of Physics 978-0-7354-0373-4/07/$23.00

harmonics. The 70-cm-long rhodium-coated bent mirror is used to focus the X-ray beam in horizontal direction and eliminate the higher harmonics of the undulator beam with a glancing angle of 3.5 mrad. The bent mirror for vertical focusing is currently not used because of the decreasing of reflectivity due to the radiation damage. The photon flux at the sample position is around 1×10^{12} photons/sec at 12.4 keV, and its size is ~70 μm (horizontal) and ~250 μm (vertical). We are planning to replace new vertical focusing bent mirror in summer, 2006. We expect that the focused beam size will be ~80 μm (vertical).

EXPERIMENTAL STATIONS:

Diffractometer

Both the BL38B1 and BL41XU are equipped with the same diffractometer, which consists of an optics unit, a goniometer, a detector base and an alignment table installing them (Fig. 1). The optics unit is composed of two quadrant slits, a beam intensity monitor, an attenuator and a beam shutter. The upstream slit is used to restrict the size of X-ray exposuring the sample, and the downstream slit is used to eliminate a parasitic scatter around the restricted beam. The beam intensity monitor is placed on the upstream side of the shutter, so that the diffractometer can be aligned to the X-ray beam without irradiating sample. The optics unit is vacuumed to reduce absorption and scattering with air. The goniometer has three translation axes (X, Y and Z) for sample centering.

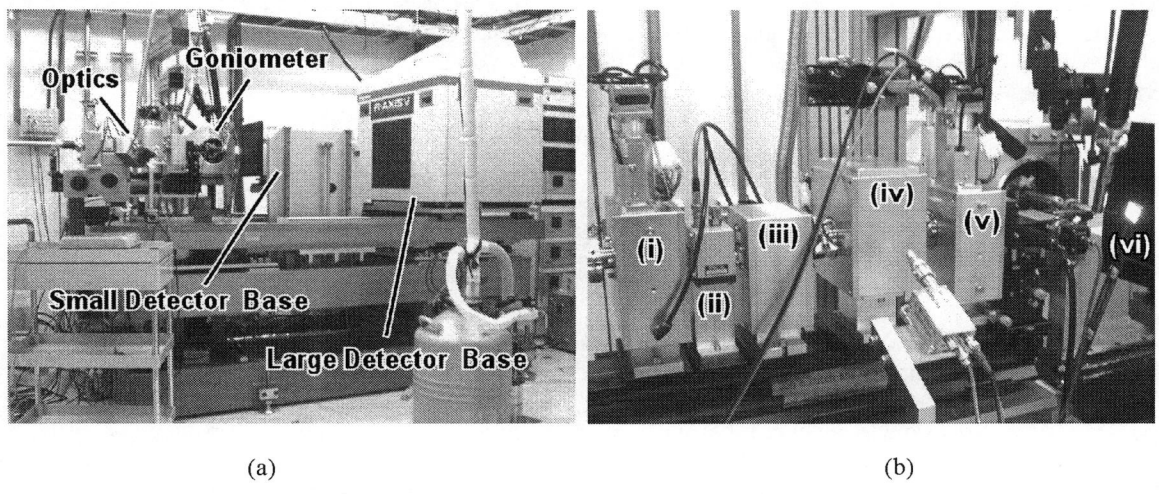

(a)   (b)

FIGURE 1. (a) Photograph of the diffractometer of BL38B1 and (b) magnified view of the optics unit. Optical components in (b) are described below: (i) and (v) are quadrant slits, (ii) is an ionization chamber as a beam intensity monitor, (iii) is a shutter, (iv) is an attenuator, and (vi) is an automatic backlight.

Two detectors, X-ray CCD (Charge Coupled Device) and IP (Imaging Plate) detectors, are installed in each beamline. Model and specification of these detectors are listed in Table 1. Our diffractometer has two detector bases (for small and large), and can be installed with both detectors. The detector base for small has a vertical translation axis. When a large size detector is used, a small size detector houses down to the table and the large size detector can move toward the goniometer without any physical interference. The Q315, installed in the BL41XU, is too large to be placed on the small detector base, therefore it is mounted on the large detector base exclusively with R-AXIS V and exchange is carried out manually.

TABLE 1. Available detectors in the BL38B1 and BL41XU.

	BL38B1		BL41XU	
	Jupiter210 (RIGAKU)	R-AXIS V (RIGAKU)	Q315 (ADSC)	R-AXIS V (RIGAKU)
Type	2×2 mosaic CCD	imaging plate	3×3 mosaic CCD	imaging plate
Sensitive Area	210 mm x 210 mm	400 mm x 400 mm	315 mm x 315 mm	400 mm x 400 mm
Pixel Dimension	4096 x 4096	4000 x 4000	6144 x 6144	4000 x 4000
Pixel Size	51 μm	100 μm	51 μm	100 μm
Readout Time	4.5 sec.	50 sec.	1.2 sec.	50 sec.
Camera Length	150 – 330 mm	170 – 950 mm	155 – 600 mm	155 – 600 mm
Ratio of Beam Time	85 ~ 90 %	10 ~ 15 %	90 ~ 95 %	5 ~ 10 %

A cryostream cooling N_2 gas-flow device (RIGAKU) is provided in both beamline. This device uses the N_2 gas generated from atmosphere and can keep the sample temperature down to 90 K. We also equip the BL41XU with a He gas-flow cryostream cooler (RIGAKU) which can keep the sample temperature down to 35 K.

Control and Data Acquisition

All components in the beamline are controlled by MADOCA (Message And Database Oriented Control Architecture) system, which is the SPring-8 standard device control system framework [5]. MADOCA brought the integration of the beamline control of beamline optics and experimental station. As the data collection software, BSS (Beamline Scheduling Software) [6] was installed to BL38B1 and BL41XU. BSS is a GUI (Graphic User Interface) based client software and can control beamline optics, diffractometer and detectors. Therefore, BSS can control all measurement condition automatically. The difference of the detector and the device composition of each beamline are hided within the software-layer of the BSS, and then users can do their experiment in the identical user interface in both beamline.

CURRENT STATUS OF BL38B1 AND BL41XU

Mail-in Data Collection System

We are now installing the mail-in data collection system at BL38B1. It is a new data collection protocol that users send crystals to SPring-8 and collect data with a help of a beamline operator. Our mail-in system makes use of automatic data collection system consist of BSS and SPACE (SPring-8 Automatic Cryo-sample Exchanger) [7]. The SPACE system is a sample changer robot with the specially designed sample pin which mounts crystal on the goniometer using screw threads. Beamline database D-Cha (Database for Crystallography with Home-lab Arrangements) has been developed to share crystal information, measurement condition and recorded diffraction data between beamlines and users' laboratory. Users can edit measurement condition and observe diffraction images using Web interface of D-Cha. All measurement conditions are determined and deposited by users-selves from home-lab. The operator start measurement after downloading measurement schedules from D-Cha to BSS. We have already succeeded in test operation of mail-in data collection between BL38B1 and several universities.

Micro Beam and Micro Crystal

To collect efficient quality data from micro-crystals, it is necessary to use micro-beam to reduce the background noise. High brilliant and low divergent beam from undulator is suitable for making such a micro beam. In the BL41XU, the micro beam was made by shaping the focused beam using the discrimination slit ((i) and (v) in Fig. 1 (b)). We have succeeded to obtain 25×25 μm² beam with 8×10^{10} photos/sec at a sample position. To handle the micro crystal, a high precision goniometer (sphere of confusion was less than ~ 2μm) and a high magnification video camera for sample observation were equipped. Using these systems, we have already collected data using micro crystals of some proteins. In the case of hen egg white lysozyme crystal whose size was less than $20 \times 20 \times 20$ μm³, we could collect a data set at 1.8 Å resolution and R_{merge} was about 8 %.

Ultra-high Resolutional Experiment

A proton and an electron play an important role to chemical reaction through biological macromolecules. In order to elucidate the reaction by X-ray crystallography, it is essential to make clear the structure in sub-atomic resolution below 0.7 Å. The BL41XU is equipped with the short wavelength X-ray by third harmonics of undulator and two-dimensional area detector with large sensitive area to collect data near 0.5 Å resolution. Using the X-ray wavelength of 0.6 Å and R-AXIS V as the detector, a sub-atomic resolution data set was collected from Endopolygalacturonase I crystal [8]. Two different condition data, in which the camera length and exposure time were optimized for high and low resolution, were collected to acquire the complete data set. Consequently, diffraction spots were visible to a resolution of 0.62 Å (Fig. 2 (a) and (b)). Intensity data were integrated and scaled, and R_{sym} of overall and outer shell (0.70-0.68 Å) were 3.2 and 30.0 %, respectively. As a result of structural refinement using SHELXL, R and R_{free} value of the obtained structure, in which 3,416 non-hydrogen atoms and 2,181 hydrogen atoms are included, is 9.72 and 10.78 %, respectively (Fig. 2 (c)).

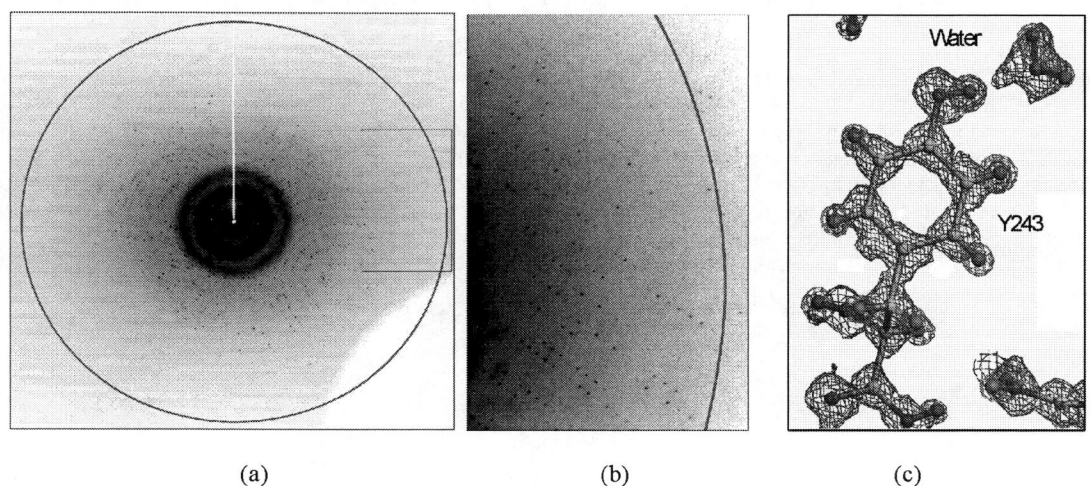

(a) (b) (c)

FIGURE 2. Diffraction pattern of endpolygalacturonase I, (a) is a whole image and (b) is a magnified image of the rounded region in (a). The circle indicates 0.7 Å resolution. (c) is the $2F_o-F_c$ map ($\sigma > 1$) and F_o-F_c map ($\sigma > 3$) around the specific residues of endopolygalacturonase I. The F_o-F_c map was calculated by omitting hydrogen atoms.

ACKNOWLEDGMENTS

The authors thank Dr. S. Muraoka (Kobe Univ.), Dr. K. Hirata (RIKEN SPring-8 Center), Dr. T. Nakatsu (Kyoto Univ.), and Prof. H. Katoh (Kyoto Univ.) for their help to collect many test datasets. We also thank the member of SPring-8 and RIKEN SPring-8 center for their cooperation and discussion.

REFERENCES

1. Tanida, H., Kikuchii, A., Miura, K., Takeshita, K., Goto, S., Shiro, Y., and Ishikawa, T., *AIP Conference Proceedings* **705**,Issue 1, 2004, pp. 651-654..
2. Kawamoto, M., Kawano, Y., and Kamiya, N., *Nucl. Instrum. Methods* 467, 1375-1379 (2001).
3. Kitamura, H., *J. Synchrotron Rad.* **7**, 121-130 (2000).
4. Yamazaki, H., Kimura, H., Kagaya, I., Yamashita, C., and Ishikawa, T., *Proc. SPIE.* 3773, 21-29 (1999).
5. Ohata, T., Nakatani, T.,. Furukawa, Y., Tamasaku, K., Ishii, M., Matsushita, T., Takeuchi, M., Tanaka, R., and Ishikawa T., *Nucl. Instrum. Methods* 467, 820-824 (2001).
6. Ueno, G., Kanda, H., Kumasaka, T., and Yamamoto, M., *J. Synchrotron Rad.* **12** 380-384 (2005).
7. Ueno, G., Hirose, R., Ida, K., Kumasaka, K., and Yamamoto, M., *J. Appl. Cryst.* **37** 867-873(2004).
8. Shimizu, T, Nakatsu, T, Miyairi, K, Okuno, T., and Kato, H., *Biochemistry* **41** 6651-6659 (2002)

Automated Sample Exchange Robots for the Structural Biology Beam Lines at the Photon Factory

Masahiko Hiraki, Shokei Watanabe, Yusuke Yamada, Naohiro Matsugaki, Noriyuki Igarashi, Yurii Gaponov and Soichi Wakatsuki

Structural Biology Research Center, Photon Factory, Institute of Materials Structure Science, High Energy Accelerator Research Organization (KEK), 1-1 Oho, Tsukuba, Ibaraki, 305-0801 Japan

Abstract. We are now developing automated sample exchange robots for high-throughput protein crystallographic experiments for onsite use at synchrotron beam lines. It is part of the fully automated robotics systems being developed at the Photon Factory, for the purposes of protein crystallization, monitoring crystal growth, harvesting and freezing crystals, mounting the crystals inside a hutch and for data collection. We have already installed the sample exchange robots based on the SSRL automated mounting system at our insertion device beam lines BL-5A and AR-NW12A at the Photon Factory. In order to reduce the time required for sample exchange further, a prototype of a double-tonged system was developed. As a result of preliminary experiments with double-tonged robots, the sample exchange time was successfully reduced from 70 seconds to 10 seconds with the exception of the time required for pre-cooling and warming up the tongs.

Keywords: automated system, high-throughput, protein crystallography, structural biology, beam line
PACS: 07.07.Tw, 87.61.Ff

INTRODUCTION

Protein crystallographic structural analyses involve many steps: protein overexpression, purification, crystallization, harvesting and freezing crystals, data collection, data processing and structure determination. In order to facilitate high-throughput structural analyses of a large number of protein crystals, we are pursuing several R&D projects aimed towards a fully automated system at the SBRC (Structural Biology Research Center) in KEK. We have, for example, developed a high-throughput protein crystallization system, and with this system have already obtained a large number of protein crystals and determined their structures. We are also developing an automated crystal harvesting system consisting of a micromanipulator and a robot hand for holding a cryo loop. As a preliminary test, we have succeeded in using this system to harvest protein crystals from crystallization drops.

X-ray protein crystallographic experiments at synchrotron beam lines are often carried out with loop-mounted crystals under cryogenic conditions. The advances in insertion devices, beam line optics and fast readout CCD detectors have reduced X-ray exposure times substantially, and consequently a major proportion of beam time is spent on sample manipulation in experimental hutches. Users need to enter the hutch, dismount an old sample, mount a cryo-loop with a new crystal and center it on the rotation axis of the goniometer before starting the next diffraction experiment. The time required for these procedures could be substantially reduced by crystal exchange robots. Several automated sample exchange systems have been developed and are used in synchrotron facilities [1-6].

We considered three factors for choosing crystal exchange robots to reduce the time required for sample exchange on our insertion device beam lines, BL-5A, BL-17A and AR-NW12. First, exchange robots should be able to handle widely used and/or commercially available cryo pins. Second, the capacity of the robots must be large enough for continuous data collection of several hundred samples (a particularly important factor for structure-based drug design). Third, they should leave ample space for manual mounting of cryo pins by users. The SAM (SSRL Automated Mounting) system [1] developed by the Macromolecular Crystallography Group at the SSRL (Stanford Synchrotron Radiation Laboratory) fulfills these requirements, and we decided to implement this system design in the structural biology beam lines at the Photon Factory with generous help from the SSRL group.

CP879, *Synchrotron Radiation Instrumentation: Ninth International Conference*,
edited by Jae-Young Choi and Seungyu Rah
© 2007 American Institute of Physics 978-0-7354-0373-4/07/$23.00

SAMPLE EXCHANGE ROBOTS

Two exchange robot systems based on the SAM system have been installed at the BL-5A and AR-NW12 beam lines at the Photon Factory (Fig. 1). Ninety-six cryo pins containing protein crystals are stored in each SSRL cassette and three cassettes can be placed in the liquid nitrogen Dewar (Fig. 2(a)). The sequence of operation of our system is the same as that of the SAM system. The procedure for mounting and dismounting a cryo pin are as follows: the robot cools the tongs in the liquid nitrogen, picks up a magnet dumbbell and pulls the required cryo pin from the cassette using the strong end of the magnet dumbbell. The robot then holds the cryo pin with the tongs, moves the tongs towards the goniometer and mounts the pin on the goniometer. The tongs are then opened and moved away from the goniometer. After an X-ray diffraction experiment, the robot dismounts the cryo pin from the goniometer using pre-cooled tongs, returns to the Dewar and puts the used cryo pin on the weak end of the magnet dumbbell (Fig. 2(b)). Then the robot grasps the magnet dumbbell and restores the used cryo pin into the cassette.

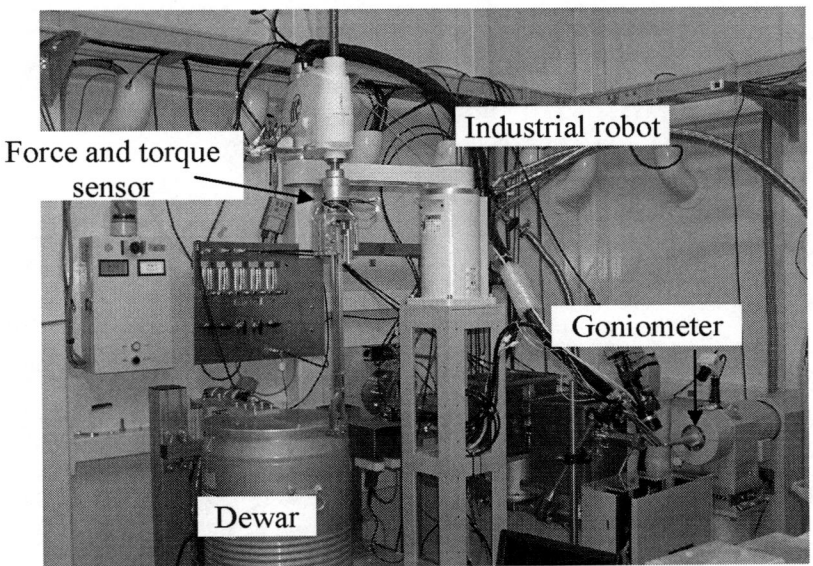

FIGURE 1. Sample Exchange Robot Installed at AR-NW12A

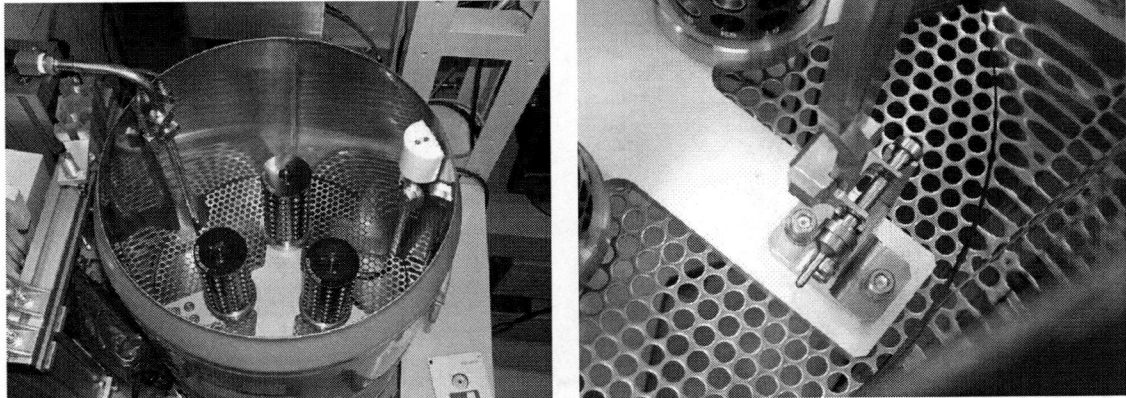

FIGURE 2. (a) Liquid Nitrogen Dewar. Three sample cassettes can be placed in the Dewar. (b) Magnet Dumbbell and Cradle. The cryo pin is attached at the weak end of the magnet dumbbell and the tongs are grasping the magnet dumbbell.

We selected for sample manipulation an industrial 4-axes robot of the same series as the SAM system. The robot is fixed on the top of a robot base that was designed to fit our beam lines. The cryo tongs hang down from the robot arm and are controlled by a pneumatic actuator (Fig. 3). To detect collision of the robot arm, a force and torque sensor is inserted between the robot arm and the pneumatic actuator. In addition, this force and torque sensor is used to calibrate the position and direction of the cassettes and the magnet dumbbell under the liquid nitrogen. Because all components, including the robot, a robot controller, a liquid nitrogen Dewar, a dryer unit and electrical devices, are implemented on the robot base, the system can be easily temporarily detached from the beam lines, for instance, during the maintenance of the beam lines.

 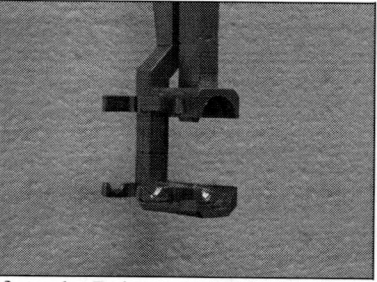 

FIGURE 3. Cryo Tongs that Hang Down from the Robot Arm. The cryo tongs are opened (center) and closed (left) by the pneumatic actuator. The cryo tongs can grasp the magnet dumbbell using side fingers (right).

REDUCTION OF EXCHANGE TIME

The sample-exchange time can be reduced by changing the sequence of handling cryo pins. As described above, after (i) dismounting the used pin, putting it on the weak end of the magnet dumbbell and (ii) restoring the pin to the cassette, (iii) the robot pulls the next cryo pin from the cassette, (iv) holds the pin attached to the strong end and mounts it on the goniometer. If the next cryo pin is put on the strong end of the magnet dumbbell before dismounting, however, the tongs can hold the next pin and move towards the goniometer immediately after putting the used pin on the weak end of the magnet dumbbell. The proposed sequence is (iii)-(i)-(iv)-(ii) instead of (i)-(ii)-(iii)-(iv). Because the movement (ii) and (iii) can be carried out in parallel with data collection and loop alignment, the periods (i) and (iv) are required for exchanging the cryo pins. Our system required about 70 seconds to execute the former sequence; 15, 20, 20 and 15 seconds are necessary for the periods (i), (ii), (iii) and (iv), respectively. By the proposed sequence, the time required for exchanging the cryo pins is decreased to 30 seconds. In the case of the proposed sequence, however, the next pin is on the strong end of the magnet dumbbell until the robot returns to the Dewar after dismounting the used pin. The used pin also remains on the weak end of the magnet dumbbell until the robot returns after mounting the next pin. These situations increase the probability that small ice formations will accumulate on the cryo loops in the Dewar.

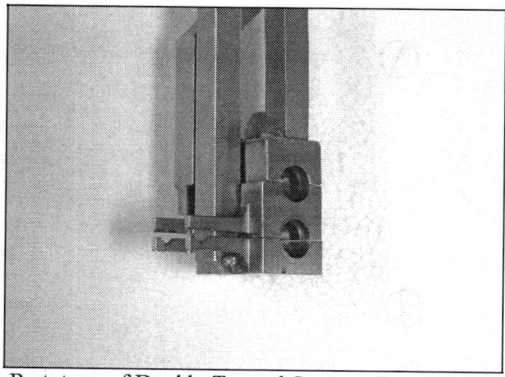 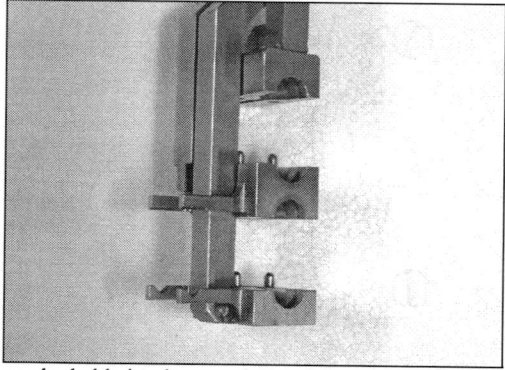

FIGURE 4. Prototype of Double Tonged System. Two cryo pins can be held simultaneously by the upper and the lower tongs. The side fingers for grasping the magnet dumbbell synchronize with the lower tongs.

The reduction of the sample-exchange time is more effective in the case of a short data collection period, as may be the case when taking several diffraction images for evaluation of the crystals, or by using a faster detector like the X-ray HARP detector [7], which is currently under development at the SBRC, and so forth. In order to reduce the number of trips between the Dewar and the goniometer by a factor of two, we modified the single tongs of the SAM system and developed a prototype of double tongs that can hold two cryo pins simultaneously (Fig. 4). The double-tonged robot moves to the goniometer holding the next cryo pin, dismounts the used cryo pin with the other tongs and mounts the next cryo pin onto the goniometer without leaving the diffractometer area. Furthermore, the pin that is prepared to be used next can be protected against small ice by the cryo tongs while the robot is waiting for the exchange command. As a result of a preliminary experiment with the double-tonged system, the cryo pin exchange time was successfully reduced to 10 seconds.

CONCLUSIONS

We developed automated sample exchange robots based on the SSRL automated mounting system and have already installed the two systems on two insertion device beam lines: BL-5A and AR-NW12A. Three SSRL cassettes with 96 cryo pins each can be placed in the liquid nitrogen Dewar and the robots can handle 288 samples continuously. The time required for dismounting and mounting samples is about 70 seconds with the exception of the time required for pre-cooling and warming up the cryo tongs. To reduce the time for exchanging samples for high-throughput operation, we proposed rearranging the sequence of handling the cryo pins and the required time was reduced to 30 seconds. In order to reduce the number of trips between the Dewar and the goniometer by a factor of two, moreover, we developed a prototype with double tongs that can hold two cryo pins simultaneously. The robot having the double tongs could exchange the samples in 10 seconds. We are currently modifying the calibration software for the position and the orientation of the cassettes in the liquid nitrogen Dewar in order to improve the stability of the system. Another exchange robot will be installed at the micro-focus beam line BL-17A that is being commissioned.

ACKNOWLEDGMENTS

The authors would like to thank Mike Soltis, Mitch Miller, Scott McPhillips and Jinhu Song, of the Macromolecular Crystallography Group at SSRL for their great help with the mechanical design and the software. We thank Masanori Kobayashi, Structural Biology Research Center, KEK-PF for his support of the experiments for data collection at the beam lines. This work was supported by the Protein 3000 project of the Ministry of Education, Culture, Sports, Science and Technology (MEXT) of Japan. This work was also supported by the programs of the Strategic International Cooperative Program and the Development of Systems and Technology for Advanced Measurement and Analysis from the Japan Science and Technology Agency (JST).

REFERENCES

1. A. E. Cohen, P. J. Ellis, M. D. Miller, A. M. Deacon and R. P. Phizackerley, *J. Applied Crystallography*, **35**, 720-726 (2002).
2. G. Ueno, R. Hirose, K. Ida, T. Kumasaka and M. Yamamoto, *J. Applied Crystallography*, **37**, 867-873 (2004).
3. G. Snell, C. Cork, R. Nordmeyer, E. Cornell, G. Meigs, D. Yegian, J. Jaklevic, J. Jin, R. C. Stevens and T. Earnest, *Structure*, **12**, 537-545 (2004).
4. W. I. Karain G. P. Bourenkov, H. Blume and H. D. Bartunik, *Acta Crystallographica*, **D58**, 1519-1522 (2002).
5. E. Pohl, U. Ristau, T. Gehrmann, D. Jahn, R. Robrahn, D. Malthan, H. Dobler and C. Hermes, *J. Synchrotron Radiation.*, **11**, 372-377 (2004).
6. J. Ohana, L. Jacquamet, J. Joly, A. Bertoni, P. Taunier, L. Michel, P. Charrault, M. Pirocchi, P. Carpentier, F. Borel, R. Kahn and J.-L. Ferrer, *J. Applied Crystallography*, **37**, 72-77 (2004).
7. High Energy Accelerator Research Organization (KEK), *Annual Report 2004*, 69-70 (2004).

Quickly Getting the Best Data from Your Macromolecular Crystals with a New Generation of Beamline Instruments

Florent Cipriani[a], Franck Felisaz[a], Bernard Lavault[a], Sandor Brockhauser[a], Raimond Ravelli[a]; Ludovic Launer[b]; Gordon Leonard[c], Michel Renier[c]

[a]EMBL-Grenoble, 6, rue Jules Horowitz, BP181 38042 Grenoble Cedex 9, France; [b]MRC-France(BM14),c/o ESRF, B.P.220, 38043 Grenoble CEDEX France; [c]ESRF, B.P. 220, 38043 Grenoble CEDEX, France.

Abstract. While routine Macromolecular x-ray (MX) crystallography has relied on well established techniques for some years all the synchrotrons around the world are improving the throughput of their MX beamlines. Third generation synchrotrons provide small intense beams that make data collection of 5-10 microns sized crystals possible. The EMBL/ESRF MX Group in Grenoble has developed a new generation of instruments to easily collect data on 10 μm size crystals in an automated environment. This work is part of the Grenoble automation program [1] that enables FedEx like crystallography using fully automated data collection and web monitored experiments. Seven ESRF beamlines and the MRC BM14 ESRF/CRG beamline are currently equipped with these latest instruments. We describe here the main features of the MD2x diffractometer family and the SC3 sample changer robot. Although the SC3 was primarily designed to increase the throughput of MX beamlines, it has also been shown to be efficient in improving the quality of the data collected. Strategies in screening a large number of crystals, selecting the best, and collecting a full data set from several re-oriented micro-crystals can now be run with minimum time and effort. The MD2x and SC3 instruments are now commercialised by the company ACCEL GmbH.

Keywords: Protein crystallography, micro crystal; diffractometer, air bearing goniometer, kappa goniometer, spine standard, datamatrix, beamline automation, sample changer, automatic crystal detection, automatic crystal centring.
PACS: 07.85.Qe

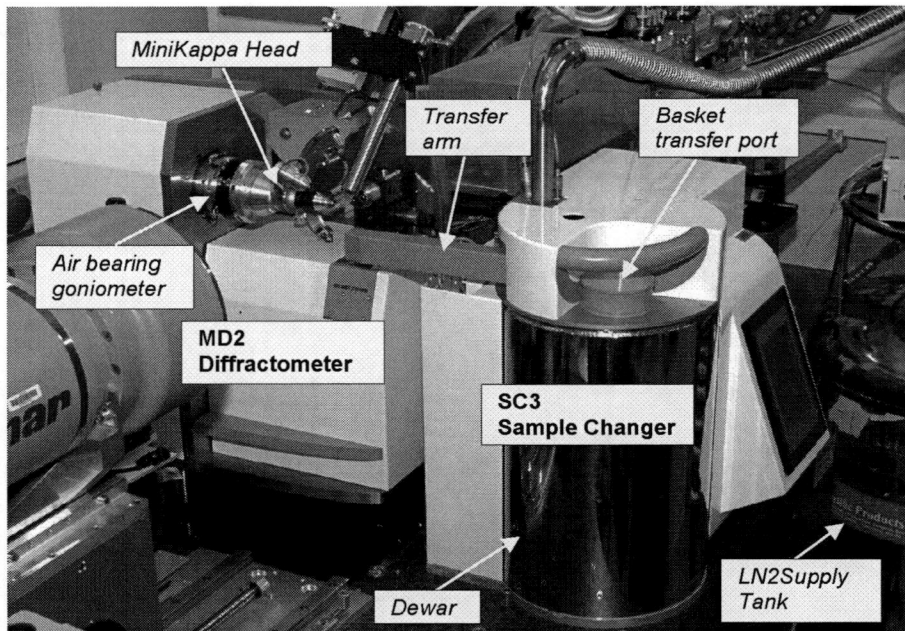

FIGURE 1. The MD2 diffractometer associated with the SC3 sample changer robot at the ESRF ID14-3 beamline.

CP879, *Synchrotron Radiation Instrumentation: Ninth International Conference*,
edited by Jae-Young Choi and Seungyu Rah
© 2007 American Institute of Physics 978-0-7354-0373-4/07/$23.00

MD2 DIFFRACTOMETER

Based on the concept of the micro-diffractometer developed for the ESRF ID13 micro-focus beamline [2], the MD2 (Fig. 1) is the most complete model of the MD2x diffractometer family. It is a turnkey instrument with integrated electronics and windows XP based control software. The MD2 was designed to allow processing routinely crystals down 10 micrometers in size. An on-axis beam-viewing video-microscope allows precise crystal-to-beam alignment. The air scattering is minimised by a capillary beamstop unit, ensuring that weak high-resolution spots can be recorded against a minimal background. An optional miniKappa goniometer head allows for the reorientation of crystals with minimum risk of collision. A motorised support for a fluorescence detector has also been integrated in the design. The MD2 can be operated from a graphical user interface (Fig. 3) or controlled remotely through a device server. Several tools are integrated for tuning the shutter synchronisation, the positioning of the beam shaping devices and to monitor the operation of the goniometer. Two crystal alignment methods are available: The 3 click crystal centring procedure and an automatic centring procedure based on the C3D software package (http://www.embl.fr/groups/instr/auto_centring/index.html). With a current success rate of about 70%, C3D makes the MD2 a good candidate for automated beamlines. A complementary technique based on UV fluorescence is being implemented to solve the most difficult cases [3].

Air Bearing Goniometer, MiniKappa head and Shutter Control

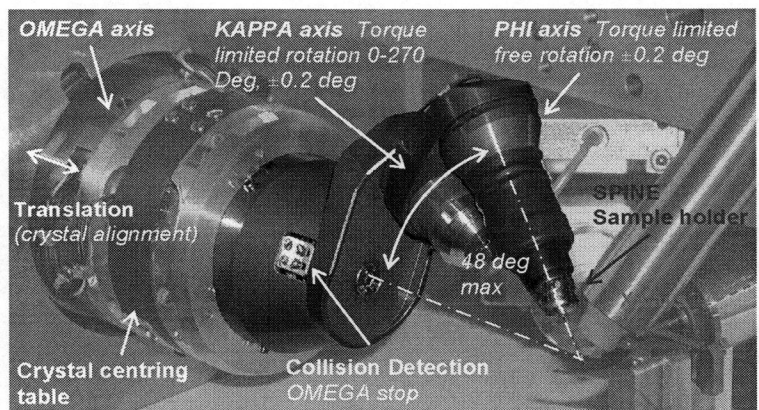
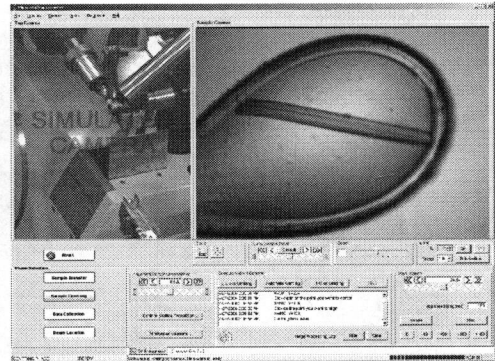

FIGURE 2. The MiniKappa Goniometer head   **FIGURE 3.** the MD2 graphical user interface

The goniometer of the MD2 is constructed around an air bearing spindle. It features a sphere of confusion of less than 3 μm diameter (typically 2), less than 1 mdeg error @20 deg/s and 230 deg/s of maximum speed. It is mounted on a XYZ stage and includes a XY crystal centring table. This permits a crystal to be centred and positioned in the beam with a micrometer resolution. A precise shutter control signal with open/close correction delays is provided. The optional miniKappa head (Fig. 2) can easily be mounted on top of the centring table of the goniometer. Its simplified design benefits from the unique SPINE sample holder length (www.spineurope.org, Protocols menu). Small crystals can be processed without any stability problems, even in fast multi-passes scans. New data collection strategies will be possible using the STAC software package that is under development and includes modules for calibration, crystal reorientation, automatic re-centring and multi-sweep strategy options. Different sample mounting methods are offered: Direct mounting using Cryo-tongs; Optional assisted mounting using a motorised transfer arc or the miniKappa goniometer head (transfer with vials filled with liquid nitrogen); robot mounting with the SC3 sample changer or other mounting robots. The later is also possible with the miniKappa head.

Beam Shaping Devices, Beam Scintillator and on Beam Axis Video Microscope

The quality of data collected from macromolecular crystals depends on the signal to noise ratio and the radiation damage incurred. This is particularly important for micro-crystals where the brilliance of the source is only useful if all the x-ray photons interact with the crystals. The best results are obtained when the beam size matches the size of the smallest crystal dimension. If the crystals are x-ray sensitive it is often best to collect at several crystal positions and from several crystals. The MD2 features all the options to efficiently collect data from micro-crystals.

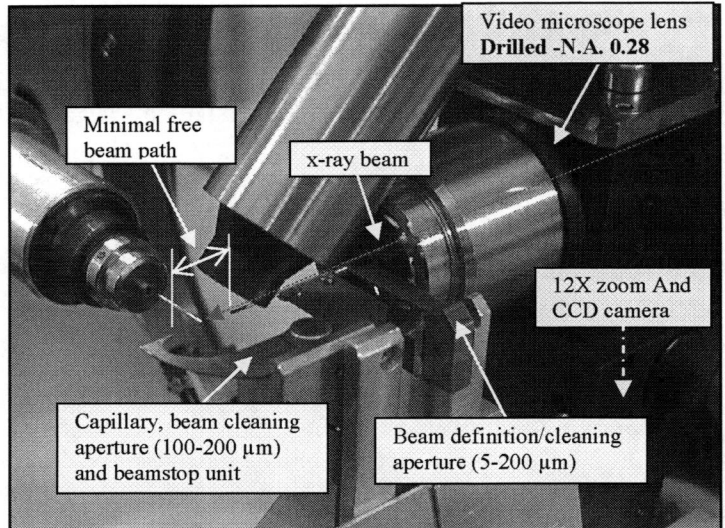

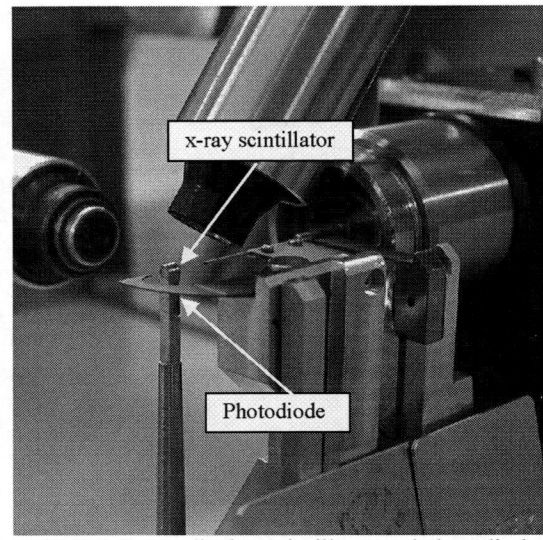

FIGURE 4. Beam shaping devices.   **FIGURE 5.** Beam display scintillator and photodiode

The beam definition aperture and the capillary beamstop unit (Fig. 4) allow a precisely shaped and low scattering x-ray beam to be delivered on the crystals. The x-ray scintillator (Fig. 5) can be remotely placed in the beam at the crystal position to verify its shape and position. Precise and parallax free alignment of the crystals is possible with the high resolution "On beam axis video microscope" (Fig. 4). The beam (Fig. 6 a, b) or the crystals (Fig 6 b, c) can be displayed with a mouse click, at any time. The photodiode (Fig. 5) can be used to adjust the position of the capillary/beamstop and the position of beam definition aperture (1D or 2D Scans), to check the stability of the beam and to tune the shutter synchronisation (Beam intensity/goniometer position plot).

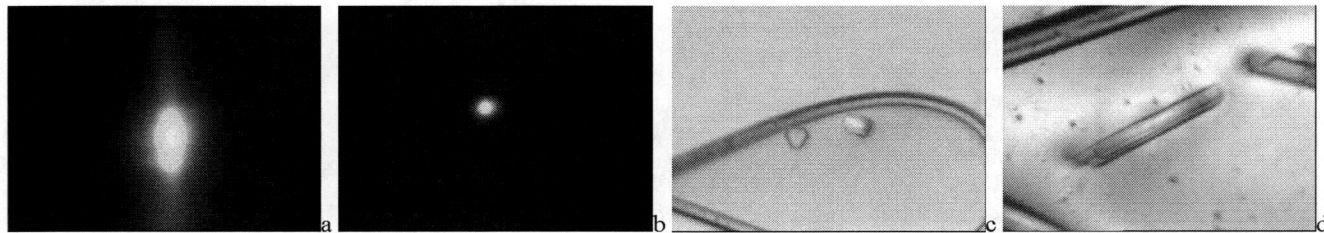

FIGURE 6. a) Direct Beam; b) Beam shaped a 10 μm; c) 12 μm crystal; d) 12 μm needle after exposure to x-rays

SC3 SAMPLE CHANGER ROBOT

The SC3 sample changer robot (Fig. 1) has been designed to automatically handle pre-frozen crystals mounted on standard SPINE sample holders. It is a very compact machine which can be operated in the presence of users without risk of injury. Compared to auto-mounters based on industrial multi axis robots, the risk of damage to other equipment is very limited. The adoption of the European SPINE standard for sample holders and vials guarantees the compatibility with sample changers over Europe, and a smooth transition from manual to automatic sample processing.

The SC3 is composed of a closed Dewar that can hold up to 50 samples placed in 5 baskets. A liquid nitrogen pump maintains a constant level of liquid nitrogen inside the sample changer. Any sample in the SC3 can be transferred in few seconds to and from a host goniometer by a compact Cartesian transfer arm. The SC3 exploits the DataMatrix codes of the SPINE sample holders to ensure safe management of the data flow. A user could "FedEx" the basket (five) in a standard CX100 transport Dewar (Fig. 8), after which the baskets are loaded manually in the sample changer. The samples can then be transferred automatically onto the host goniometer. An electro magnet on the goniometer maintains the sample holder and detects its presence. The sample holders and the baskets are identified with <u>DataMatrix</u> and clear codes. At crystal harvesting time, the sample holders can be identified using a Datamatrix reader connected to a pocket PC, itself connected to LIMS. A Datamatrix reader integrated in the SC3 ensures safe management of the data flow. The sample changer can be operated locally through a touch screen, or remotely, using the Graphical User Interface (Fig. 7) or the device server commands.

The average transfer time per sample for 50 samples is about 1'20", including basket & sample selection, Datamatrix reading, loading and unloading. In fully automatic mode, the total screening time for 50 samples is about 150 minutes (2H30'), when the SC3 is associated with a MD2 diffractometer, controlled by the ESRF beamline control program MxCuBe and a modified version of the data collection and processing software DNA [4]. The average time of 3' per sample can be broken down as follows: 2' to prepare and execute the load/unload of the sample, 50" to automatically centre the sample, and 10" seconds to generate 2 diffraction images 90 degrees apart with an exposure time of 1".

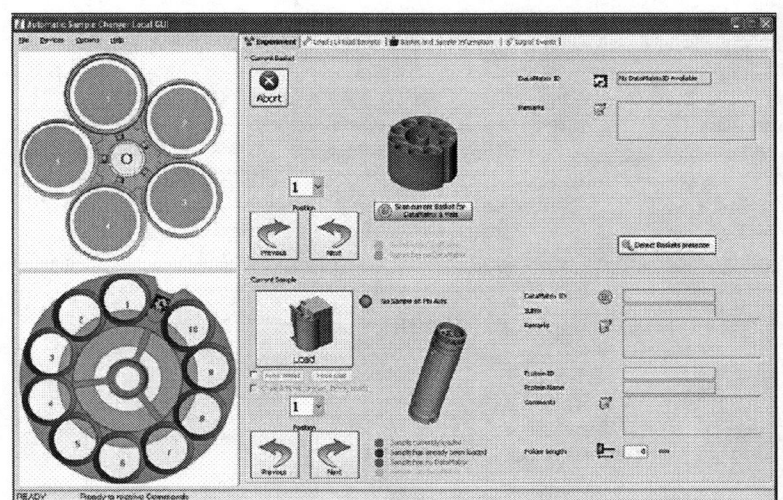

FIGURE 7. SC3 Graphical User Interface.

FIGURE 8. SC3 Tools & Consumables

ACKNOWLEDGMENTS

We would like to thank J-S.Aksoy[a] for its contribution to the development of the software of the sample changer; H.Caserotto[b], and J.Huet[a] for their work on the development of the electronics; C.Taffut[c], M.Lentini[c], M.Dallery[c], F.Di-Chiaro[c] and M.Guijarro[c] for their contribution on the construction, test and commissioning of the 8 sample changer units; J.McCarthy[c], S.McSweeney[c] for theirs scientific input and their patience in testing the instruments; S.Cusack[a], S.Larsen[c] M.A.Walsh[b] and A.Thompson[a,d], for having strongly supported these projects; and the company MAATEL for the industrialisation of the electronics of the instruments. ([d]Present address: Synchrotron SOLEIL, L'Orme des Merisiers, Saint-Aubin, 91192 GIFsur-YVETTE CEDEX, France)

REFERENCES

1. Arzt, S, Beteva, A, Cipriani, F, et al. *Automation of macromolecular crystallography beamlines,* PROG BIOPHYS MOL BIO 89 (2): 124-152 OCT 2005
2. Perrakis, A, Cipriani, F, Castagna, JC, et al., "*Protein microcrystals and the design of a microdiffractometer: current experience and plans at EMBL and ESRF/ID13*", ACTA CRYSTALLOGR D 55: 1765-1770 Part 10 OCT 1999
3. Vernede et al. *UV laser-excited fluorescence as a tool for the visualization of protein crystals mounted in loops* Acta Cryst. (2006). D62, 253–261
4. Leslie, A.G., Powell, H.R., Winter, G., Svensson, O., Spruce, D., McSweeney, S., Love, D., Kinder, S., Duke, E., Nave, C., 2002. Automation of the collection and processing of X-ray

Development of Control Applications for High-Throughput Protein Crystallography Experiments

Yurii A.Gaponov[1], Naohiro Matsugaki, Nobuo Honda, Kumiko Sasajima, Noriyuki Igarashi, Masahiko Hiraki, Yusuke Yamada, Soichi Wakatsuki

Structural Biology Research Center, Photon Factory, Institute of Materials Structure Science, High Energy Accelerator Research Organization, 1-1 Oho, Tsukuba, Ibaraki, Japan

Abstract. An integrated client-server control system (PCCS) with a unified relational database (PCDB) has been developed for high-throughput protein crystallography experiments on synchrotron beamlines. The major steps in protein crystallographic experiments (purification, crystallization, crystal harvesting, data collection, and data processing) are integrated into the software. All information necessary for performing protein crystallography experiments is stored in the PCDB database (except raw X-ray diffraction data, which is stored in the Network File Server). To allow all members of a protein crystallography group to participate in experiments, the system was developed as a multi-user system with secure network access based on TCP/IP secure UNIX sockets. Secure remote access to the system is possible from any operating system with X-terminal and SSH/X11 (Secure Shell with graphical user interface) support. Currently, the system covers the high-throughput X-ray data collection stages and is being commissioned at BL5A and NW12A (PF, PF-AR, KEK, Tsukuba, Japan).

Keywords: control system, SR, high-throughput protein crystallography, TCP/IP UNIX secure socket, database, Linux.
PACS: 07.05.DZ, 07.85.Qe

INTRODUCTION

There are several different criteria that need to be addressed during the programming of a control system for operating scientific experiments. In general, control software can be considered as an interface between the experimenter and the equipment. Therefore, it is important for the control system to support a friendly and reliable interface, which can convert experiment-related commands (top level) to equipment-related commands and operations (low level). This is particularly important for high-throughput protein crystallography experiments, which are expected to be performed in schedule mode (without the participation of the experimenter during the different stages of the experiment). It is necessary to keep all parts of the system in reliable condition and protected from any human errors or errors due to software/hardware features. Another desirable feature of a high-throughput protein crystallography experiment control system is the possibility to control an experiment remotely, thereby allowing different members of a research team to join the experiment [1,2].

To satisfy the demands described above with one stand-alone application on one computer is quite difficult; controlling several different equipment units makes the application large and difficult to modify. It is more advantageous to split the large task into smaller ones, which is the preferred choice in the field of controlling design and large-scale programming [3]. Typically, these applications are based on a data-bus with integrated CPU board computers. These systems tend to be expensive and quite complicated for extensive programming and modification. An alternative approach is Internet-based distributed systems. In a distributed system, many different computers are integrated through the network, which makes it feasible for a group of system engineers and computer scientists to simultaneously develop the control software [4].

High-throughput protein crystallography experiments comprise several stages. To perform tasks during all stages

---

[1]Contact person, email:yugru@yahoo.com

CP879, *Synchrotron Radiation Instrumentation: Ninth International Conference*,
edited by Jae-Young Choi and Seungyu Rah
© 2007 American Institute of Physics 978-0-7354-0373-4/07/$23.00

in a high-throughput mode, it is necessary to integrate all the schedules, experimental data, and results in the database to keep the control software updated according to the current experimental status and to allow one to scientifically analyze all of the related experiment data.

DATABASE

Figure 1 shows the database layout. The relational database for high-throughput protein crystallography experiments was developed using MySQL software to store all the experiment details, schedules, data, and results from data analysis. The database allows one to systematically analyze the information related to a protein crystallography experiment, which is essential to solve scientific tasks and avoid human errors and mistakes [1,2].

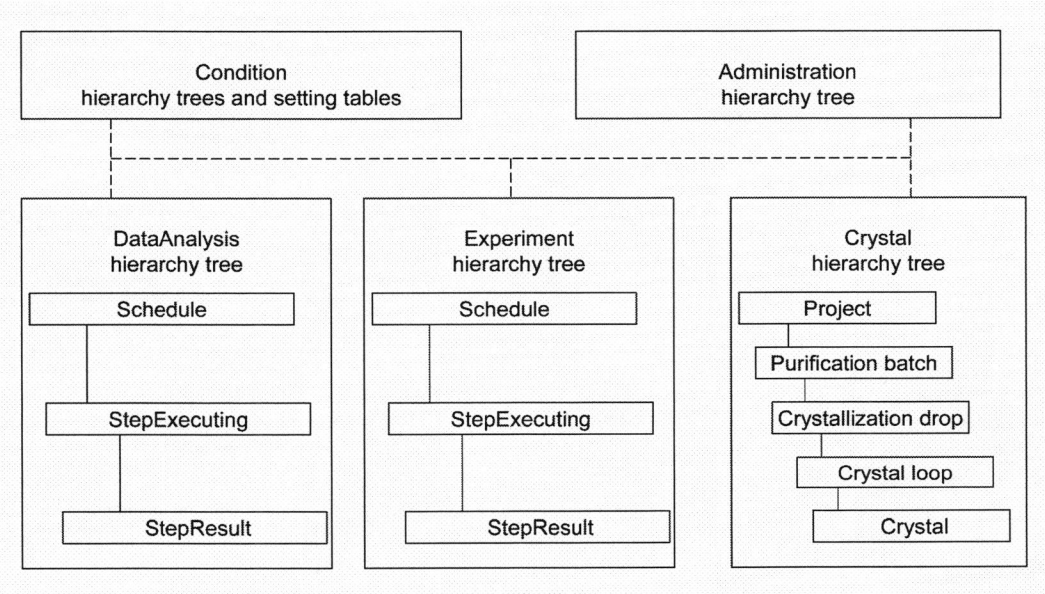

FIGURE 1. The overall scheme of the relational PCDB database.

There are several database hierarchy trees. One of these, Crystal-tree, is related to the protein crystal, and describes the information about overexpression and purification of the protein, crystallization, crystal harvesting, and crystal mounting in the crystal loop (cryo-loop). Experiment-tree is related to the X-ray experiment, and describes the details of the schedule of the experiment steps (snapshot, oscillation, MAD, and EXAFS), the progress in executing these steps, and the results. DataAnalysis-tree is related to X-ray data analysis, and describes the details of the data indexing and integrating, and data reduction procedures. Condition-tree stores the details and parameters of the experiments and data analysis. Administration-tree stores information about user accounts, groups, privileges, and storage devices.

BEAMLINE CONTROL SYSTEM

The UNIX TCP/IP socket client/server was chosen as the general communication model. To make communication secure, OpenSSL, an open-source implementation of the SSL (Secure Sockets Layer) and TLS (Transport Layer Security) protocols, is used. This model is reliable and well designed under all operating systems for inter-process/program synchronous and asynchronous communications [4, 5].

There are four types of communication interfaces between client application and the server: a) a sending command is followed by waiting for a reply; b) a receiving command is followed by a sending reply; c) a sending command or message is not followed by waiting for a reply; and d) a receiving a message or a command is not followed by sending a reply. A combination of the first two types of interfaces allows one to build synchronous network communication reliably. A combination of the last two types of interfaces allows for asynchronous network communication with a higher communication speed, mainly for exchanging status information for monitoring purposes. Our control system consists of a mixture of these two combinations.

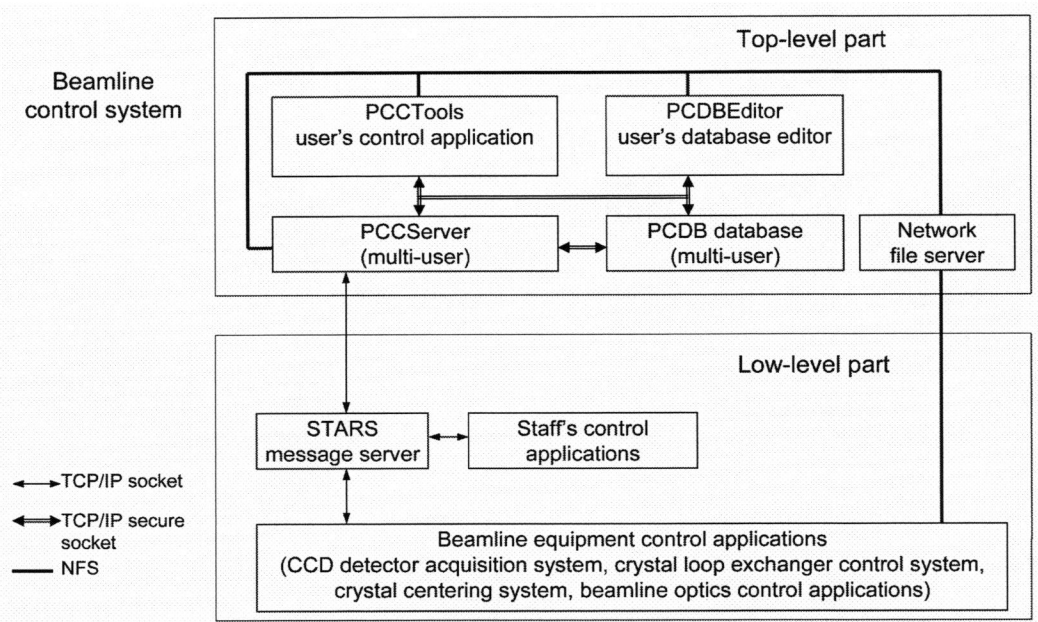

FIGURE 2. The overall scheme of the beamline control system for high-throughput protein crystallography experiments.

Figure 2 shows the overall scheme of the control system. There are two levels of control. On the top level, PCCServer, the main server module, allows the user's client application PCCTools to be connected in a multi-user mode. This server has a socket connection with PCDB database, which stores all the top-level information. The server is able to execute both X-ray data collection experiments and analysis of the experimental data by executing the commercial applications for indexing, integration, and scaling. On the top-level, control commands reflect the modes of operation of the experimental equipment, i.e., data collection, direct beam measurement, beam stopper alignment, etc. A multi-socket server, STARS, (Simple Transmission and Retrieval System, which was developed at the Photon Factory, KEK, Tsukuba, Japan) is used for communication with different equipment control systems and applications [6]. One client can be connected to STARS through several sockets to achieve any of the types of communication interfaces mentioned above.

The system has two modes of operation: schedule mode and step-by-step mode. In both modes, the experiment schedule, which includes several schedule steps, is executed. In the schedule mode, all the steps in the experiment schedule are executed at once. In the step-by-step mode, only the selected step is executed. For the schedule mode, it is necessary to create an experiment schedule in advance using the PCDBEditor or PCCTools applications. The experiment schedule will be loaded from the PCDB database and executed for continuous high-throughput experiments. For the step-by-step mode, schedule steps can only be created during the experiment using the StepWizard procedure, which was developed within the PCCTools application. This allows one to define the experimental settings using a variety of default-preset experimental parameters or to derive them from the currently opened schedule step. The CBrowser procedure was developed to create the items at all levels of Crystal-tree. All relational details of the Crystal and Condition hierarchy trees were hidden to simplify the schedule step creation procedure. For example, to select the necessary crystal, it is only necessary to select the project, purification batch, crystallization drop, crystal loop, and the crystal from the list of names. For SAD, MAD, and XAFS experiments, the EBrowser procedure was developed to define default settings by selecting the chemical element under investigation from the periodic table of the elements. For reference, EBrowser presents information about other chemical elements, possibly included some in the crystal, to warn the user about the possibility of anomalous or fluorescent signals from the other heavy atoms. Theoretical values of default-preset parameters for energy ranges are modified and updated from the experimental data of different protein crystals containing the same chemical element.

The procedure RunEngine was developed to control the execution of schedule steps and individual top/low-level commands. Control commands can be executed sequentially or, when possible, in parallel to optimize the execution of the procedure. To increase the reliability of the system, RunEngine was placed in the server side (PCCServer) allowing one to continue to control a previous experiment after restarting (for whatever reason) the PCCTools application without any interruption in the experiments. For security and reliability reasons, RunEngine may be controlled only by one connected user's application (active mode of PCCTools). All other user's applications are allowed (depending on the user's membership in the group) only to monitor the experiment and the current status of the experimental equipment (passive mode of PCCTools). The RunEngine procedure is responsible for synchronizing the experiment with the PCDB database. All created experimental data files will be associated with the individual records in the corresponding PCDB database tables. The execution status of the experiment schedule step is updated in the database to show the progress of the experiments.

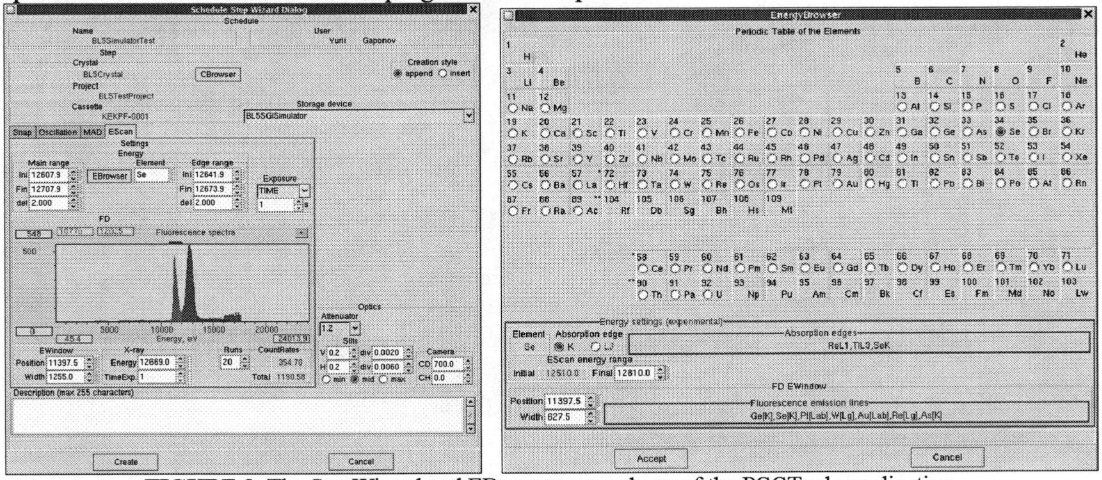

FIGURE 3. The StepWizard and EBrowser procedures of the PCCTools application.

Figure 3 represents the StepWizard and EBrowser procedures of the PCCTools application. StepWizard shows the creation of the experimental settings for the EXAFS schedule step. EBrowser offers the experimental set of parameters (energies, energy steps, and time exposure) for the Se edge EXAFS experiment.

The main operating system is Linux Red Hat 9. The STARS server, several client modules and applications for controlling the equipment, the MySQL database software, PCCServer, PCCTools, and PCDBEditor all operate under the Linux. MS Windows is also used to control the equipment.

The majority of the source code is written in C/C++ (GNU C/C++ compiler v.3.2.2). GUI modules of the system were built mainly using Glade user interface builder for GTK+ and Gnome under the Red Hat Linux 9 operating system. On the STARS server, several client modules and applications for controlling the equipment are written in Perl. Some existent client modules, connected to the STARS server, were modified to accept top-level commands. The top-level part of the control system is only connected to the STARS server. The control system is being commissioned at BL5A and NW12A (PF, PF-AR, KEK, Tsukuba, Japan). The main functions of the system, such as X-ray experiment scheduling, X-ray data collection (snapshot, oscillation and MAD experiments, and EXAFS experiment), crystal loop mounting, and crystal centering procedures (manual and automatic), were checked.

ACKNOWLEDGMENTS

This work was supported in part by Grants-in-Aid for Scientific Research from the Ministry of Education, Culture, Sports, Science, and Technology (MEXT) of Japan, Special Coordination Funds for Promoting Science and Technology, and the Protein 3000 Project of the MEXT.

REFERENCES

1. Yu. A. Gaponov, N. Igarashi, M. Hiraki, K. Sasajima, N. Matsugaki, M. Suzuki, T. Kosuge and S. Wakatsuki, J. Synchrotron Rad., **11**, 17-20 (2004).
2. Yu. A. Gaponov, N. Igarashi, M. Hiraki, K. Sasajima, N. Matsugaki, M. Suzuki, T. Kosuge and S. Wakatsuki, AIP Conference Proceedings 705, American Institute of Physics, San Francisco, California, 25-29, August, USA, 2003, pp. 1213-1216.
3. R. Pugliese, L. Gregoratti, R. Krempska, F. Bille, J. Krempasky, M. Marsi, A. Abrami, J. Synchrotron Rad, **5**, 587-589 (1998).
4. E. Abola, P. Kuhn, T. Earnest, R. C. Stevens, Nature. Struct. Biol., **7**, 973-977 (2000).
5. R. M. Sweet, J. M. Skinner, M. Cowan, Synchrotron Radiation News, **14(3)**, 5-11 (2001).
6. Kosuge, T., Saito, Y., Nigorikawa, K., Katagiri, H., Shirakawa, A., Nakajima, H., Ito, K., Kishiro, J., Kurokawa, S., Oral talk on 4th International workshop on Personal Computers and Particle Accelerator Controls, 14-17 October, 2002, Frascati (RM), Italy, http://www.lnf.infn.it/conference/pcapac2002/TALK/WE-03/WE-03_talk.pdf

Design and Construction of a High-speed Network Connecting All the Protein Crystallography Beamlines at the Photon Factory

Naohiro Matsugaki, Yusuke Yamada, Noriyuki Igarashi and Soichi Wakatsuki

Structural Biology Research Center, Photon Factory, Institute of Materials Structure Science, High Energy Accelerator Research Organization (KEK), 1-1 Oho, Tsukuba, Ibaraki 305-0801, Japan

Abstract. A private network, physically separated from the facility network, was designed and constructed which covered all the four protein crystallography beamlines at the Photon Factory (PF) and Structural Biology Research Center (SBRC). Connecting all the beamlines in the same network allows for simple authentication and a common working environment for a user who uses multiple beamlines. Giga-bit Ethernet wire-speed was achieved for the communication among the beamlines and SBRC buildings.

Keywords: protein crystallography, synchrotron beamline, network configuration
PACS: 07.85.Qe

INTRODUCTION

SBRC currently operates four protein crystallography (PX) beamlines (BL-5A, BL-6A, BL-17A, and AR-NW12A) located at two synchrotron rings, 2.5GeV PF ring and PF-Advanced Ring (PF-AR, 6.5GeV). Originally, the network equipment at the beamlines and the SBRC buildings were directly connected to the facility network. This configuration was changed by making a LAN for each beamline to hide the network from unauthorized people. However, communication among the beamlines must be done through the facility network because the beamlines were physically distant. This way of communication could be unstable depending on the traffic of the facility network. In addition to this, massive data transfer (e.g. copying diffraction images) between the PX beamlines would deteriorate the performance of the facility network. Thus a private network that was isolated physically (not virtually) and connecting all the PX beamlines was required.

A LAN that includes all the beamlines has another benefit; only one authentication procedure is necessary for a user who accesses the beamlines. This is of great use for users with remote access. If each beamline is managed on its own LAN, it is difficult to implement a simple authentication procedure in a secure manner.

PRIVATE NETWORK DESIGN

Inter-site Configuration

The cabling of the private network was designed in the same manner as that of the facility network, which was arranged in a 'star shape' with the computer center building as its center, to take advantage of the existing plant. The overview of the physical network configuration is shown in Fig. 1. Optical fiber cables were required to connect the computer center and the five sites (BL-AR-NW12, PF office bldg., BL-5A&6A, BL-17A and SBRC bldg.), which are more than 500 m away. Two pathways and two independent network switches (Catalyst 3550) from each site to the computer center were prepared for network redundancy (in case one of the switches is broken) and traffic distribution. The connection from the site BL-5A&6A was 'bonded' to double the bandwidth because the data transfer from BL-5A could be massive due to diffraction images of larger file size.

CP879, *Synchrotron Radiation Instrumentation: Ninth International Conference,*
edited by Jae-Young Choi and Seungyu Rah
© 2007 American Institute of Physics 978-0-7354-0373-4/07/$23.00

A Catalyst 6503 Layer3 switch was installed as a network gateway of the LAN. The 6503 is connected to the 3550 switches with two bonded lines. The gateway has a NAT (Network Address Transformation) function, masquerading as the hosts inside and transforming the addresses attached to the outer interface to the specified addresses corresponding to the hosts in the LAN. The other channel of the 6503 is connected to the central switch in the facility network through which packets to the facility and Internet are transmitted.

The network configuration allows data transfer among the sites with gigabit Ethernet wire-speed. Beamline users could access a large amount of data, especially diffraction images, stored even at the distant sites with little delay and instability.

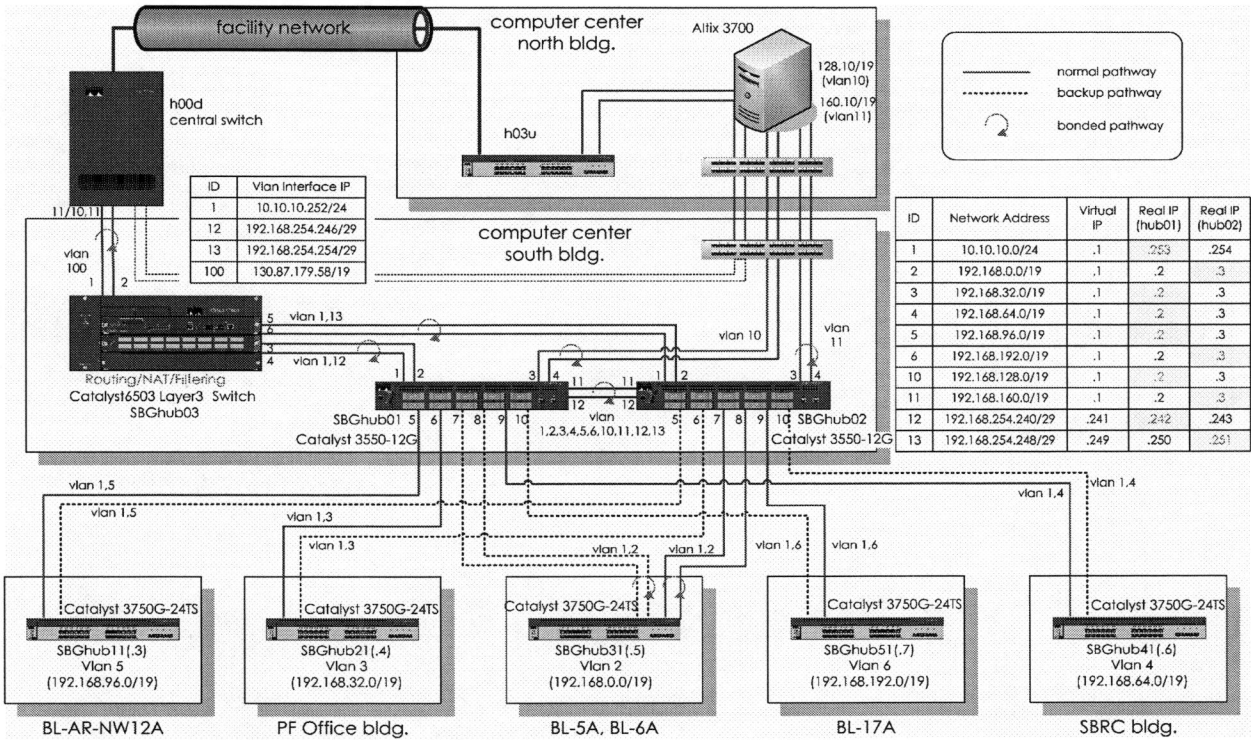

FIGURE 1. Physical configuration of the private network for protein crystallography beamlines

SGI Altix 3700 server, a single Linux box with 16 Itanium2 processors, was installed for data process and analysis. The Altix is connected directly to the two 3550 switches with two bonded lines to enlarge the bandwidth, assuming a number of simultaneous accesses from any beamlines. A RAID system (SGI TP9100) is attached to the Altix for the purpose of providing shared storage (9TB) to all the sites.

The communication speed between the Altix and the PCs at each site was tested using a ttcp program. The results are shown in Table 1. The speed for individual communication was satisfactory given the Giga-bit Ethernet wire-speed. However, when the PCs at the four sites communicated simultaneously, the speed dropped to approximately half of the cumulative speed. One of the reasons was that the Ethernet bonding implemented in the Altix was not optimized under the current package (SGI propack 3.0, Linux kernel 2.4 based). Preliminary tests showed that the performance could be improved when the Linux kernel 2.6 (SGI propack 4.0) was used.

TABLE 1. Communication speed between the Altix server and PCs at each site

	BL-5A	BL-6A	PF Office bldg.	BL-AR-NW12A	total
Individual communication	71 MB/s	75 MB/s	66 MB/s	72 MB/s	(*284 MB/s)
Simultaneous communication	34 MB/s	51 MB/s	29 MB/s	38 MB/s	151 MB/s

*cumulative value

Intra-site Configuration

Each beamline site has its own dedicated network to control beamline devices (beamline device control LAN). The beamline control is implemented using locally developed software, STARS (Simple Transmission and Retrieval System), which was originally designed for the central control system of interlock in the PF [1]. The STARS provides unified control in the heterogeneous architectural environment by using a common communication protocol. The system adopts a client-server style where client programs communicate with each other through the server process using the STARS protocol. User interface for beamline control, a STARS client running inside the users LAN, communicates with device control clients only through the STARS server process running on the gateway of the device control LAN (Fig. 2). In this way the beamline device control clients are hidden from general users.

Diffraction images from a CCD detector are stored on a local file server through another dedicated LAN. Isolation of the data transfer pathway from other networks stabilizes the performance of data acquisition. The stored images can be accessed from general users through another Ethernet channel using NFS or Samba protocol.

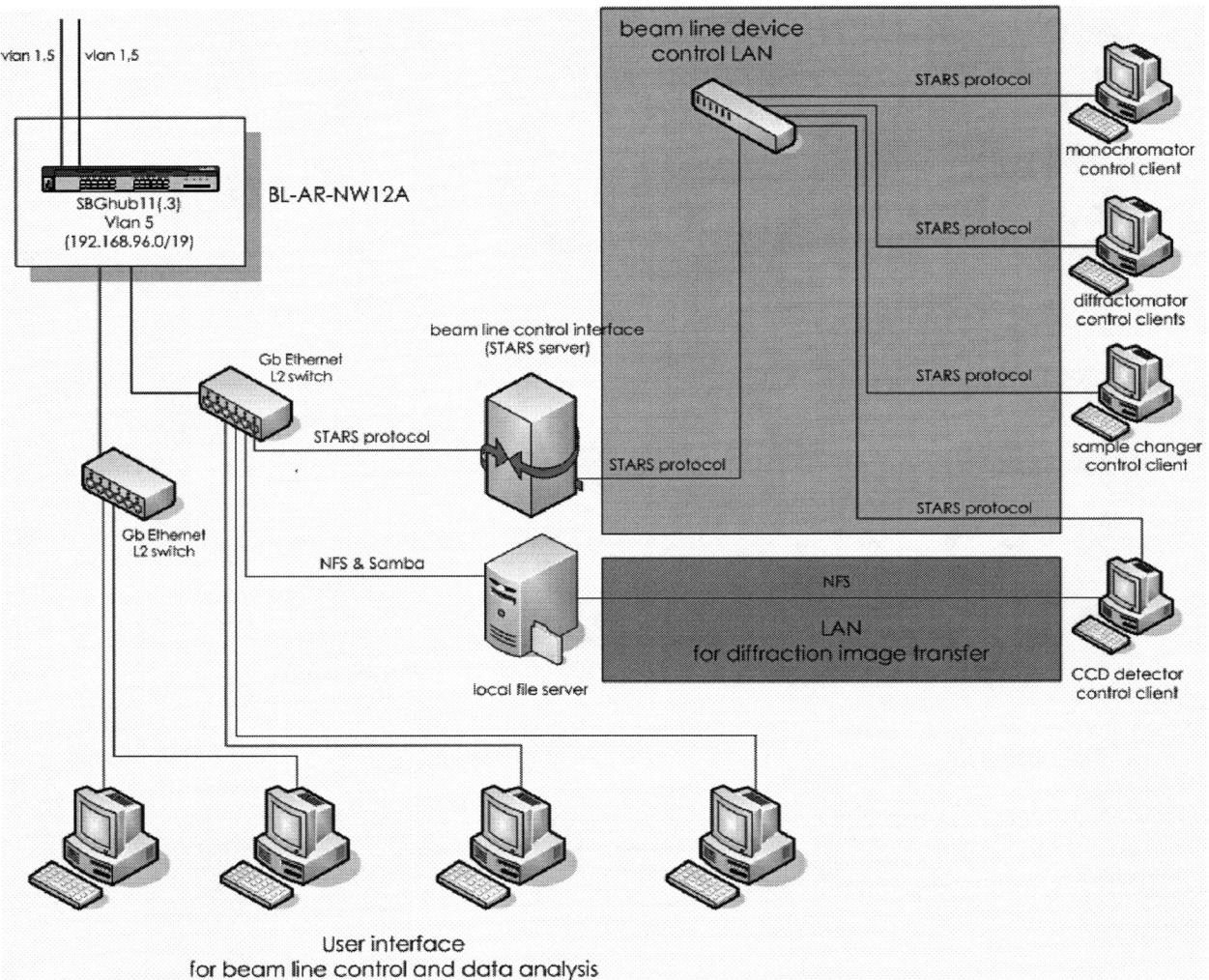

FIGURE 2. Typical network configuration of a site (BL-AR-NW12A)

AUTHENTICATION

LDAP (Lightweight Directory Access Protocol) combined with Samba was used for unified authentication between two different operating systems, Linux and Microsoft Windows. Users can log on to any PCs (of whatever

OS) with the same account. Root directories and setting files which define the working environment are shared with all the PCs. LDAP slave servers run at every sites for backup and traffic distribution; the authentication information kept in a master LDAP server is synchronized periodically.

FUTURE PERSPECTIVES

An integrated control system based on a relational database for PX experiments is under development [2]. The system, called PCCS (Protein Crystallography Control System), manages all the experimental information coming from structural biology activities, including protein purification, crystallization, X-ray data acquisition and structure solution. The interface of the PCCS runs inside the users LAN and communicates with beamline device control clients using the STARS protocol in the same manner currently used for PX experiments. The goal of the PCCS is fully automated PX experiments as well as interactive remote experiments and data analysis from the laboratories outside. Such remote access will be facilitated in future when the private LAN is connected directly to the broadband Internet, e.g. the Super SINET in Japan, instead of the facility network.

ACKNOWLEDGMENTS

The authors would like to thank the facility staff for their support and useful comments during the private network construction. The authors also express their appreciation to Toshihiro Fuji of Japan Silicon Graphics Co. for network speed tests and estimations. This work was supported in part by Grants-in Aid for Scientific research from the Ministry of Education, Culture, Sports, Science and Technology (MEXT) of Japan, Special Coordination Funds for Promoting Science and Technology, and the Protein 3000 Project of the MEXT.

REFERENCES

1. Kosuge, T., Saito, Y., Nigorikawa, K, Kawata, H., Shirakawa, A., Nakajima, H., Ito, K., Abe, I., Kishiro, J., and Kurokawa, S., PCaPAC2002 Proceedings, Frascaty, 2002.
2. Gaponov, Yu.A., Igarashi, N., Hiraki, M., Sasajima, K., Matsugaki, N., Suzuki, M., Kosuge, T. and Wakatsuki, S., J. Synchrotron Rad., 11,17-20, (2004).

X-ray Tomography and Chemical Imaging within Butterfly Wing Scales

Jian-Hua Chen, Yao-Chang Lee, Mau-Tsu Tang, and Yen-Fang Song

National Synchrotron Radiation Research Center
101 Hsin-Ann Road, Hsinchu Science Park, Hsinchu 30077, Taiwan

Abstract. The rainbow like color of butterfly wings is associated with the internal and surface structures of the wing scales. While the photonic structure of the scales is believed to diffract specific lights at different angle, there is no adequate probe directly answering the 3-D structures with sufficient spatial resolution. The NSRRC nano-transmission x-ray microscope (nTXM) with tens nanometers spatial resolution is able to image biological specimens without artifacts usually introduced in sophisticated sample staining processes. With the intrinsic deep penetration of x-rays, the nTXM is capable of nondestructively investigating the internal structures of fragile and soft samples. In this study, we imaged the structure of butterfly wing scales in 3-D view with 60 nm spatial resolution. In addition, synchrotron–radiation-based Fourier transform Infrared (FT-IR) microspectroscopy was employed to analyze the chemical components with spatial information of the butterfly wing scales. Based on the infrared spectral images, we suggest that the major components of scale structure were rich in protein and polysaccharide.

Keywords: transmission x-ray microscopy, Fourier transform infrared, infrared microspectroscopy, butterfly wing scale, photonic crystal
PACS: 07.85.Tt, 07.85.Qe, 42.79.Ci, 87.64.Je

INTRODUCTION

Photonic crystals are optical materials with periodic refractive index variations. As a result, the photonic crystals display an interesting optical behavior in which the propagation of light at certain frequencies is prohibited. The photonic crystal like structures of the butterfly wing scales are believed to reply for the flourish colors of butterfly wings [1-5]. The diffractive mechanism of the scales is expected to be resolved by the detail 3D structural information, and hopefully will lead to an innovative way for fabricating complete band gap photonic crystals. Although one can easily see the wing scales under conventional optical microscope (OM), an adequate imaging tool for revealing the 3D structures of the scales with sufficient spatial resolution is rather few. The match of the characteristic photonic-like length of the butterfly wing scales and the wavelength of reflected light (several hundreds nm), simply implies that the OM is not satisfactory of spatial resolution (sub-100 nm). The electron microscope (EM) can provide much subtle structural details, but rather in 2-D manners. In this paper, we present the tomographic x-ray microscopy with spatial resolution of 60 nm.

The NSRRC nano-transmission x-ray microscope (nTXM) was designed and constructed in late 2004 at beamline BL01B. It provides 2-D images and 3-D tomography with sub-30 nm to 60 nm spatial resolution, in the energy range 8-11 keV, and with Zernike phase contrast mode for imaging soft specimens.

The compositional analysis of the wing scales was done by synchrotron–radiation-based Fourier transform infrared microspectroscopy (FT-IMS) equipped at NSRRC infrared microscopy beamline 14A [6]. FT-IMS a combination of FT-IR spectroscopy, microscope, and semi-coherent infrared from synchrotron source, has long been applied to provide ultra-spatially resolved spectroscopic information with higher signal-to-noise ratio. When applied to biological specimens, the technique enables a correlation of chemical information to histological structures. Various biomolecular composition (proteins, membranes, nucleic acids, etc.) of biological samples contribute to a characteristic IR spectrum, which is rich in structural and functional aspects.

EXPERIMENT AND RESULTS

Figure 1 shows a schematic optical layout of the x-ray microscope. Similar to other transmission-type microscopes, incident x-rays are first condensed onto the sample and then magnified with an objective lens, to form projection images on an area detector. The detecting system consists of a scintillator converting the x-rays into visible light and forming ultimate images on a charge-coupled-detector (CCD), with a 16 bits dynamic range and 13.5 μm pixel size. The estimated photon flux density for the imaging is about 450G photons/s/0.1%BW un the energy range 5-20keV. The detail optical concept of the nTXM can be found in another article (H1-003).

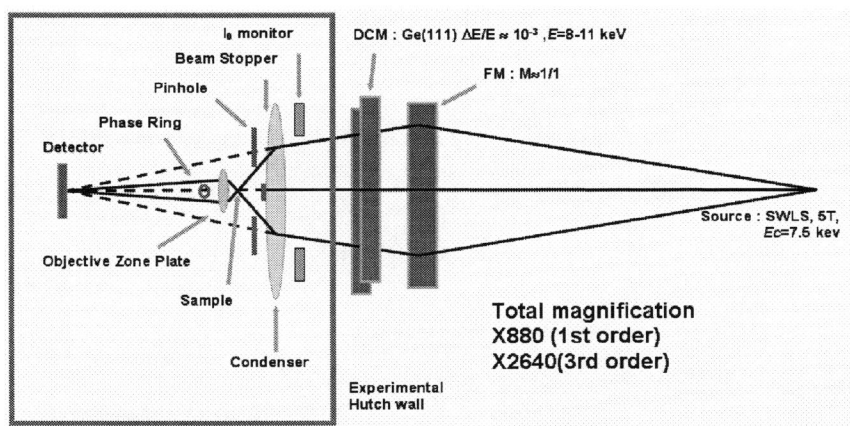

FIGURE 1. Schematic optical layout of NSRRC nano transmission x-ray microscope (NSRRC nTXM)

The wing scale was sampled from the *Papilio bianor*, a big size butterfly that can be easily found in Taiwan (Fig. 2a). The orderly arrayed wing scales can be seen with conventional optical microscope with 200 times magnification (Fig. 2b). A wing scale originally seen red was chosen for imaging. The scales were pre-aligned by an off-line OM and then positioned into the rotation center of nTXM. No any pre-chemical sample treatment is necessary. The x-ray energy was selected 8 keV for all the x-ray measurements. In order to obtain sufficient image contrast, the phase contrast mode of the microscope was acted, which was accomplished by placing a 3-micron thick gold-made phase ring at the back focal plane of the objective zone plate.

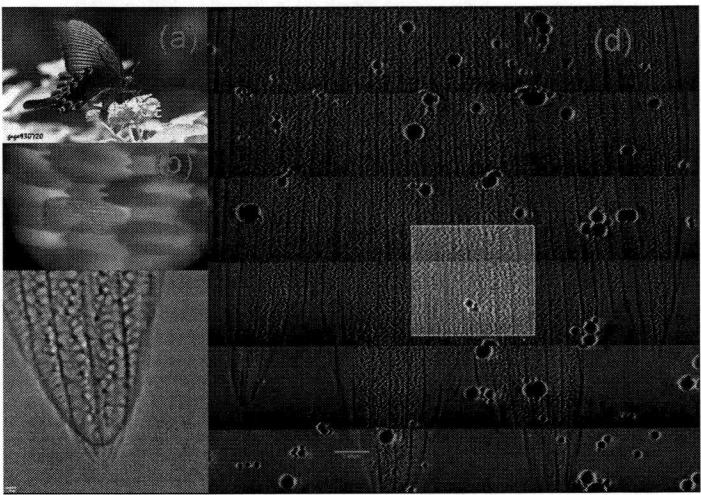

FIGURE2. The *Papilio bianor* butterfly wing scales imaged by (a) optical camera, (b) 200 times optical microscope, (c) and (d) NSRRC nTXM. The spatial resolution, filed of view and exposure time of a single exposure for the x-ray images was 60 nm, 15μm x 15μm and 30 seconds, respectively. (d) is a mosaic pattern consisting 7x7 array of single exposures. The tomographic

imaging were taken within the light square region. The gold particles was dropped on the sample as the reference point. The gold particles size were randomly distributed.

The 2-D x-ray projection images are shown in Fig. 2c and 2d. While Fig. 2c is a projection image around a tip of a wing scale, Fig. 2d exhibits a mosaic pattern consisting 7x7 array of single exposures. The field of view and exposure time for each single exposure are 15μm x 15μm and 30 seconds, respectively. In contrast to images from OM, the x-ray images visibly exhibit much detail structural information with better spatial resolution. Although the microscope can be performed with sub-30 nm spatial resolution, in this study the microscope was operated with 60 nm spatial resolution in order to reduce the radiation damage caused by x-rays and to limit the exposure time. The dark round dots shown in the x-ray images are gold particles intentionally spread on the scales surface as alignment marks when taking tomogrpahic frames.

The tomographic images were taken by azimuthally rotating the sample within ± 70 degrees with 1 degree difference. Figure 3 shows the reconstructed 3-D tomography of the wing scales. The imaged areas are shown in Fig. 2c, and 2d, respectively. Figure 3c, 3d, 3e show the virtual section images of Fig. 3b, viewing perpendicular (c, d) and parallel (e) to the incident x-ray beams. Several characteristics can be revealed from the 3-D images. The wing scales are formed in network structure as expected. We can divide the structure into two parts, the long frame-like (L) and the fine (F) structures. The long frame like structure elongating in entire scale is believed to contribute to the mechanical strength of the scales. The fine structures forming network in between the long frames, on the other hand, is believed to play the role of reflecting the light. A trial analysis to the distances between the joints of the fine structures (F) reveals a characteristic length of around 500-600 nm, close to the wavelength of red light. A further photonic bang gap calculation is in progress.

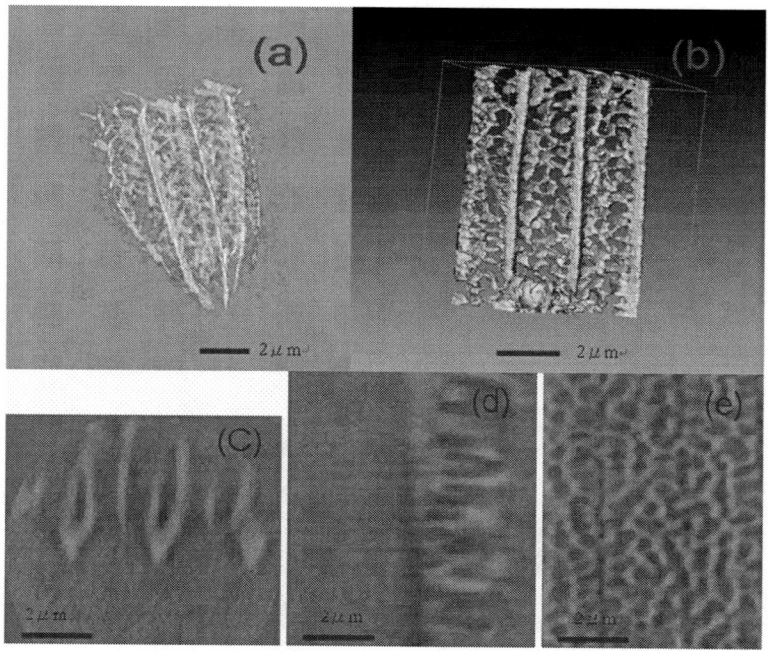

FIGURE 3. Reconstructed 3-D tomographic images of wing scales. The tomography was reconstructed based on 141 sequential 2-D frames taken with azimuth angle rotating from -70° to 70°. The (a) and (b) are reconstructed from the sampling areas of Fig. 2c and 2d, respectively. (c)-(e) are virtual sectional images for (b) perpendicular (c, d) and parallel (e) to the incident x-ray beams.

The synchrotron-radiation-based IMS was applied to analyze the chemical compositions of butterfly wing scales with spatially resolved spectral information. The infrared images and spectra of *papilio bianor* were shown in Fig. 4. The images reveal that the components of the butterfly wing scale are rich in amide, i.e. protein. The absorption peaks of amide A, amide B, amide I and amide II for protein structure were shown in the Fig. 4 (right). The positions of peaks at 1027, 1075, 1113, 1161 and 3442 cm^{-1} were assigned to the vibrational motion of four different C-O bonds and O-H stretch of chitin. In addition, the position of peak at 1263 cm^{-1} was assigned to the S=O stretching vibration. Based on these results of infrared spectra, we suggested these six characteristic absorption peaks were assigned to chondroitin sulfonic like structure of the wing scales and they were not only constructed by chitin but protein.

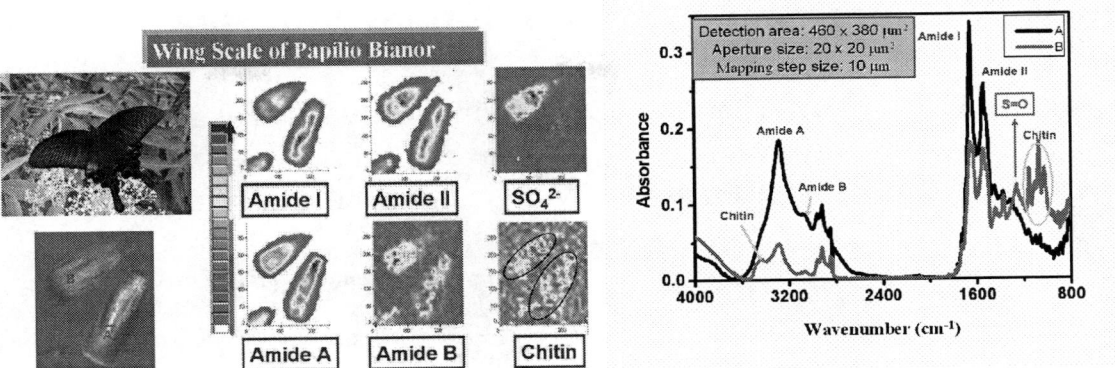

FIGURE 4. The IR spectroscopic analysis of two kinds of butterfly wing scales A and B (scale bar is 75μm). The structure reveals the signals of amide A, amide B, amide I and amide II for protein structure. The positions of peaks at 1027, 1075 1113, 1161 and 3442 cm^{-1} were the vibrational motion of four different C-O bend and O-H stretch of chitin. And the position of peaks at 1263 cm^{-1} was assigned to the S=O stretching vibration.

CONCLUSIONS

We have successfully imaged the butterfly wing scales by the newly built NSRRC-nTXM with spatial resolution 60 nm, in 2-D and 3-D manners. The capability of Zernike phase contrast of NSRRC-nTXM is crucial in the study for the wing scales almost transparent under a conventional x-ray microscope. Two characteristic internal structures can be realized in the wing scale internal structure, namely the long frame-like (L) and fine (F) structures. While the long frame-like structure may provide mechanical support to the entire scale, the fine structure may reply on the reflection of light. A trial analysis to the fine structure reveals a characteristic length closer to the wavelength of reflected light. To fully understand the role of the internal structure playing to light reflection, the photonic band gap calculation is necessary. The chemical composition inside the wing scales was analyzed by synchrotron-radiation-based Fourier transform infrared microspectroscopy. While not much surprised by the composition of proteins of the wing scales, the infrared data suggests that some minor organic elements may be related to the final dimensionalities of the wing scales.

REFERENCES

1. P. Vukusic and J. R. Sambles, *Nature* **424**, 852-855 (2003).
2. S. Yoshioka and S. Kinoshita, *Proc. Biol. Sci.* **271**, 581-587 (2004).
3. S. Kinoshita, S. Yoshioka and K. Kawagoe, *Proc. Biol. Sci.* **269**, 1417-1421 (2002).
4. S. Kinoshita and S. Yoshioka, *Chemphyschem.* **6**, 1442-1459 (2005).
5. M. Srinivasarao, *Chem. Rev.* **99**, 1935-1962 (1999).
6. G.L. Carr, J.A. Reffner, G.P. Williams, Rev. Sci. Ins. **66**, 1490-1492 (1995)

In-vivo Fluorescent X-ray CT Imaging of Mouse Brain

T. Takeda[*], J. Wu[*], Thet-Thet-Lwin[*], Q. Huo[*], N. Sunaguchi[**], T. Murakami[**],
S. Mouri[**], S. Nasukawa[**], T. Yuasa[**],
K. Hyodo[***], H. Hontani[****], M. Minami[*], and T. Akatsuka[**]

[*]Graduate School of Comprehensive Human Sciences, University of Tsukuba, Tsukuba, Ibaraki 305-8575 Japan,
[**]Faculty of Engineering, Yamagata University, Yonezawa, Yamagata 992-8510 Japan,
[***]Institute of Material Science, High Energy Accelerator Research Organization,
Tsukuba, Ibaraki 305-0801 Japan,
[****]Department of Computer Science and Engineering, Nagoya Institute of Technology,
Nagoya, Aichi 466-8555 Japan

Abstract. Using a non-radioactive iodine-127 labeled cerebral perfusion agent (I-127 IMP), fluorescent X-ray computed tomography (FXCT) clearly revealed the cross-sectional distribution of I-127 IMP in normal mouse brain in-vivo. Cerebral perfusion of cortex and basal ganglion was depicted with 1 mm in-plane spatial resolution and 0.1 mm slice thickness. Degree of cerebral perfusion in basal ganglion was about 2-fold higher than that in cortical regions. This result suggests that in-vivo cerebral perfusion imaging is realized quantitatively by FXCT at high volumetric resolution.

Keywords: In-vivo imaging, Cerebral perfusion, Fluorescent X-ray CT, Functional imaging, Molecular imaging, Small animal, Experimental study
PACS: 01.30.Cc, 07.85.Qe, 07.85.Nc, 32.50.+d, 32.30.Rj, 42.62.Be, 87.63.Lk, 87.19.Xx, 87.59.Fm

INTRODUCTION

In biomedical field, functional evaluation is quite useful to understand the process, diagnosis and treatment of various diseases [1]. Micro positron emission tomography (micro-PET) [1-3] and micro single photon emission computed tomography (micro-SPECT) [4, 5] are recently used to visualize in-vivo biochemical processes in small animal. However, the use of radionuclide agent is indispensable in PET and SPECT study, and the volumetric resolution of these techniques is limited to about 1 mm³ and 0.5 - 0.1 mm³, respectively.

Fluorescent X-ray technique using synchrotron X-ray, which is usually used to observe the surface of the object, can detect very low contents of medium or heavy trace elements with concentrations in the order of picograms [6]. To depict the distribution of specific elements inside the object without slicing procedure, fluorescent X-ray computed tomography (FXCT) with synchrotron radiation is being developed [7-10]. The FXCT could depict iodine within a phantom, and the endogenous iodine of an excised human thyroid [11-13]. Furthermore, the FXCT was applied to assess the functional information of ex-vivo brain and heart of small animals similar to autoradiogram, and we successfully observed the cerebral blood flow [14] and myocardial fatty acid metabolism [15, 16] after injecting various types of non-radioactive iodine labeled agent. Since the significant results were obtained by ex-vivo studies, in-vivo FXCT imaging was performed by using a germanium detector with high count rate capability and energy resolution [17]. Here, the experimental results of in-vivo cerebral blood flow imaging of mouse brain obtained by FXCT is described.

METHODS AND MATERIALS

The experiment was carried out at the bending-magnet beam line BLNE-5A of the Tristan accumulation ring (6.5 GeV) in Tsukuba, Japan. The photon flux rate in front of the object was approximately 9.3 x 10⁷ photons/mm²/s for

CP879, Synchrotron Radiation Instrumentation: Ninth International Conference,
edited by Jae-Young Choi and Seungyu Rah

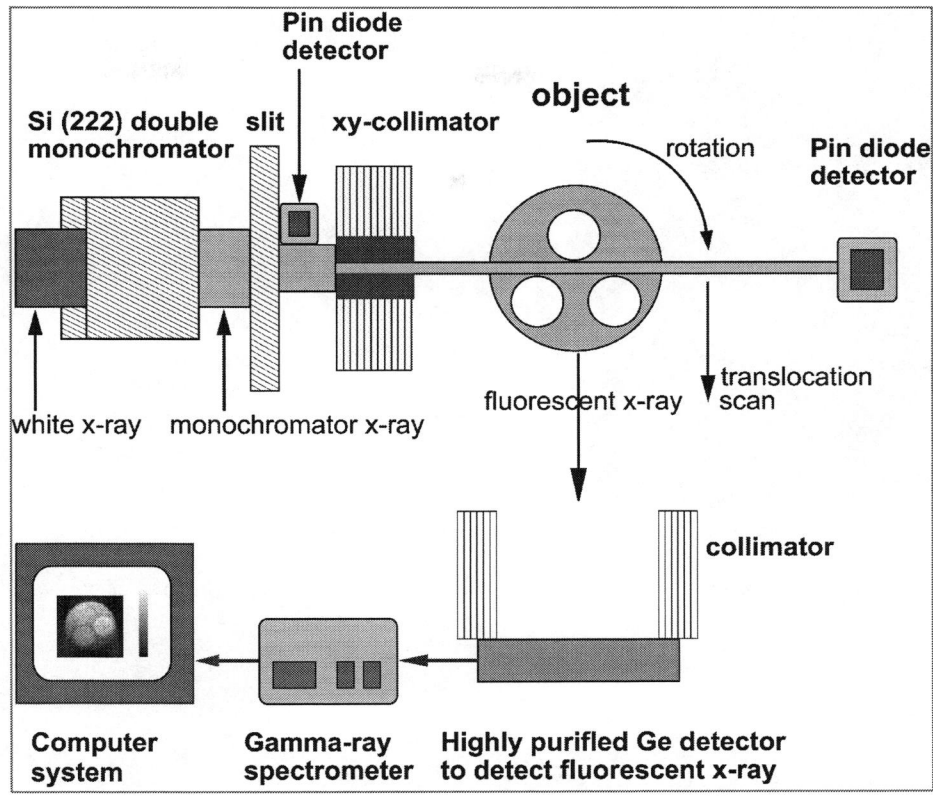

FIGURE 1. Schematic diagram of fluorescent X-ray computed tomography system.

beam current of 40 mA. FXCT system consists of a silicon (220) double crystal monochromator, an X-ray slit system, a scanning table for subject positioning, a fluorescent X-ray detector, and two pin-diode detectors for incident X-ray and transmission X-ray data (Fig. 1). The white X-ray beam was monochromatized to 37 keV X-ray energy. The monochromatic X-ray was collimated into a pencil beam (1 x 0.1 or 0.2 mm^2: horizontal and vertical direction). Fluorescent X-rays induced by incident X-ray beam were detected in a high purity germanium (HPGe) detector operating in the photon-counting mode, and the HPGe detector was oriented perpendicular to the incident monochromatic X-ray beam. The data acquisition time of the HPGe detector for each scanning step was set 5-s.

Objects were 3 mice and an acrylic phantom with three sub-holes. Since radioactive I-123 labeled N-isopropyl-p-iodoamphetamine (I-123 IMP) is popularly used to evaluate cerebral blood flow in clinical SPECT study [18], we attempted to use a non-radioactive I-127 labeled IMP (I-127 IMP) for in-vivo FXCT imaging of mouse. In-vivo FXCT imaging of mouse started 5 min after intravenous injection of I-127 IMP under the anesthetized with pentobarbital. The 20-mm in diameter acrylic phantom filled with various concentration of iodine solution was also imaged to determine the absolute iodine content within brain. Object was scanned with 1-mm translation step over target range and 6 degree rotation step over a range of 180 degrees. FXCT images were reconstructed by the algebraic method with attenuation correction for the incident beam and the emitted fluorescent X-ray using the TXCT data [19]. The TXCT image was reconstructed by using the filtered back projection method with the Shepp and Logan filter.

Our present experiment was approved by the Medical Committee for the Use of Animals in Research of the University of Tsukuba, and it conformed to the guidelines of the American Physiological Society.

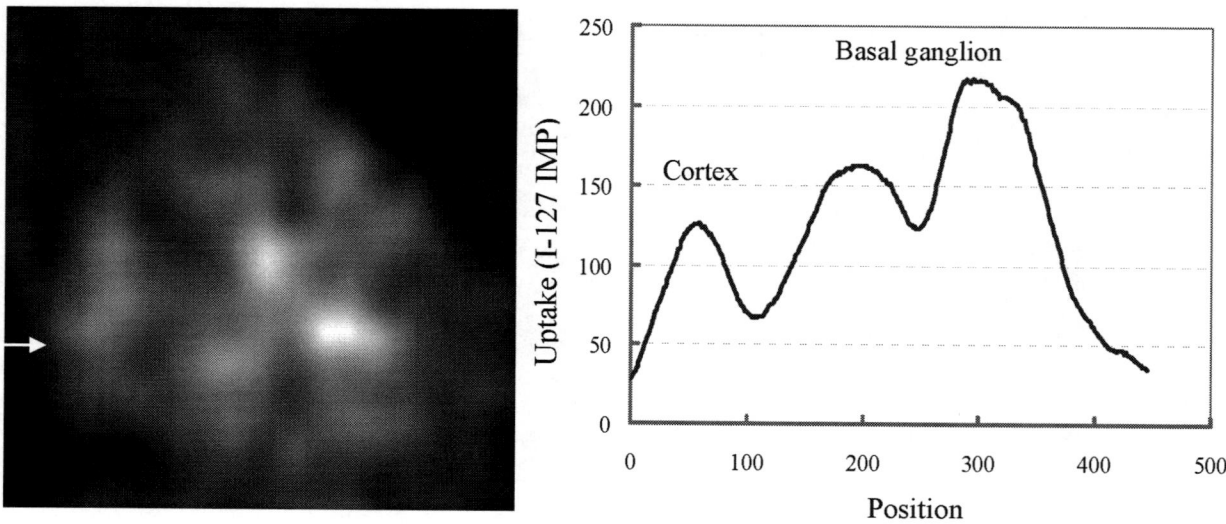

FIGURE 2. Cerebral perfusion image with I-127 IMP and its profile analysis of a mouse brain by fluorescent X-ray computed tomography.

RESULTS AND DISCUSSIONS

In-vivo cerebral perfusion of mouse was clearly imaged by FXCT at an 1 mm spatial resolution with a 0.1 or 0.2 mm slice thickness (Fig. 2). Cerebral cortex and basal ganglion were well visualized, and the cerebral perfusion in basal ganglion was about 2-fold higher than other cortical regions probably due to anesthesia. Reduced cerebral perfusion in left cerebral cortex might be caused by ischemia. In profile analysis, the degree of regional cerebral perfusion was assessed quantitatively as shown in Fig. 2. Using calibration data from the phantom, cerebral uptake of the iodine in mouse was estimated about 0.02-0.04 mg/g.

The volumetric resolution of this in-vivo FXCT image was 0.1 or 0.2 mm^3 (1-mm x 1-mm x 0.1 or 0.2-mm). This resolution was almost comparable to super micro-SPECT imaging as 0.1 mm^3 [5, 20]. Since the duration of anesthesia is limited, in this study, the spatial resolution had been restricted to 1 mm by scanning time. However, we can obtain much higher spatial resolution image of 0.5-mm with a 0.4-mm slice thickness by using the HPGe detector with much higher count rate capability. For this purpose, we are now developing high speed FXCT system. In addition, the use of non-radioactive agent is significantly suitable to perform the biomedical experiment because the preparation of drug and experiment are quite easy without radiation exposure for researchers.

Thus, in-vivo FXCT is a powerful tool to image the functional information with high spatial resolution and without the use of non radioactive labeling agents.

ACKNOWLEDGMENTS

We thank Y. Tsuchiya PhD, X. Chang PhD, and T. Kuroe MS for technical supports, Mr. K. Kobayashi for preparation of the experimental apparatus, and Nihon Medi-Physics Co., Ltd., Japan supplying I-127 IMP. This research was partially supported by a Grant-In-Aid for Scientific Research (#15390356, #15070201, #17390326, #16-04246) from the Japanese Ministry of Education, Science and Culture, Research Grant A 2117 from University of Tsukuba, and was performed under the auspices of the National Laboratory for High Energy Physics (Proposal: 2003G315, 2005G308).

REFERENCES

1.   H.R. Herschman, *Science* **302**, 605-608 (2003).
2.   A.F. Chatziioannou, *Eur. J. Nucl. Med.* **29**, 98-114 (2002).
3.   Y. Yang, Y.C. Tai, S. Siegel, D.F. Newport, B. Bai, Q. Li, R.M. Leahy and S.R. Cherry, *Phy. Med. Biol.* **49**, 2527-2545 (2004).
4.   P.D. Acton and H.F. Kung, *Nucl. Med. Biol.* **30**, 889-895 (2003).
5.   F.J. Beekman, F. Van der Have, B. Vastenhouw, A.J.A. Van der Linden, P.P. Van Pijk, J.P.H. Burbach and M.P. Smidt, *J Nucl. Med.* **46**, 1194-1200 (2005).
6.   A. Iida and Y. Gohshi, "Tracer element analysis by X-ray fluorescent". in *Handbook on Synchrotron Radiation* Vol 4, edited by S. Ebashi et al. North-Holland, Elsevier Publisher, Amsterdam 1991, pp. 307-348.
7.   T. Takeda, *Nucl. Instrum. Meth.* **A548**, 38-46 (2005).
8.   J.P. Hogan, R.A. Gonsalves and A.S. Krieger, *IEEE Trans. Nucl. Sci.* **38**, 1721-1727 (1991).
9.   T. Takeda, M. Akiba, T. Yuasa, M Kazam, A. Hoshino, Y. Watanabe, K. Hyodo, F.A Dilmanian, T. Akatsuka and Y Itai, "SPIE-The international Society for Optical Engineering Press" in *Physics of Medical Imaging 1996*. SPIE Conference Proceedings 2708, Society of Photo-Optical Instrumentation Engineers, Bellingham, WA, 1996, pp. 685-695.
10.   T. Takeda, T. Yuasa, A. Hosino, M. Akiba, A. Uchida, M. Kazama, K. Hyodo, F.A. Dilmanian, T. Akatsuka and Y. Itai, "SPIE-The international Society for Optical Engineering Press" in *Developments in X-Ray Tomography* edited by U Bonse. SPIE Conference proceedings 3149, Society of Photo-Optical Instrumentation Engineers, Bellingham, WA, 1997, pp. 160-172.
11.   G.F. Rust and J. Weigelt, *IEEE Trans.Nucl.Sci.* **45**, 75-88 (1998).
12.   T. Takeda, Q. Yu, T. Yashiro, T. Zeniya, J. Wu, Y. Hasegawa, Thet-Thet-Lwin, K. Hyodo, T. Yuasa, F.A. Dilmanian, T. Akatsuka and Y. Itai, *Nucl. Instr. Meth.* **A 467-468**, 1318-1321 (2001).
13.   T. Takeda, A. Momose, Q. Yu, T. Yuasa, F.A. Dilmanian, T. Akatsuka and Y. Itai, *Cell. Mol. Biol.* **46**, 1077-1088 (2000).
14.   J. Wu, T. Takeda, Thet-Thet-Lwin, N. Sunaguchi, T. Yuasa, T. Fukami, H. Hontani and T. Akatsuka, *Med. Imag. Tech.* **23**, 312-317 (2005).
15.   Thet-Thet-Lwin, T. Takeda, J. Wu, N. Sunaguchi, Y. Tsuchiya, T. Yuasa, F.A. Dilmanian, M. Minami and T. Akatsuka, *6th Asian-pacific Conference on Medical and Biological Engineering 2005* edited by T. Katsuhuko, et al. APCMBE & IFMBE Proceedings 8; International Federation for Medical and Biological Engineering , 2005, PA-3-34.
16.   T. Takeda, T. Zeniya, J. Wu, Q. Yu, Thet-Thet-Lwin, Y. Tsuchiya, D.V. Rao, T. Yuasa, T. Yashiro, F.A. Dilmanian, Y. Itai and T. Akstsuka, "SPIE-The international Society for Optical Engineering Press" in *Developments in X-Ray Tomography III* edited by U Bonse. SPIE Conference proceedings 4503, Society of Photo-Optical Instrumentation Engineers, Bellingham, WA, 2002, 299-311.
17.   T. Takeda, Y. Tsuchiya, T. Kuroe, T. Zeniya, J. Wu, Thet-Thet-Lwin, T. Yashiro, T. Yuasa, K. Hyodo, K. Matsumura, F.A. Dilmanian, Y. Itai and T. Akatsuka, *Synchrotron Radiation Instrumentation: Eight International Conference*, edited by T. Warwick et al. AIP Conference Procedings. American Institute of Physics, **CP705**, 2004, pp.1320-1323.
18.   H.S. Winchell, W.D. Horst, W.H. Braum, R. Oldendorf, R. Hattner and H. Parker, *J. Nucl. Med.* **21**, 947-952 (1980).
19.   T. Yuasa, M. Akiba, T. Takeda, M. Kazama, Y. Hoshino, Y. Watanabe, K. Hyodo, F.A. Dilmanian, T. Akatsuka and Y. Itai, *IEEE trans. Nucl. Sci.* **44**, 54-62 (1997).
20.   T. Takeda, J. Wu, Thet-Thet-Lwin, N. Sunaguchi, T. Yuasa, K. Hyodo, F.A. Dilmanian, M. Minami and T. Akatsuka, *International Conference on Image Processing* 2005, IEEE Conference proceedings, 2005, III: 593-596.

Images of the Rat bone, Vertebra and Test phantom Using Diffraction-Enhanced Imaging Technique with 20, 30 and 40 keV Synchrotron X-rays

Donepudi V. Rao[1], Zhong Zhong[2], Tetsuya Yuasa[3], Takao Akatsuka[3], and Tohoru Takeda[4], Giuliana Tromba[5]

[1]Department of Physics, Sir. C.R.R. (A) College, Eluru-534007., W.G. Dt., A.P., India.
[2]National Synchrotron Light Source, NSLS, Brookhaven National Laboratory, Upton, NY 11973, USA.
[3]Dept of Bio-System Engineering, Faculty of Engineering, Yamagata University, 4-3-16 Jonan, Yonezawa 992-8510, Japan
[4]Institute of Clinical Medicine, University of Tsukuba, Tsukuba, Japan
[5]Synchrotron Radiation for Medical Physics, Elettra, Trieste, Italy

Abstract. Images of rat bone of different age groups (8, 56 and 78 weeks), lumbar vertebra and calcium hydroxyapatite phantom are obtained utilizing the diffraction-enhanced imaging technique. Images obtained with DEI are of superior quality and this novel technique may be an excellent choice for better visualization of the microstructure and the embedded spongiosa. Our motivation is to develop the optimizing tomography with the use of the data obtained at multiple energies.

Keywords: Images, rat bone, vertebra, DEI, synchrotron x-rays.
PACS: 87, 87.57.-e, 87.59.Nk

MOTIVATION

X-ray imaging techniques were applied extensively to study the 3D microstructure of the bone and embedded internal features. Experimental studies are confined to the conventional techniques with the use of the tube source of X-rays. Conventional X-ray imaging technique utilizing the tube source of X-rays relies on the absorption variation in the object to give a gray scale picture. The flux emitted by the X-ray tube is considerably low and associated bremsstrahlung radiation reduces the required monochromicity. An ideal monoenergetic radiation for this purpose comes from a synchrotron source since it provides considerable monochromicity with high flux. Recently, refraction properties of X-rays turn to be more attractive advantages for imaging over the absorption properties. Refraction is orders of magnitude more sensitive, particularly for biological or low materials (Zhong 2000). The Diffraction-enhanced imaging (DEI) technique exploits these refraction for differentiating the embedded spongiosa with high collimated synchrotron X-rays [1].

With the introduction of diffraction-enhanced X-ray imaging (DEI) which is capable of rendering images with absorption, refraction and scatter rejection qualities has allowed detection of specific soft tissues based on small differences in tissue densities. This novel X-ray imaging technology carried out at NSLS, BNL has previously shown it is possible to visualize breast tissue and a number of related medical samples demonstrate that the technique has the potential to provide new information [2-9].

For the last few years, DEI is continuously and consistently upgraded with the inclusion of new optics, electronics, mechanics and mountings and software at various stages at NSLS, BNL, USA, before imaging. Improvements in image quality were carried out regularly on phantoms and biological samples. These studies suggest that, DEI is a potential, valuable and reliable tool for studying the refraction and scattering properties. Diffraction-enhanced imaging technique derives contrast from absorption, refraction and extinction. Real and imaginary parts of the refractive indices are responsible for the phase-shifts and absorption for enhanced contrast imaging. For biological samples, the real part is up to 1000 times greater than the imaginary part. In this context,

CP879, *Synchrotron Radiation Instrumentation: Ninth International Conference*,
edited by Jae-Young Choi and Seungyu Rah
© 2007 American Institute of Physics 978-0-7354-0373-4/07/$23.00

phase-sensitive imaging techniques are more efficient than conventional methods. Images are acquired using 20, 30 and 40 keV synchrotron X-rays.

In view of this, we applied this novel technique for the rat bone of different age groups and lumbar vertebra with spongiosa. The main motivation and the real thrust of this study is to utilize the DEI to image the above samples in order to produce both pure absorption and refraction images. Monochromaticity of these synchrotron X-rays and the associated DEI system, improved the contrast of the images considerably. The choice of optimum energy is chosen based on the quality of the image for better visualization and analysis.

Materials and Methods

The details of DEI technology and the associated instrumentation have been presented previously and here we provide a brief description.

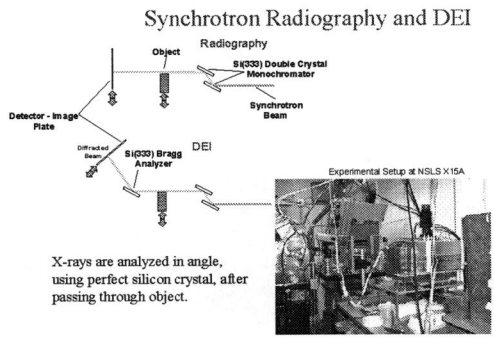

FIGURE 1. Experimental arrangement and the associated DEI system

The experimental arrangement is shown in Fig. 1. The collimated fan beam of X-rays is prepared by the Silicon [3, 3, 3] monochromator consisting of two perfect silicon crystals. Once this beam passes through the subject, a third crystal (analyzer crystal) of the same reflection index diffracts the X-rays onto radiographic film (Kodak Professional Industrie 150, Industrex SR45) or an image plate detector (Fuji HRV image plate, read out by a Fuji BAS2500 image plate reader). The distance between the X-ray source and the specimen is approximately 20 m while the distance between the specimens and the X-ray film or image plate detector is 1 m. The image of the subject is formed by scanning the subject and X-ray film at the same speed through the fan beam, in approximately opposite directions to take account of the Bragg reflection by the analyzer crystal. Because of the non-dispersive nature of the crystals, the narrow Darwin-width of the diffraction used, and the small distance between the sample and the detector, the resolution of the image obtained is limited by the resolution of the X-ray film, which is approximately 50 μm, or the pixel size of the image plate detector, which is approximately 75 μm. The Bragg condition for the analyzer crystal is met only when the incident beam makes the correct angle with the lattice planes in the crystal for a given X-ray energy. When this condition is met, the beam diffracts from the planes over a narrow range of incident angles. As the analyzer crystal is rotated about a horizontal plane, the crystal will go through a Bragg condition for diffraction and the diffracted intensity will trace out a profile or a rocking curve. The rocking curve of the analyzer in the protocol described here is roughly triangular and has peak intensity close to that of the beam striking in it. The width of this profile is typically a few micro radians (the full width at half maximum is 1.5 micro radians at an X-ray energy of 40 keV and 3.6 micro radians at 18 keV, using the Si [3, 3, 3] reflection). This narrow angular width provides the tools necessary to prepare and analyze, on the micro radian scale, the angle of X-ray beams modified by the subject while traversing it. Since the range of the angles that can be accepted by the analyzer crystal is only a few micro radians, the analyzer crystal detects the subject's X-ray scattering (ultra-small-angle-scattering) and refraction of X-rays at the micro radian level, an angular sensitivity which is not possible in conventional radiography. The X-ray intensity in the subject is therefore modulated by the scattering and refraction properties of the subject. To extract refraction information, the analyzer is typically set to the half intensity points on the low-and high angle sides of the rocking curve refereed to as -1 and +1, respectively, in the following discussion), or at the base of the rocking curve (refereed as -2 and +2, respectively, for the low- and high angle sides), while the imaging takes place. For optimal extinction (scatter rejection) sensitivity, the analyzer is typically set to the peak of the rocking curve during imaging. The reproducibility of the DEI images is maintained by monitoring the intensity of the diffracted X-rays by the analyzer just prior to imaging, to ensure that the analyzer is at the prescribed angular position.

Specimens

Bone phantom's with different calcium hydroxyapatite concentrations (10mm x 10mm, 10mm x 8mm), rat's vertebra and bone samples of different age groups (8, 56 and 78 weeks) were prepared and stored. The diameter of the cylindrical phantoms varies from 8 to 10 mm.

RESULTS AND DISCUSSION

Figures 2 to 4 shows the images of the rat bone of different age groups, calcium hydroxyapatite phantom and the rat's vertebra at 20, 30 and 40 keV. In order to visualize the interior microstructure and for sharp images we acquired the images at different energies and the choice of optimum energy is chosen based on the quality of the image. The contrast in the refraction image represents differences in intensity scattered through a particular angle. Sources of contrast in the apparent absorption image are absorption and extinction in scatter image. Extinction here means missing intensities of those X-rays that are refractive under angles substantially higher than the acceptance of the analyzer. The interior microstructure including the embedded spongiosa is sharper with high contrast. As regards to the rat's lumber vertebra, we differentiated the central portion of the vertebra at different orientations for clear visualization of the external and internal features. Expect in few orientations, majority of the images are acceptably distinct at 40 keV and the system is more sensitive and reliable at this optimum energy. The visibility of porosity is considerably large at 40 keV has mesh compared to 20 and 30 keV. The porosity of the phantom (differences in density) at 40 keV has closed mesh structure compared to 20 and 30 keV.

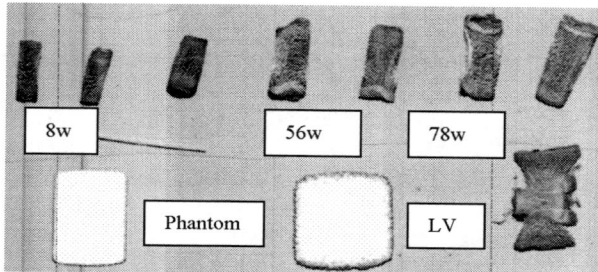

FIGURE 2. Images of the rat bone of different age groups, phantom and the vertebra (LV) with 20 keV synchrotron X-rays.

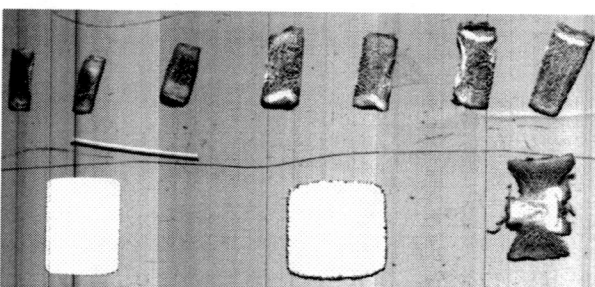

FIGURE 3. Same as in Fig. 2 but for 30 keV.

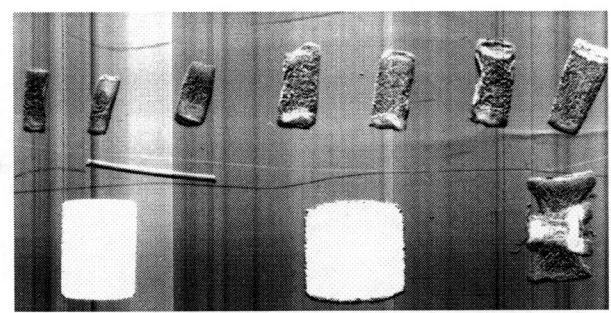

FIGURE 4. Same as in Fig. 2 but for 40 keV

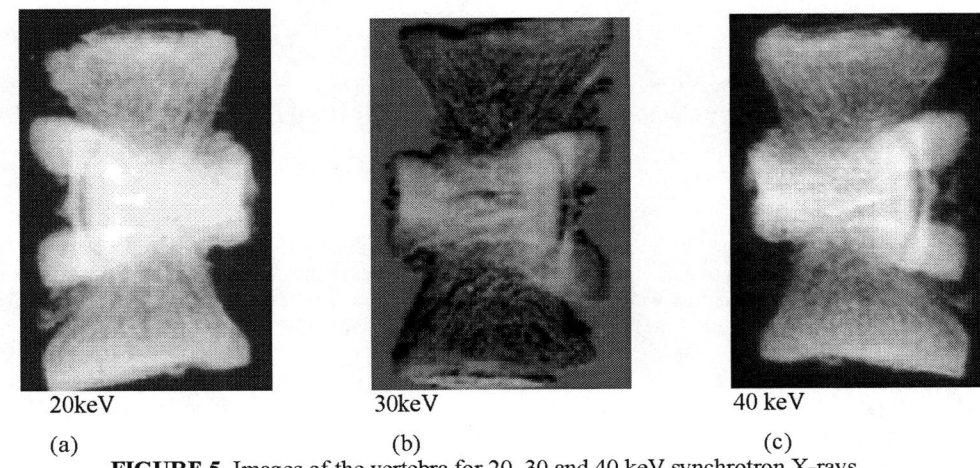

20keV   30keV   40 keV

(a)   (b)   (c)

FIGURE 5. Images of the vertebra for 20, 30 and 40 keV synchrotron X-rays

Figure 5 shows the images of vertebra obtained at 40 keV. The interior microstructure including the embedded spongiosa is sharper with high contrast. As regards to the rat's vertebra, we differentiated the central portion of the vertebra at different orientations for clear visualization of the external and internal features. Except in few orientations, the majority of the images are acceptably distinct at 40 keV and the system is more sensitive and reliable at this optimum energy. The image of the lumber vertebra is clearly visible with its associated microstructure in the middle, internal features and also the porosity.

ACKNOWLEDGEMENTS

One of the author's (DVR) undertook part of this work with a support from Department of Bio-System Engineering, Yamagata University, Yonezawa, and JSPS, Japan. In addition the potential author received per-diem from BNL and collaboration with Z. Zhong.

REFERENCES

1. Z. Zhong, W. Thomlinson, D. Chapman, and D. Sayers, *NIM A* 450, 556-567, 2000.
2. R .E. Johnston, D. Walburn, P. Pisano, C. Burns, W. C. Thomlinson, D.L. Chapman, F. Arfelli, N. Gmür, N., Z. Zhong, D. Sayers, *Radiology* 200, 659-663,1996.
3. D. Chapman, W. Thomlinson, R.E. Johnston, D. Washburn, E. Pisano, N. Gmür, Z. Zhong, R., Menk, F. Arfelli, D. Sayers, *Phys. Med. Biol.* 42, 2015-2025,1997.
4. D. Chapman, W. Thomlinson, Z. Zhong, R. E. Johnston, E. Pisano, D. Washburn, D. Sayers, C. Segre, *Synchrotron Radiat. News* 11 (2) 4, 1998.
5. D. Chapman, E. Pisano, W. Thomlinson, Z. Zhong, R. E. Johnston, D. Washburn, D. Sayers, *Breast Dis.* 10, 197-207, 1998a.
6. C. Muehleman, D. L. Chapman, K. E. Kuettner, J. Rieff, J. A. Mollenhauer, K. Massuda, and Z. Zhong, *The Anatomical Record part A* 272, 392-397, 2003.
7. O Oltulu, Z. Zhong, M. Hasnah, M. N. Wernick, and D. Chapman, *J. Phys. D: Appl. Phys.* 36, 2152-2156, 2003.
8. J. Li, Z. Zhong, R. Lidtke, K.E.Kuettner, C. Peterfy, E. Aliyeva, and C. Muehleman, *J. Ant.* 202, 463-470, 2003.
9. M. Z. Kiss, D. E. Sayers, Z. Zhong, C. Parham, and E. Pisani, *Phys. Med. Biol.* 49, 3427-3439, 2004.

Application of Synchrotron Radiation Imaging for Non-destructive Monitoring of Mouse Rheumatoid Arthritis Model

Chang-Hyuk Choi[1], Hong-Tae Kim[2], Jung-Yoon Choe[3], Jong Ki Kim[4], Hwa Shik Youn[5]

[1]Department of Orthopaedic Surgery, [2]Department of Anatomy, [3]Internal Medicine, [4]Radiology and Medical Engineering, School of Medicine, Catholic University of Deagu, 3056-6 Daemyung 4-Dong NamGu, Daegu, 705-718, Korea
[5]Pohang Acceleratory Laboratory, Pohang University of Science and Technology, Pohang, 790-784 Korea

Abstract. This study was performed to observe microstructures of the rheumatoid arthritis induced mouse feet using a synchrotron radiation beam and to compare findings with histological observations. X-ray refraction images from *ex-vivo* rheumatoid arthritis induced mouse feet were obtained with an 8KeV white (unmonochromatic) beam and 20 micron thick CsI(Tl) scintillation crystal. The visual image was magnified using a ×10 microscope objective and captured using digital CCD camera. Experiments were performed at 1B2 bending magnet beamline of the Pohang Accelerator Laboratory (PAL) in Korea. Obtained images were compared with histopathologic findings from same sample. Cartilage destruction and thickened joint capsule with joint space narrowing were clearly identified at each grade of rheumatoid model with spatial resolution of as much as 1.2 micron and these findings were directly correlated with histopathologic findings. The results suggest that x-ray microscopy study of the rheumatoid arthritis model using synchrotron radiation demonstrates the potential for clinically relevant microstructure of mouse feet without sectioning and fixation.

Keywords: Mouse, Rheumatoid arthritis, Synchrotron-microscopy, High-resolution refraction radiography
PACS: 07.85.Tt

INTRODUCTION

The mouse rheumatoid arthritis (RA) model is very useful for understanding of the pathogenesis of RA and to develop new drugs for RA. However, other imaging techniques could not directly show micro-structural changes in bone and joint without destruction of joint capsule. Radiographic examination is essential for identifying presence and progression of bone and joint pathologies including arthritis, osteoporosis and fracture. The resolution of conventional x-ray radiography limits the image in macroscopic level, although it requires comparatively little time. Microscopic examination is essential for the understanding of histopathologies of the disease. To examine bone and cartilage under a microscope, sectioning, fixation, and staining procedures are required, and require considerable processing time. However, using the third generation synchrotron radiation facility it is possible to realize a tremendous increase in radiation flux, and achieve high source coherence for a small source size. It is this source coherence that provides the superior contrast referred to as phase contrast. Synchrotron radiation (SR) is a powerful x-ray source, and a promising tool for future clinical micro-imaging. Furthermore, it allows micro-images of organs to be obtained in real time without preparation, and recent studies have achieved refraction images of mouse feet tissue with spatial resolutions of 1.2 microns. Here, we attempted to observe the rheumatoid arthritis induced mouse feet using a synchrotron radiation beam and compared findings with histological observations.

CP879, *Synchrotron Radiation Instrumentation: Ninth International Conference*,
edited by Jae-Young Choi and Seungyu Rah
© 2007 American Institute of Physics 978-0-7354-0373-4/07/$23.00

MATERIALS AND METHODS

DBA/1J mice (6 weeks of age; Jackson Laboratories, Bar Harbor, ME) are used for these studies. The animals are housed in a controlled environment and provided with standard rodent chow and water.

Bovine CII (Chondrex, Redmond, WA) is dissolved in $0.05M$ acetic acid at a concentration of 2 mg/ml by stirring overnight at 4°C. Dissolved CII is frozen at −70°C until used. Freund's complete adjuvant (CFA) is prepared by adding *Mycobacterium tuberculosis* H37Ra at a concentration of 4 mg/ml. Before injection, CII is emulsified with an equal volume of CFA. Collagen induced arthritis(CIA) is induced as follows. On day 1, mice are injected intradermally at the base of the tail with 100 μl of emulsion (containing 100 μg of CII). On day 21, a second injection of CII in CFA is administered.

Mice are evaluated daily for arthritis according to a macroscopic scoring system: 0 = no signs of arthritis, 1 = slight swelling and redness of the paw, or involvement of < 2 joints, 2= pronounced edema, involvement of > 2 joints, 3 = severe arthritis of the entire paw and digits with joint rigidity and ankylosis.

Experiments were performed at 1B2 bending magnet beamline of the Pohang Accelerator Laboratory (PAL) in Korea. Eight weeks after the immunization, an X-ray refraction image from forefeet of each grade was obtained with white (unmonochromatic) beam and CdWO4 scintillation crystal. Unit energy was set at 8 keV(= 1.5 Å), and the sample was positioned 25 meters away from the source, which was a bending magnet of the PLS. The x-ray shadow of the specimen was converted into a visual signal on the surface of a 20 micron thick CsI(Tl) scintillation crystal which was placed at distance of 5 cm from the specimen. The visual image was magnified using a ×10 microscope objective and captured using digital CCD camera. Obtained images were compared with histopathologic findings from same sample.

RESULTS

Normal joint microstructures including cartilage and subchondral bone and surrounding soft tissue structures were clearly identified with spatial resolution of as much as 1.2 micron and had good correlation with conventional light microscope findings (Fig. 1.). There were minimal changes on subchondral bone and chondrocyte with preserved joint space at lower grade. Cartilage destruction and thicken joint capsule with joint space narrowing were clearly identified at higher grades and these findings were directly correlated with histopathologic findings (Fig. 2.).

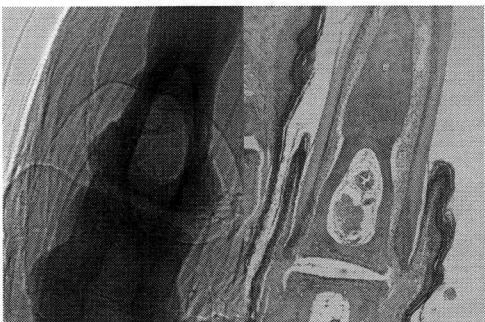

FIGURE 1. White beam X-ray image of synchrotron and corresponding light microscopic (LM) image show clear correlation in the bone and joint structures of forefoot of mouse in normal status

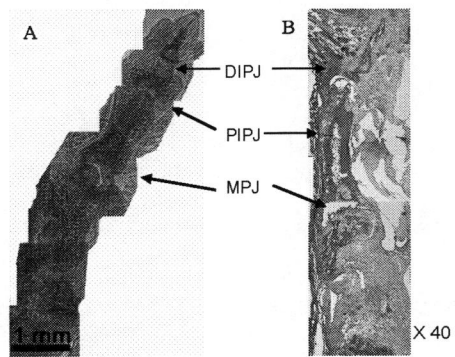

FIGURE 2. High grade RA induced mouse with SR image (A) and HE stained LM examination (B) show good relationship including cartilage, subchondral bone and joint space. The major soft tissue structures that also can be identified here, which are not visible in the conventional radiograph. DIP, distal interphalangeal joint. PIP, proximal interphalangeal joint. MP, metaphalangeal joint.

DISCUSSION

Joint induced problems such as rheumatoid arthritis, osteoarthritis, and osteoporsis have a pathology that starts in cartilage and surrounding soft tissues of the joint with various morphological alterations such as cartilage destruction, capsule and synovial thickening, joint effusion, edema of interstitial tissue, etc. Therefore it is clinically important to diagnose such pathologies with detailed microscopic imaging at an early stage of relevant structural modification. On the basis of correspondence between our SR images and histopathological findings, the present SR radiographic technique may open a new imaging modality for clinical application under the present circumstances where no other conventional radiological imaging or microscopic examinations are available for noninvasive microscopic imaging of intact tissue.

Furthermore, conventional radiological imaging such as x-ray radiography and CT are restricted to the visualization of structure and morphology, and do not provide functional information of the joint. With the introduction of novel CT technologies, our capability to evaluate image structures radiologically has exceeded general expectations and even surpasses macroscopic inspection levels. But even these advances, offering more precise visualization of normal bone and joint structure and associated pathological changes, are probably not sufficient for making therapeutic decisions. Clinicians tend to rely on clinical symptoms and pathologic changes associated with functional impairment, such as cartilage destruction and synovial proliferation, to initiate or modify treatments rather than depending on bone structure changes visible on CT scans.

Although we obtained static refraction images from resected tissue of mouse feet, we were able to produce real-time microscopic x-ray images of a rheumatoid arthritis induced model at each stage of disease progression. Thus SR imaging may also provides clinical information based on the series of such structural changes according to stage of the disease. Based on the quality of the images obtained from our *ex-vivo* model, we may expect to observe chondrocyte changes and synovial proliferation.

CONCLUSION

Although field of view and penetration depth are limited for further in vivo microscopic imaging, we are convinced that observation of clinically relevant microstructure of bone and cartilage tissue in rheumatoid arthritis model may provide potential experimental and clinical application in rheumatoid arthritis studies such as evaluation of disease progression and new drug development.

ACKNNOWLEDGEMENTS

This work was supported by the grant of Research Institute of Medical Science, Catholic University of Daegu(2005) and the Pohang Accelerator Laboratory (PAL)

REFERENCES

1. S. W. Wilkins, T. E. Gureyev, D.Gao, A.Pogany and A. W. Stevenson,. "Phase-contrast imaging using polychromatic hard X-rays". *Nature,* 1996, 384: pp. 335-338.
2. A. Snigirev, I. Snigireva, V. Kohn, S. Kuznetsov and, I. Schelokov, "X-ray phase contrast imaging with submicron resolution by using extremely asymmetrical Braqq diffractions". *Rev. Sci. Instrum,* 66, 1995, pp 54-86.
3. S. H. Jeon, H. T Kim, J. K. Kim, J. Y. Hung and H. S. Youn, "X-ray refraction imaging of the lung with histological correlation (in submission)".
4. R. Meuli, Y. Hwu, J. H. Je and G. Margaritondo, "Synchrotron radiation in radiology: radiology techniques based on synchrotron sources". *Eur Radiol,* 2004, 14: pp. 1550-1560.
5. T. Imagawa, S. Watanabe, S. Katakura, G. P. Boivin and R. Hirsch, "Gene transfer of a fibronectin peptide inhibits leukocyte recruitment and suppresses inflammation in mouse collagen-induced arthritis". *Arthritis Rheum,* 2002, 46(4): pp. 1102-8.
6. K. C. Sabino, F. A. Castro, J. C. Oliveira, S. R. Dalmau and M. G. Coelho, "Successful treatment of collagen-induced arthritis in mice with a hydroalcohol extract of seeds of Pterodon pubescens". *Phytother Res,* 1999,13(7): pp. 613-5.

7. I. Yamane, H. Hagino, T. Okano, M. Enokida, D. Yamasaki and R. Teshima, "Effect of minodronic acid (ONO-5920) on bone mineral density and arthritis in adult rats with collagen-induced arthritis". *Arthritis Rheum,* 2003, 48(6): pp. 1732-41.
8. D. M. Butler, A. M. Malfait, R. N. Maini, F. M. Brennan and M. Feldmann, "Anti-IL-12 and anti-TNF antibodies synergistically suppress the progression of murine collagen-induced arthritis". *Eur J Immunol,* 1999, 29(7): pp. 2205-12.
9. P. F. Sumariwalla, R. Gallily, S. Tchilibon, E. Fride, R. Mechoulam and M. Feldmann, "A novel synthetic, nonpsychoactive cannabinoid acid (HU-320) with antiinflammatory properties in murine collagen-induced arthritis". *Arthritis Rheum,* 2004, 50(3): pp. 985-98.

Validity of Fusion Imaging of Hamster Heart obtained by Fluorescent and Phase-Contrast X-Ray CT with Synchrotron Radiation

J. Wu[*], T. Takeda[*], Thet Thet Lwin[*], Q. Huo[*], N. Sunaguchi[**], T. Murakami[**], S. Mouri[**], S. Nasukawa[**], T. Fukami[**], T. Yuasa[**], K. Hyodo[***], H. Hontani[****], M. Minami[*], and T. Akatsuka[**]

[*]Graduate School of Comprehensive Human Sciences, University of Tsukuba, Tsukuba, Ibaraki 305-8575 Japan,
[**]Faculty of Engineering, Yamagata University, Yonezawa, Yamagata 992-8510 Japan,
[***]Institute of Material Science, High Energy Accelerator Research Organization, Tsukuba Ibaraki 305-0801, Japan,
[****]Department of Computer Science and Engineering, Nagoya Institute of Technology, Nagoya, Aichi 466-8555 Japan

Abstract. Fluorescent X-ray CT (FXCT) to depict functional information and phase-contrast X-ray CT (PCCT) to demonstrate morphological information are being developed to analyze the disease model of small animal. To understand the detailed pathological state, integration of both functional and morphological image is very useful. The feasibility of image fusion between FXCT and PCCT were examined by using ex-vivo hearts injected fatty acid metabolic agent (^{127}I-BMIPP) in normal and cardiomyopathic hamsters. Fusion images were reconstructed from each 3D image of FXCT and PCCT. ^{127}I-BMIPP distribution within the heart was clearly demonstrated by FXCT with 0.25 mm spatial resolution. The detailed morphological image was obtained by PCCT at about 0.03 mm spatial resolution. Using image integration technique, metabolic abnormality of fatty acid in cardiomyopathic myocardium was easily recognized corresponding to anatomical structures. Our study suggests that image fusion provides important biomedical information even in FXCT and PCCT imaging.

Keywords: Fusion imaging, Fluorescent X-ray CT, Phase-contrast X-ray CT, Functional imaging, Molecular imaging, Cardiomyopathy, Small animal
PACS: 01.30.Cc, 07.85.Qe, 07.85.Nc, 32.50.+d, 32.30.Rj, 42.62.Be, 87.63.Lk, 87.19.Xx, 87.59.Fm

INTRODUCTION

Micro-imaging technique for small laboratory animal model of human diseases is an important tool for study on the cause, diagnosis and treatment of diseases. To obtain the biomedical information precisely from the morphological and/or functional point of view, imaging techniques with high-contrast and high-spatial resolution are required. Fluorescent X-ray CT (FXCT) and phase-contrast X-ray CT (PCCT) by using synchrotron radiation, which can reveal functional information and morphological structures, enable to obtain the fine image with high-contrast and high-spatial resolution because of their high physical sensitivity compared to conventional absorption-contrast technique [1-8]. However, it is difficult to confirm the detailed localization of abnormal area by FXCT only. Thus, fusion of functional FXCT image and anatomical PCCT image is quite useful for well understanding of the biomedical significance. Recently, we integrated the FXCT and PCCT images of the mouse brain and confirmed the validity of the fusion image to recognize the abnormal cerebral perfusion [9]. The aim of this study is to discuss the feasibility of fusion image for the evaluation the physiological state of hamster heart.

CP879, Synchrotron Radiation Instrumentation: Ninth International Conference,
edited by Jae-Young Choi and Seungyu Rah
© 2007 American Institute of Physics 978-0-7354-0373-4/07/$23.00

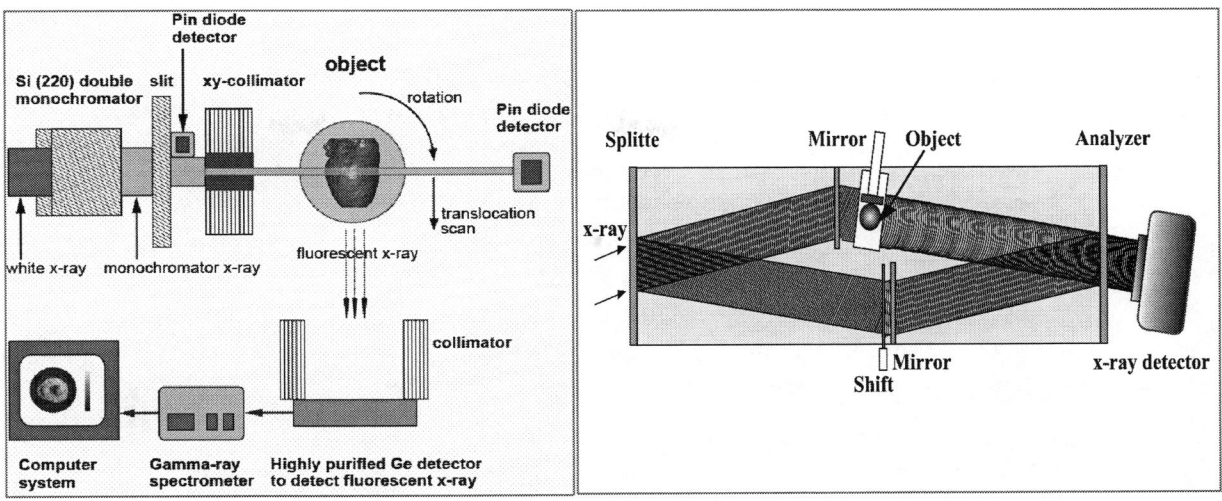

FIGURE 1. Schematic diagram of fluorescent (left) and phase-contrast (right) X-ray computed tomography system.

METHODS AND MATERIAL

Animal Preparation

Cardiomyopathic hamsters J2N-k and age matched normal J2N-n hamsters were used. Under anesthesia, the heart were extracted after 5-min intravenous injection of myocardial fatty acid metabolic agent, [127]I labeled 15-p-(iodophenyl)-3-methlpentadecanoic acid ([127]I-BMIPP). Each heart was fixed by formalin for imaging. The experiment was approved by the Medical Committee for the Use of Animals in Research of the University of Tsukuba.

Fluorescent X-ray CT Imaging

FXCT system (Left half of Fig. 1) consists of a silicon (220) double crystal monochromator, an X-ray slit system, a scanning table, a fluorescent X-ray detector, and two pin-diode for collecting incident and transmission X-ray data [5]. Fluorescent X-rays were detected by a high purity germanium (HPGe, IGRET, EG&G Ortec Ltd.) detector set placed perpendicular to the incident monochromatic X-ray beam. Monochromatic X-ray energy was set at 37keV. X-ray beam was collimated to 0.25×0.5 mm^2 pencil beam, and object was scanned in 0.25-mm translation step over the object region and 3° rotation step over 180 degrees. Data acquisition time for each scanning step was 7 sec. The experiment was performed at the bending-magnet beam line BL-NE5A of Photon Factory, High Energy Accelerator Research Organization in Tsukuba, Japan.

Phase-contrast X-ray CT Imaging

Phase-contrast X-ray CT system (Right half of Fig. 1) consists of an asymmetrically cut silicon crystal, a monolithic X-ray interferometer, a phase shifter, an object cell and an X-ray CCD camera [5, 8]. Fringe-scan technique was used and the X-ray energy was set at 35 keV by the monochromator. The field of view was 24×30 mm^2. The specimens were rotated inside the 20 mm thickness sample cells filled with formalin. Each phase map was calculated from 3 interference fringes by changing the phase of the reference beam, and each interference fringe was imaged at 5 s exposure and the number of projections was 250 over 180 degrees. The experiment was performed at a vertical wiggler beam line BL14C1 of the Photon Factory in Tsukuba, Japan.

Fusion Image Representation

Three-dimensional (3D) images were reconstructed from the trans-axial image of FXCT and PCCT respectively. We determined heart surface contour from both three-dimensional images (3D-images) after resembling for describe with the same special resolution. Two images were fused interactively on displayed images by using 3D image manipulation software (Real INTAGE). Here, one is represented by gray scale image and the other was represented by pseudo-color for easy to manipulate them under operator's decision.

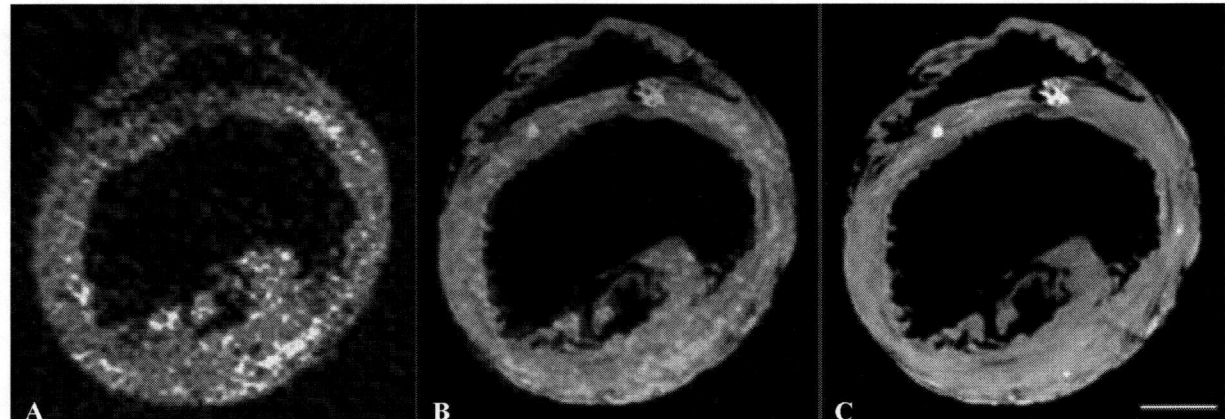

FIGURE 2. Images of a cardiomyopathic hamster obtained by fluorescent X-ray CT (FXCT) using [127]I-BMIPP myocardial fatty acid metabolic agent (A), phase-contrast X-ray CT (PCCT) (C) and fused image of FXCT and PCCT (B). Patched low BMIPP uptake was observed clearly in left ventricular wall. The defect like area was demonstrated in septum which corresponded thrombus area on phase-contrast X-ray CT image. Scale Bar = 2 mm.

RESULTS AND DISCUSSION

The extracted hearts of hamsters labeled [127]I-BMIPP were imaged by FXCT and PCCT to obtain the functional and morphological images, respectively (Fig. 2 A and C).

[127]I-BMIPP distribution within the left ventricular wall was clearly demonstrated by FXCT with 0.25 mm spatial resolution. Patched low uptake that corresponds to metabolic abnormality of the fatty acid, was observed in myocardium of J2N-k cardiomyopathic hamster (Fig. 2A), whereas normal hamster showed homogeneous distribution. These findings were the same as previously reported study with radioactive ^{123}I-BMIPP [10]. However, detailed myocardial localization was difficult to recognize in FXCT image because the FXCT image demonstrates only metabolic condition in myocardium. On the other hand, PCCT with 0.03 mm spatial resolution, clearly depicted the morphological structures such as cardiac lumen, papillary muscle, and left and right ventricular wall (Fig. 2C). Then, we integrated the FXCT image and PCCT image for hamster heart. In fusion image, fatty acid metabolic abnormality in cardiomyopathic myocardium was easily recognized corresponding to detailed anatomical structures (Fig. 2B). This procedure is very useful to understand the biochemical state in myocardium.

Our results suggested that the image fusion was a quite significant technique to evaluate the exact and quantitative myocardial function of the small animal both in FXCT and PCCT with high spatial resolution.

ACKNOWLEDGMENTS

We thank Akio Yoneyama PhD, Xiaowei Chang PhD and Yoshinori Tsuchiya PhD for technical supports, Mr. Kouzou Kobayashi for his preparation of experimental apparatus, Nihon Medi-Physics Co., Ltd. Japan for supplying ^{127}I-BMIPP, and Yukiko Kawata for help in preparing this article. This research was partially supported by Grant-In-Aid for Scientific Research (#17591244, #17390326, #17659362), Special Coordination Funds and Grant-in-Aid for Scientific Research in Priority Areas (#15070201) from the Ministry of Education, Culture, Sports, Science and Technology of the Japanese Government, and was performed under the auspices of High Energy Accelerator Research Organization (2005G308, 2005S2-001).

REFERENCES

1. T. Takeda, M. Akiba, T. Yuasa, M Kazam. A. Hoshino. Y. Watanabe. K. Hyodo, F.A Dilmanian, T. Akatsuka and Y Itai, "SPIE-The international Society for Optical Engineering Press" in *Physics of Medical Imaging 1996*. SPIE Conference Proceedings 2708, Society of Photo-Optical Instrumentation Engineers, Bellingham, WA, 1996, pp. 685-695.
2. T. Takeda, S. Matsushita, J. Wu, Q. Yu, Thet-Thet-Lwin, T. Zenia, T. Yuasa, K Hyodo, F.A. Dilmanian, T. Akatsuka and Y. Itai, "*World Publishing Corporation International Academic Publishers Press*" in Image Visualization and reconstruction edited by X.X. Zheng, et al., IEEE-EMBS Asia-Pacific Conf. on Biomedical Engineering Proceedings 1, 2000, pp 276-277.
3. T. Takeda, A. Momose, Q. Yu, T. Yuasa, F. A. Dilmanian, T. Akatsuka, Y. Itai, *Cellular & Molecular Biology* **46**, 1077-1088 (2000).
4. T. Takeda, J. Wu, A. Yoneyama, Y. Tsuchiya, Thet-Thet-Lwin, Y. Hirai, T. Kuroe, T. Yuasa, K. Hyodo, F. A. Dilmanian, T. Akatsuka, "SPIE-The international Society for Optical Engineering Press", in *Development in X-ray tomography IV*, edited by U. Bonse, SPIE Conference Proceedings 5535, Society of Photo-Optical Instrumentation Engineers, Bellingham, WA, 2004, pp. 380-391.
5. T. Takeda, *Nucl. Instrum. Meth.* **A548**, 38-46 (2005)
6. Thet-Thet-Lwin, T. Takeda, J. Wu, Sunaguchi N, Tsuchiya Y, Yuasa T, Dilmanian FA, Minami M, Akatsuka T, 2005, APCMBE Conference Proceedings, Tsukuba, 2005, PA-3-34, pp. 1-4,
7. T. Takeda, A. Momose, Y. Itai, *Acad Radiol* **2**, 799-803 (1995).
8. A. Momose, T. Takeda, Y. Itai, K. Hirano, *Nature Med.* **2**, 473-475 (1996).
9. J. Wu, T. Takeda, Thet Thet Lwin, N. Sunaguchi, T. Yuasa, F. Tadanori, H. Hontani, T. Akatsuka, *Med. Imag. Tech.* **23**, 312-317 (2005)
10. Thet-Thet-Lwin, T. Takeda, J. Wu. Y. Fumikura. K. Iida. S.Kawano, I. Yamaguchi, and Y. Itai, *Eur J Nucl Med Mol Imaging* **30**, 966-973 (2003)

On Detailed Contrast of Biomedical Object in X-ray Dark-Field Imaging

Daisuke Shimao[1], Hiroshi Sugiyama[2,3], Toshiyuki Kunisada[4], Kazuyuki Hyodo[2], Koichi Mori[1], Masami Ando[2,3,5]

[1]Department of Radiological Sciences, Ibaraki Prefectural University of Health Sciences, Ami4669-2, Inashiki, Ibaraki, 300-0394, Japan
[2]Photon Factory, institute of Materials Structure Science, High Energy Accelerator Research Organization, Oho 1-1, Tsukuba, Ibaraki 305-0801, Japan
[3]Department of Photo-Science, School of Advanced Studies, Graduate University for Advanced Studies, International Village, Hayama, Kanagawa 240-0193, Japan
[4]Department of Orthopaedic Surgery, Okayama University Graduate School of Medicine and Dentistry, Shikata-cho 2-5-1 Okayama 700-8558, Japan
[5]Research Institute for Science and technology, Tokyo University of Science, Yamasaki 2641, Noda, Chiba 278-8510, Japan

Abstract. Over the past 10 years, refraction-based X-ray imaging has been studied together with a perspective view to clinical application. X-ray Dark-Field Imaging that utilizes a Laue geometry analyzer has recently been proposed and has the proven ability to depict articular cartilage in an intact human finger. In the current study, we researched detailed image contrast using X-ray Dark-Field Imaging by observing the edge contrast of an acrylic rod as a simple case, and found differences in image contrast between the right and left edges of the rod. This effect could cause undesirable contrast in the thin articular cartilage on the head of the phalanx. To avoid overlapping with this contrast at the articular cartilage, which would lead to a wrong diagnosis, we suggest that a joint surface on which articular cartilage is located should be aligned in the same sense as the scattering vector of the Laue case analyzer crystal. Defects of articular cartilage were successfully detected under this condition. When utilized under appropriate imaging conditions, X-ray Dark-Field Imaging will be a powerful tool for the diagnosis of arthropathy, as minute changes in articular cartilage may be early-stage features of this disease.

Keywords: X-ray Dark-Field Imaging, refraction contrast, object direction, articular cartilage
PACS: 87.57.-s, 87.59.-e, 87.64.Bx

INTRODUCTION

X-rays are highly valued as a non-destructive tool for the observation of internal structure in a variety of scientific, technical, and medical fields. Especially in clinical medicine, absorption-based X-ray projection images and X-ray computed tomography are easily interpreted and are invaluable in medical diagnosis. Refraction-based X-ray imaging has been studied over the past 10 years with a perspective view to medical application [1-8]. X-ray Dark-Field Imaging (XDFI) utilizing a Laue geometry analyzer has recently been proposed [9, 10]. In this novel method, which is achieved by tuning incident X-ray energy to the thickness of an analyzer crystal, the intensity of the forward-diffracted plane wave X-rays can be suppressed to almost zero at the Bragg condition: non-refracted X-rays at an object no longer pass through in the forward diffraction direction. Higher image contrast due to refraction at an object can be expected by the XDFI; near-elimination of the background enables fine contrast to be detected. A two-dimensional field of view can be obtained with Laue geometry using a single shot because the plane wave condition required of the XDFI simultaneously produces a relatively expanded beam size incident to the object.

XDFI has been applied to imaging human articular cartilage, which has been successfully visualized even in intact objects [7, 8]. However, parasitic and unwanted fringes attached to main contrast at joint surfaces have been

CP879, *Synchrotron Radiation Instrumentation: Ninth International Conference*,
edited by Jae-Young Choi and Seungyu Rah
© 2007 American Institute of Physics 978-0-7354-0373-4/07/$23.00

observed. The relation between these fringes and the main contrast of the articular cartilage may change depending on the direction of the beam to the object. To avoid clinical misinterpretation and obtain a correct diagnosis, the precise contrast mechanism under XDFI must be revealed. In this study, the relationship between the angle of the beam incident to the object and its additional contrast is investigated for the first time.

MATERIALS AND METHODS

The experiment was performed at beamline BL14B using the vertical polarization synchrotron radiation from the 5 Tesla superconducting wave shifter of the Photon Factory in the High Energy Accelerator Research Organization, which is in operation with 2.5 GeV/ 450 mA. We used incident X-ray energy of 36.0 keV. The experimental setup is shown in Fig. 1. A Bragg case Si (440) asymmetric cut monochromator-collimator (MC) expands the horizontal size and simultaneously improves the angular divergence of the outgoing beam owing to the asymmetric factor b [11]. The factor b was calculated to be 0.02 under the conditions of this setup. A Laue case Si (440) analyzer (A[L]) with a thickness of 1.2 mm was utilized for detecting the refracted X-rays at objects. An acrylic rod with a diameter of 10 mm was chosen as a simple case to observe the contrast of its cylindrical surface, and a naked proximal phalanx head with two defect regions of articular cartilage was also chosen to assess depiction ability; this evaluation is valuable for future clinical use. This specimen had been amputated from a cadaver. Two pictures for each incident angle of the beam were acquired by changing the direction of the beam incident to the head of proximal phalanx. Both objects were soaked in water to suppress surface effects and prevent unwanted additional contrast, as significant refraction caused by large difference in refractive indices between air and object surfaces would cause serious deterioration in the image. Table 1 summarizes the densities ρ [12] and unit decrements of refractive indices δ under X-ray energy of 36.0 keV for water, acrylic resin, articular cartilage, and cortical bone. All images were stored on mammography film (Kodak Min-R 2000) without an intensifying screen in order to acquire images with high spatial resolution. The exposure time was 2 min at a ring current of approximately 400 mA.

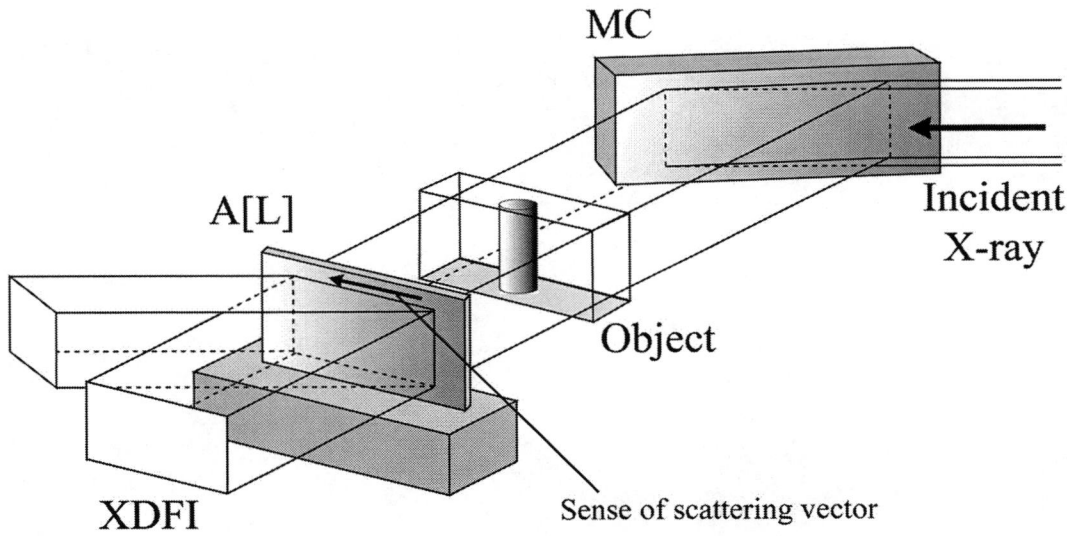

FIGURE 1. Experimental setup for XDFI constructed at BL14B of the Photon Factory.

TABLE 1. ρ and δ of materials under X-ray energy of 36.0 keV.

Material	ρ [kg/m³]	δ
Water	1000	1.60×10^{-7}
Acrylic resin	1190	1.90×10^{-7}
Articular cartilage	1100	1.76×10^{-7}
Cortical bone	1920	3.07×10^{-7}

RESULTS AND DISCUSSION

Figure 2 shows images of the acrylic rod, including a photograph of the rod (Fig. 2 (a)), an absorption contrast X-ray image (Fig. 2 (b)), and XDFI (Fig. 2 (c)) of the area outlined by the rectangle in Fig. 2 (a). All of the X-ray images appear as seen from downstream and the white area corresponds to the region of strongest X-ray intensity. The contours of the rod are depicted by XDFI in Fig. 2 (c); however, there is a complete lack of contrast in Fig. 2 (b). This result indicates that XDFI can detect the deference in δ to the degree of 3.0×10^{-8} in the case of a cylindrical surface. There are, however, differences in image contrast between the right and left sides of the acrylic rod. The right side of the rod shows a complicated contour with superimposed fringes that have a width of approximately 400 μm, while the contour on the left side is simple and clear. We consider that this effect may be caused by Pendellösung fringes that occur within the Borrmann fan in relation to the sense of the scattering vector. Figure 3 shows X-ray images of a naked proximal phalanx head with two defect regions of articular cartilage; the phalanx had been amputated from a cadaver. The absorption contrast image and XDFI obtained under the condition where the joint surface was turned to the same sense as the scattering vector are shown in Figs. 3 (a) and (b), respectively. XDFI can detect a difference in δ of 1.6×10^{-8} and depict articular cartilage and its defect areas, while absorption contrast shows only bone structure. Figures 3 (c) and (d) show the absorption contrast image and XDFI, respectively, acquired following rotation of the object by 180° from the position in which Figs. 3 (a) and (b) were obtained. The same phenomenon that was revealed in Fig. 2 (c) is observed at the articular cartilage. If the target joint surface is turned to the opposite sense, against the scatter vector, the Pendellösung fringes that occur at the boundary between water and articular cartilage are overlapped onto the articular cartilage itself. This is a serious artifact when visualizing articular cartilage in a finger with a thickness of approximately 500 μm. The setting direction of the object must be taken into account when XDFI is applied to the imaging of thin articular cartilage.

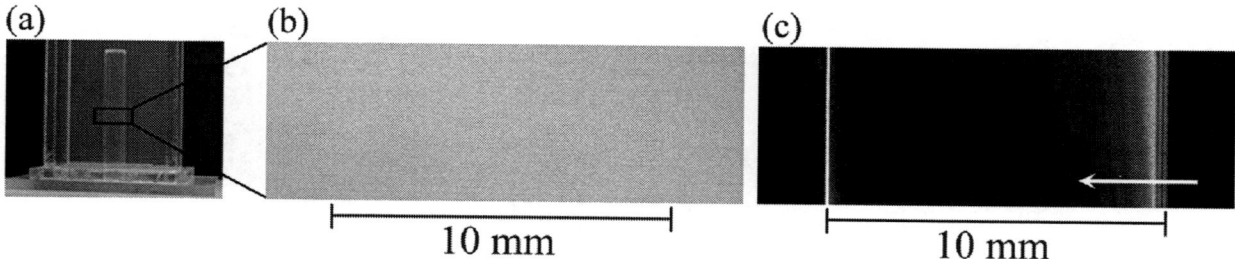

FIGURE 2. (a): photograph, (b): absorption contrast image, and (c): XDFI of acrylic rod whose diameter is 10 mm. The arrow in (c) indicates the sense of the scattering vector.

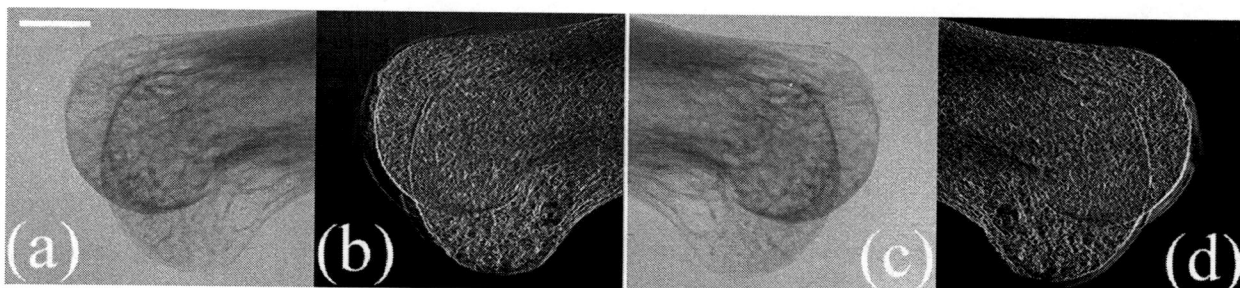

FIGURE 3. (a): absorption contrast image and (b): XDFI of the head of proximal phalanx which was set as turned to the same sense as the scattering vector. (c): absorption contrast image and (d): XDFI acquired following rotation of the object by 180° from the position in which (a) and (b) were obtained. Scale bar shown in (a) is 3 mm

CONCLUSIONS

We researched variations in the depiction ability for articular cartilage by XDFI using synchrotron X-rays in terms of the setting direction of the object. Undesirable contrast that appeared at the right side of an acrylic rod may result from Pendellösung fringes that occur within the Borrmann fan. The possibility exists that this contrast could overlap with articular cartilage, particularly where the cartilage is thin as on the phalanx, and thereby lead to misinterpretation. This situation can be avoided by considering the setting direction of the object. When thin articular cartilage on the head of a bone is to be observed, the joint surface must be directed in the same sense as the scattering vector.

ACKNOWLEDGEMENTS

This experiment was performed with approval from KEK-PF-PAC (2006G212). We acknowledge the support of a Grant-in-Aid for Scientific Research (Grant-in-Aid for Young Scientists) (#18790900) from the Japanese Ministry of Education, Culture, Sports, Science, and Technology.

REFERENCES

1. K. Mori, K. Hyodo, N. Shikano and M. Ando, *Jpn. J. Appl. Phys.* **38**, L1339-L1341 (1999).
2. F. Arfelli, V. Bonvicini, A. Bravin, G. Cantatore, E. Castelli, L. D.Palma, M. Di Michiel, M. Fabrizioli, R. Longo, R. H.Menk, A. Olivo, S. Pani, D. Pontoni, P. Poropat, M. Prest, A. Rashevsky, M. Ratti, L. Rigon, G. Tromba, A. Vacchi, E. Vallazza and F. Zanconati, *Radiology* **215**, 286-293 (2000).
3. J. Mollenhauer, M. E. Aurich, Z. Zhong, C. Muehleman, A. A. Cole, M. Hasnah, O. Oltulu, K. E.Kuettner, A. Margulis, L. D.Chapman, *Osteoarthr. Cartilage* **10**, 163-171 (2002).
4. K. Mori, N. Sekine, H. Sato, N. Shikano, D. Shimao, H. Shiwaku, K. Hyodo and K. Ohashi, *Jpn. J. Appl. Phys.* **41**, 5490-5491 (2002).
5. R. A. Lewis, C. J. Hall, A. P. Hufton, S. Evans, R. H. Menk, F. Arfelli, L. Rigon, G. Tromba, D. R. Dance, I. O. Ellis, A. Evans, E. Jacobs, S. E. Pinder and K. D. Rogers, *Brit. J. Radiol.* **76**, 301-308 (2003).
6. M. Ando, H. Sugiyama, T. Kunisada, D. Shimao, K. Takeda, H. Hashizume and H. Inoue, *Jpn. J. Appl. phys.* **43**, L1175-L1177 (2004).
7. D. Shimao, H. Sugiyama, K. Hyodo, T. Kunisada and M. Ando, *Nucl. Instrum. Meth. A* **548**, 129-134 (2005).
8. D. Shimao, H. Sugiyama, T. Kunisada and M. Ando, *Appl. Radiat. Isotopes* **64**, 868-874 (2006).
9. M. Ando, H. Sugiyama, A. Maksimenko, W. Pattanasiriwisawa, K. Hyodo and X. Zhang, *Jpn. J. Appl. Phys.* **40**, L844-L846 (2001).
10. M. Ando, K. Hyodo, H. Sugiyama, A. Maksimenko, W. Pattanasiriwisawa, K. Mori, J. Roberson, E. Rubenstein, Y. Tanaka, J. Chen, D. Xian and X. Zhang, *Jpn. J. Appl. Phys.* **41**, 4742-4749 (2002).
11. K. Kohra, *J. Phys. Soc. Jpn.* **17**, 589-590 (1962).
12. ICRU report 44. International Commission on Radiation Units and Measurement, Maryland, 1989, pp. 22.

Development of Cell Staining Technique for X-Ray Microscopy

P. Y. Tseng [a], Y. T. Shih [a], C. J. Liu [a], T. Hsu [a], C. C. Chien [a], W. H. Leng [a], K. S. Liang [b], G. C. Yin [b], F. R. Chen [b], J. H. Je [c], G. Margaritondo [d], Y. Hwu [a, b, e]

[a] Institute of Physics, Academia Sinica, Nankang, Taipei 11529, Taiwan
[b] National Synchrotron Radiation Research Center, Hsinchu 30076, Taiwan
[c] Dept. Mater. Sci. Eng., Pohang University of Science and Technology, Pohang, Korea
[d] Ecole Polytechnique Fédérale, CH-1015 Lausanne, Switzerland
[e] Institute of Optoelectronic Sciences, National Taiwan Ocean University, Keelong, Taiwan

Abstract. We report a technique for detection of sub-cellular organelles and proteins with hard x-ray microscopy. Several metals were used for enhancing contrast for x-ray microscopy. Osmium tetroxide provides an excellent stain for lipid and can delineate cell membrane. Uranyl acetate has high affinity for nucleotide and can stain nucleus. Immunolocalization of specific proteins and sub-cellular organelles was achieved by 3'3 diaminobenzidine (DAB) with nickel enhancement and nanogold-conjugated secondary antibody with silver enhancement. The x-rays emitted from synchrotron source was monochromatized by double crystal monochromator, the photon energy was fixed at 8 keV to optimize the focusing efficiency of the zone plates. The estimated resolution is about 60 nm. When compared with visible light and conventional confocal microscopy, the X-ray microscopy provides a superior resolution to both conventional optical microscopes.

Keywords: cell staining, protein localization, hard x-ray, x-ray microscopy.
PACS: 87.59.-e

INTRODUCTION

The X-ray microscopy currently provides high resolution imaging of biological specimens [1-3]. The advantage of X-ray microscopy is the high penetration which allows observation of thicker specimens. Soft X-ray microscopy at a range of photon energy between 283 and 543 eV, known as water window, enable examination of hydrated cells (up to 10 μm) without further staining [4-5]. However, the penetration depth of soft X-ray microscopy limits its application to whole cell and tissue. Although the resolution of hard X-ray microscopy at energy between 8 and 11 KeV (about 60 nm) is lower than that of soft X-ray microscopy, the hard X-ray microscopy can examine thicker specimens more than 200μm [6]. However, the natural contrast of biological specimens is poor under hard X-ray microscopy. Thus, suitable staining with heavy metals is required for enhancing contrast. To study protein and cell functions, specifically labeling proteins or cellular organelles is required for selectively tracing and marking targets. We use immunolabelling techniques and metal enhancement to label specific proteins and enhance absorption contrast under hard X-ray microscopy [7-8]. This technique can compensate for the deficit of TEM and soft X-ray microscopy and can provide a unique tool for studying biological science.

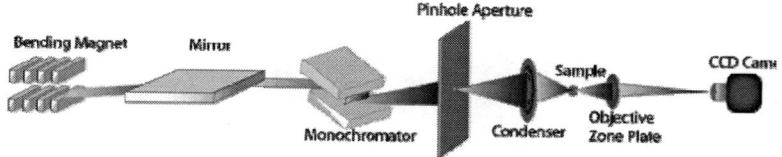

FIGURE 1. Schematic layout of hard X-ray Microscopy (TXM) at National Synchrotron Radiation Research Center (NSRRC).

CP879, *Synchrotron Radiation Instrumentation: Ninth International Conference*,
edited by Jae-Young Choi and Seungyu Rah

RAW CELLS STAINED BY OSMIUM TETROXIDE AND URANYL ACETATE

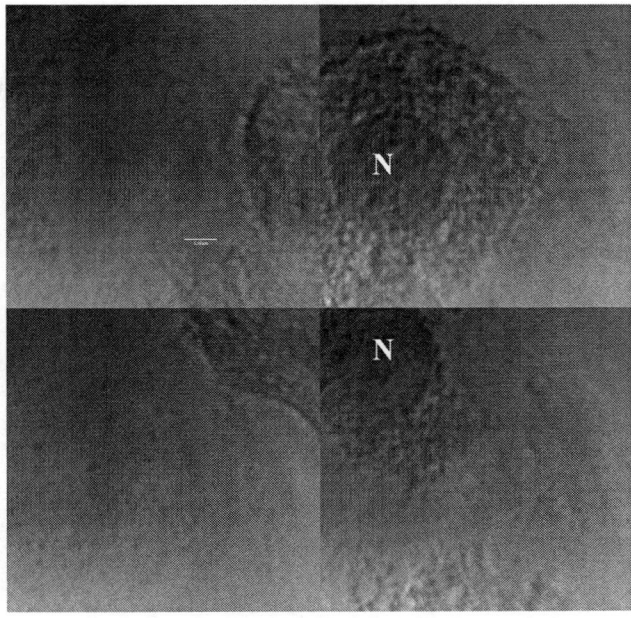

FIGURE 2. Patched image of RAW cells stained by uranyl acetate and osmium tetroxide. Uranyl acetate which has high affinity to nucleotide can stain cell nuclei. Osmium tetroxide which binds lipid can stain cell membrane and delineate cell margin. Bar = 1.56 μm, Field of View (FOV) = 30 μm.

VIMENTIN BUNDLES OF HELA CELLS WERE STAINED BY DAB WITH NICKEL ENHANCEMENT

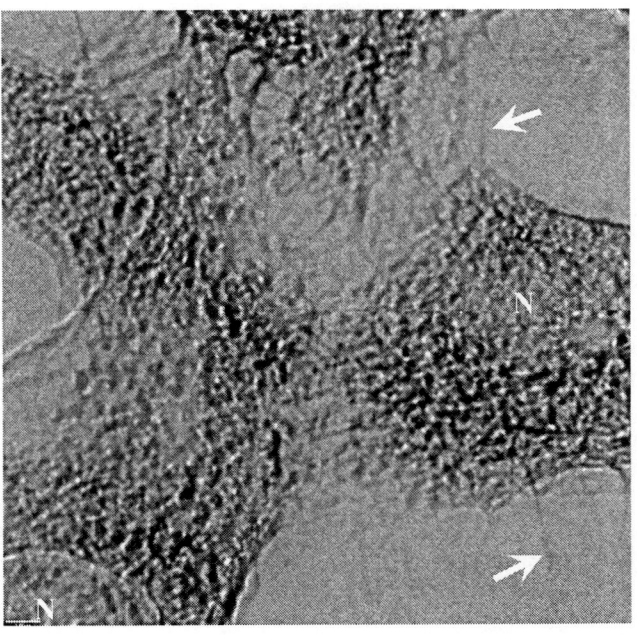

FIGURE 3. Immunolocalization of intermediate filaments vimentin of HeLa cells. Vimentin filaments were labeled by anti-vimentin antibody and stained by DAB with nickel enhancement. Arrows indicated individual vimentin bundles which diameter is less than 100 nm. N denotes nuclei. Bar = 1.56 μm, FOV = 30 μm.

MITOCHONDRIA OF HELA CELLS WERE STAINED BY NANOGOLD WITH SILVER ENHANCEMENT

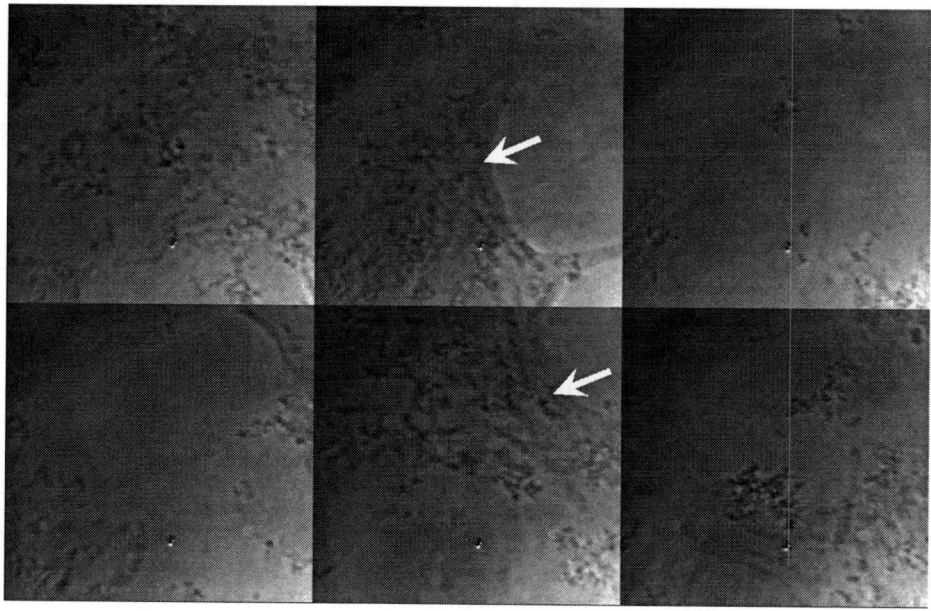

FIGURE 4. Mitochondria inside HeLa cells were immunolabeled by anti-ATPsynthase antibody and stained by nanogold-conjugated secondary antibody with silver enhancement. The dark speckles (indicated by arrows) were mitochondria. The nucleus could not be identified because mitochondria did not delineate the nucleus. Field of view of each frame is 15 μm.

TABLE 1. Brief list of properties for TEM, soft x-ray microscopy and hard x-ray microscopy

Properties	Resolution	Penetration
TEM	1 nm	1 μm
Soft X-ray microscopy	30 nm	10 μm
Hard X-ray microscopy	60 nm	200 μm

ACKNOWLEDGEMENTS

This work was supported by the Taiwan National Science Council, the Academia Sinica, Taiwan Nanoscience and Nanotechnology National Programs, the BK21 Project, the Korea Institute of Science and Technology Evaluation and Planning through the National Research Laboratory and SKORE-A projects, the Fonds National Suisse de la Recherche Scientifique, the Ecole Polytechnique Fe´de´rale de Lausanne. We thank Y. F. Song of NSRRC and W. Yun of Xradia for helps in the nano-TXM studies

REFERENCES

1. G. Margaritondo, *Elements of Synchrotron Light for Biology, Chemistry, and Medical Research,* Oxford University Press, New York, 2002.
2. Y. Hwu, W. L. Tsai, et al., *Biophys. J.* **87**, 4180-4187 (2004).
3. N. H. Chapman, J. Fu, C. Jacobsen and S. Williams, *J. Micros. Soc. Am.* **2**, 53-62 (1996).
4. W. Meyer-Ilse, C.A. Larabell, et al., *J. Microsc.* **201**, 395-403 (2001).

5. C.A. Larabell CA and M.A. Le Gros, *Mol. Biol. Cell* **15**, 957-962 (2004).
6. G. C. Yin, and K. S. Liang, et al., *Proceedings of the 8th International Conference on X-ray Microscopy,* (2005).
7. L. Scopsi, L. I. Larsson, L. Bastholm and M. H. Nielsen, *Histochemistry* **86**, 35-41 (1986).
8. M. A. Green, L. Sviland, A. J. Malcolm and A. D. Pearson, *J. Clin. Pathol.* **42** 875-80 (1989).

Phase Contrast Imaging of Biological Materials using LEBRA-PXR

T. Kuwada, Y. Hayakawa, K. Nogami, T. Sakai, T. Tanaka,
K. Hayakawa and I. Sato

Laboratory for Electron Beam Research and Application (LEBRA), Nihon University,
Narashinodai 7-24-1,Funabashi 274-8501, Japan

Abstract. Phase contrast x-ray imaging is an important technique for investigation of materials consisted of light atoms, such as soft biological tissues. The tunable monochromatic x-ray source based on Parametric X-ray Radiation (PXR), which was developed at Laboratory for Electron Beam Research and Application (LEBRA) in Nihon University, provides x-rays with a high spatial coherence which is an essential property required for phase contrast imaging. In preliminary experiment, refraction contrast images for leaf tissues of a tree and animal specimen have been obtained successfully with the LEBRA-PXR x-rays. In the imaging system, the x-ray that passed through the sample once reflects off the silicon perfect-crystal x-ray analyzer at the Bragg angle, and then enters the imaging plate. The bright-field and the dark-field phase contrast images have been obtained by infitesimal rotations of the analyzer, showing the evidence of contrast reversal. Although the conventional radiograph by absorption contrast was also taken with the LEBRA-PXR, significant differences are found between the radiograph and the phase contrast images.

Keywords: Parametric X-ray Radiation, phase contrast imaging
PACS: 87.64.Rr, 87.59.-e, 87.59.Bh

INTRODUCTION

The x-ray imaging, which provides information on internal structure without other additional destructive methods, is an important diagnostic technique in medicine, biology and material science. In particular, phase contrast imaging is utilized for observing light materials which are difficult to visualize with absorption imaging techniques. The contrast in the conventional x-ray image results only from the difference of x-ray absorption depending on the density, the thickness or the composition in the object. However, the phase contrast imaging technique can provide the additional contrast in the transmitted x-ray beam due to the phase difference resulted from inhomogeneous refractive index in the object. Therefore, phase contrast imaging offers higher contrast and spatial resolution for light materials such as biological tissues [1], though implementation of the new imaging technique requires the spatially coherent x-rays.

In the Laboratory for Electron Beam Research and Application (LEBRA) at Nihon University, a point-like x-ray source has been developed on the basis of Parametric X-ray Radiation (PXR) from a high energy electron beam [2]. The LEBRA-PXR generator using a double-crystal system provides the monochromatic x-ray beam ranging from 6 to 20 keV. The x-ray beam of LEBRA-PXR has a wide exposure field, which is restricted by the x-ray extraction window with diameter of 98mm, and has rather uniform dose in the exposure field. These characteristics of the LEBRA-PXR suggest that the PXR beam is suitable for imaging applications. In fact, the x-ray absorption images have been obtained for several materials using the LEBRA-PXR light source. Furthermore, the contrast enhancement effect has been observed at the image for the edge of an acrylic plate, which suggests that the LEBRA-PXR has a good coherency applicable for phase contrast imaging [3].

In preliminary experiment for phase contrast imaging with LEBRA-PXR, refraction contrast imaging has been tested for the x-rays which transmitted through an object and being analyzed by a perfect crystal; only the x-rays satisfying the Bragg's law for the diffraction can inject into the detector and contribute to the image formation [4]. Using the imaging technique, the refraction contrast images for a few materials have been obtained successfully.

CP879, *Synchrotron Radiation Instrumentation: Ninth International Conference,*
edited by Jae-Young Choi and Seungyu Rah
© 2007 American Institute of Physics 978-0-7354-0373-4/07/$23.00

This paper discusses on the results of the refraction contrast images for biological materials obtained with the LEBRA-PXR.

EXPERIMENTAL SETUP

The experimental setup for phase contrast imaging using the LEBRA-PXR is shown in Fig. 1, where the setup was designed in a similar manner with the refraction contrast technique using an analyzer crystal as reported by Bravin [4]. The x-ray beam is extracted through the extraction window with an inner diameter of 98 mm. Horizontal size of the beam is restricted to 30 mm by a slit placed in front the sample. The sample is directly irradiated by the x-ray beam without using a monochromator, since the LEBRA-PXR is horizontally energy resolved monochromatic x-ray source. The x-ray passed through the sample is reflected by the analyzer crystal on the goniometer when the incident angle of the x-ray satisfies the Bragg condition. Then the x-ray is detected with the imaging plate or with the ionization chamber, which depends on the experiment; imaging of the sample or measurement of the rocking curve of the analyzer crystal. The conventional radiographs are also obtained by removing the analyzer crystal and placing the imaging plate instead. The spatial resolution of the imaging plates (CrossField Co., Ltd.) is 50 μm. The x-ray images in the experiment were taken at the exposure time of 30 min for the x-rays with the energy of 14 keV.

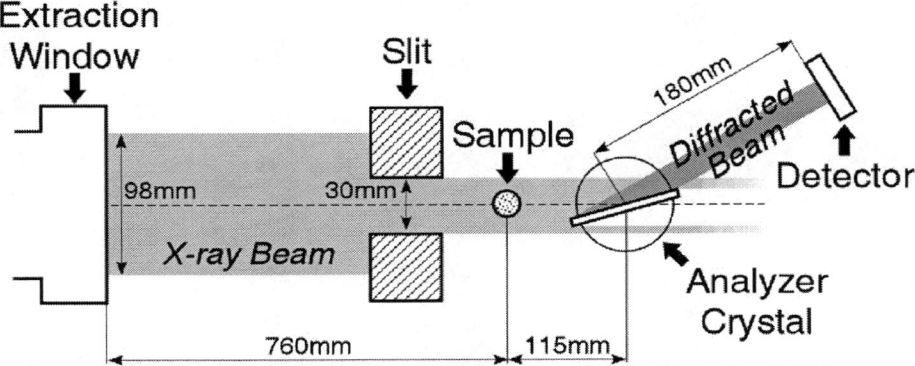

FIGURE 1. Experimental setup for refraction contrast imaging with the LEBRA-PXR. The PXR beam passes through the sample, then being diffracted by a Si (111) analyzer crystal on the goniometer. The horizontal beam size is restricted to 30 mm by the slit of plastic blocks. The diffracted beam is detected by the imaging plate or the ionization chamber.

RESULTS AND DISCUSSION

In this experiment, X-ray imaging was tested for a few biological materials by using the LEBRA-PXR. Figure 2 shows the results of imaging for the dry specimen of a web spider, *Araneus ventricosus*. Figure 2-(a) is the conventional absorption image. Figs 2-(b) to 2-(e) are the refraction contrast images, each of which were taken at the angle of the analyzer crystal specified on the rocking curve in Fig. 2-(f). Difference in the appearance of the density heterogeneities is observed among these refraction contrast images, which is obviously due to slight difference of the diffraction angle. Especially the significant result is that the contrast reversal between the bright-field image (Fig. 2-(b)) and the dark-field image (Fig. 2-(e)) is clearly observed. The contrast reversal is also obvious from the density distribution curves in Fig. 3-(f). These results suggest a high spatial coherency of the LEBRA-PXR and demonstrate the usefulness of the x-ray optics dedicated to refraction contrast imaging with the LEBRA-PXR. In this experimental setup for 14 keV x-ray beam, a relatively large imaging field of 10 x 25 mm has been obtained by detecting the diffracted beam from the whole area of the analyzer crystal with a diameter of 70 mm. However, a long exposure time is required due to low duty factor of the pulsed x-ray source which is generated by the pulsed RF electron linac with the duration of 10 - 20 μs and the repetition rate of 2 Hz.

Although absorption contrast imaging can also visualize the specimen as shown in Fig. 2-(a), refraction contrast imaging can enhance the contrast and the sharpness of the image as seen in Figs. 2-(b) to 2-(e), which has an advantage in the study on the details of the specimen structure. Figure 3 shows the magnified views of the rectangle regions indicated in Fig. 2, together with the density distribution curves along the horizontal lines indicated in the images. Comparing with the absorption contrast image in Fig. 3-(a), details of specimen structures are observed in the refraction contrast images, Fig. 3-(b) to 3-(e). For example, the distinction between the body wall (layered cuticle) and the internal structure (muscle etc.) is clearly visualized in the refraction contrast images. As shown in Fig 3-(f), the density distributions in the refraction contrast images also exhibit the presence of the body walls separately from the muscles, although the difference is not clear in the density distribution for the absorption contrast image. These results prove that refraction contrast imaging based on the LEBRA-PXR is available for visualization of light materials, which are difficult to be visualized by absorption contrast imaging, with a good spatial resolution by detecting the contrast arising not only from pure absorption but from refraction in the bright and the dark fields.

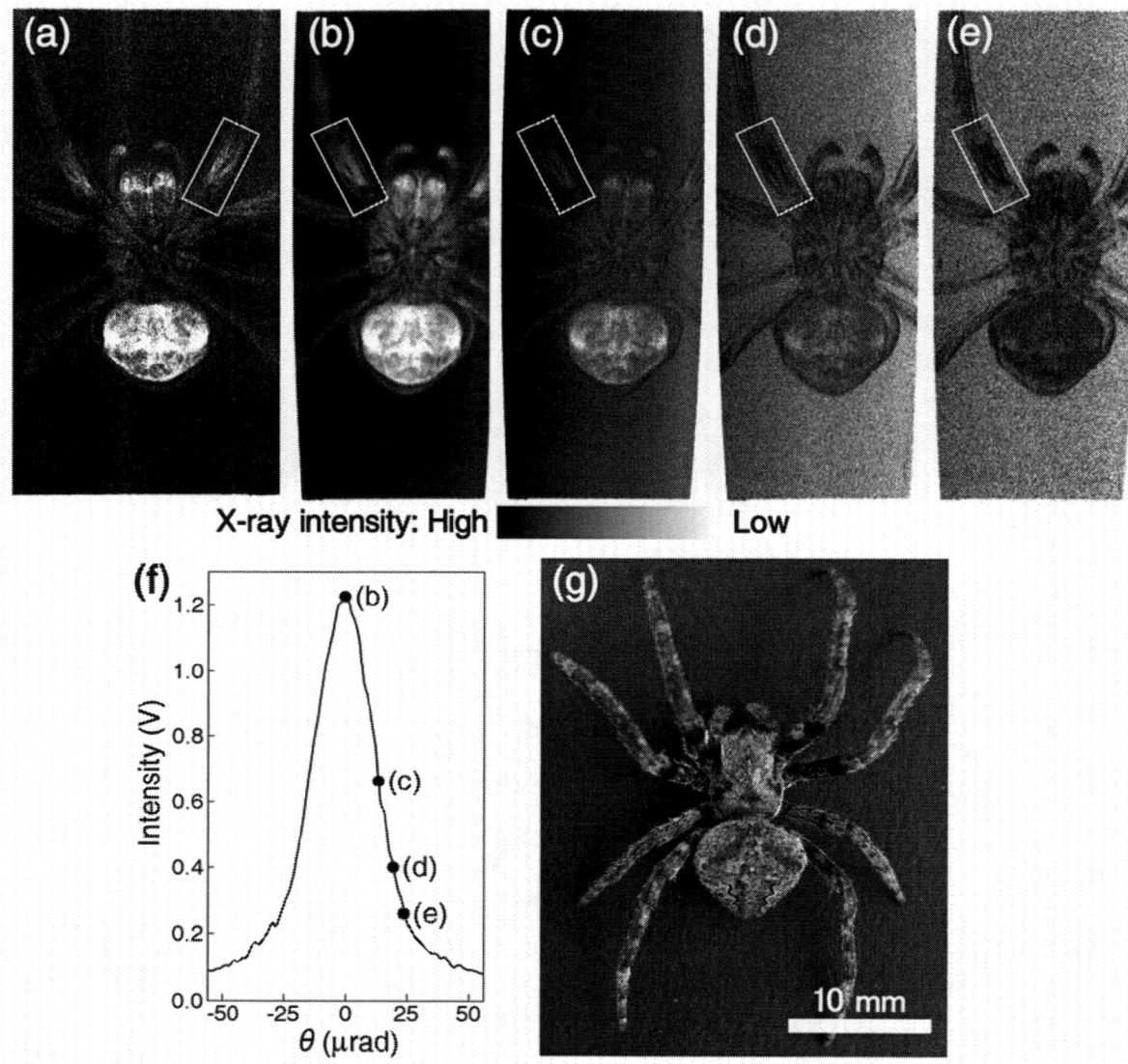

FIGURE 2. Difference of the x-ray images depending on the imaging techniques for the biological sample. (a): the conventional radiograph. (b)-(e): the refraction contrast images taken at the analyzer angles specified on the rocking curve in (f). Relative x-ray intensity (with sample) are 100%, 50%, 25 %, and 12.5% for images (b), (c), (d) and (e), respectively. Magnified view of the leg marked by the rectangle in (a)-(e) is shown in Fig. 3. (g): the dry specimen of a web spider, *Araneus ventricosus* prepared for the imaging. The body length of the spider is about 12 mm.

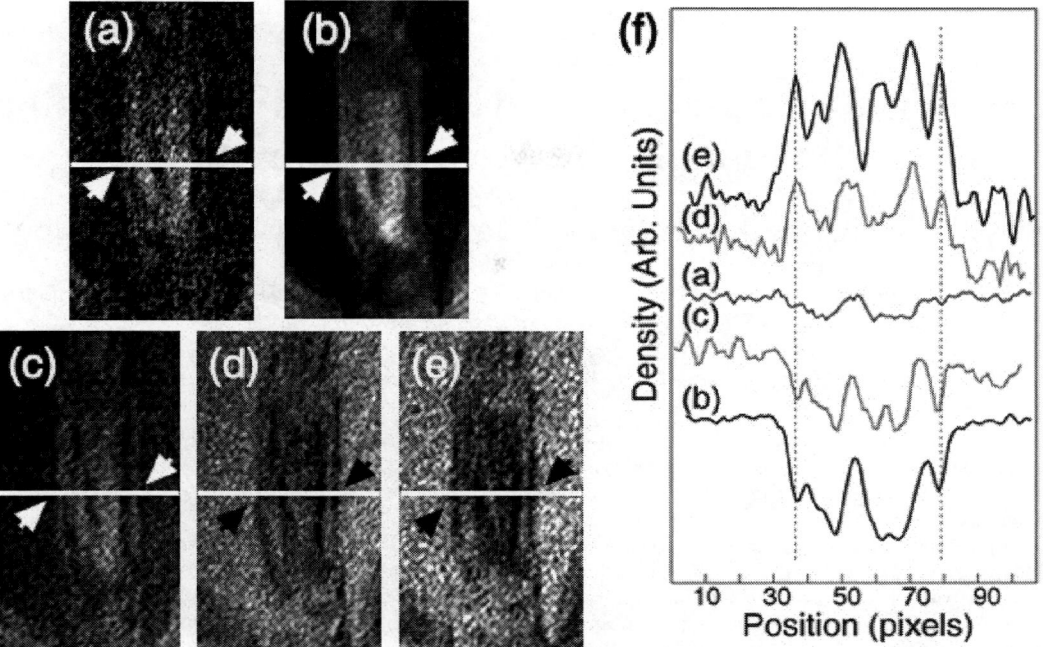

FIGURE 3. Magnified views of the images shown in Fig. 2. (a): the conventional radiograph which is mirror-reversed for comparison. (b-e): the refraction contrast images corresponding to Fig. 2-(b) to 2-(e), respectively. (f): density distribution curves along the white lines indicated in (a) to (e), which were obtained by using the R-AXIS display software (Rigaku Co.). The arrows in (a)-(e) show the body walls of the spider's leg, which is observed clearly in the refraction contrast images but not in the conventional radiograph. The body wall positions in the density curves are indicated by broken lines in (f). The structures seen between the body walls in the images are the muscles in the leg.

CONCLUSION

The refraction contrast images were obtained by using the LEBRA-PXR, which proves that the LEBRA-PXR is the spatially coherent x-ray source. These images have shown higher contrast and sharpness than the absorption contrast image, which can provide information on the detailed structure of biological samples. Successful results on phase contrast imaging in the mid-size electron linac facility may provide for the possibility of therapeutic and commercial applications of the imaging technique.

ACKNOWLEDGMENTS

The authors are grateful for rich information, experimental advices and technical supports from Dr. Y. Takahashi, Dr. A. Mori and Dr. T. Sakae, working at Nihon University. This work was supported by "Academic Frontier" Project for Private Universities: matching fund subsidy from MEXT (Ministry of Education, Culture, Sports, Science and Technology), Japan, 2000-2004 and 2005-2007.

REFERENCES

1. R. Fitzgerald, Physics Today 53, pp. 23-26 (2000).
2. Y, Hayakawa, I. Sato, K. Sato, K. Hayakawa, T. Tanaka, H. Nakazawa, K. Yokoyama, K. Kanno and T. Sakai, Proc. of 12th Symposium on Accelerator Science and Technology, Wako, Japan, 1999, p.391.
3. Y. Hayakawa, I. Sato, K. Hayakawa, T. Tanaka, A. Mori, T. Kuwada, T. Sakai, K. Nogami, K. Nakao and T. Sakae, "Status of the Parametric X-Ray Generator at LEBRA, Nihon University" (to be publlished)
4. A. Bravin, J. Phys. D 36, A24-29 (2003).

Applications of the Generalized X-ray Diffraction Enhanced Imaging in the Medical Imaging

Anton Maksimenko*, Eiko Hashimoto†, Masami Ando* and Hiroshi Sugiyama*

*Photon Factory, Institute of Materials Structure Science, High Energy Accelerator Research Organization (KEK), 1-1 Oho, Tsukuba, Ibaraki 305-0801, Japan
†Department of Photo-Science, School of Advanced Studies, Graduate University for Advanced Studies (GUAS), Shonan International Village, Hayama, Miura, Kanagawa 240-0193, Japan

Abstract. The X-ray Diffraction Enhanced Imaging (DEI) is the analyzer-based X-ray imaging technique which allows extraction of the "pure refraction" and "apparent absorption" contrasts from two images taken on the opposite sides of the rocking curve of the analyzing crystal. The refraction contrast obtained by this method shows many advantages over conventional absorption contrast. It was successfully applied in medicine, technique and other fields of science. However, information provided by the method is rather qualitative than quantitative. This happens because either side of the rocking curve of the analyzer is approximated as a straight line what limits the ranges of applicability and introduces additional error. One can easily overcome this problem considering the rocking curve as is instead of it's Taylor's expansion. This report is dedicated to the application of this idea in medical imaging and especially computed tomography based on the refraction contrast. The results obtained via both methods are presented and compared.

Keywords: Computed Tomography, X-ray imaging, Refraction contrast
PACS: 87.59.-e 07.85.Qe 07.05.Pj 87.58.Mj

INTRODUCTION

When an X-ray penetrates through an object it deflects on it's boundaries and inner nonhomogenieties. It also absorbs and scatters inside the object so that the outcoming beam holds information on the object structure in terms of intensity distribution and the refraction angle distribution. It is easy to measure the intensity by putting a detector right after the object but the measurement of the deflections are comparatively complicated procedure since the refraction coefficient for the hard X-ray is very close to unity: $\tilde{n} \equiv 1 - n \sim 10^{-5..7}$ for most substances. It leads to the very small refraction angles of the same order. The experimental systems which can visualize so small deflections usually consist of analyzing crystals or Fresnel zone plates placed between the object and detector (see Refs. [1-9] for the refraction contrast). The contrast obtained under these conditions is a mixture of absorption and refraction contrasts. In 1997th D. Chapman et al. presented Diffraction Enhanced Imaging (DEI) method[2] which can separate the absorption and refraction components of the contrast. Although the images obtained via the DEI are informative, the quantitative analysis is suspicious since the accuracy of the method is very limited while recent applications of the refraction contrast require data with higher accuracy (for example refraction-based Computed Tomography (CT) [10-14] In our recent work[15] we presented and tested extension to the DEI method. We proposed consideration of the analyzer's RC as is instead of it's Taylor approximation used in the original DEI. This new extended DEI (EDEI) method was shown to significantly improve the accuracy of the results, make the sensitivity limits wider at the same experimental conditions as well as provide additional flexibility in the experimental setup. Here we want illustrate the importance of our extension in medical imaging and refraction-based CT.

CP879, *Synchrotron Radiation Instrumentation: Ninth International Conference*,
edited by Jae-Young Choi and Seungyu Rah
© 2007 American Institute of Physics 978-0-7354-0373-4/07/$23.00

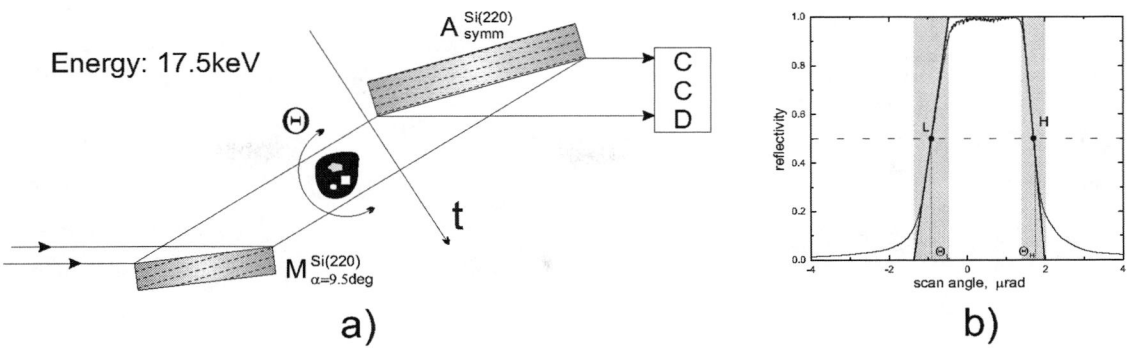

FIGURE 1.   a)Schematics of the experiment. The collimating crystal M is Si(220) with the asymmetry angle $\alpha = 9.5^{o}$, analyzer A is Si(220) with $\alpha = -9.5^{o}$. The sizes in the figure are not in scale. Angles corresponds to the photon energy $17.5keV$. b) RC of the analyzer. Full line represents experimentally measured RC, while dashed line is the two-term Taylor approximation of the RC used in the DEI method.

THEORY

Similarly to the original DEI method[2], the experimental system (see Fig.1a) consists of collimating and analyzing crystals, object placed between them and a detector after the analyzer. The images are taken on each side of the analyzers RC in points L and H like it is depicted in Fig.1b. As it was discussed in Ref.[2], the intensity diffracted by the analyzer set at a relative angle Θ_L (low-angle side) and Θ_H (high-angle side) from the Bragg angle Θ_B can be expressed as follows:

$$I_{L,H} = I_R R(\Theta_B + \Theta_{L,H} + \Delta\alpha), \tag{1}$$

where $\Delta\alpha$ is the refraction angle of the X-ray passed through the object, I_R is the apparent absorption intensity, $R(\Theta)$ is the analyzer RC shown in Fig.1b. Then the DEI utilizes two-term Taylor expansion of the RC: $R(\Theta_B + \Theta_{L,H} + \Delta\alpha) \approx R(\Theta_B + \Theta_{L,H}) + \Delta\alpha \cdot (dR/d\Theta)(\Theta_B + \Theta_{L,H})$. This converts the equations (1) to the linear system. Then the solution of this system for $\Delta\alpha$ and I_R can easily written in the explicit form (see Eqs.(6a) and (6b) in Ref.[2]). However, the Taylor approximation of the RC may lead to certain limitations of the applicability of the method and introduce unwanted error. It is illustrated in Fig.1b where the RC of the analyzer is presented together with it's Taylor approximation. In order to use the RC as is instead of it's linear approximation we can exclude member I_R from the Eqs.(1) as follows:

$$\frac{I_L - I_H}{I_L + I_H} = \frac{R_L(\Delta\alpha) - R_H(\Delta\alpha)}{R_L(\Delta\alpha) + R_H(\Delta\alpha)}, \tag{2}$$

where we used denotations $R_{L,H}(\Delta\alpha) = R(\Theta_B + \Theta_{L,H} + \Delta\alpha)$. We chose this combination of equations (1) to limit possible values of the function in the right part of eq.(2) (let us denote it as $F(\Delta\alpha)$) in the ranges [-1;1]. As it was discussed in details in Ref.[15], this function is monotonously growing in the region between minimum and maximum values what allows us, assuming that the refraction angle doesn't exceed the limits of the region, construct unambiguously determined function for calculation of $\Delta\alpha$ from input intensities:

$$\Delta\alpha = F^{-1}\left(\frac{I_L - I_H}{I_L + I_H}\right). \tag{3}$$

Once we calculated the refraction angle $\Delta\alpha$, it can be used to write down the expression for the absorption intensity I_R. It can be written in various forms. In our computations we used this one:

$$I_R = \frac{I_L + I_H}{R_H(\Delta\alpha) + R_L(\Delta\alpha)}. \tag{4}$$

EXPERIMENT AND RESULTS

The above mentioned theory was used in experiment. The x-ray beam reflected from an asymmetrical monochromator M, pass through the object and, after reflection from the analyzer A, forms contrast which is acquired by CCD-camera. Photon energy used in experiment was $17.5\ keV$. Both monochromator and analyzer used Si(2 2 0) diffraction

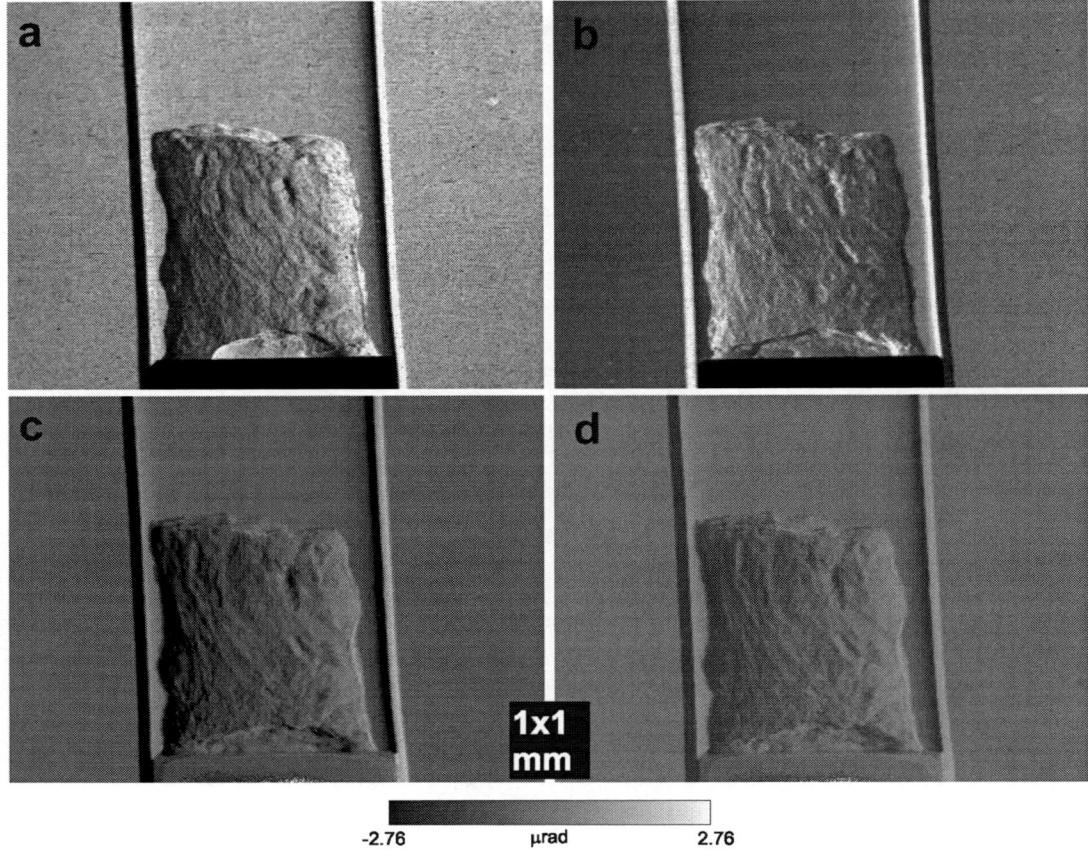

-2.76   μrad   2.76

FIGURE 2. a) and b) originally acquired images of the sample taken on the low-side and high-side of the RC respectively. c) and d) are refraction contrasts extracted via EDEI and DEI methods respectively. The gray scale in images c) and d) varies from $-2.76\mu rad$ (black) to $2.76\mu rad$ (white).

type. The monochromator was asymmetrically cut with 9.5 *deg*. At these conditions Bragg angle $\Theta_B = 10.6\ deg$. The CCD-camera used had view area of $10.0(w) \times 7.5(h)\ mm$ with 1384×1032 pixels. The choice of the symmetrically cut analyzer is not mandatory and was done reasoning from the size of the object, view area of the CCD-camera and width of the rocking curve. All experiments were performed at the vertical wiggler beamline BL14B, at the Photon Factory, KEK, Japan. The sample under investigation was the real breast cancer part of semi-cylindrical shape of approximately 3x3mm in size. Figures 2a,b are examples of the original pictures in L and H positions of the analyzer respectively. After the background correction these images serve as the input to the EDEI algorithms. The outputs are the apparent absorption contrast as well as the refraction contrast. In Ref.[15] we showed that the difference between absorption contrasts extracted via either methods is respectively small (under 10%). Therefore we will skip the description of the absorption contrast and concentrate on the refraction contrast. The refraction contrasts obtained via DEI and EDEI methods are shown in Fig.2c and d respectively. One can note that the the result of EDEI shows much more sharp and higher peaks in comparison to the DEI. It is explained by the lower sensitivity of the DEI which comes from the linear approximation of the RC (see [15] for more details). The set of images similar to depicted in Fig.2a,b taken at different rotation angles Θ are processed in accordance with the theory described in section . The result of the reconstruction process are the slices in the horizontal plane vertical to the plane of originally acquired images. The slices represents the 2D map of the refraction coefficient of the object. The example of the CT reconstruction performed using the method described in Ref.[13] can be found id Fig.3 representing the contrasts obtained via EDEI (a) and DEI (b) approaches. The pictures shows that in case of EDEI the inner structure of the sample is clearly seen. Brighter inclusions which are visible in the slice (they are magnified in the upper left corner of the image) corresponds to the calcification which are found in the cancer cell. It can be understood memorizing that the calcium is much more refractive material than the soft tissue of the body. One can also easily distinguish the milk tube structure and invasive

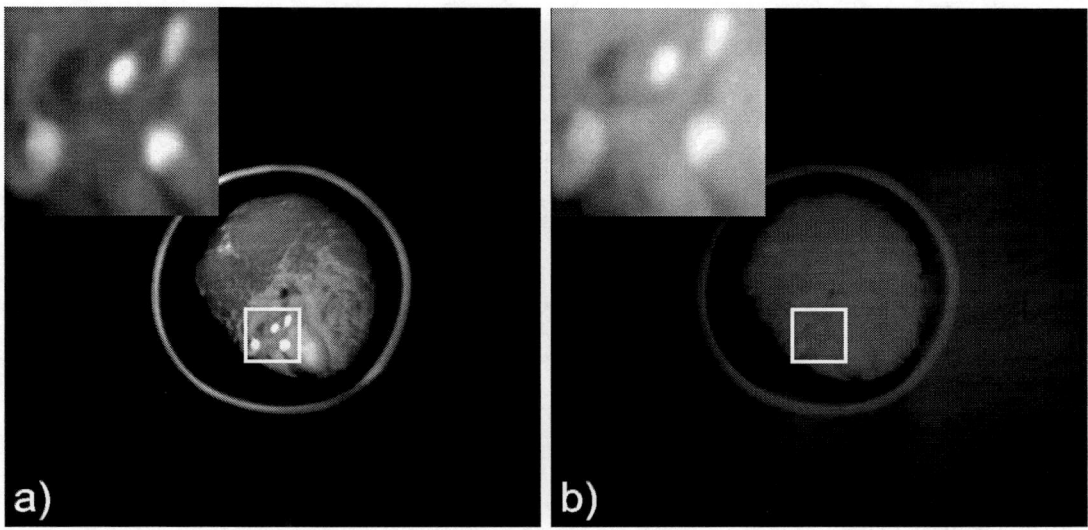

FIGURE 3. The reconstructed slice of the sample. The CT reconstruction was performed from the refraction contrast calculated via a) EDEI and b)DEI methods. The magnified fragments of the images are depicted in the upper left corners. Please note that the images themselves are shown in the same scale of gray while magnified fragments utilizes their own scale enhancements.

ductal carcinoma around it. Contrastingly to the EDEI, DEI-based reconstruction does not allow to see the inner structure so well. Although some elements are visible, the cannot be recognized the same clearly. Also the calcification inside the cancer cell are not well distinguished in the surrounding media. This is also the consequence of the linear approximation in the original DEI since it cuts out the higher refraction angles.

Finalizing all written above we can state that the EDEI method has great advantages over the original DEI. This is well seen both in the refraction-contrasts extracted from two images via either methods as well as the CT based on them shows significian improvement of the results. Therefore we expect that the EDEI method should be preferable over the original DEI in practical applications.

REFERENCES

1. V. Ingal and E. Beliaevskaya, J. Phys. D, **28**, 2314 (1995).
2. D. Chapman, W. Thomlinson, R. E. Johnston, D. Washburn, E. Pisano, N. Gmür, Z. Zhong, R. Menk, F. Arfelli, D. Sayers, Phys. Med. Biol. **42** 2015 (1997).
3. P. Cloetens, W. Ludwig, and J. Baruchel, Appl. Phys. Lett. **75**, 2912 (1999)
4. M. Ando, K. Hyodo, H. Sugiyama, A. Maksimenko, W. Pattanasiriwisawa, K. Mori, J. Roberson, E. Rubenstein, Y. Tanaka, J. Chen, D. Xian and Z. Xiaowei, Jpn. J. Appl. Phys. **41**, 4742 (2002).
5. K. Hirano, A. Maksimenko, H. Sugiyama and M. Ando, Jpn. J. Appl. Phys. **41**, L595 (2002).
6. A. Bravin, J. Phys. D: Appl. Phys. **36**, A24 (2003).
7. M. N. Wernick, O. Wirjadi, D. Chapman, Z. Zhong, N. P. Galatsanos, Y. Yang, J. G. Brankov, O. Oltulu, M. A. Anastasio and C. Muehleman, Phys. Med. **48**, 3875 (2003).
8. M. Wernick, J. Brankov, D. Chapman, Y. Yang, C. Muehleman, Z. Zong and M. A. Anastasio, Proc. of the SPIE, Developments in X-Ray Tomography IV **5535**, 369, (2004).
9. M. Ando, K. Yamasaki, F. Toyofuku, H. Sugiyama, C. Ohbayashi, G. Li, L. Pan, X. Jiang, W. Pattanasiriwisawa, D. Shimao, E. Hashimoto, T. Kimura, M. Tsuneyoshi, E. Ueno, K. Tokumori, A. Maksimenko, Y. Higashida, M. Hirano, Jpn. J. Appl. Phys. **44** L528 (2005).
10. F. A. Dilmanian, Z. Zhong, B. Ren, X. Y. Wu, L. D. Chapman, I. Orion and W. C. Thomlinson, Phys. Med. Biol. **45**, 933 (2000).
11. K. M. Pavlov, C. M. Kewish, J. R. Davis and M. J. Morgan, J. Phys. D: Appl. Phys. **34**, A168 (2001). Biol. **48** 3875 (2003).
12. I. Koyama, Y. Hamaishi and A. Momose: *AIP Conf. Proc.* **705**, 1283, (2004).
13. A. Maksimenko, M. Ando, H. Sugiyama and T. Yuasa, Appl. Phys. Lett. **86**, 124105 (2005).
14. J. G. Brankov, M. N. Wernick, Y. Yang, J. Li, C. Muehleman, Z. Zhong and M. A. Anastasio, Med. Phys. **33** (2006) 278.
15. A. Maksimenko, "Nonlinear Extension of the X-ray Diffraction Enhanced Imaging", (under review).

Hard X-ray Microscopic Imaging Of Human Breast Tissues

Sung H. Park*, Hong T. Kim*, Jong K. Kim*, Sang H. Jheon† , Hwa S. Youn¶

*College of Medicine, Catholic University of Daegu, 3056-6 Daemyung-Dong, Nam-Gu, Daegu 705-718, Korea

† College of Medicine, Seoul National University, 27 Youngun-Dong, Jongro-Gu, Seoul 110-799, Korea

¶Pohang Accelerator Laboratory, Pohang University of Science and Technology,
31 San, Hyoja-dong, Pohang, KyungBuk 790-784, Korea

Abstract. X-ray microscopy with synchrotron radiation will be a useful tool for innovation of x-ray imaging in clinical and laboratory settings. It helps us observe detailed internal structure of material samples non-invasively in air. And, it also has the potential to solve some tough problems of conventional breast imaging if it could evaluate various conditions of breast tissue effectively. A new hard x-ray microscope with a spatial resolution better than 100 nm was installed at Pohang Light Source, a third generation synchrotron radiation facility in Pohang, Korea. The x-ray energy was set at 6.95 keV, and the x-ray beam was monochromatized by W/B4C monochromator. Condenser and objective zone plates were used as x-ray lenses. Zernike phase plate next to condenser zone plate was introduced for improved contrast imaging. The image of a sample was magnified 30 times by objective zone plate and 20 times by microscope objective, respectively. After additional 10 times digital magnification, the total magnifying power was up to 6000 times in the end. Phase contrast synchrotron images of 10-μm-thick female breast tissue of the normal, fibroadenoma, fibrocystic change and carcinoma cases were obtained. By phase contrast imaging, hard x-rays enable us to observe many structures of breast tissue without sample preparations such as staining or fixation.

Keywords: x-ray microscopy, synchrotron radiation, phase contrast imaging, breast tissue and cancer.
PACS: 87.59. –e, 87.62. +n, 87.64.Bx

INTRODUCTION

Breast cancer is the 2nd leading cause of cancer death of women. In 2005, WHO estimates 1.2 million new breast cancer cases worldwide. More than 10,000 Korean women were newly diagnosed as breast cancer in 2005. Radiological imaging of breast tissues is very important for clinical and research purposes. Screening mammography plays essential role in early detection of breast cancer even though it has some limitations. Mammography with synchrotron radiation is expected to improve these limitations in clinical setting in the future. And, high-resolution synchrotron imaging of breast tissues can reveal unknown evidence of various breast conditions including breast cancer. A new x-ray microscope was installed on 1B2 beamline of Pohang Light Source, a 3rd generation synchrotron radiation facility with operating energy of 2.5 GeV at Pohang, Korea. The spatial resolution of this microscope in Zernike phase contrast mode is better than 100 nm. We got some synchrotron radiation microscopic images of tissue of normal breast, fibroadenoma, fibrocystic change and breast cancer.

EXPERIMENT

The Si (111) channel-cut crystal monochromator is about 15 m away from the source. The vertical size of the incoming white beam is 3 mm. Unit energy was set at 6.95 keV. The condenser zone plate located at 22 m from the source demagnifies the source by a factor of 8.84 at its image point and floods the sample with light. The sample was positioned 25 m away from the source. The x-ray image of a sample is magnified 30 times by the objective zone plate and converted into a visual image on the CsI (TI) scintillation crystal. This visual image is further enlarged 10 to 20 times with a microscope objective lens and captured by a full frame CCD camera.

CP879, *Synchrotron Radiation Instrumentation: Ninth International Conference,*
edited by Jae-Young Choi and Seungyu Rah
© 2007 American Institute of Physics 978-0-7354-0373-4/07/$23.00

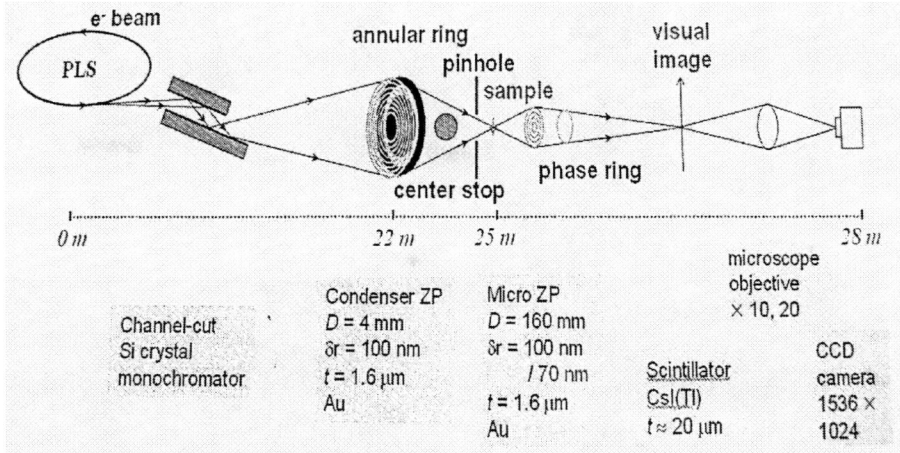

FIGURE 1. This is the layout of the 1B2 beamline optics of Pohang Light Source for phase contrast synchrotron radiation microscopy.

RESULT

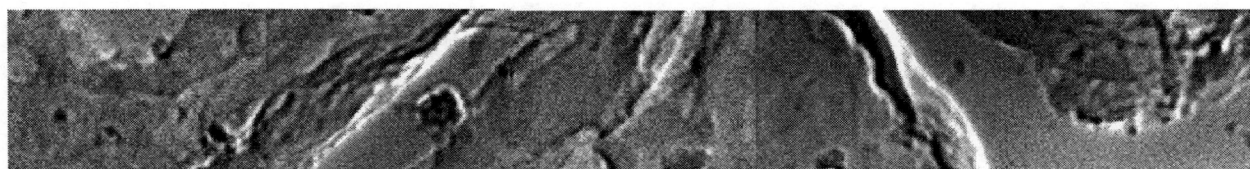

FIGURE 2. This is the monochromated phase contrast imaging of normal breast of premenopausal woman. 100 x 13 μm in size, 10 μm thick and unfixed.

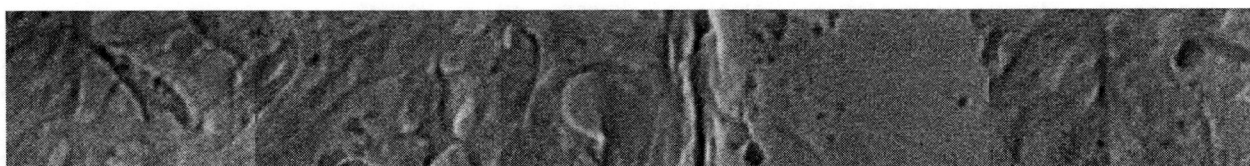

FIGURE 3. This is the monochromated phase contrast imaging of fibroadenoma of premenopausal woman. 100 x 13 μm in size, 10 μm thick and unfixed.

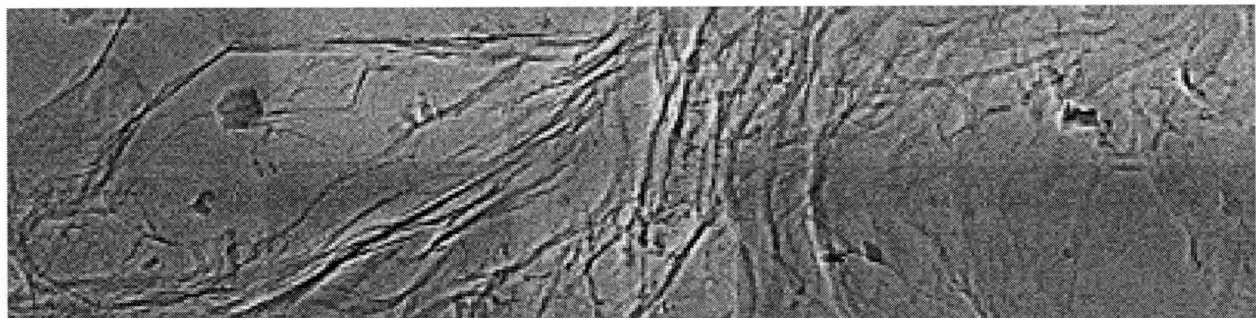

FIGURE 4. This is the monochromated phase contrast imaging of fibrocystic change of premenopausal woman. 100 x 24 μm in size, 10 μm thick and formalin-fixed.

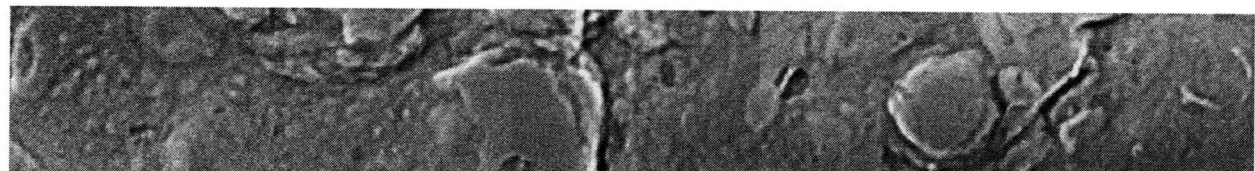

FIGURE 5. This is the monochromated phase contrast imaging of breast cancer of perimenopausal woman. 100 x 12 µm in size, 10 µm thick and unfixed.

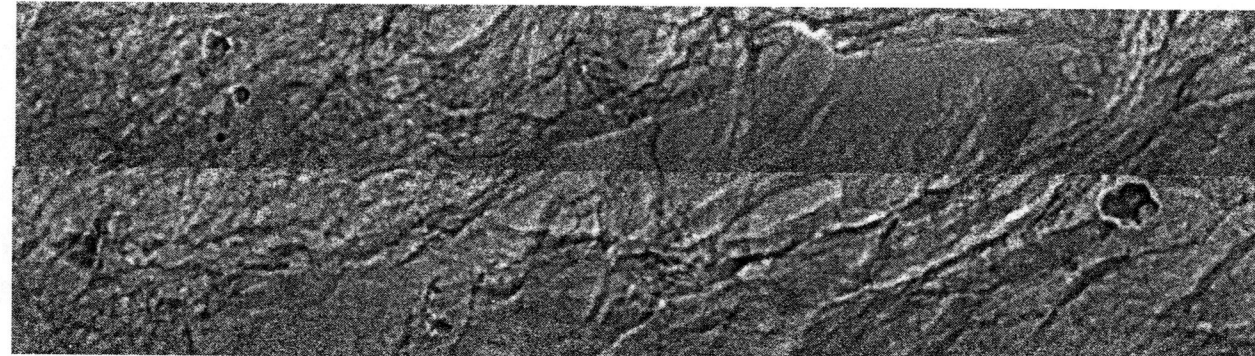

FIGURE 6. This is the monochromated phase contrast imaging of breast cancer of premenopausal woman. 100 x 22 µm in size, 10 µm thick and formalin-fixed.

CONCLUSION

Synchrotron radiation microscope images of normal breast, fibroadenoma, fibrocystic change and breast cancer tissues at better resolution than 100 nm were obtained. Monochromated beam, zone plate and Zernike phase contrast plate were used for high resolution. Synchrotron radiation enables to observe the detailed internal structures of the breast tissues without fixation or staining. From the images obtained, we can tell that x-ray microscope imaging of breast tissue with synchrotron radiation has a great potential for various settings of clinical and research purposes such as oncologic studies, early detection of cancer and adjuvant to pathologic diagnosis in the future.

REFERENCES

1. P. Suortti and W. Thomlinson, *Medical applications of synchrotron radiation*, Phys. Med. Biol. 48 (2003) R1-R35.
2. F. Arfelli, V. Bonvicini, et als. *Mammography with synchrotron radiation: phase-detection techniques*, Radiology 2000; 215:286-293.
3. U. Neuhausler, G. Schneider et als. *X-ray microscopy in Zernike phase contrast mode at 4 keV photon energy with 60 nm resolution*, J. Phys. D: Appl Phys. 36 (2003) A79-A82.
4. T. Takeda, A. Momose, et als. *Human carcinoma: early experience with phase-contrast x-ray CT with synchrotron radiation-comparative specimen study with optical microscopy*, Radiology 2000; 214:298-301.
5. E. Pagot, S. Fiedler, et als. *Quantitative comparison between two phase contrast techniques: diffraction enhanced imaging and phase propagation imaging*, Phys. Med. Biol. 50 (2005) 709-724.
6. M. Ando, K. Yamasaki, et als. *Attempt at two-dimensional mapping of x-ray fluorescence from breast cancer tissue*, Japan. J Appl. Phys: vol. 44. No. 31. 2005. pp. L998-L1001.
7. H. Youn, S. Baik, et als. *Hard x-ray microscopy with a 130 nm spatial resolution*, Rev Sci Instrum 76, 023702 (2005).
8. S. Jheon, H. Youn et als. *High-resolution x-ray refraction imaging of rat lung and histological correlations*, Micro. Research. Tech. 69. (2006).

High Quality Image of Biomedical Object by X-ray Refraction Based Contrast Computed Tomography

E. Hashimoto[1], A. Maksimenko[2], H. Sugiyama[1,2], K. Hirano[2], K. Hyodo[2], D. Shimao[3], Y. Nishino[4], T. Ishikawa[4], T. Yuasa[5], S. Ichihara[6], Y. Arai[7] and M. Ando[1,2,8]

[1] *Department of Photon-Science, School of Advanced Studies,*
Graduate University for Advanced Studies (GUAS), Shonan Village, Hayama, Kanagawa 240-0193, Japan
[2] *Photon Factory, Institute of Materials Structure Science,*
High Energy Accelerator Research Organization (KEK), 1-1 Oho, Tsukuba, Ibaraki 305-0801, Japan.
[3] *Department of Health Sciences, Ibaraki prefectural University of Health Sciences,*
4669-2Ami, Ami, Inashiki, Ibaraki, 300-0394, Japan
[4] *RIKEN Harima Institute, 1-1-1 Kouto, Mikazuki, Sayo, Hyogo, 679-5148, Japan*
[5] *Department of Bio-system Engineering, Faculty of Engineering Yamagata University,*
4-3-16 Jonan, Yonezawa, Yamagata 992-8510, Japan
[6] *Dept. of Path., Nagoya Med. Center, Nat. Hospital Organization, Naka-ku, Nagoya 460-0001, Japan,*
[7] *Matsumoto Dental University, 1980 Hirooka, Shiojiri, Nagano, Japan*
[8] *Inst. of Sci. and Tech., Tokyo Univ. of Science, Yamasaki 2641, Noda, Chiba 278-8510, Japan,*

Abstract. Recently we have developed a new Computed Tomography (CT) algorithm for refraction contrast that uses the optics of diffraction-enhanced imaging. We applied this new method to visualize soft tissue which is not visualized by the current absorption based contrast. The meaning of the contrast that appears in refraction-contrast X-ray CT images must be clarified from a biologic or anatomic point of view. It has been reported that the contrast is made with the specific gravity map with a range of approximately 10 μarc sec. However, the relationship between the contrast and biologic or anatomic findings has not been investigated, to our knowledge. We compared refraction-contrast X-ray CT images with microscopic X-ray images, and we evaluated refractive indexes of pathologic lesions on phase-contrast X-ray CT images. We focused our attenuation of breast cancer and lung cancer as samples. X-ray refraction based Computed Tomography was appeared to be a pathological ability to depict the boundary between cancer nest and normal tissue, and inner structure of the disease.

Keywords: X-ray imaging, Refraction based computed tomography, Refraction contrast, DEI, computed tomography, breast cancer,
PACS: 87.59.-e, 87.59.Ek, 87.59.Fm, 87.59.Hp, 87.61.Pk, 87.62.+n, 87.64.Bx

INTRODUCTION

Traditional X-ray [1] imaging is based on the absorption contrast of different parts of an object. This progress has been made on the improvement of detectors, and developments on source and optical elements. Refraction contrast imaging quality has been recently achieved by exploring the real component of the refraction index. Different optical arrangements and techniques, generally termed as phase-contrast radiography [2-4], diffraction-enhanced imaging (DEI) [5-7] or dark-field imaging (DFI) [8, 9] are being used to increase the density of information carried by the X-ray beam that passes through object and reaches a detector. Some groups [10-14] have attempted at clinical application to use biomedical tissue as sample. Trials to make CT reconstruction based on the refraction contrast have been done [15-19]. The properties of the refraction contrast provide certain advantages over other contrasts such as absorption and phase-shift. The refraction contrast can show tiny details of the inner structure which are invisible in other types of the X-ray imaging techniques. Another advantage of the contrast is the sensitivity to the

CP879, *Synchrotron Radiation Instrumentation: Ninth International Conference,*
edited by Jae-Young Choi and Seungyu Rah
© 2007 American Institute of Physics 978-0-7354-0373-4/07/$23.00

low Z materials. This property of the refraction contrast may be of great importance in the medical applications of the X-ray. The advantages provided by the refraction contrast allow one to expect the same advantages of the computed tomography (CT) from the refraction contrast. Although some attempts to make CT-reconstructions based on the refraction contrast are known [15-16], none of them showed faithful representation of the object under investigation. This work describes the first application of X-ray refraction-based computed tomography (CT) to a biological object.

We have developed refraction-based algorithm [20] and its application to discovery of a fine crack of object [21] unless otherwise invisible by absorption contrast. Further this was applied to successfully visualize bio-medical specimens [22-24]. The meaning of the contrast that appears in refraction-contrast X-ray CT images must be clarified from a biologic or anatomic point of view. It has been reported that the contrast is made with the specific gravity map with a range of approximately 10 marc sec. However, the relationship between the contrast and biologic or anatomic findings has not been investigated, to our knowledge. In the purpose, the ability of Refraction based contrast CT to depict breast cancer nest was tested by observing breast cancer of the human with synchrotron X-rays. Reconstructed images of breast cancer show much detail that is notes in absorption-based CT.

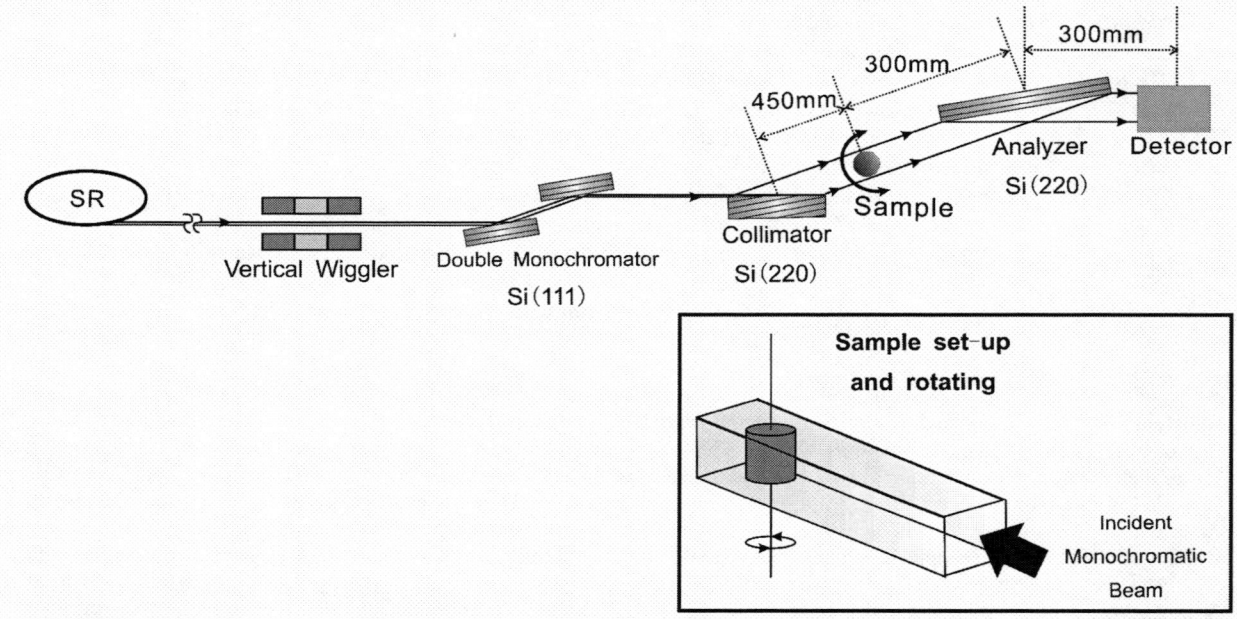

FIGURE 1. Schemes of the X-ray refraction based CT system which is consisted of DEI optics. The X-ray beam generated by a double monochromator from the vertical wiggler is introduced into the experimental hutch. An asymmetrically-cut collimator crystal has two functions to make parallelity and size of the beam with higher performance. The 2-dimensional parallel monochromatic X-ray beam which transmits through object is diffracted by an analyzer crystal that is rocked around its Bragg peak position for the data acquisition at the desired X-ray energy 17.5keV.

EXPERIMENT

X-ray refraction based contrast Computed Tomography system was contracted in BL14B at Photon Factory, Japan. Figure 1 shows the schematics of the optical set-up at 17.5keV. The X-ray beam produced by the vertical wiggler was monochromated by the double crystal monochromator Si(111) upstream, and then transported into the experiment hutch BL14B. As X-ray optics two components were installed in the hutch. One is the Si(220) 9.5°off asymmetrical cut silicon crystal to have more parallel beam tuned at the above X-ray energy 17.5 keV and to expand the beam size to fit a sample size. The second is Si(220) symmetrical cut silicon crystal for angular analysis.

The X-rays are refracted through very small angles due to the tiny variations in the refraction index. Then, we obtain $\Delta\alpha$ as total deflection angle

$$\Delta\alpha(\Theta,t)e^{i\Theta} = \int_S |\nabla\widetilde{n}|e^{i\phi}ds \tag{1}$$

where $\tilde{n}$ is refraction index as $\tilde{n} = 1 - n$, ϕ is angle between beam direction and gradient $\nabla\tilde{n}$, S is the beam pass, Θ is rotation angle and t is propagate axis of the X-ray beam. Thus, this method observes $\nabla\tilde{n}$ which vector field is.

The rocking curve is the autocorrelation function of the second crystal, and its half width is approximately 1.5 arcsec. The intensity changes are largest when the analyzer is tuned to reflect at the slope of the rocking curve in Fig. 2. The angular changes are very small, and for this reason a high resolution and stable tuning of the analyzer to the desired angular positions are essential. When a given part of the object is imaged the contrast is reversed when going from the low-angle side to the high-angle side so that the component of refraction can be extracted by a mathematical procedure [7]. The angular positions to acquire data are at the half maximum of the diffraction peak of the rocking curve of the angular analyzer.

The sample stage was settled between the first crystal and the analyzer and the rotation axis was the vertical direction to the X-ray incidence plane. The image from the sample was detected by a CCD Camera (pixel size 6.7 µm, field of view 8.7(h)mm x 6.9(v)mm, pixel number 1392×1040, Photonics Science). For reconstruction of CT image a summed projections were acquired at 900 rotational positions of the sample around 180° at each angular position of the angular analyzer.

RESULTS

Fig. 3(A) shows X-ray Micro CT image. Fig. 3(B) is cross-section of X-ray refraction based CT. We imaged in vitro a sample of human breast cancer tumors. The sample with 3.5 mm in diameter and 4.7 mm height was punched out from a specimen including calcification. In comparison Fig. 3(A) with Fig. 3(B), the refraction contrast image is visualized the inner condition of milk duct affected by the tumor. Fig. 4 shows histogram of the contrast value from each picture of Fig. 3. Fig. 4(a) is from X-ray Micro CT. Fig. 4(b) is from X-ray refraction based CT. Thus, the New CT technique can depict wide range, in other words observe not only low-Z elements but also high-Z elements. This new technique is not slave of absorption coefficient because is formed by refraction coefficient from each cell.

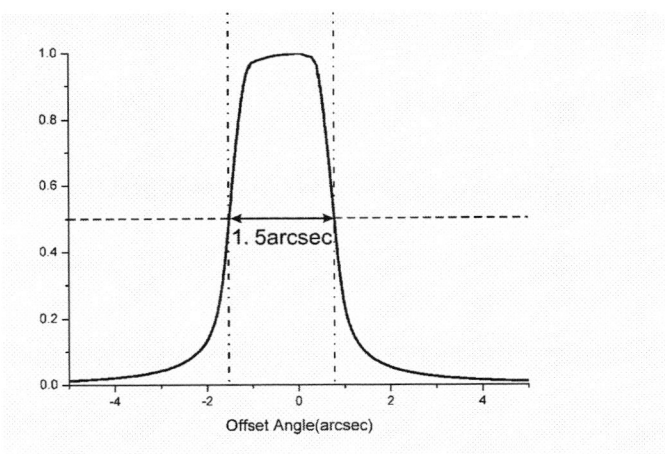

FIGURE 2. A measured reflection curve at 17.5 keV. Its half width is 1.5 arcsec. Each set of raw data for CT is acquired at the left and right steepest position of the rocking curve, namely at half maximum height of the peak.

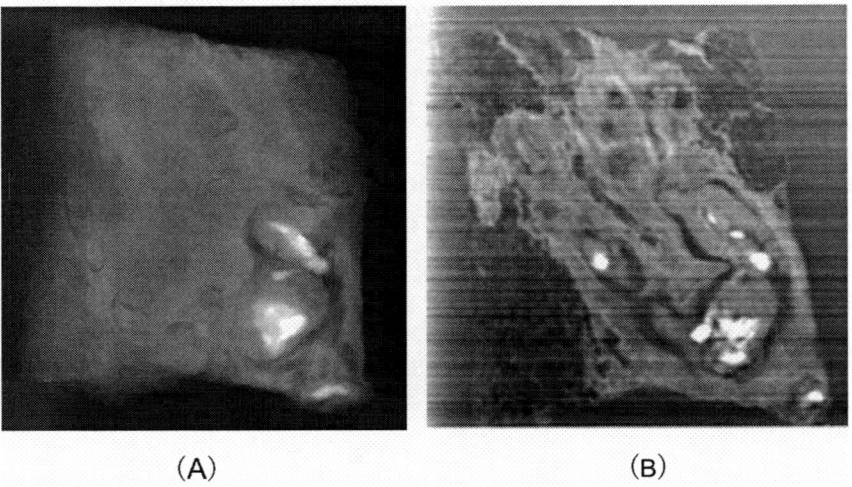

(A)   (B)

FIGURE 3. Comparison of Micro X-ray CT, reconstructed X-ray refraction based CT. It is the breast cancer tissue extracted from human. Image (A) is reconstructed image of Micro X-ray CT. (B) is refraction image by decomposed from the intensity of image detector. The image of Refraction contrast can be easy distinguished from breast cancer nest.

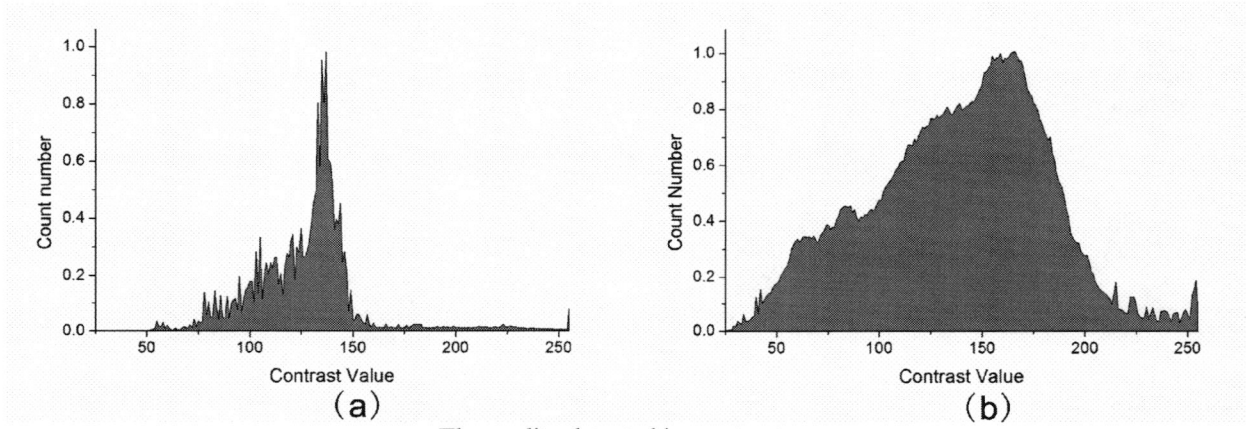

The quality detected inner structure

FIGURE 4. (a): The contrast value of Fig. 3(A). (b): The contrast value is obtain by Fig. 3(B). In comparison (a) with (b), Refraction based CT image have high performance to catch the edge of low-Z elements in detail.

CONCLUSION

The present study demonstrated successful refraction-based reconstructed images. This is partly as a result of the improved contrast. One other point to note, the method is sensitive to change in refractive index rather than the absolute value. The new method appears to be a new powerful tool for assessing diagnosis of clinical object in a future. Also in the future work, we will try to use large samples heading for clinical application.

ACKNOWLEDGMENTS

We gratefully acknowledge T. Kimura and K. Yamazaki for valuable help and advice. This experiment using synchrotron radiation was carried out under proposal number 2005PF02, 2005PF09, 2005PF13 approved by the Program Advisory Board of High Energy Accelerator Research Organization.

REFERENCES

1. W. C. Roentgen, Nature *Nature* **53**, 274-276 (1896).
2. U. Bonse, M. Hart, *Appl. Phys. Lett.*, **7**, 99-101 (1965).
3. M. Ando and S. Hosoya, *Proceedings of the 6th International Conference on X-ray Optics and Microanalysis*, edited by G. shinoda et al.,p.63. Tokyo, University of Tokyo (1972) .
4. A. Momose, *Nuclear Instruments and Methods in Physics Research,* **A 352**, 622-628 (1995).
5. E. Foerster, K. Goentz and P. Zaumseil, *Krist. Tech.*, **15**, 937-945 (1980).
6. V. N. Ingal and E. A. Beliaevskaya, *J. phys. D*, **28**, 2314-17 (1995).
7. D. Chapman, W. Thomlinson, R. E. Johnston, D. Washburn, E. Pisano, N. Gmur, Z. Zhong, R. Menk, F. Arfelli, D. Sayers,: *Phys. Me. Biol.* **42**, 2015-2025 (1997).
8. M. Ando, H. Sugiyama, A. Maksimenko, W. Pattanasiriwisawa, K. Hyodo and X. Zhang, *Jpn. J. Appl. Phys.*, **40**, 844-846(2001).
9. K. Hirano, A. Maksimenko, H. Sugiyama and M. Ando, Jpn. *J. Appl. Phys.*, **41**, 595-598 (2002)
10. E. Burattini, E. Cossu, C. D. Maggio, M. Gambaccini, P. L. Indovina, M. Marziani, M. Pocek, S. Simeoni, A. G. Simonetti, *Radiol.*, 195, 239-244 (1995) .
11. R. E. Johnston, D. Washburn, E. Pisano, C. Burns, W. C. Thomlinson, L. D. Chapman, F. Arfelli, N. F. Gmur, Z. Zhong, D. Sayers, *Radiol.,* 200, 659-663 (1996) .
12. Q. Yu, T. Takeda, K. Umetani, E. Ueno, Y. Itai, Y. Hiranaka and T. Akatsuka, J. Sych. Radiol., 6, 1148-1152 (1999) .
13. T. Takeda, A. Momose, K. Hirano, S. Haraoka, T. Watanabe, Y. Itai, *Radiol.*, 214, 298-301 (2000) .
14. T. Sera, K. Uesugi, N. Yagi, *Respir. Physiol. Neurobiol.,* 147, 51-63 (2005) .
15. F. A. Dilmanian, Z. Zhong, B. Ren, X. Y. Wu, L. D. Chapman, I. Orion and W. C. Thomlinson, *Phys. Med. Biol.*, **45**, 933 (2000).
16. Koyama I, Hamaishi Y, Momose A (2004) Phase tomography using Diffraction-enhaced imaging, AIP Conf Proc **705**: 1283-1286
17. T.Sera, K. Uesugi and N. Yagi, *Med. Phys.,* **32**, 2787-2792 (2005) .
18. S Fiedler, A Bravin, J Keyril̈ainen, M Fern ndez, P Suortti, W Thomlinson, M Tenhunen, P Virkkunen and M-L Karjalainen-Lindsberg, *Phys. Med. Biol.*, **49**, 175–188 (2004)
19. Zhi-Feng Huang et al., *Appl. Phys. Lett.*, **89**, 041124 (2006)
16. A. Maksimenko, M. Ando, H. Sugiyama, T. Yuasa, *Appl. Phys. Lett.* **86**, 124105-1~124105-3 (2005).
17. A. Maksimenko, M. Ando, H. Sugiyama, E. Hashimoto, *Jpn. J. Appl. Phys.* **44**, L633-L635 (2005).
18. Hashimoto E, Maksimenko A, Sugiyama H et al., *Zool. Sci.* (2005) (in press)
19. Hashimoto E, Maksimenko A, Sugiyama H et al., *Proc. SPIE* (2006) (in press)
20. M. Ando, A. Maksimenko, T. Yuasa, E. Hashimoto, K. Yamasaki, C. Ohbayashi, H. Sugiyama, K. Hyodo, T. Kimura, H. Esumi, T. Akatsuka, G. Li, D. Xian and E. Ueno, *Bioimages,* **13**, 1-7 (2005).

Resonance Raman Spectroscopy for In-Situ Monitoring of Radiation Damage

A. Meents[1*], R. L. Owen[1], D. Murgida[2], P. Hildebrandt[2], R. Schneider[1],
C. Pradervand[1], P. Bohler and C. Schulze-Briese[1]

[1]Swiss Light Source at PSI, 5232 Villigen PSI, Switzerland
[2]Technische Universitaet Berlin, 10623 Berlin, Germany

Abstract. Radiation induced damage of metal centres in proteins is a severe problem in X-ray structure determination. Photoreduction can lead to erroneous structural implications, and in the worst cases cause structure solution to fail. Resonance Raman (RR) spectroscopy is well suited *in-situ* monitoring of X-ray induced photoreduction. However the laser excitation needed for RR can itself cause photoreduction of the metal centres. In the present study myoglobin and rubredoxin crystals were used as model systems to assess the feasibility of using RR for this application. It is shown that at least 10-15 RR spectra per crystal can be recorded at low laser power before severe photoreduction occurs. Furthermore it is possible to collect good quality RR spectra from cryocooled protein crystals with exposure times of only a few seconds. Following extended laser illumination photoreduction is observed through the formation and decay of spectral bands as a function of dose. The experimental setup planned for integration into the SLS protein crystallography beamlines is also described. This setup should also prove to be very useful for other experimental techniques at synchrotrons where X-ray photoreduction is a problem e.g. X-ray absorption spectroscopy.

Keywords: Resonance Raman spectroscopy, macromolecular crystallography, radiation damage, photoreduction.
PACS: 87.15.-v, 87.15.Mi, 87.64.Je

INTRODUCTION

X-ray induced radiation damage is one of the limiting factors in the structure determination of biological macromolecules at 3rd generation synchrotron sources [1]. While the global effects of radiation damage can be observed directly, for example via reduction of the resolution limit or $I/\sigma(I)$, specific structural damage often only becomes apparent following structure refinement. Alternatively, radiation damage can be monitored by micro resonance Raman (RR) spectroscopy. In this technique the excitation laser line is chosen to be in resonance with an electronic transition of a chromophore in the macromolecule, typically a cofactor or an aromatic amino acid residue. The resonance condition produces a strong enhancement, ca. six orders of magnitude, solely of the Raman bands of the chromophore, providing high sensitivity and selectivity. For metalloproteins, the positions and intensities of the RR bands are particularly sensitive to the redox state, coordination pattern and spin of the metal ion and, therefore, direct markers of radiation damage [2-4].

The most efficient collection of the Raman signal is achieved in back-scattering geometry, i.e. by means of a single microscope objective. Thus, in contrast to UV/VIS spectroscopy which requires two microscope objectives [5], backcattering RR spectroscopy is ideally suited for integration into the spatially restrictive environment at typical macromolecular crystallography diffractometer.

A potential drawback of RR, however, is that high dosis of laser radiation may induce photoreduction and high dose rates may cause thermal damage of the protein crystals.

Here we report a RR study of myoglobin (Mb) and rubredoxin (Rb) single crystals that aims to establish the conditions for the future integration of a Raman set-up into the SLS beam lines.

CP879, *Synchrotron Radiation Instrumentation: Ninth International Conference*,
edited by Jae-Young Choi and Seungyu Rah
© 2007 American Institute of Physics 978-0-7354-0373-4/07/$23.00

EXPERIMENTAL

Protein crystals. Sperm whale Myoglobin (Mb) and Rubredoxin (Rb) crystals were chosen as model systems. Mb is a heme protein containing a Fe^{3+} ion in its stable oxidized state and is among the most studied proteins in RR spectroscopy. The iron is coordinated by four nitrogen atoms from the porphyrin ring and two oxygen atoms from a histidine and a water molecule. Rb is an electron transfer protein containing a single Fe^{3+} ion in a sulphur cluster. Both proteins are known for their sensitivity to photoreduction. Crystals of both compounds were grown according to standard procedures [6,7].

Resonance Raman spectroscopy. RR spectra were measured in back-scattering geometry using a confocal microscope coupled to a single stage spectrograph (Jobin Yvon, XY) equipped with a 1800 l/mm grating and liquid nitrogen cooled back illuminated CCD detector. Elastic scattering was rejected with Notch filters. The laser line was focused onto the surface of the crystal by means of a long working distance objective (20x; N.A. 0.35) resulting in a focal spot size of 10 μm diameter.

To avoid thermal degradation, crystals were cooled to 100 K using an open flow nitrogen cryostat. Mb samples were measured with the 413 nm line of cw Kr^+ laser (Coherent Innova 302) while for Rb the 514 nm line of a cw Ar^+ laser (Coherent Innova 70C) was used.

RESULTS AND DISCUSSION

Upon Soret band excitation, the RR spectra of heme proteins are dominated by the totally symmetric A_{1g} vibrational modes of the porphyrin ring. The high frequency region (ca. 1300-1700 cm^{-1}) displays the so-called marker bands that are particularly diagnostic for specific structural properties of the porphyrin such as the oxidation, spin, and ligation state of the heme iron, while the fingerprint region (ca. 200-500 cm^{-1}) reflects the interactions of the heme with the surrounding protein [4, 8, 9]. For all heme proteins the most intense RR band corresponds to the v_4 mode which appears at ca. 1360 cm^{-1} for the ferrous form and shifts to ca. 1370 cm^{-1} in the oxidised state and, therefore, constitutes the most sensitive reporter of photoreduction processes.

We have investigated the threshold of laser induced damage in Mb crystals as a representative example of a heme protein prone to photoreduction. Figure 1 shows a typical RR spectrum of a Mb crystal measured at low laser power, ca. 200 μW. Despite the relatively long laser exposure (8 minutes), the RR spectrum of the crystal is identical to that of ferric Mb in solution measured in a rotatory cuvette under ideal conditions and no indication of laser-induced damage of the protein crystal is observed.

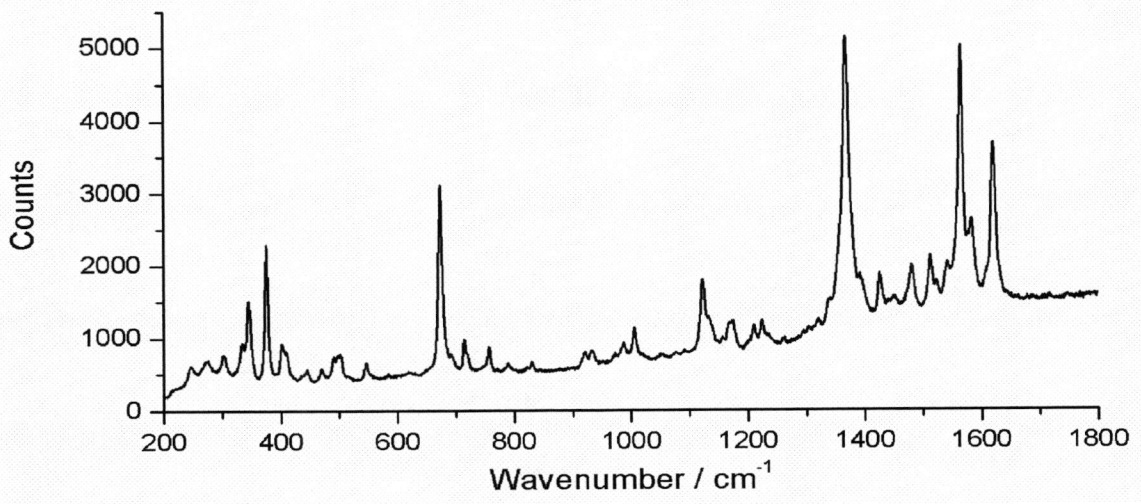

FIGURE 1. Resonance Raman spectrum from a Myoglobin crystal (excitation wavelength 414 nm). The total exposure time was 480s at a laser power of 0.23 mW.

At substantially higher laser powers we observe the accumulation of photoreduced Mb that is reflected in the RR spectra by the appearance of a shoulder in the v_4 band at 1357 cm^{-1}, as well as a shoulder in the v_3 band whose position for ferric Mb is at 1480 cm^{-1} and shifts down to 1475 cm^{-1} in the photoreduced form. This effect is exemplified in Fig. 2A, where spectra recorded at 9 mW with spectral accumulation times of 2 seconds are shown as a function of the exposure time to the laser beam.

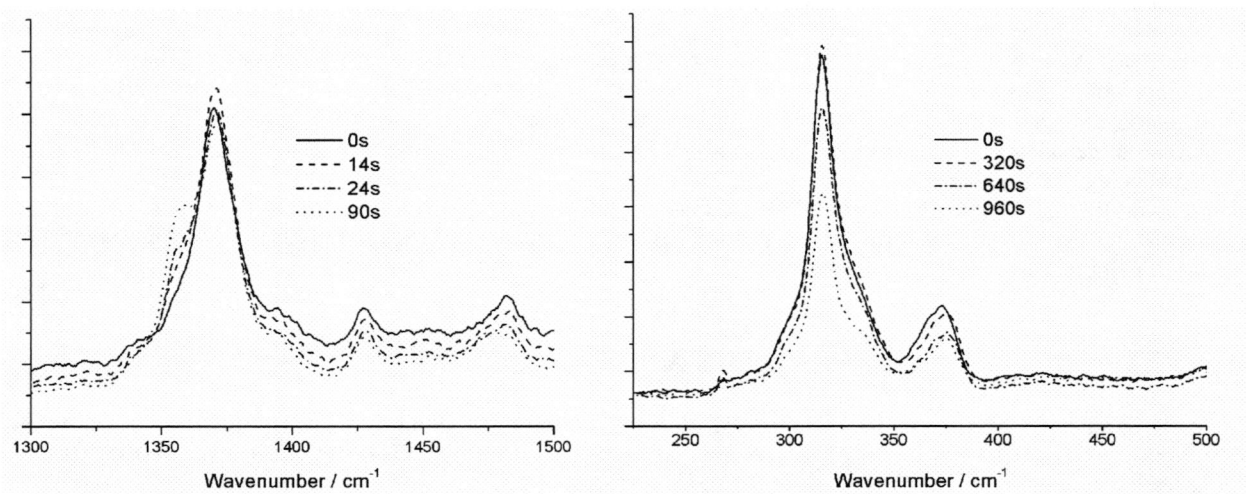

FIGURE 2. Band shifts in Mb (left) and Rb (right) spectra as a function of absorbed dose.

Similar experiments were performed for Rb crystals, as a representative example of proteins containing iron-sulphur clusters. As in the case of Mb, RR spectra do not undergo time dependent changes when measured at laser powers below 500 μW and display the typical features of fully oxidised Rb in its native state (data not shown).

At higher laser powers laser-dose dependent accumulation of photoreduced Rb becomes evident as a drop of the RR signal due to the lower extinction coefficient of the reduced form at the excitation line (Fig. 2B). For a laser power of 7 mW photoreduction of Rb starts to become evident only after irradiation times of ca. 600 seconds while RR spectra of good quality can be recorded in less than 20 seconds.

CONCLUSIONS AND FUTURE PERSPECTIVES

The test experiments show that RR is well suited to monitor the X-ray induced photo-reduction of metal centres in protein crystals. Complete RR spectra of good signal-to-noise ratio can be obtained in few seconds using sufficiently low laser powers that do not induce sample damage even after exposure times of several minutes.

Based on these results, we have started the installation of a micro RR setup at the SLS protein crystallography beamlines that is schematically shown in Fig. 3. The light source is a Kr$^+$ or Ar$^+$ laser, which is focused on the sample by the use of a microscope objective. Backscattered light is collected by the same objective and delivered to the Raman spectrometer. The objective used has a 10 x magnification and a numerical aperture of 0.28. The working distance is 33 mm. 1 mm holes were drilled in all lenses at the PSI workshop using water-cooled diamond hollow drills. Elastically scattered light is rejected by means of a notch filter placed in front of the spectrograph entrance slit. The spectrograph is an Andor Shamrock 303 equipped with an electron multiplying CCD and 3600 l/mm grating that provides a resolution of ca. 3 cm^{-1} at 413 nm excitation. The use of the on-axis geometry, with the X-ray beam passing through the drill hole in the middle of the objective, ensures that the focused laser spot probes the same sample volume which is exposed to the X-ray beam.

Installation of the RR device at the beamline will be finished by the end of July and the first experiments are scheduled for September 2006.

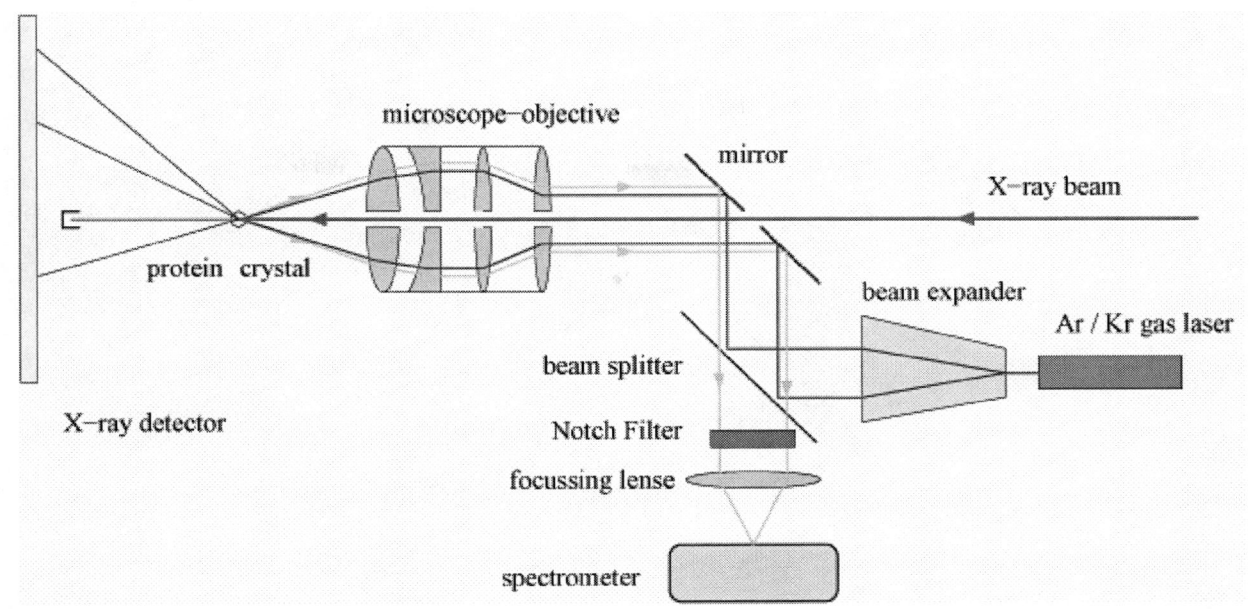

FIGURE 3. Experimental setup of a Raman microscope planned for in-situ monitoring of X-ray induced radiation damage at the SLS protein crystallography beamlines.

ACKNOWLEDGMENTS

Rubredoxin was kindly donated by Dr. Jacque Meyer from CEA Grenoble (France). The cryojet used was a kind loan from the protein crystallography team at BESSY (Germany). Special thanks to Ronald Foerster for his assistance during the experiments.

REFERENCES

1. R. L. Owen, E. Rudiño-Piñera, E. F. Garman. PNAS **103**, 4912-4917 (2006).
2. N. Engler, A. Ostermann, A. Gassmann, D. C. Lamb *et al.* Biophys J. **78**(4): 2081–2092 (2000).
3. Y. Xiao, H. Wang, S. J. George *et al.* J. Am. Chem. Soc; **127**(42) 14596 – 14606 (2005).
4. D. H. Murgida, P. Hildebrandt. Acc. Chem. Res. **37**, 854-861 (2004).
5. C.M. Wilmot, T. Sjogren, G.H. Carlsson, G.I. Berglund, J. Hajdu. Meth. Enzym. **353**, 301-318(2002).
6. J.C. Kendrew. Acta Cryst. **1**, 336 (1948).
7. Z. Dauter, L.C. Sieker, K.S. Wilson. Acta Cryst. **B48**, 42-59 (1992).
8. S. Z. Hu, I. K. Morris, J. P. Singh, K. M. Smith, T. G. Spiro. J. Am. Chem. **115**, 12446-12458 (1993).
9. S. Oellerich, H. Wackerbarth, P. Hildebrandt. J. Phys. Chem. B **106**, 6566-6580 (2002).

SRI2006 Exhibitors

Asian/Oceanic Forum for Synchrotron Radiation Research
pfwww.kek.jp/AOF2006/

ACCEL Instruments GmbH
www.accel.de

VG SCIENTA LIMITED
www.vgscienta.com

FMB Feinwerk-und Messtechnik
www.fmb-berlin.de

LEBOW COMPANY
www.lebowcompany.com

Incoatec GmbH
www.incoatec.de

VMT Co., Ltd
www.vmt.co.kr

HORIBA Jobin Yvon
www.jobinyvon.com

Oxford Danfysik
www.oxford-danfysik.com

MB Scientific AB
www.mbscientific.se

Photon Production Laboratory Ltd
www.ppl-xray.com

Hamamatsu Photonics K.K. (Modoo Tek Co., Ltd)
www.hamamatsu.com

Motion Hightech Co., Ltd
www.himotion.co.kr

TOYAMA Co., Ltd.
www.toyama-jp.com

KOHZU Precision Co., Ltd
www.kohzu.co.jp

SAES Getters Korea
www.saesgetters.com

Advanced Design Consulting USA, Inc
www.adc9001.com

XRADIA
www.xradia.com

AMERICAN MAGNETICS INC
www.americanmagnetics.com

SPECS
www.specs.de

BESTEC
www.bestec.de

Johnsen Ultravac Inc.
www.ultrahivac.com

Omicron Nano Technology GmbH
www.omicron.de

XIA LLC
www.xia.com

Carl Zeiss Optronics GmbH
www.zeiss.de

WOOSUNG VACUUM CO., LTD
www.woovac.com

SHIMADZU CORPORATION
www.shimadzu.com/opt

Rigaku Corporation
www.rigaku.co.jp
www.rigakumsc.com

Mar USA, Inc
www.mar-usa.com

NEOMAX Co., Ltd
www.neomax.co.jp

Paul Scherrer Institut
www.psi.ch

HD Technologies Inc
www.hdtechinc.com

MDC VACUUM PRODUCTS, LLC (ATC Co., Ltd)
www.mdcvacuum.com

NTT Advanced Technology Corporation
www.keytech.ntt-at.co.jp/nano/index_e.html

RAD Device Co., Ltd.
radd-yk@w6.dion.ne.jp

JTEC Corporation
www.j-tec.co.jp

Huber Diffraktionstechnik GmbH & Co. KG
www.xhuber.com

Oxford Instruments NanoScience
www.oxford-instruments.com

Varian, Inc.
www.varianinc.com
www.varian.co.kr

A

Abe, H., 463
Abel, D., 675
Abela, R., 848, 890, 1030, 1168, 1198, 1912
Ablett, J. M., 283, 1557
Abrami, A., 1895
Abramsohn, D., 1198
Adachi, H., 1691
Adachi, J., 1805
Adachi, S., 91
Adam, J. F., 1887
Adams, B., 1210
Ade, H., 505
Agafonov, A., 79
Akatsuka, T., 1944, 1948, 1956
Aki, T., 1715
Akimoto, K., 75
Akinaga, H., 1353
Akiniwa, Y., 1861
Akita, K., 1577
Akiyama, H., 1691
Al-Adwan, A., 1198
Alatas, A., 737
Alianelli, L., 836
Allier, C., 1156
Almeida, E., 1333
Alp, E. E., 894, 1073
Altissimo, M., 1486
Amemiya, K., 463
Amenitsch, H., 1287
Anders, W., 167
Anderson, E. H., 1269
Ando, A., 107, 171
Ando, M., 1899, 1960, 1972, 1979
Andrejczuk, A., 994
Androsov, V., 9, 79
Anfinrud, P., 1187
Aoki, S., 1283, 1357
Aoyagi, H., 523, 1010, 1018
Aoyagi, S., 1829
Appadoo, D., 579
Aquilanti, G., 750, 1242
Arai, N., 1129
Arai, Y., 1979
Arakawa, P., 1683
Arfelli, F., 1109, 1895
Ariake, J., 1699
Arita, M., 1597
Armelao, L., 1202
Arms, D. A., 950, 985
Arp, U., 139, 481
Asano, Y., 420, 523, 1125

Asaoka, S., 87, 1038
Ascone, I., 1883
Ashida, T., 1638
Asokan, K., 1659
Assoufid, L., 706, 1299
Attaphibal, M., 147
Attwood, D. T., 1269
Avagyan, V., 1046
Azuma, J., 623, 1218

B

Bacher, R., 175
Bächli, H., 1030
Bae, Y. S., 252
Bahou, M., 188, 603, 1462, 1503
Bahrdt, J., 315
Baik, S., 1411, 1554
Balewski, K., 175
Bando, A., 1565
Bando, H., 1899
Bang, S.-M., 1443, 1499
Bao, C. W., 1659
Barabash, R. I., 1299
Barnett, H. A., 1317
Bartelt, A., 1206
Barthelmess, M., 343
Basolo, S., 1087
Batrakov, A., 305, 396, 412
Battaile, K., 840
Baumbach, T., 301, 363
Bayat, S., 1887
Beau, J.-O., 1152
Beaud, P., 1198
Beckmann, F., 746, 804
Beesley, A., 1735
Beesley, A. M., 1739
Belkhou, R., 455
Bellamy, H., 163
Benabderrahmane, C., 359
Benn, R., 754, 1403
Benson, C., 1061
Bérar, J.-F., 1087, 1743
Berges, U., 30, 519, 583, 852, 875
Berglund, M., 34
Bergstrom, J., 1002
Berman, L., 283, 754
Bernard, F., 1156
Bernard, P., 792, 1687
Bernhard, A., 301, 363
Berruyer, G., 1887
Besson, J. C., 3

Betemps, R., 643, 848, 890
Bettles, C., 1751
Bhattacharyya, D., 671
Bianchetti, C., 163
Bianco, A., 497
Bilderback, D. H., 758, 1395
Billè, F., 683
Billette, J., 1695
bin Mahmood, S., 1512
Bissen, M., 651
Biston, M. C., 1887
Bizen, D., 1711
Bizen, T., 420
Bjeoumikhov, A., 975
Blome, C., 1254
Blomqvist, K. I., 323, 392, 408
Blumer, H., 890
Blyth, R. I. R., 473
Bocci, A., 1246
Boege, M., 388
Böge, M., 505, 1198
Bohic, S., 792
Bohler, P., 1984
Boldeman, J. W., 864, 887
Bonucci, A., 287, 432
Boonsuya, S., 58
Borchert, G., 1168
Bordessoule, M., 1087, 1137
Borel, C., 792
Borghes, R., 683
Botman, J. I. M., 79
Bouchenoire, L., 1679, 1683
Boudet, N., 1087
Boudou, C., 1887
Bowler, M. A., 627
Boye, P., 1295
Bozek, J., 509
Bradley, J., 1337
Brandin, M., 34
Brandt, G., 167
Bras, W., 1695
Bräuer, M., 343
Braun, S., 493, 1482
Bravin, A., 1887
Brefeld, W., 175
Bregant, P., 1895
Brennan, S., 1337
Breugnon, P., 1087
Brewe, D., 1202
Briquez, F., 396, 412
Broadbent, A., 615
Brochard, T., 1887
Brockhauser, S., 1928
Brönnimann, Ch., 1141
Brown, S. D., 1679, 1683
Brown, V. C., 1195

Bruce, S. L., 1845
Brüggmann, U., 907, 1755
Brum, J. A., 196
Brunelle, P., 3, 311
Bruni, C., 256
Bucaille, T., 1137
Bucourt, S., 722
Budai, J. D., 1299
Buddhakala, M., 543
Budz, P., 167
Bulgheroni, W., 388
Bulyak, E., 9, 79
Burghammer, M., 1287, 1295
Bürkmann-Gehrlein, K., 167
Burt, M., 836
Butler, L. G., 1317
Byrd, J., 1206

C

Cai, Y.-Q., 971
Cai, Z., 1065
Cai, Z.-H., 1395
Caillot, B., 1087
Cain, M. G., 1679
Caletka, D., 392, 1069
Caliebe, W., 1755
Cammarata, M., 1187
Capel, M., 832
Carré, B., 256
Casalbuoni, S., 301, 363
Casse, B. D. F., 188, 603, 1462, 1512
Castelli, E., 1895
Cauchon, G., 455, 722
Cautero, G., 683
Cautero, M., 683
Cernik, R. J., 1747
Chachai, W., 335
Chaiprapa, J., 860
Chamorro, J., 1137
Chan, C. K., 42, 62, 67
Chance, M., 675
Chang, C. C., 62, 67
Chang, C. H., 371, 376, 808, 828, 883
Chang, C. L., 1659
Chang, H. P., 13, 50, 376
Chang, J. C., 13
Chang, L. H., 46
Chang, S. H., 808, 828
Chang, S.-H., 451
Chang, S. S., 46, 981, 1507
Chang, W. D., 1466
Chao, C. H., 816
Chao, W., 1269
Chaplier, G., 1137

Fisher, M. V., 651
Flank, A.-M., 1329
Flechsig, U., 505, 643, 890
Fliegauf, R., 167
Follath, R., 513, 679
Fonne, C., 1152
Fons, P., 1309
Foran, G., 1152
Forest, E., 1341
Fourme, R., 1883
Frahm, R., 301, 1081
Frani, A., 575
Frank, K., 890
Frazer, B. H., 651
Frentrup, W., 315
Frings, P., 1695
Fritz, D. M., 1210
Fröba, M., 1755
Frommherz, U., 643
Fu, Y., 591
Fuchs, M. R., 1006
Fujimoto, H., 623
Fujisawa, M., 1117
Fujisawa, N., 1630
Fujisawa, T., 1160
Fujiwara, H., 1415, 1703
Fukami, T., 1956
Fukui, T., 240, 248
Fukumoto, K., 571
Fukushima, S., 1775
Fukushima, Y., 1470
Fukuyama, Y., 1238
Fung, H. S., 451, 563, 883
Fuoss, P., 1176
Furukakwa, Y., 954
Furukawa, Y., 523
Furuya, K., 623

G

Gang, L., 1899
Gao, C., 1849
Gao, X., 188
Gaponov, Y. A., 1924, 1932
Garbin, V., 1287
Garg, C. K., 635
Garrett, R., 1152
Garzella, D., 256
Gaupp, A., 315
Gautier, J., 489
Gawlitza, P., 493, 1482
Geandier, G., 1751
Gejo, T., 1121
Gélebart, F., 1837
Gentle, I., 887

Genzel, C., 1763
Gerlach, M., 1145
Gerson, A., 872
Ghodgaonkar, M. D., 1105
Gil, K. H., 1113
Gilbert, P. U. P. A., 651
Gillilan, R., 758
Giorgetta, J.-L., 1101
Girard, E., 1883
Giura, P., 1837
Giuressi, D., 1109
Gladkikh, P., 9, 79, 83
Glatzel, P., 1731
Glover, T. E., 1195
Godefroy, J. M., 3
Goettert, J., 163, 1759
Goo, J., 1278
Gordon, R., 1202
Gorges, B., 1531
Goto, S., 523, 718, 922, 1057, 1380
Gottwald, A., 167
Goujon, G., 1663, 1671
Goulon, J., 1663, 1671, 1675, 1707
Goulon-Ginet, C., 1663
Grabe-Celik, H., 175
Graber, T., 840
Grantham, S., 481
Green, M. A., 323
Grevtsev, V., 79
Griesebock, B., 301
Grigor'ev, Y., 79, 135
Grigoriev, M., 998
Grilli, A., 575
Grolimund, D., 1198
Groso, A., 848, 1168, 1912
Grunewaldt, J.-D., 1731
Gu, J., 188, 603
Gu, P., 603
Gu, P. D., 188, 1462
Guidi, M. C., 1246
Guilera, G., 750, 1242, 1531
Gullikson, E. H., 647
Gumprecht, L., 339
Gunjishima, I., 1550
Guo, F. Z., 1164
Guo, Q., 559
Guo, Y., 1494
Guo, Z., 1849
Guttmann, P., 1291
Gvozd', A., 79

H

Haake, U., 301
Hada, H., 1470

J

Jablonka, M., 256
Jaggi, A., 890
Jakob, B., 400
Jan, J. C., 371, 376
Jang, S. D., 272
Jang, Y. J., 1875
Janpuang, P., 58
Jansen, A. N., 989
Jaouen, N., 1663, 1671, 1675, 1707, 1825
Jark, W., 796
Jaski, Y., 1053, 1061
Jayne, R., 531, 1069
Je, J. H., 1373, 1399, 1427, 1431, 1466, 1516,
    1891, 1908, 1964
Jean, Y. C., 816
Jemian, P. R., 1833
Jeng, U., 808, 883
Jensen, J. P., 175
Jeon, D., 1443
Jeon, S. M., 1649
Jeon, S. Y., 1278
Jeong, E.-S., 1865
Jheon, S. H., 1899, 1976
Jian, L. K., 188, 1462, 1503, 1512
Jiang, D. T., 800, 1002, 1022
Jiang, J., 1451
Jiao, X., 898, 1172
Jin, Y., 163
Jinamoon, V., 264
Jinnai, H., 1369
Johnson, E., 283, 323, 408, 531, 675
Johnson, S., 1198
Jolivet, C. S., 1077
Joubert, A., 1887
Jourdain, E., 489, 639
Joyeux, D., 447
Jozwiak, C., 509
Juan, L. Z., 844
Juang, J. M., 828
Juanhuix, J., 824
Jung, M.-J., 1499
Jung, S. J., 1649
Jung, Y. G., 331
Jung, Y. K., 260
Junthong, N., 58
Jürgensen, A., 1202
Juthong, N., 147, 335

K

Kachel, T., 22, 1250
Kafle, B. P., 1809
Kageyama, T., 91

Kahn, R., 1883
Kahng, B., 1603
Kajiwara, K., 1545, 1577
Kakiuchi, T., 1419
Kakizaki, A., 543
Kalt, H., 643
Kamada, M., 559, 623, 1218
Kamakura, N., 1611
Kamali Kannangara, G. S., 1436
Kampen, Th., 1607
Kan, Y., 1494
Kanai, K., 698
Kanda, K., 694, 1423, 1817, 1853
Kane, S. R., 635
Kaneko, H., 1581
Kaneyasu, T., 1793
Kang, B. K., 260
Kang, H. S., 131, 252
Kang, L., 941, 963
Kang, T. H., 131
Kang, T.-H., 477
Kangrang, N., 264
Kannangara, G. S. K., 1642
Kanter, E. P., 1226
Karnaukhov, I., 9, 79
Karunakaran, Ch., 730
Kashiwagi, T., 710
Kaspar, J. D., 1176
Kasuga, T., 87, 91
Katagishi, K., 967, 1325
Katayanagi, H., 1809
Kato, K., 1214, 1238, 1771
Kato, M., 1121, 1129, 1821
Kato, S., 1520
Kato, Y., 694, 1423, 1853
Katoh, M., 71, 75, 192, 527
Kaulich, B., 497
Kawado, S., 1545
Kawamoto, M., 1920
Kawamura, N., 1353, 1699
Kawasaki, M., 1667
Kawata, H., 91, 1691
Kazimirov, A., 758, 1395
Kaznacheyev, K. V., 730
Keefe, L. J., 840
Kelez, N., 509, 1341
Keller, A., 388, 1198
Kennedy, B. J., 879
Kervin, J., 1683
Keyrilainen, J., 1887
Khan, S., 22, 1250
Khor, K. H., 1747
Khounsary, A., 989, 1210, 1299
Khruschev, S. V., 305
Kii, T., 240, 248
Kikuchi, M., 91

A8

Kikuchi, T., 710
Kikuta, S., 926, 930
Kikuzawa, T., 611
Kim, B. J., 477
Kim, C. K., 1113
Kim, D. E., 260, 331, 436
Kim, D. H., 1623, 1646
Kim, E.-S., 111
Kim, G., 1443, 1499
Kim, G. B., 981, 1507, 1891
Kim, G. H., 1113, 1875
Kim, H.-D., 477, 1649
Kim, H. G., 244
Kim, H. T., 1899, 1976
Kim, H.-T., 1952
Kim, H. Y., 1875
Kim, J., 1238, 1262
Kim, J. H., 477, 933, 981, 1507, 1875
Kim, J. J., 1407, 1539
Kim, J. K., 1899, 1952, 1976
Kim, J.-S., 1603
Kim, J. W., 1607
Kim, J.-Y., 477
Kim, K. J., 1875
Kim, K. W., 1875
Kim, M. G., 252
Kim, M.-G., 1222
Kim, M. J., 933, 1875
Kim, S., 1649
Kim, S. C., 244
Kim, S.-H., 1865
Kim, S. N., 933, 1875
Kim, T. C., 1603
Kim, T. K., 1187, 1262
Kim, Y., 1603
Kimura, A., 551
Kimura, H., 107, 571, 946
Kimura, S., 71, 192, 1238, 1771
Kimura, S.-I., 527, 587, 595
Kimura, T., 1899
Kinoshita, H., 1470, 1478
Kinoshita, T., 571
Kirby, N., 887
Kirimura, T., 1258
Kiriyama, K., 959
Kirz, J., 159
Kishimoto, H., 523, 718
Kishimoto, S., 1148
Kitagawa, T., 1784
Kitamura, H., 297, 355, 388, 420, 523, 571, 1010, 1018, 1034
Kitamura, T., 1010, 1180
Kitani, K., 1691
Kitchen, M. J., 1903
Kizilkaya, O., 163, 1759
Kläser, M., 301

Klaus, M., 1763
Kleemann, B. H., 659
Klein, R., 167
Klein, U., 291
Klinkhieo, S., 860
Klute, J., 175
Klysubun, P., 58, 1490
Klysubun, W., 860
Ko, I. S., 131, 244, 252, 436
Ko, J. Y. P., 1202
Kobata, M., 1407, 1539
Kobayakawa, H., 75
Kobayashi, K., 151, 1353, 1407, 1539, 1597, 1841
Kobayashi, Y., 87, 91, 95, 99, 103, 428
Koda, S., 119, 179, 184
Kohmura, Y., 1380
Kohn, V., 998
Kohno, A., 820
Koike, M., 647, 690
Koinuma, H., 1667
Kolobov, A. V., 1309
Kolokolnikov, Yu., 396
Komano, H., 1470
Komatsu, K., 1419
Kometani, R., 1423
Komorowski, P., 291
Kondo, Y., 559, 623, 1699
Kong, J. R., 188, 1462, 1512
Kong, Q., 1187, 1262
Kong, R., 1797
Konishi, H., 902, 1234, 1581
Koo, C. J., 933
Koo, K., 1443, 1499
Koo, T.-Y., 424
Korchuganov, V., 440
Koshelev, I., 840
Kostka, B., 301, 363
Kosugi, N., 192
Kotsugi, M., M., 1353
Koudobine, I., 1087
Kourinov, I., 832
Kovalyova, N., 79, 83
Koyama, A., 812
Kozhevnikov, I. V., 778
Kozin, V., 79
Kraft, P., 1141
Krämer, D., 167
Krämer, M., 852
Krasniqi, F., 1198
Krastanov, B., 683
Krebs, G. F., 159
Krempaski, J., 388
Krischel, D., 291
Krumrey, M., 1145
Krywka, C., 875

Kuan, C. K., 62, 143, 702
Kuan, C.-K., 451
Kubala, T., 651
Kubik, D., 1172
Kubota, M., 710
Küchler, M., 1295
Kuczewski, A. J., 1176
Kudo, K., 91
Kudo, T., 954, 1010
Kuetgens, U., 737
Kuk, D. H., 252
Kulesza, J., 283, 319, 323, 408, 531
Kulipanov, G. N., 234, 305
Kumar, C., 163
Kumasaka, T., 1871
Kumbaro, D., 34
Kume, T., 726
Kumigashira, H., 1569, 1667
Kunisada, T., 1960
Kuo, C. C., 13, 50, 376
Kuo, C. H., 151, 155, 367
Kuper, E. A., 305
Kuramoto, E., 1258
Kurapova, O., 1295
Kurisaki, S., 623
Kurosawa, T., 1129
Kuske, P., 167
Kusoljariyakul, K., 264
Kusukame, K., 240, 248
Kutluk, G., 1638
Kuwada, T., 123, 1841, 1968
Kwankasem, A., 58, 335
Kwon, S. J., 272
Kwon, S. Y., 1875

L

Labat, M., 256
Labiche, J.-C., 1242
Lagarde, B., 489, 607, 639, 655
Lagarde, P., 1329
Laggner, P., 1287
Lai, B., 1065, 1313, 1321, 1395
Lai, L. J., 816
Lai, T. Y., 1466
Lai, Y. H., 808
Lam, S., 1202
Lamb, R., 1642
Lambert, G., 256
Lammert, H., 667, 679, 706
Lampert, M.-O., 1152
Landahl, E. C., 950, 1210, 1226
Lange, M., 848
Lange, R., 167
Lanzara, A., 509

Lapertot, G., 1687
Lapshin, V., 79
Lara-Curzio, E., 1451
Larson, B. C., 1299
Last, A., 770
Lau, S. P., 1097
Laughon, G., 1683
Launer, L., 1928
Lavault, B., 1928
Lavender, W., 840
Lebasque, P., 3
Lebedev, A., 79
Le Duc, G., 1887
Lee, C. H., 1899
Lee, C. S., 933
Lee, D., 151, 155, 367, 1478
Lee, D. R., 1387
Lee, H., 1649
Lee, H. C., 1113
Lee, H. G., 260, 331
Lee, H.-H., 1589
Lee, H. J., 1195
Lee, H. S., 1646, 1875
Lee, H.-S., 424
Lee, J. F., 808
Lee, J. H., 1262
Lee, J.-M., 1222
Lee, J. P., 981, 1507
Lee, J. W., 188
Lee, K., 1411
Lee, P. L., 1073
Lee, S. H., 1210
Lee, S.-H., 1801
Lee, S. J., 981, 1507, 1891
Lee, T. L., 816
Lee, T.-Y., 252
Lee, W. H., 1623
Lee, W. S., 1619
Lee, W.-S., 1278
Lee, W. W., 244
Lee, Y.-C., 1940
Lee, Y. T., 1801
Lehecka, M., 283
Lekki, J., 1097
Leng, W. H., 1964
Lengeler, B., 1295
Lenke, R., 485
Leonard, G., 1928
Lerche, M., 737, 894
Le Roux, V., 3
Leson, A., 493, 1482
Lestrade, A., 3
Leuschner, A., 175
Lev, V. H., 305
Levecq, X., 722
Level, M. P., 3

Matsubara, T., 17, 327, 694
Matsuda, H., 1180
Matsudo, T., 1121
Matsugaki, N., 812, 1924, 1932, 1936
Matsui, F., 547, 1164, 1180
Matsui, S., 1423, 1853
Matsumoto, A., 1703
Matsumura, D., 1234
Matsunami, M., 1597, 1611
Matsuo, S., 623
Matsushita, T., 523, 571, 1010, 1164
Matsuyama, S., 786, 967, 1325
Mattarello, V., 575
Matteucci, M., 796, 1447
May, T., 579
McIntyre, S., 872
McKinlay, J., 615
McKinney, W., 469, 706
Medjoubi, K., 1087, 1137
Meents, A., 1984
Meessen, C., 1087
Meister, D., 848
Meng, W. J., 1451
Menk, R. H., 796, 1109
Menk, R.-H., 1895
Menouni, M., 1087
Menzel, M., 493, 1482
Méot, F., 256
Mercère, P., 455, 619, 722
Meron, M., 840
Mertin, M., 493, 497
Merz, M., 175
Meyer, D. A., 1176
Meyer-Reumers, M., 291
Mezentsev, N. A., 305, 440
Mezouar, M., 1883
Miao, H., 188
Michaelsen, C., 774
Miginsky, E. G., 305
Mikaelyan, R., 1046
Mikuljan, G., 848
Milev, A., 1436, 1642
Miller, T., 1341
Mimura, H., 786, 967, 1325, 1699
Min, C.-H., 477
Minami, M., 1944, 1956
Minkov, D., 268
Minty, M., 175
Mishina, A., 87, 95
Mitamura, T., 1861
Mitsuhashi, T., 87, 91
Mitsui, T., 17, 107, 327
Mitsuke, K., 1809, 1817
Mitzner, R., 1250
Mitzner, T., 22
Miura, K., 1638

Miura, N., 611, 1784
Miura, T., 523, 718
Miwa, D., 1407, 1597
Miyajima, T., 87, 91, 95, 99, 103
Miyake, S., 1619
Miyamoto, S., 327
Miyano, K., 1711
Miyata, H., 623
Miyauchi, H., 87, 1038
Miyawaki, J., 463
Miyayhara, T., 698
Miyazaki, H., 1821
Miyoshi, T., 812
Mizota, H., 1775
Mizuki, J., 902, 1234, 1581
Mizuno, T., 595
Mizutani, U., 75, 1821
Mocheshnikov, N., 79
Mochihashi, A., 71
Mochizuki, T., 959, 1034
Modi, M. H., 1474
Moeller, K. D., 1503
Mohr, J., 770
Momose, A., 1361, 1365, 1369
Monfraix, S., 1887
Monot, P., 256
Monroy, E., 1675
Moon, A., 611, 1784
Morand, M., 1837
Morawe, C., 764, 792
Morel, C., 1087
Moreno, T., 455, 619, 722, 1101
Morgan, M. J., 1903
Mori, A., 123, 1841
Mori, K., 1960
Mori, T., 1715
Mori, Y., 726, 786, 967, 1325
Morikawa, E., 163, 1759
Morini, P., 1246
Morishita, Y., 1121, 1129
Morita, M., 26, 268, 297
Morita, S., 75
Moritake, M., 1365
Moritomo, Y., 1238
Morris, K., 163
Morris, K. J., 1317
Moser, H. O., 188, 603, 844, 1097, 1462, 1503, 1512
Mosnier, A., 256
Motooka, T., 559
Moure, C., 1659
Mouri, S., 1944, 1956
Mueller, U., 1006
Muir, J. L., 840
Mulichak, A. M., 840
Müller, A. S., 301

Müller, P., 1145
Müller, R., 167
Muñoz, M., 750
Murakami, M., 297
Murakami, T., 1944, 1956
Murakami, Y., 1711
Murakawa, T., 1415
Muraoka, H., 1353
Murata, K., 1561
Murata, M., 1419
Murata, Y., 1419
Murayama, H., 1238
Murgida, D., 1984
Muro, T., 571, 1164, 1703
Murphy, M., 1202
Muto, T., 1117
Mytsykov, A., 9, 79

N

Nadji, A., 3
Nadolski, L., 3
Nagahashi, S., 87, 91, 95, 99, 103
Nagaoka, R., 3
Nagata, K., 820
Nagira, M., 551
Nagy, M., 730
Nahon, L., 256, 311, 447
Nakagawa, A., 1916
Nakahigashi, S., 1861
Nakai, Y., 240, 248
Nakajima, H., 543
Nakamura, A., 75
Nakamura, D., 1550
Nakamura, E., 527, 587, 595
Nakamura, M., 1711
Nakamura, N., 1117
Nakamura, S., 698
Nakamura, T., 151, 571, 1353, 1703, 1779
Nakamura, T. T., 91
Nakamura, Y., 1634
Nakanishi, H., 91
Nakanishi, K., 547, 1180, 1630
Nakano, M., 248
Nakao, H., 1711
Nakao, K., 123
Nakayama, Y., 698
Namatame, H., 551, 1597, 1638, 1788
Namba, H., 547, 1180, 1630
Namikawa, K., 1849
Namikawa, T., 1357
Namiki, T., 1283
Nandedkar, R. V., 631, 635, 1474
Nanpei, M., 547
Narayanan, S., 898, 911, 937, 1073, 1387

Nardone, M., 1695
Nariki, S., 297
Nariyama, N., 1125, 1133
Nash, P., 989
Nasiatka, J., 1206
Nasukawa, S., 1944, 1956
Nayak, M., 1474
Nazmov, V., 770
Neau, D., 163
Neisius, T., 1655
Neklyudov, I., 79
Nelles, B., 485, 497, 659, 918
Nemoz, C., 1887
Newton, M., 1531
Newton, M. A., 750, 1242
Ng, J. C., 1845
Niemann, B., 1291
Niibe, M., 1520
Nisawa, A., 1920
Nishibori, E., 1829
Nishihata, K., 698
Nishihata, Y., 1234
Nishikawa, K., 1784
Nishikawa, Y., 1369
Nishino, J., 1565
Nishino, Y., 786, 954, 967, 1325, 1376, 1407, 1597, 1979
Nishio, M., 559
Nishiyama, A., 1573
Nitta, T., 1561
Niwa, Y., 1230
Nogami, K., 123, 1968
Nogami, T., 87, 91, 95
Noguchi, T., 1779
Noh, D. Y., 1603
Noh, Y. D., 26
Nohira, H., 1597
Nohtomi, A., 1129
Noller, B. N., 1845
Nomoto, T., 1638
Nomura, M., 1230
Nozawa, Y., 547
Nutarelli, D., 256

O

Obina, T., 87, 91, 95, 428
Oda, Y., 1561
Ofuchi, H., 1565
Ogata, C., 832
Ogawa, H., 559
Ogawa, K., 547, 1630
Ogimoto, Y., 1711
Ogura, T., 1829
Oh, J. S., 252, 272, 436

Oh, S.-J., 477
Oh, T. H., 1278
Ohashi, H., 523, 706, 718, 1611
Ohata, T., 523
Ohbayashi, C., 1899
Ohgaki, H., 119, 179, 184, 240, 248
Ohishi, Y., 922
Ohkubo, K., 1715
Ohmori, H., 1573
Ohnishi, T., 1667
Ohsawa, S., 1715
Ohsawa, Y., 91
Ohshima, H., 1841
Ohta, T., 463
Ohtake, Y., 1619
Okabayashi, J., 698, 1569
Okada, K., 1121
Okada, M., 1520
Okajima, S., 179
Okajima, T., 559, 623, 820
Okajima, Y., 1234
Okawachi, N., 248
Okazaki, N., 1871
Okube, M., 1715
Okumura, K., 1214
Ollinger, C., 975
Ono, H., 946
Ono, K., 710, 1353
Ono, M., 91
Oohashi, H., 1775
Ortega, J. M., 256
Osaka, K., 1771
Oshima, M., 698, 1353, 1569, 1667
Ota, S., 1821
Otani, H., 1565
Ouchi, C., 1520
Oura, M., 523, 1034, 1121
Owen, R. L., 1984
Ozaki, T., 91, 1309

P

Pace, E., 1246
Padmore, H. A., 159, 469, 890, 1195, 1206, 1341
Paganin, D. M., 1903
Pairsuwan, W., 214
Pak, C. O., 87, 95
Palshin, V., 163
Pan, G., 963, 1849
Pan, H. B., 1615, 1626
Pang, J. W. L., 1299
Pangaud, P., 1087
Pantenburg, F. J., 1456
Parc, Y. W., 252

Park, C. D., 1113
Park, J. H., 252
Park, K. H., 260, 331, 436
Park, S., 1443, 1499
Park, S. H., 1899, 1976
Park, S.-H., 1865
Park, S. J., 260
Park, Y., 1554
Park, Y. D., 1623
Park, Y. J., 1411, 1554
Parry, D., 1751
Pascarelli, S., 750, 1242, 1531
Patelli, A., 575
Paterson, D. J., 864
Patommel, J., 1295
Paul, D. F., 1683
Paulsen, J., 473
Paulus, M., 875
Pavlov, K., 1903
Pease, M., 1759
Peev, F., 79
Peña, O., 1659
Pereira, N. R., 985
Perennes, F., 796, 1447
Perng, S. Y., 563, 702
Perng, S.-Y., 451
Peters, H. B., 539
Peterson, E., 1226
Petrosyan, A., 1046
Petrov, A., 175
Peverini, L., 778, 1349
Pflüger, J., 339, 343
Phalippou, D., 447
Pi, T.-W., 1589
Pianetta, P., 1333, 1337
Piccinini, M., 1246
Pimol, P., 58, 147
Ping, Y., 844
Piotrowski, J., 1246
Pirngruber, G., 1731
Plate, D., 1195
Plech, A., 1187
Polack, F., 447, 489, 607, 639, 655, 706, 1101
Polak, W., 1097
Polian, A., 1329
Ponchut, C., 1242
Pong, W. F., 1659
Ponwitz, D., 599
Porra, L., 1887
Pottin, B., 3
Powers, T., 1061
Pradervand, C., 1984
Prange, A., 1759
Prawanta, S., 58, 335
Prawatsri, P., 58
Preissner, C., 911, 1049, 1321

Prenting, J., 175
Prestemon, S., 509
Prinz, H., 975
Pyun, A., 1373

Q

Qi, Z., 591
Qian, H., 555
Qian, S., 706
Qiang, J., 1206
Qiu, R., 424
Quai, E., 1895
Quan, C. G., 1503
Quast, T., 22, 1250
Quinn, F. M., 627
Quitmann, C., 505

R

Raabe, J., 505, 643
Rabedeau, T., 392
Racky, B., 175
Raco, A., 575
Raghuvanshi, V. K., 631
Rah, S., 706
Rahighi, J., 208
Rahn, J., 167
Raimondi, S., 287
Rajashankar, K., 832
Rakowsky, G., 283
Rao, D. V., 1948
Raoux, S., 1341
Rappolt, M., 1287
Ravelli, R., 1928
Ravet-Krill, M.-F., 489, 1101
Regier, T., 473, 1202
Rehbein, S., 1291
Reichardt, G., 497, 667, 918, 1006
Reininger, R., 509, 535, 567, 579, 651, 667, 872
Reis, D., 1210
Ren, Y. P., 188, 1512
Renier, M., 1887, 1928
Repkov, V. V., 305
Requardt, H., 1887
Reuzayev, A., 79
Reznikova, E., 770
Rhodes, M. W., 264
Richter, M., 167
Rickers, K., 907
Riekel, C., 1295
Riekel, Ch., 1287
Rigato, V., 575

Rigon, L., 796, 1109
Rikken, G. L. J. A., 1695
Riley, M., 1845
Rimjaem, S., 264
Rivkin, L., 1198
Robin, D., 1341
Robin, D. S., 159
Robinson, A. L., 159
Rocca, F., 1349
Rodier, J. C., 447
Rodrigues, M., 1391
Rogalev, A., 1663, 1671, 1675, 1707, 1825
Rogers, G., 323, 651
Röhlsberger, R., 539
Rohrer, M., 1198
Rokvic, T., 1895
Rommeveaux, A., 706, 792
Rosenberg, R. A., 1202
Rossi, E., 1061
Rossmanith, R., 301, 363
Roulliay, M., 489
Rousseau, J., 1887
Rouvinski, E., 396
Roux, G., 396, 412
Rowen, M., 392
Roy, A., 163, 1759
Rozatian, A. S. H., 1593
Rudati, J., 1176
Rudenko, O., 359
Rueff, J.-P., 1837
Rugmai, S., 58, 335
Rujirawat, S., 58
Rüter, H. D., 737
Ryan, C. G., 864
Ryu, C. G., 1875

S

Safonova, O. V., 1731
Sahoo, G. K., 175
Sahoo, N. K., 671
Saikubo, A., 1423
Saile, V., 770
Saisut, J., 264
Saito, I., 714
Saito, N., 1121, 1129
Sakai, H., 91, 1117, 1871, 1920
Sakai, M., 587
Sakai, N., 297
Sakai, T., 123, 1968
Sakakura, S., 1775
Sakamoto, Y., 91
Sakanaka, S., 87, 91, 95
Sakata, M., 75, 1829
Sakaya, N., 1478

Takagaki, M., 1353, 1699
Takagi, H., 1419
Takahashi, H., 1569
Takahashi, K., 623, 1218
Takahashi, M., 1561
Takahashi, N., 547, 1180
Takahashi, S., 523, 1010, 1018, 1034
Takahashi, T., 71, 87, 91, 95, 107
Takahashi, Y., 1376
Takaie, Y., 726
Takaku, J., 1561
Takasaki, S., 91
Takashima, Y., 71, 75
Takata, M., 1238, 1771
Takata, Y., 1597, 1611
Takayama, Y., 698
Takeda, T., 1944, 1948, 1956
Takeda, Y., 75, 1361, 1365, 1369, 1619
Takemura, M., 1573
Takenaka, H., 647
Takeno, S., 1573
Takeshita, K., 523
Takeuchi, A., 1305, 1380
Takeuchi, M., 523, 571
Takeuchi, T., 75, 523, 587, 1821
Takumi, M., 820
Tamasaku, K., 107, 786, 922, 967, 1057,
    1325, 1407, 1597
Tamenori, Y., 1121
Tamura, N., 1751
Tan, K., 473
Tan, Y., 1026
Tanabe, T., 283
Tanaka, H., 1238, 1577
Tanaka, K., 75, 1788
Tanaka, M., 1779
Tanaka, T., 123, 297, 355, 388, 523, 559,
    1841, 1968
Tanaka, Y., 1238, 1258, 1478
Tancharakorn, S., 335
Tang, J., 1899
Tang, M. T., 883
Tang, M.-T., 1274, 1940
Tanida, H., 954, 1214
Taniguchi, M., 551, 1597, 1638, 1723, 1788
Taniguchi, S., 523, 1125
Taniguchi, Y., 1565
Tanimoto, Y., 87, 91, 95
Tanioka, K., 1091
Taniuchi, T., 1353
Tao, Y., 555
Tarawarakarn, P., 860
Tarawneh, H., 34
Tarrio, C., 481
Tatchyn, R., 79
Tavakoli, K., 311

Tavares, P. F., 196
Tay, C. J., 1503
Techert, S., 1254
Tejima, M., 91
Telegin, Yu., 9, 79
Teramoto, T., 1805
Terasawa, M., 1861
Teshima, H., 297
Thomasset, M., 706
Thomlinson, W., 1887
Thompson, P., 1679
Thompson, P. B. J., 1683
Thongbai, C., 264
Thoraud, S., 1101
Thorin, S., 34
Thornagel, R., 167
Tian, Y., 868, 1494
Tiedtke, K., 276
Tieman, B., 1172
Tischer, M., 175, 339, 343
Tischler, J. Z., 1299
Tittsworth, R. C., 163, 1317
Tobin, M., 615
Tochihara, H., 623
Tochio, T., 1775
Toellner, T. S., 1073
Tokei, Z., 1557
Tokudomi, S., 1218
Tokushima, T., 523
Tolan, M., 30, 875
Tomimasu, T., 119, 179, 184
Tominaga, J., 1309
Tonutti, M., 1895
Toomey, J., 675
Tordeux, M. A., 3
Toriumi, K., 1238
Toyoda, S., 1569
Toyoda, T., 1715
Toyokawa, H., 1141
Toyoshima, A., 710
Toyosugi, N., 26, 268
Tozawa, K., 902, 959
Tran, N., 1436, 1642
Trapananti, A., 750, 1242
Treusch, R., 667
Tromba, G., 1895, 1948
Tromp, M., 1731, 1735, 1739
Tropres, I., 1887
Trotsenko, V., 79
Tsai, H. J., 50
Tsai, H. M., 1659
Tsai, M. H., 46
Tsai, Z. D., 702
Tsakanov, V. M., 202
Tsang, K. L., 563, 808, 816, 828, 883, 971
Tsapatsaris, N., 1735, 1739

Tschentscher, Th., 1254
Tseng, C. C., 816
Tseng, P. C., 883
Tseng, P. Y., 1908, 1964
Tseng, T. C., 563, 702
Tseng, T.-C., 451
Tsubakino, H., 1861
Tsuchiya, K., 87, 91, 380, 384, 428
Tsuji, J., 623
Tsuji, N., 1691
Tsukanov, V. M., 305
Tsukihara, T., 1916
Tsukuda, N., 1258
Tsunekawa, M., 1703
Tsuru, R., 297
Tsuru, T., 1524

U

Uchida, K., 1345
Uchiyama, T., 87, 91
Udupa, D. V., 671
Ueda, A., 87, 91
Ueda, S., 1539
Ueng, T. S., 13, 143, 702
Ueno, E., 1899
Ueno, G., 1871, 1879
Ueno, K., 726
Uesugi, K., 1365, 1903
Ülkü, D., 208
Ulm, G., 167
Ulrich, J., 890
Umemori, K., 87, 91, 95
Uozumi, Y., 119
Uruga, T., 1214
Usuda, K., 1569

V

Valade, J.-P., 1687
Valentinov, A., 440
Valleau, M., 396, 412
Vanacken, J., 1695
van Bokhoven, J. A., 1731
van der Hart, A., 1295
van der Linden, P., 915, 1242, 1655
van der Meulen, M., 1333
van der Veen, F., 1198
van Vaerenbergh, P., 1349
Vaughan, G., 998
Veres, A., 1825
Vervloët, M., 447
Veteran, J., 311

Viccaro, P. J., 754
Vielitz, T., 339
Vigeolas, E., 1087
Vilaithong, T., 264
Vincze, L., 1295
Vinokurov, N., 234
Vlaicu, A. M., 1775
Vobly, P., 396, 412
Vogt, S., 1313, 1321
Vollenweider, C., 400
Volmer, M., 875
von Bohlen, A., 852
von Borany, J., 1813

W

Wada, H., 1703
Wagener, G., 1763
Wagner, A., 836
Wakabayashi, Y., 1419, 1711
Wakatsuki, S., 812, 1924, 1932, 1936
Wakita, H., 623
Wakita, T., 571, 1353
Walko, D. A., 950, 1226
Wallace, M. J., 1903
Wallén, E., 34
Wallwork, K. S., 879
Walther, H., 890
Wan, W., 1206, 1341
Wang, C. H., 1399, 1427, 1431, 1908
Wang, C. J., 151, 155, 367
Wang, Ch., 46
Wang, D., 879, 1486
Wang, D. J., 13, 563, 702
Wang, D.-J., 451
Wang, J., 451, 832, 937, 1049, 1387, 1535
Wang, M. H., 13, 50
Wang, S., 1797
Wang, X., 1030
Wang, Y., 937
Wang, Z., 1797
Wannberg, B., 1597
Wanning, A., 175
Wanzenberg, R., 175
Warren, J., 1176
Warwick, T., 159, 469, 509, 730
Watanabe, A., 1034
Watanabe, N., 1283, 1357
Watanabe, S., 946, 1924
Watanabe, T., 623, 1470, 1478
Waterman, D., 283, 323, 392, 408, 1069
Waterstradt, T., 1081
Wee, A. T. S., 188, 1097
Wei, D., 1341